DINOSAURS
THE ENCYCLOPEDIA
Supplement 2

DINOSAURS
THE ENCYCLOPEDIA
Supplement 2

by Donald F. Glut

Foreword by PHILIP J. CURRIE,
MSC, PH.D., FRSC

McFarland & Company, Inc., Publishers
Jefferson, North Carolina, and London

To Tracy Lee Ford,
"Dino Hunter" Extraordinaire,
in thanks for the continued help and support

Library of Congress Cataloguing-in-Publication Data

Glut, Donald F.
Dinosaurs: the encyclopedia : supplement two / by Donald F. Glut : foreword by Philip J. Currie.
p. cm.
Includes bibliographical references and index.

ISBN-13: 978-0-7864-1166-5
ISBN-10: 0-7864-1166-X
(library binding : 50# alkaline paper)

1. Dinosaurs—Encyclopedia. I. Title.
QE862.D5G652 2002
567.9'1'03—dc20 95-47668

British Library cataloguing data are available

Manufactured in the United States of America

McFarland & Company, Inc., Publishers
Box 611, Jefferson, North Carolina 28640
www.mcfarlandpub.com

Acknowledgments

The author thanks the following vertebrate paleontologists for their critical review of selected sections of the manuscript, for their invaluable criticisms, suggestions, and other comments, and for generally improving the text:

(For primitive saurischians and theropods) Ralph E. Molnar, Museum of Northern Arizona, Flagstaff, formerly of the Queensland Museum, Queensland; (basal sauropodomorphs and prosauropods) Peter M. Galton, College of Naturopathic Medicine, University of Bridgeport, Bridgeport, Connecticut; (sauropods) John S. McIntosh, formerly of Wesleyan University, Middletown, Connecticut; (thyreophorans) Kenneth Carpenter, Denver Museum of Natural History, Denver, Colorado; (primitive ornithischians, pachycephalosaurs, and iguanodontians excluding hadrosaurs) Hans-Dieter Sues, Royal Ontario Museum, Toronto; (hadrosaurs) Michael K. Brett-Surman, National Museum of Natural History, Smithsonian Institution, Washington, D.C., also my chief scientific advisor, who reviewed the entire work and made myriad suggestions for improving the manuscript; and (ceratopsians) Peter Dodson, School of Veterinary Medicine, University of Pennsylvania, Philadelphia.

Thanks to the following vertebrate paleontologists, who sent me reprints of articles, copies of journals, personal communications, photographs, slides, illustrations, or otherwise contributed directly to the production of this volume: John R. Bolt, John J. Flynn and William F. Simpson, The Field Museum, Chicago; Lawrence G. Barnes, Luis M. Chiappe, Samuel McCleod, J. D. Stewart, and Howell W. Thomas of the Natural History Museum of Los Angeles County; Philip J. Currie, Royal Tyrrell Museum of Palaeontology, Drumheller, Alberta; Fabio M. Dalla Vecchio, Museo Paleontologico Citadino di Monfalconel, Colloredo di Prato, Italy; James O. Farlow, Department of Geosciences, Indiana-Purdue University at Fort Wayne, Indiana; Catherine A. Forster, Department of Anatomical Sciences, Health Service Center, State University of New York, Stony Brook; Mark B. Goodwin and Robert A. Long, University of California Museum of Paleontology, Berkeley; James I. Kirkland, Division of State History and Antiquities, Utah; Giuseppe Leonardi; Spencer G. Lucas, New Mexico Museum of Natural History and Science, Albuquerque; Tim Lucas, News Services, North Carolina State University, Raleigh; James H. Madsen, Jr., DINOLAB, Salt Lake City; Dale A. Russell, North Carolina Museum of Natural Sciences, North Carolina State University, Raleigh; Paul C. Sereno and Jeffrey A. Wilson, Department of Organismal Biology and Anatomy, University of Chicago; Robert M. Sullivan, State Museum of Pennsylvania, Harrisburg; Matthew J. Wedel, Oklahoma Museum of Natural History, Norman; Thomas E. Williamson, New Mexico Museum of Natural History, Albuquerque; Donald L. Wolberg, Dinofest™; and Xu Xing, Institute of Paleoanthropology, Beijing, People's Republic of China.

Also thanks to the "paleo-artists" — including Ricardo Delgado, Brian Franczak, Mike Fredericks, Berislav Krzic, and Gregory S. Paul — who contributed dinosaur illustrations (and who retain all copyrights and any other rights to their respective illustrations, whether explicitly stated or not) to this volume.

Finally, thanks to Nina Cummings of the Photography Department, The Field Museum, Chicago; Stephen A. and Sylvia Czerkas, The Dinosaur Museum, Blanding, Utah; Allen A. and Diane E. Debus, Hell Creek Productions (*Dinosaur World*); Allen Detrich, Detrich Fossils, Great Bend, Indiana; Tim Donovan; Randy Epstein; Debra J. Faulkner and Patrick Fisher, Oklahoma Museum of Natural History, Norman; Tracy L. Ford ("Dino Hunter"); my mother, Julia Glut; Mike Fredericks, *Prehistoric Times*; Karen Kemp, North Carolina Museum of Natural Sciences, North Carolina State University; John J. Lanzendorf; Cathy McNassor, Natural History Museum of Los Angeles County; Mary Jean Odano and Toshiro Odano, Valley Anatomical Preparations, Canoga Park, California; George Olshevsky, Publications Requiring Research, San Diego; Richard C. Ryder; the Society of Vertebrate Paleontology; Susan Shaffer; Kent A. Stevens, Department of Computer and Information Science, University of Oregon, Eugene; Masahiro Tanimoto; Michael Trcic, Trcic Sculpture Studio, Sedona, Arizona; Michael Triebold, Triebold Paleontology, Woodland Park, Colorado; and Pete Von Sholly, Fossil Records.

I apologize to anyone and any institution I may have missed.

Once again, I have done my best to credit all illustrations and to obtain permissions for their use in this book, if permissions are so required. Copyrights for photographs (except for those privately taken, such as the ones shot by the author) of specimens, whether stated or not, are assumed to be held by the institution housing those specimens or by the publications in which those pictures initially appeared.

Foreword

Donald Glut's *Dinosaurs: The Encyclopedia* appeared in 1997, the same year that Academic Press published *The Encyclopedia of Dinosaurs*, edited by Kevin Padian and myself. We had been aware of each other's book projects for several years before that, and in fact, Don was kind enough to write an excellent treatment on dinosaurs in "Popular Culture" for our book. In a field with so much public demand for information, this could easily have ended up in books that heavily overlapped because the number of people working within the discipline is actually relatively low.

Fewer than 150 living people worldwide have published scientific papers on dinosaurs, and of those, only a minority of these (fewer than 50) are specifically employed to do research on these fascinating animals. But the subject matter is so extensive that there was more than enough room for two encyclopedias on this subject. And the books had fundamentally different approaches, and consequently became complementary to each other. In fact, along with *The Dinosauria*, edited by Weishampel, Osmólska and Dodson (1990), just about anything you would ever want to know about dinosaurs is available to anyone in these books.

It is so different now than when I was younger. Fifty years ago, there was virtually nothing available for a person seriously interested in dinosaurs. I had to work hard on my newspaper route to save up enough money to buy Fenton and Fenton's *Fossil Book* and Roy Chapman Andrews' *All About Dinosaurs*. These books were great works in their time, but they have been vastly eclipsed in "the information age" by books that are both more technical and more informative. And the public is ready for it with a more solid foundation of basic knowledge about dinosaurs.

There have never been so many television programs, newspaper and magazine articles, popular books, and even movies about dinosaurs. Although we do not think of movies as being educational, I had to change my opinion after the release of the movie *Jurassic Park*. The "Dinosaur Renaissance" had begun in the mid–1970s with the publication of Robert T. Bakker's article in *Scientific American* and Adrian Desmond's *Hot-Blooded Dinosaurs*. Research on dinosaurs and public interest in these animals has always been intimately linked in a mutual feedback system. As scientists learn more, the public displays more interest. When the public shows more interest, more funds and support become available, making it possible to do more research.

In spite of the ever-increasing interest in dinosaurs during the 1970s and '80s, dinosaur researchers continued to get the same kind of questions as our colleagues must have been getting since the release of the motion picture *King Kong* in 1933. But that changed overnight in 1993 with the release of *Jurassic Park*, which never made any claims to being an educational film. But both the author of the book, Michael Crichton, and the movie production crew of Steven Spielberg made more than a half-hearted effort to bring a more up to date version of our understanding of dinosaurs to the public. And overnight, the questions we paleontologists were being asked about dinosaurs took a quantum leap ahead. All of the subtle ideas introduced by *Jurassic Park*, including warm-bloodedness, complex behavior and intelligence in dinosaurs, and the dinosaurian ancestry of birds, were picked up by the public. And the level of questions we received finally pulled away from the 1930s Hollywood understanding of dinosaurs.

Along with the ready availability of information on the Internet, people are better informed than they ever were in the past about dinosaurs (and everything else, of course). And the desire for more current, more accurate, more detailed information is insatiable. Don Glut's *Dinosaurs: The Encyclopedia* has met the need for information so admirably that even I use it frequently as a first point source for information.

The author and publisher, recognizing that it is difficult to stay current with the research that is going on, made the wise decision to publish regular supplements to the main volume. Other books, in the meantime, have become out of date. In addition to being comprehensive and up-to-date, *Dinosaurs: The Encyclopedia* and its supplements are beautifully balanced — they are detailed but readable, contain a healthy dose of information on the cultural aspects of the Dinosauria, provide the keys to access more detailed sources of information, and are beautifully illustrated.

My hat is off (although I hate to admit that this is with some eagerness as I do not like wearing hats) to both Don Glut and McFarland for the vast service they have performed for both professional paleontologists and anyone interested in dinosaurs. Thank you, thank you, thank you...

PHILIP J. CURRIE, MSC, PHD, FRSC
Royal Tyrrell Museum of Palaeontology

Table of Contents

Acknowledgments v

Foreword, by Philip J. Currie, MSc, PhD, FRSC vi

Preface ix

I Introduction 1

- *Never Ending Changes* 4
- *The Mesozoic Era* 5
- *New Discoveries, Ideas and Studies* 9
- *Still Unresolved: Ectothermy or Endothermy?* 86
- *The Dinosaur-Bird Debate: Nearing a Resolution?* 94
- *Dinosaur Extinctions* 106

II Dinosaurian Systematics 113

III Dinosaurian Genera 185

IV Nomen Nudum and Excluded Genera 593

A List of Abbreviations 597

Appendix One: Displays, Sites and Attractions 601

Appendix Two: Further Reading 615

Glossary 623

Bibliography 635

Index 671

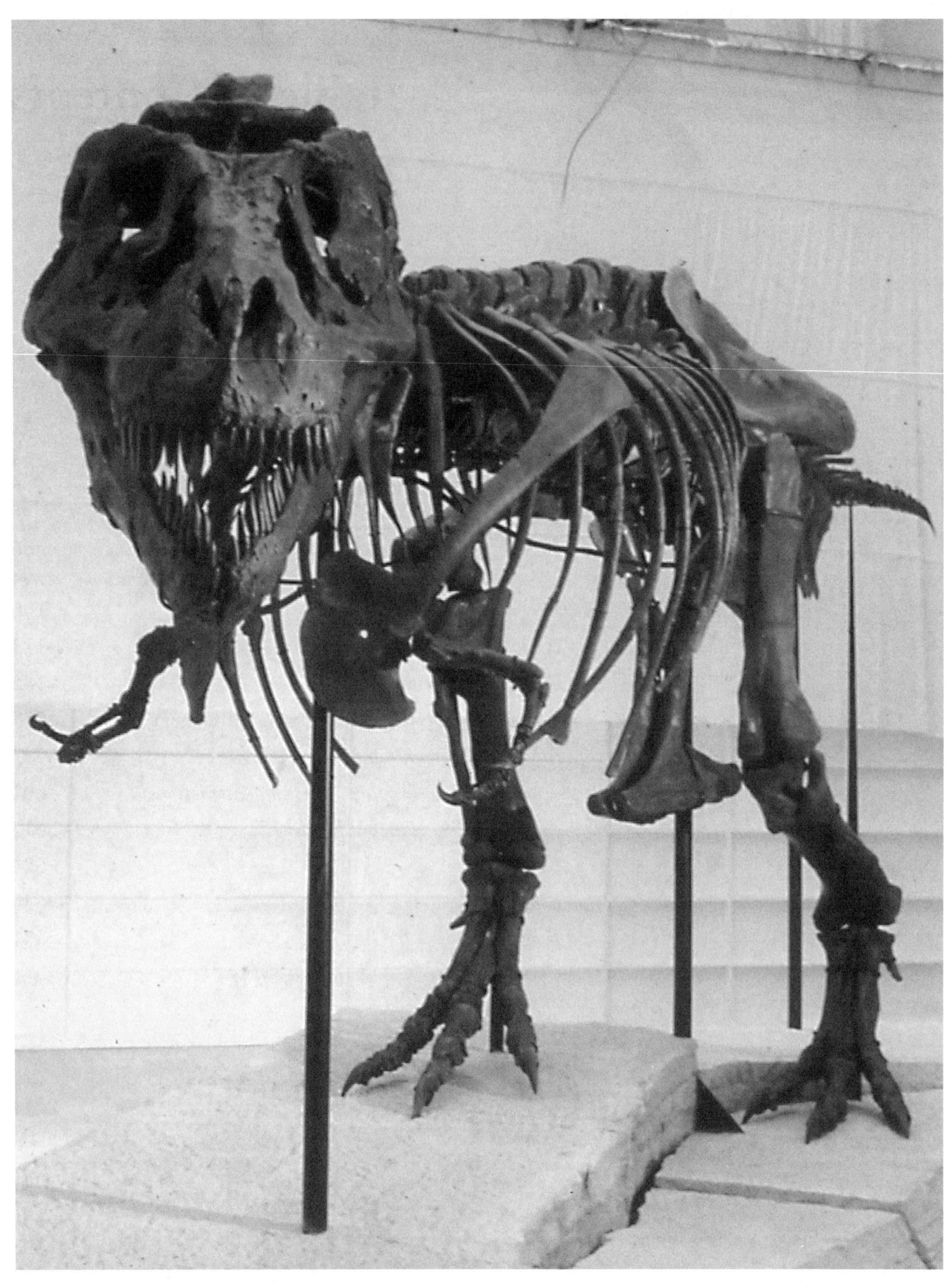

The *Tyrannosaurus rex* specimen (FMNH PR2081) known as "Sue," to date the largest (over 12 meters, or 42 feet in length), most complete (90 percent), and best-preserved *T. rex* fossil to be described. After years of delays and controversy (see *D:TE*, *S1*, and *Tyrannosaurus* entry, this volume) the mounted skeleton — one of the most famous dinosaur specimens in the world — is now on permanent display, gracing Stanley Field Hall at The Field Museum, Chicago. Each bone in this mount is cradled in an individual setting, allowing it to be removed for scientific study. Photograph by John Weinstein, courtesy The Field Museum (neg. #GN89677_47c).

Preface

"What's new?" That question is frequently asked of paleontologists by visitors to their laboratories.

Indeed, in a place where the objects of study can be millions of years old, the joke (but correct) answer to that inquiry is often, "Nothing is new."

However, "new" dinosaur fossils, ranging from some 65 million to well over 200 million years old, continue to be discovered, and through the study of these ancient wonders, our interpretation of what these extinct animals may have been like — the way they looked in life, how they behaved, and how they related to different kinds of dinosaurs and also to other groups of animals — change and evolve.

That there is so much new information concerning a subject as old as the dinosaur warrants the existence of this book.

This second supplementary book continues the intent of the first supplement (published in 1999 by McFarland & Company) — namely, to update, augment, emend, and sometimes correct material that appeared in *Dinosaurs: The Encyclopedia* (McFarland & Company, 1997), which was a compendium of data based on the original research of professional paleontologists. The original volume (referred to throughout this text as *D:TE*) was conceived as a foundation upon which all future supplementary volumes would build. The first of these subsequent volumes, *Dinosaurs: The Encyclopedia, Supplement 1* (herein referred to as *S1*, published by McFarland in 1999), built upon and hopefully enriched that flagship book. Taken together, the present volume and its two predecessors constitute a unique and hopefully ongoing experiment in publishing: To make available an open-ended, though mostly unofficial, "series" of reference books about dinosaurs that will hopefully keep the topic up to date.

The present volume, like its two predecessors, is intended both to provide a handy reference tool for professional paleontologists and students and to offer less technical information of interest to the more casual dinosaur enthusiast. It does not presume to be a substitute for the original literature, nor should it be used for formal taxonomic purposes. The author encourages the less technically-minded readers to make use of the glossary in order to understand some of the more difficult passages. Furthermore, I recommend that the reader make use of this book's bibliography and then, when possible, seek out the original technical publications.

Following the intent, format, and writing style of *D:TE* and *S1*, this book is meant to be a handy compendium of mostly new or recent information and ideas about dinosaurs. Its content was based on the original research of paleontologists, who have published the results of their often painstaking work through the accepted channels, and which became available to the present writer in time for incorporation into this text. Usually these scientists have published their work in peer-reviewed technical articles that appear in scientific journals, or, less frequently, in books written or edited by paleontologists. Some of this information may also have appeared in the form of published "abstracts" — brief, usually concise, and frequently preliminary summaries of works-in-progress — presented as talks at a scientific conference or symposium, such as the Annual Meeting of the Society of Vertebrate Paleontology (SVP). Presumably, the final, detailed results (often somewhat differing in content from what is stated in the abstract) of these early, often incomplete presentations will appear at some later date as a fully realized article. (Recently, the SVP has stated that "the technical content of the SVP sessions is not to be reported in any medium — print, electronic, or the internet — without the prior express permission of the authors," this hopefully ensuring that information heard "on the floor" or during a talk, but not included in the printed abstracts, is not prematurely made public.)

Once more it must be stressed that none of these "encyclopedia" volumes are meant to substitute for the original writings as published in the paleontological literature, and that this writer always recommends that the reader seek out the original sources whenever possible.

Paleontology is a science that depends upon what has been preserved in the fossil record (then found, collected, and prepared), then deposited in a museum certified by the American Association of Museums. Then follows the subsequent interpretation of those fossil specimens. Much of what is known to scientists or believed by them about dinosaurs, therefore, is limited by the quality (or lack thereof) of the preserved materials. New discoveries of dinosaurian fossils are regularly made throughout the world and some of these specimens represent previously unknown genera and species. Fossil specimens also continue to be recovered pertaining to already known taxa, sometimes including elements not known before, therefore leading to revised or entirely new interpretations of these taxa.

Again, this series of supplementary books is intended to be open-ended. "Facts" that are considered in this volume to be accurate and correct may well, based upon future studies and new discoveries, turn out to be somewhat (or even completely) in error, and in need of revision in some later supplement. In truth, therein lies much of the excitement and appeal of this field of paleontology, as we continue to learn and understand more and more about these former rulers of the Earth that we call dinosaurs.

Once again, I ask the reader to be tolerant in dealing with a book (or series of books) containing so many words. Errors inevitably creep into manuscripts and proof pages, no matter how many sharp editorial eyes scrutinize the text copy. In the case of errors and omissions, I urge the reader to inform me (via McFarland) so that they can be dealt with accordingly in the next supplement. Needless to add, any and all copies of original articles, photographs of specimens, artwork, *etc.*, which this writer may find useful in preparing future supplements, are always welcomed.

In order to keep this book a manageable size, I have avoided as much as possible the repeating of information (*e.g.*, background data, definitions of taxa above the level of genus, illustrations, *etc.*) that can already be found in *D:TE* and *S1*.

The content of this volume must be restricted to information published in the paleontological literature prior to the book's "going to press." This means that any material — including new genera and species, revised classifications, and so forth — not available to this writer before the editorial cut-off date imposed by the publisher will not be included. Any such data must, therefore, be saved for the intended next supplementary volume. As a book of this size, nature, and complexity can (and often does) require a year or more of editorial and production work, a certain amount of this excluded new information will, no doubt, make it into the popular press before it appears in this series.

Such delays can be unbearably frustrating for an author such as myself trying to keep the work in progress as current and correct as possible. Regarding the first supplement, imagine the disappointment in not being able to include so spectacular a discovery as *Suchomimus* (see entry, this volume), because its name and formal description were published in the paleontological literature less than three months beyond the point of "no more changes!"

For the record, then, all of the author's work on the manuscript to this volume ended on December 31, 2000.

As before, life restorations of dinosaurs featured in the "genera" section of this book are included *only* if— in my somewhat biased opinion — they represent taxa known from fossil remains sufficient to produce a reasonably accurate illustration.

Continuing the policy established in the earlier two books, out-dated life restorations — many of them highly conjectural, based upon very incomplete fossil material — are featured only when they are of particular historical, aesthetic, or other interest. Illustrations of this nature appear only in the earlier chapters in this book and in the addendum on dinosaur-related displays; none of these appear in the section on dinosaurian genera. New life restorations of dinosaurs appearing herein have been graciously supplied by artists whom I regard as among the best of a community of "paleoartists," who combine talent and craftsmanship with truthful integrity and scientific accuracy (see Acknowledgments).

DONALD F. GLUT
Burbank, California USA
December 31, 2000

I: Introduction

We have already left the twentieth century to enter not only a new century, but also a new millennium and the dinosaurs—that very special group of extinct terrestrial animals that dominated this world for more than 150 million years—continue to remain a worthy source of strong interest and fascination.

Once considered a "dead" topic (this once being the opinion of many vertebrate paleontologists, as well as other scientists, educators, and lay people), dinosaurs remain today alive in our sciences and imaginations. Indeed, our knowledge of dinosaurs continues to grow and evolve beyond the year 2000, as more fossil discoveries are made, leading to new interpretations of and perspectives on those erstwhile rulers of the Earth.

Dinosaurs have been popular with the public at least since 1852, when the first attempted life-sized reconstructions of some of them, made by sculptor Benjamin Waterhouse Hawkins under the supervision of Richard Owen, were exhibited at the Crystal Palace Grounds in Sydenham, London (McCarthy and Gilbert 1994; see also *D:TE* and *S1*). Yet it appears that now, more than ever before in human history, the public is interested in dinosaurs—not merely the popular conceptions of these animals, as presented over the decades in such media as fiction and motion pictures, but more importantly, their scientific reality. Privately published magazines (sometimes called "fanzines"), such as *Prehistoric Times* and *Dinosaur World*, bring a considerable amount of paleontological information to an audience comprised of numerous dinosaur enthusiasts who may never read a scientific paper.

Hadrosaurus foulkii (right), the first dinosaur skeleton (reconstructed in plaster, including cast bones of holotype ANSP 10005, with a skull patterned after that of an iguana lizard) to be restored and mounted, in both an earlier century and millennium, exhibited in 1894 at the Field Columbian Museum in Chicago. It is looking over the skeletons (casts from the _Catalogue of Casts of Fossils from the Principal Museums of Europe and America with Short Descriptions and Illustrations_ (1866), by Henry A. Ward of the mammals _Unitatherium_ and _Megaloceras_, among other items in the museum's collections.

Courtesy The Field Museum (neg. #CSGEO2977).

Courtesy and copyright © *The Illustrated London News.*

An example of the popularity of dinosaurs during the 1950s, this crowded and rather speculative "artificial grouping" of life in China during the ?Early Jurassic to Late Cretaceous, was created by British artist Neave Parker for the February 11, 1956 edition of *The Illustrated London News*. However, much of what is depicted in the scene sprang from the imaginations of Parker and scientific advisor Dr. W.E. Swinton rather than the sometimes very incomplete fossil evidence. From left to right: The small theropod *Sinocoelurus* (foreground), hadrosaur *Tanius*, ankylosaur *Heishanosaurus*, stegosaur *Chialingosaurus*, sauropod *Omeisaurus*, a pholidosaurid crocodile, the theropod *Szechuanosaurus*, and dinosaur of uncertain classification *Sanpasaurus* (here depicted as an ornithopod).

Beyond the fan publications, the Internet has brought these animals from our world's past into the almost futuristic world of home computers. (Once a rare tool in paleontology, computer technology now allows the dinosaur researcher "to collect and analyze data better, more easily, and more creatively than previously was possible"; see Chapman 1997 for a survey report on the use of modern technology in regards the study of dinosaurs.) Numerous "websites," "chat rooms," "bulletin boards," and other cyberspace havens allow anyone interested in dinosaurs on any level — whether scientist or layman — to post or learn new information, discuss issues, argue points and topics, buy and sell dinosaur-related items, or just get to know one another. But beware — there are hundreds of dinosaur related websites, most of them unreviewed by professional dinosaur paleontologists, many of these containing misleading, incomplete, and biased information that can mislead the layman.

Funding for dinosaur research continues to be acquired from previously untapped sources. The Jurassic Foundation, a new nonprofit society dedicated to fund international dinosaur research (in some ways, a "successor" to The Dinosaur Society [see *S1*], which unfortunately suffered its own extinction during the later 1990s), was officially launched on October 3, 1998 in connection with the Fifty-Eighth Annual Meeting of the Society of Vertebrate Paleontology (SVP) in Snowbird, Utah. This new organization was begun by Don Lessem (1999), one of the founders of The Dinosaur Society. Originally, it was financed primarily via revenues derived from "The Dinosaurs of Jurassic Park — The Lost World," a popular and quite successful traveling exhibit licensed by Universal Studios and Amblin Dreamworks Entertainment based on their 1966 blockbuster movie, *The Lost World: Jurassic Park.*

Never Ending Changes

Reflecting upon the previous two volumes, the present writer is struck by how much is now different.

Many new genera or species have been introduced over recent years. Some genera (*e.g.*, *Lisboasaurus*; see "Nomen Nudum and Excluded Genera" chapter), formerly thought to be dinosaurian, are no longer regarded as such; others (*e.g.*, *Nuthetes* and *Patagonykus*), formerly classified as nondinosaurian, are now, at least for the present, included within the Dinosauria (see *D:TE*).

Photograph by John Weinstein, courtesy The Field Museum (neg. #GEO85868_1).

The "Life Over Time" exhibit at The Field Museum, dominated by (in foreground) the skeleton (FMNH PR308; originally identified as *Gorgosaurus libratus*) of the theropod dinosaur *Daspletosaurus torosus*, posed above the hadrosaurid *Lambeosaurus lambei* (post-crania FMNH PR380; skull FMNH UC1479), and (in background) the composite skeleton (FMNH PR25112 and P27021, skull cast of CM 11162) of the giant sauropod *Apatosaurus excelsus*.

The classification of the many different kinds of dinosaurs (*e.g.*, *Lophorothon*) and the relationships between the various dinosaurian groups (*e.g.*, those within Sauropoda)—once (before the widespread use of cladistics) thought of as fairly well understood and stable, although often based upon superficial features—continue to change, sometimes quite dramatically.

Indeed, this new century and millennium promise to be truly exciting for those of us for which the many varied kinds of dinosaurs are more than a mere passive interest.

The Mesozoic Era

The span of geologic time during which the dinosaurs ruled the world is called the Mesozoic Era, currently dated by scientists as approximately 248 to 65 million years ago (although such dates have been known to change, based upon new evidence).

This long expanse of time has been subdivided into three "periods": the Triassic (245 to 208 million years ago; the earliest dinosaurs are known from the latter part of this period, reports of earlier "dinosaurs," mostly based on footprints, almost universally regarded as specious); the Jurassic (208 to 145.6 million years ago); and the Cretaceous (145.6 to 65 million years ago, after which—except for birds—all dinosaurs were extinct).

Each Mesozoic period is further divided into various chronostratigraphic "stages," designated by such (always capitalized) adjectives as "Early" and "Late" (as opposed to lithostratigraphic levels designated by [always uncapitalized] terms such as "lower" and

Photograph by Sanford Maudlin Photography, courtesy Oklahoma Museum of Natural History, University of Oklahoma.

Mounted skeleton (cast) of *Allosaurus maximus* (originally referred to the new genus *Saurophaganax*) as mounted for display in the recently opened "Ancient Life" exhibit at the Sam Noble Oklahoma Museum of Natural History.

Courtesy North Carolina Museum of Natural Sciences.

The North Carolina Museum of Natural Sciences is another new institution (opened April 7–8, 2000) featuring dinosaur exhibits. In the foreground is the skeleton of the giant theropod *Acrocanthosaurus atokensis* and in the background a life-sized model of the sauropod *Pleurocoelus nanus*. Replicas of the huge pterosaur *Quetzalcoatlus northropi* hang overhead. All three taxa are known from Texas.

One of a series of watercolor paintings by John Pemberton (Jack) Cowan depicting—quite inaccurately and in a highly romanticized style—various geological periods for the Hughes Tool Company, first used in advertisements (1948–49), then as give-away items. This Late Triassic scene portrays the prosauropod *Plateosaurus engelhardti*.

Popular art often incorrectly portrays dinosaurs from different time periods and places coexisting, contributing to the misconception that all extinct animals lived together in some vague "prehistoric world." This lively but fanciful illustration by John Pemberton (Jack) Cowan for the Hughes Tool Company depicts in very inaccurate life restorations (left) *Polacanthus* (Early Cretaceous, Europe) fighting (right) *Antrodemus* (Late Jurassic, North America).

Inaccurate romanticized and idyllic restoration of a Late Cretaceous scene dominated by the horned dinosaur *Triceratops*, painted by John Pemberton (Jack) Cowan for the Hughes Tool Company. In background are pterosaurs of the genus *Pteranodon*.

"upper"). Every stage is then subdivided into various "ages," such as "Hettangian" and "Campanian," which can also be modified by such (noncapitalized) adjectives as "early" and "late." (The Mesozoic Era and its divisions are discussed in *D:TE*.)

The following is a breakdown of the Mesozoic Era and its three periods, beginning with the oldest (Triassic). As is traditional, each period — and then the stages and ages contained within — are arranged in descending order, beginning with the most recent (Late Cretaceous; Maastrichtian). (The following datings are from the "1999 Geologic Time Scale" issued by the Geological Society of America, compiled by A. R. Palmer and John Geissman, based upon a number of recently published sources.)

Note that a new term, "Middle Cretaceous," is used in this book, but was not "officially" used in the previous supplement. Traditionally, the Cretaceous period has been formally subdivided only into "Early" and "Late" stages. In recent years, however, various workers (*e.g.,* Russell 1996) have begun to use the term "Middle Cretaceous" to include both Albian and Cenomanian ages.

In 1998, Rathkevich, while describing the new sauropod *Sonorosaurus* (see *S1*), argued that — based upon recent studies utilizing stratigraphic, radioscopic, and fossil pollen analysis — the "Middle Cretaceous" appears to be a distinct and unique transitional period in dinosaurian history. About the same time, other workers (*e.g.,* Kirkland 1998; Lucas, Kirkland and Estep 1998) began to accept and utilize this term as more than an unofficial designation.

Middle Cretaceous, therefore, is now employed as an official term in the text including the Albian and Cenomanian.

Cretaceous

LATE
MAASTRICHTIAN (71.3 to 65 million years ago)
CAMPANIAN (83.5 to 71.3 million years ago)
SANTONIAN (85.8 to 83.5 million years ago)
CONIACIAN (89 to 85.8 million years ago)
TURONIAN (93.5 to 89 million years ago)
CENOMANIAN (99 to 93 million years ago)

MIDDLE
CENOMANIAN (99 to 99.5 million years ago)
ALBIAN (121 to 99 million years ago)

EARLY
APTIAN (121 to 112 million years ago)
BARREMIAN (127 to 121 million years ago)
HAUTERIVIAN (132 to 127 million years ago)
VALANGINIAN (137 to 132 million years ago)
BERRIASIAN (144 to 137 million years ago)

Jurassic

LATE
TITHONIAN/PORTLADIAN (151 to 144 million years ago)
KIMMERIDGIAN (154 to 151 million years ago)
OXFORDIAN (159 to 154 million years ago)

MIDDLE
CALLOVIAN (164 to 159 million years ago)
BATHONIAN (169 to 164 million years ago)
BAJOCIAN (176 to 169 million years ago)
AALENIAN (180 to 176 million years ago)

EARLY
TOARCIAN (190 to 180 million years ago)
PLIENSBACHIAN (195 to 190 million years ago)
SINEMURIAN (202 to 195 million years ago)
HETTANGIAN (206 to 202 million years ago)

Triassic

LATE
"RHAETIAN" (210 to 206 million years ago)
NORIAN (221 to 210 million years ago)
CARNIAN (227 to 221 million years ago)

MIDDLE
LADINIAN (234 to 227 million years ago)
ANISIAN (242 to 234 million years ago)

EARLY
OLENEKIAN (245 to 242 million years ago)
INDUAN (248 to 245 million years ago)

New Discoveries, Ideas and Studies

Dinosaurs are mostly known from their bones, preserved over many millions of years as fossils (a word literally meaning "dug up"). The very nature of the preservation of dinosaur bone was reinvestigated by Carpenter (1998*a*). Fossil bones are preserved through a process called "permineralization." In the past, permineralization has usually been explained in terms of chemical and physical processes (*i.e.,* minerals replacing, over a long period of time, the actual bone material, thereby producing an exact "turned to stone" replacement replica). Microbes supposedly played only an accessory role in this process.

However, based on Carpenter's more recent experimental work, bacteria seems to play a much greater role in the precipitate minerals than was previously suspected, with these minerals precipitating during, or resulting from, bacterial metabolism, while the minerals produced depend upon the species of bacteria, and the kinds and abundance of ions. According to Carpenter, "The ubiquitous presence of bacteria in a decomposing carcass (dinosaur or otherwise) means that the potential agents for permineralization are present at burial." Following the decomposition of the soft tissue of the carcass, bacterial biofilms precipitate minerals throughout various canals and other spaces within the bones, with precipitation then sealing in the organic molecules, and possibly also explaining the presence of amino acids that have been reported in some dinosaur bones. As ions for bacteria must be replenished by flowing groundwater, this could also explain the very poor fossilization of dinosaurs in formations representing sand dunes. "This bacterial model for permineralization," Carpenter stated, "eliminates the need to somehow explain supersaturization of minerals in groundwater and furthermore is testable in the laboratory."

In an abstract, Carrano (1999*a*) offered a preliminary report on the phylogenetic, temporal, and scaling perspectives relating to body-size evolution in dinosaurs. As this author pointed out, body size has fundamental effects upon the design of vertebrates in general, these ranging "from imposing structural or physiological limits to relieving functional or ecological constraints." Although trends (*e.g.*, "Cope's Rule," which postulates that animals tend to become larger during their evolution) in body-size evolution have long been commonly known in vertebrate groups, Carrano's study attempted — by utilizing phylogenetic, temporal, and scaling approaches — to analyze these trends in a quantitative way.

Carrano's brief report showed the following: 1. Within a phylogenetic framework, ancestral body sizes can be reconstructed using squared-change parsimony, with each taxon compared to its ancestral state, and trends "examined at several phylogenetic levels to isolate local changes or summarize changes across the tree"; 2. body sizes mapped onto a temporal framework can reveal "an explicitly chronological pattern that is compared to the sequential pattern derived from the phylogeny"; and 3. scaling analyses can "deduce the relationship between various morphological parameters and body size, providing a means to link the two in different clades or in the group as a whole."

In conclusion, Carrano's study supported the hypothesis that dinosaurs generally increased in size during their evolution, this phenomenon being apparent phylogenetically ("more and greater increases than decreases between ancestors and descendants") as well as temporally ("maximum mean body sizes increase during the Mesozoic"); also, scaling analyses showed that size-related changes can explain many trends in the dinosaur locomotor apparatus.

Carrano (2000) subsequently published an extensive study on the subject of homoplasy and how it relates to the evolution of dinosaurian locomotion. Combining in this study both qualitative and quantitative morphologic data with a systematic survey of dinosaurian taxa, the author found that a number of hindlimb and pelvic features, all of which have potential relevance for the locomotor function of the hindlimbs, evolved parallel in several lineages of the Dinosauria. These features, all of which evolved independently within the group several times, include: 1. expanded pre- and postacetabular processes of the ilia; 2. a medially directed femoral head; and 3. an elevated lesser trochanter (all three of these changes occurring in theropods and ornithischians, the first and second in sauropods). According to Carrano (2000), they "probably reflect enlargement of several hindlimb muscles as well as a general switch in their predominant function from abduction-adduction (characteristic of sprawling limb postures) to protraction-retraction (characteristic of parasagittal or erect limb posture)."

As noted by Carrano (2000), dinosaurs exhibit various modifications — some of them related to the reduced abduction and adduction associated with a mostly parasagittal hindlimb posture — indicating the increased specialization of a permanently upright hindlimb posture and a bipedal stance. Carrano (2000) found that the three above-mentioned features "all increase the effectiveness of protraction and retraction at the expense of abduction for several major pelvic muscles." Also occurring are enlargement and functional elaboration, this resulting in four distinct Mm. iliotrochanterici (muscles on the lateral ilium of extant birds).

According to Carrano (2000), subdivision of the primitive archosaurian pelvic musculature, including at least some of the Mm. iliotrochanterici, was already underway in nonavian theropods. "Avian" characters such as this, shared not only by more basal theropods but also, acquired convergently, by other dinosaurian lineages more distantly related to birds (*i.e.,* sauropods and ornithischians), suggested to the author that the evolution of the avian hindlimb "represents a cumulative acquisition of characters, many of which were quite far removed in time and function from the origin of flight."

The significance of this study, Carrano (2000) pointed out, is that parallelism offers an insight into the possible mechanical advantages of developing such features; also, it can shed light upon evolutionary transitions in extinct taxa by showing how repeated "experiments" on an ancestral condition have led to similar results. The constraints (whether mechanical, developmental, or other) creating this evolutionary pattern remain unknown, the author noted. Their mechanical effects, however, regardless of when or where they arose, would be quite similar.

Rensberger, Mielkey, Jensen, Lee and Moorthy (1999), in an abstract, briefly reported on their study of lamellar bone — "characterized by distinctively differentiated layers" — in both coelurosaurian and ornithischian dinosaurs. Among the authors' discoveries was that the fiber bundles and thickness and continuity of individual lamillae are more regular in birds than they are in mammals. Curiously, the structural organizations in coelurosaur osteons and endosteal lamellae resembled those in birds, while examined osteonal structure in ornithischian tissues were more similar to those in mammals. According to the authors, these differences may be associated with the mechanism of osteogenesis (see Rensberger *et al.* for details).

Several brief reports have been published as abstracts comparing growth and bone development in extant vertebrates with those of dinosaurs.

In an abstract for a poster, Banks, Fenton and Hayward (2000), noting that extant birds share numerous features (*e.g.*, similar eggshell structure, bones, cardiac morphologies, nesting ecologies, and feathers; see section on dinosaurs and birds, below) with various nonavian dinosaurs, presented results of a study of specimens of 79 wild glaucous-winged gull chicks and 15 adults. This study revealed that compact bone formation parallels the increasing use of bone, and also that development of the humerus lags behind that of the femur, this reflecting the differential ontogenies of wing versus leg behavior in that group of birds. These results, the authors noted, may be compared with published dinosaur material.

In an abstract for a poster, Curry, Rogers, Castanet, Cubo and Boisard (2000) provided the first look at bone growth dynamics in extant ostriches and emus, their results intended for comparison with the bones of nonavian dinosaurs. In young ratites, Curry Rogers *et al.* observed the following primary cortical bone growth: 1. (Hatchling to two months) cortical thickness remaining constant, thereby expressing equilibrium between periosteal bone deposition and endosteal bone resorption; 2. hindlimb diaphyseal cortical bone growing rapidly, from 10 to 80 millimeters per day; 3. wing bones smaller, developing later, primary cortical growth rates from 2 to 14 millimeters per day; 4. highest growth rate corresponding to densely vascularized primary bone of reticular or laminae tissue types, growth rates decreasing as vascular density of bone diminishes; and 5. comparative and quantitative histological sample among extant taxa being among the most precise means of delimiting growth rates and ages in nonavian dinosaurs.

In a preliminary report, Erickson and Curry Rogers (2000) utilized a new research method called Developmental Mass Extrapolation (DME) to reconstruct growth curves for various kinds of dinosaurs by coupling estimates of their mass and longevity during development. Erickson and Curry Rogers' "results suggest that dinosaurs exhibit s-shaped growth curves typical of most vertebrates and that maximal growth rates among small, medium, and giant taxa are comparable to those of extant marsupials."

One of the most diverse dinosaur faunas known from the Cretaceous of Africa, discovered in 1989 in Late Cretaceous (Cenomanian, dated by Werner 1994 based on fish fauna) terrestrial sediments of Sudan (see Buffetaut and Brinkman 1990), was described in detail by Rauhut (1999). Consisting mostly of silt- to sandstones, the vertebrate-bearing beds include two lateral localities — the Wadi Milk Formation in the Wadi Abu Hashim (where most of the dinosaurian remains were recovered) northwest of Khartoum, and the Shendi Formation northwest of the city of Shendi. The fauna, the first to be found in the Cretaceous of Sudan, comprises the following 10 to 11, mostly large dinosaurian taxa, the remains of which are all housed in the Institut für Paläontologie of the Free University, Berlin:

Carcharodontosaurid theropods represented by three vertebrae (Vb-607, Vb-717, and Vb-870); possible carcharodontosaurids by a single caudal vertebra (Vb-871), the proximal end of a metatarsal (Vb-849), and a phalanx (Vb-718); a velociraptorine dromaeosaurid probably representing a new genus and species, known from a tooth (Vb-875), one phalanx (Vb-713), and five unguals (Vb-714, Vb-860, and Vb-866 to Vb-868); a possibly new taxon of a ?coelurosaurian by the proximal portion of an ungual (Vb-839); indeterminate theropods by teeth (Vb-873, Vb-874a, and Vb-874b), two fused sacral vertebrae (Vb-852) very similar to those of a juvenile tyrannosaurid specimen (MOR 533E), and isolated unidentifiable limb elements; a possible small theropod, perhaps with troodontid affinities, by a single phalanx (Vb-861); also an isolated theropod phalanx.

Dicraeosaurid sauropods represented by three

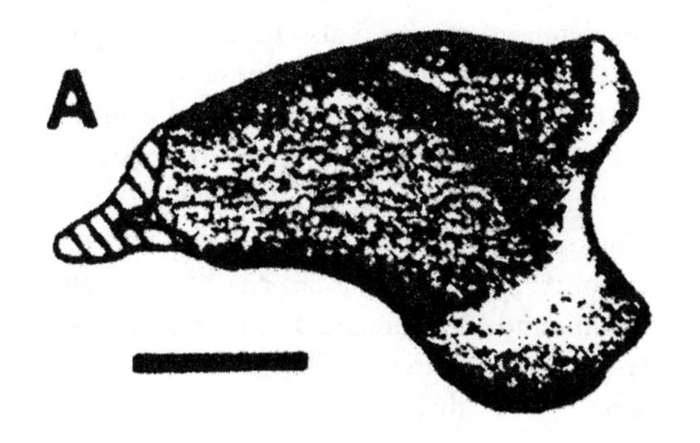

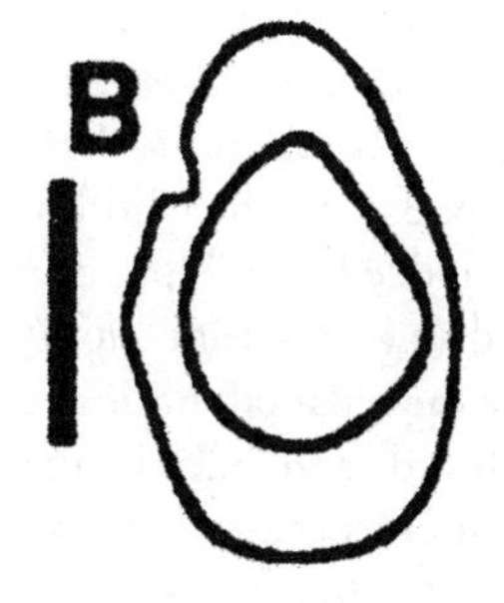

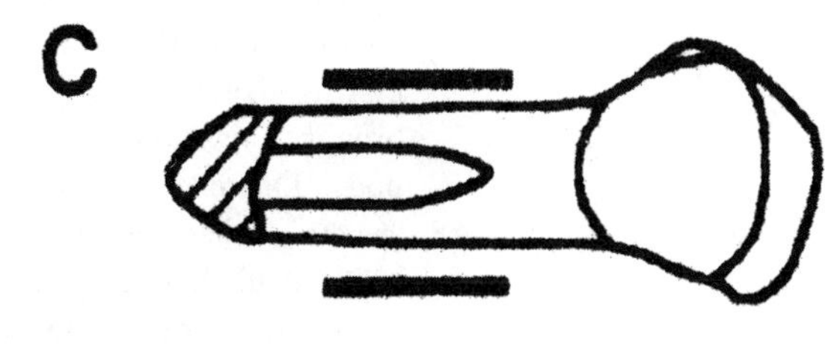

Manual ungual (Vb-839) from the Wadi Milk Formation of Sudan, possibly representing a new ?coelurosaurian theropod, in A. left lateral view, B. proximal outline, and C. ventral outline. Scale = 1 cm. (After Rauhut 1999).

caudal vertebral centra (Vb-856, Vb-857, and Vb-892), a proximal end of a left fibula (Vb-884), and the distal end of a right tibia (Vb-879); titanosaurids by 53 teeth (Vb-721, Vb-900 to Vb-912), one presacral vertebra (Vb-893), and 15 caudals (Vb-606, Vb-720, Vb-876, Vb-877, Vb-880, Vb-882, Vb-883, Vb-885, Vb-886, Vb-889, Vb-891, Vb-894, Vb-897, Vb-899, and Vb-935), the vertebrae, because of the great similarity of caudals among different titanosaurids, possibly representing more than one species; indeterminate sauropods by three vertebral centra (Vb-719, Vb-895, and Vb-896), and five limb bone fragments (Vb-845 to Vb-848, and Vb-878).

?Hypsilophodontid ornithopods represented by one tooth (Vb-936), perhaps belonging to a juvenile; ?iguanodontids, possibly *Ouranosaurus*, by two caudal vertebrae (Vb-723 and Vb-853); an indeterminate iguanodontid, referred to ?*Iguanodon* sp., by the distal end of a femur (Vb-850); and an indeterminate ornithopod closely resembling *Prosaurolophus* (personal observation of Rauhut) by a single caudal vertebra (Vb-854), similar vertebrae having been found in *Iguanodon bernissartensis*, possibly belonging to Vb-850.

Indeterminate dinosaurs represented by one tooth (Vb-722) resembling teeth in both therizinosauroids and ornithischians, a vertebra (Vb-855) possibly belonging to a sauropod with a general shape not unlike that of brachiosaurids (see Janensch 1950*a*), and the distal end of a femur (Vb-851) resembling that in theropods but exhibiting a steep angulation of the articular surface not known by Rauhut in any theropod; also some isolated phalanges.

Other specimens from the Sudan localities include various kinds of reptiles and amphibians, currently being studied by C. Werner, and numerous fossil plants (see Werner 1993, 1994; Werner and Rage 1994; Evans, Milner and Werner 1996).

Rauhut's analysis of these localities found similarities between the Wadi Milk Formation and the North American, Upper Jurassic Morrison Formation suggesting similar and rather uniform landscapes. The environment, Rauhut suggested, "was probably a semiarid savanna with some rivers, lined by dense vegetation, with abundant sauropods, less abundant theropods and rare ornithopods." Rauhut suggested that the top predators were the carcharodontosaurids, speculating that these very large theropods could have brought down both the sauropods and the ornithopods. The dromaeosaurids, if indeed they hunted in "packs" as has been suggested (*e.g.*, Ostrom 1969*b*, 1970), could have preyed upon the iguanodontids and smaller ornithopods, also on lizards and amphibians, but not the sauropods. The dicraeosaurids and titanosaurids, with their relatively short necks, may have preferred feeding on plants growing at moderate heights. The iguanodontids may have fed on low-growing plants (as suggested by Bakker 1978), thereby not competing for food with the sauropods (although their rarity in the Sudan localities could imply a more distant habitat). Finally, the hypsilophodontids probably fed on fruits and seeds, thereby not competing with the sauropods or larger ornithopods.

Also noting other dinosaur faunas (see Rauhut for examples and references), Rauhut concluded that such faunas from the Middle to Late Cretaceous of Africa seem to be characterized by the presence of carcharodontosaurids, spinosaurids, titanosaurids, dicraeosaurids, and possibly iguanodontids, most of these groups probably representing the biological inheritance from the Jurassic period.

Regarding other newly found dinosaurs, Arizona newspapers reported in August 2000 the discovery somewhere in Utah of a new dinosaur of undisclosed or unknown affinities apparently to be named "Hesperisaurus" [meaning "western lizard"]. No other pertinent information was given in these reports, however.

Carrier and Farmer (2000) published a study on the evolution of pelvic aspiration in Archosauria, that large group of vertebrates that includes, among other taxa, Dinosauria, Pterosauria, Crocodylia, and Aves. As detailed by Carrier and Farmer, participation of the pelvic girdle in the production of inspiration in vertebrates is rare. However, movements of the pelvic girdle are now known to contribute to inspiratory airflow in two groups of extant archosaurs, crocodilians (Farmer and Carrier 2000) and birds (Baumel, Wilson and Bergman 1990). This brought up the possibility that the pelvic musculoskeletal system may have been involved in the ventilation of basal archosaurs. Based upon this mechanism of pelvic aspiration in crocodilians and also in the structure of gastralia in basal archosaurs, Carrier and Farmer suggested that a parasagittally located ischiotruncus muscle, having the motor pattern seen in Recent alligators, pulled the medial aspect of the gastralia caudally, thereby aiding in the production of inspiration by increasing the volume of the abdominal cavity.

According to Carrier and Farmer, this pelvic musculoskeletal system could also have had a significant function in the ventilation of dinosaurs. As these authors pointed out, both saurischian and ornithischian dinosaurs, unlike basal archosaurs, possessed pelves distinguished thusly: 1. An ilium having a large preacetabular process; 2. reduced and dorsally relocated attachment sites for muscles basally associated with the pubis and ischium; and 3. exceptionally long and isolated pubes and ischia. These first two characters have been generally explained as a consequence of

Photograph by the author, courtesy The Field Museum.

Detail of reconstructed gastralia of the tyrannosaurid genus *Daspletosaurus torosus* (FMNH PR308, skeleton originally referred to *Gorgosaurus libratus*). According to David R. Carrier and Colleen G. Farmer (2000), such "belly ribs" may have played a role in the breathing of such theropods.

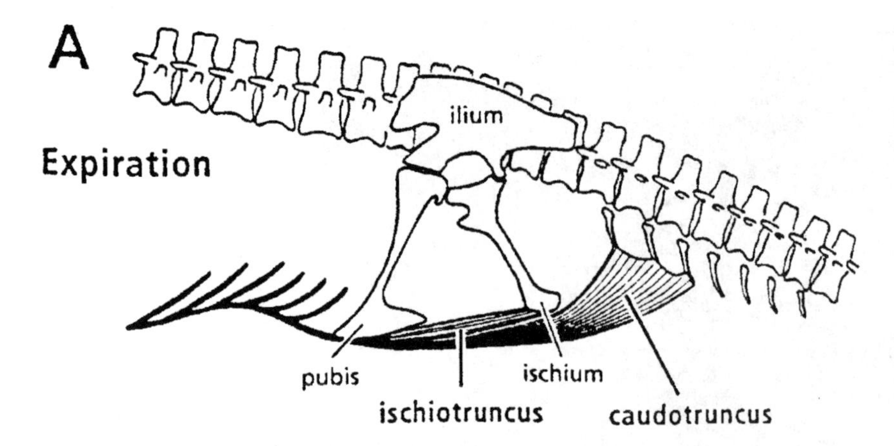

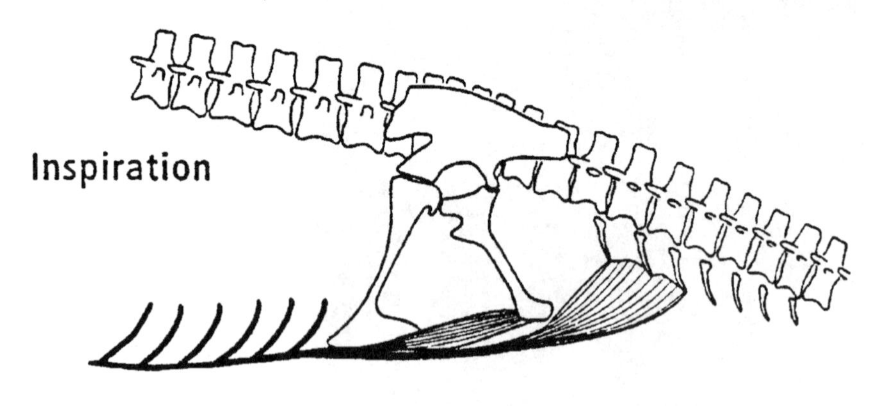

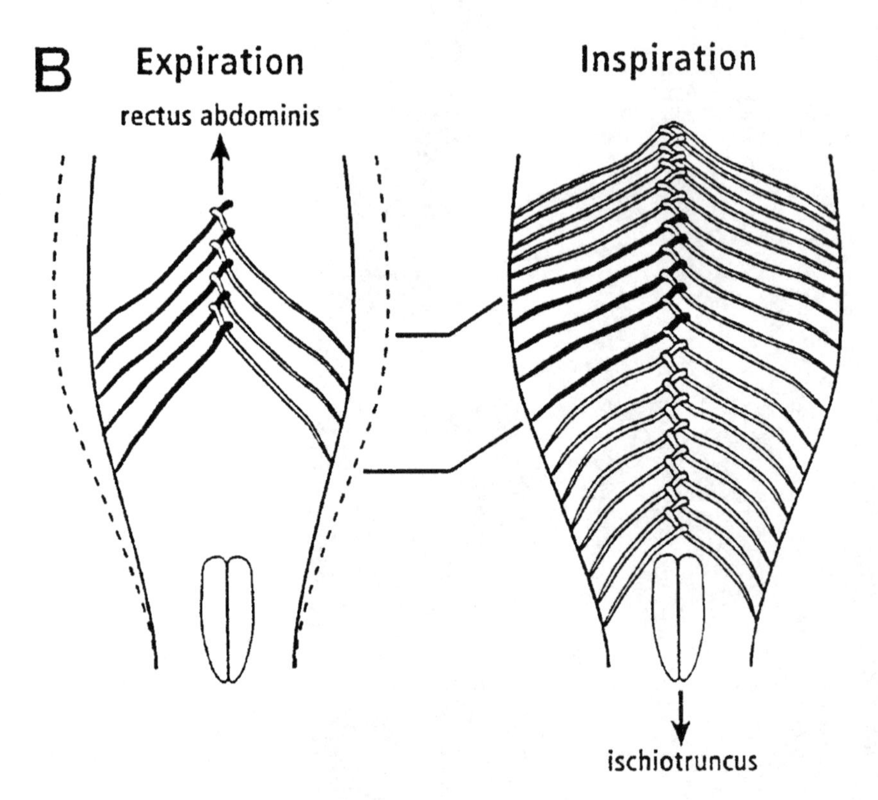

Proposed mechanism of cuirassal breathing in the theropod *Allosaurus* sp. (DNM 11541) utilizing both ischiotruncus and caudotruncus muscles, in A. lateral and B. ventral views. (After Carrier and Farmer 2000.)

the shift from a sprawling to more or less vertically oriented limb posture (*e.g.*, Romer 1923*a*, 1923*b*, 1927; Charig 1972; Walker 1977). Not yet satisfactorily explained, however, is why the pubes and ischia became so long, in some cases being only loosely attached to the ilia. Carrier and Farmer postulated that many of the modifications found in dinosaur pelves could be related to lung ventilation.

Regarding saurischians, Carrier and Farmer found the anatomy of nonavian theropods to be consistent with the mechanism of cuirassal breathing, which these authors proposed for basal archosaurs. Nonavian theropods possess well-developed, highly derived gastralia (see Claessens 1997) in the belly wall between the pubes and sternum. (Gastralia have also been reported in the basal bird *Archaeopteryx* [see Ostrom 1976*c*; Wellnhofer 1992] and also in other early avians.) In nonavian theropods, gastralia were no simple union of lateral elements at the ventral midline; rather, they crossed the midline and articulated with two gastralia from the opposite side of the body, this crossing arrangement, Carrier and Farmer noted, constituting a lever system which, through via hypaxial muscles, could have widened or narrowed the cuirassal basket (see Claessens).

As Carrier and Farmer noted, the elongate pubis terminated distally in cranial and caudal expansions called a "boot" or "foot," its ventral margin often being convex in shape, covered in cartilage, and aligned with the distal end of the elongate ischium. This alignment would have allowed an ischiotruncus muscle, originating at the distal end of the ischium, to cross the ventral surface of the "boot," reaching the gastralia, with the long pubis serving as a guide to orient the force of the ischiotruncus muscle on the gastralia.

According to Carrier and Farmer, if the ischiotruncus muscle drew the medial ends of the gastralia caudally, the lever system of the gastralia would have produced a lateral expansion of the belly. Also, the lattice-like articulation of these gastralia would have reduced the ventral expansion of cuirassal breathing that the authors proposed for basal archosaurs (L. Claessen, personal communication to Carrier and Farmer 1999), cuirassal breathing in theropod dinosaurs probably having "been a combination of mediolateral and dorsoventral expansion and contraction of the belly."

Carrier and Farmer further suggested "that the highly derived pubis and ischium of nonavian theropods are modifications of the basal archosaur condition that facilitated cuirassal breathing in animals with parasagittal limb posture."

In dromaeosaurs, the authors noted, the pelvis differs from that of other nonavian theropods in ways that would have influenced cuirassal breathing. The

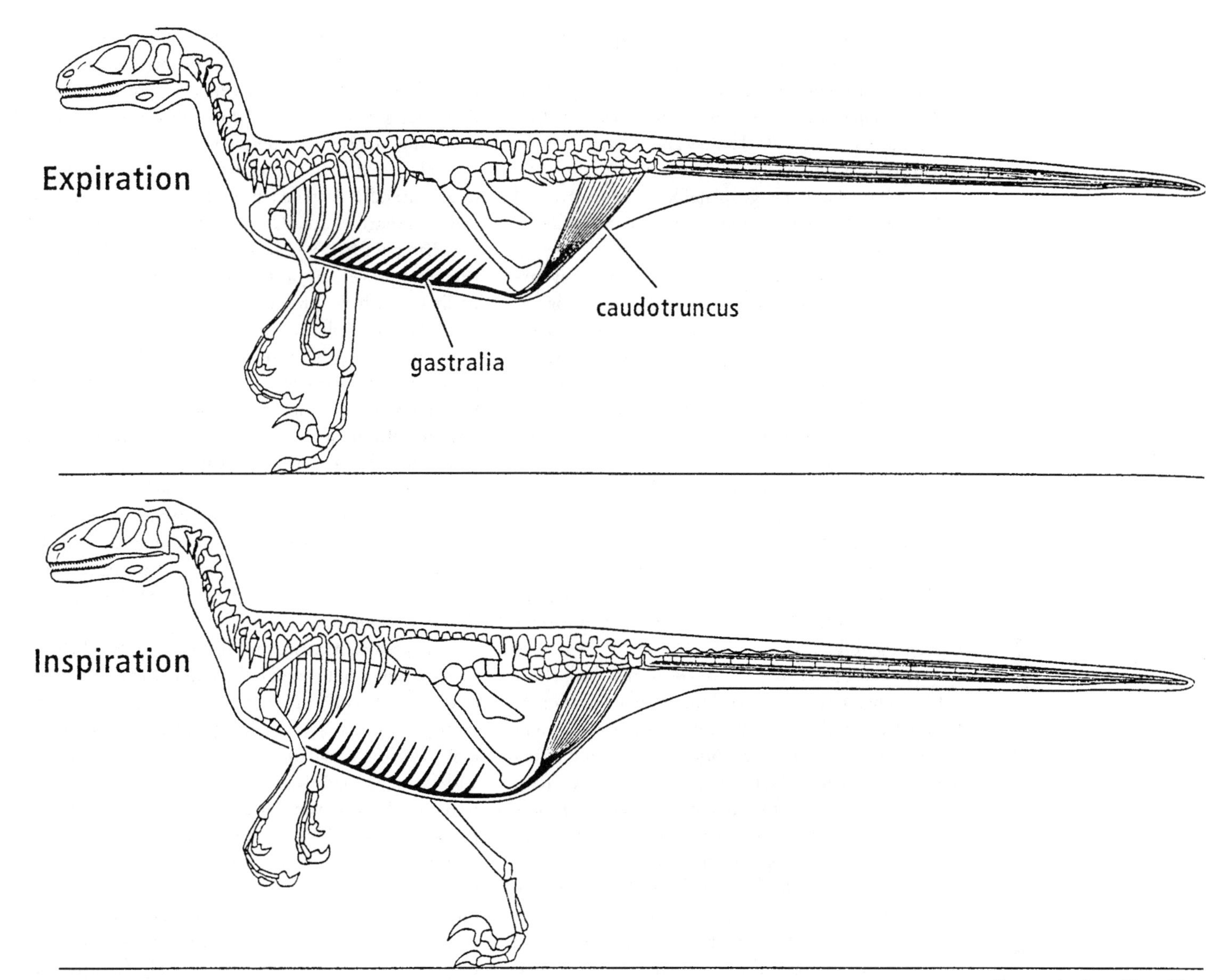

Proposed cuirassal breathing mechanism in the theropod *Deinonychus antirrhopus*, utilizing a caudotruncus muscle but not a ischiotruncus muscle. (After Carrier and Farmer 2000.)

pubes in these dinosaurs were directed caudally, not cranially, and the ischia were considerably shorter than the pubes (see Norell and Makovicky 1997). Although this arrangement does not allow for an ischiotruncus muscle, dromaeosaurs did retain a well-developed "foot" on the pubis and gastralia typical of nonavian theropods. These characters suggested to Carrier and Farmer that this group of theropods did used cuirassal breathing. This would have required, however, a caudotruncus muscle — present in Recent crocodilians although seemingly without being involved in ventilation (see Farmer and Carrier) — extending from the base of the tail to insert on the gastralia. According to Carrier and Farmer, if the tail was, in fact, a primary anchor for the muscles that expanded the belly during inspiration, simultaneous activity of the epaxial muscles would have been required, stabilizing the tail against forces exerted on the gastralia. Consequently, the early shift in Manuraptora "to retroverted pubes and relatively short ischia may have set the stage for the association of epaxial muscle activity with inspiration in Recent birds by requiring the tail to be stabilized against the ventrally directed pull of the caudotruncus muscle."

Concerning ornithischians, which lack gastralia, Carrier and Farmer noted that the pelvic structure of some of these dinosaurs seem to have been permissive of pubic and ischial kinesis. The pubis and ischium are relatively delicate elements that were loosely articulated to each other and the ilium, seemingly having played little or no role in supporting the femur at the hip joint (see Charig). This suggested to the authors that the pubis and ischium played a reduced role in generating locomotor forces, and yet these elements, rather than disappearing, acquired elaboration.

Also, a large broad and paddle-shaped anterior prepubis process evolved independently three times in the Ornithischia. It was small in primitive forms such as *Lesothosaurus* (Weishampel and Witmer 1990*a*), in basal ornithopods (Weishampel and Witmer 1990*b*), and in basal Ceratopsia (Dodson and Currie 1990; Sereno 1990). Thus, a large, platelike prepubis evolved independently in the Ceratopsia, Stegosauria, and Ornithopoda.

Carrier and Farmer noted that, in many ornithopods, the ischia were as long as, sometimes longer, than the femora and were directed almost horizontally under the tail. In ceratopsids, however, the ischia were directed more vertically and were dramatically decurved (Dodson and Currie), apparently functioning to enclose and support the posterior abdominal contents. If not a function for locomotion, then, Carrier and Farmer hypothesized that the pubis and ischium — loosely jointed and having only limited involvement in the hip joint — of some ornithischians may have been specialized for lung ventilation.

Admitting that their ideas were highly speculative and probably difficult to test, Carrier and Farmer offered the following:

In neoceratopsians like *Triceratops*, all three pelvic joints may have been kinetic, with motion at the iliopubic and ilioischial joints mostly rotational in the parasagittal plane, but motion at the punioischial joint transitional in the parasagittal plane. If correct, contraction of the rectus abdominis muscle — recognized as an expiratory muscle in crocodilians (Farmer and Carrier), mammals (De Troyer and Loring 1986), and birds (Fedde, DeWet and Kitchell 1969; Fedde 1976) — would have caused a cranial rotation of the ischium, this dorsally levering the prepubic blade of the pubis. If shortening of this muscle anteriorly rotated the ischium in ceratopsians, it would have pushed the abdominal viscera cranially into the thorax, thereby reducing the volume of the pleuroabdominal cavity. Simultaneously, dorsal rotation of the prepube would have allowed the lateral abdominal body wall to collapse medially, while caudal rotation of the ischium by the ischiocaudal or an ischioilium muscle would have resulted in a ventral rotation of the prepube that probably would have caused a lateral expansion of the abdominal wall to produce inspiratory flow. The combination of both rotations, therefore, would have increased the volume of the body cavity.

In basal ornithischians such as *Lesothosaurus*, the pubis and ischium are approximately of equal length, lying parallel and in close association with each other (Weishampel and Witmer 1990*a*). Also, the pelvic joints were so structured that the margin between these two elements (*i.e.*, the ischiopubic interosseous margin) was somewhat colinear with the iliopubic and ilioischial joints. This condition allowed for a long-axis rotation of the pubis along its long contact with the ischium, Carrier and Farmer finding it unlikely that a long-axis rotation of the pubis in such primitive dinosaurs served any function.

However, in both Stegosauria and Ornithopoda, large, laterally flattened prepubic processes evolved independently (see Galton 1990; Norman and Weishampel 1990; Weishampel and Horner 1990), the prepubic process in both groups being directed cranially beneath the preacetabular process of the ilium, so that the prepube was oriented at an acute angle to the long-axis of the pubis. Carrier and Farmer suggested that, if the pubis was rotated along its long-axis, the prepubis was swing to the side, thereby expanding the volume of the abdomen. Thus, the prepubic plates in these two groups of dinosaurs served as rotating flaps that laterally expanded the belly wall. The authors speculated that an as yet unknown muscle — one originating on the supra-acetabular crest of the ilium and inserted on the cranial margin of the prepubic process — could have produced the lateral rotation of the prepubes.

Therefore, Carrier and Farmer suggested that a mechanism of pubic and ischial kinesis evolved in derived ceratopsians, and that the prepubic plates of ornithopods and stegosaurs independently evolved lateral-rotation mechanisms. In all three cases, the authors suggested, the proposed pelvic kinesis would have aided costal ventilation in pumping air in and out of the lungs. This also implies that the above three examples of convergent or parallel ornithischian evolution also suggests a strong selection for pelvic aspiration. Comparing these ornithischians with extant crocodilians and birds, Carrier and Farmer imagined two plesiomorphic conditions strongly favoring the evolution of pelvic aspiration: If basal ornithischian dinosaurs possessed a diaphragmatic muscle similar to that of today's crocodiles, then a pelvic mechanism to expand the posterior abdomen (as has been suggested for crocodilians) would have aided in the elimination of restrictions on the return of blood. Moreover, if basal ornithischians possessed air sacs of lung diverticula extending into the lumbar region, the pelvic mechanism proposed by Carrier and Farmer "could have provided a more effective means of ventilating these lung diverticula than the more cranially located thoracic ribs" (see also the study of Claessens, Perry and Currie 1998 regarding the involvement of gastralia in theropod respiration, below).

Sereno (1999*c*) published a review on the evolution of dinosaurs, providing a new cladogram for the Dinosauria, and also proposing an expansive survey-type scenario suggesting trends occurring during the rise and fall of this entire group. Following are some of the evolutionary patterns that resulted from Sereno's (1999*c*) study:

Dinosaurs, having descended from a single common ancestor, began relatively rapidly during the Late Triassic. The earliest known dinosaurs (*e.g.*, *Eoraptor*), measured only about one meter in length. Rather than outcompeting their nondinosaurian contemporaries (*e.g.*, basal archosaurs, nonmammalian synapsids, and rhynchosaurs; see, for example, Charig 1984), these early dinosaurs seem to have taken advantage of ecological niches vacated after these other groups of animals became extinct.

With the exception of the small-bodied coelurosaurs that evolved into birds, dinosaurian radiation was both "slower in tempo and more restricted in adaptive scope than that of the therian mammals" (which, taking advantage of future vacated niches, would replace the last dinosaurs at the end of the Cretaceous period). Apart from the evolution of birds, which involved a significant decrease in body size (as well as the appearance of feathers and modifications of the skeleton for perching and flight), trends recurring in all major dinosaurian clades include an increase in body size.

Contrasted with trends seen in Cenozoic mammals, the radiation of nonavian dinosaurs was slow and constrained. Adaptive designs seen among mammalian groups—*e.g.*, gliders, burrowers, saltators, and taxa adapted to specific habitats—never evolved in nonavian dinosaurs. Among the observed phylogenetic trends recurring in dinosaurs were "incorporation of osteoderms in the skull, narial enlargement and retraction, reduction and loss of teeth, increase in neck length and number of cervicals, increase in the number of sacrals, miniaturization of the forelimb, reduction and loss of external digits in the manus, and posterior rotation of the pubis."

Comparisons of limb proportions in dinosaurian herbivores and contemporary predators suggested that, as in mammalian ungulates and their predators, pursuit predation did not significantly affect the evolution of locomotor capabilities (Janis and Wilhelm 1993; Carrano 1999*b*), large dinosaurian herbivores most often being graviportal regardless of the locomotor abilities of their carnivorous pursuers. Nor does study of the teeth of these plant eating dinosaurs during the Late Cretaceous radiation of flowering plants reveal any precise co-evolutionary pattern (Sereno 1997).

Also, dinosaurian faunas remained relatively uniform from the Late Triassic through the Jurassic, giving way to highly differentiated Cretaceous faunas. This faunal differentiation was governed mainly by three processes—vicariance (*i.e.*, "the splitting of lineages in response to geographic partitioning") and regional extinction (*i.e.*, "the disappearance from one or more geographic regions of a taxon whose former

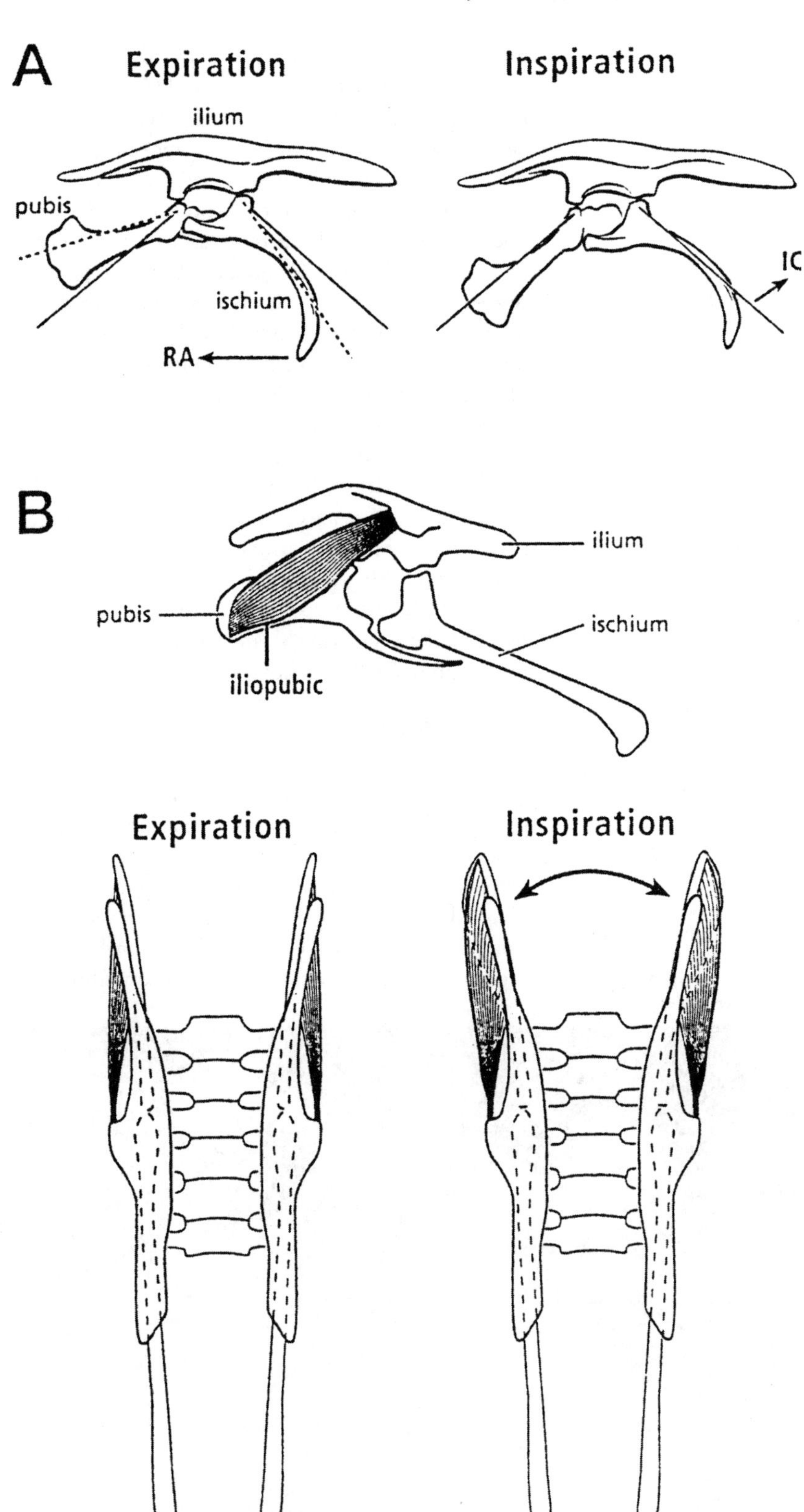

Proposed mechanisms of pubic and ischial kinesis in derived ornithischians, A. the pelvis of the ceratopsian *Triceratops*, utilizing an ischiocandalis muscle, and B. the pelvis of the hadrosaur *Corythosaurus* (illustrating the proposed mechanism of pubic kinesis in ornithopods and stegosaurs), utilizing an iliopubic muscle. (After Carrier and Farmer 2000.)

presence is clearly demonstrated by fossils"), which enhance differentiation, and dispersal (*i.e.*, "the crossing of geographic barriers"), which reduces it (see Sereno 1997). As the land masses moved apart, dinosaurian biogeography was shaped more by regional extinction (*e.g.*, "ceratosauroid" and allosauroid theropods, present on both northern and southern continents during the Jurassic and Early Cretaceous, being replaced during the Late Cretaceous in North America and Asia by large-bodied coelurosaurs [tyrannosauroids]), and intercontinental dispersal (*e.g.*, the phylogenetic patterns of spinosaurids and hadrosaurids [see Brett-Surman 1979] during the Cretaceous) than by the breakup of Pangaea.

In summation, Sereno (1999*c*) predicted that "Future discoveries are certain to yield an increasingly precise view of the history of dinosaurs and the major factors influencing their evolution."

Zhao (2000), a specialist in dinosaur eggs, recognized the egg-laying strategies of theropods and hypsilophodonts based upon the elongate fossil eggs attributable to these dinosaurian groups (see Zhao's paper for full details).

Theropods seem to display two egg-laying strategies. In the first strategy (oviraptorids), the female presumably constructed a sand mound. Then she probably laid her eggs on the slope of the mound in a concentric distribution pattern, burying them for incubation. According to Moratalla and Powell (1994), this pattern may have been an adaptation for utilizing the minimal amount of nest space without the eggs touching one another, or that the temperature varied from circle to circle, allowing for the possibility of temperature-dependent sex control.

In the second strategy, eggs laid by giant theropods — preserved with their long axis inclined slightly, arranged in the nest in an annular or arc pattern — seem to have been initially buried in sand at a certain angle of inclination (as with other elongathoiolid eggs). During fossilization, each egg would subsequently be flattened obliquely to its long axis by the pressure of the overlying sediment.

North American hypsilophodonts (*e.g.*, *Orodromeus*), Zhao noted, did not build special places to deposit their eggs, but laid them from a squatting position, forcing the eggs down vertically into the muddy soil of a lake bed (see Horner and Gorman 1988). Chinese Cretaceous hypsilophodontids, on the contrary, did construct nests in semiarid climates, possibly first loosening the sediments to make the nest, then squatting down to lay their eggs. The Chinese nests contain from six to seven eggs (of the ichnogenus *Prismatoolithus*) per nest; Horner (1984), however, reported clutches from Montana containing as many as 24 eggs per nest. Therefore, Zhao speculated that the Chinese hypsilophodonts, which generally laid their eggs in riverine sands, constructed nests more than one meter in diameter, then laid their eggs into the sand in a vertical orientation. This suggested, the author noted, that the female had produced 24 eggs which she then separated into several clutches. Furthermore, this suggested that hypsilophodonts

Skeleton (cast) of an *Oviraptor* posed "brooding" over its nest of eggs, displayed at Dinofest, held from December 8, 2000 through January 8, 2001, Navy Pier, Chicago.

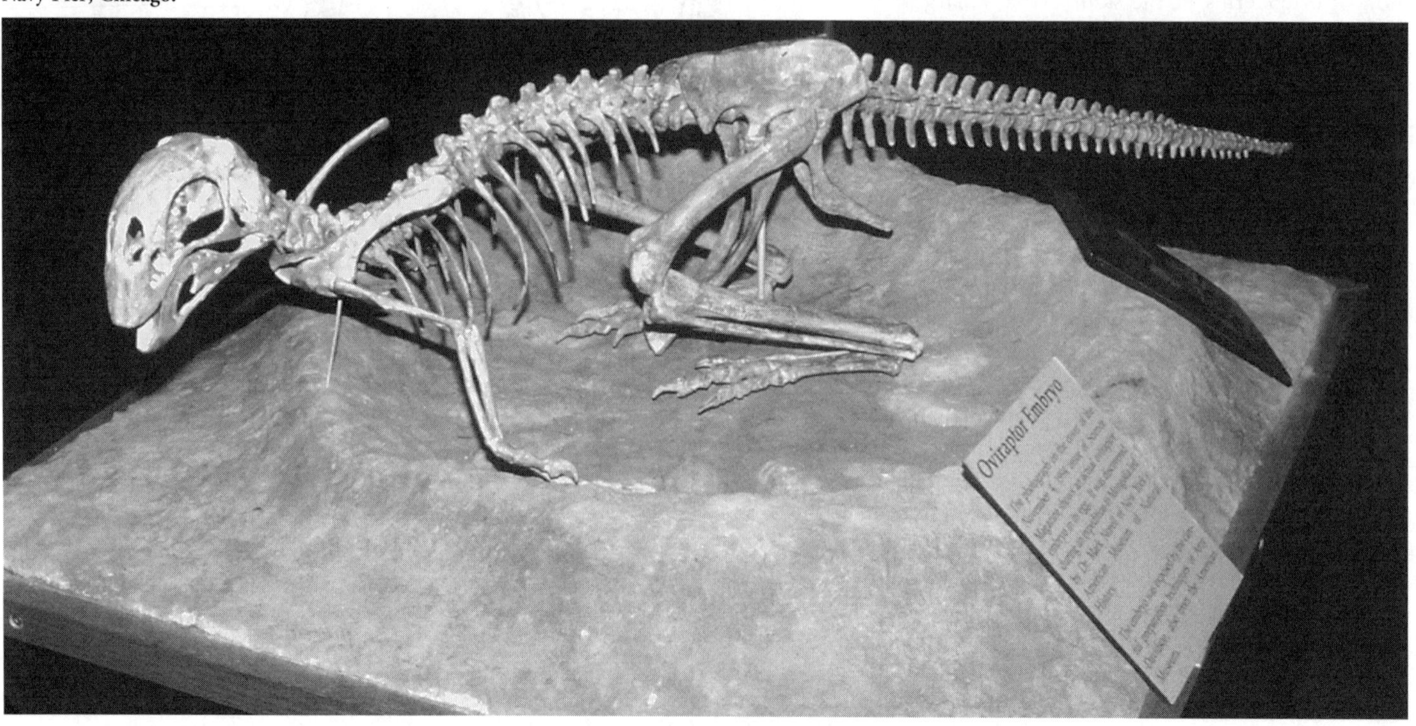

Photograph by the author.

were more intelligent than previously suspected, depositing their eggs vertically into the sediments at nesting, and also readjusting their nesting behavior as affected by different geological environments.

Evidence has been presented suggesting that at least some kinds of dinosaurs may have laid their eggs in damp environs rather than on dry land. López-Martinez, Moratalla and Sanz (2000) reported on a clutch of seven elongate (possibly hadrosaurian or titanosaurian) dinosaur eggs (of the ichnospecies *Megaloolithus siruguei*) found atop gray marls (comprising limestone, clay, and sand), presumably in its original position, in the Tremp Formation of Biscarri, Spain, these rocks representing Late Cretaceous–age tidal flats. Root casts and other plant material, plus fossils of fishes, mollusks, and crustaceans were found surrounding the marl and atop the egg clutch. Also found were iron-encrusted root traces. The authors noted that all of this evidence suggests that the eggs were laid in "water-saturated sediments in a reducing coastal environment."

According to López-Martinez *et al.*, five of the eggs were laid in a closely-spaced position and vertically oriented, the nest seemingly having been established on or near plant material. Unlike the nests of dinosaurs established on dry land, no pit or hole was established with this nest. The eggshells, measuring in thickness from 2.3 to 3.2 millimeters, display a unique design of pore canals interlacing in straight and horizontal patterns. This combination, unknown in any other dinosaur eggshell, allowed for greater shell permeability. The latter, the authors pointed out, would have been ideal for allowing the moist marsh air to hydrate the inside of the egg.

Saurischians

THEROPODS

Among the most primitive known saurischians, or "lizard-hipped dinosaurs," are the herrerasaurids, a group that is generally classified with the nonavian theropods (or simply, theropods)— mostly bipedal and carnivorous dinosaurs. In a brief report, Kischlat and Barberena (1999) presented new data on Triassic Brazilian dinosaurs known from well-preserved material recently collected from the Botucarai outcrop in Candelária City, State of the Río Grande do Sul. Among this new material (which includes two yet unstudied ?prosauropod taxa) was a "special pubis" articulated with an ischium and two vertebrae, a dorsal and a sacral. As these elements have compatible size, color, and preservation, and originate from the same region of the body, the authors considered them to belong to a single individual. According to Kischlat and Barberena, the pubis has a pronounced ambiens origin, as in *Herrerasaurus*, which Novas (1993) regarded as an autapomorphy of that genus.

Kischlat and Barberena stated that a recent study of the holotype of *Staurikosaurus*, a genus usually classified with the Herrerasauridae (see "Systematics" chapter for more recent assessments of this family), do not support the claim of Sereno and Arcucci (1994) that a trochanteric shelf is present on the femur. The authors noted that following "better preparation, this taxon should reveal differences pointing far from herrerasaurids." (Azevedo 1999 reported teeth and two isolated vertebrae, material currently under study, from the Botucarai region of Brazil, the vertebrae somewhat similar to those of the holotype of *Staurikosaurus pricei*.)

Furthermore, Kischlat and Barberena noted that as yet unpublished studies by Peter M. Galton (personal communication to Kischlat and Barberena) indicate that *Spondylosoma*, a dinosaur whose affinities are currently under question (see *D:TE*), may be a herrerasaurid, and that sacral vertebrae collected at the Botucarai locality, agree with the third sacral of that genus. The authors pointed out that "If Galton's ideas turn out to be phylogenetically tested, the name Sponlyosomatidae (Huene 1942) ... "would have precedence over Herrerasauridae (Benedetto 1973)."

Heckert and Lucas (1999) published an expansive survey regarding global correlation and chronology of Triassic dinosaurs. Accepting herrerasaurs as primitive theropods, Heckert and Lucas noted that the earliest known probable theropods — including fragmentary remains from the Chinle Formation of the United States (see Hunt and Lucas 1990), and possibly *Alwalkeria* from the Maleri Formation of India — are of early late Carnian (Otischalkian, based on the concurrence of the phytosaur *Paleorhinus*) age. Known from the same or correlative strata are the primitive ornithischians *Pisanosaurus* from the Ischigualasto Formation in Argentina and *Pekinosaurus* from the Pekin Formation in the United States, and the prosauropod *Azendohsaurus*, from the Argana series in Morocco. Therefore, dinosaurs were recognized by Heckert and Lucas as comprising "a very small, yet diverse, component of late Carnian tetrapod faunas," having already split into several major dinosaurian groups. Also, by latest Carnian times, the Theropoda had already diverged from other dinosaurs and included basal forms, herrerasaurids, and more rarely, "ceratosaurs" ("Ceratosauria" and also Neoceratosauria, the former regarded by some theropod specialists as a polyphyletic assemblage; see "Systematics" chapter, "Notes" under Neotheropoda and Ceratosauria) among various other problematic but seemingly derived forms. This divergence suggested to Heckert and Lucas that tetanurine theropods, though their remains have not yet been found in Triassic strata, should also be present at that time. By Norian time, the authors noted, while prosauropods already dominated the more terrestrial, dry ecosystems, theropods remained a minor component of all faunas. By "Rhaetian" time, following the apparent extinction of herrerasaurs, most theropod fauna consists of moderately to highly derived "ceratosaurs," including *Coelophysis* and *Liliensternus.* Abundant footprint evidence reveals that small to medium-sized theropods (1–3 meters long) were common by this time, although body fossils belonging to these theropods remain relatively rare in the fossil record.

Regarding theropods in general, a brief comparative study of the skulls of various large theropods (including "ceratosaurs," carnosaurs, and tyrannosaurs) including inferences as to such activity as feeding methods in these dinosaurs, was made by Holtz (1998):

Nontyrannosaurs were found to retain several primitive features of carnivorous archosaurs: Dorsoventrally deep, mediolaterally (oreinorostral) narrow skull, blade-like, serrated (ziphodont) teeth, and absence of ossified secondary palate, biomechanical analyses suggesting to Holtz that this design is effective for vertical slicing of food, but poorly resistant to torsional forces.

In tyrannosaurs, the following features were observed: Proportionately wider rostra, labiolingually expanded (incrassate) teeth, and well-developed ossified secondary palates, analyses by Holtz revealing this plan to be more resistant to torsional forces than the primitive condition. Also supporting the latter feeding method is the expansion of the posterior portion of the tyrannosaur skull, which increased the leverage of the neck muscles.

Holtz noted that theropods exhibit two morphologies of the infratemporal fenestrae in the dorsal surface of their skulls: The large size of this opening is achieved in some neoceratosaurs and carnosaurs by a caudal projection of the quadrate bones, this moving the jaw articulation into the region of the cervical vertebrae, thereby restricting the lateral motion of the neck. By contrast, the increased surface area (for muscle attachment) is achieved in tyrannosaurs by a forward projection of the squamosal and quadratojugal, this reducing the overlap of the jaw articulation and neck region, and, consequently, permitting greater side-to-side movement of the neck.

Speculation into the possible behavior of at least one kind of theropod dinosaur resulted from the discovery of a nest (including over a 100 well-preserved eggs) of an unknown theropod in Upper Jurassic (Tithonian) levels of Lourinhã, Portugal, reported in

Steel skull of *Tyrannosaurus rex* created by sculptor Larry Williams for an exposition in Trieste, Italy, sponsored by Stoneage.

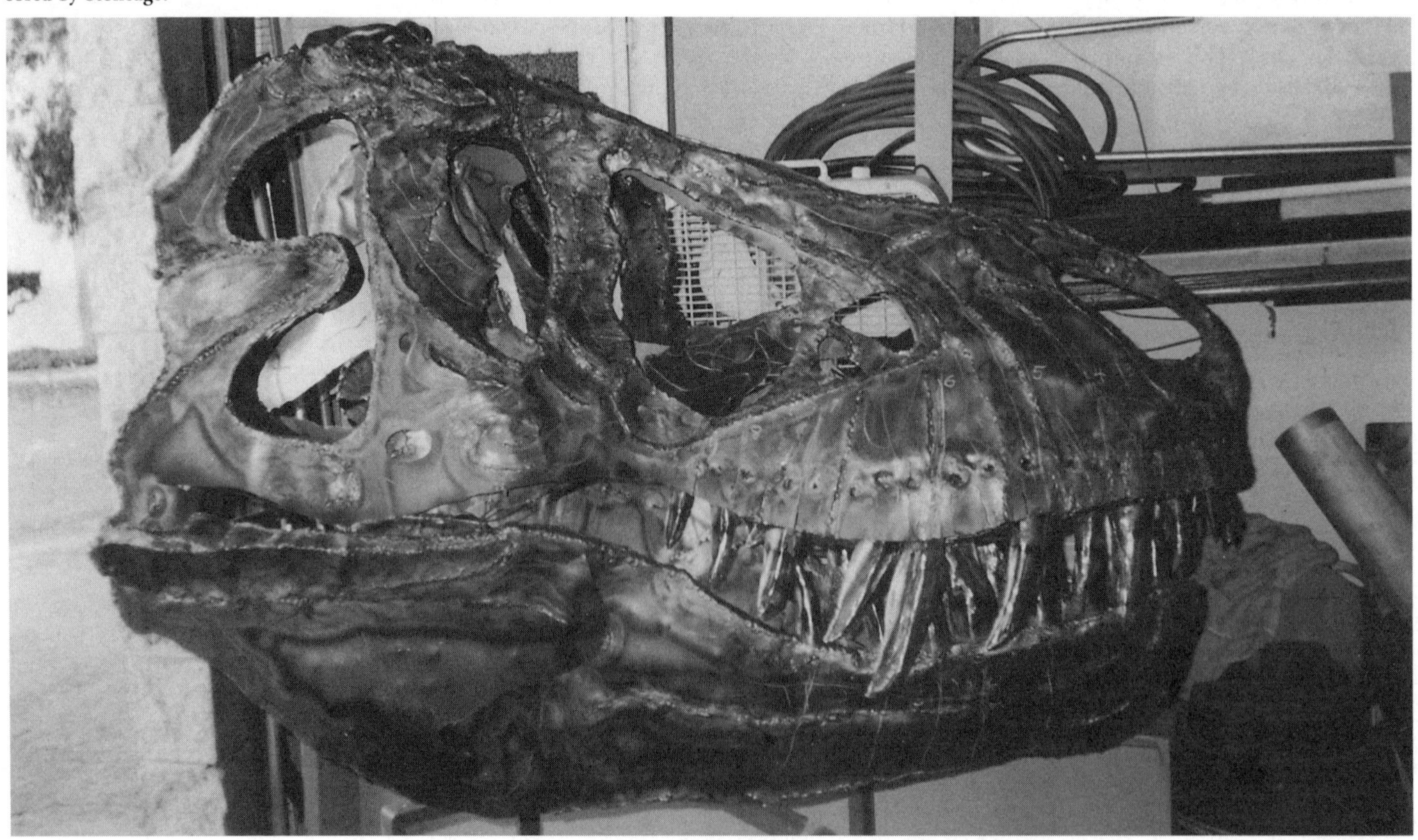

an abstract by Mateus, Taquet, Antunes, Mateus and Robeiro (1998). The eggshells were found dispersed over a large area of 11 meters in the greatest diameter, and with a high concentration in the center. Confirmation of their basal tetanuran identity was by the presence of embryonic bones in some of the eggs. Included in their brief description of these eggs, the authors stated that they are of the "Dinosauroid-prismatic obliquiprismatic type," with shells having a thickness of from 0.6 to 0.9 millimeters.

According to Mateus *et al.*, such a high concentration of eggs belonging to a single species of dinosaur, and the assumption that a single female could not have laid so many eggs during a short period of time, may suggest that this site represents a community nest. Mateus *et al.* noted that no particular nesting pattern was observed, with some of the eggs laid over others, this arrangement being "similar to that exhibited by ostrich egg-laying strategies."

The authors noted that five eggs having a microstructure similar to that in crocodiloid eggs, a jaw of the multitubercolate mammal *Kuehneodon* sp., a gastrolith, and a tooth of an adult theropod were also found at this site, and that five other such nesting sites were also found in the Lourinhã area.

Concerning theropods in general, Christiansen (1999) published a detailed analysis of allometry in 53 specimens of long bones in a large sample of adult theropods, these spanning almost the entire size range of at least 26 known species and including 13 unidentified specimens. Christainsen utilized 54 species of extant mammals — primarily ungulates and carnivorans having largely parasagittal limb kinetics — as an anatomical analog for theropod locomotion; mammals were chosen over birds due to the high diversity of the former and also "their similar limb kinematics, notably a longer, more vertically oriented femur" (see Gatesy 1990, 1991).

Theropods are very important for scaling analyses, Christainsen noted: Theropod postcranial anatomy is rather similar among taxa and all forms were apparently basically terrestrial animals, employing alternate striding during locomotion, implying that ecological specializations (*e.g.,* saltatoriality, fossoriality, and arboreality), often having profound influence on the appendicular morphology of extant mammals (*e.g.*, Casinos, Quintana and Viladieu 1993), most likely have little relevance. Also, the theropod appendicular skeleton is basically designed to support mass and accommodate locomotion. Furthermore, this group of dinosaurs spanned an estimated four orders of magnitude in mass.

Christainsen's analysis showed that theropods rather closely parallel extant mammals in long bone allometry, their skeletal allometry becoming progressively more apparent as species size increases. Large theropod forms (*i.e.*, having a body mass of more than 300 kilograms) scaled with significantly lower regression slopes than smaller theropod species. As with these mammals, the long bones of large theropod species tended to be elastically similar, while these elements in smaller species approached geometric similarity. In a similar fashion to mammals, theropods seemed to scale their limb postures to size, aligning their individual long bones more steeply to vertical as size increases, thereby "probably reducing the mass specific amount of force necessary to counteract movements about the joints."

The author pointed out that theropods and large mammals differ significantly in the relative lengths of their hindlimbs. Large theropods (*e.g.*, *Tyrannosaurus*) have retained very long hindlimbs; contrarily, large mammals capable of rapid locomotion have evolved markedly shorter (and stronger) long bones relative to body mass. This disparity between the relative hindlimb length of large theropods and large running mammals could have bearing on the continuing debate of whether or not such large bipedal dinosaurs were capable of fast locomotion. As pointed out by Christiansen, giant theropods like *Tyrannosaurus*, *Giganotosaurus*, and *Carcharodontosaurus* all possessed suites of appendicular adaptations required for rapid locomotion (*e.g.*, long limbs, well-developed appendicular muscles, feet with a reduced number of digits, hinge-like joints, *etc.*). However, the combination of this markedly cursorial anatomy, long bones scaled apparently paralleling the adaptive trends of extant mammals, and very long, but mechanically weak, hindlimbs found in these dinosaurs is not known in any living tetrapod. This suggested to Christainsen that, as had been suggested by previous authors (*e.g.*, Coombs 1978; Bakker 1986; Paul 1988*b*), that such dinosaurs "had morphological adaptations strongly suggestive of the capability of fast locomotion," but that the strength of the hindlimb bones "casts serious doubt on the capability of large theropods to include a long, suspended phase in the stride." It was Christainsen's conclusion, therefore, that giant theropods such as *Tyrannosaurus* did not move at the running speeds proposed by Bakker and Paul, but rather may have "relied more on fast progression involving long strides with brief, if any, suspended phases" (see Farlow, Smith and Robinson 1995), the long legs allowing these animals to move at quite impressive speeds.

Larsson, Sereno and Wilson (2000) performed a study centered upon the enlargement of the brain and forebrain in nonavian theropods utilizing endocasts of the carcharodontosaurid allosauroid *Carcharodontosaurus* (SGM-Din 1) and the tyrannosaurid coelurosaur *Tyrannosaurus* (AMNH 5029), two very large

Photograph by Allen A. Debus, courtesy Burpee Museum, Rockford, Illinois.

Recent findings by Per Christiansen supplements earlier studies suggesting that giant theropods such as *Tyrannosaurus rex* (cast of AMNH 5027) were capable of rapid locomotion.

Late Cretaceous forms having similar adult body mass but which differ in their phylogenetic proximity to Aves.

Larsson *et al.* observed that, despite the similar body size, endocast volumes for *Carcharodontosaurus* and *Tyrannosaurus* are quite different. The cerebral volume in *Tyrannosaurus* is about 50 percent greater than in *Carcharodontosaurus*, the percentage of total endocast volume occupied by the cerebrum, therefore, being greater in *Tyrannosaurus* (32.6 percent) than in *Carcharodontosaurus* (24.0 percent). Increase in the total volume of the brain, relative to body mass, and in cerebrum volume, relative to brain volume, also characterizes *Archaeopteryx*, extant birds, and most likely small-bodied theropods that are closely related to birds (see Russell 1969; Sues 1978; Currie 1985, 1995).

By comparing the endocasts of these two theropods, and also with endocasts of the allosaurid allosauroid *Allosaurus* and the avian *Archaeopteryx*, Larsson *et al.* concluded the following:

1. Relative cerebral and total brain volumes in *Carcharodontosaurus* (and presumably more basal theropods) are comparable to those of extant nonavian reptiles. Coelurosauria is characterized by an increase of possibly 50 percent of total endocast volume compared to body size, a disproportionate amount of this increase taking place within the cerebrum, which doubled in volume. The authors noted that this first stage of relative brain and cerebral expansion probably occurred during the Middle Jurassic (about 180 to 160 million years ago) following the divergence of that group (Sereno 1997).
2. The estimated total brain volume and relative cerebral volume in *Archaeopteryx* are intermediate between those in *Tyrannosaurus* and modern birds, this stage of brain and cerebral enlargement probably having taken place by the Late Jurassic (about 150 million years ago), most likely characterized by small manuraptorans closely related to birds (*e.g.*, *Troodon* and possibly *Caudipteryx*; see Ji, Currie, Norell and Ji 1998).
3. Additional enlargement of the brain occurred in ornithurine birds, as in *Archaeopteryx* relative brain size and the proportion of the brain including the cerebrum are less than in extant birds. The authors noted that this most recent stage of such expansion apparently occurred by earliest Cretaceous (about 140 million years ago) time at the base of Ornithurae (based upon an inspection of the skull of the bird *Confusciousornis*).

Consequently, Larsson *et al.* stated that enlargement of the brain from primitive reptilian proportions to that found in living birds "would have taken place in less than 40 million years."

Among the more recent theropod discoveries, Munter (1999), in an abstract, reported on, among other (unidentified) vertebrate fossils, two theropod specimens recovered from the upper Lower Jurassic Boca Formation, Huizachal Canyon, Mexico. One specimen consists of an articulated sacrum and pelvis, the other two fragments from the braincase. As briefly described by Munter, the pelvic specimen closely resembles that of the coelophysoid *Syntarsus*, differing from that genus in that the iliac blades contact along the midline, and are most likely contacted by the neural spines of the second, third, and probably first, sacral vertebrae.

Because the two specimens may not represent a single taxon, Munter subjected them to separate phylogenetic analyses. From these, Munter determined the pelvic specimen to represent the sister taxon of *Syntarsus*, thereby confirming its identity as a basal theropod. This classification was supported by four characters: Five sacral vertebrae; fusion of sacral ribs; the transverse breadth of acetabular portion of ischium and pubis; and expansion of ischium into acetabulum where it contacts ilium (the latter previously known only in *Syntarsus*).

Analysis by Munter of the skull pieces most parsimoniously placed the specimen either with Manuraptora or a clade comprising *Ceratosaurus* and Abelisauridae. As the paroccipital root of this specimen is pneumaticized, it could not be associated with the Coelophysoidea, a clade for which such pneumatization has not been reported. For now, according to Munter, the two specimens should not be combined to represent a single taxon. Doing so "results in it being the sister taxon of *Syntarsus*, but requires the pneumatization of the paroccipital root to be homoplasious."

Until recently, the notion of venomous theropods has been restricted to the popular media (*e.g.*, the novel *Jurassic Park* and its motion-picture adaptation). In an abstract, Rodríguez de la Rosa and Aranada-Manteca (2000) addressed the possible reality of this concept, citing a theropod tooth (FCM 06/053) collected from the Upper Cretaceous (late Campanian) outcrops of the El Gallo Formation in the State of Baja California, Mexico.

As described by these authors, the tooth is blade-like in shape, labio-lingually compressed, and small (0.95 centimeters in anteroposterior base length, 0.56 centimeters in lateral width, and about 2 centimeters in height). Although lacking its tip, the tooth seems to be sharply pointed and distally recurved. Only the bases of the denticles have been preserved, the anterior denticles being minute relative to the posterior ones.

As Rodríguez de la Rosa and Aranada-Manteca observed, the general features of this tooth are not unlike those found in other known theropods. However, the posterior denticles are located within a longitudinal groove running over the two distal thirds of the posterior carina. This structure, not yet described in any other theropod, is analogous to that found in various reptile groups adapted for envenomation (*i.e.*, producing oral toxins), these including certain kinds of snakes and lizards, both extinct and extant. Rodríguez de la Rosa and Aranada-Manteca interpreted this evidence "as the existence of a lineage of small theropods bearing teeth adapted to the conductance of oral toxins, probably related to aid in feeding strategies."

The authors noted that this record is significant, being the first its kind, both within the context of dinosaurian studies and, more importantly, under the scope of theropod biology.

Heckert, Lucas and Sullivan (2000), in a survey report on the Triassic dinosaurs of New Mexico, described in detail three theropod specimens from the Los Esteros Member of the Santa Rosa Formation, near Lamy, in Santa Fe County, north-central New Mexico. NMMNH P-25749 consists of the proximal end of a left pubis, the crushed and slightly eroded proximal end of a femur, and miscellaneous elements concreted against the pubis. The femoral head is offset and more like that of the ?coelophysoid "ceratosaur" *Eucoelophysis* (see Sullivan and Lucas 1999), and not like that of the herrerasaurid *Chindesaurus*, which is more angular and square-shaped (Long and Murry 1995), although there is no sign of a groove in the shaft as in *Eucoelophysis*. Therefore, NMMNH P-25749 conforms more with coelophysoid than herrerasaur morphology.

The second specimen (originally described briefly by Hunt and Lucas 1995) described by Heckert *et al.*, NMMNH P-25790, comprises five incomplete metatarsals, a phalanx, and shaft fragments which may or may not pertain to a single individual. Three incomplete elements — identified by the authors as the left fourth, third, and second tarsometatarsals and also metatarsals three and four — conform in size to those elements in the holotype of *Eucoelophysis baldwini*, although no autapomorphic characters were observed in NMMNH P-25790 warranting their referral to that taxon. These elements, the authors noted, support assigning of these elements to the Coelophysoidea, although the other tarsometatarsal and phalangeal fragments remain unidentifiable.

Lastly, NMMNH P-13006 includes two, as of this writing unprepared incomplete, fused sacral centra with portions of the synsacrum preserved in a concretion, the latter therefore obscured by matrix. The centra are hollow, identifying the specimen as belonging to the Theropoda. According to Heckert *at al.*, this specimen could pertain to a large coelophysoid or a herrerasaur.

Arcucci and Coria (1998), in an early report, announced the recovery of the skeleton — including a quite large (measuring 45 centimeters from tip of snout to quadrate) and almost complete articulated skull — of a new primitive theropod, identified by the authors as apparently a "ceratosaur" from the upper levels of the Upper Triassic (supposedly of Norian age, based on its faunal content) Los Colorados Formation of La Rioja, Argentina.

As briefly described by Arcucci and Coria, the extended antorbital fenestra and low laminar crest preserved on the right side of the skull, most likely represent part of a paired structure; the accessory fenestra on the anterior part of the maxilla and shape of the infratemporal opening exhibit "the advanced condition shown by *Carnotaurus* and *Sinraptor*." A preliminary analysis by Arcucci and Coria suggested a closer relationship of this theropod with more derived "ceratosaurs" than with the coelophysoids.

Sampson, Forster and Krause (1998) briefly noted that "a small ceratosaurian theropod," represented by only isolated elements, had been discovered in the Upper Cretaceous (Maastrichtian) Maevarano Formation of northwestern Madagascar (see section on sauropods, below), but provided no further details at the time.

In an abstract for a poster, Hand and Bakker (2000) reported on a new allosaurid forearm discovered in summer of 1999 at Como Bluff, Wyoming. Although belonging to a juvenile, this specimen indicates an animal that, in its adult stage, may have been substantially larger than *Allosaurus*. The bones are all exceptionally robust, and the radius and ulna are shorter than the first ungual (indicating that the arm relative to body size was probably some 25 to 30 percent shorter than in *Allosaurus*).

As briefly described by Hand and Bakker, the "first ungual is unusually long and straight but has a huge attachment site for the flexor tendon." Compared to the manus in *Allosaurus*, this one could hyper-extend; when at maximum flexion, the first digit rotated dorso-medially, becoming semi-opposable. The preserved semilunate carpal indicates that wrist movement was anteroposterior, although it moved through a limited arc when the manus was held with palmar surface medially directed.

Given the comparatively small size of this specimen, Hand and Bakker found it unlikely that such hands were used to capture prey. However, the high degree of variation observed in such short forearms implied that they must have been used for something, possibly reflecting variations in interspecific female anatomy. Perhaps, the authors speculated, they were utilized as claspers during copulation.

Sampson, Carrano and Forster (2000), in a preliminary report, announced the discovery of a new and quite unusual noasaurid theropod recently collected from the Upper Cretaceous (Maastrichtian) Maevarano Formation of Madagascar. Small-bodied (about 1.8 meters in length), this dinosaur is currently known from isolated elements representing at least 10 individuals, this material including several skull elements and together revealing about 40 percent of the postcranial skeleton.

According to Sampson *et al.* (2000), "This taxon is unique in having the most highly procumbent and strongly heterodont lower dentition of any known theropod." The authors, in briefly describing this theropod, also observed the following: First four teeth radially arrayed, enlarged first alveolus nearly horizontally oriented, first tooth, therefore, being recurved and transversely compressed.

Character analysis by Sampson *et al.* (2000) showed that this new taxon is closely related to other Late Cretaceous species, the Argentinian *Noasaurus leali* and Indian *Laevisuchus indicus*, these three taxa seemingly representing a cosmopolitan radiation that rapidly spread across the landmasses of Gondwana. As these authors pointed out, the known paleogeographic and temporal distribution (Argentina, Madagascar, and India) of this family, the Noasauridae, is congruent with that of the Abelisauridae, a clade of closely related but larger-bodied theropods. This suggests, the authors noted, that both clades dispersed in parallel. Also, the new evidence is consistent with the hypothesis that a biogeographic corridor joined South America with Indo-Madagascar.

Silva and Kellner (1999) briefly described a number of large theropod teeth collected from various Late Cretaceous localities in Brazil. These included teeth (Museu de Ciências da Terra-DMPM) from the upper part of the Bauru Group, which have a morphology somewhat distinct from those previously known from this area, and from Morro do Cambambe, north of Cuiabá (Mato Grosso). Those from the latter locality, the authors noted, have morphologies similar to the teeth of the abelisaurid *Giganotosaurus* from Argentina and carcharodontosaurid *Carcharodontosaurus* from Africa, both also from the Late Cretaceous.

Another new theropod, the same age as that reported by Coria and Salgado (1999), was announced that same year in a preliminary notice by Calvo, Rubilar and Moreno (1999). The specimen, still under preparation at the time their report was written, includes a left maxilla, dentary, cervical, dorsal, sacral (at least four), and caudal vertebrae, hemals, ilia, pubis, a left femur, left tibia, and foot bones collected from Candeleros Member of the Río Limay Formation of Añelo, Neuquén Province, Patagonia. Calvo *et al.* observed the following characters in this specimen: Cervical vertebrae having pair of pleurocoels; fusion of sacral vertebrae and ribs; fusion of astragalus and calcaneum.

According to these authors, a short laminar process of the astragalus assigns this specimen to the Tetanurae. The absence of a quadrangular symphysis and a dorsally directed femoral head, (present in the carcharodontosaurid *Giganotosaurus carolinii*) exclude the specimen from the Carcharodontosauridae; and the anterior ramus of the maxilla does not attain 70 percent of the maximum depth and the teeth are laterally flattened and serrated, this excluding it from the Spinosauridae. The ventral edge of the dentary is short and convex as in Abelisauridae. Unlike the abelisaurid *Carnotaurus sastrei*, however, the femur of the new theropod has a laminar anterior trochanter and a well-developed fourth trochanter. The astragalar condyle displays an anterior groove.

A new giant theropod, yet another contendor in the ever popular "largest meat-eating dinosaur ever to walk the Earth" category, was announced in various Associated Press newspaper accounts published in March, 2000 (see also Currie 2000). Informally described in these accounts as "a needle-nosed, razor-toothed beast," the unnamed form—apparently resembling and related to the huge and geologically older *Gigtanotosaurus*—is represented by the remains of six individuals recovered from a desert (Middle Cretaceous, about 100 million years ago) on the eastern slopes of the Andes, in Patagonia. The remains were found in 1995 by a local shepherd. Two years later, excavation of the bones began under the direction of Rodolfo A. Coria and Philip J. Currie. This new theropod was heavier built than *Tyrannosaurus rex* and has slightly shorter legs. The skull is long and narrow with scissors-like jaws. The total estimated length of the animal is 45 feet (more than 13 meters).

Currie (2000) further reported that, as technicians at the Museo Municipal "Carmen Funes" in Argentina began to prepare the collected bones of this new theropod, Coria realized that elements of three skeletons had actually been recovered, and the site, therefore, constituted a bonebed. Digging was therefore resumed at the site from 1998 to 2000, the quarry consequently extended to become the largest dinosaur quarry yet excavated in Argentina. To date of this writing, six individuals of this dinosaur have been collected, together representing about 90 percent of the skeleton.

In the past, giant theropods were generally regarded as solitary animals; the number of specimens of this new taxon, however, suggest otherwise. As noted by Currie, large carnivorous animals are relatively rare animals in their environment, generally comprising about 5 percent of the dinosaur faunas. Therefore, as herbivorous dinosaurs comprise approximately 95 percent, these enjoy a greater chance of being preserved as fossils. The Argentinean bonebed, however, consists entirely of theropods. Currie pointed out that chance might be invoked to explain the presence of these six theropods at a single locality; however, the likelihood of this being the case, according to mathematical probability, is 64 million to one.

As Currie noted, something obviously brought these dinosaurs to the same place to die and be buried. The condition of the bones studied and the rocks in which they were imbedded indicate that all of these individuals died at approximately the same time and were buried rather quickly after death. "This strongly suggests," Currie stated, "that they died together at the same time, which in turn suggests that they were living together just before they died." As the specimens recovered represent individuals of vastly different sizes (the smallest being about 5 meters or less than 17 feet long, the largest more than 11 meters) and ages, Currie theorized that these theropods had not come together merely to breed and lay eggs, but that they were hunting in a pack.

Although to date this new theropod has not yet been formally named or described, some of the fossil material belonging to it was exhibited that March at the Riverfront Arts Center in Wilmington, Delaware.

Coria, Chiappe and Dingus (2000) reported yet another new abelisaur from the Anacleto Member of the Río Colorado Formation, at Auca Mahuevo, northwestern Patagonia, Argentina, this dinosaur represented by a very well-preserved specimen including most of an articulated skeleton (including fully articulated forelimbs, hindlimbs, and most of the tail) and traces of soft tissue around the pelvis. The specimen was found in finely grained sediments indicating shallow water and lacustrine deposits. This interpretation based the well-developed laminae on the specimen and also the presence of microinvertebrates.

As briefly described by Coria *et al.* (2000), this theropod is very similar to although, about 30 percent smaller than, the abelisaur *Carnotaurus sastrei*. Indications that the Auca Mahuevo specimen represents a new taxon include the following: Rostrum and antorbital fenestra longer and lower; maxillary fenestra laterally exposed; skull with swells rather than horns; coracoid process reduced; humerus slender, craniocaudally compressed, condyles well defined; hemal canals open. According to the authors, the presence of various derived characters (*e.g.*, cervical vertebrae with cranially projected epipophyses, proximal and middle caudal vertebrae having hypantrum-hyposphene articulations, frontal protuberances, coracoid broad, humeral head large and hemispherical, ulna and radius extremely short) suggest that this taxon is the sister taxon of *Carnotaurus.*

Coria *et al.* (2000) noted that this new abelisaur offers additional osteological data that clarifies significant aspects of the postcranial anatomy of *Carnotaurus* and also, to some degree, that of other abelisaurs.

Martill, Frey, Sues and Cruickshank (2000) described some well-preserved and uncrushed remains belonging to an unknown coelurosaurian theropod, possibly a compsognathid, recovered by a local fossil collector from (precise locality not known) the Chapada do Araripe, in the Romualdo Member of the ?Lower Cretaceous (?Albian, based on palynological studies by Pons, Berthou and Campos 1990) Santana Formation (see Martill and Wilby in Martill 1993), Pernambuco, northeastern Brazil. The specimen (SMNK 2349 PAL) comprises well-preserved, associated, and uncrushed postcranial remains including vertebrae (the last dorsal vertebra with an attached rib, one dorsosacral), a nearly complete sacrum (two sacral vertebrae and a partial third), several incomplete gastralia, the pelvic girdle (with almost complete ilia, ischia, and pubes mostly preserved in natural articulation), and partial hindlimbs (including incomplete femora and proximal ends of the right tibia and fibula); most remarkably, however, it also preserves rare dinosaurian soft tissues interpreted as a portion of the lithified intestinal tract.

According to Martill *et al.*, the pelvic girdle of this specimen "is noteworthy for the bilaterally asymmetrical development of various bony features." As observed by these authors, an oval obturator foramen is present on the subacetabular portion of the right pubis, although this feature is absent on the left pubis.

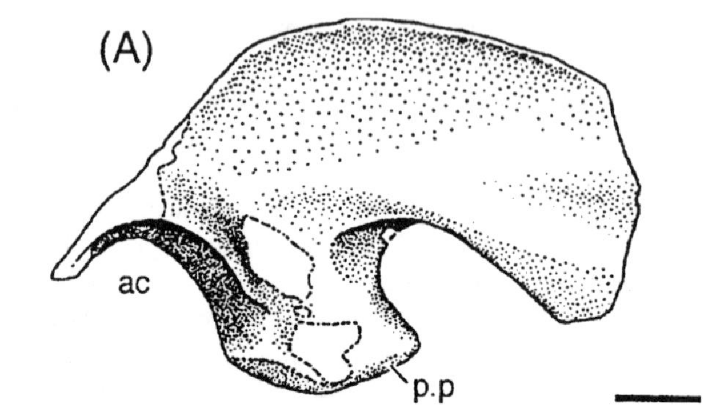

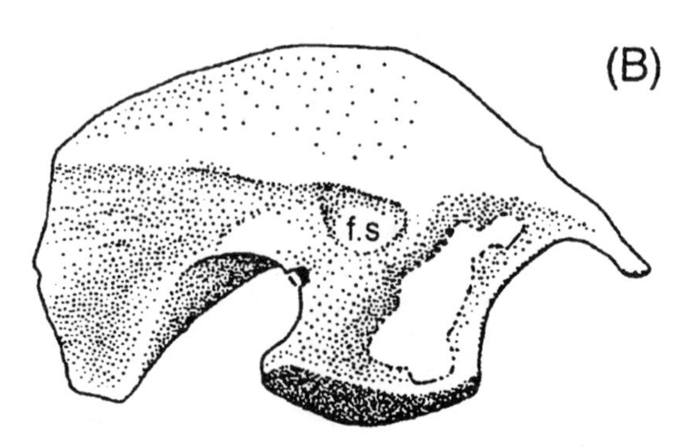

Incomplete right ilium (SMNK 2349 PAL) of unnamed coelurosaurian theropod in A. lateral and B. medial views. Scale = 1 cm. (After Martill, Frey, Sues and Cruickshank 1999.)

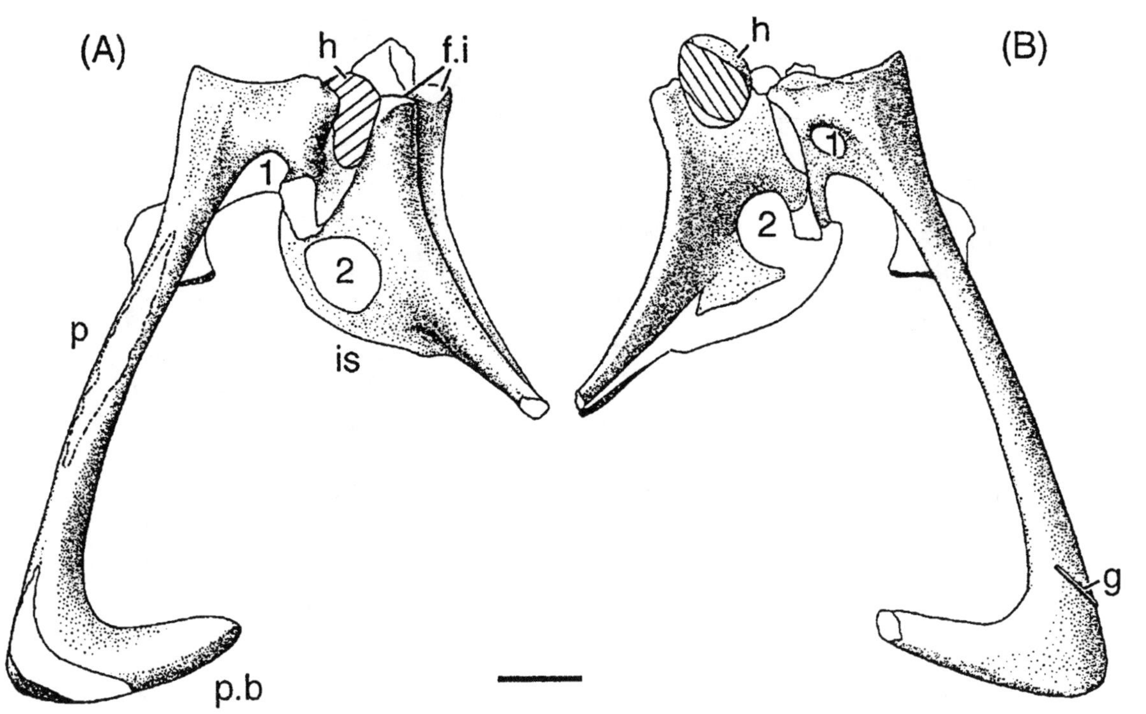

Unnamed coelurosaurian theropod (SMNK 2349 PAL), A. left pubis and ischium and B. right pubis and ischium, lateral views. Scale = 2 cm. (After Martill, Frey, Sues and Cruickshank 1999.)

Ventral portion of pelvic girdle (SMNK 2349 PAL) of unnamed coelurosaurian theropod in A. anterior and B. posterior views, part of intestinal infill (int) having been removed for thin-sectioning. Scale = 2 cm. (After Martill, Frey, Sues and Cruickshank 1999.)

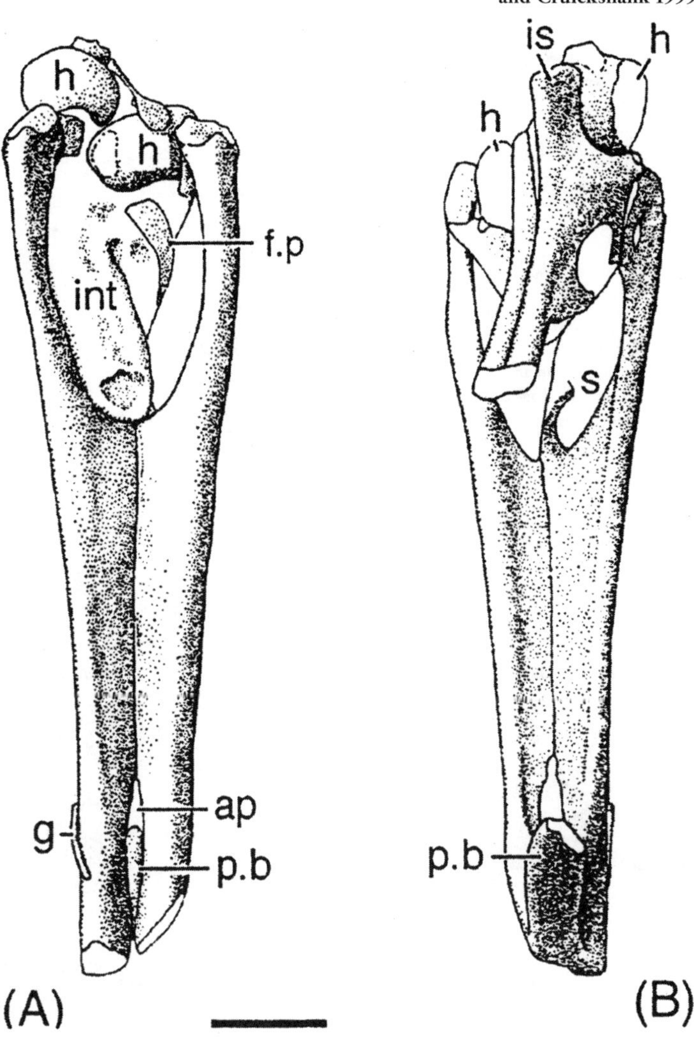

The subacetabular portion of the right ischium exhibits a ventral notch; on the left ischium, a large foramen occupies this location. There is a perforation at the base of the right transverse process of the dorsocacral vertebra, and part of the iliosacral ligament seems to be ossified on the right transverse process of the vertebra. Martill *et al.* noted that some of the features present on only one side of this specimen may be diagnostic at high taxonomic levels among Theropoda (see, for example, Holtz 1994). An obturator foramen was considered by Holtz to be a synapomorphy of "Ceratosauria" and is plesiomorphic for Dinosauria. However, Martill *et al.* interpreted the asymmetrical development of features of the pelvic girdle as probably representing individual variation, therefore having no diagnostic utility.

Interpreted by Martill *et al.* as the intestinal tract, a blunt and cone-shaped structure can be seen in the matrix, located behind the pubis in the middle between the ventral surface of the sacrum and distal end of the pubis, the cone tapering towards the pubis and extending posterodorsally. This structure, as originally preserved prior removal for cross-sectioning, measured 27 millimeters in length and 17 millimeters in width. It has a spongy texture with small, irregular, and calcite-lined vacuities, this being interpreted by the authors as representing the lithified intestinal infill. The outer layer is smooth and dense, possibly representing the fossilized intestinal tube itself.

The course of the intestine, the authors noted, is virtually the same as that observed in the dinosaur *Scipionyx* (Dal Sasso and Signore 1998*a*, 1998*b*; see

Right femur (SMNK 2349 PAL) of unnamed coelurosaurian theropod in A. anterior, B. medial, C. lateral, and D. posterior views. Length = 165 mm. (After Martill, Frey, Sues and Cruickshank 1999.)

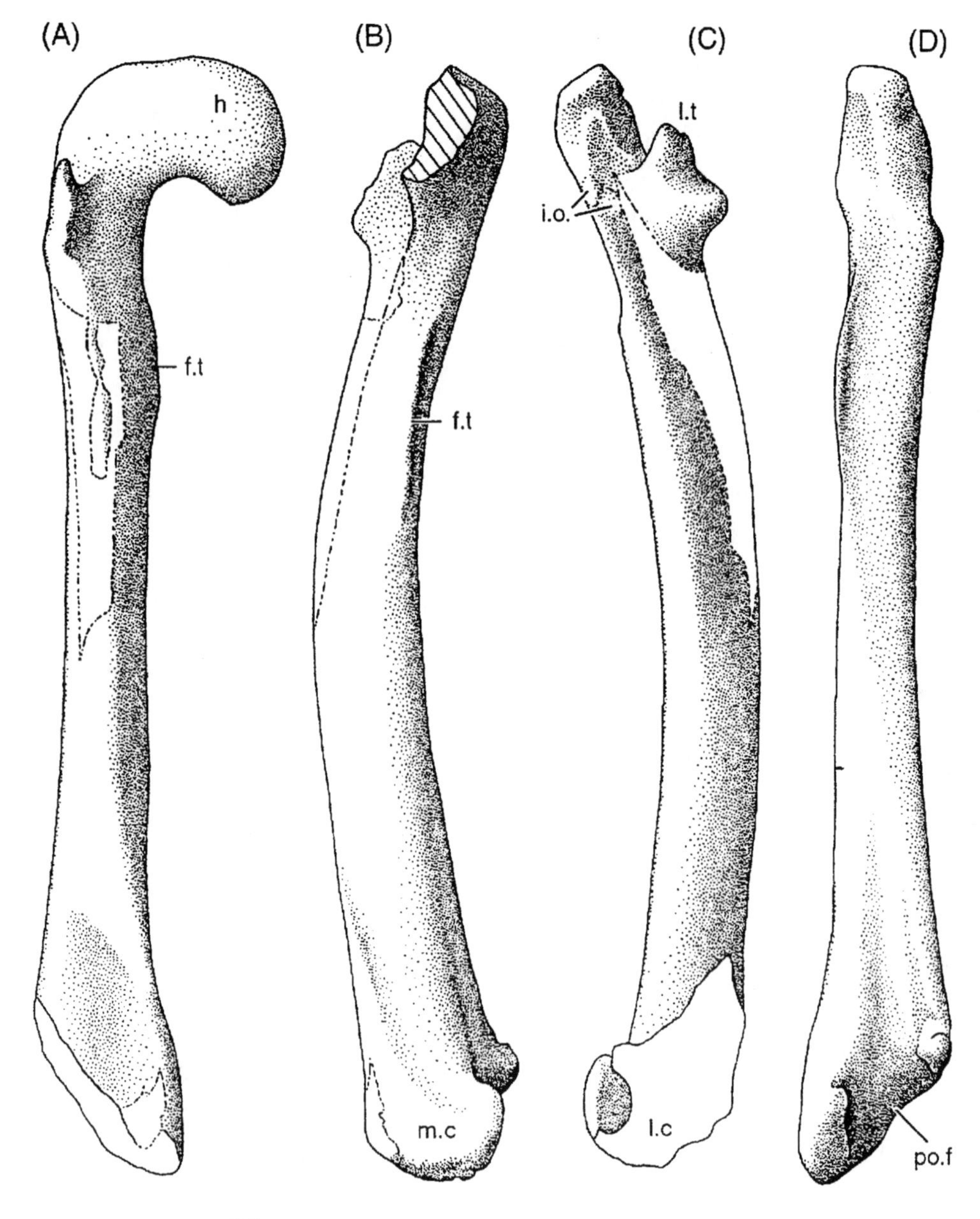

Partial sacrum (SMNK 2349 PAL) of unnamed coelurosaurian theropod in A. right lateral and B. dorsal views. Scale = 2 cm. (After Martill, Frey, Sues and Cruickshank 1999.)

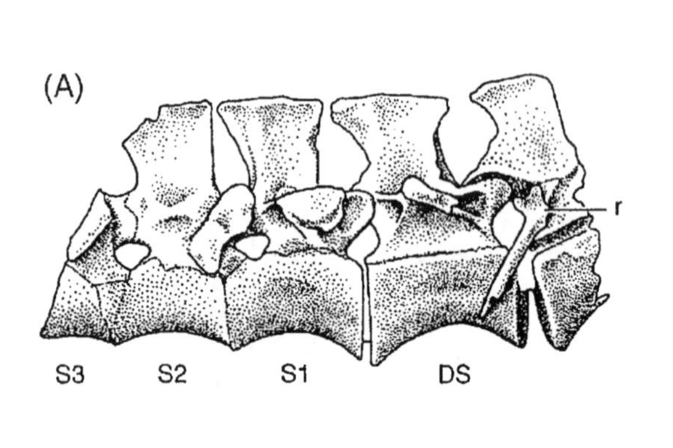

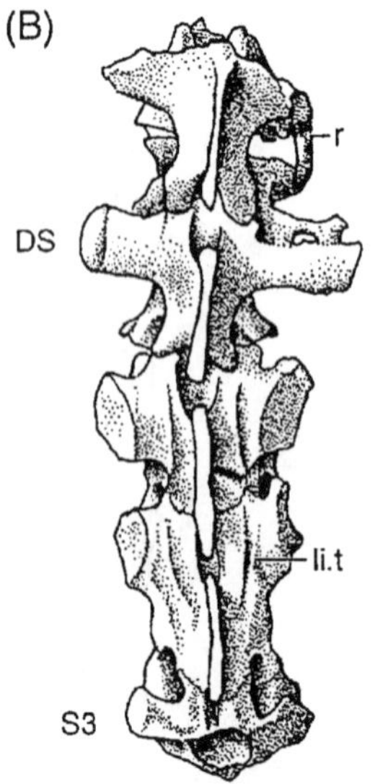

entry, also *S1*). Furthermore, it extends well ventral to the vertebral column and not in the dorsal abdominal cavity, as reconstructed by Ruben, Dal Sasso, Geist, Hillenius, Jones and Signore (1999) for *Scipionyx* by analogy to extant crocodilians. As pointed out by Martill *et al.*, unlike the uncrushed specimen from Brazil, the material reconstructed by Ruben *et al.* is crushed and flattened, thereby providing source material that cannot be considered reliable for determining the configuration of the intestinal tract.

Indeed, Martill *et al.* reached a conclusion contrary to that of Ruben *et al.* Martill *et al.* observed that, posterior to the pubic apron, a large vacuity, partially lined with small calcite crystals, was originally preserved in the matrix. Destroyed during preparation (although casts were first made), this vacuity had "extended from the dorsal aspect of the pubic boot dorsally along the concave posterior surface of the pubic apron and terminated at the distal end of the proximal end of the pubis"; posteriorly "it extended half way towards the ischia." Among various hypotheses considered by the authors to explain this vacuity, the most plausible was that it was once occupied by a postpubic air sac extending "into the space between the posterior surface of the pubic shaft, the dorsal aspect of the pubic boot, and the ischia." Martill *et al.* speculated that such an air sac "could have been ventilated by a dorsal pneumatic duct passing through the left side of the gap between the sacrum and pubes and a ventral one passing through the distal opening the pubic apron." (For further discussions, pro and con, regarding such air sacs in nonavian dinosaurs, see section on dinosaurs and birds, below).

As the neurocentral sutures can still be seen on all preserved vertebrae, Martill *et al.* identified SMNK 2349 PAL (by analogy to modern crocodilians; see Brochu 1996) as an immature individual. Based on dimensions of the holotype of the compsognathid *Compsognathus longipes* (Ostrom 1978) to estimate the size of the missing bones, the authors suggested that the coelurosaur from Brazil stood approximately 60 centimeters high at the hips. From dimensions of the existing bones, Martill *et al.* deduced that the body was very narrow across the pelvic region.

Martill *et al.* referred SMNK 2349 PAL to the Theropoda based upon the following features: Preacetabular portion of iliac blade large, extending far anteriorly; curvature of femoral shaft; and thin-walled limb bones. Referral to the Tetanurae was supported by the following: Lesser trochanter of femur "wing-like"; and large pubic boot. Referral to the Coelurosauria was based on the following: Absence of distal "foot" on ischium; and ischium less than two-thirds length of pubis.

As observed by Martill *et al.*, the pubis of the Brazilian specimen resembles closely in size and shape that of *Coelurus* (Marsh 1884*a*) and that of *Compsognathus* (Bidar, Demay and Thomel 1972). Also, SMNK 2349 PAL exhibits two features regarded by Chen, Dong and Zhen (1998) as diagnostic for Compsognathidae: Fan-shaped neural spines of dorsal vertebrae; and limited anterior expansion of pubic boot. However, too little information was presently available regarding most small theropods of the Late Jurassic and Early Cretaceous for the authors to evaluate more fully the phylogenetic importance of these similarities.

Martill *et al.* could find no coelurosaurian in South America or anywhere else displaying a similar combination of characters as seen in SMNK 2349 PAL, suggesting that the specimen represents a new taxon. In view of the incompleteness of this material, however, the authors refrained from proposing a new genus and species to embrace it.

SMNK 2349 PAL is significant in representing the first coelurian theropod known from the Romualdo Member of the Santana Formation of Brazil.

Naish (1999*b*) described for the first time a robust, partial right tibia (BMNH R9385) that had been collected during the nineteenth century from

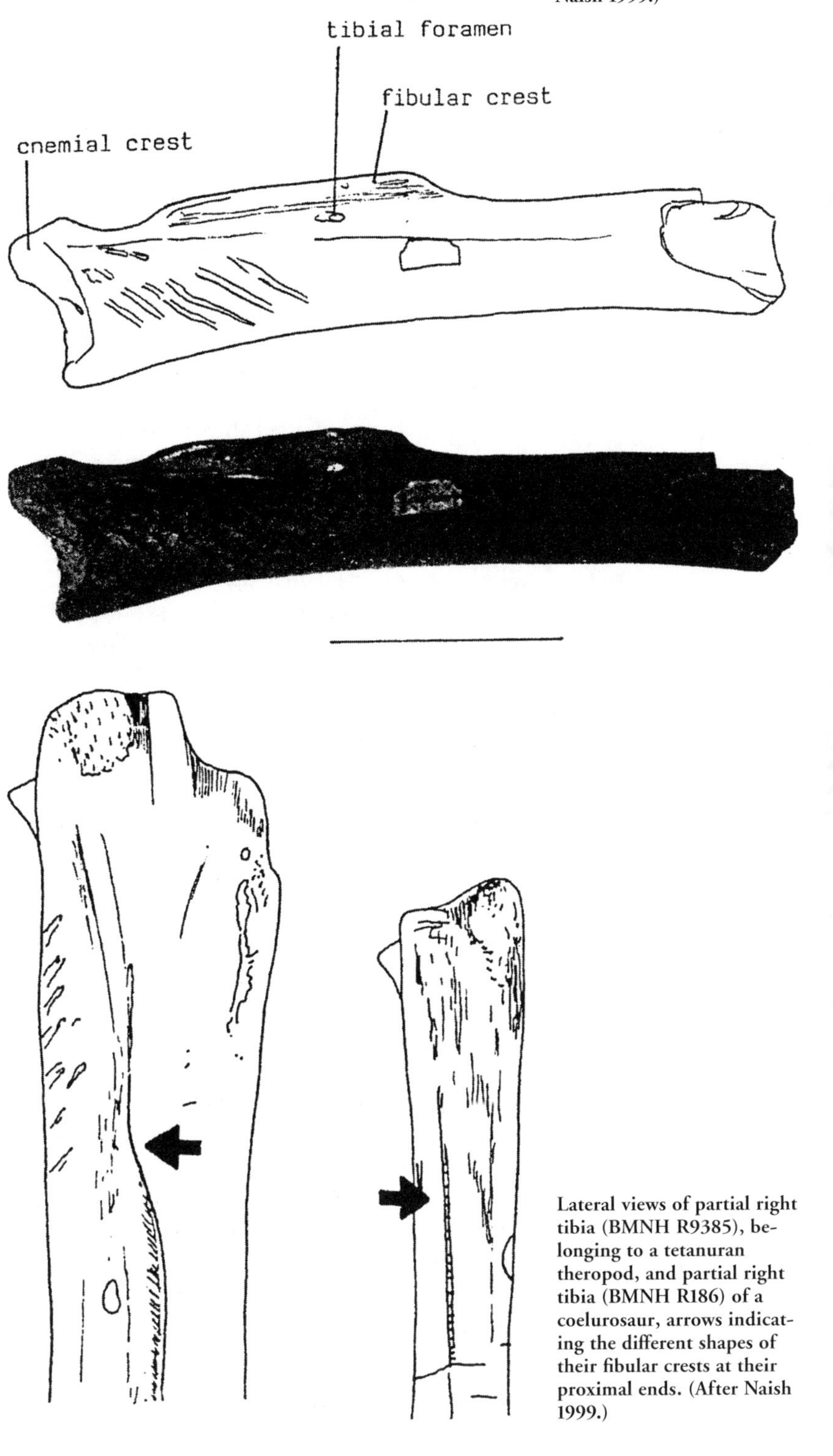

Partial right tibia (BMNH R9385) of an unnamed tetanuran theropod in caudal view. Scale = 50 mm. (After Naish 1999.)

Lateral views of partial right tibia (BMNH R9385), belonging to a tetanuran theropod, and partial right tibia (BMNH R186) of a coelurosaur, arrows indicating the different shapes of their fibular crests at their proximal ends. (After Naish 1999.)

presumably the Hastings Beds (Lower Cretaceous: Berriasian–Valanginian), Wealden Group, at Hastings, Sussex, England. According to Naish, no secondary source reviews the specimen's history (except for a mention by Brookes 1997 in her unpublished master's thesis and briefly by Naish 1998*a*), although various notes and letters suggest it was first obtained by an unidentifiable medical doctor and was originally part of the Dawson Collection prior to its acquisition by the British Museum (Natural History), now The Natural History Museum, London. This stratigraphic position, the author noted, distinguishes BMNH R9385, along with various indeterminate large theropods, as one of the oldest theropods from the Wealden Group.

The specimen is 160 millimeters as preserved, but is missing both proximal articulatory surface and distal end, the total length estimated to be about 250 millimeters. In proportions, the specimen is similar to the tibia of *Deinonychus antirropus*, and has a tibia of about 300 millimeters in length (see Ostrom 1969*b*). As *D. antirropus* has an adult size of 3 meters (about 10.5 feet), Naish suggested that the specimen from Hastings may represent a dinosaur of comparable size.

As observed by Naish, BMNH R9385 belongs to a theropod as identified by both the prominent cnemial crest and the sharp-edged fibular crest, the latter proposed by Holtz (1994) as a character of the Tetanurae. (Brookes had also concluded that BMNH 9385 belonged to a tetanuran.) The tibial foramen of the specimen is located adjacent to a point within the distal third of the crest. Previous authors (*e.g.*, Osmólska, Roniewicz and Barsbold 1972; Kirkland, Gaston and Burge 1993) have noted that this feature — the tibial foramen higher up the tibial shaft — in other theropods may be characteristic of the Coelurosauria, a foramen positioned lower down possibly characteristic of more primitive theropods.

Comparing BMNH R9385 with other known Wealden theropods (*i.e.*, the holotypes of *Neovenator salerii* and *Baryonyx walkeri*, plus BMNH R186, an isolated coelurosaurian tibia that is smaller, more gracile, and geologically younger than BMNH R9385), Naish found that BMNH R9385 belonged neither to *N. salerii* nor BMNH R186. In the latter specimen, the tibial foramen is positioned more proximally relative to the fibular crest, a condition tentatively regarded as a coelurosaurian condition and unlike that of BMNH R9385. The possibility remains however, that BMNH R9385 may be referrable to *Baryonyx*, although the preserved tibia of *B. walkeri* is crushed (Charig and Milner 1997), thereby making a meaningful comparison impossible. According to Naish, if BMNH 9385 belongs to a baryonychine (=baryonychid of his usage), it represents a smaller and stratigraphically older form than *B. walkeri*. The author could only identify BMNH 9385 with certainty as a tetanuran theropod.

Also, Naish identified a series of parallel tracts on the caudal surface of BMNH R9385 as marks made by serrated teeth (see also Naish 1998*a*), this — the first reported occurrence of theropod tooth marks on a Wealdon theropod bone — interpreted as "evidence for intraspecific aggression, cannibalism, predation by another species, or opportunistic scavenging."

Naish (2000) subsequently described a well-preserved, complete left theropod femur (IMIWG 6214), collected in November, 1998, by K. Minter from the Lower Plant Debris Bed in the Wessex Formation (Wealden Group, Lower Cretaceous), between Compton and Hanover Point, on the Isle of Wight, England (this specimen first mentioned briefly by Naish 1998*b*). Naish (2000) observed that the entire area of the caput and crista trochanteris and the proximal part of the anterior trochanter is heavily pitted, suggesting that the specimen could pertain to a juvenile. However, as some other juvenile and even embryonic theropods possess entirely ossified limb bones (see Currie and Peng 1993; Norell, Clark, Dashzeveg, Barsbold, Chiappe, Davidson, McKenna, Perle and Novacek 1994), the pitting in MIWG 6214 may be pathological. If that is the correct interpretation, the small size of the specimen (123 millimeters in length as measured along the lateral condyle) would indicate that the animal represented by MIWG 6214, if an adult, is possibly one of the smallest known nonavian theropods.

As observed by Naish (2000), MIWG 6214 is very distinctive, particularly in the morphology of its anterior trochanter. Significantly, this trochanter "is consistently the same height (3 mm) along its entire 21 mm length, an unusual feature that invites comparison with the anterior trochanters of other theropods."

Naish (2000) identified the femur as dinosaurian based on its prominent anterior trochanter, presence of a deflected caput, and lack of mediolateral curvature. It is not a nonthyreophoran ornithischian, as indicated by its ridge-like fourth trochanter, not thyreophoran, as shown by its craniodorsal curvature, and not from a small or juvenile sauropod, as indicted by the craniodorsal curvature and prominent anterior trochanter. Therefore, MIWG 6214 can only be from a theropod.

However, Naish (2000) was unable to refer this specimen to any theropod for which femora are known. Compared to other Wealden Group theropods, MIWG 6214 is notably distinct and "may be potentially important in indicating further theropod diversity within the Wealden Group fauna, as it does not seem to represent the same taxon as do the other Wealden Group theropods for which femora are known."

As MIWG 6214 lacks the common tetanuran feature of an extensor groove, Naish (2000) questioned whether or not the specimen belonged to the Tetanurae. Noting that an extensor groove is absent in the alvarezsaurid *Mononykus olecrans* (Perle, Chiappe, Barsbold, Clark and Novell, *et al.* 1994; Novas 1996*b*), and present in less derived members of the Alvarezsauridae (Novas 1996*b*, 1997*c*), this could represent an autapomorphic loss of this character within that clade. The presence or absence of the extensor groove, then, may have ontogenetic significance (it is absent in the juvenile oviraptorosaur *Microvenator*; P. J. Makovicky, personal communication to Naish). The absence of this groove in the Isle of Wight specimen, therefore, is unreliable as a character of phylogenetic importance.

Naish (2000) noted that the lack of a "wing-like" anterior trochanter (a character unique to Tetanurae; see Gauthier 1986) may indicate that MIWG 6214 does not belong to a tetanuran. However, as other features mentioned above suggest that this specimen is either juvenile or pathological, it was deemed unwise by the author to regard this feature as unequivocal. Furthermore, if MIWG 6214 is a "non-neoceratosaurian not-tetanuran theropod," its presence in the Wealden Group (and in the Lower Cretaceous) would be highly significant, all other such theropods being known only from the Late Triassic to Early Jurassic. Naish (2000) further noted that a number of tetanuran groups had modified the morphology of their anterior trochanter, deviating from the "wing-like" shape. Comparisons with the Isle of Wight specimen provide no useful information.

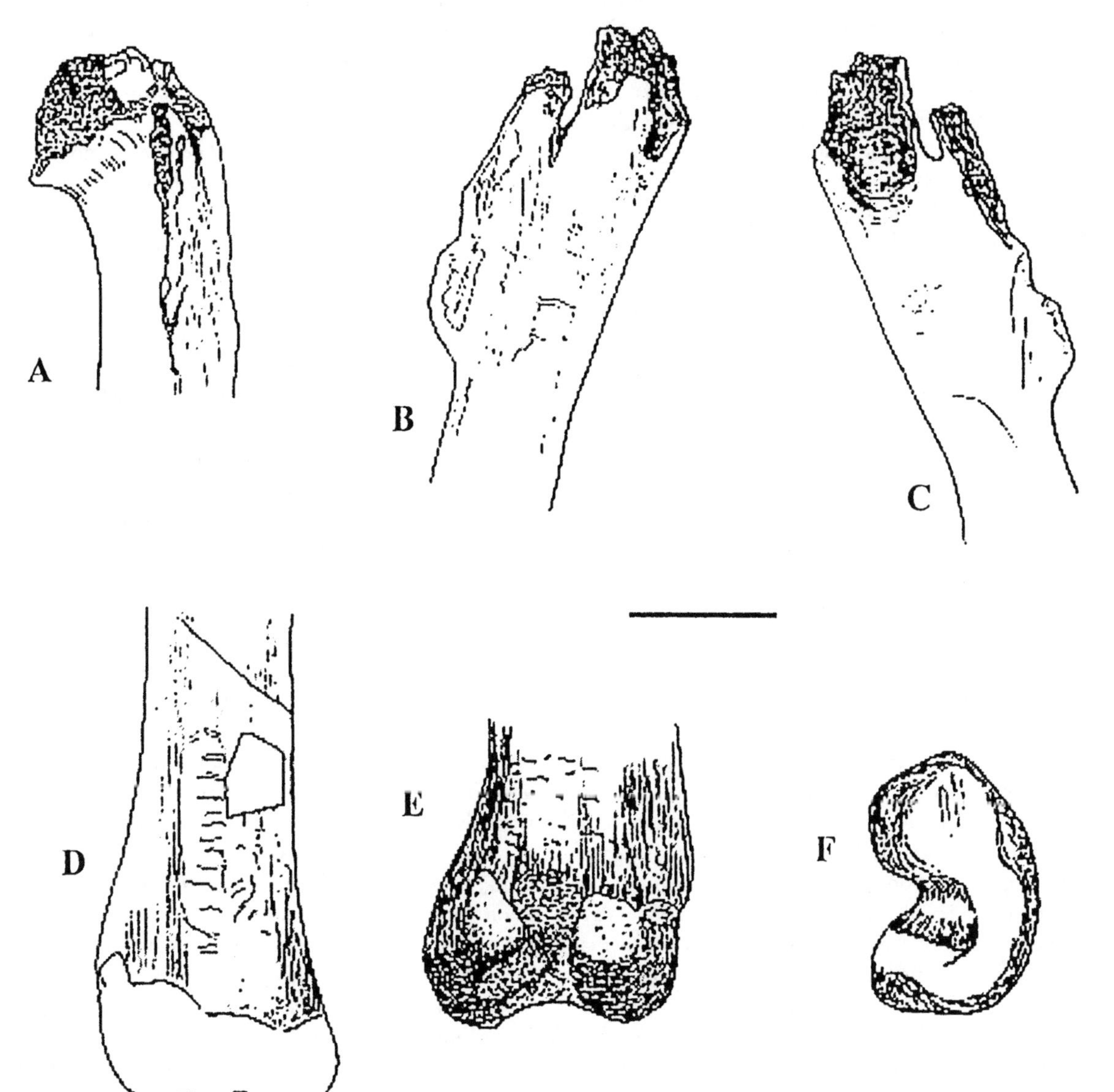

Proximal end of MIWG 6214, femur representing a small theropod from the Lower Cretaceous of the Isle of Wight, in A. cranial, B. lateral, and C. medial views; distal end in D. cranial, E. caudal, and F. distal views. Scale = 1 cm. (After Naish 2000.)

As observed by Naish (2000), MIWG 6214, compared with other theropods, most closely resembles *Microvenator* from the Cloverly Formation of Montana in the general proportions of the femur. There are also differences: The crista trochanteris of *Microvenator* is "broadly separated from the head by a saddle-shaped sulcus" (Makovicky and Sues 1998). Naish (2000) found no indication of such a sulcus in the Isle of Wight specimen, adding that "this area is broken and partially reconstructed." Also in MIWG 6214, it did not seem possible for the anterior trochanter to have projected dorsally like the tall, finger-shaped crest seen in *Microvenator* (see Ostrom 1970; Makovicky and Sues). The absence of a fourth trochanter in *Microvenator* (see Ostrom; Makovicky and Sues), Naish (2000) noted, might preclude a relationship between that genus and MIWG 6214.

Naish (2000) acknowledged that the differences between MIGW 6214 and *Microvenator* are difficult to evaluate, such features in theropods being susceptible to much individual variation. However, the femora in both forms resemble one another. Therefore, "future discoveries may demonstrate the presence of [*Microvenator*], or of a closely related one, in the Wealden Group. Its precise affinities unresolved, Naish (2000) regarded the Isle of Wight femur as an indeterminate member of the Theropoda.

In an abstract, Alcober, Sereno, Larsson, Martinez and Varricchio (1998) reported on the partial skeleton (and mentioned at least one referred specimen) of a new and still unnamed medium-sized theropod discovered in exposures of the Middle Cretaceous (approximately Cenomanian) Río Colorado Formation of Mandoza Province, Argentina. Alcober *et al.* briefly described the skull of this specimen as exhibiting a large canal close to the base of the quadrate, a structure previously not seen in nonavian theropods, suggesting an avian-like course for the pneumatic siphonium. The postcranial skeleton is distinguished by marked camellate pneumacity, and pneumatic cavities invading the centra and neural arches of the vertebrae and apparently extending into the furculum and ilium. A referred astragalus has proportions similar to that in coelurosaurs and an ascending process that is taller than that in any other known allosauroid.

Alcober *et al.* identified the specimen as a late-surviving carcharodontosaurid based on the following synapomorphies of the Carcharodontosauridae: Extreme axial pneumacity; coossification of some medial gastralia as V-shaped struts; and hypertrophied pubic foot. The authors found this new form to have a sister-taxon relationship with *Giganotosaurus* and *Carcharodontosauus*, which is "consistent with an hypothesis of continent-level vicariance during the Early Cretaceous."

Sadleir (1998) briefly reported on the recovery of more than 100 theropod teeth from the Kem Kem beds of southeastern Morocco, these specimens shedding light on a community of Middle Cretaceous (Cenomanian) African carnivorous dinosaurs. The specimens were acquired by surface collection and screening large amounts of sediment from numerous localities. The most common large-bodied forms represented by these teeth were carcharodontosaurids and perhaps spinosaurids (the latter being difficult to distinguish from crocodilian teeth, as both lack serrations), with dromaeosaurids possibly included among the small-bodied forms, and potential juveniles found with the microfossil specimens.

In a somewhat related study, Sadleir and Chapman (1999) briefly described a computer method for identifying isolated theropod teeth, the latter of which are very abundant in the fossil record, often providing the only evidence of a particular taxon. Teeth used in this study included (but were not exclusive to) those belonging to spinosaurids, carcharodontosaurids, and dromaeosaurids.

Sadleir and Chaplan applied geometric morphometric methods to analyze variation in these teeth, both within individuals and within and across taxa, combined coordinate data for tooth base and tip points with tooth outlines, providing input from the entire morphology, and utilized different perspectives for each tooth. From this data, Sadleir and Chapman were able "to construct an empirical morphospace of tooth shape" wherein teeth could be plotted, thereby "identifying morphological, temporal and taxonomic trends." Adding to this information, axes for serration sizes and densities offered more discrimination between shapes and, therefore, the taxa. Unoccupied morphospace areas were explained by the authors in terms of "geometric or functional impossibilities, or, in some cases, are simply unexplored by theropods." In the future, Sadleir and Chapman hope to utilize this technique in identifying less complete, isolated teeth, and also apply it to nontheropod taxa.

In an abstract, Novas, Martínez, Valais and Ambriosio (1999) reported a tooth pertaining to the Carcharodontosauridae from the Mata Amarilla Formation (Upper Cretaceous; Turonian), at a locality bordering the Shehuen River, in the province of Santa Cruz, Patagonia. The specimen was found in association with remains of enormous sauropods and also the teeth of other vertebrate animals. The tooth, missing the root and base of the crown, was briefly described by Novas *et al.* as 105 millimeters in length, 20 millimeters in width, and 45 millimeters anteroposteriorly, with estimated total length and width of 18 by 15 centimeters. The tooth is laterally compressed, with curved margins and serrations, and small denticles. The denticles comprise a series of ridges that are concave toward their bases, this trait being characteristic of carcharodontosaurids; also, slight perpendicular undulations at the based of the crown, a character seen in Carcharodontosauridae and also the abelisaurid *Abelisaurus*. This find is significant in constituting the first known occurrence of the Carcharodontosauridae in this area during Turonian times.

Eberth, Currie, Coria, Garrido and Zonneveld (2000) reported briefly on a recently discovered carcharodontosaurid bonebed in exposures of the Huicul Member of the Middle Cretaceous (Cenomanian) Río Limay Formation in Neuquén, Argentina. The bonebed consists of isolated bones, representing at least six individuals, scattered throughout the base of a multistoried sheet-sandstone that indicates a channel origin. Taphonomic information and the fact that just a single taxon is represented at this site indicated to the authors a mass death of these theropods, also suggesting "some form of pre-mortem community or social structure." Eberth *et al.* did not speculate as to the cause of death.

A number of tyrannosaurid specimens, some of which had been collected decades ago, were reported on for the first time by Carr and Williamson in their comprehensive review of the Tyrannosauridae from New Mexico. These include the following: A large pedal phalanx (AMNH 5882) recovered from the Naashoibito Member of the Kirtland Formation (see section of "Paleocene Dinosaurs," below) of the San Juan Basin, New Mexico, collector and date not recorded; an incomplete skeleton (USNM 365551, not catalogued until 1986) collected during the 1920s or 1930s, including a femur (about 665 millimeters long), right tibia, a metatarsal, and an incomplete pubis, the relatively small size of this specimen indicating a juvenile individual; a femur (NMMNH P-25083), measuring 993 millimeters in length, and a tibia (NMMNH P-25085), 883 millimeters, collected in 1980 from the De-na-zin Member of the Kirtland Formation, the disproportionate lengths of these bones suggesting two individuals; a partial skull and skeleton (NMMNH P-27470) collected by Thomas E. Williamson from the De-na-zin Member of the Kirtland Formation in the Bisti-De-na-zin Wilderness Area, including the rostral end of a left dentary, the pubic peduncle with the acetabular margin of the left ilium, a caudal neural arch, and a caudal centrum, also (collected at a later date) a femur (NMMNH P-25083) associated with some partial caudal vertebrae (NMMNH P-22722), all of this material possibly belonging to one individual; and a specimen (NMMNH P-7199) referrable to *Tyrannosaurus rex* (see *Tyrannosaurus* entry).

Carr and Williamson also reported for the first time that a tooth (AMNH 2479) from the Judith River Formation of Montana, identified correctly by these authors as belonging to an indeterminate tyrannosaurid, is included in the type series of the hadrosaurid *Dysganus encaustus* (see *Dysganus* entry, *D:TE*).

New insights into the feeding behaviour of tyrannosaurids, a group which includes some of the largest carnivorous dinosaurs, as well as that of other kinds of theropods, were recently gleaned from examinations of various fossil bones bearing teeth marks. Jacobsen (1999) briefly reported on tooth-serration marks found on a dentary belonging to an articulated specimen of the theropod *Saurornitholestes*, collected from the Upper Cretaceous (Campanian) Judith River Group of Alberta, Canada. These serrations match those previously found on other bones (*e.g.*, hadrosaurids, ceratopsians, and other tyrannosaurids) which were attributed to the denticles of tyrannosaurids. According to Jacobsen, analysis of these marks suggest that they were inflicted as the result of hunting or scavenging. They are significant in suggesting "that large theropods at least occasionally included small theropods in their diet."

In another brief report, Jacobsen and Ryan (1999) commented on remains of the hadrosaurid *Edmontosaurus*, collected from a bonebed in the Late Cretaceous (lower Maastrichtian) Horseshoe Canyon Formation of Alberta. Some of this material is disarticulated, many elements apparently having come from associated partial skeletons. A good number of these bones have been marked by theropod teeth as punctures or tooth-serration drag patterns. According to Jacobsen and Ryan, many of these tooth marks can be attributed to the tyrannosaurid *Albertosaurus* (the second most common taxon in this bonebed, represented by more than 130 teeth), while those made by small theropods can be referred to *Saurornitholestes*. The great number of tooth marks indicated to the authors that the hadrosaurid remains were manipulated after death by the theropods, possibly constituting evidence "suggestive of scavenging where theropods shed teeth while feeding on an abundance of prey carcasses" (although the hadrosaurs could also have been "manipulated" after being killed by a theropod; R. M. Molnar, personal communication 2000).

Mongelli, Varricchio and Borkowski (1999), in an abstract, discussed wear surfaces and breakage patterns seen on 386 isolated tyrannosaurid premaxillary and lateral teeth collected from the Upper Cretaceous (Campanian) Two Medicine Formation and Judith River Formation of Montana. When compared, the studied specimens fell into three subjective wear classes: 1. little or no wear, 2. slight to moderate wear, and 3. heavy wear. Class 3 teeth also displayed smooth,

Photograph by the author, courtesy Natural History Museum of Los Angeles County.

Skeleton (LACM 7244/ 23844, most of postcrania cast from RTMP specimen) of *Tyrannosaurus rex*, the largest known tyrannosaurid theropod species, mounted for the Natural History Museum of Los Angeles County's "Dueling Dinosaurs" exhibit (see *Tyrannosaurus* entry, *S1*).

almost flat, wear facets, these teeth having broken while in the animal's mouth, and flatted and smoothed during post-break usage. Damaged tyrannosaurid teeth were also compared by these authors with reported damaged teeth in extant carnivores.

According to Mongelli *et al.*, tyrannosaurid premaxillary teeth display the most frequent heavy wear (class 3), while the lateral teeth commonly show light to moderate wear (class 2). Of the examined specimens, 47 percent of the premaxillary teeth, but only 8.4 percent of the lateral teeth, showed smooth, nearly flat wear facets (indicating post-break usage). Several hypotheses were put forth to explain this disparity, "including taphonomic processes, differential rates of tooth replacement, greater stresses or usage [of premaxillary teeth] than lateral teeth, and less resistance to torsional forces than lateral teeth," the authors finding the latter two to be best supported.

The lower jaw in theropods comprises two segments often described as loosely hinged. The anterior segment includes the dentary, splenial, and supradentary bones, the posterior, the surangular, angular, coronoid, prearticular and articular (see Bakker, Williams and Currie 1988). The intramandibular joint (a synapomorphy of Theropoda; Currie 1997), moveable in many theropods, unites these two segments.

In the past, various authors have suggested that in tyrannosaurids these segments were rather loosely joined. Hence, the jaws were relatively weak. Hurum and Currie (2000) performed a study regarding the issue of the crushing bite of tyrannosaurids, their research based upon newly prepared lower-jaw specimens of the tyrannosaurids *Gorgosaurus libratus*, *Tyrannosaurus rex*, and *Tyrannosaurus bataar* (*Tarbosaurus bataar* of their terminology). In this study the authors described in detail for the first time the intramandibular jaw joint and reported on the discovery of the first entirely preserved and fused supradentary/coronoid bones in these taxa.

As described by Hurun and Currie, the supradentary and coronoid bones of the lower jaw in these taxa are fused. This fusion of the two elements crosses the intramandibular jaw joint, thereby restricting its

Photograph by the author, courtesy The Field Museum.

Skull (FMNH PR308, originally referred to *Gorgosaurus libratus*) of the tyrannosaurid theropod *Daspletosaurus torosus*. Recent comparative studies of tyrannosaurid skulls were made independently by paleontologists including Thomas R. Holtz, Jr. and Thomas D. Carr.

movement. This, the authors noted, lead to a lower jaw constituting a secondary specialization for a crushing bite.

Chin, Eberth and Sloboda (1999) reported in an abstract a highly fractured theropod coprolite containing abundant bone fragments and also, more rare, exceptionally well-preserved soft tissue. The phosphaic specimen — attributed to a tyrannosaur because of its large size (64 × 17 × 11 centimeters), composition, and paleontological context — was recovered from the Dinosaur Park Formation (Upper Cretaceous) of Alberta. The specimen includes numerous small, three-dimensional soft-tissue impressions, the most common patterns being striated soft tissues sometimes exhibiting cord-like or fluted morphologies, suggesting undigested muscle tissues. The abundance of pulverized primary fibrolamelar bone preserved in the specimen suggested to the authors that the soft tissues belonged to a subadult dinosaur.

According to Chin *et al.*, the undigested soft-tissue material "suggests a short gut-residence time and may reflect gorging behavior by the theropod." Preservation of this material would have necessitated "rapid lithification [=mineralization] and minimal subsequent diagenetic recrystallization."

Manabe (1999) described an isolated, small premaxillary tooth (IBEF VP 001) belonging to a tyrannosaurid, collected in 1996 by Masatoshi Okura from the Early Cretaceous Jobu Formation, Itoshiro Subgroup, Tetori Group, in Hayashi-dani, Izumi Village, Fukui Prefecture, Japan. In describing the specimen, Manabe observed that the tooth has a well-developed D-shaped cross section. Tooth crown height is 11 millimeters, anteroposterior basal length 4.5 millimeters, and tooth basal width 3.8 millimeters. There are four denticles per one millimeter on the carinae, these being significantly larger than comparably sized tyrannosaurid teeth from the Judith River Formation.

As noted by Manabe, this specimen complements *Sinotyrannus* (see *D:TE*), a primitive tyrannosaurid known only from postcranial remains recovered from the Early Cretaceous of Asia. IBEF VP 001, therefore, constitutes additional evidence supporting the hypothesis that the Tyrannosauridae originated in Asia during the Early Cretaceous, migrating to North America when the two continents were joined by a land connection (*e.g.*, see Buffetaut, Suteethorn and Tong 1996). This interpretation, the author noted, is consistent with the earliest previous Asian fossil record of the Tyrannosauridae, *Alectrosaurus olseni* from the ?Cenomanian Iren Dabasu Formation.

Manabe could not ascertain whether IBEF VP 001 belongs to a juvenile or an adult representing a small species. The specimen differs significantly from typical Late Cretaceous tyrannosaurid premaxillary teeth in lacking a median ridge on the posterior surface of the crown. As the premaxillary teeth of the primitive tyrannosaurids *Aublysodon* and *Alectrosaurus* are not serrated, perhaps the premaxillary teeth of earlier tyrannosaurids were also not serrated. IBEF VP 001 has distinctive serrations, however, which help to polarize this character and weakly supports the monophyly of the tyrannosaurid subfamily Aublysodontinae. Furthermore, Manabe noted, the small sizes of the more primitive Asian theropods are consistent with the cladistic analysis of Holtz (1994), wherein tyrannosaurs are regarded as a derived coelurosaurian group that achieved very large size only during the Late Cretaceous.

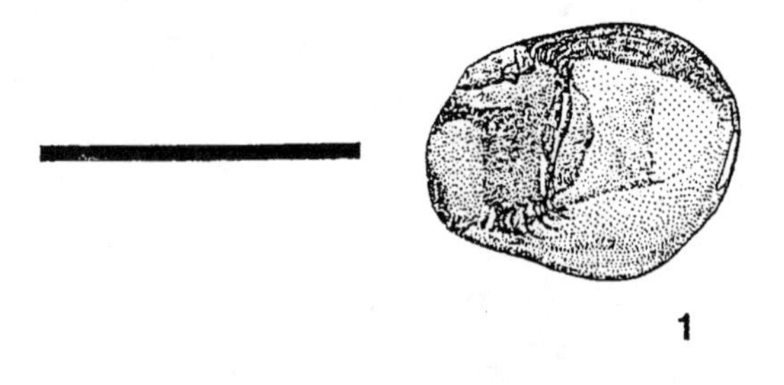

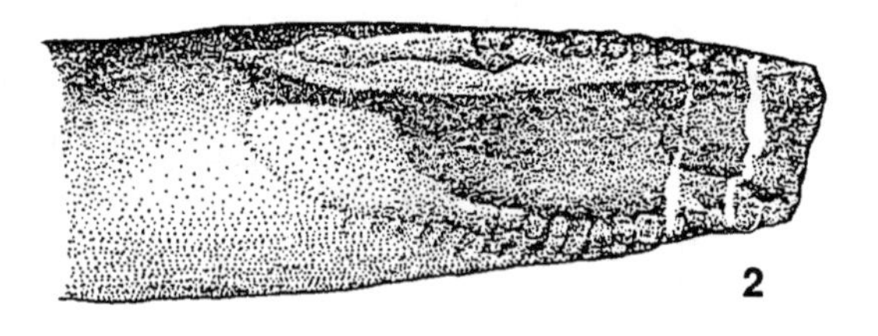

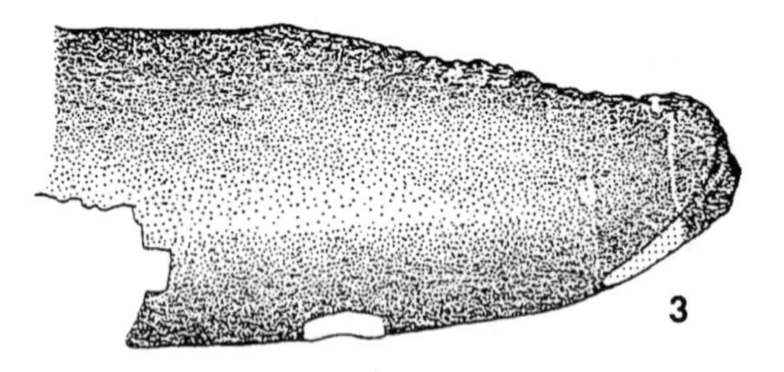

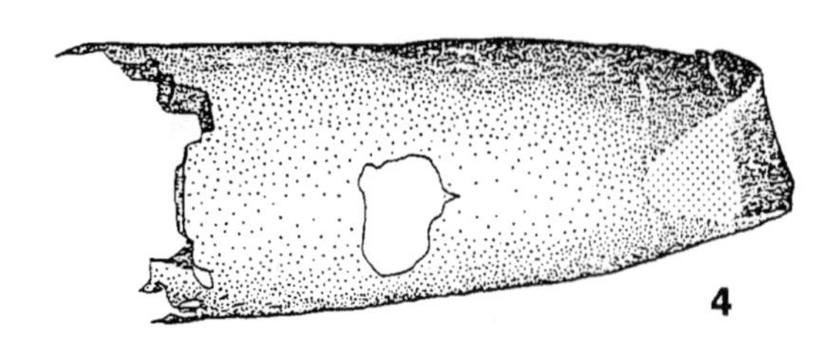

IBEF VP 001, premaxillary tooth of a primitive Asian tyrannosaurid in 1. occlusal, 2. posterior, 3. lateral, and 4. anterior views. Scale = 5mm. (After Manabe 1999.)

Manabe cited Carpenter's (1997*b*) observation that, among the adaptations in becoming top predators, tyrannosaurids increased their bite size by thickening the width of their teeth, especially the premaxillary teeth. Such thickening in IBEF VP 001 is already apparent, suggesting that this adaptation had already started during the Early Cretaceous, and implying that this modification could have predated the increase in body size in the Tyrannosauridae.

Longrich (2000), in an abstract, presented the unusual hypothesis that alvarezsaurids may have been myrmecophagous (ant eating) theropods. Comparing the alvarezsaurid skeleton to various ant- and termite-eating vertebrates (including the aardvark, aardwolf, anteaters, armadillos, echidnas, numbat, pangolins, sloth bear, horned toads, and the snake *Typhlops*), Longrich found these dinosaurs to show features that are common among members of this guild, especially anteaters and pangolins. These features include the following: Short, robust forelimbs with hypertrophied muscular attachments and in-levers that concentrate force through a strong, compact manus and powerful pick-like claw; typical myrmecophagous features as elongate snout, slender mandible, reduced dentition, possibly reduced jaw musculature, and planar jaw hinge; in mononykines, dorsal vertebrae opisthocoelous, neural spines posteriorly forked, dorsals lacking hyposphene-hypantra articulations, zygapophyses and intraspinous ligament attachments aligning with rotational axes of vertebrae (these features possibly allowing extensive dorsal and lateral flexion of the spine); dorsally displaced ribs improving leverage of epaxial muscles during dorsoflexion, rib cage stabilized by having parapophyses and diapophyses in the same horizontal plane, the back perhaps functioning "as a proximal extension of the digging arm, working in concert with it for an up-and-in ripping motion to tear open insect nests."

As further noted by Longrich, myrmeocophay seems to be consistent with the insect fossil record. Also, this "may explain the high rate of skeletal evolution in alvarezsaurs, which has hindered study of their [phylogenetic] relationships."

A new ornithomimid specimen (IGM 100/987), reported by Makovicky and Norell (1998), was discovered at Ukhaa Tolgod (Upper Cretaceous), in the southwestern Gobi Desert of Mongolia, constituting the first ornithomimid occurrence in Djadokhta-like beds and, therefore, adding "yet another taxon to an already astounding diversity of theropod dinosaurs from this locality" (see Norell, Clark and Chiappe 1996). The fragmentary specimen — including the occipital portion of a braincase, several cervical vertebrae, and a cervicodorsal vertebra — was collected in summer of 1993 by the Mongolian Academy of Sciences–American Museum of Natural History expeditions to that locality. As the vertebrae show closure of neurocentral sutures but the braincase is not fused, the individual represented by this specimen probably had not yet attained full maturity. (The remains of a significantly smaller troodontid theropod, to be described later by Mark A. Norell, Peter J. Makovicky and James M. Clark, were also recovered at this time from the same locality.)

Makovicky and Norell briefly described the broken and slightly distorted braincase thusly: The braincase exhibits a high degree of pneumatization. All three tympanic recesses have been preserved, the posterior recess being large, opening into the middle ear as in most other nonavian coelurosaurs. Although the dorsal recess has not been preserved, a groove connects the middle ear to the anterior face of the paroccipital process where this recess is generally found in other advanced theropods.

Autapomorphies of the Ornithomimidae observed by Makovicky and Norell preserved in this specimen include the following: Expansive pneumatization of the basioccipital-exoccipital area dorsal to the basal tubera; and a large depression of the posterior face of the quadrate shaft, the hollowness of this shaft indicating that the quadrate may also have been pneumatic.

As observed by the authors, IGM 100/987 differs in various subtle features from preserved cranial material referred to other advanced ornithomimids. In IGM 100/987, the basal tubera are shallow and widely separated, with a small tubercle in the center of the cleft between them (these tubera are deeper, separated by a narrow cleft in the North American species *Struthiomimus altus* (AMNH 5355) also proportionally deep in the large Mongolian species *Gallimimus bullatus*, although the extent of their separation is unknown). Unlike the condition in *S. altus* (AMNH 5355), there is no midline ridge on the supraoccipital in either IGM 100/987 or *G. bullatus* (IGM 100/12). The quadrate shaft in IGM 100/987 is proportionally more narrow with a more developed posterior fossa than in *G. bullatus* (IGM 100/12), the quadrate fossa also well developed in *Dromiceiomimus samueli* (ROM 840) and *Ornithomimus edmontonicus* (ROM 851). Such differences, Makovicky and Norell pointed out, clearly indicate that IGM 100/987 is distinct from any known ornithomimid taxon.

At present, advanced ornithomimid taxa are primarily diagnosed based upon limb-bone proportions and the form of the manual unguals, elements not preserved in IGM 100/987, while braincase anatomy remains to date undescribed or unknown in many ornithomimid taxa. Therefore, Makovicky and Norell found it impossible to refer the Ukhaa Tolgod specimen to any particular ornithomimid taxon. A more precise taxonomic placement, the authors noted, must depend upon the discovery of additional materials and, most importantly, "a reevaluation of ornithomimid phylogeny based on discrete osteological characters."

Hasegawa, Manabe, Kase, Nakajima and Takakuwa (1999) described an ornithomimid vertebral centrum (GMNH-PV-028) lacking a neural spine, diapophyses, and transverse processes. The specimen was discovered by Shuichi Nakajima and Tomoki Kase in January, 1981, in the brackish (see Matsukawa 1983; Kase 1984) shale of the Sebayashi Formation (Late Barremian to Aptian) in Nakazato-mura, near Kagahara, Nakazato Village, Sanchu Terrane, Tano County, Gunma Prefecture, Japan. (The shale has also yielded abundant disarticulated bivalves, gastropods, and fragments of fossil plants.) The specimen is notable as the first ornithomimid found in Japan and, if body size can be estimated from this single bone, the largest ornithomimid yet known.

Although they did not name this specimen, Hasegawa *et al.* diagnosed it as follows: Highly elongated; spool-like; amphicoelous with pleurocoels. As preserved, the centrum has the following dimensions: Approximately 110 millimeters long; maximum width 80 millimeters; maximum height 75 millimeters; maximum width of neural canal 20 millimeters.

Hasegawa *et al.* identified this specimen as an ornithomimosaur based on its elongated, spool-like shape and as either a dorsal, sacral, or anterior caudal vertebra by its lack of parapophyses and distinct chevron facets. The shape and amphicoelous articular surfaces suggest that the specimen pertains to the Ornithomimidae. Comparing it to elements in other ornithomimosaurs (*e.g.*, *Gallimimus*), the authors determined that its amphicoelous condition and deep pleurocoels identify the specimen as the 13th dorsal or first, second, third, or fourth sacral.

Hasegawa *et al.* found this specimen to resemble most closely the only slightly smaller 13th dorsal vertebra of the ornithomimid *Gallimimus bullatus*. As these authors observed, the flat nature of the ventral surface of the specimen is almost identical in morphology to that Mongolian species (the flat ventral surface developing only on the 13th dorsal or first sacral vertebra in *G. bullatus*).

Kaneko (2000) reported briefly on a new and to

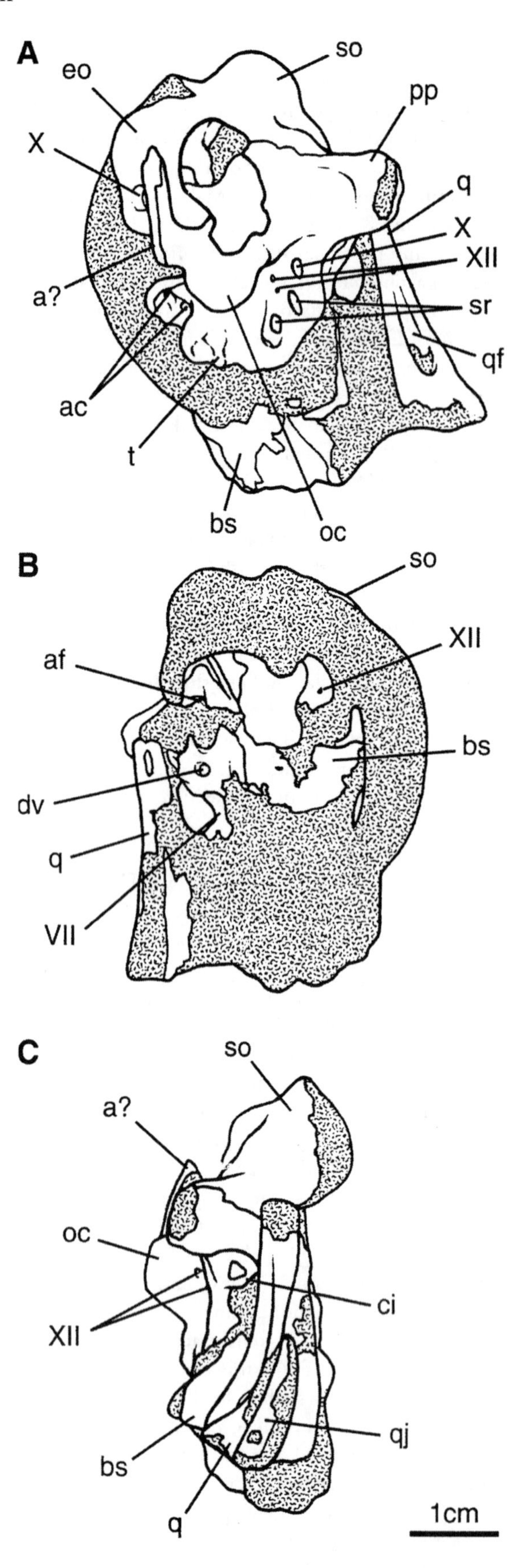

Braincase (IGM 100/987) of an indeterminate ornithomimid from the Ukhaa Tolgod locality in the Southeastern Gobi Desert, Mongolia, in A. posterior, B. anterior, and C. lateral views. (After Makovicky and Norell 1998.)

date undescribed ornithomimid from Thailand. Apparently to be named "*Ginnareeminus*," this new genus is represented by several vertebrae and a metatarsal collected from the Sao Khua Formation in Phu Wiang. Reportedly, the metatarsal of this specimen displays the arctometatarsalian condition.

Koboyashi, Lu, Dong, Barsbold, Azuma and Tomida (1999) presented evidence suggesting that at least some kinds of ornithomimid theropods were herbivorous. As reported by Koboyashi *et al.*, a dozen well-articulated ornithomimid skeletons, ranging in ontogeny from juvenile to adult, were discovered in 1997 in the Ulansuhai Formation (Upper Cretaceous) of China. Within each of these skeletons, preserved inside the ribcage and attached to the medial surface of the articulated dorsal ribs and gastralia, was a gastrolith mass (the latter previously found only in the birdlike genus *Caudipteryx* (see entries, *S1* and this volume) among nonavian theropods). As the authors pointed out, the occurrence and characteristics of such masses indicate that these toothless theropods, as do modern birds, may have possessed gizzards and used grit to grind up plant matter.

As described by Koboyashi *et al.*, these gastroliths — found in the same region of each specimen, generally in the area near the middle dorsal vertebrae — are mostly composed of silicate grains with no bony elements. One isolated gastrolith mass, seemingly belonging to an adult individual (15 × 11 × 3 centimeters) based on the size of its rib impressions, occupies a greater total volume than those of juveniles (10 × 7.5 × 2.5 centimeters), this being "indicative of a body-size to total gastrolith volume like that observed in crocodilians" (see Cott 1961). Most of these gastrolith grains measuring less than one millimeter in diameter vary in shape from angular to very angular; those between one and two millimeters are mainly subangular. Excluding the subangular to angular grains, grains average at about 2.41 millimeters, fitting the predicted one based on values in modern birds.

Koboyashi *et al.* concluded that the large number of gastrolith grains found in these ornithomimid specimens "are consistent with a herbivorous diet and the possession of a muscular ventricular stomach, or gizzard, like that found in modern herbivorous birds."

Gastroliths had previously been reported in the small theropod *Caudipteryx* (see *S1*). According to Koboyashi *et al.*, as *Caudipteryx* or the bird clade and the Ornithomimidae are not closely related taxa, the presence of these masses within the skeletons of the Chinese ornithomimids suggests that they are the result of convergence, herbivory apparently having evolved in the Theropoda more than once.

An impact fracture in a right fourth metatarsal (SMP VP-971) belonging to an indeterminate ornithomimid was described by Sullivan, Tanke and Rothschild (2000). This unusually well-preserved specimen was discovered in 1997 by Robert M. Sullivan in the De-na-zin Member of the upper Kirtland Formation (late Campanian), in the San Juan Basin, New Mexico, in the same general locality that had recently yielded the tibia of a juvenile *Ornithomimus antiquus* previously described by Sullivan (1997; see *S1*). Associated with this specimen were two unidentified bone fragments, one of which may be the greater portion of the medial shaft of the third metatarsal.

Sullivan *et al.* could not refer SMP VP-971 to any particular ornithomimid genus. However, although this specimen is shorter, it compares quite favorably with the right fourth metatarsal of an incomplete skeleton (ROM 41966) identified only as "Ornithomimidae indet." The authors, therefore, regarded SMP VP-971 as an indeterminate member of the Ornithomimidae. As the bone is comparatively small and its surface "is slightly roughened with a wood-like long-grained texture" (*sensu* Sampson, Ryan and Tanke 1997) Sullivan *et al.* extrapolated that the bone probably represents an immature individual.

As observed by Sullivan *et al.*, SMP VP-971 is notable for the distinct, partially remodeled fracture borne near its distal end. Sullivan *et al.* interpreted this pathology "as a comminuted [reduced to powder or pulverized] impact fracture, " which is significant as (to the best of the authors' knowledge) the first such fracture ever reported in any kind of dinosaur. X-ray examination of the fracture by the authors revealed a foreshortening of the bone with a slight malalignment (angulation) related to a healed comminuted impact or stove fracture. Moreover, the partially healed splint can be plainly seen in the distal part of the fracture. This kind of healed fracture, the authors pointed out, is easily distinguishable from the postmortem, taphonomy-related fractures caused by the pressures of sediment, the latter type — often observed to have affected the hollow-shafted long bones in both small and large theropods (D. Tanke, field observation) — being located just proximal to the fracture sight.

Sullivan *et al.* speculated that this injury was probably the result of some traumatic event involving a relatively low impact with the substrate (*e.g.*, "the dinosaur landing on uneven ground, during a running or jumping episode"). The partially remodeled nature of this bone illustrates that this ornithomimid survived the event that resulted in the break. Furthermore, "the smooth, raised interdigitated edge, or callus, clearly indicates that this phenomenon occurred in life, and that the affected individual survived for an indeterminate length of time, possibly up to several weeks duration."

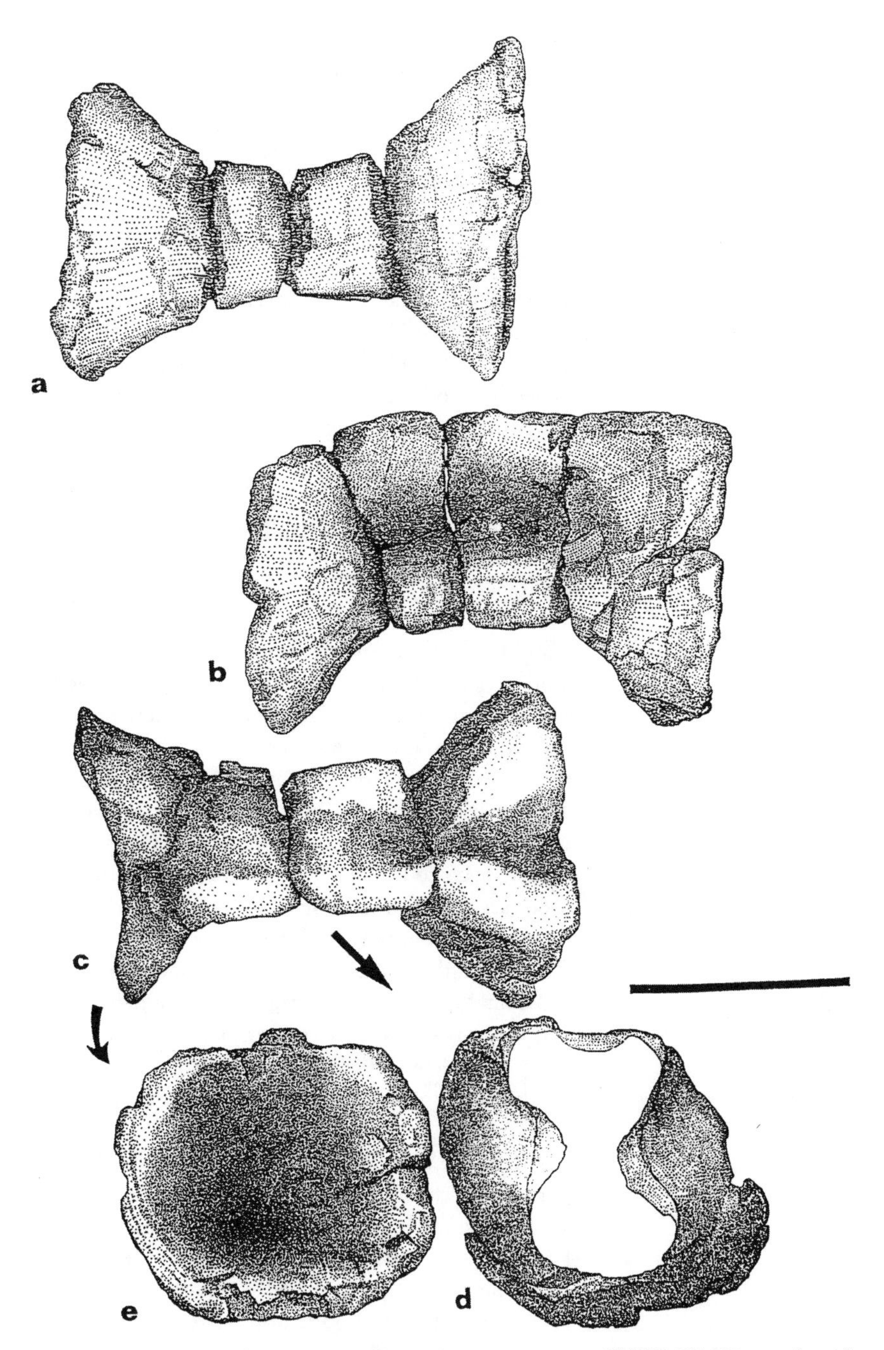

GMNH-PV-028, vertebra of a very large ornithomimid from Japan, in a. ventral, b. lateral, c. dorsal, d. cross sectional, and e. posterior views. Scale = 5 cm. (After Hasegawa, Manabe, Kase, Nakajima and Takakuwa 1999.)

Weishampel, Fastovsky, Watabe, Barsbold and Tsogtbataar (2000), in an abstract, briefly reported on three partial, yet articulated, oviraptorosaur embryos recently collected from a single nesting site in the Nemegt Formation (?early Maastrichtian) at Bugin Tsav, southwestern Gobi area of Mongolia, by members of the Hayashibara Museum of Natural Sciences–Mongolian Paleontological Center Joint Paleontological Expeditions. The embryos were preserved in eggs of the family Elongatoolithidae with lineartuburculate external ornamentation. As noted by Weishampel *et al.*, these embryonic remains "indicate for the first time asynchronous, within-nest development and, presumably therefore, asynchronous hatching." They are also, the authors pointed out, the first known fossil embryos of Nemegt age.

Clark, Norell and Chiappe (1999) reported on yet another oviraptorid skeleton found preserved atop a nest of oviraptorid eggs in a posture suggesting that taken by both extinct and extant birds while brooding their nests. The specimen, the partial skeleton of an adult individual, was discovered in the Upper Cretaceous rocks of Djadokhta Formation, in Ukhaa Tolgod, Mongolia. The find marked the fourth among 17 skeletons recovered from this formation that were preserved in this "brooding" type position (see *Oviraptor* entries, *D:TE* and *S1* for more information on other specimens, *S1* and p. 18, this volume, for photographs).

The seemingly undisturbed skeleton was preserved with the body centered above the nest, its thorax intact. Clark *et al.* briefly described this skeleton as follows: Three pairs of ribs articulate with the costal margin of the sternum (as opposed to two pairs in more basal theropods, but fewer than in all known birds except some moas [*Dinornis*]). Posterior to this margin is a lateral process on each plate, analagous to the lateral xiphoid process of the sternum in birds. The anterior thoracic ribs have one ossified ventral segment like that in extant birds, unlike the two segments in crocodilians. The ribs have free uncinate processes, resembling those in *Velociraptor* and the extinct birds *Chaoyangia* and *Hesperornis*.

The authors noted that this new specimen offers additional evidence suggesting that brooding was a consistent behavior of oviraptorids; furthermore, as the skeleton contacts the eggs directly, "the nest was not buried, unlike in crocodylians but as in avians primitively."

Later, Clark, Norell and Barsbold (2000) reported in an abstract that two new oviraptorid species are represented by the material recently found in the Djadokhta Formation at Ukhaa Tolgod.

The first species reported by Clark *et al.* (2000) is represented by two adults on nests (including the specimen previously described previously by Clark, Norell and Chiappe), an exceptionally well-preserved skull, and an embryonic skeleton. As briefly described by these authors, the skull of this species has an anterodorsally sloping occiput, a parietal that extends further anteriorly, and a vertical ascending process of the maxilla. The cervical vertebrae are elongate, possibly having a functional relationship to those specializations of the skull. There is an epipterygoid, with a small, delicate coronoid demonstrating that the former was not entirely absent in oviraptorids. Among the recovered specimens is one pertaining to a relatively large individual, its skull having a midline length of 17.2 centimeters. The authors noted that the second specimen found sitting on its nest, like the one described earlier by Clark *et al.* (1999), was preserved in a birdlike position, its body contacting the eggs, its arms spreading around the perimeter of the nest.

The second new species reported by Clark *et al.* (2000), represented by three almost complete articulated skeletons, most closely resembles *Conchoraptor*. The former differs from *Conchoraptor*, however, in possessing fused nasals and a more horizontal narial opening. It differs from all other known oviraptorids in that metacarpal III is reduced proximally and does not contact the carpus.

The authors further noted that the exceptional preservation of this new material sheds light upon the anatomy of the poorly-preserved type specimen of *Oviraptor philoceratops*, allowing for a new description of this specimen and a revised diagnosis of this species (to be published).

Triebold, Nuss and Nuss (2000) announced in a brief preliminary report a new North American oviraptorid known from two skeletons including disarticulated skull elements, the postcrania complete save for the distal caudal vertebrae and the metatarsals. The specimens were found in October, 1998, by fossil hunter Fred Nuss of Nuss Fossils approximately 50 meters apart on private ranchland in an exposure of the Hell Creek Formation in Harding County, southwest of Buffalo, South Dakota. In June of the following year, the specimens were identified as oviraptorid by Michael Triebold of Triebold Paleontology, Inc. Although about the same size, one skeleton is gracile and the other robust, suggesting to the authors sexual dimorphism.

As briefly described by Triebold *et al.*, the edentulous skull bears a large nasal crest; front limbs measure one meter in length with manual ungual phalanges attaining 18.4 centimeters in length. Standing almost two meters tall at the hips, these skeletons, according to the authors, "are dramatically larger than any other known oviraptor."

Frankfurt and Chiappe (1999) reported on an isolated, relatively small, and almost complete cervical vertebra (MACN-622) recently recovered from a small quarry in the middle section of the Lecho Formation (Upper Cretaceous) in the Estancia El Brete, in the province of Salta in northwestern Argentina. The specimen was associated with remains of the sauropod *Saltasaurus loricatus*, the theropod *Noasaurus leali*, various indeterminate theropod teeth (see Bonaparte and Powell 1980), and some enantiornithine birds (Walker 1981; Chiappe 1993). The authors described this specimen as bearing pneumatic foramina (a theropod feature; see Gauthier 1986) and epipophyses, the latter not yet known in any other nonavian theropod.

Photograph courtesy Michael Triebold.

Skeleton of an unnamed and undescribed giant (more than 7 feet tall, slightly less than 9 feet in length) oviraptoroid. This mount is a composite comprising the remains of two individuals discovered by Fred Nuss of Nuss Fossils, Otis, Kansas, in Harding County, South Dakota, with preparation and casting by Triebold Paleontology, Inc., Woodland Park, Colorado. It represents at least 90 percent of the skeleton (missing elements based upon material from the Sandy Quarry in the Hell Creek Formation, near Buffalo, South Dakota, with little conjecture).

Performing a cladistic analysis, Frankfurt and Chiappe found that MACN-622 is most closely allied with the caenagnathid *Chirostenotes* and oviraptorids, with therizinosaurids and a new taxon represented by a single vertebra from Como Bluff, Wyoming (see Makovicky 1997, and "Introduction," *S1*) as successive outgroups. It was Frankfurt and Chiappe's conclusion that MACN-622 may represent a member of the Oviraptorosauria or a close relative of that clade. If correct, their assessment offers new and stronger evidence supporting the once weakly supported hypothesis that this group of dinosaurs lived in Gondwana during Cretaceous times.

Frankfurt and Chiappe stated that MACN-622 probably represents a new theropod taxon, being different from all described cervical remains from South America. However, the authors postponed proposing a new name for this taxon until additional materials might be recovered.

Manabe, Barrett and Isaji (2000) reported one of the earliest known members of the Therizinosauroidea, an unnamed form known only from a single manual ungual from the Kuwajima Formation (Early Cretaceous, Valanginian or Hauterivian) of Shiramine, Ishikawa, Japan. As observed by Manabe *et al.*, this specimen has a pronounced lip posterodorsal to the articular surface of the ungual (a synapomorphy of the unnamed clade Oviraptorosauria plus Therizinosauroidea).

In an abstract for a poster, Sankey and Brinkman (2000) briefly reported on more than 1,700 theropod teeth recovered via intensive screening of microverte-

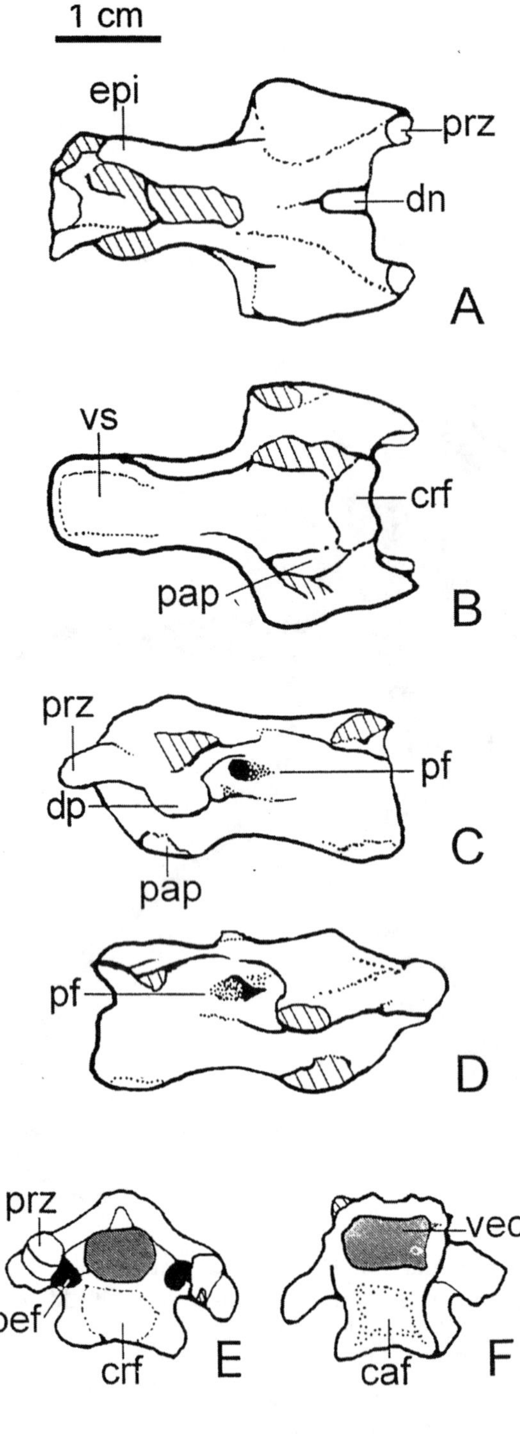

Cervical vertebra (MACN-622) of a possible oviraptorosaur from El Brete, in A. dorsal, B. ventral, C. left lateral, D. right lateral, E. cranial, and F. caudal views. (After Frankfurt and Chiappe 1999.)

brate sites from the Upper Cretaceous (Campanian) Judith River Group of Alberta. The material was recovered during the past 15 years by the Royal Tyrrell Museum of Palaeontology. Study of these teeth revealed several new taxa or morphotypes, these including the following: One new morphotype of the dromaeosaurine genus *Dromaeosaurus*, three new morphotypes of the dromaeosaurid *Saurornitholestes*, two of the dromaeosaurid *Paronychodon*, and one new species of the caenagnathid *Richardoestesia* (now the generally accepted spelling; see *D:TE*). Also recognized by Sankey and Brinkman from this study is that *Saurornithoides* is a valid genus. Fossil bird teeth were also recognized from this collected material.

As Sankey and Brinkman pointed out, understanding the reasons for the high diversity in this rich and well-studied unit, and also for the possible decrease in diversity in later Maastrichtian times, is essential in attempting to explain the extinction of theropods at the K-T (Cretaceous–Tertiary) boundary.

Gishlick (2000), in an abstract, evaluated the climbing capability of basal manuraptoran theropods. Functional studies of the forelimbs of *Deinonychus* suggested to that author that basal manuraptorans were not suited to a scansorial life style. As Gishlick pointed out, the wrist in scansorial animals flexes in a dorsal to ventral plane when propelling the body forward, moving the hands parallel to the direction of body motion. However, in manuraptorans, the semilunate carpal limits the wrist, preventing it from moving in a dorsal-ventral plane. The elbow flexes in a dorsal to ventral plane and the semilunate carpal in the lateral plane, unfolding the manus into the forearm. Consequently, the forearm moves perpendicular to the body, pulling the hands toward the body or pushing the hands away from it.

According to Gishlick, 1. the semilunate carpal and resultant restricted wrist motion would be inefficient for vertical climbing, as there would be a tendency to pull the animal into, rather than along, the surface it was climbing; 2. lateral folding of the wrist would twist the claws out of the surface to which the animal was supposedly clinging; and 3. the contortions required of the body to accommodate the structural constraints of the forelimb would make climbing utilizing four limbs awkward.

Gishlick's study concluded that, while basal manuraptorans may have been able to climb trees, they were not adept quadrupedal climbers; also, that scansorial habits did not constitute the primary biological role of the manuraptoran forelimb.

Anthony R. Fiorillo (1998), a paleontologist who has published a number of studies of the teeth of various kinds of dinosaurs (see *D:TE*, *S1*), presented a new study on the microwear observed on a total of 20 teeth belonging to various theropods (*Dromaeosaurus albertensis*, *Saurornitholestes langstoni*, *Troodon formosus*, *Richardoestesia gilmorei*, and also a large unnamed taxon informally referred to as "Theropod 'A'"). The specimens were all yielded by quarries from the approximate middle third of the Judith Rover Formation of south-central Montana.

As Fiorillo noted, the only wear patterns observed

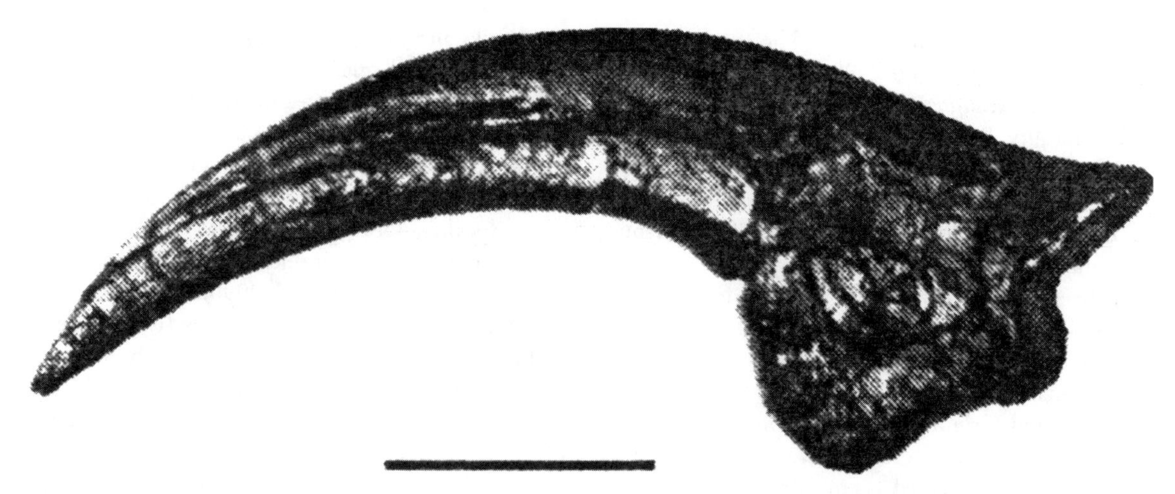

Manual ungual of a new therizinosauroid from the Kuwajima Formation of Japan. Scale = 5 millimeters. (After Manabe, Barrett and Isaji 2000.)

on any of these teeth — examined with a scanning electron microscope with 200×–600× magnification — were fine scratches, the pattern extending from the wear facet of each tooth to regions close to the facet. Consistent with modern-day analogues that consume soft food items, this pattern suggested to Fiorillo that either these theropods were eating meat but not bone, or that one or more of them was omnivorous and consuming soft plant material. According to that author, the limited number of theropods in this ecosystem may be indicative of an environment with poor resources for carnivorous dinosaurs.

Mongelli and Varricchio (1998), in an abstract, reported the collection of 86 theropod teeth from the Lower Two Medicine Formation (Upper Cretaceous: Campanian) of northwestern Montana, over three field seasons. In addition to teeth identified as belonging to *Dromaeosaurus*, *Saurornitholestes*, *Richardoestesia*, tyrannosaurids, and one tooth indeterminate, the assemblage included five teeth apparently pertaining to a new velociraptorine taxon. As briefly described by these authors, the latter is substantially larger than those of any other known Judithian velociraptorine.

In a preliminary report published as an abstract, Xu (2000) reported on a theropod tail section pre-

Skull (cast of holotype AMNH 5356 without plaster reconstruction; see *D:TE*) of *Dromaeosaurus albertensis*. New studies are underway by paleontologists including Anthony R. Fiorillo, Anthony Mongelli, Jr. and David J. Varricchio regarding the teeth of this theropod and other carnivorous dinosaurs.

Photograph by the author, courtesy Royal Tyrrell Museum/Alberta Community Development.

serving filamentous integument. The specimen was collected in the summer of 1999 from the same quarry that yielded *Caudipteryx* in the Yixian Formation, Sihetun locality, Beipiao, in western Liaoning, China. It belongs to a juvenile individual representing a new species of manuraptoran theropod, as indicated by a suite of unique characters not found in other manuraptorans. According to Xu, this specimen "is extremely similar to early birds in a few features but resembles oviraptorosaurs, dromaeosaurids and troodontids in other features."

Remains of small theropods have been reported by Csiki and Grigorescu (1998) from the Densus-Ciula Formation at Valioara. Two teeth, originally mixed with and regarded as crocodilian teeth (MAFI V.12685, in the depository of the Hungarian Geological Survey), were identified as possible theropod teeth by the authors in 1994. Their "laterally compressed, pointed, distally recurved shape and serrated carinae" are suggestive of small manuraptorans. From the same area were collected two frontals (MAFI v.1 3528), misidentified by Nopcsa as the hadrosaurid *Telmatosaurus* (see entry), which Jianu and Weishampel (1997) identified as a possible undetermined arctometatarsalian theropod.

During the late 1970s, more remains of small theropods were discovered in the Hateg Basin, including "coelurosaurian" teeth from different localities within the Sinpetru Formation (see Grigorescu 1984*a*, 1984*b*; Grigorescu, Hartenberger, Radulescu, Samson and Sudre 1985), and, a small, partial skull roof collected in 1992 from the Sibisel Valley, identified by Weishampel and Jianu (1996) as that of a dromaeosaurid apparently closely related to *Saurornitholestes langstoni* (see below).

Excavations in the Hateg Basin carried out after 1978 produced additional remains of small theropods: shed teeth being collected between 1979 and 1996 from various localities of the Densus-Ciula (Valioara) and Sinpetru (Sibisel Valley) formations. These included laterally compressed, pointed and sharply recurved teeth (FGGUB R.1271, R.1321, R.1322, R.1428, R.1430, and R.1582) identified as belonging to velociraptorine dromaeosaurids; conical, laterally compressed (though less so than those identified as "velociraptorine" or "eurynchodont") teeth (FGGUB R.1318, (?)R.1319, R.1320, and MAFI v.12685a,b) identified as "troodontid-like"; a small, elongated, strongly recurved, and pointed tooth (FGGUB R.1431) closely resembling those of *Euronychodon portucalensis*; and a partial laterally compressed, pointed, and curved distally and slightly lingually, indeterminate tooth (FGGUB R.1583) suggestive of the primitive ornithomimosaur *Pelecanimimus*, plus recently collected specimens from the Fantanele microvertebrate site at Valioara.

Csiki and Grigorescu also reassessed other previously described small theropod remains, including those referred to the derived manuraptoran *Eolopteryx* and noncoelurosaurian tetanuran *Bradycneme* (=*Heptasteornis*) (see entries). The small, slender tibia FGGUB R.252, labeled "theropod," collected at the Tustea locality, does not belong to a theropod, the authors noted, lacking as it does a crest for the articulation of the fibula (see Benton 1990).

Based on the studied materials (isolated teeth and limb fragments were treated separately, as no associated remains have been collected), Csiki and Grigorescu discovered an unexpected diversity at the top of the food chain of the Hateg Basin fauna (see also Jianu and Weishampel). Study of the small isolated teeth led the authors to identify at least three and perhaps four different taxa — a velociraptorine dromaeosaurid, a "troodontid-like" small theropod, an "euronychodont," and possibly another having affinitities with primitive ornithomimosaurs. Other remains suggested possible clades represented by the small *Elopteryx* and *Bradycneme*, while additional material, a distal femur (FGGUB R.351) referred by Grigorescu and Kessler (1981) to *Elopteryx*, could represent a small abelisaurid.

Csiki and Grigorescu observed that the recently described Mongolian ?avetheropod, *Bagaraatan ostromi* (see *S1*), has some of the unusual features present in the isolated theropod hindlimb material from the Hateg Basin. Possibly, then, some of these elements from Romania belong to a single taxon closely related to *B. ostromi*. If true, this constitutes supporting evidence for Asian affinities of the Hateg fauna (previously suggested by Csiki 1995).

As the authors noted, the most striking aspects of the Hetig Basin theropod assemblage were its previously unsuspected diversity, and (while all other known contemporaneous faunas seem to have had large theropods at the top of their food chains) the absence of any large "top" theropod, the latter case distinguished as the first ever reported for Late Cretaceous faunas. The authors suggested that this phenomenon is probably linked to the restricted, insular, small-area habitat of the Hateg fauna.

Makovicky, Norell and Clark (1999) reported a new troodontid theropod discovered by the Mongolian Academy of Sciences–American Museum of Natural History expeditions at the Ukhaa Tolghod locality, Peoples Republic of Mongolia. Representing both adult and juvenile specimens, some of this material is preserved in three dimensions. The new taxon was identified by these authors based on its display of the following apomorphies of the Troodontidae: Enlarged otosphenoidal crest delimiting large "lateral depression"; dentary having lateral groove for nutrient foramina; and distal caudal vertebrae having a groove

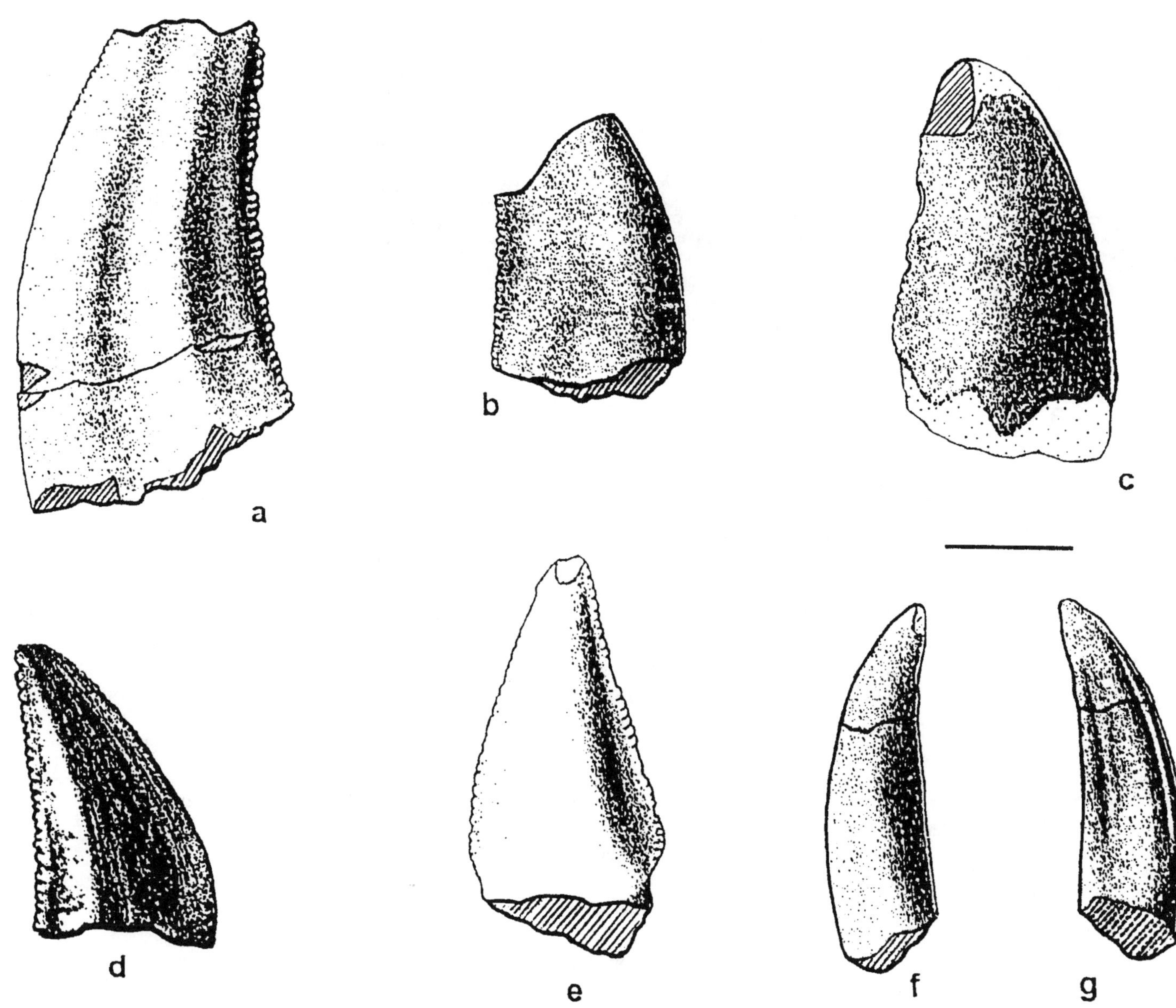

Small theropod teeth from the Hateg Basin: Velociraptorine dromaeosaurid, a. FGGUB R.1428, b. FGGUB R.1430, c. FGGUB R.1322, d. FGGUB R.1580, all labial view; "troodontid-like" theropod, e. FGGUB R.1320 (?anterior dentary tooth), cf. *Euronychodon* (FGGUB R.1431), in f. labial and g. lingual views. Scale = 3 mm. (After Csiki and Grigorescu 1998.)

along dorsal midline. It is unique among known troodontids in possessing the following characters: Teeth unserrated; antorbital and accessory fenestrae separated by depressed interfenestral bar. Troodontid anatomical details seen in this material for the first time, thanks to the preservation, includes the presences of a secondary palate and complex sinus system in the snout.

According to a preliminary analysis performed by Makovicky *et al.*, this new taxon seems to be closely related to such species as *Troodon formosus* and *Saurornithoides mongoliensis*. Although "the fragmentary nature of most troodontid taxa makes any conclusions tentative," this new Mongolian find "underscores that the known radiation of troodontids occurred mainly in Central Asia."

Novas, Apesteguia, Pol and Cambiaso (1999) briefly reported on a probable troodontid — if that, the first known troodontisaurid in the Southern Hemisphere — represented from postcranial remains, including a complete hind foot, collected from the Portezuelo Formation (Upper Cretaceous; Turnonian–Coniacian), in the Río Nequén Supergroup, of Patagonia. The unnamed new taxon was tentatively referred by Novas *et al.* (1999) to the Troodontidae because it exhibits a combination of traits observed only in this group, including the following: Arctometatarsalian condition of the metatarsus; second digit of pes having raptorial phalange; articulation between phalanges 1 and 2 of second pedal digit tending towards hyperextension; and claw of second digit of pes strongly recurved and laterally compressed. However, the Patagonian form differs from troodontids of the Northern Hemisphere in exhibiting characters (not

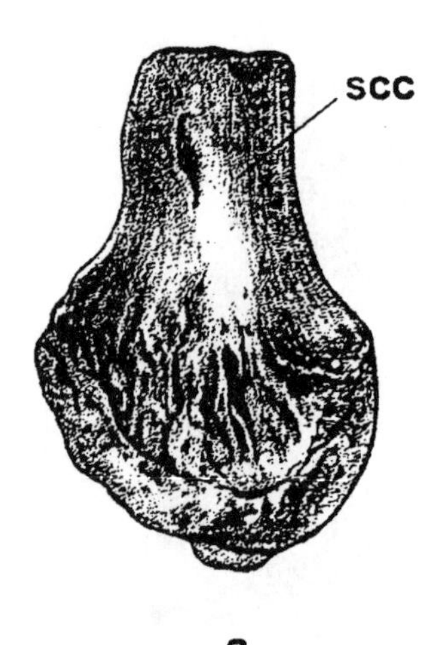

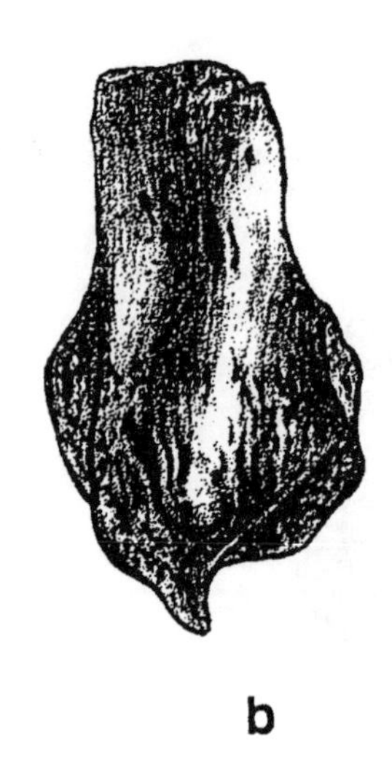

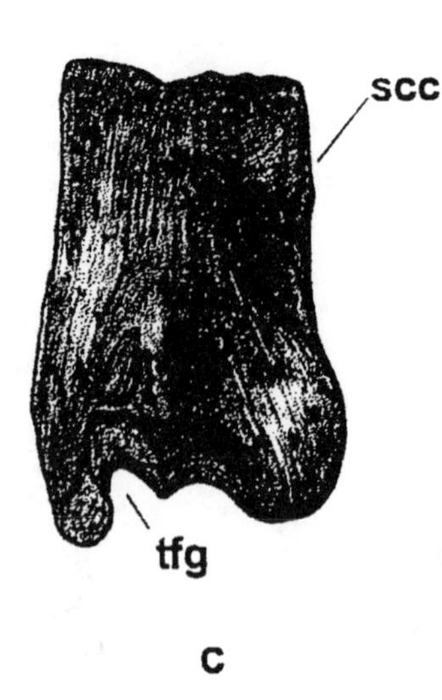

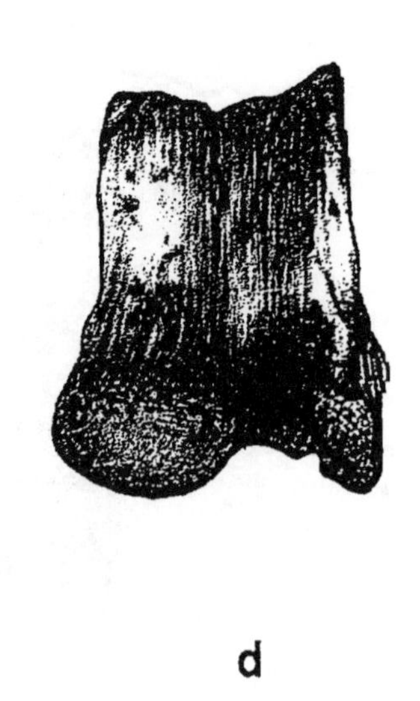

Distal femur (FGGUB R.351) possibly belonging to a abelisaurid, in a. medial, b. lateral, c. caudal, and d. cranial views. Scale = 10 mm. (After Csiki and Grigorescu 1998.)

yet stated by the authors in this preliminary report) that warrant its referral to a new taxon.

In an abstract, Martinez, Lamanna, Smith, Casal and Luna (1999) reported on a new theropod based on two specimens from the Middle Cretaceous (?Cenomanian) Bajo Barreal Formation of Chubut, Patagonia. The first and larger specimen consists of associated remains including a large but incomplete ungual phalanx of digit II of the left pes. Resembling the same ungual in the slightly geologically younger *Megaraptor* (see *S1*), the Patagonian ungual was briefly described by the authors as follows: Transversely compressed, with grooved lateral and medial surfaces, medial groove more nearly dorsal than lateral; flexor tubercle dorsoventrally low, articular surface keeled; metatarsal distally robust, considerably more massive than metatarsal III in *Megaraptor* (suggesting heavier pedal construction).

The previously discovered second specimen comprises a dorsal vertebra, three caudal centra, a complete rib, partial metatarsal II, and manual and pedal unguals. According to Martinez *et al.*, the foot elements in this and also the above specimen have the same morphology, suggesting that they belong to a single taxon. The dorsal vertebra shares some features (*e.g.*, high neural arch with three large and deep caudal pneumatic fossae, widely-spaced suboval articular facet, low neural spine, marked hapidocoel, short and low-angled centrum with small pleurocoel) with the dromaeosaurid *Deinonychus*; the second metatarsal, however, lacks the deep distal groove found in dromaeosaurids. The phylogenetic relationships of this theropod must, therefore, remain uncertain until additional material is recovered.

Currie and Varricchio (2000) briefly reported on a new and relatively small dromaeosaurid theropod, based on a partial skull discovered in the Horseshoe Canyon Formation (Upper Cretaceous; uppermost Campanian to lowermost Maastrichtian) of Alberta, Canada. As briefly described by Currie and Varricchio, the skull differs from that of *Bambiraptor* (see entry), *Saurornitholestes*, and *Velociraptor* in having a short, deep face, and teeth that "are more strongly inclined towards the throat than they are in most other dromaeosaurids." Cladistic analysis suggested that this new form may belong to an independent lineage tracing its origins back to the Early Cretaceous *Deinonychus*.

Additionally, Currie and Varricchio mentioned a specimen from the Two Medicine Formation of Montana that could represent another new genus.

Is Spinosuchus *a Theropod?*

Spinosuchus (deriving from the Latin *spina* for "spine" and Greek *souchos* for "crocodile"), named and described by German paleontologist Friedrich von Huene in 1932, was listed in *D:TE* among the "excluded genera." Recently, however, its status as a possible theropod dinosaur has been reconsidered.

During the summer of 1921, fossil collector and paleontologist Ermin C. Case discovered the major part of a presacral vertebral column (UMMP 7507) preserved in a light cream-colored clay in the Tecovas Member of the Upper Triassic (late Tuvalian, a substage of the Carnian) Dockum Formation of Crosby County, a few miles north of Cedar Mountain, in west Texas. The specimen, found lying in natural position, comprised 22 somewhat damaged (probably eight cervical and 14 dorsal) vertebrae. The bones had been somewhat crushed before permineralization commenced. At the time of their discovery, the vertebrae were hailed as "the largest series of Dinosaur vertebrae yet found in a free condition in the Triassic of North America." Also found in these same beds were various new taxa of phytosaurs and stegocephalians ("plate-headed" tetrapods) (Case 1922).

Case, in a preliminary description, referred the remains to the theropod *Coelophysis* sp., this original material representing an animal probably measuring about 2.5 meters (less than 8.5 feet) in length. (Case also referred other material from the Tecovas Member of the Dockum Formation of Texas to the Dinosauria. These specimens include a left femur [UMMP 3396] which, according to Hunt, Lucas, Heckert, Sullivan and Lockley 1998, belongs to an aetosaur, possibly *Desmatosuchus*; also various other specimens; see *Caseosaurus* and *Coelophysis* entries; see also Heckert, Lucas and Sullivan 2000. As also pointed out by Heckert *et al.*, a specimen previously reported as dinosaurian by Case 1916 from the San Pedro Arroyo Formation, near Socorro, New Mexico, is actually the femur of a phytosaur). Later, in addition to the vertebral material, Case (1927) also referred to *Coeplophysis* sp. an ilium (UMMP 8870) and a basicranium (UMMP 7473).

In a more detailed description following additional preparation of the material, Case (1927) observed that the vertebral centra are somewhat low, the cervical column inclining upward. Case (1927) noted that the "obviously striking thing about the vertebral column is the presence of elevated neural spines," some of them measuring as much as three times the length of a centrum. As later suggested by other workers (*e.g.*, Long and Murry 1995), these spines possibly constituted the basis of a low dorsal "sail" or hump.

The above remains were later redescribed by Huene (1932), who, believing them to represent a primitive theropod of the family "Podokesauridae" [=Coelophysoidea], though one different from *Coelophysis*, referred them to the new genus and species *Spinosuchus caseanus*. At the same time, Huene designated the vertebral column as the holotype (this specimen now mounted in plaster as a half-relief exhibit at the University of Michigan Museum of Paleontology, in Ann Arbor).

Since the middle 1970s (see Zhang 1975), most paleontologists have regarded *Spinosuchus* as a nondinosaurian archosaur (*e.g.*, Padian 1986; Murry and Long 1989). Chatterjee (1985) recognized the braincase referred to *S. caseanus* as belonging to the rauisuchian *Postosuchus kirkpatricki*. Long and Murry (1995) identified the referred ilium as that of a herrerasaurid, removing it from *Spinosuchus* and

***Spinosuchus caseanus*, UMMP 7507, holotype partial vertebral column (entire specimen not shown in photograph). Are these the remains of a theropod dinosaur?**

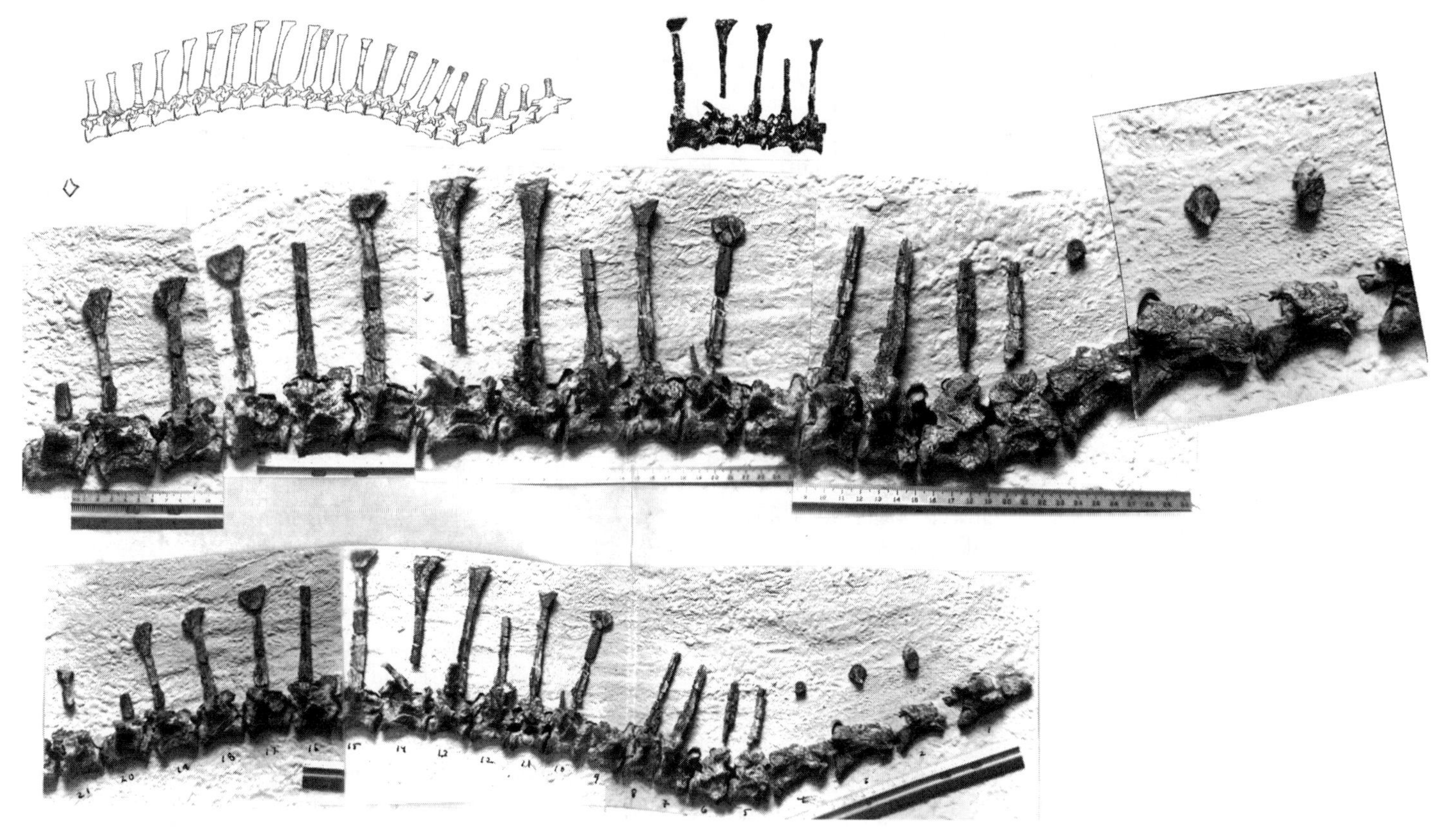

Photograph by Robert A. Long and Samuel P. Welles, courtesy University of Michigan Museum of Paleontology.

assigning it to the new taxon *Chindesaurus bryansmalli* (see *Caseosaurus* entry). Consequently, *Spinosuchus* was included by the present writer among the "excluded genera" in *D:TE.*

The dinosaurian or nondinosaurian status of *Spinosuchus* has recently come again under investigation. Hunt *et al.*, in a survey of Late Triassic dinosaurs from the western United States, reassessed the genus following recent analyses within Dinosauria (*e.g.*, Novas 1993; Sereno and Novas 1993). Hunt *et al.* noted the following: The hollowness of the vertebral centra (Case 1922) suggest that *Spinosuchus* could be a theropod after all; however, hollow centra are also known in pterosaurs (which *Spinosuchus* clearly is not), although it has not yet been demonstrated whether or not this feature may also be present in other archosaurian groups. *Spinosuchus* might share with herrerasaurids spine "tables" on the posterior dorsal vertebrae, the summit of the neural spine in UMMP 7507's 22nd vertebra being markedly expanded compared to the others in the series. However, this latter feature, in these authors' opinion, may not constitute a valid character given the atypical neural spines. Other possible dinosaurian synapomorphies (*e.g.*, postaxial cervical epipophyses, hyposphene-hypantrum: see Novas; Sereno and Novas) were not visible due to poor preservation or the enclosure of the holotype in plaster.

For the above reasons, Hunt *et al.* tentatively assigned *Spinosuchus* to "cf. Theropoda *incertae sedis*," based upon the hollow centra of its vertebrae. However, Case (1927) had considered an isolated bone found with the vertebral column to represent a (by inference postaxial) cervical intercentrum belonging to *Spinosuchus*. If the latter identification and association can be shown as correct, Hunt *et al.* noted, then *Spinosuchus* is not a dinosaur but a very derived and unusual proterosuchid-grade archosaur (Benton and Clark 1988). According to Hunt *et al.*, recent discoveries of dorsal vertebrae having long neural spines (private collection) indicate that *Spinosuchus* may also be present in the Adamanian fauna of the Los Esteros Member of the Santa Rosa Formation, in Santa Fe County, central New Mexico.

In a later abstract, Richards (1999) again questioned the possible dinosaurian status of *Spinosuchus*. Richards acknowledged that various characters of the holotype (*e.g.*, cervical centra having the form of a parallelogram in side view, well-developed anterior and posterior centrodiapophyseal laminae in pectoral region, supporting laminae of zygapophyses, diapophyses, and neural spines) resemble apomorphic vertebral characters of the Ornithodira, Dinosauromorpha, and Saurischia, but noted that at least one of these has also been observed in different basal archosauromorphs and proterosuchid archosauriforms. Lacking any cranial, mandibular, or appendicular skeletal elements, the type material cannot, stated Richards, be diagnosed as a crown group archosaur or archosauriform. Also, deformation of the bones, missing portions of elements, and artifacts of reconstruction "allow for enough latitude to cast doubt on the dinosaurian status of *Spinosuchus*."

Richards further pointed out that *Spinosuchus* and *Trilophosaurus*, a (nondinosaurian) euryapsid reptile, "share the development of paramedian interzygapophyseal plates between the postzygapophyses of the cervical vertebrae," and that fragmentary remains from the Late Permian Kawinga Formation of Tanganyika, previously assigned to the Proterosuchidae, could represent an animal closely related to *Spinosuchus*.

Therefore, the current volume continues to regard *Spinosuchus* as a nondinosaurian genus, pending, of course, any discovery of additional materials that might again alter this assessment.

SAUROPODOMORPHS

Sauropodomorphs comprised the earlier and more primitive prosauropods, a group that included both small and also the first large dinosaurs, and the sauropods, an assemblage including the largest tetrapods of all time.

A major study on the evolution of presacral vertebrae in this group of dinosaurs was presented by Bonaparte (1999*b*), who contended that these vertebrae represent "a rich source of information for understanding the evolution and systematics of sauropodomorphs." According to Bonaparte's study, three morphological types of vertebrae — a condition later developed further by sauropodomorphs — are present in the basal dinosauromorph *Marasuchus*.

Among prosauropods, the melanorosaurid *Riojasaurus* exhibits the most primitive condition in the number and organization of cervical vertebrae, having nine cervicals, the first five showing the cervical morphological type, the next four the dorsal morphological type; the plateosaurid *Plateosaurus* is more derived than *Riojasaurus* in characters of the neck.

Among primitive sauropods, the vertebrae are more derived than typical prosauropod vertebrae but show more primitive characters than the vertebrae in "cetiosaurid" sauropods (the "cetiosaurid" type most likely representing the ancestral condition). Bonaparte hypothesized that "cetiosaurid" type presacral vertebrae may correspond to an evolutionary stage that allowed for gigantism to occur in these dinosaurs, triggering the adaptive radiation seen in

North America, Asia, and Africa during the Late Jurassic.

As Bonaparte observed, the presacrals of diplodocid sauropods exhibit clear relationships to those of "cetiosaurids," however the number of cervicals is higher and the opisthocoely of their centra are more developed. Important differences in the cervical vertebrae of *Apatosaurus*, compared to those of *Diplodocus*, *Camarasaurus*, and other sauropods, suggested to the author that that genus warrants its own family, Apatosauridae (see *Apatosaurus* entry, *S1*). Presacrals of the brachiosaurid type, as seen in *Brachiosaurus brancai*, show characters that are more advanced than in *Diplodocus* and *Patagosaurus*, including opisthocoely extending to the last dorsal.

The prescral vertebrae of *Haplocanthosaurus* show a distinct morphology, a criterion for which Bonaparte referred this genus to its own new family, the Haplocanthosauridae. Re-examination by Bonaparte of camarasaurid type presacral vertebrae found them to be more derived in several features than those of *Diplodocus*, not primitive as previously interpreted. Dicraeosaurid type presacrals possibly evolved independently of other presacral types. Characters in brachiosaurids that are more primitive than in *Diplodocus* imply an origin from a more primitive condition than that exemplified in "cetiosaurids."

Several morphological types of presacral vertebrae were observed in titanosaurs. Among members of the Titanosauridae, various distinctive features were noted by Bonaparte, including "a unique design of the infrapostzygapophyseal construction, the reduced distance between the centrum and the zygapophyses, the near absence of neural spines." *Malawisaurus*, *Andesaurus*, and *Argentinosaurus* were found to be more primitive than titanosaurids.

Prosauropods

Among the recent discoveries of new prosauropods, the more primitive sauropodomorph group, are specimens of very ancient forms collected by a team from The Field Museum. Flynn, Parrish, Rakotosamimanana, Simpson, Whately and Wyss (1999), in a report on a ?Middle to Late Triassic (about 225 to 230 million years ago) vertebrate fauna from basal sites of the Isalo II, in the region east of Sakahara, in the Morondava Basin of Madagascar, announced these discoveries, which included two new (as of this writing unnamed) prosauropod taxa. Flynn *et al.* dated the Iasalo II to be of Triassic age, as Buffetaut had estimated in 1983; they did not, however, necessarily accept the latter's suggestion that these rocks were as old as Middle Triassic. (Other material yielded by the basal Isalo II sites included the remains of rhyncosaurs, a sphenodontian, a kanneymeriid dicynodont, and eucynodonts, constituting the first co-occurrence of dinosaurs and eucynodonts ever recorded.)

According to Flynn *et al.*, one of the prosauropods — known from several well-preserved maxillae and dentaries (UA collection, including UA 60603, a right maxilla, and UA 60604, a partial left dentary, both temporarily exhibited at The Field Museum) — resembles *Azendohsaurus* from the Argana Formation (Carnian) of Morocco (see Gauffrey 1993*c*). The authors referred this material to the prosauropod based upon the following synapomorphies: Anterior dentary curving downward; dorsal process of maxilla robust, its base located in anterior third of bone; and medial face of maxilla with series of small nutrient foramina.

This new prosauropod shares the following derived characters with *Azendohsaurus*: Prominent longitudinal keel on medial face of maxillary; fossa on medial face of maxilla posterior to dorsal fossa; consistent presence of neck between crown and root of teeth; and anteroposterior expansion of crowns always beginning at their bases. Differences between the Madagascar taxon and *Azendohsaurus* include: maxillary tooth count of 11 to 13 in new taxon (15 or 16 in *Azendohsaurus*); prominent wear facets present on maxillary teeth (absent in *Azendohsaurus*); and

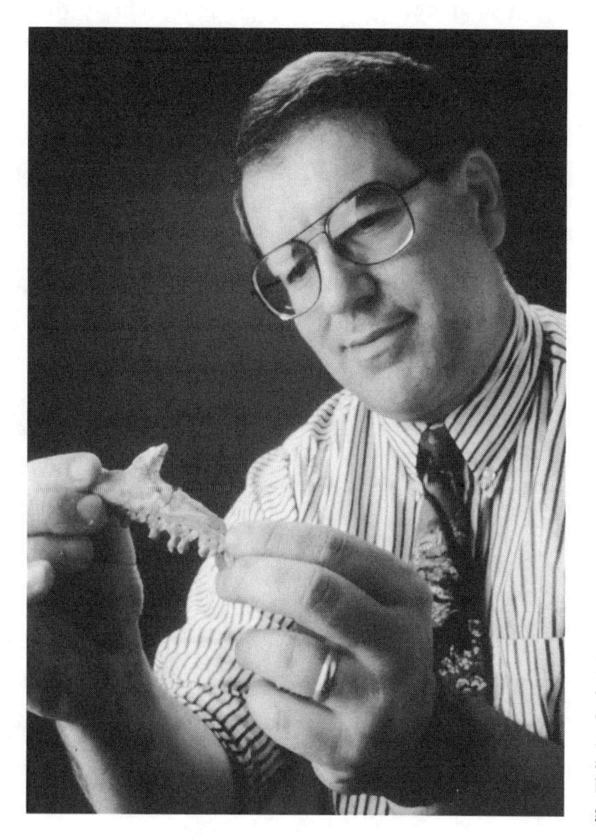

Photograph by John Weinstein, courtesy The Field Museum (neg. #GEO86239.2C).

Paleontologist John J. Flynn of The Field Museum geology department with the right maxilla (UA 16030) of a new prosauropod from the basal Isalo II beds of Madagascar.

Photograph by John Weinstein, courtesy The Field Museum (neg. #GEO86110_1C).

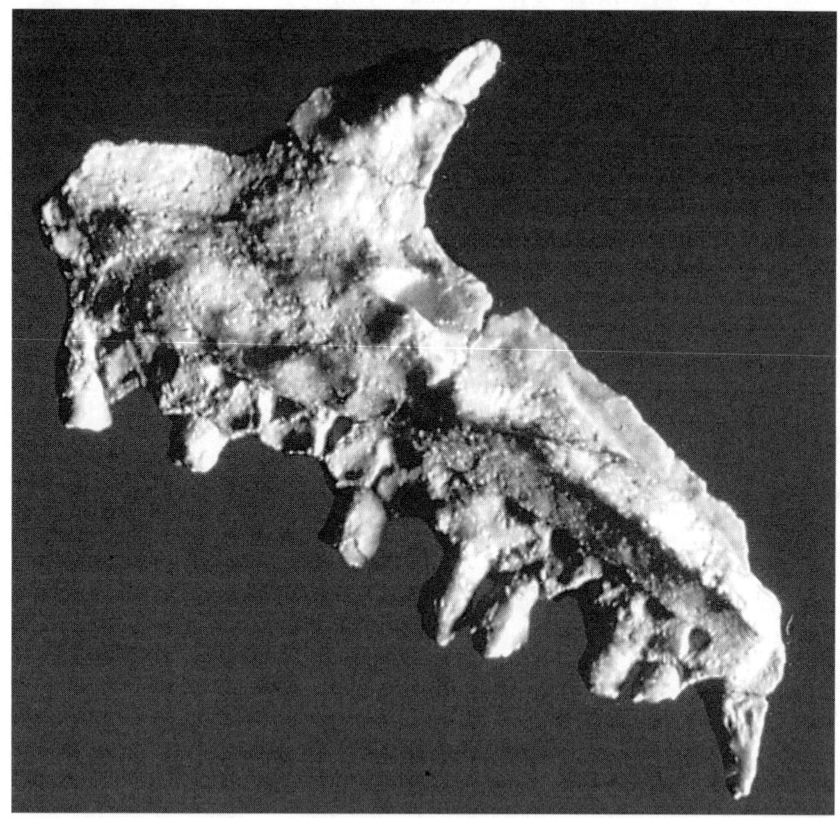

Right maxilla (UA 60603) of an unnamed prosauropod recovered from the basal Isalo II beds of Madagascar.

attenuate, caudoventrally projecting posterior process present on maxilla.

The authors reserved assigning a name to this material pending the anticipated collection of more complete specimens.

The second new prosauropod taxon displays a tooth morphology ("elongate crowns and tightly packed teeth lacking expanded crowns")—typical of most other prosauropods, including *Sellosaurus* and *Thecodontosaurus*—that distinguishes it from both the other Isalo II prosauropod and from *Azendohsaurus*.

As pointed out by Flynn *et al.*, the basal Isalo II dinosaurs are most interesting because of their great antiquity, the horizon's age now better constrained biostratigraphically, allowing time-related correlation to other land-based faunas worldwide. Since rhynchosaurs and dicynodonts are not known elsewhere after the Carnian, and prosauropods and sphenodontians are not known elsewhere before that time, this suggests "an undifferentiated Carnian age at a minimum and an approximate temporal correlation to the [middle or late Carnian] Ischigualasto Formation of Argentina" from which the most complete early dinosaurs have been recovered.

Because the basal Isalo II fauna is at least of Carnian age, the authors noted that "previously undocumented co-occurrences provide evidence that is either older than other dinosaur-bearing faunas worldwide or represents a unique Carnian fauna preserving the latest occurrences of several other taxa" (*e.g.*, eucynodonts). Therefore, the basal Isalo II beds seem to hold either the youngest known record of various nondinosaurian groups (*e.g.,* eucynodonts), or (more likely, based upon additional evidence relating to nondinosaurian forms) the remains of dinosaurs that are older than those from the Ischigualasto Formation (see Langer, Abdala, Richter and Benton 1999 and *Saturnalia* entry).

Photograph by John Weinstein, courtesy The Field Museum (neg. #GEO86236.4C).

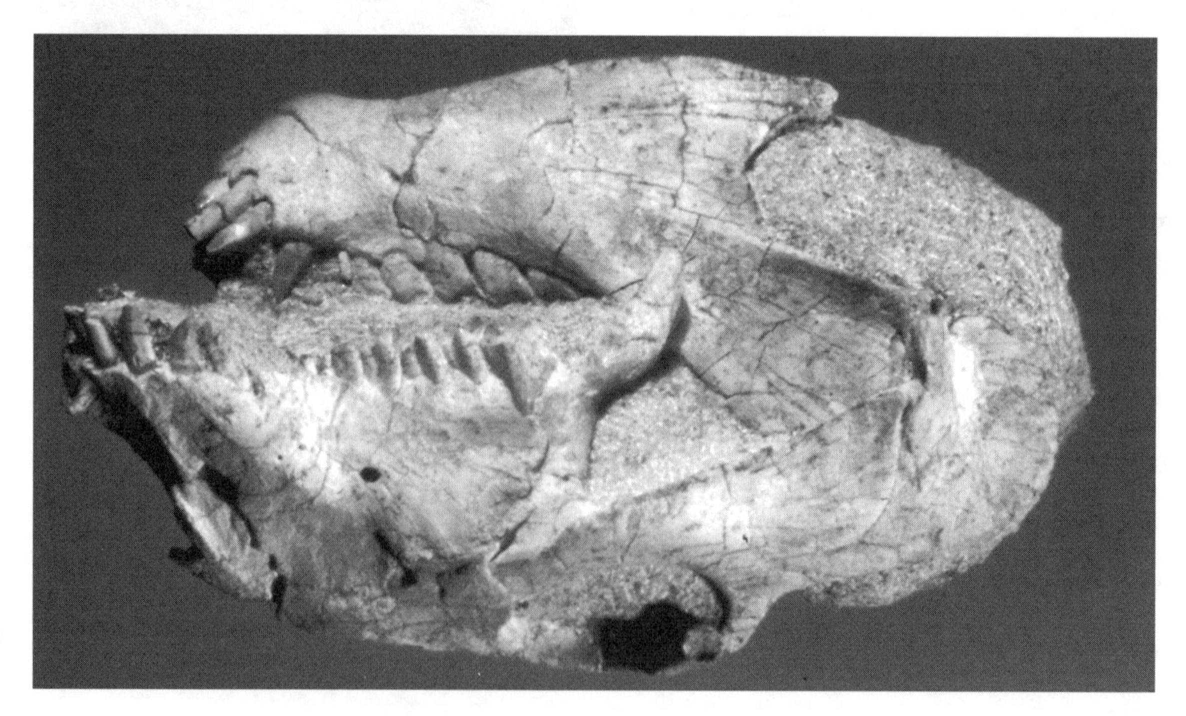

Partial left dentary (UA 60604) of an unnamed prosauropod recovered from the basal Isalo II beds of Madagascar.

Kellner, Azevedo, Rosa, Boelter and Leal (1999), and also Azevedo, Rosa, Boelter and Leal (1999), briefly reported on another new prosauropod from Brazil. The almost complete skeleton (housed at the Laboratório de Estratigrafia e Paleobiologia, Department de Geosciências/UFSM, catalogued as 11069) was recovered from fluvial sandstones in the Upper Triassic Caturrita Formation, Água Negra locality, near the city of Santa Maria, State of Río Grande do Sul, southern Brazil. Although still being prepared (by technicians and researchers of UFSM and of the Setor de Paleovertebrados of the Departmento de Geologia and Paleontologia of the Museu Nacional/UFRJ), some information has already been derived from this specimen, mostly pertaining to the skull.

As described by Kellner *et al.*, the teeth are typically prosauropod in form, being small and spatulate, with marginal dentitions measuring from 2 to 3 millimeters. They are closely spaced and slightly asymmetrical, the posterior teeth positioned obliquely relative to the tooth row, anterior teeth gradually shifting to a more parallel position. There are three, or more likely, four premaxillary teeth, an unknown number of maxillary teeth, and 20 teeth in each mandibular ramus. The morphology of these teeth, the authors noted, differs from that of thecodontosaurids, plateosaurids, and most melanorosaurids.

Kellner *et al.* described the frontal bones of the skull as long, extending posteriorly, forming a small laterodorsally oriented process with the parietal (this condition not reported in other known prosauropods); the posterior portion of the frontal shows a deep depression. The thin, subvertical nasal process of the maxilla resembles that of *Massospondylus*. The jaw articulation is ventrally offset below the tooth rows, although not as much as in plateosaurids. Also, the anterior portion of the dentary exhibits a complex morphology, including ridges and depressions, which are, to date unique to this prosauropod. As noted by Kellner *et al.*, other material pertaining to this dinosaur includes vertebrae, ribs, gastralia, incomplete pectoral girdles, a scapula, humerus, radius, ulna, and carpal elements.

Although the relationships of this Brazilian dinosaur have yet to be determined, the above unique combination of characters suggested to Kellner *et al.* that the specimen represents a new prosauropod taxon. The first South American occurrence of *Massospondylus*—a genus known mainly from South Africa, but also in North America—was reported by Martinez (1999). This new material, collected from the Lower Jurassic Cañon del Colorado Formation in the province of San Juan, northern Argentina, represents several different-sized individuals and includes a skull, lower jaw, partial vertebral column, pelvic girdle, and a partial forelimb and hindlimb. The elements show little signs of preburial weathering; some bones are articulated. According to Martinez, the presence of the genus in this formation indicates that these sediments are of Hettangian to Pliensbachian age, constituting "the first record of Lower Jurassic continental deposits with dinosaurs in South America."

Prosauropods may have been a key factor in the origin of high browsing, at least in part due to their size. As pointed out by Parrish (1998), prosauropods, which made their debut during the Late Triassic (Late Carnian), were the first tetrapods on this planet able to feed off vegetation much more than one meter off the ground. Indeed, the general prosauropod shape or bauplan was new among vertebrates, the characteristic elongate neck measuring about 25 percent as long as the whole body, and subequal in length to the distance between the pectoral and pelvic girdles.

Parrish postulated that the elongated prosauropod neck of the smaller (and comparatively rare) genera like *Azendohsaurus* and *Thecodontosaurus* would have had little ecological impact. More profound, however, would have been the impact of later, much bigger and more common forms such as the European *Plateosaurus* (Late Triassic, Norian), and the African and North American *Massospondylus* and the Asian *Lufengosaurus* (Early Jurassic, Hettangian–Pleinsbachian). The zygapophyseal geometry of the cervical vertebrae in such large prosauropod forms allowed "significant dorsiflexion such that, even in a quadrupedal posture, these dinosaurs had a feeding envelope roughly two meters higher than any of their contemporaries." Parrish concluded that this ability to access otherwise untapped sources of vegetation could have been instrumental in the relatively rapid spread of large sauropodomorphs during the Late Triassic and Early Jurassic.

Galton and Upchurch (2000), in an abstract (for a detailed study to be published in the second edition of the book *The Dinosauria*, currently in press), offered preliminary findings regarding their study of prosauropod vertebrae. Galton and Upchurch stated the following: in *Plateosaurus* (see entry), sacral vertebra 1, the proximal part of the diapophysis is wide, rounded distally, and laterally directed, the anterodistal part being narrow and anterolaterally directed. The diapophysis in sacral 2 is posterolaterally directed, while that in sacral 3 merges imperceptibly with the rib. Sacral ribs 1 and 2 are large, deep, and distally fused ventrally, forming a somewhat V-shaped surface for the ilium. The second sacral rib is winglike and has a shallow S-shape in left side view. As

Photograph by the author, courtesy Natural History Museum of Los Angeles County.

Skeleton of the prosauropod *Lufengosaurus hueni* (specimen originally referred to *L. magnus*, part of a temporary exhibition of Chinese dinosaurs at the Natural History Museum of Los Angeles County in 1989). J. Michael Parrish (1998) has shown how the long neck was a great advantage over this dinosaur's plant-eating contemporaries.

Galton and Upchurch pointed out, these vertebrae correspond to those of Triassic archosaurs (*e.g.*, Dinosauriformes [*Marasuchus*], Suchia, Aetosauria [*Stagonolepis*], and Rauisuchia [*Postosuchus*]) having a reptilian sacrum comprising two vertebrae. In large *Plateosaurus* individuals, the authors noted, the distally wide but shallow third sacral rib fuses only with the dorsal part of rib 2, forming a sacricostal yoke. Sacral vertebrae 1 and 2 in other prosauropods are posterior, while sacral 3 is a modified dorsal vertebra.

According to a cladistic analysis performed by Galton and Upchurch, utilizing 135 characters and 16 genera, the sacral states in this group do not represent a dichotomy. Rather, in some taxa (*Thecodontosaurus*, *Saturnalia*, *Ammosaurus*, *Melanorosaurus*, and also the sister group Sauropoda), the condition of two sacral vertebrae plus a caudosacral vertebra is plesiomorphic for Sauropodomorpha; a dorsosacral vertebra plus two sacrals (in *Riojasaurus*, *Jingshanosaurus*, *Massospondylus*, *Yunnanosaurus*, ?*Sellosaurus*, "*Gyposaurus*" *sinensis* [=*Massospondylus*], and *Lufengosaurus*) is the derived condition that occurred more than once; and two sacral vertebrae plus a caudosacral (in *Euskelosaurus*, *Sellosaurus*, and *Plateosaurus*) represents apomorphic reversals to the plesiomorphic condition.

According to Galton and Upchurch, the above changes most likely "represent several homeotic transformations or 'frame shifts,' both anteriorly and posteriorly, that are dependent on the anatomy during vertebral development of early morphogenesis (involving segmentation genes) and the development of character identity (involving homeotic genes)."

Sauropods

Information regarding the Sauropoda, a group comprising those sometimes gigantic sauropodomorphs basically known for their large bodies, long necks, and small heads, continues to surface.

Paik (2000) discovered bore holes in various sauropod bones recovered from the Hasandong Formation (Cretaceous) in Korea. The bones were found in what was once a floodplain. Among these specimens, imbedded in a sandy mudstone, was a sauropod scapula measuring approximately 1.5 meters in length. The specimen appeared to have suffered weathering before the boring commenced. Directly below this bone was found a number of cylindrical burrows containing chips of sauropod bone.

That author identified the maker of both the bore holes and the burrows as the dermestid beetle, a kind of insect living today (the kind maintained in museums for the purpose of devouring the soft tissues of dead animals, reducing the carcasses to a skeletal state). According to Paik, these insects apparently preferred the larger sauropod bones, the smaller ones, including a rib fragment, showing no signs of boring.

More than three decades ago, Simmons (1965) referred to the prosauropod *Yunnanosaurus robustus* a left maxilla with uninterrupted teeth present in 10 of the 16 alveoli (FMNH CUP [Catholic University of Peking Collection] 2042) from the Dark Red Beds of the Lower Lufeng Formation (Early Jurassic: Sinemurian–Pliensbachian; see Luo and Wu 1994, 1995), Ta Ti, Yunnan Province, People's Republic of China.

Barrett (1999) more recently described this specimen in detail, reidentifying it as belonging to a primitive (though unnamed) sauropod. According to Barrett, FMNH CUP 2042 exhibits features indicative of both sauropodomorph groups. The caudal ramus of the maxilla is quite dorsoventrally shallow and elongate, as in prosauropods (see Galton 1984*a*), and the tooth morphology (spatulate, labiolingually compressed, laterally convex) resembles that in *Yunnanosaurus* (Barrett 1998*b*). However, a preponderence of other features — lateral plate (Upchurch 1995, 1998), absence of sheet of bone backing rostral end of antorbital fenestra (Upchurch 1998), tooth crowns with lingual ridges (Upchurch 1993), lingual surface of tooth crowns concave (Upchurch 1998), and tooth enamel wrinkled (Wilson and Sereno 1998) — support this specimen's referral to the Sauropoda.

As FMNH CUP 2042 shares no convincing synapomorphies with members of any of the major sauropod clades, Barrett (1999) was not able to assess the phylogenetic placement of this specimen. However, Barrett (1999) noted that the teeth in this specimen differ from all other sauropod teeth known from the Lower Lufeng Formation, *e.g.*, FMNH CUP 2051, which is almost cylindrical in cross section, and lacks a lingual ridge and wrinkled enamel (see Barrett 1998; Wilson and Sereno), raising the possibility of two sauropod genera being present in the Lufeng Basin during Lower Lufeng times. According to Barrett (1999), this combination of prosauropod-like and sauropod features in FMNH CUP 2042 may eventually prove to have some phylogenetic significance.

Salgado (1999) published a preliminary study on the macroevolution of the "Diplodocimorpha" (a clade introduced by Calvo and Salgado in 1995 to include *Rebbachisaurus tessonei*, Diplodocidae, and all descendants of their common ancestor; see "Systematics" chapter, *S1* and this volume, "Notes" under "Diplodocoidea" and "Diplodocidae"), that author examining "the role that spaciotemporal changes in development might have played in the evolution of the Diplodocimorpa," and then speculating "on their possible adaptive meaning." In this study, Salgado regarded the family Diplodocidae as comprising two subfamilies, the "Diplodocinae" (large to very large forms, with very long necks and tails) and the "Dicraeosaurinae" (relatively smaller forms, having shorter necks and tails, but disproportionately tall neural spines) (these two taxa generally regarded as families, the Diplodocidae and Dicraeosauridae; see "Systematics" chapter).

From an evolutionary perspective, Salgado noted, two of the most interesting aspects of "diplodocimorphs" is their unusual skull — having a ventrally displaced occiput and confluent nares situated atop the skull — and the extreme bifurcation of the neural spines of the cervical and dorsal vertebrae. Salgado's study, focusing upon the morphological changes in mature specimens, attempted to explain "the processes that may have affected the developmental trajectory of these structures."

Traditionally, the author noted, two distinct, basic kinds of sauropod skulls have been recognized — the relatively longer, so-called *Diplodocus*-type skull, interpreted by Huene (1927) as "highly modified" or "transformed," and the more box-like *Camarasaurus*-type skull, regarded as more "generalized" (Huene 1927) and "primitive" (Huene 1929). Salgado and Calvo (1997) later demonstrated that Huene's perceptions were basically correct — that the features used to define the *Diplodocus*-type skull are synapomorphies of the "Diplodocimorpha," while those characterizing the *Camarasaurus* type are plesiomorphic for the Sauropoda. Salgado and Calvo theorized that the migration of the external nares to a high position on the adult "diplodocimorph" skull may "have resulted from the strong inclination of the braincase as a whole." Mainly, these authors argued that the reorientation of the gap for the olfactory tracts, which faces dorsally (unlike anterodorsally as in other dinosaurs), could have impelled the skull's olfactory chamber and narial openings, thereby modifying the gap's relative position. Furthermore, Salgado and Calvo suggested that most of the synapomorphies of the "diplodocimorph" skull resulted from a single event, "the coordinated reorientation of the antero-posterior axis of the skull with respect to the neck axis."

Salgado's study suggested that this morphological transformation could have been promoted by

Photograph by the author, courtesy Carnegie Museum of Natural History.

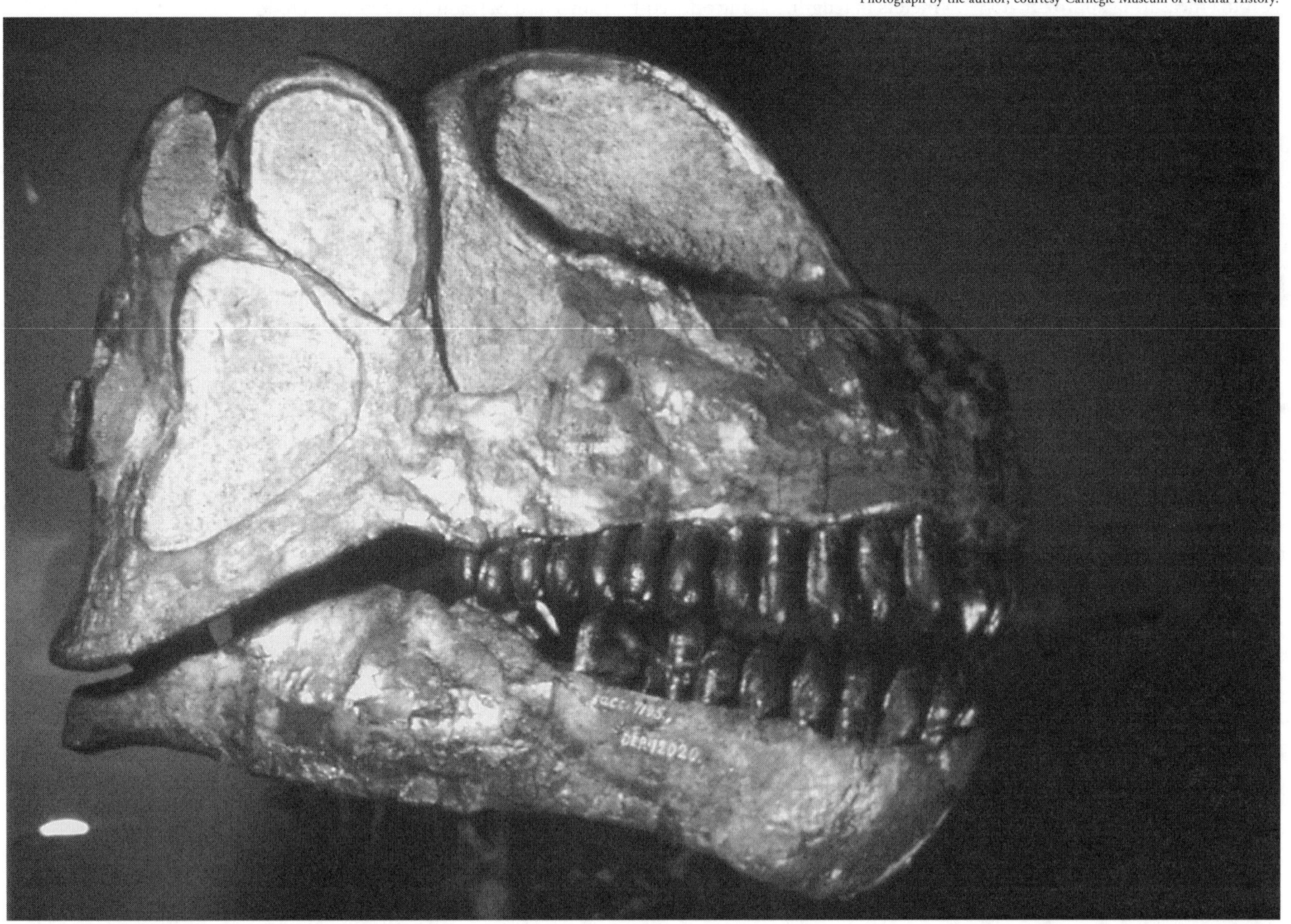

The short, box-like skull (CM 12020) of the camarasaurid sauropod *Camarasaurus lentus*, collected at Dinosaur National Monument, in right lateral view. Prior to the late 1970s, casts of this specimen were ubiquitous in museum displays incorrectly attached to mounted skeletons of the diplodocid *Apatosaurus*.

spatio-temporal changes during ontogeny. The author pointed out that in most vertebrate embryos, the antero-posterior axis of the head is almost perpendicular to the axis of the trunk; gradually during ontogeny, the foramen magnum displaces posteriorly, the head and the trunk thereby coming to lie in the same plane (see Gould 1977). Conversely, the basic embryonic condition was apparently retained in "diplodomorphs" (as well as in apes and humans) throughout their lives.

Salgado determined that dissociated heterochrony seems to have been the dominant mode of evolution in the "Diplodocimorpha." The group arose by deep modification of the morphology of the skull. This may have begun via an heterotopic change during an early developmental stage involving the relationship between the vomer and maxilla, eventually leading to subsequent development of the skull. In the diplodocids *Diplodocus* and *Apatosaurus*, the vomers are tightly clasped between two bony shelves of the maxilla, the palatine processes (see McIntosh and Berman 1975), these structures not known in any other sauropod genus. Based on their location and spacial relationships, these palatine processes and the anterior portion of the premaxillary processes, found in other sauropodomorphs, are most likely homologous. Salgado postulated that, during ontogeny, the incipient palatine shelves of the "diplodocimorph" ancestor collapsed the membranous vomers laterally (rather than joining each other along the midline, as in *Camarasaurus* type skulls), this hypothetical heterotopy possibly significantly affecting the development of the skull. Among these effects, the choanae could have remained in an anterior position, having not accompanied the migration of the external nares; also, the opening of the preantorbital fenestra and atypical shape of the anterior third of *Diplodocus* and *Apatosaurus* skulls could be associated with the early clasping of the vomers.

Concerning the slender, peglike teeth found in *Diplodocus* and *Apatosaurus*, Salgado noted that such teeth are also found in juveniles of at least two broad-toothed sauropodomorphs, the (juvenile) prosauropod *Mussaurus patagonicus* (see Bonaparte and Vince

1979) and the "cetiosaurid" *Patagosaurus fariasi* (Bonaparte 1986*c*), suggesting that peglike teeth may persist in fully mature "diplodocimorphs" as a result of heterochrony. Perhaps the new architecture of skull also led to further changes in the jaw muscles, allowing "diplodocimorphs" to undertake new forms of feeding (see Barrett and Upchurch 1994) which aided in attaining a novel lifestyle. Salgado speculated that the possession of confluent, retracted nares, and also a ventral occipital condyle that permitted rapid rotation of the head, may have been advantages employed in limiting predatory effects.

As Salgado noted, the very high dorsal and caudal neural spines (three times the height of the centra) are "diplodocimorph" characters that are probably the result of local acceleration or hypermorphosis (see McKinney and McNamara 1991). Furthermore, the comparatively short humerus in these dinosaurs may have evolved via local neoteny (*e.g.*, a decrease in cartilege growth), a change that "probably contributed to constrain the potential feeding niches to be occupied by diplodocimorphs."

A probable heterochronic trait identified by Salgado in the skulls of diplodocids (*e.g.*, *Diplodocus*, *Apatosaurus*, *Dicraeosaurus*, and *Amargasaurus*) was the squamosal not contacting the quadratojugal, the infratemporal bar consequently remaining incomplete, this possibly explained by paedomorphosis.

The extreme bifurcation of the vertebral neural spines has usually been explained as providing a space for a continuous "ligamentum nuchae" that served to support the neck (*e.g.*, see Powell 1986; McIntosh 1990). Given the probably adaptive significance of this feature, Salgado proposed a heterochronic explanation. Salgado pointed out that in vertebrate fetuses, "a cartilaginous neural arch is formed from two centers of chondrification"; theoretically, if the early termination of growth delays the fusion of both cartilaginous germs (local progenesis or post-displacement), or the growth rate of the germs becomes lower (local neoteny), a paedomorphic bifid neural spine may result. In the diplodocid ancestor, a fusion defect may have affected all presacral (excluding the axis), sacral, and anterior caudal neural arches during an early ontogenetic stage. "Hypothetically," Salgado stated, "the primarily double spines closed differentially during ontogeny, from bottom to top, following a cranial progression, but in the anterior dorsal and cervical vertebrae of mature individuals the spine remained divided perhaps as result of local neoteny," with the posterior dorsal, sacral, and anterior caudal neural spines, therefore, remaining entirely closed.

According to Salgado, heterochrony may also have played a crucial role in the origin and evolution of the "Dicraeosaurinae." Synapomorphies of this group interpreted to be the result of heterochronies include the existence of a parietal and post–parietal fenestra in adult individuals. However, the long basipterygoid processes, reduced supratemporal fenestra, and fused frontals seem to be peramorphic, possibly having arisen by local acceleration. The downward extension of the descending wing of the frontal (see Salgado and Calvo 1992) may also be a peramomorphic trait, related to the reduction of the supratemporal fenestra.

The smaller size of "dicraeosaurines" is sometimes associated with global progenesis (*e.g.*, see McKinney and McNamara) rather than neoteny. However, the partition of the neural spines reaching a maximum degree in "dicraeosaurines" (McIntosh 1990) was interpreted by Salgado as paedomorphic, possibly casually associated with the comparatively smaller size of these sauropods. The relatively taller anterior caudal neural spines, however, might be explained by local growth acceleration, hypermorphosis, or predisplacement. Progressive lengthening of these spines in a sequence comprising the oldest species *Dicraeosaurus hansemanni*, the younger referred species *"D." (=Amargasaurus) sattleri*, and the geologically youngest species *Amargasaurus cazui*, suggested to Salgado a disassociated heterochronoline, with the lengthening of neural spines interpreted as forming a peramorphocline and the opening of the neural spines forming a paedomorphocline.

Salgado's study concluded that the Diplodocidae arose mainly via the process of paedomorphosis, evidenced by both the shortening of the forelimbs and severe bifurcation of the neural spines. Within this family, two different heterochronic processes were involved — peramorphosis driving "diplodocine" evolution (resulting in large body size and higher number of vertebrae) and paedomorphosis dominating "dicraeosaurine" evolution (relatively small body side and progressive extension of bifurcation degree in neural spines).

Bilbey, Hall and Hall (2000), in an abstract for a poster, briefly reported on a new haplocanthosaurid from the lower Morrison Formation of northeastern Utah. The discovery site is on land administered by the United States Bureau of Land Management (BLM) less than seven kilometers (four miles) away from the Carnegie Quarry at Dinosaur National Monument and about 92 meters (300 feet) lower stratigraphically. The specimen was "preserved in a sandbar of interbedded fluvial sandstone and conglomeratic beds of the Salt Lake Wash Member of the Morrison Formation along the pluging anticlinal nose of Split Mountain." It was found on June 21, 1999, when a ditching unit impacted a sauropod dorsal vertebra and several metacarpals during construction of the Williams Rocky Mountain Pipeline. Subsequent excavation of the right-of-way over a slightly more than one-month period yielded more than 60 percent of the solitary skeleton. Many of the bones were found in articulation, with only minor structural distortion, and with minor to moderate damage to articular surfaces.

As Bilbey *et al.* pointed out, of particular importance was the collection of most of the appendicular skeleton, which is poorly known in other haplocanthosaurid specimens found at other sites in the Western United States. According to preliminary analyses by these authors, this specimen represents an older adult individual having a well-fused scapulocoracoid, although its stature is relatively small, being approximately 2.25 meters (7.5 feet) in height at the pelvis.

Foster, Curtice and Pagnac (1998) briefly reported on the remains of a new sauropod — a humerus and metatarsal collected in 1962 for what is now named The Field Museum — from the southern Black Hills, in Fall River County, South Dakota. Geologic evidence acquired from the type locality in 1997 compared with the matrix in The Field Museum material suggests that the latter was probably recovered from the Lower Cretaceous Lakota Formation and, as such, constitutes the first sauropod fossils identified from that formation in the Black Hills. According to Foster *et al.*, this material probably belongs to a camarasaurid or titanosaurid.

Remains of a new, primitive sauropod apparently related to the titanosauriformes was reported by Martinez (1998) in a short report. The material comprises a well-preserved, articulated skull and cervical series found in the Upper Cretaceous Bajo Barreal Formation of Chubut, Central Patagonia.

As briefly described by Martinez, the skull is a flattened structure possessing 15 thick maxillary and premaxillary teeth and 11 dentary teeth. The single nasal opening in the specimen exhibits signs of a *Brachiosaurus*-like nasal arch, and there the snout reveals traces of a prenasal depression. Three rod-like ossified tendons are present along both sides of the neck, two of which are alike with a 1-centimeter diameter, and a smooth, brown-colored surface, the third having a 0.3-millimeter diameters diameter, and a rugose, black surface. Examination of these rods revealed a dense concentration of Haversian systems.

According to Martinez, this unnamed form shares with *Nemegtosaurus* a mandibular symphysis of the dentary that is perpendicular to the long axis of the lower jaw; with *Nemegtosaurus* and *Diplodocus*, a ventrally directed occipital condyle; with *Diplodocus*, a ventrally oriented lacrimal; with *Nemegtosaurus* and *Brachiosaurus*, teeth having sharply inclined wear facets; with *Brachiosaurus* and *Camarasaurus*, teeth along the maxilla and dentary, each being in contact with two teeth of the opposite mandible, wide and dorsally visible supratemporal fenestrae, a jugal that does not participate in the margin of the preorbital fenestral, a surangular with an elevated coronoid, a quadrate having a shallowly excavated posterior face, and a dentary with a curved symphyseal ramus.

Kirkland, Aguillon-Martinez, Hernández-Rivera and Tidwell (2000), in an abstract for a poster, briefly reported on a proximal caudal vertebra belonging to a brachiosaurid, collected from the Late Cretaceous (late Campanian) Cerro del Pueblo Formation of Coahuila, Mexico. The specimen was described by these authors thusly: no chevron facets (identifying it as a first or second caudal vertebra); centrum lacking pleurocoels (therefore it does not pertain to the Titanosauridae, generally regarded as the only sauropods occurring in North America during the Late Cretaceous; see below); centrum short, wider than tall (160 millimeters in height, 179 millimeters in width, and 82 millimeters in length), with anteriorly located (incomplete) neural arch; neural canal circular, measuring 25 millimeters in width; caudal ribs angling back (as in brachiosaurids); centrum naturally flexed from side to side, aligned with swept-back caudal ribs, faces of centra almost parallel with each other (suggesting good lateral flexion); and caudal ribs with well-developed laminae connecting them with neural arch.

According to Kirkland *et al.*, this specimen, although unique, compares most favorably with taxa assigned to the basal Titanosauriformes close to or included in the Brachiosauridae. As none of the Lower Cretaceous South American titanosaurids are as primitive as the one represented by this caudal vertebra, the specimen supports the hypothesis that there was no "sauropod hiatus" in North America during the Late Cretaceous. Furthermore, the authors pointed out, nontitanosaurid sauropod lineages from Early Cretaceous North America may have survived into Late Cretaceous times in the relatively arid environments postulated for the Cretaceous refugia of Mexico.

Titanosaurs, in fact once among the more neglected and least understood of sauropod groups, continue to receive more attention; and as new titanosaur material is collected and studied, the more data becomes known about this quite common assemblage.

Wedel (2000), in an abstract for a poster, reported on a new, moderate-sized sauropod collected from the recently discovered Wolf Creek quarry, located in an outcrop of fine-grained claystone, in Unit V of the Early–Middle Cretaceous (Aptian–Albian) Cloverly Formation of Montana. This dinosaur is represented by disarticulated elements including humeri, tibiae, an ischium, other fragmentary pelvic elements, one metatarsal, and numerous partial vertebrae and vertebral fragments. Wedel noted that the general morphology of both the axial and appendicular elements suggests affinities with the Titanosauria, although it could not be confirmed at the time of his report. An almost complete, uncrushed cervical vertebra indicates that this sauropod had a relatively short neck. As this vertebra contrasts sharply with YPM 5294, a very elongate cervical centrum found in Unit VII, Wedel noted that it demonstrates the presence of at least two sauropod taxa in the Cloverly Formation. (Wedel reported that remains of a velociraptorine theropod, probably *Deinonychus antirrhopus*, represented by one tooth, at least one carnosaur based on a large tooth and caudal vertebra, and the crocodilian *Goniopholis* and chelonian *Naomichleys*, represented by teeth and scutes, have been collected from this quarry. Also found at this site were numerous small [2 to 10 centimeters in diameter], bony plates, roughly hexagonal or pentagonal, pitted on both faces, not corresponding well with previously described titanosaurian or ankylosaurian osteoderms.)

Britt, Scheetz, McIntosh and Stadtman (1998) briefly reported on the remains of at least a dozen titanosaurid sauropod individuals (the collected remains representing virtually all skeletal elements, including partial crania) recovered from the Dalton Wells quarry, located at the base of the Lower Cretaceous Cedar Mountain Formation, near Moab, Utah.

As observed by these authors, this unnamed genus exhibits the following titanosaurid characters: all vertebrae having low neural spines; caudal vertebral centra strongly procoelous; cranial-most caudal centrum biconvex (biconvex caudal centra known in dinosaurs only among neosauropods); large craniocaudally elongate sternals; and ulna with well-developed olecranon process.

Derived characters of this genus noted by Britt *et al.* include the following: basisphenoid with caudally directed tubera; neural spines of mid- and posterior caudal vertebrae shallowly bifid; high neural arch peduncles; cervical ribs deeply bifid; humerus with robust deltopectoral crest; and humero-femoral ratio of 0.8.

As pointed out by Britt *et al.*, the tooth crowns of this titanosaurid are spatulate and large, measuring up to 60 millimeters in length and 35 millimeters in width. The ilium of this new form is comparatively primitive to that of most known titanosaurids, as the anterior blade is not ventrally reflected; however, the blade is laterally reflected at an approximately 60-degree angle relative to the parasagittal plane. The authors concluded that "The presence of an Early Cretaceous titanosaurid in North America argues against the long-standing hypothesis that titanosaurs did not invade North America until the Late Cretaceous."

Dinosaur specimens from Chile are quite rare, most of them consisting of isolated bones. In an abstract, Vargas, Kellner, Diaz, Rubilar and Soares (2000) reported a new titanosaurid found in 1999 by a joint expedition of the Museo Nacional de Historia Natural (Santiago) and the Museo Nacional (Río de Janeiro), in collaboration with the Universidad Catolica del Norte (Antofagasta), in a copper mine north of the town of Calama and east of El Abra, in the Atacama Desert in northern Chile. The sauropod, represented by an incomplete skeleton, was found in a red conglomerate layer, overlain by reddish sandstones, in a sedimentary unit presumed to be of Late Cretaceous age.

As reported by Vargas *et al.*, the remains of this titanosaurid collected and identified thus far include incomplete ilia and ischia, dorsal and caudal vertebrae, ribs, an incomplete humerus, and an incomplete femur and proximal articulation of the tibia. Distortion of some of these elements seems to have taken place during the process of fossilization. Although the bones were not articulated, they were found in association, presumably belonging to a single individual. Most of this specimen had not yet been prepared to date of Vargas *et al.*'s report; however, the procoelic condition of the distal caudal vertebrae identified the sauropod as a titanosaurid.

As noted by the authors, this find is significant as the most complete dinosaur yet discovered in Chile; also, it suggests that the Atacama Desert has a high potential for yielding additional fossil vertebrate remains.

Calvo (2000), in an abstract on dinosaurian remains recovered by the Museum of Geology and Paleontology of the National University of Comahue from the Río Neuquén and Río Colorado formations, on the coast of Los Barreales Lake (Upper Cretaceous), Neuquén, Patagonia, Argentina, reported the following titanosaurid specimens: MUCPv-204, collected in 1990, comprising 13 caudal vertebrae, two femora, a scapula, and a humerus representing a primitive form; MUCPv-304, several articulated dorsal vertebrae collected in February, 1999; MUCPv-302, two dorsal vertebrae collected in April, 1999; MUCPv-303, an incomplete specimen representing a very large individual including a scapula, ribs, and dorsal and caudal vertebrae, collected in February–March, 2000; and MUCPv-304, a specimen representing a giant individual including five middle cervical vertebrae (one having a length of 95 centimeters and height of 85 centimeters) and other unspecified material, the triangular neural spine and general morphology indicating a titanosaurine. (Calvo also reported from this locality a complete tetanuran humerus [MUCPv-275] found in 1996, and an almost complete and articulated theropod skeleton [MUCPv-301] found in April, 1999, which, based upon preliminary field determination, seemingly does not belong either to the Abelisauridae or Carcharodontosauridae.) Most of the above material, Calvo noted, is either unpublished or unprepared, still in protective jackets.

The first articulated distal tail segment belonging to a Gondwanan titanosaur, and the first specimen which offers definite information concerning the supposed presence of a "whiplash" tail in this group of sauropod dinosaurs (which in the past, by association, has sometimes inferred affinities between the Titanosauria and Diplodocoidea), was described by Wilson, Martinez and Alcober (1999). The unnamed specimen (MCNA-PV 3136) consists of an isolated distal caudal series comprising 10 articulated vertebrae. It was collected in 1996 by a joint University of San Juan–Museo Juan Cornelio Moyano de Mendoza–University of Chicago expedition to the Río Colorado Formation (Middle Cretaceous: ?Cenomanian) at Cañadon Amarillo, Department of Maragüe, Mendoza Province, Argentina.

Wilson *et al* described these vertebrae as comparatively abbreviated, becoming shorter distally. The vertebrae are procoelous, becoming progressively biconvex in the distal region of the tail (biconvex caudals constituting a salient tail specialization that became modified in different ways within the various sauropod subgroups). Caudal 10 has a gently rounded posterior surface lacking an apex, this confirming the presence of biconvex distal caudal vertebrae in titanosaurs. In short, the authors noted that MCNA-PV 3136 probably represents a tail segment that possessed a relatively short, archless tip.

Because these procoelous caudal vertebrae become biconvex in the distal part of the tail (a condition also seen in the saltasaurids *Neuquensaurus*, *Opisthocoelicaudia*, and *"Titanosaurus" (=Laplatasaurus) araukanicus*, but also found in diplodocoids and *Cetiosauriscus*), the authors referred the new specimen to the titanosaurian family Saltasauridae. The distribution of this biconvex condition on a cladogram of sauropod relationships suggested to Wilson *et al.* "two equally parsimonious interpretations for the evolution of biconvex distal caudal vertebrae in neosauropods": 1. That this condition represents a neosauropod synapomorphy with a reversal in the genus *Camarasaurus*, and 2. that the condition was acquired independently in diplodocoids and saltasaurids.

Regarding the question of whether or not titanosaurs possessed tapering "whiplash" tails, which might have been used as a weapon to ward off predators or to produce loud warning or signatory noises—such tails currently known unambiguously only in the diplodocids *Apatosaurus* and *Diplodocus*—Wilson *et al.* defined this kind of specialized tail as one comprising a sequence of at least 30 elongate, biconvex distal caudal vertebral centra. This complex feature involves: 1. changes in the morphology of the articular face of the centrum (platycoelous or biconvex), 2. in the centrum's shape (short or long), and 3. in the number of vertebrae (less than 10 or at least 30) in the series. From a cladistic analysis of various sauropods, the authors found biconvex distal caudal vertebrae to be more common among this group of dinosaurs than the more specialized elongate caudal centra or series of at least 30 biconvex vertebrae required for a "whiplash" tail. Thus, the presence of biconvex distal caudal vertebrae constitutes neither a useful criterion for inferring a "whiplash" tail, nor one for supporting unambiguously presumed affinities between diplodocids and titanosaurs.

As noted by the authors, the short series of abbreviate biconvex caudal vertebrae in the Mendoza tail segment illustrates that the tail of titanosaurs is very distinct from that of *Diplodocus* and *Apatosaurus*, casting serious doubt that titanosaurs also possessed elongate "whiplash" tails. Wilson *et al.* concluded that a short, quite flexible tail tip evolved with the appearance of the earliest neosauropods of the Middle Jurassic, which eventually developing into the highly specialized "whiplash" tail of diplodocids by the Late Jurassic.

Curtice (1998) reported that characters were scored for more than 1,600 articulated and disarticulated sauropod vertebrae. This analysis was performed in an attempt to understand more clearly "the range and degree of vertebral variation and its effects upon the creation and scoring of phylogenetic characters." The scored characters were as follows: degree and shape of procoely; morphology of neural spine bifurcation; size, shape, and depth of pneumatic angle and ventral sulci; position of neural arch on centrum; declination angle on neural spine; and development degree of all laminae.

The results of Curtice's analysis showed artificial variation can be created through the oversimplification of vertebral characters. Present-or-absent characters were shown to be surprisingly susceptible to such variation, their presence or absence on only some vertebrae requiring the selection of a "correct" character state upon which not all researchers may agree. Curtice suggested that the likelihood of artificial variation can be lessened by utilizing more precise character definitions (*e.g.*, left- versus right-side differences within one individual, amphicoelous caudal vertebrae within "advanced" titanosaurids, anteriorly placed caudal neural arches on diplodocids). Characters that vary with respect to their position in the vertebral

Photograph by the author, courtesy American Museum of Natural History.

Composite skeleton (AMNH 460, 222, 339, and 592) of the diplodocid *Apatosaurus excelsus*, one of the small number of sauropods known unequivocally to have possessed a highly specialized "whiplash" tail.

column "receive special consideration due to their potential impact on phylogenetic analyses which incorporate disarticulated vertebrae."

As a first step towards interpreting the phylogenetic significance of sauropod vertebral laminae — those "numerous bony struts that connect the costovertebral and intervertebral articulations, centrum, and neural spine of the presacral, sacral, and anterior caudal vertebra" — Wilson (1999*b*) offered a new standardized nomenclature for these laminae. Unlike earlier nomenclatures (Osborn 1899; Janensch 1929*b*), the one proposed by Wilson utilizes morphological landmarks connected by the laminae, and also provides the same name for serial homologues.

Wilson identified, defined, and described in detail 19 different neural arch laminae (no single vertebra possessing all of them) based upon morphological landmarks. These laminae were divided and placed into four distinct regional categories which Wilson named diapophyseal laminae, parapophyseal laminae, zygapophyseal laminae, and spinal laminae, which are subdivided into additional categories. Wilson also provided a corresponding set of abbreviations. Thus, diapophysial laminae (di) includes the anterior centrodiapophyseal lamina (acdl), posterior centrodiapophyseal lamina (pcdl), prezygapophyseal lamina (prdl), postzygapophyseal lamina (podl), spinodiapophyseal lamina (spdl), and paradiapophyseal lamina (ppdl); parapophyseal laminae (pl) includes the anterior centroparapophyseal lamina (acpl), posterior centroparapophyseal lamina (pcpl), and prezygoparapophyseal lamina (prpl); zygapophyseal laminae (pol, prl) includes the centroprezygapophyseal lamina (cprl), spinoprezygapophyseal lamina (sprl), centropostzygapophyseal lamina (cpol), spinopostzygapophyseal lamina (spol), and intrapostzygapophyseal lamina (tpol); and spinal laminae (sl) includes the prespinal lamina (prsl) and postspinal lamina (posl).

According to Wilson, five diapophyseal laminae and six zygapophyseal characterize the presacral vertebrae of saurischians in general, this basic pattern having been established early in the evolution of dinosaurs by the Late Triassic. According to Wilson, the Sauropoda, among all saurischian groups, possess the best-developed vertebral laminae. Vertebral laminae in this clade were found to be fairly conservative, varying mostly at low taxonomic levels. The primitive sauropod *Barapasaurus* and all more derived taxa (*i.e.*, *Omeisaurus* and Neosauropoda) are characterized by the presence of an additional spinodiapophyseal and three parapophyseal laminae. Also present in these dinosaurs are "divided" laminae (zygapophyseal laminae divided into medial and lateral sublaminae) and "composite" laminae ("divided" laminae joining other laminae forming unique combinations). Diplodocids are characterized by the unique presence of dispophyseal laminae in the caudal vertebrae, also by divided zygapophyseal laminae in the dorsal vertebrae.

Given the presence of vertebral laminae in small theropod dinosaurs predating the earliest known sauropods by some 25 million years, Wilson suggested that these structures did not, contrary to earlier opinions, originate as a means of lessening the weighty loads imposed on the axial columns of these giant dinosaurs. However, the increased number and complexity of these laminae, Wilson speculated, "may reflect their increasing importance as structural supports in sauropod dinosaurs."

Skull material pertaining to the Titanosauridae has always been rare, with titanosaurid skulls traditionally (and incorrectly) being restored resembling the elongated skull of *Diplodocus*; however, more recently collected fossil evidence has revealed that titanosaurid skulls probably look more like the short, high skulls of sauropods such as *Camarasaurus* and *Brachiosaurus* than those of diplodocids (see *S1*).

Gomani (1998) briefly reported on cranial material (including a basicranium, quadrate, jugal, partial maxilla, premaxilla, pterygoids, an ectopterygoid, a dentary, and a splenial) belonging to the titanosaur *Malawisaurus* (now regarded as a titanosaurid), and also the dentary belonging to a smaller titanosaurid taxon from Malawi, East Africa. As Gomani observed, this new material suggests that the nasals of titanosaurids are less retracted than in *Diplodocus*, the rostrum short, cranium high, tooth row anteriorly restricted or extending along half of dentary, and teeth primitively slightly broad or cylindrical.

Dodson, Krause, Forster, Sampson and Ravoavy (1998*b*) formally described three isolated (FMNH PR 2021 and PR 2052, and UA 8675) and eight small (UA 8676) associated titanosaurid osteoderms, collected in 1996 (briefly reported and described in an abstract by Dodson *et al.* 1998*a*; see *D:TE* for details), from four separate sites in the Upper Cretaceous (?Campanian) Maevarano Formation, near the village of Berivotra, in the Mahajanga Basin of northwestern Madagascar.

As noted by Dodson *et al.* (1998*a*, 1998*b*), this discovery confirms and extends Depéret's (1896) suggestions of more than a century ago that titanosaurid sauropods, including *Titanosaurus madagascarensis*, bore armor, these ideas having been published almost a century before any sauropods were known to have been armored.

Dodson *et al.* (1998*b*) pointed out that, despite a century of intermittent collecting, titanosaurid osteoderms are rare finds in Madagascar. The authors offered a number of explanations for this: 1. Malagasy titanosaurids may have been sparsely armored; 2. despite their robust morphology, the porpous bone texture of some osteoderms (including the largest preserved, FMNH PR 2021) could have made them especially susceptible to weathering and fragmentation; 3. titanosaurid osteoderms are often difficult to recognize in the field, particularly when fragmented; and 4. since two titanosaurid species are known in the Maevarano Formation (see Krause and Hartman 1996), only one may have been armored. As titanosaurids comprise the more abundant of Malagasy dinosaur remains (including specimens found in bonebeds), and because isolated ankylosaur osteoderms are common in Upper Cretaceous rocks in western North America, Dodson *et al.* (1998*b*) found the first of the above postulations to be the most likely.

Based upon the size and variety of these osteoderms, Dodson *et al.* (1998*b*) suggested that the armor may been rather sparse and varied over different areas of the body. The features of vertical sides and nutrient canals entering and exiting from those sides on FSL 92827 and UA 8675 suggested to the authors that these kinds of osteoderms lay adjacent to others, with vascular canals passing between them. Larger osteoderms taper toward the periphery, this suggesting that they did not abut other osteoderms, but instead lay either isolated or were at least removed from direct contact with other osteoderms. Two specimens, USA 8676 and FMNH PR 2052, consist of small nodules of bone that could have filled the spaces between larger osteoderms, or possibly "occupied the periphery of an osteoderm field." Furthermore, the size of some of the larger osteoderms suggested that the dermis in which they were imbedded approximated, in places, 7 centimeters in thickness.

In a subsequent brief report, Sampson, Forster and Krause (1998) reported that at least two titanosaurid species are now known from the Maevarano Formation of Madagascar, represented by one almost complete skeleton and four partial skeletons, all of modest size, plus two partial skulls that offer important new insights into titanosaur morphology.

González Riga (1999), in an abstract, made the preliminary announcement of new titanosaurian specimens discovered in 1998 in the Upper Cretaceous (Coniacian–Santonian) Río Colorado Formation, Arroyo Seco locality, province of Mendoza, Argentina.

The material, belonging to two adults and one juvenile individual, seem to represent two species of Titanosauria.

One species, known from some articulated caudal vertebrae and numerous disarticulated bones pertaining to the limbs, offers a combination of characters that suggest a new genus and the first to be described from Mendoza. This as yet unnamed taxon was assigned by González Riga to the Titanosauria, but excluded from the Titanosauridae because it lacks pronounced pleurocoels in middle and posterior caudal vertebrae. Additional characters such as the lower development of the neural spines in the middle caudal vertebrae, were interpreted by the author as plesiomorphic.

The other species represented by this material differs from ore basal titanosaurs like *Andesaurus* in a number of apomorphic features (including anterior caudals with pronounced pleurocoels; and postzygapophyses located in midanterior of midcaudal centra) suggesting a closer relationship to more derived titanosaurian forms such as *Titanosaurus* and *Aeolosaurus*. Also, González noted, the relatively slender limbs and the posterior caudal vertebrae with pronounced pleurocoels suggest possible affinities with titanosaurids.

Calvo, Moreno, and Rubilar (1999), in an abstract, briefly reported on a new titanosaurid — represented by four articulated dorsal vertebrae, at least one posterior dorsal articulated with neural arches, and some ribs — collected from Los Barreales lake in the Upper Cretaceous Río Colorado Formation of Neuquén Province, Argentina, establishing the presence of these sauropods in this area. Calvo *et al.* referred this material to the Titanosauria based on the following characters: posterior trunk vertebrae having ventrally widened, slightly bifurcated infradiapophyseal lamina; centro-parapophyseal lamina; and eye-shaped pleurocoels. It was referred to the Titanosauridae based on the absence of a hyposphene-hypantrum. The presence of diapopostzygapophyseal laminae in the anterior trunk vertebrae separates this material from *Opisthocoelicaudia* and an indeterminate titanosaurine from Brazil.

It is no longer speculation that sauropods laid eggs and now there is also evidence suggesting that at least some kinds of sauropods may have cared for their young. The first reliable evidence of sauropod (?titanosaur) embryos was reported in an abstract by Chiappe, Jackson, Dingus, Grellet-Tinner and Coria (1999). The embryos were contained in fossil megaloolithid eggs found in 1997 at the vast (extending several kilometers) Auca Mahuevo site in the Late Cretaceous Río Colorado Formation of northwestern Patagonia, Argentina. As related by Chiappe *et al.*, the eggs are highly concentrated at this site which includes four stratigraphic layers. In the third layer, 30 eggs were mapped in an area of 600 square meters, while more than 200 eggs were discovered distributed over a 30 square-meter quarry. To date, embryonic bones were found in least one-third of the examined *in situ* eggs; embryonic skin was discovered in dozens of egg fragments in this same layer.

According to Chiappe *et al.*, the eggs' distribution

Photograph by the author, courtesy Mesa Southwest Museum.

Now it is certain that sauropods laid eggs, as did other kinds of dinosaurs, with evidence suggesting that at least some of these very large dinosaurs also cared for their young. This cluster of 1.7- to 1.8-liter sauropod eggs — part of the "Great Russian Dinosaurs Exhibition" — are from the Upper Cretaceous (Campanian) Barun Goyot Formation, Peoples' Republic of Mongolia.

Photograph by the author, courtesy Houston Museum of Natural Science.

Diplodocus hayi **subadult skeleton (composite including holotype CM 662, also CM 94, skull (cast) of *D. carnegii*, CM 1161). Recent studies by John Martin, Valérie Martin-Rolland, and Ebergard (Dino) Frey (1998), and Kent A. Stevens and J. Michael Parrish (1999), have demonstrated that the neck of this sauropod could not be raised to the height shown in this exhibit (and also in numerous life restorations).**

and the characteristics of the sediments in which they were found preclude any significant transport after the eggs were laid. Therefore, the Auca Mahuevo site is especially important, "clarifying the reproductive biology of sauropod dinosaurs and providing uncontroversial evidence of their gregarious nesting behavior."

Martin, Martin-Rolland and Frey (1998) stated that the mechanics of sauropod necks are still poorly understood. Two reasons for this problem are that relatively few (approximately seven or eight) sauropod taxa have necks known well enough for their posture and mobility to be reconstructed fully and reliably, and that no living animal offers a perfect anatomical or mechanical for the very long necks of sauropods.

Earlier reconstructions of skeletons and life restorations of sauropods tended to depict these animals with their necks held out horizontally or sloping downwards, with the heads held close to the ground. Among these illustrations is the widely reproduced (sometimes subsequently modified) skeletal reconstruction of *Diplodocus carnegii* published by John Bell Hatcher in 1901. As pointed out by Martin *et al.*, and also by Stevens and Parrish (1999; see below), more recent life restorations regularly depict these dinosaurs with the neck held almost vertically, or in an aesthetically satisfying "S" curve, suggesting that the sauropods were high browsers or could fight by "neck whipping" (*e.g.*, Bakker 1986; Paul 1987; see also Czerkas and Glut 1982 and *D:TE* for examples). However, some modern workers (*e.g.,* Coombs 1973; Dodson 1990*a*) have argued that sauropod necks were incapable of such "artistic" maneuverings due to their complex intervertebral articulation systems. This would imply that the earlier reconstructions are correct.

Martin *et al.* addressed this issue by applying to sauropod necks the principles of biomechanics, including only those taxa in which enough of the necks have been preserved so that "the limits of mobility imposed by the zygapophyseal and central articulation structures may be calculated." The authors used as an analogy for the sauropod neck a single beam supported at only one end (*e.g.*, a crane jib), which has its upper part stressed by tension while the lower part is under compression. Most such beams are braced to

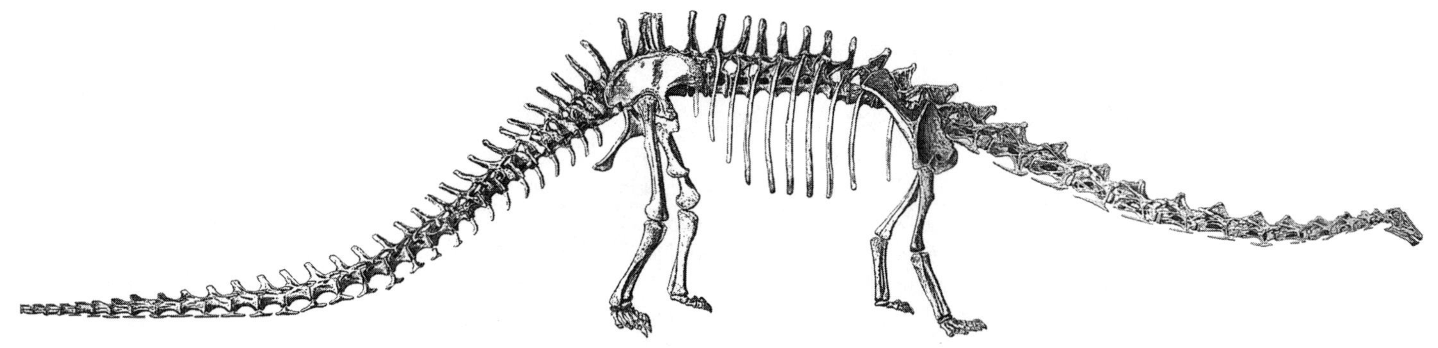

John Bell Hatcher's original skeletal reconstruction of *Diplodocus carnegii*. Current studies by John Martin, Valérie Martin-Rolland, and Ebergard (Dino) Frey (1998), and Kent A. Stevens and J. Michael Parrish (1999), indicate that the depiction of this gigantic sauropod — with its neck sloping downward and head held close to the ground — is reasonably accurate. (After Hatcher 1901.)

distribute and reduce these stresses in one of three ways — by cross-sectional design, internal struts, or external braces.

As these authors observed, sauropod necks are basically similar in their structure to these beams. However, unlike such beams, the sauropod neck comprises a series of individual segments — the cervical vertebrae — which must be free to move against its neighbors. Consequently, "the tensile and compressive stresses must be redistributed away from the intersegmental articulations by bracing systems running the whole length of the structure."

As in beam-like structures (*e.g.*, some bridges), sauropod necks must be braced above and below incorporating various dorsal and ventral bracing systems. Dorsal, tensile-bracing elements almost surely consisted of unmineralized connective tissues, primarily ligaments, but also muscles and tendons, which would have provided hydraulic stabilization for other structures but would have not aided in maintaining the neck in its "normal" or "rest" position. The ventral system comprised the long forward and backward directed cervical ribs in conjunction with associated soft-tissue structures. In order to deal with the extremely complex stresses imposed on the neck by gravity, the inertia of a long swinging beam and by the muscle-to-bone reactions on a dynamic, flexible beam, the dorsal and ventral systems had to evolve so that the construction is versatile. Therefore, these bracing systems were more or less combined in the necks, becoming either mainly dorsally braced, ventrally braced, or in a more equitable mixture of the two.

Based upon their anatomy, sauropod necks, Martin *et al.* observed, can be grouped into three basic categories: 1. Dorsally braced — seemingly the least common condition among known Sauropoda, include taxa (*e.g.*, *Apatosaurus* and *Dicraeosaurus*) having high neural spines, typically bifurcated to increase the volume of the suspension ligaments, all of these structures implying large mass that restricted the number of segments and length of the neck; 2. ventrally braced — including taxa (especially the Chinese genera *Mamenchisaurus*, *Euhelopus*, and *Omeisaurus*) in which the overlapping (and sometimes very long) cervical ribs were bound together by connective tissues and bundles of short-fibered muscles (as in some crocodilians; see Frey 1988); and 3. dual systems — most known sauropods (*e.g.*, *Diplodocus*, *Cetiosaurus*, and *Haplocanthosaurus*), in which the neural spines and, to a lesser extent, the transverse processes, were of moderate size and the cervical ribs of moderate length.

As Martin *et al.* noted, *Camarasaurus* (see entry, also Parrish and Stevens 1998; see below) basically fits into the second of the above categories, but is a special case. The first 11 cervical neural spines are modest and unforked, and the cervical rib processes are extremely elongated (up to three segments), which indicates ventral bracing. However, the vertebrae themselves are short and wide, the neural arches high, and the zygapophyses widely separated, features indicative of a very powerful although inflexible neck.

Regarding *Brachiosaurus* (see entry) — which is often restored as an elegant, upwardly directed, giraffe-like animal with its very long neck raised high and almost vertically disposed — Martin *et al.* emphasized that the neck in this genus is not fully known and that early descriptions (Janensch 1914, 1929*a*) of the bones of *Brachiosaurus brancai* are equivocal, with lengths of only a few limb bones given and that the mounted skeleton at the Humboldt Museum für Naturkunde of *B. brancai* (see *D:TE* for photograph), upon which many life restorations have been based, is a composite including elements from other individuals and also plaster modeling and other species referred to this genus (*e.g.*, *B. altithorax* Riggs 1903 and ?*B. atalaiensis* Lapparent and Zbyszewski 1957) seem to possess front and hindlimbs of approximately the same length. Martin *et al.* concluded that there is no direct evidence that *Brachiosaurus* had a neck posture different from other sauropods and that this genus was a "four-square" sauropod, "higher in front than most genera (those with limb proportions closer to 0.8:1), but not a dinosaurian giraffe."

Concerning the other sauropods studied which

Courtesy Kent A. Stevens.

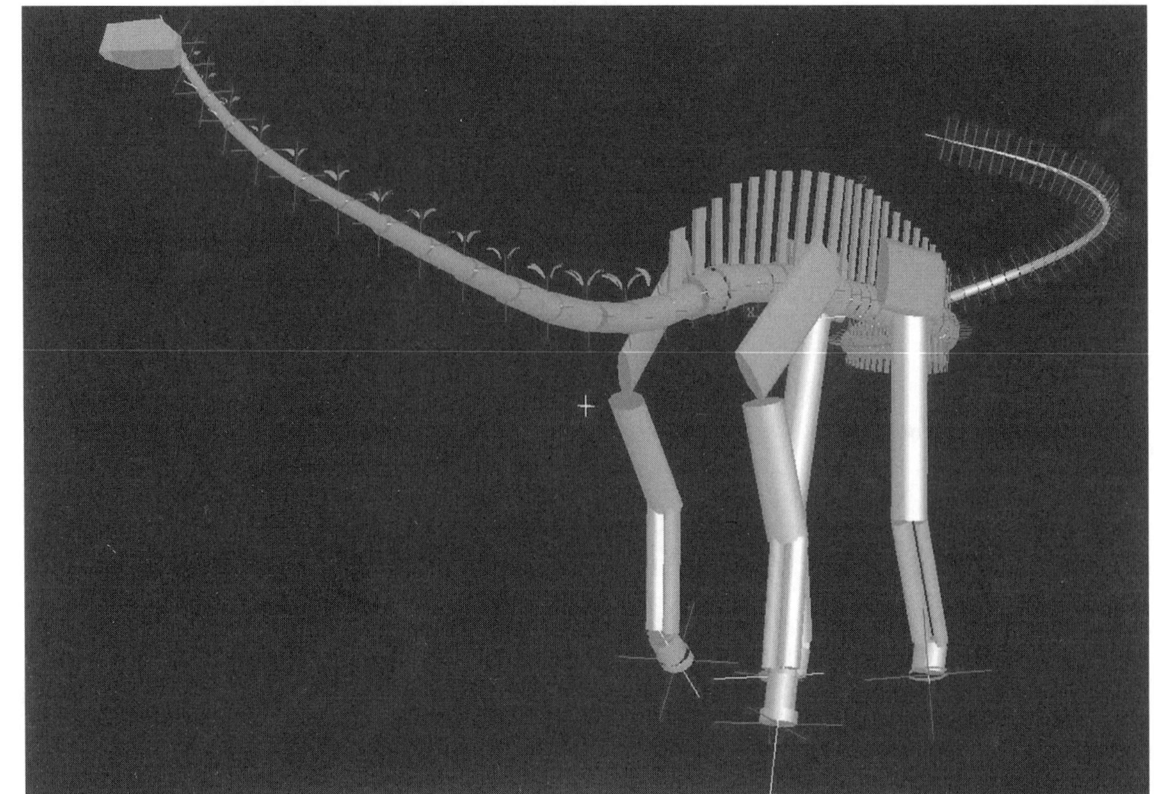

Computer generated model (front left view) of the skeleton of *Diplodocus carnegii* (based on CM 84) with the neck in neutral (undeflected) and dorsoflexion positions (the rest of the pose and viewpoint held constant).

Courtesy Kent. A. Stevens.

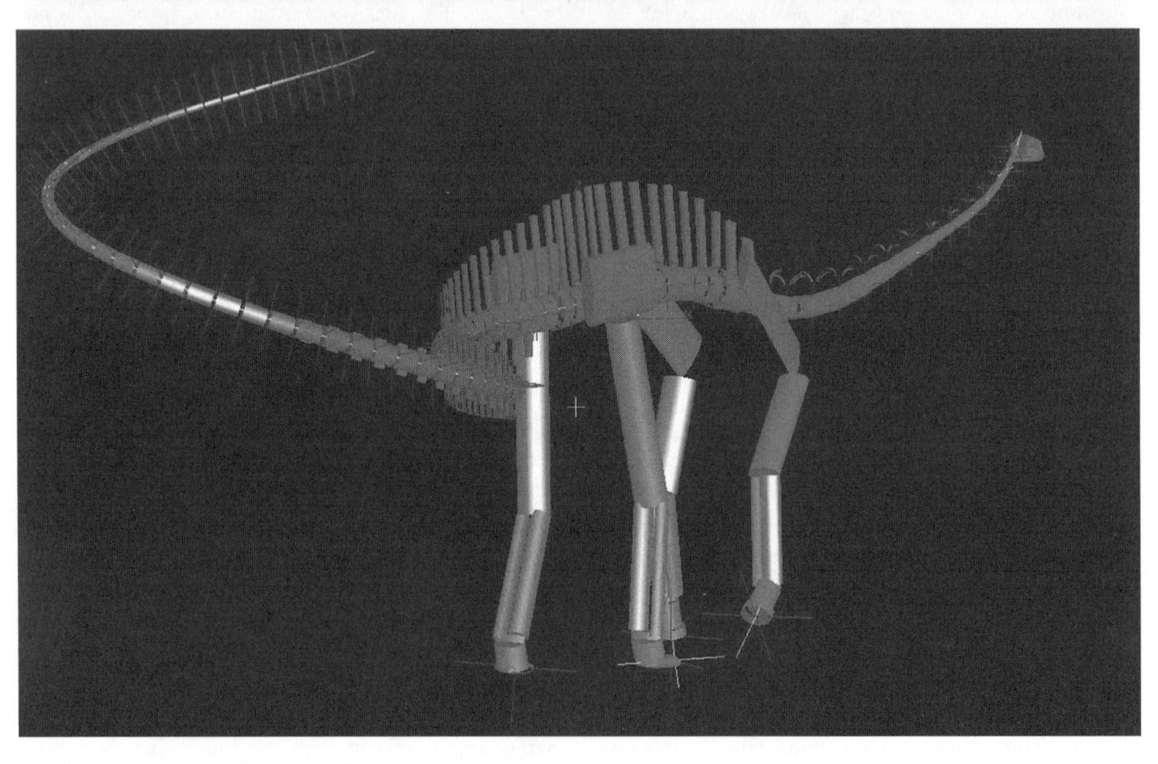

Computer generated model (rear left view) of the skeleton of *Diplodocus carnegii* (based on CM 84) with the neck in maximum dorsoflexion.

had sufficiently known necks, Martin *et al.* concluded that the necks of these animals — regardless of the mobility-controlling systems — were relatively or very rigid structures, habitually carried with the ventral aspect "down" with respect to gravity — "in other words as beams, not masts."

In a separate line of research on basically the same theme, Stevens and Parrish (1999, following earlier reports appearing in nontechnical publications (see *S1*), utilized a computer to generate articulated digital reconstructions in order to produce data indicating the neck posture and suggesting the feeding

Photograph by the author, courtesy The Field Museum.

Reconstructed skeletal cast of *Brachiosaurus altithorax* (including cast of holotype FMNH PR25107) mounted by the PAST company outside The Field Museum. Recent studies have indicated that the neck orientation shown in this mount was not possible. This is the museum's second cast of this skeleton; the first, formerly displayed inside the building (see *D:TE, S1*), is currently on exhibit at Chicago's O'Hare airport.

habits of two Late Jurassic sauropods, the more massive *Apatosaurus* and more gracile and longer-necked *Diplodocus* (based on mounted skeletons at the Carnegie Museum of Natural History, see *D:TE* for photographs and published accounts and illustrations). Stevens and Parrish chose this cyber-route of investigation rather than attempting direct manipulation of the fossil bones, as the originals are too heavy, awkward, and fragile to arrange manually and also because postdepositional distortion often prevents correct articulation.

From this study, Stevens and Parrish determined the "neutral" poses for both of these dinosaurs (the Carnegie mounted skeletons sharing an equal number of cervical vertebrae and having shorter forelimbs than hindlimbs, but differing in build and vertebral form), wherein the paired articular facets of the postzygapophyses of each cervical were centered over the facets of the prezygapophyses of its posteriorly adjacent neighbor. In this position, the muscles and ligaments controlling the movements of the neck were relaxed, with a minimization of the stresses on the joint capsules connecting the paired zygapophyses. In this pose, the necks of both *Apatosaurus* and *Diplodocus* were seen to be remarkably straight, the downward tilt of the spine in the shoulder region bringing their heads to just above ground level. At or near this pose, the authors noted, the ligament supporting the neck would be taut, holding the neck in a horizontal, cantilevered position with the need for much muscular exertion.

Given the postventroventral inclinations of the occipital condyles of both genera, the heads — when the necks were in this relaxed pose — were bent downward almost vertically, not unlike the orientation seen in some living, ground-feeding, ungulate mammals. Also in these genera, the extreme dorsal placement of the orbits, accompanied by a curved narrowing of the frontals between the orbits, would have allowed effective anterior vision even when the head was angled downward.

Stevens and Parrish found that the necks of both *Apatosaurus* and *Diplodocus* were less flexible than usually depicted in life restorations. Of the two genera, *Apatosaurus* had the more flexible neck, capable of curling back to form a tight S-curve or U-curve, with its head held out laterally over 4 meters. The neck of *Diplodocus*, though capable of less angular deflection laterally, was comparatively longer than that of *Apatosaurus*, thereby resulting in similar total lateral deflection of about 4 meters. *Diplodocus*, however, was barely capable of raising its head above back level.

As noted by Stevens and Parrish, *Apatosaurus* and *Diplodocus* were found to be quite similar in their skull shape, dentition and neck structure, suggesting that they may have fed on the same kinds of vegetation. The main difference between the neck posture of these two genera was in dorsiflexion. *Apatosaurus* could feed at a maximum height of 6 meters above ground level, *Diplodocus* at about 4 meters. Furthermore, based upon the zygapophyseal design of these genera, *Apatosaurus* had a feeding envelope that was, in anterior aspect, subrectangular (high and low feeding at lateral extremes). The envelope of *Diplodocus* was more diamond-shaped, enjoying less of vertical range allowed at the lateral extremes and less elevation. The necks of both genera were surprisingly found to be capable of more ventriflexion than dorsiflexion, attaining heights of at least 1.5 meters below ground level. According to these authors, this extreme ventral flexibility in both *Apatosaurus* and *Diplodocus* would have allowed a diplodocid standing at the edge of a lake or river to extend its head outward and downward to feed on high-nutrition vegetation found on or under the water.

Regarding the theory that diplodocids could rear up tripodally to feed off tall trees, Stevens and Parrish noted the following: the upper canopy aborescent plants of the Late Jurassic would have mostly consisted of conifers and ginkgoes, neither of which would have provided substantial nourishment for such large herbivores; high-nutrition vegetation (including ferns, seed ferns, algae, cycadeoids, and horsetails) was available to sauropods at the lower-canopy levels, where great neck length and the ability to flex it ventrally and transverse a broad lateral arc would have been an advantage; and the presence of a near-horizontal neck renders moot the problem of supplying blood to the brain if the neck were raised almost vertically. "Rather than flexing their necks like dinosaurian counterparts of giraffes or swans," the authors concluded, "they appear to have fed more like giant, long-necked bovids" (see also *Camarasaurus* entry).

These recent computer-based studies regarding the flexibility of sauropod necks have not gone without criticism. In their paper describing the new Cretaceous sauropod genera *Jobaria* and *Nigersaurus* (see entries), Sereno, Beck, Dutheil, Larrson, Lyon, Moussa, Sadleir, Sidor, Varrichio, Wilson and Wilson (1999) commented upon such recent work (*e.g.*, Stevens and Parrish 1999). According Sereno *et al.*, living long-necked herbivores (*e.g.*, the dromedary camel, *Camelus domesticus*), despite having relatively flat zygapophyseal articular surfaces, "routinely engage in extreme neck postures." The necks of sauropods, these authors suggested, may also have been in general "more flexible than inferred from computer modeling."

Seymour and Lillywhite (2000) performed their own study of sauropod hearts, neck posture, and metabolic intensity, reevaluating the problem that a tall, upright neck posture requires an exceptionally high arterial blood pressure and an enormous heart (see Poupa and Ostadal 1969; Hohnke 1973; Seymour 1976; Lillywhite 1991). Seymour and Lillywhite's study reconsidered this problem taking into account new data on the scaling of heart-muscle stress, considering various proposed solutions to this problem and evaluating a new solution proposed by the authors linking heart size with metabolic intensity.

Based on this study (see paper for details), Seymour and Lillywhite proposed that the stroke volumes of sauropods were smaller than previously assumed and that they required smaller hearts that either had to beat faster or reduce the rate of blood flows. The authors concluded that it was unlikely that large sauropods carried their necks upright, although such a posture may have been possible if these dinosaurs possessed metabolic rates substantially lower than those expected for endotherms. The authors noted that ectothermy for sauropods is further suggested by the uninsulated skin found on well-developed embryos (see Chiappe, Coria, Dingus, Jackson, Chinsamy and Fox 1998). Furthermore, high metabolic rates and upright neck postures seem to have been mutually exclusive in sauropods regardless of their metabolism. If genera such as *Apatosaurus* and *Diplodocus* were unable to raise their heads high because of the structure of their cervical vertebrae (*e.g.*, Stevens and Parrish 1999), they avoided high blood pressure and may have had either low or high metabolic rates. Moreover, if genera such as *Brachiosaurus* held their heads high (*e.g.*, Gunga, Kirsch, Rittweger, Röcker, Clarke, Albertz, Wiedmann, Mokry, Suthau, Wehr, Heinrich and Schultze 1999), such dinosaurs were most likely ectothermic (see section on ectothermy versus endothermy, below).

Work has continued by Paul (1998) concerning the somewhat controversial hypothesis that sauropods could rear up on their hind legs. This ability, the author noted, required a skeleton that displayed the adaptations and strength necessary for standing on two legs.

Paul pointed out that small-tailed elephants, though not designed for bipedal standing, sometimes rear up to feed. Fossil trackways show that giant sloths could walk bipedally; their massive dorsal vertebrae and stout tails with their protective "sled-chevrons" would have allowed a tripodal stance. Sauropods possessed large tails, vertebral columns stronger than those of elephants and approaching or exceeding those of giant sloths and always had hindlimbs that were stronger than their forelimbs. Based on these features, Paul concluded that all sauropods should have been able to stand bipedally more easily than elephants and that most of them were seemingly adapted to do so on a regular basis.

Paul also observed the following basic types of sauropod rearing: 1. "Cetiosaurs" — caudal and pelvic adaptations for rearing minimal; 2. shunosaurs and diplodocids — sled-chevrons protected by prop-like tail, a normal pelvic orientation preventing the hindlimbs to function when the body was erect, this favoring an immobile tripodal posture; 3. camarasaurs — absence of sled-chevrons suggesting the tail was kept off the ground, a retroverted pelvis permitting slow bipedal walking while feeding; and 4. mamenchisaurs and omeisaurs — sled-chevrons and retroverted pelves allowing both tripodal rearing up and also slow bipedal walking.

The evolution of the manus shape and the antebrachium in sauropod dinosaurs was a topic of study briefly reported on in an abstract by Bonnan (1999*b*). As pointed out by Bonnan, the sauropod manus has a unique structure among dinosaurs distinguished as a columnar ring of metacarpals that probably evolved in that way to support the great weight of these animals. However, prior to this study, the mechanism by which a comparatively flat saurischian manus evolved into this U-shaped forms had not been formally investigated.

As Bonnan noted, evidence from sauropod trackways indicates that the manus was pronated as it made contact with the substrate with digit III facing anteriorly. In most dinosaurs, pronation of the manus was accomplished by rotation of the proximal radius about the ulna, the radius consequently crossing over the ulna. However, in most sauropods, the radius cannot rotate across the ulna as these elements are interlocked at their proximal ends.

Also, the olecranon process in most sauropods is very reduced and flattened. If this process is oriented posteriorly, Bonnan stated, then the ulna lies posterior and the radius anterior, this configuration rotating the manus laterally so that digit I (rather than III) faces anteriorly, an orientation that does not agree with known manus tracks.

Based on examination and scale modeling of the forelimbs in two sauropod groups, the Camarasauridae and Diplodocidae, Bonnan proposed the following: 1. The manus was constrained by carpal, radius, and ulna morphology, the direction of flexion and extension thereby passing through digit III (not digits I and V); 2. to achieve pronation of the manus, the radius was medial and the ulna lateral; and 3. the ulna supported most of the humerus with its medial arm and flattened olecranon process, while the lateral arm of the ulna and radius articulated respectively with the lateral and medial condyles of the humerus.

The above observations suggested to Bonnan that "the radius initially crossed the ulna but, early in sauropod evolution, the entire radius migrated medially,

Photograph by the author, courtesy Peabody Museum of Natural History, Yale University.

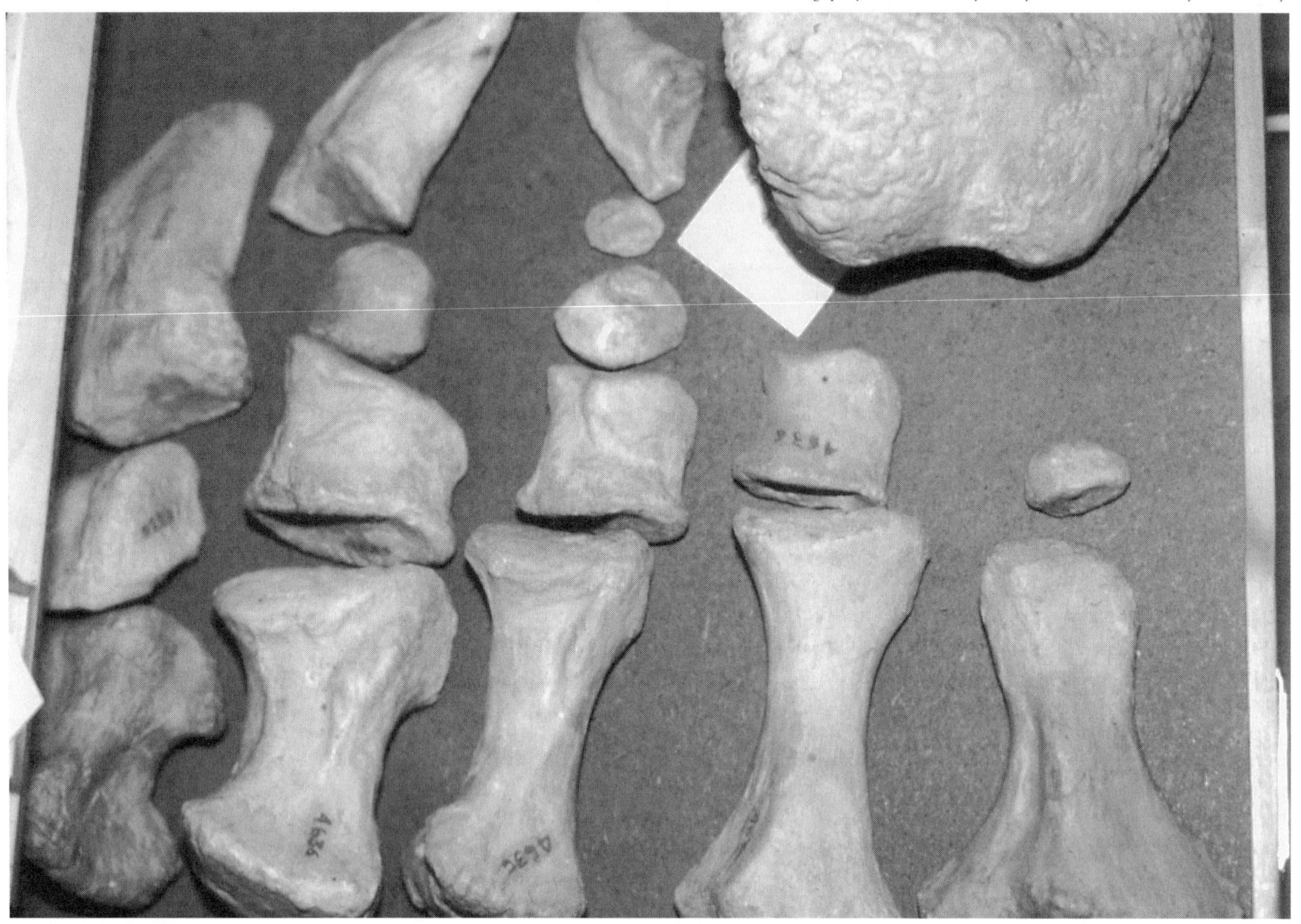

Manus of the diplodocid *Apatosaurus excelsus* (YPM 4636, cast of CMNH 89, the most perfect front foot of this genus yet collected). Matthew F. Bonnan recently published a brief report on the evolution of the manus in sauropod dinosaurs, focusing upon diplodocids and camarasaurids.

perhaps to reduce the strain of a crossed antebrachium under heavy compression," with the U-shaped manus perhaps forming as a result of the first digit following along as the radius migrated.

Ornithischians

Almost a century ago, ornithischian dinosaurs—all of which were plant eaters—were restored as possessing cheeks. Richard Swnn Lull, for example, suggested as early as 1908 that the ceratopsian *Triceratops* may have had cheeks. More than 30 years later, in 1940, Barnum Brown and Erich Maren Schlaikjer argued to the contrary regarding another ceratopsian, *Protoceratops*. According to the latter authors, the buccinator, the facial muscle present in mammals and required for cheek activity, is not found in reptiles. Thus, following Brown and Schlaikjer's study, it became more fashionable to portray these plant eating animals without cheeks.

In 1973, Peter M. Galton, in a study based on Lull's earlier research, published his own conclusions that ornithischians actually did have cheeks. Galton's (1973*a*) hypothesis was founded upon an anatomical feature seen in most ornithischian skulls—an edentulous area external to both the maxillary and dentary tooth rows, roofed by a prominent horizontal maxillary ridge and floored by a massive dentary. Galton suggested that this ridge served as a place where, in life, muscles were attached for a structure—bordering a space or vestibule that would have received chewed food that was moved lateral to the lower teeth—somewhat analagous in function to the muscular cheeks in mammals. According to Galton, ornithischian dinosaurs, possessing such a cheek-like structure, could retain food in their mouth while chewing; furthermore, a long and slender tongue could scoop up any food that had ended up in this space and return it to the mouth for chewing.

Since its original presentation in the literature, Galton's model has been the one mostly accepted. In fact, only recently has this interpretation been seriously challenged.

For example, Papp and Witmer (1998) pointed

Courtesy The Field Museum (neg. #GEO81446).

Skeleton (FMNH PR12991) of *Protoceratops andrewsi* with eggs possibly belonging to this dinosaur recovered in 1926 during the Third Asiatic Expedition of the American Museum of Natural History) to Mongolia. The idea that ornithischians had cheeks has recently come into question. In 1940, Barnum Brown and Erich Maren Schlaikjer expressed their opinion that dinosaurs such as this did not have cheeks.

out that neither birds nor crocodilians (the closest living relatives of dinosaurs) possess cheeks. These authors also noted that the feature of medially inset dentition, the most frequently cited osteological correlate for cheeks in ornithischians, is not present in mammals; furthermore, the muscular cheeks theorized for ornithischians could not have functioned like a mammalian buccinator.

Papp and Witmer offered the alternative hypothesis that the buccal region in ornithischians could be associated with the horny beak occurring on the premaxillary and predentary bones, as the beaks of some ornithischians display osteological similarities to those seen in birds and turtles. Noting the distinctiveness of ornithischians, the buccal fossae of which do not closely resemble those of any extant animal group, Papp and Witmer stated that these dinosaurs "may well have totally novel structures without any modern analog, confounding our attempts to reconstruct their soft tissues without confidence," further suggesting that "alternatives to mammalian-style cheeks should be pursued."

Czerkas (1998) also questioned the earlier models of ornithischians possessing mammal-like cheeks. Comparing the jaws of dinosaurs with those of extant reptiles, Czerkas concluded that certain criteria used to indicate the presence of cheeks (*e.g.,* teeth inset, longitudinal dorsal shelf laterally adjacent to tooth row, and prominent lateral ridge on outside of jaw) are also seen in some extant lizards and are, consequently, invalid. This further suggested to Czerkas "that the abraded teeth of derived ornithischians did not contribute to a true mammal-like mastication process, but instead may have played a more limited function of primarily cutting." Rather than teeth, Czerkas proposed that ornithischians had mouths with a "beak in front, followed by either typically reptilian lips, or a further development of an extended beak along the sides of the jaws."

With the exception of stegosaurs, all ornithischian dinosaurs possess ossified tendons, bony strand-like structures that reinforced their vertebrae. However, until recently, little research has been done on these structures, even though they have been known to scientists since the early days of paleontology. McNeil (1999), in an abstract, briefly published on his examination of ossified tendons from various ornithischian taxa — primarily the Ankylosaurida, Hadrosauridae, and Neoceratopsia. The studied remains were collected from many Late Cretaceous localities, including monogeneric bonebeds, in southern Alberta, some of this material originating from articulated specimens to ensure correct identification.

Following histological examination of thin sections of this material using transmitted light, McNeil found that all of these tendons possessed a similar, basic structure. An avascular outer layer of lamellated bone surrounds primary, fibro-lamellar bone. Secondary Haversian reconstruction heavily replaces the primary bone, which may grade into a cancellous center. Importantly, McNeil also discovered that the bone microstructure in these tendons significantly differs — mainly in size, shape, and the arrangement of secondary Haversian canals — between the clades, thereby allowing identification based only on a fragment of ossified tendon.

As ossified tendon fragments are relatively common, they can now, utilizing McNeil's new method of identification, be associated with particular ornithischian groups. This method "may also provide new insight into phylogenetic relationships among the ornithischian dinosaurs."

Organ (2000), in an abstract, reported on his own study of ossified tendons in ornithischian dinosaurs utilizing engineering statistics and computer models. Organ's analysis indicated that both trellised patterns — parallel and crisscrossed — of ossified tendons passively served to elevate and stiffen the tail, not laterally as previously suggested, but dorsoventrally. Such stiffening, the author proposed, would reduce deflection of the tail resulting from protraction of the hindlimb and contraction of the m. caudofemoralis longus. Organ hypothesized that this resistance to tail deflection played "a large role in locomotion due to the increase in locomotion due to the increase in force transfer to the retracting hindlimb.

THYREOPHORANS

Barrett (1998*a*), in an abstract, published a brief report on possible feeding mechanisms — until now, only poorly understood — in thyreophoran (including armored and plated) dinosaurs, based upon his study of craniodental material belonging to the primitive thyreophoran *Scelidosaurus*, and also various stegosaurs and ankylosaurs (see also abstract of Rybcznski and Vickaryous 1998).

These features were noted by Barrett: *Scelidosaurus* teeth are heavily worn, their wear patterns distinctive, the high-angled wear facets being on the lingual crown surfaces of the maxillary teeth and labial surfaces of the dentary teeth. Wear facets often extend beyond the mesial and distal margins of the tooth crowns over the crown apex. Sometimes the apical wear facets on opposing maxillary and dentary teeth can be matched. These observations may suggest that the tooth crowns, during jaw closure, met in opposition to each other rather than interlocking. However, opposing wear facets could not be identified in some tooth positions.

The foregoing observations indicated to Barrett that some of the wear may have been caused by tooth-food rather than tooth-tooth contact. As the bases of the dentary tooth crowns are labially expanded in mesial view, Barrett suggested "that the dentary teeth form a broad basin which allows puncture-crushing of food items by the maxillary teeth," possibly explaining "the random distribution of wear facets along the tooth row."

Barrett added that similar patterns of tooth wear were also observed in some stegosaurs and ankylosaurs, and suggested that the "development of paired crescentic cingula on the teeth of the ankylosaurs *Texasetes* and *Priodontognathus* may represent a refinement of this puncture-crushing mechanism."

Ankylosaurs

Carpenter (1998*b*), in reexamining the Ankylosauria for the then in-progress Tree of Life Project, cited a small number of "odds and ends" points concerning ankylosaurs, including the following:

1. The holotype of *Hylaeosaurus armatus* includes an incomplete skull including a bowed left quadrate attached to the paroccipital process, posterior portions of both mandibles, and an upper right horn resembling that in *Pawpawsaurus*; 2. in *Polacanthus foxii*, the perceived lesser trochanter of one femur is a matrix-filled crushed zone, the crushing giving another a midshaft "kink" and making it superficially resemble the femur of the distinct genus *Hoplitosaurus* (see also Coombs 1995; also *D:TE*), the sometimes referred-to "tail-club" in *Polacanthus* actually being the transition zone between the short anterior caudal vertebrae and more elongate distal caudals; 3. the neck region of the holotype of *Scolosaurus cutleri* [usually considered to be a junior synonym of *Euoplocephalus tutus*; see *Euoplocephalus* entry, *D:TE*] had been damaged by erosion before discovery, resulting in the loss of most of the keel on the second cervical ring, giving the armor the appearance of large and flat discs, this character no longer being a criterion for separating *Dyoplosaurus* (=*Scolosaurus* [both genera usually regarded as junior synonyms of *Euoplocephalus*]) from *Euoplocephalus*; 4. the lateral temporal fenestra in *Silvisaurus* is visible in side view; the surface of most of its cranial armor was lost during preparation, its pattern therefore indeterminate; and, 5. preliminary phylogenetic analysis of *Minmi* places this genus as a sister taxon to the Nodosauridae plus Ankylosauridae.

Carpenter and Kirkland (1998) published a preliminary review of Lower and Middle Cretaceous, North American ankylosaurs, including *Hoplitosaurus marshi*, *Nodosaurus textilis*, *Pawpawsaurus campbelli*, *Priconodon crassus*, *Sauropelta edwardsi*, *Stegopelta landerensis*, *Silvisaurus condrayi*, *Texasetes pleurohalio*, *Gastonia burgei*, and an unnamed nodosaurid and ankylosaurid (see individual genus entries for the authors' revised or emended diagnoses for most of these taxa).

In this study, Carpenter and Kirkland also corrected two errors in terminology that, in the past, had been applied to the scapula and proximal end of the femur, these regions being taxonomically diagnostic in ankylosaurian dinosaurs:

1. The lateral surface of the scapula in many nodosaurid ankylosaurs has a ridge which has been referred to as the acromion (*e.g.*, Russell 1940), pseudoacromion (*e.g.*, Carpenter, Dilkes and Weishampel 1995), and scapular spine (*e.g.*, Coombs 1995). According to Carpenter and Kirkland, this feature is primitively present in the basal thyreophoran *Scelidosaurus* as a swelling along the dorsal margin of the scapula and remains there in ankylosaurids (see Coombs 1978*a*). In nonavian theropods and birds, the homologous structure, referred to as the acromion, supports the furcula (*e.g.*, Makovicky and Currie 1996); in mammals, it supports the clavicle (Gregory 1918). As ankylosaurs possess neither furcula nor clavicle, this structure is probably "the acromion, which in nodosaurids, has migrated ventrally onto the lateral surface of the scapula." As previously suggested by Coombs (1978*b*), the M. scapulohumeralis cranialis (a muscle) originated on the acromion; however, according to Carpenter and Kirkland, based upon the topography in birds, the M. scapulotriceps most likely originated along the dorsal margin of the acromion, instead of posterior to the glenoid.
2. Previously, the dorsal region lateral to the condylar head of the femur had been identified (see Carpenter *et al.* 1995, following Gilmore 1914 for *Hoplitosaurus*) as the greater trochanter, this location coinciding with that identified by Marsh for the stegosaurid thyreophoran *Stegosaurus*. According to Carpenter and Kirkland, however, that identification is incorrect, as first noted by Gregory. The latter author pointed out that the greater trochanter in reptiles (including Sauropoda) is found on the outer side of the femoral shaft somewhat near the proximal end; in ornithischians, the proximal surface of what has incorrectly been called the greater trochanter was covered by bursa, resembling the smooth surface above the greater trochanter found in the ostrich. Examination by Carpenter and Kirkland of a domestic chicken revealed "cartilage capping the rugose dorsal surface of the femur, and that the greater trochanter is located on the lateral surface of the femoral shaft." Separating the greater trochanter from the femoral head is the ridge-like crista trochanteris, this ridge in ankylosaurs sometimes prominent and well ossified (as in *Sauropelta*), sometimes not. The greater trochanter in birds serves as the insertion point for the heads of three muscles identified by Coombs (1979) in ankylosaurs — M. iliotrochantericus anteriorly, the M. iliofemoralis externus medially, and the M. ischiofemoralis posteriorly.

Furthermore, Carpenter and Kirkland observed that, topographically, "the insertion for the M. iliotrochantericus caudalis, at the antero-dorsal most part of the greater trochanter, coincides with what has been referred to as the lesser trochanter in ankylosaurs" (*e.g.*, Gilmore 1914; Coombs 1995; Pereda-Suberbiola 1994). As Gregory had already noted, however, the true lesser trochanter is medially found below the femoral head, serving as the insertion point for the M. pubo-ischio-femoralis, having migrated from its primitive location in early reptiles to its derived position on the medial side of the femur, in mammals. Consequently, stated Carpenter and Kirkland, "the lesser trochanter of ankylosaurs (and dinosaurs in general) is not homologous with the lesser trochanter of primitive reptiles and mammals." Rowe (1989) referred to this structure in the small theropod *Syntarsus* as the anterior trochanter, an identification embraced by Carpenter and Kirkland.

Penkalski (1998), in an early report previewing a forthcoming systematic analysis of the Ankylosauridae from the Late Campanian of North America), this family, the author pointing out, being more diverse than is generally realized. At the same time, Penkalski briefly reported on a new ankylosaurid genus found in the Late Campanian Two Medicine Formation of Montana.

Incorporating skull measurements from this unnamed taxon, as well as those from other known ankylosaurids taxa, Penkalski performed a morphometric study (one that has come under criticism by some workers [K. Carpenter, personal communication 2000]) that revealed several clusters. Taking into account that one of more clusters could be explained by sexual dimorphism, Penkalski found at least two species present. Clusters were corroborated by analyses of skull and postcranial characters, dentition, and armor morphology. Four species — *Euoplocephalus tutus*, *Dyoplosaurus acutosquameus*, *Scolosaurus cutleri* [the latter two taxa generally accepted as junior synonyms of *E. tutus*], and the Two Medicine form — were determined by this study.

Several unusual, perhaps primitive, features were observed by Penkalski in the Two Medicine ankylosaurid. They include a relatively large humerus, six closely-set armor scutes in the first cervical half-ring and nodosaur-like cervical spikes on the possibly second half-ring.

From this preliminary study, Penkalski noted that the new form could be allied with "*Dyoplosaurus*," a juvenile form which possesses a comparatively small tail club and regose armor that tends to form high-keeled scutes; that *Euoplocephalus* has a skull that is trapezoidal in dorsal view, six scutes in the first half-ring, and sharply keeled armor; and that "*Scolosaurus*" has a skull [most of which is missing in the holotype of *Scolosaurus cutleri*] with a more restricted snout, four amorphous scutes in the first half-ring, and rugose armor that tends to form conical scutes. Specimens having large plates, Penkalsli speculated, might represent another species.

ORNITHOPODS

Coria (1999), in a review of the Ornithopoda from the Neuquén Group (?Middle to ?Late Cretaceous) of Patagonia, Argentina, mentioned various previously unreported indeterminate ornithopods, including one represented by part the proximal half of a tibial shaft (MCF-PVPH-165) collected from the Río Neuquén Formation. Although the specimen was too fragmentary for phylogenetic evaluation, its great size (41 centimeters) "indicates the presence of a large iguanodontian form." Other indeterminate ornithopod remains from the same formation, reported by Coria, include a partial right femur (MCF-PVH-73), fragments of a proximal and distal tibia (MCF-PVPH-165), and several ungual phalanges.

Hypsilophodonts and Other Primitive Forms

New insights regarding hypsilophodonts — an assemblage of rather small and primitive bipedal ornithischians — were presented by Rich and Vickers-Rich (1999) in their report on the Hypsilophodontidae of southeastern Australia, all of these taxa being represented by rather meager and sometimes poorly-preserved specimens consisting mostly of isolated bones and teeth.

Earlier, Rich and Vickers-Rich (1989) had divided the femora of all known Australian hypsilophodontids into four basic groups (see *Fulgurotherium* entry, "Notes," for details, *D:TE*): 1. Femora assigned to the type species *Leaellynasaura amicagraphica*; 2. those designated Victorian Hypsilophodontid Type 1; 3. those referred to the type species *Atlascoposaurus loadsi*; and 4. those designated Victorian Hypsilophodontid Type 2. New details observed by Rich and Vickers-Rich following the collection of additional Australian hypsilophodontid specimens over the past decade, and also observations made of other nonhypsilophodontid specimens, suggested that the boundaries originally recognized by them in 1989 "may not be realistic."

During the interim between 1989 and 1999, Vickers and Vickers-Rich made observations relating to this issue. A study was performed by these authors of a collection of 15 beautifully preserved femora (HMN collection) belonging to *Dryosaurus lettowvorbecki*, a small, bipedal and primitive iguanodontian ornithopod superficially similar to hypsilophodontids. All of these specimens were recovered from a single site — an Upper Jurassic quarry in the Tendaguru Hills, west of Lind, Tanzania, representing "a monospecific mass death assemblage where animals of all sizes are thought to have perished during widespread drought conditions" (see Heinrich, Ruff and Weishampel 1993). Vickers and Vickers-Rich were impressed by the striking uniformity of shape of this material, despite that the linear dimensions of the smallest specimen are only one-third that of the largest. "Essentially," these authors observed, "the smallest femora are miniatures of the largest with little apparent proportional differences." This suggested to them that allometric changes within a single small ornithopod species can mostly be discounted, save perhaps for the bones of hatchlings (identified by a lesser degree of ossification).

As noted by Rich and Vickers-Rich, the sample size of the Australian specimens is very small due to their incompleteness and also the distortion of their shape resulting from the high pressures imposed upon them during burial, resulting in two things making their analysis difficult: 1. Linear quantitative measurements are of little use (the only measurements of any meaning and consistency being those of the crudest attributes, *e.g.*, femoral length) and 2. shape differences may either have interspecific significance or be the result of postmortem distortion. Therefore, as suggested by the *D. lettowverbecki* remains, any possible intraspecific variation in the Australian material, many of these elements having been extensively crushed, has probably been obscured.

Consequently, taking into consideration the collection of additional Australian specimens and also their observations regarding the great range of morphological variation in *D. lettowverbecki*, Rich and Vickers-Rich revised their assessment of the Australian femora. In 1989, Rich and Rich had included in their diagnosis of *L. amicagraphic* that the femur is characterized as being anteroposteriorly compressed; at the same time, Victorian Hypsilophodontid Femur Type 1 was characterized by them as having a mediolaterally compressed distal end. However, reexamination of one more recently collected specimen (NMV P18596) — a femur in which the distal end is oriented to suggest that it appears intermediate between the two categories — cast doubt on that division. Also pointing out that all of the above morphologies are known only from a single locality (Dinosaur Cove in the Albian Eumerella Formation), Rich and Vickers-Rich, therefore, found it "appropriate to combine the material originally assigned to *L. amicagraphica* with that assigned to Victorian Hypsilophodontid Type 1"; similarly, specimens assigned to *Fulgurotherium australe* and Victorian Hypsilophodontid Femur Type 2 may also be combined. However, a total of 39 (including those recovered since 1987) femora belonging to this grouping — the most abundant femoral type, occurring in both the Aptian Wonthaggi Formation and Albian Eumerella Formation — are currently known. Based upon what is now understood of femoral variation, these "could belong to one highly variable species or, on the other hand, represent three or four different species or even genera."

The discovery of a new hypsilophodontid taxon, represented by a partial skull from the Early Cretaceous (Valganinian–Hauterivian) Tetori Group of central Honshu, Japan, was reported in an abstract by Barrett and Manabe (2000). The specimen includes much of the facial region, a partial left dentary, and bones of the skull roof and palate. Based upon a preliminary study of this material, this dinosaur seems to be more similar to the Australian taxa *Atlascoposaurus* and *Leaellynasaura* than to *Hypsilophodon*. (Other ornithopods represented at this locality include the fragmentary skull remains of a large iguanodontian [see below] having teeth very similar to those of *Iguanodon*. Nonornithischian remains (represented by teeth and ungual claws) from this locality include theropods [tyrannosaurid, velociraptorine, and member of the unnamed clade of Oviraptorosauria plus Therizinosauroidea] and sauropods [a basal titanosaurian or euhelopodid and a ?nemegtosaurid]).

As noted by Barrett and Manabe, the above discoveries supplement knowledge of East Asian paleobiogeography during Early Cretaceous times and offer new information on the Jurassic/Cretaceous terrestrial transition, current data suggesting "that Eastern Asia was a centre of origin and diversification for many dinosaur clades as well as a refugium."

Iguanodonts

Iguanodontian ornithopods comprise a diverse group of mostly bipedal ornithischian dinosaurs. Kobayashi and Azuma (1999) reported, in an abstract, on a presently unnamed iguanodontian, remains of which were found during two field periods (1988–1993 and 1996–1998) by the Fukui Prefectural Museum in freshwater deposits of the Lower Cretaceous (Barremian–Aptian) Kitadani Formation, Tetori Group, Fukui Prefecture, Japan. These remains consist of skull parts belonging to at least three individuals, ranging from juvenile to subadult. (Other materials collected at this locality include a modest-sized theropod and an almost complete goniopholidid crocodyliform.)

As briefly described by Kobayashi and Azuma, this new taxon displays a mosaic of both plesiomorphic and derived characters, differing from other known iguanodontians in a number of ways (*e.g.*, rostrum narrow; no embayment of lacrimal process of maxilla; high lacrimal process; jugal process without caudolateral projection; paraquadratic foramen; dentary without diastema; surangular with second foramen). Some [not yet disclosed] features, however, indicate affinities with iguanodontid dinosaurs (*e.g.*, *Iguanodon*, *Altirhinus*, and *Ouranosaurus*). Therefore, according to a preliminary phylogenetic analysis by the authors utilizing cranial characters, the new form from Japan is closely related to iguanodontid-grade ornithopods, but this analysis indicates that Iguanodontidae is a paraphyletic assemblage.

According to Kobayashi and Azuma, the occurrence of this new iguanodontian taxon, as well as fragmentary remains of iguanodontid-grade dinosaurs previously reported from other Lower Cretaceous Japanese formations, could indicate a dispersal of iguanodontid-grade dinosaurs into eastern Asia by Valanginian times, plus a maximum temporal range in Japan from Valanginian to Aptian times.

"Anabisetia saldiviai," a new dryomorph iguanodontian to be named and formally described by Coria and Calvo (to date in review), was first mentioned by Coria (1999) in a review of the ornithopods from the Upper Cretaceous Río Limay Formation, Nequén group, of Patagonia, Argentina, collected at the same stratigraphic level of claystone deposits derived from floodplain and paleosol environments. The new taxon is known from the very well-preserved remains of at least five individuals, some of which are articulated or semiarticulated and vary only slightly in size. The skeletons represent a small (2 meters or about 6.8 feet in length) dinosaur with strong hindlimbs.

Coria diagnosed "A. saldivai" as follows: fifth metacarpal flattened; scapula having very strong acromion process; preacetabular process of ilium longed than 50 percent of total length of ilium; preacetabular process of ilium (when articulated) reaching more anterior point than prepubic process; ischiatic shaft proximally triangular, distally quadrangular in cross section; fibular-astragalur contact of tarsus.

According to Coria, an interesting feature in this dinosaur is the pes, which "has a conspicuous, though reduced, metatarsal I which bears digit I." This toe is formed by two phalanges, the second of which is an ungual with a degree of reduction similar to that in such basal dryomorph genera as *Dryosaurus* and *Camptosaurus*. Other features of the Dryomorpha observed by Coria in "Anabisetia" include "a laterally compressed prepubic process, a proximally placed obturator process and anteriorly oriented ischial foot."

Based on cladistic analysis, Coria found "Anabisetia" to belong within the Iguanodontia because of its possession of such iguanodontian synapomorphies as the following: dentary with parallel margins;

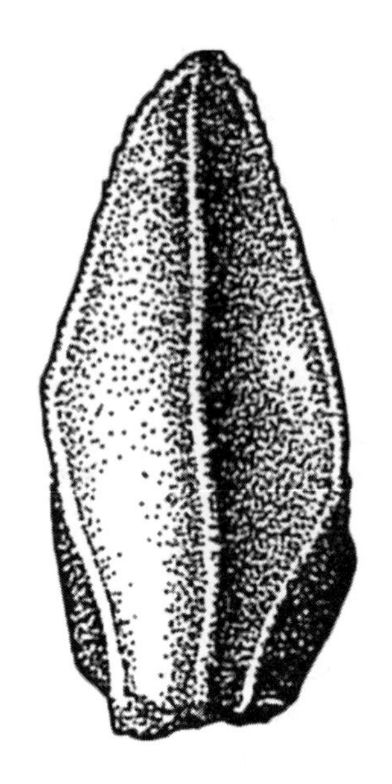

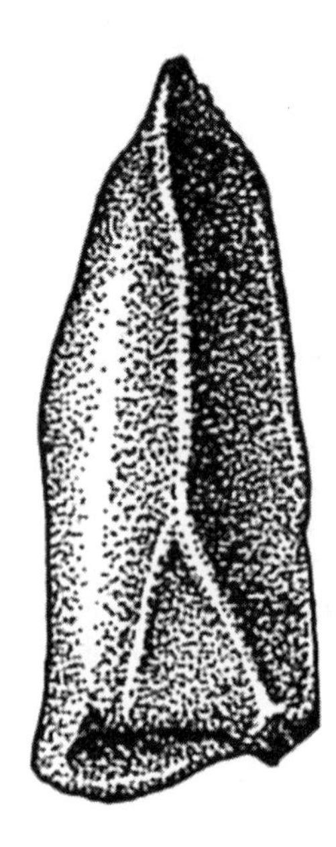

Hadrosaur tooth (MLP 98-I-10-1) from Antarctica in (left) occlusal and (right) lateral views. Scale = 10 mm. (After Case, Martin, Chaney, Reguero, Marenssi, Santillana, and Woodburne 2000.)

sinuous dorsal margin of iliac blade; femur having deep anterior condylar groove; and condylid slightly overlapping posterior intercondylar groove. Euiguanodontian synapomorphies exhibited by "Anabisetia" include these: maxillary teeth having strong primary ridge; brevis shelf well developed; and metatarsal I reduced. According to Coria, this new genus is more derived than *Gasparina* (see *S1*), having an ischial shaft that is rounded in cross section and a "footed" distal ischial shaft.

Coria also reported a fragment of a proximal femur collected recently from the same outcrops (but not the same quarry) that yielded the "Anabisetia" material. In this specimen, unlike "Anabisetia," the greater and lesser trochanters are fused, this feature being an autapomorphy proposed for the primitive euiguanodontian *Gasparinisaura* (see Coria and Salgado 1996). According to Coria, this feature could indicate the cohabitation of two different kinds of ornithopods at this locality, including a *Gasparinisaura*-related form, "suggesting that the biocron of this form is more extensive than thought, and that the split between *Gasparinisaura* and "Anabisetia" may have occurred earlier than the Cenomanian."

Buffetaut and Suteethorn (1998) briefly announced the discovery of the remains of iguanodontid iguanodontian dinosaurs from two localities in the Lower–Middle Cretaceous (Aptian–Albian) Khok Kruat Formation of northeastern Thailand. This area previously yielded material belonging to the primitive ceratopsian *Psittacosaurus* (see *D:TE*) and also indeterminate theropods. The ornithopod material—which, the authors observed, is indistinguishable from *Iguanodon*—includes a quadrate, numerous teeth (most of them shed and worn), vertebrae, and a femur. The relatively small size of most of these remains indicated to the authors that they belong to juveniles.

According to Buffetaut and Suteethorn, these iguanodontid remains "complement the postcranial elements found by Hoffet in the 1930s in an equivalent of the Khok Kruat Formation in Laos (and misidentified by him as hadrosaurs)." Furthermore, this discovery confirms that iguanodontids lived in South East Asia at the end of the Early Cretaceous that a significant faunal exchange took place there during the Early Cretaceous. As the older (?Valanginian–Hauterivian) Sao Khua Formation preserves much dinosaurian material but no ornithopods, it is implied that iguanodontids arrived, presumably from Europe, in South East Asia relatively late during the Early Cretaceous, a scenario possibly also true for Central Asia. Buffetaut and Suteethorn commented that "reports of undoubted iguanodontids from earlier parts of the Early Cretaceous in Japan are not easy to reconcile with the Thai and Central Asian record."

Xu, Zhao, Dong and Huang (1999) announced in an abstract the discovery in spring 1994 of a partial iguanodontian skeleton from the Late Cretaceous Sangpin Formation of Nanyang Basin, southeastern Henan Province, China, a site from which thousands of dinosaur eggs were also collected. The specimen is celebrated as the first dinosaur skeleton known from this area. Its estimated length, as reconstructed by the authors, is about 4.5 meters (more than feet).

According to Xu *et al.*, the skeleton represents a new genus and species distinguished from other known iguanodontians by the following apomorphies:

metacarpals II to IV partially fused; metacarpal II more than 90 percent length of subequal metacarpals III and IV; metacarpal IV more robust than others, with a very thick distal end; and phalanx IV-1 of manus wider transversely than proximally.

It is similar to hadrosaurids in the following features: caudal neural spines longer than corresponding chevrons; deltopectoral crest having angular ventral margin; carpus reduced; metacarpals slender; loss of metacarpal I; femur straight; anterior intercondylar groove tunnel-like; transverse widening of proximal tibial head, extending onto diapophysis; distal tarsals lost; and metatarsal I lost.

Xu *et al.* observed that this new form is more derived than hadrosaurids in the following features: only six sacral vertebrae; centra of anterior 17 caudal vertebrae presenting rounded or rectangular outline in cranial aspect; and astragalus rectangular in anterior aspect with weak ascending process. Phylogenetic analysis by Xu *et al.* suggested that the new iguanodontian lies outside of Hadrosauridae and is more derived than *Probactrosaurus.*

Hadrosaurs

One of the earliest hadrosaurs, or duckbilled dinosaurs, known to date was described by Casanovas, Pereda-Suberbiola, Santafé and Weishampel (1999). This new form, as yet unnamed, is known only from a lower left jaw preserving most of the dentary bone and part of the tooth battery. The specimen was collected from the Fontllonga site in the Tremp Formation (uppermost Maastrichtian; see Galbrun, Feist, Colombo, Rocchia and Tambareau 1993) in the Noguera region, southern limb of the Ager syncline, province of Lleida, southern Pyrenees of Catalonia, between the villages of Camarasca and La Baronia de Sant Oisme, on the Iberian peninsula. Stratigraphically, this specimen is distinguished as one of the youngest dinosaur bones yet collected in Europe.

To understand the phylogenetic relationships of the "Fontllonga hadrosaurid," Casanovas *et al.* performed a cladistic analysis based upon eight ornithopod taxa (ingroups including *Iguanodon* sp., *Ouranosaurus nigeriensis*, *Probactrosaurus gobiensis*, *Bactrosaurus johnsoni*, *Telmatosaurus transsylvanicus*, Hadrosaurinae, Lambeosaurinae, with *Hypsilophodon foxii*, *Camptosaurus dispar*, and *Rhabdodon robustus* used as outgroups), 17 mandibular and dentary tooth characters coded. Results of this analysis showed that the "Fontllonga hadrosaurid" is more derived than *Telmatosaurus,* making it a member of the Euhadrosauria, a clade sharing the following dentary tooth-related synapomorphies: miniaturization of dentary teeth; dentary teeth not recurved; and marginal denticles of dentary crowns not supported by subsidiary ridges. However, the "Fontllonga hadrosaurid" lacks some of the derived features that unite the subfamilies Hadrosaurinae and Lambeosaurinae. Therefore, the specimen from Catalonia was regarded by Casanovas *et al.* as a basal euhadrosaurian outside of both Hadrosauria and Lambeosaurinae. This study, Casanovas *et al.* noted, provided new information concerning the diversity of European hadrosaurids at the end of the Cretaceous.

The first hadrosaur specimen (MLP 98-I-10-1; cast, SDSM C641) ever found in Antarctica was described by Case, Martin, Chaney, Reguero, Marenssi, Santillana and Woodburne (2000). The specimen, a lower cheek tooth, was collected from Upper Cretaceous (late Maastrichtian) marine sediments of the Sandwich Bluff Member of the Lopez de Bertodano Formation, "Reptile Horizon," on the eastern flank of Sandwich Bluff, on Vega Island, Antarctic Peninsula. The tooth was recovered during a joint United States–Argentinean field expedition. As described by Case *et al.*, the tooth is definitely hadrosaurian in form, closely resembling (*i.e.,* lozenge-shaped enamel face with median carina, broad vertical wear facet on opposite side of tooth from enamel face), because its crown morphology is very similar to that of North American duck-billed dinosaurs, particularly *Edmontosaurus.* Various features (*e.g.*, tooth relatively symmetrical, rather short, with coarse, poorly formed denticles) of MLP 98-I-10-1 suggested to the authors that the specimen could belong to the hadrosaurinae. However, the broken root precluded determination of the crown-root angulation, a characteristic that can be utilized in separating the Hadrosaurinae and Lambeosaurinae (*e.g.*, Lull and Wright 1942; Horner 1990).

According to the authors, this dinosaur is certainly an immigrant into Gondwana from North America. Additionally, as duckbilled dinosaurs are also known from South America, the discovery of the Antarctic form supports an earlier hypothesis that an isthmus joined Antarctica with South America during Late Cretaceous times allowing for faunal exchange (see Molnar 1989).

Dinosaur material from Mexico is rare. In an abstract, Westgate, Brown, Cope and Pittman (2000) reported remains of hadrosaurian limb bones and vertebrae found littering the outcrop surface of the Late Cretaceous Aguja Formation in Chihuanhua, Mexico, near Big Bend National Park. (Other taxa reported from this site by Westgate *et al.* include large and small carnosaurs, fishes [gar, bowfin, and shark], soft-shelled turtles, and oysters.)

The first dinosaur specimen found in New Mexico preserving fossilized skin impressions in association with skeletal remains was reported by Anderson,

Photograph by the author.

Skeleton (cast of GDF 300) of *Ouranosaurus nigeriensis*, a long-spined iguanodontid from Niger, Africa, exhibited at Dinofest (2000–2001), held at Chicago's Navy Pier.

Lucas, Barrick, Heckert and Basabilvazo (1998). In 1990, George T. Basabilvazo spotted fragmentary fossil bone material and unidentified integument impression preserved within sandstone float of the Ringbone Formation (Upper Cretaceous) in the Little Hatchet Mountains east of the town of Playas, in Grant County, southwest New Mexico. Three years later, Adrian P. Hunt and Spencer G. Lucas (1993) identified this material as possibly having vertebrate affinities and having a morphology comparable with chondrychthian skin impressions like *Petrodus* sp. The actual specimen (NMMNH P-2611) was collected in January 1996. It represents the distal-tail region of an indeterminate hadrosaur and includes 20 articulated caudal vertebral centra, fragments of several other disarticulated centra, one chevron, ossified tendons, skin impressions, and unidentified bone fragments; several float blocks also contain fragmentary, poorly preserved neural spines. The integument, comprising six discrete patches, was preserved in positive and negative relief between two beds of very finely grained sandstone, which were interpreted by these authors as a fluvio-lacustrine facies package.

Anderson *et al.* described the skin impressions as characterized by predominantly apical, circular to ovate tubercles. Excluding some isolated larger limpet-like forms, the tubercles in the tail region are of approximately the same size, ranging from 3–12 millimeters and 10–16 millimeters on the short and long axes, respectively. Virtually every examined tubercle is ornamented with radiating ridges and grooves that meet at their apex, their number and complexity increasing in direct proportion to tubercle size.

Although NMMNH P-2611 could not be referred to any particular hadrosaurid genus, the authors noted that the size, shape, and ornamentation of the isolated limpet-like tubercles were similar in some ways to those reported in hadrosaurine "gryposaurs" and in the lambeosaurine *Corythosaurus casuarius* (see *Gryposaurus* and *Corythosaurus* entries for information and various authors, *D:TE*).

As noted by Anderson *et al.*, the specific type of ornamentation in the integument of the Ringbone hadrosaurid is not known in the skin of any extant reptile, bird, or mammal; however, the tubercular morphology of NMMH P-2611 resembles that of the lizard *Heloderma*, although the individual tubercles of this modern reptile lack ornamentation. As speculated by the authors, a possible advantage of the radial sculpturing in the Ringbone specimen "would be

Photograph by the author, courtesy American Museum of Natural History.

Detail of ossified tendons in the hip and tail region of a skeleton of the duckbilled dinosaur *Corythosaurus casuarius*. Such bony, strand-like structures have recently been studied by Paul E. McNeil (1999).

more efficient heat exchange across the skin-air interface." These features possibly also represent "a form of structural reinforcement of the integument" offering additional surface resistance to damage (see also Lucas, Heckert and Sullivan 2000 for a review of and comments on NMMNH P-2611).

Exceptionally well-preserved hadrosaurian skin impressions, from body areas for which such impressions had not yet been described, were reported in an abstract for a poster by Hernández-Rivera and Delgado-de Jesús (2000). These impressions pertain to at least two hadrosaurian individuals collected from the Cerro de los Dinosaurios locality in the Cerro del Pueblo Formation of southeastern Coahuila, Mexico. The integumentary material includes positive skin impressions of the limb, tail, and hip areas, plus negative impressions from other areas of the bodies.

Preliminary analysis by the authors showed that the positive relief surfaces are preserved mostly at the base of a mudstone, while the negative impressions occur along the top of the mudstone underlying the skeletal remains. Hernández-Rivera and Delgado-de Jesús noted that additional skin impressions, not associated with any skeletal material, were found in a nearby quarry.

Weishampel, Mulder, Dortangs, Jagt, Jianu, Kuypers, Peeters and Schulp (1999) described isolated unnamed hadrosaurid remains recovered from various outcrops of the Gulpen Formation (Lanaye Member) and Maastricht Formation (Valkenburg, Gronsveld, Schiepersberg, Emael, Nekum, and Meerssen Members), in the Maastrichtian type area of the Netherlands and Belgium. This material includes two maxillary teeth (original in E. Croimans Colln collection [cast NHMM 1999012] and NHMM 1997274) of unresolved relationships within the Hadrosauridae, a dentary tooth (NHMM RD [R. W. Dortangs Collection] 214) apparently belonging to the Euhadrosauria, a right metatarsal III from the CBR-Romontbos quarry in the Valkenburg Member of the Maastricht Formation, representing a large but relatively gracile hadrosaurid, and a fragmentary ?right humerus (TM 11253) perhaps lacking an extensor intercondylar groove (a potentially diagnostic feature for this form), this specimen possibly belonging to a nonlambeosaurine hadrosaurid.

TM 11253, Weishampel *et al.* related, has had a convoluted history. Collected in the nineteenth century, it was originally recorded by Winkler (1865), in

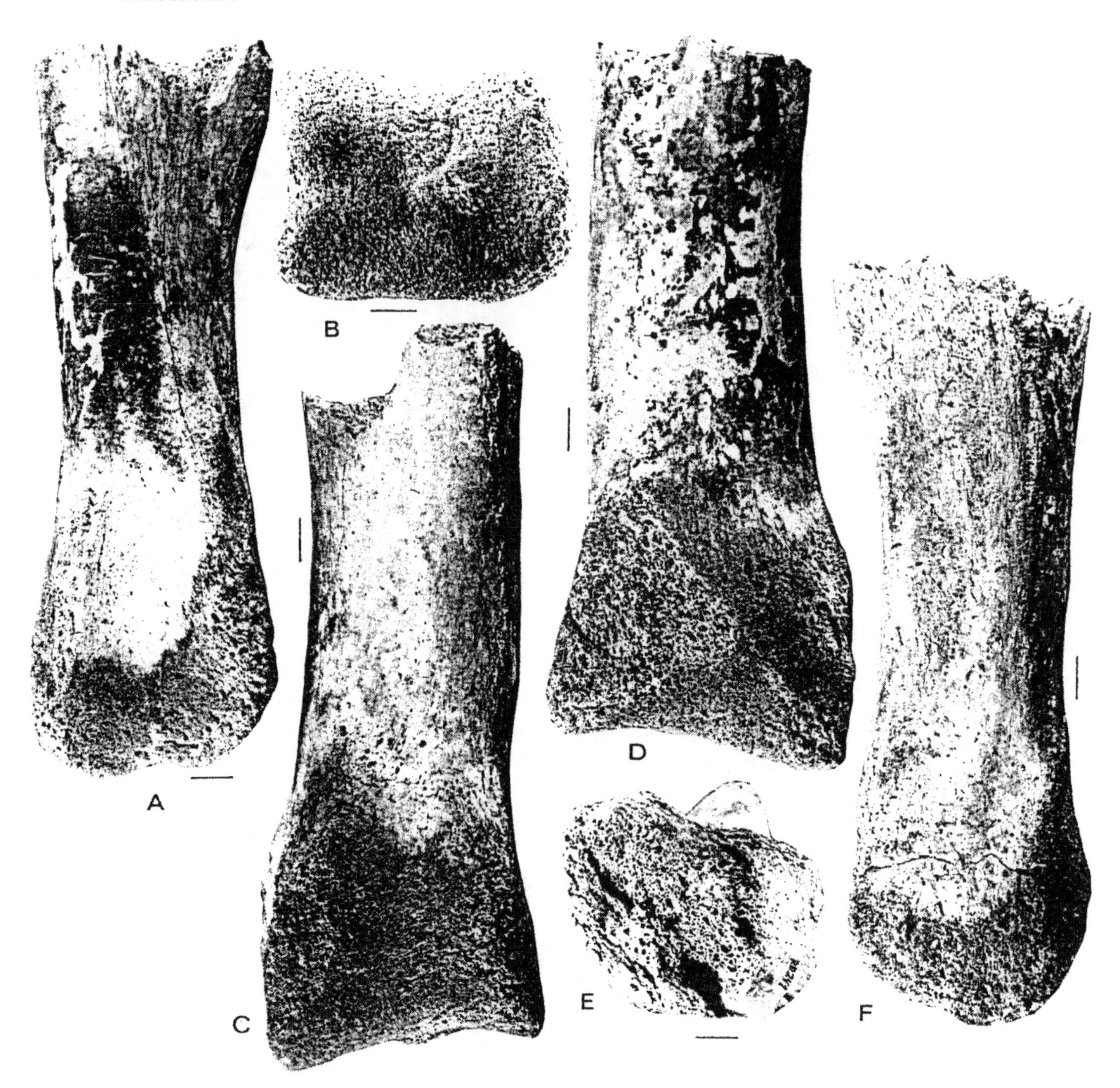

Hadrosaurid right metatarsal III (NHMM RD 241) in various views. Scale = 19 cm. (After Weishampel, Mulder, Dortangs, Jagt, Jianu, Kuypers, Peeters and Schulp 1999.)

his catalogue of specimens in the Teylers Museum, as the humerus of the giant marine lizard *Mosasaurus camperi*. W. E. Swinton in a 1955 letter (lost) to C. O. van Regteren Altena, then curator of the Teylers Museum, opined that the specimen could be the humerus of *Megalosaurus* [now *Betasuchus*] *bredai*, a species based on a single fragmentary right femur, representing the only theropod known from the type Maastrichtian strata. According to Weishampel *et al.*, these hadrosaurid remains, all collected *in situ*, seem to be "associated with regressive phases (shallowing) during which riverine input may be considered to have been more important, or with the onset of transgressive pulses or with actual highstand periods, during which neighbouring (low relief?) land areas were inundated." Except for *B. bredai*, all dinosaurian fossils collected from the type Maastrichtian belong to hadrosaurids, the new material suggesting that more than a single taxon is represented.

Apesteguía and Cambiaso (1999) briefly reported on the first hadrosaurid remains found in the Upper Cretaceous (Campanian–Maastrichtian) Paso del Sapo

Formation ("Senoniano lacustre" locality) of Chubut, Argentina. The material, representing at least two individuals, includes various vertebrae, coracoids, bones of the pes, phalanges, teeth, and other elements. Preliminary examination of these remains by the authors revealed similarities to the unnamed hadrosaurine "gryposaur" described by Bonaparte, Franchi, Powell and Sepúlveda (1984) as "*Kritosaurus*" *australis* (see *Gryposaurus* entries, *D:TE* and *S1*), a taxon from the Los Alamitos Formation of Río Negro, Patagonia (see *D:TE*, *S1*), particularly in the general morphology of the coracoids and foot elements. However, the bones apparently differ from the Patagonian taxon enough to warrant referring the Chubut hadrosaurid to a new species. Apesteguía and Cambiaso did not, at this time, specify those differences.

González, Riga and Casadío (2000) reported the first recorded specimen (MPHN-Pv 01-30) of a dinosaur (a possibly hadrosaurine hadrosaurid) from La Pampa Province, Argentina, including an incomplete posterior cervical, anterior, middle, and posterior dorsal, sacral, and anterior and middle caudal vertebrae, a left scapula and coracoid, right femur, and two right pedal phalanges. It was collected from sandy facies of litoral deposits, influenced by river flow and tides, in the lowest levels of the Upper Cretaceous (upper Campanian to lower Maastrichtian) Allen Formation, Grupo Malargüe, Islas Malvinas, department of Puelén. The authors described the caudal centra as anteriorly hexagonal, this feature regarded as a synapomorphy of the Hadrosauridae. The dorsal neural arches possess strongly upward oriented diapophyses and relatively short neural spines, suggesting that MPHN-Ov 01 may be referable to the subfamily Hadrosaurinae.

The authors noted that the specimen is both similar to and yet different from "*Kritosaurus*" *australis*. The taxa are especially similar in the form of the cervical, dorsal, and caudal vertebrae. The La Pampa specimen differs in details of the postzygapophysis in some cervical vertebrae, the presence of a concavity at the base of the transverse process of the dorsal neural arch, the position of the postzygapophysis in the anterior caudal vertebrae, and the straighter dorsal border of the distal end of the scapula. These differences suggest that MPHN-Pv 01 represents a distinct genus, although a more exact systematic assignment was difficult to make given the fragmentary state of the specimen. This record of the Hadrosauridae in South America is significant, adding new evidence on the immigration of this group of dinosaurs from North America near the end of the Cretaceous period. Its record from the Allen Formation and equivalent units of northern Patagonia also suggest the existence of a land connection between both North America and South America during the Campanian.

Dinosaur specimens of any kind found in Italy are extremely rare. Vene (1998), in the Italian magazine *Airone*, reported on a fine, complete, and articulated skeleton (housed at the Museo di storia naturale di Trieste) belonging to an as yet undescribed hadrosaur discovered in 1993 at the Villaggio del Pescatore, near Trieste (Duino), Italy. To date of this writing, this second dinosaur known from Italy (nicknamed

Partial remains (MPHN-Pv 1-30) of the unnamed hadrosaurid from La Pampa, Argentina: A. (MPHN Pv 26) coracoid, left lateral view; B. (MPHN Pv 14) distal portion of scapula, lateral view; C. (MPHN Pv 20) distal portion of left femur, posterior view; D. (MPHN Pv 27) pedal phalanx of digit III, dorsal view; and E. (MPHN Pv 34) pedal ungual phalanx, dorsal view. Scale = 5 cm. (After González Riga and Casadóo 2000.)

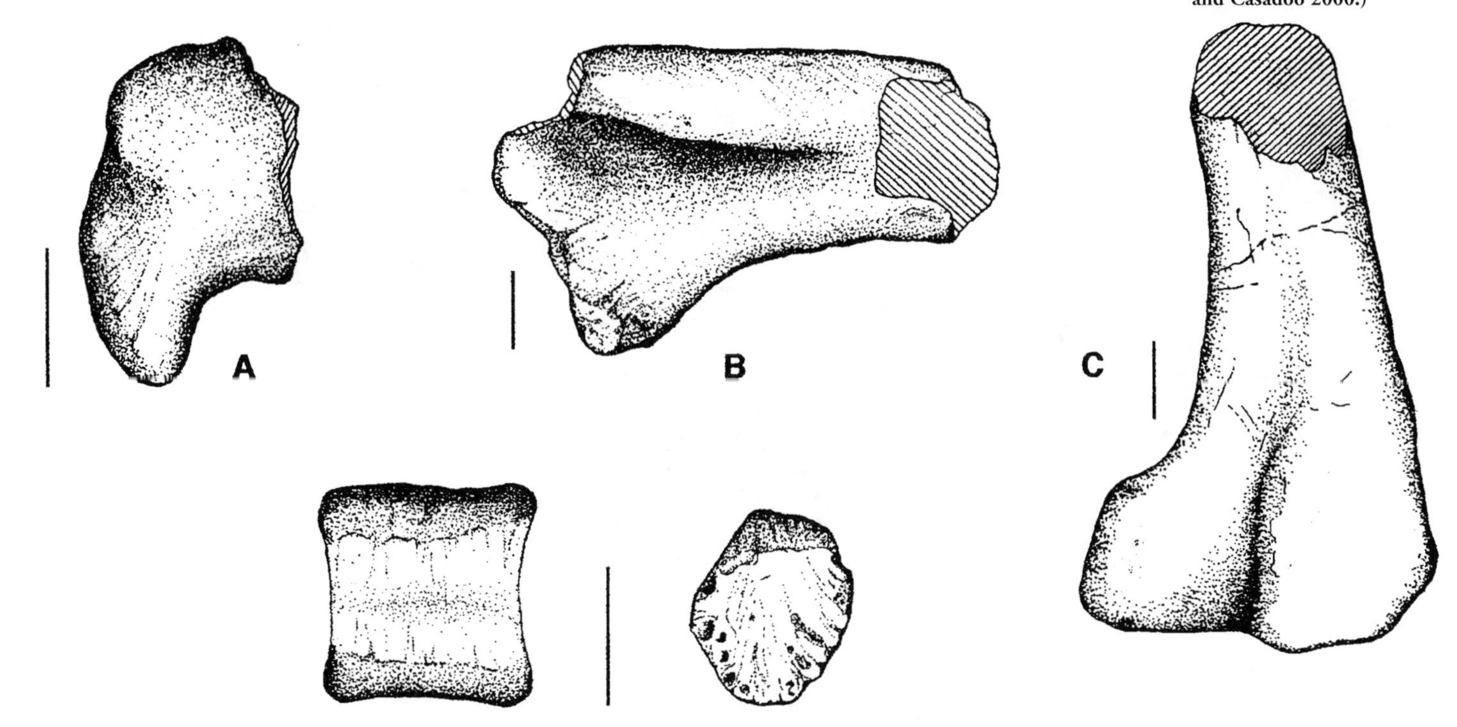

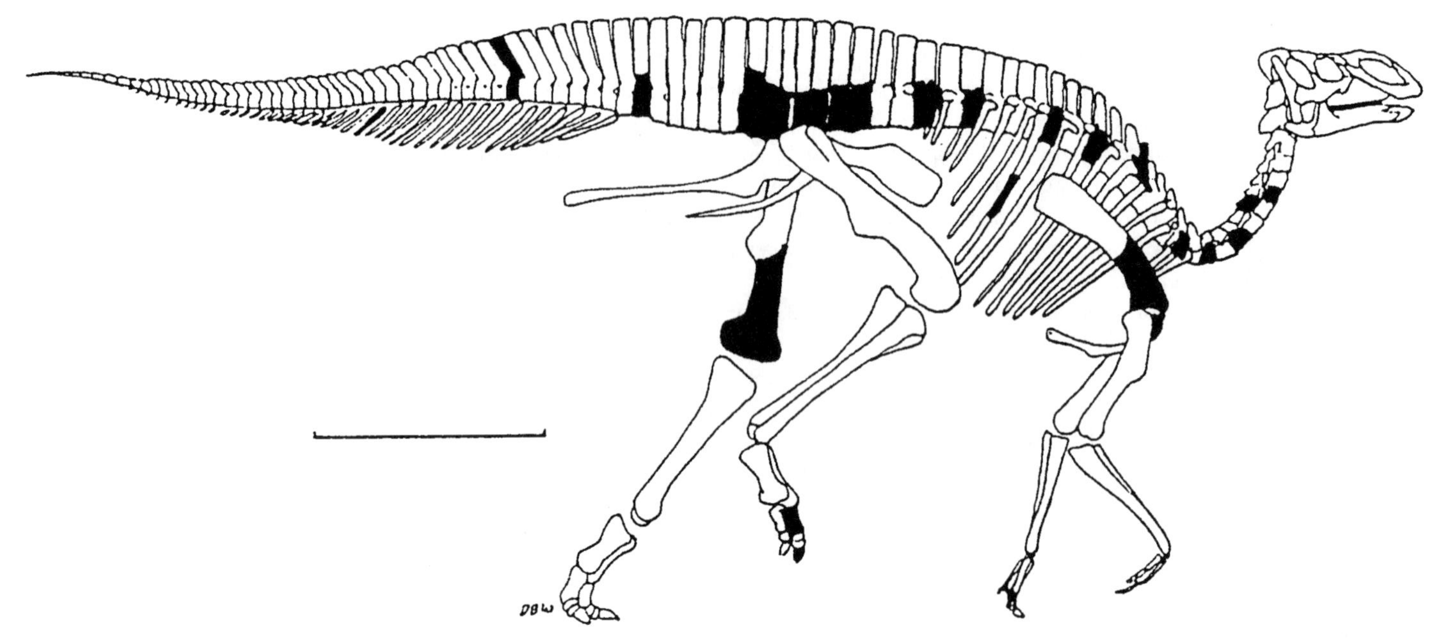

Partial remains (MPHN Pv 01-30) of the hadrosaurid from La Pampa, shaded black on the skeleton of *Kritosaurus* (after Norman 1985). Scale = 1 m. (After González Riga and Casadóo 2000.)

Complete articulated skeleton (Museo di storia naturale di Trieste collection) of the undescribed "gryposaur" nicknamed "Antonio." (After Vene 2000.)

"Antonio") is still being prepared in acid. The skull is crushed and lacks a crest. Reportedly, the material represents a "gryposaur" measuring some 4 meters (almost 14 feet) in length. A national committee will decide who will describe this specimen in detail.

Due to the derived specializations of their teeth and bodies for herbivory, hadrosaurs have often been portrayed as analogous in lifestyle to modern ungulates. This idea was rigorously tested by Carrano, Janis and Sepkoski (1999), who compared several morphological features correlated with diet, habitat preference, and sexual dimorphism (see Dodson 1975) that are displayed in both ungulates and duckbilled dinosaurs. Entering into their statistical analyses were numerous specimens representing a variety of hadrosaurine (*Anatotitan copei*, *Edmontosaurus regalis*, *E. annectens* "*E. edmontoni*" *[=E. annectens]*, *E. saskatchewanensis*, "*Gryposaurus*" *[="Kritosaurus"] incurvimanus*, *Prosaurolophus maximus*, and *Saurolophus osborni*) and lambeosaurine (*Corythosaurus casuarius*, *Hypacrosaurus altispinus*, *H. stebingeri*, and *Lambeosaurus lambei*) taxa.

The Cretaceous terrestrial habitats include analogs to today's open and closed ("open" and "closed" being relative terms depending on the size of the

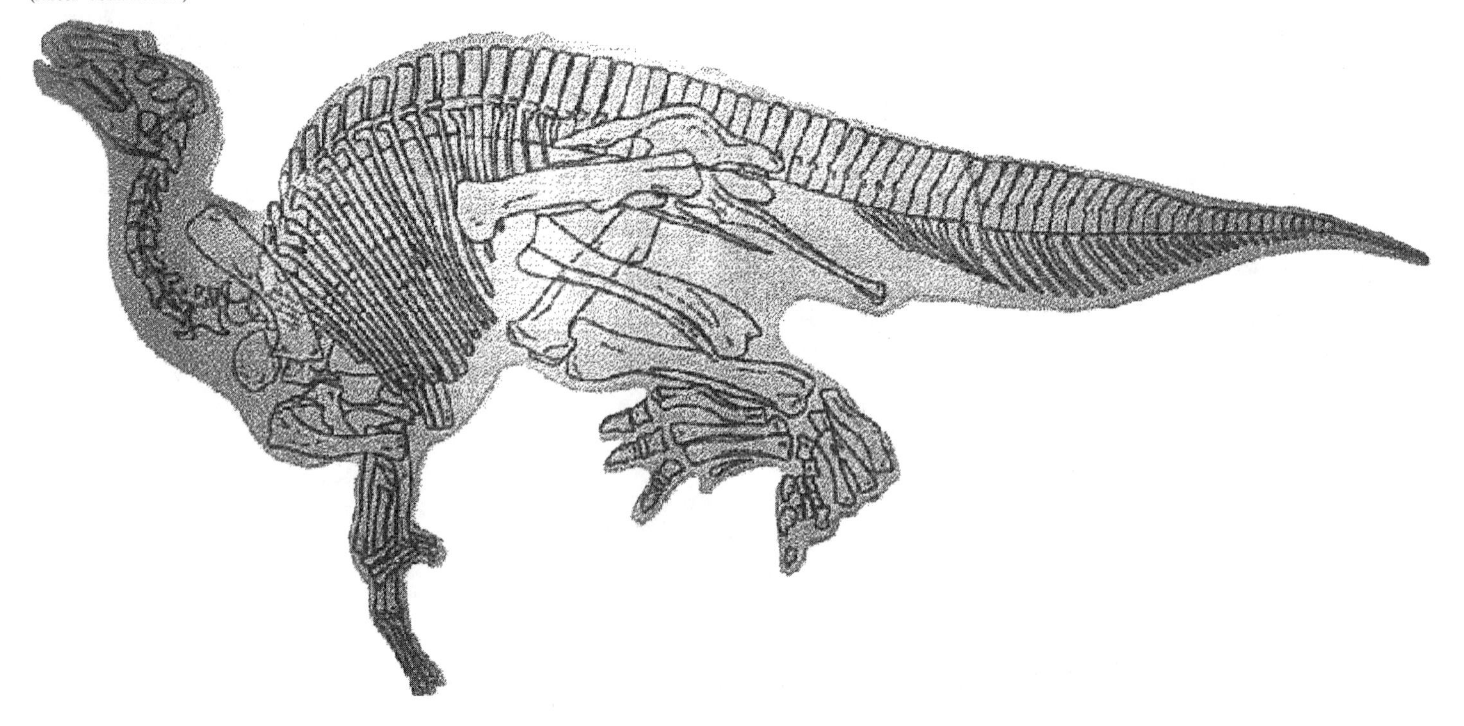

specific animals involved) habitats, although the specific floral composition differed markedly (see Carrano *et al.* for references). Cretaceous woodlands may have been more open than most modern forest environments, but hadrosaurs were generally larger than most extant forest ungulates. As the distribution of hadrosaurian remains shows that these dinosaurs apparently inhabited both open and closed environments, it may be inferred "that they were subject to the same basic physical and biological constraints as ungulates in similar habitats, which in turn might be manifested as similar morphological and behavioral adaptations."

Results of Carrano *et al.*'s experiments showed that hadrosaurines, which are monomorphic, exhibit the same wider muzzles and longer distal limbs as seen in open-habitat, less selective-feeding, monomorphic modern ungulates. However, Lambeosaurines, which are dimorphic, exhibit the narrower muzzles and shorter distal limbs characterizing the more selective feeders that inhabit closed habitats. Hadrosaurines were, therefore, interpreted to be more open-habitat animals and, as indicated by their lack of pronounced sexual dimorphism, more gregarious than lambeosaurines. Lambeosaurines were inferred to be closed-habitat animals and, as indicated by their marked dimorphism, more territorial than hadrosaurines. Both hadrosaurines and lambeosaurines were found to parallel late Cenozoic and Recent ungulates in being large, common, and diverse, also paralleling these mammals in aspects of behavioral ecology related to habitat preference. Although the lack of grasses precluded grazing *per se* during the Cretaceous, the authors noted that herbivory on high-fiber, open-habitat vegetation was a viable option. Carrano *et al.* concluded that, while terrestrial ecosystems of the late Mesozoic and late Cenozoic were very different in terms of both fauna and flora, "there may be universal constraints on the ecological roles played by large herbivores, resulting in convergence in morphology (and, by implication, behavioral ecology) between groups as taxonomically distinct as dinosaurs and mammals."

Addressing the possibility that incomplete data (due to incomplete fossil materials, inadequate population samples, or other paleontological biases) may significantly affect the outcome of such analyses, Carrano *et al.* performed the same analyses using the hadrosaur dataset as a template to degrade artificially the (previously complete) ungulate datasets. Nevertheless, the results proved consistent with the authors' earlier results.

Marginocephalians

PACHYCEPHALOSAURS

The first certain remains of a pachycephalosaur — one of a group of "dome-headed," bipedal dinosaurs belonging to the larger group Marginocephalia — from New Mexico was reported by Williamson and Sealey (1999) in an abstract. The partial skull (NMMNH P-27403) was recovered from the Upper Cretaceous (late Campanian) De-Na-Zin Member of the Kirtland Formation in the San Juan Basin. It includes an almost complete frontoparietal bone, right supraorbital II, right postorbital and postfrontal, right squamosal, much of the basicranium, and the interorbital area.

Through phylogenetic analysis, Williamson and Sealey discovered the specimen seems to be closest to the Mongolian, late Campanian species *Prenocephale*. It possesses many unique derived "*Stygimoloch* + Pachycephalosaurinae" characters (*e.g.*, high dome, closure of supratemporal fenestrae, anteroventral rotation of occiput); lacks posterior and lateral shelves consisting of marginal roofing bones, as in "Pachycephalosaurinae"; and resembles *Prenocephale* in having a single row of tubercles along the lateral and posterior edges of the squamosal, postorbital, and postfrontal. The new specimen differs from *Prenocephale*, however, in its lack of inflation of the lateral roofing bones; also, in the more ventrally directed posterolateral corners of the squamosals, the width across the squamosals, therefore, being less than that of the skull as measured above the orbits.

CT scans of the specimen revealed to the authors internal details not previously reported in dome-headed dinosaurs. These included several anteroposteriorly oriented vacuities, possibly vascular canals, located dorsolaterally to the endocranial cavity close to the base of the dome and four to five regularly-shaped laminae of denser bone traversing a comparatively low bone-density region between the endocranial cavity and the roof of the dome.

Experiments have continued regarding the putative (although once commonly accepted) theory that pachycephalosaurs butted heads in intraspecific contests. Although evidence keeps surfacing that seems to negate this popular and dramatic scenario, the notion is still presented as fact in books about dinosaurs, with illustrations of pachycephalosaurus banging their heads together (usually in even more unlikely head-on confrontations) still ubiquitous in such publications (see *D:TE*, *Pachycephalosaurus*, *Stegoceras*, and *Stygimoloch* entries; also *Stygimoloch* entry, *S1*).

Horner and Goodwin (1998) briefly reported on their thin-sectioning of three pachycephalosaur frontal-parietal domes, representing different species

Referred partial skull (MPM 8111), in left lateral view, of the pachycephalosaur *Stygimoloch spinifer*. Such so-called "bone-headed" dinosaurs were once believed to have butted heads in intraspecific contests to attract mates or establish dominance. The theory has recently come into serious doubt regarding this dinosaurian group. Scale = 5 cm. (After Goodwin, Buchholtz and Johnson 1998, courtesy Mark B. Goodwin.)

and apparently three stages of ontogeny (two juveniles and one adults), in order to test this proposed kind of behavior. In their histological study, the authors noted that vascularization decreased with increased size of the dome, which supports the ontogeny hypothesis. The domes possessed from two to three histological zones, designated I, II, and III, respectively, from ventral to dorsal. The adult dome was found to be minimally vascularized, particularly in zones II and III; contrarily, the juveniles had much vascularization, indicating the very rapid growth of periosteal bone.

The most striking histological feature observed by Horner and Goodwin in each specimen was the presence of numerous large Sharpey's fibers that were oriented perpendicularly to the surface of the dome. These fibers indicated that, in life, each dome possessed an attached tissue structure such as a keratinous cover. In the adult specimen, abundant fibers in a matrix of sparse vascularization suggested that at least the outer zone III was formed by some tendinous ossification instead of periosteal deposition, this further suggesting the possession of an osteodermic attachment.

Horner and Goodwin's study revealed no histological evidence of alteration by trauma, and, therefore, none supporting the head-butting scenario.

CERATOPSIANS

Among dinosaurs, discoveries of new taxa of ceratopsians, or "horned dinosaurs" (as opposed to theropods, ornithopods, sauropods, *etc.*), are relatively rare. Nevertheless, new ceratopsian discoveries continue to be made.

Ryan, Russell, Getty, Eberth and Brinkman (1999), in an abstract, reported the finding of a new centrosaurine ceratopsid recently in the only known bonebed in the Oldman Formation of Dinosaur Provincial Park, Alberta, Canada. This new taxon, most likely closely related to *Centrosaurus*, is distinguished by its unique parietal ornamentation, which Ryan *et al.* briefly described as follows: ornamentation consisting of at least three short, blunted, procurving hooks on both sides of the parietal midline; these hooks developing from fused epoccipital processes, the open sutures of which can be clearly observed on smaller specimens; each hook closely associated with a mass of small dermal ossifications, the latter developing to and between the fused epoccipitals; dermal ossifications only weakly fused to the parietal, variably developing as low beads or taller (more than 100 millimeters) narrow spines that tend to fuse along their length.

As pointed out by Ryan *et al.*, based upon bonebed data, *Centrosaurus* is found in rocks older than this unnamed form, restricted to lower sediments of the Dinosaur Park Formation of Alberta, while *Styracosaurus* is segregated to yet a higher stratigraphic level. It could not be determined, however, if "this succession of three centrosaurines is the result of the presence of a single evolving lineage or of shifts in biogeographical distributional patterns."

The elaborate, outermost chambers of the nasal vestibule or cavity in ceratopsids have played important roles in phylogenetic reconstructions of these horned dinosaurs. They have also been implicated in a number of hypotheses regarding their behavior. In an abstract, Sampson and Witmer (1999) presented their preliminary findings on this unusual narial anatomy, noting that, as in hadrosaurs and sauropods,

The ornithischian dinosaur *Triceratops horridus*, here depicted without cheeks (see pages 68–69) faces off against its classic saurischian rival *Tyrannosaurus rex*. This pencil drawing was made on brown paper by artist Charles R. Knight during the late 1920s prior to his mural for the Field Museum of Natural History (see *D:TE* for reproduction of finished mural).

the outermost chamber in ceratopsids is much enlarged and exhibits a suite of derived features.

From their detailed study of numerous skulls belonging to both ceratopsid clades (the Centrosaurinae and Chasmosaurinae), Sampson and Witmer gleaned significant new insights into narial morphology and evolution, including neurovascular pathways and certain structural homologues of both groups. Most surprising was "strong evidence that all ceratopsids possessed a unique, previously unrecognized form of paranasal pneumaticity." This, the authors noted, accounted for much of the variation seen in chasmosaurines. As observed by Sampson and Witmer, skulls belonging to both clades have a large, ventral pneumatic cavity located mostly within the ascending maxillary ramus. In chamosaurines, this condition is more elaborate, with a pneumatic diverticulum passing rostrally along the inner surface of the maxillary ascending ramus, invading the premaxilla and finally emerging within the nasal vestibule. In the chasmosaurine *Triceratops*, this condition is most derived, with pneumatic passageways connecting rostral and caudal portions of the vestibule. According to Sampson and Witmer, comparisons of these features

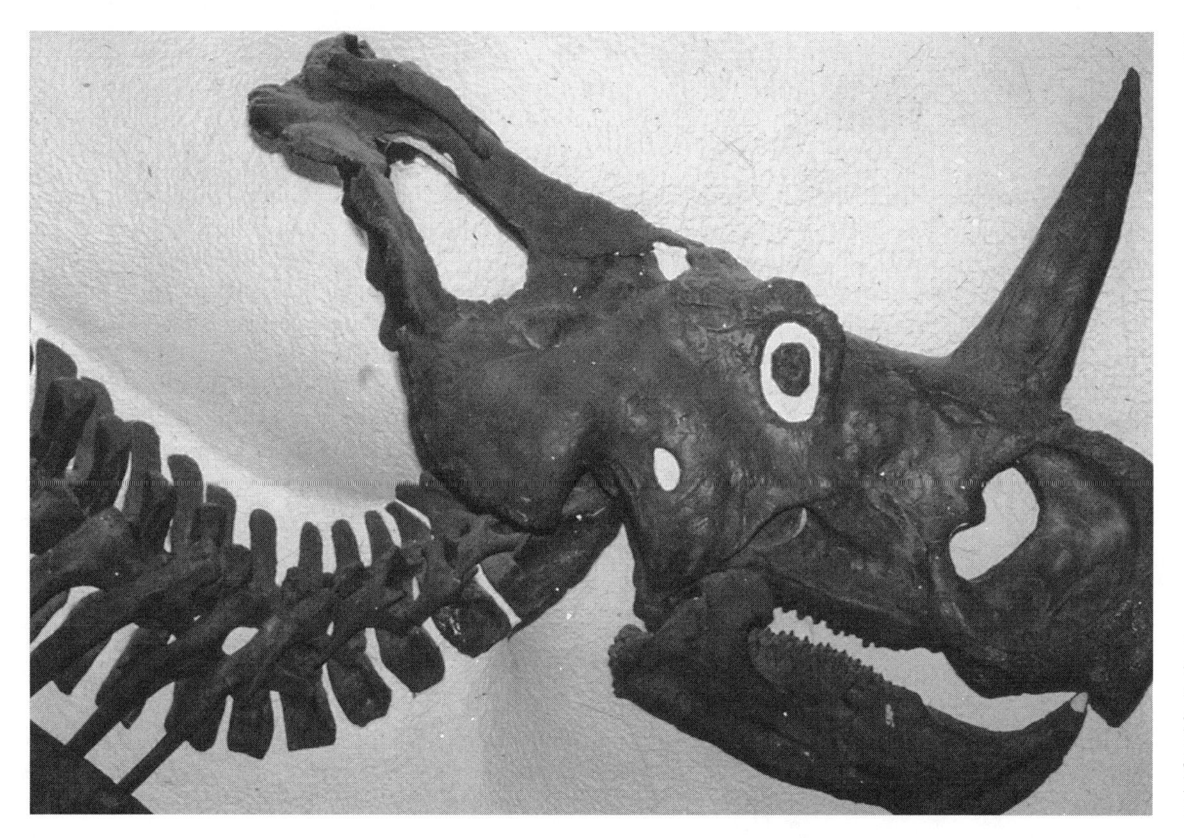

Photograph by the author, courtesy Raymond M. Alf Museum.

***Centrosaurus nasicornis*, cast of AMNH 5351. The narial anatomy of dinosaurs such as this has recently been studied in detail by Scott D. Sampson and Lawrence M. Witmer (1999).**

A. Multiview reconstruction of the limbs of *Triceratops* or *Torosaurus* based primarily on *Triceratops horridus* skeleton USNM 4842, shown walking out a probably ceratopsid trackway (from Lockley and Hunt 1995). anteriormost manus print about to be impressed (modified after Paul 1991); B., C., trackways of walking (B) *Diceros* (black rhinoceros) and (C) *Connochaetes* (wildebeest) (modified after Bird 1987). Scale = 1 m. (After Paul and Christiansen 2000.)

with birds and crocodilians offer "the basis for inferring soft tissues as well as assessing higher order functions relating to the nasal cavity of ceratopsids."

Paul and Christiansen (2000) probed in detail the ongoing issue of the posture of the forelimbs in neoceratopsian dinosaurs. As these authors pointed out, this long standing debate has been unnecessarily polarized, forcing a choice between two rather extreme hypotheses: 1. The forelimbs of Ceratopsids (as traditionally restored) were oriented in a more lizard-like sprawling posture, suggested as early as 1905 by Gilmore, this implying that such dinosaurs were incapable of running at high speed, or 2. the forelimbs were restored in an elephantine, near-columnar posture. Neither of these interpretations, the authors contended, is correct.

According to Paul and Christiansen's study, some of the misconceptions regarding the forelimb posture of ceratopsid dinosaurs stem from various errors in the reconstruction and mountain of skeletons, particularly in the rib and vertebral articulations, rather than in the forelimb itself. When correctly reconstructed, the authors noted, a mounted ceratopsid skeleton should match the evidence of probable ceratopsid trackways (*e.g.*, Lockley 1986, 1991; others cited by Paul and Christiansen), revealing "that the hands [of the ceratopsid] were placed directly beneath the glenoids, and that manual impressions were directed laterally, not medially as in sprawling reptiles." Furthermore, the impressions of hindfeet "are medial to the manual impressions, owing to the slightly averted elbow and to the asymmetrical distal femoral condyles, which directed the crus slightly medially."

In their reconstructions of ceratopsid skeletons, Paul and Christainsen found the best extant analogies for both forelimb and hindlimb posture to be large mammals, particularly the rhinoceros, with the forelimbs slightly flexed and elbows directed somewhat posteriorly. As the authors observed, these mammals and the horned dinosaurs "share a number of

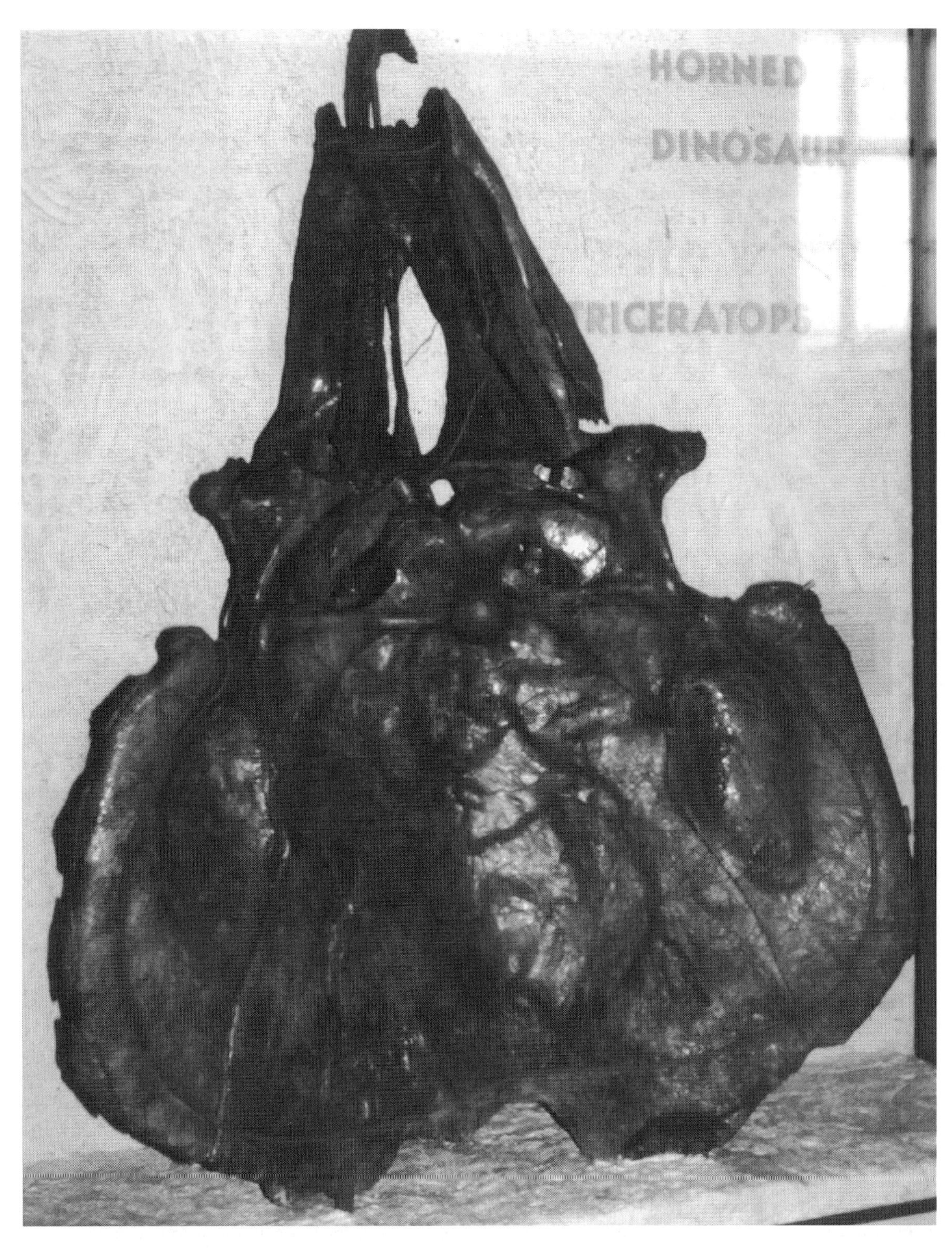

Skull (AMNH 970) of *Triceratops horridus*, originally referred to *T. serratus*. John Anton has proposed that the ceratopsian crest functioned to amplify and absorb sound.

Photograph by the author, courtesy Natural History Museum of Los Angeles County.

convergent anatomical resemblances probably forced upon them by the need to support a large body mass while retaining the ability to perform true running with a suspended phase."

Regarding the top running speeds which larger ceratopsids could attain, Paul and Christainsen found it likely, although not certain, that these dinosaurs were able to reach a full gallop (Christainsen and Paul, in press), exceeding the speed of elephants and probably approximating that of rhinoceroses.

In the past, various explanations have been proposed for the possible function of the frill in ceratopsian dinosaurs, including defense, thermal venting, sexual ostentation, and identification. Anton (2000), in an abstract, offered yet another theory, this one involving sound, explaining the primary function of the crest may have been an acoustic device. According to Anton's idea, the crest's reflection and absorption of acoustic energy "would theoretically increase the amount of sound available for translation to the auditory system and naturally result in enhanced hearing capability."

Anton tested this theory first using models, observing the reflection of water waves off scaled-down ceratopsian crests (this serving as an analog revealing how sound waves interacted with crests) and then experimentally, measuring acoustic energy reflected off a cast of a crest referred to *Triceratops* sp. Anton's experiments indicated that sound interacted with these crests in a way predicted by physical laws. The crest of *Triceratops* sp. apparently amplified lower frequencies by about three times relative to a control and absorbed higher frequencies so that little (under six percent) or no sound amplification was detected.

Enhanced hearing sensitivity, Anton noted, would be an advantage to these dinosaurs in communicating with one another over greater distance and in the detection of stealthy predators. That, plus a relatively large interaureal distance (in some cases about one meter), would be effective in determining from which direction a sound originated.

Still Unresolved: Ectothermy or Endothermy?

Arguments, both pro and con, continue to appear in the paleontological literature (also in popular books and via the news media), as scientists attempt to solve the complicated problem of whether or not some (or all) dinosaurs were cold-blooded (ectothermic), warm-blooded (endothermic), or something in between.

In recent years, some of this discussion has centered around the speculative topic of respiration in dinosaurs. As pointed out by Hengst (1998), a persistent supply of oxygen is required for sustained activity, with a greater capacity to breath supporting higher performance limits. A small amount of the blood made available through breathing is utilized by physically related systems (*e.g.*, circulatory and cardiovascular). Continuing work on the possible breathing methods in theropods, Hengst, in a brief report discussing the limitations offered by muscle on the breathing process in these carnivorous dinosaurs, considered two proposed models of respiration — costal breathing (involving the ribs, as in birds and mammals) and hepatic-piston breathing (utilizing a diaphragm to ventilate a bellows-like septate lung, as in crocodiles).

Hengst explained that the total amount of air delivered to lung surfaces is affected by contractile velocity, length, and frequency, with excesses of any of these possibly resulting in muscle fatigue, especially when thoracic compliance is low or thoracic viscosity high (as with the hepatic-piston model). Data including that resulting from flouroscopic studies of respiratory muscle action in alligators (tested at different temperatures over a range of sizes), rats, and opposums were applied both to costal and hepatic-piston models for theropod breathing in order to bracket the highest and lowest physiological limits. According to Hengst, "Costal breathing favors a mammalian type respiratory system using high frequency-small volumes, while hepatic piston breathing favors low frequency, with high volumes." Analysis of the convergence of various clades on a common thoracic morphology compared well with the hepatic-piston models; high abdominal compliance in theropods would have resulted in respiratory support for higher levels of activity.

Geist and Hillenius (1998) addressed recent arguments focusing upon the supposed lack or presence of nasal respiratory turbinates (or RT) in dinosaurian taxa and how they relate to metabolism (see *S1*, "Introduction"). As noted by Geist and Hillenius, the presence of RT in mammals and birds has casually been linked to high routine rates of ventilation and resting levels of oxygen consumption, and, therefore, been used as evidence for endothermy. Conversely, the absence of RT in extinct reptiles is a probable sign of low ventilatory rates related to ectothermic metabolic physiology.

Computerized tomographic or CT scans by Geist and Hillenius of several Late Cretaceous theropods revealed relatively narrow nasal cavities that, in their opinion, were inadequate to have housed respiratory turbinates; furthermore, new information concerning the skull morphology of several sauropods, theropods, and ornithischians confirmed to the authors the general lack of RT in these taxa.

In regard to the manuraptoran theropod *Scipionyx* (see entry, *S1* and this volume), Geist and Hillenius found that its amazingly preserved type specimen "provides compelling new evidence for the existence of a hepatic-piston diaphragm in theropod dinosaurs." The morphology and orientation of the preserved liver, trachea, and colon were closely comparable to the anatomy of those features in extant crocodilians, while contrasting sharply with those in birds. According to these authors, reliance on a hepatic-piston mechanism in *Scipionyx* most likely "constrained theropods from developing an avian-style lung/air sac respiratory system."

In an overview of recent perspectives on the respiratory physiology of dinosaurs, Ruben, Jones and

Photograph by Jerry Van Wyngarden, courtesy Museum of Western Colorado.

Crushed fossil egg, possibly laid by the theropod ***Allosaurus***, from the Young Locality (Upper Jurassic), Dominguez Canyon, Delta County, Colorado. Eggs are now being used as evidence in hypotheses regarding dinosaurian metabolism.

Geist (1998) acknowledged that deciphering the physiology of these animals remains by necessity speculative, with determination of their metabolic status especially problematical. However, many workers have advocated the now popular notion that dinosaurs ("... among the most distinctive and successful of all land vertebrates"), like modern birds and mammals, were probably endothermic, thereby "providing a model supposedly essential to the interpretation of these animals as having led particularly active, interesting lives." In contrast, the more traditional idea that dinosaurs were ectothermic animals is generally, "albeit mistakenly, associated with unintelligent brutes leading sluggish, sedentary lives."

Again emphasizing that respiratory turbinates are present in virtually all birds and mammals, serving to alleviate various problems that would otherwise be associated with endothermy, Ruben *et al.* concluded the following:

1. The apparent lack of respiratory turbinates in dinosaurs "provides the first causally linked fossil evidence that these animals were unlikely to have maintained metabolic rates equivalent to those of modern endotherms."

2. The dinosaurs' close relatives, the earliest birds (*e.g., Archaeopteryx* and enantiornithines), apparently had not attained endothermy (see Ruben 1996). Lines of arrested growth (LAGs) in long bones of enantiornithine birds suggest variable body temperature, a pattern of thermoregulation usually inconsistent with endothermy (Chinsamy, Chiappe and Dodson 1994). Also, analysis of the anatomy and physiology of modern birds suggests that flight may have evolved before the evolution of avian endothermy (Randolph 1994).

3. The dinosaurs' metabolic status probably reveals less about their lifestyle than is usually presumed. Though seemingly not possessing the basic anatomical attributes generally associated with endothermy in modern animals, dinosaurs did not have to be similar in lifestyle to most modern, temperate-latitude reptiles. Considering the mild Mesozoic climates, most dinosaurs were probably

large enough to have been homeothermic without being endothermic (Spotila, O'Connor and Paladino 1991). Therefore, even if dinosaurs were fully ectothermic, had these animals "possessed aerobic metabolic capabilities and predatory habits equivalent to those of some large, modern, tropical-latitude lizards (*e.g.*, *Varanus*), they may well have maintained large home ranges, actively pursued and killed large prey, and defended themselves fiercely when cornered" (Auffenberg 1978).

Witmer and Sampson (1999) also focused on earlier discussions of respiratory turbinates, or nasal conchae, and the cross-sectional area of the nasal cavity in dinosaurs, confronting previous workers' conclusions that the nasal tube in theropods was too narrow to house the conchae required to support endothermic rates of ventillation. Witmer and Sampson identified enlarged nasal cavities in a variety of dinosaurian clades including ankylosaurids, ceratopsids, and other ornithischians; sauropods, ceratopsids, and hadrosaurines were found to possess apomorphically enormous nasal vestibules. In at least ankylosaurids and centrosaurines, the authors identified structures explained as probably supports for nasal conchae. Also, in sauropods and hadrosaurines, bony narial structures were identified that "had the effect of increasing mucosal surface area within the vestibule."

Skull of the hadrosaurid species *Edmontosaurus annectens* collected from the Upper Cretaceous of Wyoming. Lawrence J. Witmer and Scott D. Sampson (1999) have identified much enlarged nasal vestibules in duckbilled dinosaurs, evidence suggesting that hadrosaurids may have been warm-blooded.

According to these authors, the above data, combined with new information "that these narial regions were served by an apomorphically rich blood supply, perhaps with masses of cavernous tissue," are consistent with inferring mucosal elaboration in these dinosaurian groups.

Witmer and Sampson cautiously stated that their data did not contradict Ruben *et al.*'s conclusions regarding theropods, nor did their findings necessarily indicate that the studied dinosaurs were endothermic. However, the derived narial apparatus—*i.e.,* "enlarged vestibule, elaborated mucosal area, increased vascularization, and, in some, cochea"—observed in the above dinosaurian groups "bears the hallmarks of a well developed counter-current exchanger, a device with a variety of physiological implications, including "an effect on water economy (which could be coupled to endothermy) ... also heat exchange associated with brain cooling, extrapulmonary gas exchange, and mechanical effects on air flow." Such devices in large animals, the authors noted, "is consistent with the scaling of thermal phenomena."

Claessens, Perry and Currie (1998) have continued their investigations into the possible respiratory capabilities of nonavian theropods (see "Introduction," *S1*), reaching rather different conclusions than

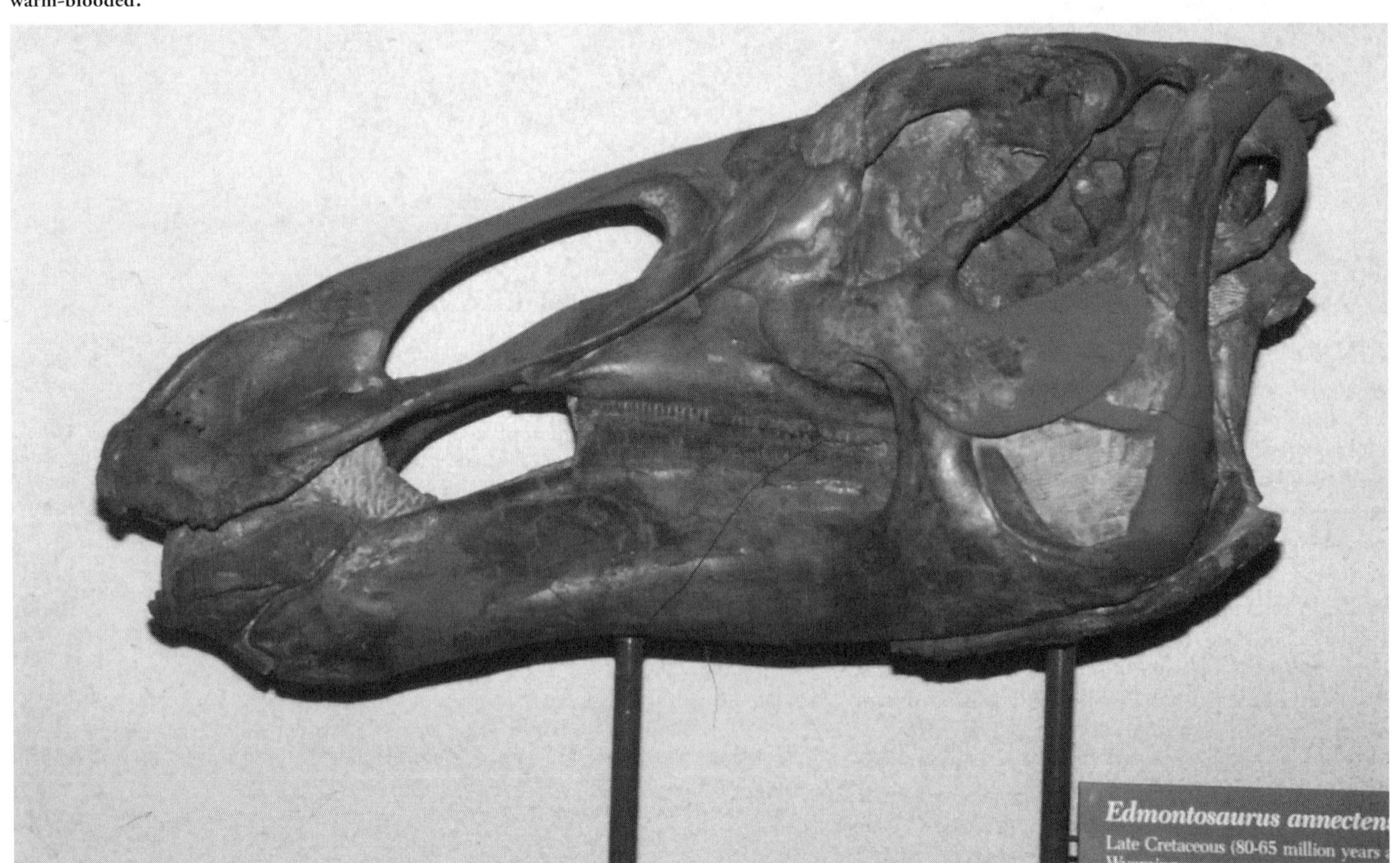

Photograph by the author, courtesy Royal Tyrell Museum/Alberta Community Development.

the above authors. In a brief report, in which comparative anatomy was utilized in reconstructing theropod respiration, Claessens *et al.* pointed out the following: theropod ribs have large areas for the attachment of intercostal muscles, which possess widely separated heads that define particular planes of movement; at least one theropod genus, the dromaeosaur *Velociraptor*, is known to have well-developed uncinate processes (also known in birds; see Clark, Norell and Chiappe 1998, section on birds below, for an oviraptorid specimen preserved with these processes); and kinetic gastralia, which are jointed in a characteristic interlocking pattern, could have further aided in aspiration breathing.

According to Claessens *et al.*, it is possible to determine at least the probability of a birdlike or reptile-like respiratory system in nonavian theropods based on skeletal analysis, because of unique functional correlates of the avian and reptilian lung systems. Saurischian dinosaurs, like birds, have pneumatic vertebrae, suggesting that their lungs were attached dorsally. Comparison with extant reptiles shows that attached lungs are partitioned heterogeneously, while some advanced theropods are known to have developed high-compliance pulmonary diverticula that approached those of cursorial paleognathous birds. Furthermore, lungs in saurischian dinosaurs were probably multichambered, given the large size of some members of this group.

Claessens *et al.* concluded that "the combination of costal and gastralial movement would have sufficed to support a constant, high aerobic level" in theropods. The authors also speculated, "Although it is possible that unidirectional air flow in the dorsal part of the lung became established late in the theropod-avian transition, the anatomical prerequisites for avian degree cross-current gas exchange are already present in the crocodilian lung, and thus predate the origin of dinosaurs."

Claessens *et al.* disavowed, upon skeletal evidence, the claims by other workers that theropods possessed a hepatic-piston pump as seen in crocodilians. (See also Carrier and Farmer 2000, who, in their analyses of dinosaurian ventilation, found no evidence to support that theropods utilized a hepatic-piston pump in ventilation.)

Taking a different approach to addressing the above issues, Norton (1998), in a brief report, stated that "the constraints placed on putative dinosaur lung structure by the taphonomy of nest sites and the physical characteristics of eggshells cannot be ignored."

As explained by Norton, eggs buried in sand or mounds of vegetation (like those of crocodilians) usually "have high conductances to maximize gas exchange with little risk of desiccation," while those laid

Photograph by the author, courtesy Natural History Museum of Los Angeles County

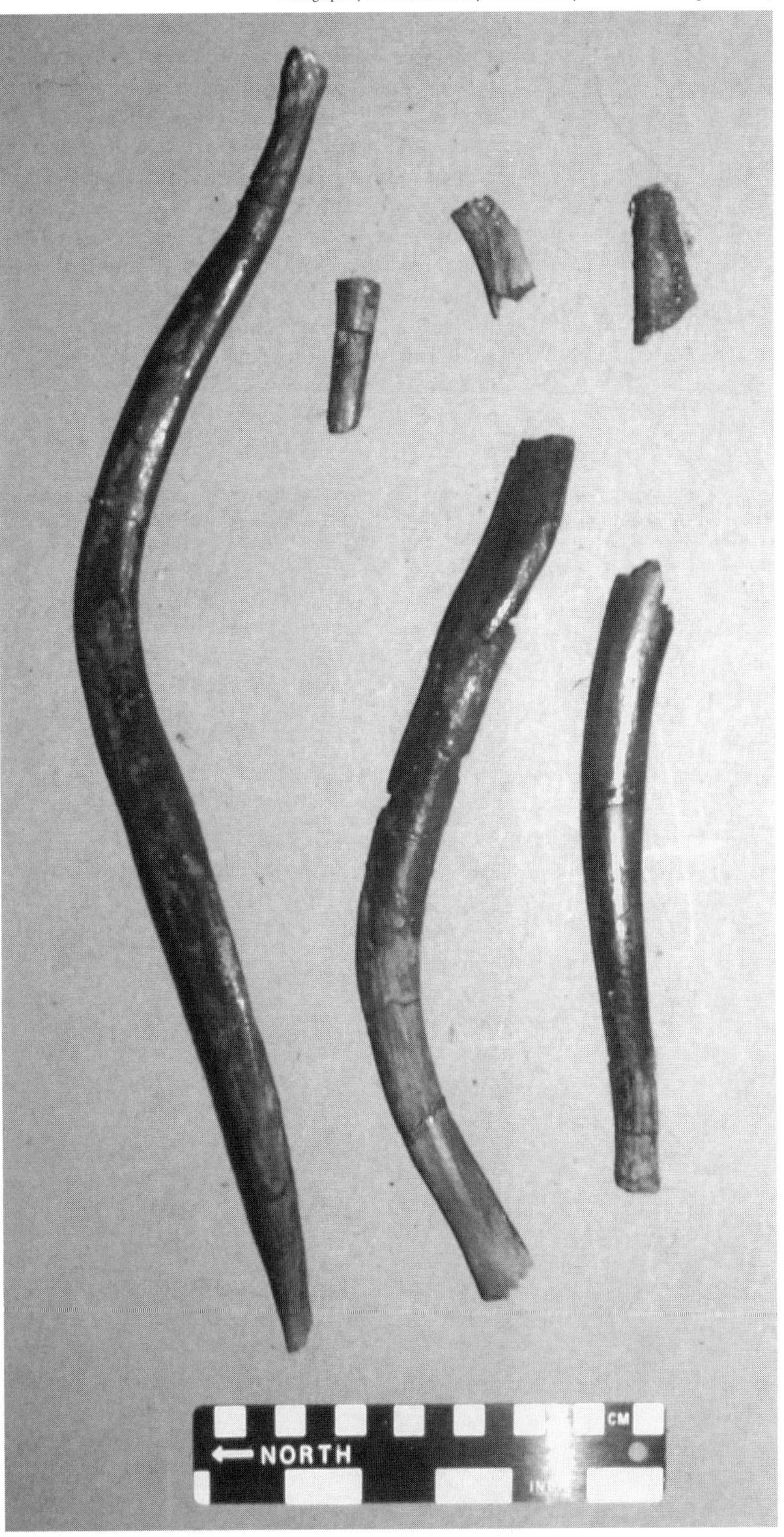

Gastralia or "belly ribs" belonging to a specimen (LACM 23844) of *Tyrannosaurus rex*. Leon P. A. M. Claessens, Steven F. Perry and Philip J. Currie have performed a study wherein kinetic gastralia, present in at least some advanced theropods, may have had a function enhancing respiration.

in open nests exposed to the air "have conductances low enough to prevent excessive desiccation but high enough to preserve oxygenation." Partial desiccation of the egg and an air pocket are necessary for the development of an avian-style parabronchial lung having air capillaries. The avian embryo ventilates its lungs and absorbs fluid from its air capillaries via internal systems. The eggs of crocodilians, in covered nests, generally remain well hydrated and do not develop air pockets. These conditions are consistent with the model for the crocodilian lung.

Norton suggested that a highly effective avian gas exchange apparatus could not have developed in dinosaurs if their eggs were laid in sand or vegetation; consequently, "Evidence of eggs laid in open nests would support the presence of avian-style lungs in dinosaurs, and the parental brooding behavior suggested by recent fossil finds [see *Oviraptor* entry; also, entries in *D:TE* and *S1*] would reflect attempts to control nest humidity as well as temperature" (see also Norton 1999, *Deinonychus* entry).

Horner, Padian and Ricqlès (1999), in an abstract, published on the osteohistology of some embryonic and perinatal archosaurs (including birds, nonavian dinosaurs, and other reptiles) and the phylogenetic and behavioral consequences these results have for dinosaurs (this study having implications relating dinosaurs to birds). According to these authors, a strong evolutionary signal can be seen in the distribution of tissues and the patterns in the shafts and ends of these bones.

The embryonic bones of basal reptiles and basal

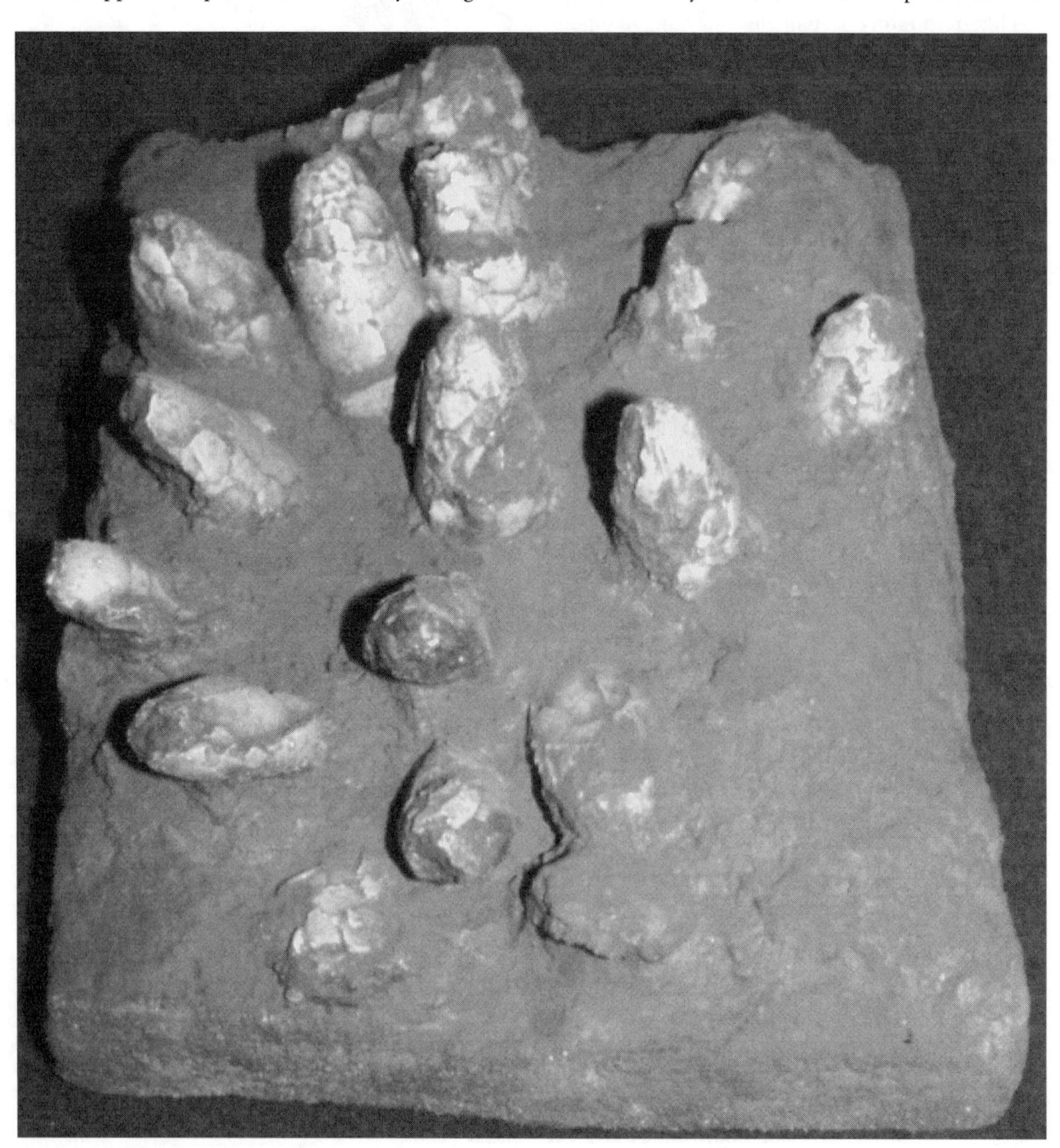

Fossil eggs attributed to the theropod *Oviraptor*, formerly assigned to the ceratopsian dinosaur *Protoceratops*.

Photograph by the author, courtesy Natural History Museum of Los Angeles County.

archosaurs, Horner *et al.* noted, possess thin-walled cortices and large, subdivided marrow cavities. In basal reptiles, these cortices are poorly vascularized, osteocytes are common though not organized, and there is no evidence of fibro-lamellar tissue; in basal archosaurs, the bones display increased vascularization, organized osteocytes, and some fibro-lamellar tissue; and in dinosaurs, vascularization and cortical thickness further increase, osteocyte lacunae are better organized and more abundant, and fibro-lamella tissue is abundant.

According to Horner *et al.*, metaphyseal morphology varies as new features are acquired in derived groups. For example, the cartilage cone is persistent among other reptiles; in ornithischian dinosaurs, however, this is entirely calcified before erosion by marrow processes. Cartilage canals, found in dinosaurs, are absent in basal archosaurs. Also in dinosaurs, a thickened calcified hypertrophy is a sign of accelerated bone growth.

The variation seen in nonavian dinosaurs is also found in birds, wherein it is associated with life history strategies (including the adults caring for their young). Horner *et al.* concluded that this variation, in conjunction with independent taphonic evidence from dinosaur nesting sites, strengthens "the hypothesis that variations in bone growth strategies in dinosaurs reflect different life histories, including parental care."

In another abstract, Padian, Horner and Ricqlès (1999) briefly reported on their study relating dinosaurian growth rates to the evolution of life-style strategies. Like other tetrapods, dinosaurs grew more quickly during their early ontogeny, their growth slowing and eventually ceasing in adulthood. According to Padian *et al.*, the skeletal tissues of small basal dinosaurs (*e.g.*, the primitive thyreophoran *Scutellosaurus*) and pterosaurs reveal evidence indicative of slow growth rates, comparable more to young, quickly growing crocodiles than to saurischians and the more derived ornithischians (*e.g.*, hadrosaurs). Therefore, the authors postulated that, based on ontogenetic and phylogenetic patterns in dinosaur bone tissues, large size — which developed independently in several lineages — can often be correlated with strategies of rapid growth that are reflected in the characteristic highly vascularized, fibro-lamellar bone tissues that comprise most of the bone cortex.

Padian *et al.* suggested that the ability of some dinosaurs to grow to large adult size could have been an adaptive advantage, with the smaller lineages retaining high metabolic rates and growth rates for reasons pertaining to other strategies of life history. In basal dinosaurs and pterosaurs, low vascularity and the low growth rates inferred from that "may or may not be good indicators of thermometabolic regime by themselves," as some endotherms have low growth rates and present few osteohistological features indicating endothermy. According to Padian *et al.*, the relating of life history strategies in dinosaurs to growth rates and body sizes will be better comprehended "as more histological studies place these data into phylogenetic and ontogenetic contexts."

Barrick, Russell and Showers (1998) performed a study involving the possible metabolisms of dinosaurs utilizing oxygen isotopes to decipher thermoregulatory patterns in several dinosaurian groups. Barrick *et al.* pointed out that isotopic variability within bones reflects temperature variations that took place during bone growth and such variability between bones reflects mean temperature differences between different regions of the skeleton. In endothermic animals, mean body water isotopic values varies with body size mostly due to shifts in mass specific metabolic rates — recorded in adults through bone remodeling — with increasing body size. In ectotherms, because of low metabolic rates, there are no shifts.

As stated by Barrick *et al.*, shifts in mean oxygen isotope values from remains of various dinosaur taxa spanning a large range of body sizes can provide evidence of metabolic strategies. Material examined consisted of fossil remains collected from mass-death assemblages, eliminating problems of time and environment mixing. From this study, Barrick *et al.* concluded that dinosaurs seem "to have several thermoregulatory strategies and intermediate metabolic rates." Among the taxa studied, hadrosaurs apparently had the highest metabolic rates. Hadrosaurs lived during the Cretaceous, in environments wherein "many dinosaurs were able to maintain homeothermy and the benefits of endothermy with greatly reduced food requirements." A hadrosaur weighting 4 tons, the authors calculated, "had the same food requirements as an 800 kg bison whereas a similarly sized true ectotherm would have had the food requirement of a 60 kg goat."

Briefly, Hicks and Farmer (1998) addressed the issue that the lung morphology of ectothermic amniotes has been offered as evidence "to 1. limit rates of oxygen consumption (VO2) due to diffusion barriers, constraining the evolution of endothermy and 2. disprove ab avian-theropod relationship" (see below). According to these authors, both of the above suggestions are incorrect, as indicated by morphological, physiological, and phylogenetic evidence."

Hicks and Farmer disagreed with the above suggestions, pointing out the following:

1. Oxygen-consumption rates in some ectotherms, such as varanid lizards, can actually be higher than those

Courtesy Sinclair Oil Corporation.

Full-scale model of *Corythosaurus casuarius* made by Louis Paul Jones Studies as it originally appeared in the "Sinclair's Dinoland" exhibit at the New York's World's Fair (1964), held at Flushing Meadows, Long Island. An adult hadrosaur of about this size may have eaten approximately as much food per day as an adult bison.

of some equivalently-sized endotherms while running at the same speeds. Lung morphology does not constrain endothermy. Endotherms take in comparatively more oxygen because of greater ventilation rates and the greater respiratory surface area of the lung. Furthermore, in both ectotherms and endotherms the diffusion barrier is approximately the same.

2. Phylogenetic and embryological evidence shows that the multicameral septate lung of numerous amniotes (including crocodilians) is the primitive condition; also, according to morphological data, "this lung could be modified through elaboration of the dorsal surface area to produce the avian parabronchial lung."

Additional direct evidence (if correctly interpreted) supporting the hypothesis that dinosaurs were warm-blooded was offered by Fisher, Russell, Stoskopf, Barrick, Hammer and Kuzmitz (2000) in their description of what appeared to be the first heart ever found preserved in a dinosaur skeleton. CT scans of this putative heart revealed a four-chambered structure, similar to that found only in birds and mammals. Such a heart suggests a relatively high metabolism and is associated with endothermic rather than ectothermic animals (see *Thescelosaurus* entry for details).

Jensen, Sharp and Lucas (2000; see their paper for more in-depth details) approached the question of whether theropod dinosaurs were ectothermic or endothermic utilizing evidence gleaned from a new technique in oxygen isotope geochemistry. Jensen *et al.*'s study focused upon data derived from the tooth enamel of dinosaurs (metabolism uncertain) and crocodiles (known ectotherms) living at approximately the same time, the appropriate values of an animal's tooth enamel constituting a function of its body temperature.

The authors analyzed populations of theropod and crocodilian teeth recovered from both the Kirtland Formation, San Juan Basin, New Mexico, and the Judith River Formation, Alberta Basin, Alberta. In order to separate the possible ambiguities associated with giganothermy, teeth from these localities pertaining to theropods varying greatly in size (*Saurornitholestes* and

Paleontologist Dale A. Russell with a specimen (NCSM 15728) of *Thescelosaurus neglectus* preserving what appears to be the heart, constituting new evidence for endothermy in non-saurischian dinosaurs (see *Thescelosaurus* entry)

Photograph by Jim Page, courtesy North Carolina Museum of Natural Sciences, North Carolina State University.

Albertosaurus) were analyzed. Results of Jensen *et al.*'s analyses supported two important conclusions concerning the metabolism and behavior of theropods: 1. The oxygen isotope "values of small and large theropods are very similar, indicating comparable metabolisms for both groups and effectively eliminating gigantothermy as a possible thermoregulatory strategy" and 2. limited latitudinal variability in these values for these dinosaurs (but not crocodiles) "indicates that tooth formation for theropods occurred over a wide range of latitudes, implying that theropods were migratory and likely endothermic."

In an abstract, Van Valkenburgh and Molnar (2000) took yet another approach in a study that had implications regarding the possible metabolism of theropods only. Van Valkenburgh and Molnar hypothesized that, although bipedal dinosaurian predators were very different morphologically from the largest quadrupedal mammalian carnivores in their respective ecosystems, some of the same processes that shape mammalian predators and their communities were probably important to theropod predators also. Exploring this, the authors compared the predatory adaptations of theropod dinosaurs and mammalian carnivores, focusing mainly upon aspects of their feeding morphology (*i.e.*, skulls, jaws, and teeth), but also examining suites of sympatric species or "ecological guilds" of predatory theropods and mammals, with emphasis on species richness and body size distribution within those guilds.

As all sampled theropods seem to have been hypercarnivorous, Van Valkenburgh and Molnar's morphological comparisons showed reduced trophic diversity among theropods relative to mammalian carnivores. No clear analogs among theropods were found by the authors of felids, canids, and hyaenids. Furthermore, theropods were found to parallel canids more so than felids in skull proportions, while all theropods examined seem to have possessed weaker jaws than carnivorans.

The authors found it surprising, given the apparent trophic similarity of theropods and their large body sizes, that that species's richness of theropod guilds exceeded that observed in mammalian carnivore guilds. Finding the apparent ecological overpacking of the theropod guilds difficult to understand with mammalian energetic requirements assumed for theropods, the results of Van Valkenburgh and Molnar's study "suggest that theropods had reduced metabolic rates relative to mammals and were not endothermic."

The Dinosaur-Bird Debate: Nearing a Resolution?

At The Florida Symposium on Dinosaur Bird Evolution, held in Fort Lauderdale in April, 2000, paleontologist Kevin Padian stated the following in an abstract: "The 'debate' about whether birds evolved from dinosaurs is no longer scientific and it has not been for some years. Opponents to this hypothesis have no alternative hypotheses of their own, have not tested any hypotheses using standard phylogenetic methods, have not proposed alternate methods, and ignore 90% of the evidence in favor of dinosaurian origin." Padian further posited that this debate is kept alive only in the popular press, as reporters do not understand scientific methods and standards of evidence.

Indeed, the once very heated debate over the possible relationships of dinosaurs to birds seems to have noticeably tempered, as more birdlike (see *Bambiraptor* entry) and feathered specimens representing different nonavian theropod clades are discovered and described (including *Beipiaosaurus* and *Sinornithosaurus*; see entries), and with most vertebrate paleontologists now to accepting that birds evolved directly from theropod dinosaurs. Consequently, many scientists now unequivocally state that birds — both extinct and extant — *are* theropods. Nevertheless, studies continue to be made relating to this once more controversial topic, the majority of them in support of the theropod-bird direct connection.

Brochu and Norell (2000) addressed the issue (often raised by various workers who maintain that Aves has an origin outside of Dinosauria; see *D:TE* and *S1*) of a perceived temporal paradox regarding bird origins — namely, that, as preserved in the fossil record, primitive birds (*e.g.*, the Late Jurassic *Archaeopteryx*) predated by millions of years the most birdlike of nonavian theropods (*e.g.,* the Early Cretaceous dromaeosaurids). The authors pointed out that this argument 1. "assumes that stratigraphic order should overturn a well-corroborated phylogeny for a group with a fragmentary fossil record," 2. "confuses the concepts of 'ancestor' and 'sister taxon,'" and 3. "ignores or dismisses fragmentary remains that may close the stratigraphic gap between the first birds and their closest nonavian theropod relatives."

Tackling this issue empirically, Brochu and Norell performed the first study of this apparent disparity in which measures of relative stratigraphic congruence to phylogenetic hypotheses were applied to contemporary hypotheses of archosauromorph relationships,

Photograph by the author.

Life-sized sculpture by Stephen A. Czerkas and Sylvia Czerkas depicting *Deinonychus antirrhopus* with feathers, exhibited at Dinofest (2000–2001), Chicago. Although feathers have not yet been found with any dromaeosaurid specimen, evidence of related taxa suggests that such birdlike theropods may indeed have been feathered.

and with the placement of birds conforming to alternative hypotheses of avian origins. These authors found that, by simply removing birds from Theropoda altogether and placing them into one of several other relationships proposed in the literature—(*e.g.*, birds as the sister group to the Crocodylomorpha (Walker 1972, 1977; Martin 1991), *Euparkaria* (Welman 1995*b*), *Scleromochlus* (Feduccia 1996), or Dinosauromorpha (Tarsitano and Hecht 1980)—the node joining bird to its sister taxon is consistent with stratigraphy and local congruence improved. However, based on that analytic approach, such arrangements did not result in trees that fit stratigraphy better. Contrarily, global congruence did not increase and could, in fact, decrease, as new, previously-nonexisting inconsistencies were introduced. Brochu and Norell further noted that, according to this analysis, any placement for birds outside Dinosauria requires more missing fossil record than a placement within that clade, and that a nontheropod origin necessitates a much greater temporal gap (or about 55 million years) between the first birds and any putative nondinosaurian ancestor. Brochu and Norell, therefore, concluded "that stratigraphy presents no significant paradox to the congruent hypotheses supported by multiple independent character data sets."

Continuing their work regarding theropod limb proportions (see *S1*), Middleton and Gatesy (2000) examined in morphospace the relationship between the design and function of theropod forelimbs across the 230 million year history of their evolution. Middleton and Gatesy's study assessed the disparity in the forelimbs of various kinds of theropods (including birds) based on the contributions of the three main limb elements in the forelimb and the functional evolution of these limbs. The authors postulated that obligate bipedalism in theropods freed these animals' forelimbs from terrestrial constraints, resulting in morphological diversity. However, the rules governing the size and shape of theropod forelimbs have, until now, been largely unexplored.

Hypothesizing that five factors (see below) may have been involved in the evolution of theropod forelimb design and utilizing ternary diagrams (those with three variables, in this case for humerus, radius, and carpometacarpus), Middleton and Gatesy found that theropods can be grouped into the following categories:

1. Predatory (*e.g.*, ornithomimids and compsognathids)— those theropods that used their forelimbs for predation possess similarly proportioned forelimbs, despite sometimes striking differences in the absolute sizes of these animals. All known theropods in this category possess functionally three-fingered, grasping, or raking limbs.
2. Reduced (*e.g.*, tyrannosaurids and abelisaurids)— theropods in which forelimb length is significantly reduced compared to the hindlimbs. This group is characterized by humeri that are proportionally longer than in nonavian theropods retaining predatory forelimbs, although the forelimbs in this group are relatively shorter.
3. Flying (*e.g.*, dromaeosaurids and other theropods regarded as most closely related to birds)— nonavian theropods that retain predatory forelimbs and share similar, though not identical, proportions with the forelimbs of primitive birds (*e.g.*, *Archaeopteryx*). Indeed, only slight proportional differences separate basal birds from nonavian manuratorans, the origin of flight apparently not entailing significant forelimb reproportioning.
4. Wing-propelled diving (*e.g.*, some birds such as penguins and diving petrels)— birds that have proportionately longer humeri than do most birds. Accompanying the radiation of birds was an increase in disparity among taxa.
5. Flightless (*e.g.*, ostriches)— birds in which reduction of the distal wing elements relative to total forelimb length results in a higher humeral proportion. Flightlessness has evolved numerous times among avian families, with each transition accompanied by increased relative humeral length resulting from comparatively short distal limb elements.

According to Middleton and Gatesy, as ratites converge on nonavian forelimb proportions, primitively and secondarily flightless theropods occupy the same region of their diagram; therefore, although theropod limb proportions are functionally diagnostic in some instances, they should not be used solely in diagnoses for controversial theropods like *Mononykus*.

The authors proposed five biomechanical and developmental factors that may have influenced the evolution of limb proportions in theropods:

1. Limb folding— interference during bipedal locomotion between limbs possibly reduced by tucking the forelimbs against the body.
2. Limb inertia—flapping wings benefiting from minimized limb inertia.
3. Spacial access— forelimbs during predation accessing the region around the shoulder for grasping and seizing prey.
4. Minimum proportions— the absolute lengths of the forearm bones perhaps restricted to the amount of bone required to form complete joints on each end; minimum bone lengths possibly determined by the functional demands of muscle.
5. Developmental pathways— changes during embryonic development and growth possibly leading "to adult flightless birds with proportions resembling a juvenile stage of their flying ancestor" (*i.e.*, paedomorphosis).

Kenworthy, Chapman, Holtz and Sadleir (2000) performed a study utilizing morphometric models to compare the claws of dinosaurs with those of other reptiles, extant birds, and mammals. More than 250 claw examples were digitized, the data for each representative group of animals then plotted in morphospace, with similarities and differences among those groups noted and tested. Preliminary results of this study, reported in an abstract, showed that the claws of the theropod manus are most similar to those of modern birds of prey, possibly indicating a manipulation of prey role for the forelimb in nonavian theropods. However, the authors noted, the pedal claws in dinosaurs are quite different from those in extant birds, which is attributed by them to the degree of curvature in the claws of the hindfoot and also possibly related to the extreme digitigrade condition of and the lack of a basal "footpad" in the pes of nonavian theropods.

Modern birds are pestered by parasites such as mites, ticks, fleas, and lice, and the same may have been true for ancient birds and feathered theropods. In examining a well-preserved, Early Cretaceous fossil feather (NSM collection) preserved in a rock slab from the Nova Olinda Member of the Crato Formation, Martill and Davis (1998) discovered in the specimen what they interpreted to be direct evidence for the presence of mites in ancient birds and feathered nonavian theropods. Identified by the authors as a caudal feather, the specimen was described by them as long, slender, slightly asymmetric, measuring 85 millimeters in length and 11 millimeters in width.

Martill and Davis found clinging to the barbs and shaft of the specimen more than 100 "hollow, subspherical structures" measuring 68–75 micrometers in diameter, scattered about the feather but not in the surrounding matrix. These structures, the authors observed, resemble eggs of the small mites called acari. Considering the possibility that these structures could represent spores or pollen, Martill and Davis noted the following: spores and pollen are usually organic in composition, while the structures in the NMS specimen were preserved as "limonitic replacements" and several of these structures exhibited tiny circular openings that differ morphologically from those in pollen. The authors reinforced their conclusions by referring to a louse egg found in Oligocene amber.

The presence of mites in the Crato Formation

specimen indicate a long evolutionary history for these parasites and suggests that early birds and feathered nonavian theropods, like modern birds, scratched to alleviate the torment inflicted by such pests.

The origin of feathers ("a complex evolutionary novelty characterized by structural diversity and complex hierarchical development") was addressed in a brief report by Prum (2000). Rather than focusing on speculative functional explanations, Prum instead turned his attention to the hierarchical details of feather development, offering information supporting "an explicit model of origin and diversification of feather morphology."

Prum's model showed feathers originating with the evolution of the first feather follicle, "a cylindrical epidermal invagination around the base of a dermal papilla." Later follicle and feather morphologies then "evolved through a series of stages of increasing complexity in follicle structure and follicular developmental mechanisms." Prum's model predicted the original feather to be an undifferentiated cylinder that was succeeded by a tuft of unbranched barbs. With the subsequent origin of rachis and barbules, the bipinnate feather evolved, eventually followed by the pennaceous feather possessing a closed vane displaying additional structural diversity. According to Prum, the filamentous integumental structures seen in the coelurosaurs *Sinosauropteryx* and *Beipiaosaurus* are consistent with his model's predictions of early feather morphology. The author cautioned, however, that additional research is required to determine if these structures evolved from follicles and are homologous with the feathers of birds.

Accepting that birds did evolve from a theropod ancestor, Wagner and Gauthier (1998) addressed the more than century-old problem of attempting to explain how the theropod manus, known to comprise digits I, II, and III, developed into the avian manus, generally accepted as comprising digits II, III, and IV (see also *S1*, section on birds). Complicating this issue is the fact that anatomical and functional evidence unequivocally supports the I-II-III-digits hypothesis in theropods (including *Archaeopteryx*). Embryological evidence, however, suggests that the digits of the bird manus are really II, III, and IV, digits I and V being thought to have been lost during ontogeny through "condensation," with the first digit (the one most likely to be lost via condensation during ontogeny) never appearing at all. The authors noted that the conflict between the paleontological and embryological data is real, but that it cannot be resolved by rejecting one kind of data, or by assuming that birds are not theropods.

Wagner and Gauthier proposed (see original paper for full details) a provocative theoretical explanation which they referred to as their "frame-shift hypothesis." The authors suggested "that the theropod lineage leading toward birds faced a conflict between two evolutionary constraints: a functional necessity to retain the inner three fingers — especially the thumb — and an opposing developmental constraint favoring loss of the last-formed and most anterior condensation." Simply stated, the developmental properties responsible for digits I through III are "shifted" during ontogeny onto embryonic precartilaginous condensations and as a result, digit II becomes digit I, and so forth. From this perspective, the paleontological and embryological data do not conflict.

As stated by Wagner and Gauthier, novelties arising early in a life cycle can profoundly influence later development, "which may explain how shared development and genetic pathways can so often succeed at providing congruent patterns of synapomorphy." However, the authors cautioned, as such novelties can occur at any point in ontogeny, one must be careful "to avoid mistaking the homology of developmental precursors for the homology of the adult characters themselves."

Wagner and Gauthier's "frame-shift" theory came under fire in a commentary written by Alan Feduccia, a paleontologist well known for not accepting the theropod-to-bird scenario, who stated the following objections: 1. There is no evidence supporting any substantial morphological change in theropod hands to indicate any kind of shift during their evolution; 2. the similarity between the hands of theropods and *Archaeopteryx* are overemphasized in drawings and the semilunate wrist bone (a feature used to link dinosaurs and birds) may not be homologous; and 3. in bird development, the fore- and hindlimbs display the same pattern of highly conserved development. Therefore, a "frame shift," if there was one, would have to occur in the forelimb only.

Feduccia concluded his commentary by stating that other versions of the frame-shift hypothesis will be required "to explain such problems as the transformation of teeth and tooth replacement, the transformation of a dinosaurian septate, hepatic-piston breathing system to a bird flow-through lung (see below), the complete abandonment of a balanced seesaw body plan to the avian model, and the reelongation of already shortened forelimbs, to mention a few." According to that author, the greatest form of "special pleading," however, must be invoked to explain flight originating from the "ground up" (requiring both small size and high places), why superficially birdlike theropods are found in the fossil record 30–80 million years after the appearance of *Archaeopteryx*, and why Triassic theropods do not exhibit any birdlike features.

Longrich (1999) restudied two synapomorphies of Manuraptora, which also have bearings on the theropod-bird hypothesis. The first of these is the semilunate carpal, comprising a carpal capping metacarpals I and part of II (which permits extensive flexion of the wrist, coupled with rotation) and, fused to it, a second carpal that is expanded to cap II in dorsal view. As noted by Longrich, semilunate carpals are known in dromaeosaurids, troodontids, oviraptorids, *Caudipteryx*, *Protarchaeopteryx*, and birds [reported by Chure 1999 also in *Allosaurus*; see entry]; they are not found in *Scipionyx*, ornithomimids, nor tyrannosaurids. Unfused semilunate carpals, perhaps a therizinosaur synapomorphy, are present in *Alxasaurus* and *Therizinosaurus*. Unfused medial portions of right and left semilunates are preserved in the type specimen of *Coelurus*. This lack of fusion is possibly related to immaturity. As in modern birds, the arc of the semilunate in the holotype of *Oviraptor* is extended ventrolaterally by a separate carpal over metacarpal III.

The other synapomorphy is a crest — possibly appearing during ontogeny, possessed only by some manuraptorans (*e.g.*, oviraptorids, therizinosaurs, troodontids, *Caudipteryx*, and birds) — formed by complete fusion of the "greater" and "lesser" trochanters. According to Longrich, the so-called "lesser trochanter" perceived in some of these taxa, and possibly persisting in extant birds, is actually the accessory trochanter.

In his monograph on *Protoavis* (see entry) and the early evolution of birds, Chatterjee (1999) addressed the issue of flight and how it relates to the various known feathered theropods. Subscribing to the scenario of flight originating from the "trees down" (see also *Caudipteryx* entry), Chatterjee (1999) postulated that the protofeathers covering the bodies of these chicken-sized, lightly built animals helped to trap body heat in the cooler environment of the trees, arboreal life in three-dimensional space consequently promoting the enlarging of the cerebellum and increased visual acuity.

Chatterjee (1999) recognized four stages of avian flight — "gliding flight, undulating flight, horizontal flight, and maneuvering flight" — based on phylogeny, paleoecology, and functional anatomy. *Sinosauropteryx* (see *S1*), the author speculated, utilized its long, stout hallux with recurved claw for grasping branches, while the hair-like proto-feathers on its forelimbs, body, and tail offered a large surface area to catch air, slow descent, and cushion landing during parachuting. In manuraptorans, this adaptation for climbing was refined with the evolution of longer forelimbs, swivel joints in the wrists, a proportionately longer third phalanx in relation to the second in manual digit II, and a stiff tail. According to Chatterjee (1999), the pectoral and supracoracoideus muscles, anchored by the furcula and sternum, were originally utilized in climbing, but subsequently co-opted for flight. The symmetrical remiges and retrices of *Caudipteryx* and *Protarchaeopteryx* served to extend airborne time, flatten the gliding path, and aid in landing. Bounding flight was pioneered by *Archaeopteryx* which, with its asymmetric flight feathers, flew by alternating flapping and gliding; *Protoavis* (which Chatterjee regards as avian; see entry), the author suggested, was capable of "horizontal flight" and could take off from the "ground up."

Additional thoughts regarding an arboreal origin for avian flight was offered in an abstract by Chatterjee and Templin (2000). Utilizing the "trees down" model as a framework, Chatterjee and Templin phylogenetically interpreted the evolution of avian flight "by analyzing the sequence of modifications of coelurosaurian forelimbs, feathers, and tails," resulting in the following proposed scenario:

In eluding predators by climbing trees, small coelurosaurs became opportunistically arboreal, exploiting new resources. This climbing adaptation led to longer forelimbs and recurved claws. Life in the trees in three-dimensional space would have promoted increases in brain size and also in visual acuity. The cooler climate in the trees — due to greater shade, increased wind, and lack of heat reflection from the ground — could have prompted the evolution of insulatory downy feathers (as in *Sinosauropteryx*), which were later co-opted for parachuting. Such structures also decreased the rate of descent during parachuting and offered a cushion to minimize the forces of impact when landing. In *Caudipteryx* and *Protarchaeopteryx*, symmetrical contour feathers have evolved in the hands and tail allowing prolonged gliding without muscular effort, which increased maneuverability and retarding landing. In dromaeosaurs, theropods generally regarded as the closest to birds, climbing ability was enhanced with the development of a powerful shoulder girdle, biceps tubercle, elongated coracoid, laterally facing glenoid, ossified sternum, swivel wrist joint, and retroverted pubis. Freed from femoral retraction, the stiffened tail became a supporting prop during climbing. With the acquisition of the swivel wrist joint, "the forelimbs developed an automatic linkage system between the elbow and the wrist joints, as in birds," the result being that the forelimbs, no longer functioning for catching and eating prey, were primarily used for climbing and gliding. Furthermore, the authors postulated that the sequence of forelimb movements during climbing presaged the "up-and-forward" and "down-and-backward" flight stroke. Gliding having been perfected, "flapping flight appeared in *Archaeopteryx* with the development of asymmetric contour feathers that enabled prolonged flight." The tail in this stage became an integral tool for controlling pitch and increase lift. Chatterjee and Templin concluded that "For *Archaeopteryx*, takeoff from a perch would have been more efficient and cost effective than from the ground."

Gatesy (2000), in an abstract, also discussed flight evolution, noting that the origin of the flight stroke is regarded as the "central problem" of this issue and pointing out the difficulty in comparing the complex three-dimensional forelimb motion among birds and other tetrapods. Gatesy cited two major problems in making such comparisons: 1. Kinetic descriptions may vary depending on their frame of reference, as wing/forelimb motion can be homologized with respect to the external environment, the long axis of the body, phases of the locomotor cycle, and the shoulder girdle components; and 2. lack of appreciation for the basis of motion.

As Gatesy explained, the trajectory of a limb "is the net result of intrinsic neuromechanical factors interacting with extrinsic loads from the substrate, prey, or air flow," the relative influence of these forces (rarely considered when limb trajectories are compared) evolving. According to this author, the loading patterns of forelimbs underwent at least two major transitions on the line from basal archosaur to extant bird. Muscular forces in primitive quadrupedal forms were produced to counteract ground reaction forces during locomotion. The forelimb in basal theropods was freed from terrestrial constraints, with forelimb motion resulting "from muscles interacting with inertial loads of prey." Finally, aerodynamic forces supplanted these forces, becoming "the primary extrinsic determinant of wing motion." Gatesy concluded that this spatial and functional perspective on the homology of forelimb motion "has implications for evaluating different scenarios of flight stroke evolution, including a predatory strike, a vertical climbing cycle, a terrestrial thrust-generating flap, and gliding."

(For a theory regarding the origins of flight viewed in the context of predation, see *Bambiraptor* entry, and Garstka, Marsic, Carroll, Heffelfinger, Lyson and Ng 2000.)

In an abstract, Burnham and Zhou (1999) compared the furcula in various groups of extinct and extant birds and dinosaurs, in the latter case particularly a furcula preserved in the not then named *Bambiraptor* (see entry). The furcula in modern birds, these authors noted, exhibits shape and flexibility, a U-shaped furcula with dorsal ends recurved caudally being mostly associated with flight. In both *Archaeopteryx* and *Confusciusornis*, the furcula is boomerang-shaped, limited to a single plane, and cranio-caudally compressed along the full length of this bone. The furcula in *Archaeopteryx*, *Confusciusornis*, and enantiornithines is grooved on the caudal side along its entire length. The U-shaped furcula of *Bambiraptor* is very similar to that of *Archaeopteryx* and *Confusciusornis*. Unlike the V-shaped furcula of *Velociraptor*, which is round in cross section, the furcula in *Bambiraptor feinbergi* "is cranio-caudally flattened and caudally grooved along its full length" as is in those two avian genera. Consequently, Burnham and Zhou concluded, *Bambiraptor* seems to be more similar to early birds than to other known dinosaurs.

Accepting that birds descended directly from theropod dinosaurs, Rubilar, Vargas and Lemus (2000), in an abstract for a poster, compared this evolution to the skeletal development of the chick (*Gallus gallus*) and a lizard (*Liolaemus gravenhorsti*). The author utilized cleared and stained embryos of different stages. Rubilar *et al.* found striking resemblances in the development of the chick to nonavian dinosaurs and to the dinosaur-to-bird transition, dinosaurian features observed in the chick but not in the lizard including the following: small first pedal digit in high position, at about midheight of metatarsi of remaining digits, apparently oriented in same direction as those digits; infraprezygapophyseal laminae in neck posteriorly inclined (as in large tetanuran theropods); inward torsion of innermost finger (digit I) before fusion of digits II and III; furcula having wide, v-shaped angle; and cone-shaped "protopygostile" of four to five vertebrae, neural arches inclining anteriorly (as in oviraptorids). These changes were observed by Rubilar *et al.* in chick limb proportions, occurring as in the transition of dinosaurs to birds: scapula and manus lengthening; forelimb proportions at six to seven days very similar to large tetanurans (*e.g.*, *Allosaurus*), at nine to 10 days to manuraptorans; reduction and fusion occurring after establishment of basically dinosaurian configuration (*e.g.*, fibula as long as tibia, tarsi and metatarsi unfused); and long feathers appearing on manus and on pygostyle (as in *Caudipteryx zoui*).

Britt (2000), in an abstract, pointed out that the only living tetrapods having lungs partitioned into exchange (lung core) and pump (air sac) components are birds, which also constitute the only extant taxon wherein the lungs invade the postcranial skeleton. This kind of system exceeds in respiratory effectiveness that of mammals several times over. In both birds and saurischian dinosaurs, diverticula of the air sacs invade the postcranial skeleton. Only coelurosaurian theropods, among all the proposed avian ancestors, possess these highly derived avian-grade postcranial pneumatic characters. This, to Britt, constituted "the strongest osteological evidence of the theropod-bird connection."

Photograph by the author, courtesy The Field Museum.

Skeletons of two individuals of *Confuciusornis sanctus*, the earliest known bird (Late Jurassic, about 120 million years ago) capable of flight. This species, displayed in National Geographic Society's traveling exhibit "China's Feathered Dinosaurs," is known from hundreds of specimens found in the Chaomidianzi Formation near Sihetun, a village in northeastern China.

According to Britt, the presence of pneumatic vertebrae and ribs in coelurosaurians suggest that the lungs of even the earliest dinosaurs were partitioned, demonstrating that partitioning preceded sternal ventilation, a character found in derived birds. Pneumatic theropod vertebrae, the author noted, comprise three types of conditions: 1. Fossate, the least complex, consisting of external pneumatic fossae on the centra and neural arches (as in *Coelophysis*); 2. camarate, consisting of thick-walled vertebrae filled with a small number of large pneumatic chambers in the centra that open to the surface by way of large foramina, the neural arch being pneumatized externally by deep fossae separated by laminae (as in *Sinraptor*); and 3. camellate, the most complex, consisting of thin-walled vertebrae filled with many interconnected alveoli communicating with the surface by way of small foramina, the neural arch being internally pneumaticized resulting in "swollen" apophyses, often combined with external pneumatic features (as in tyrannosaurids). Britt concluded that the appearance of the camellate condition — an autapomorphy of Coelurosauria, including extant birds — "may correlate with the development of the parabronchial lung."

Chiappe and Grellet-Tinner (2000) published a brief report offering new evidence regarding the evolution of archosauriform eggshell structure and its relationship to the origin of birds. Fossil egg and eggshell, these authors noted, carry phylogenetically valuable data and are appropriate for cladistic analysis, this conclusion "corroborated by the congruence found between morphology-based phylogenetic hypotheses of paleognathous birds and a recent cladistic analysis of these birds based on eggshell microstructural characters."

According to Chiappe and Grellet-Tinner, known theropod eggshells, unlike those of neosauropods and ornithopods, possess in radial view two macrostructural layers, expressed by an abrupt or gradual boundary between layers one (mammillary) and two (squamatic). Also, radiating acicular mammillae (a synapomorphy of both *Deinonychus antirrhopus* and oviraptorids) is unequivocally associated with the abrupt-boundary state, the remaining theropods "united by having eggshells with a gradual interlayer boundary and blocky calcite mammillae."

As Chiappe and Grellet-Tinner noted, the advent of avian eggshells is marked by the accretion of a third layer (for neognathous birds) and even a fourth (as in some paleognaths), the origin of these zones possibly found within Aviale (not including Aves). Although these authors cautioned that their speculation does not exclude a nonavian theropod origin, they concluded that "eggshell character optimization compared with already existing cladograms proves most parsimonious and supports the theropodan hypothesis of bird origins."

(See also *Triceratops* entry for the study by Marsic, Carroll, Heffelfinger, Lyson, Ng and Garstka 2000 involving possible DNA in the bones of this ceratopsian dinosaur and the implications regarding a dinosaurian ancestry for birds.)

In summation, Gregory S. Paul (2000) listed the following as among the lines of evidence supporting the hypothesis that birds descended from derived predatory dinosaurs: 1. Cladistic analyses; 2. skeletons of basal birds and advanced nonavian theropods sharing numerous derived features including those of the cranial sinuses, the palate, braincase, pelvis, and fore- and hindlimbs that are not seen in other potential bird ancestors; 3. in theropods, the progressive evolution of a preavian pulmonary air-sac complex; 4. theropods with simple feathers and birdlike egg shell microstructures; 5. the dromaeosaur *Sinornithosaurus* having climbing adaptations supporting the idea that avian flight evolved among partly arboreal theropods; and 6. dromaeosaurs, troodonts, caudipterygians, oviraptorosaurs, therizinosaurs, and protarchaeopterygians having "ossified uncinate processes, sternal ribs, hinged sternocostal articulations, horizontal scapula blades, sharply reflexed coracoids that articulate via a hinge joint with large, platelike sterna, folding arms with a semilunate carpal block, tails that are either very reduced or are similar to those of pterosaurs and urvogels, and symmetrical to secondarily flightless birds."

As Paul (2000) noted, some of the above-mentioned avian features, though not present in *Archaeopteryx*, are common in secondarily flightless birds. Also, some of the above-mentioned theropods display avian features not seen in archaeopterygiforms. For example, the small and urvogel-like *Sinornithosurus* possesses a very large sternal plate. This implied the possibility "that some or all such dinosaurs were closer to modern birds than the original bird."

Among the more spectacular of recent discoveries supporting the hypothesis that birds descended directly from theropod dinosaurs is the partial skeleton of *Nomingia gobiensis*, a Mongolian oviraptorosaur having a short tail culminating in a pygostyle, a feature designed to hold tail feathers and, until now, known only in birds (see *Nomingia* entry).

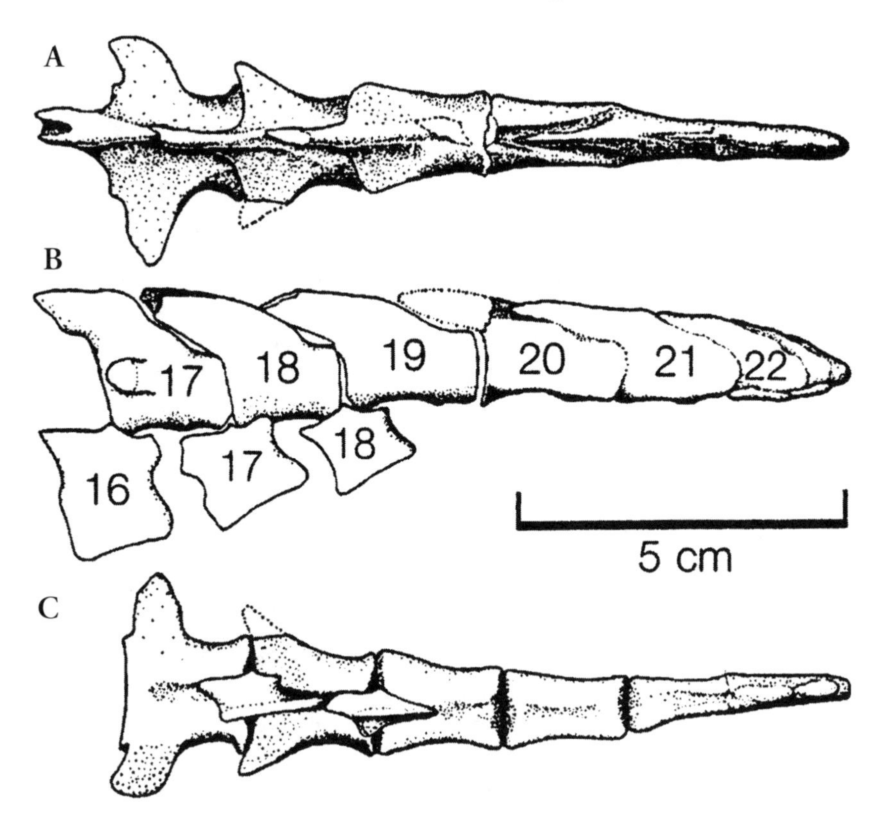

GIN 100/119, holotype of *Nomingia gobiensis*, showing pygostyle in A. dorsal, B. left lateral, and C. ventral views. See *Nomingia* entry for photographs. (After Barsbold, Currie, Myhrvold, Osmólska, Tsogtbaatar and Watabe 2000.)

The Vocal Minority

A long list of characters now seem to link birds with theropod dinosaurs. Dissentors, although in the minority, remain among paleontologists.

Among the comparatively few recent studies that deny a direct theropod-bird relationship was one published by Ruben, Dal Sasso, Geist, Hillenius, Jones and Signore (1999). Ruben *et al.* utilized ultraviolet light to examine closely the soft tissues preserved in fossils of two small theropods, *Scipionyx* from Italy and *Sinosauropteryx* from China, primarily the lungs and position of the colon in the former specimen which the authors compared with those of extant birds and terrestrial animals.

Ruben *et al.* (1999) discovered that the position of the posteriorly situated colon in *Scipionyx* is similar to that seen in today's mammals and crocodiles, but unlike that in birds, wherein the colon is located in the midabdominal region of the body. Also, the liver in *Scipionyx* is located anterior to the large intestine as in crocodilians and mammals. In crocodilians, the liver figures importantly in the function of the hepatic-piston type lungs, driven by the reptile's diaphragm during breathing. The preserved soft tissues in *Sinosauropteryx*, found to be similar to those in *Scipionyx*, were revealed to possess a subdivided visceral cavity and lung ventilation aided by a diaphragm.

Therefore, the authors concluded that these dinosaurs may have lived lives not unlike those of modern crocodilians, spending much of their time in relative inactivity, utilizing bursts of energy when attacking prey. This model, in the authors' opinion, seriously questions the hypotheses that dinosaurs were endothermic and that birds have a theropod ancestry. (It also calls into question any kind of archosaurian ancestry for birds, as there is no evidence that any other archosaurs had other types of visceral cavities or ventillation mechanisms; R. M. Molnar, personal communication 2000).

Geist (1999), in a brief report, considered the aforementioned "ground up" explanations for the origins of avian flight to be unsupportable "by most theoretical aerodynamic and energetic models, or chronologically appropriate fossils," finding the alternative or "trees down" scenario more tenable. The latter hypothesis, argued Geist, derives the earliest birds from within the adaptive radiation of small, arboreal, Triassic archosaurians.

According to Geist, examination of two Late Triassic nondinosaurian reptiles — the "avimorph" *Cosesaurus* and the quite birdlike *Megalancosaurus*, incorrectly described in the past as aquatic archosauromorphs — suggest a number of arboreal/aerial adaptations; these, along with *Longisquama*, a gliding archosaurian reptile of the Triassic period, "provide compelling morphological alternatives to a dinosaurian ancestry of birds."

However, in a later abstract, Hartman (2000) also considered the "ground up" hypothesis of flight origins, presenting a model for the evolution of aerodynamic surfaces within the context of a cursorial life style. Hartman's model suggested that some effects, such as drag (traditionally regarded as a hindrance to cursorial locomotion), "actually enhance maneuverability when employed selectively towards a common aerodynamic goal." Furthermore, lengthening these surfaces offers both the pitch control and stability demanded by hunting behavior, ending "in short duration leaps onto prey, as is commonly attributed to dromaeosaur and troodontid thereopods."

Citing that recent finds, such as the primitive bird *Rahoavis* (see *S1*), offer evidence that wing and tail feathers may have first evolved in a terrestrial context, Hartman concluded that terrestrial cursors were not restricted by any barrier to their evolution of aerodynamic surfaces.

A rather controversial paper, which suggests that the origins of birds can be found in animals other than theropod dinosaurs, was published by Jones, Ruben, Martin, Kurochkin, Feduccia, Maderson, Hillenius, Geist and Alifanov (2000). Jones *et al.* reexamined the most complete specimen (PIN 2584/4) of *Longisquama insignis*, an unusual small, Late Triassic (Norian), gliding reptile whose remains were collected from a lake bed in what is today Kyrgystan in Central Asia, and first described by Sharov (1970). What makes this reptile so spectacular is the unique feature of a half dozen or more pairs or vane-like integumentary appendages, or "plumes," sprouting high from its back, structures that have been traditionally interpreted as elongate scales.

As recounted in a follow up article by Stokstad (2000*b*), this specimen has, since it was originally described, remained housed at the Paleontological Institute, Academy of Science, in Moscow, where few paleontologists from the West had the opportunity to examine it. In 1999, while touring the United States as part of a privately funded fossil show, the specimen was noticed by John Ruben of Oregon State University in Corvallis and his graduate student Terry D. Jones. The fossil was subsequently taken to Larry D. Martin's laboratory at the University of Kansas, Lawrence, where it was examined by various workers including Martin, Ruben, Jones, and Alan Feduccia of the University of North Carolina, Chapel Hill.

In their joint paper on this specimen, Jones *et al.* interpreted (as had Sharov) *Longisquama* as an archosaur. The dorsal integumentary appendages, however, were interpreted by these authors as closely resembling the feathers of birds. According to these authors, these structures in *Longisquama* have a central shaft with narrow ribs extending out to the edges, which are comparable to the barbs on a feather. Furthermore, the shafts have wide, tubular bases near the spine resembling the hollow calamus of the feathers in modern birds.

Pointing out that *Archaeopteryx* "possessed a complete plumage of flight feathers that differed little from those of many extant birds," Jones *et al.* interpreted *Longisquama* as an archosaur possessing nonavian feathers probably homologous to the feathers of birds. If this latter interpretation is correct, the authors concluded, these integumentary appendages "may provide insight into an evolutionary grade through which feathers passed almost 75 million years before *Archaeopteryx* and perhaps before the origin of Aves itself."

According to Stokstad's report, Ruben and his coauthors, outside of the report's pages, have touted *Longisquama* as "an ideal bird ancestor." This conclusion quickly elicited a number of criticisms from various paleontologists, which Stokstad relayed:

Kevin Padian of the University of California, Berkeley, objected to Jones *et al.* simply stating that *Longisquama* is birdlike, without showing that it is more closely related to birds than something else. John Mereck and Thomas R. Holtz, Jr., both of the

University of Maryland, College Park, as well as Padian, objected to the lack of cladistic analysis, as well as the lack of falsifications of existing cladograms supporting a theropod origin for birds. Richard O. Prum, curator of ornithology at the University of Kansas's Museum of Natural History, pointed out a number of details of the integumentary structures of the specimen suggesting that they may be ribbed membranes rather than feathers (*e.g.*, edges fused rather than frayed, as in feathers: ribs near their nases extending from the shaft backward toward the body, instead of toward the tip as does a feather barb and one plume crossing another showing the clear imprint of the one underneath, something a feather would not do. Hans-Dieter Sues of the Royal Ontario Museum in Toronto noted that *Longisquama* might not even be an archosaur, noting that the cracked skull of the specimen makes it impossible to identify two diagnostic features of the Archosauria (openings anterior to the orbit and in the lower jaw), the vast group of vertebrates to which all birds and nonavian dinosaurs belong.

Feduccia (2000), in a brief report from his own perspective, restated some of his previously published, "red flag" objections to the theropod-to-bird scenario and the perception of birds as living dinosaurs, mainly: 1. *Velociraptor*, the most superficially birdlike theropod [although the recently described *Bambiraptor* seems to be a more birdlike dromaeosaurid; other more birdlike taxa, such as *Caudipteryx*, were regarded by Feduccia 1999*a* as birds, see below], occurs some 80 million years after *Archaeopteryx*; 2. the issue is clouded by various cladistic analyses that are either incorrect or suspect; 3. the theropod-bird cladogram should be tested by falsification with individual key characters (*e.g.*, manual digits, method of lung ventillation, and the condition of the tarsus), not by competing cladograms; and 4. embryological evidence reveals the digits of the avian manus to be II-III-IV, while paleontological evidence shows the theropod's to be I-II-III.

"Controversial" Taxa

Some workers (*e.g.*, Olshevsky 2000) continue to list *Archaeopteryx* among the nonavian Theropoda. However, according to fossil bird and pterosaur authority Peter Wellnhofer (2000), despite the numerous primitive, theropod-like osteological characters found in *Archaeopteryx*, some structures of the skull and especially the ulnar abduction in the wrist can be diagnosed as typically avian. All seven recovered specimens of include flight feathers, in varying states of completeness and preservation, that are mostly compatible with those of modern birds. Additionally, as fossil bird specialist Larry D. Martin (2000) observed, the hand and wrist of *Archaeopteryx* are consistent in their unique skeletal anatomy and soft tissues with the hand of birds — in which the hand is used, not for grasping prey, but in climbing and flight.

The phylogenetic status of *Unenlagia comahuensis*, the so-called "half bird" — a genus, based on very fragmentary material, originally described by Novas and Puerta (1997) as an unusual nonavian theropod, but subsequently regarded as avian by other workers (*e.g.*, Forster, Sampson, Chiappe and Krause 1998; see "Introduction," *S1* for details) — was readdressed by Norell and Makovicky in a paper describing newly revealed features of the postcranial anatomy of the dromaeosaurid *Velociraptor* (see entry).

As observed by Norell and Makovicky, some of the features shared by *U. comahuensis* and birds are also found in *Velociraptor mongoliensis*, including "derived characters of the pelvis, the dorsal vertebrae, and the shoulder girdle." These authors noted that the pelves of *U. comahuensis*, *Archaeopteryx lithographica*, and dromaeosaurids all possess a pubic apron that is formed by pubes that expand distally, forming a concave dorsal surface that is proximal to the pubic boot. The preserved left ilium of *U. comahuensis* is quite similar to that of both *V. mongoliensis* and *A. lithographica*, having an antilium with a laterally deflected ventral margin, which expands onto the pubic peduncle. The preserved dorsal vertebrae of *U. comahuensis* share with dromaeosaurids two features not found in basal birds, these being the possession of "stalked" parapophyses that project laterally from the Centrum (see Ostrom 1969*a*, 1969*b*) and a mediolateral expansion of the tip of the neural spine in posterior dorsals (a condition also seen in troodontids). Furthermore, the centra of the thoracic vertebrae of some velociraptorine dromaeosaurids (*e.g.*, *Deinonychus antirrhopus*, *Saurornitholestes langstoni*) have pneumatic foramina. Regarding the pectoral girdle, *V. mongoliensis* shows a glenoid process that is similar to, though not as extreme as, that of *U. comahuensis*. Also, other characters of the pectoral elements that generate a more horizontal and dorsal position of the scapula are seen in *V. mongoliensis*. Finally, a possible synapomorphy shared by *V. mongoliensis* and *U. comahuensis* is the pronounced curvature of the scapula near the glenoid (this curvature being less pronounced in more primitive nonavian theropods, wherein the scapular blade appears to be straighter in dorsal aspect).

For these reasons, Norell and Makovicky concluded that, pending a more complete description, the status of *U. comahuensis* "as a phylogenetically intermediate link between nonavialian dinosaurs and avialians needs to be examined within a larger phylogenetic context, one that includes *Velociraptor mongoliensis* and related taxa."

Fernando E. Novas (2000), one of the authors who named and originally described *Unenlagia comahuensis*, readdressed this taxon, focusing upon some details of its ilium relating to the acquisition of avian characteristics. According to Novas (2000), the pelvic girdle of this species is seemingly anatomically intermediate in various respects between that of dromaeosaurids and primitive birds (*e.g.*, *Rahonavis*, *Archaeopteryx*, and *Confusciusornis*). In the ilium of *U. comahuensis*, an avian processus supratrochantericus is quite prominent and triangular and is associated with a ridge that runs posterodorsally, above the acetabulum, on the lateral surface of the ilium. Therefore, the ilium in this species is more derived towards the avian condition than that of dromaeosaurids (*e.g.*, *Deinonychus*, *Saurornitholestes*, and unnamed form IGM 100/985), in which a precursor of this process is expressed as a slight transverse enlargement of the posterodorsal iliac margin.

As Novas (2000) observed, the preacetabular blade of *Unenlagia* is rounded and cranially extended and resembles that found in early birds. In less derived coelurosaurs, the cranial margin of the ilium is straight (*e.g.*, *Tyrannosaurus* and most dromaeosaurs) or notched (*e.g.*, oviraptorosaurs and ornithomimids). Also in *Unenlagia*, the cuppedicus fossa is elongate (as in *Rahonavis*); it is craniocaudally shorter in dromaeosaurids and less derived tetanurans.

According to Novas (2000), recent claims suggesting that dromaeosaurids are more derived than *Unenlagia* in the loss of a medial antiliac shelf for the M. cuppedicus are unfounded, as "such a crest is present and prominent in *Deinonychus* (AMNH 3015), *Sauronitholestes* (MOR 660), the unnamed dromaeosaurid (IGM 100/985), and *Rahonavis* (UA 5658)." Therefore, the author concluded, the above-cited morphological features "strengthen the hypothesis that *Unenlagia* is closer to birds than are dromaeosaurids."

Mononykus had previously been regarded by most authors as a bird (see *D:TE* and *S1* for pro and con arguments regarding the avian or nonavian status of this genus). Based upon recent studies (*e.g.*, Sereno 1999*a*), however, *Mononykus* as well as other genera nestled in the family Alvarezsauridae (*e.g.*, *Alvarezsaurus*, *Patagonykus*, *Shuvuuia*, and *Parvicursor*, see individual entries), previously considered to be birds, are herein treated as nonavian theropods (for details, see "Systematics" chapter, "Notes" under Alvarezsauridae and Mononykinae).

Oviraptorosaurs Interpreted as Flightless Birds

Similarities found by Elźanowski (1999) in the structure of the jaws and palate in birds and in the Oviraptorosauria (a well established clade of toothless manuraptoran theropods known from the Late Cretaceous of Asia and North America, a group not particularly close phylogenetically to birds) raised a potentially explosive possibility with far-reaching consequences — that oviraptors are not nonavian theropods, as they have always been classified, but rather "the earliest known flightless birds."

Basing his study on a skull (ZPAL MgD-1/95) referred to *Oviraptor* sp., which had never previously been described in detail, Elźanowski observed that the medial wall of the caudal sinus in the oviraptorid skull is similar in shape and location to a vertical, median or paramedian element (Elźanowski and Wellnhofer 1995) in the upper jaw of the fifth skeleton of *Archaeopteryx* (see "Introduction," section on birds, *S1*). He also observed that oviraptorosaurids, therizinosauroids, ornithomimosaurs, caenagnathids (see Sues 1997*a*), and birds have a palatine possessing a maxillary process that much exceeds in length the rostromedial vomeral process, overlapping ventrally the maxillary palatal shelf. In oviraptorids, therizinosauroids, birds, and probably caenagnathids, the palatine has a broad pterygoid (caudal) wing. Also, the palatine in ornithomimosaurs has a dorsal, transversely oriented process located near the lacrimal (see Osmólska, Roniewicz and Barsbold 1972), comparable in position to a prominent transverse crest in *Archaeopteryx* (see Elźanowski and Wellnhofer) and three transverse crests in the caenagnathid *Chirostenotes* (see Sues). The ectopterygoid connects the lacrimal to the palatine, as does the uncinate (lacrimopalatine) that is widespread among neornithine birds, and differing from the ectopterygoid in other known theropods including therizinosauroids and *Archaeopteryx*. In oviraptorids and caenagnathids, the pterygoid has a basal process for articulating with the cranial base. In *Archaeopteryx*, this basal process being poorly developed; it may, Elźanowski suggested, constitute a synapomorphy joining the oviraptorosaurs (including oviraptorids and caenagnathids) with the ornithurines. The quadrate has a backward-bent head in oviraptorids (see Maryańska and Osmólska 1997), the therizinosaurid *Erlikosaurus*, ornithomimosaurs, troodontids, and *Gobipteryx* and most other ornithurine birds, while in dromaeosaurids and most remaining theropods the quadrate is straight in caudal profile, the quadrate socket in the squamosal opening more or less ventrally; only the oviraptorid quadrate agrees with that of ornithurines in possessing "an otic capitulum for the articulation with the braincase, the pterygoid articular surface approaching the medialmandibular condyle, and a distinct (but shallow) quadratojugal cotyla (Maryańska and Osmólska 1997)." The articular surface for the quadrate is expanded both laterally and medially in both oviraptorids and caenagnathids, these projections in oviraptorosaurs being similar to those seen in *Gobipteryx*. In oviraptorids and caenagnathids, at least in *Garudimimus* among ornthiomimosaurs (see Barsbold and Osmólska 1990), and some birds, the prearticular continues far rostrally as a straight bony rod. Among theropods, the ends of the mandibular rami are known to be fused only in oviraptorosaurs and the smallest specimens of caenagnathids (Currie, Godfrey and Nessov 1993), a condition also found in most birds, which possibly constitutes another synapomorphy for oviraptorosaurs and ornithurines. An intraramal joint (including articulations between dentary and surangular and between splenial and angular), allowing mediolateral mobility within each mandibular ramus, is present in most theropods (including dromaeosaurids), but absent in ornithomimosaurs, oviraptorosaurs, *Erlikosaurus*, and primitive birds (including *Archaeopteryx*, *Gobipteryx*, and probably *Confuciousornis*). Most birds and oviraptorids possess a thin, rod-like jugal bar formed largely of the quadratojugal; in *Archaeopteryx* and remaining theropods, this bar is a robust slat.

Regarding the lack of teeth in oviraptorosaurs, in more advanced ornithomimosaurs, and in most birds, Elźanowski considered this condition to be the result of convergence, noting that the reduction of teeth is a common theme during vertebrate evolution.

Convergence, however, could not explain the similar morphologies of the skulls of ornithomimosaurs, oviraptorosaurs, therizinosauroids, and birds. Consequently, Elźanowski's study suggested that, contrary to most recently proposed phylogenetic reconstructions based primarily on postcranial evidence, these clades are the closest theropodan relatives of birds. This study also found no specific avian similarities in the jaws and palate of dromaeosaurids, a group of manuraptoran theropods now usually classified as most closely related to birds.

It was reasonable, therefore, for Elźanowski to suspect that the unique cranial similarities of birds, ornithimomimosaurs, and therizinosauroids are synapomorphic, thereby indicative of a clade comprising these taxa and probably also including the Troodontidae. This controversial new arrangement, the author noted, would agree with a phylogeny proposed earlier by Russell and Dong (1993*a*; see "Systematics" chapter and *Alxasaurus* entry, *D:TE*), who separated tetanuran theropods into two groups — one including dromaeosaurs and carnosaurs, the other ornithomimosaurs, troodonts, therizinosauroids, and oviraptorosaurs.

Intriguingly, Elźanowski's study suggested that oviraptorosaurs branched off after *Archaeopteryx*,

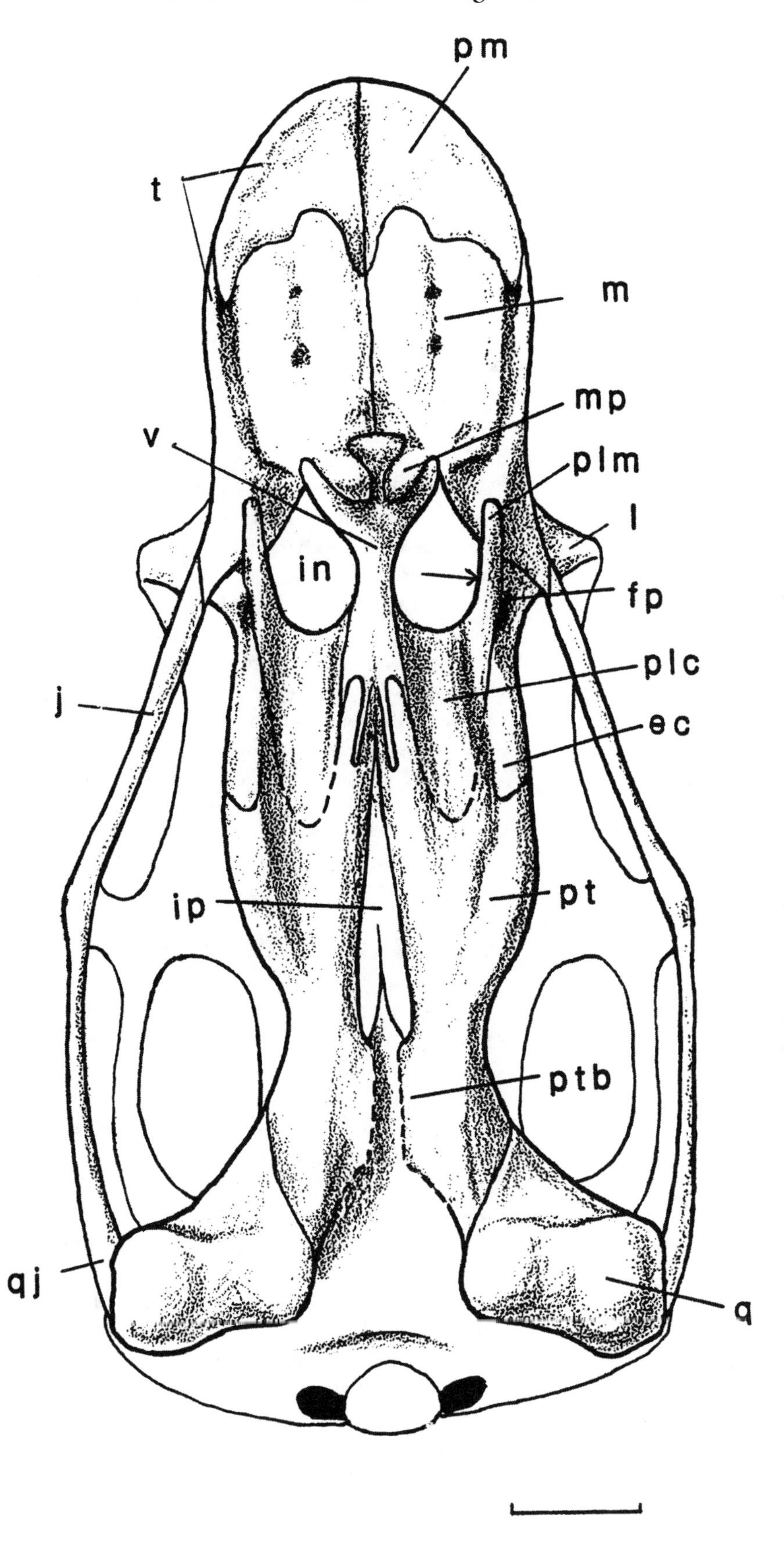

Reconstruction of the bony palate of the skull (ZPAL MgD-1/95) of the oviraptorid *Oviraptor* sp. A recent study by Andrzej Elźanowski suggests that oviraptorosaurs may not be nonavian theropods but rather early flightless birds. Scale = 10 mm. (After Elźanowski 1999.)

representing the earliest known flightless birds. Save for the elongate forelimbs (shortened in flightless forms), the postcranial skeleton of *Archaeopteryx* shows no any avian traits not found in oviraptorids (see Barsbold 1983*a*, 1983*b*); therefore, the author speculated, "if flightlessness had evolved at a stage of avian evolution close to *Archaeopteryx*, this would be extremely difficult to distinguish from the primary flightlessness of the theropods."

Elźanowski pointed out that, while ornithomimosaurs are clearly theropodan, as evidenced by cranial details, oviraptorosaurs have always been difficult to classify. The unusual skulls of oviraptorosaurs have made comparison with other theropods difficult and evidence for their affinities has mostly come from comparisons of postcranial skeletons. Indeed, the only consensus among workers regarding this theropod group is that the Oviraptorosauria belongs in Coelurosauria.

Naturally, so controversial a conclusion as Elźanowski's will have to be discussed and evaluated by other paleontologists before a final decision is reached. If, in fact, this assessment of the Ornithomimosauria proves to be correct, then a number of well-known taxa—*e.g.*, *Oviraptor*, *Ingenia*, *Chirostenotes*, and others—will need to be removed from the list of dinosaurian genera and relegated to "excluded genera" status, while another major reorganization of the higher taxa will be in order.

Dinosaur Extinctions

The mystery of the extinction of all nonavian dinosaurs at the end of the Cretaceous period remains a source of fascination, and clues to solving this intriguing puzzle continue to surface.

New evidence supporting the hypothesis that worldwide fires may have destroyed almost 18 to 24 percent of the Earth's biomass at the culmination of the Cretaceous was presented by Arinobu, Ishiwatari, Kaiho and Lamolda (1999). Arinobu *et al.* reported on the discovery of a one- to two-meter wide layer of red mudstone in a very well-preserved part of the K-T (Cretaceous–Tertiary) boundary in Caravaca, Spain. As noted by these authors, the layer—dubbed by them the "fallout lamina"—is located at the base of boundary and contains iridium, an element generally found in meteorites but which can also constitute evidence of ancient volcanic activity.

High-resolution analyses by Arinobu *et al.* indicated that the iridium layer was formed instantaneously. Dating the entire stratum in which the layer was found revealed an age ranging from the last 6,800 years of the Cretaceous to the first 29,000 of the succeeding Tertiary. Analysis by the authors of unweathered rock slices just above the this boundary revealed a composition of mostly large amounts of pyrene with phenanthrene, flouranthene, and methylpyrene in a "highly precondensed" form, compounds that form only at high temperatures. Based upon the amount of carbon produced, Arinobu *et al.* suggested that these fires burned on a global scale.

Some scientists (*e.g.*, Clemens and Archibald 1980; MacLeod, Rawson, Forey, Banner, BouDagher-Fadel, Brown, Burnett, Chambers, Culver, Evans, Jeffrey, Kaminski, Lord, Milner, Milner, Morris, Owen, Rosen, Smith, Taylor, Urquhardt and Young 1997; Hickey 1981; Williams 1994; Sarjeant 1996), who postulate that the extinction of dinosaurs at the end of the Cretaceous was gradual rather than catastrophic or comparatively sudden, support their position by referring to a three-meter or 10-foot gap in dinosaur fossils at the top of the Hell Creek Formation (uppermost Maastrichtian) in western North America. According to these researchers, dinosaur fossils are either rare in or entirely absent from the three-meter interval of sedimentary rocks just below the geological horizon interpreted by other workers as indicative of an impact with the Earth by an enormous extraterrestrial bolide (Alvarez, Alvarez, Asaro, and Michel 1980; see below). Consequently, this paucity of dinosaurian remains implied that the last dinosaurs had already become extinct (see Clemens and Archibald; Hickey; Sarjeant) or that they had significantly decreased in numbers (MacLeod *at al*; Williams; Archibald 1996) before the collision occurred, the impact at best serving as the *coup de grace* in wiping out the last of this group of animals.

Sheehan, Fastovsky, Barreto and Hoffman (2000) investigated this reported gap and then related that the interval was originally reported as a "gap" in the upper two to three meters of the Hell Creek Formation in northeastern Montana by Clemens and Archibald. Hickey subsequently found this "gap" to be present throughout western North America. After dinosaur bones (*e.g.*, see Sheehan, Fastovsky, Hoffman, Berghaus and Gabriel 1991) and tracks (*e.g.*, Pillmore, Lockley, Fleming and Johnson 1994) were eventually reported from the "gap," it was regarded as an interval in which fossils, though quite rare, were nevertheless present. Other workers (Galton 1974; Bohor, Triplehorn, Nichols and Millard 1987; Rigby and Rigby 1990; Sheehan *et al.*; Lockley 1991; Pilimore *et al.*), following that original report of Clemens and Archibald, have also cited evidence of dinosaurs in the uppermost three meters of the formation. According to Sheehan *et al.* (1991), this gap, despite rigorous documentation of low dinosaur abundance in the interval (see Fastovsky and Sheehan 1997), has continued to be used as important evidence in support of the gradual argument, accepted by some workers — Williams, Sarjeant, Archibald, and MacLeod *et al.* — as established fact.

To test whether or not dinosaur abundance was declining in this three-meter interval, Sheehan *et al.* (2000) examined this abundance and compared the results with dinosaur abundance in comparable lower intervals of the Hell Creek Formation. Materials used in this study had been collected by survey crews during a search in the badlands near Marmath, North Dakota, and Glendive, Montana, for changes in dinosaur communities through the formation (see Sheehan *et al.* 1991; White, Fastovsky and Sheehan 1998). The survey had been structured to determine if these communities in the lower, middle, and upper thirds of the formation were degrading during deposition of the formation. Given that a skeleton comprises numerous bones, only the absolute minimum number of dinosaur individuals that could be represented at a single locality were incorporated into this study. For example, if a right and left femur belonging to *Triceratops* were found, a single individual was recorded; if two complete femora were found, two *Triceratops* individuals were recorded.

Sheehan *et al.* (2000)'s study brought in 113 separate bones representing a minimum of 38 individual dinosaurs from the upper three-meter interval of the Hell Creek Formation. The authors identified 2,324 separate dinosaur bones representing at least 985 individual dinosaurs collected from 1.5-meter intervals in the upper 60 meters of the formation in Montana and the upper 36 meters in North Dakota. Their data revealed that the numbers of fossils in the upper three meters of the formation were less than those in intervals having the most abundant fossils, but, conversely, that the upper intervals are not among those with the least abundant fossils. Sheehan *et al.* (2000) also discovered that (*contra* Williams) dinosaur remains are not evenly distributed throughout the formation; rather, intervals "with abundant fossils reflect localized pockets of abundant fossils from place to place in the formation" (see Fastovsky 1987; White *et al.*).

Comparing dinosaur abundance in the upper three meters of the Hell Creek Formation to that in the underlying 36 meters in Montana and North Dakota, Sheehan *et al.* (2000) found that dinosaur fossils in the former occur in numbers comparable to those in the latter. (Sheehan *et al.* did not study the three-meter "gap" in northeastern Montana first reported by Clemens and Archibald, noting that because they found no gap in either of the two examined regions, a gap in northeastern Montana, if present, would "not have much to contribute to the

Mounted skeleton (cast of AMNH 5116, 5039, 5045, and 5033) of the giant ceratopsid *Triceratops horridus*, one of the last nonavian dinosaurs to become extinct. A growing body of evidence indicates that an impact from some kind of gigantic celestial body occurred at the termination of the Cretaceous period. However, what effect such a collision may have had upon the dinosaurs' existence or demise remains open to debate.

Photograph by the author, courtesy Boston Museum of Science.

extinction debate"). From these results, Sheehan *et al.* (2000) therefore concluded that there is evidence for a three-meter gap at the top of the Hell Creek Formation and that the distribution of dinosaurs in the uppermost interval of the formation is consistent with the scenario of "an abrupt extinction event that coincided with a bolide impact."

More on the Catastrophic Scenarios

A growing body of evidence, some of it originating from outside the field of paleontology, now further supports the hypothesis that a celestial impact (*e.g.*, that of an asteroid, meteorite, or comet) did occur at the end of the Cretaceous period and that, at least in some ways, the forces released by that impact played a salient role in the vanishing of the last dinosaurs (see *D:TE* and *S1* for information and conflicting theories).

Frank T. Kyte (1998) of the Institute of Geophysics and Planetary Physics, University of California, Los Angeles, described a fragment of a meteorite, measuring 2.5 millimeters across and dredged from a deep sea drilling core off the floor of the North Pacific Ocean at the Cretaceous–Tertiary boundary. The drilling core was located just below a clay deposit containing iridium. The deposit also contains shocked quartz grains, evidence for such an impact. The fragment itself is mostly made up of hematite, enriched with chromium and gold. Apparently retained in the specimen are "primary textures" indicative of meteorites, although these may have been geochemically changed since the time of deposition.

Kyte inferred that this tiny piece of rock, discovered after an intensive two-year search, may indeed "be a piece of the projectile responsible for the Chicxulub crater [the alleged impact site, located off the coast of Yucatan; see *S1*]."

Geochemical and petrographic analyses showed that the impacting object from which this fragment originated was not the porous aggregate type of interplanetary dust regarded as typical of comets, but rather the typical metal and sulfide rich carbonaceous chondrite indicative of asteroids. Differences between this ancient impactor and more recent ones could be explained by weathering action in the Pacific area during a 65 million-year time span.

According to Kyte's calculations, an oblique angle of impact would have reduced the fraction of the asteroid exposed to temperatures higher than its own melting point, favoring the preservation of its fragments preserved at various sites corresponding with the K-T (or Cretaceous–Tertiary) boundary.

A. Shukolyukov of Scripps Institution of Oceanography and G. W. Lugmair of the University of California, San Diego, performed research connecting the K-T boundary with the cosmic-origin mass-extinction scenario, but which may, in one way at least, contradict the findings of Kyte.

Utilizing state of the art techniques incorporating a mass spectrometric detection system, Shukolyukov and Lugmair (1998) discovered the presence of a chromium isotope in mineral samples collected from the K-T boundary at Stevens Klint, Denmark, and Caravaca, Spain. As the isotopic composition of chromium detected in these samples differs from the ratios known on Earth, the authors deduced their origin to be extraterrestrial.

According to these author's findings, however, the chromium ratios found at the K-T boundary — once the required fractional corrections are made in their calculations — are identical to those (within the range of experimental uncertainty) found in the class of meteorites referred to as carbonaceous condrites, which are believed to have a composition similar to that of comets. *Contra* Kyte, then, Shukolyukov and Lugmair theorized that it was probably a comet and not an asteroid that brought about the demise of the dinosaurs at the end of the Cretaceous.

Accepting that catastrophic events lead to the extinctions at the Cretaceous–Tertiary boundary (while at the same time addressing the question of why only some plant and animal taxa of the Late Cretaceous went extinct), Retallack (2000) pointed out in an abstract that acid is one of the consequences of such events — "nitric acid from atmospheric shock by bollides and from burning of trees, sulfuric acid from volcanic aerosols and from impact vaporization of evaporites, hydrochloric acid from volcanic aerosols, and carbonic acid from the carbon dioxide of volcanoes and fires." The author noted that the amounts of acid entered into the environment at that time can be determined by chemoassay (chemical reactants), pedoassay (paleosols), and bioassay (selective extinctions). Soil samples were examined for their acid content from Montana and India.

According to Retallack's assessments, the acid content in the environment during that time would have allowed the survival of flora including coccolithophores, foraminifera, and dinoflagellates, and also fishes and amphibians, while mollusks in Montana would have suffered strong extinctions. Also, although angiosperms and conifers survived in Montana and South Dakota, acidification may have resulted in heavy extinctions among evergreen angiosperms. Browning of vegetation via acidification "would have been difficult for herbivorous dinosaurs and their large predators, but less problematic for small insectivorous and detritivorous birds and mammals."

Paleocene Dinosaurs?

Over the years, various reports have been made by paleontologists supposedly documenting the presence of dinosaur fossils in Paleocene-age rocks (see *D:TE*). If correctly interpreted, such finds would extend the occurrence of at least some kinds of dinosaurs beyond the termination of the Cretaceous period. They would cast doubts on the orthodox hypothesis that dinosaurs, as a group, suffered mass extinctions at the Cretaceous–Tertiary boundary and also weaken the hypothesis that dinosaurian extinctions were the result of some catastrophic event such as an impact by a gigantic extraterrestrial bolide.

Lucas and Sullivan (2000*a*) recently addressed the possibility that some Late Cretaceous dinosaurian taxa may have survived into the early Cenozoic Era, or Age of Mammals in their study concentrating on the San Juan Basin of New Mexico. As noted by these authors, new mapping, lithostratigraphy, and the collection of vertebrate fossils at Betonnie Tsosie Wash — a locality first mentioned by Sinclair and Granger (1914), apparently unaware of dinosaur fossils in Cretaceous Tsosie outcrops, as early Paleocene (Puercan, a very early part of this geologic period, equivalent to part of the Danian age), subsequently mapped with Cretaceous strata included by some authors (*e.g.*, Reeside 1924) — in the west-central San Juan Basin, allows for new interpretations of the K-T boundary at that locality and throughout the San Juan Basin. Exposed at the Betonnie Tsosie Wash, in ascending order, is the following stratigraphic section: the Kirtland Formation (including the Hunter Wash, Farmington, De-na-zin, and Naashoibito members), Ojo Alamo Sandstone, and Nacimiento Formation.

As noted by Lucas and Sullivan, the Betonnie Tsosie Wash vertebrate fossils are known only from the Kirtland and Nacimiento formations, none being known from the Ojo Alamo Sandstone at the Betonnie Tsosie Wash regardless of earlier claims (*e.g.*, Rigby and Lucas 1977; Lucas and Rigby 1979; Rigby 1981). Four stratigraphically distinct vertebrate-fossil-bearing intervals were recognized by Lucas and Sullivan at the Betonnie Tsosie Wash locality:

1. The Farmington Member-one locality has yielded diverse fossil remains including those of hadrosaurian and other kinds of dinosaurs.
2. Several localities in the De-na-zin Member, between the Farmington and Naashoibito members, has yielded isolated bones mostly pertaining to hadrosaurs.
3. The outcrop belt of the Naashoibito Member has produced isolated turtle and numerous dinosaur bones, most significantly ceratopsid frill fragments (SMP VP-1243, 1245, and 1246), a hadrosaurid vertebra (SMP VP-1247), an almost complete tyrannosaurid right femur (SMP VP-1113), and an incomplete sauropod femur referable to *Alamosaurus sanjuanensis* (see *Alamosaurus* entry).
4. The lower part of Nacimiento Formation (see Williamson 1996) has yielded mammal remains of Puercan age.

Lucas and Sullivan found some strata regarded by various previous workers (*e.g.*, Rigby and Lucas; Lucas and Rigby; Rigby) as the "Paleocene-age" lower Ojo Alamo Sandstone actually to be a sandstone-dominated lithofacies of the Upper Cretaceous Naashoibito Member. In other words, at some locations the lithostratigraphic unit known as the Ojo Alamo Sandstone consists not only of rocks of Paleocene age, as has generally been assumed, but comprises two intervals, a lower Late Cretaceous interval and an upper early Paleocene interval. This stratigraphic reassignment, the authors pointed out, has basinwide implications for the interpretation of the K-T boundary.

Therefore, regardless of persisting claims to the contrary (*e.g.*, Fassett, Lucas, Zielinski and Budahn 2000; Fassett and Lucas 2000; see below), Lucas and Sullivan found "no *proven* records of Paleocene dinosaurs in the San Juan Basin, despite continuing claims to the contrary" (*e.g.*, Fassett *et al.* 2000; Fassett and Lucas); also, "Many of the supposedly reworked dinosaur bones in the lower part of the Ojo Alamo Sandstone, claimed by some to be Paleocene dinosaurs, probably come from this lower interval and are of Late Cretaceous age."

Although Lucas and Sullivan posited that there are no *proven* records of Paleocene dinosaurs in the San Juan Basin, Fassett and Lucas offered new (and seemingly conflicting) evidence supporting the possibility of Paleocene dinosaurs in the Ojo Alamo Sandstone (see also Lucas, Heckert and Sullivan 2000).

Fassett and Lucas dated the Ojo Alamo Sandstone as of Paleocene age based on the following evidence: 1. The lowermost part of the Nacimiento Formation, overlying and intertonguing with the Ojo Alamo Sandstone in the Ojo Alamo type area in the southern part of the San Juan Basin, contains Puercan mammalian fossils; 2. plant megafossils from Ojo Alamo are indicators of Tertiary age; and 3. pollen-productive rock samples from Ojo Alamo at several San Juan Basin localities have all yielded Paleocene-age palynomorph assemblages but none of Cretaceous age.

The evidence presented by Fassett and Lucas comprised both the first reported discovery of Paleocene-age palynomorphs from Ojo Alamo Sandstone outcrops in the northern part of the basin and also the hard evidence of a dinosaur bone. The rocks containing the pollen were recovered "from a coaly, carbonaceous shale bed 120m above the base of the Ojo Alamo and 3m below the level of a large hadrosaur femur."

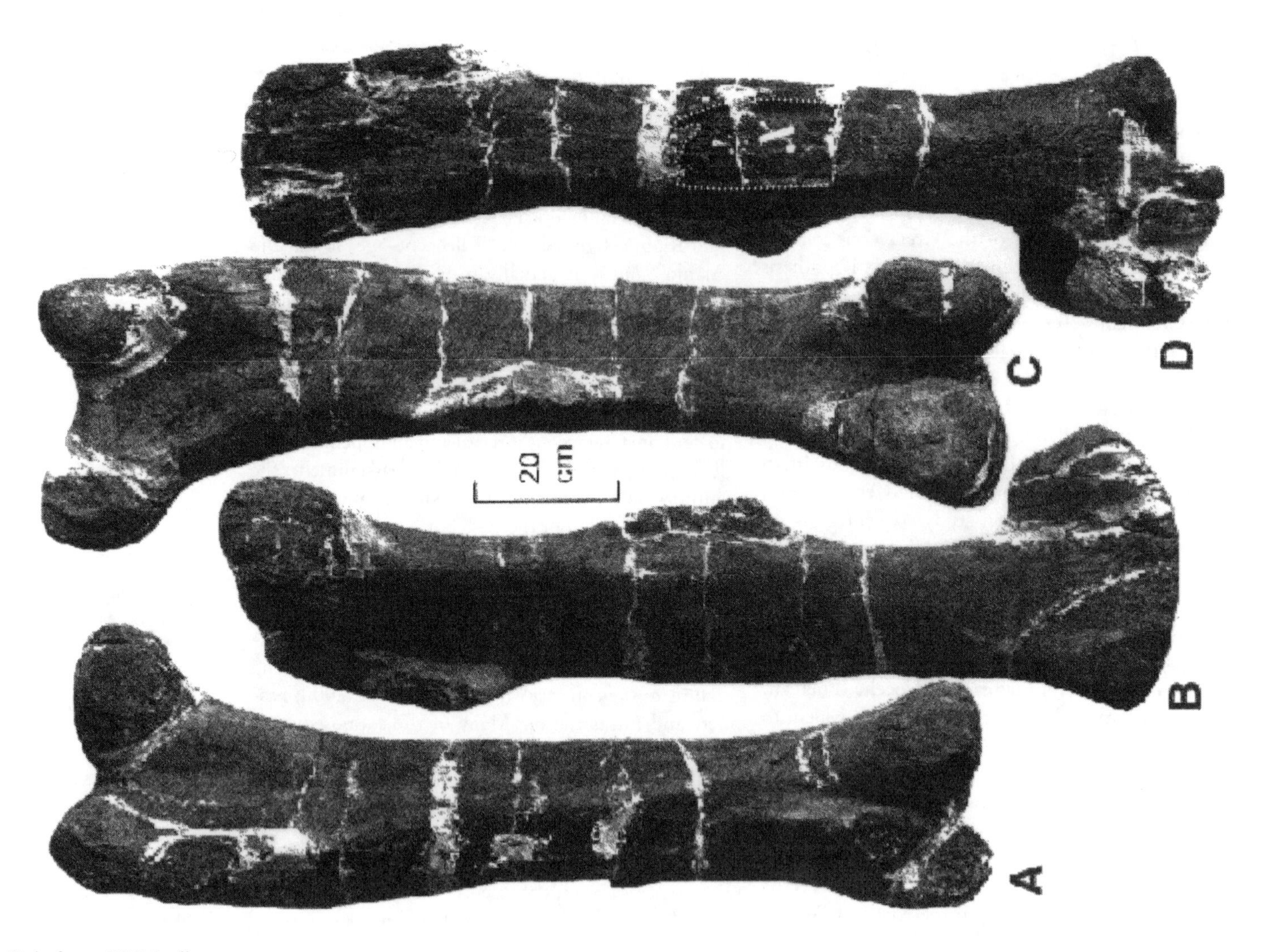

Right femur (UNM collection) of a hadrosaurian dinosaur from the Ojo Alamo Sandstone, San Juan River site, San Juan Basin, New Mexico, in A. anterior, B. medial, C. posterior, and D. lateral views. Palynmorph evidence suggests that this dinosaur may have lived during the early Paleocene period, after the culmination of the Cretaceous. (After Fassett and Lucas 2000.)

This right femur (now on display at the Geology Museum, University of New Mexico, UNM collection), discovered in 1983 in the bluffs south of the San Juan River approximately 5 kilometers southwest of Farmington, New Mexico, is significant as the first dinosaur bone found in the Ojo Alamo Sandstone in the northern part of the basin. The specimen was collected in the summer of that year by a field party led by Michael O'Neill of the Farmington District Office of the Bureau of Land Management (BLM) and with Sid Ash, David A. Thomas, and Brad Peterson participating, under the auspices of the New Mexico Bureau of Mines and Mineral Resources, Socorro. The first photograph of the specimen was published in 1987 by Fassett, Lucas and O'Neill.

As measured by Fassett and Lucas, the bone has a maximum length of 1310 millimeters, maximum proximal width of 370 millimeters, and maximum distal width of 330 millimeters. The authors could not identify the specimen beyond the level of family. Citing Lull and Wright's (1942) classic monograph on North American duckbilled dinosaurs, Fassett and Lucas noted that the Ojo Alamo specimen is about 8 to 12 percent longer than the longest femora (belonging to *Kritosaurus* and *Saurolophus*) listed by those authors; however, it is shorter than the femur of the Chinese *Shantungosaurus giganteus* (1650 millimeters long; Hu 1973), the largest hadrosaur known.

Although the bone had been excavated from a cobble-to-boulder conglomerate sandstone, the specimen exhibits no signs of abrasion by erosion, indicating to Fassett and Lucas that burial occurred directly after the animal died and that mineralization of the bone occurred rather early, leading the authors to the opinion that the femur had not been reworked from underlying Cretaceous-age strata.

The presence of this bone in rocks dated as Paleocene in age certainly offers problems, although one of the authors (S. G. Lucas, personal communication 2000) felt that the information concerning it be published. As of this writing, Fassett and Lucas have submitted dinosaur bone samples from both the Ojo Alamo Sandstone and the uppermost Kirtland Formation at several sites in the southern San Juan Basin for geochemical analysis by U. S. Geological Survey geochemists Robert Zielinski and James Budahn, Denver. Preliminary results indicate the distinct geochemical differences can be gleaned from Paleocene, Ojo Alamo Sandstone, and the Cretaceous Kirtland Formation trace element assemblages (Fassett *et al.* 2000). Fassett and Lucas stated that the ultimate results of these analyses may finally answer "lay to rest any questions regarding the reworking of Cretaceous dinosaur bones from the Kirtland Formation into the Paleocene Ojo Alamo Sandstone."

On the basis of the data currently available, Fassett and Lucas could only conclude that "some dinosaurs in the San Juan Basin survived the 'terminal' end-Cretaceous asteroid impact event only to become extinct a few hundred thousand years (at most) later, in earliest Paleocene time."

II: Dinosaurian Systematics

The following breakdown of the Dinosauria is built upon those that appeared previously in both *D:TE* and then *S1*. As in those earlier volumes, this arrangement is the present writer's conservative attempt to organize the various taxa above the genus level into a convenient and usable system, based upon recently published data, most notably the phylogeny proposed by Sereno (1999*c*). Some of the following choices, as to which phylogenies to follow, were subjective, based upon the author's opinions after weighing the published evidence. Needless to add, much of what follows will undoubtedly change as future supplementary volumes in this series are issued.

As stated in the first two books, this breakdown is not the author's attempt to propose or sanction any "official" organization of the higher taxa. Lacking the luxury of having a complete fossil record that accurately reveals all of the dinosaurian taxa that ever existed, it can never be possible to produce an entirely stable classification of the Dinosauria. Indeed, as pointed out by editors James O. Farlow and Michael K. Brett-Surman in their book *The Complete Dinosaur* (1997, p. 62), "Unfortunately, stability can also reflect consensus due to the lack of an adequate fossil record — or a stagnation of research."

Cladograms are not included herein, as one specialist's cladogram can (and often does) differ from someone else's, sometimes in the extreme. As in *D:TE* and *S1*, the following attempt at organizing a phylogeny of the Dinosauria is consequently highly tentative and subject to future changes.

Well-known published monotypic groups (*e.g.*, the family Dryptosauridae, which includes but a single genus, *Dryptosaurus*) are listed according to the name of the higher taxon, although many systematists — to avoid redundancy, particularly on cladograms — simply list the genus. The terms "family," "subfamily," and "superfamily" have been retained for use in the text of this book, although some modern workers have abandoned these designations — and even moreso, such familiar terms as "suborder," "infraorder," and others, once commonly used in Linnaen classification — altogether. A family is identified by the suffix "-idae," a subfamily by "-inae," and a superfamily by "oidea." Taxa preceeded by a question mark are usually considered to be valid (*e.g.*, HERRERASAURIDAE and *Cryolophosaurus*), although their phylogenetic position is currently uncertain. Those taxa set in quotation marks (*e.g.*, "MEGALOSAURIDAE") are in dispute or under discussion or for some other reason may eventually be found invalid. Erring on the side of completion, included among the listed genera (*e.g.*, *Leipanosaurus*) are *nomina dubia*, that is taxa regarded as having dubious validity, some of which may actually be referrable to other taxa.

DINOSAURIA

- I. SAURISCHIA
  - GUAIBASAURIDAE
  - THEROPODA
    - ?*Eoraptor*
    - ?HERRERASAURIDAE
    - NEOTHEROPODA
      - "CERATOSAURIA"
        - COELOPHYSOIDEA
          - COELOPHYSIDAE
          - DILOPHOSAURIDAE
        - NEOCERATOSAURIA
          - CERATOSAURIDAE
          - ABELISAUROIDEA
            - *Ilokelesia*
            - ?NOASAURIDAE
            - ABELISAURIDAE
      - TETANURAE
        - SPINOSAUROIDEA
          - "MEGALOSAURIDAE"
          - EUSTREPTOSPONDYLIDAE
          - SPINOSAURIDAE
            - BARYONYCHINAE
            - SPINOSAURINAE

SPINOSAURINAE (cont.)
AVETHEROPODA
CARNOSAURIA
?*Cryolophosaurus*
Monolophosaurus
ALLOSAUROIDEA
?*Piatnizkysaurus*
?*Becklespinax*
?*Szechuanosaurus*
SINRAPTORIDAE
ALLOSAURIDAE
?CARCHARODONTOSAURIDAE
COELUROSAURIA
COMPSOGNATHIDAE
?DRYPTOSAURIDAE
MANURAPTORIFORMES
ARCTOMETATARSALIA
TYRANNOSAURIDAE
?AUBLYSODONTINAE
TYRANNOSAURINAE
ORNITHOMIMOSAURIA
?ALVAREZSAURIDAE
MONONYKINAE
Patagonykus
Alvarezsaurus
HARPYMIMIDAE
GARUDIMIMIDAE
ORNITHOMIMIDAE
OVIRAPTOROSAURIA
?*Microvenator*
OVIRAPTORIDAE
OVIRAPTORINAE
INGENIINAE
CAENAGNATHIDAE
THERIZINOSAUROIDEA
Beipiaosaurus
ALXASAURIDAE
THERIZINOSAURIDAE
MANURAPTORA
?*Ornitholestes*
DEINONYCHOSAURIA
Protarchaeopteryx
?CAUDIPTERIDAE
TROODONTIDAE
EUMANURAPTORA
DROMAEOSAURIDAE
DROMAEOSAURINAE
VELOCIRAPTORINAE
Sinornthosaurus
Bambiraptor
?AVIALE [AVES]
SAUROPODOMORPHA
PROSAUROPODA
Saturnalia
THECODONTOSAURIDAE

        THECODONTOSAURIDAE (cont.)
        ANCHISAURIDAE
        MASSOSPONDYLIDAE
        YUNNANOSAURIDAE
        PLATEOSAURIDAE
        MELANOROSAURIDAE
        BLIKANASAURIDAE
        ?*Riojasaurus*
    SAUROPODA
        VULCANODONTIDAE
        EUSAUROPODA
            BARAPASAURIDAE
            "CETIOSAURIDAE"
            Jobaria
            NEOSAUROPODA
                Cetiosauriscus
                DIPLODOCOIDEA
                    Amphicoelias
                    ?NEMEGTOSAURIDAE
                    REBBACHISAURIDAE
                    DICRAEOSAURIDAE
                    DIPLODOCIDAE
                    ?*Antarctosaurus*
                MACRONARIA
                    ?HAPLOCANTHOSAURIDAE
                    CAMARASAUROMORPHA
                        CAMARASAURIDAE
                        ?TENDAGURIIDAE
                        TITANOSAURIFORMES
                            BRACHIOSAURIDAE
                            SOMPHOSPONDYLI
                                EUHELOPODIDAE
                                    SHUNOSAURINAE
                                    EUHELOPODINAE
                                TITANOSAURIA
                                    ANDESAURIDAE
                                    Malawisaurus
                                    Argentinosaurus
                                    TITANOSAURIDAE
                                    EUTITANOSAURIA
                                        SALTASAURIDAE
                                        [unnamed clade including *Lirainasaurus*]
II. ORNITHISCHIA
    Pisanosaurus
    Technosaurus
    Lesothosaurus
    FABROSAURIDAE
    GENASAURIA
        THYREOPHORA
            Scutellosaurus
            Emausaurus
            THYREOPHOROIDEA
                Scelidosaurus
                EURYPODA
                    STEGOSAURIA

STEGOSAURIA (cont.)
HUAYNAGOSAURIDAE
STEGOSAURIDAE
ANKYLOSAURIA
NODOSAURIDAE
Minmi
ANKYLOSAURIDAE
POLACANTHINAE
SHAMOSAURINAE
?*Tsagantegia*
[unnamed clade including *Saichania* plus *Tarchia* plus *Nodocephalosaurus*]
[unnamed clade including *Euoplocephalus* plus *Ankylosaurus*]
ANKYLOSAURINAE
CERAPODA
ORNITHOPODA
HETERODONTOSAURIDAE
EUORNITHOPODA
HYPSILOPHODONTIDAE
HYPSILOPHODONTINAE
THESCELOSAURINAE
IGUANODONTIA
?*Tenontosaurus*
?*Muttaburrasaurus*
Rhabdodon
EUIGUANODONTIA
Gasparinisaura
DRYOMORPHA
DRYOSAURIDAE
ANKYLOPOLLEXIA
CAMPTOSAURIDAE
[unnamed clade]
IGUANODONTIDAE
HADROSAUROIDEA
?*Bactrosaurus*
?[unnamed clade including *Eolambia* plus
?*Probactrosaurus*]
HADROSAURIDAE
?*Protohadros*
Telmatosaurus
EUHADROSAURIA
"Fontllonga" hadrosaurid
HADROSAURINAE
LAMBEOSAURINAE
MARGINOCEPHALIA
Stenopelix
PACHYCEPHALOSAURIA
HOMALOCEPHALIDAE
PACHYCEPHALOSAURIDAE
CERATOPSIA
?*Chaoyangsaurus*
PSITTACOSAURIDAE
NEOCERATOPSIA
Asiaceratops
?*Microceratops*
?*Graciliceratops*

?*Graciliceratops* (cont.)
PROTOCERATOPSIDAE
Montanoceratops
CERATOPSOMORPHA
Zuniceratops
?*Turanoceratops*
CERATOPSOIDEA
Leptoceratops
CERATOPSIDAE
CENTROSAURINAE
CHASMOSAURINAE

Explanations of Higher Taxa of the Dinosauria

In the original *Dinosaurs: The Encyclopedia*, the chapter on dinosaurian systematics included a section explaining each higher taxon above the level of genus. These explanatory passages included information regarding the clade's author, any junior synonyms referred to it along with their authors, the most recently available diagnosis or definition of that taxon, a generalized description of representatives of the group, the time span and geographic distribution known for it (from either fossil bones or trace fossils, *e.g.*, tracks, and usually excluding "ghost" lineages, *i.e.*, the theorized geological range extensions of taxa before their earliest known occurrence), and, finally, the lower-level taxa (*e.g.*, genera) that have been referred to that taxon.

As that section was quite lengthy and much of its data remained basically the same, a similar section—in order to avoid repetition and to conserve on space—was not included in the first supplementary volume. However, much has changed in the way of dinosaurian systematics since *D:TE* went to press. Many new fossil discoveries have been made, and many new genera and species have been named and described since the publication of that book and also *S1*. The knowledge gleaned from these more recent discoveries has led to a number of reinterpretations by paleontologists of some of these higher clades (*e.g.*, Spinosauridae, see *S1*) since the publication of the first book and also to the introduction of a number of various new groups (*e.g.*, Titanosauriformes, see *S1*).

It seemed that an updated version of that explanatory section was due, although in a slightly "streamlined" (*e.g.*, excluding original authors, junior synonyms, etc.) but, in some ways, expanded form. What follows below is an assemblage of the higher taxa, listed according to their order of appearance in the breakdown, along with pertinent information. The reader is encouraged to compare this section with its quite very different counterpart in the first volume.

Already named genera (excluding those generally regarded as junior synonyms) belonging to higher taxa are listed by themselves after each of the following groups in alphabetical order, all of which have been discussed in their respective entries in the first two books or the present volume. However, unnamed genera belonging to these taxa (*e.g.*, "Herreasaurid A" and "Herrerasaurid B" mentioned in the "Notes" under this volume's *Chindesaurus* entry), even though they may have been figured or briefly described in the literature, are not listed.

Dinosauria

Node-based definition: Including Saurischia, Ornithischia, and all descendants of their common ancestor [with *Eoraptor* and Herrerasauridae accepted here as saurischian theropods rather than basal Dinosauria] (Novas 1996*a*); or [if accepting all birds as theropod dinosaurs], all descendants of the most recent common ancestor of *Triceratops* and modern birds [Aves of Gauthier 1986; Neornithes of other authors] (Holtz, personal communication 2000).

Diagnosis: Temporal musculature extended anterio-medially onto skull roof; cervical vertebrae with epipophyses; deltopectoral crest distally projected; manual digit IV having three or fewer phalanges; perforate acetabulum; brevis shelf on lateroventral side of postacetabular blade of ilium; ischium having slender shaft and ventral keel (obturator process) restricted to proximal third of bone; reduction of tuberosity that laterally bounds ligament of femoral head; femur with prominent anterior (lesser) trochanter; tibia overlapping anteroproximally and posteriorly ascending process of astragalus; calcaneum having concave proximal articular surface (for reception of distal fibular end); distal fourth tarsal proximodistally depressed and triangular-shaped in proximal view (Novas 1996*a*, emended after Galton 1999).

Photograph by John Weinstein, courtesy The Field Museum (neg. #GEO85871-53).

Reconstructed skeleton (including casts of PVSJ 407, 53, and 373) and life sculpture (by Stephen A. Czerkas) of the primitive carnivorous saurischian *Herrerasaurus ischigualastensis*, usually classified as a theropod.

General description: Very small to gigantic terrestrial ornithodiran archosaurs, with erect gait, much variety, various diets, all apparently egg-laying.

Age: Late Triassic to Late Cretaceous.

Geographic distribution: Worldwide.

Taxa: Saurischia plus Ornithischia.

Dinosauria *incertae sedis*: *Aliwalia*, *Alwalkeria*, ?*Macrodontophion*, *Sanpasaurus*, *Teyuwasu*, *Tichosteus*.

Notes: Bonaparte, Ferigolo and Ribeiro (1999) observed that the acetabulum of the primitive saurischian *Guaibasaurus candelariensis* is either only slightly open or entirely closed (see *Guaibasaurus* entry). If the latter interpretation is correct and the condition of the acetabulum in this species is not an artifact of preservation [and if the acetabulum is not open in adults; M. K. Brett-Surman, personal communication 2000], then the diagnosis of the Dinosauria should be emended accordingly.

Novas (1996*a*) had included the feature of at least two sacral vertebrae (sacrals 1 and two, plus an additional dorsosacral vertebra) as a synapomorphy of Dinosauria. Galton (1999), however, identified two distinct morphs of sacra in the prosauropod *Sellosaurus gracilis*, one including three sacral vertebrae and the other only two, this discrepancy interpreted by Galton as a sexual dimorphism. According to Galton, a sacrum with a dorsosacral vertebra plus two sacrals cannot be used to characterize the Dinosauria. Furthermore, a maximum of two reptilian sacral vertebrae is plesiomorphic for Dinosauria, present in Herrerasauridae, in *S. gracilis*, and also in some (but not all) other prosauropods (see *Sellosaurus* entry for more details).

Saurischia

Node-based definition: Modern birds [accepted as members of Theropoda] and all taxa closer to modern birds than to *Triceratops* (Gauthier 1986; Holtz, personal communication 2000).

Diagnosis: Skull with subnarial foramen; wedge-shaped ascending process of astragalus (Sereno 1993); temporal musculature extending onto frontal; quadratojugal overlapping laterally onto caudal process of jugal; elongate caudal cervical vertebrae resulting in relatively longer neck than in other archosaurs; axial postzygapophyses set lateral to prezygapophyses; epipophyses present on cranial cervical postzygapophyses; accessory intervertebral articulations (hyposphene-hypantrum) present in dorsal vertebrae;

manus over 45 percent length of humerus; manus distinctly asymmetrical, digit II longest; proximal ends of metacarpals IV and V lying on palmar surfaces of manual digits III and IV, respectively; pollex heavy, with very broad metacarpal (Gauthier 1986).

General description: Very small to gigantic, much variety, bipedal, semibipedal, and quadrupedal forms, with various diets (including the only carnivorous forms), skull and pelvis relatively remodeled as compared to ornithischians.

Age: Late Triassic to Late Cretaceous.

Geographic distribution: Worldwide.

Taxa: Guaibisauridae plus Theropoda plus Sauropodamorpha.

Saurischia *incertae sedis*: ?*Nyasasaurus* [*nomen nudum*].

GUAIBISAURIDAE

Diagnosis: Dorsal vertebrae with parapophyseal-prezygapophyseal lamina and fossa below it; infrazygapophyseal lamina not divided, pneumatic cavities in dorsal neural arches, hyposphene-hypantrum; centra of sacral vertebrae with disparity in size; scapula slender, expanded distally; iliac part of acetabulum only slightly reduced, very much developed crista supracetabularis; ischia and pubes elongated; femora with small anterior trochanter, lacks trochanteric shelf; calcaneum transversel narrow with pronounced ventromedial process; metatarsal V reduced, lacking phalanges (Bonaparte, Ferigolo and Ribeiro 1999).

General description: Small, very primitive, bipedal forms, probably carnivorous, apparently ancestral both to theropods and sauropodomorphs.

Age: Late Triassic.

Geographic distribution: South America.

Genus: *Guaibasaurus.*

Note: This family was introduced by Bonaparte, Ferigolo and Ribeiro in their description of the primitive saurischian *Guaibasaurus* (see entry). Bonaparte *et al.* determined that guaibasaurids comprise a group of saurischians that may have been ancestral to the Theropoda and Sauropodomorpha. In their analysis, the Herrerasauridae was excluded from the Theropoda.

THEROPODA (EXCLUDING AVES)

Stem-based definition: All saurischians more closely related to birds than to Sauropodomorpha (Currie 1997).

Diagnosis: Reduced overlap of dentary onto base of postdentary bases, reduced mandibular symphysis; lacrimal exposed on skull roof; maxilla with accessory fenestra; vomers rostrally fused; expanded ectopterygoid with ventral fossa; first intercentrum having large fossa and small odontoid notch; second intercentrum with broad, crescent-shaped fossa for reception of first intercentrum; pleurocoel in presacral vertebrae; tail with transition point having marked changes in form of articular processes; distal first carpal enlarged, overlaps bases of first and second metacarpals; manual digit I absent or reduced to vestige; manual digit IV absent or reduced; penultimate manual phalanges elongate; manual digit III with short first and second phalanges; manual unguals enlarged, sharply pointed, strongly recurved, with enlarged flexor tubercles; ilium with long preacetabular process; caudal part of ilium with pronounced brevis fossa; femur cranially convex; fibula closely appressed to tibia, articulated with tibial crest; metatarsus narrow, elongate; pedal digit IV reduced; pedal digit V represented by very strong reduced metatarsal; metatarsal I reduced, not contacting tarsus, attached halfway or further down side of metatarsal II; long bones thin-walled, hollow (Benton 1990, based upon other authors, including Gauthier 1986 and Osmólska 1990).

General description: Very small to giant, usually slender and bipedal, with small to very large head, either possessing sharp teeth or partially or entirely toothless, most of them with reduced (often greatly so) forelimbs, birdlike feet, the first digit of which is reduced and spurlike, some forms feathered or possessing "protofeathers," most genera carnivorous, with also insectivorous, omnivorous, and, most rare, herbivorous forms, the dinosaurian group most likely to have been endothermic, some more derived forms possessing feather-like structures, proto-feathers or feathers.

Age: Late Triassic to Late Cretaceous.

Geographic distribution: Worldwide.

Taxa: ?*Eoraptor* plus ?Herrerasauridae plus Neotheropoda.

Theropoda *incertae sedis*: *Archaeornithoides, Asiamerica, Betasuchus, Calamosaurus, Calamospondylus, Chuandongocoelurus, Coeuluroides, Dandakosaurus, Deinocheirus, Dolicosuchus, Embasaurus, Halticosaurus, Inosaurus, Itemerus, Jubbulpuria,* ?*Likoelesaurus* [*nomen nudum*], *Lukousaurus, Ozraptor, Podokesaurus, Prodeinodon, Protoavis, Pterospondylus, Rapator,* ?*Saltopus, Sinocoelurus, Sinosaurus, Teinurosaurus, Thecoelurus, Thecospondylus, Tugulusaurus,* ?*Velocipes, Wyleia,* ?*Zatomus.*

Notes: This document basically follows Padian, Hutchinson and Holtz's (1997 [abstract], 1999) recently published work in its organization of the Theropoda and in its choice of nomenclature used for the various theropod groups and also the definitions of those names. Padian *et al.* (1999), taking into consideration various other recent studies (*e.g.*, Currie and Padian 1997; Sereno 1997, 1998, 1999*c*) dealing with theropod classification, proposed that standardized stem-based and node-based terms be used for the

Courtesy The Field Museum (neg. #GEO81689).

Detail of a mounted skeleton (FMNH PR308) of the theropod dinosaur *Daspletosaurus torosus* showing the typically saurischian (or "lizard hipped") pelvis.

major theropod groups currently recognized by phylogenetic analyses as relatively stable. Their proposed terminology was "intended to resolve conflict among names applied to the same clade, or sometimes applied without indication of whether the taxon is node-based or stem-based." Only one new name, the node-based Eumanuraptora (see below), was introduced in this study. Based on apparently consensus conclusions, Padian *et al.* (1999) proposed a "'bare bones' classification of nodes and stems," their cladogram excluding taxa (*e.g.*, Abelisauridae, Spinosauroidea, Therizinosauridae, etc.), the phylogenetic position of which these authors regarded as debatable.

Padian *et al.* (1999) further suggested that 1. node- and stem-based taxa are preferable to apomorphy-based taxa because the former are more stable; 2. names extant in the literature should be conserved when possible, although their basis may sometimes require adjustment (*e.g.*, changing stems to nodes, nodes to stems, or converting apomorphy-based taxa to stems and nodes); 3. terms used by other authors can be adjusted to reflect common use or priority; 4. names should generally be based on well-resolved phylogenies; and 5. workers should refrain from introducing new names for theropodan nodes and stems pending the better resolution of the phylogenetic relationships of theropodan sub-clades. Padian *et al.* (1999) recognized that their proposals "will not solve all potential problems in theropod phylogeny," but may offer to them at least a partial solution.

Sereno (1999*b*), however, subsequently criticized Padian *et al.* (1999), as well as other recent authors (*e.g.*, Holtz 1994; Padian 1997*b*) of recently revised dinosaurian phylogenies, pointing out how they differ in several major ways — rationale, priority, and historical interpretation — from phylogenies offered by Sereno (1997, 1998, 1999*c*).

According to Sereno (1999*b*), phylogenies frequently lack in such areas as accuracy and consistency regarding phylogenetic definitions and in the selection of definitional types or reference taxa (rationale); in utilizing a higher taxon simply because of its date of publication, while disregarding the current and often very different definition of that taxon (priority); and in attributing accurate information, including correct phylogenetic definitions and the mention of reference taxa, to cited authors (historical interpretation). A goal of phylogenetic systematics, Sereno (1999*b*) concluded, should be a consensus in taxonomy, which is "best achieved by careful historical interpretation, judicious selection of reference taxa, and an effective configuration of taxonomic definitions."

HERRERASAURIDAE

Diagnosis: Epipophyses of cervical vertebrae prong-shaped; trochanteric shelf on proximal femur; humerus almost 50 percent length of femur; metacarpals I–III with deep extensor pits; metacarpal IV strongly reduced, having fewer than two phalanges; metacarpal V strongly reduced, without phalanges; splenial posterior process spoon-shaped, horizontally placed below angular; angular anteriorly hooked; lacrimal exposed on skull roof; prezygapophyses of distal caudal vertebrae elongate; penultimate manual phalanges elongate; manual unguals of digits 2 and 3 enlarged, compressed, sharply pointed, strongly recurved, with enlarged flexor tubercles; pubis distally enlarged (Novas 1997*b*).

General description: Small (1–5 meters long), obligatorily bipedal, large head, typical theropod teeth, short neck, rather long forelimbs (half length of hindlimbs), elongate hands, first through third finger having large, trenchant claws, relatively long thighs, carnivorous.

Age: Late Triassic.

Geographic distribution: South America.

Genera: *Caseosaurus, Chindesaurus, Herrerasaurus, ?Spondylosoma, Staurikosaurus.*

Notes: In the past, some workers (see *D:TE* and *S1* for references) have regarded the Herrerasauridae as a clade made up of basal saurischians, primitive theropods, or even archosaurs excluded from the Dinosauria. In the previous supplementary volume, the Herrerasauridae was placed under Theropoda as a basal taxon within that group, a position sustained by Sereno (1999*c*) in his most recent cladogram of the Dinosauria.

However, in performing a cladistic analysis of the basal sauropodomorph *Saturnalia*, including in this study *Herrerasaurus* and *Staurikosaurus*, Langer, Abdala, Richter and Benton (1999) got two opposing results, both of which disagree with recent phylogenetic propositions that have placed these latter genera within Theropoda. Langer *et al.* found that either 1. *Herrerasaurus* and *Staurikosaurus* are nondinosaurian genera, forming together a monophyletic Herrerasauridae that is the sister group of Dinosauria, or that 2. these genera constitute successive sister taxa of Theropoda plus Sauropodomorpha. The authors pointed out, however, that these results could be biased, as they took into account only morphological features available for *Saturnalia*, these excluding those of the skull and hand.

Bonaparte, Ferigolo and Ribeiro (1999), in their analysis of the primitive genus *Guaibasaurus*, found that the Herrerasauridae "appears to a sterile lineage [outside of Theropoda] with precocious derived characters."

More recently, Kuznetsov and Snnikov (2000), in their study on the function of a perforated acetabulum in archosaurs and birds, regarded, without giving evidence supporting their opinion, that the herrerasaurids (including *Herrerasaurus* and *Staurikosaurus*) are not dinosaurs but rather advanced dinosauromorph "thecodonts."

In the present volume, therefore, the Herrerasauridae occupies a tentative position within Theropoda, although this placement may change again based upon future analyses.

NEOTHEROPODA

Stem-based definition: All theropods excluding *Eoraptor* and Herrerasauridae [if these two taxa are accepted as theropods] (Padian, Hutchinson and Holtz 1997).

General description: Most theropod forms (excluding the most primitive theropods) very small to giant.

Age: Late Triassic to Late Cretaceous.

Geographic distribution: Worldwide.

Taxa: "Ceratosauria" plus Tetanurae.

Neotheropoda *incertae sedis*: *Genyodectes.*

Note: Following Holtz (1994), most workers divided the Neotheropoda into the Ceratosauria plus Tetanurae, with the Ceratosauria broken down into the clades Coelophysoidea and Neoceratosauria. However, this conception of the Neotheropoda is currently being questioned (see "Notes" under "Ceratosauria"). For example, for more than half a decade Currie (1995*b*) has argued that the Ceratosauria is a paraphyletic clade definable only by plesiomorphic characters if it includes *Ceratosaurus*, a genus which that author has interpreted to be the sister taxon to the

Carnosauria (see *D:TE, S1*). Also, Currie (1997) found the Neoceratosauria to be only weakly supported. According to Currie (1997), the perceived synapomorphies of this group concerning the premaxilla and the quadrate, the latter resulting in the large infratemporal fenestra, may be related to size rather than having diagnostic significance. Another purported synapomorphy, the femoral head directed anteromedially rather than medially (as in tetanurans), is a feature common to all theropods that have been classified as "ceratosaurians."

"CERATOSAURIA"

Stem-based definition: All neotheropods closer to *Ceratosaurus* than to birds (Padian, Hutchinson and Holtz 1999).

Diagnosis: Cervical vertebrae having two pairs of pleurocoels; perforation of pubis plate by two fenestrae; fusion of sacral vertebra and ribs (in adults); fusion of astragalus and calcaneum (in adults); flange of distal end of fibula that flares medially to overlap ascending process of astragalus anteriorly (Hutchinson 1997).

General description: Small to very large, skull lacking maxillary fenestration, hand with four fingers and fourth digit shorter than the others (primitive features), some forms possessing cranial crests or horns.

Age: Late Triassic to Late Cretaceous.

Geographic distribution: North America, South America, Europe, Asia, Africa.

Taxa: Coelophysoidea plus Neoceratosauria.

"Ceratosauria" *incertae sedis*: *Eucoelophysis*, *Genusaurus*, *Sarcosaurus*.

Notes: Most workers have long accepted "Ceratosauria" (the present writer's quotes) as a valid taxon. Currie (1995*b*), however, regards "Ceratosauria" as definable only by plesiomorphic characters if it includes *Ceratosaurus*, a genus regarded by the author as the sister taxon to the Carnosauria (see *D:TE, S1*; also, see Currie's 1997 review of the "Ceratosauria").

Rauhut (1998; see *Elaphrosaurus* entry), in a brief report, found "Ceratosauria" to be a paraphyletic assemblage, "with Abelisauroidea and *Ceratosaurus* representing successively closer outgroups to Tetanurae." Britt and Sampson (1999), in performing their own preliminary phylogenetic reanalysis of the "Ceratosauria"—aided in their study by new abelisaurid specimens, reexamination of basal tetanurans, and no taxa above the level of genus—also found it to be a paraphyletic assemblage. Therefore, "Ceratosauria" was rejected by them as a valid clade, with the taxa (*i.e.*, Coelophysoidea, Abelisauroidea, and Ceratosauridae) formerly included therein positioned serially outside of Tetanurae. Britt and Sampson noted that, if their conclusions are correct, then 1. a major reconsideration of theropod evolution is in order; 2. long abelisaurid and tetanuran ghost lineages are reduced, these taxa being better sampled than previously suspected; and 3. abelisaurid radiation is no longer a Cretaceous counterpart of earlier coelo-

Skull (USNM collection) of a baby *Coelophysis bauri*, a coelophysoid "ceratosaur," collected from the Ghost Ranch quarry in New Mexico.

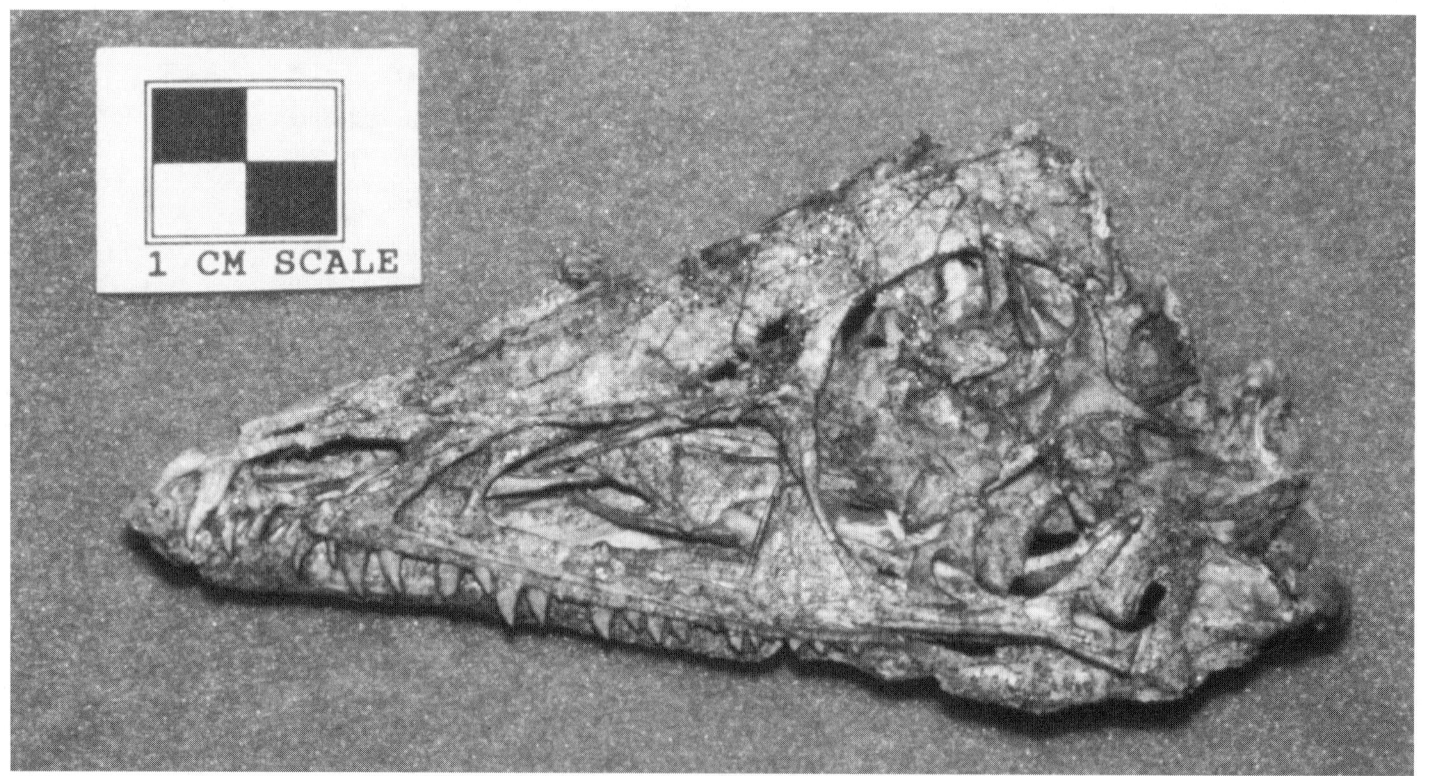

Photograph by Michael K. Brett-Surman, courtesy National Museum of Natural History, Smithsonian Institution.

Photograph by the author.

Reconstructed holotype (MWC 1) skull cast by DINOLAB of the ceratosaurid *Ceratosaurus magnicornis*, exhibited at Dinofest (2000–2001), Navy Pier, Chicago.

physoid evolution. Furthermore, their findings should "also have significant implications for patterns of homoplasy, biogeography, and body size changes during theropod evolution."

More recently, Madsen and Welles (2000), in their monograph on *Ceratosaurus*, accepted "Ceratosauria" as a valid assemblage. These authors — the main focus of their paper being a revised osteology of *Ceratosaurus*— did not, however, define or diagnose this group, nor did they address recent cladistic studies by other workers (*e.g.*, Padian, Hutchinson and Holtz 1999) including those who consider the "Ceratosauria" to be polyphyletic (see above). Madsen and Welles divided "Ceratosauria" into the superfamilies Podokesauroidea (see "Notes" under Coelophysidae, below) and Neoceratosauroidea. They then divided Podokesauroidea into two families, the Podokesauridae and Dilophosauridae, and these, respectively, into the (redundant) subfamilies Podokesaurinae and Dilophosaurinae. Only one family, Ceratosauridae, was (redundantly) included by them in Neoceratosauroidea.

Subsequently, Britt, Chure, Holtz, Miles and Stadtman (2000), in an abstract considering the phylogenetic affinities of *Ceratosaurus* (see entry), based in part upon newly collected specimens, found through phylogenetic analyses (PAUP and NONA) that *Ceratosaurus* (and Ceratosauridae) remains the sister taxon to the Abelisauridae within Neoceratosauria. However, these authors noted, "tetanurine-like features in *Ceratosaurus* and abelisaurs significantly weaken the union of coelophysoids and neoceratosaurs," with trees only two steps longer than the most parsimonious ones placing the latter "closer to Tetanurae than to Coelophysoidea."

COELOPHYSOIDEA

Stem-based definition: All "ceratosaurs" closer to *Coelophysis* than to *Ceratosaurus* (Padian, Hutchinson and Holtz 1999).

Diagnosis: Unambiguous synapomorphies including subnarial gap (indicating possibly mobile premaxilla-maxilla joint); reduction of axial parapophysis; loss of diapophysis; loss of axial pleurocoel (Holtz 1994).

General description: Small to medium-sized, lightly built, gracile, rather elongated head, some more specialized forms bearing cranial crests, moderately long neck, short and slender forelimb with four-fingered hand, foot birdlike with three functional toes.

Age: Late Triassic to Late Jurassic.

Geographic distribution: North America, Europe, Africa, ?Asia.

Taxa: Coelophysidae plus Dilophosauridae.

Notes: In *Predatory Dinosaurs of the World*, Paul (1988*b*) introduced the family Coelophysidae, a taxon (founded upon the genus *Coelophysis*) which has since been accepted by most workers. Colbert (1964*a*) had previously synonymized *Podokesaurus* and *Coelophysis*, but later (Colbert 1989) acknowledged that this synonymy could not be adequately demonstrated, as the holotype and only specimen of the type species *Podokesaurus holyokensis* had been destroyed.

Following Colbert (1989), Madsen and Welles (2000) reinstated the family name Podokesauridae (Huene 1914), noting that it has priority over Coelophysidae. Consequently, Madsen and Welles replaced Coelophysoidea with Podokesauroidea, which in turn included the subfamily Podokesaurinae. However, various paleontologists (*e.g.*, Olsen 1980; Padian 1986; Rowe and Gauthier 1990; Norman 1990) have shown that *P. holyokensis* cannot be positively classified below Theropoda *incertae sedis*. Therefore, this document retains the name Coelophysoidea for this group of theropods (see *Podokesaurus* entry, *D:TE* for more details) and also the familial name Coelophysidae.

COELOPHYSIDAE

Diagnosis: Unambiguous synapomorphies including pronounced subnarial gap (indicating possible mobile premaxilla-maxilla joint); reduced axial parapophysis; loss of diapophysis; loss of axial pleurocoels (Holtz 1994).

General description: Small forms.

Age: Late Triassic to Early Jurassic.

Geographic distribution: North America, Europe, Africa.

Genera: *Camposaurus*, *Coelophysis*, ?*Gojirasaurus*, *Liliensternus*, *Syntarsus*.

Coelophysoidea *incertae sedis*: ?*Procompsognathus*.

DILOPHOSAURIDAE

Diagnosis: Medium-sized; skull high; parassagital crest; jugal large, deep; parapophyses on short stems; supraacetabular crest of ilium broad, notched anteriorly (Madsen and Welles 2000).

General description: Medium-sized, bearing cranial crests.

Age: Early Jurassic.

Geographic distribution: North America, Asia.

Genus: *Dilophosaurus*.

Note: The above diagnosis by Madsen and Welles (2000) was for the subfamily Dilophosaurinae. As this was the only subfamily included in their family Dilophosauridae, the diagnosis is used here for that family.

NEOCERATOSAURIA

Stem-based definition: All "ceratosaurs" closer to *Ceratosaurus* than to *Coelophysis* (Padian, Hutchinson and Holtz 1999).

Diagnosis: Unambiguous synapomorphies including femoral head directed anteromedially; premaxilla very deep subnarially; parietal projected dorsally; infratemporal fenestra very large; articulation of quadrate projected deeply posteroventrally (Holtz 1994).

General description: Small to large-sized, massively built, large head, short neck, short forelimbs, heads bearing horns or crests.

Age: Late Jurassic to Late Cretaceous. Geographic distribution: North America, South America, Europe, Asia, Africa.

Taxa: Ceratosauridae plus Abelisauroidea.

Note: According to Currie (1997), Neoceratosauria remains only weakly supported; the perceived synapomorphies regarding the premaxilla and the quadrate, the latter resulting in the large infratemporal fenestra, may be related to size rather than having diagnostic significance. Furthermore, another purported synapomorphy, the femoral head directed anteromedially rather than medially (as in tetanurans), is a feature common to all "ceratosaurians."

CERATOSAURIDAE

Diagnosis: [No modern diagnosis published.]

General description: Large forms with median nasal horns, bony protuberances along midline of back and probably tail, fused ankles.

Age: Late Jurassic to ?Early Cretaceous.

Geographic distribution: North America, ?Europe, Africa, ?Asia.

Genus: *Ceratosaurus*.

Note: Madsen and Welles (2000) tentatively included in the Ceratosauridae the European genus *Proceratosaurus* but gave no reasons for this referral. Most workers, however, regard this genus as a coelurosaur of uncertain affinities (see *D:TE*).

ABELISAUROIDEA

Diagnosis: [None currently accepted.]

General description: Most "ceratosaurians," small to medium-sized, heads with horns or crests.

Age: Middle to Late Cretaceous.

Geographic distribution: South America, Asia, Africa.

Taxa: *Ilokelesia* plus ?Noasauridae plus Abelisauridae.

Note: According to Rowe, Tykowski and Hutchinson (1997), in a review of the "Ceratosauria," the characters used by Holtz (1994; see *D:TE*) to diagnose

Abelisauroidea are homoplastic with many tetanturine theropods; consequently, the monophyly of the Abelisauridae should be regarded with caution. These authors also noted that fragmentary taxa including *Noasaurus* may also belong to the Abelisauridae (see also Currie 1997).

NOASAURIDAE

Partial diagnosis: Quadrate and quadratojugal unfused; also [unstated] differences between maxilla and cervical ribs of noasaurids and those of *Abelisaurus* and *Carnotaurus* (Bonaparte 1996*b*).

General description: Small to medium-sized forms.

Age: Early to Late Cretaceous.

Geographic distribution: South America.

Genera: *Laevisuchus*, ?*Ligabueino*, *Noasaurus*.

Note: Some paleontologists (*e.g.*, Currie 1997) regard *Noasaurus*, the fragmentary type genus of the family Noasauridae, as a possible abelisaurid.

ABELISAURIDAE

Node-based definition: *Carnotaurus*, *Abelisaurus*, and all descendants of their most recent common ancestor (Padian, Hutchinson and Holtz 1999, following Bonaparte and Novas 1985).

Diagnosis: Premaxilla cranicaudally short and deep; snout dorsoventrally deep at level of narial openings; frontals dorsoventrally thickened, resulting in dorsal bulking; paired hornlike structures or domelike prominence; posterior surface of basioccipital wide, smooth below occipital condyle; dentary short, having convex ventral margin; loose contacts among dentary, splenial, and postdentary bones (Novas 1997*a*).

General description: Large, with a high and broad head, rather short face, large eyes, and cranial adornment in the form of horns or crests, very short forelimbs.

Age: Middle to Late Cretaceous.

Geographic distribution: South America, Asia, Africa.

Genera: *Abelisaurus*, *Carnotaurus*, ?*Elaphrosaurus*, *Indosaurus*, *Indosuchus*, *Majungatholus*, *Tarascosaurus*, ?*Velocisaurus*, ?*Xenotarsosaurus*.

Note: Currie (1997) found the monopoly of Abelisauridae to be weakly supported, noting that, of all the genera referred to this family, only *Carnotaurus* was (then) known from a relatively complete skeleton and that, "To date, only phenetic resemblances [those arising in response to environmental stimuli] have been offered in support of the recognition of Abelisauridae."

TETANURAE

Stem-based definition: All theropods closer to birds than to *Ceratosaurus* (Holtz and Padian 1995).

Diagnosis: Skull with increased pneumaticity; manus reduced to three digits; fibula reduced, clasped by tibia; anteriorly placed horizontal groove on astragalar condyles; maxillary tooth row restricted to position anterior to orbit; ischium with obturator notch (Hutchinson and Padian 1997*b*, based on Gauthier 1986, Holtz 1994, Sereno, Wilson, Larsson, Dutheil and Sues 1994).

General description: Small to giant, wide range of diversity, posterior part of tail having reduced flexibility.

Age: ?Late Triassic to Late Cretaceous.

Geographic distribution: Worldwide.

Taxa: Spinosauroidea plus Avetheropoda.

Tetanurae *incertae sedis*: *Bruhathkayosaurus*, *Chilantaisaurus*, *Coelurus*, *Dryptosauroides*, *Gasosaurus*, *Iliosuchus*, *Kaijiangosaurus*, *Kelmayisaurus*, *Labocania*, *Orthogoniosaurus*, *Ornithomimoides*, ?*Proceratosaurus*, *Shanshanosaurus*, ?*Unquillosaurus*, *Walgettosuchus*.

SPINOSAUROIDEA

Diagnosis: Jaws elongate, especially in prenarial region; greater or lesser tendency (greater in upper than lower jaw) to develop terminal rosette; lower jaw turned upwards at extreme anterior end, constricted transversely just posterior to it; increase from five to seven premaxillary teeth; teeth showing reduction in compression of whole tooth in 1. labio-lingual direction, 2. recurvature of crown, and 3. size of denticles on anterior and posterior carinae (Charig and Milner 1997).

General description: Large to giant, highly derived forms, head with narrow jaws, more teeth than in other theropods, some genera having elongated neural spines, some with relatively long necks, some possibly quadrupedal.

Age: Early to Late Cretaceous.

Geographic distribution: Africa, Europe, South America, ?Asia.

Taxa: "Megalosauridae" plus Eustreptospondylidae plus Spinosauridae.

"MEGALOSAURIDAE"

Diagnosis: [No modern diagnosis published; see *D:TE*, "Systematics" chapter.]

General description: Now usually regarded as an unnatural grouping of medium-sized to large forms apparently superficially resembling allosaurs, head with long jaws, double-edged teeth, very short forelimbs with three-fingered hands, toes with large talons.

Age: ?Early Jurassic to Early Cretaceous.

Geographic distribution: North America, Europe, Africa, Asia.

Genera: *Afrovenator*, *Altispinax*, *Edmarka*, *Magnosaurus*, *Megalosaurus*, ?*Metriacanthosaurus*, ?*Piveteausaurus*, *Poekilopleuron*, *Torvosaurus*, ?*Wakinosaurus*, ?*Xuanhanosaurus*.

Skeletal cast of the abelisaurid theropod *Carnotaurus sastrei*, based on holotype MACN-CH 894.

Photograph by the author, courtesy Natural History Museum of Los Angeles County Museum.

Note: R. E. Molnar (personal communication 2000) does not agree with even this tentative referral of *Wakinosaurus* and *Xuanhanosaurus* to the "Megalosauridae" (see individual entries, *D:TE*, for other opinions). Also, Molnar regards *Pivetausaurus* as a probable junior synonym of *Poekilopleuron*.

EUSTREPTOSPONDLYDAE

Diagnosis: Skull low, long; frontals long; postfrontals shifted forward; dorsal vertebrae having long centra, more opisthocoelous than in "megalosaurids" and allosaurids; pleurocoels more extensively developed (Paul 1988*b*).

Photograph by the author.

Reconstructed skull (cast) by DINOLAB of the giant "megalosaurid" *Torvosaurus tanneri*, exhibited at Dinofest (2000–2001, held at Chicago's Navy Pier.

General description: Medium-sized, rather lightly-built forms, some having long heads with low snouts, some with rather high neural spines.

Age: Middle to Late Jurassic.

Geographic distribution: Europe.

Genus: *Eustreptospondylus*.

Note: The genus *Becklespinax* was previously classified with the Eustreptospondylidae, an assignment not accepted by R. E. Molnar (personal communication 2000). As the vertebrae in this genus most closely match those of the ?allosauroid *Piatnitzkysaurus* (see Olshevsky 1991), *Becklespinax* is herein tentatively listed as an allosauroid close to that genus.

SPINOSAURIDAE

Diagnosis: Teeth having sub-circular transverse sections; external nares displaced backwards; seven premaxillary teeth; anterior portion of lower margin of upper jaw concave, conforming to convexity of dentaries; anterior portion of upper and lower jaws laterally expanded (Kellner and Campos 1996).

General description: Large to giant forms, with long-snouted head resembling that of a crocodilian, teeth more numerous than in other theropods, some forms with relatively long neck, three-fingered hand, some possessing elongated neural spines that apparently supported a "hump" or possibly "sail," apparently piscivorous, some possibly semi- or fully quadrupedal.

Age: Early to Late Cretaceous.

Geographic range: Africa, South America, ?Asia.

Taxa: Baryonychinae plus Spinosaurinae.

Note: Taquet and Russell (1998) accepted three spinosaur families — Spinosauridae (Stromer 1915), Baryonychidae (Charig and Milner 1986), and Irritatoridae (Martill, Cruickshank, Frey, Small and Clarke 1996) — commenting that the "morphology of spinosaurs is too poorly understood at present to attempt to assess the validity of the latter two."

BARYONYCHINAE

Stem-based definition: All spinosaurids more closely related to *Baryonyx* than to *Spinosaurus* (Sereno, Beck, Dutheil, Gado, Larsson, Lyon, Marcot, Rauhut, Sadleir, Sidor, Varricchio, Wilson and Wilson 1998).

General description: Medium-sized to giant

Photograph by Allen A. Debus.

Skeleton (cast, reconstructed from holotype UC DBA 1, missing elements from *Allosaurus fragilis*) of the "megalosaurid" theropod *Afrovenator abakensis* in 1994, under the direction of Paul C. Sereno, temporarily displayed at the Harold Washington Library Center in Chicago.

forms, hand with especially large claw, more primitive than spinosaurines.

Age: Early to ?Late Cretaceous.

Geographic distribution: Europe, Africa.

Genera: *Baryonyx, Suchomimus.*

SPINOSAURINAE

Stem-based definition: All spinosaurids more closely related to *Spinosaurus* than to *Baryonyx* (Sereno, Beck, Dutheil, Gado, Larsson, Lyon, Marcot, Rauhut, Sadleir, Sidor, Varricchio, Wilson and Wilson 1998).

General description: Small to giant forms.

Age: Middle to Late Cretaceous.

Geographic distribution: Africa, South America, ?Asia.

Genera: *Irritator, ?Siamosaurus, Spinosaurus.*

AVETHEROPODA

Node-based definition: The most recent common ancestor of birds and *Allosaurus* and all descendants of that ancestor (Padian, Hutchinson and Holtz 1999).

Diagnosis: Unambiguous synapomorphies including loss of obturator foramen; proximally situated lesser trochanter; basal half of metacarpal I closely appressed to II; cnemial process arising from lateral surface of tibial shaft; pronounced pubic "boot"; coracoid tapers posteriorly; premaxillary symphysis U-shaped; premaxillary tooth crowns asymmetrical in cross-section (Holtz 1994).

General description: Very small to giant, wide range of diversity and specialization.

Age: ?Late Triassic to Late Cretaceous.

Geographic distribution: North America, South America, Europe, Africa, Australia, Asia, ?Antarctica [see *D:TE*, p. 48, footnote under AVETHEROPODA].

Taxa: Carnosauria plus Coelurosauria.

Avetheropoda *incertae sedis*: *Antrodemus, Bagaraatan, Bradycneme*

CARNOSAURIA

Stem-based definition: All avetheropods closer to *Allosaurus* than to birds (Padian, Hutchinson and Holtz 1999).

Diagnosis: Unambiguous characters including: horizontal groove across anterior face of condyles; cervical vertebrae strongly opisthocoelous; pleurocoels on cervical centra, sometimes on anterior dorsals; adductor fossa on distomedial surface of femoral shaft; sharply defined mediodistal crest; ridge for cruciate ligaments in flexor groove of femur (Azuma and Currie 2000).

Photograph by the author, courtesy University Museum, Oxford.

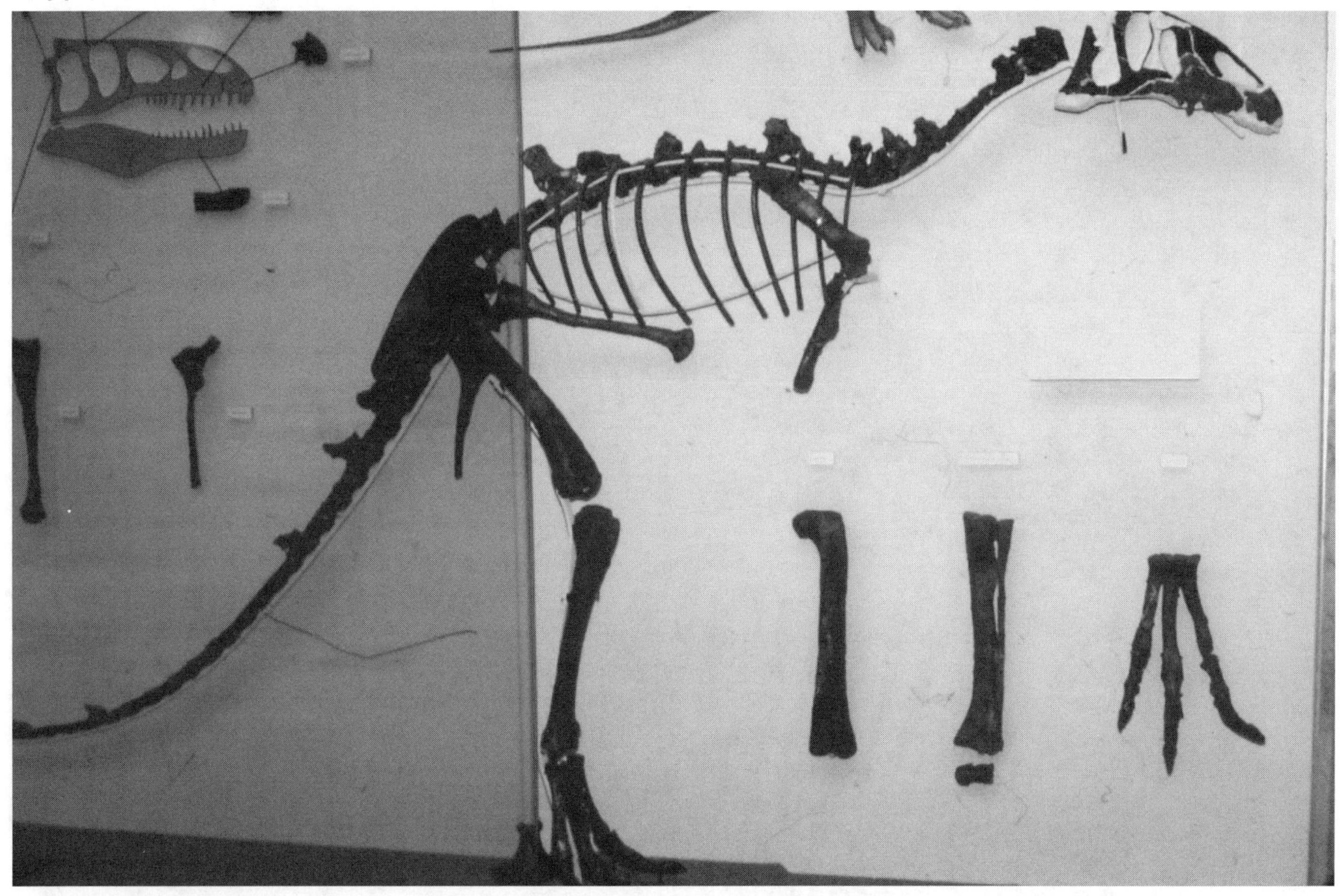

Subadult incomplete skeleton (OUM J13558, holotype of *Megalosaurus cuvieri*) of the eustreptospondylid tetanuran *Eustreptospondylus oxoniensis*.

General description: Large to giant, some heavily built, heads large, forelimbs reduced, forelimbs powerfully constructed in some taxa.

Age: Middle Jurassic to Late Cretaceous.

Geographic distribution: ?Worldwide.

Taxa: ?*Cryolophosaurus* plus *Monolophosaurus* plus Allosauroidea.

ALLOSAUROIDEA

Node-based definition: Including *Allosaurus*, *Sinraptor*, and all descendants of their most recent common ancestor (Padian, Hutchison and Holtz 1999).

Diagnosis: Nasal participating in antorbital fossa; flange-shaped lacrimal process on palatine; basioccipital excluded from basal tubera; articular having pendant medial process (Sereno, Wilson, Larsson, Dutheil and Sues 1994).

General description: Large to giant, head large, neck short, short forelimbs with three fingers and trenchant claws.

Age: Late Jurassic to Late Cretaceous.

Geographic distribution: North America, South America, Europe, Africa, Asia, Australia.

Taxa: ?*Piatnizkysaurus* plus ?*Becklespinax* plus ?*Szechuanosaurus* plus Sinraptoridae plus Allosauridae plus ?Carcharodontosauridae.

Allosauroidea *incertae sedis*: *Neovenator*.

Note: Holtz (1995) regarded *Piatnizkysaurus* and *Szechuanosaurus* as possible allosauroids outside of Allosauridae.

SINRAPTORIDAE

Stem-based definition: *Sinraptor* and all allosauroids [excluding carcharodontosaurids] closer to it than to *Allosaurus* (Padian, Hutchison and Holtz 1999).

Diagnosis: Allosauroids more primitive than allosaurids in that the quadrate is more elongate, paroccipital processes not as strongly down-turned, and distal end of scapula not expanded; autapomorphies including: presence of two or more sets of accessory openings in antorbital fossa connected to maxillary sinus; low lacrimal horn; ventral margin of axial intercentrum rotated above ventral margin of axial centrum (Currie and Zhao 1993*a*).

General description: Large, with relatively low "lacrimal horns."

Age: Late Jurassic.

Geographic distribution: Asia.

Photograph by the author.

Skeleton of the giant sinraptorid allosauroid *Yangchuanosaurus shangyouensis* mounted at Dinofest in 1998.

Genera: *Sinraptor*, ?*Lourinhanosaurus*, *Yangchuanosaurus*.

ALLOSAURIDAE

Stem-based definition: *Allosaurus* and all allosauroids [excluding carcharodontosaurids] closer to it than to *Sinraptor* (Padian, Hutchison and Holtz 1999).

Diagnosis: Unambiguous synapomorphies including pubic "boot" longer anteriorly that posteriorly, triangular (apex posterior) in ventral aspect (Holtz 1994).

General description: Large- to giant-sized, more derived than other sinraptoroids, with prominent lacrimal "horns."

Age: Late Jurassic to Late Cretaceous.

Geographic distribution: North America, Asia.

Genera: *Acrocanthosaurus*, *Allosaurus*, ?*Compsosuchus*, *Fukuiraptor*, ?*Valdoraptor*.

CARCHARODONTOSAURIDAE

Diagnosis: Large allosauroids having pleurocoelous proximal caudal vertebrae (Rauhut 1995).

General description: Large to giant forms, including some of the largest known theropods.

Age: Early to Late Cretaceous.

Geographic distribution: North America, South America, Africa.

Genera: ?*Bahariasaurus*, *Carcharodontosaurus*, *Giganotosaurus*.

Note: Currie and Carpenter (2000), in performing a cladistic analysis to determine the phylogenetic placement of the genus *Acrocanthosaurus*, could not determine whether or not the Carcharodontosauridae—including the genera *Giganotosaurus* and *Carcharodontosaurus*—should be included within the Allosauroidea.

COELUROSAURIA

Stem-based definition: All avetheropods closer to birds than to *Allosaurus* (Padian, Hutchinson and Holtz 1999).

Diagnosis: Unambiguous synapomorphies including ischium only two-thirds or less length of pubis; loss of ischial "foot"; expanded circular orbit;

ischium having triangular obturator process; ascending process of astragalus greater than one-fourth length of epipodium; 15 or fewer caudal vertebrae with transverse processes (Holtz 1994).

General description: Very small to giant, with large eyes, birdlike feet, much variation, including carvorous, insectivorous, possibly herbivorous and omnivorous forms, some forms possessing protofeathers or feathers.

Age: ?Late Triassic to Late Cretaceous.

Geographic distribution: North America, South America, Europe, Asia, Africa.

Taxa: Compsognathidae plus ?Dryptosauridae plus Manuraptoriformes.

Coelurosauria *incertae sedis*: *Deltadromeus*, ?*Diplotomodon*, *Kakuru*, *Marshosaurus*, ?*Megaraptor*, *Ngexisaurus* [*nomen nudum*], *Nqwebasaurus*.

Notes: Holtz (1999), in a preliminary report utilizing a "new data matrix, culled from a much larger analysis," performed a new phylogenetic analysis of the Coelurosauria which yielded a single most parsimonious tree in which tyrannosaurs is the sister taxa to manuraptorans, troodontids the sister taxon to other manuraptorans, and therizinosauroids sister taxon to oviraptorosaurs. Differences between this analysis and other previously proposed analyses were the results of "different character choice and coding, and in part to the effect on character distribution by other taxa not included in this study."

Kakuru, formerly classified as Theropoda *incertae sedis*, was regarded by Molnar (personal communication 2000) as a coelurosaur.

COMPSOGNATHIDAE

Diagnosis: Premaxillary teeth unserrated, maxillary teeth serrated; neural spines of dorsal vertebrae fan-shaped; caudal transverse processes rudimentary or absent; powerful manual phalanx I-1 (diameter of shaft greater than that of radius); public boot having limited anterior expansion; ischium with prominent obturator process (Chen, Dong and Zhen 1998).

General description: Very small forms, hand with at least two and possibly three functional fingers, with third finger reduced, at least some forms possessing fibrous structures that have been interpreted by some paleontologists as "protofeathers."

Age: Late Jurassic to ?Early Cretaceous.

Geographic distribution: Europe, Asia.

Genera: *Aristosuchus*, *Compsognathus*, *Sinosauropteryx*.

Reconstructed skeleton of *Deltadromeus agilis* (including cast of holotype SGM-Din 2), a coelurosaurian theropod from Africa, prepared and mounted under the direction of paleontologist Paul C. Sereno at the Crerar Library, University of Chicago.

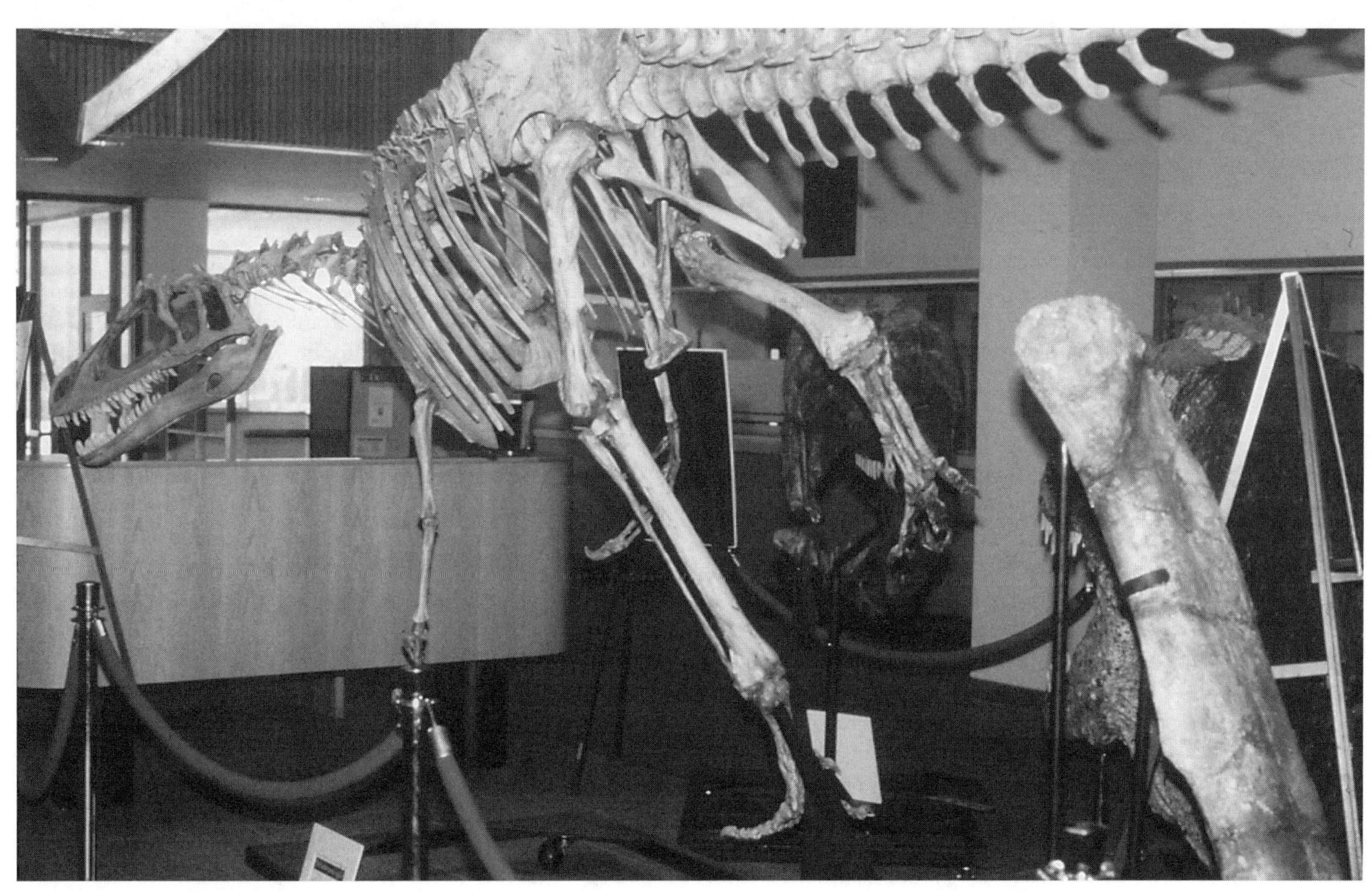

Photograph by the author, courtesy University of Chicago.

DRYPTOSAURIDAE

Diagnosis: Maxillary teeth having gaps between serrations equal to half width of a serration; blood grooves ["perpendicular to the longitudinal axis of the tooth"] of maxillary teeth not extending onto crown; dentary slender; humerus with well-developed deltopectoral crest occupying proximal one-third of shaft; manus ungual I-2 very large relative to phalanx I-1, (exceeding *Baryonyx* in ungual/humeral ratio); phalanx I-1 straight; other manual phalanges relatively straight; hindlimb gracile; lesser trochanter lower than femoral head, spike-like; femur lacking deep intercondylar groove anteriorly; no deep groove separating cnemial crest and fibular condyle; astragalus having bulbous medial condyle; ascending process of astragalus probably tall, centrally located; metatarsal not of arctometatarsal type (Carpenter, Russell, Baird and Denton 1997).

General description: Very large, relatively large forelimbs, large talon-like claws.

Age: Late Cretaceous.

Geographic distribution: North America.

Genus: *Dryptosaurus*.

MANURAPTORIFORMES

Node-based definition: *Ornithomimus*, birds, and all descendants of their most recent common recent ancestor (Holtz 1996).

Diagnosis: Ulna bowed posteriorly; metacarpal III long, slender; posterodorsal margin of ilium curving ventrally in lateral aspect; zygapophyses of cervical vertebrae flexed; jugal located on rim of antorbital fenestra; metacarpal I one-third length of II (Holtz 1994, based on Gauthier 1986).

General description: Very small to very large birdlike theropods, much variety, at least some forms possessing feathers.

Age: ?Late Triassic to Late Cretaceous.

Geographic distribution: North America, South America, Europe, Asia, Africa.

Taxa: Arctometatarsalia plus Oviraptorosauria plus Therizinosauroidea plus Manuraptora.

Manuraptoriformes *incertae sedis*: ?*Nedcolbertia*, ?*Santanaraptor*, *Scipionyx*.

Notes: Phylogenetic analysis by Makovicky and Sues (1998) found that a sister-taxon relationship between Oviraptorosauria and Therizinosauroidea is supported by the following unambiguous cranial and postcranial apomorphies: premaxilla edentulous; basipterygoid processes reduced; symphyseal region of dentary inflected medially toward midline; distal caudal vertebrae short; posterodorsal "lip" on proximal end of manual unguals; and preacetabular portion of ilium deep.

According to these authors, ambiguous apomorphies linking these taxa include the following: no accessory maxillary fenestra; lateral depression in middle ear region on braincase; cervical vertebrae having two pairs of pneumatic foramina; all presacral vertebrae pneumatized; sacral centra decreasing in width toward posterior end of series; manual phalanx III-3 shorter than III-1; broad fossa for origin of M. cuppedicus; pubic boot anteriorly expanded.

This book continues to use the spelling "Manuraptora" (and, consequently, "Manuraptoriformes" and Eumanuraptora), which was corrected in 1997 from the older "Maniraptora" (hence, "Manuraptoriformes") by Charig and Milner (see *S1*). However, most authors continue to use the incorrectly derived spellings.

ARCTOMETATARSALIA

Stem-based definition: *Ornithomimus* and all coelurosaurs closer to that genus than to birds (Holtz 1996).

Diagnosis: Unambiguous synapomorphies including tibia and metatarsus elongate; metatarsals deeper anteroposteriorly than mediolaterally; arctometatarsus; metatarsus gracile; reversal of flexed cervical zygapophyses (Holtz 1994).

General description: Small to giant, feet birdlike, with the upper end of the third metatarsal "pinched," includes toothed and toothless forms.

Age: Late Cretaceous.

Geographic distribution: North America, Asia.

Taxa: Tyrannosauridae plus Ornithomimosauria.

Arctometatarsalia *incertae sedis*: *Avimimus*, *Stokesosaurus*.

TYRANNOSAURIDAE

Diagnosis: Head large relative to body size; neck, trunk, and forelimbs short; nasals very rugose, often fused along midline, constricted between lacrimals; frontals excluded from orbits by lacrimals; jugal pierced by large foramen; premaxillary teeth D-shaped in cross-section, carina along posteromedial and posterolateral margins; vomer with diamond-shaped process between maxillae; ectopterygoid with large ventral opening; surangular foramen large; centra of cervical vertebrae slightly opisthocoelous, anteroposteriorly compressed, broad; scapula long, slender; humerus with weakly- to moderately well-developed deltopectoral crest; manus reduced to two functional digits (II and III), digit IV deduced to splint; pubis having well-developed anterior "foot"; distal end of ischium not expanded; astragalus with broad, tall ascending ramus (Carpenter 1992).

General description: Small- to giant-sized forms (including some of the largest carnivorous dinosaurs, although perhaps larger nontyrannosaurid genera have been described), heads very large proportionally, forelimbs extremely small though muscular and apparently strong in some forms, hand with two functional fingers, a splintlike third digit present in some less derived genera, most known forms having serrated premaxillary teeth.

Age: Early to Late Cretaceous.

Geographic distribution: North America, Asia.

Taxa: ?Aublysodontinae plus Tyrannosaurinae.

Tyrannosauridae *incertae sedis*: *Shanyangosaurus*, ?*Siamotyrannus*.

AUBLYSODONTINAE

Diagnosis: Premaxillary incisiform teeth lacking serrations (Holtz 1997).

General description: Small to medium-sized forms, only known significant differences from other tyrannosaurids being the possession of unserrated incisiform teeth.

Age: Late Cretaceous.

Geographic distribution: North America, Asia.

Genera: *Alectrosaurus*, *Aublysodon*, *Stygivenator*.

TYRANNOSAURINAE

Diagnosis: Tyrannosaurids having serrated premaxillary teeth (Holtz 1997).

General description: The majority of tyrannosaurids, distinguished from aublysodontines by the possession of serrated teeth, large to giant.

Genera: *Albertosaurus*, *Alioramus*, ?*Chingkankousaurus*, *Daspletosaurus*, *Deinodon*, *Gorgosaurus*, *Tyrannosaurus*.

ORNITHOMIMOSAURIA

Node-based definition: ?*Pelecanimimus*, *Ornithomimus*, and all descendants of their most recent common ancestor (Padian, Hutchison and Holtz 1999).

Diagnosis: Enlarged beak-like premaxilla; secondary palate formed by premaxilla and maxilla; reduced lower temporal fenestra; humerus lightly constructed, deltopectoral crest reduced; manus long, with subequal metacarpals and digits; unguals less trenchant; metatarsus elongated, about 80 percent length of tibia; pedal digits short, stout (Gauthier 1986).

General description: "Ostrich-like dinosaurs," small to rather large, lightly built, superficially resembling modern ground birds, with long snout, beak-like and usually toothless jaws (teeth in more primitive forms), toothless jaws probably covered by

Mounted skeleton of the large tyrannosaurid *Gorgosaurus libratus*.

Photograph by the author, courtesy Royal Tyrrell Museum/Alberta Community Development.

Courtesy The Field Museum (neg. #GEO81671).

Chief Preparator Orville L. Gilpin and assistant Cameron Gifford in 1956 mounting a skeleton (FMNH PR308; originally referred to *Gorgosaurus libratus*) of the giant tyrannosaurid *Daspletosaurus torusus*.

Photograph by the author, courtesy American Museum of Natural History.

Skeletons of the tyrannosaurid theropod *Tyrannosaurus rex* (AMNH 5027), in background, the diplodocid sauropod *Apatosaurus excelsus* (composite: AMNH 460, 222, 339, and 592), remounted for the American Museum of Natural History's cladistically arranged "Hall of Saurischian Dinosaurs," which opened in 1995.

horny sheath, eyes large, neck long, forelimbs relatively long (about half length of hindlimbs), nonraptorial hand with three long fingers, slender hindlimbs, thigh shorter than shin, feet with from three to four toes, possibly insectivorous, herbivorous, or omnivorous, cursorial (among fleetest of dinosaurs)."

Age: ?Late Jurassic to Late Cretaceous.

Geographic distribution: North America, Asia.

Taxa: ?Alvarezsauridae plus Harpymimidae plus Garudimimidae plus Ornithomimidae.

ALVAREZSAURIDAE

Node-based definition: *Patagonykus*, *Alvarezsaurus*, *Mononykus* and all the descendants of their most recent common ancestor (Novas 1996*b*).

Diagnosis: Unambiguous synapomorphies including cervical vertebrae having craniocaudally short and dorsoventrally low neural spines; sacral and caudal vertebrae procoelous; caudal sacral centra transversely compressed; hemal arches of proximal caudal vertebrae dorsoventrally elongate; ungual phalanx of manual digit I stout, robust; pubic peduncle of ilium slender; pubic peduncle projecting cranioventrally; absence of fossa for M. cuppedicus; supraacetabular crest extending cranially above pubic peduncle; ambiguous traits including postacetabular blade of ilium with caudolaterally oriented brevis shelf and ventrally curved medial flange; proximal femur without "posterior" trochanter (Novas 1996*b*).

General description: Small to medium-sized, birdlike, superficially resembling ornithomimids, short forelimbs, hand in some forms reduced to a single robust finger, apparently predators of small animals and insects, possibly omnivorous or herbivorous.

Age: Late Cretaceous.

Geographic distribution: South America, Asia.

Taxa: Mononykinae plus *Patagonykus* plus *Alvarezsaurus*.

Note: *Alvarezsaurus* was originally classified as a birdlike nonavian theropod (see entry, *D:TE*). Subsequently, *Alvarezsaurus* and its family Alvarezsauridae (tentatively included here) were removed from the Dinosauria and reclassified as Late Cretaceous birds (see *S1*). More recently, Sereno (1999*a*, 1999*c*) questioned placing the Alvarezsauridae into Aves, based on recent cladistic analyses (*e.g.*, Novas 1996*b*; Chiappe, Norell and Clark 1996), as "(1) ingroups did not include any nonavian taxa or (2) available character evidence uniting alvarezsaurids and nonavian taxa was not considered"; furthermore, character evidence uniting alvarezsaurids and birds more advanced than *Archaeopteryx* has now been undermined by the discovery of basal alvarezsaurids (*e.g.*, *Alvarezsaurus* and *Patagonykus*) and basal ornithurine birds (*e.g.*, *Confusciusornis*), taxa which introduce important homoplasy.

According to Sereno's (1999*a*) new analysis, the "remaining alvarezsaurid-bird synapomorphies are overwhelmed by character evidence linking alvarezsaurids and ornithomimids" (among other unstated characters, peculiar dorsoventrally flattened internarial bar; enlarged prefrontal having broad orbital flanges; extension of dentary tooth row posterior to that of maxilla; alveoli having incomplete septa; metacarpals I–III having extensive shaft-to-shaft contact; metacarpal I long, phalanx 1 of manual digit I with dorsomedial extensor tubercle and paired flexor flanges; manual unguals having flattened ventral surfaces and distally displaced flexor tubercles; iliac blade meeting in midline; and other unstated characters).

Sereno (1999*a*, 1999*c*) noted that additional features also favor placing the Alvarezsauridae near the base of Coelurosauria within Ornithomimosauria, including the primitive form of the pectoral girdle, forelimb, tarsus, and pes (among other unstated features, prominent scapular acromion, crescent-shaped coracoid posterior process, posteroventrally facing glenoid, lack of edge-to-edge coracosternal contact, strong ulnar olecranon, distinct tibial malleoli, low ascending process of acetabulum, unreversed pedal digit I).

MONONYKINAE

Node-based definition: *Mononykus*, *Shuvuuia*, and *Parvicursor*, and all the descendants of their most recent common ancestor (Chiappe, Norell and Clark 1998).

Diagnosis: [See "Notes" below.]

General description: Small to medium-sized, hand, in at least some forms, with one robust digit.

Age: Late Cretaceous.

Geographic distribution: Asia.

Genera: *Mononykus*, *Parvicursor*, *Shuvuuia*.

Notes: The Mononykinae was proposed by Chiappe, Norell and Clark (1998) in their paper describing the new alvarezsaurid genus *Shuvuuia*. Chiappe diagnosed this subfamily in supplementary material to that paper. That diagnosis, however, is already dated and an emended diagnosis has not yet been proposed (L. M. Chiappe, personal communication 2000).

In *D:TE* and *S1*, various authors were cited offering evidence that suggested either that the genus *Mononykus* was a bird or a nonavian theropod. Considering the evidence on both sides of this debate, the present writer originally opted to regard *Mononykus* as avian. The more recent analysis of Sereno (1999*a*, 1999*c*), however, seems to suggest that the family Alvarezsauridae—which includes Mononykinae, which in turn includes *Mononykus* and other related genera generally classified as birds—constitutes a clade of very birdlike nonavian theropods. For the present pending possible future arguments that could sway this debate back toward a strict avian interpretation of Mononykinae, this taxon is herein regarded as a group of nonavian theropods.

HARPYMIMIDAE

Diagnosis: Front of dentary having six small, blunt, cylindrical teeth; humerus short, about length of skull, almost length of scapula; metacarpal I slightly more than half length of II; metacarpal III longest (Barsbold and Osmólska 1990).

General description: Very primitive, the only known ornithomimosaurs known to have teeth (blunt and cylindrical), hand comparatively large, feet probably three-toed, probably capable of swift locomotion.

Age: Middle Cretaceous.

Geographic distribution: Asia.

Genus: *Harpymimus*.

GARUDIMIMIDAE

Diagnosis: Dorsal and sacral vertebrae lacking pleurocoels; ilium shorter than pubis; metatarsal III strongly narrowed proximally, visible throughout extensor face of metatarsus, entirely separating metatarsals II and IV; proximal phalanx of pedal digit II as long as in III; pedal digit III slightly longer than IV (Barsbold and Osmólska 1990).

General description: Primitive, with four-toed foot that retains short first toe, capable of moderately swift locomotion.

Age: Late Cretaceous.

Geographic distribution: Asia.

Genus: *Garudimimus*.

ORNITHOMIMIDAE

Diagnosis: Femur distinguished from that of *Elmisaurus* (Caenagnathidae) by having more proximally situated base of lesser trochanter (femora of Harpymimidae and Garudimimidae not described) (Rich and Rich 1994).

General description: Medium to rather large size, fingers subequal in length, foot without first toe, third toe longer than second and fourth, probably fastest locomotion among all ornithimomimosaurs, some forms apparently herbivorous.

Age: Early to Late Cretaceous.

Geographic distribution: North America, Asia.

Photograph by the author, courtesy National Museum of Natural History, Smithsonian Institution.

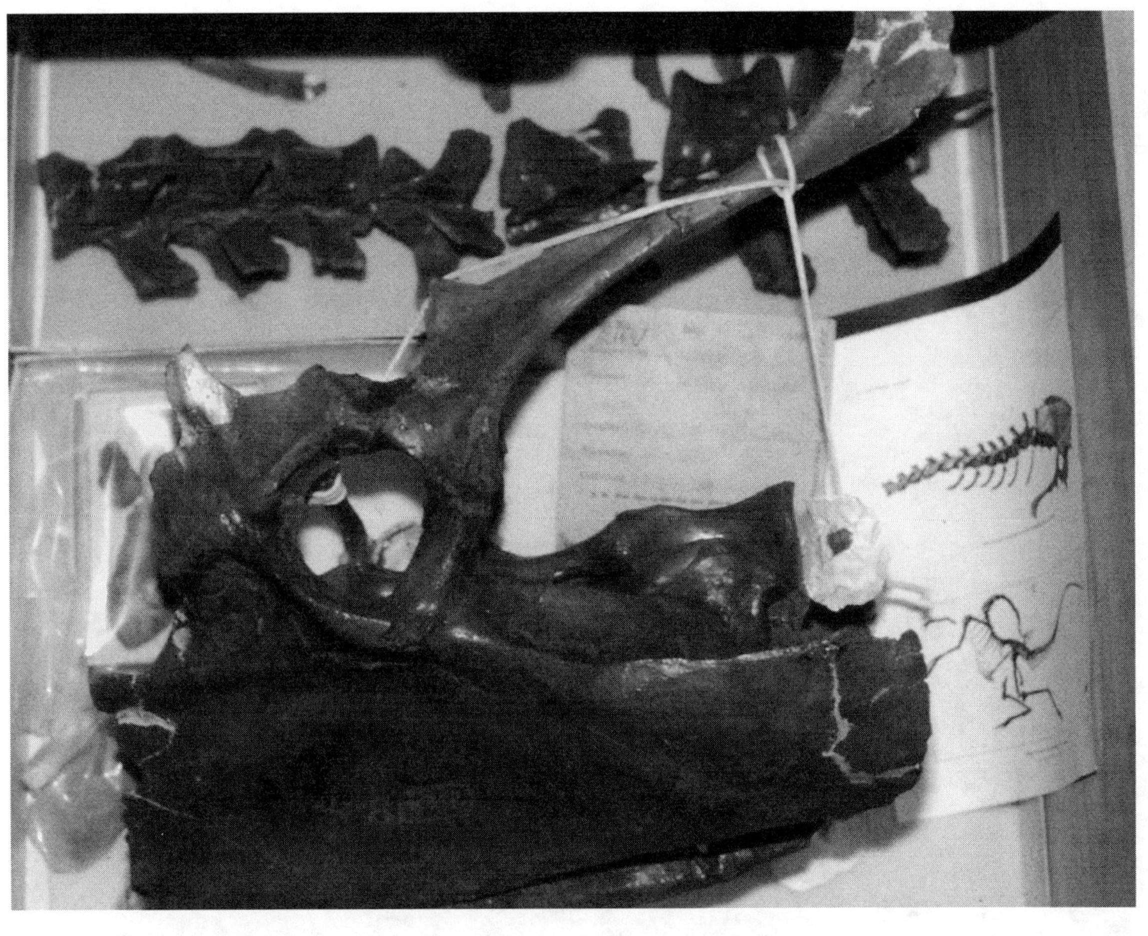

Holotype (USNM 930) of the ornithomimid *Ornithomimus altus*, which was referred to the genus *Struthiomimus*. The specimen includes partial pelvic elements, a right hindlimb, and toes of the left foot.

Genera: *Anserimimus, Archaeornithomimus, Dromiceiomimus, Gallimimus, Ornithomimus, ?Pelecanimimus, Struthiomimus.*

OVIRAPTOROSAURIA

Node-based definition: *Oviraptor*, *Chirostenotes*, and all descendants of their most recent common ancestor (Sues 1997*a*).

Diagnosis: Skull deep, strongly fenestrated; snout short; toothless jaws forming broad base for horny beak; medial process on articular of mandible; large external mandibular fossa; anterodorsal margin of mandible arched; dentary with two long posterior processes (Barsbold 1997).

General description: Mostly small-sized, also large, lightly built forms, head toothless with beak, neck moderately long, hand three-fingered, possibly egg-eaters, omnivorous or herbivorous, some forms apparently feathered at least in the tail region.

Age: Late Cretaceous.

Geographic distribution: Asia, North America.

Taxa: ?*Microvenator* plus Oviraptoridae plus Caenagnathidae.

Oviraptorosauria *incertae sedis*: *Nomingia*.

Note: In their analysis of *Microvenator* (see entry), Makovicky and Sues (1998) determined that this genus is probably a member of the Oviraptorosauria or a sister taxon to that group.

OVIRAPTORIDAE

Stem-based definition: All oviraptorian taxa closer to *Oviraptor* than to *Chirostenotes* (Padian, Hutchison and Holtz 1999).

Diagnosis: Skull deep, with shortened snout; mandible short, deep; dentary having short, concave rostral portion thickened at symphysis; two widely separated caudal dentary processes bordering vertically enlarged external mandibular fenestrae ventrally and dorsally; external mandibular fenestra subdivided by spinelike rostral process of surangular; palate with pair of median, toothlike processes on maxillae (Barsbold, Maryańska and Ósmólska 1990).

General description: Small to large, head with short jaws, with parrot-like beak, relatively large braincase, frontlimbs very long (similar to those in ornithomimosaurs, but differing in that claws are more sharply curved and first finger is somewhat longer than other two fingers).

Age: Late Cretaceous.

Geographic distribution: Asia.

Taxa: Oviraptorinae plus Ingeniinae.

Photograph by the author, courtesy Royal Tyrrell Museum/Alberta Community Development.

Skeleton (cast of AMNH 5339) of the ornithomimid ornithomimosaur *Struthiomimus altus.* One of the most complete ornithomimid specimens collected, it was found at Red Deer River, near Steveville, Alberta, Canada (see *D:TE*).

OVIRAPTORINAE

Diagnosis: Manus with comparatively long, slender digits; digit I shortest; penultimate phalanges longest; unguals strongly curved, with strong flexor tubercles and dorsoposterior "lips" (Barsbold 1981).

General description: More gracile than ingeneniines, cranial crest in some forms, behavior including birdlike brooding.

Age: Late Cretaceous.

Geographic distribution: Asia.

Genera: *Conchoraptor, Oviraptor.*

INGENIINAE

Diagnosis: Manual digit I longest, very strong; digits II and III much shorter, thinner due to strongly reduced phalanges, especially penultimate ones, which are shorter that preceding phalanx; unguals of digits II and III short, weakly curved, with small flexor tubercles; unguals without "lip" (Barsbold 1981).

General description: More robust than oviraptorines, head apparently without crest, relatively smaller claws on second and third fingers.

Age: Late Cretaceous.

Geographic distribution: Asia.

Genus: *Ingenia.*

CAENAGNATHIDAE

Node-based definition: *Chirostenotes pergracilis* and *C. elegans*, an unnamed caenagnathid (upper Maastrichtian of North and South Dakota, United States; Currie, Godfrey and Nessov 1994, Triebold and Russell 1995), *Elmisaurus rarus, Caenagnathasia martinsoni*, and their most recent common ancestor (Sues 1997*a*).

Diagnosis: (Based on *C. pergracilis*) antorbital fossa on maxilla with pronounced rim; medial wall of antorbital fossa extensive, without accessory antorbital fenestra; mandible with long, low, undivided external mandibular fenestra; braincase much deeper than long, having distinctly verticalized basicranial region; otic region having deep but anteroposteriorly narrow lateral depression; manual digit III longer than I, with very slender phalanges; synsacrum comprising six co-ossified vertebrae with pneumatic foramina; proximal end of metatarsal III distinctly constricted, pinched between II and IV (Sues 1997*a*).

Photograph by the author, courtesy Mesa Southwest Museum.

Anterior part of skeleton of the oviraptorid theropod *Ingenia yanshini* collected in the Gobi Desert, Kulsan, Mongolia, exhibited in "The Great Russian Dinosaurs Exhibition" (1996), here presented at the Mesa Southwest Museum, Mesa, Arizona. This Late Cretaceous taxon is known from at least six specimens.

General description: More primitive than oviraptorids, small, lightly built, with gracile hands and large feet (heads unknown).

Age: Late Cretaceous.

Geographic distribution: North America, Asia.

Genera: *Caenagnathasia*, *Chirostenotes*, *Elmisaurus*, *Richardoestesia*.

THERIZINOSAUROIDEA

Diagnosis: Ala from jugal embracing ventrolateral corner of lateral temporal fenestra; heavy crest rising from anterior third of dentary dorsoposteriorly toward anterodorsal end of surangular; teeth becoming larger anteriorly; cervical vertebrae with widely spaced zygapophyses, zygapophyseal pedicels well spaced, zygapophyseal extremities defining shape intermediate between right-angled "X" in dorsal outline; shaft of posterior cervical ribs bifurcating anteriorly; neural arches of anterior dorsal vertebrae elevated so that distance between arch suture and level of zygapophyseal articulations are greater than three-fourths height of centrum; shaft of dorsal ribs straight, oriented at right angle to capitular ramus; rectangular buttress from metacarpal underlying metacarpal II; manual unguals flat-sided, deep proximally; rim of ilium dorsal to sacral transverse processes strongly elevated; ilium with reduced pubic peduncle; ischium strap-like, laterally bowed; femur with prominent posterior trochanter, crest on femur proximally situated; proximomedial fossa on fibula shallow or absent; metatarsus less than one-third length of tibia; facet on metatarsal I for articulation with metatarsal II enlarged, deeply concave; first digit of first toe (and entire first toe) slightly enlarged, so that I-1 is longer than IV-1 (Russell and Dong 1993*a*).

General description: Medium-sized to large, somewhat superficially resembling both prosauropods, with small head, but with beak probably covered with horny sheath, followed by toothless area, which is followed by series of rather small cheek teeth, neck relatively long and lightly constructed, arms atypically long for theropods, hands with three digits, fingers very long with claws, hindlimbs stout, feet broad with claws, tail short, apparently herbivorous.

Age: Late Cretaceous.

Geographic distribution: Asia.

Taxa: *Beipiaosaurus* plus Alxasauridae plus Therizinosauridae.

ALXASAURIDAE

Diagnosis: Approximately 40 dentary teeth; teeth in symphyseal region of dentary; ribs not fused to cervical centra; ligament pits often well developed in manual phalanges; ilium not greatly shortened anteroposteriorly; preacetabular ala moderately expanded; ungual shorter than or subequal to first phalanx in pedal digits II–IV (Russell and Dong 1993*a*).

General description: Superficially resembling moderately large prosauropod, with small head, relatively long neck, long arms, hand with unusually long fingers, short tail, overall look and posture suggesting somewhat a modern sloth.

Age: Late Cretaceous.

Geographic distribution: Asia.

Genus: *Alxasaurus.*

THERIZINOSAURIDAE

Diagnosis: Skull and mandible shallow, long, toothless rostrally; external nares greatly elongate; palate highly vaulted, with elongate, caudally shifted vomers and palatines, rostrally reduced pterygoids; well-developed premaxillary-maxillary secondary palate; basicranium and ear region strongly enlarged, pneumaticized; mandible downwardly curved rostrally; six firmly coalesced sacral vertebrae, with long transverse processes and sacral ribs; humerus having strongly expanded proximal and distal ends; pelvis opisthopubic, ilia separated from each other; ilium with deep and long preacetabular process, with pointed cranioventral extremity that flares outward at right angle to sagittal plane; very short postacetabular process with strong, knob-like caudolateral protuberance; astragalus with tall ascending process, reduced astragalar condyles, partly covering distal end of tibia; pedal claws comparatively large, narrow, decurved (Barsbold and Maryańska 1990).

General description: Medium to large, forelimbs relatively large, claws of hand narrow but not strongly curved, some forms possessing feathers.

Age: Late Cretaceous.

Geographic distribution: Asia.

Genera: *Enigmosaurus, Erlikosaurus, Nanshiungosaurus, Segnosaurus, Therizinosaurus.*

MANURAPTORA

Stem-based definition: All theropods closer to birds than to ornithomimids (Holtz 1996).

Diagnosis: [None presently accepted.]

General description: Small- to large-sized bird-like forms, including all extinct and extant birds.

Age: ?Late Triassic to Late Cretaceous (continuing to present, if including birds).

Mounted skeleton (cast of reconstructed holotype AMNH 619) of the possible manuraptoran *Ornitholestes hermanni*. According to Lowell Dingus, Eugene S. Gaffney, Mark A. Norell and Scott D. Sampson (1998), evidence of a nasal "horn" sometimes perceived in this specimen (see *D:TE*) "is probably a result of crushing down during fossilization."

Photograph by the author, courtesy Royal Tyrrell Museum/Alberta Community Development.

Geographic distribution: North America, Asia (worldwide, if including birds).

Taxa: ?*Ornitholestes* plus Deinonychosauria.

Manuraptora *incertae sedis*: *Elopteryx*.

DEINONYCHOSAURIA

Stem-based definition: All taxa closer to *Deinonychus* than to Aves (Padian, Hutchinson and Holtz 1997).

Diagnosis: [None presently accepted.]

General description: Birdlike forms (and birds), brains relatively large, some forms possessing feathers.

Age: ?Late–Triassic to Late Cretaceous (continuing to present, if including birds).

Geographic distribution: North America, Asia (worldwide, if including birds).

Taxa: *Protarchaeopteryx* plus ?Caudipteridae plus Troodontidae plus Eumanuraptora.

CAUDIPTERIDAE

Diagnosis: Distinctive features including tail having less than 25 caudal vertebrae, comparatively shorter than tail in known oviraptorids and dromaeosaurids; ischium with extremely large triangular obturator process, less developed than in dromaeosaurids and oviraptorids; midshaft of third metatarsal significantly medio-laterally compressed; digit IV shorter than II; pes with reversed halux, differing from halux in other known theropods (*e.g.*, Norell and Makovicky 1997; Clark, Norell and Makovicky 1999) (Zhou and Wang 2000).

General description: Small, birdlike, feathered, cursorial, herbivorous.

Age: Early Cretaceous.

Geographic distribution: Asia.

Genus: *Caudipteryx*.

Note: Zhou and Wang (2000), in describing the new species *Caudipteryx asianensis*, referred the genus *Caudipteryx* (see entry) to its own family Caudipteridae. These authors suggested that this family could be the sister taxon to Ornithomimosauria or a separate lineage somewhere between Oviraptosauria and Dromaeosauridae. This latter suggestion is closer to Ji, Currie, Norell and Ji's (1998) original proposal that *Protarchaeopteryx* and *Caudipteryx* combined to form an unnamed manuraptoran clade that was the sister group to the Velociraptorinae. The placement of Caudipteridae in the present list, therefore, is only tentative.

TROODONTIDAE

Diagnosis: Skull long, with narrow snout, large endocranial cavity; external naris bounded by maxilla caudoventrally; basioccipital and periotic sinus systems; parasphenoid capsule bulbous; braincase with large lateral depression on lateral wall, containing enlarged middle ear cavity; basipterygoid process enlarged, hollow; mandible having rostrally tapering dentary that laterally exposes row of foramina located within groove beneath alveolar margin; up to 25 premaxillary and maxillary teeth, 35 dentary teeth; all teeth small, recurved, spaced closely together; distal and often mesial edges of teeth with large, hooked denticles pointing toward tip of crown; distal and mesial denticles subequal in size; calcaneum thin, fused with astragalus; metatarsus long to very long, with longitudinal, proximal trough along extensor face of metatarsal III; metatarsal III only slightly longer than IV, reduced to splinter for about half proximal length; proximal end of metatarsal III hidden on extensor side by tightly adhering ends of metatarsals II and IV; distal end of metatarsal III having tongue-like articular surface that extends proximocaudally; metatarsal IV most robust, occupying more than half width of metatarsus in caudal view (Osmólska and Barsbold 1990); all six sacral vertebrae fully anklosed; first two sacrals with pneumatic foramina; ventral surface of sacrals flattened, shallow medial groove developing on second sacral, very pronounced on sacrals 3–6; neural platform (Howse and Milner 1993).

General description: Small, with long, narrow head, relatively largest brain of all dinosaur groups, long, slender, and delicate jaws, small and closely-spaced teeth, very long and slender hindlimbs, foot with first (innermost) and second toes enlarged, digit II largest, foot claws sometimes straight.

Age: Late Jurassic to Late Cretaceous.

Geographic distribution: North America, Asia.

Genera: *Borogovia*, *Byronosaurus*, *Koparian*, ?*Ornithodesmus*, *Saurornithoides*, *Tochisaurus*, *Troodon*.

Note: Norell, Makovicky and Clark (2000*a*), in performing a cladistic analysis of the Troodontidae (excluding *Borogovia*, *Koparion*, and *Tochisaurus*; see *Byronosaurus* entry), found *Byronosaurus jaffei* to lie in a clade including *Troodon formosus*, *Saurornithoides mongoliensis*, and *S. junior*, the sister group to this clade being *Sinornithoides youngi*, with the base of the troodontid tree occupied by an unnamed Mongolian taxon (IGM 100/44) (see Barsbold, Osmólska and Kurzanov 1987).

EUMANURAPTORA

Node-based definition: Including deinonychosaurs and birds and all descendants of their common ancestor (Padian, Hutchison and Holtz 1999).

Diagnosis: [None published.]

General description: Birdlike theropods (and birds).

Age: ?Late Triassic to Early to Late Cretaceous (continuing to present, if including birds).

Geographic distribution: North America, Asia.

Taxa: Dromaeosauridae plus ?Aviale [Aves = birds].

Notes: Padian, Hutchinson and Holtz (1999) introduced this node-based taxon, essentially replacing the stem-based Paraves introduced by Sereno (1997), because it "represents an important juncture in theropod evolution (the most recent common ancestors of avialians and their deinonychosaurian relatives), and deserves a term facilitating reference to it."

Holtz (1994) had formerly regarded the Troodontidae — united by Gauthier (1986) with Dromaeosauridae into the grouping Deinonychosauria — as a sister taxon to the Ornithimomimidae (see *D:TE*, *S1*), this later assessment thereby making Deinonychosauria virtually redundant with Dromaeosauridae. Based upon more recent analyses, however, Padian, Hutchison and Holtz (1999) regarded Gauthier's original assessment as probably correct. Therefore, Deinonychosauria, a taxon that was not used in *S1*, is herein reinstated.

DROMAEOSAURIDAE

Stem-based definition: *Deinonychus* and all maniraptorans closer to it than to birds (Holtz 1996).

Diagnosis: Lacrimal slender, T-shaped; frontal process of postorbital upturned, meeting pronounced and posteroventrally inclined postorbital process of frontal; T-shaped quadratojugal different from L-shaped ones in most theropods; pronounced ventrolateral process of squamosal sutured to top of paroccipital process, extending conspicuously lateral to intertemporal bar and quadratojugal; distal end of paroccipital process twisted to face posterodorsally; pneumatopore in anterior surface of paroccipital process opening into pneumatic sinus within this process; splenial with relatively large, triangular process exposed on lateral surface of mandible between dentary and angular; relatively large, triangular internal mandibular fenestra (infra–Meckelian fenestra); retroarticular process broad, shallow, shelf-like, with vertical, columnar process rising from posteromedial corner; interdental plates in premaxilla, maxilla, and dentary fused to each other and to margins of jaws; pubis presumably retroverted; specialized (sickle-clawed) pedal digit II (Currie 1995*a*, modified by Barsbold and Osmólska 1999).

General description: Rather small to large, lightly built, large head with narrow snout, moderately deep and straight lower jaw, backwardly curved teeth, unusually long forelimbs, hand with very mobile wrist, three long fingers with large curved claws, foot with second toe developed into highly raptorial "sickle" claw, tail long and stiff, apparently fairly active and (by dinosaurian standards) relatively intelligent, some forms possessing feathers.

Age: Early to Late Cretaceous.

Geographic distribution: North America, Asia.

Taxa: Dromaeosaurinae plus Velociraptorinae plus *Sinornithosaurus* plus *Bambiraptor*.

Dromaeosauridae *incertae sedis*: ?*Euronychodon*, *Hulsanpes*, *Microraptor*, ?*Nuthetes*, *Paronychodon*, ?*Phaedrolosaurus*, *Pyroraptor*, *Saurornitholestes*, *Utahraptor*, *Variraptor*.

Notes: The number of characters in Currie's diagnosis of the Dromaeosauridae was reduced by Barsbold and Osmólska (1999), based on their recent study of numerous specimens of *Velociraptor mongoliensis* (see *Velociraptor* entry for details).

Norell, Makovicky and Clark (2000*b*), in a brief review of the Dromaeosauridae, noted the relative rarity of this group, for which good fossils are few and most taxa are very incompletely known, and the paucity of information relating to many aspects of dromaeosaurid anatomy and intraspecific variability. Norell *et al.* (in preparation) will publish a revised diagnosis of the Dromaeosauridae, review the anatomy of this group, discuss variation among its described members, and also describe several new forms recovered during the Mongolian Academy of Sciences-American Museum of Natural History Paleontological Expeditions. According to Norell *et al.* (2000*b*), of particular significance is intraspecific variation in features of the hindlimb, pelvis, vertebrae, and basicranium, these differences having special importance regarding the relationships of dromaeosaurids, either alone or with the troodontids, with Avialae. These authors will propose a new technique for examining the stability of relationships where much data are missing.

DROMAEOSAURINAE

Diagnosis: Fewer maxillary teeth (nine) than in other dromaeosaurids; anterior carina of maxillary tooth close to midline of tooth near tip, but not far from tip, tooth twists toward lingual surface; differs from velociraptorines in following features: premaxilla deeper, thicker; quadratojugal stouter; top of frontal flatter, margin of supratemporal fossa not as pronounced; postorbital process of frontal more sharply demarcated from dorsomedial orbital margin; posteromedial process of palatine more slender; anterior and posterior tooth denticles of subequal size (Currie 1995*a*).

General description: Head more massive than in velociraptorines, serrations on front and back of teeth the same size, premaxillary teeth about equal in size.

Age: Late Cretaceous.

Geographic distribution: North America, ?Asia.

Genera: ?*Achillobator*, ?*Adasaurus*, *Dromaeosaurus*.

Photograph by the author, courtesy Royal Tyrrell Museum/Alberta Community Development.

Mounted skeleton (cast of holotype AMNH 5356, reconstructed) of the dromaeosaurine dromaeosaurid *Dromaeosaurus albertensis*.

VELOCIRAPTORINAE

Diagnosis: Denticles on anterior carina of maxillary and dentary teeth significantly smaller than posterior denticles; second premaxillary tooth significantly larger than third and fourth premaxillary teeth; nasal depressed in lateral view (Paul 1988*b*) (this bone unknown in *Dromaeosaurus*) (Currie 1995*a*).

General description: More gracile and agile than dromaeosaurines, denticles on back of teeth much larger than those on front, second premaxillary tooth the largest.

Age: Early to Late Cretaceous.

Geographic distribution: North America, Asia.

Genera: *Deinonychus*, *Koreanosaurus* [*nomen nudum*], *Sinornithosaurus*, *Velociraptor*.

SAUROPODOMORPHA

Stem-based definition: Prosauropoda plus Sauropoda, and all saurischians closer to them than to birds (Upchurch 1997*b*).

Diagnosis: Derived characters including teeth with lanceolate crowns; anteroposterior elongation of cervical vertebrae; trunk region long relative to hindlimb length; astragalus having ascending process that keys into tibia, the latter with matching descending process (see Gauthier 1986) (Upchurch 1997*b*).

General description: Small to gigantic, bipedal, semibipedal, and quadrupedal forms, with small head, long neck, heavy bones, stocky hindlimbs, five-fingered forefoot, five-toed hindfoot, all forms apparently herbivorous.

Age: Late Triassic to Late Cretaceous.

Geographic distribution: Worldwide.

Taxa: Prosauropoda plus Sauropoda.

Sauropodomorpha *incertae sedis*: *Dachungosaurus* [*nomen nudum*], ?*Halticosaurus*, ?*Thotobolosaurus* [*nomen nudum*].

PROSAUROPODA

Stem-based definition: Thecodontosauridae, Anchisauridae, Plateosauridae, Melanorosauridae, and all sauropodomorphs closer to them than to Sauropoda (Upchurch 1997*a*).

Diagnosis: Skull approximately half length of femur; jaw articulation slightly below level of maxillary tooth row; teeth small, homodont or weakly homodont, spatulate, with coarse marginal serrations; manual digit I bearing twisted first phalanx and enormous, trenchant ungual medially directed when hyperextended; digits II and III of subequal length, with small, slightly recurved ungual phalanges; digits IV and V reduced, lacking ungual phalanges; typical phalangeal formula of 2-3-4-(3 or 4)-3; bladelike distal parts of pubes forming broad, flat apron; fifth pedal digit vestigial (Galton 1990*a*); femur having longitudinal crest proximal to lateral condyle; lesser trochanter a weak ripple proximodistally lying on latero-anterior surface, main part of trochanter below level of femoral head (Gauffre 1993*a*).

General description: Small to large, bipedal, facultatively bipedal, or quadrupedal, with small head, leaf-shaped teeth, large body, elongate neck, first finger of forefoot powerfully developed, fourth and fifth fingers reduced and clawless, fifth toe of hindfoot very small.

Age: Late Triassic to Early Jurassic.

Geographic distribution: Worldwide.

Taxa: *Saturnalia* plus Thecodontosauridae plus Anchisauridae plus Massospondylidae plus Yunnanosauridae plus Plateosauridae plus Melanorosauridae plus Blikanasauridae plus ?*Riojasaurus*.

Prosauropoda *incertae sedis*: ?*Arctosaurus*, *Fulengia*, *Tawasaurus*.

Notes: Most modern workers (*e.g.*, Sereno 1989, 1997, 1998; 1999*c*; Galton 1990; Gauffre 1995; Wilson and Sereno 1998) have accepted the Prosauropoda as a monophyletic taxon within Sauropodomorpha rather than as a series of outgroups to Sauropoda (*e.g.*, Colbert 1964*b*; Charig, Attridge and Crompton 1965; Gauthier 1986; Benton 1990). In 1990, Galton divided this group into the suite of families listed above, most of these (*e.g.*, Anchisauridae) including only one genus.

More recently, however, Benton, Juul, Storrs and Galton (2000), in performing a cladistic analysis, found that the Prosauropoda is probably a valid taxon within Sauropodomorpha, but that this conclusion is weakly supported. Benton *et al.*'s study did not resolve relationships within the Prosauropoda, although *Thecodontosaurus* (see entry) was determined to be the outgroup to all other prosauropods (including *Saturnalia*), which in turn is the sister group to the remaining prosauropods; *Plateosaurus* and *Euskelosaurus* were found to be closely-paired genera. Various taxa were omitted from this analysis including dubious forms, juveniles (*Mussaurus*), the Chinese genera *Yunnanosaurus* and *Lufengosaurus* ("since it is not clear whether the two genera are truly distinct or not, nor which of the material from the Lufeng Formation belongs to which of the two taxa"), and *Riojasaurus* (see note under Melanorosauridae).

THECODONTOSAURIDAE

Diagnosis: Derived characters of relatively long retroarticular process of mandible, closure of interpterygoid vacuity (Galton 1990*a*).

General description: Presumably the most primitive forms, with small, lightly-built body, relatively large skull, larger eyes and smaller nostrils than most other forms, short tooth rows, jaws designed to accommodate more resistant food, posture erect, only fully bipedal prosauropods, at least subcursorial.

Age: Late Triassic.

Geographic distribution: Europe, Africa.

Genera: ?*Azendohsaurus*, *Thecodontosaurus*

ANCHISAURIDAE

Diagnosis: Derived characters of large basisphenoid tubera projecting farther ventrally than very small basipterygoid process, proportionally elongate cranial process of ilium, ventral emargination of proximal part of pubis, and size reduction of first pedal ungual so that it is smaller than second (Galton 1990*a*).

General description: Presumably primitive with small body, comparatively large head with larger eyes and smaller nostrils than more advanced prosauropods, short tooth rows, lightly built with slender limbs, mostly bipedal, at least subcursorial.

Age: Early Jurassic.

Geographic distribution: North America.

Genus: *Anchisaurus*.

MASSOSPONDYLIDAE

Diagnosis: Derived character of centrally located, almost vertical dorsal process of maxilla (Galton 1990*a*).

General description: Medium-sized, lightly-built, with relatively small head and jaws, bipedal.

Age: Early Jurassic.

Geographic distribution: Africa, North America, South America.

Genus: *Massospondylus*.

YUNNANOSAURIDAE

Diagnosis: Maxillary and dentary teeth weakly spatulate, noticeably asymmetrical, apices slightly medially directed; a few coarse apically directed marginal denticles (Galton 1990*a*).

General description: Large, head tall and narrow, snout short, teeth resembling those of some sauropods.

Age: Late Jurassic.

Geographic distribution: Asia.

Genus: *Yunnanosaurus*.

Note: See "Notes" under Prosauropoda (above) regarding *Yunnanosaurus*.

PLATEOSAURIDAE

Diagnosis: Jaw articulation ventrally offset, well below level of dentary tooth row (Galton 1990*a*).

General description: Medium- to large-sized, head proportionately small, long tooth rows, possibly having rather primitive cheeks, neck relatively long, limbs stout, some forms heavily built, facultatively bipedal, at least subcursorial to slow-moving.

Age: Late Triassic to Early Jurassic.

Geographic distribution: North America, South America, Europe, Asia, Africa.

Genera: *Ammosaurus, Coloradisaurus, Euskelosaurus, Jingshanosaurus, ?Lufengosaurus, Mussaurus, Plateosaurus, Sellosaurus, Yimenosaurus.*

Note: See "Notes" under Prosauropoda (above) regarding *Lufengosaurus.*

MELANOROSAURIDAE

Diagnosis: Ilium with ischial peduncle shorter than pubic peduncle (Galton 1990*a*); latero-medial width of femur greater than antero-posterior width (width measured at middle length of femur) (Gauffre 1993*a*).

General description: Large-sized, heavily built, somewhat sauropod-like, with tendency toward quadrupedal locomotion, some fully quadrupedal, slow moving, presumably among more advanced forms.

Age: Late Triassic to ?Early Jurassic.

Geographic distribution: Africa, Europe, South America.

Genera: *Camelotia, Lessemsaurus, Melanorosaurus.*

Note: *Riojasaurus* has been usually classified among the melanorosaurids. However, Benton, Juuls, Storrs and Galton (2000), in performing their cladistic analysis of various sauropodomorph groups, found this genus to be "a 'rogue' taxon in certain ways" ... not exhibiting "any close alliance with any of the other prosauropods." Its inclusion in the Melanorosauridae was based upon its large size and proportions. Furthermore, Benton *et al.* found that the traditional view of the Melanorosauridae as a well-defined group close to Sauropoda (*e.g.*, Gauthier 1986; Benton 1990) can no longer, based on recent cladistic analyses, be supported.

BLIKANASAURIDAE

Diagnosis: Maximum proximal width of tibia 48 percent of maximum length, long axes of proximal and distal ends at 60 degrees, distally posterior process relatively deep, robust; distal end of fibula with large, ventromedially directed articular surface; astragalus having relatively high, prominent ascending process; calcaneum small; metatarsal I short but very robust; metatarsals II–IV subequal in length, III longest; ratio of maximum proximal width to maximum length for metatarsals I–IV is 0.91, 0.72, 0.53, 0.53, and 0.74, respectively (Galton and Heerden 1998).

General description: Apparently medium-sized, stockily built, hindfoot with backwardly directed "spur," possibly quadrupedal, apparently slow moving.

Age: Late Triassic.

Geographic distribution: Africa.

Genus: *Blikanasaurus.*

SAUROPODA

Stem-based definition: Sauropodomorphs more closely related to *Saltasaurus* than to the prosauropod *Plateosaurus* (Wilson and Sereno 1998).

Diagnosis: Obligatory quadrupedal posture with columnar orientation of principal limb bones (humerus, radius-ulna, femur, tibia-fibula); four or more sacral vertebrae (one caudosacral added); humerus having low deltopectoral crest; olecranon absent; proximal end of ulna triradiate; distal end of radius subrectangular, posterior margin for ulna flattened; ilium having low ischial peduncle; ischial blade equal to or longer than pubic blade; ischial shafts having dorsoventrally flattened distal ends; cross-section of femur eliptical (with transverse long axis); fourth trochanter of femur reduced to low crest or ridge; astragalus lacking anterior fossa and foramina at base of ascending process; distal tarsals 3 and 4 absent (no ossified distal tarsals); proximal ends of metatarsals I and V subequal in area to that of metatarsals II and IV; metatarsal V 70 percent or more of length of IV; ungual of pedal digit I enlarged, deep, narrow (sickle-shaped) (Wilson and Sereno 1998).

General description: Medium-sized to gigantic (including largest tetrapods of all time), quadrupedal (though some may have been capable of rearing up on hind legs), with small head, smallest brain-to-body size of all dinosaurs, long to extremely long neck, strong upright limbs, broad feet, five toes on both front and hind feet, first toe of forefoot having claw, tail short to extremely long, mostly terrestrial, least likely of all dinosaurian groups to be endothermic (possibly homeothermic).

Age: Late Triassic to Late Cretaceous.

Geographic distribution: North America, South America, Europe, Asia, Africa, Australia, ?Antarctica.

Taxa: Vulcanodontidae plus Eusauropoda.

Sauropoda *incertae sedis*: *Aepisaurus, Asiatosaurus, Atlantosaurus, Austrosaurus, Chinshakiangosaurus* [*nomen nudum*], *Clasmodosaurus, Damalasaurus* [*nomen nudum*], *Gongxianosaurus, Isanosaurus, ?Kotasaurus, Kunmingosaurus* [*nomen nudum*], *Lancanjiangosaurus* [*nomen nudum*], *Macrurosaurus, Microdontosaurus* [*nomen nudum*], *Mongolosaurus, Qinlingosaurus, Ultrasaurus.*

Notes: Wilson (1998), in a preliminary report on

a new phylogenetic analysis of the Sauropoda, concluded that early sauropod skull synapomorphies include retracted nares and overlapping teeth having V-shaped wear facets. Enlarged nares characterize the Macronaria, dorsally-directed nares are a synapomorphy of Diplodcoidea. Loss of tooth overlap, V-shaped wear facets, and broad crowns took place independently within the Neosauropoda. Axial specializations of basal sauropods include a complex system of vertebral laminae, pleurocoels, and incorporation of trunk vertebrae into the neck and sacrum. Independent pleurocoel and laminae reduction and multiple neck lengthenings characterize later sauropod evolution. Evolution of a digitigrade manus in the Neosauropoda and a "wide guage" limb posture in titanosaurs succeeded the early acquisition of a quadrupedal, semiplantigrade stance.

According to Wilson, this study reveals "a complex pattern of character evolution marked by early acquisition of herbivorous and locomotor adaptations followed by independent modifications of each within sauropod subclades." The results of this analysis suggests "the parallel evolution of narrow crowns and bifid neural spines, and convergent neck lengthing events."

VULCANODONTIDAE

Diagnosis: Teeth with coarse denticles on both edges; presacral vertebrae without pleurocoels; sacrum narrow, with four functional sacral vertebrae; sacrocostal yoke not contributing to acetabulum; caudal vertebrae deeply furrowed beneath with strong chevron facets; pubes primitive, forming apron (as in prosauropods); ischium noticeably longer than pubis; forelimb relatively long, femur slender; calcaneum present; metatarsals elongated (McIntosh 1990).

General description: Primitive, large, with sauropod type limbs, forelimbs almost as long as hindlimbs, also prosauropod type features including an enlarged thumb claw on first toe.

Age: Late Triassic to Early Jurassic.

Geographic distribution: Africa, Asia.

Genera: ?*Ohmdenosaurus*, *Vulcanodon*.

EUSAUROPODA

Stem-based definition: Sauropods more closely related to *Saltasaurus* than to *Vulcanodon* (Wilson and Sereno 1998).

Diagnosis: External nares retracted posterodorsally; snout with stepped narial margin; absence of antorbital fossa; maxillary border of external naris long; absence of anterior process of prefrontal; absence of squamosal-quadratojugal contact; anterior ramus of quadratojugal elongate, distally expanded; infraorbital region of cranium shortened anteroposteriorly; supratemporal region of cranium shortened anteroposteriorly; supratemporal fossa broadly expanded laterally; quadrate shaft having elongate posterior fossa; lateral ramus of palatine narrow; maximum depth of anterior end of dentary ramus approximately 150 percent minimum depth of ramus; tooth rows broadly arched anteriorly; tooth crowns spatulate; enamel having wrinkled texture; crown overlap; crown-to-crown occlusion precise; V-shaped wear facets (interdigitating occlusion); minimum of 13 cervical vertebrae; cervical centra opisthocoelous; mid-cervical neural arches tall (height greater than posterior face of centrum); dorsal neural spines broader transversely than anteroposteriorly; distal chevrons having anterior and posterior processes; block-shaped carpals; phalanges on manual digits II and III reduced (II-ungual, III-3 and ungual absent; manual phalangeal formula 2-2-2-2-2 or lower); manual phalanges (except for unguals) broader than long; iliac blade having semicircular dorsal margin and expanded preacetabular process; pubic apron canted posteromedially; tibia with laterally projecting cnemial crest; posteroventral process of tibia reduced; fibula having lateral trochanter; length of metatarsal III 25 percent or less that of tibia; minimum shaft width of metatarsal I greater than that of metatarsals II–IV; metatarsals with spreading configuration; nonterminal phalanges of pedal digits short; ungual of pedal digit I longer than metatarsal I; penultimate phalanges of pedal digits II–IV rudimentary or absent; unguals of pedal digits II–III sickle-shaped; ungual of pedal digit IV rudimentary or absent (Wilson and Sereno 1998).

General description: The majority of sauropod taxa, excluding the most primitive forms, moderate size to giant, with considerable variation.

Age: Early Jurassic to Late Cretaceous.

Geographic distribution: North America, South America, Europe, Asia, Australia.

Taxa: Barapasauridae plus "Cetiosauridae" plus *Jobaria* plus Neosauropoda.

BARAPASAURIDAE

Diagnosis: Autapomorphies including large cavities inside neural spines of dorsal vertebrae; depressions on lateral sides of vertebral centra (Halstead and Halstead 1981).

General description: Primitive (the earliest known eusauropods, retaining some prosauropod features), large, with spoon-shaped teeth, slender limbs.

Age: Early Jurassic.

Geographic distribution: Asia.

Genus: *Barapasaurus*.

"CETIOSAURIDAE"

Diagnosis: Primitive, medium to large; skull relatively low; external nares in dorso-anterior area of skull; orbits broad in midposterior part of skull; lower jaw comparatively slender; teeth medium sized,

Photograph by the author.

Skull (cast, based on MNN TIG 3-5) of the primitive sauropod *Jobaria tiguidensis* on display at Dinofest (2000–2001), Navy Pier, Chicago.

relatively more teeth in upper than lower jaw; neck not greatly elongate, approximately up to 25 presacral vertebrae; pectoral and pelvic girdles solidly constructed; ilium low; front legs two-thirds to three-fifths length of hind legs; radius three-quarters length of humerus; tibia over half femoral length (Zhang and Chen 1996).

General description: Primitive, some with exceptionally long necks.

Age: Early to Late Jurassic.

Geographic distribution: North America, South America, Europe, Asia, Africa.

Genera: *Amygdalodon*, *Cardiodon*, *Cetiosaurus*, *Chuanjiesaurus*, ?*Patagosaurus*, ?*Protognathosaurus*, *Rhoetosaurus*, *Tehuelchesaurus*.

Note: A number of workers, having come to regard this taxon as a paraphyletic grouping of taxa based upon plesiomorphic characters, have suggested that the "Cetiosauridae" be abandoned (see "Systematics" chapter, *S1*). Recently, however, in describing the new "cetiosaurid" *Techuelchesaurus* (see entry), Rich, Vickers-Rich, Gimenéz, Cúneo, Puerta and Vacca (1999) retained this family as diagnosed as follows by Bonaparte (1986*a*) and McIntosh (1990): Coracoid roughly circular in outline; humerus to femur ratio low relative to that in Brachiosauridae; deltoid crest weakly developed; femora lacking medial deflection in anterior aspect; preacetabular process of ilium not outwardly inclined; posterior dorsal vertebrae opisthocoelous; diapophyses on dorsal vertebrae directed upward and outwards; caudal vertebrae only slightly procoelous.

In the present document, "Cetiosauridae" is therefore retained only provisionally.

NEOSAUROPODA

Node-based definition: *Diplodocus*, *Saltasaurus*, their common ancestor, and all its descendants (Wilson and Sereno 1998).

Skeletal cast of the sauropod dinosaur ***Jobaria tiguidensis*** (adult) displayed at Dinofest (2000–2001), Navy Pier, Chicago.

Photograph by the author.

Diagnosis: Preantorbital fenestra developed; ventral process of postorbital broader transversely than anteroposteriorly; lack of jugal-ectopterygoid contact; closed external mandibular fenestra; tooth crowns lacking denticles; two or fewer carpal elements; metacarpals bound, with long intermetacarpal articular surfaces; metacarpus having U-shaped proximal articular surface (270-degree arc); ilia preacetabular process laterally divergent; proximal end of tibia subcircular; astragalus having ascending process extending to posterior margin; astragalus wedge-shaped in anterior view (Wilson and Sereno 1998).

General description: Most sauropods except primitive forms.

Age: Late Jurassic to Late Cretaceous.

Geographic distribution: North America, South America, Europe, Africa, Asia.

Taxa: *Cetiosauriscus* plus Diplodocoidea plus Macronaria.

Neosauropoda *incertae sedis*: *Agustinia*, *Dystrophaeus*, *Huabeisaurus*.

Note: Wilson, Martinez and Alcober (1999) removed *Cetiosauriscus*, a genus possessing biconvex caudal vertebrae, from the Diplodocoidea, regarding it as a basal neosauropod. This assessment was based on a cladogram (modified from Wilson and Sereno 1998) of neosauropod relationships indicating the distribution of biconvex caudal vertebrae, elongate biconvex caudals, and a series of at least 30 biconvex caudals (see "Introduction," section on sauropodomorphs).

DIPLODOCOIDEA

Stem-based definition: Neosauropods more closely related to *Diplodocus* than to *Saltasaurus* (Wilson and Sereno 1998).

Diagnosis: Dentary with ventrally projecting "chin" and transversely narrow symphysis; tooth crowns subcylindrical; narrow subcylindrical basipterygoid process (length at least four times basal diameter) that project anteroventrally or anteriorly; atlantal intercentrum with anteroventrally expanded occipital fossa; cervical ribs shorter than their respective centra; arches of dorsal and caudal vertebrae tall (more than two and one half times length of dorsoventral centrum height); whiplash tail (at least 30 elongate, biconvex posterior caudal vertebrae) (Wilson and Sereno 1998).

General description: Gigantic, slender to massively built forms, some forms with extremely long and also relatively shorter neck, head rather long, forelimbs shorter than hindlimbs, some extremely long (including longest of all terrestrial animals).

Age: Late Jurassic to ?Late Cretaceous.

Geographic distribution: North America, South America, Europe, Asia, Africa.

Taxa: *Amphicoelias* plus ?Nemegtosauridae plus Rebbachisauridae plus Dicraeosauridae plus Diplodocidae.

Diplodocoidea *incertae sedis*: *Histriasaurus*.

Note: Salgado (1999) continues to use the taxon Diplodocimorpha, which Calvo and Salgado (1995) proposed to include *Rebbachisaurus* and Diplodocidae, and diagnosed by the following characters: teeth pencil-like; anterior extension of quadratojugal located beyond anterior border of orbit; basipterygoid processes forwardly directed; basioccipital condyle downwardly directed; quadrate inclined posterodorsally; infratemporal fenestra oval or slit-shaped; narial opening located above orbit; "whip-lash" tail; neural arch in posterior dorsals three times higher than centra; neural arch in caudals at least 15 times higher than centra; caudals with wing-like transverse processes; ratio of humerus to femur less than 0.70.

NEMEGTOSAURIDAE

Diagnosis: Elongate preorbital region (Barrett and Upchurch 1994); at least one genus (*Nemegtosaurus*) having fully retracted, dorsally facing external nares (Upchurch 1994).

General description: Skulls resembling those of dicraeosaurids.

Age: Late Cretaceous.

Geographic range: Asia.

Genera: *Nemegtosaurus*, ?*Phuwiangosaurus*, *Quaesitosaurus*.

Notes: Although the genera *Nemegtosaurus* and *Quasitosaurus* (possibly a junior synonym of *Nemegtosaurus*) had recently been reclassified as titanosaurs (see "Systematics" chapter and *Nemegtosaurus* entry, *S1*), Upchurch (1999) presented evidence suggesting that these genera — both known only from isolated skulls, making it difficult to assess their affinities — belong in their own family of Mongolian sauropods, the Nemegtosauridae, closely related to other diplodocoids, and that this new group belongs, along with Dicraeosauridae and Diplodocidae, in the "superfamily" Diplodocoidea (see *Nemegtosaurus* entry, this volume).

However, in a more recent paper redescribing *Phuwiangosaurus* (see entry), Martin, Suteethorn and Buffetaut (1999) regarded this genus (classified as a titanosaur by Upchurch 1998) as a titanosaur, with the Nemegtosauridae. According to these authors, the phylogenetic position of the Nemegtosauridae may require reappraisal pending the description of skull material belonging to *Phuwiangosaurus*.

REBBACHISAURIDAE

Diagnosis: [Not yet published; see Wilson's master's thesis, 1999.]

General description: Medium-sized to large sauropods, neural spines of dorsal vertebrae undivided, vertebrae of some taxa with high neural spines, teeth at least in some taxa much more numerous than in other sauropod groups.

Age: Middle to Late Cretaceous.

Geographic range: South America, Africa.

Genera: *Nigersaurus, Rayososaurus, Rebbachisaurus.*

DICRAEOSAURIDAE

Partial diagnosis: Dorsal vertebrae distinct in lack of pleurocoels, apparently secondarily cancelled; proportionally small centra; few nonbifurcated dorsal neural spines paddle-shaped; cervical, dorsal, and sacral vertebrae very derived; girdles rather primitive; limb bones blunt (Bonaparte 1996*a*).

General description: Medium-sized, similar to diplodocids, especially in the form of the head and teeth, but with relatively shorter neck and rather high back, some forms possessing extremely long neural spines.

Age: Late Jurassic to Late Cretaceous.

Geographic distribution: Africa, South America, Asia.

Genera: *Amargasaurus, Dicraeosaurus, ?Dyslocosaurus.*

DIPLODOCIDAE

Diagnosis: Skull relatively long; quadrate inclined rostroventrally; external nares opening dorsally; basipterygoid process slender, elongate; mandible light; teeth slender, peglike; proximal caudal vertebrae with procoelous centra and wing-like transverse processes; middle caudal vertebrae having forked chevrons; ischia with expanded distal ends, meeting side by side (McIntosh 1990).

General description: Very large to giant, both massive and more slender forms, very long neck, very small head, with small, slender, pencil-like at the front of the jaws only, some forms having tails ending in a "whiplash," at least one form possessing row of cone-shaped epidermal spines down tail (possibly also on neck and back).

Age: Middle Jurassic to Early Cretaceous.

Genera: *Apatosaurus, Barosaurus, Dinheirosaurus, Diplodocus, Eobrontosaurus, Lourinhasaurus, Megacervixosaurus* [*nomen nudum*], *Seismosaurus, Supersaurus.*

Note: Salgado (1999) continues to divide the Diplodocidae into two subfamilies, the Diplodocinae and Dicraeosaurinae. Salgado diagnosed the Diplo-

Reconstructed skeleton MACN-N 15 of the dicraeosaurid *Amargasaurus cazaui* (skull based on *Dicraeosaurus*) mounted at the Museo Argentino de Ciencias Naturales, Buenos Aires. This unusual sauropod is noted for its large dorsal spines.

Photograph by Brian Franczak, courtesy José F. Bonaparte and Museo Argentino de Ciencias Naturales.

Courtesy Travel Division, South Dakota Department of Highways, Pierre, SD.

Though inaccurate by modern standards, this life-sized concrete-and-steel *Apatosaurus excelsus*—made through the Works Progress Administration (WPA) in 1936 under the supervision of local sculptor Emmett A. Sullivan, based on both a mural by Charles R. Knight at the Field Museum of Natural History and a full-scale animated figure at the 1933 Chicago World's Fair — remains a popular tourist attraction atop its hilltop home at Dinosaur Park, Rapid City, South Dakota. This and other figures in the park, made to "perpetuate the facts of history," constitute the first permanent display of life-sized dinosaurs in the United States (see "Dinosaur Displays" chapter).

docinae by the following snapomorphies: cervical neural spine bifurcated; anterior dorsal neural spine bifurcated; anterior dorsals with hyposphene-hypantrum; anterior caudal vertebrae slightly procoelous; midcaudal chevrons having fore- and aft-directed processes; hemal canal closed in anterior caudals; ratio of maximum width of distal scapular blade to minimum width of scapular blade less than or equal to 2; ischia with expanded distal end; wide dorsoventral contact of both distal ends of ischia; lateral projection of distal articulation of metatarsal I.

Salgado diagnosed the Dicraeosaurinae by the following synapomorphies: parietal and postparietal fenestrae; long basipterygoid processes; small (less than 30 degrees) angle between two basipterygoid processes; tetrahedron-like supraoccipital crest; supratemporal fenestra smaller than foramen magnum; frontal bones fused; posterior rim of infratemporal fenestra in front of posterior rim of orbit; absence of pleurocavities in cervical and dorsal vertebrae; bifurcation reaching at least sixth presacral vertebra; transverse processes of dorsal vertebrae projecting dorsolaterally at 45-degree angle; tall anterior caudal arch-neural spines.

MACRONARIA

Stem-based definition: Neosauropods more closely related to *Saltasaurus* than to *Diplodocus* (Wilson and Sereno 1998).

Diagnosis: Preantorbital fenestra developed; ventral process of postorbital broader transversely than anteroposteriorly; lack of jugal-ectopterygoid contact; external mandibular fenestra closed; tooth crowns lacking denticles; two or fewer carpal elements; metacarpals bound, having long intermetacarpal articular surfaces; metacarpus having U-shaped proximal articular surface (270-degree arc); ilia preacetabular process laterally divergent; proximal end of tibia subcircular; astragalus with ascending process extending to posterior margin; astragalus wedge-shaped in anterior view (Wilson and Sereno 1998).

General description: Very large to gigantic, massive and relatively slender, relatively larger and more boxlike heads than diplodocoids, forelimbs relatively longer than in diplodocoids.

Age: Late Jurassic to Late Cretaceous.

Geographic distribution: North America, South America, Europe, Asia, Africa.

Taxa: ?Haplocanthosauridae plus Camarasauromorpha.

HAPLOCANTHOSAURIDAE

Diagnosis: Fourth and fifth presacral vertebrae constituting cervical-dorsal transition, remaining cervicals and dorsals more or less consistent in form with one another (Bonaparte 1999*b*).

General description: Basically similar to camarasaurids (see below), differing largely in the form of the presacral vertebrae.

Age: Late Jurassic.

Geographic distribution: United States.

Genus: *Haplocanthosaurus.*

CAMARASAUROMORPHA

Node-based definition: The most recent common ancestor of Camarasauridae and Titanosauriformes and all of its descendants (Salgado, Coria and Calvo 1997).

Diagnosis: Maximum diameter of naris greater than maximum diameter or orbit; quadrate fossa deep; depth of surangular more than twice that of angular; posterior trunk and sacral centra opisthocoelous (from Salgado, Coria and Calvo 1997); metacarpals relatively long (from Salgado *et al.*), longest metacarpal 45 percent longer than radius, metacarpal I longer than metacarpal IV; ischium with dorsoventrally expanded pubic articulation (from Salgado *et al.*) (Wilson and Sereno 1998).

General description: Macronarians more advanced than *Haplocanthosaurus*, "characterized by an enlarged coranoid process on the lower jaw" (Wilson and Sereno 1998).

Age: Late Jurassic to Late Cretaceous.

Geographic distribution: North America, South America, Europe, Asia, Africa.

Taxa: Camarasauridae plus ?Tendaguriidae plus Titanosauriformes.

CAMARASAURIDAE

Diagnosis: Basipterygoid processes short, sturdy;

Referred right forelimb and foot (left) and articulated dorsal vertebrae (right), IVPP V. 11121, of the euhelopodid camarasauromorph sauropod, *Hudiesaurus sinojapanorum*, on display at Dinofest in 1998.

Photograph by the author.

jugal excluded from lower rim of skull; midpresacral vertebrae having U-shaped divided spine (McIntosh 1990).

General description: Rather stocky, relatively medium-to giant-sized, skull massively constructed, head relatively large and boxlike, short snout, nostrils anterior to eyes, heavy, thick-rooted spatulate teeth extending along jaw margins, forelimbs slightly longer than forelimbs, back horizontally oriented, tail long to relatively short.

Age: Late Jurassic to Late Cretaceous.

Geographic distribution: North America, Europe, Asia.

Genera: *Abrosaurus, Aragosaurus, Camarasaurus, ?Janenschia, Nurosaurus.*

Camarasauridae *incertae sedis*: *Neosodon.*

TENDAGURIIDAE

Diagnosis: Approximately 20 meters long; anterior dorsal vertebrae opisthocoelous, with very low, anteroposteriorly laminar neural spines not rising above surrounding area of neural arch; neural spine not distinct corpus of bone, but integrated within osseous structures around it, continuous with transverse processes; tranverse process elongated with dorsal depression lateral to prezygapophyses; deep anterior cavities on anterior surface of transverse process, shallow cavities on posterior surface; well-developed infradiapophyseal laminae; strong, thick epipophyses near uppermost border of vertebrae, these connected to transverse processes (Bonaparte, Heinrich and Wild 2000).

General description: Gigantic, apparently somewhat similar to camarasaurids.

Age: ?Late Jurassic.

Geographic distribution: Africa.

Genus: *Tendaguria.*

TITANOSAURIFORMES

Node-based definition: *Brachiosaurus, Saltasaurus*, their common ancestor, and all of its descendants (Wilson and Sereno 1998).

Diagnosis: Pterygoid lacking dorsomedially oriented basipterygoid hook; dorsal ribs with pneumatic cavities; distal condyle of metacaral I undivided, with

Skeleton (cast) of *Camarasaurus supremus*, prepared by DINOLAB for the New Mexico Museum of Natural History and Science in Albuquerque. Recent studies have shown that the neck of this sauropod was probably quite flexible.

Photograph by the author, courtesy New Mexico Museum of Natural History and Science.

reduced articular surface; iliac preacetabular process semicircular; proximal one-third of femoral shaft deflected medially (Wilson and Sereno 1998).

General description: Very large to gigantic, some having longer forelimbs than hindlimbs.

Age: Late Jurassic to Late Cretaceous.

Geographic distribution: North America, South America, Europe, Africa, Asia.

Taxa: Brachiosauridae plus Somphospondyli.

BRACHIOSAURIDAE

Stem-based definition: Titanosauriforms more closely related to *Brachiosaurus* than to *Saltasaurus* (Wilson and Sereno 1998).

Diagnosis: Synapomorphies including subrectangular muzzle (twice as long as deep); elongate cervical centra (attaining maximum of seven times as long as deep); centra with deep accessory depressions; humerus elongate (subequal to femur in length); humerus with prominent deltopectoral crest (Wilson and Sereno 1998).

General description: Gigantic, with short, low snout, nostrils sometimes raised above eyes, teeth broad and spatulate, very long neck, front legs almost as long as, or longer than, hind legs, relatively short tail.

Age: Late Jurassic to Late Cretaceous.

Geographic distribution: North America, South America, Europe, Africa, Asia.

Genera: *Astrodon*, ?*Atlasaurus*, *Bothriospondylus*, *Brachiosaurus*, *Cedarosaurus*, *Chondrosteosaurus*, *Dystylosaurus*, ?*Gigantosaurus*, *Ischyrosaurus*, *Klamelisaurus*, *Lapparentosaurus*, *Morinosaurus*, *Oplosaurus*, *Pelorosaurus*, *Pleurocoelus*, *Sauroposeidon*, *Sonorasaurus*, *Volkeimeria*.

Note: Until recently, brachiosaurids have not been known from rocks more recent than of Early Cretaceous age. Kirkland, Aguillon-Martinez, Hernández-Rivera and Tidwell (2000; see "Introduction," section on sauropods) have described a brachiosaurid caudal vertebra from the Upper Cretaceous of Mexico.

SOMPHOSPONDYLIDI

Diagnosis: Titanosauriforms more closely related to *Saltasaurus* than to *Brachiosaurus* (Wilson and Sereno 1998).

Diagnosis; Cervical neural arches with reduced lamination; presacral vertebrae composed of spongy bone; anterior and middorsal neural spines posteroventrally inclined; six sacral vertebrae (one dorsosacral vertebra added); scapular glenoid deflected medially (Wilson and Sereno 1998).

General description: Similar to brachiosaurids, but generally less massive.

Age: Late Jurassic to Late Cretaceous.

Geographic distribution: North America, South America, Asia, Europe, Africa.

Taxa: Euhelopodidae plus Titanosauria.

EUHELOPODIDAE

Diagnosis: Teeth spatulate; neural spines of cervical vertebrae low; chevrons of caudal vertebrae bifurcated; fourth trochanter of femur located at about midshaft; tail of some genera terminating in bony, club-like mass (Martin-Rolland 1999).

General description: Similar to camarasaurids, very long neck, relatively high skull with spatulate teeth, forelimbs approximately equal to or subequal to hindlimbs, some genera having tail clubs.

Age: Early to Late Jurassic.

Geographic distribution: Asia.

Taxa: Shunosaurinae plus Euhelopodinae.

Euhelopodidae *incertae sedis*: *Chiayüsaurus*, *Hudiesaurus*.

Note: Martin-Rolland (1999), in a review and revision of the sauropods of China, regarded all Chinese sauropods as belonging to the family Euhelopodidae, a group endemic to Asia during the Jurassic period. Martin-Rolland accepted two euhelopodid subfamilies, the Shunosaurinae (including the genera *Bellusaurus*, *Datousaurus*, *Shunosaurus*, and *Zizhongosaurus*) and the Euhelopodinae (including *Euhelopus*, *Mamenchisaurus*, *Omeisaurus*, and *Zigongosaurus*).

SHUNOSAURINAE

Diagnosis: Cervical vertebrae less elongated than in Euhelopodinae (Martin-Rolland 1999).

General description: More primitive forms, some bearing tail clubs.

Age: Early to Middle Jurassic.

Geographic distribution: Asia.

Genera: *Bellusaurus*, *Datousaurus*, *Shunosaurus*, *Zizhongosaurus*.

EUHELOPODINAE

Diagnosis: Cervical vertebrae very elongate, fortified transversely; cervical ribs extremely long; neural spines of dorsal vertebrae relatively low; bifurcation of neural spines of posterior cervical and anterior dorsal vertebrae (when present) weak (Martin-Rolland 1999).

General description: More advanced forms apparently lacking tail clubs.

Age: Late Jurassic.

Geographic distribution: Asia.

Genera: *Euhelopus*, *Mamenchisaurus*, *Omeisaurus*, *Zigongosaurus*.

TITANOSAURIA

Stem-based definition: Titanosauriforms more closely related to *Saltasaurus* than to *Brachiosaurus* or *Euhelopus* (Wilson and Sereno 1997).

Diagnosis: Anterior caudal centra with prominently convex (hemispherical) posterior face; anterior and middle caudal vertebrae having prespinal and postspinal laminae; sternal plates with strongly concave lateral margins; ulna with prominent olecranon process; carpals unossified or absent; ischium longer than pubis (Wilson and Sereno 1998).

General description: Rather primitive, slender to heavy, medium-sized to gigantic, with wide, sloping head, peglike teeth, frontlimbs about three-fourths length of hindlimbs, long "whiplash" tail, some or possibly all forms possessing dermal armor.

Age: Early to Late Cretaceous.

Geographic distribution: North America, South America, Africa, Europe, Asia.

Taxa: Andesauridae plus *Malawisaurus* plus *Argentinosaurus* plus Titanosauridae plus Eutitanosauria.

Titanosauria *incertae sedis*: ?*Campylodoniscus*, *Tangvayosaurus*, *Titanosaurus*.

Note: This division of the Titanosauria basically follows the cladogram of Wilson, Martinez and Alcober (1999), which was modified from Wilson and Sereno (1998). According to Wilson *et al.* (1999), three species of *Titanosaurus*—*T. indicus*, *T. colberti*, and *T.* [=*Laplatosaurus*] *araukanicus*—may not all belong in the same genus.

ANDESAURIDAE

Diagnosis: Titanosaurs with small pleurocoels located in anterior region of ovoid depression; two or three deep, triangular depressions on lateral surface of neural arch; neural arch and neural spine of dorsal vertebrae higher than in Titanosauridae; hyposphene-hypantrum very developed, some genera with extra articular surfaces; middle and distal caudal vertebrae with short prezygapophyses (Bonaparte and Coria 1993).

General description: More primitive than "Senonian" titanosaurids, back relatively taller, much diversification among genera.

Age: ?Late Jurassic to Late Cretaceous.

Geographic distribution: South America, ?Africa.

Genera: *Andesaurus*, ?*Epachthosaurus*.

Note: See "Note" under Eutitanosauria (below).

TITANOSAURIDAE

Diagnosis: Most derived characters including transversely expanded ischium and strongly procoelous anterior caudal vertebrae (those with caudal ribs or transverse processes [see "Notes" below]); in more derived forms, middle and posterior caudals also procoelous; teeth narrow, flattened in more primitive forms, may be more pencil-like in derived forms; sternal plates robust; neural spines of cervical vertebrae undivided, cervical ribs extending beyond centrum to overlap following vertebra (both primitive characters); teeth not limited to anterior portion of jaw, at least primitively; external nares far anterior (Jacobs, Winkler, Downs and Gomani 1993).

General description: Medium-sized to gigantic, variety of tooth forms, teeth small and slender, superficially resembling teeth in diplodocids.

Age: Late Jurassic to Late Cretaceous.

Geographic distribution: North America, South America, Africa, Europe, Asia.

Genera: *Aegyptosaurus*, *Aeolosaurus*, *Alamosaurus*, *Ampelosaurus*, *Chubutisaurus*, *Hypselosaurus*, *Iuticosaurus*, *Jainosaurus*, *Loricosaurus*, *Magyarosaurus*, *Pellegrinisaurus*.

Note: This family, for many years a kind of "catch basin" for numerous sauropod genera, is in need of revision and may not be valid. In fact, *Titanosaurus*, the genus for which this family was named, is now regarded as Titanosauria *incertae sedis*. Some taxa once included in this group were referred by Wilson, Martinez and Alcober (1999) to the new family Saltisauridae (see below).

Bonaparte, Heinrich and Wild (2000) questioned the taxonomic value of procoelous anterior caudal vertebrae, a character generally regarded as diagnostic for the Titanosauridae, pointing out that this has also been identified in various other nontitanosaurid taxa (*e.g.*, *Bellusaurus*, *Janenschia*, and *Mamenchisaurus*). These authors proposed that procoelous anterior caudals "represent an adaptation of the structure of the tail which was probably developed [independently] after the middle Late Jurassic in several different groups of sauropods"; that this condition is "associated with anterior dorsal vertebrae possessing double (bifurcated) or single neural spines." In the Titanosauridae, Bonaparte *et al.* noted, all caudal vertebrae are procoelous. Furthermore, titanosaurids are "characterized by a certain type of cervical and dorsal vertebrae "which are associated with procoelous caudals in most parts of the tail."

EUTITANOSAURIA

Node-based definition: The most recent common ancestor of *Saltasaurus*, *Argyrosaurus*, *Lirainosaurus*, plus the "Peirópolis titanosaur" and all its descendants (Sanz, Powell, Le Loeuff, Martinez and Suberbiola 1999).

Diagnosis: Synapomorphies including absence of hyposphene-hypantrum articulation in posterior dorsal vertebrae; reduced fourth trochanter of femur; presence of osteoderms (verified only in *Saltasaurus* and *Lirainosaurus*) Sanz, Powell, Le Loeuff, Martinez and Suberbiola 1999).

Generalized descriptions: Some or all forms armored.

Age: Late Cretaceous.

Geographic distribution: South America, Asia.

Taxa: Saltasauridae plus unnamed clade (including

Lirainosaurus and the titanosaur from Peirópolis, Brazil (Powell 1987).

Note: Other recent analyses of the Sauropoda have been proposed following that of Wilson, Martinez and Alcobar (1999). The most comprehensive of these, differing dramatically in some ways from Wilson *et al.*'s phylogeny, was published by Sanz, Powell, Le Loeuff, Martinez and Suberbiola (1999) in their description and analysis of the titanosaur *Lirainosaurus* (see entry). Sanz *et al.* retained the group Titanosauroidea (*sensu* Upchurch 1995; see *S1*) in their analysis, utilized *Patagosaurus* for outgroup comparison, and used as their ingroup *Haplocanthosaurus priscus*, *Andesaurus*, *Opisthocoelicaudia*, *Saltasaurus loricatus*, *Argyrosaurus*, *Epacthosaurus*, and the "Peirópolis titanosaur." Excluded from the analysis by these authors were *Aeolosaurus*, *Alamosaurus*, *Ampelosaurus*, *Magyarosaurus*, and such poorly known forms as *Titanosaurus*, *Antarctosaurus*, and *Neuquensaurus*.

The analysis of Sanz *et al.*—which may or may not be adopted by other workers—resulted in a single tree indicating to them the following: *Haplocanthosaurus* plus *Andesaurus* comprise a clade within "Titanosauroidea" (but outside of Titanosauria), *Opisthocoelicaudia* is a "titanosauroid" and close sister taxon of Titanosauria, and Titanosauria includes the genus *Epacthosaurus* plus a new taxon, Eutitanosauria. The latter group includes two clades—one joining *Argyrosaurus* plus *Saltasaurus*, the second made up of *Lirainosaurus* plus the "Peirópolis titanosaur."

SALTASAURIDAE

Partial diagnosis: Centra of distal caudal vertebrae procoelous (Wilson, Martinez and Alcober 1999).

General description: Medium to large-sized, some or all genera possessing dermal armor in the form of bony plates or ossicles imbedded in skin.

Age: Late Cretaceous.

Geographic distribution: South America, Asia.

Genera: *Argyrosaurus*, *Neuquensaurus*, *Opisthocoelicaudia*, *Rocasaurus*, *Saltasaurus*.

Note: The new family Saltasauridae was introduced by Wilson, Martinez and Alcober (1999) based upon caudal vertebrae morphology. (In 1986, Powell had proposed the new name Saltasaurinae for a subfamily of the Titanosauridae. This taxon is still being used by various South American workers, *e.g.* Salgado and Azpilicueta 2000.)

Ornithischia

Node-based definition: *Triceratops* and all taxa closer to Triceratops than to modern birds (Padian and May 1993).

Diagnosis: Synapomorphies including premaxilla with toothless and roughened (except in *Technosaurus*) rostral tip; palatal process of premaxilla horizontal or broadly arched; large lateral process of premaxilla excluding maxilla from margin of external naris; antorbital fenestra reduced or absent; ventral margin of antorbital paralleling maxillary tooth row; palpabral in orbit; prefrontal with long caudal ramus overlapping frontal; subrectangular quadratojugal located behind infratemporal fenestra; quadrate massive, elongate; predentary bone at front of mandible; dorsal border of coronoid eminence formed by dentary; mandibular condyle set below tooth row; upper and lower jaws having buccal emargination suggesting possession of pseudo-cheeks; cheek teeth with low triangular crowns having well-developed cingulum beneath; cheek teeth with crowns having low, bulbous base, margins with enlarged denticles; maxillary and dentary teeth with overlapping adjacent crowns; maxillary and dentary teeth not recurved; maximum tooth size near middle of maxillary and dentary tooth rows; at least five true sacral vertebrae in adults; no gastralia; ossified tendons (epaxial) at least above sacral region; pelvis opisthopibic, pubis with small prepubic process; lateral swelling of ischial tuberosity of ilium; iliac blade with preacetabular process and deep caudal process; pubis with obturator notch instead of foramen, formed between pubis and ischium; distal pubic and ischial symphyses; pubic and ischial symphyses only at their distal ends; femur with pendant fourth trochanter and finger-like lesser trochanter; pedal digit V reduced to small metacarpal with no phalanges (Benton 1990, based on Maryańska and Osmólska 1984, 1985, Norman 1984*a*, 1984*b*, Sereno 1984, 1986, and Gauthier 1986).

General description: Herbivorous, small to giant-sized, bipedal, semibipedal, and quadrupedal forms, much variety, head more remodeled than in Saurischia, gut larger than in saurischians, most forms with head having presumably horn-covered beak, most possessing cheek teeth, some forms with cranial crests, horns, or other ornamentation, some with dermal armor.

Age: Late Triassic to Late Cretaceous.

Geographic distribution: Worldwide.

Taxa: *Pisanosaurus* plus *Technosaurus* plus *Lesothosaurus* plus Fabrosauridae plus Genasauria.

Ornithischia *incertae sedis*: *Alocodon*, *Galtonia*, *Gongbusaurus*, *Lucianosaurus*, *Notoceratops*, ?*Revueltosaurus*, ?*Taveirosaurus*, *Tecovasaurus*, *Trimucrodon*, *Xiaosaurus*.

FABROSAURIDAE

Diagnosis: Unequivocal synapomorphies of lacrimal inserting into narrow slot in apex of maxilla;

mandible with peculiarly salient finger-like retroarticular process; particularly short forelimb, approximately 40 percent length of hindlimb; ilium with supra-acetabular flange over anterior half of acetabulum; posterior process of ilium with distinctive brevis shelf that first turns medially, then downwards; ischium with dorsal groove on proximal shaft; reduced pedal digit I, having splintlike shaft of metatarsal I and ungual extending just beyond send of metatarsal II (Peng 1997).

General description: Primitive, small and lightly built, bipedal and cursorial, unarmored, head triangular in shape, eyes at side of head, having narrow snout and possibly possessing beak, neck and trunk very short neck, long tail comprising almost half of body length, limbs slender, forelimbs about 40 percent length of hindlimbs, with very small hands.

Age: Late Triassic to Early Jurassic.

Geographic distribution: Africa, Asia.

Genera: *Agilisaurus*, *Fabrosaurus*.

GENASAURIA

Diagnosis: Cheeks; maxillary dentition offset medially; mandibular symphysis spout-shaped; coronoid process moderate; premaxilla with edentulous anterior portion; antorbital fossa with entire margin sharply defined, or extended as secondary lateral wall that encloses fossa; external mandibular foramen relatively smaller; pubic peduncle of ilium relatively less robust than inschial peduncle (Sereno 1986).

General description: All except very primitive ornithischians, all having cheek teeth and true cheeks, much variety.

Age: Early Jurassic to Late Cretaceous.

Geographic distribution: Worldwide.

Taxa: Thyreophora plus Cerapoda.

Genasauria *incertae sedis*: *Onychosaurus*.

THYREOPHORA

Diagnosis: Jugal orbital bar with transversely broad orbital rim, transversely broader than dorsoventrally tall; medial portion of quadrate condyle much more robust than lateral portion; dorsal body with parasagittal row of keeled scutes, lateral rows of low keels scutes (Sereno 1986, emended by Rosenbaum and Padian 2000).

General description: Small to very large, mostly quadrupedal but including more primitive bipedal or semibipedal forms, all possessing body armor of spikes or plates.

Age: Early Jurassic to Late Cretaceous.

Geographic distribution: Worldwide.

Taxa: *Scutellosaurus* plus *Emausaurus* plus Thyreophoroidea.

Thyreophora *incertae sedis*: *Lusitanosaurus*, ?*Tatisaurus*.

Note: The feature relating to the quadrate condyle in the above diagnosis, previously regarded by Sereno, as a synapomorphy of Thyreophoroidea, was also identified by Rosenbaum in *Scutellosaurus* (see entry).

THYREOPHOROIDEA

Diagnosis: Dentary tooth row having sinuous curve; supraorbital completely incorporated into orbit; basisphenoid markedly shortened in length relative to basioccipital; median palatal keel in midpalate, palatines and especially vomers vertically tall (Sereno 1986, modified by Rosenbaum and Padian 2000).

General description: Small to very large forms.

Age: Early Jurassic to Late Cretaceous.

Geographic distribution: Worldwide.

Taxa: *Scelidosaurus* plus Eurypoda.

Note: Rosenbaum and Padian noted that one character (quadrate condyle more robust medially than laterally) originally used by Sereno to diagnose Thyreophoroidea is also present in the basal thyreophoran *Scutellosaurus*. The latter two characters in Sereno's diagnosis could not be assessed in *Scutellosaurus lawleri* as those portions of the skull are not preserved in the described material (see entry).

EURYPODA

Diagnosis: Skull having two lateral supraorbitals forming dorsal margin of orbit; quadrate condyle strongly angled ventromedially; quadrate shaft not laterally distinct from pterygoid ramus; absence of otic notch between quadrate and paroccipital process; vertical median portion of pterygoid palatal ramus developed posteriorly, median palatal keel extending to posterior end of palate; exoccipital border of foramen magnum indented to form short recess on each side, supraoccipital and dorsal portion of border overhanging recess from above, occipital condyle floors recess from below; dentary with symphyseal portion very slender relative to ramus at midlength; atlas with neural arches fused to atlar intercentrum; parallel dorsal and ventral borders of scapular, distal expansion of blade minor or absent; femur having lesser trochanter entirely fused to greater trochanter; ilium with preacetabular process directed about 40 degrees lateral to axial column, postacetabular process relatively shorter; ischial blade without distal expansion and ventromedial slant; pedal digit IV with no more than four phalanges; metacarpals and metatarsals relatively short (Sereno 1986).

General description: "Plated" and "armored" dinosaurs, heavily built and slow moving, larger than more primitive thyreophorans.

Age: Early Jurassic to Late Cretaceous.

Geographic distribution: Worldwide.

Taxa: Stegosauria plus Ankylosauria.

STEGOSAURIA

Stem-based definition: All thyreophoran ornithischians closer to *Stegosaurus* than to *Ankylosaurus* (Galton 1997*a*).

Diagnosis: Quadrupedal, armored, derived cranial characters of large oval fossa on pterygoid ramus of quadrate and subrectangular, platelike quadrate head; derived postcranial characters; dorsal neural arch pedicles at least 1.5 times height of centrum; mid-dorsal vertebrae with transverse processes 50–60 degrees above horizontal; anterior dorsal vertebrae with spacious neural canal equal to more than half diameter of face of centrum; distal caudal centra subquadrate; absence of ossified epaxial tendons; acromion broad, platelike; triceps tubercile prominent, humerus with descending ridge; proximal carpals large, blocklike; intermedium and ulnare coossified; distal carpals absent; acetabular surface of pubis oval, laterally directed; prepubic process more than half length of postpubic process; metatarsal I and its digit absent; pedal digit III with only three phalanges (one phalanx lost); prominent osteoderms angling slightly away from sagittal plane, grading from short, erect plates anteriorly to longer posterodorsally angling spines posteriorly; proportionately narrow terminal spine pair projecting beyond last caudal vertebra; characteristic parascapular spine projecting posterolaterally from shoulder (Sereno and Dong 1992).

General description: Medium to very large, quadrupedal, with proportionally small, rather long head, brains comparatively smallest among all dinosaurs (except sauropods), short neck, short and massive forelimbs, limbs and feet elephant-like, front feet with five short, robust toes (at least two terminating in hooflike unguals), hind feet with short but massive and stout toes (digits I and apparently V either lost, vestigial or cartilaginous), two rows of back-directed (probably horn-covered) dermal plates or spines, plates usually paired, some forms with shoulder spines, plates grading from smaller to larger plates to spines, plates having one or more theorized functions (*e.g.* thermoregulation, increasing size of body outline, meager defense, intraspecific display, etc.), terminal spines covering most of tail, apparently used

Mounted skeleton of the *Stegosaurus ungulatus* (DINOLAB cast), the most common stegosaurid species.

Photograph by the author, courtesy New Mexico Museum of Natural History and Science, Albuquerque, New Mexico.

as defensive weapons when tail was swung horizontally.

Age: Middle Jurassic to Late Cretaceous.

Geographic distribution: North America, ?South America, Europe, Africa, Asia, ?Australia.

Taxa: Huayangosauridae plus Stegosauridae.

HUAYANGOSAURIDAE

Diagnosis: Cranial dorsal ribs with prominent hamularis process; distal end of pubic rod much enlarged (Dong, Tang and Zhou 1982).

General description: More primitive than Stegosauridae, medium-sized, possessing premaxillary teeth, apparently having two rows of small, paired dorsal plates, shoulder and dorsal spines.

Age: Middle Jurassic.

Geographic distribution: Asia.

Genera: *Huayangosaurus*, *Regnosaurus*.

STEGOSAURIDAE

Diagnosis: Skull with orbit caudal to maxillary tooth row; femorohumeral ratio of at least 145 percent (modified from Sereno 1986), coracoid higher than wide, width slightly more than half ventral part of scapula; scapula with prominent acromial process; ilium with prominent antitrochanter, reduced ischial peduncle (Galton 1990*b*).

General description: More derived than huayangosaurids, mostly large-sized, plates small to very large, arranged in two paired or alternating rows, spines on posterior region of back or tail, four to eight tail spikes.

Age: Middle Jurassic to Late Cretaceous.

Geographic distribution: North America, Africa, Europe, Asia.

Genera: *Changdusaurus* [*nomen nudum*], *Chialingosaurus*, *Craterosaurus*, *Dacentrurus*, *Kentrosaurus*, *Lexovisaurus*, *Monkonosaurus*, *Paranthodon*, *Stegosaurus*, *Tuojiangosaurus*, *Wuerhosaurus*, *Yingshanosaurus* [*nomen nudum*].

ANKYLOSAURIA

Stem-based definition: All thyreophoran ornithischians closer to *Ankylosaurus* than to *Stegosaurus* (Carpenter 1997*a*).

Diagnosis: Cranium low, flat, rear of skull wider than high; antorbital and supratemporal fenestrae closed; (in adults) sutures between cranial bones of skull roof obliterated; maxilla having deep, dorsally arched cheek emargination; passage between space

Life restoration of *Huayangosaurus taibaii*, a huayangosaurid stegosaur from China, painted by Berislav Krzic.

Photograph by the author, courtesy American Museum of Natural History.

Partial skeleton (AMNH 5665) of the nodosaurid ankylosaur *Edmontonia rugosidens*, left lateral view. A distinctive feature of nodosaurids is the lack of a tail club.

above palate and that below braincase closed by pterygoid; accessory antorbital ossification(s) and postocular shelf partially or completely enclosing orbital cavity; bony median septum extending from ventral surface of skull roof into palate as fused vomers; quadratojugal contacting postorbital; quadrate with dorsoventrally narrow pterygoid process; quadrate slanting rostroventrally from underside of squamosal; mandible with coossified keeled plate along ventrolateral margin; ribs of posterior dorsal vertebrae tending to fuse to vertebrae at centrum and along transverse process, may also underlie and fuse to expanded preacetabular segment of ilium; at least three posterior dorsal vertebrae fusing to sacrals, forming presacral rod, ribs of former contacting preacetabular process of ilium; neural and hemal spines of distal caudal vertebrae elongating along axis of tail, contact between adjacent hemal spines; ilium rotated into horizontal plane; acetabulum closed secondarily; pubis small, quite short extension of shaft adjacent to ischium; pubis almost entirely excluded from acetabulum; no prepubic process; body covered dorsally by armor plates of three-four shapes (including flat, oval to rectangular plates with keel, ridge or short spine externally; larger armor plates commonly symmetrically arranged in transverse rows) (Coombs and Maryańska 1990).

General description: "Armored dinosaurs," stocky, quadrupedal, small- to large-sized, with short legs, forelimbs about two-thirds to three-fourths length of hindlimbs, feet broad and short, primitive forms with five toes on hind feet, more advanced forms having four toes, head broad, carried low, covered by mosaic of armor plates, teeth small, leaf-shaped, body armor usually consisting of small rounded plates arranged in parallel strips, forming continuous bony shield on back, tail making up about half of body length.

Age: Early Jurassic to Late Cretaceous.

Geographic distribution: North America, South America, Europe, Asia, Australia, Antarctica.

Taxa: Nodosauridae plus *Minmi* plus Ankylosauridae.

Ankylosauria *incertae sedis*: *Brachypodosaurus*, *Danubiosaurus*, ?*Peishansaurus*, *Sauroplites*, *Stegosaurides*.

Note: Hill (1999), after performing a phylogenetic analysis utilizing 37 cranial characters in 19 thyreophoran taxa, found that the traditional division of the Ankylosauria into Nodosauridae and Ankylosauridae is strongly supported (uniting characters including closure of lateral temporal fenestra, vertically oriented quadrate, etc.) with the problematic genus *Minmi* positioned as the sister taxon to the latter family. The genera *Gastonia*, *Gargoyleosaurus*, and *Nodocephalosaurus* were found by Hill to "represent three successive sister taxa to the Ankylosauridae plus *Minmi* clade," which is supported by characters of the quadrate and cranial armor; exact placement of these taxa, however, must await the discovery of additional postcranial remains.

In a review of ankylosaurian osteoderms and a preliminary review of ankylosaurian armor, Ford (2000) proposed that such armor is diagnostic (see Coombs and Demere 1996) at various taxonomic levels, including that of subfamily (although K. Carpenter, personal communication 2000, is of the opinion that these osteoderms are not diagnostic enough to warrant the erection of new taxa). Therefore, Ford introduced a number of new subfamilies to both the Nodosauridae and Ankylosauridae, while also reinstating various previously proposed, but subsequently abandoned, taxa (including genera now regarded as junior synonyms, *e.g.*, *Scolosaurus*).

Ford retained Polacanthidae as a valid family under which he grouped the genera *Gastonia*, *Hoplitosaurus*, *Hylaeosaurus*, *Mymoorapelta*, "*Polacanthoides*" [=*Polacanthus*], and *Polacanthus*. Under Nodosauridae, that author included the subfamilies Edmontoniinae Russell 1940 (including "*Chassternbergia*" [=*Edmontonia*] and *Edmontonia*), Nodosaurinae Nopcsa 1923 (*Nodosaurus*), Panoplosaurinae Nopcsa 1929 (*Panoplosaurus*), Ford's new taxon Sauropeltinae (*Sauropelta* and *Silvisaurus*),

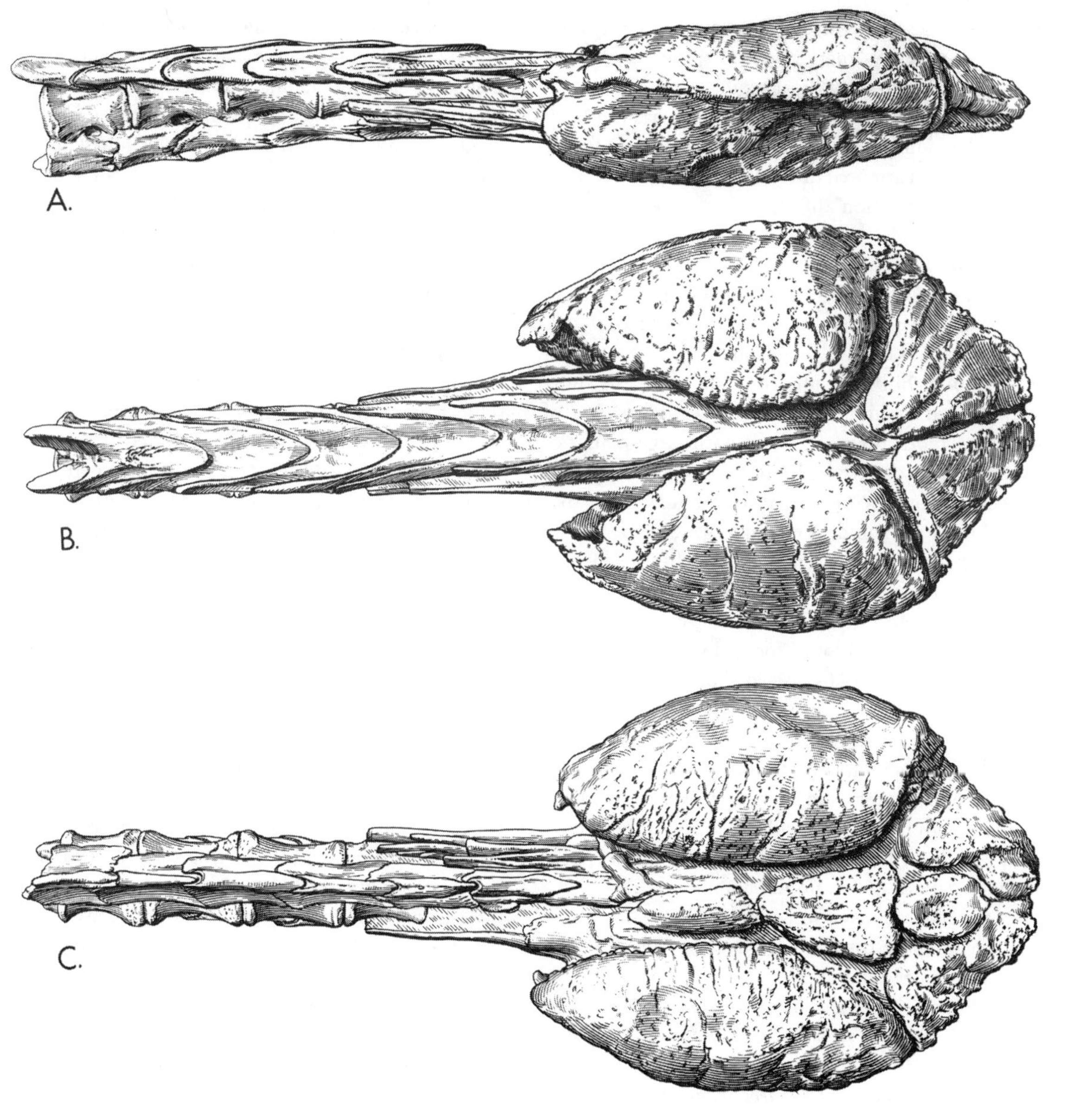

Distal caudal vertebrae and tail club of (AMNH 5214) of the large ankylosaurid ankylosaur species *Ankylosaurus magniventris* in A. left lateral, B. dorsal, and C. ventral views. Possession of a tail club is a diagnostic feature of the family Ankylosauridae. (After Coombs 1978.)

and Struthiosaurinae Nopcsa 1929 (*Struthiosaurus*). Under Ankylosauridae, Ford included Ankylosaurinae Brown 1908 (*Ankylosaurus*, "*Dyoplosaurus*" [=*Euoplocephalus*], *Euoplocephalus*, "*Scolosaurus*" [=*Euoplocephalus*], Syrmosaurinae Maleev 1952 (*Amtosaurus*, *Maleevus*, *Nodocephalosaurus*, *Pinacosaurus*, *Saichania*, *Shanxia*, *Talarurus*, *Tarchia*, and "*Tianzhenosaurus*" [=*Shanxia*]), Ford's new taxon Stegopeltinae (*Glyptodontopelta* and *Stegopelta*), and Shamosaurinae Tumanova 1983 (*Shamosaurus* and *Tsangantegia*).

Various other ankylosaurian genera were given the status of *incertae sedis*.

Of the new families introduced by him, Ford only supplied a diagnosis for the Sauropeltinae: nodosaurs having square cervical rings; medial scute round, flat, primary and secondary being posteriorly pointing spines; tranverse bands of armor across body; pelvic shield large, with round scutes intertwined with asymmetrical scutes not forming rosettes; large, posteriorly directed primary pectoral spine; small postorbital scute and large jugal scute.

Ford's system for the Ankylosauria has not been adopted and may be rejected by Carpenter in a future publication on armored dinosaurs (K. Carpenter, personal communication 2000).

NODOSAURIDAE

Diagnosis: Palate hourglass-shaped; basipterygoid processes consisting of a pair of rounded, rugose stubs; occipital condyle hemispherical, composed of basioccipital only, set off from braincase on short neck, angled about 50 degrees downward from line of maxillary tooth row; skull roof with large plate between orbits, rostrocaudally narrow plate along caudal edge of snout rostral to orbits; scapular spine displaced toward glenoid; coracoid large, craniocaudally long relative to dorsoventral width; ischium ventrally flexed at midlength (Coombs and Maryańska, modified from Sereno 1986).

General description: More primitive than ankylosaurids, small- to large-sized, body more upright than in ankylosaurids, head relatively longer, rather pear-shaped with narrow beak, with rounded corners, "hornless," single large plate on top center of skull, pairs of plates anterior to that plate over snout, head fortified with bony plates, side openings of head covered with bony plates, back of body sometimes bearing large lateral spikes, no tail club.

Age: Late Jurassic to Late Cretaceous.

Geographic distribution: North America, Europe, Australia, Antarctica.

Genera: *Acanthopholis*, *Animantarx*, *Anoplosaurus*, *Crataeomus*, ?*Cryptodraco*, *Dracopelta*, *Edmontonia*, *Hierosaurus*, ?*Lametasaurus*, *Leipsanosaurus*, *Niobrarasaurus*, *Nodosaurus*, *Palaeoscincus*, *Panoplosaurus*, *Pawpawsaurus*, ?*Priconodon*, *Priodontognathus*, *Sarcolestes*, *Sauropelta*, *Silvisaurus*, *Struthiosaurus*, *Texasetes*.

ANKYLOSAURIDAE

Diagnosis: Skull with maximum width equal to or greater than length; snout arching above level of postorbital skull roof; premaxillary septum dividing external nares, ventral or lateral opening leading into maxillary sinus; caudal process of premaxilla along margin of beak extending lateral to most mesial teeth; premaxillary palate wider than long; no ridge separating premaxillary palate from lateral maxillary shelf; premaxillary beak having edge continuous with lateral edge of maxillary shelf; secondary palate complex, twisting respiratory passage into vertical S-shaped bend; paired sinuses in premaxilla, maxilla, and nasal; postorbital shelf comprising postorbital and jugal bones extending farther medially and ventrally than in nodosaurids; near-horizontal epipterygoids contacting pterygoid and prootic; quadratojugal dermal plate prominent, wedge-shaped, projecting caudolaterally; infratemporal fenestra, paroccipital process, quadratojugal, and all but quadrate condyle hidden in side view by united quadratojugal and squamosal dermal plates; each lateral supraorbital element having sharp lateral rim and low dorsal prominence; flat lateral supraorbital margin above orbit; skull roof covered by numerous small scutes (including paired premaxillary scutes, single median nasal scute, two scutes between orbit and external nares laterally, scutes on each supraorbital element, single scute on squamosal and on quadratojugal dermal plates); scute pattern poorly defined in supraorbital region; coronoid process lower, more rounded than in nodosaurids; centra of distal caudal vertebrae partially or completely fused, with elongate, dorsoventrally broad prezygapophyses and elongate postzygapophyses united to form single dorsoventrally flattened, tongue-like process; distal caudal vertebrae having hemal arches with zygapophysis-like overlapping processes and elongate bases that contact, forming fully encased hemal canal; sternal plates fused; deltopectoral crest and transverse axis through distal humeral condyles in same plane; ilium with short postacetabular process; ischium almost vertical below acetabulum; pubis reduced to very small bone fused to ilium and ischium; fourth trochanter on distal half of femur; tail [except in polacanthines; see below] terminating with club comprising large pair of lateral plates and two smaller terminal plates; distal caudal vertebrae surrounded by ossified tendons (Coombs and Maryańska 1990, modified from Sereno 1986).

General description: Large, body more robust than in nodosaurids, head triangular with wide beak, entire surface of skull (in most forms) covered in a mosaic of small armor elements, triangular armor

Photograph by the author.

Skeleton (cast) of a recently discovered small (to date unnamed and undescribed) ankylosaurid exhibited at Dinofest (2000–2001), Navy Pier, Chicago.

plates suggesting small "horns" present at upper and lower corners of head, armor including lateral spikes (when present), some forms with sacral shield, most forms having tail terminating in a club (most likely used for defense) or incipient club.

Age: Middle Jurassic to Late Cretaceous.

Geographic distribution: North America, Asia.

Taxa: Shamosaurinae plus Polacanthinae plus ?*Tsagantegia* plus unnamed clade (including *Saichania* and *Tarchia* plus *Nodocephalosaurus*) plus unnamed clade (including *Euoplocephalus* and *Ankylosaurus*) plus Ankylosaurinae.

Ankylosauridae *incertae sedis*: *Pinacosaurus*.

POLACANTHINAE

Diagnosis: Ankylosaurid-like skull, almost straight, with parallel tooth rows, long basipterygoid processes; scapula having well-developed acromium flange arising from dorsal margin of scapula, ischia with ventral flexion at midlength as in Nodosauridae; armor including sacral shield of fused (not sutured) elements; elongate, posteriorly grooved shoulder spines; large, elongate, laterally directed, symmetrically-hollow-based, triangular caudal plates extending down the tail (Kirkland 1998*a*).

General description: Primitive ankylosaurids, with long head, exhibiting some features reminiscent of the nodosaurids, shield over sacral area of the body comprising fused elements, some or all genera possessing an incipient tail club.

Age: Late Jurassic to Late Cretaceous.

Geographic distribution: North America, Europe.

Genera: *Gargoyleosaurus*, *Gastonia*, *Hoplitosaurus*, *Hylaeosaurus*, *Mymooropelta*, *Polacanthus*.

Note: This taxon may be raised to family status (K. Carpenter, personal communication 2000).

SHAMOSAURINAE

Diagnosis: [No updated diagnosis currently published.]

General description: Large, differing slightly from other more advanced ankylosaurids in form of armor and other respects.

Age: Early Cretaceous.

Geographic distribution: North America, Asia.

Genera: ?*Glyptodontopelta, Shamosaurus, Stegopelta.*

[UNNAMED CLADE]

"Diagnosis": Possession of bulbous, polygon-shaped osteoderms (Sullivan 1998).

General description: Large, heavily armored forms.

Age: Late Cretaceous.

Geographic distribution: Asia.

Genera: *Nodocephalosaurus, Saichania, Tarchia.*

[UNNAMED CLADE]

Diagnosis: [None published.]

General description: Large, heavily armored forms.

Age: Late Cretaceous.

Geographic distribution: North America.

Genera: *Ankylosaurus, Euoplocephalus.*

ANKYLOSAURINAE

Diagnosis: [No modern diagnosis published.]

General description: Large, more derived, heavily armored ankylosaurids.

Age: Late Cretaceous.

Geographic distribution: North America, Asia.

Genera: ?*Amtosaurus*, ?*Heishansaurus*, ?*Maleevus*, *Sangonghesaurus* [*nomen nudum*], ?*Shanxia*, *Talarurus*, ?*Tianchisaurus*.

CERAPODA

Diagnosis: Spout-shaped mandibular symphysis; entire margin of antorbital fossa sharply defined, or extending as lateral wall enclosing fossa (Benton 1990).

General description: All primitive to advance ornithopods, dome-headed and horned ornithischians, having thicker enamel on one side of upper and lower cheek teeth.

Age: Early Jurassic to Late Cretaceous.

Geographic distribution: North America, South America, Europe, Asia, Africa, Australia.

Taxa: Ornithopoda plus Marginocephalia.

ORNITHOPODA

Diagnosis: Premaxillary teeth (if present) on level lower than maxillary teeth; jaw joint lower than tooth rows so that jaws come together like nutcrackers; premaxilla with process extending caudally toward orbit; femur with very large fourth trochanter (for attachment of caudifemoralis muscle group) (Fastovsky and Weishampel 1996).

General description: Most diverse of dinosaurian herbivorous groups, small to very large, smaller forms bipedal, larger forms bipedal to semibipedal to quadrupedal, unarmored, the first plant-eating dinosaurs possessing multiple tooth rows, cheek pouches, and true mastication, when present premaxillary teeth are on lower level than maxillary teeth, jaw joint lower than tooth row resulting in "nutcracker-like" biting action, jaws specialized for grinding soft vegetation, the first bipedal dinosaurian herbivores occupying complete range of size.

Age: Early Jurassic to Late Cretaceous.

Geographic distribution: Worldwide.

Taxa: Heterodontosauridae plus Euornithopoda.

Ornithopoda *incertae sedis*: *Drinker, Loncosaurus, Othnielia.*

HETERODONTOSAURIDAE

Diagnosis: Cheek teeth high-crowned; crowns chisel-shaped, denticles restricted to apical-most third of crown; caniniform tooth in both premaxilla and dentary (Weishampel and Witmer 1990*b*).

General description: Primitive forms, among smallest and earliest of ornithopods, predentary probably covered with horn, canine-like teeth present, cheek teeth closely packed to form solid dental battery, arms relatively long, hands large.

Age: Early Jurassic.

Geographic distribution: North America, Africa, Asia.

Genera: *Abrictosaurus, Dianchungosaurus*, ?*Echinodon, Geranosaurus, Heterodontosaurus, Lanasaurus, Lycorhinus*, ?*Oshanosaurus* [*nomen nudum*].

EUORNITHOPODA

Diagnosis: Antorbital fenestra smaller in area than antorbital fossa it overlies; sinuous lower margin to jugal arch; space separating lower edge of quadratojugal from lateral margin of jaw articulation; pleurokinetic hinge (lateral rotation of cheek teeth developed); anterior process of pubis comparable in length to (or exceeding in length) anterior process of ilium; medioventral edge of shaft of ischium with discrete, finger-shaped obturator process; shallow anterior intercondylar groove on distal end of femur; ?angle between anterior and posterior pubic ram less than 100 degrees (Norman 1998*a*).

General description: Most ornithopods excluding most primitive forms, small to giant, mostly bipedal.

Age: Middle Jurassic to Late Cretaceous.

Geographic distribution: Worldwide.

Taxa: Hypsilophodontidae plus Iguanodontia.

HYPSILOPHODONTIDAE

Stem-based definition: All ornithopods closer to *Hypsilophodon* than to *Iguanodon* (Sues 1997*b*).

Diagnosis: Maxilla with rostral process fitting into groove on posteromedial aspect of premaxilla; frontal with lateral peg fitting into socket on medial

Photograph courtesy Steve Hutt and the Museum Isle of Wight Geology.

Skeletal cast prepared by Steve Hutt of the primitive ornithopod *Hypsilophodon foxi*. The validity of the Hypsilophodontidae, a long accepted family based upon the genus *Hypsilophodon*, has recently come into question.

surface of postorbital; cingulum on dentary teeth; scapula shorter than, or as long as, humerus; prepubic process rod-like in shape, generally wider mediolaterally than deep dorsoventrally; ossification of sternal segments of more anterior dorsal ribs; sheath of epaxial and hypaxial ossified tendons surrounding distal part of tail (Sues and Norman 1990; Weishampel and Heinrich 1992).

General description: Small- to medium-sized, lightly-built, facultatively and possibly obligately bipedal, with comparatively heavy hind legs, head with narrow snout, chisel-shaped cheek teeth, primitive features including teeth in premaxilla, the first ornithopod group to have widespread, almost global distribution, formerly (though erroneously) believed to have been arboreal.

Age: Middle Jurassic to Late Cretaceous.

Geographic distribution: North America, Europe, Asia, Australia, Antarctica.

Taxa: Hypsilophodontinae plus Thescelosaurinae.

Note: Scheetz (1998) reported briefly on a new phylogenetic analysis of basal ornithopods. Analyzing 20 ornithopod taxa, incorporating 76 cranial and 48 postcranial characters, Scheetz found that the long traditional family "Hypsilophodontidae is a pectinate grade of dinosaurs with a concommitment trend in increased size and herbivorous efficiency," with widespread distribution of small ornithopod taxa taking place before the Late Jurassic.

Scheetz's study reached the following conclusions: *Gasparinisaura* fills an evolutionary gap between *Hypsilophodon* and *Tenontosaurus*; *Tenontosaurus* and *Rhabdodon* (see entry) should be placed between what were traditionally termed Hypsilophodontidae and Iguanodontidae; the terms "Hypsilophodonts" and "Iguanodonts" should not be used for taxa occurring before the taxa for which these group-names are derived; also, most major herbivorous adaptations were present within ornithopods before the first occurrence of angiosperms in North America, negating claims of an evolutionary radiation.

Similar conclusions were reached by Winkler, Murry and Jacobs (1998), published in a brief report on an abundant sample of fossils representing a population of a new, small ornithopod dinosaur, recovered from the Early Cretaceous Proctor Lake locality in Comanche County, Texas. This form exhibits features that have been used to diagnose the Hypsilophodontidae, which, the authors pointed out, has generally grouped together most small primitive ornithopods of the Middle Jurassic to Late Cretaceous. Although the new form "preserves individual variation in other characters that are often used to diagnose this family," it also possesses "attributes used to exclude its members from the Iguanodontia." Winkler *et al.* discovered a mosaic of characters in this ornithopod, including polymorphism in the manual phalangeal formula and the shape of the prepubis.

For these reasons, Winkler *et al.* suggested that the "Hypsilophodontidae" be removed from phylogenetic considerations, that at least some of its members, including the genus *Hypsilophodon* upon which this family was erected, be best regarded "as an array of sister taxa to the Iguanodontia," and that, given the glimpse of an evolutionary transition afforded by the Proctor Lake assemblage, the concept of the Iguanodontia also reevaluated.

More recently, in a review on the hypsilophodontids of southeastern Australia (in which the new genus *Qantassaurus* was introduced (see entry), Rich and Rich (1999) accepted the Hypsilophodontidae as a valid family.

HYPSILOPHODONTINAE

Diagnosis: Skull roof having narrow interorbital width; frontal relatively narrow transversely; ventral margin of occipital condyle, basal tubera, and pterygoid processes lie in plane sloping approximately 35 degrees anteroventrally (Sereno 1986; Weishampel and Heinrich 1992).

General description: More primitive forms.

Age: Late Jurassic to Late Cretaceous.

Geographic distribution: North America, Europe, Asia, Australia, Antarctica.

Genera: *Atlascoposaurus, Fulgurotherium, Hypsilophodon, Laosaurus, Leaellynsaura, Nanosaurus, Notohypsilophodon, Orodromeus, Othnielia, Parksosaurus, Phyllodon, Qantassaurus, ?Siluosaurus, Yandusaurus, Zephyrosaurus.*

THESCELOSAURINAE

Diagnosis: Cheek teeth with thicker enameled side (buccal for maxillary and lingual for dentary teeth) bearing series of secondary ridges medially and distally to form two converging circles on either side of central ridge; posterior half of ventral ridge of jugal offset ventrally, covered ventrolaterally with obliquely inclined ridges (see Galton 1997*c*); femur longer than tibia (Galton 1997*b*).

General description: More advanced forms.

Age: ?Late Jurassic to Late Cretaceous.

Geographic distribution: North America, ?Europe.

Genera: *Bugenasaura, Thescelosaurus.*

IGUANODONTIA

Diagnosis: Digit I of manus having conical, pointed, strongly divergent ungual; elongate and ligamentously bound ("bundled") metacarpals II–IV; metacarpals IV subequal in length to III, both considerably longer than metacarpal II; metacarpal II and its first phalanx approximately equal in length to III; metacarpal II proximal to III and IV, projecting into recess on distal surface of carpus; digits II and III with flattened, distally rounded ungual phalanges; sternal bone having caudolateral process; dorsoventrally expanded distal end of anterior pubic process; posterior pubic process shorter than ischium; pedal ungual phalanges of digits II–IV flattened, broad; lattice-like ossified tendons flanking dorsal, sacral, and proximal caudal neural spines (Norman 1998*a*).

General description: Medium- to giant-sized, mostly bipedal (larger forms apparently facultatively bipedal or semiquadrupedal), no premaxillary teeth.

Age: Late Jurassic to Late Cretaceous.

Geographic distribution: Worldwide.

Taxa: ?*Tenontosaurus* plus ?*Muttaburrasaurus* plus ?*Rhabdodon* plus Euiguanodontia.

Iguanodontia *incertae sedis*: *Callovosaurus, Craspedodon, Lophorhothon.*

EUIGUANODONTIA

Stem-based definition: *Gasparinisaura* plus Dryomorpha, but excluding *Tenontosaurus* and all other ornithopods (Coria and Salgado 1996).

Diagnosis: Unequivocal synapomorphies including jugal-postorbital articulation facing laterally; maxillary teeth having prominent lateral primary ridge; well-developed brevis shelf; metatarsal I reduced or absent (Coria and Salgado 1996.)

General description: Small to giant, slightly more advanced iguanodontians.

Age: Late Jurassic to Late Cretaceous.

Geographic distribution: North America, South America, Europe, Asia, Australia, Africa, ?Antarctica.

Taxa: *Gasparinisaura* plus Dryomorpha.

DRYOMORPHA

Diagnosis: Predentary with paired ventrolateral process; maxilla with lateral and medial rostral processes separated by oblique channel for caudolateral premaxillary process; teeth supported also by bone of alveolous, molded around each root and crown; development of specialized premaxilla-maxilla hinge as part of functional evolution of pleurokinesis; channel-like anterior intercondylar groove on femur; prominent articular head developed on posterior side of proximal end of humerus; quadratojugal reduced to small element separating jugal and quadrate; quadrate having notchlike embayment in its lateral wing, forming caudal margin of quadrate (paraquadratic) foramen; close-packed teeth in maxillary and dentary dentitions; width across orbital region of skull roof greater than that across occipital region; ischial shaft arched ventrally along its length; development of small distal boot with flat medial surface on ischium; obturator process situated proximally on shaft of ischium (within 25 percent of total length of ischium) (Norman 1998*a*).

General description: Primitive but more advanced than hypsilophodontians, small, bipedal, quite long tail.

Age: Middle Jurassic to Late Cretaceous.

Geographic distribution: North America, South America, Europe, Asia, Africa, ?Antarctica.

Taxa: Dryosauridae plus Ankylopollexia.

DRYOSAURIDAE

Diagnosis: Skull with premaxilla not enclosing external naris dorsally; ilium with wide bevis shelf; distal articular end of femur having deep extensor groove; femur with deep pit (for insertion of M. caudifemoralis) developed at base of fourth trochanter; metatarsal I vestigial (Sues and Norman 1990).

General description: Small and lightly built, unspecialized, somewhat resembling hypsilophodontids, with sharp-ridged cheek teeth, most likely obligatively bipedal, the last ornithopods with relatively very short arms, hand five-fingered, long and stiffened tail making up more than half total length of animal, the first ornithopods to weigh over 100 kilograms (about 245 pounds).

Age: Late Jurassic to Early Cretaceous.

Geographic distribution: North America, Africa.

Genera: *Dryosaurus*, *Kangnasaurus*, *Valdosaurus*.

ANKYLOPOLLEXIA

Stem-based definition: Iguanodontians more advanced than Dryosauridae (Sereno 1986).

Diagnosis: Differs from more primitive ornithopods in dentary tooth crowns having lingual surface with distally offset and reduced primary ridge separated by shallow, vertical trough from low, broad secondary ridge (small number of tertiary ridges developed from base of marginal denticles); carpals partially fused into two blocks, one associated with distal radius, one with distal ulna; manual ungual modified into spur; transversely flattened and broadened prepubic process; rhamboidal latticework of ossified tendons (Norman and Weishampel 1990, based on Sereno 1986).

General description: Medium-sized to giant iguanodontians.

Age: Middle Jurassic to Late Cretaceous.

Geographic distribution: North America, South America, Europe, Asia, Africa.

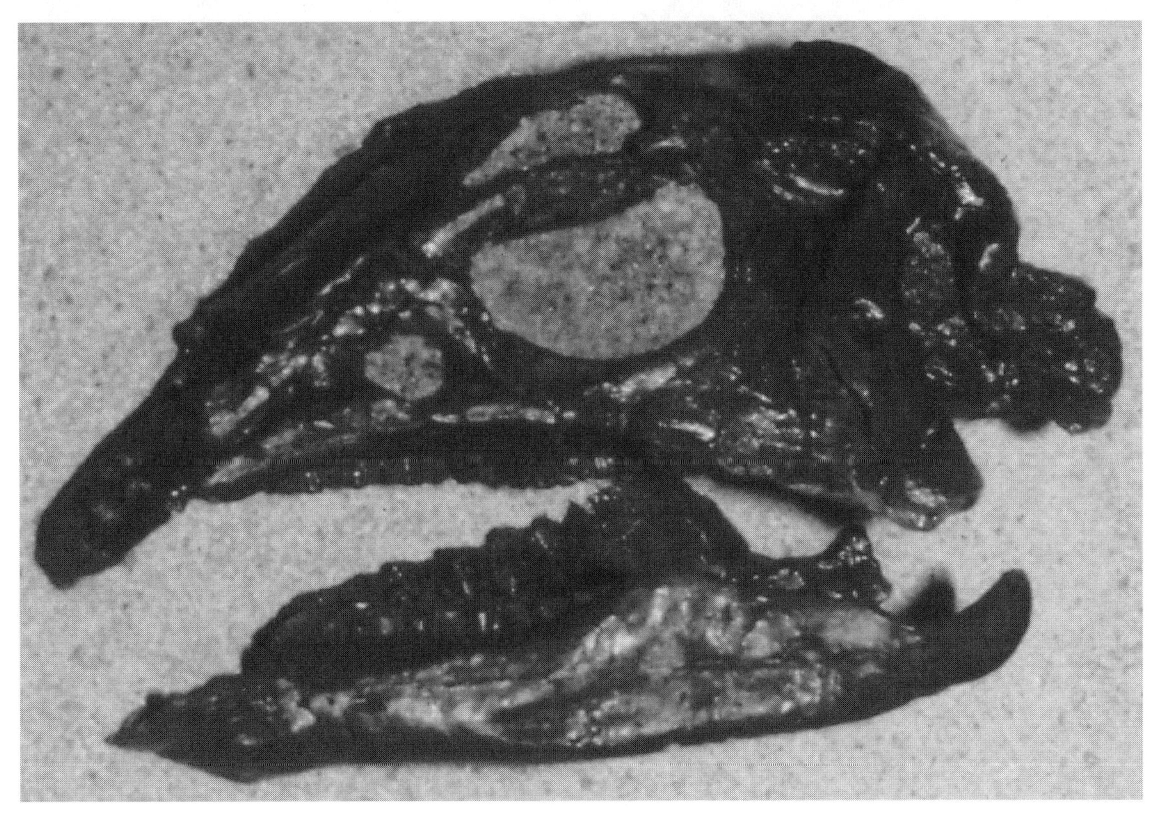

Skull (cast) of the dryosaurid ornithopod *Dryosaurus altus*.

Photograph by the author, courtesy Royal Tyrrell Museum/Alberta Community Development.

Taxa: Camptosauridae plus unnamed clade (including Iguanodontidae plus Hadrosauroidea).

CAMPTOSAURIDAE

Diagnosis: Prominent primary ridge on labial surface of maxillary crowns; inrolling of enamel on lower half of mesial edge of crown (in lingual view), forming oblique cingulum; maxillary crowns narrower and more lanceolate (in labial view) than dentary crowns (in lingual view); heavy ossification of carpus (metacarpal I shortened, incorporated obliquely into carpus; consequent partial fusion of modified carpus into two blocks — metacarpal 1 + radiale + intermedium + adjacent distal carpals, and ulnare + distal carpals 4 and 5 — these articulating via broad concave facets with distal end of radius and ulna); short, divergent digit I of manus and subconical manus digit I ungual (Sereno 1986, regarding *Camptosaurus*); digits II and III dominant, ending in dorsoventrally flattened unguals; digits displaying ability to hyperextend and well-defined articular relationships (indicating use in weight support/locomotion); manus digit III having lost one phalanx (Norman 1998*a*).

Photograph by the author, courtesy National Museum of Natural History, Smithsonian Institution.

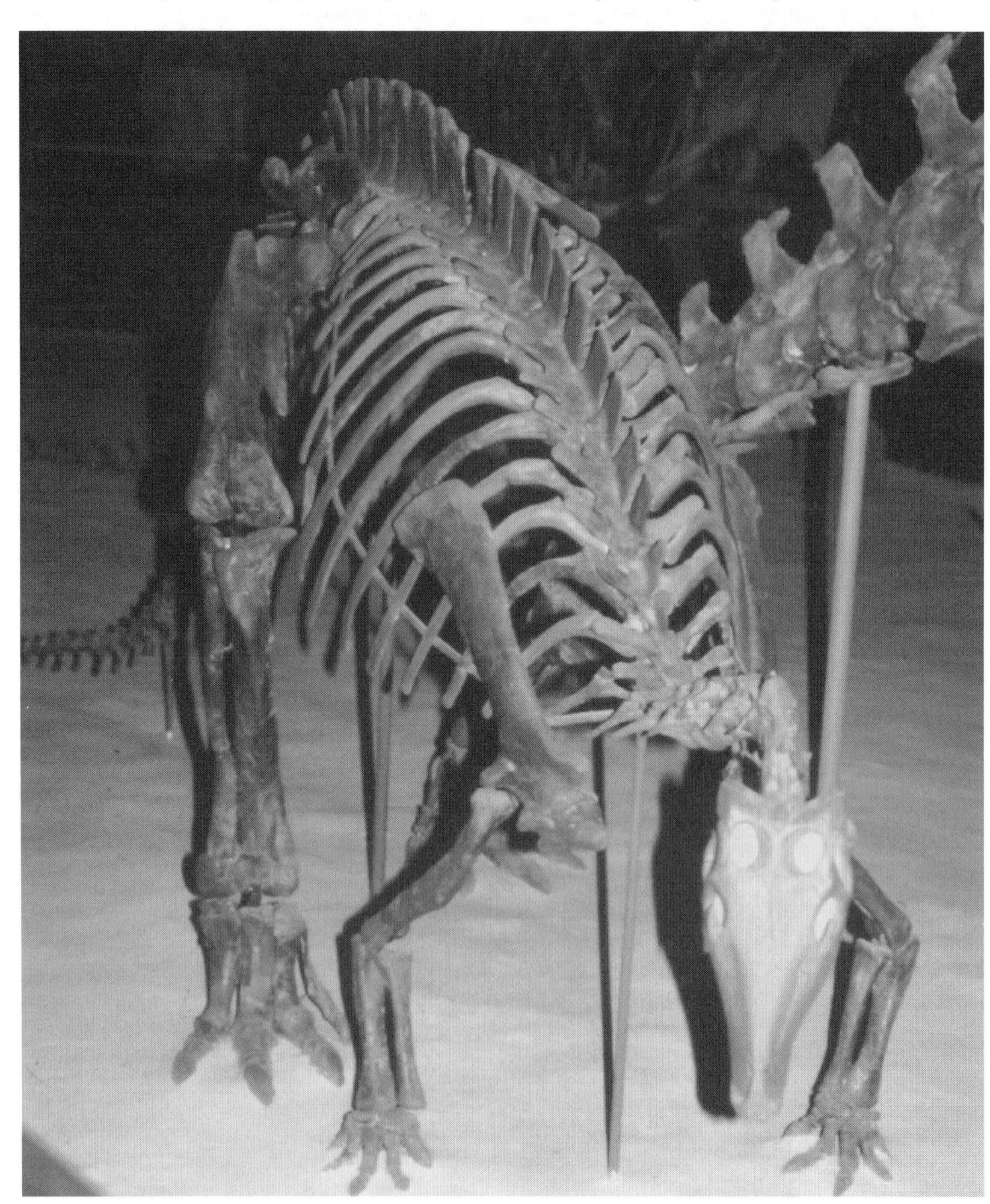

Holotype (USNM 4282) skeleton of *C. browni* the camptosaur *Camptosaurus* now referred to *C. dispar* mounted by Charles Whitney Gilmore in a quadrupedal pose, "an attitude which it is believed was often assumed" (Gilmore 1909). This nearly complete specimen was collected by Othniel Charles Marsh from the Morrison Formation, in Albany County, east of Como Bluff, Wyoming.

General description: More advanced than dryosaurids, medium to semilarge, rather robust, bipedal to semibipedal, wide hips, forelimbs comparatively longer than in dryosaurids, forelimbs much shorter than hindlimbs, hands with large wrist and short, spur-like first finger, hindlimbs thick, feet broad and stocky, tail without tall spines, the first heavily built ornithopods measuring over 3 meters in length, the first with noticeably elongated muzzle, the first with two functional tooth rows in both upper and lower jaws, teeth set above each other in staggered pattern to form a single chewing unit.

Age: Middle to Late Jurassic to ?Early Cretaceous.

Geographic distribution: North America, Europe.

Genus: *Camptosaurus*.

[UNNAMED CLADE]

Diagnosis: Synapomorphies including supraoccipital excluded from margin of foramen magnum; sternal bones having rod-shaped, caudolaterally directed processes; ulna with well-developed olecranon process; prepubis dorsoventrally expanded, prepubic blade and neck distinctly separated; caudal ramus of pubis reduced in length; femur almost straight in lateral view (possible reversion in *Telmatosaurus transsylvanicus*) (Godefroit, Dong, Bultynck, Hong and Feng 1998).

General description: Medium to very large-sized forms.

Age: Late Jurassic to Late Cretaceous.

Geographic distribution: Europe, North America, South America, Africa, Asia, Antarctica.

Taxa: Iguanodontidae plus Hadrosauroidea.

IGUANODONTIDAE

Diagnosis: Dentary teeth broad, asymmetrical, having distally off-set but relatively low primary ridge, separated by shallow trough from parallel secondary ridge which meets apex of crown at a mesial "shoulder" (a variable number of parallel tertiary ridges running down crown surface, usually taking their origin from bases of marginal denticles); dentary and maxillary teeth interlocking through shape of their adjacent crowns (little evidence of cementum locking crowns and roots of teeth into battery); lacrimal locked on to fingerlike process at apex of maxilla, extensively overlain by premaxilla and prefrontal; palpebral 1 (supraorbital) with large base-plate articulating loosely against prefrontal alone; palpebral curving upward and backward, following line of orbital margin (forming equivalent of a prominent brow ridge) terminating near postorbital and sometimes followed by loose (rarely preserved) accessory palepebral; jugomaxillar suture complex, consists of finger-like caudolateral process of maxilla, which slots into large recess of medioventral surface of jugal (the two meeting in externally sinuous sutural line); ungual phalanx of manus digit I hypertrophied as long, medially directed, robust conical spike, articulating against its metacarpal via flattened, disclike first phalanx (this ungual may be subequal to, or greater in length to digit II of manus); manus digit V showing evidence of hyperphalangy, diverging strongly from digits II–IV; carpal bones and metacarpal 1 fused indistinguishably into two bone blocks (Norman 1998*a*).

General description: Medium to large, head long with low snout (broader than in camptosaurs), horny beak, numerous grinding teeth, large shoulders and forelimbs, second and third fingers blunt and hooflike, thumb developed into "spike" in some or all forms, hindlimbs and feet broad.

Age: Late Jurassic to Early Cretaceous.

Geographic distribution: Europe, North America, Africa, Asia, ?South America.

Genera: *Altirhinus*, *Bihariosaurus*, *Iguanodon*, *Lurdusaurus*, *Ouranosaurus*.

HADROSAUROIDEA

Diagnosis: Iguanodontia characterized by more than one replacement tooth in each tooth position; crowns firmly cemented together, forming rigid tooth battery; articular surface of occipital condyle not steeply inclined downwards; basipterygoid processes long, slender, projecting more ventrally than occipital condyle; presence of fontanelles on skull of younger specimens; surangular foramen absent; maxillary teeth miniaturized, bearing single large carina; preacetabular process of ilium very deflected ventrally; developed antitrochanter of ilium; obturator foramen of pubis fully open; transverse widening of proximal head of tibia extending distally (Godefroit, Dong, Bultynck, Li and Feng 1998).

General description: Medium-sized, somewhat resembling but more primitive than hadrosaurids.

Age: Late Cretaceous.

Geographic distribution: Asia, Europe, Antarctica.

Taxa: ?*Bactrosaurus* plus ?[Unnamed clade including *Eolambia* plus ?*Probactrosaurus*] plus Hadrosauridae.

Hadrosauroidea *incertae sedis*: *Claosaurus*, *Gilmoreosaurus*, *Mandschurosaurus*, *Secernosaurus*, *Tanius*.

HADROSAURIDAE

Node-based definition: Euhadrosauria plus *Telmatosaurus*, *Secernosaurus*, their common ancestor, and all its descendents (Weishampel, Norman and Grigorescu 1993; see "Note" below).

Diagnosis: Synapomorphies including reduction of denticulations on oral margin of premaxillary and

Life restoration of the Mongolian, "bulbous nosed" iguanodontid *Altirhinus kurzanovi*, painted by Berislav Krzic.

predentary bones; narrow, subhemispherical mandibular condyle; caudal border of skull not indented; medial rami of paired squamosals almost in contact; diastema in mandible (convergently evolved in *Ouranosaurus nigeriensis*); coronoid process of dentary forwardly inclined; dentary teeth miniaturized; dentary teeth not recurved; dentary teeth with large single carina; odontid process free from axis; angular deltopectoral crest (convergently evolved in *Bactrosaurus johnsoni*) (Godefroit, Dong, Bultynck, Hong and Feng 1998).

General description: Similar to, but more advanced than, iguanodontids, includes all hadrosaurids (duckbilled dinosaurs), medium- to giant-sized, with three vertical tooth rows having hundreds of interlocking teeth forming tooth battery.

Age: ?Early to Late Cretaceous.

Geographic distribution: North America, South America, Europe, Asia, Antarctica.

Taxa: ?*Protohadros* plus *Telmatosaurus* plus ?*Gilmoreosaurus* plus *Secernosaurus* plus Euhadrosauria.

Note: As pointed out by M. K. Brett-Surman (personal communication 2000), the Hadrosauridae has traditionally been defined as the clade including Hadrosaurinae plus Lambeosaurinae and its common ancestor. The classification scheme was confused, however, by introducing Euhadrosauria at the wrong level.

EUHADROSAURIA

Node-based definition: Crown-based clade including Hadrosaurinae plus Lambeosaurinae, their common ancestor, and all of its descendants (Weishampel, Norman and Grigorescu 1993).

Diagnosis: Absence of denticulate oral margin of premaxilla; quadrate with narrow mandibular condyle; dentary teeth narrow; dentary teeth with large single carina; angular deltopectoral crest (Weishampel *et al.* 1993).

General description: Large to giant, body plan generally quite similar to iguanodontids (no "spike" thumb), head with "duckbills" formed by flat, broad jaws, some forms having solid or hollow cranial crests, possibly some with dorsal dermal "frill," forelimbs

Photograph by the author, courtesy Natural History Museum of Los Angeles County.

Mounted skeleton (cast of IVP AS V251) of the lambeosaurine hadrosaurid *Tsintaosaurus spinorhinus*. The narrow, vertically oriented "crest" shown in earlier mounts (for photograph, see *Tsintaosaurus* entry, *D:TE*) of this species has recently been brought down to a horizontal orientation and reinterpreted as a nasal.

relatively longer than iguanodontids with long forearms, fingers covered in fleshy "mitten," fingers and toes hooved, this group comprising the most widespread, diverse, and (with the ceratopsinae) most numerous of Late Cretaceous large-bodied herbivores.

Age: ?Early to Late Cretaceous.

Geographic distribution: North America, South America, Europe, Asia, Antarctica.

Taxa: "Fontllonga hadrosaurid" plus Hadrosaurinae plus Lambeosaurinae.

Photograph by the author, courtesy Natural History Museum of Los Angeles County.

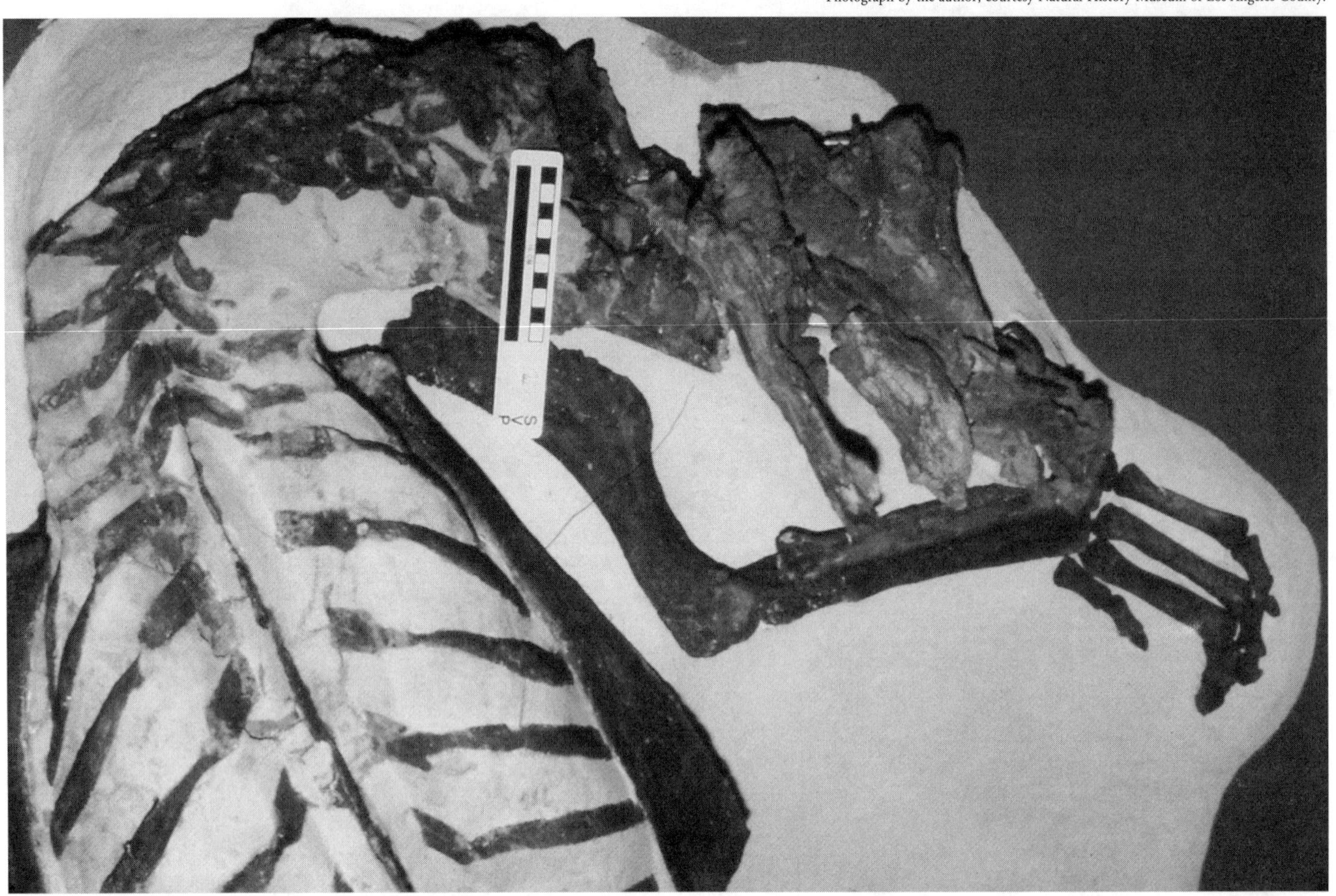

Skeleton (LACM 7233/ 23504), juvenile individual, of the hadrosaurine *Edmontosaurus annectens*, collected by Harley J. Garbani from the Hell Creek Formation, Garfield County, Montana.

Hadrosauridae *incertae sedis*: *Cionodon*, *Diclonius*, *Hypsibema*, *Microhadrosaurus*, *Ornithotarsus*, ?*Orthomerus*, *Pneumatoarthrus*, *Thespesius*.

HADROSAURINAE

Diagnosis: Premaxillary lip reflected; external naris enlarged, up to 40 percent of basal skull length (reversed in *Maiasaura peeblesorum*); maxilla symmetrical in lateral view; circumnarial depression extending up onto nasal; angle between crown and root of dentary teeth less than 130 degrees (evolved convergently in *Iguanodon*); groove on ventral side of sacrum (homoplastic character [M. K. Brett-Surman, personal communication 2000] convergent in *Iguanodon*); ischial shaft tapering distally (Godefroit, Dong, Bultynck, Hong and Feng 1998, including characters from Kirkland 1998*b*, Sereno 1986, Brett-Surman 1989, Weishampel and Horner 1990, Weishampel, Norman and Grigorescu 1993).

General description: Generally large (including giant) but more gracile forms, with comparatively longer head than lambeosaurines, either "flat-headed" or with solid cranial crest.

Age: Late Cretaceous.

Geographic distribution: North America, South America, Europe, Asia.

Genera: *Anantotitan*, *Aralosaurus*, *Brachylophosaurus*, *Edmontosaurus*, *Gryposaurus*, *Hadrosaurus*, *Kritosaurus*, *Maiasaura*, *Prosaurolophus*, *Saurolophus*, *Shantungosaurus*.

Note: In their phylogenetic analysis of the lambeosaurine genus *Nipponosaurus*, published as an abstract, Suzuki, Weishampel and Minoura (2000) found that the Hadrosaurinae collapses, with *Prosaurolophus* and *Saurolophus* positioned as outgroups of Lambeosaurinae. This potentially controversial conclusion, which differs dramatically from earlier (and traditional) studies, resulted from "the addition of many newly-found postcranial characters and revision of former hadrosaurine synapomorphies." Seemingly, this result shows "that hadrosaurid postcrania are conservative and therefore retain a transitional state in their characters." Details of this analysis are presumably to be published at a later date, after which time the conclusion of Suzuki *et al.* may be adopted. For the present, however, this document conservatively retains the generally accepted division of the Hadrosauridae into two subfamilies, Hadrosaurinae and Lambeosaurinae.

Photograph by the author, courtesy Royal Tyrrell Museum/Alberta Community Development.

Mounted skeleton of the medium-sized pachycephalosaur *Stegoceras validum*. The hypothesis that such dome-headed dinosaurs engaged in head-butting behavior has been largely discounted in recent years.

LAMBEOSAURINAE

Diagnosis: Frontal entirely excluded from orbital rim; parietal shortened (evolved convergently in *Ouranosaurus nigeriensis*); lacrimal laterally "lapped" by maxilla; maxillary shelf developed; hollow supranarial crest (expansion of nasal apparatus); external naris completely surrounded by premaxilla; jugal with truncated and rounded rostral process; absence of premaxillary foramen; caudal neural spines very tall; forelimb robust (Godefroit, Dong, Bultynck, Hong and Feng 1998, with additional characters from Kirkland 1998*b*, based on characters proposed by Sereno 1986, Brett-Surman 1989, Weishampel and Horner 1990, Weishampel, Norman and Grigorescu 1993).

General description: More robust forms, head relatively shorter than hadrosaurines, head adorned with hollow crest, body taller in region of back because of more elongate sacral neural spines, forelimbs shorter, but stouter, than in hadrosaurines.

Age: ?Early to Late Cretaceous.

Geographic distribution: North America, Europe, Asia.

Genera: *Amurosaurus*, ?*Barsboldia*, *Charonosaurus*, *Corythosaurus*, *Hypacrosaurus*, ?*Jaxartosaurus*, *Lambeosaurus*, *Nipponosaurus*, *Pararhabdodon*, *Parasaurolophus*, *Tsintaosaurus*.

MARGINOCEPHALIA

Node-based definition: Ceratopsia and Pachy-

Photograph by the author, courtesy The Academy of Natural Sciences of Philadelphia.

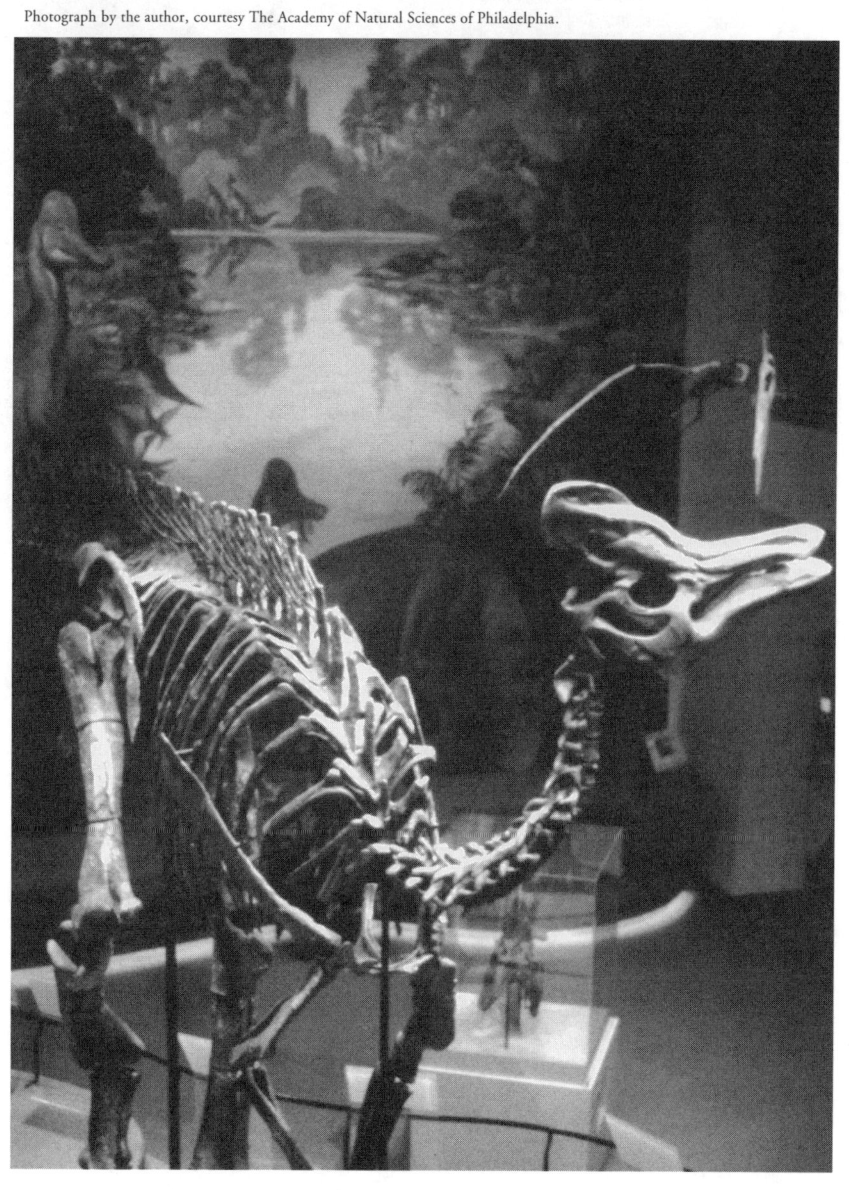

Lambeosaurine hadrosaurid skeleton labeled *Corythosaurus casuarius*, as remounted in 1984 by paleontologist Kenneth Carpenter for the "Discovering Dinosaurs" exhibit at The Academy of Natural Sciences of Philadelphia. The mount comprises the postcrania (ANSP 16969) of an indeterminate lambeosaurine referred to *Corythosaurus* sp. and skull (cast of ROM 776, holotype of *C. intermedius*, female) of *C. casuarius*. This skeleton was originally mounted in 1957 with an incorrect posture for the Academy's former Hall of Earth History.

Courtesy The Field Museum (neg. #GEO81554).

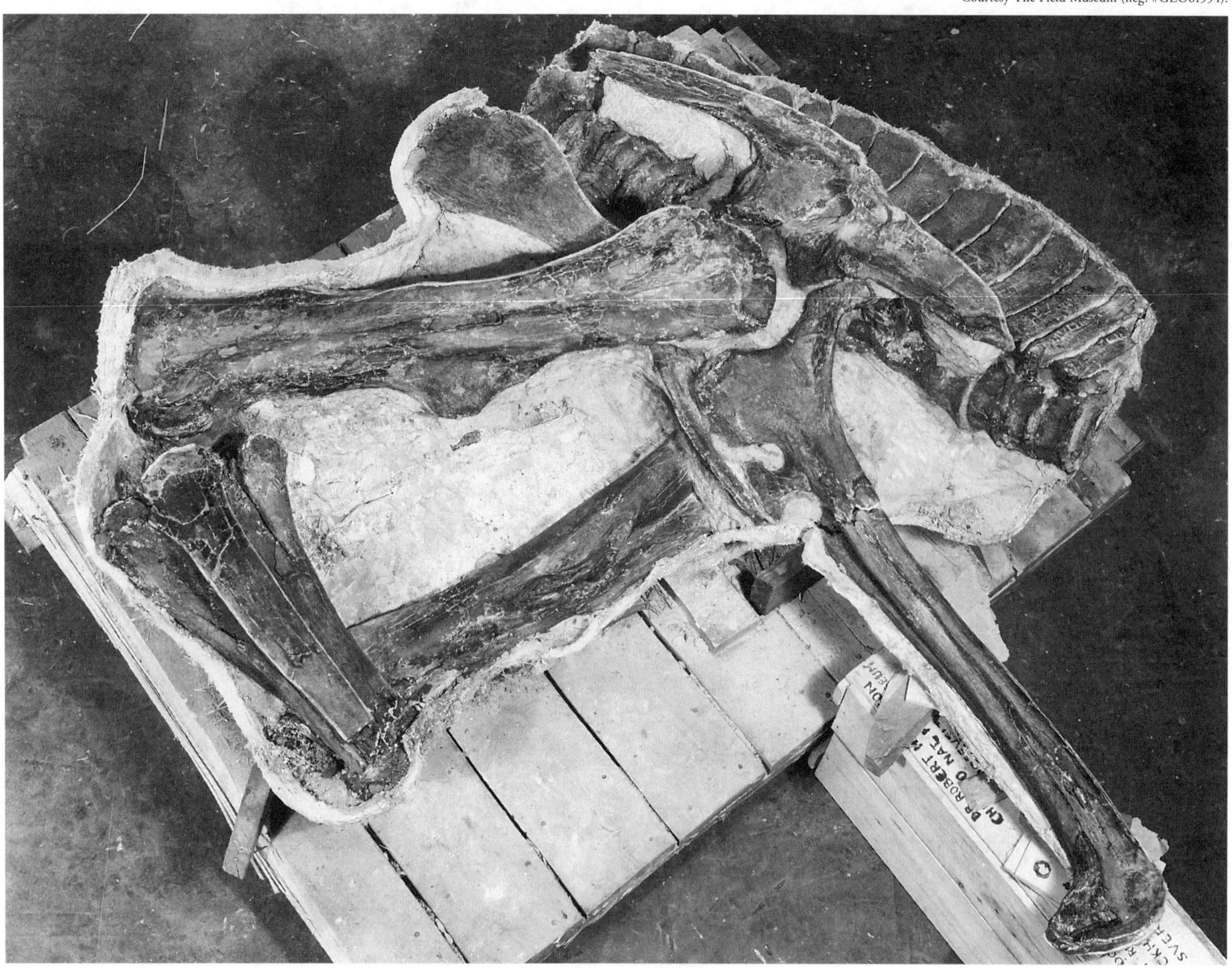

Detail of skeleton (FMNH PR380) of the ornithopod dinosaur *Lambeosaurus lambei* showing the typical ornithischian (or "bird hipped") pelvis.

cephalosauria and all descendants of their most recent common ancestor (Anonymous 1997).

Diagnosis: Ischium lacking obturator process; derived characters including formation of parietosquamosal shelf (or incipient frill) overhanging occiput; pubis reduced, lacking symphysis (Sereno 1986); greatest separation of acetabula than dorsal borders of ilia (Maryańska and Osmólska 1985; Dodson 1990*b*).

General description: Small to giant, bipedal and secondarily quadrupedal, most forms in Late Cretaceous, unique among dinosaurs in that skull roof overhangs occiput region of the skull as a narrow shelf, group including "dome-headed," "parrot-beaked," "frilled," and horned ornithischians.

Age: Early to Late Cretaceous.

Geographic distribution: North America, Europe, Asia, ?Australia.

Taxa: *Stenopelix* plus Pachycephalosauria plus Ceratopsia.

PACHYCEPHALOSAURIA

Diagnosis: Skull roof thickened, table-like or domed; jugal and quadratojugal strongly extended ventrally toward articular surface of quadrate; jugal meeting quadrate ventral to quadratojugal; orbit having coossified rostral and medial walls; orbital roof with two supraorbitals, contact between first supraorbital and nasal excluding prefrontal-lacrimal contact; basicranial region strongly shortened sagittally, basal tubera thin and platelike, contact between prootic-basisphenoid plate and quadrate ramus of pyterygoid; squamosal broadly expanded on occiput; external surfaces of cranial bones strongly ornamented, prominent osteoderms on squamosal rim; dorsal and caudal vertebrae with ridge-and-groove articulation between zygapophyses; sacral and proximal caudal vertebrae with long ribs; forelimb about 25 percent length of hindlimb; humerus slightly bowed and twisted, with rudimentary deltopectoral crest; ilium with cranially

Courtesy North Carolina Museum of Natural Sciences.

Skeleton of the hadrosaurine hadrosaurid *Edmontosaurus annectens* mounted protecting its eggs and hatchlings from attack.

broad and horizontal preacetabular process, medial flange on postacetabular process; pubis reduced, excluded from acetabulum; ischium with long dorsoventrally flattened cranial peduncle that contacts pubic peduncle of ilium and sacral ribs (instead of pubis); middle portion of tail having multiple rows of fusiform ossified tendons (Maryańska 1990).

General description: Small- to medium-sized, bipedal, head with relatively short facial region, skull roof thickened (sometimes developed into dome), teeth small and leaf-like, bony ornamentation of varying degrees on snout and back of skull, stiff tail, sometimes referred to as "thick-headed" or "dome-headed" dinosaurs, thickened head formerly believed (though no longer by most paleontologists) to have functioned in head-butting.

Age: Early to Late Cretaceous.

Geographic distribution: North America, Asia, Europe.

Taxa: Homalocephalidae plus Pachycephalosauridae.

Pachycephalosauridae *incertae sedis*: *Micropachycephalosaurus*, ?*Tianchungosaurus* [*nomen nudum*].

HOMALOCEPHALIDAE

Diagnosis: Derived character of flat, table-like skull roof; relatively large, primitive supratemporal fenestra (Maryańska 1990).

General description: Small- to medium-sized, "flat-headed" forms, skull roof somewhat evenly thickened and with flat dorsal surface.

Age: Late Cretaceous.

Geographic distribution: Asia.

Genera: *Goyocephale*, *Homalocephale*, *Wannanosaurus*.

Note: Sereno (1986; see also 2000, for his recent review of the Marginocephalia) regarded the genus *Stegoceras* as bridging a gap between flat-headed and fully-domed pachycephalosaurs, this observation calling into question the monophyly of the Homalocephalidae. As noted by Sullivan (2000*b*), with the referral of some specimens originally referred to

Courtesy Royal Ontario Museum

Skeleton of a "gryposaur," holotype (ROM 4614) of the hadrosaurine hadrosaurid *"Kritosaurus" incurvimanus.*

Stegoceras to the genus *Prenocephale* (see entry), "the apparent transition from the flat-headed pachycephalosaurs to the more fully-domed forms is not a simple linear relationship," but rather "a more complex scenario whose phylogenetic history is unresolved." Sullivan (2000*b*, 2000*c*) further pointed out that flat-headed specimens assigned to *Ornatotholus browni* pertain to *Stegoceras validum* and may represent a sexual dimorph of the latter species (see *Stegoceras* entry), such morphs within a single species thereby invalidating the Homalocephalidae as a clade.

PACHYCEPHALOSAURIDAE

Diagnosis: More or less extensive dome-like thickening of frontals and parietals (Sues and Galton 1987).

General description: Small- to medium-sized, "dome-headed" forms, domes thicker than in homalocephalids and varying in size among genera, some possessing large cranial spike-like projections.

Age: Early to Late Cretaceous.

Geographic distribution: North America, Asia, Europe.

Genera: *Gravitholus, Pachycephalosaurus, Prenocephale, Stegoceras, Stygimoloch, Tylocephale, Yaverlandia.*

Note: Sullivan (2000) noted that Sereno (2000) stated that *Stygimoloch* and *Pachycephalosaurus* lack a parietosquamosal shelf, although Goodwin, Bucholtz and Johnson (1998) had figured a *Stygimoloch spinifer* specimen (MPM 8111; see *S2* for additional photographs) having a distinct, though rather abbreviated, parietosquamosal shelf, the latter serving as a platform for the attachment of the large nodes characteristic of this genus. Therefore, Sullivan found it conceivable that *Stegoceras, Stygimoloch,* and *Pachycephalosaurus* form a clade (the parietosquamosal shelf being well developed in *Stegoceras*, less so in *Sygimoloch*, and, not substantiated, absent or almost so in *Pachycephalosaurus*), although synapomorphies among these genera must be identified to support such a clade.

CERATOPSIA

Stem-based definition: All Marginocephalia closer to Ceratopsidae than to Pachycephalosauria (Dodson 1998).

Diagnosis: Head triangular in dorsal view; median rostral bone on anterior snout overlying premaxilla; snout tall, with relatively tall premaxillae; external nares high on snout; better developed biplanar lateral surface on jugal; maxilla proportionately tall (at least two-thirds as tall as long); premaxilla vaulted (deep, transversely arched); mandibular symphysis immobile, symphysis broad, strong union between dentary rami and predentary; jugal orbital bar broader dorsoventrally than posterior ramus below laterotemporal fenestra; parietal transversely broad, overhangs most of occipital margin (Sereno 1986).

General description: Small to giant, some forms bipedal but most quadrupedal, head triangular in top

Courtesy Royal Ontario Museum.

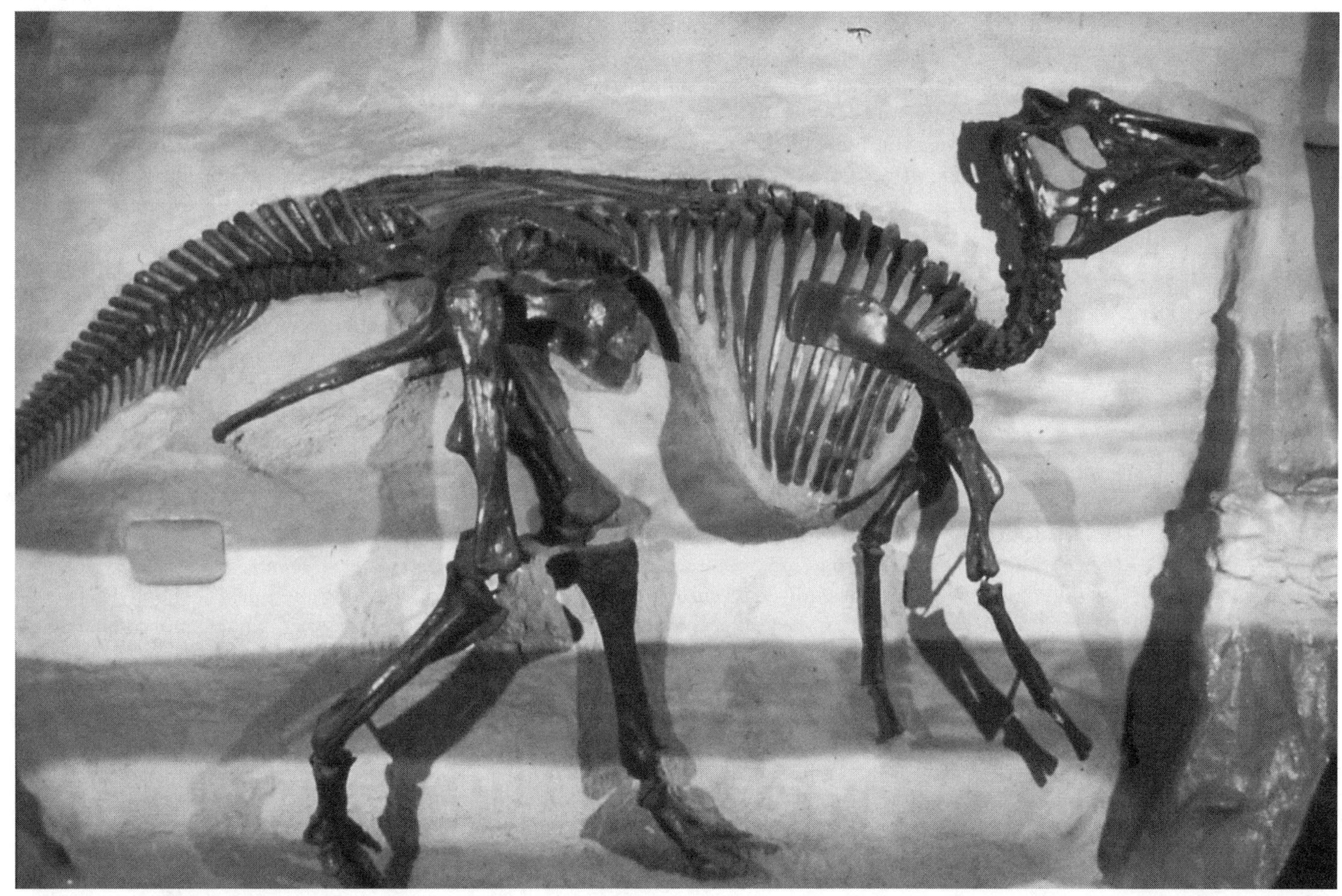

Skeleton (ROM 787) of the hadrosaurine hadrosaurid *Prosaurolophus maximus.*

view, snout high, hooked, parrot-like beak probably horn-covered in life, head usually having large parietal frill and facial horns, front feet with five toes.

Age: ?Middle Jurassic to Late Cretaceous.

Geographic distribution: North America, Asia, ?South America, ?Australia.

Taxa: ?*Chaoyangsaurus* plus Psittacosauridae plus Neoceratopsia.

Ceratopsia *incertae sedis*: *Claorhynchus*, *Trachodon*.

PSITTACOSAURIDAE

Diagnosis: Short preorbital skull segment (equal to less than 40 percent skull length); external naris very high on snout; nasal extending rostroventrally below external naris, establishing contact with rostral; extremely broad caudolateral premaxillary process separating maxilla from external naris by wide margin, extending dorsally to form tall, parrot-like rostrum; premaxilla, maxilla, lacrimal, and jugal sutures converging to point on snout; absence of antorbital fenestra and antorbital fossa; unossified gap in wall of lacrimal canal; eminence on rim of buccal emargination of maxilla near junction with jugal; pterygoid mandibular ramus elongate; dentary crown with bulbous primary ridge; manual digit IV with one simplified phalanx; absence of manual digit V (Sereno 1990).

General description: Small, facultatively bipedal, with tall, parrot-like beak, snout comparatively shorter than in other known dinosaur, rather long forelimbs, hand with four fingers, hindfoot having four functional toes, superficially resembling primitive ornithopods in general body proportions.

Age: Late Jurassic [see Shúan and Haichen 1998] to Middle Cretaceous.

Geographic distribution: Asia.

Genus: *Psittacosaurus*.

NEOCERATOPSIA

Diagnosis: Head very large relative to body; rostral and predentary sharply keeled, terminating in point; quadrate sloping rostrally; cranial frill prominently broad, confluent with supratemporal fossa; exoccipitals excluding basioccipital from foramen magnum; predentary with bifurcate caudoventral process and laterally sloping triturating [crushing] surface; teeth, in buccal aspect, with ovate crowns; maxillary teeth having prominent buccan primary ridge; one-three or four fused cervical vertebrae, forming syncervical; ischium gently decurved (Sereno 1986).

General description: More advanced than psittacosaurids, small to giant, mostly quadrupedal, head very large relative to body size, with prominent beak probably covered in life with horny material, head frills of varying shapes and sizes, some frills with bony processes, larger forms with horns.

Age: Middle to Late Cretaceous.

Geographic distribution: North America, Asia, ?South America, ?Australia.

Taxa: *Asiaceratops* plus ?*Microceratops* plus ?*Graciliceratops* plus Protoceratopsidae plus *Montanoceratops* plus Ceratopsomorpha.

Neoceratopsia *incertae sedis*: *Arstanosaurus*, *Kulceratops*.

Note: You and Dodson (2000), in a preliminary report published as an abstract, performed their own cladistic analysis of the Neoceratopsia. The one most parsimonious tree resulting from this study indicated that basal neoceratopsians diverged into two clades before the Late Cretaceous, the Protoceratopsidae (including *Archaeoceratops*, *Bagaceratops*, *Breviceratops*, and *Protoceratops*) and the Ceratopsoidea (whose basal member is *Leptoceratops*). You and Dodson pointed out that, to date, all protoceratopsids are known only from Asia and all ceratopsoids only from North America. This phylogeny, therefore, has "the virtue of enhancing the biogeographic coherence of the Neoceratopsia."

PROTOCERATOPSIDAE

Diagnosis: Rostrum narrow; premaxilla with concave ventral edge; naris small, elliptical, high-positioned; squamosal caudodorsally extended; frill fenestrated (You and Dodson 2000).

General description: Small, relatively primitive, mostly quadrupedal, basically hornless (though, in adults, region above snout and eyes can be raised and roughened, with rudimentary nasal horn sometimes present), neck frill somewhat developed (but not to extent as in ceratopsids), some forms possibly capable of rapid and bipedal locomotion.

Age: Late Cretaceous.

Geographic distribution: Asia.

Taxa: *Archaeoceratops*, *Bagaceratops*, *Breviceratops*, *Protoceratops*, *Udanoceratops*.

Note: The long-presumed monophyletic status of the Protoceratopsidae — which, until comparatively recently, had generally included the genera *Kulceratops*, *Leptoceratops*, *Microceratops*, and *Montanoceratops*— came under challenge during the last couple of decades (*e.g.*, Sereno 1984, 1986), with some workers (*e.g.*, Ryan and Currie 1996; Sampson, Ryan and

Mounted skeleton (adult and juvenile) at Dinofest (1998) of *Protoceratops andrewsi*. The Protoceratopsidae, the family to which this species has generally been referred, is now regarded by some vertebrate paleontologists as a paraphyletic assemblage.

Photograph by the author.

Tanke 1997; Foster and Sereno 1997) preferring to refer informally to taxa usually assigned to this group as "protoceratopsians." Chinnery and Weishampel (1998), following a cladistic analysis of this group, found the Protoceratopsidae to be paraphyletic unless restricted to include only the genera *Protoceratops*, *Leptoceratops*, and *Udanoceratops* (see *Montanoceratops* entry).

More recently, You and Dodson (2000; see "Note" under "Neoceratopsia," above), in performing their own cladistic analysis, accepted Protoceratopsidae as a valid family including *Archaeoceratops*, *Bagaceratops*, *Breviceratops*, and *Protoceratops*. In this preliminary report (full details presumably to be published at a later date), the authors did not mention the Ceratopsomorpha.

CERATOPSOMORPHA

Diagnosis: Ceratopsians with paired orbital brow horns ancestrally (Wolfe and Kirkland 1998).

General description: Primitive forms possessing brow horns, small, with single-rooted or double-rooted teeth.

Age: Late Cretaceous.

Geographic distribution: North America, Asia.

Taxa: *Zuniceratops* plus ?*Turanoceratops* plus Ceratopsoidea.

CERATOPSOIDEA

Stem-based definition: All neoceratopsians closer to *Leptoceratops* than to *Protoceratops* (You and Dodson 2000).

Diagnosis: Rostral bones wide; ventral edges of premaxillae convex; loss of premaxillary teeth; nares round, low-positioned; downward processes of squamosals; caudal margins of frills round (You and Dodson 2000).

General description: Small to giant forms, semi-bipedal to fully quadrupedal, most taxa bearing prominent horns and frills.

Age: Late Cretaceous.

Geographic distribution: North America, ?South America.

Taxa: *Leptoceratops* plus Ceratopsidae.

CERATOPSIDAE

Diagnosis: Head large, with prominent parietosquamosal frill, and variably, yet often, strongly-developed nasal or postorbital horns; external nares enlarged, set in narial fossae; prefrontals meeting on midline to separate nasal from frontals; frontals folding to form secondary skull roof expressed as frontal fontanelle; infratemporal fenestrae reduced; squamosal with strong postquadrate expansion; free border of squamosal often ornamented with or without epidermal ossifications; broad contact of jugal and squamosal above infratemporal fenestra; quadrate with pocket to receive wing of pterygoid; chonae enlarged; palatine reduced, longitudinal process lost; reduced ectopterygoid, eliminated from exposure on palate; palatine foramen lost; eustachian canal impressed on pterygoids; supraoccipital excluded from foramen magnum by exoccipitals, which form two-thirds of occipital condyle; one-third of condyle formed by basioccipital; dentary with expanded apex of coronoid process; dental batteries; teeth with split roots; tendency to add epidermal ossifications to squamosal and parietal; centra of cervical vertebrae expanded in width to support head; usually ten fused sacral vertebrae; humerus having robust deltopectoral crest; ilium with everted dorsal border; ischium decurved; femur considerably longer than tibia; ungual blunt, rounded, hooflike (Dodson and Currie 1990).

General description: Relatively small to giant, quadrupedal, with prominent parietal frill sometimes developed to very large size, horns of varying sizes (probably used for defense or in intraspecific contests), cutting edges of teeth arranged almost vertically resulting in "scissors-like" shearing action, jaws adapted for tougher-fiber vegetation, having largest brains compared to body size of all quadrupedal dinosaurs, bodies very similar among genera, forelimbs slightly shorter than hindlimbs, hind feet short, toes on front and hind feet spreading, hooflike.

Age: Late Cretaceous.

Geographic distribution: North America, ?South America.

Taxa: Centrosaurinae plus Chasmosaurinae.

Ceratopsidae *incertae sedis*: *Dysganus*.

Notes: Lehman (1990) divided the Ceratopsidae into the subfamilies Centrosaurinae and Chasmosaurinae (see *D:TE*), based mostly on cranial characters, but also on some differences in their postcranial bones.

More recently, Chinnery and Chapman (1998) published a brief preliminary report on their analysis, incorporating a combination of morphometric methods, of the ceratopsid postcranial skeleton. This study, based on linear measurements of ceratopsid girdle and limb elements, confirmed some, though not all, of Lehman's conclusions. Chinnery and Chapman found a high degree of variation throughout the Ceratopsidae, mostly within Chasmosaurinae. Most significant differences between Centrosaurinae and Chasmosaurinae were observed in the coracoid, ilium, pubis, ulna, femur, fibula, and especially the ischium. A landmark-based analysis supported the above study, and also revealed such important (but otherwise not apparent utilizing standard linear dimensions) trends as proximal and distal limb-bone orientation and the shape of the dorsal iliac border.

According to Chinnery and Chapman, results of the above study "confirm the applicability of morphometric methods in documenting and describing the taxonomic differences between the postcrania of these two groups." Also, they "support the utility of combining analyses based on conventional and geometric methods of morphometrics."

CENTROSAURINAE

Diagnosis: Large, with short, deep facial region (preorbital length/height = 1.2 to 2.4); no inter-premaxillary fossa; nasal bones with finger-like processes projecting into narial apertures; premaxilla and predentary bones with wide laterally included shearing surfaces; nasal horncore, when present, large, formed primarily by upgrowth of nasal bones; supraorbital horns poorly developed; no postfrontal foramen; postfrontal fontanelle walled primarily by large frontal bones; prefrontal bones large; cranial frill short (0.54–1.00 basal skull length); squamosal quadrangular (length/height about 1.0); first sacral rib expanded ventrally, forming triangular web of bone; ventral surface of sacrum with poorly developed longitudinal channel; ulna with reduced olecranon process; ischium relatively straight (Lehman 1990).

General description: Generally less derived than chasmosaurines, relatively small to large forms, face relatively short and high, nose horn usually longer than in chasmosaurines and formed from extension of nasal, shorter to nonexistent brow horns than in chasmosaurines, parietal frill usually (possibly always) fenestrated, frill usually shorter than length of basal part of head, frill generally having scalloped border with epoccipital processes sometimes present, frill more derived in degree of ornamentation than in chasmosaurines.

Age: Late Cretaceous.

Geographic distribution: North America.

Genera: *Achelousaurus*, ?*Avaceratops*, *Brachyceratops*, *Centrosaurus*, *Einiosaurus*, *Monoclonius*, *Pachyrhinosaurus*, *Styracosaurus*.

CHASMOSAURINAE

Diagnosis: Large to giant, head with long, low facial region (preorbital length/height = 1.4–3.0); inter-premaxillary fossae; premaxilla and predentary with horizontal or poorly developed lateral cutting flange; nasal horncore, when present, small, formed in part by separate ossification; supraorbital horns more prominent than in centrosaurines; postfrontal foramen; postfrontal fontanelle walled mostly by postorbital bones; frontal bones reduced; prefrontal bones

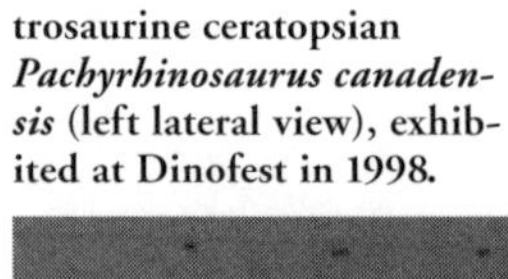

Skull (cast) of the centrosaurine ceratopsian *Pachyrhinosaurus canadensis* (left lateral view), exhibited at Dinofest in 1998.

Photograph by the author.

Courtesy Browarny Photographs, Ltd. (neg. #10-62-21).

Fanciful life-sized model of the centrosaurine ceratopsian *Styracosaurus albertensis* at the Calgary Zoological Society's dinosaur park, one of a suite comprising the first such permanent outdoor exhibit in North America.

small; cranial frill long (0.94 to 1.70 basal skull length, excluding *Triceratops* [see "Note" below]); squamosal bones triangular (length/height = 2.0 to 3.5); ventral surface of sacrum with prominent longitudinal channel; ulna having large olecranon process; ischium strongly curved (Lehman 1990, modified slightly following Penkalski and Dodson 1999).

General description: More derived than centrosaurines, large- to giant-sized, head with long and low facial region, pronounced beak, jaws very powerful, nose horn (when present) shorter than in centrosaurines and formed from bone separate from nasal, paired brow horns longer than in centrosaurines (see "Note" below) than in centrosaurines, with long usually fenestrated (solid in *Triceratops* [see "Note" below]) frill that is longer than basal part of head, frill border having more prominent epoccipital processes than in centrosaurines.

Age: Late Cretaceous.

Geographic distribution: North America.

Genera: *Agathaumas, Anchiceratops, Arrhinoceratops, Ceratops, Chasmosaurus, Diceratops, Pentaceratops, Polyonax, Torosaurus, Triceratops.*

Note: In the past, prominent brow horns have been used in identifying ceratopsid specimens as members of the subfamily Chasmosaurinae. More recently, in redescribing the basal ceratopsid genus *Avaceratops* (see entry for additional details), Penkalski and Dodson (1990) discussed the previously unaddressed possibility that postorbital horns might actually be a primitive rather than derived condition within the Ceratopsidae.

Penkalski and Dodson noted that at least once, postorbital horns evolved from larger to smaller size (from *Chasmosaurus mariscalensis* to *C. belli*; see Forster, Sereno, Evans and Rowe 1993). This evolution may also have occurred in the Centrosaurinae, as such horns are entirely lost in *Pachyrhinosaurus*, the most derived known member of that clade. They also noted that prominent brow horns are present in a basal nonchasmosaurine ceratopsid skull (MOR 692) which Penkalski and Dodson tentatively referred to *Avaceratops*.

Also, type material belonging to the basal ceratopsomorph *Turanoceratops* (see Nessov 1993, 1995) includes, among other elements, a straight, rounded bone measuring about 16 centimeters in length, identified as a partial nose horn; however, it also includes

Photograph by Robert A. Long and Samuel P. Welles (neg. #73/534-1), reproduced with permission of Canadian Museum of Nature, Ottawa, Canada.

Close view of the skull of *Chasmosaurus belli*, part of a mounted skeleton (plesiotype CMN 2245) of this chasmosaurine ceratopsid dinosaur.

a 9-centimeter in length "additional horn" that appears to be the tip of a slightly recurved horncore. Consequently, Penkalski and Dodson raised the possibility that *Turanoceratops* (and perhaps also the primitive *Zuniceratops*, which has prominent brow horns) might also be an early ceratopsid with postorbital horns. If correct, this would constitute additional evidence that brow horns are a basal ceratopsid character.

In performing their cladistic analysis on *Avaceratops*, Penkalski and Dodson found this genus and also MOR 692 (which was separately entered into the analysis) to be the most primitive known ceratopsid(s), which further suggests that large postorbital horns could be a basal condition within the Ceratopsidae. A previously conducted cluster analysis performed by Dodson (1993) had already revealed that the highly derived *Triceratops* (which, among all known ceratopsian genera, most closely resembles *Avaceratops*; see entry), if brow horns were not taken into account, grouped with the Centrosaurinae instead of the Chasmosaurinae. Penkalski and Dodson concluded that "*Avaceratops* is the least derived known ceratopsid, and hence is close to the ancestry of *Triceratops* and the Chasmosaurinae." This assessment implies that *Avaceratops* and *Triceratops*— and, by association, various closely related taxa (*e.g.*, *Diceratops*)— could belong in their own clade existing outside of Chasmosaurinae.

Some ceratopsian specialists (*e.g.*, P. J. Currie, personal communication 2000) have found the above new assessment reasonable, but others have not. In the estimation of C. A. Forster (personal communication 2000), "*Triceratops* and *Diceratops* fit snugly into the Chasmosaurinae." Furthermore, dividing out these taxa would make the Chasmosaurinae a paraphyletic group. According to Forster, these genera "really aren't that different from, say, *Torosaurus*, and fit nicely into Chasmosaurinae without difficulty."

III. Dinosaurian Genera

The following section consists of an alphabetically arranged compilation of dinosaurian genera, with information about the genus and its species. Some of these entries are new to this volume. Other genus entries are revised from entries previously published in the original *Dinosaurs: The Encyclopedia* (*D:TE*) and *Dinosaurs: The Encyclopedia, Supplement 1* (*S1*).

Entries marked with a dagger (†)—*e.g., Montanoceratops*— identify those which appeared in one or both of the earlier volumes, but in this book offer new, revised, or corrected data, new species, or new, or updated, or otherwise different illustrations. Within each dagger-marked entry, basic introductory information (*e.g.*, that relating to classification, age, diagnosis, *etc.*) is included only when it is new or has been revised. Information presented in this volume that conflicts with statements appearing in either *D:TE* or *S1* replaces that which was published earlier. Most photographs and drawings of type specimens and referred fossil material, and also life restorations relevant to dagger-marked entries, will be found in either of the previous books.

Revised or new diagnoses, definitions, and generalized descriptions and explanations of higher taxa, with broader applications to each genus or species, can be found in the preceding "Systematics" chapter; unchanged data relevant to those topics will be found in *D:TE* or *S1*.

The following genera (excluding taxa which were recently named and described but are already regarded as junior synonyms, *e.g.*, *Gondwanatitan*) have their own entries which are new to this volume:

Achillobator, Amurosaurus, Atlasaurus, Bambiraptor, Beipiaosaurus, Byronosaurus, Camposaurus, Caseosaurus, Cedarosaurus, Charonosaurus, Chuanjiesaurus, Cristatusaurus, Dinheirosaurus, Eucoelophysis, Fukuiraptor, Glyptodonopelta, Gongxianosaurus, Graciliceratops, Histriasaurus, Huabeisaurus, Isanosaurus, Jeholosaurus, Jobaria, Lessemsaurus, Lirainosaurus, Lourinhanosaurus, Lourinhasaurus, Lurdusaurus, Microraptor, Mononykus, Nigersaurus, Nipponosaurus, Nodocephalosaurus, Nomingia, Notohypsilophodon, Nqwebasaurus, Ozraptor, Parvicursor, Patagonykus, Protohadros, Pyroraptor, Qantassaurus, Qinlingosaurus, Rocasaurus, Santanaraptor, Saturnalia, Sauroposeidon, Shanyangosaurus, Shuvuuia, Sinornithosaurus, Stygivenator, Suchomimus, Tangvayosaurus, Tehuelchesaurus, Variraptor, Yimenosaurus.

The following genera (this list excluding taxa now regarded as junior synonyms, *e.g., Angaturama*) had their own entries in *D:TE* or *S1*, but also have entries in this volume:

Acanthopholis, Acrocanthosaurus, Aeolosaurus, Agustiana, Alamosaurus, Albertosaurus, Aliwalia, Allosaurus, Alvarezsaurus, Amargasaurus, Andesaurus, Anoplosaurus, Antarctosaurus, Antrodemus, Apatosaurus, Archaeoceratops, Aristosuchus, Astrodon, Aublysodon, Avaceratops, Avimimus, Bactrosaurus, Baryonyx, Blikanasaurus, Brachiosaurus, Brachylophosaurus, Bradycneme, Bugensaura, Camarasaurus, Camelotia, Carcharodontosaurus, Caudipteryx, Centrosaurus, Ceratosaurus, Chaoyangsaurus, Chasmosaurus, Chiayüsaurus, Chilantaisaurus, Chindesaurus, Coelophysis, Coelurus, Compsognathus, Dacentrurus, Daspletosaurus, Deinonychus, Dilophosaurus, Diplodocus, Dryptosauroides, Echinodon, Edmontonia, Edmontosaurus, Elaphrosaurus, Elopteryx, Eolambia, Epachthosaurus, Euhelopus, Euoplocephalus, Euskelosaurus, Gallimimus, Guaibasaurus, Heterodontosaurus, Hoplitosaurus, Hypacrosaurus, Iguanodon, Ilokelesia, Indosaurus, Irritator, Janenschia, Kritosaurus, Leptoceratops, Lophorhothon, Magyarosaurus, Maiasaura, Majungatholus, Malawisaurus, Mamenchisaurus, Megalosaurus, Microvenator, Minmi, Montanoceratops, Nedcolbertia, Nemegtosaurus, Nodosaurus, Nuthetes, Omeisaurus, Othnielia, Pachycephalosaurus, Pachyrhinosaurus, Parasaurolophus, Pawpawsaurus, Pelecanimimus, Pentaceratops, Phuwiangosaurus, Pinacosaurus, Plateosaurus, Prenocephale, Priconodon, Probactrosaurus, Protoceratops, Psittacosaurus, Rhabdodon, Saltasaurus, Saurolophus, Sauropelta, Saurornitholestes, Scipionyx, Scutellosaurus, Seismosaurus, Sellosaurus, Shamosaurus, Shanxia, Silvisaurus, Sinosauropteryx, Sonorosaurus, Spinosaurus, Stegoceras, Stegopelta, Stegosaurus, Styracosaurus, Syntarsus, Tarchia, Telmatosaurus, Tendaguria, Texasetes, Thecodontosaurus, Thescelosaurus, Triceratops, Troodon, Tyrannosaurus, Ultrasaurus, Velociraptor, Zigongosaurus, and *Zuniceratops*.

Note that *Alvarezsaurus, Avalonianus, Elopteryx, Mononykus, Nuthetes,* and *Patagonykus*, included in this book as dinosaurian taxa, were listed in one or both of the previous volumes as nondinosaurian and, therefore, were previously included among the groupings of "excluded genera"; *Ultrasaurus* was listed in *D:TE* among the *nomina nuda*.

†**ACANTHOPHOLIS** Huxley 1867 [*nomen dubium*]

Type species: *A. horridus* Huxley 1867 [*nomen dubium*].

Other species: ?*A. eucercus* Seeley 1869 [*nomen dubium*], *A. macrocercus* [in part] Seeley 1869 [*nomen dubium*], *A. platypus* [in part] Seeley 1869 [*nomen dubium*], *A. stereocercus* [in part] Seeley 1869 [*nomen dubium*].

Age: Middle Cretaceous (Albian–Cenomanian).

Comments: *Acanthopholis*, one of the earliest-named ankylosaurian genera, was founded in 1867

Acanthopholis horridus **syntypes: BGS GSM 109057, basicranium, A. right lateral and B. anterior views, BGS GSM 109045, tooth, C. ?lingual and B. ?labial views, E. BGS GSM 109047, left cervical spine, dorsal view, BGS GSM 109050, right cervical spine, F. dorsal and G. ventral views, and H. BMNH 44581, referred caudal vertebra, anter view. Scale = .60 cm. (After Suberbiola and Barrett 1999.)**

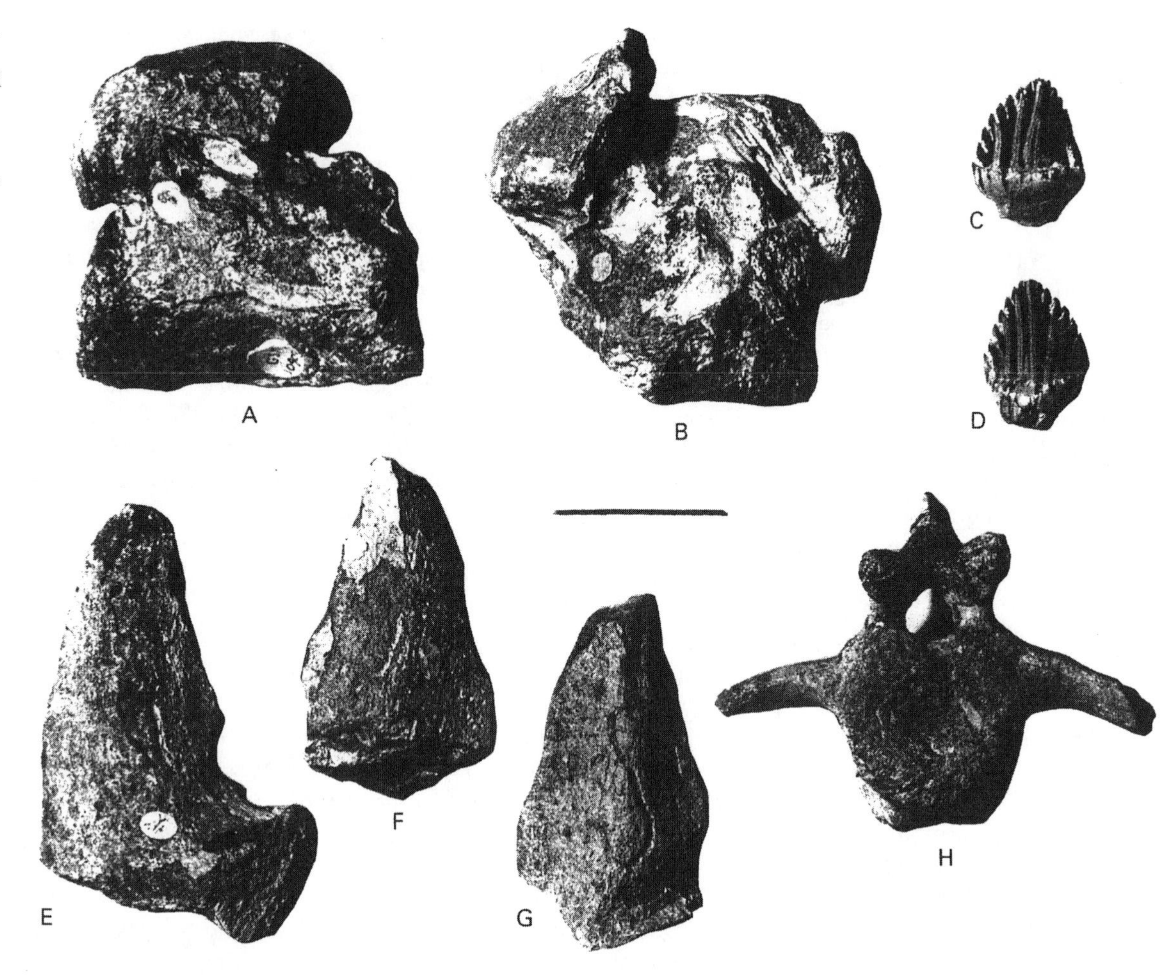

upon fossil remains by Thomas H. Huxley. The pioneer paleontologist named the type species of this new genus *A. horridus*. Later, Harry Govier Seeley referred various other species to *Acanthopholis—A. macrocerus*, *A. platypus*, and *A. stereocercus* in 1869, and *A. eucercus* in 1879 — each one founded upon scrappy fossil remains, and long regarded as *nomina dubia*. In more recent decades, ankylosaur specialist Walter P. Coombs, Jr. (1971, 1978) considered the genus *Acanthopholis* to be a *nomen dubium*, having been founded upon undiagnostic nodosaurid remains. In a later review of the Ankylosauria, however, Coombs and Maryańska (1990) provisionally listed *A. horridus* as a valid "small nodosaurid" taxon (mostly based "on tooth structure and presence of tall conical spines and keeled plates that are flat or slightly convex medially"), but pointed out that it could be a *nomen dubium*. (See *D:TE*, *Acanthopholis* entry, for more details on the genus and its species.)

More recently, Suberbiola and Barrett (1999), in a detailed systematic review of ankylosaurian remains recovered from the Albian–Ceonomanian of southeastern England, reconsidered the status of *Acanthopholis horridus*.

Suberbiola and Barrett reassessed the following materials assigned to the type species: BGS GSM 109057, a fragmentary basicranium; BGS GSM 109045–109046 (in part) and 109051 (in part), three isolated teeth; BGS GSM 109053, a dorsal vertebra (doubtfully ankylosaurian, according to the authors); and BGS GSM 109046–109052, 10955–10956, dermal armor (spines and scutes). Other elements mentioned by Huxley, including a dorsal vertebra, fragmentary humerus, and numerous scutes, could not be located by the authors for their study. Other specimens (BMNH 33581, 47234, and 49917) later acquired by the British Museum (Natural History) [now The Natural History Museum, London] and referred to *A. horridus*, seemingly representing a number of individuals, were also used in this study, although their origins were not entirely clear to the authors.

According to Suberbiola and Barrett, the original syntypes (BGS GSM 109057, basicranium, 109045, tooth, 109047, left cervical spine, and 109050, right cervical spine) of *A. horridus* might pertain to a single individual, Huxley having suggested that "as they all came from one small area, they probably belonged to one animal"; however, lack of accurate field information prevents confirming this hypothesis, while two vertebrae from this grouping are doubtfully ankylosaurian.

Noting that comparing the type material of *A. horridus* to other nodosaurid taxa is difficult due to the paucity of the former's material, Suberbiola and Barrett found only a few skeletal elements suitable for detailed comparison. The damaged occipital condyle of *A. horridus* is oriented similarly to that observed in *Silvisaurus condrayi* and also in an undetermined juvenile nodosaurid from Texas that may be referred provisionally to *Texastes* (see Jacobs, Winkler, Murry and Maurice 1994); its cervical spines are quite comparable to the anterolaterally directed cervical spines in both *Hylaeosaurus* and *Edmontonia*; the remaining elements cannot be distinguished from those of other known nodosaurids. Furthermore, none of the available type material of *A. horridus* exhibits diagnostic characteristics. For these reasons, Suberbiola and Barrett concurred with earlier opinions (*e.g.*, Coombs 1978) that *Acanthopholis horridus* should be regarded as a *nomen dubium* and indeterminate nodosaurid.

Many more remains, based upon very incomplete or fragmentary specimens from the Cambridge Greensand (upper Albian) (see Rawson, Currey, Dilley, Hancock, Kennedy, Neale, Wood and Worssam 1978), have been referred to *Acanthopholis*, some of these originally described as new species of this genus or as new genera (see *D:TE*). All diagnostic ankylosaurian remains collected from Europe to date have been referred to the Nodosauridae, but Suberbiola and Barrett took the more conservative approach in evaluating the following taxa. All of these taxa were regarded (in part, based on the ankylosaurian material) by Coombs and Maryańska as junior synonyms of *A. horridus*; Suberbiola and Barrett found them to be *nomina dubia*, representing indeterminate members of the Ankylosauria:

A. macrocercus Seeley 1869 (referred by Seeley 1879 to the new genus and species *Sygongosaurus macrocercus*) was established on composite remains (SMC B55570–55609) — a cervical, nine dorsal, four sacral, and five caudal vertebrae, a fragmentary humerus and tibia, four metatarsals, two phalanges, 11 osteoderms, and three unidentified bones, most of the vertebral material (SMC B55571–55579, B55582–55586) belonging to an ornithopod — found near Cambridge, Cambridgeshire, some of this material not found for study. The armor, however, can be assigned to the Ankylosauria.

A. platypus Seeley 1869 was based on composite remains (SMC B55449–55461) including six caudal centra, a chevron, five articulated metapodials, and one phalanx from the village of Bottisham, east of Cambridge, the metapodials (SMC B55449–55453), metarsals I–V, later referred by Seeley (1876) to the sauropod *Macrurosaurus* (see entry, *D:TE*).

A. stereocercus Seeley 1869 was based on composite remains (SMC B55558–55569) including 11 fragmentary vertebrae and a dermal plate, recovered near Cambridge. Later, Seeley (1879) referred some of this material to *Anoplosaurus* as the new species *A. major*, retaining two dorsal vertebrae, five caudals, and a fragmentary spine (SMC B55558–55560, B55562, B55566–55569) as *Acanthopholis stereocercus*, some of the vertebrae (SMC B55558–55559) possibly representing ornithopods (see Brinkmann 1988) while others are probably ankylosaurian.

A. eucercus Seeley 1879, based on six caudal centra (SMC B55552–55557), were regarded by Seeley (1879) as closely resembling those of *A. horridus*, by Nopcsa (1923) as ornithopod-like, and Coombs and Maryańska as doubtfully ankylosaurian. While acknowledging that the caudals appear to belong to an ornithischian dinosaur, Suberbiola and Barrett found them to be undiagnostic and too fragmentary for a precise identification.

In addition to the above cited remains, Suberbiola and Barrett described and commented upon numerous other ankylosaurian materials, much of it previously undescribed, from the Cambridge Greensand and housed in the Sedgwick Museum, University of Cambridge. Although some of these specimens have, in the past, been referred to *Acanthopholis* (*e.g.*, Coombs 1978), they exhibit no diagnostic characters, are sometimes from uncertain localities, should be considered *nomina dubia*, and were regarded by the authors as simply indeterminate Ankylosauria or Nodosauridae. Some of these additional fossils included names informally supplied but never published by Seeley, hand-written on labels accompanying the material. Labeled "Acanthopholis hughessi," SMC B55463–55490 includes six dorsal and seven caudal vertebrae, a transverse process, four metapodials, three phalanges, and seven dermal plates, this assemblage including indeterminate ornithopod (SMC B55463–55467, dorsal vertebrae) and ankylosaurian remains. "Acanthopholis keepingi," SMC B55493, catalogued by Seeley (1869), includes five dorsal and six caudal centra (SMC B55493, a caudal, could not be found by Suberbiola and Barrett), three transverse processes, a rib, two chevrons, one metatarsal, two phalanges, a fragmentary ?ilium, 13 dermal plates, and two unidentified elements. This seems to be a composite specimen, SMC B55494 and B55503 (dorsal and caudal vertebrae) belonging to an indeterminate ornithopod, SMC B55509 (limb fragment) a turtle, and an ankylosaur, the various pieces apparently mixed at the time of collection.

Key references: Brinkmann (1988); Coombs (1971, 1978); Coombs and Maryańska (1990); Huxley (1867); Nopcsa (1923); Seeley (1869, 1776, 1879); Suberbiola and Barrett (1999).

ACHILLOBATOR Perle, Norell and Clark 1999

Saurischia: Theropoda: Neotheropoda: Tetanurae: Avetheropoda: Coelurosauria: Manuraptoriformes: Manuraptora: Deinonychosauria: Eumanuraptora: Dromaeosauridae: Dromaeosaurinae.

Name derivation: Latin "Achilles [hero of Homer's *Iliad*]" + Mongolian *bator* = "hero."

Type species: *A. giganticus* Perle, Norell and Clark 1999.

Other species: [None.]

Occurrence: Bayan Shireh Formation, Dorno Gobi Aimak (Sain Shand), Burkhant, Mongolia.

Age: Late Cretaceous (?Santonian–?Campanian).

Known material/holotype: Fragmentary skeleton including left maxilla with nine well-preserved teeth and two empty alveolae, left femur and tibia, metatarsals II and III, isolated unguals and phalanges, unguals of right pes, pubes, well-preserved right ilium, right ischium, fragments of caudal vertebrae and ribs.

Diagnosis of genus (as for type species): Gigantic manuraptoran with large head; hindlimbs very stout, massive, comparatively short; forelimbs probably enlarged (as indicated by well-preserved radius); pes of medium (shortened) length; penultimate and ungual phalanges of digit II robust, especially proximo-ventral process of the former; metacarpal III long, irregular in shape, with slender shaft; skull having large semicircular to oval first antorbital fenestra, vertically oriented; second and third antorbital fenestrae subvertically oriented; 11 maxillary teeth; all teeth having anterior and posterior serrated margins;

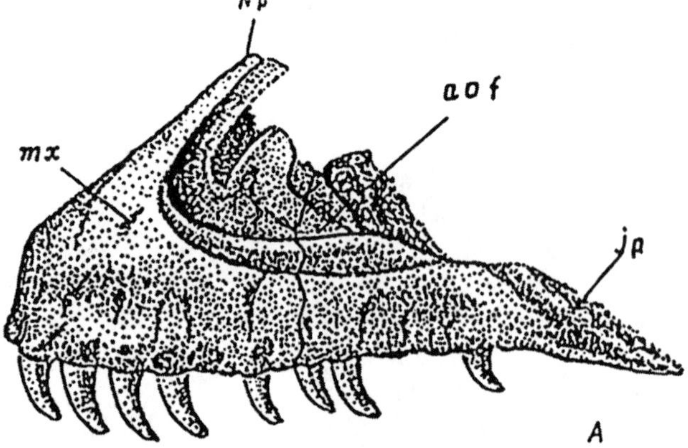

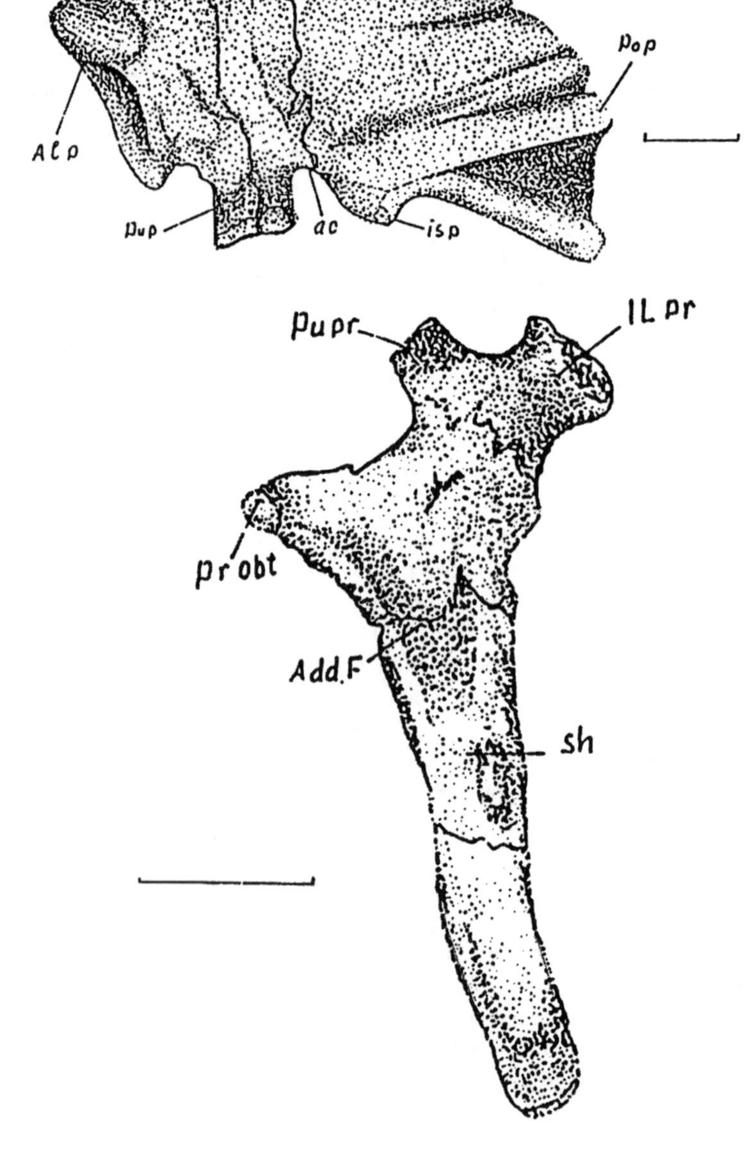

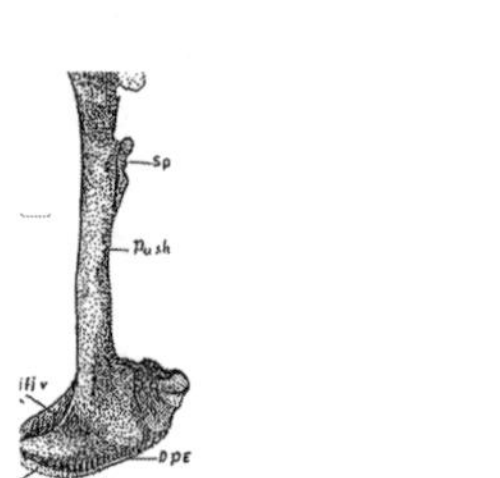

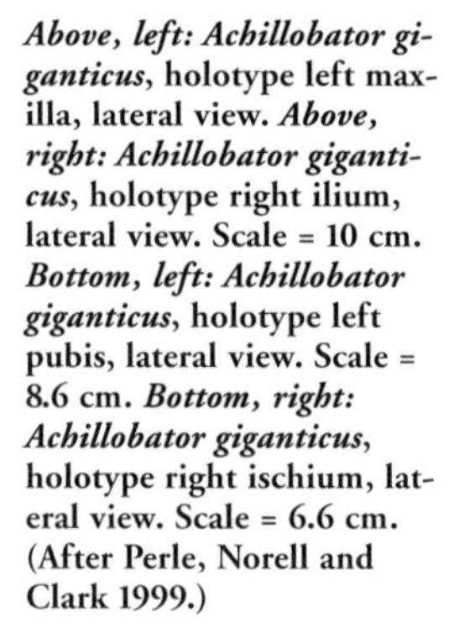
***Above, left: Achillobator giganticus*, holotype left maxilla, lateral view. *Above, right: Achillobator giganticus*, holotype right ilium, lateral view. Scale = 10 cm. *Bottom, left: Achillobator giganticus*, holotype left pubis, lateral view. Scale = 8.6 cm. *Bottom, right: Achillobator giganticus*, holotype right ischium, lateral view. Scale = 6.6 cm. (After Perle, Norell and Clark 1999.)**

denticles of posterior serrations larger than anterior ones; cervical vertebrae short, massive, sharply angled; caudal vertebrae long, platycoelous; caudal vertebrae having extremely long, rod-like prezygapophysial processes; chevrons very much elongated into long paired, double bony rods extending forward; ischium having large triangular obturator process located on proximal half of ischial shaft; pubis long, very stout, with long antero-posteriorly directed distal expansion, as in carnosaurs (Perle, Norell and Clark 1999).

Comments: The very large "dromaeosaur" *Achillobator giganticus* was founded upon a fragmentary associated skeleton collected during the Mongolian-Russian Paleontological Expedition in 1989 by Mr. Namsarai, assistant paleontologist at the Mongolian Museum of Natural History, at a single sight in the Burkhant locality, in outcrops of the Bayan Shireh Formation (lowermost Upper Cretaceous), at Burkhant, southwest of Dzun Bayan village and southwest of Khongii Tsav, in the southeastern Gobi Desert, south central Mongolia. In adjacent localities of the Bayan Shireh Formation the remains of crocodilians (Efimov 1988), sauropods, pinacosaur-like ankylosaurs (Tumanova, in press), and hadrosaurs (Perle, Norell and Clark (1999) were collected.

According to Perle *et al.*, *Achillobator* is notable for its size, which is almost three times that of the large dromaeosaurid *Deinonychus antirrhopus*.

Perle *et al.* found it difficult to refer *Achillobator* precisely to any particular group pending a phylogenetic revision of the "dromaeosaurs." The genus was regarded as a "dromaeosaur" based on characters including a highly modified, second digit of the hind foot. It seems to be closely related to *Dromaeosaurus albertensis*, differing mainly in the prepubic constructions of the pelvis (as in carnosaurs) and having caudal vertebrae with anterior prezygapophyses spanning adjacent vertebrae. Although it was beyond the scope of their present study to discuss "dromaeosaur" relationships as a whole, the authors commented that the character conditions in this new genus and species "have broad implications for the study of dromaeosaur phylogeny and perhaps the monophyly of this group."

Note: Mark A. Norell and James M. Clark, in a note, stated that their paper authored jointly with Altangerel Perle *et al*, originally intended for publication in *American Museum Novitates*, instead appeared "in Mongolia in an extremely preliminary form without the knowledge of the junior authors, and without a planned comparative analysis." Neither Norell nor Clark contributed to a section included in the paper dealing with "Habits and Affinities of Dromaeosaurian Dinosaurs"—which included discussions and speculations on such disparate topics as posture, locomotion, predation, and the origins of flight—before publication.

Key reference: Perle, Norell and Clark (1999).

†ACROCANTHOSAURUS

Saurischia: Theropoda: Neotheropoda: Tetanurae: Avetheropoda: Carnosauria: Allosauroidea: Allosauridae.

Diagnosis: Large theropod having elongate neural spines that are more than 2.5 times corresponding presacral, sacral, and proximal caudal lengths of centra; lacrimal contacting postorbital; supraoccipital expanding on either side of midline, protruding as double boss behind nuchal crest; pleurocoelous fossae and foramina pronounced on all presacral and sacral vertebrae; cervical neural spines having triangular anterior processes inserting into depressions beneath overhanging processes on preceding neural spines; accessory transverse processes on midcaudal vertebrae (Currie and Carpenter 2000).

Comments: Among the largest known theropods is *Acrocanthosaurus atokensis*. The most complete skeleton (OMNH 10168) belonging to this type species, which includes the only known complete skull, was found in the Antlers Formation (Trinity Group) of McCurtain County, Oklahoma.

As chronicled by Currie and Carpenter, parts—including two posterior cervical or anterior dorsal centra, an ischial fragment, and the distal end of a femur—of this specimen were collected by the [Sam Noble] Oklahoma Museum of Natural History. However, most of the specimen—including the skull (see *S1* for photograph), about two dozen vertebrae, some ribs, chevrons, most of the front limbs, pelvic fragments, parts of both femora and tibiae, and most of the foot bones—was recovered by Cephis Hall and Sid Love. Allen Graffham of Geological Enterprises, Inc., later acquired the unprepared specimen and then arranged for its preparation by the Black Hills Institute of Geological Research. The skeleton was completely prepared by the summer of 1996 and, the next year, it went to the North Carolina State Museum of Natural Sciences (acquiring the new catalog number of NCSM 14345) where it is currently on display.

Currie and Carpenter noted that, as mounted by the Black Hills Institute, the skeleton of *Acrocanthosaurus* measures 11.5 meters (approximately 38 feet) in length, ranking this theropod as smaller than *Tyrannosaurus rex* and *Giganotosaurus carolinii* (Coria and Salgado 1995), but larger than the biggest known specimens of such genera as *Carcharodontosaurus* (Stromer 1931), *Allosaurus fragilis*, *A. maximus* ("*Saurophaganax*"; see Chure 1996), *Suchomimus* (Sereno, Beck, Dutheil, Gado, Larsson, Lyon, Marcot, Rauhut,

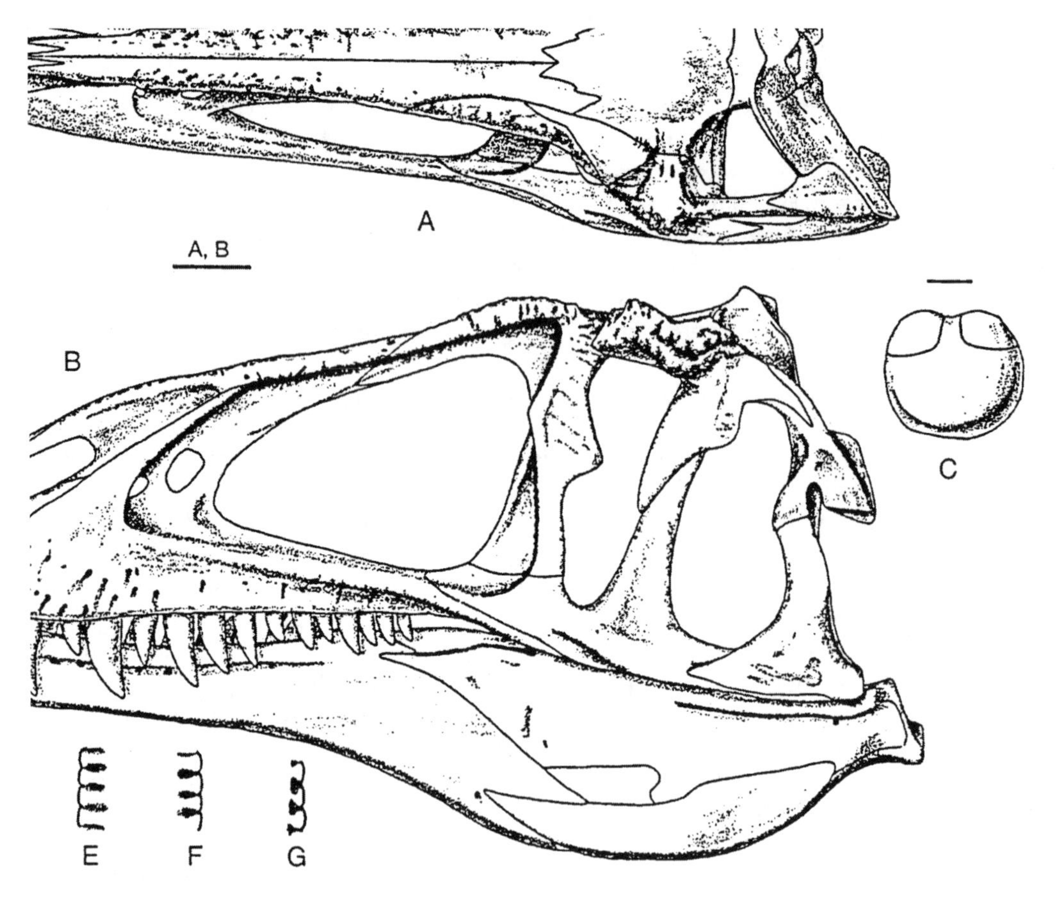

***Acrocanthosaurus atokensis*, NCSM 14345, reconstruction of skull in A. dorsal and B. left lateral views, C. occipital condyle, D.-G. denticles of maxillary teeth. Scale bars = (A, B) 10 cm. and (C) 2 cm. (After Currie and Carpenter 2000.)**

Sadleir, Sidor, Varricchio, Wilson and Wilson 1998), and *Tyrannosaurus bataar* (=*Tarbosaurus* of Currie's usage in Currie and Carpenter [K. Carpenter, personal communication 2000]) [although the type skull of *T. bataar* is as large as that of *T. rex*; R. E. Molnar, personal communication]. The length of *Acrocanthosaurus* compares with that of *T. rex* because, as with *Allosaurus*, its vertebrae are relatively longer when compared with the lengths of the skull or femur.

Utilizing the formula for estimating body weight based upon the circumference of the femur (Anderson, Hall-Martin and Russell (1985), Currie and Carpenter suggested that the weight of the individual represented by NCSM 14345 was about 4.16 metric tonnes (more than 4.5 tons), heavier than most specimens of *Allosaurus* and *T. bataar*, and lighter than *T. rex*, *Giganotosaurus*, and *Carcharodontosaurus*.

Currie and Carpenter rediagnosed the genus (and type species) and described the entire skeleton of *Acrocanthosaurus* in detail, focusing primarily upon new information gleaned from NCSM 14345 and especially pertaining to the skull, pectoral girdle, forelimb, and, to a lesser degree, the hindlimb.

Stovall and Langston (1950), in their original description of *Acrocanthosaurus*, classified this genus among the Allosauridae, an assessment accepted by various subsequent authors (*e.g.*, Madsen 1976; see *D:TE*). More recently, Sereno, Dutheil, Iarochene, Larsson, Lyon, Magwene, Sidor, Varracchio and Wilson (1996) referred *Acrocanthosaurus* to the Carcharodontosauridae, a conclusion subsequently also reached in a more thorough analysis, including new information gleaned from additional material, by Harris (1998; see *S1*).

Currie and Carpenter, however, found *Acrocanthosaurus* to compare in overall morphology most favorably with *Allosaurus*, with few characters useable in uniting the genus with *Carcharodontosaurus* (see Currie and Carpenter for detailed evaluations of the characters supposedly uniting *Acrocanthosaurus* with the Carcharodontosauridae). In performing their own cladistic analysis, Currie and Carpenter found *Allosaurus* to be the theropod closest to *Acrocanthosaurus*; *Giganotosaurus* and *Carcharodontosaurus* were determined to form a second clade, its inclusion within the Allosauroidea, however, undetermined.

Courtesy North Carolina Museum of Natural Sciences.

***Acrocanthosaurus atokensis*, reconstructed skeleton (including cast of NCSM 14345), the "Terror of the South" exhibit at the North Carolina Museum of Natural Sciences. The skeleton, the only one of *Acrocanthosaurus* on display in the world, measures approximately 40 feet in length.**

Copyright Kazuhiko Sano, courtesy North Carolina Museum of Natural Sciences.

Painting by Kazuhiko Sano, originally published in *Scientific American* magazine, depicting a confrontation on an Early Cretaceous mudflat between the theropod *Acrocanthosaurus atokensis* and sauropod *Pleurocoelus nanus.*

Key references: Anderson, Hall-Martin and Russell (1985); Chure (1996); Coria and Salgado (1995); Currie and Carpenter (2000); Harris (1998); Madsen (1976); Sereno, Beck, Dutheil, Gado, Larsson, Lyon, Marcot, Rauhut, Sadleir, Sidor, Varricchio, Wilson and Wilson (1998); Sereno, Dutheil, Iarochene, Larsson, Lyon, Magwene, Sidor, Varracchio and Wilson (1996); Stoval and Langston (1950); Stromer (1931).

†**AEOLOSAURUS**—(=*Eolosaurus, Gondwanatitan*)
Saurischia: Sauropodomorpha: Sauropoda: Eusauropoda: Neosauropoda: Macronaria: Camarasauromorpha: Titanosauriformes: Somphospondyli: Titanosauria: Titanosauridae.

Comments: New South American material has been referred to *Aeolosaurus*, a genus founded upon postcranial remains from the Allen Formation of Argentina (see *D:TE*).

Bertini and Santucci (2000), in an abstract for a poster, briefly reported on an as yet undescribed caudal vertebral series collected from the Upper Cretaceous Adamantina Formation of Northern São Paulo State, in southeastern Brazil. According to these authors, the specimen is characterized by an upwardly projecting and forwardly directed neural spine and a neural arch positioned anteriorly on the centrum, characters which are also present in *Aeolosaurus*.

In 1999, Kellner and Azevado named and described an incomplete skeleton (MN 4111-V)—including two partial cervical, seven dorsal, six sacral, 24 caudal (some articulated), and some unidentified vertebrae, the proximal part of a left scapula, an incomplete left ilium, the middle portion of the pubes, incomplete ischia, humeri, tibiae, parts of several ribs, and several unidentified fragments—as *Gondwanatitan faustoi*. As related by Kellner and Azevado, in 1983 Yoshitoshi Myzobuchi found some fossil bones on his property located in the Bauru Group (Upper Cretaceous; Campanian, dated as between Santonian and Maastrichtian by Fernandes and Coimbra 1996), near the city of Alvares Machado, State of São Paulo, Brazil. Fausto Luiz de Souza Cunha, a vertebrate paleontologist then curator of the Museu Nacional/Universidade Federal do Rio de Janeiro, was informed of the find and recognized these remains as belonging to a titanosaurid sauropod. Dr. Cunha and his crew excavated the bones from 1984 through 1986. Collected from the same outcrop of the "Sitio Myzobuchi," these remains apparently represent a single individual. The material was subsequently partially prepared and briefly mentioned (Cunha and Suarez 1985; Cunha, Régo and Capilla 1987). In 1997, preparation of the fossils was resumed. Eventually they were formally named and described as a new titanosaurian genus and species, *Gondwanatitan faustoi* (Kellner and Avezedo 1999).

Kellner and Avezado diagnosed *G. faustoi* as follows: Titanosaur having "heart-shaped" distal articulation surface of proximal and midcaudal vertebrae; humerus with very well-developed, medially curved deltopectoral crest; anterior part of proximal articulation of tibia projecting dorsally; cnemial crest only slightly curved laterally (Kellner and Avezedo 1999). Based upon the recovered material, Kellner and Azevedo determined that this was a relatively small sauropod, measuring approximately 6 to 7 meters (about 21 to 24 feet) in length.

Studies of titanosaurs, this new taxon included, have always been hampered by a paucity of diagnostic specimens. Nevertheless, Kellner and Azevedo referred their material to the Titanosauria based upon its possession of three titanosaurian synapomorphies (following Salgado, Coria and Calvo 1997)—dorsal vertebrae with elongated ("eye-shaped") pleurocoels; posterior dorsal vertebrae having widened, slightly forked infradiapophyseal laminae; and anterior caudal vertebrae procoelic. Further cladistic analysis by these authors determined that this taxon is not a member of the Saltasauridae (=Saltasaurinae of their usage); also, it can be separated from basal titanosaurs such as *Malawisaurus* and *Andesaurus*. However, Kellner and Azevedo observed that *G. faustoi* shares at least one major feature—anterior caudal and anterior midcaudal vertebrae having strongly anteriorly directed neural spines—with the titanosaur *Aeolosaurus*.

Aeolosaurus rionegrinus, reconstructed holotype (MN 4111-V) incomplete skeleton of _Gondwanatitan faustoi_, preserved parts indicated. Length of animal = approximately 6–7 m. (After Kellner and Azevedo 1999.)

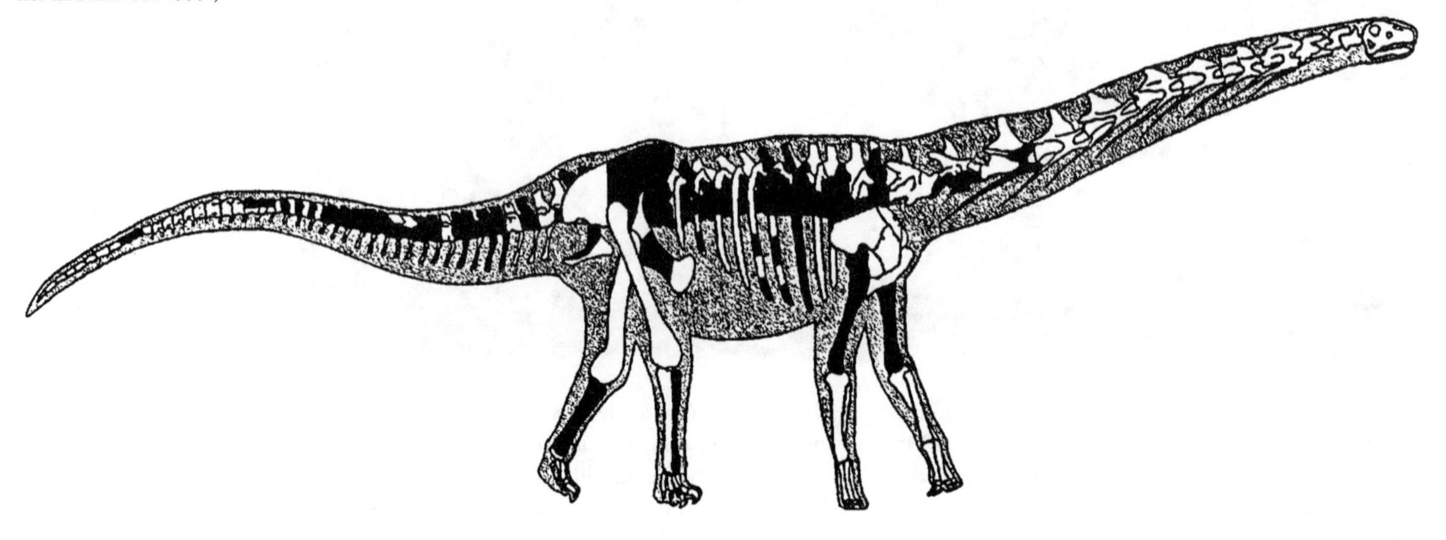

This suggested to the authors that *Gondwanatitan* and *Aelosaurus* could form a monophyletic clade, although such could not be determined given the fossil materials available.

More recently, Bertini suggested that *Gondwanatitan* and *Aeolosaurus* are congeneric, based upon diagnostic features in their respective anterior caudal vertebrae and also unspecified appendicular characters (a more comprehensive comparison of these taxa presumably to follow later in a detailed paper).

Kellner and Azededo commented that, given the wide distribution of titanosaurids in Brazil and the lack of ornithischian material (Kellner 1996), "it appears that sauropods constituted the major herbivore group in the vertebrate fauna during the Cretaceous in this part of Gondwana."

Bertini stated that the presence of *Aeolosaurus* in Brazil suggests Campanian to Maastrichtian ages for the Adamantina Formation and the Marília Formation (from which an undescribed titanosaurid vertebral series has recently been collected).

According to Bertini, the Adamantina Formation of Western São Paulo State has yielded articulated caudal vertebrae with characters atypical for titanosaurids (*e.g.*, slightly backwardly directed neural spine having strong expansion, especially on superior portion; prezygapophysis short, with wide articular faces). Titanosaurids from Brazil can be divided into the following associations: 1. Western São Paulo State, taxa characterized by procoelic and a laterally expanded neural spine, 2. Northern São Paulo State, by slightly developed procoelic caudal vertebrae, and 3. Western Minas Gerais State, by procoelic caudals and extended zygapophyseal laminae. Bertini further noted that titanosaurids from Southeastern Brazil exhibit both endemic and immigrant clades.

Key references: Bertini (2000); Cunha, Rego and Capilla (1987); Cunha and Suarez (1985); Kellner (1996); Kellner and Azevedo (1999), Salgado, Covia and Calvo (1997).

†**AGROSAURUS** [*nomen dubium*]—(See *Thecodontosaurus.*)

†**AGUSTINIA** Bonaparte 1999

Saurischia: Sauropodomorpha: Sauropoda: Eusauropoda: Neosauropoda *incertae sedis.*

Name derivation: "Agustin [Martinelli]."

Type species: *A. ligabuei* (Bonaparte 1998).

Other species: [None.]

Occurrence: Lohan Cura Formation, Neuquén-Province, Argentina.

Age: Early Cretaceous (Aptian).

Known material/holotype: MCF-PVPH-110, incomplete series of 18 fragmentary, articulated vertebrae (three incomplete dorsals, six incomplete sacrals, 10 incomplete caudals), nine dermal ossifications (eight preserved in sequence, two isolated), almost complete right tibia and fibula, five articulated metatarsals.

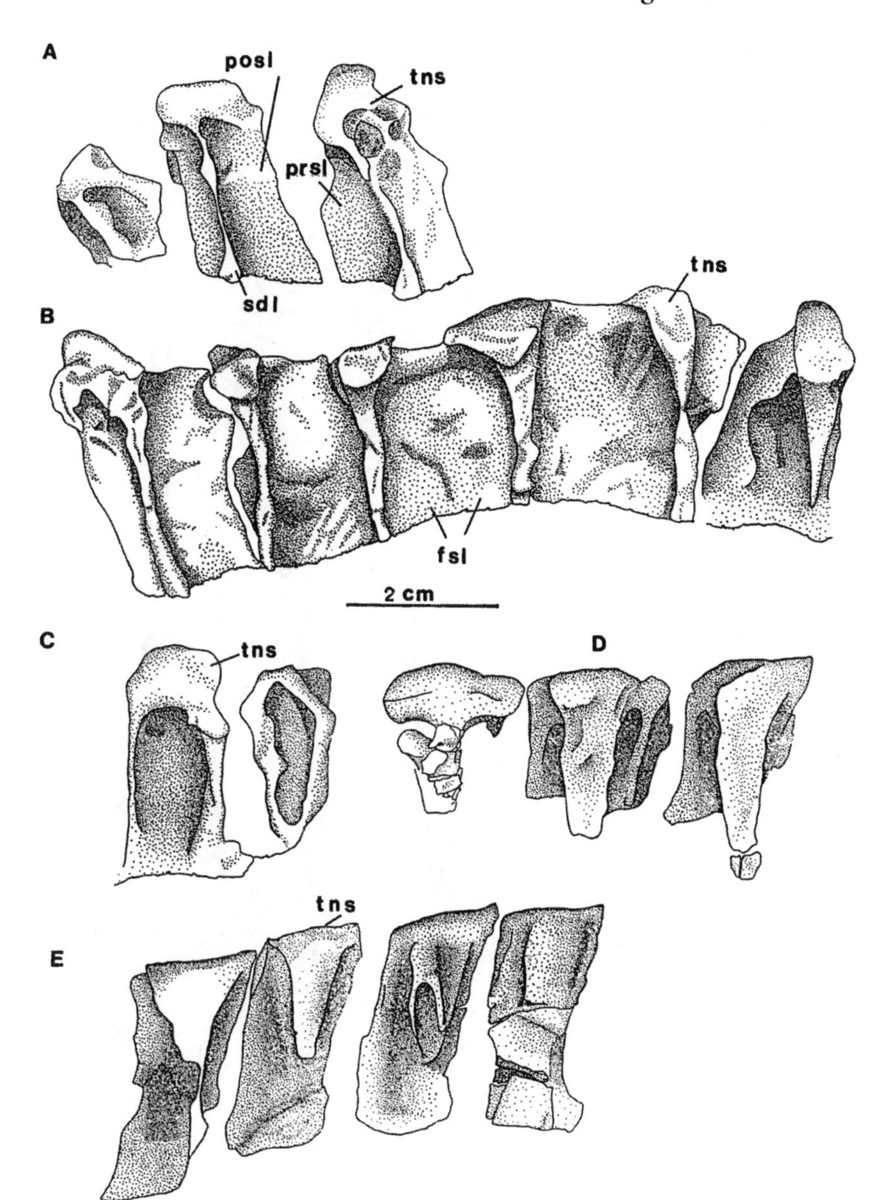

***Agustinia ligabuei*, MCF-PVPH-110, holotype (lateral view), A. last three incomplete dorsal neural arches, B. six incomplete sacral neural arches, C. incomplete neural arches of caudals 1 and 2, D. and E. sequence of seven incomplete neural arches of articulated caudal vertebrae. (After Bonaparte 1999.)**

Diagnosis of genus (as for type species): Sauropod with summit of neural spines expanded in last dorsal, all sacral, and three anterior caudal vertebrae; various types of osteoderms articulating on summits of neural spines — a. Type 1, unpaired, leaf shaped, b. Type 2, laminar, transversely wide, with lateral projections, and c. Type 3 and Type 4, elongate, flat, or cylindrical, dorsolaterally projecting; proximal portion of fibula having pronounced posterior projection; tibia with bend internally directed, bounding cnemial crest; metatarsals of the type found in titanosaurs (Bonaparte 1999*a*).

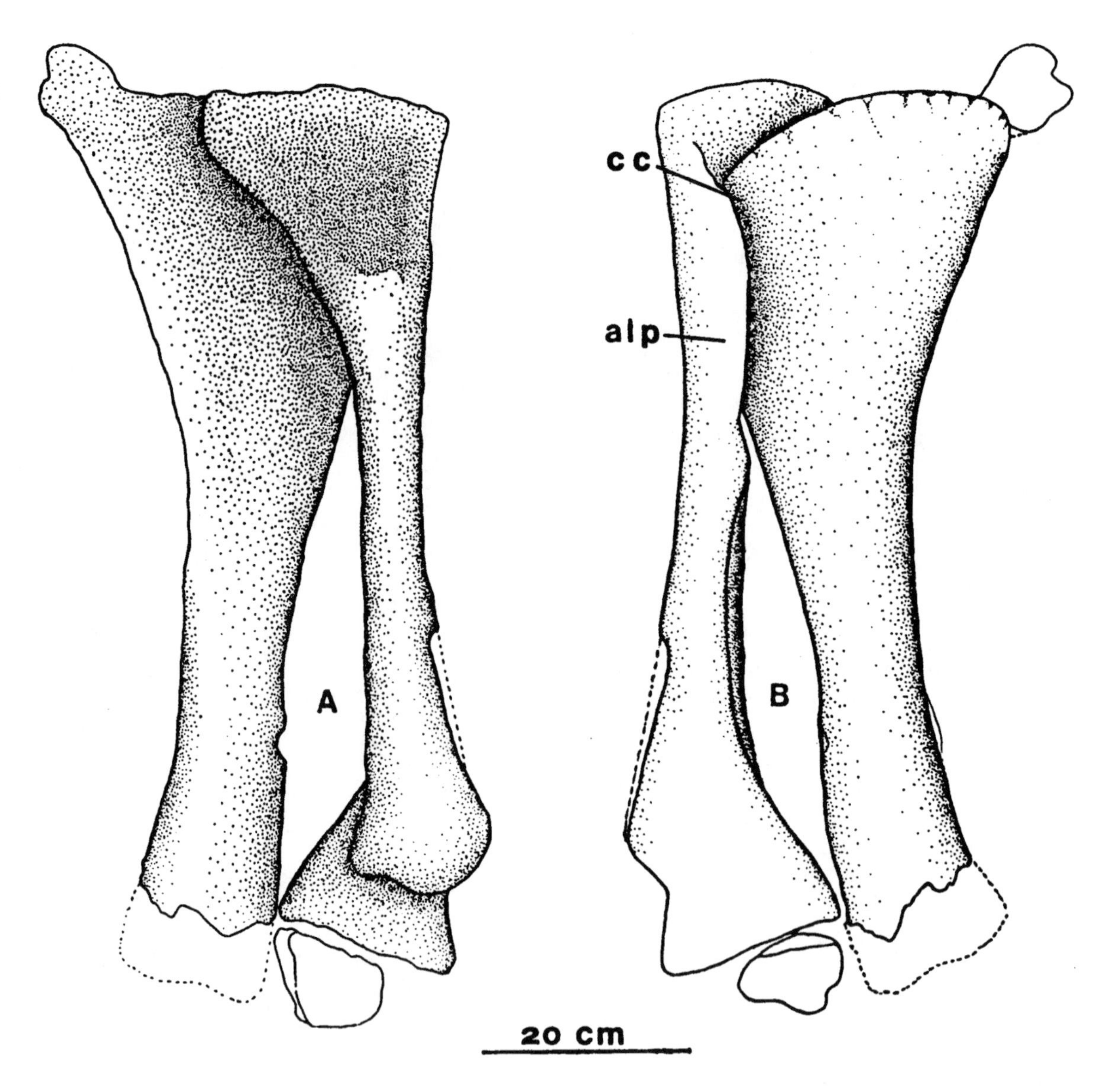

***Agustinia ligabuei*, MCF-PVPH-110, holotype right tibia and fibula in A. lateral and B. internal views. (After Bonaparte 1999.)**

Comments: Another in the growing number of armored sauropods, the type species now named *Agustinia ligabuei* was founded upon partial postcranial remains (MCF-PVPH-110) collected from the upper part of the Lohan Cura Formation (Leanza and Hugo 1996), near Arroyo China Muerta, on the north side of Cerro El León, west of the town of Picún Leufú, in southern Neuquén Province, Argentina. The material was discovered by student Agustin Martinelli, a member of a paleontological field team from the Museo Argentino de Ciencias Naturales of Buenos Aires, which had dispatched the team to the area during January and February in 1996 and 1997 (Bonaparte 1999*a*).

The type species was first announced and briefly described (but not figured) by Bonaparte (1998) in an abstract as *Augustia ligabuei*, a name which proved to be preoccupied (see *S1* for details).

As noted by Bonaparte (1999*a*), eight of the osteoderms are rather complete and well preserved, six of them having been found "in natural sequence, parallel to one another, and following the line of the posterior dorsal and sacral vertebrae." All of these, the author observed, consist of a large body (probably occupying a more external position), also one or two smaller and thicker pieces that were connected to the larger bodies and to the tops of the vertebrae by soft tissue. In effect, *Agustinia* seems to have been quite striking in appearance, the anterior region of the back and possibly also the neck adorned by a row of prominent platelike ossifications, this row grading into two parallel rows of taller spikes (some measuring as long as from 50 to 75 centimeters), projecting laterodorsally along the posterior region of the back.

In his earlier description, Bonaparte (1998) was unable to assign this taxon to any particular sauropod group, owing to the fragmentary nature of the preserved type material; the author did, however, speculate

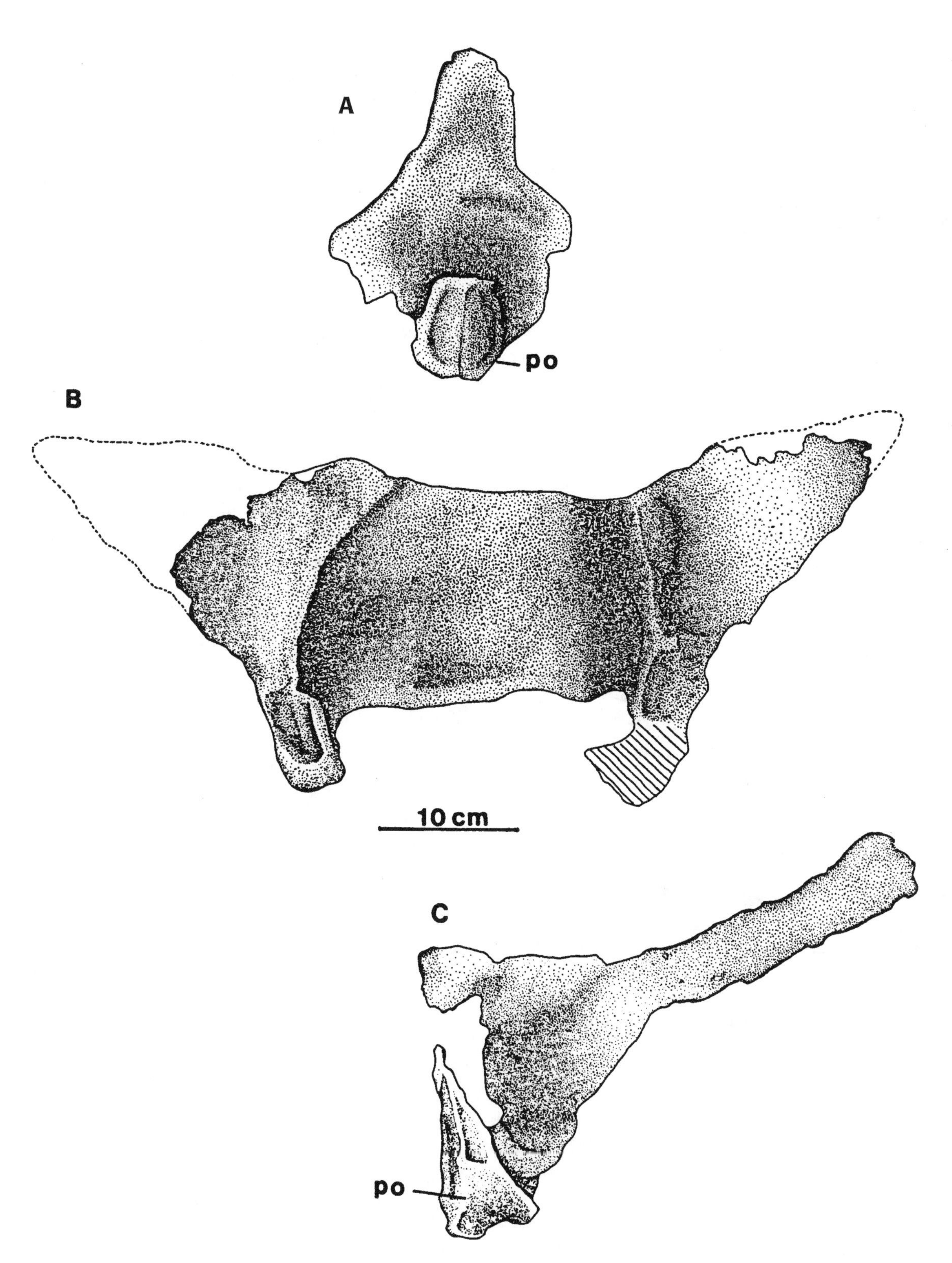

***Agustinia ligabuei*, MCF-PVPH-110, holotype osteoderms, A. Type 1, B. Type 2, and C. Type 3. (After Bonaparte 1999.)**

that this taxon might belong to a new family of Aptian-age South American sauropods.

Subsequently, Bonaparte (1999*a*) identified MCF-PVPH-110 as a member of the Sauropoda (rather than of the Theropoda or Ornithischia) on the basis of morphological similarities between the preserved vertebrae, tibia, fibula, and metatarsals with those elements in other sauropods. Based on the morphology of the neural spines, having four laminae at right angles, that author assessed that *Agustinia* is more similar to the Rebbachisauridae than to the Titanosauridae, the projection of the laminae being different in the latter group, and the prespinal lamina larger (the reversal of the condition in *Agustinia*). Bonaparte (1999*a*) proposed

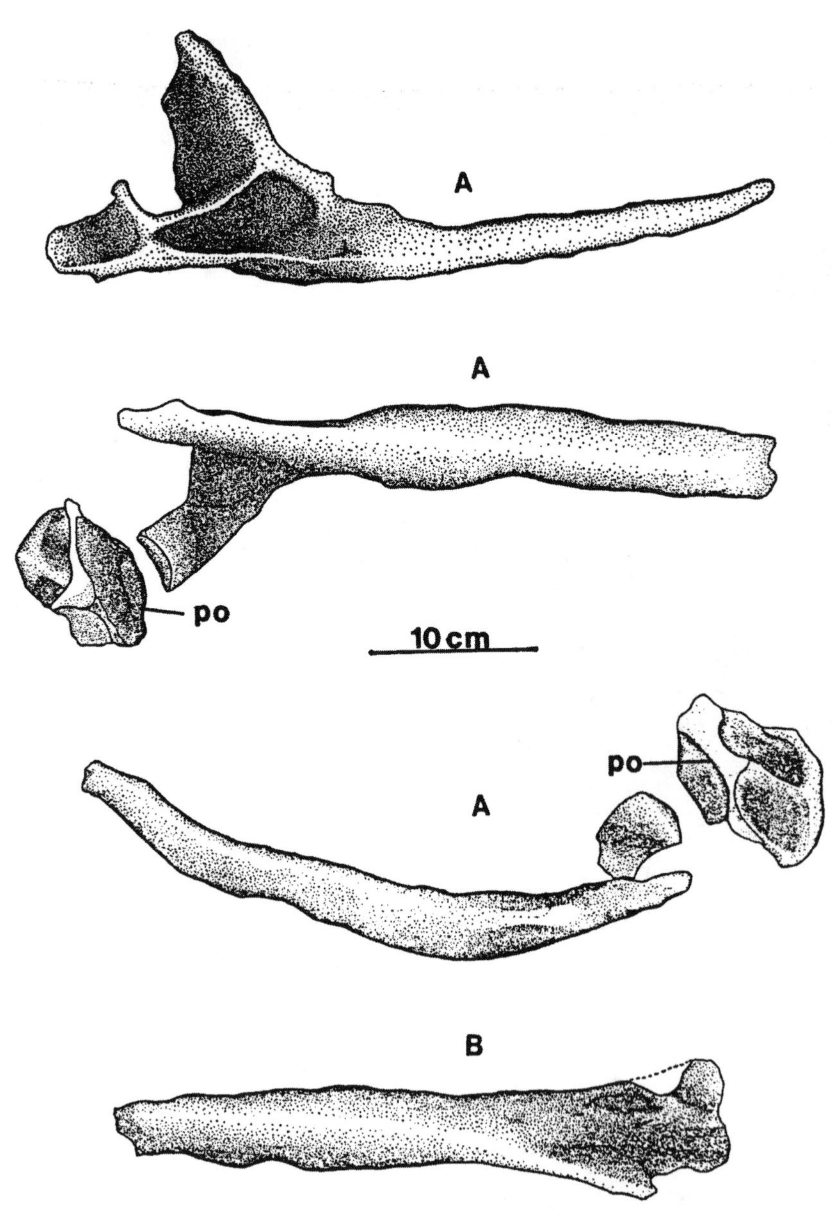

***Agustinia ligabuei*, MCF-PVPH-110, holotype osteoderms, A. Type 3 and B. Type 4. (After Bonaparte 1999.)**

that "possibly the presence of such heterogenous types of osteoderms and the notably expanded top of the neural spines may be enough characters to propose a new family of Sauropod," which he named Angustinidae. However, that proposal was not firmly stated; nor did the author suggest how the Agustinidae might relate phylogenetically to other higher taxa within the Sauropoda.

Bonaparte (1999*a*) diagnosed the Angustinidae as follows: Large sauropods having three or more types of dermal ossifications located along vertebral column above neural spines, including the following: a. unpaired, transversely narrow dermal plates, b. unpaired, transversely broad dermal plates, and c. paired elongated dermal plates.

Key references: Bonaparte (1998, 1999*a*).

†ALAMOSAURUS

Saurischia: Sauropodomorpha: Sauropoda: Eusauropoda: Neosauropoda: Macronaria: Camarasauromorpha: Titanosauriformes: Somphospondyli: Titanosauria: Titanosauridae.

Age: Late Cretaceous (late Campanian–late Maastrichtian).

Comments: *Alamosaurus sanjuanensis* (see *D:TE*)— a titanosaurid taxon founded by Charles Whitney Gilmore in 1922 upon an incomplete left scapula (USNM 10486; see *D:TE* for details and illustration) from the San Juan Basin in New Mexico — remains distinguished as the geologically youngest known North American sauropod. Specimens of this sauropod have been recovered from Upper Cretaceous rocks in Wyoming, Utah, Texas, and New Mexico. Indeed, as pointed out by Lucas and Sullivan (2000*b*) in their review of the material referred to this taxon from New Mexico's San Juan Basin, all sauropod remains collected from the San Juan Basin subsequent to Gilmore's naming of this genus have been referred to *A. sanjuanensis* (*e.g.*, Mateer 1976; Kues, Lehman and Rigby 1980; Lehman 1981; Lucas, Mateer, Hunt and O'Neill 1987; Lucas and Hunt 1989; Sullivan and Lucas 2000*a*) while other North American Late Cretaceous sauropods have been referred either to *Alamosaurus* or to the Titanosauridae.

Lucas and Sullivan described and illustrated, for the first time, additional remains from the San Juan Basin that had been collected in recent years and are referrable to *A. sanjuanensis*. These include an incomplete femur (SMP VP-1138, originally cited collection by Lehman 1985) recovered in 1998 by the State Museum of Pennsylvania from the Naashoibito Member of the Kirtland Formation in Hunter, Wash. Lucas and Sullivan also reported another sauropod femur, discovered by these authors in 1999 while prospecting the Upper Cretaceous strata of Betonnie Tsosie Wash in the Kirtland Formation (Lucas and Sullivan 2000*a*).

As Lucas and Hunt (2000*b*) observed, Gilmore's comparison of the holotype scapula well differentiates *A. sanjuanensis* from such sauropods as *Amphicoelias*, *Apatosaurus*, *Brachiosaurus*, *Camarasaurus*, *Diplodocus*, and *Haplocanthosaurus*. However, separating this scapula from that of some other titanosaurids such as the South American *Laplatasaurus araukanicus* (Huene 1929) and *Saltasaurus loricatus* (Bonaparte and Powell 1980) proves more difficult. In fact, Lucas and Hunt (2000*b*) found it impossible to diagnose *Alamosaurus* from either of those species based only on its holotype scapula. "Therefore," the authors stated, "strict application of the type concept indicates that *Alamosaurus sanjuanensis* is a *nomen vanum* (*nomen dubium*)." According to Lucas and

Photograph by the author, courtesy National Museum of Natural History, Smithsonian Institution.

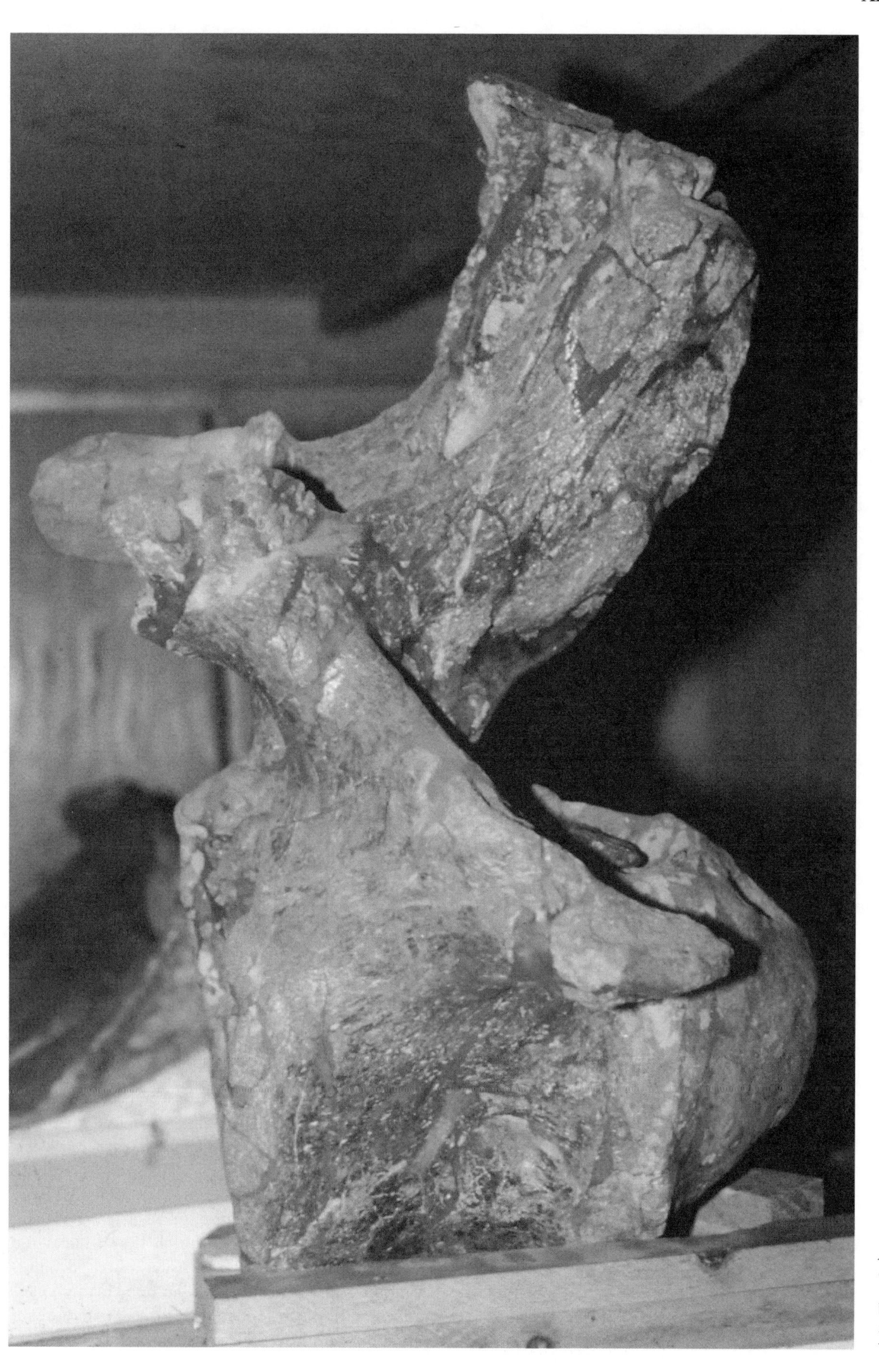

***Alamosaurus sanjuanensis*, caudal vertebra number 2 (USNM 15560), described by Charles Whitney Gilmore (1946*b*), seen in lateral view.**

Sullivan (2000*b*), at least one titanosaurid taxon is represented by the sauropod specimens collected to date in the San Juan Basin, but the type species *A. sanjuanensis* was established upon an inadequate holotype. In the future, pending the discovery of a potentially diagnostic specimen, a neotype for this species may be designated. For the present, however, Lucas and Sullivan (2000*b*) found it "useful to continue using the name *Alamosaurus* as a form genus [term coined by Spencer G. Lucas] for all North American Late Cretaceous sauropods" (see also Sullivan and Lucas 2000*a*).

Generally *Alamosaurus* has been considered to be of Lancian (late Maastrichtian) age (*e.g.*, Hunt and Lucas 1992; Lehman 1980; Lucas 1980; Lucas, Mateer, Hunt and O'Neill).

It has long been believed that *Alamosaurus* first appeared in North America following a so-called "sauropod hiatus," a term coined by Lucas and Hunt for the absence of sauropods on that continent between Cenomanian and late Maastrichtian time. Lucas and Hunt argued that this absence resulted from the extinction of North American sauropods at approximately the end of the Albian. Presumably, not until the late Maastrichtian did sauropods (*i.e.*, titanosaurs) return to that continent, probably immigrating from South America.

As already noted, *Alamosaurus* was regarded by Sullivan and Lucas and Lucas and Sullivan (2000*b*) as a "form genus" to which these authors provisionally refer all Late Cretaceous sauropod remains from New Mexico's San Juan Basin. Thus, to *A. sanjuanensis*, Sullivan and Lucas referred an incomplete sauropod tooth (SMP VP-1097) collected in 1998 by Robert M. Sullivan from the De-na-zin Member of the Kirtland Formation of the San Juan Basin, New Mexico. Although the holotype of *A. sanjuanensis* lacks skull material (see Gilmore), these authors referred the tooth to that taxon because of the "form genus" status of *Alamosaurus*, and also based on an earlier report on teeth referred to this genus published by Kues, Lehman and Rigby. Also, Sullivan and Lucas referred to *A. sanjuanensis* previously unidentified and uncatalogued caudal vertebrae (NMMNH P-28741, P-29722, and P-29723) collected in 1984 by a University of New Mexico field party from the same stratigraphic interval. Sullivan and Lucas described these specimens in detail (see original paper).

This material is especially significant, Sullivan and Lucas (see also Lucas and Sullivan 2000*b*) pointed out, because of its age. Radioisotopic dating of two air-fall ash beds in the De-na-zin Member of the Kirtland Formation indicate a late Campanian age for these specimens (see Obradovich 1993; Fassett and Steiner 1997). The presence of the dinosaurs *Parasaurolophus*, *Pentaceratops*, and *Albertosaurus* in the Campanian Fruitland Formation is consistent with dating the De-na-zin Member as of late Campanian age (see Sullivan 1999; Sullivan and Williamson 1999). Therefore, the above newly described material, being of late Campanian age, shortens the "sauropod hiatus." Consequently, these new records undermine the traditional utilization of *Alamosaurus* as an index fossil for Lancian time (see also Lucas and Sullivan 2000*b*), thereby also calling into question the age of the geologically younger Naashiobito Member of the Kirtland Formation. As pointed out by Lucas, Heckert and Sullivan (2000), the Naashiobato Member has been generally regarded as of Maastrichtian age (*e.g.*, Lehman 1981), based on the presence of *Torosaurus*, *Tyrannosaurus*, and *Alamosaurus*. However, it has also been regarded as late Campanian (Sullivan 1987), based upon the inadequacy of the material pertaining to those Lancian-age indicators, and, less likely, Paleocene (*e.g.*, Fassett 1982; Fassett and Steiner 1997), based on an unpublished record of presumed Paleocene-age pollen from the uppermost De-Na-Zin Member (J. E. Fassett, personal communication to Lucas *et al.* 2000).

Sullivan and Lucas further noted that the above specimens do not constitute the first documentation of a late Campanian North American sauropod. McCord (1997) had previously reported a titanosaurid caudal vertebra from the Fort Crittenden Formation in the Santa Rita Mountains of Arizona, this formation being a correlative of (possibly the same lithostone as) the Ringbone Formation in southwestern New Mexico (see Lucas, Kues and Gonzaplez-León 1995). An ash bed at the base of the Hidalgo Formation, which immediately overlies the Ringbone Formation in the Little Hatchet Mountains of southwestern New Mexico, has been dated as 71.44 ± 0.19 million years old (see Lawton, Basabilcazo, Hodgson, Wilson, Mack, McIntosh, Lucas and Kietzke 1993). This suggested to Sullivan and Lucas that both underlying Ringbone Formation and also its Arizona equivalent, the Fort Crittenden Formation, are also of late Campanian age.

Therefore, according to Sullivan and Lucas and Lucas and Sullivan (2000*b*), the age of the new *Alamosaurus* material and the Arizona titanosaurid material indicate that, if sauropods were indeed extirpated in North America by the end of the Early Cretaceous, they returned — apparently immigrating from South America or Asia — by late Campanian time. As *Alamosaurus* seems to be related to the Asian titanosaur *Opisthocoelicaudia* (see Upchurch 1998), Lucas and Sullivan (2000*b*) noted, "an Asian sauropod immigration event to terminate the sauropod hiatus is consistent with the introduction of other Asian dinosaur taxa to North America during Campanian time."

Key references: Bonaparte and Powell (1980); Fassett (1982); Fassett and Steiner (1997); Gilmore (1922); Huene (1929); Hunt and Lucas (1992); Kues, Lehman and Rigby (1980); Lawton, Basabilcazo, Hodgson, Wilson, Mack, McIntosh, Lucas and Kietzke (1993); Lehman (1980, 1981, 1985); Lucas (1980); Lucas and Hunt (1989); Lucas, Kues and González-León (1995); Lucas, Mateer, Hunt and O'Neill (1987); Lucas and Sullivan (2000*a*, 2000*b*); Mateer (1987); McCord (1997); Obradovich (1993); Sullivan (1987, 1999); Sullivan and Lucas (2000*a*); Sullivan and Williamson (1999); Upchurch (1998).

†ALBERTOSAURUS

Comments: Philip J. Currie (2000) presented evidence, gleaned from fossil specimens collected from a rediscovered bonebed, that the tyrannosaurid *Albertosaurus* may have been a pack hunter.

According to Currie's report, in 1910, Barnum Brown of the American Museum of Natural History had originally found this site in Alberta, Canada. While floating on a raft down the Red Deer River, Brown and his assistant Peter Kaisan spotted an *Albertosaurus* skeleton encased in "ironstone" on a shoulder of rock located some 50 meters above the river. To Brown and Kaisan's surprise, the bones of more than one individual were preserved at this site. After collecting some of this material over a period of three weeks, they moved on downriver, never returning to this site. Although Brown later discussed this site in various nontechnical articles, he never addressed its significance.

Having read some of these articles during the late 1970s, Currie became fascinated by the fact that as many as a half dozen tyrannosaurs could have been preserved in a single quarry and pondered the significance of this. In 1996, following a number of failed attempts to locate the quarry, Currie examined the material collected there by Brown in 1910. Currie found that Brown had collected only parts — mostly feet — of nine specimens of *Albertosaurus*. This suggested to Currie that Brown had been "high-grading," that is collecting elements indicating that numerous specimens belonging to this one genus were recovered from a single locality.

Studying the field notes written by Brown and Kaisan and the correspondence between Brown and his boss Henry Fairfield Osborn, and examining photographs taken in 1910, Currie launched an expedition to the area co-sponsored by the Royal Tyrrell Museum of Palaeontology and the Dinamation International Society. Following Brown's journey down the Red Deer River, Currie eventually found the long abandoned site, marked by the presence of numerous *Albertosaurus* bones and teeth.

The site was subsequently worked in the summers of 1998, 1999, and 2000 by field teams from the Tyrrell Museum. *Albertosaurus* remains were recovered representing individuals of all ages that died *en masse* approximately 70 million years ago. This evidence suggested to Currie that *Albertosaurus* individuals, at least part of the time, lived together and hunted in packs. Currie pointed out that this hypothesis correlates with footprint evidence from the Peace River Canyon of British Columbia showing several individuals of large theropods traveling together. It also correlates with the discovery of bones belonging to adult and juvenile *Tyrannosaurus rex* individuals collected from the quarry that yielded the *T. rex* specimen known as "Sue" (see *Tyrannosaurus* entries, *D:TE*, *S1*, and this volume).

According to Currie, bits of charcoal mixed in with the *Albertosaurus* bones may suggest that these dinosaurs were killed during a forest fire.

In a comprehensive summary and review of the record of the Tyrannosauridae of New Mexico, Carr and Williamson (2000) addressed previously published claims of *Albertosaurus* remains being found in New Mexico. As recounted by these authors, Lucas, Mateer, Hunt and O'Neill (1987) had referred four tyrannosaurid teeth (originally UNM FKK-077, 078, 079, and 080, now catalogued as NMMNH) to *Albertosaurus* sp. and one (originally UNM FKK-76, now NMMNH P-13000) to cf. *Tyrannosaurus rex*. All of these specimens, according to Lucas *et al.*, were collected from the Naashoibito Member of the Kirtland Formation. As pointed out by Carr and Williamson, however, it is now clear that only the tooth referred to cf. *T. rex* was recovered from the Naashoibito Member, the teeth referred to *Albertosaurus* sp. having been collected from the De-na-zin Member.

According to Carr and Williamson, Lucas *et al.*'s referral of the majority of these teeth was based on tooth size and the number of denticles (nine to 11) per five millimeters. At the same time, these authors, without offering diagnostic characters to support their opinion, referred with "almost" certainty other specimens from the San Juan Basin of New Mexico to *Albertosaurus*, these including an incomplete skeleton (OMNH 10131; referred by Lehman and Carpenter (1990) to *Aublysodon* cf. *A. mirandus*, and by Carr and Williamson to *Daspletosaurus* sp.; see *Daspletosaurus* entry) collected in 1940 by J. W. Stovall and Donald E. Savage from the Kirtland Formation, a dentary (USNM 8346; referred by Gilmore 1935 to *Gorgosaurus*, regarded by Carr and Williamson as an indeterminate tyrannosaurid) collected by J. B. Reeside from the vicinity of the Ojo Alamo Sandstone (see Gilmore 1916*b*); and a privately owned metatarsal IV (also regarded by Carr and Williamson as an indeterminate tyrannosaurid), collected by Ronald P. Ratkevich from the Naashoibito Member of the Kirtland Formation and reported by Lehman (1981) as ?*Albertosaurus* sp.

According to Carr and Williamson, however, all such referrals of tyrannosaurid material from New Mexico to *Albertosaurus* are unsubstantiated, as no positive occurrences of this genus have been recorded from that state.

Key references: Carr and Williamson (2000); Currie (2000); Gilmore (1916*b*); Lehman (1981); Lehman and Carpenter (1990); Lucas, Mateer, Hunt and O'Neill (1987).

†ALIWALIA

Age: Late Triassic (?early Carnian).

Known material: Partial femur, ?partial cranial remains.

Cotypes: NMW 1886-XV-39 and NMW 1876-VII-B124, both ends of femur.

Comments: *Aliwalia rex* (see *D:TE*), a poorly known dinosaur of uncertain affinities, was founded upon the proximal and distal ends of a large femur (NMW 1886-XV-39 and NMW 1876-VII-B124, respectively), which, if complete, would have measured approximately 900–1000 millimeters in length. The type material was recovered from the lower Elliot Formation (Upper Triassic, Carnian, possibly early Carnian; see Gauffre 1993*b* and *Blikanasaurus* entry) of South Africa, probably from the same excavation site (Barnard's Spruit near Aliwal North) as the lectotype of the prosauropod *Euskelosaurus browni* (Galton and Heerden 1998).

To *A. rex*, Galton and Heerden tentatively referred a large (approximately 420 millimeters long) left premaxilla and maxilla (BMNH R3301) with carnivorous replacement teeth in place, recovered from the same excavation in the lower Elliot Formation at Barnard's Spruit. This specimen was originally described by Seeley (1894), and later by Charig, Attridge and Crompton (1965), the latter authors noting that it gave little indication as to the animal's affinities.

According to Galton and Heerden, BMNH R3301 is not a prosauropod, and therefore, clearly not referrable to *Euskelosaurus*. The specimen lacks any derived characters diagnostic of rauisuchian "thecodontians" or the early ?theropod *Herrerasaurus*, but is of appropriate size to go with the type femur of *A. rex*.

Although Galton (1985*a*) had originally referred *Aliwalia* to the "Herrerasauria" [=Herrerasauridae], Sues (1990), in a study of *Staurikosaurus* and the

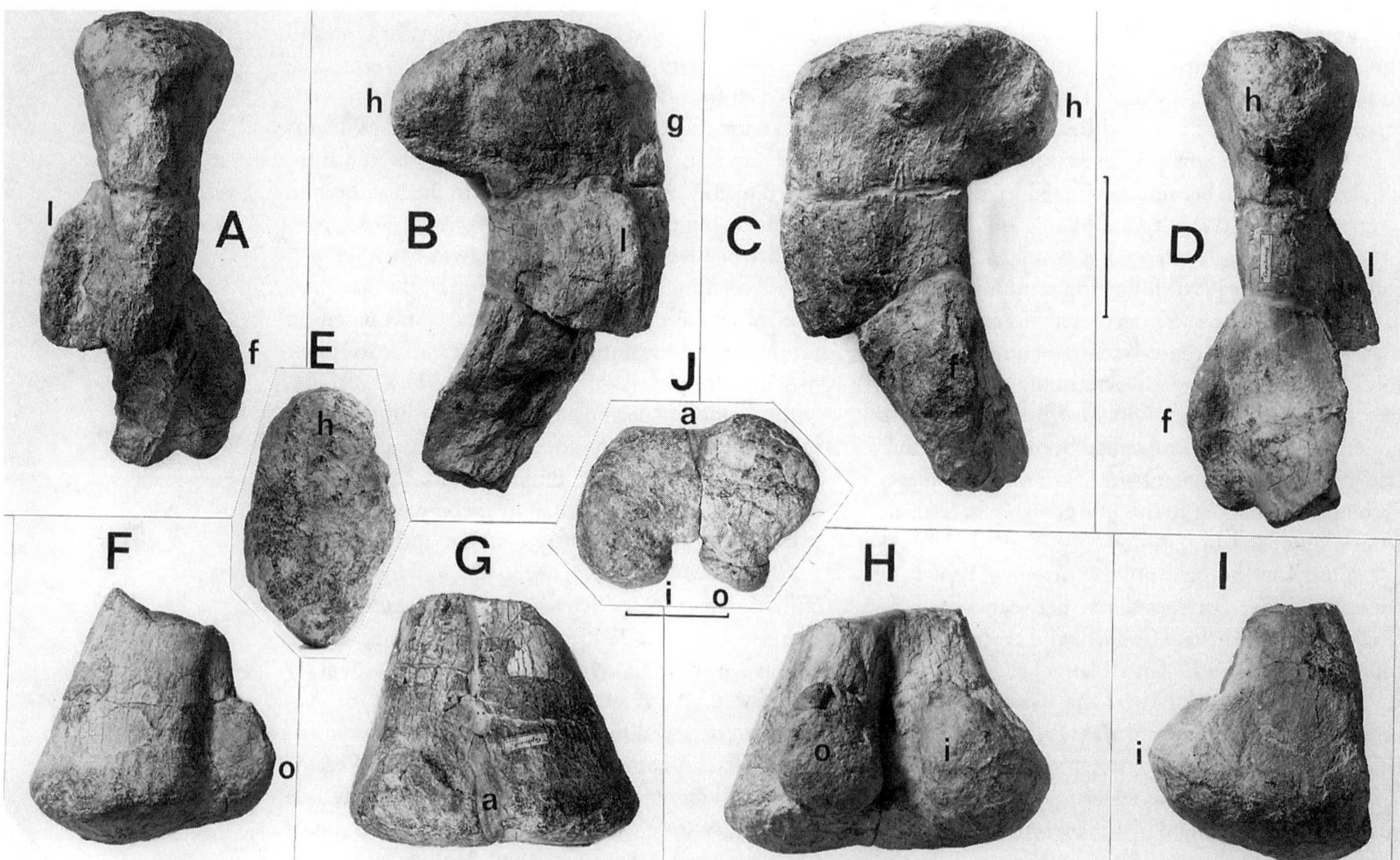

***Aliwalia rex*, cotypes, incomplete femur originally described as ?*Euskelosaurus* and included in the paralectotype of *E. browni*: NMW 1889-XV-39, proximal end in A. lateral, B. anterior, C. posterior, D. medial, and E. proximal views; NMW 1876-VII-B124, distal end in F. lateral, G. anterior, H. posterior, I. medial, and J. distal views. Scale = 10 cm. (After Galton and Heerden 1998.)**

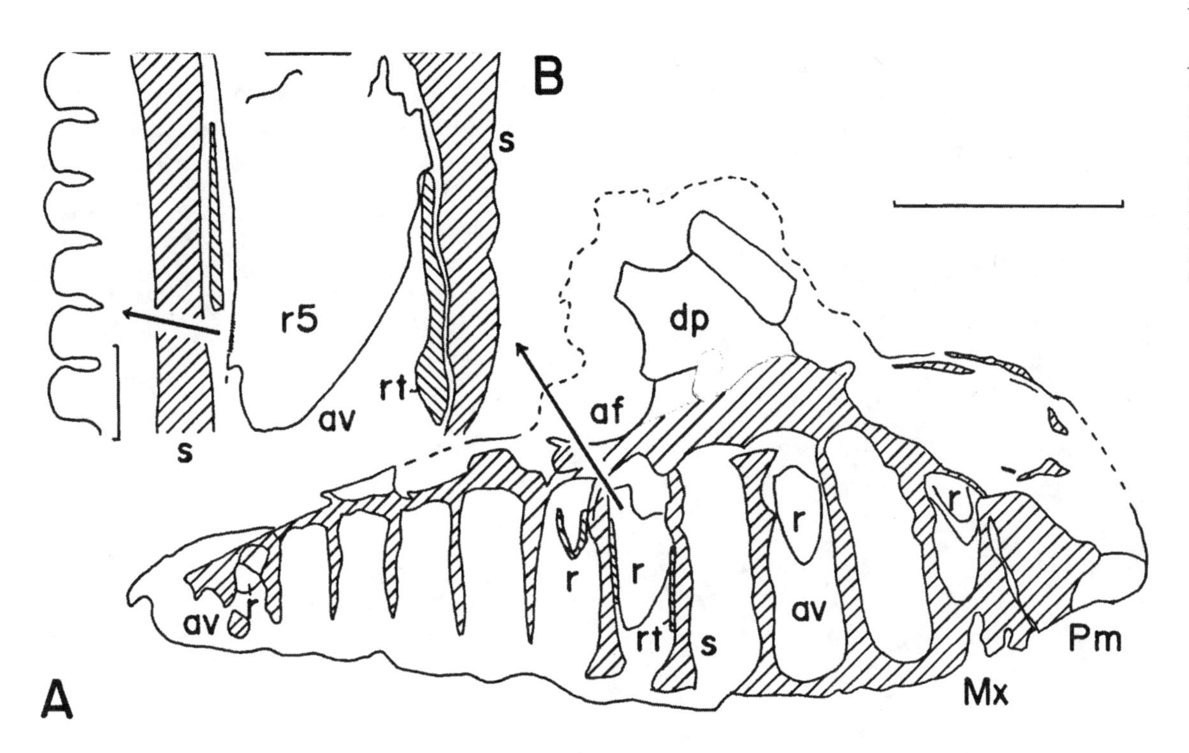

***Aliwalia rex*, BMNH R3301, tentatively referred left maxilla and premaxilla in medial view, A. photograph and explanatory drawing, B. replacement tooth in alveolus 5 showing detail of posterior serrations. Scale = (A) 10 cm. and (B) 1 mm. (After Galton and Heerden 1998.)**

Herrerasauridae, regarded the genus as Dinosauria *incertae sedis*. Agreeing with Sues' reassessment, Galton and Heerden noted that, in particular, "the femur of *Aliwalia* lacks a prominent trochanteric shelf passing posteriorly from the lesser trochanter that is much larger than that of [the herrerasaurid genus] *Herrerasaurus*."

Key references: Charig, Attridge and Crompton (1965); Galton (1985*a*); Galton and Heerden (1998); Seeley (1894); Sues (1990).

†**ALLOSAURUS** (=*Apatodon, Creosaurus, Epanterias, Hypsirophus* [in part], *Labrosaurus, Saurophaganax*; =?*Antrodemus*)

Comments: *Allosaurus* remains one of the best known of all theropod genera, represented by myriad specimens. Previously known with certainty only from the Late Jurassic Morrison Formation of North America, the type species *Allosaurus fragilis* is now also celebrated as the first known "intercontinental dinosaur species," and the first diagnostic Jurassic theropod found on the Iberian Peninsula (earlier putative reports of *Allosaurus* in Africa, Europe, and Australia [see note below] having been based upon undiagnostic specimens).

In 1999, Pérez-Moreno, Chure, Pires, da Silva, Santos, Dantas, Póvoas, Cachão, Sanz, and Galopim Carvalho redescribed a specimen (MNHNUL/AND.001, Andres collection) found in 1988 in the Upper Jurassic (upper Kimmeridgian–Tithonoian) Lourinhã Formation (Lusitanian Basin)—a site mostly

Allosaurus fragilis **mounted skeleton (CM 11844, skull cast of UUVP 6000, forelimbs cast from a USNM specimen). This relatively complete skeleton, mounted in 1938, was collected between 1913 and 1915 by Earl Douglass and his field crew.**

Photograph by the author, courtesy Carnegie Museum of Natural History.

represented by micaceous sandstones, fine sands, and stilts — near the village of Andre's, District of Leira, east of Lisbon, in west-central Portugal. The specimen consists of a partial skeleton including the ventral articular end of the right quadrate, several vertebrae and chevrons, several dorsal ribs and gastralia, an incomplete pelvis, parts of the hindlimbs, and a few indeterminate fragments. The material compares quite favorably with corresponding elements in both described and undescribed specimens of *A. fragilis*. The

Photograph by the author.

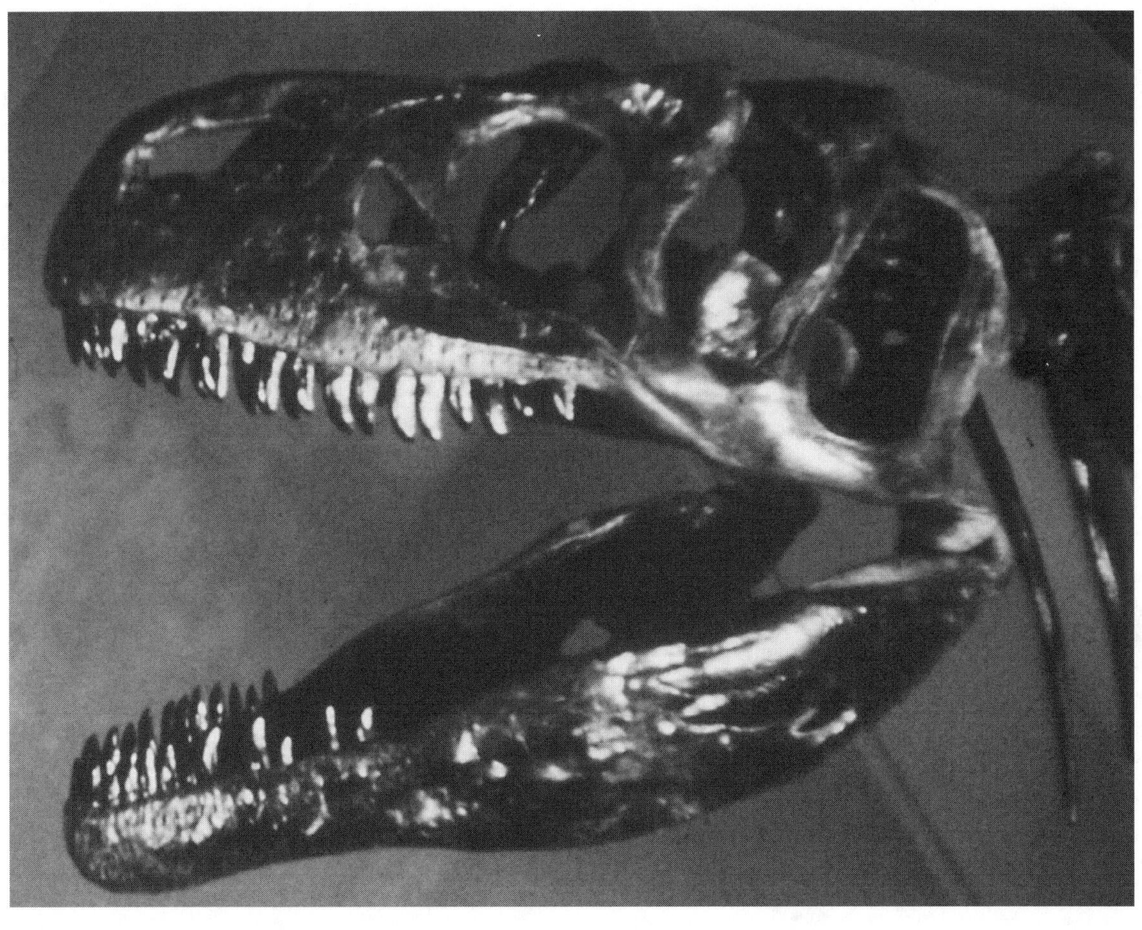

Skull of *Allosaurus fragilis*. Recent studies were performed focusing on the skull of this large carnosaur.

ischium of both the Portuguese specimen and that designated DINO 11541 (the latter representing a possibly new species currently under study by Daniel J. Chure), from Dinosaur National Monument, show a very large obturator process, which is a diagnostic character of *Allosaurus* (see Madsen 1976).

The occurrence of *A. fragilis* in Portugal, the authors noted, suggests that this species was also probably present in the eastern United States, an area from which rocks of Mid- and Late Jurassic age are unfortunately absent (see Weishampel and Young 1996). It also implies that an Atlantic land connection or "bridge" existed during the Late Jurassic period between the North American continent and Eurasia (as suggested by previous authors, *e.g.*, Galton 1980*a*, 1980*b*; and Galton and Powell 1980). This idea conflicts dramatically with earlier paleogeographic reconstructions that depict a fully open North Atlantic sea.

In an abstract, Bybee and Smith (1999) brought attention to an "unusual allosaurid skull" (BYU collection) preserved with some postcranial remains originally referred to *A. fragilis*. The specimen was reported during the 1970s to James A. Jensen by the Hinkle Family from an isolated Morrison Formation site in eastern Utah, adjacent to Rabbit Valley, Colorado, and later collected.

As observed by Bybee and Smith, this skull exhibits various unusual characteristics that set it apart from other *Allosaurus* remains recovered from the Dry Mesa Dinosaur Quarry in Colorado and the famed Cleveland-Lloyd Dinosaur Quarry in Utah (a site which may, based on a recent analysis of bone orientation utilizing the quantitative method of vector summation, represent either or both a "fluvial concentration" or "predator trap"; see McGee 1999 for details). According to these authors, the animal represented by the BYU specimen was more robust and heavily ossified than the complete skull (UUVP 6000, the neotype of *A. fragilis*) displayed at Dinosaur National Monument (see *D:TE* for photograph of this specimen). Among these characteristics are the following: Sphenethmoid ossified; cultriform process very thin; basisphenoid recess posteriorly oriented; cranial nerve foraminae coalesced at base of occipital condyle into single, large foramen; well-developed longitudinal trough; orientation of paroccipital processes more horizontal than in Dry Mesa and Cleveland-Lloyd *Allosaurus* specimens.

According to Bybee and Smith, the differing characters in the BYU specimen cannot be attributed to allometry, as the braincases of the Dry Mesa and Dinosaur National Monument *Allosaurus* specimens

Photograph by the American Museum of Natural History Photo Studio. Courtesy Department of Library Services, American Museum of Natural History (neg. #338592).

AMNH 5753, one of the most photographed of all the many recovered skeletons (skull cast of another specimen) of *Allosaurus fragilis*. Originally part of the Edward Drinker Cope collection, this specimen was purchased by the American Museum of Natural History in 1899 and prepared and mounted by 1908. Posed over a partial skeleton (AMNH 222) of the sauropod *Apatosaurus*, this mount was the subject of a famous painting by Charles R. Knight (see Czerkas and Glut 1982). AMNH 222 was one of the American Museum's first dinosaur specimens, collected in 1897 or 1898 by Henry Fairfield Osborn and Barnum Brown at Como Bluff, Wyoming.

are similar in size. Therefore, these differences "probably reflect population or systematic variation."

In an abstract, Smith, Bybee and Anderson (1999) briefly discussed their updated analysis of cranial kinetics in *Allosaurus*. Unlike earlier interpretations (which proposed a two-part skull, the tooth-bearing elements rotating dorsoventrally about a prokinetic joint relative to the braincase), this hypothesis offered a more elaborate mechanism. This new model — based on UUVP 6000, an almost complete skull recovered from an isolated site in eastern Utah, and disarticulated remains (BYU collection) from the Cleveland-Lloyd and Dry Mesa quarries — was "summarized as a posterior expansion of the skull associated with an anterior extension of the maxilla."

Bybee and Smith found in *Allosaurus* that, as a condition of this hypothesis, a complete postorbital bar does not preclude extensive kinetic mechanism; the streptostylic jaw mechanism (found in most, perhaps all theropods) is well developed; and opening and closing the jaws involved more than simple abduction and adduction, an independent clockwise rotation occurring as the mandibles opened, thereby bringing the dentary teeth in line more with the maxillary teeth.

Recently, Smith (1999) performed morphometric analyses on samples of limb, manus, pes, pectoral- and pelvic-girdle, and many cranial elements (*e.g.*, premaxilla, maxilla, ectopterygoid, lacrimal, and postorbital) referred to the type species *Allosaurus fragilis*. It was that author's goal to determine and define the patterns of size-related variation (*i.e.*, the first principal component) present within the genus *Allosaurus*. In performing this study, Smith divided the skeleton into five major regions — the cranium, girdles, limbs, manus, and pes. Studied materials included specimens recovered from the Cleveland-Lloyd Dinosaur Quarry, Utah, Dinosaur National Monument, Utah, Dry Mesa quarry, Colorado, Cañon City, Colorado, and Como Bluff, Wyoming. Measurements from these elements were converted by Smith to natural logarithms, then, applying a principal components analysis, extracted from a convariance matrix (for details, see Smith 1998; also, *S1*).

Based on this study, Smith observed the following in *Allosaurus*: the cranium is the region most susceptible to variability in shape, exceeding (as expected) that of the postcranial skeleton. The percentage of noted variability in the cranium attributable to size ranges from 53 percent (lacrimal) to 84 percent (premaxilla). The humerus, radius, and ulna remain moderately conservative in shape. The pectoral girdle (the second most variable region), however, represents a separate allometric region, with only 64 percent of the observed variation in the scapula explained by size, 71 percent of the variation in the coracoid apparently being the result of the first principle component. The manual phalanges have a consistently greater shape influence on total variation than do the pedal phalanges, which is possibly explained by greater functional and environmental influences on the foot than on the hand. Also, the very high proportion of variation in manual digit I can be explained by size changes, while the proportion of variation in manual digits II and III attributed only to size change successively declines, this pattern explained as resulting from increasing functional influence on the outside of the hand. No pattern defines the pelvic girdle as a whole, although considerably more variation attributed to shape can be seen in the ilium than in the comparatively conservative ischium, and the girdles are more prone to shape change than the limbs. In the hindlimb, the proportion of variation explained by size change only declines until the metatarsals, which are highly shape-variable. The 76 percent proportion of variation in the astragalus, explained by size, is similar to that seen in the metatarsals. The pes exhibits the most conservative region of the skeleton, with 90–95 percent of the variation within the phalanges attributed to ontogeny.

As pointed out by Smith, all the Dry Mesa material clusters together away from the other specimens. If consistent, this pattern might suggest the presence of a second species in that quarry. However, as only the astragalus was found to display this pattern, it was not interpreted as having any systematic significance. Smith noted that the above study and its "resulting allometric patterns will have implications for theropod species identification and the effects of allometry on changes in bone morphology and variation."

Chure (1999), in an abstract, briefly reported on a complete, articulated forelimb, including two proximal (ulnare and intermedium) and two distal carpals, belonging to a subadult and virtually complete *Allosaurus* recovered at Dinosaur National Monument. This specimen (DNM collection) is important in better comprehending the evolution of the semilunate carpal, historically an important synapomorphy in deriving birds from theropod dinosaurs. The semilunate carpal had previously been listed among the synapomorphies of the Tetanurae; however, information derived by Chure from the new DNM specimen has revealed that several steps were involved in the evolution of this birdlike feature.

As described by Chure, the DNM forelimb lacks an ulnare (either lost or unossified). Distal carpal 1 possesses a well-developed trochlear surface and constitutes a semilunate carpal. The second distal carpal is reduced but not fused to the semilunate carpal. Also examining an ontogentic series of 25 *Allosaurus* distal carpals collected from the Cleveland-Lloyd Dinosaur Quarry, Chure observed a greatly reduced distal carpal 2 that, even in the smallest specimens, fuses to the semilunate carpal.

Photograph by Patrick Fisher, courtesy Oklahoma Museum of Natural History, University of Oklahoma.

Mounted skeleton (cast)—including casts of OMNH 01123 (holotype), 01338, 01737, 01425, 01708, 01370, 01935, originally assigned to the new genus *Saurophaganax*— of *Allosaurus maximus.*

Chure noted that the primitive condition of disk-shaped carpals is present in the herrerasaurid *Herrerasaurus* and the abelisaurid *Carnotaurus*. Distal carpals 1 and 2 are fused in the coelophysoids *Coelophysis* and *Syntarsus*; however, the resulting carpal in these taxa is flat, lacks a trochlea, and is not a semilunate carpal. This fusion is considered by Chure to be homoplastic with the same condition in Manuraptora.

According to Chure, the semilunate carpal in *Allosaurus* mostly consists of distal carpal 1, which is restricted to the end of metacarpal 1. Among manuraptorans, therizinosauroids possess subequal and unfused distal carpals 1 and 2, these together forming a trochlea; however, in other manuraptorans, the fused distal carpals 1 and 2 form the semilunate carpal, covering metacarpals I and II. This fusion, Chure proposed, may have taken place more than once depending upon the phylogenetic position of Therizinosauroidea within Manuraptora. The primitive condition of unfused, disk-shaped carpals is found in the manuraptoran group Ornithomimidae. This, Chure postulated, is either a heterochronic effect or the semilunate carpal in *Allosaurus* is homoplastic with that of manuraptorans; hence, a semilunate carpal is not a tetanuran synapomorphy.

Chure (2000*a*), in another abstract, presented new data on the morphology of the gastral basket in the same *Allosaurus* skeleton, correcting a number of previously published misconceptions. Chure (2000*a*) observed the following: 18 rows of gastralia; two gastralia present on each side of a row (as in other

Photograph by the author, courtesy National Museum of Natural History, Smithsonian Institution.

USNM 2315, holotype left dentary of *Labrosaurus ferox* in medial view (see *D:TE* for lateral view). James H. Madsen, Jr. regards this specimen as "an aberrant and damaged" specimen of *Allosaurus fragilis*.

theropods); right medial element overlapping the left ventrally; no medial V-shaped element; rostral row of gastralia consisting only of medial elements that, unlike the rest of the basket, do not overlap; rather, the medial ends are expanded, meeting one another at a right angle, this angle opening rostrally.

This specimen — the central topic of Chure's Ph.D. dissertation — will be made a new species of *Allosaurus* by Chure, named in honor of *Allosaurus* authority James H. Madsen, Jr. (D. J. Chure, personal communication 2000).

Chure (2000*a*) also reexamined a disarticulated gastral basket in the pathologic *Allosaurus* specimen (USNM 4734), which Gilmore had described in 1920 and which is mounted at the National Museum of Natural History. As observed by Chure (2000*a*), the medial V-shaped element identified by Gilmore is really a furcula (like that also seen in the DNM specimen). Gilmore counted seven gastrial per row. However, as Chure (2000*a*) pointed out, this animal during life received a severe though nonfatal blow to the right side of the basket, resulting in transverse breaks in several medial bones. In healing, some of the breaks fused, and others formed movable joints. Consequently, Gilmore interpreted these segments as separate elements. Additional injuries in the gastralia of USNM 4734, noted by Chure (2000), include "long axis green breaks with no signs of healing," which occurred when the animal died or in postmortem.

Foster and Chure (1999) published a brief report on hindlimb proportion allometry in juvenile to adult *Allosaurus* individuals.

As these authors pointed out, juvenile theropods are rare in the Morrison Formation, most juveniles of *Allosaurus* represented by disarticulated remains from the Cleveland-Lloyd Dinosaur Quarry. These include the smallest representative yet found of this taxon, a femur measuring approximately 250 millimeters in length. Because these juvenile remains were not articulated, positive association of elements from a specific age category could not be made, nor could allometric differences in limb proportions between age categories be examined.

The most complete juvenile *Allosaurus* yet found consists of a partial but associated specimen (SDSM 30510, with a femur length of 274 millimeters) recovered from the Black Hills of northeastern Wyoming. According to Foster and Chure, this specimen

Opposite: Allosaurus atrox **depicted as a pack feeder by artist Gregory S. Paul, the animals shown here with bloated bellies, the youngsters partly covered in down feathers. The flora includes cycads and araucarian conifer trees.**

offers insights into the relative hindlimb proportions in this genus relative to larger individuals (femur lengths up to 108 millimeters).

Based on examination of the above materials, Foster and Chure determined that *Allosaurus* juveniles, compared to adults, have relatively longer legs, that the tibia and metatarsals are comparatively longer in juveniles than adults, and that the femur in juveniles is longer relative to the ilium. Furthermore, in juveniles, the combined length of femur, tibia, and metatarsal is 33 percent greater, compared to the length of the ilium, to that in adults.

Based on the above findings, Foster and Chure proposed that *Allosaurus* juveniles were probably more cursorial than the subcursorial subadult and adult individuals, having proportions basically equivalent to those of modern cursorial quadrupedal animals and the limb proportions, lost in *Allosaurus* adults, of cursorial small adult theropods (*e.g.*, *Coelophysis*).

A separate study of femoral ontogeny in *Allosaurus fragilis* was performed by Loewen and Sampson (2000), based upon a juvenile partial skeleton from the Cleveland-Lloyd Dinosaur Quarry, and including a comprehensive reanalysis of other *A. specimens* in the Cleveland-Lloyd collection, and also including extant components (*e.g.*, ostrich).

Subjecting this material to gross morphological and histological study, and also cross-sectional geometry, and applying principles of beam theory to assess quantitatively ontogenetic parameters, Loewen and Sampson found that the femur undergoes 1. changes in muscle insertions, 2. allometric increases in the thickness of cortical bone, and 3. shifts in cross-sectional geometry, the latter manifested as a major decrease in circularity indicating that loading regimes differed at various stages of growth.

Breithaupt, Chure and Southwell (1999) commented on the most famous of all *Allosaurus* skeletons (AMNH 5753; see *D:TE* for additional photograph), that of *A. fragilis* mounted at the American Museum of Natural History in 1907.

As recounted by Breithaupt *et al.*, this skeleton, upon its discovery one of the most complete theropod specimens yet found, was collected in October 1879 by F. F. Hubbel for Edward Drinker Cope from the latter's Quarry 3 at Como Bluff, Wyoming. Apparently unaware of the completeness of the specimen, Cope never removed it from the packing crates. The crated specimen was purchased by the American Museum in 1899, although it was not until four years later that the crates were at last opened and its ancient treasures revealed.

AMNH 5753 is distinguished as the first freestanding mount of a theropod skeleton. It was mounted under the direction of Professor Henry Fairfield Osborn, then curator of the museum's Department of Vertebrate Paleontology. At the time, this pose was rather innovative, having the *Allosaurus* bent over and apparently scavenging a carcass of *Apatosaurus* found at Como Bluff in 1897. Photographs of these two skeletons have, over the years, been published in numerous articles and books about dinosaurs; the exhibit was also the inspiration for at least two classic life-restoration paintings by artist Charles R. Knight (see Czerkas and Glut 1982).

Curiously, Breithaupt *et al.* pointed out, AMNH 5753, its fame and exposure notwithstanding, has never been formally described. A series of plates of the specimen were indeed prepared by E. S. Christman under the direction of Walter Granger for an intended publication. Unfortunately, the latter was never printed because of the publication of Charles Whitney Gilmore's (1920) classic theropod monograph which focused mainly on *Allosaurus*.

Chure (2000*b*), in a subsequent abstract, addressed the bypass in the fossil record of the Morrison Formation in favor of *Allosaurus* specimens. As that author noted, theropods other than *Allosaurus* are rare in this formation, often known from one quarry. New Morrison Formation records from Utah, Colorado, Wyoming, and South Dakota of the theropods *Ceratosaurus*, *Coelurus*, *Stokesosaurus*, and *Torvosaurus*, including juvenile and adult specimens, are all of single individuals. Such discoveries, Chure (2000*b*) noted, support the concept of a Morrison Formation theropod community dominated by *Allosaurus*, with that genus comprising from 60 to 70 percent of all known theropod specimens. This pattern, the author pointed out, can be observed in the Cleveland-Lloyd Dinosaur Quarry and continues across all stratigraphic levels and depositional environments.

Chure (2000*b*) was unable to explain this pattern of relative abundance of genera with one genus dominating, noting that 1. the rare taxa are in the same size range as *Allosaurus* and should, therefore, have an equal potential for fossilization, 2. the geographic and stratigraphic patterns do not suggest ecological segregation, and 3. all of these theropod genera are found in the predator trap at the Cleveland-Lloyd site. Chure (2000*b*), therefore, suggested that this pattern probably reflects the biological reality of the theropods in the Morrison Formation. According to that author, the above-mentioned new specimens plus other as yet undisclosed paleontological and geological information "shows that giant theropods are stratigraphically widespread in the Morrison Formation and that biozonation based on body size or stratigraphic occurrence of theropod taxa is unwarranted."

Rayfield (1999), in an abstract, briefly reported on her use of computer technology (Finite Element

Analysis) and CT data to generate a three-dimensional virtual image of an *A. fragilis* skull (MOR 693) for use in detecting stress and strain patterns within the skull of this species. Following the estimation of total adductor muscle force, Rayfield calculated the theoretical bite force, applying this to appropriate tooth-bearing areas of the skull.

Rayfield's resulting data mimicked the stress and strain patterns during biting or feeding activity. Compressive stresses were observed in the region of the impact teeth, maxilla, and skull roof; tensile stresses were observed along the ventral edge of the skull. According to Rayfield, this study documents high-stress intensity areas of the *A. fragilis* skull, exposes morphological characters involved in the resistance of strain-induced deformation of the skull, and reveals "the possibilities and limitations of FEA in producing structurally viable virtual models of fossil vertebrates."

In a subsequent abstract, Rayfield (2000), again utilizing the computer model of the *A. fragilis* skull, determined that the static biting ability of this dinosaur was relatively poor, and that muscle-only-generated bite forces are small, with the lower jaw susceptible to bending. Conversely, the skull is well designed for impact, the cranium being capable of withstanding loads many times larger than those experienced during static biting and yet operating within reasonable safety factors. Rayfield (2000) noted that the geometry of the skull transmitted stresses from the tooth row to the skull roof. There these stresses were dissipated and absorbed by soft tissue, while stresses also looped around fenestrae.

Rayfield (2000) also noted that the interaction of kinetic joints in the *A. fragilis* skull is complex, with reduced stress in one area possibly countered by increased stress in another.

Results of Rayfield's (2000) study implied "that *Allosaurus* employed a high impact attack strategy," with feeding involving little bone crushing and consisting "of ripping flesh, aided by laterally compressed slashing teeth, neck retraction and the stress absorbing capacity of soft tissues and kinetic joints."

Madsen and Welles (2000), in their monograph on *Ceratosaurus* (see entry), reviewed in detail various species referred to *Labrosaurus*, a genus usually regarded as a junior synonym of *Allosaurus* (see *D:TE*):

The type species *Labrosaurus lucaris* was established by Marsh (1879) on vertebrae and forelimbs (YPM 1931; this specimen never figured) from Reed's Yale Peabody Museum Quarry 3, at Como Bluff, Wyoming, Marsh (1978) having originally described this material as *Allosaurus lucaris*. Marsh (1879) diagnosed *L. lucaris* thusly: Smaller than *A. fragilis*; cervical vertebrae short, opisthocoelous, dorsal vertebrae moderately so; all vertebrae with very large centrocoels connecting with exterior by small foramina on each side; dorsal spines high, not showing zygosphenal articulation; forelimbs very small, humerus curved and with large crest. Although Hay (1902), then Ostrom and McIntosh (1966), referred *L. lucaris* to the genus *Antrodemus* (usually regarded as a junior synonym of *Allosaurus*; see *Antrodemus* entry), Madsen and Welles found this species to be not diagnostic, regarding it as a *nomen vanum*. Described by Marsh (1884*b*), *L. ferox* was founded on a left dentary (USNM 2315; see *D:TE* for photograph) from Felch's Quarry 1, Garden Park, Colorado. The specimen is edentulous anteriorly and highly arched ventrally. Marsh (1884*b*) perceived the teeth to be more triangular than those in other species. Hay (1908) pointed out that the condition of the teeth could not be verified as the crowns were missing and the roots buried under matrix. According to Madsen and Welles (after a study of USNM 2315 by James H. Madsen, Jr.), this specimen "is an aberrant, pathological left dentary of *Allosaurus fragilis*," the downcurved, posterior third explained by over-preparation and postmortem distortion. Although rare, these authors noted, the edentulous section is a condition also seen in the holotype (UUVP 158) left dentary of *Ceratosaurus dentisulcatus*.

A pelvis labeled "*Labrosaurus fragilis*" was published by Marsh (1896) in his classic monograph on North American dinosaurs. As Madsen and Welles noted, Hay (1908) had pointed out "that this was a slip of the pen for *Labrosaurus ferox*."

Labrosaurus medius (Marsh 1888), based on material from the Jurassic of Potomac, Maryland, was referred by Matthew and Brown (1922) to the genus *Dryptosaurus* and the questionable species ?*D. medius*. *L. stechowi* Janensch 1920 and *L. sulcatus* were both referred by Madsen and Welles to *Ceratosaurus* sp. (see *Ceratosaurus* entry.)

Note: In 1981, Ralph E. Molnar, Timothy F. Flannery and Thomas H. V. Rich described an incomplete left astragalus (NMV P150070), identified by them as *Allosaurus* sp., which had been found in Lower Cretaceous rocks of Victoria, Australia. Until then, *Allosaurus* was only known from the Late Jurassic of North America (and possibly Africa). Consequently, this specimen was used as evidence supporting the hypothesis that Australia had become a place of refuge for certain late-surviving, terrestrial vertebrates.

Although not all workers agreed with the above identification of NMV P150070) (*e.g.*, Welles 1983), Molnar *et al.* (1983) maintained that the specimen indeed represented the well-known genus (see *D:TE* for details).

A decade and a half later, Chure (1998*b*) reexamined this fossil in light of the recent discovery of other basal tetanuran theropods. Chure found that some of the synapomorphies reportedly shared by the Australian form and *Allosaurus* actually have a wider distribution among tetanurans, and that other presumed synapomorphies are founded upon misinterpretations of the morphology of NMV P150070. According to Chure, the Australian specimen represents neither *Allosaurus* nor the Allosauridae, although it could belong to the Allosauroidea, a clade that persisted at least into the Cenomanian. Therefore, Chure concluded, this specimen can no longer be used to support "the current concept that Australia served as a refugium for terrestrial vertebrates that went extinct elsewhere at the end of the Jurassic."

More recently, Azuma and Currie (2000) pointed out that the morphology of the Australian specimen is similar to that of the Asian genus *Fukuiraptor* (see entry). As noted by these authors, Welles had presented 19 morphological differences between NMV P150070 and the astragalus of *Allosaurus*. According to Azuma and Currie, the differences between NMV P150070 and *Allosaurus* cited by Welles (1983) are also those that exist between *Fukuiraptor* and *Allosaurus*, which included the following: higher ascending process in NMV P150070; distinct vertical ridge marking medial margin of fibular contact with ascending process; deeper upper horizontal groove with a sharper anterior edge above condyles; shallower lower groove across condyles; and stronger crescentic groove on posterior surface of ascending process.

Azuma and Currie noted that the *Allosaurus* autapomorphies used by Molnar *et al.* are also found in *Fukuiraptor* and other carnosaurs, the morphology of NMV P150070 therefore suggesting a possible relationship with *Fukuiraptor*.

Key references: Azuma and Currie (2000); Breithaupt, Chure and Southwell (1999); Bybee and Smith (1999); Chure (1998*b*, 1999, 2000*a*, 2000*b*); Czerkas and Glut (1982); Galton (1980*a*, 1980*b*); Galton and Powell (1980); Gilmore (1920); Hay (1902, 1908); Janensch (1920); Madsen (1976); Madsen and Welles (2000); Marsh (1878, 1879, 1884*b*, 1888, 1896); Matthew and Brown (1922); McGee (1999); Molnar, Flannery and Rich (1981, 1983); Ostrom and McIntosh (1966); Perez-Moreno, Chure, Pires, da Silva, Santos, Dantas, Póvoas, Cachão, Sanz and Gallopim de Carvalho (1999); Rayfield (1999, 2000); Smith (1998, 1999); Smith, Bybee and Anderson (1999); Weishampel and Young (1996); Welles (1983).

†ALVAREZSAURUS

Saurischia: Theropoda: Neotheropoda: Tetanurae: Avetheropoda: Coelurosauria: Manuraptoriformes: Arctometatarsalia: ?Ornithomimosauria: Alvarezsauridae.

Diagnosis: Autapomorphies including cervical vertebrae having amphicoelous centra; postzygapophyses of cervical vertebrae dorsoventrally flattened, paddle-shaped in dorsal aspect, with pair of strong craniocaudal ridges; distal caudal vertebrae more than 20 percent length of proximal caudals; scapular blade slender, reduced; ungual phalanx of digit I ventrally keeled (Novas 1996*b*).

Comments: The genus *Alvarezsaurus*, based on a single partial skeleton (MUCPv 54), was originally described by Bonaparte (1991*a*) as a small nonavian theropod (see *D:TE*).

Later, Bonaparte (1996*b*), in a review of Cretaceous tetrapods from Argentina, noted the superficial similarities between *Alvarezsaurus* and Laurasian ornithomimids, emphasizing several significant anatomical differences: although the cervical, dorsal, and sacral vertebrae of *Alvarezsaurus* are generally similar to those of ornithomimids, the caudal vertebrae, especially the central ones, "are strikingly different," being quite low, without neural spines, procoelous, and possessing wing-like (aliform) processes. The morphology of the sacrum, ilium, and proximal portion of the femur suggested the similar organization of these elements in the ornithomimids *Gallimimus* and *Struthiomimus*, while the very different morphology of the astragalus and calcaneum may be plesiomorphic. In *Alvarezsaurus*, the foot is considerably less derived than in ornithomimids, with metatarsal III not reduced, and with its entire dorsal surface exposed. The small pectoral girdle also contrasts with the condition in ornithiomimids, exhibiting a number of unique characters (*e.g.*, dorsal border almost straight without acromial expansion; ventral border having marked curvature towards glenoid; glenoid more posteriorly projected), its size suggesting a much reduced girdle and forelimb.

Bonaparte (1996*b*), using ornithomimids as a model, postulated that *Alvarezsaurus* was not a predator. According to that author, the pedal ungual phalanges suggest that these were not aggressive organs. Bonaparte (1996*b*) interpreted *Alvarezsaurus* as probably representing an herbivorous, vicariant adaptive type to the ornithomimids of Laurasia, but with a different evolutionary history—"an endemic group of theropods which evolved in Gondwana, isolated from the Laurasian supercontinent, eventually from forms like *Compsognathus* or *Ornitholestes* of possible worldwide distribution during the Jurassic."

Novas (1996*b*), in a review of the Alvarezsauridae, pointed out a number of problems with Bonaparte's (1991) original diagnosis of the type species, *Alavreszsaurus calvoi*, some of the cited characters (*e.g.*, cervical pleurocoels, 5–6 sacrals, ilium low and long, unfused metatarsals and tarsals, astragalus with wide condyles, metatarsal III narrower in caudal aspect relative to remaining metatarsals, metatarsal IV greater proximally than others) being widely distributed within Tetanurae; some (*e.g.*, cranial sacrals with slight axial depression on ventral surface, illium with postacetabular blade [process] greater than preacetabular blade) are features widely and unevenly distributed among several nonavialian and avialian taxa; others (*e.g.*, neural spines vestigial in cervical and cranial dorsal vertebrae, caudal sacrals with narrow ventral margin) are also present in *Mononykus*; and another (small size) is basically irrelevant, the holotype representing an immature individual.

Most workers (*e.g.*, Chiappe 1996; Chiappe, Norell and Clark 1996; Novas) regard *Alvarezsaurus* (and also the family to which Bonaparte referred it, the Alvarezsauridae) as avian (see *S1*). In the previous volume, therefore, *Alvarezsaurus* was relegated to the section on "excluded genera."

However, more recently, Sereno (1999*a*) found through cladistic analysis that alvarezsaurids seem not to be birds after all, and that this group is better understood when interpreted as specialized Cretaceous ornithomimosaurs (see "Systematics" chapter "Note" under Alvarezsauridae). For those reasons, *Alvarezsaurus* is hereby tentatively reinstated among this listing of dinosaurian genera.

Key references: Bonaparte (1991*b*, 1996*b*); Chiappe (1996*a*); Chiappe, Norell and Clark (1996); Novas (1996*b*); Sereno (1999*a*).

†AMARGASAURUS

New species: *A. sattleri* (Janensch 1929).

Comments: The elongation of the neural spines of the cervical and dorsal vertebrae of diplodocoid dinosaurs reached the extreme in the Patagonian genus *Amargasaurus*, wherein the bifurcated cervical spines attained a length of four times that of the centra (see *D:TE*, *S1*).

In a study on macroevolution in "diplodocimorph" sauropods, Salgado (1999) suggested that these spines in *Amargasaurus* might be explained by Salgado as the result of both natural selection and local acceleration. The author also addressed Salgado and Bonaparte's (1991) earlier skeletal reconstruction of this dinosaur, in which the neck was held high with the head in an almost horizontal position (see *S1*).

Salgado published a new reconstruction of the skeleton of the type species, *Amargasaurus cazaui*, showing that the elongated neck spines made it impossible for this dinosaur to carry its head as did other diplodocoids (=diplodocids of Salgado's terminology). While *Diplodocus* could carry its head in an almost horizontal position because the neck was mechanically designed to assume the shape of an "exponential curve," *Amargasaurus* could only carry its head with its anteroposterior axis almost vertical, the first cervical vertebra being nearly horizontal and the neck forming a "sigmoid curve." Therefore, Salgado and Bonaparte's earlier reconstruction, in which the neck is raised higher and the head held almost horizontally, was shown to be inaccurate.

Also, Salgado noted that one of the two species of the African genus *Dicraeosaurus* (see *D:TE*), the lesser known referred taxon *Dicraeosaurus sattleri*, shares a number of apomorphies with *A. cazaui*, including the following: height of posterior dorsal neural spines twice height centrum plus neural arch (measured up to hyposphen); bifurcation processes reaching at least fifth presacral. Because of these shared characters, Salgado proposed that *D. sattleri* be regarded as a second species of *Amargasaurus*, the implied new binomen being *A. sattleri*.

Key references: Salgado (1999); Salgado and Bonaparte (1991).

AMUROSAURUS Bolotsky and Kurzanov 1991

Ornithischia: Genasauria: Cerapoda: Ornithopoda: Euornithopoda: Iguanodontia: Euiguanodontia: Dryomorpha: Ankylopollexia: Hadrosauroidea: Hadrosauridae: Euhadrosauria: Lambeosaurinae.

Name derivation: "Amur [region of Russia]" + Greek *sauros* = "lizard."

Type species: *A. riabini* Bolotsky and Kurzanov 1991.

Other species: [None.]

Occurrence: Blagoveschensk, Kundur, Russian Amur region of northeastern Asia.

Age: Late Cretaceous (late Maastrichtian).

Known material: Various specimens.

Holotype: Partial skull.

Diagnosis of genus (as for type species): [None yet published.]

Comments: The new type *Amurosaurus riabini* was named and only briefly described in 1991 in an abstract by Bolotsky and Kurzanov, based on a partial skull collected from a bonebed in the Blagoveschensk locality in the Amur region of northeastern Asia. This large lambeosaurine comprises more than 90 percent of the vertebrate fossils collected at this site, the other material representing theropod, ankylosaurian, crocodile, and turtle specimens. Unfortunately, the paper in which these authors introduce this taxon is quite rare and has not been seen by most workers in the United States (the present writer included).

More recently, Godefroit, Zan and Jin (2000) cited *A. riabini* in their paper describing the new lambeosaurine *Charonosaurus jiayinensis* (see *Charonosaurus* entry). Godefroit *et al.* observed that *A. riabini* differs from *C. jiayinensis* in many ways, including the following: shorter frontal platform suggesting a different kind of supracranial crest; well-developed sagittal crest; longer paroccipital processes; longer parietal; and especially well-separated squamosals.

According to Godefroit *et al.*, numerous new specimens apparently belonging to *A. riabini* have been found at Blagoveschensk and Kundur in the Amur region.

As noted by these authors, the *A. riabini* material is currently under preparation by Y. L. Bolotsky and Pascal Godefroit.

Key references: Bolotsky and Kurzanov (1991); Godefroit, Zan and Jin (2000).

†ANASAZISAURUS—(See *Kritosaurus.*)

†ANDESAURUS

Saurischia: Sauropodomorpha: Sauropoda: Eusauropoda: Neosauropoda: Macronaria: Camarasauromorpha: Titanosauriformes: Somphospondyli: Titanosauria: Andesauridae.

Diagnosis of genus: Ischium having long and laminar process (Sanz, Powell, Le Loeuff, Martinez and Suberbiola 1999).

Comment: Although *Andesaurus* has usually been classified as a titanosaur (see *D:TE*), Sanz, Powell, Le Loeuff Martinez, and Suberbiola (1999) recently reassessed it phylogenetic position, placing it with *Haplocanthosaurus* outside of Titanosauria (see "Systematics" chapter, "Note" under "Eutitanosauria").

Key reference: Sanz, Powell, Le Loeuff, Martinez and Suberbiola (1999).

†ANGATURAMA—(See *Irritator.*)

†ANOPLOSAURUS Seeley 1879

Ornithischia: Genasauria: Thyreophora: Thyreophoridea: Eurypoda: Ankylosauria: Nodosauridae.

Type species: *A. curtonotus* Seeley 1897.

Other species: *A. major* [in part] (Seeley 1869) [*nomen dubium*].

Occurrence: ?Upper Gault Clay or ?Cambridge Greensand, Cambridge, Cambridgeshire, Reach, England.

Age: Early Cretaceous (upper Albian) (Rawson, Currey, Dilley, Hancock, Kennedy, Neale, Wood and Worssam 1978).

Lectotype: SMC B55731, partial right scapula.

Diagnosis of genus (as for type species): Small nodosaurid distinguished by thumblike acromion process placed centrally on scapula; coracoid almost trapezoidal, wider anteroventrally than posterodorsally; trochanter fingerlike, separated laterally from femoral shaft by shallow but short cleft (Suberbiola and Barrett 1999).

Comments: The armored genus *Anoplosaurus* (see entry, *D:TE*), founded upon a number of skull and postcranial elements, was originally described by early paleontologist Harry Govier Seeley in 1879 as a rather small, "semi-erect" dinosaur. Zittel (1893) interpreted the genus to be a "scelidosaurid"; later, Hennig (1915) classified it as a genus of armored dinosaur [=ankylosaur].

Nopcsa, in a 1923 report on British dinosaurs, removed some of these remains (including a dentary, scapula, coracoid, and femur) from *Anoplosaurus* and transferred them to *Acanthopholis* (see entries, *D:TE* and this volume), a genus of nodosaurid ankylosaur, referring the remaining material (most likely because

***Amargasaurus cazaui*, new skeletal reconstruction showing corrected orientation of the neck and head. (After Salgado 1999.)**

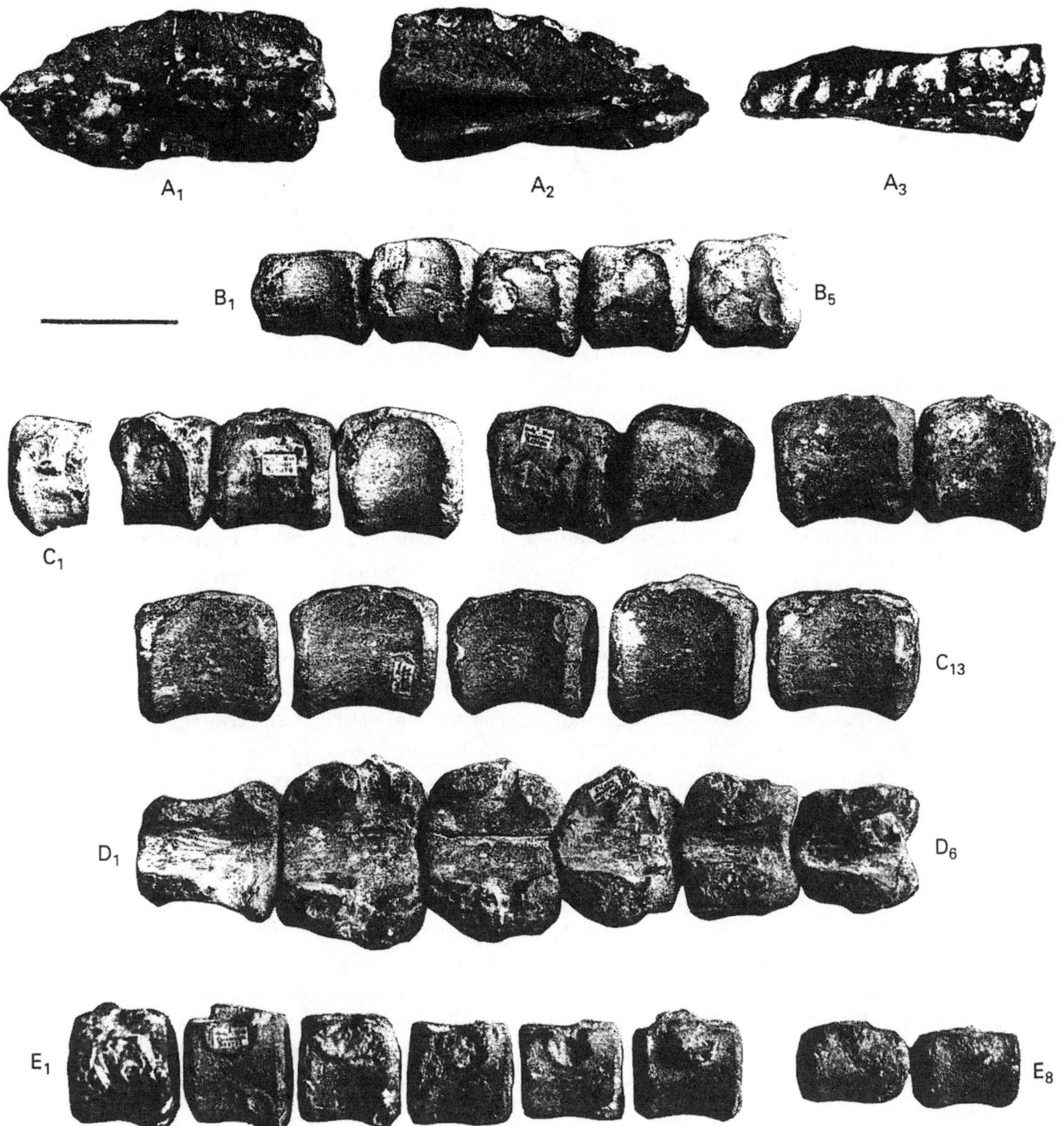

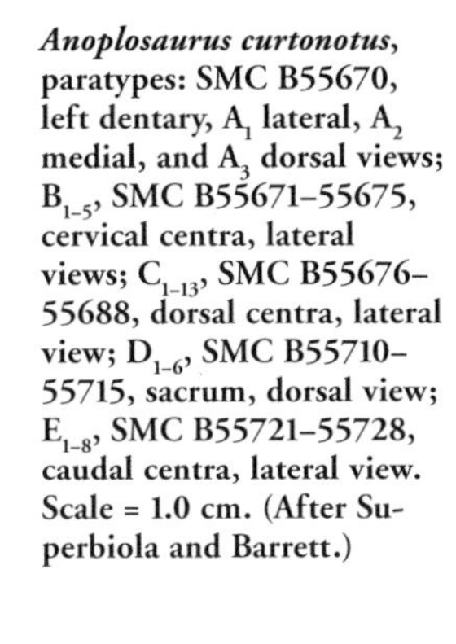

***Anoplosaurus curtonotus*, paratypes: SMC B55670, left dentary, A_1 lateral, A_2 medial, and A_3 dorsal views; B_{1-5}, SMC B55671–55675, cervical centra, lateral views; C_{1-13}, SMC B55676–55688, dorsal centra, lateral view; D_{1-6}, SMC B55710–55715, sacrum, dorsal view; E_{1-8}, SMC B55721–55728, caudal centra, lateral view. Scale = 1.0 cm. (After Superbiola and Barrett.)**

it lacked associated dermal plates) to the ornithopod family Camptosauridae. Over the years, subsequent workers (*e.g.*, Coombs 1971; Galton 1983) have tended to accept Baron Nopcsa's removal of some of the original fossils from *Anoplosaurus*, regarding the retained material as belonging to the Ornithopoda, and referring it either to the Camptosauridae (*e.g.*, Huene 1956) or Iguanodontidae (Kuhn 1936, 1964; Steel 1969), or to an indeterminate member of the Iguanodontia (Brinkmann 1988; Norman and Weishampel 1990).

More recently, however, Suberbiola and Barrett (1999), in a review of ankylosaurian dinosaur specimens from the Albian–Cenomanian of England, reassessed the systematic status of the type species *Anoplosaurus curtonotus*. The authors based this study on the lectotype specimen (SMC B55731, an incomplete right scapula), and also the following paralectotypes: SMC B55670, partial left dentary; SMC B55671–55675, five cervical centra; SMC B55676–55688, 13 dorsal centra; SMC B55689–55694, isolated neural arches; SMC B55695–55696, dorsal ribs; SMC B55697–55709, rib fragments; SMC B55710–55715, six unfused sacral centra; SMC B55716–55720, sacral ribs; SMC B55733, ?metacarpal; SMC B55734–55735, fragmentary left femur; SMC B55736, fragmentary left tibia; SMC B55738–55741, ?metatarsals; SMC B55742, phalanx.

Suberbiola and Barrett concluded that *A. curtonotus* is not an ornithopod, but indeed a nodosaurid ankylosaur, based mostly upon the following features (not present in ornithopods): 1. sinuous tooth row characteristic of thyreophorans (Sereno 1986); 2. vertebral count (probably seven–eight cervicals, 13 dorsal, six sacrals, and indeterminate number of caudals) comparable to ankylosaurs, especially nodosaurids (Carpenter 1984; Coombs and Maryańska 1990); 3. articular faces of cervical centra acoelous (or amphyplatyan) to slightly amphicoelous, as in ankylosaurs (Coombs 1978; Coombs and Maryańska 1990); 4. anterior area of sacrum longer than wide, as are sacrocaudals, midsacrals comparatively shorter and wider (condition opposite from that of the ornithopod *Iguanodon*; Norman 1986), sacrum very similar in general form to those of ankylosaurs like *Polacanthus* (Blows 1987; Pereda-Suberbiola 1994); 5. scapula with well-developed pseudo-acromial process displaced towards glenoid, and with small pre-spinous fossa anterior to scapular spine, features interpreted as derived conditions for nodosaurids (Coombs 1978; Coombs and Maryańska 1990); 6. coracoid, though incomplete, apparently relatively larger relative to size of scapula than in ornithopods, a nodosaurid feature (Coombs 1978; Coombs and Maryańska 1990); 7. humerus (regarded by Nopcsa as camptosaur-like, but without giving supporting evidence), though fragmentary, straight, similar to ankylosaurian humeri (Coombs 1978); and 8. femur displaying nodosaurid condition of lesser trochanter separated from greater trochanter by cleft deeper laterally than medially (Galton 1983).

As pointed out by Suberbiola and Barrett, a number of other nodosaurid taxa are known from the same age range as *Anoplosaurus*, especially in North America—*Sauropelta edwardsi*, *Silvisaurus condrayi*, *Nodosaurus textilis*, *Texastes pleurohalio*, *Pawpawsaurus campbelli*, and the undetermined juvenile from Texas provisionally referred to *Texastes* (see Jacobs, Winkler, Murry and Maurice 1994).

Suberbiola and Barrett theorized that the unexpected absence of armor—dermal armor being the most conspicuous ankylosaurian feature (Coombs 1978; Coombs and Maryańska 1990)—in the type material of *Anoplosaurus* may bear ontogenetic significance. (Jacobs *et al.* and Galton 1982 had observed the seeming absence of osteoderms in both juvenile nodosaurids and stegosaurs, respectively; more recently, however, Philip J. Currie has noted two cervical rings in baby [one meter in length] *Pinacosaurus* specimens; C. Carpenter, personal communication 2000.) Consequently, Suberbiola and Barrett suggested that the lack of armor in *Anoplosaurus* may be explained by the juvenile status of the type material. The authors noted that, based on observations by Galton (1982) and Coombs (1986), various features of the type material of *A. curtonotus* support this interpretation, including the following: absence of fusion between centra and neural arches; absence of fusion between sacral ribs and centra; articular faces of centra less expanded than in large vertebrae; coracoid and scapula not coossified; and surfaces of long bones smooth. Furthermore, other features (*e.g.*, lack of fusion of splenial to lower jaw, unusual form of coracoid foramen, and small size of specimens) may also be regarded as juvenile features (see Jacobs *et al.*; Pereda-Suberbiola, Astibia and Buffetaut 1995).

The authors noted that it was tempting to refer other ankylosaurian remains from the Cambridge Greensand and other localities to *Anoplosaurus* based on proximity of localities, similar ages, and observed low range of variability. This might provide additional skeletal elements to compare with other taxa. However, given the doubtful associations of this material, none of which displays characteristics diagnostic for *Anoplosaurus*, they conservatively limited their comparisons to the type material.

Suberbiola and Barrett interpreted *Anoplosaurus* to be a rather primitive nodosaurid, as indicated by these features (some shared with other primitive nodosaurids such as *Silvisaurus*): anterior extent of dentary tooth row; lack of differentiation of humeral head from shaft; four sacral vertebrae; and distinct lesser trochanter on femur. It is distinct from and more derived than *Hylaeosaurus* and *Polacanthus* in that its scapula has a prominent acromion process and small pre-spinous fossa (the latter feature also present in the apparently more derived *Struthiosaurus*). Therefore, *Anoplosaurus* seems to be a taxon intermediate between such very primitive genera as *Hylaeosaurus* and *Polacanthus* and various more derived nodosaurids including *Struthiosaurus*, *Sauropelta*, *Edmontonia*, and *Panoplosaurus*. According to these authors, the conservative character of *Anoplosaurus* relative to contemporary North American nodosaurids may be explained in terms of geographical isolation of European Albian faunas (see Doré 1991; Russell 1995).

A second *Anoplosaurus* species, *A. major*, was erected by Seeley (1869) founded upon a cervical vertebra and three caudal centra (SMC B55561, B55563–55565) that Seeley (1879) removed from the original specimens of *Acanthopholis stereocercus* (see *Acanthopholis* entries, *D:TE* and this volume), collected from the Cambridge Greensand, near Cambridge, in Cambridgeshire. As observed by Suberbiola and Barrett, the cervical centrum of this specimen is amphicoelous and resembles more those of ankylosaurs than those of ornithopods; the very elongate caudal centra (SMC B55563–55565), though, may belong to an ornithopod. Therefore, as had Norman and Weishampel, the authors regarded *A. major* as a *nomen dubium* and an indeterminate iguanodontian, based upon composite remains.

Key references: Blows (1987); Brinkmann (1988); Carpenter (1984); Coombs (1971, 1978, 1986); Coombs and Maryańska (1990); Doré (1991); Galton (1982, 1983); Hennig (1915); Huene (1956); Jacobs, Winkler, Murry and Maurice (1994); Nopcsa (1923); Norman (1986); Norman and Weishampel (1990); Pereda-Suberbiola (1994); Pereda-Suberbiola, Astibia and Buffetaut (1995); Russell (1995); Seeley (1869, 1879); Sereno (1986); Suberbiola and Barrett (1999); Zittel (1893).

†ANTARCTOSAURUS

Saurischia: Sauropodomorpha: Sauropoda: Eusauropoda: Neosauropoda: ?Diplodocoidea.

Comments: The placement of the giant South American type species *Antarctosaurus wichmannianus*— generally classified with the Titanosauridae— has been problematic. Huene (1929) originally reconstructed the partial holotype skull belonging to the holotype (MACN 6904) of this taxon after that of the diplodocid *Diplodocus*. In more recent years, the type specimen has been regarded as comprising a diplodocoid skull and titanosaur postcranial remains (McIntosh 1990), as representing a diplodocoid (Jacobs, Winkler, Downs and Gomani 1993), and as representing a titanosaur (Powell 1986; Bonaparte 1996*b*; Salgado and Calvo 1997) (see *Antarctosaurus* entries, *D:TE* and *S1*, for details and illustrations).

More recently, Upchurch (1999), noting that earlier studies of this taxon have been primarily based upon individual characters, conducted the first full phylogenetic analysis of *A. wichmannianus*. This analysis demonstrated that the most parsimonious placement for *A. wichmannianus* is within the Diplodocoidea, as the sister-taxon to the Nemegtosauridae, although this relationship is only weakly supported. As Upchurch pointed out, the skull material included in MACN 6904 exhibits several synapomorphies of the Diplodocoidea (*e.g.*, mandible subrectangular in dorsal view; tooth row terminating rostral to antorbital fenestra); allegedly associated postcrania, however, possessing various "titanosauroid" features (*e.g.*, biconvex first caudal vertebra; radius to third metacarpal ratio greater than 0.45; dorsoventrally elongate muscle scar on lateral surface of fibula).

Upchurch concluded that, based on the following observations, MACN 6904 is really a composite of remains representing several possibly nonspecific sauropod individuals:

1. Maximum transverse widths of the occiput

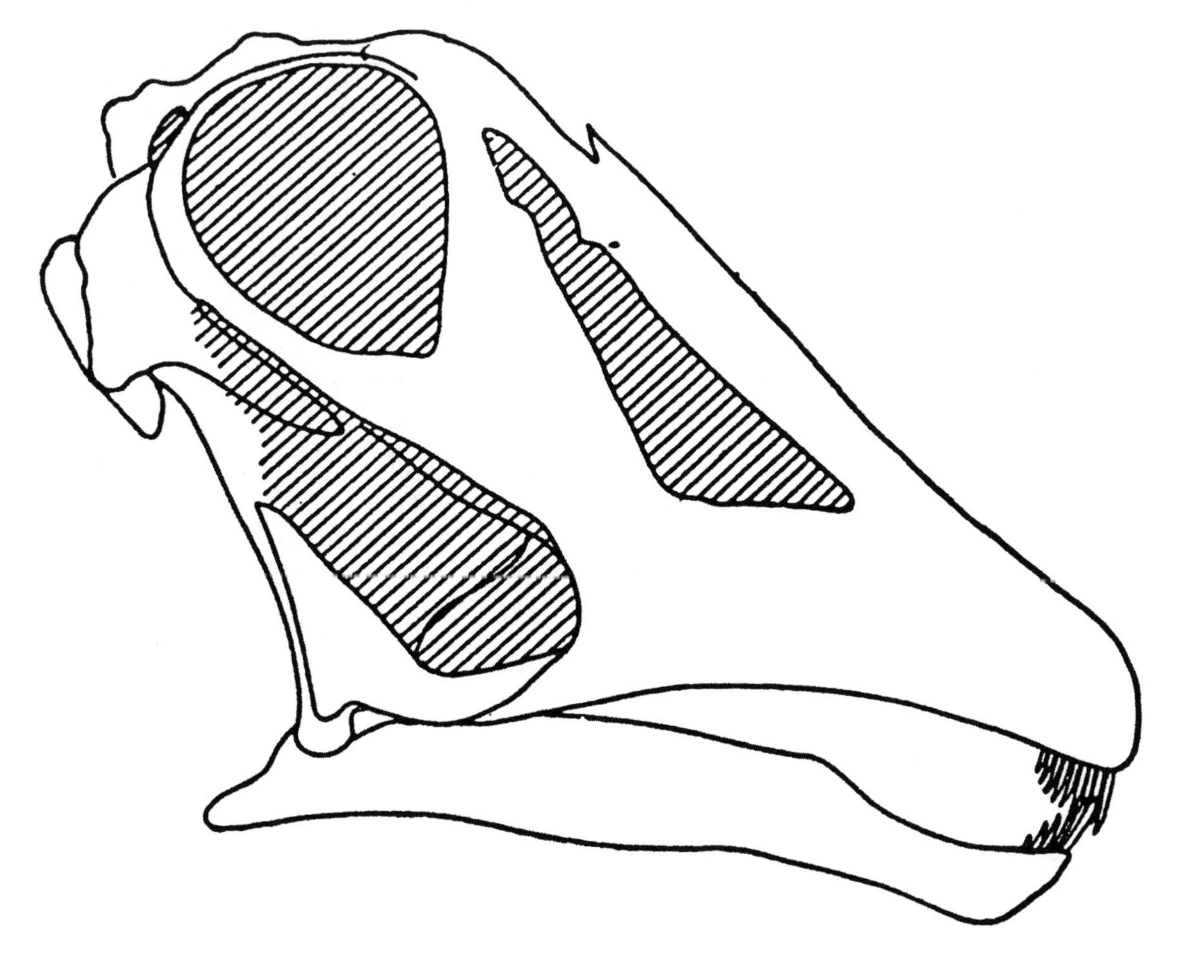

Friedrich von Huene's original inaccurate reconstruction (see *S1* for modern reconstruction) of the holotype skull (MACN 6904) of *Antarctosaurus wichmannianus*, based upon his assumption that it resembled that of *Diplodocus*. (After Huene 1929.)

and mandible are 20 and 26 centimeters respectively, as measured by Huene, while in almost all other dinosaurs, sauropods included, the transverse width of the snout is less than or at most equal to the width of the occiput. While noting that the snout of *Antarctosaurus* could have been apomorphically wide, Upchurch found it more likely that the mandible and snout pertain to different individuals.

2. The only known vertebrae assigned to this species are a poorly-preserved caudal cervical, a first caudal (which Huene believed could belong to *Laplatasaurus*), and series of middle caudals (which Huene suggested were not part of the holotype); consequently, no presacral vertebrae exist that can offer a link between the cranial and postcranial remains. Upchurch further pointed out, given the usual pattern of preservation, that sauropod skull material is normally found either in isolation, with a series of cervical vertebrae, or as part of a skeleton including a number of presacral vertebrae and limb material.

While acknowledging that the presence of both diplodocoid and "titanosauroid" apomorphies could reflect a true aspect of sauropod evolutionary history (*i.e.*, a diplodocoid clade independently acquiring various titanosaur postcranial apomorphies), Upchurch found it more plausible that MACN 6904 is composed of elements pertaining to different taxa. Furthermore, if skull elements of the holotype do not belong to the same individual, the characters in the mandible are consistent with the assignment of *A. wichmannianus* with the Diplodocoidea.

Key references: Bonaparte (1996*b*); Huene (1929); Powell (1986); Salgado and Calvo (1997); Upchurch (1999).

Photograph by the author, courtesy National Museum of Natural History, Smithsonian Institution.

Antrodemus valens **(USNM 218), holotype incomplete caudal vertebra of *"Poicilopleuron" valens*.**

†**ANTRODEMUS** Leidy 1873 [*nomen dubium*]—(=?*Allosaurus*, ?*Marshosaurus*, ?*Stokesesaurus*)

Saurischia: Theropoda: Neotheropoda: Tetanurae: Avetheropoda *incertae sedis*.

Name derivation: Greek *antron* (or Latin *antrum*) = "cave [or 'cavity in bone']" + Greek *demas* = "body."

Type species: *A. valens* (Leidy 1870) [*nomen dubium*].

Other species: [None.]

Occurrence: Morrison Formation, Colorado, United States.

Age: Late Jurassic (Kimmeridgian–Tithonian).

Known material/holotype: USNM 218, posterior half of sixth caudal vertebra.

Diagnosis of genus: [No modern diagnosis published.]

Comments: In 1870, Joseph Leidy of the Academy of Natural Sciences of Philadelphia, referred an incomplete theropod caudal centrum (USNM 218) to the European genus "*Poicilopleuron*" (Leidy's misspelling of *Poekilopleuron*) as the new species *P. valens*. The specimen had been found by Ferdinand Vandiveer Hayden in the Morrison Formation of Grand County, Middle Park, Colorado. Leidy offered no other stratigraphic data, other than stating that it had come from Middle Park, and that Hayden believed it to be Cretaceous in age.

Leidy diagnosed USNM 218 by only one observed character — medullary cavity of centrum divided into smaller recesses by trabeculae — noting that, if other characters not included in his diagnosis were ever deemed significant, this taxon should be referred to a new genus, for which he proposed the name *Antrodemus*.

In 1873, Leidy published a figure of the vertebra labeled *Antrodemus*, although his text continued to refer to the specimen by its original name.

Later, Charles Whitney Gilmore (1920), observing that the specimen now known as *Antrodemus valens* closely resembled the caudal vertebrae in the better known *Allosaurus fragilis* Marsh 1877, referred *A. valens* to that species. However, as the name *Antrodemus* had been first published four years earlier than *Allosaurus*, it clearly had priority. Thus, during the 1950s, the name *Antrodemus* began to appear in various texts (*e.g.*, *Osteology of the Reptiles* by Alfred Sherwood Romer, 1956, *The Dinosaurs* by W. E. Swinton, 1970), including popular books (*e.g.*, *The Dinosaur Quarry* by Good, White and Stucker, 1958), when referring to specimens and life restorations pertaining to *Allosaurus*. Finally, James H. Madsen, Jr. (1976), in his classic monograph on *Allosaurus*, demonstrated that *A. valens* is a dubious taxon (see below) and that the name *Allosaurus* is clearly the valid name for that better known theropod (see *Allosaurus* entry, *D:TE*).

Madsen argued that Leidy (1870) provided only broad geographic reference and no precise stratigraphic information, that Leidy had obtained the specimen third hand, and that USNM 218 is but questionably diagnostic at the genus level and indeterminate at the species level. According to Madsen, *A. valens* could not be unequivocally referred to *A. fragilis*.

However, in the approximately 80 years following Gilmore's synonymy of these two taxa, *Antrodemus valens* has been generally regarded as a junior synonym of *Allosaurus fragilis* (*e.g.*, Molnar, Kurzanov and Dong 1990; see also *D:TE, S1*). However, there are problems with this alleged synonymy. Indeed, during those years, this referral has become more clouded, as other large theropods (*e.g.*, the carnosaurian *Marshosaurus* and arctometatarsalian *Stokesesaurus*, see respective entries, *D:TE*; also other new allosaurid taxa, "Introduction"), with caudal vertebrae indistinguishable from USNM 218, have been found in the Morrison Formation. As aptly noted by Olshevsky (2000) on his checklist of dinosaurian species, "This genus is probably a senior synonym of *Allosaurus*, but it is best isolated as indeterminate until comparative work on all the Morrison allosaurid specimens is completed." Therefore, *Antrodemus* is herein regarded as an avetheropod of otherwise uncertain affinities.

Key references: Gilmore (1920); Good, White and Stucker (1958); Leidy (1870, 1873); Madsen (1976); Marsh (1877); Olshevsky (2000); Romer (1956); Swinton (1970).

†**APATOSAURUS**—(=*Brontosaurus*; =?*Atlantosaurus*).

Comments: *Apatosaurus*, a gigantic diplodocid sauropod possessing a very long "whiplash" tail, continues to be a topic of study.

Recently, physicist Nathan P. Myhrvold and paleontologist Philip J. Currie (1997) published their findings regarding the capabilities of diplodocids—their study mostly focused upon *Apatosaurus*, among the most well known of all dinosaurian genera—to "crack" their geometrically scaled tails in the fashion of a bullwhip, thereby creating a supersonic boom (see "Introduction," *S1*, for additional information). Utilizing computer simulations, primarily of the tail of the referred species *Apatosaurus louisae* (CM 3018), Myhrvold and Currie showed that this taxon (and apparently other diplodocids, like the genera *Diplodocus* and *Barosaurus*) and also other kinds of sauropods (*e.g.*, dicraeosaurids and *Mamenchisaurus*) possessed the fundamental potential to employ their tails in producing a sound analagous to a bullwhip's "crack."

Tentative confirmation of this capability was illustrated by several osteologic features that served to stiffen the tail in the fashion of a bullwhip. Such features, the authors noted, include the progressively increased length of the caudal centra between positions 18 and 25 of the vertebral series (consistent with an adaptation to stresses generated by this kind of tail movement); and also the coossification of these caudal vertebrae (formerly attributed to damage resulting from contact with the ground when the heavy animal reared up on its hind legs, using its tail for support) due to diffuse idiopathic skeletal hyperostosis (or DISH; *e.g.*, Rothschild and Berman 1991; see also *D:TE*). This latter condition had been found in approximately half the specimens studied and possibly having sexually dimorphic significance.

As suggested by Myhrvold and Currie, the lengthening of the vertebrae in the region of caudals 18 through 25 seems to be "an evolutionary adaptation to the need for increased stiffness and reinforcement in this area," such stiffness further supplemented in some individuals by vertebral fusion (DISH). The authors further speculated that the long tail of *Apatosaurus* and related sauropods may have extended beyond the terminal caudal vertebra in the fashion of a "popper," this supposed extension presumably comprising a combination of skin, tendon, and keratin. Simulations incorporating such a "popper" generated supersonic motion and generally required less torque to accomplish this. The authors speculated that a broad range of physical conditions allowed for the possibility of supersonic motion in *Apatosaurus* and other dinosaurs having geometrically scaled tails.

According to Myhrvold and Currie, the production of such loud noises may have been used for defense, communication, intraspecific rivalry, or courtship. Comparisons with the club-ended tails of the Asian sauropods *Shunosaurus lii* and *Omeisaurus tianfuensis* indicated that the whiplash tail of diplodocids was well adapted as a noisemaker but not as a direct-impact weapon.

The authors cautioned that the arguments and simulations offered in this study were "at best only circumstantial evidence," any such behavioral characteristics (*e.g.*, tail cracking) being very difficult to

Courtesy The Field Museum (neg. #CSGEO23972).

Partial skeleton (FMNH PR25112) of *Apatosaurus excelsus*, collected by the Museum Paleontological Expedition of 1901 in the Grand River Valley, Fruita, Colorado, being mounted ca. 1902 for display at the Field Columbian Museum.

prove from direct fossil evidence; however, based on the biomechanical structure of diplodocid tails, the fundamental potential for supersonic motion was plausible. In conclusion, Myhvold and Currie found "it pleasing to think that the first residents of Earth to exceed the sound barrier were not humans."

Once extremely rare, skulls of *Apatosaurus* continue to be found and studied. Previously, Connely (1997) briefly reported on the first almost complete skull with jaws (TATE 099, assigned to *A. excelsus*; see Bakker 1998*a*) of this genus collected from an Upper Jurassic quarry of Como Bluff, Wyoming (see *Apatosaurus* entry, *S1*).

In a subsequent abstract, Connely and Hawley (1998) briefly reported on their proposed reconstruction, based upon this well-preserved skull, of the jaw musculature and other soft cranial tissues of *Apatosaurus*. Fortunately, the bones were slightly disarticulated. Therefore, individual cranial elements could be removed for study, permitting a detailed examination of the skull's inner framework. Following preparation and casting of each bone (at the Tate Geological Museum, Casper College), Connely and Hawley then fitted them together, thereby revealing details of this sauropod's skull anatomy. Noting preserved suture lines and muscle attachments, and combining their observations with modern analogues, Connely and Hawley were able to reconstruct some of the soft tissues in the *Apatosaurus* skull. Most significant, the authors stated, is the jaw musculature, this reconstruction suggesting "that *Apatosaurus* may have used its jaws in a front to back sliding motion to aid in the cropping or biting techniques used in feeding."

Curry (1998, 1999) made a study of sexual maturity and longevity in sauropods, particularly the genus *Apatosaurus*. As Curry (1999) pointed out, these

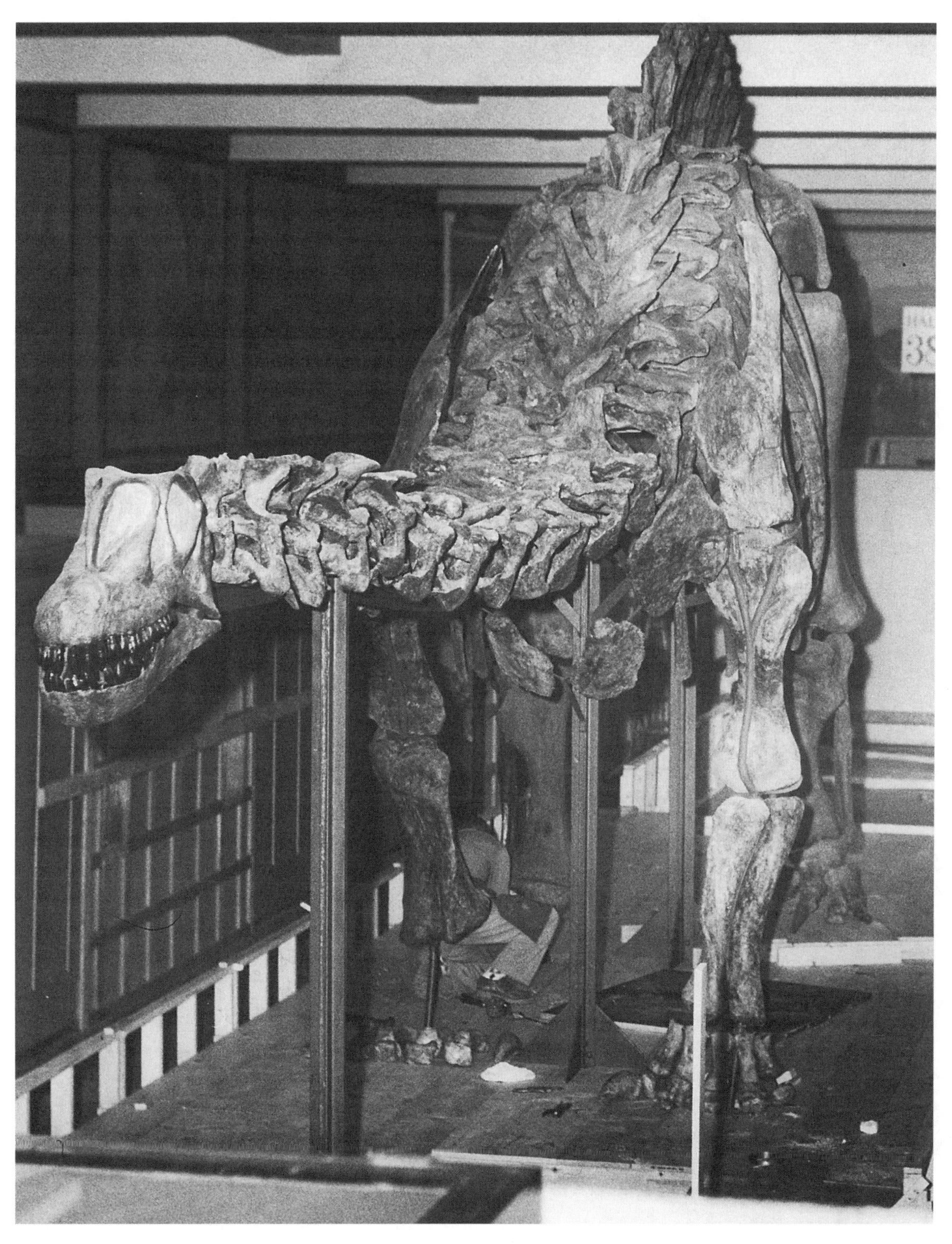

Courtesy The Field Museum (neg. #GEO84054).

***Apatosaurus excelsus*, composite skeleton (anterior part, FMNH PR27021, posterior part FMNH PR25112) during its 1958 remounting under the direction of Chief Preparator Orville L. Gilpin, still sporting a skull (FMNH PR26230, cast of CM 12020) of *Camarasaurus lentus*.**

topics, which have long intrigued paleontologists, have been difficult to address. Nevertheless, questions regarding the life history strategies of these gigantic animals continue to be asked and their possible answers debated: "Did sauropods grow indeterminately? Did sauropods grow at constant rates? Did they experience regular cycles of relative growth rate variation? How long did it take for them to reach adult size?" The key to some of these questions, Curry (1999) stated, may be in the bone histology of these giant dinosaurs.

Traditionally, sauropods like *Apatosaurus*, based upon the maximum growth rates in living reptiles, have been imagined as relatively slowly growing, long-lived animals, reaching sexual maturity at approximately age 60 and adult sizes as late as 120. According

Photograph by John Weinstein, courtesy The Field Museum (neg. #GEO85868-11).

***Apatosaurus excelsus*, composite skeleton (anterior part, FMNH PR27021, posterior part FMNH PR25112, skull cast of CM 11162) as remounted for The Field Museum's "Life Over Time" exhibit.**

to Curry (1998, 1999), however, bone tissue organization in *Apatosaurus*, as revealed through histological study, indicates that extant mammals and birds are better analogues than reptiles for examining sauropod ontogeny.

Curry (1999) presented a detailed ontogenetic, histologic analysis of *Apatosaurus* utilizing three appendicular elements (radius, ulna, and scapula) among four age classes—Age Class 1 (early juveniles), Age Class 2 (late juveniles), Age Class 3 (subadults), and Age Class 4 (adult). The study revealed, for the first time, histological differences among *Apatosaurus* elements throughout ontogenetic development. Materials examined by Curry (1999) included two scapulae, three radii, and four ulnae (BYU 641) referred to *Apatosaurus* sp., collected from the Cactus Park bonebed

Photograph by Sanford Maudlin Photography, courtesy Sam Noble Oklahoma Museum of Natural History, University of Oklahoma.

(Right foreground) the largest ***Apatosaurus excelsus*** yet known, this mounted skeleton (cast) measures almost 27 meters (90 feet) and has a femur about 2 meters (almost 7 feet) long. (Left background) a mounted skeleton (cast) of ***Allosaurus maximus.***

in the Upper Jurassic Brushy Basin Member of the Morrison Formation of Colorado (this site, yielding disarticulated *Apatosaurus* remains at various ontogenetic stages, possibly representing "a single event mass death assemblage" [K. Stadtman, personal communication to Curry, 1995]); also, "to extend ontogenetic coverage," an adult *Apatosaurus excelsus* specimen (UW 15556) mounted at the University of Wyoming's Geological Museum.

This study indicated to Curry (1999) that the microstructure of the radii, ulnae, and scapulae of *Apatosaurus* underwent three distinct phases or stages of osteogenesis or bone growth:

Growth Stage A, including Age Classes 1 and 2 ("primary deposition of fibro-lamellar bone, laminar to plexiform vascularity, no annuli of LAGS, Haversian bone only at endosteal margin"). This sustained, relatively rapid growth pattern, occurring in individuals up to 73 percent adult size, constitutes the bulk of osteogenesis in the genus.

Growth Stage B, including Age Class 3 ("Haversian reconstruction throughout cortex, primary fibro-lamellar bone for ~75 percent of cortical deposition, parallel-fibered bone at periosteal surface, decrease in vascularity to primarily longitudinal post–LAGs"). This pattern occurs in *Apatosaurus* individuals up to 91 percent adult size.

Growth Stage C, including Age Class 4 ("dense Haversian tissue extensively developed, primary deposition of lamellar bone at periosteal surface, periosteal bone is vascular, accretionary bone deposits at periosteal margin, periosteal bone marked with multiple annuli and LAGs"). This pattern affects only in radii, ulnae, and scapulae in adult *Apatosaurus* individuals.

These phases indicated to Curry (1998, 1999) that sustained, rapid bone growth occurred in *Apatosaurus* until late in this dinosaur's ontogeny, followed by a gradually declining growth rate, with probable plateauing at about age 8 to 10, and reaching maximum size in adulthood.

Photograph by the author, courtesy The Field Museum.

Detail of a skull (cast of CM 11162, referred to *Apatosaurus*, probably belonging to the species *A. louisae*) mounted on the composite skeleton of *Apatosaurus excelsus* displayed at The Field Museum. Skulls of this genus — and sauropods in general — are extremely rare.

Histological examination of periosteal bone in *Apatosaurus* scapulae indicated that, contrary to earlier hypotheses, prolonged ontogenies (or basically advanced ages) were not necessary for these animals to attain great body size. Cyclic deposition of periosteal bone (assumed to be seasonal or yearly events) in *Apatosaurus* scapulae, and apposition of laminar bone in scapulae, radii, and ulnae belonging to this genus, indicated that large subadult sizes were attained at approximately eight to 10 years of age. Appositional rates determined experimentally by Curry (1998, 1999) in extant birds and mammals and applied to *Apatosaurus* yielded minimal longevity rates of seven to 10 years. The results of these experiments suggested that sustained rapid growth occurred in dinosaurs like *Apatosaurus*, hatchlings attained half size within four to five years, with maximum growth achieved at approximately 10 years, this estimated growth rate being on a par with that occurring in modern ducks.

Regarding the question of longevity, Curry (1998, 1999) noted that, while gigantic sauropods such as *Apatosaurus* may have lived for at least a century, such extended life spans were not necessary for these dinosaurs to attain extremely large body size. *Apatosaurus* attained a maximum size, and probably plateaued," Curry (1999) concluded, "well within the temporal limitations associated with maximum size acquisition in extant mammals and birds."

Notes: In a preliminary report, Bakker (1998*b*) described the cranial/cervical articulations in a complete, articulated "apatosaurine" (informally used by Bakker 1998*a* for all "massive limbed, wide necked-wide tailed diplodocids") skull and neck from the Upper Jurassic of Como Bluff, Wyoming.

Bakker (1998*b*) observed the following features in this specimen: air pathway leading from throat to external nostril surrounded by unusual basket-work of joints; first two cervical ribs possessing movable joints with their centra, all remaining ribs fused to their vertebral bodies; maxilla and quadratojugal weakly joined, vertical hinge line permitting outward flexture; lacrimal/jugal joint a horizontal hinge; all cheek bones below orbit less than half the thickness of those bones in sauropods with the same skull length in other

Apatosaurus louisae **as restored by artist Brian Franczek.**

families, cheek and neck ribs therefore making up a vibratory shelf.

According Bakker (1998*b*), the large channels for arteries and nerves observed in the "apatosaurine" skull indicate that thick dermal tissue was present on the anterior half of the snout, but did not extend posteriorly onto the vibratory shelf. As layers of heavy soft tissue on the outer posterior surfaces would have dampened vibrations, this suggested to Bakker that "Much of the hitherto puzzling anatomy of diplodocids may well be explained by adaptations for shaping low frequency sound."

Salgado (1999), in a study on the macroevolution of the "Diplodimorpha," reassessed the status of the juvenile specimen (CM 566) originally named *Elosaurus parvus* by Peterson and Gilmore (1902), and which McIntosh (1990) and other authors have regarded as a possible juvenile *Apatosaurus excelsus*. However, as noted by Salgado, the cervical and dorsal spines of this specimen are singular, not bifurcated; therefore, as shown by Salgado's study, if this specimen belongs to *Apatosaurus*, the spines should have forked during postnatal development. As this scenario is not in accordance with Salgado's hypothesis on the origin of bifid spines (see "Introduction," section on sauropods), that author regarded *E. parvus* as not a juvenile diplodocid, but rather an immature *Haplocanthosaurus* sp., *Haplocanthosaurus* being a genus having unforked neural spines when fully grown.

Erratum: The *Camarasaurus lentus* skull (cast of CM 12020) formerly displayed on The Field Museum mounted skeleton of *Apatosaurus excelsus* (see photographs in *D:TE*) bears the catalogue number of FMNH PR26230.

Key references: Bakker (1998*a*, 1998*b*); Connely (1997); Connely and Hawley (1998); Curry (1998, 1999); McIntosh (1990); Myhrvold and Currie (1997); Peterson and Gilmore (1902); Rothschild and Berman (1991); Salgado (1999).

†ARCHAEOCERATOPS

Ornithischia: Genusauria: Cerapoda: Marginocephalia: Ceratopsia: Neoceratopsia: Protoceratopsidae.

Comments: Among the most primitive of the neoceratopsian dinosaurs is *Archaeoceratops oshimai*, a type species founded upon a partial skeleton from the Mazonoshan area of northwest China (see entry, *S1*),

In an abstract briefly reporting on various dinosaurs from the late Early Cretaceous of the Mazonoshan area, You, Dodson, Dong and Azuma (1999) noted the following features in *A. oshimai*: skull showing no trace of nasal horn; squamosal not extending beyond quadrate cotylus; orbit comparatively largest among all known neoceratopsians; infratemporal fenestra narrow rostrocaudally; palpebral prominent, fused to prefrontal; and three (preserved) maxillary teeth.

Autapomorphic features in this species observed by You *et al.* include the following: delicate and pendant rostral bone; triangular antorbital fossa; and modest pustulose ornamentation covering much of lateral surface of jugal. The frill is very much abbreviated.

Performing a preliminary cladistic analysis, You *et al.* found *Archaeoceratops*, displaying several neoceratopsian synapomorphies, to be the sister taxon to the Neoceratopsia, occupying "a crucial basal position in the phylogeny of all ceratopsians." More recently, You and Dodson, following a cladistic analysis of the Neoceratopsia, referred *Archaeoceratops* to the Protoceratopsidae (see "Systematics" chapter).

Key references: You and Dodson (2000); You, Dodson, Dong and Auma (1999).

ARCHAEORAPTOR Sloan 1999 —(See *Microraptor.*)

Name derivation: Greek *archaio* = "ancient" + Latin *raptor* = "thief."

Type species: *A. liaoningensis* Sloan 1999.

†ARISTOSUCHUS Seeley 1887

Saurischia: Theropoda: Neotheropoda: Tetanurae: Avetheropoda: Coelurosauria: Compsognathidae.

Name derivation: Greek *aristo* = "best" + Greek *souchos* = "crocodile."

Type species: *A. pusillus* Seeley 1887.

Other species: [None.]

Occurrence: Wealden, Isle of Wight, England.

Age: Early Cretaceous (upper Barremian).

Known material/holotype: BMNH R178, pelvis.

Diagnosis of genus (as for type species): [No modern diagnosis published.]

Comments: For more than a century, the small theropod type species *Aristosuchus pusillus* has generally been regarded by most authors as a junior synonym of the type species *Calamospondylus oweni*.

A. pusillus had been founded upon remains (BMNH R178) collected from the Wealden Group of the Isle of Wight, England, and originally named *Poiklopleuron pusillus* by Sir Richard Owen in 1876. According to tradition, Owen, believing that this material, as well as that referred by Fox to *C. oweni*, belonged to "*Poikilopleuron*" [=*Poekilopleuron*], a large theropod which he incorrectly interpreted to be crocodilian, referred it all to the new species *P. pusillus*. Seeley (1887), subsequently finding some of Owen's diagnostic criteria for that assessment to be invalid (while also questioning the reference of the sacral vertebrae, part of the holotype of *C. oweni*, to this taxon), referred these remains to the new genus *Aristosuchus* (for more details, see *Calamospondylus* entry, *D:TE*).

Thus, *Calamospondylus* and *Artistosuchus* have been regarded as congeneric, with the name *Aristosuchus* having priority. Later, Norman (1990) pointed out that William Fox's publication of the original taxon did, in fact, have priority, the name *Calamospondylus* having correctly been accompanied by a valid albeit poor description (hence, its listing under *Calamospondylus* and not *Aristosuchus* in *D:TE*).

In a yet more recent brief report, Naish (1999*a*), who had been investigating the above taxa and the fossil materials relating to them, noted that only one recorded specimen ("the compsognathid pelvis BMNH R178") can correctly be associated with these alleged synonyms. According to Naish's research, discrepancies in the original descriptions of *C. oweni* and *A. pusillus* suggest that two specimens were involved; also, "in previously overlooked correspondence," Fox stated that these "two specimens were different." Consequently, Owen's description of BMNH R178 was not of *C. oweni*, *Aristosuchus pusillus*, therefore, being the valid name for this specimen. Naish (2000) later regarded *Aristosuchus* as "a Wessex Formation compsognathid."

Erratum: In *D:TE*, *Calamospondylus* entry, the holotype of *C. oweni* was numbered as BMNH R178. This catalog number is actually that of the type specimen of *A. pusillus*. The holotype of *C. oweni* comprises two dorsal, five sacral, and two caudal vertebrae, the fused distal portions of the pubes, and one ungual phalanx. The whereabouts of this material, Naish noted, is unknown.

Key references: Fox (1866); Naish (1999*a*, 2000); Norman (1990); Owen (1876); Seeley (1887).

†ASTRODON Johnston 1859 [*nomen dubium*]—(=?*Pleurocoelus*)

Saurischia: Sauropodomorpha: Sauropoda: Eusauropoda: Neosauropoda: Macronaria: Camarasauromorpha: Titanosauriformes: Brachiosauridae.

Name derivation: Greek *astros* = "star" + Greek *odous* = "tooth."
Type species: *A. johnstoni* Leidy 1865 [*nomen dubium*].
Other species: [None.]
Occurrence: Arundel Formation, Maryland, United States.
Age: Early Cretaceous (Aptian).
Known material/holotype: YPM 789, tooth.

Diagnosis of genus (as for type species): [None published.]

Comments: The genus *Astrodon* was founded upon a single tooth (YPM 789; see *D:TE* for illustration) belonging to an adult sauropod. The specimen was found by state Geologist Philip T. Tyson in late November, 1858, on the property of J. D. Latchford in the Arundel Formation of Muirkirk, Maryland. According to research by Peter M. Krantz (1998), this location was substantiated in late 1996 following the inadvertent acquisition of Latchford's property in December of 1995 by the Maryland-National Capitol Park and Planning Commissions. (As the property was adjacent to an active dinosaur fossil site, that company sought it for the location of a new dinosaur park.) Discovery of parts of the Latchford house substantiated the property as the type locality for the tooth.

Christopher Johnston (1859) first reported (but did not describe) this tooth in a dentistry journal. Johnston named the tooth *Astrodon* but did not propose a specific name. Leidy (1865) later erected the type species *Astrodon johnstoni* and described the tooth as spoon-shaped.

In 1903, Hatcher, observing close similarities between the holotype tooth of *A. johnstoni* and teeth of the more completely represented but juvenile sauropod *Pleurocoelus* (see *D:TE, S1*), while also pointing out that all related material collected by him was yielded by iron ore beds at about or possibly the same horizon and locality, suggested that *Astrodon* and *Pleurocoelus* were congeneric. Indeed, most subsequent authors have followed Hatcher's suggestion, regarding *Astrodon* as synonymous with *Pleurocoelus*. Romer (1967), in his benchmark book *Vertebrate Paleontology*, for example, accepted this synonymy, retaining the earlier published name *Astrodon*, while most subsequent workers, considering *Astrodon* to be a dubious taxon, have preferred *Pleurocoelus*, even though the latter was published at a later date (Marsh 1888*b*).

Astrodon was referred to *Pleurocoelus* in both *D:TE* and *S1*, not only because this synonymy had become generally adopted in the paleontological literature, but also because a number of paleontologists (personal communication) had advised that I utilize that synonymy. Sauropods from the Cretaceous of North American are rare to begin with, and no other sauropod is known from the deposits that yielded both *Astrodon* and *Pleurocoelus*. However, there are problems with this referral. As sauropod authority John S. McIntosh (1990) stated in his review of the Sauropoda (wherein he listed these two forms as possibly congeneric), "It is not unlikely that the tooth named *Astrodon johnstoni* (Leidy 1865) represents an adult form of *Pleurocoelus*, but this remains to be established."

More recently, Olshevsky (2000), in his annotated checklist of dinosaurian species, listed *Astrodon johnstoni* as a genus and species apart from *Pleurocoelus*. Pending future discoveries which can demonstrate unequivocally that the two genera are the same, this document now, at least tentatively, lists *Astrodon* as a genus separate from *Pleurocoelus*.

Note: The species *Astrodon pusillus* has been referred by Galton and Boine (1980) to the stegosaur *Dacentrurus armatus*.

Key references: Galton and Boine (1980); Hatcher (1903); Johnston (1859); Krantz (1998); Leidy (1865); Marsh (1888*b*); McIntosh (1990); Olsveksky (2000); Romer (1967).

ATLASAURUS Monbaron, Russell and Taquet 1999
Saurischia: Sauropodomorpha: Sauropoda: Eusauropoda: Neosauropoda: Macronaria: Camarasauromorpha: Titanosauriformes: ?Brachiosauridae.
Name derivation: "Atlas [mountain chain from Morocco; also, the mythological giant]" + Greek *sauros* = "lizard."
Type species: *A. imelakei* Monbaron, Russell and Taquet 1999.
Other species: [None.]
Occurrence: Tilgougguit Formation, Azilal, Morocco.
Age: Middle Jurassic (Bathonian–Callovian).
Known material/holotype: Musée des sciences de la Terre de Rabat, Morocco collection, nearly complete skeleton.

Diagnosis of genus (as for type species): Autapomorphies including supratemporal fenestra twice as wide as long (140 by 70 millimeters), not visible in lateral view; combined width of paroccipital processes 48 percent of estimated mandibular length; paroccipital processes extending horizontally at almost right angles to long axis of skull; mandibular symphysis and dentary very shallow (symphyseal depth 116 percent minimum depth of dentary, probably reversal from primitive state in Sauropoda); length of humerus 65 percent of estimated length of dorsal vertebrae series; length of ulna exceeding that of tibia by approximately 115 percent (Monbaron, Russell and Taquet 1999).

Comments: The genus *Atlasaurus* was based on an almost complete skeleton with skull (first mentioned but not named by Monbaron and Taquet 1981, then Monbaron 1983) collected from the Wawmda locality in the Tilougguit Formation, Azilal province,

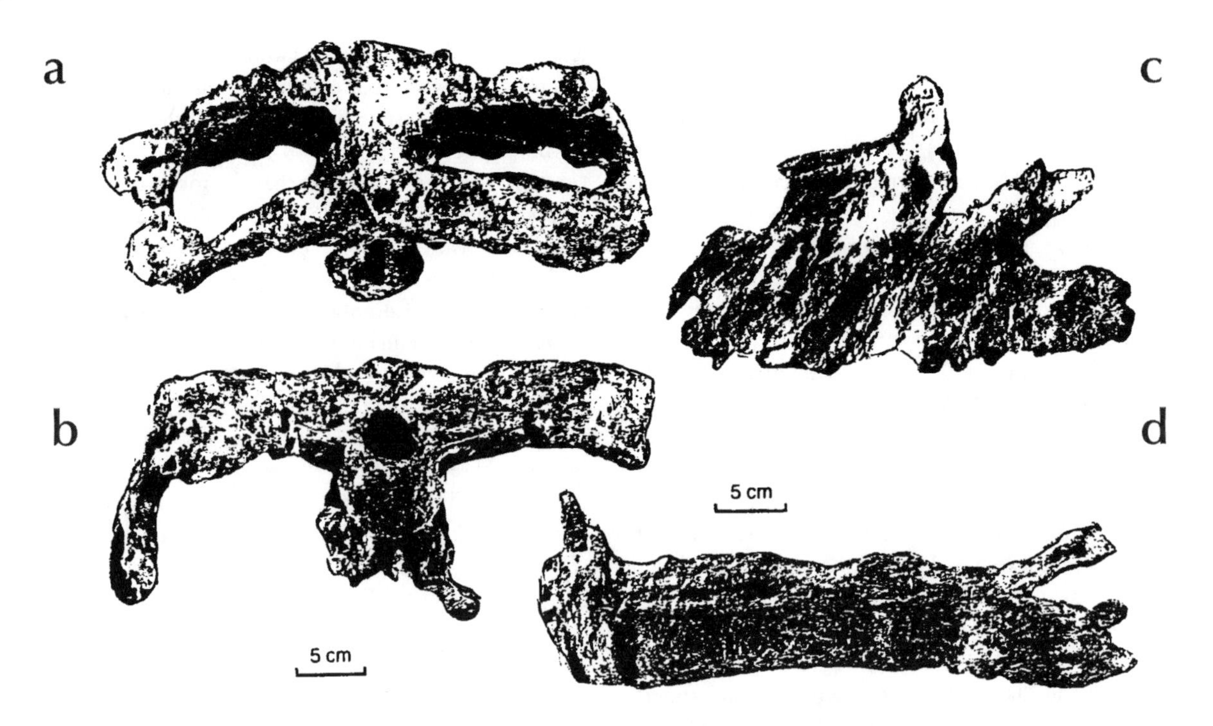

***Atlasaurus imelakei*, holotype braincase in A. dorsal and B. posterior views, C. left maxilla, lateral view, and D. left dentary, lateral view. (After Monbaron, Russell and Taquet 1999.)**

High Atlas of Morocco (see Monbaron and Taquet) in rocks of Bathonian–Callovian (see Jenny, Le Marrec and Monbaron 1981) age (Monbaron, Russell and Taquet 1999).

According to Monbaron *et al.*, the holotype of the type species *Atlasaurus imelakei* is distinguished as one of the relatively few almost complete sauropod skeletons that have been recovered, especially one including a skull; also, it is among the oldest relatively complete sauropod skeletons known, about 15 million years older than the classic Late Jurassic sauropod assemblages of North America and Tanzania. The relative completeness of the specimen suggested to the authors "that the animal was borne away and drowned in a flood, then caught in an oxbow or meander and rapidly covered by fluvial sediments mixed with vegetation," thereby being partially protected from dismemberment and scattering (see Monbaron and Taquet). Found in association with the specimen were a number of theropod teeth.

Monbaron *et al.* only briefly described the type specimen (which will be more fully described elsewhere) as representing a moderately large, adult sauropod. Compared to the estimated total length of the dorsal column in the related genus *Brachiosaurus*, *A. imelakei* has a relatively large skull (though not as large as that of *Camarasaurus*, short neck, long tail (not as long as in *Apatosaurus*), and very long limbs. Based on the minimum midshaft circumferences of the humerus (565 millimeters) and femur (690 millimeters), and following a mathematical formula devised by Anderson, Hall-Martin and Russell (1985), the authors estimated the adult animal to have weighed approximately 22.5 metric tonnes (about 26 tons).

As observed by Monbaron *et al.*, the holotype of *A. imelakei* exhibits various plesiomorphic features — *e.g.*, relatively short neck; undivided neural spines in cervical and preserved anterior dorsal vertebrae; metacarpals I–III facing anteriorly as in *Omeisaurus*, IV–V sharply rotated posteriorly and medially around lateral edge of forefoot, the metacarpus consequently being only incipiently columnar — that identify this species as a rather primitive sauropod.

Compared with other primitive sauropods, *Atlasaurus* was found by the authors to be positioned within the Camarasauromorpha, as indicated by various features — *e.g.*, major rami of quadratojugal embracing 90-degree angle or less; metacarpals relatively long; dorsal pleurocoels invaginated, extending to posterior end of presacral vertebral column. Close similarities between the vertebral column and limbs suggested to Monbaron *et al.* "a closer affinity to *Brachiosaurus* than to any other known sauropod." (*Atlasaurus* is about 15 million years older than the larger and more derived *Brachiosaurus*, from the Upper Jurassic of both East Africa and North America.)

As pointed out by Monbaron *et al.*, the evidence of fossil bones and footprints representing various other taxa suggests that a major early radiation of sauropods took place in Morocco and in China during the Middle Jurassic. This radiation coincides with an increase in size, with levels of atmospheric carbon dioxide during the Middle Jurassic (see Berner 1997), and perhaps higher levels of plant productivity. The Atlantic rift having not, at that time, yet separated the

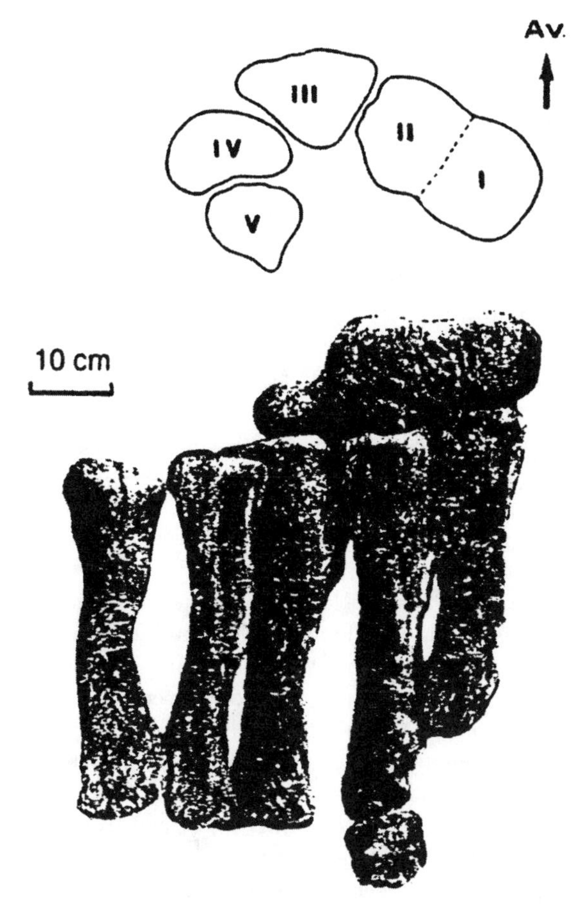

***Atlasaurus imelakei*, holotype carpals and metacarpals of left manus, posterior view (metacarpals of left manus, posterior view (metacarpal V preserved upside down). After Monbaron, Russell and Taquet 1999.)**

coasts of southeastern North America and West Africa (see Gradstein, Jansa, Srivastava, Williamson, Carter and Stam 1990), Moroccan sauropods most likely resembled some of those inhabiting the western side of this rift in southeastern North America. Therefore, the authors stated, "the Early and Middle Jurassic of Morocco potentially hold information of fundamental importance in resolving the early history of the sauropodan giants of the Mesozoic."

Key reference: Anderson, Hall-Martin and Russell (1985); Jenny, Le Marrec and Monbaron (1981); Monbaron (1983); Monbaron, Russell and Taquet (1999); Monbaron and Taquet (1981).

†**AUBLYSODON** [*nomen dubium*]—(=?*Stygivenator*)

Saurischia: Theropoda: Neotheropoda: Tetanurae: Avetheropoda; Coelurosauria: Manuraptoriformes: Arctometatarsalia: Tyrannosauridae: ?Aublysodontinae.

Diagnosis of genus (as for type species): Teeth D-shaped in cross section, lacking serrations (Carpenter 1982).

Comments: The type species *Aublysodon mirandus* (see *D:TE* for further details) was established upon teeth described in 1868 by Joseph Leidy of The Academy of Natural Sciences of Philadelphia. The lectotype tooth (ANSP 9335, designated as such by Carpenter 1982), is unserrated and therefore distinct from the teeth of most tyrannosaurids known from skeletal remains.

Over the years, a number of specimens, most of them consisting of only teeth, have been referred to *Aublysodon*. Carpenter referred numerous teeth from the Lance Formation and Hell Creek Formation to *Aublysodon*, which led him to the conclusion that *Aublysodon* is a valid genus. Without offering compelling evidence, Paul (1988*b*), in his book *Predatory Dinosaurs of the World*, referred to *Aublysodon* the Chinese theropod *Shanshanosaurus houyanshanensis* and also, for a juvenile partial tyrannosaurid skull (LACM 28471) from the Hell Creek Formation of Montana, erected the new species *A. molnaris*. Subsequently, Molnar and Carpenter (1989) removed LACM 28471 from its own species status and, along with more teeth, referred the specimen to *Aublysodon* cf. *A. mirandus*. Later, Lehman and Carpenter (1990) referred to *Aublysodon* cf. *A. mirandus* a partial skeleton and skull (OMNH 10131) from the Kirtland Formation of New Mexico. More recently, Olshevsky, Ford and Yamamoto (1995*a*, 1995*b*) referred LACM 28471 to the new genus *Stygivenator* (see entry) and Carr and Williamson (2000) referred OMNH 10131 to *Daspletosaurus* sp. (see *Daspletosaurus* entry).

Additional teeth have been referred to *Aublysodon*, including specimens from the Dinosaur Park Formation of Alberta, while morphologically identical teeth, referrable to the genus *Alectrosaurus*, were reported from the Iren Dabasu Formation in the People's Republic of China (Currie, Rigby and Sloan 1990). Eaton, Diem, Archibald, Schierup and Monk (1999) referred teeth from a Santonian-age locality in southwestern Utah to ?*Aublysodon* sp. Similarly, Eaton, Cifelli, Hutchison, Kirkland and Parrish (1999) referred to cf. *Aublysodon* sp. teeth from the middle to late Turonian Smoky Hollow Member of the Straight Cliffs Formation of Utah (Parrish 1999).

Recognizing OMNH 10131 as an adult *Daspletosaurus* specimen and regarding LACM 28471 as an indeterminate juvenile tyrannosaurid (apparently unaware at the time of their writing that Olshevsky *et al.* were referring this specimen to a new genus), Carr and Williamson considered *Aublysodon* to be a *nomen dubium*.

Therefore, Molnar and Carpenter's diagnosis of *Aublysodon*, based upon remains other than teeth, is no longer applicable.

Key references: Carpenter (1982); Carr and

Williamson (2000); Currie, Rigby and Sloan (1990); Eaton, Cifelli, Hutchison, Kirkland and Parrish (1999); Eaton, Diem, Archibald, Schierup and Monk (1999); Lehman and Carpenter (1990); Molnar (1987); Molnar and Carpenter (1989); Olshevsky, Ford and Yamamoto (1995*a*, 1995*b*); Parrish (1999); Paul (1988*b*).

†**AUGUSTIA** Bonaparte 1998—(Preoccupied; see *Agustinia.*)

†AVACERATOPS

Ornithischia: Genusauria: Cerapoda: Marginocephalia: Ceratopsia: Neoceratopsia: Ceratopsomorpha: Ceratopsoidea: Ceratopsidae: ?Centrosaurinae.

New material: Skull elements, ?partial skull, ?adult.

Diagnosis: Small ceratopsid having short, fan-shaped frill; no sagittal indentation on caudal border of frill; sagittal scallop on caudal border of parietal, with four or five scallops to either side; no parietal fenestrae; squamosal robust relative to parietal; dorsal border of squamosal straight; rostral portion of squamosal relatively long; pronounced bump on lateral surface of squamosal above quadrate arch, one or two lesser bumps located rostrally; postorbital horns ?moderately large; naso-premaxillary process protruding into caudal border of external naris; premaxilla with laterally inclined cutting edge that is bulbous caudoventrally, jutting well below maxillary tooth row; dentary robust, having ?21 teeth; humerus with conservative deltopectoral crest; olecranon process weakly developed; pedal unguals tapered (Penkalski and Dodson 1999).

Comments: The most primitive known ceratopsid species (see Penkalski (1993), *Avaceratops lammersi* was named and first described by Peter Dodson in 1986, this taxon based on an almost complete juvenile skeleton (ANSP 5800; see *D:TE* for photograph of the reconstructed skeleton mounted at The Academy of Natural Sciences of Philadelphia) from the Judith River Formation of south-central Montana. Since that date, additional materials referred to this taxon have been collected from the same locality, including an isolated left squamosal (USNM 4802) and a left squamosal (USNM 2415) and, provisionally referred to *Avaceratops*, a partial small, but probably adult, skull (MOR 692). These newly recovered materials have provided information that allowed the publication of a revised diagnosis and detailed description of the genus and species (Penkalski and Dodson 1999).

As the holotype of *A. lammersi* represents an immature animal, the adult MOR 692 could not be referred by Penkalski and Dodson with certainty to *Avaceratops*. Nevertheless, the authors noted the following similarities between this skull and that of ANSP 5800: parietals without hooks or spikes and apparently lacking fenestrae; pronounced squamosal bump; preserved portion of maxilla exhibiting centrosaurine morphology, large brow horns apparently chasmosaurine; skull in overall shape neither centrosaurine nor chasmosaurine (see below). Penkalski and Dodson concluded that MOR 692 must either represent a new genus or else an adult *Avaceratops*. Taking a conservative approach, the authors tentatively referred MOR 692 to the already established taxon.

According to Penkalski and Dodson, the type specimen of *Avaceratops* clearly represents an immature animal, unfused skull elements and the striated periosteal bone texture indicating that the individual was still growing at its time of death. Nevertheless, various lines of evidence in this specimen show that

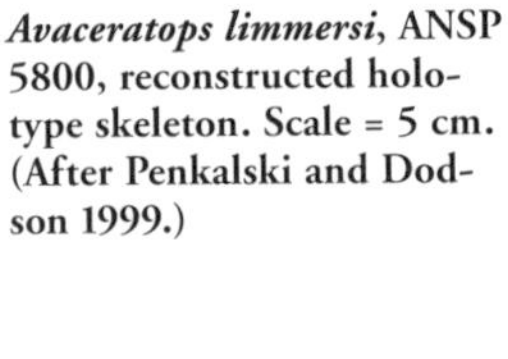

***Avaceratops limmersi*, ANSP 5800, reconstructed holotype skeleton. Scale = 5 cm. (After Penkalski and Dodson 1999.)**

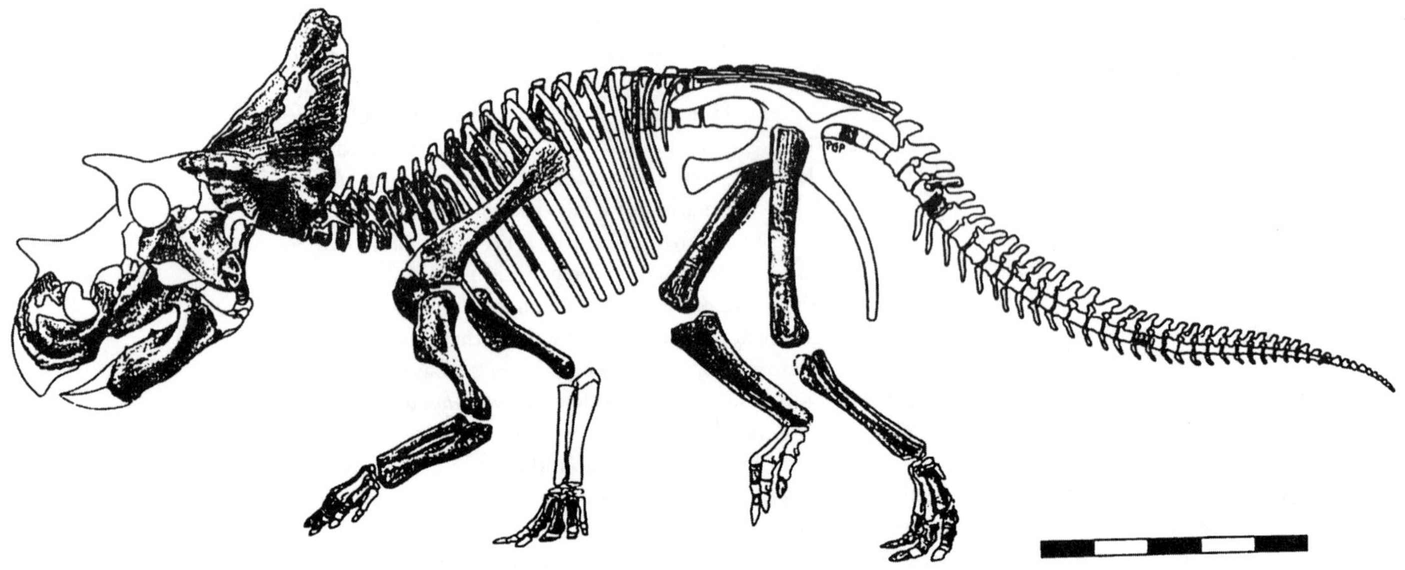

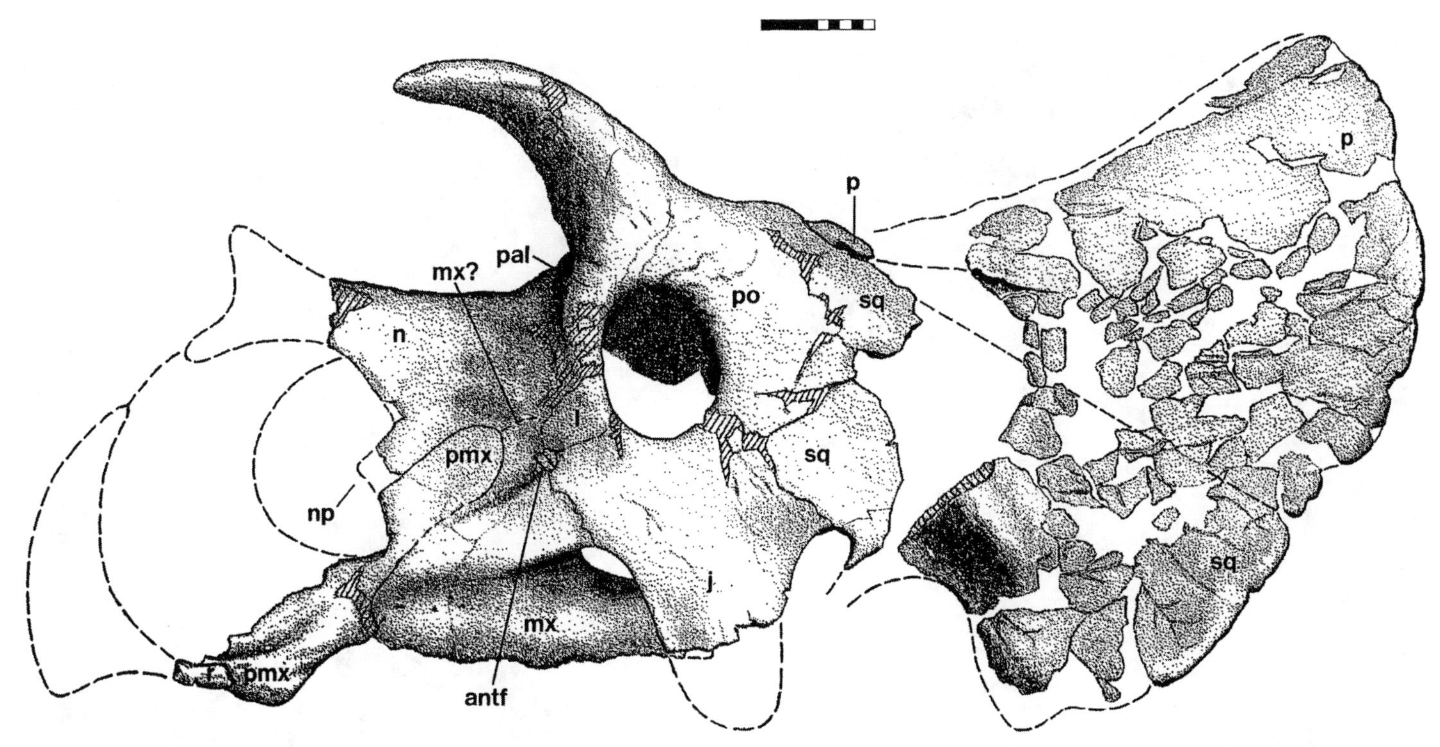

***Avaceratops limmersi*, MOR 692, tentatively referred skull. Scale = 10 cm. (After Penkalski and Dodson 1999.)**

the adult size of this dinosaur was smaller than that of other ceratopsids:

1. In the holotype, the occipital condyle is entirely fused at a relatively small diameter.

2. The pedal unguals are shaped more like those of some protoceratopsids than larger ceratopsids.

3. The olecranon process is weakly developed, the deltopectoral crest not as distally expanded as in other ceratopsids, and the limb bones are very graceful, indicating a small, lightly-built animal

4. The holotype snout is not as short or high, relatively, as in known juvenile horned dinosaurs (*e.g.*, *Protoceratops* and *Brachyceratops*)

5. Though small, MOR 692 is most likely an adult skull, as evidenced by its extensive fusion and sculpturing.

6. One of the referred squamosals (USNM 4802) is approximately 1.55 times as large as that of the holotype and seems to be close in size to that reconstructed on MOR 692. Scaled up taking into account these squamosals and measured through the vertebral column, ANSP 5800 would have an adult length of about 4.2 meters (approximately 14 feet), distinguishing this *Avaceratops* as the smallest known ceratopsid. (Adult centrosaurines and *Chasmosaurus belli*, the smallest known chasmosaurine, have a length of approximately 5 meters.)

Dodson originally referred *Avaceratops* to the ceratopsid subfamily Centrosaurinae, and Dodson and Currie (1990) regarded the genus as a quite primitive member of that group. Penkalski and Dodson, however, reassessed the systematic position of this genus, taking into account the newly collected specimens and noting that the two most diagnostic bones in ceratopsid dinosaurs, respectively, are the squamosal and the parietal.

As observed by Penkalski and Dodson, the squamosal of *Avaceratops* differs from the squamosals of all other known ceratopsids in being short, somewhat triangular in outline, with both rostral and caudal halves of equal length, and with a uniquely very straight dorsal border. As Dodson (1986) previously noted, the squamosal is comparatively long for a centrosaurine, estimated to be about 86 percent of the length of the parietal. (This ratio ranges from about 60 to 75 percent in centrosaurines and about 80 to 100 percent in chasmosaurines.)

Penkalski and Dodson noted that in all centrosaurines the squamosals are roughly quadrangular (see Sampson, Ryan and Tanke 1997) or polygonal in shape, yet never rectangular. This general shape is due to a "step-up" (present even in very small juveniles) along the dorsal border where this bone articulates with the parietal, matching a flange jutting outward at the same point along the lateral edge of the parietal. Chasmosaurine squamosals are narrowly triangular or scythe-shaped in outline (Dodson 1991), with a smooth dorsal border that curves upward without a step. The long axis of the centrosaurine squamosal is directed caudoventrally; however, the more highly derived chasmosaurine squamosal has two axes, the first directed caudoventrally, the other caudodorsally

Photograph by the author.

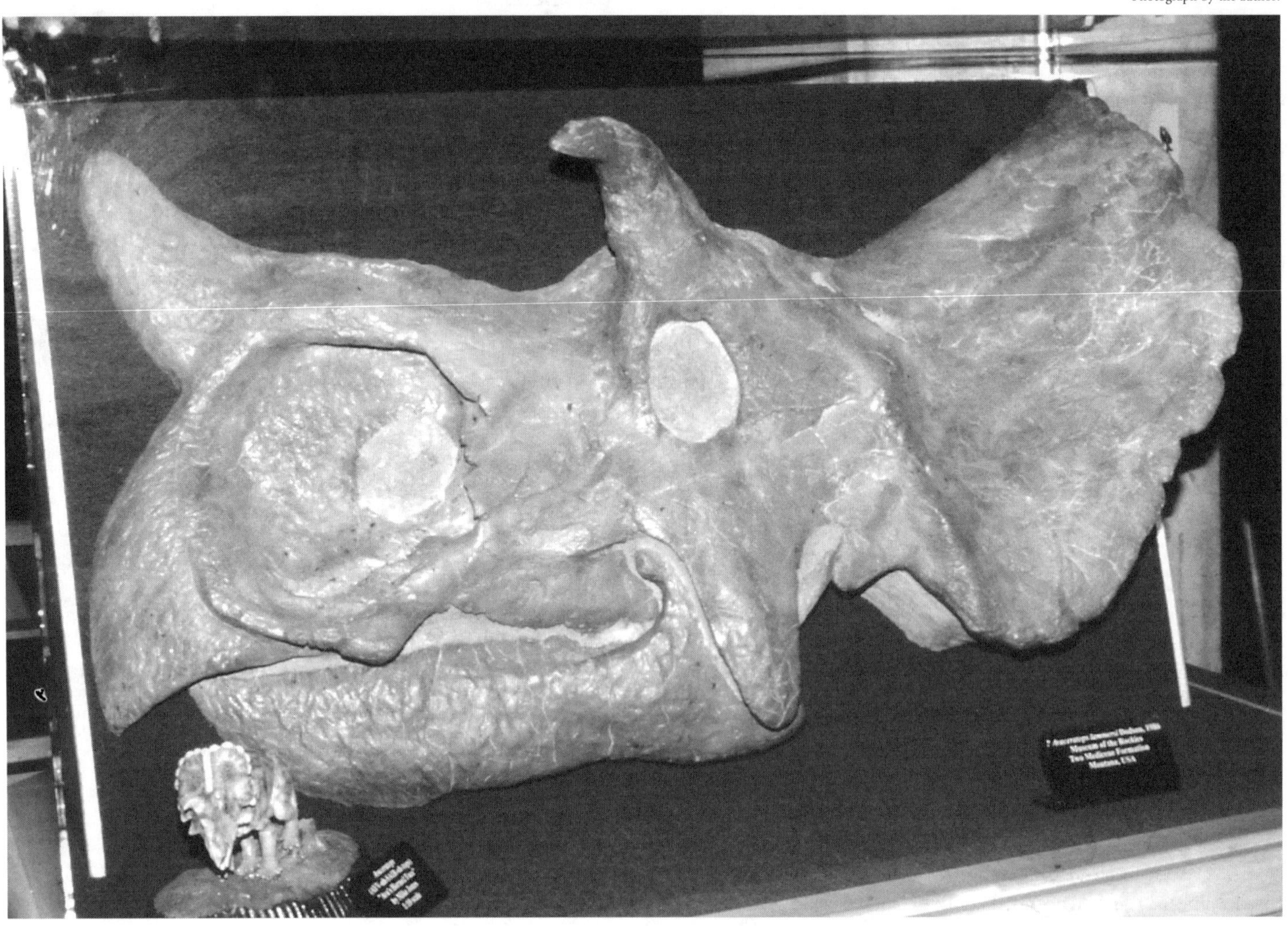

Reconstructed skull (cast of MOR 692) tentatively referred to *Avaceratops lammersi*, exhibited at Dinofest (2000–2001), held at Navy Pier, Chicago.

(Dodson 1993). In *Avaceratops*, the squamosal resembles that more typical of centrosaurines, but, as in chasmosaurines, the rostral half is comparatively longer and the dorsal border smooth.

As earlier observed by Sampson *et al.*, the parietal is the most diagnostic element within the Centrosaurinae. In adults, the parietal undergoes major ontogenetic changes (see Gilmore 1939; Sampson 1990; Dodson and Currie 1988; Lehman 1990). In juveniles, Penkalski and Dodson noted, the parietal, "because of its grossly different morphology, may not be useful in identifying specimens and may even be misleading." Penkalski and Dodson observed that the parietal of *Avaceratops* is unique in being seemingly solid rather than fenestrated, scalloped sagittally, possessing a relatively smooth border, and lacking any substantial ontogenetic developments (*e.g.*, hooks, processes, epoccipitals), and that it most closely resembles the solid parietal of *Triceratops* (the only derived ceratopsid known to have an unfenestrated crest).

A cladistic analysis performed on *Avaceratops* by Penkalski and Dodson, assuming this genus to be a centrosaurine, including six well-known centrosaurine genera (*Monoclonius*, *Centrosaurus*, *Styracosaurus*, *Einiosaurus*, *Achelosaurus*, and *Pachyrhinosaurus*), and using Chasmosaurinae as the first outgroup, produced unexpected results:

Avaceratops grouped not with the six centrosaurines, but fell between *Montanoceratops* and the Chasmosaurinae. This placement is well supported by various unambiguous characters (*e.g.*, parietal lacking caudosagittal indentation and apparently lacking parietal fenestrae, relatively long rostral half on squamosal, relatively small adult size). Furthermore, the pointed pedal unguals of *Avaceratops* seemed intermediate between the claw-like unguals of protoceratopsids and the rounded, hooflike ones of other horned dinosaurs.

Previous authors (Sereno 1986; Dodson and Currie 1990) had accepted an unfenestrated parietal as a basal feature for "protoceratopsids" (*e.g.*, including *Leptoceratops*, no longer classified as a protoceratopsid; see entry). Therefore, Penkalski and Dodson

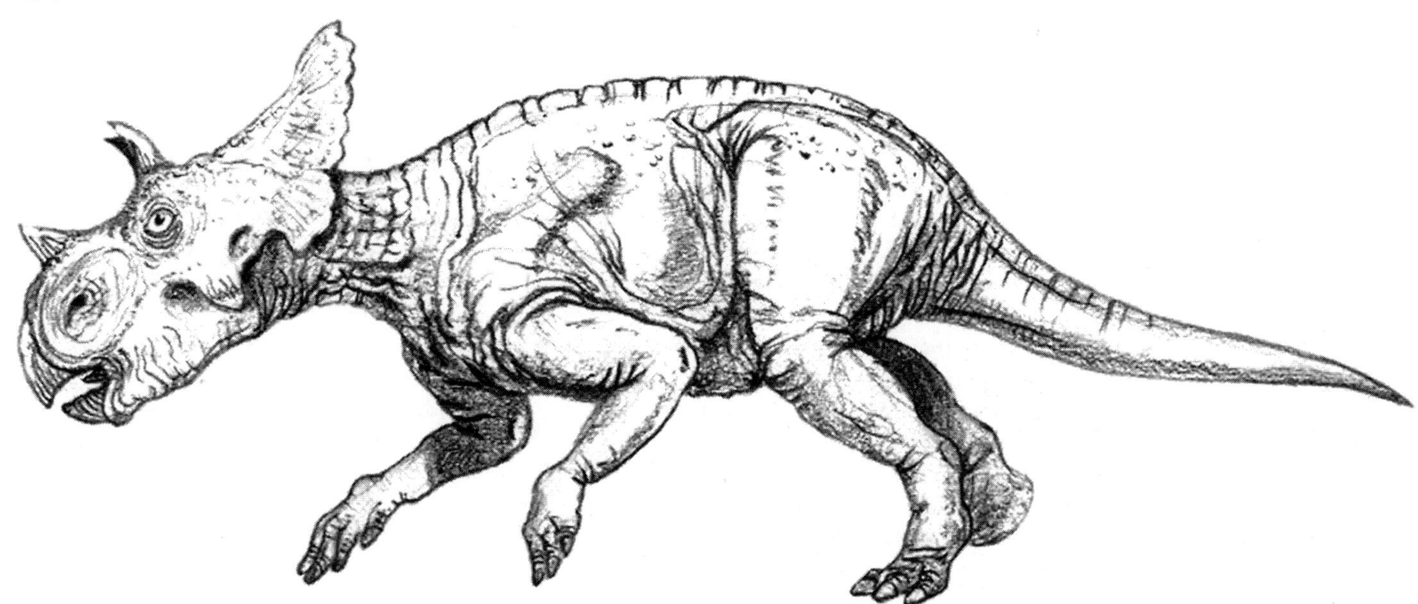

Life restoration of *Avaceratops limmersi* by Mike Fredericks.

postulated that the seemingly solid parietal of *Avaceratops* might represent a basal condition within the Ceratopsidae. Also, prominent brow horns — long regarded as diagnostic of the Chasmosaurinae — are present in the nonchasmosaurine skull MOR 692, suggesting to Penkalski and Dodson that this feature alone can no longer be used as a criterion for referring a specimen to the Chasmosaurinae. As certain key elements are missing, the holotype of *Avaceratops* and MOR 692 were entered separately in Penkalski and Dodson's analysis; still, MOR 692 grouped with ANSP 5800 as the most primitive known ceratopsid(s). Consequently, the presence of large brow horns might constitute evidence for their basal nature among ceratopsid dinosaurs (see "Systematics" chapter, "Notes" under "Chasmosaurinae").

According to Penkalski and Dodson, *Avaceratops*, whether or not possessing prominent brow horns, displays both unique and basal characters. While more primitive, the squamosal seems to exhibit characters of both the Centrosaurinae and Chasmosaurinae; the comparatively long rostral half is like that in protoceratopsids but unique to ceratopsids, the parietal seems to have been "either solid or had unusually small fenestrae," and its caudal border has the uncommon feature of a sagittal scallop, suggesting just the basically unadorned solid crest expected in an ancestral ceratopsid.

Clearly not a chasmosaurine, *Avaceratops* was found by Penkalski and Dodson to exhibit a number of traits (*e.g.*, face short and relatively high, naso-premaxillary process, general shape of premaxilla, and short frill) that could reasonably include this genus within the Centrosaurinae; however, it also differs significantly from all other members of that clade, the short face and frill possibly being symplesiomorphies. Penkalski and Dodson, therefore, preferred to regard *Avaceratops* not as "an aberrant centrosaurine" but rather a basal ceratopsid close to the ancestry of *Triceratops* and the Chasmosaurinae.

Penkalski and Dodson's revised assessment as to the phylogengtic placement of *Avaceratops* has not, been accepted by all specialists in ceratopsian dinosaurs. For example, Catherine A. Forster (personal communication 2000) sees *Avaceratops* as a doubtless centrosaurine and, as a juvenile, one difficult to define. Forster heartily disagreed with Penkalski and Dodson's assignment of the poor "flattened MOR specimen to *Avaceratops*," perceiving no characters linking the two.

Erratum: In *D:TE*, it was incorrectly stated that *Avaceratops* was collected from the Dinosaur Park Formation (="Judith River Formation").

Key references: Dodson (1986, 1991, 1993); Dodson and Currie (1988, 1990); Gilmore (1939); Lehman (1990); Penkalski (1993); Penkalski and Dodson (1999), Sampson (1990); Sampson, Ryan and Tanke (1997); Sereno (1986).

†**AVALONIANUS** Kuhn 1961—(See *Megalosaurus.*)

Name derivation: *Avalonia* [Celtic mythology: "Avalon," an island paradise in the western seas where heroes such as King Arthur went to their death] + Latin *ianus* = "of" or "belonging to."

Type species: *A. sanfordi* (Seeley 1898).

†AVIMIMUS

Comments: *Avimimus portentous*, a type species that was treated as a nonavian theropod in both *D:TE* and *S1*, has proven to be a rather controversial taxon in recent years.

This turkey-sized animal, based on several incomplete skeletons, was named and originally described in 1981 by Seriozha M. Kurzanov as a very birdlike theropod which he referred to the new monotypic family Avimimidae (see *D:TE*). Kurzanov pointed out that *Avimimus* was, in some ways, more birdlike (*e.g.*, large orbit, toothless jaws but with toothlike denticulations in premaxilla, orbit and lower temporal opening not confluent with development of typical horizontal jugal bar, *etc.*) than even *Archaeopteryx*, possessing a number of avian characters not found in the latter genus. Various other authors (*e.g.*, Thulborn 1984; Molnar 1985; Paul 1988*b*) subsequently agreed that this genus is more avian than *Archaeopteryx*.

Chatterjee (1991), when first describing *Protoavis* (see entries, *D:TE*, *S1*, and this volume), regarded *Avimimus* as avian and belonging to the order Avimimiformes, listing the following apomorphic characters in this genus: ascending process of quadratojugal lost (as in other birds), so that quadrate forms posterior margin of lower temporal fenestra; (unlike other birds) quadrate fused to squamosal; teeth in premaxilla, absent in maxilla and dentary. Most subsequent workers (*e.g.*, Norman 1990), however, have continued to classify *Avimimus* as a rather unusual, although very birdlike, nonavian theropod. Recently, Dyke and Thorley (1998) stated "it is evident that *Avimimus* exhibits several derived characteristics shared between birds and theropods, such as the structure of the forelimbs."

Later, Chatterjee (1999) again addressed the status of *Avimimus*, based on his examination of a cast of a well-preserved pelvic girdle and hindlimbs of this genus provided to him by Dr. Kurzanov. As observed by Chatterjee (1999), *Avimimus* is an unusual animal displaying "a curious mixture" of both plesiomorphic features (*e.g.*, ilium with brevis shelf, pubes propubic with large "foot," shaft of femur with fourth trochanter) and highly derived features (*e.g.*, skull more avian than *Archaeopteryx*, cervical vertebrae with well-developed hypapophyses, pubes fused, distal condyles of humerus carinate-shaped as in modern birds). Therefore, Chatterjee (1999), utilizing cranial synapomorphies, "placed *Avimimus* within Aves and as the sister-group of *Protoavis* and higher birds" (see also Chatterjee 1991).

Avimumus portentosus, referred skeleton from the Djadokta Formation, Gobi Desert, People's Republic of Mongolia.

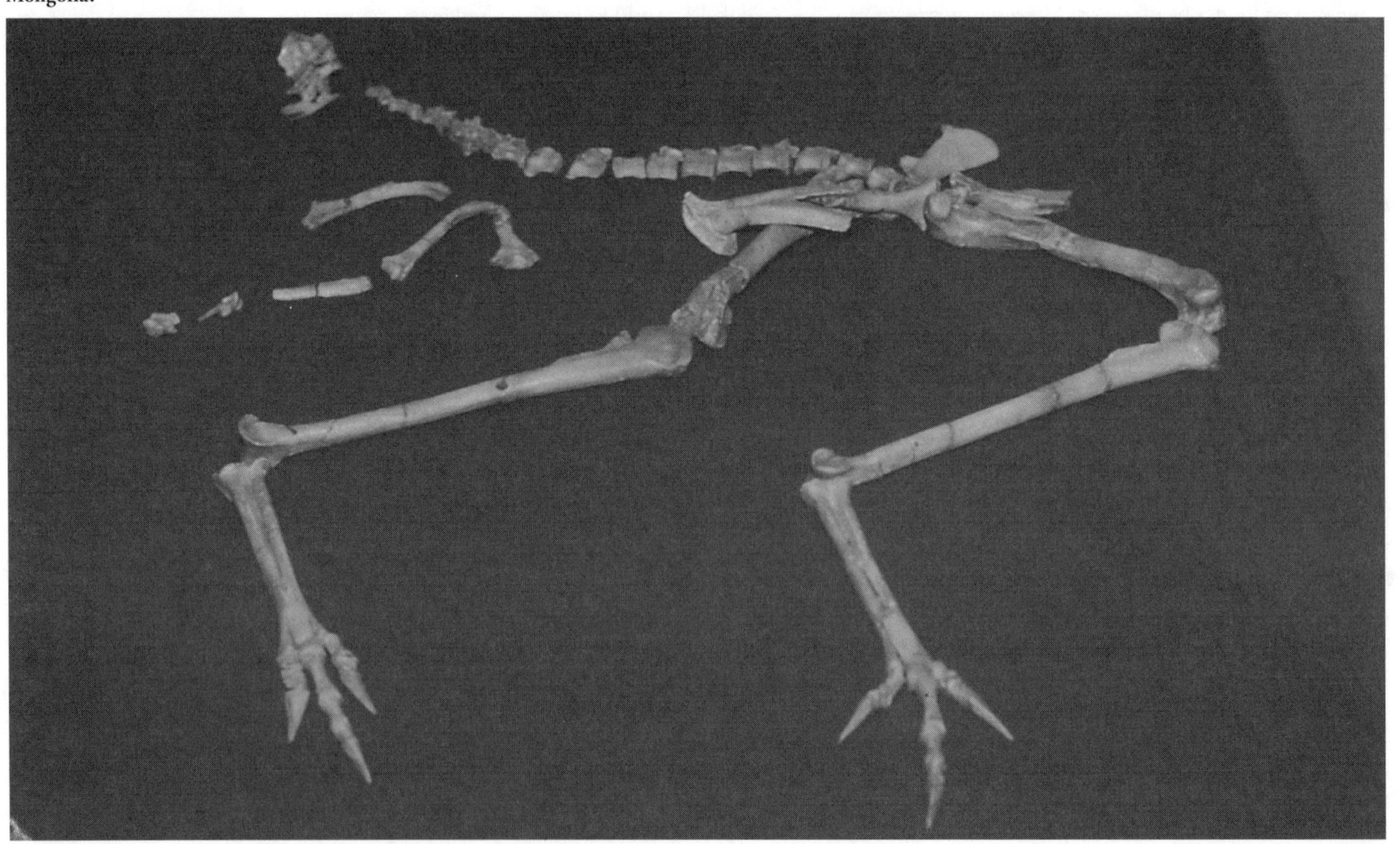

Photograph by the author, courtesy Mesa Southwest Museum.

More recently, Watabe, Weishampel, Barsbold, Tsogthbataar and Suzuki (2000), in an abstract, reported on an almost complete skeleton with the skull of *A. portentosus* found in 1996 by a team of the Hayashibara Museum of Natural Sciences–Mongolian Paleontological Center Joint Paleontological Expeditions, in the Upper Cretaceous beds of Shar Tsav, Eastern Gobi region, Mongolia. As these authors noted, the specimen preserves many anatomical elements not included in the holotype (PIN 3907/1), such as the premaxilla, an articulated forelimb (scapulocoracoid, humerus, radius, ulna, and fused carpometacarpal), and an articulated tail.

The specimen was described by Watabe *et al.* as follows: Premaxilla very narrow, with depression for narial fossa; premaxillary teeth small; preserved cervical and dorsal vertebrae with well-developed hypapophysis; posterior margin of ulna with very sharp and thin keel (crista); radius very narrow, rod-like; carpometacarpal completely fused; scapula narrow but elongate; coracoid large; metatarsus showing very advanced arctometatarsalian condition; unguals of three pedal digits claw-like, straight, long axis of unguals of digits II and IV angled laterally; preserved caudal vertebrae anteroposteriorly elongated, no evidence of pygostyle.

According to Watabe *et al.*, the discovery of this specimen "confirms the validity of the original description by Kurzanov" of this taxon as a nonavian theropod, its avian-like features showing "birdlike adaptations in a small Late Cretaceous theropod that inhabited heterogenous environments."

Also, the authors speculated that several thousand very small, bipedal, tridactyl footprints, occurring as gregarious trackways at the Shar Tsav locality, may be attributable to *A. portentosus*.

Key references: Chatterjee (1991, 1999); Dyke and Thorley (1998); Kurzanov (1981); Molnar (1985); Norman (1990); Paul 1988*b*); Thulborn (1984); Watabe, Weishampel, Barsbold, Tsogthbataar and Suzuki (2000).

†BACTROSAURUS

Ornithischia: Genasauria: Cerapoda: Ornithopoda: Euornithopoda: Iguanodontia: Euiguanodontia: Dryomorpha: Ankylopollexia: Hadrosauroidea *incertae sedis*.

Age: Middle Cretaceous (?Cenomanian).

Diagnosis (as for type species): Nonhadrosaurid hadrosauroid characterized by (characters autapomorphic of *Bactrosaurus johnsoni*) long, rounded rostral process of jugal, bearing very elongated and excavated maxillary facet; postcotyloid process of squamosal markedly curved backwards; club-shaped neural spines of most posterior dorsal vertebrae in fully ossified adult specimens [a growth feature; M. K. Brett-Surman 2000]; distal blade of scapula widening very regularly toward its end, ratio "length of the scapula/maximal width of the distal blade" less than 3.5; humerus with prominent and angular deltopectoral crest; (among characters distinguishing *B. johnsoni* from *Gilmoreosaurus mongoliensis*) rather narrow caudal process of jugal; greatly expanded prepubic blade [Brett-Surman 1979, 1989]; preacetabular process of ilium very deflected ventrally; ischial shaft very thick, ischial foot greatly expanded; femur perfectly straight in lateral view; ungual phalanges thick and truncated in adults (Godefroit, Dong, Bultynck, Hong and Feng 1998).

Comments: In 1933, Charles Whitney Gilmore of the United States National Museum, Smithsonian Institution, named and described the new type species *Bactrosaurus johnsoni*, remains of which were discovered in 1922 and 1933 in the Iren Dabashu Formation of Inner Mongolia. Gilmore believed this taxon to be the most primitive known lambeosaurine hadrosaurid, apparently displaying an odd mixture of hadrosaurine (*e.g.*, a flat head) and lambeosaurine (*e.g.*, a hollow narial crest) features while also exhibiting various more primitive iguanodontid characters. This assessment of *B. johnsoni* as a primitive member of the Lambeosaurinae has been followed by most subsequent authors (*e.g.*, Steel 1969; Brett-Surman 1979; Maryańska and Osmólska 1981; Weishampel and Horner 1986) (see *D:TE*).

However, a new interpretation concerning the phylogenetic position of this dinosaur was proposed following the more recent collection of new remains pertaining to *B. johnson*. As reported by Godefroit, *et al.*, in 1995, the Sino-Belgian Dinosaur Expedition — a joint project of the Inner Mongolian Museum at Hohhot and Institut Royal des Sciences Naturelle de Belgique, Brussels, Belgium — discovered a rich, virtually monospecific bonebed in the Upper Cretaceous Iren Dabasu Formation northeast of Erenhot, Inner Mongolia, People's Republic of China. The site has yielded the scattered, though sometimes articulated, remains, totalling several hundred skeletal elements and mostly vertebrae, pelvic elements, ribs, and limb bones, representing at least four individuals belonging to *Bactrosaurus johnsoni*. (The only nonspecific specimen recovered from this bonebed thus far has been a single ornithomimid caudal vertebra.)

Specimens from this bonebed, regarded as "topotypes," are collectively catalogued as SMDE 95E5, and most of the elements are housed at the Inner Mongolian Museum. Part of this material is temporarily stored at Institut Royal des Sciences Naturelle, which will receive casts of this collection

Drawing of *Bactrosaurus johnsoni* bonebed discovered by the Sino-Belgian Dinosaur Expedition in 1995, near Erenhot, Inner Mongolia. (After Godefroit, Dong, Bultynck, Hong and Feng 1998.)

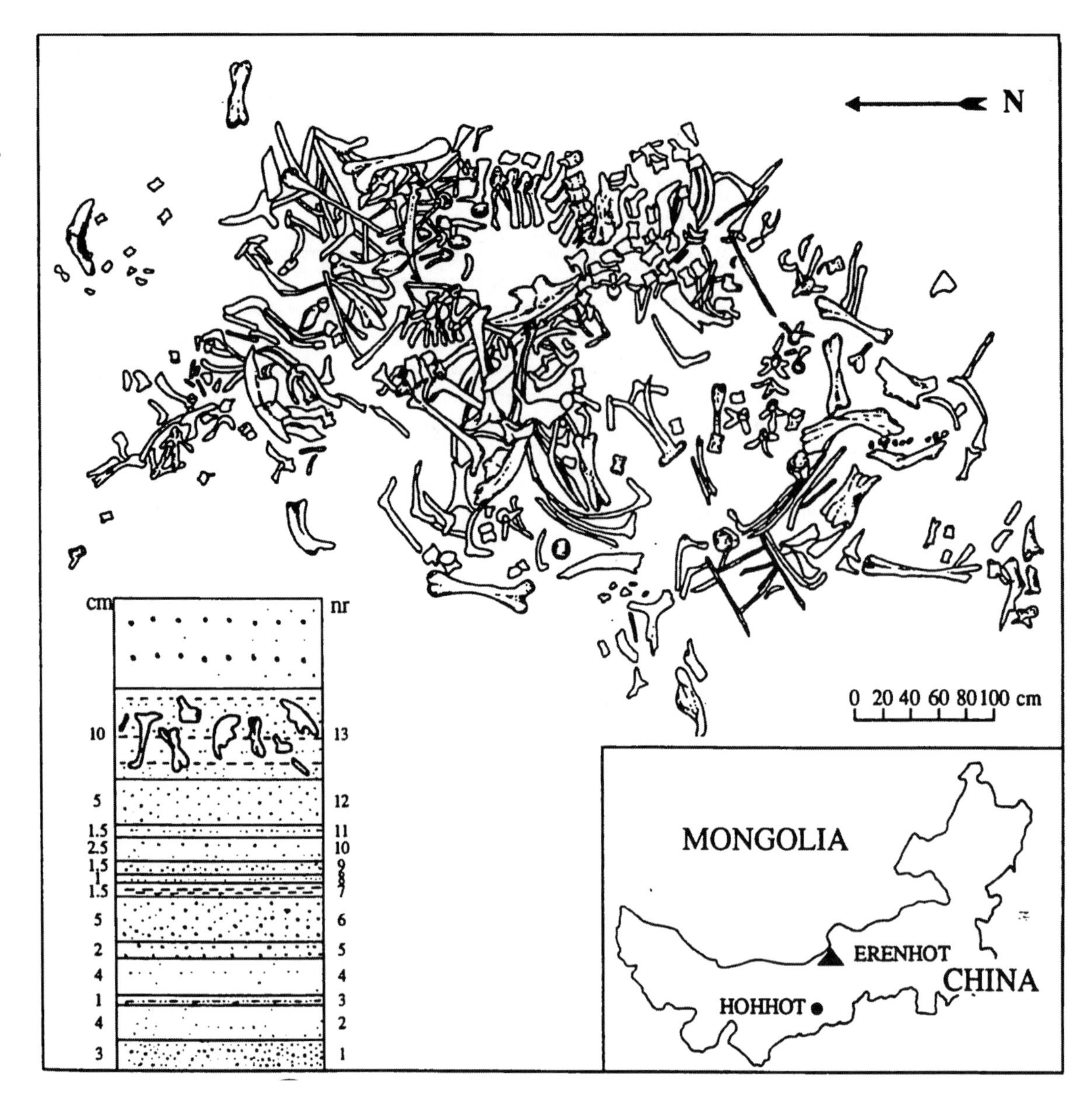

registered as IRSNB IG28522. These remains include the following: partial braincases, supraoccipital, parietals, frontals, premaxillae, maxillae, nasals, jugals, quadrates, squamosals, lacrimal, prefrontals, postorbital, supraorbitals, predentaries, dentaries, surangulars, isolated teeth, isolated and articulated vertebrae, sacra, ribs, scapulae, coracoids, sternals, humeri, radius, ulna, metacarpals, ilia, ischia, pubes, femur, fibula, astragalus, metacarpals, and phalanges.

In addition to this new material, Godefroit *et al.* included in their analysis of *B. johnsoni* the holotype (AMNH 6553), comprising dentary, maxillary, and other skull fragments, 10 dorsal, seven sacral, and 36 caudal vertebrae, a left scapula, left sternal, pubes, ischia, a left femur, fibula, a complete left and part of the foot (all of these elements associated based on their size); and the paratype specimens AMNH 6353, 6356, 6370, 6372–3, 6375, 6379–80, 6384–6, 6388–98, 6501, 6553, 6574–5, 6577–8, and 6580–7.

Addressing the issue that hadrosaurine and lambeosaurine remains may have been mixed at Gilmore's original bonebed site, Godefroit *et al.* pointed out that the same apparent "mixing" was also seen at the SBDE bonebed. Indeed, examination of all the material found at this locality revealed to the authors that all the skeletal elements — represented by at least four specimens, and belonging to different-sized individuals — are quite similar to one another, suggesting that only one species is present at this site. Consequently, the authors found it highly unlikely that two different localities would yield an identical fortuitous combining of identical skeletal elements from two different hadrosaurid taxa, their most logical conclusion being that both bone-beds are monospecific (a view not shared by all workers; M. K. Brett-Surman, personal communication 2000).

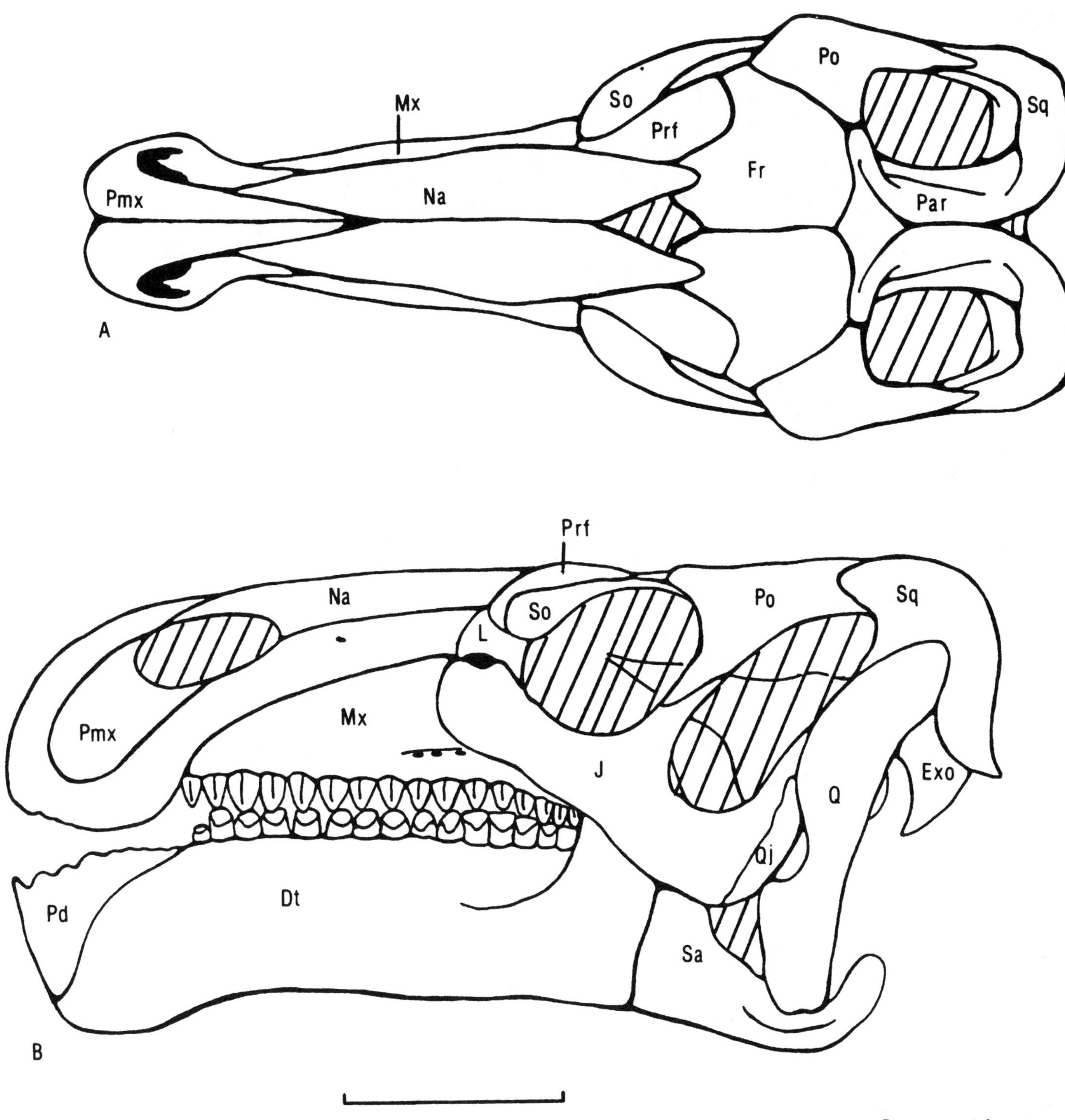

***Bactrosaurus johnsoni*, reconstruction of the skull in A. dorsal and B. left lateral views. Scale = 10 cm. (After Godefroit, Dong, Bultynck, Hong and Feng 1998.)**

Based upon a cladistic analysis involving the above specimens, Godefroit *et al.* noted that *B. johnsoni* shares a list of plesiomorphic characters with both hadrosaurid subfamilies, the Hadrosaurinae and Lambeosaurinae, while possessing some characters that seem to have evolved convergently in all three of these taxa, possibly explaining the apparent "mixing" that has intrigued and confused workers over the years.

These characters (in addition to those cited in their emended diagnosis, above) were regarded by the authors as plesiomorphic in *B. johnsoni*: (Distinguishing this species from the early hadrosauroid *Gilmoreosaurus mongoliensis*, also from the Iren Dabasu Formation) maxilla markedly asymmetrical in lateral view; about 20 maxillary and dentary tooth positions in adults; dentary downturned rostrally; (distinguishing this species from the basal hadrosaurid

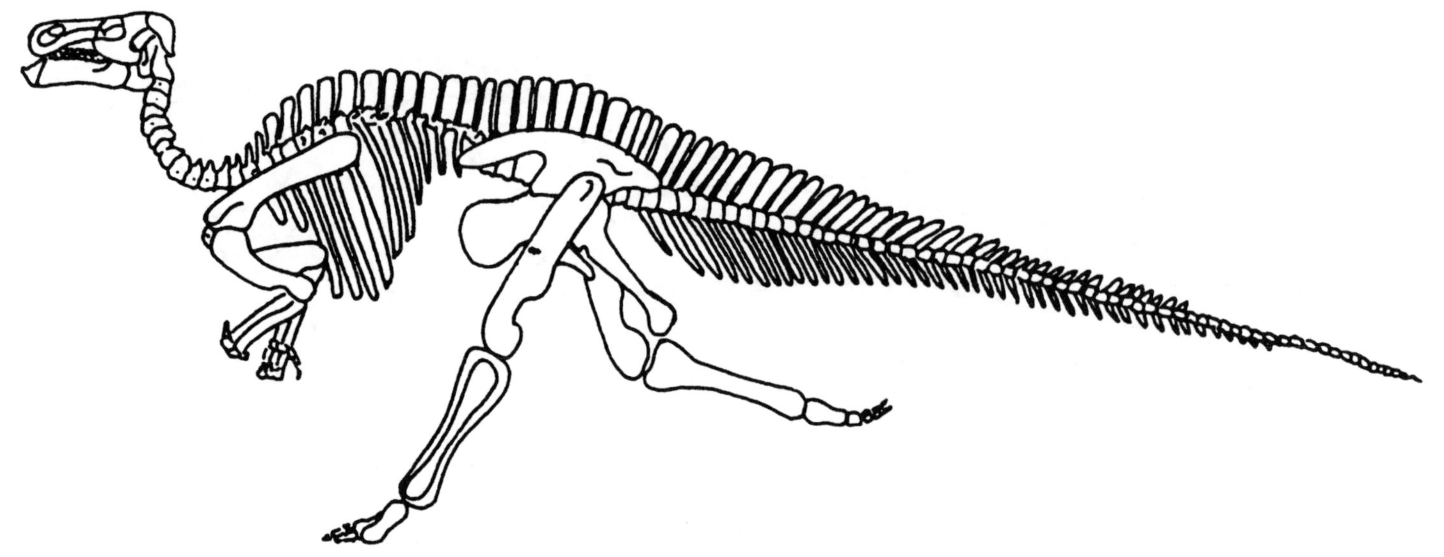

***Bactrosaurus johnsoni*, reconstruction of skeleton in bipedal pose. (After Godefroit, Dong, Bultynck, Hong and Feng 1998, modified after Paul *see in* Brett-Surman 1989.)**

Telmatosaurus transsylvanicus, from the late Maastrichtian of Romania) maxillary/dentary tooth count (see above); laterally positioned antorbital fenestra; paraquadrate foramen developed; and dentary teeth with one caudal secondary ridge.

Therefore, Godefroit *et al.* concluded that *B. johnsoni* is neither a lambeosaurine nor a hadrosaurid, but rather appears to occupy a basal position within a new iguanodontian clade, which they named Hadrosauroidea (see "Systematics" chapter).

Also, Godefroit *et al.* questioned previous datings for the Iren Dabasu Formation, noting that their morphologic and cladistic study of *B. johnsoni* are more compatible with an early Middle Cretaceous (Cenomanian) age for this geological unit. According to these authors, *B. johnsoni* is more primitive than any known typical Campanian–Maastrichtian hadrosaurid, possessing many apparently plesiomorphic features such as those found in Early Cretaceous iguanodontids. A Turonian–early Campanian dating of the Iren Dabasu Formation, therefore, would imply a phylogenetic continuity ("Iguanodontidae—'Probactrosaurs'—'Bactrosaurs'—Hadrosauridae") gap of at least 5 million years, a separation eliminated by a Cenomanian age for this unit. The proposed basal position proposed for *B. johnsoni*, the authors noted, is in accord with this presumed early stratigraphic distribution.

Key references: Brett-Surman (1979, 1989); Gilmore (1933*b*); Godefroit, Dong, Bultynck, Hong and Feng (1998); Maryańska and Osmólska (1981); Steel (1969); Weishampel and Horner (1986).

BAMBIRAPTOR Burnham, Derstler, Currie, Bakker, Zhou and Ostrom 2000

Saurischia: Theropoda: Neotheropoda: Tetanurae: Avetheropoda: Coelurosauria: Manuraptoriformes: Manuraptora: Deinonychosauria: Eumanuraptora: Dromaeosauridae.

Name derivation: "Bambi" [nickname (short for "bambino") for the fossil coined by the Linster family] + Latin *raptor* = "thief."

Type species: *B. feinbergi* Burnham, Derstler, Currie, Bakker, Zhou and Ostrom 2000.

Other species: [None.]

Occurrence: Two Medicine Formation, Montana, United States.

Age: Late Cretaceous (Campanian).

Known material: Skeleton, miscellaneous remains, subadult and adult.

Holotype: FIP 001, nearly complete partially articulated skeleton, subadult.

Diagnosis of genus (as for type species): Jugal having row of foramina along ventral margin; scapula having large, medially directed acromion; scapulocoracoid suture short and distinct; coracoid with neck or peduncle forming part of glenoid; absence of coracoid foramen; 13 dentary teeth, nine maxillary teeth; ratio of humerus plus ulna to femur large (1.68); pubis having distal shaft and booth rotated posterodorsally; ischium having small proximal dorsal process; femur strongly recurved laterally and posteriorly. (Burnham, Derstler, Currie, Bakker, Zhou and Ostrom 2000).

Comments: *Bambiraptor*, a small and very birdlike genus, was founded upon a more than 90 percent complete and partially articulated skeleton (FIP 001) discovered in September 1993 by Wes Linster in a siltstone layer of the Two Medicine Formation, about two-thirds of the way above the unit, near Bynum, Montana. Nicknamed "Bambi," the specimen was in such an excellent state of preservation that the animal's posture could be reconstructed with some confidence.

Bambiraptor feinbergi, FIP 001, holotype skull and lower jaw elements in left lateral view. Scale = 5 cm. (After Burnham, Derstler, Currie, Bakker, Zhou and Ostrom 2000.)

Scattered within the same deposit were remains (FIP 002-036) apparently belonging to other individuals of the same species, most of these pertaining to animals about one third larger than the holotype. These bones suggest that little anatomical change occurred in this taxon during ontogeny. Associated with the holotype were remains of hadrosaurs (presumably *Maiasaura peeblesorum*), at least three species of tyrannosaurids, two type of eggshell fragments, and carbonized scraps of wood. (Burnham, Derstler, Currie, Bakker, Zhou and Ostrom 2000).

Discovery of the holotype was first mentioned in an early report by Burnham, Derstler and Linster (1997), who found it to closely resemble the dromaeosaurid *Saurornitholestes langstoni*, but more conservatively referred the specimen to *Velociraptor* cf. *V. langstoni* (see *Velociraptor* entry, *S1*). Later, this specimen was figured and briefly discussed by Feduccia (1999*b*) utilizing information supplied to him by David A. Burnham. Demonstrating that FIP 001 differs from all other known theropods including *S. langstoni*, Burnham *et al.* (2000) erected for this material the new genus and species *Bambiraptor feinbergi* and offered a preliminary description of the material.

Burnham *et al.* (2000) identified *B. feinbergi* as a member of the Dromaeosauridae based on its possession of such dromaeosaurid synapomorphies as the enlarged, retractable second pedal ungual, the rod-stiffened tail, and the T-shaped lacrimal.

In comparing *B. feinbergi* with various other related theropod taxa, Burnham *et al.* (2000) distinguished this species from the almost contemporaneos *S. langstoni* by its relatively long, anteriorly tapering frontal. The authors noted that, although this feature could be exaggerated because of the immaturity of the type specimen, it is probably not entirely an ontogenetic difference, as the frontals in the type specimens of both *B. feinbergi* and *S. langstoni* are almost the same length, while the orbital rim in the former species is twice the length of that in the latter. The limbs and axial skeleton of *B. feinbergi* resemble those of the dromaeosaurid *Deionychus antirrhopus*, possibly reflecting the quality of information based on these two taxa. The authors noted that *B. feinbergi* shares many features with *Sinornithosaurus milleni*, most notably the *Archaeopteryx*-like furcula and the sternal plates, although *S. milleni* and *B. feinbergi* differ in other ways (*e.g.*, the latter species lacking a pubic boot, having a long obturator process, a skull longer than the femur, and proportionally shorter arms).

The following more obvious birdlike features identified in *B. feinbergi* were briefly discussed by Burnham *et al.* (2000):

1. The furcula (the wishbone, "one of the traditional hallmarks of a bird") in *B. feinbergi* is unusual in having thick, curved arms, a rounded interior angle, and lacking a hypocledium. Only one nonavian theropod, *Sinornithosaurus*, and some early birds (*e.g.*, *Archaeopteryx* and *Confuciusornis*) were known by the authors to possess this type of wishbone, other

Bambiraptor feinbergi, FIP 001, holotype pelvic girdle in left lateral view. Scale = 5 cm. (After Burnham, Derstler, Currie, Bakker, Zhou and Ostrom 2000.)

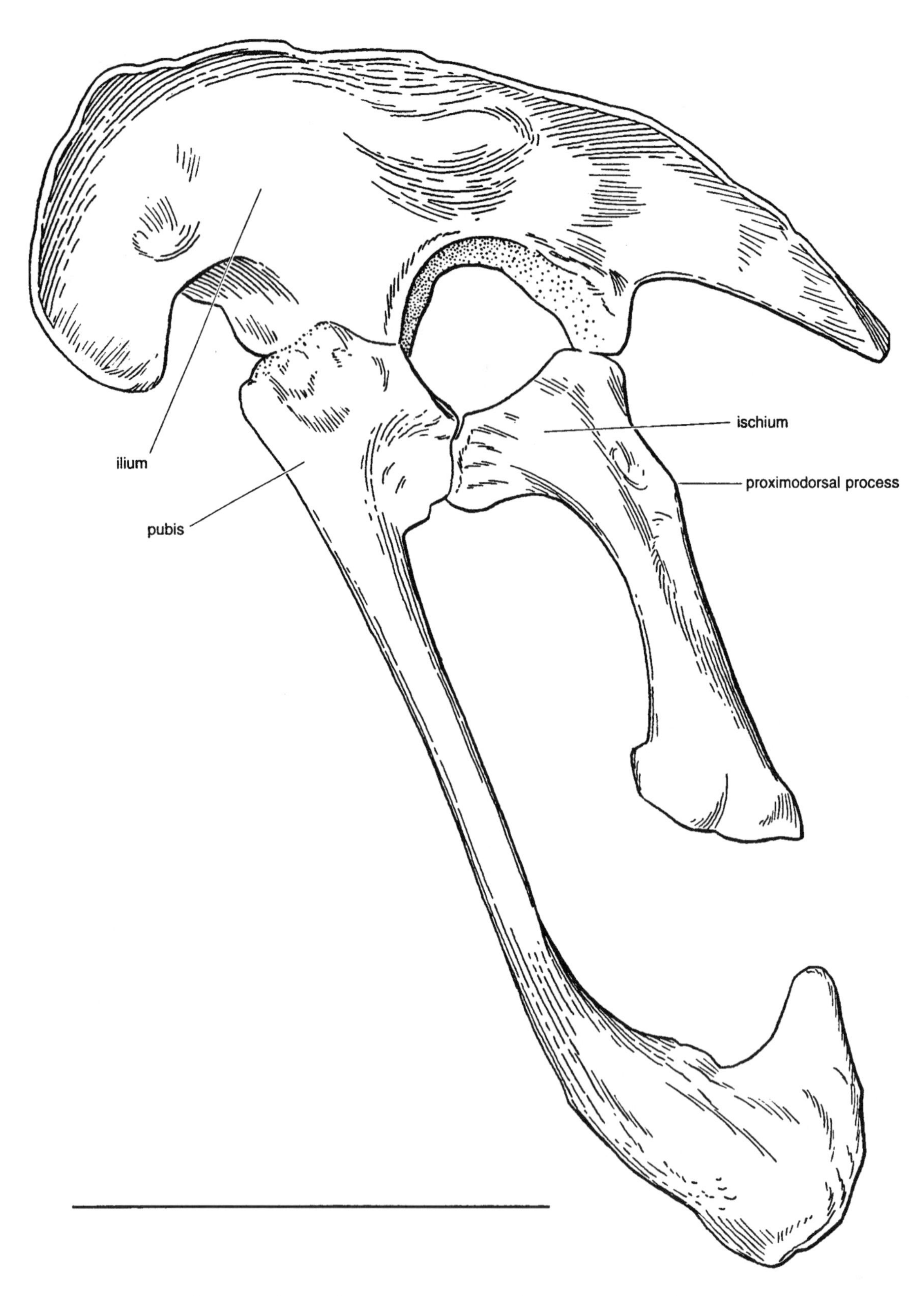

theropods having a furcula with straight arms and a less rounded interior angle, most birds having more slender arms and a large platelike hypocleidium (see also Burnham and Zho).

2. A large ossified sternum, produced by two ossifications, is found in birds (in which these ossifications are fused) and *B. feinbergi* (not fused). Similar ossified plates are also being found in the manuraptorans *Velociraptor*, *Sinornithosaurus*, and *Oviraptor*.

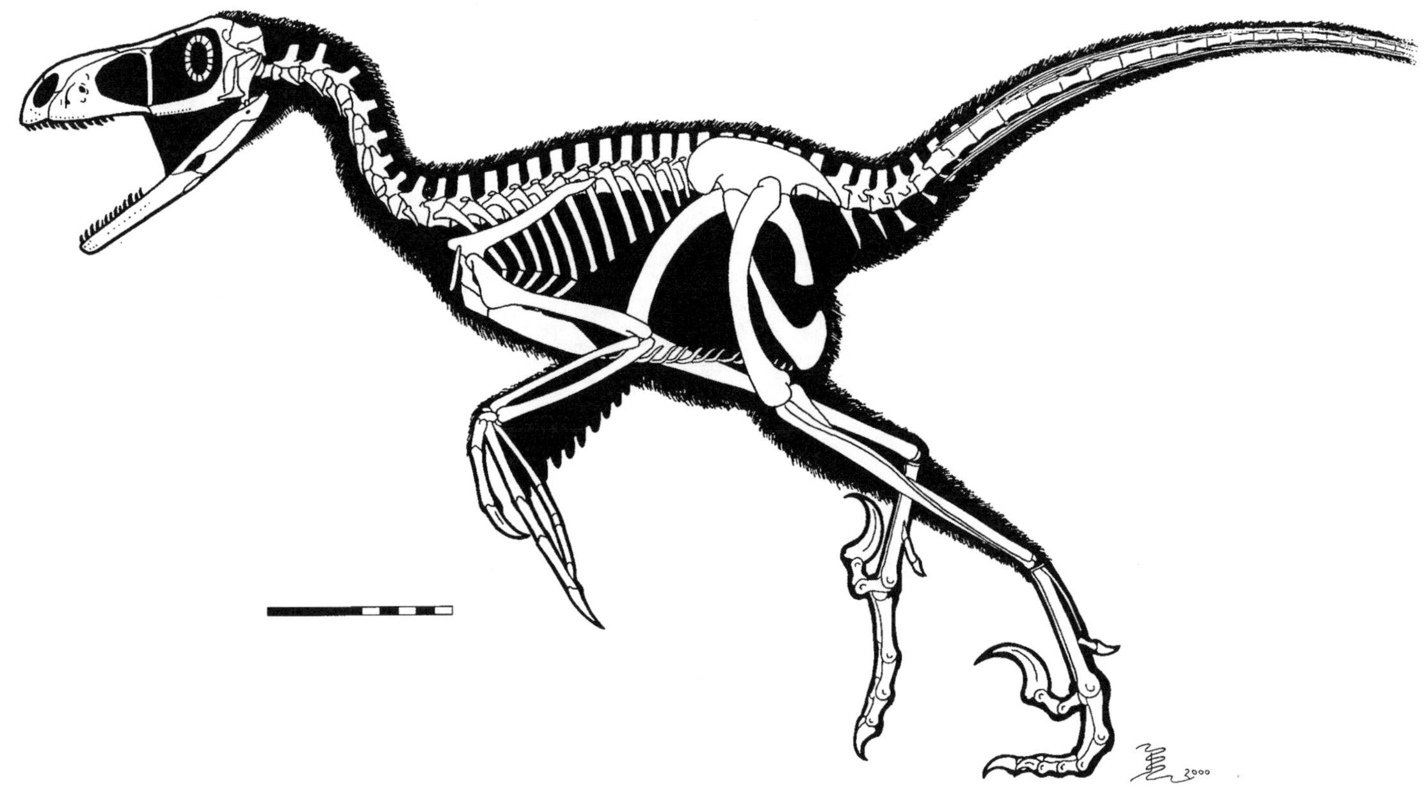

Bambiraptor feinbergi, FIP 001, holotype skeleton reconstruction with hypothetical feather-like integument. (Scale = 10 cm.) (After Burnham, Derstler, Currie, Bakker, Zhou and Ostrom 2000.)

3. The glenoid in *B. feinbergi* faces more laterally than in any other known theropod, a laterally oriented glenoid seemingly constituting a preadaptation for avian flight (see Novas and Puerta 1997).

4. An elongate coronoid process, perhaps the most notable feature of *B. feinbergi*, is similar to the condition in most modern birds.

5. A prominent free acromion is present in *B. feinbergi*, as in the primitive birds *Unenlagia* and bird *Rahonavis*, and modern birds. In birds this condition contributes to the construction of the triosseal canal.

6. A biceps tubercle is large in *B. feinbergi*, many small theropods (*e.g.*, *Gallimimus bullatus*), and in birds (here termed an acrocoracoid, forming part of the triosseal canal; see Ostrom 1976*a*).

7. The elongated arms and manus in *B. feinbergi* are proportionally longer than in any other known nonavian theropod, approaching the proportions in *Archaeopteryx* and other early birds.

8. A large semilunate carpal, allowing for the development of a laterally folding wrist, is found in *B. feinbergi*, most manuraptoran theropods, birds, and *Allosaurus* (see Chure 1999).

9. A proximodorsal process on the ischium is found in *Bambiraptor*, *Unenlagia*, and *Rahonavis*, also in early birds like *Archaeopteryx*, *Confuciusornis*, and *Cathayornis*.

10. A retroverted pubis, a condition seen in *Bambiraptor* and all other dromaeosaurs, is also observed in all modern birds.

11. A splintlike fibula is found in dromaeosaurids, including *Bambiraptor*, and birds.

12. Pneumatic cervical and dorsal vertebrae, associated with the kind of respiratory system found in birds, are observed in *B. feinbergi*, similar to those in droaeosaurids and birds.

Gartska, Marsic, Carroll, Heffelfinger, Lyson and Ng (2000), in a preliminary report, observed that *B. feinbergi* exhibits specialized structures of the breast, shoulder, and forelimb that are consistent with avian structures. These structures in *B. feinbergi* suggested to Gartska *et al.* that this dinosaur exhibited a mode of predation whereby both forelimb and hindlimb were utilized in a coordinated, opposing manner.

In Gartska *et al.*'s model of the manuraptoran forelimb, the manus rotates "on its semilunar carpals against a rigid curved surface presented by the distal radius and ulna." The digits of the manus are parallel, arranged in an avian fashion with the second digit (the alular digit in birds) most distant from the radius and ulna with the manus in a relaxed, folded position. The authors theorized that, during extension of the forelimb, the claws of the manus "would have swung upward through a rising arc, placing their cutting action in an opposing direction to the cutting of the downwardly curved hindlimb sickle-claws." The

***Bambiraptor feinbergi*, restored with feather-like integument by Mike Fredericks.**

presence of the furcula cushioned the shoulders of *Bambiraptor* for rapid forelimb extension, although limiting mediolateral movement of the coracoids along the sternal plates. As Gartska *et al.* concluded, "This model places the necessary structural and functional components of powered flight in another adaptive context; that of predatory behavior."

Although direct evidence of feathers or similar structures was not found with the specimen, Derstler and Burnham (2000) stated that "Superimposing available integument information on our cladogram indicates that *Bambiraptor feinbergi* may have had both dinofuzz and vaned feathers."

According to Derstler and Burhnam, cladistic analysis of this species, several early birds, and other theropods suggest it to be the sister taxon of *Archaeopteryx* and subsequent birds, thereby distinguishing *Bambiraptor* as the known dinosaur closest to birds. This analysis also implies that the dromaeosaurid gestault (*e.g.*, "killer claw," T-shaped lacrimal, rod-stiffened tail) may not represent synapomorphies of the Dromaeosauridae, but that these features are rather shared with and were further modified in early birds.

The holotype skeleton of *B. feinbergi* was unveiled at The Florida Symposium on Dinosaur Bird Evolution in April, 2000, hosted by the Graves Museum of Archaeology and Natural History. It was subsequently displayed at the Florida Institute of Paleontology; casts were exhibited at the University of Kansas Natural History Museum and Biodiversity Research Center (KUVP 129737) and the Science Museum of Minnesota (SMM P99.3.1c).

Key references: Burnham, Derstler, Currie, Bakker, Zhou and Ostrom (2000); Burnham, Derstler and Linster (1997); Burnham and Zhou (1999); Chure (1999); Derstler and Burhnam (2000); Feduccia (1999*b*); Garstka, Marsic, Carroll, Heffelfinger, Lyson and Ng (2000); Novas and Puerta (1997); Ostrom (1976*a*).

†BARYONYX

Saurischia: Theropoda: Neotheropoda: Tetanurae: Avetheropoda: Spinosauroidea: Spinosauridae: Baryonychinae.

Diagnosis (as for type species): Spinosaurid characterized by fused nasals having median crest terminating posteriorly in cruciate process, solid subrectangular lacrimal horn, marked transverse constriction of sacral or anterior caudal centra, well-formed pen-and-notch articulation between scapula and coracoid, everted distal margin of pubic blade, and very shallow fibular fossa (Sereno, Beck, Dutheil, Gado, Larsson, Lyon, Marcot, Rauhut, Sadleir, Sidor, Varricchio, Wilson and Wilson 1998).

Comments: Information continues to surface regarding *Baryonyx*, an unusual spinosaurid theropod from England noted for (among other features) its very large manual claw.

In naming and describing another spinosaurid genus, *Suchomimus* (see entry), Sereno, Beck, Dutheil, Gado, Larsson, Lyon, Marcot, Rauhut, Sadleir, Sidor, Varricchio, Wilson and Wilson (1998) offered alternative identifications for several cranial elements in the holotype (BMNH R9951) of the British spinosaurid, *Baryonyx walkeri* (see *D:TE*). Sereno *et al.* interpreted bones originally identified in this specimen as the left postorbital, left jugal, right atlantal neural arch, and left angular as, respectively, the posterior portion of the right surangular, the right prearticular, the central body of the left pterygoid, and the right angular (see *Suchomimus* entry for skull reconstruction by Sereno *et al.* incorporating these reidentified elements). A plate-shaped element previously identified as a vomer was reidentified by Sereno *et al.* as the anteromedial process of the maxillae; the supposed deeper proportions of the occiput was regarded as an artifact of unnatural displacement of the quadrate.

Consequently, the cranium of *Baryonyx* was interpreted to be long, low, and narrow, similar to that of *Suchomimus*.

Key reference: Sereno, Beck, Dutheil, Gado, Larsson, Lyon, Marcot, Rauhut, Sadleir, Sidor, Varricchio, Wilson and Wilson (1998).

BEIPIAOSAURUS Xu, Tang and Wang 1999

Saurischia: Theropoda: Neotheropoda: Tetanurae: Avetheropoda: Coelurosauria: Manuraptoriformes: Therizinosauroidea.

Name derivation: "Beipiao [city near locality where type specimen was found]" + Greek *sauros* = "lizard."

Type species: *B. inexpectus* Xu, Tang and Wang 1999.

Other species: [None.]

Occurrence: Yixian Formation, Liaoning, China.

Age: ?Late Jurassic (Kimmeridgian) or ?Early Cretaceous (Valanginian).

Known material/holotype: IVPP V11559, incomplete partially articulated skeleton, feather-like structures.

Diagnosis of genus (as for type species): Differing from other therizinosauroids in having teeth with shorter, more bulbous crowns, larger skull, tridactyl pes with splintlike proximal first metatarsal, shallow anterior iliac process, long manus (10 percent longer than femur), long tibia (275 millimeters greater than 265 millimeters of femur), elongated lateral articular surface on palmar side of manual phalanx I-1, and proximally compressed metatarsals III and IV (Xu, Tang and Wang 1999).

Comments: Distinguished as the first therizinosauroid genus known to have possessed feather-like structures, *Beipiaosaurus* was founded upon a partial skeleton (IVPP V11559) found and collected in 1996 by farmer Li Yinxian from the Sihetun locality—which previously yielded fine specimens of *Sinosauropteryx*, *Caudipteryx*, *Protarchaeopteryx*, and Early Cretaceous birds (see *S1*)—in the lower part of the Yixian Formation, near Beipiao, in Liaoning, China. (The lower part of the Yixian Formation was dated as late Early Cretaceous by Swisher, Wang, Wang, Xu and Wang 1998, based on the evidence of radiometric dating [see also Wang, Wang, Wang and Xu 2000].) Jiang, Chen, Cao and Komatsu (2000), based upon invertebrate faunas in this and other localities, suggested a Late Jurassic [Kimmeridgian] to Early Cretaceous [Valanginian] age. See also Manabe, Barrett and Isaji (2000) for comments regarding the interpretation of this formation as a refugium, whereby some typically Late Jurassic lineages survived into the Early Cretaceous.)

The holotype, believed to belong to a single individual, includes a partial right dentary with teeth, a right postorbital, right parietal, right ?nasal, right prootic, several cervical and dorsal vertebrae, an incomplete caudal vertebra, incomplete ribs, a partial scapula, coracoids, furcula, partial humerus, radius, and ulna, almost complete manus, partial ilium, pubis, and ischium, complete right femur, right tibia, and right fibula, incomplete left femur, tibia, and fibula, and incomplete right pes, some elements represented by impressions, and feather-like integumentary structures (Xu, Tang and Wang 1999).

According to Xu, Tang and Wang, the type species *Beipiaosaurus inexpectus* represents the largest theropod yet discovered in the Yixian Formation, having an estimated length of 2.2 meters (about 6.75 feet), its comparatively large skull possessing a jaw measuring 65 percent of the femoral length.

In performing a phyogenetic analysis, Xu *et al.* found this new genus to be a basal taxon within Therizinosauroidea, displaying such primitive features as a relatively large head, tridactyl foot, and a tibia with a fibular crest. Pelvic elements quite similar to those

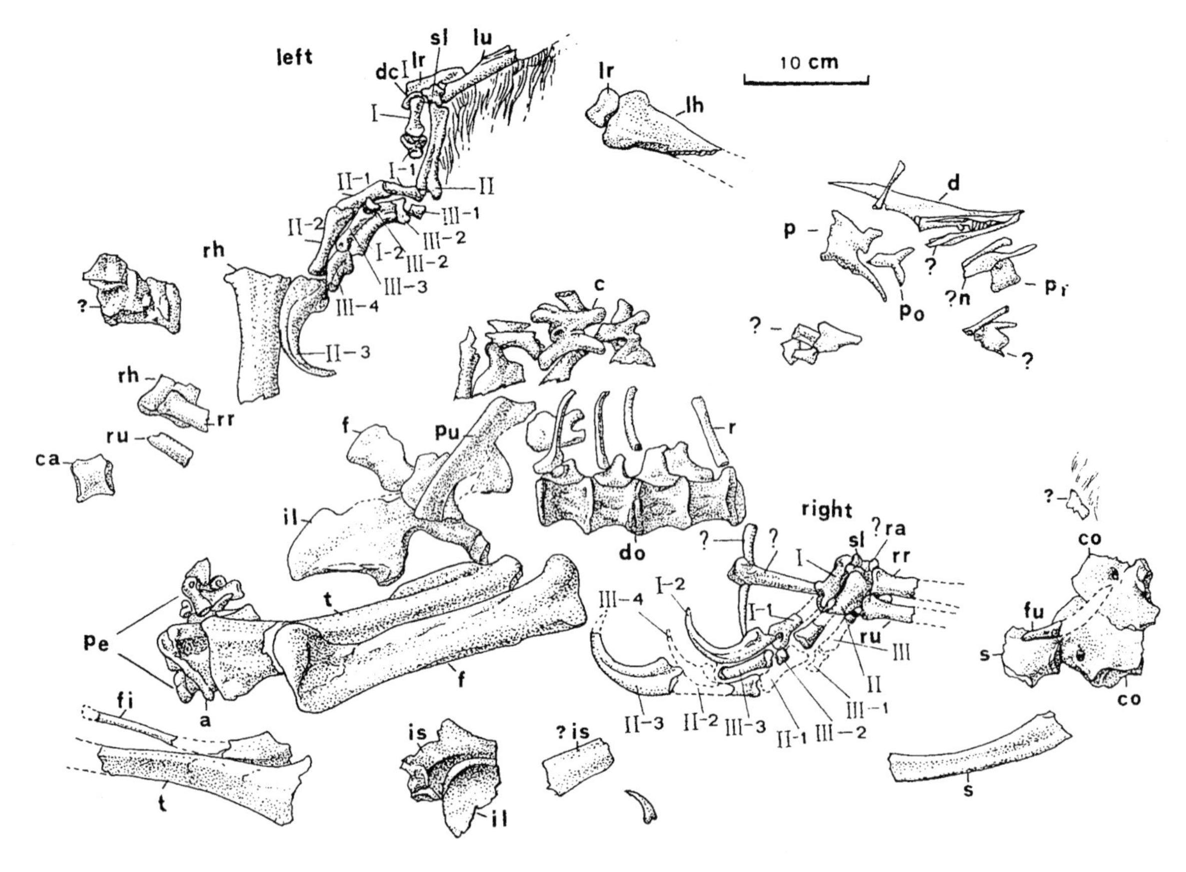

***Beipaiosaurus neglectus*, IVPP V11559, holotype partial skeleton with feather-like integumentary structures. (After Xu, Tang and Wang 1999.)**

seen in other coelurosaurian theropods (*e.g.*, ilium parallelogram-shaped as in dromaeosaurids and basal birds; posteroventral margin of ilium deflected laterally at right angle to vertical ramus, with shallow brevis fossa, as in coelurosaurians; pubic peduncle longer than ischiadic peduncle, similar to those of dromaeosaurids and *Archaeopteryx*) indicate that the Therizinosauroidea belongs in the Coelurosauria. Various derived characters of therizinosauroids (excluding *Beipiaosaurus*) were most parsimoniously interpreted as having evolved convergently with some other dinosaurian groups, particularly sauropodomorphs, these unusual theropods having "re-evolved a robust first digit in which the proximal end of metatarsal I articulates with the tarsals."

Preserved with the type specimen were large patches of filamentous integumentary structures found in close association with the ulna, radius, femur, tibia, and pectoral elements. As described by Xu *et al.*, these structures resemble the so-called "protofeathers" seen in the compsognathid *Sinosauropteryx* in their parallel arrangement, differing in their contact with the ulna and in their greater length. The authors rejected the possibility that these structures are preserved muscle fibres or frayed collagen. The integumentary structures are most dense close to the bone. Most of them measure approximately 50 millimeters in length, the longest being up to 70 millimeters. Some of these structures possess shallow and faint median grooves, perhaps indicating collapsed hollow cores, with indications of branching distal ends (as in *Sinosauropteryx*).

The presence of such structures in both compsognathids and therizinosauroids indicated to the authors that there could be a broader distribution of similar structures among various theropod groups, thus supporting the hypothesis "that these simple integumentary filaments may represent an intermediate evolutionary stage to the more complex feathers of *Protarchaeopteryx*, *Caudipteryx* and more derived Avialae." The absence of such structures in most theropod fossils could be explained by the lack of such excellent preservation as is found in the Yixian Formation. According to Xu *et al.*, this suggests "that feathers preceeded flight, because both therizinosauroids and compsognathids apparently could not fly and did not descend from flying animals."

Key references: Jiang, Chen, Cao and Komatsu (2000); Swisher, Wang, Wang, Xu and Wang (1998); Xu, Tang and Wang (1999); Wang, Wang, Wang and Xu (2000).

†BLIKANASAURUS

Age: Late Triassic (Carnian or ?early Carnian).

Diagnosis of genus (as for type species): Tibia

having maximum proximal width 48 percent of maximum length, long axes of proximal and distal ends at 60 degrees, distally posterior process relatively deep, robust; distal end of fibula having a large, ventromedially directed articular surface; astragalus with relatively high, prominent ascending process; calcaneum small; metatarsal I short but very robust; metatarsals II–IV subequal in length, III being the longest; ratio of maximum proximal width to maximum length for metatarsals I–IV is 0.91, 0.72, 0.53, 0.53, and 0.74, respectively (Galton and Heerden 1998).

Comments: In 1985, Peter M. Galton and Jacques Van Heerden named and briefly described *Blikanasaurus cromptoni*, an early (Late Triassic), heavily-built, quadrupedal prosauropod which they placed in its own monotypic family, Blikanasauridae (see *D:TE*). More than a decade later in a well-illustrated paper, Galton and Heerden (1998) fully described and evaluated the holotype (SAM K403, a left lower hindlimb including tibia, fibula, a tarsus, and a pes, and missing two or three phalanges), and thus far the only known specimen of *B. cromptoni*, the authors at the same time slightly emending their original diagnosis of the type species.

As chronicled in detail by Galton and Heerden (1998), the type specimen was found in January 1962 by Dr. Christopher E. Gow (a member of the British-South African Museum Expedition to the Herschel area and Lesotho). The discovery was made in the so-called "Passage Beds" at the base of the lower Elliot Formation, Stormberg Group, Karoo Supergroup (Upper Triassic, Carnian, possibly lower Carnian; redated by Gauffre 1993*b*, see below), at a site northeast of the trading store at Blikana, in the Herschel district of the Eastern Cape (from a region once called Transkei), South Africa. Following preparation by immersion in 12 percent acetic acid, the specimen was tentatively (and incorrectly) referred by A. W. Crompton and M. L. Wapenaar (in an unpublished description) to the prosauropod *Melanorosaurus*, after which various other authors informally referred to it as the "Blikana dinosaur" or "Blikana melanorosaurid" (*e.g.,* Charig, Attridge and Crompton 1965, who briefly described but did not figure the specimen).

Galton and Heerden (1998) noted that, in past years, various authors (*e.g.,* Walker 1969, Benton 1993) have claimed that prosauropod specimens have been found predating a Carnian age, particularly from the Middle Triassic of England. As later stated by Galton and Walker (1996), however, these specimens are not prosauropod, nor even dinosaurian, but pertain to other kinds of reptiles.

Gauffre (1993*b*) redated the lower Elliot Formation as of Carnian or ?early Carnian (rather than Carnian/Norian, as proposed by Galton 1990, or Norian, as suggested by Benton 1993) age based on the presence of traversodontid cynodonts and rauisuchid "thecodontians," considered to be of Carnian age by Shubin and Sues (1991). Therefore, Galton and Heerden (1998) noted that the fauna of South Africa's lower Elliot Formation includes the earliest prosauropods known from articulated specimens and also the first radiation of the Prosauropoda, the latter represented by three families and the earliest genus recorded for each family — the Plateosauridae (*Euskelosaurus*), the Melanorosauridae (*Melanorosaurus*), and the Blikanasauridae (*Blikanasaurus*). Probably these genera also represent the first record of the Prosauropoda as a group.

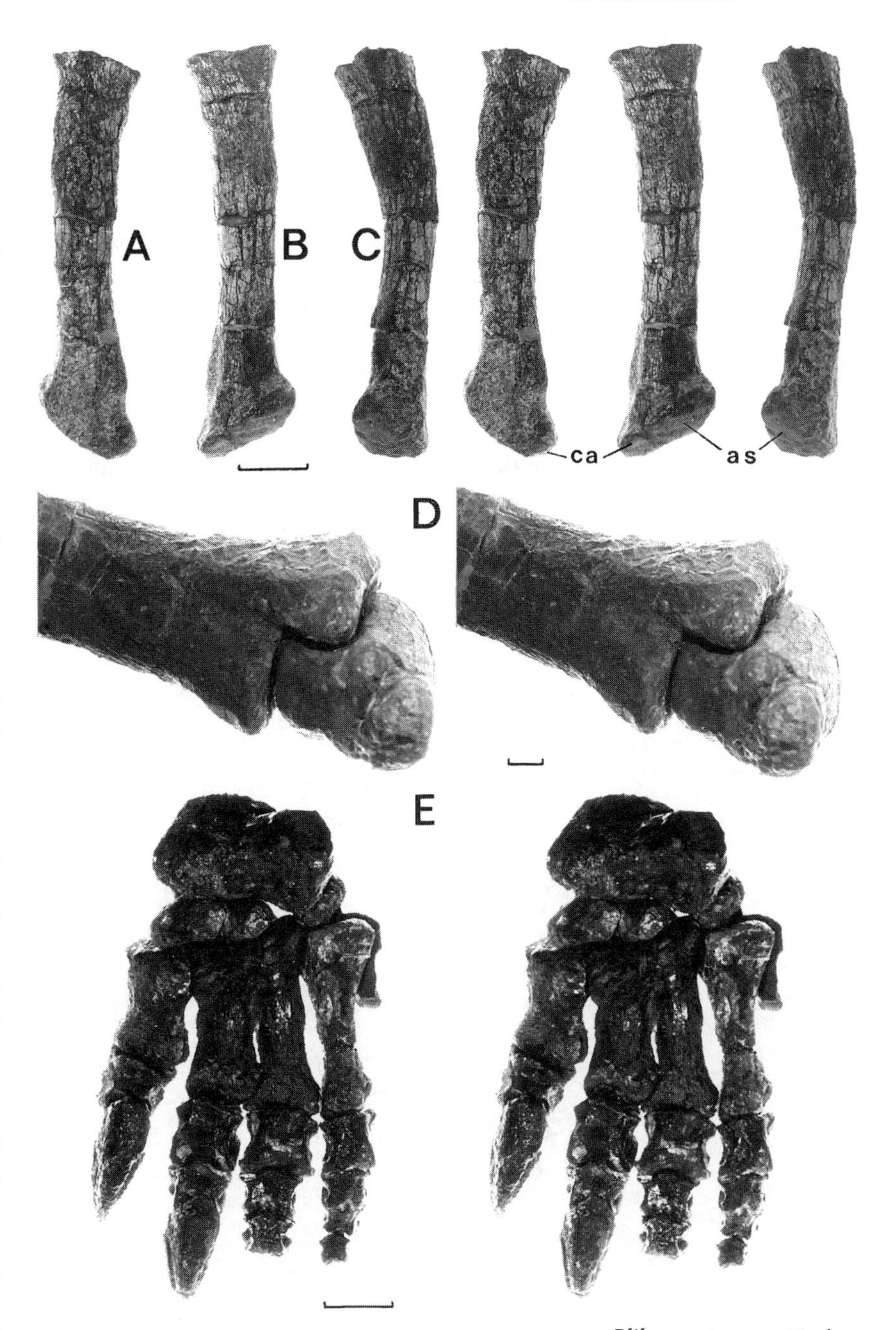

***Blikanasaurus cromptoni*, SAM K403, stereo photos of holotype left fibula (lacking proximal part) in A. lateral, B. medial, and C. anterior views; D. distal end of left tibia, astragalus, and calcaneum, lateral view; and E. left proximal tarsals and pes, anterior view. Scale (A–C, E) = 5 cm. and (D) 1 cm. (After Galton and Heerden 1998.)**

Key references: Benton (1993); Charig, Attridge and Crompton (1996); Galton (1990); Galton and Heerden (1985, 1998); Galton and Walker (1996); Walker (1969).

†**BRACHIOSAURUS**—(=*Giraffatitan*; =?*Oplosaurus*, ?*Pelorosaurus*)

Comments: Among the largest and most recognizable of sauropods is the gigantic *Brachiosaurus*.

In 1988, Gregory S. Paul separated the African species of *Brachiosaurus*, *B. brancai*, named by Werner Janensch (1914*a*), represented by specimens recovered from the Tendaguru Beds (Late Jurassic: Kimmeridgian–Tithonian) of Tanzania, and the North American species, *B. altithorax* (Riggs 1903), from the Morrison Formation of Colorado and Utah. Paul's division of these taxa was based on differences in the vertebral columns which he interpreted as significant. For *B. brancai*, Paul introduced the new subgenus name of *Giraffatitan*, the resulting combination thereby becoming *Brachiosaurus* (*Giraffatitan*) *brancai* (see *D:TE* for details).

Olshevsky (1991), in his list of dinosaurian taxa, later raised Paul's new taxon to the full status of genus, and this subsequent combination became *Giraffatitan brancai*. During the following years, neither Paul's nor Olshevsky's proposed renamings have been adopted.

In a subsequent list published in 2000, Olshevsky retained the generic name of *Giraffatitan* for the African species, although sauropod specialist John S. McIntosh (personal communication to Olshevsky) "does not consider the differences generically significant." *Giraffatitan*, therefore, is regarded in this document as a junior synonym of *Brachiosaurus*. Olshevsky added that a skull from the Morrison Formation of North America, identified as a brachiosaurid (K. Carpenter, personal communication to Olshevsky), "differs significantly in morphology from that attributed to *Giraffatitan brancai* and, if it indeed is a *Brachiosaurus* skull, further, supports separating the African genus from the North American." As of this writing, this skull is on exhibit at the Denver Museum of Natural History.

A series of technical articles discussing various topics relating to the Tendaguru sauropods—*Brachiosaurus*, *Barosaurus*, and *Dicraeosaurus*, but primarily *Brachiosaurus*, appeared in a single journal published in 1999.

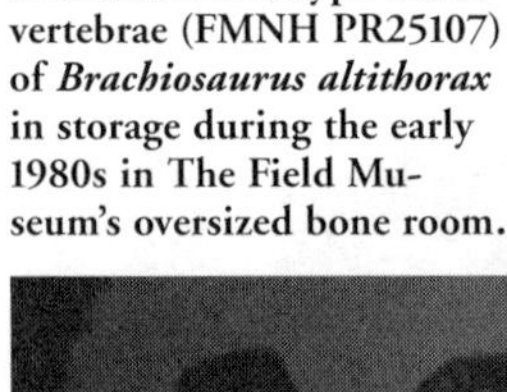

Articulated holotype dorsal vertebrae (FMNH PR25107) of *Brachiosaurus altithorax* in storage during the early 1980s in The Field Museum's oversized bone room.

Photograph by the author, courtesy The Field Museum.

Photograph by the author, courtesy The Field Museum.

Skeleton (cast) of *Brachiosaurus altithorax* (including cast of holotype FMNH PR25107) mounted by the PAST company, based somewhat on the more completely known species *B. brancai*, displayed outside The Field Museum.

Christian, Müller, Christian and Preuschoft (1999) presented their study on reconstructing the limb movements and speed estimates in large dinosaurs — including *Brachiosaurus* and other Tendaguru taxa — based on data from fossil trackways, but more importantly from detailed analyses of limb movements in walking Asian elephants and giraffes. Speed estimates based on stride frequencies from trackways were found to be less reliable for such dinosaurs and other terrestrial vertebrates than those based upon relations among stride length, hip height, and speed (see Alexander 1976), or those of maximum speed based on calculations of bone length (Alexander 1978, 1996; Farlow, Smith and Robinson 1995). Christian *et al.* concluded that sauropods such as *Brachiosaurus*, and most likely other large dinosaurs, usually walked at rather low speeds comparable to the low walking speeds of humans, although under certain circumstances these dinosaurs may have been capable of walking fast, economically using a mechanism similar to that of large terrestrial mammals, shifting the energy from trunk to limbs by accelerating the hip and shoulder joints.

The posture of the limbs of *Brachiosaurus* has been a topic of dispute for decades. Originally, Janensch (1950*c*) reconstructed the skeleton of *B. brancai* with the forelimbs in a semiextended orientation. This interpretation was subsequently perpetuated in the literature and also in popular books in no small way, via an often reproduced painting (based on the composite skeleton, the largest mounted dinosaur skeleton in the world, in the Museum für Naturkunde der Humboldt-Universität zu Berlin; see *D:TE* for photograph) by artist Zdeněk Burian which first appeared in the book *Prehistoric Animals*, by Josef Augusta (1960; see also *D:TE*). More recent skeletal reconstructions and life restorations, however, have generally depicted the forelimbs of this giant in a fully erect position (*e.g.*, Charig 1979; Alexander 1985, 1989; Bakker 1986; Paul 1987, 1988*a*; Fastovsky and Weishampel 1996).

Christian, Heinrich and Golder (1999) utilized biomechanical models to analyze the posture and mechanics of the forelimbs of *B. brancai*. Christain *et al.* stated that an important argument favoring the extended posture hypothesis is the very high body weight in dinosaurs such as *Brachiosaurus*. "In geometrically similar shaped animals of different size,

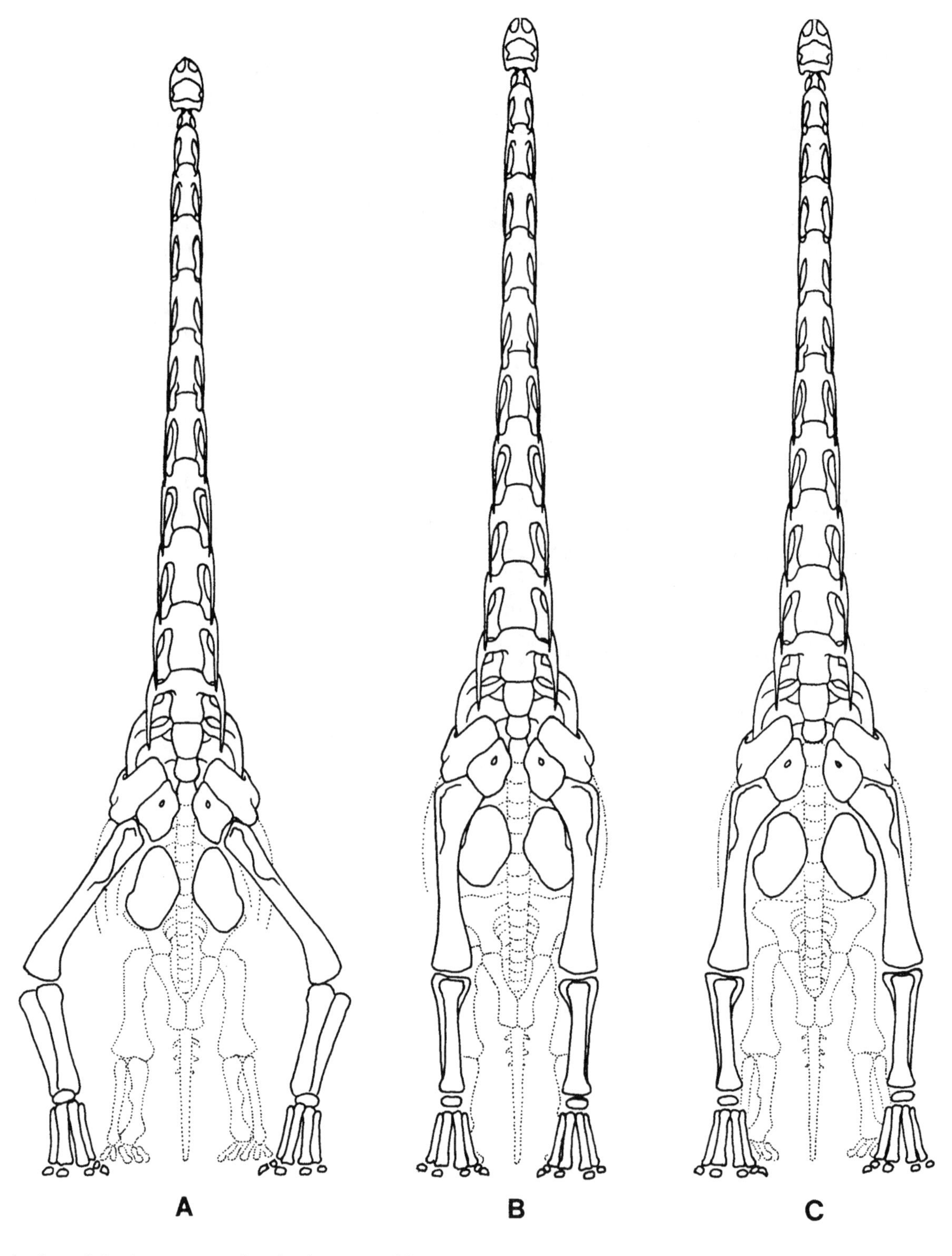

Skeleton of ***Brachiosaurus brancai*** **reconstructed with A. a semi-extended forelimb posture according to Janensch (1950*c*), B. an extended forelimb posture with the bones in line, and C. an extended forelimb posture with the elbow joints slightly flexed. Andreas Christian, Wolf-Dieter Heinrich and Werner Golder (1999) found the latter two reconstructions to seem reasonable. (After Christian, Heinrich and Golder 1999).**

body weight increases to the third power of linear dimensions (*e.g.*, body length) whereas maximum muscle forces and strength of bones, tendons and ligaments increase only with the square of linear dimensions." Thus, larger animals are generally less forceful with respect to body weight than small animals. This implies that vertebrates of different size may also have a different limb posture. The limbs of larger vertebrates tend to be more extended, thereby reducing the lever arms of external forces (*e.g.*, McMahon and Bonner 1983; see Christian *et al.* 1999 for additional references). Also, narrow sauropod fossil trackways show that limb movements of such dinosaurs were restricted to parasagittal planes (see Thulborn 1990; Lockley 1991; Lockley and Hunt 1995).

Body dimensions were derived by Christian *et*

al. (1999) from the mounted skeleton in Berlin. Estimates of the fraction of body mass borne by the shoulder joints were based on estimates of the mass distribution in the body of this dinosaur (total volume of the body of the Berlin skeleton estimated at 74.4 square meters) published earlier by Gunga *et al.* (1995). Christian *et al.* (1999) corrected various overestimations by Gunga *et al.* (1995), taking into account the proposed sizes of the lungs and pneumatic cavities, and also the dimensions of the commercial model of *Brachiosaurus* produced by The Natural History Museum (London). The neck was oriented vertically, as proposed earlier by Christian and Heinrich.

The resultant model indicated to Christian *et al.* (1999) a total body mass of about 63,000 kilograms for the Berlin *Brachiosaurus* skeleton, with approximately 20,000 kilograms of that total carried on the shoulder joints. Assuming the lungs to have been carried in the frontal part of the trunk, the authors found the center of gravity of the rest of the body to be located nearer the hip joints than shoulder joints. Consequently, considerably less than half the total body weight was carried on the shoulder joints.

Christain *et al.* (1999) cited a study by Golder and Christian (1999) in which CT-analyses of the humeri of several Tendaguru sauropods revealed "that in all of these dinosaurs forces were transmitted mainly through the broadened middle fractions of the proximal epiphyses of the humeri." Also, the dimensions of the long bones of the forelimb (Christiansen 1997) and thick walls of the shaft in these bones "indicate an approximately axial loading of the long forelimb bones." These results, Christain *et al.* (1999) noted, are in accordance with the hypothesis of fully erect forelimbs, the humerus in line with the radius and ulna.

Regarding a walking *Brachiosaurus*, however, Christian *et al.* cited evidence offered by Christainsen indicating very restricted movements in the forelimbs and very little displacements in the pectoral girdle under load. The authors pointed out that Christainsen's concept is critical as related to peak forces occurring in the locomotor apparatus, especially in the cartilage of the joints. Christian *et al.* (1999) suggested that, to reduce peak forces when walking forward at a constant, relatively rapid speed, either the forelimbs had to be flexed to some degree at the elbows during the middle of the support phase, or the supposedly rigid pectoral girdle must have indeed allowed some mobility, so that vertical shifts of the shoulder joint relative to the trunk were possible, thereby reducing peak forces acting on the limbs.

According to Christain *et al.* (1999), the locomotor capability of *Brachiosaurus* was quite restricted, its overall construction being related mainly to the extreme task of high browsing. The maximum forward-walking speed of this dinosaur, as estimated for the Berlin mounted skeleton, is 4.5 meters per second or 16.2 kilometers per hour (see Alexander 1976, 1989; Thulborn; Christain, Müller, Christain and Preuschoft 1999). Tasks other than standing and walking forward at constant speed—*e.g.*, reproduction, defense, and feeding—would have been strenuous, with sharp changes of direction being impossible for the entire body and also for such large body parts as the neck and limbs. Furthermore, all involving "considerable fractions of the total body mass must have been slow and smooth, giving the moving animal a somewhat fluid appearance."

However, the authors noted, some mobility in both pairs of limbs were necessary for *Brachiosaurus* to perform such basic tasks as turning to the side or accelerating. Indeed, forward and backward accelerations (see Christainsen), as well as torques required to turn the body to the side, must have been mostly produced by the hindlimbs. Because its body mass was concentrated nearer the center of gravity, resulting in a reduced moment of inertia about the vertical axis through the center of gravity, *Brachiosaurus* could probably turn more easily than could sauropods having a more horizontal neck orientation and very long tail (*e.g.*, *Diplodocus*). Christian *et al.* (1999) envisioned *Brachiosaurus* with one hindlimb pushing backward and the other forward, thereby creating the torque required to turn the massive body to the side, possibly aided by the sideways movement of the relatively short tail, with the forelimbs possibly contributing some additional sideways forces.

Christain *et al.* (1999) calculated that at least 8,400 kilograms of muscle mass was necessary to lift the *Brachiosaurus* body from a lying to standing position, this being equivalent to a human rising from a prone to standing position bearing a backpack of 150 kilograms mass (about 340 pounds). The authors noted that if *Brachiosaurus* ever did lie down, rising must have been an incredible feat. Christain *et al.* (1999) speculated that, if this sauropod accomplished this, it may have employed a special technique to reduce peak forces in rising the hindlimbs possibly flexing first during lying down, the forelimbs then extending, thereby reducing the load on the hindlimbs, and the tail finally serving as a fifth limb.

Given their findings, Christain *et al.* found it doubtful that *Brachiosaurus* could rear up on its hind legs.

Noting that sauropods "present exceptional challenges in understanding their biology because of their exceptionally large body size, Sander (1999, 2000) conducted paleohistological studies of growth strategies based on the long bones of four Tendaguru sauropod taxa, *Barosaurus*, *Brachiosaurus*, *Janenschia*, and *Dicraeosaurus*. Thin sections of bone were cut from small cores drilled into a growth series of humeri and femora at midshaft of these genera. Also incorporated into this study were growth records of single bones.

Regarding the individual Tendaguru sauropod genera, Sander's (1999, 2000) studies concluded the following:

1. In *Barosaurus*, a growth series beginning with the smallest humerus (43 centimeters in length) to the largest adult humerus (99 centimeters) documents a uniform strategy of continuous growth rates — the highest following hatching, the rate decreasing in juveniles, and then plateauing from midontogeny until death (the latter apparently reflecting sexual maturity, correlated with a humeral length of from 50–60 centimeters). Also, *Barosaurus* displays two distinctive types of bone histology — the more than twice as common type A specimens, characterized by a cortex of fibrolamellar bone that is little affected, even in large individuals, by Haversian remodelling, and type B specimens, which show numerous regularly spaced LAGS ("lines of arrested growth"). These different patterns of bone growth may indicate sexual dimorphism, the fast growth in type A possibly representing the male because of competition for mates, type B perhaps representing females which, not facing this pressure, could have afforded slower, uninterrupted growth.

2. In *Brachiosaurus*, the growth pattern is similar to that of *Barosaurus*, except growth was apparently interrupted periodically following sexual maturity (indicated by red lines in the bone sections).

3. In *Dicraeosaurus* (known from fewer specimens, all of them representing adults, and a smaller range in size than the above taxa), the growth was apparently cyclical. Sexual maturity in this genus seems to have been reached when the femur attained a length of approximately 80 percent maximum size.

4. In *Janenschia* (see entry), the rarest of the Tendaguru sauropods (only one specimen, a large femur, being available to Sander for study), growth seems to have been relatively rapid and uninterrupted at first, becoming cyclical later during life.

Sander (1999) speculated that the greatly different growth strategies could be linked to different adult body sizes in these taxa, Haversian being most important in the largest genus, *Brachiosaurus*, and least in the smallest, *Dicraeosaurus*. It could also be related to the reactions of these individual taxa to a seasonal climate.

These growth series suggested to Sander (1999) that all four of these genera "are characterized by continued growth after sexual maturity but that growth was determinate." In *Barosaurus* type A specimens, sexual maturity seems to have been attained when the animal reached approximately 70 percent its maximum size, while in *Brachiosaurus* sexual maturity was reached at 40 percent maximum size (see also Sander 2000). Compared to the former genus, *Brachiosaurus* attained its enormous size "by prolonging the phase of fast adult growth" (a pattern similarly observed in the giant crocodilian *Deinosuchus*; see Erickson and Brochu 1999), this perhaps necessary to maintain viable populations. If *Brachiosaurus* and *Barosaurus* grew at approximately the same rate as the microstructure of their bones seems to indicate, they also attained sexual maturity at about the same age, as their long bones were approximately the same size at sexual maturity. Sander (1999) suggested that, as 20 years appears to be the top limit to age at sexual maturity in sauropods (based on demographic simulations; see Dunham, Overall, Porter and Forster (1989), "*Brachiosaurus* may have had to mature relatively earlier than *Barosaurus* to avoid this limit."

Sander (1999) determined that the dominance of fibrolamellar bone in the studied samples indicated that these dinosaurs grew at rates comparable to those seen in modern mammals, the growth pattern of the Tendaguru sauropods thereby combining typically reptilian traits of continued growth after sexual maturity with traits typical of mammals. Furthermore, considerable variation was possible within the life history framework of the Tenadaguru sauropods, in keeping with the somewhat different bauplan of these dinosaurs and the different ecological niches they seem to have inhabited.

In addition to gleaning information related to the growth of the Tendaguru sauropods, Sander (2000) also demonstrated that the histology of the long bones of these four genera, particularly in the humerus and femur, allow taxon recognition, especially if bones of the same size are compared. Sander (2000) noted that all of the Tendaguru sauropod long bones exhibit a peculiar kind of growth mark called the "polish line," defined by that author "as a growth line in fibrolamellar bone that is visible in polished section but not in thin section." As Sander (2000) described polish lines, they follow the fabric of primary bone, generally in the middle of a lamina of fibrolamellar bone, and represent former bone surfaces, thereby being growth markers rather than diagnostic artifacts. The various Tendaguru sauropods can be identified by the degree and nature of the bone remodeling and also by the presence and spacing of their distinctive polish lines (see Sander 2000 for details).

Gunga, Kirsch, Rittweger, Röcker, Clarke, Albertz, Wiedmann, Mokry, Suthau, Wehr, Heinrich and Schultze (1999) published their study to calculate the body size and body volume distribution in *Brachiosaurus* and *Dicraeosaurus*.

As noted by these authors, allometric equations are often based on an animal's total body mass, the latter particularly important in determining numerous physiological functions (see Peters 1983; Schmidt-Nielsen 1984, 1997; Withers 1992). However, body-mass estimates for *Brachiosaurus* (14.9 to 102 metric tonnes) and *Dicraeosaurus* (10 to 40 metric tonnes) have been widely divergent (see Peczkis 1994). Such estimates were reached based upon projections from models of the fleshed-out animals or circumferential measurements of the humerus and femur (*e.g.*, Colbert 1962; Lambert 1983; Anderson, Hall-Martin and Russell 1985; Paul 1988*a*; Alexander 1989; see also *D:TE*), these methods being dependent on the accuracy or inaccuracy of the model, and also upon the enlargement factor, whereby even the slightest errors can lead to differences having a multiplication factor of from 10 to 50.

Gunga *et al.* noted that 1. calculating body mass utilizing measurements of the humerus or femur is affected when bones examined belong to different specimens; 2. bone growth in dinosaurs is not yet completely understood (see Reid 1984*a*, 1984*c*; Sander 1999, 2000); 3. the tremendous strains on the limbs, as occurs during walking and braking, also play important roles in the growth of bone; 4. bones having an otherwise equal circumference and cross-sectional surface can vary, their corticalis, for example, becoming thickened overproportionally to increase stability; and 5. the envelope describing the range of body masses for mammals is much less than that likely for sauropods.

In order to determine the precise body mass and volume distribution in these dinosaurs, Gunga *et al.* employed two methods. First, they applied classical three-dimensional stereophotogrammetry (see earlier work in this area by Gunga, Kirsch, Baartz, Röcker, Heinrich, Lisowski, Wiedemann and Albertz 1995; Widermann, Suthau and Albertz 1999; see also a similar study by Henderson 1999) to the skeletons of *Brachiosaurus brancai* (mounted in 1937) and *Dicraeosaurus hansemanni* (mounted in 1930) exhibited at the Berlin museum. They also used a newly developed laser scanner technique designed for large-scale objects (see Wiedermann *et al.*). Although both of these skeletons are composites, each one mostly comprises bones belonging to a single individual.

Mounting points of reference on these skeletons, Gunga *et al.* (1995) made a three-dimensional reconstruction of the skeletons using stereo projectors (see Gunga *et al.* 1995; Wiedermann and Wehr 1998; Wiedermann *et al.*). Dimensions of the skeletons were then determined, those for *Brachiosaurus* by classical stereophotogrammetry and for *Dicraeosaurus* by laser-scanning line by line. With these photographic/electronic data as a basis, Gunga *et al.* (1999) utilized scaling equations to estimate the size of organ systems.

From this study, Gunga *et al.* (1999) calculated the body mass of *Brachiosaurus* to be 74.4 metric tonnes (approximately 83 tons), close to Colbert's estimate of 66 to 77 metric tonnes. The thorax was found to occupy almost 74 percent of the entire body volume, with the limbs adding at least 80 percent. The neck was found to occupy 15 percent of the entire body volume and the tail only 4.4 percent, this imbalance suggesting to these authors that the tail could not counterbalance the neck and head and that, consequently, the neck was held vertically rather than horizontally, moreso even than generally restored (Christian and Heinrich 1988; *contra* Martin, Martin-Rolland and Frey 1998); Parrish and Stevens 1998; Stevens and Parrish 1999).

Gunga *et al.* (1999) calculated the body mass of the smaller and relatively shorter-necked *Dicraeosaurus* to be 12.8 metric tonnes (less than 14 tons), lower than Janensch's (1950*b*) estimate of 10 to as much as 40 metric tonnes. Also, the head of this genus is 20 times larger compared to body volume than in *Brachiosaurus*. Together, the head and neck comprise 6.4 percent of the total body volume and are closer to being balanced by the tail (8.5 percent), suggesting a horizontal orientation of the neck.

Given the disparity in body masses and utilizing allometric formulas given for endotherms, Gunga *et al.* (1999) deduced different sets of physiological data for these two sauropods, with *Brachiosaurus* assumed to have six times the lung volume of *Dicraeosaurus* and a difference in blood volume in the same range. Among thc conclusions reached by Gunga *et al.* (1999) from this study are the following:

The skeleton of *B. brancai* in the Berlin museum represents an animal not yet fully grown (see Sander 1999). Based on a femur (1.86 meters in length, 0.8 meters in circumference, 158.5 square centimeters in cross-sectional area at middiapophysis) comparable in size with those of the Berlin mount, Gunga *et al* (1999). calculated them maximum strength in compression for this bone to be from about 180 to 190 metric tonnes.

Fossil tracks of sauropods indicate that these kinds of dinosaurs, like elephants, walked with their feet near the midline beneath the body (see Norman 1991; Lambert 1993), this posture being the most practical for gigantic animals walking on dry land as the force of gravity is directed straight to the broad, rounded feet. The fore- and hindfeet of *Brachiosaurus*, were measured by Gunga *et al.* (1999) at 2.20 and 2.8, respectively. Therefore, the authors assumed for *Brachiosaurus* "a very slow locomotion due to its body mass, or also due to the limitations exerted by the thermoregulatory and circulatory systems."

Assuming a metabolism similar to that of endotherms because of its fast growing rates (see Sander 1999, 2000) and mass homeothermy because of large body size (Spotila 1980; Hotton 1980), Gunga *et al.* (1999) applied allometric equations based on data relating to mammals to calculate apparent organ sizes and physiological parameters (see Peters; Withers; Schmidt-Nielsen 1984, 1997). Gunga *et al.* (1999) estimated the lung volume of *Brachiosaurus* to have been about 6000 litres, its tidal volume about 520 litres, and its respiratory rate approximately three breaths per minute. Oxygen consumption at rest was calculated as about 50 litres per minute. The amount of ingested food would probably have been approximately 350 kilograms, less than one percent of the total body weight, 70 percent of the plant intake being water, 30 percent representing the dry mass, 50 percent of the latter actually being absorbed by the gastro-intestinal tract (as in extant, very large herbivores; see Owen-Smith 1988).

Accepting reconstructions of *Brachiosaurus* with a vertically oriented neck, Gunga *et al.* (1999) estimated its blood volume to be about 3600 litres, the weight of the presumably classical four-chambered heart at least 386 kilograms, with a stroke volume of about 17.4 litres, and a heart frequency on 14.6 minutes^{-1}, this, in the authors' opinion, being the only plausible reconstruction of this dinosaur's cardiovascular system. Because of the hydrostatic load of blood at the extremities, Gunga *et al.* (1999) theorized that *Brachiosaurus* must have possessed some kind of "oedema prevention mechanism," not unlike the venous valves, muscular venous pumps, very long connective tissue, and thickened basal membrane observed in the extremities of giraffes and gazelles (Hargens, Millard, Petterson and Johansen 1987; Withers). Furthermore, Gunga *et al.* (1999) pointed out that the thoracic-abdominal cavity in both *Brachiosaurus* and *Dicraeosaurus* appear large enough to accommodate these organs and also a gastro-intestinal tract, the latter actually being relatively small in such megaherbivores as elephants (Owen-Smith).

In conclusion, Gunga *et al.* (1999) found sauropods such as *Brachiosaurus* and *Dicraeosaurus* to be highly specialized in regards their metabolism, circulation, and temperature regulation, with even the most minor environmental changes involving climate, food, or fluid intake having catastrophic consequences for their survival.

The study by Perry and Reuter (1999) focused upon the possible lung structure (fossilized lungs tissue having never been found) of *Brachiosaurus*. Perry and Reuter noted that, being "an archosaur, *Brachiosaurus* presumably had multichambered lungs with an asymmetrical (monopodial) branching pattern and tubular, arching chambers. Therefore, invoking Baron Georges Cuvier's principle (introduced during the early nineteenth century) of utilizing recent analogues in reconstructing the soft tissues of fossil organisms, Perry and Reuter used as a starting point the lungs of extant archosaurs and the closest living relatives of dinosaurs, crocodilians and birds.

The resultant hypothetical model for *Brachiosaurus* was a lung consisting of four rows of chambers (as in crocodilians and birds), each chamber radiating independently from an intrapulmonary bronchus. Cranially, this latter structure is reinforced by cartilage; caudally, the reinforcement of cartilage diminishes, the radiating pattern of tubular chambers becoming irregular.

Perry and Reuter noted that extensive pneumatic spaces in the dorsal vertebrae and also the ribs of *Brachiosaurus* indicate a dorsal and lateral attachment of the lungs to the body wall, which is indicative of a heterogenous lung structure. (Reptiles either possess lungs with heterogenously or homogeneously distributed parenchyma, or gas exchange tissue; in homogeneously structured lungs, the parenchyma are intrinsically stable, often being freely suspended in the body cavity.) Similar spaces were also found in the cervical vertebrae, suggesting that the oesophagus part of the alimentary canal was surrounded by air spaces, as the trachea must have lain either lateral or ventral to the oesophagus while the pneumatic bones laid dorsal to it. The pneumatic spaces, combined with gigantic size, suggested to Perry and Reuter that the dorsal margin of the lungs was strongly attached to the body wall, thereby limiting lung inhalation in this region.

The authors suggested that the cavernous, ventral portions of the very large lungs of *Brachiosaurus*, as in the lungs of snakes (see Lillywhite 1987), may have been entirely devoid of gas-exchange capillaries. If true, these ventral portions of the lungs were postulated by Perry and Reuter to be cavernous air sac-like chambers that served to store and convect dorsally-concentrated gas. This large lung volume resulted "in a low total body density, thus reducing the weight on land." Also, Perry and Reuter noted that a large lung volume would be an advantage considering the extremely long neck in this sauropod. "Since the resistance to flow in tubular structures is inversely proportional to the fourth power of the diameter and is only directly proportional to the length, it is energetically advantageous to have a wide-bore trachea combined with a large-volume lung to reduce the deleterious effects of rebreathing a large trocheal volume."

According to Perry and Reuter, a heterogenous lung structure for *Brachiosaurus* would result in a low work of breathing. Large lungs require less work to inflate, with less physical work being necessary to move one millilitre of air in a large lung than in a small one. However, in lungs of dissimilar structure, heterogenous lungs have better compliance—*i.e.*, the ease of lung inflation—than homogenous lungs. Consequently, "if *Brachiosaurus* had highly heterogenous lungs its work of breathing could have been extremely low for two reasons: 1) because the lungs are very large and 2) because they were heterogenous."

Perry and Reuter found it unlikely that the ancestral archosaurian hepatic piston, proposed by some recent workers (see "Introduction," section on ectothermy versus endothemy), existed in *Brachiosaurus*. According to these authors, "The work required to move the liver and other abdominal viscera with each breath would have been enormous and the same end could have been reached by costal breathing, with the sac-like regions of the lungs disposed laterally and ventrally beneath the ribs." Maintaining a relatively low breathing frequency, given the great weight of the rib cage, would have been energetically advantageous, this hypothesis consistent with large, heterogenous lungs.

The avian lung, the authors noted, is characterized by a very efficient cross-current gas exchange model (see Scheid and Piper 1970); while no evidence compellingly demonstrates that large dinosaurs like *Brachiosaurus* maintained a high metabolic rate, it would certainly be of selective advantage to a large sauropod to have an efficient respiratory system. Extracting a large amount of oxygen with every breath would have made fewer breaths necessary, resulting in the conservation of metabolic energy. Additionally, even modest levels of aerobic activity could lead to problems with heat loss. Perry and Reuter concluded that the cross-current model for *Brachiosaurus* "may result in increased efficiency of heat loss and therefore could have been instrumental in the attainment of extreme body size in a subtropical environment."

Heinrich (1999) published the first major study on the taphonomy of *B. brancai* as well as the other dinosaurs from the Tendaguru Beds of Tanzania. As noted by this author, available information on this topic is somewhat limited, most of it envisioning a catastrophic demise for these dinosaurs. Janensch (1914*b*, 1914*c*), for example, had proposed that the "Tendaguru Saurien Beds" were of marginal-marine origin, suggesting that the animals had become trapped and were killed in the "mireholes" of mud of lagoons temporarily exposed by ebbing tides (see also Hennig 1912*a*, 1912*b*), the carcasses then floating and decomposing in shallow water before burial. Hennig (1912*a*, 1925) and Janensch (1914*b*, 1914*c*) both explained bonebeds mostly consisting of the disarticulated remains of the Tendaguru stegosaur *Kentrosaurus* or ornithopod *Dryosaurus* as the result of sudden, catastrophic, mass-mortality events. Other authors (*e.g.*, Abel 1927; Kitchin 1929; Parkinson 1930; Colbert 1968) have proposed an estuarine environment for the area now known as Tendaguru, with the carcasses of these dinosaurs washed in by rivers, then transported downstream to be buried near the mouth of rivers flowing into the sea. Contrary to those interpretations, Reck (1925) argued against catastrophic mortality for these dinosaurs, instead interpreting the depositional environment of the Tendaguru Beds as saline marshes.

Heinrich's detailed taphonomic study was based upon previously unpublished field sketches (drawn in pencil and ink by Werner Janensch, Edwin Hennig, and Hans Reck, most of these sketches documenting the dinosaur bone assemblages from the Upper Saurian Bed) in the archive of the Museum für Naturkunde der Humboldt-Universität zu Berlin, made by the German Tendaguru Expedition (1909–1913). Heinrich pointed out that these drawings—documenting bone assemblages of sauropods and ornithopods from the Middle Saurian Bed, Upper Saurian Bed, and Transitional Sands above the *Trigonia smeei* Bed—reveal details of the preservation characteristics of the Tendaguru dinosaurs and shed light on their taphonomy.

The invertebrate fossil record of the Tendaguru Beds—including freshwater gastropods (*e.g.*, Dietrich 1914; Hennig 1914*c*), *Asmussia* (=*Estheria*; see Janensch 1933), and limnic to brackish ostracods (Schudak, Schudak and Pietrzeniuk 1999)—suggested to Heinrich a brackish to limnic depositional environment for the Middle and Upper Saurian Bed (*e.g.*, Dietrich; Hennig 1914*a*, 1914*b*; Schudak 1999; Schudack *et al.*). The co-occurrence of dinosaur bones and marine to brackish invertebrates (*e.g., Trigonia* and *Nerinea*; see Janensch 1914*d*, 1961) shows that the depositional environment for the Transitional Sands was shallow marine to brackish. Also, the presence of *Asmussia* (=*Estheria*) in the Upper Saurian Bed could indicate periods of drought (see Janensch 1933; Russell, Béland and McIntosh 1980).

One incomplete skeleton of *B. brancai* from Tendaguru "Site S" was found with both the left humerus and tibia in an upright position, suggesting to Heinrich "that the animal became mired in soft mud and died laying on its belly," *in situ* decay of the soft tissues indicated by a series of articulated cervical vertebrae and by parallel-oriented dorsal ribs lying upon each other. Most of the bones, save for the cervical vertebrae, were concentrated near these left front-limb bones, suggesting that dispersal of the bones was minimal and that the place of death was probably the place of final deposition.

As the author noted, the stages of disarticulation of these remains range from incomplete skeletons to solitary bones, which argues strongly for decomposition and postmortem transport of the carcasses prior to burial. Accumulations of sauropod bones (*B. brancai*, ?*Barosaurus africanus*, and *Amargasaurus sattleri* [formerly *Dicraeosaurus sattleri*]) are dominated by adult individuals, with juveniles either rare or missing. The fact that bones occurred in different superimposed dinosaur-bearing horizons (*i.e.*, Saurian Beds, Transitional Sands) indicated to Heinrich that the remains were accumulated over a long span of Late Jurassic time, the majority of these accumulations most likely being attritional rather than catastrophic, probably the result of year-to-year bone imput due to short-term intervals of normal death events caused by starvation, disease, weakness, old age, and seasonal drought.

Note: Burge, Bird, Britt, Chure and Scheetz (2000) reported brachiosaurid remains recently collected from the Price River II Quarry — a site that has also yielded the remains of a large ankylosaur, ornithopods, theropod teeth, and, more rarely, turtle fragments — at the base of the Ruby Ranch Member of the Early Cretaceous (Barremian) Cedar Mountain Formation near Price, Utah. Representing at least four individuals, the Brachiosaurid remains recovered thus far include disarticulated long bones, metapodials including a complete set of elongate metatarsals, vertebrae, and pelvic elements. Brachiosaurid characters noted by the authors in this material include the following: pubic peduncle elongate; puboischial contact broad; vertebrae camellate; cervical vertebrae having slender centra; caudal vertebrae having short spines and persistent backwardly directed ribs. According to Burge *et al.*, most of these bones "are surprisingly similar to those of *Brachiosaurus*, the primary difference being the relatively robust humerus, which has a 0.35 length to minimum diameter ratio.

As pointed out by Burge *et al.*, the Early Cretaceous North American sauropod fauna differs strikingly from that of the Late Jurassic. Cedar Mountain Formation forms are considerably smaller (less than 14 meters in length) than the giants in the Upper Jurassic Morrison Formation. Three sauropod genera (a titanosaurid, brachiosaurid, and camarasaurid) are known from the basal member of the Cedar Mountain Formation, while brachiosaurids constitute the only sauropod taxa yet known from the geologically younger Poison Strip Sandstone and Ruby Ranch members of that formation, and small teeth tentatively referred to the Brachiosauridae comprise the only sauropod material from the Mussentuchit Member. Burge *et al.* further noted that, while Morrison Formation brachiosaurids are extremely rare, these dinosaurs become more abundant numerically, becoming the dominant sauropod group in the Cedar Mountain Formation. According to the authors, this pattern of relatively small size and numerical abundance is consistent for the Brachiosauridae across most of Early Cretaceous North America.

Key references: Abel (1927); Alexander (1976, 1978, 1985, 1989, 1996); Bakker (1986); Burge, Bird, Britt, Chure and Scheetz (2000) Charig (1979); Christian and Heinrich (1998); Christian, Heinrich and Golder (1999); Christian, Müller, Christian and Preuschoft (1999); Christainsen (1997); Colbert (1962, 1968); Dietrich (1914); Dunham, Overall, Porter and Forster (1989); Farlow, Smith and Robinson (1995); Fastovsky and Weishampel (1996); Golder and Christian (1999); Gunga, Kirsch, Baartz, Röcker, Heinrich, Lisowski, Wiedemann and Albertz (1995); Gunga, Kirsch, Rittweger, Röcker, Clarke, Albertz, Wiedmann, Mokry, Suthau, Wehr, Heinrich and Schultze (1999); Hargens, Millard, Petterson and Johansen (1987); Heinrich (1999); Hennig (1912*a*, 1912*b*, 1914*a*, 1914*b*, 1925); Hotton (1980); Janensch (1914*a*, 1914*b*, 1914*c*, 1914*d*, 1933, 1950*b*, 1950*c*, 1961); Kitchin (1929); Lambert (1993); Lillywhite (1987); Lockley (1991); Lockley and Hunt (1995); Martin, Martin-Rolland and Frey (1998); McMahon and Bonner (1983); Norman (1991); Olshevsky (1991, 2000); Owen-Smith (1988); Parkinson (1930); Parrish and Stevens (1998); Paul (1987, 1988*a*); Peckzis (1994); Peters (1983); Reck (1925); Riggs (1903); Russell, Béland and McIntosh (1980); Sander (1999, 2000); Schmidt-Nielsen (1984, 1997); Schudak (1999); Schudak, Schudak and Pietrzeniuk (1999); Spotila (1980); Stevens and Parrish (1998); Thulborn (1990); Wiedermann, Suthau and Albertz (1999); Wiedermann and Wehr (1998); Withers (1992).

†BRACHYLOPHOSAURUS

Ornithischia: Genasauria: Cerapoda: Ornithopoda: Euornithopoda: Iguanodontia: Euiguanodontia: Dryomorpha: Ankylopollexia: Hadrosauroidea: Hadrosauridae: Euhadrosauria: Hadrosaurinae.

Comments: Among the rarer hadrosaurid genera is the crested *Bracylophosaurus*, previously known only from several specimens.

Ancell, Harmon and Horner (1998) reported, in a recent abstract, the discovery of a nearly complete, articulated skeleton of this duckbilled dinosaur (see *D:TE*) in the Upper Cretaceous Judith River Formation of Montana. The skeleton was found retaining its three-dimensional architecture that reveals the structure of both pectoral and pelvic girdles. Stapes and ossified tendons were preserved in place.

As noted by Ancell *et al.*, this specimen is especially interesting in regards its taphony, as the complete and articulated skeleton of *Lepisosteus* (a garfish

Photograph by the author, courtesy Royal Tyrrell Museum/Alberta Community Development.

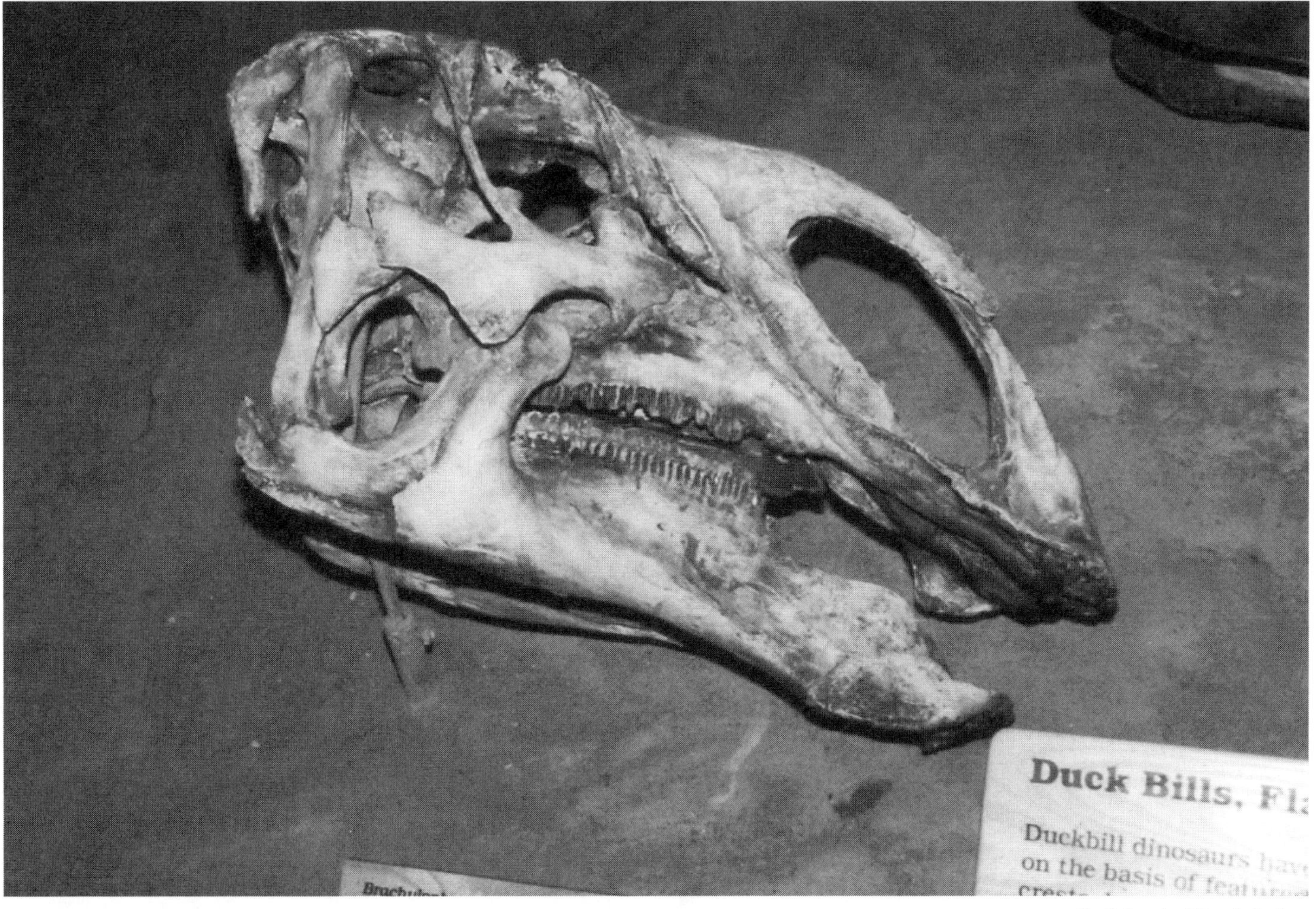

Skull (cast) of ***Brachylophosaurus canadensis.***

measuring 80 centimeters in length, with fins and scales seemingly in place) was preserved in the hadrosaur's stomach region, ventral to the pubes but within the area that would have made up the body. Both dinosaur and fish skeletons were preserved in uncemented sandstone that also encased various isolated bones belonging to other dinosaurian taxa, as well as numerous leaf and twig fossils.

The discovery did not indicate that this *Brachylophosaurus* individual was a fish-eater, however. A taphonomic study of the specimen suggested to Ancell *et al.* that the gar had invaded the stomach region of the *Brachylophosaurus* shortly after the dinosaur died, after which both animals were rapidly and entirely covered and trapped by sand.

Hanna, LaRock and Horner (1999) briefly reported on fossil remains collected in 1998 from a bonebed in the terrestrial Upper Cretaceous Judith River Formation near Malta, in northeastern Montana, this site to date having yielded more than 650 elements representing at least three dinosaurian and several nondinosaurian taxa, with about 90 percent of the material referred to *Brachylophosaurus* sp. Ten of the approximately 450 bones examined by the authors were found to be pathological, six belonging to *Brachylophosaurus* (including a pedal phalanx, metacarpal, and two caudal and two dorsal vertebrae), the others (a fragment and three caudal vertebrae) not yet assignable. Abnormalities will later be described and classified regarding possible origins of the pathologies; continued documentation of pathological elements from this bonebed "will help define the frequency and anatomical location of brachylophosaur abnormalities."

In an abstract for a poster, Prieto-Marquez (2000) noted a suite of postcranial features in specimens of the referred species *Brachylophosaurus goodwini* collected from the above-mentioned bonebed. Material examined by Prieto-Marquez included a complete, articulated adult skeleton (MOR 794), plus over 1,200 specimens representing almost all of the postcranial elements and the skull, some of these remains showing both individual and ontogenetic variation.

Prieto-Marquez observed the following in these remains (the features of the coracoid, carpus, and phalangeal count of the manus described here in the Hadrosauridae for the first time): axial skeleton containing 13 mediolaterally broad cervical and 18 dorsal vertebrae, sacrum with nine fused vertebrae; appendicular skeleton including coracoid with anterior and ventral processes, scapula with caudally expanded blade, pubis with ellipsoidal preacetabular blade, very elongated ulna and fibula; carpus comprising tiny pebble-like bone and much larger tetrahedal bone; phalangeal formula of manus of 0-3-?2-3-4; pedal unguals possessing plantar keel (a synapomorphy uniting *B. goodwini* with the hadrosaurine *Maiasaura peeblesorum*).

As noted by Prieto-Marquez, the following features support the inclusion of *B. goodwini* in the Hadrosaurinae: ischium unfolded; deltopectoral crest projecting less than diameter of humeral shaft; and three caudal-most sacral vertebrae grooved. These postcranial features support the affinity of this taxon with *M. peeblesorum*.

The author further noted that the observed variability in the thickness of the scapular neck argues against regarding this character as diagnostic for the Hadrosaurinae; rather, it seems to be dependent upon age and individual variation.

Key reference: Ancell, Harmon and Horner (1998); Hanna, LaRock and Horner (1999); Prieto-Marquez (2000).

†BRADYCNEME (=*Heptasteornis*; =?*Elopteryx*)

Saurischia: Theropoda: Neotheropoda: Tetanurae: Avetheropoda *incertae sedis*.

Occurrence: Sinpetru Formation, Transylvania, Romania.

Known material: Distal tibiotarsi.

Comments: In 1913, Charles W. Andrews included in the holotype (BMNH A.1234) of *Elopteryx nopcsai* the distal part of a left tibiotarsus. At the same time, Andrews referred to this new genus and species various fragmentary specimens. These included the distal parts of two tibiotarsi (BMNH A.1588 and A.1528). All of this material was collected from outcrops of the Sinpetru Formation in the Sibisel Valley (D. B. Weishampel, personal communication to Csiki and Grigorescu 1998), Hateg Basin, Transylvania (other details regarding their precise locality are unknown).

Later, Harrison and Walker (1975) separated the distal left tibiotarsus from the type specimen of *Elopteryx*, assigning it the new catalog number of BMNH A.4358. These authors designated BMNH A.1588 to be the holotype of a new genus and species, *Bradycneme draculae*, and referred both BMNH A.4358 and A.1528 to another new genus and species, *Heptasteornis andrewsi*, with BMNH A.4358 named as the holotype. Although both of these new taxa were originally described by Harrison and Walker as

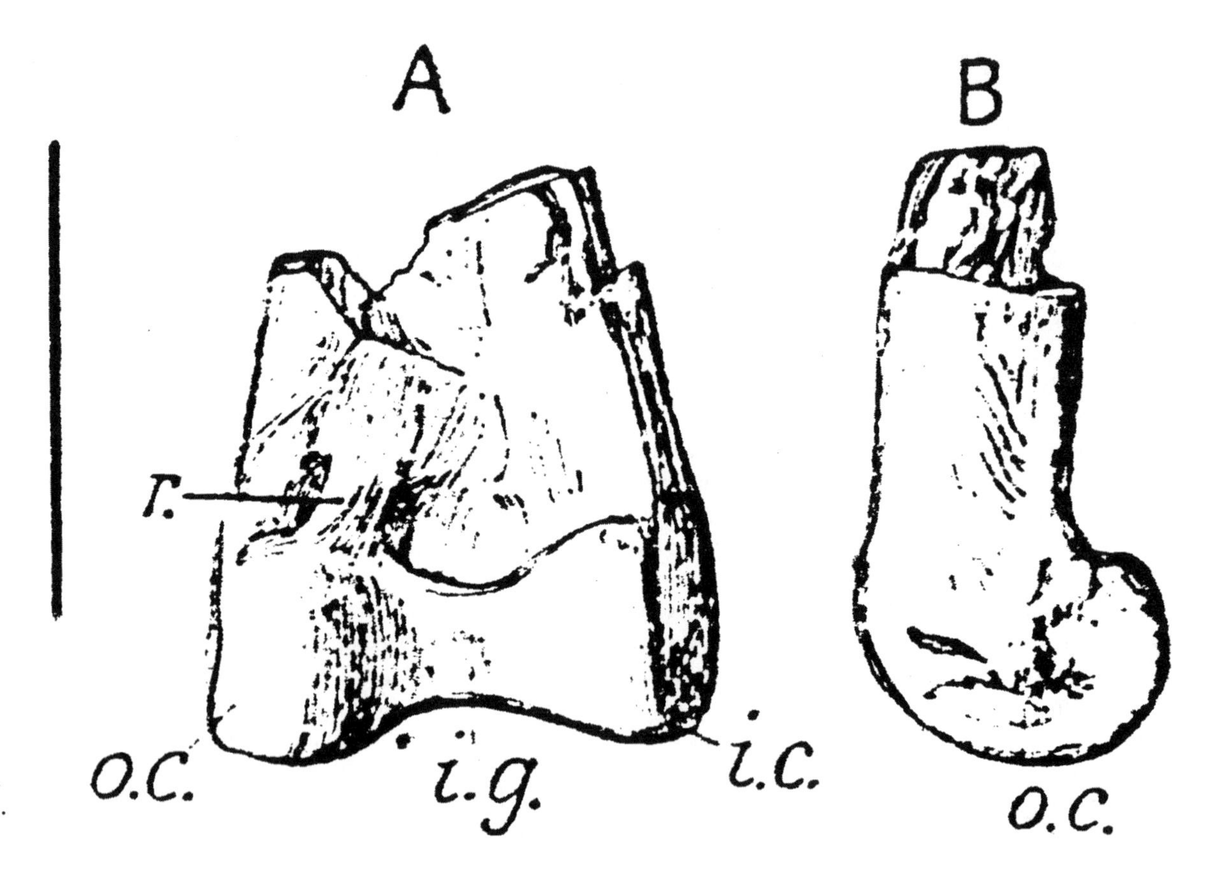

***Bradycneme draculae*, BMNH R.1588, holotype right distal tibiotarsus, in A. cranial and B. right lateral views. Scale = 1.33 cm. (After Andrews 1913.)**

strigiform birds, a number of subsequent workers (*e.g.*, Elżanowski 1983; Grigorescu 1984*a*; Norman 1985; Osmólska 1987; Paul 1988*b*) noted their dinosaurian affinities. Osmólska and Barsbold (1990), in their review of the Troodontidae, regarded *B. draculae* as possibly belonging to that clade (see *Bradycneme* and *Heptasteornis* entries, *D:TE*). Le Loeuff (1992), and Le Loeuff and Buffetaut (1991), later proposed that *Bradycneme* is an "elopterygine" dromaeosaurid (Le Loeuff suggesting that the tibiotarsi are similar to those of *Deinonychus*), and that both this genus and *Heptasteornus* are junior synonyms of *Elopteryx*; and Howse and Milner (1991) regarded *Heptasteornis* as a troodontid.

More recently, Csiki and Grigorescu (1998) agreed that *Heptasteornis* should be synonymized with *Bradycneme*, but basically disagreed that these taxa should be referred to *Elopteryx*, nor are they closely related.

As observed by these authors, the tibiotarsi of *Bradycneme* display the following derived characters in common with the Tetanurae (Sereno, Dutheil, Iarochene, Larsson, Lyon, Magwene, Sidor, Varracchio and Wilson 1996): astragular ascending process platelike, relatively well developed; astragalar condyles having cranial transverse groove (this latter feature a tetanuran synapomorphy excluding this taxon from Manuraptora; see Sereno *et al.*; Holtz 1994).

Compared with other tetanurans, Csiki and Grigorescu noted, *Bradycneme* seems to have the most in common with troodontids, comparable to them in the following features: more or less well-fused tibiotarsus (also a feature of *Avimimus* and derived birds); broad, cranio-caudally flattened shape of distal tibiotarsi (see Russell 1969; Barsbold 1974; Osmólska 1987), showing well-pronounced, cranially projected condyles separated cranially and distally by deep intercondylar groove that becomes very narrow caudally; calcaneum fused to astragalus (Russell and Dong 1993*a*) or missing (Currie and Peng 1993); lateral condyle extending more distally than medial condyle; condyles slightly divergent cranially; and lateral condyle triangular in distal view, with cranially projecting "lip" and narrow laterocranial groove for articulation with fibula.

Csiki and Grigorescu (*contra* Le Loeuff) perceived no similarity between the tibiotarsi assigned to *Bradycneme* and those of *Deinonychus*. The authors concluded that *Bradycneme* is a small theropod sharing most characters with troodontids, although it exhibits one definite synapomorphy that excludes it from the derived coelurosaurians. Regarding Le Loeuff's synonymizing of *Bradycneme* with *Elopteryx*, the authors commented that this seems to preclude referring *Bradycneme* to the Manuraptora, a clade to which *Elopteryx* most likely belongs.

Therefore, the systematic placement of *B. draculae* could not, at present, be more accurately determined.

Erratum: In *D:TE*, the holotype of *H. andrewsi* was incorrectly numbered BMNH A.4359.

Key references: Andrews (1913); Csiki and Grigorescu (1998); Elżanowski (1983); Grigorescu (1984*a*); Harrison and Walker (1975); Howse and Milner (1993); Le Loeuff (1992); Le Loeuff and Buffetaut (1991); Norman (1985); Osmólska (1987); Osmólska and Barsbold (1990); Paul (1988*b*).

†BUGENASAURA

Ornithischia: Genasauria: Cerapoda: Ornithopoda: Euornithopoda: Hypsilophodontidae: Thescelosaurinae.

Diagnosis: Cheek tooth row very deeply recessed, with deep, massive dentary and very prominent overhanging ridge (with braided appearance) on ventral part of maxilla, distal end of elongate palpebral (= supraorbital) truncated with ridges medially (presumably for accessory palpebral) (Galton 1999*b*).

Comments: In 1995, Peter M. Galton named and described the small ornithopod *Bugenasaura infernalis*, founded upon a specimen (SDSM 7210) comprising an incomplete skull, four incomplete dorsal vertebrae, and two manual phalanges found in 1972 by Kenneth H. Oson and Arland Jacobson in the Hell Creek Formation of South Dakota. The specimen had originally been described by Morris (1976) as ?*Thescelosaurus* sp. and was subsequently cited by Sues and Norman (1990) as an unnamed hypsilophodontiod. Later, Galton (1995) provided an original diagnosis and a brief description of the holotype skull (see *D:TE*). Four years later, Galton (1999*b*) published a full description of this skull and proposed a new diagnosis for the genus and type species.

Galton (1999*b*) assigned *Bugenasaura* to the Hypsilophodontidae based on the following cranial hypsilophodontid synapomorphies: 1. Maxilla having anterior process that fits into groove of premaxilla (see Sues and Norman); 2. presence of ridges culminating in marginal denticles on heavily enameled surface of cheek teeth (see Weishampel and Heinrich 1992); and 3. presence of cingulum on dentary teeth (Weishampel and Heinrich).

Galton (1999*b*) accepted Sternberg's (1940) division of the Hypsilophodontidae into two subfamilies, the Hypsilophodontinae and the more derived Thescelosaurinae. Galton (1999*b*) found *Bugenasaura* to be allied with and the sister taxon of the less derived *Thescelosaurus*, these genera comprising the Thescelosaurinae, based upon the following synapomorphies: 1. Cheek teeth having thicker enameled side

***Bugenasaura infernalis*, holotype skull (SDSM 7210) in left lateral view. Scale = 1 cm. (After Galton 1999*b*.)**

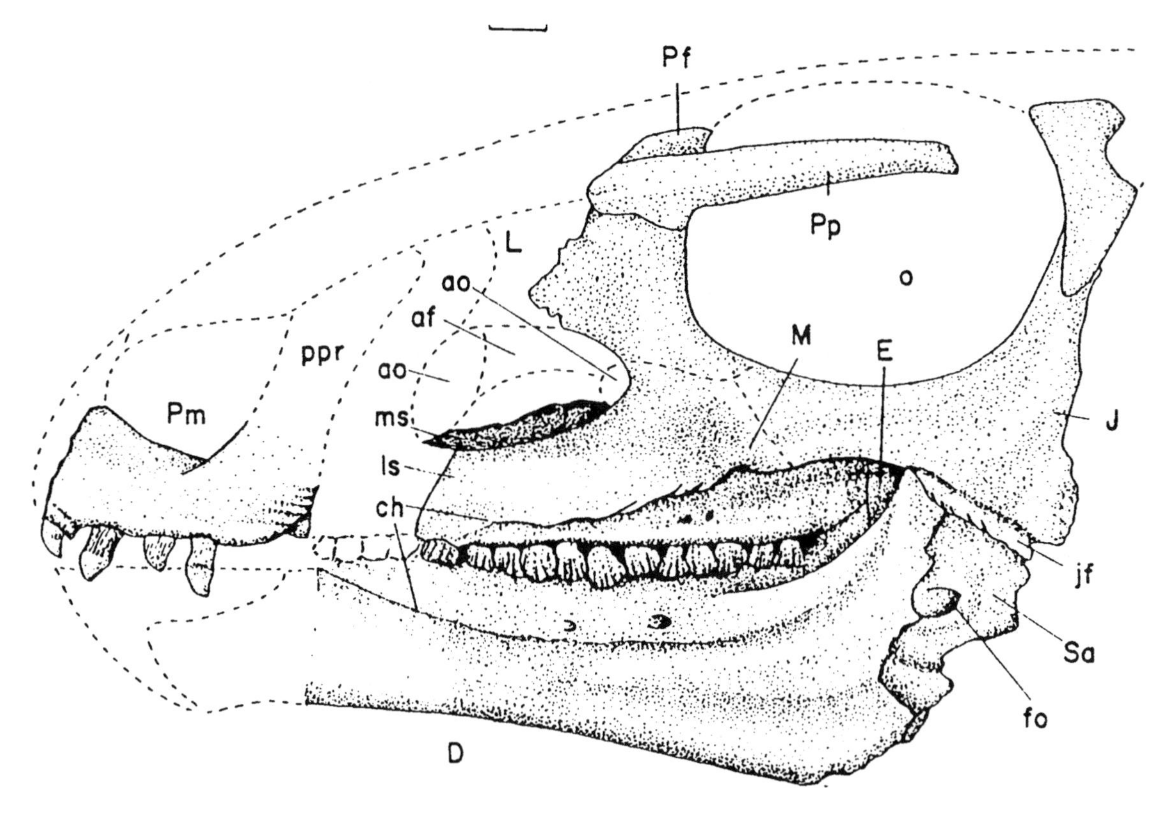

(buccal for maxillary and lingual for dentary teeth) bearing series of secondary ridges medially and distally forming two converging circles on either side of central ridge; 2. posterior half of ventral ridge of jugal offset ventrally, covered ventrolaterally with obliquely inclined ridges, as in *Thescelosaurus* (see Galton 1997*c*); femur longer than tibia.

As noted by Galton (1999*b*), an argument could be made (*e.g.*, Peng 1992) supporting an affinity of *Bugenasaura* with the Chinese genus *Agilisaurus* (regarded by Peng as a fabrosaurid, based upon derived characters shared with the African genus *Lesothosaurus*, and by Padian 1997*a* and Sues 1997*b* as a hypsilophodontid). This possible affinity was based upon various characters that might be interpreted as synapomorphies, including the following: 1. Maxilla having prominent ridges compared to other hypsilophodontines; and 2. deep dentary having prominent lateral ridge compared to other hypsilophodontines. However, Galton (1999*b*) pointed out, the cheek teeth of *Agilisaurus* are quite simple, their denticles not supported by ridges on the well-enameled surface, and the proportions of the hindlimbs, as in other hypsilophodontines, are subcursorial. Given the similarities between *Agilisauurus* and *Lesothosaurus*, the form of the teeth in the former genus, and also the close

***Bugenasaura infernalis*, referred right dentary tooth (YPM 8098) in A. lingual, B. distal, C. buccal, and D. mesial views. Scale = 1 mm. (After Galton 1999*b*.)**

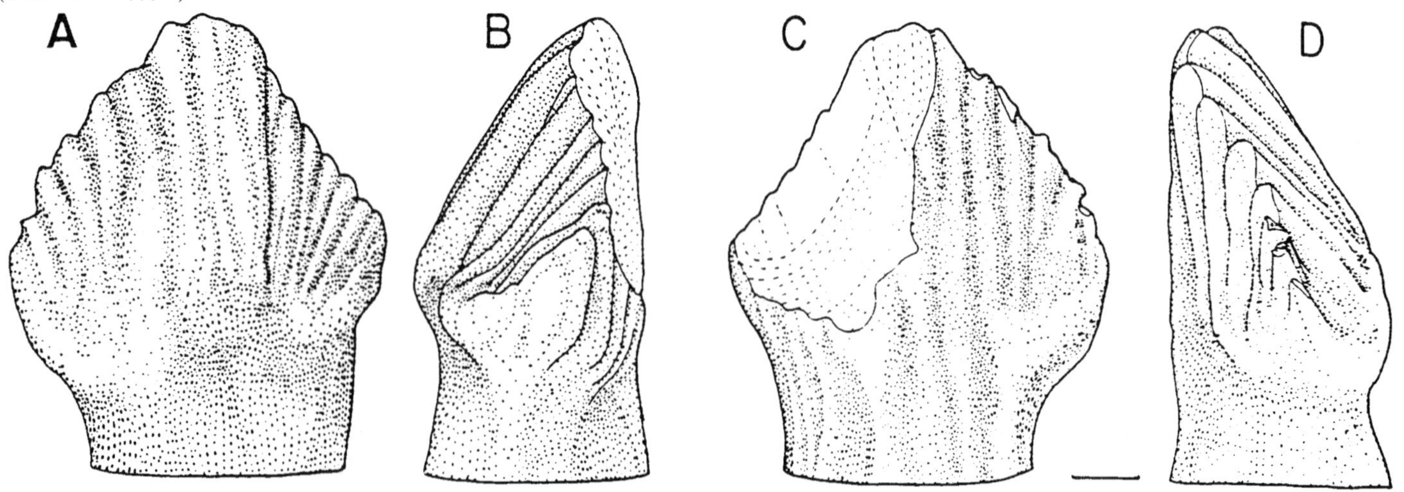

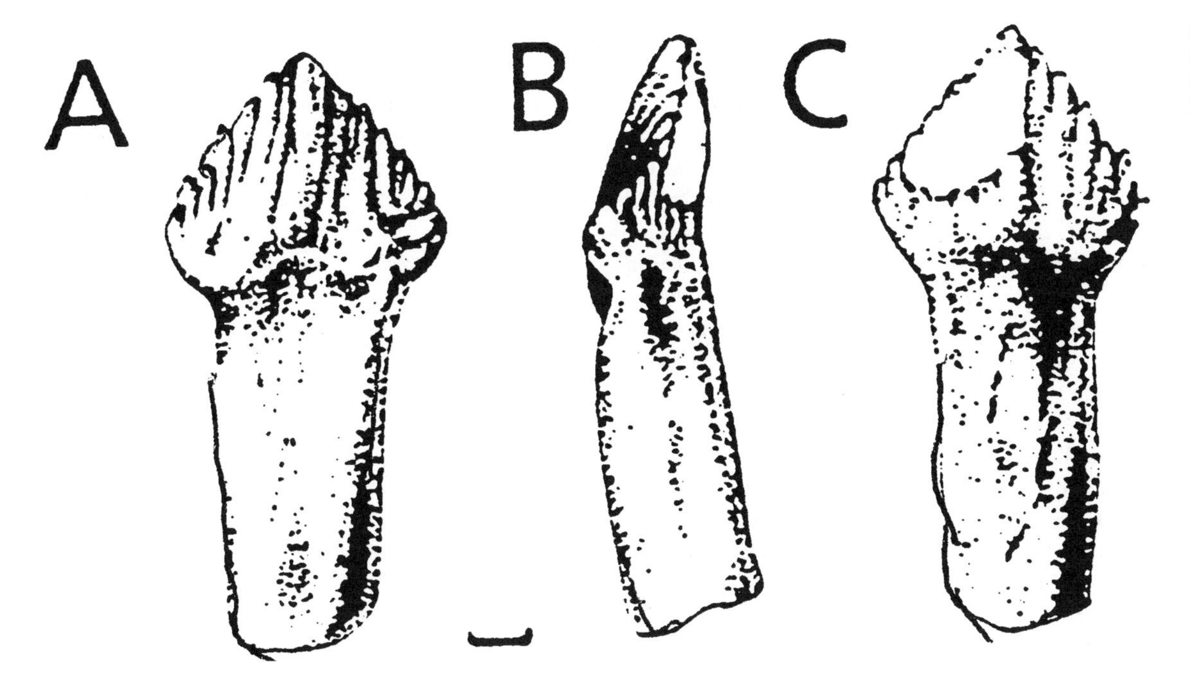

Right dentary tooth (UCMP 46911) referred to cf. *Bugenasaura*. Scale = 1 mm. (After Galton 1999*b*.)

morphological similarities between the cheek teeth, jugal, and the hindlimb proportions of *Bugenasaura* and *Thescelosaurus*, Galton (1999*b*) found it "more reasonable to regard the massiveness of the maxillary and dentary ridges of *Agilisaurus* as a parallel development" in that genus. The retention in both *Bugenasaura* and *Agilisaurus* of a continuous premaxillary tooth arcade anteriorly and an elongate palpebral were interpreted by Galton (1999*b*) to be plesiomorphic for the Ornithischia.

Galton (1999*b*) interpreted one character, the reduction of the marginal denticles on the premaxillary teeth, as homoplastic, having developed independently in other hypsilophodontids. In *Bugenasaura* and *Othnielia*, that author observed, these denticles are only marginal; in the basal ornithischian *Lesothosaurus* they are more numerous (see Sereno 1991) and in *Hypsilophodon*, these denticles are more numerous, thereby being plesiomorphic for Ornithischia; in *Drinker* they are very asymmetrical, unlike the simple cones in other hypsilophodontids.

Nonautapomorphic characters noted by Galton (1999*b*) separating *Bugenasaura* from other known hypsilophodontids include the following: 1. Teeth probably continuing to anterior end of premaxillae where tooth rows form continuous arcade; and 2. long palpebral (100 percent of maximum anteroposterior length of orbit; Weishampel and Heinrich).

Galton (1999*b*) also described a hypsilophodontid right dentary tooth (UCMP 49611), referred to cf. *Bugenasaura* and originally previously described by Galton (1975, 1980*b*), presumably from the Upper Jurassic Kimmeridge Clay of Weymouth, England (see Galton 1999*b* for details regarding the questionable provenance data associated with this specimen). According to that author, a few differences distinguish this specimen from the holotype dentary teeth of *B. infernalis*, including crown shape (more triangular in UCMP 49611), and the comparatively shorter and less prominent form of the vertical edges on the distal part of the buccal surface and on the lingual surface of the crown. As these differences are also consistent with the smaller teeth of the right dentary of the holotype, they are not a function of their positions within the tooth row. More likely, Galton (1999*b*) suggested, these differences may represent ontogenetic changes, with UCMP 49611 representing a larger tooth from a smaller and probably younger individual.

Key references: Galton (1975, 1980*b*, 1995, 1997*c*), 1999*b*); Morris (1976); Padian (1997*a*); Peng (1992); Sereno (1991); Sternberg (1940); Sues (1997*b*); Sues and Norman (1990); Weishampel and Heinrich (1992).

BYRONOSAURUS Norell, Makovicky and Clark 2000

Saurischia: Theropoda: Neotheropoda: Tetanurae: Avetheropoda: Coelurosauria: Manuraptoriformes: Manuraptora: Deinonychosauria: Troodontidae.

Name derivation: "Byron [Jaffe]" + Greek *sauros* = "lizard."

Type species: *B. jaffei* Norell, Makovicky and Clark 2000

Other species: [None.]

Occurrence: Ukhaa Tolgod, Mongolia.

Age: Late Cretaceous.

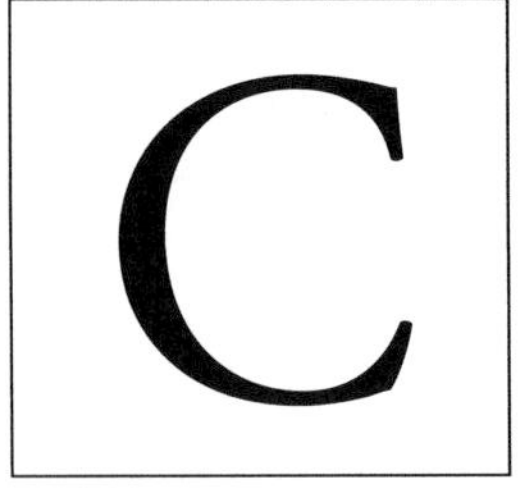

Known material: Two fragmentary skulls, partial postcranial remains, adult.

Holotype: IGM 100/983, fragmentary skull with partial postcrania.

Diagnosis of genus (as for type species): Differs from all other known troodontid species based on the following derived characters: teeth lacking serrations on anterior and posterior carinae (also found in Spinosauridae, *Pelacanimimus*, and some basal Avialae); presence of large interfenestral bar that is not inset medially from lateral surface of maxilla; connection between nasal passage and antorbital fenestra through interfenestral bar; horizontal groove on maxilla adjacent and parallel to tooth row, containing small foramina (Norell, Makovicky and Clark 2000).

Comments: The very birdlike theropod *Byronosaurus jaffei* was founded upon a specimen (IGM 100/983) including a fragmentary skull (one of the best preserved troodontid skulls ever recovered), postcranial remains collected as float, plus additional remains collected in subsequent field seasons (1994 and 1995), during the Mongolian Academy of Sciences–American Museum of Natural History expeditions to the Gobi Desert of Mongolia. A referred specimen (IGM 100/984), collected on July 15, 1996 from a locality called Bolor's Hill about five kilometers from Ukhaa Tolgod, consists of a fragmentary skull that preserves some elements not found on the holotype. This new taxon is significant in being the seventh troodontid known from Central Asia, its discovery underscoring the extensive diversity of the family in Asia (Norell, Makovicky and Clark 2000).

The authors found especially interesting in *B. jaffei* the unserrated teeth, which are very similar to the teeth found in basal avialans (*e.g.*, *Archaeopteryx lithographica*). This feature is of particular note as most known theropod taxa possess serrated teeth (see Feduccia 1996). Also of special note, a secondary palate is present, formed by extensive palatal shelves meeting the vomer on the midline, a condition also reported in the dromaeosaurid *Velociraptor mongoliensis* (Norell and Makovicky 1998) and the ornithomimosaur *Garudimimus brevipes* (Barsbold 1983*a*).

Norell *et al.* referred *Byronosaurus* to the family Troodontidae on the basis of the combination of the following derived shared characters: numerous tightly packed teeth on anterior part of tooth row (Currie 1987); maxilla participating in posterior margin of nares (Barsbold and Osmólska 1990) (also present in therizinosaurs, *e.g.*, *Erlikosaurus*); large, prominent ostosphenial crest (Currie and Zhao 1994); dentaries roughly triangular in lateral view, mental foramina in deep groove (Currie 1987); and lacrimal having elongate anterior process dorsal to antorbital fossa.

The authors performed a cladistic analysis of the Troodontidae — including *B. jaffei*, *Troodon formosus*, *Saurornithoides mongoliensis*, *S. junior*, *Sinornithoides youngi*, plus an unnamed Early Cretaceous form (IGM 100/44) (see Barsbold, Osmólska and Kurzanov 1987) from Mongolia — to determine the phylogenetic position of *Byronosaurus*. Excluded from this analysis were *Borogovia gracilicrus*, *Koparion douglassi*, and *Tochisaurus nemegtensis*, as their codings were redundant with various other taxa, and the latter two taxa cannot be diagnosed by autapomorphies. *Ornithodesmus cluniculus* and *Bradycneme draculai*, sometimes incorrectly referred to the Troodontidae (see Norell and Makovicky 1997), were also excluded. Included as outgroups were the taxa Dromaeosauridae, Oviraptoridae, Avialae, and Ornithomimosauria. From this analysis, *Byronosaurus* was found to be positioned within a clade including both species of *Saurornithoides* and the North American *Troodon formosus*, the sister group of this clade being *S. youngi* from the Early Cretaceous of Central Asia.

Key references: Barsbold (1983*a*); Barsold and Osmólska (1990); Barsbold, Osmólska and Kurzanov (1987); Currie (1987*a*); Currie and Zhao (1994); Feduccia (1996); Norell and Makovicky (1997, 1998); Norell, Makovicky and Clark (2000).

†CAMARASAURUS

Comments: Much is known about the large sauropod, *Camarasaurus*, thanks to the recovery of a number of fine specimens that have been referred to this genus.

One of the most complete and perfectly preserved dinosaur skeletons ever found is that of a juvenile individual of the camarasaurid sauropod species *Camarasaurus lentus*. The articulated specimen (CM 11338), collected at the Carnegie Quarry (now Dinosaur National Monument) in Utah, was described by Gilmore (1925). As related by Ashworth (1996), the skeleton was displayed at the Carnegie Museum of Natural History as a panel mount, keeping "its original position, except that the tail was straightened out, a few displaced bones were rearticulated, and the missing left ilium was provided from another specimen...," and a sternal plate, found next to the tail, "was placed beneath the neck in the final display (see *D:TE* for a photograph of the original specimen as displayed at the Carnegie Museum of Natural History).

Like many articulated dinosaur skeletons, CM 11338 was found preserved in a so-called "death pose." That is, it exhibits a post-mortem deformation wherein the series of cervical vertebrae has been flexed dorsally (most likely due to dewatering and contraction of the nuchal ligament), the result being that the

Photograph by the author, courtesy Natural History Museum of Los Angeles County.

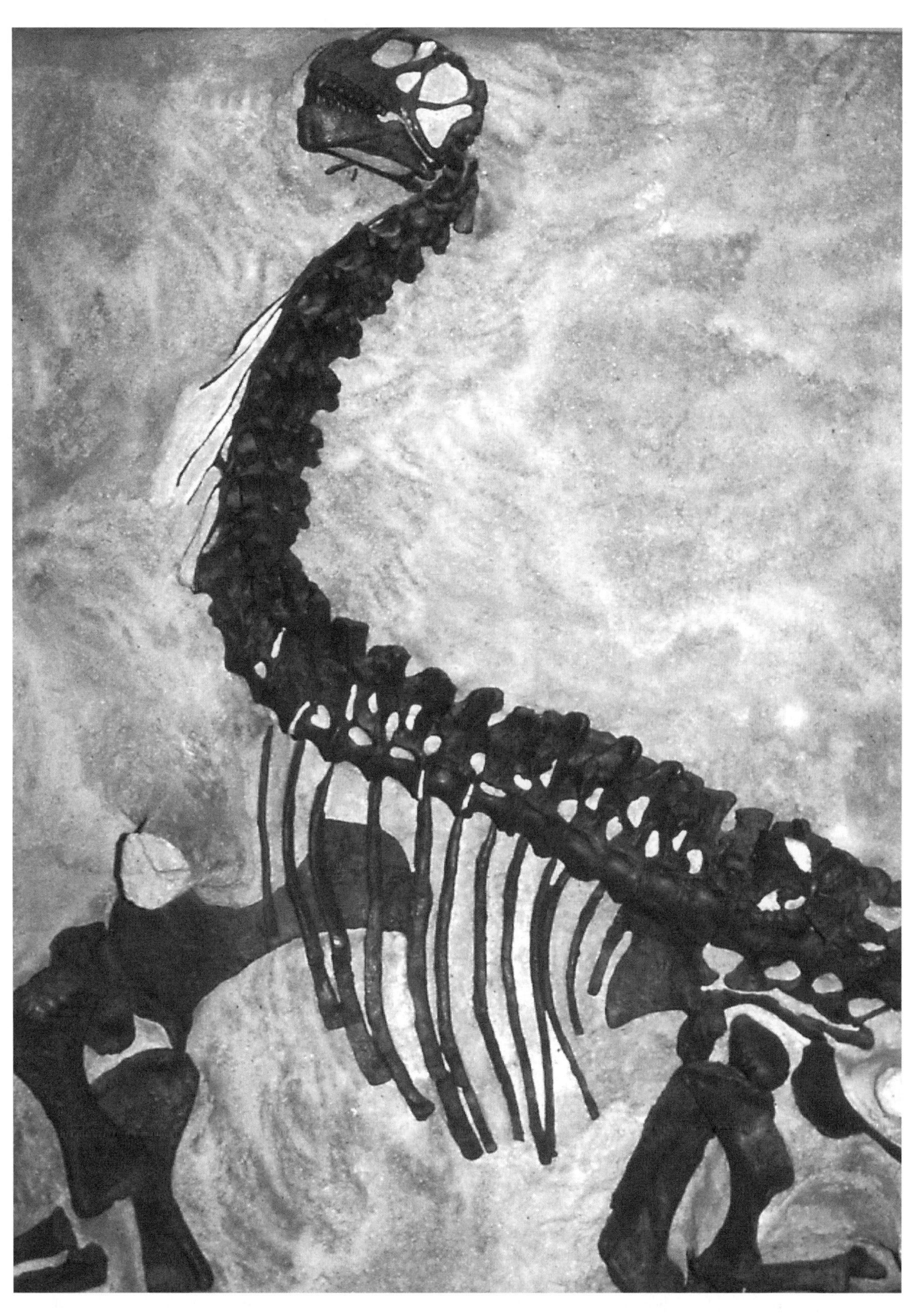

Articulated, virtually complete skeleton (LACM cast of CM 11338) of *Camarasaurus lentus* (juvenile) as found in postmortem "death pose." Utilizing computer graphics, J. Michael Parrish and Kent A. Stevens have refigured this specimen in its original neutral pose.

vertebrae are no longer in articulation, and the neck is bent backwards.

J. Michael Parrish and Kent A. Stevens have continued experiments regarding sauropod anatomy, their work this time concerning CM 11338, utilizing three-dimensional computer graphics software called DinoMorph (see *S1*). Manipulating images of the CM 11338 elements, both manually and via computer,

Photograph by the author, courtesy Royal Tyrrell Museum/Alberta Community Development.

Skeleton of an adult *Camarasaurus supremus* mounted with its neck in vertical position, this orientation not possible according to various recent studies by John Martin, Valérie Martin-Rolland, and Eberhard (Dino) Frey (1989; see also Introduction) and J. Michael Parrish and Kent A. Stevens (1998).

Parrish and Stevens (1998) restored the skeleton to its undistorted neutral pose. Results of this exercise included the revelation that this sauropod was capable of greater neck flexibility than in the longer-necked diplodocids, but that it could not achieve the severe dorsiflexion of the head as seen in the death pose (see also "Introduction" for more on these authors' studies of sauropod neck posture).

Wilhite and Curtice (1998) investigated ontogenetic variation in sauropods, citing the problem that ontogeny affects the morphology of growing sauropod individuals. Focusing upon *Camarasaurus*, examining limb-proportion changes at numerous ontogenetic stages in this genus, Wilhite and Curtice found limb bones to grow isometrically, exhibiting only slight proportional changes over a wide range of increments in size; early analyses by the authors of other sauropod genera supported this conclusion throughout the Sauropoda.

Wilhite and Curtice noted the following characters of sauropod vertebrae that were affected by ontogeny: development of the bifurcation and laminae of the neural spine; depth of the ventral sulcus; the size, depth, and shape of pneumatic fossae; and the order of neurocentral fusion. The authors noted that this latter feature could be of phylogenetic importance, as in diplodocids and titanosaurids fusion commences in the distal caudal vertebrae and proceeds anteriorly, while in *Dicraeosaurus* the dorsal neural arches are the last to fuse to their centra.

Wilhite and Curtice cautioned against including taxa known only from juvenile specimens in phylogenetic analyses, as the often severe morphological changes undergone by vertebrae during ontogeny can seriously impact such analyses.

Lucas and Heckert (2000), in a survey of the Jurassic dinosaurs of New Mexico, reported on a caudal vertebra referred to *Camarasurus* sp., collected from the Summerville Formation in the Hagan basin of north-central New Mexico (see Hunt and Lucas 1993*a*; Lucas, Anderson and Pigman 1995), constituting one of the stratigraphically lowest, and therefore oldest, occurrences of sauropod body fossils in North America. According to Lucas and Heckert, the

Courtesy The Field Museum (neg. #CSGEO8793).

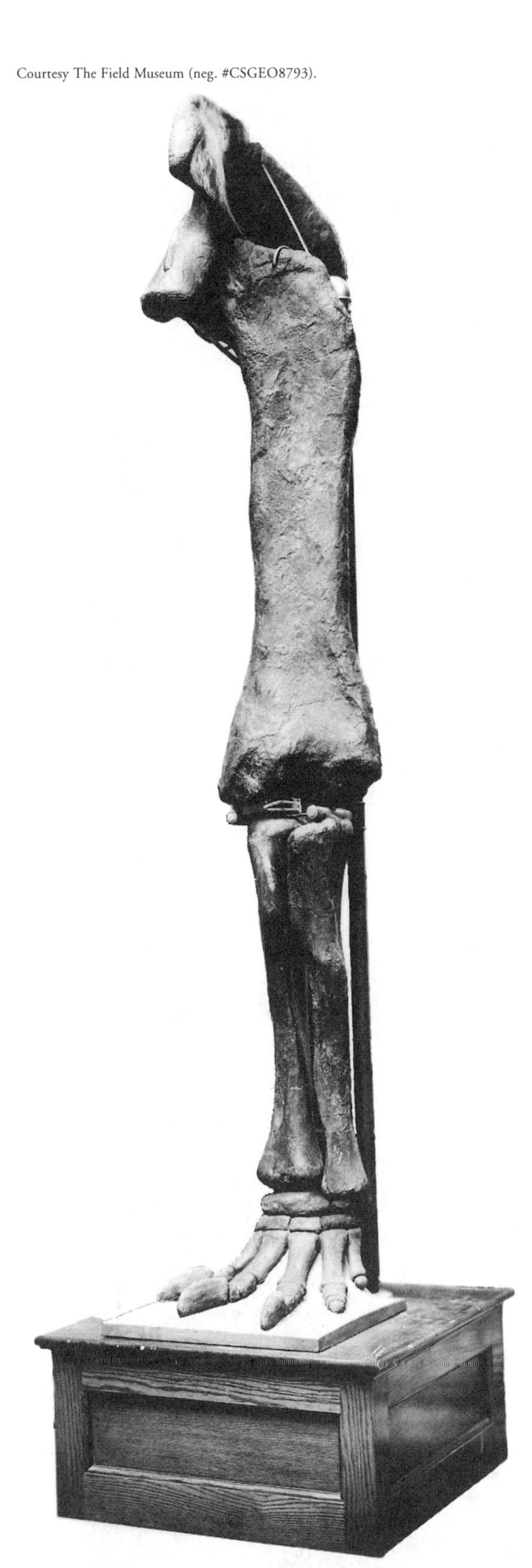

Camarasaurus grandis **left hind limb (originally referred to *Morosaurus impar*), as exhibited during the early 1900s at the Field Columbian Museum. In 1998, Ray Wilhite and Brian David Curtice published on ontogenetic variation in sauropods based upon limb proportions of dinosaurs such as *Camarasaurus*.**

occurrence of this genus supports a Late Jurassic age for the upper part of this formation.

Key references: Ashworth (1996); Gilmore (1925); Hunt and Lucas (1993*a*); Lucas, Anderson and Pigman (1995); Lucas and Heckert (2000); Parrish and Stevens (1998); Wilhite and Curtice (1998).

†CAMELOTIA

Name derivation: "Camelot" [seat of King Arthur's court in Avalon, probably in Somerset, England].

Diagnosis: Large melanorosaurid in which height of centra of posterior dorsal and anterior caudal vertebrae exceed length; proximal end of tibia only slightly expanded anteroposteriorly relative to shaft (Galton 1998).

Comments: In 1989, Peter M. Galton, in a study of the saurischian dinosaurs of the Upper Triassic of England, published a detailed study of the holotype of *Camelotia borealis*, a large prosauropod dinosaur which that author had named in 1985 (see *D:TE*).

To *C. borealis*, Galton (1998) referred all the prosauropod remains recovered from a small quarry at Wedmore Hill ("Rhaetic," Upper Triassic) in the parish of Wedmore in the Vale of Glastonbury, Somerset, these fossils originally reported by Sanford (1894). These included the following, all probably pertaining to a single individual: BMNH R2870, left femur, R2871, proximal part of left tibia, R2872, ungual phalanx of right pedal digit III, phalanx 3 of right pedal digit III, phalanx 3 of right pedal digit IV, R2873, dorsal vertebrae 14 and 15, R2877, dorsal vertebra, R2876, caudal vertebra 6, all described by Seeley (1989); BMNH R2874, incomplete proximal part of left dorsal rib, incomplete proximal part of right pubis, distal end of right ischium, R2876, two distal caudal vertebrae, (mislaid) caudal 2 or 3, (mislaid) distal caudal, R2878, distal portion of ?left fibula, described by Huene (1907–08); also, BMNH R2878, incomplete neural arch of anterior dorsal vertebra, and R2878, centrum of posterior cervical vertebra (see also *Megalosaurus* entry).

Wellnhofer (1993) had suggested that *C. borealis* could be a junior synonym of the plateosaurid prosauropod *Plateosaurus engelhardti*. This assessment was based upon similarities between the femur of *C. borealis* and that of an approximately 40 percent complete specimen, representing a large individual referred to *P. engelhardti*, collected from a bonebed in the Feurletten (uppermost Middle Keuper or Upper Triassic: Middle Norian) in Ellingen, in Middle Franconiam in Bavaria. Galton (1998*a*, 1998*b*, 2000) commented that, as described by Wellnhofer, the Ellingen specimen differs from the many *Plateosaurus* specimens recovered from Trössingen, Nordwürttemberg

Camelotia borealis, holotype elements: a. BMNH R2870, left femur, anterior view; c. BMNH R2873b, dorsal vertebra, right lateral view; d.–e. BMNH R2877, dorsal vertebra 4, posterior and right lateral views; f. BMNH R2876a, caudal vertebra, right lateral view; h. BMNH R2872a, ungual of phalanx III, right pes, proximal view; and i. BMNH R2872a–b, ungual of phalanx and phalanx 3, digit III, right pes, medial view. Scale = 10 cm. (a–c, f) and 1 cm. (g–i, j, k). (After Seeley 1898.)

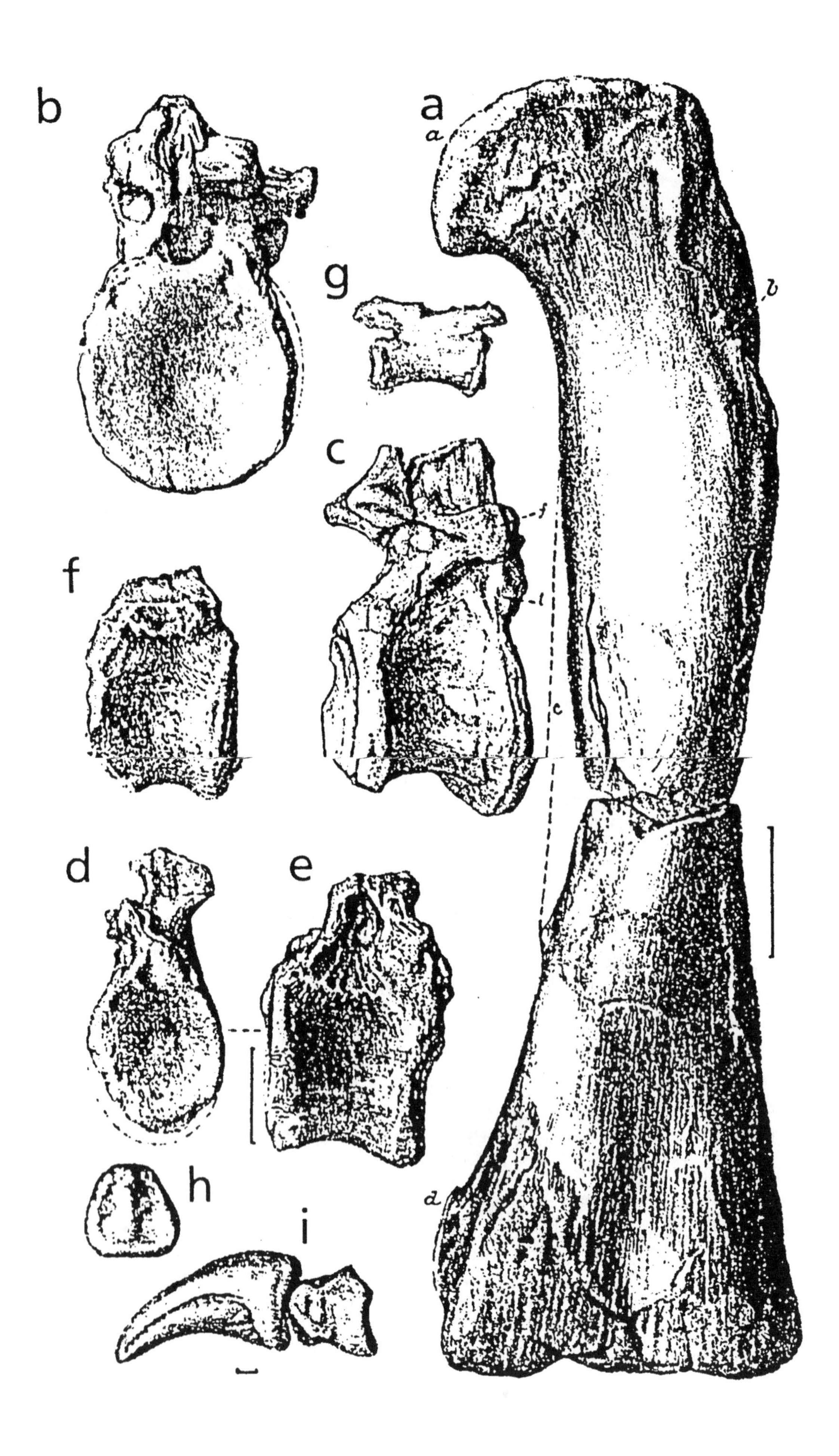

(see *Plateosaurus* entry, *D:TE*) in being more massively built, and having a more robust femur and relatively shorter metatarsals. Further addressing this possibility, Galton (1998) noted that the holotype femur of *C. borealis* corresponds more closely in its midshaft ratio to the melanorosaurid *Melanorosaurus readi* than to *P. engelhardti*. Furthermore, *C. borealis* displays two synapomorphies of the Melanorosauridae: the lesser trochanter is almost sheet-like, and the transverse width of the shaft at midlength exceeds the antero-posterior width. Also, the centra of the most anterior caudal vertebrae are tall but short, as in *Melanorosaurus*, which, according to Galton (1998), is probably the sister taxon of *Camelotia*. The Ellingen prosauropod, Galton (1998) suggested, may represent another valid species of *Plateosaurus* in western Europe, this idea remaining for the present a topic for future discussion.

Key references: Galton (1985*d*, 1998*a*, 1998*b*, 2000); Huene (1907–08); Sanford (1894); Seeley (1898); Wellnhofer (1993).

CAMPOSAURUS Hunt, Lucas, Heckert, Sullivan and Lockley 1998 [*nomen dubium*]—(=?*Coelophysis*, ?*Podokesaurus*, ?*Syntarsus*)

Saurischia: Theropoda: Neotheropoda: "Ceratosauria": Coelophysoidea: Coelophysidae.

Name derivation: "[Charles] Camp" + Greek *sauros* = "lizard."

Type species: *C. arizonensis* Hunt, Lucas, Heckert, Sullivan and Lockley 1998 [*nomen dubium*].

Other species: [None.]

Occurrence: Bluewater Creek Formation, Arizona, United States.

Age: Late Triassic (late Carnian).

Known material: Miscellaneous vertebral and hindlimb elements, possibly other postcranial remains.

Holotype: UCMP 34498, associated right and left distal hindlimb bones, with fused astragalus and calcanea.

Diagnosis of genus (as for type species): Theropod differing from all "ceratosaurs" except *Coelophysis* [=*Rioarribasaurus* of the authors' usage] and *Syntarsus* in possessing tarsals fused to tibia and fibula, differing from *Coelophysis* and *Syntarsus* in that the ventral margin of astragalus is horizontal rather than deeply concave in anterior/posterior view (Long and Murry 1995), further differing from *Coelophysis* in that the fibula was in contact with tibia visible in anterior view for distance equal to at least one and a half times the width of astragalus plus calcaneum (Hunt, Lucas, Heckert, Sullivan and Lockley 1998).

Comments: The genus *Camposaurus* was founded upon numerous dissociated theropod postcranial elements of similar size, with no duplication of elements (figured by Lucas, Hunt and Long 1992; see also Long and Murry 1995, therein figured with corrected specimen numbers). The holotype consists of associated, fused right and left tibia-fibula-astragalae-calcanea (UCMP 34498) collected by Charles Camp from the *Placerias* quarry in the Upper Triassic/late Carnian Bluewater Creek Formation (Chinle Group) of Apache County, Arizona (see Lucas *et al*; Long and Murry). Paratypes from the same locality include an uncatalogued proximal right femur (originally cited by Long and Murry as belonging to an indeterminate "ceratosaur"), a partial sacrum preserving portions of four fused centra (figured by Lucas, Hunt and Murry 1992; Long and Murry), a dorsal centrum (figured by Long and Murry), and three incomplete sacra (all of these specimens uncatalogued, in the UCMP collection); also a dorsal centrum (MNA collection, figured by Long and Murry). Because the above material revealed no duplication of elements, and as all specimens are of comparable size, it was most parsimonious to assume that they belong to a single individual (Hunt, Lucas, Heckert, Sullivan and Lockley 1998).

According to Hunt *et al.*, the number of sacral vertebrae in *Camposaurus* and the presence of a high ascending process preclude this taxon's assignment to the Herrerasauridae (see Novas 1993), a group of very primitive Late Triassic theropods. The possession of four fused sacral vertebrae and the fusion of the distal limb elements suggested to the authors that the holotype represents a "ceratosaur" (*sensu* Rowe and Gauthier 1990). As Hunt *et al.* emphasized, the paratypes of the type species *C. arizonensis* offer additional characters (*e.g.*, femoral head more rectangular and dorsal centra more waisted in ventral aspect) separating this taxon from both *Coelophysis* and *Syntarsus* (see entries).

Additional theropod remains have been recovered from the *Placerias* quarry, including a proximal left pubis and some dorsal centra (Long and Murry). According to Hunt *et al.*, these specimens, representing a smaller individual than that to which the above material belongs, cannot be referred with certainty to *Camposaurus*.

Downs (2000), in a comparative study of the theropods *Coelophysis bauri* and *Syntarsus rhodesiensis*, commented that Hunt *et al.*, in citing a paper by Padian (1986) on *Coelophysis*, had contrasted the type material of *C. arizonensis* with inaccurate drawings by Lois Darling of *C. bauri* skeletal elements published in Colbert's (1989) monograph on *Coelophysis*. According to Downs, "*Camposaurus* appears to fall easily within the range of variation seen in the Ghost Ranch sample of *Coelophysis* and is consequently a *nomen dubium*."

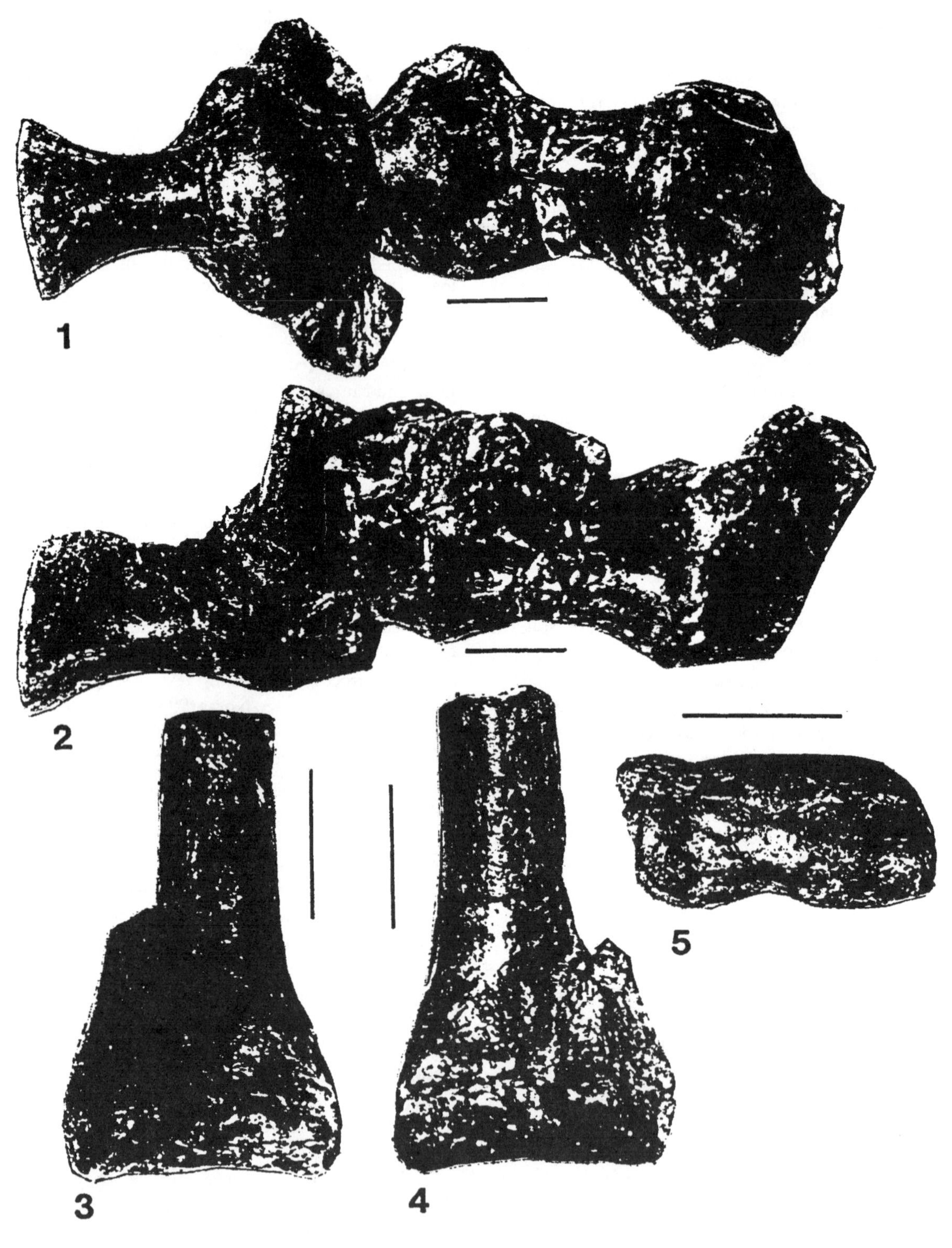

***Camposaurus arizonensis*, paratype (uncatalogued UCMP specimen; field number A269cg2) partial sacrum, in 1 ventral and 2. lateral views; holotype (UCMP 34498) fused right tibia/fibula/astragalus/calcaneum, 3. posterior, 4. anterior, and 5. distal views. Scale = 1 cm. (After Hunt, Lucas, Heckert, Sullivan and Lockley 1998).**

Note: Hunt *et al.* reassessed additional remains referred to various dinosaurian taxa collected from the *Placerias* quarry. As noted by these authors, specimens referred by Murry and Long (1989) and Long and Murry (1995) to *Chindesaurus bryansmalli* (see *Chindesaurus* entry) represent an indeterminate herrerasaurid.

According to Hunt *et al.*, a single tooth (MNA V3690) described by Tannenbaum (1983) as belonging to the Prosauropoda, actually belongs to an ornithischian, so identified because it is not symmetrical in anterior view. The tooth resembles one referred earlier by Hunt and Lucas (1994) to the primitive ornithischian *Lucianosaurus* and, along with two other teeth (MNA V3682 and V3683), by Kaye and Padian (1994) to *Revueltosaurus*. However, it differs from *Revueltosaurus* in such features as "(1) much smaller size,

(2) much finer denticulations; and (3) more pointed apex." Hunt *et al.* interpreted all three of these MNA specimens as representing indeterminate ornithischians.

Teeth identified by Tannenbaum, Murry and Long, and Long and Murry as prosauropod were found by Hunt *et al.* to belong to the ornithischian *Tecovasaurus murryi*; a partial tibia (UCMP A269/25793, figured by Long and Murry) exhibits no prosauropod synapomorphies (Novas 1989) and may, according to Hunt *et al.*, represent a primitive theropod.

Key references: Colbert (1989); Downs (2000); Hunt, Lucas, Heckert, Sullivan and Lockley (1998); Kaye and Padiau (1994); Long and Murry (1995); Lucas, Hunt and Long (1992); Novas (1993); Padian (1986); Rowe and Gauthier (1990); Tannenbaum (1983).

†CARCHARODONTOSAURUS (=*Sigilmassasaurus*)

Saurischia: Theropoda: Neotheropoda: Tetanurae: Avetheropoda: Carnosauria: ?Allosauroidea: Carcharodontosauridae.

Age: Middle Cretaceous (Albian–Cenomanian).

Comments: *Carcharodontosaurus*, once a rather poorly known theropod taxon, is now represented by more and quite substantial specimens. Some of these fossils were originally referred to other taxa, and have only recently been identified as belonging to this very large ?allosauroid genus.

In 1934, Ernst Stromer described postcranial remains collected from the Baharija Formation, Baharija oasis, Egypt, as belonging to what he believed to be a new species of the giant theropod *Spinosaurus*. Stromer unofficially referred to this form as "*Spinosaurus* B" (see *D:TE*). Significantly, remains of this so-called "*Spinosaurus* B" were collected from the same locality that would later yield a partial skeleton of *Carcharodontosaurus saharicus* (see Sereno, Dutheil, Iarochene, Larsson, Lyon, Magwene, Sidor, Varricchio and Wilson 1996).

Over a half century later, Russell (1996) erected a new genus and species, *Sigilmassasaurus brevicollis*, upon isolated vertebrae recovered from Cenomanian-age rocks in Morocco. Russell also referred Stromer's "*Spinosaurus* B" material to this new taxon (see *Sigilmassasaurus* entry, *S1*).

Carcharodontosaurus saharicus sculpture by Garfield Minott and reconstructed skull (including cast of SGM-Din 1), both prepared under the direction of Paul C. Sereno and displayed at the Crerar Library, University of Chicago.

Photograph by the author, courtesy University of Chicago.

More recently, Sereno, Beck, Dutheil, Gado, Larsson, Lyon, Marcot, Rauhut, Sadleir, Sidor, Varricchio, Wilson and Wilson (1998) reassessed all of the above-mentioned material. Sereno *et al.* pointed out that the remains designated "*Spinosaurus* B" overlap the partial skeleton of *C. saharicus*. Also, these authors questioned Russell's distinguishing *S. brevocollis* from *C. sarahicus* on the basis of the proportions of a single vertebral centrum. Therefore, Sereno *et al.* (1998) referred *S. brevicollis* to *C. saharicus*, at the same time provisionally referring to the latter species all carcharodontosaur remains collected from rocks of Albian and Cenomanian age in northern Africa.

Lamanna, Smith, Lacovara, Dodson and Attiya (2000), in an abstract, reported the collection of additional fossils comprising more than five tons of vertebrate, invertebrate, and plant material from the Baharija Oasis in 2000 by a joint University of Pennsylvania-Egyptian Geological Museum team. Among these remains are theropod vertebrae collected from Gebel Harra, which according to Lamanna *et al.*, probably belong in the holotype of *C. saharicus*. (Other taxa found at during this field expedition include a possible abelisaur represented by shed tooth crowns, other theropods [including *Spinosaurus*] represented by isolated postcrania and teeth, sauropods represented by an associated partial skeleton pertaining to a very large genus, the fragmentary skeleton of a small form, plus several isolated specimens, crocodyliforms, the squamate *Simoliophis*, turtles, and several genera of fishes.)

Key references: Lamanna, Smith, Lacovara, Dodson and Attiya (2000); Russell (1996); Sereno, Beck, Dutheil, Gado, Larsson, Lyon, Marcot, Rauhut, Sadleir, Sidor, Varricchio, Wilson and Wilson (1998); Sereno, Dutheil, Iarochene, Larsson, Lyon, Magwene, Sidor, Varracchio and Wilson 1996); Stromer (1934).

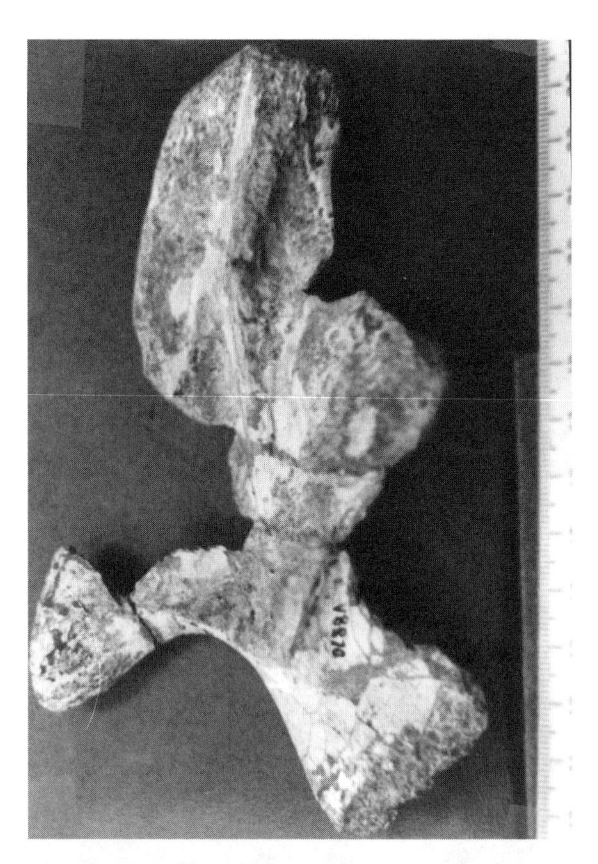

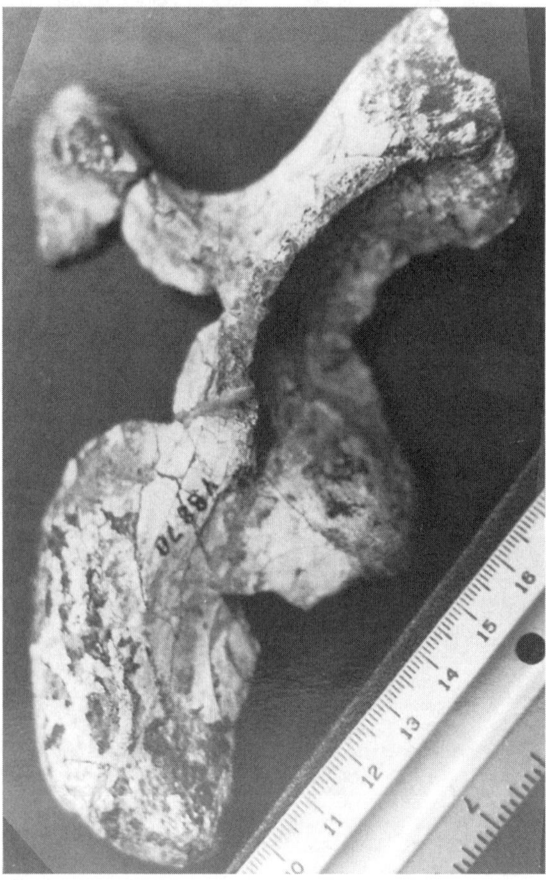

***Right, top: Caseosaurus crosbyensis*, UMMP 8870, holotype right ilium originally referred by E. C. Case to *Coelophysis* sp., outer view. *Right, bottom: Caseosaurus crosbyensis*, UMMP 8870, holotype right ilium originally referred by E. C. Case to *Coelophysis* sp., inner view.**

Photographs by Robert A. Long and Samuel P. Welles, courtesy University of Michigan Museum of Paleontology.

CASEOSAURUS Hunt, Lucas, Heckert, Sullivan and Lockley 1998

Saurischia: ?Theropoda: Herrerasauridae.

Name derivation: "[Ermin C.] Case" + Greek *sauros* = "lizard."

Type species: *C. crosbyensis* Hunt, Lucas, Heckert, Sullivan and Lockley 1998.

Other species: [None.]

Occurrence: Dockum Formation, Texas, United States.

Age: Late Triassic (late Carnian/"late Tuvalian" [substage of the Carnian]).

Known material/holotype: UMMP 8870, right ilium.

Diagnosis of genus (as for type species): "Herrerasaur" having ilium differing from that of *Staurikosaurus* and *Herrerasaurus* in possessing elongate and dorsoventrally narrow posterior blade, narrow ridge on lateral margin extending from anterodorsal margin of acetabulum to anterodorsal anterior spine (*sensu* Novas 1993), highly reduced brevis fossa, and semicircular margin dorsal to acetabulum; differs from *Chindesaurus bryansmalli* in possessing less deep brevis shelf that does not extend to posterior margin, lateral longitudinal ridge (for sacral ridge articulation) placed more ventrally, and much thinner (less than half) posterior blade in dorsal view (Hunt, Lucas, Heckert, Sullivan and Lockley 1998).

Comments: The genus *Caseosaurus* was founded upon a right ilium (UMMP 8870) discovered by E. C. Case in the Upper Triassic (late Tuvalian, a substage of the Carnian) Tecovas Member of the Dockum Formation (exact locality unknown) of Crosby County, Texas. Case originally figured this specimen as *Coelophysis* sp., after which Long and Murry (1995) referred it to the primitive herrerasaurid *Chindesaurus bryansmalli* (see *Chindesaurus* entry, *D:TE*).

In reassessing UMMP 8870, Hunt, Lucas, Heckert, Sullivan and Lockley (1998) observed that this specimen is identified as dinosaurian by its brevis shelf and medially opened acetabular wall. The ilium bears a strong resemblance to that of the herrerasaurids *Staurikosaurus* and *Herrerasaurus* in being anteroposteriorly short and dorsoventrally high, and having a partially perforate acetabulum. Also, the brevis shelf is reduced to a slight ridge, and the brevis fossa is small, features indicative of herrerasaurids (Novas 1993).

According to Hunt *et al.*, UMMP 8870 differs from *Staurikosaurus* and *Herrerasaurus* in possessing the following: 1. elongate, dorsoventrally narrow posterior blade; 2. narrow ridge on the lateral margin, extending from the anterodorsal margin of the acetabulum to the anterodorsal anterior spine (*sensu* Novas 1993); 3. highly reduced brevis fossa; and 4. semicircular margin dorsal to the acetabulum.

Hunt *et al.* noted that Long and Murry had assigned this ilium to *C. bryansmalli*, although the holotype of the latter (PEFO 10395) includes only two small fragments of the ilium — a fragment of the pubic process and a portion of the left posterior iliac blade — which the latter authors regarded as indistinguishable from UMMP 8870. Although the pubic process exhibited no diagnostic features, the fragment of iliac blade allowed Hunt *et al.* to compare UMMP 8870 with PEFO 10395. They noted that the posterior iliac blade of PEFO 10395 differs from that of the Texas specimen in having a brevis shelf that extends to the posterior margin, a more distally located lateral longitudinal ridge (for sacral rib articulation), and a much thicker (over twice) posterior blade in dorsal aspect.

Deciding that UMMP 8870 clearly represents a herrerasaurid but not *Chindesaurus*, Hunt *et al.* referred the ilium to the new genus and species *Caseosaurus crosbyensis*, at the same time accepting the Herrerasauridae as an assemblage of basal theropods.

Key references: Case (1927); Hunt, Lucas, Heckert, Sullivan and Lockley (1998); Long and Murry (1995).

†CAUDIPTERYX

Saurischia: Theropoda: Neotheropoda: Tetanurae: Avetheropoda: Coelurosauria: Manuraptoriformes: Manuraptora: ?Deinonychosauria: Caudipteridae.

New species: *C. dongi* Zhou and Wang 2000.

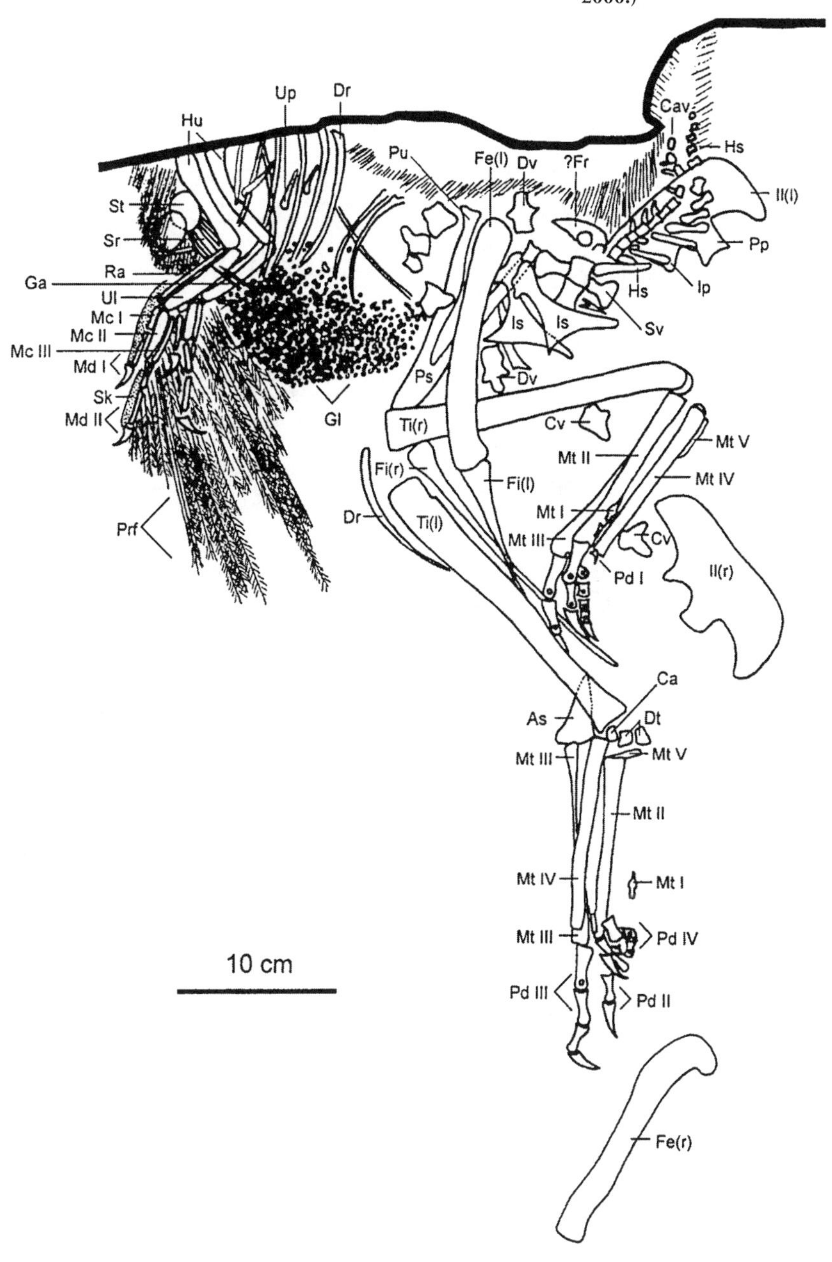

***Caudipteryx dongi*, IVPP V 12344, holotype skeleton. (After Zhou and Wang 2000.)**

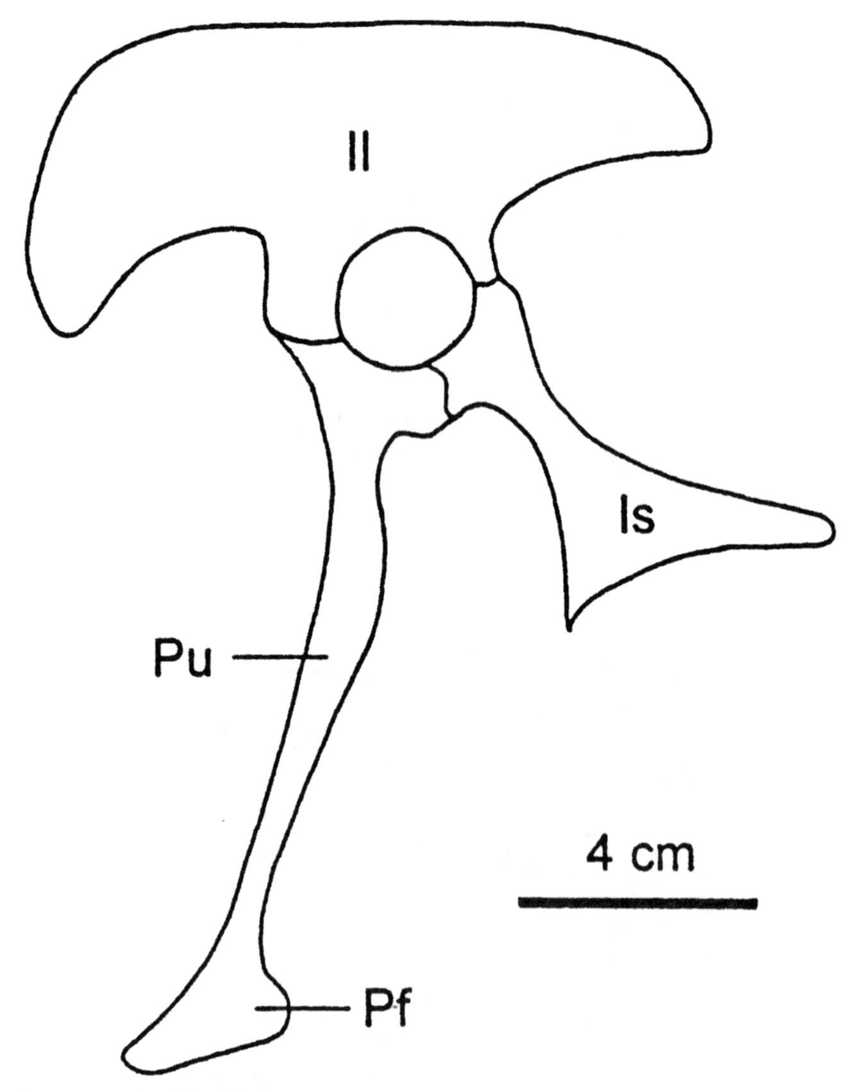

Reconstructed pelvis of the very birdlike dinosaur *Caudipteryx* in left lateral view. (After Zhou and Wang 2000.)

Occurrence of *C. dongi*: Yixian Formation, Liaoning Province, People's Republic of China.

Age of *C. dongi*: ?Late Jurassic (Kimmeridgian) or ?Early Cretaceous (Valangian).

Known material/holotype of *C. dongi*: IVPP V 12344, incomplete postcranial skeleton including complete forelimbs, hindlimbs, and pelves, feather impressions.

Diagnosis of *C. dongi*: Sternum small, femur to sternum ratio of 6.0 (4.1 in *C. zoui*); metacarpal I to II ratio of 0.45 (0.40 in *C. zoui*); ischium short; ilium long (Zhou and Wang 2000).

Comments: One of the more important and spectacular dinosaur discoveries of recent years has been *Caudipteryx* (see *S1*; also "Introduction," section on birds, this volume), a small, feathered, and very birdlike theropod based on two almost complete and partially articulated skeletons, first reported and briefly described by Ji, Currie, Norell and Ji in 1998.

Zhou and Wang (2000) subsequently reported that, in the summer of 1998, a field crew of the Institute of Vertebrate Paleontology and Paleoanthropology of the Chinese Academy of Sciences recovered an incomplete skeleton (IVPP V 12344) lacking the skull, but with the feathers well preserved and the bones in better articulation than in the original two specimens. The new skeleton was found in layer six at the ?Late Jurassic–Early Cretaceous (Jiang, Chen, Cao and Komatsu 2000) Zhangijagou locality in the Yixian Formation, near the Sihetun locality in Beipiao, in Liaong Province, northeastern China. Representing a larger individual than the type species, *C. zoui*, Zhou and Wang referred it to the new species *C. dongi*.

Compared with the type species, *C. dongi* is a larger animal, with a relatively longer first metacarpal, although the sternum (length of 25 millimeters) is longer in *C. zoui* (36 millimeters). The authors reported that a recently collected, as of yet uncatalogued specimen of *C. zoui*, though also smaller than the holotype of *C. dongi*, has a sternum measuring 30 millimeters in length. This relatively long metacarpal I and short sternum indicated to Zhou and Wang that *C. dongi* may be a slightly more primitive species than *C. zoui*.

As discussed by Zhou and Wang, the holotype of *C. dongi* reveals details of the *Caudipteryx* skeleton not preserved in the collected *C. zoui* specimens. For example (*contra* Ji *et al.*), the pubis in *C. dongi* is clearly antero-ventrally oriented, as in oviraptorids and most saurischian dinosaurs, not backwardly as in dromaeosaurids and birds. The pubic peduncle has a flat ventral margin in side view, unlike the concave margin in dromaeosaurids and the convex margin in oviraptorids. The fibula is just slightly shorter than the tibia, and probably contacted the calcaneum as in oviraptorids and most theropods. The fifth metatarsal is a slender rod-like element as in *Deinonychus*, oviraptorids, *Archaeopteryx*, and *Confuciusornis*. Uncinate processes, resembling in general shape those of oviraptorids (see Clark *et al.*) and the early ornithurine bird *Chaoyangia* (Hou and Zhang 1993), articulate with dorsal ribs near their midshaft. Pedal digit III is the longest and IV is shorter than II. As in birds, the hallux of the pes is reversed, while the ungual is shorter than those of the other digits. This reversed hallux suggested to Zhou and Wang "that the ancestors of *Caudipteryx* had obtained perching or grasping capability, supposedly a prerequisite for the origin of avian flight, therefore supporting the arboreal hypothesis of the origin of avian flight" (see also "Introduction," section birds and the origin of flight).

Zhou and Wang observed that the longest primary feather preserved in IVPP P 12344 is longer than the femur. However, the barbules could not be perceived in the features in this specimen, perhaps, the authors speculated, because they had not yet developed in this genus. Therefore, this feather in *Caudipteryx* was most likely more primitive than feathers in birds, an alternative explanation being that it was secondarily reduced.

Caudipteryx and also *Protarchaeopteryx* (Ji and Ji 1996; see *S1*) are known from well-preserved skeletons including feather impressions collected in Liaong Province, the latter genus having originally been described its their authors as a bird (later reclassified as a dinosaur). Indeed, most vertebrate paleontologists have accepted the classification of these taxa as nonavian theropods.

However, Feduccia (1999*a*) briefly offered a dissenting opinion that both of these taxa are not, in fact, theropod dinosaurs, but rather birds more primitive than *Archaeopteryx*. Feduccia criticized that half of the characters used to diagnose these genera are primitive and should not have been used, and half are not present in the preserved material, with but two or three characters remaining that cannot be properly interpreted due to the crushed condition of the skulls. According to that author, "*Caudipteryx* exhibits a suite of features that show it to be a secondarily flightless bird, a Mesozoic kiwi, including a protopygostyle (fused tail vertebrae), an avian occiput, reduced fibula, wing feathers attached as in archaic birds, *etc.*" As no examples of theropod skin shows feathers, Feduccia regarded the idea of feathered dinosaurs as a myth. As it would have been beyond the scope of his commentary to do so, Feduccia did not back up the above claims with a detailed analysis of these two taxa.

Similarly, Ruben and Jones (2000) suggested that *Caudipteryx* could be a bird. Briefly these authors pointed out that bipedal dinosaurs and cursorial birds exhibit distinctly different cursorial mechanisms. In cursorial birds, Ruben and Jones stated, stride generation centers on parasagittal rotation of the lower hindlimb, while in bipedal dinosaurs the entire hindlimb seems to have been involved in stride generation, these variations being causally and tightly linked to the proportionately longer hindlimbs in cursorial birds. *Caudipteryx*, based on previously undescribed criteria, was apparently the only known theropod to have utilized the avian running style, raising "questions about character matrices used previously to assign *Caudipteryx* to Theropoda," while also calling attention to "the degree to which continued misinterpretation of the Mesozoic avian pelvis has contributed to current views of theropod-bird relationships."

Norell, Ji and Ji (1998), in a brief report on the anatomy of both *Protarchaeopteryx* and *Caudipteryx*, stated that these genera "are morphologically and phylogenetically intermediate between the compsognathid *Sinosauropteryx* ... and the primitive bird *Confuciusornis*," and that, based on cladistic analysis, the former two taxa should remain outside of Aviale.

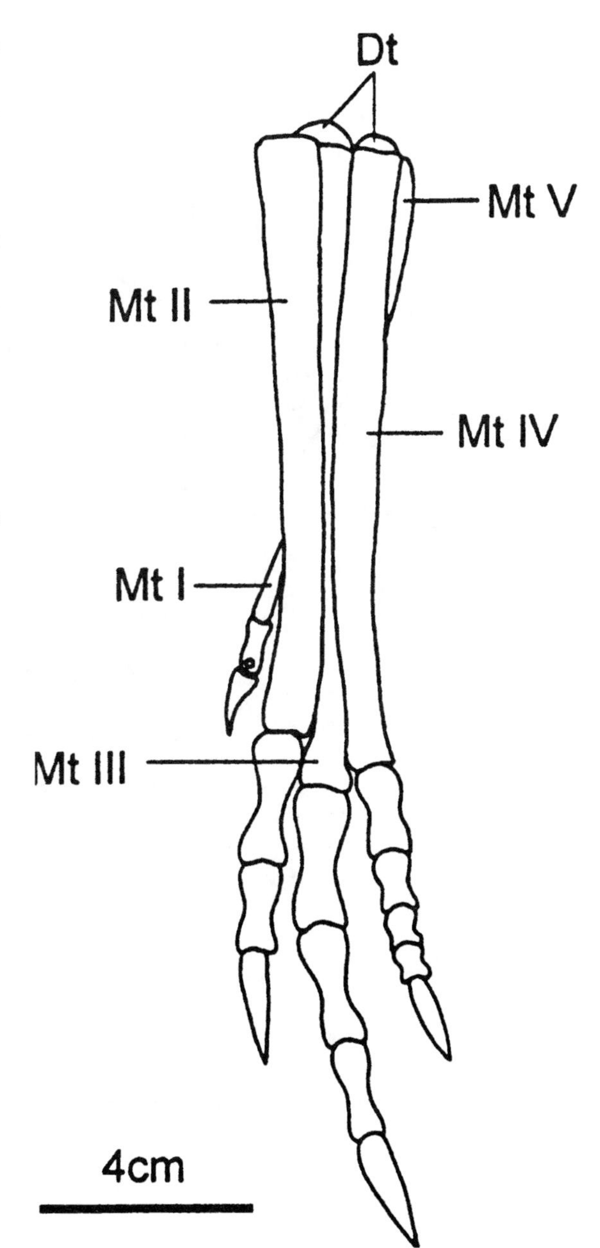

Reconstructed left hindlimb of *Caudipteryx* in dorsal view. (After Zhou and Wang 2000.)

Currie (1999), in a later abstract, stated that new information on the skeletal anatomy of *Sinosauropteryx*, *Protarchaeopteryx*, and *Caudipteryx* "confirms that all three are nonavian theropods representing distinct stages of coelurosaurian evolution." All three of these genera, the author posited, are insulated by simple integumentary structures; *Protarchaeopteryx* also possesses added retrices to its pelage and *Caudipteryx* has well-developed remiges and rectrices. According to Currie, the phylogenetic positions of these theropods suggest that many other coelurosaurs were also feathered.

Subsequently, Wellnhofer (2000) commented that *Caudipteryx* has relatively shorter feathered arms

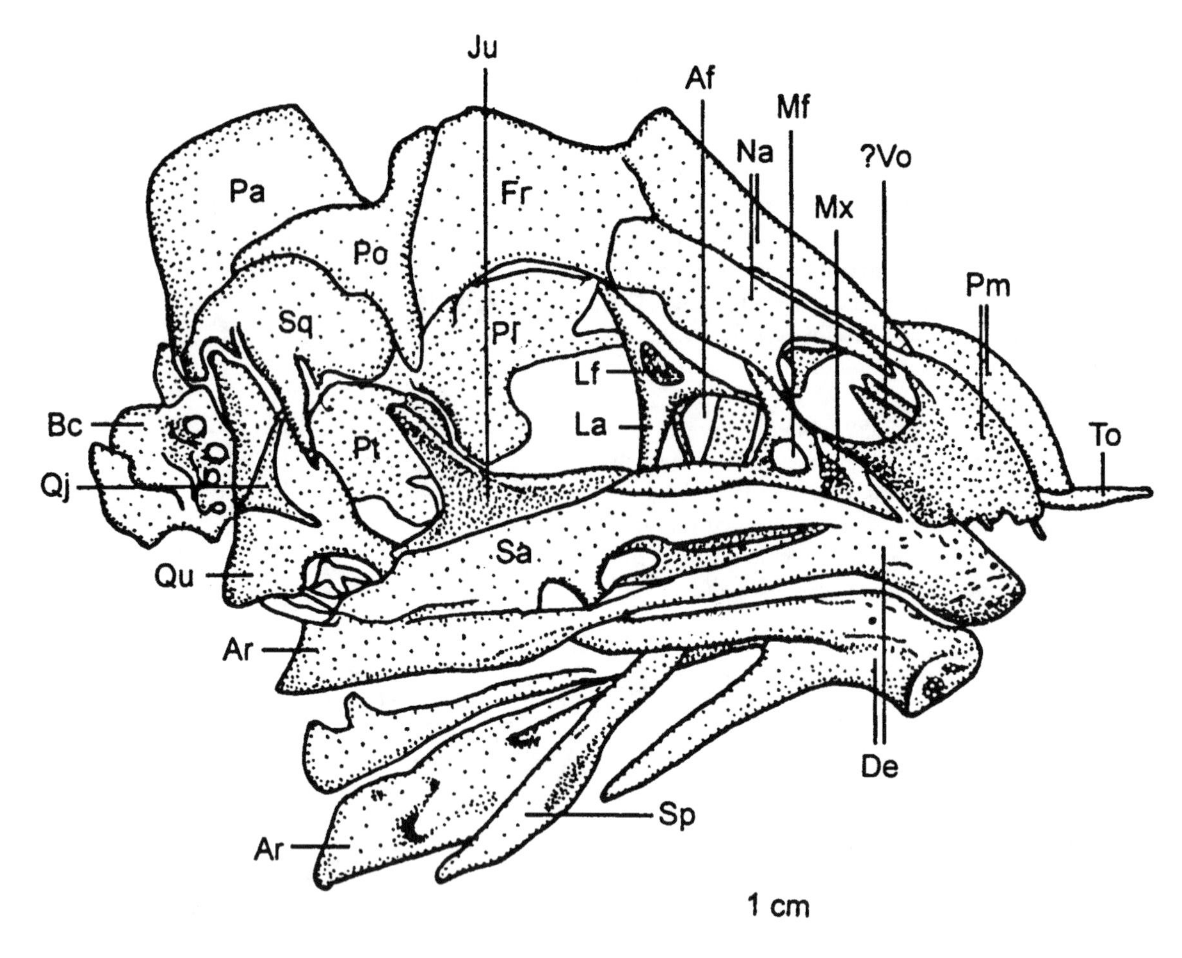

***Caudipteryx zoui*, referred skull (BPM 0001). (After Zhou, Wang, Zhang and Xu 2000.)**

than *Archaeopteryx* and lived approximately 25 million years later. However, the precise classification of animals like *Caudipteryx* either as feathered theropods or secondarily flightless birds, pending a detailed analysis of the skeleton and feathers, remains a problem that may never be unequivocally resolved. Moreover, Wellnhofer added, this question "does not invalidate the supposed close relationship between theropods and *Archaeopteryx*.

In their own study, Zhou and Wang addressed the question of whether *Caudipteryx* was indeed a feathered theropod or, as had been recently suggested, a flightless bird. Fortunately, IVPP V 12344 preserved salient information which permitted Zhou and Wang to make the following observations:

Caudipteryx possesses several avian characters that are usually not found in dinosaurs, including: 1. hallux reversed; 2. slender or absent iliac fossa for M. cuppedicus (as in *Archaeopteryx* and *Confusciusornis*; large and deep in dromaeosaurs; see Norell and Makovicky); 3. ribs with uncinate processes (present in oviraptorids and advanced birds, absent in most theropods and early birds); 4. teeth constricted at base of crown (typical of all tooth birds and a small number of theropods; X. Xu, personal communication to Zhou and Wang); and 5. tail (relatively shorter than that of *Archaeopteryx*) and tibia of about equal length.

However, *Caudipteryx* also displays the following suite of characters that are similar to those of manuraptoran dinosaurs and are more primitive than in birds: 1. jugal short with high dorsal ascending process; 2. hemal spines more developed than in *Archaeopteryx* (see Wellnhofer 1974); 3. distal end of scapula expanded; 4. angle between scapula and coracoid much greater than 90 degrees; 5. sternal plates unfused; 6. ratio of length of metacarpal I to II about 0.45 (only about 0.36 in *Confusciusornis* and 0.33 in *Archaeopteryx*); 7. first phalanx of manual digit I extending distally to middle of first phalanx of manual digit II; 8. pubic peduncle not extending posteriorly to below middle of acetabulum; 9. ischium with large obturate process; 10. pubis antero-ventrally directed; 11. pubic symphysis at least 50 percent length of pubis (near or less than one third length of pubis in *Archaeopteryx* and *Confusciusornis*, symphysis reduced further in more advanced birds); 12. astragalus and calcaneum separated from tibia; 13. absence of pretibial bone (present in birds; see Martin, Stewart and Whetstone 1980); 14. midshaft of metatarsal III laterally compressed; 15. tarsometatarsus unfused; and 16. forelimb much shorter than hindlimb.

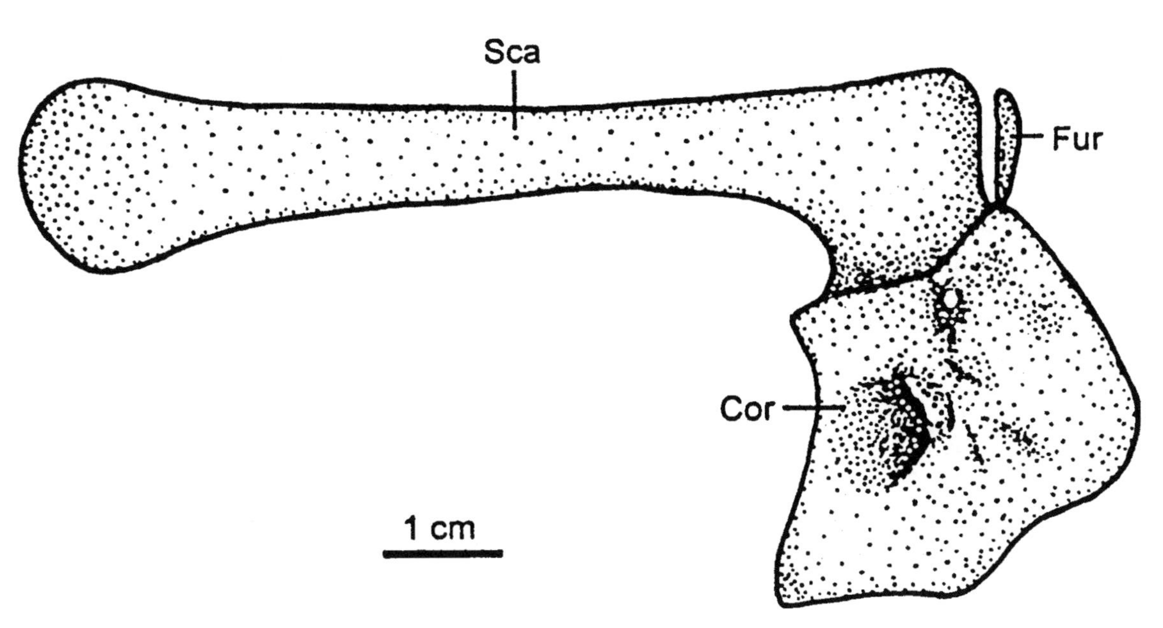

***Caudipteryx zoui*, referred pectoral girdle (BPM 0001). (After Zhou, Wang, Zhang and Xu 2000.)**

Zhou and Wang noted that arguing for an avian state for *Caudipteryx* would necessitate the reversal of all of the above-listed 16 characters—including the propubic pelvis—from its presumed bird ancestors. However, when interpreted as a feathered dinosaur, the lesser number of avian characters can be explained either by convergence or by their appearance prior to the origin of birds. Evoking parsimony, the authors found the second of the above hypotheses to be more likely, also noting that, in recent years, some characters (*e.g.*, the furcula and opisthopubic pelvis) formerly regarded as strictly avian have been found in many theropod dinosaurs (see Sereno 1999*c*).

Originally, *Caudipteryx* was regarded by Ji *et al.* as the closest relative to birds, an interpretation that was later challenged by various other paleontologists (*e.g.*, Sereno) who generally associated it with oviraptorids.

As observed by Zhou and Wang, *Caudipteryx* has (as in dromaeosaurids, troodontids, and birds) a shortened first metacarpal that is less than half the length of the second. The fused furcula in this genus is also

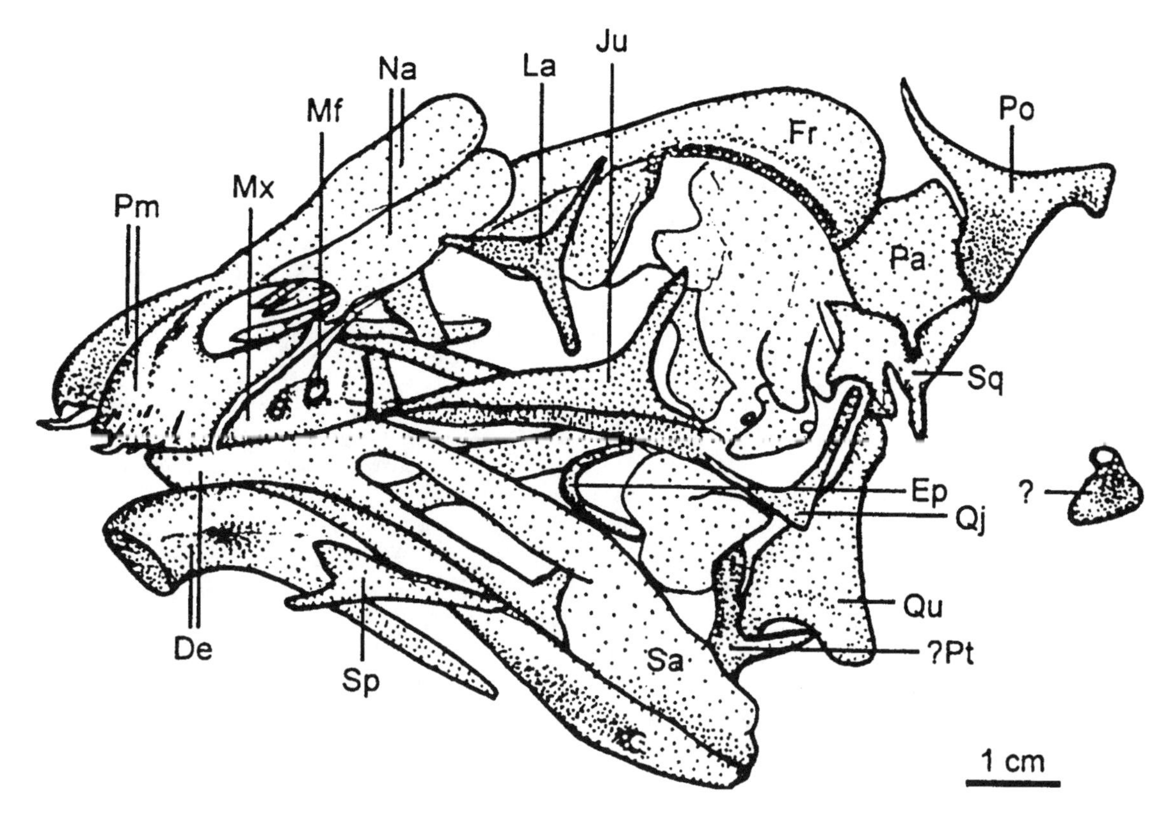

***Caudipteryx* sp., skull (IVPP V 12430). (After Zhou, Wang, Zhang and Xu 2000.)**

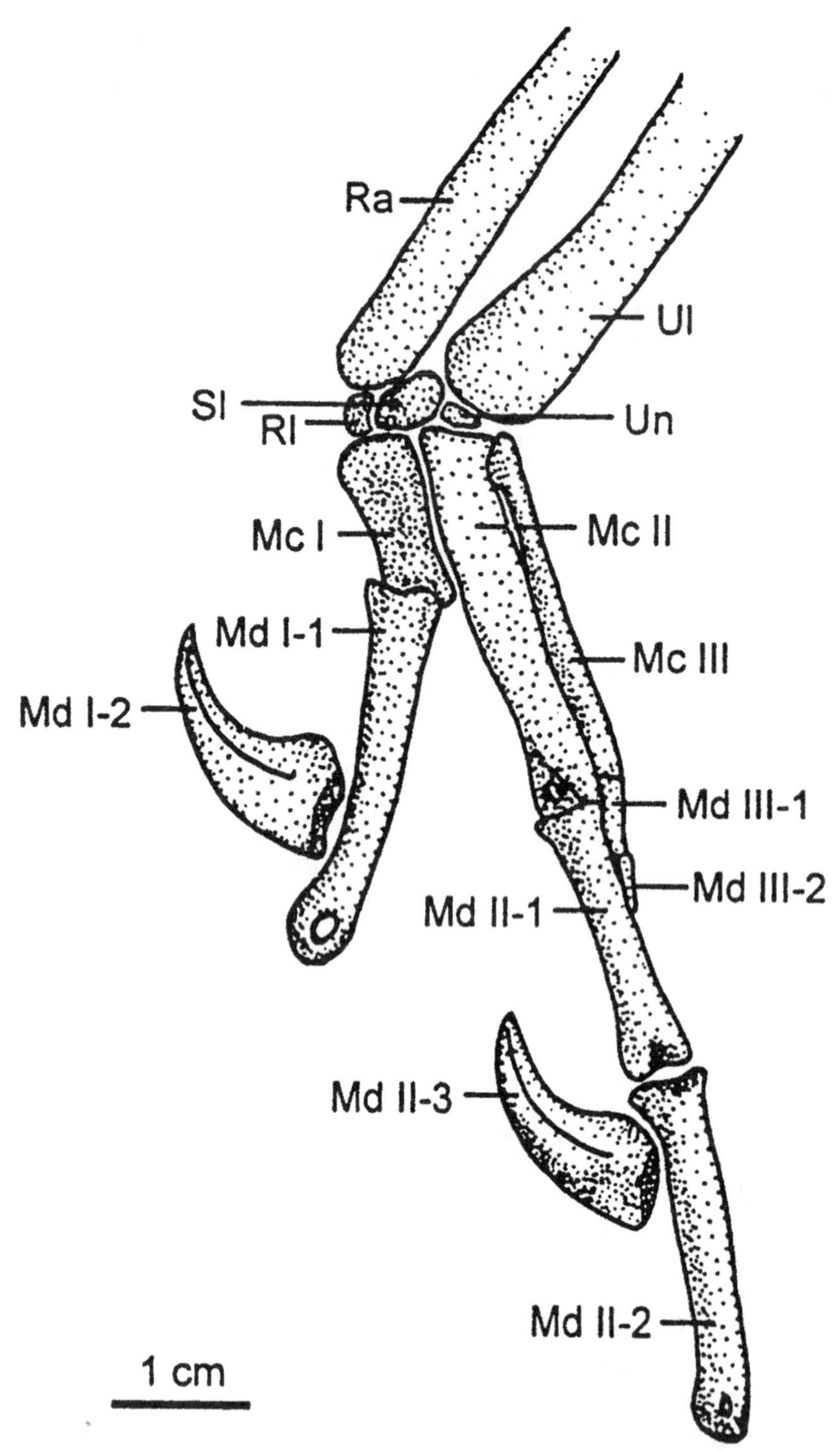

***Caudipteryx* sp., left manus (IVPP V 12430). (After Zhou, Wang, Zhang and Xu 2000.)**

known in oviraptorids (see Norell, Makovicky and Clark 1997) dromaeosaurids, and birds. The pelvis, in having an antero-ventrally oriented footed pubis, resembles that in oviraptorids more than that in dromaeosaurids; the reduced iliac fossa (large and deep in dromaeosaurs) for the M. cuppedicus is similar to that in oviraptorids and birds.

Zhou and Wang observed that *Caudipteryx* shares with oviraptorids uncinate processes (not yet found in dromaeosaurids), a short, high skull, and the loss of teeth. As in dromaeosaurs, manual digit I is less robust than in oviraptorids. Also, as in dromaeosaurs and birds, the pubic peduncle is much deeper than the ischiadic peduncle (pubic peduncle almost as deep as ischiadic peduncle in oviraptorids).

However, many distinctive characters present in *Caudipteryx* precluded its referral by Zhou and Wang either to Oviraptorosauria or Dromaeosauridae, including the following: tail shorter than in all known oviraptorids and dromaeosaurids, possessing fewer than 25 caudal vertebrae; ischium having extremely large triangular obturator process, less developed than in oviraptorids and dromaeosaurids; midshaft of third metatarsal significantly medio-laterally compressed (third metatarsal in oviraptorids and *Sinornithosaurus* narrowing slightly at proximal end [X. Xu, personal communication to Zhou and Wang]; in birds and some dromaeosaurs [Norell and Makovicky], third metatarsal not changing in width much from proximal to distal end [Chiappe 1996]). Also, digit IV is shorter than digit II, and the hallux of the pes is reversed, the latter condition unknown in other theropods (Norell and Makovicky; Clark, Norell and Makovicky 1999).

Zhou and Wang, therefore, concluded 1. that dromaeosaurids are most likely the closest dinosaurian relatives to birds (Padian and Chiappe 1998; Xu, Wang and Wu 1999; Sereno); 2. both *Caudipteryx* and Oviraptoridae are too specialized (*e.g.*, large leg to arm ratio and medio-laterally compressed metatarsal III) to be considered ancestral to birds, although both taxa independently developed various avian characters (*e.g.*, uncinate processes and reduction of teeth); 3. as certain features of *Caudipteryx* are more similar to dromaeosaurids and birds than oviraptorids, this genus could constitute the sister group to Oviraptorosauria or a separate lineage between Oviraptorosauria and Dromaeosauridae; and 4. that the odd combination of characters displayed by *Caudipteryx* warrants its referral to its own family, which Zhou and Wang named Caudipteridae (see "Systematics" chapter).

Zhou and Wang speculated as to the habits and lifestyle of this dinosaur. *Caudipteryx*, the authors postulated, unlike the majority of theropod dinosaurs, may not have been a meat eater. The maxilla and dentary of *Caudipteryx* are toothless (as in most ornithomimids). Well-preserved gastroliths (stones almost never reported in theropods [see Kobayashi, Lü, and Dong 1999]) were found with both specimens of *C. dongi*. The neck of *Caudipteryx* is long, comprising at least 10 cervical vertebrae (a recently found herbivorous oviraptorid having 13; J. Lü, personal communication to Zhou and Wang). These features collectively suggested to Zhou and Wang that *Caudi-*

pteryx, like ornithomimids and oviraptorids, may have been an eater of plants. Furthermore, *Caudipteryx* lacks the "killer claw" found in dromaeosaurids and troodontids suggestive of predation. Additionally, the unguals of the forelimbs are relatively small, and the forelimbs' long wing feathers would have prevented this animal from being predatory.

According to Zhou and Wang, the extreme shortening of the forelimbs in *Caudipteryx* may reflect the reduction of the use of the forelimbs, while the long "wing feathers" most likely had a display purpose.

The first phalanx of the foot of *Caudipteryx* does not extend to the distal end of metatarsal II (similar to the condition in the dromaeosaurid *Deinonychus*). This feature, the authors noted, along with the high position of the hallux-metatarsal II articulation and the comparatively short, less curved ungual of the hallux, is indicative of cursoriality. The opposable hallux was interpreted as a character from retained ancestors having a perching or grasping habit. That *Caudipteryx* was terrestrial was evidenced by pedal digit IV being shorter than digit II (digit IV longer than II in the arboreal *Archaeopteryx* and *Confuciusornis*). Furthermore, the phalangeal proportions of the foot of *Caudipteryx* are in keeping with fast cursorial animals in which the proximal phalanges are considerably longer than the second or third in digits II and III. Zhou and Wang concluded that the small, lightly built body, the shortened tail, tail, and compact metatarsals are also consistent with the idea that *Caudipteryx* was "a fast, cursorial herbivorous animal."

Jones, Farlow, Ruben, Henderson and Hillenius (2000) focused upon *Caudipteryx* in a study of cursiorality in bipedal archosaurs.

As noted by these authors, modern avians possess markedly foreshortened tails and, to provide stability during powered flight, their body mass is centered anteriorly, far from the pelvis and close to the wings (Pennycuick 1986; also others cited by Jones *et al.*), posing potential balance problems for running birds. Cursorial birds compensate for this by maintaining the femur subhorizontally, its distal end located anteriorly near the animal's center of mass, stride generation stemming mostly from parasagittal rotation of the lower hindlimb about the knee joint (Gatesy 1990; also others cited). In almost all bipedal cursorial dinosaurs, by contrast, the center of mass is located near the hip joint, the entire hindlimb rotating during stride generation (Alexander; also others cited). The authors demonstrated that these contrasting styles of fast running are closely related to longer relative hindlimb length in cursorial birds than in bipedal dinosaurs.

Jones *et al.* observed that in *Caudipteryx*, surprisingly, the mechanism of cursoriality is quite different from that of other bipedal dinosaurs. The relative total hindlimb proportions, contrasting sharply with those of other such dinosaurs, are indistinguishable from those in modern birds. Also, the tail is very short — relatively, the shortest tail of all known nonavian theropods. Thus, *Caudipteryx* probably had its center of mass located anteriorly as in birds rather than posteriorly as in other dinosaurs (*e.g.*, *Deinonychus*). Consequently, the authors deduced, *Caudipteryx* must have had a cursorial-birdlike rather than dinosaur-like running style.

As pointed out by Jones *et al.*, the above observations, while offering valuable clues as to the lifestyle of this dinosaur, also have implications regarding its taxonomic relationships. The authors further suggested "that the anatomical uniqueness of *Caudipteryx* must be consistent with one of the following":

1. *Caudipteryx* is an unusual theropod, descended from cursorial ancestors that had abandoned dinosaurian locomotion, assuming the unique morphology and running style of cursorial birds. As the authors pointed out, this alternative is supported by cladistic analysis that indicates the genus to be a coelurosaurid theropod (Ji *et al.*; Sereno). However, these analyses did not take into account the cursorial-birdlike locomotory characters of this genus; also, "it is difficult to construct a hypothesis in which the terrestrial theropod ancestors of *Caudipteryx* might have switched to a specialized, running style with an anterior centre of mass, resembling that of cursorial birds, when they were already adept cursors with a posterior centre of mass."

2. *Caudipteryx* is a theropod that had derived from flighted ancestors (Paul 1988*b*). Jones *et al.* noted that, although there is no evidence to support this hypothesis, its possibility cannot be dismissed.

3. *Caudipteryx* is not a feathered, nonavian theropod, but rather a post-*Archaeopteryx*, secondarily flightless bird (this idea consistent with earlier work by Elźanowski 1983). Considering the problems associated with the first and second theory, Jones *et al.* suggested that the third deserves closer scrutiny, finding "it a striking coincidence that the only unambiguously feathered theropod was the only known theropod likely to have utilized locomotory mechanisms identical to those of cursorial birds."

Additional information regarding *Caudipteryx* was reported by Zhou, Wang, Zhang and Xu (2000) from two almost complete, but slightly smaller, specimens belonging to this genus, and with almost completely articulated postcrania — the first (BPM 0001), referred to *C. zoui*, the second (IVPP V 12430), referred to *Caudipteryx* sp.

Zhou *et al.* observed the following new information regarding the skeletal anatomy of this genus:

nasal opening of skull larger than antorbital fenestra; digital formula of 2-3-2 as in advanced birds; loss of ungual of manual digit III; tail comprising 22 unfused caudal vertebrae; approximately 12 cervical and only nine dorsal vertebrae; confirmation of pubis being antero-ventrally (rather than posteriorly) oriented, fibula contacting calcaneum, and quadratojugal contacting both squamosal and quadrate; teeth restricted (in known specimens) to premaxilla; scapula expanded at distal end, coracoid with prominent biceps tubercle, possessing elliptical supracoracoid foramen; confirmation that metatarsal I articulates with postro-medial surface of II, hallux at least partially reversed (suggesting possible arboreal capability); manual digit III reduced, offering additional evidence that three manual digits in birds and theropods is homologous (although the reduction of manual digit III appeared numerous times during the history of dinosaurs and birds).

The authors concluded that, although *Caudipteryx* is currently classified as a nonavian dinosaur, the above newly described characters suggest "that its phylogenetic position remains a debatable issue."

Key references: Chiappe (1996*b*); Clark, Norell and Makovicky (1999); Currie (1999); Elźanowski (1983); Feduccia (1999*a*); Hou and Zhang (1993); Ji, Currie, Norell and Ji (1998); Ji and Ji (1996); Jones, Farlow, Ruben, Henderson and Hillenius (2000); Kobayashi, Lü and Dong (1999); Martin, Stewart and Whetstone (1980); Norell, Makovicky and Clark (1997); Padian and Chiappe (1998); Paul (1988*b*); Ruben and Jones (2000); Sereno (1999*c*); Wellnhofer (1974, 2000); Xu, Wang and Wu (1999); Zhou and Wang (2000); Zhou, Wang, Zhang and Xu (2000).

CEDAROSAURUS Tidwell, Carpenter and Brooks 1999

Saurischia: Sauropodomorpha: Sauropoda: Eusauropoda: Neosauropoda: Macronaria: Camarasauromorpha: Brachiosauridae.

Name derivation: "Cedar [Mountain Formation]" + Greek *sauros* = "lizard."

Type species: *C. weiskopfae* Tidwell, Carpenter and Brooks 1999.

Other species: [None.]

Occurrence: Cedar Mountain Formation, Utah, United States.

Age: Early Cretaceous (Barremian).

Known material/holotype: DMNH 39045, eight articulated dorsal vertebrae, 25 caudal vertebrae, several chevrons, proximal portions of right and left scapulae, right and left sternal plates, right humerus, radius, and ulna, metacarpal IV, right pubis, partial left pubis, proximal portions of both ischia, partial left femur, right femur, right tibia, three metatarsals, phalanx, three unguals, ribs, numerous gastroliths.

Diagnosis of genus (as for type species): Medium-sized brachiosaurid having anterior caudal

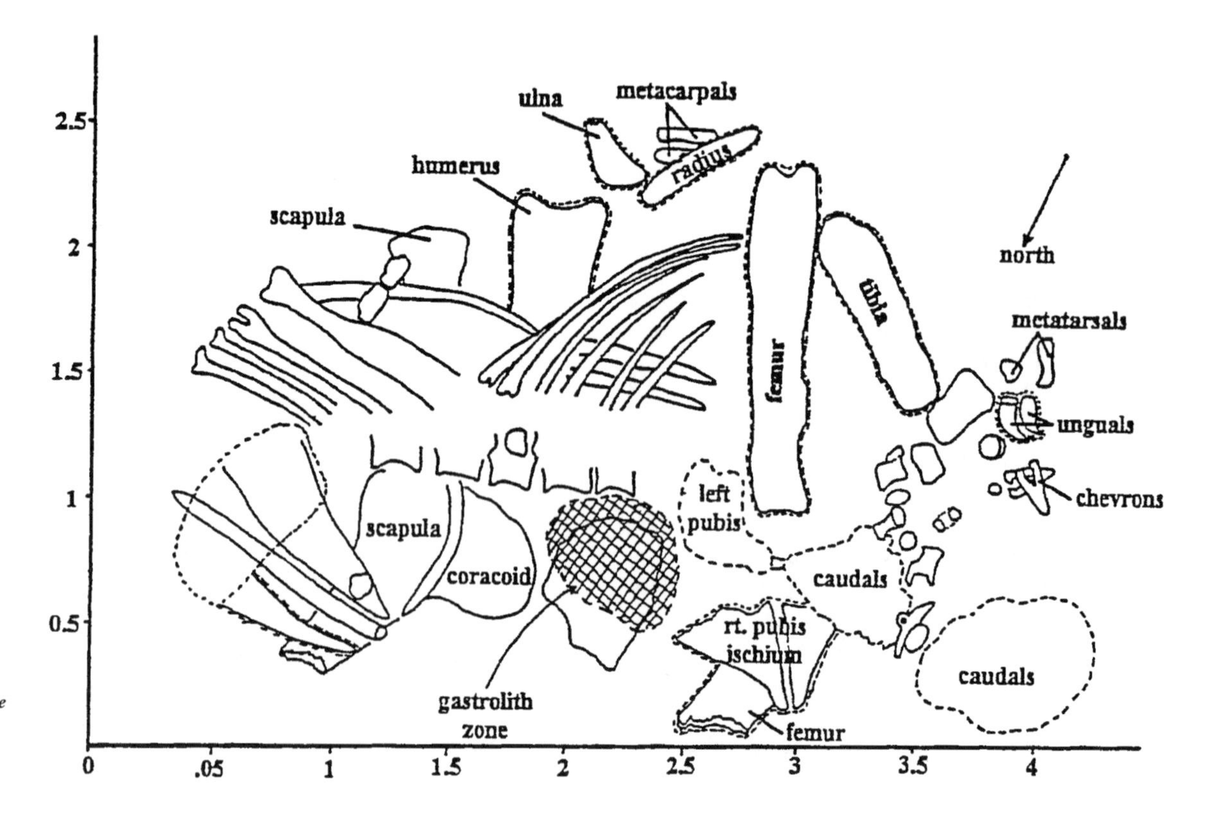

Quarry map showing the holotype (DMNH 39045) of *Cedarosaurus weiskopfae* as found. (After Tidwell, Carpenter and Brooks 1998.)

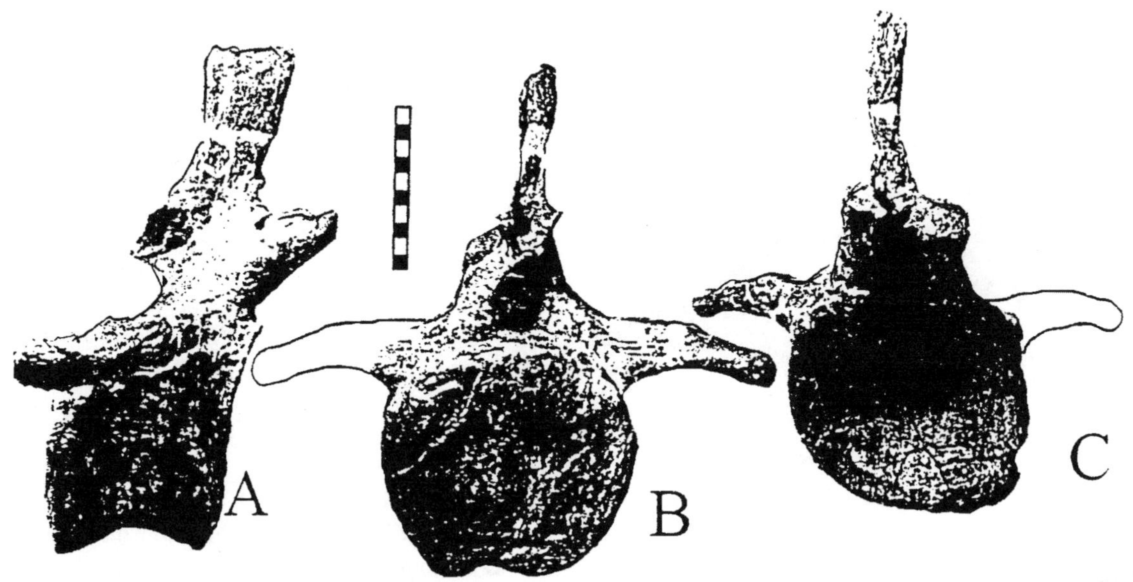

***Cedarosaurus weiskopfae*, holotype (DMNH 39045) anterior caudal vertebra in A. lateral, B. posterior, and C. anterior views. (After Tidwell, Carpenter and Brooks 1998.)**

vertebrae possessing deeply concave anterior faces (in contrast to *Brachiosaurus*, *Sonorosaurus*, and *Pleurocoelus nanus*, and which lack well-developed hypophenes, in contrast to SMU 61732; see "Note," below); posterior faces of anterior caudal vertebrae flat to concave, lacking posterior ball of titanosaurs; mid-caudal vertebrae having sharp ridge extending axially along sides of neural arch; deltopectoral crest of humerus placed closer to midshaft than in *Brachiosaurus*, *P. nanus*, or SMU 61732; humerus/femoral ratio of .98, similar to *Brachiosaurus*, higher than *P. nanus*; humerus proportionally more robust than in *Brachiosaurus*; radius very slender, with two ridges beginning at midshaft, one extending along lower posterior side, the other curving along lateral side to terminate in prominent rugosity near distal end (ridge absent in *Brachiosaurus*); distal end of radius subrectangular and flat (oval and lower along lateral ventral side in *Pleurocoelus*); ulna having prominent posterior condyle similar to *Pelorosaurus* and unlike *Brachiosaurus*, with distinct groove separating distal condyle from lateral wall (in contrast to *Pleurocoelus* and SMU 61732) (Tidwell, Carpenter and Brooks 1999).

Comments: The most completely known Early Cretaceous brachiosaurid, *Cedarosaurus* was founded upon a partial, semiarticulated postcranial skeleton (DMNH 39045) discovered by Denver Museum of Natural History volunteer Billy Kinneer in a hard, maroon mudstone in a quarry (opened by the Denver Museum in 1996) located in the Cat Member of the Cedar Mountain Formation of Grand County, eastern Utah. The skeleton, some of which had apparently eroded away, was found in a ventrally prone position, the scapulocoracoids divergent from each other, the dorsal vertebrae lying articulated on their right side upon the scapulocoracoids. The specimen

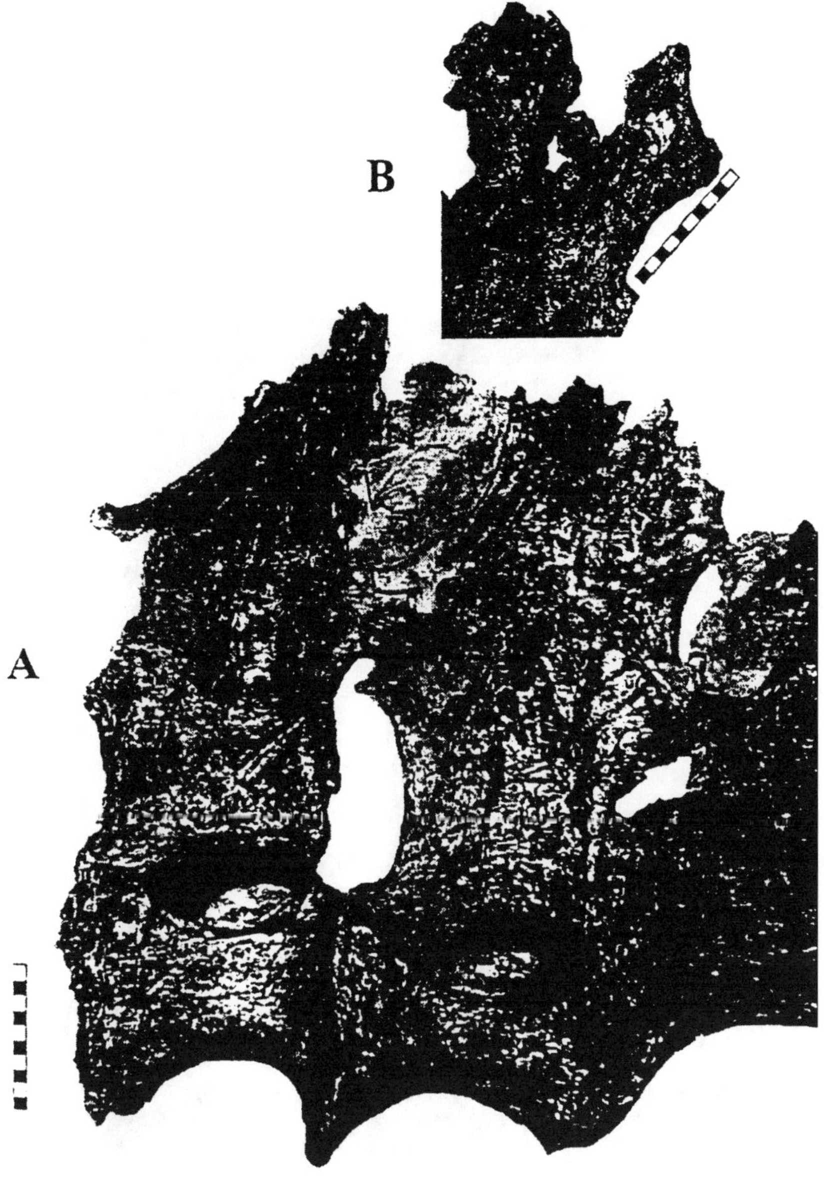

***Cedarosaurus weiskopfae*, holotype (DMNH 39045) thirteenth caudal vertebra in A. posterior, B. anterior, and C. lateral views. Scale = 10 cm. (After Tidwell, Carpenter and Brooks 1998.)**

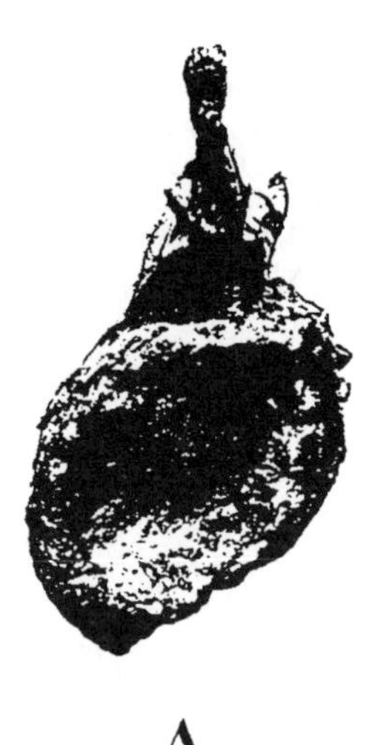

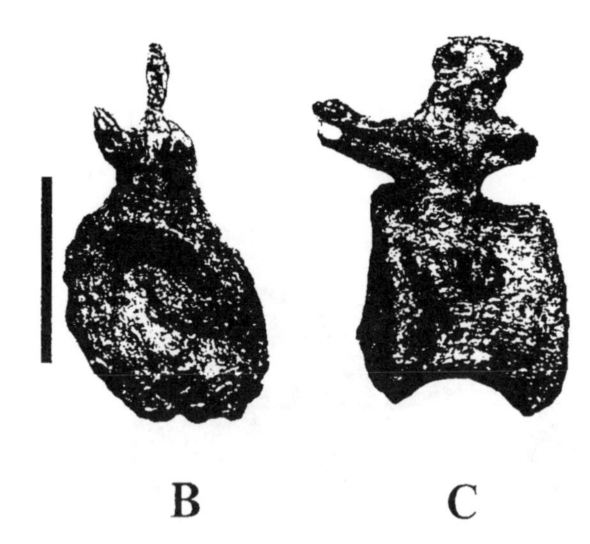

***Cedarosaurus weiskopfae*, holotype (DMNH 39045) dorsal vertebrae, lateral view. (After Tidwell, Carpenter and Brooks 1998.)**

was excavated over a two-year period by a team including Kenneth Carpenter, Virginia A. Tidwell, and William Brooks (Tidwell, Carpenter and Brooks 1998).

Included in the holotype were a large number of gastroliths found in approximately the belly region, this, as noted by Sander and Carpenter (1998), indicating that this dinosaur had remained at least partially intact following burial. Tidwell *et al.* observed that many of the elements "show extensive erosive damage along a single somewhat irregular plane." Fragments of bone on the damaged surfaces of these elements suggested that the skeleton was first only partially buried under overbank sediments, subjecting the exposed parts to trampling and weathering which essentially destroyed the left side of most vertebrae, the distal ends of the scapulae, and the lateral side of the right femur. Later, the eroded surface of the specimen was also buried by overbank sediments.

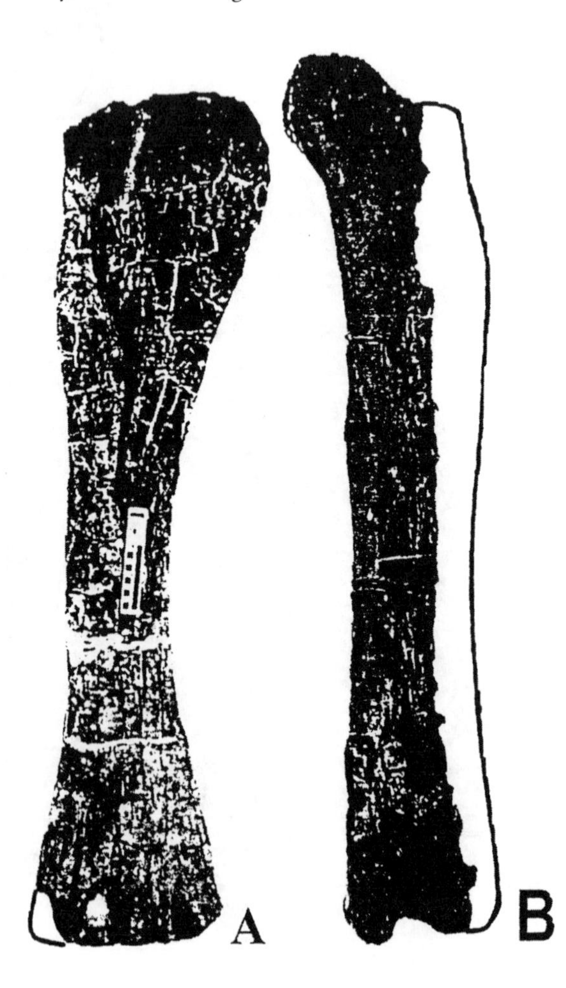

***Cedarosaurus weiskopfae*, holotype (DMNH 39045) A. right humerus in anterior view, B. right femur in posterior view. (After Tidwell, Carpenter and Brooks 1998.)**

The authors referred *Cedarosaurus* to the Brachiosauridae based upon the following features: humerus almost equal in length to femur; centra of dorsal vertebrae having extended length; neural arches of caudal vertebrae occupying almost entire length of centra in anterior caudals, posterior edge of neural arch not extending more than three-fourths length of centra beginning approximately at caudal 11, only about two-thirds length of centrum on last caudal present; and long, slender metacarpals. Although DMNH 39045 represents a relatively small member of the Brachiosauridae, the authors identified it as "a fully adult individual, as evidenced by the fusion of the scapulae to coracoids and neural arches to centra."

As pointed out by Sander and Carpenter, the gastroliths (originally reported as belonging to an unnamed camarasaurid) found associated with *Cedarosaurus* constitute the first gastrolith set found *in situ* in the Yellow Cat Member of the Cedar Mountain Formation of eastern Utah. This set of at least 70 stones, found in the pelvic region, were partially supported in matrix, some being supported on edge and some found in stone-to-stone or stone-to-bone contacts, these orientations suggesting to Sander and Carpenter that they were deposited while still inside the soft tissue of the carcass.

As described by Sander and Carpenter, the gastroliths range in size and mass from 0.6–166 cubic centimeters and 1.7–406 grams. Their gastrolith surfaces are mostly polished. Low-energy conditions and seeming initial containment in soft tissue suggested to the authors that this set is most likely complete.

Sander and Carpenter observed that these stones are drably colored and come in a wide variety of shapes, suggesting low selectivity for these characteristics by the animal that consumed them. As they range in composition from hard chert to fragile sandstone and siltstone, the usually accepted ball-mill model for the sauropod gut may be incorrect, indicating to the authors that these stones may have served some other function.

Note: Until recently, as pointed out by Tidwell *et al.*, remains of Cretaceous North American sauropods have been restricted to the derived, Upper Cretaceous (Maastrichtian) titanosaur *Alamosaurus*, while widely scattered remains from the Lower Cretaceous have been referred to the brachiosaurid *Pleurocoelus*. This latter material includes several specimens (SMU 61732), from the Lower Cretaceous of Texas, which Langston (1974) originally referred to *Pleurocoelus* based on similarities of the distal caudal vertebrae (*e.g.*, slender. spool-shaped, centra amphyplatyan, neural arch occupying cranial half of centrum).

However, as Tidwell *et al.* noted, distal caudals such as these are common to all brachiosaurids as well as to many titanosaurids. Furthermore, significant differences between the anterior caudals of *Pleurocoelus* and the remains from Texas demonstrate that these are two distinct genera. The centra in *Pleurocoelus* are amphiplatyan while those of SMU 61732 are strongly concave, the latter resembling those of the brachiosaurid *Cedarosaurus*.

The presence of prominent hyposphenes or hyposphenal ridges were reported by Upchurch (1998) in several sauropod taxa, and Langston (also McIntosh 1990) considered hyposphenes to have major importance in diagnosing sauropod taxa. As noted by Tidwell *et al.*, the anterior caudal neural spine (USNM 5650) of the type species *Pleuroelus nanus* shows little indication of a hyposphene; however, prominent hyposphenes were reported by Langston in the anterior to midcaudals of SMU 61732.

In *Cedarosaurus*, only the eighth caudal shows a small hyposphene; the postcygapophysese in all other caudals of that genus extend down to the neural canal, with little room remaining for hyposphenes. Interlocking hypantra are present on the prezygapophyses of SMU 61732, but not on those of *Cedarosaurus*.

Tidwell *et al.* concluded, therefore, that SMU 61732 is clearly not referrable to *Pleurocoelus*, based on the presence of concave anterior faces and of hyposphenes on the anterior caudals. Furthermore, other workers (see Salgado, Calvo and Coria 1995; Gomani, Jacobs and Winkler 1998) have suggested that the Texas material may be more closely related to the Titanosauridae than to the Brachiosauridae. The relationship of SMU 61732 to *Cedarosaurus*, however, remains in doubt pending the preparation of more of the SMU material.

Key References: Gomani, Jacobs and Winkler (1998); Langston (1974); McIntosh (1990); Salgado, Calvo and Coria (1995); Tidwell, Carpenter and Brooks (1998); Sander and Carpenter (1998); Tidwell, Carpenter and Brooks (1999).

†CENTROSAURUS—(=*Eucentrosaurus*; =?*Monoclonius*)

Ornithischia: Genusauria: Cerapoda: Marginocephalia: Ceratopsia: Neoceratopsia: Ceratopsomorpha: Ceratopsoidea: Ceratopsidae: Centrosaurinae.

Comment: Chure and McIntosh (1989), in their bibliography of the Dinosauria, stated that the name *Centrosaurus*, proposed by Lawrence M. Lambe in 1904, was preoccupied (having been proposed by Fitzinger in 1843 for a lizard). Thus, they suggested the new name *Eucentrosaurus* for this well-known centrosaurine ceratopsian. More recently, however, extensive research through the literature by Ben Creisler and George Olshevsky indicates that *Centrosaurus* is indeed the valid named for this horned dinosaur (G. Olshevsky, personal communication 2000).

As noted by Olshevsky, the generic name *Centrosaurus* (Fitzinger 1843) was first published as a junior synonym of the "horned toad" *Phrynosoma*. According to the ICZN (International Code of Zoological Nomenclature) rules in effect when Chure & McIntosh proposed *Eucentrosaurus*, Fitzinger's name, to preoccupy *Centrosaurus* (Lambe 1904), must have been used as an available generic name, not merely as a junior synonym, before 1961. Creisler and Olshevsky's search found no such usage. Furthermore, the prevailing usage during the 20th century has been the name *Centrosaurus* for the dinosaur described by Lambe, not for Fitzinger's lizard. Moreover, the new ICZN code published in 2000 "strongly supports maintaining prevailing usage and suppressing forgotten senior homonyms that threaten nomenclatural stability in this manner." Therefore, the name *Centrosaurus* is valid for Lambe's ceratopsian, Fitzinger's taxon is a *nomen oblitum* not preoccupying *Centrosaurus*, and the name proposed by Chure and McIntosh is a junior objective synonym of *Centrosaurus*.

Key references: Chure and McIntosh (1989); Lambe (1904).

†CERATOSAURUS

Saurischia: Theropoda: Neotheropoda: "Ceratosauria": Neoceratosauria: Ceratosauridae.

Type species: *C. nasicornis*

New species: *C. magnicornis* Madsen and Welles (2000)

New species: *C. dentisulcatus* Madsen and Welles (2000).

Other species: [None.]

Occurrence of new species: Morrison Formation, Colorado, Utah, United States.

Age of new species: Late Jurassic.

Known material of two new species: Numerous specimens including two partial skeletons, ?isolated dermal ossicles, juvenile, subadult, and adult.

Diagnosis of *C. magnicornis*: Differs from *Ceratosaurus nasicornis* in being more massive; having longer, lower skull (height to length ratio of 40 versus 47); anterior border of premaxilla straighter;

Photograph by the author, courtesy National Museum of Natural History, Smithsonian Institution.

***Ceratosaurus nasicornis*, UNSM 4735, holotype skeleton. Until the recent discovery of a juvenile specimen from Bone Cabin Quarry West in Wyoming, this was the smallest known individual of *Ceratosaurus*.**

maxilla longer (412 versus 360 millimeters); anterior edge of maxilla almost vertical (versus anterior dip of about 50 degrees); lower border of maxilla more convex, its lateral face more deeply impressed by recess, nasal process having deep maxillary vacuity; upper edge of maxilla, below antorbital fenestra, dipping only 15 degrees posteriorly (versus 25 degrees), anterior edge of maxilla lower at front of fenestra; nasal horn core longer, lower; teeth longer, stouter, especially posteriorly; lacrimal more massive ventrally; quadrate having much longer and lower articular surface, its pillar more concave posteriorly; dentary much more concave dorsally and convex ventrally, chin much more rounded with the bone more massive, becoming 148 millimeters high at surangular contact at 546 millimeters from chin (versus 92 millimeters at 320 millimeters); 11 (possibly 12) alveoli (versus 15); sixth cervical vertebra 80 millimeters long (versus 65 millimeters), its neural spine much higher (145 versus 120 millimeters) and longer anteroposteriorly (52 versus 150 millimeters); table slanting up more steeply posteriorly; posterior chonos much shorter; diapophysis much higher above parapophysis; stout epipophysis; femur 630 millimeters long (versus 620 millimeters), its head 120 millimeters broad (versus 150 millimeters), distal end of each 135 millimeters broad; shaft 75 millimeters broad below trochanteric shelf (versus 52 millimeters) and much straighter; tibia 520 millimeters long (versus 555 millimeters), tibia to femur ratio 83 (versus 90), turberosity not so well developed; proximal diameters 135 and 189 millimeters, distal diameters 132 and 140 millimeters; astragalar facet similar, but dorsal process of astragalus completely fills facet; calcaneum broader anteriorly, occupying 43 percent of astragalocalcanean breadth (versus 28 percent), suture running dorsolaterally; calcaneum broader in lateral view (Madsen and Welles 2000).

Diagnosis of *C. dentisulcatus*: Differs from *C. nasicornis* in larger size; subnarial border of premaxilla arched, almost horizontal (versus straight, dipping forward); nasal process lower; body of premaxilla longer, with several large foramina; maxilla more massive, alveolar border more concave, recess more pronounced; posterior edge of nasal process rising more steeply; front of antorbital fenestra more open; three large openings into body of maxilla at front of maxillary recess and base of nasal process; 12 alveoli (versus 15); teeth more massive, more strongly recurved; dentary more massive, more upturned from sixth tooth forward; dentary teeth more massive, only 11 (versus 15); atlas-axis 100 millimeters long (versus 84); odontid more prominent, axial centrum much shorter,

Photograph by the author, courtesy National Museum of Natural History, Smithsonian Institution.

its ventral edge less downcurved, its spine higher and very much shorter anteroposteriorly, its anterior edge dipping 70 degrees anteriorly (versus 20 degrees); edge of table of axial centrum much steeper and straighter, no prezygapophysis, epipophysis extending far behind spine; third centrum shorter, its ends nearly vertical, spine shorter and almost vertical (versus strongly recurved), its epipophysis very much larger; tibia longer (394 versus 544 millimeters), more massive; tuberosity and shaft heavier; astragalar overhang dipping only 5 degrees (versus 28 degrees); distal end of tibia 165 millimeters broad (versus 140 millimeters); weak horizontal groove across front of astragalus, its dorsal process ossified; fibula longer (564 versus 502 millimeters); upper end of fibula vertical in lateral view (versus dipping 70 degrees anteriorly), broader; tibial flange dipping posteriorly (versus anteriorly), its upper edge (versus lower edge) projecting anteriorly; distal end of tibia broader (81 versus 53 millimeters), convex, evenly rounded (versus truncated); differs from *C. magnicornis* in having much more massive premaxilla; muzzle more rounded, nasal process curving back more sharply, making snout and naris longer; lower border of naris convex behind an anterior concavity (versus smoothly concave); maxilla about same length, but with steeper nasal process having straight posterior border, so that antorbital fenestra has more open anterior border; three additional openings into maxilla at front of recess and base of nasal process; lateral face of maxilla higher above first three teeth; lower half of anterior border of maxilla dipping steeply posteriorly (versus anteriorly); alveolar border slightly more convex; foramina in row above alveolar border larger, groove below them deeper; anterior three maxillary teeth more recurved, posterior teeth more vertical; fifth cervical vertebra with similarly downset centrum, but is shorter; transverse process ending with diapophysis higher above parapophysis; table shorter; epipophysis much smaller; spine lower, shorter, shoulder weaker; humerus longer (333 versus 292 millimeters), much more massive; head of humerus dipping only 10 degrees laterally (versus 20 degrees), anteroventral process of greater tuberosity much larger; deltoid tuberosity 93 millimeters long (versus 84 millimeters), and thinner; shaft of humerus 63 millimeters broad (versus 35 millimeters) (Madsen and Welles 2000).

Comments: *Ceratosaurus* (see *D:TE*), a large,

***Ceratosaurus nasicornis*, USNM 4735, holotype skeleton, close-up detailing the dermal ossifications preserved with this specimen (see *D:TE*).**

Photograph by the author, courtesy Museum of Western Colorado.

***Ceratosaurus magnicornis*, MWC 1, holotype skull on display at the original location of the Dinosaur Valley museum in Grand Junction, Colorado.**

nasal-horned theropod of the Late Jurassic, was named and first described by Othniel Charles Marsh in 1884. The genus was recently reassessed by Madsen and Welles (2000) in a monograph intended to supplement the landmark review of the type species *Ceratosaurus nasicornis* Marsh 1884*b* (as well as various other theropods) published 80 years earlier by Charles Whitney Gilmore (1920). Madsen and Welles did not rediagnose or redescribed the genus or the type species, *Ceratosaurus nasicornis*, apparently accepting Gilmore's work, a paper "basic to any study of the genus *Ceratosaurus*." In their same study, however, the authors introduced two new species of *Ceratosaurus*—*C. magnicornis* and *C. dentisculatus*—each founded upon more recently collected remains.

As related by Madsen and Welles, *C. magnicornis* was based on an articulated, almost complete skeleton (MWC 1) that was collected in 1976 by the Museum of Western Colorado from the lower part of the Brushy Basin Member of the Morrison Formation, in the Fruita Paleontological Area, south of Grand Junction, near Fruita, Ute Meridian, Mesa County, western Colorado.

C. dentisuclcatus was based on a less complete specimen (UMNH 5278, the individual bones having separate catalog numbers; material first mentioned by Madsen and Stokes (1963) consisting of disarticulated skeletal elements from the Brushy Basin Member of the Morrison Formation and its contact with the Cedar Mountain Formation, at the Cleveland-Lloyd Dinosaur Quarry, in Emory County, east-central Utah. The type specimen of this latter species, the authors pointed out, represents the largest "ceratosaur" yet on record. Isolated amorphous dermal ossicles (UUVP 15, 80, 93, 102, 113, 208, 236, 351, 361, 367, 433, 469, 507, 677, and 6788), found at the Cleveland-Lloyd Dinosaur Quarry (and similar to those found with the holotype of *C. nasicornis*; see *D:TE*), may pertain to *C. dentisulcatus*.

Based upon *C. nasicornis* and these two new species, Madsen and Welles offered a revised osteology of the genus *Ceratosaurus*.

Madsen and Welles also addressed the validity of various specimens that have, over the years, been referred to *Ceratosaurus*:

Ceratosaurus roechlingi, described by Janensch (1925*a*), was based on a left quadrate and a left tibia (see *D:TE* for illustration) collected from the Upper Saurian bed of Tendaguru, Tanzania, East Africa. As the quadrate is generally diagnostic, Madsen and

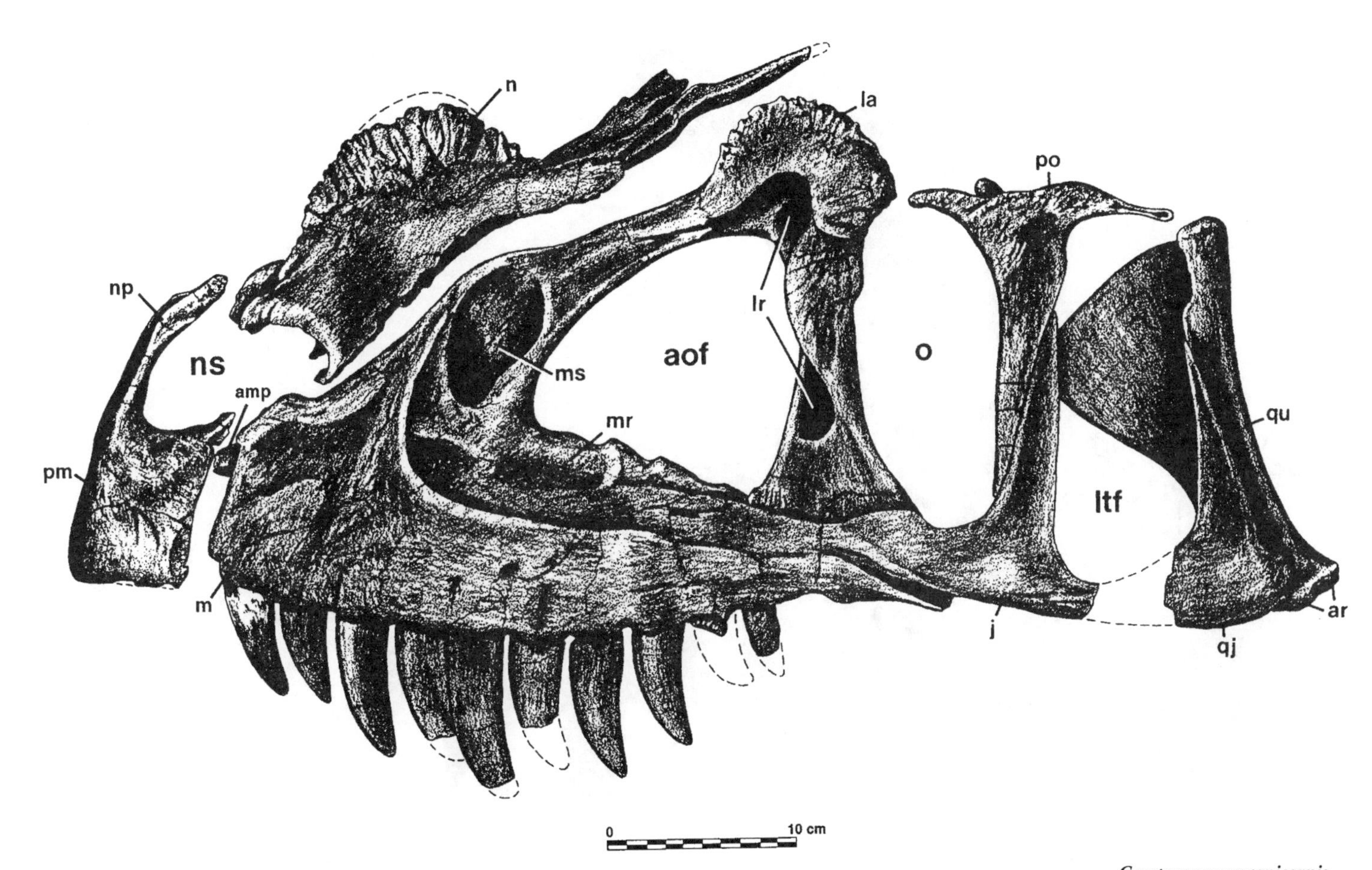

Ceratosaurus magnicornis, MWC 1, holotype skull including premaxilla (reversed right premaxilla), maxilla, paired nasals, lacrimal, jugal, postorbital, quadratojugal, and quadrate, left lateral view. (After Madsen and Welles 2000.)

Welles designated it to be the type specimen. The quadrate measures 130 millimeters broad distally (versus 78 millimeters for *C. magnicornis*). The specimen indicates a theropod nearly twice the size of *C. magnicornis*. Unable to distinguish this material at the species level, Madsen and Welles designated it *Ceratosaurus* sp.

Earlier, Janensch (1920) described a number of scattered theropod teeth from Tendaguru as *Labrosaurus* (?) *stechowi*. As noted by Madsen and Welles, these serrated, recurved, and longitudinally grooved teeth are like those in the known species of *Certaosaurus* and, with the skeletal material referred to "*C. roechlingi*," confirm the presence of a large *Ceratosaurus* in this fauna. As these teeth are probably synonymous with that species, Madsen and Welles referred them to *Ceratosaurus* sp. Also from Tendaguru, Janensch (1920) described three small dorsal vertebrae representing a juvenile, plus a middorsal and a posterior dorsal vertebra from the same locality. Madsen and Welles, noting that these remains are not diagnostic at the specific level, agreed with Janensch's (1920) referral of them to *Ceratosaurus* sp.

Madsen and Welles also referred the following remains to *Ceratosaurus* sp.: *Labrosaurus sulcatus*, based on a tooth from the Morrison Formation of Colorado (see Hay 1930), figured by (but not illustrated by) Marsh (1896), described by Hay (1908) as 30 millimeters high, base 12.5 by 12 millimeters, identified by Madsen as an anterior ?dentary tooth of *Ceratosaurus*; a right premaxilla (DNM 972) collected at Dinosaur National Monument, this specimen first identified by White (1964); a tooth with weak grooves on its lateral surface, originally named *Megalosaurus meriani* [*nomen dubium*] by Greppin (1870) and later referred to *Labrosaurus meriani* by Janensch (1920), from the Upper Jurassic Vigulla beds of the Malm, near Mouteir, Savoy, in the Bern Jura, Switzerland; teeth from the Mygatt-Moore Quarry (also known as the Rabbit Valley Quarry), Morrison Formation, Colorado, identified as *Ceratosaurus* sp. by James I. Kirkland (personal communication to Madsen and Welles 1992); an undescribed specimen (BYUVP 13024) including a complete, ossified left scapula and coracoid, axial and appendicular elements, from the Dry Mesa Quarry, near Delta, Colorado (see Britt 1991); a specimen (BYUVP 12893) comprising a complete cranium, partial nasal horn cores, distal fragment of a quadrate, seven fragmented dorsal vertebrae, and incomplete pelvic bones (ischium and pubis), recovered in 1992 from the Agate Basin area on the west flank of the San Rafael Swell southeast of Moore, Emery

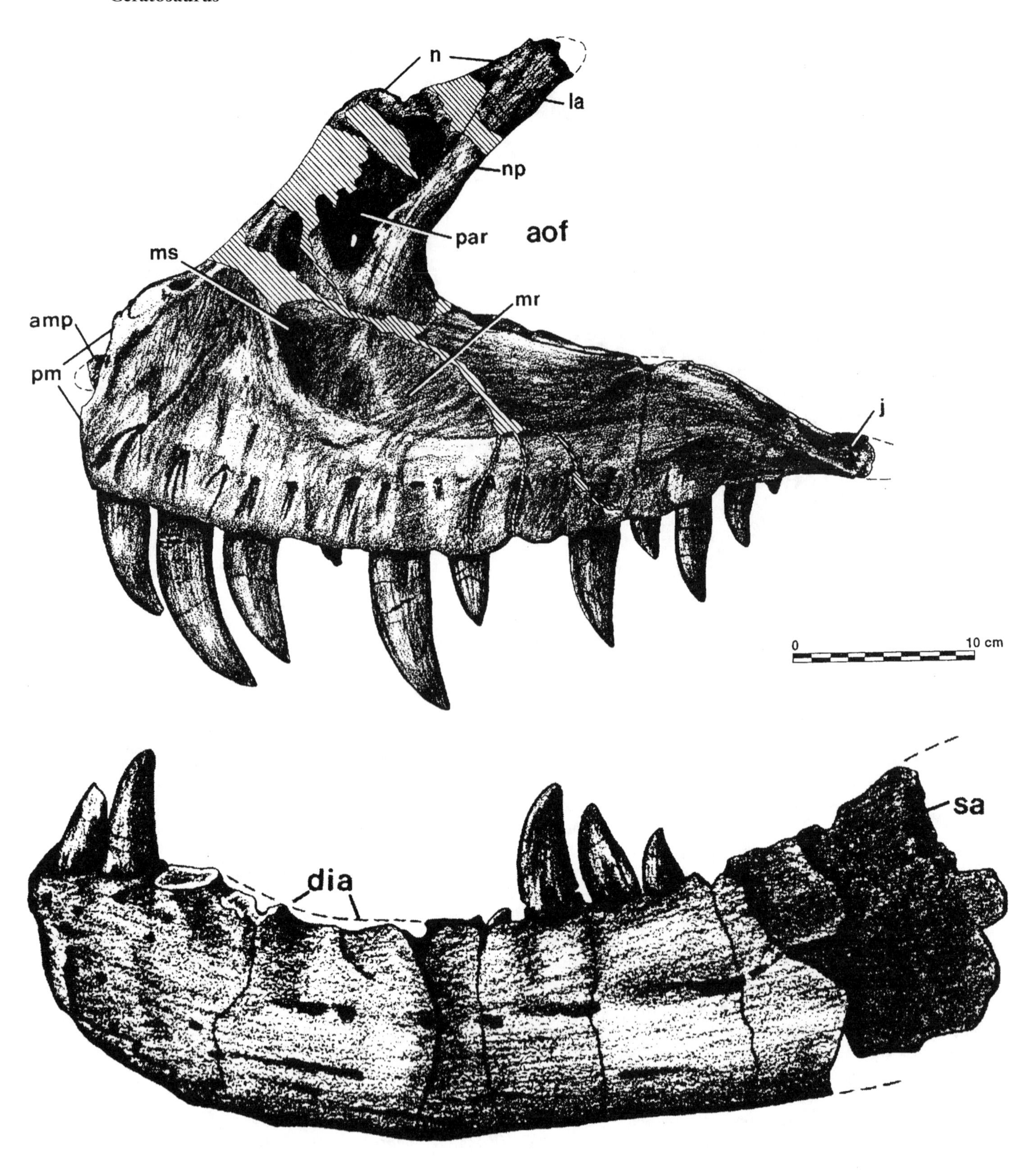

Top: Ceratosaurus dentisulcatus, UMNH 5278, holotype left maxilla, lateral view. (*Bottom: Ceratosaurus dentisulcatus*, UMNH 5278, holotype left dentary, lateral view. (After Madsen and Welles 2000.)

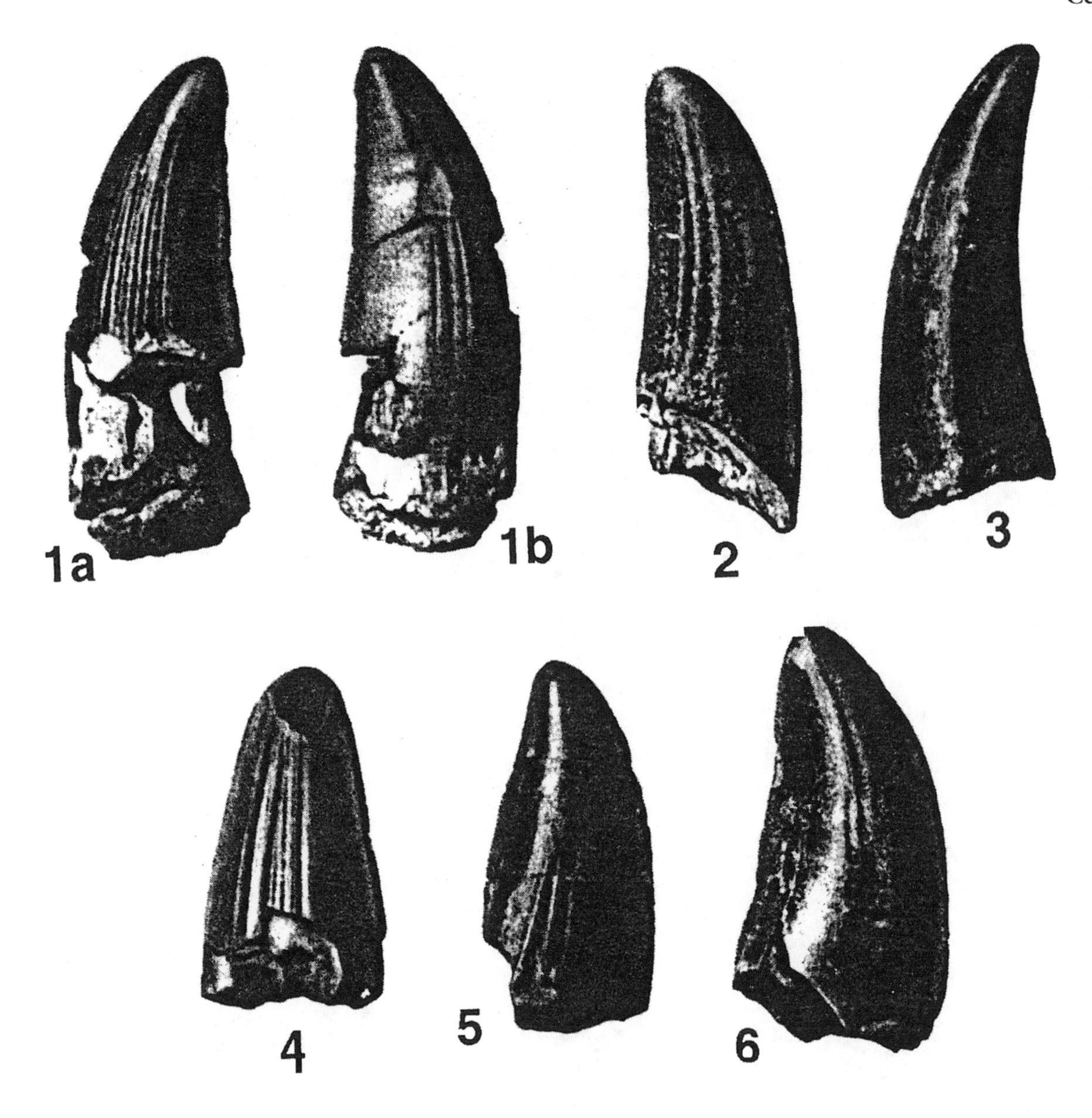

Teeth originally described and figured by German paleontologist Werner Janensch (1925) as *Labrosaurus* (?) *stechowi*, referred by James H. Madsen, Jr. and Samuel P. Welles (2000) to *Ceratosaurus* sp.

County, Utah, representing one of the largest *Ceratosaurus* skeletons yet found (to be described by Britt, Stadtman, Chure and Madsen (B. Brooks, personal communication to Madsen and Welles 1999); and a juvenile skeleton (see Britt, Miles, Cloward and Madsen 1999, below). Possibly referrable to *Ceratosaurus* sp. is a large, fused scapula and coracoid reported by Robert T. Bakker (personal communication to Madsen and Welles, 1991) from Quarry Nine at Como Bluff, Wyoming, a site where a single *Ceratosaurus* tooth had been collected.

Until recently, *Ceratosaurus* has been mostly known from adult specimens (see *D:TE*). The discovery of an associated juvenile specimen of *Ceratosaurus* was reported in an abstract and presented by Britt *et al.* (1999). The specimen — preserved with the partially articulated skull lodged against the pelvis, in the classic "death pose" — comprises a complete skull and about 30 percent of the postcrania. It was collected from fluvial sandstones of the Upper Jurassic Morrison Formation at Bone Cabin Quarry West (west of the famous original Bone Cabin Quarry), in Albany County, Wyoming.

As briefly described by Britt *et al.*, the skull is about 34 percent smaller than that of the holotype of *C. nasicornis*, USNM 4735, which until the present discovery had been the smallest representative of *Ceratosaurus* known. Juvenile features identified by the authors in the Bone Cabin Quarry West specimen include "open neurocentra sutures, minimal fusion of ischia and pubes, unfused vomers, and small horn size." The specimen is unique in being the only *Ceratosaurus* known in which the halves of the nasal horn are not coossified.

According to Britt *et al.*, this juvenile specimen reveals autapomorphies of *Ceratosaurus* not previously

Head of the horned theropod *Ceratosaurus nasicornis* as restored by Gregory S. Paul.

observed, "including a large foramen that extends dorsoventrally between the distal ends of the pubes, and a peg-and-socket articulation between the ilium and ventral pelvic elements" (the peduncles of the ilium bearing prominent "pegs" that insert into appropriate sockets in the pubis and ischium).

In a subsequent abstract for a poster, Britt, Chure, Holtz, Miles and Stadtman (2000) noted the following significant ontogenetic changes in the juvenile, subadult, and adult *Ceratosaurus* specimens from Bone Cabin Quarry and the San Rafael Swell: distal ends of pubes changing from small, nonexpanded terminus (juveniles) to robust terminus with caudal extension (adults); ilium long (juveniles), lacking dorsoventral expansion of cranial blade (present in older individuals); nasal and lacrimal horns undergoing expansion from subadult to adults stages (suggesting USNM 4735 to be subadult). Unusual features observed by these authors include the following: bundles of ossified tendons covering lateral surfaces of neural spines of sacral vertebrae; quadrate having caudal extension.

Note: Stovall (1938) described various bones (including teeth, metatarsals, a cervical vertebra, and a dorsal vertebra) suggesting to him an animal like *Ceratosaurus*. As observed by Madsen and Welles, however, the tooth, though serrated and recurved, is not longitudinally grooved as in *Ceratosaurus*. The cervical and dorsal vertebra, though large enough (89 and 124 millimeters long, respectively) to belong to a large theropod, are indeterminate. Thus, Madsen and Welles referred this material to "Theropoda indet."

Megalosaurus (?) *ingens* Janensch 1920, sometimes referred to *Ceratosaurus ingens* (see Rowe and Gauthier 1990), was based on the tooth of a gigantic theropod from the Upper Saurian level in Tendaguru. According to Madsen and Welles, this tooth seems to be too big to pertain to any known *Ceratosaurus* species.

Erratum: The photograph of the holotype skull of *Ceratosaurus magnicornis* appeared in *S1* incorrectly labeled *C. nasicornis*.

Key references: Britt (1991); Britt, Chure, Holtz, Miles and Stadtman (2000); Britt, Miles, Cloward and Madsen (1999); Gilmore (1920); Greppin (1870); Hay (1908, 1930); Janensch (1920, 1925*a*); Madsen and Stokes (1963); Madsen and Welles (2000); Marsh (1884*b*, 1896); Rowe and Gauthier (1990); Stovall (1938); White (1964).

†**CHAOYANGSAURUS** Zhao, Cheng and Xu 1999 — (= *Chaoyoungosaurus*)

Ornithischia: Genasauria: Cerapoda: Marginocephalia: Ceratopsia.

Name derivation: "Chaoyang [Area of Liaoning Province]" + Greek *sauros* = "lizard."

Type species: *C. youngi* Zhao, Cheng and Xu 1999.

Other species: [None.]

Occurrence: Tuchengzi Formation, Liaoning Province, China.

Age: Middle (Bathonian) or Late Jurassic.

Known material/holotype: IGCAGS V371, dorsal part of skull, almost complete mandible, complete axis, isolated cervical vertebra, five articulated cervical vertebrae, fragmentary humerus and scapula.

Diagnosis of genus (as for type species): Ceratopsian having weakly-developed and smooth jugal boss; quadratojugal overlapping posterior side of quadrate shaft; lack of broad lateral surface of quadrate shaft; quadrate shaft with convex posteroventral margin; coronoid process rather low, with planar top; ridge between planar lateral and ventral surface of angular (Zhao, Cheng and Xu 1999).

Comments: The type species now named *Chaoyangsaurus youngi* is distinguished as the earliest ceratopsian dinosaur found to date. It was based upon a partial skull and postcranial remains (IGCAGS V371) collected in 1976 by Cheng Zhengwu from the Ershijiazi locality of the Tuchuengzi Formation, in the Chaoyang Area of Liaoning Province, in northeastern China. Although the correct date for the Tuchengzi Formation is currently in dispute (Middle or Late Jurassic), recent work by Wang (1998) suggests a late Middle Jurassic. This assessment is based on the occurrence of *Mesolimnadia*, *Sentestheria*, and *Pseudograpta*, conchostracan invertebrates that can be correlated with nonmarine conchostracans from Bathonian-age rocks in northwest Scotland (Zhao, Cheng and Xu 1999).

The type specimen was originally mentioned with the spelling *Chaoyoungosaurus* by Zhao (1983), who used the genus as the basis for his newly proposed superfamily Chaoyoungosauridea, regarded by that author as a clade of basal ceratopsians or "proto-psittacosaurids" that had evolved from genera like or similar to *Heterodontosaurus*. Zhao characterized the Chaoyoungosauroidea by the presence of a massive, strongly-developed jugal arch, three-lobed teeth having low crowns, and weak, caniniform premaxillary teeth. Later, in an article appearing in the book *The Jurassic in China*, Zhao (1985) proposed the binomen *Chaoyoungosaurus liaosiensis* for this material. However, as the specimen was neither sufficiently diagnosed or described in these publications, the type species *C. liasiensis* was relegated to the status of a *nomen nudum* (see *D:TE*, "Nomen Nudum and Excluded Genera" chapter).

As described by Zhao *et al.*, *Chaoyangsaurus* is a small ceratopsian, the preserved length (from tip of snout to back of quadrate) of its skull measuring only about 140 millimeters (approximately 5.4 inches).

Zhao *et al.* identified *C. youngi* as a member of the Ceratopsia based on its possession of the following ceratopsian synapomorphies: rostral bone; jugal flaring well beyond skull roof; deep jugal infraorbital ramus; and wide predentary ventral process. Originally considered to be a basal ceratopsian by Zhao (1983), the genus was more recently reinterpreted by Sereno (1997) as possibly a basal neoceratopsian. However, Zhao *et al.* pointed out that the rostral bone in *Chaoyangsaurus*— poorly defined in the holotype and largely fused to the premaxilla — is psittacosaur-like (thin dorsally and broad ventrally in anterior aspect, forming transversely convex shield capping triangular surface on conjoined premaxillae) rather than neoceratopsian-like (transversely narrow, the rostral bone being strongly compressed).

As observed by Zhao *et al.*, *C. youngi* is unique in exhibiting a combination of primitive ornithischian features and various derived psittacosaurid and neoceratopsian features. This species resembles psittacosaurs in possessing tall, quite short snout and a dorsally external naris, and neoceratopsians in the pointed anterior end of its mandible. *C. youngi* exhibits the

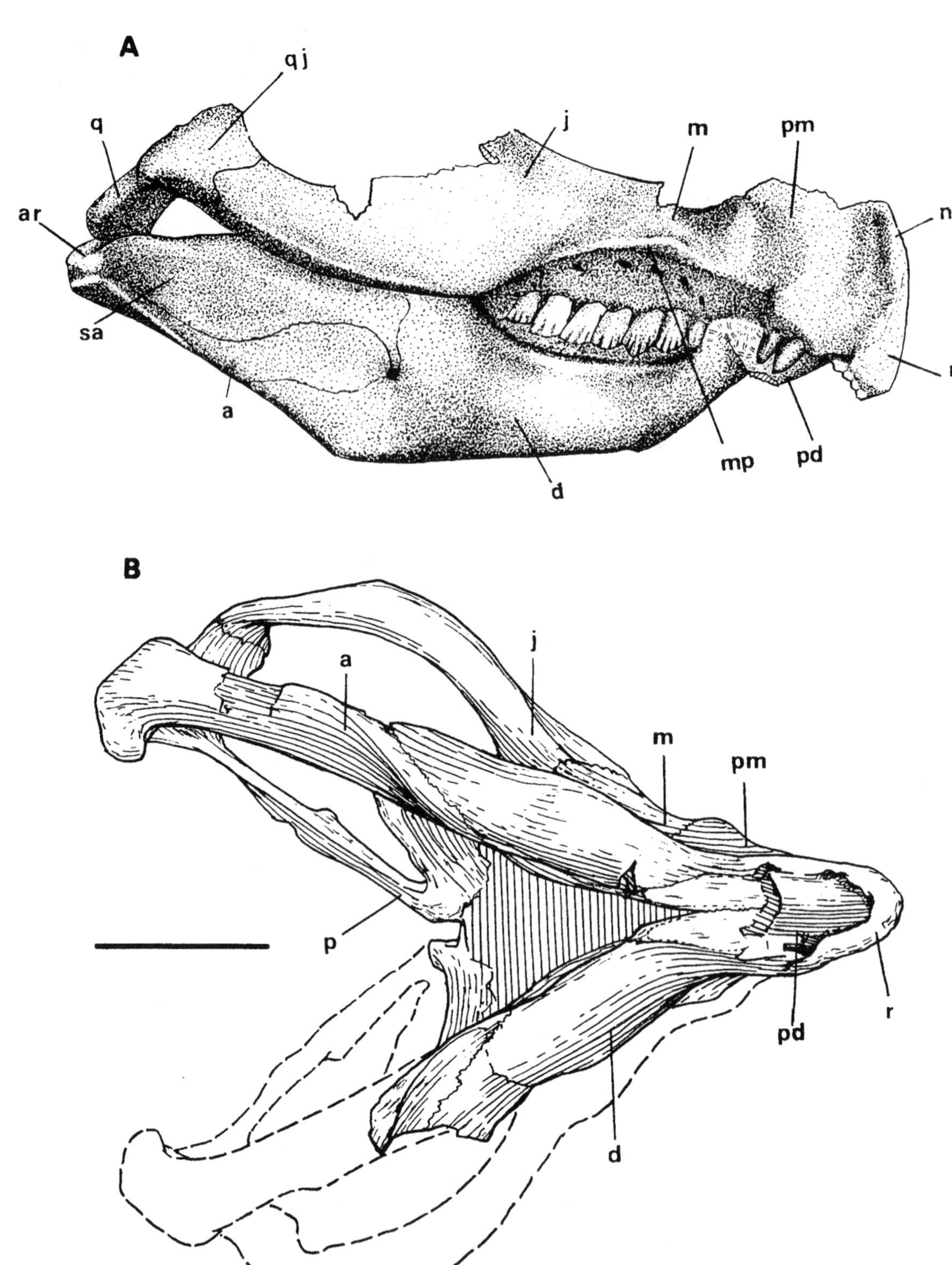

***Chaoyangsaurus youngi*, IGCAGS V371, holotype partial skull in A. right lateral and B. ventral views. Scale = 3 cm. (After Zhao, Cheng and Xu 1999.)**

plesiomorphic feature of a jugal boss, as seen in the possibly related, more primitive ornithischian *Heterodontosaurus*, but lacks a crest on the lateral surface of the jugal as in the latter genus. This lack of a crest separates *C. youngi* from all other known ceratopsians, wherein the lateral aspect of the jugal is divided by a dorsoventral crest into rostral and caudal surfaces (see Sereno 1990). As in other nonceratopsian ornithischians, the ventral margin of the jugal in *C. youngi* forms a gentle arc; in ceratopsians excluding *C. youngi*, the ventral margin of the jugal is angular due to the abrupt inward deflection of the caudal ventral margin.

In attempting to determine the systematic position of *C. youngi*, Zhao *et al.* performed a cladistic analysis, utilizing character data offered by Sereno (1986, 1990), and restricting the ingroups to *Psittacosaurus*, *Chaoyangsaurus*, *Archaeoceratops*, *Leptoceratops*, and Ceratopsidae. This analysis was not able to resolve the position of *Chaoyangsaurus* unequivocally within Ceratopsia, producing five trees, three of which placed the genus as the sister taxon to *Psittacosaurus* (supported by four synapomorphies), one as the sister taxon to *Psittacosaurus* plus Neoceratopsia (five synapomorphies), and one indicating that it belongs within Neoceratopsia (three synapomorphies).

Of the above hypotheses, Zhao *et al.* found the second to be the most acceptable, the clade of *Psittacosaurus* plus Neoceratopsia being supported by the following characters: premaxillary palate vaulted; maxilla dorsoventrally high; dorsoventral crest on lateral surface of jugal; ventral margin of jugal angular; and neural spine of axis anteroposteriorly long. This assessment, the authors noted, was only tentative, pending the acquisition of additional material.

Errata: In *D:TE*, it was incorrectly stated that Zhao (1983) originally regarded this genus as a pachycephalosaur. Also, the reference for Zhao (1985) was inaccurately cited as Zhao (1986).

Key references: Sereno (1986, 1990, 1997); Zhao (1983, 1985); Zhao, Cheng and Xu (1999).

CHAOYOUNGOSAURUS Zhao 1985 [*nomen nudum*]—(See *Chaoyangsaurus*.)

CHARONOSAURUS Godefroit, Zan and Jin 2000

Ornithischia: Genasauria: Cerapoda: Ornithopoda: Euornithopoda: Iguanodontia: Euiguanodontia: Dryomorpha: Ankylopollexia: Hadrosauroidea: Hadrosauridae: Euhadrosauria: Lambeosaurinae.

Name derivation: "Charon [from Greek mythology, the ferryman who conveyed the dead to Hades over the river Styx]" + Greek *sauros* = "lizard."

Type species: *C. jiayinensis* Godefroit, Zan and Jin 2000.

Other species: [None.]

Occurrence: Yuliangze Formation, Heilongjiang, People's Republic of China.

Age: Late Cretaceous (late Maastrichtian).

Known material: Partial skull, various postcranial remains, juvenile and adult.

Holotype: CUST V-V1251-57, partial skull.

Diagnosis of genus (as for type species): Dorsal surface of parietal regularly rounded, without sagittal crest; lateral side of squamosal almost completely covered by caudal ramus of postorbital; paroccipital and postcotyloid processes very low, extending only at midheight of foramen magnum; alar process of basisphenoid prominent, symmetrical; tricipital crest developed on humerus; forearm elongated, slender, with ratio of total length/maximal width of proximal head greater than 6.3 for ulna and 6.6 for radius; preacetabular process of ilium more elongated than in other known hadrosaurids; ratio of ilium length to preacetabular length greater than 2.1 (Godefroit, Zan and Jin 2000).

Comments: The genus *Charonosaurus* was founded upon a partial skull (CUST J-V1251-57) collected from a large bonebed in the Yuliangze Formation (dated as late Maastrichtian, based on the abundance of *Wodehouseia spinata* plant fossils), on the southern banks of the Amur (Heilongjiang) River, west of Jiayin Village, in Heilongjiang Province, People's Republic of China. Material from the same locality referred to the type species, *C. jiayinensis*, includes CUST J-lll and J-V, and GMH Hlj-16, 77, 87, 101, 140, 143, 144, 178, 195, 196, 207, 278, A10, A12, and "magnus" [the material in these specimens, including postcranial elements, not specified] (Godefroit, Zan and Jin 2000).

According to Godefroit *et al.*, the disarticulated remains of many kinds of animals (including theropods, ankylosaurians, crocodiles, and turtles) are mixed together in this bonebed, the apophyses and neural arches are broken off, and the long bones indicate a preferential direction, this all suggesting that the carcasses accumulated in a fluvial environment with relatively slight currents. Most of these remains comprised the bones of lambeosaurine dinosaurs. Although various ontogenetic stages, from juvenile to adult, were represented at this site, the lambeosaurine sample is particularly homogenous, indicating the presence of only one lambeosaurine taxon. The numerous theropod resorbed teeth at this site indicated to the authors that the lambeosaurines were either killed by predators along the river or that their carcasses had been scavenged.

C. jiayinensis was a large hadrosaurid, Godefroit *et al.* noting that the femur attained a length of 135 centimeters (about 51 inches), longer than any known hadrosaur other than *Shantungosaurus giganteus*.

Godefroit *et al.* observed the following lambeosaurine synapomorphies in *Charonosurus*: parietal shortened (length to minimal width ratio greater than 2.0), completely excluded from occipital surface by squamosals; frontal forming broad excavated platform as floor for hollow crest, excluded from orbital rim by postorbital-prefrontal joint; vertical caudal portion of prefrontal; elevated lateral side of squamosal; rounded and symmetrical rostral process of jugal; sinuous median carina of dentary teeth; and very developed (in length and width) deltopectoral crest of humerus extending down below midpoint of bone.

Performing a phylogenetic analysis (excluding

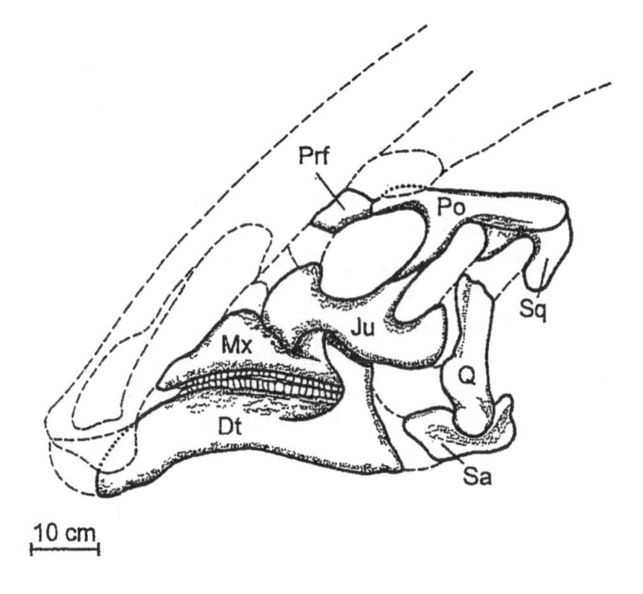

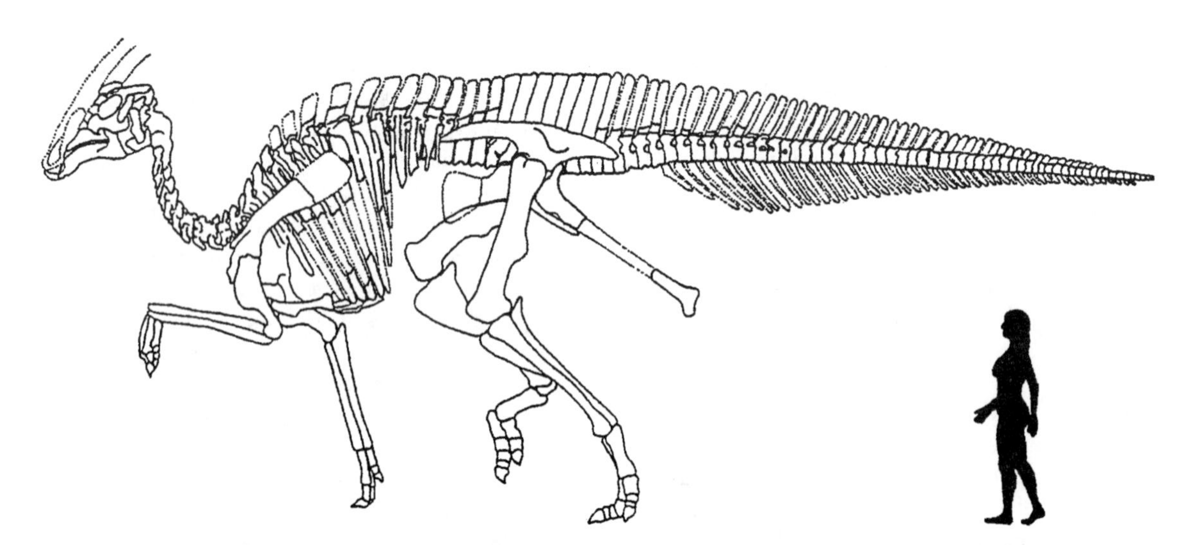

Charonosaurus jiayinensis **(top) CUST J-V1251-57, holotype skull, in left lateral view, and (bottom) reconstruction of skeleton based upon holotype and referred specimens. (After Godefroit, Zan and Jin 2000.)**

such controversial genera as *Tsintaosaurus* and *Eolambia*), the authors obtained a single most parsimonious cladogram in which *Charonosaurus* and *Parasaurolophus* are closely united based "on the frontal platform extending backwards above the rostral part of the parietal and supratemporal fenestra, the very enlarged club-shaped distal end of the fibula, and the equilateral cranial ascending process of the astragalus."

As Godefroit *et al.* pointed out, the late Maastrichtian dinosaur faunas now known in northeastern Asia are dominated by lambeosaurine hadrosaurids (comprising over 90 percent of the fossils collected thus far at Jiayin, and also the sites at Blagoveschensk and Kudur, in the Russian Amur region), while the presence of this subfamily in the Maastrichtian of North America has not yet been unequivocally substantiated (see Sullivan and Williamson 1999). This disparity suggested to the authors "that different kinds of dinosaur communities existed outside North America just before the K/T crisis," and that future research into these Asian and also other localities could perhaps lead to the clarification of the extinction pattern of nonavian dinosaurs at the end of the Cretaceous period.

Key references: Godefroit, Zan and Jin (2000); Sullivan and Williamson (1999).

†CHASMOSAURUS

Ornithischia: Genusauria: Cerapoda: Marginocephalia: Ceratopsia: Neoceratopsia: Ceratopsomorpha: Ceratopsoidea: Ceratopsidae: Chasmosaurinae.

Type species: *C. belli* (Lambe 1914).

Other species: *C. mariscalensis* Lehman 1989.

Comments: *Chasmosaurus*, a chasmosaurine ceratopsian dinosaur genus having a large, fenestrated

Life restoration of *Chasmosaurus belli*, the type species of the genus *Chasmosaurus*, painted by Berislav Krzic.

frill and relatively short horns, has had a number of species referred to it over the years (see *D:TE*).

In 1995, Stephen J. Godfrey and Robert B. Holmes recognized as valid species of the genus *Chasmosaurus* both the type species, *C. belli*, and a referred species, *C. russelli*. The criteria used by these authors to distinguish these two species included the following: 1. Posterior parietal margin is almost straight, with little or no posteromedial emargination, in *C. belli*, this margin is broadly arched on either side of a median emargination in *C. russelli*; 2. large triangular epoccipitals at posterolateral corners of parietals with other parietal epoccipitals much smaller in *C. belli*; three epoccipitals of subequal size in *C. russelli*; and 3. lateral bars of parietal completely enclosing parietal fenestrae in *C. belli*; these bars reduced and not entirely enclosing fenestrae in all but one specimen (AMNH 5656) in *C. russelli* (see *Chasmosaurus* entry, *D:TE*).

More recently, in a paper mostly devoted to *Pentaceratops* (see *Pentaceratops* and *Triceratops* entries for additional related comments), Thomas M. Lehman (1998), a specialist in ceratopsian dinosaurs, also discussed *Chasmosaurus* and variation in this genus. Following a comparison of seven *Chasmosaurus* specimens (CMN 8803, 2280, and 491, RTMP 83.25.1, AMNH 5402, YPM 2016, and ROM 843), Lehman demonstrated that none of the above characters used by Godfrey and Holmes exhibit discrete states useful in distinguishing species, but are, in fact, "end members of a gradational spectrum of variation." As with *Pentaceratops* and also *Triceratops* (see entry), Holmes could draw no clear morphologic, geographic, or stratigraphic separation between the end-member morphotypes in the type species.

Holmes, Shepherd and Forster (1999), in an abstract, reported a chasmosaurine specimen recovered from the Dinosaur Park Formation southeast of Irvine, Alberta. The specimen — preserved in an upright position, limbs folded underneath the body, the postcranial skeleton especially well articulated — is largely complete, basically missing parts of the snout, left cheek, and most of the tail. The completeness of the skeleton, the authors pointed out, provides [not yet disclosed] "information on the relative positioning of limbs and pectoral girdle."

Based on their preliminary analysis, Holmes *et al.* referred this specimen to *Chasmosaurus*, noting that a

Photograph by the author, courtesy Royal Tyrrell Museum/Alberta Community Development.

Skull of the chasmosaurine horned dinosaur *Chasmosaurus belli.*

number of striking apomorphies (*e.g.*, "very broad, square frill with a straight posterior parietal bar bearing conspicuous anteriorly curled epoccipitals, and small transversely oval and posteriorly positioned parietal fenestrae") suggest a much greater anatomical variation for this genus than previously suspected.

Hernández-Rivera and Jesús (1999) briefly reported on a juvenile ceratopsid specimen, tentatively referred to *Chasmosaurus*, found in 1999 during a field season of the Cretaceous Dinosaur Project of the State of Coahuila, in the Upper Cretaceous (latest Campanian) Cerro del Pueblo Formation (Difunta Group), at Rincón Colorado, west of the town of Saltillo, in Coahuila, Mexico. The specimen includes skull fragments, vertebrae, limb bones, and various postcranial elements. Most unexpectedly, it also includes a nearly complete endocast of the brain cavity missing only the medulla.

As described by these authors, this natural brain cast measures 6.8 centimeters in length and 2.5 centimeters in width; identified on it were the cerebrum, optic lobes, cerebellum, dorsal venous sinus, venous canal, and both trigeminal and abducens nerves. Compared to described *Triceratops* endocasts, the specimen from Mexico is about half the size, with far more developed and conspicuous optic lobes, abducens nerves located more posteriorly relative to the germinal nerves, and less distance between the trigeminal nerves and optic lobes. According to Hernández-Rivera and Jesus, the differences between the Mexican endocast and those belonging to *Triceratops* may have generic significance.

Key references: Godfrey and Holmes (1995); Hernández-Rivera and Jesús (1999); Holmes, Shepherd and Forster (1999); Lehman (1998).

†CHIAYÜSAURUS

Saurischia: Sauropodomorpha: Sauropoda: Eusauropoda: Neosauropoda: Macronaria: Camarasauromorpha: Titanosauriformes: Somphospondyli: Euhelopodidae.

Type species: *C. lacustris* Bohlin 1953.

New species: *C. asianensis* Lee, Yang and Park 1997.

Known material/holotype of *C. asianensis*: KPE 8001, tooth.

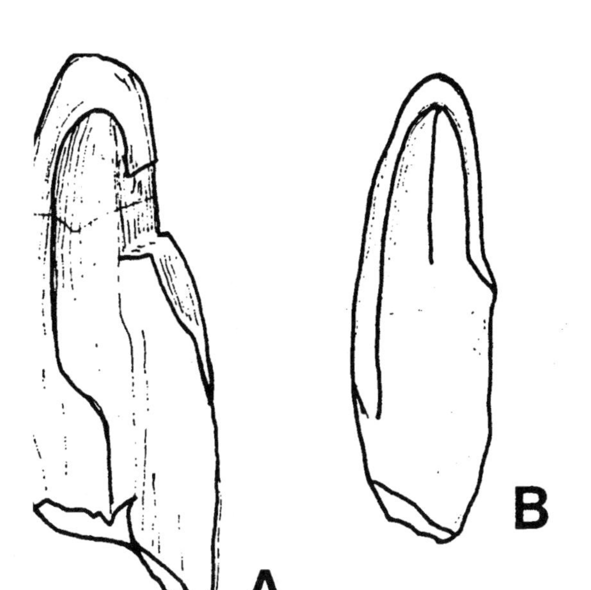

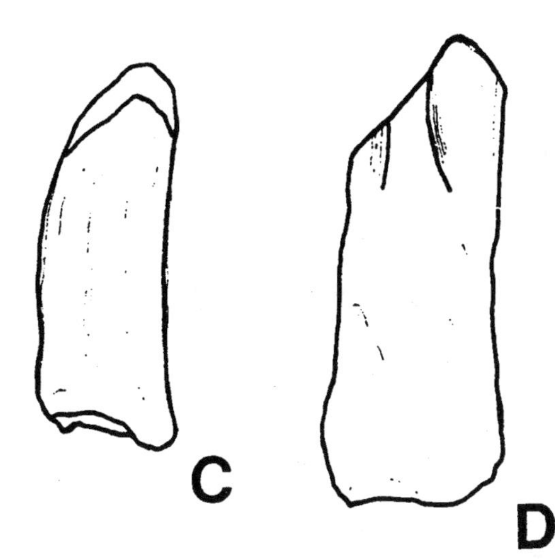

A. Tooth designated Species A. (aff. *Chiayüsaurus*), B. holotype tooth (KPE 8001) of *Chiayüsaurus asianensis*, C. ?Titanosaurid tooth (KPE 8002), and D. ?camarasaurid tooth (KPE 8003), lingual views. Scale = 2 cm. (After Lee, Yang and Park 1997.)

Occurrence: Xinminbao Group, Chia-yükuan, Hui-Hui-Pu, Gansu Province, Huoyanshan Formation, Turpan, Xinjiang, China; Hasandong Formation, Jinju, Gyeongsang Province, Korea.

Age: Late Jurassic–Early Cretaceous.

Diagnosis of *C. lacustris*: Tooth spatulate, lingual concavity occupying distal half of height, proximal half broadly convex; external side strongly convex, forming broad ridge bordered by very narrow, thinner edge anteriorly, broader posteriorly; wear surfaces inclined almost vertically (Lee, Yang and Park 1997, modified from Sun, Li, Ye, Dong and Hou 1992).

Diagnosis of *C. asianensis*: Tooth showing prominent U-shaped vertical wear surfaces along lingual margins of crown; narrow median ridge extending from apical tip to base within lingual concavity (Lee, Yang and Park 1997).

Comments: The type species *Chiayüsaurus* (see *D:TE*) is known only from teeth. As noted by Lee, Yang and Park (1997), the age of the holotype tooth — collected from the Xinminbao Group, Chia-yü-kuan, in Gansu Province, China, and described by Birger Bohlin in 1953 — is uncertain, probably dating as either Late Jurassic (see Sun *et al.* 1992) or Early Cretaceous (Shen 1981). Lee *et al.* identified this tooth as belonging to the Euhelopodidae (see also Park, Yang and Currie 2000 for this assessment).

Lee *et al.* referred a new species to this genus, *Chiayüsaurus asianensis*, founded upon a single tooth (KPE 8001) collected in 1989 by Seong-Young Yang at the Yusuri site near Jinju, in the Hasandong Formation, Gyeongsangnam-do (Gyeongsang Supergroup), Korea. As measured by these authors, the tooth as preserved is 46 millimeters in height.

According to Lee *et al.*, KPE 8001 "is clearly duplicated" in a tooth described by Bohlin as "Species A. (aff. *Chiayüsaurus*)." Bohlin believed that this and the holotype tooth possibly belonged to the same species. Their differences, Lee *et al.* noted, could be attributed to individual variation (such variation having also been reported in other kinds of sauropods, *e.g.*, Janensch 1935–1936; Osborn and Mook 1921; Wiman 1929), the larger Species A. (aff. *Chiayüsaurus*) representing an anterior tooth and the smaller *C. lacustris*, a posterior tooth. Russell and Zheng (1993) regarded *C. lacustris* as a *nomen vanum*, finding the holotype tooth of this taxon to be indistinguishable from teeth of *Mamenchisaurus*. However, as pointed out by Lee *et al.*, the anterior teeth of *Mamenchisaurus*, unlike the tooth of *C. asianensis*, have a median ridge that is limited to the apical portion lingually. Lee *et al.* concluded that *C. asianensis* is not the anterior tooth of *C. lacustris* and the species are distinct.

As the genus *Chiayüsaurus* is now known from two Asian countries, the authors stated, the species *C. asianensis* is significant as representing "the first dinosaurian fauna to imply the biogeographic affinities between China and Korea during the Early Cretaceous."

Lee *et al.* also described a possible titanosaurid tooth (KPE 8002) collected by Yang in 1989 and a possible camarasaurid tooth (KPE 8003) collected by Yoshikazu Hasegawa, both specimens from the Yusuri site. These specimens, together with the new material referred to *Chiayüsaurus*, indicate that at least three families of sauropods lived in Korea.

Note: Park, Yang and Currie (2000), in addition to describing KPE 8001, identified other dinosaurian teeth from the Hasandog Formation of Korea. These include an isolated tooth (KPE 8003) from Yusuri, Jinju, measuring 44 millimeters in height, having a medial ridge on the lingual surface similar to that of *Camarasaurus supermus* (see Osborn and Mook), this specimen tentatively referred to the Camarasauridae; three isolated teeth (KPE 8002, 8006, and 8007) from Yusuri and Guryangri, Sepomyeon, measuring 34.6, 17+, and 16+ millimeters in height, respectively, slender and peglike as in Diplodocidae and Titanosauridae, but without the circular cross section as in diplodocids (see Marsh 1884*a*; Holland 1924; Berman and McIntosh 1978), therefore tentatively referred to the Titanosauridae; and two isolated teeth (KPE 8004 and 8005) from Sumunri, Hadongun, and Tusuri, Jingu, measuring 38+ and 37.5+ millimeters in height, resembling teeth of the theropod *Allosaurus* and referred to the Allosauridae. As noted by Park *et al.*, these teeth were identified utilizing enamel microstructures as taxonomic parameters, which could possibly have utility as diagnostic features at the genus level.

Key references: Berman and McIntosh (1978); Bohlin (1953); Holland (1924); Janensch (1935–1936); Lee, Yang and Park (1997); Marsh (1884*a*); Osborn and Mook (1921); Park, Yang and Currie (2000); Russell and Zheng (1993); Shen (1981); Sun, Li, Ye, Dong and Hou 1992; Wiman (1929).

†CHILANTAISAURUS

Saurischia: Theropoda: Neotheropoda: Tetanurae *incertae sedis.*

Type species: *C. tashuikouensis* Hu 1964.

Other species: [None.]

Occurrence: [Unnamed formation], Nei Mongol, Zizhiqu, People's Republic of China.

Comments: Previously classified as an allosaurid (see *D:TE*) but also sometimes as a "megalosaurid" or spinosauroid, the genus *Chilantaisaurus*—a large Asian theropod distinguished by long and massive forelimbs, with atypically large claws — was reassessed by Chure (1998*a*) in a brief report.

As recounted by Chure, *Chilantaisaurus* originally included two species — the type species, *C. tashuikouensis*, known from fore- and hindlimb, and pelvic remains (IVP AS 2884), and a referred species, *C. maortuensis*, based on partial skull and vertebral remains (IVP AS 2885), with no elements overlapping between the two (see *D:TE* for photographs of some of these elements). Both species were named and described by Hu Show-yung in 1964.

Reexamining the skull material referred to *C. maortuensis*, Chure discovered a number of "unusual features," including the following: only 12 maxillary teeth; a large pneumatic chamber in the nasals; a small invasion of the frontals by the M. pseudotemporalis; a sagittal crest made up of the frontals and parietals; very small fused maxillary interdental plates that can only be differentiated from the maxilla by their texture; absence of a groove for the interdental artery; and a highly pneumaticized, rostrocaudally shortened basicranium.

According to Chure's reevaluation, the second species shares synapomorphies with *Labocania anomala*, dromaeosaurs, troodontids, and tyrannosaurs. In Chure's opinion, while earlier assessments of the type species may be correct, "*C.*" *moaortuensis* should be considered to be the type species of "a new genus of large [manuraptoran] coelurosaur, with an estimated skull length of 70 cm."

Erratum: In *D:TE*, the diagnosis for *C. maorutensis* was incorrectly identified as being that of the type species, *C. tashuikouensis.*

Key references: Chure (1998*a*); Hu (1964).

†CHINDESAURUS

Saurischia: ?Theropoda: Herrerasauridae.

Occurrence: Chinle Formation, Arizona, "PreTecovas" and Tecovas Formation, Dockum Group, Texas, United States.

Comments: One of the most primitive of known dinosaurs is the rather small genus *Chindesaurus*, definite remains of which have been found in Arizona and Texas, although other occurrences have been incorrectly reported (see *D:TE*).

Murry and Long (1989) and Long and Murry (1995) stated that the type species *Chindesaurus bryansmalli* also occurs at the *Placerias* quarry, located in the Upper Triassic/late Carnian Bluewater Creek Formation (Chinle Group) of Apache County, Arizona, represented there by five dorsal centra. However, according to Hunt, Lucas, Heckert, Sullivan and Lockley (1998), in their report on dinosaurs of the Late Triassic western United States, these remains, though belonging to a herrerasaurid, are generically indeterminate.

Hunt *et al.* were surprised that Long and Murry did not mention the teeth that were included in the holotype (PEFO 10395) of *C. bryansmalli*. As described for the first time by Hunt *et al.*, the most complete tooth, presumably belonging to this species, is laterally compressed and recurved with a concave posterior margin. The tooth measures 22 millimeters high (as preserved, the tip being broken) and has a basal length of approximately 12.5 millimeters. Both anterior and posterior margins are finely serrated, having about five serrations per millimeter.

Heckert *et al.* subsequently reported the proximal end of a left femur (NMMNH P-4126) from the Bull Canyon Formation in Bull County, and also commented on specimen NMMH P-4415. Noting that

Photograph by the author, courtesy Mesa Southwest Museum.

The primitive dinosaur ***Chindesaurus bryansmalli*,** cast of holotype partial skeleton displayed with preserved bones in their probable placement at the Mesa Southwest Museum, Mesa, Arizona.

these specimens are, in part, characterized by a squared-off femoral head having a pronounced medial offset, these authors found them readily comparable to the type specimen of *C. bryansmalli*.

Also, Heckert *et al.* reported on isolated vertebral centra from the Bull Canyon Formation that may represent both fragmentary herrerasaurids and coelophysoids. Of these, vertebrae catalogued as NMMNH P-4882 could represent a member of either of these clades, the larger of the two — being amphicoelous, hourglass-shaped, and with shallow pleurocoels on both sides — possibly being a sacral vertebra of *C. bryansmalli*. Additionally, the striated outer edge of the articular surfaces of this vertebra conform to the holotype sacral of *C. bryansmalli* (see Long and Murry). According to Heckert *et al.*, the second vertebra — differing from that of *Chindesaurus* in not being hourglass-shaped, having a centrum that is taller than wide (wider than tall in *C. bryansmalli*) and laterally constricted, and lacking pleurocoels — is apparently a caudal vertebra belonging to a coelophysoid.

An incomplete dorsal vertebra (NMMNH P-16844) belonging to a theropod dinosaur, also from the Bull Canyon Formation, conforms generally to the holotype of *C. bryansmalli*; however, the articular surfaces of both ends, the authors pointed out, are slightly taller than wide (unlike the condition in *C. bryansmalli*, wherein these surfaces are about as wide as they are tall).

Chindesaurus was recognized by Hunt *et al.* and Heckert *et al.* as a member of the Herrerasauridae, this family accepted by those authors as a group of basal theropods (see "Systematics" chapter).

Note: Hunt *et al.* reported that a total of eight dinosaurian taxa from the Bull Canyon Formation of New Mexico have been described in Adrian P. Hunt's unpublished dissertation, but will later be described elsewhere (see Hunt 1994, 1995). The largest of these new forms are herrerasaurids, which Hunt informally designated "Herrerasaurid A" and "Herrerasaurid B." According to Hunt *et al.*, "Herrerasaurid A" is known from several partial skeletons (including UCM 47221, NMMNH P-17134, and P-17258), and "Herrerasaurid B" is represented by a partial skeleton (NMMNH P-4569) and various isolated elements which Long and Murry had previously referred to the Prosauropoda. A smaller taxon, presently referred to as "Herrerasaurid C" (regarded by Heckert, Lucas and Sullivan 2000 as a herrerasaur, but not presently diagnostic as the genus level) was the basis for the identification by Murry and Long, and then Long and Murry, of *C. bryansmalli* from New Mexico. NMMNH P-4415, P-16656, P-17325, and P-22494, all referred by Long and Murry to *C. bryansmalli*, pertain to the "ceratosaur" *Gojirasaurus*, a possible theropod and an indeterminate herrerasaurid, respectively (see Hunt 1994; Hunt *et al.*; Heckert *et al.*).

According to Hunt *et al.*, the above specimens

can only establish the presence of *Chindesaurus*-sized herrerasaurids in the Bull Canyon Formation of east-central New Mexico.

Hunt *et al.* reported that other material found in the Bull Canyon Formation includes a partial skeleton (NMMNH P-17375) representing a smaller theropod, and an isolated premaxilla (UCM 52081) establishing the presence of the putative "theropod" *Shuvosaurus inexpectus* (which has been reclassified as nondinosaurian, possibly a rauisuchian; see Murry and Long).

An illium described by Case (1927), and referred by Long (in Murry 1989; see also Long and Murry) to *C. bryansmalli*, was referred by Hunt *et al.* to the new genus and species *Caseosaurus crosbyensis* (see *Caseosaurus* entry).

Key references: Case (1927); Heckert, Lucas and Sullivan (2000); Hunt (1994, 1995); Hunt, Lucas, Heckert, Sullivan and Lockley (1998); Long and Murry (1995); Murry and Long (1989).

CHUANJIESAURUS Fang, Pang, Lu, Zhang, Pan, Wang, Li and Cheng 2000

Saurischia: Sauropodomorpha: Sauropoda: Eusauropoda: "Cetiosauridae."

Name derivation: Chinese "Chuanjie" + Greek *sauros* = "lizard."

Type species: *C. anaensis* Fang, Pang, Lu, Zhang, Pan, Wang, Li and Cheng 2000.

Other species: [None.]

Occurrence: Ána, Chuanjie, Lufeng, Yunnan Province, China.

Age: Middle Jurassic.

Known material: At least one humerus, scapula, series of at least 12 (perhaps as many as 16) caudals vertebrae with chevrons, two or three smaller long bones, some ribs.

Comments: The new type species *Chuanjiesaurus ánaensis* was named and briefly described by Fang, Pang, Lu, Zhang, Pan, Wang, Li and Cheng (2000) in a paper on the stratigraphy of the Jurassic of Lufeng, China, based upon partial postcranial remains. To date of this writing, an English translation of this paper has not yet been made available. Although a diagnosis was included in this paper, a description and figures of the type material were not.

Apparently, this taxon represents a relatively small sauropod.

Note: According to the ICZN, a species name cannot have an apostrophe; therefore, the species should be written *Chuanjiesaurus anaensis* (G. Olshevsky, personal communication 2000).

Key reference: Fang, Pang, Lu, Zhang, Pan, Wang, Li and Cheng (2000).

†COELOPHYSIS—(=*Longosaurus*, *Rioarribasaurus*; =?*Camposaurus*, ?*Podokesaurus*, ?*Syntarsus*)

Saurischia: Theropoda: Neotheropoda: "Ceratosauria": Coelophysoidea: Coelophysidae.

Type species: *C. bauri* (Cope 1887).

Other species: *C. longicollis* (Cope 1887) [*nomen dubium*], *C. willistoni* (Cope 1889) [*nomen dubium*].

Comments: Studies continue regarding *Coelophysis*, a small, primitive theropod known only from Upper Triassic rocks of North America.

In the previous two volumes, conflicting opinions were cited concerning the taxonomic validity of this genus, with some paleontologists positing that the generic name *Coelophysis* should be reserved for the various specimens first described by Edward Drinker Cope in 1877 (as new species of the genus *Coelurus*), with the newer name *Rioarribosaurus* designating the virtually countless theropod specimens collected from the Whitaker Quarry at Ghost Ranch (*e.g.*, Colbert 1989). Although the controversy over this dinosaur seemed to have been resolved when the International Commission on Zoological Nomenclature (ICZN)—after evaluating numerous unsolicited pro and con opinions from various workers—ruled that the better known name *Copelophysis* was indeed valid for the Ghost Ranch specimens (see *S1*), questions continue to surface concerning the taxonomy of this well-known theropod.

In a survey of western North American dinosaurs of the Late Triassic, Hunt, Lucas, Heckert, Sullivan and Lockley (1998) stressed that the original syntypes (AMNH 2701-2708) of the various species of *Coelophysis* (see *Coelophysis* and *Rioarribasaurus* entries, *D:TE* and *S1*) were collected in the 1880s by David Baldwin from at least two localities in the Petrified Forest Formation, Rio Arriba County, New Mexico (see Colbert; Hunt and Lucas 1991; Sullivan, Lucas, Heckert and Hunt 1996). Six other Petrified Forest Formation localities have also yielded remains attributed to *Coelophysis*. More than a century after Baldwin collected those original specimens, apparent topotypes of *Coelophysis*, found at about the same stratigraphic level and of similar preservational quality to the syntypes, were described by Sullivan *et al.* as morphologically indistinguishable from the syntypes.

According to Hunt *et al.*, all of the above specimens seem to differ from the neotype (AMNH 7224) specimen of *Coelophysis bauri* which was collected from the famous Ghost Ranch quarry, in the geologically younger (Apachean) Rock Point Formation (Hunt and Lucas; Sullivan *et al.*) (not the Petrified Forest Formation, as previously stated by other authors). Hunt *et al.*, therefore, concluded that the syntypes of the type species *C. bauri* could represent a different taxon from the theropod known from so

Photograph by the author.

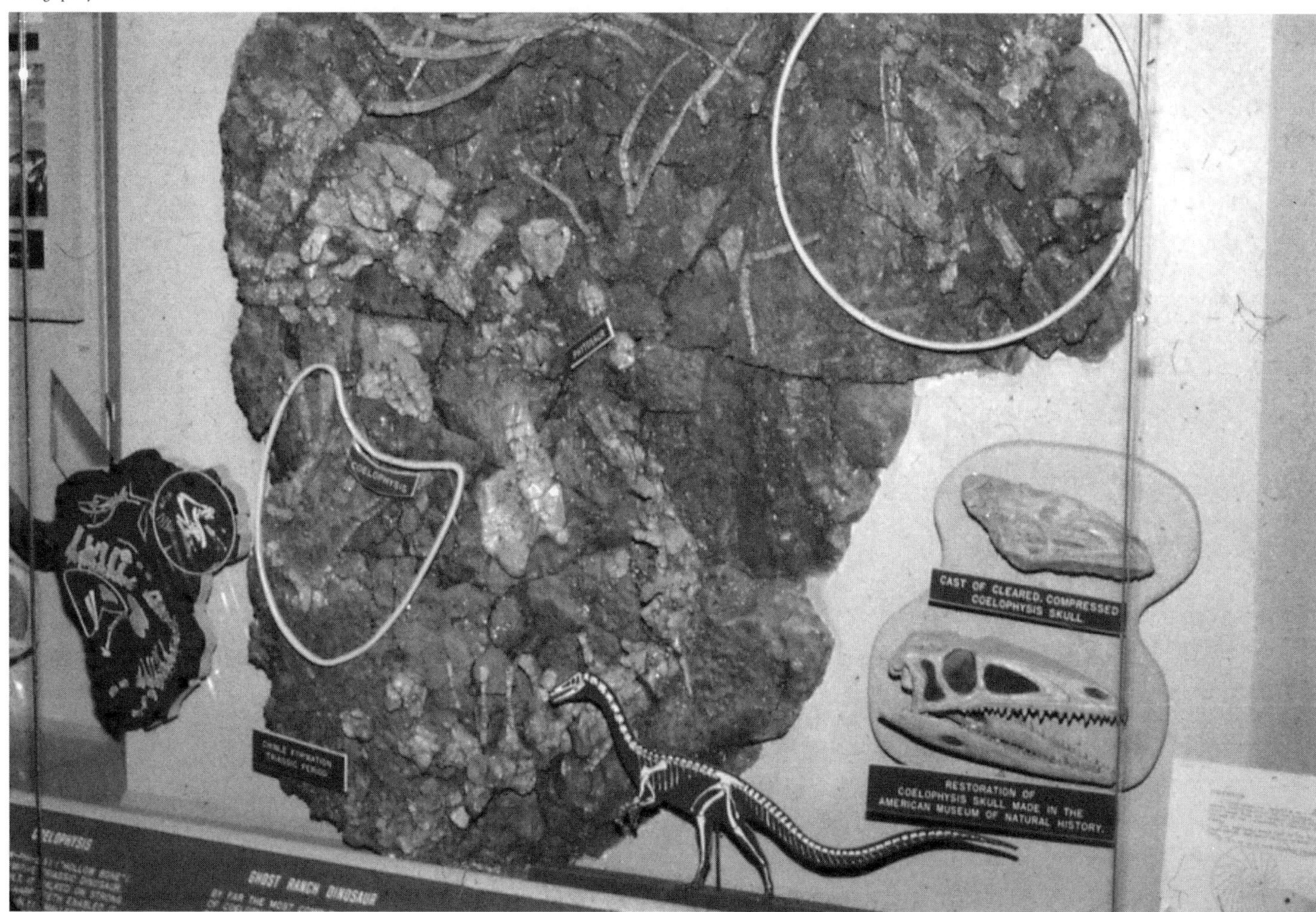

One of the many Whitaker Quarry blocks containing fossils of the small theropod *Coelophysis bauri*, this one exhibited at a visitors center near Ghost Ranch, New Mexico.

many specimens found at the Ghost Ranch site (see *Eucoelophysis* entry).

Padian (1986) referred to *C. bauri* a fragmentary skeleton from the Dinosaur Hill locality (also called Bolt quarry and Lacey Point), Painted Desert Member (the upper dinosaur-bearing unit) of the Petrified Forest Formation, at Petrified Forest National Park. According to Hunt and Lucas, and Hunt *et al.*, this specimen seems to differ from AMNH 7224 and associated Ghost Ranch skeletons in possessing a well-developed obturator foramen (although Robert M. Sullivan has identified the latter feature in the neotype); it is similar to the related genus *Syntarsus* (see below; also see *D:TE*) and apparently identical to an undescribed specimen (NMMNH P-4415) from the Bull Canyon Formation of east-central New Mexico. Also similar to Padian's specimen, in the possession of a well-developed obturator foramen (a plesiomorphic character), is a specimen referred by Cope (1887) to *Coelophysis* from the "Revueltian" Petrified Forest Formation of New Mexico.

Regarding the Ghost Ranch specimens, Hunt *et al.* noted that the most important dinosaur body fossils from rocks of Apachean age are those from this quarry. According to these authors, only some of the theropods represented at this site are *C. bauri* (Colbert 1989), this taxon having an obturator foramen, but possibly lacking the pubic fenestrae and pubio-ischiac plate seen in *Syntarsus*. Indeed, some Ghost Ranch specimens possess two pubic foramina, incipient skull crests, and fused tarsals indistinguishable, except for their smaller size, from such features in *S. kayentakatae*, the North American species of *Syntarsus*.

Modern technology has again contributed to our understanding of dinosaurs, in this instance Late Triassic theropods such as *Coelophysis*. In a preliminary report published as an abstract, Gatesy and Middleton (1998) wrote about their utilization of computers and three-dimensional animation ("a powerful technique for analyzing complex movements, which are difficult to create, compare, and communicate using standard 2-D images") to create a visualized simulation of the moving foot of *Coelophysis*. As the authors noted, the hindlimbs of theropods comprise at least 17 linked segments of bone which undergo cyclical

Photograph by Robert A. Long and Samuel P. Welles.

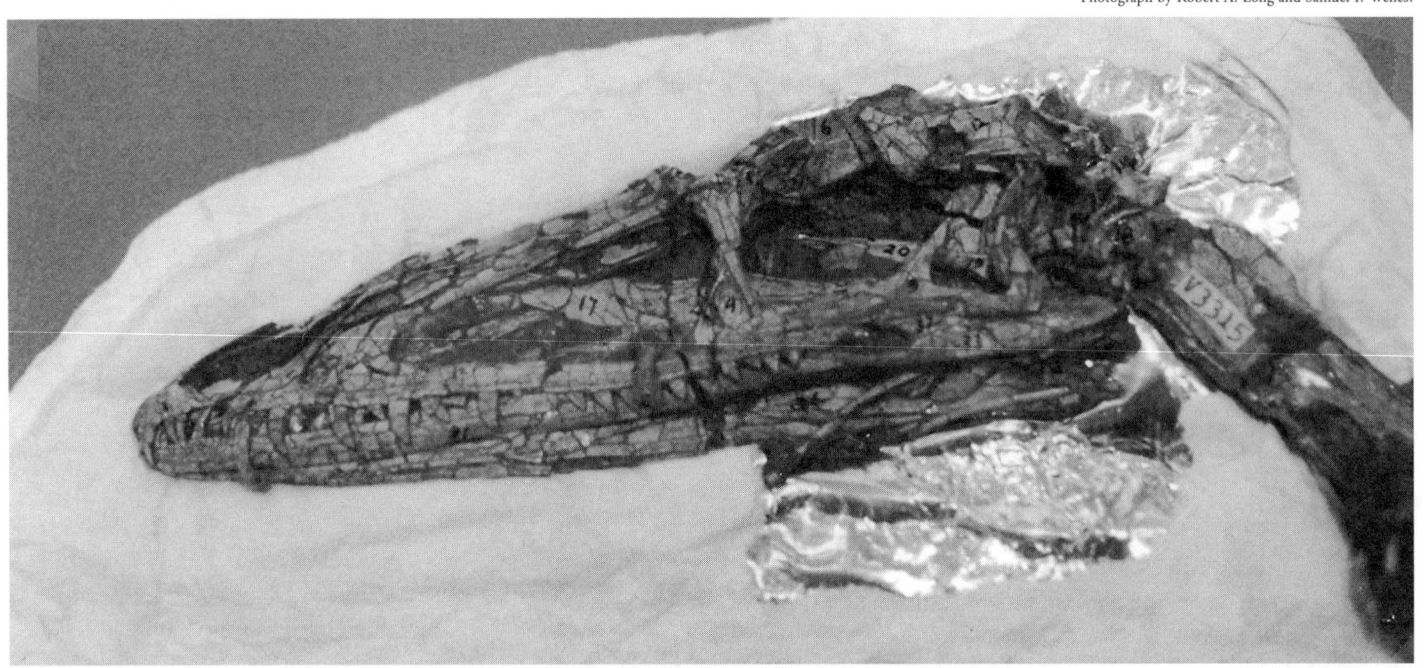

One of numerous skulls (left lateral view) recovered from the Ghost Ranch quarry and referred by Edwin H. Colbert to *Coelophysis bauri.*

movements during walking, this morphological, spacial, and temporal complexity being well suited to the tools of 3-D animation.

Gatesy and Middleton produced a computer-generated "puppet," based upon the hind-limb bones of *Coelophysis*, and controlled by six inverse kinematic constraints. This "puppet" was then rotoscoped (a motion-picture and video process by which an animated image is traced over the real image of a person, animal, or object) onto video footage of a walking turkey. The image was subsequently modified in accordance with Late Triassic fossil footprints (presumably made by a theropod like *Coelophysis*) from East Greenland and various data gleaned from those impressions. Results of this experiment "confirmed that Triassic theropod feet underwent toe convergence as in the turkey, but differed from this bird in hallucal orientation and the timing of metatarsal rotation."

Detailed data from this study were subsequently formally presented in a joint paper authored by Gatesy, Middleton, Jenkins and Shubin (1999). Gatesy *et al.* found that the hallux (or first digit) of the foot was slightly flexed when the track-making theropod slid down into the mud that preserved the prints; that the hallux was not reversed as it is in modern birds that use this toe for perching; and that the deeper the substrate, the more pronounced was the imprint of the metatarsals, this suggesting that dinosaurs such as these track-makers consistently maintained a 25- to 30-degree continual metatarsal slope even while walking through mud. The authors further determined that the method of locomotion of these theropods involved a hip extension similar to that of walking crocodilians as opposed to the knee extension employed by modern birds.

In an abstract, Smith, Andersen, Larsson and Bybee (2000) briefly reported on their CT scan of a recently prepared, well-preserved skull (CMNH collection) of *C. bauri* from Ghost Ranch, this study intended to extract the braincase and clarify the internal skull anatomy of a basal theropod. The specimen selected for this study is less distorted than most collected *Coelophysis* skulls and preserves an undistorted occipital region.

Scanning was executed in Chicago by Charles Smith of BIR Laboratories, resulting in about 600 transverse sections taken 0.5 millimeters apart. The complete skull and first three cervical vertebrae were imaged.

The scan resulted in good separation between bone and matrix in most cranial regions except for the skull roof, bones from that area being filled in by a laser scan, the anatomy of other internal areas able to be reconstructed. Smith *et al.* compared the braincase and endocranium anatomy of *Coelophysis* with the related form from Zimbabwe, *Syntarsus rhodesiensis* (see below). Results of this comparison will presumably be published in detail at some later date.

A long-standing debate in the paleontological literature has been whether or not *Coelophysis* and *Syntarsus* are congeneric (see "*Rioarribasaurus*" [=*Coelophysis*] and *Syntarsus* entries, *D:TE* for pro and con arguments). Indeed, Gregory S. Paul referred *Syntarsus* to *Coelophysis* in 1993, although the majority of subsequent authors opted to maintain generic separation of these taxa. This question was recently addressed

From *Predatory Dinosaurs of the World* (1988), copyright © Gregory S. Paul.

Coelophysis bauri **as restored by Gregory S. Paul. Previously only speculation, feather-like integumentary structures such as those shown are now known in some small theropods, although none have yet been found in this taxon.**

once again by Alex Downs (2000), who, starting in early 1987, undertook preparation work on a large Carnegie Museum of Natural History block (field number C-3-82) of Ghost Ranch *Coelophysis* specimens as part of a fossil preparation exhibit at the National Museum of Natural History.

Downs, who at the time of his study did not have access to the actual fossil material of *Syntarsus rhodesiensis*, acquired skeletal figures of this species (from Raath's doctoral dissertation of 1977). Downs learned that much of the *Coelophysis* material collected during the 1940s and prepared at the American Museum of Natural History in the 1950s had been prepared mechanically, without the help of a microscope (E. H. Colbert and C. Sorenson, Jr., personal communication to Downs), although "the use of a good stereo-dissecting microscope is essential to minimize the damage to these delicate fossils." Furthermore, Downs noted that the description of *Coelophysis* in Colbert's 1989 monograph was mostly based upon specimens collected in the 1940s at Ghost Ranch, many of which had suffered damage from weathering, rendering them difficult to interpret, and also that the line drawings by Lois Darling (see Padian 1996) of *C. bauri* bones conflicted in many places with the description written by Colbert.

In comparing the illustrations of *Syntarsus* with the specimens in the *Coelophysis* block and noting various points made earlier by Paul, Downs observed that the morphologies of both genera matched in almost every instance. Downs noted various cranial differences between the two forms. For example, *S. rhodesiensis* possesses a nasal fenestra whereas *C. bauri* does not. (Noting the otherwise remarkable match between the Ghost Ranch and Zimbabwe theropods, Downs expressed doubt that the African *Syntarsus* actually possesses a nasal fenestra, although this hypothesis cannot be tested, the author noted, until articulated skulls of *S. rhodesiensis* may be discovered.) Also, the median ridge of the anterior process of the lacrimal in *Coelophysis* does not show the slight lateral "bowing" described for *Syntarsus*. However, both genera share a number of features. In *C. bauri*, the morphology of the bones that would border the nasal fenestra (if present) is very similar to that described in *S. rhodesiensis*. Allowing for ontogenetic changes, the morphology of the nasal of a juvenile *C. bauri* specimen (CM collection) is quite similar to that element figured by Raath for *R. rhodesiensis*. The lacrimal in Raath's figure of *S. rhodesiensis* is virtually identical to those observed in the skulls from Ghost Ranch. The maxillae of both genera possess interdental plates. Furthermore (*contra* Colbert), *Coelophysis*, as evidenced by "one of the finest *Coelophysis* skulls in any collection" (CM 31374), possesses a vaulted palate as seen in *Syntarsus*.

Postcranially, regarding the pelvic girdle, an opening in the proximal pubis of *S. rhodesiensis* (Raath 1969, 1977) is homologous with the obturator foramen in *C. bauri*. Both taxa have a similar antitrochanter on the caudal margin of the acetabulum (see Raath 1969, 1977 for *Syntarsus*; Rowe and Gauthier 1990 for *Coelophysis* and *Syntarsus*). Also, Downs noted, the pubes, ilia, and ischia are fused together (*contra* Colbert) as they are in *Syntarsus* (Raath 1977). In general, the pelvic anatomy of both theropods (see Rowe and Gauthier) are virtually identical.

According to Downs, the femur of *C. bauri* (*contra* Colbert), is typically theropod, having a strongly inturned head, and, like that of *S. rhodesiensis*, has three condyles on its distal end; moreover, all of the *Coelophysis* femora examined by Downs displays characters (*e.g.*, lesser trochanter block-like, greater trochanter rugose) that conform to Raath's "robust morph" of *S. rhodesiensis*. Likewise, the fibula and tibula of both taxa were indistinguishable to Downs (*contra* Darling's illustrations in Colbert's monograph). Also, the astragalus and calcaneum in both *Coelophysis* and *Syntarsus* are fused together, forming a single unit in the ankle. In both forms, the astragalus has a short, anterior ascending process.

Downs noted that the anterior margins of the transverse processes of the anterior dorsal vertebrae are "swept back" in both *Coelophysis* and *Syntarsus*, and, in both forms, the transverse processes of the dorsal vertebrae grade into a delta shape. Finally, Downs identified what appeared to him to be a furcula, a birdlike element previously identified in both *S. rhodesiensis* (Raath 1977) and in the North American species *S. kayentakatae* (Tykoski 1998).

Downs acknowledged that he did not have the opportunity to study the actual specimens of *S. rhodesiensis* personally. However, the author found the illustrations prepared for Raath's dissertation, which Downs assessed to be mostly accurate, to depict *Coelophysis bauri* more precisely than do the drawings and descriptions in Colbert's monograph. "Barring the most unlikely coincidences," Downs stated, "*C. bauri* and *S. rhodesiensis* must be remarkably similar," differing only in various minor details (*e.g.*, neck length, proximal and distal hindlimb proportions, and proximal caudal anatomy). Downs suggested that "These differences do not justify a generic separation." However, the author did not, in the present writer's opinion, formally refer *S. rhodesiensis* to the genus *Coelophysis* (as a new species *C. rhodesiensis*); nor did he comment upon the species *S. kayentakatae*, having not seen the fossils (A. Downs, personal communication 2000).

As Downs' study mainly focused upon the genus *Coelophysis* and did not compellingly, in the present writer's opinion, demonstrate that *Coelophysis* and *Syntarsus* are synonymous, this document, at least for now, retains these taxa as separate genera.

Note: Kundrát (1998), in an abstract, briefly commented on various earlier reports by Gerard Gierlinski (see "Note," *Dilophosaurus* entry, *S1*; also, Sabath and Gierlinski 1998, below) regarding an integumentary impression of feather-like features of tracks assigned to the ichnospecies *Eubrontes miniscułus* Hay 1902, which were originally described in 1858 by Edward Hithcock. This trace fossil (AC 1/7) recorded a moment, from the Early Jurassic of Massachusetts, when the medium-sized theropod trackmaker (believed by Hitchcock to be a marsupial, interpreted by subsequent workers as an ornithischian) had rested in a sitting position, with subdominal, ischiadic, and two autopodial imprints preserved as the tail was held off the ground.

As stated by Kundrát, the diagnostically important distal pubic print, seen inside the subdominal imprint, was impressed through the skin by a rod-like, slightly swollen distal end of the pubic shaft. As this condition is known in both *Coelophysis* and *Syntarsus*, the trackmaker was probably a coelophysoid theropod. Sabath and Gierlinski also identified AC 1/7 as a propubic theropod, as the pubic imprint is separated from the ischiadic imprint.

As verified by Kundrát, the trackmaker left detailed impressions of soft tissues on both sides of its subabdominal imprint, probably consisting "of parallelly arranged barbs (comprising incipiently developed barbules in their lower parts) branched alternately from the basal stem zone," and measuring about 10 millimeters in length by 2 millimeters in width. As these structures predate true avian feathers, Kundrát suggested the new termed "dinosaur feather" for them, diagnosed by the brush-like structure (suitable for insulation), and the absence of any avian feather aerodynamic quality. Kundrát commented that the discovery of these structures offers more evidence supporting the hypothesis that birds had their origin in the Theropoda, that feathers are not structures unique to birds, and that this find also suggests an endothermic affinity of the physiology of coelophysoids.

Sabath and Gierlinski (1998) pointed out that AC 1/7 predates by some 50 million years (Early Liassic Turner Falls System) the first advanced avian feathers (*i.e.,* those with vanes), which begin in the fossil record with *Archaeopteryx* in the Late Jurassic. These authors stated that experiments with AC 1/7 reveal that these brush-like (rather than pennate, or advanced feather-like) imprints closely resemble the semiplumes (intermediaries between down and contour feathers) of extant birds, and that similar structures are present near the tibia of the Berlin *Archaeopteryx* specimen, and associated with the recently described Chinese compsognathid, *Sinosauropteryx primus*.

According to Sabath and Gierlinski, semi-plume feathers effectively serve insulation functions without a protective layer of contour feathers. In modern birds, the continuous intergradation between down, semiplumes, and contour feathers "may suggest that non-avian semiplumes differentiated into avian contour feathers and down, simultaneously specializing in aerodynamic and insulating functions, respectively."

From the Tecovas Member of the Dockum Formation (Upper Triassic: late Carnian) in west Texas, Case (1927, 1932) referred various dinosaur-like caudal vertebrae (UMMP 7277, 9805, and 13670) and isolated teeth (*e.g.*, UMMP 2680, 13766, and 13765) to *Coelophysis* sp. or the Dinosauria. According to Hunt *et al.*, these vertebrae exhibit no synapomorphies of Dinosauria, poor preservation of the material precluding the evaluation of any potential dinosaurian characters (*e.g.*, elongate prezygapophyses; see Novas 1993). All the teeth are serrated and laterally compressed, and some are slightly recurved (see Case 1932), and could, Hunt *et al.* noted, belong to theropods, rauisuchians, or a heterodont phytosaur.

Murry and Long (1989) referred isolated teeth from the *Placerias* quarry, in the Upper Triassic/late Carnian Bluewater Creek Formation (Chinle Group) of Apache County, Arizona to "cf. *Coelophysis*," representing "probably some variety of thecodont or primitive saurischian." Hunt *et al.* found these teeth to be indeterminate.

Key references: Case (1927, 1932); Colbert (1989); Cope (1887); Downs (2000); Gatesy and Middleton (1998); Gatesy, Middleton, Jenkins and Shubin (1999); Hay (1902); Hitchcock (1858); Hunt and Lucas (1991); Hunt, Lucas, Heckert, Sullivan and Lockley (1998); Kundrát (1998); Long and Murry (1989); Padian (1986, 1996); Paul (1993); Raath (1966, 1977); Rowe and Gauthier (1990); Sabath and Gierlinski (1998); Smith, Andersen, Larsson and Bybee (2000); Sullivan, Lucas, Heckert and Hunt (1996); Tykoski (1998).

†COELURUS

Saurischia: Theropoda: Neotheropoda: Tetanurae: Avetherapoda: Manuraptoriformes: Manuraptora *incertae sedis*.

Comments: An important new, partial skeleton belonging to the small theropod type species *Coelurus fragilis* (see *D:TE*), collected from the Upper Jurassic Morrison Formation, west of Bone Cabin Quarry, in Albany County, Wyoming, was briefly announced in an abstract by Miles, Carpenter and Cloward (1998). Recovered material includes the first skull material ever found belonging to this species, and also a complete pectoral girdle, both forelimbs, an entire right hindlimb, a left femur, the proximal halves of the left tibia and fibula, 14 vertebrae, ribs, three chevrons, the distal halves of both pubes, and possibly one sacral centrum.

As briefly described by Miles *et al.*, this specimen is nearly twice the size of the holotype (YPM 1991) of *C. fragilis*, although both specimens are immature, as fusion has not taken place between the neural arches and centra.

Miles *et al.* noted that the recently collected specimen is significant in offering new data supporting the taxonomic separation of *C. fragilis* and the Late Jurassic manuraptoriform *Ornitholestes hermanni*. The latter is known from an adult skeleton (see *Ornitholestes* entry, *D:TE*) approximately half the size of the new *C. fragilis* specimen, but in which much fusion can be seen in the skull and postcranial skeleton. Also, the skull of *C. fragilis* seems to be dorso-ventrally taller and anterior-posteriorly shorter than that of *Ornitholestes*.

Although, in the past, *Coelurus* has been classified as Tetanurae *incertae sedis*, Miles *et al.* pointed out the following salient Manuraptoran features: Scapula long, strap-like; ulna bowed; manus long and slender; (small) carpal semilunate; tibia longer than femur; anterior (=lesser) trochanter low, square; elongated prezygapophyses on distal caudals and cervicals.

Key reference: Miles, Carpenter and Cloward (1998).

†COMPSOGNATHUS

Comments: The hand of this small theropod has never been entirely understood due to incomplete preservation and the partial disarticulation of the manus in the holotype specimen (BSP 1563; see *D:TE* for photograph), this leading to confusion as to whether the manus possessed two or more digits. According to Gauthier and Gishlick (2000) in a brief report, most of this confusion has stemmed from the position of the three long bones, often interpreted as metacarpals I through III, at the base of the left manus.

Recently, Gauthier and Gishlick reexamined this specimen, discovering that the element usually identified as metacarpal I is actually the first phalanx of digit I, bearing the asymmetrical proximal articular facets typical of other theropods. That identification is supported by an articulated digit positioned the same way in the right manus, the base of which is obscured beneath the left ulna. The authors observed that that digit bears an ungual phalanx articulated to a long phalanx, thereby exhibiting a phalangeal formula correct for a first digit.

Gauthier and Gishlick identified a "mystery bone" preserved above the skull as metacarpal I of the left manus. They briefly described this element as block-like in shape, having a flat proximal articular surface and two asymmetrical distal articular condyles typical of a first metacarpal, these condyles matching those on the element identified by these authors as a first phalanx.

It was the conclusion of Gauthier and Gishlick, therefore, that *Compsognathus* possessed a manus similar to that of other theropods except that metacarpal

Skull of a new small theropod similar to (and possibly representing a new though more primitive species) of *Compsognathus*. Scale = 2 cm. (After Viohl 1999.)

I is abnormally short rather than unusually long as generally depicted. The manus has a minimum of two functional digits, with metacarpal III of the third digit reduced, and with no phalanges preserved in the type specimen that could belong to a third digit. However, as the impression of the distal end of that third metacarpal seems to bear a possible collateral ligament pit, the presence of a third functional digit is feasible. There is no evidence that the manus was four-fingered.

Gauthier and Gishlick concluded that *Compsognathus* possessed a manus similar in many ways to the hands of other basal coelurosaurs, especially those of *Scipionyx*, *Sinosauropteryx*, and Tyrannosauridae.

Note: Viohl (1999) briefly described a new and as yet unnamed small coelurosaurian theropod species from Upper Jurassic (late Kimmeridgian) strata near Schamhaupteen, district of Eichstätt, in Bavaria, Germany. (Also known from this area are a variety of fossil forms, including nano- and microfauna, invertebrates, fishes, turtles, rhyncocephalians, coprolites, regurgitates, and numerous land plants.)

The theropod specimen, preserved in a slab, was recovered in July, 1998, by brothers Hans and Klaus-Dieter Weiss, volunteer workers for the Jura-Museum. It includes at least a partial though excellently preserved skull and also some cervical vertebrae. As strong silicifation of the matrix had made the rock harder than steel, preparation of the fossil was extremely difficult. Nevertheless, Jura-Museum preparator Pino Völkl managed to uncover more of the skull and glue on the displaced elements.

As related by Viohl, uncovering the skull allowed for the unambiguous identification of the specimen as a theropod. Unlike *Compsognathus longipes*, however, a tiny theropod known from the Solnhofen Lithographic Limestone in Bavaria, the unnamed form is especially remarkable for its very large teeth. In morphology, however, the teeth of this new form are virtually identical to those of *C. longipes* (P. Wellnhofer, personal communication to Viohl). The posterior edges of the tooth crowns are serrated, only the long premaxillary teeth being smooth, and their tips are strongly bent backwards. This suggested to Viohl that this new form could be a new species of *Compsognathus* possibly ancestral to *C. longipes*.

From the geologic and other fossil evidence at this locality, Viohl deduced that this new theropod probably lived on a nearby island where it hunted other vertebrates. Based on the size of its teeth, the author inferred that this dinosaur's spectrum of prey could have included animals larger than those hunted by *C. longipes* (remains of the small lizard *Bavarisaurus*, having been found in the stomach area of the type specimen of this species; see *D:TE*). Viohl further speculated that this dinosaur was most likely swept away by a torrent, washed during a rainy season into the basin, drowning and sinking to the seafloor, there rapidly being covered by a bacterial mat and thereby preserved.

A more detailed description and the naming of this new theropod will be published following the complete preparation and further study of the specimen.

Key reference: Gauthier and Gishlick (2000); Viohl (1999).

CRISTATUSAURUS Taquet and Russell 1998 [*nomen dubium*].

Saurischia: Theropoda: Neotheropoda: Tetanurae: Avetheropoda: Spinosauroidea: Spinosauridae: Baryonychinae.

Name derivation: Latin *cristatus* = "crested" + Greek *sauros* = "lizard."

Type species: *C. lapparenti* Taquet and Russell 1998 [*nomen dubium*].

Other species: [None.]

Occurrence: Elrhaz Formation, Gadoufaoua, Niger.

Age: Early Cretaceous (Aptian).

Known material: Partial skull and postcranial remains.

Holotype: MNHN GDF 366, both premaxillae, portion of right maxilla and dentary, subadult.

Diagnosis of genus (as for type species): Differs from *Spinosaurus ?maroccanus* in the following: Premaxilla short, strongly "hooked" in lateral aspect, increasing in height posteriorly; dorsal surface narrowing into crest posteriorly; anteriormost premaxillary alveolus relatively large, lateral alveoli uniformly closely spaced; maxillary and dentary teeth laterally compressed in cross section, maxillary teeth vertically oriented posteriorly; dentary slightly constricted vertically in midsection, alveoli closely spaced; maxillary and dentary teeth having finely serrated carinae (Taquet and Russell 1998).

Comments: The genus *Cristatusaurus* was established upon partial skull material (HMNH GAD 366) including premaxillae, a part of a right maxilla (the most complete yet known for any spinosaurid), and a dentary recovered from the Elrhaz Formation (Aptian) of Gadoufaoua, south of the Aïr Massif, in northern Niger. The nonfused suture separating the premaxillae is indicative of an immature individual. Referred remains from the same locality includes fused premaxillae and dorsal vertebrae numbers 357–359 and 361 (HMNH GDF 365) (Taquet and Russell 1998).

***Cristatusaurus lapparenti*, MNHN GDF 366, holotype (top) anterior portion of premaxillae, left lateral view, (middle) portion of right maxilla, medial view, and (bottom) portion of right dentary, medial view. (After Taquet and Russell 1998.)**

While not named until 1998, the above material had already been figured and described by Taquet (1984), and later by Kellner and Campos (1996).

In comparing the *Cristatusaurus* remains to those of other spinosaurids, Taquet and Russell observed that the holotype dentary fragment shows an anterior constriction that is similar to but less marked than in *Spinosaurus aegyptiacus* and *S. ?maroccanus*. The base of a spine preserved on a neural arch of an anterior dorsal vertebra (mentioned by Taquet 1976) is less robust than its counterpart of similar dimensions in *S. maroccanus*, this suggesting to Taquet and Russell that the neural spines were less elevated in *C. lapparenti*.

Although Taquet and Russell found *C. lapparenti* to differ from *Baryonyx walkeri* by the "brevirostrine condition of premaxilla," Charig and Milner (1997) (see also Sereno, Beck, Dutheil, Gado, Larsson, Lyon, Marcot, Rauhut, Sadleir, Sidor, Varricchio, Wilson and Wilson 1998) found these premaxillae to possess no distinguishing features. Charig and Milner attributed

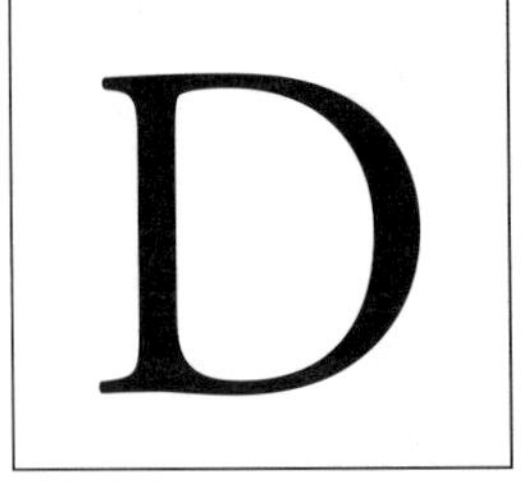

these premaxillae to an indeterminate species of *Baryonyx*; Sereno *et al.* regarded *C. lapparenti* as a *nomen dubium*.

Taquet and Russell also separated *Cristatusaurus* from both *Irritator* and "*Angaturama*" (=*Irritator*; see entry) by the presence of serrations on the teeth.

Key references: Charig and Milner (1997); Kellner and Campos (1996); Sereno, Beck, Dutheil, Gado, Larsson, Lyon, Marcot, Rauhut, Sadleir, Sidor, Varricchio, Wilson and Wilson (1998); Taquet (1976, 1984); Taquet and Russell (1998*b*).

†DACENTRURUS

Ornithischia: Genasauria: Thyreophora: Thyreophoridea: Eurypoda: Stegosauria: Stegosauridae.

Comment: Casanovas-Cladellas, Santafé-Llopis, Santisteban Bové and Pereda Suberbiola (1999), in a review of all stegosaurian remains collected from the possibly Late Jurassic (Purbeck facies; previously dated as Early Cretaceous) Aras de Alpuente area (Losilla and Cerrito del Olmo localities) of Valencia, Spain, described a new femur (designated CO-II-1) which they referred to the stegosaurid *Dacentrurus armatus*.

This femur, the authors noted, is morphologically comparable to one figured by Zbyszewski (1946; see also Lapparent and Zbyszewski 1957) from Baleal (Peniche, Portugal) (see *D:TE*), which they referred to *Omosaurus armatus* (=*Dacentrurus armatus*; see Galton 1985*f*). The identification of this specimen as *D. armatus*, therefore, expands the distribution of this species in Europe.

Note: In a brief report on the dinosaurs of the Late Jurassic of Portugal, Mateus (1999) pointed out that the alleged "egg," described by Lapparent and Zbyszewski as belonging to *Dacentrurus armatus* (see "note" in *Dacentrurus* entry, *D:TE*, wherein the reference was incorrectly cited as 1955), "is not an egg but a geological structure."

Key references: Casanovas-Cladellas, Santafé-Llopis, Santisteban Bové and Pereda Suberbiola (1999); Galton (1985*f*); Lapparent and Zbyszewski (1957); Mateus (1999); Zbyszewski (1946).

†DASPLETOSAURUS

Comments: There is no doubting that theropods such as the giant tyrannosaurid *Daspletosaurus* were carnivores that fed upon prey of considerable size. Until recently, however, the exact nature of that prey has been basically limited to evidence derived from the analysis of coprolite contents and from tooth marks found on bones.

The first gut contents for a tyrannosaurid were reported by Varricchio (1999) in a *Daspletosaurus* skeleton recently recovered from the Upper Cretaceous (Campanian) Two Medicine Formation of western Montana. Associated with these skeletal remains were the acid-etched vertebrae and fragmentary dentary belonging to a juvenile hadrosaurid. As Varricchio noted, the corrosion displayed on the surface of these hadrosaurid elements matches the kind produced by stomach acids and digestive enzymes in a wide variety of living animals.

Based upon the above observations, combined with gut contents of other taxa and also tooth-marked bone studies, Varricchio suggested that *Daspletosaurus* (and most carnivorous dinosaurs) ingested and digested food in a manner similar to crocodilians and birds, "employing a two-part stomach with [an] enzyme-producing proventriculus followed by a thick-walled muscular gizzard," this two-part stomach apparently a synapomorphy of Archosauria.

Further studies have been made of the excellent specimen FMNH PR308, a well-preserved partial tyrannosaurid skeleton with incomplete skull excavated in 1914 from rocks of the Dinosaur Park Formation (75.5 to 74.5 million years ago) of Alberta, Canada (see *S1* for background and historical details; see *D:TE* and *S1* for additional photographs of this specimen). The specimen was originally referred to *Gorgosaurus libratus* (see Matthew and Brown 1923), subsequently transferred to the genus *Albertosaurus* (Russell 1970*b*), and officially referred by tyrannosaurid specialist Thomas David Carr (1999*b*, 2000) to the type species *Daspletosaurus torosus*.

In the first installment of a two-part article appearing in the privately published journal *Dinosaur World*, Carr (1999*a*) detailed the following:

The specimen, originally housed at the American Museum of Natural History (catalogued as AMNH 5434), was first mentioned by American Museum paleontologists William Diller Matthew and Barnum Brown in a brief notice published by that institution in 1923. The specimen lacks parts of the skull, both feet, and portions of the tail. Representing one of the largest tyrannosaurid specimens yet recovered from the Dinosaur Park Formation, the skull is estimated (if complete) to measure approximately 1050 millimeters (more than 38 inches) in length. As measured by Matthew and Brown, the skull's muzzle, through the last tooth row, measures 310 millimeters deep; the back of the skull is 425 millimeters (almost 16 inches) from the nuchal crest to the mandibular condyle; and the lower jaw totals 1015 millimeters (about 37 inches) in length.

Carr (1999*a*) noted that the skull of FMNH PR308 to this day still bears the original American Museum number, written in black ink on the left jugal. Inexplicably, it also displays "AMNH 5336"

Photograph by John Weinstein, courtesy The Field Museum (neg. #GEO85818).

Skull (FMNH PR308, originally referred to *Gorgosaurus libratus*) of *Daspletosaurus torosus* on display separate from the mounted reconstructed skeleton.

inscribed in red ink on the left pterygoid, this being the catalog number given a skull of *Gorgosaurus libratus* [=*Albertosaurus libratus* of Carr's usage] found by Brown, now displayed in the American Museum's Hall of Saurischian Dinosaurs.

Also, Carr (1999*a*) pointed out that a photograph of the specimen included in Matthew and Brown's paper indicated that the skull was intended to be a "showpiece" intended more for display at the American Museum than for scientific study. Consequently, much of the skull had already been restored in plaster (see also Carr 1999*b*), sometimes concealing diagnostic features (see Russell). It was this quite heavily restored specimen that was eventually purchased in 1954 by the Chicago Natural History Museum (now The Field Museum) and, after two years of mounting, displayed at its new and permanent home.

According to Carr's (1999*a*) close examination of the skull of FMNH PR308, the palate, the actual part seen in ventral view only, has been restored in plaster after that of a specimen (AMNH 5336) of *G. libratus*. Surprisingly, most of the facial part of the skull proved to be a plaster reconstruction, "which provided an explanation for the enigmatic nature of the skull" (see *S1* for opinions in this regard of Bakker, Currie and Williams 1988; also Paul 1988*b*). Much of the restoration of this mature skull was revealed to be plaster sculpturing based upon smaller and less mature specimens of *G. libratus* (at least four individuals) in the American Museum collections. All of the premaxillary and maxillary teeth proved to be plaster

Skull (FMNH PR308, originally referred to *Gorgosaurus libratus*) in dorsal view of *Daspletosaurus torosus*.

Courtesy The Field Museum (neg. #GEO81591).

models too small for their respective alveoli. Such key features of the skull as "both premaxillae, both lacrimal horns, the midregion of both maxillae, the rostral end of the right jugal, and most of each postorbital horn" were modeled in plaster. Plaster smeared onto the dorsal surface of the prefrontal and frontals, obscuring important details. Also plaster are the "matrix" filling in the antorbital fenestra, the maxillary fenetra, and the orbit. The unrestored part of the palate, the adductor chamber, and the braincase are complete and in articulation, although unprepared in some places, generally to accommodate the mounting techniques. Much of the matrix has also remained unprepared, thereby offering an internal support for the articulated facial elements.

Damage observed by Carr in the skull of FMNH PR308 includes possible tooth marks (trough-like gouges in the left nasal; a large, deep cavity in the dorsal surface of the left frontal) apparently made by scavenging tyrannosaurids. Four parallel raised scars on the outer surface of the left surangular suggest "a glancing scrape from the premaxillary teeth of another tyrannosaurid," but may also have come from "a snap from a crocodile, accidental impalement on a splintered bone, or even a scrape from a fall." Also possibly indicating an injury, Carr noted, is a large, burr-like rugosity that interrupts the dorsal row of foramina on the left nasal.

In a detailed study of craniofacial ontogeny in tyrannosaurids, Carr (1999*b*) stated that, morphologically, the skull of FMNH PR308 is indistinguishable from the morphotype represented by the type species *Dapletosaurus torosus*.

In the second part of his *Dinosaur World* article, Carr (2000*a*) noted that the skull of FMNH PR308 — as figured in a half-tone illustration published by Dale A. Russell, in his landmark study of as North American tyrannosaurids, as exemplary of *Albertosaurus libratus*— was reconstructed, incorporating other material belonging to *A. libratus* (*e.g.*, AMNH 5336, including the prepared palatal bones). Carr (2000*a*) pointed out that this illustration, therefore, became "an idealized portrait" of *A. libratus*, although there was no published acknowledgement that Russell's illustration was, in fact, a composite. Ironically, Carr (2000*a*) added, this illustration appeared in the same publication in which Russell named and diagnosed the new genus *Daspletosaurus*.

Carr (1999*b*, 2000*a*) noted the following *Daspletosaurus* characters seen in FMNH PR308 — some of them, the author speculated in the later article, perhaps later to be found in other tyrannosaurids or not as distinct as they currently seem to be, or possibly explained by individual variation): Maxilla thick, alveolar deepened, lateral surface sculpture pronounced

Chief Preparator Orville L. Gilpin in 1956 adding "finishing touches" to the skeleton (FMNH PR308, nicknamed "Gorgy," and originally identified as *Gorgosaurus libratus*) now referred by Thomas D. Carr to *Daspletosaurus torosus.*

and incised by deep neurovascular sulci; nasals dished between lacrimals; caudal margin of jugal pneumatic foramen resorbed, exposing secondary fossa usually concealed from lateral view, giving recess a round appearance; rounded strut separating secondary fossa from antorbital fossa, depressing side of snout, deep pocket depressing lateral surface of postorbital process of jugal; cornual process of jugal prominent, articular surface of quadratojugal arising ahead of midlength of ventral process; maxillary process deep through level of jugal pneumatic recess; (in lateral view) tip of jugal process of postorbital ending above floor of

Skeleton (FMNH PR308, originally referred to *Gorgosaurus libratus*), of *Daspletosaurus torosus* as first mounted in 1956 in Stanley Field Hall in the [then named] Chicago Natural History Museum.

Courtesy The Field Museum (neg. #GN79225).

orbital fenestra and having low suborbital process; (in dorsal view) frontals as wide as long, notch for lacrimal rostrocaudally short and wide; frontals sloping medioventrally, contacting each other; prefrontal wide proximally; (in caudal view) portions of supraoccipital-occipital suture coalesced; caudal surface of basioccipital rugose, pit-like beneath occipital condyle; basisphenoid recess toward its ceiling; muscular scar on caudolateral process of extopterygoid broad, swollen, flaring ventrally, flanking medial pneumatic foramen.

According to Carr (2000*a*), several characters found in FMNH PR308 correspond to those used by Russell to diagnose *Daspletosaurus*, these including the following: Nasals slightly constricted between lacrimals; anteroexternal edge of antorbital ramus of lacrimal not continued across lateral surface of jugal; and ventral process of ectopterygoid inflated, having large ventrally opening sinus. However, as revealed The Field Museum specimen, some of Russell's stated characters do not diagnose this genus (*e.g.*, "there are two basisphenoid pneumatic foramina, not one," and also "the angular extends caudal to the caudal surangular foramen, not stopping beneath it" [see Carr 1999*b*]).

Casts of the skull of FMNH PR308 have been placed on exhibit at the Field Station of the Royal Tyrrell Museum of Palaeontology in Dinosaur Provincial Park and at Chicago's O'Hare Airport.

In an abstract, Carr and Williamson (1999) briefly reported on a well-preserved, disarticulated skull and skeleton (NMMNH P-25049) of a tyrannosaurid that shows affinities to *Daspletosaurus*. Recovered from the Farmington Member of the Kirtland Formation (late Campanian), in the San Juan Basin of New Mexico, the specimen is distinguished as the most complete Late Cretaceous theropod from the southern Rocky Mountain region, including an incomplete skull with jaws, most caudal vertebrae, a shoulder girdle, a partial ilium, partial forearm, and the right hindlimb. The specimen represents an immature individual, as indicated by size-independent morphological characters, open cranial and postcranial sutures, a partially ossified scapula, proportions of the hindlimb, and the animal's overall size (estimated to be about 1.7 meters or approximately 5.8 feet tall at the hips, and weighing some 278 kilograms or slightly more than 600 pounds).

As briefly described by Carr and Williamson, NMMNH P-25049 is unusual among tyrannosaurids in that the nuchal crest of the skull is rostrally everted, almost roofing the dorsotemporal fenestra. An exceptionally wide adductor chamber is "indicated by a broad and flat dorsal skull roof that forms a shallow dorsotemporal fossa, a foreshortened midregion of the braincase, and a steeply declined dorsal margin of the orbital process of the quadrate. Revealed in the specimen are previously undescribed contact surfaces in the tyrannosaurid skull (*e.g.*, plug-in-socket contact between parietal and frontals); also a large pneumatic foramen in the medial surface of the prootic, this indicating "that a pneumatic diverticulum gained entry into the rostral portion of the bone, inflating it into a tuberous and swollen condition" (see Carr 1999*b*; also see *Tyrannosaurus* entry).

As stated by Carr and Williamson (2000) in their review of the tyrannosaurids of New Mexico, these authors (1999, in preparation) will refer NMMNH P-25049 to a new species of *Daspletosaurus*. The presence of this new species, Carr and Williamson noted, "represents another dinosaur together with lambeosaurine hadrosaurids [*Parasaurolophus tubicen*] (Wiman 1931; Sullivan and Williamson 1996, 1999), hadrosaurine hadrosaurids [*Kritosaurus navajovius*] (Brown 1910), chasmosaurine ceratopsids [*Pentaceratops sternbergii*] (Osborn 1923), pachycephalosaurids [as of yet unnamed] (Williamson and Sealey 1999), and ankylosaurs [*Nodocephalosaurus kirtlandensis*] (Sullivan 1999) from the late Campanian Kirtland Formation of New Mexico," this distinct fauna implying "that there was a northern and southern faunal province among late Campanian dinosaurs in western North America because it overlaps in time with the Dinosaur Park Formation of Alberta and the upper Two Medicine Formation of Montana (Lehman 1978, 1997)."

Carr and Williamson considered a partial skull and skeleton (OMNH 10131) which Lehman and Carpenter (1990) had referred to *Aublysodon* cf. *A. mirandus* (see *D:TE*, *Aublysodon* entry for details and illustrations). Lehman and Carpenter had concluded that the specimen, collected in June, 1940 by W. J. Stovall and Donald E. Savage of the University of Oklahoma, was recovered from probably low in the Kirtland Formation, most likely the Ahshi-sle-pah Wash. Carr and Williamson noted that the specimen includes an incomplete left postorbital, the frontoparietal region, a partial left dentary, partial femora, the distal portion of a right tibia, an incomplete metatarsal II, an incomplete metatarsal III, plus (apparently missing) an ilium and a pubis.

Carr and Williamson disagreed with Lehman and Carpenter's referral of this specimen to *Aublysodon*, pointing out that the associated premaxillary tooth is denticulate (*contra* Lehman and Carpenter), and that the supposedly unique form of the distal end of the tibia is actually an artifact of erosion and also preparation. Furthermore, the cornual process of the postorbital of OMNH 10131 consists, as it does in other specimens of *Daspletosaurus*, "of a reduced dorsal ridge and a hypertrophied caudoventral boss with a convex caudal margin." As this feature is diagnostic among the species of Tyrannosauridae, Carr and Williamson referred OMNH 10131 to *Daspletosaurus* sp.

Erratum: In *D:TE* it was incorrectly stated that the original mounting of FMNH PR308 included a cast (numbered PR85769) of the specimen's skull, also that the real skull was given its own number (the latter); and in *S1*, it was stated that the original skull is now attached to the skeleton as currently mounted at The Field Museum. Actually, the skull and postcranial skeleton bear one and the same FMNH catalogue number, PR308; the cast replica was never given its own catalogue number. The real skull was included in the mount only briefly when the skeleton was first displayed in 1956; in subsequent years, the skeleton has borne the cast skull. Today, the actual reconstructed skull of PR308 is on exhibit in a display case near the mounted skeleton where it can be viewed up close and is accessible for study.

Key references: Bakker, Currie and Williams (1988); Carr (1999*a*, 1999*b*, 2000*a*, 2000*b*); Carr and Williamson (1999, 2000); Lehman (1987, 1997); Lehman and Carpenter (1990); Matthew and Brown (1923); Paul (1988*b*); Russell (1970*b*); Varricchio (1999).

†DEINONYCHUS

Saurischia: Theropoda: Neotheropoda: Tetanurae: Avetheropoda: Coelurosauria: Manuraptoriformes: Manuraptora: Deinonychosauria: Eumanuraptora: Dromaeosauridae: Velociraptorinae.

Age: Early to Middle Cretaceous (Aptian–Albian).

Comments: Since this fascinating dinosaur was first named and described in 1969 by John H. Ostrom of Yale University's Peabody Museum of Natural History, *Deinonychus antirrophus*— a medium-sized "sickle-clawed" predacious theropod now well known to the public in addition to scientists — has only been represented by reasonably complete and unequivocal specimens found in the Cloverly Formation (Lower–Middle Cretaceous: Aptian–Albian) of Montana and Wyoming (see *D:TE*, *S1*). Recently, however, Brinkman, Cifelli and Czaplewski (1998) reported the first occurrence of this type species from the Antlers Formation of Atoka County, Oklahoma.

In the opinion of Brinkman *et al.*, the Antlers Formation, based upon the presence of certain invertebrate fossils, is probably of Aptian–Albian age, almost contemporary with the Cloverly Formation (although lack of sufficient vertebrate faunas precludes a precise comparison between these two units), and laterally equivalent to the Trinity Group in Texas. Based on the somewhat equivocal information provided by its vertebrate fauna — including *Deinonychus*, the large ornithopod *Tenontosaurus* (and the presumed prey of this theropod), and triconodontid mammals — the Antlers Formation (and also the Trinity Group) seems to be geologically older, at least in part, than the Cloverly Formation (this assessment being in agreement with Jacobs, Winkler and Murry 1991; *contra* Ostrom 1970; Clemens, Lillegraven, Lindsay and Simpson 1979, who believed the reverse).

The Antlers Formation specimen (OMNH 50268) consists of an incomplete, partially articulated skeleton that was found on the grounds of the Howard McCleod Correctional Center, southwest of the town of Antlers. The skeleton was preserved in the proximity of several presumably shed *Deinonychus* teeth associated with at least four partial skeletons (OMNH 16563, 50269, 51091, and 53781, all of which were at least partially articulated) of *Tenontosaurus*. According to Brinkman *et al.*, various other reports of isolated dromaeosaurid bones and teeth attributed to *Deinonychus* from presumed Lower Cretaceous deposits of Arizona (see Thayer and Ratkevich 1996), the Trinity Group of Texas (Thurmond 1974; Winkler, Murry, Jacobs, Downs, Branch and Trudel 1988; Gallup 1989; Jacobs 1995), the Cedar Mountain Formation of Utah (Pomes 1988; Kirkland and Parrish 1995; Kirkland 1996), and the Potomac Group of Maryland (T. R. Lipka, personal communication to Brinkman *et al.* 1997) cannot be evaluated, as none of the claimed material has yet been described or figured in detail.

To date of this writing only partially prepared, OMNH 50268 represents a somewhat smaller individual than those typical of the Cloverly Formation belonging to *D. antirrhopus*. The specimen represents every major region of the skeleton, some features not preserved in any of the described Cloverly specimens. It includes 26 teeth, the best example yet found of the occiput (to be described in detail by L. M. Witmer and W. D. Maxwell in a forthcoming monograph on the *Deinonychus* cranium) of this species, a partial right prootic, parts of two middorsal ribs, an almost complete proximal chevron, most of what seems to be the fifth sacral vertebra, an exceptionally well-preserved right coracoid, poorly-preserved portions of the right humerus and femur, parts of left and right metacarpals I and II, part of the left and most of the right manual phalanges I-1, most of left manual phalanx I-2 (ungual), portions of left manual phalanges II-1 and II-2, a complete left manual phalanx III-2 and right manual phalanx III-3, a portion of left metatarsal I, most of right pedal phalanx II-2 (the "sickle claw"), portions of left pedal phalanges III-3 and IV-3, and a complete pedal phalanx IV-4. The authors noted that, when fully prepared, the Oklahoma specimen promises to provide skeletal elements thus

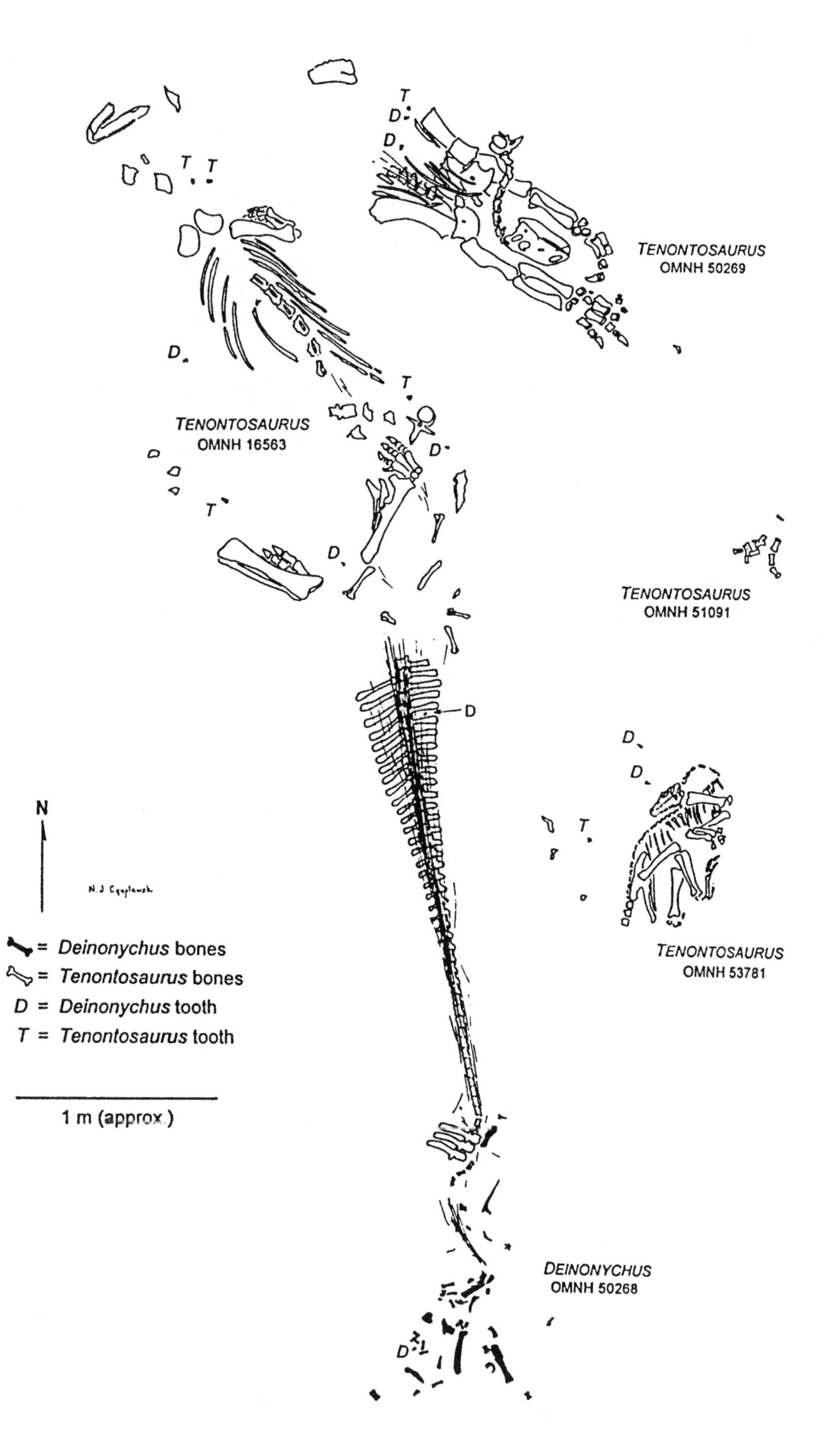

Quarry sketch map showing association of ***Deinonychus antirrhopus*** (OMNH 50268) skeleton with those of *Tenontosaurus* (OMNH 16563, 50269, 51091, and 53781) at the Antlers Formation locality, Atoka County, Oklahoma. (After Brinkman, Cifelli and Czaplewski 1998.)

Photograph by the author, courtesy Museum of Comparative Zoology, Harvard University.

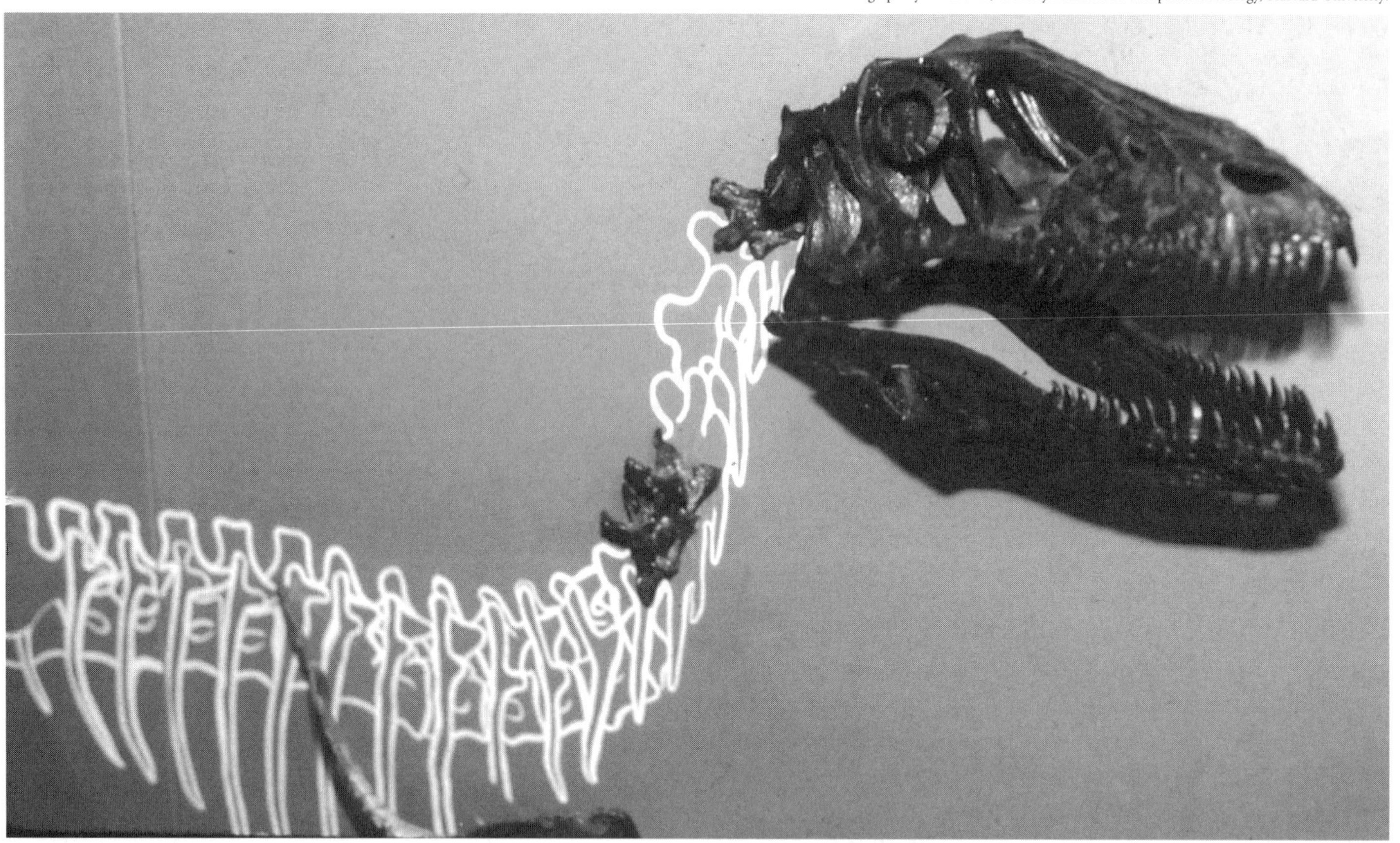

***Deinonychus antirrhopus*, cast of MCZ 4371, a partial skeleton discovered by Steven Orzack, collected in July 1974, in the Cloverly Formation (Lower Cretaceous) of southern Montana by an expedition for Harvard University led by Farish A. Jenkins.**

far not known in this taxon and significant new information.

Brinkman *et al.* compared OMNH 50268 with the holotype and hypodigm specimens of *D. antirrhophus* housed at the Peabody Museum, supplementing those comparisons with published accounts (Ostrom 1969*a*, 1976*b*) of *Deinonychus* specimens at Harvard University's Museum of Comparative Zoology and the American Museum of Natural History. The authors found that the Antlers Formation specimen compares favorably with skeletal elements diagnostic of the larger and presumably more mature Clovery Formation taxon, differing only in its smaller size and in various subtle ways. These differences in OMNH 50268 — seen in such elements as the occiput (sutures not entirely fused), coracoid (differing in its smaller size, the shape of its supracoracoid foramen, the development of its biceps tubecle, sternal process, and glenoid, the relative thickness of its continuously rounded cranial and ventral borders, and the development of its medial concavity), metapodials and phalanges (shorter, less dorsoventrally deep, and less mediolaterally broad), and manual elements (relatively shorter than most pedal elements) — are generally associated with immaturity. Therefore, Brinkman *et al.* did not erect a new species for the reception of the specimen from Oklahoma, but regarded its smaller size and slight differences as features having ontogenetic rather than taxonomic significance, OMNH 50268 interpreted by these authors as representing a subadult *D. antirrhopus* individual.

As Brinkman *et al.* noted, the Oklahoma discovery constitutes only the second known occurrence to preserve substantial skeletal remains of both *Deinonychus* and *Tenontosaurus* (the other being in the Cloverly Formation of Montana). This association of the two taxa, along with the presence of what seem to be shed *Deinonychus* teeth, further documents the predator-prey relationship between *Deinonychus* and *Tenontosaurus*. Also, the occurrence of both genera in both the Antlers Formation and Cloverly Formation offers the most compelling biostratigraphic link yet between these units, indicating "broad geographic distribution for at least some taxa of terrestrial vertebrates in the Early Cretaceous of North America."

Norton (1999), in a preliminary report, briefly explained the use of computer technology in creating a model for rib motion in theropod dinosaurs, specifically *Deinonychus*. According to Norton, as soft-tissue physiology in extinct animals may best be understood via functional analyses of skeletons, theropod lung function might best be understood through investigating ventilatory movements of the rib cage. *Deinonychus* was selected for this project "because its body mass and rib cage dimensions are comparable both to those of modern predators and to representative of the dinosaur extant phylogenetic bracket, allowing comparisons of rib dynamics among creatures of similar size, lifestyle, or thoracic structure." Through such comparisons, published allometric relationships among mass, metabolism, and respiratory parameters may be confidently used combined with the results of rib motion analysis.

The computer-generated *Deinonychus* model used for reference a combination of published drawings, photographs of mounted skeletons, and first-hand examinations of fossil specimens. The finished model was capable of describing the movement of a rib through any arc in three-dimensional space. Furthermore, it permitted the hinge axis produced by the capitulum-tuberculum articulation of the rib with the ventral body and transverse process to be tilted at any angle away from the vertical, this accommodating morphological variability among the dorsal vertebrae. As Norton predicted, by eventually combining the motions of an entire set of theropod ribs, the computer-generated "model will provide both patterns of dimensional change in different regions of the rib cage and overall volume changes during ventilation, allowing predictions regarding the ability of the theropod thorax to ventilate an avian style parabronchial lung with its associated air sacs."

Key references: Brinkman, Cifelli and Czaplewski (1998); Clemens, Lillegraven, Lindsay and Simpson (1979); Gallup (1989); Jacobs (1995); Jacobs, Winkler and Murry (1991); Kirkland (1996); Kirkland and Parrish (1995); Norton (1999); Ostrom (1969*a*, 1970, 1976*b*); Pomes (1988); Thayer and Ratkevich (1996); Thurmond (1974); Winkler, Murry, Jacobs, Downs, Branch and Trudel (1988).

†DILOPHOSAURUS

Saurischia: Theropoda: Neotheropoda: "Ceratosauria": Coelophysoidea: Dilophosauridae.

Comments: Madsen and Welles (2000), in their monograph on the "ceratosaur" *Ceratosaurus*, pointed out errors made by Rowe and Gauthier (1990) regarding the skull of the type species *Dilophosaurus wetherilli*.

Rowe and Gauthier had stated that the junction between the premaxilla and maxilla in this species is firm. According to Madsen and Welles, however, articulated specimens (UCMP 37303 and 77270) reveal "that the posteromedial process of the premaxilla is transversely concave and rides smoothly over the anteromedial process of the maxilla, forming a rather weak connection," not interdigitating with the latter process, but rather forming an arched cover over it. The anteromedial process "interlocks with its opposite by slightly arched anteroposterior tongues and

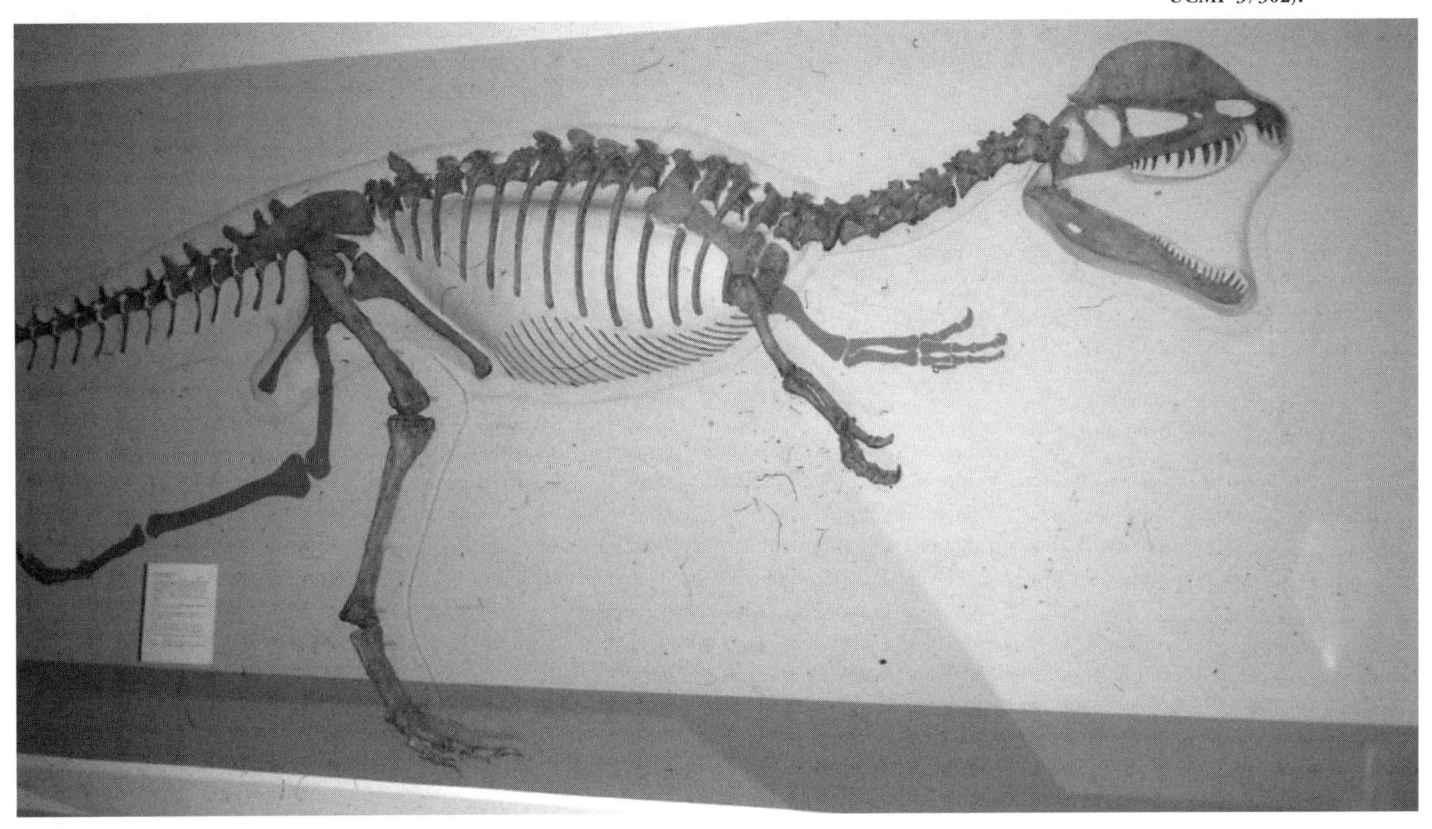

Dilophosaurus wetherilli **skeleton (LACM 4462/118118, cast of holotype UCMP 37302).**

Photograph by the author, courtesy Natural History Museum of Los Angeles County Museum.

Photograph by the author.

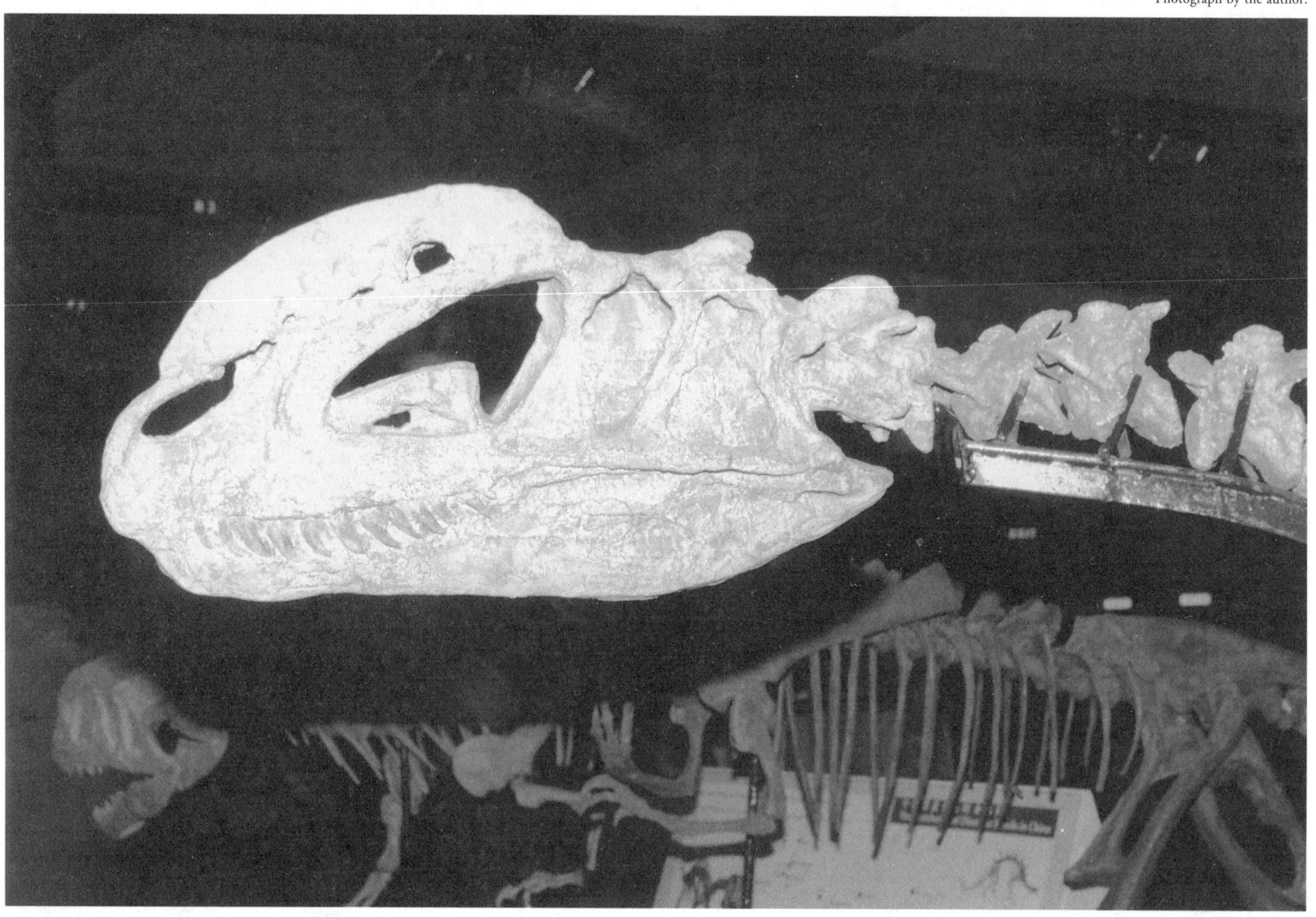

***"Dilophosaurus" sinensis*, holotype skeleton (KMV 8701) mounted at the third Dinofest event (the "World's Fair of Dinosaurs") held in Philadelphia, Pennsylvania in 1998, close-up look at the skull in left side view. The generic status of this theropod has recently come into doubt.**

grooves." Furthermore, the nasal process of the premaxilla rides smoothly over the nasal, another indication of a weak attachment. Therefore, Madsen and Welles concluded, "the premaxilla is loosely attached to both the maxilla and nasal."

Madsen and Welles also addressed the suggestion of Rowe and Gauthier that the subnarial gap and pit accommodated a large dentary tooth, pointing out that no such tooth is in evidence in either UCMP 37303 or 77270.

The Chinese theropod *"Dilophosaurus" sinensis* was referred to the genus *Dilophosaurus* (see entry, *D:TE* and *S1*), primarily based on its possession of distinctive paired nasocranial crests that resembled in form those in the North American species *D. wetherilli*. Recently, Lamanna, Smith, You, Holtz and Dodson (1998) reassessed the Chinese taxon, taking note of numerous apparent peculiarities in its cranial and postcranial anatomy that have cast doubt upon this species' taxonomic and systematic placement.

As pointed out by Lamanna *et al.*, ornamentation such as these crests, once believed to be uniquely found in *Dilophosaurus*, is now also known in the North American species *Syntarsus kayentakatae*. Future research may, in fact, reveal that the feature of paired crests is sporadically distributed within the Coelophysoidea, thereby being of limited use in assessing interrelationships within this group.

Unstated differences regarding "premaxillary shape and tooth count, size and position of the external mandibular fenestra, shape of the infratemporal fesestra and squamosal, and a derived antorbital tooth row" in *D. wetherilli* suggested to the authors that the Chinese and American species do not belong to the same genus.

As noted by R. E. Molnar (personal communication 2000), there are — based upon counts of different specimens (mostly skeletons) in published photographs — two *Dilophosaurus*-like taxa in southwestern China: *D. sinensis* is not *Dilophosaurus*, but the second is more like *Dilophosaurus* and may belong to that genus.)

Key references: Lamanna, Smith, You, Holtz and Dodson (1998); Madsen and Welles (2000); Rowe and Gauthier (1990).

Head of the double-crested theropod *Dilophosaurus wetherilli* shown in front and side views, restored by Gregory S. Paul.

DINHEIROSAURUS Bonaparte and Mateus 1999—(=?*Lourinhasaurus*)

Saurischia: Sauropodomorpha: Sauropoda: Eusauropoda: Neosauropoda: Diplodocoidea.

Name derivation: "[Praia de Porto] Dinheiro [locality where holotype was collected]" + Greek *sauros* = "lizard."

Type species: *D. lourinhanensis* Bonaparte and Mateus 1999.

Other species: [None.]

Occurrence: Camadas de Alcobaca Formation, Lourinhã, Portugal.

Age: Late Jurassic (late Kimmeridgian).

Known material/holotype: ML414, two incomplete cervical and nine rather complete, articulated dorsal vertebrae, seven fragmented centra, some incomplete neural arches, 12 dorsal ribs, very fragmentary appendicular bones, gastroliths.

Diagnosis of genus (as for type species): Posterior cervical and dorsal vertebrae of *Diplodocus* type, with bifurcated neural spines, but only one well-defined infraparapophyseal lamina present from dorsal 4 backwards; cited lamina in mid- and posterior dorsals obliquely directed from parapophysis to ventroposterior corner of neural arch; complex structure derived from hyposphene making accessory articulation, exposed in lateral view; lower section of neural arch of mid- and posterior dorsal vertebrae dorsoventrally shorter than in *Diplodocus* (Bonaparte and Mateus 1999).

Comments: The genus *Dinheirosaurus* was based upon an articulated series of cervical and dorsal vertebrae, ribs, and fragmentary appendicular bones (ML414), plus almost a hundred gastroliths, found in 1987 (discovery first announced by Dantas, Sanz and Gallopim de Carvalho 1992) by Carlos Anunciacào, an assistant at the Museum da Lourinhã, in the upper section (dated as of Late Kimmeridgian age by Helmdach 1973–74, based on the occurrence of brackish-water ostracods; see also Manupella 1996) of the Camadas de Alcobaca Formation (="Formacão da Lourinhã), at Porto Dinheiro, near the top of the coastal cliff, in the Lusitanian Basin of Lourinhã, in central-eastern Portugal. The type specimen—originally referred by Dantas, Sanz, Silva, Ortega, Santos and Cachao (1998) to *Lourinhasaurus alenquerensis*—was collected from 1987 to 1992 by a team including (from the Museu da Lourinhã) Anunciacão, H. Mateus, I. Mateus, O. Mateus, and V. Ribeiro; (Museu de Historia Natural da Universidade de Lisboa) P. Dantas, C. Marquez da Silva, M. Cachão, and V.

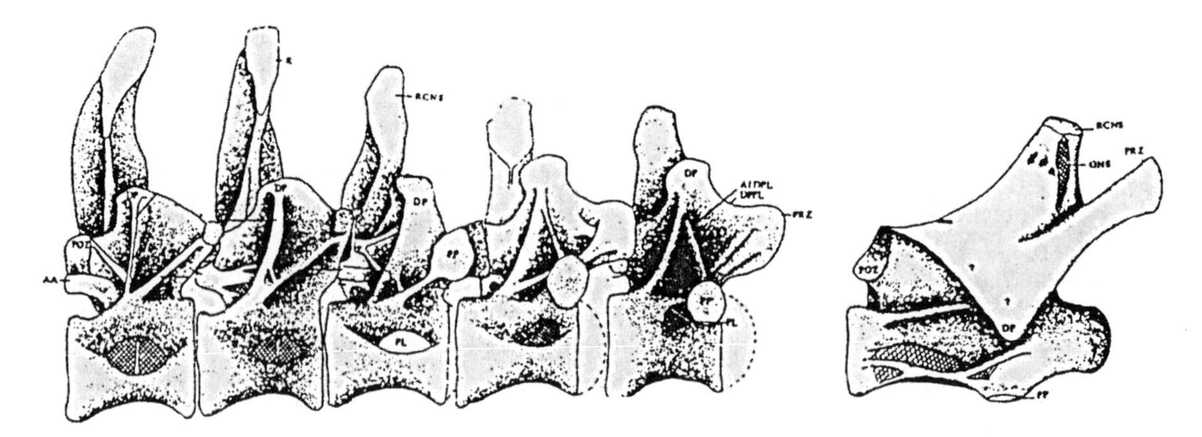

***Dinheirosaurus lourinhanensis*, ML414, holotype 14th cervical vertebra (right), and sequence of dorsal vertebrae numbers 3 to 7 in right lateral view. (After Bonaparte and Mateus 1999).**

Santos; and (Universidade de Salamanca) F. Ortega and S. Tudanca (Bonaparte and Mateus 1999).

Bonaparte and Mateus questioned the assignment by Dantas *et al.* to *L. alenquerensis*, pointing out that "there is no anatomical evidence for such a procedure because the taxonomical identification of both specimens is impossible," neither of the two specimens preserving corresponding elements fundamental for identification. However, pending the discovery of more complete materials, Bonaparte and Mateus acknowledged that this new taxon might someday be reinterpreted as to its relationships with other sauropods or as a possible synonym or some other sauropod—including *Lourinhasaurus*—from the Late Jurassic of Portugal.

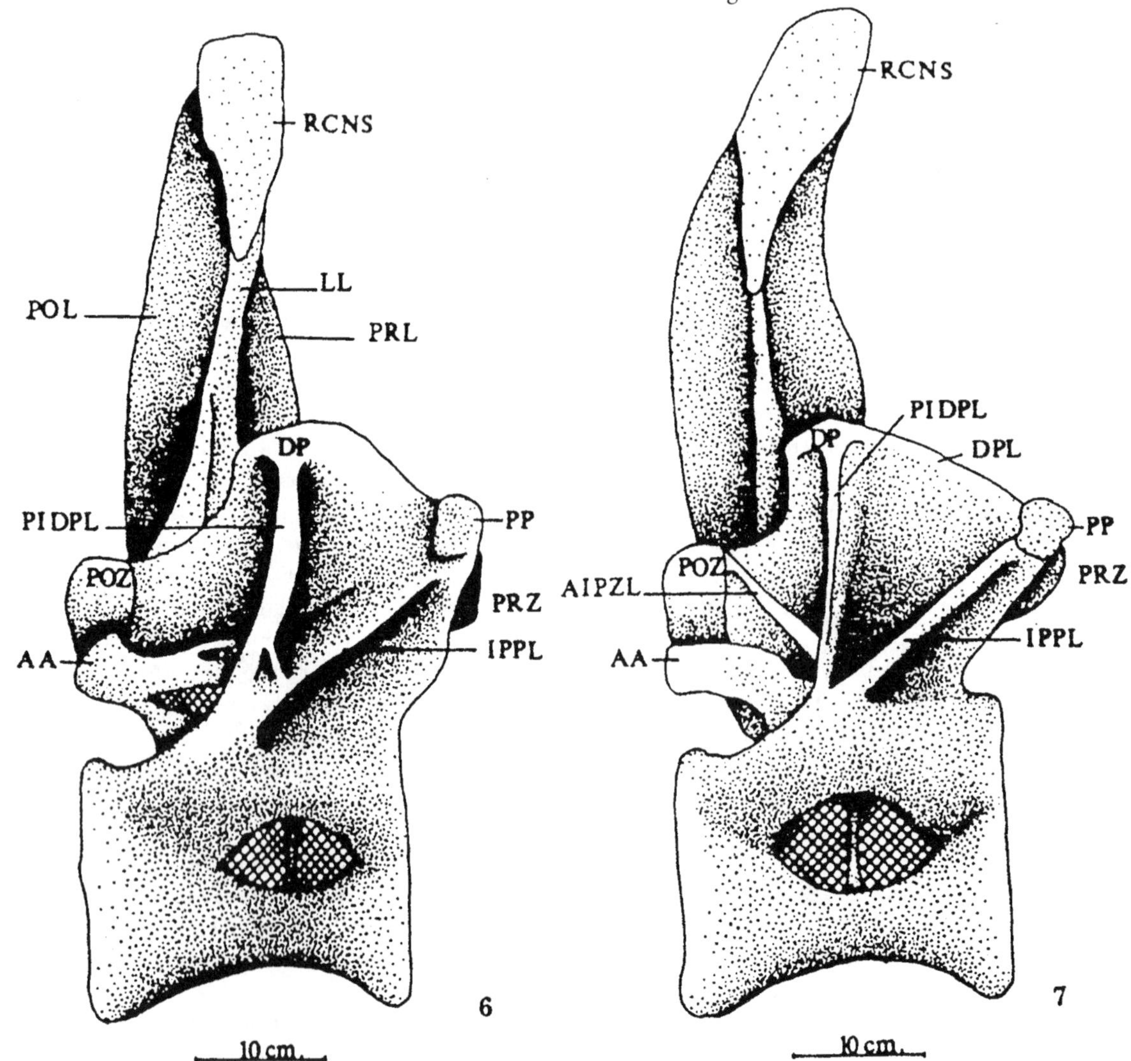

***Dinheirosaurus lourinhanensis*, ML414, sixth dorsal vertebra (reconstructed), right lateral view. (After Bonaparte and Mateus 1999.)**

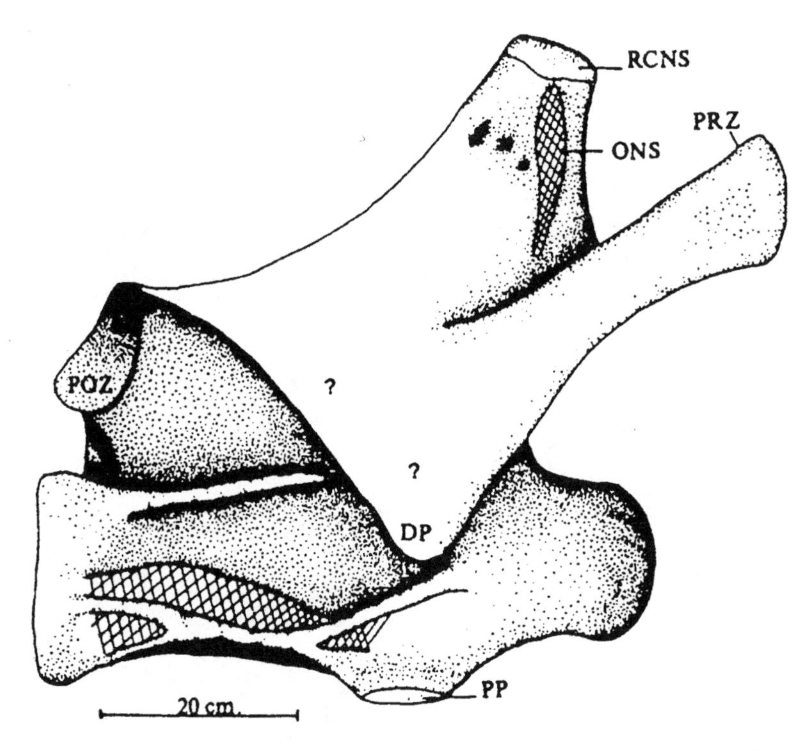

***Dinheirosaurus lourinhanensis*, ML414, holotype 14th cervical vertebra, right lateral view. (After Bonaparte and Mateus 1999.)**

In comparing *Dinheirosaurus* with other sauropods, Bonaparte and Mateus noted that the different morphology of the presacral vertebrae in this genus preclude its referral to either *Camarasaurus* or *Apatosaurus*. Most closely these vertebrae in *Dinheirosaurus* resemble those of *Diplodocus*. However, the following derived characters seen in *Dinheirosaurus* imply generic differences between these two taxa: 1. Presence of a well-defined paired structure originating from the posteroventral area of the neural arch, making an accessory articulation between two vertebrae, derived from the hyposphene and contacting with the hypantrum; and 2. lower section of neural arch dorsoventrally lower than in *Diplodocus*. Therefore, the organization of the dorsal neural arch in *Dinheirosaurus* seems to be more derived than in *Diplodocus*, while the dorsoventral development of the lower portion of the arch is more derived in *Diplodocus*.

Bonaparte and Mateus speculated that the geography of Portugal, isolated as it was from North America by water in the Late Jurassic during which time the faunal exchanges with other continents was interrupted, gave rise to the distinct characters exhibited in *Dinheirosaurus*.

Key references: Bonaparte and Mateus (1999); Dantas, Sanz and Gallopim de Carvalho (1992); Dantas, Sanz, Silva, Ortega, Santos and Cachao (1998).

DINOTYRANNUS Olshevsky, Ford and Yamamoto 1995—(See *Tyrannosaurus*.)

Name derivation: Latinized *dino* (from Greek *deinos*) = "terrible" + Latinized *tyrannus* (from Greek *tyrannos*) = "tyrant."

Type species: *D. megagracilis* (Paul 1988).

†DIPLODOCUS—(=?*Dystrophaeus*, ?*Seismosaurus*)

?New species: *D. hallorum* (Gillette 1991) [see "Note" below].

Comments: In the past, the absence of an ossified calcaneum has been suggested as a synapomorphy of the Diplodocidae, the sauropod family founded upon the gigantic yet comparatively slender sauropod genus *Diplodocus*. McIntosh, Brett-Surman and Farlow (1997) considered the absence of the calcaneum to be a character of diplodocids. Other workers have regarded the absence of the calcaneum as a possible synapomorphy within the Diplodocidae (Wilson and Sereno 1998) or as a synapomorphy of that family (Upchurch 1995, 1998).

Indeed, although ossified calcanea—in the form of a fist-sized, globe-shaped element, with all surfaces very rugose and pitted (McIntosh 1990; Bonnan 2000, personal observation)—have been reported in various sauropod genera (*Vulcanodon*, *Barapasaurus*, *Shunosaurus*, *Brachiosaurus*, *Neuquensaurus* [as "*Titanosaurus*" Huene], *Euhelopus*, *Mamenchisaurus*, *Antarctosaurus*, though questionably, and *Camarasaurus*; see Huene 1929; Raath 1972; Cooper 1984; McIntosh; McIntosh *et al.*; Wilson and Sereno), this feature had never been observed in any diplodocids (McIntosh; Wilson and Sereno; Upchurch 1995, 1998). As noted by Bonnan, however, since all other known saurischians possessed ossified calcanea, the lack of this feature in diplodocids would be unprecedented.

Testing this idea, Bonnan examined the fibula, astragalus, and disarticulated pes belonging to a specimen (CM 30767)—collected at Dinosaur National Monument between 1909 and 1922 by Earl Douglass, this specimen originally part of field specimen 175, which seemed to have comprised the remains of more than one individual (J. S. McIntosh, personal communication to Bonnan), the pes prepared at the

Courtesy Houston Museum of Natural Science.

Skeleton of the diplodocid sauropod *Diplodocus hayi*, including holotype CM 662, referred material CM 94 (collected from Dinosaur National Monument), and skull cast of *D. carnegii* (CM 1161), mounted under the direction of Wann Langston, Jr.

Carnegie Museum of Natural History in 1936 — referred to *Diplodocus* sp. and housed in the Carnegie Museum.

As Bonnan pointed out, the pes of CM 30767 is diplodocid, this determination "based in part on the distinctive, posteriorly facing tab on the lateral edge of metatarsal I, a character recognized as unique to Diplodocidae" (see McIntosh; Upchurch 1998). Also, the astragalus is wide anteroposteriorly and has a pyramid-shaped medial apex, this condition known in diplodocids but absent in camarasaurids (Bonnan 1999*a*). Furthermore, this specimen was assigned to *Diplodocus* by John S. McIntosh (personal communication to Bonnan; also, Bonnan 2000, personal observation) based on the more slender nature of the metatarsals and overall similarity of the pes to that of *Diplodocus carnegie* and other species of *Diplodocus*.

Bonnan (2000) found, in the same drawer that contained the pes and astragalus, a small, fist-sized bone marked CM 30767 but lacking identification. Examining this element and comparing it with the foot bones of other sauropods, Bonnan (2000) determined that it was indeed an ossified calcaneum. This discovery suggested to that author that at least one member of the diplodocidae possessed an ossified calcaneum. In describing this element, Bonnan (2000) observed that it is morphologically similar to the calcanea of *Camarasaurus*. Bonnan (2000) noted that the element could be an aberrant phalanx or carpal, or, not having been found in articulation with the astragalus or fibula, it might not belong to CM 30767, but to a different sauropod, perhaps a camarasaurid.

However, as Bonnan (2000) pointed out, the phalanges of sauropods (and other saurischians) have distinct cranial, lateral, medial, and caudal surfaces, plus well-defined proximal and distal articular surfaces; nor do any known sauropod phalanges have highly rugose surfaces as those seen in the CM 30767 element. Bonnan (2000) further noted that the known carpals of camarasaurids comprise two or possibly three disc-shaped elements that articulated tightly with the proximal articular surfaces of the metacarpals (Osborn 1904; Bonnan 2000, personal observation). Also, the proximal and distal articular surfaces of these

Photograph by the author, courtesy Carnegie Museum of Natural History.

Partial skeleton (CM 3452) of *Diplodocus carnegii* including skull and several cervical vertebrae, displayed as found.

elements are flat and smooth, while in the CM 30767 element they are rugose. Additionally, the CM 30767 bone is pentagonal in proximal outline, not disc-shaped as in camarasaurids, while its ventral surface is convex, not flattened and comparatively smooth as in the carpals of camarasaurids. Noting the possibility that this element could represent some unknown or undescribed ossified tarsal or carpal, but considering all of the above observations and also its morphological similarity to calcanea of *Camarasaurus*, Bonnan (2000) found it "unlikely that Element 30767 was a carpal or aberrant phalanx."

Further addressing the assignment of the CM 30767 element to the Diplodocidae, while noting that the sauropod assemblage from which this element was excavated at Dinosaur National Monument also contains several camarasaurids (Dodson 1990*a*), Bonnan (2000) cited the following:

1. The pes of CM 30767 is complete except for the missing phalanges II-2 and V-1, even the smallest phalanges of III-2 and III-3 having been preserved. Although it is not known if the pes was articulated when found, its complete nature suggests it was not significantly disarticulated. Therefore, although the pes was separated from the hindlimb, the individual elements of the foot remained almost completely intact.

2. The material given to which Earl Douglass gave the field number of 175 contained only diplodocid remains. It is possible that camarasaurid material contaminated the assemblage, although it is unlikely that only one such element would be included with the otherwise strictly diplodocid remains.

3. Even if the calcaneum is not a part of CM 30767, the closest source for the element would have been another diplodocid, being morphologically similar to other known sauropod calcanea, different from those of *Camarasaurus*, and having been found in close proximity to a diplodocid pes.

Therefore, Bonnan (2000) deduced that the CM 30767 element is a calcaneum belonging to a diplodocid, probably of *Diplodocus* itself.

According to Bonnan (2000), an ossified calcaneum in a diplodocid dinosaur has significant functional implications, providing various advantages to the foot. An ossified calcaneum may have contributed stability to the pes during locomotion, and also to the correct alignment and insertion of the plantar musculature.

Bonnan (2000) concluded that, while the systematic affinities of CM 30767 may be contested, the distinctive morphology and relatively rare preservation of the sauropod calcaneum "suggest that the presence of an ossified calcaneum in sauropods should not be used as a character state, nor as a synapomorphy of the Diplodocidae."

Note: Various authors have suggested that the type species *Seismosaurus halli* (see *Seismosaurus* entry, this volume and *D:TE*) is invalid and represents a new species of *Diplodocus*, named *D. hallorum* (Gillette 1991) (see Curtice 1996; Lucas 2000; Lucas and Heckert 2000). This assessment may be correct, although a comparative study demonstrating this proposed synonymy of *Seismosaurus* and *Diplodocus* and the referral of this species has not yet been published.

Erratum: Elizabeth A. Hill, in number 175 (February 1999) of the *Society of Vertebrate Paleontology New Bulletin*, announced that the famous and often cited 1898 newspaper article depicting a reared-up *Diplodocus* peering into an upper window of the New York Life Building was not, contrary to common belief, published in the *New York Journal* (see *D:TE*). A recent discovery in the Vertebrate Paleontology archives of the Carnegie Museum of Natural History has revealed that the article actually appeared in the *New York World*.

Key references: Bonnan (1999*a*, 2000); Cooper (1984); Curtice (1996); Dodson (1990*a*); Gillette (1991); Huene (1929); Lucas (2000); Lucas and Heckert (2000); McIntosh (1990); McIntosh, Brett-Surman and Farlow (1997); Osborn (1904); Raath (1972); Upchurch (1995, 1998); Wilson and Sereno (1998).

†DRYPTOSAUROIDES

Erratum: In *D:TE*, the entry for this genus incorrectly noted that the Lameta Formation, from which the holotype of *Dryptosauroides grandis* was collected, is of Coniacian–Santonian age. Actually it is of middle to late Maastrichtian age.

†ECHINODON

Ornithischia: Cerapoda: Ornithopoda: ?Heterodontosauridae.

Comments: Among the more lesser known and poorly understood dinosaurian genera has, for more than a century, been *Echinodon*.

Founded upon two isolated fragmentary dentaries and two maxillae (BMNH 48209-23) recovered from the Purbeck Limestone Formation (Berriasian) of Dorset, England, the type species *Echinodon becklesii* (see *Echinodon* entry, *D:TE*) was first described as a lizard by the yet to be knighted Richard Owen in 1861. Other fragmentary material was subsequently referred to this taxon, representing several individuals. Better preparation of the original fossil material has revealed anatomical details identifying it as pertaining to an ornithischian dinosaur. Later workers have, over the years, regarded this taxon as a fabrosaurid, heterodontosaurid, or possibly an armored dinosaur.

In 1879, Owen referred some dermal objects, which he described as "granicones," to his already established genus *Nuthetes* (see entry), which he had classified as a lacertilian. Owen (1879) believed that these objects were akin to the armor-like scutes in some modern lizards. Galton (1986*b*) suggested that these "granicones" could indeed represent dermal armor. Consequently, Galton's interpretation supported the referral of *Echinodon* to the Thyreophora (an assessment followed by other workers, *e.g.*, Coombs, Weishampel and Witmer 1990 in their review of basal Thyreophora).

In an abstract, however, Norman and Barrett (1999), noted that, while various thyreophorans (*e.g.*, tooth and caudal vertebra of a nodosaurid, indeterminate thyreophoran teeth previously referred to ?*Stegosaurus*) have been found in the Purbeck Limestone of Dorset, and as a fragmentary thyreophoran (?stegosaur) skeleton has been reported from the Purbeck Limestone of Buckinghamshire, there seems to be no real basis for the referral of this material to the Thyreophora, as no associated *Echinodon* remains were found with these "granicones." Therefore, Norman and Barrett found it most reasonable to regard *E. becklesii* as Ornithischia *incertae sedis*.

Barrett (1999*a*), in a more recently published abstract, again addressed the phylogenetic status of *E. becklesii*, stating that alleged fabrosaurid and thyreophoran apomorphies found in this taxon are primitive for Ornithischia and, therefore, of little use in determining its systematic position. Additionally, the enigmatic so-called "granicones" cannot with certainty be referred to *E. becklesii* based upon current data and, consequently, cannot be used as evidence for referral of the genus to the Thyreophora.

According to Barrett, a detail reassessment of the collected material suggests that *Echinodon* may represent a late-surviving heterodontosaurid; however, "the apomorphies used to support this assignment (lack of replacement foramina, disposition of the denticles on maxillary and dentary tooth crowns) differ from those suggested by previous authors."

Note: Norman and Barrett addressed the poor representation of dinosaur fossils in the Purbeck Limestone Formation. As noted by these authors, the Purbeck environment seems to be transitional between the earlier, mostly marine facies of the Portland

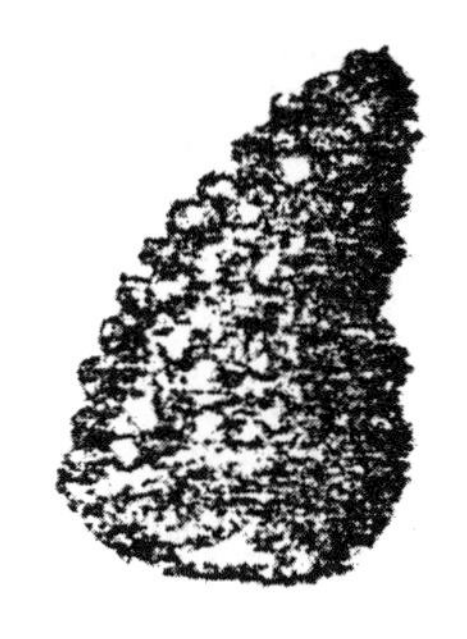

Armor scutes of *Echinodon becklesii*, originally referred by Richard Owen to the theropod *Nuthetes destructor*. (After Owen 1879.)

Formation and later freshwater-terrestrial dominated sequence represented by the Wealden. This suggested to them that the Purbeck environment of the Wessex Basin was probably quite variable: "on land generally hot, semiarid conditions prevailed, occasionally becoming quite hostile to larger terrestrial organisms during prolonged droughts." Possibly, this environment was more conducive to inhabitation by smaller and more resilient kinds of animals, particularly lepidosaurs and mammals, "with dinosaurs representing comparative rare 'passers by' and leaving a consequently impoverished fossil record locally at least."

Key references: Barrett (1999*a*); Coombs, Weishampel and Witmer (1990); Galton (1986*b*); Norman and Barrett (1999); Owen (1861, 1879).

†EDMONTONIA

Ornithischia: Genasauria: Thyreophora: Thyreophoridea: Eurypoda: Ankylosauria: Nodosauridae.
New species: *E. australis* Ford 2000 [*nomen dubium*].
Occurrence of *E. australis*: Kirtland Formation, New Mexico, United States.
Age of *E. australis*: Late Cretaceous (Maastrichtian).
Known material of *E. australis*: Four armor scutes.
Holotype of *E. australis*: NMMNH P-25063, pair of cervical medial scutes.

Diagnosis of *E. australis*: Medial cervical scutes of ?second cervical band almost square, with rounded lateral edges with low ridge running down middle or near middle of scute ending in short conical point; more similar to *E. longiceps* than to *E. rugosidens* in being nearly square with conical point [see "Comments," below] (Ford 2000).

Comments: Ford (2000), in a review of ankylosaurian armor from New Mexico and a preliminary review of ankylosaurian armor in general, denoted a well-preserved pair of scutes (NMMNH P-26063), collected from the Kirtland Formation, San Juan Basin, New Mexico, as the holotype of a new species of *Edmontonia* (see *D:TE*), which he named *E. australis*. To this species Ford also referred a well-preserved right medial cervical scute (NMMNH P-27450) and a primary cervical scute (NMMNH P-27420) presumably from the same locality.

As described by Ford, the two holotype scutes attach at the midline but are not coossified. The right (complete) scute measures about 22.3 centimeters in length and 17.3 centimeters in width, the ridge being 8.7 centimeters from the medial edge. The left (missing its posterior edge) scute measures 18.3 centimeters in length and 6.2 centimeters in width, its "ridge running nearly down the middle" and not ending in a point.

In his diagnosis of *E. australis*, Ford used the name *Chassterbergia rugosidens* instead of *Edmontonia rugosidens* (as used above), the name of the type species. The name *Chassternbergia* (see *D:TE* for more details) was coined by Robert T. Bakker (1988) to designate what he regarded as a subgenus of *Edmontonia*, Bakker's new combination becoming *Edmontonia* (*Chassternbergia*) *rugosidens*, based on "long-headed" specimens referred to *E. rugosidens*. This new combination was not adopted by the majority of workers and officially rejected by ankylosaur specialist Kenneth Carpenter (1990). Ford, however, followed Olshevsky (1991), who tends to be more of a "splitter" than a "lumper" when compiling lists of dinosaurian genera and species, in accepting *Chassternbergia* as a valid genus.

Ford found it reasonable to erect new taxa at the specific level based upon osteoderms, citing Coombs and Demere (1996), who were of the opinion that 1. "if individual osteoderms of a distinctive type are known from a single taxon, then they are diagnostic," and 2. "if a specimen shows osteoderms *in situ*, then the arrangement may be diagnostic."

Carpenter (personal communication, 2000), however, criticized Ford's assessment of the osteoderms that Ford referred to *E. australis*. Carpenter pointed out that, although ankylosaurian armor is more diagnostic than Coombs originally thought in his pre–1990 papers, it is not as diagnostic as Ford believes; if it were, virtually every specimen could be interpreted as a new taxon. According to Carpenter (who had not seen the holotype of *E. australis*), the two scutes are so asymmetrical and of different sizes

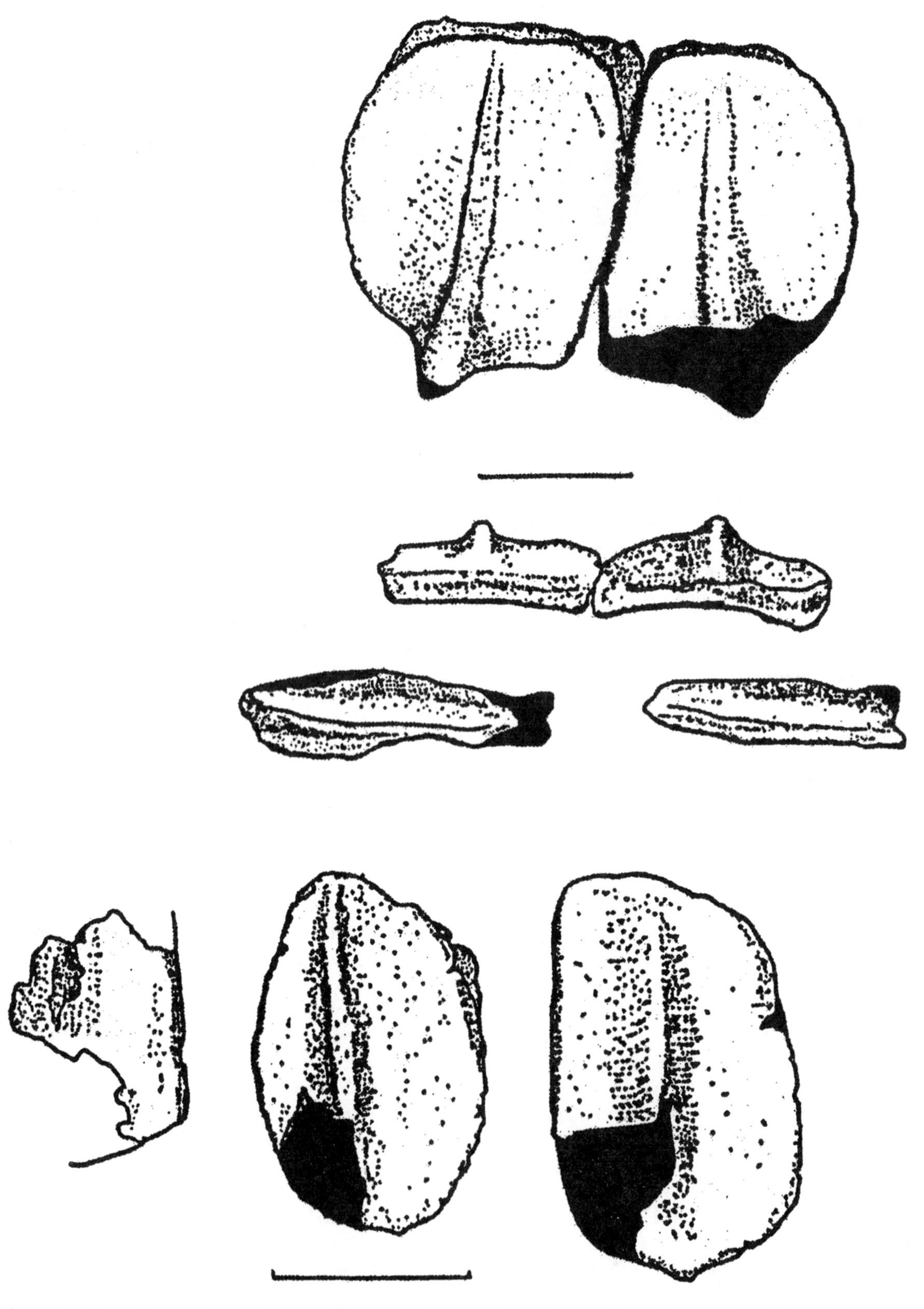

***Edmontonia australis*, NMNH 25063, holotype osteoderms in (from top to bottom) dorsal, cranial, and lateral views. Scale = 10 cm. (After Ford 2000.)**

that probably only one of them is actually a median scute, the larger possibly being a left median and the other a right lateral scute. Also, while Ford added to the tips of the two plates in his figuring of the material, these scutes are complete and, therefore, not elongated as shown. In a future publication, Carpenter will officially designate *E. australis* to be a *nomen dubium*.

Note: In an abstract for a poster, Watabe, Nakaya and Nakahara (2000) reported on a mostly well-preserved nodosaurid skeleton excavated from the Upper Cretaceous (Maastrichtian) Lance Formation in Greasewood Creek, Wyoming. The specimen, housed in the Hayashibara Museum of Natural Sciences in Okayama, Japan, includes a poorly-preserved skull, well-preserved lower jaws, a series of cervical and dorsal vertebrae, partial caudal vertebrae, forelimb and hindlimb complexes, and apparently an abundant representation of dermal scutes (including three cervical-thoracic armor rings, plus other isolated scutes of varying shape and size).

Watabe *et al.* observed that the tooth morphology (*e.g.*, large size, development of cingulum) and scute morphology (*e.g.*, bifurcated lateral spine of third armor ring) suggests referral of this specimen to *Edmontonia rugosidens*. Also, the variation pattern of the cervical to dorsal vertebrae is similar to other kinds of quadrupedal dinosaurs, such as the ceratopsian *Chasmosaurus*.

As briefly described by Watabe *et al.*, the main lateral spine on the third armor ring is ventrally rather than anterolaterally directed, this condition reflecting the function of the large lateral shoulder spine. Such ventrally oriented spines, the author suggested, would not have been an obstacle as the animal walked through bushy environments.

Furthermore, Watabe *et al.* noted that various pathological traces on the scutes of this specimen—"classified as chips, pits, and melting"—were most likely the result of predatory attack or disease.

Key references: Bakker (1988); Carpenter (1990); Coombs and Demere (1996); Ford (2000); Olshevsky (1991); Watabe, Nakaya and Nakahara (2000).

†EDMONTOSAURUS

Ornithischia: Genasauria: Cerapoda: Ornithopoda: Euornithopoda: Iguanodontia: Euiguanodontia: Dryomorpha: Ankylopollexia: Hadrosauroidea: Hadrosauridae: Euhadrosauria: Hadrosaurinae.

Note: Michael Triebold, a fossil collector and dealer in fossil specimens and casts, in an interview conducted by Tony Campagna for the fan publication *Prehistoric Times*, reported on a virtually complete skeleton with an almost complete skull of the duck-billed dinosaur *Edmontosaurus annectens*, discovered in the fall of 1994 "on private ranchland" in South Dakota, and collected the next spring. The skeleton was found semiarticulated "with a large section of the vertebral column intact and the limbs close to their life positions." The skull was found approximately in place, the predentary and one lower jaw in perfect position.

Triebold stated that the skeleton is gracile, measuring 25 feet [about 7.4 meters] in length. The skull, for which only the right quadrato-jugal required reconstruction and in which the entire tooth battery is preserved, is 34 inches [about 90 centimeters] long.

Most surprising, Triebold noticed, during preparation in the laboratory, that the outer surfaces of both jaws bore the teeth marks of small theropod dinosaurs and that these marks mirrored each other. Triebold also observed that the marks on one dentary had begun to heal.

From the evidence, Triebold envisioned a scenario out of time in which a small theropod (or group of theropods) attacked the herbivorous dinosaur by

Skeleton (cast) of *Edmontosaurus annectens* from Triebold Paleontology, Inc., Woodland Park, Colorado.

Courtesy Michael Triebold.

Courtesy Michael Triebold.

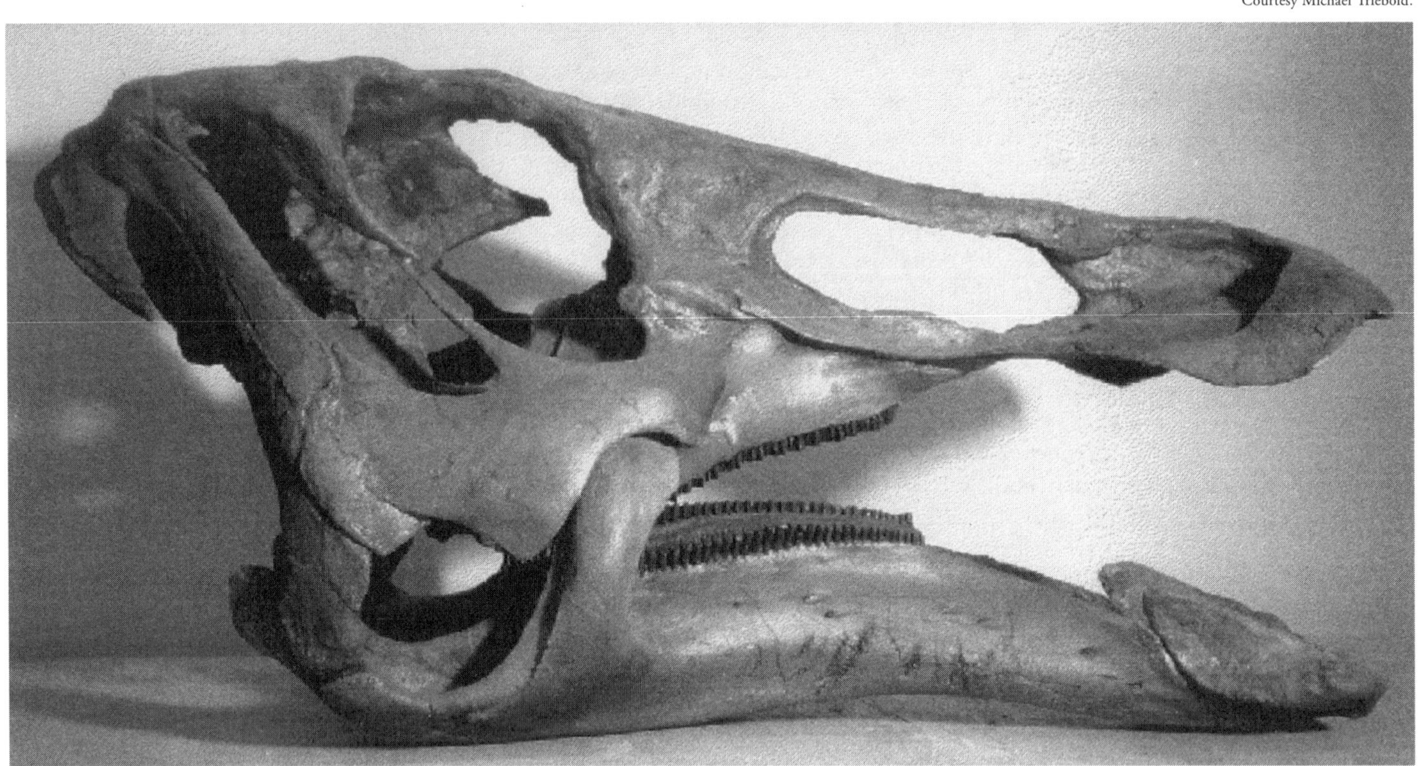

Above: **Skull (cast) of** ***Edmontosaurus annectens*****, original specimen collected by Michael Triebold.** ***Right:*** **Lower jaw of the skull** ***Edmontosaurus annectens*** **showing bite marks made by a small theropod dinosaur.**

Courtesy Michael Triebold.

seizing the throat in the fashion of many modern carnivores. The *Edmontosaurus* "survived the attack," Triebold speculated, "but the doomed animal was mortally wounded. With its throat horribly mangled, it either could not eat, or a nasty infection set in, progressively and inexorably sealing its fate. After a couple weeks, the huge creature finally succumbed to its injuries, and on a warm river delta near the edge of the sea, it finally gave up the ghost, separated from us not by a great distance, but only by time, as it breathed [its] last breath, was entombed by a spring flood, and revealed for human eyes to see 66,000 [millennia] later."

A cast of this specimen is now housed at the North Carolina State Museum of Natural Sciences in Raleigh.

Key reference: Campagna (2000).

†ELAPHROSAURUS

Saurischia: Theropoda: Neotheropoda: "Ceratosauria": Neoceratosauria: Abelisauroidea: ?Abelisauridae.

Comments: Work continues regarding the classification of *Elaphrosaurus bambergi*, a medium-sized theropod from the Late Jurassic of Tanzania, particularly regarding its role in the evolution of the carnivorous dinosaurs. Originally, this type species — founded upon an incomplete skeleton lacking a skull — was described by Janensch (1920) as a "Megalosaurid." More recent studies, however, have placed this taxon with other theropod groups (see entry, *D:TE*; also "Systematics" chapter, *S1*, for opinions by various authors, *e.g.*, Holtz 1995).

As pointed out by Rauhut (1998), in a preliminary report, *E. bambergi* has often been regarded as a primitive member of the Ornithomimosauria, this assessment based primarily on appendicular characters. However, this taxon actually combines primitive vertebral characters with ornithomimosaur-like limbs. According to Rauhut (as already suggested by other recent authors), analysis of *E. bambergi*, mostly founded upon characters in the sacrum and forelimbs, indicates a relationship of this species with the family Abelisauridae. Rauhut, therefore, interpreted *Elaphrosaurus* as representing a Gondwanan appearance of the "ostrich dinosaur" morphology, without having any closer relationship with the Ornithomimosauria.

Rauhut acknowledged the importance of *Elaphrosaurus* within the framework of theropod evolution, with the Abelisauridae usually nestled within "Ceratosauria," thereby forming a part of the first major theropod radiation. *Elaphrosaurus*, however, exhibits only a few characters usually considered to be synapomorphies of the "Ceratosauria," the new analysis by Rauhut indicating the latter to be a paraphyletic group (see "Systematics" chapter).

Key references: Holtz (1995); Janensch (1920); Rauhut (1998).

†ELOPTERYX Andrews 1913 (=?*Bradycneme*)

Saurischia: Theropoda: Neotheropoda: Tetanurae: Avetheropoda: Coelurosauria: Manuraptoriformes: Manuraptora *incertae sedis*.

Name derivation: Greek *helos* (also *elo*) = "marsh" + Greek *pteryx* = "wing."

Type species: *E. nopcsai* Andrews 1913.

Other species: [None.]

Occurrence: Sinpetru Formation, Transylvania, Romania; Fox-Amphoux and Roques-Hautes, Provence, France.

Age: Late Cretaceous (latest Maastrichtian).

Known material: ?Various fragmentary postcranial elements.

Holotype: BMNH A.1234, proximal femoral fragment.

Diagnosis of genus (as for type species): [No modern diagnosis published.]

Comments: In 1913, Charles W. Andrews

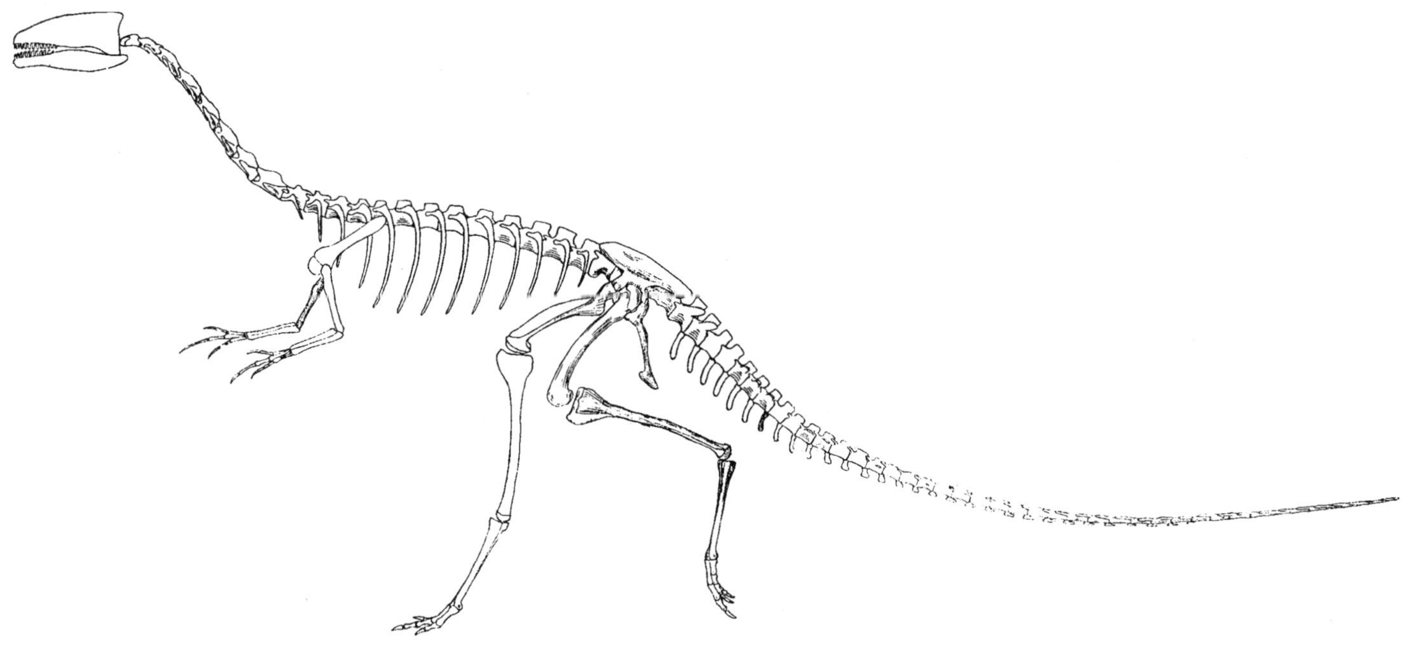

Paleontologist Werner Janensch's original reconstruction of the holotype postcranial skeleton (HMN collection) of the primitive theropod *Elaphrosaurus bambergi*. (After Janensch 1920.)

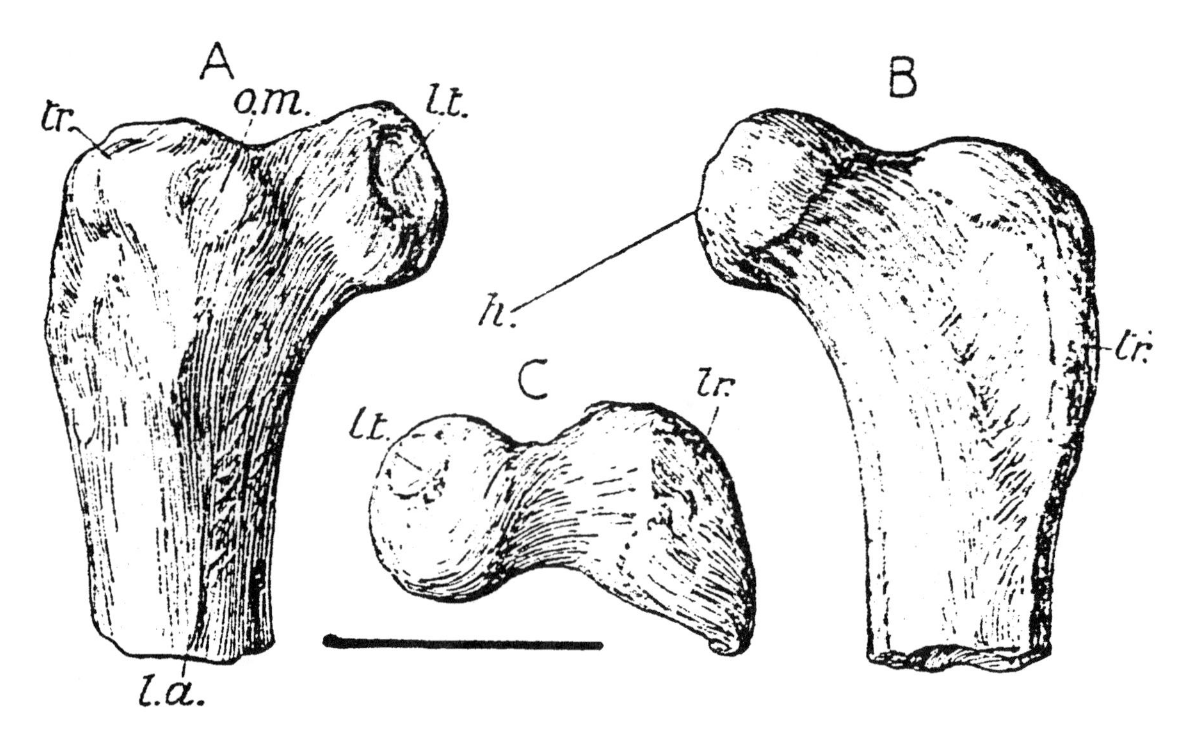

***Elopteryx nopcsai*, BMNH R.1234, holotype proximal left femur, in A. caudal, B. cranial, and C. dorsal views. Scale = 1.33 cm. (After Andrews 1913.)**

described in detail a new genus and species, *Elopteryx nopcsai*, which he believed to be a pelecaniform bird. Andrews established this taxon on a well-preserved proximal fragment of a femur and the distal part of a left tibiotarsus (BMNH A.1234) collected by Baron Franz Nopcsa, apparently during the late 1800s, from the Sinpetru Formation, in the Sibisel Valley in the Hateg Basin of Transylvania, Romania (then in Hungary). Nopcsa, the author noted, also believed these remains to be avian.

At the same time, Andrews referred to *E. nopcsai* and described other fairly well-preserved specimens, including another proximal femur (BMNH A.1235), and two additional distal tibiotarsi (BMNH A.1588 and A.1528). All that is known about these specimens is that they also were found in the Sibisel Valley (D. B. Weishampel, personal communication to Csiki and Grigorescu 1998).

Although Andrews did not diagnose this taxon, that author noted, in comparing the above material with various extant bird groups, that *E. nopcsai* most closely resembles "the Steganopodes, *e.g.,* the cormorant (*Phalaecrocorax*)," the similar features being "(1) the form of the great trochanter, especially the strong forward prominence of its antero-external angle, (2) the position and depth of the muscle impressions on the outer face of the trochanter, (3) the fact that the summit of the head rises above the trochanter, (4) the large size of the pit for the *ligamentum teres.*" Also, Andrews observed that the arrangement of muscular ridges, at least those in the anterior portion of the shaft, closely resembles that in steganopodians." Andrews concluded that *E. nopcsai* is a bird about the size of a pelican, but different from any bird previously known.

Harrison and Walker (1975) subsequently removed the distal left tibiotarsus from BMNH A.1234, giving it the new catalog number BMNH A.4358, and designating it the holotype of a new genus and species, *Heptasteornis andrewsi.* These authors also referred BMNH A.1528 to this new taxon, while establishing BMNH A.1588 as the holotype of another new genus and species, *Bradycneme draculae.* Both of these new taxa were interpreted as avian by their authors (see *Bradycneme* entries, *D:TE* and this volume).

Later, Grigorescu and Kessler (1981) referred to *E. nopcsai* a distal left femoral fragment (FGGUB R.351) (incorrectly interpreted by Le Loeuff in 1992 to be a distal tibiotarsus).

Le Loeuff assigned yet more remains from the Hateg Basin of Transylvania to *E. nopcsai* (which Grigorescu and Kessler regarded as a dromaeosaurid theropod), these including the following: The specimens described by Harrison and Walker; caudal vertebrae (FGGUB R.70 and R.71) and a proximal ulna and distal tibia (FGGUB R.72 and R.73), described by Grigorescu (1984*b*); paired frontals (MAFI v.13528), described by Jianu and Weishampel (1997); and a partial skull roof with associated frontal (MCDRD 454) and parietals (MCDRD 254), described by Weishampel and Jianu (1996) as a dromaeosaurid closely related to *Saurornitholestes.*

Also, Le Loeuff, Buffetaut, Mechin and Mechin-Salessy (1992) tentatively referred to the Dromaeosauridae and described various specimens collected by Patrick Mechin and Annie Mechin-Salessy from two Lower Rognacian (probably late Campanian–early Maastrichtian; see Westphal and Durand 1990) localities — the Fox-Amphoux (Var) syncline and Roques-Hautes (Bouches-du-Rhone) — in the Aix-en-Provence basin, these constituting the first birdlike theropod remains from southern France. This material, "reminiscent of the Transylvanian theropods," belonging to the Mechin collection, included (from Fox-Amphoux) a left femur (number 203), and (from Roques-Hautes) a cervico-dorsal vertebra and a sacro-caudal vertebra. Le Loeuff *et al.* suggested that these specimens could all belong to the genus *Elopteryx*, though probably representing a new species. The vertebrae, however, were subsequently referred by Le Loeuff and Buffetaut (1998) to the new genus *Variraptor* (see entry).

Over the years, *Elopteryx* has been classified as both a bird and a theropod dinosaur *e.g.*, Grigorscu; Paul 1988*b*), although most workers have favored the avian interpretation (see "excluded genera," *D:TE*). Le Loeuff, also Le Loeuff and Buffetaut (1991), and Le Loeuff *et al.*, for example, considered this genus to be an "elopterygine" dromaeosaurid, with *Bradycneme* and *Heptasteornis* regarded as junior synonyms.

More recently, Csiki and Grigorescu (1998), treating with all of the above specimens as separate items, identified the holotype proximal femoral fragments of *E. nopcsai* as representing a manuraptoran theropod. According to the authors, this material shares with the Tetanurae the following synapomorphies — mediodorsally inclined femoral head (Pérez-Moreno, Sanz, Sudre and Sigé 1993); lesser trochanter laterally displaced, adjacent and posterior to greater trochanter, projecting above proximal margin of femoral head (Rowe and Gauthier 1990; Pérez-Moreno *et al.*); absence of trochantic shelf (Pérez-Moreno); with the Coelurosauria the following characters — tip of lesser trochanter level with greater trochanter proximally; fourth trochanter weak or absent (Pérez-Moreno *et al.*; Sereno, Dutheil, Iarochene, Larsson, Lyon, Magwene, Sidor, Varrichio and Wilson 1996).

Csiki and Grigorescu observed that within the Coelurosauria, *E. nopcsai* can be characterized by the following features: Confluence of lesser trochanter with greater trochanter (Benton 1990), a condition shared with troodontids (*e.g.*, Russell 1969; Currie and Peng 1993; Russell and Dong 1993*a*), dromaeosaurids (Le Loeuff and Buffetaut 1998; Le Loeuff, Buffetaut, Mechin and Mechin-Salessy), and *Mononykus olecranon* (Perle, Chiappe, Barsbold, Clarke and Norell 1994) (though the continuous plate of bone formed by the trochanters is oblique as in dromaeosaurids rather than perpendicular to the long axis of the femoral head as in troodontids; Currie and Peng); well-developed posterior trochanter, a condition shared with dromaeosaurids (Ostrom 1990), troodontids (Le Loeuff; Currie and Peng; Russell and Dong), *Avimimus portentosus* (Norman 1990), *Archaeopteryx lithographica* and enantiornithine birds (Chiappe and Calvo 1994); fourth trochanter very reduced or absent, this condition shared with troodontids, dromaeosaurids, caenagnathids [=elmisaurids of Csiki and Grigorescu's usage] (Currie 1990), *Microvenator celer* (Norman), and *Mononykus* (Perle *et al.*).

Furthermore, the authors pointed out that, as *Elopteryx* lacks certain synapomorphies of Ornithothoraces (a group of birds including all modern birds) or other higher clades of birds, it cannot be classified with the Neornithes as Andrews had originally proposed. Csiki and Grigorescu concluded that *E. nopcsai*, as known only from the femoral fragments included in the holotype, is a derived manuraptoran theropod close to either dromaeosaurids, troodontids, or *Avimimus*, possibly representing a new taxon within Manuraptora (Le Loeuff *et al.*'s "elopterygines").

Notes: Regarding other specimens that have been referred to *E. nopcsai*, Csiki and Grigorescu pointed out the following:

Distal femur FGGUB R.351 lacks several characteristic avian features listed by Chiappe and Calvo, but exhibits features indicative of a close relationship to "ceratosaurs" (possibly not a valid group; see "Systematics" chapter): Absence of well-marked cranial intercondylar groove on distal femur (presence of this groove regarded as a synapomorphy of Tetanurae; Pérez-Moreno *et al.*; Novas 1992*b*); presence of distinctively deep groove at base of crista tibiofibularis (Rowe and Gauthier; listed as synamporphy of "Ceratosauria" by Pérez-Moreno *et al.*; Holtz 1994); presence of nonelliptical muscle scar on craniodistal region of femur (character shared with the Neoceratosauria and some nontetanurans (Pérez-Moreno *et al.*); and presence of craniomedial crest (=supracondylar crest) originating above medial (tibial) condyle (reported in abelisaurids; Le Loeuff and Buffetaut 1991). Autapomorphies of this specimen include a triangle-shaped, depressed popliteal area bordered by the converging ascending ridges of the articular condyles, and a cranially extended medial condyle. Therefore, FGGUB R.351 could represent a small abelisaurid.

The paired frontals MAFI v.13528, partial skull roof with frontal MCDRD 454, and parietals MCDRD 254 are dromaeosaurid as the original authors described them. However, Csiki and Grigorescu pointed out that if *Eloperyx* does prove to be a dromaeosaurid, it seems to exhibit characters that deny its referral to the taxon represented by this material.

Lacking more complete, associated material, it

was not possible for Csiki and Grigorescu to recognize whether the other disarticulated theropod remains referred to *E. nopcsai* from the Hateg Basin belongs to one taxon or more.

Key references: Andrews (1913); Csiki and Grigorescu (1998); Grigorescu (1984*b*); Grigorescu and Kessler (1981); Harrison and Walker (1975); Jianu and Weishampel (1975); Le Loeuff (1992); Le Loeuff and Buffetaut (1991, 1998); Le Loeuff, Buffetaut, Mechin and Mechin-Salessy (1992); Paul (1988*b*); Weishampel and Jianu (1996).

†ELOSAURUS—(See *Haplocanthosaurus, D:TE, S1*)
Name derivation: Greek *helos* (also *elos*) = "marsh" + Greek *sauros* = "lizard."

†EOLAMBIA

Ornithischia: Genasauria: Cerapoda: Ornithopoda: Euornithopoda: Iguanodontia: Dryomorpha: Ankylopollexia: ?Hadrosauroidea.

Comments: The type species *Eolambia caroljonesa* was originally described by James I. Kirkland in 1998 as a basal lambeosaurine lacking a crest, Kirkland using cranial and postcranial synapomorphies in its diagnosis (see *S1*).

More recently, Head (1999), after examining paratypic material belonging to this taxon, observed "none of the previously reported cranial synapomorphies uniting *Eolambia* and Lambeosaurinae," and found postcranial characteristics to be "dubious with regard to polarity and distribution." Furthermore, *Eolambia* was found to retain characters plesiomorphic with the Iguanodontia (*e.g.*, seven sacral vertebrae, lack of iliac trochanter, open femoral anterior intercondylar groove [both regarded by M. K. Brett-Surman, personal communication 2000, as possibly invalid features], primitive surangular, ?transversely-oriented antorbital fenestra). Phylogenetic analysis performed by Head on *Eolambia* resulted in the exclusion of this genus from any definition of Hadrosauridae, this indicating that *Eolambia* "is a *Probactrosaurus*-grade iguanodontian." Head also determined that *Eolambia* plus *Probactrosaurus* form a clade "weakly supported by the mutual possession of a shortened posterior skull roof"; furthermore, the materials referred to *Eolambia* could represent a chimera of two nonhadrosaurid taxa.

Reevaluation of this genus "demonstrates a North American representation of the evolutionary heritage leading to hadrosaurids near the beginning of the Late Cretaceous." According to Head, the limited geographic distance and brief minimum divergence time separating *Eolambia* from higher taxa suggest that the duckbilled dinosaurs evolved *in situ* in the North American early Late Cretaceous. On the other hand, similar trends perceived in Asia "suggest either rapid migration of the evolutionary grade leading to hadrosaurids between North America and Asia, or large-scale morphological convergence not detected by phylogenetic analysis."

Kirkland (personal communication to T. L. Ford, 2000), however, maintains that *Eolambia* is indeed a lambeosaurine and that new, as of this writing undescribed material will support this assessment.

Key reference: Head (1999); Kirkland (1998*b*).

†EPACHTHOSAURUS

Saurischia: Sauropodomorpha: Sauropoda: Eusauropoda: Neosauropoda: Macronaria: Camarasauromorpha: Titanosauriformes: Somphospondyli: Titanosauria: ?Andesauridae.

Diagnosis of genus: Characterized by U-shaped groove on surface of posterior condyle (Sanz, Powell, Le Loeuff, Martinez and Suberbiola 1999).

Key reference: Sanz, Powell, Le Loeuff, Martinez and Suberbiola (1999).

EUCENTROSAURUS Chure and McIntosh 1989—(See *Centrosaurus.*)
Name derivation: Greek *eu* = "good/true" + *Centrosaurus.*
Type species: *E. apertus* (Lambe 1904.)

EUCOELOPHYSIS Sullivan and Lucas 1999

Saurischia: Theropoda: Neotheropoda: "Ceratosauria": ?Coelophysoidea.
Name derivation: Greek *eu* = "good/true" + *Coelophysis.*
Type species: *E. baldwini* Sullivan and Lucas 1999.
Other species: [None.]
Occurrence: Petrified Forest Formation, New Mexico, United States.
Age: Late Triassic (Norian ["Revueltian"]).
Known material: Various postcranial remains, two individuals.
Holotype: NMMNH P-22298, postcrania including two dorsal and four incomplete caudal vertebrae, incomplete left scapulacoracoid, almost complete right pubis, partial right ischium, fragment of ?ilium, portions of both femora, greater proximal portion of right tibia, complete metatarsal III, incomplete metatarsals II and IV, phalanges, unidentified bone fragments, subadult.

Diagnosis of genus (as for type species): "Ceratosaur" distinguished from all other known

Photograph courtesy Robert M. Sullivan.

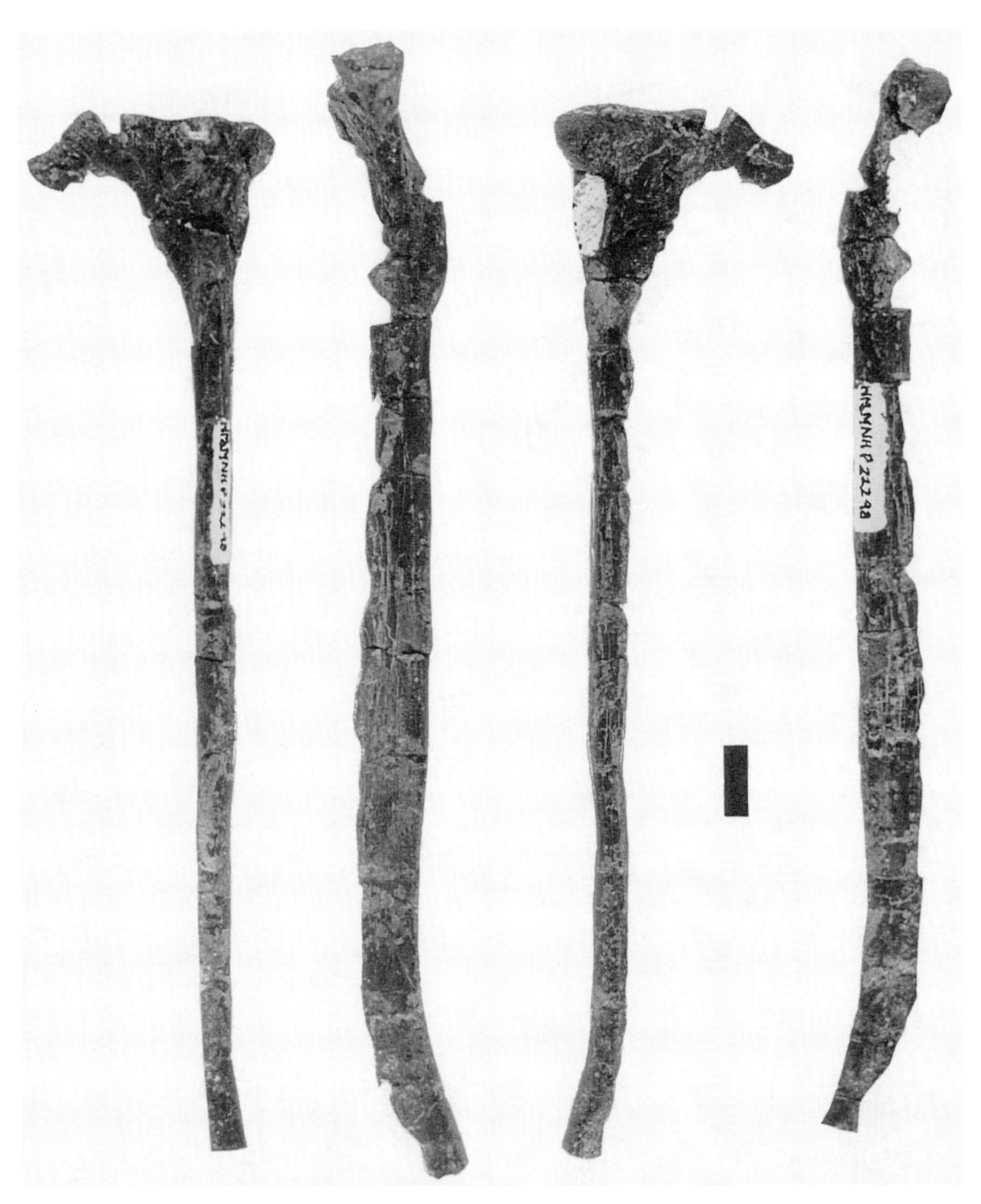

***Eucoelophysis baldwini*, NMMNH P-22298, holotype right pubis. Scale = 1 cm.**

"ceratosaurs" [Ceratosauria no longer a valid taxon; see "Systematics" chapter] by ischio-acetabular groove in proximal end of pubis [between ischial facet and acetabular facet], and sulcus incised into proximal articular surface of femur; differs from *Coelophysis bauri* and *Syntarsus rhodesiensis* in lacking well-developed posterior femoral concave notch below femoral head, from *C. bauri* in that tibia has large (two-thirds length of tibia) appressed surface on its lateral side (distally) for contact with fibula, lacks fibular crest (crista fibularis) (Sullivan and Lucas 1999).

Comments: Informally known as the "Orphan Mesa theropod," the new type species *Eocoelophysis baldwini* was founded on an incomplete appendicular skeleton (NMMNH P-22298) including an incomplete articulated right leg. The specimen was discovered in 1983 — associated with a small number of phytosaur bones and some unidentifiable elements — by Robert Cross along the northeast base of a topographical feature known locally as Orphan Mesa (see Sullivan 1993; Sullivan, Lucas, Heckert and Hunt 1996), in the Ghost Ranch Quadrangle in the upper

Drawing courtesy Robert M. Sullivan. (After Sullivan and Lucas 1999.)

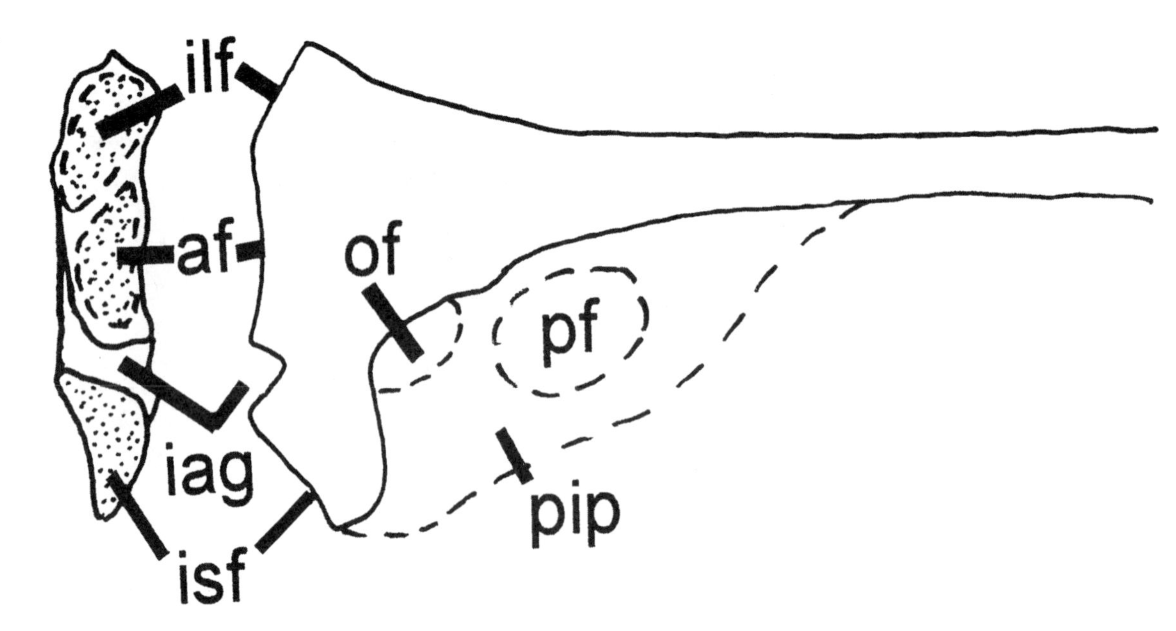

***Eucoelophysis baldwini*, NMMNH P-22298, holotype right pubis with pubio-ischiac plate restored; "iag" = ischio-acetabular groove.**

part of the Petrified Forest Formation, Chinle Group, Rio Arriba County, in north-central New Mexico (see Colbert 1989). Ten years later, fragmentary remains of other theropods, including isolated vertebrae, the proximal end of a left tibia, and the distal portion of a left scapulocorocoid, were recovered around other the western, eastern, and southern sides of the base of Orphan Mesa. These fragmentary specimens compare superficially in both morphology and preservation with the original type material (originally referred to *Coelurus*; see *Coelophysis* and *Rioarribasaurus* entries, *D:TE*; also *Coelophysis* entries, *S1* and this volume) that established the genus *Coelophysis*, collected in 1881 by David Baldwin for Edward Drinker Cope. However, these specimens lack apomorphic characters and, therefore, cannot be diagnosed (Sullivan and Lucas 1999).

Previously, Sullivan *et al.* (1996) showed that the Orphan Mesa area is the same Arroyo Secco locality where Baldwin recovered the type material that came to be known as *Coelophysis*. It is not, as demonstrated by these authors, the geologically younger (Rock Point Formation) Whitaker (Ghost Ranch) quarry which yielded what is now the neotype (AMNH 2706) of the type species *Coelophysis bauri* (the specimen designated as such by a ruling of the International Commission on Zoological Nomenclature (ICZN); see *Coelophysis* entry, *S1*), which is now the type locality for this genus, and is famous for its many articulated skeletons belonging to this small theropod.

As pointed out by Sullivan *et al.*, the decision of the ICZN left Cope's original "Arroyo Secco" specimens without a genus and species name; nor can they be given such because of their undiagnostic nature (*e.g.*, Hunt and Lucas 1991; Sullivan 1993, 1994, 1995). One of these specimens, however — a nearly complete right pubis (AMNH 2706), formerly referred to *Coelophysis longicollis* and later to *C. bauri*— could be

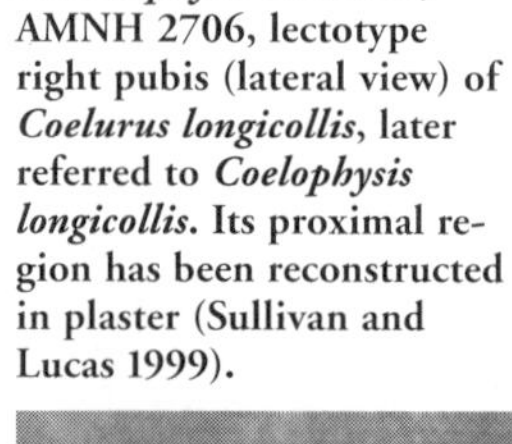

***Eucoelophysis baldwini*, AMNH 2706, lectotype right pubis (lateral view) of *Coelurus longicollis*, later referred to *Coelophysis longicollis*. Its proximal region has been reconstructed in plaster (Sullivan and Lucas 1999).**

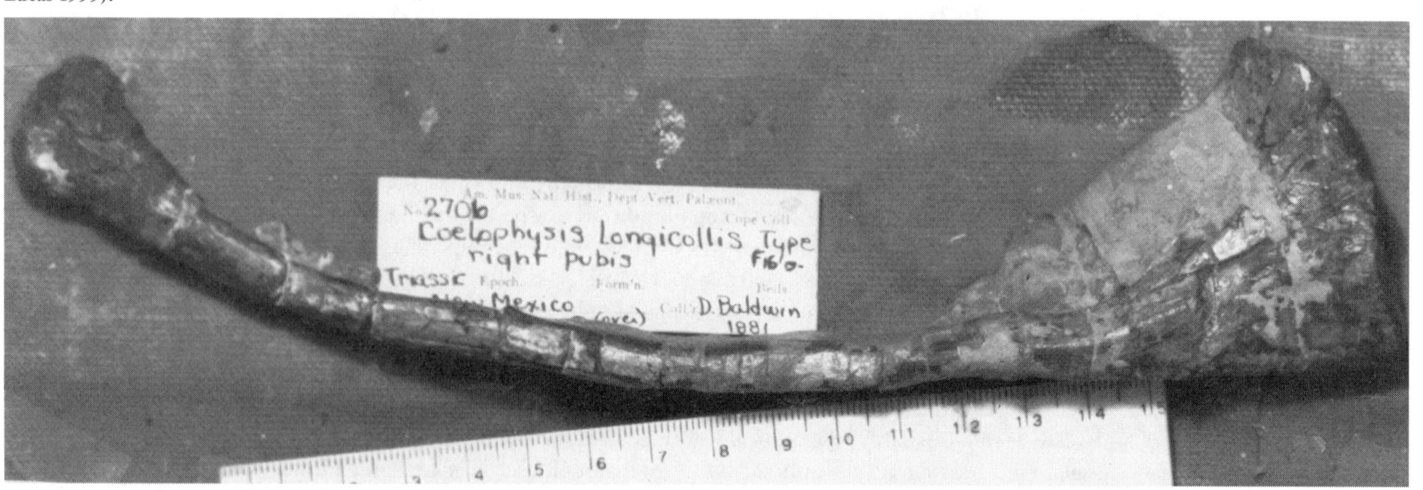

Photograph by Robert A. Long and Samuel P. Welles, courtesy American Museum of Natural History.

Photograph by Robert A. Long and Samuel P. Welles, courtesy American Museum of Natural History.

"Coelophysis bauri," AMNH 2722, lectotype four fused sacral vertebrae joined with pubic process of ilium, originally referred to the genus *Coelurus.* This specimen is not diagnostic, but could belong to *Eucoelophysis baldwini.* It was one of the various specimens collected by David Baldwin and described by Edward Drinker Cope.

referred with confidence by Sullivan and Lucas to *E. baldwini*. Sullivan and Lucas noted that Cope's other "*Coelophysis*" specimens—referred to the species *C. longicollis* and *C. willistoni*—could belong to *Eucoelophysis*, being "ceratosaurian" and originating from the same general locality and stratigraphic level, although a positive identification of these remains could not be demonstrated.

Sullivan and Lucas assigned *Eucoelophysis baldwini* to the "Ceratosauria" based upon several features (*e.g.*, triangular and posteriorly directed transverse processes of dorsal vertebrae; presence of prominent trochanteric shelf on lesser trochanter of femur; possibly large pubic fenestra).

The authors envisioned *Eucocoelophysis* as a theropod much smaller and more gracile than *Ceratosaurus* and *Dilophosaurus*, most similar to *Coelophysis* and *Syntarsus*. Lack of a skull prevented Sullivan and Lucas from placing the new genus in any "ceratosaurian" clade.

Heckert, Zeigler, Lucas, Rinehart and Harris (2000) (see also Heckert, Lucas, Rhinehart and Harris 2000) described a number of skull and postcranial elements collected from the Upper Triassic ("Revueltian") Snyder quarry (the second-most productive Chinle theropod locality, discovered in 1998 by Mark Snyder), Arroyo Seco, about eight kilometers northeast of the type locality of *Eucoelophysis*, stratigraphically

Photograph by Robert A. Long and Samuel P. Welles, courtesy American Museum of Natural History.

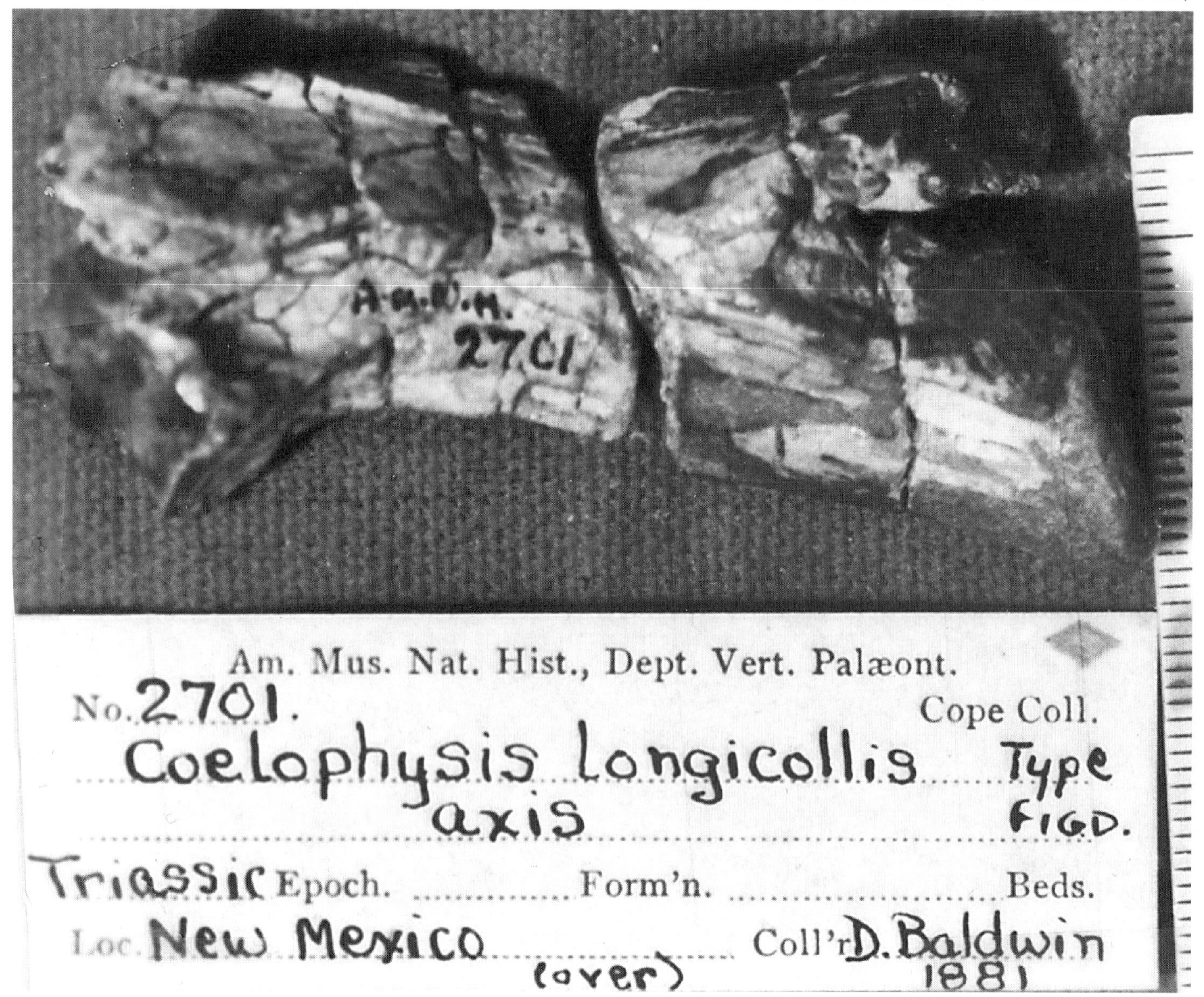

***Coelophysis longicollis*, AMNH 2701, lectotype cervical vertebra, left lateral view, originally referred to the genus *Coelurus*. The specimen is not diagnostic, but could belong to *Eucoelophysis baldwini*.**

high in the badlands of the Petrified Forest Formation near Ghost Ranch. Three field seasons of excavations yielded the relatively well-preserved remains (NMMNH collection) of perhaps as few as four individuals representing two taxa of different sizes. Other fossil remains recovered from this quarry include invertebrates (*e.g.*, conchostracans, bivalves, and decapod crustaceans), semionotid and colobonotid fishes, metoposaurid amphibians, and archosaurs (*e.g.*, phytosaurs and aetosaurs), and possible cynodonts.

Most of the theropod elements — including an incomplete skull, lower jaws, two articulated cervical vertebrae, one dorsal vertebra, an incomplete sacrum, incomplete left scapulocoracoid, a fragmentary right ilium, an almost complete right ischium, two femora, three tibiae, an astragalus, and numerous metapodials, phalanges, and ribs — pertain to the smaller taxon. Heckert, Zeigler *et al.* referred this material to the "Ceratosauria," and more precisely to the Coelophysoidea, based upon such features as a "subnarial gap, heterodont premaxilla, gracile limb bones, and numerous other features."

According to these authors, the smaller theropod, in its general morphology, most closely resembles *Eucoelophysis*, especially in details of the scapulocoracoid, ischium, and tibia (particularly, the surface is strongly appressed, contacting the fibula). However, a number of differences were noted between the Snyder quarry material and that genus. In the former, the glenoid of the scapulocoracoid is more curved and the scapula, as a whole, is more gracile (although the authors pointed out that this feature could be an artifact

Photograph by Robert A. Lon and Samuel P. Welles, courtesy American Museum of Natural History.

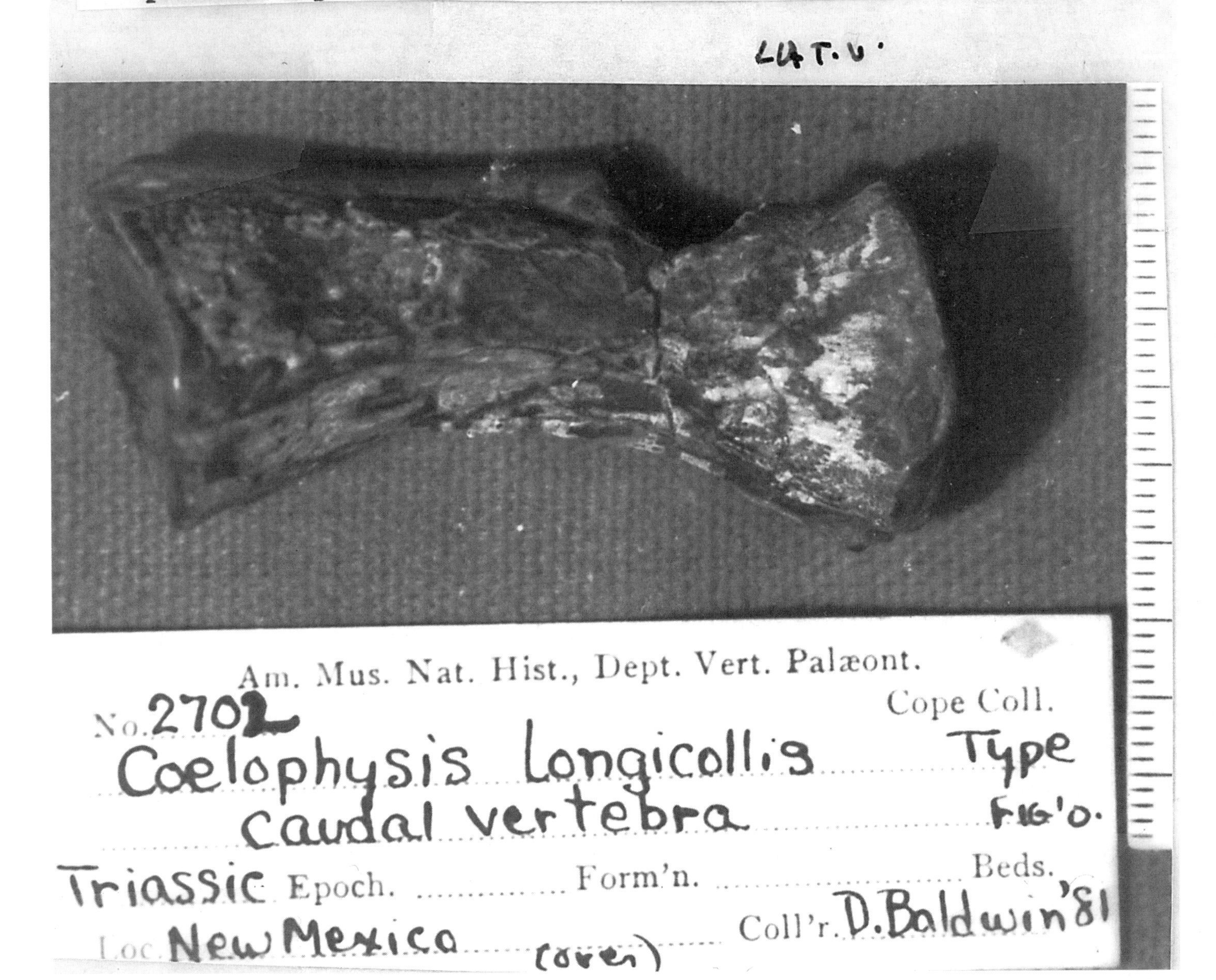

***Coelophysis longicollis*, AMNH 2702, lectotype caudal vertebra, originally referred to *Coelurus longicollis*. The specimen is not diagnostic but could belong to *Eucoelophysis baldwini*.**

relating to the larger size of the *Eucoelophysis* holotype); also, the coracoid possesses a strong glenoid pillar and is more expanded than in the coelophysoids *Eucoelophysis*, *Coelophysis*, and *Syntarsus*. The Snyder quarry femora lack the proximal groove or sulcus found in *Eucoelophysis* and, unlike the condition in that genus, the lesser trochanter is a prominent subtriangular ridge (similar to the "robust" morphs of the species of *Syntarsus*; see Raath 1977; Tykoski 1998). Heckert, Zeigler *et al.* proposed that both the similarities to *Eucoelophysis* combined with the differences between these theropods could warrant the erection of a second species of that genus, although some of these differences, especially those in the femur, could have sexual significance (see Raath). Therefore, the authors tentatively referred this material to *Eucoelophysis* sp.

The larger Snyder quarry theropod is represented only by a fused tibia-fibula-astragalus-calcaneum. Heckert, Zeigler *et al.* found this specimen too fragmentary to identify positively, while noting that it is surely unique among all known "ceratosaurs" (see also Heckert, Lucas and Sullivan 2000).

As stated by Heckert, Zeigler *et al.*, the Snyder quarry theropods are particularly significant for the

Photograph by Robert A. Long and Samuel P. Welles, courtesy American Museum of Natural History.

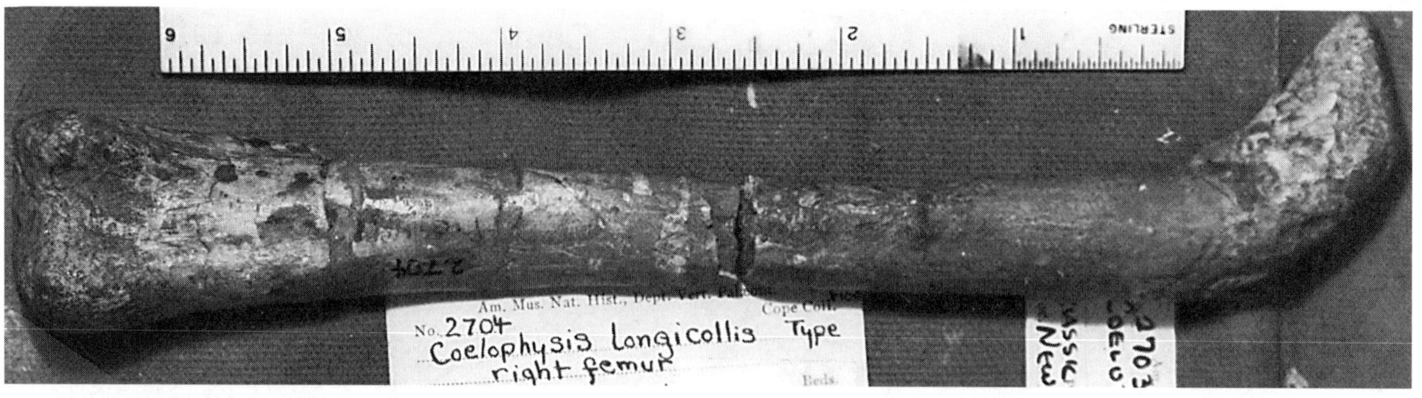

***Coelophysis longicollis*, AMNH 2704, lectotype right femur, posterior view, originally referred to *Coelurus longicollis*. The specimen is not diagnostic, but could belong to *Eucoelophysis baldwini*.**

following reasons: 1. Dinosaurs are relatively rare among Chinle fauna; 2. these are among the best-preserved early coelophysoids; 3. their stratigraphic position, roughly correlative of the *E. baldwini* type locality, enhances understanding of the superposition of Upper Triassic dinosaur faunas; and, 4. all of the above combine to improve understanding of Late Triassic dinosaur evolution.

As further noted by Heckert, Zeigler *et al.*, the Snyder quarry theropod fossils demonstrate that, by "Revueltian" time, "ceratosaurs," have evolved into a body plan "classically identified as coelophysoid," a design so successful that it would remain basically the same in such later taxa as *Coelophysis* and *Syntarsus rhodesiensis*, the latter surviving as a species until almost 185 million years ago. "With the possible exception of some of the more plesiomorphic titanosaurid sauropods," the authors stated, "no other group of dinosaurs has retained such a consistent body form for such a considerable (approximately 30 myr) span of time."

Key references: Colbert (1989); Heckert, Lucas, Rhinehart and Harris 2000); Heckert, Lucas and Sullivan (2000); Heckert, Zeigler, Lucas, Rinehart and Harris (2000); Hunt and Lucas (1991); Raath (1977); Sullivan (1993, 1994, 1995); Sullivan and Lucas (1999); Sullivan, Lucas, Heckert and Hunt (1996); Tykoski (1998).

†**EUHELOPUS**—(=*Helopus* Wiman 1929, *Tienshanosaurus*; =?*Asiatosaurus*)

Saurischia: Sauropodomorpha: Sauropoda: Eusauropoda: Neosauropoda: Macronaria: Camarasauromorpha: Titanosauriformes: Somphospondyli: Euhelopodidae: Euhelopodinae.

Comments: In a review of the sauropods of China, Valérie Martin-Rolland referred to *Euhelopus* the genus *Tienshanosaurus*, a genus informally described during the 1920s and later officially named and described by Yang [Young] (1937), based upon two cotype postcranial skeletons (IVP AS 40002 and 40003; see *D:TE* for information and illustrations).

Martin-Rolland proposed this synonymy based upon the following similarities between the two taxa:

***Coelophysis longicollis*, AMNH 2705, incomplete right ilium. This specimen, originally referred to *Coelurus longicollis*, was apparently once considered to be a lectotype for *C. bauri* (see *D:TE*). The specimen is not diagnostic, but could belong to *Eucoelophysis baldwini*.**

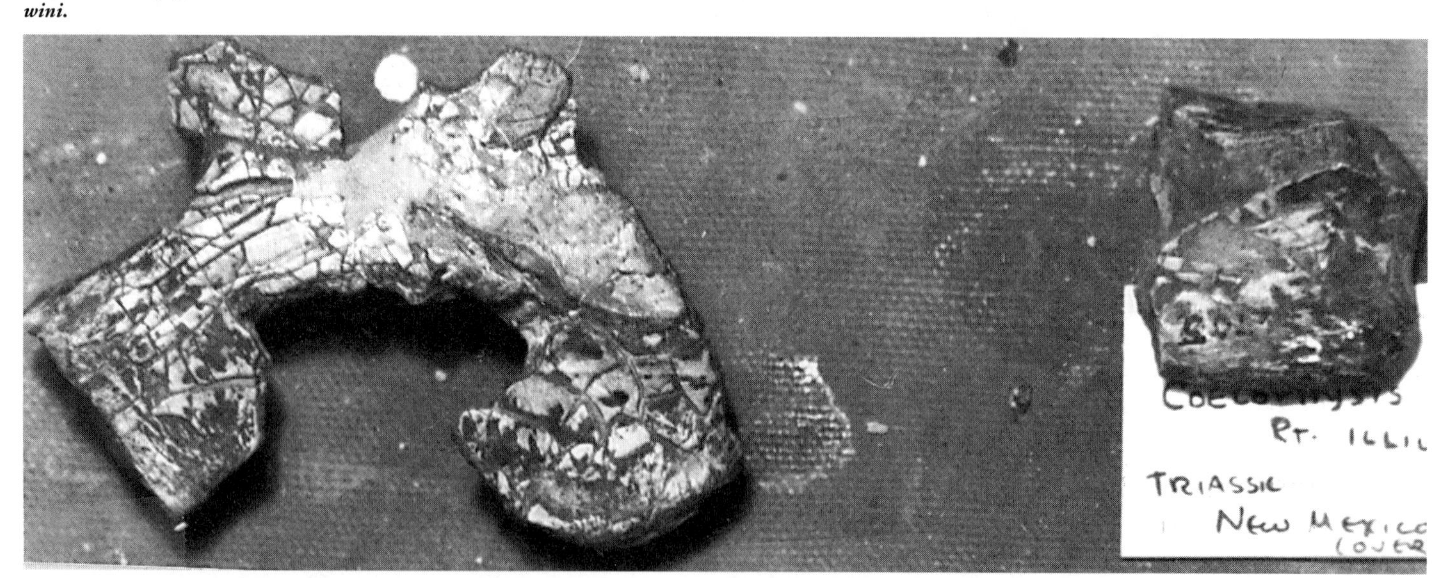

Photograph by Robert A. Long and Samuel P. Welles, courtesy American Museum of Natural History.

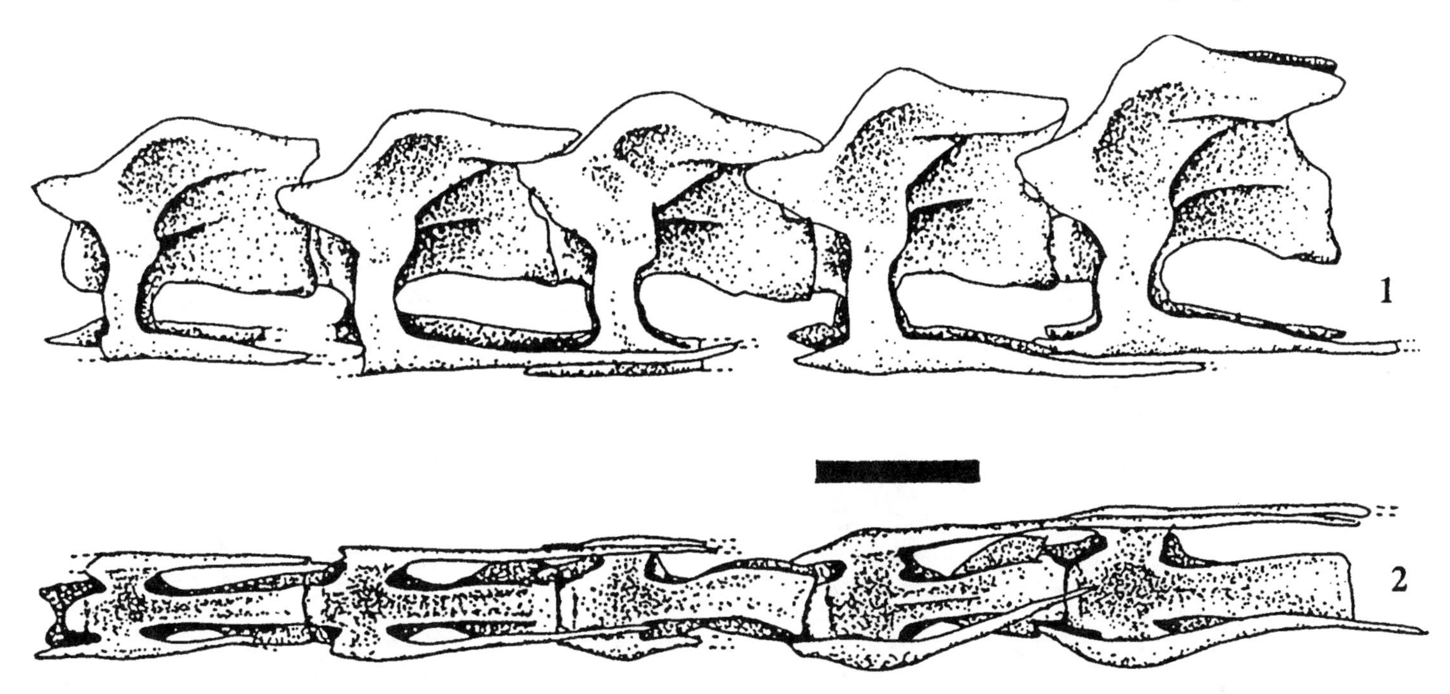

***Euhelopus zdanskyi*, cervical vertebrae in left 1. lateral and 2. ventral views. Scale = 20 cm. (After Wiman 1929.)**

Dorsal vertebrae exhibiting the same characters in both taxa; scapula displaying similar constriction at lower portion of blade, therefore being very close to ventral protuberance; cervical vertebrae of *Tienshanosaurus*, although fragmentary, comparable in their development to those of *Euhelopus*.

However, sauropod specialist J. McIntosh (personal communication 2000) is not convinced of this synonomy. As noted by McIntosh, the type material of *Tienshanosaurus* has been destroyed and the only extant collected material pertaining to this genus is a left scapula (see *D:TE*) that was found in 1930. In his opinion, although *Tienshanosaurus* and *Euhelopus* may very well be congeneric, uniting these taxa into a single genus is premature until additional materials are collected. Therefore, Martin-Rolland's synonomy of *Tienshanosaurus* with *Euhelopus* is herein regarded as provisional and may later be reversed pending future discoveries.

Key references: Martin-Rolland (1999), Wiman (1929); Young [Yang] (1937).

†EUOPLOCEPHALUS

Ornithischia: Genasauria: Thyreophora: Thyreophoroidea: Eurypoda: Ankylosauria: Ankylosauridae.

Comments: New studies have been conducted concerning the skull of the large, Late Cretaceous ankylosaurid *Euoplocephalus tutus*.

One line of research was performed by Rybczynski and Vickaryous regarding the jaw movements of this species, the results of which were briefly reported. This study challenged earlier research that suggested that thyreophorans (and pachycephalosaurs) utilized a simple orthal (or up and down) movement of the lower jaws during the powerstroke, and were, consequently, "orthal pulplers," that is employers of a mastication technique that implies minimal oral processing.

Examination by Rybczynski and Vickaryous of an *in situ* specimen of *E. tutus* revealed that tooth wear facets are continuous between adjacent teeth; also, examination of these surfaces via scanning electron microscopy revealed numerous microscopic striations that are roughly parallel with the tooth row. Combining this evidence with morphological data, the authors produced a more complex model of jaw action in this armored dinosaur, their new model describing "a bilateral powerstroke that is predominantly retractive, with a small transverse component."

As noted by these authors, *E. tutus* is distinguished as the first thyreophoran taxon to be diagnosed having a complex jaw action, this study (and perhaps similar studies to be made of other ornithischian "orthal pulpers") possibly providing "the basis for inferences of diet and niche separation in these forms."

With *E. tutus* as a basis of study, Vickaryous, Russell, Currie, Carpenter and Kirkland (1998) addressed the issue of the presence of cranial sculpturing along the dorsal and lateral surfaces of ankylosaur skulls, this feature having long been regarded as diagnostic for the Ankylosauria (see "Systematics" chapter). As noted by these authors, two different hypotheses have been offered to explain this sulpturing (the first most commonly accepted): 1. Overlying osteoderms have coosified with the skull roof; and 2.

Life restoration of the armored dinosaur *Euoplocephalus tutus* by artist Brian Franczek.

sculpturing is derived from the elaboration of the cranial periosteum.

Based upon fossil evidence from *E. tutus*, and also comparative data from living, nondinosaurian taxa exhibiting a similar condition, Vickaryous *et al.* found the first hypothesis best supported. Their proposed developmental model "involves the formation of osteoderms within the dermis under the influence of overlying epidermal structures," this, the authors further pointing out, also explaining the "bony cheeks" present in some ankylosaurian taxa.

However, subadult specimens of *E. tutus* and another ankylosaur, *Pinacosaurus grangeri*, "indicate that the bosses located at the posterior corners of the skull and over the orbits underwent a separate developmental history, involving elaboration of the periosteum." Thus, Vickaryous *et al.* noted, both of the above proposed developmental mechanisms may have been involved, to some extent, in this sulpturing of the skull.

Key references: Rybczynski and Vickaryous (1998); Vickaryous, Russell, Currie, Carpenter and Kirkland (1998).

†EUSKELOSAURUS

Age: Late Triassic (Carnian or ?early Carnian).

Comments: *Euskelosaurus* is the "common prosauropod dinosaur from the lower Elliot Formation" (Upper Triassic, ?early Carnian; see Gauffre 1993*b* and *Blikanasaurus* entry, this volume) of Africa (Galton and Heerden 1998). It is a large though relatively gracile prosauropod (compared to more massive genera such as *Plateosaurus*) known from numerous fragmentary skeletons and isolated postcranial remains (see *D:TE* for information and figures).

Galton and Heerden chronicled in detail the history of early discoveries and descriptions of remains attributed to the type species, *Euskelosaurus browni*, a brief summary of which follows: In 1894, Harry Govier Seeley wrote that Alfred Brown had sent four separate collections "of the same animal" of *Euskelosaurus browni*, the first London shipment having been described by Huxley (1866), the Paris shipment by Fischer (1879), and the third London shipment now by Seeley (who was unable to locate the second London shipment). As little information was included on the shipment and subsequent (BMNH) collection

labels, there has been controversy regarding the type locality that yielded the fossils included in these shipments.

According to Broom (1911), Brown had informed him in a note that all of these collections shared the same point of origin. Huxley stated that Brown had written to him that the specimens were obtained in the division of Aliwal North, some 30 miles distant from the town of Aliwal North. Seeley, however, gave the locality as Barnard's Spruit, 15 miles south of Aliwal North. Haughton (1924) then stated that the location where the fossils were recovered was almost surely the Kraai River. Years later, Kitching and Raath (1984) combined these various localities—"Aliwal North-Kraai River" and "Albert (Burgersdorp)-Bernard's Spruit (Ezelsklip)." In relating how he and Brown had gone off seeking *Euskelosaurus* remains, Seeley (quoted by Cooper 1980) implied that the locality was not the Kraai River. Therefore, Galton and Heerden deduced that, as Seeley seems to have visited the type locality of *E. browni* with Brown, the "Barnard's Spirit" lectotype locality is the more logical choice.

In a brief report published as an abstract, Welman (1998) used the braincase of *Euskelosaurus* as the basis for a study on the origins of dinosaurs (the first bonafide dinosaurs being known only from Upper Triassic rocks). As Welman pointed out, the braincase of *Euskelosaurus* (see Welman 1995*a*) "is unique among the archosauromorphs and may represent the most primitive condition in any dinosaur," this being most "evident from the condition of the median eustachian system in the basicranium" of this genus.

As observed by Welman (1998), the median eustachian system in more primitive archosauromorphs (*e.g., Proterosuchus* and *Erythrosuchus*) "consists of paired tubes that enter between the basioccipital and basiphenoid-basiparasphenoid and connect the endocranium with the pharynx." In some early prosauropods (*e.g., Plateosaurus* and *Massospondylus*) and primitive theropods (*e.g., Syntarsus*) "a much more transformed eustachian system," similar to that in extant and extinct crocodiles (Walker 1990), can be seen (Welman 1995*a*). The transformation between the early archosauromorph system and more advanced median eustachian system seems to have occurred "by a process of a gradual increase in size of the eustachian tubes between the anterior and posterior parts of the braincase and a subsequent rearrangement of these parts, which resulted in the formation of the advanced archosaurian pattern."

According to Welman (1998), the entirely opened up condition of the Eustachian tubes in the skull of *Euskelosaurus* seems to represent a transitional stage between the conditions present in more primitive archosauromorphs and those in some of the earliest dinosaurs. "Such an opening up and reclosing in a common ancestor," the author proposed, "would explain the complicated enclosed path of the palatine ramus of the facial nerve by the posterolateral process of the basiparasphenoid, which is apparently present in more advanced dinosaurs."

Note: According to Galton and Heerden, the only fossil footprints from the lower Elliot Formation that can be attributed to prosauropods are those named *Tetrasauropus unguiferus* Ellenberger 1970 (see Olsen and Galton 1984).

Key references: Broom (1911); Cooper (1980); Fischer (1879); Galton and Heerden (1998); Gauffre (1993*b*); Haughton (1924); Huxley (1866); Kitching and Raath (1984); Olsen and Galton (1984); Seeley (1894); Walker (1990); Welman (1995*a*, 1998).

FUKUIRAPTOR Azuma and Currie 2000

Saurischia: Theropoda: Neotheropoda: Tetanuurae: Avetheropoda: Allosauroidea: Allosauridae.

Name derivation: "Fukui [Prefecture]" + Latin *raptor* = "thief."

Type species: *F. kitadaniensis* Azuma and Currie 2000.

Other species: [None.]

Occurrence: Kitadani Formation, Fukui Prefecture, Japan.

Age: Middle Cretaceous (Albian).

Known material: Partial skeleton, tooth, possibly other postcranial elements.

Holotype: FPMN 97122, associated partial skeleton including left maxillary fragment (FPMN 9712201), right dentary fragment (FPMN 9712202), four isolated teeth (FPMN 9712203, 9712204, 9712205, 9712206), dorsal centrum (FPMN 9712207), distal caudal (FPMN 9712208), humeri (FPMN 96082443, 9712209), right ulna (FPMN 9712210), right manual ungual I-2 (FPMN 9712211), right manual phalanx II-2 (FPMN 9712212), left manual ungual II-3 (FPMN 9712213), fragments of right ilium (FPMN 9712214), pubis (FPMN 9712215 shaft, and possibly 9712216), ischia (FPMN 9712217, 9712218), right femur (FPMN 9712219), proximal half of right tibia (FPMN 9712220), distal half of right fibula (FPMN 97080206), right astragalus (FPMN 9712221), right metatarsal I (FPMN 9712222), right metatarsal II (FPMN 9712223), right metatarsal III (FPMN 9712224), pedal phalanges left I-1 (FPMN 9712225), right III-1 (FPMN 9712226), right III-2 (FPMN 9712227), and right IV-2 (FPMN 9712228), immature individual.

Diagnosis of genus (as for type species): Medium-sized carnosaur distinguished by the following autapomorphies: interdental plates in maxilla and

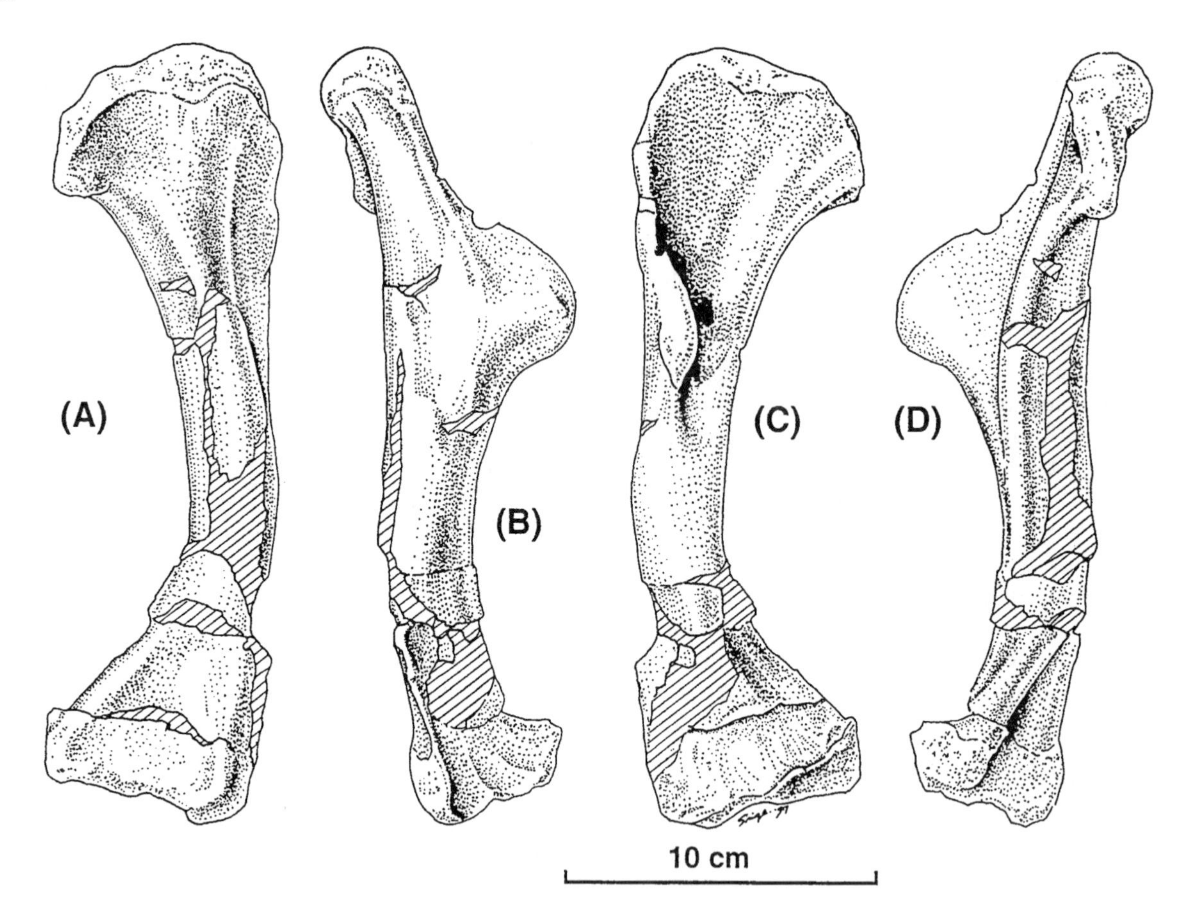

***Fukuiraptor kitadanensis*, FPMN 9712209, holotype left humerus in A. dorsal, B. lateral, C. ventral, and D. medial views. (After Azuma and Currie 2000.)**

dentary fused (convergent with dromaeosaurids); narrow dentary; teeth narrow, blade-like, with oblique blood grooves (convergent with tyrannosaurids); 0.92 ratio of ulna to humerus (significantly higher than any other carnosaurs except *Xuanhanosaurus*); hands relatively larger with better-developed unguals than in *Allosaurus*; pubic peduncle lateromedially wider than anteroposteriorly long; ascending process relatively taller than in other known carnosaurs (Azuma and Currie 2000).

Comments: The genus *Fukuiraptor* was established on a partial well-preserved, associated skeleton (FPMN 97122) found in one small area of the Kitadani quarry in the Middle Cretaceous (Albian) Kitadani Formation (Akaiwa subgroup, Tetori Group) on the Sugiyama River, in the northern part of the city of Katsuyama, in Fukui Prefecture, Hokuriku region of central Japan (Azuma and Currie 2000).

As related by Azuma and Currie, the initial find was made kin 1993, when a right manual ungual (I-2), right astragalus, and right metatarsal III were discovered in a small area of this quarry and recognized as belonging to an associated skeleton. This discovery subsequently led to the reopening of that section of the quarry in 1996 and 1997, during which time numerous additional theropod bones were recovered. Specimens from the same quarry referred to the type species *Fukuiraptor kitadaniensis* include a fragment of a left maxilla with two complete and one partial alveoli (FPMN 9712229), a partial dentary with three alveoli (FPMN 9712230), isolated shed teeth (FPMN 9712231, 9712232, 9712233, 9712234, 9712235, 9712236, 9712237, 9712238, and ?9712239), a tentatively referred cervical centrum (FPMN 9712240), a partial cervical neural arch (FPMN 9712241) that fits onto the centrum (even although it was collected from a different part of the quarry, and which may belonging to the same individual), and a nearly complete dorsal neural arch (FPMN 9712242) that may belong to this genus. Also, a theropod coracoid (FPMN 9712243), recovered from the same level as the main concentration of theropod bones but found five meters away in a different part of the quarry, is the correct size to belong to the holotype. All of these remains may pertain to a single individual (Azuma and Currie 2000).

Azuma and Currie noted that the lengths of all of the collected bones indicate that this individual dinosaur, in life, measured about 4.2 meters (over 14 feet) long. The hands, as described by these authors, are relatively large and armed with strongly curved, sharp claws, these features suggesting *Fukuiraptor* was an active predator. The authors estimated the weight of this theropod to have been about 175 kilograms (approximately 450 pounds).

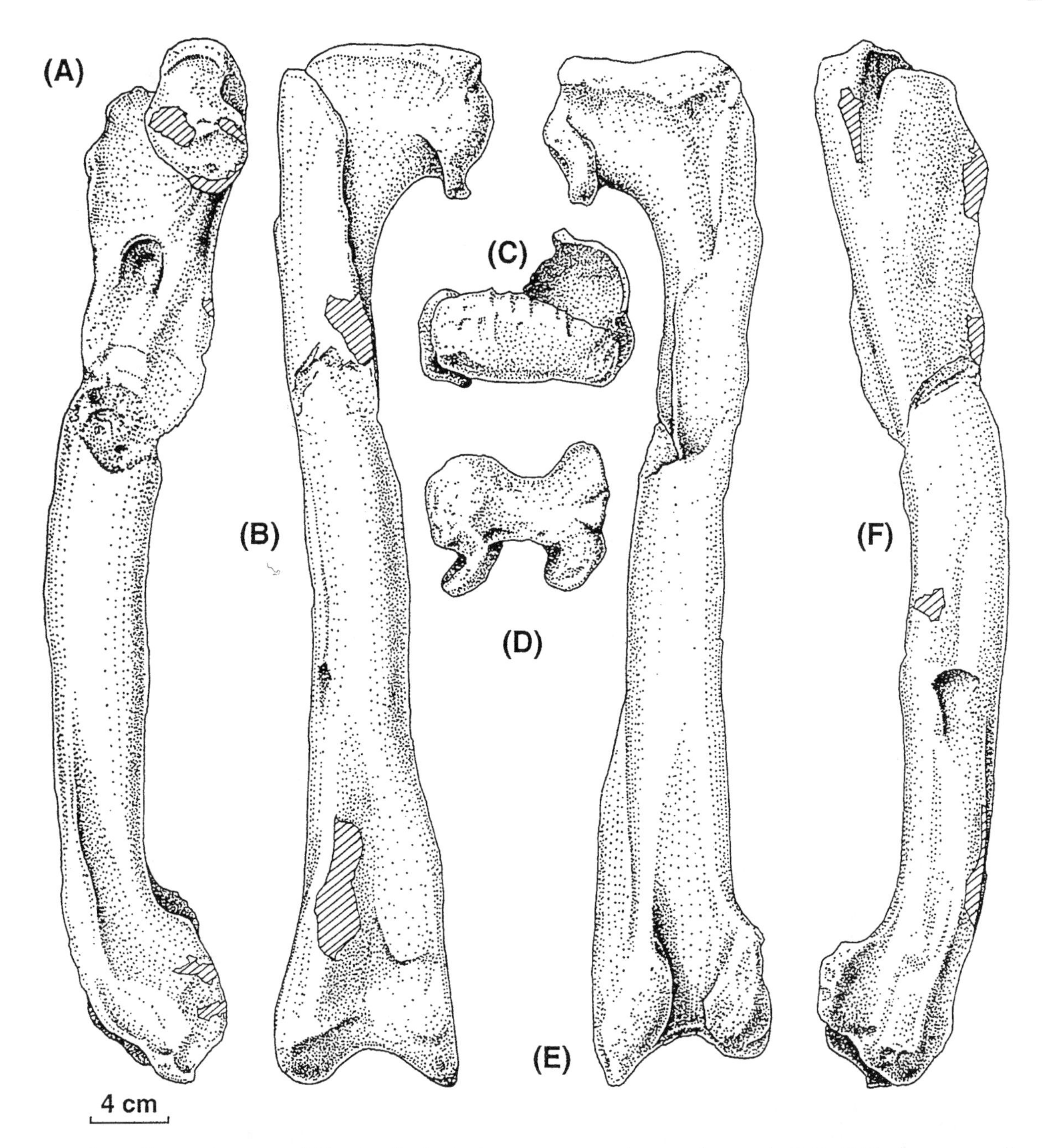

***Fukuiraptor kitadanensis*, FPMN 9712219, holotype right femur in A. medial, B. anterior, C. proximal, D. distal, E. posterior, and F. lateral views. (After Azuma and Currie 2000.)**

According to Azuma and Currie, *F. kitadaniensis* is definitely a carnosaur, but possesses various coelurosaurian characters. The dental shelves of the maxilla and dentary are narrow and the interdental plates have fused to each other and to the jaw margin, these characteristics having been previously reported only in dromaeosaurids. Based on these features and on the unguals, the specimen was originally referred by Azuma and Currie (1995) to the Dromaeosauridae (see *D:TE* and *S1*).

In all other characters, Azuma and Currie (2000) noted, *Fukuiraptor* more closely resembles advanced carnosaurs (humerus long and slender; ulna shorter than humerus, stout, powerfully built; manual phalanx II-2 relatively shorter than in *Deinonychus*, two preserved manual unguals (I-2, II-3) almost as well developed, both powerfully built and strongly curved, most reminiscent of carnosaurs like *Allosaurus*; fragment of ilium with an anteroposteriorly short pubic, apparently much larger than ischial peduncle; femur that of a carnosaur, comparing well with advanced allosauroids, *e.g., Allosaurus* and *Sinraptor*, femoral head perpendicular to shaft, deep cleft separating aliform lesser trochanter, fourth trochanter well developed, distal end with broad, shallow extensor groove; tibia apparently at least as long as femur.

Phylogenetic analysis by Azuma and Currie — including 32 cranial and dental characters and 78 postcranial characters, 59 of which could be coded for *Fukuiraptor*— indicated that *Fukuiraptor* is a basal allosauroid.

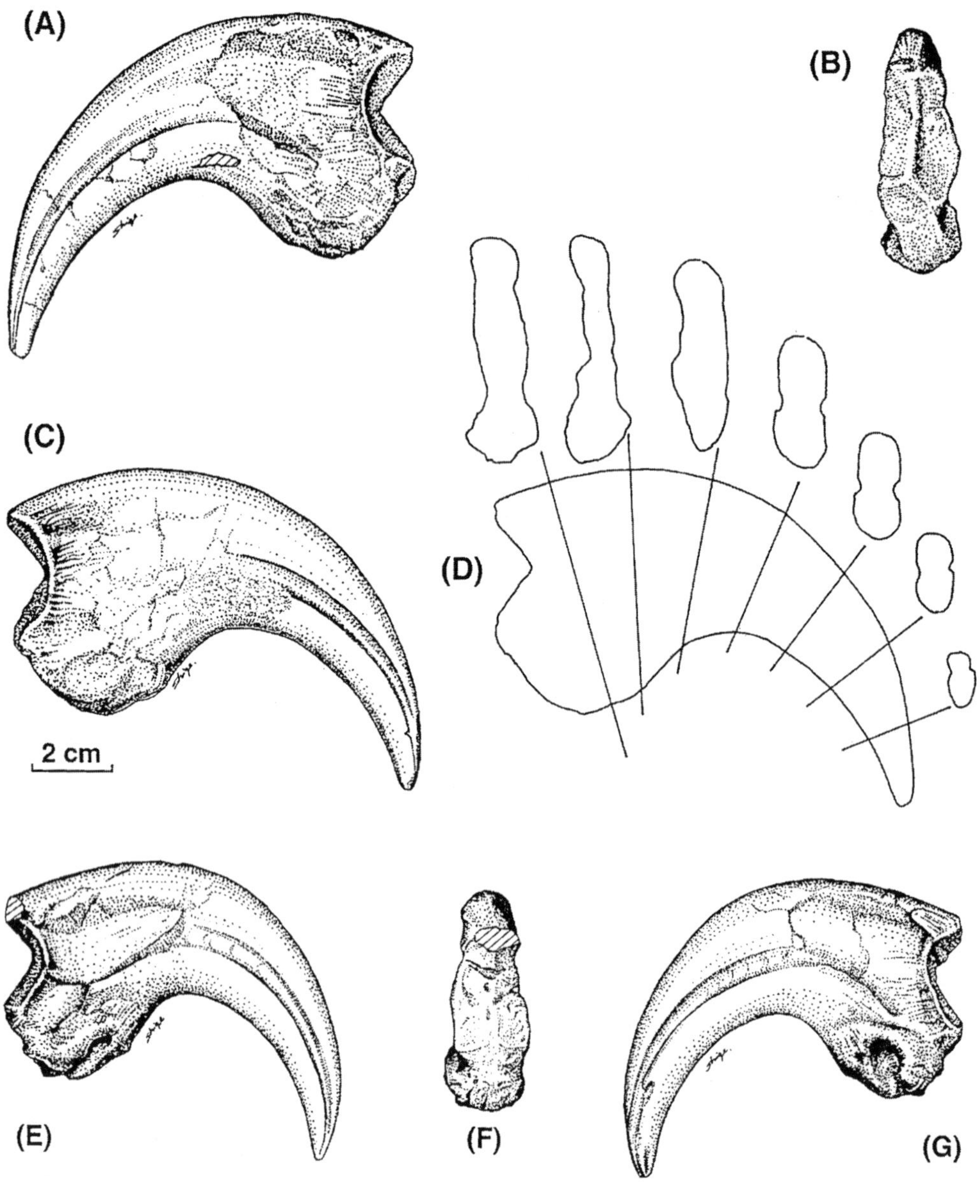

Fukuiraptor kitadanensis, holotype manual unguals, FPMN 9712211 (right I-2) in A. medial, B. posterior, and C. lateral views, and D. cross sections, FPMN 9712213 (left II-3) in E. lateral and G. medial views. (After Azuma and Currie 2000.)

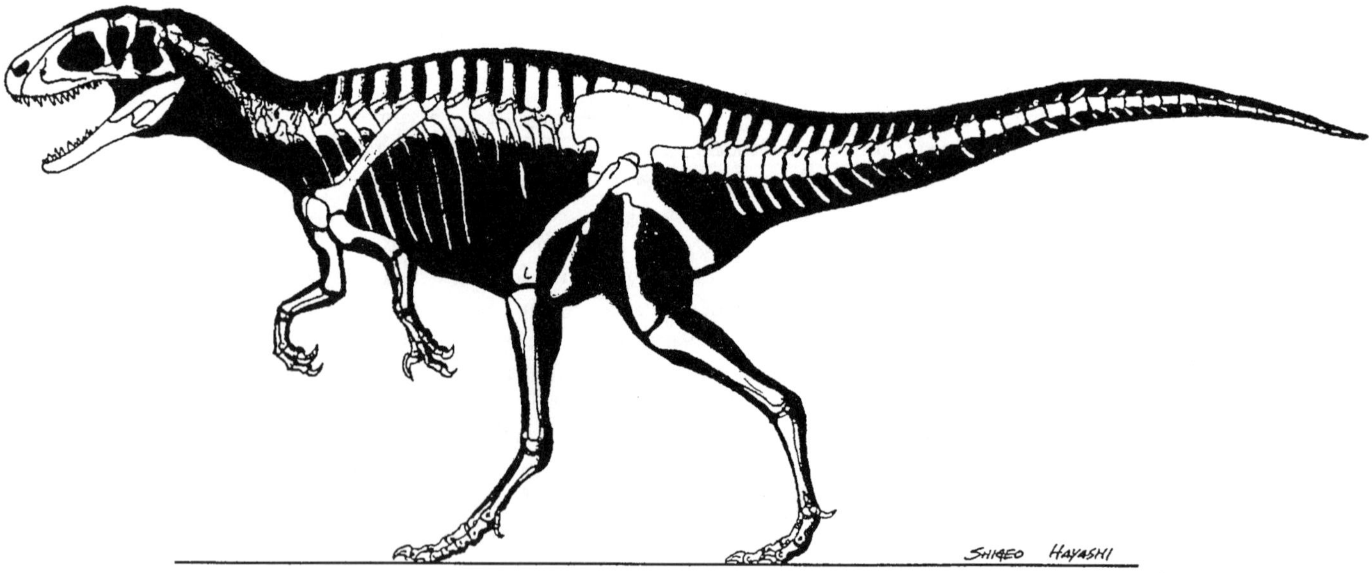

Fukuiraptor kitadanensis, reconstructed skeleton based upon holotype and referred specimens. Illustration by Shigeo Hayashi. (After Azuma and Currie 2000.)

G

Note: A fossil tooth from the Kitadani Formation of Katsuyama City, Japan, unofficially referred to as "Kitadanisaurus" in several Japanese publications, originally considered by Dong, Hasegawa and Azuma (1990, in the book *The Age of Dinosaurs in Japan and China*) to be a dromaeosaurid (see also Lambert 1990, *The Dinosaur Data Book*), may be *Fukuiraptor*. Possibly referable to that genus is a large, indeterminate large allosaurid informally named "Katsuyamasaurus," known only from an ulna and caudal vertebra. Also, an indeterminate large theropod from the Early Cretaceous of Shiramine Village, Japan, based on two teeth and informally referred to as "Kagasaurus," might also be referable to *Fukuiraptor* (G. Olshevsky, personal communication 2000).

Key references: Azuma and Currie (1995, 2000); Dong, Hasegawa and Azuma (1990); Lambert (1990).

†GALLIMIMUS

Saurischia: Theropoda: Neotheropoda: Tetanurae: Avetheropoda: Coelurosauria: Manuraptoriformes: Arctometatarsalia: Ornithomimosauria: Ornithomimidae.

Erratum: In *D:TE* it was incorrectly stated that the genus *Gallimimus* is also known from Naran Bulak, which is actually a locality of Paleocene–Eocene age.

†GASTONIA

Comments: This highly ornate, extensively-armored polacanthine ankylosaur — known from abundant skeletal remains collected from the Lower Cretaceous Yellow Cat Member of the Cedar Mountain Formation of Utah (see *S1*)— has now been made

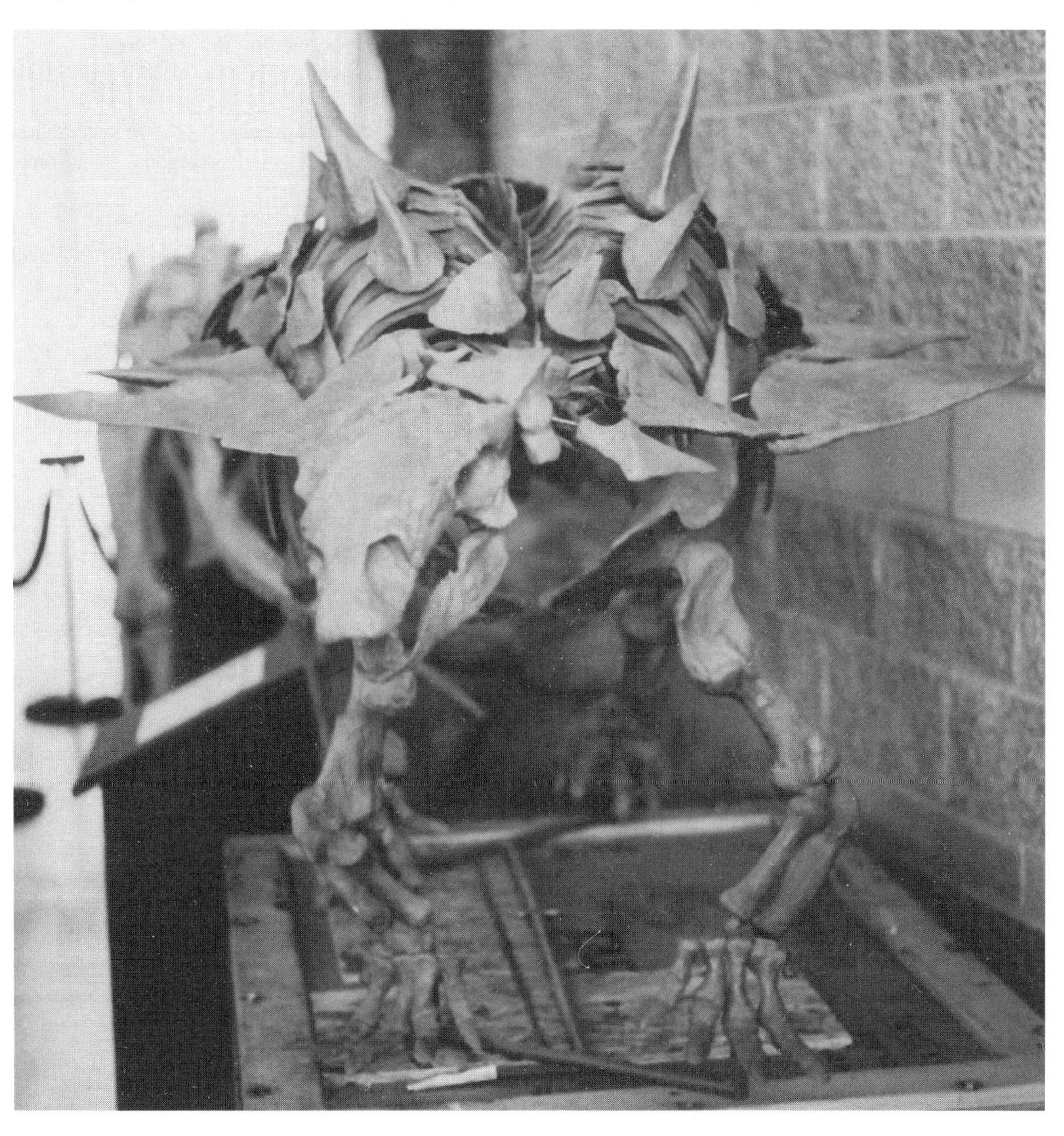

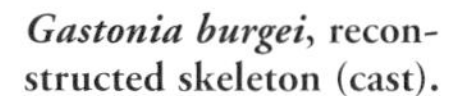

***Gastonia burgei*, reconstructed skeleton (cast).**

Photograph by Allen A. Debus.

accessible to the public via a free-standing skeletal mount.

As related in an abstract by Burge, Gaston, Kirkland and Carpenter (1998), every element belonging to this dinosaur, except the lower jaw, is well represented. Although some of these remains are preserved in excellent condition (including the skull), most of the bones were badly crushed. Therefore, in preparing this mount, a good number of elements (*e.g.,* vertebrae and the sacrum) required sculpting; others (*e.g.,* some spines and plates) were inflated from hollow casts of crushed elements. Armor arrangement was patterned after that of related ankylosaurs.

Burge *et al.* stated that "The robust skull, cervical rings, lateral and dorsal spines, sacral shield and lateral caudal plates make this the most impressive ankylosaur ever mounted."

The mounted skeleton of *Gastonia* made its debut appearance at the Fifty-Eighth Annual Meeting of the Society of Vertebrate Paleontology, held from September 30 to October 3, 1998.

Key reference: Burge, Gaston, Kirkland and Carpenter (1998).

GIRAFFATITAN Olshevsky 1991—(See *Brachiosaurus.*)

Name derivation: "Giraffe" + "Titan [one of a group of Greek divinities]."

Type species: *G. brancai* (Janensch 1914).

GLYPTODONTOPELTA Ford 2000 [*nomen dubium*]—(=?*Stegopelta*)

Ornithischia: Genasauria: Thyreophora: Thyreophoridea: Eurypoda: Ankylosauria: Ankylosauridae: ?Shamosaurinae.

Name derivation: "Glyptodon [armadillo-like mammal of the Pleistocene period]" + Greek *pelta* = "shield."

Type species: *G. mimus* Ford 2000 [*nomen dubium*].

Other species: [None.]

Occurrence: Kirtland Shale, New Mexico, United States.

Age: Late Cretaceous (Maastrichtian).

Known material: Numerous armor scutes.

Holotype: USNM 8610, section of coossified pelvic osteroderms.

Diagnosis of genus (as for type species): Scutes large, asymmetrical, irregularly hexogonal, pentagonal,

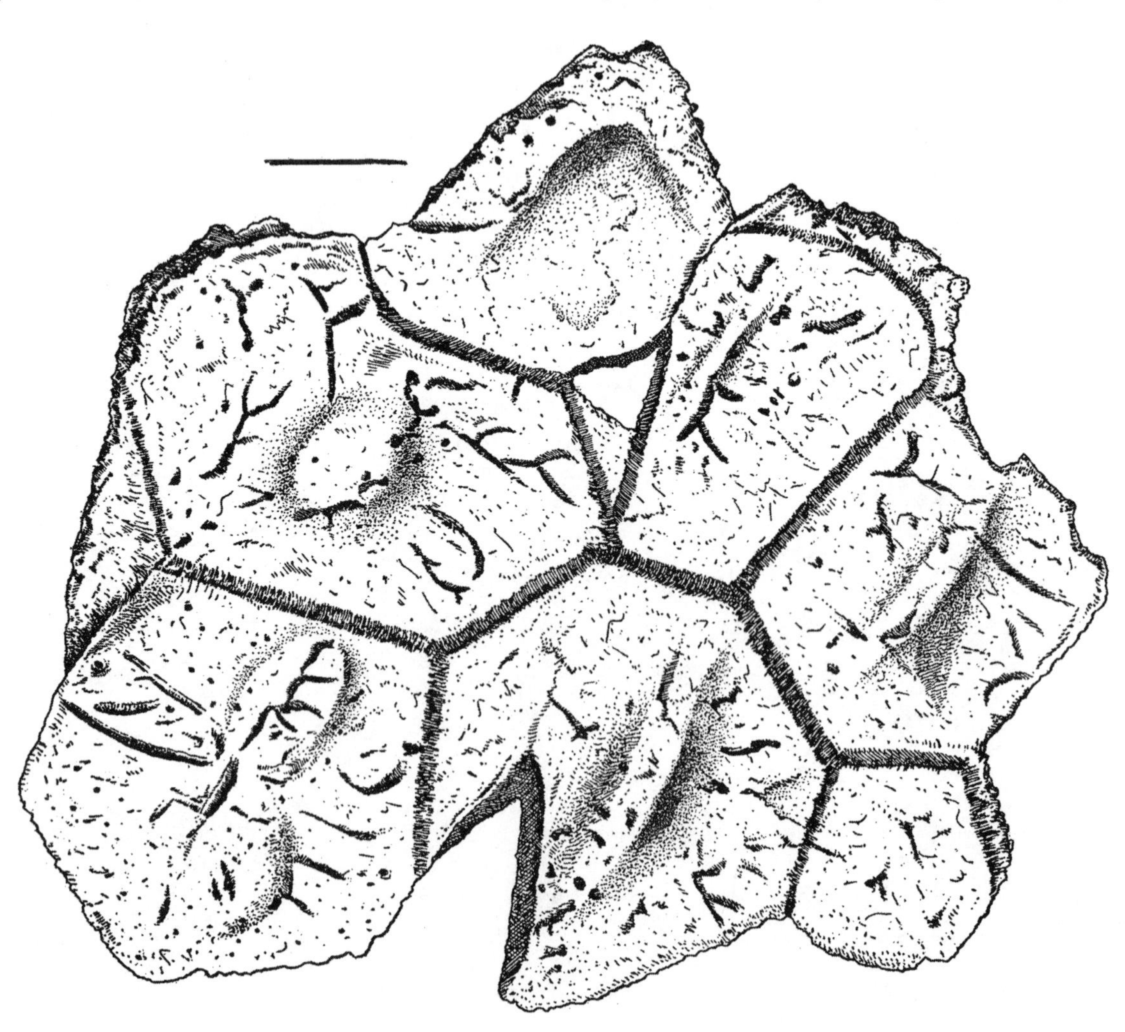

***Glyptodontopelta mimus*, holotype joined osteoderms. Scale = 4 cm. (After Gilmore 1919.)**

or quadrilateral, with low or flat peaks abutting each other, set in mosaic pattern, scutes forming solid shield over ilia; scutes flat or with short ridge, abutting each other so as to leave no room for small ossicles or rosettes around scutes (Ford 2000).

Comments: The genus *Glyptodontopelta* was established by Ford (2000) on a section of pelvic scutes (USNM 8610, first figured and described by Gilmore 1919) collected by the Reeside party from the Naashoibito Member of the Kirtland Shale (formerly the Ojo Alamo Formation) at Barrel Springs Arroyo, southwest of Ojo Alamo, San Juan County, New Mexico.

Gilmore very briefly described these scutes as comparatively thick, flattened, and of various sizes and shapes. Some of them have low dorsal ridges or asymmetrically placed carinae. Gilmore noted that several of these scutes have smooth rounded edges, indicating "that they existed as independent plates," although most of them possesses sutural edges, showing "that they were closely united into a carapace-like covering." Gilmore surmised that these scutes collectively formed a pelvic shield such as that found in *Polacanthus* and *Stegopelta*.

Gilmore did not name this material, but referred it to the abandoned family Scelidosauridae.

Specimens referred by Ford to his new type species *Glyptopelta mimus* include USNM 8611, from the same locality; and, from other Naashoibito Member localities, NMMNH P-14266, P-27849, P-27405, P-22654, and P14266, comprising 38 scute fragments in various states of preservation, NMMNH P-27849, a fragmentary "pup-tent" scute and a fragment of a pelvic scute, NMMNH P-27405, two fragmentary "pup-tent" scutes and several fragmentary scutes, and NMMNH P-22654, including two fragmentary pelvic scutes.

Ford named *Glyptodontopelta* based upon its pelvic shield's resemblance to the shields of glyptodonts. The author noted that the fused osteroderms in this new genus are very similar to that found in the latest Albian- or earliest Cenomanian-age *Stegopelta* (see entry) and also the late Campanian SDNHM 33909 (a specimen to be named and described as a new genus and species by Ford and Kirkland, in press), this suggesting to him that these forms could be congeneric, despite their chronological separation. Ford observed that "the pelvic osteoderms of SDNHM 33909 lack the ridge and flatness of *Glyptodontopelta*, and some osteoderms have a more pointed center area," although this different pattern could be attributed to individual variation.

In erecting this new genus and species only upon armor scutes, Ford cited Coombs and Demere (1996), who opined that 1. "if individual osteoderms of a distinctive type are known from a single taxon, then they are diagnostic," and 2. "if a specimen shows osteoderms *in situ*, then the arrangement may be diagnostic."

Ankylosaur specialist Kenneth Carpenter (personal communication, 2000), however, stated that, although the armor of these dinosaurs is more diagnostic than Coombs originally believed in his papers published before 1990, it is not as diagnostic as Ford believes. If that were the case, almost every specimen of ankylosaurian armor could be interpreted as a new taxon. Carpenter pointed out that Ford's diagnosis for *Glyptodontopelta* is the same as that for Ford's new subfamily Stegopeltinae (see "Systematics" chapter), the author, therefore, effectively admitting that *Glyptodontopelta* is a synonym of *Stegopelta*. Also, Ford states, but offers no evidence, that the specimens listed in his "hypodigm" (actually paratypes) belong to his new genus, although some of these specimens could just as readily belong to *Nodocephalosaurus* or some other ankylosaur. Carpenter, in a future publication about armored dinosaurs, will designate *E. australis* to be a *nomen dubium*.

Key references: Coombs and Demere (1996); Ford (2000); Gilmore (1919).

GONDWANATITAN Kellner and Azevado 1999 — (See *Aeolosaurus*.)

Name derivation: "Gondwana [land mass once uniting all southern continents and India]" + "Titan [one of a group of Greek divinities]."

Type species: Kellner and Avezedo 1999.

GONGXIANOSAURUS He, Wang, Liu, Zhou, Liu, Cai and Dai 1998

Saurischia: Sauropodomorpha: Sauropoda *incertae sedis*.

Name derivation: "Gongxian [County]" + Greek *sauros* = "lizard."

Type species: *G. shibeiensis* He, Wang, Liu, Zhou, Liu, Cai and Dai 1998.

Other species: [None.]

Occurrence: Ziliujing Formation, Sichuan, People's Republic of China.

Age: Early Jurassic.

Known material: [To date] three postcranial specimens, adult, ?juvenile.

Holotype: Incomplete partially articulated postcranial skeleton, adult.

Diagnosis of genus (as for type species): Early, primitive sauropod of moderate to large size (estimated length of about 14 meters or 48 feet); premaxilla on posterodorsal region with well-developed

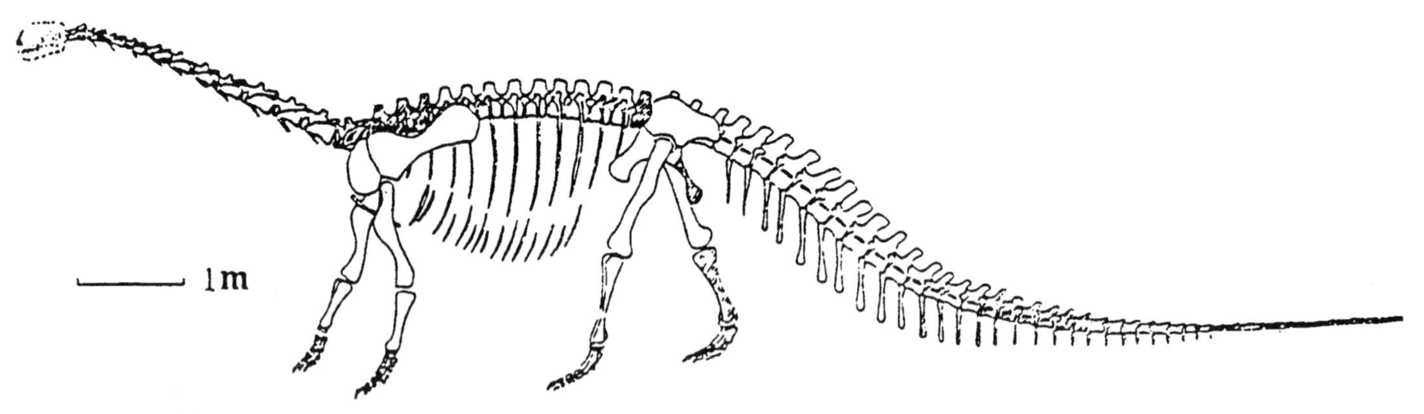

***Gongxianosaurus shibeiensis*, reconstructed skeleton. (After He, Wang, Liu, Zhou, Liu, Cai and Dai 1998.)**

posterior process in addition to anterodorsal process; three to four premaxillary teeth, mandibular dentition relatively long with more than 12 teeth (estimated at 14–16) in sequence; dental morphology typically spoon-shaped, lacking either anterior or posterior serrations; lingual and labial surfaces with relatively thick longitudinal striations, lacking attenuating folds, lingual surface lacking medial ridge; cervical vertebrae amphiplatyan, morphology simple, with relatively small neural spines that are rectangular in lateral view; cervical ribs particularly short, only two-thirds length of centrum; dorsal vertebrae amphiplatyan to very weakly amphicoelous, medially constricted, maintaining simple rectangula spine that is longer than high; diapophyses almost horizontal, with two nearly parallel transversely extended ventral lamina or spines (resembling ventral diapophyseal plates); three fused ?sacral vertebrae, very slightly constricted with relatively narrow neural spines; anterior sacral vertebra amphiplatyan to amphicoelous, internal structure of centra compact and lacking conspicuous honeycombed fabric, pleurocoels absent and supporting laminar structure is undeveloped; anterior caudal vertebrae amphiplatyan, with broad, plate-shaped neural spines that are distinctly extended, broad anteroposteriorly, thin dorsoventrally; caudal vertebrae without bifurcated hemal arch; scapula thin, plate-shaped, with broad distal end and strongly constricted medial shaft, its anterior margin curving in arc posteriorly; scapulocoracoid contact particularly straight, no coracoid foramen; sternal outline an elongated oval; humerus of typically sauropod morphology, with relatively well-developed deltopectoral crest; ulna 60 percent length of humerus, forelimb 70–75 percent length of hindlimb; pubic peduncle especially short, almost equal in length to ischiac peduncle; iliac crest high, conspicuously narrow posteriorly; pubic shaft broad, anterior margins of pubes completely fused; femur relatively short but robust, head not conspicuously set off from shaft, lesser trochanter absent; hindlimb articulated with astragalus, calcaneum, and two distal metatarsals; pes formula 2.3.4.5.1, ungual phalanges on digits I–IV; digit and ungual morphology typically sauropodomorph (He, Wang, Liu, Zhou, Liu, Cai and Dai 1998).

Comments: The genus *Gongxianosaurus* was founded upon a well-preserved incomplete skeleton belonging to a rather primitive sauropod, discovered in 1979 by Fengyun Zhou and associates from the 202nd Corps of the Sichuan Geological Survey. The original specimen was found in dark purple silty mudstones of the middle to upper portion of the Dongyumiao Member of the Lower Jurassic Ziliunging Formation (this member directly overlying the Zhenzhuchong Member of the lower Ziliujing Formation, which correlates to the Lower Lufeng Formation), near Hongshacun Hamlet, Shibeixiang Village, in Gongxian County, south Sichuan, China. Recovered that same year from this site were additional specimens, two of them partially articulated and presumably belonging to the new type species, *Gongxianosaurus shibeiensis* (although two species or even more than one higher-level taxa are may be represented here), but also other taxa including theropods. Of the sauropod skeletons, two were of about the same size and were designated the holotype and paratype. As these specimens were not entirely articulated, it was difficult to determine if some of the elements belonged to one or the other. A third and smaller specimen could represent a juvenile individual. Among the elements represented by these specimens are two right premaxillae, almost complete right and left mandibles, numerous isolated teeth, incomplete and disarticulated cervical, dorsal, sacral, and caudal vertebrae, a complete pectoral girdle, a forelimb, pelvic girdle, hindlimb, hind foot, and isolated sternal elements, ribs, and gastralia (He, Wang, Liu, Zhou, Liu, Cai and Dai 1998).

Although more material relating to *G. shibeiensis* has yet to be collected, prepared, and studied, preliminary analyses by *et al.* indicate that this site offers a new Early Jurassic faunal complex — with *Gongxianosaurus* thus far being the only identified representative — apparently postdating the Early Jurassic

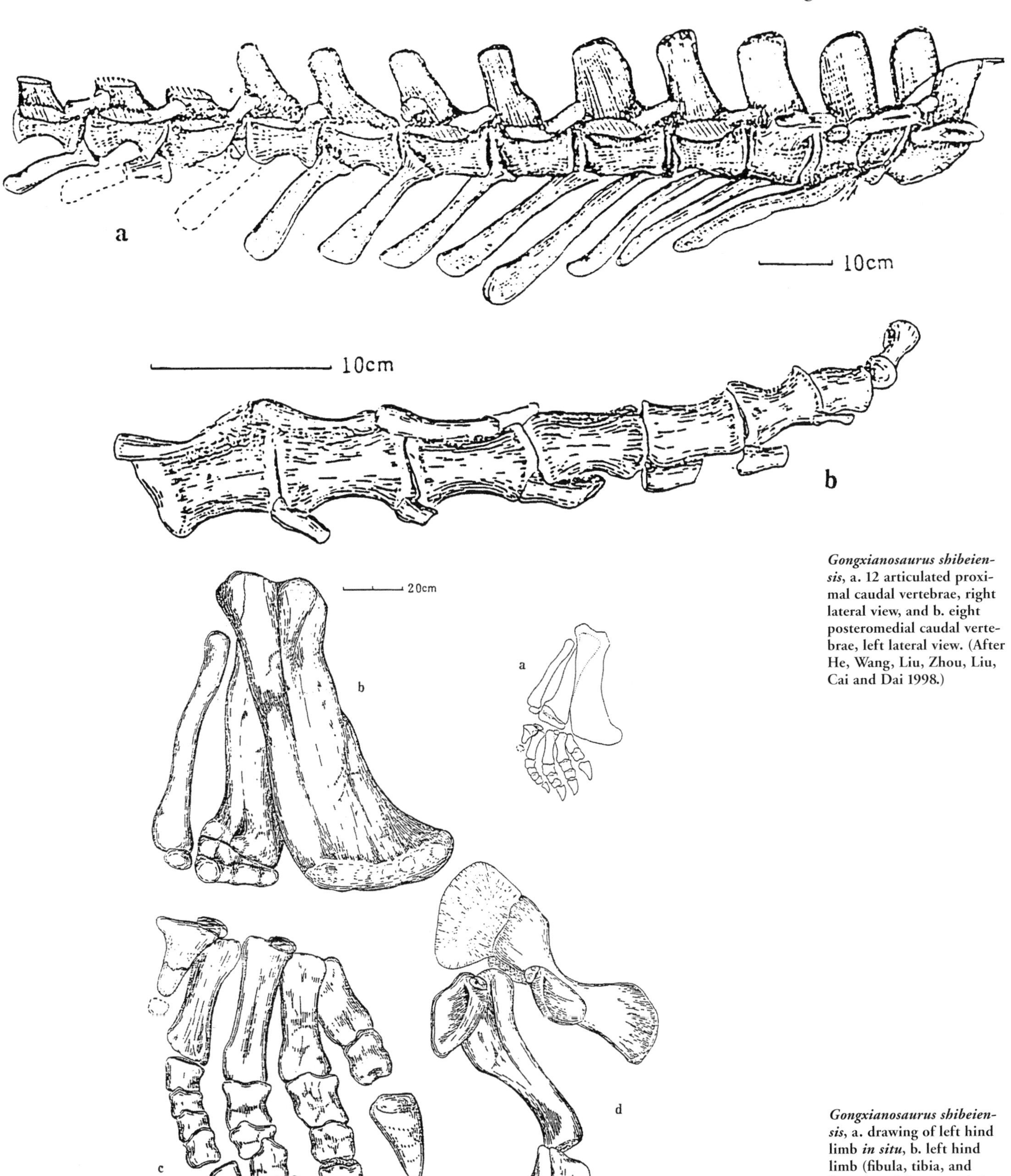

Gongxianosaurus shibeiensis, a. 12 articulated proximal caudal vertebrae, right lateral view, and b. eight posteromedial caudal vertebrae, left lateral view. (After He, Wang, Liu, Zhou, Liu, Cai and Dai 1998.)

Gongxianosaurus shibeiensis, a. drawing of left hind limb *in situ*, b. left hind limb (fibula, tibia, and femur), c. left pes, and d. left shoulder girdle in general articulation with sternal plates and forelimb. (After He, Wang, Liu, Zhou, Liu, Cai and Dai 1998.)

Lufengosaurus Fauna of Lufeng, Yunnan Province, and predating the *Shunosaurus* Fauna of Dashanpu, Sichuan Province.

As the He *et al.* noted, *G. shibeiensis* exhibits numerous plesiomorphic characters seen in prosauropods such as *Lufengosaurus* and *Jinshanosaurus*, these including the following: The well-developed posterior process of premaxilla; the dense internal structure of the anterior sacral vertebrae; the absence of pneumatocoels; the absence of opisthocoelous centra; the unbifid neural spines; gastralia; the low sacral centra count; the unbifurcated caudal hemal arches; an ankle joint preserving the calcaneum; an astragalus articulated to two distal tarsals; and the pedal formula. Characters shared with the prosauropods of Yunnan Province include the following: The extremely short cervical ribs; the prosauropod-like morphology of the scapula; and the almost rectangular and vertical diapophyses.

However, the authors referred this new taxon to the Sauropoda based on the following: As in sauropods, the teeth are extremely large and spoon-shaped. The forelimb is very long, unlike the condition in prosauropods. Although the pes resembles the prosauropod foot, the phalanges are thick and short, lacking proximolateral ligament depressions, and the unguals lack dorsolateral longitudinal grooves that more closely suggest the condition in sauropods. Finally the size is greater than that of any known prosauropod.

The authors concluded that *Gongxianosaurus* is an early and primitive sauropod that shares a number of symplesiomorphies with the Prosauropoda, especially taxa from Southwest China, this establishing the new taxon's genesis in that area. Consequently, "further in-depth research on this genus will have extreme significance toward understanding the origin of the Sauropoda."

He *et al.* also commented that *Gongxianosaurus* differs in numerous significant ways from such Middle to Late Jurassic sauropods from Sichuan, such as *Mamenchisaurus*, *Omeisaurus*, and *Shunosaurus*, these differences to be discussed in detail at some later date. Furthermore, although the fragmentary primitive sauropod taxa as the Indian *Barapasaurus* and *Kotasaurus*, the African *Vulcanodon*, and Australian *Rhoetosaurus* all display many plesiomorphic characters, these taxa are more typical of Sauropoda and quite distinct from *Gongxianosaurus*, which exhibits more symplesiomorphic characters with prosauropods. Within an evolutionary context, therefore, *Gongxianosaurus* was "recognized as an ancestral/descendant taxon."

Key reference: He, Wang, Liu, Zhou, Liu, Cai and Dai (1998).

GRACILICERATOPS Sereno 2000

Ornithischia: Genusauria: Cerapoda: Marginocephalia: Ceratopsia: Neoceratopsia.

Name derivation: Latin *gracilis* = "slender" + Greek *keratos* = "horn" + Greek *ops* = "face."

Type species: *G. mongoliensis* Sereno 2000.

Other species: [None.]

Occurrence: Mongolia.

Age: Late Cretaceous.

Known material/holotype: Partial skeleton.

Diagnosis of genus (as for type species): [Not yet available to the present writer.]

Comments: The type species *Graciliceratops mongoliensis* was erected by Sereno (2000), based upon a partial skeleton from Mongolia previously referred to *Microceratops gobiensis*. Reportedly, Sereno redescribed the type specimen of the latter as an indeterminate juvenile, thereby making *Microceratops* a *nomen dubium*. (To date of this writing, the present author has not seen Sereno's article, published in a volume edited by Benton, Shishkin, Unwin and Kurochkin.)

Key reference: Sereno (2000).

†GUAIBASAURUS

Saurischia: Guaibassauridae.

Name derivation: "[Rio] Guaiba [Hydrographic Basin]" + Greek *sauros* = "lizard."

Type species: *G. candelariensis* Bonaparte, Ferigolo and Ribeiro 1999.

Occurrence: Caturrita Formation, Rio Grand du Sol, Brazil.

Age: Late Triassic (?Carnian).

Known material: Two partial skeletons.

Holotype: MCN-PV 2355, partial skeleton comprising five incomplete centra of dorsal vertebrae, three incomplete dorsal neural arches, two incomplete sacral centra, 10 almost complete articulated caudal vertebrae with hemal arches, five incomplete dorsal ribs, incomplete left scapula, coracoids, fragment of left ilium, almost complete right ilium, almost complete pubes, incomplete ischia, incomplete femora, incomplete tibiae and fibulae, feet including fragmentary tarsals (except for complete right calcaneum).

Diagnosis of genus (as for type species): Distinguished from all other known dinosaurs by a combination of primitive and derived characters, including the following: midanterior neural arches of dorsal vertebrae having large pneumatic cavities; parapophyseal-prezygapophyseal lamina with fossa below it; infradiapophyseal lamina undivided; hyposphenum-hypantrum articulation; sacral centrum 1 having double the volume of sacral centrum 2; scapula slender with distal expansion; acetabular area of ilium closed or nearly closed, with only incipient reduction of its medial wall; ilium having dorsoventrally deep posterior projection, extremely large crista supracetabularis, and well-developed fossa brevis; ischium elongated, slender, triangular in cross section, possessing long symphysis, with dorsoventrally flat distal expansion; laminar and elongated pubes with pronounced proximolateral trochanter; femur with small anterior trochanter; lateroposterior ridge above level of anterior trochanter, absence of trochanteric shelf; tibia almost as long as femur, distal end subquadrangular in cross section, posteroventral projection well defined; astragalus large, with well-defined dorsal process; small, narrow calcaneum having pronounced ventromedial process; two distal tarsals; reduced metatarsal V, lacking phalanges; metatarsal I unreduced (Bonaparte, Ferigolo and Ribeiro 1999).

Comments: This very primitive, early Late Triassic dinosaur — collected by a field team of the Museu de Ciências Naturais da Fundacão Zoobotânica do Rio Grande do Sul, from the Caturrita Formation (see Andreis, Bossi and Montando 1980) of Rio Grand du Sol in southern Brazil — was first announced in an abstract by Bonaparte and Ferigolo (1998) as the new type species *Guaibasaurus candelariai* (originally misspelled, due to a printing error, as *G. dandelariai*). In that preliminary report (see *S1* for additional information), however, the authors did not diagnose or formally describe this taxon; nor were figures published of the holotype material.

Subsequently, Bonaparte, Ferigolo and Ribeiro (1999), in a more detailed preliminary report, formally described *Guaibasaurus candelariensis* (the spelling corrected), identifying the holotype specimen as MCN-PV 2355. An incomplete left tibia and fibula, articulated

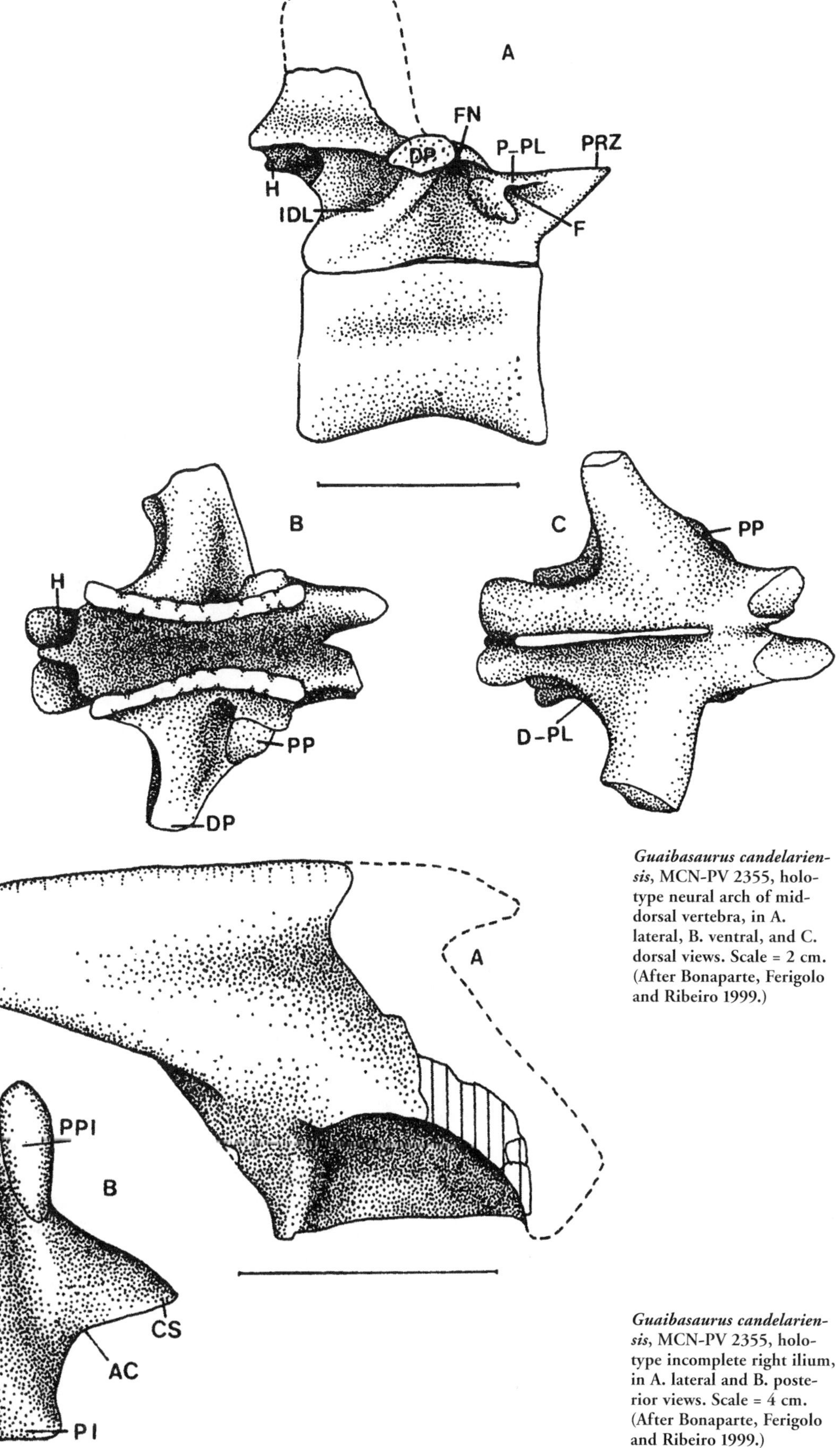

***Guaibasaurus candelariensis*, MCN-PV 2355, holotype neural arch of mid-dorsal vertebra, in A. lateral, B. ventral, and C. dorsal views. Scale = 2 cm. (After Bonaparte, Ferigolo and Ribeiro 1999.)**

***Guaibasaurus candelariensis*, MCN-PV 2355, holotype incomplete right ilium, in A. lateral and B. posterior views. Scale = 4 cm. (After Bonaparte, Ferigolo and Ribeiro 1999.)**

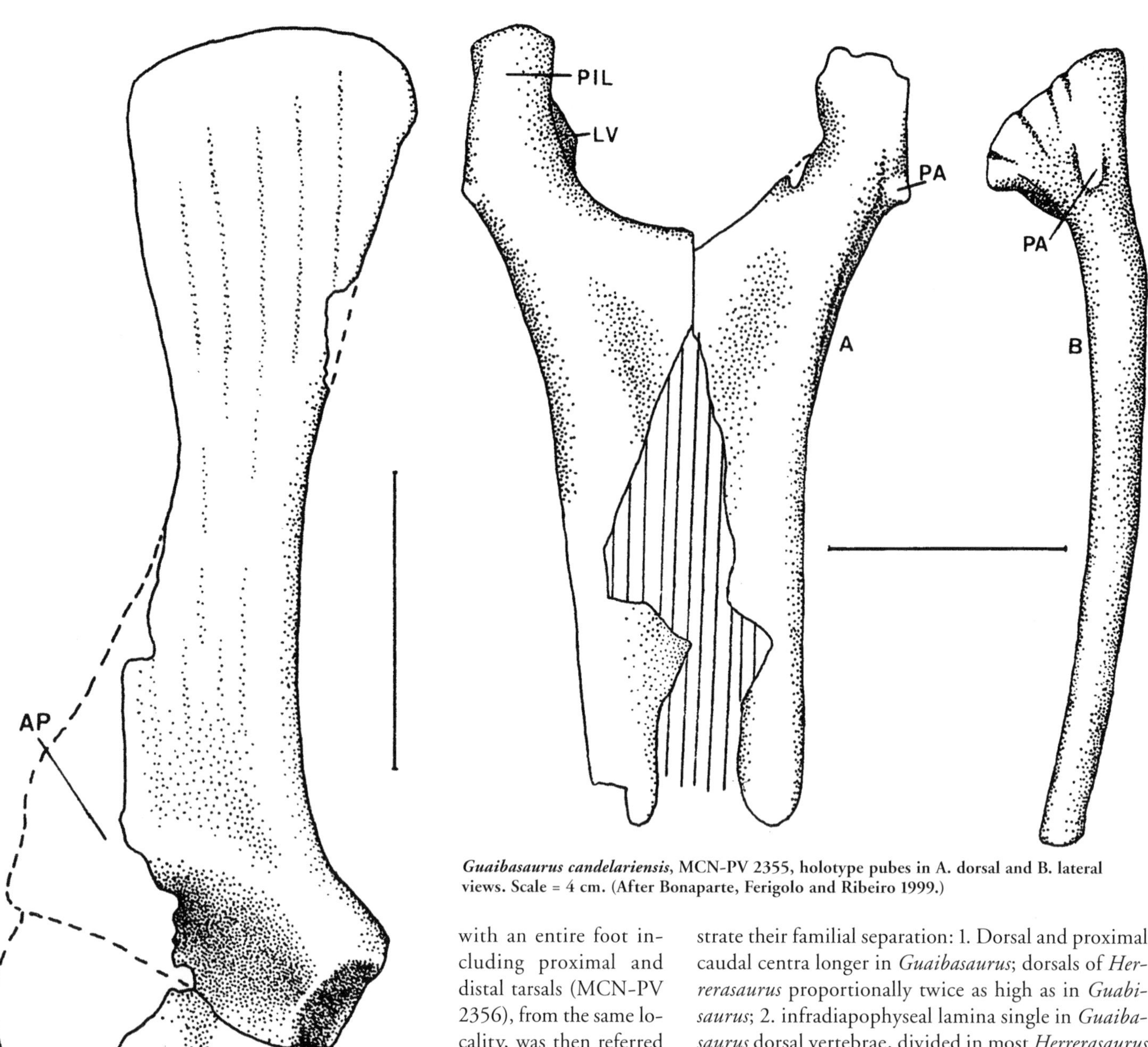

Guaibasaurus candelariensis, MCN-PV 2355, holotype pubes in A. dorsal and B. lateral views. Scale = 4 cm. (After Bonaparte, Ferigolo and Ribeiro 1999.)

Guaibasaurus candelariensis, MCN-PV 2355, holotype incomplete left scapula and coracoid, lateral view. Scale = 4 cm. (After Bonaparte, Ferigolo and Ribeiro 1999.)

with an entire foot including proximal and distal tarsals (MCN-PV 2356), from the same locality, was then referred by Bonaparte *et al.* to the type species.

As observed by Bonaparte *et al.*, the diagnostic features of *G. candelariensis* indicate that this species is a very primitive saurischian, more than even herrerasaurids (*e.g.*, *Herrerasaurus* and *Staurikosaurus*), which the authors regarded as carnivorous "basal saurischians" rather than primitive theropods (see "Systematics" chapter), as well as more primitive than prosauropods.

Bonaparte *et al.* enumerated the following morphological differences between *Guaibasaurus* and the Herrerasauridae which, the authors noted, demonstrate their familial separation: 1. Dorsal and proximal caudal centra longer in *Guaibasaurus*; dorsals of *Herrerasaurus* proportionally twice as high as in *Guabisaurus*; 2. infradiapophyseal lamina single in *Guaibasaurus* dorsal vertebrae, divided in most *Herrerasaurus* dorsals; 3. scapula narrow above acromial area and from there laterally expanded in *Guiabasaurus*, dorsoventrally narrow in *Herrerasaurus*; 4. postacetabular process of ilium dorsoventrally higher in *Herrerasaurus*; 5. acetabulum in *Herrerasaurus* and *Staurikosaurus* with medial wall of ilium more open than in *Guaibasaurus*; 6. crista supracetabular is more developed in *Guaibasaurus*; 7. derived features of pubis with ventral torsion, "foot," and transversely narrow distal end lacking in *Guaibasaurus*; 8. pubis with process for attachment of ambiens more developed than in *Herrerasaurus*; 9. proximal dorsal torsion of ischia of *Herrerasaurus* forming a distinct angle, forming a gentle curvature in *Guaibasaurus* (and in *Staurikosaurus*); 10. shafts of ischia elongate, triangular in cross section, with distal expansion in *Guaibasaurus*,

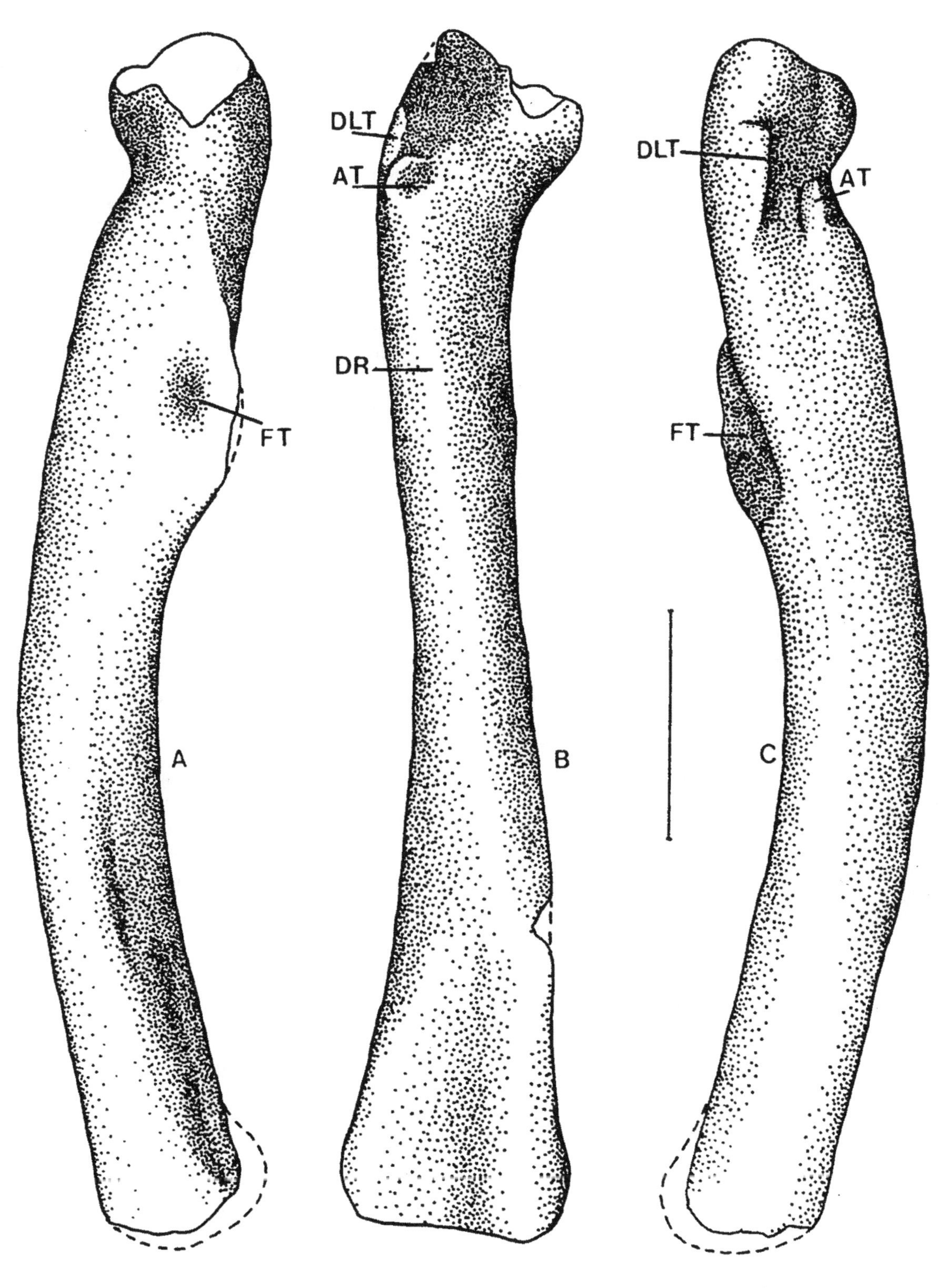

***Guaibasaurus candelariensis*, MCN-PV 2355, holotype incomplete right femur, in A. medial, B. dorsal, and C. lateral views. Scale = 4 cm. (After Bonaparte, Ferigolo and Ribeiro 1999.)**

shorter and without expansion in *Herrerasaurus*, shafts of ischia laminar in *Staurikosaurus* (see Colbert 1970); 11. femur of *Herrerasaurus* bearing well-defined trochanteric shelf, lacking in *Guaibasaurus*, femur in *Staurikosaurus* proportionally slender, with more defined anterior trochanter, slightly lower in position than in *Guaibasaurus*; 12. calcaneum of *Guaibasaurus* having larger medial projection than in *Herrerasaurus*; and 13. metatarsal I of *Guaibasaurus* not reduced as in *Herrerasaurus*, metatarsal V more reduced than in *Herrerasaurus*.

The following differences were cited by Bonaparte *et al.* indicating that *Guaibasaurus* and the primitive theropod *Coelophysis* do not belong to the same

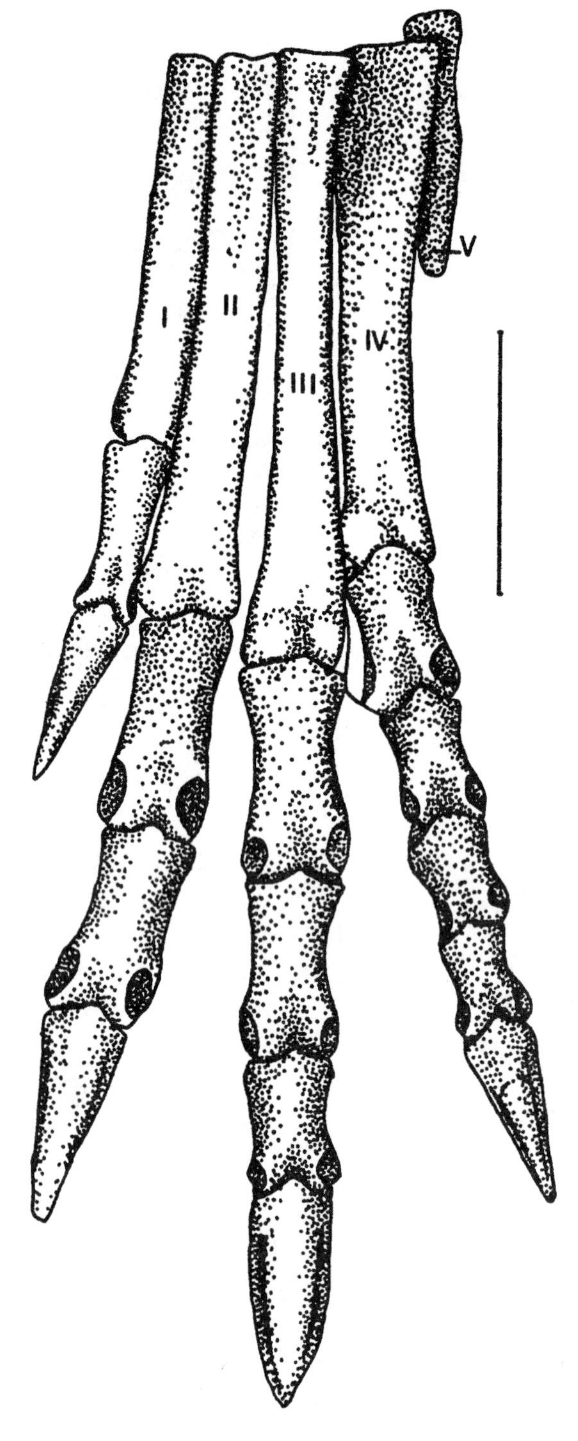

***Guaibasaurus candelariensis*, MCN-PV 2355, holotype left pes, dorsal view. Scale = 4 cm. (After Bonaparte, Ferigolo and Ribeiro 1999.)**

family: 1. Acetabulum closed or almost so in *Guaibasaurus*, entirely open in *Coelophysis*; 2. pubis of *Guaibasaurus* wide, that of *Coelophysis* transversely narrow; 3. astragalocalcaneum fused in *Coelophysis*, not so in *Guaibasaurus*; 4. proximal ends of metatarsals II and metatarsal IV separated from one another by III in *Guaibasaurus*, proximal superposition of II and IV over III in *Coelophysis*, and metatarsal I unreduced in *Guaibasaurus*, extremely reduced in *Coelophysis*.

Because of the above differences, Bonaparte *et al.* found it justifiable to refer *Guaibasaurus* to its own family, the Guaibisauridae (see "Systematics" chapter), diagnosed by its possession both of characters more primitive than in those of the Herrerasauridae and more derived than in the nondinosaurian ornithodiran *Marasuchus*. The authors regarded this family as including dinosaurs yet more primitive than herrerosaurids, until now the most primitive unequivocal dinosaurs known.

Comparing the Guaibsauridae with the Prosauropoda, Bonaparte *et al.* noted that these two groups differ in various significant features of the dorsal vertebrae; also, unlike the conditions in prosauropods, *Guaibasaurus* possesses a large crista supracetabularis, a femur with a more proximally positioned fourth trochanter, and a dorsal crest extending down from the anterior trochanter.

The following characters were listed by Bonaparte *et al.* supporting these authors' interpretation that *Guaibasaurus* may belong to a group ancestral to the Sauropodomorpha: 1. Ischium elongate, massive, with long symphysis and distal expansion; 2. pubis elongate, straight, with lateroproximal process for ambiens muscle (present in *Sellosaurus* ["*Efraasia*"] (see Galton 1986*a*) and *Riojasaurus* (Bonaparte 1972); 3. femur with moderately developed anterior trochanter; 4. dorsoventrally elongated fourth trochanter, with medial depression for muscle attachment; 5. femur straight in anterior aspect; 6. posterior side of distal section of tibia positioned more laterally than anterior side, resulting in oblique direction of internal side of tibia; 7. calcaneum with reduced ventromedial projection; 8. metatarsal I not reduced; and 9. metatarsal V reduced to half length of IV.

Given the very primitive nature of *Guaibasaurus*, Bonaparte *et al.* suggested that the diagnostic characters used to define the Saurischia —*e.g.*, medially open acetabulum — should be reconsidered. This portion of the acetabulum, almost closed in *Guaibasaurus* (as preserved), is incipiently to moderately open in *Herrerasaurus*, *Staurikosaurus*, and in melanorosaurid prosauropods. Unfortunately, only one incomplete ilium is available in the type material of *Guaibasaurus*; therefore, a comparison cannot be made to determine if the condition observed in this specimen is actually the case.

Bonaparte *et al.* concluded that, if *Guaibasaurus* and its family — with characters more primitive than those of herrerasaurids and more derived than in *Marasuchus*— are part of a group that is ancestral to Sauropodomorpha and also more primitive than the Herrerasauridae, then "its inferred position as the ancestral group of both Sauropodomorpha and Theropoda is a reasonable possibility." No reason was seen

by these authors to link the Guaibasauridae with the Ornithischia.

Although there was no evidence to prove that *Guaibasaurus* was a carnivorous animal, Bonaparte *et al.* suggested that this dinosaur probably was, noting that all of the most primitive known saurischians were meat-eaters. Additionally, the authors pointed out that various characters found in *Guaibasaurus* (*e.g.*, the high position of the fourth trochanter, the dorsal crest of the femur running distally from the anterior trochanter, the subquadrangular cross section of the distal end of the tibia) correspond well to the known Triassic theropods.

Key references: Bonaparte and Ferigolo (1998); Bonaparte, Ferigolo and Ribeiro (1999); Colbert (1970); Galton (1986*a*).

†**HEPTASTEORNIS**—(See *Bradycneme.*)

†**HETERODONTOSAURUS**

Comments: Until now, the Heterodontosauridae—a family of small and primitive ornithopods possessing derived cranial and dental features, and exemplified by the tusked genus *Heterodontosaurus* (see *D:TE*)—has only been known in Africa and North America. Now this group is also represented in South America.

Baez and Marsicano (1998) reported that a weathered, left posterior maxillary fragment, belonging to "a relatively advanced heterodontosaurian," had been recently found in an apparently Upper Triassic (probably Norian) outcrop of the El Tranquilo Group, in Santa Cruz Province, in southern Patagonia, Argentina. The specimen includes three preserved maxillary teeth, and also an isolated caniniform possessing mesial and lateral serrations.

As briefly described by Baez and Marsicano, the maxillary teeth "bear flat wear facets, and are columnar

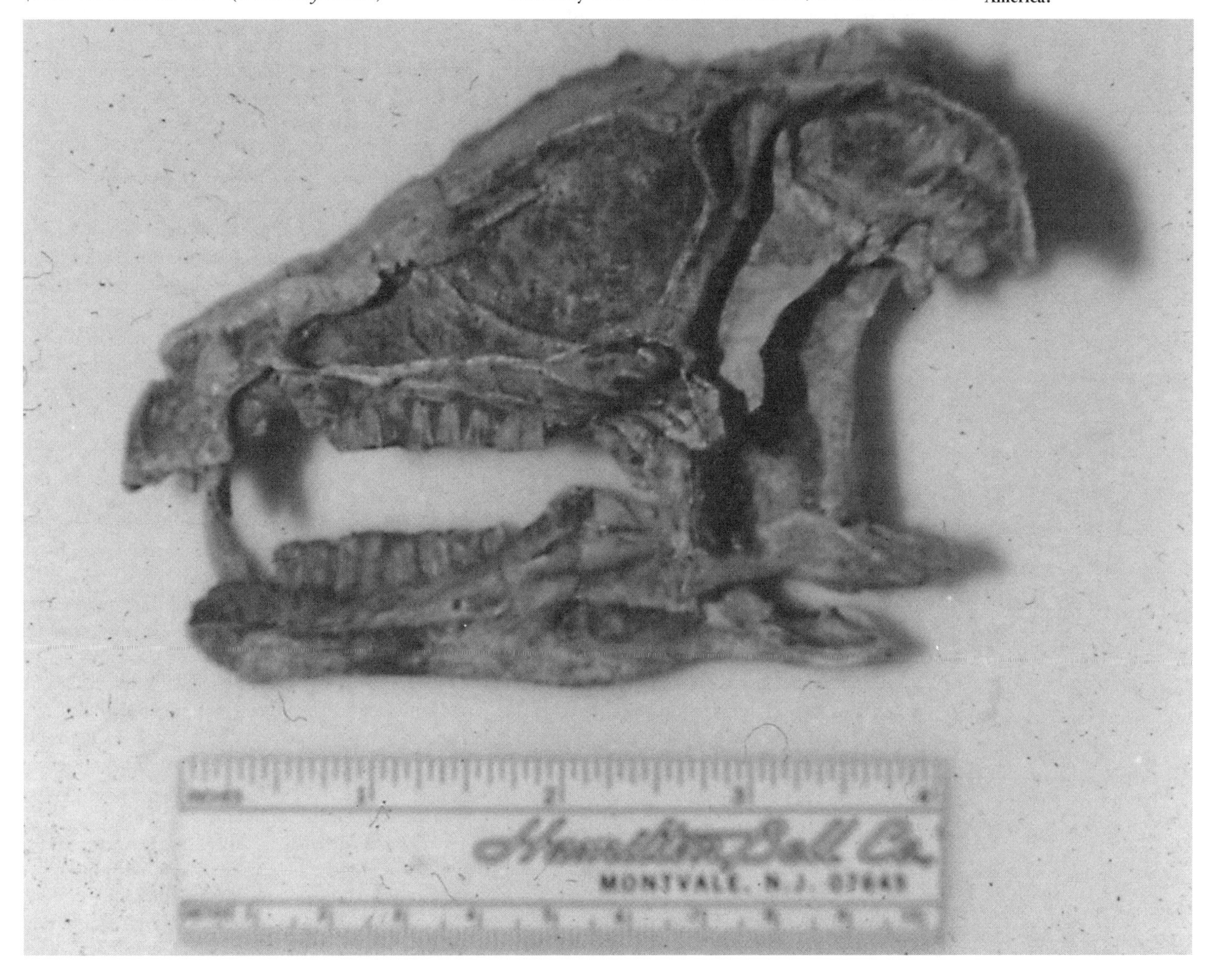

***Heterodontosaurus tucki*, skull belongng to skeleton SAM K1332, this species known from South Africa. Remains of the genus *Heterodontosaurus* have now also been found in South America.**

and closely packed," and the "mesial and distal surfaces of the crown are in broad contact," the latter feature regarded as a derived character supporting "a sister-group relationship of *Heterodontosaurus* and *Lycorhinus*." As in *Heterodontosaurus*, the South American form has teeth in which no cingulum or constriction separates the crown and root, while the wear facets of adjoining teeth form one continuous surface. This suggested to the authors that the South American "heterodontosaur" is a relatively advanced form that is either congeneric with *Heterodontosaurus* or a distinct but closely related taxon.

As pointed out by Baez and Marsicano, this discovery extends the temporal range of the most derived "heterodontosaurians." This, plus consideration of the phylogeny of ornithischians, indicated to the authors "that a more extensive phyletic diversification of these dinosaurs was taking place in the Late Triassic than that documented by actual records."

Key reference: Baez and Marsicano (1998).

HISTRIASAURUS Dalla Vecchia 1998

Saurischia: Sauropodomorpha: Sauropoda: Eusauropoda: Neosauropoda: Diplodocoidea *incertae sedis*.

Name derivation: Latin *Histria* = "Istria [Croatia]" + Greek *sauros* = "lizard."

Type species: *H. boscarolli* Dalla Vecchia 1998.

Other species: [None.]

Occurrence: Croatia.

Age: Early Cretaceous (upper Hauterivian–lower Barremian).

Known material/holotype: WN-V6 [temporary number], almost complete posterior dorsal vertebra.

Diagnosis of genus (as for type species): Combined presence in dorsal vertebrae of hyposphene-hypanium complex, well-developed outer suprazygapophysial lamina running parallel to axis of vertebral spine, and high (approximately 45 degrees) inclination of long diapophyses (Dalla Vecchia 1998).

Comments: The new type species *Histriasaurus boscarollii* was established upon a dorsal vertebra (WN-V6; not yet officially catalogued, but to be stored at the Museum of the Municipality of Bale) collected from Upper Cretaceous limestone outcrops of the Adriatic sea bottom (also the beach), a site discovered by Dario Boscarolli on the southwestern Istrian peninsula at Kolone, near Bale/Valle village, northwestern Croatia.

The holotype of *H. boscarollii* numbered among a collection of approximately 200, mostly fragmentary specimens comprising the first dinosaur bones ever found in the Adriatic sea bottom. These remains were first reported by Boscarolli, Laprocina, Tentor, Tunis and Venturini (1993), then briefly discussed by subsequent workers (*e.g.*, Dalla Vecchio 1994*a*). While some of these bones were subsequently identified as those of sauropods (Dalla Vecchio 1994*b*, 1997*a*, 1997*b*, 1997*c*; Dalla Vecchio and Tarlao 1995), none was described in any detail until Dalla Vecchia's comprehensive study of the material published in 1998.

As noted by Dalla Vecchio (1998), all of these remains were recovered "from the same outcrop, from the same stratigraphic level, and probably from the same bed." They may, the author speculated, belong to the same taxon; but as they were collected randomly as scattered fragments, they cannot be referred with certainty to the same taxon (see "note" below). Due to the difficulties involved, no attempts have yet been made to collect the many bones that remain embedded in the rock below sea level, although this site, Dalla Vecchio (1998) noted, has the potential of becoming one of the richest and most interesting localities of present Europe.

According to Dalla Vecchia (1998), WN-V6 represents a "diplodocimorph" [=diplodocoid] quite probably related to *Rebbachisaurus*, only more primitive in the presence of the hyposphene-hypantrum in the dorsal vertebrae.

Note: Dalla Vecchio (1998) also described a sampling of the many other retrieved specimens (none of which has been referred to *Histriasaurus*) from this site, the material including a complete cervical vertebra, almost complete cervical centrum, fragments of possible cervical ribs, three incomplete dorsal vertebrae, five incomplete caudal vertebrae, parts of caudal neural spines, one chevron, the distal portion of a femur, the proximal portion of a tibia, and other fragments. The greatly different sizes and morphological differences of these specimens suggest that the material represents several individuals and at least two taxa, both a large and small form.

Regarding this unidentified material, Dalla Vecchio (1998) observed that the complete caudal vertebra (WN-V3) and the anterior to midcaudal vertebrae (MPCM-V14, MPCM-V15, and WN-V4) seem to be closest in their morphology to the Brachiosauridae; the proximal cervical centrum (WN-V1) resembles those of *Chondrosteosaurus* (see *D:TE*); a caudal neural spine (MPCM-V9) resembles the spines in Camarasauridae; and the dorsal vertebrae (MPCM-V1, MPCM-V3) possess a very short neural arch, a well-developed laminar complex, and various other characters suggesting assignment to basal Titanosauriformes and possibly "Diplodocimorpha."

The above remains, as well as fragmentary elements apparently belonging to nonsauropod dinosaurs, were prepared for further study by Fabio M. Dalla Vecchio at the Museo Paleontològico Cittadino of Monfalcone (Gorizia).

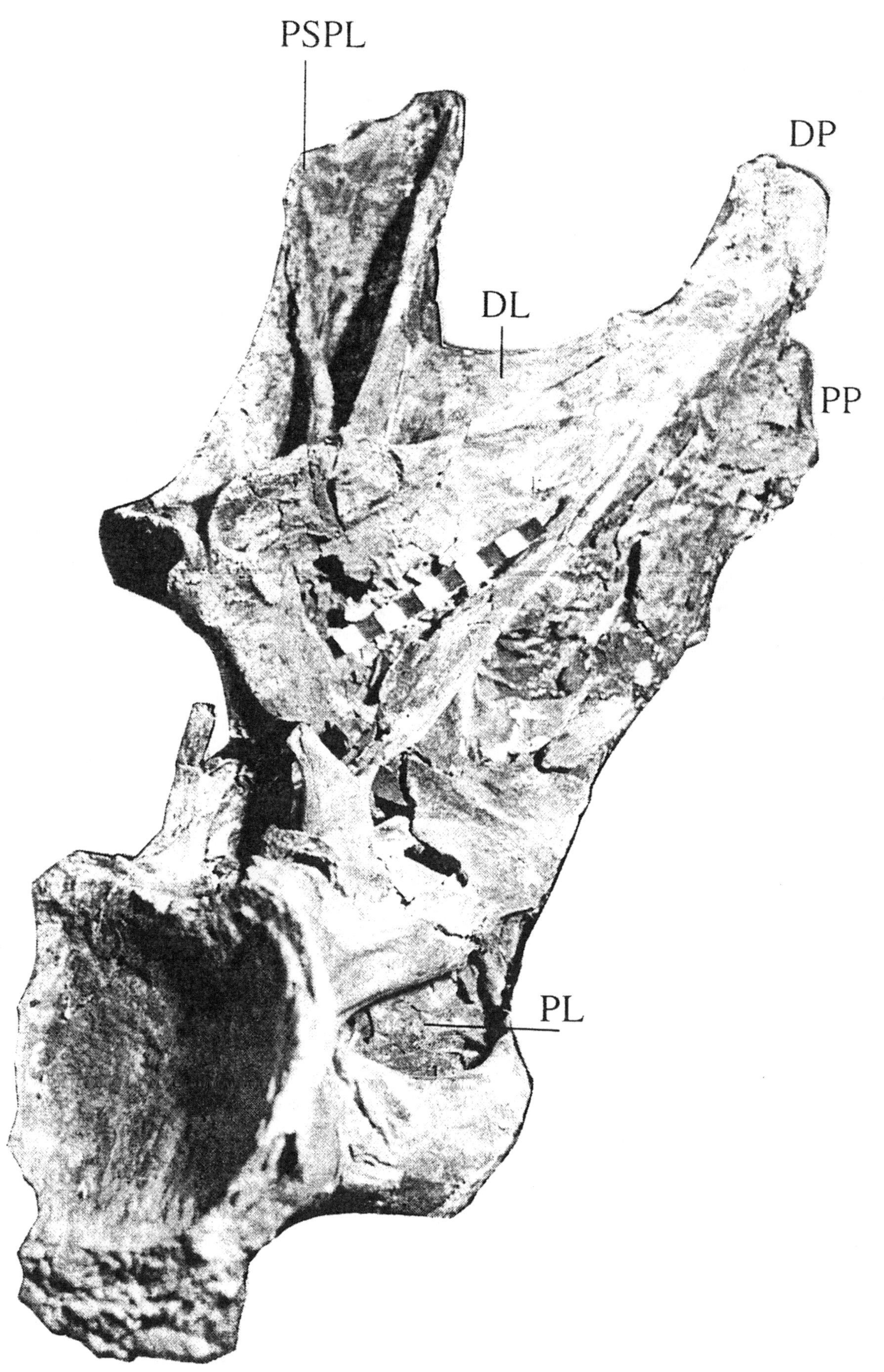

Histriasaurus boscarolli, WN-V6, holotype posterior dorsal vertebra, postero-lateral view. (After Dalla Vecchia 1999.)

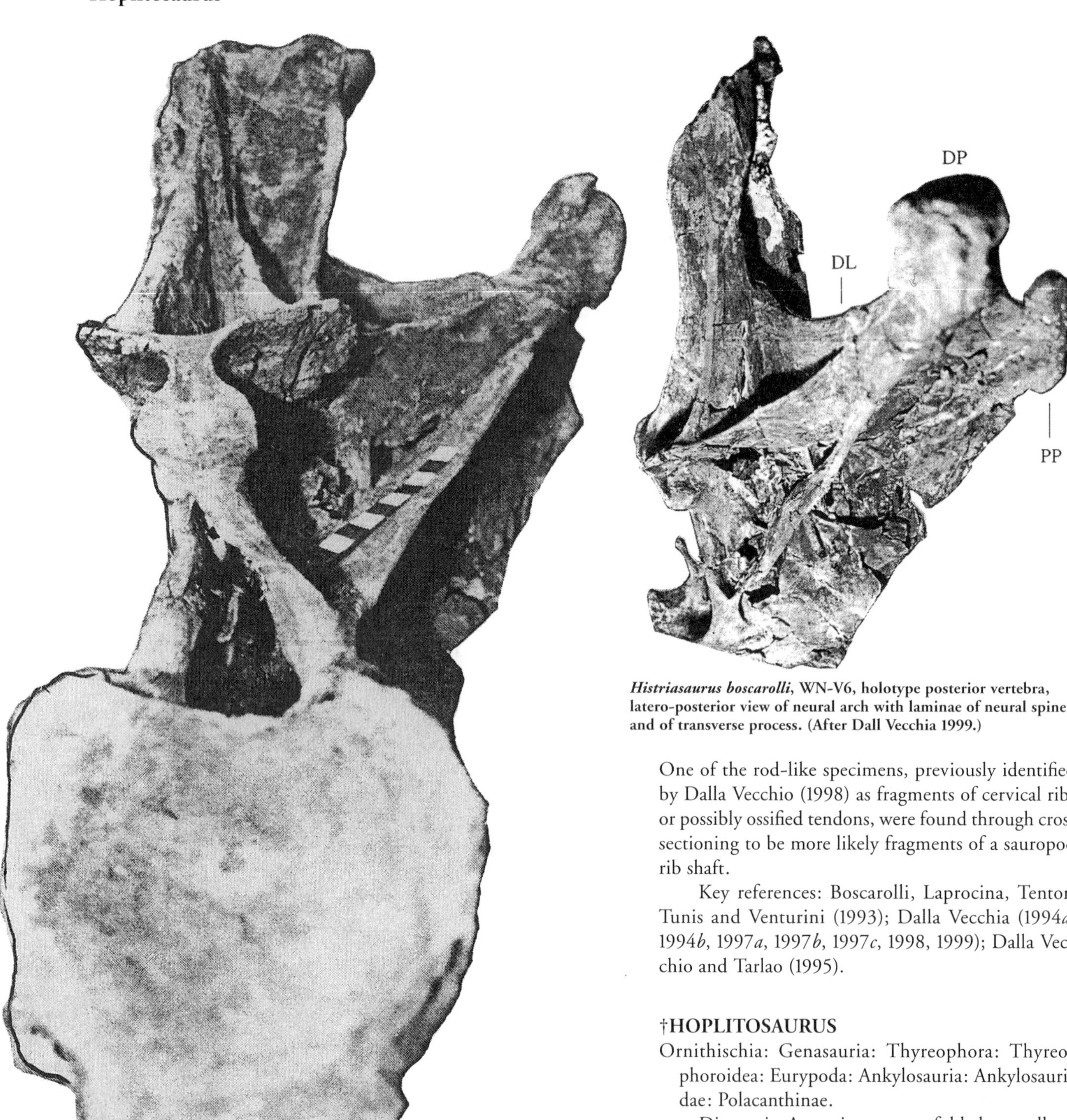

Histriasaurus boscarolli, WN-V6, holotype posterior vertebra, latero-posterior view of neural arch with laminae of neural spine and of transverse process. (After Dall Vecchia 1999.)

Histriasaurus boscarolli, WN-V6, holotype posterior dorsal vertebra, postero-ventral view. (After Dalla Vecchia 1999.)

As a completion to his 1998 paper, Dalla Vecchio (1999) subsequently published a photographic survey of the specimens from Istria, noting that they seem to represent a large-size titanosauriform, "diplodocimorph" (=diplodocoid), and a possible camarasaurid. One of the rod-like specimens, previously identified by Dalla Vecchio (1998) as fragments of cervical ribs or possibly ossified tendons, were found through cross sectioning to be more likely fragments of a sauropod rib shaft.

Key references: Boscarolli, Laprocina, Tentor, Tunis and Venturini (1993); Dalla Vecchia (1994*a*, 1994*b*, 1997*a*, 1997*b*, 1997*c*, 1998, 1999); Dalla Vecchio and Tarlao (1995).

†HOPLITOSAURUS

Ornithischia: Genasauria: Thyreophora: Thyreophoroidea: Eurypoda: Ankylosauria: Ankylosauridae: Polacanthinae.

Diagnosis: Acromion process folded ventrally as in *Hylaeosaurus* and *Gastonia* (not ridged as in nodosaurids); femoral head angled laterally steeply, more than in other known nodosaurids; crista trochanteris not well developed as in *Sauropelta*, *Cryptodraco*, or *Struthiosaurus*; greater trochanter expanded laterally more than in *Polacanthus* or *Niobrarasaurus*; anterior trochanter well developed, separated from shaft as a spike (ridge on femoral shaft in *Niobrarasaurus*, *Gastonia*, and *Sauropelta*, small spike in *Cryptodraco*); armor including posteriorly grooved cervical spine as in *Mymoorapelta*, *Gastonia*, and *Polacanthus*, wide-based spined plates as in *Polacanthus* (Carpenter and Kirkland 1998).

Comments: *Hoplitosaurus*— previously regarded by some workers as synonymous with *Polacanthus* (see *D:TE*, *Hoplitosaurus* entries for pro and con arguments), based upon similarities between the humeri, femora, and some of the armor — was accepted as a valid genus and also redescribed by Carpenter and Kirkland (1998) in their review of Lower and Middle Cretaceous ankylosaurs from North America.

According to Carpenter and Kirkland, reexamination of the holotype material —(USNM 4752, a partial postcranial skeleton including armor, from the Lakota Formation of Custer County, South Dakota — casts doubt on that synonymy.

Carpenter and Kirkland noted that features (*e.g.*, low position of deltopectoral crest) cited by Pereda-Suberbiola (1994) shared between the partial humerus of *Hoplitosaurus* and a humerus referred by Pereda-Suberbiola to *Polacanthus* are also seen in other ankylosaurs (*e.g.*, *Gastonia*). The feature of a separate anterior trochanter of the femur, which Pereda-Suberbiola stated was present in both *Hoplitosaurus* and *Polacanthus*, can be seen only in a single femur of *Polacanthus*, while preparation of that area revealed "that the separation between the anterior trochanter and greater trochanter is actually a crushed, matrix-filled zone formed during compaction of the femur." A protruding fourth trochanter was another feature used by Pereda-Suberbiola to synonymize the two genera, a feature, Carpenter and Kirkland pointed out, "only occurs on one femur and corresponds to a compact fracture."

Based on the above elimination of features, Carpenter and Kirkland accepted as valid genera *Hoplitosaurus* and *Polacanthus*. Also, these authors acknowledged the similarity of some of their armor (especially the spined plates and grooved spines), this suggesting a close relationship between these genera.

Originally, Carpenter, Kirkland, Miles, Cloward and Burge (1996) referred *Hoplitosaurus* to the subfamily Polacanthinae based on similarities in the armor of this genus with that of other polacanthine taxa, and also by Blows' (1987) observation of an incipient tail club in USNM 4752. Reexamination of this specimen, however, revealed to Carpenter and Kirkland that this perceived "club" is really "a segment of about six posterior caudals from the middle section of the tail," this segment including "the transition to the elongated caudals of the posterior part of the tail." The authors did not speculate as to how many caudals distal to the tail club were missing.

Key references: Blows (1987); Carpenter and Kirkland (1998); Carpenter, Kirkland, Miles, Cloward and Burge (1996); Pereda-Suberbiola (1994).

HUABEISAURUS Pang and Cheng 2000—(=?*Tangvayosaurus*)

Saurischia: Sauropodomorpha: Sauropoda: Eusauropoda: Neosauropoda *incertae sedis*.

Name derivation: Chinese "Huabei" [phonetic annotation of the Chinese characters for "North China"] + Greek *sauros* = "lizard."

Type species: *H. allocotus* Pang and Cheng 2000.

Other species: [None.]

Occurrence: Huiquanpu Formation, Shanxi Province, People's Republic of China.

Age: Late Cretaceous.

Known material: Postcranial remains, teeth.

Holotype: HBV-2000, two teeth, almost complete postcranial skeleton including cervical, dorsal, sacral, and caudal vertebrae, ribs, pectoral and pelvic girdles, left humerus and radius, left and right femora, tibiae, and fibulae, vertebrae partially articulated.

Diagnosis of genus (as for type species): Large sauropod about 20 meters (almost 70 feet) in length and 5 meters (almost 17 feet) in height; teeth strong, peglike, tooth crown high; ratio of 3:1 between length of tooth crown and length of tooth; cervical vertebrae opisthocoelous, with wide-flat ventral flute, pleurocoels wide, deep, their neural spines bifurcated; dorsal vertebrae opisthocoelous, with deep, oval-shaped pleurocoels, their neural spines high, unbifurcated; five sacral vertebrae, their spines combined into a plate; caudal vertebrae amphicoelous, their neural arches located on anterior half of centrum, spines unbifurcated; scapula long, its distal dilation not pronounced; humerus robust, slightly flattened proximally; ilium very large, long, fan-shaped, pubic peduncle located slightly anterior of middle part of lower edge of ilium, ischic peduncle undeveloped; pubis broad, almost platelike, with distinctly expanded proximal portion; ischium yoke-like with short, flat shaft; femur straight, long, narrow, flat, having well-developed, oblate head, fourth trochanter dilated and located on posterior inner lateral edge of slightly upper portion of midshaft; tibia and fibula long, flat; length ratio of 0.78:1 between humerus and femur, that of 0.77:1 between radius and humerus, 0.75:1 between tibia and femur (Pang and Cheng 2000).

Comments: The gigantic sauropod genus *Huabeisaurus* was founded upon a well-preserved, partial postcranial skeleton plus two teeth (HBV-20001) discovered in 1983 by Pang Qiqing and Cheng Zhengwu in the Upper Cretaceous (see below) Huiquanpu Formation, on the northeastern slope of Kangdailiang Hill, Zhaojiagou Town, Tianzhen County, Shanxi Province, People's Republic of China. The initial discovery was that of 12 articulated caudal vertebrae. Five subsequent excavations from 1989 to 1994 (reported

Huabeisaurus allocotus, HBV-2000, holotype tooth in 1. lingual and labial views, mid-cervical vertebra in 3. left lateral and 4. anterior views, posterior cervical vertebra in 5. left lateral and 6. anterior views, dorsal vertebra in 7. left lateral and 8. posterior views, anterior caudal vertebra in 9. posterior and 10. right lateral views, mid-caudal vertebra in 11. posterior and 12. right lateral views, chevron of anterior caudal vertebra in 13. right lateral and posterior views, 15. right scapula, internal view, and right coracoid in 16. internal and 17. external views. (After Pang and Cheng 2000.)

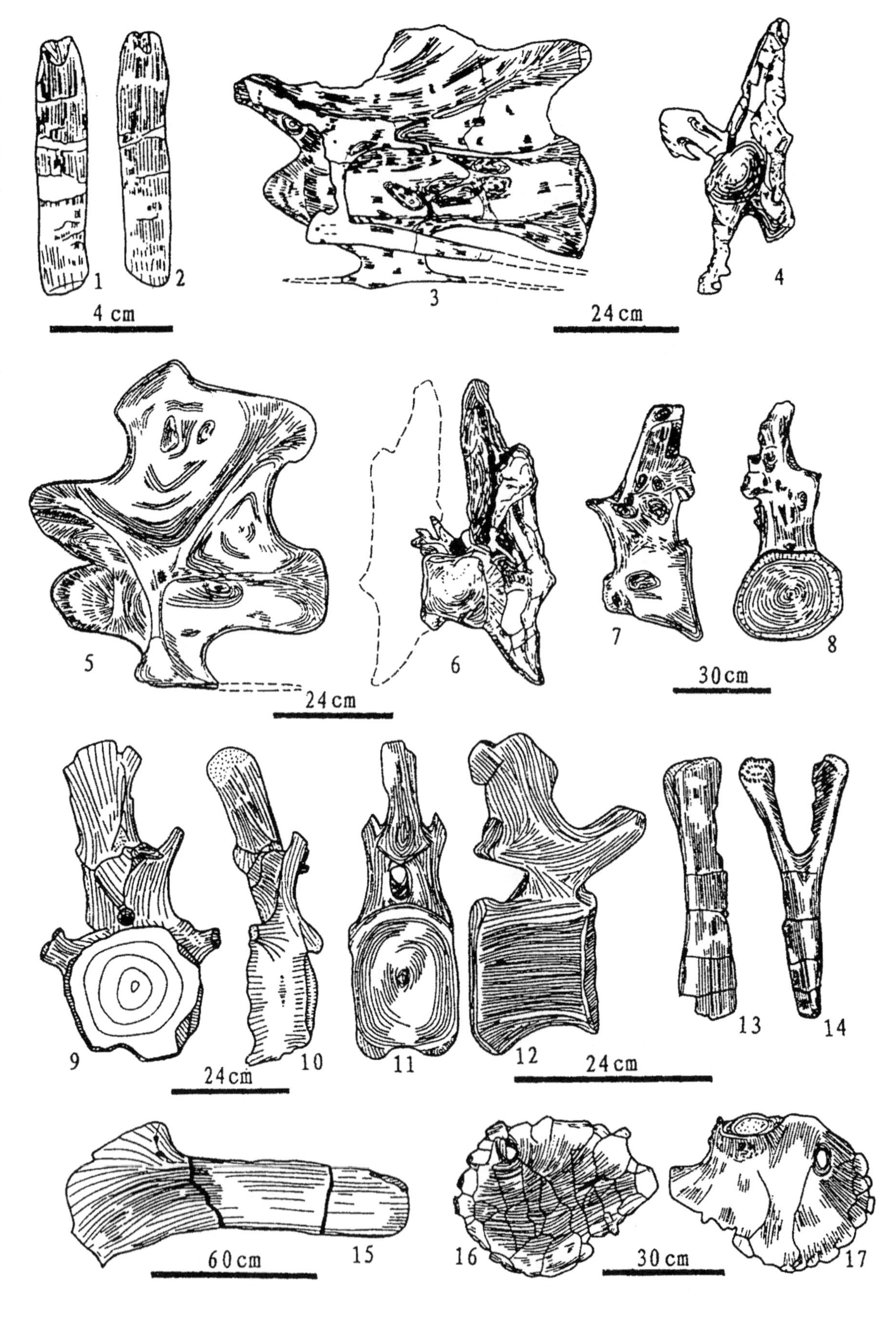

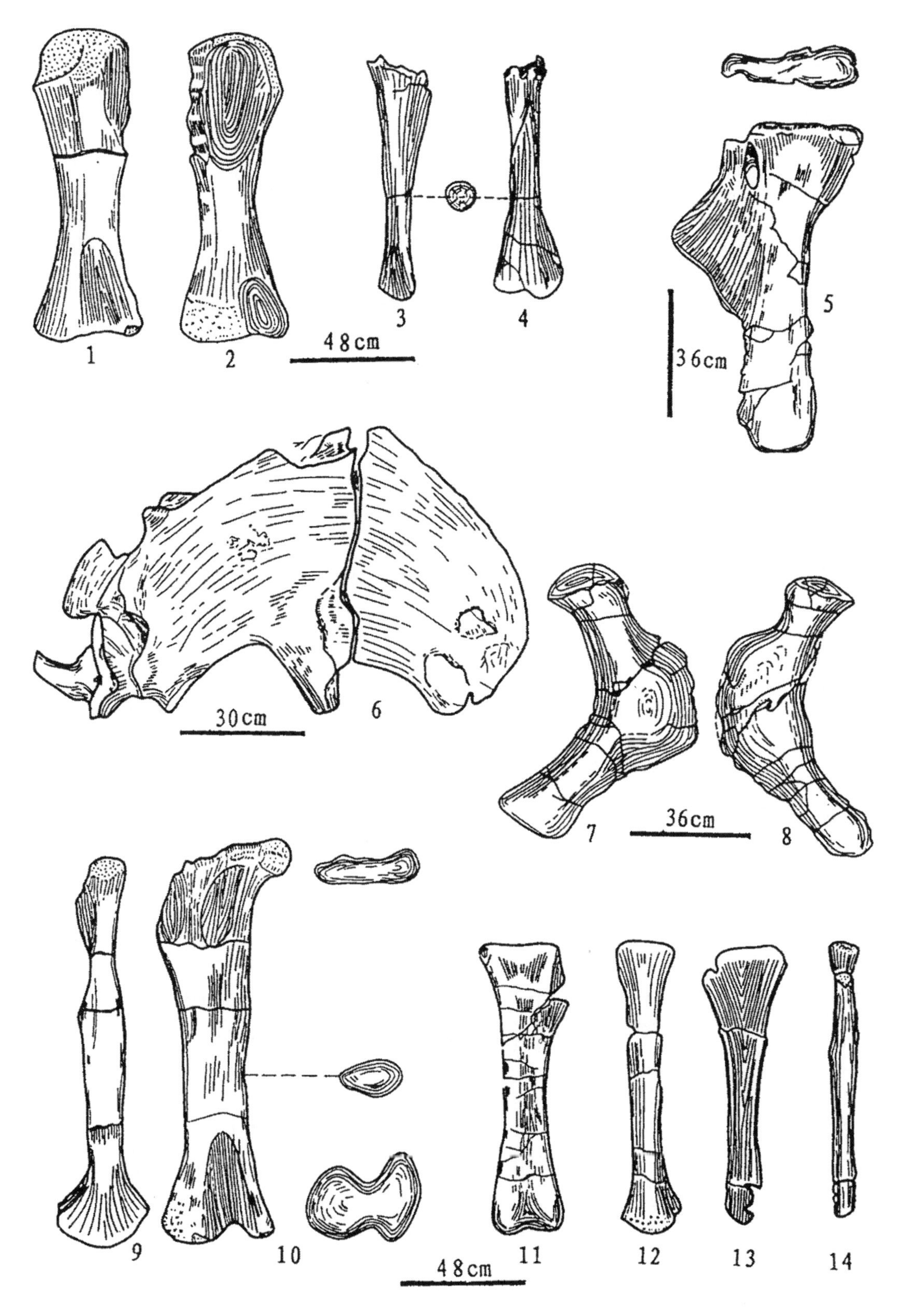

***Huabeisaurus allocotus*, HBV-2000, left humerus in 1. posterior and 2. anterior views, left radius in 3. posterior and 4. internal views, 5. right pubis in external view, 6. right ilium, external lateral view, 7. left and 8. right ischia in internal view, left femur in 9. internal and 10. posterior views, right tibia in 11. posterior and 12. internal views, and left fibula in 13. internal and anterior views. (After Pang and Cheng 2000.)**

by Pang, Cheng, Yang, Xie, Zhu and Luo 1996) yielded 2,300 pieces of dinosaur bone, including a paratype (HBV-20002), an almost complete left humerus belonging to the type species *Huabeisaurus allocotus*, plus remains of other kinds of dinosaurs (see Cheng and Pang 1996; Pang and Cheng 1998) including theropods, ankylosaurids (see *Shanxia* entry), and hadrosaurids (Pang and Cheng 2000).

As noted by Pang and Cheng (2000), the suite of distinguishing characters of *H. allocotus—i.e.*, peglike teeth, deep and regular pleurocoels of dorsal vertebrae, high, unbifurcated neural spines, amphicoelous caudal vertebrae having neural arches in their anterior parts and unbifurcated chevrons, undeveloped ischic peduncle, yoke-shaped ischium, and long, narrow, and flattened limb bones — separate this taxon from the Vulcanodontidae, Cetiosauridae, Brachiosauridae, and Camarasauridae. However, this species is similar to the Diplodocidae and Titanosauridae (families which these authors regarded as belonging to the superfamily Homalosauropoidea; see "Systematics" chapter for a different and more generally accepted assessment of the relationships between these two families) in the possession of peglike teeth.

Comparing *Huabeisaurus* to the diplodocoid family Nemegtosauridae, Pang and Cheng (2000) pointed out that the ratio of length of tooth crown to that of root is 3:1 in the former taxon and 2:1 in the latter; also, the teeth of *Huabeisaurus* "are robust, a little flat in the upper part of the tooth crown, with short ridges at the two edges and without plumb grooves in the concave lingual facet." These differences, the authors stated, separate *Huabeisaurus* from the Nemegtosauridae.

For the above reasons, Pang and Cheng (2000) proposed that *Huabeisaurus* be placed in its own family, the Huabeisauridae (same diagnosis as for the genus), which they in turn assigned to the "Homalosauropoidea." According to these authors, Huabeisauridae differ from Diplodociae and Titanosauridae in various ways. In Diplodocidae, the teeth are slender, the anterior dorsal vertebrae have bifurcated neural spines, and the middle chevrons are forked distally; in Huabeisauridae the teeth are robust, the dorsal vertebrae have unbifurcated spines, and caudal vertebrae have singular spines. In Titanosauridae, the dorsal vertebrae have irregularly shaped pleurocoels, and the middle caudal vertebrae are procoelous, the distal end of their centrum having a prominent ball; in Huabeisauridae, the cervical vertebrae have bifurcated neural spines, the dorsal vertebrae have deep, regularly shaped pleurocoels, and the caudal vertebrae are amphicoelous (McIntosh 1990).

According to Pang and Cheng (2000), *Huabeisaurus* seems to be more closely related to the Titanosauridae, particularly *Titanosaurus* (Huene and Matley 1933). It is also rather similar to the titanosaurid *Alamosaurus* (see McIntosh) in characters of the scapula, ischium, and femur, although in the latter genus the scapula is distinctly dilated, the humerus remarkably shrunken in the middle portion, the ischium is short with a strong iliac process, and the caudal vertebrae are procoelous.

As Pang and Cheng's (2000) interpretation of the relationships between the Diplodocidae, Nemegtosauridae, and Titanosauridae markedly differs from other generally accepted, recently published schemes (*e.g.*, Wilson and Sereno 1998), this book regards, at least for the present, both *Huabeisaurus* and its family Huabeisauridae as Neosauropoda *incertae sedis.*

Pang and Cheng (2000) observed that the species *Titanosaurus falloti* Hoffet 1942 is similar to *Huabeisaurus* in having amphicoelous caudal vertebrae and a relatively flat femur. The authors agreed with McIntosh that *T. falloti* belongs to a new genus possibly also representing a new family. Agreeing with McIntosh's appraisal but apparently not cognizant that this species had already been referred by Allain, Taquet, Battail, Dejax, Richir, Véran, Limon-Duparcmeur, Vacant, Mateus, Sayarath, Khenthavong and Phouyavong (1999) to the new genus, *Tangvayosaurus* (see entry), Pang and Cheng suggested that this taxon be referred to *Huabeisaurus.* If, however, *Huabeisaurus* and *Tangvayosaurus* do later prove to be congeneric, then the latter name, having been published earlier and thereby having priority, will be the valid one.

Huabeisaurus exhibits various characters representing the latest stages of the evolution of the Sauropoda, these including the following: teeth peglike, without edge denticles; deep pleurocoels in cervical and dorsal vertebrae, neural arches and spines bearing well-developed pits and ridges; five sacral vertebrae with combined, plate-shaped spines; ilium with pubic peduncle located almost in middle and with undeveloped ischic peduncle; femur with oblate-shaped head, fourth trochanter weak and located in median-upper portion of bone; and shaft of limb bones hollow. These characters, Pang and Cheng noted, are very different from those found in primitive Late Jurassic to Early Cretaceous sauropods, but are at the same level of evolution as the comparatively advanced sauropods of the Late Cretaceous from around the world, such as *Titanosaurus* (Huene and Matley) and *Saltasaurus* (Bonaparte and Powell 1980). Consequently — and also because ankylosaurids (Pang and Cheng 1998), hadrosaurids, ostracods, and charophytes have been found in these rocks — Pang and Cheng found it advisable to assign a Late Cretaceous age to beds that yielded *Huabeisaurus.*

As Pang and Cheng (2000) pointed out, *Huabeisaurus allocotus* is a significant species, being the first very late sauropod found in China based on relatively complete material, and also an important discovery on a global basis, further study of this material potentially leading "to the understanding of the distribution, classification, systematic evolution, migration and extinction of the sauropod dinosaurs, as well as the Cretaceous palaeoclimate and palaeogeography."

Key references: Allain, Taquet, Battail, Dejax, Richir, Véran, Limon-Duparcmeur, Vacant, Mateus,

Sayarath, Khenthavong and Phouyavong (1999); Bonaparte and Powell (1980); Cheng and Peng (1996); Hoffet (1942); Huene and Matley (1933); McIntosh (1990); Pang and Cheng (1998, 2000); Pang, Cheng, Yang, Xie, Zhu and Luo (1996); Wilson and Sereno (1998).

†HYPACROSAURUS

Ornithischia: Genasauria: Cerapoda: Ornithopoda: Euornithopoda: Iguanodontia: Euiguanodontia: Dryomorpha: Ankylopollexia: Hadrosauroidea: Hadrosauridae: Euhadrosauria: Lambeosaurinae.

Comments: The long bones of the holotype specimen (MOR 549) of *Hypacrosaurus stebingeri*—the second species erected for this long-spined, crested duckbilled dinosaur (see *D:TE*) collected from the Jason's Giant Site in Montana, were histologically sectioned and studied. In an abstract, Cooper, Lamm and Horner (1998) noted that lines of arrested growth (LAGs), believed to be deposited annually, were observed in the sectioned bones. Examination of three consecutive LAGs allowed for calculations involving the change in bone diameter as periosteal bone was deposited along the midshaft, the possible bone length at the time the LAG was deposited, the body length, and also the change in body mass between each LAG. The results of this study were compared with an alligator from the Love Site in Florida.

According to Cooper *et al.*, the amount of bone deposited between each LAG was greater in the alligator than in *H. stebingeri* within the initial stages of development, after which the LAGs are closer together, with little new bone deposited. Also, the diameter-to-bone length ratio in the alligator was about half that in the hadrosaur. From these results, the authors inferred that *H. stebingeri* "deposited more mass between each LAG throughout its development" than did the alligator.

Continuing this line of research, Cooper and Horner (1999), in another similar report, briefly discussed a femur belonging to the holotype of *H. stebingeri*. Histological sectioning of this bone also revealed LAGs. The circumference of the femur was determined at each LAG. Utilizing an allometric equation, Cooper and Horner were able to calculate the overall body mass at each measured circumference, and utilizing a monomolecular growth equation, the age of the animal at each mass. A growth curve was then generated for *H. stebingeri*, incorporating the histologically-determined measurements and also a number of whole-bone femurs of various sizes belonging to this taxon.

In comparing the results of this growth curve to growth sequences from ratite birds and a male alligator, Cooper and Horner found that "*Hypacrosaurus stebingeri* sustained a growth rate proportionally similar to that of ratites for several years which was greater than the growth rate of the alligator."

Key references: Cooper and Horner (1999); Cooper, Lamm and Horner (1998).

†IGUANODON—(=*Heterosaurus*, *Hikanodon*, *Sphenospondylus*, *Therosaurus*, *Vectisaurus*)

Type species: *I. bernissartensis* Boulenger 1881 *see in* Beneden 1881.

Lectotype: IRSNB 1534, almost complete, articulated skeleton.

Comments: The traditional story of how the first remains (teeth) of the ornithopod dinosaur *Iguanodon* were supposedly discovered as been told many times over. According to that original account, these fossils were found in the Wealden of Tilgate Forest, in Sussex, by Mary Ann Mantell while her physician husband, Dr. Gideon Algernon Mantell, was visiting a patient. Although Gideon Mantell originally believed the teeth to be mammalian, he eventually recognized

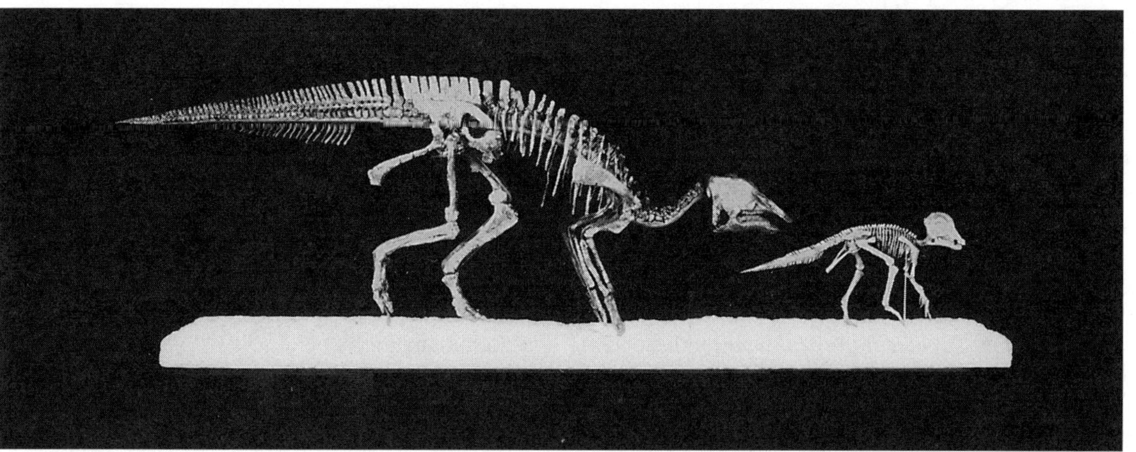

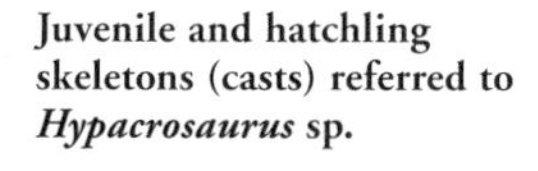

Juvenile and hatchling skeletons (casts) referred to *Hypacrosaurus* sp.

Photograph by J. Beckett. Courtesy Department of Library Services, American Museum of Natural History (neg. #2A21928).

Photograph by the author, courtesy The Natural History Museum, London.

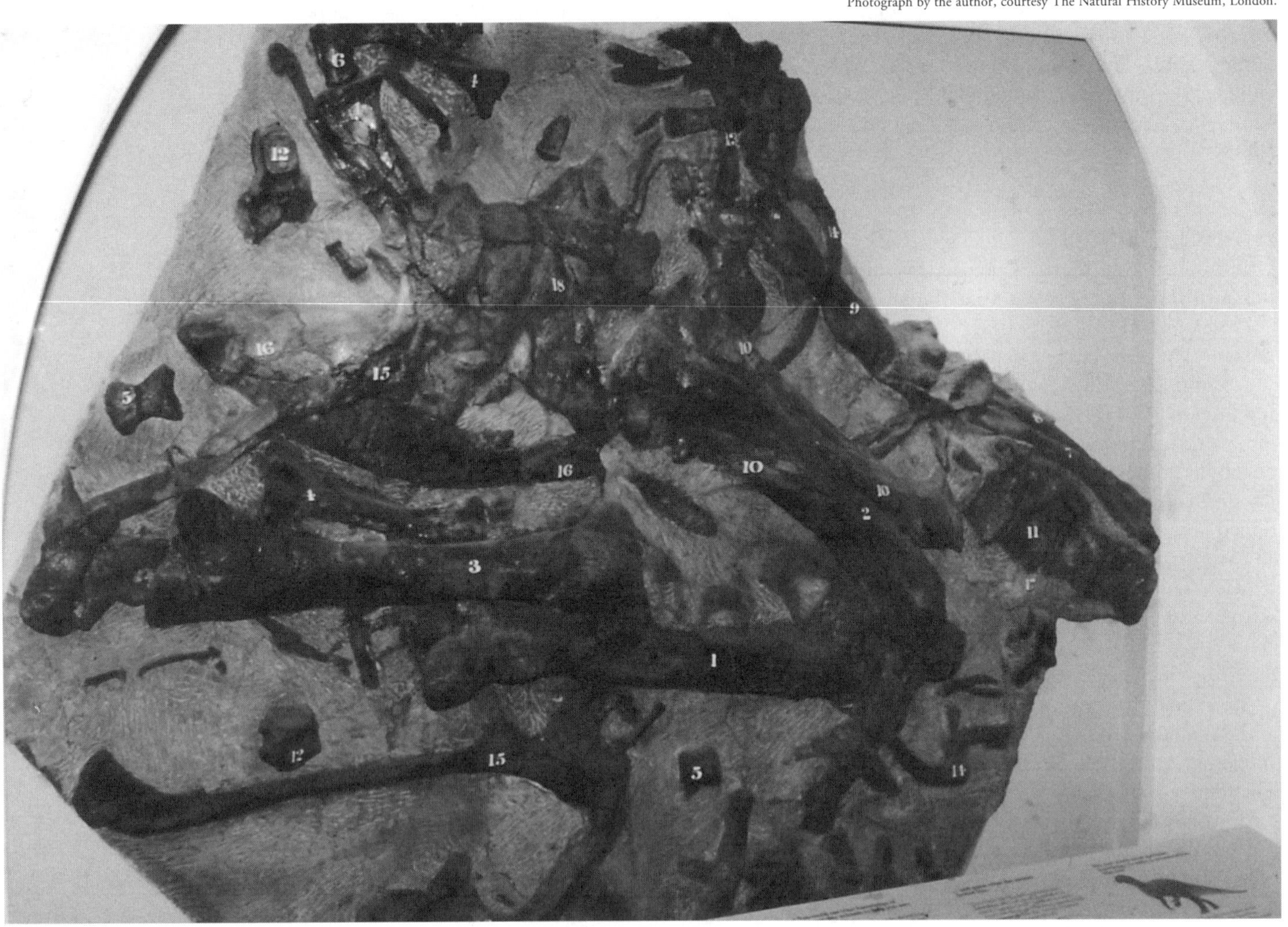

The so-called "Maidstone" *Iguanodon bernissartensis*, originally referred to *I. mantelli*, discovered in 1834 in Kent, England.

them as reptilian, having a resemblance to the much smaller teeth of a modern iguana lizard (see *D:TE*).

More recently, various researchers (*e.g.*, Dean 1995, 1999; Buffetaut 1999) have assembled data that seem to present a more accurate illustration as to how *Iguanodon*— historically significant as the second dinosaur to be named and described—was discovered and finally recognized for what it is.

According to Buffetaut, the details of the discovery of *Iguanodon* and Mantell's eventual realization as to what its fossils represented, have long remained unclear. This is mostly because Mantell's own various accounts have presented conflicting information, and also because subsequent authors have augmented those accounts with "fictitious embellishments." Therefore, building upon Dean's research concerning Mantell and the discovery of *Iguanodon*, and including as evidence previously unpublished annotations by Mantell to a copy of his 1824 publication "Outlines of the Natural History of the Environs of Lewes" sent to Alexandre Brongniart, Buffetaut reconstructed in detail the following scenario:

In 1822, teeth that would later be named *Iguanodon* were discovered in Tilgate Forest, some of these specimens having been found by Mantell's wife. At that time, Mantell suspected that the teeth were those of a very large herbivorous reptile, an idea that was largely met with skepticism when he presented some of his specimens to the Geological Society in London in June, 1822. Only a Dr. Wollaston supported Mantell's opinion and encouraged him to continue his researches in the direction that he had discovered a previously unknown plant eating reptile. Among the dissenters, the eminent geologist William Buckland, who two years later would describe the theropod dinosaur *Megalosaurus*, believed Mantell's fossils to belong to a fish or some "diluvial" mammal such as a whale. More influenced by Buckland's opinions, Mantell began to think of the mysterious Tilgate animal in mammalian terms.

By February of 1824, influenced by reports of other very large fossil reptiles (dinosaurs and plesiosaurs) from Tilgate Forest, Mantell gradually began to sway from idea that his specimens belonged to a

Iguanodon

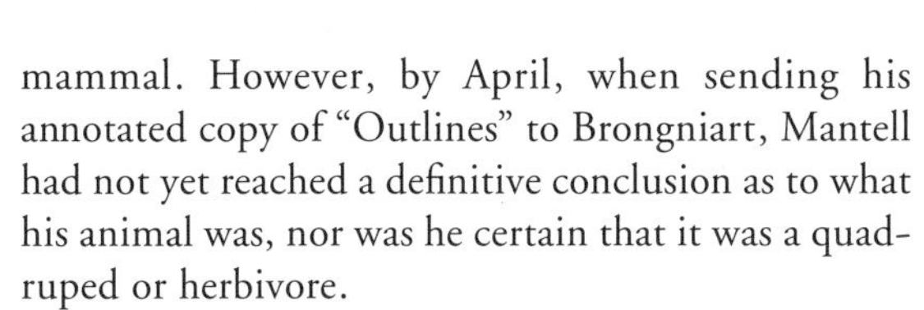

Skeletal reconstruction of *Iguanodon bernissartensis* by French paleontologist Louis Dollo in 1883. Though the kangaroo-like pose is outmoded, the anatomy in this illustration is quite correct, this picture being historically significant as the first published accurate illustration of this dinosaur's skeleton.

mammal. However, by April, when sending his annotated copy of "Outlines" to Brongniart, Mantell had not yet reached a definitive conclusion as to what his animal was, nor was he certain that it was a quadruped or herbivore.

In August or September of that year, Mantell made his famous visit to the Hunterian Museum at the University of Glasgow. There, Samuel Stutchbury showed him the teeth of an iguana. Noting the resemblances of the teeth to those of the Tilgate animal, Mantell was at last convinced that his fossils belonged to some kind of giant reptile.

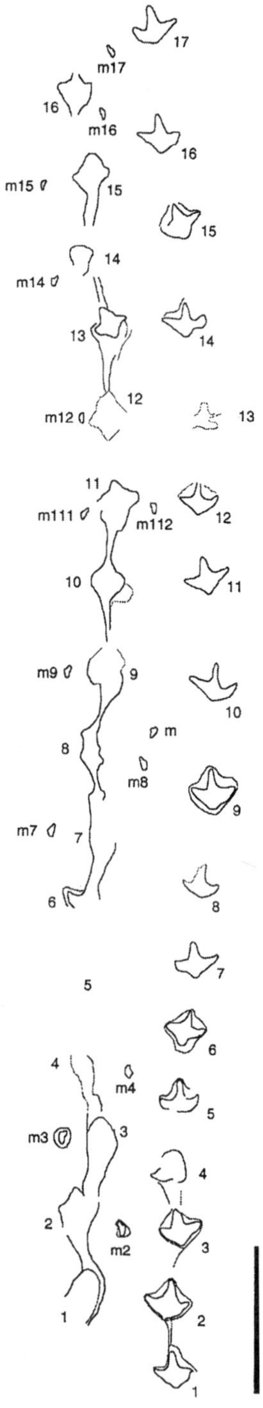

Backtracking, in June of 1823, Mantell had sent his fossils to Paris for examination by comparative anatomist Baron Georges Cuvier, the "Father of Paleontology," whose pronouncements almost always went unchallenged. From 1923 to 1824, Cuvier also mistook them for the remains of mammals, either a large rhinoceros or some other unknown quadruped from the Tilgate sandstone. This assessment, Buffetaut noted, also influenced Mantell and "led his British correspondent on the wrong track, and this may have delayed the recognition of the true nature of *Iguanodon* for a few months." Not until Mantell sent Cuvier additional worn and unworn teeth, probably in spring of 1824, did the Baron alter his original opinion. In June of that year Cuvier wrote to Mantell that the latter's original idea of the teeth being reptilian seemed to be correct.

Buffetaut concluded, "Whether the crucial factor was Cuvier's letter or the comparison with the teeth of the iguana at the Hunterian Museum, or simply the discovery of more *Iguanodon* teeth showing various degrees of wear, is uncertain; all may have played a part [in Mantell's final identification of the specimens]." On February 10, 1825, Mantell formally published his name for the fossil teeth, *Iguanodon*.

Norman and Barrett (1999), in an abstract, commented upon *Iguanodon hoggi*, a small species based upon an isolated right dentary, collected from the Purbeck Limestone Formation of the Isle of Purbeck, England, and figured by Richard Owen in 1874.

As related by Norman and Barrett, this specimen has, until relatively recently, been only partially visible as it was embedded in a small block of limestone. Acid preparation of the specimen at The Natural History Museum, London, however, has removed the matrix, revealing an almost complete and intact, only slightly crushed specimen with dentition intact. According to these authors, this taxon is distinguished from *I. fittoni* and *I. dawsoni*, younger species from the Lower Wealden of the Weald Basin, by the shape of the coronoid process, structure of the tooth row, and morphology of the individual teeth. Comparisons with other ornithopods revealed to Norman and Barrett a close relationship of *I. hoggi* with the Late Jurassic North American and British species of *Camptosaurus*.

In a brief preliminary report, Norman (1999) pointed out the poor resolution of the anatomically diagnostic and stratigraphic "limits" of *Iguanodon*. According to Norman (1999), a revision of Berriasian-age (pre–Wealden) fossil remains from Great Britain "calls into question the validity of the attribution of this material to the genus *Iguanodon*," and also other ornithopod remains collected from the early Wealden deposits of Southern England and from North America. Norman's (1999) yet to be completed study, which has already inspired a broader view of Late Jurassic–Early Cretaceous, European and North American medium to large-sized ornithopods, will attempt a synthesis resulting from the detailed comparative anatomy and stratigraphic and geographic distribution of these animals, "with a view to resolving, in a consistent manner, the taxonomy and systematics of these relatively ubiquitous and (apparently) well-known dinosaurs."

On January 13, 1997, the International Commission on Zoological Nomenclature (ICZN) received from the late Dr. Alan J. Charig and Sandra D. Chapman of The Natural History Museum, London, an application to designate *Iguanodon bernissartensis* as the new type species of the genus *Iguanodon* (replacing the original type species, *I. anglicus* Holl 1829, and to designate IRSNB 1534, a nearly complete skeleton mounted at the Institut Royal des Sciences Naturelle de Belgique in Brussels, Belgium, as the lectotype. The ICZN accordingly sent notices of their application (Case 3037) to various scientific journals (see *S1* for details; also, see *D:TE* and *S1* for photographs of mounted casts of IRSNB 1543).

Comments by vertebrate paleontologists supporting Charig and Chapman's application were subsequently published in the *Bulletin of Zoological Nomenclature*— Professor David B. Norman and Dr. Angela C. Milner in *BZN* 55: 172 (September 1998); and Dr. Paul M. Barrett and Dr. Kenneth Carpenter in *BZN* 55: 239–240 (December 1998). Dr. Hans-Dieter Sues (*BZN* 55: 240–241) agreed with designating IRSNB 1534 the lectotype of *I. bernissartensis*, but opposed designating that taxon as the type species of *Iguanodon*.

On September 1, 1999, the Commission members were invited to vote on Case 3037. At the close of the voting period on December 1, 1999, all of the Commission members participating in this matter voted in the affirmative. The final ICZN ruling (Opinion 1947), published in *BZN* 57 (1) and dated March, 2000, was as follows:

1. All previous fixations of type species for the genus *Iguanodon* were abandoned, with *Iguanodon bernissartensis* designated as the type species, and the skeleton catalogued as IRSNB 1534 designated as the lectotype;

2. the name *Iguanodon* was placed on the Official List of Generic Names in Zoology; and

3. the name *bernissartensis*, as published in the binomen *Iguanodon bernissartensis* and defined by the lectotype IRSNB 1534, was placed on the Official List of Specific Names in Zoology.

Note: In life restorations, the genus *Iguanodon* has usually been reconstructed in a normal, quadrupedal walking pose with the forelimbs placed slightly

Two of the three Purbeck trackways apparently made by an iguanodontid, possibly *Iguanodon*. The trackway on the left shows manus as well as pes impressions, that on the right showing pes impressions only. Scale = 1 m. (After Wright 1999.)

nearer to the midline than the hindlimbs, the dorsal surface of the manus facing forwards. This familiar image — relating to iguanodontids in general and, perhaps, *Iguanodon* in particular — has been recently challenged by Wright (1999) after a study of three series of large dinosaur footprints discovered in rocks of the Intermarine Member of the Purbeck Limestone Group (Lower Cretaceous) Dorset, southern England. Discovered in summer, 1961, the trackways were first reported on by Charig and Newman (1962). Some of the undamaged tracks were excavated during from June to July, 1963, by The Natural History Museum (then the British Museum [Natural History]), one trackway (BMNH R 8643) eventually being put on permanent display outside the building.

These Purbeck trackways preserve impressions of three-toed hind feet; what were interpreted as manus impressions seem to have been made by three digits set closely together, the smallest at the back or outside, the largest in the middle. This latter design is consistent with the manus structure of all iguanodontids, in which metacarpals II, III, and IV were bound together by ligaments, the joints of the phalanges allowing the digits to hyperextend, thereby providing a stable walking surface for quadrupedal locomotion (see Norman 1980; Norman and Weishampel 1990). Although these trackways cannot be referred with certainty to any dinosaurian species, they are, Wright observed, most consistent — in terms of morphology, age, and geographic occurrence — with those of *Iguanodon*, and more closely with the species *I. atherfieldensis* than *I. bernissartensis*.

According to Wright, the generally accepted posture is problematic, as it would have necessitated an unnatural rotation of the radius around the ulna, this leading to distortion and dislocation of the joints at the wrist or elbow. The Purbeck trackway evidence, however, allows for a different interpretation. As noted by Wright, the manus impressions are oriented with the palmar (or bottom) surfaces directed inward toward the midline of the trackway, digit I (which was not used in walking) therefore facing anteriorly, not medially as has usually been depicted. In this model, there is no unnatural twisting of the lower forelimb bones. The posture indicated by the trackways is more compatible with what is known of the skeletal anatomy of iguanodontids. Furthermore, the foot emplacement seen in these ichnites is typical of a facultative quadruped.

As observed by Wright, the forefeet impressions of these trackways lie in two lines on either side of the line formed by the hindfeet impressions. Thus, when the animal walked on all fours, the forefeet were placed in a wider trackway than were the hindfeet, although not in a lizard-like sprawling forelimb posture. In bipedal animals, the hindfeet — the weight-bearing limbs — must be placed directly under the body midline to maintain balance. The forelimbs of the Purbeck trackmaker, not required for weight-bearing and balance during walking, did not have to operate under such restraints. Wright noted that the angulation pattern in the Purbeck manus prints is about equal to the width of the shoulder girdle of *I. atherfieldensis* (see Norman 1986), these relative proportions suggesting "that the arms would have been held in a vertical position when the forefeet were in contact with the ground."

(See *D:TE* for more information about ichnites attributed to *Iguanodon*.)

Key references: Beneden (1881); Buffetaut (1999); Charig and Newman (1962); Dean (1995, 1999); Holl (1829); Mantell (1824, 1825); Norman (1980, 1986, 1999); Norman and Barrett (1999); Norman and Weishampel (1990); Owen (1874); Wright (1999).

ILOKELESIA Coria and Salgado 2000

Saurischia: Theropoda: Neotheropoda: "Ceratosauria": Neoceratosauria: Abelisauroidea.

Type species: *I. aguadagrandensis* Coria and Salgado 2000.

Other species: [None.]

Occurrence: Río Limay Formation, Patagonia, Argentina.

Age: Middle Cretaceous (Albian–Cenomanian).

Known material/holotype: Postorbital, quadrate, occipital condyle, two proximal cervical vertebrae, a posterior dorsal vertebra, five midcaudal vertebrae articulated with chevrons, one tibial shaft, pedal phalanges.

Diagnosis of genus (as for type species): Quadrate with reduced lateral condyle, posterior border of articular surface formed completely by medial condyle; cervical vertebrae having very reduced diapopostzygapophysial laminae; dorsal vertebrae having ventrally concave infraparapophysial laminae and ventrally oriented parapophysis; dorsals lacking pleurocoels; midcaudal vertebrae having distally expanded transverse processes bearing anteriorly and posteriorly projecting processes; and distal edge of caudal transverse processes showing gently sigmoid profile that is convex anteriorly, concave posteriorly. (Coria and Salgado 1999).

Comments: In a preliminary report published as an abstract, Coria and Salgado (1999) announced the discovery of a primitive abelisaurid based upon fragmentary remains collected from the Río Limay Formation (Upper Cretaceous; Albian–Cenomanian) of Patagonia, Argentina.

This dinosaur was found by Coria and Salgado

to be the most plesiomorphic known abelisaurid taxon, sharing with both Abelisauridae and Noasauridae the following characters: Supraorbital ossification of postorbital; intraorbital projection of jugal process of postorbital; quadrate with reduced lateral condyle; cervical vertebrae having reduced neural spines; prezygo-epipophysial laminae well developed; and caudals with distally expanded transverse processes.

However, the new theropod could not be referred by the authors to either abelisauroid family, retaining as it does the following features: jugal process of postorbital that is perpendicular to its horizontal branch; quadrate foramen; and cervical epipophyses without anterior projections.

Coria and Salgado (2000) subsequently named this theropod *Ilokelesia aguadagrandensis* and formally described the material in a paper (published in the volume on theropods edited by Perez-Moreno, Holtz, Sanz and Moratalla 2000) not, to date of this writing, yet available to the present writer.

Key references: Coria and Salgado (1999, 2000).

†INDOSAURUS

Saurischia: Theropoda: Neotheropoda: Abelisauroidea: ?Abelisauridae.

Comments: The large Indian abelisaurid *Indosaurus* (see *D:TE, S1*) figured in a study, published as a brief report by Chatterjee (1998*a*), regarding plate techtonics:

During the Mesozoic Era, the central element in the composition and later fragmentation of Gondwana — a "southern continent" that began to form at the end of the Triassic period during the break-up of the super-continent Pangaea — is today the subcontinent of India. Chatterjee reevaluated the largely accepted plate tectonic model that India remained an island continent during the Jurassic and Cretaceous periods after splitting off from Africa and Antarctica. If this model were true, Chatterjee noted, and India had, in fact, stayed isolated for over a 100 million years, then its long period of isolation would have produced a highly endemic biota.

However, as Chatterjee observed, "Indian dinosaurs do not show the evolutionary effects of isolation and endemism," some Indian dinosaurs exhibiting rather close similarities to those on other continents (*e.g.*, the resemblance of the Indian theropod *Indosaurus* to South American abelisaurids; see *Indosaurus* entries, *D:TE* and *S1*). Contrary to the older scenarios, as suggested by recent paleogeographic reconstructions, several biogeographic corridors must have connected India with other Gondwana land masses during most of the Jurassic and Cretaceous, this permitting dinosaurs to enter and leave India. Chatterjee suggested that, when India collided with a peninsular extension of East Africa during the Late Cretaceous, a land-bridge was created that allowed for the passage of dinosaurs from Africa to India.

Key reference: Chatterjee (1998*a*).

†IRRITATOR—(=*Angaturama*)

Saurischia: Theropoda: Neotheropoda: Tetanurae: Avetheropoda: Spinosauroidea: Spinosauridae: Spinosaurinae.

Comments: The genus *Irritator* now seems to include another recently described taxon, *Angaturama limai.*

In 1996, Alexander W. A. Kellner and Diogenese de A. Campos named and described *A. limai*, a new genus and species founded on a partial skull (GP/2T-5) recovered from the Santana Formation of northeastern Brazil (see *Angaturama* and *Baryonyx* entries, *S1*). However, following its publication, various workers have questioned the validity of this taxon.

Kellner (1996), for example, suggested that the two type species *Irritator challengeri* and *A. limai* could very well be congeneric. However, Kellner was reluctant to refer the latter genus to the former due to what he considered to be a lack of sufficient corresponding skull material for comparison.

Later, Charig and Milner (1997) observed that *Angaturama* shares the following features with the spinosaurid *Irritator*: Elongated jaws; external nares set far back; and a number of tooth characters unique to spinosaurids (see *Irritator* entry, *S1*). It was the opinion of Charig and Milner authors that *Angaturama* is probably congeneric with the earlier named *Irritator.*

More recently, in reviewing various spinosaurids, Sereno, Beck, Dutheil, Gado, Larsson, Lyon, Marcot, Rauhut, Sadleir, Sidor, Varricchio, Wilson and Wilson (1998) referred *Angaturama* to the genus *Irritator.*

Sues, Frey and Martill (1999), in an abstract, briefly reported these newly observed anatomical details concerning the spinosaurid skull following detailed preparation of the holotype (SMNS 58022; see *D:TE* for illustration) of *I. challengeri*: The elongated snout is quite narrow transversely, but deep dorsoventrally. The nasal and frontal bones form a median crest (but no evidence was found for the originally claimed parietal crest). The long axes of the large antorbital fenestra and orbit are posterodorsally inclined. The braincase is deep dorsoventrally, short anteroposteriorly. The crowns of the conical teeth are nearly straight, have distinct, unserrated carinae, and are deeply set in the jaws.

The above features indicated to Sues *et al.* that *Irritator* possessed jaws that could close rapidly and forcefully upon prey items of limited size. Considering the idea that spinosaurids were fish eaters, these authors did not rule out the possibility, but noted that "the transversely narrow and dorsoventrally deep (oreinirostral) snout differs significantly from the tubular (platyrostral) snout in piscivorous crocodyliform archosaurs."

Key references: Kellner (1996); Kellner and Campos (1996); Sereno, Beck, Dutheil, Gado, Larsson, Lyon, Marcot, Rauhut, Sadleir, Sidor, Varricchio, Wilson and Wilson (1998); Sues, Frey and Martill (1999).

ISANOSAURUS Buffetaut, Suteethorn, Cuny, Tong, Le Loeuff, Khansubha and Sutee Jongautchariyakul 2000

Saurischia: Sauropodomorpha: Sauropoda *incertae sedis.*

Name derivation: *Isan* [local name for northeastern Thailand] + Greek *sauros* = "lizard."

Type species: *I. attavipachi* Buffetaut, Suteethorn, Cuny, Tong, Le Loeuff, Khansubha and Sutee Jongautchariyakul 2000.

Other species: [None.]

Occurrence: Nam Phong Formation, Chaiyaphum Province, Thailand.

Age: Late Triassic (late Norian or "Rhaetian").

Known material/holotype: CHA4, associated remains including one cervical vertebra, one dorsal and six caudal vertebral centra, neural arch of posterior dorsal vertebra, two chevrons, fragmentary ribs, right sternal plate, right scapula, left femur, ?subadult.

Diagnosis of genus (as for type species): Primitive sauropod having robust femur bearing prominent, S-shaped fourth trochanter tapering to point, located in proximal half of bone (Buffetaut, Suteethorn, Cuny, Tong, Le Loeuff, Khansubha and Sutee Jongautchariyakul 2000).

Comments: Distinguished as the earliest known sauropod dinosaur and the first pre–Jurassic sauropod known from evidence other than fossil footprints (*e.g.*, Ellenberger and Ellenberger 1958), the type species *Isanosaurus attavipachi* was based on a partial postcranial skeleton (CH4) apparently belonging to a single individual, recovered from a natural outcrop of dark red sandstones of the fluviatile Nam Phong Formation (dated as late Norian or "Rhaetian," based on vertebrate fauna and palynoflora; see Racey, Love and Polachan 1996) at Phu Nok Khian, near the village of Ban Non Thaworn, in Chaiyaphum Province, on the Khorat Plateau, in northeastern Thailand. Apparently the specimen had been largely eroded away before its discovery (Buffetaut, Suteethorn, Cuny, Tong, Le Loeuff, Khansubha and Sutee Jongautchariyakul 2000).

Buffetaut *et al.* measured the holotype femur (CH4-1) of *I. attavipachi* as 76 centimeters in length, the entire length of the animal estimated by the authors as about 6.5 meters (approximately 22 feet). This relatively small size, plus the unfused neurocentral sutures of the vertebrae, indicate that this individual was not fully grown at the time of death.

The authors identified the Thai specimen as a primitive sauropod rather than an advanced prosauropod based on the following features: Regarding the vertebrae, a short cervical centrum (CH4-3) is markedly opisthocoelous, in contrast to the amphicoelous centra of prosauropods or the anteriorly flat centra of the Early Jurassic sauropod *Gongxianosaurus* (see He, Wang, Liu, Zhou, Liu, Cai and Dai 1998) and shows a strong median ridge as in primitive sauropods. The sides of the centrum are deeply concave (such depression also occurring on a posterior dorsal centrum), not excavated by pleurocoels as in more advance sauropods, these vertebrae thereby resembling those of the primitive sauropods *Barapasaurus tagorei* (Jain, Kutty Roychowdhury and Chatterjee 1979) and *Shunosaurus lii* (Zhang 1988). A probably posterior dorsal neural arch (CH4-7) is very tall, unlike the relatively low neural spines of prosauropods but like those in some later sauropods (He *et al.*), and longer (rostrocaudally) than broad (transversely) as in primitive sauropods. Consequently, *Isanosaurus* is less advanced than the primitive *B. tagorei* and *Zizhongosaurus chuanchengensis* (Dong, Zhou and Zhang 1983), in which this spine is transversely broadened, and more like such Middle Jurassic sauropods as *S. lii* (Zhang 1988), *Lapparentosaurus madagascariensis* and *Volkheimeria chubutensis* (see Bonaparte 1999*b*; Upchurch 1988) in which the dorsal neural spines are laterally flattened. Additional features found in sauropods (see Bonaparte) but not prosauropods include incipient posterolateral laminae and a ridge that extends from the transverse base of the spine. The scapula, although incomplete, resembles that of *Shunosaurus*, having the primitive sauropod feature (see Upchurch) of a moderate, dorsally rounded proximal expansion and also a slight distal expansion.

Buffetaut noted that the robust femur has a straight, craniocaudally flattened, sauropod-like shaft, with no sign of the S-shaped curvature usually seen in prosauropods. The articular head of the femur is well defined and dorsomedially oriented (more hook-shaped in prosauropods, even in large plateosaurids and melanorosaurids; see Galton 1985*d*; Gauffre 1993*a*; Wellnhofer 1993). Unlike the condition in

Drawings by Haiyan Tong.

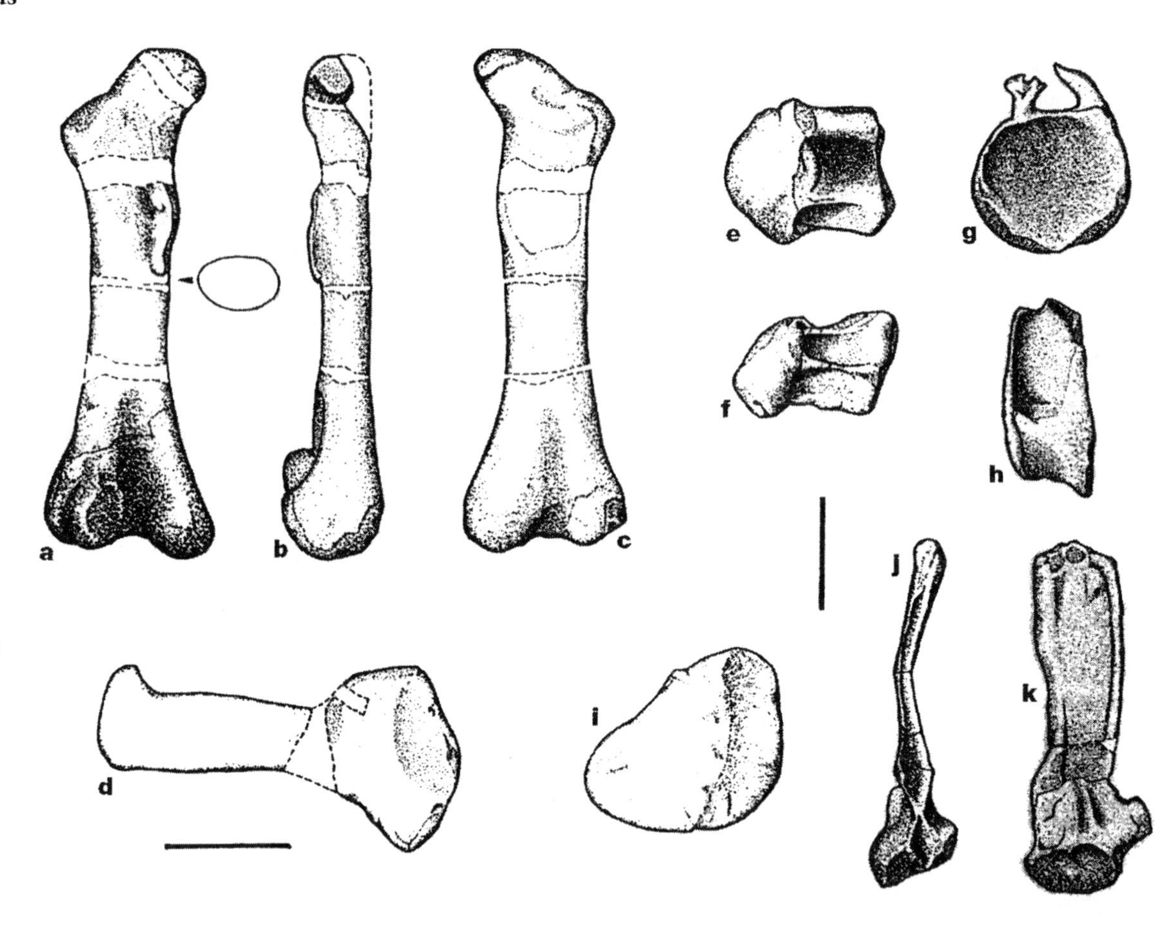

***Isanosaurus attavipachi*, CH4, holotype, left femur (CH4-1) in a. posterior (with cross section at level of distal end of fourth trochanter), b. medial, and c. anterior views, d. right scapula in lateral view, centrum of cervical vertebrae (CH4-3) in e. left lateral and f. ventral views, centrum of posterior dorsal vertebra (CH4-6) in g. posterior and h. left lateral views, i. right sternal plate in external view, neural arch of posterior dorsal vertebra (CH4-7) in j. posterior and k. right lateral views. Scale bars = (horizontal for a–d) 20 cm. and (vertical for e–k) 10 cm. (After Buffetaut, Suteethorn, Cuny, Tong, Le Loeuff, Khansubha and Sutee Jongautchariyakul 2000.)**

prosauropods and *Vulcanodon* (Cooper 1984), there is no evidence of a lesser trochanter. The fourth trochanter in *Isanosaurus* is more prominent than in *Barapasaurus*, *Vulcanodon*, and *Shunosaurus* and is not wing-shaped as in prosauropods. It ends distally in an acute, slightly hook-shaped tip suggesting that in *Barapasaurus* and *Vulcanodon*. The unusual shape of the fourth trochanter, different from the condition in other sauropods, was regarded by Buffetaut as an autapomorphic character of *Isanosaurus*, which otherwise exhibits features that are plesiomorphic for Sauropoda. Also, the "strongly expanded distal end of the femur shows massive condyles, a well developed ectepicondyle (not usually seen in prosauropods), and no longitudinal crest proximal to the lateral condyle (unlike prosauropods" (see Gauffre).

Noting the primitive features exhibited by *I. attavipachi*, Buffetaut *et al.* found them not to be suggestive of the Prosauropoda, but rather illustrative of an early stage in the evolution of characters that would be more developed in more advanced members of the Sauropoda. As the interrelationships of other known primitive sauropods remain uncertain, and because comparisons between *Isanosaurus* and those sauropod genera are difficult to make, the authors referred that genus to Sauropoda *incertae sedis*.

As pointed out by Buffetaut *et al.*, the discovery of this dinosaur is important, substantiating theoretical predictions that sauropods had already appeared by Late Triassic time and that the period of their evolution was longer than previously known. It demonstrates that the earliest sauropods coexisted with another group of large-bodied sauropodomorphs, the melanorosaurid prosauropods.

Also, the facts that Thailand and China were joined during the Late Triassic and the earliest known sauropod is an Asian form, and also that *Vulcanodon* documents the presence of sauropods in Africa at the beginning of the Jurassic, all suggest "that by the time of the Triassic–Jurassic boundary, sauropods already had a vast geographical distribution, doubtless made possible by Pangaeoan palaeogeographical conditions."

Key references: Bonaparte (1999*b*); Buffetaut, Suteethorn, Cuny, Tong, Le Loeuff, Khansubha and Sutee Jongautchariyakul (2000); Cooper (1984); Dong, Zhou and Zhang (1983); Ellenberger and Ellenberger (1958); Galton (1985*d*); Gauffre (1993*a*); He, Wang, Liu, Zhou, Liu, Cai and Dai (1998); Jain, Kutty Roy-chowdhury and Chatterjee (1979); Racey, Love and Polachan (1996); Upchurch (1998); Wellnhofer (1993); Zhang (1988).

†JANENSCHIA—(=?*Tendaguria*)

Saurischia: Sauropodomorpha: Sauropoda: Eusauropoda: Neosauropoda: Macronaria: Camarasauromorpha: ?Camarasauridae.

Occurrence: Upper Tendaguru Beds, Mtwara, Tanzania, East Africa.

Age: ?Late Jurassic.

Known material: Postcranial remains including, limb and vertebral elements, representing at least four individuals.

Holotype: SMNS 12144, postcranial remains including distal section of femur, tibia, fibula, astragalus, metatarsals I–V and corresponding phalanges.

Diagnosis (as for type species): Sauropod 15 to 20 meters (about 52 to 68 feet) in length, with tibia bearing two cnemial trochanters, one internal on line of cnemial crest, second in more lateral plane and separated from former by well-defined depression; previous character associated with dorsoventrally low but transversely broad astragalus, with posterior indentation making medial half anteroposteriorly distinct from lateral half of astragalus; pes rather primitive, with well-developed metatarsals I–IV and ungual phalanges in digits I–III; metatarsal V derived in being unusually smaller than metatarsal IV; metatarsal V enlarged proximally (Bonaparte, Heinrich and Wild 2000).

Comments: In 1908, Professor Eberhard Fraas of Stuttgart described the remains of sauropod dinosaurs found the previous year and collected during the German Tendaguru expedition (1909–1913) in the Upper Tendaguru Beds near Tendaguru Hill, in what is now Mtwara, East Africa. Upon an well-preserved, incomplete hindlimb (SMNS 12144; see *D:TE* for photograph) Fraas erected the new genus *Gigantosaurus*, referring to it two species, *Gigantosaurus africanus* (the type species) and *G. robustus*. As the generic name was found to be preoccupied (Seeley 1869), Sternfield (1911) later renamed Fraas' material *Tornieria*, revising the spelling of the type species to *Tornieria africana* and the referred species to *T. robusta*. Next, Janensch (1922) referred *T. africana* to the species *Barosaurus africanus*, the referred species thereby being left without a valid generic name. More than half a century later, Wild (1991) referred this available species to a new genus which he named *Janenschia* (see *D:TE*).

As material referred to *Tornieria* sp. was recorded by Raath and McIntosh (1987) from the Kadzi Formation of Zimbabwe, and because the dinosaur bed of this formation seems to be equivalent to the marine *Trigonia smeei* Bed of Tendaguru, both units were tentatively dated by these authors as Late Jurassic.

In 2000, Bonaparte, Heinrich and Wild published a review of the type species *Janenschia robusta* in which these authors rediagnosed the type species and described in detail all the material referrable to this taxon.

Bonaparte *et al.* noted that, although Fraas (1925, 1929*a*, 1961) had referred various other Tendaguru specimens including caudal vertebrae to *J. robusta*, only the material with comparable parts of the skeleton can be referred with confidence to that species. These specimens included the following: MB. R2095.1-13, two tibiae (only one figured by Janensch in 1961), two fibulae (one almost complete), two left astragali, an almost complete left humerus (similar to that of *Camarasaurus*), two radii, and a poorly-preserved left manual ungual, from the Upper Saurian Bed (material originally cited by Janensch 1929*a*; figured by Janensch 1961); MB.R2096.1, a right femur from the Upper Transitional Sands (Janensch 1961); MB.R2096.2, a left radius (smaller than MB.R.2095.12) from the Upper Transitional Sands (Janensch 1961); and the tentatively assigned MB.R.2093.1-12, an almost complete right manus resembling that of *Camarasaurus*, originally referred by Janensch (1922) to *Gigantosaurus robustus*) and later (1961) to *Torneria robusta*, collected near the type locality of *J. robusta*.

Also, Bonaparte *et al.* noted that a significant character of the manus of *J. robusta* is the general proportions of the metacarpals, which are robust, proportionally short and thick, and have a length to width ratio of about 3:1.

While admitting that comparative material is quite incomplete, Bonaparte *et al.* noted that the hindlimb in the holotype of *J. robusta* allows some indications of possible relationships with other sauropods. In the past, *Janenschia* has been regarded as a possible titanosaurid. However, anatomical differences between the hindlimb of this species and those in *Brachiosaurus brancai*, *Dicraeosaurus hansemanni*, *Mamenchisaurus hochuanensis*, *Diplodocus carnegii*, ?*Barosaurus africanus*, and *Saltasaurus loricatus* suggested that *J. robusta* is not related to any of these taxa; hence, it is not referrable to the Brachiosauridae, Dicraeosauridae, Euhelopodidae (=Mamenchisauridae of their terminology), Diplodocidae, or Titanosauridae.

On the other hand, Bonaparte *et al.* observed a number of similarities between the hindlimb of *J. robusta* and that of the camarasaurid *Camarasaurus supremus* (see Osborn and Mook 1921). The tibia of *J. robusta* resembles that of *C. supremus* in "the lateral bending of the cnemial crest, the lateral expansion of the lower shaft, and the flat anterior surface of the lower shaft," although the lateral process of the cnemial crest, present in *Janenschia*, is not in *Camarasaurus*. The fibula in both taxa are similar, although in *Camarasaurus* there is a more pronounced concavity along the distal shaft of the internal face of this

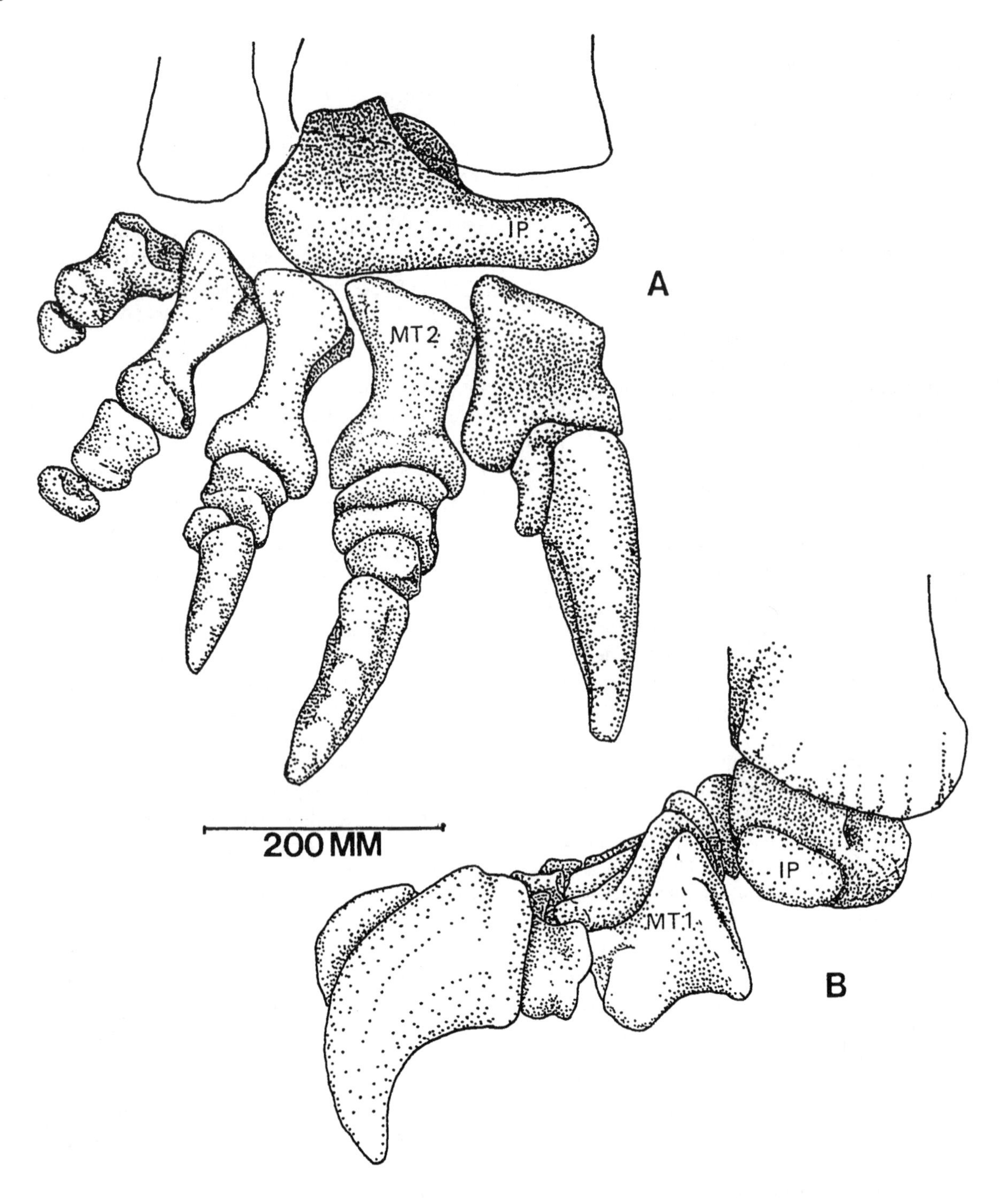

***Janenschia robusta*, SMNS 12144, holotype right foot and astragalus of *Gigantosaurus robustus* in A. dorsal and B. medial views. (After Bonaparte, Heinrich and Wild 2000.)**

bone. The astragalus in both taxa are similar in their very small proportions, the medial half being flat, the anterior border almost straight, the posterior border strongly angular, the latter "narrowing the medial half of the astragalus anteroposteriorly into a sort of internal process." The third metatarsal (the only element available to the authors for comparison) of *Camarasaurus* corresponded favorably in morphology with that of *Janenschia*. Both taxa also compared well in the ratio between the lengths of the tibia and humerus.

Bonaparte *et al.* suggested that the astragalus in *Brachiosaurus*, *Diplodocus*, and *Dicraeosaurus* exhibits the pelsiomorphic condition of being proportionally high laterally, gradually diminishing medially in both height and width; it is wedge-shaped, internally low and narrow, laterally high and quadrangular. This bone in *Janenschia* and *Camarasaurus* is low medially, flat dorsally, narrow anteroposteriorly; its posterior border is indented, thickening internally to form a medial process, the external half having a more conservative shape.

The authors, therefore, concluded that a close relationship exists between *Janenschia* and *Camarasaurus*, possibly at the family level with both genera belonging to the Camarsauridae, or, less likely, to different but related families.

Note: Bonaparte *et al.* reassessed other material from Tendaguru which, in the past, had been referred to *Janenschia*.

Janensch (1929*a*) had referred an almost complete tail (MB.R.2091.1-30) missing the distal end, comprising 30 articulated vertebrae and including two incomplete chevrons, to "*Gigantosaurus robustus*." In their detailed description of this material, Bonaparte *et al.* noted that the anterior eight vertebrae in this series were affected on the left side by weathering. The vertebrae in this series exhibits decreasing procoely posteriorly; the first six vertebrae are strongly procoelous, the succeeding four slightly procoelous, the next three almost amphiplatyan (anterior side of the centrum concave), and the last 17 gently amphicoelous. As this material could not be compared with that of *Janenschia*, it was not referred to that genus.

MB.R.2094, a larger isolated caudal vertebra from Obolello, near Tendaguru Hill, referred by Janensch (1929*a*) to *Giganotosaurus*, likewise could not be assigned to *Janenschia*. According to Bonaparte *et al.*, this specimen could pertain to a more adult individual of the same species represented by MB.R.2091.1-30, exhibiting different proportions and a more pronounced procoely.

Key references: Bonaparte, Heinrich and Wild (2000); Fraas (1908); Janensch (1922, 1925*b*, 1929*a*, 1961); Osborn and Mook (1921); Raath and McIntosh (1987); Seeley (1869); Sternfield (1911); Wild (1991).

JEHOLOSAURUS—Xu, Wang and You 2000

Ornithischia: Genasauria: Cerapoda: Ornithopoda.

Name derivation: "Jehol [old geographical name for western Liaoning and northern Hebei, China]" + Greek *sauros* = "lizard."

Type species: *J. shangyuanensis* Xu, Wang and You 2000.

Other species: [None.]

Occurrence: Yixian Formation, Liaoning Province, China.

Age: ?Late Jurassic (Kimmeridgian) or ?Early Cretaceous (Valanginian).

Known material: Two skulls, incomplete postcrania remains.

Holotype: IVPP V 12529, almost complete skull, partial postcrania.

Diagnosis of genus (as for type species): Small ornithopod differing from other known ornithischians in combination of the following primitive and derived characters: six premaxillary teeth; few foramina on dorsal surface of nasal; large quadrate foramen on lateral side of quadratojugal; predentary approximately 1.5 times length of main body of premaxilla; lack of external mandibular fenestra; absence of anterior intercondylar groove; metatarsals not in same plane; pedal phalanx III-4 longer than other phalanges of pedal digit III (Xu, Wang and You 2000).

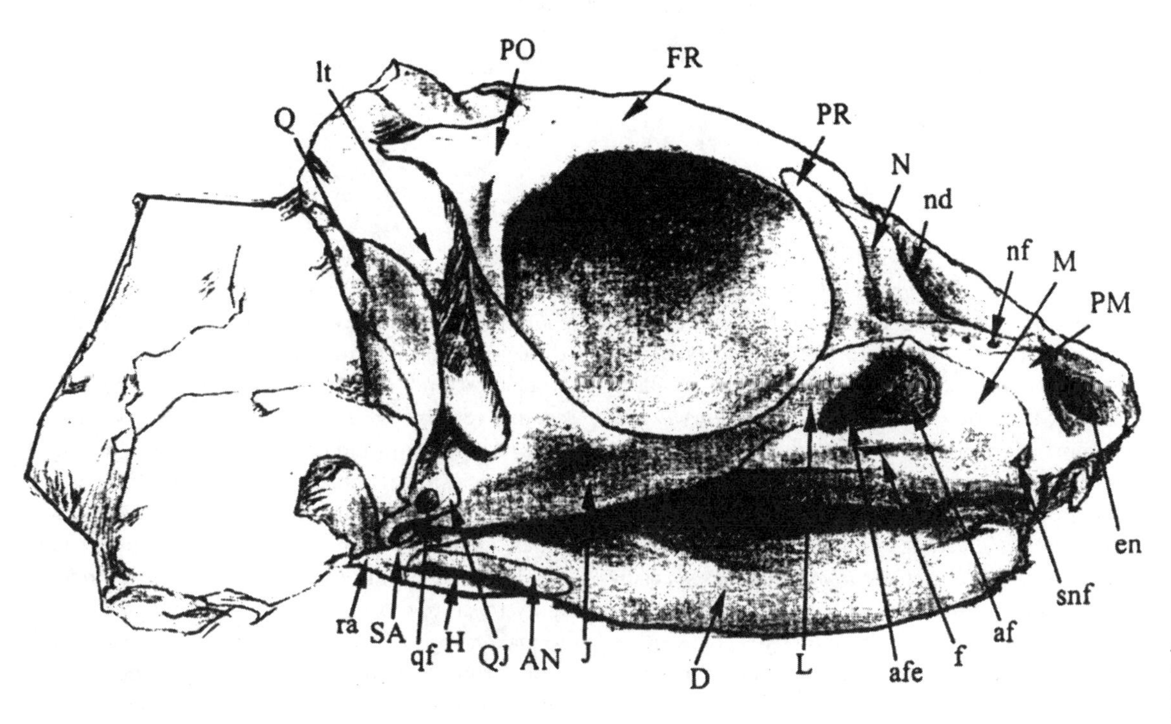

Jeholosaurus shangyuanensis, IVPP V 12530, referred skull, right lateral view. Scale = 10 mm. (After Xu, Wang and You 2000.)

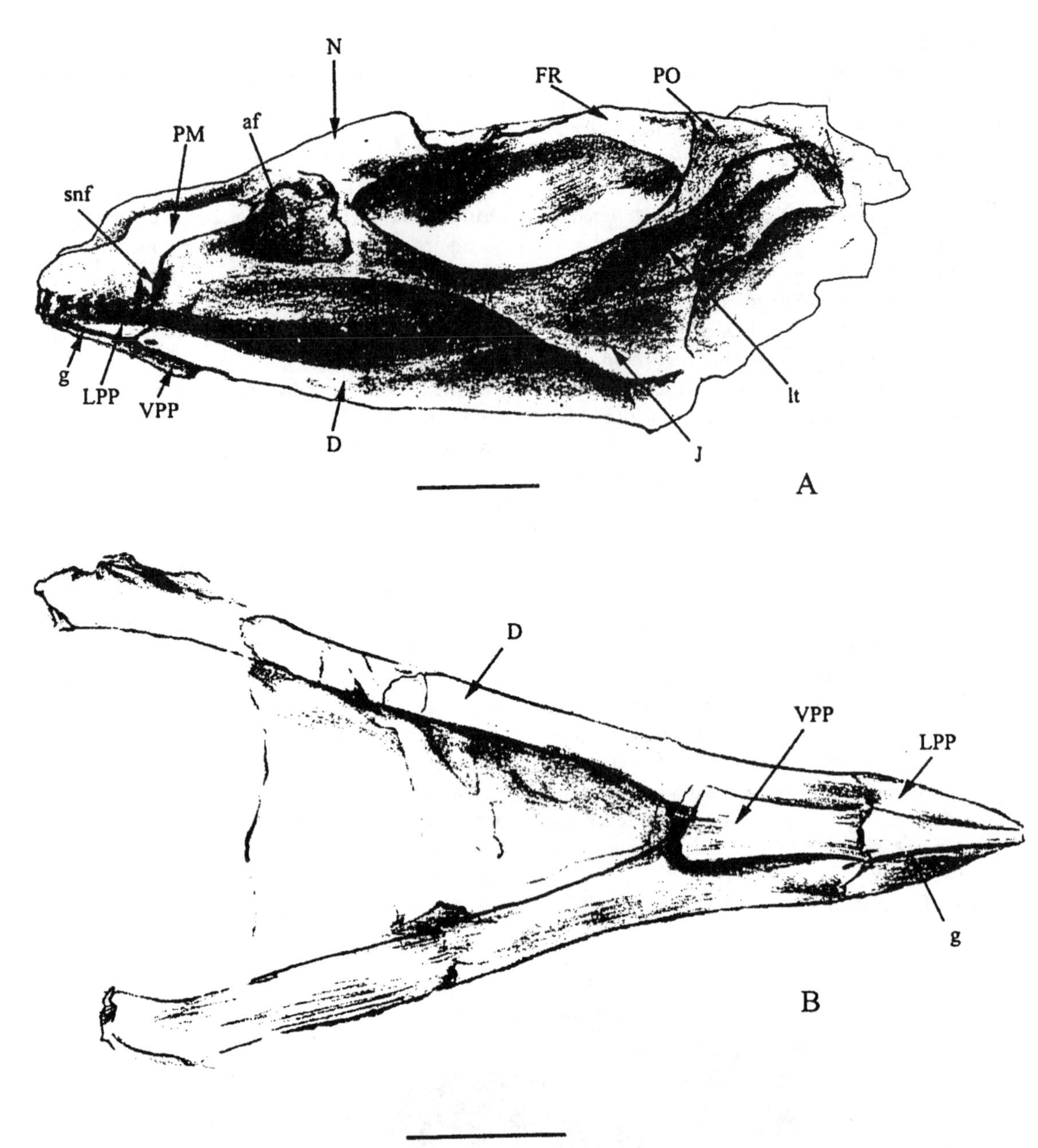

***Jeholosaurus shangyuanensis*, IVPP V 12529, holotype A. skull, left lateral view, and B. mandible, dorsal view. Scale = 10 mm. (After Xu, Wang and You 2000.)**

Comments: The genus *Jeholosaurus* was based upon two specimens collected from the Lujiatum Locality of the Yixian Formation in Lianing Province, Peoples Republic of China. The holotype (IVPP V 12529) of the type species *Jeholosaurus shangyuanensis* comprises a nearly complete skull and partial postcranial remains; a referred specimen (IVPP V 12530) consists of an almost complete skull and some cervical vertebrae (Xu, Wang and You 2000).

As noted by Xu *et al.*, *Jeholosaurus* is especially interesting in its possession of both primitive and derived features. With euronithopods this genus shares the following synapomorphies: Small antorbital fenestra; quadrate foramen located on lateral aspect of quadratojugal; large quadrate foramen; and absence of external mandibular fenestra (Sereno 1999*c*). Also, the morphology of the proximal portion of the femur is similar to derived ornithopods in the finger-like lesser trochanter and anteroposteriorly wide greater trochanter (Sereno 1986). The long predentary suggests that *Jeholosaurus* could have an immobile mandibular symphsis as in ceratopsians (Sereno 1986). However, *Jeholosaurus* lacks various ornithopod autapomorphies, including the following: Presence of premaxilla-lacrimal; jaw articulation ventrally offset to maxillary tooth row, and premaxillary tooth row ventral to maxillary tooth row (Sereno 1999*c*). *Jeholosaurus* is more primitive than other ornithopods and marginocephalians in the following features: Possession of six premaxillary teeth (Sereno 1986; Weishampel

1990), short edentulous anterior portion of premaxilla, and short diastema between premaxillary and maxillary teeth.

Xu *et al.* found it notable that *Jeholosaurus* is similar in some ways to some primitive ornithischians from China, a longitudinal fossil also being present along the midline of the nasals in *Agilisaurus louderbacki* (Peng 1997) and *Yandusaurus multidens* (He and Cai 1984), the latter two taxas also sharing a somewhat dumbbell-shaped lower temporal fenestra (possibly a plesiomorphic feature also observed in *Lesothosaurus*). Also, *Jeholosaurus* shares with *Xiaosaurus* a long pedal phalanx III-4. These similarities, the authors noted, suggest that the above Chinese taxa could form a monophyletic group, although such an assessment could not yet be confirmed due to insufficient evidence.

Key reference: He and Cai (1984); Peng (1997); Sereno (1986, 1999*c*); Weishampel (1990); Xu, Wang and You (2000).

JENGHIZKHAN Olshevsky, Ford and Yamamoto 1995—(See *Tyrannosaurus.*)

Name derivation: Latinization of [Mongol tyrant] "Ghengis Khan."

Type species: *J. bataar* (Maleev 1955).

JOBARIA Sereno, Beck, Dutheil, Larrson, Lyon, Moussa, Sadleir, Sidor, Varrichio, Wilson and Wilson 1999

Saurischia: Sauropodomorpha: Sauropoda: Eusauropoda.

Name derivation: Tamacheck *Jobar* [mythical creature] + Greek *ia* = "pertaining to."

Type species: *J. tiguidensis* Sereno, Beck, Dutheil, Larrson, Lyon, Moussa, Sadleir, Sidor, Varrichio, Wilson and Wilson 1999.

Other species: [None.]

Occurrence: Tiourarén Formation, Taméret, Fako, Tawachi, other localities, Republic of Niger.

Age: Early Cretaceous (Hauterivian).

Known material: Several partial skeletons including skulls, gastralia, isolated bones, adult and subadult, all of the skeleton represented.

Holotype: MNN TIG3, partial articulated skeleton including axis, fore- and hindlimbs, pubes, most caudal vertebrae.

Diagnosis of genus (as for type species): Eusauropod characterized by cervical prezygapophyses having accessory anterior process, cervical neural arches with deep fossa between centropostzygapophyseal and intrapostzygapophyseal laminae, dorsal prezygapophyses with broad pendant flange, dorsal

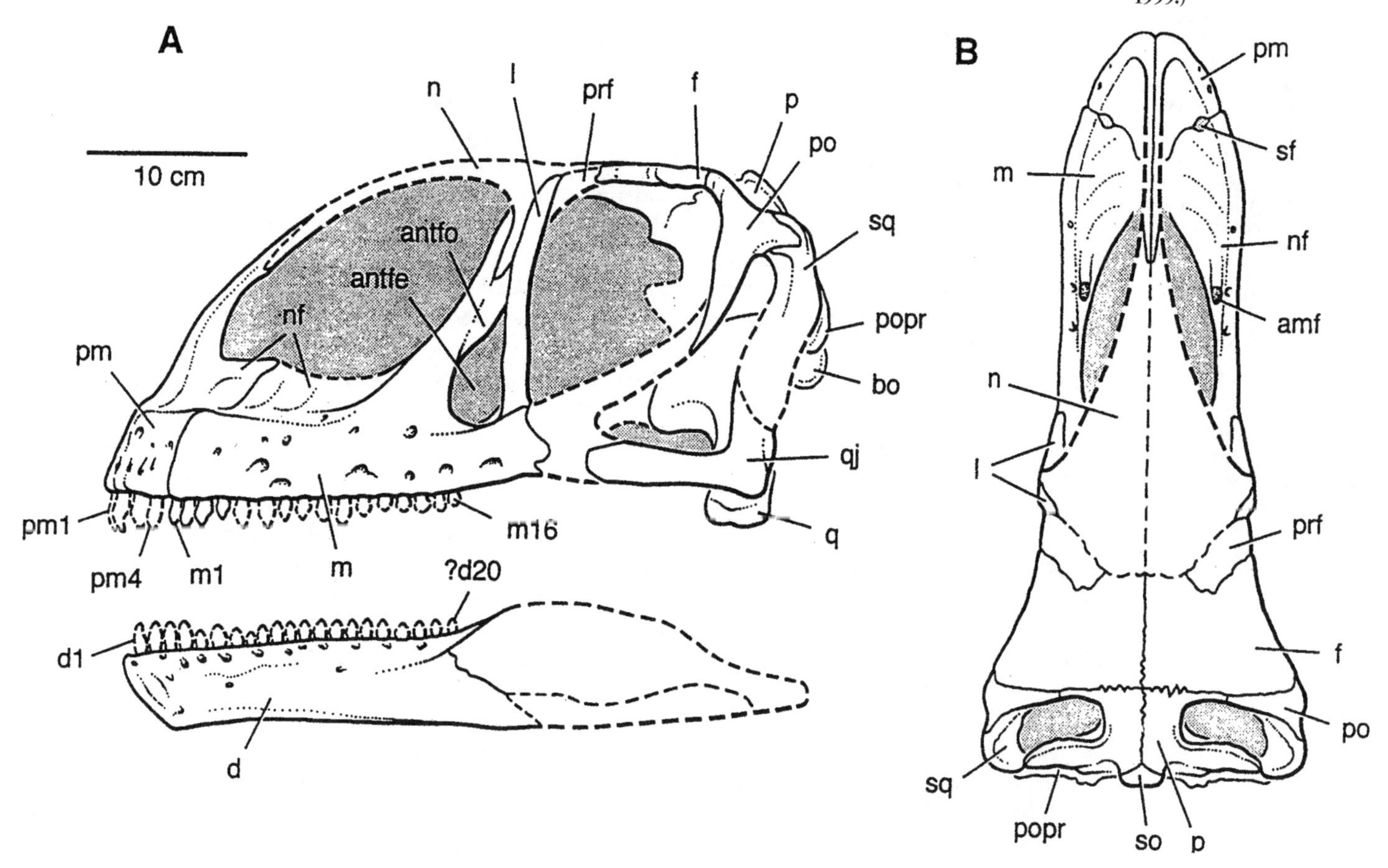

***Jobaria tiguidensis*, skull based on MNN TIG 3-5, including holotype, in A. left lateral and B. dorsal views. (After Sereno, Beck, Dutheil, Larrson, Lyon, Moussa, Sadleir, Sidor, Varrichio, Wilson and Wilson 1999.)**

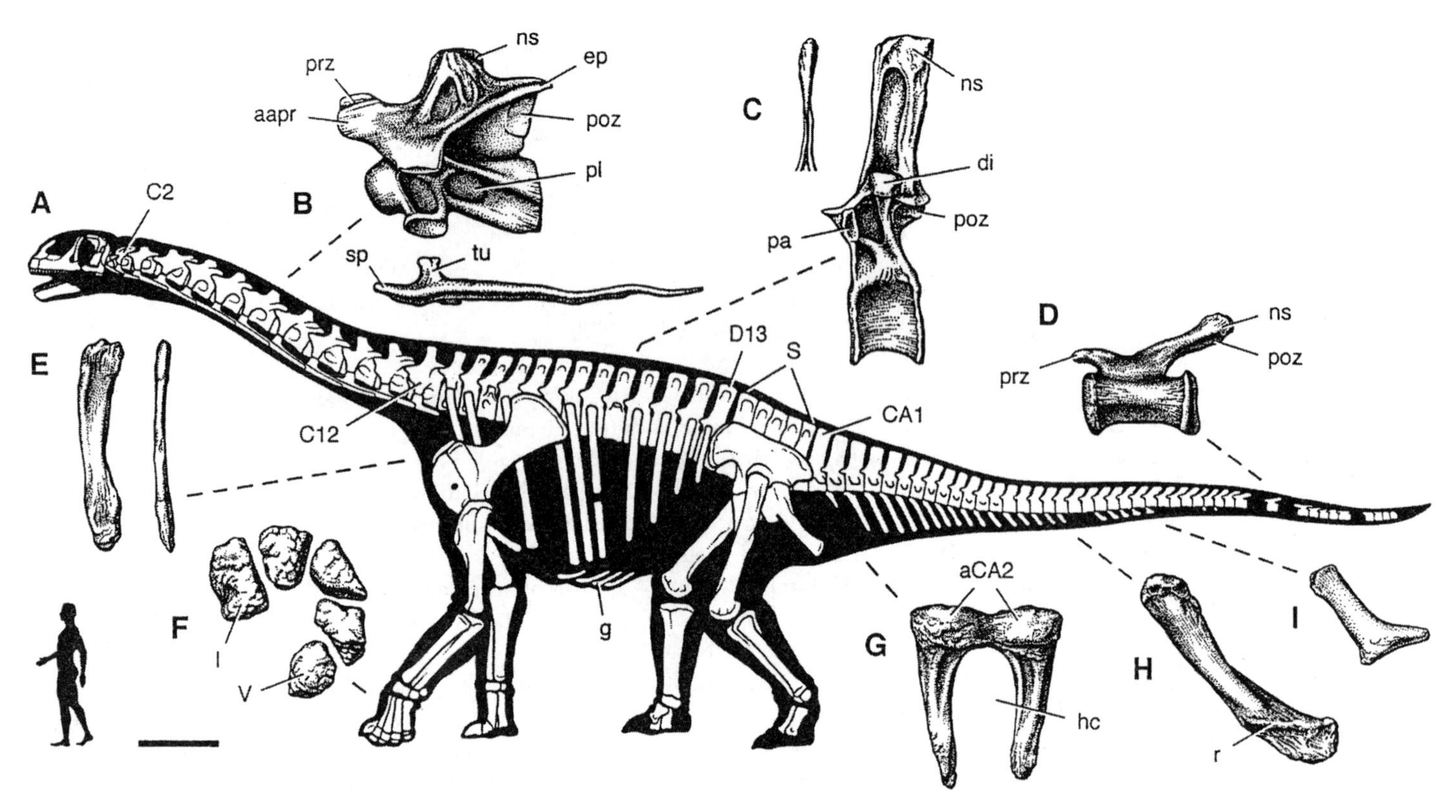

***Jobaria tiguidensis*, approximately 90 percent complete skeletal reconstruction, illustrating A. preserved bones, B. cervical vertebra 7 and aassociated rib (left lateral view), C. dorsal vertebra 9 (left lateral view) and spine (posterior view), D. caudal vertebra 35 (left lateral view), E. right clavicle (anterior view, ventral end toward top), F. right metacarpus (proximal view, anterior side toward top), G. chevron 1 (anterior view), H. chevron 16 (lateral view), and I. chevron 24 (left lateral view). Scale for A. = 1 m, B. and C. = 20 cm., D., E., G., H. and I. = 10 cm., and F. = 15 cm. (After Sereno, Beck, Dutheil, Larrson, Lyon, Moussa, Sadleir, Sidor, Varrichio, Wilson and Wilson 1999.)**

neural arches with fossa between parapophyses and diapophyses, anterior caudal neural spines with cervical ribs 3 to 6 having accessory anterior process, U-shaped first chevron, and midcaudal chevrons with rugose ridge across distal end of blade (Sereno, Beck, Dutheil, Larrson, Lyon, Moussa, Sadleir, Sidor, Varrichio, Wilson and Wilson 1999).

Comments: An unusual sauropod (and one that contributes important new information considering the evolution of the Sauropoda in Africa) and the most primitive Cretaceous sauropod known, the genus *Jobaria* was based on a number of incomplete skeletons and isolated bones collected by a University of Chicago field team led by Paul C. Sereno at several clay-rich localities in the "Neocomian" (=Hauterivian) Tiourarén Formation of central Niger, in northern Africa. The team had been alerted to the bones by Taureg tribesmen who attributed them to a mythical creature called a "Jobar." The holotype consists of a partial articulated skeleton (MNN TIG3). Referred remains include several partial skeletons and isolated elements from Fako, Tawechi, and other localities near the base of the Falaise de Tiguidi, this material including a partial adult skull and skeletons (MNN TIG4 and TIG5), a relatively complete subadult skeleton (MNN TIG6), and a subadult braincase (MNN TIG7). The type species *J. tiguidensis* proved to be the most abundant terrestrial vertebrate in the formation, the remains of no other large-bodied herbivore being recovered from these sites. Other dinosaurs represented at these sites included the theropod *Afrovenator* and two smaller-bodied theropods of uncertain affinities (Sereno, Beck, Dutheil, Larrson, Lyon, Moussa, Sadleir, Sidor, Varrichio, Wilson and Wilson 1999).

Although not yet named, the new dinosaur was first mentioned — very briefly described as a broad-toothed sauropod — although not named, by Sereno, Wilson, Larsson, Duthell and Sues (1994) in their paper naming and describing *Afrovenator.*

Comparing *Jobaria* with other sauropods, Sereno *et al.* (1999) observed that the adult skull is approximately the length of that of a subadult *Camarasaurus*, the latter possessing substantially longer limb bones. Unlike macronarians such as *Brachiosaurus*, the external naris of *Jobaria* is especially large, the snout anteriorly short. At least 20 teeth are present in the upper and lower jaws; the tooth crowns are spatulate, possessing a variable number of marginal denticles.

As described by Sereno *et al.* (1999), this new genus has a relatively short neck comprising only 12 vertebrae. One specimen was preserved with the articulated neck "in a fully dorsi-flexed, C-shaped posture, suggesting that the cervical column had considerably more latitude in the sagittal plane than suggested recently for some other sauropods" (see Stevens and Parrish 1999). Limb proportions are primitive, the forelimb not being as elongate compared to the hindlimb as in *Brachiosaurus*, the manus, relative to the forearm, comparatively shorter than in *Camarasaurus* and other macronarians.

Reconstructed skeleton (cast by the PAST Company) of the adult *Jobaria tiguidensis* on exhibit at Dinofest (2000–2001), held at Navy Pier, Chicago. This skeleton measures more than 17.5 meters (approximately 60 feet) in length.

Photograph by the author.

Photograph by the author.

Reconstructed skeletal casts by the PAST Company of the theropod *Afrovenator abakensis* and a juvenile of the sauropod *Jobaria tiguidensis*, the latter measuring almost 12 meters (about 40 feet) long, displayed at Dinofest (2000–2001), Navy Pier, Chicago.

The authors reconstructed the skeleton (see figure) of *J. tiguidensis* as measuring 18 meters (more than 60 feet) in length, estimating the maximum adult length to be approximately 21 meters (about 66 feet).

Sereno *et al.* (1999) noted that, despite its age, "*Jobaria* is strikingly primitive," as indicated by such features as "the abbreviated snout, terminally positioned nares, lower number of cervical vertebrae, simple neural spines," plus numerous additional features. This primitive condition suggested to the authors that *Jobaria* lies outside the radiation of the Neosauropoda, a group which includes all other known Cretaceous sauropods. Therefore, this genus "represents an unknown lineage of broad-toothed sauropods that had diverged by the Middle Jurassic some 30 to 40 million years earlier," a time interval during which few skeletal changes took place.

The discovery of *Jobaria* and also *Nigersaurus* (see entry), a recently described basal diplodocoid from Niger, offered to the authors a framework for better understanding the history and evolution of the Sauropoda in Africa during the Cretaceous. According to Sereno *et al.* (1999), a minimum of three sauropod lineages survived in Africa into that period: 1. A primitive broad-toothed group (*e.g.*, *Jobaria*), which flourished during the Hauterivian but is not known from younger rocks; 2. rebbachisaurids (*e.g.*, *Nigersaurus*), known from the Aptian–Albian through the Cenomanian, though not from younger rocks; and 3. basal titanosaurs (*e.g.*, *Malawisaurus*), known only rarely from Middle to Early Cretaceous rocks, at least through the Cenomanian.

Performing a quantitative analysis of skeletal changes among nonavian dinosaurian taxa in general over time, Sereno *et al.* (1999) found that the rates of these changes in sauropods and other major dinosaurian groups are highly variable, with a slow rate of evolutionary change apparently characterizing the lineage to which *Jobaria* belongs.

A reconstructed cast skeleton of *J. tiguidensis* was mounted by the PAST company (Prehistoric Animal Structures, Inc.) of Alberta, under Sereno's direction, for temporary display in November, 1999 at the National Geographic Society in Washington, D.C. Another PAST cast — accompanied by one of 27-foot long, contemporaneous predator *Afrovenator abakensis*—

Photograph by Allen A. Debus.

Reconstruction of the head and neck of ***Jobaria tiguidensis*** **by Gary Staab, in the temporary "Dinosaur Giants" exhibit at Chicago's Navy Pier. The opposite side is a life restoration (this exhibit suggested by dinosaur art collector John Lanzendorf).**

went on temporary display as part of "Dinosaur Giants," from January 14 to March 19, 2000 an exhibition created by Project Exploration at the Crystal Gardens, Navy Pier, Chicago (see Debus and Debus 2000 for more details on this exhibit).

Featured in the above display was a life-sized restoration of the head and neck of *Jobaria* revealing the muscular system, pneumatic air sacs, and skin. As reported in an abstract for a poster by Sanders, Wedel, Sereno and Staab (2000), soft tissues in this reconstruction were based upon dissections of ratite birds (*Struthio camellus* and *Rhea americana*). As Sanders *et al.* pointed out, most osteological characters of the cervical vertebrae of sauropods are homologous with those of birds, indicating that the origins and insertions of most muscles are basically the same in both of these groups. The authors noted that, although "the pneumatic characters in the vertebrae of *Jobaria* are less extensive and complex than those of either derived sauropods or birds, large pneumatic fossae are present on every cervical vertebra except the atlas." Avian vertebrae are pneumaticized by diverticulae that arise from pulmonary air sacs in the body cavity. A similar system, Sanders *et al.* suggested, may have also been present in the body cavity of this sauropod.

Notes: Prior to the issuance of the joint paper in which *Jobaria* was officially named and described, various newspaper articles — published on November 12, 1999, in such papers as the *Chicago Sun-Times* and *Chicago Tribune*— announced the dinosaur's discovery. Paul Sereno was quoted stating that one juvenile *Jobaria* specimen shows healed teeth marks across the ribs from teeth quite like those of *Afrovenator*, this suggesting that the individual died in an attack or was scavenged after death.

In another newspaper article, co-authored by Paul C. Sereno and Gabrielle H. Lyon, published in the November 20, 2000 edition of the *Chicago Sun-Times*, the authors reported on a subsequent expedition to the *Jobaria* locality. During this expedition, Sereno and Lyon stated, a shoulder blade, front limb, neck, plus a 10-foot-long hind limb of *Jobaria* were found. The latter, the writers reported, includes the hind foot, the only part of the skeleton that had previously been unknown in this dinosaur.

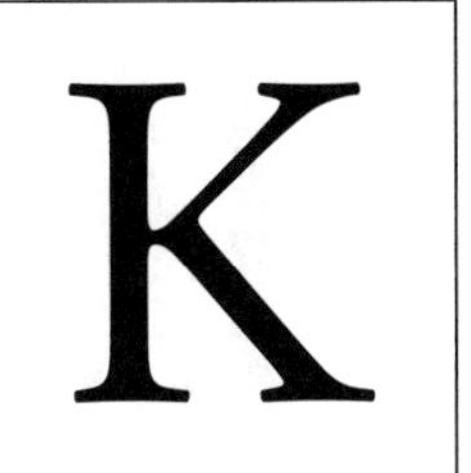

(Also reported by Sereno and Lyon in their article, found near these additional *Jobaria* remains, were bones and small, bony plates with keels, belonging to a small ornithischian. This six-foot-long skeleton, to be described in 2001, is the first of its kind found in Africa and seems to be "a distant relative of our North American armored dinosaurs.")

Key references: Debus and Debus (2000); Sanders, Wedel, Sereno and Staab (2000); Sereno, Beck, Dutheil, Larrson, Lyon, Moussa, Sadleir, Sidor, Varrichio, Wilson and Wilson (1999); Sereno, Wilson, Larsson, Duthell and Sues (1994); Stevens and Parrish (1999).

†**KRITOSAURUS**—(=*Anasazisaurus, Naashoibitosaurus*; =?*Hadrosaurus*)

Ornithischia: Genasauria: Cerapoda: Ornithopoda: Euornithopoda: Iguanodontia: Euiguanodontia: Dryomorpha: Ankylopollexia: Hadrosauroidea: Hadrosauridae: Euhadrosauria: Hadrosaurinae.

Diagnosis: Differs from all other known hadrosaurines by possessing nasal arch that extends above orbits and (in adults) posteriorly to between orbits, nasals bifurcated posteriorly to accept median tongue of frontal and overlapped by extensive medial wings of prefrontals so that it appears in dorsal view that transverse width of nasals is narrower than transverse width of prefrontals, and circumnarial depression that, at its posterior extent, extends from near dorsal margin of nasals laterally onto prefrontals and lacrimals (Williamson 2000).

Comments: The status of *Kritosaurus* as a valid genus has long been under debate, but may at last have been resolved by Thomas A. Williamson (2000).

In brief, as related by Williamson, the type species *Kritosaurus navajovius* was established by American Museum of Natural History collector Barnum Brown (1910) on an incomplete skull (AMNH 5799) from the Ojo Alamo Sandstone, San Juan Basin, New Mexico (see *D:TE* for details and photograph of the specimen). Lull and Wright (1942), in their classic review of North American hadrosaurs, regarded *Kritosaurus* as a senior subjective synonym of *Gryposaurus* (see *D:TE* and *S1*). Baird and Horner (1977) considered *Kritosaurus* to be a junior synonym of *Hadrosaurus*, even though no cranial material is known for the latter genus. Later, Brett-Surman (1989) split these two genera again, preferring to retain *Hadrosaurus* as a valid genus pending the possible discovery of cranial material belonging to this taxon.

Weishampel and Horner (1990) then regarded *Gryposaurus* as a valid genus, but found the holotype of *K. navajovius* to be probably too incomplete to have diagnostic value. Subsequently, Horner (1992), stating that the more complete and diagnostic specimens NMMNH P-16106 (collected by David D. Gillette and David A. Thomas) and BYU 12950 (collected by Brooks Britt), both from the Kirtland Formation of New Mexico, are almost identical to AMNH 5799 and quite different from *Gryposaurus*, revealing characters indicating *K. navajovius* to be valid. Therefore, Horned retained *K. navajovius* but noted that its holotype is missing the majority of diagnostic features (*e.g.*, nasals, anterior prefrontals, lacrimals, anterior jugals, premaxillae, portions of maxillae).

Hunt and Lucas (1992) at first assessed *Kritosaurus* to be a valid genus distinguished from most other hadrosaurines by various features (*e.g.*, skull with short, high lateral profile, lack of posterior extension of nasals dorsal to orbit, much larger lateral temporal fenestra, differences in the ilium [although AMNH 5799 does not include the ilium]) and also a subjective senior synonym of *Gryposaurus*. Later, Hunt and Lucas (1993) regarded *N. novajovious* as a *nomen dubium*, as the holotype does not exhibit diagnostic characters. At the same time, these authors referred the two skulls previously mentioned by Horner to their own new taxa on the basis of differences in the morphology of the nasal area, and also because each were collected from different stratigraphic units with different faunas. Thus, NMMNH P-16106 became the holotype of the new genus and species *Naashoibitosaurus ostromi* and BYU (12950) that of *Anasazisaurus horneri* (see entries in *D:TE* for details and illustrations).

The holotype of *K. novajovious*—heavily restored in plaster by Brown after the more complete holotype skull (CMN 2278) of *Gryposaurus notabilis*—is on display in a glass case at the American Museum of Natural History and inaccessible. Nevertheless, Williamson was able to observe sufficient characters in AMNH 5799 to diagnose the specimen at both the generic and specific levels. Additionally, Williamson noted that AMNH 5799 also offers details of the atlas, a part of the holotype specimen that has never been illustrated.

Regarding the new taxa erected by Hunt and Lucas (1993), Williamson cited the diagnoses for these taxa as proposed by Hunt and Lucas (1993) and compared their distinguishing characters with those of *Kritosaurus navajovius*.

Williamson found sufficient morphological information preserved in the holotype skulls of *K.*

Opposite: **Sculpture of *Jobaria tiguidensis* by artist Michael Trcic made for the "Dinosaur Giants" exhibition (2000), Navy Pier, Chicago. Photograph by Allen A. Debus.**

navajovius and *A. horneri* "to indicate that both specimens share a similar crest morphology and that this morphology differs from that observed in all other hadrosaurs." Conversely, Williamson stated that it "cannot be demonstrated that AMNH 5799 and BYU 12950 differ sufficiently to justify recognition of a separate genus and species for BYU 12950." According to Williamson, the diagnosis of *N. ostromi* "does not allow *Naashoibitosaurus* to be distinguished from *Kritosaurus navajovius* (as demonstrated by the holotype AMNH 5799 and the referred specimen BYU 12950; holotype of *Anasazisaurus horneri*)."

Williamson further criticized distinguishing taxa on anything but morphological ground. Nevertheless, regarding Hunt and Lucas' (1993) claim that the holotypes of *A. horneri* and *N. ostromi* "are from different stratigraphic units whose faunas and superposition show them to be of different ages," Williamson showed that these specimens plus the holotype of *K. navajovius* "are all from the same stratigraphic horizon and are part of similar fauna." Consequently, Williamson regarded both *A. horneri* and *N. ostromi* as junior synonyms of *K. navajovius*.

Key references: Baird and Horner (1977); Brett-Surman (1989); Brown (1910); Horner (1992); Hunt and Lucas (1992, 1993); Lull and Wright (1942); Weishampel and Horner (1990); Williamson (2000).

†LEPTOCERATOPS

Ornithischia: Genasauria: Cerapoda: Marginocephalia: Ceratopsia: Neoceratopsia: Ceratopsomorpha: Ceratopsoidea.

Comments: The first bonebed occurrence of a basal neoceratopsian — in this case, the primitive ceratopsian dinosaur *Leptoceratops* (see entry, *D:TE*) — was reported by Chinnery and Trexler from the Two Medicine Formation (Upper Cretaceous: late Campanian) of Montana. Six subadult specimens have been found in this accumulation. According to the authors, the material is complete enough, preserving almost every skeletal element, to represent a species distinct from the type species *Leptoceratops gracilis*. Additionally, these remains provide new details about one aspect — "the internal structure of the skull and the full extent of sutural elements between skull elements" —

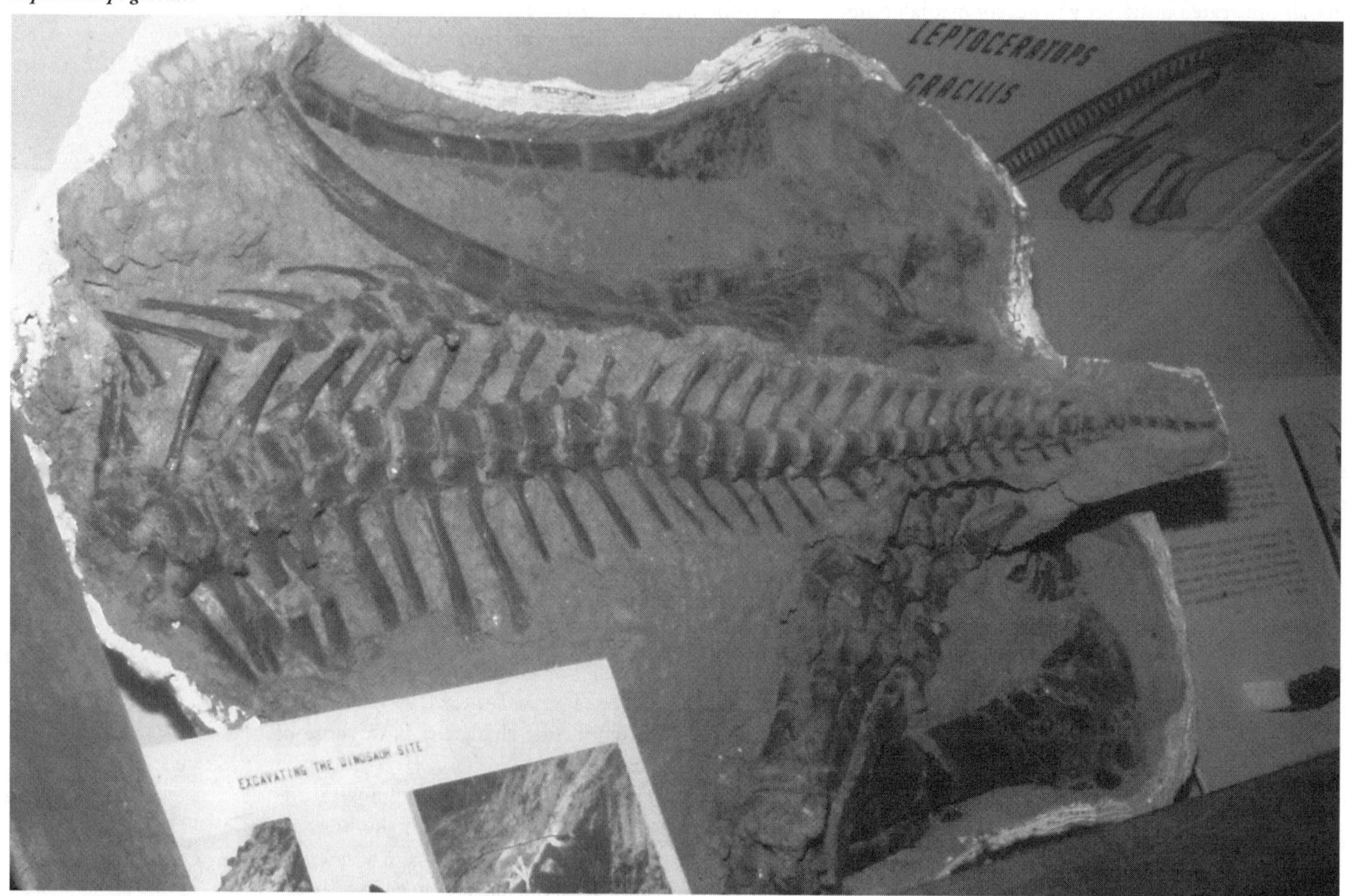

Partial skeleton of a primitive horned dinosaur referred to the type species *Leptoceratops gracilis*.

Photograph by the author, courtesy Princeton Museum of Natural History, Princeton University.

Photograph by the author.

Skull (cast) of *Leptoceratops gracilis* exhibited at Dinofest (2000–2001), held at Navy Pier, Chicago.

of *Leptoceratops*, as well as most neoceratopsians, previously unknown.

Chinnery *et al.* identified these remains as belonging to *Leptoceratops* based upon the following morphological similarities with the holotype (AMNH 5205) and plesiotypes (CMN 8887 and 8888, and 8889) of *L. gracilis*: Face sloping, no nasal horn; frill short; ventral dentary very curved; vertical-notch tooth-wear pattern; unguals tapering; and numerous other unspecified traits.

Nonontogenetic differences seen in the bonebed specimens that argued for referring the material to a new species include the following: Posterior nasal and anterior width across frontals very constricted; very deep frontal depression; coronoid process of dentary exhibiting distinct notch not found on the type species; angular very unusual, having expanded anterior end; metatarsals longer than in type species.

This new species of *Leptoceratops* has yet to be formally named and described.

According to Chinnery and Trexler, two fragmentary specimens from the Two Medicine Formation of Montana, described by Gilmore (1939) as *Leptoceratops* sp., could be sufficiently different from *L. gracilis* to represent a separate species; however, this material is too meager for such an assessment to be made.

Leptoceratops has generally been classified as a protoceratopsid. However, in a recent phylogenetic analysis of the Neoceratopsia, a brief preliminary report of which was published in an abstract, You and Dodson assessed this genus to be a basal member of the superfamily Ceratopsoidea, positioned closer to Ceratopsidae than to Protoceratopsidae.

Key references: Chinnery and Trexler (1999); Gilmore (1939); You and Dodson (2000).

LESSEMSAURUS Bonparte 1999

Saurischia: Prosauropoda: Melanorosauridae.

Name derivation: "[Don] Lessem" + Greek *sauros* = "lizard."

Type species: *L. sauropoides* Bonaparte 1999.

Other species: [None.]

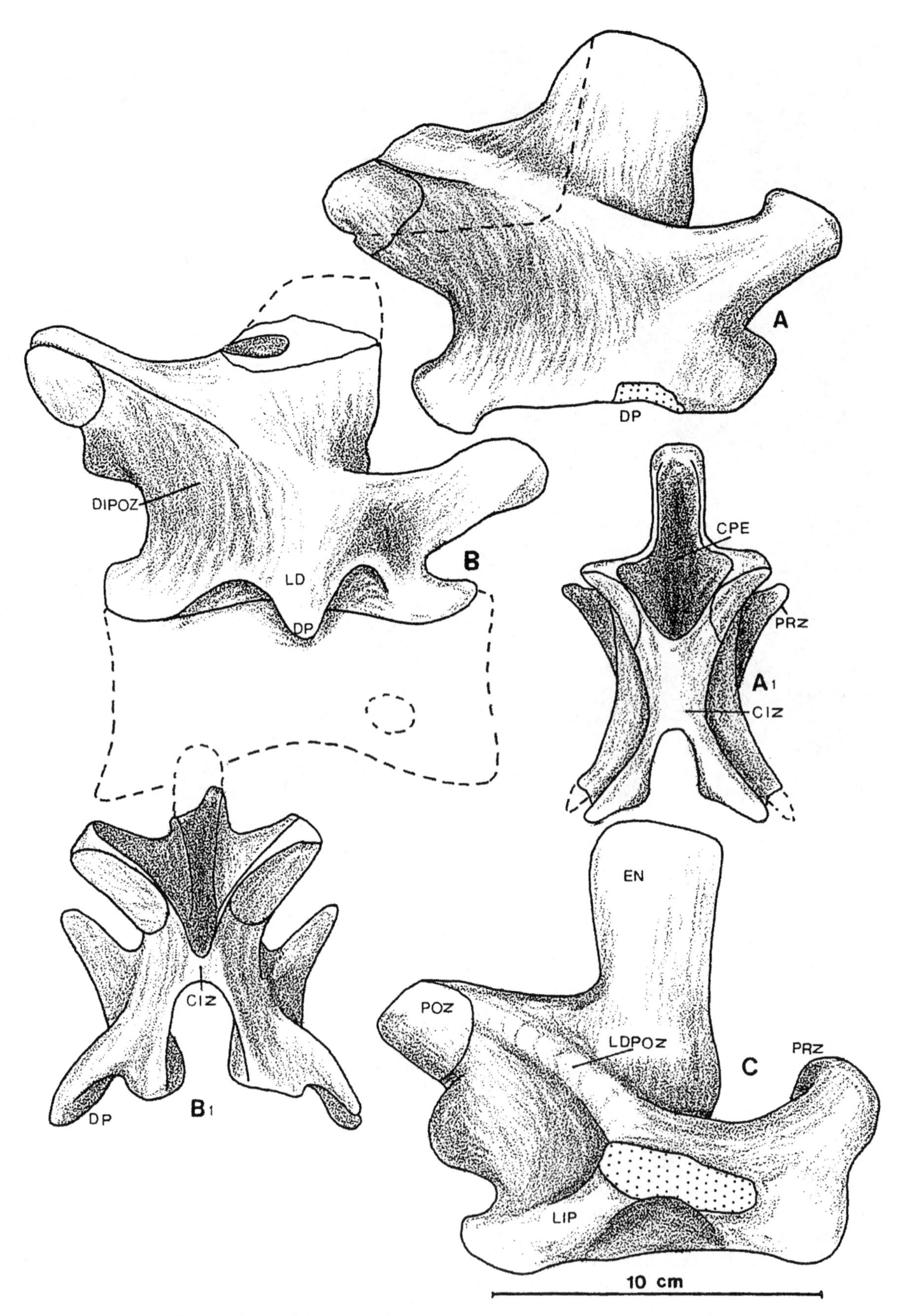

Lessemsaurus sauropoides, PVL 4822-1, holotype vertebrae, A. and A1., neural arch of cervical vertebra ?7, lateral and posterior views, B. and B1., neural arch of cervical ?8, lateral and posterior views, C. neural arch of dorsal 1, lateral view. (After Bonaparte 1999*b*.)

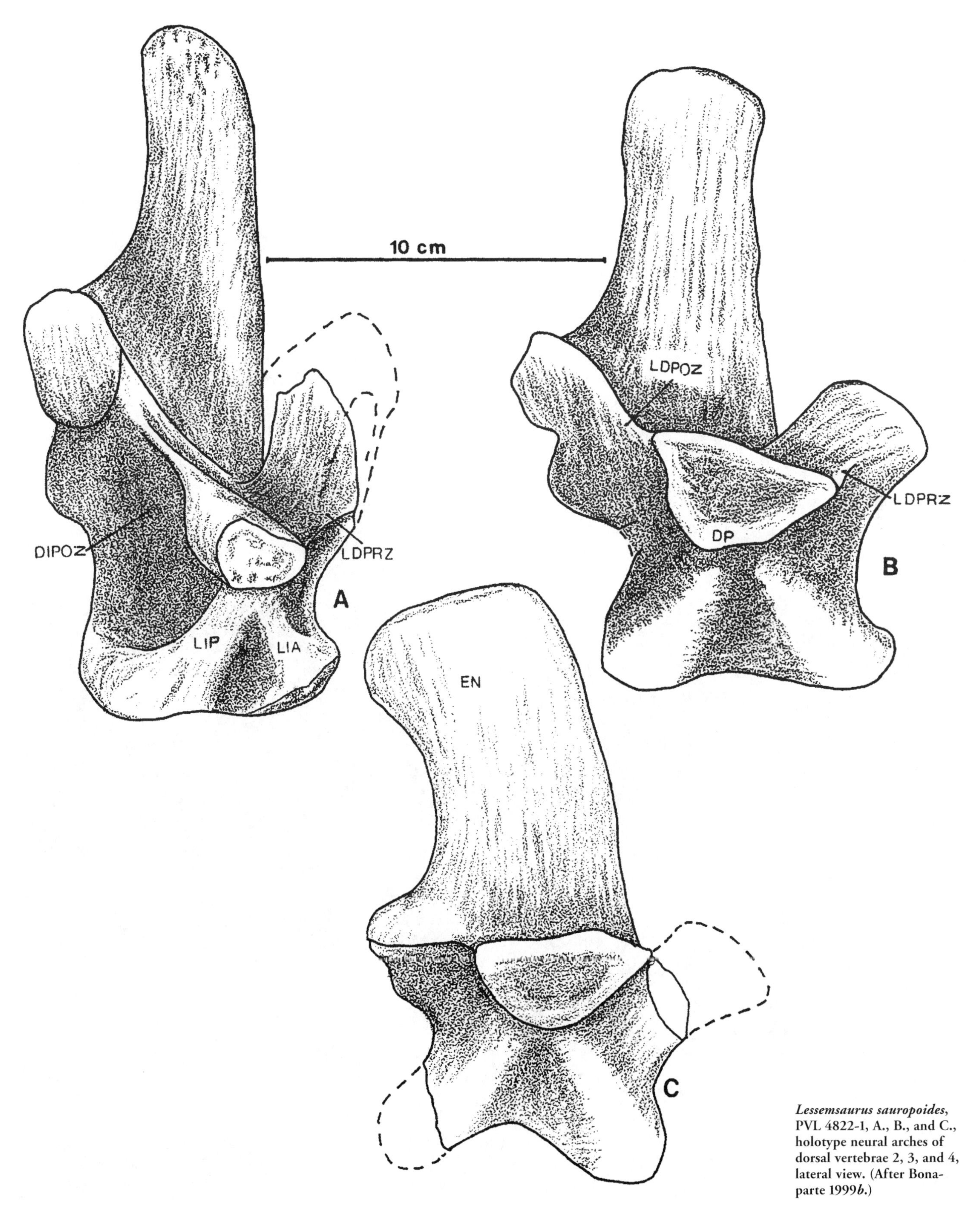

Lessemsaurus sauropoides, PVL 4822-1, A., B., and C., holotype neural arches of dorsal vertebrae 2, 3, and 4, lateral view. (After Bonaparte 1999*b*.)

Occurrence: Los Colorados Formation, Paraje La Esquina, Argentina.

Age: Late Triassic (upper Norian).

Known material: Vertebrae and possibly associated postcranial remains, perhaps representing more than one individual.

Holotype: PVL 4822-1, vertebrae represented by three neural arches, three cervical centra, 14 dorsal neural arches.

Diagnosis of genus (as for type species): Prosauropod larger than *Riojasaurus*; middle and posterior cervical vertebrae and dorsal vertebrae having higher neural arches than in other known prosauropods; infrapostzygapophyseal constriction of posterior cervical vertebrae, postspinal fossa pronounced, infrapostzygapophyseal depression deep; middle and posterior dorsal vertebrae having neural spines much higher than wide, ratio of 1.5–2 (height) to 1 (width) (Bonaparte 1999*b*).

Comments: The type species *Lessemsaurus sauropodoides* was founded upon partial postcranial remains (PVL 4822-1), mostly consisting of vertebrae, from the Los Colorados Formation (Upper Cretaceous) of Paraje La Esquina, Independencia department, province of La Rioja, Argentina (Bonaparte 1999*b*).

Key reference: Bonaparte (1999*b*).

LIRAINOSAURUS Sanz, Powell, Le Loeuff, Martińez and Suberbiola 1999

Saurischia: Sauropodomorpha: Sauropoda: Eusauropoda: Neosauropoda: Macronaria: Camarasauromorpha: Titanosauriformes: Somphospondyli: Titanosauria: Eutitanosauria.

Name derivation: Basque *lirain* = "slender" + Greek *sauros* = "lizard."

Type species: *L. astibiae* Sanz, Powell, Le Loeuff, Martińez and Suberbiola 1999.

Other species: [None.]

Occurrence: Laño outcrop, Condado de Treviño, Spain.

Age: Late Cretaceous (?upper Campanian).

Known material: Various cranial and postcranial remains, teeth, ?dermal scutes, several individuals, adult and juvenile.

Holotype: MCNA 7458, anterior caudal vertebra.

Diagnosis of genus (as for type species): Characterized by autapomorphies including restricted articular condyle in posterior caudal vertebrae, sagittal condylar groove in posterior articular surface of distal caudals, lamina in interzygapophyseal fossa in anterior caudals, spinopostzygapophyseal structure not posteriorly projected in posterior caudals,

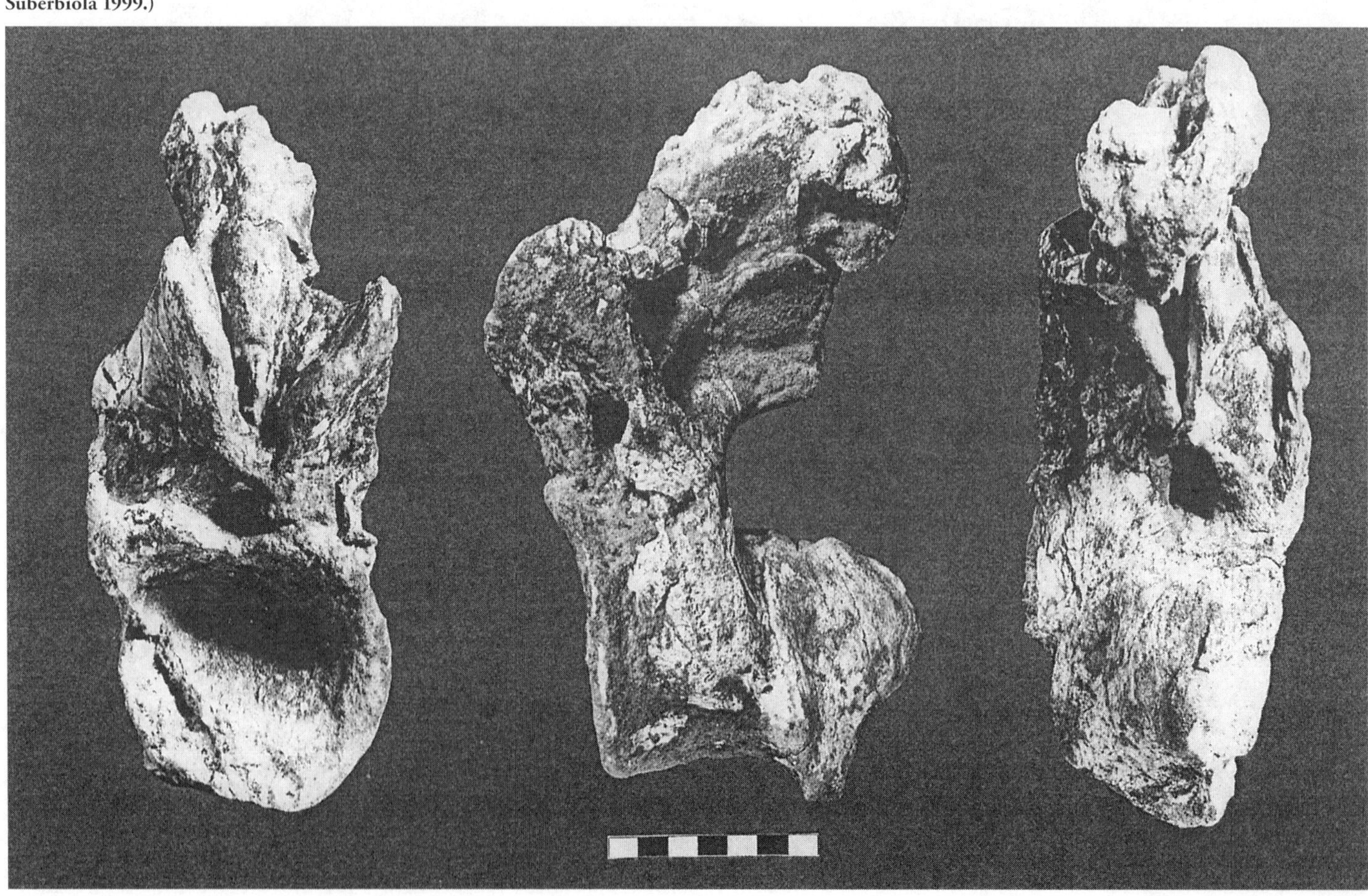

Lirainosaurus astibiae, MCNA 7458, holotype (?first) caudal vertebra in anterior, lateral, and posterior views. Scale in centimeters. (After Sanz, Powell, Le Loeuff, Martińez and Suberbiola 1999.)

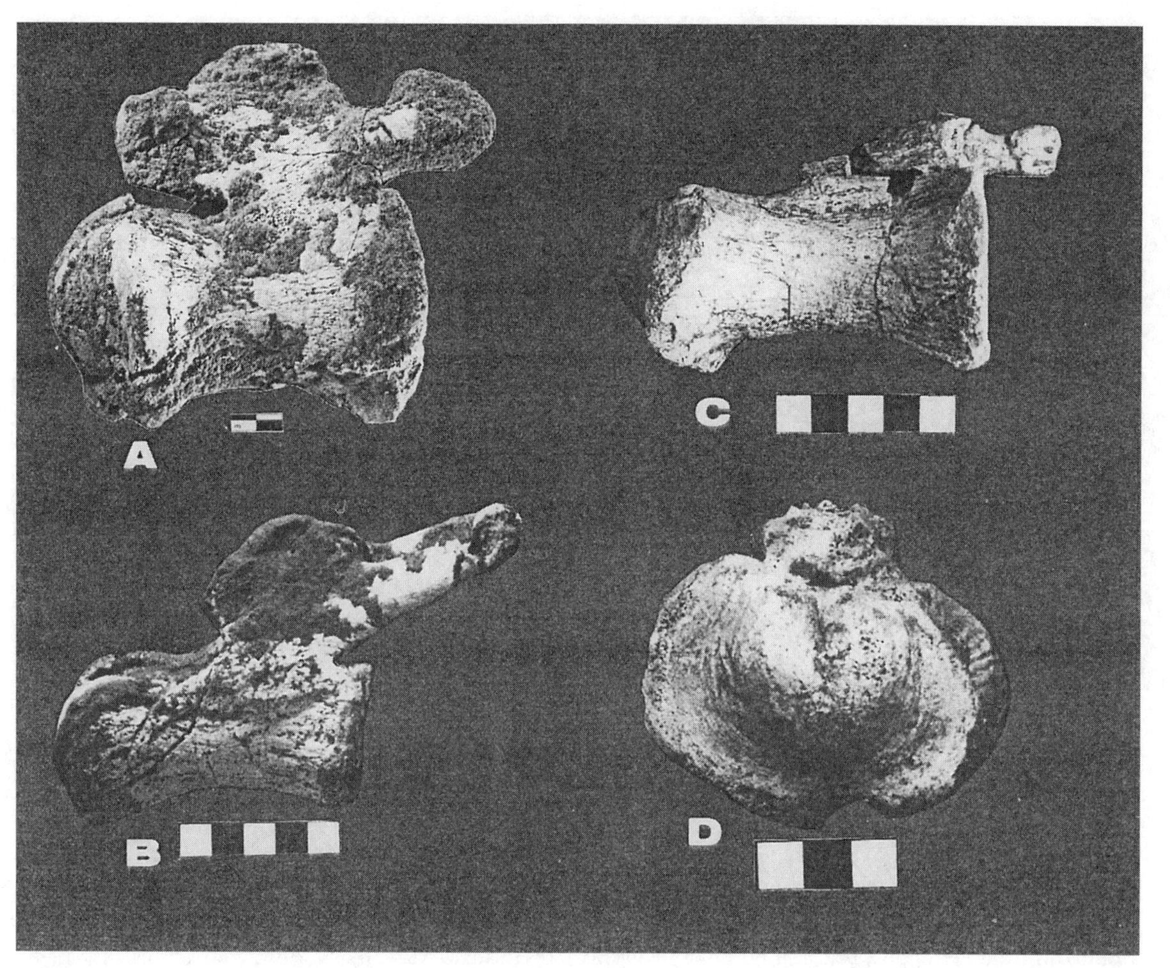

Lirainosaurus astibiae, A. MCNA 7457, mid-caudal vertebra (lateral view), B. MCNA 1812, distal caudal vertebra (lateral view), C. and D., MCNA 7452, distal caudal view (lateral and posterior views, respectively). Scale in centimeters. (After Sanz, Powell, Le Loeuff, Martińez and Suberbiola 1999.)

lateroanterior process on sternal plate, and ridge in ventral margin of medial side of scapular blade; also characterized by combination of synapomorphies including: teeth peglike, axial keel in centrum of dorsal vertebrae, developed and clearly defined pleurocoelic cavity in middle and posterior dorsals, horizontal surface at end of diapophysis in posterior dorsals, absence of hyposphene-hypantrum articulation in posterior dorsals, spinodiapophyseal lamina in posterior dorsals, acuminate posterior margin in pleurocoelous outline in anterior and middorsal vertebrae, spinodiapophyseal accessory lamina in posterior dorsals, low neurapophysis in dorsals, parapophysis located in vertical plane near that of diapophysis in posterior dorsals, prespinal lamina having basal branching poorly developed in mid- and posterior caudal vertebrae, cancellous tissue in presacral vertebrae, developed posterior condyle in anterior caudals, depression under prezygapophysis in anterior caudals, developed prespinal and postspinal laminae in anterior caudals, hemapophysial ridges in midcaudals, sternal plate with anteroventral ridge, median prominence near dorsal margin of sapular blade, anteromedial coracoid outline straight, coracoidal foramen located near coracoscapular suture and dorsal margin of scapula, humerus with developed and medially twisted deltopectoral crest, humerus with posterior supracondylar ridges, depression at base of pubic pedicile of ilium, femur with proximal buttress, femur with reduced fourth trochanter, presence of osteoderms (Sanz, Powell, Le Loeuff, Martińez and Suberbiola 1999).

Comments: The genus *Larainosaurus* was founded upon an anterior caudal vertebra (probably the first in the series), excavated from the Laño quarry (Upper Cretaceous: lower Maastrichtian or, more likely, upper Campanian), in the Condado de Treviño (within Alava Province, in north-central Spain, by a field team led by Dr. Humberto Astibia. Paratype specimens from the same locality referred to the type species *L. astibiae* include a skull fragment (MCNA 7439), isolated different-sized teeth (MCNA 7440–7441), nine dorsal vertebrae (MCNA 7442–7450), seven caudal vertebrae (MCNA 7451–7457, 1812), a scapula (MCNA 7459), coracoid (MCNA 7460), sternal plate (MCNA 7461), four humeri (MCNA 7462–7465), fragment of an ilium (MCNA 7466), fragment of a pubis (MCNA 7467), three femora (MCNA 7468, 3160, 7470), a tibia (MCNA 7471), fibula (MCNA 7472), and (provisionally referred to this species) two dermal scutes (MCNA 7473–7474). To date, all of

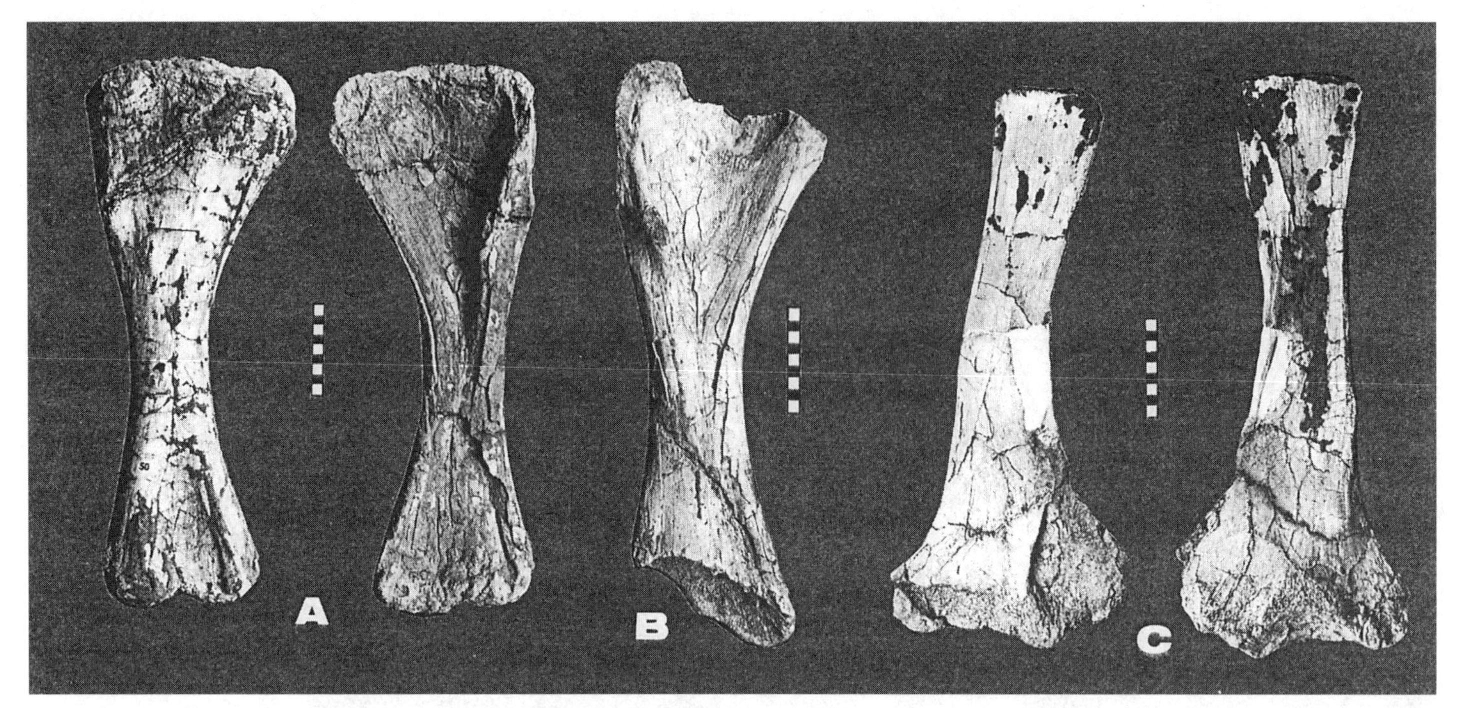

***Lirainosaurus astibiae*, A. MCNA 7463, left humerus (posterior and anterior views, respectively), B. MCNA 7462, right humerus (anterior view), C. MCNA 7459, left scapula (lateral and medial views, respectively). Scale in centimeters. (After Sanz, Powell, Le Loeuff, Martiñez and Suberbiola 1999.)**

these remains are provisionally housed at the Museo Nacional de Ciencias Naturales de Alvala, Vitoria-Gasteiz, in Madrid (Sanz, Powell, Le Loeuff, Martinez and Suberbiola 1999).

In performing a systematic analysis to determine the phylogenetic relationships of *Lirainosaurus*, Sanz *et al.* found that this genus plus an unnamed titanosaur from Peirópolis, Brazil, reported by Powell (1987), form a clade (unnamed) belonging to a newly proposed titanosaurian group which they named Eutitanosauria (see "Systematics" chapter, "Note" under "Eutitanosauria").

Key references: Powell (1987); Sanz, Powell, Le Loeuff, Martiñez and Suberbiola (1999).

†LOPHORHOTHON

Ornithischia: Genasauria: Cerapoda: Ornithopoda: Euornithopoda: Iguanodontia *incertae sedis*.

Comments: The ornithopod genus *Lophorhothon*, based upon a single specimen (FMNH PR27383) from Alabama, has, since its original description by Wann Langston, Jr. in 1960, been classified with the Hadrosauridae (see *D:TE*). More recently, as reported in an abstract by Lamb (1998), a second specimen was collected belonging to this genus, this one preserving numerous postcranial elements not represented or only poorly so in the holotype.

Lamb observed the following suite of characters in both specimens: One replacement tooth per position; less than eight sacral vertebrae; sacrum ventrally ridged; ischium narrow, lacking distal boot; scapula dorsally straight-edged, with medio-laterally compressed glenoid fossa; dentary teeth having prominent primary median ridge and smaller subsidiary ridges. (The second, third, and fifth of these characters are regarded as questionable by hadrosaur specialist M. K. Brett-Surman, personal communication 2000.)

As suggested by Lamb, this combination of characters precludes referring *Lophorhothon* to the Hadrosauridae. More likely, this genus represents "a late-surviving iguanodontian, possibly allied with *Tenontosaurus*," the general resemblance of its skeleton to hadrosaurine hadrosaurs attributed to homoplasy.

According to Lamb, the fact that this genus is only known from the Gulf Coast reflects "the geographic isolation of the Appalachian region from the Cordilleran-Asian ornithopod faunas due to the Western Interior Seaway."

Key references: Lamb (1998); Langston (1960).

LOURINHANOSAURUS Mateus 1998

Saurischia: Theropoda: Neotheropoda: Tetanurae: Avetheropoda: Carnosauria: Allosauroidea: ?Sinraptoridae.

Name derivation: "Lourinhã [Formation]" + Greek *sauros* = "lizard."

Type species: *L. antunesi* Mateus 1998.

Other species: [None.]

Occurrence: Lourinhã Formation, Peralta, Portugal.

Age: Late Jurassic (upper Kimmeridgian–lower Tithonian).

Known material/holotype: ML370, partial skeleton including vertebrae (six cervicals with six ribs, five dorsals with ribs, five sacrals, 14 caudals with

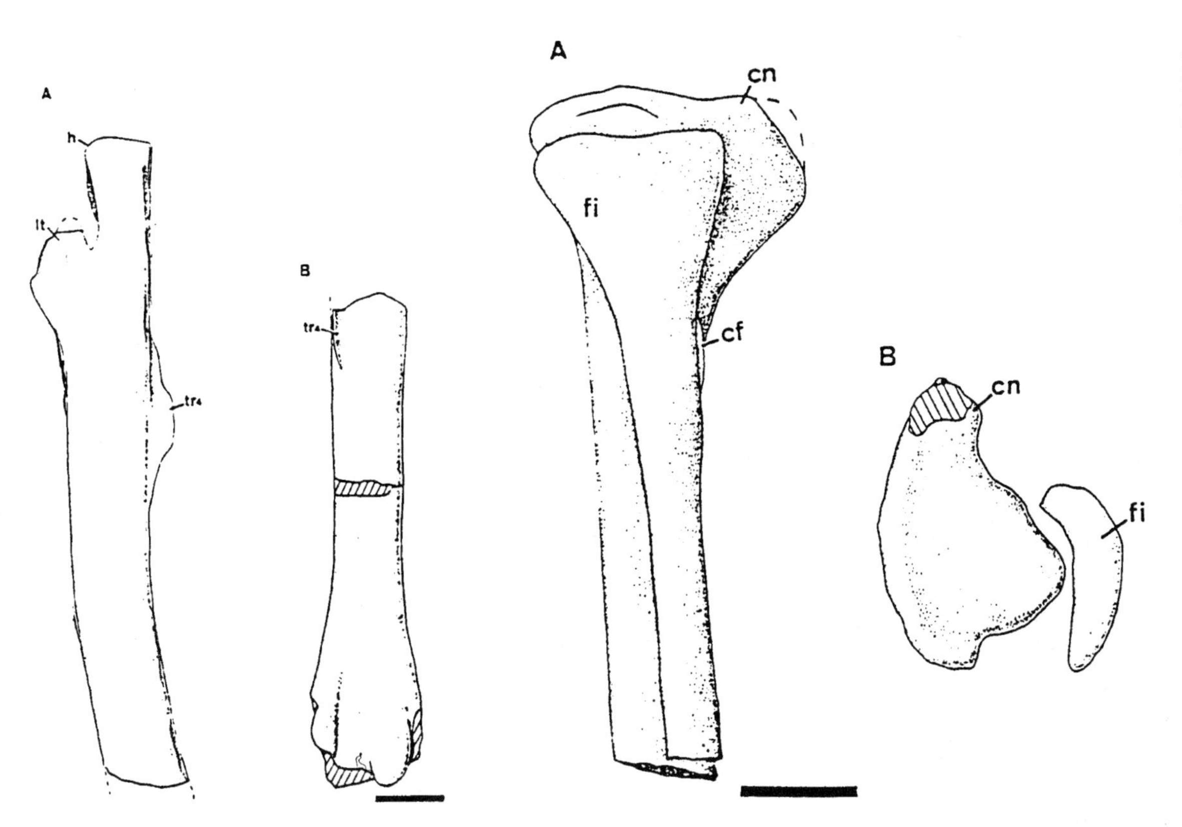

***Lourinhanosaurus antunesi*, ML370, holotype (left) a. left femur in lateral view and b. right femur in posterior view; (right) right tibia and fibula, A. lateral and B. proximal views. Scales = 5 cm. (After Mateus 1998.)**

chevrons), ilia, proximal parts of pubes and ischia, femora (left femur missing tibial and fibular condyles, proximal end of right femur lacking), proximal part of articulated right tibia and fibula, proximal end of one metatarsus, ?associated 32 gastroliths.

Diagnosis of genus (as for type species): Differs from all other known allosauroids in the following: all vertebral centra longer than tall; neural spines of anterior caudal vertebrae with well-developed spikelike anterior process; pubic blade perforated by large vertical ellipsoidal foramen; lesser trochanter (in lateral view) well separated from main body axis of femur (Mateus 1998).

Comments: The new type species *Lourinhanosaurus atunesi* was founded upon the incomplete, partially articulated remains of a single individual (ML370), discovered by farmer Luis Mateus in gray, micaceous, fine sandstone of the Lourinhã Formation, Subral Unit (see Manupella 1998) at Peralta, northwest of Lisbon, in central-eastern Portugal. The specimen was found lying on its right side and shows no signs of deformation (Mateus 1998).

Mateus classified *Lourinhanosaurus* with the Allosauroidea based on the following allosauroid synapomorphies: Transverse processes of middle caudal vertebrae situated near middle of centrum; pubic peduncle of ischium with slender "neck"; ischium (in lateral view) having trapezoidal obturator process; small notch at distal end of ischial obturator process; aliform lesser trochanter.

In comparing *Lourinhanosaurus* to various other theropod taxa, Mateus observed that this genus is distinct from *Neovenator*, the only allosaurid known to date from Europe, in that the last dorsal vertebra is nonpleurocoelous, and also by the position of the fourth trochanter on the posterior surface of the femoral shaft. Although several features (*e.g.*, trapezoidal ischial obturator process, wing-shaped lesser trochanter; see Madsen 1976) of the new genus can be found in *Allosaurus*, *Lourinhanosaurus* differs from the latter in other features (*e.g.*, proportionally longer vertebrae, tibia with less-developed cnemial crest, other diagnostic characters).

Mateus noted that the pubic blade gap in *Lourinhanosaurus* may be the result of an enlargement of the obturator foramen, as in *Piatnitkysaurus*, *Torvosaurus*, *Carnotaurus*, *Monolophosaurus* (see Zhao and Currie 1993*a*), and the sinraptorid allosauroid *Yangchuanosaurus shangyouensis* (Dong, Zhou and Zhang 1983). this hypothesis taking into account that the latter two taxa are allosaurids (following Holtz 1996; *Monolophosaurus* is now generally interpreted as positioned outside of Allosauroidea, see "Systematics" chapter). As the sinraptorid *Sinraptor* exhibits a quite similar though smaller opening which forms an obturator notch, the author suggested that *Lourinhanosaurus* is an allosauroid that is more primitive than allosaurids, probably a member of the Sinraptoridae and, therefore, possibly the only member of this clade thus far known from Europe.

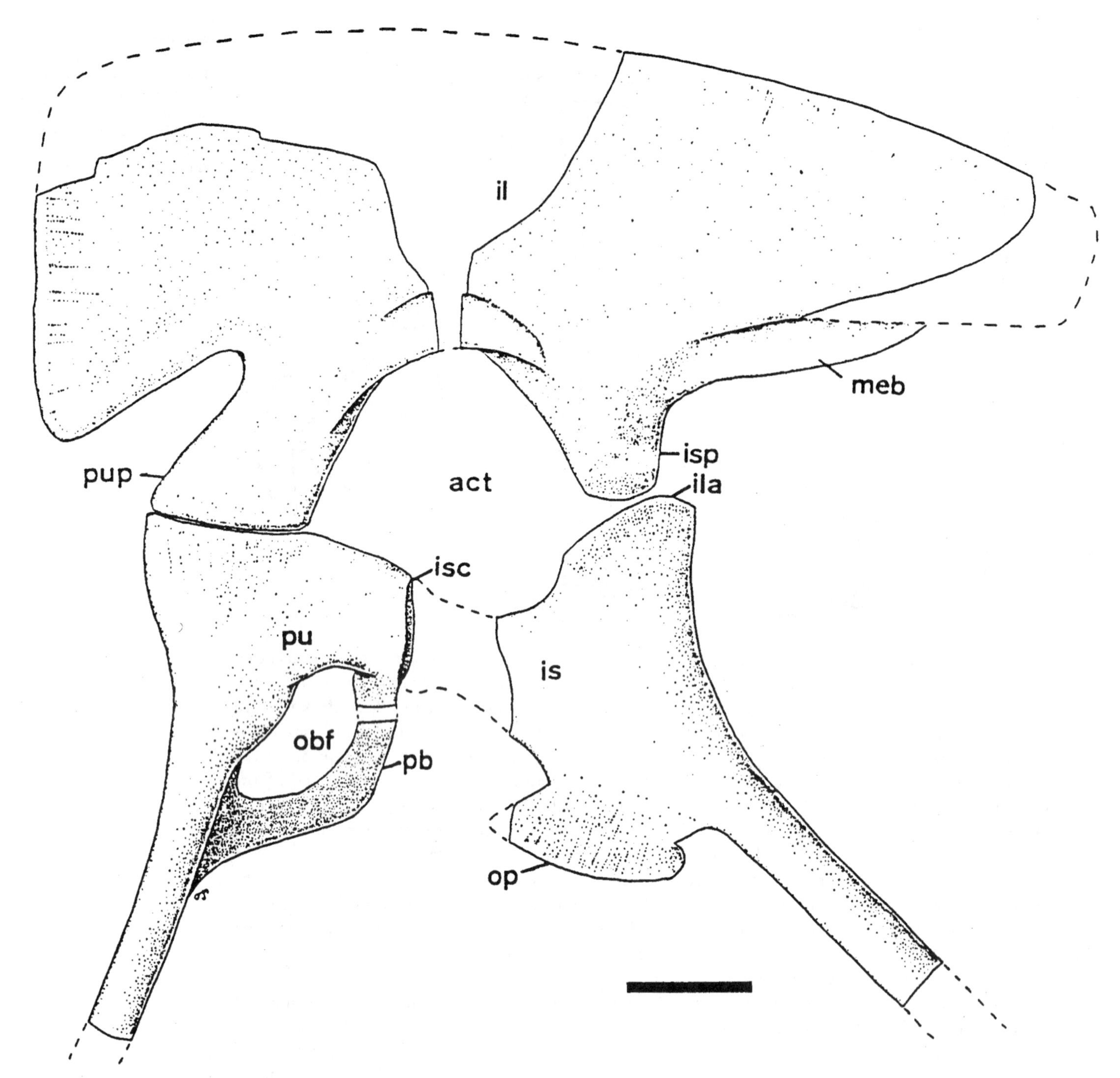

***Lourinhanosaurus antunesi*, ML370, holotype reconstructed pelvic girdle, ilium reconstructed based on features of both ilia. Scale = 5 cm. (After Mateus 1998.)**

A total of 32 gastroliths and the imprints of three more were found apparently associated with this specimen, these, according to Mateus, constituting the first gastroliths known in a nonavian theropod (although gastroliths had been recently reported in the birdlike theropod *Caudipteryx*; see *S2*). The gastroliths were found in the rib cage of the specimen, below the eleventh dorsal vertebra. The largest of these pebbles measures 22 millimeters in length. Found near them were three small bone fragments that appear to be food remains. Mateus interpreted that "The high number, concentration and relative size of the gastroliths suggest that they belong to this specimen," and that they were not swallowed when eating the other dinosaur.

Key references: Dong, Zhou and Zhang (1983); Holtz (1996); Madsen (1976); Manupella (1998); Mateus (1998); Zhao and Currie (1993*a*).

LOURINHASAURUS Dantas, Sanz, Silva, Ortega, Santos and Cachao 1998 — (=?*Dinheirosaurus*)

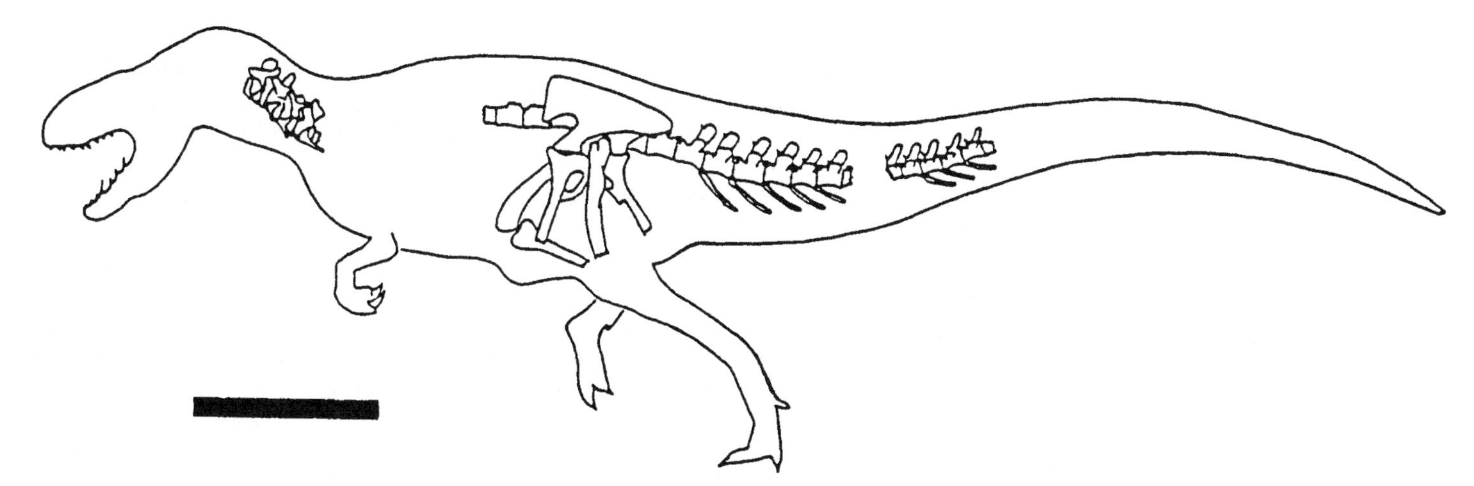

***Lourinhanosaurus antunesi*, ML370, holotype partial skeleton. Scale = 0.5 cm. (After Mateus 1998.)**

Saurischia: Sauropodomorpha: Sauropoda: Eusauropoda: Neosauropoda: Diplodocoidea: Diplodocidae.

Name derivation: "Lourinhã [Formation] + Greek *sauros* = "lizard."

Type species: *L. alenquerensis* (Lapparent and Zbyszewski 1957).

Other species: [None.]

Occurrence: Lourinhã Formation, Lourinhã, Portugal.

Age: Late Jurassic (late Kimmeridgian–early Tithonian).

Known material: Various postcranial remains.

Holotype: Museu Geológico do Instituto Geológico e Mineiro (*ex* Servicos Geológicos de Portugal), Coleccão da Jazida do Moinho do Carmo, Alenquer, partial, somewhat articulated postcranial skeleton including 26 vertebrae (some cervicals, most dorsals, the sacrals, a few anterior caudals), almost complete shoulder girdle (scapula and associated coracoid), part of pelvic girdle (left ilium, ischia, and pubes), forelimb elements (humeri, right radius and ulna, left carpal elements including portion of metacarpal III, phalange of digit II of left forelimb), and hindlimb elements (femora, left tibia and fibula, left astragalus, and partial calcaneum).

Diagnosis of genus (as for type species): Large sauropod having dorsal vertebrae distinguished by the following features: neural apophyses relatively high, vertical, divided; supradiapophyseal lamina much reduced; suprapostzygapophyseal lamina modestly small; postspinal lamina much divided; prespinal lamina great; lower portion of neural arch extended to cover entirety of vertebral centrum; consistency of characters in lateral surface of lower part of neural arch; one articulated divided accessory in lower region, lateral to neural arch which "supports" lateral expansion of hypoephene; infraparadiapophyseal lamina varying from slight (in dorsal 1) to large (in posterior dorsals); vertebral centra proportionately high compared to total height of vertebrae; vertebrae opisthocoelous (including posterior cervicals); pleurocoels moderately middle, eliptical; posterior cervicals long on ventral surface); first dorsal vertebra slightly marked, next five dorsals not marked (Dantas, Sanz, Silva, Ortega, Santos and Cachao 1998).

Comments: The new type species *Lourinhasaurus alenquerensis*— based on an incomplete, partially articulated postcranial skeleton lacking vertebral arches, excavated from the Lourinhã Formation, Praia de Porto Dinheiro outcrop, Lusitanian Basin, Lourinhã, in central-eastern Portugal — was originally named *Apatosaurus alenquerensis* and described by Lapparent and Zbyszewski (1957). In 1990, McIntosh tentatively assigned this species to the genus *Camarasaurus* based on such characters as the number of dorsal vertebrae, the opisthocoelian condition of the posterior dorsals, the ratio of humerus to femur, and the expanded distal end of the scapula.

Additional remains from the same locality and later referred to this taxon were reported by Dantas, Sanz and Gallopim de Carvalho (1992) and Dantas, Sanz, Silva, Ortega, Santos and Cachao (1998). Including details not preserved in the holotype, this specimen consisted of nine neural apophyses (some of them somewhat eroded), 12 vertebrae (dorsals and posterior cervicals), very incomplete remains of both girdles, various indeterminate bone fragments, and possibly some isolated teeth.

McIntosh, Miller, Stadtman and Gillette (1996) later suggested that *C. alenquerensis* might represent a new genus. Wilson and Sereno (1998), assuming that the slender ischial shaft of this species resembles that of *Camarasaurus*, subsequently also interpreted this species as probably belonging to that genus (see *D:TE* for illustration of holotype ilium and pubis).

The taxon was reevaluated by José F. Bonaparte

and P. M. Dantas in a study begun in 1993, results of which were published by Dantas *et al.* (1998), who commented that the "diagnostic characters present in the specimen assign it to *Apatosaurus alenquerensis*," although analysis of those characters, primarily those found in the first seven dorsal vertebrae, suggest referral of this material to a new genus. These authors therein proposed the new generic name *Lourihasaurus* to embrace the material.

Note: Incomplete postcranial sauropod remains from Praia de Porto Dinheiro, assigned by Dantas *et al.* to *L. alenquerensis*, were subsequently referred by Bonaparte and Mateus (1999) to the new genus *Dinheirosaurus* (see entry).

Key references: Dantas, Sanz and Gallopim de Carvalho (1992); Dantas, Sanz, Silva, Ortega, Santos and Cachao (1998); Lapparent and Zbyszewski (1957); McIntosh (1990); McIntosh, Miller, Stadtman and Gillette (1996).

LURDUSAURUS Taquet and Russell 1999

Ornithischia: Genasauria: Cerapoda: Ornithopoda; Euornithopoda: Iguanodontia: Euiguanodontia: Dryomorpha: Iguanodontidae.

Name derivation: Latin *lurdus* = "heavy [both in weight and significance]" + Greek *sauros* = "lizard."

Type species: *L. arenatus* Buffetaut and Russell 1999.

Other species: [None.]

Occurrence: Elrhaz Formation, Niger.

Age: Middle Cretaceous (Aptian).

Known material: Skeleton, cranial elements, at least two individuals.

Holotype: MNHN GDF 1700, almost complete skeleton.

Diagnosis of genus (as for type species): Iguanodontid (as indicated by coosified carpalia on hallux) of unusually massive proportions, exhibiting the following diagnostic characters: endocranial axis and long axis of occipital condyle approximately parallel to long axis of skull; quadrate short (about twice average length of dorsal centra); cervical centra and dorsal centra approximately same length; well-developed hypapophyses on anteroventral margin of posterior cervical centra; height of posterior dorsal centra approximately 40 percent of total vertebral height; axial tendons evidently unossified; radius about 54 percent as long as humerus; length of metacarpals about 17 percent length of humerus; length of hallux ungual more than half that of radius; length of tibia 84 percent that of femur (Taquet and Russell 1999).

Comments: The genus *Lurdusaurus*, a large, short-spined iguanodont distinguished by its massive bones, was founded upon a nearly complete skeleton (MNHN GDF 1700), the scattered elements of which were collected in 1965 from "Camp des deux arbres," southeast of Elhraz, at Gisement de Gadoufaoua — a site exceptional for its abundance and quality of dinosaur skeletons, and the world's leading equatorial site in its number and variety of ornithopods (see Taquet 1976, 1994*a*) — in the upper part of the Elhraz Formation, Tégama Series, Sahara Desert, Niger. The specimen, first reported by Taquet (1967), was found near the type skeleton of the long-spined iguanodontid *Ouranosaurus nigeriensis* (see *D:TE*). Taquet (1976) at first referred to this dinosaur as "Iguanodontide trapu" in his monograph on *O. nigeriensis*; later, the new form became the subject of an unpublished dissertation by Chabli (1988). The specimen was later briefly described in an abstract by Taquet and Russell (1998*a*). Additional specimens referred to the type species, *L. arenatus*, include a dentary fragment (MNHN GDF 43G) and a right corocoid (MNHN GDF 381) (Taquet and Russell 1999).

Originally, Taquet (1976) briefly diagnosed this as yet unnamed form as follows: Forelimbs extremely robust, with radius and ulna very short but extremely thick, hallux with short spike; vertebrae enormous, possessing particularly short neural spines, carpal block very massive; metacarpals very short and very large (of graviportal type), hindlimbs short and massive.

According to Taquet and Russell (1999), the holotype skeleton of *L. arenatus* belongs to a mature individual measuring almost 9 meters (approximately 30 feet) from snout to tip of tail, standing about 2 meters (about 6.8 feet) high at the hips, and weighing in life some 5.5 metric tonnes (about 61 tons).

The skull of *L. arenatus*, as described by Taquet and Russell (1999), is relatively small, measuring approximately 83.3 centimeters in length, about 20 centimeters wide at the snout, and about 30 centimeters wide at its posterior end. In life, the body would have been quite bulky. The neck is comparatively longer than that of other iguanodonts. The pelvis is powerfully constructed, more so than in *Iguanodon bernissartensis*. The back of the animal, as suggested by the heavy, diverging anterior alae of the ilia and horizontally-oriented posterior thoracic ribs, seems to have been rather flat, measuring about one meter across. The powerfully built forelimb, with its manus possessing a heavy carpal block and large thumb claw, reminded the authors of a mace-and-chain, obviously more suited to defend the animal than to rapid quadrupedal locomotion. The tail is relatively short (more than 4 meters in length). The

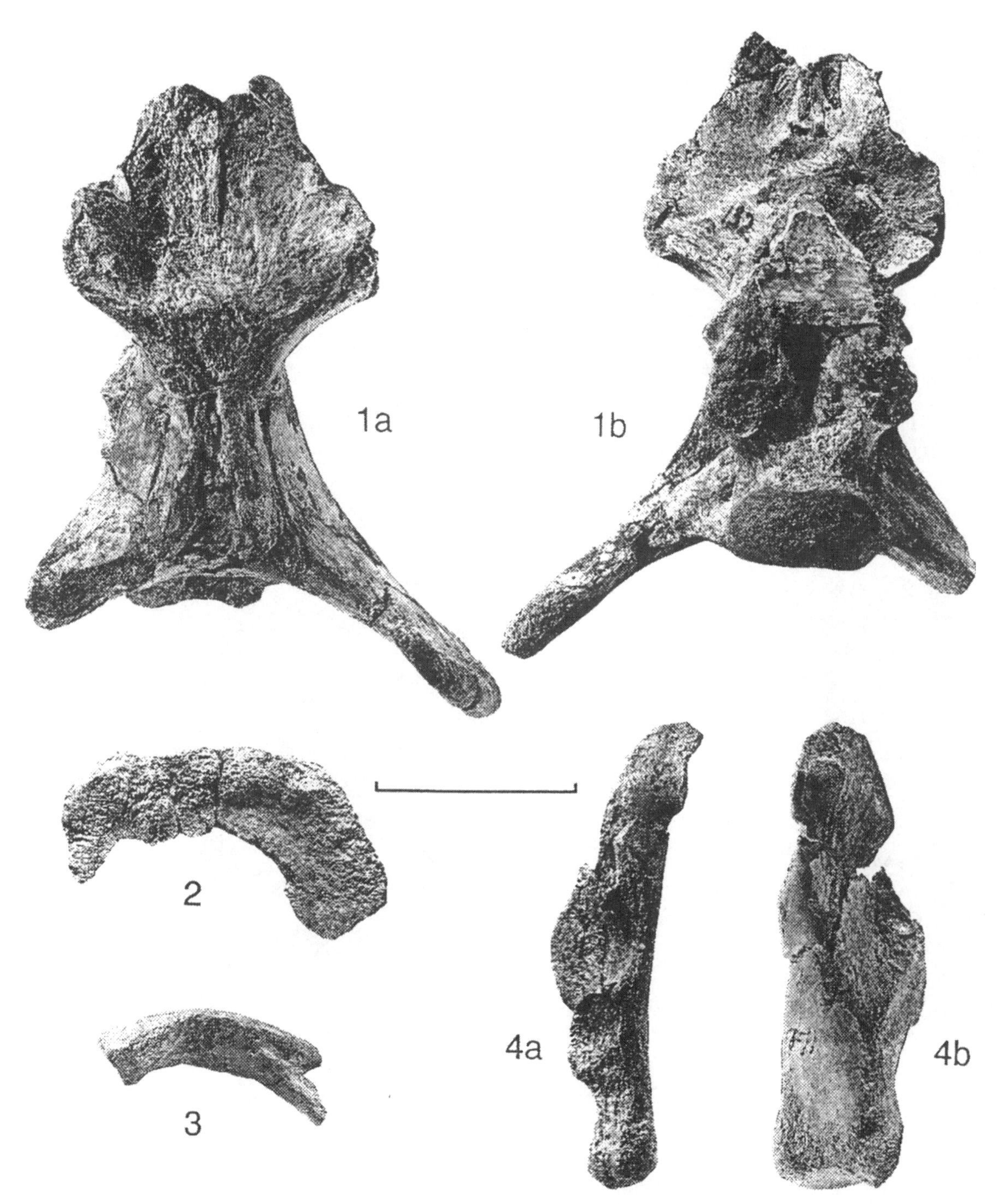

***Lurdusaurus arenatus*, MNHN GDF 1700, holotype skull roof in 1a. dorsal and 1b. ventral views, 2. right premaxilla, dorsal view, 3. predentary, dorsal view, and left quadrate, 4a. lateral and 4b. posterior views. Scale = 10 cm. (After Taquet and Russell 1999.)**

femur is short and massive, resembling the femur in ceratopsians in its slight medial curvature, and that of sauropods in its anteroposterior flattening. When the animal was posed quadrupedally, its circular chest would clear the ground by only some 70 centimeters.

Compared with *O. nigeriensis*, this "ponderous iguanodont" (as the authors informally dubbed the new taxon) differs in the robust nature of its skeletal elements and in the proportions of its various diagnostic skeletal structures. Although Taquet and Russell (1999) discussed numerous details of the skeletal anatomy of *L. arenatus* that indicate a close relationship with Early Cretaceous European iguanodontids (*i.e., Iguanodon atherfieldensis* and *I. bernissartensis*), the new Saharan form differs from all known European iguanodonts in a number of features. Primary among these differences are its massive proportions, dorsoventrally flattened skull, the well-developed cervical hypapophyses, short dorsal spines, short limbs, and the powerfully developed hallux.

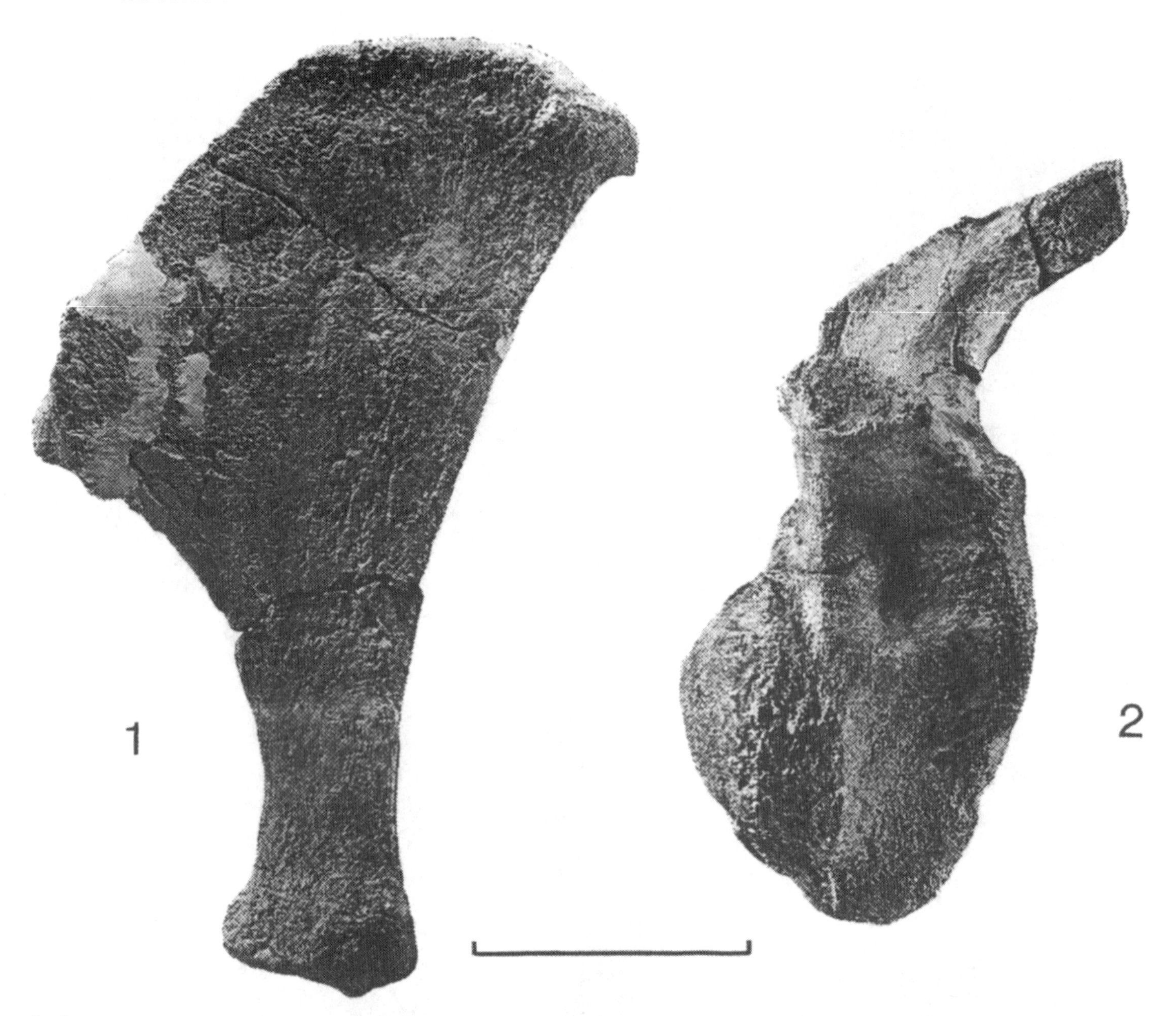

***Lurdusaurus arenatus*, MNHN GDF 1700, holotype 1. left sternal bone, ventral view, and 2. cervical vertebra, left lateral view. Scale = 10 cm. (After Taquet and Russell 1999.)**

Based on the following features, Taquet and Russell (1999) interpreted *Lurdusaurus* to be a genus among intermediate and higher iguanodontians: Presence of posterolaterally directed process on sternum; distal expansion of pubis (though not extreme); reduction of posterior process of pubis; and strongly opisthocoelous cervical and anterior dorsal vertebrae. The authors excluded *Lurdusaurus* from the Hadrosauridae based on the following: Presence of first digit of manus; well-ossified carpal block; and open intercondylar groove on femur.

The authors found *Lurdusaurus* to be a somewhat ambiguous genus in its possession of a femoral fourth trochanter that is almost pendant, as in lower iguanodontians, but also a large number (from 12 to 14) of cervical vertebrae, as in hadrosaurids. However, based on current data on ornithopod evolution, Taquet and Russell (1999) referred *Lurdusaurus* to the family Iguanodontidae, although regarding it as an atypical genus, its bulky body, rather long neck, massively constructed forelimb and thumb spike, and comparatively short hindlimbs and tail suggesting habits and physical and constitutional characteristics different from those of other iguanodontids.

In all, Taquet and Russell (1999) noted that *Lurdusaurus*, with its squat posture, might have superficially resembled an ankylosaur; however, its relatively small head, circular chest, powerful and heavily clawed forelimbs, transversely flattened femoral shaft, and massive bones strikingly suggest "extinct giant ground sloths in general body form."

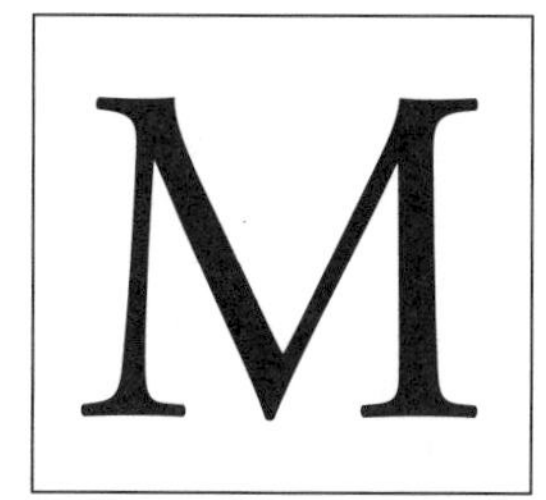

Key references: Chabli (1988); Taquet (1967, 1976, 1994*a*); Taquet and Russell (1998*a*, 1999).

†MAGYAROSAURUS

Comments: In 1914, Franz Baron Nopcsa interpreted the sauropod that Huene (1932) would later name *Magyarosaurus*— a comparatively small titanosaurid from the Late Cretaceous of Transylvania — as a dwarf dinosaur. Nopcsa interpreted the Transylvanian region of the Late Cretaceous as an island (based on geological evidence and the identification of members of the fauna identified as dwarfs), and had viewed dwarfing as resulting from insular habitation affecting body size (see *D:TE*, *S1*).

Based on the above claims (and following Upchurch's 1998 phylogenetic analysis of the Sauropoda), Jianu and Weishampel (1999) published "a preliminary re-evaluation of Nopcsa's identification of dwarfing in *Magyarosaurus* and its heterochronic consequences," this reevaluation taking the following two approaches: "a regression analysis of humeral data as a means of establishing patterns in body size among closely related sauropods, and an optimization analysis of humeral data onto sauropod phylogeny to evaluate evolutionary trends within the clade." For comparative and phylogenetic purposes, this study was based upon 20 sauropod species distributed among 14 neosauropod genera (principally titanosaurids, brachiosaurids, and camarasurids, grouped together by Upchurch into the clade Brachiosauria) utilizing length and midshaft mediolateral humeral width of presumed adult forms (the humerus being one of the most common available elements to the authors), and of ontogenetic samples representing postnatal to adult individuals.

For their regression analysis of intra- and interspecific differences in humeral form, Jianu and Weishampel first divided their samples into those believed to represent adults and those that could be assembled into an ontogenetic series, this age classing based upon previously published statements, and also on joint-surface maturity and overall body size. Adding *Magyarosaurus* humeral specimens to this analysis, Jianu and Weishampel's data showed that the Transylvanian sauropod is more similar to juvenile than adult "brachiosaurians." From this result, the authors concluded the following: *Magyarosaurus*, among adult neosauropods, seems to be represented only by the smallest individuals; humeri of this genus appear to be more similar to those of subadults of other taxa;

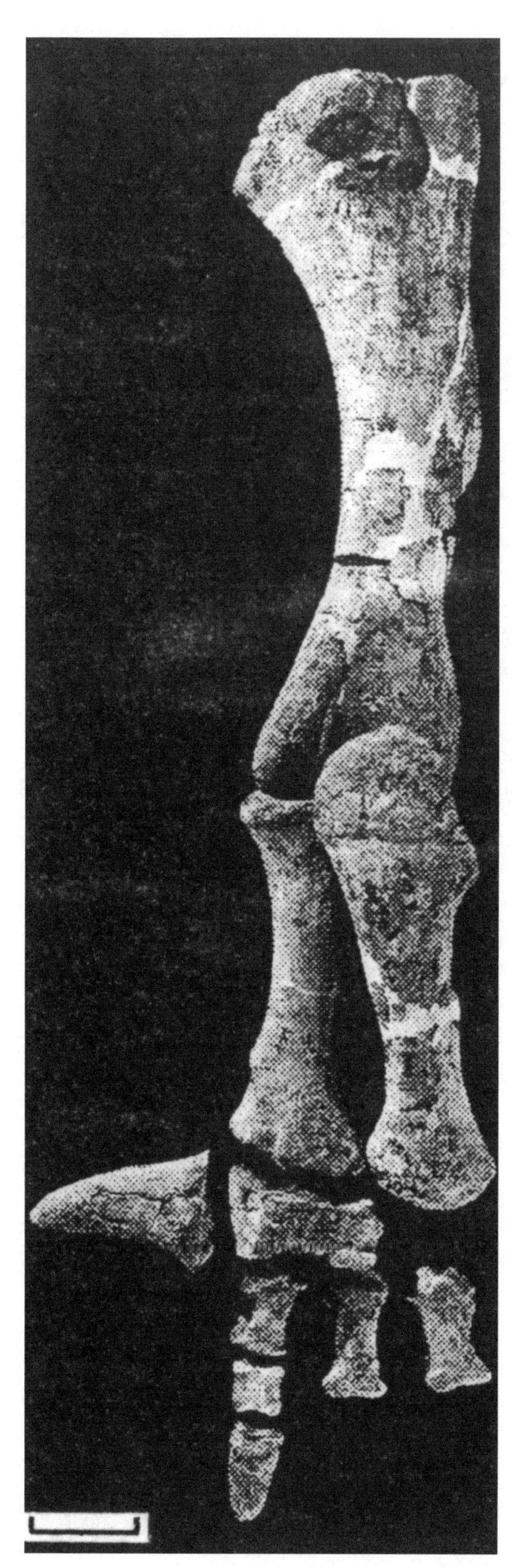

***Lurdusaurus arenatus*, MNHN GDF 1700, holotype left forelimb, posterior view. Scale = 10 cm. (After Taquet and Russell 1999.)**

FGUUB R.1410, osteoderm referred to *Magyarosaurus dacus* in a. externa and b. lateral views. Scale = 2 cm. (After Csiki 1999.)

and the "juvenile" morphology of this genus may constitute dwarfism by paedomorphosis.

To determine how these regressions may reflect more clearly their evolutionary context, Jianu and Weishampel presented optimism analyses (*i.e.*, those that begin with an explicit phylogenetic hypothesis and data to be optimized onto them; see Farris 1970) of the form of the humerus within the "Titanosauroidea" (see "Systematics" chapter). From these analyses, Jianu and Weishampel found that the humerus of *Magyarosaurus* shows an apomorphic shift to a small, juvenilised morphology, the same apparently also true for *Ampelosaurus*. This could possibly suggest a close relationship of these two genera based on some degree of dwarfing, although the authors declined to make such a claim at that point, noting that it would be based on lack of phylogenetic resolution of these genera, and also that of many other "titanosauroid" forms. However, from both of the above approaches, Jianu and Weishampel concluded that the small size of *Magyarosaurus* seems to represent an apomorphic shift in body size, the possible result of dwarfing by heterochronic paedomorphosis.

Jianu and Weishampel also noted that, although all of the species referred by Huene to *Magyarosaurus* have been consolidated into one, *M. dacus* (see Weishampel, Grigorescu and Norman 1991; Le Loeuff 1993), "this situation clearly requires further taxonomic and systematic research" (Weishampel, Jianu and Csiki, in preparation).

Continuing the work of Jianu and Weishampel, Mussell and Weishampel (2000), in an abstract for a poster, further tested on a larger scale the hypothesis that *Magyarosaurus* was a dwarf sauropod. Mussell and Weishampel also considered measurements of the humerus and femur, their focus being on articular surfaces and bone robusticity, compared among six sauropod families (Vulcanodontidae, "Cetiosauridae," Diplodocidae, Camarasauridae, Brachisauridae, and Titanosauridae). Measurements were selected for their biological significance, with regression analyses performed on the data to detect adult and ontogenetic growth series, and nonparametric statistical tests employed to elucidate significant differences. The authors concluded that, while *Magyarosaurus* does seem to be a dwarf sauropod, this condition does not seem to have been caused by progenesis.

Csiki (1999) described a comparatively small, "peculiar osteoderm" (FGGUB R.1410), referred to *M. dacus*, which was found in 1995 in proximity of some scattered titanosaurid bones in the late Maastrichtian of the "La Cărare" locality near Sînpetru village, Sibisel Valley, Hateg Basin, Hunedoara County, Romania. The specimen, the third report on osteoderms in European titanosaurids, demonstrates that dermal armor in these sauropods had a wide distribution during the Late Cretaceous.

Csiki observed, among other details, that the osteoderm has an elongated, somewhat elliptical outline, with two distinct regions — a "scute" or "bulb" region (see Le Loeuff, Buffetaut, Cavin, Martin, Martin and Tong 1994) and a "root" region. The "scute" region measures 80 millimeters in length and 68 millimeters in width. It consists of a low, rather rounded central cone (42 mm in height) surrounded by a partially broken off shelf. If complete, the scute may have had an almost circular outline. The "root" region, which underlies the scute, has an elliptical to triangular shape. Its outline is mostly rounded at one end below the scute, narrowing to a triangular, pointed opposite end. The root's long axis apparently diverges from that of the scute by a 30-degree angle.

Until the discovery of FGGUB R.1410, no armor had been reported in *M. dacus*. Csiki pointed out that this osteoderm does not compare favorably with that of crocodilians or nodosaurids, the only other known reptilian groups bearing armor from the Hateg Basin. It does, however, share the following characters with osteoderms referred to titanosaurs known from South America and Africa: Outline oval to slightly circular; cingulum at the margins of the scute; cone-shaped external surface with rugosities of a somewhat radiated "groove and ridge" pattern; conspicuous woven bone on the internal face; presence of internal ducts. Except for minor differences, the Romanian specimen is also quite similar to an osteoderm (FMNH PR2021) from Malagasy described by Dodson, Krause, Forster, Sampson and Ravoavy (1998*b*).

Based on the similarities of FGGUB R.1410 with the osteoderms of other titanosaurids, plus the co-occurrence of titanosaurid bones at the locality site, Csiki referred the specimen to *M. dacus*.

Csiki addressed the relatively small size of this osteoderm. Its diameter is just 8 centimeters; however, FMNH has a diameter of 17 centimeters, while some other specimens are yet considerably larger. The author noted that this size difference might be explained by the phenomenon of dwarfism among the Hateg dinosaurs.

Following Dodson *et al.*, Csiki viewed this specimen as being embedded alone in the dermis. Also, being asymmetric, the osteoderm was most likely displaced from the midline of the body, perhaps being part of a longitudinal row of scutes. Noting that this osteoderm was reminiscent of a specimen for which Sanz and Buscalioni (1987) proposed a sacral position, and that it also seems less suitable for shoulder defense being lower and more rounded than a specimen located in the shoulder region, Csiki very tentatively found a sacro-dorsal region of the back to be a possible location for FGGUB R.1410.

Key references: Csiki (1999); Dodson, Krause, Forster, Sampson and Ravoavy (1998*b*); Farris (1970); Huene (1932); Jianu and Weishampel (1999); Le Loeuff (1993); Le Loeuff, Buffetaut, Cavin, Martin, Martin and Tong (1994); Mussell and Weishampel (2000); Nopcsa (1914); Sanz and Buscalioni (1987); Weishampel, Grigorescu and Norman (1991); Upchurch (1998).

†MAIASAURA

Ornithischia: Genasauria: Cerapoda: Ornithopoda: Euornithopoda: Iguanodontia: Euiguanodontia: Dryomorpha: Ankylopollexia: Hadrosauroidea: Hadrosauridae: Euhadrosauria: Hadrosaurinae.

Comments: Schmitt, Horner, Laws and Jackson (1998) briefly reported on their sedimentologic and taphonomic analyses of a bonebed pertaining to *Maiasaura peeblesorum*— the celebrated "good mother lizard" (see *D:TE*, *S1*) — containing thousands of skeletal elements belonging to this species and representing a life assemblage ranging from juveniles through adults, in the Two Medicine Formation of the Willow Creek anticline, in northwest Montana. Damage to the specimens included breakage patterns indicating to the authors transport before fossilization; also, the bones showed no signs of weathering or burning, with only slight indications of boring or scavenging. Schmitt *et al.* offered a number of possible causes for this flow event, these including the movement of volcanic ash, rainfall, and flooding.

Earlier, Deeming and Unwin (1993), and then Dodson (1995), had suggested that two groups of very young hadrosaurines found within the confines of possible nests — referred by Horner and Makela (1979) to *M. peeblesorum* and, having worn teeth, interpreted by them as post hatchlings that had already begun to feed — were actually embryos. This latter interpretation was largely based on the later discovery of embryonic lambeosaurines, referred by Horner and Currie (1994) to *Hypacrosaurus stebingeri*, possessing worn teeth. Deeming and Unwin suggested that the *Maiasaura* nestlings may originally have been underestimated by a factor of four; Dodson further asked how such closely related genera as *Maiasaura* and *Hypacrosaurus* could have hatchlings of such disparate sizes (see *D:TE* and *S1* for details). Horner (1999) addressed these critics' questions, which he found to be

Photograph by the author.

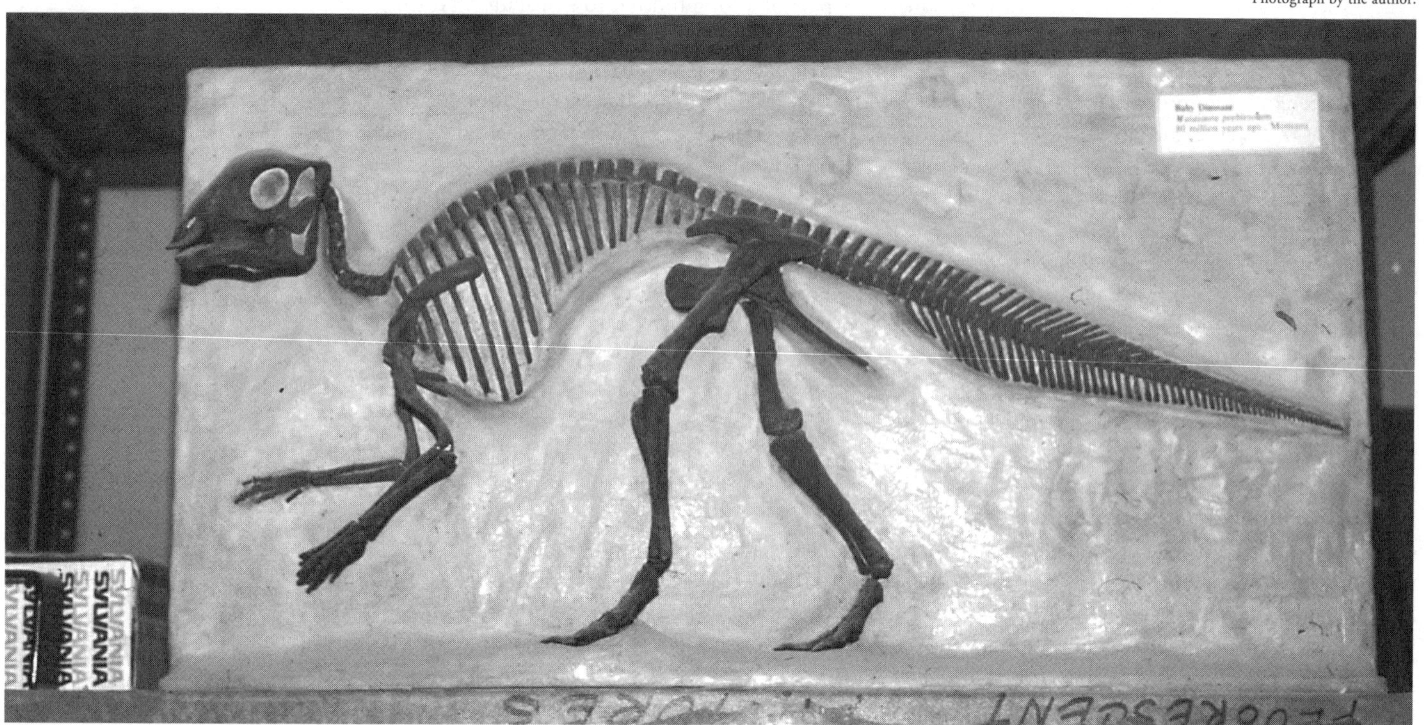

***Maiasaura peeblesorum*, skeletal reconstruction (cast) of nestling by Stephen A. Czerkas and Sylvia J. Czerkas.**

"well-founded," while reporting on two new occurrences of dinosaur eggs possessing embryos, and representative clutches, representing both the Hadrosaurinae and Lambeosaurinae.

The hadrosaurine eggs — each clutch occurring in two layers — were collected from the Two Medicine Formation on the Willow Creek Anticline, Teton County, Montana. They were referred by Horner to *Maiasaura* based upon embryonic remains found within some of the eggs, and because "all hadrosaurian skeletal elements of any ontogenetic stage, found in strata of the Willow Creek Anticline, have characters consistent with the Hadrosaurinae; also, "the skeletal elements identifiable to genus are consistent with *Maiasaura peeblesorum*." Thickness of the shells is variable, ranging from 0.75 to 1.5 millimeters. One clutch (MOR 281) included at least 16 unhatched eggs, none of them seemingly including embryonic remains. As described by Horner, these eggs seem to be spherical in shape (although they may have been somewhat distorted by crushing), with a maximum diameter of 12 centimeters and (if spherical) a volume of about 900 millileters. The asymmetric egg shape originally predicted by Horner (1984) was, therefore, incorrect, and the predicted size not underestimated but overestimated.

Another partial hadrosaurine clutch (MOR 244) includes associated embryonic material. Removed skeletal elements belonging to at least two individuals were identified by Horner as hadrosaurine "based on the midshaft position of the equalateral triangular fourth trochanter of the femur" and also "the pentagonal shapes of the mid and distal caudal vertebrae" (see Weishampel and Horner 1990). Femora in these individuals were measured by Horner (1999) as from 3.5 to 4.0 centimeters in length.

The lambeosaurine eggs — with a clutch occurring in a single layer — were collected from the Judith River Formation in central Montana (see Clouse and Horner 1993), some clutches referred to the Lambeosaurinae based upon embryonic remains (Horner and Clouse 1998). These eggs appear to be spherical in shape. One clutch (MOR 770) includes 22 unhatched eggs, individual eggs measuring about 20 centimeters in diameter, eggshell thickness from 0.8 to 1.50 millimeters, volume estimated at approximately 4,190 millileters. Immature embryonic remains extracted from one egg were identified as hadrosaurid; all hadrosaur skeletal elements recovered from these nesting horizons possess characteristics consistent with the Lambeosaurinae (see Weishampel and Horner). Horner (1999) noted that the Judith River lambeosaurine eggs are fairly consistent in size, predicted shape, and surface patterns with the eggs referred by Horner and Currie to *H. stebingeri*.

Based upon this embryonic material from Montana, Horner (1999) concluded that hadrosaurid dinosaurs laid eggs of spherical (or possibly sub-spherical) shape (see also Yang [Young] 1959); and that there was considerable difference in egg size between hadrosaurines and lambeosaurines, this probably reflecting "either phylogenetic differences or different life history strategies" (see Weishampel and Horner; Horner 1994). Noting that among birds "larger eggs correlate with longer incubation periods" (see Ricklefs and Stark 1998), Horner (1999) speculated that a similar situation may have been the case with *Maiasaura*: Apparently living in a dry upland environment (see Lorenz and Gavin 1984) in which a briefer incubation period may have been an advantage, *Maiasaura* laid small eggs that yielded smaller hatchlings than the larger eggs of lambeosaurs.

Horner, Ricqlès, and Padian (2000) published on the histology of long bones of *M. peeblesorum*, a study, based on an ontogenetic series of skeletal elements, revealing certain new insights pertaining to the growth dynamics and physiology of this duck-billed dinosaur. The authors began their study with the generalization that four principal factors — ontogeny, phylogeny, mechanics, and environment — determine the type and form of hard tissues deposited at any given time in vertebrate skeletons (although other factors, including chance, illness, injury, starvation, and individual variation can also affect bone formation).

As noted by Horner *et al.*, the use of bone histology to test some of the recently proposed hypotheses regarding growth, physiology, behavior, and ecology in extinct animals requires the study of as many parts of the skeleton as is practical through as many allometric stages as possible. The unprecedented abundance of skeletal material pertaining to *M. peeblesorum*, ranging from embryos in eggs to adults and representing countless individuals, has made such a study possible. Horner *et al.* based their study on numerous specimens derived from nesting grounds, bonebeds, and isolated skeletons collected from the upper middle part of the Two Medicine Formation (Upper Cretaceous; middle Campanian) of Montana. The histological samples were derived from bones recovered from 1978 through 1995 by field teams from Princeton University and Montana State University; specimens designated as adult, subadult, and late juvenile by Horner *et al.* were collected from a catastrophic bonebed in the Willow Creek Anticline of Montana. The specimens studied were of discrete, nonoverlapping size classes, thereby representing age-classes of a single population (see Schmitt, Horner, Laws and Jackson 1998).

Horner *et al.* found six distinct allometric stages in *M. peeblesorum*, each recognized on the basis of both relative bone size and on patterns of histological differences that were observed during this study. These stages included 1. small nestlings (overall length of approximately 45 centimeters), 2. large nestlings (about 90 centimeters), 3. early juveniles (about 120 centimeters), 4. late juveniles (about 3.5 meters), 5. subadults (about 5 meters), and 6. adults (not necessarily fully grown at about 7 meters).

The following observations were made by the authors: The earliest stages are generally marked by spongy bone matrix having large vascular canals. Cortical bone, through growth, differentiates into fibrolamellar bone which tends to become more regularly layered through growth in the outer cortex. Epiphyseal pads of calcified cartilage tend to become thin, while Haversian osteons and endosteal laminar bone spread at a rapid rate. Lines of arrested growth (LAGs) begin appearing regularly by the subadult stage, although there are no signs that growth slowed before or following LAG deposition. Resorption lines and substantial Haversian substitution begin to appear at the subadult stage, and the external cortex of some bones shows a lamellar-zonal structure, this indicating imminent cessation of bone growth.

Horner *et al.* concluded that growth rates in the genus *Maiasaura* changed ontogenetically. Comparing the above patterns with rates of apposition of similar bone tissues in living amniote animals, these authors deduced that young *Maiasaura* nestlings seem to have grown at very high rates, and that individuals grew at high and moderately high rates during nestling, juvenile, and sub-adult stages, with growth rates in adults (from about 7 to 9 meters in length) decreasing to low to very low. Using living animals for comparison, the authors calculated that a *Maiasaura* hatchling attained nestling size in approximately one to two months, juvenile size in one to two years (these estimates comparable to those in living ratite birds), and adult size in from six to no more than eight years.

The histological tissues, patterns, and inferred growth rates of the bones of *Maiasaura* were found by Horner *et al.* to be "completely different from those of living nonavian reptiles, generally similar to those of most other dinosaurs and pterosaurs for which data are available, and much like those of extant birds and mammals." Therefore, the authors concluded that *Maiasaura* did not grow like modern nonavian reptiles, and consequently cautioned that the latter cannot be utilized as informative analogs for most aspects of dinosaurian growth, nor physiology to the extent that growth rates reflect metabolism.

Key references: Clouse and Horner (1993); Deeming and Unwin (1993); Dodson (1995); Horner (1984; 1999); Horner and Clouse (1998); Horner and Currie (1994); Horner and Makela (1979); Horner, Ricqlès and Padian (2000); Lorenz and Gavin (1984); Ricqlès, Horner and Padian (1998); Schmitt, Horner, Laws and Jackson (1998); Weishampel and Horner (1990); Yang [Young] (1959).

†MAJUNGATHOLUS

Saurischia: Theropoda: Neotheropoda: "Ceratosauria": Neoceratosauria: Abelisauroidea: Abelisauridae.

Comments: O'Connor and Sampson briefly reported additional information concerning two recently collected, remarkably well-preserved specimens of this large, short-faced, and "domed" theropod from Upper Cretaceous (Campanian) rocks of northwestern Madagascar (see *S1*)—one specimen (UA 8678) including an almost complete skull, 25 associated caudal vertebrae, and numerous chevrons; the other, 23 presacral vertebrae, a partial sacrum, and five proximal caudals, found articulated or in close association, and also a partial skull, cervical ribs, and complete left ilium.

From this newly recovered material, O'Connor and Sampson observed that the vertebrae of *M. atopus* exhibit several synapomorphies of the "Ceratosauria" (including transverse processes of dorsal vertebrae triangular in dorsal view); shares several derived characters with the Abelisauridae, especially *Carnotaurus* (*e.g.*, postaxial cervical vertebrae having elongate epipophyses; dorsals with laterally expanded,

Skull (FMNH PR2100, cast) of the "dome-headed" abelisaurid theropod *Majungatholus atopus* in right three-quarter view.

Photograph by the author, courtesy The Field Museum.

buttressed parapophyses). CT scanning and examination of breaks in the fossil material revealed to the authors that the presacral column is made up of vertebrae of the camerate type (that is, having large internal chambers). Apparently, the cervical ribs are unique in their possession of "multiple, hypertrophied (diameter > 10 mm) pneumatic foramina proximally and bifurcate shafts distally."

Key reference: O'Connor and Sampson (1998).

†MALAWISAURUS

Saurischia: Sauropodomorpha: Sauropoda: Eusauropoda: Neosauropoda: Macronaria: Camarasauromorpha: Titanosauriformes: Somphospondyli: Titanosauria.

Comments: Winkler, Gomani and Jacobs (2000) performed a taphonomic investigation of CD-9, the most productive quarry in the Dinosaur Beds of northern Malawi, Africa, most remains from this site pertaining to the sauropod *Malawisaurus* (see *D:TE, 1*) (other remains including those of fishes, terrestrial invertebrates, small tetrapods, and theropod dinosaurs). The *Malawisaurus* include semiarticulated and isolated skeletal elements representing at least two individuals and all regions of the body.

Based upon patterns of bone orientation and articulation, Winkler *et al.* inferred that the *Malawisaurus* carcasses lay decomposing on land for as long as two months before burial, within or near the dry river channel.

As pointed out by these authors, the most likely segments of the sauropod skeleton to remain intact during decomposition were the neck, tail, and feet. The skull, dorsal vertebrae, and upper limbs seem to be most susceptible to disarticulation. The dorsal vertebrae probably did not disarticulate early after death as they may often have been temporarily suspended above the ground by the rib cage, tumbling downward during decomposition. Also, the authors noted, the dorsal vertebrae would be near active microbial decay of the abdomen, hastening their disarticulation.

Winkler *et al.* compared this bonebed with other sauropod burial sites, including an Early Cretaceous sauropod (presumably *Pleurocoelus*) quarry (Jones Ranch, Hood County, Texas; see Winkler, Jacobs and Murry 1997; Gomani, Jacobs and Winkler 1999) in the Twin Mountains Formation near Glen Rose, Texas. At both sites, the authors noted, the event that caused the death of these sauropods seems to have taken place prior to the depositional event, while the mortality event apparently occurred close to the eventual burial site. That semiarid climates have been interpreted for both of these sites suggested that water was most likely a limiting resource at times and that drought was likely, playing a role in the mortality of these animals. Therefore, it is probable that these dinosaurs, seeking out water, seasonally entered dry stream beds, thus being more likely to be caught in floods. Because of the terrestrial depositional setting, however, the authors found it unlikely that these dinosaurs died while trying to cross flooding rivers.

Key references: Gomani, Jacobs and Winkler (1999); Winkler, Gomani and Jacobs (2000); Winkler, Jacobs and Murry (1997).

†MAMENCHISAURUS

Saurischia: Sauropodomorpha: Sauropoda: Eusauropoda: Neosauropoda: Macronaria: Camarasauromorpha: Titanosauriformes: Somphospondyli: Euhelopodidae: Euhelopodinae.

Type species: *M. constructus* Yang [Young] 1954.

Other species: *M. hochuanensis* Yang [Young] and Chao 1972, *M. jingyanensis* Zhang and Li 1996, *M. sinocanadorum* Russell and Zheng 1993, *M. youngsi* Zhang and Chen 1996.

Comments: An extremely long-necked sauropod previously known only from Upper Jurassic rocks of Sichuan Province, China, the genus *Mamenchisaurus* now seems also to have been found in Upper Jurassic rocks of Yunnan Province.

In 1999, Guan, Rigby and Zheng briefly reported on a new dinosaur assemblage, including new material tentatively identified as *Mamenchisaurus* c.f., recovered from the small structural Chuanjie Basin — located southwest of the Lufeng Formation, between the Chuxiong Basin and Lufeng Basin (both Lower Jurassic) — of Yunnan. This site, named Lao Chang Quing and located southwest of the town of Lufeng and north of Chuanjie, has to date of this writing yielded remains representing at least four individuals of this dinosaur, also portions (including 18 articulated vertebrae, an ilium, and a pubis) of a theropod dinosaur, and several turtles.

As noted by Guan *et al.*, the apparent *Mamenchisaurus* material, well-preserved and only moderately weathered, includes three complete scapulae, three forelimbs (comprising humeri, radii, ulnae, a complete set of metacarpals, and an ungual), complete pubes, femora, tibiae, 16 articulated cervical vertebrae, dorsal and sacral vertebrae, four sets of articulated caudal vertebrae of varying numbers, two ilia, left and right ischia, three complete scapulae, and a scapulocoracoid (articulated with the humerus, radius, and ulna mentioned above).

According to these authors, the amount of limb material will permit the calculations of valid forelimb to hindlimb ratios in this genus, while the fossils

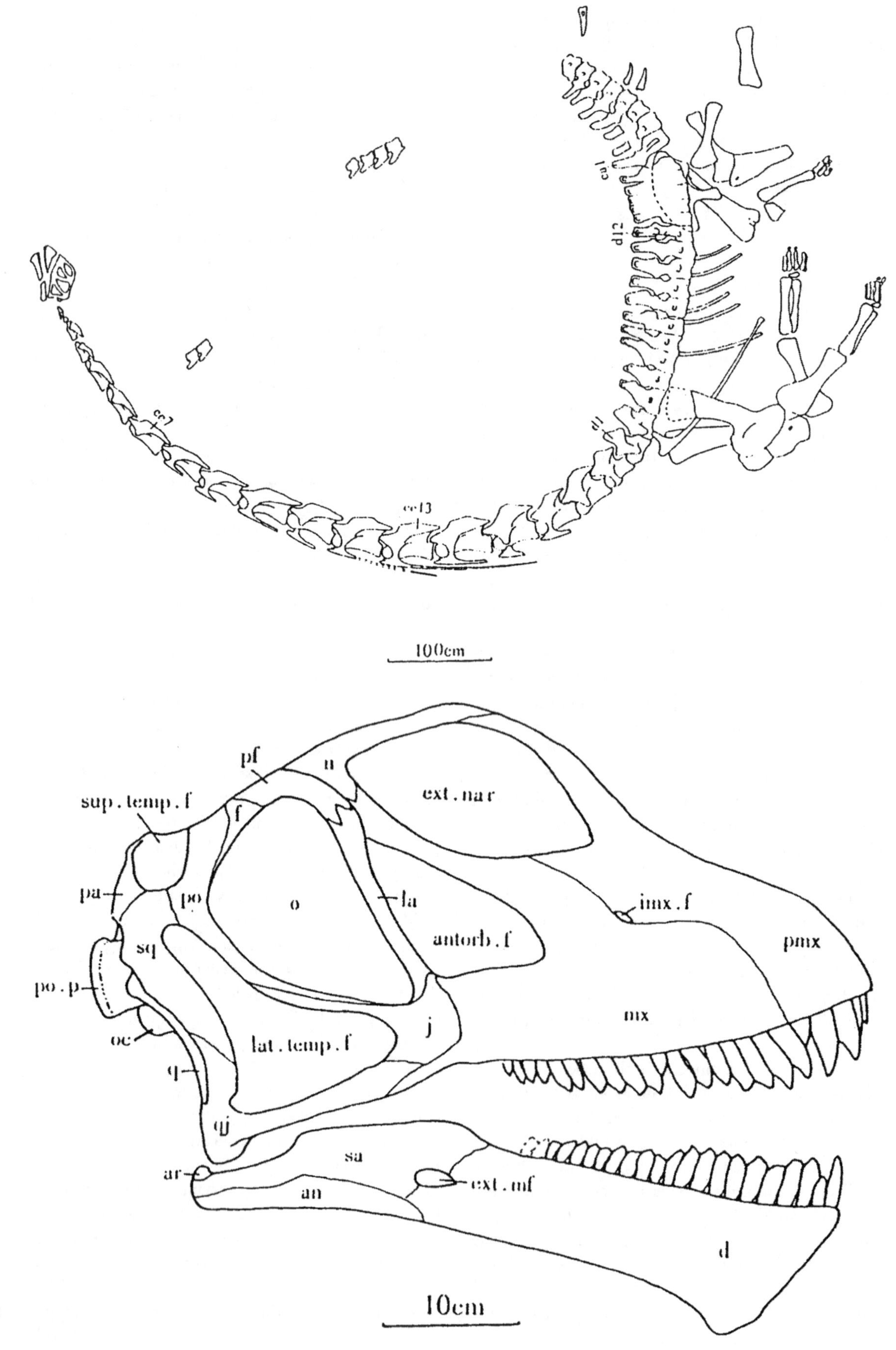

***Mamenchisaurus youngi*, holotype skeleton as found. (After Pi, Ouyang and Ye 1996.)**

***Mamenchisaurus youngi*, holotype skull in right lateral view. (After Pi, Ouyang and Ye 1996.)**

should also "contribute substantially to our understanding of *Mamenchisaurus* morphology and to our knowledge of Upper Jurassic faunal distribution in China."

A new, large species of *Mamenchisaurus*, originally reported by Zhang and Chen (1996; see *S1*) with the spelling *Mamenchisaurus youngsi*, was described by Pi, Ouyang and Ye (1996). This species was founded on nearly complete, largely articulated skeleton, including a well-preserved skull, collected by Ouyang and colleagues from the Zigong Dinosaur Museum. An English translation of Pi *et al.*'s paper describing this species was not available at this time of this writing. However, Zhang and Chen reported that the skull and teeth are very similar to those of *Mamenchisaurus jingyanensis*, its neck comprises 19 cervical vertebrae, and the cervicals, dorsals, caudals, and chevrons are exactly like those of *M. hochuanensis*.

Key reference: Guan, Rigby and Zheng (1999); Pi, Ouyang and Ye (1996); Russell and Zheng (1993); Yang [Young] (1954); Yang [Young] and Chow (1972); Zhang and Chen (1996); Zhang and Li (1996).

†**MEGALOSAURUS**—(=*Avalonianus, Picrodon*; =?*Torvosaurus*).

Saurischia: Theropoda: Neotheropoda: Tetanurae: Spinosauroidea: "Megalosauridae."

Comments: For more than a century, the name *Megalosaurus* has been assigned to fossil specimens pertaining to various large theropod individuals found primarily in Europe, but also known from other continents. Indeed, numerous species have been referred to this genus over the years, some of them now known to belong to other genera, not all of which are theropod. One entry in this lengthy list is the tentatively named ?*Megalosaurus cambrensis*, a species having a rather lengthy history, to which two other genera, *Avalonianus* (a genus sometimes classified as nondinosaurian or "thecodontian") and *Picrodon*, were tentatively referred in 1998 by Peter M. Galton (see *D:TE*, *Megalosaurus* and *Picrodon* entries, also "excluded genera").

Galton (1998), in a study of the saurischians of the Upper Triassic of England, related in much detail the involved and rather convoluted taxonomic history of ?*M. cambrensis*, as well the closely related histories of "*Avalonianus*" and "*Picrodon*." The highlights of Galton's report are as follows:

During the early 1890s, the first fossil belonging to a large dinosaur from the "Rhaetic" (Upper Triassic) of England were recovered from a quarry at Wedmore Hill, in the parish of Wedmore in the Vale of Glastonbury, Somerset. Material collected from this site included teeth (to be catalogued as BMNH R2869, R2875), also various postcranial remains including a hindlimb, and some dorsal and caudal vertebrae (see *Camelotia* entry). The originally recovered

Miscellaneous bones assigned to the genus *Megalosaurus* on display at the Dinosaur Museum in Dorchester, Dorset, England.

Photograph by the author, courtesy Dinosaur Museum.

Photograph by the author, courtesy Geological Sciences Museum.

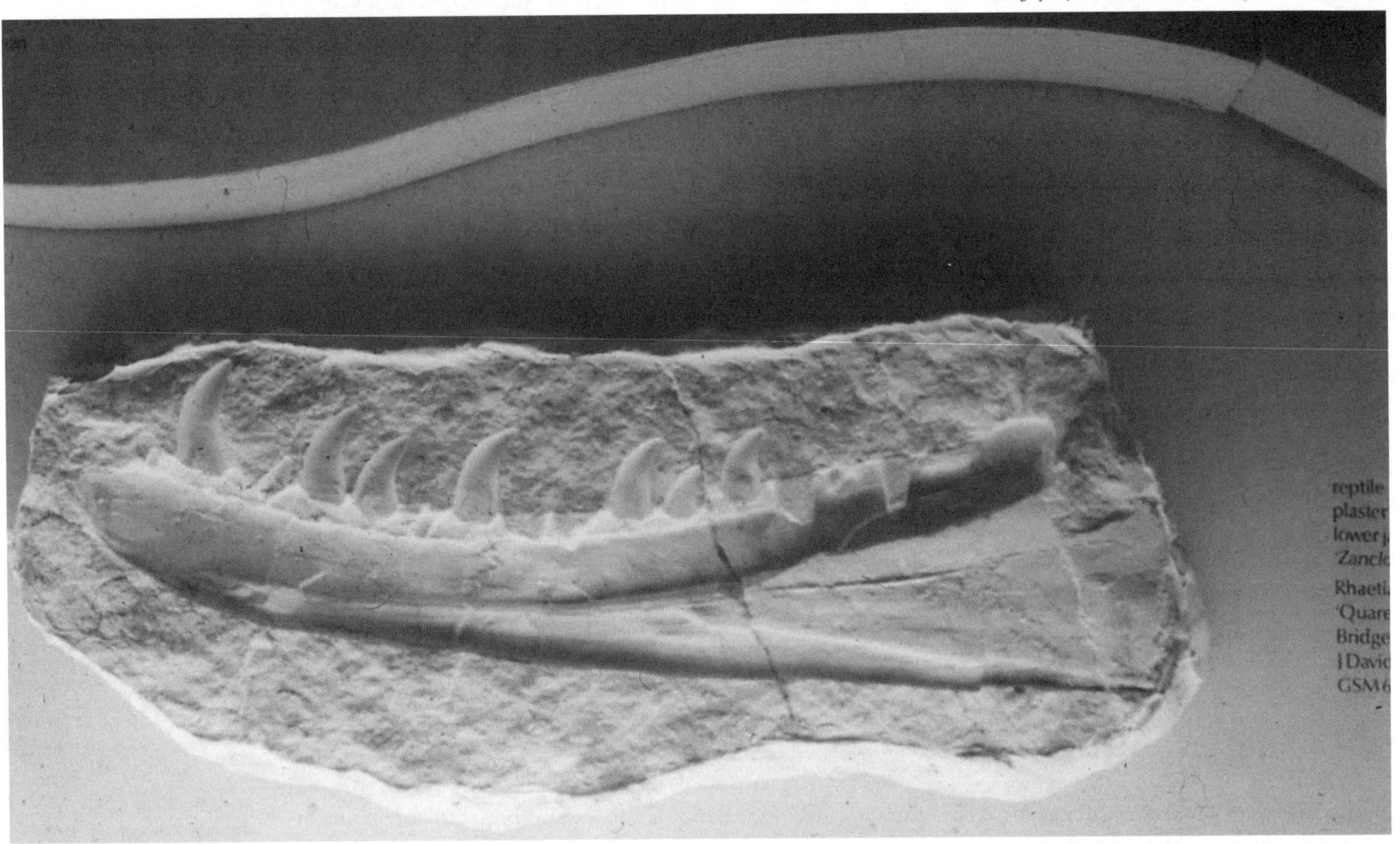

?*Megalosaurus cambrensis*, holotype natural mold of left lower jaw of *Zanclodon cambrensis* (Geological Sciences Museum, now in collections of the British Geological Survey, Keyworth).

specimen (BMNH R2869) consisted of an isolated large dentary tooth, which was briefly described by Sanford (1894). In his report, Sanford stated that, near the spot where the limb bones and vertebrae were discovered, portions of the lower jaw "and an entire tooth were found, the root of which I fear has crumbled into powder; but the crown is still nearly perfect."

In 1898, Seeley described all of the dinosaurian remains from this quarry, referring them to the theropod group "Carnosauria," and erecting for them two new taxa—*Avalonia sanfordi*, founded on the tooth first described by Sanford, and *Picrodon herveyi*, upon another isolated though slightly smaller dentary tooth (BMNH R2875). In the same paper, Seeley referred to *A. sanfordi* the hind-limb bones and larger dorsal vertebrae, and to *P. herveyi* the smaller dorsals and caudal vertebrae. Seeley differentiated these two type species based upon perceived distinctive features of the holotype teeth, finding the tooth of *Picrodon* to exhibit oblique serrations which he believed were of diagnostic significance. Concerning the tooth named *Avalonia*, Seeley wrote that "Mr. Sanford states that the root of the tooth crumbled, and that portions of the lower jaw [see below] were found."

The following year, Newton (1899) noted that the slight differences between Seeley's genera were "barely perceptible." Not only did this author regard these taxa as congeneric, he also believed that the two holotypes represented different parts of a single skeleton. Newton pointed out that the anterior edge of the *Avalonia* tooth was quite worn, which suggests, as Galton pointed out, that it was probably originally serrated near the apex.

At the same time, Newton erected the new type species *Zanclodon cambrensis*, based on the natural mold of a left lower jaw (on exhibit at the Geology Museum, London) from the "Rhaetic" of Glamorganshire, Wales. Newton made the observation that the holotype teeth of both *A. sanfordi* and *P. herveyi* were very similar to those of *Z. cambrensis*. (In 1974, Waldman observed that *Z. cambrensis* is strikingly similar to that of *Megalosaurus hesperis*, a species based on skull material from the Middle Jurassic of Dorset; in their 1990 review of the Carnosauria, Molnar, Kurzanov and Dong tentatively referred *Z. cambrensis* to *Megalosaurus* as the species *?M. cambrensis*, at the same time noting the resemblance of this species to *M. hesperis*; see *D:TE* for details.)

In various papers published in the early part of the 20th century, Huene (1905*b*, 1907–08, 1932, 1956) expressed his opinion that all of the dinosaurian remains from the Wedmore Hill quarry named either *Avalonia* or *Picrodon* represented one nontheropod

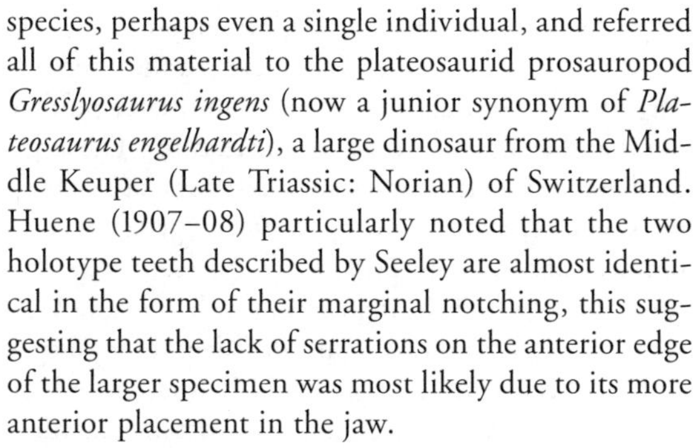

species, perhaps even a single individual, and referred all of this material to the plateosaurid prosauropod *Gresslyosaurus ingens* (now a junior synonym of *Plateosaurus engelhardti*), a large dinosaur from the Middle Keuper (Late Triassic: Norian) of Switzerland. Huene (1907–08) particularly noted that the two holotype teeth described by Seeley are almost identical in the form of their marginal notching, this suggesting that the lack of serrations on the anterior edge of the larger specimen was most likely due to its more anterior placement in the jaw.

More than half a century after the genus was named, Kuhn (1961) pointed out that the generic name *Avalonia* was preoccupied (Walcott 1889) and renamed it *Avalonianus*.

In 1965, Charig, Attridge and Crompton also interpreted the material described by Seeley as "*Avalonia*" as belonging not to a theropod, but a prosauropod. As the bones were considerably larger than those of any known plateosaurid, Charig *et al.* referred them to the family Melanorosauridae. Later, Steel (1970), following the division of the postcranial remains first proposed by Seeley, considered the tooth and dorsal vertebrae assigned to *Picrodon* to be of "carnosaurian" origins, but kept the tooth, limb bones, and larger vertebrae referred to *A. sanfordi* in the Melanorosauridae. Charig (1971) subsequently referred all the postcranial material from Wedmore Hill (including BMNH R2874, incomplete proximal part of left dorsal rib, proximal part of right pubis, distal end of right ischium, R2876, distal caudal vertebra and caudal 2 or 3, R2878, distal part of ?left fibula) described by Huene (1907–08) to the Melanorosauridae.

More recently, Galton noted that the statements of Sanford and Seeley regarding the original condition of the holotype of *A. sanfordi* imply that the tooth was found *in situ* in the portion of lower jaw. From a box containing pieces of broken bone, Galton produced a handwritten note, presumably written by Sanford, stating that two jaws were discovered *in situ*, one of which "fell to pieces and into mud as soon as found." According to Galton's investigation of this note, the almost perfect unworn tooth crown, found *in situ* in the first of these jaws and mentioned by Sanford, seems to have actually been the holotype tooth of *Picrodon* and not (*contra* Seeley) that of "*Avalonia*."

Citing a study by Currie, Rigby, and Sloan (1990) on theropod teeth, Galton pointed out that theropod teeth with roots are found in dentigerous bones, but almost never in isolation. A "perfect worn crown found separate in the clay" described in the above-mentioned note was the holotype tooth of *A. sanfordi*, the apical parts of the anterior and posterior ridges of this specimen displaying wear on their outer or labial face agreeing with the kind of slight wear de-

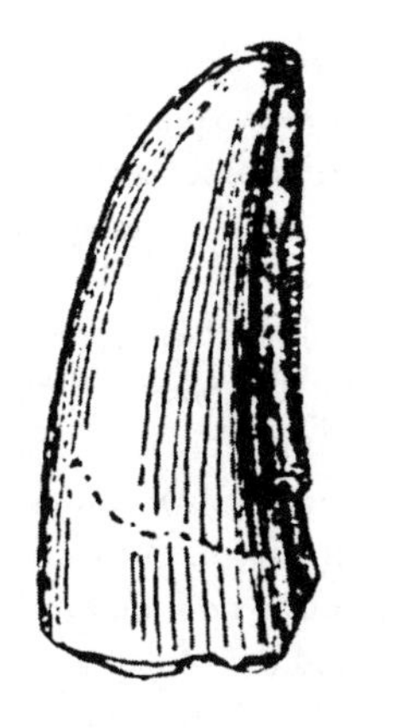

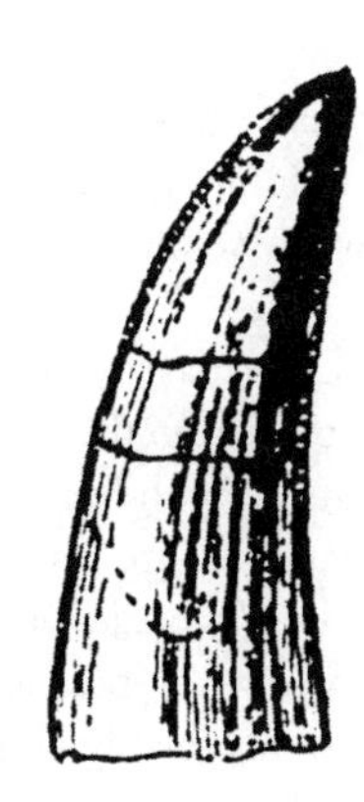

?*Megalosaurus cambrensis* dentary teeth: (Left) BMNH R2869, holotype tooth of *Avalonia sandfordi* and (right) BMNH R2875, holotype tooth of *Picrodon herveyi*. Scale = 1 mm. (After Seeley 1898.)

scribed for shed tyrannosaur teeth from the Upper Cretaceous of western Canada (see Farlow and Brinkman 1994). Galton deduced that probably this tooth was "loose in the jaw so that it could have fallen out of one of the two jaws during preservation or, alternatively, it was shed while *Avalonianus* fed on the carcass of either [the prosauropod represented by the postcranial remains from Wedmore Hill] *Camelotia* (see entry, *D:TE* and this volume) or *Picrodon*."

While referring the postcranial remains from Wedmore to *Camelotia*, Galton retained the two teeth in the Theropoda. Galton regarded both *A. sanfordi* and *P. herveyi* as *nomina dubia*, the minor differences between their holotypes being less than are found in teeth from different areas of the jaws of the same species of the theropod *Dromaeosaurus albertensis* from the Upper Cretaceous of Alberta, Canada (see Currie *et al.*). Therefore, Galton tentatively referred *A. sanfordi* and *P. herveyi* to ?*Megalosaurus cambrensis*.

Notes: Csiki and Grigorescu (1998) published a detailed study in which they reviewed various small theropods — known from both already described specimens and also new materials — from Upper Cretaceous (latest Maastrichtian) rocks of Transylvania in, western Romania, primarily taxa from the Hateg Basin, a region where remains of herbivorous dinosaurs are relatively common, although those of theropods — mostly teeth and hindlimb elements — are scarce.

The problematic species ?*Megalosaurus hungaricus* (see *Megalosaurus* entry, *D:TE*) was based on two small, isolated teeth (MAFI Ob. 3106) possibly representing a dwarf theropod, found at Nagybáród (now Bordod, Bihor County) (see Nopcsa 1915) and, according to a label once accompanying the specimen, from the "Gosau coal outcrops near Nagyabáród locality." Based on its "Gosau coal" (lower part of the Senonian sequence) origin, however, this taxon probably belongs to a stratigraphically older and yet unknown assemblage different from the Hateg Basin fauna.

Other taxa reviewed by Csiki and Grigorescu are all known from the Hateg Basin: The alleged first "top predator" to be identified from the Hateg Basin, *Megalosaurus* sp. was founded upon two caudal centra from Valioara (Densus-Ciula Formation), and interpreted by Nopcsa as representing an indeterminate carnosaur possibly closely related to ?*M. pannoniensis* or ?*M. hungaricus*, both of which were based on teeth. As these remains are quite similar to midposterior caudal vertebrae of the titanosaurid *Magyarosaurus dacus*, Csiki and Grigorescu tentatively referred the *Megalosaurus* sp. material to that sauropod taxon.

Mateus (1999), in a brief review of dinosaurs from Upper Jurassic of Portugal, stated that several bones referred to *Megalosaurus insignis* (see *Megalosaurus* entry, *D:TE*) belong to *Lourinhanosaurus* (see entry), and that all *Megalosaurus* spp. remains from Portugal should be considered *nomina dubia*.

Galton (1999*b*) described theropod teeth (UCMP 49612) catalogued by Samuel P. Welles (personal communication to Galton) as *Megalosaurus* sp., collected from the Kimmeridge Clay (Upper Jurassic) of Weymouth, England. Galton (1975) referred these teeth to *Megalosaurus* sp., although Sues (1980) later pointed out that they display no diagnostic features. The larger of these teeth may represent a different taxon from the smaller teeth. According to Welles' personal communications to Galton (1999*b*), these teeth were presumably collected during the 1940s, before the University of California Museum of Paleontology began active collecting in the type Lance Formation in 1956 (see Clemens 1963), although other details regarding their collection are unknown.

Key references: Charig (1971); Charig, Attridge and Crompton (1965); Csiki and Grigorescu (1998); Currie, Rigby and Sloan (1990); Farlow and Brinkman (1994); Galton (1975, 1998, Galton 1997*b*); Huene (1905*b*, 1907–08, 1932, 1956); Kuhn 1961; Mateus (1999); Molnar, Kurzanov and Dong (1990); Newton (1899); Nopcsa (1915); Sanford (1894); Seeley (1898); Steel (1970); Sues (1980); Waldman (1974).

MICRORAPTOR Xu, Zhou and Wang 2000 (=*Archaeoraptor*)

Saurischia: Theropoda: Neotheropoda: Tetanurae: Avetheropoda: Coelurosauria: Manuraptoriformes:

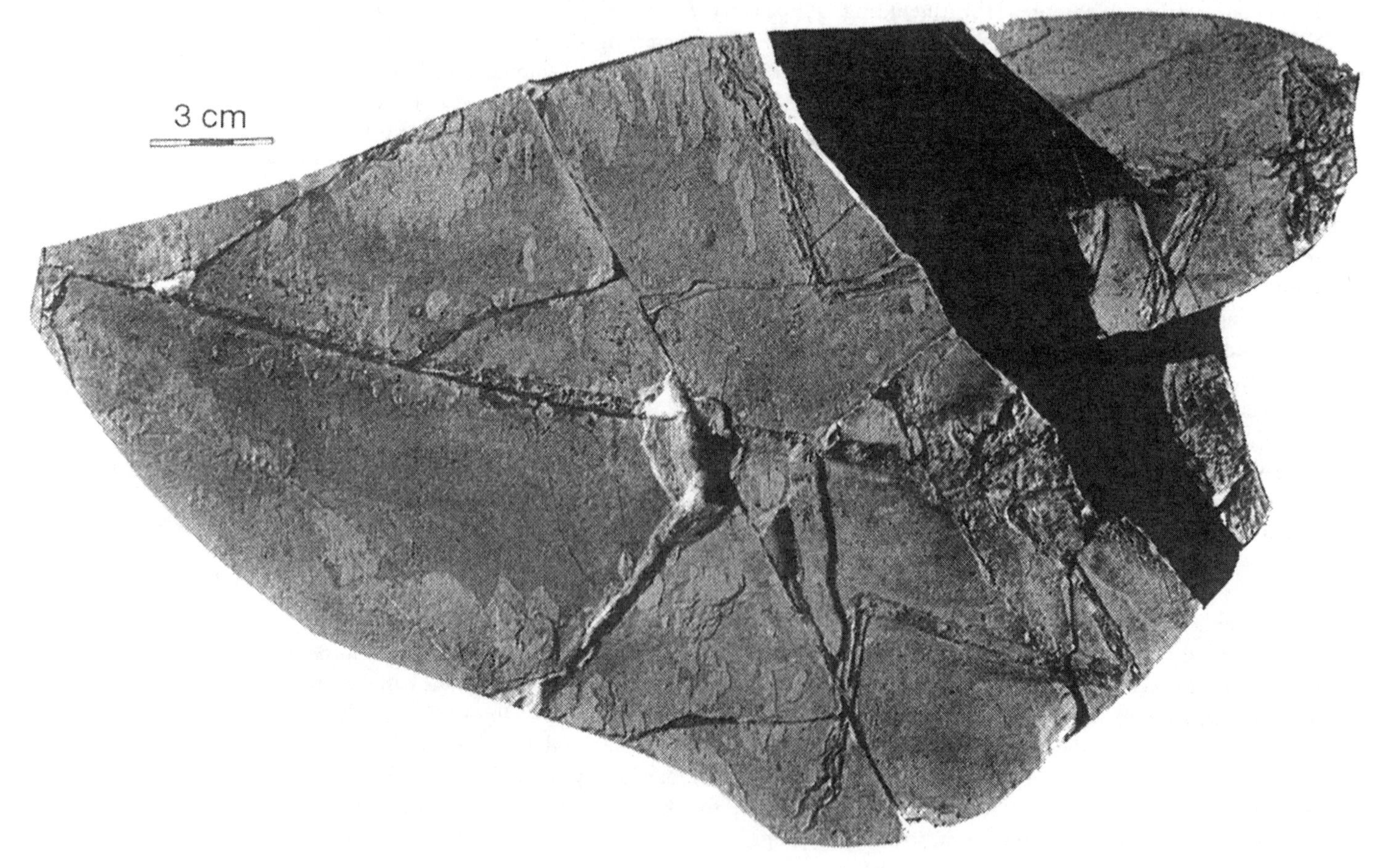

***Microraptor zhaoianus*, IVPP V12330, holotype skeleton. Scale = 3 cm. (After Xu, Zhou and Wang 2000.)**

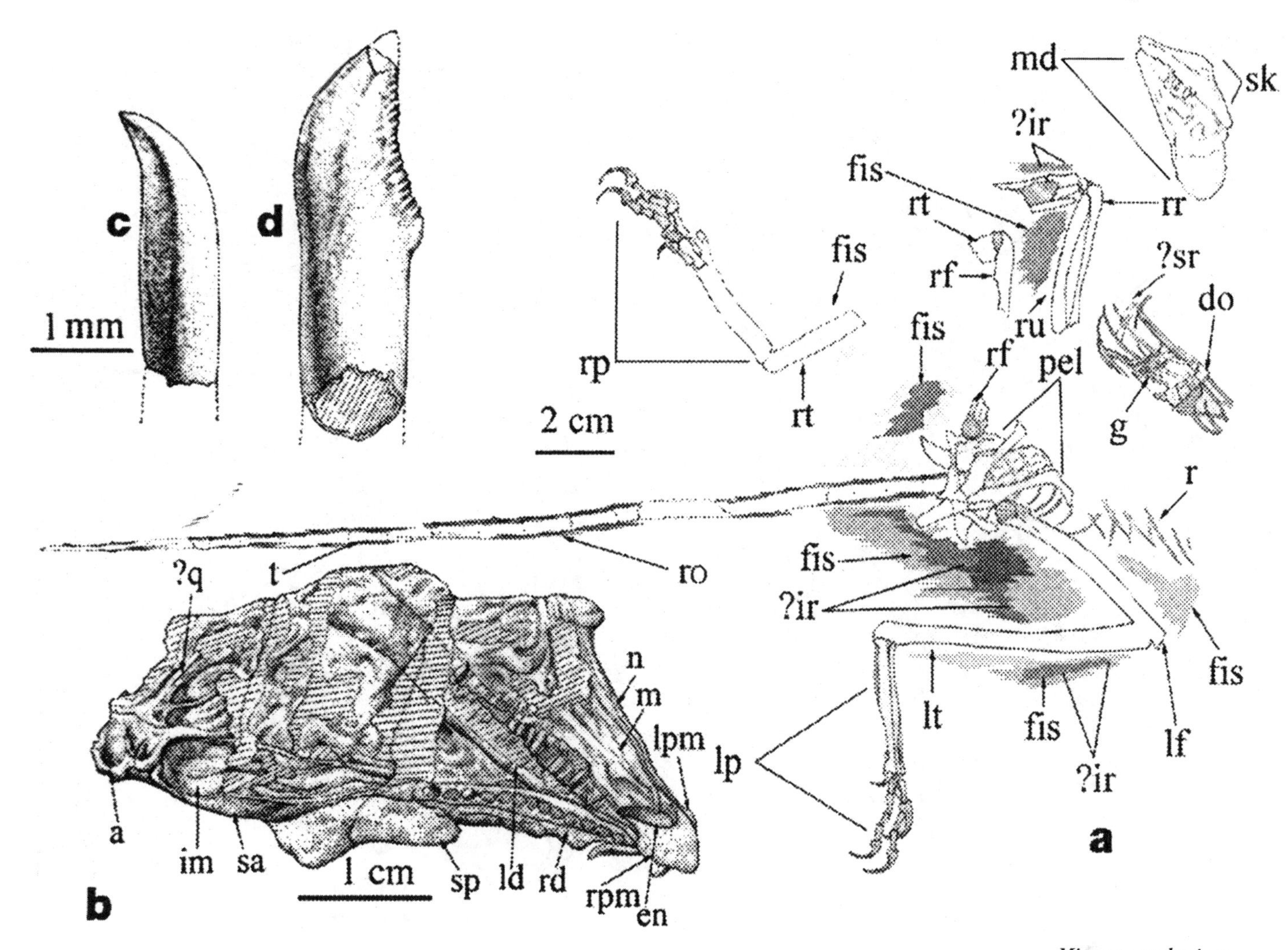

***Microraptor zhaoianus*, IVPP V12330, holotype, a. outline of skeleton (broken lines indicating elements preserved as impressions). b. skull with lower jaws, c. anterior dentary tooth (medial view), and d. posterior dentary tooth (lateral view). (After Xu, Zhou and Wang 2000.)**

Manuraptora: Deinonychosauria: Eumanuraptora: Dromaeosauridae *incertae sedis*.

Name derivation: Greek *mikros* = "small" + Latin *raptor* = "thief."

Type species: *M. zhaoianus* Xu, Zhou and Wang 2000.

Other species: [None.]

Occurrence: Jiufotang Formation, Liaoning, China.

Age: Late Cretaceous.

Known material/holotype: IVPP V12330, articulated partial skeleton preserved in main and counter slabs, missing middle portion of body, including skull with at least teeth, five fused sacral vertebrae, 24 to 25 caudal vertebrae, sternal ribs with uncinate processes, pubis, ischium, radius, ulna, carpals, femur, tibia, articulated foot, metatarsals, integument.

Diagnosis of genus (as for type species): Distinguished from all other known dromaeosaurids in the following: all teeth lacking anterior serrations; posterior teeth having basal constriction between crown and root; middle caudal vertebrae approximately three to four times length of anterior dorsal vertebrae; femur with accessory crest at base of lesser trochanter; tail having less than 26 vertebrae; pedal ungual slender and strongly recurved, with prominent flexor tubercle (Xu, Zhou and Wang 2000).

Comments: Until recently, no mature nonavian theropod had ever been found smaller than the primitive bird *Archaeopteryx*. Xu, Zhou and Wang (2000) reported on the first such nonavian theropod, *Microraptor zhaoianus*, a very small dromaeosaurid (in fact, the smallest adult theropod of any kind found to date). This type species was based on a partial skeleton (IVPP V12330) from the Lower Cretaceous Jiufotang Formation (underlain by the Yixian Formation) of Xiasanjiazi, Chaoyong County, western Liaoning, China. (The holotype comprises the counter slab to the posterior section of the composite specimen formerly known as *Archaeoraptor*; see below).

Xu *et al.* described *M. zhaoianus* as having an extremely short body with an estimated trunk length (based on the assumption that *Microraptor* possessed 13 dorsal vertebrae and that, as in other known dromaeosaurids, the cervicodorsal vertebrae are relatively long compared to the other dorsals) of approximately

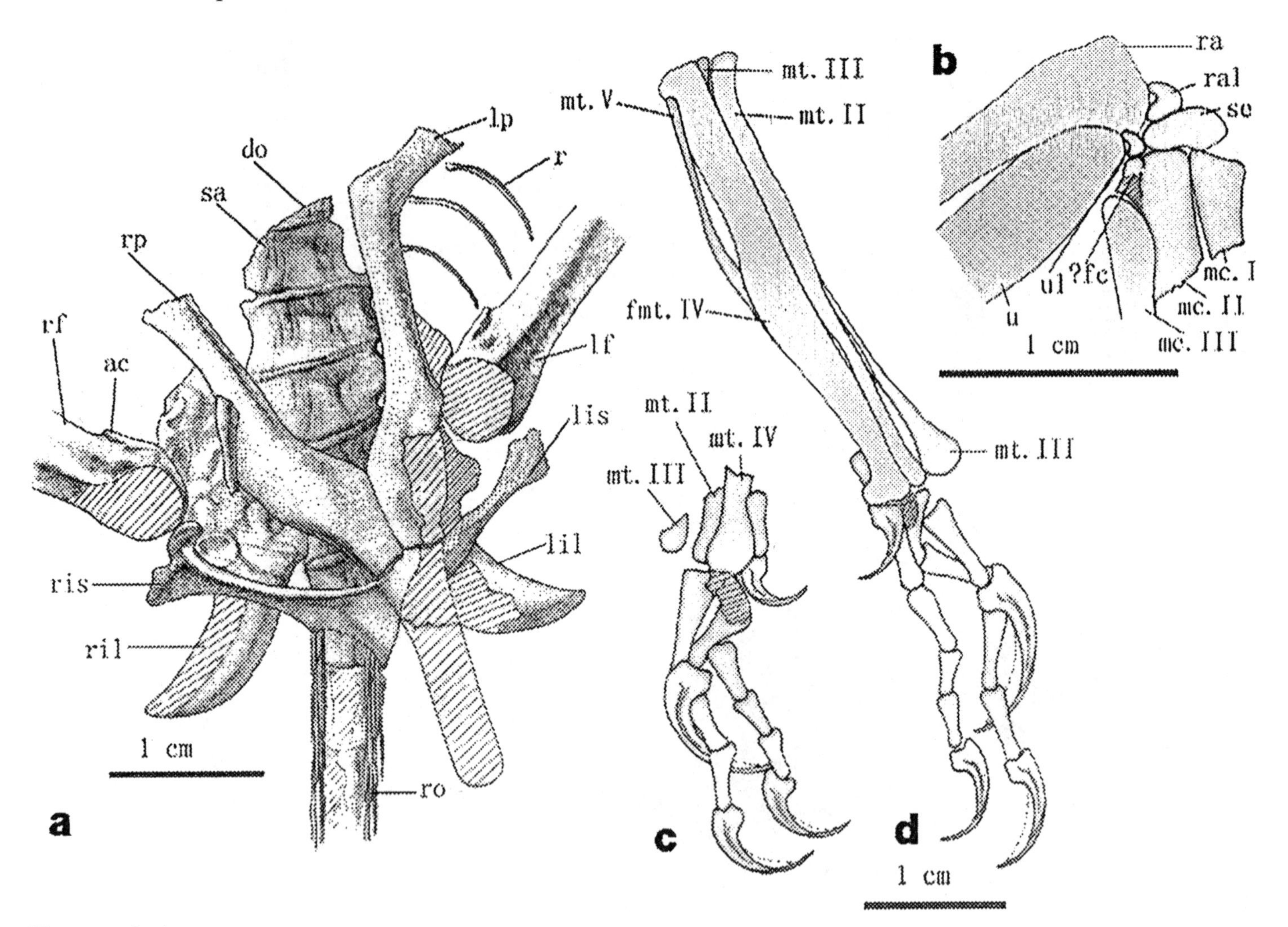

***Microraptor zhaoianus*, IVPP V12330, holotype, a. pelvic girdle (ventral view), b. right forelimb (ventral view), c. right pes (posterior view), and d. left pes (posterior view). (After Xu, Zhou and Wang 2000.)**

47 millimeters. There are at least 15 dentary teeth, which are closely packed, as in troodontids. The anterior (including premaxillary) teeth are more recurved and laterally compressed than the posterior teeth, as in most other theropods, although they lack serrations on anterior and posterior carinae, as in birds and in some nonavian theropods. The posterior teeth have the birdlike feature of a less compressed crown and a constriction between root and crown. The ischium, the authors observed, has a posterior process as in the nonavian theropod *Sinornithosaurus* and the birds *Unenlagia* and *Rahonavis*; also, a large, distally positioned obturator process, as in *Sinornithosaurus* and *Rahonavis*, is present. The metatarsals are of the arctometatarsalian condition as in troodontids.

Large integumentary patches preserved around the skeleton *in situ* are similar to such integument found with early-bird specimens from the same locality. As noted by Xu *et al.*, these structures are best preserved near the femur, running almost perpendicular to that element. The authors described them as long (averaging from 25 to 30 millimeters), narrow, and with a feather-like contour. Those structures along the tibia and in the hip area are shorter. According to Xu *et al.*, some of these integumentary impressions "contain a structure similar to that of a rachis, suggesting that true feathers may have been present among dromaeosaurids."

Performing a cladistic analysis of *Microraptor* utilizing 13 taxa and 89 characters, Xu *et al.* referred this genus to the Dromaeosauridae based on at least 18 synapomorphies, primarily the following: Ossified caudal rods extending length of prezygapophyses and chevrons; longitudinal ridge along posteromedial surface of metatarsal IV; second pedal claw apparently larger than pedal phalanx II.

Xu *et al.* found *Microraptor* to be less derived than other known dromaeosaurids, lacking the following synapomorphies of the Dromaeosauridae: Teeth with denticles on anterior and posterior carinae; all teeth laterally compressed, recurved; constriction between crown and root absent on all teeth; absence of maxilla contribution to naris; number of caudal vertebrae more than 30; femoral neck; tibia less than

115 percent length of femur; and internal mandibular fenestra triangular, relatively large.

The authors found this genus to be a basal dromaeosaurid displaying various salient similarities to troodontids. The more birdlike teeth of *Microraptor*, the *Rahonavis*-like ischium, and the small number of caudal vertebrae distinguish this new genus as unique among all known dromaeosaurids. These birdlike characters, the authors noted, were therefore "more widespread and retained in the lineage leading to birds and in the basal droaeosaur *Microraptor*, but were subsequently lost (through reversal to a more primitive state) in more derived dromaeosaurids."

Xu *et al.* interpreted IVPP V12330, despite its very small size, as representing a mature animal as evidenced by the fusion of the sacral vertebrae, the pubic symphysis, partial fusion of the last dorsal vertebra to the synsacrum, the relatively small sacrum (0.36 sacrum to femur length ratio), the relatively small skull (0.85 skull to femur length ratio), the fine serrations on the tooth crown, the well-developed accessory crest on the femur, and the well-ossified cortical bone. The authors stated that finding this dinosaur eliminates the size disparity between the earliest birds and their nearest nonavian theropod relatives (all previously reported adult theropods being larger than *Archaeopteryx*). At the same time, the small size of this new dinosaur "suggests that nonavian thereopods were pre-adapted for the origin of flight." Furthermore, the discovery of such avian features in nonavian theropods offers new insights for studying the palaeoecology of some birdlike theropod dinosaurs. Furthermore, these features augment our understanding of the apparent morphological transition from nonavian theropods to birds.

According to the authors, the almost completely articulated foot displays features such as the distally positioned digit I, slender and recurved pedal claws, and elongated penultimate phalanges, comparable to those of arboreal birds, suggesting that *Microraptor* was a tree climber. However, as this dinosaur retains an arctometatarsalian foot, *Microraptor* seems to have evolved from a cursorial ancestor.

The paper in which Xu *et al.* formally named and described *Microraptor zhaoianus* did not, however, mention the controversy over the specimen from which this taxon, at least in part, was derived.

The binomen *Archaeoraptor liaoningensis*, the intended name of a supposed feathered dromaeosaurid from Liaoning Province, was first published by Christopher P. Sloan in the November, 1999 issue of the magazine *National Geographic*. As the name appeared in italics without quotation marks and without an accompanying diagnosis or description, it technically became upon publication a *nomen nudum*. According to Sloan's article, this new type species clearly represented a "missing link" between dinosaurs and birds and was distinguished as the first known nonavian theropod capable of flight.

The purported type specimen — an almost complete skeleton possessing distinct feather impressions (some hair-like protofeathers as in *Sinosauropteryx*, others long and broad suggesting flight feathers), preserved in a rock slab — had recently been acquired at the Tucson Gem, Mineral, and Fossil Show in Arizona (via purchase — reportedly for $80,000— by a private benefactor) by Stephen A. Czerkas and wife Sylvia J. Czerkas of the Dinosaur Museum in Blanding, Utah. The fact that this was a vertebrate fossil brought out of China allegedly illegally added to the controversy that would soon focus upon it.

According to this article, details of the specimen brought out by ultraviolet light and CT scans [performed by Tim Rowe of the University of Texas, Austin] showed that the forelimbs, including arms, shoulders, wrists, and hands were quite birdlike. Preliminary examinations seemed to indicate that, with its particular birdlike features, *Archaeoraptor* "was a better flier than *Archaeopteryx*, the earliest known bird." After completion of studies of the specimen at the Dinosaur Museum, the fossil was to be returned to its country of origin. In the *National Geographic* notice was a two-page photograph of the skeleton, also a life restoration sculpted by Stephen Czerkas, who at some later date was to name and describe the specimen formally. Unfortunately, most scientists in the paleontological community were not allowed access to the specimen for study, the spectacular skeleton reserved for its intended future full-color unveiling in the pages of *National Geographic*.

On October 20, 1999, the specimen was displayed to the public by Czerkas, Philip J. Currie of the Royal Tyrrell Museum of Palaeotology in Drumheller, Alberta, and Xu Xing of the Institute of Vertebrate Paleontology and Paleoanthropology in Beijing, at a press conference held at the National Geographic Society in Washington, D.C. The first real look at the specimen by a large number of paleontologists was afforded during the 1999 annual meeting of the Society of Vertebrate Paleontology, held from October 20–23 in Denver, Colorado. Once the photographs were finally revealed and evaluated, rumors commenced circulating among SVP antendees that the specimen was a composite. Some paleontologists were concerned that the bones connecting the tail to the body were missing and that the slab showed telltale evidence of reworking. Within a short time, the specimen's authenticity was under both question and suspicion. Opinions began appearing over the Internet, some suggesting that the unique specimen of *A. liaoningensis* was actually an accidentally or intentionally made chimera comprising the remains of two different taxa.

Storrs L. Olson, curator of birds at the National Museum of Natural History, Washington D.C., urged *National Geographic* not to run the article about *Archaeoraptor*, mainly because the specimen had not yet been fully evaluated and formally described in a peer-reviewed paper. Two papers written about the fossil were submitted to the journals *Science* and *Nature*; however, these were summarily rejected without peer review, in part because the fossil discussed in the articles was controversial, as if for no other reason, it had been smuggled out of its native country.

In December, 1999, following the issuance of the *National Geographic* report, the specimen was publicly exhibited at the National Geographic Society's Explorers Hall in Washington, D. C. along with various other fossils of Mesozoic birds and feathered dinosaurs from China. Newspapers (*e.g.*, the *Chicago Sun-Times*) quoted Currie attesting that the specimen was that of a dinosaur capable of flying, its long, stiff tail probably helping it maneuver, but that the animal was probably not a very good flier.

The controversy was finally resolved by Xu, as reported by Monastersky (2000), in an article published in *Science News*. Xu, while examining a dromaeosaurid skeleton in a private collection (now IVPP V12330) in China, had realized that the tail of that specimen was identical, although "reversed," to that in the *Archaeoraptor* fossil. (Sometimes, when rocks containing fossils are split, they break into "mirror images" of one another.) Apparently some unknown party, presumably in China, had created a chimera by joining a dromaeosaurid tail section to a fossil of a bird. Xu subsequently reported his observations that the *Archaeoraptor* material represented a composite in a letter published in March, 2000 issue of *National Geographic*. In an editorial response to that missive, *National Geographic* stated that CT scans of the specimen seem to corroborate Xu's interpretation, promising that "Results of the Society-funded examination of *Archaeoraptor* and details of new techniques that revealed anomalies in the fossil's reconstruction will be published as soon as the studies are completed."

Currie and other workers stressed to the media that this embarrassing situation in no way casts doubts on or otherwise affects existing evidence supporting the hypothesis that birds are ancestral to theropod dinosaurs; furthermore, the tail section seems to represent a new kind of dromaeosaurid — the first known member of the group to have been feathered — which may later be described by Xu.

As pointed out in an article by Holden (2000*a*) in the journal *Science*, the *Archaeoraptor* situation has resulted in two publishable specimens, one avian, the other nonavian dinosaurian. Reportedly, Czerkas and Xu intend to co-author a joint paper on the avian part, which Czerkas (who, contrary to Xu, preferred referring to as *Archaeoraptor*) presented at The Florida Symposium on Dinosaur Bird Evolution in Fort Lauderdale as "A New Toothed Bird from China." In his abstract, Czerkas (2000) noted that the numerous teeth resemble those of *Archaeopteryx*; the furcula is longer and more deeply U-shaped than in that genus; and the highly derived shoulder girdle and development of the wings are indicative of a strong ability to fly.

For full details on the saga of *Archaeoraptor*, see Lewis M. Simon's (2000) report in *National Geographic*.

Olson (2000), in the April, 2000 issue of *Backbone*, the "Newsletter of the Department of Vertebrate Zoology, National Museum of Natural History" (which Olson sent out "to nearly 50 individuals, libraries, and museums in 12 countries to establish its date of publication), noted that the *National Geographic* article, in publishing the name *Archaeoraptor liaoningensis* along with a photograph of the specimen, had made the name "available for purposes of nomenclature." Furthermore, Olson, pointing out that the dromaeosaurid part of the composite *Archaeoraptor* specimen had counterparts in IVPP V12330, designated the latter as the lectotype of *Archaeoraptor liaoningensis* Sloan, the posterior part of the composite skeleton, consequently, also belonging to that proposed type species. "This means," Olson stated, "that the name now goes with the dinosaur and may be expunged from the literature of avian paleontology," the avian parts of the composite thereby left for proper naming and description.

Olson's proposed naming of IVPP V12330 has also not been without controversy. Various members in the paleontological community questioned the validity of establishing a taxon in what is basically an "in house" newsletter; another concern was that Olson had not described the type material. The article by Xu *et al.*, naming *Microraptor*, based upon the counter slab to the *Archaeoraptor* posterior section, makes no reference to Olson's article or his proposal to retain the original name. Xu (X. Xu, personal communication 2000), in fact, after consulting with other paleontologists knowledgeable in the rules of scientific nomenclature, decided that *Archaeoraptor* is an invalid name. Therefore, pending a possible future ruling to the contrary by the International Commission on Zoological Nomenclature (ICZN), the name *Microraptor* is accepted in this document as the valid one for IVPP V12330.

Note: Xu also has a second dromaeosaurid specimen with an "*Archaeoraptor*"-like tail (X. Xu, personal communication 2000.)

Key references: Holden (2000*a*); Monastersky (2000); Olson (2000); Simon (2000); Sloan (1999); Xu, Zhou and Wang (2000).

†MICROVENATOR

Saurischia: Theropoda: Neotheropoda: Tetanurae: Avetheropoda: Coelurosauria: Manuraptoriformes: ?Oviraptorosauria.

Known material/holotype: AMNH 3041, partial skeleton including several cranial bones, parts of at least 23 vertebrae (cervicals, dorsals, sacrals, and caudals), four ribs, left coracoid, left humerus, radius and ulna, four bones of manus, fragments of ilia, both pubes, femora and tibiae, partial left fibula, left astragalus, two pedal bones, several indeterminate bones or fragments.

Age: Early to Middle Cretaceous (Aptian–Albian).

Diagnosis (as for type species): Autapomorphies including dorsal and caudal centra that are distinctly wider than high and accessory crest (for insertion of M. iliotrochantericus anterior) at base of lesser trochanter of femur (Makovicky and Sues 1998).

Comments: In 1933, an American Museum of Natural History field team led by museum paleontologist and collector Barnum Brown recovered a partial skeleton (AMNH 3041; see *D:TE* for photographs) belonging to a small manuraptoran theropod from the Cloverly Formation of south-central Montana. Brown, who had intended eventually to name and describe this specimen, informally called it "Megadontosaurus ferox," a name which he utilized in public lectures, to identify numerous preliminary skeletal reconstructions drawn under his supervision, on photographs of the fossil material, and, for a while, on specimens exhibited at the American Museum (see Chure and McIntosh 1989). AMNH 3041 was not formally named and described until decades later, when John H. Ostrom of Yale University's Peabody Museum of Natural History, in his 1970 monograph on the stratigraphy and fossil vertebrates of the Clovery Formation of Montana and Wyoming, made it the holotype of a new genus and species, *Microvenator celer* (see *D:TE*).

Later, Peter J. Makovicky and Hans-Dieter Sues, in a short preliminary report published in 1997, briefly

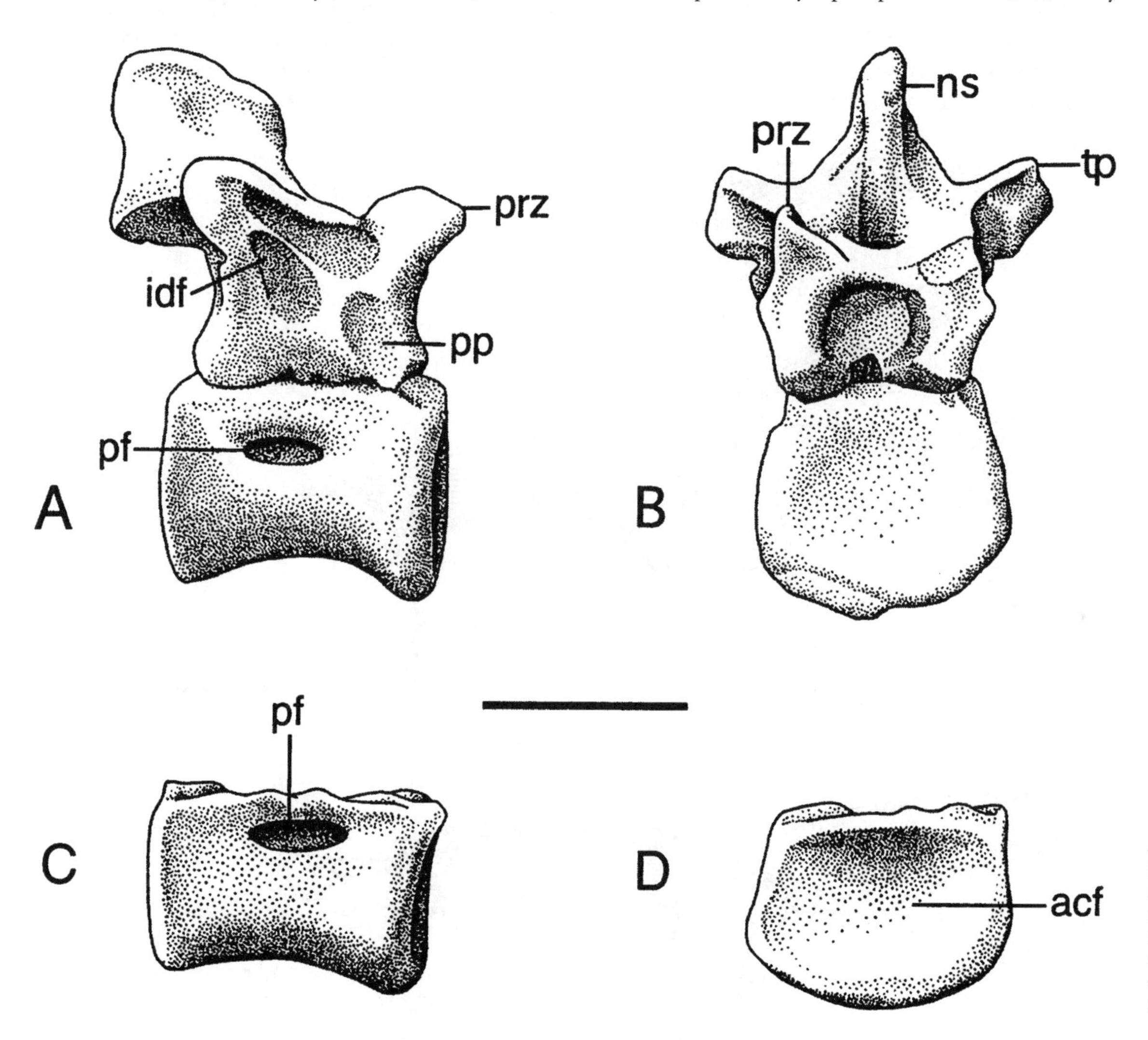

***Microvenator celer*, AMNH 3041, holotype posterior dorsal vertebra in A. lateral and B. anterior views, and isolated dorsal centrum in C. lateral and D. anterior views. Scale = 1 cm. (After Makovicky and Sues 1998.)**

Microvenator celer, AMNH 3041, holotype caudal vertebra in A. lateral and B. anterior views. Scale = 5 mm. (After Makovicky and Sues 1998.)

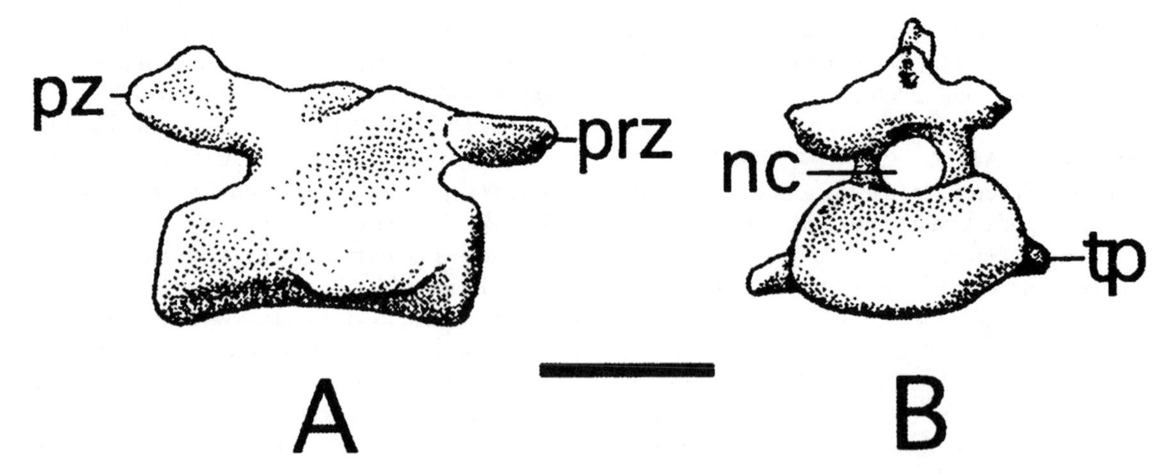

commented on *M. celer* (see *S1*). In 1998, Makovicky and Sues published their subsequent detailed redescription of the species, including a new diagnosis and emphasizing details relating to the phylogenetic relationships of *M celer*. Also, for the first time, these authors published some of the original drawings prepared for Brown's never published study "because they show the limb bones prior to extensive restoration."

This new study by Makovicky and Sues (1998) was based upon their examination of the holotype. (Ostrom had tentatively referred to *M. celer* a small theropod tooth [YPM 5366]; however, Makovicky and Sues (1998) pointed out, AMNH 3041 does not include any unambiguously associated teeth, therefore this identification cannot be verified.)

AMNH 3041 was interpreted by Makovicky and Sues (1998) as representing a juvenile animal based upon several postcranial features, these including the following: (Most significant) cervical, dorsal, sacral, and anterior caudal vertebra lacking fusion between their centra and neural arches; and no apparent fusion between scapula and coracoid. Possible additional

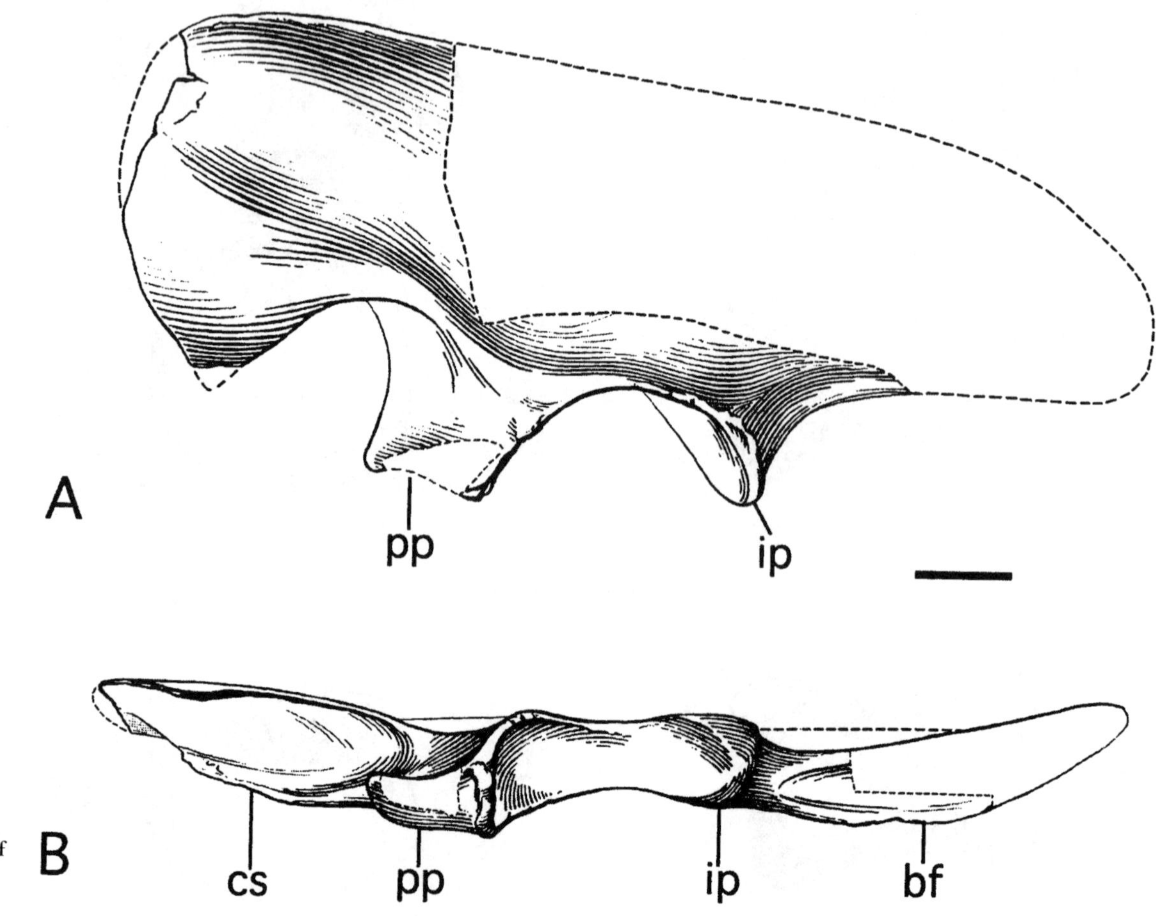

Microvenator celer, AMNH 3041, holotype left ilium in A. lateral and B. ventral views. Scale = 1 cm. This is one of the previously unpublished drawings prepared under the direction of Barnum Brown. (After Makovicky and Sues 1998.)

features indicating immaturity include these: Depression on posteromedial aspect of femoral shaft rather than crestlike fourth trochanter (trochanters increasing in size during ontogeny; see Horner and Weishampel 1988); presence of paired depressions instead of invasive perforations on cervical centra; and cervical centra having distinct nutrient foramina, separated from pneumatic structure by wall of bone (see Britt 1993).

Makovicky and Sues found *M. celer* to possess a suite of at least four features that distinguish this species from all other known theropods: 1. Dorsal centra wider than high; 2. caudal centra wider than high; 3. femur having accessory trochanteric crest; and 4. deep concavity on proximomedial part of pubis.

Ostrom had originally referred *Microvenator* to the Coeluridae. This assessment was based upon the degree of hollowness in the bones (observed by Ostrom to be similar to that in *Compsognathus* and *Ornitholestes*), high degrees of vertebral pneumacity (as in *Coelurus*), and to overall small body size. Subsequent authors have regarded this genus as a paraphyletic taxon lacking any diagnostic characters and belonging to the Manuraptora (Gauthier 1986); as a possible basal caenagnathid (Currie and Russell 1988); and as a possible oviraptorid (Russell and Dong 1993*a*).

Makovicky and Sues (1998) performed their own phyologenetic analysis of *Microvenator*, which determined the following:

Microvenator could be a member of the Oviraptorosauria, the genus possessing these four unambiguous synapomorphies of that group: Dentary (and probably all other jaw bones) edentulous; neural spines of cervical vertebrae short, centered on neural arch, giving neural arches X-shaped appearance in dorsal aspect; dorsal vertebrae with short, wide transverse processes; and pubic shaft anteriorly concave in lateral aspect. However, the following five ambiguous apomorphies support a sister-group relationship between *Microvenator* and Oviraptorosauria: Propubic condition (this character apparently diagnostic for the Oviraptorosauria); paroccipital processes pendant, deflected ventrally; mandible with anteroposteriorly elongate, shallow articular facet; retroarticular process long and tapering; and sacrum including six or more vertebrae. This relationship is also supported by the following ambiguous apomorphies: Prenasal-nasal contact excluding maxilla from margin of external narial fenestra; quadrate pneumatic (a character state known only in Oviraptoridae; see Maryańska and Osmólska 1997); dorsal centra wider than high; and no distinct transition point in tail; also, broad fossa for origin of M. cuppedicus. One character — dentary having widely divergent posterior rami — supports a possible sister-taxon relationship between *Microvenator* and the family Oviraptoridae. Also, at least two ambiguous characters — all sacral vertebrae pneumatic; ischium less than two-thirds length of pubis — are potentially diagnostic of an oviraptorosaurian, excluding *Microvenator*.

The above analysis of Makovicky and Sues (1998) indicated that *Microvenator* is not nested within other manuraptoran groups (*e.g.*, Caenagnathidae). However, given that 47 percent of the characters could not be scored due to the incompleteness of the holotype material, these authors predicted that "any future discoveries of additional material are likely to affect the topology of the relationships found in our analysis." For the present, then, Makovicky and Sues (1998) concluded that *M. celer* is most likely either the earliest known member of the Oviraptorosauria or the earliest oviraptorosaur-like genus represented by diagnostic skeletal remains.

Key references: Britt (1993); Chure and McIntosh (1989); Currie and Russell (1988); Gauthier (1986); Horner and Weishampel (1998); Makovicky and Sues (1997, 1998); Maryańska and Osmólska (1997); Ostrom (1970); Russell and Dong (1993*a*).

†MINMI

Ornithischia: Genasauria: Thyreophora: Thyreophoroidea: Eurypoda: Ankylosauria.

Comments: Gut contents or cololites, which are rarely preserved as fossils, provide the most reliable identification of food items to specific taxa. Molnar and Clifford (2000) described gut contents in the small, Early Cretaceous Australian ankylosaur *Minmi*, this constituting "the first report of a cololite for any thyreophoran dinosaur, and the most reliable report of dietary items for any specific herbivorous dinosaur.

The gut contents were discovered associated with an almost complete, articulated skeleton (QM F18101) collected by the Queensland Museum in 1990 from the Toolebuc Formation (Middle Cretaceous, Albian) in north-central Queensland and referred to *Minmi* sp. (see Molnar 1996; also*S1*).

As recounted by Molnar and Clifford, the specimen — preserved upside down — was originally enclosed mostly in what then appeared to be a single, large calcareous nodule. A small variegated patch of material was found as the modular matrix was being removed. This, upon closer examination, was found to consist of short fragments of plant debris which were interpreted to be the contents of the animal's digestive tract. Subsequent examination found additional plant material between this patch and the right antacetbular process. No indications of gastroliths were found with the specimen.

The cololite mass was found lying apparently in

a flate "smear," 13 millimeters the left of a dorsal (apparently the seventh) vertebra and 25 millimeters dorsal to the ventral margin of the centrum; it is ventral to the ventral margin of the nearest rib, below the level of the elements of the manus. As noted by Molnar and Clifford, this position does not correspond to the gut contents of extant animals, but is consistent with the upside-down position of the skeleton.

About 10 percent of the surface of the cololite consists of plant remains. Under examination using a binocular microscope, the remains showed little detail. They did, however, exhibit some vascular structure, this indicating that they represent vascular tissue in various stages of decomposition.

Molnar and Clifford addressed the possibility that this plant matter was introduced into the animal's body after burial, brought therein by flowing water. The authors found this scenario unlikely, pointing out that QM F18101 was buried in a marine environment. This "would imply that the plant material either was introduced before or during transportation of the carcass to its eventual burial site, but not flushed out or otherwise removed during the process of transport, or that it was introduced after the carcass had come up to rest on the sea floor, more than 100 km from the nearest inferred shoreline (Molnar 1996). Unlikely was the possibility that plant fragments introduced randomly from the sea would all be of about the same size, with no evidence of complete leaves or larger pieces, all brought together into one, relatively small mass. Furthermore, plant matter was found in a plausible position for a part of the digestive tract.

As pointed out by the authors, this discovery is significant for several implications: 1. It verifies that at least some ankylosaurs were herbivorous; 2. the material is consistent with the usually accepted assessment that these dinosaurs ate soft vegetation (in this case, foliage, small stems, and fruiting bodies); 3. the abrupt, perpendicular ends of the vascular bundles shows they were cut by teeth or a beak (not or not only reduced by the grinding of gastroliths); 4. the sizes of the pieces suggest the presence of cheeks (without which it would not have been possible for the mouth to hold the pieces while reducing them to such small segments); and, 5. the sizes suggest that these dinosaurs may have been selective browsers, the shortness of the segments seemingly reflecting the lengths in which they were cut by a series of bites rather than chewing. Therefore, Molnar and Clifford found it possible dinosaurs such as *Minmi* "could nip off small pieces from specific plants, or of specific parts of plants."

Key references: Molnar (1996); Molnar and Clifford (2000).

MONONYKUS Perle, Norell, Chiappe and Clark 1993

Saurischia: Theropoda: Neotheropoda: Tetanurae: Avetheropoda: Coelurosauria: Manuraptoriformes: Arctometatarsalia: ?Ornithomimosauria: Alvarezsauridae: Mononykinae.

Name derivation: Greek *monos* = "single" + Greek *onyx* = "claw."

Type species: *M. olecrans* Perle, Norell, Chiappe and Clark 1993.

Other species: [None.]

Occurrence: Nemegt Formation, Bugin Tsav, Tugrugeen Shireh, Djadokhta Formation, Bayn Dzak, Mongolian People's Republic, Iren Dabasu Formation, China.

Age: Late Cretaceous (?Campanian–?Maastrichtian).

Known material: Numerous specimens including incomplete skeletons.

Holotype: MGI 107/6), incomplete skeleton including posterior part of skull, most precaudal vertebrae, fore- and hindlimbs, thoracic girdle, portions of ilium and pubis.

Diagnosis of genus (as for type species): Cervical vertebrae lacking pleurocoels; sulcus caroticus in cervical vertebrae; presacral vertebrae with diapophyses and parapophyses occupying same level; dorsal vertebrae lacking hyposphene-hypantrum, postzygapophyses lateroventrally oriented; cranial dorsal vertebrae transversely compressed; caudal dorsal vertebrae strongly procoelous; last sacral vertebra having extreme transverse compression; coracoidal shaft elliptical in lateral aspect, craniocaudally long and dorsoventrally low; coracoid transversely flat; sternum having thick carina; radius with extensive articular surface for ulna; first phalanx of digit I of manus with very prominent proximocaudal process; pubis oriented caudoventrally; absence of pubic foot; ischium strongly reduced; femoral distal condyles transversely expanded, almost confluent below popliteal fossa; tibia having accessory (medial) cnemial crest; outer malleolus of distal tibia craniocaudally thick; astragalocalcaneum having deep intercondylar groove; ascending process of astragalus laterally displaced; femoral trochanteric crest; fibula not articulating with tarsus (Novas 1996*b*)

Comments: A very unusual and rather controversial taxon in regards its phylogenetic position, the type species *Mononykus olecrans* was described (originally spelled *Mononychus olecranus*) by Perle, Norell, Chiappe and Clark (1993) based on two specimens from the Upper Cretaceous of the Gobi Desert, Mongolian People's Republic. The holotype is an incomplete skeleton (MGI 107/6) collected in 1987 during the Mongolian-Russian Paleontological expedition.

Additional remains referred to the type species

***Mononykus olecrans*, reconstructed skeleton based on various specimens including holotype MGI 107/6. Originally classified as a dinosaur-like bird, this alvarezsaurid species may rather be a nonavian theropod. (After Chiappe, Norell and Clark 1996, modified from Perle, Norell, Chiappe and Clark 1993.)**

collected at these localities include specimens 100/975 and 100/977, with other referred specimens subsequently assigned by Chiappe, Norell and Clark (1998) to the new genus *Shuvuuia* (see entry); also, discovered recently in the collections of the American Museum of Natural History (see Norell, Chiappe and Clark 1993), specimens recovered during the Central Asiatic Expedition of the American Museum of Natural History of the 1920s, led by Roy Chapman Andrews and Walter Granger, from the Djadokhta Formation at Bayn Dzak and the Iren Dabasu Formation in northern China (see *D:TE* and *S1* for photographs of the cast skeleton mounted at the American Museum of Natural History).

Mononkyus was referred by Perle *et al.* to the Metornithes, a newly introduced group of birds diagnosed by these authors as follows: (Unambiguous characters) large, longitudinally oriented rectangular sternum; ossified sternal carina; ilium with prominent antitrochanter; femoral trochanteric crest undivided; fibula not reaching tarsus; (ambiguous characters owing to lack of data of conflicting characters in closely related taxa) carpometacarpus; pubic apices not in contact; vertebral foramen wide; distal condyles of femora nearly confluent below popliteal fossa; pubis lacking foot (present in *Sinornis*).

The authors originally diagnosed the type species thusly: (Unambiguous characters) maxilla edentulous (convergent with Aves); humerus with pronounced pectoral crest; humerus with single distal condyle; pronounced ventral tubercle of humerus; shafts of ulna and radius extremely short; olecranon process of ulna very long; short, massive carpometacarpus, quadrangular without intermetacarpal space; manus digit I much larger than digits II and III; claw of manus digit I robust; coracoid not expanded ventrally (reversal to primitive manuraptoran condition); thick sternal carina; one posterior dorsal vertebra biconvex, synsacrum procoelous (convergent with *Patagopteryx*); zygapophyses, costal fossa, and transverse process of anterior dorsal vertebrae on same level; posterior synsacral centra sharply keeled; hemal arches elongate; ischium extremely slender; medial cnemial crest on tibiotarsus (convergent with Ornithurae); metatarsal III limited to distal third, triangular in cross section (convergent with several nonavialian theropods); (ambiguous characters) absence of pubic foot.

Life restoration of *Mononykus olecrans* by Mike Fredericks.

At the same time, the authors diagnosed Ornithothoraces, an avian clade including all modern birds, as follows: (Unambiguous characters) coracoid strut-like; scapulocoracoid articulation well below shoulder end of coracoid; coracoid forming sharp angle with scapula at level of glenoid cavity; scapula tapering caudally; humerus shorter than ulna; shaft of ulnar much thicker than that of radius; (ambiguous characters) absence of ventral ribs (gastralia); furcula U-shaped (interclavicular angle less than 90 degrees).

Perle *et al.* described "*Mononychus*" as a flightless bird closer to modern forms than is *Archaeopteryx*, but a quite bizarre avian indeed, being larger than most basal birds, with extremely short forelimbs terminating in a robust hand possessing a single hypertrophied digit. The authors noted that the highly modified forelimb in this new genus is similar to that of burrowing animals; at the same time, however, this feature seems to be incongruous with the long, gracile hindlimbs, suggesting that this animal was probably not a digger.

Later, Norell *et al.* (introducing the new spelling *Mononykus olecranus*) pointed out such features shared with modern birds as a breastbone with a strong median keel, a fibula reduced to a small spike, and fused wrist bones. However, primitive features found also in theropods included the presence of teeth, a long tail, and foot bones separate from each other.

Performing a phylogenetic analysis, Chiappe, Norell and Clark (1996) subsequently referred *Mononykus*, as well as the South American genera *Alvarezsaurus* and *Patagonykus* (see respective entries) to the family Alvarezsauridae, regarded by these authors as the sister taxon to Ornithothoraces, and closer to the group Neornithines than is *Archaeopteryx*.

Although the original authors as well as other workers have continued to posit the avian status of *M. olecrans* (*e.g.*, Perle, Norell, Chiappe, Barsbold, Clark and Norell 1994), not all paleontologists (*e.g.*, Patterson 1993; Feduccia 1994; Martin 1995, 1997; Martin and Rinaldi 1994; Ostrom 1994; Wellnhofer 1994) have agreed with that assessment, instead regarding this taxon as a theropod dinosaur. Among these dissenting views, Martin (1995) (a fossil bird specialist who has long rejected a theropod ancestry for birds), and Martin and Rinaldi, further argued that *Mononykus* is a really a theropod dinosaur related to ornithomimids (see *S1*, section on birds for various pros and cons regarding this issue).

In a paper discussing the differences between dinosaurs and birds, Martin (1997) criticized that previous studies had only compared *Mononykus* to birds and not dinosaurs. Martin (1997) pointed out that *Mononykus* is actually more like dinosaurs in various respects — including having a fully improved posture and a body balanced on the femoral head, with a long tail serving as a counterweight, this arrangement unlike that in any other known bird including *Archaeopteryx*. Based on Martin's (1997) own analysis, *Mononykus* was found to share with ornithomimids a number of synapomorphies, these including: Wrist reduced; metacarpal I enlarged; and astragalus enlarged, becoming the only functional proximal tarsal in the mesotarsal joint. Martin (1997) concluded, therefore, that "*Mononykus* is probably an ornithomimid."

The question of whether *Mononykus* should be classified as a birdlike dinosaur or a dinosaur-like bird has been hotly debated since this genus was originally described. Some more recent interpretations of this taxon have probably been influenced by such discoveries as what appears to be a growing list of feathered Asian theropods, these further blurring the distinctions between dinosaurs and birds. In a brief reported on a new cladistic study, Sereno (1999*a*) found the Alvarezsauridae to be better understood as a theropod clade close to the base of Coelurosauria and nestled within Ornithomimosauria (see "Systematics" chapter for details) than as an assemblage of Cretaceous birds. Consequently, *Mononykus*, pending strong new evidence to the contrary, is herein tentatively included in this compilation of dinosaurian genera.

In 1998, Chiappe *et al.* used *Mononykus* as the basis for a new subfamily of alvarezsaurids, the Mononykinae, to which they also referred the closely related genera *Shuvuuia* and *Parvicursor* (see "Systematics" chapter).

Key references: Chiappe, Norell and Clark (1996, 1998); Norell, Chiappe and Clark (1993); Feduccia (1993); Martin (1991, 1995, 1997); Martin and Rinaldi (1994); Novas (1996*b*); Ostrom (1994); Patterson (1994); Perle, Chiappe, Barsbold, Clark and Norell (1994); Perle, Norell, Chiappe and Clark (1993); Sereno (1999*a*); Wellnhofer (1994).

†MONTANOCERATOPS

Ornithischia: Genasauria: Cerapoda: Marginocephalia: Ceratopsia: Neoceratopsia.

Material: Almost complete skeleton and fragmentary

Mounted skeleton (cast) of the small horned dinosaur *Montanoceratops cerorhynchus* mounted at Dinofest (1998), in Philadelphia. Once classified as a protoceratopsid, this species has been reassessed by paleontologists Brenda J. Chinnery and David B. Weishampel to be the most advanced basal neoceratopsian and the sister taxon to the Ceratopsidae.

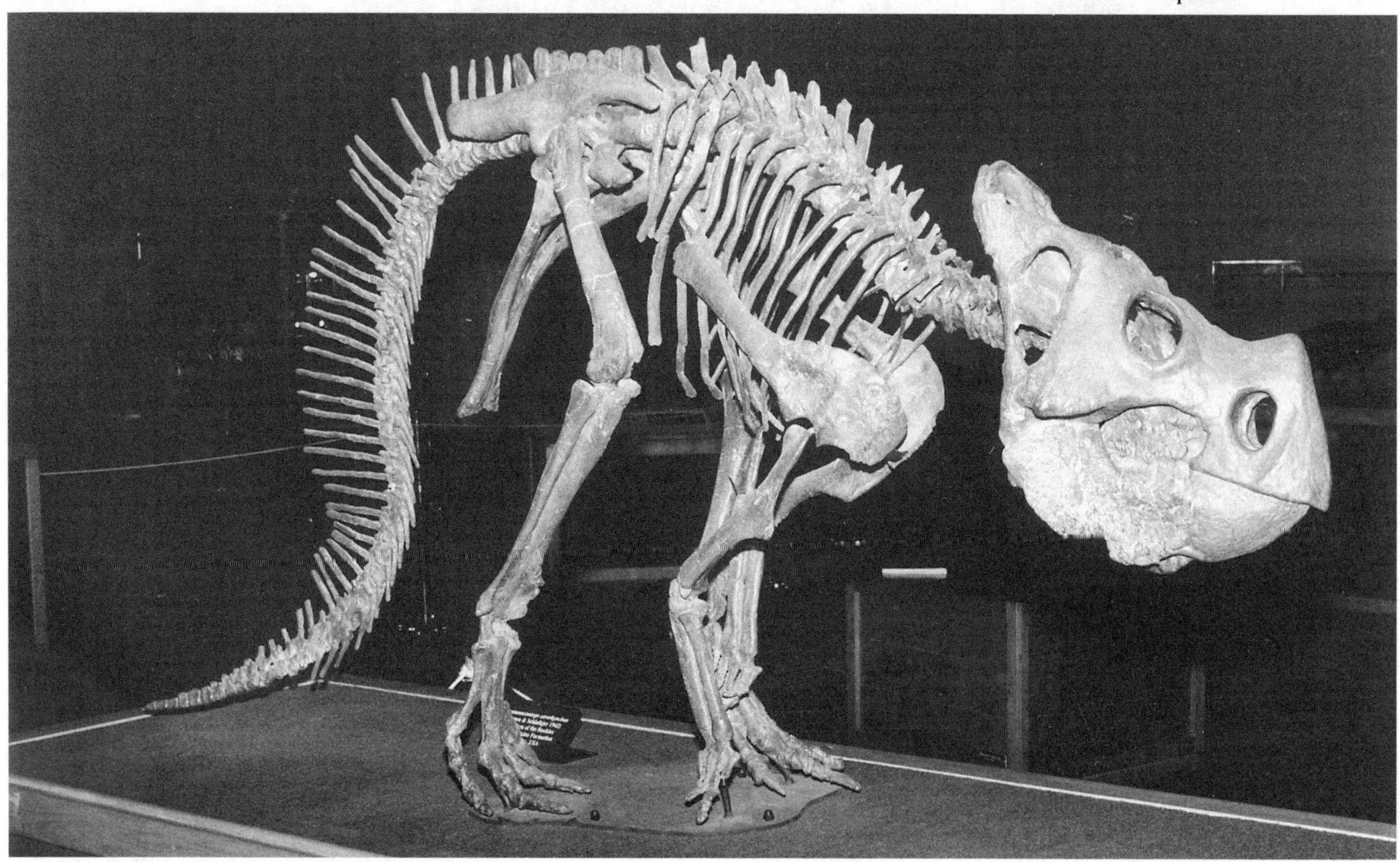

Photograph by the author.

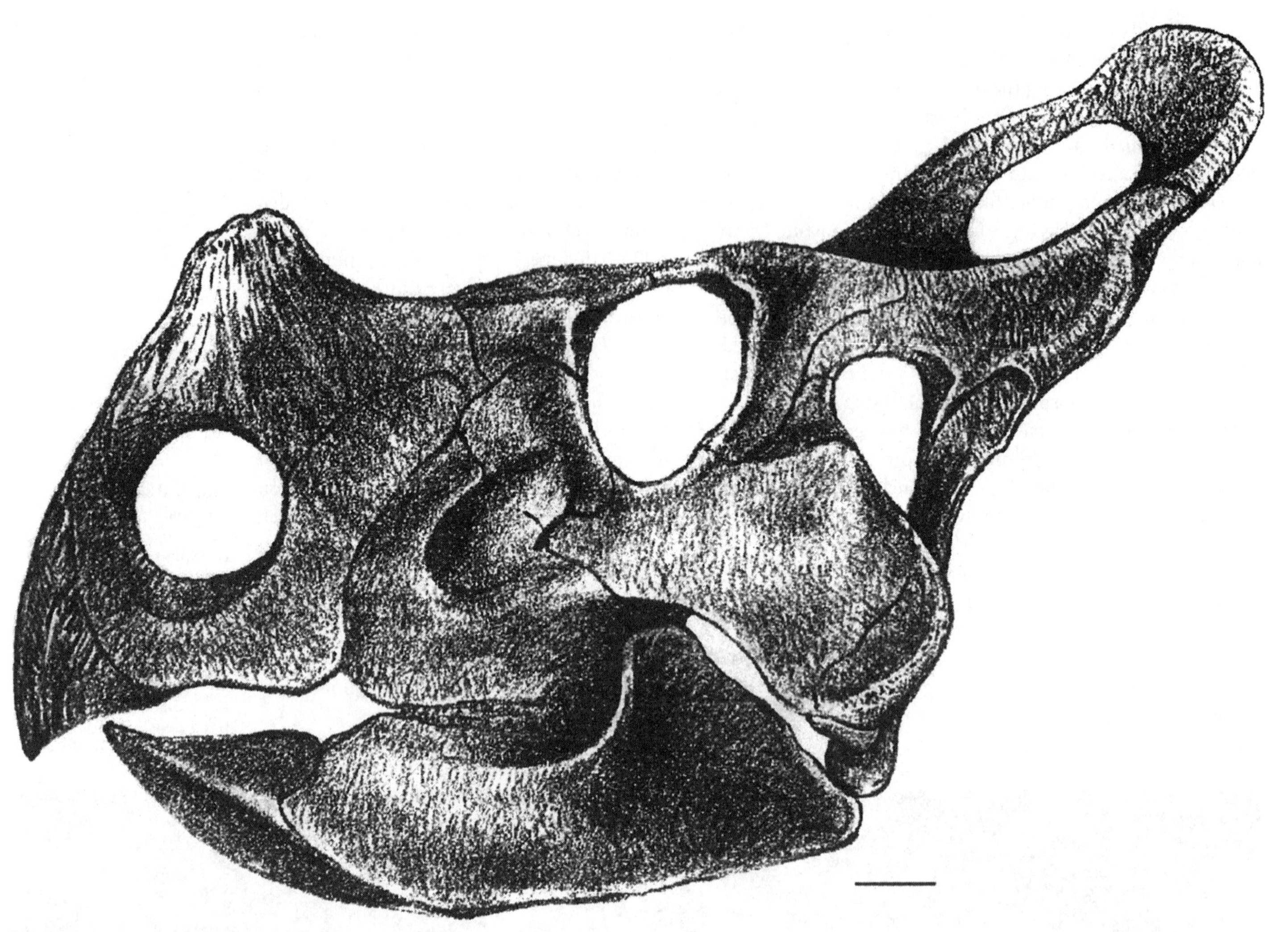

***Montanoceratops cerorrhynchus*, reconstruction of skull based on both MOR 542 and AMNH 5464, left lateral view. Scale = 3 cm. (After Chinnery and Weishampel 1998.)**

skull, partial skull with partial skeleton (postcrania including three cervical vertebrae, incomplete scapulae, clavicals, sternal plates, left ilium and ischium, right pubis, right manus, right hindlimb including incomplete femur, tibia, fibula, almost complete right pes), subadults.

Holotype: AMNH 5465, partial skull (including fragments of nasal, prefrontal, postorbital, squamosal, quadrate, and jugal) with jaws, almost complete postcrania, subadult.

Diagnosis: Rostral basisphenoid reduced; basipterygoid processes caudally curved; ovate antorbital fossa; mandibular teeth having unique buccal protruvision; third and fourth cervical vertebrae short; ilium partially everted (Chinnery and Weishampel 1998).

Comments: Remains of primitive neoceratopsians have been relatively rare in North America, with remains of the most derived basal neoceratopsian, *Montanoceratops* (see *D:TE*)—a small to medium-sized genus with a rudimentary nose horn and a short, fenestrated crest—limited to but a single specimen.

New information about the type species *Montanoceratops cerorhynchus* (originally described by Brown and Schlaikjer 1942 as a second species of *Leptoceratops*) has resulted from a recent study of an incomplete skeleton with a fragmentary skull (MOR 542). These remains, representing an individual about two-thirds the size of the holotype skeleton (AMNH 5464), were discovered by David B. Weishampel in 1986 in Upper Cretaceous (early Maastrichtian) St. Mary Formation at Little Lake Coulee, north of Cutbank, Montana, presumably just a few meters away from the discovery site of the type specimen. Referral of MOR 542 to *Montanoceratops* was founded upon its resemblance to the holotype, particularly the mandible, cervical and dorsal vertebrae, pelvis, tibia, and fibula. Importantly, MOR 542 included material (most notably the posterior half of the braincase, the pelvic gridle, and manus) that permitted a more complete diagnosis of this taxon, and also resolved uncertainties regarding its systematic position in the Neoceratopsia (Chinnery and Weishampel 1998).

After the collection of the holotype specimen in

1916 by Barnum Brown of the American Museum of Natural History, this species was regarded by Sternberg (1951) (who referred it to the new genus *Montanoceratops*) as intermediate between the Asian species *Protoceratops andrewsi* and the North American *Leptoceratops gracilis*. As Chinnery and Weishampel pointed out, the holotype skeleton — still identified as *Leptoceratops cerorhynchus* — was mounted for display at the American Museum with the skull reconstructed from the preserved fragments, and with the missing parts cast from elements (including portions of the forelimbs, ribs, caudal vertebrae, and pes) of *L. gracilis*.

For about three and one half decades, *Montanoceratops* has most often been regarded as belonging to the Protoceratopsidae, a family to which a number of primitive neoceratopsians had, over the years, been referred. More recently, however, with cladistic methods becoming the norm in paleontological classification, problems have arisen regarding the assessment of *Montanoceratops* as a protoceratopside, as well as concerning the validity of this family itself.

Sereno (1984, 1986), in his benchwork cladistic reanalysis and restructuring of the Ornithischia, questioned the acceptance of Protoceratopsidae as a monophyletic assemblage of primitive neoceratopsians. Sereno regarded *Montanoceratops* not as a protoceratopsid, but the sister taxon to Ceratopsidae.

Following their detailed description of MOR 542, Chinnery and Weishampel performed a cladistic analysis of *Montanoceratops*, excluding from their study the fragmentary genus *Kulceratops*, but including all other known basal neoceratopsians, four North American ceratopsids, plus *Psittacosaurus* and Pachycephalosauria. For the sake "of attaining the most complete record possible of relationships and migration events among ceratopsians," the authors included data on genera (particularly *Asiaceratops*, *Microceratops*, and *Udanoceratops*) known only from very fragmentary specimens. The result of their analysis yielded three most parsimonious, node-based trees, the consensus tree indicating the following:

Node 1: The clade Neoceratopsia, as defined by Sereno (1986), is supported by these characters — 1. presence of frill; 2. ventral border of mandible either quite curved of straight (rather than slightly curved); 3. (in adults) fusion of first three cervical vertebrae; and 4. parietal fenestra.

Node 2: An unnamed clade including Protoceratopsidae and an unresolved trichotomy (see numbers 3 and 5, below, respectively) is strongly supported by these characters — 1. lateral upper and medial lower teeth with enamel; 2. orbit with angular rostrodorsal rim; 3. antorbital fossa large; and 4. paroccipital processes at least 40 percent length of basal skull.

Node 3: Protoceratopsidae, including *Protoceratops*, *Leptoceratops*, and *Udanoceratops*, is supported by the following — 1. jaw articulation position at or above tooth row; 2. ventral border of mandible highly curved; and 3. glenoid fossa formed primarily of coracoid.

Node 4: An unnamed clade including *Leptoceratops* and *Udanoceratops* share these characters — 1. maxillary tooth row curved; and 2. vertical-notch wear pattern. Together, these genera have a sister-group relationship with *Protoceratops*.

Node 5: The unresolved trichotomy including *Breviceratops*, *Bagaceratops*, and a clade including *Montanoceratops* and Ceratopsidae, share the following — 1. Nasal horn; 2. predentary having less than 67 percent length of dentary; 3. articulation between dentary and prearticular; 4. contact above or beside infratemporal fenestra between squamosal and jugal; and 5. at least eight sacral vertebrae; iliac blade partially everted.

Node 6: *Montanoceratops* is the sister taxon of the family Ceratopsidae, with the superfamily Ceratopsoidea (Sereno 1986) supported by these characters — 1. Small infratemporal fenestra; 2. large nares; 3. supraoccipital excluded from foramen magnum; 4. more circular intercentrum; and 5. (in adults) complete fusion of cervicals 1–3.

Node 7: The Ceratopsidae (as the authors expected) is the most strongly supported of all neoceratopsian groups.

As determined from Chinnery and Weishampel's analysis, Protoceratopsidae is monophyletic when restricted to *Protoceratops*, *Leptoceratops*, and *Udanoceratops*, while *Montanoceratop* has a sister-group relationship with the Ceratopsidae.

Adding the concept of "ghost lineages" to their study, Chinnery and Weishampel also concluded the following:

1. Ghost lineages are long for most neoceratopsian genera because of the early stratigraphic occurrence of the basal Asian forms *Microceratops* and *Asiaceratops*, this implying much information on this clade remains missing, especially in Asia.

2. There were at least two neoceratopsian Asia-to-North America migrations — one of the ancestors of the North American *Leptoceratops*, another of the common ancestor of the North American *Montanoceratops* and Ceratopsidae (see also Forster and Sereno 1997). The biogeographic distribution of *Leptoceratops* (belonging to the clade including *Microceratops* and *Asiaceratops*) probably required a dispersial event via Beringia (a land bridge joining eastern Asia with western North America) that took place before the late Maastrichtian. Following that migration, all cladogenic events among neoceratopsians probably occurred in Asia until the divergence of *Montanoceratops*

and Ceratopsidae. As these taxa have a North American distribution, the authors suggested that their common ancestor most likely also migrated from Asia.

3. If the very fragmentary Asian neoceratopsian *Turanoceratops* is, in fact, a valid taxon, that shares similarities with the Centrosaurinae (see Nessov 1995), then this genus is nested within Ceratopsidae, this suggesting that there was also a third such migration.

Key references: Brown and Schleikjer (1942); Chinnery and Weishampel (1998); Forster and Sereno (1997); Nessov (1995); Sereno (1984, 1986); Sternberg (1951).

NAASHOIBITOSAURUS—(See *Kritosaurus.*)

†NEDCOLBERTIA

Comment: *Nedcolbertia* (see *S1*), a small, lightly-built coelurosaur previously known only from North America, may also be represented in Europe. Naish (2000) tentatively referred to cf. *Nedcolbertia* the distal end of a left femur (BMNH R36539) collected from the Weald Clay of Sussex. As described by Naish, the clearly tetanuran specimen possesses an extensor groove, a quite sharp-edged mediodistal crest, and very pronounced, posteriorly-directed protuberances on the distal condyles, those on the lateral condyle being markedly square-shaped in side view.

Naish observed that BMNH R36539 is very similar to the distal femur of *Nedcolbertia.*

Key reference: Naish (2000).

†NEMEGTOSAURUS (=?*Quaesitosaurus*)

Saurischia: Sauropodomorpha: Sauropoda: Eusauropoda: Neosauropoda: Diplodocoidea: Nemegtosauridae.

Known material: Skull, tooth.

Comments: The systematic placement of two Mongolian sauropod type species, *Nemegtosaurus mongoliensis* and *Quaesitosaurus orientalis*— the latter regarded by Wilson (1997) as a junior synonym of the former taxon — has been controversial in recent years.

Aleksander Nowiński (1971), who named and first described *N. mongoliensis*, referred this species to the Dicraeosaurinae (=Dicraeosauridae of Upchurch 1995), a taxon then accepted a subfamily of the Diplodocidae. In like manner, Kurzanov and Bannikov (1983), who named and first described *Q. orientalis*, assigned their new species to the Dicraeosaurinae. (See *D:TE* for details on both of these taxa, as well as figures of the holotype skulls.)

Paul Upchurch (1994, 1995), a specialist in sauropodomorph dinosaurs, referred both of these genera and species to a new family, Nemegtosauridae, and joined that taxon with Dicraeosauridae and Diplodocidae to form the superfamily Diplodocoidea. However, subsequent workers (*e.g.*, Salgado and Calvo 1997) have made strong cases suggesting that the affinities of these dinosaurs were rather with the "Titanosauroidea" (see "Systematics" chapter and *Nemegtosaurus* entry, *S1*).

More recently, Upchurch (1999) published detailed cladistic analyses of these nemegtosaurids, evaluating them as both possible "titanosauroids" and also diplodocoids, and critically examining the conflicting character states that support the two hypotheses.

The most parsimonious result of Upchurch's (1999) study favored inclusion of the Nemegtosauridae within the Diplodocidae, this conclusion supported by the following unique synapomorphies: Premaxilla with transversely narrow rostral end, elongate rostrocaudally; subnarial foramen elongated along premaxilla-maxilla suture; in caudal view, caudal margin of external naris lying caudal to rostral tip of prefrontal; in external view, vomerine processes of maxilla hidden by modified premaxilla; loss of intercoronoid; in dorsal view, mandible subrectangular; and teeth restricted to rostral end of jaws.

Contrarily, only two derived character states — orientation of mandibular symphysis (dependent on the inclusion of *Antarctosaurus wichmannianus*; see below) and orientation of tooth crowns — were found to be unique to nemegtosaurids and one or more "titanosauroid" taxa.

Upchurch's (1999) findings supported the hypothesis that the Nemegtosauridae is a sister-taxon to a clade including *Rebbachisaurus*, dicraeosaurids, diplodocids, and that Nemegtosauridae may be the sister taxon to *A. wichmannianus*. The latter species, however, may have been founded on a holotype comprising remains of different sauropod taxa (see *Antarctosaurus* entry), thereby making it unreliable as an operational taxonomic unit.

Upchurch (1999) further noted that, if his interpretations of the skulls of *Nemegtosaurus* and *Quaesitosaurus* are correct, they constitute evidence that some diplodocoid forms survived to the end of the Cretaceous (a time from which most collected sauropod remains pertain to titanosaurs), diplodocoids thereby having a long ghost lineage extending back into the Late Jurassic; that, while basal "titanosauroids" most likely had skulls resembling those of broad-nostrilled camarasauromorphs (*e.g.*, camarasaurs and brachiosaurs), this resemblance is the result of convergence, with the degree of convergence between higher "titanosauroid" and diplodocid skulls perhaps greater than previously suspected; and that diplodocoid cranial apomorphic features were acquired gradually, and do not necessarily form a single complex character.

Note: Previous reports of postcranial remains referred to *Nemegtosaurus* (*e.g.*, *D:TE*) are questionable (J. S. McIntosh, personal communication 2000).

Key references: Kurzanov and Bannikov (1983); Nowiński (1971); Upchurch (1994, 1995, 1999); Wilson (1997).

NIGERSAURUS Sereno, Beck, Dutheil, Larsson, Lyon, Moussa, Sadleir, Sidor, Varrichio, Wilson and Wilson 1999

Saurischia: Sauropodomorpha: Sauropoda: Eusauropoda: Diplodocoidea: Rebbachisauridae.

Name derivation: "[Republique du] Niger" + Greek *sauros* = "lizard."

Type species: *N. taqueti* Sereno, Beck, Dutheil, Larrson, Lyon, Moussa, Sadleir, Sidor, Varrichio, Wilson and Wilson 1999

Other species: [None.]

Occurrence: Various localities, Tégama Group, Gadoufaoua, Niger.

Age: Early to Middle Cretaceous (Aptian–Albian).

Known material: Several partial skeletons.

Holotype: MNN GAD512, partial articulated skeleton including partial skull, neck, scapula, fore- and hindlimbs.

Diagnosis of genus (as for type species): Rebbachisaurid distinguished by the following characters: elongate frontal (much narrower than long) having marked cerebral fossa on frontal; upper and lower tooth rows transversely oriented; extension of tooth rows lateral to plane of principal ramus of lower jaw; tooth crowns having prominent medial and lateral marginal ridges; paired pneumatic spaces at base of neural spines of dorsal vertebrae; prominent rugosity on medial aspect of base of scapular blade (Sereno, Beck, Dutheil, Larsson, Lyon, Moussa, Sadleir, Sidor, Varrichio, Wilson and Wilson 1999).

Comments: *Nigersaurus* is among the most common vertebrates in the Gadoufaoua exposures and also one of the smallest known sauropods (maximum body length of approximately 15 meters or about 52 feet).

The new genus — originally described (but not named) by Phillipe Taquet in 1976 as a dicraeosaurid having affinities with the titanosaurs — was founded upon a partial articulated skeleton (MNN GAD512) from the "niveau des innocents" locality in the terrestrial Tégama Group, (Aptian–Albian, about 115 to 105 million years ago), Gadoufaoua region, Republic of Niger, in northern Africa; specimens referred to the type species *Nigersaurus taqueti* from other Gadoufaoua localities include several partial skeletons and

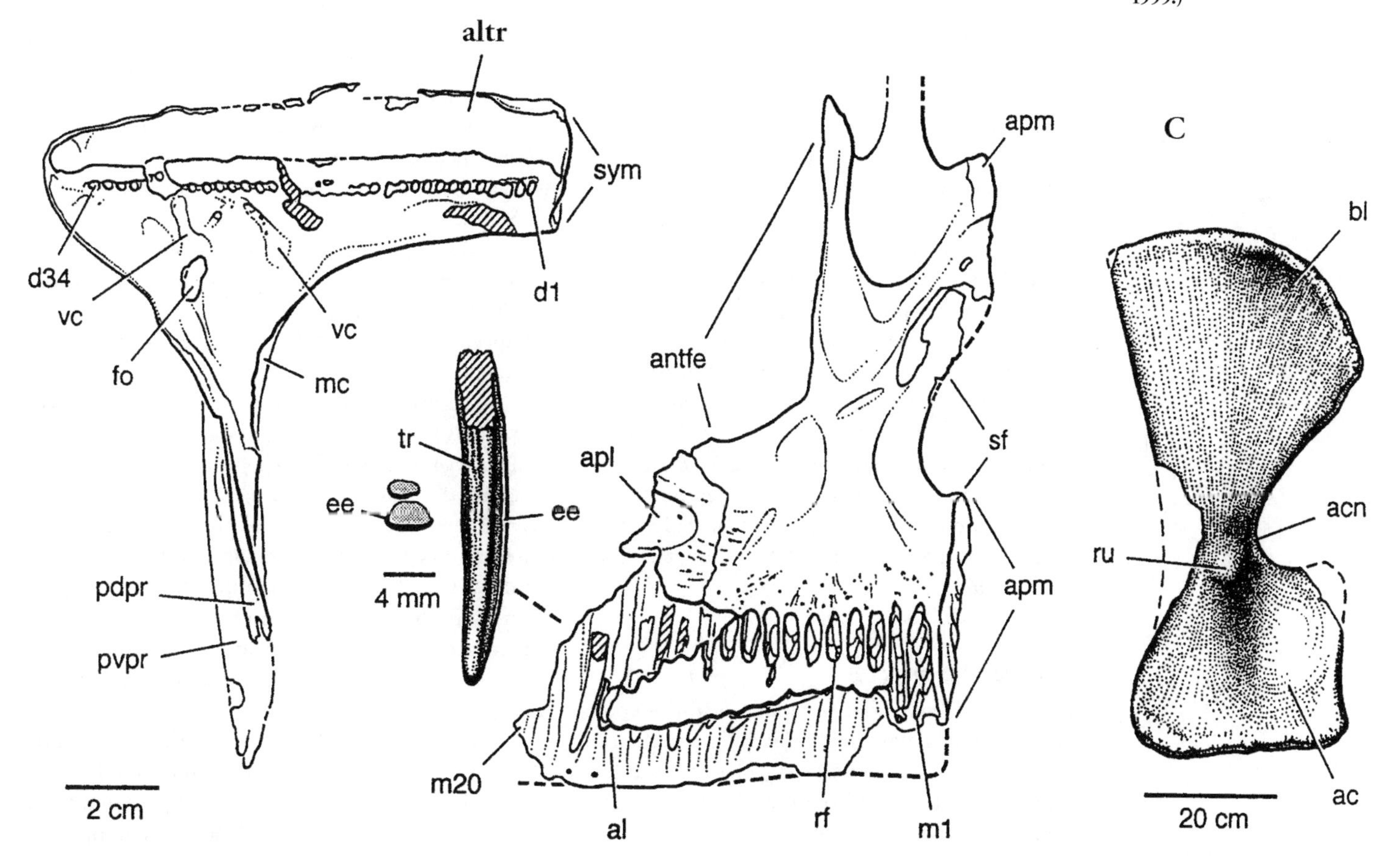

***Nigersaurus taqueti*, MNN GAD512, holotype C. left dentary (dorsal view), B. ventral half of left maxilla (posterior view), and C. left scapula (medial view). Enamel in crown cross section shown as black, dentine as gray shading. Scale bar for A. also applies to B., excluding magnified view of teeth. (After Sereno, Beck, Dutheil, Larrson, Lyon, Moussa, Sadleir, Sidor, Varrichio, Wilson and Wilson 1999.)**

some isolated elements. The fossil-bearing exposures of the Tégama Group in this region (corresponding with the upper part of the Elrhaz Formation and lower part of the Echkar Formation), made up almost completely of cross-bedded fluvial sandstones, have in the past also yielded remains of the spinosaurid *Suchomimus*, the ornithischians *Ouranosaurus* and *Lurdosaurus*, and a nameless though diagnostic titanosaurian sauropod (Sereno, Beck, Dutheil, Larsson, Lyon, Moussa, Sadleir, Sidor, Varrichio, Wilson and Wilson 1999).

In describing *Nigersaurus*, Sereno *et al.* observed that the external nares of the skull are retracted, as in dicraeosaurids and diplodocids. The skull displays features unknown in other Sauropoda, these including the shape of the snout, particularly at the dentary; also, features of the tooth rows, which are strictly transversely oriented, extending lateral to the plane of the lower jaw. The dentary of *Nigersaurus* most closely resembles a specimen from the Cenomanian of Argentina, referred by Huene (1929) to *Antarctosaurus*, in the extreme width of its symphyseal ramus. The enamel of the teeth is many times thicker on the lateral side of the crown than on the medial side, this condition previously only known in advanced ornithischian dinosaurs. There are 34 dentary and at least 20 maxillary teeth, triple the number in *Diplodocus*; including the replacement teeth stacked within the dentary and maxilla, the jaws of *Nigersaurus* would have possessed more than 600 teeth in all.

As described by Sereno *et al.*, the scapular blade, with its broad U-shaped notch at the base of the acromial process, resembles those of the Argentinian (Albian–Cenomanian) *Rayososaurus* and Moroccan (Cenomanian) *Rebbachisaurus*.

Because of the above mentioned similarities, Sereno *et al.* regarded *Nigersaurus* and the dentary attributed to *Antarctosaurus* as apparently "the earliest and latest known representatives, respectively, of this basal clade of diplodocoids, the Rebbachisauridae (Wilson 1999). Furthermore, the establishment of this family in Africa provides information regarding the evolutionary history of the Sauropoda on that continent during the Cretaceous period (see *Jobaria* entry).

Key references: Huene (1929); Sereno, Beck, Dutheil, Larsson, Lyon, Moussa, Sadleir, Sidor, Varrichio, Wilson and Wilson (1999); Taquet (1976); Wilson (1999*a*).

†NIPPONOSAURUS

Ornithischia: Genasauria: Cerapoda: Ornithopoda: Euornithopoda: Iguanodontia: Euiguanodontia: Dryomorpha: Ankylopollexia: Hadrosauroidea: Hadrosauridae: Euhadrosauria: Hadrosauridae: Lambeosaurinae.

Comments: *Nipponosaurus sachalinensis*, one of the relatively few dinosaurs known from Japan (and the first dinosaur discovered in that country), has been a rather neglected and enigmatic taxon since it was first described by Takumi Nagao in 1936. Recently, Suzuki, Weishampel and Minoura (2000) reexamined this type species, known only from isolated skull and postcranial elements representing at least five individuals and founded upon an incomplete skeleton (see *D:TE*).

In a preliminary report published as an abstract for a poster, based on a new study of the holotype, Suzuki *et al.* stated that the type specimen of the genus *Nipponosaurus* represents a subadult individual exhibiting several (unspecified at this time) diagnostic characters.

Performing a phylogenetic analysis utilizing 90 characters for 23 ingroup and three outgroup taxa, Suzuki *et al.* concluded that *Nipponosaurus* is a lambeosaurine, forming a clade with *Hypacrosaurus altispinus*.

Key references: Nagao (1936); Suzuki, Weishampel and Minoura (2000).

NODOCEPHALOSAURUS Sullivan 1999

Ornithischia: Genasauria: Thyreophora: Thyreophoroidea: Eurypoda: Ankylosauria: Ankylosauridae.

Name derivation: Latin *nodus* = "knob/swelling" + Greek *cephalo* = "head + Greek *sauros* = "lizard."

Type species: *N. kirtlandensis* Sullivan 1999.

Other species: [None.]

Occurrence: Kirtland Formation, New Mexico, United States.

Age: Late Cretaceous (upper Campanian).

Known material: Incomplete skull material, ?miscellaneous indeterminate remains.

Holotype: SMP VP-900, incomplete skull, cranial fragments.

Diagnosis of genus (as for type species): Medium-sized ankylosaurid differing from North American *Ankylosaurus* and *Euoplocephalus* and Asian *Pinacosaurus*, *Shamosaurus*, *Talarurus*, and *Tsagantegia* in having semiinflated to bulbous polygonal osteoderms fused to nasal, frontal, and supraorbital regions of skull; differs from all Asian and North American ankylosaurids in having prominent quadratojugal protuberance directed anteroventrally and prominent post-maxillary/lacrimal ridge (osteoderm) (Sullivan 1999).

Comments: In August, 1996, the partial skull (SMP VP-900) of an ankylosaurid was found weathering out of gray mudstone in the Upper Cretaceous De-na-zin Member of the Kirtland Formation, in the

Courtesy Robert M. Sullivan.

***Nodocephalosaurus kirtlandensis*, SMP VP-900, holotype partial skull, left lateral view.**

San Juan Basin, Alamo Mesa East Quadrangle, San Juan County, New Mexico (several hundred meters west of a recently discovered site that yielded the skull of the hadrosaur *Parasaurolophus*; see Sullivan and Williamson 1996). The specimen, consisting of the left side of a crushed and fractured skull, included the basicranium, left squamosal, quadratojugal protuberances, and associated cranial fragments. It is distinguished as "the first specimen of its kind to come from New Mexico" (Sullivan 1999).

Enough distinctive osteological features — not present in other known ankylosaurids, particularly *Ankylosaurus* and *Euoplocephalus* [including *Dyoplosaurus*]) — were preserved in SMP VP-900 for Sullivan to identify it as representing a new genus and species, which he named *Nodocephalosaurus kirtlandensis*. According to the author, the presence of the bulbous osteoderms ally this new form to the Asian ankylosaurids *Saichania* and *Tarchia*, uniting these three taxa in a single clade within the Ankylosauridae, a group that had already been proposed by Kirkland (1998*a*) (see "Systematics" chapter, *S1*). Sullivan, as had Kirkland, refrained from naming this new clade and assessing its precise phylogenetic placement owing to the incomplete nature of SMP VP-900. Furthermore, Sullivan further stated that all of the characters used to diagnose all of the ankylosaurid species are in need of rigorous review.

Noting that the age of the Kirtland Formation has been the source of debate (*e.g.*, Clemens 1973; Lehman 1981; Lucas, Mateer, Hunt and O'Neill 1987), Sullivan suggested that the similarity of *Nodocephalosaurus* to *Saichania* and *Tarchia* could indicate, from an evolutionary standpoint, that are all of approximately the same geologic age. The upper part of the Kirtland Formation, which yielded the holotype of *N. kirtlandensis*, has been most recently dated as 73 million years ago (Fassett and Steiner 1997). This date is comparable to the middle late Campanian (Obradovich 1993); also, it is consistent with the earliest relative dates (middle–late Campanian) proposed for the Barun Goyot Formation, where *Saichania* was collected, and the Red beds of the Khermeen Tsav, which yielded *Tarchia* (Maryańska 1977; Jerzykiewicz and

Russell 1991). Therefore, Sullivan stated, the similarity in both age and morphology of these three ankylosaurid taxa offers further support for a paleogeographic contiguity between North America and Asia during middle to late Campanian time, this hypothesis already having been suggested by Jerzykiewicz and Russell.

Sullivan mentioned other ankylosaurian specimens (incorrectly referred earlier to the geologically younger nodosaurid *Panoplosaurus* and ankylosaurid *Euoplocephalus*) collected over past years from the San Juan Basin that could pertain to *N. kirtlandensis*, although this material is too incomplete and nondiagnostic for exact identification. These include: A right humerus (USNM 8360) collected by J. B. Reeside, Jr. from the De-na-zin member of the Kirtland Formation, identified as ankylosaurid by Barnum Brown and reported by Gilmore (1916*a*), later referred by Lehman to *Euoplocephalus*; a left scapula (USNM 8571), referred to ?*Panoplosaurus*, and dermal plates (USNM 8610), these specimens, almost certainly from the Naashoibito Member of the Kirtland Formation, also reported by Gilmore; isolated dermal plates (USNM 8611) from the Naashoibtio Member; miscellaneous vertebrae, scutes, and a terminal phalanx (NMMNH P-22981), plus other isolated dermal plates and scute fragments in the NMMNH collections (Hunt and Lucas 1993*b*); and an incomplete dermal armor scute from the Kirtland Formation reported by Sullivan and Williamson, housed at the Paleontological Museum, University of Upsalla.

As Sullivan noted, until the discovery of *Nodocephalosaurus* "our understanding of the taxonomic diversity of Late Cretaceous ankylosaurids in the San Juan Basin has been based largely on the parsimonious assumption that the only known North American Late Cretaceous taxa were *Euoplocephalus* and *Ankylosaurus*." However, the discovery of this more primitive genus, and one closer to *Saichania* and *Tarchia* than to the North American forms, forces a reassessment of the relationships of ankylosaur specimens collected previously from this area.

Key references: Gilmore (1916*a*); Hunt and Lucas (1993*b*); Jerzykiewicz and Russell (1991); Kirkland (1998*a*); Sullivan (1999); Sullivan and Williamson (1996).

†NODOSAURUS

Ornithischia: Genasauria: Thyreophora: Thyreophoridea: Eurypoda: Ankylosauria: Nodosauridae.

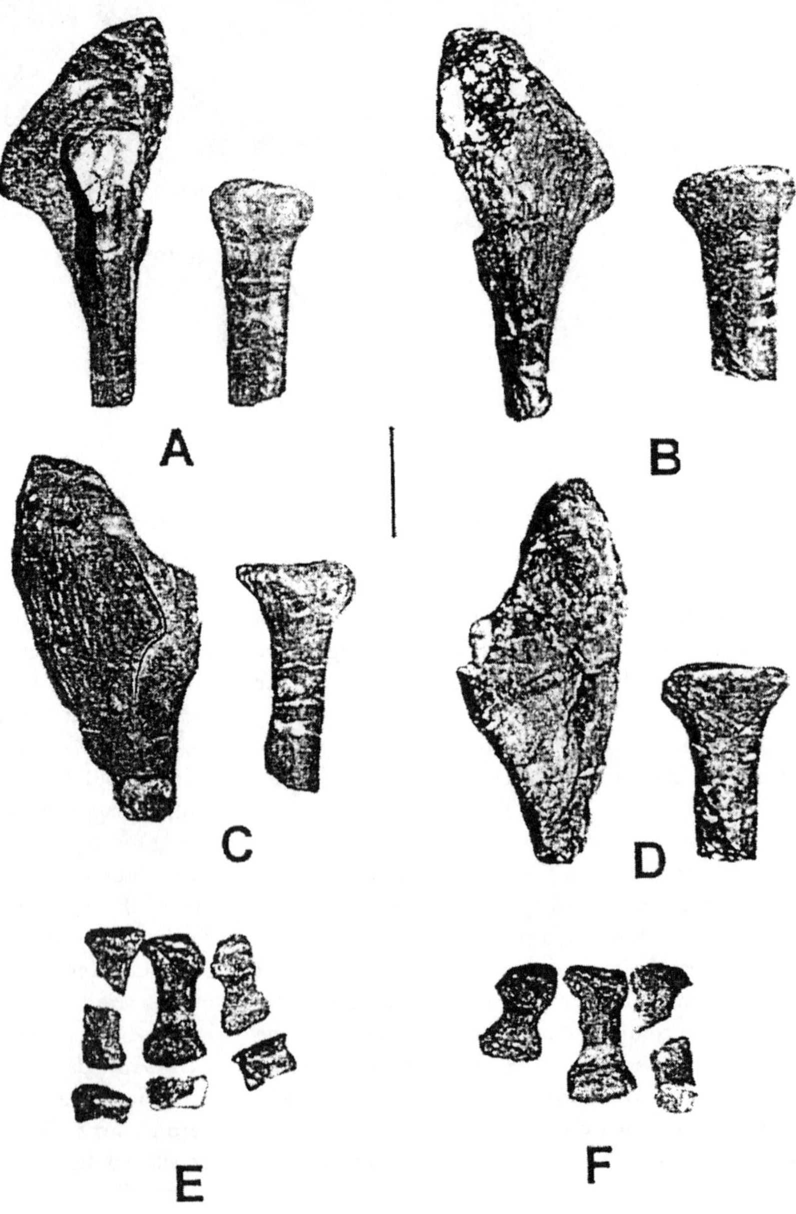

Nodosaurus textilis, YPM 1815, holotype right ulna and radius in A. anterior, B. posterior, C. lateral, and D. medial views; partial right manus, E. metacarpals I–III and phalanges in anterior view, F. metacarpals in posterior view. Scale = 10 cm. (After Carpenter and Kirkland 1998.)

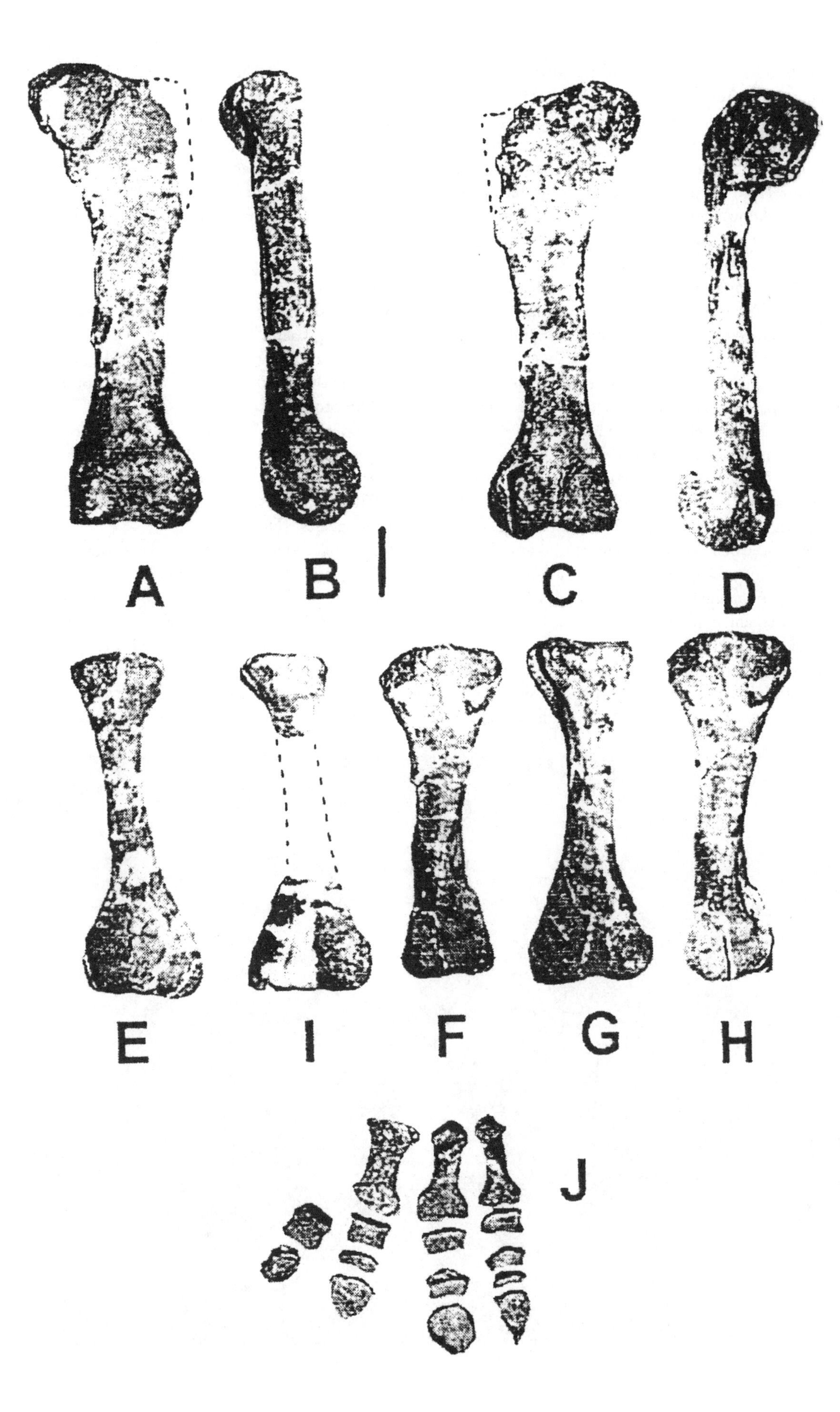

Nodosaurus textilis, YPM 1815, holotype left femur in A. anterior, B. lateral, C. posterior, and D. medial views; left tibia in E. anterior, F. lateral, G. posterior, and H. medial views; I. right tibia in anterior view; J. partial pes in anterior view. Scale = 10 cm. (After Carpenter and Kirkland 1998.)

Age: Middle Cretaceous (middle Cenomanian).

Occurrence: Frontier Formation, Wyoming, Niobrara Chalk Formation, Kansas, United States.

Diagnosis: Femoral head sloping laterally; tibia having slender shaft, distal end not sharply flared as in *Sauropelta*; posterior dorsal armor differing from all other known nodosaurids in having pair of midline rectangular scutes with raised central dome, transverse bands of subrectangular scutes with domed center, alternating with transverse bands of smaller, flat, square scutes; sacral armor comprising flat, somewhat hexagonal, coossified scutes (Carpenter and Kirkland 1998).

Comments: *Nodosaurus textilis* was founded upon a partial postcranial skeleton (YPM 1815) collected from the Belle Fourche Member of the Frontier Formation of Albany County, Wyoming (see *D:TE*). The Frontier Formation (Dunbar 1944) in Wyoming was revised and reidentified by Merewether, Cobban and Cavanaugh (1983) and Merewether (1983) as the Belle Fourche member of that formation; based upon invertebrate stratigraphy noted by those authors, *Nodosaurus* is distinguished as "one of the few Cenomanian dinosaurs known world-wide" (Carpenter and Kirkland 1998).

In redescribing this taxon, Carpenter and Kirkland addressed Lull's (1921) hypothesis that *Nodosaurus* and *Hoplitosaurus*, both of them fragmentary forms, might be congeneric. This suggestion seemed doubtful to Carpenter and Kirkland, who pointed out "the lack of similarity between the armor of the two forms"; also, *Hoplitosaurus* is probably Barremian, whereas *Nodosaurus* is middle Cenomanian, a difference of about 25 million years.

Key references: Carpenter and Kirkland (1998); Lull (1921).

NOMINGIA Barsbold, Osmólska, Watabe, Currie and Tsogtbaatar 2000

Saurischia: Theropoda: Neotheropoda: Tetanurae: Avetheropoda: Coelurosauria: Manuraptoriformes: Oviraptorosauria.

Name derivation: "Nomingiin [Gobi, part of Gobi Desert near the type locality]."

Type species: *N. gobiensis* Barsbold, Osmólska, Watabe, Currie and Tsogtbaatar 2000.

Other species: [None.]

Occurrence: Beds of Bugeen Tsav, Bayankhongor Province, Gobi Desert, Mongolia.

Age: Late Cretaceous (?upper Campanian, ?lower Maastrichtian, or ?middle Maastrichtian).

Known material/holotype: GIN 100/119 (formerly GIN 940824), incomplete postcranial skeleton including continuous series of vertebrae (13 presacrals, five synsacrals, 24 caudals with most chevrons

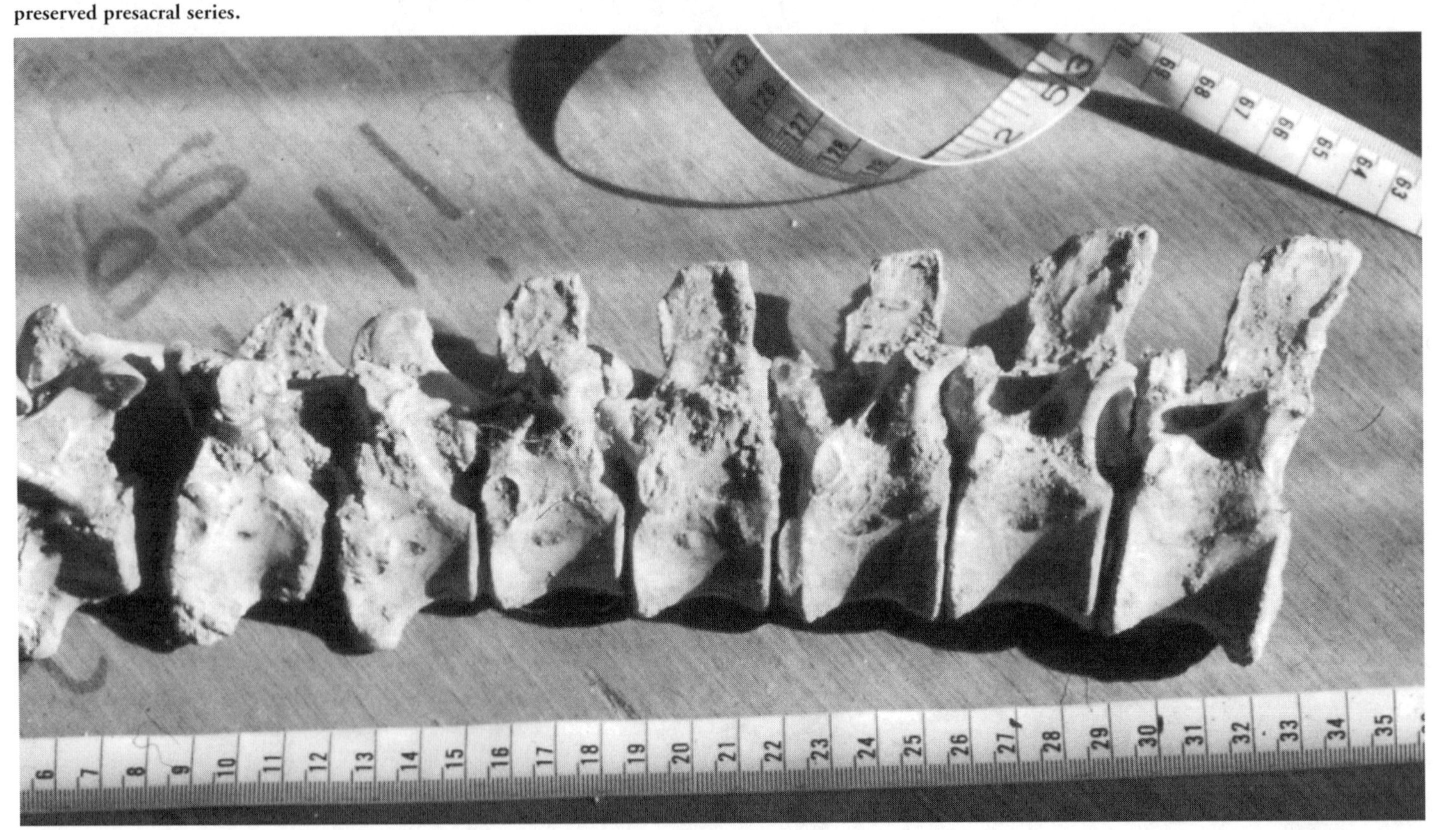

***Nomingia gobiensis*, GIN 100/119, holotype 10 anteriormost vertebrae from the preserved presacral series.**

Courtesy Philip J. Currie.

Courtesy Philip J. Currie.

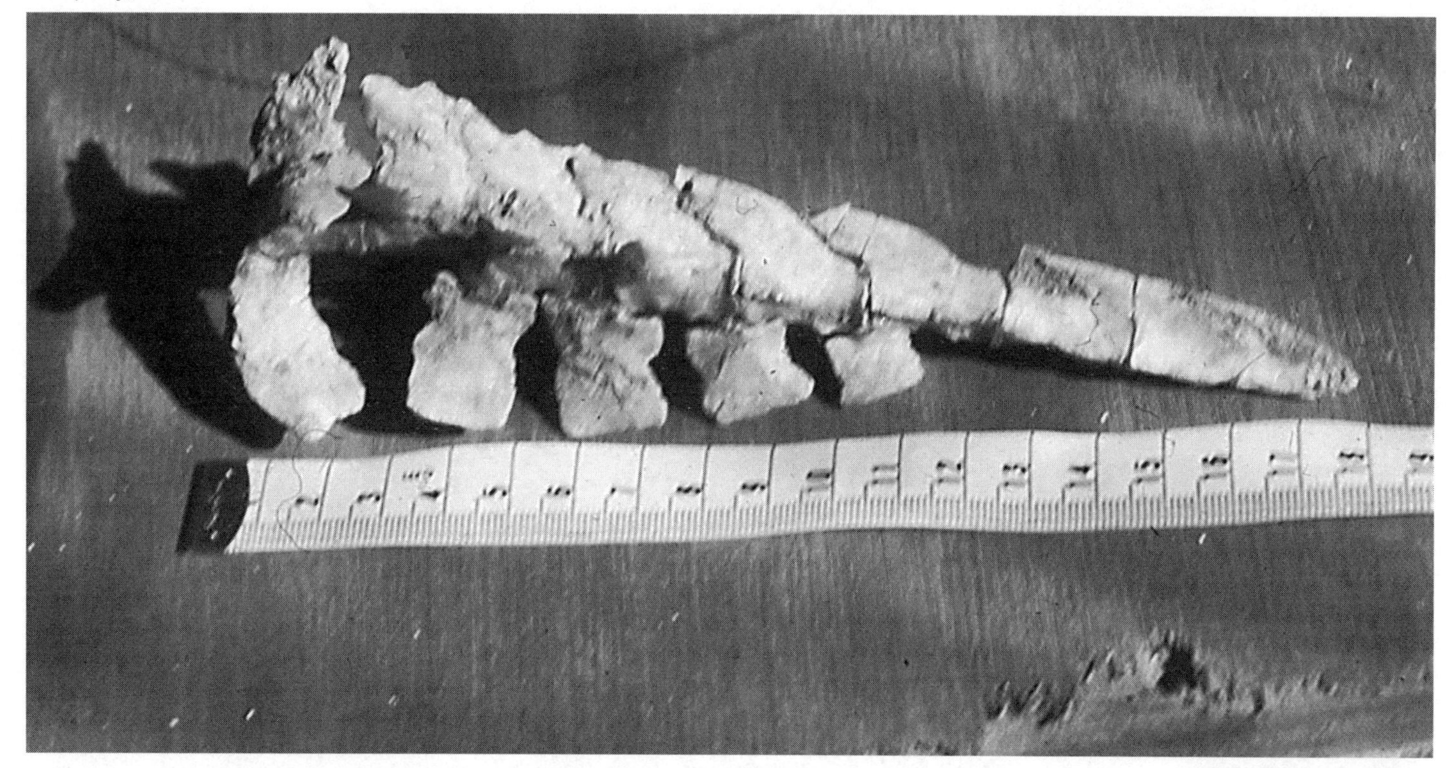

***Nomingia gobiensis*, GIN 100/119, holotype caudal vertebrae, the last five of which are fused into a pygostyle.**

articulated, complete pelvis, 10 fragmentary thoracic ribs, several disarticulated gastralia, left femur, tibiae and fibulae.

Diagnosis of genus (as for type species): Differs from all other known oviraptorosaurs in having tail comprising only 24 vertebrae, last five caudal vertebrae fused into pygostyle; chevrons wide, dorsoventrally elongate on most caudals, except on pygostyle; pelvis weakly propubic, pubic axis at angle of approximately 30 degrees to vertical; dorsal margin of ilium convex along preacetabular process, horizontal along postacetabular process; preacetabular process about 25 percent longer than postacetabular process; postacetabular process equally deep along its entire length; iliac process of pubis twice length of respective process of ilium (Barsbold, Osmólska, Watabe, Currie and Tsogtbaatar 2000).

Comments: The genus *Nomingia* constitutes another recent discovery providing support for the hypothesis that birds descended from theropod dinosaurs.

The type species *Nomingia gobiensis* was founded upon an incomplete postcranial skeleton (GIN 100/119) discovered in 1994 by the Mongolian–Japanese Palaeontological Expedition in a bluish-white, fine sandstone layer of the Beds of Bugeen Tsav (equivalent of the Nemegt Formation; dated as ?upper Campanian and ?lower Maastrichtian by Gradziński, Kielan-Jaworowska, and Maryańska 1997; as ?middle Maastrichtian by Jerzykiewicz and Russell 1991), at Bugin Tsav, Bayankhongor Province, Gobi Desert, Mongolia (Barsbold, Osmólska, Watabe, Currie and Tsogtbaatar 2000).

The discovery was first announced, then as an unnamed oviraptorosaur, in an article published in *National Geographic* magazine (see Sloan 1999). The piece was accompanied by a photograph of the posterior part of the specimen. What made this specimen so special, Sloan noted, was that the tail terminated in a pygostyle.

Still unnamed, the specimen (then bearing the catalog designation of GIN 940824) was subsequently given a preliminary description by Barsbold, Currie, Myhrvold, Osmólska, Tsogbaatar and Watabe (2000). As noted by these authors, the specimen has a relatively shorter tail than other known oviraptorosaurs (or any other nonavian theropods). The last five caudal vertebrae are coossified into the birdlike feature of a pygostyle, a condition otherwise found only in birds designed to hold tail feathers, sometimes serving as a flight-control device.

The new genus and species were formally named and fully described in a subsequent paper by Barsbold, Osmólska, Watabe, Currie and Tsogtbaatar (2000).

The specimen was identified by Barsbold, Currie *et al.* as an oviraptorosaur based upon such features as pneumatic proximal caudal centra, short distal centra, dorsally close ilia, and vertical pubis. Barsbold, Osmólska *et al.* could not, however, presently

Photograph Philip J. Currie.

***Nomingia gobiensis*, GIN 100/119, holotype left ilium with three articulated posterior dorsal vertebrae and sacrum, lateral view.**

refer *Nomingia* to any oviraptorosaurian family. Comparing the type specimen to other theropod taxa, Barsbold, Currie *et al.* and Barsbold Osmólska *et al.* noted that, like the caenagnathid *Chirostenotes* (but unlike the oviraptorid *Ingenia*, a genus known from the same site), the preacetabular blade of the ilium is higher than the postacetabular region. The tail is shorter than that of the oviraptorid *Oviraptor* (MGI 100/42), and the preacetabular iliac blade is lower with a less pronounced dorsal curvature. The pubis has a straighter shaft than that of *Ingenia*; the ventrally directed pubis is 171 percent of the length of the ischium (148 percent in *Oviraptor*, 121 percent in *Ingenia*, 221 percent in *Chirostenotes*). As in *Chirostenotes*, the ischium curves strongly posteriorly, while similarities of the ilium and ischium with those elements in North American members of the Caenagnathidae, a family also known in Asia, suggest it may be referable to that clade. "Unfortunately," Barsbold, Osmólska *et al.* stated, "Caenagnathidae are poorly known, and species assigned to that taxon are based on incomplete specimens, which severely restricts any further comparisons of *N. gobiensis* with the caenagnathid taxa."

Barsbold, Currie *et al.* (see also Barsbold, Osmólska *et al.*) described the pygostyle as follows: A comparatively straight, tapering mass of fused bones; caudal vertebrae 15 to 24 differing in morphology from those elements in other oviraptorosaurs, with coossification between the twentieth to twenty-first caudals limited to the dorsal parts of the centra, the ventral margins remaining separate; neural arches of the first centra in the pygostyle not fused (the condition seen in birds); other intervertebral articulations completely fused; at least two fused hemal arches; the number of coossified vertebrae the same as the number of caudals supporting rectrices (tail feathers) in *Caudipteryx* and modern birds. The authors rejected the possibility that the pygostyle was pathologic, noting that the caudal vertebrae "undergo a steady reduction in dimensions from the base to the end of the tail," and that no elements exhibit "rough surface texture, asymmetry, growth interruption or any indication of trauma" (Barsbold, Currie *et al.*).

Barsbold, Currie *et al.* further noted that, as oviraptors are not phylogenetically placed especially close to birds (see "Systematics" chapter; see also discussion on oviraptorosaurs as possible avians, "Introduction"),

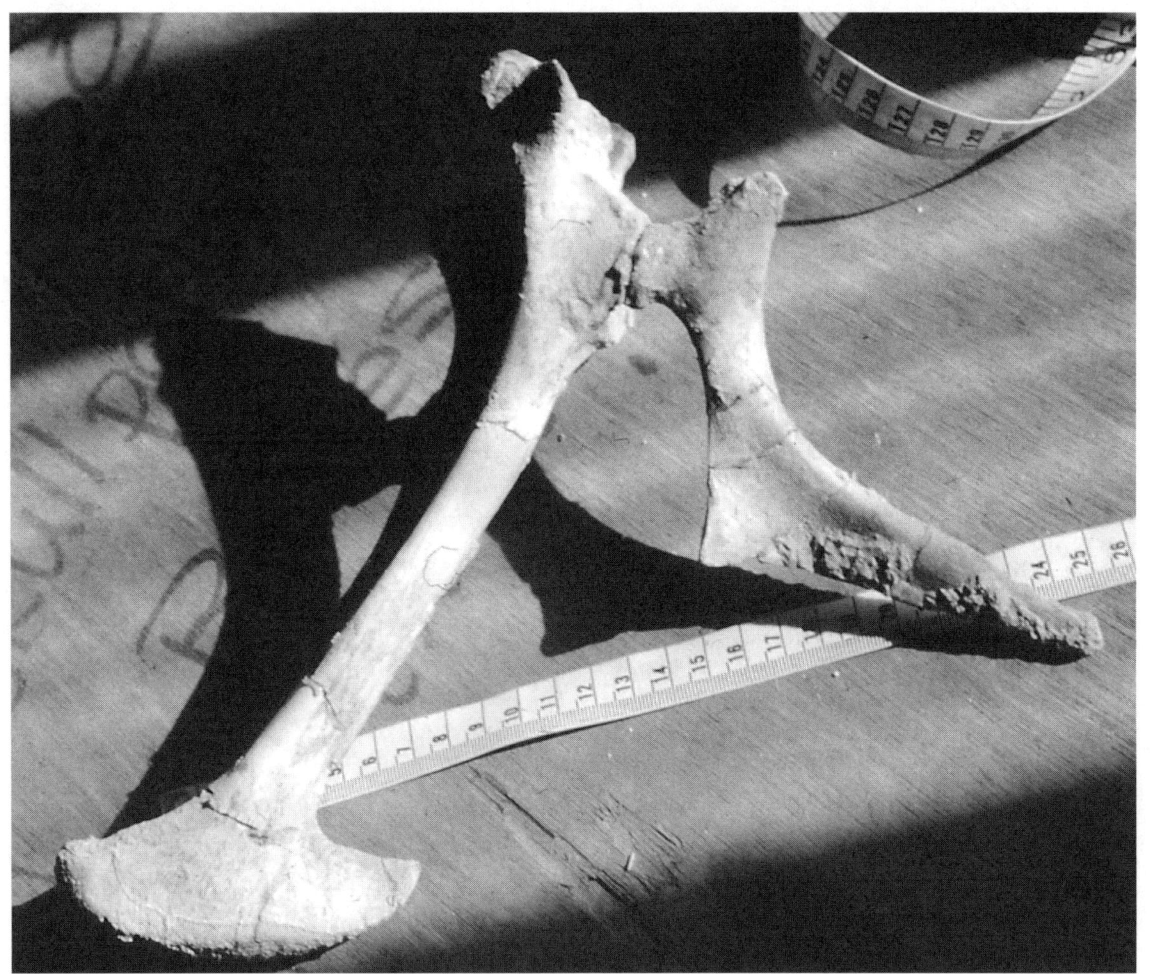

***Nomingia gobiensis*, GIN 100/119, holotype right articulated pubis and ischium.**

Left and below: Courtesy Philip J. Currie.

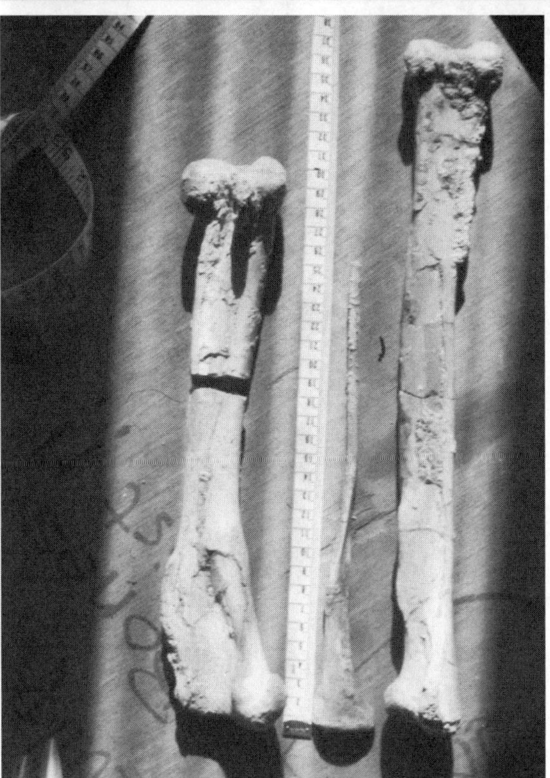

***Nomingia gobiensis* (left) left femur in posterolateral view and (right) right tibiotarsus in posterior view.**

the short tail and fused caudal vertebrae of this new taxon are not assumed to be holologous with avian tails. However, as feathers are now known in some nonavian theropods, this specimen may have possessed rectrices similar to those seen in *Caudipteryx* and *Protarchaeopteryx* specimens. Furthermore, the birdlike genus *Caudipteryx* (see entry) is most likely a basal orviraptorosaur, as evidenced by its "edentulous, beak-like dentary that is deeply invaded posteriorly by the enlarged external mandibular fenestra, a posteriorly concave ischium, short caudal centra, and a short tail." Rather than a pygostyle, the terminal caudal vertebrae in this genus apparently form a stiffened rod. The short tail and coossified caudals of this specimen may indicate a fan of elongate tail feathers (rectrices) in oviraptorosaurs more derived than in *Caudipteryx*.

According to Barsbold, Currie *et al.*, pygostyle-like structures may have evolved independently at least three times in theropod dinosaurs, "although the presence of rectices in two of these taxa suggests a functional association."

Key references: Barsbold, Currie, Myhrvold, Osmólska, Tsogbaatar and Watabe (2000); Barsbold, Osmólska, Watabe, Currie and Tsogtbaatar (2000); Gradziński, Jaworowska and Maryańska (1997); Jerzykiewicz and Russell (1991); Sloan (1999).

NOTOHYPSILOPHODON Martínez 1998

Ornithischia: Genasauria: Cerapoda: Ornithopoda: Euornithopoda: Hypsilophodontidae: Hypsilophodontinae.

Name derivation: Greek *notos* = "south [for South America]" + *Hypsilophodon*.

Type species: *N. comodorensis* Martínez 1998.

Other species: [None.]

Occurrence: Bajo Barreal Formation.

Age: Late Cretaceous (?Cenomanian).

Known material/holotype: UNPSJB—PV 942, partial skeleton including four cervical, seven dorsal, four sacral, and six caudal vertebrae, fragments of four ribs, sacrals, partial left scapula, right coracoid, right humerus, both ulnae, partial left femur, a right tibia, partial left tibia, a left fibula, partial right fibula, right astragalus, left calcaneum, and 13 pedal phalanges with three unguals.

Diagnosis of genus (as for type species): Humerus without deltopectoral crest; tibia with anteromedial bulge; fibula abruptly narrowed at distal middle of shaft; astragalus with proximal surface in two levels, calcaneum with strong posterodistal projection, ungual phalanges with flat ventral surfaces (Martínez 1998*b*).

Comments: The type species *Notohypsilophodon comodorensis* is significant as the first hypsilophodontid known from South America. It was founded upon a partial postcranial skeleton (UNPSJB—PV 942)

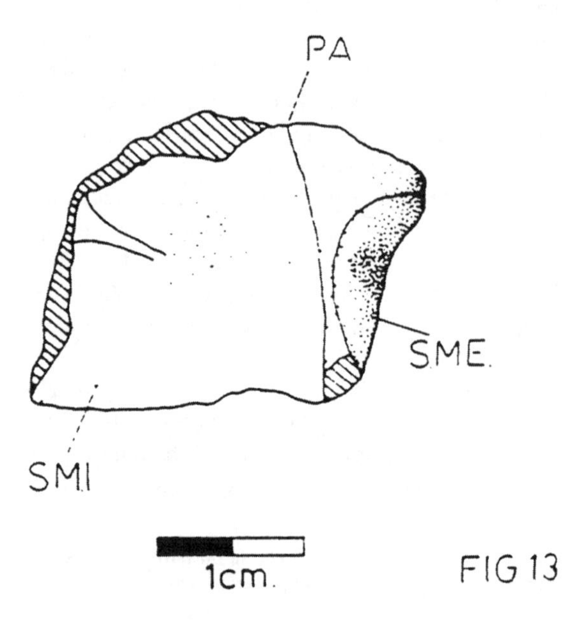

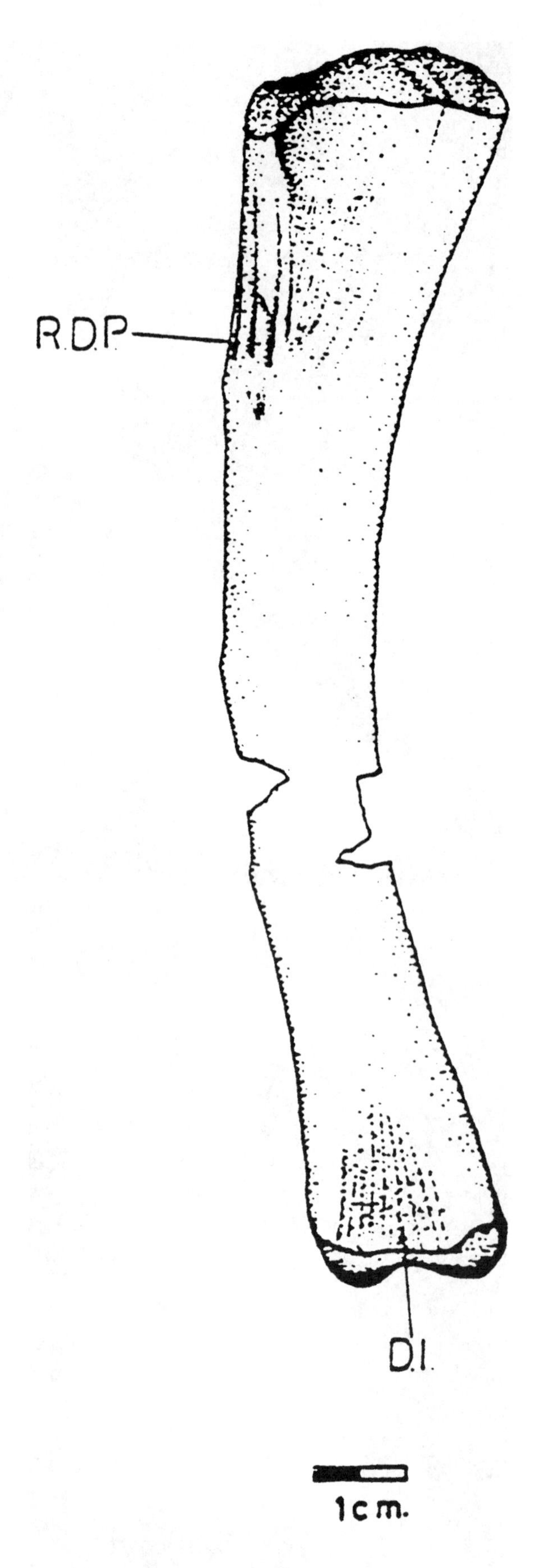

***Left: Notohypsilophodon comodorensis*, UNPSJB—PV 942, holotype right astragalus in proximal view. (After Martinez 1998*b*.) *Right: Notohypsilophodon comodorensis*, UNPSJB—PV 942, holotype right humerus in anterior view. (After Martinez 1999*b*).**

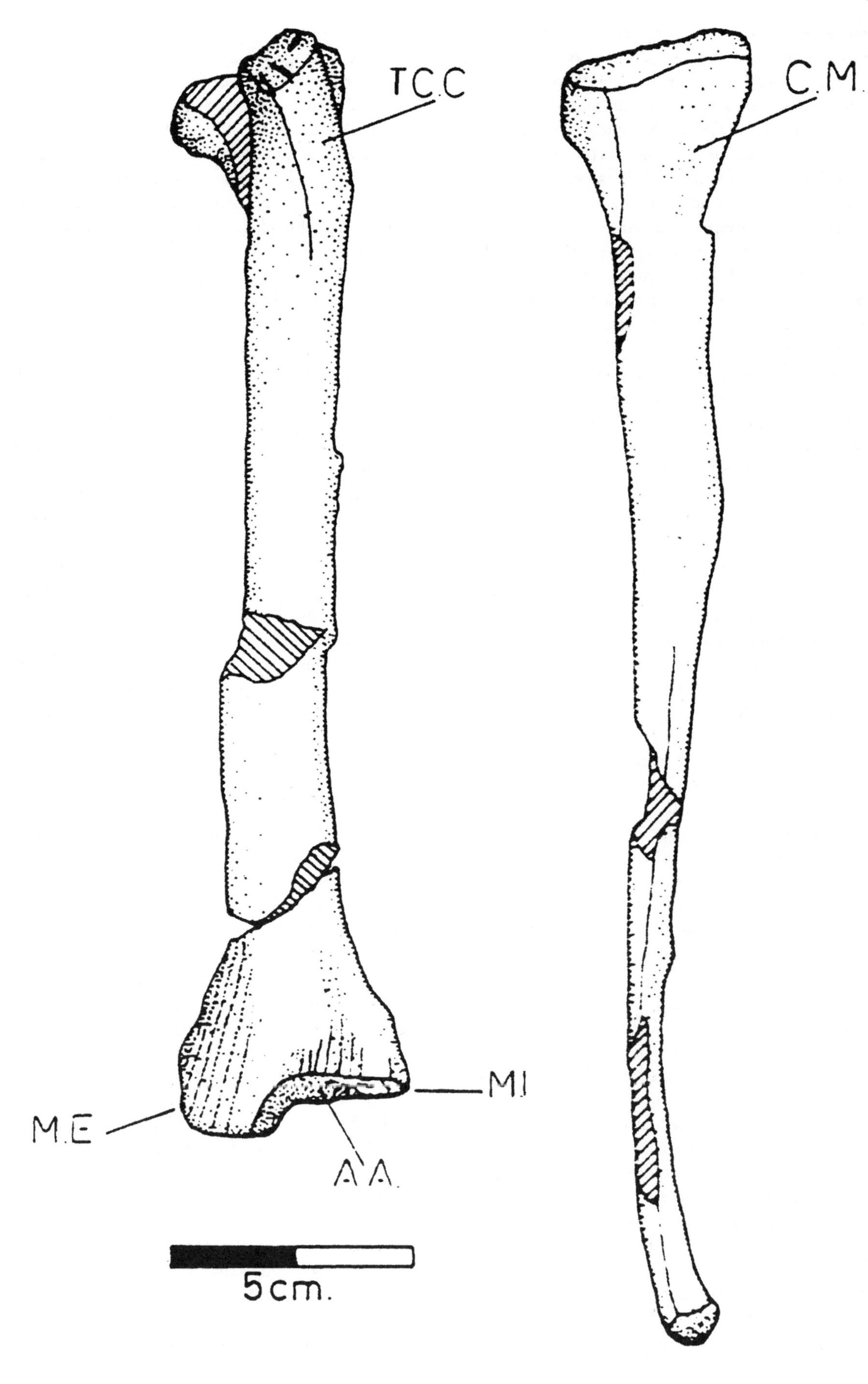

Left: Notohypsilophodon comodorensis, **UNPSJB—PV 942, holotype left fibula in medial view. (After Martinez 1998*b*.)** ***Right: Notohypsilophodon comodorensis***, **UNPSJB—PV 942, holotype right tibia in anterior view. (After Martinez 1998*b*.)**

recovered from the San Jorge Basin, Barreal Formation, southern Chubut, Argentina (Martínez 1998*b*).

This small dinosaur first mentioned by Coria (1999), the name set in quotation marks, in a review of ornithopods from the Neuquén Group. In that paper, Coria diagnosed the monotypic *N. comodorensis* as follows: Humerus caudally twisted; femur with reduced lesser trochanter; femur with reduced ischatic canal, lacking intercondylar groove.

Martínez referred *N. comodorensis* to the Hypsilophodontidae based upon the following synapomorphies: (Pertaining to the femur) ischiatic groove; anterior trochanter well below greater trochanter; shallow intertrochantric cleft, absence of extensor groove in distal end; also, humerus having caudal flexture at level of deltopectoral region. No iguanodontian features were perceived by Martínez in the type specimen.

Key references: Coria (1999); Martínez (1998*b*).

NQWEBASAURUS de Klerk, Forster, Sampson, Chinsamy and Ross 2000

Saurischia: Theropoda: Neotheropoda: Tetanurae: Avetheropoda: Coelurosauria *incertae sedis.*

Name derivation: Xhosa *Nqweba* [name for Kirkwood region] + Greek *sauros* = "lizard."

Type species: de Klerk, Forster, Sampson, Chinsamy and Ross 2000.

Other species: [None.]

Occurrence: Kirkwood Formation, Eastern Cape Province, South Africa.

Age: Early Cretaceous (Berrisian–Valanginian).

Known material/holotype: AM 6040, fragmentary skull, almost complete, articulated postcranial skeleton, late juvenile or subadult.

Diagnosis of genus (as for type species): Basal coelurosaurian theropod having the following autapomorphies: ginglymus of metacarpal I very robust and asymmetrical, having hypertrophied articular surfaces and greatly enlarged lateral (ulnar) condyle; manual ungual phalanx of digit I elongate (length = four times proximal depth), mediolaterally compressed; fibular shaft reduced distally to thin splint; metatarsal IV reduced in width to approximately half width of metatarsal III (de Klerk, Forster, Sampson, Chinsamy and Ross 2000).

Comments: Distinguished as one of the most completely known and best-preserved Cretaceous African theropods yet described, the type species *Nqwebasaurus thwazi* was based on a nearly complete skeleton, discovered in 1996 by Callum F. Ross and William J. de Klerk during a joint Albany Museum–State University of New York expedition, in coarse, red-brown mudstone in the unnamed uppermost member of the Lower Cretaceous Kirkwood Formation, Uitenhage Group, Algoa Basin, west of the village of Kirkwood, in Eastern Cape Province, South Africa (de Klerk, Forster, Sampson, Chinsamy and Ross 2000; see also McMillan 1999). This specimen, de Klerk *et al.* pointed out, is "the first well preserved and nearly complete taxon to be described from the Kirkwood Formation," providing "important temporal and geographic range extensions for coelurosaurians in Gondwana."

Found with the holotype skeleton, preserved scattered around the abdominal region, were at least a dozen small, round to ellipsoidal stones (maximum diameter of 5.2 to 14.5 millimeters, average of 8.0 millimeters). As observed by de Klerk *et al.*, these stones are of diverse lithologies, although most of them consist of fine-grained quartzites. The latter are smooth and polished, while the remaining pebbles of softer rock type appear etched. No small, polished, or etched pebbles have otherwise been found in the matrix preserving the specimen and none are commonly found in the mudstone facies of the Kirkwood Formation. Consequently, the authors interpreted these stones as gastroliths, extremely rare in theropod dinosaurs, presumably utilized in the mechanical breakdown of ingested food.

Addressing the ontogenetic age of the *N. thwazi* type specimen, de Klerk *et al.* noted that, while neurocentral sutures in the cervical vertebrae appear to be fused, the neural arches and centra are disarticulated on the preserved dorsal and caudal vertebrae, this suggesting an immature individual. Also, examination of transverse histological thin sections cut from the midshafts of the femur and tibia were examined under a polarizing microscope (see Chinsamy and Raath 1992). This examination revealed at least three growth lines comprising zones, annuli, and rest lines in the c compact bone (Chinsamy 1997). The cycles observed in this specimen suggested to the authors "that AM 6040 has undergone some period of growth and is not a young juvenile." However, other observed features (*e.g.*, absence of inner circumferential lamella and any secondary reconstruction) indicate that the specimen represents a subadult or late juvenile. Nevertheless, even the fully adult *Nqwebasaurus* was a small dinosaur, the individual represented by AM 6040 standing only about 30 centimeters (approximately 11.5 inches) high at the hips.

According to de Klerk *et al.*, *Nqwebasaurus* exhibits various characters that identify this genus as a tetanuran theropod, these including the following: Elongate, three-fingered manus; reduced trochanteric shelf; tall, broad ascending process of astragalus; reduced distal portion of fibula; metatarsal III wedge-shaped at midshaft; and highly reduced, distally positioned metatarsal I (Gauthier 1986; Holtz 1994, 2000; Sereno, Dutheil, Iarochne, Larsson, Lyon, Magwene, Sidor, Varricchio and Wilson 1996; Sereno 1999*c*). Within Tetanurae, the genus possesses the following characters which support its referral to the Coelurosauria: Centra of cervical vertebrae amphyplatyan; greater trochanter of femur cleft from femoral head; proximal end of fibula at least 75 percent proximal width of tibia; and head of femur set off at over 90 degrees from shaft (head distally directed) (Holtz 2000). Finally, *Nqwebasaurus* also displays the derived state for the following characters distributed among basal coelurosaurians: Articular surface of cervical vertebrae more broad than deep, with kidney-shaped articular surfaces; length of midcervical centra approximately twice diameter of anterior articular face; loss of transverse groove across anterior face of astragalus (see also Sereno); cervical vertebrae with flexed prezygapophses; length ratio of radius to humerus greater than 76 percent (exactly 76 percent in *Nqwebasaurus*); manual ungual region palmar to ungual groove subequal in width to region dorsal to ungual groove; manual ungual blade-like in cross section, more than three times as deep as wide; and metacarpal III shorter than II (Holtz 2000).

As de Klerk *et al.* pointed out, *Nqwebasaurus* is presently the oldest known Gondwanan coelurosaurian, predating previously described coelurosaurian taxa by about 50 million years, thereby documenting the presence of the Coelurosauria on the Gondwana "supercontinent" at the beginning of the Cretaceous, long before the main phase of fragmentation of that great landmass. This occurrence in southern Africa also supports the hypothesis that the Coelurosauria could have been distributed globally by the Late Jurassic (see Sereno *et al.*). Furthermore, the presence of this genus at the southern tip of Africa substantiates the postulation that coelurosaurians may have been widely dispersed throughout Gondwana before continental isolation. This idea, the authors noted, is consistent with the recent announcement of a therizinosaur—seemingly the oldest known coelurosaur—from the Lower Jurassic of China (see Zhao and Xu 1998), this report, if correct, suggesting that the origins of the Coelurosauria predate even the breakup of Pangaea.

Although *Nqwebasaurus* can be used to support Late Jurassic cosmopolitanism of the Coelurosauria, it cannot, de Klerk *et al.* emphasized, verify this hypothesis without additional evidence: "Whether the Gondwanan coelurosaurian legacy reflects a separate southern radiation of this clade, periodic immigration, extensive and continuous interchange with Laurasia, or some combination of these factors, can only be tested through finer resolution of coelurosaurian phylogeny and additional occurrences. Although older coelurosaurians may be known from Laurasia, the paucity of Jurassic dinosaurs from Africa and elsewhere in Gondwana cautions against a priori assumptions of a Laurasian origination and diversification for Coelurosauria."

Nqwebasaurus thawzi, AM 6040, holotype skeleton in right lateral view. Scale = 5 cm. (After de Klerk, Forster, Sampson, Chinsamy and Ross 2000.)

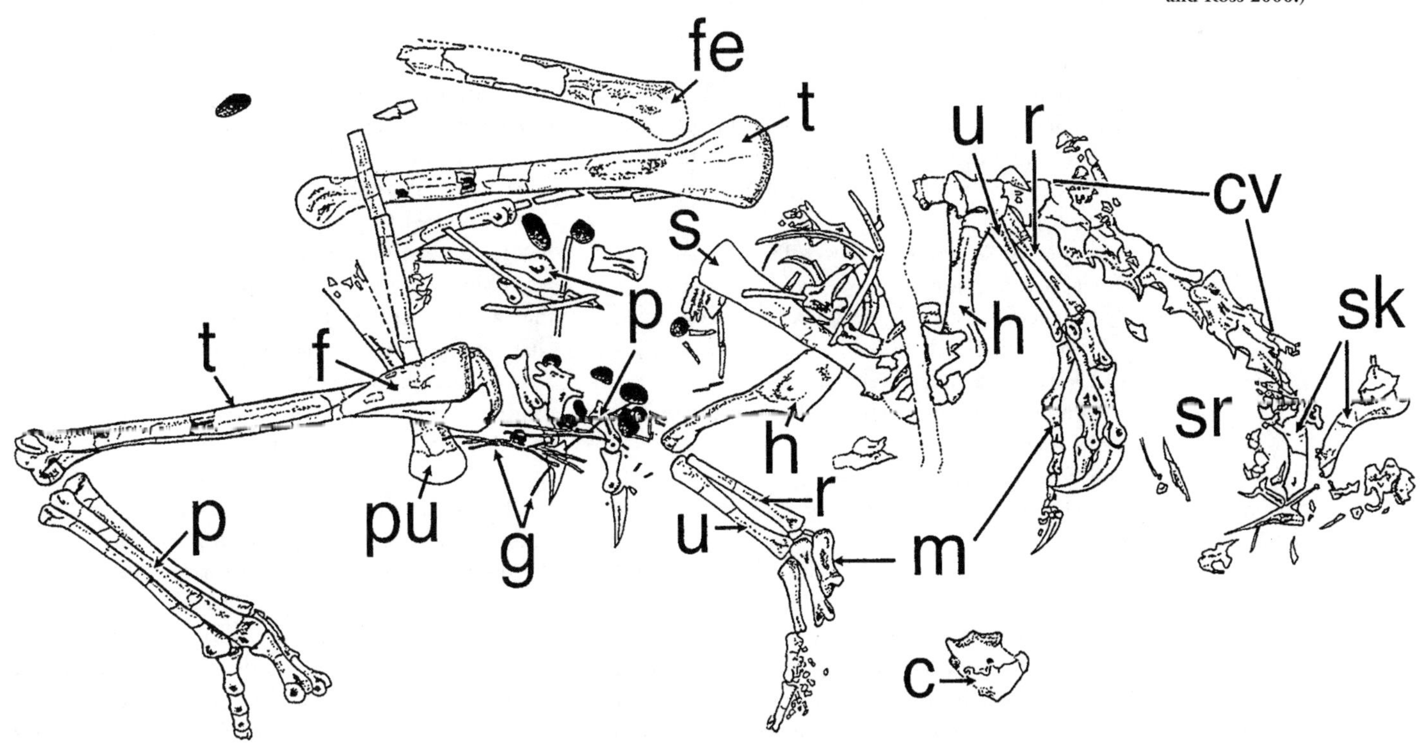

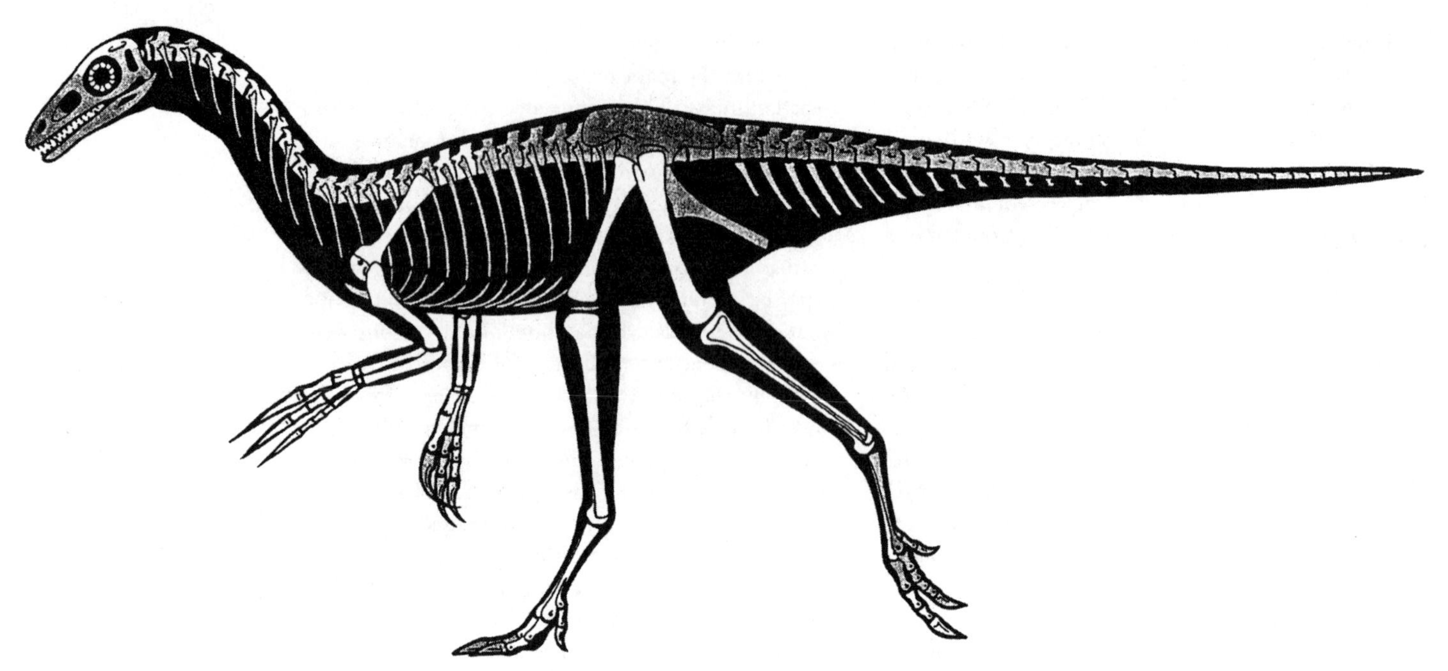

***Nqwebasaurus thawzi*, reconstructed skeleton based on holotype AM 6040. (After de Klerk, Forster, Sampson, Shinsamy and Ross 2000.)**

Key references: Chinsamy (1997); Chinsamy and Raath (1992); de Klerk, Forster, Sampson, Chinsamy and Ross (2000); Gauthier (1986); Holtz (1994, 2000); McMillan (1999); Sereno (1999*c*); Sereno, Dutheil, Iarochne, Larsson, Lyon, Magwene, Sidor, Varricchio and Wilson (1996); Zhao and Xu (1998).

†**NUTHETES** Owen 1854

Saurischia: Theropoda: Neotheropoda: Tetanurae: Avetheropoda: Coelurosauria: Manuraptoriformes: Manuraptora: Deinonychosauria: Eumanuraptora: ?Dromaeosauridae.

Name derivation: Greek abbreviation of *nouthetetes* = "one who warns, a monitor [lizard, in reference to the teeth's resemblance to those of modern varanid lizards]."

Type species: *N. destructor* Owen 1854.

Other species: [None.]

Occurrence: Purbeck Limestone Formation, Isle of Purbeck, England; Wiltshire and Sussex, England.

Age: Early Cretaceous (Berriasian).

Known material: Jaw, possibly other jaw and postcranial remains.

Holotype: Portion of left dentary.

Diagnosis of genus (as for type species): [No modern diagnosis published].

Comments: The only theropod known from the Isle of Purbeck, England, *Nuthetes* was named and first described by Richard Owen in 1854, based on a small, fragmentary left dentary with teeth, collected by Charles Wilcox of Swanage, Dorsetshire, from Bed k93 (see Austen 1852). Owen believed that the specimen belonged to a carnivorous or insectivorous lacertilian — more precisely, "a Pleurodont Lizard, allied to the Monitors of the modern genus *Varanus*" the size of *V. crocodilinus*, "or great Land Monitor of India." According to the author, these teeth were adapted "for piercing, cutting, and lacerating its prey."

Owen described the fragment as measuring 1.5 inches in length. Some of the features (none of them of diagnostic value today) noted in Owen's original description were that the ramus is compressed; the "exterior surface of the bone is smooth and polished, but impressed by very thin, longitudinal linear markings, and performated by nervous or vascular foramina along the alveolar wall: it is traversed near the lower dentary from the angular piece in the jaw of *Varanus*"; and the enamelled tooth crowns "are moderately long, slender, compressed, pointed, slightly recurved, and with a well-marked serrated margin both before and behind." The author noted that the teeth resemble, in miniature, those of *Megalosaurus* in that they are thickest anteriorly; they differ, however, in "having the inner alveolar ridge of the jaw not more developed than in the modern *Varani*, and in not exhibiting any rudiments of alveolar divisions."

Also, Owen tentatively referred to the type species, *N. destructor*, a tibia and fibula, also collected by Wilcox, from the Feather Quarry on the Isle of Purbeck. Later, Owen (1879) additional material from the "Feather-bed Marl" to *Nuthetes*, including dentary fragments, isolated teeth, and phosphatic "granicones" from the Samuel H. Beckles Collection, the latter interpreted by that author as dermal bones or armor similar to the scutes of some modern lizards. Owen (1879) described these structures as "ossiciles of a conical shape, the cone showing various degrees of elevation, with a granulate surface, the base being flat and smooth, or faintly and minutely pitted."

Additional remains referred to this genus, recovered from Wiltshire and Sussex, were later reported by Delair (1958).

Classifying *Nuthetes* has been equivocal since the genus was originally described. For example, Swinton (1934) regarded *Nuthetes* as a rather small or juvenile theropod dinosaur, classifying it among the megalosaurs, an assessment that was accepted by some subsequent workers (*e.g.*, Chure and McIntosh 1989). Delair identified the enigmatic "granicones" as pertaining to armored dinosaurs. Galton (1986*b*) later referred these to the ornithischian *Echinodon* (see entry, *D:TE* and this volume; see also Coombs, Weishampel and Witmer 1990). Steel (1970) listed *Nuthetes* as a junior synonym of *Megalosaurus*. Molnar, Kurzanov and Dong (1990) listed *Nuthetes* as Carnosauria *nomen dubium*.

Other workers, however, have favored Owen's original assessment that *Nuthetes* is a lizard. White (1972), in his catalog of dinosaurian genera, regarded *Nuthetes* as "non Carnosaur" and belonging to the Lacertilia. Likewise, Olshevsky (1991) excluded *N. destructor* from his list of dinosaurian taxa, noting that "The type specimen of this species was originally (and correctly) described as lacertilian, but the species has occasionally been classified as a small theropod." Following the opinion of these latter authors, *Nuthetes* was listed in the "excluded genera" section of *D:TE*.

More recently in an abstract, Milner (1999), accepting this genus' status as belonging to the Theropoda, noted that the "dentary fragments of *Nuthetes* do not show well-developed interdental plates on the lingual margin of the tooth row, as they are coossified and fused to the dentary margin," a condition characteristic of dromaeosaurids. Isolated teeth referred to *Nuthetes* match the few preserved in the holotype fragments, these being compressed, strongly recurved, and with denticulated mesial and distal carinae. Some theropod teeth exhibiting characters having taxonomic value at family levels, Milner determined that "data obtained for *Nuthetes* suggests that it may be referable to the Dromaeosauridae," this conclusion being consistent with the stratigraphical range of dromaeosaurids.

Note: According to Milner, a large isolated

Fig. 1.

Fig. 2.

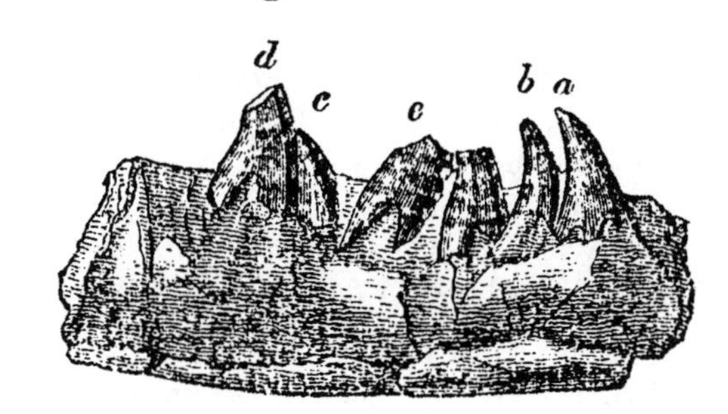

Fig. 3.

Fig. 4.

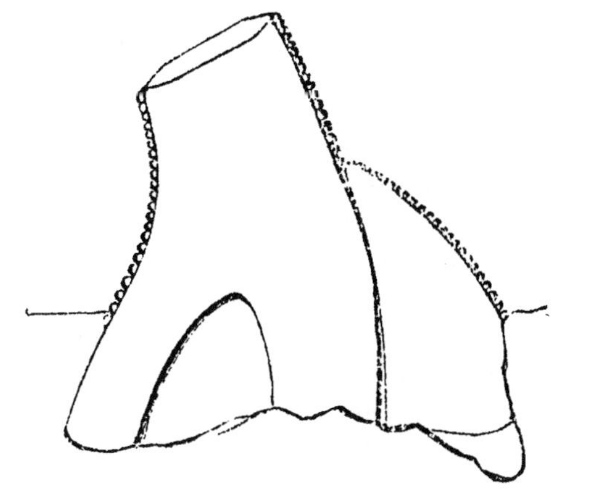

***Nuthetes destructor*, holotype incomplete left dentary in (Fig. 1) external and (Fig. 2) internal views, (Fig. 3) "end" view (scale = 2 cm.), and (Fig. 4) magnified view of two teeth. (After Owen 1854.)**

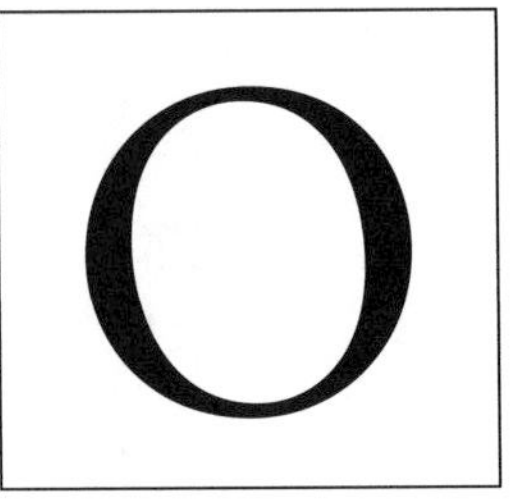

theropod tood crown collected from an unknown Purbeck locality, and referred by Lydekker (1890) to "*Megalosaurus dunkeri*" [=*Becklespinax altispinax*], very similar to isolated and possibly indeterminate tooth crowns from the Purbeck Limestone of Aylesbury, Buckinghamshire, England, may represent one large theropod in this area.

Key references: Austen (1852); Chure and McIntosh (1989); Delair (1958); Galton (1986*b*); Lydekker (1890); Milner (1999); Molnar, Kurzanov and Dong (1990); Olshevsky (1991); Owen (1854, 1879); Steel (1970); White (1972).

†OMEISAURUS

Saurischia: Sauropodomorpha: Sauropoda: Eusauropoda: Neosauropoda: Macronaria: Camarasauromorpha: Titanosauriformes: Somphospondyli: Euhelopodidae: Euhelopodinae.

Type species: *O. junghsiensis* Yang [Young] 1939.

Other species: *O. tianfuensis* He, Li, Cai and Gao 1984, ?*O. luoquanensis* He, Li and Cai 1998.

Comments: In a review of the sauropods of China, Martin-Rolland regarded the fragmentary species ?*Omeisaurus changshouensis* as a junior synonym of the type species *O. junghsiensis*. Martin-Rolland noted that material belonging to the former taxon does not appear to be distinct enough to warrant its retention in a separate species.

That author referred *Omeisaurus* to the Euhelopodidae, proposing that all known Chinese sauropods belong in that family, which, according to Martin-Rolland, was endemic to Asia during the Jurassic.

Key references: He, Li, Cai and Gao (1984); Martin-Rolland (1999); Yang [Young] (1939, 1958).

†ORNATOTHOLUS—(See *Stegoceras*.)

†OTHNIELIA

Ornithischia: Genasauria: Cerapoda: Ornithopoda: Euornithopoda: Hypsilophodontidae: Hypsilophodontinae.

Note: Additional material that may belong to the small, Late Jurassic dinosaur *Othnielia* (see *D:TE*) was reported in an abstract by Crouse and Carpenter (1999). These remains consist of a poorly preserved partial skeleton — comprising a frontal, most of the (semiarticulated) dorsal vertebrae, some ribs, and several long bones including a femur, humerus, and partial tibia and fibula — representing a "toddler ornithopod," preserved in several blocks of maroon sandstone recovered from the Morrison Formation of Garden Park, Colorado.

According to the authors, the lack of sufficient diagnostic characters makes it difficult to identify this specimen with certainty. However, it may belong to *Othnielia* based on the following features: Absence of pit near fourth trochanter of femur; shape of proximal end of humerus; and ratio of length of transverse processes to centrum height on vertebra.

As the neural arches are not fused, the bones are quite small (femur length of about 8.5 centimeters), and the ends of the long bones are spongy and not completely formed, Crouse and Carpenter suggested that the specimen may represent a juvenile animal, though one larger than a hatchling. The straight shaft of the humerus and the pronounced emargination of the frontal above the orbit, the authors noted, sets this specimen apart from larger juvenile skeletons, these features probably due to ontogeny.

Key reference: Crouse and Carpenter (1999).

OZRAPTOR Long and Molnar 1998

Theropoda *incertae sedis*.

Name derivation: Colloquial "Oz [from the novels of L. Frank Baum, once popular in Australia, used as slang for "Aus," which is short for Australia; R. E. Molnar, personal communication 2000] + Latin *raptor* = "thief."

Type species: *O. subotaii* Long and Molnar 1998.

Other species: [None.]

Occurrence: Colalura Sandstone, Western Australia.

Age: Middle Jurassic (Bajocian).

Known material/holotype: UWA 82469, distal end of left tibia.

Diagnosis of genus (as for type species): Small theropod having distal end of tibia with well-defined, high rectangular facet for ascending process of astragalus set into anterior surface of tibia, this facet having relatively straight dorsal margin and distinct centrally located vertical ridge; medial malleolus weakly developed (Long and Molnar 1998).

Comments: *Ozrapotor subotaii* is distinguished as the first Jurassic theropod known from Australia, and also the first Western Australian dinosaur to be formally named (excluding ichnotaxa).

This type species was established on a partial left tibia (UWA 82469; transferred to the collections of the Western Australian Museum, Perth) discovered in 1966 by Scotch College students Steven Hincliffe, Peter Peebles, Robert Coldwell, and Trevor Robinson in the Colalura Sandstone, at the Bringo Railway Cutting, east of Geraldton, Western Australia. Casts of the specimen, still embedded in matrix, were sent to the British Museum (Natural History) [now The Natural History Museum, London], where Alan J. Charig at first tentatively identified one of the casts as the bone of a turtle. Following more recent preparation of the original material, the specimen was correctly identified by John A.

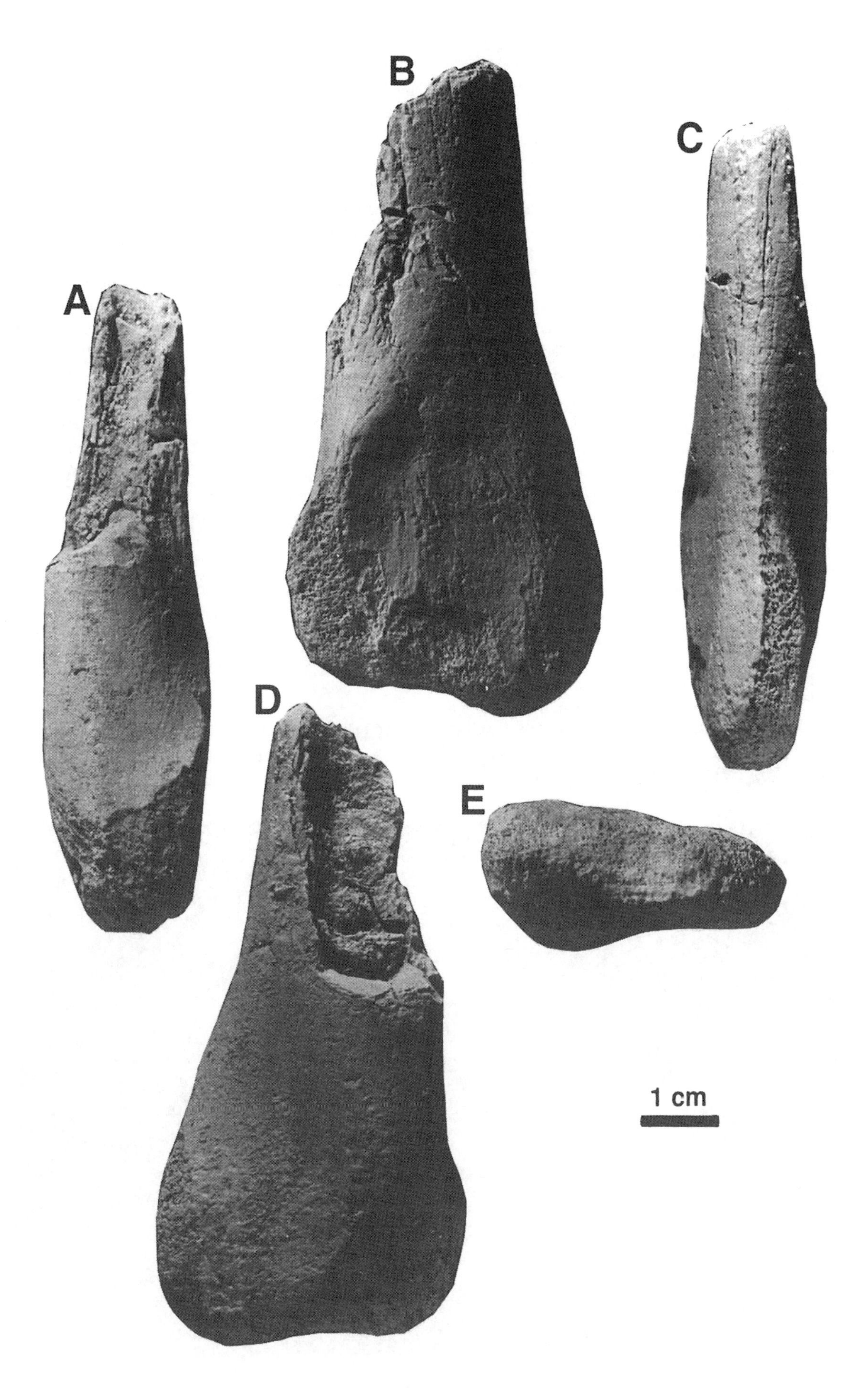

***Ozraptor subotai*, UWA 82469, holotype distal end of left tibia in A. lateral, B. anterior, C. posterior, D. medial, and E. ventral views. (After Long and Molnar 1998.)**

Long of the Western Australian Museum as belonging to a theropod dinosaur (Long and Molnar 1998).

UWA 82469 was referred by Long and Molnar to the Theropoda based on criteria identifying the morphology of the astragalus or its corresponding facet on the tibia in theropod dinosaurs (see Welles and Long 1974; Molnar and Pledge 1980; Molnar, Flannery and Rich 1981; Paul 1988*b*; Molnar, Angriman and Gasparini 1996). As preserved, the specimen was not substantial enough for Long and Molnar to refer it to any particular theropod group. However, the authors noted that, this being the first Australian Jurassic theropod from bone material recorded, "it is unlikely to be confused with any described existing dinosaur skeletal remains." The closest theropod in age and geographic proximity to the Australian specimen is *Cryolophosaurus elliotti*, an Early Jurassic form in which (assuming that the postcranial remains have been correctly associated with the holotype skull) the astragalus and calcaneum are fused (J. Long, personal observation 1996). Comparing UWA 82469 with numerous Late Triassic to Early–Middle Jurassic theropods, the authors concluded that "based on the relative size and almost square shape of the dorsal process of the astragalus in *Ozraptor*, that it has no intimate relationship with any theropods so far described." Following Welles and Long, who showed that the ascending process of the theropod astragalus is unique, Long and Molnar stated that in no known theropod other than that represented by UWA 82469 "is the astragalus developed as an almost rectangular dorsal process with a straight dorsal margin."

Long and Molnar postulated that the functional morphology of the shape of the astragalus might reveal something about the nature of theropod lifestyles, those having high astragalar dorsal processes being the more agile and often small to medium-sized forms (*e.g.*, ornithomimosaurs, oviraptorids, caenagnathids, dromaeosaurids, and some smaller tetanurans), but also including some large forms (*e.g.*, tyrannosaurids). *Ozraptor* is relatively more derived, for its age in terms of astragalar shape and possible locomotory implications (similar in these respects to the carnosaur *Allosaurus*), than other contemporary Early to Middle Jurassic theropods (*e.g.*, the "ceratosaur" *Dilophosaurus*), none of the latter having an astragalas possessing a high, broad ascending process. Consequently, Long and Molnar speculated "that the high, broad ascending process may be (partially) 'locked' in place by the central ridge, suggesting resistance to stresses at the ankle." The authors further suggested that, because this morphology of ascending processes seemingly correlated with small, agile theropods, *Ozraptor* was an agile dinosaur.

Key reference: Long and Molnar (1998); Molnar, Angriman and Gasparini (1996); Molnar and Pledge (1980); Molnar, Flannery and Rich (1981); Paul (1988*b*); Welles and Long (1974).

†PACHYCEPHALOSAURUS

Note: In both *D:TE* and *S1*, photographs of the pachycephalosaurid specimen collected by Michael

Photograph by the author.

Skull (cast) of *Pachycephalosaurus wyomingensis* collected by Michael Triebold from the "Sandy Quarry," Hell Creek Formation, near Buffalo, South Dakota, which has been identified as both *Stygimoloch* and *Pachycephalosaurus*. Although this specimen was labeled as *Stygimoloch* in *D:TE* and *S1*, Triebold (personal communication 2000) has informally identified it as belonging to *Pachycephalosaurus wyomingensis.*

Life restoration by Berislav Krzic of *Pachycephalosaurus wyomingensis* from the "Sandy Quarry."

Tribold from the Sandy Quarry, in the Hell Creek Formation, near Buffalo, South Dakota, were identified as *Stygimoloch*. Although this skeleton has not yet been formally described, Triebold (personal communication 2000) is now convinced that it pertains to the type species *Pachycephalosaurus wyominensis*.

†PACHYRHINOSAURUS

Ornithischia: Genusauria: Cerapoda: Marginocephalia: Ceratopsia: Neoceratopsia: Ceratopsomorpha: Ceratopsoidea: Ceratopsidae: Centrosaurinae.

Comment: What may be the earliest record yet of the ceratopsian dinosaur *Pachyrhinosaurus* was reported by May and Gangloff (1999) in an abstract in which they discussed a recently defined bonebed. The new site, defined in 1998, is located in the Upper Cretaceous Prince Creek Formation, along the west bank of the Colville River, north of Kilak Creek and south of Ocean Point, National Petroleum Reserve, Alaska.

As related by these authors, all bones collected at this site exhibit some degree of pre-burial weathering and dessication. Thus far, bones recovered both *in situ* and as "float" indicate the presence of at least four tax — tyrannosaurid and dromaeosaurid theropods, and hadrosaurids, but primarily the ceratopsian dinosaur *Pachyrhinosaurus*, the latter contributing 90 percent of these remains, all of these elements of subadult proportions.

Key reference: May and Gangloff (1999).

†PARASAUROLOPHUS

Ornithischia: Genasauria: Cerapoda: Ornithopoda: Euornithopoda: Iguanodontia: Euiguanodontia: Dryomorpha: Ankylopollexia: Hadrosauroidea: Hadrosauridae: Euhadrosauria: Lambeosaurinae.

Known material: Six skulls, ?partial skull material representing at least four individuals, adults, subadults, juvenile.

Diagnosis of genus: Lambeosaurine differing from all other hadrosaurs in having tubular, slightly curved narial crest composed mostly of premaxillae; prefrontals lapping onto ventral lateral surface of premaxillae, where skull roof and narial crest meet; frontals arising from posterior base of narial crest, extending posteriorly, where they articulate with nasals at point before farthest extent of squamosals; nasals confined to ventral margin of anterior part of narial crest, extending short distance beyond posteriormost extent of frontals; premaxillae forming lower border of external naris, extending posteriorly, each flanked by lateral groove below (separating premaxilla from maxilla), terminating just beyond dorsal apex of maxilla (Sullivan and Williamson 1999).

Diagnosis of *P. walkeri*: Long-crested, differing from *P. tubicen* in having narial crest consisting of two dorsal tubes that extend posteriorly, forming U-shaped bend at apex of narial crest returning along diverticula; dorsal and ventral tracks separated by lateral groove on each side of narial crest; differing from "*P. cyrtocristatus*" [see below] in having elongated narial crest, more than twice length (Sullivan and Williamson 1999).

Diagnosis of *P. tubicen*: Long-crested, differing from *P. walkeri* in having longer and internally more complex narial crest consisting of pair of dorsal tubes that extend from external naris and terminate posteriorly, forming apex of narial crest below two parts of tubes that extend from behind external narial openings (and are presumably contiguous with them), forming U-bend posteriorly where they coalesce into single pair of ventral tubes and return anteriorly, forming ventral margin of narial crest where they rise slightly above nasal and frontal region, fusing with ventral tubes of lateral diverticulae above before entering choana; lateral diverticulae paired, arising anteriorly (presumably communicating with external nares), extending posteriorly, forming tight U-bend and returning ventrally above ventral tubes, coalescing with ventral tubes anteriorly before entering choana; dorsal and dorsolateral exterior surface of narial crest bearing anastomosing furrows; (Sullivan and Williamson 1999).

Comments: "Perhaps the most beloved, and the most profusely illustrated, hadrosaurid dinosaur is *Parasaurolophus*," pronounced paleontologists Robert M. Sullivan and Thomas E. Williamson in the opening paragraph of their 1999 monograph, wherein they thoroughly revised and redescribed this taxon and

Courtesy Thomas E. Williamson and the New Mexico Museum of Natural History and Science.

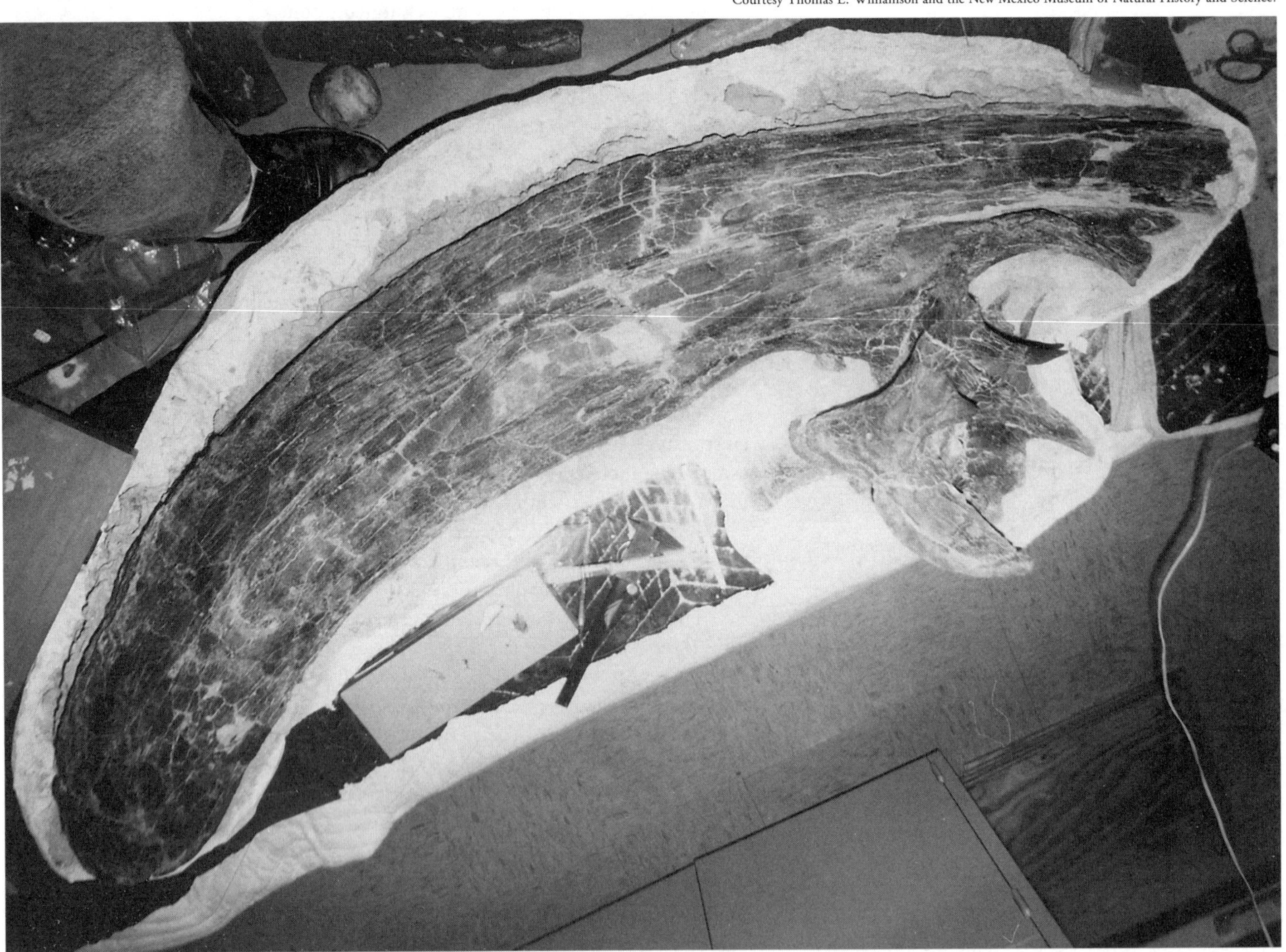

***Parasaurolophus tubicen*, NMMNH P-25100, referred almost complete narial crest, skull roof, and basicranium.**

reassessed its three species — the long-crested *Parasaurolophus walkeri* and *P. tubicen*, and short-crested *P. cyrtocristatus* (see below; see also *D:TE* for photographs of the holotypes) — and also described an important new skull belonging to the species *P. tubicen* (see Sullivan and Williamson 1996).

Sullivan discovered this specimen, an almost complete skull with lower jaw (NMMNH P-25100), on August 17, 1995 on a tributary of Hunter Wash, in the De-na-zin Member of the Upper Cretaceous Kirtland Formation (late Campanian), Bisti/De-na-zin Wilderness, San Juan Basin, San Juan County, New Mexico. The specimen was collected in August, 1995 by a team comprising Ray Geiser, Pete Reser, Warren Slade, Sullivan, Michael Tipping, and Williamson. Including the left jugal, left maxilla, skull roof (lacrimals, prefrontals, frontals, nasals, and nearly complete narial crest [premaxillae]), basicranium, left dentary, left angular, left surangular, and left articular, NMMNH P-25100 compares readily to the holotype (PMU.R1250) of this species, and is distinguished as the second most complete *Parasaurolophus* skull and second definitive skull of *P. tubicen* (Sullivan and Williamson 1999).

As Sullivan and Williamson (1999) noted, the excellent state of preservation of NMMNH P-25100 permitted, for the first time, the discrimination of sutures defining the bones of the upper skull and narial regions, thereby settling long-standing debate regarding the interpretation of the pattern of the cranial sutures.

Through computer tomographic analysis of the internal structure of the narial crest of NMMNH P-25100, Sullivan and Williamson (1999) discovered a previously unknown, extremely complex network of tubes consisting of the following: 1. A dorsal pair extending from the external naris to the tip of the crest, there terminating in paired chambers; 2. two pairs extending from the external naris to the wall of the terminal chamber, U-bend and return anteriorly, coalescing into one pair of tubes, and coalescing with a pair of inner tubes (lateral diverticulae); and 3. an

Photograph by the author, courtesy The Field Museum.

Parasaurolophus tubicen*, reconstruction (cast) of holotype partial skull (FM P27393) of *Parasaurolophus cyrtocristatus.

innermost pair (lateral diverticulae) extending from the anterior part of the skull under the two pairs of tubes, posteriorly forming a tight U-shaped bend, returning ventrally to coalesce with the ventralmost tubes and descend into the choana.

In their reconsideration of *Parasaurolophus*, Sullivan and Williamson (1999) then accepted all three of its named species as valid (see below).

To *P. cyrtocristatus* the authors referred BYU 2467, an incomplete narial crest comprising premaxillae, posterior portions of the prefrontals, frontals, nasals, and many unidentified fragments probably belonging to the narial crest and skull regions, this specimen having previously been referred by Weishampel and Jensen (1979) to *Parasaurolophus* sp., collected from the Kaiparowits Formation of Utah (see *D:TE*). Sullivan and Willaimson (1999) noted that Weishampel and Jensen were in error regarding certain details of their description of BYU 2467 "due to the misidentification of the narial tracts and the failure to recognize the associated skull roofing bones; also, this specimen is quite similar to the holotype (FMNH P27393) of *P. cyrtocristatus*, the osteological differences between them due to ontogenetic or individual variation.

Also to this species, Sullivan and Williamson (1999) referred UCMP 143270 (see below), an undescribed specimen currently under study by J. Howard Hutchinson. The authors pointed out that the short-crested skull was found in the same horizon of the Kaiparowits Formation as BYU 2467 (see below). Although only about 66 percent the size of the earlier collected specimen, UCMP 143270 has an external appearance almost identical to that of the holotype of this species, this suggesting to the authors "that the narial crest does not change significantly during ontogeny and negates the notion that the short-crested morphs are juveniles of the long-crested species."

Sullivan and Williamson (1999) discussed other specimens possibly belonging to *Parasaurolophus*. The first of these (MNA Pl.529), originally referred by Mateer (1981) by its locality number (MNA V529) to *Parasaurolophus* sp., was a nearly complete right maxilla, with partial right and incomplete left palatine, from the upper part of the Fruitland Formation (Upper Cretaceous: middle–late Campanian) of San Juan County. In describing this specimen, Sullivan and Williamson (1999) noted that the jugal is smaller and has a slightly different shape to that of *P. tubicen* (based on NMMNH P-25100). The maxillary

Photograph by the author, courtesy Museum of Paleontology, University of California, Berkeley.

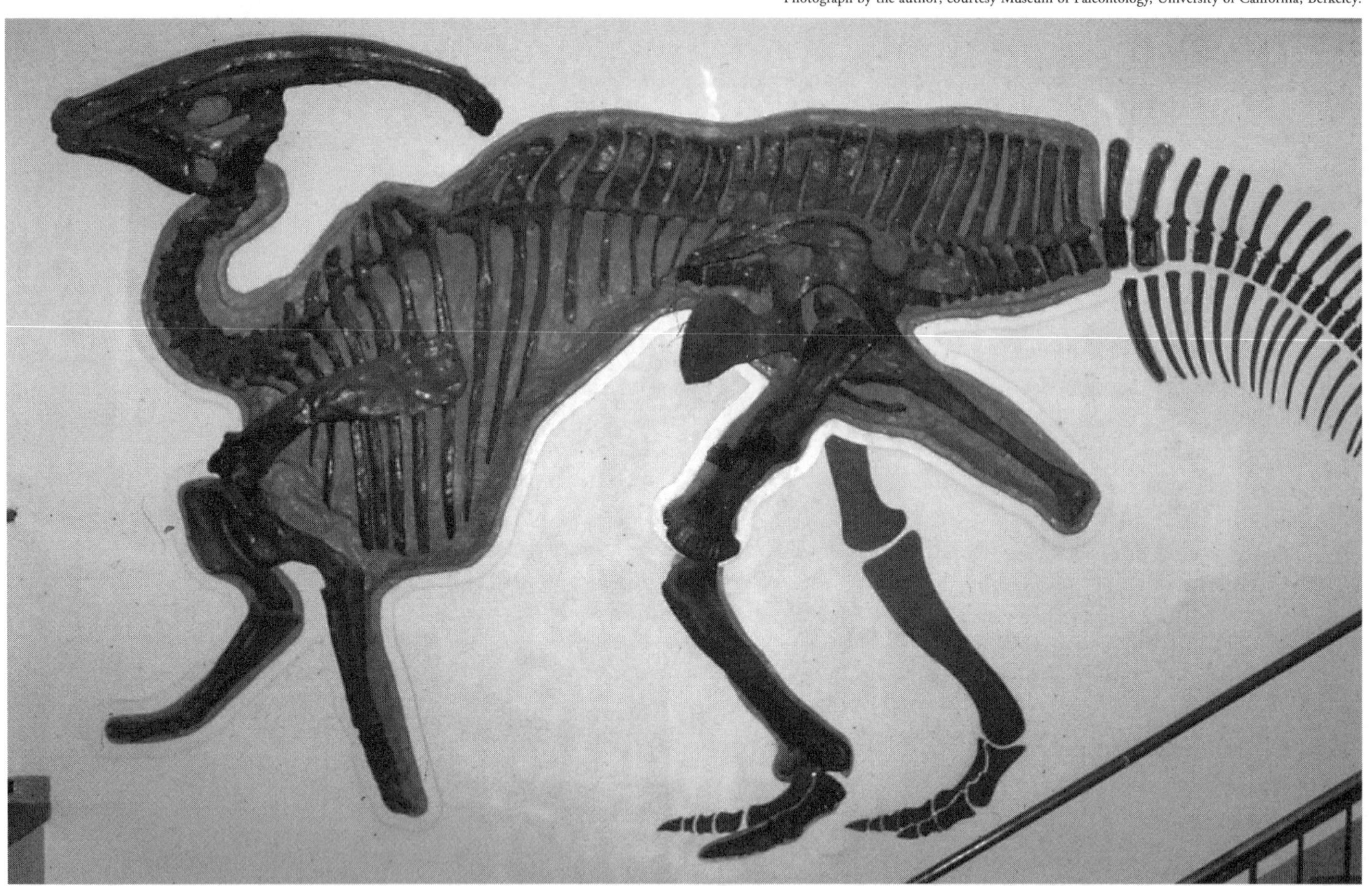

Parasaurolophus walkeri **skeleton (cast of holotype ROM 764).**

processes is clearly lambeosaurine (Heaton 1972; Horner 1990), and the maxilla is comparable in size and shape to that of NMMNH P-25100. However, finding no diagnostic characters allowing for an unambiguous assignment to the genus, Sullivan and Williamson (1999) provisionally referred MNA Pl.529 to *Parasaurolophus* sp.

Another specimen (NMMNH P-27286) consists of a nearly complete right dentary with teeth discovered in 1966 by Warren Slade in the De-na-zin Member of the Kirtland Formation (Upper Cretaceous: late Campanian), San Juan County. The dentary, belonging to a subadult individual, is much smaller than that of NMMNH P-25100 and about 89 percent the size of the holotype of *P. cyrtocristatus* (FMNH PR27393). In describing this specimen, Sullivan and Williamson (1999) observed that the shape of the coronoid differs from that of NMMNH P-25100 and resembles that of another lambeosaurine, *Corythosaurus*. However, as *Parasaurolophus* is the only lambeosaurine known to date from San Juan Basin strata, the authors questionably referred this specimen to *Parasaurolophus* sp.

A third specimen (USNM 13492) includes an almost complete articulated tail, incomplete left ilium, and a femur, collected in 1929 by N. H. Boss from the Kirtland Formation (Upper Cretaceous: late Campanian) of San Juan County. The specimen was first reported by Lull and Wright (1942) in their classic monograph on North American hadrosaurs. Although Lull and Wright had included an incomplete right maxilla with teeth in USNM 13492, a maxilla was never associated with the specimen (M. K. Brett-Surman, personal communication to Sullivan and Williamson, 1997). Sullivan and Williamson (1999) regarded this specimen as cf. *Parasaurolophus* sp.

Sullivan and Williamson (1999) reassessed the stratigraphic occurrence of the three *Parasaurolophus* species, pointing out that earlier interpretations had been "based on less-than-certain biostratigraphic horizons." Based upon more recent advances in the understanding of the stratigraphy since the discovery of the three holotypes of these species, especially in those units in southern Alberta and northwestern New Mexico, Sullivan and Williamson (1999) discussed in detail this issue, concluding the following:

1. The holotype of *P. cyrtocristatus* seems to have been recovered from the Coal Creek region of the present day Fossil Forest in the upper Fruitland Formation or, more likely, the lower Kirtland (Hunter Wash

Member) Formation (see Hunt and Lucas 1992) of New Mexico, dated between 75.5 and 74.5 million years ago (Fassett and Steiner 1997), this being slightly younger than the date given for the holotype of *P. walkeri* (D. A. Eberth, personal communication to Sullivan, 1997).

2. The holotype of *P. walkeri* was collected from the Dinosaur Park Formation (formerly Oldman Formation, in part, equal to the Belly River beds, in part), of Alberta, Canada, these strata dated between 76 and 74 million years ago (Eberth and Hamblin 1993).

3. The holotype of *P. tubicen* (this specimen bearing "the same distinctive black preservation as the new specimen [NMMNH P-25100]), assuming that it was recovered from the De-na-zin Member of the Kirtland Formation of New Mexico, has been dated 73.37 and 73.04 million years ago.

4. The relative age of the lower third of the Kaiparowits Formation of New Mexico, from which two referred *P. cyrtocristatus* have been collected, has been interpreted as of Campanian (="Judithian") age, although the latter has not been unequivocally established.

The authors concluded that the above paleogeographic and stratigraphic data confirm that the three species of *Parasaurolophus* did not coexist. It therefore did not seem likely to them that the short-crested and long-crested morphs represent ontogenetic stages of a single species; moreover, the holotypes of these species seemed to belong to fully mature individuals. Also, these data strongly seemed to weaken earlier suggestions (*e.g.*, Hopson 1975, who did not consider the possibility of two separate taxa) that long-crested forms of this genus indicate males and short-crested morphs females (see below).

Sullivan and Williamson (1999) proposed the alternative explanation that the two morphs reflect a phyletic sequence including three valid species of *Parasaurolophus*. The apparently older *P. cyrtocristatus* was determined to be the sister taxon to younger *P. walkeri* and then youngest *P. tubicen*, and also a structural ancestor to the long-crested forms represented by these two species. Of these three species, *P. tubicen* was found to possess the most complex narial crest, distinguishing it as the most derived of the taxa.

Comparing this genus to other lambeosaurines, the authors noted that *Parasaurolophus* could be derived from *Hypacrosaurus*, as the "narial crest of the former compared to the latter is not all that different with respect to the juxtaposition of the narial crest bones." The authors suggested that the highly derived narial crest in *Parasaurolophus* arose from a common ancestor similar to *Hypacrosaurus*. Furthermore, the cranial features of *Parasaurolophus*, particularly the narial crest, distinguish this genus as the most derived of all known lambeosaurines (*contra* Weishampel and Horner 1990).

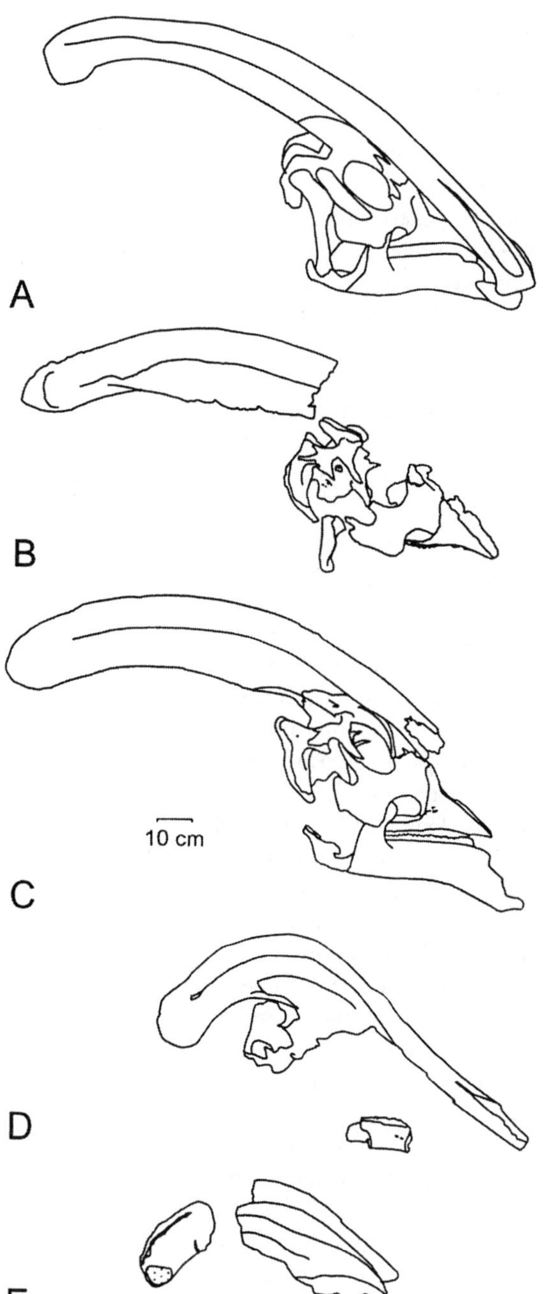

***Parasaurolophus* skull specimens: A. ROM 768, holotype of *P. walkeri*; B. PMU.R1250, holotype of *P. tubicen*; C. NMMNH P-25100, referred to *P. tubicen*; D. FMNH PR27393, holotype of *P. cyrtocristatus*, referred to *P. tubicen*; and E. BYU 2467, referred to *P. tubicen*. (After Sullivan and Williamson 1999.)**

More recently, however, Williamson (2000), in an extensive review of the Hadrosauridae from the San Juan Basin of New Mexico, reassessed the above conclusions reached by Sullivan and Williamson concerning *Parasaurolophus* and its species. According to Williamson, those authors had argued, without justification, that the type specimen of *P. cyrtocristatus* represents an adult individual and that this species is the plesiomorphic sister taxon to *P. tubicen* and *P. walkeri*. Williamson referred to other lambeosaurine taxa, for which relatively large samplings of skulls are known, and also to the Dodson's (1975) study of lambeosaurine crests from the Dinosaur Park Formation of

Alberta. Dodson's analyses resulted in a number of conclusions including that the crest in lambeosaurine hadrosaurids is highly variable, and that this variability can be ascribed both to ontogenetic changes and sexual dimorphism.

Dodson's assessment of these crests, Williamson noted, suggested that *P. cyrtocristatus* and short-crested specimens referred to *Parasaurolophus* actually represent juveniles or females of either *P. walkeri* or *P. tubicen*. Furthermore, based upon the growth in other lambeosaurines documented by Dodson, it was logical to assume that hatchlings of these two species lacked crests and that their crests developed ontogenetically, with a short-crested stage of transitional development to be expected. Presumably, then, Williamson suggested, "females of either *P. walkeri* and *P. tubicen* (assuming that these both represent male specimens) would have a somewhat shorter crest and might more closely resemble juvenile males."

Addressing the stage of maturity of the holotype of *P. cyrtocristatus*, Williamson noted that this specimen has been assumed to represent an adult individual based solely upon its postcranial size (Ostrom 1963; Hopson; Sullivan and Williamson), which, among hadrosaurs, is not a good estimator for relative age. As Dodson had suggested, the crests in other lambeosaurines did not fully develop until complete maturity had been reached; also, because dinosaurs did not experience determinate growth, their adult body size could display a high degree in variation. Williamson further noted that, "based on relative skull size, *P. tubicen* is significantly larger than *P. walkeri*, and if '*P. cyrtocristatus*' represents a subadult of *P. tubicen*, it might have a larger postcranial skeleton than *P. walkeri* even at an earlier ontogenetic stage."

Also, Williamson pointed out that the dentary of FMNH P27393 is significantly smaller than ROM 768, the holotype of *P. walkeri* (23 percent smaller in depth at the midpoint; see Sullivan and Williamson), and NMMNH P-25100 (28 percent; Sullivan and Williamson). If the latter specimen represents an adult individual, Williamson noted, then FMNH P27393, based on the comparative size of its dentary, is only 72 percent adult size.

As to Sullivan and Williamson's rejection of the hypothesis that short-crested *Parasaurolophus* specimens represent juveniles or females, because long- and short-crested morphs have not been found in the same stratigraphic horizon and consequently cannot be shown to be sympatric, Williamson stated that the small sampling of these specimens renders them statistically insignificant. Furthermore, specimens of *P. tubicen* and *P. cyrtocristatus* are separated by only about 1.1 to 2.5 million years (Sullivan and Williamson), a span of time certainly within the expected range for a single dinosaur species.

Given the above reasons, Williamson found it "most parsimonious to regard *P. cyrtocristatus* as a subjective junior synonym of *P. tubicen*, representing either a subadult or a female of that species." Although Williamson acknowledged that some *Parasaurolophus* specimens having short crests could represent a less derived species of this genus (a position taken by Sullivan and Williamson), that author found it "more likely that the short-crested specimens of *Parasaurolophus* represent juveniles or females of either *P. tubicen* or *P. walkeri*.

Regarding BYU 2467 and UCMP 143270, which Sullivan and Williamson had referred to *C. cyrtocristatus*, Williamson noted that it is not possible to refer juvenile *Parasaurolophus* specimens to either adult species of that genus. However, the additional complexity inside the crest of BYU 2467 suggests that it may be similar to *P. tubicen*, a reason for which that author tentatively referred this specimen to that species. However, the internal structure of the crest of UCMP 143270 has yet to be described.

Sullivan and Bennett (2000), and also Bennett and Sullivan (2000; abstract for a poster), referred to *Parasaurolophus* sp. a juvenile specimen (SMP VP-1090) discovered in the summer of 1998 by R. M. Sullivan in the Upper Cretaceous Fruitland Formation in Ah-shi-sle-pah Wash (in the Ah-shi-sle-pah Wilderness Study Area; this area not yet mapped, but see Hunt and Lucas 1992 for illustrations of the dinosaur-bearing outcrops of this area), San Juan Basin, and collected on August 5 of that year by Sullivan, Fred Widmann, and Michael Lamirata. The specimen comprises an almost complete right dentary with teeth, both surangulars, a possibly left quadratojugal, both jugals, a left quadrate, possibly left postorbital, plus miscellaneous unidentified bone fragments.

As Sullivan and Bennett, and Bennett and Sullivan, observed, some features (*e.g.*, strong ventral deflection of anterior part of dentary, tooth crown margins having denticles on upper part of tooth, short edentulous rostral section of dentary, anteriorly bowed quadrate) indicate the specimen's affinities with the Lambeosaurinae. As observed by these authors, the teeth are slightly elongate and have slightly sinuous carinae, as reported in *P. tubicen*, although some of the preserved teeth exhibit very faint, sinuous secondary ridges as found in juvenile hadrosaurines.

Although the specimen lacks a distinctive crest, other features — the teeth and morphology of the jugal being almost identical to NMMNH P-25100 — identify it as belonging to the genus *Parasaurolophus*. To a lesser degree, the morphology of the dentary, the surangular, and quadrate are also consistent with referral of this specimen to *Parasaurolophus* although, the authors pointed out, these similarities are arguably only indicative of the subfamily.

As measured by Sullivan and Bennett (also Bennett and Sullivan), the dentary (including the coronoid process) measures 270 millimeters in length and 83 millimeters in depth, being about one-third the length of the adult NMMNH P-25100. Finding that SMP VP-1090 falls between the size ranges of small and large juvenile hadrosaurs (see Horner, de Ricqlès and Padian 2000), the authors estimated that it represents an individual from one to two years old.

Sullivan and Bennett noted that SMP VP-1090 is significant as the first unequivocal juvenile *Parasaurolophus* specimen recovered from New Mexico.

Over the years, numerous theories have been offered regarding possible secondary functions of the narial crest (*i.e.*, besides its principle function of allowing the air flow from the external naris to the lungs) in *Parasaurolophus*, some of these including visual display (*e.g.*, Hopson, plus others cited by Sullivan and Williamson 1999), thermoregulation (Wheeler 1978), and acoustic resonance (Hopson; Weishampel 1981, 1997) (see *D:TE* for additional hypotheses).

Recent research was conducted concerning the possible function(s) of the crest in *P. tubicen*, including that continued by Diegert and Williamson (1998, abstract; paper in preparation to date of this writing) following earlier work by Sullivan and Williamson (1996; see *S1*). As briefly reported by Diegert and Williamson, the respiratory tract of this species was reconstructed mainly through CT scans of NMMNH P-25100, the specimen previously studied by Sullivan and Williamson (1996), in which the crest is almost complete and only slightly distorted. The resulting model allowed the authors to obtain acoustic predictions for this species, both with various presumed vocal organs and without such an organ; it also predicted steady-state frequency spectra and depicted specific vocalizations. The crest proved to be multiply resonant, the main shortest path from internal to external nares being a simple sinus passage of approximately 2.9 meters.

Diegert and Williamson found that, lacking a vocal organ, the crest of this species "can be excited by an 'air-jet reed' formed by exhaling air over the internal nares and out the mouth." With the external nares shut, that "2.9-meter sinus passage would resonate at about 30 Hz." However, the "actual branching sinus anatomy produces a complicated dependence of crest input impedence on frequency, comprising an irregular series of impedance peaks (resonant frequencies) and valleys."

In their monograph on *Parasaurolophus*, however, Sullivan and Williamson (1999) disagreed with each other regarding the various proposed secondary functions of the narial crest.

Of these, Sullivan found the acoustic resonance hypothesis to be the least plausible, noting that (*contra* Weishampel 1981) acoustic resonance is not the same as vocalization ("sounds produced in the pharynx and emitted through the oral cavity as produced by many vertebrates"), and that vocalization does not depend on resonating properties of the lambeosaurine narial crest. Sullivan also cautioned that modeling of the narial crest by some artificial means (*e.g.*, computer imaging) does not necessarily demonstrate function. For example, if the crest evolved for the purpose of visual display and related in acoustic resonance, the latter is a by-product of crest development. It was that author's conclusion that the most likely secondary function of the narial crest of this genus was visual display for species, not gender recognition, as had been suggested by Dodson and also Hopson, Sullivan stating that "The ability to discriminate species by cranial profile would be advantageous given that the postcranial profile of all hadrosaurs is nearly the same."

Williamson, on the other hand, interpreted the crest as serving as both a visual display and an acoustic display structure, primarily for intraspecific recognition. This author noted that, although the crests of lambeosaurines have been interpreted as serving as sexual display devices (*e.g.*, Hopson), there does not seem to be a structural reason for these crests to contain elaborate extensions of the respiratory tracts. In fact, some lambeosaurine genera (*e.g.*, *Corythosaurus*, *Lambeosaurus*, and *Hypacrosaurus*) have crests that are mostly solid, while also containing extensions of these tracts, this suggesting that the elaboration of the respiratory tracts served as some "special" function of crest development rather than being a mere "by-product." According to Williamson, the pattern of crest growth observed in lambeosaurines may help establish their function. Dodson had concluded that the crests in lambeosaurines did not become apparent until about 35 percent of maximum body size had been attained and did not fully develop until maturity, this suggesting that structures seem to have served, at least in part, a sexual display function. Growth within the crests of the enclosed narial passageways of certain lambeosaurines also exhibit a delayed expression (see Weishampel 1981), this possibly also indicative of a secondary sexual character. However, as this change alters significantly the narial passageways within the crest, it results in a relatively insignificant change in the crest's external appearance, this suggesting that the addition of these passageways is not related to visual display. Williamson concluded that the development of the lateral diverticulae could "significantly alter the potential acoustical resonance properties of the crest" (see Weishampel 1981), "and therefore lends some support to the idea that the crest served, at least in part, as a resonating structure for acoustic display."

Key references: Bennett and Sullivan (2000); Diegert and Williamson (1998); Dodson (1975); Hopson (1975); Horner, de Ricqlès and Padian (2000); Hunt and Lucas (1992); Lull and Wright (1942); Mateer (1981); Ostrom (1963); Sullivan and Bennett (2000); Sullivan and Williamson (1996, 1999); Weishampel 1981, 1997); Weishampel and Horner (1990); Weishampel and Jensen (1979); Wheeler (1978); Williamson (2000).

PARVICURSOR Karhu and Rautian 1995

Saurischia: Theropoda: Neotheropoda: Tetanurae: Avetheropoda: Coelurosauria: Manuraptoriformes: Arctometatarsalia: ?Ornithomimosauria: Alvarezsauridae: Mononykinae.

Name derivation: Latin *parvus* = "small" + Latin *cursor* = "runner."

Type species: *P. remotus* Karhu and Rautian 1995.

Other species: [None.]

Occurrence: Barun Goyot Formation, Southern Gobi Aimak, Mongolia.

Age: Late Cretaceous (middle or late Campanian).

Known material/holotype: PIN 4487/25, incomplete skeleton including fragments of dorsal, sacral, and caudal vertebrae, partial pelvis, right hindlimb (missing digits I and III), disarticulated elements of left hindlimb.

Diagnosis of genus (as for type species): Crista ventralis developed only on first sacral vertebra; well-pronounced cranial curving of femur; femur S-shaped in anterior view; length of metatarsal III about one-fourth that of total tarsometatarsus (Karhu and Rautian 1995).

Comments: The genus *Parvicursor* was based on an incomplete skeleton (PIN 4487/25) collected in 1992, during a Joint Russian–Mongolian Paleontological Expedition, from the Khulsan locality of the Baron Goyot Formation, in the Southern Gobi Aimak, southern Mongolia. Originally determined to be a bird, the specimen was subsequently reidentified as representing a very small, lightly-built nonavian theropod (Karhu and Rautian 1995).

Despite its size and general gracility, the holotype of the type species *Parvicursor remotus* was judged by Karhu and Rautian as probably not belonging to a young individual, this assessment based on the following observations: Complete fusion of vertebral arch and centrum, also of two (of three) sacral vertebrae (see Gauthier 1986); complete sutural obliteration in region of acetabular opening; relatively small size of acetabular foramen; fusion between proximal tarsals and metatarsals; smooth articular surface of vertebral centra and zygapophyses, also of acetabular cavity.

In comparing *Parvicursor* to other taxa, Karhu and Rautian found it to be closest in morphology to *Mononykus* (see entry), another genus from the Late Cretaceous of southern Mongolia, sharing with it a number of traits (*e.g.*, complete wedging of proximal

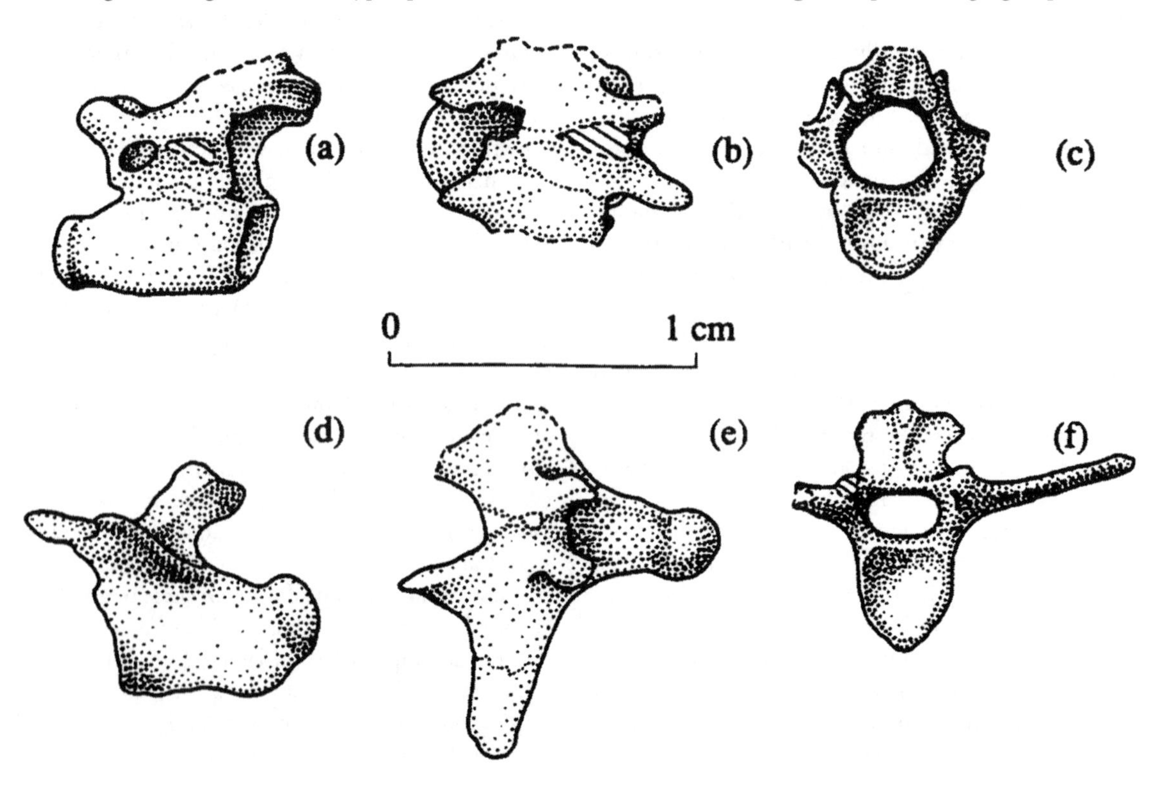

***Parvicursor remotus*, PIN 4487/25, holotype last presacral vertebra in a. lateral, b. dorsal, and c. posterior views, and first three caudal vertebrae in d. lateral, e. dorsal, and f. anterior views. (After Karhu and Rautian 1995.)**

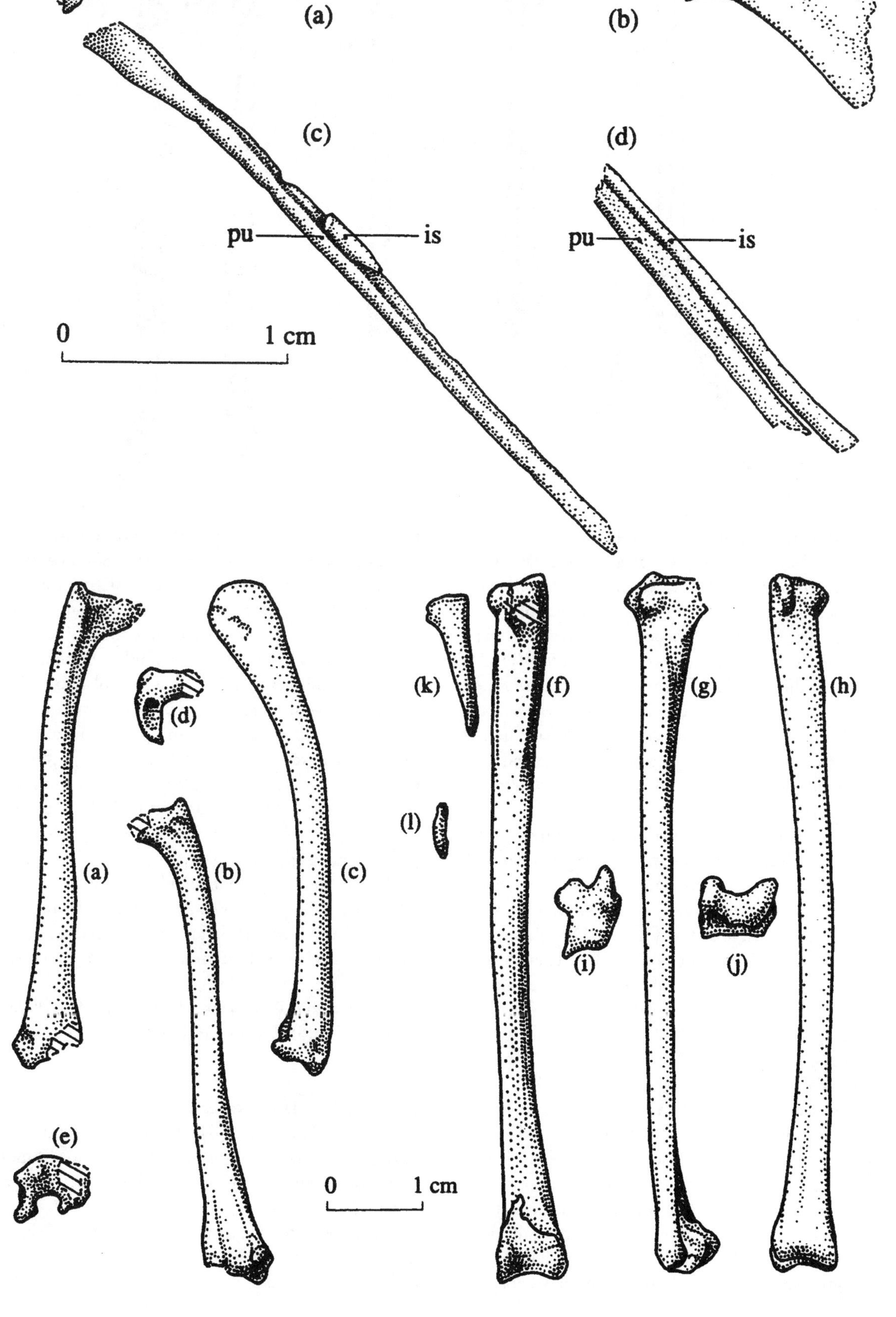

Parvicursor remotus, PIN 4487/25, holotype left ilium in a. lateral and b. dorsal views, c. partial left pubis and ischium in lateral view, and d. distal parts of right pubis and ischium in medial view. (After Karhu and Rautian 1995.)

Parvicursor remotus, PIN 4487/25, holotype right femur in a. anterior, b. posterior, c. lateral, d. proximal, and e. distal views, right tibiotarsus in f. anterior, g. lateral, h. posterior, i. proximal, and j. distal views, and right fibula in k. lateral, and l. proximal views. (After Karhu and Rautian 1995.)

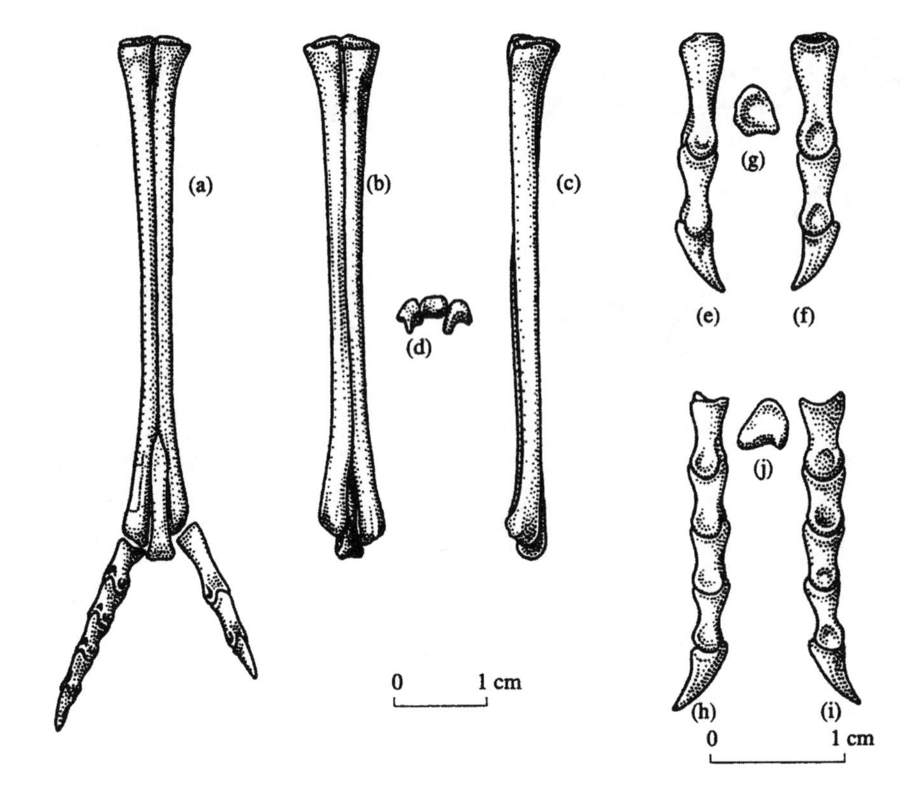

***Parvicursor remotus*, PIN 4487/25, holotype right tarsometatarsus in a. dorsal view of pedal digits II and IV, b. plantar, c. lateral, and d. distal views, pedal digit II in e. lateral and f. medial views, g. articular cotyle of proximal phalanx, pedal digit IV in h. lateral and i. medial views, and j. articular cotyle of proximal phalanx. (After Karhu and Rautian 1995.)**

portion of metatarsal III combined with unfused proximal epiphyses of metatarsals II and IV). However, *Parvicursor* also differs from *Mononykus* in various ways, these including the structure of the vertebrae (last dorsal opisthocoelous, more vertically positioned costal fossae, frontally expanded vertebral centra), the complete fusion between the pelvic elements in the region of the acetabulum, the substantially smaller relative size of the acetabular foramen, the femur lacking a fourth trochanter, the wider position of the femoral condyles, and the comparatively longer tibiotarsus.

Consequently, Karhu and Rautian proposed that the type species *P. remotus*, though very similar to *M. olecranus*, was distinctive enough to warrant its own new manuraptoran family, the Parvicursoridae, which the authors diagnosed as follows: Dorsal vertebrae (including last) opisthocoelous; free caudal vertebra procoelous; absence of additional articulations (hyposphen-hypantrum) of dorsal vertebrae; dorsal and caudal vertebrae without pleurocoels; postacetabular portion of ilium in parafrontal position; ilium strongly expanded laterally directly caudal to acetabulum; narrow, long pubis and ischium closely drawn together, posteroventrally oriented; pubis and ischium not forming distal symphyses with contralateral bones; pelvic elements entirely fused in acetabular region; acetabular foramen relatively small, less than one-half of diameter of acetabular cavity; major and minor trochanters fused in single stout crest, femoral trochanter; absence of fourth trochanter; basis of collum femoris in extreme caudoproximal position on medial side of proximal femoral epiphysis; femoral condyles deeply divided by fossa poplitea, which is caudally wide, distally opened; lateral side of distal femoral epiphysis bearing epicondylus lateralis; tibiotarsus about 1.5 times length of femur; fibula represented by very short rudiment having wide mediolaterally flattened epiphysis; tarsometatarsus longer than half of tibiotarsus; metatarsal I retaining only distal rudiment; diaphyses of metatarsals II and IV approximately equally developed; metatarsal III completely wedged proximally, not reaching half of tarsometatarsal length; trochleae metatarsus not ginglymoid; pedal digit II supports slightly longer digit IV; ungual phalanx of digit II proportionally developed.

Later, Chiappe, Norell and Clark (1998) referred *Parvicursor* to a new subfamily of alvarezsaurids, the Mononykinae (see "Systematics" chapter and *Mononykus* entry).

Key references: Chiappe, Norell and Clark (1998); Gauthier (1986); Karhu and Rautian (1995).

PATAGONYKUS Novas 1996

Saurischia: Theropoda: Neotheropoda: Tetanurae: Avetheropoda: Coelurosauria: Manuraptoriformes: Arctometatarsalia: ?Ornithomimosauria: Alvarezsauridae.

Name derivation: "Patagonia" + Greek *onyx* = "claw."

Type species: *P. puertai* Novas 1996.

Other species: [None.]

Occurrence: Río Neuquén Formation, Neuquén-Province, Patagonia, Argentina.

Age: Late Cretaceous (?Turonian).

Known material/holotype: PVPH 37, two incomplete dorsal vertebrae, incomplete sacrum, two proximal and two distal caudal vertebrae, incomplete coracoids, proximal and distal ends of both humeri, right proximal portions of ulna and radius, articulated carpometacarpus and first phalanx of digit I of right manus, incomplete ungual phalanx probably corresponding to digit I, portions of ilia, proximal ends of ischia, portions of pubes, proximal and distal portions of right and distal end of left femora, proximal and distal ends of tibiae fused with proximal tarsals, metatarsals II and III fused to distal tarsals III, several pedal phalanges.

Diagnosis of genus (as for type species): Postzygapophyses in dorsal vertebrae having ventrally curved, tongue-shaped lateral margin; dorsal, sacral, and caudal vertebrae with bulge on caudal base of neural arch; humeral articular facet of coracoid transversely narrow; internal tuberosity of humerus subcylindrical, wider at its extremity than in its base; humeral entepicondyle conical, strongly projected medially; first phalanx of manual digit I having proximomedial hook-like processes; ectocondylar tuber of femur rectangular in distal aspect (Novas 1996*b*).

Comments: The genus *Patagonykus* was founded upon partial postcranial remains (PVPH 37) collected from the Portezuelo Member of the Río Neuquén-Formation, Sierra del Portezuelo, west of Plaza Huincul City, in northwest Patagonia, Argentina (Novas 1996*b*).

Novas, after performing a phylogenetic analysis of the type species *Patagonykus puertai*, determined that it belongs in the Alvarezsauridae, a family which he (and other workers; see systematics section, *S2* and this volume) interpreted as a family of Mesozoic birds and the sister taxon of the avian group Ornithothoraces. However, although the Alvarezsauridae has subsequently been regarded as an avian assemblage, Sereno (1999*a*) recently reinterpreted this taxon as more likely a theropod group within Ornithomimosauria (see "Systematics" chapter); therefore, *Patagonykus* is tentatively regarded herein, at least for the present, as a nonavian theropod dinosaur.

Comparing *Patagonykus* with other known alvarezsaurids, Novas found that this genus and the geologically younger Asian taxon *Mononykus* share 17 characters which are absent or unknown in *Alvarezsaurus*. This suggested that *Patagonykus* and *Mononykus* are more closely related to one another than either is to *Alvarezsaurus*— despite the close geographic and stratigraphic provenances of the Patagonian forms — and, consequently, may form their own clade.

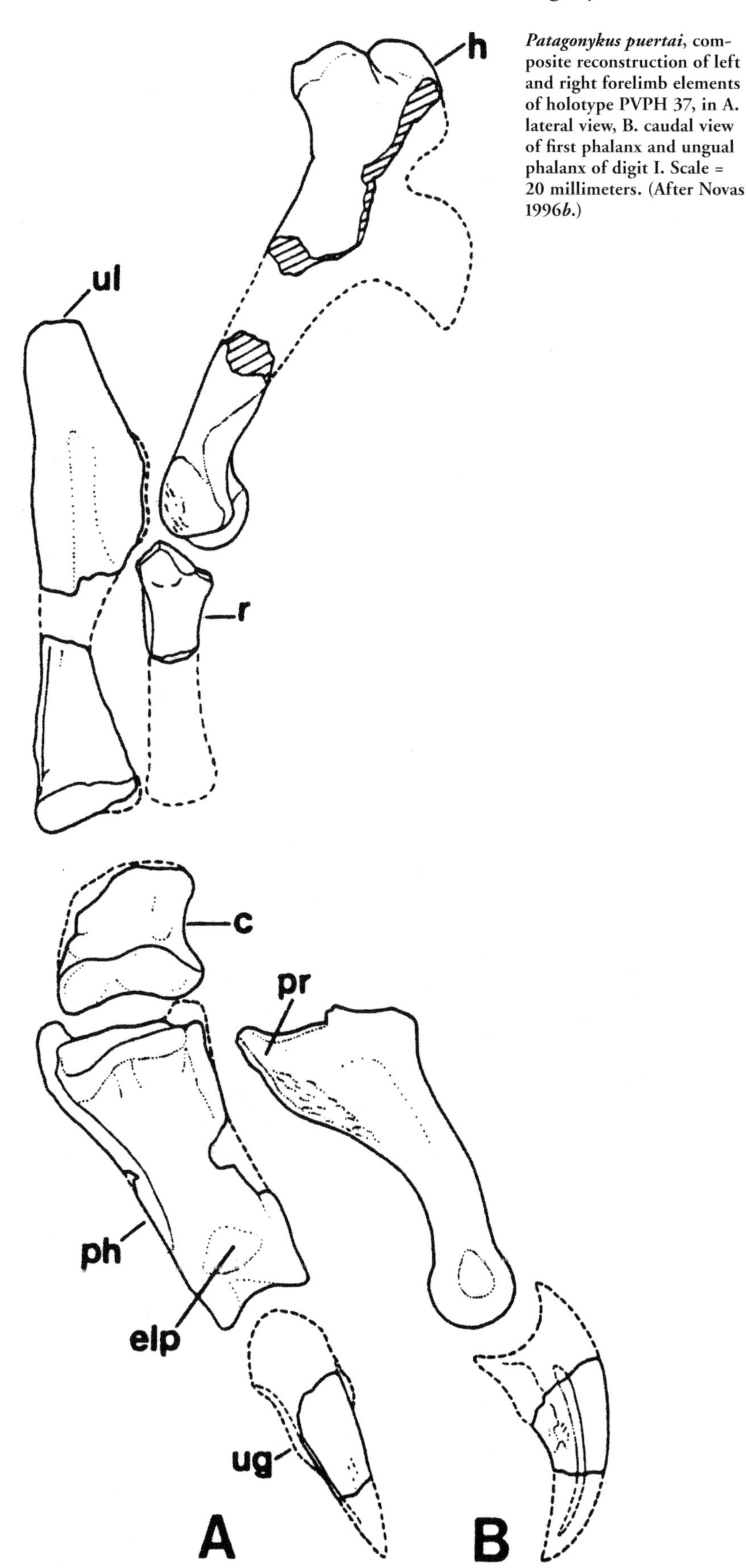

***Patagonykus puertai*, composite reconstruction of left and right forelimb elements of holotype PVPH 37, in A. lateral view, B. caudal view of first phalanx and ungual phalanx of digit I. Scale = 20 millimeters. (After Novas 1996*b*.)**

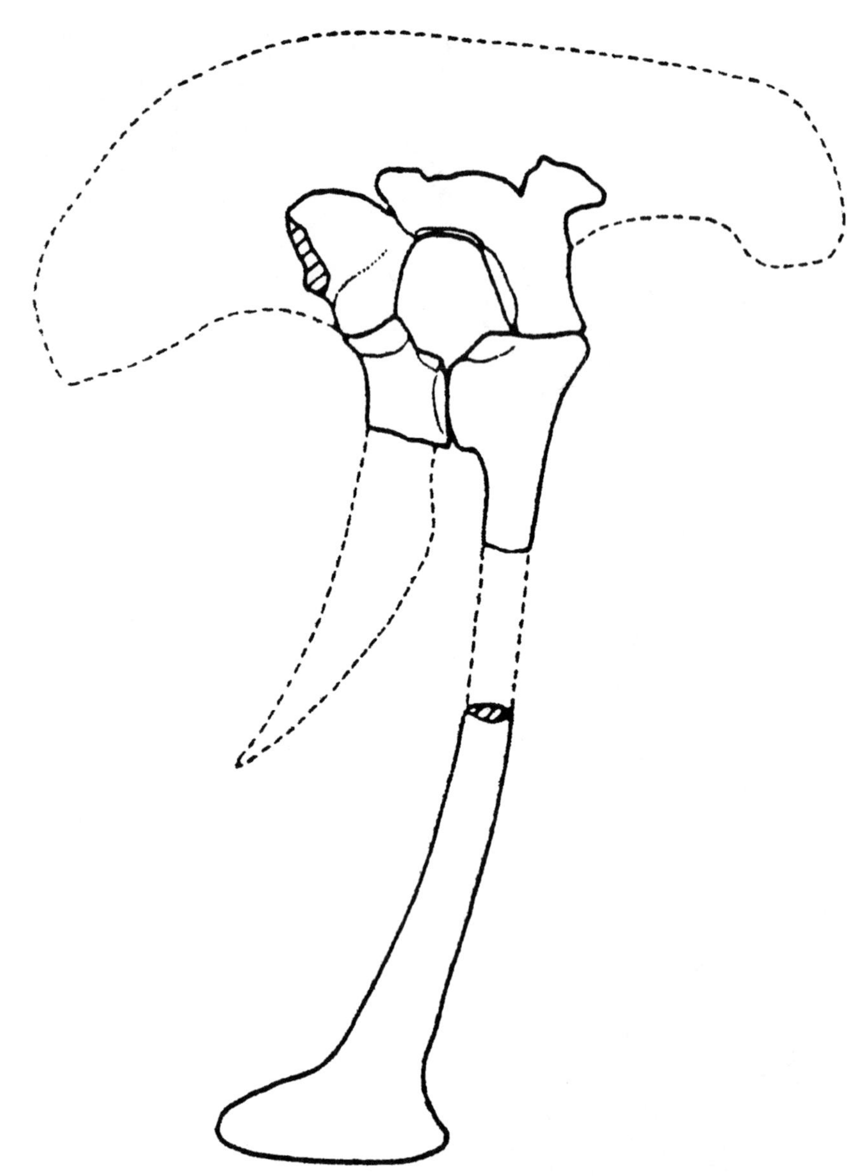

***Patagonykus puertai*, composite reconstruction of pelvis based on left and right bones of pelvis of holotype PVPH 37. (After Novas 1996*b*.)**

Characters shared by *Patagonykus* and *Mononykus* include: Caudal articular surface of centra of last sacral and proximal caudal vertebrae strongly spherical; sacral vertebrae ventrally keeled; femur with fourth trochanter; supracetabular crest; posterior dorsal vertebrae procoelous; coracoid without bicipital tubercle; forelimbs less than 20 percent length of hindlimbs; humeral head having major transverse axis ventrolaterally inclined relative to longitudinal axis of humerus, internal tuberosity proximally projected; humerus with single distal condyle, ulna and radius tightly appressed proximally, forming cup-like articular surface for humerus; strongly-developed olecranon process; ulnar caudal margin straight; carpometacarpus massive, short, quadrangular; digit I larger than remaining digits of manus; phalanx 1 on manual digit I showing B-shaped proximal articular surface, hook-like proximomedial processes, symmetrical distal ginglymus, deep extensor ligamentary pit; medial condyle of femur transversely wide, distally flat; ectocondylar tuber projected caudally, well behind medial distal condyle; fibular condyle of femur conical, projected distally.

Novas pointed out that the condition of most apomorphies uniting *Patagonykus* and *Mononykus* are unknown in *Alvarezsaurus* due to the fragmentary nature of its available material; therefore, some of these uniting characters may eventually prove to be synapomorphic for a more encompassing Alvarezsauridae.

Also, Novas addressed the apparent inconsistency that, in their greater age, the Patagonian alvarezsaurids *Patagonykus* and *Alvarezsaurus* are more primitive than the Asian *Mononykus*, while *Patagonykus* and *Mononykus* share various derived characters that suggest a closer relationship between them than either to *Alvarezsaurus*. This suggested to Novas that both "*Patagonykus* and *Mononykus* are descendants from a common ancestor not shared with *Alvarezsaurus*."

Novas further discussed the issue of the occurrence of the Alvarezsauridae in both Mongolia and Patagonia, this being puzzling mostly because the Alvarezsauridae is only one of two taxa (the other being the sauropod group Titanosauridae) which are known among the very different Late Cretaceous faunas of South America and Asia. In fact, various previous workers (*e.g.*, Bonaparte 1986*b*; Bonaparte and Kielan-Jaworoska 1987; Russell 1993) had pointed out the many dissimilarities between Gondwanan and Laurasian taxa; these differences were attributed mainly to the separation of both super-continents over 70 to 80 million years spanning most of the Cretaceous period, during which time a number of dinosaurian taxa evolved separately.

Novas proposed two possible alternative hypotheses to explain the shared presence of the Alvarezsauridae in South America and Asia:

1. Alvarezsaurid species from both continents were vicariant taxa descended from an ancestral species that was widely distributed over northern and southern landmasses, this having taken place before the emplacement of major barriers to overland dispersal in Gondwana and Laurasia (see Lillegraven, Kraus and Brown 1979). Accepting this interpretation, the origin of alvarezsaurids must be traced back to Valanginian times when the two supercontinents were closer together (Scotese and Golonka 1992).

2. *Patagonykus* and *Alvarezsaurus* were endemic Gondwanan taxa that evolved in isolation during Cenomanian and Santonian times. Accepting this explanation, these two genera may be interpreted to be later emigrants to Asia, possibly via North America during the Campanian, when land connections existed between Gondwana and Laurasia. According to

Novas, this explanation agrees with the currently available fossil evidence of alvarezsaurids, the paleogeographic reconstructions (*e.g.*, Lillegraven *et al.*; Scotese and Golonka), and paleobiogeographic interpretations the evolution of Gondwanan vertebrate faunas in general (*e.g.*, Bonaparte; Bonaparte and Kielan-Jawarowska).

According to Novas (see Novas for further details and related references), the current South American and Asian documentation of alvarezsaurids suggests that the Alvarezsauridae successfully occupied a wide range of environmental conditions, these ranging from desertic (as indicated by sedimentology of the Djadokhta Formation) to more humid conditions (suggested by the fluvial deposits of the Nemegt, Río Neuquén and Río Colorado formations).

Key references: Bonaparte (1986*b*); Bonaparte and Kielan-Jawarowska (1987); Novas (1996*b*); Sereno (1999*a*).

†PAWPAWSAURUS (=*?Texasetes*)

Ornithischia: Genasauria: Thyreophora: Thyreophoridea: Eurypoda: Ankylosauria: Nodosauridae.

Diagnosis: Postorbital scute in dorsal view more prominent than in *Edmontonia*, *Panoplosaurus*, *Sauropelta*, *Silvisaurus*, and *Struthiosaurus*; muzzle parallel-sided (not tapering as in *Edmontonia*, *Panoplosaurus*, *Sauropelta*, and *Silvisaurus*); orbits protruding laterally (unlike *Edmontonia*, *Panoplosaurus*, *Sauropelta*, *Silvisaurus*, and *Struthiosaurus*); temporal notch of skull roof more prominent than in *Edmontonia*, *Panoplosaurus*, *Sauropelta*, *Silvisaurus*, or *Struthiosaurus*; jugal horn in lateral view moderate in size, rounded as in *Silvisaurus* and *Niobrarasaurus* (unlike small, round horn in *Edmontonia* or large scute in *Sauropelta*); in lateral view, premaxillaries with teeth as in *Silvisaurus* (not edentulous as in *Edmontonia* and *Panoplosaurus*); vomer keeled as in *Edmontonia*, *Niobrarasaurus*, and *Silvisaurus* (not swollen and grooved as in *Panoplosaurus*); premaxillary palate ("scoop") wide (width = 88 percent length) as in *Panoplosaurus* and *Edmontonia longiceps* (width = about 85 percent length) (unlike long and narrow in *Silviaurus* [width = 66 percent length] or very wide as in *Edmontonia rugosidens* [width = 110 percent length]; palate not strongly hourglass-shaped; maxillary tooth rows strongly diverging posteriorly as in *E. rugosidens* and *Panoplosaurus*; pterygoid plates very long as in *Silvisaurus* (short and more vertical in *Edmontonia* and *Panoplosaurus*) (Carpenter and Kirkland 1998).

Comments: The type species *Pawpawsaurus campbelli* (see *D:TE*) was redescribed by Carpenter and Kirkland (1998) in their review of Lower and Middle Cretaceous ankylosaurs from North America.

Carpenter and Kirkland found Lee's (1996) original diagnosis of this armored dinosaur to be inadequate, noting that it was "brief and not comparative." According to these authors, two of the character's used in Lee's diagnosis (U-shaped "prevomer" ridge and "prevomer" oval foramen) and their use is questionable; the "prevomer" ridge is not oriented symmetrically along the midline, as Lee had stated, but rather "slightly towards the left side of the skull, suggesting that perhaps this structure is abnormal; and "both the ridge and foramen are topographically located within the palatal portion of the premaxilla, not the "premover" structure identified by Gilmore (1930)."

In 1997, Lee referred a humerus (no catalogue number given) to *Pawpawsaurus* (without justification according to Carpenter and Kirkland), pointing out that this bones is "strongly twisted." This element, the authors noted, is not the same humerus having a straight shaft which Lee (1996) had previously referred to *Pawpawsaurus*. These two morphs of humeri indicated to Carpenter and Kirkland that two nodosaurid taxa are present in the Paw Paw Formation. Carpenter were not able to determine if the holotype skull of *P. campbelli* is that of *Texasetes*. However, the twisted-shaft humerus apparently does not belong to *Texasetes*, which possesses a humerus with a straight shaft. For the present, then, the authors urged caution about arbitrarily assigning postcranial remains to *Pawpawsaurus* (see *Texasetes* entry, "Note").

Key references: Carpenter and Kirkland (1998); Gilmore (1930); Lee (1996, 1997).

†PELECANIMIMUS

Saurischia: Theropoda: Neotheropoda: ?Tetanurae: ?Avetheropoda: ?Coelurosauria: ?Manuraptoriformes: ?Arctometatarsalia: ?Ornithomimosauria: ?Ornithomimidae.

Comments: Previously regarded as a primitive orinithomimosaur (see *D:TE*) and a primitive ornithomimid (*S1*), this small theropod dinosaur from the Hauterivian–Barremian of Spain was recently reassessed by Taquet and Russell (1998) in their report on spinosaurids from the Early Cretaceous of the Sahara.

As noted by these authors, *Pelecanimimus* displays a number of unusual autapomorphies, *e.g.*, more than 200 teeth, premaxillary and anterior teeth "D" shaped in cross section (other teeth being laterally compressed), and manus resembling that in ornithomimids (Pérez-Moreno, Sanz, Buscalloni, Moratalla, Ortega, and Rasskin-Gutman 1994).

However, Taquet and Russell pointed out various features of *Pelecanimimus* otherwise known only in spinosaurs: Premaxilla containing seven teeth; median longitudinal ridge in the temporal region

(Briggs, Wilby, Pérez-Moreno, Sanz and Martínez, 1997); jugal not reaching antorbital fenestra (Martill, Cruickshank, Frey, Small and Clarke 1996); maxillary teeth larger than dentary teeth, dentary teeth atypically numerous, skull narrow, shallow and having elongated facial region (Charig and Milner 1990, 1997); and absence of interdental plates (Buffetaut 1989).

For these reasons, Taquet and Russell suggested that the genus *Pelecanimimus* should be reassessed as a possible "ornithomimid mimic" allied with spinosaurs. If such an evaluation can be demonstrated, then, the authors noted, the diversity of spinosaurs is much greater than previously suspected.

Key references: Briggs, Wilby, Pérez-Moreno, Sanz and Martínez (1997); Buffetaut (1989); Charig and Milner (1990, 1997); Martill, Cruickshank, Frey, Small and Clarke (1996); Pérez-Moreno, Sanz, Buscalloni, Moratalla, Ortega, and Rasskin-Gutman (1994); Taquet and Russell (1998).

†PENTACERATOPS

Ornithischia: Genusauria: Cerapoda: Marginocephalia: Ceratopsia: Neoceratopsia: Ceratopsomorpha: Ceratopsoidea: Ceratopsidae: Chasmosaurinae.

New material: Incomplete skeleton and skull.

Type species: *P. sternbergi* Osborn 1923.

Other species: [None.]

Diagnosis of genus (as for type species): Large chasmosaurine having long, straight, narrow squamosals with eight to 10 pronounced marginal undulations with or without attached epoccipitals, slender strap-like parietal with indented medial posterior margin and large elongate fenestrae, two very large triangular posterior parietal epoccipitals and pair of upturned epoccipitals on midline dorsal surface of posterior part of parietal, large curved epijugal horncores with posteriorly directed jugals, large anteriorly curved supraorbital horncores arising directly over orbits, and deep face with finger-like posterior process of premaxillary on nasal (Lehman 1998).

Comments: The first accurate impression of the body proportions of the giant horned dinosaur *Pentaceratops sternbergi* was afforded by a well-preserved and mostly undistorted exceptionally large skeleton representing most elements, and including an incomplete skull (OMNH 1065). The specimen comprised an almost complete but disarticulated postcranial skeleton lacking the pubes, many vertebrae, the metapodials, and phalanges; the skull, having partially articulated jaws, lacks most of the frill and left side of the face. It was collected in July 1941 by Wann Langston, Jr., Donald E. Savage, and J. W. Stovall for the University of Oklahoma from San Juan County in northwestern New Mexico. Although the precise locality of the find was not recorded, Langston (1989), in reviewing field notes and automobile logs, found the site to be in the headwaters of Coal Creek, and therefore in the Fruitland Formation or Kirtland Shale exposures. Thus far, OMNH 1065 constitutes the only specimen of *P. sternbergi* found with the skull associated with most of the postcranial remains. Only recently, however, has this gigantic specimen been fully prepared for study (Lehman 1998).

As measured by Lehman, the complete skull of this specimen would have been almost 3 meters (about 10 feet) in length, therefore distinguishing OMNH 10165 as the largest known ceratopsian individual, possessing the largest skull known to date for any tetrapod. The entire rostrum to tail-tip length of the skeleton, as reconstructed by Lehman, would be about 6.8 meters (almost 22 feet).

Lehman described in detail the intricate narial structure of the skull of *Pentaceratops* and those of other chasmosaurines, this structure including the development of a short secondary hard palate. Noting that ankylosaurs and hadrosaurs also possess complex respiration-related structures, and pointing out that these ornithischian groups along with ceratopsians constitute the most derived of reptilian herbivores, Lehman hypothesized that the complex narial structure developed in all three groups may be somehow related to advanced reptilian herbivory.

Weishampel (1984) had suggested that the elaborate dental batteries and buccal pouches in ceratopsian dinosaurs indicate that extensive mastication of vegetation occurred in the mouth, as in hadrosaurs. According to Lehman, ceratopsians must therefore have fed by shearing, chopping, or crushing their food prior to swallowing, instead of bolting their food, or relying upon a gizzard or fermentation to macerate plant matter. The secondary palate was "likely extended with soft palate to deliver inspired air into the pharynx behind the shearing mechanism and allow continued feeding without interruption in breathing." Lehman interpreted this intricate narial structure, as well as the similar adaptations in ankylosaurs and hadrosaurs, as possibly "a consequence of an advanced level of reptilian herbivory."

As previously stated by Lehman (1996), *P. sternbergi*, though among the largest of ceratopsians, is also one of the least derived chasmosaurines. In comparing OMNH 10165 to other specimens of *Pentaceratops* (including PMU.R268, a nearly complete postcranial skeleton assigned by Wiman in 1933 to *Pentaceratops*, this referral accepted by Lehman 1998) and also those of *Triceratops* and *Chasmosaurus*, Lehman (1998) saw that such features as the well-developed sinuses in the supraorbital horns, and the shape of the limb (*e.g.*,

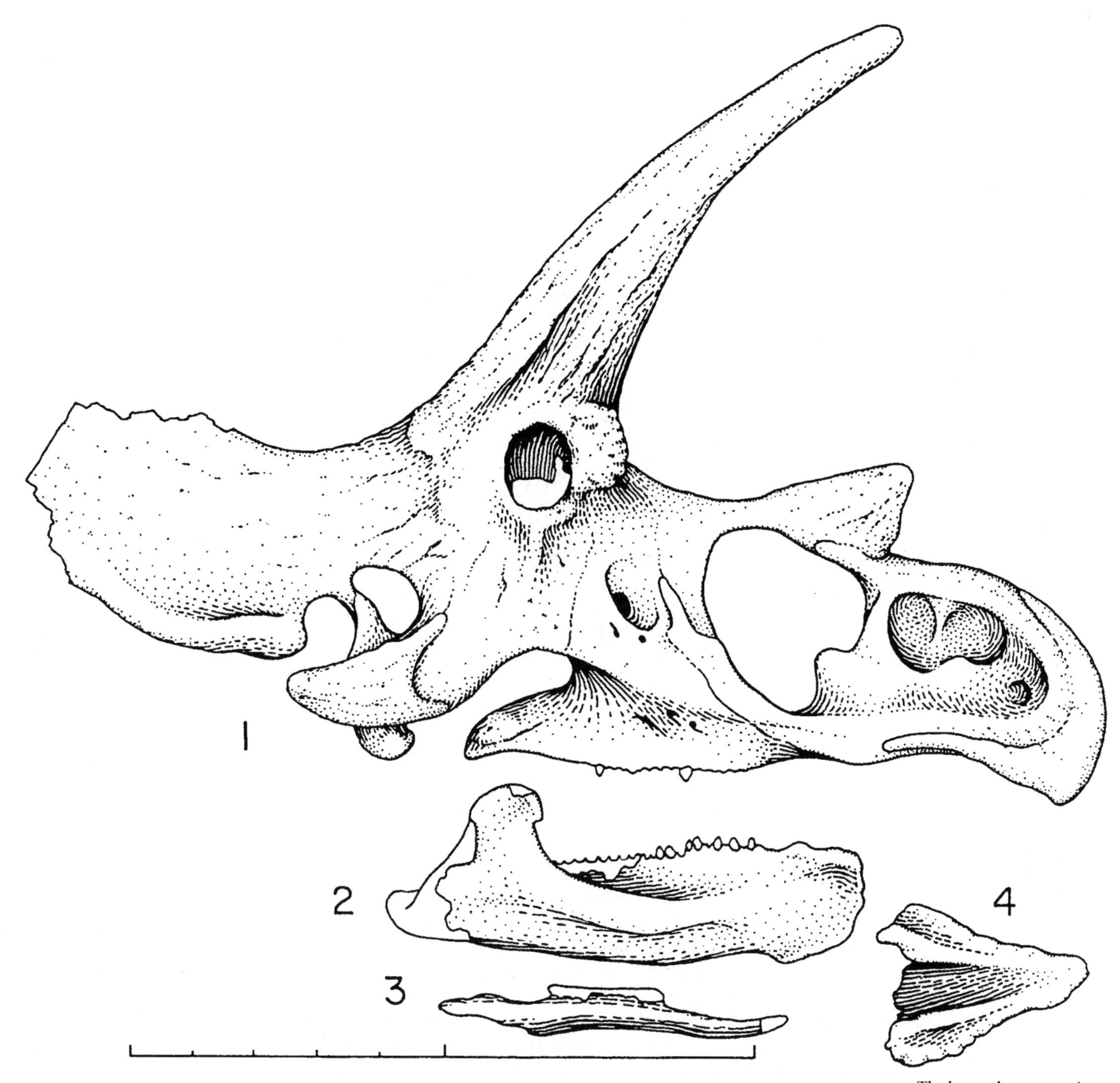

The largest known specimen (OMNH 10165) of *Pentaceratops sternbergi*, including 1. skull, 2. right dentary, and 3. splenial (right lateral view), and 4. predentary (ventral or oral view). Scale = 1 m. (After Lehman 1998).

femur robust and erect) and girdle (*e.g.,* ilium relatively broad, with less sinuous lateral margin) elements, more closely resemble the highly derived condition seen in *Triceratops* than in *Chasmosaurus* or other *Pentaceratops* individuals. This suggested to Lehman (1998) that characters traditionally used in phylogenetic analyses of ceratopsids are actually size related. However, although the limb bones in *Pentaceratops* are similar in size and morphology to those of *Triceratops,* the limb proportions in the former genus are closer to those found in more primitive chasmosaurine taxa such as *Anchiceratops* and *Chasmosaurus*. This suggested to Lehman (1998) that, although large body size seems to have evolved comparatively early among chasmosaurine ceratopsids, more primitive limb ratios were retained in *Pentaceratops.*

Lehman (1998) also observed among different *Pentaceratops* specimens (including OMNH 10165) — all remains of this genus having been collected from a small geographic area and narrow stratigraphic interval (the Upper Cretaceous Fruitland Formation and Kirtland Shale in the San Juan Basin of New Mexico — much variation in the features of the elongate rostrum, elaborate premaxillary structure, and complex

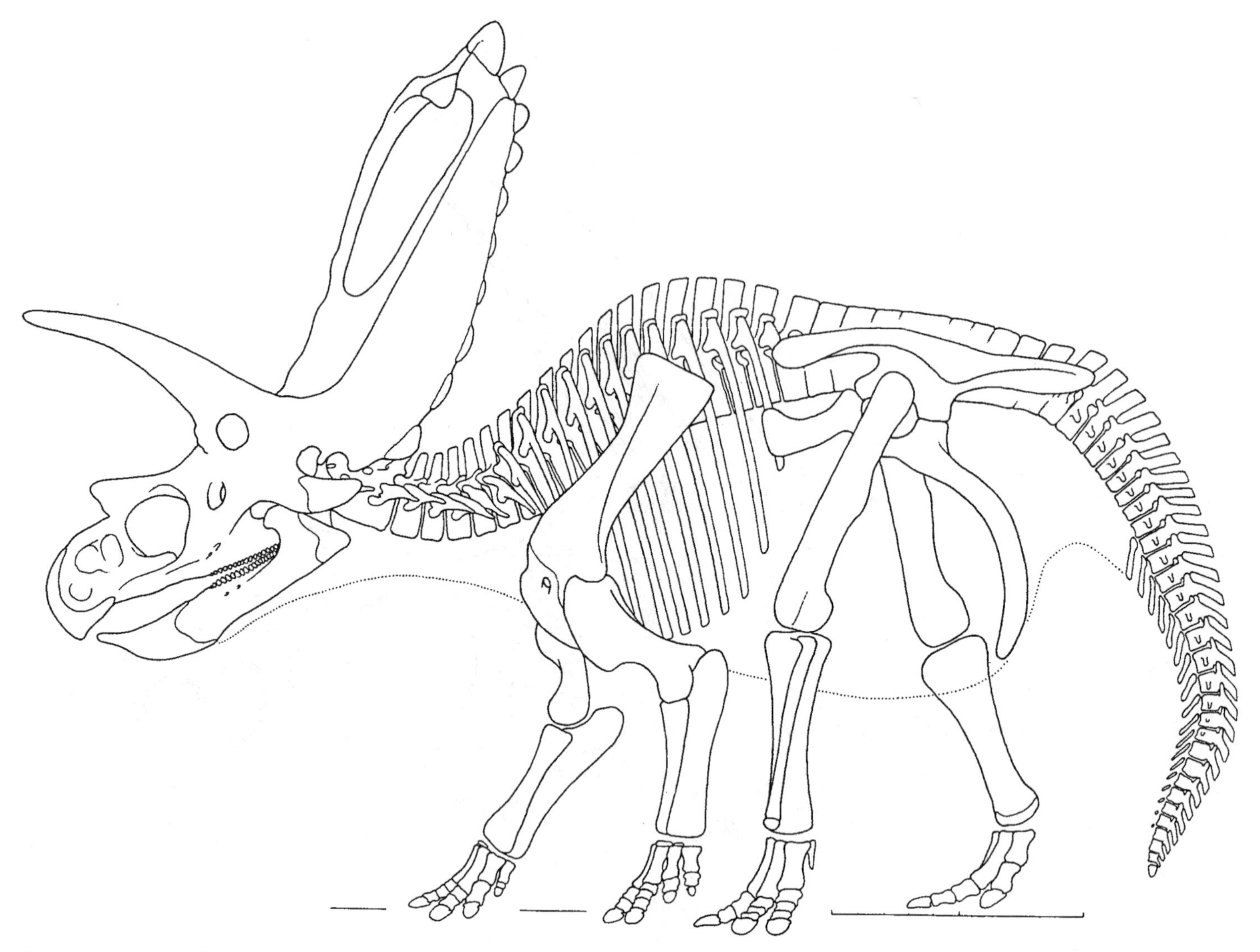

***Pentaceratops sternbergi*, reconstructed skeleton (OMNH 10165, missing elements based on PMU. R268). Scale = 1 m. (After Lehman 1998).**

frontal and cornual sinuses. This indicated to the author that these features, rather than being used (as they have been in the past) as criteria for providing species-level discrimination in chasmosaurines, may really demonstrate a high level of intraspecific sexual and ontogenetic variability. As this assessment seems also likely true for *Triceratops* (see entry) and *Chasmosaurus*, it suggested to Lehman (1998) that, although individual ceratopsian populations display high individual variability, lengthy periods of evolutionary stasis persisted during the time recorded by each species range zone, and that most ceratopsian genera were monospecific and limited to a small geographic area.

As chasmosaurine taxonomy is based completely on cranial characters, with useful specimens limited to essentially complete skulls, it was Lehman's (1998) contention that perceived morphological differences among specimens of even relatively well-known ceratopsids "will typically not pass rigorous tests for statistical significance because of the small sample sizes"; consequently, a "purely intuitive evaluation of the variation among specimens is as appropriate as biometric analysis of such small samples or reliance on a few presumably diagnostic characters."

Note: Lehman (1998) pointed out that, according to Article 32 of the International Code of Zoological Nomenclature, misspellings because of improper latinization of taxonomic names should be corrected. Osborn (1923) had originally spelled the type species name for *Pentaceratops* as "*sternbergii.*" As pointed out by Lehman (1998), this specific name "should have been spelled *sternbergi* because the root is the personal named 'Sternberg' and requires only one 'i' to make it genitive singular (D. Prothero, personal commun., 1995)."

Key references: Lehman (1996, 1998); Langston (1989); Osborn (1923); Weishampel (1984); Wiman (1933).

†PHUWIANGOSAURUS

Saurischia: Sauropodomorpha: Sauropoda: Eusauro-

Photograph by the author, courtesy Museum of Northern Arizona.

Skull (MNA P.1747) of *Pentaceratops sternbergi* from New Mexico imbedded in rock matrix.

poda: Neosauropoda: ?Diplodocoidea: ?Nemegtosauridae.

Age: Early Cretaceous.

Comments: *Phuwiangosaurus* is a medium-sized sauropod that was based upon abundant fossil material collected from Sao Khua Formation (Early Cretaceous, based on palynological evidence; see Racey, Goodall, Love, Polachan and Jones 1994) in the Phu Wiang area of northeastern Thailand, and described in 1994 by Martin, Buffetaut and Suteethorn (see *D:TE*).

In 1999, Martin, Suteethorn and Buffetaut published a new detailed description of the type species *Phuwiangosaurus sirindhornae*, this description augmented mostly by specimens referable to this species from the Phu Pha Ngo and Ban Nong Mek sites in Kalasin Province. Both sites were discovered in 1991, the Phu Pha Ngo site located near a Buddhist temple. Some 40 identifiable bones, first thought to be the remains of elephants, were excavated during the construction of this temple. A subsequent excavation at this site yielded fossil representing at least three individuals.

Importantly, a major new locality discovered in November, 1994, near the temple at Wat Sakawan (Sahan Sakhan, Kalasin Province) yielded several partly articulated skeletons (to date consisting of more than 600 bones), including numerous teeth and elements of the upper jaw (see Suteethorn, Martin, Buffetaut, Triamwi Chanon and Chaimanee 1995; Buffetaut and Suteethorn 1999).

Originally, Martin *et al.* (1994) assumed that *Phuwiangosaurus* might be related to Chinese Jurassic sauropods belonging to the Euhelopodidae, this assessment based in part on biogeographical considerations, and also because the Sao Khua Formation was originally believed to be Jurassic in age rather than Early Cretaceous. In their more recent study, Martin *et al.* (1999) found that the skeleton of *Phuwiangosaurus* does not, in fact, resemble those of euhelopodids. As these authors observed, the cervical vertebrae in those Chinese genera exhibit a trend toward a strong transversal compression; in *Phuwiangosaurus*, the flattening occurs dorsoventrally, resulting in quite broad cervical vertebrae. Furthermore, the bifurcation

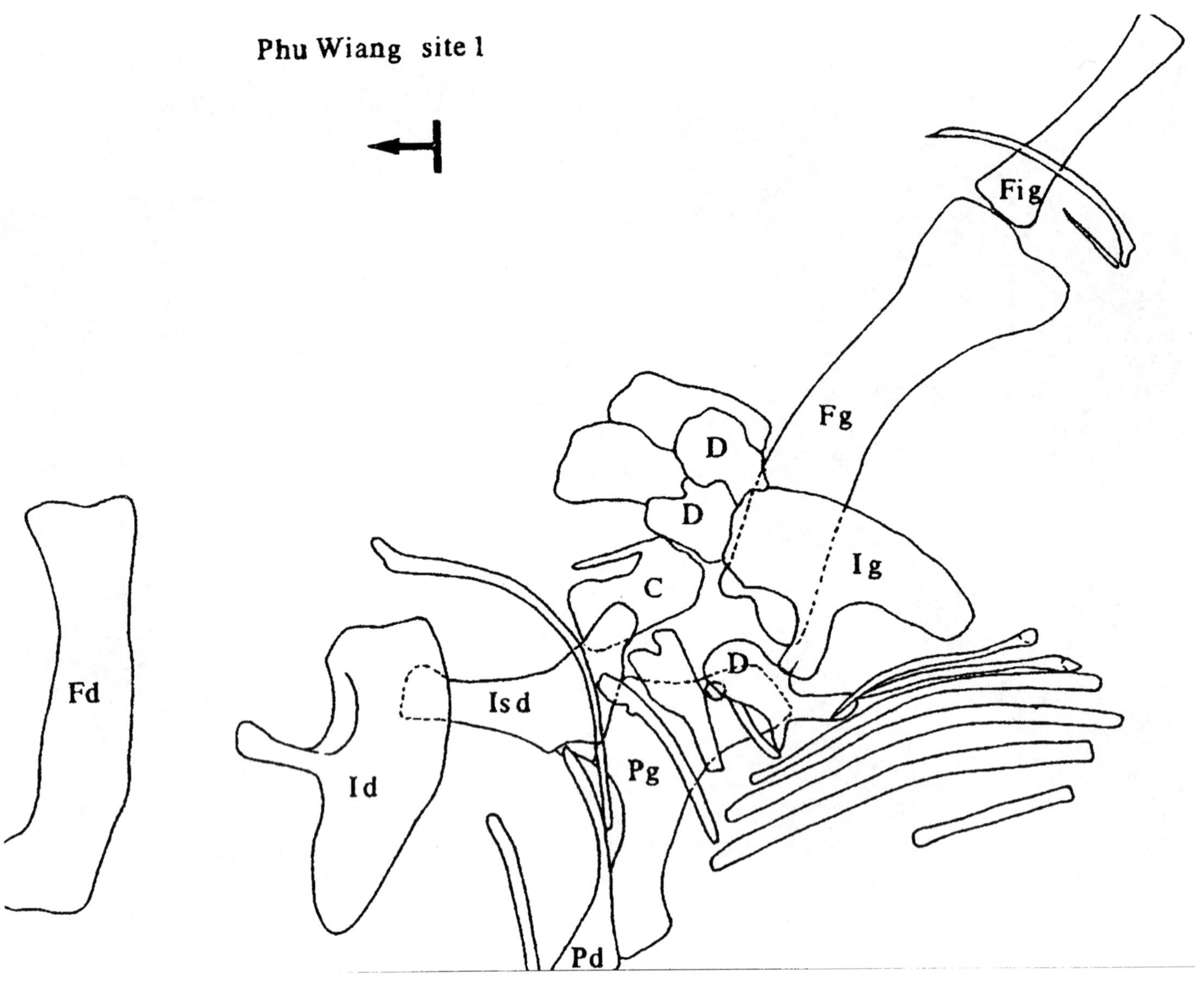

***Phuwiangosaurus strindhornae*, P.W.1-1 to P.W. 1-21, holotype skeleton as found at site P.W.1. Figure by Varavudth Suteethorn. (After Martin, Suteethorn and Buffetaut 1999.)**

of the last cervical is very shallow in euhelopodids yet very deeply marked in *Phuwiangosaurus*. Also, at that time, comparing the genus to other sauropod groups, Martin *et al.* (1994) were unable to assign it to any definite family.

Later, Martin *et al.* (1999), though not yet attempting to discuss at length the phylogenetic affinities of *Phuwiangosaurus*, stated that the few recently recovered "teeth and cranial elements clearly suggest affinities with the Nemegtosauridae." As observed by these authors, the slender teeth, with their lanceolate crowns, are very different from the broader, spoon-shaped teeth of euhelopodids, but closely resemble the teeth of *Nemegtosaurus*, a nemegtosaurid from the Late Cretaceous of Mongolia and China. This suggested that "*Phuwiangosaurus* may be close to the ancestry of *Nemegtosaurus* and related Late Cretaceous forms" (Buffetaut and Suteethorn). Other cranial elements, including a braincase recently found at a new site in Kalasin Province, seem to "confirm the idea that the Thai form belongs to the Nemegtosauridae." Assignment of *Phuwiangosaurus* to this family, therefore, has the potential to "shed new light on the evolutionary history of the Cretaceous sauropods of Asia."

Martin *et al.* addressed the fact that, in cladograms of the Sauropoda published by Upchurch (1998), *Phuwiangosaurus* consistently clustered with "titanosauroids," leading that author to place this genus in the "titanosauroid" family Andesauridae (see "Systematics" chapter). Therefore, "further comparisons of *Phuwiangosaurus* with the titanosaurs will be needed, in conjunction with a reappraisal of the phylogenetic position of the Nemegtosauridae, once the cranial material of *Phuwiangosaurus* is described."

Key references: Buffetaut and Suteethorn (1999); Martin, Buffetaut and Suteethorn (1994); Martin, Suteethorn and Buffetaut (1999); Suteethorn, Martin,

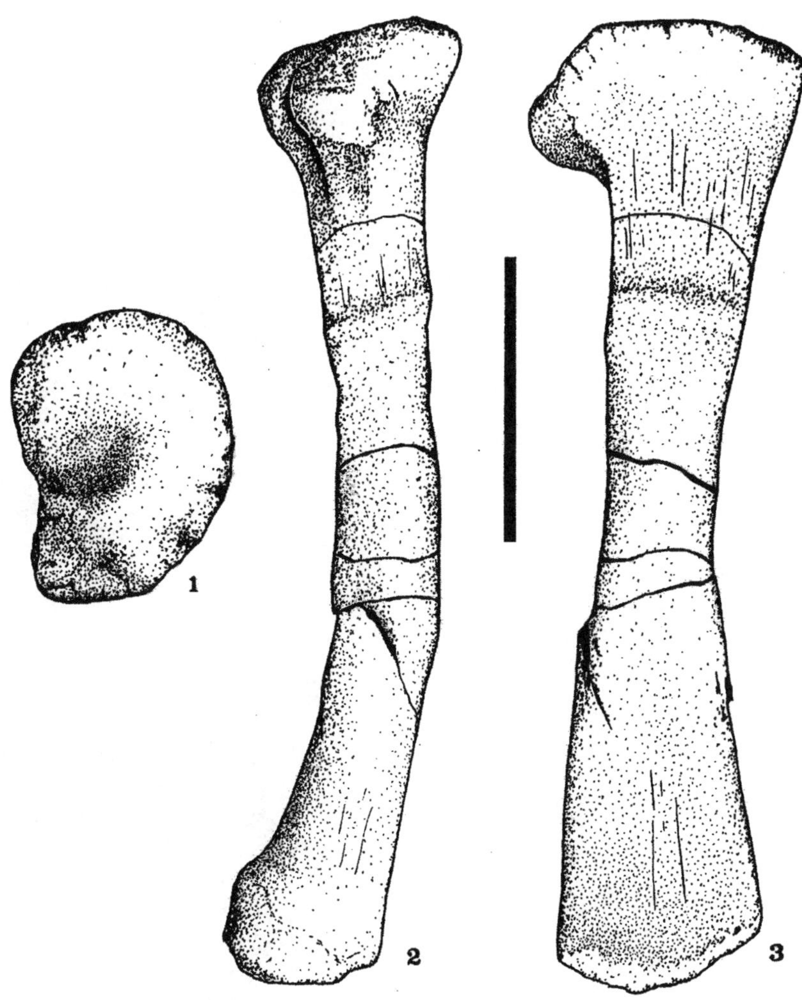

***Phuwiangosaurus strindhornae*, referred left radius (P.W.K.1-36) in 1. proximal, 2. medial, and 3. cranial views. Scale = 10 cm. (After Martin, Suteethorn and Buffetaut 1999.)**

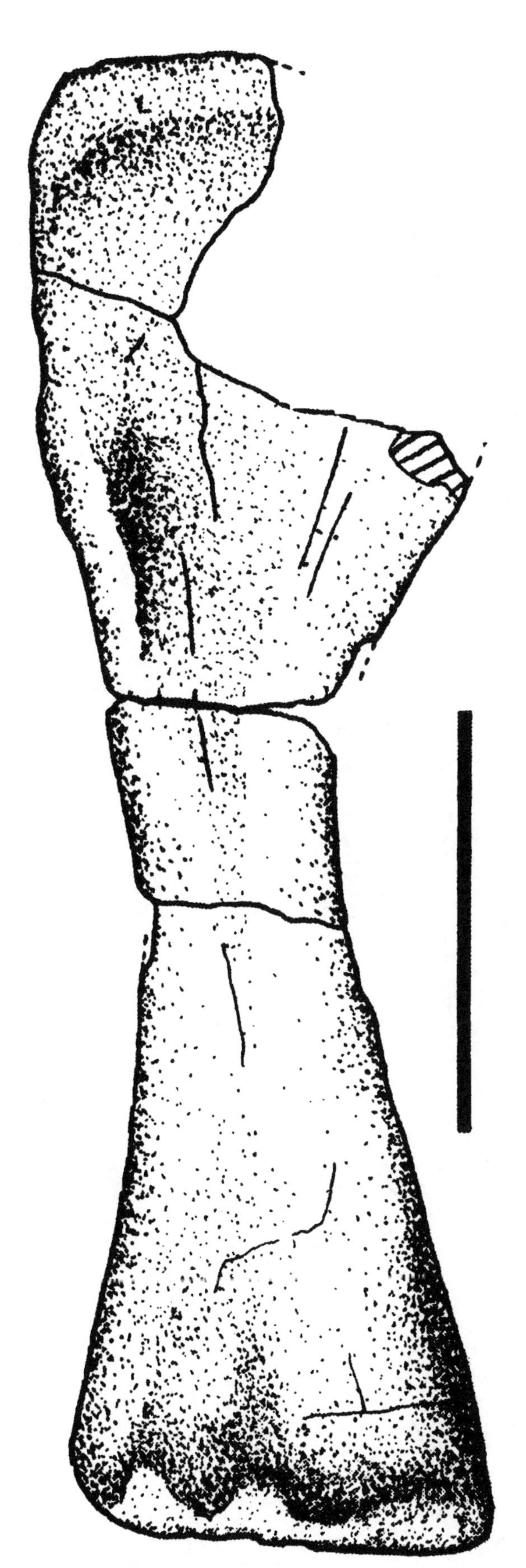

Buffetaut, Triamwi Chanon and Chaimanee (1995); Upchurch (1998).

†**PICRODON** (See *Megalosaurus*.)

Name derivation: Greek *pikros* = "sharp" + Greek *odous* (also *odon*) = "tooth."

Type species: *P. herveyi*.

†**PINACOSAURUS**

Ornithischia: Genasauria: Thyreophora: Thyreophoridea: Eurypoda: Ankylosauria.

Type species: *P. grangeri* Gilmore 1933.

New species: *P. mephistocephalus* Godefroit, Pereda Suberbiola and Dong 1999.

Occurrence of *P. mephistocephalus*: Bayan Mandahu Formation, Inner Mongolia, People's Republic of China.

Age of *P. mephistocephalus*: Late Cretaceous (Campanian).

***Phuwiangosaurus strindhornae*, referred right humerus (P.W.K.1-28) in anterior view. Scale = 20 cm. (After Martin, Suteethorn and Buffetaut 1999.)**

Pinacosaurus mephistocephalus, IMM 96BM3/1, ?subadult holotype skull in (top) rostral, (middle) dorsal, and (bottom) left lateral views. (After Godefroit, Pereda Suberbiola, Li and Dong (1999).

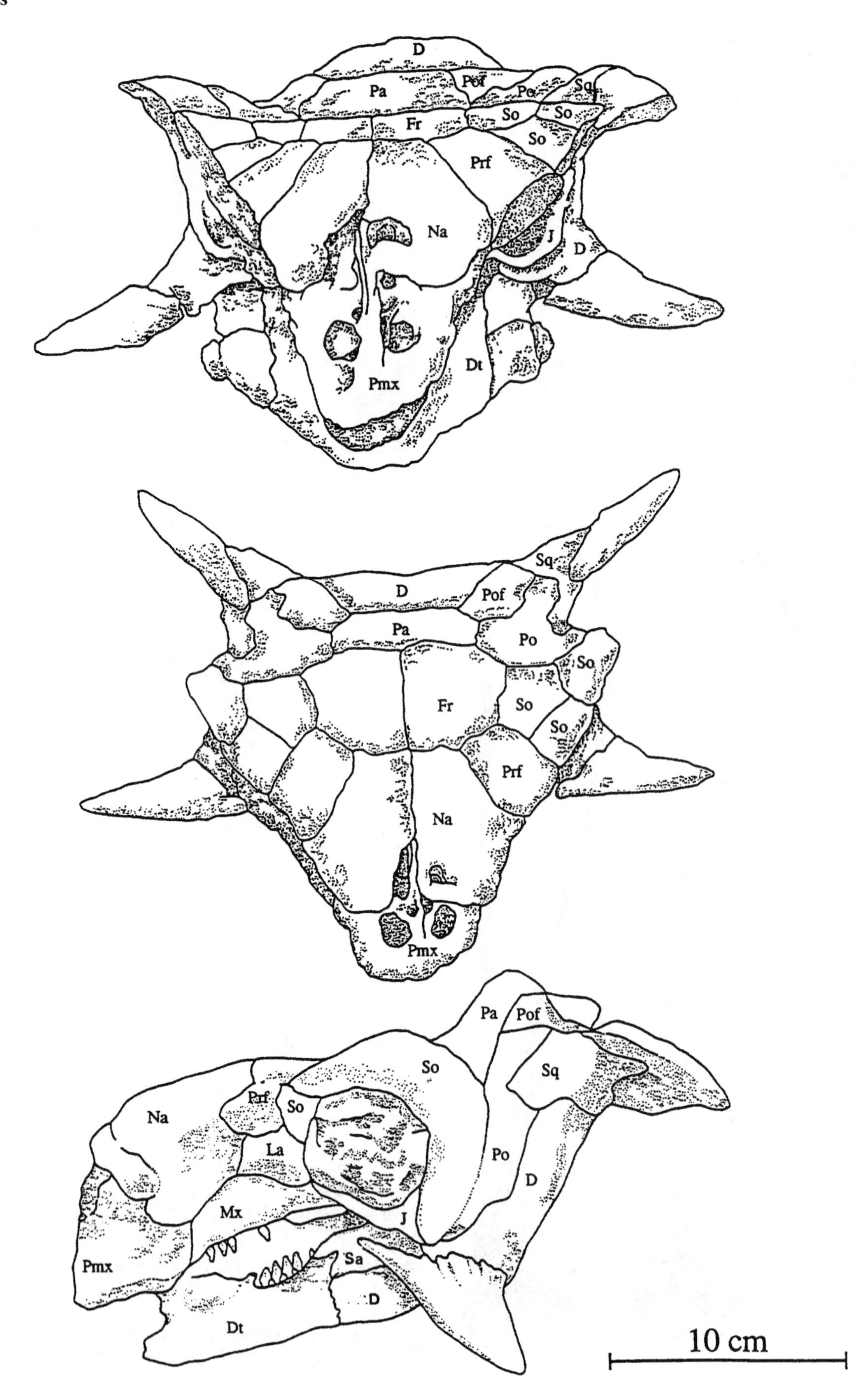

Life restoration of *Pinacosaurus grangeri*, the type species of the genus *Pinacosaurus*, painted by Berislav Krzic.

Known material/holotype of *P. mephistocephalus*: IMM 96BM3/1, almost complete, articulated skeleton including skull with lower jaws, much of the postcranial skeleton, cervical armor and tail club, ?subadult.

Diagnosis of genus: Premaxilla excavated by pair of large, rounded "gland openings"; dentary low, subrectangular; small, edged ossification on angular area of mandible reaching level of penultimate alveolus; neural arches on caudalmost cervical vertebrae completely reduced; ilium with strongly divergent preacetabular process; two cervical half rings comprising three to four fused elements, including oval and low-keeled dorsomedial elements and pointing triangular spines projected ventrolaterally; lightened dermal armor, seemingly formed by two longitudinal series of small oval scutes parallel to vertebral column along dorsal region (Godefroit, Pereda Suberbiola and Dong 1999).

Diagnosis of *P. mephistocephalus*: Two pairs of premaxillary foramina leading to premaxillary sinuses; rostrally facing "gland" opening; external nares visible only in dorsal view; orbits circular, as high as long, laterally oriented; no caudodorsal premaxillary process protruding between maxilla and nasal; lacrimal square shaped; parietal much shorter than frontal; wide frontoparietal process of postorbital; deep frontoparietal depression; scapula relatively short, robust (ratio of proximal width/length of scapula = 0.36 in holotype), with well-developed acromial process; deltopectoral crest well-developed, extending down shaft, terminating distal to midlength of humerus; proximal articular head of radius strongly expanded both medio-laterally and cranio-caudally (Godefroit, Pereda Suberbiola and Dong 1999).

Comments: *Pinacosaurus*, a medium-sized armored dinosaur, is a relatively common taxon found in the Upper Cretaceous rocks of the Gobi Basin (Maryańska 1977; Jerzykiewicz and Russell 1991), the type species of which, *Pinacosaurus grangeri*, was first described by Charles Whitney Gilmore in 1933 (see *D:TE*).

A new species of *Pinacosaurus*, *P. mephistocephalus*, was founded on a well-preserved, nearly complete skeleton (IMM 96BM3/1) from red sandstones in the badlands of the Bayan Mandahu Formation, on the northern flank of the Lang Shan (Wolf Mountains), in Inner Mongolia (site discovered in 1978 by members of the Inner Mongolia Museum at Hohhot; see

Dong, Currie and Russell 1989; site excavated since 1987 by the Sino-Canadian Dinosaur Project; see Dong 1993; worked more recently by the Sino-Belgian Dinosaur Expedition; see Godefroit, Dong, Bultynick, Li and Feng 1998). The skeleton was found lying in a natural position, the cervical armor and tail club in place, the limb bones doubled beneath the body (as with other dinosaur skeletons, *e.g.*, *Protoceratops*, found in the same beds; see Jerzykiewicz, Currie, Eberth, Johnston, Koster and Zheng 1993) (Godefroit, Pereda Suberbiola and Dong 1999).

In addition to describing this new species, Godefroit *et al.* (1999) also proposed a new diagnosis for *P. grangeri*. Following Maryańska (1971, 1977) and Coombs and Maryańska (1990), Godefroit *et al.* (1999) primarily based their diagnosis of the type species on a nearly complete skull and associated postcranial skeleton of a juvenile individual (ZPAL MgD-II/1, the best preserved specimen of this taxon), in addition to the crushed adult holotype skull (AMNH 6523), both specimens having been collected from the Djadokhta Formation of Bayn Dzak, Mongolia. Their diagnosis of *P. mephistocephalus* was based mainly on comparisons with ZPAL MgD-II/1. As Godefroit *et al.* (1999) remarked, both IMM 96BM3/1 and ZPAL MgD-II/1 represent different ontogenetic stages. Therefore, if the diagnostic value of the first three characters in their diagnoses of these species is unequivocal, the possibility exists that some of the other characters could reflect ontogenetic variation.

Godefroit *et al.* (1999) identified IMM 96BM3/1 as belonging to the genus *Pinacosaurus*, as the new species shares with *P. grangeri* the following apomorphies, most of them listed as characters in their diagnosis of the genus: Paired "gland" openings on premaxilla; mandible slender, with roughly parallel dorsal and ventral margins of dentary (Maryańska 1977 explained this morphology by ontogentic changes, although Godefroit *et al.* 1999 found a taxonomic significance to be more plausible); small edged ossification on angular area of mandible reaching level of penultimate alveolus; complete reduction of neural arches on caudalmost cervical vertebrae; strongly divergent preacetabular process of ilium; and lightened dorsal armor.

The authors justified the erection of a new species of *Pinacosaurus* by citing the following differences in the skeleton of *P. mephistocephalus* that separate it from *P. grangeri*: Two pairs of premaxillary foramen leading to premaxillary sinuses (one pair known in *P. grangeri*); "gland" opening rostrally located (ventrolaterally in *P. grangeri*); external nares visible only in dorsal view (ventral view in *P. grangeri*); orbit circular, as high as long, and laterally oriented (oval, longer that high, and ventrolaterally oriented in *P. grangeri*); no caudodorsal process of premaxilla protruding between maxilla and nasal (this process present in *P. grangeri*); lacrimal square-shaped (subrectangular, more elongated in *P. grangeri*); parietal much shorter than frontal (both subequal in length in *P. grangeri*); frontoparietal processes of postorbital wider than in *P. grangeri*; frontoparietal depression parallel to caudal margin of skull roof much deeper than in *P. grangeri*; scapula shorter, less convex externally, with better-developed acromial process than in *P. grangeri*; deltopectoral crest well-developed, extending down below shaft to terminate distal to midlength of humerus (less developed in *P. grangeri*, terminating at about level of humeral midlength); proximal articular end of radius strongly expanded medio-laterally and cranio-caudally (very narrow in *P. grangeri*, especially in ZPAL MgD-II/1.

Godefroit *et al.* (1999) interpreted the above morphological differences in the external nares to be unequivocal specific characters. However, the authors noted that, although the remaining characters may also have specific significance, they might reflect ontogenetic variation, the validity of their taxonomic status pending detailed study of more *Pinacosaurus* material from Bayan Mandahu.

Pinacosaurus was originally classified by Gilmore as a member of the Nodosauridae, after which Coombs, in his reassessment of the ankylosaurian families, referred the genus to the Ankylosauridae (see *D:TE*). Godefroit *et al.* (1999) accepted *Pinacosaurus* as a member of that family based on the following ankylosaurid synapomorphies (see Coombs 1978*a*; Sereno 1986; Coombs and Maryańska 1990) of the skull and postcrania of the new species: Maximum skull width greater than length; snout arches above level of postorbital skull roof; external nares divided by horizontal premaxillary septum; complex secondary palate; sinus-complex developed in premaxillary-maxillary area; prominent wedge-shaped, posterolaterally projecting quadratojugal horns; infratemporal fenestra and quadratojugal hidden in lateral aspect by quadratojugal and squamosal dermal ossifications; sharp lateral rim and low dorsal prominence for each supraorbital element; flat lateral supraorbital margin above orbit; coronoid process very low; two cervical half-rings of armor; postacetabular process of ilium short; distal caudal vertebrae surrounded by ossified tendons; terminal tail-club.

The skeleton of *P. mephistocephalus* offers new information regarding the ontogenetic changes in ankylosaurs. Godefroit *et al.* (1999) identified the type specimen of this species as an immature individual. The authors noted that, in all fully-ossified adult ankylosaurs (including AMNH 6523; see Gilmore), the sutural boundaries of the cranial bones are entirely obliterated due to the fusion of armor plates or the addition of secondary epidermal bone (see Coombs and Maryańska). In both IMM 96BM3/1 and ZPAL MgD-II/1 (see Maryańska 1971, 1977), however, the sutural patterns of the skull are well known. Among the features supporting the immature nature of IMM 96BM3/1 is the lack of fusion between the skull roof and the occiput.

IMM 96BM3/1, unlike ZPAL MgD-II/1, possesses well-developed quadratojugal and squamosal ossifications that form prominent horns. Noting that the development and orientation of these horns probably show a significant ontogenetic variability and perhaps also constitute a sexual dimorphism, Godefroit *et al.* (1999) found it difficult to evaluate their taxonomic value given the data currently available.

The following ontogenetic characters of the postcrania indicated to Godefroit *et al.* (1999) that the holotype of *P. mephistocephalus* is probably a subadult rather than a juvenile (see Maryańska 1977; Galton 1982; Coombs 1986): Neural arches of vertebrae fused to centra; dorsal ribs fused to centra; sacral ribs firmly fused to centra and to ilia; very well-developed ossified tendons in dorsal area and along distal portion of tail; tail-club well developed.

Godefroit *et al.* (1999) considered other material that has been referred to *Pinacosaurus*. Postcranial remains collected by the Sino-Swedish Expedition in 1923 (Wiman 1929) outside the Gobi Basin from the Wangshi Group of Shandong Province were referred by Buffetaut (1995) to *Pinacosaurus* cf. *grangeri*. Other postcranial bones collected in 1932 by the Sino-Swedish Expedition from the Alashan Desert of northern Ningxia Province were originally described by Yang [Young] (1935) as *Pinacosaurus nighsiensis* and later referred to *P. grangeri* by Maryańska (1977). As no specific characters can be seen in this material, Godefroit *et al.* referred these taxa to *Pinacosaurus* sp.

As noted by Godefroit *et al.* (1999), the type species *P. grangeri*, thus far, is only known to occur in the Djadokhta Formation, while *P. mephistocephalus* is known only from the Bayan Mandahu Formation, the distance separating these two units being approximately 350 kilometers. Additionally, preliminary studies of the morphology of *Protoceratops* skulls from both of these formations also reveal specific differences between the faunas. This suggested to the authors the possibility that these differences could indicate local speciation processes between geographically isolated populations, and may also reflect slightly distinct ecological conditions (this also suggested by the abundance of trionychoid turtles at Bayan Mandahu; see Brinkman and Peng 1996). Godefroit *et al.* (1999) cautioned, however, that "it is not certain that both formations are strictly synchronous, as no chronostratigraphic scheme based on palynofloral, radioscopic or palaeomagnetic data has been devised for Late Cretaceous continental sequences in the Gobi Basin."

Key references: Buffetaut (1995); Coombs (1978*a*; 1986); Coombs and Maryańska (1990); Galton (1982); Gilmore (1933*a*); Godefroit, Dong, Bultynick, Li and Feng (1998); Gilmore (1933); Godefroit, Pereda Suberbiola and Dong (1999); Jerzykiewicz, Currie, Eberth, Johnston, Koster and Zheng (1993); Jerzykiewicz and Russell (1991); Maryańska (1971, 1977); Sereno (1986); Yang [Young] (1935).

†PLATEOSAURUS

Type species: *P. engelhardti* Meyer 1837.

Other species: *P. erlenbergiensis* Huene 1907–08 [*nomen dubium*], *P. longiceps* Jaekel 1914, *P. plieningeri* (Huene 1905) [*nomen dubium*], ?*P. poligniensis* (Pidancet and Chopard 1862) [*nomen dubium*], *P. reinigeri* Huene 1905 [*nomen dubium*], *P. robustus* (Huene 1907–08) [*nomen dubium*].

Diagnosis of *P. engelhardti*: Autapomorphies including first sacral rib originating from complete length of centrum, second sacral rib originating from posterior 75 percent of centrum; distal part of femur straight in anterodorsal views (Galton 2000).

Comments: In 2000, prosauropod specialist Peter M. Galton published another installment in his series of papers on the large genus *Plateosaurus* (see *D:TE* and *S1*). Galton, in this study, stated that *Plateosaurus* is distinguished as "the fifth oldest valid genus of dinosaur ... the oldest one from outside of England ... one of the best known Triassic dinosaurs, the best known from Europe, the best known genus of prosauropod, and the basis for the family Prosauropoda ... the oldest genus of prosauropod after *Thecodontosaurus* Riley and Stutchbury, 1836, with *P. engelhardti* Meyer 1855 as the oldest species (cf. *T. antiquus* Morris, 1843)." Indeed, so many *Plateosaurus* specimens were found in southern Germany that they were nicknamed "der Schwäbische Lindwurm," meaning "the Schwabian Dragon" (Quenstedt 1865).

The purpose of Galton's latest paper on this dinosaur was to offer a well-illustrated description of the syntype specimens of the type species *Plateosaurus engelhardti*—which had not been done since the descriptions published by Meyer (1837) and Huene (1907–08)—and also to discuss other European prosauropod specimens having (as does an undescribed specimen of *P. engelhardti* from Ellingen, near Nürnberg, in Bavaria, Germany; Wellnhofer 1993) a distally straight femur (see Galton 1998, 2000, and *S1* for details). (In a subsequent paper in this series, Galton, in press, b, will discuss the validity and synonymies of the other species referred to *Plateosaurus*.)

The syntypes of *P. engelhardti* (referred to briefly

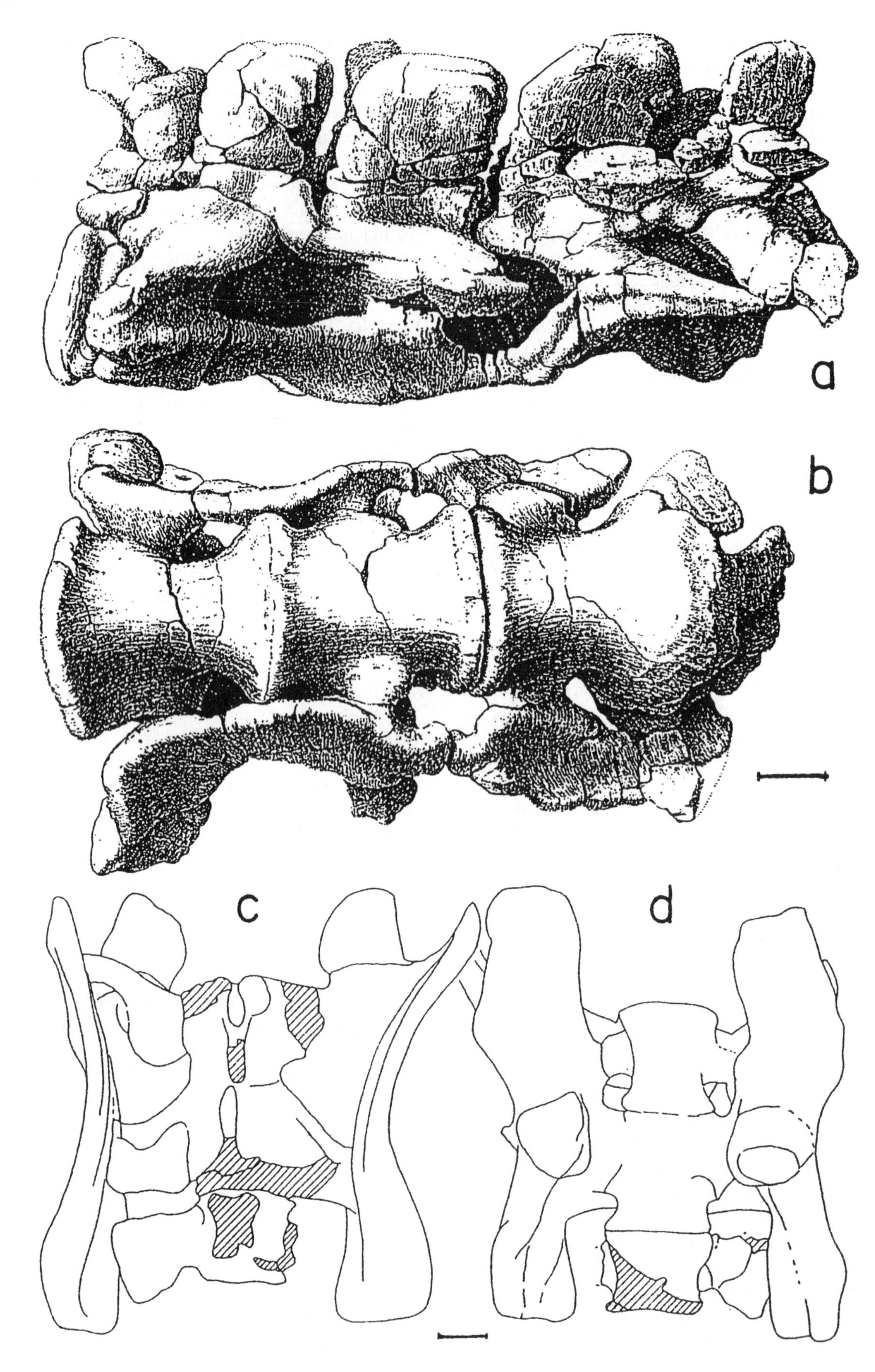

***Plateosaurus einigeri*, SMNS 80664, holotype sacrum and incomplete first caudal vertebra, in a. left lateral and b. ventral views; SMNS 53537, holotype sacrum with ilia in c. dorsal and d. ventra; views. Scale = approximately 5 centimeters. (After Huene 1907–08 and Huene 1905, respectively, reproduced in Galton 2000.)**

by Meyer 1839, described by Meyer 1855 and Huene), described in detail by Galton in his 2000 paper, comprise just a few fragmentary specimens all recovered from the *Plateosaurus*-Konglomerat ("*Zanclodon*-Breccie") of the Feurletten ("*Zanclodon*"-Letten," equivalent to Knollenmergel, upper Middle Keuper, upper Norian [see Gwinner 1980] of Heroldsberg, near Nürnberg, Middle Franconia, Bavaria [see Urlichs 1966, 1968]), in southern Germany. These specimens include UE 561 and 562, dorsal vertebrae, the former being more complete, another UE specimen from the same site referred by Huene to a large, generically indeterminate plateosaurid, and UE 557, a large encrusted posterior dorsal vertebra not mentioned by Huene; a rib fragment listed by Huene and three pieces of rib (one catalogued as UE 554, figured by Probst and Windolf 1993; UE 552, a somewhat vertically compressed sacrum; UE 550, 558, and 559, caudal vertebrae, UE 558 included in the holotype by Meyer (1855), identified by Huene as an anterior caudal vertebra belonging to a generically indeterminate plateosaurid, reidentified by Galton (2000) as a slightly incomplete neural arch of an anterior caudal, UE 559 not described by either Meyer or Huene; UE 556, the distal part of an apparently straight left femur, described by Meyer (1855) as a single piece and by Huene as two (see *D:TE* for figure); and UE 556, the left tibia (see Probst and Windolf for color photograph; also *D:TE* for figure).

The original syntypes also included UE 549, a flat bone identified by Meyer (1855) as a posterior cranial element but described by him as a sternum. This element, later removed from the holotype by Galton (1984*a*) because it does not resemble any bone in *Plateosaurus* or in any other known prosauropod, was reidentified by him as a dermal scute. Huene had also listed as part of the holotype another bone that he identified as a fragment of a pubis. Meyer (1855) later listed this element as a middle hand or middle foot bone (elements not listed by Huene). As noted by Galton (2000), this bone (not seen by Galton) is not the distal end of a prosauropod pubis or ischium, but is probably the proximal end of a metopodial. The concave proximal end of this element does not match the metapodials in any referred *Plateosaurus* skeletons from the Trössingen quarry in Germany, wherein the proximal surface is convex (as in other prosauropods). Consequently, Galton (2000) removed this bone from the syntypes of *P. engelhardti*, suggesting that it is probably the proximal end of the left metatarsal IV of

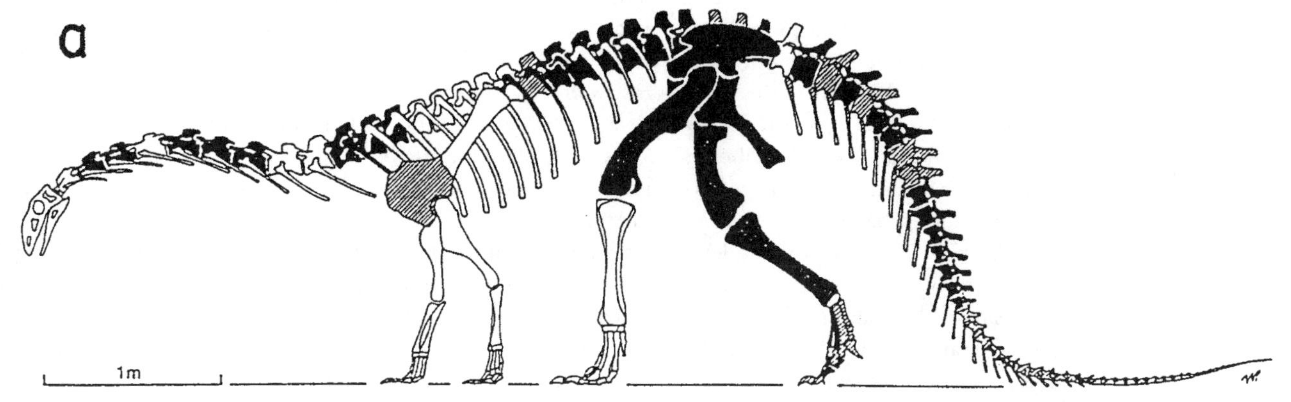

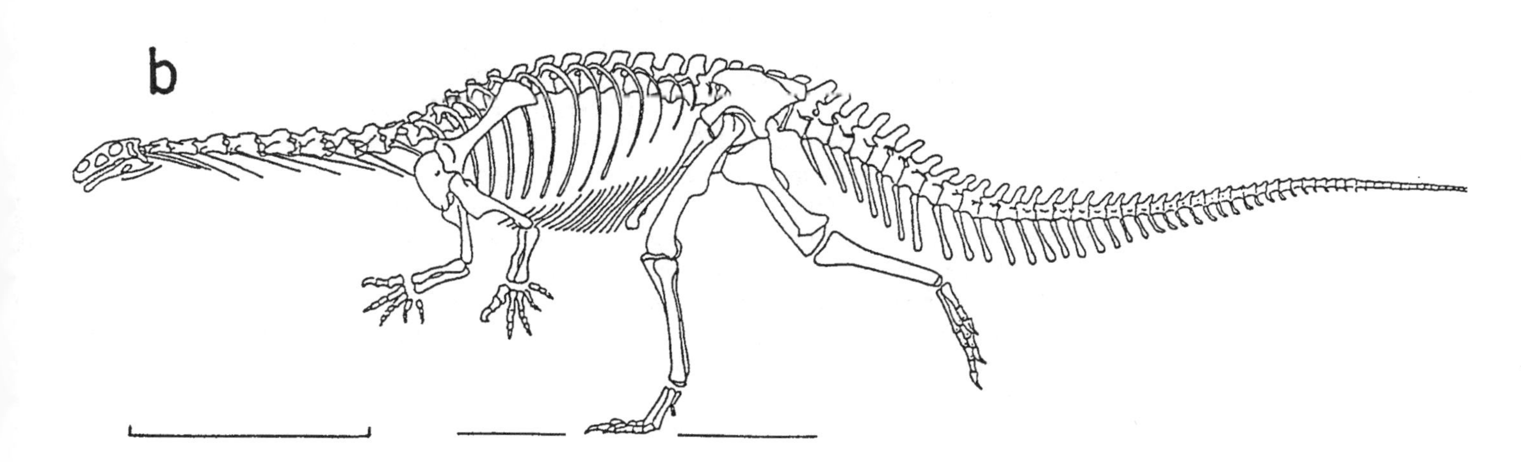

Reconstructed skeletons referred to *Plateosaurus engelhardti*: a. BSP 1962 XLVI from Ellingen (from Wellnhofer 1993), a fully quadrupedal animal, and b. SMNS 13200, holotype of *Plateosaurus fraasianus* (=*P. engelhardti*) Huene 1932 from Galton 1990, modified after Huene 1926, a facultatively bipedal animal. Scale = 1 m. (After Galton 2000.)

a theropod such as *Lilienstrnus liliensterni*, wherein the proximal end is concave (see Huene 1934).

Reviewing the other European prosauropod specimens with distally straight femora, Galton (2000) noted the following:

All other prosauropod bones from the Feuerletten of Franconia, in Bavaria, especially those found in a region east and southeast of Nürnberg, are generically indeterminate (see Gümbel 1865; Blanckenhorn 1898; Huene; Stromer 1909; Dehm 1935; Markthaler 1937; Urlichs 1966, 1968; Probst and Windolf).

Zanclodon bavaricus Fraas 1894 (*see in* Sandberger 1894) was briefly described but not figured by Sandberger from a well boring in the Upper Keuper of Altenstein near Marolsweisach near Würzburg, Lower Frankonia. The holotype (UW collection, not seen by Galton) comprises fragments of a seventh cervical and middle caudal vertebra (cervical rib not located by Huene) and the distal end of a tibia. Although Huene considered this material to be referrable to *P. engelhardti*, Galton (2000) regarded it as a *nomen nudum*.

As noted by Galton (2000), most femora attributed to *Plateosaurus*, including those from Trössingen and those analyzed by Weishampel and Chapman (1990; see *S1*), are gently S-shaped, with the distal portion gently curved in both anterior and posterior views. However, additional prosauropod specimens are known from Europe, several of these having been referred by Huene (1907–08) to *Gresslyosaurus robustus* (a species subsequently referred to by Huene 1932 as *Plateosaurus robustus*) and described as possessing a distally straight femur. These specimens — plus an additional specimen referred to *Gresslyosaurus* which, although the form of the femur is indeterminate, may differ from the typical Trössingen prosauropod form — were considered by Galton (2000) based on locality:

From Franconia, UE 556, the holotype of *P. engelhardti* from near Heroldberg, includes a femur, the distal part of which appeared to Galton (2000) as if it were originally straight in anteroposterior view; and BSP 1962 XLVI, the massive specimen from Ellingen described by Wellnhofer (1993; see *S1* for additional details), includes femora, both of which are distally straight.

From Württemberg, UT B, the holotype of *Gresslyosaurus robustus*, includes several vertebrae (mostly centra) from all regions, also parts of the manus, pelvic girdle, and hindlimb, recovered from the Rothen Graben (upper Middle Keuper, upper Norian; see Brenner 1973), near Bebenhausen, near Tübingen (precise locality uncertain). Huene (1907–08) reconstructed the femur that is represented by these pieces as relatively straight distally in posterior view, although as Galton (2000) pointed out, this outline was most likely based on a specimen from Wedmore Hill that Huene (1907–08) had referred to the type species, *Gresslyosaurus ingens*, and which Galton (1985*d*) later referred to the new genus and species *Camelotia borealis* (see below). Galton (1985*c*, 1990), who previously accepted *P. robustus* as a junior synonym of *P. engelhardti*, subsequently regarded it as a *nomen dubium* based upon the incomplete preservation of the sacrum and femur (Galton 2000).

Other specimens from Würtemberg were by described by Huene (1932) and referred by him to "*G. robustus*." SMNS 13200 a+c consists of a partial skeleton comprising dorsal vertebra 13 through caudal vertebra 7, plus a complete pelvic girdle and right hindlimb, from the lower bonebed at Trössingen. Huene (1932) figured the femur of this specimen as relatively straight in posterior view, although this bone is actually of the typical, gently sigmoid Trössingen-type (R. Wild, personal communication to Galton 2000), the sacrum also being of the typical Trössingen type (Huene 1932; Galton, in press, b). SMNS 13200d, an entire left hindlimb described but not figured by Huene (1932), possibly with humerus SMNS 132001, includes a femur preserved and restored so poorly that it appears straight in posterior view (R. Wild, personal communication to Galton 2000). UT III, comprising ischia and a complete right hindlimb, includes a femur that, as examined by Galton (2000), proved to be distally curved in both anterior and posterior views.

From Halberstadt, Thuringia, in central Germany, HMN Fund XVIII is a large specimen (femur length estimated at approximately 1000 millimeters), which Jaekel (1914) described as a nonplateosaurid and Huene (1932) referred to *Gresslyosaurus ingens*. Only the distal portion of the femur is preserved in this specimen. HMN Fund 45.2, a specimen referred by Huene (1932) to *Plateosaurus erlenbergiensis*, includes a right femur that was figured as relatively straight in posterior view. As noted by Galton (2000), however, this bone could not be found, nor are any femora that are straight in anteroposterior views known to be housed in the HMN collection.

From Poligny, France, POL 75 consists of numerous bones originally referred to *Plateosaurus polignienis* that Huene (1907–08) designated *Gresslyosaurus* cf. *plieningeri*. Considered by Huene (1907–1908) to represent a single individual, these elements include the neural arch of an anterior dorsal vertebra (POL 1), an incomplete right scapula (POL), a compressed right femur (POL 75), two fibulae, the right lacking its distal end (POL 37), and three pedal phalanges (see Huene 1907–08 and Galton 1998*a* for figures). As Galton (2000) observed, the distal end of the femur is straight in anterior and posterior views. Also, while being only slightly longer than the lectotype left femur of *P. poligniensis* (821 as opposed to 772 millimeters), POL 75 is much more massive, some of which is possibly the result of crushing. However, Galton (2000) further noted that the lectotype right femur of *P. poligniensis* is also crushed and, consequently, appears to be more massive. Nevertheless, in posterior view, the bone in this specimen is still S-shaped, its distal portion not straight in anteroposterior views, thus indicating the presence of two distinct morphs. According to Galton (2000), the slender though larger *Thecodontosaurus*-like form from Vilette near Arbois (see Gaudry 1890; for figures, see also Galton 1998*a*; Huene 1907–08) could also belong to the above taxon with the straight femur.

From England, BMNH R2870, the holotype of *Camelotia borealis* (Galton 1985*d*) includes a complete left femur which is basically straight in anterior and posterior views (see *Camelotia* entries, *D:TE* and this volume, for details, and this volume for Wellnhofer's suggestion that *C. borealis* might be congeneric and conspecific with *P. engelhardti*).

Evaluating the above specimens, Galton (2000) cited Wellnhofer's suggestion that the species *P. engelhardti* be restricted to the more massive, stratigraphically younger *Plateosaurus* material from the Frankonian Feueletten (including the Ellingen material) and from the Knollenmergel (upper Norian; see Brenner) of Stuttgart-Dagerloch, the latter specimens described by Huene (1907–08) as, respectively, the new species *Plateosaurus reinigeri* (see below) and *Gresslyosaurus plieningeri*; nor should the geologically older Trössingen plateosaurids, in which even the larger individuals are more gracile with more slender femora and comparatively longer metatarsals, nor the forms from France and Switzerland, be referred to the type species.

As Galton (2000) pointed out, however, Wellnhofer offered no evidence except stratigraphy supporting his grouping of the type species with the Ellingen specimens rather than the Trössingen material described by Huene (1926, 1932). Galton (2000) noted that Huene (1907–08) had already described the main part of the holotype tibia of *P. engelhardti* as having a rather slim structure. Morphometric analysis by Weishampel and Chapman found the holotype femur, although incomplete, to group with the smaller femoral morph from Trössingen. However, Chapman (in Wellnhofer 1993) found a taxonomic separation between the Ellingen femora referred to *P. engelhardti* and the Trössingen specimens. Based on the structure of the tibia and femur, Galton (2000) noted that, instead of grouping the holotype of *P. engelhardti* with the more massive Ellingen material, it might seem more reasonable to group it with the slender Trössingen material, the Ellingen remains, therefore, perhaps being more comparable to the larger, generically indeterminate plateosaurid represented four vertebrae collected from the type locality at Heroldsberg (see above).

However, as Galton (2000) showed, the second sacral rib in *P. engelhardti* rib displays the autapomorphic condition as it originates from the posterior 75 percent of the centrum (see also Galton and Upchurch 2000; and also "Introduction," section on prosauropods). In the Trössingen material (as well as in the more primitive prosauropods *Sellosaurus*, *Ammosaurus*, *Anchisaurus*, *Massospondylus*, *Lufengosaurus*, and *Yunnanosaurus*; see Galton 1999), the plesiomorphic condition was observed, whereby this rib originates from the anterior 75 percent. Additionally, the distal end of the femur in *P. engelhardti* seems to be rather straight in both anterior and posterior views, while in the Trössingen specimens the femur is curved. Consequently, Galton (2000) accepted Wellnhofer's separation of *P. engelhardti* from the Trössingen material.

It has already been mentioned that Wellnhofer grouped the stratigraphically younger plateosaurid species of Stuttgart-Degerloch with the Ellingen material as *P. engelhardti*. The species *Plateosaurus reinigeri* was founded on a reasonably complete postcranial skeleton (SMNM 53537) missing some cervical vertebrae, the tail, and portions of the scapula, forearm, manus, and pes, originally described by Plieninger (1852) as "erste Stuttgarter Skelett," later referred by Huene (1905*a*) to the new species of *Plateosaurus*. *P. plieningeri*, established on a partial skeleton comprising 17 vertebrae including a complete sacrum, some ribs, both ilia, a pubis, and fragments of the hindlimbs, was first described by Plieninger as the "zeweite Stuttgarter Skelett," and later as *Gresslyosaurus plieningeri* by Huene (1907–1908).

As noted by Galton (2000), Wellnhofer compared the left femur of *P. reinigeri* with the Ellingen femur, both of which are straight distally. However, the head of that belonging to *P. reinigeri* was figured as directed somewhat anteromedially; if directed medially, Galton (2000) pointed out, then the distal end is oriented slightly posterolaterally as in the Trössingen specimens. As Huene (1907–08) had reported earlier, both femora had been broken during excavation and restored to display an unnatural curve. Weishampel and Chapmen found the femur of *P. reinigeri* to be more similar to that of the Trössingen specimens. Galton (2000) observed that the hindlimb and pelvis of the type specimens of both these species do not seem to be more massive than the ones from Trössingen. Also, the obturator foramen of the pubis, open in the Ellingen material (see Wellnhofer), was most likely closed in *P. reinigeri* (see Plieninger) and *P. plieningeri* (see Plieninger; Huene 1907–08). Based on similarities between the Trössingen skeletons and *P. reinigeri* and *P. plieningeri*, Galton (1985*e*, 1990) previously regarded these species as junior synonyms of *P. englehardti*. However, as the sacra of *P. reinigeri* and *P. plieningeri* are of the Trössingen type, not *P. engelhardi*, Galton (in press, b) will again separate these taxa from the type species, although both of these species will be designated *nomina dubia* (P. M. Galton, personal communication, 2000).

Galton (2000) also readdressed the issue of the posture of *Plateosaurus*, noting that Huene (1907–08, 1926) had reconstructed this dinosaur in a bipedal pose, although Galton (1971, 1976, 1990) later suggested that the uniformity of hand structure in prosauropod dinosaurs (see Galton and Cluver 1976; Galton in press c) relates to "the development of an enormous and trenchant first ungual that was used while bipedal for offense or defense in a group that was only facultatively bipedal." According to Galton (2000), while dinosaurs such as *Plateosaurus* were walking on all fours with the manual digits in full extension, the weight of the anterior part of the body was taken by the second through fourth digits, primarily digits II and III. During quadrupedal locomotion, that enormous first claw was held entirely off the ground, its lateral surface only touching the ground when it was soft or irregular, and with the claw point never suffering damage (see Galton 1971). As suggested earlier by Galton (1971), the hindlimb to trunk ratio for *Ammosaurus*, *Anchisaurus*, and *Plateosaurus* is intermediate between those for undoubtedly bipedal and fully quadrupedal dinosaurs (see Galton 1970), this interpretation being in agreement with a fossil trackway from the Navajo Sandstone of Arizona named *Navahopus falcipollex*, attributed to a plateosaurid prosauropod walking on all fours (see Baird 1980). Galton (2000) noted that the basic adaptations of digit I are found in the habitually bipedal, basal prosauropod *Thecodontosaurus* (Gauthier 1986; Galton 1990; Benton, Juuls, Storrs and Galton 2000), so they must have been present before the Prosauropoda became facultatively bipedal or entirely quadrupedal. The retention of this basic plan for the manus for a span of 25 million years "shows that it met the basic functions of weight support, while walking quadrupedally, and defense while bipedal, with or without the propping action of the tail" (Galton 2000; see also in press c).

As Wellnhofer had observed, the ratio of hindlimb to trunk in the Ellingen prosauropod is approximately 1.06. Also, the centra of the proximal caudal vertebrae possess ventrally converging articular surfaces, forming a 12- to 20-degree angle, imparting a circular curvature to the base of the tail. This indicated that the natural orientation of the tail was curving downwards rather than being held rather straight and horizontal (as restored by Weishampel and Westphal 1986).

According to Wellnhofer (also quoted by Galton 2000), this condition of the tail "alone suggests a quadrupedal stance and gait, even if we do not know the size of the forelimbs. However, this is not in contrast to the assumption that *Plateosaurus* could assume a tripodal feeding posture in order to browse at higher levels, assisted by the relatively long neck. On the other hand, the morphology of the caudal vertebrae does not support an even facultatively bipedal running gait of *Pateosaurus*. It was an obligatory quadruped."

However, Wellnhofer's assessment of the posture of *Plateosaurus* seems to be only partially correct. As Galton (2000) pointed out, reconstructions of *Plateosaurus* running bipedally with the tail stiffly lifted well clear of the ground (*e.g.*, Weishampel and Westphal; Galton 1990) were based on the almost complete holotype skeleton (SMNS 13200) of *P. fraasianus* from Trössingen (see Huene 1926). The pose was adapted with only slight modifications from Huene's (1926) first reconstruction, wherein the tail was drawn straight. Likewise, the tail was positioned straight in the skeleton as originally mounted, as well as in mounted casts of this original specimen.

Galton's (2000) personal observations of this skeleton confirmed, however, that the proximal caudal vertebrae do not have the marked angulation of the centra as reported by Wellnhofer in the Ellingen specimen, nor is this condition true for other mounted skeletons based upon associated articulated specimens from Trössingen. Galton (2000) further noted that the Trössingen quarry diagram depicts specimen UT 1 with the tail preserved in the straight and extended orientation. Therefore, Galton (2000) concluded (assuming the angling of the proximal caudal centra not to be an artifact of preservation) that the more massive Ellingen *Plateosaurus engelhardti* was "an obligatory quadruped," as suggested by Wellnhofer, while those individuals from Trössingen were facultatively bipedal.

Wellnhofer had cited a number of features in the Ellingen specimen that separate it from the Trössingen specimens. These include the following: A more robust and massively built form, with even larger individuals more lightly built; more slender femora and comparatively longer metatarsals; an open obturator foramen of the pubis; the ends of the third through seventh caudal vertebrae markedly angled. These distinctions led to the conclusion by Galton (2000) that the Ellingen and Trössingen *Plateosaurus* specimens represent separate species, the former the more derived in these characters, the more lightly built taxon from Trössingen, as well as those from the Knollenmergel of Halberstadt and from equivalent beds at Poligny, France, and Fric, Switzerland, yet to be discussed (Galton, in press b).

Considering the possibility that the Ellingen specimen could be referrable to the type species of *Plateosaurus*, Galton (2000) pointed out that Wellnhofer did not sufficiently compare the former to the syntypes of *P. engelhardti*. As Galton (2000) noted, morphometric differences in the femora of the Ellingen and Trössingen forms could indicate their taxonomic separation (Chapman in Wellnhofer 1993). However, the syntype femur of *P. engelhardti* plots morphometrically with Trössingen material (see Weishampel and Chapman), thus indicating a separation of the Ellingen material from the type species. Galton (2000) suggested that the slender femur and tibia in *P. engelhardti* and their massive counterparts in the Ellingen specimen might indicate the presence of two distinct species. On the other hand, this difference could also represent a sexual dimorphism, with the more massive form probably representing the female (see Galton 1999 for discussion, also in press b). A final judgement on this issue, Galton (2000) noted, must await a future detailed description of the Ellingen material, including the smaller specimens.

Galton (personal communication, 2000) will accept *Plateosaurus longiceps* (=*P. fraasianus* Huene 1932) as a valid species and plans to apply to the International Commission of Zoological Nomenclature (ICZN) to suppress the name *Plateosaurus erlenbergiensisis*, all additional species that have been referred to the genus *Plateosaurus* other than *P. engelhardti* being regarded by Galton as either *nomina dubia* or *nomina nuda*.

Key references: Baird (1980); Benton, Juuls, Storrs and Galton 2000); Blanckenhorn (1898); Brenner (1973); Dehm (1935); Fraas (1894) *see in* Sandberger (1894); Galton (1970, 1971, 1976, 1984*a*, 1985*a*, 1985*c*, 1985*e*, 1998, 1990, 1998*a*, 1999, 2000, in press b, c); Galton and Cluver (1976); Galton and Upchurch (2000); Gaudry (1890); Gauthier (1986); Gümbel (1865); Gwinner (1980); Huene (1905*a*, 1907–08, 1926, 1932, 1934); Jaekel (1914); Markthaler (1937); Meyer (1837, 1839, 1855); Morris (1843); Pidancet and Chopard (1862); Plieninger (1852); Probst and Windolf (1993); Quenstedt (1865); Riley and Stutchbury (1836); Stromer (1909); Urlichs (1966, 1968); Weishampel and Chapman (1990); Weishampel and Westphal (1998); Wellnhofer (1993).

†PRENOCEPHALE

New species: *P. edmontonensis* (Brown and Schlaikjer 1943).

New species: *P. brevis* (Lambe 1918).

Occurrence of new species: Edmonton Formation, Oldman Formation, Dinosaur Park Formation, Alberta, Canada, Kirtland Formation, New Mexico, United States.

Age: Late Cretaceous (Campanian).

Known material: Complete and incomplete frontoparietal domes, numerous individuals.

Holotype of *P. edmontonensis*: CMN [formerly GSC and NMC] 8830, complete frontoparietal dome.

Holotype of *P. brevis*: CMN 1423, complete frontoparietal.

Diagnosis of genus (as for type species): Nearly fully-domed pachycephalosaur, differing from *Stegoceras validum* in these features: frontonasl boss well developed; absence of supratemporal fenestrae; absence of parietosquamosal shelf; retention of small, discrete linear row of nodes (tubercles) on posterior and lateral edges of squamosals; lateral corner node on squamosal (below linear row); differing from *Stygimoloch* in that frontoparietal dome is not narrow, but more rounded; squamosals reduced, not forming thickened posterior squamosal shelf, lacking distinctive hypertrophied nodes (low-angle horns) associated with squamosals; distinguished from *Pachycephalosaurus* in lacking well-developed nodes forming clusters along posterior portion of squamosals and by smaller size (Sullivan 2000*b*).

Diagnosis of *P. edmontonensis*: Differs from *P. prenes* and *P. brevis* in these features: three prominent nodes (tubercles) in row on occipital part of squamosal, fourth (medial-most) node straddling posterior part of parietal and left and right squamosals; nodes larger relative to those in *P. prenes*; angle of linear row to horizontal high (about 32 degrees); incipient corner node; peripheral frontoparietal elements fully incorporated into dome forming contiguous surface with frontoparietal (Sullivan 2000*b*).

Diagnosis of *P. brevis*: Differs from *P. prenes* and *P. edmontonensis* in having medial-most (fourth) nodes located entirely on parietal; anterior portion of frontoparietal sloped; frontals forming three lobes (two lateral, one medial) (Sullivan 2000*b*).

Comments: *Prenocephale*, a rare genus of dome-headed ornithischian dinosaur previously known only from a single species (*Prenocephale prenes*) and a single specimen — an almost perfect skull (ZPAL MgD-I/104) from the Nemegt Formation, Mongolian People's Republic, Asia — is now, according to Robert M. Sullivan (2000*b*), also represented in North America (Maryańska and Osmólska 1974; see *D:TE*).

Sullivan, in examining various pachycephalosaurid specimens that had previously been referred to the genus *Stegoceras*, determined that many of these actually pertain to *Prenocephale*. Indeed, Sullivan referred some of these specimens to two new species of

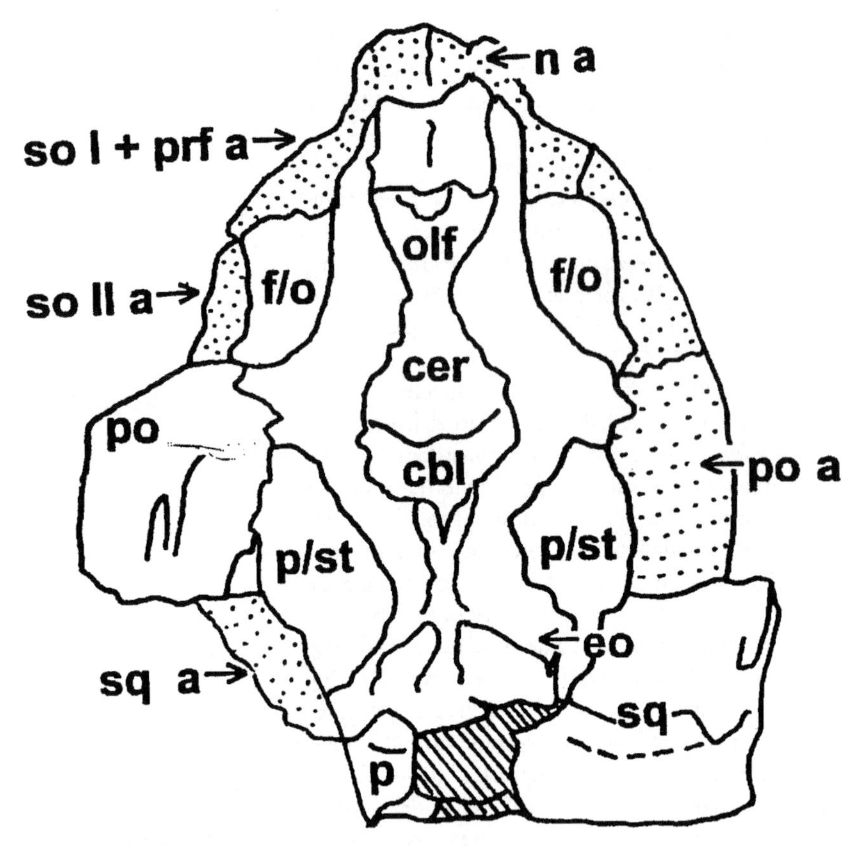

***Prenocephale edmontonensis*, RTMP 87.113.3, referred incomplete skull (anterior view). Scale = 5 cm. (After Sullivan 2000.)**

Prenocephale, which he named *P. edmontonensis* and *P. brevis*. At the same time, that author offered a new and revised diagnosis of the genus and the type species, *P. prenes*. Sullivan found the presumably primitive trait of the retention on the squamosals — four to five in a linear row — of discrete nodes to be significant in the genus. It was Sullivan's opinion that the presence of these nodes in specimens referred to *P. edmontonensis* and *P. brevis*, as well as in one specimen (NMMNH P-27403) from New Mexico, precludes referral of that material to *Stegoceras* (see below).

Sullivan founded the new species *Prenocephale edmontonensis* upon a frontoparietal dome (CMN 8830) collected from the Edmonton Formation, opposite the mouth of Big Valley Creek below Scollard Ferry, Red Deer River, Alberta. This specimen was originally described by Brown and Schlaikjer (1943) as *Troodon edmontonensis*, then renamed *Stegoceras edmontonensis* by Sternberg (1945), and later again renamed *Stegoceras edmontonense* by Sues and Galton (1987). To this species Sullivan referred various incomplete frontoparietal domes, described completely for the first time by that author, including CMN 8831 and 8832 (paratypes of *T. edmontonensis*), from the Edmonton Formation of Alberta and originally described by Brown and Schlaikjer, LACM 64000, from the Hell Creek Formation of Montana, referred by Wall and Galton (1979) to "*Stegoceras* species undetermined," and RTMP 87.113.3, also from the Hell Creek Formation of Montana, briefly described by Giffin (1989). To cf. *Prenocephale edmontonensis* Sullivan referred an incomplete, weathered frontoparietal (LACM 15345) from the Hell Creek Formation of Montana, previously referred by Sues and Galton and later by Giffin to "*Stegoceras*" *edmontonense*.

In one of the above specimens (RTMP 87.113.3), Sullivan observed a depression which the author interpreted to be a puncture wound, this suggesting a predation event. In comparing this opening with the tip of an albertosaur/daspletosaur tooth, Sullivan noted that the tooth fits well within that depression, with sufficient room around the tip of the tooth for soft tissue. A number of tiny depressions and grooves within the base of the depression were interpreted by the author as corresponding to repeated occlusal events inflicted during predation. "The size of the depression, its geometry and basal characters," Sullivan stated, "strongly suggest that this is in fact a bite mark."

Sullivan established the species *Prenocephale brevis* on a complete frontoparietal (CMN 1423) from the Oldman Formation of Alberta, originally referred by Lambe (1918) to *Stegoceras brevis*, and later by Sues and Galton to *Stegoceras breve*. Frontoparietal domes from the same locality previously referred to this species include the plesiotypes CMN 121 and 194, CMN 193 (incomplete), and CMN 8819 (almost complete). Additional frontoparietal domes newly referred by Sullivan to *P. brevis* include RTMP 85.36.265, from the Oldman Formation of Alberta, RTMP 91.36.265, 99.55.122, and 2000.12.01, the latter three all from the Dinosaur Park Formation of Alberta.

Sullivan rejected the possibility that the specimens referred to this species represent juvenile individuals or are sexual dimorphs (see Brown and Schlaikjer), pointing out that all *P. brevis* specimens display the characters cited in the diagnosis and significantly differ from specimens of *Stegoceras validum* (Sullivan 2000, in preparation). Reaching the same conclusion, Sternberg had noted that in *P. brevis* the parietals sharply turn down, while in both adult and immature specimens of *S. validum* parietals extend posteriorly, forming a horizontal shelf over the occiput. As the squamosals are not known in *P. brevis* and the nature of this "horizontal crest" cannot, therefore, be substantiated, Sullivan speculated that this species lacked a distinct shelf (Sullivan, in preparation).

Sullivan addressed the possibility that *P. brevis* belongs in *Stegoceras* rather than *Prenocephale*, but noted the following: *Stegoceras* is largely characterized by a well-developed parietosquamosal shelf, opened supratemporal fenestrae, and other features (Sullivan

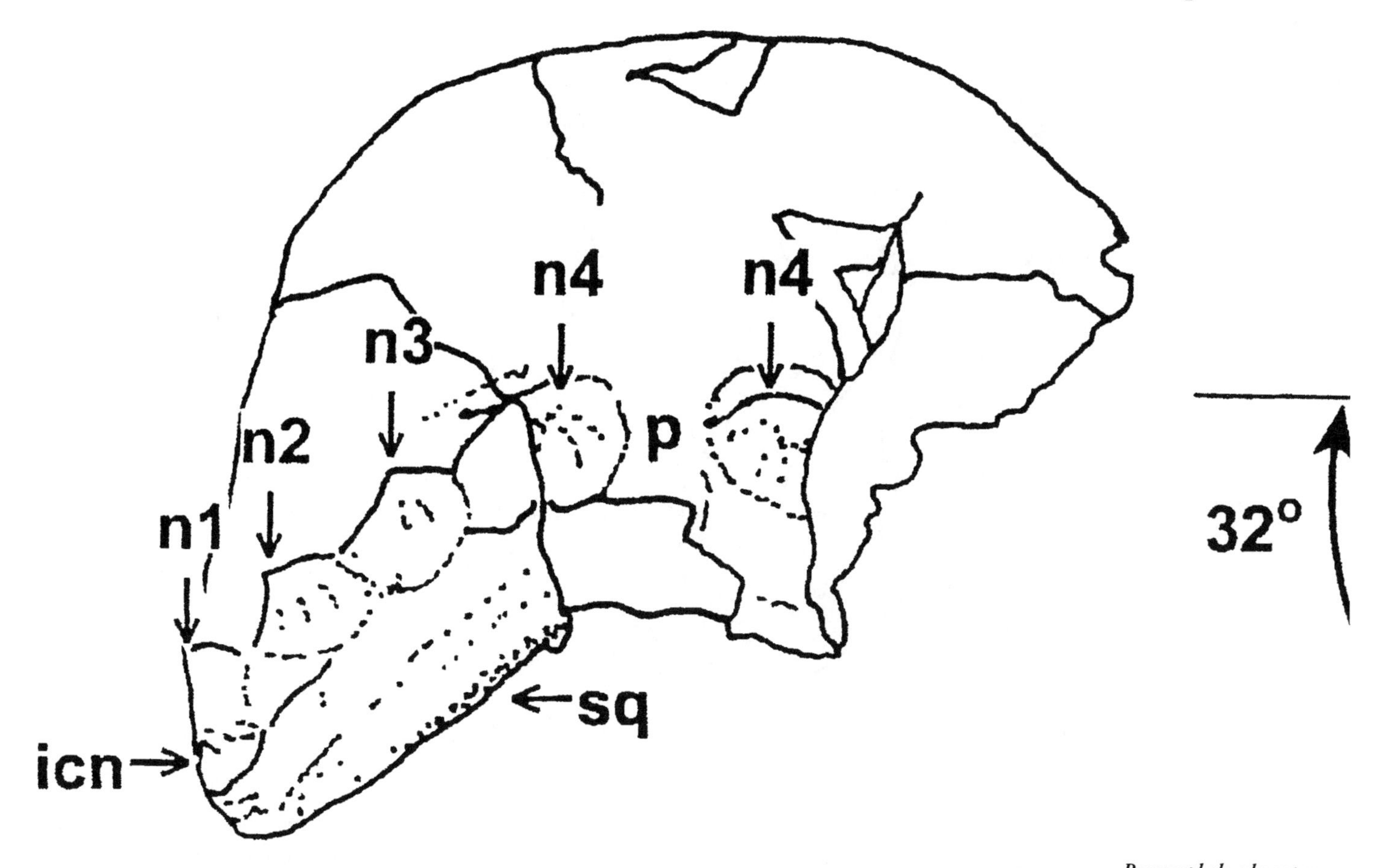

***Prenocephale edmontonensis*, RTMP 87.113.3, referred incomplete skull (posterior view), drawing showing location of primary nodes numbers 1 to 4. Scale = 5 cm. (After Sullivan 2000.)**

2000, in preparation). A more parsimonious assignment to *Prenocephale* is based on the lack of a parietosquamosal shelf, closed supratemporal fenestra, and the presence of medial nodes located on the down-turned area of the parietal. Sullivan pointed out that the presence of these nodes as well as other features, which are consistent among all known specimens of this species, plainly indicates a relationship with *P. edmontonensis*. Furthermore, the small size of this species may suggest a sister taxon relationship with the larger *P. edmontonensis*.

In addition to the above two newly proposed species, Sullivan mentioned four pachycephalosaur specimens from the Kirtland Formation, San Juan County, New Mexico. Two of these specimens, SMP VP-1084 and the slightly smaller NMMNH P-27403, were collected from the De-na-zin Member of the Kirtland Formation. Both of these specimens were regarded by Sullivan as "*Prenocephale* species indeterminate."

Although Sullivan did not include a cladistic analysis in his study of *Prenocephale* and its three species, he did point out the following: *Prenocephale* is, in part, characterized by peripheral cranial elements (*e.g.*, supraorbitals I plus prefrontals, supraorbitals II, postorbitals, and squamosals) that are completely or almost completely incorporated into the frontoparietal dome. The genus is distinct from the more primitive *Stegoceras validum*, which is characterized, in part, by a well-developed parietosquamosal shelf extending posteriorly and with open supratemporal fenestrae (Sullivan 2000, in preparation), this shelf being ornamented with clusters of incipient nodes similar in arrangement to (but not as pronounced as) that of the hypertrophied nodes seen in *Stygimoloch* and *Pachycephalosaurus*. The presence of discrete nodes on the squamosals of *P. edmontonensis*, *P. bervis*, and the indeterminate species NMMNH P-27403, plus additional diagnostic features, therefore suggested to Sullivan that *Prenocephale* belongs to a clade distinct from the one formed by *Stegoceras*, *Stygimoloch*, and *Pachycephalosaurus*.

Sullivan suggested that the presence of this genus in both the Campanian and Maastrichtian dinosaur faunas of North America supports the hypothesis of dinosaur migration between Asia and North America. Although it has generally been assumed that the migration of dinosaurs during this time was mostly from Asia to North America, Sullivan's interpretation was based on the known occurrence of dinosaurian families from pre–middle Campanian rocks.

Key references: Brown and Schlaikjer (1943); Giffin (1989); Lambe (1918); Maryańska and Osmólska (1974); Sternberg (1945); Sues and Galton (1987); Sullivan (2000*b*); Wall and Galton (1979).

Prenocephale sp. indet., SMP VP-1084, frontoparietal dome in ventral view. Scale = 5 cm. (After Sullivan 2000.)

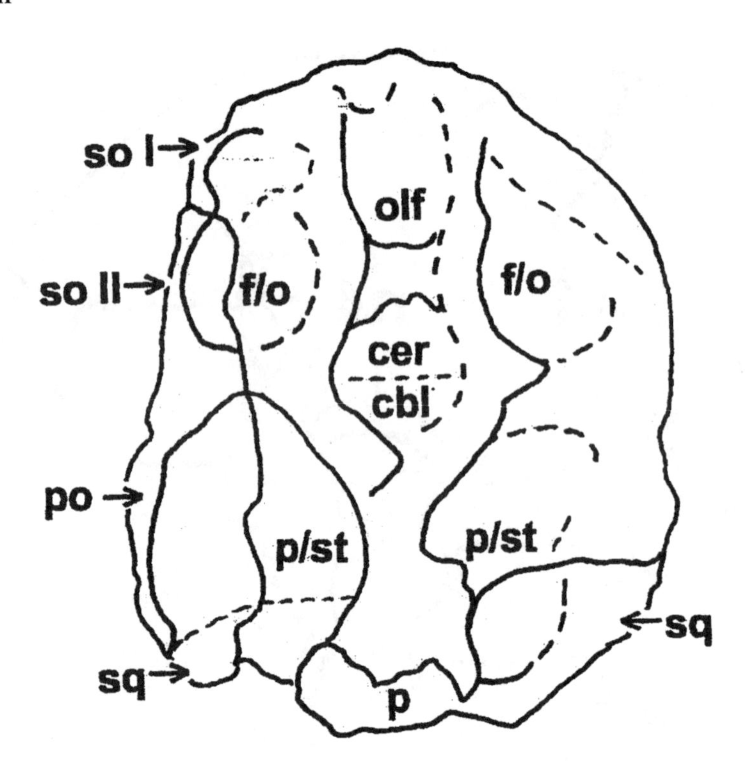

†PRICONODON

Ornithischia: Genasauria: Thyreophora: Thyreophoroidea: Eurypoda: Ankylosauria: ?Nodosauridae.

Name derivation: Greek *prion* = "saw" + Greek *odous* = "tooth."

Diagnosis: Teeth differing from those of all other known nodosaurids in their very large size; cingulum more prominent that in *Sauropelta*, less than in *Panoplosaurus*, *Edmontonia*, or *Silvisaurus*; denticles not extending much onto crown face (unlike *Edmontonia*, *Panoplosaurus*, *Silvisaurus*, or *Stegopelta*; cingulum lacking sharp shelf of *Panaplosaurus*, *Edmontonia*, *Silvisaurus*, or *Stegopelta* (Carpenter and Kirkland 1998).

Comments: *Priconodon crassus* (see *D:TE*)—a type species known mostly from isolated fossil teeth collected in Maryland—was redescribed by Carpenter and Kirkland (1998) in their review of Lower and Middle Cretaceous ankylosaurs from North America.

Carpenter and Kirkland based their revised diagnosis of this taxon upon the holotype, USNM 2135, a worn tooth from the Arundel Formation (Aptian–Albian boundary; Doyle and Hickey 1976) of Muirkirk, Maryland; various isolated teeth (USNM 8437–8441, 337984, 437985, 442458, 442551, 451960, 466057, and 481142), almost all being abraded and lacking enamel, having been discovered by Peter Kranz, Tom Lipka, and Bob Wirst in generally the same area as the holotype; and a tentatively referred left tibia (USNM 9154) from Coffee Mine, Muirkirk. The latter specimen—measuring 31.3 centimeters in length, 13.1 centimeters in width distally, and having a minimum circumference of 20.6 centimeters—shares the stoutness observed in other nodosaurid taxa.

Following Lull (1911), Carpenter and Kirkland tentatively accepted the assignment of *Priconodon* to the Nodosauridae. In attempting to explain the lack of armor found with all specimens of *P. crassus*, Carpenter and Kirkland noted that they either "are mistaken and *Priconodon* is not an ankylosaur, or," noting that fossils of any kind are rare in the Arundel Formation, "that it is simply fortuitous that armor has not yet been found."

Key references: Carpenter and Kirkland (1998); Lull (1911).

†PROBACTROSAURUS

Ornithischia: Genasauria: Cerapoda: Ornithopoda: Euornithopoda: Iguanodontia: Euiguanodontia: Dryomorpha: Ankylopollexia: ?Hadrosauroidea.

Type species: *P. gobiensis* Rozhdeskvensky 1966.

Other species: ?*P. alashanicus* Rozhdestvensky 1966, *P. mazongshanensis* Lu 1997.

Age: Middle Cretaceous (Albian–Cenomanian).

Comments: Relatively little work has been done on *Probactrosaurus* since Anatoly Konstantinovich Rozdestvensky named and first described this important Asian ornithopod in 1966. Originally, Rozdestvensky referred two species to this genus, *P. gobiensis* and *P. alashanicus*; more than 30 years later, Lu (1997) erected the new species *P. mazongshanensis* (see *Probactrosaurus* entries, *D:TE* and *S1*).

Photograph by the author, courtesy Natural History Museum of Los Angeles County.

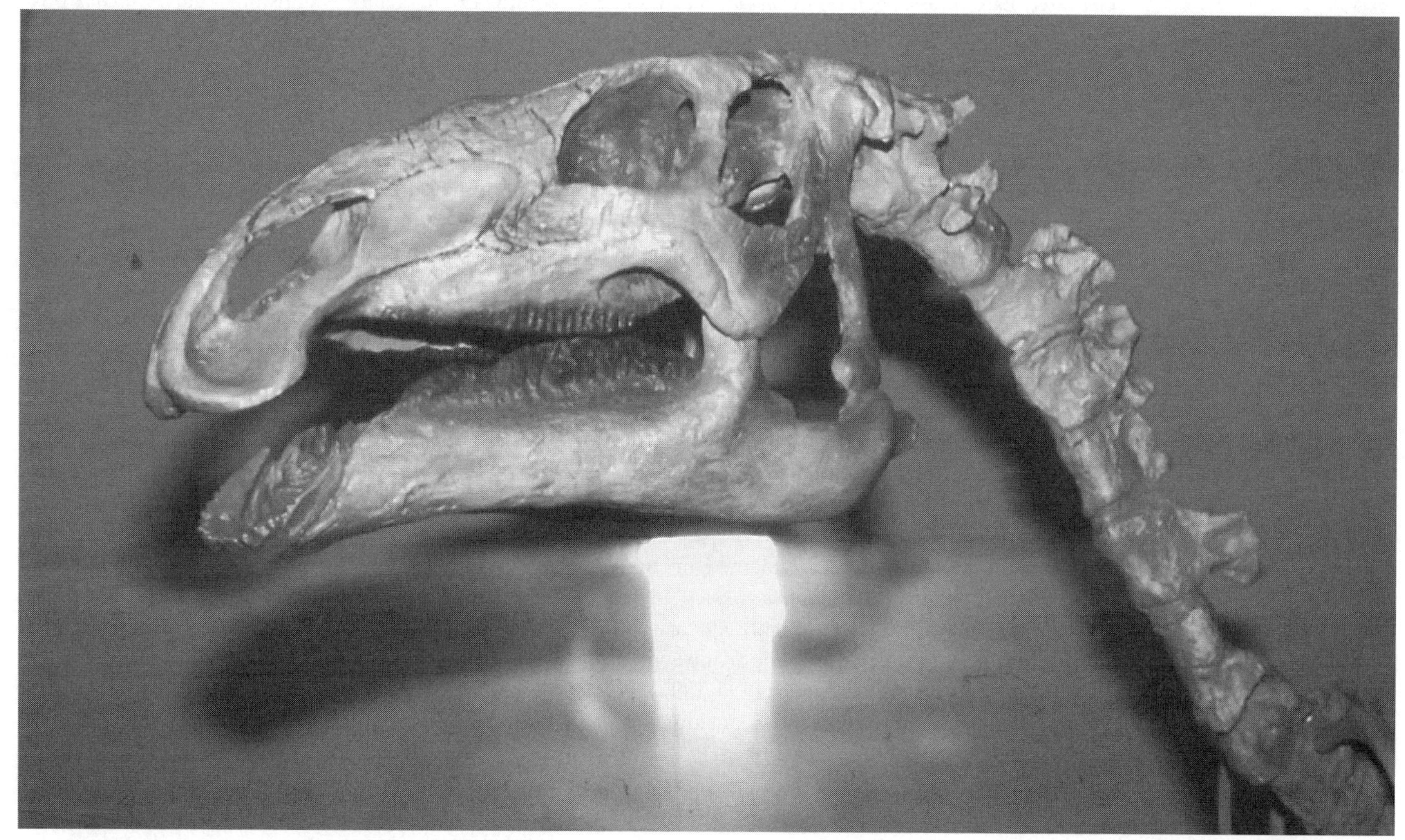

Close view of *Probactrosaurus gobiensis* skeleton (cast), original specimen collected from the Alashan Desert (Early Cretaceous/ Aptian-Albian) of Inner Mongolia, China, displayed in 1999 as part of "The Great Russian Dinosaurs Exhibition," here presented at the Natural History Museum of Los Angeles County.

Norman (1998*b*) briefly redescribed the original *Probactrosaurus* material described by Rozdestvensky and housed in the Paleontological Institute, Academy of Science, Moscow. While some of the type and referred material belonging to *P. alashanicus* could not be located, Norman found the distinction differentiating this taxon from the type species to be apparently "highly subjective."

As observed by Norman, *P. gobiensis* is a very conservative ornithopod of medium size (5 to 7 meters, or about 17 to 24 feet in length), exhibiting such features as miniature maxillary teeth, mesiodistally compressed dentary-tooth crowns, a hadrosaur-like dental battery, a small, distinctive ungual "thumb-spike," and slender metacarpals.

Norman argued the following: *Probactrosaurus* represents a primitive genus that is basal to the Hadrosauridae; hadrosaurids are derived from unspecialized medium-sized ornithopods of Middle Cretaceous (Albian–Cenomanian) Asia, such as *Probactrosaurus*; and homoplasy is common among mid–Cretaceous ornithopods. According to Norman, key anatomical innovations in the origin of the Hadrosauridae include miniaturization of the maxillary teeth, possession of a full dental battery, forearm lengthening, reduced ossification of the carpus, straightening of the femoral shaft, and enclosure of the anterior intercondylar canal [a growth feature; M. K. Brett-Surman, personal communication 2000].

More recently, Casanovas, Pereda-Suberbiola, Santafé and Weishampel (1999), in performing their own phylogenetic analyses of various ornithopods, found that *Probactrosaurus* shares at least two synapomorphies of the Hadrosauroidea — three or more dentary teeth per tooth position (*Probactrosaurus* showing a possible third replacement tooth; see Rozhdestvensky 1967), and teeth firmly cemented together to form a compact dental battery. Pending a full revision of *Probactrosaurus*, Casanovas *et al.* tentatively regarded this genus as a basal hadrosauroid.

Key references: Casanovas, Pereda-Suberbiola, Santafé and Weishampel (1999); Lu (1997); Norman (1998*b*); Rozdestvensky (1966, 1967).

†PROTOAVIS

Comments: Still a controversial taxon (see *D:TE, S1*), *Protoavis texensis* was founded upon two fragmentary skeletons from the Upper Dockum Formation (Upper Triassic) of Texas, and originally described in 1991 by Sankar Chatterjee as the earliest known (Late Triassic/early Norian) bird in the fossil record. Indeed, evaluating *Protoavis* as a bird would push back avian antiquity significantly, predating the celebrated Late Jurassic genus *Archaeopteryx* by some 75 million years and therefore dethroning the latter from the unique status it has enjoyed for well over a century. However, this interpretation of the *Protoavis* material was widely challenged (*e.g.*, Currie and Zhao 1993; Wellnhofer 1994; Chiappe 1995) and even vehemently criticized (Ostrom 1987, 1991, 1996) by various theropod and fossil bird specialists who reinterpreted these fossils as representing a nonavian theropod.

In 1997, Luis M. Chiappe, a specialist in fossil birds, stated the following objections to Chatterjee's identification of *Protoavis* as a bird: "The available material of *Protoavis* is fragmentary and its association in the two specimens alleged by Chatterjee is not clear. In addition, some of the elements regarded as avian (*e.g.*, furcula and carinate sternum) can be alternatively interpreted as something else."

Nevertheless, Chatterjee has remained undaunted in his original position that *Protoavis* does, in fact, represent the oldest known bird, subsequently providing additional arguments and supporting evidence to that interpretation (see Chatterjee 1995, 1997). Following Chatterjee's interpretation, various other workers (*e.g.*, Peters 1994; Kurochkin 1995) have stated that they had recognized the avian traits of this genus, placing it nearer to the ancestry of modern birds than *Archaeopteryx*.

In 1998, Chattejee countered Professor Ostrom's opinions, stating that they were "flawed and misleading," and noting that the specimens of *Protoavis*, despite their fragmentary nature, "offer a wealth of anatomical information which is not available from most of the Mesozoic birds." Chatterjee (1998*b*), while pointing out that Ostrom did not cite any evidence to support his objections and did not analyze the morphological data of *Protoavis*, cited the following as Ostrom's apparently major criticisms: 1. The fragmentary type material of *Protoavis* could represent a mixture of bones including those of pterosaurs, squamates, crocodilians, and rhynchocephalians; and, 2. except for "a keeled breast bone and a possible strut-like coracoid," the bones of *Protoavis* do not show many avian characters.

According to Chatterjee (1998*b*), the specimens of *Protoavis*, unlike the two-dimensionally preserved specimens of other Mesozoic birds (including *Archaeopteryx*), are essentially preserved in three dimensions and are undistorted, thereby providing critical anatomical information. Regarding Ostrom's question of the type material of *Protoavis* perhaps representing mixed assemblages, Chatterjee (1998*b*) countered that this possibility is supported neither by taphonomy nor comparative anatomy.

Performing a cladistic analysis, Chatterjee (1998*b*) found 59 characters establishing that *Protoavis* possesses features of Dinosauria, Saurischia, Theropoda, Tetanurae, Coelurosauria, Manuraptora, and Aves in a nested hierarchy, this presumably negating Ostrom's suggestion that the type material may be a chimera including various nondinosaurian bones. Chatterjee (1998) pointed out that this conclusion was also supported by the fact that different parts of the *Protoavis* skeleton exhibit a morphologic integrity and not a mixture of different taxa.

Chatterjee (1998*b*) found *Protoavis* to have acquired all of the evolutionary novelties that define primitive birds like *Archaeopteryx*, and to possess the following diagnostic avian characters: Absence of prefrontal bone; nasal process of frontal extremely narrow; teeth having unserrated crowns; presence of posterior maxillary sinus; less than 25 caudal vertebrae; preacetabular process of ilium more than 40 percent length of entire ilium; fusion of astragalus and calcaneum; pes anisodactyl. Chatterjee (1998*b*) defined Aves as encompassing *Archaeopteryx*, Ornithurae [=some Mesozoic and all modern birds] plus their descendants, with *Protoavis* regarded as having acquired all the avian attributes.

Protoavis was found by Chatterjee (1998*b*) to be more derived than *Archaeopteryx* and deeply imbedded within Ornithothoraces (an avian clade including all birds except *Archaeopteryx*) based on its possession of the following characters: Reduction of tooth count from condition in *Archaeopteryx*; absence of ectopterygoid; absence of maxillary fenestra; anterior cervical vertebrae having ventral hypapophysis; anterior cervical vertebrae heterocoelous; cervical neural spine atrophied; large neural canal in dorsal vertebrae; flexible articulation between scapula and coracoid; coracoid strut-like, subtriangular; acrocoracoid process on coracoid; caudal end of scapula tapering; sternum keeled; furcula U-shaped, forming acute interclavicular angle; distal condyles of humerus pronounced on cranial aspect; radiale large; fusion of ilium and ischium; absence of iliac brevis fossa; small ischiadic peduncle on ilium; ischium lacking distal symphysis; cranial cnemial crest on tibia.

From his analysis (scoring 66 cranial and postcranial characters among four taxa of Mesozoic birds, and utilizing dromaeosaurs such as *Velociraptor* as the outgroup), Chatterjee (1998*b*) generated a single most parsimonious tree, with *Archaeopteryx* found by him to be the sister taxon of Ornithothoraces (including *Protoavis*). This analysis suggested that *Protoavis* is the basal member of Ornithothoraces and the sister taxon to the avian group Pygostylia, the latter comprising two lineages, Enantiornithes and Ornithurae. In conclusion, Chatterjee (1998*b*) remained adament that

Protoavis texensis **skeleton reconstructed as a bird. (After Chatterjee 1999.)**

Protoavis is indeed a bird, and one closer to the ancestry of modern birds than *Archaeopteryx*.

Chatterjee (1999) subsequently published a major redescription of *Protoavis* in which he again affirmed the avian status of this genus (and also *Avimimus* and *Mononykus*). In this study, the author published the following emended diagnosis of the type species, *P. texensis*: Ornithothoracine bird (*sensu* Chiappe and Calvo 1994) having reduced dentition, enormous, frontally placed orbits; lateral (tympanic) Eustachian foramen relatively large, without parasphenoid cover; axis having anterior hypapophysis; scapula pneumatic, with short acromial process; proximal and distal expansions of coracoid at oblique plane, coracoidal sulcus for sternal articulation; sternal bearing V-shaped articular surface for coracoid; possible quill knobs on metacarpals II and III; anteroventral process of ilium highly pronounced, twisted; strongly developed renal fossa bounded by ilioischiadic pila; large nutrient foramen on distal femur; tibial crest overhanging on fibular shaft.

Some paleontologists have accepted Chatterjee's assessment regarding the phylogenetic standing of *Protoavis*. Regarding both *Avimimus* and *Protoavis* as birds rather than nonavian theropods, Dyke and Thorley (1998) performed a cladistic analysis to elucidate the interrelationships of these genera and Mesozoic birds. Parsimony analysis of a data-set consisting of all well-known Mesozoic "birds," including *Avimimus* and *Protoavis*, resulted in a total of 92 most parsimonious trees. Utilizing an alternative consensus method to summarize the phylogenetic relationships common to all of these trees indicated to them that including both *Avimimus* and *Protoavis* in no way affected these relationships among Mesozoic birds, and that the placement of these two genera within this phylogeny is highly variable. Three possible nonexclusive explanations were given for the latter:

Composite skeleton (including holotype TTU P 9200) of *Protoavis texensis*, regarded by some paleontologists as a birdlike theropod, restored here as a bird by artist Michael W. Nikell in 1993. (Scale = 5 cm.) (After Chatterjee 1999.)

1. Lack of data purporting to these "rogue" taxa; 2. homoplasy between the taxa used in this analysis; and 3. incorrect coding of character states for *Avimimus* and *Protoavis* (suggested by Chiappe 1995 and then Ostrom 1996).

Since the issuance of Chatterjee's 1999 paper on *Protoavis*, little has yet been published by dinosaur paleontologists, at least in any great detail, either refuting or accepting the supposed avian status of this genus. Sereno (1999*c*), for example, briefly noted, in a systematic study of the Dinosauria, that it has been suggested that the material assigned to *Protoavis* is a composite including material belonging to several nonavian theropod species.

More recently, Thomas R. Holtz, Jr. (personal communication 2000), a specialist in nonavian theropods, stated that most dinosaur experts who have seen the *Protoavis* material agree with his own observations, namely:

1. The fossil material is not as complete as restored by Chatterjee, the skull, for instance, comprising only a braincase and a few fragments of facial bones.

2. The postcranial material positively contains some nondinosaurian remains, probably rauisuchian bones.

3. The cervical vertebrae of *Protoavis* are very birdlike, but they are also quite similar to those of Triassic drepanosaurids [a group of Triassic nondinosaurian reptiles), particularly the genus *Megalancosaurus*.

4. The bones appearing to be nonavian theropod (*e.g.*, the skull and some limb bones) are not particularly different from those of coelophysoids.

Therefore, *Protoavis* is still retained here as a nonavian theropod.

Key references: Chatterjee (1991, 1995, 1997, 1998*b*, 1999); Chiappe (1995, 1997); Chiappe and Calvo (1994); Currie and Zhao (1993); Dyke and Thorley (1998); Kurochkin (1995); Ostrom (1987, 1991, 1996); Peters (1994); Sereno (1999*c*).

†PROTOCERATOPS

Ornithischia: Genusauria: Cerapoda: Marginocephalia: Ceratopsia: Neoceratopsia: Protoceratopsidae.

Comments: A close-knit assemblage of 15 *Protoceratops* hatchlings was reported in an abstract by Weishampel, Fastovsky, Watabe, Barsbold and Tsogtbataar (2000). The specimens were recently collected by members of the Hayashibara Museum of Natural Sciences-Mongolian Paleontological Center Joint Paleontological Expeditions from eolian sandstones of the Upper Cretaceous (Campanian) Tugrikin Shireh in the central Gobi region of Mongolia.

As briefly described by Weishampel *et al.*, most of these individuals are preserved complete and uncrushed, including skulls and postcranial skeletons, total length from tip of beak to end of tail averaging at about 16 centimeters. As preserved *in situ*, the orientation of these specimens corresponds to the visible part of the bowl-like depression of the nest, "crowded at one end, often placed one above another, and oriented in a common direction to the southeast." Weishampel *et al.* suggested that "This orientation faced the hatchlings away from the prevailing wind," and that these individuals "were buried alive in a rapid, catastrophic sandstorm."

As no eggshell was found within the nest, and because the hatchlings were found in abundance and excellently preserved, Weishampel *et al.* deduced that these dinosaurs were nest residents and that this taxon seems to have been altricial.

Key reference: Weishampel, Fastovsky, Watabe, Barsbold and Tsogtbataar (2000).

PROTOHADROS Head 1998

Ornithischia: Genasauria: Cerapoda: Ornithopoda: Euornithopoda: Iguanodontia: Euiguanodontia: Dryomorpha: Ankylopollexia: ?Hadrosauroidea: ?Hadrosauridae.

Name derivation: Greek *proto* = "first" + Greek *hadros* = "sturdy [for Hadrosauridae, indicating phylogenetic status]."

Type species: *H. byrdi* Head 1998.

Other species: [None.]

Occurrence: Woodbine Formation, Texas, United States.

Age: Middle Cretaceous (mid–Cenomanian).

Known material: Disarticulated skull, isolated postcrania, teeth.

Holotype: SMU 74582, nearly complete skull, isolated postcrania including a right atlantal neural arch, rib fragments, and manual phalanx.

Diagnosis of genus (as for type species): Differs from other known hadrosaurids in possession of ventrally deflected muzzle including massive predentary and robust, rostrally expanded and deflected dentary; also recognized by possession of strongly bilobate jugal-maxillary articulation, elongate maxillary rostral processes, and following combination of primitive characteristics: retention of jugal-ectopterygoid contact, short jugal-postorbital articulation, ventral orientation of angular, and retention of surangular foramen (Head 1998).

Comments: Originally described as the oldest definitive hadrosaurid genus, *Protohadros* was founded upon an almost complete skull (the most complete hadrosaurid skull known from before the Turonian; see Weishampel 1990; Weishampel and Horner 1990) and some postcranial remains (SMU 74582), discovered in 1994 by Gary Byrd of the Dallas Paleontological Society in the Woodbine Formation (Middle Cretaceous/mid–Cenomanian, Kennedy and Cobban 1990; 95 million years ago, Obradovich 1993), in the township of Flower Mound, Denton County, Texas. Material referred to the type species *P. byrdi* from the same locality includes a maxillary tooth morph and dentary tooth morph (SMU 74638 and 73571, respectively) reported by Lee in 1995, who had also reported footprints and isolated postcrania establishing the presence of hadrosaurids in the Woodbine Formation (Head 1998).

Head performed a phylogenetic analysis of *Protohadros* utilizing its primitive cranial characters, while including characters and character states from recent analyses of hadrosaurids and more primitive iguanodontians (see Head for details). Noting that this genus can be taxonomically defined as both the most derived iguanodontian (see Norman and Weishampel 1990) and the most basal hadrosaurid, Head concluded that *Protohadros*— possessing the fewest hadrosaurid synapomorphies — is a member of the Hadrosauridae, occupying the most basal position within this clade. This inclusion, however, would require a redefinition of the Hadrosauridae "to relocate certain synapomorphies to more exclusive nested groups, a situation that is likely to be repeated as additional primitive taxa are discovered." Head further suggested that some characteristics previously regarded as derived for the Hadrosauridae (*e.g.*, mandibular diastema and dorsoventrally expanded jugal) and which show a discontinuous distribution in his analysis are perhaps "indicators of a wide variety of functional ecological niches occupied by hadrosaurids" rather than having phylogenetic significance. Regardless of

PROTOCERATOPS ANDREWSI
Pronunciation: Pro-toe-CER-a-tops
Locality
Age
Size
Meaning of Name
Classification

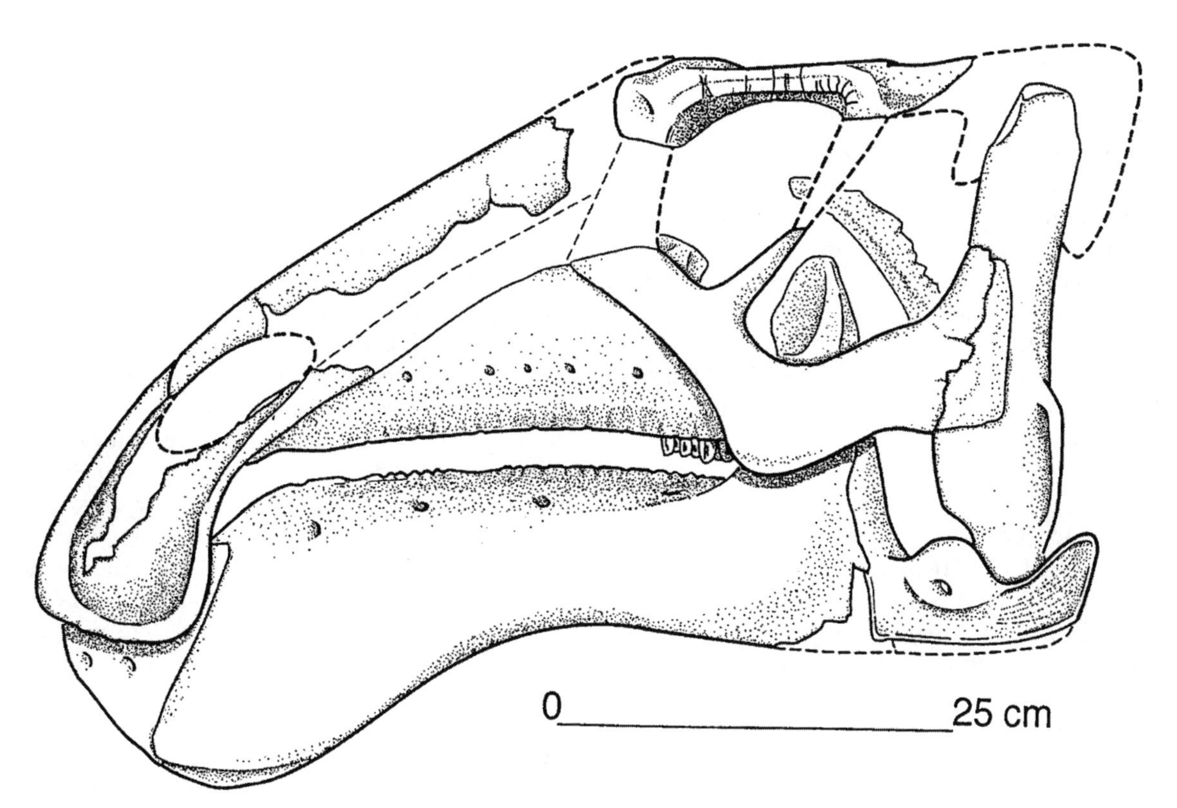

***Protohadros byrdi*, SMU 74582, holotype skull (reconstructed), left lateral view. (After Head 1998.)**

its basal position, *Protohadros* exhibits a number of features regarded as diagnostic of later hadrosaurids (*e.g.*, expanded premaxillary beak, compact maxillary teeth, and expanded pleurokinetic hinge; see below). As pointed out by Head, the discovery of this new taxon offers an opportunity to test the established hypothesis of the evolution of the Hadrosauridae, and also provides new information regarding the early development and history of this clade of dinosaurs.

More recently, Casanovas, Pereda-Suberbiola, Santafé and Weishampel (1999) commented that *Protohadros* could either be a derived iguanodont (*sensu* Norman and Weishampel 1990) or the most basal hadrosaurid, exhibiting some synapomorphies of the Hadrosauridae (dentary alveoli forming parallel-sided vertical furrows, and distal secondary ridges on dentary crowns) but lacking one (more than 29 dentary tooth positions, 28 in *Protohadros*). These authors concurred with Head that, if *Protohadros* is a member of the Hadrosauridae, then this clade should be redefined.

Head noted that the phylogenetic and stratigraphic positions of this new genus allows for an examination of adaptations for cranial kinesis in hadrosaurid skulls. Among ornithopods, hadrosaurids possessed the most derived form of such kinesis, comprising both expanded pleurokinesis (Norman 1984*a*) and streptostyly (Weishampel 1984), including the rotation of the quadrate at the glenoid and evolution of a flexible palatine-pterygoid hinge, some (but not all) of these adaptations being found in *Protohadros*. In the skull of *Protohadros*, the combination of a derived maxillopalatal unit and a primitive quadrate reveal a level of cranial kinesis that seems to be intermediate between primitive iguanodontians and other hadrosaurids. According to Head, some features of the hadrosaurid skull — many of which were not included in his phylogenetic analysis, due either to lack of preservation or documentation — reveal "an evolutionary sequence of adaptations in which modifications for increased masticatory ability of the maxillary region precede the evolution of both streptostylic cranial kinesis and a more derived mandible in Hadrosauridae."

The massive, uniquely deflected rostral dentary and strongly denticulate predentary, combined with the occurrence of this dinosaur in the deltaic environment of the Woodbine Formation, suggested to Head that *Protohadros* was exclusively a low-browsing feeder that could have used its muzzle as a scoop for acquiring aquatic plants in swampy areas.

Although remains of early hadrosaurids are scarce and ambiguous, this clade has generally been thought of as originating in Asia. However, this prevalent hypothesis has been challenged by the discovery of

***Opposite:* Adult and hatchling skeletons (casts) of *Protoceratops andrewsi* from the Gobi desert, southern People's Republic of Mongolia, part of "The Great Russian Dinosaurs Exhibition," here presented at the Natural History Museum of Los Angeles County. Photograph by the author, courtesy Natural History Museum of Los Angeles County.**

Protohadros in the Cenomanian of North America. Head proposed that, given the age and location of *Protohadros*, combined with other early hadrosaurid occurrences (*e.g.*, see *Eolambia* entry, *S1*), the Hadrosauridae may have originally been endemic to North America, with short-term regressive phases in eustatic sea levels permitting pre–Turonian dispersals across ephemeral seaways such as the Western Interior Seaway and Bering Straits.

Key references: Casanovas, Pereda-Suberbiola, Santafé and Weishampel (1999); Head (1998); Norman (1984*a*); Norman and Weishampel (1990); Weishampel (1984).

†PSITTACOSAURUS

Age: ?Late Jurassic (Kimmeridgian) or Early to Middle Cretaceous (Aptian–Albian).

Comments: Another specimen of the small, primitive, and rather common ceratopsian dinosaur *Psittacosaurus* has recently been discovered and was described by Xu and Wang (1998). The fossil, an almost complete though badly crushed skeleton, was collected by a local farmer from the black shale of the lower part of the Yixian Formation of Sihetun, Beipiao City, Liaong Province, China, a locality known for a diverse fauna including the "feathered" theropods *Sinosauropteryx prima* and *Protarchaeopteryx robusta*, the pterosaur *Eosipterus*, the birds *Confusiusornis sanctus* and *Liaoningornis longiditrus*, and amphibians of the group Anura (personal observation of the authors). The specimen (designated L0001) is housed in the collection of Liang Shikuan.

Xu and Wang referred the new specimen to *Psittacosaurus* because of various features including the following: Naris small, highly situated; eminence on rim of buccal emargination of very caudal maxilla; developed jugal horn; bulbous primary ridge on dentary crown; absence of manual digit V. Some features —*e.g.*, diapophysese of neural arch straplike and extremely caudally positioned; ulna with unexpanded distal end, proximal end of ulna more than twice as wide as distal end; distal end of radius strongly expanded; metacarpal I relatively broad — are not known in other species of *Psittacosaurus*, although these differences may, the authors pointed out, be due to the distortion of the specimen. Other features —*e.g.*, dorsoanteriorly directed crest on rim of buccal emargination of maxilla, enamel distributed almost equally on opposing sides of cheek teeth, and semicircular (rather than crescent shaped) sternum — distinguish L0001 from other known psittacosaur species.

The authors observed numerous morphological differences (*e.g.*, a comparatively weakly developed jugal horn in L0001) between the new material and other previously described psittacosaur species —*Psittacosaurus mongoliensis*, *P. sinensis*, *P. xinjiangensis*, *P. meileyingensis*, *P. neimongoliensis*, *P. ordosensis*, and *P.*

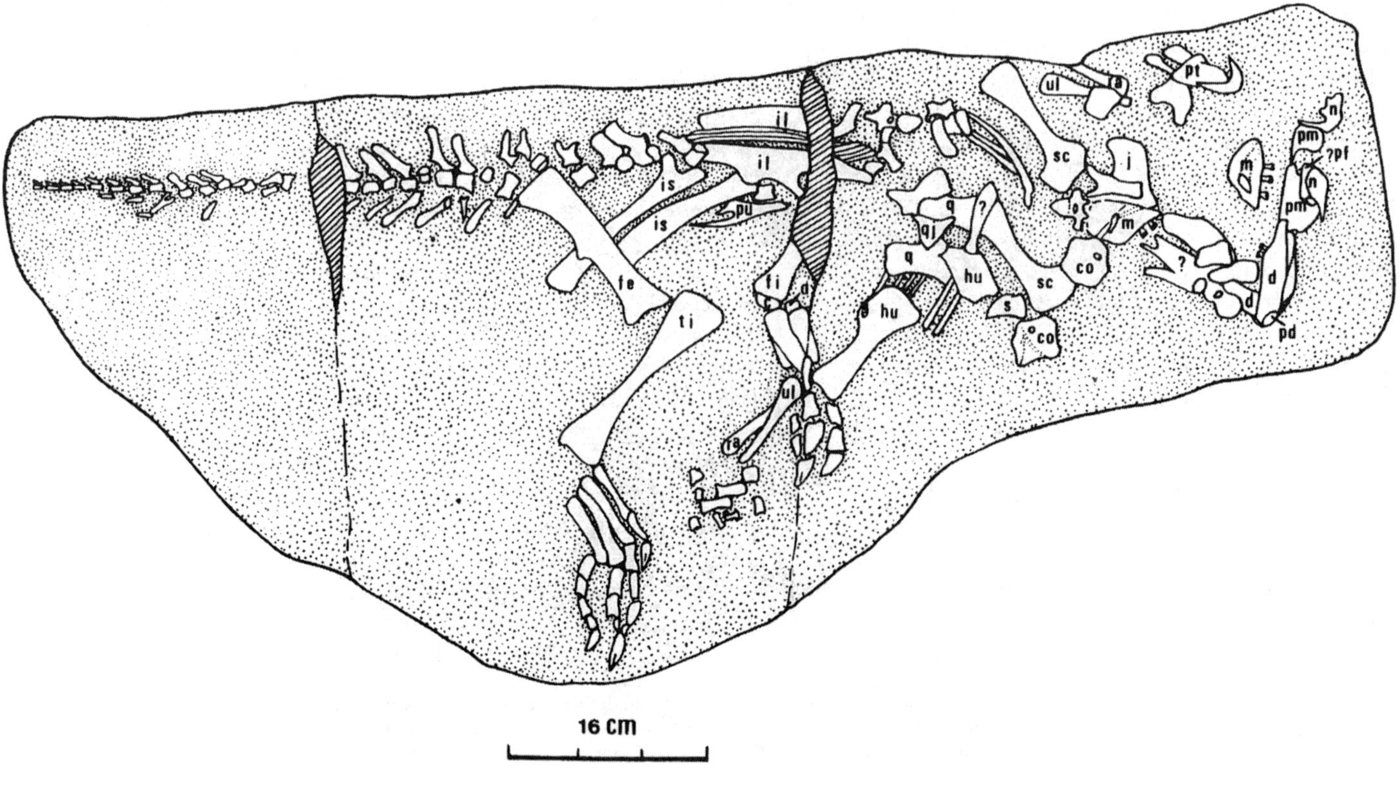

Psittacosaurus **sp., skeleton (L0001) of an apparently more primitive psittacosaurid, in right lateral view. (After Xu and Wang 1998.)**

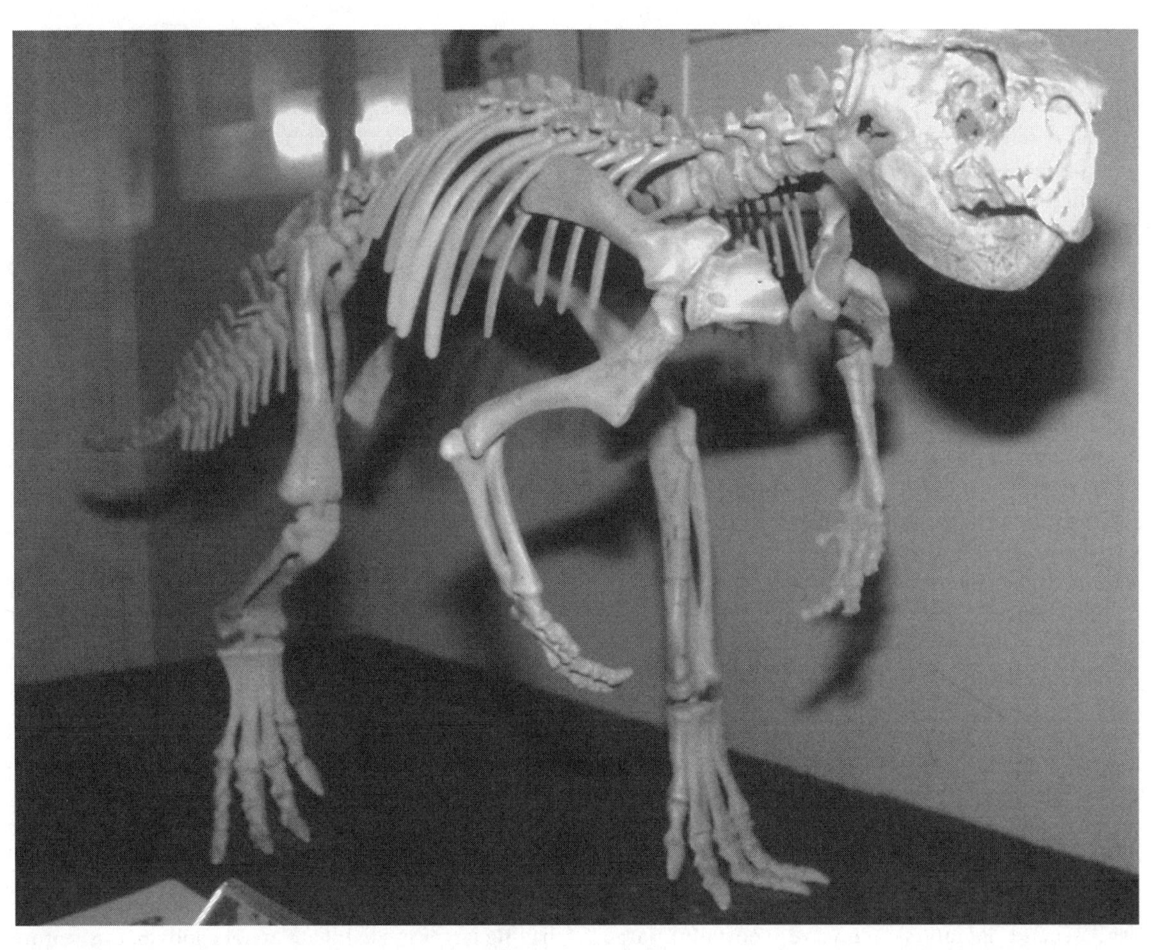

Photograph by the author, courtesy Natural History Museum of Los Angeles County.

***Psittacosaurus mongoliensis*, cast of skeleton from the Barunbayan Formation, Gobi Desert, southwestern People's Republic of Mongolia, part of the traveling "Great Russian Dinosaurs Exhibition."**

mazongshanensis— but, because of the poor preservation of the material, identified few autapomorphies on L0001. Therefore, Xu and Wang tentatively referred the new specimen to *Psittacosaurus* sp. rather than assign to any new species.

Recently there has been a considerable amount of controversy over the age of the Yixian Formation rocks (see *Sinosauropteryx* entries, *S1* and this volume). As pointed out by Xu and Wang, there is a consensus that all known *Psittacosaurus* localities (in Mongolia, China, Siberia, Japan, and Thailand) are of late Early to Middle Cretaceous (Aptian–Albian) age. This would, therefore, suggest that the psittacosaur remains found at Siheton, Liaong are also no older than Early Cretaceous. However, based upon some primitive features (*e.g.*, apparently shorter rostral process of nasal, weakly developed jugal horns, diapophysis more strap-like than rod-like) the authors observed in L0001, the Yixian Formation may be earlier than other *Psittacosaurus* localities. Recently, Jiang, Chen, Cao and Komatsu (2000), based upon invertebrate fossils, suggested an age for this locality close to the Jurassic–Cretaceous boundary. More recently, Wang, Wang, Wang and Xu, in an abstract, argued for an Early Cretaceous age (see *Sinosauropteryx* entry).

Erickson and Tumanova (2000) published a study of the growth curve in *Psittacosaurus mongoliensis* based upon the histology of major long bones (all PIN specimens) of this species recovered from the Khuktseka Svita of the People's Republic of Mongolia, the authors noting that the skeleton of this dinosaur "undergoes substantial histological modification during ontogeny in association with longitudinal growth, shape changes, reproductive activity, and fatigue repair." Specimens were examined ranging in developmental stages from juvenile to adult. Results of this study suggested, as shown on a graph, that 1. *P. mongoliensis* had an S-shaped growth curve—"best described using a logistic (or sigmoidal) equation"—as found in most extant vertebrates, and that 2. the maximum growth rates in this taxon exceeded those of extant reptiles and marsupials, but were slower than most birds and other warm-blooded taxa.

Note: Shu'an and Haichen (1998) described recently discovered psittacosaurid skin impressions recovered from Sihetun Village, Beipiao City, in western Liaoning Province, China, these being significant as the first tangible example of fossil skin reported pertaining to this group of dinosaurs. The authors described the scales on these impressions as small, polygonal, their edges not overlapping, their sizes and shapes varying slightly in different areas of the body.

According to Shu'an and Haichen, psittacosaurids arose early during the Late Jurassic, flourishing in the Early Cretaceous.

Key references: Erickson and Tumanova (2000); Jiang, Chen, Cao and Komatsu (2000); Shu'an and Haichen (1998); Wang, Wang, Wang and Xu (2000); Xu and Wang (1998).

PYRORAPTOR Allain and Taquet 2000

Saurischia: Theropoda: Neotheropoda: Tetanurae: Avetheropoda: Coelurosauria: Manuraptoriformes: Manuraptora: Deinonychosauria: Eumanuraptora: Dromaeosauridae.

Name derivation: Greek *pyros* = "fire" + Latin *raptor* = "thief."

Type species: *P. olympius* Allain and Taquet 2000.

Other species: [None.]

Occurrence: Fluvio-lacustrine sandstone, Bouches-du-Rhône, France.

Age: Late Cretaceous (upper Campanian–lower Maastrichtian).

Known material: Bones of front and hindfeet, teeth, vertebrae.

Holotype: MNHN B0001, complete ungual phalanx of left second digit of pes.

Diagnosis of genus (as for type species): Small dromaeosaurid characterized by deep depression on lateral face of bowed ulna below its proximal extremity; second metatarsal concave ventrally; large, strongly curved ungual claw; metatarsal II distally grooved, asymmetrical; metatarsal II subequal in length to ulna; tooth serrations restricted to anterior carina (Allain and Taquet 2000).

Comments: Dromaeosaurid remains have been rarely found in Europe and none of these have offered substantial taxonomic information. The genus *Pyroraptor* is distinguished as the first member of the Dromaeosauridae from Europe to provide good taxonomic data and diagnostic characters for this theropod family (Allain and Taquet 2000).

As recounted by Allain and Taquet, the type species *Pyroraptor olympius* was founded upon an ungual phalanx (MNHN B0001) collected from the fluvio-lacustrine sandstones at La Boucharde, southeast of Trets, in France. Paratype specimens of *P. olympius* from the same locality include (MNHN B0002) a second phalanx of a right pedal digit II, (B0003) a left metatarsal II, (B0004) a complete ungual phalanx of a right pedal digit II, (B0005) a right ulna, and (B0014 and B0015) teeth. Referred specimens include (MNHN B0006, B0007, B0008, and B0009) pedal phalangeal elements, (B0011) a manual phalanx, (B0012) the distal end of metacarpal I, (B0013) a right radius, (B0016) an anterior caudal vertebra, and (B0017) a dorsal vertebra.

This new taxon occurs associated with the remains of the ornithopod *Rhabdodon priscus*, nodosaurid, titanosaurid, and ceratopsian dinosaurs, dinosaur eggshells, fragments of turtle shells (*Dortoka* sp., *Solemys* sp., *Polysternon* sp., and an as of yet unnamed chelydroid turtle), and the bones of an alligatoroid, this fauna as a whole representing, the authors noted, a typical continental vertebrate assemblage of Late Cretaceous, western Europe.

Allain and Taquet identified *P. olympius* as a dromaeosaurid by numerous synapomorphic features: Elements of second toe highly modified for predation, shape of claw more reminiscent of dromaeosaurids than troodontids (see Currie and Peng 1993); second toe laterally flattened, ventrally sharp, curved through 140-degree arch (Ostrom 1969*b*); strongly developed flexor tubercule located at proximal end of ungual phalanx as in dromaeosaurids (Ostrom; Barsbold 1983*a*); second phalanx of digit II laterally compressed as in dromaeosaurids (compressed proximodorsally in most troodontids) (Osmólska 1987; Russell and Dong 1993*b*; Currie and Peng); proximal end of that bone extended ventrally into projection or heel (see Ostrom), its distal end having deeply grooved ginglymus extending further ventrally than dorsally in proximal portion (see Ostrom), and deep, subcircular fossae for ligament attachment; distal end of metatarsal II deeply grooved, very asymmetrical ginglymus having larger medial than lateral condyle, this feature known only in dromaeosaurids (Norell and Makovicky 1997); metatarsal II almost same length as ulna, as in dromaeosaurids (metatarsal II sometimes 1.5 or more times length of ulna in troodontids; see Russell and Dong).

Due to the paucity of the material, it was not possible for Allain and Taquet to compare *Pyroraptor* adequately with other dromaeosaurids. However, in describing this new taxon, the authors noted that the claw of left digit II, being more trenchant ventrally and less recurved, more resembles that of the dromaeosaurid *Saurornitholestes langstoni* (Sues 1978; Currie 1995) than of *Deinonychus antirrhopus*. As measured by Allain and Taquet, the complete ungual phalanx of this digit is 66 millimeters in length on its dorsal margin, 18 millimeters high, and 7 millimeters wide at its proximal end.

Various authors (Ostrom 1990; Le Loeuff and Buffetaut 1998) had previously suggested that the Dromaeosauridae originated either in North America or Euromerica during the Early Cretaceous, a hypothesis not supported by recently discovered Early Cretaceous dromaeosaurid material from China (Xu, Wang and Wu 1999). Regardless of their place of origin, Allain and Buffetaut assessed the dromaeosaurids of Europe to be "remnants of a generalized Late Jurassic

or Early Cretaceous fauna which have evolved on their own, after the late Cretaceous isolation of southern Europe."

Key references: Allain and Taquet (2000); Barsbold (1983*a*); Le Loeuff and Buffetaut (1998); Currie (1995); Currie and Peng (1993); Osmólska 1987; Ostrom (1969*b*, 1990); Russell and Dong (1993*b*); Sues (1978); Xu, Wang and Wu (1999).

QANTASSAURUS Rich and Vickers-Rich 1999

Ornithischia: Genasauria: Cerapoda: Ornithopoda: Euornithopoda: Hypsilophodontidae: Hypsilophodontinae.

Name derivation: "QANTAS [airlines] + Greek *sauros* = "lizard."

Type species: *Q. intrepidus* Rich and Vickers-Rich 1999

Other species: [None.]

Occurrence: Wonthaggi Formation, Victoria, Australia.

Age: Middle Cretaceous (Aptian).

Known material: Dentaries and dentary fragment, three individuals.

Holotype: NMV P199075, complete left dentary.

Diagnosis of genus (as for type species): Only 10 cheek teeth present in foreshortened dentary having anteriorly converging dorsal and ventral margins (Rich and Vickers-Rich 1999).

Comments: Most of the so-called "polar dinosaurs" known from southeastern Australia are primarily represented by isolated bones and teeth, which have been referred to a group of small ornithishians commonly called hypsilophodontids. Another among this relatively rare sampling of dinosaurs is *Qantassaurus*. The genus was founded upon a left dentary (NMV P199075) preserving all alveolar sockets, with at least one tooth present in seven of the 10 alveolar positions. The specimen was collected from the Flat Rocks fossil vertebrate site, in the Wonthaggi Formation, Strzelecki Range, on the shore platform of Bunurong Marine Park, Victoria, in southeastern Australia. Specimens referred to the type species *Qantassaurus intrepidus* from the same locality include a complete left dentary (NMV P198962) preserving all alveolar sockets with four teeth present, and a fragment of a right dentary (NMV P199087) missing the symphysial region and much of the mandible posterior to the most posterior alveolus (Rich and Vickers-Rich 1999).

As Rich and Vickers-Rich observed, the individual cheek teeth preserved in NMV P199075 are quite similar to those of another Victorian hypsilophodontid, *Atlascoposaurus loadsi* (see Rich and Rich 1989; also *D:TE*). In both taxa these teeth are characterized mainly by a strong primary ridge and prominent secondary ridges. As the authors pointed out, Victorian hypsilophodontids having cheek teeth with that pattern can be grouped into two distinct categories — "those which have elongated dentaries with their dorsal and ventral margins parallel (the advanced condition [see Weishampel and Heinrich 1992]) and foreshortened forms with fewer cheek teeth and the two margins converging anteriorly." The new genus and species *Q. intrepidus*, embracing these latter forms, was proposed by Rich and Vickers-Rich to recognize this difference adequately. Because *A. loadsi* was

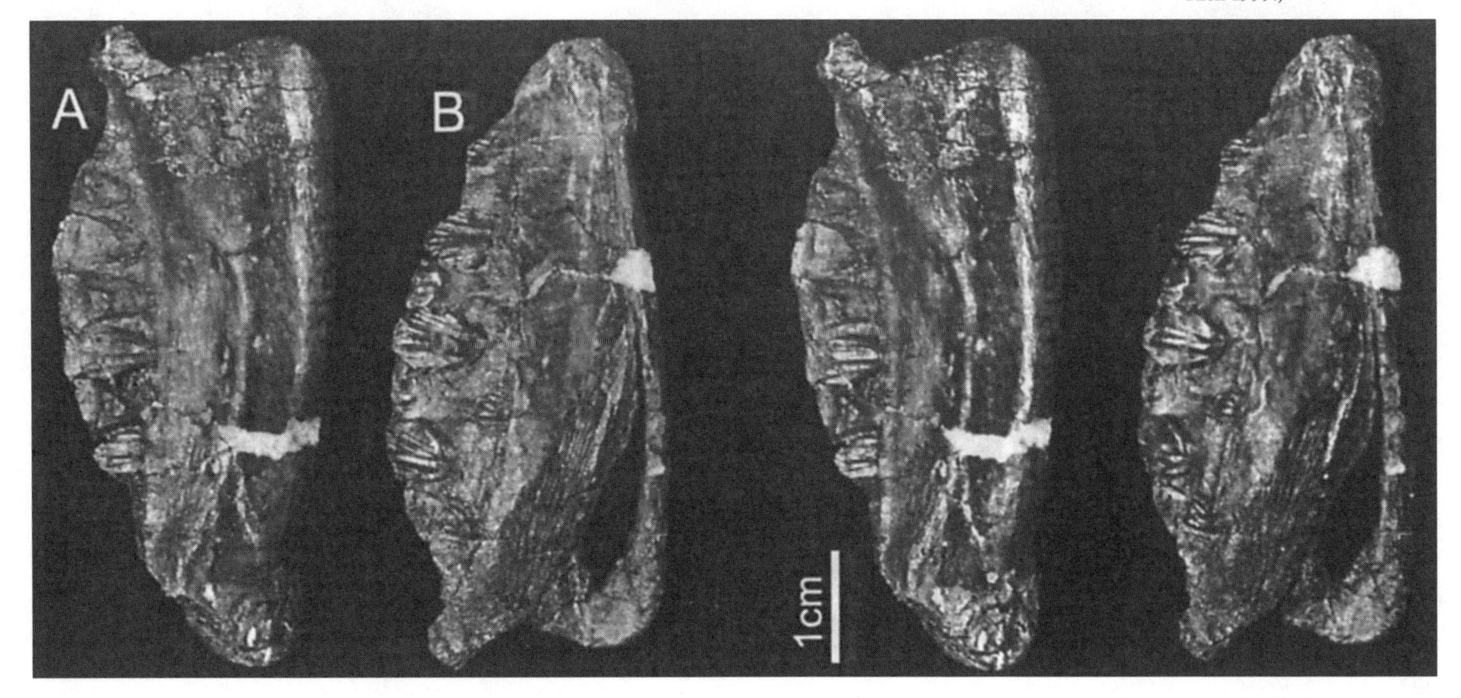

***Qantassaurus intrepidus*, NMV P199075, holotype left dentary, in A. lingual and b. lateral stereo views. (After Rich and Vickers-Rich 1999.)**

founded upon a fragment of a maxilla, it was not possible for the authors to determine that this holotype did not belong to an animal having a lower jaw like that of *Q. intrepidus*, rather than the elongated morph with more than 10 cheek teeth. However, seven hypsilophodontid jaws falling into the first of the above categories, with parallel dorsal and ventral margins and having one cheek tooth or more, have been recovered from the Eumerella Formation which yielded the type specimen of *A. loadsi*; in but one only possibly hypsilophodontid specimen from that formation, an edentulous jaw fragment (NMV P185854), these margins converge anteriorly. Therefore, Rich and Vickers-Rich found it reasonable to presume that *A. loadsi* had jaws of the first category, and that *Q. intrepidus*, with jaws of the second category, is a valid taxon.

Key references: Rich and Rich (1989); Rich and Vickers-Rich (1999); Weishampel and Heinrich (1992).

QINLINGOSAURUS Xue, Zhang and Bi 1996 *see in* Xue, Zhang, Bi, Yue and Chen 1996

Saurischia: Sauropodomorpha: Sauropoda *incertae sedis.*

Name derivation: "Qingling [Mountains]" + Greek *sauros* = "lizard."

Type species: *Q. luonanensis* Xue, Zhang and Bi Xue, Zhang and Bi 1996 *see in* Xue, Zhang, Bi, Yue and Chen 1996.

Other species: [None.]

Occurrence: Shanyang Formation, Qinling Mountains region, China.

Age: Late Cretaceous.

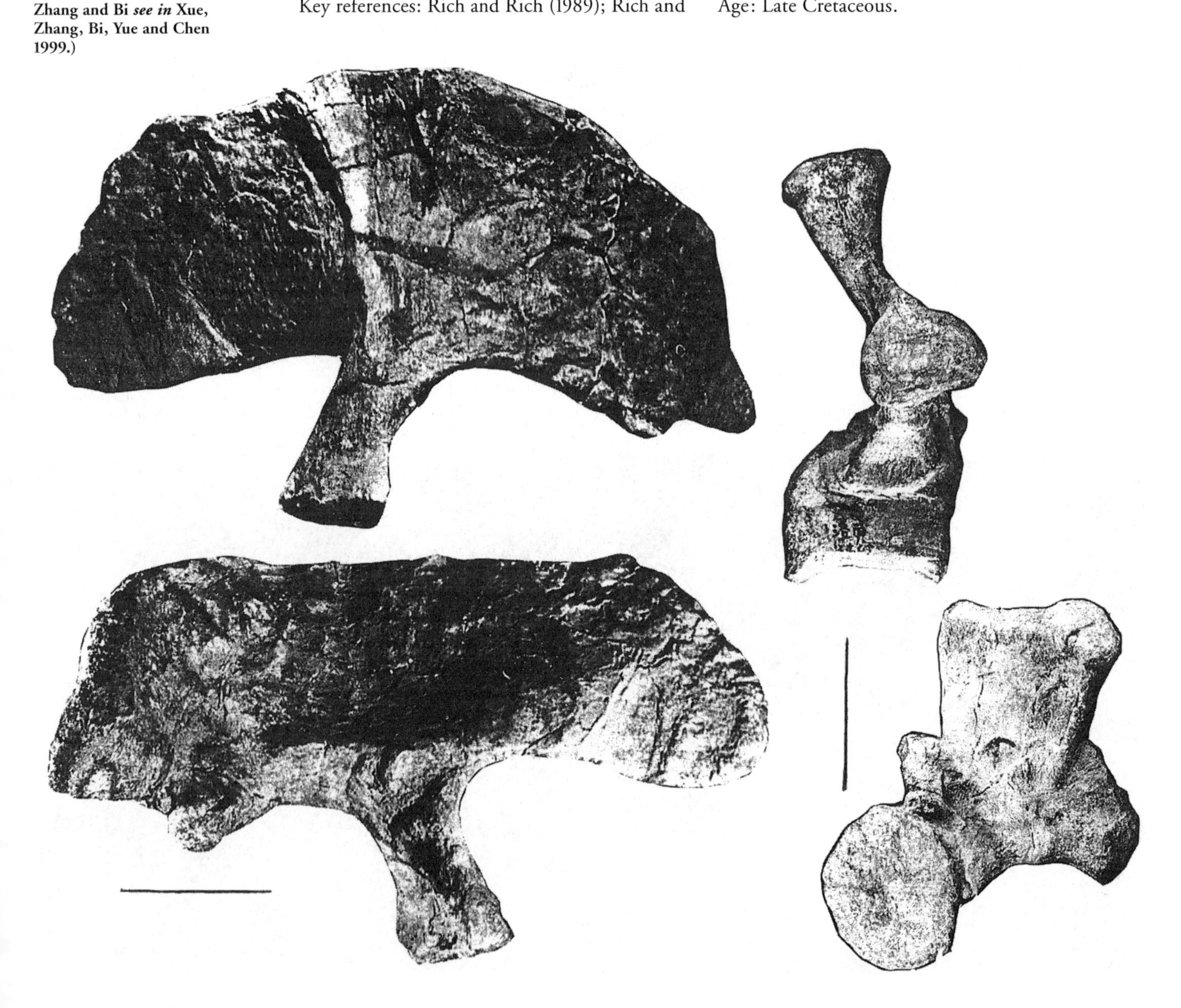

Qinglingosaurus luonanensis, **(left) holotype right ilium (NWUV1112.1; scale = 17 cm.), lateral and inner views, (right) holotype vertebrae (NWUV1112.2; scale = 16 cm.). (After Xue, Zhang and Bi *see in* Xue, Zhang, Bi, Yue and Chen 1999.)**

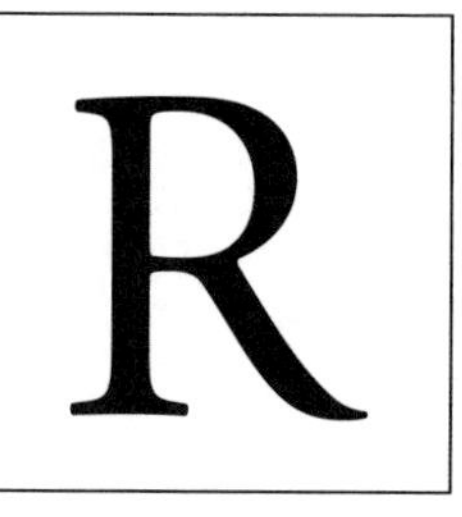

Known material/holotype: NWUV1112.1-2, complete right ilium, fragment of ischium, three vertebrae.

Diagnosis of genus (as for type species): Sauropod having ilium that is thin and in long, narrow rhomboid form, anterior and posterior sides mostly curving outward, posterior side much longer than anterior side; pubic peduncle long, strong, located at center of bone plate; ischiac peduncle small; acetabulum large (Xue, Zhang and Bi 1996).

Comments: The type species *Qinlingosaurus luonanensis* was founded upon very incomplete postcranial remains (NWUV1112.1-2) recovered from the Shanyang Formation in the eastern part of the Qinling Mountains, in central China. This new taxon was only briefly discussed in a volume edited by Xue, Zhang, Bi, Yue and Chen (1996) about development and environmental changes of the intermontane basins in the eastern part of the Qinling Mountains (Xue, Zhang and Bi 1996).

Key reference: Xue, Zhang and Bi 1996 *see in* Xue, Zhang, Bi, Yue and Chen (1996).

†RHABDODON

Ornithischia: Genasauria: Cerapoda: Ornithopoda: Euornithopoda: ?Iguanodontia.

Diagnosis of genus (as for type species): Maxillary teeth having parallel ridges without prominent primary ridges; dentary teeth having distal prominent primary ridge; enamel restricted to lateral side of maxillary teeth and medial side of dentary teeth; sacrum comprising six sacral vertebrae and one sacrodorsal vertebra; sacral neural spines fused; scapula having strongly widened distal border, cranial and caudal margins concave; coracoid with prominent sternal process, coracoid foramen closed to glenoid cavity; prepubic blade long, straight, laterally flattened, with closed obturator foramen; postpubic blade long, curved, dorsoventrally flattened; ischium blade straight, laterally flattened, with widened distal end and obturator process located on proximal half of ischial shaft; femur having prominent but nonpendant fourth trochanter and proximo-lateral bulge on caudal margin; femur longer than tibia (Garcia, Pincemaille, Vianey-Liaud, Marandat, Lorenz, Cheylan, Cappetta, Michaux and Sudre 1999).

Comments: *Rhabdodon* is a rather conservative euornithopod genus, remains of which were first collected from the Sânpetru Formation in the Hateg Basin of Transylvania, in western Romania (see *D:TE*).

Because *Rhabdodon* has been found in almost all well-known Late Cretaceous terrestrial faunas of Europe, and due to its relevance both to euornithopod systematics and island biogeography, Weishampel, Jianu, Csiki and Norman (1998) subjected this taxon to a numerical phylogenetic analysis, the results of which were presented in a preliminary report.

Utilizing 13 taxa and 32 characters, Weishampel *et al.* produced a cladogram in which *Rhabdodon* was "positioned as a basal iguanodontian euornithopod in an unresolved polytomy with *Tenontosaurus*, *Muttaburrasaurus*, and Euiguanodontia." *Rhabdodon* also possesses numerous (unspecified) autapomorphies, a greater array than in any other member of the Hauteg fauna (see below).

Although the type species, *Rhabdodon priscus*, is known from numerous specimens (mostly disarticulate cranial and postcranial remains representing juveniles to adults), it was only recently that a nearly complete skeleton has been available for study. That specimen (NHM.Aix collection) was discovered in

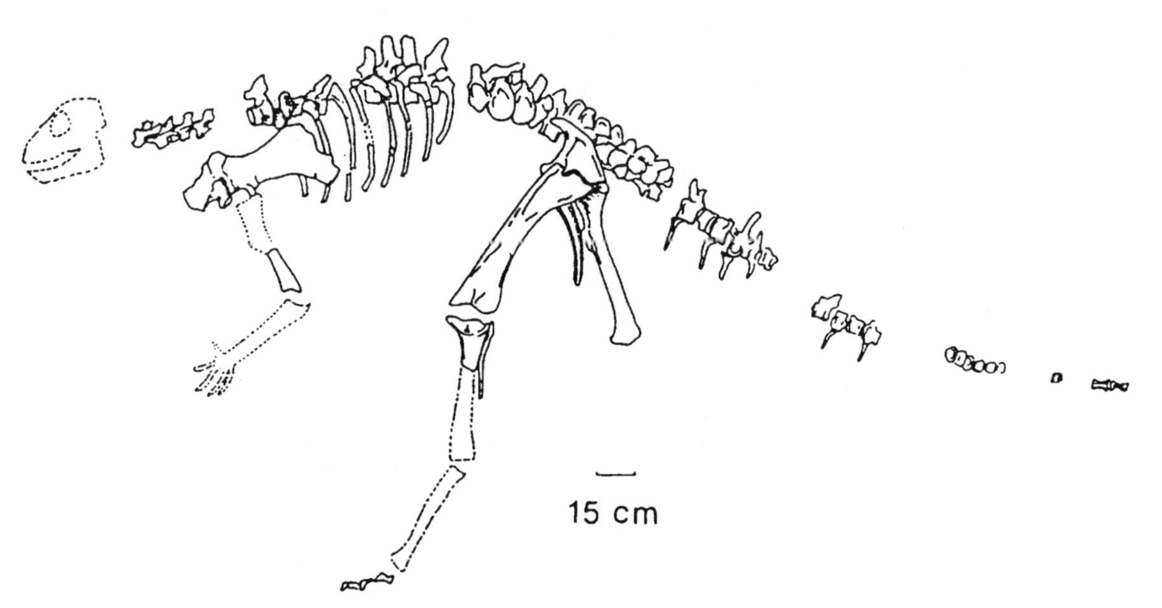

Drawing of the prepared skeleton (NHM.Axis) of *Rhabdodon priscus* from Vitrolles-Couperigne, Provence, France. (After Garcia, Pincemaille, Vianey-Liaud, Marandat, Lorenz, Cheylan, Cappetta, Michaux and Sudre 1999).

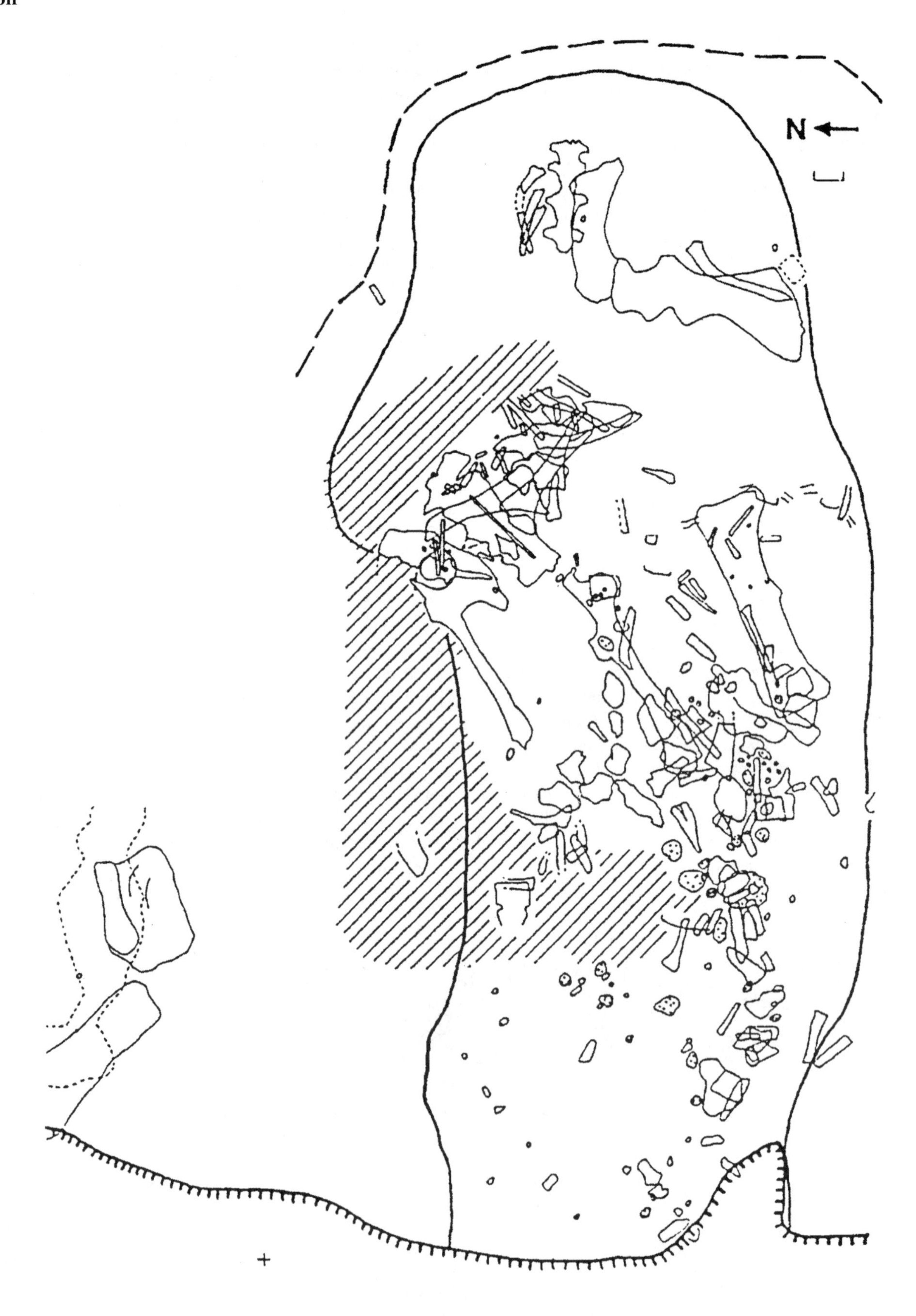

Skeleton (NHM.Axis) of ***Rhabdodon priscus*** as found in the main locality at Vitrolles-Couperigne, Provence, France. (After Garcia, Pincemaille, Vianey-Liaud, Marandat, Lorenz, Cheylan, Cappetta, Michaux and Sudre 1999).

1995 by Edgar Lorenz in the gray marls level (early Maastrichtian; see Westphal and Durand 1990), Vitrolles-Couperigne, Provence, in southern France. The specimen was subsequently excavated by a team of paleontologists from the Natural History Museum of Aix-en-Provence and the Institut des sciences de l'évolution de Montepellier. It lacks only the skull, the forelimbs, and various other elements (Garcia, Pincemaille, Vianey-Liaud, Marandat, Lorenz, Cheylan, Cappetta, Michaux and Sudre 1999).

Garcia *et al.* proposed a new diagnosis for *R. priscus* based primarily on this recently collected skeleton, with supplementary data from the lectotype (MNPL 30, a partial left dentary first described by Matheron 1869) and various paralectotypes (MNPL 31, a partial right dentary; MNPL 34, a dorsal vertebra; MNPL 36, three fused sacral vertebrae; MNPL A, a posterior caudal vertebra; a partial right femur identified by Matheron as a humerus; MNPL ?60, half of the proximal border of a right femur; and MNPL 51, a left ulna).

Garcia *et al.* found *R. priscus* to measure approximately 6 meters (about 21 feet) in length. According to Garcia *et al.*, this dinosaur was surely capable of some bipedal locomotion, although the strong bones are suggestive of frequent locomotion on all fours. Furthermore, "the morphology of the sacrum illustrates this sturdiness with sacral vertebrae fused at the level of the vertebral body and the neural spine."

Garcia *et al.* provided their own systematic analysis of *R. priscus*, noting that the following features permit the placement of this taxon near the clade Iguanodontia (see Sereno 1986): Heterogenous enamel distribution (*i.e.*, restricted to lateral side of maxillary teeth and to medial side of dentary teeth); ventral and dorsal margins of dentary parallel; femur very robust, 4th trochanter quite prominent and nonpendant (as in iguanodontids and hadrosaurids); morphology of prepubic blade, laterally compressed with closed obturator process, seemingly relating this species to *Tenontosaurus tilleti* (see Forster 1990).

Key references: Forster (1990); Garcia, Pincemaille, Vianey-Liaud, Marandat, Lorenz, Cheylan, Cappetta, Michaux and Sudre (1999); Matheron (1869); Sereno (1986); Weishampel, Jianu, Csiki and Norman (1998); Westphal and Durand (1990).

†**RHODANOSAURUS** [*nomen dubium*]. See *Struthiosaurus* (*D:TE*, *S1*. K. Carpenter, personal communication 2000)

ROCASAURUS Salgado and Azpilicueta 2000

Saurischia: Sauropodomorpha: Sauropoda: Eusauropoda: Neosauropoda: Macronaria: Camarasauromorpha: Titanosauriformes: Somphospondyli: Titanosauria: Eutitanosauria: Saltasauridae.

Name derivation: "Roca [city named in honor of General Roca, near the Salitral Moreno locality]" + Greek *sauros* = "lizard."

Type species: *R. muniozi* Salgado and Azpilicueta 2000.

Other species: [None.]

Occurrence: Allen Formation, Patagonia, Argentina.

Age: Late Cretaceous (Maastrichtian or Campanian–Maastrichtian).

Known material: Postcranial remains representing a number of individuals, juvenile and adult.

Holotype: MPCA-Pv 46, vertebral elements comprising a cervical centrum, cervical neural arch, two dorsal centra, three dorsal neural arches, two sacral neural arches, a midcaudal and midposterior caudal vertebra, ischia, left pubis, left ilium, fragment of right ilium, left femur, juvenile.

Diagnosis of genus (as for type species): "Saltasaurine" characterized by distal expansion on lateral margin of pubis; ischium with amplified lamina; caudal vertebrae with pronounced ventral depression divided by longitudinal septum, posterior articulation depressed, ventrally extended forward (Salgado and Azpilicueta 2000).

Comments: *Rocasaurus muniozi* was founded upon partial postcranial remains (MPCA-Pv 46) collected from the lower member of the Upper Cretaceous (Maastrichtian, Ballent 1980, or Campanian–Maastrichtian, Uliana and Dellapé 1981) Allen Formation (see Powell 1986) at Salitral Moreno, near the city of General Roca, in the province of Rio Negro, Patagonia, Argentina. Material from the same locality referred to the type species *R. muniozi* includes three anterior caudal centra (MPCA-Pv 47 to 49), a midcaudal centrum (MPCA-Pv 49), and six distal caudal centra (MPCA-Pv 51 to 56). Also from that locality, referred to *Rocasaurus* sp., was an anterior caudal vertebra (MPCA-Pv 57), a midcaudal (MPCA-Pv 58), and a posterior caudal (MPCA-Pv 59) (Salgado and Azpilicueta 2000).

Salgado and Azpilicueta referred *R. muniozi* to the "Saltasaurinae" (Powell 1986, 1992) [=Saltasauridae of Wilson, Martinez and Alcober 1999; see "Systematics" chapter], noting that the closest relationships of this taxon are with the titanosaur *Saltasaurus loricatus*, both species sharing caudal vertebrae bearing a ventral depression divided by a longitudinal septum. The record of this sauropod group, Salgado and Azpilicueta noted, is presently restricted to the Upper Cretaceous of Argentina, including three distinct taxa—*S. loricatus*, *Neuquensaurus australis*, and *R. munionzi*.

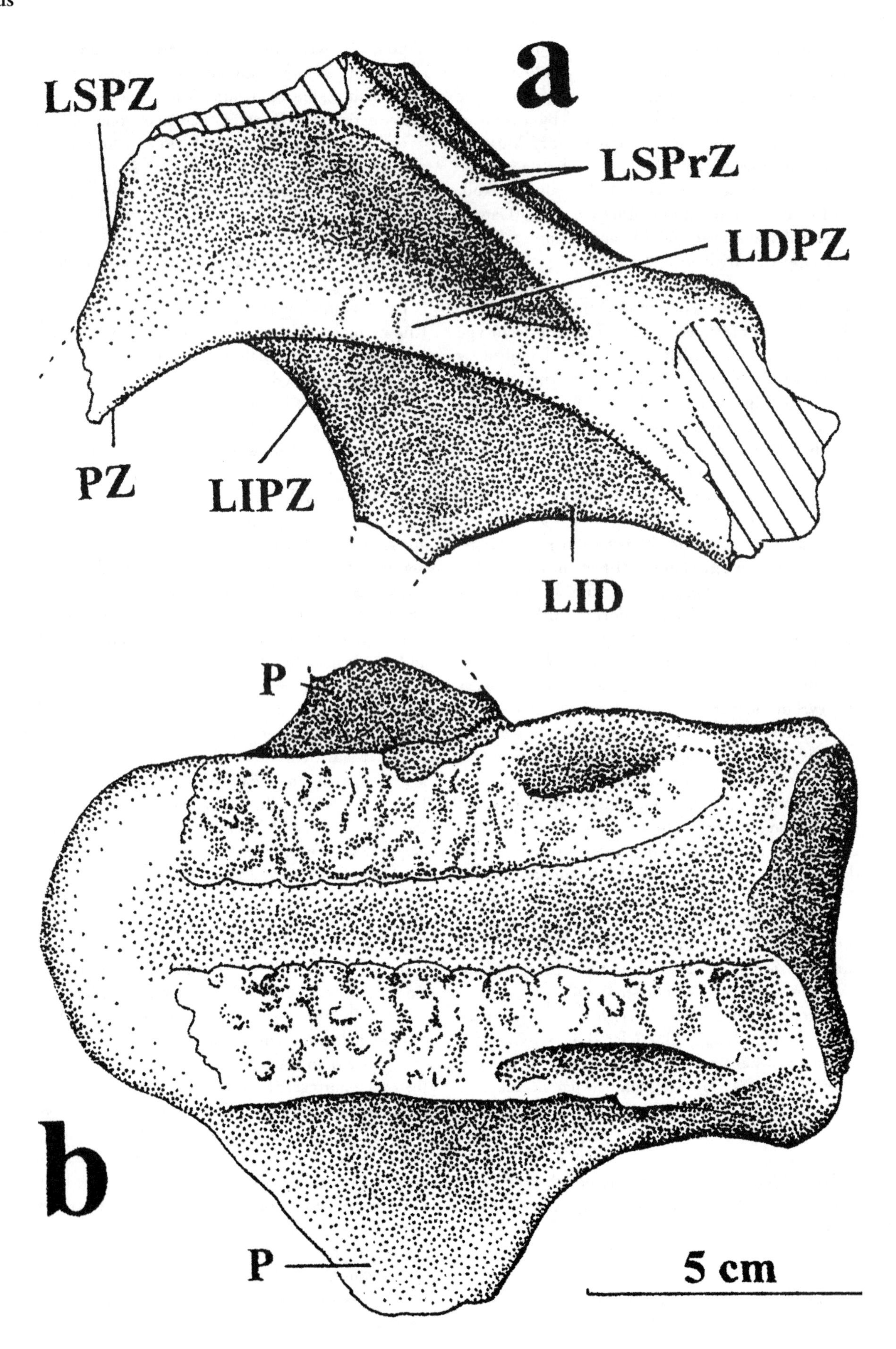

Rocasaurus muniozi, MPCA-Pv 46, holotype a. cervical neural arch, lateral view, and b. cervical centrum, dorsal view. (After Salgado and Azpilicueta 2000.)

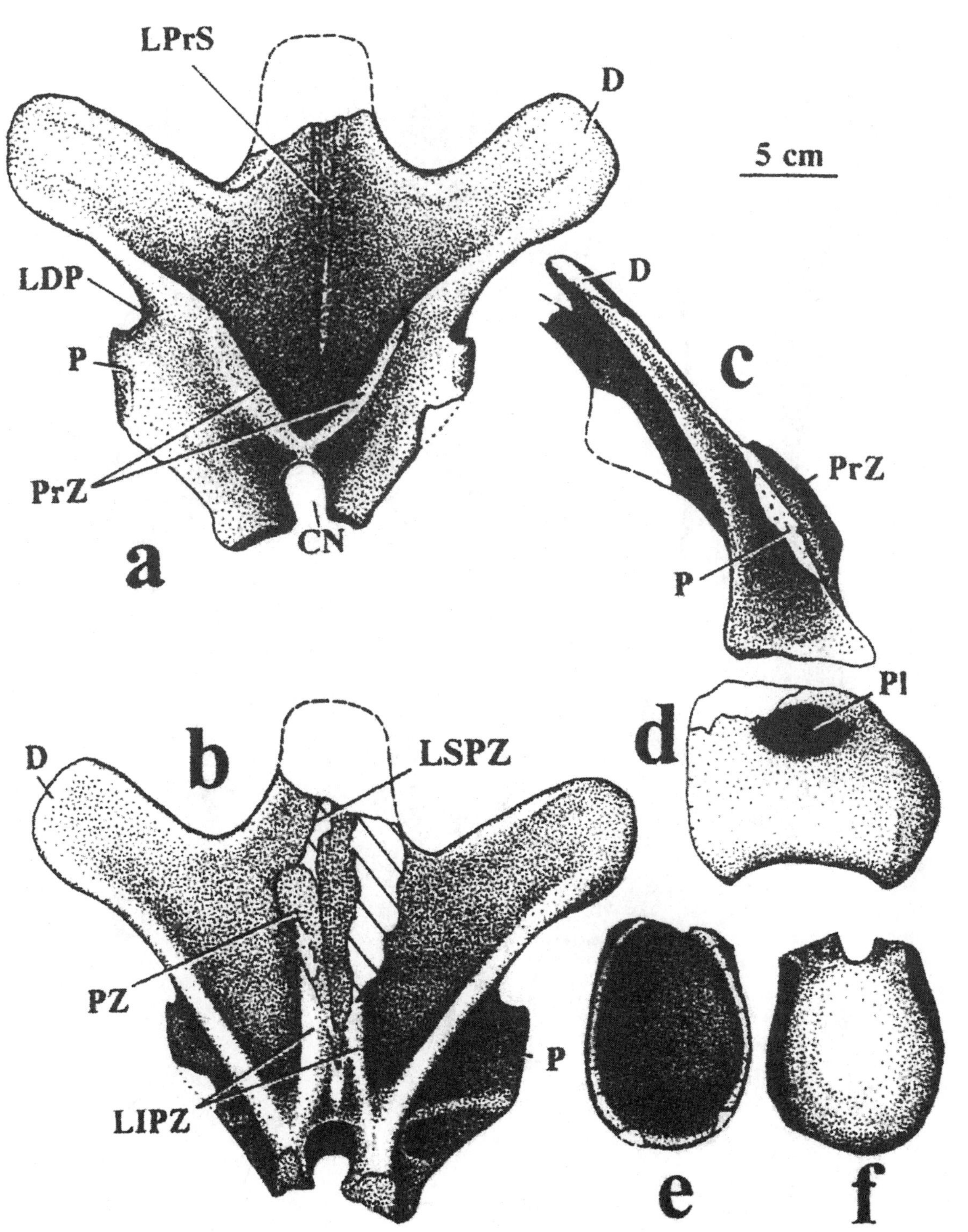

***Rocasaurus muniozi*, MPCA-Pv 46, holotype middorsal neural arch in a. anterior, b. posterior, and c. lateral views; dorsal centrum in d. lateral, e. posterior, and f. anterior views. (After Salgado and Azpilicueta 2000.)**

Key references: Ballent (1980); Powell (1986, 1992); Salgado and Azpilicueta (2000); Uliana and Dellapé (1981); Wilson, Martinez and Alcober (1999).

†SALTASAURUS

Saurischia: Sauropodomorpha: Sauropoda: Eusauropoda: Neosauropoda: Macronaria: Camarasauromorpha: Titanosauriformes: Somphospondyli: Titanosauria: Eutitanosauria: Saltasauridae.

Diagnosis of genus: Autapomorphies including double infraprezygapophyseal laminae in first dorsals; cancellous osseous tissue in presacral and anterior caudal vertebrae; axial ventral crest between hemapophyseal ridges in centra of midcaudal vertebrae; perpendicular angle between plane including greatest proximal dimension of tibia regarding that of distal region (Sanz, Powell, Le Loeuff, Martinez and Suberbiola 1999).

Rocasaurus muniozi, MPCA-Pv 46, holotype a. left femur, lateral view, b. left ilium, lateral view, c. right ischium, lateral view, and d. left pubis, medial view. (After Salgado and Azpilicueta 2000.)

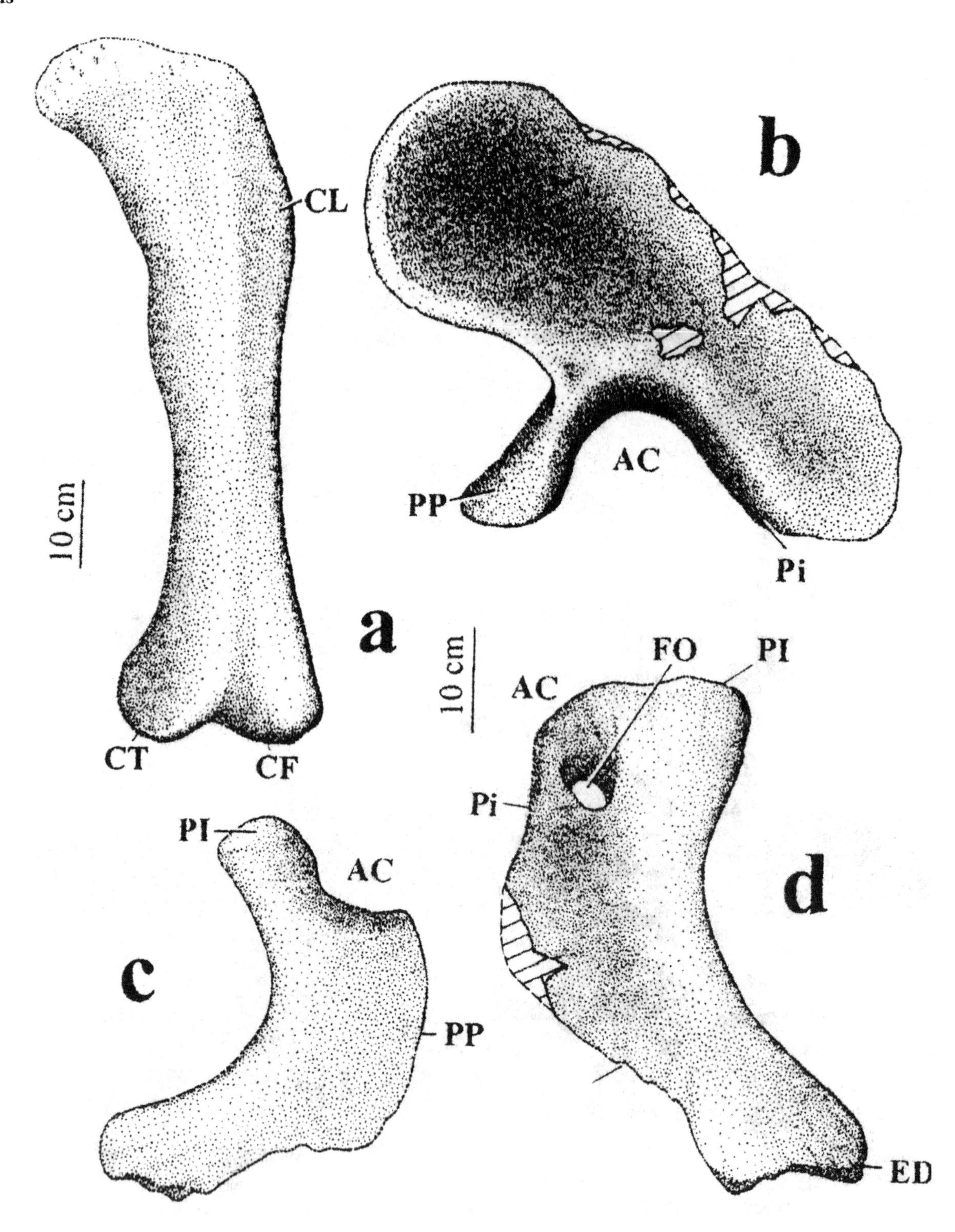

Comment: Sanz, Powell, Le Loeuff, Martinez and Suberbiola (1999) recently placed *Saltasaurus* and the genus *Lirainosaurus* into the new titanosaurian clade Eutitanosauria (see "Systematics" chapter).

Notes: Although, in the past, a number of fossil egg specimens have been assigned to various sauropoda taxa, primarily because of the formers' size, no positive identification could be made of those specimens due to the lack of such evidence as diagnostic embryonic remains within the eggs. Recently, however, the first reliable identification of egg specimens with sauropods was possible following a discovery made in South America by a research expedition led by Luis M. Chiappe, Rodolfo A. Coria, Lowell Dingus, Frankie Jackson, Anusuya Chinsamy, and Marilyn Fox. Their discovery is also significant in that it

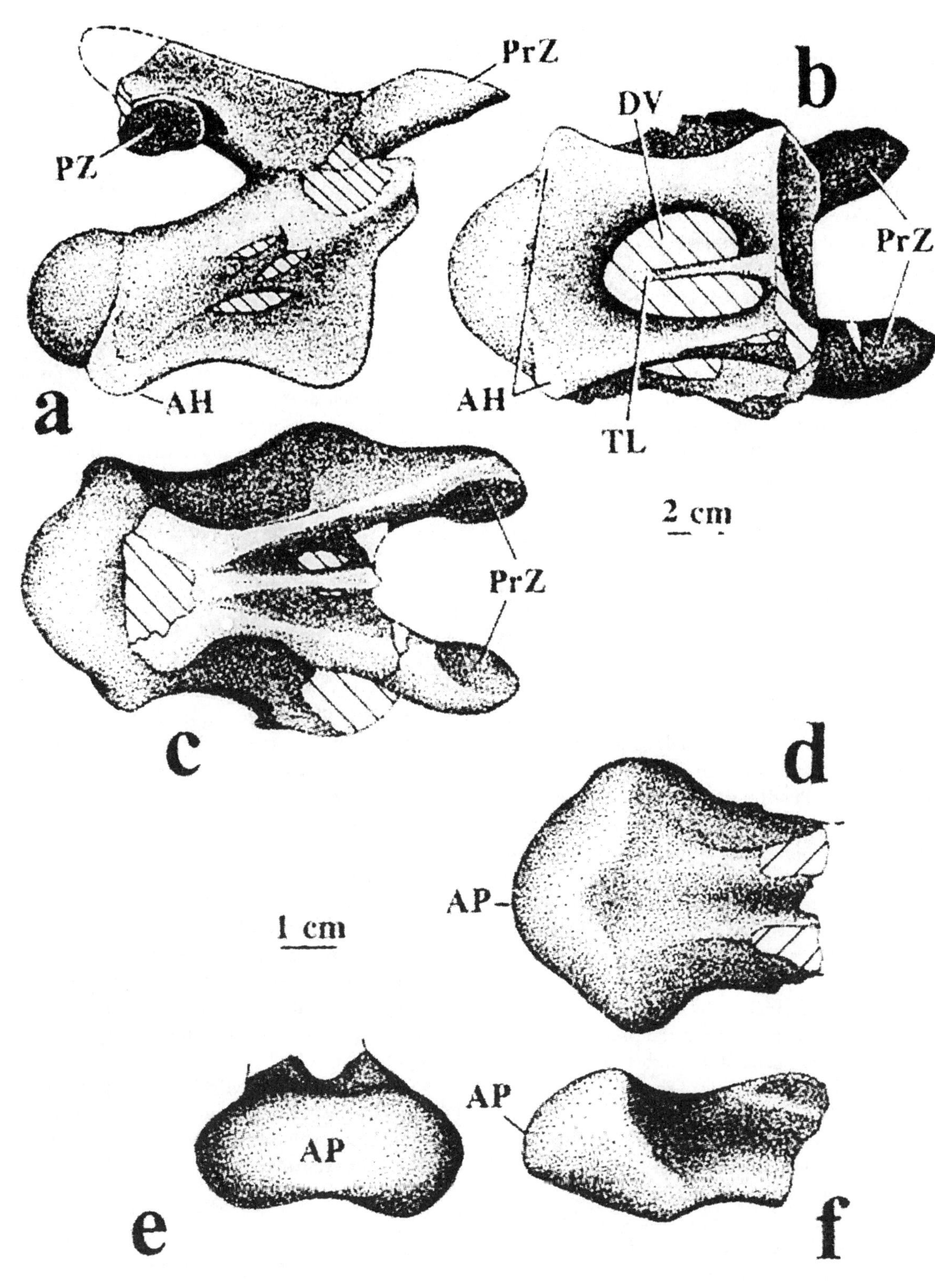

Rocasaurus **sp., MPCA-Pv 58, midcaudal vertebra in a. lateral, b. ventral, c. dorsal views; *Rocasaurus muniozi*, MPCA-Pv 56, referred posterior caudal vertebra in d. dorsal, e. posterior, and f. lateral views. (After Salgado and Azpilicueta 2000.)**

countradicts earlier speculations (see Bakker 1986) that sauropods were viviparous.

In 1998, Chiappe, Coria, Dingus, Jackson, Chinsamy and Fox reported on and described the first definitive, nonavian dinosaur embryos from Gondwana. The specimens, comprising thousands of megaloolithid eggs, some including embryonic remains, were collected over an area covering more than one square kilometer, at the Auca Mahuevo nesting ground in the Upper Cretaceous stage of the Río Colorado Formation of Patagonia, Argentina. Some of the embryonic specimens include large patches of fossil skin casts in addition to bones, these distinguished as the first unequivocal portions of integument ever reported for nonavian dinosaur embryos. The identical morphology of the eggs containing the embryonic remains indicated to the authors that these specimens all belong to the same sauropod species.

Chiappe *et al.* noted that cranial elements preserved in at least two of the specimens (PVPH-112 and PVPH-113) allowed a confident taxonomic identification of the embryos.

As described by Chiappe *et al.*, the left postorbital of PVPH-112 displays "an inverted L-shape with a long, tapering jugal process, a depressed and rostrally expanded anterior process, and a short, blunt squamosal process," similar to the postorbital of sauropods (*e.g.*, McIntosh 1990); the thin, elongate jugal process is considered to be a diagnostic feature of the Sauropoda (Gauthier 1986). The frontals of the visible skull roof of PVPH-113 are slightly wider than long, this feature also regarded as a synapomorphy of the Sauropoda (Gauthier).

The teeth are pencil-like, with straight margins that taper gradually towards the apex of the crown. The crowns are formed by smooth enamel without denticles or primary ridges, this feature regarded as a synapomorphy of the Neosauropoda (see Wilson and Sereno 1998). Pencil-like teeth constitute a feature found in diplodocid, dicraeosaurid, and titanosaurian sauropods, while the dental morphology of the Auca Mahievo specimens most favorably compares with the chisel-like teeth of titanosaurs. One tooth (belonging to specimen PVPH-112) shows a wear facet, the presence of such facets of embryonic individuals considered to be a synapomorphy of the Eusauropoda (Wilson and Sereno). The wear facet in PVPH-112 has an angle of 33 degrees with respect to the main axis of the tooth, this angle, Chiappe *et al.* noted, being typical of titanosaurs but smaller than that known for diplodocids.

Petrographic analysis by Chiappe *et al.* of the preserved integument revealed the presence of clay, quartz, and a carbonate mineral, these results showing that the patches are casts of the original skin.

The authors described the general pattern of the preserved skin patches as comprising "round, nonoverlapping, tubercle-like scales of subequal size" approximately 300 millimeter in diameter. One sample (PVPH-126) "contains a stripe formed by three rows of larger scales." The subrectangular scales of the central row are twice the size (about 800 millimeters) of the scales of the lateral rows (about 400 millimeters). Although the exact anatomical position of these patches could not be established, the authors speculated that the "stripe" probably ran along the embryo's back. Specimen PVPH-130 includes scales displaying a rosette pattern; the round central scale (approximately 780 millimeters in diameter) is encircled by ten smaller scales with about one-half the diameter of the former. The scale pattern of the South American specimens was found to be tuberculated, the nonoverlapping pattern of scales and rosettes comparing well with other collected specimens of dinosaur skin (see Czerkas 1994).

As observed by Chiappe *et al.*, the scale patterns of the Auca Mahuevo specimens are similar to the mosaic pattern seen on small osteoderms associated with the bones of *Saltasaurus loricatus*, an armored sauropod from the Late Cretaceous of Argentina. However, no indications of distinct osteoderms were found in thin sections of the embryonic specimens. Perhaps then, the authors suggested, osteoderms in sauropods such as *Saltasaurus* appeared only after hatching (a pattern of development also known in various extant reptiles, *e.g.,* the cordylid lizard *Cordylus cataphractus* and the Nile crocodile *Crocodylus niloticus*), with development occurring during ontogeny, most likely "developing in a one-to-one correspondence with the scales."

The fractured and rather compressed eggs, as described by Chiappe *et al.*, are round, about 13–15 centimeters in diameter, and were apparently spherical to subspherical in shape, with shell thickness varying from 1.00 to 1.78 millimeters and an average thickness of 1.40 millimeters. Some of the less weathered specimens show ornamentation consisting of compact, domed tubercles; in some specimens, the latter coalesce to form short ridges.

Based upon their observations of the Auca Mahuevo specimens, Chiappe *et al.* stated that the limited osteological evidence permitted no positive taxonomic identification beyond Sauropoda, although dental morphology nests them within Neosauropoda. Furthermore, as titanosaurian neosauropods are the only sauropods known from the Río Colorado Formation and the only Late Cretaceous forms having pencil-like teeth, the embryos are most likely titanosaurs.

Key references: Bakker (1986); Chiappe, Coria, Dingus, Jackson, Chinsamy and Fox (1998); Czerkas (1994); Gauthier (1986); McIntosh (1990); Sanz, Powell, Le Loeuff, Martinez and Suberbiola (1999); Wilson and Sereno (1998).

SANTANARAPTOR Kellner 1999

Saurischia: Theropoda: Neotheropoda: Tetanurae: Avetheropoda: Coelurosauria: ?Manuraptoriformes.

Name derivation: "Santana [Formation]" + Latin *raptor* = "thief."

Type species: *S. placidus* Kellner 1999.

Other species: [None.]

Occurrence: Santana Formation, Ceara State, Brazil.

Age: Middle Cretaceous (Albian).

Known material/holotype: MN 4802-V, both (almost complete) ischia, complete right hindlimb and incomplete left hindlimb (both femora preserved), three caudal vertebrae, unidentified elements, soft tissue.

Diagnosis of genus (as for type species): Coelurosaur with the following autapomorphies: ischial obturator notch very large (25 to 30 percent length of ischium); foramen on base of lesser trochanter; very well-developed sulcus on head of femur; fibular trochlea triangular in shape, constricted at base (Kellner 1999).

Comments: The genus *Santanaraptor* was established upon a well-preserved, partial postcranial skeleton (MN 4802-V; cast MCT 1502-R) found in a calcareous nodule in an outcrop of the Romualdo Member in the upper lithostratigraphic unit of the Santana Formation (Beurlen 1971), near the town of Santana do Cariri, in Ceara State, northeastern Brazil. Preserved at this same site were ostracods and a number of fish scales. Thus far, six dinosaur specimens are known from the Romualdo Member of the formation, two of which have been named (the spinosaurids *Irritator challengeri* and *Angaturama limai* [the latter now considered a junior synonym of *I. challengeri*; see *Irritator* entry]) (Kellner 1999).

To date of Kellner's original publication on this

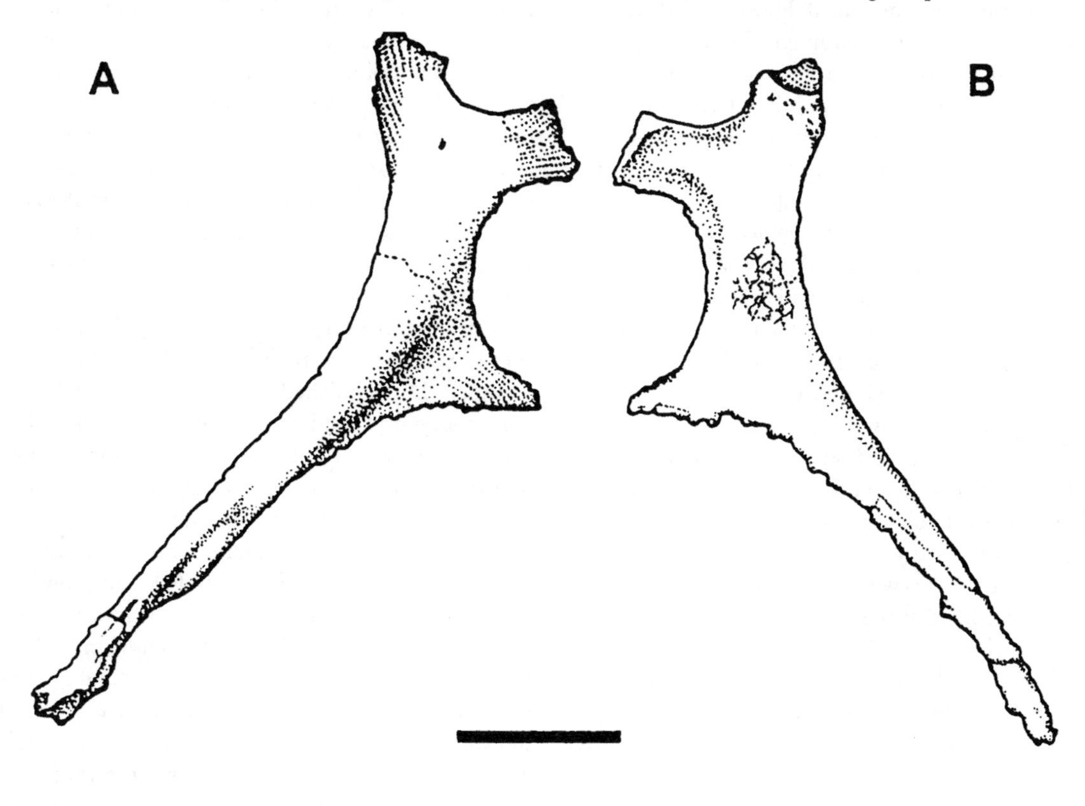

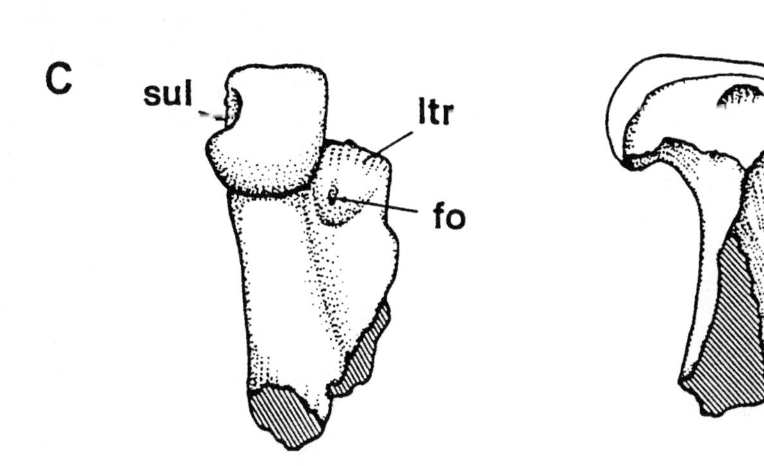

***Santanaraptor placidus*, MN 4802-V, holotype right ischium in A. lateral and B. medial views, and proximal portion of left femur in C. anterior and D. medial views. Scale = 20 mm. (After Kellner 1999.)**

dinosaur, the holotype of the type species, *S. placidus*, had only been partially prepared. Therefore, Kellner only briefly described this taxon, stating that a more detailed description will be published at a later date. The author also noted that the material represents a rather small animal, the shaft of the femur measuring only 13 millimeters in diameter.

As observed by Kellner (and based on Sereno's 1999 phylogenetic breakdown of the Dinosauria), *Santanaraptor* lacks any "ceratosaurian" features, but exhibits at least one coelurosaurian synapomorphy — the triangle-shaped oburator process. Also, the unusual morphology of the obturator notch — U-shaped, with slightly diverging sides — suggests affinities with the Manuraptoriformes. Kellner noted, however, that relating *Santanaraptor* to this group can only be confirmed following the full preparation of the type specimen.

In a subsequent report published as an abstract, Kellner (2000) offered new information regarding *S. placidus*. The type specimen, the author noted, also includes associated preserved soft tissue comprising dermis (lacking feather-like filaments) and epidermis, preserved in a calcareous nodule typical from the Romualdo Member of the Santana Formation.

Kellner (2000) also noted the following features not reported in his original description of the type specimen, some of which may be incorporated into a later emended diagnosis of *S. placidus*: Femur gracile, hollow, curved; deep cleft separating blade-like lesser trochanter from greater trochanter; fibular trochlea well developed; fibula very thin, laterally compressed, elongated, with sharp flange at distal portion near articulation; astragalus with large, blade-like ascending process, dorsomedial edge crossing medial margin of tibia at high angle; calcaneum small, higher than wide, occupying less than 10 percent maximum width of astragalus; metatarsals long, reaching approximately 70 percent of length of femur; metatarsal III slightly larger, not pinching out proximally; metatarsals II to IV about same size.

Key references: Beurlen (1971); Kellner (1999, 2000); Sereno (1999).

SATURNALIA Langer, Abdala, Richter and Benton 1999

Saurischia: Sauropodomorpha: Prosauropoda.

Name derivation: Latin *Saturnalia* = "[equivalent of] carnival."

Type species: Langer, Abdala, Richter and Benton 1999.

Other species: [None.]

Occurrence: Santa Maria Formation, Rio Grande do Sul, Brazil; ?Pebbly Arkose Formation, Zimbabwe, Africa.

Age: Late Triassic (Carnian).

Known material: Partial postcranial skeletons, three individuals.

Holotype: MCP 3844-PV, partial, semiarticulated postcranial skeleton including most of presacral vertebrae, both sides of pectoral girdle, right humerus, partial right ulna, both sides of pelvic girdle with sacral series, left femur, most of right hindlimb.

Diagnosis of genus (as for type species): Gracile, measuring 1.5 meters (approximately 5.2 feet) in length; skull length about one third that of femur; teeth lanceolate; deltopectoral crest extending for more than 50 percent of humeral length; anterior blade of ilium short, pointed; acetabulum not fully opened; proximal portion of pubes deep, lateral border of pubis robust and triangular in distal aspect; femur subequal in length to tibia; proximal portion of femur having well-developed trochanteric shelf; ascending process of astragalus broad, laterally elongated (Langer, Abdala, Richter and Benton 1999).

Comments: Until the recent discovery of *Saturnalia tupiniquim*, sauropodomorphs of Carnian age were represented only by fragmentary remains found in North America and Africa, with large, heavily-built forms restricted to rocks some 10 million years younger. Distinguished as the oldest well known sauropodomorph and the best known Carnian herbivorous dinosaur, this type species was founded upon well-preserved, incomplete postcrania (MCP 3844-PV) collected from private land in the Alemoa Member of the Santa Maria Formation, on the outskirts of Santa Maria, in state of Rio Grande do Sul, southern Brazil. Paratype specimens recovered from the same locality consist of MCP 3845-PV, comprising a skeleton including the natural cast of a mandibular ramus with 16 teeth, some articulated trunk vertebrae, both sides of the pectoral girdle, and most of the right hindlimb, and MCP 3846-PV, a skeleton (not completely prepared), which includes, among other elements, some trunk vertebrae, and a partial tibia and foot (Langer, Abdala, Richter and Benton 1999).

In describing *S. tupiniquim*, Langer *et al.* deduced that the length of the skull, unknown in the collected materials, can be roughly estimated based on the partial mandibular ramus of the paratype specimen MCP 3845-PV. From this fragment, probably making up most of the dentary, the authors projected the entire length of the head to have a maximum length of 100 millimeters. Langer *et al.* described the teeth as like those of prosauropods, "being small, lanceolate and coarsely serrated (Galton 1990*a*)."

The vertebral column displays similarities with basal sauropodomorphs — *e.g.*, penultimate cervical vertebra much longer than trunk vertebrae, vertebra

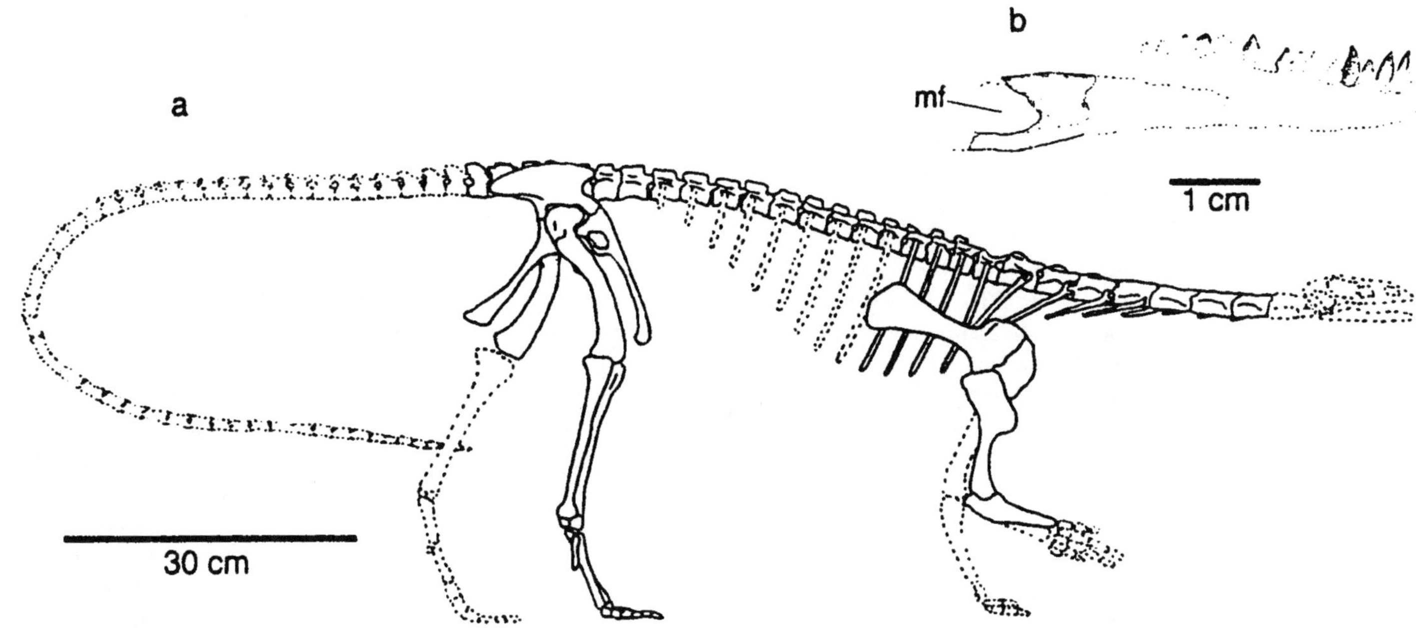

***Saturnalia tupiniquim*, a. skeletal reconstruction based on preserved parts of holotype MCP 3884-PV (full lines); b. paratype MCP 3845-PV, natural cast (dashed lines) and preserved parts (full line) of mandibular ramus. (After Langer, Abdala, Richter and Benton 1999.)**

of caudal origin added to sacrum (Galton 1973*b*, 1990*a*). The humerus, with a deltopectoral crest that extends for about half the bone length, is also like that of sauropodomorphs (Galton 1990*a*). The only partially opened acetabulum is a plesimorphic feature that suggested to Langer *et al.* that the complete opening of this structure may have occurred more than once during the evolution of dinosaurs. According to Langer *et al.*, a cladistic analysis of *Saturnalia*, based upon its preserved remains, supports its referral to the Sauropodomorpha, while its overall morphology indicates that this genus is the most basal sauropodomorph yet known. This position, the authors noted, corresponds to the greater age of this genus. However, a more recent cladistic analysis by Galton and Upchurch (in press), to be published in the second edition of the book *The Dinosauria* (in press; see also Galton and Upchurch 2000), utilizing 135 characters, places *Saturnalia* well into the Prosauropoda (P. M. Galton, personal communication 2000).

From the slender tibia, fibula, and metatarsals, *Saturnalia* was judged by Langer *et al.* to be less robust than most basal prosauropods [=sauropodomorphs of their usage] (see Galton 1990*a*). According to those authors, this gracile morphology, apparently typical of basal representatives of the other major dinosaurian lineages (see Bonaparte 1976; Sereno, Forster, Rogers and Monetta 1993), supports the hypothesis of a common origin for most of these groups from lightly-built, basal ornithodirans of the Middle Triassic (Gauthier 1986; Novas 1996*a*).

Langer *et al.* noted that, until the discovery of *Saturnalia*, the occurrence of "basal sauropodomorphs" from a time equivalent in age to the earliest known carnivorous dinosaurs was expected but only suggested, based on an unnamed form represented by a proximal femoral fragment from the Pebbly Arkose Formation of Zimbabwe (see Raath, Oesterlen and Kitching 1992; Raath 1996), and also the fragmentary *Azendohsaurus laaroussii* from the Argana Formation of Morocco (Gauffre 1993). Combined, however, these records now offer evidence that supports a widespread distribution of sauropodomorphs during the late Carnian. Furthermore, the authors observed that the morphology of the Zimbabwean femoral fragment "is almost indistinguishable from that of *Saturnalia*, and it would not be surprising if they belonged in the same taxon." (According to prosauropod specialist Peter M. Galton, personal communication 2000, this comparison cannot adequately be made, the Zimbabwean material being indeterminate.)

Key references: Bonaparte (1976); Galton (1973*b*, 1990*a*); Galton and Upchurch (2000); Gauffre (1993); Gauthier (1986); Langer, Abdala, Richter and Benton (1999); Novas (1996*a*); Raath, Oesterlen and Kitching (1992); Raath (1996); Sereno, Forster, Rogers and Monetta (1993).

†SAUROLOPHUS

Ornithischia: Genasauria: Cerapoda: Ornithopoda: Euornithopoda: Iguanodontia: Euiguanodontia: Dryomorpha: Ankylopollexia: Hadrosauroidea: Hadrosauridae: Euhadrosauria: Hadrosaurinae.

Note: Ishigaki (1999) reported the discovery of more than 600 large (from 25 to 155 centimeters in length), tridactyl fossil footprints found in Upper Cretaceous rocks at Bugeen Tsav and Gurilin Tsav, in the western Gobi Desert, Mongolia. These ornithopod tracks, discovered by the Hayshsibara Museum of Natural Science–Mongolian Paleontological Center Joint Paleontological Expedition Team between 1995 and 1998 have been attributed to the large duckbilled, solid-crested dinosaur *Saurolophus*, based on the abundance of bones belonging to that genus found in the same region.

According to Ishigaki, more than 15,000 dinosaur footprints from 11 different Mongolian Gobi localities were found during this expedition. Other groups of dinosaurs represented by these tracks included theropods and a probable ankylosaurid. Additional materials from the same beds included skeletal remains of dinosaurs, crocodiles, turtles, and fish, also dinosaur eggs, mollusks, and "petrified wood."

Key reference: Ishigaki (1999).

†SAUROPELTA

Ornithischia: Genasauria: Thyreophora: Thyreophoridea: Eurypoda: Ankylosauria: Nodosauridae.

Age: Early to Middle Cretaceous (Aptian–Albian).

Diagnosis (for skull): Skull armor not well defined as in *Edmontonia*, *Panoplosaurus*, and *Pawpawsaurus*; snout apparently tapering (not parallel-sided as in *Panoplosaurus*); orbital rim not projecting as in *Pawpawsaurus*; skull roof flat (not domed as in *Pawpawsaurus*, *Silvisaurus*, and *Struthiosaurus*); lateral temporal notch on skull roof more developed than in *Panoplosaurus* and *Edmontonia*, less than in *Pawpawsaurus*; postorbital scute small, rounded (large in *Pawpawsaurus*); jugal scute large, rounded (moderate-sized, rounded in *Pawpawsaurus*, *Niobrarasaurus*, and *Silvisaurus*, round horn in *Edmontonia*) (Carpenter and Kirkland 1998).

Comments: Although Ostrom (1970) described in detail *Sauropelta edwardsi* (see *D:TE*)—a large nodosaurid type species founded upon a partial skeleton (AMNH 3032) from the Cloverly Formation (Aptian–Albian) of Big Horn County, Montana—he did not illustrate the skull. More recently, Carpenter and Kirkland (1998) published a revised diagnosis of the skull of this taxon, at the same time offering multiple views of the most complete skull (AMNH 3035), also the postorbital region of another skull (YPM 5499), and a braincase (YPM 5178).

Carpenter and Kirkland described in detail the skull of *Sauropelta* noting that it is large, measuring 35 centimeters (about 13.5 inches) wide across the postorbital scutes. Individual cranial scutes are not well-defined. The loss of such scutes in *Edmontonia* had been assumed by Carpenter (1990) to indicate old age; however, additional preparation by Carpenter of one *Edmontonia* specimen (DMNH 468) exposed "a thin layer of concretionary material masking a typical *Edmontonia* pattern (Carpenter, unpublished)." The possibility exists, therefore, "that a thin layer of concretionary material obscures these features in the *Sauropelta* skull, or that preparation obliturated these sutures." Ostrom had inferred that *Sauropelta* possessed premaxillary teeth. This could not be confirmed by Carpenter and Kirkland, who pointed out that the skull described by Ostrom had been cemented to its base, thus preventing examination of the palate.

In 1984, Carpenter believed that the cervical armor in *Sauropelta* was arranged in a single paired row with dorso-posteriorly projecting spines (see *D:TE*). However, further examination by Carpenter and Kirkland of the *Sauropelta* material—including three more recently discovered pairs of large plates apparently associated with the articulated cervical armor (AMNH 3035)—suggested that the neck armor also consisted of laterally projecting scutes. This interpretation was supported by a partial cervical ring (YPM 5178) including a dorso-posteriorly-projecting spine coosified with a laterally-projecting spine.

Carpenter and Kirkland noted "that the lateral spines along the sides of the body and anterior part of the tail are probably much larger than illustrated by Carpenter (1984)." Additionally, the holotype includes two large lateral scutes having a posteriorly directed spine (see Ostrom), while another specimen (YPM 5490) consists of a large lateral plate having a paired or bifurcated spine. Comparing these elements with the articulated caudal armor of referred specimen AMNH 3036 (see *D:TE* for additional photograph), Carpenter and Kirkland suggested that these plates were most likely located at the base of the tail. Although such large plates overlapping several caudal vertebrae would have restricted the lateral flexibility of a tail which seems to have functioned as a weapon, "a less flexible tail would provide the caudo-femoralis with a more stable platform for retraction of the hindlimb."

Contrary to earlier reports, *Sauropelta* is not known from the Cedar Mountain Formation of Utah (K. Carpenter, personal communication 2000).

Notes: For reasons already stated by Carpenter in 1984, Carpenter and Kirkland reaffirmed their belief that *Sauropelta* is the trackmaker of fossil footprints named *Tetrapodosaurus borealis* (see "Notes," *Sauropelta* entry, *D:TE*).

Parsons and Parsons (2000) reported, in an abstract for a poster, the recovery of a ?nodosaurid skull from the purplish red mudstone layers of the basal portion of Unit VII of the Cloverly Formation, Middle Dome Area, of Wheatland County, Montana. The authors tentatively referred this specimen to the Nodosauridae based upon the following features: Large, open lateral temporal fenestra; considerable separation between paroccipital processes and postorbital horn

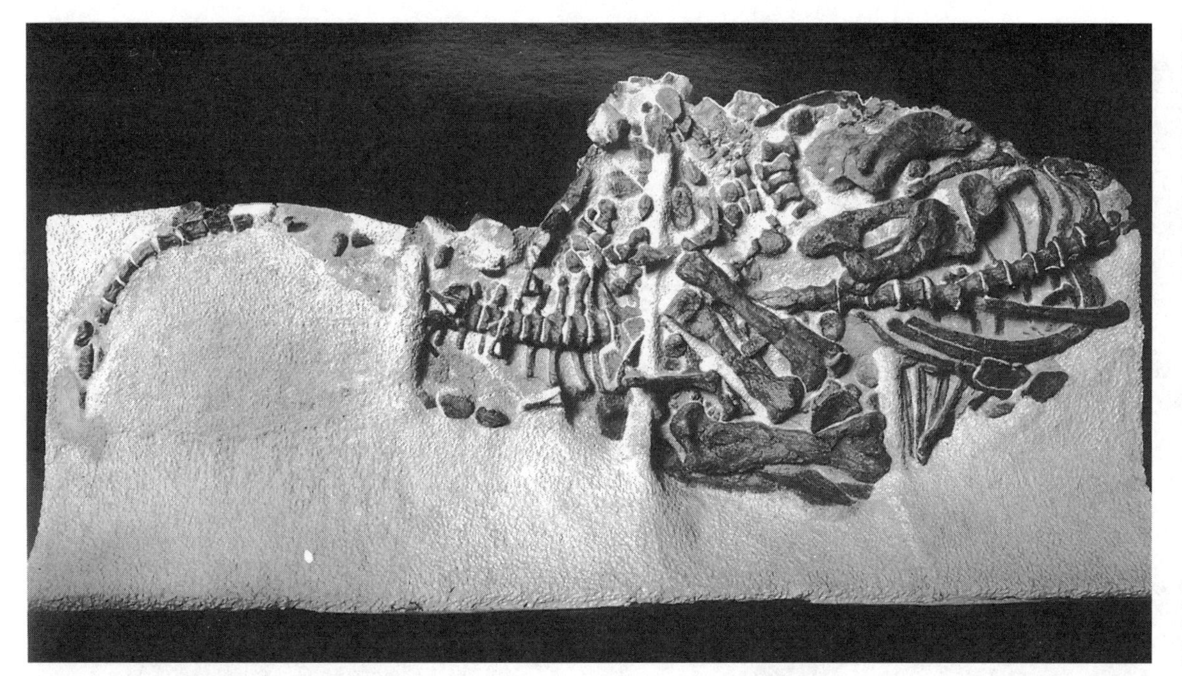

Skeleton (AMNH 3036) of the armored dinosaur *Sauropelta edwardsi* collected from Upper Cretaceous rocks south of Billings, Montana. For an opposite-side view of this more than four-meters long specimen showing the dermal armor, see *Sauropelta* entry, *D:TE.*

Photograph by the American Museum of Natural History Photo Studio. Courtesy Department of Library Services, American Museum of Natural History (neg. #314803).

(due to expanded temporal region); and lack of any degree of overhanging in supraoccipital region. According to Parsons and Parsons, this specimen possesses some interesting phylogenetic characters, including the following: Distinct jugal horn; expanded (not notched) temporal region; and prominent postorbital horn (other structures still being prepared for examination). A distinct, rather large bony orbital capsule has also been exposed in the specimen.

As *Sauropelta edwardsi* is not currently known from good skull material, Parsons and Parsons were unable to compare the new specimen with that taxon.

Key references: Carpenter (1984, 1990); Carpenter and Kirkland (1998); Ostrom (1970); Parsons and Parsons (2000).

SAUROPOSEIDON Wedel, Cifelli and Sanders 2000

Saurischia: Sauropodomorpha: Sauropoda: Eusauropoda: Neosauropoda: Macronaria: Camarasauromorpha: Titanosauriformes: Brachiosauridae.

Name derivation: Greek *sauros* = "lizard" + Poseidon [in Greek mythology, god of earthquakes].

Type species: *S. proteles* Wedel, Cifelli and Sanders 2000

Other species: [None.]

Occurrence: Antlers Formation, Oklahoma, United States.

Age: Early to Middle Cretaceous (Aptian–Albian).

Known material/holotype: OMNH 53062, articulated cervical vertebrae 5 to 8, cervical ribs preserved in place.

Diagnosis of genus (as for type species): Cervical centra extremely elongate; centrum length over five times height of posterior centrum; differing from all other sauropods in possessing well-defined centroparapophyseal laminae that extend to posterior ends of centra, diapophyses located approximately one third of centrum length behind anterior condyles, deeply excavated neural spines that are perforate in anterior cervicals, and hypertrophied central pneumatic fossae that extend posteriorly to cotyles; neural spines occupying anterior nine-tenths of centra, not bifurcated; cervical ribs slender, elongated, with long, robust anterior processes extending nearly to anterior condyles; total length of each cervical rib equalling or exceeding length of three centra (Wedel, Cifelli and Sanders 2000*a*).

Comments: The new taxon *Sauroposeidon proteles*— a sauropod larger than the gigantic *Brachiosaurus*— was founded upon a series of articulated cervical vertebrae (OMNH 53062) collected by Oklahoma Museum of Natural History crews in May and August of 1994 from a claystone outcrop of the Antlers Formation of Atoka County in southeast Oklahoma (Wedel, Cifelli and Sanders 2000*a*).

Comparing *Sauroposeidon* to various other sauropod taxa, Wedel *et al.* (2000*a*) found it to be most similar to the brachiosaurid *Brachiosaurus*, sharing with that genus a number of synapomorphies including the following: 1. Elongate cervical centra and long cervical ribs (independently derived in other sauropod lineages, but definitely separating advanced Brachiosauridae from the Titanosauridae); 2. the derived camellate pattern of internal pneumatic structure (this feature having evolved at least three times — reported in some basal titanosaurs (M. Webel, unpublished

Courtesy Matthew J. Wedel.

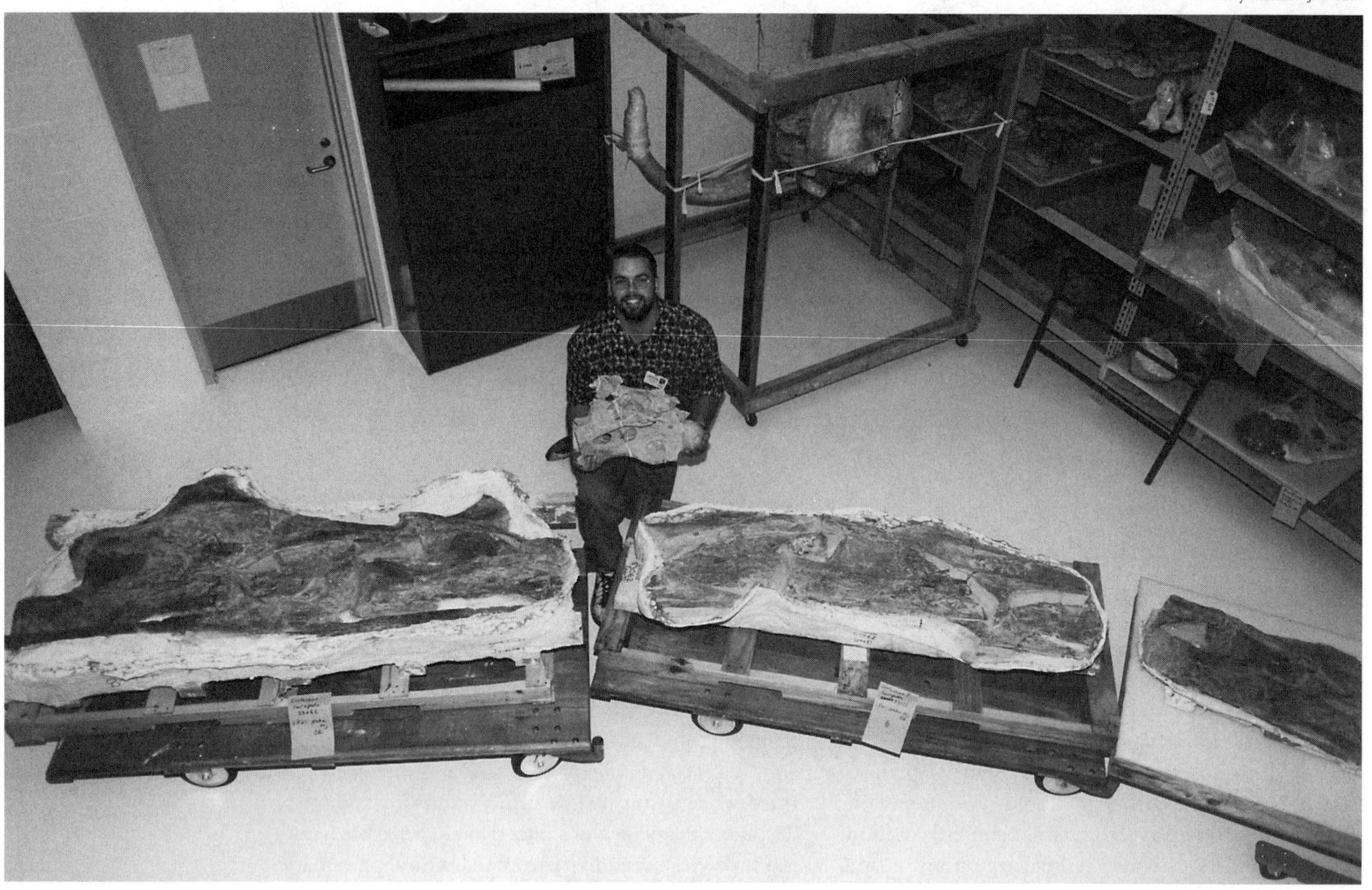

Paleontologist Matthew J. Wedel, holding for comparison a cervical (probably the sixth) vertebra of *Apatosaurus*, with holotype cervical vertebrae 5 to 8 of *Sauroposeidon proteles*.

data), *Brachiosaurus* (see Britt 1993; Upchurch 1998) and *Diplodocus* (see Britt)— as a weight-saving adaptation correlated with neck elongation); and 3. a mid-cervical transition point in neural spine height and morphology. Therefore, the authors referred *Sauroposeidon* to the Brachiosauridae.

Wedel *et al.* (2000*a*) also observed similarities between *Sauroposeidon* and a juvenile sauropod reported by Ostrom (1970) represented by a poorly preserved cervical vertebra (YPM 5294) from the Cloverly Formation (Lower–Middle Cretaceous; Aptian–Albian) of Montana, this suggesting that the latter may represent a young *Sauroposeidon* or a closely related taxon. As described by Ostrom, YPM 5294 has not undergone fusion, has a centrum length of 470 millimeters, and an uncrushed centrum height of 90 millimeters. Wedel *et al.* (2000*a*) noted that the 5.2 length-to-diameter ratio of the Montana vertebra closely approximates the proportions of *Sauroposeidon*; also, as in *Sauroposeidon*, the juvenile specimen possesses long, thin centroparapophyseal laminae that extend posteriorly from the parapophyses approximately halfway to the posterior end of the centrum. According to these authors, "YPM demonstrates that the distinctive vertebral proportions seen in *Sauroposeidon* can be achieved at a relatively early age, and that the development of centroparapophyseal laminae in some long-necked taxa predates fusion of the neural elements and may be an ontogenetically stable feature."

North American Cretaceous sauropods are relatively rare. In the past, much of the sauropod remains collected from the Lower Cretaceous rocks on that continent have been traditionally referred to *Pleurocoelus* (see *D:TE, S1*) or *Astrodon* (see entry), a poorly understood genus founded upon four disarticulated vertebrae from a very young animal. As shown by Webel *et al.*, however, both *Sauroposeidon* and YPM 5294, given the gross proportional differences between their vertebrae, can be excluded from *Pleurocoelus*. Furthermore, with the discovery of this new genus, the assignment of any sauropod remains from the Aptian–Albian of North America to *Pleurocoelus* should be reexamined.

According to Wedel *et al.* (2000*a*), *Sauroposdeidon* is notable for lightening the vertebrae of the neck to an extreme degree without sacrificing structural integrity. Indeed, the "intersection of vertical septa of bone, such as the medium septum of the centrum, with bony laminae extending laterally to support the

Courtesy Matthew J. Wedel, revised from Wedel, Cifelli and Sanders 2000*a*.

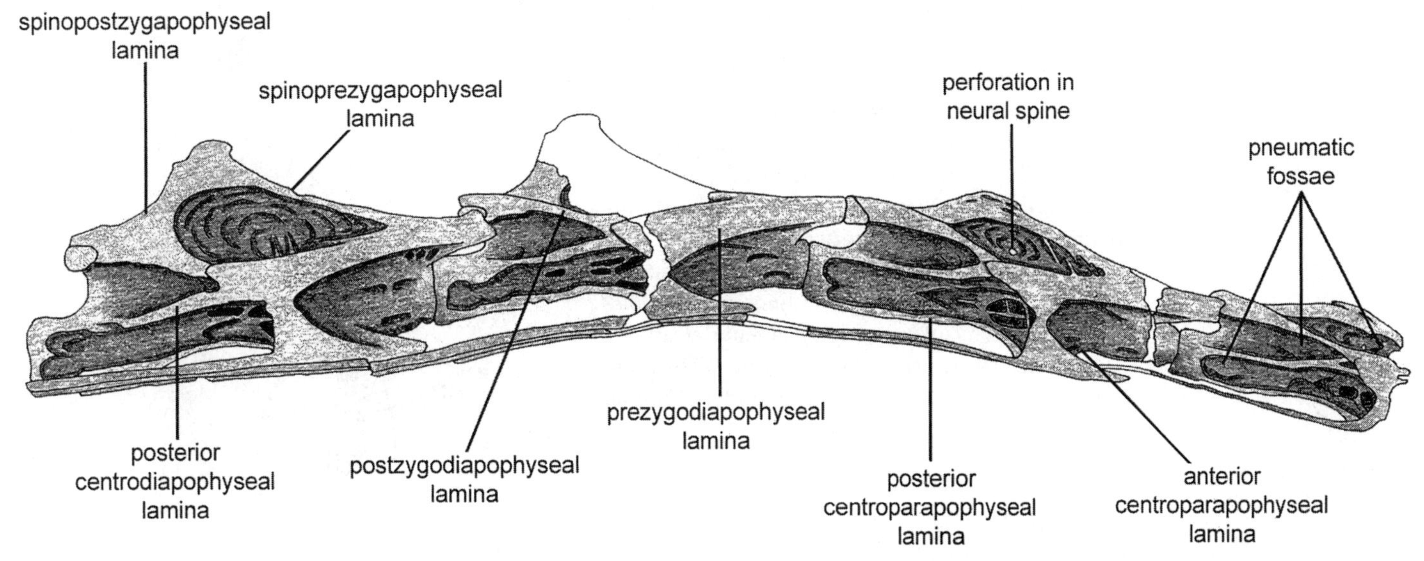

***Sauroposeidon proteles*, OMNH 53062, holotype articulated cervical vertebrae 5 to 8, lateral view, bottom drawing showing bone restored to missing areas. Scale = 1 m.**

diapophyses and ventrolaterally to support the parapophyses (centroparapophyseal laminae) gives the vertebra a longitudinal structure similar to that of an I-beam … and allows for great mechanical strength with little cross-sectional area."

Wedel *et al.* (2000*a*) further noted the biomechanical significance of the midcervical transition point in both *Sauroposeidon* and *Brachiosaurus*. The high neural spines of the posterior cervical vertebrae would have enhanced the leverage of the dorsal muscles

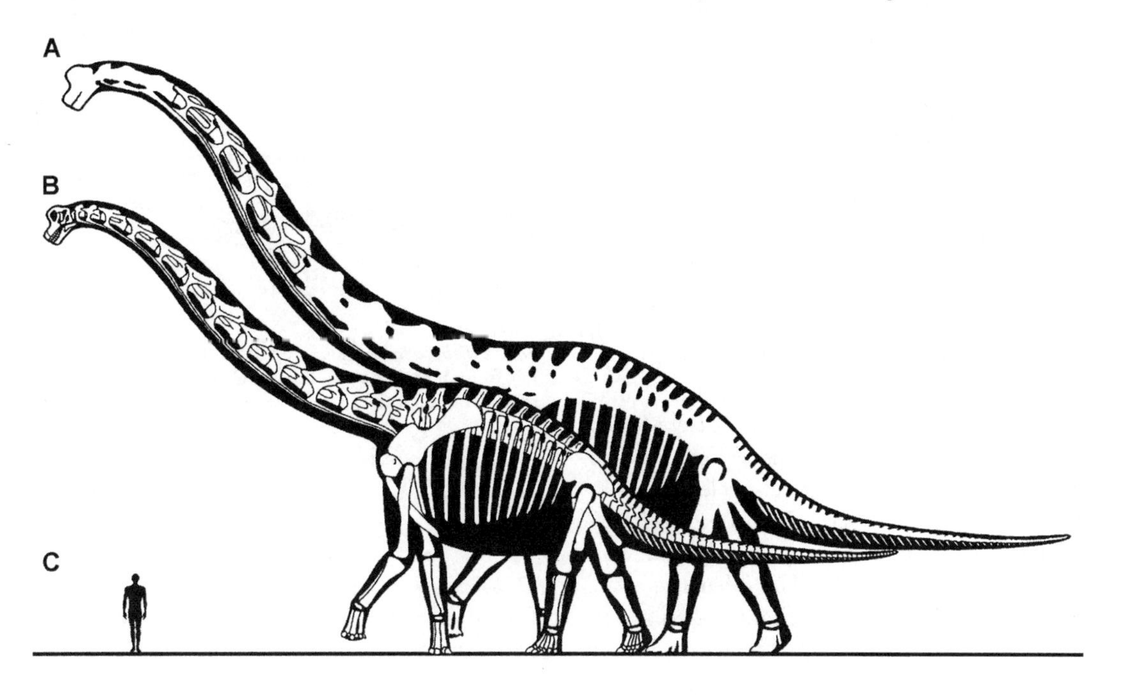

A. *Sauroposeidon proteles* (hypothetical reconstruction based on *Brachiosaurus brancai*) and B. *Brachiosaurus brancai* skeletons and C. *Homo sapiens* shown to indicate scale, the human figure being 1.8 meters tall. (After Wedel 2000*a*.)

supporting the neck, the more slender morphology of the anterior cervicals increasing the flexibility of the head and distal region of the neck. The transition point, marking the change from a more upright to more horizontal posture in the distal third of the neck, would have given a shallow S-curved to the neck when in neutral pose. *Sauroposeidon* seems "to be the last of the giant North American sauropods and represents the culmination of brachiosaurid trends towards lengthening and lightening the neck."

Wedel *et al.* (2000*b*) subsequently published an osteology of *S. proteles* (not available to date of this writing).

Note: Prior to the official publication of this new type species, reports on the dinosaur's discovery appeared in various newspaper articles, including one published in the November 3, 1999 edition of the *Chicago Sun-Times*. According to these articles, the dinosaur's remains were first spotted by a dog handler at a state prison in Atoka County. The remains supposedly indicated "one of the biggest [dinosaurs] ever discovered: a 60-ton, six-story high giraffelike creature."

Key references: Britt (1993); Ostrom (1970); Upchurch (1998); Wedel, Cifelli and Sanders (2000*a*, 2000*b*).

†SAURORNITHOLESTES

Saurischia: Theropoda: Neotheropoda: Tetanurae: Avetheropoda: Coelurosauria: Manuraptoriformes: Manuraptora: Deinonychosauria: Eumanuraptora: Dromaeosauridae.

Occurrence: Dinosaur Park Formation, Alberta, Kirtland Formation, New Mexico, United States.

Known material: Numerous partial skulls consisting mostly of frontals, partial postcrania.

Holotype: RTMP 74.10.5, partial skull including frontals, left pterygoid, two teeth, postcrania including two vertebrae, gastralia, fragments of thoracic ribs, a metacarpal, several phalanges, all unguals of left manus, three isolated prezygapophyses, some very fragmentary indeterminable bones.

Comments: Sullivan and Lucas (2000*b*) described a recently discovered left frontal (SMP VP 1270), referred to the genus *Saurornitholestes*, collected from the Upper Cretaceous De-na-zin Member of the Kirtland Formation, at the head of the east branch of Hunter Wash (southeast), in San Juan County, New Mexico.

The specimen differs only slightly from the holotype (RTMP 74.10.5) of the type species *Saurornitholestes langstoni*, from Alberta. In SMP PV-1270 the anterior portion of the frontal is slightly more constricted than in RTMP 74.10.5. This difference was regarded by the authors as insignificant and within the range of intraspecific variation. Therefore, Sullivan and Lucas confidently referred SMP VP-1270 to *S. langstoni*. Sullivan and Lucas referred it to that species.

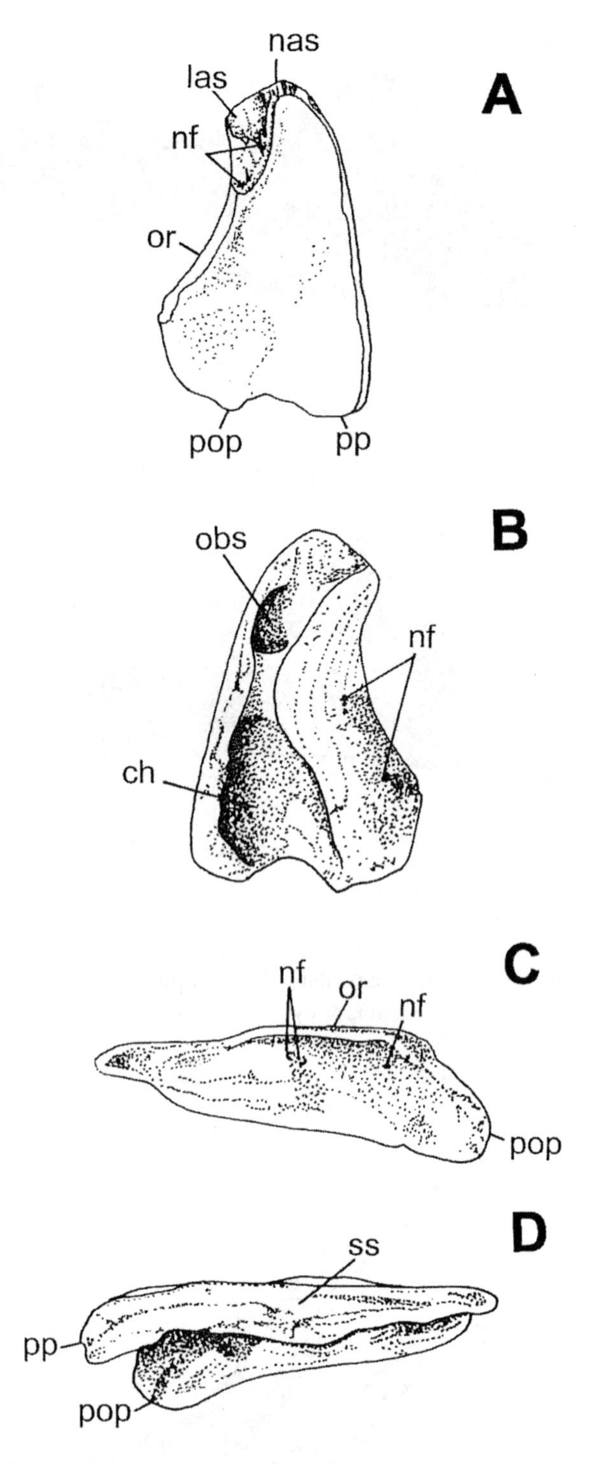

***Saurornitholestes langstoni*, SMP VP-1270, almost complete left frontal, in A. dorsal, B. ventral, C. left lateral, and D. medial views. Scale = 1 cm. (After Sullivan and Lucas 2000.)**

As observed by Sullivan and Lucas, SMP VP-1270 exhibits features that clearly distinguish the frontal of *S. langstoni* from all other known theropods, these including the following: 1.Triangular in shape, not bowel- or basin-shaped (dorsally, lacking frontoparietal crest; 2. projecting posteroventrally to meet parietal; 3. postorbital process having concave dorsal surface behind sigmoidal slope break that runs medially forming dorsal edge of parietal process; and 4. posterior concavity around supratemporal fenestra (Sues 1978; Currie 1987*b*).

Sullivan and Lucas compared the New Mexico specimen with a specimen (IGM 100/976) of the Mongolian dromaeosaurid *Velociraptor mongiliensis*, the latter preserving both frontals and including a partial postcranial skeleton (Norell and Makovicky 1999; see *Velociraptor* entry). Compared with the frontals of *V. mongoliensis*, the authors found those of *S. langstoni* 1. to be more lightly constructed or gracile and 2. subrectangular in shape, 3. to possess a weakly-developed olfactory bulb surface depression (ventrally) and 4. a weakly-developed cerebral hemisphere depression (ventrally), and 5. to have a lower angle of orbital rim (ventrally) and 6. a weakly developed posterior lateral wing. *Saurornitholestes*, Sullivan and Lucas noted, is also generically distinct from *Bambiraptor feinbergi* (see *Bambiraptor* entry).

As Sullivan and Lucas pointed out, radioisotopic dating of the De-na-zin Member of the Kirtland Formation has produced numbers indicating a Campanian age (see Obradovich 1993; Fassett and Steiner 1997). As Sullivan and Lucas (2000*a*) noted, several other De-na-zin dinosaurian genera — mostly *Albertosaurus*, *Parasaurolophus*, and *Pentceratops*— are known also from the Campanian Fruitland Formation, this being consistent with ascribing a late Campanian age to the De-na-zin.

Key references: Currie (1987*b*); Fassett and Steiner (1997); Obradovich (1993); Sues (1978); Sullivan and Lucas (2000*b*).

†SCIPIONYX

Comments: In 1998, Cristiano Dal Sasso and Marco Signore named and described the manuraptoriform theropod, *Scipionyx samniticus*, thus far the only dinosaur found in Italy, and one of the relatively few theropod genera for which a juvenile is known. The nearly complete type specimen (Soprintendenza Archeologica collection, Salernoa) is also notable — as well as unique — for its remarkably preserved internal organs, the fossilization of its soft tissues due to exceptionally conservative depositional conditions (see *S1*).

Subsequent to the original description of this dinosaur, Dal Sasso and Signore (1998*b*) published a brief report on *S. samniticus*, noting that the outstandingly preserved "muscles and internal organs, such as whole intestine, the pectoral musculature, windpipe rings, horny claws, liver remains, and *m. caudifermoralis* fibres, allows paleobiological inferences until now inconceivable." As further noted by these authors, paleobiogeographical data support the hypothesis that *Scipionyx* "evolved independently on isolated, emersed lands in the Cretaceous Central Tethys Sea."

More recently, Galliano and Signore (1999), in pointing out the rarity of very young theropods and the paucity of knowledge regarding the possibility of parental care in these kinds of dinosaurs, noted that some insights into the latter issue might be derived from the holotype *S. samniticus*, owing to its completeness and excellent state of preservation.

From a preliminary analysis of the type specimen, Galliano and Signore deduced that *Scipionyx* was capable of hunting its own live food, and that sustained predatory activity could be sustained, as evidenced by the specimen's bone articulations and teeth. The authors further surmised that small theropods such as *Scipionyx* "were precocial hatchlings, and could hunt small preys, while possibly staying with parents to acquire more hunting tactics, like modern big cats."

Ruben, Dal Sasso, Geist, Hillenius, Jones and Signore (1999) reconstructed the intestinal tract of *Scipionyx* in the dorsal cavity as an analogy to present-day crocodilians, this interpretation supposedly offering evidence that this dinosaur and, by inference, other nonavian theropods, lacked dorsal air sacs. This interpretation was criticized, however, by Martill, Frey, Sues and Cruikshank (1999) in their paper describing an unnamed coleurosaurian theropod from Brazil in which what was interpreted by them as a lithified intestinal tract was preserved (see Introduction). According to Martill *et al.*, the original material of *Scipionyx* is crushed flat, therefore being unreliable in preserving the original three-dimensional configuration of the viscera.

Key references: Dal Sasso and Signore (1998*a*, 1998*b*); Galliano and Signore (1999); Martill, Frey, Sues and Cruickshank (1999); Ruben, Dal Sasso, Geist, Hillenius, Jones and Signore (1999).

†SCUTELLOSAURUS

Comments: The small armored dinosaur *Scutellosaurus lawleri*— based on the holotype (MNA P1.175) and paratype (MNA P1.1752) skeletons from the Kayenta Formation (Lower Jurassic) of northern Arizona, the type species named and first described in 1981 by Edwin H. Colbert — has, until recently, been only incompletely known (see *D:TE* for details). As noted by Rosenbaum and Padian (2000), this taxon is especially important for its placement as a basal

Photograph by the author, courtesy Museum of Northern Arizona.

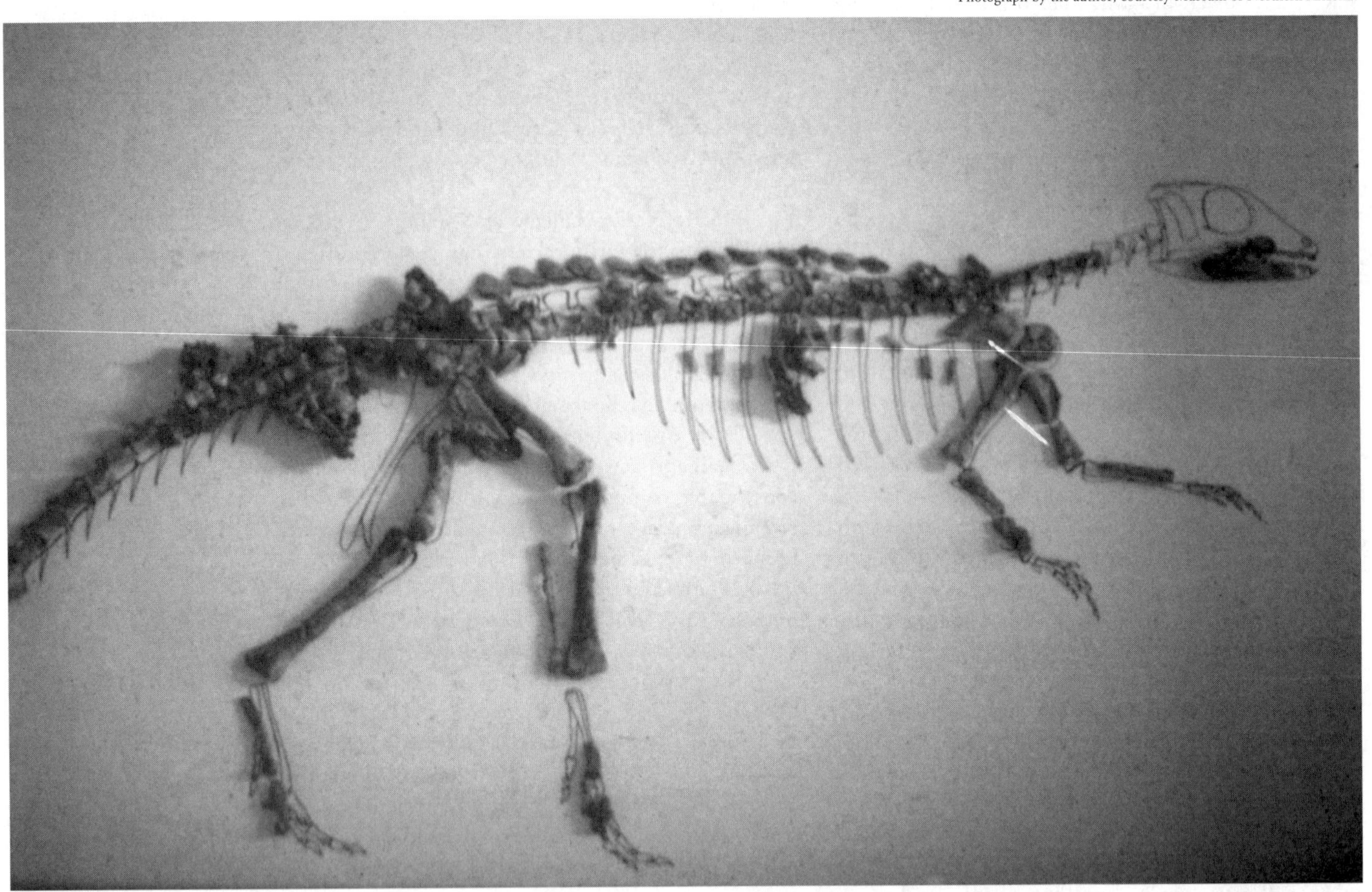

***Scutellosaurus lawleri*, holotype incomplete skeleton (MNA P1.175) originally described in 1981 by Edwin H. Colbert.**

member of the Thyreophora and also in being one of the earliest comparatively well known members of the Ornithischia, information about *S. lawleri* therefore having the potential to "elucidate the origins and history of both of these groups."

As related by Rosenbaum and Padian, additional materials from Arizona's Kayenta Formation of *S. lawleri*, specimens not utilized by Colbert in his description of this dinosaur, have been recovered. These include as of yet undescribed specimens collected during the late 1970s and early 1980s by Harvard University, and in 1997 by a joint Harvard-University of Texas team (F. A. Jenkins, Jr., personal communication to Rosenbaum and Padian), and also specimens collected by James M. Clark in 1981 and 1983 for the University of California Museum of Paleontology at Berkeley.

Clark's specimens were all recovered from localities on Navajo Nation lands along the Adeei Ecchii Cliffs in northern Arizona. The specimens, comprising sparse fragments representing from four to six individuals, consist of the following: UCMP 130580, the most complete (though mostly fragmented) specimen, including portions of both frontals, both quadrates, off-center portion of a parietal, right jugal, dentaries, 33 complete vertebral (five cervical, eight thoracic, two sacral, and 18 caudal) centra, many rib fragments, incomplete scapula, fragmented and possibly distorted humeri, fragments of shafts of both ischia, crushed femora, proximal ends of fibulae, well-preserved left astragalus, well-preserved metatarsals from both feet, three proximal phalanges, two ungual phalanges, and five fragments of a pes, and a great number of osteoderms; UCMP 170829, fragments (some distorted, but having comparative value), including a small fragment of ilium, distorted distal end of femur; also, UCMP 130581, UCMP 175166, UCMP 175167, and UCMP 17568. Described for the first time by Rosenbaum and Padian, these specimens revealed numerous details not previously known regarding the anatomy of *S. lawleri*. Of special importance, these authors noted, were portions of the skull, ankle, forearm, and pelvis that were either missing or only poorly preserved in the holotype of *S. laweri*.

As observed by Rosenbaum and Padian, *S. laweri* represents, in many ways, a good example of a basal, generalized ornithischian. However, the cranial material preserved in UCMP 130580, although fragmentary, displays a similarity to the skull of the basal thyerophoran *Emausaurus*, especially in the morphology

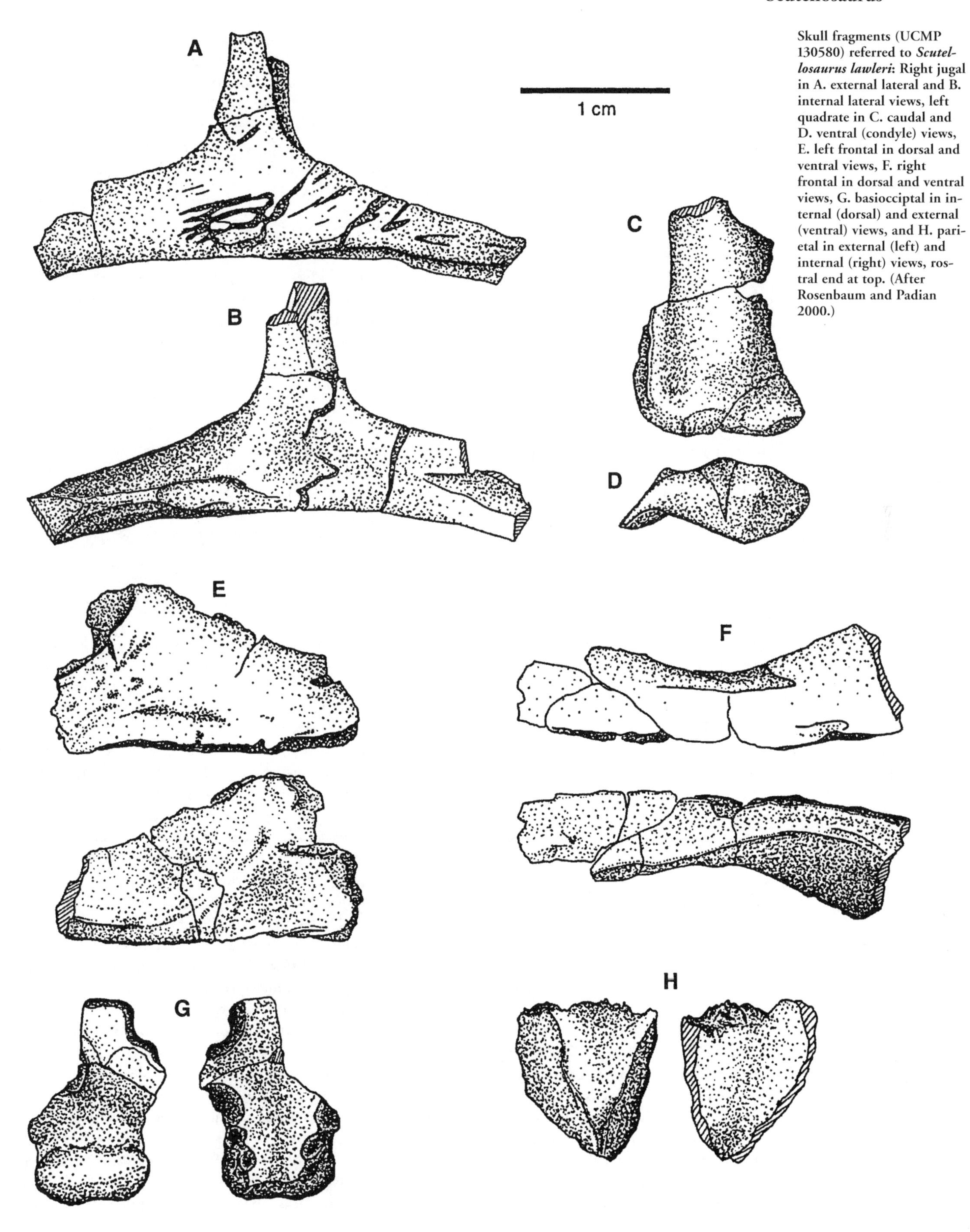

Skull fragments (UCMP 130580) referred to ***Scutellosaurus lawleri***: Right jugal in A. external lateral and B. internal lateral views, left quadrate in C. caudal and D. ventral (condyle) views, E. left frontal in dorsal and ventral views, F. right frontal in dorsal and ventral views, G. basioccipital in internal (dorsal) and external (ventral) views, and H. parietal in external (left) and internal (right) views, rostral end at top. (After Rosenbaum and Padian 2000.)

of the jugals. In both *Scutellosaurus* and *Emausaurus*, the jugal is an inverted T-shape; and the vertical process of the jugal in *Scutellosaurus* has a groove on its rostral side which, as in *Emausaurus* (see Haubold 1990), probably held another bone in place.

Therefore, Rosenbaum and Padian accepted the present classification of *Scutellosaurus* as a basal thyreophoran. The authors pointed out that, while this genus is smaller and lacks the more developed armor characteristic of derived thyreophorans (*e.g.*, *Scelidosaurus*, Stegosauria, and Ankylosauria), it possesses osteoderms and also exhibits cranial features similar to *Emausaurus*, these uniting *Scutellosaurus* with the rest of Thyreophora.

In previous descriptions (*e.g.*, Colbert) of *Scutellosaurus*, the length of the forearm could only be estimated. Colbert supposed that the ratio of forelimb to hindlimb was 0.58, suggesting that this dinosaur was an obligate quadruped that occasionally walked on its hind legs. As observed by Rosenbaum and Padian, however, UCMP 130580, which is only slightly smaller than the holotype, possesses a substantially shorter radius than that estimated by Colbert, indicating a more accurate ratio of about 0.52. Also, the forelimbs of *S. lawleri* are of slight construction when compared to the hindlimbs, and they lack substantial or elaborated structures for the attachment of muscles. This all implied to Rosenbaum and Padian that *Scuttelosaurus* was not an obligate quadruped and may not have extensively utilized its forelimbs in locomotion. Furthermore, the very long tail was most likely used to counterbalance the animal on its hindlimbs (see also Colbert).

Addressing Colbert's question as to whether the relatively small holotype of *S. lawleri* belonged to an adult or juvenile animal, Rosenbaum and Padian noted that that specimen and UCMP 130580 are approximately the same size, while both are significantly smaller than other collected specimens. Furthermore, neurocentral sutures in vertebrae in both specimens are fused only in the caudal series. Although this constitutes a juvenile feature in many tetrapods, such sutures are also lacking in the larger *S. lawleri* specimens (this feature also lacking in adults of the primitive ornithischian *Lesothosaurus diagnosticus*; see Thulborn 1972).

According to Rosenbaum and Padian, preliminary histological studies of *S. lawleri* supports the conclusion that this dinosaur grew rather slowly compared to hadrosaurs such as *Maiasaura* and *Hypacrosaurus*, theropods, and most sauropods (Horner *et al.*, in press; Ricqlès *et al.*, in press; Padian *et al.*, unpublished data). The fact that some *Scutellosaurus* specimens are approximately 25 percent larger than others may indicate that growth was relatively slow during this dinosaur's life, maybe because its adult size was also small. Although sexual dimorphism might also explain this size disparity between specimens, Rosenbaum and Padian cautioned that both hypotheses must await the discovery of additional specimens.

Key references: Colbert (1981); Haubold (1990); Rosenbaum and Padian (2000); Thulborn (1972).

†SEISMOSAURUS—(=?*Diplodocus*)

Comments: Originally hailed as the longest known dinosaur, a gigantic sauropod boasting an estimated length of 39 to 52 meters (135 to 180 feet; see Gillette 1991, 1994), and presumably known from both skeletal remains and gastroliths (Gillette, Betchel and Betchel 1990), the type species *Seismosaurus halli* (see *D:TE*) has recently come under question as to its validity, size and the materials referred to it.

Recently, Lucas and Heckert (2000) briefly addressed this taxon again from various viewpoints, stating that *Seismosaurus* has been misrepresented in the following ways: 1. As a valid genus distinct from *Diplodocus*, 2. having an overestimated length, and 3. having been found with gastroliths associated with its skeletal remains.

Regarding the validity of *S. halli*, sauropod specialist Brian David Curtice (1996) had previously questioned the taxonomic validity of this genus and species. Lucas and Heckert, and also Lucas (2000), agreed with Curtice (personal communication to the authors, and oral communication to Lucas, both 2000) that *Seismosaurus* is not a valid genus, but that its material represents a new species of *Diplodocus* (see entry), which all of these authors referred to as *Diplodocus hallorum* (Gillette 1991). This assessment may well be correct. However, pending the publication of a detailed comparison of *Seismosaurus* and *Diplodocus*, this document prefers retaining, at least for the present, the name *Seismosaurus* as the valid name for this giant from New Mexico. Regarding the length of this dinosaur, Lucas and Heckert, and Lucas, noted that Paul (1988, 1994) had isometrically scaled the length more accurately at 33 meters (about 110) feet. Furthermore, Curtice had pointed out that the longest caudal vertebra of *S. halli* is just 20 millimeters longer than the longest known caudal of *Diplodocus*, while its pelvis is actually smaller than that of the diplodocid *Supersaurus*.

Lucas mainly addressed in detail the issue of more than 240 polished chert and quartzite stones, ranging from one to four inches in diameter, found near and, to a lesser degree, associated with the type material (NMMNH 3690) of *S. halli* and which were originally described as gastroliths (see *D:TE*). As noted by Lucas, however, authenticated gastroliths in

Photograph by the author, courtesy New Mexico Museum of Natural History and Science.

***Seismosaurus halli*, NMMNH 3690, holotype caudal vertebra 20. Some paleontologists (*e.g.*, Brian David Curtice, Spencer G. Lucas, and Andrew B. Heckert) have recently considered this taxon to be a junior synonym of *Diplodocus*, representing a new species *Diplodocus hallorum*.**

dinosaurs are rare (known mainly in the prosauropod *Massospondylus* and the primitive ceratopsian *Psittacosaurus*), and those associated with sauropods are virtually unknown (the evidence for gastroliths in the Argentinian *Rebbachisaurus*, as reported by Calvo and Salgado 1995, having been based on uncompelling evidence). Lucas stated that previously published claims (*e.g.*, Manley 1989, 1991*a*, 1991*b*) that a high degree of polish is diagnostic of dinosaur gastroliths are unsubstantiated "because high polish has not been demonstrated as distinctive of *bona fide* dinosaur gastroliths." Furthermore, Lucas stated that Darby and Ojakangas (1980) had previously concluded that high polish is not a distinctive characteristic of gastroliths, particularly those found in plesiosaurs; and also that Whittle and Onorato (2000) argued that true gastroliths are not characterized by a high polish, but rather when viewed under high magnification, a highly pitted and rilled surface.

Concerning the supposed "gastroliths" of *Seis-*

Photograph by the author.

Reconstructed skeletal cast of *Seismosaurus halli*, exhibited at Dinofest International (2000–2001), Navy Pier, Chicago. "The World's Largest Museum Specimen," this display was prepared over a four-year period by Dinosauria International, Inc., based upon the holotype material and also specimens of *Diplodocus*. The mount measures almost 40 meters (137.5 feet long) and about 20 meters (22.5 feet) tall at the hips.

mosaurus specifically, Lucas listed the following objections to that interpretation of these stones:

1. Only the stones found next to the pelvis — not in the position of a crop or gizzard — are directly associated with this dinosaur's skeleton, and these were interpreted by Lucas "as sedimentary clasts [cobbles composed of sedimentary minerals] trapped by the bone in a northeasterly-flowing current."

2. Gillette's (1994) argument that these stones are not sedimentary clasts was based upon a misunderstanding of the fluvial processes, the sizes, shapes, and configuration of these stones in sediment attesting to "their origin as sedimentary clasts (conglomerate) in a channel-lag deposit."

3. Manley's (1991*a*, 1991*b*, 1993) arguments that such stones can be identified as gastroliths by their high polish were based upon spurious comparisons.

4. No compelling evidence for gastroliths in sauropod dinosaurs has yet been published.

Researching earlier published accounts of gastroliths in dinosaurs, Lucas found most of them to be nonscientific anecdotes. In the case of *Seismosaurus*, the stones associated with its type material exhibit a wide range of polish, and resemble similar stones from the Morrison Formation that were not associated with skeletal remains. The polished stones found associated with *Seismosaurus* were explained by Lucas, rather than stones swallowed by a dinosaur for the purpose of grinding its food, as cobbles deposited near the skeleton by a stream.

Note: The first mounted skeleton of *Seismosaurus*— a cast with the missing elements restored after those of other diplodocids — went on display in December, 2000 at Dinofest International, held at Navy Pier in Chicago.

Key references: Calvo and Salgado (1995); Curtice (1996); Darby and Ojakangas (1980); Gillette (1991, 1994); Gillette, Betchel and Betchel (1990); Lucas (2000); Lucas and Heckert (2000); Manley (1989, 1991*a*, 1991*b*, 1993); Paul (1988, 1994); Whittle and Onorato (2000).

†SELLOSAURUS

Holotype: SMNS 5715, partial postcranial skeleton.

Comments: A medium-sized prosauropod from the Upper Triassic of Germany, the type (and only valid) species *Sellosaurus gracilis* is known from numerous partial skeletons and partial skulls representing juvenile to adult individuals (see *D:TE* for details and illustrations of the type specimen).

In 1999, Peter M. Galton, a specialist in pro-

Courtesy Peter M. Galton.

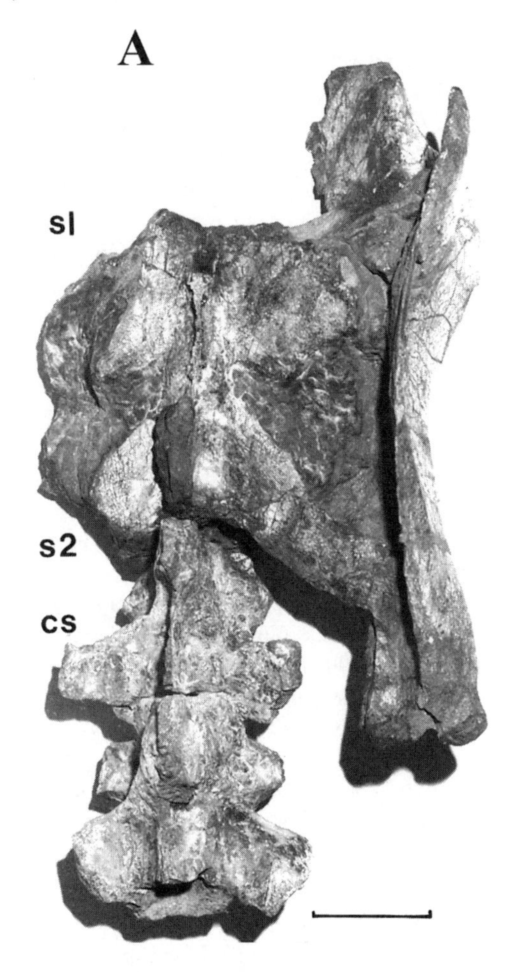

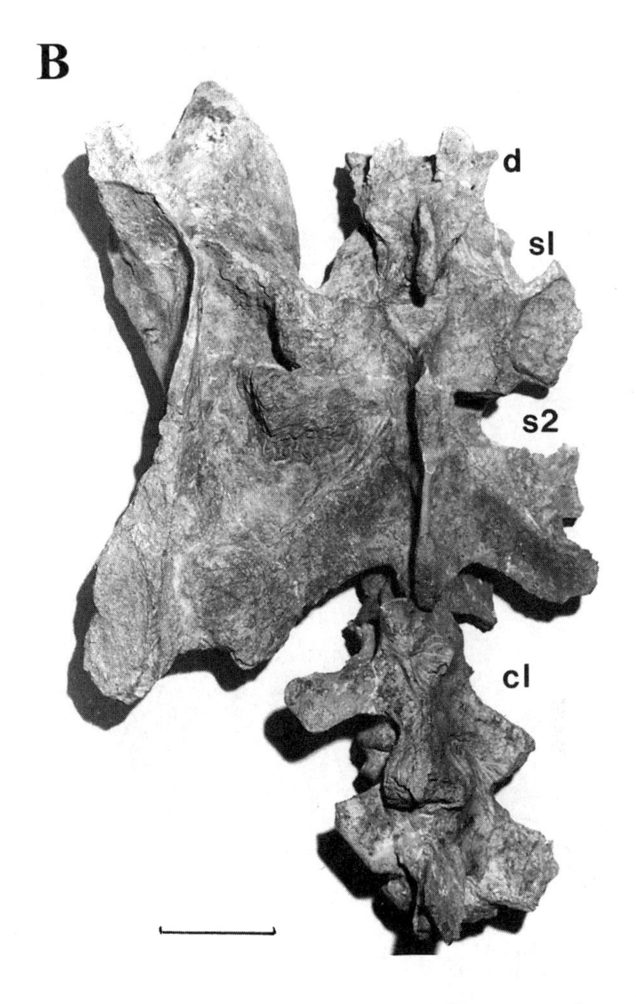

***Sellosaurus gracilis*, A. SMNS 5715, sacrum with three vertebrae and right ilium and B. SMNS 17928, sacrum with two vertebrae and left ilium, dorsal views.**

sauropods (as well as other groups of dinosaurs), published a paper in which he reinterpreted the sacra in various specimens referred to *S. gracilis*. Galton observed that the sacra in this species can be divided into two distinct groups. One group, comprising specimens with sacra possessing two vertebrae (sacrals 1 and 2), include the following (all of these recovered from the Lower Stubensandstein, Ochsenbach, Stromberg, Nordwürttemberg): SMNS 17928 (described previously by Galton 1984*b* as having three sacral vertebrae, that author then believing that the sacrum had been displaced posteriorly relative to the ilium during preservation); SMNS 14881; SMNS 12684 (the vertebrae incorrectly identified by Huene 1932 as sacrals 2 and 3); and SMNS 12667 (the juvenile holotype of *Palaeosaurus diagnosticus* Huene 1932, designated by Galton 1973*b* the type species of the new genus *Efraasia*, subsequently referred by Galton 1985*b* to *S. gracilis*). The second group, made up specimens with sacra having three vertebrae (sacrals 1 and 2, plus a caudosacral vertebra, or a caudal vertebra that incorporates into the sacrum), includes the following: SMNS 5715, the holotype of *S. gracilis*, from the Middle Stubensandstein, Heslach, Stuttgart; from the Middle Stubensandstein, Pfaffenhofen, Stromberg, SMNS 12217, an isolated caudosacral vertebra, and SMNS 12669, almost complete sacrum; and, from Aishem near Rottweil, a specimen in the UT collection representing a large individual (figured by Huene 1908 as ?*Teratosaurus suevicus*).

As noted by Galton (1999*a*), almost all other known prosauropods have three sacral vertebrae, the exceptions being the larger and more derived *Melanorosaurus* (see Heerden and Galton 1997), from the Upper Triassic of South Africa, and *Massospondylus* (see Cooper 1981), from the Lower Jurassic of southern Africa, both of which have four. As in some individuals of *S. gracilis*, a caudosacral vertebra is also present in various other prosauropod species, in some primitive sauropods, and probably also in the theropods *Dilophosaurus* and *Allosaurus*. Some prosauropods, however, have a dorsosacral as the third sacral vertebra.

The possession of at least three sacral vertebrae has long been regarded as diagnostic of the Dinosauria (*e.g.*, Bonaparte 1976, Gauthier 1986 [at the level of

Courtesy Peter M. Galton.

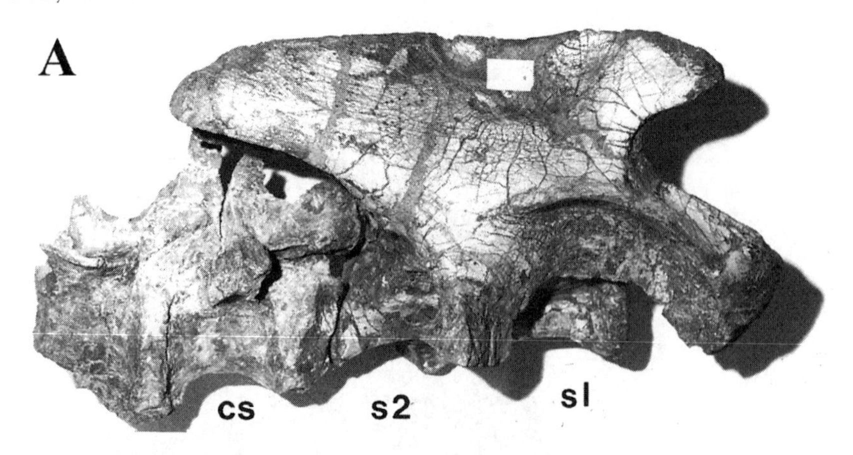

***Sellosaurus gracilis*, A. SMNS 5715, sacrum with three vertebrae and right ilium and B. SMNN 17928, sacrum with two vertebrae and left ilium, lateral views.**

Ornithodira], Benton 1990, Fastovsky and Weishampel 1996, Novas 1997*b*, Sereno 1997, other authors). Novas (1992*a*) had suggested that three sacrals represented a synapomorphy of Dinosauria, while the plesiomorphic character in *Herrerasaurus* supported the outgroup position of that genus; later, Sereno and Novas (1992) proposed that in the common dinosaurian ancestor the number of sacral vertebrae increased from two to three by the addition of a dorsal vertebra, *i.e.,* a "dorsosacral" (see also Sereno, Forster, Rogers, and Monetta 1993; Sereno 1997), although in herrerasaurids one segment was apomorphically lost (see Novas 1993, 1996*a*, 1997*b*). However, Galton's (1999*a*) recognition that some specimens of *S. gracilis* possess only two sacral vertebrae while others have three, negates this one feature used to characterize the Dinosauria. In other words, as two reptilian sacral vertebrae are present in Herrerasauridae and some *S. gracilis* individuals, this feature is plesiomorphic for Dinosauria; consequently, a sacrum including a dorsosacral plus sacrals 1 and 2 cannot be used in the diagnosis of the Dinosauria, nor is the addition of a third vertebra, a dorsal, a synapomorphy of that group (see "Systematics" chapter).

Galton (1999*a*) interpreted the fact that some individuals of *S. gracilis* possess the reptilian count of two sacral vertebrae and others have three (including a caudosacral vertebra), while the rest of their postcranial skeletons are quite similar, as a sexual dimorphism. According to Galton (1999*a*), the morphs with two sacrals probably indicate male individuals, those with three identify females. This assessment, the author noted, is in accordance with various earlier interpretations of dinosaurs, primarily theropods, suggesting that the more robust morph indicates a female and the more gracile morph a male (*e.g.*, see Raath 1977; Carpenter 1990*a*; Larson 1994). A robust pelvis offers more room for the passage of eggs; furthermore, "the gracile morph has a longer chevron bone at the base of the tail and the extra chevron of male crocodiles functions to anchor the penis retractor muscles (Larson 1994)."

A detailed description of the sacra of *S. gracilis* is currently in preparation by Galton.

Key references: Benton (1990); Bonaparte (1976); Carpenter (1990*a*); Cooper (1981); Fastovsky and Weishampel (1996); Galton (1973*b*, 1984*b*, 1985*b*, 1999*a*); Gauthier (1986); Heerden and Galton (1997); Huene (1908, 1932); Larson (1994); Novas (1993, 1996*a*, 1997*b*); Raath (1977); Sereno (1997); Sereno, Forster, Rogers, and Monetta (1993); Sereno and Novas (1992).

†SHAMOSAURUS

Ornithischia: Genasauria: Thyreophora: Thyreophoroidea: Eurypoda: Ankylosauria: Ankylosauridae: Shamosaurinae.

Comments: *Shamosaurus scutatus* is a typical member of the family Ankylosauridae, according to Tatyana A. Tumanova, who erected this Early Cretaceous genus and species from the Kukhteskaya Svita, Dornogov, Ovorkhangai, Mongolian People's Republic, in 1983 (see *D:TE*). However, as Tumanova (1985) pointed out, *Shamosaurus* is also, "to a certain extent intermediate between the two families of ankylosaurs." As later specified by Tumanova (1998), *S. scutatus* shares various features with the family Nodosauridae, these including the following: Narrow premaxillary beak; quadrate condyle visible in lateral aspect; occipital condyle round and vertically oriented.

Tumanova (1998) observed various (though mostly unspecified) by him features in the evolutionary transformations of the Asian ankylosaurid lineage leading from *Shamosaurus* to *Tarchia*: Quadratojugal and squamosal dermal plates becoming horn-like in form; articular condyle somewhat hidden in side view; change in width of premaxillary beak; change in position of orbits; differences in articulations of basipterygoids; shortening of tooth rows.

According to Tumanova (1998), recent examination of the lower-jaw morphology of *S. scutatus* and also new material from Khongil, another Lower Cretaceous locality in Mongolia, reveals differences in the length, height, and morphology of the coronoid and articular processes, these new data expanding what is known "of the direction of evolutionary modifications and serving as the base for new morpho-functional reconstruction."

Note: Carpenter and Kirkland (1998), in their review of Lower and Middle Cretaceous ankylosaurs from North America, briefly reported on an as yet unnamed *Shamosaurus*-like ankylosaurid (to be named and described by Carpenter, Kirkland and Burge, in preparation), known from at least three individuals represented by two partial skulls and partial skeletons, recovered from the same quarry in the Ruby Ranch Member (Albian; Kirkland, Britt, Burge, Carpenter, Cifelli, DeCourten, Eaton, Hasiotis and Lawton 1997) of the Cedar Mountain Formation of Carbon County, Utah. As briefly described by Carpenter and Kirkland, the least disarticulated of the skulls displays the same boxy profile as *S. scutatus.* It differs from *Shamosaurus*, however, in the following features: Pterygoids much more elongated posteriorly; occipital condyle neck longer; hemispherical occipital condyle facing posteriorly. Also, the maxillary teeth have a poorly developed cingulum, a feature lacking in *Shamosaurus* teeth (undescribed IVPP specimen).

Key references: Carpenter and Kirkland (1998); Kirkland, Britt, Burge, Carpenter, Cifelli, DeCourten, Eaton, Hasiotis and Lawton (1997); Tumanova (1983, 1985, 1998).

†**SHANXIA**—(= *Tianzhenosaurus*)

Ornithischia: Genasauria: Thyreophora: Thyreophoroidea: Eurypoda: Ankylosauria: Ankylosauridae: Ankylosaurinae.

Comments: Ankylosaurid taxa are rare in China. Recently, the Chinese form *Shanxia tianzhenensis* (see *S1*), from the Huiquanpu Formation (Upper Cretaceous) of Shanxi Province, People's Republic of China, was introduced by Barrett, You, Upchurch and Burton (1998).

The same month and year that *S. tianzhenensis* was named and described, Pang and Cheng (1998), unaware of that taxon, introduced the new genus and species *Tianzhenosaurus youngi. T. youngi* was based on various specimens including three skulls, a partial mandible, an almost complete disarticulated postcranial skeleton, numerous dermal scutes, and a tail club recovered from the Upper Cretaceous (age based on associated hadrosaurid, a possible Late Cretaceous titanosaurid, and some nondinosaurian Late Cretaceous fossils from the same locality) Huiquanpu Formation at Kangdailian, near Zhaojiagou Village, Tianzhen County, Shanxi. The first of these specimens to be found consisted of a dozen articulated vertebrae discovered in 1983 by Pang Qiqing and Cheng Zhengwu. Successive excavations at this site resulted in the recovery of more than 3,300 fossil specimens (see Pang, Cheng, Yang, Xie, Zhu and Luo 1996), representing a fauna including this new ankylosaurid, and also a sauropod, theropod, and hadrosaurid (Cheng and Pang 1996). Of the ankylosasaurid remains, a well-preserved, nearly complete skull (HBV-10001) was designated the holotype of *T. youngi*; an incomplete right mandible (HBV-10002) and an almost entire postcranial skeleton (HBV-1003) were named paratypes (Pang and Cheng 1998).

T. youngi was diagnosed by Pang and Cheng as follows: Skull low, flat, medium-sized, having shape of isoceles triangle [in dorsal view]; skull roof covered with irregular dermal bony tubercles; premaxilla relatively long; orbit small, surrounded by dermal bony ring; narial opening horizontally elongate, septomaxilla not separate narial openings; maxillary tooth rows almost parallel, slightly convergent posteriorly; basicranial part short; maxiloturbinal situated laterally in middle part of palatal vault; occipital region almost vertical, with narrow, high occipital condyle not extending beyond posterior margin of skull roof; opisthotic extended lateroventrally as curved process; mandible deep, having convex ventral border and no dermal covering on lateral surface; tooth crowns having basal cingulum labially, swollen base, and well-developed middle ridge lingually; cervical centrum short, amphicoelous; centrum of dorsal vertebrae forming presacral rod, succeeded by fusion of four true sacral vertebrae and first caudal vertebra; anterior caudal vertebrae short, thick, posterior caudals narrow and elongate, ending with tail club; scapula roughly triangular and platelike; both ends of humerus moderately expanded (not twisted); femur thick, without fourth trochanter; tarso-metatarsal and digital bones ankylosaur-type.

Pang and Cheng assigned their new genus to the family Ankylosauridae based on the following characters: Skull short, wider than long; skull roof covered with dermal scutes and nodes; occipital region wider than high; orbit small. Comparing it with other ankylosaurid genera (including *Pinacosaurus* and the North American genus *Ankylosaurus*), Pang and Cheng observed that their new genus is closest in morphology to *Saichania.* Both taxa have skulls with the general shape of an isosceles triangle. In both, the orbit is located at the midposterior part of the skull, the occipital condyle does not extend beyond the posterior edge of the skull roof, and the skull roof is

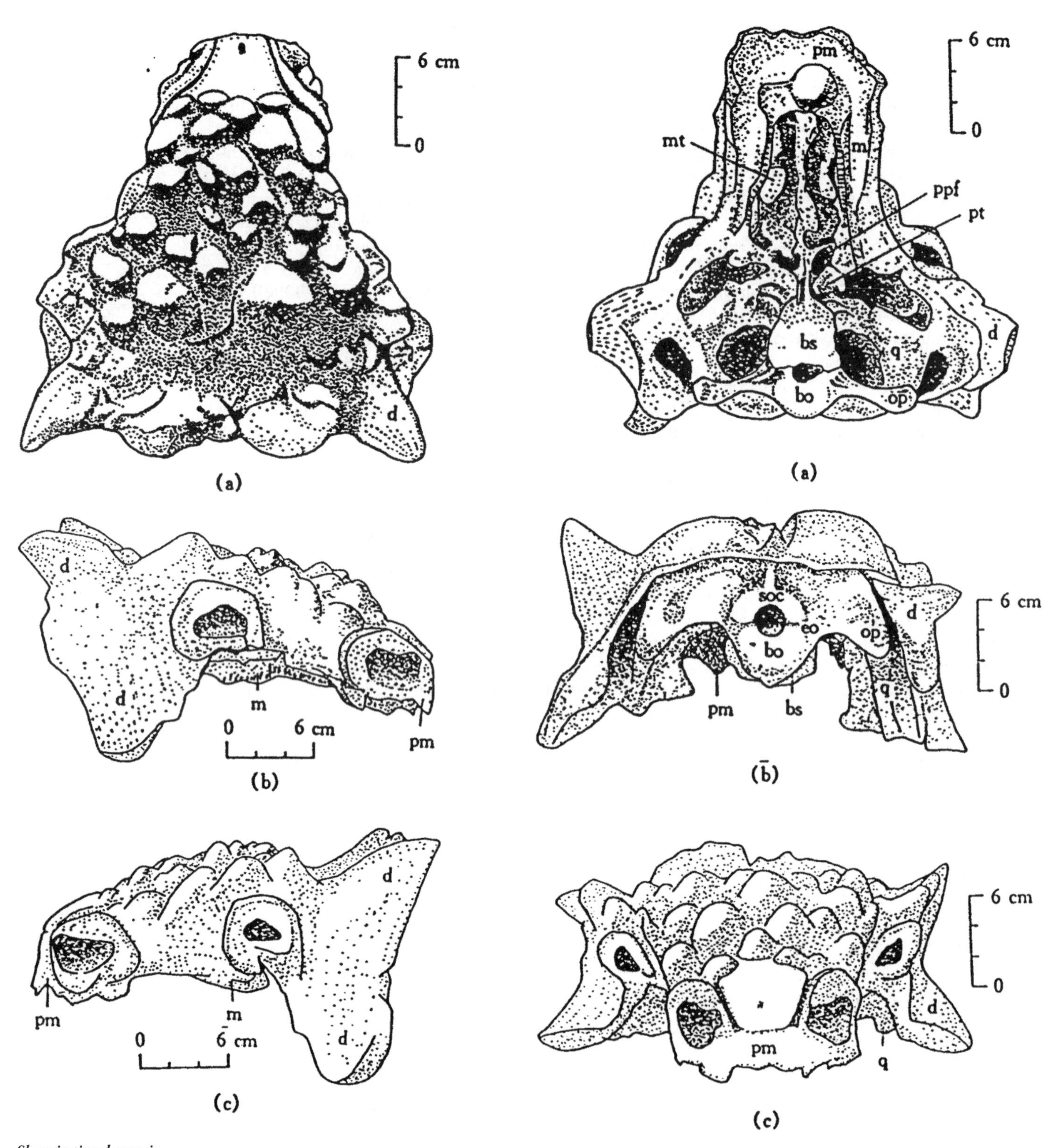

***Shanxia tianzhenensis*, HBV-10001, holotype skull of *Tianzhenosaurus youngi* in (left) a. dorsal, b. right lateral, c. left lateral, and (right) a. palatal, b. occipital, and c. anterior views. (After Pang and Cheng 1998.)**

covered with dermal plates and bony nobs. According to Pang and Cheng, the two forms differ in their skulls' shape (*e.g., Tianzhenosaurus* with a longer premaxilla and, therefore, a long snout, *Saichania* having a short snout) and size (*Tianzhenosaurus* 275 millimeters long, with a maximum width of 294 millimeters, *Saichania* 455 millimeters long, with a maximum width of 480 millimeters). *Tianzhenosaurus* differs from *Saichania* in having maxillary tooth rows that are almost parallel, with a slightly narrower posterior width than snout width, and having a "vertical postemporal region with a narrow and vertically expanded occipital condyle." In a detailed comparison between these two genera, the authors also noted important differences in the structures of the ornamentation of the dermal roofing (cranial dermal tubercles in *Tianzhenosaurus* differing in shape, irregularly arranged; in *Saichania*, polygonal in shape, symmetrically arranged).

Following the original publications of both *Shanxia* and *Tianzhenosaurus*, Sullivan (1999), in a paper naming and describing the new ankylosaurid *Nodocephalosaurus* (see entry), suggested that *S. tianzhenensis* should be regarded as a *nomen dubium* based on two points: 1. Squamosal "horn" shape — the single autapomorphy by which *Shanxia* was originally diagnosed — is highly variable within the genus *Euoplocephalus* (see also Coombs 1971); and 2. as both *S. tianzhenensis* and *T. youngi* come from the same formation in the same country, they most likely are synonymous, while *Tianzhenosaurus* was regarded as a junior synonym of *Saichania*.

The above points were later addressed by Upchurch and Barrett (2000), who defended the validity of *Shanxia tianzhenensis* by challenging Sullivan's arguments, while also identifying two previously features distinguishing *Shanxia* from *Saichania*:

1. Upchurch and Barrett accepted that the shape of the squamosal "horn" varies within the genus *Euoplocephalus* (Coombs; also K. Carpenter, personal communication to Upchurch and Barrett). Sullivan, however, did not offer data regarding the extent of this variation in other ankylosaurid genera, thereby not demonstrating that the squamosal "horn" morphology of *Shanxia* falls within the variation range of this feature in *Saichania*. Furthermore, Barrett (1998*a*), in examining squamosal "horn" morphology in various ankylosaurid genera (*Ankylosaurus*, *Saichania*, *Tarchia*, *Talarurus*, *Shamosaurus*, *Tsagantegia*, *Pinacosaurus*, and *Euoplocephalus*), had already showed that this condition in *Shanxia* is unique. Therefore, in Upchurch and Barrett's opinion, *Shanxia* is not a dubious taxon.

2. Stratigraphic and geographic range similarities constitute little more than circumstantial evidence in supporting the synonymy of taxa and cannot be regarded as a reliable taxonomic criterion without supporting anatomical evidence. Furthermore, many dinosaurian faunas are known containing two or more valid ankylosaurian genera (*e.g.*, see Weishampel 1990; Jerzykiewicz and Russell 1991).

3. Two additional character states distinguish *Shanxia* from *Saichania*: In *Saichania*, the humerus has a large processus medalis humeri (see Maryańska 1977), a feature not present in *Shanxia* (see Barrett *et al.*) (the humerus of *Tianzhenosaurus* having not yet been figured or described in detail for comparison). Also in *Saichania*, the posterior border of the skull roof is characterized by two large domed subrectangula osteoderms located between the squamosal "horns" (see Maryańska); these are not present in *Shanxia*, although similar ossifications are found in *Tianzhenosaurus* (see Pang and Cheng).

Upchurch and Barrett also confronted the following:

It was suggested (J. I. Kirkland, personal communication to Upchurch and Barrett) that the dorsal-view slenderness of the squamosal "horn" in *Shanxia* is an artifact of preparation, due to an excess of plaster used in restoring the posterolateral part of the holotype skull roof. As noted by Upchurch and Barrett, however, plaster was only applied to the ventral surface of the skull roof, the skull roof in dorsal view forming a single continuous surface, with no evidence suggesting that it was artificially widened or lengthened by the incorporation of plaster.

Also, it was suggested (K. Carpenter, personal communication to Upchurch and Barrett) that apparent differences between the preservation of the skull and postcranial elements of *Shanxia* may indicate that these elements do not belong to the same individual. However, Upchurch and Barrett pointed out that the preservation of the braincase and postcrania is identical; that it is uninformative to compare the dermal armor preservation with that of the cranial and postcranial elements, because of the very different fabrics of these elements; and vertebrate paleontologists from the Institute of Vertebrate Paleontology and Paleoanthropology, who collected the type specimen, verified that this material was associated when discovered (You Hailu, personal communication to Upchurch and Barrett).

Finally, Upchurch and Barrett addressed the possibility of *Shanxia* and *Tianzhenosaurus* being synonymous. It was the opinion of these authors that, if this synonymy proves to be valid, the name *Shanxia* would have priority. Upchurch and Barrett pointed out that both names were published in June, 1998, with Barrett *et al.*'s publication appearing on June 15 of that month, and Pang and Cheng's work appearing in a journal that lacked a specific date. Therefore, according to International Commission on Zoological Nomenclature rules, Pang and Cheng's paper is deemed to have appeared at the end of the month, therefore being published after *Shanxia*.

The authors agreed with Sullivan's assessment that *Tianzhenosaurus* could be synonymous with *Saichania*, however the morphology of the squamosal "horn" and lack of large osteoderms from the posterior margin of the skull roof indicated to Upchurch and Barrett that *Tianzhenosaurus* and *Shanxia* are distinct genera.

In a response to Upchurch and Barrett, Sullivan (2000*a*) countered with the following points:

1. The differences in squamosal "horn" morphology among the taxa cited by Upchurch and Barrett are mostly known from one or two specimens in which individual variation is unknown; therefore, each seems to be different or unique in its own way. Only in *Euoplocephalus*, for which 14 skulls had been studied (see Combs), is sample size large enough to demonstrate the taxonomic insignificance of the observed variation. Based on this known variation in *Euoplocephalus*, therefore, there is no intrinsic reason to suggest that the morphology of the squamosal "horn" is unique for *Shanxia*. Sullivan (2000*a*) further noted that he had examined the squamosal "horn" morphology in numerous ankylosaurid specimens (see Sullivan 1999, Appendix 1, for list), but that Upchurch and Barrett "did not list material examined in their paper, nor does it appear that they examined any material first hand, other than their specimen, based on their published acknowledgements." Sullivan (2000*a*) concluded that "their study was conducted by reviewing the literature," Upchurch and Barrett's statement that they had examined "a wide variety of material," therefore, unsubstantiated.

2. Although not definitive, the fact that *Shanxia* and *Tianzhenosaurus* originate from the same lithologic unit at least supports their synonymy. Although it is possible to have more than one ankylosaurid from the same stratum and stratigraphic correspondence is no substitute for morphology, given the poor quality of the holotype of *S. tianzhensis* and based on parsimony, it seemed "more likely that both *Shanxia* and *Tianzhenosaurus* are the same taxon."

3. As the morphology of the squamosal "horn" is the only character Barrett *et al.* used to diagnose *Shanxia*, the morphology of the humerus of *Shanxia* has no bearing in comparing this genus with *Tianzhenosaurus*. At any rate, the humerus in the latter genus has not yet been described. Furthermore, dermal ossifications are highly variable in ankylosaurid skulls (see Coombs).

Regarding Upchurch and Barrett's comments on nomenclature, Sullivan (2000*a*) discovered that the publication date for Pang and Cheng's paper naming and describing *Tianzhenosaurus* was no later than June 1, 1998. Therefore, in Sullivan's (2000) estimation, if *Tianzhenosaurus* and *Shanxia* prove to be synonymous the former name has priority.

While a definite decision regarding this nomenclature issue has yet to be made, *Tianzhenosaurus youngi* will be formally referred to *Shaxia tianzhenensis* in a joint paper by Carpenter, Maryańska and Weishampel (in preparation).

Key references: Barrett (1998*a*); Barrett, You, Upchurch and Burton (1998); Coombs (1971); Jerzykiewicz and Russell (1991); Maryańska (1977); Pang and Cheng (1998); Pang, Cheng, Yang, Xie, Zhu and Luo (1996); Sullivan (1999, 2000); Upchurch and Barrett (2000); Weishampel (1999)

SHANYANGOSAURUS Xue, Zhang and Bi 1996 *see in* Xue, Zhang, Bi, Yue and Chen 1996

Saurischia: Theropoda: Neotheropoda: Tetanurae: Avetheropoda: Coelurosauria: Manuraptora: Arctometatarsalia: Tyrannosauridae.

Name derivation: "Shanyang [Formation]" + Greek *sauros* = "lizard."

Type species: *S. niupanggouensis* Xue, Zhang and Bi 1996 *see in* Xue, Zhang, Bi, Yue and Chen 1996.

Other species: [None.]

Occurrence: Shanyang Formation, Qinling Mountains region, China.

Age: Late Cretaceous.

Known material/holotype: NWUV111.2, 4, 5, 6, 8, 9, 10 and 11), left scapula, humeri, scarum, right femur, right tibia, left and right fourth metatarsal, partial phalange, ungual.

Diagnosis of genus (as for type species): Small theropod; limb bones slender, hollow; femur short, curved, lacking fourth trochanter; posterior intercondylar furrow deep, wide; tibia slender, with crooked twist and long cnemial crest lacking furrow on distal portion; claw narrow, sharp; ribs having horizontal hooks (Xue, Zhang and Bi 1996).

Comments: The type species *Shanyangosaurus niupanggouensis* was founded upon partial postcranial remains (NWUV111.2, 4, 5, 6, 8, 9, 10 and 11) collected from the Shanyang Formation, in the eastern part of the Qinling Mountains, in central China. It was only briefly discussed in a volume edited by Xue, Zhang, Bi, Yue and Chen (1996) about development and environmental changes of the intermontane basins in the eastern part of the Qinling Mountains (Xue, Zhang and Bi 1996).

Key reference: Xue, Zhang and Bi 1996 *see in* Xue, Zhang, Bi, Yue and Chen (1996).

SHUVUUIA Chiappe, Norell and Clark 1998

Saurischia: Theropoda: Neotheropoda: Tetanurae: Avetheropoda: Coelurosauria: Manuraptoriformes: Arctometatarsalia: ?Ornithomimosauria: Alvarezsauridae: Mononykinae.

Name derivation: Mongolian *shuvuu* = "bird."

Type species: *S. deserti* Chiappe, Norell and Clark 1998.

Other species: [None.]

Occurrence: ?Djadokhta Formation, Ukhaa Tolgod, Tugrugeen Shireh, South Gobi Aimak, Mongolia.

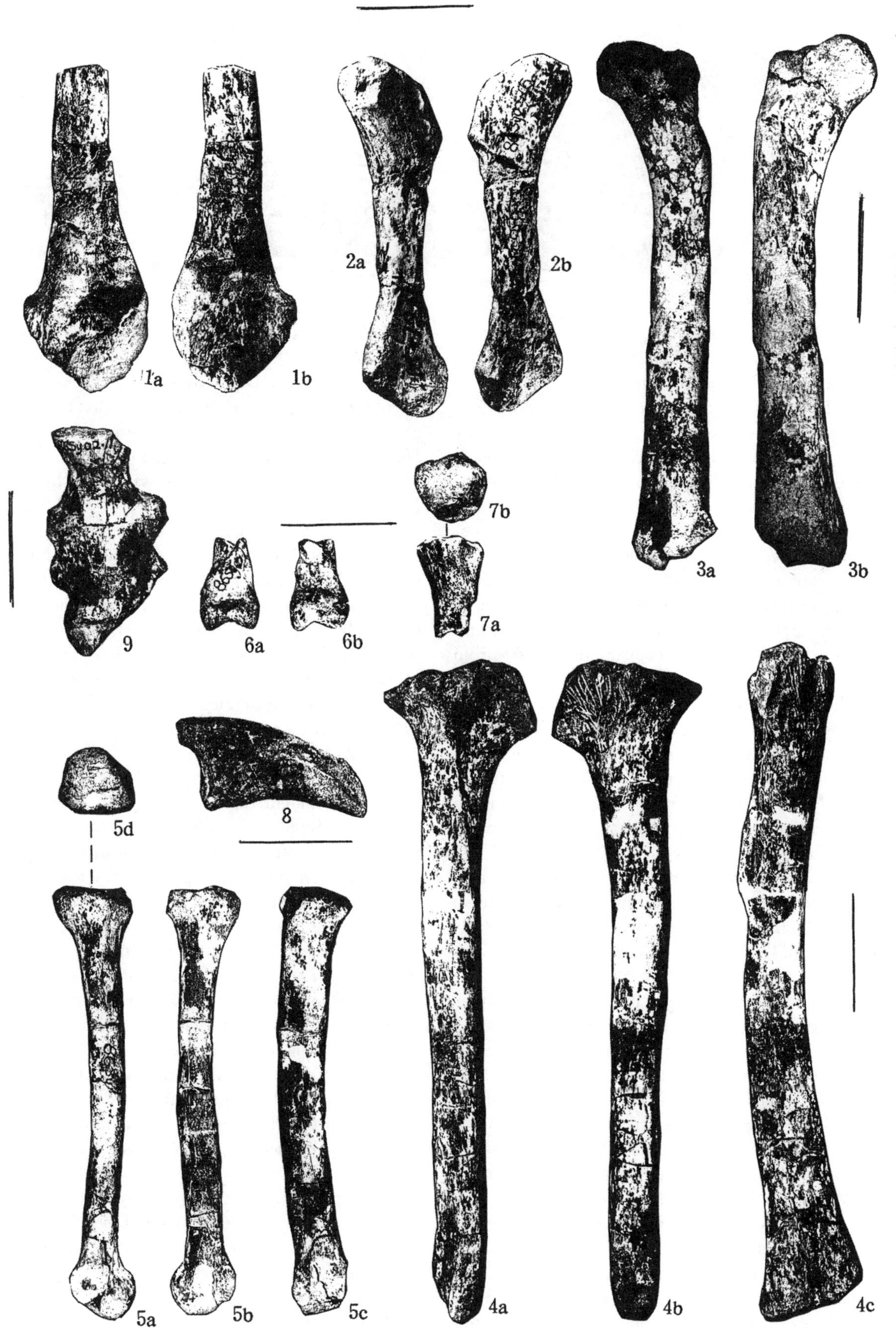

***Shanyangosaurus niupang-gouensis*, holotype 1. left and 2. right humeri (NWUV111.11 and NMUV111.2; scale = 4 cm.), 3. right femur (NWUV1111.4; scale = 6 cm.), 4. right tibia (NWUV1111.5; scale = 6 cm.), 5 left scapula (NWUV1111.6; scale = 4 cm.), 6.–7. partial metatarsals (NWUV1111.8 and 9; scale = 4 cm.), 8. ungual (NWUV1111.10; scale = 2 cm.), and 9. (NWUV1111.11; scale = 4 cm.) sacrum. (After Xue, Zhang and Bi *see in* Xue, Zhang, Bi, Yue and Chen 1999.)**

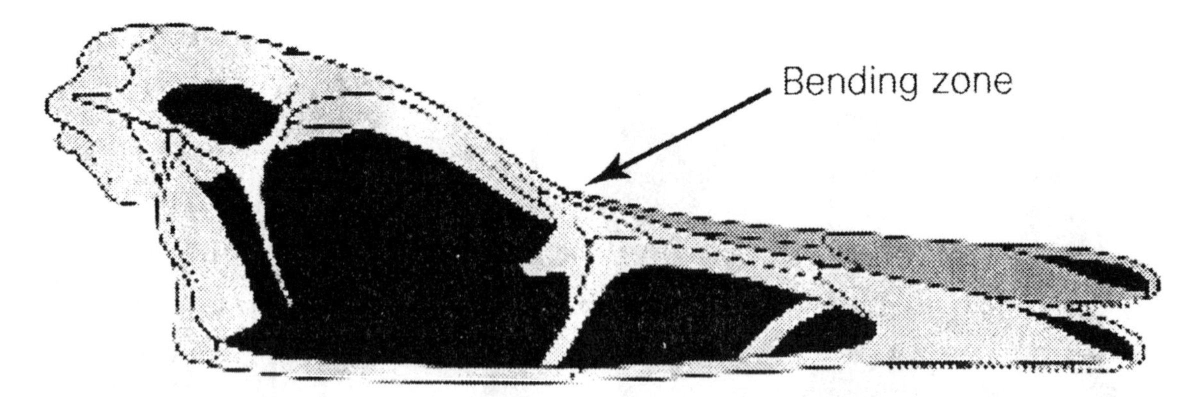

***Shuvuuia deserti*, reconstruction of skull showing its kinetic capacity for raising the snout independent of the braincase. (After Chiappe, Norell and Clark 1998.)**

Age: Late Cretaceous (?Campanian).
Known material: Various skull and postcranial remains.
Holotype: MGI 100/975, skull.

Diagnosis of genus (as for type species): Mononykine distinguished from *Mononykus olcranus* by less compressed cervical centra having large pneumatic foramina, humeral deltopectoral crest that is continuous with its head, pubis with subcircular cross section, femoral and tibiotarsal shafts bowed latero-medially, and less excavated medial margin of ascending process of astragalus; less coossification of proximal tarsals to tibia and among metacarpals in specimens of comparable size to holotype of *M. olecranus*, suggesting lower rate of bone coossification; differs from *Parvicursor remotus* in lacking ventral keel in most rostrally located synsacral vertebrae, and in less coossification between proximal tarsals and tibia; differs from all other alvarezsaurids by the autapomorphy of sharp ridge on medial margin of distal tibiotarsus; autapomorphic characters of the skull including articulation between quadrate and postorbital, elongated basipterygoid processes, numerous teeth, and hypertrophied prefrontal/ectethmoid (some of these characters possibly diagnosing a more inclusive taxon, pending future discovery of additional alvarezsaurid skull material (Chiappe, Norell and Clark 1998).

Comments: Tentatively included here as a nonavian theropod, the genus *Shuvuuia* was founded upon a beautifully-preserved skull (MGI 100/975) collected by joint expeditions of the American Museum of Natural History and the Mongolian Academy of Sciences from Late Cretaceous rocks of the Gobi Desert in Mongolia. Other specimens from South Gobi Aimak referred to the type species *S. deserti* include MGI N 100/99 and MGI 100/977, a skull with jaws, from Ukhaa Tolgod, and MGI 100/99 from Tugrugeen Shireh. Two additional specimens — MGI N 100/99, recovered in 1992 from the Upper Cretaceous (?Campanian) Tugrugeen Shireh by the Mongolian-American Museum Expedition, comprising a skull fragment and articulated postcranial skeleton lacking cervical vertebrae, thoracic girdle, and proximal forelimb bones, plus MGI 100/1001, a well-preserved skull with right jaw from Ukhaa Tolgod — were previously referred by Chiappe *et al.* (1996) to the closely related genus *Mononykus* (see entry). Material assigned to *Shuvuuia* is significant in that it includes the first skull ever found belonging to alvarezsaurids (Chiappe, Norell and Clark 1998).

In their detailed description of *S. desertis*, Chiappe *et al.* (1998) observed that the skull is delicately constructed, elongate, and with large orbits and terminal tear-shaped nares. As in enantiornithine birds, the maxilla accounts for most of the lateral surface of the snout. Not preserved is the tooth-bearing margin of the premaxilla; however, the maxilla possesses numerous tiny and unserrated teeth indistinguishable from those of *Mononykus*.

According to Chiappe *et al.* (1998), the configuration of the temporal region of the skull of *Shuvuuia* reveals an apparent capability for some movement, probably prokinetic (prokinesis being the primitive type of kinesis found in birds). Flexion areas were located at the junction of the upper jaw and the neurocranium, indicating that the snout could elevate independently of the braincase. The snout appears to have moved as a single unit, this indicated by the "thinning of the jugal (bending zone) just caudal to its lacrimal contact and the loose connection between the frontals and the preorbital bones (nasal and prefrontals/ectethmoids)," this interpretation "supported by the absence of a continuous naso-orbital septum." The authors speculated that their interpretation of the design of the skull of *Shuvuuia* provides evidence "supporting the theory that prokinesis was a primitive type of kinesis."

Chiappe *et al.* (1998) originally described the type species *Shuvuuia deserti* as a bird (see *S1*, "Introduction," section on dinosaurs and birds), noting that it shares the following cranial characters with birds but not with any known nonavian theropods: Absence of postorbital-jugal contact; (nonsutured) moveable joint between quadratojugal and quadrate; articulation of quadrate with braincase separate; foramen magnum disproportionately large relative to occipital condyle. According to the authors, *Shuvuuia* shares with *Archaeopteryx* (but not with Velociraptorinae) various skull characters including these: Lack of squamosal-quadratojugal contact; coronoid in mandible; triradiate palatine; caudal tympanic recess confluent with columellar recess; tooth crowns unserrated.

Although the Alvarezsauridae was previously regarded as a stem-group avian family (see *S1*), it has more recently been reinterpreted by Sereno (1999*a*) as probably a clade of birdlike nonavian theropods (see *Alvarezsaurus* entry and "Systematics" chapter).

Key references: Chiappe, Norell and Clark (1996, 1998); Sereno (1999*a*).

†SIGILMASSAURUS—(See *Carcharodontosaurus.*)

†SILVISAURUS

Ornithischia: Genasauria: Thyreophora: Thyreophoridea: Eurypoda: Ankylosauria: Nodosauridae.

Diagnosis: (Cranial) Skull in dorsal view having greatest width across

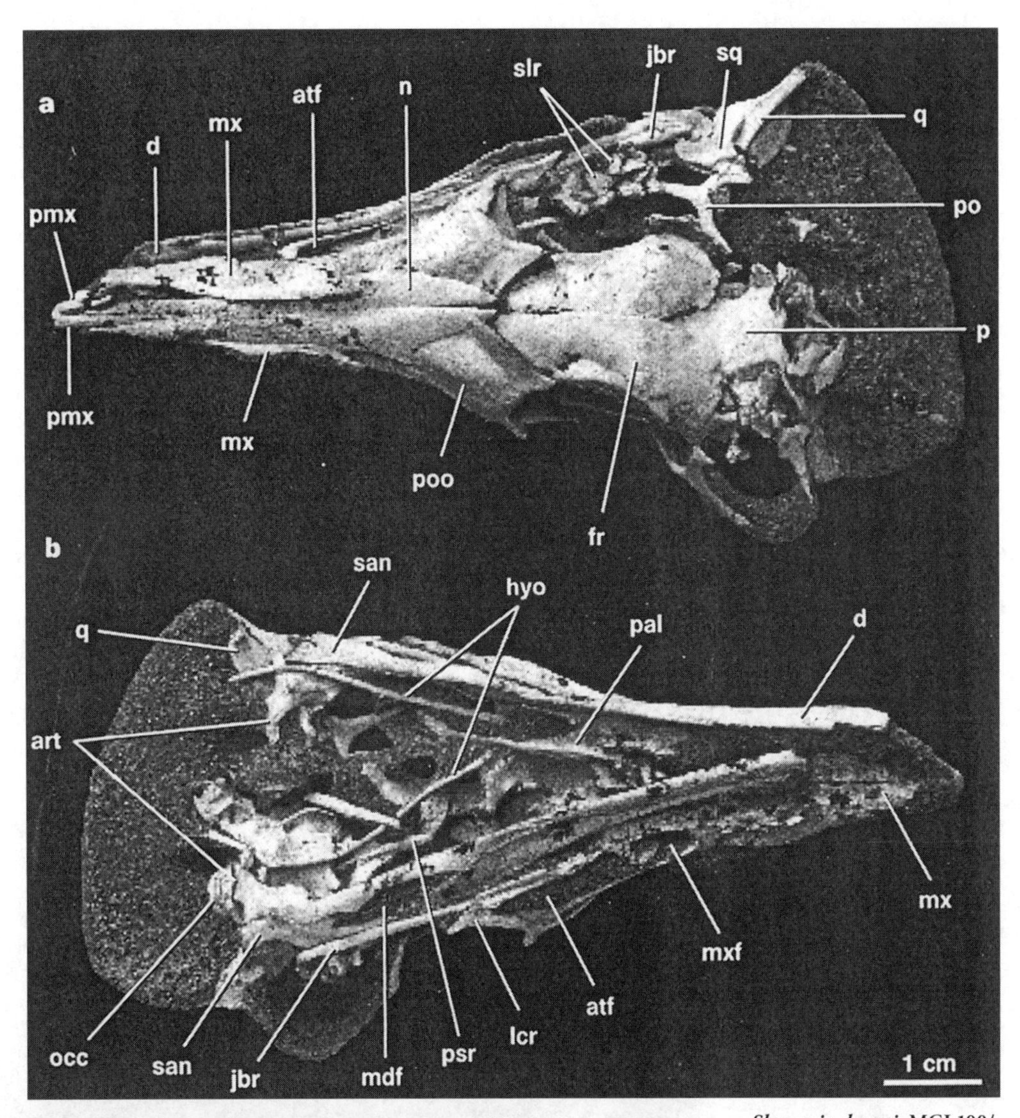

Shuvuuia deserti, MGI 100/977, referred skull in a. dorsal and b. ventral views. (After Chiappe, Norell and Clark 1998.)

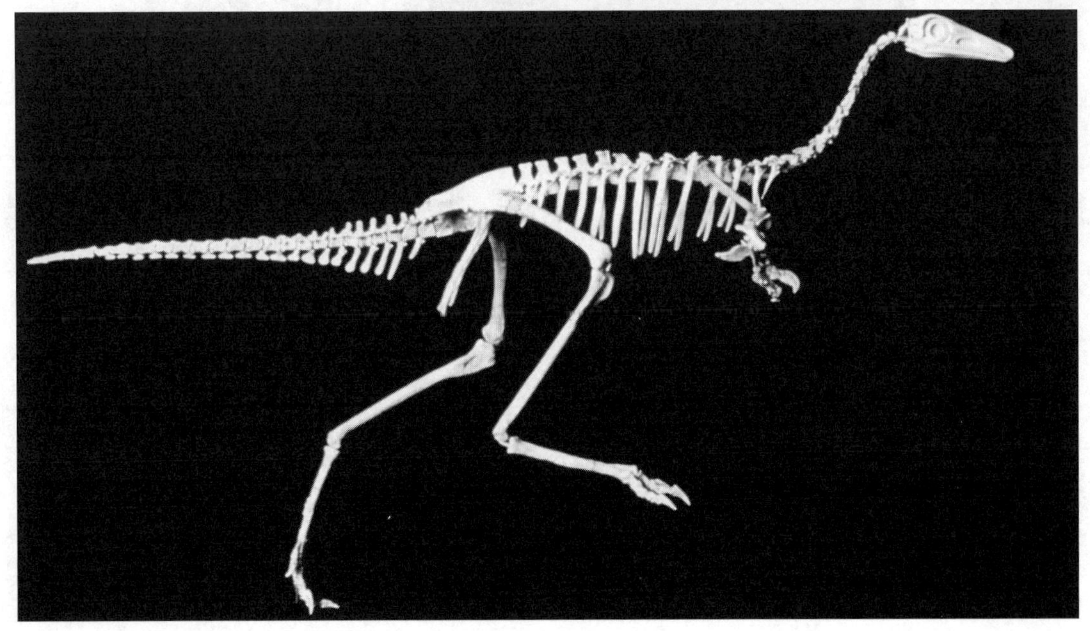

Courtesy Louis M. Chiappe, Mark A. Norell and American Museum of Natural History.

Composite skeleton (cast) of *Shuvuuia deserti.*

postorbital region as in all nodosaurids; postorbital scute more prominent than in *Edmontonia*, *Panoplosaurus*, and *Struthiosaurus*, less than in *Pawpawsaurus* or *Sauropelta*; muzzle more tapering in dorsal view than in *Edmontonia* and *Panoplosaurus* (not parallel-sided as in *Pawpawsaurus*); lateral temporal notch of skull roof more prominent than in *Panoplosaurus* and *Edmontonia*, not as prominent as in *Pawpawsaurus*; prominent node-like scute posterodorsal to external nares; jugal scute moderate-sized and rounded in lateral view as in *Pawpawsaurus* and *Niobrarasaurus* (small, rounded horn in *Edmontonia*, large scute in *Sauropelta*); premaxillaries in ventral view having teeth as in *Pawpawsaurus* (edentulous in *Edmontonia* and *Panoplosaurus*); vomer keeled as in *Edmontonia* and *Niobrarasaurus* (swollen and grooved in *Panoplosaurus*); premaxillary palate ("scoop") very long and narrow (width = ~66 percent length, unlike condition in *Panoplosaurus* in which width = ~88 percent length, *Edmontonia longiceps* in which width = ~85 percent length, *E. rugosidens* = ~110 percent length, or *Pawpawsaurus* (~92 percent length); palate not strongly hourglass-shaped; maxillary tooth rows most similar to *E. longiceps*, being narrow anteriorly (not diverging posteriorly to extreme amount as in *Pawpawsaurus*, *E. rugosidens*, and *Panoplosaurus*; pterygoid plates very long as in *Pawpawsaurus* (short, more vertical in *Edmontonia* and *Panoplosaurus*; posterolateral corners of pterygoids folded ventrally, forming pocket, this condition unknown in other nodosaurids; (postcranial) cervical and dorsal neural spines blade-like (expanded in *Edmontonia*); mid- and posterior cervical centra most similar to *Sauropelta* (shorter in *Edmontonia*, longer in *Struthiosaurus*); dorsal centra not heart-shaped as in *Edmontonia*; synsacrum lacking groove or paired ridge seen in *Gastonia*, *Edmontonia*, *Niobrarasaurus*, and *Nodosaurus*; cervical armor having keeled spines (Carpenter and Kirkland 1998).

Comments: *Silvisaurus condrayi* was redescribed in detail by Carpenter and Kirkland (1998) in their review of Lower and Middle Cretaceous ankylosaurs from North America, following reexamination of the holotype (KUVP 10296), an almost complete skull and partial postcrania from the Dakota Formation of Ottawa County, Kansas (see *D:TE*).

Skull of *Silvisaurus condrayi*, a medium-sized nodosaurid ankylosaur from the Early Cretaceous of Kansas, in left lateral view.

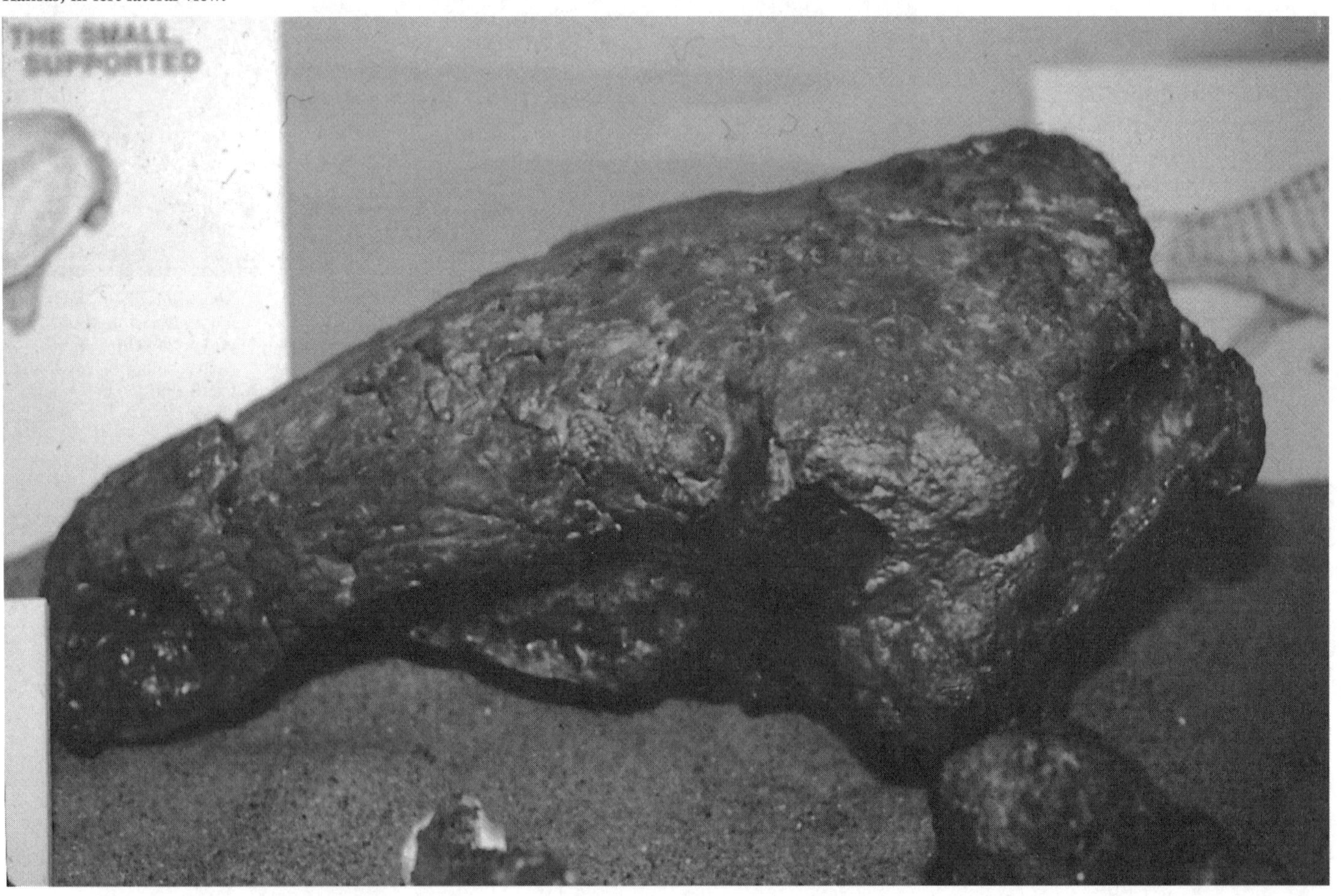

Photograph by the author, courtesy Museum of Natural History, University of Kansas.

Carpenter and Kirkland commented on a number of inaccuracies stated in Theodore H. Eaton Jr.'s (1960) original description of this type species. As noted by these authors, damage during collection and preparation resulted in loss of much of the surface bone of the holotype skull. Bone along the midline of the snout suffered damage greater than that acknowledged by Eaton. The shallow median groove extending anteriorly on the snout, described by Eaton, is in part due to the loss of bone which exposes the suture between the nasals. The external nares are large anteriorly and most likely not restricted by armor and beak as extensively as Eaton had suggested. *Contra* Eaton, the lateral temporal fenestrae are visible in side view, and the antorbital fenestra is not visible through the palate. The pterygoid shelves in the choanae of the palate, rather than extending anteriorly along the vomers as Eaton had illustrated, terminate opposite the anterior margin of the orbits (as in *Pawpawsaurus*). The teeth are not as slender as figured by Eaton. The anterior part of the tooth row does not taper down to the symphysis as gradually as figured by Eaton, but exhibits the condition more similar to that of *Edmontonia*, but with a longer tooth row (see Gilmore 1930).

As pointed out by Carpenter and Kirkland, the neural arches of *Silvisaurus* are not as tall as Eaton had figured them. The partial synsacrum does not correspond with Eaton's illustrations. These authors' examination of the fused sacrals showed "three true sacrals with sacral ribs on the left side, two fused caudals posteriorly, and one dorsal anteriorly" rather than the six sacrals and five or six fused dorsal figured by Eaton. According to Carpenter and Kirkland, as Eaton had reported, there could have been an additional five or six fused dorsal vertebrae, resulting in a synsacrum of 11 or 12 vertebrae. Finally, as Ostrom (1970) had pointed out, the single sternal plate (resembling that of a hadrosaur more than the diamond-shaped sternal on ankylosaurs; see Maryańska 1977), was misidentified by Eaton as a pubis.

Carpenter and Kirkland identified the following kinds of armor in the type specimen of *Silvisaurus*: 1. Three partial cervical rings from the left side of the neck, somewhat similar to the partial fused cervical ring in *Sauropelta*, differing in being composed of two rather than three keeled scutes fused together, the most lateral scute having a low keel; 2. a long ?shoulder spine, apparently from the left side (based on its angled base); and 3. sacral armor possibly represented by a single flat, rectangular scute (based on the flat, hexagonal scutes described by Lull 1921 in *Nodosaurus*).

Key references: Carpenter and Kirkland (1998); Eaton (1960); Gilmore (1930); Lull (1921); Maryańska (1977); Ostrom (1970).

SINORNITHOSAURUS Xu, Wang and Wu 1999

Saurischia: Theropoda: Neotheropoda: Tetanurae: Avetheropoda: Coelurosauria: Manuraptoriformes: Manuraptora: Deinonychosauria: Eumanuraptora: Dromaeosauridae.

Name derivation: Greek *Sinai* = "China" + Greek "ornis" = "bird" + Greek *sauros* = "lizard."

Type species: *S. millenii* Xu, Wang and Wu 1999

Other species: [None.]

Occurrence: Yixian Formation, Liaong Province, China.

Age: ?Early Jurassic (Kimmeridgian) or ?Early Cretaceous (Valangian).

Known material/holotype: IVPP V12811, disarticulated skeleton with skull, integumentary filaments.

Diagnosis of genus (as for type species): Differs from other known dromaeosaurids in presence of ornament-like pits and ridges on antorbital surface of antorbital fossa; posterolateral process of parietal turning sharply posteriorly; dentary bifurcated posteriorly; premaxillary teeth unserrated; coracoid with supracoracoid fenestra; manual phalanx III-1 more than twice length of phalanx III-2; pronounced tubercle near midshaft of pubis; ischium with posterodorsal process; metatarsal III partially arctometatarsalian (Xu, Wang and Wu 1999).

Comments: The number of dinosaurs possessing feathers and feather-like structures increased to five with the discovery of the new theropod dromaeosaurid type species *Sinornithosaurus millenii*. Announced in the October, 1999 issue of *National Geographic* magazine (which included a photograph of a model life restoration of the animal, depicted with a coat of feathers), this genus is distinguished as the first dromaeosaurid known to have a filamentous integument. The genus was founded on the skeleton (IVPP V12811) of a small dromaeosaurid (skull approximately 13 centimeters or about 5 inches long) collected from the probably Early Cretaceous lower (Chaomidianzi; see Ji, Currie, Norell and Ji 1998) Yixian Formation, Jehol Group of Shihetun, western Liaoning, China. *Sinornithosaurus* is currently the best represented (and probably earliest) dromaeosaurid known to date, its type specimen revealing details that add significantly to our knowledge of dromaeosaurid anatomy (Xu, Wang and Wu 1999).

Although the skeleton is small, it was identified by the authors as representing an adult animal, this evidenced by the coossified sacral vertebrae and partially fused astragalus and calcaneum.

Xu *et al.* referred *Sinornithosaurus* to the Dromaeosauridae based upon the following shared derived features: Lacrimal T-shaped; supratemporal fossa large, with strongly sinusoidally curved anterior frontal margin; quadratojugal T-shaped; widely open

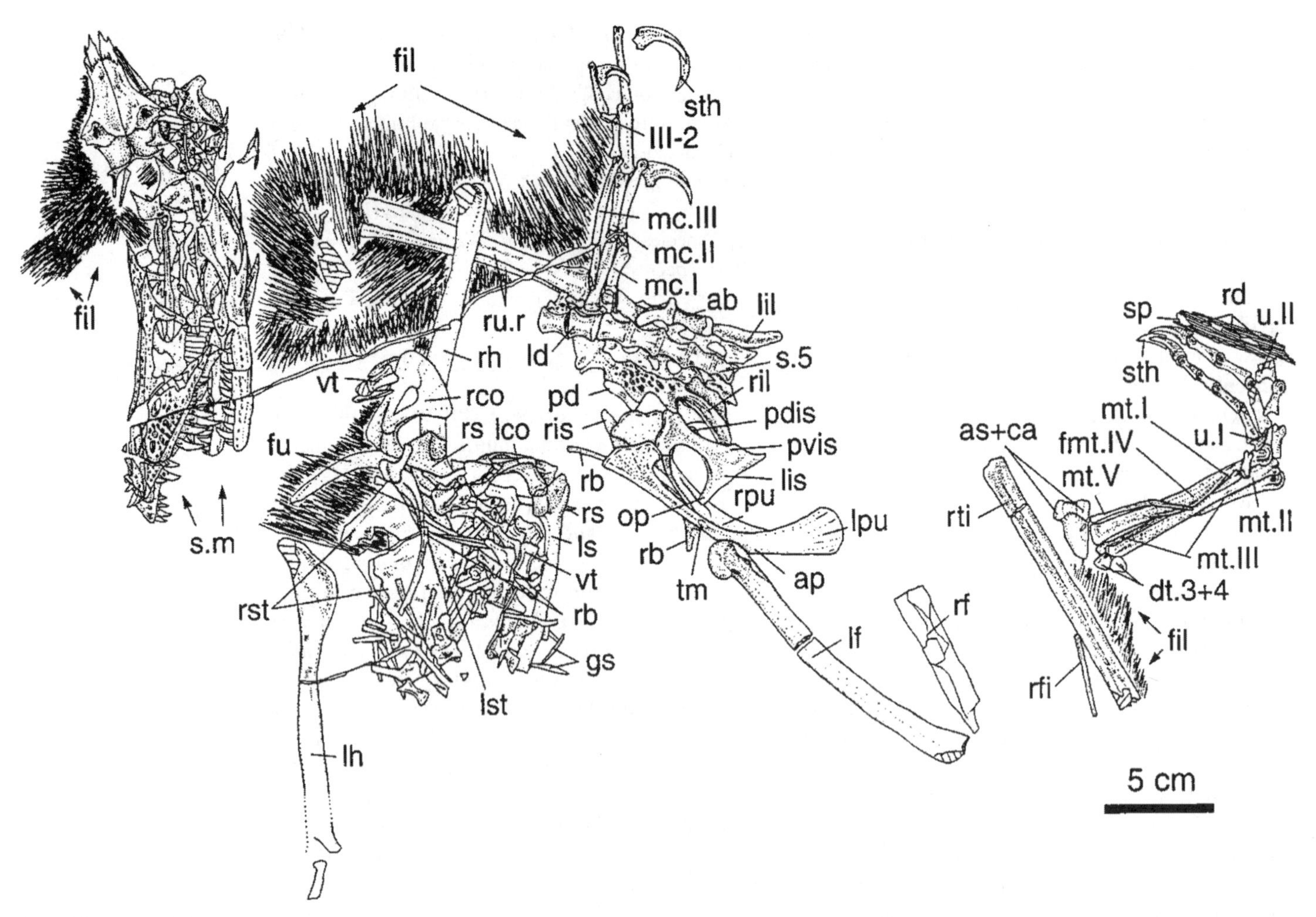

***Sinornithosaurus milleni*, IVPP V12811, holotype skeleton with filamentous integument. (After Xu, Wang and Wu 1999.)**

fenestra between quadratojugal and quadrate; dentary having subparallel dorsal and ventral margins; ossified caudal rods increasing lengths of prezygapophyses and chevrons.

Sinornithosaurus was described by Xu *et al.* as the most birdlike of all known nonavian theropods. The shoulder girdle and forelimb of this genus are very similar to those of early birds; the articulated left scapula and coracoid, as in *Archaeopteryx*, form an angle of less than 90 degrees, and the scapula is shorter than the humerus (67%) and ulna (77%), forming most of the laterally facing glenoid. Indeed, the length of the forelimb — estimated to be approximately 80 percent as long as the hindlimb — is greater relative to that of all other known nonavian theropods.

The authors observed that *Sinornithosaurus* is also more birdlike than other nonavian theropods in the following derived features of the pelvic girdle and hindlimb: Pubic peduncle of ilium broader than acetabulum (see Forster, Sampson, Chiappe and Krause 1998); acetabulum open, tending to close off medially (Martin 1991; Novas and Puerta 1997); pubis posteroventrally directed, bearing short symphysis (less than half the bone's length), distal end cup-like (Norell and Makovicky 1998); ischium short, platelike, less than half length of pubis, indicating, as in *Velociraptor* (see Norell and Makovicky) and birds (Forster *et al.*) absence of ischial symphysis; thin shaft of fibula (about one seventh of tibial diameter; see Forster *et al.*).

Additionally, the shoulder girdle of *Sinornithosaurus* is similar to that of *Archaeopteryx*, having a laterally facing glenoid indicating an avian mode of forelimb movement (elevation and relative rotation and adduction). This modification, the authors speculated, may also have been the case for *Deinonychus* and other dromaeosaurids for which complete pectoral girdles are not known. This altered orientation of the shoulder girdle may have permitted a wider and more birdlike range of motion at the glenohumeral joint. As *Sinornithosaurus* and other dromaeosaurids were bipedal, cursorial terrestrial animals closely related to birds, this suggested to Xu *et al.* that "the anatomical modification of their shoulder girdle supports a cursorial origin for avian flight" (see Gauthier and Padian 1985; Ostrom 1986).

Among the more spectacular aspects of IVPP P12811 is the layer of integumentary filaments —

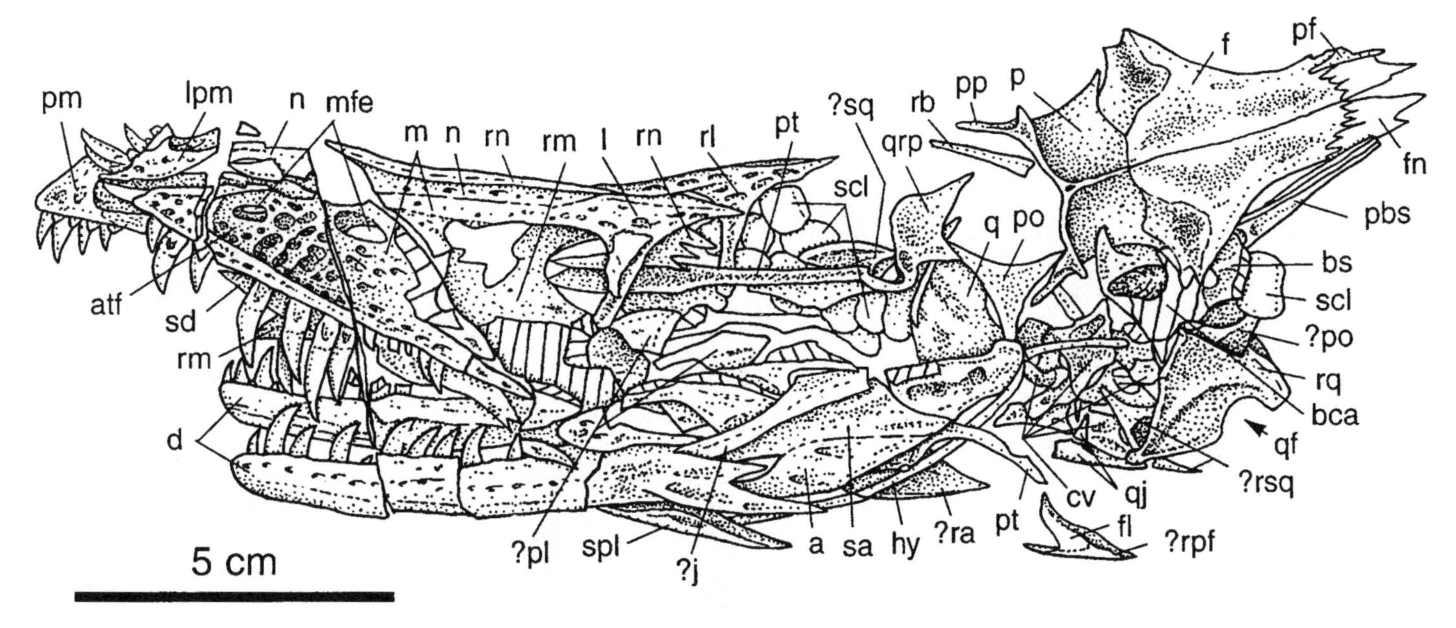

Sinornithosaurus milleni, IVPP V12811, holotype skull with lower jaws. (After Xu, Wang and Wu 1999.)

preserved as patches beneath or near most of the bones, no longer their original positions due to postmortem displacement—that apparently covered the body. As described by Xu *et al.*, these filaments (then only partially exposed) generally reach 40 millimeters in length, with those preserved nearer the postcranial bones apparently longer than those around the skull, those around the tibia relatively sparse. The filaments differ only slightly from those known in some other nonavian theropods (*e.g., Caudipteryx*) or the pulmaceous feathers of the primitive bird *Confuciusornis* from the same locality. The authors could not determine, however, if these structures in *Sinornithosaurus* were in life rectrix-like feathers, as in *Caudipteryx* and *Protarchaeopteryx*, or remix-like structures, as observed on the arm of *Caudipteryx*. Furthermore, the broad distribution of these structures across the body further indicates that "they do not represent internal collagenous fibres for skin support in semiaquatic animals [Geist, Jones and Ruben 1997 suggested this regarding *Sinosauropteryx*; see entry, *S1*], but integumentary derivations of terrestrial animals," possibly serving to insulate the body for maintaining heat (see Ji *et al.*).

Xu *et al.* performed a phylogenetic analysis of *Sinornithosaurus*, incorporating into it all of the theropod taxa that have previously been regarded as closely related to birds. Characters relating to feathers and integumentary filaments were not included in their analysis, as the authors could not determine if their absence in the majority of other derived theropods is real or merely a preservational artifact. From this analysis, the authors concluded that *Sinornithosaurus* is a basal dromaeosaurid; Dromaeosauridae and birds (Aves=Avialae) are more closely related to one another than either is to Troodontidae; and *Protarchaeopteryx* and *Caudipteryx* are more distant from birds than is Troodontidae. Although this study suggested to Xu *et al.* that true feathers may have been present in dromaeosaurids, the authors stated that "the validity of this interpretation cannot be confirmed until more direct evidence is available."

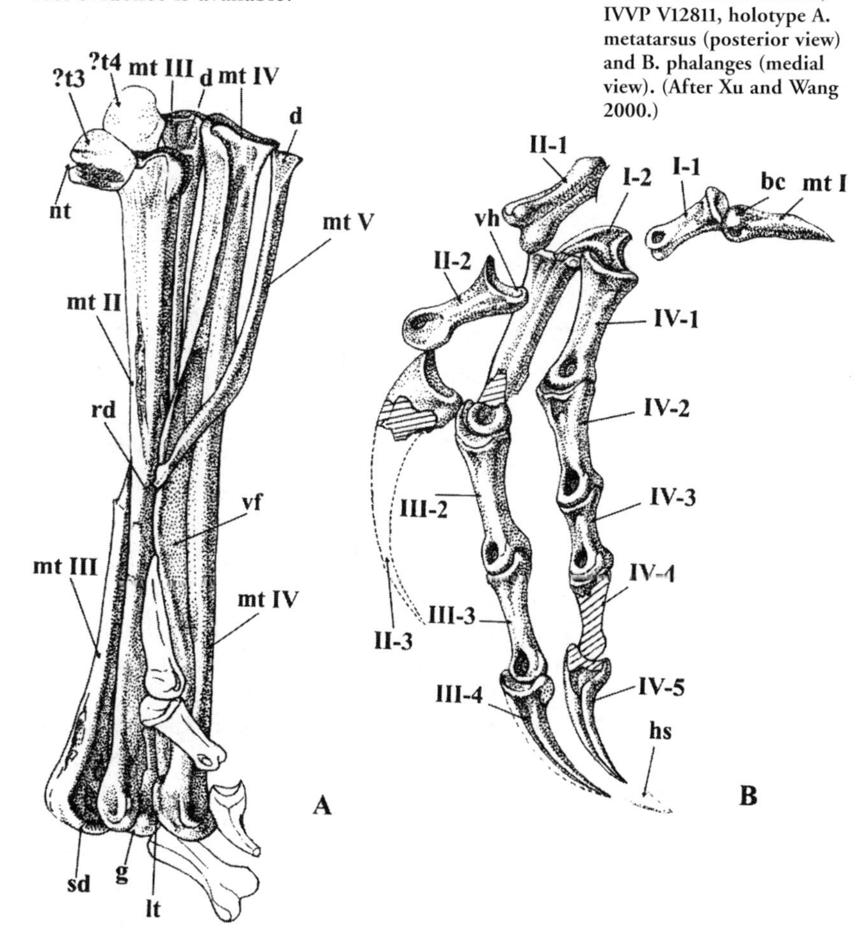

Sinornithosaurus milleni, IVVP V12811, holotype A. metatarsus (posterior view) and B. phalanges (medial view). (After Xu and Wang 2000.)

Life restoration of *Sinornithosaurus milleni* with feathers by Mike Fredericks.

Subsequent to the description of *Sinornithosaurus*, Xu and Wang (2000) described in detail the pes of the holotype, finding that, although this genus is a dromaeosaurid, its foot is troodontid-like. (In the past, the pes of Dromaeosauridae and Troodontidae were regarded as fundamentally different, the specialized pedal digit considered to have convergently evolved in both of these groups; see Osmólska 1990).

Xu and Wang observed the following in the foot of *Sinornithosaurus*: 1. Metatarsus relatively long and slender (in this character, *Sinornithosaurus* being closer to troodontids than to dromaeosaurids); 2. distal articulation of metatarsal I ball-like; 3. ginglymus of phalanx II-1 moderately elevated (in this feature, more similar to troodontids; extending significantly above and below shaft in other dromaeosaurids); 4. proximoventral heel of phalanx II-2 prominent (uniquely shared with troodontids, dromaeosaurids, and *Rahonavis*, but less developed than in other dromaeosaurids); 5. phalanx II-2 moderately constricted (in this feature, more similar to troodontids; distal facet of this phalanx extending well above and below shaft in other dromaeosaurids); 6. over-sized, strongly recurved ungual digit II (intermediate in size between troodontids and dromaeosaurids); 7. metatarsal III proximally compressed, wedged between adjacent metatarsals (similar to troodontids); 8. distal articulation of metatarsal III not glymoid (in this character, more similar to troodontids and most other theropods; ginglymus in other dromaeosaurids); 9. metatarsal III covering anterior surfaces of metatarsals II and IV distally (all three more or less parallel in same plane in other dromaeosaurids); and 10. metatarsal IV almost as long as III, more robust than other metatarsals (in this feature, more similar to troodontids than dromaeosaurids).

Considering the basal position of *Sinornithosaurus* within Dromaeosauridae, the previously thought primitive metatarsus seen in other derived dromaeosaurids may represent a reversal to the primitive condition, perhaps explained as synapomorphies for more derived dromaeosaurids. Furthermore, Xu and Wang noted, this troodontid-like pes in a basal dromaeosaurid could offer additional evidence for a monophyletic Deinonychosauria including Troodontidae and Dromaeosauridae. However, this hypothesis remains to be tested by a thorough phylogenetic analysis of the Coelurosauria.

Key references: Forster, Sampson, Chiappe and Krause (1998); Gauthier and Padian (1985); Geist, Jones and Ruben (1997); Ji, Currie, Norell and Ji (1998); Martin (1991); Norell and Makovicky (1998); Novas and Puerta (1997); Osmólska (1990); Ostrom (1986); Xu and Wang (2000); Xu, Wang and Wu (1999).

†SINOSAUROPTERYX

Age: ?Late Jurassic (Kimmeridgian) to ?Early Cretaceous (Valanginian).

Comments: One of the most important dinosaur finds of recent years has been *Sinosauropteryx*, a small compsognathid theropod known from several specimens recovered from a quarry in the lower part of the Yixian Formation, in the Sihetun area of northeastern China. The dinosaur was distinguished by a very long tail, yet more significantly what appear to be impressions of "protofeathers" (see *S1*).

Not only have these feather like structures remained a topic of controversy since this dinosaur's remains were first described, so also has the age of the rocks from which the specimens (as well as those of two other very birdlike nonavian theropods, *Caudipteryx* and *Protarchaeopteryx*) were extracted. The consensus has been that these fossils were found in rocks of either Upper Jurassic or Lower Cretaceous age. Most opinions have tended to favor the geologically younger dating.

Recently, Ji (1998) briefly reported that remains of two new pterosaurs (one a rhamphorhynchoid, the other belonging to the genus *Eosipterus*) had been found in the Yixian Formation in this area. Because 1. these pterosaurs were recovered from the horizon 2–3 meters higher than the *Confuciusornis* beds, 2. rhamphorhynchoids are not known beyond the Late Jurassic, and 3. *Eosipterus* is similar to certain Late Jurassic European forms, Ji interpreted the correct age for the above-mentioned taxa also to be Late Jurassic.

Contrarily, Smith, You and Dodson (1998) stated in a brief report that both recent American and Canadian radiometric dating of volcanics from the Yixian Formation, near and at the site where *Sinosauropteryx* was found, and also several Chinese palynomorph datings, consistently find Early Cretaceous ages for these

***Sinosauropteryx prima*, NIGP 127586, counterpart of holotype skeleton GMV 2123, displayed in the "China's Feathered Dinosaurs" exhibit from National Geographic Society.**

Photograph by the author, courtesy The Field Museum.

Photograph by the author, courtesy The Field Museum.

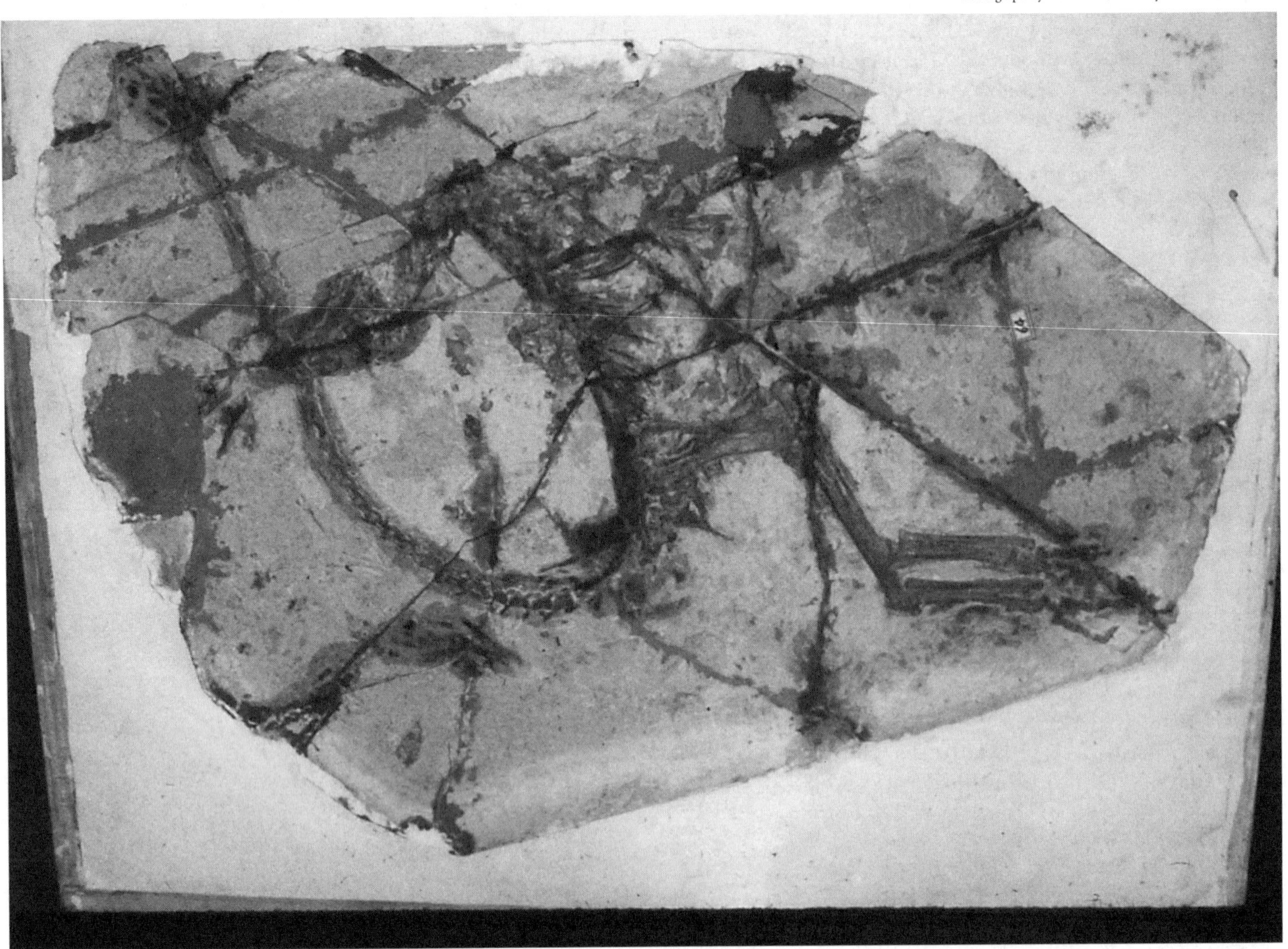

***Sinosauropteryx prima*, NIGP 127587, referred specimen preserving stomach contents and apparent eggs, on display at National Geographic Society's "China's Feathered Dinosaurs" exhibit.**

rocks. According to Smith *et al.*, resistance to the Early Cretaceous datings seems to be based "on the grounds that the Sihetun igneous rocks might be intrusives emplaced after sedimentation ceased," while "our structural work at Sihetun and petrographic work done on the volcanics indicate that the rocks formed from extrusive flows, not intrusions."

In a report on the stratigraphic sequence and vertebrate-bearing beds of the lower part of the Yixian Formation in Sihetun and neighboring areas, Wang, Wang, Wang, Xu, Tang, Zhang and Hu (1998) noted that vertebrate fossils (including *Sinosauropteryx*) recovered from the third and geologically youngest member of this formation represent all of the major vertebrate groups, including reptiles, birds, amphibians, and fish. (The first and oldest member overlies the Tuchengzi Formation, which is of Late Jurassic age.) This rich record of diverse fossil vertebrates "recorded a great evolutionary event and mass mortality event of vertebrates near the Jurassic–Cretaceous boundary." Wang *et al.* interpreted the lithostratigraphic sequence of the lower part of the Yixian Formation as revealing the development of the sedimentary basin from alluvial facies to deep, shallow, and coastal lacustrine facies. Tuffite, tuffaceous sandstone, and gravity flow event sediments, above and below the vertebrate-bearing beds in Yixian Formation lacustrine strata, indicated to the authors "that the lake was a pan-basin with wide and deep water in some periods, and volcanism had greatly affected the sedimentation in the basin."

In a later abstract, Wang, Wang, Wang, and Xu (2000) stated that, based upon recently excavated fossils (the birds *Confusciousornis*, *Liaoningornis*, and *Eoenantiornis*, theropods *Sinosauropteryx*, *Protarchaeopteryx*, *Sinornithosaurus*, and *Beipiaosaurus*, ceratopsian *Psittacosaurus*, pterosaurs *Eosipterus* and *Dendrorhynchoides*, frogs *Liaobatrachus* and *Callobatrachus*, and mammals *Zhangheotherium* and *Jeholodens*) showing derived characters respect to comparable taxa, the Jehol Group is likely of Early Cretaceous age, this being consistent with radiometric dating results of the

Photograph by the author, courtesy The Field Museum.

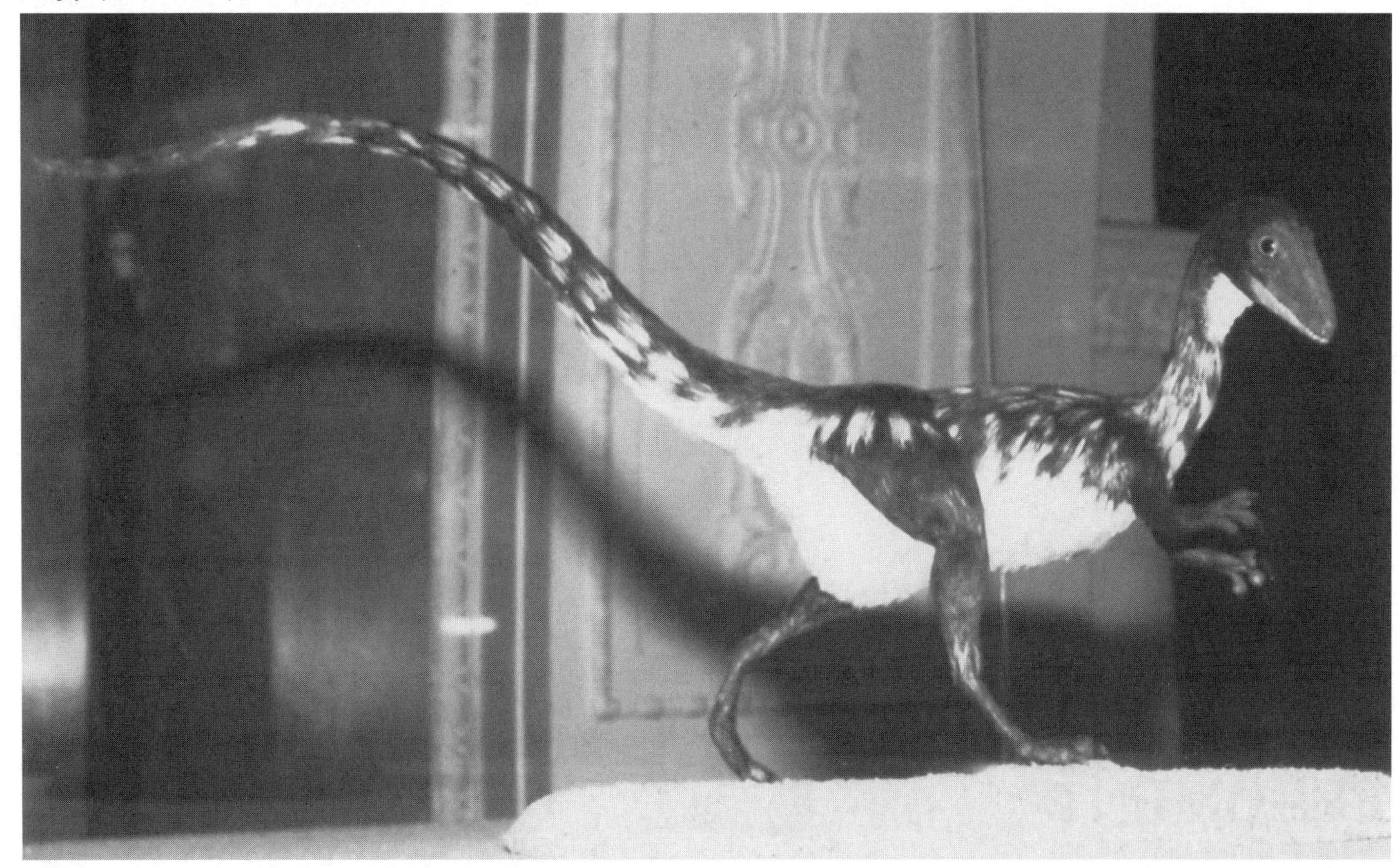

Sculpture of the nonavian theropod *Sinosauropteryx prima* restored with "protofeathers" by artist Brian Cooley for National Geographic Society's "China's Feathered Dinosaurs" exhibit.

Yixian Formation ranging from 133 to 120 million years ago.

(See also †*Psittacosaurus* entry, this volume, regarding the possible age of the Yixian Formation.)

Key references: Ji (1998); Smith, You and Dodson (1998); Wang, Wang, Wang, Xu, Tang, Zhang and Hu (1998); Wang, Wang, Wang and Xu (2000)

†SONOROSAURUS—(=?*Brachiosaurus*)

Occurrence: Turney Ranch Formation, Arizona, ?Cloverly Formation, Wyoming, Montana, United States.

Age: Middle Cretaceous (Albian–Cenomanian).

Known material/holotype: ASDM-500, left tibia, proximal end of right ?tibia, left and right radii, left ulna, complete ?left metacarpal I, complete metacarpal II, proximal and distal ends of metacarpal ?III, distal half and marginal elements of left ilium, distal one-third of ischium, acetabulum, medial portions of pubis and ?ischium, distal half of left femur, midshaft of right ulna, left ?tibia, left fibula, partial right ?fibula, metatarsals I, II, IV, and V, phalanges I-I, II-I, IV-I, II-III, ungual digits I and II, first, ?second, and ?third dorsal vertebrae, caudal vertebrae approximately 5–12, chevron fragments, dorsal and cervical ribs, gastroliths.

Comments: The type species *Sonorosaurus thompsoni*, as pointed out by Curtice (2000), is distinguished as the first brachiosaurid known from Arizona and is also the most completely known North American sauropod of Albian–Cenomanian age yet recovered.

As chronicled by Curtice, the type specimen, a partial postcranial skeleton (ASDM-500), was collected over a five-year period by the geologic staff of the Arizona-Sonora Desert Museum from the Turney Ranch Formation in Arizona (see *S1*). Following its discovery, the specimen underwent a series of conflicting identifications. Early reports suggested that the remains, based on their size and temporal position, belonged to a large hadrosaur (see Thayer and Ratkevich 1995). Later, at the DinoFest 2 symposium held in Tempe, Arizona, in 1997, Ratkevich suggested that the remains could belong to a therizinosaur (R. D. McCord, personal communication 1999). The material was subsequently examined at that conference by Brian Curtice and John S. McIntosh, who correctly identified it as a brachiosaurid sauropod. Following this identification, various talk shows, local newspapers, and *Discover* magazine reported that an almost complete skull belonging to this dinosaur had also been recovered (see Ratkevich 1997), this eventually proving to be a distorted vertebra.

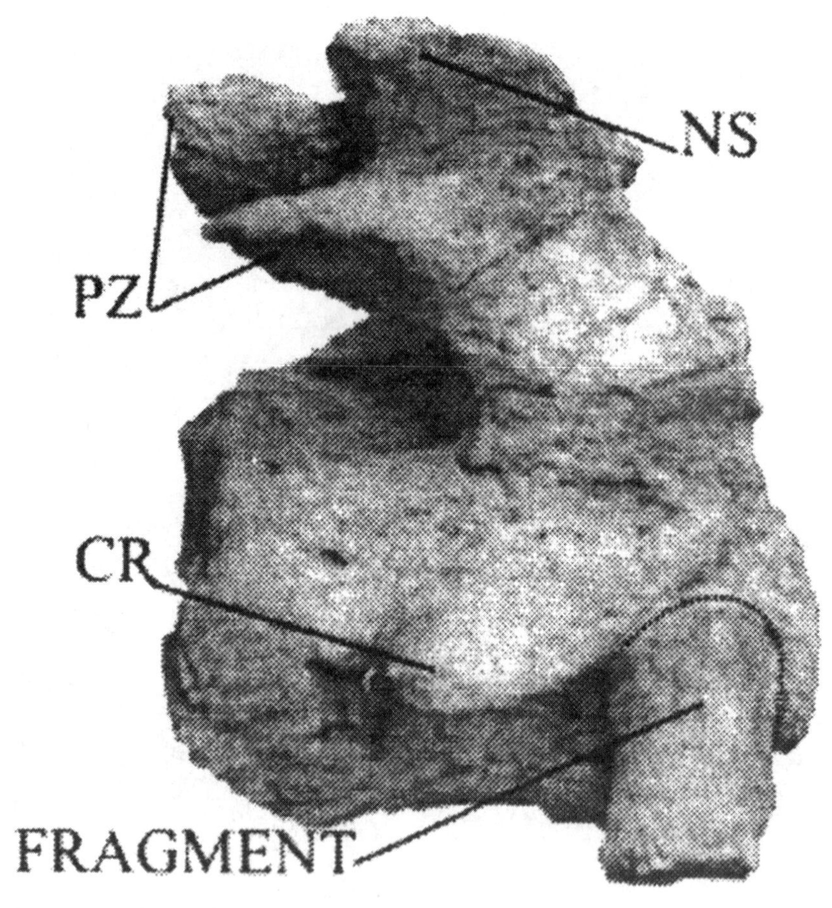

Sinorosaurus thompsoni, ASMD-500, holotype dorsal vertebra, right lateral view. (After Curtice 2000.)

Following the original description of *S. thompsoni* by Ratkevich (1998), Curtice published a detailed description of the axial skeleton. As no neurocentral sutures were found on all the preserved axial elements, Curtice (following the criteria set down by Brochu 1996, based on a study of extant crocodilians) determined that ASDM-500, although representing a smaller brachiosaurid than *Brachiosaurus*, is not a juvenile. In describing these remains, Curtice noted that the chevron facets of the caudal vertebrae are especially large, approaching in the most extreme examples those of *Haplocanthosaurus*. The preserved bases of these facets suggested to Curtice that the chevrons possess open hemal arches; however, personal examination by that author of the chevrons of *Brachiosaurus brancai* housed in the Museum für Naturkunde der Humboldt, Berlin, revealed that this character is quite variable.

Comparing *Sonorosaurus* with other sauropods, Curtice noted that some elements (YPM 5104, 5116, 5199) recovered from the Cloverly Formation of Wyoming and Montana (see Ostrom 1970), referred to *Astrodon*, bear a striking resemblance to *Sonorosaurus*, particularly with their stout amphiplatyan centra, swept back caudal ribs, and large chevron facets." However, as the Cloverly remains include an abundance of prezygapophyses but no postzygapophyses (preserved in *Sonorosaurus*), a complete comparison with *Sonorosaurus* was not possible. Furthermore, the Cloverly's apparently multiple sauropod taxa made "sweeping generalities difficult" beyond a tentative referral of those specimens to this genus. Also, disarticulated sauropod material from the Dalton Wells quarry, Cedar Mountain Formation (Lower Cretaceous), near Moab, Utah (see Britt, Scheetz, McIntosh and Stadtman 1998; also "Introduction," section on sauropods) includes caudal vertebrae resembling those of *Sonorosaurus*.

According to Curtice, the only fact differentiating

Sonorosaurus thompsoni, ASMD-500, holotype caudal vertebra, right lateral view. (After Curtice 2000.)

Sonorosaurus from *Brachiosaurus* is its temporal location and ASDM-500, if discovered in the Morrison Formation, would surely have been labeled cf. *Brachiosaurus.* Curtice noted that it would be unprecedented for a sauropod genus to have survived for 40 million years. Therefore, the author felt that — despite the fact that the axial elements of ASDM-500 and *Brachiosaurus* do not differ in any meaningful way — it was important to retain the name *Sonorosaurus* "because it is a single associated specimen with many skeletal elements existing for comparison," and also "[because] one cannot cladistically differentiate it from *Brachiosaurus* does not a *Brachiosaurus* it make."

Key references: Britt, Scheetz, McIntosh and Stadtman (1998); Brochu (1996); Curtice (2000); Ostrom (1970); Ratkevich (1997, 1998); Thayer and Ratkevich (1995).

†SPINOSAURUS

Saurischia: Theropoda: Neotheropoda: Tetanurae: Avetheropoda: Spinosauroidea: Spinosauridae: Spinosaurinae.

Type species: *S. aegyptiacus* Stromer 1915.

Other species: *S. ?maroccanus* Russell 1996 [*nomen dubium*].

Age: Middle Cretaceous (Albian–Cenomanian).

Diagnosis of *S. ?maroccanus*: Differs from *Cristatusaurus lapparenti* in the following: premaxilla bulbous, slightly "hooked" in profile, decreasing in height posteriorly; posterodorsal surface rounded in cross section; anteriormost premaxillary alveolus relatively small, lateral alveoli grouped into two pairs; maxillary and dentary teeth circular in cross section, maxillary teeth procumbent anteriorly and posteriorly; dentary markedly constricted vertically in midsection, alveoli grouped in pairs; maxillary and dentary teeth having smooth carinae (Taquet and Russell 1998).

Comments: In 1996, Dale A. Russell named and described a second species of the giant, long-spined theropod *Spinosaurus*, which he named *Spinosaurus maroccanus*. This latter taxon, described by Russell as a relatively long-necked species, was based upon material — including a caudal vertebra (CMN 50791, the holotype), dentary fragments (CMN 50832), cervical vertebrae (CMN 41768), and a dorsal neural arch (CMN 50813) — collected in southern Morocco, and also material from Algeria (Taquet and Russell 1998) (see *S1*).

Subsequently, Taquet and Russell (1998) referred to this new species additional material collected from Gara Samani (of Albian age) on the northwestern edge of the Tademaït, in the Algerian Sahara. These specimens include a rostrum containing both premaxillae,

Spinosaurus ?maroccanus, MNHN SAM 124, referred rostrum with premaxillae, maxillae, and vomers, in a. dorsal, b. ventral, and c. left lateral views. (After Taquet and Russell 1998.)

maxillae, vomers, and fragments from the midshaft of the right dentary (MNHN SAM 124) (pertaining to a mature individual, as evidenced by the closed though visible sutures separating the premaxillae); a fragment of a premaxilla (MNHN SAM 125); centra of two cervical vertebrae (MNHN SAM 126 and 127); and a neural arch of a dorsal vertebra (MNHN SAM 128).

Originally, Russell differentiated *S. maroccanus* from the type species *S. aegyptiacus* based "on the greater length of the midcervical vertebrae relative to the height of the posterior articular facet"; however, Taquet and Russell's later diagnosis differentiated the Moroccan species from a new spinosaurid taxon, *Cristatusaurus lapparenti*, "based upon the anterior portion of a skull and fragment of the right dentary from Gara Samani" (see *Cristatusaurus* entry).

In describing MNHN SAM 124, Taquet and Russell suggested that the structure of the rostrum probably reflects conditions in other spinosaurs (*i.e.*, articulated premaxillae and maxillae of about equal width across rostrum at their widest extent, constricted laterally in area of avleolar margin between them containing premaxillary-maxillary suture; seven premaxillary teeth).

According to Taquet and Russell, the dentary of *S. maroccanus*, at least what is known of it, seems to be similar to that of the type species. *Contra* Buffetaut's (1992) opinion, the authors deduced that the neck of *Spinosaurus* forms an upward curve when articulated, this evidenced in the central facets of the vertebra from Gara Samani and those from Morocco (see Russell 1996).

More recently, however, the specific validity of *S. maroccanus* has been challenged. In a paper in which they named and described the new spinosaurid *Suchomimus* (see entry), Sereno, Beck, Dutheil, Gado, Larsson, Lyon, Marcot, Rauhut, Sadleir, Sidor, Varricchio, Wilson and Wilson (1998) reevaluated the *Spinosaurus* materials from Morocco and Algeria. In their opinion, Russell's criteria (primarily the proportions of an isolated cervical centrum) for his referral of the Moroccan material to a new species were questionable (*i.e., S. ?maroccanus*). Therefore, regarding *S. maroccanus* as a *nomen dubium*, Sereno *et al.* provisionally referred all spinosaur found in Albian- to Cenomanian-age rocks in northern Africa to belong to *S. aegyptiacus*.

Note: Sereno *et al.* reinterpreted material originally identified by Stromer (1934) as "*Spinosaurus* B" to belonging to the allosauroid theropod *Carcharodontosaurus* (see entry).

Key references: Russell (1996); Sereno, Beck, Dutheil, Gado, Larsson, Lyon, Marcot, Rauhut, Sadleir, Sidor, Varricchio, Wilson and Wilson (1998); Stromer (1915, 1934); Taquet and Russell (1998).

†STEGOCERAS

Type species: *S. validum* Lambe 1902.

Other species: [None.]

Comments: *Stegoceras*, as pointed out by pachycephalosaur specialist Robert M. Sullivan, is distinguished as "one of the most studied, yet poorly understood, dinosaurs ever since its recognition," having been originally confused with the theropod *Troodon*. In fact, numerous species have been assigned to this genus over the years, most of which have been subsequently either synonymized with the type species *S. validum* or referred to new genera (*e.g., Gravitholus, Ornatotholus* and *Prenocephale*) (see *D:TE*).

In an abstract for a poster, Sullivan (2000) stated the following: *Stegoceras validum* is a primitive pachycephalosaur characterized by a well-developed squamosal and open supratemporal fenestra. *Ornatotholus browni* Galton and Sues 1983 (see *D:TE*), which has a parietal identical to that of *S. validum*, was consequently regarded as a subjective junior synonym of the latter. Specimens represented by a flat frontal were recognized by Sullivan as a sexual (perhaps female) dimorph of the type species. "*Stegoceras*" *lambei* Sternberg 1945, the oldest North American species, represents a new taxon. The species "*Gravitholus*" [=*Troodon*] *sternbergi* (Brown and Schlaikjer 1943) and *G. albertae* Wall and Galton 1979 are conspecific, *Gravitholus sternbergi* being a new combination. Previously, Sullivan (2000*b*) referred "*S.*" *breve* Lambe 1918 and "*S.*" *edmontonense* Brown and Schlaikjer 1943 to the genus *Prenocephale* (see entry). This taxonomic revision, Sullivan (2000*c*) noted, "demonstrates that small to medium size pachycephalosaurs were rather diverse during the late Campanian (Judithian).

Presumably more details regarding the above poster study will be published by Sullivan at a later date.

Key references: Brown and Schlaikjer (1943); Galton and Sues (1983); Lambe (1902, 1918); Sternberg (1945); Sullivan (2000*b*, 2000*c*); Wall and Galton (1979).

†STEGOPELTA Williston 1905—(=?*Glyptodontopelta*)

Ornithischia: Genasauria: Thyreophora: Thyreophoroidea: Eurypoda: Ankylosauria: Ankylosauridae: Shamosaurinae.

Name derivation: Greek *stegos* = "roofed" + Latin *pelta* = "shield."

Type species: *S. landerensis* Williston 1905.

Other species: [None.]

Occurrence: Frontier Formation, Wyoming, United States.

Age: Middle Cretaceous (Albian or Cenomanian).

Photograph by the author, courtesy The Field Museum.

***Stegopelta landerensis*, FMNH UR88, holotype armor and partial sacrum.**

Known material/holotype: FMNH UR88 [formerly Walker Museum collection, University of Chicago], fragmentary skeleton comprising ?maxillary with three partial alveoli, ?skull fragments, seven cervical and two dorsal vertebrae, parts of synsacrum, proximal caudal centrum, distal caudal centrum, parts of scapulae, both humeral heads, proximal end of left ulna, proximal ends of radii, parts of ilia, distal end of tibia, metacarpal, metatarsal, bifurcated ?shoulder spine, half of cervical neck ring, scutes.

Diagnosis: Axis uniquely very long, slender, strongly compressed laterally; mid- and posterior cervical centra having paired deep fossa separated by horizontal ridge for capitulum of cervical rib as in *Texasetes* (unlike other nodosaurids); neural arch of dorsal vertebrae flush with or overhanging anterior articular surface (inset from articular face in all other known nodosaurids); dorsal centrum cylindrical as in *Struthiosaurus* and *Mymoorapelta* (strongly constricted in *Sauropelta*, *Gastonia*, *Polacanthus*, or *Edmontonia*); acromion process centrally located high on scapular blade as in *Panoplosaurus* (directed toward glenoid in *Sauropelta*); pelvic armor comprising closely placed hexagonal scutes (alternating transverse rows of oval, keeled scutes in *Edmontonia*, tranverse rows of circular disks forming rosettes with smaller hexagonal scutes in *Sauropelta*, irregularly spaced keeled oval or domed circular scutes surrounded by smaller hexagonal scutes in *Polacanthus*, large flat hexagonal scutes in *Nodosaurus*) (Carpenter and Kirkland 1998).

Comments: The type species *Stegopelta landerensis* was based upon very incomplete skeletal remains (FMNH UR88; see *D:TE*, *Nodosaurus* entry for additional photograph of armor) collected from the brown sandy clay at the base of the Belle Fourche Member (latest Albian or earliest Cenomanian; Merewether 1983) of the Frontier Formation of Fremont County, in Benton, near Lander, Wyoming (precise type locality lost; William F. Simpson, personal communication to Carpenter and Kirkland 1998). As related by Moodie (1910), the first fragments belonging to this specimen "had washed and rolled down the slope and had been trodden in the mud by cattle so that when first discovered it was discarded as a worthless plesiosaur specimen." Not until University of Chicago paleontologist Samuel Wendell Williston who, while cleaning up some of these fragments "came across some typical stegosaurian teeth," was the significance of this discovery realized and the rest of the specimen collected.

The type specimen, then in the collection of the Walker Museum, University of Chicago, was first briefly described (but not diagnosed) by Williston

Photograph by the author, courtesy The Field Museum.

***Stegopelta landerensis*, FMNH UR88, holotype armor.**

(1905). As ankylosaurs were poorly known at that time (indeed, the genus *Ankylosaurus* and its family Akylosauridae would not be named by Barnum Brown until three years later), Williston regarded *Stegopelta* as a genus allied to, but only about half the size, of *Stegosaurus*. Williston wrote that the new genus "is peculiar in having a heavy bony carapace, two inches or more in thickness ... covered with, and for the most part firmly united to, a mosaic of pentagonal dermal bony plates, much like those of *Glyptodon* ... [the plates] about four inches in diameter, scrobiculate and somewhat elevated in the middle ... [the animal] was evidently covered with a dermal shield, and probably each eminence bore a more or less elongated horny spine."

In 1910, Moodie described *Stegopelta* in greater detail, although he, like Williston, did not diagnose the genus or species. Moodie also regarded *Stegopelta* as stegosaurian, although he cited similarities between this genus and the British armored dinosaur *Polacanthus*, and also compared it with Brown's (1908) recently described *Ankylosaurus*. In quoting his predecessor (see Williston 1908), Moodie wrote, "Indeed, Doctor Williston says, '*Ankylosaurus* is either very closely allied to or identical with *Stegopelta* Williston, a genus overlooked by Mr. Brown.'"

Walter J. Coombs, Jr. (1978*a*), in his groundbreaking reorganization of the families of Ankylosauria, regarded *Stegopelta* as a nodosaurid Ankylosaur and a junior synonym of *Nodosaurus* (see *D:TE*). More recently, however, Carpenter and Kirkland, in the 1998 review of Lower and Middle Cretaceous ankylosaurs from North America, accepted *Stegopelta* as a valid nodosaurid genus. Also, Carpenter and Kirkland were the first authors to diagnose formally the type species.

In redescribing this taxon, Carpenter and Kirkland pointed out that Moodie had made various errors in his redescription of *Stegopelta*. For example, the supposed right pubis identified by Moodie is actually the left scapula, identified as such by the long, curved blade and acromion process (now lost). Included in the holotype are seven cervical vertebrae not reported by Moodie. These elements resemble those of *Texasetes* but are unlike those of most other known ankylosaurs, this suggesting a close affinity between the two genera. The dorsal vertebra is somewhat different from Moodie's illustration. Contrary to Moodie's

Photograph by the author, courtesy The Field Museum.

***Stegopelta landerensis*, FMNH UR88, holotype armor.**

implication, the armor — in cross section, seen overlying the ilium but separated from it by matrix — was apparently not fused directly to the surface of the ilium. Some of the material belonging to the type specimen is not as complete as figured by Moodie; other material could not be found.

Key references: Brown (1908); Carpenter and Kirkland (1998); Coombs (1978*a*); Moodie (1910); Williston (1905, 1908).

†STEGOSAURUS

Ornithischia: Genasauria: Thyreophora: Thyreophoridea: Eurypoda: Stegosauria: Stegosauridae.

Comments: Among the best known and popular dinosaurs among lay people is the plated *Stegosaurus*; and perhaps the most familiar life depiction of this genus is in the conflict scenario, with *Stegosaurus* portrayed utilizing its spiked tail as a defensive weapon against an attacking predator. Such dramatic, life and death-related visualizations are not founded upon mere speculation, but are sometimes based upon hard evidence preserved in the fossil record.

In a recent study, McWhinney, Rothschild and Carpenter (1998) briefly reported on their examination of a sampling of 51 *Stegosaurus* dermal tail spines (or "spikes"). As stated by these authors, "Evidence of trauma-induced bone fractures in individuals who survived show alterations in the bony structure seen in endosteal new bone formation (callus)," these alterations producing "various degrees of bone resorption, bone formation and neurosis," and with no callus or other such alterations occurring after death.

Out of the sampling of dermal spines examined, four, or about 8 percent, exhibited trauma. Of these spines, two showed the effects of post-traumatic chronic osteomyelitis, a bone disease. According to McWhinney *et al.*, the pathological dermal spines showed remodeled growth indicating that each individual had survived following a traumatic encounter resulting in the breaking off of the spike.

The authors briefly described the outer cortical

Copy by J. Beckett, courtesy Department of Library Services, American Museum of Natural History (neg. #313965).

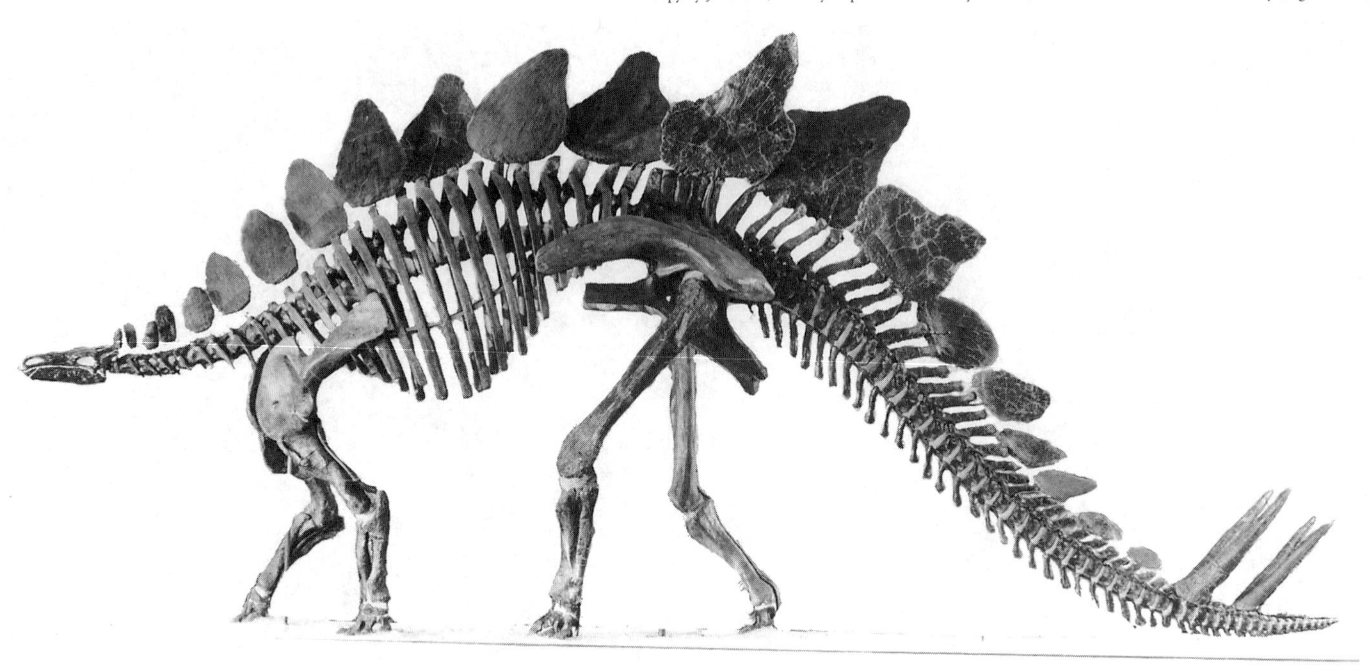

Skeleton (AMNH 650) of *Stegosaurus stenops* found by Peter Kaisen in 1901 at Bone Cabin Quarry, Wyoming, during the American Museum of Natural History's first major dig. A relatively incomplete specimen, the missing elements from other skeletons filled in the gaps when it was first mounted in 1932.

surface of the two spines as showing "deposits of extraneous bone and prominent areas of filigree bone." Drainage areas, or sinus cavities, for the excavation of pyogenic material, were observed by the authors at or close to the broken ends of the spikes. The osteomyelitis in these spikes "was a direct response to the traumatic open fractures." These fractures—which permitted direct access for bacteria to enter the wound—had a higher rate of disturbances during healing because of the increased frequency of infection.

Main, Padian and Horner (2000), in an abstract, addressed the possible functions of osteoderms in *Stegosaurus* and other archosaurs, particularly those in various thyreophoran dinosaurs. Main *et al.* performed a histologic study of the plates and spikes of *Stegosaurus*, and also the scutes of other thyreophorans and various crocodilians and their relatives.

Main *et al.*'s study resulted in the following observations: 1. The mostly flat scutes of crododiles, phytosaurs, and aetosaurs grow by periosteal deposition and external and internal modeling; 2. scutes of the basal thyreophoran *Scutellosaurus* grew in a quite similar fashion; and 3. the mostly flat parasagittal plates of stegosaurs grew mainly by basal osteogenesis, with some lateral periosteal deposition and extensive remodeling internally, these plates, consequently, evolving by hypertrophic growth of the parasagittal dorsal keel of the basal thyreophoran scutes.

The authors questioned the hypothesis that the plates of *Stegosaurus*, as evidenced by their medullary "pipes" and surface grooves, had primarily a thermoregulatory function (see *D:TE*). As Main *et al.* pointed out, other stegosaurian genera possessed a variety of plate and spike designs; therefore, the design of the plates in *Stegosaurus* seem neither optimal nor necessary for the regulation of heat. As observed by these authors, the internal "pipes" and external grooves are also sometimes present in various kinds of broad, flat bones (*e.g.*, the frills of ceratopsians and the antlers and horns of artiodactyls) that are not utilized in thermoregulation. Main *et al.* suggested that these features may actually be constructional artifacts reflecting the processes and modes of bone growth. Furthermore, "Variation in the cranial ornamentation of other ornithischian dinosaur groups suggest that display and species recognition were plausible and broadly distributed functions of these structures."

Note: Kenneth Carpenter and Peter M. Galton will show that *Stegosaurus ungulatus* does not possess eight to 10 tail spikes, as this species is commonly depicted in skeletal mounts (as the composite skeleton mounted at Yale University's Peabody Museum of Natural History) and various life restorations (K. Carpenter, personal communication 2000).

Key references: Main, Padian and Horner (2000); McWhinney, Rothschild and Carpenter (1998).

STYGIVENATOR Olshevsky, Ford and Yamamoto 1995 (=?*Aublysodon*, ?*Tyrannosaurus*) [*nomen dubium*]

Saurischia: Theropoda: Neotheropoda: Tetanurae: Avetheropoda; Coelurosauria: Manuraptoriformes: Arctometatarsalia: Tyrannosauridae: ?Aublysodontinae.

Photograph by the author, courtesy Natural History Museum of Los Angeles County.

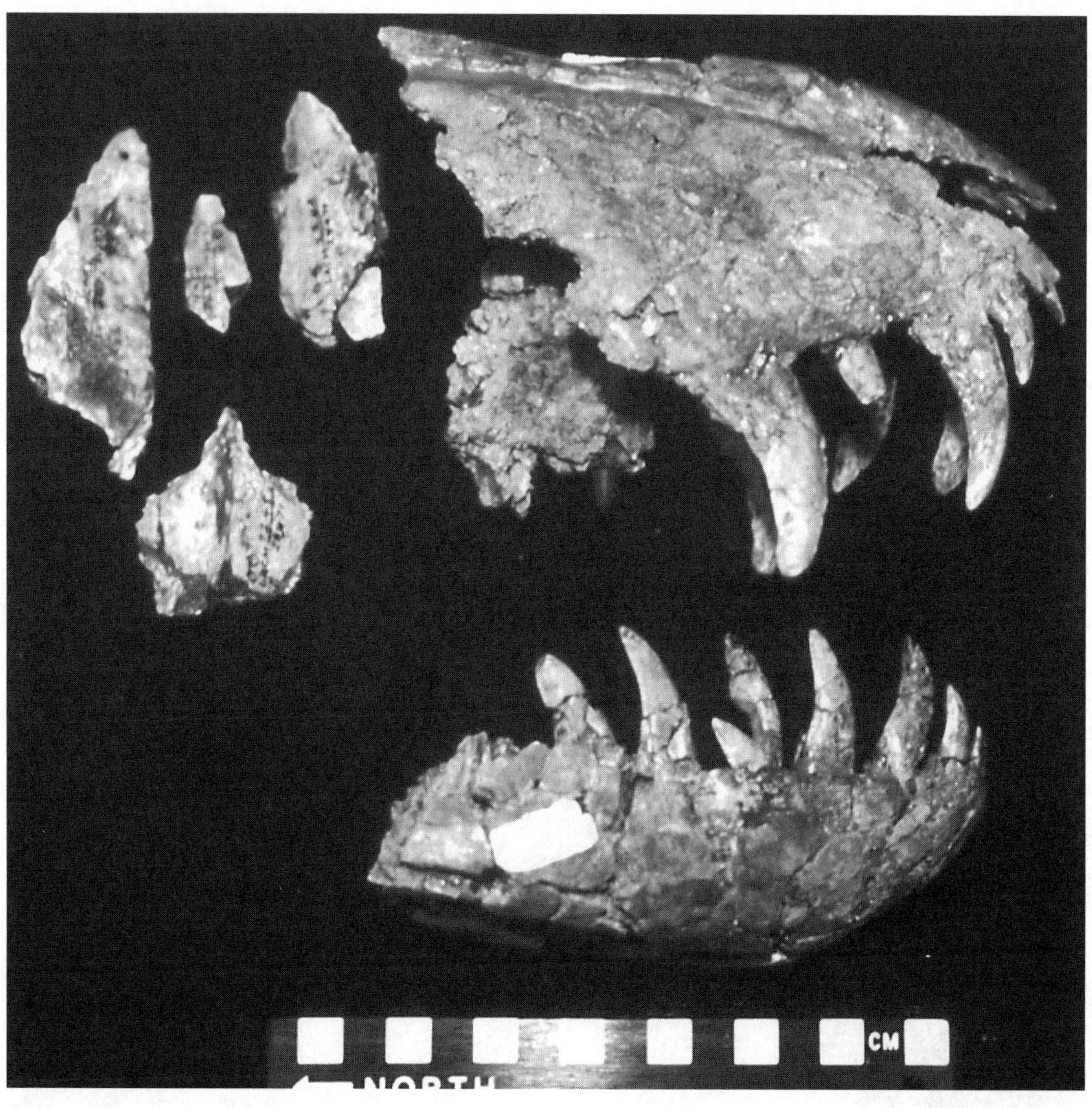

***Stygivenator molnari*, holotype partial skull (LACM 28471), in right lateral view (see *D:TE* for left lateral view), including anterior part of snout and lower jaws, and incomplete disarticulated bones of the skull roof.**

Name derivation: Latinized form of "Styx [one of three rivers of Hades, referring to the Hell Creek Formation]" + Latin *venator* = "hunter."

Type species: *S. molnari* Olshevsky, Ford and Yamamoto 1995 [*nomen dubium*].

Other species: ?*S. amplus* (Marsh 1892) [*nomen dubium*], ?*S. cristatus* Marsh 1892 [*nomen dubium*].

Occurrence: Hell Creek Formation, ?Judith River Formation, Montana, ?Two Medicine Formation, Montana, ?Lance Formation, Wyoming, United States, ?Judith River Formation, Alberta.

Age: Late Cretaceous (late Campanian–Maastrichtian).

Known material: Partial skull, ?isolated teeth.

Holotype: LACM 28471, anterior part of skull, juvenile.

Diagnosis of genus (as for type species): (Including characters from Molnar 1978) "Shanshanosaurine" [=Aublysodontine] characterized by several long anterior maxillary and dentary teeth; height of largest anterior maxillary tooth equal to or surpassing depth of dentary at its location (so that when jaws closed completely, tips of anterior maxillary teeth would extend slightly below ventral margin of snout); anterior premaxillary tooth considerably narrower in lateral aspect than lectotype tooth of *Aublysodon mirandus* and somewhat smaller; tooth row of anterior portion of dentary, including first three dentary teeth, elevated above level of remainder of tooth row, anterior dentary teeth procumbent (Olshevsky 2000).

Comments: In 1978, Ralph E. Molnar, then at the University of South Wales, described the fragmentary skull (LACM 28471) of a moderate-sized, juvenile theropod collected during the summer of 1966 by Harley G. Garbani for the Museum of Natural History, Los Angeles County (now the Natural History Museum of Los Angeles County). Garbani had found the specimen — consisting of the anterior end of the

Photograph by the author, courtesy Natural History Museum of Los Angeles County.

Holotype partial skull (LACM 28471) of the "Jordan theropod," *Stygivenator molnari*, as formerly exhibited at the Natural History Museum of Los Angeles County, then identified as a juvenile *Tyrannosaurus rex*.

snout and some incomplete, disarticulated bones from the skull roof— in the Hell Creek Formation on the ranch of F. S. McKeever, in Garfield County, Montana, near the town of Jordan. Preferring at that time not to assign the specimen to any particular genus, Molnar referred to it simply as the "Jordan theropod" (see Molnar 1978). Shortly after the collection and preparation of the specimen, the Los Angeles County museum placed it on public exhibition, the label originally identifying it as a juvenile *Tyrannosaurus rex*. (Although the specimen had not been on exhibit for a number of years, the museum is, as of this writing, has further prepared and also cast the specimen for display in its dinosaur hall.)

Molnar observed the following characters in the "Jordan theropod": Premaxillary teeth D-shaped in cross section; frontals thin dorsoventrally, with no indication of recessed areas associated with supratemporal fenestrae; maxillae dorsoventrally low; dentary showing no anterior ascent of alveolar margin.

LACM 28471 was identified by Molnar as a juvenile based on such features as its small size, the parietal being more widely separated than in specimens of adult tyrannosaurids, and the dentary being shallower relative to the length of the maxilla than in any known tyrannosaurid.

Although Molnar did not yet assign the specimen to any particular theropod group, he noted then that certain characters suggested that LACM 28471 might be the remains of a large dromaeosaurid. Molnar further noted that the snout in this specimen is longer, relative to height, than that of any known tyrannosaurid, while the maxilla most strongly resembles that of an immature *Gorgosaurus* (AMNH 5664) (see *Aublysodon* entry, *D:TE*, for more details concerning this specimen).

Later, in his book *Predatory Dinosaurs of the World*, Paul (1988*b*) referred LACM 28471 to the enigmatic genus *Aublysodon* as the new species *A. molnaris*. The lectotype of the type species *Aublysodon mirandus* consists of a single, nonserrated tooth (ANSP 9335), D-shaped in cross section, recovered from the Hell Creek Formation of Montana, named and described by Joseph Leidy in 1868 (see *D:TE* for details). Paul differentiated his newly proposed species by its relatively larger size and on not yet described remains from the Judith River Formation; at the same time, that author referred the Asian theropod *Shanshanosaurus houyanshanensis* (regarded by Molnar 1990 as not referrable to a taxon above the level of family) to *Aublysodon*.

Molnar and Carpenter (1989) subsequently agreed with Paul's assignment of the "Jordan theropod" to *Aublysodon*, but, as the undescribed material Paul mentioned was not available for comparison, these authors more conservatively referred it to *Aublysodon* cf. *A. mirandus*, also noting that *S. houyanshanensis* could not be evaluated without examination of its material. In this same paper, Molnar and Carpenter referred to *Aublysodon* numerous additional teeth (see below) from the Upper Cretaceous of Montana, Wyoming, New Mexico, Colorado, and Alberta. The following year, Lehman and Carpenter (1990) referred to *Aublysodon* cf. *A. mirandus* a partial tyrannosaurid skeleton (OMNH 10131) recovered from the Kirtland Shale of New Mexico (see *D:TE* for more details on the above).

Since its referral by Molnar and Carpenter to *Aublysodon*, the "Jordan theropod" has been kept in that genus. Recently, however, this referral of LACM 28471 has been contested. Olshevsky, Ford and Yamamoto 1995*a*, 1995*b*, in the ninth and tenth issues of the Japanese periodical *Kyoryugaki Saizensen* [*Dino-Frontline*], referred this specimen to a new genus, *Stygivenator*, at the same emending the specific name, the resulting combination being *Stygivenator molnari*. Unfortunately, *Kyoryugaki Saizensen* was published only in Japanese, a language that most dinosaur researches do not read.

Later, in Olshevsky's publication *Mesozoic Meanderings* number 3, Olshevsky (2000) explained his reasons for erecting the new taxon. Olshevsky pointed out that isolated teeth resembling the type tooth of *Aublysodon*, although uncommon, have been found in several western North American Campanian and Maastrichtian horizons (see Molnar and Carpenter); similar teeth are also found in the Asian forms *Shanshanosaurus* (Olshevsky agreeing with Paul's assessment of this genus) and *Alectrosaurus* (Dong 1977; P. J. Currie, personal communication to Olshevsky; see also *Aublysodon* entry). Without giving explanatory details, he then noted that two western North American theropod specimens that surely represent distinct genera have been referred to *Aublysodon*, one the "Jordan theropod," the other OMNH 10131. Olshevsky further stated that, upon reading Molnar's 1978 paper on this specimen, he "realized that this specimen represented a distinct theropod genus."

As to the genus *Aublysodon*, Olshevsky, of the opinion that its type specimen could not be distinguished between two different genera, regarded this genus as a *nomen dubium*. (*Aublysodon* was also declared a *nomen dubium* by Carr and Williamson 2000 in their review of the Tyrannosauridae from New Mexico.) As to the generic name (meaning "hunter from the River Styx"), Olshevsky explained that, in correspondence, Molnar had once mentioned that he had had the name *Stygivenator* in mind before publishing his 1978 paper, but did not feel confident at that time to propose the new name.

Molnar and Carpenter had referred to *A. mirandus* various species from the United States and Canada—*A. amplus* Marsh 1892, ?*A. lateralis* Cope 1876, and *A. cristatus* Marsh 1892—all based on teeth (see *Aublysodon* entry, *D:TE*, for specimen numbers and localities). Of those, Olshevsky *et al.* tentatively referred *A. amplus* (YPM 296) to ?*Stygivenator amplus* and *A. cristatus* (YPM 297) to ?*S. cristatus*.

Naming new tyrannosaurid genera based upon a single, very incomplete, and also juvenile specimen is dangerous (see, *e.g.*, Carr 1996; Carr and Essner 1997), particularly when that specimen originates from a horizon as late as the Massastrichtian wherein the niche of dominant carnivore is occupied by such sizable taxa as *Tyrannosaurus* and also the enigmatic *Aublysodon*. However, theropod specialist Thomas R. Holtz, Jr. (personal communication 2000) has stated "I think [*Stygivenator*] has a good chance at legitimacy."

Kenneth Carpenter (personal communication 2000) added that he has "long suspected that [*Aublysodon*] is not a valid taxon, mostly because of the variation seen in theropod teeth in general." Therefore, this document tentatively accepts — for the present — Olshevsky *et al.*'s proposal that LACM 28741 represents a valid genus and that *Aublysodon*, although possibly synonymous with *Stygivenator*, should be regarded as a *nomen dubium*.

Subsequent to the publication of the name *Stygivenator*, Carr and Williamson, seemingly unaware of Olshevsky's work, found LACM 28471 to differ in no significant way from other known tyrannosaurids. Supposedly primitive features listed by Molnar and Carpenter (*e.g.*, muzzle long and low, frontals longer than wide, nasals smooth, frontoparietal suture wedge-like) are, the authors noted, typical for juvenile theropods (see Carr 1999*b*). Regarding Molnar and Carpenter's diagnosis of *Aublysodon* incorporating data from LACM 28471, Carr and Williamson stated that "the acute angle of the rostroventral and alveolar borders of the dentary is typical of the shallow bone in juvenile tyrannosaurids." Also typical of tyrannosaurids in general are such features as "the ligually placed carina of the first dentary tooth, different extents of the mesial and distal denticles along the tooth crown, and relatively finer mesial than distal denticles." Although Carr and Williamson were uncertain as to the significance of the "step" in the alveolar border of the dentary at the third alveolus, they noted that this region of the margin is typically convex in this family of dinosaurs, this "step" possibly reflecting the change from a concave to convex margin."

Carr and Williamson further noted that the lack of denticles on the associated premaxillary tooth of this specimen, the generally small size of teeth referred to "*Aublysodon*" (Currie, Rigby and Sloan 1990), plus the widespread geographic and temporal occurrences of teeth at least tentatively referred to that genus (Eaton, Cifelli, Hutchison, Kirkland and Parrish 1999; Eaton, Diem, Archibald, Schierup and Monk 1999; Parrish 1999; see *Aublysodon* entry), "strongly suggests that a lack of denticles reflects relative development, as in other coelurosaurs such as velocoraptorine dromaeosaurids." In the estimation of Carr and Williamson and pending their own examination of the specimen, LACM 28471 represents an indeterminate juvenile tyrannosaurid.

Molnar (personal communication 2000), however, believes that *Stygivenator* and *Aublysodon* are synonymous; Currie (personal communication 2001) believes LACM 28471 may represent a juvenile *Tyrannosaurus*.

Key references: Carr (1996, 1999*b*); Carr and Essner (1997); Carr and Williamson (2000); Cope (1876); Currie, Rigby and Sloan (1990); Eaton, Cifelli, Hutchison, Kirkland and Parrish (1999); Eaton, Diem, Archibald, Schierup and Monk (1999); Lehman and Carpenter (1990); Leidy (1868); Marsh (1892); Molnar (1978); Molnar and Carpenter (1989); Olshevsky (2000); Olshevsky, Ford and Yamamoto (1995*a*, 1995*b*); Paul (1988*b*); Parrish (1999).

†STYRACOSAURUS

Ornithischia: Genusauria: Cerapoda: Marginocephalia: Ceratopsia: Neoceratopsia: Ceratopsomorpha: Ceratopsoidea: Ceratopsidae: Centrosaurinae.

Occurrence: Dinosaur Park Formation, Alberta, Canada.

Holotype: CMN [formerly GSC, subsequently NMC] 344, almost complete skull missing lower jaws.

Comments: *Styracosaurus*, a large centrosaurine ceratopsid distinguished by its long nasal horn and spiked frill, is known from relatively few confirmed specimens, these including an almost complete skull (CMN 344), plus some partial skulls, some undescribed, and other material, all found in the upper portion of what is now the Dinosaur Park Formation (formerly Oldman Formation), in Dinosaur Provincial Park, Alberta.

The first subadult specimen of *Styracosaurus* was reported by Ryan, Eberth and Russell, who pointed out the difficulty in referring nonadult centrosaurine specimens to any particular genus because of the similarity of their ontogenetic development. This reported specimen, an almost complete skull, was also discovered in the upper Dinosaur Park Formation in Dinosaur Provincial Park.

The skull, which had originally been referred to *Centrosaurus* sp., was identified by Ryan *et al.* as subadult "by the mixed long-grained and mottled bone texture preserved on the ventral surface of the posterior parietal frill and the presence of unfused epoccipitals." It was recognized as belonging to *Styracosaurus* by the partially developed spike-like, lateral processes on the right side of the frill, which most closely resemble putative subadult parietal spikes from *Styracosaurus albertensis*, but differ in shape and size from the parietal processes of adult and subadult specimens of the related genus, *Centrosaurus*.

According to Ryan *et al.*, earlier reports of *Styracosaurus* from the Two Medicine Formation cannot be verified due to the incompleteness of the fossil material, and also because two other centrosaurines bearing parietal spikes (*Achelousaurus* and *Einiosaurus*) are now also known from that formation. This material, as well as supposed *Centrosaurus* remains from Montana, therefore, can only be diagnosed with confidence at the subfamily level (see *Centrosaurus* and *Styracosaurus* entries, *D:TE*).

Key reference: Ryan, Eberth and Russell (1998).

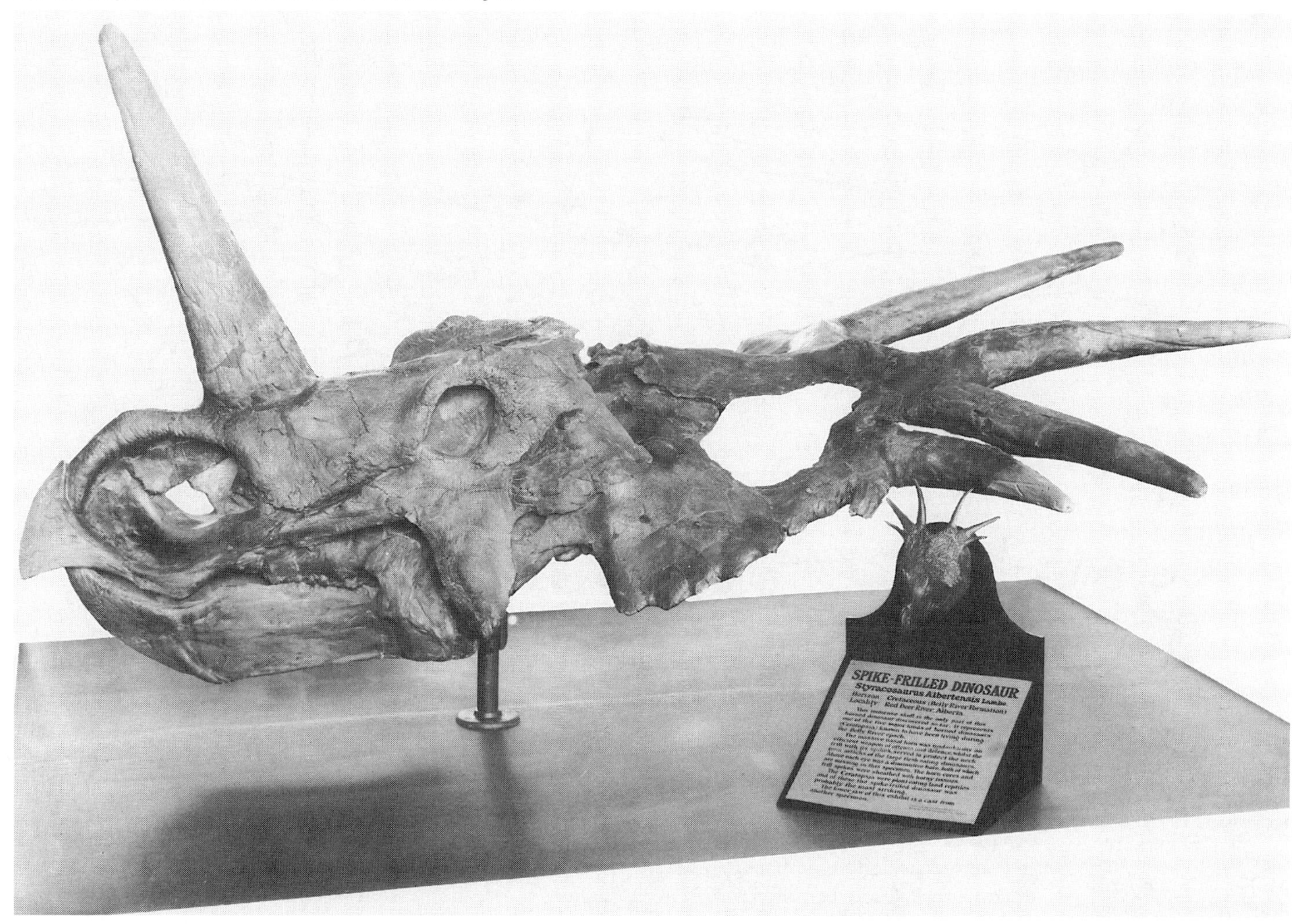

***Styracosaurus albertensis*, CMN 344, holotype skull. The lower jaw was reconstructed based upon another ceratopsian specimen.**

SUCHOMIMUS Sereno, Beck, Dutheil, Gado, Larsson, Lyon, Marcot, Rauhut, Sadleir, Sidor, Varricchio, Wilson and Wilson 1998

Saurischia: Theropoda: Neotheropoda: Tetanurae: Spinosauroidea: Spinosauridae: Baryonychinae.

Name derivation: Greek *souchos* = "crocodile" + Greek *mimos* = "mimic."

Type species: *S. tenerensis* Sereno, Beck, Dutheil, Gado, Larsson, Lyon, Marcot, Rauhut, Sadleir, Sidor, Varricchio, Wilson and Wilson 1998.

Other species: [None.]

Occurrence: Elrhaz Formation, Tégama Group, Niger.

Age: Early Cretaceous (Aptian).

Known material: Incomplete skeleton, various cranial and axial remains, teeth, miscellaneous elements.

Holotype: MNN GDF500, partial skeleton and skull.

Diagnosis of genus (as for type species): Spinosaurid characterized by elongate posterolateral premaxillary process that nearly excludes maxilla from external naris; broadened and heightened posterior dorsal, sacral, and anterior caudal neural spines; robust humeral tuberosities; hypertrophied ulnar olecranon process, offset from humeral articulation; and hook-shaped radial ectepicondyle (Sereno, Beck, Dutheil, Gado, Larsson, Lyon, Marcot, Rauhut, Sadleir, Sidor, Varricchio, Wilson and Wilson 1998).

Comments: New insights regarding the Spinosauridae — until relatively recently, a rather poorly known group of sometimes long-spined theropods, with heads somewhat resembling those of crocodilians — resulted from the discovery of the new genus *Suchomimus*.

The large genus — approaching *Tyrannosaurus rex* in size — was founded upon an incomplete disarticulated skeleton with partial skull (MNN GDF500), recovered from Lower Cretaceous Elrhaz Formation, in rocks exposed in low outcrops and dune fields in the Ténéré Desert of central Niger. Material referred to the type species, *S. tenerensis*, includes articulated premaxillae and maxillae (MNN GDF501), a right quadrate (MNN GDF502), partial dentaries (MNN GDF503, GHF504, and GDF505), axis (MNN GDF506), a posterior cervical 9141 vertebra (MNN GDF507), a posterior dorsal vertebra (MNN GDF508),

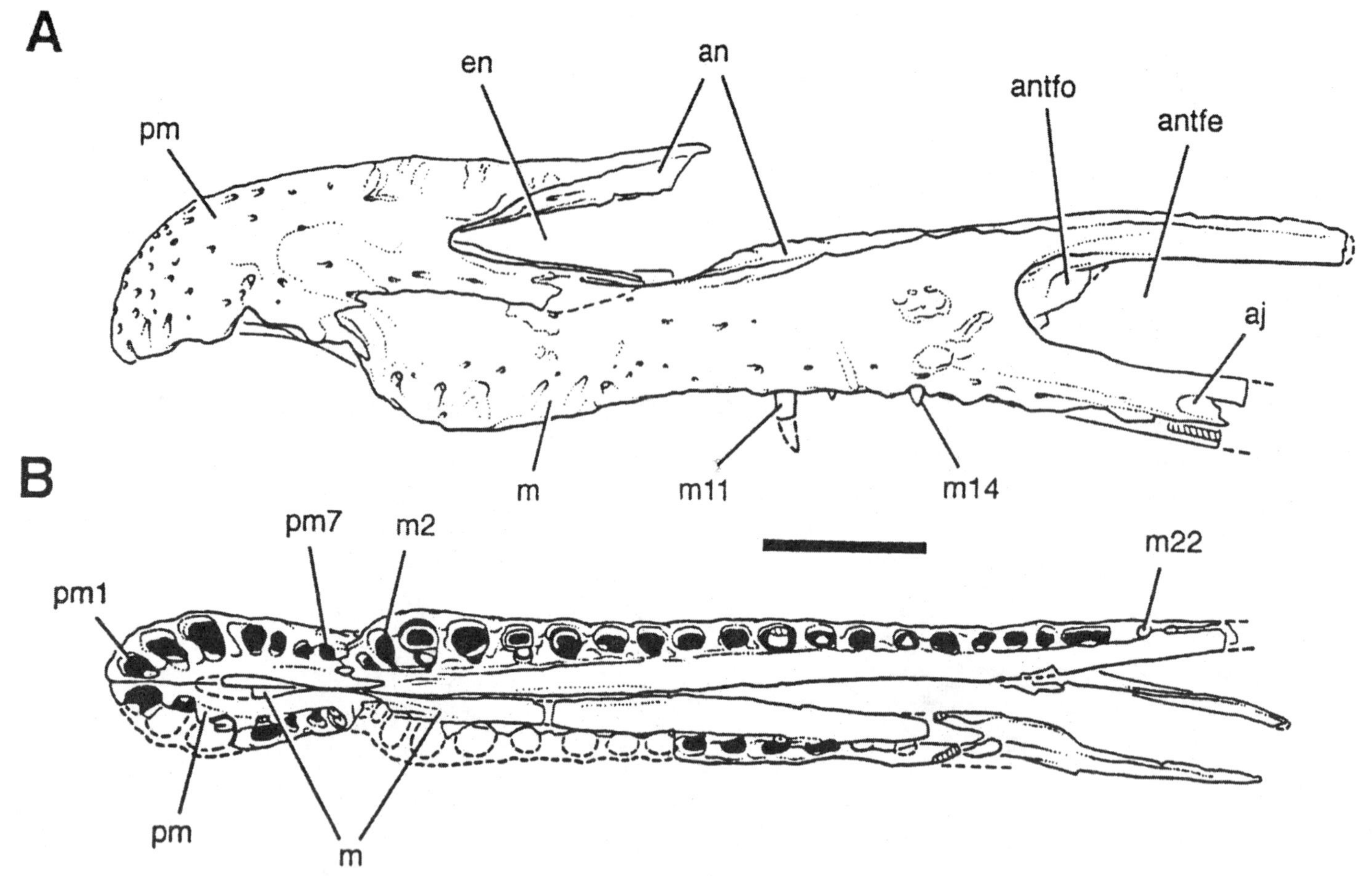

Suchomimus tenerensis, MNN GDF501, holotype skull, A. articulated premaxillae and maxillae, left lateral (reversed from right) and B. ventral views. Scale = 10 cm. (After Sereno, Beck, Dutheil, Gado, Larsson, Lyon, Marcot, Rauhut, Sadleir, Sidor, Varricchio, Wilson and Wilson 1998).

two caudal vertebrae (MNN GDF510 and GDF511), and numerous additional bones and teeth (Sereno, Beck, Dutheil, Gado, Larsson, Lyon, Marcot, Rauhut, Sadleir, Sidor, Varricchio, Wilson and Wilson 1998).

The partial skull of the holotype specimen includes an articulated snout, which, Sereno *et al* noted, is remarkable for its long, low, and narrow proportions, the elongation due to the hypertrophy of both the premaxilla and the anterior ramus of the maxilla. These cranial remains show that the skull of spinosaurid theropods is much lower, narrower, and longer than shown in earlier reconstructions, the snout being particularly narrow in dorsal view.

As described by Sereno *et al.*, the neural spines rapidly increase in height in the middorsal vertebrae, forming what the authors interpreted as "a low median sail that is deepest over the sacral vertebrae." The morphology of these vertebrae (only incipiently developed in the British spinosaurid *Baryonox*) is distinct from that of the Egyptian *Spinosaurus*, "in which the much deeper sail arches to an apex over the middorsal vertebrae" (see respective entries, *D:TE* and *S1*).

The forearm of *Suchomimus*, the authors observed, "is remarkably stout," with an especially robust third manual digit.

Phylogenetic analysis by Sereno, Dutheil, Iarochene, Larsson, Lyon, Magwene, Sidor, Varracchio and Wilson (1996), and Sereno *et al* (1998) (see "Systematics" chapter) joined spinosaurids with "megalosaurids" [=torvosaurids of their usage], positioning this clade, the Spinosauroidea, as the sister-taxon to Avetheropoda [=Neotetanurae of their usage]. According to Sereno *et al.* (1998), derived features shared by spinosaurids and "megalosaurids" (*e.g.*, short forearms, enlarged ungual of manual digit I) seem to have evolved by Middle Jurassic times; furthermore, the hook-shaped coracoid that characterizes *Suchomimus* and *Baryonyx*, but is not found in "megalosaurids," must either have evolved convergently in avetheropods and spinosaurids, or originated as a tetanuran synapomorphy later lost in "megalosaurids."

Sereno *et al.* (1998) interpreted some of the derived in spinosaurids as features related to fish-eating, these primarily including the following: An unusually long snout with a long secondary palate; a terminal rosette of teeth in both the upper and lower jaws; subcylindrical, spaced tooth crowns; posteriorly displaced external nares; and a ventrally located basipterygoid articulation.

These authors' analysis suggested that the Spinosauridae comprises two new clades — the Baryonychinae (including *Baryonyx* and *Suchomimus*) and

Suchomimus tenerensis, MNN GDF501, skeletal reconstruction. Scale = 1 meter. (After Sereno, Beck, Dutheil, Gado, Larsson, Lyon, Marcot, Rauhut, Sadleir, Sidor, Varricchio, Wilson and Wilson 1998).

the Spinosaurinae (including *Spinosaurus* and *Irritator*), which diverged before the Barremian. Though distinct, *Suchomimus* and *Baryonyx* are closely related genera, as indicated by various derived features including the increased number and small size of dentary teeth posterior to the terminal rosette, and the deeply keeled anterior dorsal vertebrae. Sereno *et al.* pointed out that other similarities between these genera remain unknown for the present, as they have not been preserved in other known spinosaurids. *Spinosaurus* and *Irritator* are joined based upon the straight unserrated tooth crowns, the small first premaxillary tooth, and increased spacing of the teeth in both upper and lower jaws (see "Systematics" chapter).

As recounted by Sereno *et al.* (1998), it was usually accepted that the geographic distribution and relationships of the Spinosauridae matched the general

Suchomimus tenerensis, reconstructed skeleton (including cast of holotype MNN GDF501), prepared and mounted by PAST under the direction of palentologist Paul C. Sereno.

Photograph by the author, courtesy Chicago Children's Museum.

Photograph by the author, courtesy Chicago Children's Museum.

Skeletal reconstruction of *Suchomimus tenerensis*, elements prepared in fiberglass by Paul C. Sereno and students and mounted by the PAST company (Prehistoric Animal Structures, Inc.) of Alberta, Canada, under the direction of Sereno, including cast of holotype MNN GDF501 (skull partially restored after *Baryonyx walkeri*). This skeleton was displayed in December 1998 at the Chicago Children's Museum, Navy Pier as the "Colossal Claw" Dinosaur Discovery Exhibit.

pattern of continental break-up during the latter half of the Mesozoic Era, this idea having been explained by large-scale vicariance. In this scenario, the split between *Baryonyx* and the "southern spinosaurines" might be ascribed "to the opening of the Tethyan seaway between Laurasia and Gondwanaland, and the divergence among spinosaurines could be the result of the subsequent opening of the Atlantic Ocean between South America and Africa."

However, the occurrence of *Suchomimus* (a dinosaur more closely related to the European *Baryonyx* than to the African *Spinosaurus*) in the mid–Cretaceous of Africa complicates the above scenario. Therefore, Sereno *et al.* (1998)— assuming their ascertained phylogentic relationships of these taxa to be correct, and accepting the above-stated rifting sequence between the continental areas — proposed an alternative hypothesis that most parsimoniously accounts for the

Life restoration of *Suchomimus tenerensis* prepared for this book by artist Ricardo Delgado.

distribution of the four spinosaurid taxa. According to this new scenario, "spinosaurids may have had a distribution across Pangaea that was split by the opening of the Tethys; baryonychines evolved from the north (Europe, or Laurasia), and spinosaurines evolved on the southern landmass (South America and Africa, or Gondwanaland)," with a single Europe-to-Africa dispersal event explaining the African presence of *Suchomimus*. As noted by Sereno *et al.*, the phylogenetic and biogeographic relationships of this genus and other spinosaurids offer additional evidence of an Early Cretaceous faunal dispersal across the Tethyan seaway.

A reconstructed skeletal cast of *S. tenerensis*— the bones prepared in fiberglass by Sereno and his University of Chicago students, the missing elements sculpted by them, the skeleton then mounted by the PAST company of Drumheller, Alberta, Canada — went on public display in December, 1998 at the Chicago Children's Museum as the "Colossal Claw" Dinosaur Discovery Exhibit. As reported by Paul C. Sereno and Gabriel H. Lyon in an article published in the *Chicago Sun-Times* (November 10, 2000), another skeletal cast of this dinosaur went on outdoor display at the Flamme de la Paix in 2000 in the desert oasis of Agadez, in the Sahara.

Note: Sereno *et al.* (1998) reported that other fossils found at this locality include plants, invertebrates, and vertebrates (including a new long-snouted basal crocodyloid and an azhdarchid pterosaur); among the vertebrates, dinosaurs are represented by at least two additional theropods (an indeterminate tetanuran known mostly from teeth, plus a small basal coelurosaur), two sauropods (a high-spined basal diplodocid, formerly referred by Taquet 1976 to the Dicraeosaurinae, and a rare titanosaur), and three ornithopods (the dryosaur *Valdosaurus nigeriensis*, the common "*Iguanodon* trapu" [see Chablis 1988], and the high-spined *Ouranosaurus nigeriensis*).

Key references: Chablis (1988); Sereno, Beck, Dutheil, Gado, Larsson, Lyon, Marcot, Rauhut, Sadleir, Sidor, Varricchio, Wilson and Wilson (1998); Sereno, Dutheil, Iarochene, Larsson, Lyon, Magwene, Sidor, Varracchio and Wilson (1996); Taquet (1976).

†SYNTARSUS (=?*Coelophysis*)

Saurischia: Theropoda: Neotheropoda: "Ceratosauria": Coelophysoidea: Coelophysidae.

New material: Partial snout, partial pelvis.

Comments: The small "ceratosaurian" type species *Syntarsus rhodesiensis*, until recently, has only been known in South Africa from isolated fragmentary postcranial remains, these including pieces of femora and tibiae, plus a few vertebrae and isolated bones of the hind foot (see *D:TE*). Munyikwa and Raath (2000) have now described in detail additional South African material belonging to this taxon consisting of a well-preserved partial snout (BP/1/5278) and also a partial pelvis and sacrum (BP/1/5246). Both specimens were found in the Ladybrand District of Free State Province by J. W. Kitching, the snout in 1981 on the Paradys farm, the pelvis in 1985 on land between the adjoining Welbedacht and Edelweiss farms.

According to Munyikwa and Raath, comparison of the snout with corresponding parts of the skull of other "ceratosaurs" (*e.g.*, *Ceratosaurus nasicornis*, *Dilophosaurus wetherilli*, *Liliensternus liliensterni*, *Coelophysis bauri*, *Syntarsus rhodesiensis*, and *S. kayentakatae*) reveals a close agreement between its characters and those of "ceratosaurs" generally (see Raath 1977; Welles 1984; Colbert 1989; Rowe 1989; Benton 1990; Rowe and Gauthier 1990), and of *Coelophysis* and *Syntarsus* particularly. The subnarial gap between the premaxilla and maxilla unite BP/1/5278 with members of the "ceratosaurian" sub-clade Coelophysoidea (see Rowe; Holtz 1994). The above features plus the depression of the antorbital fossa anterior to the antorbital

T

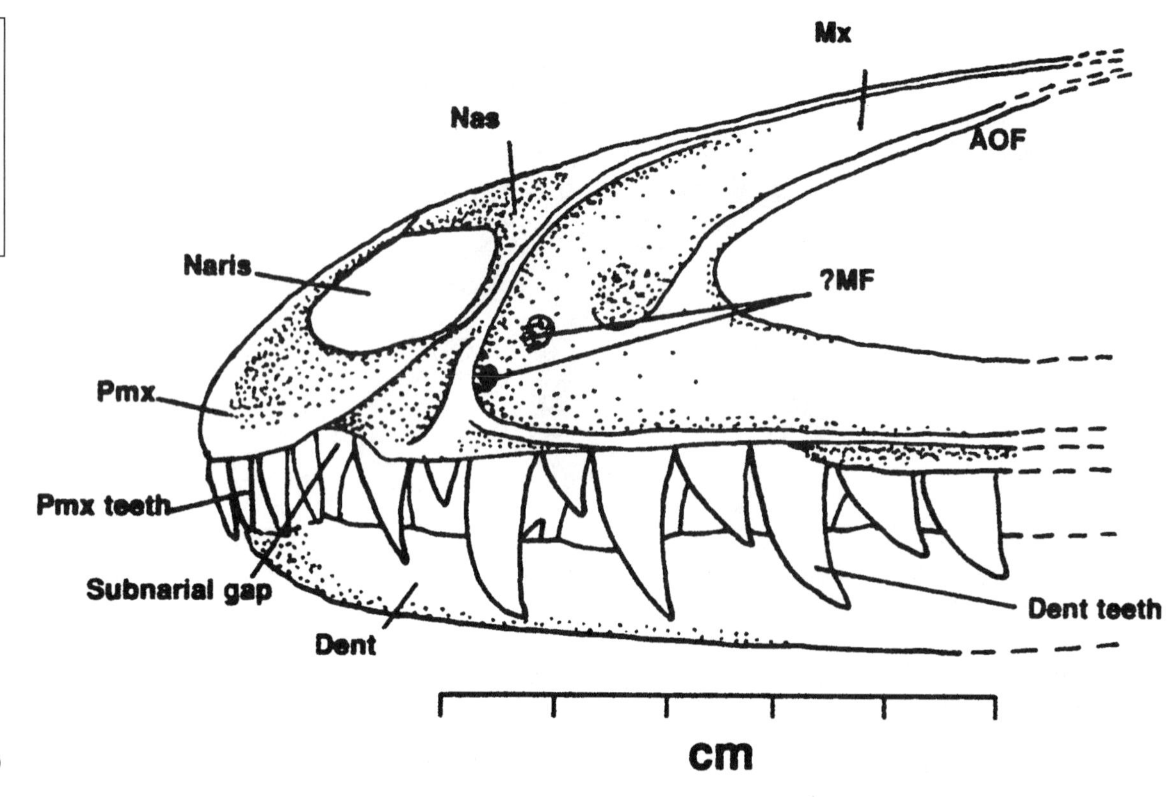

***Syntarsus rhodesiensis*, referred snout (BP/1/5278) from South Africa. (After Munyikwa and Raath 1999.)**

fenestra joins the South African form with the *Syntarsus* material from Zimbabe (see *D:TE*), the only perceived differences between them being the lack of serrations on the premaxillary teeth from South Africa. This condition, the authors noted, could be due to wear and, therefore, have no diagnostic significance.

BP/1/5246, the authors noted, though significantly smaller than the pelvis of the holotype of *S. rhodesiensis*, is, particularly in ventral aspect, basically identical morphologically.

For the above reason, Munyikwa and Raath confidently referred these South African specimens to *Syntarsus rhodesiensis*.

Recently, Downs (2000), in a comparative study of *S. rhodesiensis* and *Coelophysis bauri*, suggested that these two species belonged in the same genus, with the name *Coelophysis* having priority. Downs did not, however, formalize this proposed synonymy (see *Coelophysis* entry).

Key references: Benton (1990); Colbert (1989); Downs (2000); Holtz (1994); Munyikwa and Raath (1999); Raath (1977); Rowe (1989); Rowe and Gauthier (1990); Welles (1984).

TANGVAYOSAURUS Allain, Taquet, Battail, Dejax, Richir, Véran, Limon-Duparcmeur, Vacant, Mateus, Sayarath, Khenthavong and Phouyavong 1999 — (=?*Huabeisaurus*)

Saurischia: Sauropodomorpha: Sauropoda: Eusauropoda: Neosauropoda: Macronaria: Camarasauromorpha: Titanosauriformes: Somphospondyli: Titanosauria *incertae sedis*.

Name derivation: "Tang Vay [type locality]" + Greek *sauros* = "lizard."

Type species: *T. hoffeti* Allain, Taquet, Battail, Dejax, Richir, Véran, Limon-Duparcmeur, Vacant, Mateus, Sayarath, Khenthavong and Phouyavong 1999.

Other species: [None.]

Occurrence: Grès Supérieur Formation, Savannakhet Province, Laos.

Age: Early to Middle Cretaceous (Aptian–Albian).

Known material: Two partly articulated skeletons, partial postcranial remains, two or three individuals.

Holotype: TV4-1 to TV4-36 (housed at the Musée des Dinosaures, Savannakhet, Laos), partial skeleton, articulated.

Diagnosis of genus (as for type species): Titanosaur measuring 15 meters (approximately 51 feet) in length; posterior dorsal vertebrae having unforked neural spines; neural arches in mid- and posterior caudal centra anteriorly positioned; caudal vertebrae amphicoelous; puboischial contact deep dorsoventrally; ischial distal shafts dorsoventrally flattened, almost coplanar; ilium to pubis length ratio less than 0.9; proximal one-third of femoral shaft medially deflected; fibular condyle located medially relative to lateral margin of femoral shaft (Allain, Taquet, Battail,

Dejax, Richir, Véran, Limon-Duparcmeur, Vacant, Mateus, Sayarath, Khenthavong and Phouyavong 1999).

Comments: The genus *Tangvayosaurus* was founded upon two partial skeletons excavated from red siltstone and mudstone deposits (related to flood-plain deposits) in two localities — designated Tang Vay 2 and Tang Vay 4 (see Taquet, Battail and Dejax 1992; Taquet, Battail, Dejax, Richir and Véran 1995) — at the top of the Grès Supérieur Formation (dated as Aptian–Albian by Kobayashi 1963, 1968, based on the presence of fresh-water pelecypods of the superfamily Trigonioidacea), near Ban Tang Vay, in Savannakhet Province, R.D.P., Laos. The holotype skeleton (TV4-1 to TV4-36), collected from Tang Vay 4, includes a complete both pubes, ischia, posterior dorsal vertebrae, one of the most anterior caudal vertebrae, dorsal ribs, and the distal end of a humerus. The referred specimen (not used in the diagnosis of the type species *T. hoffeti*, but for comparisons with other material) from Tang Vay 2 includes 38 caudal vertebrae (see Taquet), chevrons, a metacarpal, cervical vertebra, distal end of femur, a tibia (see Taquet), a fibula, astragalus, and three metatarsals and nine pedal phalanges of the left hindlimb. Both of these specimens are housed at the Musée des Dinosaures in Savannakhet (Allain, Taquet, Battail, Dejax, Richir, Véran, Limon-Duparcmeur, Vacant, Mateus, Sayarath, Khenthavong and Phouyavong 1999).

Earlier, Hoffet (1942) had described material including an almost complete robust femur, the distal end of another femur, two proximal heads of femora, and some amphicoelous caudal vertebrae recovered from the Grès Supérieur Formation, near Muong Phalane, in Laos. Hoffet referred these remains to the genus *Titanosaurus*, erecting for them the new species *Titanosaurus falloti*. McIntosh (1990, in his review of the Sauropoda, suggested that this material could belong to a genus other than *Titanosaurus* and possibly to a family other than the Titanosauridae. Allain *et al.*, noting that the femur referred by Hoffet to *T. falloti* and also that a large caudal vertebra referred by Hoffet (1944) to the hadrosaurid *Mandschurosaurus laoensis*, resemble those found at Tang Vay 4, referred these remains to the new genus as *Tangvayosaurus* sp. (see *Huabeisaurus* entry for another assessment of the *Titanosaurus falloti* material).

As Allain *et al.* observed, *Tangvayosaurus* exhibits close affinities with the Late Jurassic–Early Cretaceous sauropod species *Phuwiangosaurus sirindhornae*, although in the latter is different in various features (*e.g.*, more open angle between axis of shaft and ischial border of pubis, caudal border of ischial shaft well marked and less regular, puboischial contact dorsoventrally shorter).

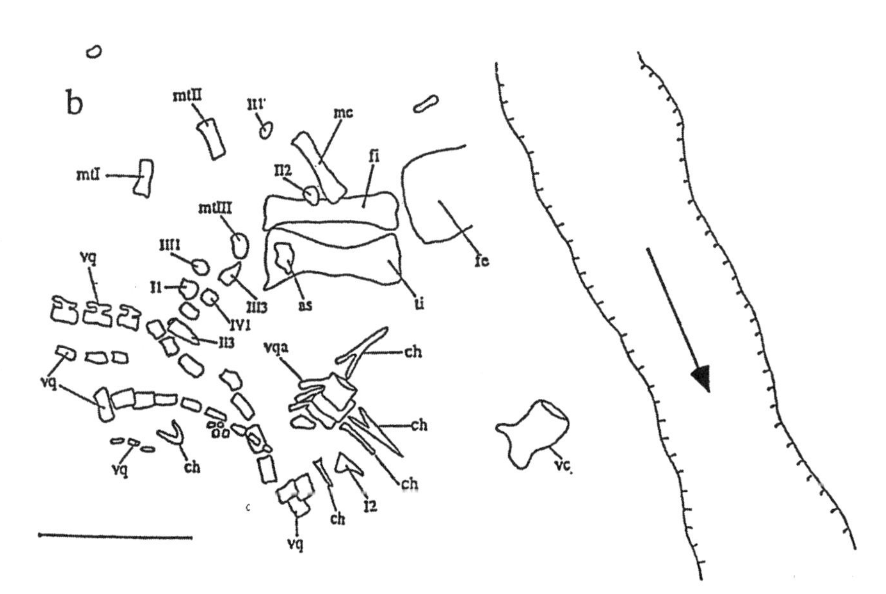

***Tangvayosaurus hoffeti*, plan of excavation surface at (a) Tang Vay 4, showing holotype skeleton as found, and (b) Tang Vay 2, showing referred specimen. Scale = 1 m. (After Allain, Taquet, Battail, Dejax, Richir, Véran, Limon-Duparcmeur, Vacant, Mateus, Sayarath, Khenthavong and Phouyavong 1999.)**

Following the phylogeny of the Sauropoda proposed by Wilson and Sereno (1998), Allain *et al.* found *Tangvayosaurus* to exhibit a derived feature of the Titanosauriformes (femur with proximal one-third of its shaft medially deflected). *Tangvayosaurus* and *Phuwiangosaurus* were found by Allain *et al.* to share a single derived character with the Titanosauria (pubis

***Tangvayosaurus hoffeti*, holotype a. TV4-3, left ischium (dorsomedial view), b. TV4-5, right pubis (ventrolateral view), c. TV4-1, right femur (posterior view), referred d. TV2-40, left fibula (anterior view), e. TV2-39, left tibia (posterior view), and f. TV2-6, caudal vertebra (lateral view). (After Allain, Taquet, Battail, Dejax, Richir, Véran, Limon-Duparcmeur, Vacant, Mateus, Sayarath, Khenthavong and Phouyavong 1999.)**

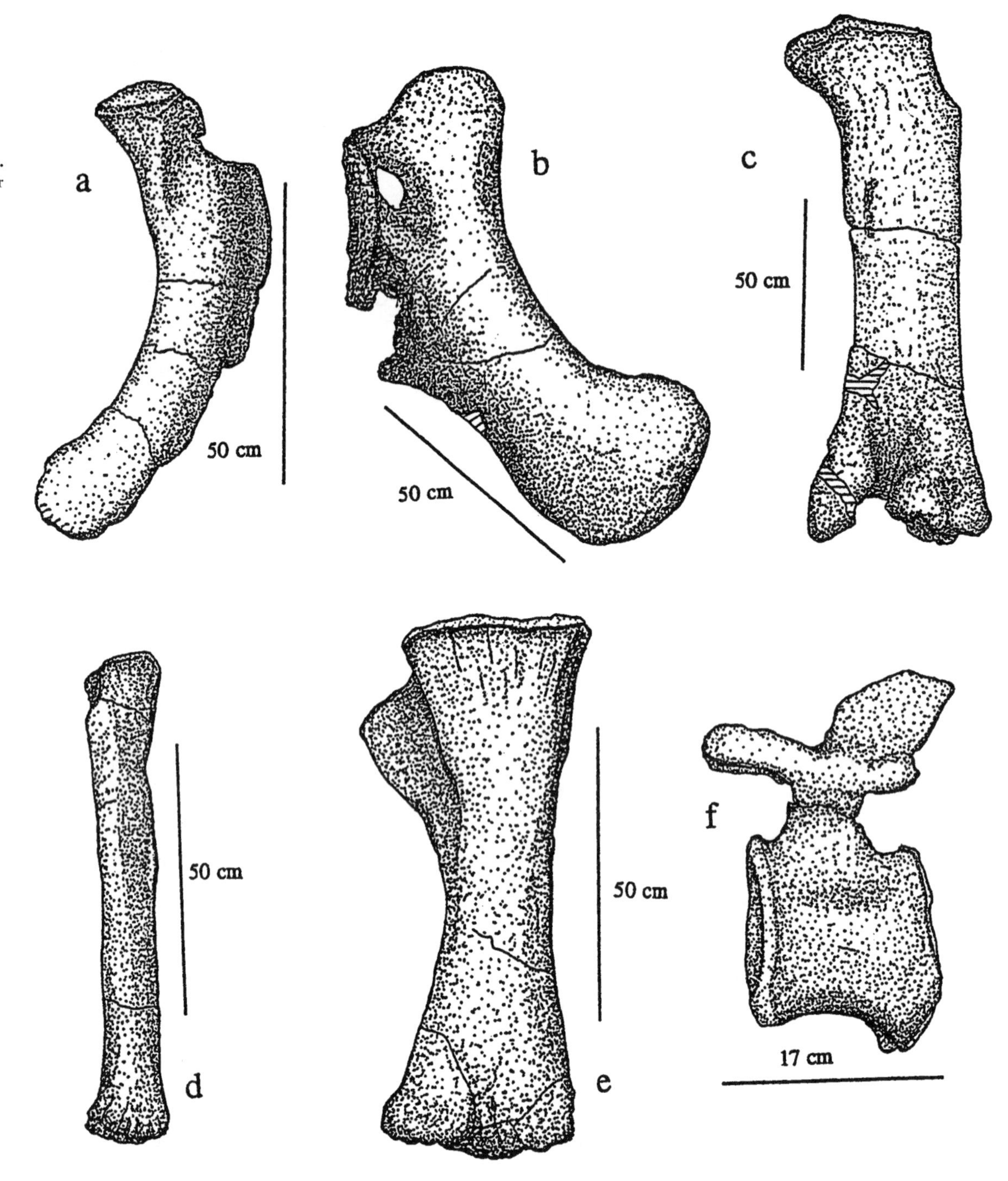

significantly longer than ischium), thus supporting the hypothesis that both genera are more closely related to titanosaurs than to the Brachiosauridae or the Nemegtosauridae (see Buffetaut and Suteethorn 1999). Consequently, *Tangvayosaurus* and *Phuwiangosaurus* were regarded by these authors "as the oldest and the most plesiomorphic titanosaurs, morphologically distant of their South American cousins," and establishing the occurrence of primitive titanosaurs in Southeast Asia during the Early and Middle Cretaceous.

Key references: Allain, Taquet, Battail, Dejax, Richir, Véran, Limon-Duparcmeur, Vacant, Mateus, Sayarath, Khenthavong and Phouyavong (1999); Buffetaut and Suteethorn (1999); Hoffet (1942, 1944); McIntosh (1990); Taquet (1994*b*); Wilson and Sereno (1998).

†TARCHIA

Ornithischia: Genasauria: Thyreophora: Thyreophoridea: Eurypoda: Ankylosauria: Ankylosauridae.

Erratum: In *D:TE*, it was apparently incorrectly stated that *Tarchia gigantea* specimen PIN 551-29, the holotype of *Dyoplosaurus giganteus*, was collected from Altan Ula II. According to Tatyana A. Tumanova, this specimen was collected from Nemegt. As pointed out by Tim Donovan (personal communication 2000), Tumanova may be a more accurate source of information in this matter as a Russian paleontological team collected the specimen.

TEHUELCHESAURUS Rich, Vickers-Rich, Giménéz, Cúneo, Puerta and Vacca 1999

Saurischia: Sauropodomorpha: Sauropoda: Eusauropoda: "Cetiosauridae."

Name derivation: "Tehuelche [Indians, inhabitants of area where this dinosaur was found]" + Greek *sauros* = "lizard."

Type species: *T. benitezii* Rich, Vickers-Rich, Giménéz, Cúneo, Puerta and Vacca 1999

Other species: [None.]

Occurrence: Cañadón Asfalto Formation, Chubut Province, Patagonia, Argentina.

Age: Early Middle Jurassic (?Callovian).

Known material/holotype: MPEF-PV 1125, partially articulated postcrania including right scapulocoracoid, left humerus, radius and ulna, femora, left ischium, distal part of right ischium, left pubis, partial right pubis, partial ilium, 10 partial dorsal vertebrae, at least two partial sacral vertebrae, one caudal vertebra, dorsal rib fragments, skin impressions.

Diagnosis of genus (as for type species): Distinguished from *Omeisaurus* by 1. subequal measurements on coracoid of a. distance from scapular surface to antero-ventral border and b. distance from glenoid surface perpendicular to it across to opposite side (rather than "b" being markedly greater than "a"), 2. anteroposterior length of distal end of humerus only slightly less than mediolateral width (rather than anteroposteriorly compressed), 3. radius and ulna stouter (*i.e.,* comparative widths greater relative to their lengths), 4. anterior border of distal shaft of pubis deeply concave (rather than modestly concave), and 5. expansion of acetabular region of ischium noticeably deeper dorsoventrally; distinguished from *Patagosaurus* and *Barapasaurus* by all (rather than just anterior dorsal) vertebrae having pseudopleurocoels and being opisthocoelous (rather than posterior dorsal being amphiplatyan), centra having only modest depressions; pseudopleurocoels (*sensu* Bonaparte 1986*a*) deep lateral depressions in centra, centra having pseudopleurocoelous condition (unlike true pleurocoels) without internal chambers within centrum (Rich, Vickers-Rich, Gimenéz, Cúneo, Puerta and Vacca 1999).

Comments: The genus *Tehuelchesaurus* was established on a specimen (MPEF-PV 1125) comprising approximately 50 percent of the postcranial skeleton, discovered about 1980 by Aldo Benitez (who had been prospecting for uranium for the Argentinian National Commission of Atomic Energy, of CNEA) at the Fernañdez Estancia locality in the Cañadón Asfalto Formation (regarded by some workers as early Middle Jurassic [Callovian], based on radiometric dating of the Pampa de Agnia Formation, the underlying volcanic unit; see Turner 1983) in western Chubut Province, Patagonia, Argentina. It was recovered during the summer of 1994-1995 by a field team led by Pablo Puerta, Raul Vacca, and Olga Giménez from the Museo Paleontològico Egidio Feruglio, Trelew, Chubut Province, Argentina. The specimen represents a single individual and constitutes the only vertebrate material found at this site, save for several isolated

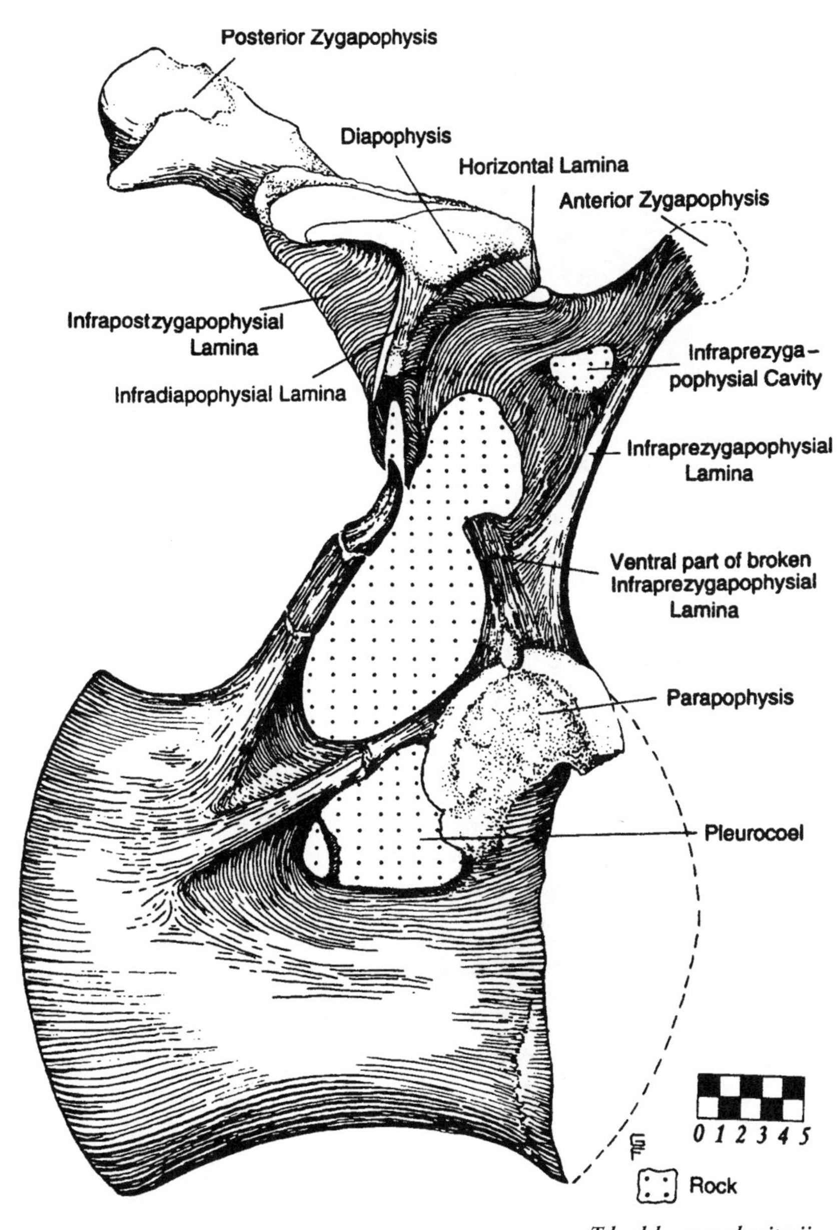

***Tehuelchesaurus benitezii*, MPEF-PV 1125, holotype dorsal vertebra (restored, based principally on dorsal 2). (After Rich, Vickers-Rich, Gimenéz, Cúneo, Puerta and Vacca 1999.)**

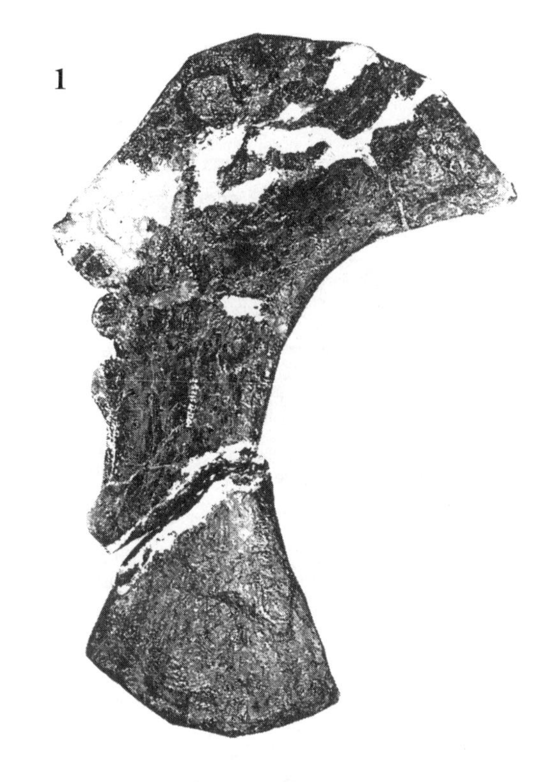

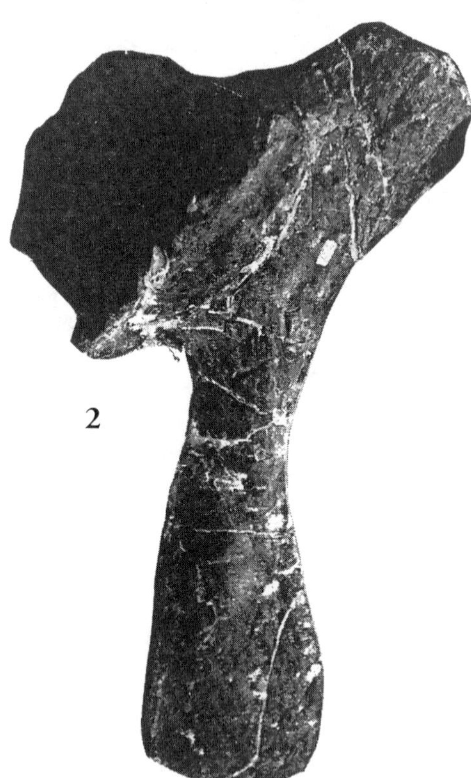

Tehuelchesaurus benitezii, MPEF-PV 1125, holotype 1. right pubis and 2. right ischium, lateral views. (After Rich, Vickers-Rich, Giменéz, Cúneo, Puerta and Vacca 1999.)

theropod teeth (Rich, Vickers-Rich, Gimenéz, Cúneo, Puerta and Vacca 1999).

As noted by Rich *et al.*, the presence of carbonates and gypsum within the Cañadón Asfalto Formation suggests sometimes semiarid conditions where lakes sometimes dried up entirely; the banding of shales implies seasonal variations possibly controlled by varying amounts of rainfall. Fossil plant fragments recovered from the Fernañdez Estancia locality all belong to ferns and conifers, this suggesting a plant community consisting "of a coniferous tree stratum and a fern understorey," this assemblage belonging to a forest near the lake in which these fossils were eventually buried in sediments close to shore.

Rich *et al.* referred *Tehuelchesaurus* to the "Cetiosauridae," a family currently abandoned by some workers (see "Systematics" chapter, *S1*, also this volume). The authors made this referral based upon characters found in the type specimen of *Tehuelchesaurus* that conform to those used by Bonaparte (1986*a*) and McIntosh (1990) to diagnose the "Cetiosauridae," these including the following: Ratio of humerus to femur length low (.73) relative to that in Brachiosauridae; deltoid crest weakly developed; ilium without any indication of transverse blade; dorsal vertebrae opisthocoelous; caudal vertebra platycoelous.

Tehuelchesaurus numbers among the few sauropods known for which fossil skin impressions have been preserved. Two of the best preserved samples, as described by Rich *et al.*, show that the skin of this dinosaur "was covered by flattened scales, hexagonal in outline, not imbricated, showing a rosette-like pattern (Gimenéz 1996)." Later, in an abstract for a poster, Gimenéz (2000) further noted that these skin impressions pertain to different parts of the body, fragments of the thorax, and a large impression from the region of the scapula. The larger samples are covered by tubercular scales having two different patterns — scales almost semismooth and hexagonal in outline (measuring 3 centimeters by 2.5 centimeters) alternating with smaller samples (2 centimeters by 1.5 centimeters) in small, nonoverlapping rosette patterns. Smaller integument samples, rhomboidal in outline, range in size from 1 to 3 millimeters in diameter and 2 to 4 millimeters in length.

Of all known sauropods, *Tehuelchesaurus benitezii*

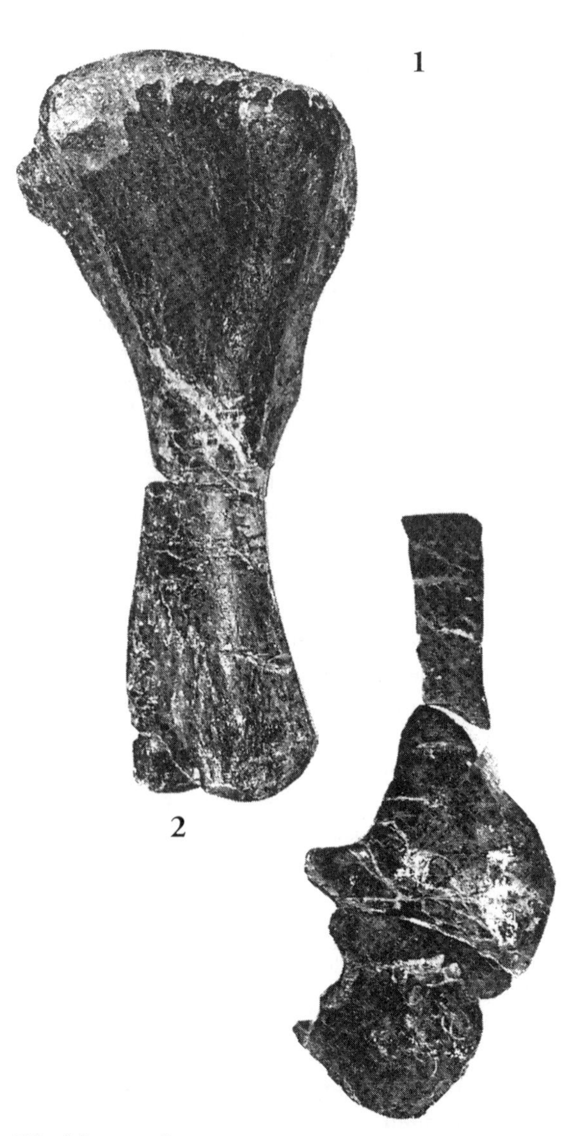

Tehuelchesaurus benitezii, MPEF-PV 1125, holotype 1. left humerus (anterior view) and 2. scapulocoracoid (lateral view). Length of scapula = 1750 mm. (After Rich, Vickers-Rich, Gimenéz, Cúneo, Puerta and Vacca 1999.)

most closely resembles in morphology the Asian species *Omeisaurus tianfuensis*, especially in the very similar structure of their dorsal vertebral centra, and to a lesser degree, the general variation in their girdle and limb elements. The presence of *T. benitezii* in Argentina and *O. tianfuensis* in China — contemporaneous sauropods separated that are very similar to one another, though so widely separated geographically — was regarded by Rich *et al.* as evidence of a mid–Jurassic interchange of land-dwelling vertebrates between central Asia and the rest of the world as already suggested by other authors (*e.g.*, Rage 1988; Sereno 1997). In noting that Russell (1993) had previously recognized two distinct Middle Jurassic biogeographic provinces, Central Asia and Neopangaea (a giant land mass including all major land masses other than Central Asia), Rich *et al.* commented that the close relationship existing between the South American and Chinese species "is an exception to the generalization of Russell."

In December, 1997, the holotype specimen went on display in the former temporary exhibition area of the Museo Paleontològico Egidio Feruglio.

Key references: Bonaparte (1986*a*); Gimenéz (1996, 2000); McIntosh (1990); Page (1988); Rich, Vickers-Rich, Gimenéz, Cúneo, Puerta and Vacca (1999); Russell (1993); Sereno (1997).

†TELMATOSAURUS

Ornithischia: Genasauria: Cerapoda: Ornithopoda: Euornithopoda: Iguanodontia: Euiguanodontia: Dryomorpha: Ankylopollexia: Hadrosauroidea: Hadrosauridae.

Comments: The genus *Telmatosaurus* was first described by Nopcsa (1900) as a hadrosaurid dinosaur which he originally named *Limnosaurus* (see *D:TE*). As noted by Casanovas, Pereda-Suberbiola, Santafé and Weishampel (1999), this genus is significant as "the first of the early named forms to contribute to Cope's taxon Hadrosauridae," erected in 1869 to embrace a number of mostly poorly known taxa, the majority of these today regarded as *nomina dubia* (see Weishampel and Horner 1990).

In recent years, the status of *Telmatosaurus* as a hadrosaurid has been challenged. Forster (1997) regarded *Telmatosaurus* as a derived iguanodontian outside of Hadrosauridae, while Norman (1998*a*) subsequently classified this genus as a member of the Euhadrosauria (see "Systematics" chapter, *S1*).

According to Godefroit, Dong, Bultynck, Hong and Feng (1998), who erected the superfamily Hadrosauroidea, the type species *Telmatosaurus transsylvanicus* shares with Hadrosauridae the following synapomorphies not seen in *Bactrosaurus johnsoni*: Migration of antorbital foramen along upper reaches of premaxillary articular surface of maxilla; more than 30 maxillary and 29 dentary tooth positions [a growth feature; M. K. Brett-Surman, personal communication 2000]; absence of paraquadtare foramen; and absence of caudal secondary ridge on dentary teeth.

More recently, Casanovas *et al.* performed a new cladistic analysis of various ornithopod taxa (see "Introduction," section on hadrosaurs). These authors concluded, as had Nopcsa, that *Telmatosaurus* is neither an euhadrosaurian nor a derived iguanodontian, but a basal member of the Hadrosauridae outside of this taxon's two subfamilies.

Key references: Casanovas, Pereda-Suberbiola, Santafé and Weishampel (1999); Cope (1869); Forster (1997); Godefroit, Dong, Bultynck, Hong and Feng (1998); Nopcsa (1900); Norman (1998*a*); Weishampel and Horner (1990).

TENDAGURIA Bonaparte, Heinrich and Wild 2000—(=?*Janenschia*).

Saurischia: Sauropodomorpha: Sauropoda: Eusauropoda: Neosauropoda: Macronaria: Camarasauromorpha: ?Tendaguriidae.

Name derivation: "Tendaguru [fossil locality in Tanzania]."

Type species: *T. tanzaniensis* Bonaparte, Heinrich and Wild 2000.

Other species: [None.]

Occurrence: Tendaguru Beds, Mtwara, Tanzania, East Africa.

Age: ?Late Jurassic.

Known material: Vertebrae, possibly other remains.

Holotype: MB.R.2092.1-2, NB4, NB5, two (originally articulated) anterior dorsal vertebrae, ?sacrum and ?ilium.

Diagnosis of genus (as for type species): Sauropod approximately 20 meters (almost 70 feet) in length; anterior dorsal vertebrae opisthocoelous, with very low, anteroposteriorly laminar neural spines which do not rise above surrounding area of neural arch; neural spine not distinct corpus of bone, but integrated within osseous structures around it, continuous with transverse processes; tranverse process elongated with dorsal depression lateral to prezygapophyses; deep anterior cavities on anterior surface of transverse process, shallow cavities on posterior surface; well-developed infradiapophyseal laminae; strong, thick epipophyses near uppermost border of vertebrae, these connected to transverse processes (Bonaparte, Heinrich and Wild 2000).

Comments: The genus *Tendaguria* was established on two rather well-preserved dorsal vertebrae (MB.R.2092.1-2, NB4 and NB5), originally reported by Janensch (1925*b*, 1929*a*), recovered from the

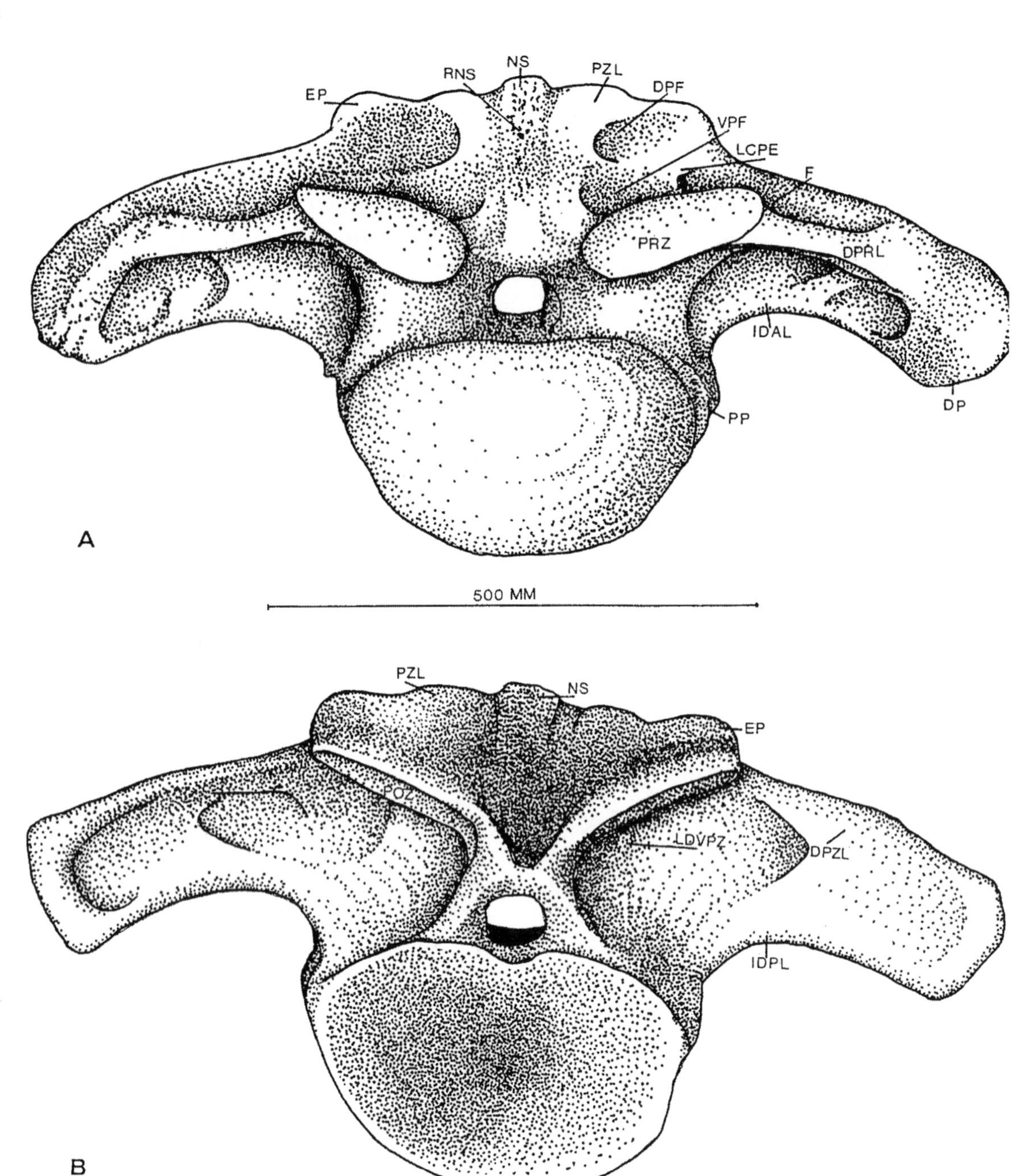

***Tendaguria tanzaniensis*, MB.R.2092.1, NB4, the more anterior of two holotype anterior dorsal vertebrae, in A. anterior and B. posterior views. (After Bonaparte, Heinrich and Wild 2000.)**

Nambango site, Tendaguru Beds (?Upper Jurassic; see Raath and McIntosh 1987; also see *Janenschia* entry), southeast of Tendaguru Hill, Mtwara, Tanzania. Probably associated with these vertebrae were also a sacrum and ilium mentioned by Janensch (1929*a*), although this material could not be found in the paleontological collection of the Museum für Naturkunde in Berlin. Referred to the type species, *Tendaguria tanzaniensis*, was a vertebra (MB.R.209.31; possibly the eighth or ninth cervical, figured with only a cursory description by Janensch 1929*a* as *Gigantosaurus robustus*) (Bonaparte, Heinrich and Wild 2000).

According to Bonaparte *et al.*, these vertebrae most likely belong to the same individual and succeeded one another, as in both preservation and size they correspond well (although an intermediate vertebra may have been located between them). Their relative position was based upon these features: 1. Relative position of parapophysis; 2. distance between prezygapophysis and postzygapophysis; 3. relatively anterior position of prezygapophysis; 4. position of postzygapophysis relative to centrum; 5. fossae on dorsal surface of transverse processes; and 6. orientation of posterior diapophysial lamina.

As Bonparte *et al.* observed, the anterior dorsal vertebrae in *Tendaguria* seem to be more highly derived

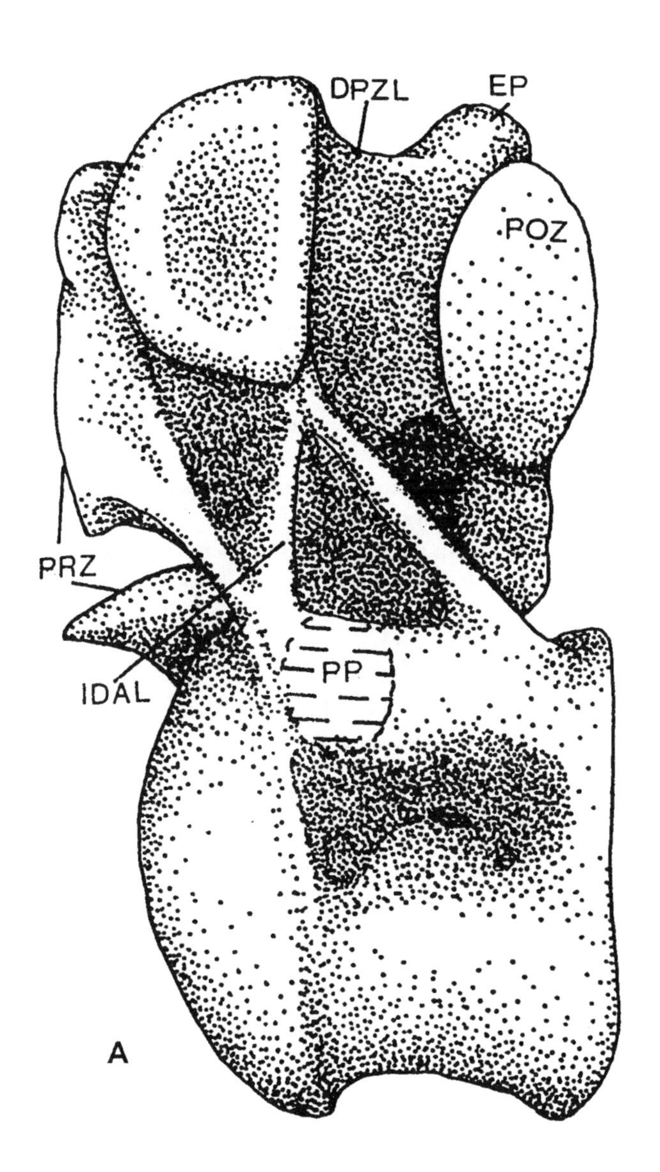

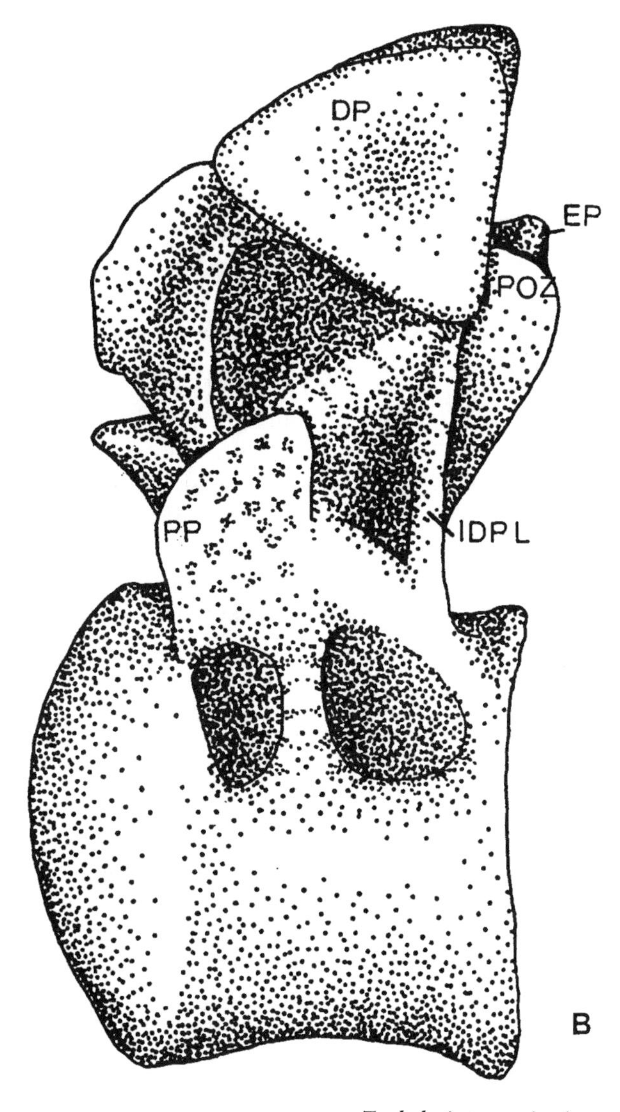

Tendaduria tanzaniensis, MB.R.2092.1, A. NB4 and B. NB5, holotype anterior dorsal vertebrae in left lateral views. (After Bonaparte, Heinrich and Wild 2000.)

than those in other known sauropods. Apparently the extremely reduced neural spines were associated with a unique shifting of the axial musculature to the dorsal surface of the transverse processes, this musculature normally located lateral to the neural spine. This condition probably resulted from more precise control of the head and more sophisticated movements of the neck, the latter being longer and heavier than in more primitive sauropods. Taxa for which the cervicodorsal transitional region of the vertebral series is known, wherein the neural spines are low, include titanosaurs, *Haplocanthosaurus*, *Tendaguria*, and possibly brachiosaurids (see Janensch 1950*a*).

In comparing this new genus and species with other sauropods, Bonaparte *et al.* found that the cervical vertebra referred to *Tendaguria* is distinct from that of all other known taxa. In certain significant aspects of the neural arch, it most closely resembles that of *Camarasaurus*; it differs, however, in the form of the centrum and in the system of pleurocoelous cavities in the neural arch and on the lateral side of the centrum. The dorsal vertebrae in these two genera are quite different. While the neural spine in *Tendaguria* is singular, that in *Camarasaurus* is bifurcated and complemented by the expanded distal end of the transverse processes (absent in *Tendaguria*). Furthermore, *Tendaguru* is more derived than *Camarasaurus* in having greatly reduced neural spines, well-developed epipophyses, and more fossae in the anterior surface of the transverse processes.

According to Bonaparte *et al.*, the new genus seems to be a unique sauropod related most closely to *Camarasaurus*, although the above cited differences between these two genera indicate that they do not belong in the same family. Consequently, Bonaparte *et al.* found it justified to refer *Tendaguria* to its own new family, the Tendaguriidae.

Bonaparte *et al.* noted that the *T. tanzaniensis* type material was possibly found in the Upper Saurian Bed not far from the type locality of *Janenschia* (see entry). The authors acknowledged that future discoveries at Tendaguru could demonstrate that *Tendaguria* and *Janenschia* are congeneric. For the present, however, such a synonym cannot be made, as the holotype vertebrae of *T. tanzaniensis* cannot be compared with the hindlimb of *J. robusta*.

Key references: Bonaparte, Heinrich and Wild (2000); Janensch (1925*b*, 1929*a*, 1950*a*); Raath and McIntosh (1987).

†TEXASETES—(=?*Pawpawsaurus*)

Ornithischia: Genasauria: Thyreophora: Thyreophoridea: Eurypoda: Ankylosauria: Nodosauridae.

Diagnosis: Acromion process apparently a ridge angled towards glenoid as in *Sauropelta* (centrally located in *Panoplosaurus* and *Silvisaurus*); deltopectoral crest small (large, projecting forward in *Sauropelta*); unla proportionally more slender and straighter than in *Niobrarasaurus*, *Panoplosaurus*, or *Sauropelta*; anterior trochanter a small, vertical ridge on femur shaft; oblique ridge on greater trochanter (Carpenter and Kirkland 1998).

Comments: *Texasetes*, a nodosaurid genus erected by Coombs (1995) founded upon a holotype broken into many pieces (see *D:TE* for photographs), has been redescribed by Carpenter and Kirkland (1998) in their review of Lower and Middle Cretaceous ankylosaurs from North America. Noting that many of these pieces have since been assembled into whole bones or larger fragments, Carpenter and Kirkland focused their comments on corrections or addenda to Coombs.

According to Carpenter and Kirkland, the knob on the acromion process of the scapula extends much further beyond its base than implied by Coombs. The deltopectoral crests of the humeri do not project forward much (unlike the condition in *Sauropelta*; see Coombs 1978*b*), and are small and more rounded than in *Sauropelta*. The humeral shaft appears to be straight. The olecranon, as Coombs (1995) had predicted, is nearly 25 percent the length of the ulna (one third the length in *Mymoorapelta*). The ulnar shaft is more slender and straight, and the olecranon less bulbous, than in *Sauropelta* and *Mymoorapelta*.

The ilia are flat and the post-acetabular process short, the dorsal surface of the ilium apparently not closely appressed to the armor (as in *Stegopelta*). The shaft of a partial left ischium (identified by Carpenter and Kirkland among the holotype fragments) tapers distally and seems to possess an anterior bend or "kink" similar to that seen in *Edmontonia*, *Polacanthus*, and *Sauropelta*.

An oblique ridge can be seen on the greater trochanter of the femur, this process most likely separating the M. iliotrochantericus and M. iliofemoralis externus. The crista trochanteris is not well developed and the anterior trochanter is a small vertical ridge (not separated from the head as in *Hoplitosaurus*).

Note: Carpenter and Kirkland also reported on an unnamed nodosaurid, based upon a specimen (CEUM 6228) including a partial skull and right mandible, a cervical and three dorsal vertebrae, ribs, both scapulocoracoids, a sternal plate fragment, a right humerus, left ilium with ischium, and left femur from the Upper Mussentuchit Member (Cenomanian) of the Cedar Mountain Formation of Utah. The specimen was discovered using a modified scintillometer in exposures where no bone had been exposed (Jones and Burge 1995).

According to Carpenter and Kirkland, this specimen shares several features in common with *Texasetes* (*e.g.*, size, oblique ridge on greater trochanter) but differs from *Pawpawsaurus* (*e.g.*, in lacking prominent scutes, especially at postorbital corners). As in *Pawpawsaurus*, the coracoid is nearly half as long as the scapulocoracoid. The deltopectoral crest is small, though not rounded as in *Texasetes*. The anterior portion of the iliac blade is ventrally folded suggesting the condition in *Mymoorapelta* (see Kirkland and Carpenter 1994), although this appearance may be an artifact of preservation. The ischium, in its tapered shape and forwardly kinked shaft, resembles that of *Texasetes*.

Key references: Carpenter and Kirkland (1998); Coombs (1978*b*, 1995); Jones and Burge (1995); Kirkland and Carpenter (1994).

TEYUWASU Kischlat 1999

Dinosauria *incertae sedis*.

Name derivation: Tupi *teyu* = "lizard" + Tupi *wásu* = "big."

Type species: *T. barberenai* Kischlat 1999.

Other species: [None.]

Occurrence: Rio Grande do Sul, Brazil.

Age: Late Triassic.

Known material/holotype: München Universität, 1933L 53–54, right femur and tibia.

Diagnosis of genus (as for type species):

Comments: The genus *Teyuwasu* was named and briefly described in an abstract by Kischlat (1999), founded upon two appressed "osteoderms" and tentatively referred additional material, the latter including a femur and tibia (München Universität, 1933L 53–54), probably (based on "the preservation, size and fitting articular surfaces") from the same specimen.

These remains — collected at the beginning of the twentieth century at Sanga Grande, Alemôa, Santa

Maria, State of Rio Grande do Sul, Brazil — were originally named *Hoplitosaurus* by Huene (1938) and described as a nondinosaurian erythrosuchid archosaur. When that name proved to be preoccupied (Lucas 1902), Huene (1942) renamed it *Hoplitosuchus*.

Recently, in reexamining the above material, Kischlat observed that the referred long bone elements represent a dinosaur, as the "proximal end of the tibia has the cnemial crest (as defined by Novas, 1996*a*) and the distal end has the helicoidal rounded surface, resembling *Marasuchus* and *Herrerasaurus*." However, the distal process of the tibia is so developed as to encompass the ascending process of the astragalus, and the femur has but two parallel ridges proximodistally rather than a trochanteric shelf.

As pointed out by Kischlat, these remains have been altered taphonomically and need better preparation; nevertheless, "they represent robust dinosaurian form."

Key references: Huene (1938, 1942); Kischlat (1999); Lucas (1902); Novas (1996*a*).

†**THECODONTOSAURUS** Riley and Stutchbury 1836 — (=*Agrosaurus*)

Age: Late Triassic (?late Carnian or ?early Norian).

Diagnosis: Small, gracile prosauropod (up to 2 meters), mainly distinguished by absence of derived characters seen in other prosauropods; autapomorphies including elongate basipterygoid processes in braincase; dentary less than half length of mandible; and posterior process of iliac blade subquadratric (Benton, Juul, Storrs and Galton 2000).

Comments: The small sauropodomorph *Thecodontosaurus* is historically significant as one of the first substantial reptiles to be described from the Triassic period and also the fourth dinosaurian genus to be named from Great Britain, founded upon a dentary (originally BCM [Bristol City Museum] 1) collected from a limestone quarry of Durdham Down, Clifton, Bristol, in southwest England (see *D:TE*). Recently, *Thecodontosaurus* was fully reevaluated in a paper co-authored by Benton, Juul, Storrs and Galton (2000), the initial work of which was performed by Lars Juul in 1992–93 as part of his Candidatus Scientiae thesis for the University of Copenhagen (supervised by Michael J. Benton).

As chronicled in detail by Benton *et al.*, the dentary upon which *Thecodontosaurus* was established was collected in 1834 by Dr. Henry Riley, a local surgeon and medical school teacher, and Mr. Samuel Stutchbury, curator of the Bristol Institution. The "saurian" find was first announced in various anonymous reports published that same year and in 1835, seven years before dinosaurs were recognized as a group (Owen 1842). A fuller account was published the following year by Riley and Stutchbury. Although not including a figure of the dentary, this report described the specimen and referred it to a new genus, *Thecodontosaurus*. In the same paper, Riley and Stutchbury also announced *Palaeosaurus*, referring to this new genus two species, *P. cylindricum* [*sic*] and *P. Platyodon* [*sic*] (see *D:TE*), characterizing the genus only briefly and the species not at all. Subsequently, Riley and Stutchbury (1837) mentioned this latter

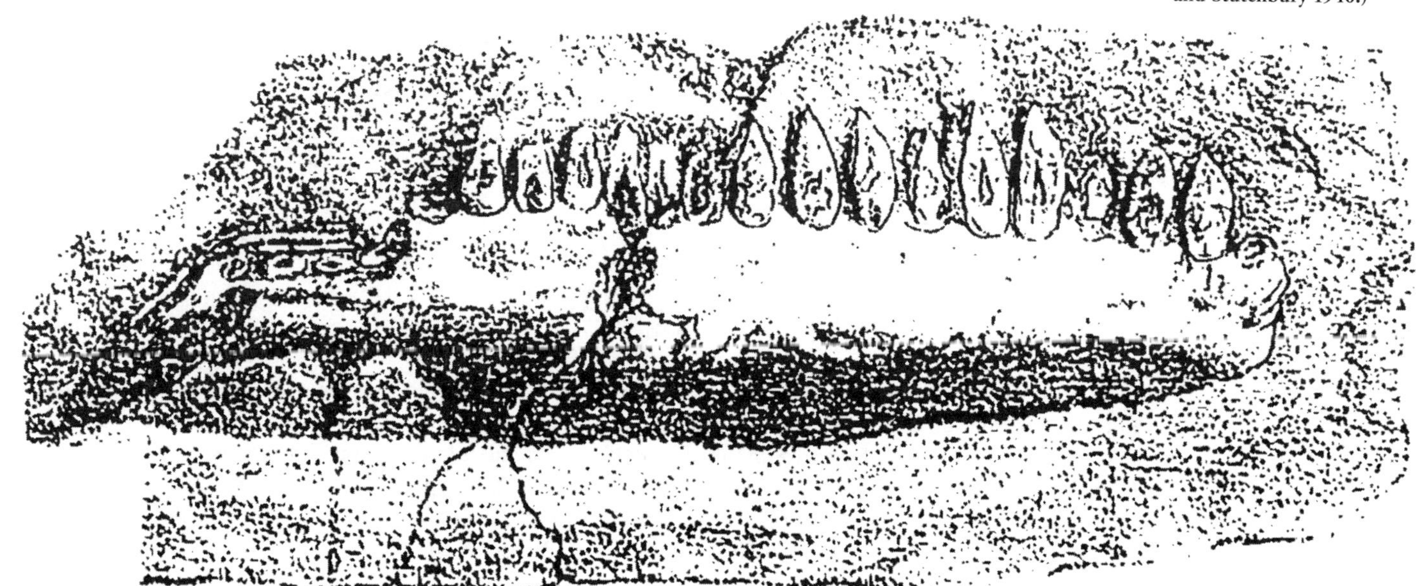

***Thecodontosaurus antiquus*, "BCM 1," holotype right dentary with teeth. (After Benton, Juul, Storrs and Galton 2000, from Riley and Stutchbury 1940.)**

***Thecodontosaurus antiquus*, YPM 2195, articulated left arm and manus bones, interpretive drawing. (After Benton, Juul, Stoors and Galton 2000.)**

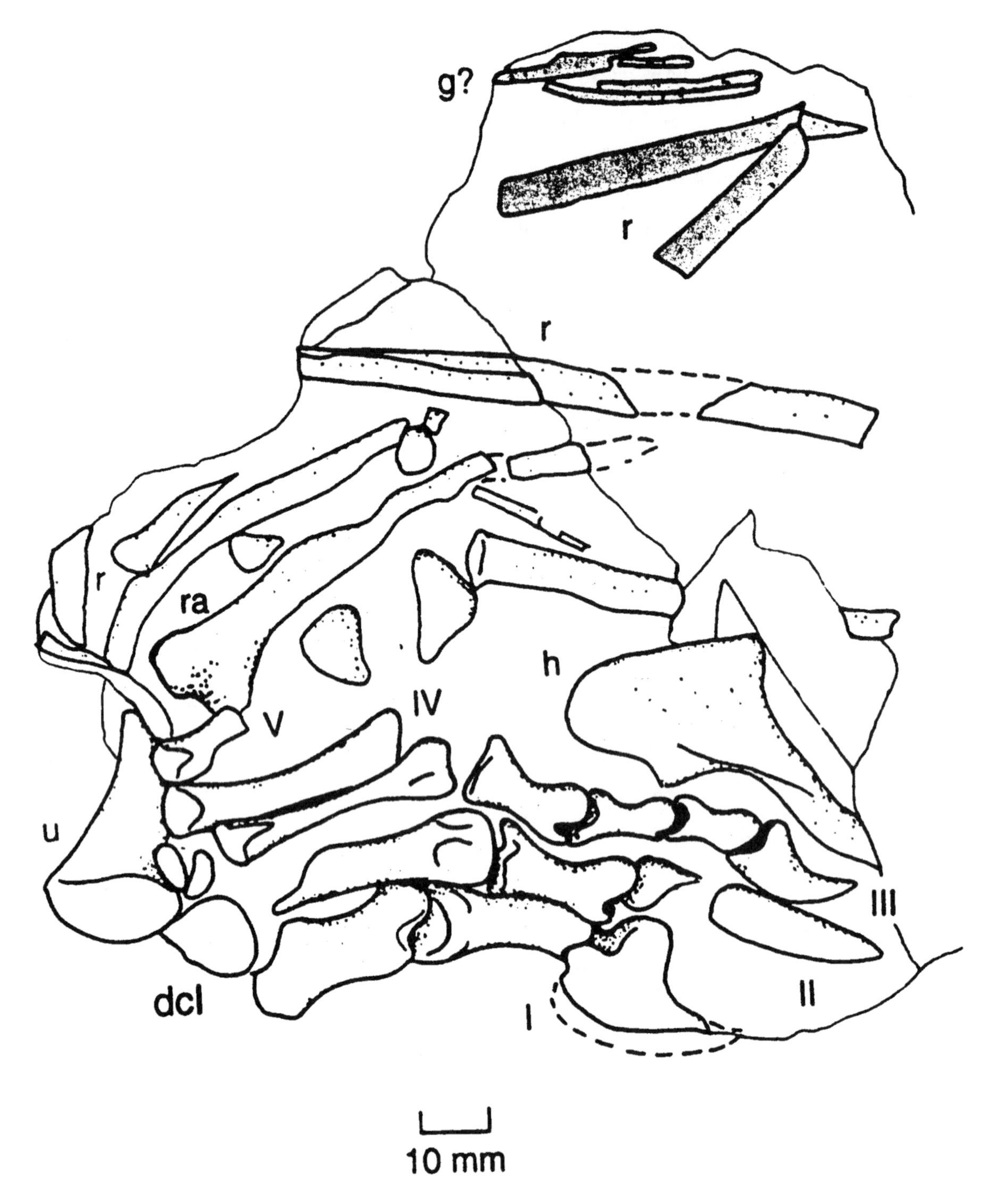

genus again, but with the name altered to *Paleosaurus* and the two species now corrected to *P. cylindrodon* and *P. platyodon*. Later, Riley and Stutchbury (1840) fully described *Thecodontosaurus*, confirming the indication from their original report that the type specimen consisted of a right dentary containing 21 teeth. In that same paper, Riley and Stutchbury sufficiently figured and described both *P. cylindrodon* and *P. platyodon*, each based on a single tooth, while also describing numerous postcranial remains that were not clearly referred to either *Thecodontosaurus* or *Palaeosaurus*. In all their reports, Riley and Stutchbury named only the genus *Thecodontosaurus*.

The type species *Thecodontosaurus antiquus* was finally named by Morris (1843) in his Catalogue of British Fossils. Years later, Huxley (1870), after examining the fossil specimens from Durdham Down, recognized that *Thecodontosaurus* was a dinosaur. Eventually, the type specimen of *T. antiquus* would be lost when, in November, 1940, a German bombing raid destroyed the Bristol City Museum's original geology gallery. (As *Palaeosaurus* would prove to be preoccupied, this genus would be renamed *Palaeosauriscus* by Kuhn in 1959.)

Thecodontosaurus has generally come to be regarded as "the most primitive reasonably complete prosauropod" (see Gauthier 1986; Benton 1990; Galton 1990). As pointed out by Benton *et al.*, however, such assessments have not been solely based on the material referrable to *T. antiquus*, but also on much smaller and probably juvenile material from south Wales referred by Kermack (1984) to *Thecodontosaurus* sp. According to Benton *et al.*, although it has often been assumed that all of the original material referred to *T. antiquus* was destroyed during the World War II bombing, several drawers of topotype material of this species yet remain in Bristol collection. Comprising 184 specimens, most of this material was collected during the 1830s; many of these specimens have been figured, some by Riley and Stutchbury (1840). More topotype material of *T. antiquus*— some of it distributed during the 1880s by Edward Wilson, then Curator at the Bristol Museum — is still housed in The Natural History Museum, London, the Peabody Museum of Natural History, and The Academy of Natural Sciences of Philadelphia.

The redescription of *Thecodontosaurus* by Benton *et al.* was primarily based upon the following specimens, all of which were recovered from the Durdham Down locality at Quarry Steps, Clifton, Bristol, England: Left dentary (BRSMG C4529 [formerly BCM 2], designated the neotype by Galton 1985*b*; the only survivor of four dentaries originally collected at this site); partial occiput and braincase (YPM 2192); four cervical vertebrae (BRSMG Ca7567, YPM 2192, 2195); 19 partial dorsal vertebrae (ANSP 9861, 9865, BRSMG C4533, Cb4153–Cb4156, Cb4163a, b, Cb4174a, b, Cb4182, Cb4197, Cb4221, Cb4293, Cb4714, YPM 2192); two sacral vertebrae (YPM 2192); 19 caudal vertebrae (ANSP 9854, 9875, BMNH 49984, R1534, R1550, BRSMG C4532a, b, Ca7473–Ca7475, Ca7507, Ca7510, Cb4164a, b, Cb4166, Cb4276, YPM 2192, 2193); 16 unidentified vertebrae (ANSP 9855, 9858, BMNH R1533, R1535, BRSMG C4301, Ca7475, Cb4157, Cb4167, Cb4171, Cb4174a, b, Cb4178, Cb4280, Cb4283, Cb4151, Cb4152, Cb4301); a neural spine (ANSP 9857); three chevrons (BRSMG C4532c, C4534, Ca7510); at least 43 ribs and rib fragments (BMNH R1536a, b, R1537a–c, R1538, BRSMG C4528, Ca7466, Cb4168a, Cb4169, Cb4170, Cb4172–Cb4174, Cb4194–Cb4196a, b, Cb4200a–e, Cb4206, Cb4212–Cb4213, Cb4218, Cb4227, Cb4234, Cb4255–Cb4256, Cb4285–Cb4286, Cb4297–Cb4298, Cb4300, Cb4528, Cb4714, YPM 2192; both scapulae and articulated forelimb (YPM 2195); right scapular blade (BRSMG Ca7481) and three ?scapula fragments (ANSP 9865, 9867, BRSMG Cb4216); 13 partial humeri (ANSP 9871, 9880, BMNH R1541–1543, BRSMG Cb4189, Cb4201, Cb4209, Cb4219, Cb4265–Cb4266, Cb4284, YPM 2192; radius (BRSMG Ca7504); right ulna (BRSMG Ca7486); fragmentary ?ulna (ANSP 9860); four ?metacarpals (BRSMG Ca7482, Cb4174, Cb4179); two ?manual phalanges (BRSMG Ca7485, Ca4187); two right ilia (ANSP 9870, BMNH R1539); three fragmentary ilia (BMNH R1540, BRSMG Cb4180–Cb4181); left pubis (BRSMG Cb4267); two distal fragments of ischia (YPM 2192) and ?proximal end of ischium (ANSP 9870); 23 partial femora (ANSP 9854, 9874, BMNH R1544–R1545, BRSMG C4530, Ca7456, Ca7481, Ca7490, Ca7494–Ca7495, Cb4176–Cb4177, Cb4183–Cb4184, Cb4190, Cb4207, Cb4228, Cb4243, Cb4259, Cb4266, Cb4269, Cb4288–Cb4289, YPM 2192; 12 partial tibiae (BMNH 49984, R1546, BRSMG C4531, Ca7495, Cb4185–Cb4186, Cb4188–Cb4189, YPM 2192); right fibula (BRSMG Ca7497) and five partial fibulae (ASNP 9863, 9867, 9869, 9872, YPM 2192); three unidentified limb bones (BMNH R1547–R1549); right metatarsal III (BRSMG Ca7451a) and indeterminate metatarsal

Courtesy Peter M. Galton.

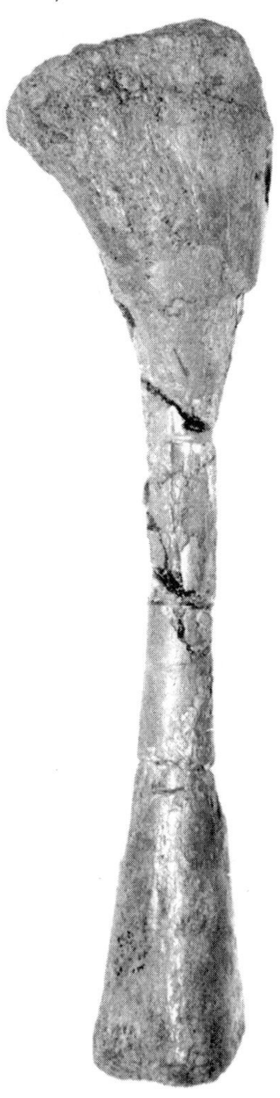

***Thecodontosaurus antiquus*, BMNH 49984, syntype proximal end of right tibia of *Agrosaurus macgillivrayi*, lateral view.**

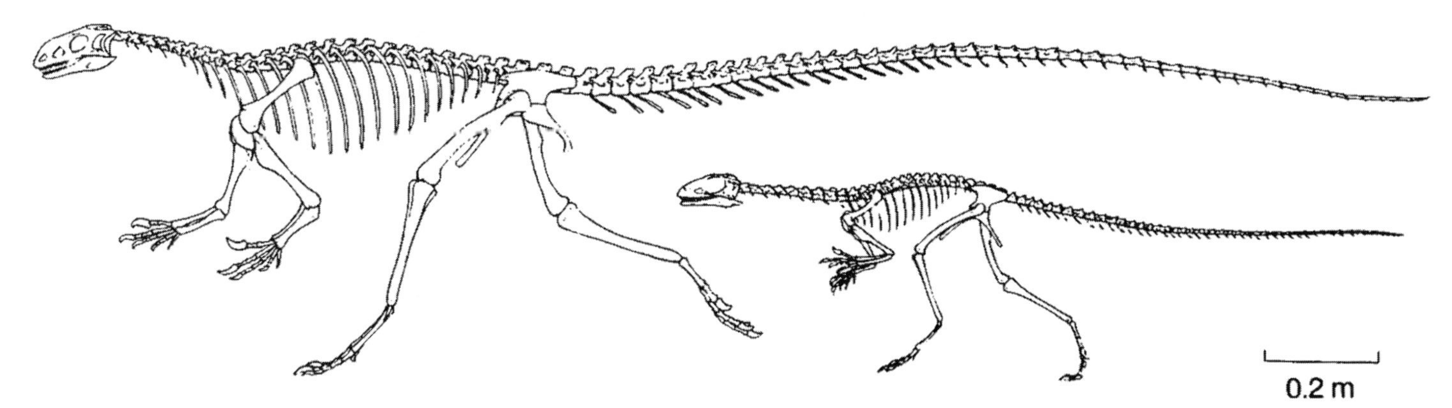

Reconstructed skeletons of an adult and juvenile *Thecodontosaurus antiquus*, the adult based on the Durdham Down specimens and a reconstruction by Huene (1932), the juvenile taken from Kermack (1984), drawing by John Sibbick. (After Benton, Juul, Storrs and Galton 2000.)

Courtesy Peter M. Galton.

***Thecodontosaurus antiquus*, BMNH 49984, syntype left tibia of *Agrosaurus macgillivrayi*, medial view.**

(BRSMG Ca7499); six ?pedal phalanges (BMNH R1552–R1553, BRSMG Ca7501, YPM 2192); three ungual phalanges (BMNH 49984, ANSP 9861, BRSMG Ca7451b). It incorporated other catalogued and uncatalogued material documented for the first time (see Benton *et al.*, Appendix I); most postcranial remains referred by Huene (1902, 1908) to the ?phytosaur *Rileya bristolensis* [*nomen dubium*]; and some material referred to *Palaeosauriscus platyodon* and *P. cylindrodon* (the holotype teeth of which, according to Benton *et al.*, are nondinosaurian; see *Palaeosauriscus* entry, "Nomen Nudum and Excluded Genera" chapter).

Regarded by Benton *et al.* as a junior synonym of *T. antiquus* is the type species *Agrosaurus macgillivarayi*, from the ?Late Triassic of Durdham Down. Once believed to represent Australia's oldest known dinosaur, this taxon was known only from the syntype specimen (BMNH 49984), comprising an isolated tooth, a distal caudal vertebra, an ungual phalanx, the distal end of the right radius, a left tibia, and the proximal part of a the right tibia. This material was originally reported by Harry Govier Seeley in 1891 as having been recovered from the northeast coast of Australia (see *D:TE*).

Recently, Vickers-Rich, Rich, McNamara and Milner (1999) reassessed this rare taxon as to its age and phylogenetic placement. As retold in detail by these authors (for full account, see also Jukes 1847), the fossils were found by John MacGillivray in 1844 during a mission of four British ships, headed by H.M.S. *Fly*, sent to construct a beacon on Raine Island off the eastern side of Cape York Peninsula, Australia. When Seeley (1891) named and described this material almost half a century later, the only locality information, quoted by that author, was the handwritten label: "*Fly*, 1844. Jn. Macgillaray, from the N.E. coast of Australia." Seeley believed that the material was most similar to dinosaurs known to him of Late Triassic to Early Jurassic age. More recent workers (*e.g.*, Galton 1990; Molnar 1991) have basically followed Seeley's assessment, regarding "*A. macgillivrayi*" as belonging to a group of dinosaurs restricted to that time, the Prosauropoda.

Vickers-Rich *et al.* related that MacGillivray was a naturalist with the *Fly* expedition who collected animals for Edward Smith Stanley, the 13th Earl of Derby (Ralph 1993). Unfortunately, MacGillivray's notebooks for the expedition have never been located; nor did he mention this specimen in the book he authored about another expedition to the same area several years later (MacGillivray 1852).

In 1993, air reconnaissance over the northeastern coast of Australia revealed that the only Mesozoic outcrops that do not appear to be deeply weathered in this area are the Helby Beds, which are exposed on the east coast of Cape York Peninsula. Based on the known location of H.M.S. *Fly* in 1844 and also consulting other now available information about the northeast coast of Australia, the Helby Beds — which would have been easily approachable for the expedition's voyagers — were therefore considered to be the most plausible source for the type specimen of "*A. macgillivrayi.*"

A modern expedition, chronicled in detail by Vickers-Rich *et al.*, was dispatched to the Helby area in July, 1995, its members including Patricia Vickers-Rich (Monash Science Center, Monash University,

Clayton, Victoria), Thomas R. Rich (Museum Victoria, Melbourne, Victoria), Gregory C. McNamara (Albert Kersten Geocentre, Broken Hill, New South Wales), Angela C. Milner (The Natural History Museum, London), Timothy Hamley (University of Queensland), Leaellyn Rich (Ormond College, University of Melbourne), Lucinda Hann (Wesley College, Glen Waveryly, Victoria), Lesley Kool (Monash University), Alan Fraser (Johns Hopkins University, Maryland), Pablo Puerta (Museo Paleontologico Egidio Feruglio, Trelew, Argentina), and Fan Jun Hang (Institute of Vertebrate Paleontology and Paleoanthropology, Beijing, China). Thanks to exceptionally favorable weather and sea conditions denied to members of the harsher 1844 expedition, these new searchers were able to examine all the Helby outcrops "for as long as necessary to establish with the highest probability possible that fossil bones similar in preservation to those of *Agrosaurus macgillivrayi* either do not occur at all in that unit of if so, are extremely rare."

After examining first hand this possible source area over a span of six days, Vickers-Rich *et al.* later determined in the laboratory at Monash University that the rock in that unit was quite unlike that which was associated with the type specimen of "*A. macgillivrayi.*" Furthermore, gross and trace-element comparisons between possible fossil bone fragments (QM F35957 and F35958, found by Vickers-Rich) from the Helby Beds and the "*A. macgillivrayi*" material revealed these to be very different in preservation. However, Vickers-Rich *et al.* found BMNH 49984 to be quite comparable to fossil bone recovered from the Magnesian Conglomerate at Durdham Down which had yielded remains of the prosauropod *T. antiquus*, and was already a well-known site when the holotype of "*A. macgillivrayi*" was purchased in 1879, as noted by Seeley, by the British Museum (Natural History) [now The Natural History Museum, London].

It was pointed out by the authors that, in 1906, Friedrich von Huene had already observed that the rock which yielded the holotype of "*A. macgillivrayi*" was reminiscent of that at Durdham Down. Huene (1906) had also remarked upon the close similarity of the latter to *T. antiquus*, noting that they differed only in that "*A. macgillivrayi*" possessed a more slender tibia and radius. Because of this close similarity, Huene (1906) referred the presumably Australian taxon to the genus *Thecodontosaurus* (see below) as the new combination *T. macgillivrayi*. Therefore, it was the conclusion of Vickers-Rich *et al.* that the holotype of "*A. macgillivrayi*" most likely did not originate from northeastern Australia, but was collected at Durdham Down, the original label associated with the specimen being in error; that this taxon was not, obviously, Australia's oldest dinosaur; and that, as foreshadowed by Huene (1906), this species is a probable junior synonym of *T. antiquus*.

Courtesy Peter M. Galton.

***Thecodontosaurus antiquus*, BMNH 49984, syntype distal end of right radius (on bloc of matrix before acid preparation) of *Agrosaurus macgillivrayi*, anterior view.**

Galton (in press a), based upon proportions of the distal end of the tibia, will formally refer *Agrosaurus* to *Thecodontosaurus antiquus*.

Both Seeley (1895) and Huene (1908*a*) had recognized two morphs — one robust, the other gracile — in the *Thecodontosaurus* postcranial material from Durdham Down, which these authors interpreted to represent at least two distinct species. Benton *et al.* also recognized these two clearly distinguished morphs, noting that the robust bones described by Huene (1908*a*) include a scapula, humerus, three femora, a tibia, and fibula. Benton *et al.* observed the following: The scapula (BRSMG Ca7481) is larger, shorter, less curved, and its narrowest part is comparatively more distally located than in the gracile morph. The humerus is relatively broader proximally, has a thicker shaft, the deltopectoral crest extends further distally but projects less anteriorly. The femora (*e.g.*, BRSMG Ca7456) have a rather smaller lesser trochanter, larger and stronger fourth trochanter, a straighter distal shaft, and transversely broader distal end. Also, the tibia has a relatively longer shaft, although the proximal end is more narrow transversely and the condyles are shaped differently than in the gracile morph.

According to Benton *et al.*, this variation in postcranial bones could be explained, following Seeley (1895) and Huene (1908*a*), by specific differences; however, they might also be the result of sexual dimorphism (see Galton 1997*b*), allometric variation, or individual variation. The idea that two species are represented — one robust, the other gracile — was ruled out, as the only character distinguishing the two forms is one of relative proportions. Allometry was rejected as both morphs are otherwise basically the same size. Individual variation was deemed unlikely as only two morphs rather than a continuum of forms are seen in the Durdham Downs specimens. Of these explanations, sexual dimorphism proved to be the most acceptable. A distinction into robust and gracile forms is a common distinction of sexual dimorphism in vertebrates. Benton *et al.* arrived at no definite conclusion as to whether the robust morph indicates a male or female. Males, the authors noted, are often larger for reasons of display or combat, while females can be larger for reasons related to the rigors of egg production. (For Galton's 1999 discussion of this topic, wherein that author suggests that the robust morph is indicative of a female, see *Sellosaurus* entry.)

In 1984, Kermack reported a small dinosaur specimen collected from a fissure deposit in the Pant-y-ffynnon Quarry (probably Norian; see Benton and Spencer 1995), South Glamorgan, South Wales, which she identified as a possible juvenile *Thecodontosaurus* sp. The remains were sufficient enough for Kermack to reconstruct the skull and skeleton.

As Benton *et al.* observed, the Welsh specimen is indeed a juvenile. The cervical centra and neural arches are not fused, and some of the long bones exhibit rugose articular surface (indicating that the cartilege had not yet ossified). The skull and orbit are comparatively large, and the overall length (about one meter) is less than half that of an average-sized adult *Thecodontosaurus* (about 2.5 meters; see below).

While some parts (*e.g.*, much of the skull and foot) of the Welsh skeleton have not been preserved in the Durdham Down material, other parts (*e.g.*, forearm and manus, dorsal vertebrae, parts of the pelvis and femur) are missing in the juvenile specimen. Nevertheless, the corresponding elements in the Welsh and British material compare most favorably.

As the specimen from Wales displays all three diagnostic characters of *T. antiquus* (*i.e.*, elongate basipterygoid processes, relatively short dentary, squared posterior process of ilium), Benton *et al.* accepted its assignment to *Thecodontosaurus*, but noted that the "comparative materials are insufficient to decide whether they belong to *T. antiquus* or to another species."

Based on the above specimens, Benton *et al.* offered a new reconstruction of the skeleton of the adult *T. antiquus*, based somewhat on the reasonably accurate reconstructions published by Huene (1908*a*, 1932) and also upon examination of the Durdham Down specimens. Conjectural in the new reconstruction were the skull, pes, and tail. The skull was restored using *Plateosaurus* as a general model, with the neotype dentary (70 millimeters in length) indicating a total lower-jaw length of perhaps 140 millimeters. The vertebral count and size of the individual vertebrae were based on those found in *Plateosaurus* and other known prosauropods (see Galton 1990), resulting in an adult skeleton measuring about 2.5 meters in length, more than half (1.5 millimeters) of which is tail. The limbs were reconstructed based on various collected elements. Measurements of the shoulder girdle and forelimbs were based on those of recovered specimens, including the articulated forelimb (YPM 2195). The pelvis was based upon isolated remains, the hind foot on those of other prosauropods.

As stated above, *Thecodontosaurus* is usually classified as a prosauropod dinosaur occupying a basal position within Prosauropoda (*e.g.*, Gauthier 1986; Galton 1990). In assessing the phylogenetic placement of this genus, however, Benton *et al.* found it necessary "to widen the ingroup to include all sauropodomorphs ... since the nature of the 'Prosauropoda' is disputed" (although most recent workers [*e.g.*, Sereno 1989, 1997, 1998; Galton 1990; Gauffre 1995; Wilson and Sereno 1998] have regarded this group as monophyletic).

A rigorous phylogenetic analysis was performed by Benton *et al.* including a number of reasonably completely known prosauropod taxa, but excluding some taxa which may or may not be distinct (*e.g.*, *Lufengosaurus* and *Yunnanosaurus*). Outgroups comprised the basal avemetatarsalian *Scleromochlus*, sister taxon to Ornithodira (see Benton 1990), and the non-sauropodomorph *Herrerasaurus*. Four sauropod (*Vulcanodon*, *Barapasaurus*, *Shunosaurus*, and *Brachiosaurus*) genera were included simply to represent the crown group Sauropoda. This analysis indicated that Prosauropoda is probably a valid clade within Sauropodomorph, but that support for this conclusion is weak. Relationships within the Prosauropoda were not resolved, although strong pairings were found of *Plateosaurus* and *Euskelosaurus*. *Thecodontosaurus* was found to be the outgroup of all other prosauropods including *Saturnalia* (see "Systematics" chapter, "Notes" under Prosauropoda), which, in turn, is the sister group to the remaining prosauropods.

Erratum: In *D:TE*, *Palaeosauriscus* entry, the type species name was incorrectly given as *P. cylindrodon* rather than *P. platyodon*.

Key references: [Anonymous] (1834, 1835); Benton (1990); Benton, Juul, Storrs and Galton (2000); Benton and Spencer (1995); Galton (1985*b*, 1990, 1997*b*, 1999); Gauffre (1995); Gauthier (1986); Huene (1902, 1906, 1908*a*, 1908*b*, 1932); Huxley (1870); Jukes (1847); Kermack (1984); Kuhn (1959); Molnar (1991); Morris (1943); Owen (1842); Ralph (1993); Riley and Stutchbury (1836, 1837, 1840); Seeley (1891, 1895); Sereno (1989, 1997, 1998); Vickers-Rich, Rich, McNamara and Milner (1999); Wilson and Sereno (1998).

†THESCELOSAURUS

Ornithischia: Genasauria: Cerapoda: Ornithopoda: Euornithopoda: Hypsilophodontidae: Thescelosaurinae.

Comments: Soft tissues are rarely preserved in dinosaur specimens, especially those inside the body; and not until the discovery of a specimen referred to the small ornithischian *Thescelosaurus* has what appears to be a heart been preserved (Fisher, Russell, Stoskopf, Barrick, Hammer and Kuzmitz 2000).

According to the report by Fisher *et al.*, the putative heart was found in a concretion noticed by Michael Hammer, a fossil preparator and paleontological investigator, and Dr. Andrew Kuzmitz, a family practitioner and amateur paleontologist, in the skeleton (NCSM 15728) of a small ornithischian dinosaur which he had found by Hammer's son Jeff in 1993 in sandstone beds of the Upper Cretaceous Hell Creek Formation of Harding County, near Buffalo, in northwestern South Dakota. The extremities and left side of the skeleton having been lost to erosion, the specimen is otherwise comparatively well preserved, containing much evidence of tissues not usually preserved, including sternal ribs and cartilaginous plates attached to the caudal surfaces of thoracic ribs. It is also significant in being the first *Thescelosaurus* specimen in which a complete skull has been preserved.

Aware of the specimen's potential, the senior Hammer took it to Kuzmitz, at Ashland Community Hospital, for reconstructive CT scanning. The resulting images convinced Hammer and Kuzmitz that the concretion in the chest cavity was a heart.

As calculated by Fisher *et al.*, the 197-millimeter circumference of the femur suggests a body weight of 300 kilograms (about 660 pounds), and the 468-millimeter femur length a total body length of 3.9 meters (more than 13 feet). Most closely, the specimen resembles the hypsilophodontid *Thescelosaurus*. (The

Photograph by Jim Page, courtesy North Carolina Museum of Natural Sciences, North Carolina State University.

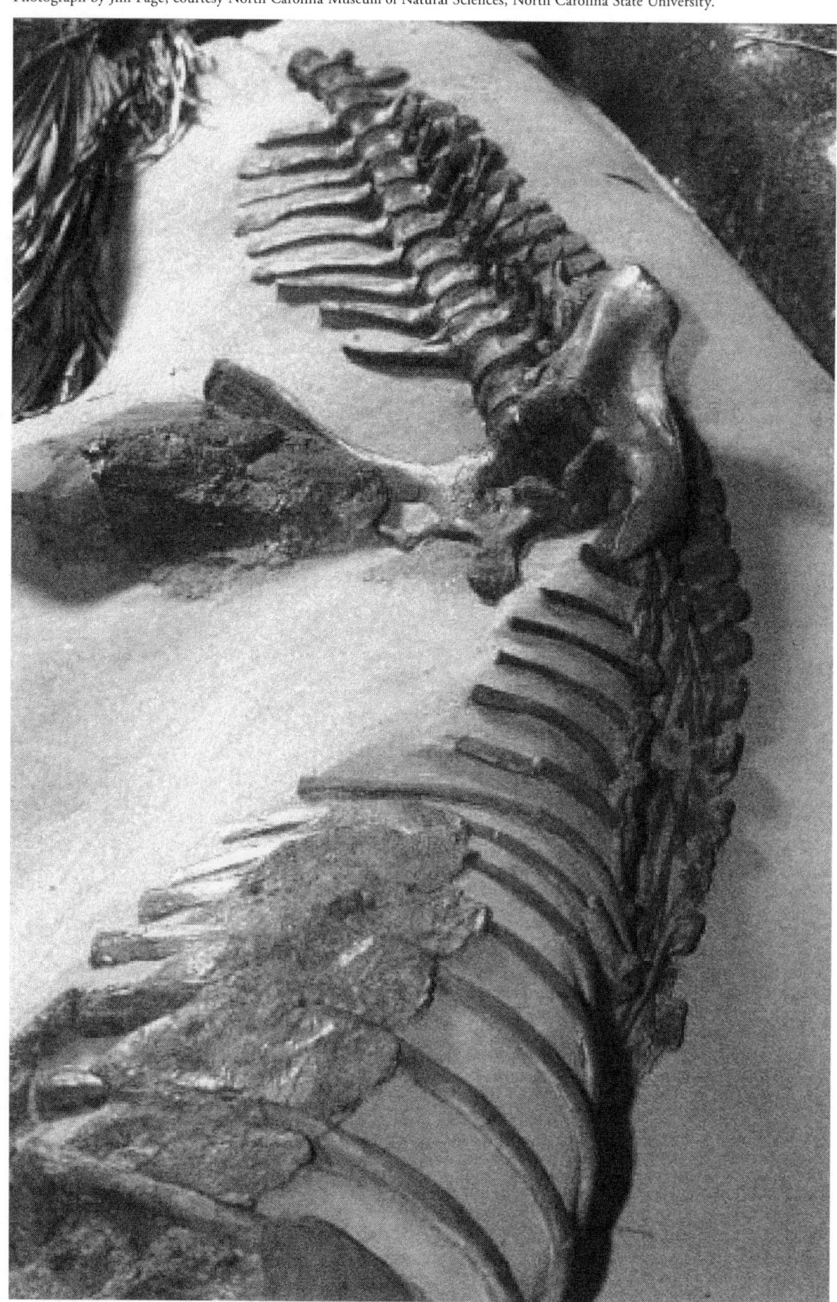

Skeleton (NCSM 15728), nicknamed "Willo," referred to *Thescelosaurus neglectus*, displayed as found in sandstone matrix.

authors noted that the gracile dentary of this specimen separates it from fragmentary remains belonging to the closely related and sympatric genus *Bugenasaura*.) The authors believed that the specimen belongs to the type species, *Thescelosaurus neglectus*.

The ferruginous or rust-colored concretion — the kind of rocky material usually discarded when fossils are prepared — was found in the thoracic cavity beneath the skeleton's upper ribcage. To determine whether this concretion contained internal structures, the somewhat compressed specimen was subjected to CT scanning by Dr. Michael Stoskopf, professor of wildlife and aquatic medicine and environmental

Photograph by Jim Page, courtesy North Carolina Museum of Natural Sciences, North Carolina State University.

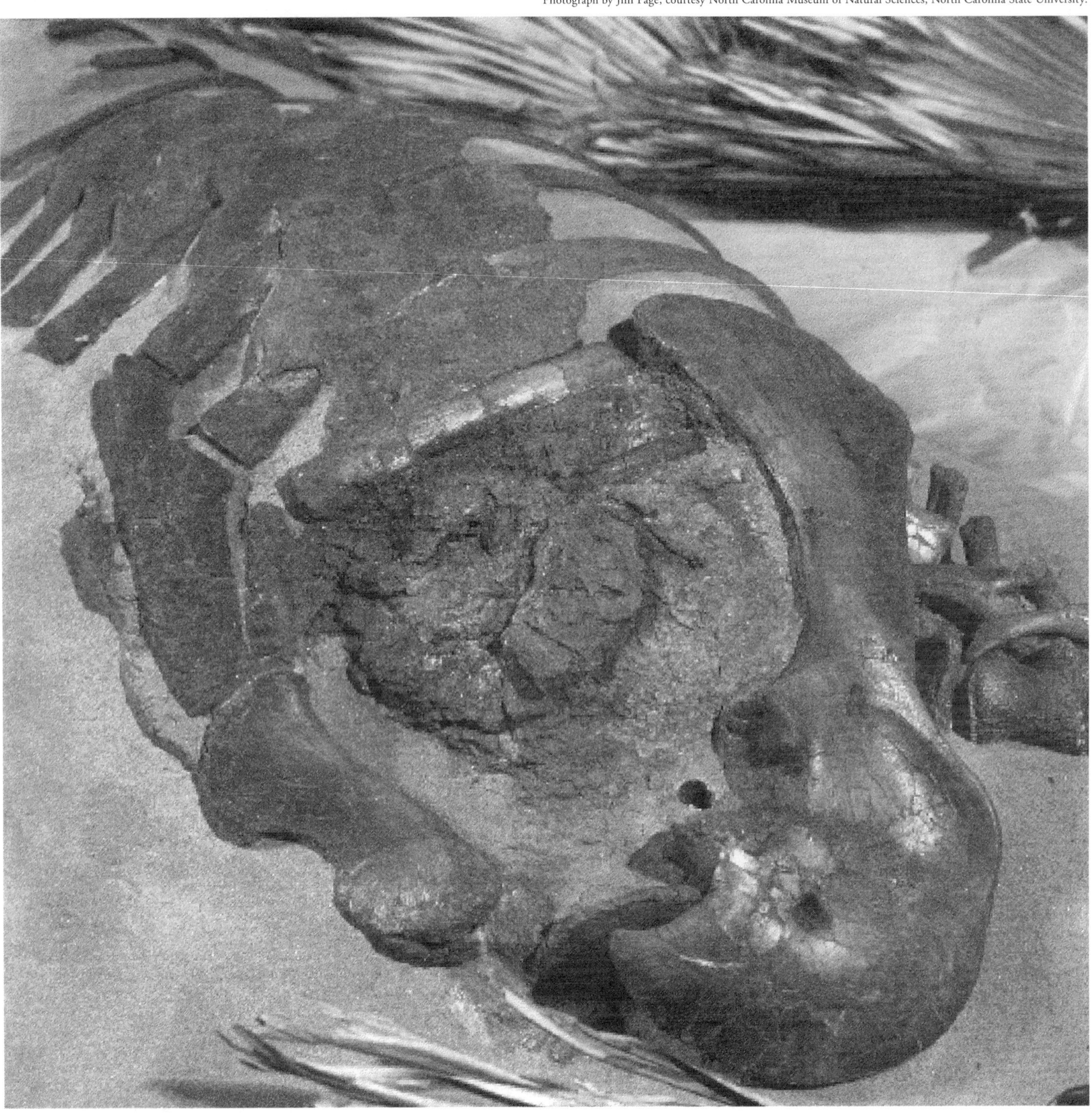

Apparent heart (center circular area) preserved in a skeleton (NCSM 15728) referred to *Thescelosaurus neglectus*.

toxicology at North Carolina State University, and Dr. Dale A. Russell, director of the Center for the Exploration of the Dinosaurian World, a joint project of the North Carolina Museum of Natural Sciences and North Carolina State University, Raleigh, aided by imaging specialists at the university's College of Veterinary Medicine. The individual "slices" were realigned utilizing a special software program, transforming the two-dimensional images into several three-dimensional images that could be studied on a computer screen.

According to Fisher *et al.*, shape patterns and radiodensities revealed "structures that are suggestive of a four-chambered heart and a single systemic aorta." Not reliably discernible in the specimen were the atria, usually very thin-walled and often collapsing in death, and the small vessels expected to be found in the region of the base of the heart. While soft tissues are

Courtesy North Carolina Museum of Natural Sciences, North Carolina State University.

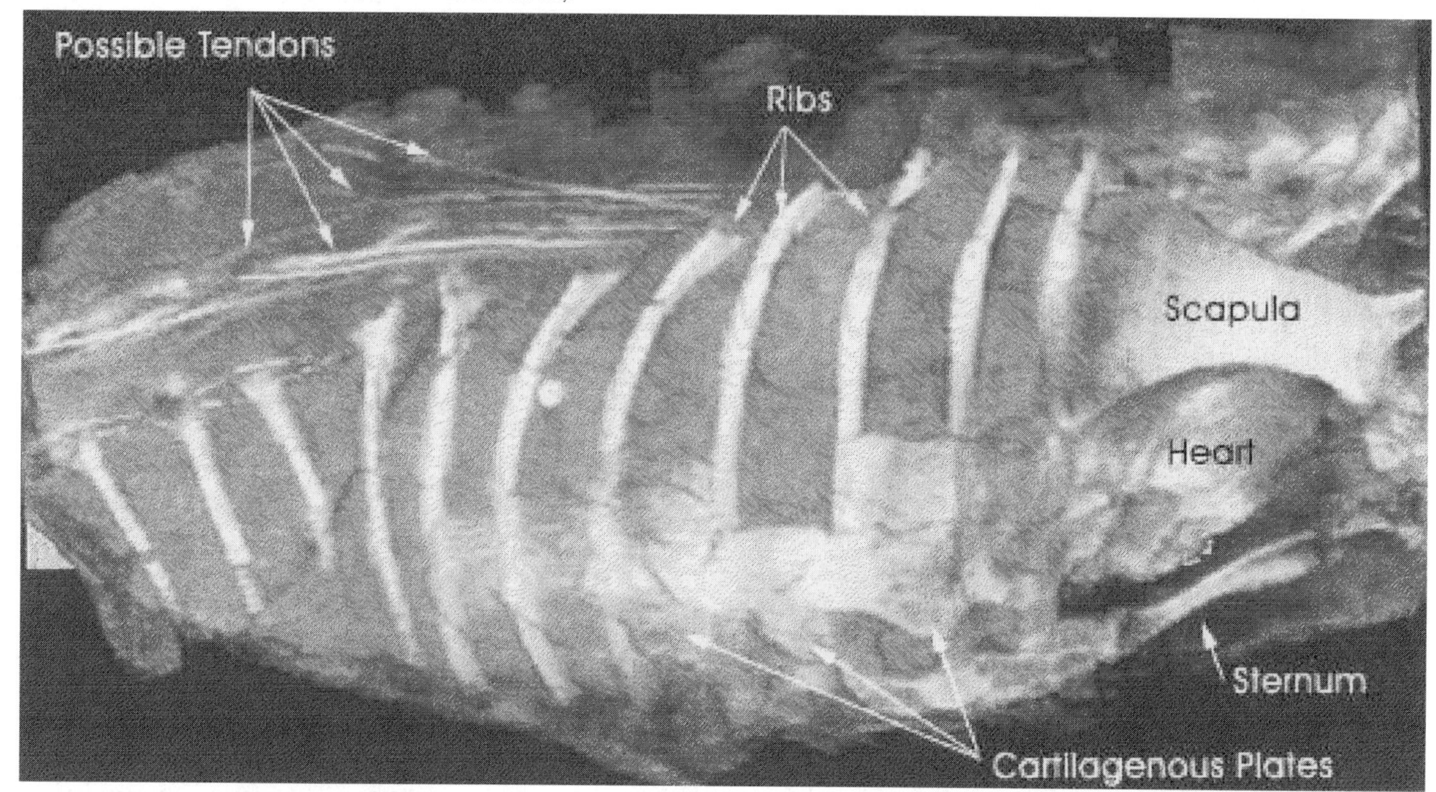

Three-dimensional imaging by Paul E. Fischer, North Carolina State University College of Veterinary Medicine Biomedical Imaging Resource Facility, of the skeleton (NCSM 15728) referred to *Thescelosaurus neglectus* showing what appears to be the preserved heart.

rarely preserved in dinosaur specimens, Fisher *et al.* theorized that chemical reactions between the blood and iron of the heart and minerals in the ground water preserved the organ's shape. (The presence of iron in the specimen's apparent heart was confirmed by X-ray diffraction analyses conducted by Dr. Reese E. Barrick, a North Carolina State University paleontologist, and graduate student William Straight.)

Among all animal groups, only birds and mammals possess a four-chambered heart, with two completely separated ventricles and one systemic aorta ensuring that only completely oxygenated blood is distributed to the body, these modifications having been correlated with higher metabolic rates than those occurring in modern reptiles (see Reid 1997). Therefore, the anatomy of the putative heart of NCSM 15728, being more like that of a bird or mammal than of living reptiles, suggested to Fisher *et al.* that this dinosaur's metabolic rate would have been more like that of an endotherm than an ectotherm.

There now seems to be evidence of an advanced heart with a single systemic aorta in at least one dinosaurian clade (hypsilophontids). As a similar kind of heart is known in birds, generally regarded as having derived from theropods, Fisher *et al.* noted that "it might be concluded that ancestral dinosaurs also possessed an advanced heart (thus making the attribute a synapomorphy for dinosaurs)." However, pointing out the more than 150 million years separating ancestral dinosaurs from NCSM 15728, the authors were "uncertain that the effects of long-term parallel selection and evolution on the cardiovascular system were negligible," this leaving open the question of whether "high metabolic rates and advanced hearts arose once or more than once among dinosaurs."

Russell was quoted on the free website maintained by the museum and university for this fossil: "The finding suggests the dinosaur's circulatory system was more advanced than that of reptiles, and supports the hypothesis that dinosaurs were warm-blooded... This challenges some of our most fundamental theories about how and when dinosaurs evolved."

According to Russell's website comments and to those by the original research paraphrased in other publications (*e.g.*, see Reed 2000), the specimen's soft tissues may have been preserved by saponification, a process by which soft tissues are converted into a soap-like substance when they are submerged in wet, oxygen-free environments, this allowing the soft parts to petrify rather than decompose.

In 2000, NCSM 15728, nicknamed "Willo" (after the wife of the rancher upon whose land the 66 million year-old fossil was discovered) was displayed at the Prehistoric North Carolina Exhibit at the North Carolina State Museum of Natural Sciences. Due to the specimen's importance and fragile condition, it

Courtesy North Carolina Museum of Natural Sciences, North Carolina State University.

"Willow," an ornithischian about the size of a short-legged pony and referred to *Thescelosaurus neglectus*, as restored by artist Ed Heck.

was presented in its original posture, still imbedded in the sandstone matrix.

Following the publication of Fisher *et al.*'s paper describing this specimen, gossip began to circulate, both verbal and in print, stating that the "heart" concretion had been misinterpreted (see Dalton 2000 for examples). Extensive subsequent testing, however, seems at this time to verify the authors' original interpretation of the concretion, and they stand by that interpretation (D. A. Russell, T. Lucas, and K. Kemp, personal communication, 2000).

Key references: (Dalton 2000); Fisher, Russell, Stoskopf, Barrick, Hammer and Kuzmitz (2000); Reed (2000); Reid (1997).

TIANZHENOSAURUS Pang and Cheng 1998— (See *Shanxia*)

Name derivation: "Tianzhen [County]" + Greek *sauros* = "lizard."

Type species: *T. youngi* Pang and Cheng 1998.

Key reference: Pang and Cheng (1998).

†TIENSHANOSAURUS (See *Euhelopus.*)

†TRICERATOPS

Ornithischia: Genusauria: Cerapoda: Marginocephalia: Ceratopsia: Neoceratopsia: Ceratopsomorpha: Ceratopsoidea: Ceratopsidae: Chasmosaurinae.

Type species: *T. horridus* (Marsh 1889).

Other species: [Excluding taxa regarded as *nomina dubia* (see *D:TE*), none.]

Comments: The well-known genus *Triceratops*—a giant ceratopsid famous for its three horns and non-fenestrated frill—once included 16 species.

In 1986, paleontologists John H. Ostrom and Peter Wellnhofer referred all of the species formerly regarded as valid to the type species, *Triceratops horridus*, noting that the perceived differences between these taxa were due to variation among individuals within that single species, and regarding the remaining species as dubious taxa. Subsequently, Catherine A. Forster (1996), using cladistics and morphometrics, separated the genus into what she considered to be two valid species, *T. horridus* and *T. prorsus*. Forster characterized *T. horridus* skull as having a long rostrum, an open frontal (="postfrontal") fontanelle, and comparatively long supraorbital brow horncores; *T. prorsus* specimens were identified as having a short rostrum, closed frontal fontanelle, and relatively short horns (see *Triceratops* entry, *D:TE*).

Later, Happ and Morrow (1996), in a study based upon an analysis of the orientation of *Triceratops* nasal horncores, and then Farke (1997), using bivariate plot analysis concerning sexual dimosphism and individual variation in *Triceratops*, concluded, as had Forster, that the genus contained two valid species, *T. horridus* and *T. prorsus* (see *DI*).

More recently, Lehman (1998), in a paper focusing on the genus *Pentaceratops* (see entry for additional related comments) but also discussing other chasmosaurine taxa, noted that ceratopsians known from a sufficient number of specimens (*e.g., Triceratops*, *Chasmosaurus*, and *Pentaceratops*) display a considerable amount of individual, ontogenetic, and sexual variability, with the relative elongation of the rostrum and the morphology and angle of the nasal horncore apparently "common modes of variation in many or all ceratopsian species." Lehman pointed out that this kind of variation, used in the past (and by Forster) as a criterion to identify species in horned dinosaurs, is now widely regarded as having little use in species-level ceratopsian taxonomy (*e.g.,* Lehman 1990; Dodson and Currie 1990; Ostrom and Wellnhofer 1990; Godfery and Holmes 1995).

Lehman (1998) noted, after a comparison of

Photograph by the author, courtesy University of Colorado Museum, University of Colorado, Boulder.

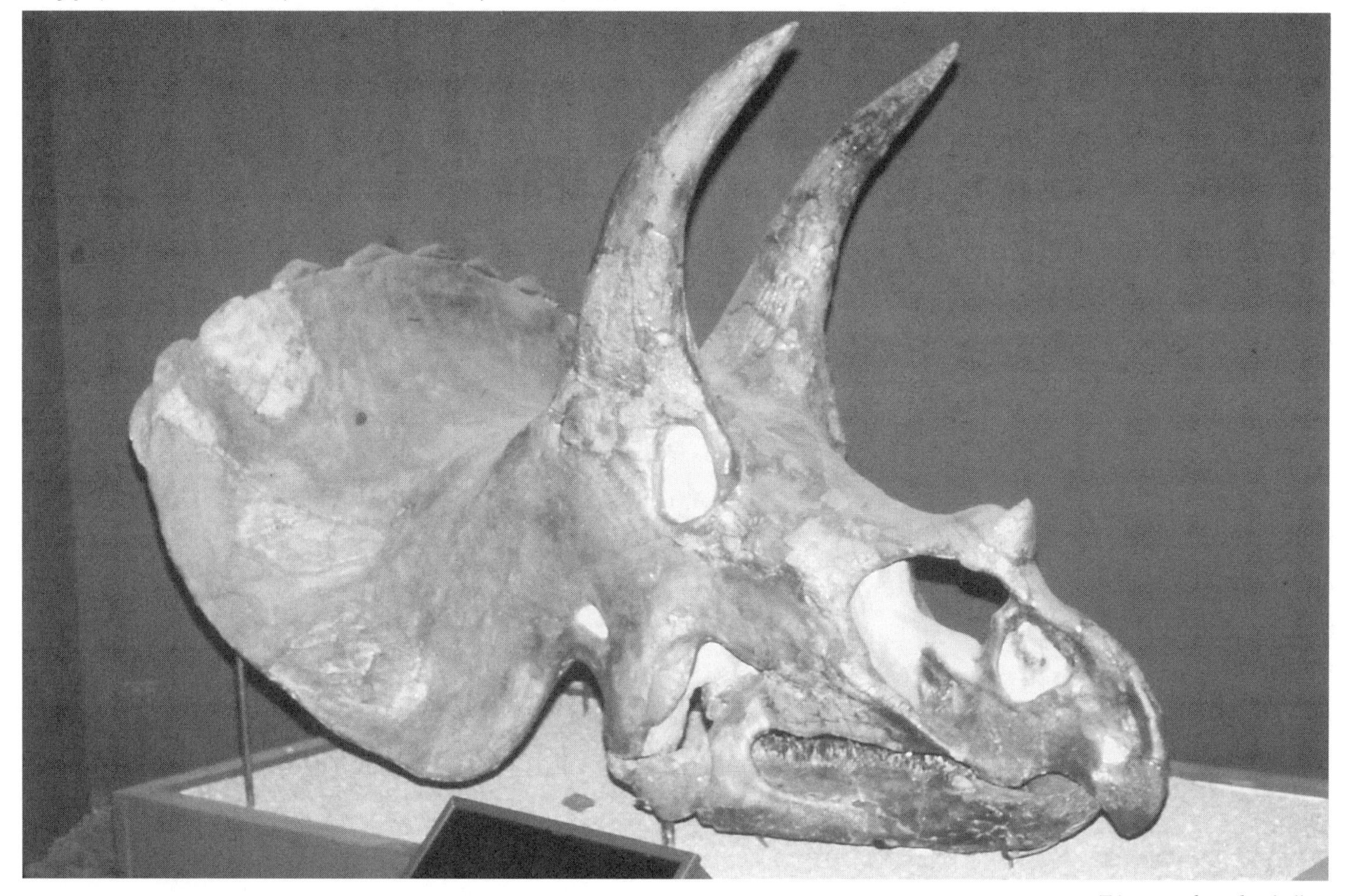

Triceratops horridus **skull (originally referred to *T. calicornis*). According to paleontologists John H. Ostrom, Peter Wellnhofer, and Thomas M. Lehman, this is the only valid species of *Triceratops*.**

Triceratops, *Chasmosaurus*, and *Pentaceratops* specimens, that 1. "the shape of the rostrom and nasal horncore spans a gradational spectrum variation," 2. "two end-members do not represent discrete morphotypes," 3. (*contra* Forster) there is no distinct separation between these two morphotypes (*e.g.,* YPM 1821, the holotype of *T. horridus*, possessing relatively short horns; referred specimen YPM 1823 having a closed fontanelle), and 4. there is no stratigraphic or geographic separation between these two morphotypes (most *Triceratops* specimens having been found separated by no more than a few hundred kilometers), a situation quite unlikely if these morphotypes represent distinct species of such large animals.

It was Lehman's (1998) opinion, based upon the above observations, that an alternative idea proposed by Forster — that the two morphotypes of *Triceratops* may represent sexual dimorphism — is more likely than that they indicate distinct species. Variation in horncore length and supraorbital horncore orientation is continuous (the latter cited by Lehman 1990 as a criterion for sexual dimorphish — erect indicating males, forwardly directed indicating females). Lehman observed that *Triceratops* specimens having the most elongate rostrum (*e.g.,* USNM 2412 and 4828, YPM 1823) also have the most erect brow horncores, while those with the shortest rostrum (YPM 1822 and 1834) have forwardly directed brow horns, this showing that some correlation exists between the morphotypes suggested by both Lehman (1990) and Forster. Therefore, Lehman (1998) found it most parsimonious to conclude that, as determined earlier by Ostrom and Wellnhofer (1990), that *T. horridus* is the only valid species of *Triceratops*, this taxon exhibiting "a gradational spectrum of variation in length of the rostrum and length and orientation of the nasal supraorbital horncores," the end members of this spectrum sometimes exhibiting the most pronounced sexual character states.

As recounted in *D:TE*, *Triceratops* was once classified by Lull (1933) — long before the widespread use of cladistics in performing phylogenetic studies — among the so-called "short-frilled" ceratopsians (a group of genera now grouped within the ceratopsid subfamily Centrosaurinae), despite its having long postorbital or brow horns and a short nasal horn, features indicative of "long-frilled" forms. This, however, resulted in a conspicuous gap in the fossil record

Photograph by the author, courtesy University of Colorado Museum, University of Colorado, Boulder.

Triceratops horridus **skull (originally referred to *T. calicornis*), showing detail of the epoccipital processes of the parietal frill.**

between *Triceratops* and taxa that were more primitive than this genus yet also geologically younger, and also leaving unexplained the pronounced horn-length disparity between *Triceratops* and these other taxa. Almost 60 years later, Dodson (1990), utilizing modern phylogenetic methods, referred *Triceratops* to the Chasmosaurinae.

More recently, however, Penkalski and Dodson (1990), in a study primarily focusing upon the small, basal ceratopsid *Avaceratops* (see entry; also see "Systematics" chapter, "Note" under "Chasmosaurinae"), found via cladistic methodology that the presence of long brow horns alone — one of the criteria for placing *Triceratops* within the Chasmosaurinae — no longer constitutes sufficient grounds for identifying a ceratopsian specimen as chasmosaurine. Furthermore, Penkalski and Dodson observed that the rather primitive Campanian genus *Avaceratops* and the highly derived Maastrichtian genus *Triceratops* seem to share a suite of similarities, including relative postorbital horn length and an unfenestrated parietal; indeed among all known taxa, *Triceratops* resembles *Avaceratops* more than any other horned dinosaur. In an earlier cluster analysis of ceratopsians, Dodson (1993) found that *Triceratops*— if brow horns were not included — grouped not with the chasmosaurines, but rather with the centrosaurines. Although it was not in the scope of Penkalsi and Dodson's study to reanalyze the entire Neoceratopsia, their findings imply that *Triceratops* (and, by association, various genera possibly synonymous with *Triceratops*; see *D:TE*) may not be a chasmosaurine but could, with *Avaceratops*, form a clade outside of both the Centrosaurinae and the Chasmosaurinae.

A variety of functions have been offered over the years for the parietal frill of *Triceratops*, including protection, intraspecific display, and muscle attachment (see *D:TE*), any, all, or some combination of them perhaps being correct. In a recent study, Barrick, Stoskopf, Marcot, Russell and Showers (1998), while not discussing these other possible functions, addressed the hypothesis that the frill (and also the horncores) may also have had thermoregulatory functions. According to these authors, body temperature variability in fossil vertebrates can be measured by the oxygen isotopic composition of bone phosphate, with intrabone and interbone variations indicating heat flow within an individual and having use in establishing

Photograph by the author, courtesy Natural History Museum of Los Angeles County.

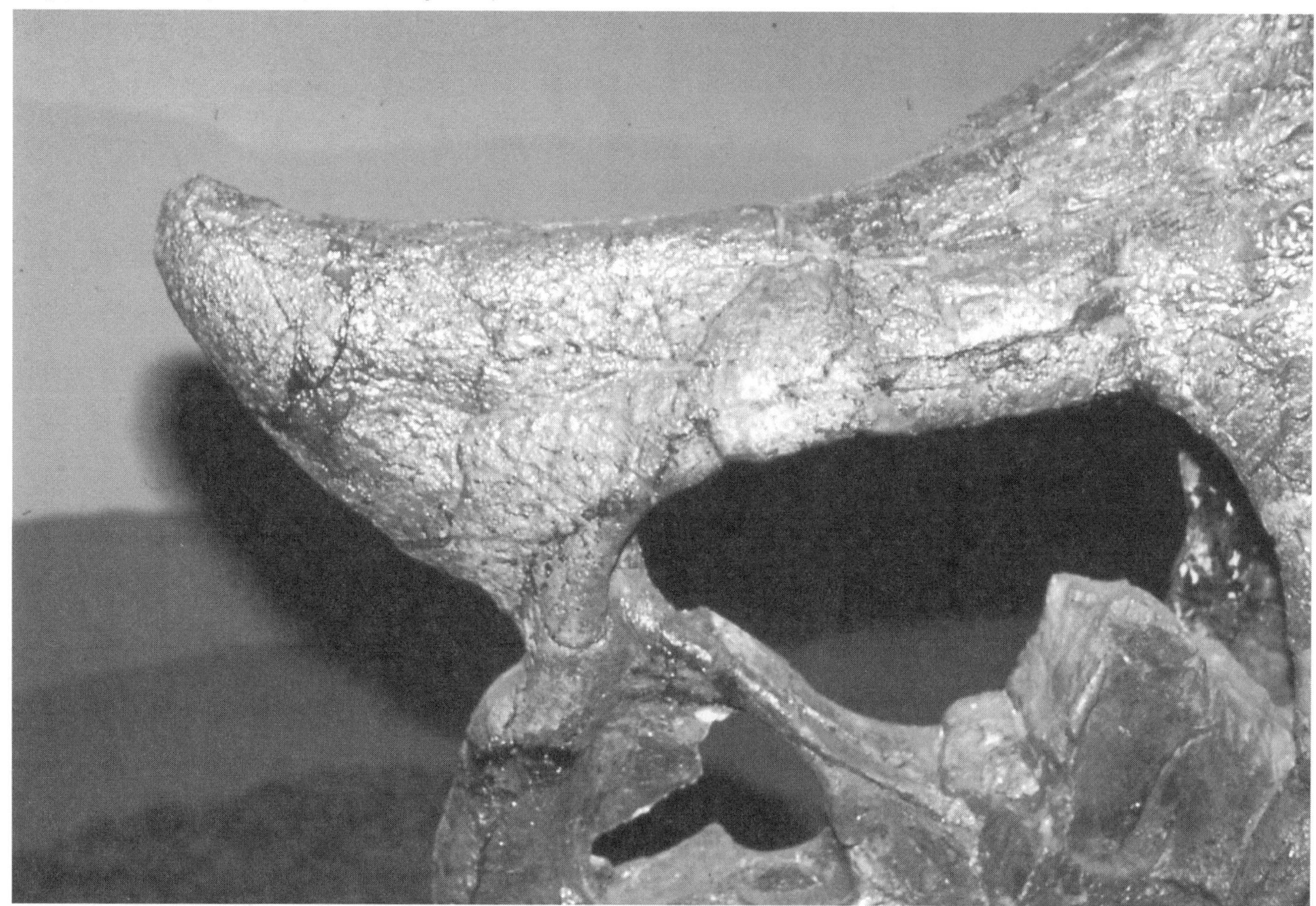

Detail of the nasal horn core of a skull (LACM 7207/59049) of *Triceratops horridus* (this specimen formerly referred to *T. prorsus* based on the orientation and the relatively greater length of this horn; see *D:TE* and *S1*).

thermoregulatory strategies (*e.g.*, homeothermy and regional heterothermy).

Applying this method to 108 samples from 18 bones of a well-preserved skeleton of *Triceratops* (CNM [formerly NMC] 34824), Barrick *et al.* achieved results suggesting a quite high and uniform heat flow through the parietal frill. This could have maintained mean frill temperatures of between 0–4 degrees centigrade below the temperature of the body core, with heat flow to the postorbital horn cores being considerably more variable. The authors posited that the very large frill, because of surface characters that enhanced convective transfer by increasing the flow of air across its surface (its dorsal surface area accounting for some 10 percent of the animal's entire surface area), functioned as an efficient heat exchanger. By turning its head or facing the wind, a *Triceratops* could use its frill somewhat in the ways suggested for the plates of *Stegosaurs* or the dorsal sail of the pelycosaur *Edaphosaurus*. According to Barrick *et al.*, the relatively constant bone temperature revealed by their study suggests the "frill's specialized importance in standard temperature control and evidence of either a very stable environmental temperature or body temperature or both."

Data pertaining to the horncores showed that heat flow to the postorbital horncore is considerably more variable than that of the frill. Combined with the low isotopic variability of the rest of the *Triceratops* skeleton, these horncore data suggest a more stable body temperature (*i.e.*, homeothermy) for this dinosaur.

Barrick *et al.* concluded that the frill and horns of *Triceratops*, regardless of their other possible functions, were indeed thermoregulatory structures. The frill was interpreted as serving a fairly consistent role in the regulation of overall body temperature; on the other hand, the horecores, not unlike the role of the horns of modern bovids, may have been important in the stabilization of brain temperature at extreme temperatures. Trends "of minor increases in isotopic variability and cooler temperatures distally in the frill" suggested that *Triceratops* possessed a fine control of the transfer of forced convective heat to the environment. The authors found their evidence for thermal control and the dumping of heat to be "more consistent

Photograph by the author, courtesy Natural History Museum of Los Angeles County.

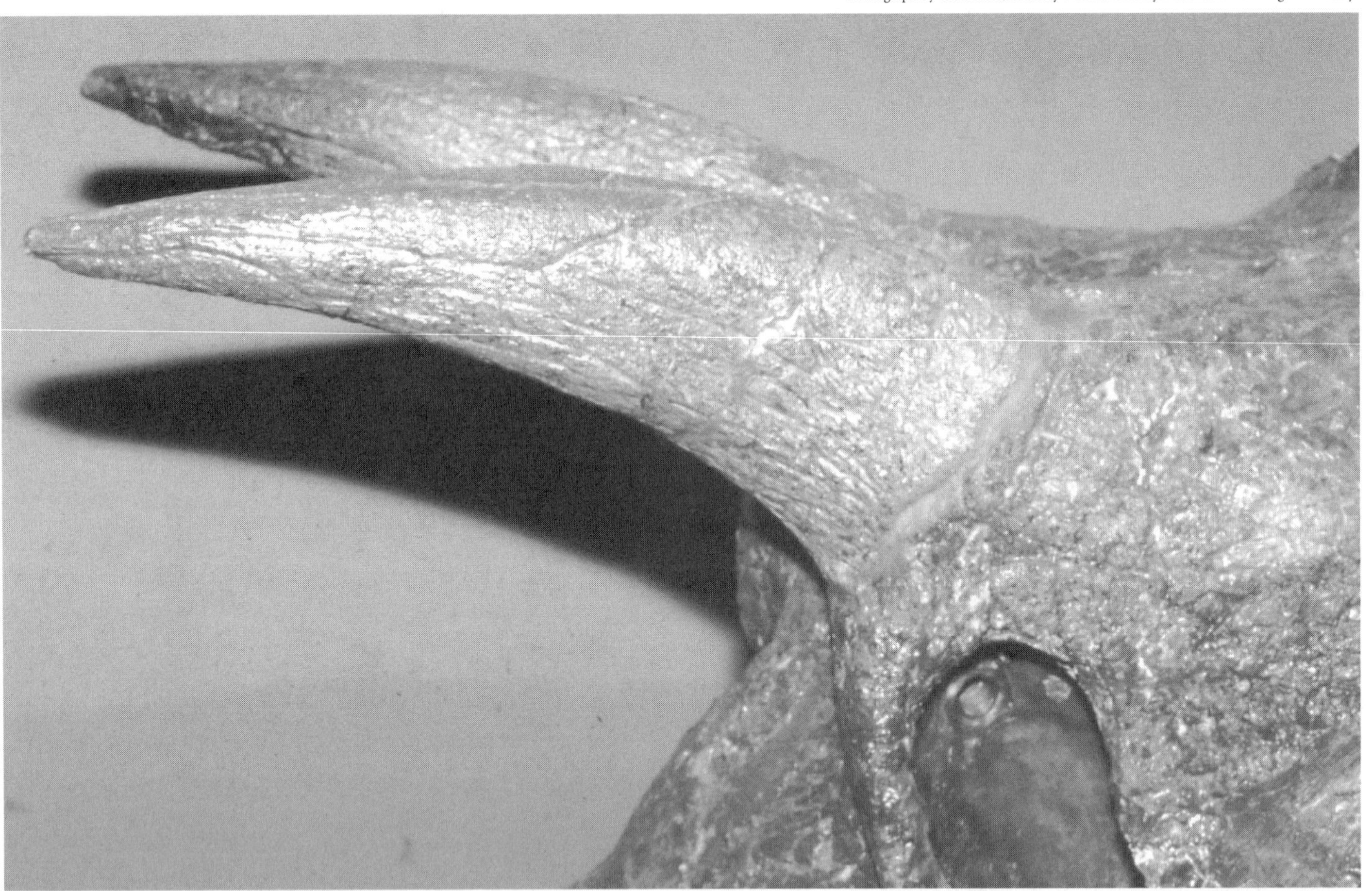

Detail of skull (LACM 7207/59049) of *Triceratops horridus* (this specimen formerly referred to *T. prorsus*). Although relatively short, these erect supraorbital horn cores, according to Thomas M. Lehman, may indicate the male sex of this individual.

with a somewhat elevated intermediate metabolism for *Triceratops* rather than simple bradymetabolic gigantothermy."

A number of brief reports have been published concerning the composite skeleton (USNM 4842) of *Triceratops horridus* that, for almost a century, had been mounted at the National Museum of Natural History, Smithsonian Institution (see *D:TE* for additional information and photograph; see Gilmore 1905 for details on the mounting of this skeleton).

Brett-Surman, Jabo, Kroehler and Parrish (1999) reported on the disassembly, restoration, and future remounting of various fossil skeletons exhibited at the National Museum, emphasizing USNM 4842 (the *Triceratops* Rescue Project). As chronicled by Brett-Surman *et al.*, this specimen (nicknamed "Mr. T." in recent years after the motion picture and television personality)—originally referred to *T. prorsus*—is distinguished as the first *Triceratops* skeleton ever put on display. Including bones from several *Triceratops* individuals, borrowed cast bones from related species, sculpted elements, and even hadrosaur foot bones (Deck 2000), the skeleton was originally mounted in 1905 for the then-named United States National Museum (with casts of this specimen subsequently appearing at various other institutions and events, *e.g.*, at the Smithsonian displays in the United States Government Building at the 1901 Pan-American Exposition, held in Buffalo, New York, and the Government Building at the Louisiana Purchase Exposition, the 1904 World's Fair in St. Louis, Missouri). The skeleton was dismantled following the discovery that real bone displayed without protective cases in a public environment, despite cleaning and standard attempts at preservation, eventually deteriorates due to vibration and changes in humidity. Damage documented on USNM 4842 include pyrite disease discovered on the articular joint surfaces after the specimen was taken apart (similar damage being found on the National Museum's mounted skeletons of *Stegosaurus* and the mammoth *Mammut*).

The implication of these damaging effects on fossil bone could mean, the authors speculated, "that all nonencased fullmounts on display may have a limited lifespan, and that cast material may be the only solution for display, especially for specimens collected in the 1800s."

In a rather new and "high tech" kind of project regarding *Triceratops*, Chapman, Andersen and Jabo (1999), and Andersen, Chapman, Kenny and Larsson

Photograph by John Weinstein, courtesy The Field Museum (neg. #GEO85828).

Skeleton of ***Triceratops horridus*** **(cast of composite including AMNH 5116, 5033, 5039, 5044, and apparently some *Torosaurus* elements) mounted before the classic mural painted by Charles R. Knight depicting this dinosaur's encounter with *Tyrannosaurus rex*. According to Barrick, Stoskopf, Marcot, Russell and Showers (1998), the frill and postorbital horns of *Triceratops* seem to have had thermoregulatory functions more than serving as a defensive shield against such giant predators.**

(1999), briefly reported on their use of digital data to recreate virtually USNM 4842, work on this project following the recent disassembly the mounted skeleton. Sufficient elements were surface scanned into a computer, with missing or insufficient elements recreated via mirror imagery based upon bilateral counterparts or scaling up or down related elements. This allowed the authors to assemble a complete, virtual skeleton. The computer-generated skeleton provided the first opportunity to animate a dinosaur's movements virtually based on accurate, scanned, anatomical information; also, Chapman *et al.* noted, "to produce full-sized prototypes of individual bones that have been cast to use in the new mount, and produce accurate, reduced versions of the huge skull or other bones, and a full 1/6 size skeleton." (The first reproduction of this virtual skeleton was published in the "New and Improved" feature in the December 2000 issue of *National Geographic* magazine.)

Andersen *et al.* briefly described their use of the virtual skeleton to animate with accuracy the step cycle of an adult *Triceratops* individual. Also incorporated into this project were images of a trackway attributed to large ceratopsian dinosaurs approximately the size of USNM 4842, these indicating the positions of manus and pes during contact with the ground. The above information was then transferred into the animation software. Excursions of forelimb and hindlimb joints were based upon anatomical evidence, and, using large-bodied extant quadrupeds as models, "the limbs were optimized to as much of a parasagital and a columnar posture as possible."

The results of this experiment proved to be somewhat unexpected for the authors: As quadrupeds require equal hand and foot excursions during locomotion, the scapula must rotate caudally with each advancing step to permit the forelimb to reach its

Courtesy Richard C. Ryder.

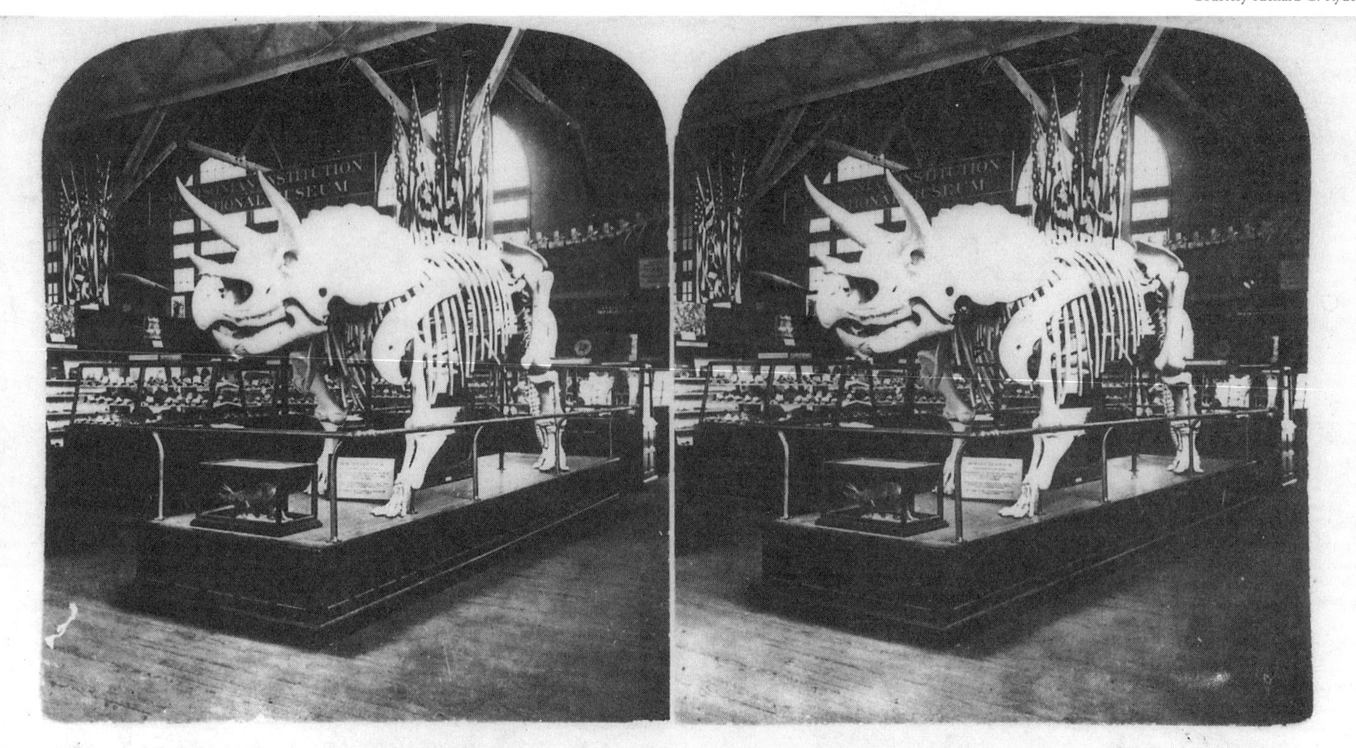

Stereograph of the skeleton of *Triceratops horridus*, plaster cast of USNM 4842 (a mount incorporating hadrosaurid hindfeet), as displayed in the Smithsonian exhibit in the United States Government Building at the Pan-American Exposition (1901), Buffalo, New York.

designated position; but during the last phase of contact between pes and ground, the scapula must rotate cranially, extending the forelimb back far enough to match the information of the trackway. Despite having a long femur and tibia, the *Triceratops* hindlimb was found to be constrained to an acute arc.

The above project, Andersen *et al.* pointed out, "sets the stage to accurately add details such as appendicular muscle origins and insertions, optimal muscle excursions, axial skeleton motion, and flesh reconstructions."

As reported in an abstract by Deck (2000), a cast (USNM 500000) of the Smithsonian skeleton — the bones recreated utilizing computer-aided analyses and prototyping of the original skeletal materials, the bones recreated via a computer-driven milling process — has now been mounted at the National Museum in an updated articulation and a more correct and dynamic pose (see also Deck 1999; Jabo, Kroehler, Andersen and Chapman 1999). This new mount includes a cast skull that is approximately 15 percent larger than that on the original mount. The correct feet have replaced the hadrosaur feet originally included in the former mount. Also, the left humerus, ilium, and scapula — formerly either sculpted or represented by elements of the wrong size — are now mirror-imaged computer models of the original right-side bones.

Happ and Morrow (1999), in an abstract, announced the discovery of a predentary belonging to a young (not a hatchling) ceratopsid — probably *Triceratops*— from the Hell Creek Formation (Upper Cretaceous) in Garfield County, near Jordan, Montana. If scaled up nonallometrically to adult size, the authors noted, it would suggest a total body length of nine feet.

As described by Happ and Morrow (1999), the predentary has a sharp, triangularly bony protuberance on its distal end, distinguished from the predentary itself by an indented groove. When the subadult predentary is scaled up so that its outline is superimposed over that of an adult *Triceratops*, the bony process of the former extends significantly away from the adult outline.

Citing for comparison the two most promical extant outgroups, reptiles and birds, Happ and Morrow (1999) pointed out that both possess an egg tooth on the upper or lower mandible that is utilized in cutting through the egg shell membranes and shell during hatching, most "egg teeth" being lost or resorbed within a few days of hatching. It was the interpretation of these authors that the extention on the predentary of this ceratopsid was an "egg tooth" used to extricate the dinosaur from its egg, and that this structure was probably retained by the animal for an extended time.

In another abstract, Happ and Morrow (2000) gave a preliminary report of apparent soft tissue in a skull (SUP9713) of *Triceratops* from the Hell Creek Formation near Jordan, Montana. Belonging to a

Courtesy Richard C. Ryder.

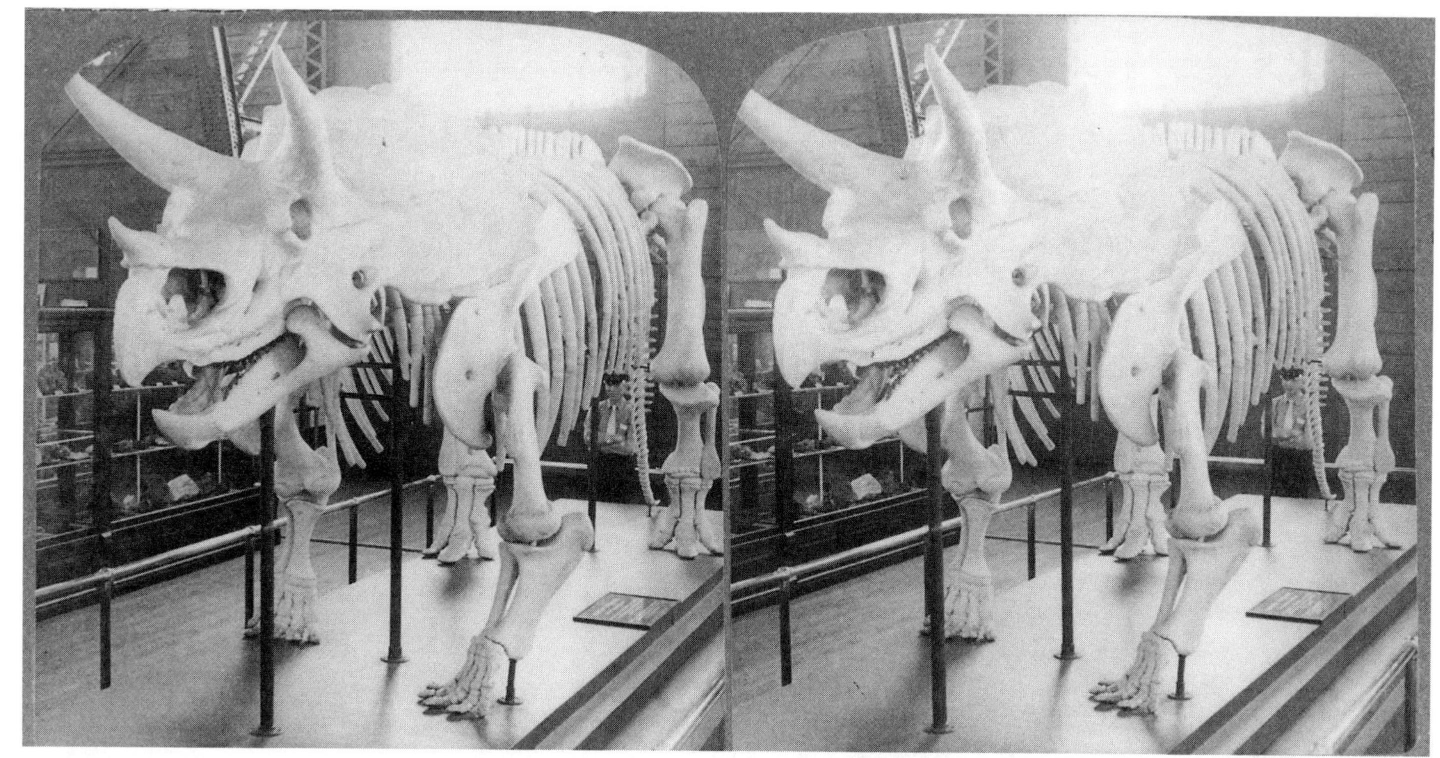

Stereograph of the skeleton of *Triceratops horridus*, plaster cast of USNM 4842, part of the Smithsonian display in the United States Government Building, the 1904 World's Fair, the Louisiana Purchase Exposition, in St. Louis, Missouri.

mature individual, the skull is distinguished as bearing the largest nasal horn core yet measured. More interesting than the length of this horn, however, was the mineralized layer removed from the left supraorbital horn core. As briefly described by Happ and Morrow (2000), this mass grades from 7 to 33 millimeters in thickness and is distinct in composition from both the bony horn core or the surrounding clay matrix.

As Happ and Morrow (2000) observed, the underside of this layer, intermixed with bone matrix, is marked by a mass of traces of blood vessels averaging 12 longitudinal vessels per lateral surface width of 100 millimeters. These traces are mineralized with ferruginous, hollow tubular structures that show fine detail.

According to Happ and Morrow (2000), arteries branch and narrow into arterioles and oppositely oriented structures display venules that branch into veins; arteries and veins range from 2.3 to 5.8 millimeters in diameter, arterioles and venules from 0.16 to 1.3 millimeters; veins and arteries are similar in number; cross sectioning of vessels reveals thick walls surrounding a sickle-shaped opening partially hidden by a globule mass, a matching cross-sectional morphology also found in blood vessels of mammals and birds suffering from advanced atherosclerosis.

Happ and Morrow (2000) also found evidence of soft tissue in the braincase of SUP9713, solid tubular structures passing through the exoccipital exit through the jugular foramen, identified by these authors as the combined exit for nerves IX, X, and XI, and perhaps also the jugular vein. In the typmanic cavity was found a solid ovate structure.

When dinosaur-bird relationships are discussed they almost always involve the saurischian group Theropoda (see "Introduction," sections on bird-dinosaur relationships, *D:TE*, *S1*, and this volume). However, Marsic, Carroll, Heffelfinger, Lyson, Ng and Gartska (2000), all of the University of Alabama in Huntsville, reported in an abstract the discovery of what may be an unexpected link between dinosaurs and avians in bones belonging to one of the least bird-like of all dinosaurian taxa, *Triceratops*. As noted by Holden (2000*b*), this team was aided in their research by NASA and also scientists from the Russian Academy of Sciences.

According to Marsic *et al.*, small fragments of DNA can survive in fossil bone millions of years old and their sequences can reveal phylogenetic information (although most fossils do not offer materials suitable for genetic analysis). A very poorly mineralized partial skeleton of *Triceratops* was uncovered by the authors in the Hell Creek Formation of Slope County, North Dakota. A stretch of what seems to be mitochondrial DNA was then extracted from one compact and one cancellous bone of the centra of two dorsal vertebrae and a rib fragment, belonging to this specimen. Due to the poor mineralization of these bones, Marsic *et al.* were able to get a 130-based pair sequence

Two *Triceratops horridus* individuals are depicted in this painting by Brian Franczak.

from this material. The sequence was then matched against samples from 28 animals, including 13 species of birds. The results were indeed surprising.

According to Marsic *et al.*, the putative DNA sequence fragment of the mitochondrial 12S rRNA gene displays a high (>90%) homology to avian sequences. As related by Holden, the *Triceratops* sequence made a 100 percent match with the DNA of a turkey and at least a 94.5 percent match with other kinds of birds. Taking into consideration the possibility of contamination from turkey sandwiches consumed either in the field or laboratory, the team tested and found no turkey DNA in turtle bones, dirt, or burlap from the same site that yielded the *Triceratops* remains.

Holden quoted William R. Garstka as stating that, "at this point, I remain quite skeptical of our own work. We would expect this kind of result from a theropod, but here we're talking *Triceratops*."

Key references: Andersen, Chapman, Kenny and Larsson (1999); Barrick, Stoskopf, Marcot, Russell and Showers (1998); Brett-Surman, Jabo, Kroehler and Parrish (1999); Chapman, Andersen and Jabo (1990); Deck (1999, 2000); Dodson (1993); Farke (1997); Forster (1996); Gilmore (1905); Happ and Morrow (1996, 1999, 2000); Holden (2000*b*); Jabo, Kroehler, Andersen and Chapman (1999); Lehman (1998); Marsh (1889); Lull (1933); Marsic, Carroll, Heffelfinger, Lyson, Ng and Gartska (2000); Ostrom and Wellnhofer (1986); Penkalski and Dodson (1999).

†TROODON

Saurischia: Theropoda: Neotheropoda: Tetanurae: Avetheropoda: Coelurosauria: Manuraptoriformes: Manuraptora: Deinonychosauria: Troodontidae.

Comments: In 1999, Varricchio, Jackson and Trueman further described and interpreted an unusual and unique trace specimen (MOR 963) pertaining to the 3-meter-long, 50-kilogram carnivorous dinosaur *Troodon formosus*— one of the best preserved dinosaur nests yet found, and containing a clutch of 24 eggs (see Varricchio, Jackson, Borkowski and Horner 1997, also *Troodon* entry, *S1*, for early reports), one clutch containing embryos (see Varrichio and Jackson 2000*b*, abstract for a poster). The trace was found at the upper surface of the Egg Mountain site, in the Upper Cretaceous (Campanian) Two Medicine Formation of Teton County, Montana.

Varricchio *et al.* (1997, 1999) assigned this trace to *T. formosus* based upon the shape and symmetry of the nest and clutch, and referred the eggs to the oospecies *Prismatoolithus levis* based on their arrangement, orientation, outer surface, and microstructure (Zelenitsky and Hills 1996).

As described by Varricchio *et al.* (1999), the nest consists of a shallow, bowl-shaped depression with an inner area measuring approximately one meter square and surrounded by a distinct raised rim. The nest is distinguished by two lithologies — a hard micritic limestone that forms the depression and the rim, and also a softer calcareous mudstone filling the depression and covering the nest and eggs. Asymmetrical and elongate, the eggs stand almost vertically, pointed end down, leaning toward the center of the clutch with upper portions near or in contact with one another. The eggs are tightly packed in the center of the nest, roughly forming an ellipse. Both nest and clutch share a bilateral symmetry about a north-south axis.

Varricchio *et al.* (1999; see also Varrichio and Jackson 2000*a*) interpreted MOR 963 as representing brooding and nesting behavior in *Troodon*. Apparently the adult *Troodon* constructed its nest by digging a shallow bowl-shaped depression in floodplain soils, utilizing loosened earth to raise a surrounding rim. As indicated by their spacing, eggs were deposited within the loosened soil two at a time in a vertical (or nearly so) position at daily or greater intervals (Varrichio and Jackson 2000*b*). If only one *Troodon* female was involved in a single nest, clutch formation would probably have required from about 11 days to three weeks. As incubation via body heat commenced, the adult may have compacted or mounded sediments around the clutch, thereby tilting the eggs toward the center of the clutch.

Varrichio *et al.* (1999) found the size, symmetry, and arrangement of the nest and clutch, and also the low organic carbon of the mudstone, to suggest active brooding behavior in the adult dinosaur (no evidence existing to suggest incubation with vegetative cover). Evidence was not found suggesting that the young remained in the nest for any significant length of time after hatching. Varrichio and Jackson (2000*a*) later commented that the pattern of egg-laying, delayed incubation, and brooding in *Troodon* differs only slightly from the behavior of exant birds (*e.g.*, ratites, tinamous, galliforms, and anseriforms). From MOR 963, Varricchio *et al.* (1997, 1999) deduced that "*Troodon* exhibits a combination of reproductive features primitively shared with crocodilians (some burial of eggs, prohibition of egg rotation), progressively shared with birds (open nest, exposed eggs, incubation by a brooding adult), and possibly unique to troodontids (steeply-inclined eggs with associated developmental requirements)." Thus, *Troodon* and other coelurosaurs possessed both reproductive characters inherited from more primitive ancestors (crocodilians) and those shared with their modern descendants (birds).

Varrichio and Jackson (2000*a*) noted that, with two eggs produced at a time, simultaneous hatching would require arrested embryonic development during the days in which the eggs were laid. Furthermore, the reproductive behavior of this genus "implies that ambient temperatures only maintained embryonic statis and that incubation and adult body temperatures were significantly higher than that of the environment," elevated body temperatures joined with increasing egg size possibly explaining the loss of egg retention in such nonavian coelurosaurs. According to Varrichio and Jackson (2000*b*), the pattern of egg-laying evident in *Troodon* seems to have evolved within Theropoda, possibly in Coelurosauria, the extant avian reproductive behavior of this dinosaur also implying a greater physiological similarity to modern birds.

Fiorillo and Gangloff (2000), in an abstract, reported on 70 theropod teeth collected from six different localities in the Kogosukruk Tongue of the Upper Cretaceous Prince Creek Formation of the North Slope of Alaska. As the authors observed, this assemblage reveals slightly less diversity compared to teeth assemblages from the slightly older Judith River Formation of south-central Montana, Aguja Formation of west Texas, and Hell Creek Formation of eastern Montana. Furthermore, the Alaskan theropod teeth, combined with data of herbivorous dinosaurs from more southerly faunas, show that Cretaceous diversity decreases as latitude increases.

Dominating these assemblages of Alaskan teeth were specimens attributed to *Troodon*. This dominance, the authors suggested, can be attributed to a faunal adaptation of this dinosaur "to low light conditions while over-wintering at a high paleolatitude."

Key reference: Fiorillo and Gangloff (2000); Varrichio and Jackson (2000*a*, 2000*b*); Varricchio, Jackson, Borkowski and Horner (1997); Varricchio, Jackson and Trueman (1999); Zelenitsky and Hills (1996).

†TYRANNOSAURUS (=*Dynamosaurus, Jenghiskhan, Manospondylus, Nanotyrannus, Tarbosaurus*; =?*Aublysodon*, ?*Stygivenator*)

Comments: Tyrannosaurids have been known for about a century, the family Tyrannosauridae having been founded upon the genus *Tyrannosaurus*, its largest known member. Over the years a number of new tyrannosaurid genera and species have been named and described, some of them regarded as "dwarf" taxa, although their validity has often been questioned. Recently, at least some of the questions concerning tyrannosaurid taxonomy, including those relating to *Tyrannosaurus* itself, may have been answered.

Photograph by the author, courtesy American Museum of Natural History.

Skeleton (AMNH 5027) of *Tyrannosaurus rex* mounted in the American Museum of Natural History's "Hall of Saurischian Dinosaurs," this specimen being one of the most famous dinosaur fossils ever collected.

Tyrannosaurid specialist Thomas David Carr has continued his research focusing upon ontogeny in this group of carnivorous dinosaurs. However, Carr's work has also pertained to taxonomy, involving such genera as *Tyrannosaurus* (see *Daspletosaurus*, *Gorgosaurus*, and *Tyrannosaurus* entries, *S1*). One of Carr's (1999*b*) studies was intended "to clarify aspects of tyrannosaurid taxonomy and investigate the supposed phenomenon of dwarfism in the clade."

Carr's computer-assisted phylogenetic character analysis included the genus *Tyrannosaurus*, but focused primarily on the more primitive but smaller tyrannosaurid species, *Gorgosaurus libratus* [=*Albertosaurus libratus* of his usage, Carr agreeing with Russell 1970*b* that *Albertosaurus* and *Gorgosaurus* are congeneric], a comparatively well-represented taxon for which a wide range of growth stages is known. In analyzing the ontogenetically variable morphological characters of *G. libratus*, Carr reconstructed the ontogenetic trajectory of the craniofacial skeleton of this species. Via this reconstruction, the author produced a sequence of progressive growth changes wherein nascent ontogenetic characters were seen as equivalent to primitive phylogenetic characters.

Included in Carr's study were the description of the taxonomic differences between tyrannosaurid taxa and the evaluation of the taxonomic status of the type species *Nanotyrannus lancensis* (see Carr 2000*b* for a detailed history of the discovery, reconstruction, taxonomic history, and various published studies of this specimen, CMNH 7541), the only complete skull of a juvenile tyrannosaurid yet collected, and *Maleevosaurus novojilovi* (both species originally referred to the genus *Gorgosaurus*; see *Nanotyrannus* and *Maleevosaurus* entries, *D:TE*). Carr (1996) suggested that the latter two species, previously described earlier as dwarf yet valid taxa, were really juveniles of the North American species *Tyrannosaurus rex* and the Mongolian *Tyrannosaurus bataar*, respectively. (Carr accepted Carpenter's 1992 referral of the Asian species *Tarbosaurus bataar* back to the genus *Tyrannosaurus*; see entry, *D:TE* and *S1*; however, see also Larson 1999, below). In a later report on CMNH 7541, published in *Dinosaur World*, Carr (2000*b*) pointed out that earlier descriptions (*e.g.*, Gilmore 1946; Russell 1970*b*) of the holotype skull (CMNH 7541) included features (*e.g.*, the supposed lacrimal horn) not present on the

Photograph by the author, courtesy Cleveland Museum of Natural History.

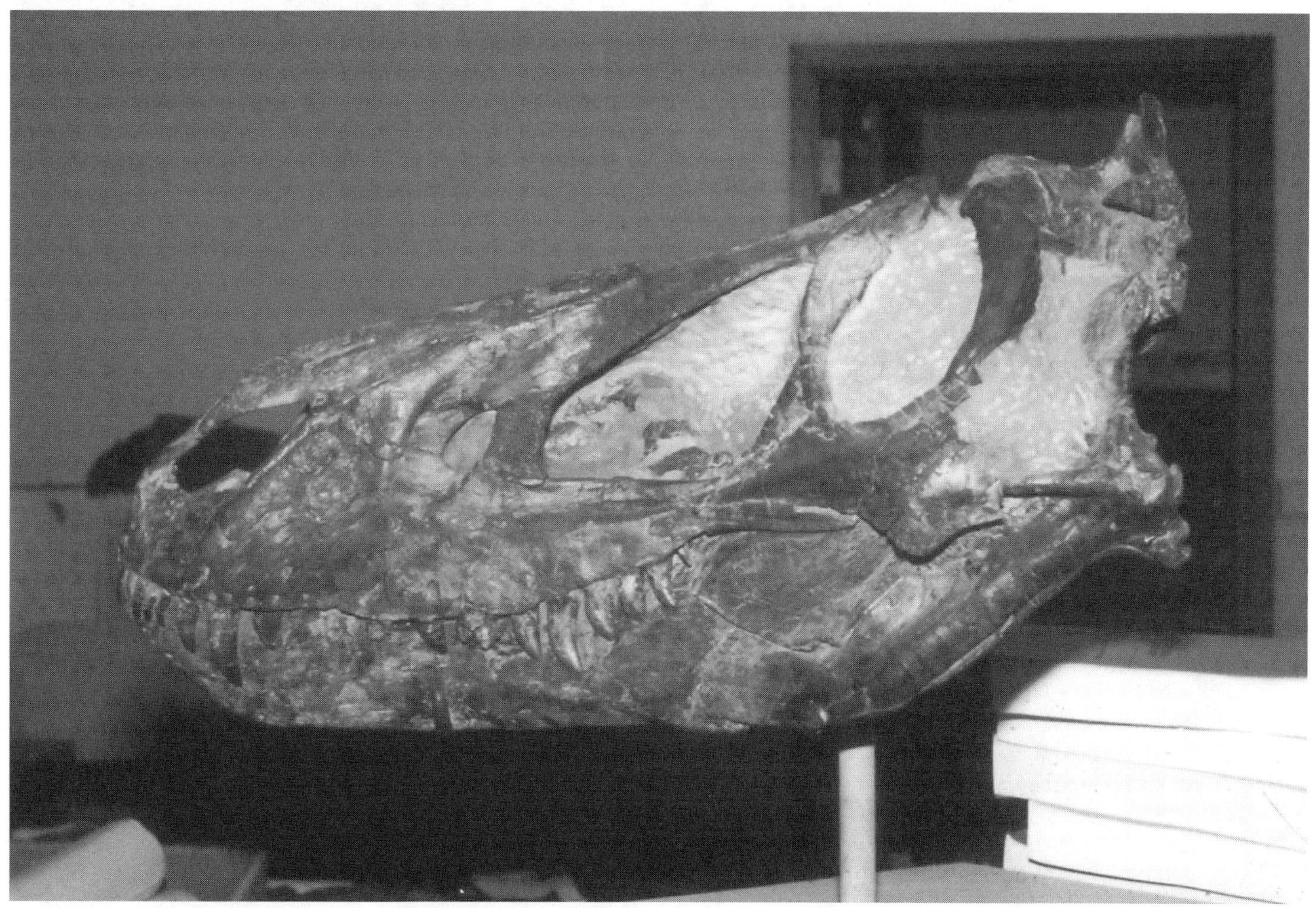

Holotype skull (CM 7541) of *Gorgosaurus lancensis*, a species subsequently named *Nanotyrannus lancensis*, but regarded by tyrannosaurid specialist Thomas D. Carr and other (but not all) paleontologists as a juvenile *Tyrannosaurus rex*.

genuine fossil, but which had actually been molded in the plaster reconstruction of its presumed missing parts.

Carr's (1999*b*) analysis — based on observations and measurements of 46 tyrannosaurid specimens in the collections of North American institutions, plus figures and photographs of other specimens published in the scientific literature, with skulls ranging from 400 millimeters to more than 1,050 millimeters in length — produced an ontogenetic tree, proceeding from early to late ontogeny, that could be "used as a reference to evaluate the relative maturity of individual specimens of other taxa."

Carr (1999*b*) pointed out that young individuals of Ornithodira (a group to which all dinosaurs belong; see *D:TE*) "are characterized by striated cortical [*i.e.*, outer surface] bone that follows the direction of growth" (see Bennett 1993; Sampson 1993). Unfortunately, not all of the available tyrannosaurid skulls — especially those on public display — could be properly examined for the observation of their bone texture. Indeed, many of the finer details of such specimens could not be documented sufficiently, obscured as they were behind their glass display cases or beneath layers of paint or consolidant.

Nevertheless, based upon the observed optimal distribution of changes of discrete morphological characters, Carr (1999*b*) was able to divide the ontogeny of *G. libratus* into three stages of specimens:

The least mature "Stage 1" specimens possess striated cortical bone grain and show nascent ontogenetic characters (see below).

"Stage 2" specimens show development of the homologous features that exemplified the former stage (*e.g.*, large marginal maxillary neurovascular foramina, depressed interfenestral strut, oblique caudal lacrimal suture of jugal, postorbital located dorsal to orbit floor, spheroid occiptal condyle, deep surangular, large, asymmetrical caudal surangular foramen, lacrimal cornual process with one apex, dorsolateral lamina of lacrimal that is deep as antorbital fossa, ventrally oriented and wide oval scar of basisphenoid, deep scar ventrolateral to glenoid of surangular) and also sometimes nascent "Stage 1" features).

"Stage 3" specimens display further development of the typical "Stage 2" features and additional

Courtesy Royal Tyrrell Museum/ Alberta Community Development.

Detail of *Tyrannosaurus rex* skeleton (RTMP 81.6.1)—discovered in 1946 by Charles M. Sternberg in central Alberta, Canada—mounted at the Royal Tyrrell Museum of Palaeontology. This is a "Stage 4" tyrannosaurid skull, as designated by paleontologist Thomas D. Carr in 1999.

Photograph by the author, Natural History Museum of Los Angeles County.

Life-sized bronzed *Tyrannosaurus rex* and *Triceratops horridus* figures based on the "Dueling Dinosaurs" skeletal exhibit inside the Natural History Museum of Los Angeles County (see *S1*), rendered in bronze by Douglas Van Howd of Sierra Sculpture.

development or retention of those of "Stage 1" (*e.g.*, antorbital fenestra with height approaching length, expansive rostrolateral surface of maxilla, convex rostral margin of rostroventral lamina of lacrimal, deep maxillary process of jugal, elongate rostral process of dorsal ramus of lacrimal). In this stage, all small features of "Stage 1" are transformed, while large features of both previous stages may remain unmodified.

An additional ontogenetic stage, dubbed "Stage 4," is represented only by adult specimens of the larger tyrannosaurids *Tyrannosaurus* and *Daspletosaurus*, these exhibiting homologous ontogenetic features developed beyond those of "Stage 3" specimens of *G. libratus*.

Carr's (1999*b*) ontogenetic tree revealed a general congruence between skull length and growth stage in *G. libratus*. A general decrease was observed in the number of maxillary teeth (from 15 to 13 tooth positions, as also seen in *T. rex*), suggesting that this pattern may be plesiomorphic for the Tyrannosauridae. However, some topological resolution was lost in the tree when the number of teeth was used as an ontogenetic character.

As Carr (1999*b*) pointed out, evidence for ontogeny in the skull and lower jaws of *G. libratus* "provides parameters by which similar variation in other tyrannosaurid crania may be inferred," the inference of a pattern of such change in one taxon, consequently, requiring verification in another. Therefore, craniofacial changes from an early to late stage of ontogeny may constitute "an alternative hypothesis to interpretations of tyrannosaurid diversity in which sympatric tyrannosaurid taxa are seen as comprising a giant and dwarf, or more lightly built taxon" (see below).

Craniofacial ontogeny in *T. rex*, Carr (1999*b*) observed, is characterized by an overall shift from a gracile early to a robust late growth morphotype, this process including a number of changes:

Rostral maxillary and dentary teeth become conical in shape, with expansion and deepening of the maxillary and dentary alveolar processes. The tooth row become reoriented rostrodorsally, teeth becoming procumbent (see Bakker, Williams and Currie 1877). The maxilla loses from three to four teeth from the rostral end of the tooth row, the rostral teeth undergoing

Photograph by the author, courtesy Natural History Museum of Los Angeles County.

Tyrannosaurus bataar **skeleton (cast), formerly known as *Tarbosaurus bataar*, for which George Olshevsky has proposed the new generic name *Jenghizkhan*. From the traveling "Great Russian Dinosaur Exhibition."**

the most change (tooth counts, the author cautioned, apparently subject to ontogenetic and individual variation, thereby constituting unreliable criteria for distinguishing taxa).

The skull of *Tyrannosaurus rex*, constructed for forceful biting (see the detailed study of Molnar 1973), undergoes other changes during ontogeny related to biting. The dorsotemporal fossa becomes deeply excavated, reflecting the hypertrophied adductor musculature. To accommodate this change, the entire skull becomes modified, alterations including a more rostral orientation of the orbits. Accommodating the increased biting force, the muzzle and jaws become deep, contacts between bones reinforced by peg-and-socket structures. The facial part of the skull becomes buttressed (*e.g.*, by the strut-like rostral margin of the antorbital fossa, passing to the columnar dorsum of the snout) to deliver and absorb the forces resulting from this more powerful biting capability. Premaxillary teeth become fewer in number during ontogeny, this "indicating less specialized grasping function characteristic of the remainder of the rostral maxillary tooth row."

The craniofacial air sac system (*e.g.*, see Witmer 1987) also affected the bone structure of the skull. The antorbital air sac, which rested within the antorbital fenestra and antorbital fossa, sent diverticula into the ectopterygoid, palatine, lacrimal, jugal, and maxilla, these invading the bones more fully as the animal attained maturity, expanding the bones and sinuses. Thus, combined with changes resulting from enlarged dentition, more mature specimens of *T. rex* have a "swollen-faced" appearance relative to earlier ontogeny specimens. Carr (1999*b*) speculated that the "swollen" bones probably "had greater cross-sectional strength than the strap-like bones of smaller animals, a morphological shift that would be important for taking live prey."

Lastly, pneumatic features of the *T. rex* skull also reflect ontogenetic change. Some "Stage 4" *T. rex* specimens (holotype CM 9380, LACM 23844, and UCMP 118742) possess "maxillary fenestrae that are rostrodorsally deep and extend medial to the rostral margin of the antorbital fossa to a greater degree than in other specimens (*e.g.*, the "Stage 4" specimens AMNH 5027, BHI 2033, RTMP 81.6.1, MOR 555).

Photograph by the author, courtesy Natural History Museum of Los Angeles County.

Fragmentary skull elements (LACM 23845) of a subadult *Tyrannosaurus rex* individual (regarded by George Olshevsky as representing a new genus and species, ***Dinotyrannus megagracilis***); (top) right surangular, (middle) right and left angulars, and (bottom) right dentary.

Also, the interfenestral strut is thin in the late ontogeny specimens, with additional pneumatic foramina present at the apex (CM 9380) or base (LACM 23844) of this strut.

From the above observations, Carr (1999*b*) concluded that the feeding behaviors of *T. rex* individuals of different ages were constrained by the distinct structural patterns of their respective ontogeny, which in older animals appear to be specializations for grasping and holding live prey, or dismembering large carcasses. Conversely, these ontogenetic changes can be related to the increased sized of the skull, greater biting forces, and larger prey, with small and large individuals perhaps differing "in foraging strategy, consumption technique, or prey type."

Based on the results of this study, *Nanotyrannus lancensis* and *Maleevosaurus novojilovi* were again found by Carr (1999*b*) to be early ontogenetic individuals of *T. rex* and *T. bataar*.

According to Carr (1999*b*), the holotype (CNMH 7541) of *N. lancensis*— originally described in a posthumously published article by Gilmore (1946*a*) as a new species of *Gorgosaurus*, the specimen having been collected from the Hell Creek Formation of Montana (see *Nanotyrannus* entry, *D:TE*)—was heavily restored in plaster. Because of this restoration, Gilmore erred in his report on supposed sutural fusion in the specimen (an indication of a relatively mature individual), an error repeated by subsequent authors (*e.g.*, Russell 1970*b*; Bakker *et al.* 1988; Paul 1988*b*). Actually, except for fusion of the intranasal and intraparietal sutures (typical of "Stage 1" in *G. libratus*), Carr (1999*b*) could discern no evidence of sutural fusion in this specimen. (The supposed fusion between prefrontal and parietal, and between parietal and frontal, were criteria used by Bakker *et al.* for referring CMNH 7541 to the new genus *Nanotyrannus*, interpreted by them to be a mature although dwarf tyrannosaurid; nor did Carr in his personal observations find such alleged fusions in any other tyrannosaurid skull.)

As observed by Carr (1999*b*), striated cortical bone grain, the presence of which is a sign of immaturity, can be found in CMNH 7541 on the antorbital fossa of the maxilla and lacrimal, the lateral surface of the vomer, dentary, surangular, angular, palatine,

Photograph by the author, courtesy Natural History Museum of Los Angeles County.

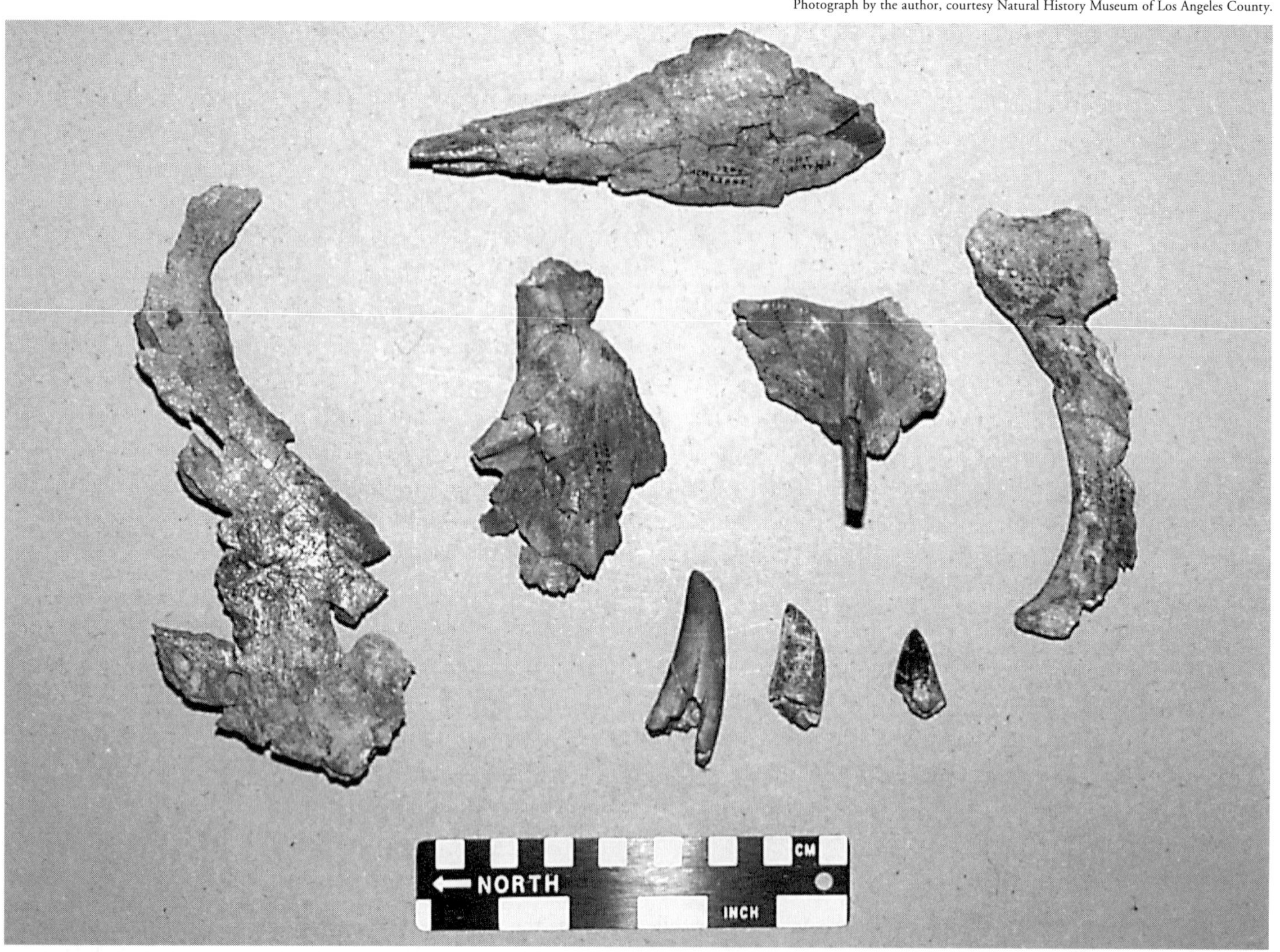

Skull elements (LACM 23845) of a subadult *Tyrannosaurus rex* individual (renamed *Dinotyrannus megagracilis* by George Olshevsky); (top) right lacrimal, (left) fragmentary left maxilla, (middle) left pterygoid, ?supraoccipital crest, ?palatine, and (bottom) teeth.

jugal, the ventral process of the maxilla, quadratojugal process of the squamosal, the squamosal ramus of the postorbital, the rostral surface of the supraoccipital crest of the parietal, the medial surface of the prearticular and splenial, the caudal margin of the quadratojugal, the caudal surface of the quadrate, and the dorsal surface of the frontal and nasal.

Consequently, Carr (1999*b*) identified CMNH 7541 as a "Stage 1" skull specimen, agreeing in morphological structure with juvenile "Stage 1" skulls of *G. libratus*. Agreeing with previous workers (*e.g.*, Rozhdesvensky 1965, who suggested that this skull may be referrable to *T. rex*, based on his observations of *T. bataar* material; see also Carpenter), Carr (1999*b*) noted that CMNH 7541 and fossils referred to *T. rex* share the following 13 characters: Nasal processes of premaxillae tightly appressed throughout their length; jugal with restricted exposure within antorbital fenestra; antorbital fossa reaching nasal suture caudodorsally; jugal pneumatic recess transversely broad; elongate frontal sagittal crest; basal tubers short, strongly divergent; caudal occipital plate rostroventrally oriented; shallow subcondylar recess; rostroventrally-deep basisphenoid plate, rostrocaudally-restricted basisphenoid recess; inflated ectopterygoid; rostral plate of surangular strongly convex; snout transversely narrow, broad temporal region relative to other tyrannosaurids (see Carpenter); and mandible deep relative to other tyrannosaurids.

M. novojilovi was first described by Maleev (1955) as a new species of the genus *Gorgosaurus*, based on a nearly complete skeleton (PIN 552-2) with partial skull (see *D:TE* for photograph) from the Nemegt Formation, Mongolian People's Republic. Rozhdostvensky subsequently identified this specimen as a juvenile of the Asian genus "*Tarbosaurus*" (=*Tyrannosaurus*). Later, Carpenter designated PIN 552-2 the type of a new genus which he named *Maleevosaurus* (see entry, *D:TE*), noting various features (*e.g.*, small maxillary fenestra, premaxillary fenestra not visible in side view, large antorbital fenestra, low lacrimal horn, low postorbital cornual process, shallow maxilla,

Photograph by John Weinstein, courtesy The Field Museum (neg. #GN89671_53c).

Skeleton of the *Tyrannosaurus rex* skeleton (FMNH PR2081) known as "Sue," mounted by Phil Fraley Productions, Inc., prior to its unveiling in May 2000 at The Field Museum, Chicago.

slender jugal) which he interpreted as falling outside the range of variation observed in *T. bataar* and *G. libratus* juveniles.

Based upon Maleev's (1974) published figure, Carr (1999*b*) observed that the skull of PIN 552-2 exhibits features (*e.g.*, delicate nasal with slotted maxilla suture, long antorbital fenestra, round maxillary fenestra located midway between interfenestral strut and rostral margin of antorbital fossa, premaxillary fenestra not recessed, rostral margin of antorbital fossa not overlapped by lateral surface, maxilla with shallow alveolar process; also (from Carpenter) shallow, delicate jugal, low postorbital cornual process, shallow dentary) consistent with the early ontogeny of *G. libratus*. Other features (no lacrimal cornual process, relatively large maxillary fenestra, patch-like postorbital cornual process) are consistent with late ontogeny *T. bataar* specimens. The incompleteness of PIN 552-2 allowed only limited diagnostic information to be derived from Maleev's (1974) illustration. Therefore, "in the absence of apomorphies and on geographic and stratigraphic grounds," Carr (1999*b*) deemed it

Photograph by Allen A. Debus, courtesy The Field Musuem.

Mounted skeleton (FMNH PR2081) of the *Tyrannosaurus rex* called "Sue" making its public debut at its new and final home, The Field Museum, on May 17, 2000. The skull seen here is a cast, too heavy for mounting on the skeleton, the real skull being displayed elsewhere in the museum.

most parsimonious to regard this specimen, as had Rozhdestvensky, as representing a young *T. bataar* individual.

Carr (1999*b*) concluded that craniofacial ontogeny in tyrannosaurids follows a conservative pattern that can be seen across taxa, the recognition of this variation reducing the number of sympatric taxa, and with taxa previously regarded as dwarf tyrannosaurids now identified as immature individuals of known species. In an additional note, the author pointed out that early ontogeny tyrannosaurid skulls tend to resemble in their proportions the adult skulls of small theropods, this phenomenon (possibly attributed to heterochrony), which may have misled some workers in the past, further complicating the correct identification of taxa.

Among the authors who have not accepted *Tyrannosaurus bataar* and "*Nanotyrannus lancensis*" as referrable to the genus *Tyrannosaurus* is George Olshevsky (2000). In issues number 9 and 10 of the Japanese periodical *Kyoryugaki Saizensen* [*Dino-Frontline*], Olshevsky, Ford and Yamamoto (1995*a*, 1995*b*) referred *T. bataar* to a new genus, *Jenghizkhan*. At the same time he erected the new North American genus *Dinotyrannus* to embrace a single subadult specimen (LACM 23845) that Molnar (1980) first tentatively referred to *Albertosaurus* cf. *A. lancensis* (see below). Unfortunately, *Kyoryugaki Saizensen* publications had little distribution outside of Japan, while relatively few workers who did obtain copies of them could read the Japanese text. Later, Olshevsky's (2000), in his own publication *Mesozoic Meanderings* number 3, described these new taxa again (but in English).

Regarding the taxon "*Tarbosaurus*," Olshevsky (2000) addressed two questions:

1. Why do most specimens of the subadult Mongolian species fall into a "subadult" size class (8 to 10 meters long), sometimes referred to as "*Tarbosaurus efremovi*," with but one specimen (PIN 551-1, an incomplete skull and at least two associated cervical and two dorsal vertebrae) known in the supposed "adult" size class (13 to 14 meters long) named "*Tarbosaurus bataar*" (the opposite of the situation prevailing in North America)?; and,

2. How did the common ancestor of both supposed species of *Tyrannosaurus* cross from eastern Asia

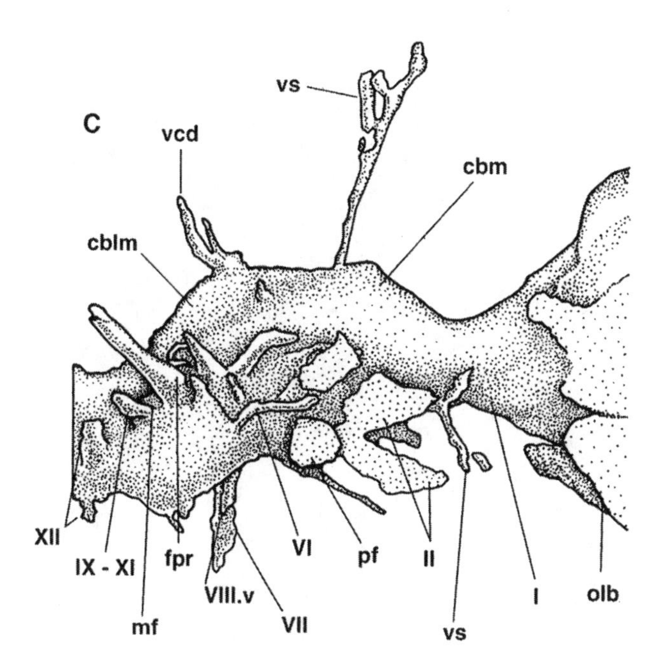

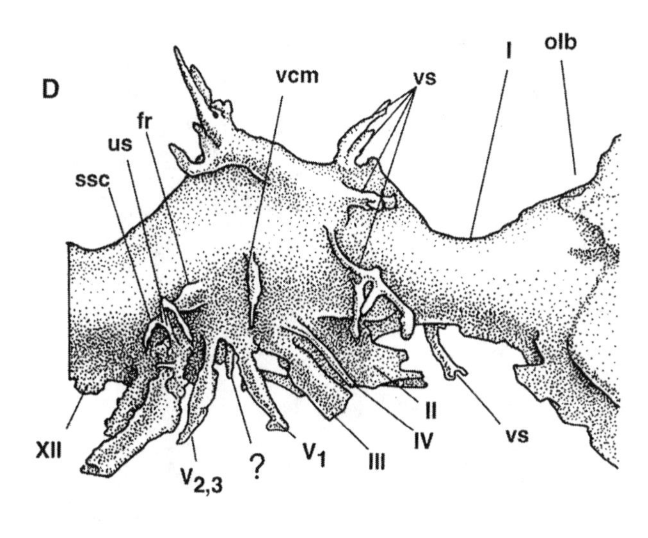

Computer-generated endocast for FMNH PR2081, the *Tyrannosaurus rex* specimen known as "Sue," in A. ventrolateral and B. dorsolateral views. (After Brochu 2000.)

into western North America (or vice versa) at a time when the Bering land bridge and the contentental land masses of the world were inundated with epicontinental seas?

Assuming that no faunal interchange occurred (and noting that much evidence supporting such an interchange originates from cladograms in which Asiatic forms sometimes appear in mostly American clades), Olshevsky proposed that, despite anatomical similarities and shared apomorphies, *Tyrannosaurus* and "*Tarbosaurus*" evolved independently of each other and are, therefore, distinct genera. Olshevsky distinguished *Tyrannosaurus* from these Asian forms "by tooth size and shape, dentary tooth count, skull width across the jugals and the occiput, narrowness of frontals and nasals, inclination angle of the occipital condyle and direction of the orbits, relative shape of the muzzle, coracoid shape, ratio of humerus length to femur, and ratio of height to centrum diameter of the dorsal vertebrae." In Olshevsky, Ford and Yamamoto (1995*a*, 1995*b*) these authors introduced the tribes Tarbosaurini and Tyrannosaurini to embrace the Asian and North American forms, respectively (these terms not adopted).

Olshevsky diagnosed his genus *Jenghizkhan* as follows: Differs from other members of the "Tarbosaurini" in larger size (adults at least 13 meters in length) and massively constructed, rugose vertical skull elements, namely, the postorbital bar and lacrimal-jugal bar; lacrimal and postorbital meeting above orbit, creating almost continuous circumorbital flange from approximately middle of vertical ramus of lacrimal around to the suborbital tuberosity ("postorbital bar") on vertical ramus of postorbital, frontal thereby excluded from orbital rim, although a notch remains at dorsal apex where lacrimal contacts postorbital rugosity; tallest neural spines relative to centrum diameter of any "tarbosaurinid"; anterior dorsal vertebrae with pointed neural spines in lateral aspect; parapophyses very well developed (possibly an ontogenetic feature due to maturity of holotype); distinguished from *Tyrannosaurus* by greater relative length and slenderness of muzzle and dentaries, and significantly greater dentary tooth count (15 to 16 teeth in *Jenghizkhan*, 13 in *Tyrannosaurus*); maxillary and dentary teeth generally more laterally compressed than thick teeth of *Tyrannosaurus*; suborbital tuboroisty not pendant as in *Tyrannosaurus*; occiput seemingly narrower than in *Tyrannosaurus*; occipital condyle apparently directed less posteroventrally than in *Tyrannosaurus*.

LACM 23845, upon which Olshevsky *et al.* founded the type species *Dinotyrannus megagracilis*, consists of a poorly preserved, mostly incomplete skeleton recovered during the summers of 1967 to 1969 by a field team of the Los Angeles County Museum of Natural History (now the Natural History Museum of Los Angeles County) led by J. Reed McDonald, in the Hell Creek Formation, L. D. Engdahl Ranch, Garfield County, Montana. The specimen includes a partial skull (nasals lacking posteriormost portions), pieces of maxillae, the horizontal ramus of a right lacrimal (lacking a "horn"), both nearly complete prefrontals, both nearly complete frontals, the anteromedial portions of a parietal, most of the quadrate processes of both pterygoids, a quadratojugal, central portions of the right dentary and of both angulars and prearticulars, most of the right and fragments of the left surangulars, the left ulna (lacking the radial and olecranal edges of the proximal articular surface), metacarpal II, a manual ungual phalanx,

Photograph by the author.

Cast of the *Tyrannosaurus rex* skeleton nicknamed "Stan" on display at Dinofest (2000–2001), Chicago. The original specimen, at least 60 percent complete, was found by Stan Sacrison in 1987 in the Hell Creek Formation of Harding County, South Dakota, and collected in 1992 by the Black Hills Institute of Geological Research.

the proximal two-thirds of the left femur, proximal half of the left tibia, complete right fibula, the right astragalus (first described by Welles and Long 1974), right metatarsal II and the distal half of III, and 11 phalanges of the right pes (Molnar 1980).

Molnar interpreted LACM 23845 as seemingly "unique among tyrannosaurids" in the subdued olecranon process of the ulna, the angulate proximal articulation of the manual claw, and the absence of a lacrimal horn. As originally described by Molnar, the specimen differs from that of *T. rex* in the generally more slender proportions of the postcranial elements, and in the form of some of the skull elements (most notably, posteriormost portion of nasals in LACM 23845 seemingly less markedly narrowing than in *T. rex*; lacrimal-nasal articulation restricted to lower half of medial face of lacrimal in LACM 23845, occupying three-quarters of height of that face in *T. rex*; upper surface of quadratojugal-quadrate articulation elpitical in outline and smooth in LACM 23845, polygonal and sutural in *T. rex*, anterior portion of that surface lacking concavity in LACM 23845, markedly concave in *T. rex*; dorsal margin of sagittal crest reaching to flattened dorsal margin of supraoccipital crest posteriorly in LACM 23845, dorsal margin of sagittal crest not extending so far dorsally in *T. rex*). Although LACM 23845 does not exhibit tightly interlocking cranial sutures and has more slender proportions than *Tyrannosaurus*, Molnar did not interpret this specimen to be a juvenile of that genus for the following reasons: 1. The supraoccipital alae are well developed; 2. the form of the prefrontal-frontal and quadratojugal-quadrate articulations differ markedly from those of *Tyrannosaurus*; and 3. the form of the proximal portion of the fibula differs from that of *Tyrannosaurus* (more slender in LACM 23845, anterior margin less convex in lateral aspect, dorsal margin more nearly flat).

Molnar found LACM 23845 to resemble most closely *Albertosaurus*, further suggesting that, "largely on stratigraphic grounds," this specimen might represent an ontogenetically older individual of the juvenile "*Albertosaurus*" *lancensis*, also from the Hell Creek of Montana. Thus, Molnar tentatively referred this specimen to *Albertosaurus* cf. *A. lancensis*. Later, Paul (1988*b*), in his book *Predatory Dinosaurs of the World*, proposed a new species for LACM 23845, *Albertosaurus megacracilis*, a taxon which did not become largely adopted. As the type specimen of *Gorgosaurus lancensis* (and the subsequently renamed

Photograph by Don Waller, courtesy Natural History Museum of Los Angeles County.

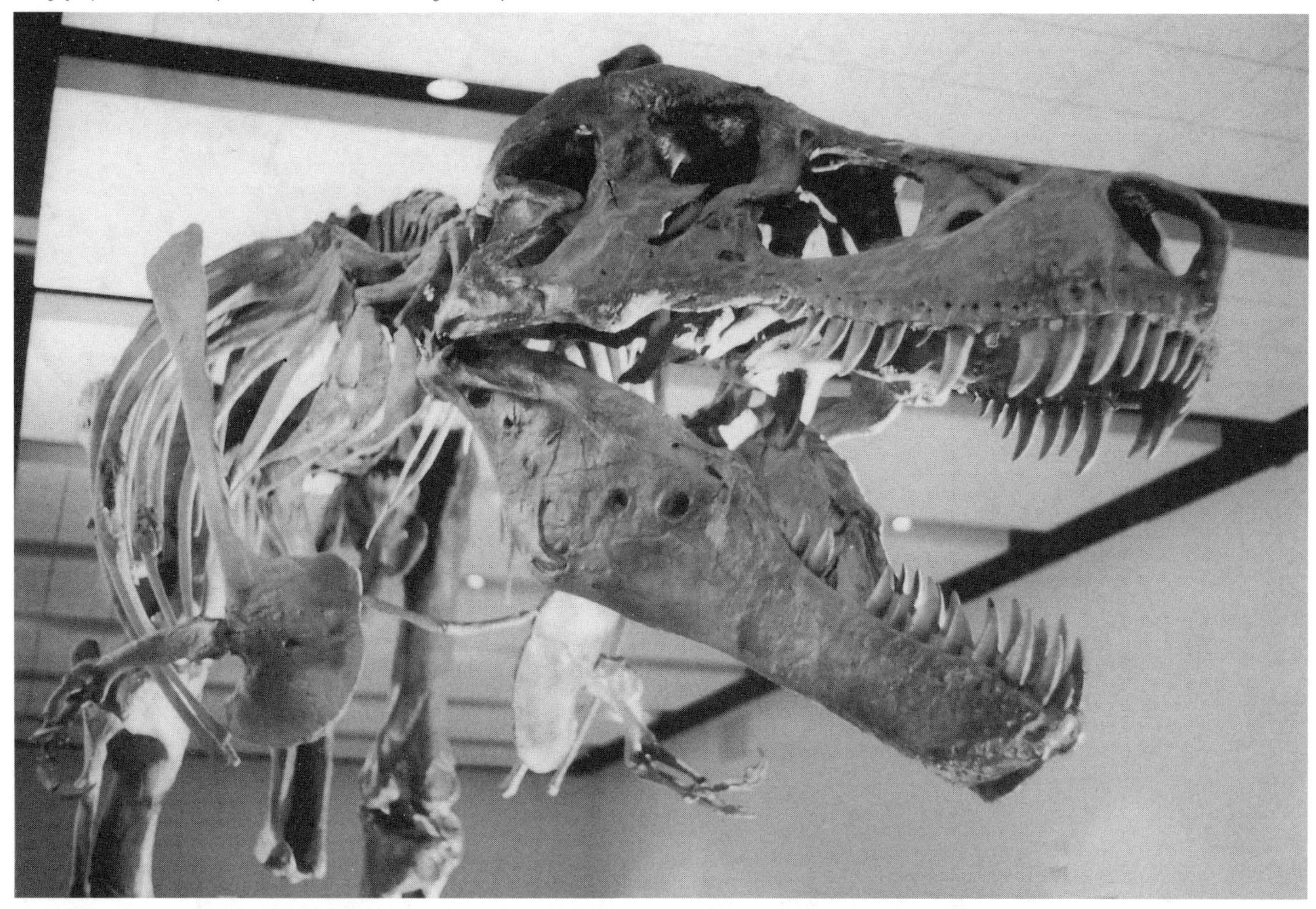

One of several traveling cast reproductions of the *Tyrannosaurus rex* skeleton called "Sue," this one on temporary display at the Natural History Museum of Los Angeles County in December 2000.

Nanotyrannus lancensis) has been referred to *Tyrannosaurus*, LACM 23845, by inference, was also referred to *Tyrannosaurus*.

Olshevsky posited that because "the presence of a lacrimal horn diagnoses the genus *Albertosaurus*, [LACM 23845] cannot belong to that genus." Consequently, Olshevsky referred this specimen to a new taxon retaining Paul's specific name, the new combination becoming *Dinotyrannus megagracilis*.

However, Olshevsky *et al.*'s two new taxa have not been generally adopted. Russell (1970*b*), in his landmark paper on North American tyrannosaurids, had found the absence of a horn on the lacrimal to be common in both *T. rex* and *T. bataar*. More recently, theropod authority Philip J. Currie (personal communication 2000) has not yet seen any characters to suggest that there is more than one tyrannosaurid disguised under the name *Tarbosaurus*. According to Currie, the perceived differences between specimens fit under the amount of expected variation for a single animal, this automatically suggesting "that the most we could squeeze out of *Tarbosaurus* is a new species (not a new genus)." Theropod specialist Thomas R. Holtz, Jr. (personal communication 2000) commented that "*Jenghizkan* is almost certainly just an older individual of the same species as '*Tarbosaurus efremovi*', although "*Tarbosaurus*" *bataar* shares many features with *T. rex* not found in other tyrannosaurids, while clearly having distinct features of its own. Holtz found no feature in LACM 23845 indicating that it is not a young *T. rex*. According to Holtz, even Gregory S. Paul, who named this specimen *Albertosaurus megagracilis*, now considers this specimen to be a young *T. rex*. It was also the opinion of Ralph E. Molnar (personal communication 2000), also a theropod specialist, that *Jenghizkahn bataar* is synonymous with *Tyrannosaurus bataar*, possibly even *T. rex*. "I've seen the type material and my feeling is that were it found in Wyoming, no one would think twice about assigning it to *T. rex*." Regarding LACM 23845, Molnar further wrote, "I am not convinced anyone has read my paper carefully. This is a large tyrannosaurid with what SEEMS to have been a kinetic skull. *T. rex* certainly didn't have cranial kinesis."

Carr and Williamson (2000), in their extensive review of the Tyrannosauridae of New Mexico, stated

Courtesy Allen Detrich and Detrich Fossils, Great Bend, Kansas.

Fossil dealer Allen Detrich with the skull of the *Tyrannosaurus rex* specimen (nicknamed "Mr. Z") at Buffalo, South Dakota.

that, pending personal examination of the specimen, they conservatively regard LACM 23845 as representing "a subadult *T. rex* on the basis of its relative size, the narrow frontal process of the nasals (a character that typifies *Daspletosaurus* and *Tyrannosaurus*), and the absence of a lacrimal cornual process, a character specific to *Tyrannosaurus*." Furthermore, these authors considered "*T. rex* to be the only tyrannosaurid present in late Maastrichtian sediments of western North America."

Therefore, following Carpenter (1992), Carr (1996), and Carr and Williamson, and considering the above mentioned opinions of Currie, Holtz, and Paul, and noting that "*D. megagracilis*" is known only from a single, very incomplete, and apparently subadult specimen, "*Jenghizkan*" and "*Dinotyrannus*" are herein regarded as junior synonyms of *Tyrannosaurus bataar* and *Tyrannosaurus rex*, respectively.

In an abstract, Larson (1999) reported that several newly recovered specimens, and also some previously collected specimens, of the horned dinosaur *Triceratops* provide new information regarding the feeding habits of both adult and juvenile (=*Nanotyrannus lancensis* of Larson's usage) individuals of *Tyrannosaurus rex*. The predator's identity, that author noted, was revealed by the presence of shed teeth, the method of feeding by the morphology of those teeth.

Larson pointed out that only complete bones are found at (adult) *T. rex* feeding sites. This suggests that the dinosaur disassembled a carcass by punching through the skin, muscle, and bone with its bullet-shaped teeth, the strong, offset jaws subsequently sheering through flesh and bone, the kinetic skull lessening the damage to its own teeth and bones.

However, one *Triceratops* specimen comprised only complete bones, although the forelimbs and tail were missing, and was found with over 20 shed teeth belonging to a juvenile *T. rex* (or "*Nanotyrannus*"). This suggested to Larson that the smaller predator was a "pack hunter" that utilized its blade-like teeth for slashing flesh from bone, cutting away and carrying off individual parts of the prey animal.

Because of these different feeding strategies, and also the discovery of (an as yet undescribed) "baby" *T. rex*, Larson concluded, contrary to the findings of Thomas Carr, that "*N.*" *lancensis* are distinct species.

The above findings, however, may offer clues regarding the lengthy and ongoing debates as to whether *Tyrannosaurus* and other tyrannosaurids were active predators or opportunistic scavengers (for various pro and con opinions, as well as authors' references, see discussions in *Tyrannosaurus* entry, *D:TE* and *S1*).

Clearly, *Tyrannosaurus* adults were designed for tearing apart the carcasses of large animals (*e.g.*, hadrosaurs and ceratopsians), while the smaller juveniles of this genus, not unlike adult dromaeosaurs and other similar-sized theropods, seem to have been active predators that possibly hunted in groups. Indeed, the recent discovery of the juvenile *T. rex* specimen known as "Tinker" (see "Notes" below) revealed teeth that were developed as far as in adults, implying that even individuals of this size (35 to 41 feet long) had the required equipment for hunting prey (see Carr 2000*b*).

Among the frequently stated arguments favoring a scavenging feeding strategy have been that carnivores like *Tyrannosaurus* adults were too large and massively built for active predation (*e.g.*, Lambe 1917, discussing the relatively smaller tyrannosaurid *Gorgosaurus*); that the femur of this dinosaur, unlike that of typically fast-running theropods, is about equal in length — not shorter — than the tibia, this indicating a slower-moving animal (Horner and Lessem 1994); and that a *Tyrannosaurus*-sized adult animal pursuing prey at top speeds (estimated up to 45 miles per hour; *e.g.*, Bakker 1986) would risk injury when the running stopped (Farlow, Smith and Robinson 1995; but see Sweeney 2000, below).

Tyrannosaurus is particularly notable because of its great size and mass, adults known to have attained lengths of more than 12 meters (40 feet) and weights of more than 5 metric tonnes (approximately 6 tons). In growing to such size, however, *Tyrannosaurus* underwent an ontogenetic series of changes from hatchling to adult, with the smaller and younger individuals no doubt capable of occupying various environmental niches that might otherwise have been taken over by different kinds of predacious adult theropods. Regarding the proportions of the hindlimb bones, Foster and Chure (1999) recently found that, at least in the very large theropod genus *Allosaurus* (see entry), the tibia and also the entire hind leg are relatively longer in juveniles than in adults, this suggesting that juveniles were better adapted for swift locomotion. Perhaps, then, *Tyrannosaurus* was both an active hunter and a scavenger — a feisty predator during its earlier ontogeny, more a scavenger after the much larger adult basically "grew out" of the more energy-expending life style of its youth. Moreso, perhaps young *Tyrannosaurus* individuals, working collectively, performed the more initiating work of bringing down larger prey, not only for themselves, but also for the benefit of the elders.

Starkov (2000), in an abstract for a poster, noted that Campanian and early Masstrichtian tyrannosaurids did not exceed 10 meters (about 34 feet) in length and 5 tons in weight. During late Maastrichtian times, however, the North American species *Tyrannosaurus rex* was substantially larger, reaching a length of 12 meters (over 40 feet) and weighing more than 6 tons. Starkov postulated that the size increase in these theropods may relate to qualitative changes of the composition of taxa, specifically the reappearance of sauropods which directed the natural selection to the size increase in tyrannosaurids. Starkov speculated that, as a reaction to the presence of sauropods, usual prey forms (*e.g.*, the hadrosaur *Edmontosaurus*, the ceratopsians *Triceratops* and *Torosaurus*) are larger (by about 15 to 20 percent) than Campanian–early Maastrichtian forms. The author further speculated that the presence of *T. bataar* in Asia during Santonian–early Maastrichtian time (earlier than *T. rex* in North America) could have been due to the simultaneous presence of *Nemegtosaurus* and its sauropod relatives.

Sweeney (2000), in an abstract for a poster, addressed the issue of whether or not *T. rex* was built for speed, utilizing CT scans and stereo X-rays to study the structure of the distal phalanges of this species. Sweeney's study revealed "a biomechanical optimization of strength while minimizing mass of the bony elements," detailed as follows: The broad articular surface of the distal phalanges is a spun solid mass comprising two half back to back arches, this creating a curved I-beam shape that wraps around the distal end of the phalange. In turn, this structure produces a maximum surface for joint contact, while at the same time minimizing mass. The epiphysis transitions to and is supported on a circular column of bone that evenly transfers force to the diapophysis. The distal diapophysis is dome-shaped or a spun arch, this shape causing the force to be transferred to the cortex of the phalangeal diapophysis.

According to Sweeney, this arrangement results

in the maximizing of the strength of the bony elements of each phalange and also the minimizing of the mass of the distal epiphysis, the latter resulting in the "dimples" found in the sides of the metatarsals and phalanges. Because bone is denser than other tissues, the distal mass of the limb element is minimized. As noted by Sweeney, minimizing the mass of the limb element reduces inertia distally, that reducing the energy needed to accelerate that element, the latter, in turn, maximizing the acceleration for a given expenditure of energy.

Sweeney concluded that the efficiency of the above arrangement "suggests that acceleration and speed were important forces in the evolution of the distal phalangeal elements of the *Tyrannosaurus rex* foot."

Carr and Williamson reported on a *T. rex* specimen (NMMNH P-7199) collected in May, 1998 by Williamson above the base of the Naashoibito Member of the Kirtland Formation in the San Juan Basin of New Mexico. The badly weathered yet associated specimen includes a partial left dentary, fragments of teeth, and a partial vertebra. Carr and Williamson pointed out that teeth of *T. rex* are identifiable by the size of their denticles, being significantly larger, regarding crown length or density in adults, than in other tyrannosaurids. Therefore, NMMNH P-7199 was referred by Carr and Williamson to *T. rex* based upon "its large apical and midheight denticles, 7.5 denticles per 5 mm and 8.5 denticles per 5 mm, respectively."

At the same time, Carr and Williamson disagreed with the suggestion of Lehman and Carpenter (1990) that the specimens NMMNH P-3698 (identified by Gillette, Wolberg and Hunt 1986 as *T. rex*), a partial skeleton from the upper Maastrichtian Hall Lake Member of the McRae Formation, New Mexico, and RTMMP-41436-1, a maxilla from the late Maastrichtian Javelina Formation of Texas (referred by Lawson 1976 to *T. rex*), represent a new genus of large theropod.

As observed by Carr and Williamson, the dentary, teeth, palatine, prearticular, and hemal arch of NMMNH P-3698 agree in terms of size and morphology to other specimens of *T. rex* (*e.g.*, AMNH 5027 [see *D:TE* for additional photographs] and 5029, BHI 3033, and SDSM 12047 [see *D:TE* for photograph]); also, as in other specimens (*e.g.*, BHI 3033 and BMNH R7994), the palatine is much inflated and the dentary possesses 13 alveoli.

Lawson had referred RTMM-414356-1 to *T. rex* based upon such features as the general proportions of the maxilla, the rostral position of the maxillary fenestra, and the narrow interfenestral strut. In addition to these, Carr and Williamson noted other features that, to them, positively refer this specimen to *T. rex*, including "the position of the maxillary fenestra immediately above the ventral margin of the antorbital fossa and the shallow dermal surface above the antorbital fossa."

In an abstract, Hutchinson (2000) addressed the question "How did a large theropod such as *Tyrannosaurus rex* stand or move?" by mapping onto a cladogram character data from bones and soft tissues of extant Reptilia. Hutchinson's analysis resulted in the least speculative reconstruction of the thigh muscles of this dinosaur, differing largely from earlier reconstructions while supporting some inferences from previous studies. The author noted that some aspects of his reconstruction (*e.g.*, anatomy of the flexor cruris muscle) are equivocal, while many others (*e.g.*, insertion of M. pubo-ischio-femoralis externus 1 plus 2) are unequivocal.

Utilizing this reconstruction in a computer model, Hutchison tested various hypotheses regarding hindlimb function in *Tyrannosaurus*. The author found that "Estimates of the active volume of hip extensor musculature appear sufficient (across a wide range of hindlimb joint angles) to counter the flexor hip joint movement induced by the ground reaction force." The knee joint, however, is extremely sensitive to such angles. Therefore, a more columnar stance would require active knee flexors, while a more crouched stance would need active knee extensors. Stabilizing the knee in the more columnar stance could be achieved by a reasonable estimate of the volume of the potentially active flexor muscle. Such stabilization in a crouched stance during a run would require a very large volume of active knee extensor muscles. Hutchinson concluded that inferring that *Tyrannosaurus* was capable of running in a crouched stance requires much speculation.

The *Tyrannosaurus rex* specimen called "Sue" (FMNH PR2081; see *D:TE* and *S1*, also Fiffer 2000, for details regarding the discovery, collection, and acquisition of this celebrated skeleton; see *S1* for a photograph of the skull prior to its acquisition by The Field Museum)— thus far, the largest, most complete, and best preserved *T. rex* specimen described — has at last been, after two years, fully prepared and mounted for display in The Field Museum's grand Stanley Field Hall, being unveiled to the public on May 17, 2000 (see Debus and Brusatte for details on the unveiling and presence of "Sue" at the Field Museum).

Webster (2000), in an article published in *National Geographic*, related how the 67 million-year old, 41 feet long (over 13 feet high at the hips) skeleton was mounted by Phil Fraley Productions, Inc. in Trenton New Jersey. It was decided to mount "Sue" on a base that mimics shattered rock. According to designer Phil

Fraley, "the specimen itself will be crouched and turning slightly, like it was in the midst of eating when something — maybe the viewer — startled it," this creating a tense relationship between "Sue" and the viewer.

The specimen (sex not yet determined, regardless of its nickname) was approximately 90 percent complete, the missing pieces being cast, sometimes as "mirror-image replicas" of preserved bones (the cast bones painted reddish brown to differentiate them from the deep brown of the true fossil material). As the somewhat distorted skull and jaws, weighing almost 750, was too heavy for the mount, it was exhibited separately in a glass case. A cast of the skull, the distortions slightly corrected, was attached to the mount. The actual fossil bones were mounted in a fashion allowing any one of the elements to be easily removed for study.

Webster noted that one of the more unusual features of the "Sue" skeleton are two caudal vertebrae — "a knobby mass of calcified clumps" (see "Notes," below). As The Field Museum's Chief Preparator William F. Simpson stated to Webster, "They show Sue was a living animal once. It had injuries or diseases or infections that left it damaged. It shows that Sue is not only an example of *T. rex* as a species but was an individual animal too — and one that spent its days with a backache."

As Webster noted, "Sue" — now the mascot of The Field Museum — will be the subject of various spin-off projects, including fiction and nonfiction books, and a seven-movement *Tyrannosaurus Sue: A Cretaceous Concerto*, composed by Bruce Adolphe. Casts of the "Sue" skeleton will go on permanent display at Disney's Animal Kingdom near Orlando, Florida, and in a pair of traveling exhibits sponsored by McDonald's.

In the first technical paper concerning "Sue" published subsequent to the specimen's acquisition by The Field Museum, Christopher A. Brochu (2000*a*), the chief researcher of the specimen, reported on a high-resolution computed tomographic (CT) analysis (see Stokstad 2000*a*) that had recently been performed on the skull, which was scanned at Boeing's Santa Susana Field Laboratory. In the past, endocasts had been produced by Osborn (1912) via hemisection for *T. rex*, from the skull of the smaller specimen AMNH 5029), and also by Maleev (1965) for *T. bataar*. The more technically advanced work by Brochu allowed for the production of a computer-generated endocast that revealed obscured objects — internal skull details until now inaccessible in intact tyrannosaurid skulls — before they were manually exposed. These details, Brochu related, include the ossified medial wall of the maxillary antrum and the internal morphology of the lacrymal, jugal, and ectopterygoid sinuses.

The endocranial cavity, Brochu found, is 28.2 centimeters in length. The olfactory tract, including bulbs, extends anteriorly for about 20 centimeters; the cerebral portion measures about 9 centimeters wide and 7 centimeters deep.

Brochu noted that this superior digital endocast of *T. rex* preserves the pathway of nerves throughout the braincase and the internal details of the otic capsule. Traced laterally were the divergent channels for the ophthalmic and maxillo-mandibular branches of the trigeminal nerve; also, the sixth cranial nerve, which had not until now been observed in other tyrannosaurid endocasts. Although smaller than the cerebellum, the forebrain is enlarged relative to the forebrains of more basal theropods.

The presence of a large olfactory nerve was confirmed by the endocast, and olfactory bulbs, greatly enlarged (each bulb about 1.5 times as wide as the cerebral portion of the endocast) compared to those of other nonavian theropods, were revealed. These suggested that an emphasized sense of smell was an important factor in the life of this genus. The author cautioned, however, that ecological conclusions derived from this observation would be premature, pointing out that while some variation in olfactory bulb size in birds may be associated with finding food, other behaviors (*e.g.*, mate location) have also been inferred (*e.g.*, Cobb 1960; Bang and Cobb 1968). Brochu further stated that, pending the availability of detailed information about the comparative size of these bulbs in other gigantic theropods, such as *Giganotosaurus* and *Carcharodontosaurus*, "we cannot rule out an allometric explanation for the large size of the olfactory bulbs in *T. rex*."

In a subsequent abstract, Brochu (2000*b*) briefly reported on the postcranial axial morphology of FMNH PR2081. Brochu (2000*b*) noted that this specimen is inferred to represent a mature individual based upon the indistinct neurocentral sutures throughout the vertebral column; however, the cervical ribs are not fused to their respective vertebrae. Elements preserved in the specimen previously unknown in tyrannosaurids and other derived theropods include a proatlas arch, a rib on the most posterior trunk vertebra, and a chevron between the first two caudals (the apparent absence of this element leading to previous suggestions that FMNH PR2081) is female; see "Notes" below).

Brochu (2000*b*) noted that the axial skeleton of this specimen is extensively pneumaticized, most cervical ribs and the centra and neural arches of all vertebrae, posteriorly through the fourth sacral, having pneumatopores, the tail, however, probably being apneumatic. Pneumatic openings in the exoccipitals possibly communicated with counterparts in the atlas-axis complex. The preserved gastral basket includes four ossifications per mediolateral row. Medial elements in the anteriormost segments are extensively fused, which would have limited the mobility of the gastral complex. Various healed fractures in the trunk ribs and fused caudal vertebrae do not appear to have resulted from fracture. Brochu (2000*b*) further noted that exostotic bone in the fused caudal vertebrae grew around muscular bands, preserving a "natural mold" of the musculature of the tail (see "Notes" below).

Brochu will later publish a monograph on *T. rex* in which he fully describes the "Sue" specimen.

Notes: In their article on "Sue," Debus and Brusatte related as yet unpublished ideas spoken by Brochu during his lecture given at The Field Museum on May 20, 2000. Brochu stated that, among reptiles, the only groups in which the females are larger than the males are turtles and snakes, the opposite condition being true from most reptiles and birds. Based on his examination of modern birds in the museum's collection, Brochu found that, among carnivorous birds, sexual dimorphism decreases, particularly in ground-hunting predators. This observation casts some doubt on the conclusion reached earlier by Larson (1994) that male tyrannosaurids were smaller than the females (see *D:TE*). At this point in time, Brochu was unable to determine positively the gender of the "Sue" specimen.

In his lecture, Brochu also contested claims made by Larson (1997) in both the literature and the media (*e.g.*, an Associated Press article), namely that 1. "Sue" had suffered from a broken left fibula, which subsequently healed, this suggesting that the animal managed to survive only through the intercession of "her" mate, and 2. perforation in "Sue's" lower jaw are bite marks made by another *Tyrannosaurus rex* which inevitable took "her" life (see *S1*). First, examination by Brochu discovered that there was no offsetting of the left fibula, which would have occurred had the leg been broken and then healed. Furthermore, the left fibula differs from the right structurally. Additionally, the dinosaur's left leg was never broken. Second, Brochu and his Field Museum colleagues now agree that the lower jaw's alleged "bite marks"—which do not match up with the teeth of any known animal—are, in reality, the result of an infection. Furthermore, these marks exhibit healing and regrowth, indicating that, even if they were bite marks, they did not cause "Sue's" death.

Observing that the mysterious "clumps" on the caudal vertebrae (see above) were bilaterally symmetrical, Brochu interpreted this mass as a mold of the dinosaur's tail muscles. Brochu was paraphrased as stating that, during an infection, spongy bone grew around the tail vertebrae in this *Tyrannosaurus*. In running out of room, they continued to grow around the tail muscles, thereby creating a cast of these muscles. (Because of its scientific value, this cast was not included on "Sue's" mounted skeleton, but was instead put on display in The Field Museum's "Life Over Time" exhibit.

New *Tyrannosaurus* specimens continue to be discovered and collected by amateurs.

In 1999, newspaper accounts (*e.g.*, that appearing in the *Chicago Sun-Times*) told the story of fossil hunter and dealer Allen Detrich's (of Detrich Fossils, "Specializing in Rare and Spectacular Cretaceous Fossils," in Great Bend, Indiana) attempts to sell a fine specimen of that most well known of all giant carnivorous dinosaurs, *Tyrannosaurus rex*. The partially articulated specimen (owned by Detrich and Fred J. Nuss of Fred J. Nuss Fossils), nicknamed "Mr. Z" by Detrich and advertised in the April 23 *Wall Street Journal* as the "Largest Male & Best Skull" yet collected, was recently found on a farm in the Hell Creek Formation (Upper Maastrichtian) of Buffalo, in Harding County, South Dakota.

Paleontologist Dale A. Russell, in preliminary notes (including a description) on the specimen supplied to Detrich, observed that the skull is unusually well preserved, with only minor distortion, originally buried by sediment with the palate facing upwards. Partial disarticulation of some of the elements suggests that the soft tissues had not yet entirely decomposed when the skeleton was first dismembered by flowing water. Interestingly, partly healed punctures in the upper rim of the left lacrimal and the lateral surface of the left surangular indicate an attack by another *Tyrannosaurus*. CT scan analysis of a tibial fragment belonging to the specimen revealed additional pathology, a hairline fracture into which infection had set in and over which bone grew. In a letter to Detrich, Russell commented, "Personally (and I'm not the doctor), I would visualize [a] scabby protruding bump beneath the skin from which pus was draining, and (like a sore thumb) protrusion was repeatedly being bumped."

By Detrich's estimates, the skeleton is 70 percent complete (although the Black Hills Institute of Geological Research, Inc. has reduced that figure by 20 percent). The skull measures more than 140 centimeters (54–56 inches) in length. The femur is more than 1360 meters (over 4.5 feet) long. Length of the entire skeleton was estimated to be 11.1 to 12.6 meters (approximately 37–41 feet).

Influenced by the recent sale of the female *T. rex* called "Sue" to The Field Museum for $8.3 million, Detrich has asked $20 million for "Mr. Z." As reported

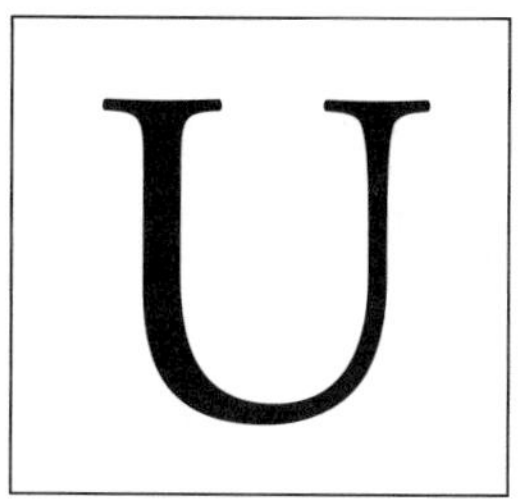

on January 18, 2000 by the Reuters news service, the specimen was subsequently placed on auction via the Internet.

Although fossil tracks attributed to *T. rex* have been previously reported (see *D:TE*), two additional suites of tracks were briefly reported on by Caneer (1999). The ichnites, found after six years of searching in the Upper Cretaceous Lower Coal Zone of the Raton Basin in southern Colorado and northern New Mexico, are consistent with *T. rex*. One set of tracks, discovered in the northern part of the northern part of the basin in Colorado, was interpreted by Caneer as that of a *T. rex* walking in a normal fashion. The New Mexico track, however, is most unusual, preserving, in addition to the hind foot, impressions of the forearms and hands. Caneer interpreted these tracks as recording a preserved moment in time during which a *T. rex* "was rising from a prone position."

On January 1, 1999, the Associated Press reported on an almost complete (about 70 to 90 percent) skeleton of a juvenile *T. rex* collected in summer, 1998, north of Belle Fourche, South Dakota. The site also yielded well-preserved fossil crocodile teeth, clams, snails, nuts, leaves, and stems, evidence suggesting an ox-bow pond environment. Associated acid-etched and flattened hadrosaur bones may have been swallowed and digested. The discovery was made by a team led by private fossil collector Mike Farrell of Houston Texas.

This *T. rex* skeleton was identified as a juvenile because of the unfused vertebrae. Nicknamed "Tinker," the fossil, still mostly encased in rock, is currently undergoing preparation at a laboratory owned by amateur fossil collector Ron Frithuif west of San Antonio, Texas. This specimen, the first recovered juvenile *T. rex* this complete, should offer important evidence regarding both ontogeny and taxonomy in the genus *Tyrannosaurus*.

Paleontologist Robert T. Bakker was reported stating that the juvenile was "quite gangly, particularly long in the shin and ankle," this being "a pleasant confirmation of earlier work." As "the jaws are 100 percent adult" possessing "massive bone-crushing teeth," Bakker suggested "that a juvenile *T. rex* ate an adult diet, even though it doesn't appear strong enough to wrangle large prey to the ground."

Paleontologist John R. Horner, in an interview with Tony Campagna published in *Prehistoric Times* number 45 (2000/2001), stated that five additional *T. rex* skeletons had been found in 1999 in the Hell Creek Formation. One of these specimens is reportedly about 15 percent larger than "Sue." As stated by Horner, *T. rex*, once a quite rare dinosaur in terms of the number of recovered specimens, now "appears to be much more common than previously thought ... more commonly found than edmontosaurs (I'm working on an explanation)." To date of this interview, one of these specimens has been collected, the remaining four to be excavated in 2001.

Erratum: The pedal element (CMN [formerly NMC] 9950), reported by Sternberg (1946) and referred by Langston (1965) to *Tyrannosaurus*, was collected from the Scollard Formation, not the Horseshoe Canyon Formation (W. Langston, Jr., personal communication to T. Donovan).

Key references: Bakker, Williams and Currie (1988); Barry and Cobb (1968); Bennett (1963); Brochu (2000*a*, 2000*b*); Caneer (1999); Carpenter (1992); Carr (1996, 1999*b*); Carr and Williamson (2000); Cobb (1960); Debus and Brusatte (2000); Fiffer (2000); Gillette, Wolberg and Hunt (1986); Gilmore (1946*a*); Hutchinson (2000); Lambe (1917); Langston (1965); Larson (1994, 1997, 1999); Lawson (1976); Lehman and Carpenter (1990); Maleev (1955, 1965, 1974); Molnar (1973, 1980); Olshevsky (2000); Olshevsky, Ford and Yamamoto (1995*a*, 1995*b*); Osborn (1912); Paul (1988*b*); Rozhdestvensky (1965); Russell (1970*b*); Sampson (1993); Sternberg (1946); Stokstad (2000*b*); Sweeney (2000); Webster (2000); Welles and Long (1974); Witmer (1987).

†**ULTRASAURUS** Kim 1983 [*nomen dubium*]

Saurischia: Sauropodomorpha: Sauropoda *incertae sedis*.

Name derivation: Latin *ultra* = "beyond" + Greek *sauros* = "lizard."

Type species: *U. tabriensis* Kim 1983 [*nomen dubium*].

Other species: [None.]

Occurrence: Gugyedong Formation, Gyeongsangbuk-do, South Korea.

Age: Middle Cretaceous (Aptian).

Known material/holotype: DGBU-1973, proximal portion of left humerus.

Diagnosis of genus (as for type species): [None published.]

Comments: Dinosaur remains from Korea of any kind are rare and of relatively recent discovery. In 1981, Haang Mook Kim reported the first dinosaur specimen — an incomplete limb bone (DGBU-1973), housed in the Department of Geology, Kyungpook National University, Taegu, Korea — ever found in that country. The specimen had been recovered in 1977 by K. H. Chang from a cliff in the Gugyedong Formation (Aptian; see Chang, Seung and Park 1983), Gyeongsang Supergroup, southwest of Tabri Station near Bongham Pass, Geumseong-myeon, Euiseong-gun, Gyeongshangbuk-do, South Korea (Lee, Yang and Park 2000).

Kim (1981) originally misidentified this specimen as the proximal end of a right ulna which he referred,

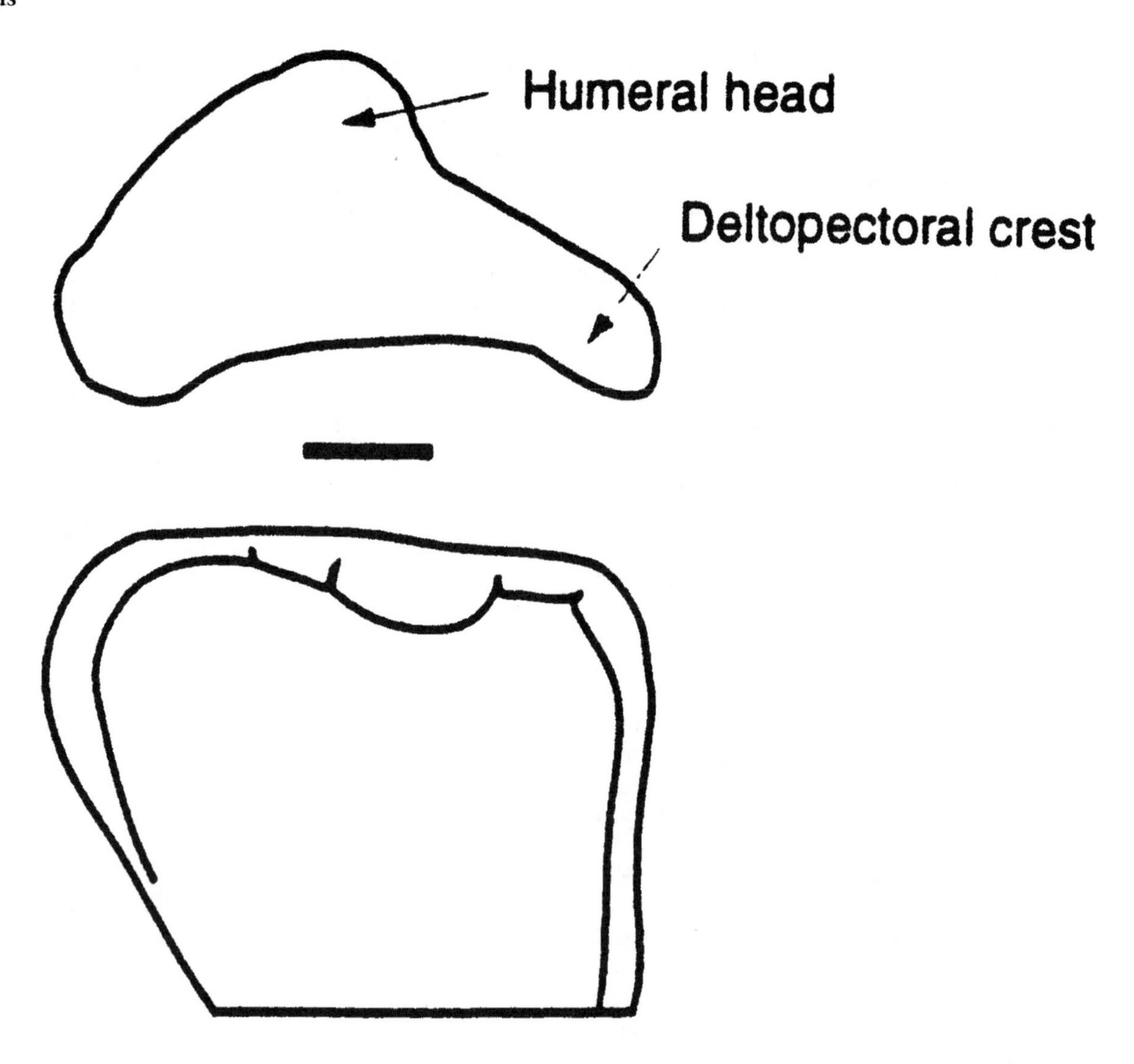

***Ultrasaurus tabriensis*, DGBU-1973, holotype proximal part of left humerus. Scale = 10 cm. (After Lee, Yang and Park 1997.)**

because of its great size, to the Sauropoda and also to the sauropod family Brachiosauridae. Kim believed that the specimen indicated a sauropod that was considerably larger than *Supersaurus* (see Jensen 1985), a North American genus that, in 1981, had not yet been formally named or described. Subsequently, Chang *et al.* reinterpreted this specimen as part of a sauropod femur or tibia.

In a 1983 paper, Kim — seemingly unaware that James A. Jensen had been informally using the name "*Ultrasaurus*" in various publications for a North American sauropod and would officially name and describe that dinosaur in 1985 (see *Ultrasauros* entry, *D:TE*; *Supersaurus* entry, *S1*— formally referred the specimen to its own new genus and species, which he named *Ultrasaurus tabriensis*. This proved to be somewhat problematic as that author issued two almost identical variants of the publication — the first, apparently an offprint, not including a specific name, the second differing from the other by naming *U. tabriensis*. Although Kim's *Ultrasaurus* clearly had priority over Jensen's genus, it was regarded by the present writer (see *D:TE*) as a *nomen nudum*.

Lee *et al.*, however, correctly identified DGBU-1973 as a partial left humerus and designated *U. tabriensis* as a *nomen dubium*.

Although the specimen presented no diagnostic features, Lee *et al.* described it as follows: Shaft constricted with expanded proximal end; transverse distance of proximal end 435 millimeters in length; cranial face of humerus broadly concave; proximal end slightly rounded mediolaterally, its surface quite rugose; head prominently expanded caudally, occupying medial shaft of proximal end; maximum craniocaudal distance of head located midway between medial and lateral borders; medial corner of proximal end making approximately 60-degree angle contrasted with lateral corner, forming approximately 100-degree angle; prominently developed deltopectoral crest, extending along lateral edge from proximal end to midshaft.

Although it cannot be properly defined, *Ultra-*

V

saurus is significant in establishing the occurrence of sauropod dinosaurs in Korea during the Cretaceous period.

Key references: Chang, Seung and Park (1983); Jensen (1985); Kim (1981, 1983); Lee, Yang and Park (1997).

VARIRAPTOR Le Loeuff and Buffetaut 1998 [*nomen dubium*]

Saurischia: Theropoda: Neotheropoda: Tetanurae: Avetheropoda: Coelurosauria: Manuraptoriformes: Manuraptora: Deinonychosauria: Eumanuraptora: Dromaeosauridae.

Name derivation: French *Var* [name of river and administrative department] + Latin *raptor* = "thief."

Type species: *V. mechinorum* Le Loeuff and Buffetaut 1998 [*nomen dubium*].

Other species: [None.]

Occurrence: Grès à Reptiles Formation, La Bastide Neuve, France.

Age: Late Cretaceous (late Campanian–early Masstrichtian).

Known material: Various postcranial remains, mostly vertebrae and limb elements.

Holotype: MDE-D168 and MDE-D169, articulated posterior dorsal vertebra and sacrum, respectively.

Diagnosis of genus (as for type species): Cervicodorsal vertebrae with prominent epipophysis and very well-developed hypapophysis; cervico-dorsals bearing two pleurocoels; cervico-dorsals to last dorsal bearing hyposphen-hypantrum articulation; centra shortening from anterior to posterior dorsals; sacrum consisting of five coossified sacral vertebrae; sacrocaudal vertebra having trapezoidal centrum; transverse process of sacrocaudal aliform; humerus with well-developed deltopectoral crest and internal tubercle, also bearing strongly developed medial tubercle (Le Loeuff and Buffetaut 1998).

Comments: The genus *Variraptor* was founded upon an articulated posterior dorsal vertebra (MDE-D168) and sacrum (MDE-169), collected by Patrick and Annie Méchin from the Upper Cretaceous (late Campanian–early Maastrichtian; see Buffetaut and Le Loeuff 1991) Grès à Reptiles Formation, La Bastide Neuve (Fox-Amphoux, Var) in southern France (Le Loeuff and Buffetaut 1998).

Material referred by Le Loeuff and Buffetaut to the type species, *V. mechinorum*, includes the following: A right humerus (MDE-D158), a right femur (Méchin Collection; cast MDE-D49), a cervico-dorsal vertebra (Méchin Collection; cast MDE-D01), and a sacrocaudal vertebra (Méchin Collection; cast MDE-D49). The latter two specimens, found at another locality, Roques-Hautes (Bouches-du-Rhône Department, Grès à Reptiles Formation, Campanian–Maastricthian), were previously referred by Le Loueff, Buffetaut, Mechin and Mechin-Salessy (1992) to an undetermined dromaeosaurid (see *Elopteryx* entry). The authors noted that the fifth sacral vertebra of MDE-D169 has exactly the same shape as the sacrocaudal from Roques-Hautes (see Le Loeuff *et al.*), and both specimens most likely belong to the same species. Consequently, Le Loeuff and Buffetaut used the Roques-Hautes specimens to supplement their diagnosis of *V. mechinorum*

Le Loeuff and Buffetaut identified the theropod remains from the Grè à Reptiles Formation as belonging to the Dromaeosauridae based on the following: The prominent epipophysis and well-developed

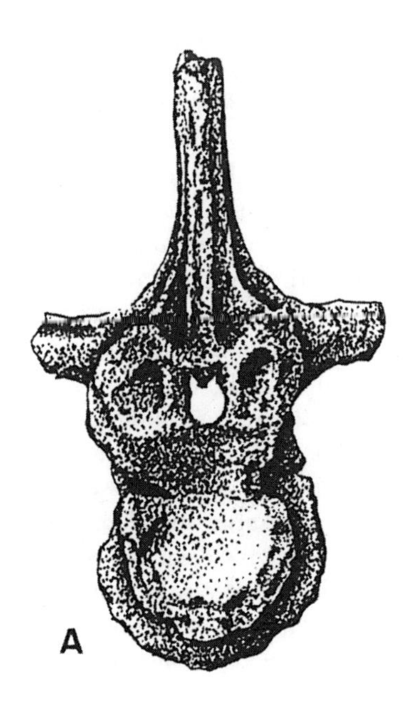

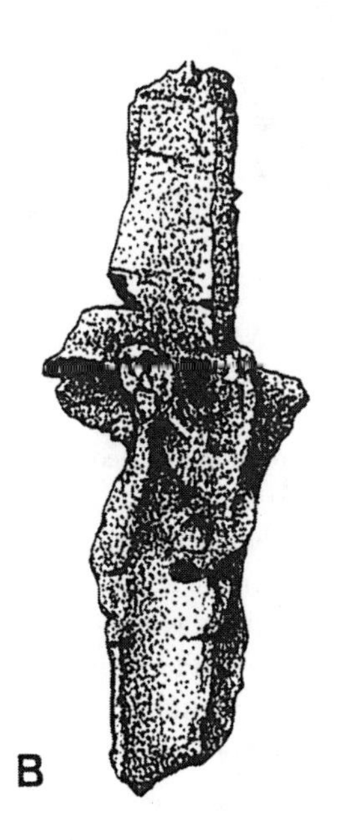

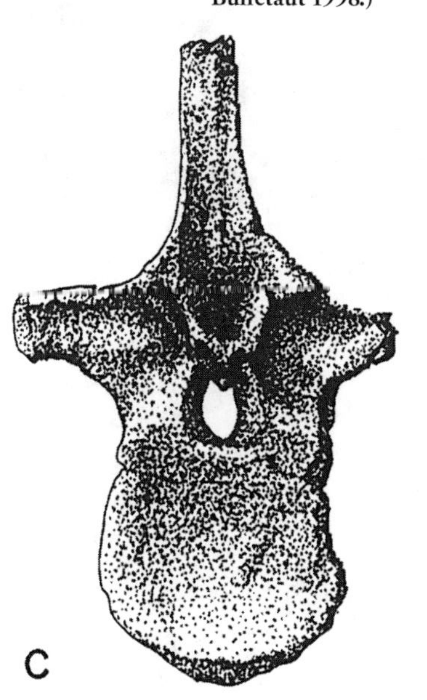

***Variraptor mechinorum*, MDE-D168, holotype last dorsal vertebra in A. cranial, B. lateral, and C. caudal views. Scale = 20 mm. (After Le Loeuff and Buffetaut 1998.)**

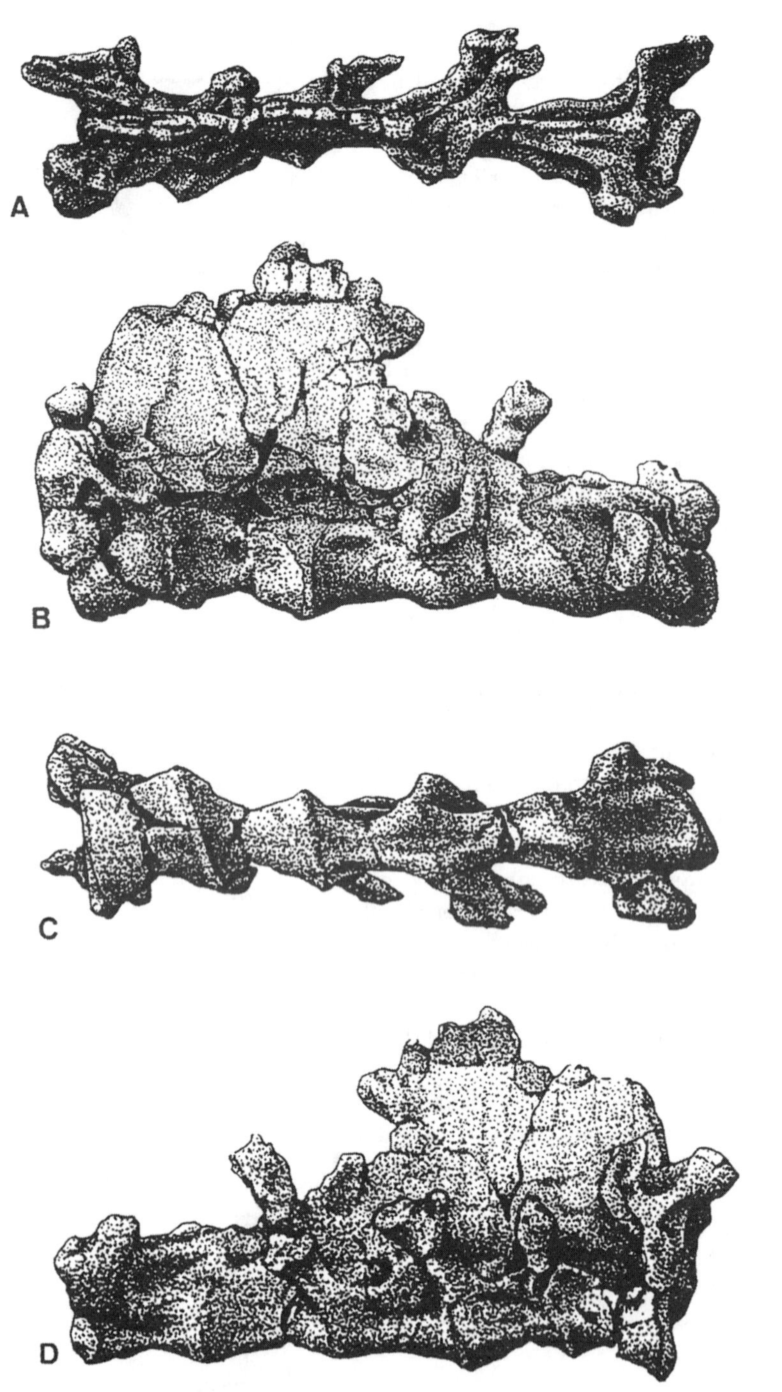

***Variraptor mechinorum*, MDE-D169, holotype sacrum in A. dorsal, B. left lateral, C. ventral, and D. medial views. Scale = 20 mm. (After Le Loeuff and Buffetaut 1998.)**

hypapophysis on the cervico-dorsals are synapomorphies of the Manuraptora (Gauthier 1986). Within Manuraptora, only the Dromaeosauridae seem to have retained five sacral vertebrae (although this is currently not certain; *e.g.*, see Ostrom 1976, 1990). Other dromaeosaurid synapomorphies seen in the French material include: Sacrocaudals having same trapezoidal centrum as anterior caudals of *Deinonychus antirrhopus* (Ostrom 1969*b*); last dorsal from Bastide Neuve, with its hyposphene-hypantrum articulation, reminiscent of that of *Deinonychus*, although medial tubercle is much stronger.

The authors attempted to compare *V. mechinorum* with various other described dromaeosaurids, based on the presently available published data arranged chronologically from oldest to youngest, these taxa including the following: *D. antirrhopus*, Cloverly Formation (Aptian–Albian), Wyoming and Montana; *Saurornitholestes langstoni*, Judith River Formation (Campanian), Alberta, Canada; *Velociraptor mongoliensis*, Djadochta Formation (Campanian), Mongolia; *Dromaeosaurus albertensis*, Judith River Formation, Alberta; *Hulsanpes perlei*, Barun Goyut Formation, Mongolia (Campanian); *Adasaurus mongoliensis*, Nemegt Formation, Mongolia (Campanian or Maastrichtian) (see respective genus entries, *D:TE* and *S1*); and also *Elopteryx nopcsai* (known only from a femur and metatarsals from late Maastrichtian Romania, formerly regarded as avian; see *Elopteryx* entry).

Of the above taxa, only *D. antirrhopus* is known from enough adequately described corresponding material for comparison. As Le Loeuff and Buffetaut observed, *D antirrhopus* significantly resembles *V. mechinorum* in the following: Epipophysis and hypophysis of cervicodorsal; five coossified sacrals; trapezoidal sacrocaudal centrum; well-developed deltopectoral crest and internal tubercle of humerus. The main differences between these taxa, saliant enough to warrant the authors' erection of a new genus and species, are the two pleurocoels of the cervico-dorsals and very strongly-developed medial tubercle of the humerus of *V. mechinorum.*

From a paleogeographical perspective, Le Loeuff and Buffetaut found the occurrence of the Dromaeosauridae in Europe at the end of the Cretaceous to be not unexpected. Their presence was considered to indicate "vestigial taxa from an Early Cretaceous Euramerican paleobioprovince, which broke up during the Aptian following transgression by epicontinental seas." The authors concluded that *V. mechinorum* seems to represent an isolated European lineage of dromaeosaurids that evolved in Euramerica from an Early Cretaceous ancestral stock, the latter dispersing into Asia by way of a land connection across the Bering region that seems to have existed between North America and Asia during the late Early Cretaceous (Russell 1993), a time when both Europe and North America had possibly become distinct paleobioprovinces (see Le Loeuff 1991; 1998).

Allain and Taquet (2000), in describing another new dromaeosaurid from France, *Pyroraptor* (see entry), recognized no distinguishing features on the holotype of *V. mechinorum*, and also pointed out that most of Le Loeuff and Buffetaut's diagnosis of this taxon was based on a cervico-dorsal vertebra found in a different locality, Roques-Hautes. Therefore, Allain and Taquet regarded this species as a *nomen dubium.*

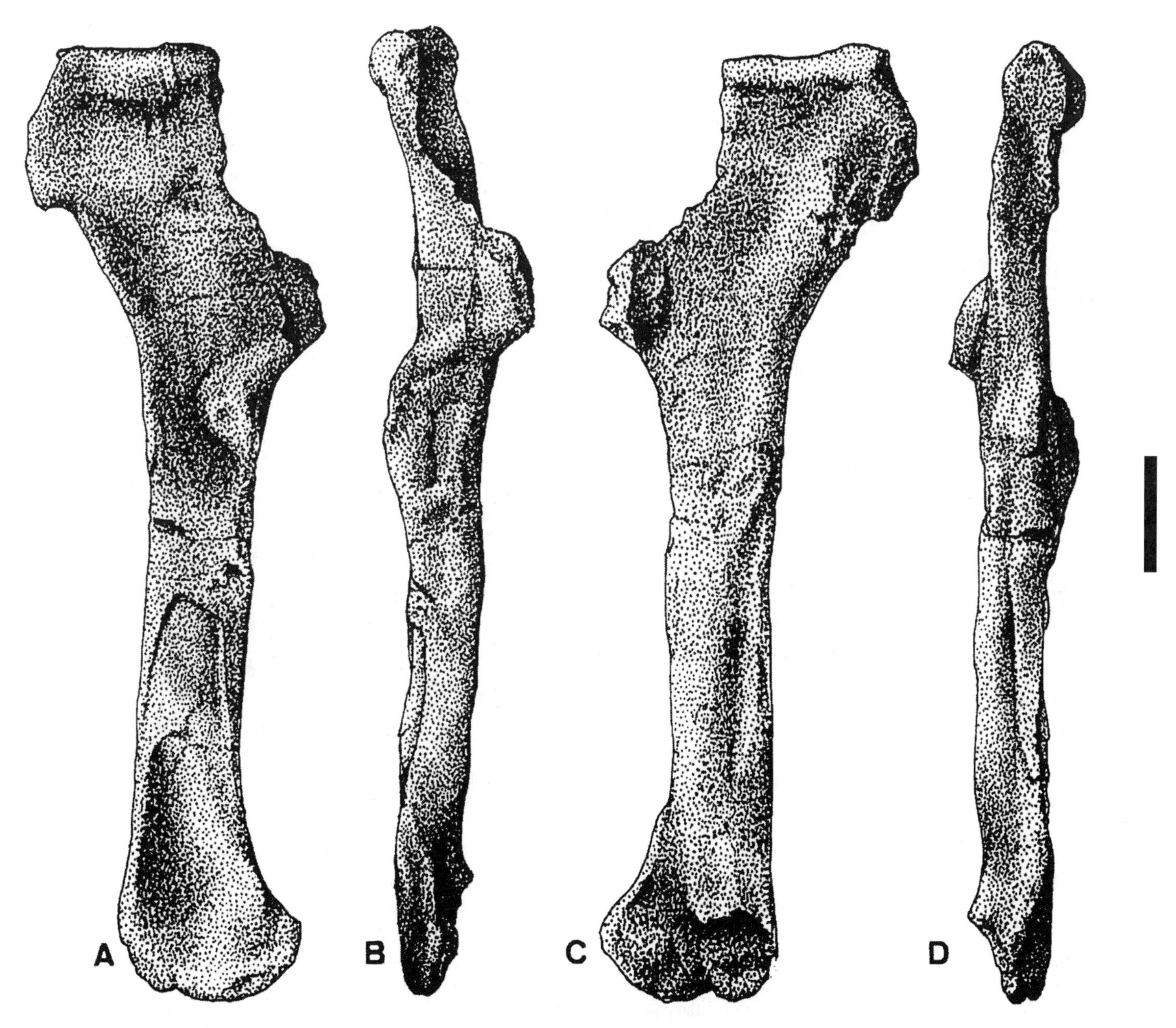

Variraptor mechinorum, MDE D-158, referred right humerus in A. caudal, B. lateral, C. cranial, and D. medial views. Scale = 20 mm. (After Le Loeuff and Buffetaut 1998.)

Key references: Allain and Taquet (2000); Gauthier (1986); Le Loeuff and Buffetaut (1998); Le Loeuff, Buffetaut, Mechin and Mechin-Salessy (1992); Ostrom (1969*b*); Russell (1993).

†VELOCIRAPTOR

Saurischia: Theropoda: Neotheropoda: Tetanurae: Avetheropoda: Coelurosauria: Manuraptoriformes: Manuraptora: Deinonychosauria: Eumanuraptora: Dromaeosauridae: Velociraptorinae.

Species: *V. mongoliensis* Osborn 1924.

Other species: [None.]

Known material: Numerous skulls and skeletons, some complete, adult and juvenile.

Diagnosis of genus (as for type species): (Based on skull characters which, due to the lack of skull data for some dromaeosaurid taxa, constitute equivocal synapomorphies of *V. mongoliensis*) skull shallow, snout long, preorbital length making up 60 percent of total skull length (estimated at approximately 50 percent in *Dromaeosaurus albertensis* and *Deinonychus antirrhopus*); supratemporal fossa and fenestra subcircular, bound by laterally convex supratemporal arcade (elongate, with straight arcade in *D. antirrhopus* (shape of fossa unknown in other dromaeosaurids); frontal long, almost four times longer than wide across orbital portion, almost four times as long as parietal (wider in *Saurornitholestes langstoni*, frontal length no more than three times width across orbital portion; frontal shorter in *D. albertensis*, about twice as long as wide; parietal/frontal length ratio unknown in these

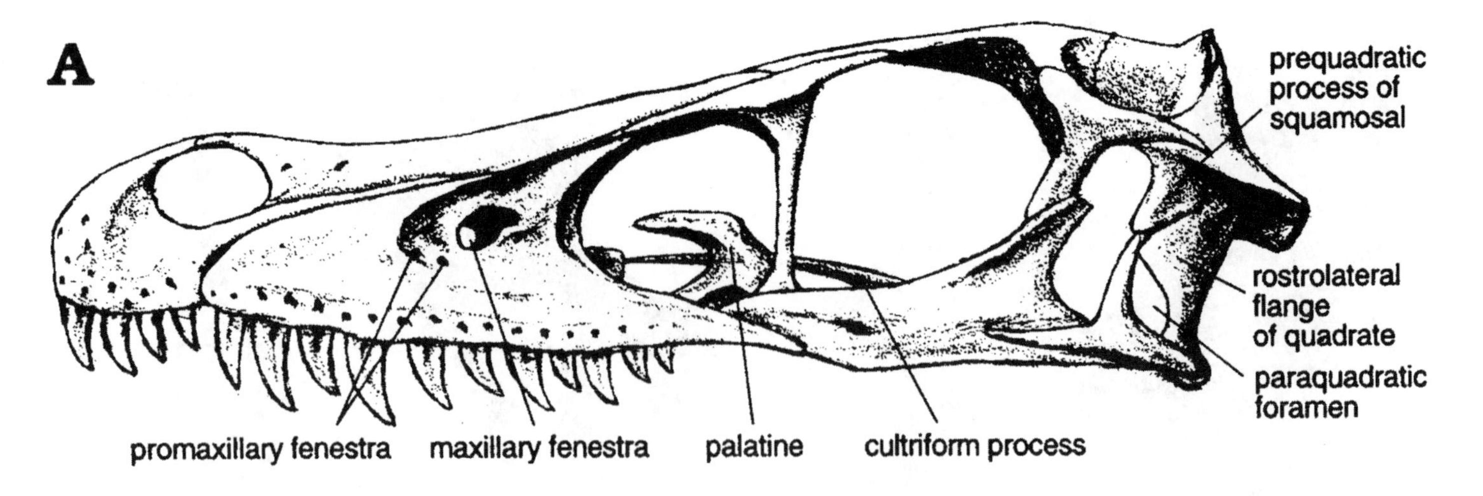

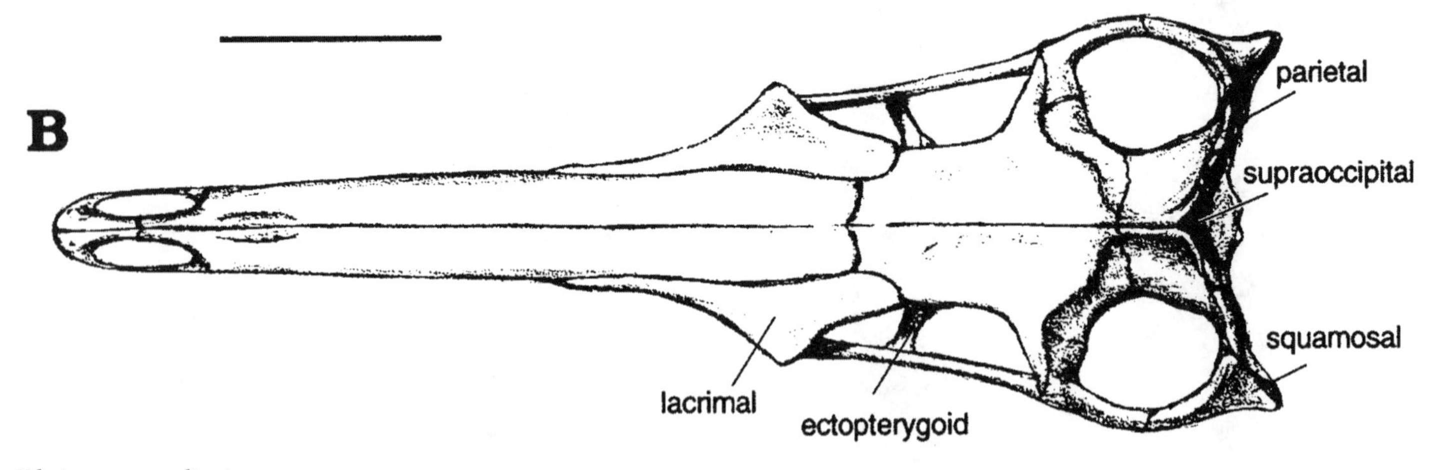

***Velociraptor mongoliensis*, skull reconstruction in A. left lateral and B. dorsal views, drawn by Karol Sabath. Scale = 4 cm. (After Barsbold and Osmólska 1999.)**

species; frontal in *D. antirrhopus* only three times longer than wide, twice as long as parietal [L. M. Witmer, personal communication to Barsbold and Osmólska 1999]; rostral border on internal antorbital fenestra broadly rounded (subrectangular in *D. antirrhopus*, shape unknown in other dromaeosaurids); premaxilla having long maxillary process reaching well beyond caudal margin of external naris (in *D. antirrhopus*, not extending beyond naris, unknown or incomplete in other dromaeosaurids); nasal depressed, deepest just behind external naris (not depressed in *D. antirrhopus*, unknown in other dromaeosaurids); maxilla having longitudinal ridge dorsal to row of neurovascular foramina, these arranged in one row (no ridge in *D. antirrhopus*, maxilla unknown in other dromaeosaurids); no separate prefrontal (separate in *D. antirrhopus*, probably separate in *D. albertensis*); dentary very shallow, depth comprising one-eighth to one-seventh length, ventral margin convex (dentary relatively deeper and with straight ventral margin in other dromaeosaurids); first and second premaxillary teeth larger than third and fourth (Barsbold and Osmólska 1999).

Comments: Once numbering among the lesser of the known theropods, *Velociraptor*— a relatively small, active genus discovered in Mongolia more than 70 years ago, but now also reported from the United States (see entries, *D:TE* and *S1*)— is currently one of the most popular of dinosaurs among lay people. More recently, this genus has become better known to scientists from a wealth of newly recovered, well-preserved cranial and postcranial specimens belonging to the Asian type species *Velociraptor mongoliensis*. Thanks to the recovery of these materials, *Velociraptor* is presently the best known dromaeosaurid genus in terms of most complete and most numerous collected specimens. Data derived from such specimens have allowed paleontologists to supplement earlier published descriptions of this intriguing and important species.

In an abstract, Norell and Makovicky (1998) pointed out that skeletons (especially the skulls) of dromaeosaurids in general are often reconstructed incorrectly due to the incompleteness of specimens. Few of the skulls had been found in an undistorted condition, this problem having led to various misunderstandings regarding the origins of birds, functional

From *Predatory Dinosaurs of the World* (1988), copyright © Gregory S. Paul.

Pair of *Velociraptor mongoliensis*, shown resting after their last meal, as restored by Gregory S. Paul. Although feature-like structures have not yet been found in this species, they are now known in other kinds of theropod dinosaurs.

morphology, and even the physiology of dromaeosaurids.

Norell and Makovicky reported on new *Velociraptor* material, collected during a recent Mongolian Academy of Sciences–American Museum of Natural History expeditions, that greatly added to our understanding of these theropods.

As briefly described by Norell and Makovicky, the skull of *Velociraptor* possesses an extremely thin muzzle, secondary palate, and orbits directed to allow for stereoscopic vision. Several unstated pneumatic cranial features can, for the first time, allow for comparisons with skulls of avialans. The authors also observed a number of postcranial features that "are remarkably similar to avialan." These features include an opisthopubic pelvis having a pubis showing a hypopubic cup and large pubic apron; a reduced cupedicus fossa; and ischia that do not fuse into a symphysis. The pectoral girdle includes a birdlike furcula, sternal plates having attached ossified sternal ribs, a laterally facing glenoid, and an L-shaped coracoid connecting to the anterior surface of the sternal apparatus.

More recently, Barsbold and Osmólska (1999) published a detailed new description of the skull of *Velociraptor*, including a revised diagnosis of the genus and type species. The authors based this study on a number of *V. mongoliensis* skulls belonging to relatively complete skeletons collected from Upper Cretaceous sandstone deposits of the Mongolian Gobi.

These include the following:

MGI [formerly GIN] 100/25 (see Kielan-Jaworowska and Barsbold 1972; Gradziński, Kielan-Jaworowska and Maryańska 1977) was recovered from the Djadokhta Formation (?early Campanian; see Kielan-Jaworowska and Hurum 1997), Tugrikin-Shire, Omnogov, Mongolia, during the Polish-Mongolian Palaeontological Expeditions. This specimen includes two skeletons, one of *V. mongoliensis*, the other of the primitive ceratopsian *Protoceratops andrewsi*, both preserved in what has been interpreted as a combat position (the famed "Fighting Dinosaurs"; see *D:TE* and *S1* for information and photographs).

PIN 3143/8, consisting of a nearly complete skull with left mandibular ramus and lacking the right temporal region, was collected at Tugrikin-Shire during the Soviet-Mongolian Palaeontological Expeditions.

MGI [formerly GIN] 100/24, collected at Tugrikin-Shire during an expedition to southern Mongolia (Barsbold 1983), consists of an almost complete, articulated although dorsoventrally compressed skull, both mandibular rami, and several fragmentary postcranial elements.

MGI [formerly GIN] 100/2000, recovered during the Mongolian-Japanese Palaeontological Expeditions from Tugrikin-Shire, comprises the complete skeleton of a young individual, the skull including a mandible.

Finally, ZPAL MgD-1/97 (see Kielan-Jaworowska and Barsbold 1972; Gradziński, Kielan-Jaworowska and Maryańska 1977), from the Barun Goyot Formation (?late Campanian; see Jaworowska and Hurum), Khulsan, Bayankhongor, Mongolia, includes the left rostral half of the skull, lacking the premaxilla and tips of the vomers.

Not included in this study was the abundant *Velociraptor* material recently collected in Mongolia by the Mongolian Academy of Sciences–American Museum of Natural History Expeditions (see Norell and Makovicky 1998, 1999); the postcrania preserved with the above mentioned specimens will be described at a later date by Barsbold and Osmólska.

Barsbold and Osmólska addressed an earlier suggestion by Paul (1988*b*; see *D:TE*), who determined that *Velociraptor* is generically a senior synonym of the North American genus *Deinonychus* (and also *Saurornitholestes*). This assessment was subsequently rejected by most workers (*e.g.,* Ostrom 1990), and more recently by Witmer and Maxwell (1996; see *Deinonychus* entry, *S1*), who listed various significant differences between these two taxa based in part upon more complete recently collected materials belonging to *D. antirrhopus*. As pointed out by Barsbold and Osmólska, these new data, as well as that included in a redescription of *Dromaeosaurus albertensis* published by Currie (1995), have made invalid some of the criteria used in Paul's synonymy. In fact, Barsbold and Osmólska's study has increased farther the number of differences separating *V. mongoliensis* and *D. antirrhopus*. Although the skulls of these taxa are similar to one another, that of *V. mongoliensis* differs from *D. antirrhopus* in various ways (as cited in their above new diagnosis).

Currie had listed a suite of autapomorphies of the Dromaeosauridae among the skull and mandible. However, based in part on their study of *Velociraptor*, Barsbold and Osmólska found several of these characters to be equivocal, these including the following: Long, shallow maxillary process of premaxilla (widespread feature among theropods); absence of ventrally extended pterygoid flange (missing in many coelurosaurs including ornithomimids, troodontids, and oviraptorids); palatine-ectopterygoid contact (also in at least ornithomimids and oviraptorids); and caudal tympanic recess (tetanuran synapomorphy; see Clark, Perle and Norell 1994). Apomorphic characters not present in *Velociraptor*, and therefore not regarded by Barsbold and Osmólska as dromaeosaurid synapomorphies, include these: Basipterygoid processes relatively short, not extending ventrally beyond level of basal tubera (extending well below nasal tubera in *Velociraptor*); frontal-lacrimal contact slot-like (in *Velociraptor*, no vertical slot on this contact, lacrimal extensively overlapping frontal dorsally); and tall, labiolingually thin dentary (dentary thick relative to height in *Velociraptor*).

In 1997, Norell and Makovicky had described selected features of the dromaeosaurid skeleton based on specimens recently collected during the Mongolian Academy of Sciences–American Museum of Natural History Expeditions to the Gobi Desert (see "Note," *Velociraptor* entry, *S1*). Two years later, Norell and Makovicky expanded their original description, concentrating on poorly known aspects of the skeleton of *V. mongoliensis*.

This description focused primarily upon three specimens collected from the Djadoktha Formation. They comprise a partial skeleton of *V. mongoliensis* with fragmentary skull (IGM 100/976) found *in situ* in 1991 at Tugrugeen Shireh (see Norell, Clark and Chiappe 1993), recovered over two field seasons; a partial skeleton (IGM 100/986) found *in situ* with various cranial and postcranial fragments at the nearby Chimney Buttes locality (found during the 1993 field season), collected as float in 1993; and a skeleton with a well-preserved skull (IGM 100/982), collected *in situ* in 1995 at the Flaming Cliffs (Bayn Dzak). These and other previously described specimens (*e.g.*, IGM 100/985) seems to represent "the largest assemblage of articulated dromaeosaurid postcrania yet described," preserving many skeletal elements until now unknown or only poorly described (*e.g.*, pectoral girdle, pelvis and sacrum, posterior dorsal and cervical vertebrae, hindlimb, and manus).

In describing the postcranial skeleton of *Velociraptor*, Norell and Makovicky (1999) particularly noted features that closely resembled characters found in basal birds, including *Archaeopteryx lithographica*. These include a pectoral girdle displaying a scapula lying in a subhorizontal position relative to the dorsal vertebral column, sternal plates that articulate with

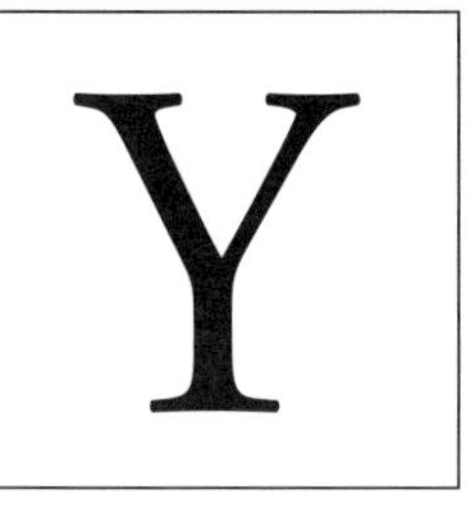

the coracoids, a furcula, and a pelvic girdle with a reduced antiliac shelf.

Norell and Makovicky (1999) addressed in detail "several common misconceptions regarding aspects of dromaeosaurid anatomy that are relevant to understanding the origin of Avialae," for example, that the pelvis of nonavian dinosaurs is different from that of basal birds such as *Archaeopteryx* (see Ruben, Jones, Geist and Hillenius 1997). Norell and Makovicky (1999) criticized various recently published reconstructions (*e.g.*, Ruben *et al.*) of the pelvis of *A. lithographica*, which, they stated, incorrectly reconstruct the pubis in an overly opisthopubic position. Based on the new specimens of *V. mongoliensis*, however, the pelves of that species and *A. lithographica*, except for various minor details and the orientation of the pubis, were found to be "exceedingly similar."

Norell and Makovicky (1999) noted that (*contra* Martin 1991, 1995) the pectoral girdle of *V. mongoliensis* shares with *A. lithographica* several derived characters, these including the large, curved, anteriorly facing coracoids. Furthermore, the similarity between the shoulder girdles of *A. lithographica*, *V. mongoliensis*, and the fragmentary remains of the primitive bird *Unenlagia comahuensis* suggested to the authors that all three of these animals were capable of similar ranges of motion. This mobility, plus a folding mechanism in the wrist distally as a primitive character for Manuraptora, "indicates that these features originated phylogenetically before the origin of powered flight in avialians."

Key references: Barsbold (1983*a*); Barsbold and Osmólska (1999); Clark, Norell and Chiappe (1998, 1999); Clark, Perle and Norell (1994); Currie (1995); Gradziński, Kielan-Jaworowska and Maryańska (1977); Kielan-Jaworowska and Barsbold (1972); Kielan-Jaworowska and Hurum (1997); Martin (1991, 1995); Norell, Clark and Chiappe (1993); Norell and Makovicky (1997, 1998, 1999); Osborn (1924); Ostrom (1990); Paul (1988*b*); Ruben, Jones, Geist and Hillenius (1997); Witmer and Maxwell (1996).

YIMENOSAURUS Bai *see in* Bai, Yang and Wang 1990

Saurischia: Sauropodomorpha: Prosauropoda: Plateosauridae.

Name derivation: "Yimen [County]" + Greek *sauros* = "lizard."

Type species: *Y. youngi* Bai *see in* Bai, Yang and Wang 1990

Other species: [None.]

Occurrence: Fengjahe Formation, Yunnan Province, China.

Age: Early Jurassic.

Known material: Several partial skeletons.

Holotype: YXV 8701, damaged partial skeleton including disarticulated skull and mandible, atlas, axis, fourth and eight cervical vertebrae, four articulated middle dorsal vertebrae and several fragmentary dorsals, several incomplete ribs, three complete sacral vertebrae, relatively complete medial section of caudal series, several fragmentary caudal vertebrae, complete anterior hemal arch, relatively complete ilium, relatively complete ischia, complete femur.

Diagnosis of genus (as for type species): Moderate-sized to large prosauropod, possibly attaining nine meters in length; skull height to length proportions moderate, skull approximately 1.65 times as long as high (including mandible), cranium of relatively delicate construction, elements comprising skull long and gracile; external nares elliptical, relatively large; premaxilla and ascending process of maxilla relatively well developed, ventral dental margin rather shortened anteroposteriorly; mandible relatively delicately constructed, medial portion of ramus relatively high, mandibular fenestra well developed, articular process lying ventral to plane of dentition, articular well developed but comparatively weak; relatively numerous teeth (four premaxillary, 17–18 maxillary, and 21–23 dentary); teeth compact, crowns relatively high, slightly spoon-shaped, with gently convex labial margin, gently concave lingual margin, medial ridge not well developed, all teeth conspicuously striated, anterior

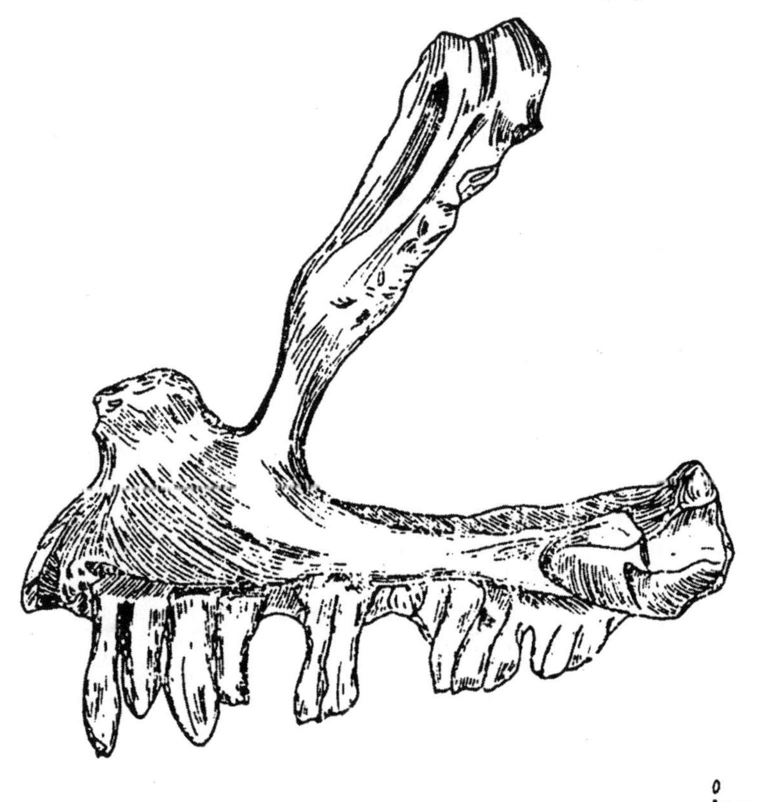

***Yimenosaurus youngi*, YXV 8701, holotype left maxilla, lateral view. Scale = 2 cm. (After Bai, Yang and Wang 1990.)**

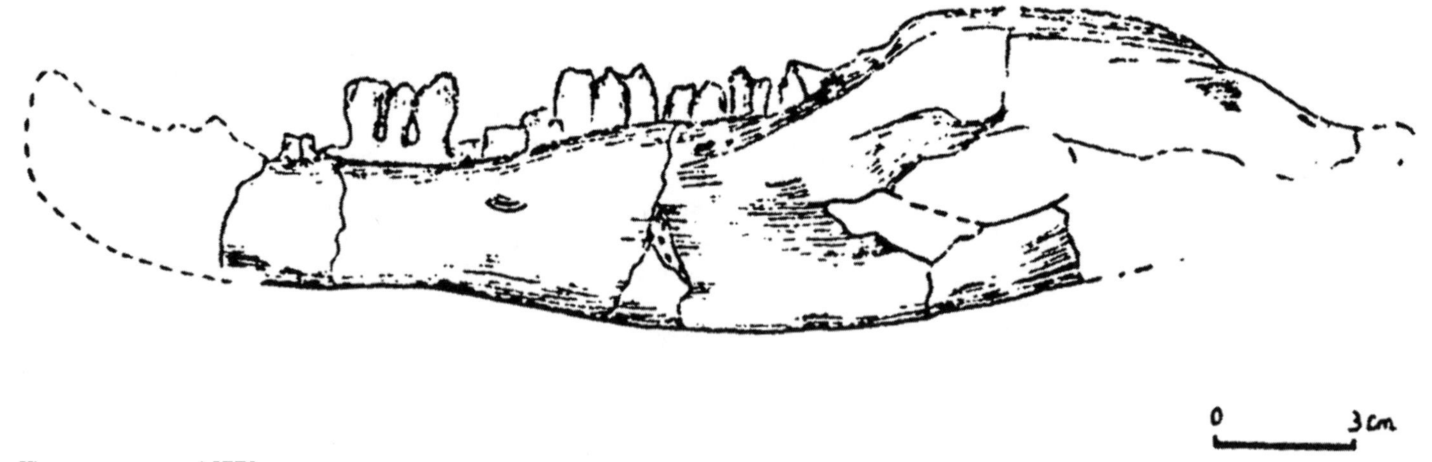

***Yimenosaurus youngi*, YXV 8701, holotype left dentary, lateral view. Scale = 3 cm. (After Bai, Yang and Wang 1990.)**

and posterior margins of teeth having well-developed denticles; presacral vertebral endochondral construction compact, all vertebrae (except atlas and axis) amphicoelous; proatlas associated with elongated intercentrum, neural arch, and two well-developed lobes, postzygapophyses relatively well developed; cervical vertebrae relatively elongated, middle cervicals rather reduced, pleurocoels undeveloped, conspicuous ventral keel present, neural arch and spine low, simply constructed; dorsal vertebrae relatively long, rather laterally compressed, low neural arches, spines shaped like thin plated crests; three amphyplatyan sacral vertebrae, with strong centra that support low, thinly walled neural arches and spines, ribs and zygapophyses fused to form robust sacriliac yoke that supports proximal end of coracoid; anterior caudal vertebrae short, robust, amphicoelous; neural arches and spines relatively low, diapophyses comparatively well developed, hemal arches particularly well developed, anterior hemal arches spinous and particularly robust with club-shaped distal ends; scapula straight, thick, with intense medial curvature and relatively narrow anterior oblique depression; coracoid short, with thick margin, proximal end of coracoid contacting anterior scapula to form broad and spacious glenoid fossa; humerus short (three-fifths length of femur), thick, lacking curvature, with well-developed medial head and strong deltopectoral crest; pelvis robust, ilium relatively low maintaining well-developed pubic and ischiac peduncles; pubis and ischium long, obturator foramen large, open, pubis with relatively large and inflated distal crest (for enhanced muscle attachment); ischium relatively short, thick, possessing well-developed proximal puboischiatic plate for contact with pubis, distal ends relatively elongated, expanded, fused, with coarsened crest (for enhanced muscle attachment); femur robust, with conspicuous sinuous curvature, shaft subcircular in cross section with thickened walls, absence of distinct neck between head and shaft, well-developed greater trochanter, lesser trochanter inconspicuous, fourth trochanter well developed and located posterodorsally at midshaft, distal condyles not exceptionally defined, intercondylar notch shallow; tibia and fibula almost equal in length, tibia two-thirds length of femur; astragalus massive, two proximal articular facets relatively well developed, distal astragalar process well developed, mediolateral breadth of foot exceeding anteroposterior length; calcaneum small, thick, with long, spherical proximal end; pes formula of 1-2-3-4-1; digit V reduced, ungual on digit I robust, unguals on remaining digits II, III, and IV weakened, reduced (Bai *see in* Bai, Yang and Wang 1990).

Comments: The type species *Yimenosaurus youngi* was founded upon several almost complete skeletons including YXV 8701 (holotype) and YXV 8702 (paratype), collected from a silty mudstone of the Fengjahe Formation at Jiaojiadian, in Yimen County, Yuxi Prefecture, Yimen Basin, southern China. The paratype specimen, larger than the holotype, comprises a damaged skull and a partial postcranial skeleton including five somewhat damaged medial and posterior cervical vertebrae, five articulated dorsals, three relatively complete sacrals, six relatively complete, articulated anterior caudals, a relatively complete scapulocoracoid, relatively complete ilia, pubes, and ischia, complete femora, complete tibiae, fibulae, astragali, and calcaneae, and a complete left pes (Bai *see in* Bai, Yang and Wang 1990).

As related by Bai *et al.*, the specimens, found at the surface, had been subjected to intense weathering. Therefore, the specimens were easily damaged during excavation and preparation. Additionally, some postmortem disarticulation took place before preservation, a slight amount of compressional distortion had affected the skull, while some of the cranial elements (*e.g.*, occipital portion, parietal, squamosal, left frontal, prefrontal, and jugal) were lost to weathering. Thus, "a certain amount of restoration" was required that lead to difficulties in diagnosing the material.

Z

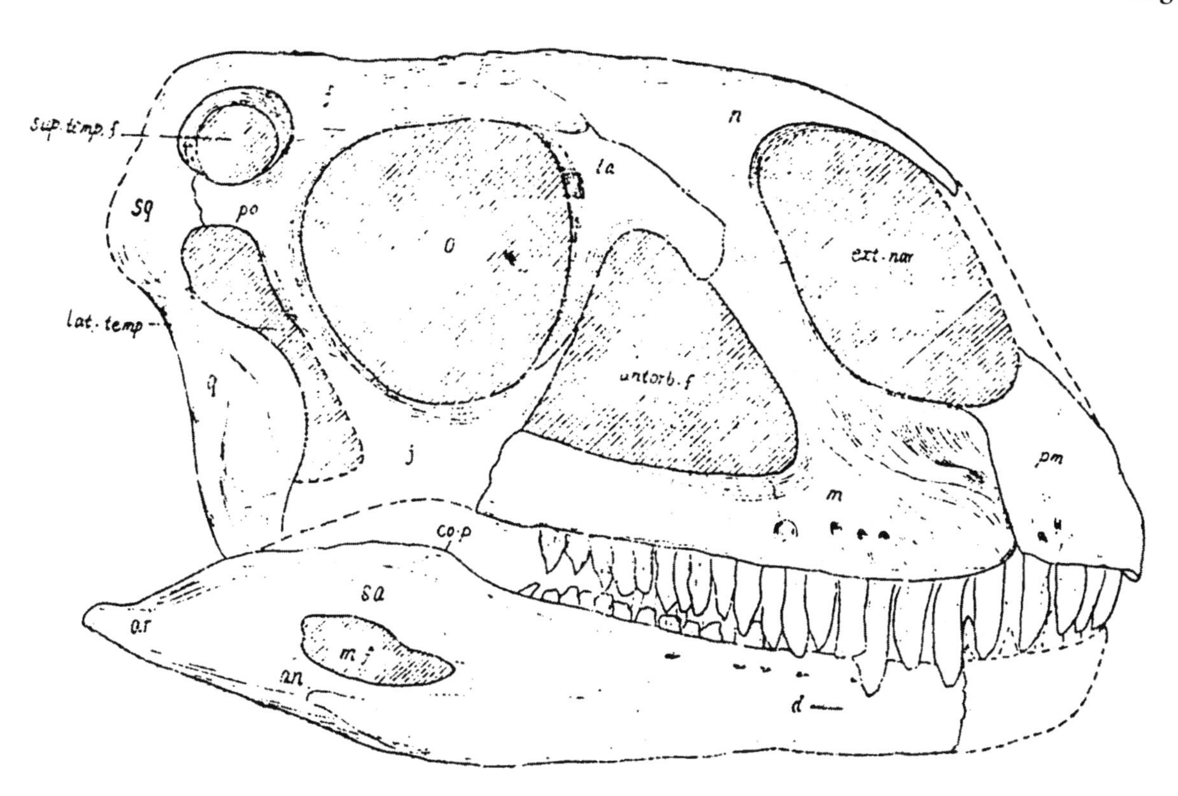

***Yimenosaurus youngi*, reconstructed skull including holotype YXV 8701. (After Bai, Yang and Wang 1990.)**

Bai described the skull as unusually short and deep, having a maxilla similar in outline to that of primitive sauropods, and with spoon-shaped teeth.

Comparing *Yimenosaurus* with other prosauropods, the author assigned this new genus to the family Plateosauridae. As in plateosaurids, the Yimmen prosauropod is quite large, the collected specimens ranging from more than six meters (about 21 feet) to over nine meters (more than 30m feet) in length. Also, *Yimenosaurus* displays the following plateosaurid features: Skeletal elements (except for the comparatively delicately constructed skull) rather robust; teeth relatively long; upper premaxillary tooth count of four, maxillary count 17–18, mandibular count about 23.

Key reference: Bai, Yang and Wang (1990).

†**ZIGONGOSAURUS** Hou, Chao and Chu 1976

Saurischia: Sauropodomorpha: Sauropoda: Eusauropoda: Neosauropoda: Macronaria: Titanosauriformes: Somphospondyli: Euhelopodidae: Euhelopodinae.

Name derivation: "Zigong" [city in China] + Greek *sauros* = "lizard."

Type species: *Z. fuxiensis* Hou, Chao and Chu 1976.

Other species: [None.]

Occurrence: Shaximiao Formation, Zigong, People's Republic of China.

Age: Middle to early Late Jurassic.

Known material: Cranial and postcranial remains representing numerous individuals.

Holotype: CV 00261, fragment of mandible with teeth, maxilla, basioccipital.

Diagnosis of genus (as for type species): Medium- to large-sized sauropod; skull relatively high; snout moderately developed; external nares paired, located at anterior part of skull; occipital portion broad; supratemporal fenestra relatively large; teeth medium-sized, spatulate; four premaxillary, 12–14 maxillary, and 15–17 dentary teeth; ?17 cervicals, having opisthocoelous centra with rather large pleurocoels, lamellar structure of neural spines developed; 12 or 13 dorsal vertebrae, anterior neural spines weakly bifurcated; four sacral vertebrae, with three fused anterior neural spines; first and second caudal vertebrae with fan-like flat caudal ribs; chevrons unforked; scapula long, thin, extended at proximal end; coracoid comparatively round; sternum oval in shape; ilium high, with pubic peduncle at middle part of ilium; limb bones relatively flat (Zhang and Chen 1996).

Comments: The new type species *Zigongosaurus fuxiensis*— prepared by Hou L. H., Zhou Xinjin, and Chao S. W., based upon the remains of many individuals (J. S. McIntosh, personal communication 1987)—was founded upon partial cranial material (CV 00261; see *D:TE* for figures) collected from the upper Shaximiao Formation in the Sichuan Basin, Wuijiaban, Zigong, People's Republic of China (Hou, Chao and Chu 1972).

While noting that some *Zigongosaurus* postcranial

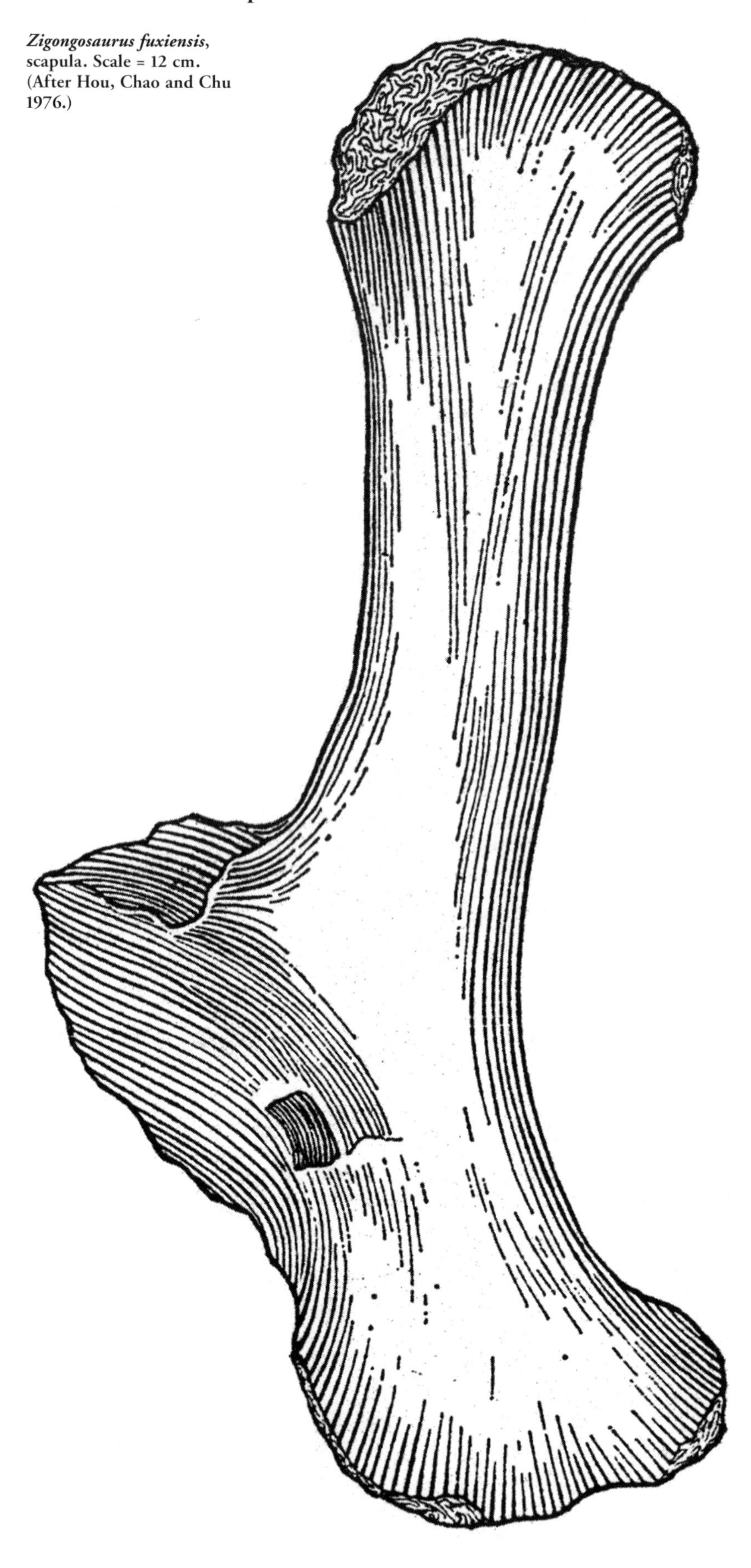

***Zigongosaurus fuxiensis*, scapula. Scale = 12 cm. (After Hou, Chao and Chu 1976.)**

material resembles that of *Omeisaurus*— primarily the rather large pleurocoels of the presacral vertebrae and the development of the laminae of the neural spines — Hou *et al.* kept the two genera separate. As observed by these authors, *Zigongosaurus* differs from *Omeisaurus* in that the anterior causal vertebrae possess neural spines that are not very distinct, and the anterior caudal vertebrae are procoelous.

Since this genus was named, the taxon has had a rather checkered history (see *D:TE*, *S1*). Dong, Zhou and Zhang (1983) erected a new species of *Omeisaurus*, *O. fuxiensis*, including the material assigned to *Zigongosaurus* and also additional incomplete skull bones, at the same time abandoning the latter generic name. However, both *Z. fuxiensis* and *O. fuxiensis* were distinguished as separate taxa based upon their own type specimens.

McIntosh (1990), in his comprehensive review of the Sauropoda, regarded *Z. fuxiensis* as another species of *Omeisaurus*, *O. junghsiensis*. Later, Zhang and Chen (1996) determined that the stratigraphic position of *Z. fuxiensis* is higher than either that of *Omeisaurus tianfuensis* or *O. junghsiensis*, yet lower than that of two of the species of *Mamenchisaurus*, *M. hochuanensis* and *M. jingyanensis*. Consequently, Zhang and Chen noted, the age of *Z. fuxiensis* must be between Middle and early Late Jurassic in age. Also, these authors pointed out that the posterior cervical and anterior dorsal neural spines of *Z. fuxiensis* are weakly bifurcated. Therefore, this species represents a taxon more advanced than *Omeisaurus* and more primitive than *Mamenchisaurus*. Noting that *Z. fuxiensis* is very similar to *Mamenchisaurus*, Zhang and Chen further suggested that this species be referred to *Mamenchisaurus* as the new species *M. fuxiensis*.

More recently, however, Martin-Rolland (1999), in her review of Chinese sauropods, reinstated *Zigongosaurus* as a valid genus. According to Martin-Rolland, *Zigongosaurus* differs sufficiently from *Mamenchisaurus* in the weak degree of bifurcation in the neural spines. That author agreed with Hou *et al.* in their reasons for separating *Zigongosaurus* from *Omeisaurus*.

Key references: Dong, Zhou and Zhang (1983); Hou, Chao and Chu (1976); Martin-Rolland (1999); McIntosh (1990); Zhang and Chen (1996).

ZUNICERATOPS

Comments: In 1998, Douglas G. Wolfe and James I. Kirkland described an important new neoceratopsian, *Zuniceratops christopheri*, uniquely distinguished as the most primitive known ceratopsian having brow horns (see *S1*).

At the time of publication of the Wolfe and

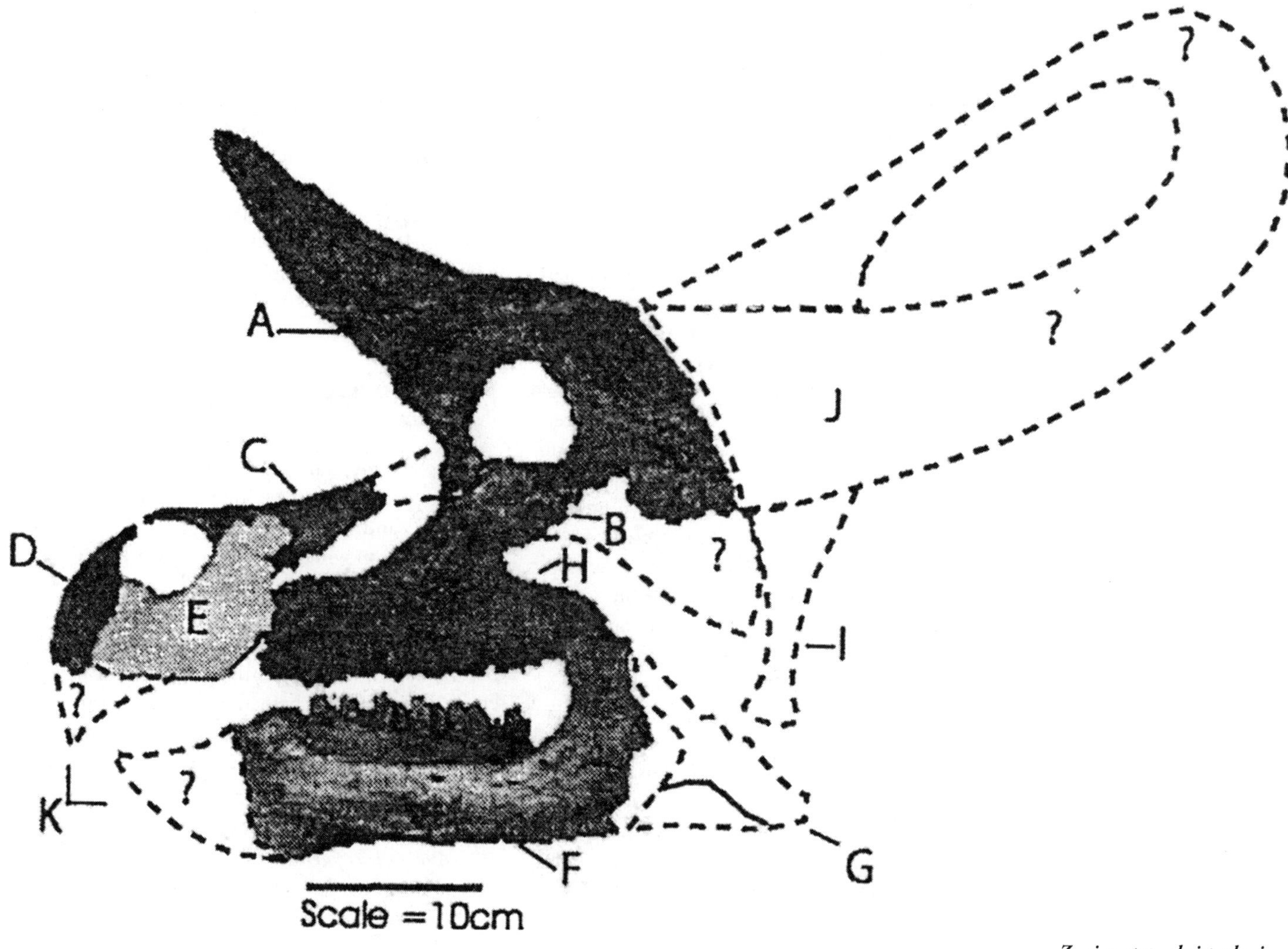

Zuniceratops christopheri, preliminary skull reconstruction by Douglas G. Wolfe based upon currently prepared elements. (After Wolfe 2000.)

Kirkland's original description of this type species, a newly discovered bonebed in the Moreno Hill Formation, in the Zuni Basin of western New Mexico, yielded additional elements belonging to this dinosaur, including a well-preserved maxilla and horncore. Also collected from this site was a specimen (MSM P2106) which the authors tentatively identified as a squamosal of *Z. christopheri*.

In a later publication on *Z. christopheri* and this bonebed, Wolfe (2000) reported the recovery of at least five disarticulated individuals of this dinosaur, "validating the principal components of the *Zuniceratops* holotype (MSM P2101), while providing important, previously missing elements." (Wolfe also reported the recovery of numerous elements representing a new theropod, this material revealing that MSM P2106 does not belong to *Zuniceratops* but rather to a new therizinosauroid taxon, as of this writing under preparation and description.)

Wolfe noted that additional *Zuniceratops* material collected from this bonebed and prepared since 1998 include four relatively prepared left dentaries, several relatively complete and partial maxillae, two premaxillae, two nasals, more horncores, two well-preserved scapulae, a well-preserved ischium, a coronoid larger than that in the holotype, a humerus, femur, sacral assemblage, vertebrae, and ribs.

The author observed that there is no evidence of a nasal horn in any of these recovered specimens. A large, well-preserved postorbital horncore, which retains the sutured palpebral, lacrimal, and jugal borders of the orbit, apparently matches the well-preserved left maxilla, thereby offering significant information on the relative position of some of the skull elements. Preserved frill fragments indicate that the frill of *Zuniceratops* was most likely relatively thin, probably fenestrated, and lacking epoccipital processes. Wolfe further noted that, although partial limb elements associated with the holotype suggest that *Zuniceratops* was a relatively gracile animal, the ischium is robust and recurved as in some later ceratopsids.

According to Wolfe, apparent variation seen in comparable elements (*e.g.*, four left dentaries) from a number of *Zuniceratops* individuals provides evidence of ontogenetic and perhaps sexual variation in this taxon. For example, the number of tooth rows, between the smallest and largest dentary specimens recovered, was estimated to range from 16 to 20. The comparable range of variation suggested to Wolfe that the holotype of *Z. christopheri* probably represents an older juvenile-young adult or possibly the more sexual dimorpoh. Differences between the specimens also illustrates subtle burial deformation that could otherwise be mistaken for morphology. "Most importantly," Wolfe stated, "the associated skull elements provide the opportunity to create a preliminary composite reconstruction of the *Zuniceratops* skull, suggesting a relatively gracile animal exhibiting many ceratopsid characteristics, with prominent brow horns but no nasal horn."

In summary, Wolfe listed the following skull characters currently identifying *Zuniceratops*: Exoccipital process of braincase expanded (as in ceratopsids), occipital condyle prominent, spherical; postorbital brow horns prominent, laterally compressed, slightly recurved, lengthening during ontogeny; postorbital area thickened, exhibiting intracranial cavities; nasal smooth, no evidence of nasal horn; nares relatively large and sub-ovate, formed by relatively simple premaxilla and naval (neither centrosaurine nor chasmosaurine in character); antorbital fenestra relatively large, formed by maxilla/nasal; two sub-parallel ridges extending along relatively rectangular lateral surface of dentary; single-rooted teeth extending caudally along dentary to caudal edge of coronoid; coronoid prominent, with slight (if any) rostral-dorsal projection; single-rooted teeth exhibiting vertical shear wear pattern in both dentary and maxilla; approximately 16 to 20 teeth (number increasing with age) in one or two replacement galleries; partial resorption of worn tooth roots prior to shed resembling incipient "ceratopsid" condition; frill thin, probably having fenestrae, probably lacking emargination.

Future publications, Wolfe noted, following complete preparation of the collected *Zuniceratops* remains, will offer a more complete description of consistent characters and compare this taxon within and between the Neoceratopsia. Furthermore, characters currently identified in this genus, its Turnonian age (see Wolfe and Kirkland), as well as information pertaining to other neoceratopsians (*e.g.*, Williamson 1997; Chinnery and Weishampel 1998; Chinnery, Lipka, Kirkland, Parrish and Brett-Surman 1998; Penkalski and Dodson 1999) indicate "that traditional views of ceratopsian diversification may require substantial revision. In particular, the presence of prominent brow horns in *Zuniceratops*, combined with the absence of a nasal horn, raises questions regarding expected relationships between early North American ceratopsids."

Key references: Chinnery, Lipka, Kirkland, Parrish and Brett-Surman (1998); Chinnery and Weishampel (1998); Williamson (1997); Wolfe (2000); Wolfe and Kirkland (1998).

Final Note

As already stated elsewhere in the text, much time can pass between a book's completion and going "to press" and its actual publication. In the case of this volume, in which the editorial work is immense, nine months have elapsed between the point of "no more changes" and the printing and binding of the finished book. (My official cut-off date was December 31, 2000.) Consequently, a number of new genera and species were named and described in the paleontological literature too late for inclusion. These include the following (although there will almost certainly be more):

Aucasaurus garridoi (abelisaurid theropod)
Bienosaurus lufengensis (scelidosaurid)
Citipati osmolskae (oviraptorid theropod)
Draconyx loureiroi (camptosaurid ornithopod)
Eotyrannus lengi (tyrannosauroid theropod)
Eshanosaurus deguchiianus (therizinosauroid theropod)
Khaan mckennai (oviraptorid theropod)
Ilokelesia aguadagranensis ("abelisaurian" theropod)
Losillasaurus giganteus (diplodocoid sauropod)
Masiakasaurus knopfleri (?noasaurid abelisauroid)
Nothronychus mckinleyi (therizinosaurid theropod)
Paralititan stromeri (titanosaurid sauropod)
Planicoxa venenica (?iguanodontid ornithopod)
Quilmesaurus curriei (theropod)
Rapetosaurus krausei (titanosaurid sauropod)
Richardoestesia isosceles (new species)
Ruehleia bedheimensis (plateosaurid prosauropod)
Venenosaurus dicrocei (titanosauriform sauropod)

IV. Nomen Nudum and Excluded Genera

A *nomen nudum* constitutes a usually doubtful genus that has appeared (printed in italics) in the literature (*e.g.*, in faunal lists, notes on taxa to be described, popular and technical books, popular articles and reports, *etc.*), but without proper description following the rules for establishing a new taxon (*e.g.*, without designating a holotype). Such a taxon is, therefore, considered to be *nomen nudum*, or a "naked name" (plural, *nomina nuda*). A *nomen nudum* may, however, be formally described at a future date, sometimes with the name altered in some way or altogether changed.

NUROSAURUS Dong 1992 [*nomen nudum*]

Saurischia: Sauropodomorpha: Sauropoda: Eusauropoda: Neosauropoda: Macronaria: Camarasauromorpha: Camarasauridae.

Name derivation: "[Oagan] Nur [Salt Mine]" + Greek *sauros* = "lizard."

Type species: *N. qaganensis* Dong 1992.

Occurrence: Qagganur Formation, southeast of Erenhot, Inner Mongolia.

Age: Early Cretaceous.

Known material/holotype: Skeleton.

Comments: Formerly given its own entry in *D:TE* (see for information and photograph of the mounted, mostly reconstructed skeleton), on the assumption that the formal description of this taxon was "in press" (Dong and Li), subsequently regarded by Martin-Rolland (1999) as, for the present, a *nomen nudum*.

Key references: Dong (1992); Martin-Rolland (1999).

Both *Dinosaurs: The Encyclopedia* and *D:TE, Supplement 1* included lists of genera that were, at one time or another, classified as dinosaurs but are now generally regarded as belonging to other nondinosaurian groups. This supplement offers more such removed taxa, including some which were omitted from the previous listings (see George Olshevsky's *Mesozoic Meanderings* number 3 for more information on excluded taxa).

Interestingly, as additional and sometimes more complete fossil materials are collected, prepared, and studied, and as the results of new phylogenetic analyses are published in the paleontological literature, interpretations of various taxa can sometimes change. Consequently, some of the "excluded genera" listed in either of the previous volumes have now been awarded dinosaurian status, at least tentatively, and have been accordingly given their own entries in the "Dinosaurian Genera" chapter of this book, these including *Alvarezsaurus*, *Avalonianus*, *Elopteryx*, *Mononykus*, *Nuthetes*, and *Patagonykus*. Of these, *Avalonianus* (a junior synonym of *Megalosaurus*) had been formerly interpreted as a possible "thecodontian," *Nuthetes* a possible lepidosaur, the others believed to be birds (but now tentatively accepted as nonavian dinosaurs).

As before, taxa formerly regarded as dinosaurs continue to be reassessed and reidentified as something else. Sometimes new classifications can occur quite rapidly. In one case, a genus (*Spinosuchus*; see "Introduction") passed from nondinosaurian status to dinosaurian, then back to nondinosaurian status, *during* the writing of the present supplement.

The following new list comprises additional genera considered, at least for the present by a majority of workers, to be nondinosaurian:

ACTIOSAURUS Sauvage 1882 [*nomen dubium*]

?Ichthyosaur.

BELODON Meyer 1842

Phytosaur.

BRACHYTAENIUS Meyer 1842

Crododilian.

CENTEMODON Lea 1856

Phytosaur.

KUSZHOLIA Nessov 1992

Palaeognathan bird.

LAORNIS Marsh 1870

?Charadriiform bird.

PALAEOSAURISCUS Kuhn 1959 [*nomen dubium*]

Archosaur, possibly phytosaurian.

Note: This genus and its species, *P. cylindrodon* and *P. platyodon*, were based on mostly undiagnostic teeth that can be identified no more precisely than Archosauria *incertae sedis* (see Benton, Juul, Storrs and Galton 2000), although the holotype tooth of *P. platyodon* seems to that of a phytosaur (A. Hungerbühler, personal communication to Benton *et al.*). Some of the materials referred to this genus have been referred to *Thecodontosaurus* (see entry, *D:TE* and this volume) and *Sellosaurus* (see entry, *D:TE*).

PALAEOSAURUS Riley and Stutchbury 1836 [*nomen dubium*]—(Preoccupied, Saint-Hillaire 1831; see *Palaeosauriscus.*)

PAREIASAURUS Owen 1876
Pareiasaurid cotylosur.

PONEROSTEUS Olshevsky 2000
Note: Olshevsky (2000) proposed the new generic name *Ponerosteus* (meaning "bad bone" or "poor bone," the name coined by Ben Creisler [personal communication to Olshevsky] referring to the quality of the type specimen) for a possible marrow-cavity cast of a partial tibia from the Cenomanian of Holubitz, Bohemia, originally referred by Fritsch [Fric] (1878) to a new *Iguanodon* species, *I. exogyrarum.* Fritsch (1905) later referred this species to a new genus, *Procerosaurus.* Chure and McIntosh (1989) regarded *I. exogyrarum* as nondinosaurian. Norman and Weishampel (1990) referred to it as "indeterminant scrap." As *Procerosaurus* was preoccupied (Huene 1902), Olshevsky suggested that a new name was warranted.

PROTOROSAURUS Meyer 1830
Nondinosaurian archosauromorph.

RILEYASUCHUS Kuhn 1961
Indeterminate archosaur, ?phytosaur.

SQUALODON Grateloup 1840
Cetacean.

TERMATOSAURUS Meyer and Plieninger 1844
Indeterminate, ?plesiosaurian.

TRIBELESODON Bassani 1886
Pterosaur.

A List of Abbreviations

The following abbreviations, which are used in this book, refer to museums and other institutions in which fossil specimens are housed:

AC Pratt Museum, Amherst College, Amherst, Massachusetts, United States
AM Albany Museum, Grahamstown, South Africa
AMNH American Museum of Natural History, New York, New York, United States
ANSP The Academy of Natural Sciences of Philadelphia, Philadelphia, Pennsylvania, United States
ASDM Arizona-Sonora Desert Museum, Tucson, Arizona, United States
BGS GSM British Geological Survey, Keyworth, Nottingham, England
BHI Black Hills Institute of Geological Research, Rapid City, South Dakota, United States
BMNH The Natural History Museum, London (formerly British Museum [Natural History]), London, England
BP Bernard Price Institute for Palaeontological Research, University of the Witwatersrand, Johannesburg, South Africa
BRSMG Bristol City Museum and Art Galleries, Bristol, England
BSP Bayerische Staatssammlung für Paläontologie und Historische Geologie, München, Germany
BYU [also **BYUVP**] Brigham Young University Vertebrate Paleontology, Provo, Utah, United States
CEUM College of Eastern Utah Prehistoric Museum (also known as Prehistoric Museum), Price, Utah
CHA Palaeontological collection, Department of Mineral Resources, Thailand
CM Carnegie Museum of Natural History, Philadelphia, Pennsylvania, United States
CMN Canadian Museum of Nature (formerly National Museum of Canada [NMC]), Ottawa, Canada
CMNH Cleveland Museum of Natural History, Cleveland, Ohio, United States
CUST Changchun University of Sciences and Technology, Changchun, People's Republic of China
CV Municipal Museum of Chunking, People's Republic of China
DGBU Department of Geology, Pusan National University, Pusan, Republic of Korea
DMNH Denver Museum of Natural History, Denver, Colorado, United States
DNM Dinosaur National Monument, Jensen, Utah, United States
FCM Facultad de Ciencias Marinas, Ensenada, B.C., Mexico
FGGUB Facultatea de Geologie si Geofisica, Universitatea Bucuresti, Bucharest, Hungary
FIP Florida Institute of Paleontology, Dania Beach, Florida, United States
FMNH The Field Museum (formerly Field Museum of Natural History; Chicago Natural History Museum), Chicago, Illinois, United States
FPMN Fukui Prefectural Museum, Fukui, Japan
FSL Faculté des Sciences, Université Claude Bernard, Lyon, France
GIN see MGI
GMH Geological Museum of Heilongjang Province, Harbin, People's Republic of China
GMNH Gunma Museum of Natural History, Gumna, Japan
GP Universidade de São Paulo, São Paulo, Brazil
GSM see BGS GSM
HBV Hebei College, Chinese Academy of Geological Sciences, Beijing, People's Republic of China
HMN Museum für Naturkunde der Humboldt-Univerität zu Berlin, Berlin, Germany
IBEF Izumi Village Board of Education, Fukui Prefecture, Japan
IGCAGS Institute of Geology, Chinese Academy of Geological Science, Beijing, People's Republic of China
IGM [also **GI**] Mongolian Museum of Natural History (formerly Geological Institute Section of Palaeontology and Stratigraphy), The Academy of Sciences of the Mongolian People's Republic Geological Institute, Ulan Bator
IMM Inner Mongolia Museum, Hohhot, Inner Mongolia, People's Republic of China
IRSN Institut Royal des Sciences Naturelle de Belgique, Brussels, Belgium
IVP [also **IVPP**] Institute of Vertebrate Paleontology and Paleoanthropology, Academia Sinica, Beijing, People's Republic of China
KPE Kyungpook Earth, Kyungpook National University, Taegu, Korea
KUVP Kansas University, Vertebrate Paleontology, Lawrence, Kansas, United States
LACM Natural History Museum of Los Angeles County (also known as Los Angeles County Museum of Natural History), Los Angeles, California, United States
MACN Museo Argentino de Ciencias Naturales, Buenos Aires, Argentina
MAFI Magyar Allami Foldtani

Intezet, Budapest, Hungary

MB Museum für Naturkunde der Humboldt-Universität zu Berlin, Institut für Paläontologie, Berlin, Germany

MCDRD Muzeul Civilizat Dacie si Romane (formerly Deva County Museum), Deva, Romania

MCF "Carmen Funes" Museum, Plaza Huincul, Neuquén Province, Argentina

MCN Museu de Ciências Naturais da Fundação Zoobotânica do Rio Grande do Sul, Brazil

MCNA Museo Nacional de Ciencias Naturales de Alava, Madrid, Spain

MCP Museu de Ciências e Tecnologia, Pontificia Universidade Católica do Rio Grande do Sul, Porto Alegre, Brazil

MCT Museu de Ciências da Terra, Departmento Nacional da Producão Mineral, Rio de Janeiro, Brazil

MDE Musée des Dinosaures, Espéraza, France

MGI Mongolian Geological Institute (formerly Geological Institute Section of Palaeontology and Stratigraphy), Ulan-Bator, Mongolian People's Republic

MIWG Museum of Isle of Wight Geology, Sandown, Isle of Wight, England

ML Museum of Lourinhã, Lourinhã, Portugal

MLP Museo de La Plata, La Plata, Argentina

MN Museu Nacional, Universidade Federal do Rio de Janeiro, Brazil

MNA Museum of Northern Arizona, Flagstaff, Arizona, United States

MNHN Musée National d'Histoire Naturelle, Paris, France

MNHNUL Museu Nacional de História Natural [and] Departamento de Geologia e Centro de Geologia da Universidade de Lisboa, Lisbon, Portugal

MNN Musée National du Niger, Niamey, Republic of Niger

MNPL Natural History Museum of Marseilles, Bouches-du-Rhône, France

MOR Museum of the Rockies, Bozeman, Montana, United States

MPCA Museo de Ciencias Naturales, Universidad Nacional del Comahue, Buenos Aires, Neuquén, Argentina (see also MUCP)

MPEF Museo Paleontológico Egidio Feruglio, Chubut Province, Argentina

MPHN Museo Provincial de Historia Natural de La Pampa, La Pampa, Argentina

MSM Mesa Southwest Museum, Mesa, Arizona, United States

MUCP Museo de Ciencias Naturales de la Universidad Nacional del Comahue, Neuquén, Argentina

MWC Museum of Western Colorado, Grand Junction, Colorado, United States

NCSM North Carolina State Museum of Natural Sciences, North Carolina State University, Raleigh, North Carolina, United States

NHM Natural History Museum, Aix-en-Provence, France

NHMM Natuurhistorisch Museum Maastricht, Netherlands

NMMNH New Mexico Museum of Natural History and Science, Albuquerque, New Mexico, United States

NMC see CMN

NMV National Museum of Victoria, Milbourne, Australia

NSM National Science Museum, Tokyo, Japan

NWUV Northwestern University, Xian, Shaanxi, People's Republic of China

OMNH Sam Noble Oklahoma Museum of Natural History, University of Oklahoma, Norman, Oklahoma, United States

OUM Oxford University Museum, Oxford, England

PEFO Petrified Forest National Park, Arizona, North America

PIN Paleontological Institute, Academy of Science, Moscow, Russia

PMU Paleontological Institute of Uppsala University, Sweden

POL Musée de Poligny [specimens now housed in the Musée Archaeologique de Lons-le-Saunier, Jura, France]

PVPH Paleontología de Vertebrados, Museo del Neuquén, Argentina

P.W. Paleontological collection, Department of Mineral Resources, Bangkok, Thailand

QM Queensland Museum, Queensland, Australia

ROM Royal Ontario Museum, Toronto, Canada

RTMP [formerly TMP] Royal Tyrrell Museum of Palaeontology, Drumheller, Alberta, Canada

SDNHM San Diego Natural History Museum, San Diego, California, United States

SDSM Museum of Geology, South Dakota School of Mines and Technology, Rapid City, South Dakota, United States

SAM South African Museum, Cape Town, South Africa

SGM Ministère de l'Energie et des Mines, Rabat, Morocco

SMC Sedgwick Museum, University of Cambridge, Cambridge, England

SMM Science Museum of Minnesota, St. Paul, Minnesota, United States

SMNK Staatliches Museum für Naturkunde Karlsruhe, Karlsruhe, Germany

SMNS Staatliches Museum für Naturkunde, Stuttgart, Germany

SMP State Museum of Pennsylvania, Harrisburg, Pennsylvania, United States

SMU Southern Methodist University, Shuler Museum of Paleontology, Dallas, Texas, North America

SUP Shenandoah University, Winchester, Virginia, United States

TATE Tate Geological Museum, Casper College, Casper, Wyoming, United States

TM Teylers Museum, Haarlem, Netherlands
UA Université d'Antananarivo, Antananarivo, Madagascar
UC DBA University of Chicago, Department of Biology and Anatomy, Chicago, Illinois, United States
UCM University of Colorado Museum, Boulder, Colorado, United States
UCMP University of California Museum of Paleontology, Berkeley, California, United States
UE Universität Erlangen, Institut für Geologie und Mineralogie, Erlangen, Germany
UMMP University of Michigan Museum of Paleontology, Ann Arbor, Michigan, United States
UNM University of New Mexico, Albuquerque, United States (collection now at NMMNH)
UMNH Utah Museum of Natural History, University of Utah, Salt Lake City, Utah, United States
UNPSJB Universidad Nacional de la Patagonia, "San Juan Bosco," Argentina
USNM National Museum of Natural History (formerly United States National Museum), Smithsonian Institution, Washington, D.C., United States
UT Eberhard-Karis-Universität Tübingen, Institut und Museum für Geologie und Paläontologie, Tübingen, Germany
UUVP University of Utah, Vertebrate Paleontology Collection, Salt Lake City, Utah, United States
UW Geological Museum, University of Wyoming, Laramie, Wyoming, United States; also, Univeritát Tübingen, Institut und Museum für Geologie, Tübingen, Germany
UWA University of Western Australia, Perth, Western Australia, Australia
YPM Yale Peabody Museum of Natural History, Yale University, New Haven, Connecticut, United States
YXV Yuxi Regional Academy of Yunnan Province, Yunnan, People's Republic of China
ZPAL Instytut Paleobiologii (also known as Institute of Paleobiology), Polish Academy of Sciences, Warsaw, Poland

Appendix One: Displays, Sites and Attractions

The following is a list of places of interest around the world relating to dinosaurs and their Mesozoic world which are readily available to the general public.

This compilation includes places — usually located out of doors–where full-sized or mechanized figures of dinosaurs are displayed, where fossil bones and traces can be viewed *in situ*, sites from which dinosaur and other Mesozoic fossils have been collected, and so forth. The list excludes most museums (these having been listed in Appendix 2 of *S1*), mentioning or repeating such institutions only when an item of interest is displayed outside of the museum building (*e.g.*, the *Stegosaurus* outside the Cranbrook Institute of Science); it also excludes Mesozoic-related sites primarily known for taxa other than dinosaurs (*e.g.*, Icthyosaur State Park). Furthermore, this list includes only places that are of educational value, excluding such tourist attractions as "Flintstones Villages" (which feature fanciful life-sized dinosaur statues), caves (boasting Mesozoic-age rock formations), and fossil and dinosaur stores (which sometimes offer dinosaur displays).

The list is first broken down into alphabetically arranged sets based on territory of origin. Places of interest are then listed within their respective final grouping in alphabetical order according to the name of the attraction (when there is one), not necessarily the much larger attraction to which it belongs. For example, "Kingdom of the Dinosaurs," a ride at Knott's Berry Farm theme park, is listed under the former name rather than the latter; and the bronzed sculptures standing outside the Natural History Museum of Los Angeles County Museum are listed under "Dueling Dinosaurs," the actual name of the tableau, not the name of the institution.

When two addresses are given, the first is the actual location of the place; the second (which may indicate a different town or city) is the official mailing address from which additional information regarding it can be obtained. In some cases, only the mailing address is given.

Note that the majority of the listed "dinosaur parks" (*i.e.* permanent outdoor displays of usually life-sized figures) are privately owned, presenting models built by independent contractors often having little or no knowledge of (or interest in) paleontology, or reflecting ideas of their owners. Many of these parks were erected decades ago, when dinosaurs were still believed to be sluggish, tail-dragging lizard-like reptiles. Consequently, the dinosaur figures in these parks often tend to be somewhat inaccurately designed, given what is known today about dinosaurs. They are, however, educational in a number of ways, offering factual and useful information about these animals and their environments. Sometimes such displays are important for their place in history alone (*e.g.*, the statues at Crystal Palace, London, and Dinosaur Park, Rapid City).

For obvious reasons, not listed are the numerous temporary dinosaur attractions, such as the Dinofest events and the numerous displays of life-sized mechanical figures manufactured and distributed by such companies as Dinamation International Corporation and Kokoro Company, Ltd. (see *S1*).

Finally, note that, given the enormous popularity of dinosaurs in today's world, this rather subjectively compiled list neither pretends nor attempts to be all-inclusive.

Brazil

ARARAQUARA
Ouro District
São Paulo
City where dinosaur tracks preserved in sidewalks.

FRANCA
São Paulo
City where dinosaur tracks preserved in sidewalks.

PIAU LOCALITY
Bacino Rio do Peixe
Sousa, Paraíba
Early Cretaceous dinosaur footprints *in situ.*

RIFAINA
São Paulo
City where dinosaur tracks preserved in sidewalks.

Canada

THE CALGARY ZOO PREHISTORIC PARK
Calgary Zoological Society
Calgary
Life-sized figures in a badlands setting including over 100 species of plants suggesting Western Canada's subtropical past.

Note: This 6.7-acred display includes modern reconstructions of dinosaurs (including copies of those at

Dinosaur World, Colwyn Bay, Wales, and other places), as well as copies made by the Louis Paul Jonas Studios, in Churchtown, New York, of their figures for the 1964-65 New York World's Fair. Still on display is "Dinny," the fanciful larger-than-life *Apatosaurus* made for the zoo's original prehistoric-animals displays, which opened at the zoo's St. George's Island in 1935.

DINOSAUR PARK MINIATURE GOLF
4946 Clifton Hill
Niagara Falls, Ontario L2E 6S8
Miniature golf course featuring life-sized dinosaur models.

DINOSAUR PROVINCIAL PARK
P.O. Box 60
Patricia, Alberta TOJ 2KO
Vast Upper Cretaceous locality from which numerous dinosaur and other vertebrate fossils have been and are still being recovered, including complete skeletons.
Note: This locality — where field work is still being done — has been worked by the Sternberg family, Barnum Brown, and other famous fossil collectors.

DINOSAUR VALLEY
Wasaga Beach, Ontario LOL 2PO
Life-sized figures in a natural setting.

LE MONDE PREHISTORIQUE (also known as PREHISTORIC WORLD)
Upper Canada Road
Morrisburg, Ontario KOC 1XO
Life-sized figures in a natural setting.

PREHISTORIC PARKS
South Dinosaur Trail
(1 kilometer from City Centre)
P.O. Box 2686
Drumheller, Alberta TOJ OYO
Life-sized figures exhibited in a rugged, naturally rock-hewn setting.
Note: Other life-sized dinosaur statues are exhibited in various places in the town of Drumheller.

ROYAL TYRRELL MUSEUM OF PALAEONTOLOGY
P.O. Box 7500
Drumheller, Alberta TOJ OYO
Vast badlands outside the museum, a site for ongoing collection of Upper Cretaceous dinosaur and other fossils.

Cuba

VALLE DE LA PREHISTORIA
Baconao National Park, Santiago de Cuba
Apparently the world's largest park featuring life-sized figures of prehistoric animals, including at least 243 sculptures representing at least 59 species (Paleozoic, Mesozoic, and Cenozoic taxa), many of them dinosaurs, based on Zdenek Burian's paintings in Zdenek Spinar's book *Life Before Man*.

Germany

DINOSAURIERFRELLICHTMUSEUM (also known as DINOSAURIERPARK)
Muchehagen (northwest of Hanover)
Life-sized figures.
Note: The figures in this park are quite accurate, based upon current scientific data. They were made by Polish sculptor Krzyszstof Kuchino.

HAMBURG ZOO
Stellingen (near Hamburg)
Life-sized figures.
Note: These were the first fairly accurate life-sized dinosaur statues made. Inspired by the figures at Crystal Palace, they are the work of German sculptor Josef Franz Pallenberg, created around 1909 for Carl Hagenbeck's "Tiergarten," the first zoo without bars.

KLEINWELKA DINOSAUR PARK
Kleinwelka
Life-sized figures sculpted by Franz Gruss.

Great Britain

CHILTON CHINE
Isle of Wight, England
Fossilized *in situ* tracks (Lower Cretaceous) attributed to *Iguanodon* and *Megalosaurus*.

THE CRYSTAL PALACE
Sydenham, London SE20, England
Life-sized figures.
Note: Features the suite of life-sized models of dinosaurs and other prehistoric animals made by English sculptor Benjamin Waterhouse Hawkins under the direction of pioneer paleontologist Richard Owen and unveiled in 1854. These figures, historically distinguished as the first full-scale dinosaur models ever made, were constructed of brick, cement and plaster over iron frameworks. Though in many ways highly fanciful, they reflect what was known and believed about these animals during the Victorian era (see *D:TE, S1*).

THE DINOSAUR PARK
Blackgang Chine, Isle of Wight, England
Life-sized figures.
Note: This park, featuring numerous fiberglass-reinforced plastic figures, was instituted in 1972.

DINOSAUR WORLD
Eirias Park
Colwyn Bay, Wales
The largest collection of life-sized dinosaur figures in the British Isles.

Courtesy German Consulate, Los Angeles (negative #3706).

***Ceratosaurus* attacking *Stegosaurus*, full-scale models by sculptor Josef Franz Pallenberg exhibited at the Hamburg Zoo (Carl Hagenbeck's Tiergarten) in Stellingen, Germany.**

Note: These figures are fairly accurate. Copies can be seen in other parks, such as The Calgary Zoo Prehistoric Park.

FOSSIL BEACH
Charmouth (near Lyme Regis), Dorset, England
Site abundant with Middle Jurassic fossils, mostly ammonites.

GULLIVER'S KINGDOM
Matlock Bath
Derbyshire, England
Life-sized figures.

ISLE OF PURBECK
England
In situ fossil tracks from the Purbeck Limestone (Lower Cretaceous), attributed to *Iguanodon* and *Megalosaurus*.

Italy

PARCO ZOO DEL GARDA (PARCO DEI DINOSAURI)
Bussolengo, Pastrengo (VR)
Life-sized figures.
Note: These statues are fairly accurate.

People's Republic of China

HEILONGTAN-ZHANGIAWA-DAWA
Lufeng, Yunnan
Prosauropod skeletons *in situ* near the villages of Heilongtan-Zhangiawa-Dawa.

Poland

THE VALLEY OF DINOSAURS
Silesian Park of Culture and Recreation
Chorzow, Silesia
Life-sized figures of Mongolian dinosaurs.
Note: The original replicas were made by Silesian sculptors from the Katowice Art Workshop (including Henryk Fudali, Waldemar Madej, and Josef Sawicki) featuring dinosaurs collected in the Gobi Desert during the Polish-Mongolian Paleontological Expeditions of 1963–1971 and first exhibited in 1975. The scenario for the park was prepared by

Courtesy Zofia Kielan-Jaworowska.

Full-sized statue of the ankylosaurid *Saichania chulsanensis* at the Valley of Dinosaurs, Silesian Park of Culture and Recreation, in Chorzow, Poland. Phylogenetic analyses by James I. Kirkland (1998) and Robert M. Sullivan (1998) suggest that this species, the very similar *Tarchia gigantea*, and also a new unnamed genus and species, form their own clade within Ankylosauridae.

paleontologist Teresa Maryańska of the Museum of the Earth, Warsaw. The figures were made after consultation with such noted paleontologists as Zofia Kielan-Jaworowska, Halszka Osmólska, and Magdalena Borsuk-Bialynicka. The original figures have since deteriorated and are no longer on display. Newer, permanent figures have subsequently been installed.

Queensland

LAKE QUARRY ENVIRONMENTAL PARK
120 km southwest of Winton
Central west Queensland
c/o Queensland National Parks and Wildlife Service
Queensland
Fossilized dinosaur tracks *in situ.*

South Korea

SAMCHEANPO
Southern coast of South Korea
Dinosaur tracks *in situ.*

Spain

LA RIOJA PROVINCE
North-central Spain
Dinosaur tracks *in situ.*

RIBADESELLA
Northern Spain
Dinosaur tracks *in situ.*

Taiwan

NATIONAL MUSEUM OF NATURAL HISTORY
Taichung

Life-sized statues by sculptor David A. Thomas of the dinosaurs *Tyrannosaurus rex* and *Pentaceratops sternbergii.*

Thailand

Phu Wiang National Dinosaur Park
Phu Wiang

Preserving some of the first dinosaur dig sites found in Thailand (1976), one including casts of the bones of the sauropod *Phuwiangosaurus* displayed as found; also features a number of life-sized figures of dinosaurs.

United States

Academy of Natural Sciences of Philadelphia
19th and Parkway
Philadelphia, Pennsylvania 19103

Life-sized *Deinonychus* figures outside the building.

Badlands National Monument
Drainages of the White, Bad, and Cheyenne Rivers, southwestern South Dakota
c/o Black Hills, Badlands & Lakes Association
900 Jackson Boulevard
Rapid City, South Dakota 57702

Vast, mostly barren locality yielding numerous fossils (primarily Cenozoic mammals), but also vertebrates (*e.g.*, dinosaurs and mosasaurs) from the Late Cretaceous.

Note: In 1938, President Franklin Delano Roosevelt established this haven for fossil collecting (formerly the White River Badlands) as a National Monument.

Barton Cove Footprint Quarry
c/o Greenfield County Chamber of Commerce
395 Main Street
Greenfield, Massachusetts 01301

Fossil tracks *in situ.*

Black Hills
Black Hills, Badlands & Lakes Association
900 Jackson Boulevard
Rapid City, South Dakota 57702

Includes Dinosaur Park (Rapid City).

Courtesy Ted Hustead and Wall Drug Store.

Sculptor Emmett A. Sullivan's Wall Drug Store *Apatosaurus* in 1968 being moved to its final destination, near the northern "wall" of Badlands National Monument.

BLACK HILLS PETRIFIED FOREST
One mile off Interstate 90
Piedmont Route, P.O. Box 766
Piedmont, South Dakota 57769

A 22-acre site in which fossil trees of Early Cretaceous age can be seen *in situ*; includes a small museum.

BROOKFIELD ZOO
8400 West 31
Brookfield, Illinois 60513

Life-sized Louis Paul Jonas Studios *Edmontosaurus* figure.

Note: This is the original (or a copy) figure made for "Sinclair's Dinoland" exhibit at the 1964 New York World's Fair.

CHAMBER OF COMMERCE
Clayton-Union County Chamber of Commerce and Tourist Information Center
1101 South First Street (Highway 87)
P.O. Box 476
Clayton, New Mexico 88415

Life-sized figures exhibited outside the Chamber of Commerce headquarters/tourist information office.

Note: These models were constructed by Santa Fe, New Mexico roofing contractor Larry Wilson.

CLAYTON LAKE STATE PARK
Rural Route Box 20
Seneca, New Mexico 88437

In situ 100 million year old dinosaur, pterosaur, and crocodile footprints.

CLEVELAND-LLOYD DINOSAUR QUARRY
c/o Bureau of Land Management
P.O. Drawer A.B.
900 North 700 East
Price, Utah 84501

Within "The Dinosaur Triangle," the site of countless specimens of dinosaus and other fossil vertebrate from the Upper Jurassic Morrison Formation, primarily such dinosaurian genera as *Allosaurus, Ceratosaurus, Apatosaurus, Diplodocus, Barosaurus, Stegosaurus, Dryosaurus,* and *Camptosaurus.*

CLEVELAND MUSEUM OF NATURAL HISTORY
One Wade Oval Drive
University Circle
Cleveland, Ohio 44106

Life-sized Louis Paul Jonas Studios *Stegosaurus* figure outside the building.

Note: This is a copy of an original figure made for "Sinclair's Dinoland" exhibit at the 1964 New York World's Fair.

COLORADO NATIONAL MONUMENT
P.O. Box 438
Fruita, Colorado 81521

Locality for Upper Jurassic dinosaur fossils (including *Allosaurus, Apatosaurus*, and *Stegosaurus*); visitor center displaying a cast of the holotype humerus of Elmer S. Riggs' gigantic sauropod *Brachiosaurus altithorax.*

COUNTY SEAT AND COURTHOUSE TRACKS
421 North Main
Nashville, Arkansas 72204

Fossilized dinosaur footprints *in situ.*

CRANBROOK INSTITUTE OF SCIENCE
500 Lone Pine Road
P.O. Box 801
Bloomfield Hills, Michigan 48013

Life-sized Louis Paul Jonas Studios *Stegosaurus* figure outside the building.

Note: This is a copy of an original figure made for "Sinclair's Dinoland" exhibit at the 1964-65 New York World's Fair.

DEAD HORSE POINT STATE PARK
P.O. Box 609
Moab, Utah 83511

DINAMATION INTERNATIONAL CORPORATION
9560 Jeronimo Drive
Irvine, California 92618

Designs, manufactures, distributes, and displays robotic dinosaurs and other extinct creatures, most of them life-sized.

Note: While its main facility is not open to the general public, Dinamation exhibits its mechanical prehistoric creatures throughout the world, sometimes on a permanent, but usually a temporary basis.

DINAMATION INTERNATIONAL SOCIETY
550 Jurassic Court
Fruita, Colorado 81521

Tours led by noted paleontologists to dinosaur dig sites in Colorado, New Mexico, Utah, also Mexico, South America and Asia; discovery, collection, and preparation of fossils in the field and laboratory.

DINOLAND U.S.A.
Animal World
Walt Disney World
Orlando, Florida
P.O. Box 1000
Lake Buena Vista, Florida 32830

Life-sized "Audio-Animatronic" figures in the "Dinoland" section of the theme park; also a cast of the *Tyrannosaurus rex* skeleton popularly known as "Sue."

DINOSAUR FLATS
4381 FM 2673
Canyon Lake, Texas 75960

Courtesy Sinclair Oil Corporation.

Louis Paul Jonas studio's life-sized *Stegosaurus* model prior to its shipment to the "Sinclair Dinoland" exhibit, 1964 New York World's Fair. Copies of this model can now be seen at a number of locations.

Dinosaur Footprints Reservation
c/o Trustees of Reservations
Western Regional Office
P.O. Box 792
Stockbridge, Massachusetts 01261

Dinosaur tracks *in situ.*

DINOSAUR GARDENS
Interstate 10
San Gorgonio Pass (west of Palm Springs)
Cabazon, California

Giant figures of *Apatosaurus* and *Tyrannosaurus.*

Note: These oversized figures ("Dinney," the *Apatosaurus* measuring 150 feet long, the *Tyrannosaurus* 90 feet tall) were made during the 1970s by Claude Kenneth Bell. They stand near the The Wheel Inn truck stop.

DINOSAUR GARDENS
Utah Field House of Natural History
235 East Main Street
Vernal, Utah 84078
Life-sized figures.

Notes: These fairly accurate fiberglass figures were made by Utah sculptor Elbert H. Porter and assistants. Originally displayed in Draper, Utah, and then in other locations, they made their debut behind the Vernal Field House in 1977.

Also displayed on the mueum grounds are a cast (nicknamed "Dippy") of the Carnegie Museum of Natural History's original *Diplodocus carnegii* skeleton, a group tableau of life-sized cement models of *Ceratosaurus, Stegosaurus,* and *Camarasaurus* made by sculptor Millard Fillmore Malin, and a full-scale *Utahraptor* sculpted by David A. Thomas.

The town of Vernal (the upper anchor of "The Dinosaur Triangle") bills itself as the "Dinosaur Capital of the World." The area in and around Vernal is a kind of outdoor dinosaur display presenting life-

Photograph by the author.

Created by Claude Kenneth Bell, this gigantic "old-fashioned" *Apatosaurus* (still sporting a *Camarasaurus*-type head), at Dinosaur Gardens, in Cabazon, can be seen by drivers along Interstate 10 in California's San Gorgonio Pass.

sized figures, most notably very dated versions of *Tyrannosaurus* and *Allosaurus*. (Some dinosaur figures previously displayed at Vernal have been relocated to the town of Dinosaur, Colorado.)

DINOSAUR GARDENS
Highway 59
P.O. Box 98
Moscow, Texas 75960

Life-sized figures in a natural setting, enhanced by sound effects.

DINOSAUR GARDENS PREHISTORICAL ZOO (also known as DOMKE'S GARDENS)
11160 U.S. 23 South (south of Alpena)
Ossineke, Michigan 49766

Life-sized figures in a swampy setting.

Note: The figures were made by a "Mr. Domke." The paleontology presented is combined with Christian beliefs.

DINOSAUR HILL
Highway 340
Fruita, Colorado
c/o Museum of Western Colorado
P.O. Box 20000-5020
Grand Junction, Colorado 81502-5020

Marked by a bronze plaque, the quarry where paleontologist Elmer S. Riggs discovered the skeleton of *Apatosaurus excelsus* on exhibit at The Field Museum.

DINOSAUR LAND
Shenandoah Valley (between Winchester and Front Royal)
White Post, Virginia

Life-sized figures.

DINOSAUR NATIONAL MONUMENT
P.O. Box 128
Jensen, Utah 84035

An 80-acre stretch of land including parts of Utah and Colorado in the Morrison Formation.

Note: Located within "The Dinosaur Triangle," this area was declared a National Monument in 1915 by President Woodrow Wilson. It has yielded the

Courtesy State of Utah Division of Parks and Recreation.

Top: **Southern Utah sculptor Elbert H. Porter poses in the late 1970s with a pair of his fiberglass *Protoceratops* creations. These figures now reside in the Dinosaur Gardens behind the Utah Field House of Natural History in Vernal, Utah. *Bottom:* Sculpted by Millard Fillmore Malin, these full-scale cement statues of *Ceratosaurus, Stegosaurus,* and *Camarasaurus* stand in front of the Utah Field House of Natural History in Vernal, Utah.**

world's finest collection of Late Jurassic vertebrate fossils, including such "classic" dinosaurs as *Allosaurus, Ceratosaurus, Apatosaurus, Diplodocus, Barosaurus, Brachiosaurus, Camarasaurus, Stegosaurus, Dryosaurus,* and *Camptosaurus*. Work continues to be done at this site.

DINOSAUR PARK
940 Skyline Drive
Rapid City, South Dakota 57702

Life-sized figures on a hilltop.

Note: Created in 1936 as a WPA (Works Progress Administration) project atop a hill in the Black Hills, these concrete and steel figures are distinguished as constituting the first permanent outdoor display of dinosaurs in the United States. Most of them were made by Frank Lockhart and George McGraw under the direction of local sculptor Emmett A. Sullivan, based upon various paintings by Charles R. Knight at what is now The Field Museum, and by full-sized mechanical dinosaurs made for the "Sinclair Dinosaur Exhibit" at the 1933 Chicago World's Fair. On July 21, 1990, Dinosaur Park was placed on the Register of Historic Places. A number of dinosaur footprints have been found near this attraction.

DINOSAUR QUARRIES
P.O. Box 36
Kenton, Oklahoma 73946

DINOSAUR QUARRY
Dinosaur National Monument
P.O. Box 128
Jensen, Utah 84035

Morrison Formation (Upper Jurassic) quarry where over 2,000 dinosaur and other vertebrate fossil bones can be seen *in situ*.

Note: This display (dubbed the "Dinosaur Ledge"

Courtesy Travel Division, South Dakota Department of Highways, Pierre, South Dakota.

Sculptor Emmett A. Sullivan's life-sized *Triceratops horridus* and *Tyrannosaurus rex* statues at Dinosaur Park, Rapid City, South Dakota. These figures were based directly on both a painting by Charles R. Knight (see *D:TE* and *S1*) and a Sinclair Refining Company display at the 1933 Chicago World's Fair (see *S1*). By the time this photograph was taken, the *T. rex* figure had lost its teeth and fingers to the elements and vandalism.

or "Great Wall") opened in 1958 at Dinosaur National Monument. Exhibited outside the quarry enclosure is a Louis Paul Jonas Studios life-sized *Stegosaurus* model.

DINOSAUR RIDGE
16381 West Alameda Parkway
c/o Morrison Natural History Society
P.O. Box 654
Morrison, Colorado 80465
Locality for Upper Jurassic (Morrison Formation) footprints and bones.

DINOSAUR STATE PARK
West Street
Rocky Hill, Connecticut 06067
Late Triassic–Early Jurassic dinosaur footprints exhibited outside the main building.

DINOSAUR TRACK SITE
Cimarron Heritage Center
P.O. Box 1146
Boise City, Oklahoma 73933

DINOSAUR TRACKS
c/o Chamber of Commerce
P.O. Box 126
Hondo, Texas 78043

THE DINOSAUR TRIANGLE
Vernal, Utah/Price, Utah/Grand Junction, Colorado
c/o Castle Country Travel Council
P.O. Box 1037
Price, Utah 84501

DINOSAUR VALLEY
c/o Museum of Western Colorado
P.O. Box 20000-5020
Grand Junction, Colorado 81502-5020
One of the world's richest deposits of Upper Jurassic (Morrison Formation) dinosaur fossils, including bones and footprints.
Note: The southern anchor point of the "Dinosaur Triangle," Dinosaur Valley operates under the auspices of the Museum of Western Colorado.

DINOSAUR VALLEY STATE PARK
P.O. Box 396
Glen Rose, Texas 76043
Dinosaur footprints *in situ* in the Glen Rose Formation (Lower Cretaceous) limestone, including those left by a giant theropod apparently stalking a sauropod, possibly *Acrocanthosaurus* and *Pleurocoelus*, respectively; also life-sized dinosaur figures (originally made by Louis Paul Jonas Studios for "Sinclair's Dinoland" exhibit, 1964-65 New York World's Fair).

DINOSAUR WORLD
5154 Harvey Tew Road
Plant City, Florida 33565
Approximately 100 life-sized dinosaur and other Mesozoic reptile figures.

DUELING DINOSAURS
Los Angeles County Museum of Natural History
900 Exposition Boulevard
Los Angeles, California 90007
On museum grounds, life-sized figures of *Tyrannosaurus rex* and *Triceratops horridus*.
Note: Completed in 1998, these bronze statues — based on mounted skeletons on exhibit inside the museum — were made by Douglas Van Howd of Sierra Sculpture.

EPCOT CENTER
Walt Disney World
Orlando, Florida
P.O. Box 1000
Lake Buena Vista, Florida 32830
Life-sized "Audio-Animatronic" figures.

The FARWELL DINOSAUR PARK (formerly LAND OF KONG DINOSAUR PARK)
Highway 187 (near Beaver Dam)
8 miles west of Eureka Springs, Arkansas
Life-sized figures.
Note: The design of a few of these more than 100 figures was influenced by the statues at Dinosaur Park, Rapid City, South Dakota.

GARDEN PARK FOSSIL AREA
c/o Garden Park Paleontological Society
P.O. Box 313
Cañon City, Colorado 81215
Items of interest include a working laboratory and a full-scale *Allosaurus* model made by sculptor David A. Thomas.

GLEN CANYON NATIONAL RECREATION AREA
Visitor Center
P.O. Box 1597
Page, Arizona
Early Jurassic locality from which dinosaurs and other fossil vertebrates have been collected.

GRAND CANYON CAVERNS (also known as DINOSAUR CAVERNS)
(between Sligman and Kingman)
Old Highway 66
P.O. Box 108
Peach Springs, Arizona 86434
Life-sized *Tyrannosaurus rex* figures standing outside the entrance to the 345 million year old Grand Canyon Caverns.

ISLANDS OF ADVENTURE
Universal Studios Escape
Orlando, Florida

Courtesy Sinclair Oil Corporation.

Apatosaurus **model under construction at the Louis Paul Jonas Studios in upstate New York prior to its appearance at "Sinclair's Dinoland" exhibit at the 1964 New York World's Fair. This figure can now be seen at Dinosaur Valley State Park, Glen Rose, Texas. Seated above is Louis Paul Jonas.**

Life-sized figures, activity centers, other activities.

Note: Based on the motion picture *Jurassic Park* (1993), this attraction, which opened in 1998, treats the dinosaurs as if they are real animals and not mechanical creatures displayed at a theme park.

JURASSIC PARK—THE RIDE
Universal Studios Hollywood
Universal City, California 91608

Life-sized animatronic figures.

Notes: Opening in 1996 and based on the 1993 blockbuster motion picture *Jurassic Park*, this is essentially a boat-ride attraction enhanced by mechanical creatures. Although the dinosaur figures (made by the Sarcos company) are reasonably accurate, some have been "embellished" (*e.g.*, *Dilophosaurus* equipped with flaring frills and poison saliva) to conform with those seen in the movie.

Fanciful mechanical dinosaurs also figure into

the movie-based "Back to the Future" ride at the Universal theme parks.

KINGDOM OF THE DINOSAURS
Knott's Berry Farm
Buena Park, California
Life-sized mechanical figures.
Note: Part of an amusement park ride which opened in 1988, these dinosaurs are fairly accurate by modern standards. They were created by Bill Novey and Joe Garlington of the Art Technology company; Kevin Nadeau designed the sound effects.

LURAY REPTILE CENTER
Dinosaur Park & Petting Zoo
1087 US Highway 221 West
Luray, Virginia 22835
Life-sized dinosaur figures, also live reptiles and birds.

MILL CANYON DINOSAUR TRAIL
c/o Bureau of Land Management
Grand Resource Area
885 South Sand Flats Road
Moab, Utah 84532

MOENAVE DINOSAUR TRACKS
West of Tuba City (en route to the village of Moenave)
Northwestern Arizona
In situ concentration of dinosaur footprints from the Kayenta Formation (Lower Jurassic), including theropods and ornithischians.

MUSEUM OF THE ROCKIES
Montana State University
600 West Kagy Boulevard
Bozeman, Montana 59717
Outside the museum, a pair of full-scale models of *Maiasaura peeblesorum* made by sculptor David A. Thomas.

MYSTERY HILL & PREHISTORIC FOREST
Route 3
Marblehead, Ohio 43440
Life-sized animated figures in an adventure scenario.
Note: This park was created by James Q. Sidwell.

NEW MEXICO MUSEUM OF NATURAL HISTORY AND SCIENCE
1801 Mountain Road NW
Albuquerque, New Mexico 87104
Displayed outside, life-sized bronze figures of *Pentaceratops sternbergii* (nicknamed "Spike") and *Gorgosaurus libratus* made during the 1980s by sculptor David A. Thomas.

PAINTED DESERT
(East of the Little Colorado River)
East-central Arizona
Site of numerous Upper Triassic fossil discoveries including dinosaurs, other vertebrates, and invertebrates.
Note: The Painted Desert includes Petrified Forest National Park.

PETRIFIED CREATURES MUSEUM OF NATURAL HISTORY
U.S. Route 20
Richfield, New York
P.O. Box 751
Cooperstown North, New York 13439
Life-sized figures displayed outdoors.

PETRIFIED FOREST NATIONAL PARK
P.O. Box 2217
Petrified Forest National Park, Arizona 86028
Expansive site of countless specimens of giant fossilized logs (commonly referred to as "petrified wood") from the Upper Triassic Chinle Formation left *in situ*, a vast locality from which a number of early dinosaur, phytosaur, and other vertebrate specimens have been collected; includes museums.
Note: This grand south-central Arizona site, part of the Painted Desert comprising more than 52,500 acres, was declared a National Forest in 1906 by President Theodore Roosevelt.

PINE BUTTE SWAMP PRESERVE
HC 58, P.O. Box 34B
Choteau, Montana 59442

POTASH ROAD DINOSAUR TRACKS
c/o Bureau of Land Management
Grand Resource Area
885 South Sand Flats Road
Moab, Utah 84532

POWDER HILL DINOSAUR PARK
c/o Chamber of Commerce
393 Main Street
Middlefield, Connecticut 06455

PREHISTORIC FOREST (also known as DINOSAUR LAND)
U.S. 12 (¼ mile west of Hayes State Park)
Irish Hills
Onsted, Michigan
Life-sized animated figures seen in a simulated trip back through time.
Note: This park was created by James Q. Sidwell. At least some of its figures seem to be copies of those at Dinosaur Land at White Post, Virginia.

PREHISTORIC GARDENS
36848 Highway 101 South
Port Orford, Oregon 97465

Life-sized figures.

Note: Open since the late 1950s, this park — set in a rain forest — was the creation of E. V. (Ernie) Nelson.

PRIMEVAL WORLD
Disneyland® Park
P.O. Box 3232
Anaheim, California 92803

Life-sized mechanical figures.

Note: These "Audio-Animatronic" dinosaurs and other moving prehistoric-creature models were made for the Magic Skyway ride inside the Ford Motor Company Pavilion, a drive-through attraction at the 1964 New York World's Fair. The figures were designed and built by technicians at WED (Walter Elias Disney) Enterprises. Though well made and realistic, they are dated in some ways, based on older paleontological information.

RABBIT VALLEY TRAIL THROUGH TIME
c/o Museum of Western Colorado
P.O. Box 20000-5020
Grand Junction, Colorado 81502-5020

Located on public land administered by the Bureau of Land Management, in the southern part of "The Dinosaur Triangle," includes a 1.5-mile "Trail Through Time," an "outdoor museum" featuring *in situ* Upper Jurassic (Morrison Formation) fossils.

RED FLEET RESERVOIR STATE PARK
Steinaker Lake North 4334
Vernal, Utah 84078

RIGGS HILL
c/o Museum of Western Colorado
P.O. Box 20000-5020
Grand Junction, Colorado 81502-5020

Marked by a bronze plaque, the quarry where paleontologist Elmer S. Riggs found remains of *Allosaurus*, *Camarasaurus*, *Stegosaurus*, and the holotype of the giant sauropod *Brachiosaurus altithorax*.

SAUROPOD TRACKSITE
c/o Bureau of Land Management
Grand Resource Area
885 South Sand Flats Road
Moab, Utah 84532

UNCLE BEAZLEY
National Museum of Natural History
Smithsonian Institution
Washington, D.C.

Life-sized *Triceratops* figure.

Note: Displayed outside the natural history museum since 1967 is the *Triceratops horridus* model — its mechanical head no longer in operation — made by the Louis Paul Jonas Studios. This is a copy of the one Jonas made earlier for "Sinclair's Dinoland" exhibit at the 1964 New York World's Fair.

UNIVERSITY OF WYOMING GEOLOGICAL MUSEUM
Geology Department
University of Wyoming
Laramie, Wyoming 82070

Life-sized *Tyrannosaurus rex* figure outside the museum.

VALLEY ANATOMICAL PREPARATIONS
9520 Owensmouth Avenue
Unit 5
Chatsworth, California 91311

Resin casts of dinosaur and other fossil specimens for purchase or rent, includes a display room.

WALL DRUG DINOSAUR
Main Street
Wall, South Dakota 57790

Life-sized *Apatosaurus* figure.

Note: This 80-foot model, standing outside the entrance of Ted Hustead's famous Wall Drug Store, at the northern extension of the "wall" of Badlands National Monument, was designed during the late 1950s under the supervision of Emmett A. Sullivan, creator of the original dinosaur statues at Rapid City's Dinosaur Park. Other dinosaur-related attractions can be found within the Wall Drug mall.

WARNER VALLEY TRACKSITE
c/o Bureau of Land Management
Dixie Resource Area
255 North Bluff Street
Saint George, Utah 84770

Fossil tracks *in situ*.

THE WYOMING DINOSAUR CENTER
P.O. Box 868
Thermopolis, Wyoming 82443

"Dig-for-a-Day Adventures" including museum, preparation lab, "Dig-Site Tours."

Appendix Two: Further Reading

Following is a selective list of popular and semipopular books and booklets about dinosaurs (or about dinosaurs as well as other forms of prehistoric life) that should be of interest and value both to the casual and advanced reader. Many of these titles are now out of print; these can sometimes be obtained from stores dealing in collector's or used books, via certain Internet websites, or found in public, private, and school libraries, as well as other appropriate sources.

It must be stressed, however, that books on this list published before the so-called "Dinosaur Renaissance" that began in the mid–1970s are generally out of date in any of a number of ways, sometimes very much so. Some of these older books are valuable nonetheless, either for their historical or biographical information, or as interesting reflections of how dinosaurs were interpreted during earlier decades, for their photographs of important or interesting specimens or their artwork, or because they were authored by a scientist of note. Indeed, even some of the more recent books may contain their own share of inaccuracies, influenced as they are by an author's subjective opinions, personal biases, or agendas. Therefore, the reader is advised to use discretion when evaluating any of the books in this compilation.

This list does not include journal and other periodical articles, monographs, most (but not all) symposium volumes, most strictly technical books, most juvenile or children's books, books about paleontology, geology, zoology, or other general subjects in which the material pertaining to dinosaurs is minimal or inconsequential, books primarily about other kinds of Mesozoic life (plants, invertebrates, fish, therapsids, "thecodontians," pterosaurs, birds, marine reptiles, mammals, *etc.*), "books" in the CD-ROM format, books mostly about dinosaurs in the media and popular arts, dinosaur-related collectibles and memorabilia, and, of course, fiction.

For references cited specifically in the text of this supplement, consult the general bibliography in the back of this book.

[Anonymous], *Chicago's Dinosaurs*. "Written and produced by Peter R. Crane, Nina M. Cummings, Marlene H. Donnelly, Ron Dorfman, John J. Flynn, Laura D. Gates, Lori L. Grove, Zbigniew T. Jastrzebski, Olivier C. Rieppel, Clara R. Simpson, William F. Simpson, Nancy E. Walsh, and John S. Weinstein." Chicago: Field Museum of Natural History, 1994.

[Anonymous]. *The Dinosaur Journal*. Illustrated by Helen I. Driggs. Philadelphia: Running Press, 1988.

[Anonymous]. *Dinosaur Provincial Park: World Heritage Site*. Acknowledgements: Deborah and Robert Enns, Sandra Leckie, and John Walper; special thanks to Philip J. Currie. Alberta, Canada: Wildland Publishing, [no date; 1979?].

[Anonymous]. *Dinosaurs from China*. Victoria, Melbourne, Australia: Council of the National Museum of Victoria, 1982.

[Anonymous]. *Discovering Dinosaurs: A Compelling New Look at an Ancient Subject*. Philadelphia: The Academy of Natural Sciences of Philadelphia, [do date; 1983].

[Anonymous]. *Encyclopedia of Dinosaurs*. Contributing writers: Brooks Britt, Kenneth Carpenter, Catherine A. Forster, David D. Gillette, Mark A. Norell, George Olshevsky, J. Michael Parrish. and David B. Weishampel. New York: Beekman House (Publications International), distributed by Crown Publishers, 1990.

[Anonymous, "by the Editors of Time-Life Books"]. *Life Before Man*. Editor-in-Chief: Hedley Donovan. New York: Time-Life Books, 1972.

[Anonymous]. *Ranger Rick's Dinosaur Book*. Illustrations from Robert Byrd, Alex Ebel, Walter Furgoson, Biruta Akerbergs, John D. Dawson, Mark Hallett, Eleanor M. Kish, Charles R. Knight, and Rudolph F. Zallinger. National Wildlife Federation, 1984. For young readers.

Abel, Othenio. *Geschichte und Methode der Rekonstruktion vorzeitischer Wirbeltiere*. Jena, 1925.

_____. *Das Reich der Tiere*. Berlin: Im Deutscher Verlag, 1939.

Adams, W. H. Davenport. *Life in the Primeval World* (founded on S. Meunier's *Les Animaux D'Autrefois*). Edinburgh and New York: T. Nelson and Sons, Paternoster Row, 1872.

Alexander, R. McN. *Dynamics of Dinosaurs and Other Extinct Giants*. New York: Columbia University Press, 1989.

Allen, Tom, Jane D. Allen, and Savanah Waring Walker.

Dinosaur Days in Texas. Dallas: Hendrick-Long Publishing Company, 1989.

Alvarez, W. Walter. *T. rex and the Crater of Doom*. Princeton, New Jersey: Princeton University Press, 1997.

Andrews, Roy Chapman. *All About Dinosaurs*. Illustrated by Thomas W. Voter. New York: Random House, 1953. For young readers.

_____. *Exploring with Andrews*. New York: C. P. Putnam's Sons, 1938.

_____. *In the Days of the Dinosaurs*. Illustrated by Jean Zallinger. New York: Random House, 1959. For Young Readers.

Araki, Kazunari. *The Dinosaur Sculptures*. Japan: Model Graphix, 1987.

Archibald, J. David. *Dinosaur Extinction and the End of an Era: What the Fossils Say*. New York: Columbia University Press, 1996.

Ash, Sidney. *Petrified Forest: The Story Behind the Scenery* [revised edition]. [Petrified Forest National Park, Hollbrook, Arizona]: Petrified Forest Museum Association, 1985.

Ash, Sidney R., and David D. May. *Petrified Forest: The Story Behind the Scenery*. Petrified Forest National Park, Hollbrook, Arizona: Petrified Forest Museum Association, 1969.

Ashworth, William B., Jr. *Paper Dinosaurs, 1824–1969: An Exhibition of Original Publications from the Collections of the Linda Hall Library*. Kansas City, Missouri: Linda Hall Library, 1996. Photographs by Ashworth, Exhibit Preparation by Bruce Bradley.

Augusta, Josef. *Prehistoric Animals*. Illustrated by Zdeněk Burian. Translated by Greta Hort. London: Paul Hamlyn, 1960.

Averett, Walter, editor. *Paleontology and Geology of the Dinosaur Triangle*. Grand Junction, Colorado: Museum of Western Colorado, 1987.

Bakker, Robert T. *The Dinosaur Heresies: New Theories Unlocking the Mystery of the Dinosaurs and Their Extinction*. Illustrated by Bakker. New York: William Morrow and Company, 1986.

Barnett, Lincoln, and the Editorial Staff of *Life* magazine. *The World We Live In*. New York: Time Incorporated, 1955. Reprints from the "World We Live In" series of articles published in *Life* magazine, illustrated by reproductions of paintings made by Rudolf F. Zallinger for the Peabody Museum of Natural History.

Bausum, Ann. *Dragon Bones and Dinosaur Eggs*. Washington, D.C.: National Geographic Press, 2000. The account of Roy Chapman Andrews and the American Museum of Natural History expeditions to Mongolia.

Benton, Michael J. *Dinosaurs*. Illustrated by James Field, others. London: Kingfisher Chambers, 1998.

_____. *The Kingfisher Pocket Book of Dinosaurs*. London: Kingfisher Books, 1984. (Reprinted in 1984 as *The Dinosaur Encyclopedia* by Simon and Schuster [a Wanderer Book], New York.

_____. *On the Trail of Dinosaurs*. London: Eagle Editions, 1989.

_____. *The Penguin Historical Atlas of the Dinosaurs*. London: Penguin Books, 1996.

_____. *Vertebrate Paleontology*. London: Unwin Hyman, 1990.

Bird, Roland T. *Bones for Barnum Brown: Adventures of a Dinosaur Hunter*. V. Theodore Schreiber, editor, foreword by Edwin H. Colbert, introduction by James O. Farlow. Fort Worth, Texas: Texas Christian University Press, 1985.

Bishop, Nic. *Digging for Bird-Dinosaurs: An Expedition to Madagascar*. "Scientist in the Field" Series. New York: Houghton Miffin Co., 2000.

Blows, William T. *Reptiles on the Rocks*. Isle of Wight, England: Isle of Wight County Council, 1978.

Bonaparte, José F. *Dinosaurios de America del Sur*. Buenos Aires: Museo Argentino de Ciencias Naturales "Bernardino Rivadaria," 1996.

_____. *Los Dinosaurios de la Patagonia Argentina*. Illustrated by Jorge Blanco. Buenos Aires: Museo Argentino de Ciencias Naturales "Bernardino Rivadaria," 1998.

Breed, William J. *The Age of Dinosaurs in Northern Arizona*. Drawings by Barton A. Wright. Flagstaff: Museum of Northern Arizona, 1968.

Brett-Surman, Michael K., and Thomas R. Holtz, Jr. *The World of Dinosaurs*. Introduction by Jack Horner, main illustrations (designed for United States Postal Service) by James Gurney. Shelton, Connecticut: The Greenwich Workshop Press, 1998.

Brochu, Christopher A., John A. Long, John D. Scanlon, and Paul Willis. *The Time-Life Guides: Dinosaurs*. Consulting Editor: Michael K. Brett-Surman. New York: Time-Life Books (Time- Life, Inc.), 2000.

Calagrande, John, and Larry Felder. *In the Presence of Dinosaurs*. Foreword by Jack Horner. Illustrated by Felder. New York: Time-Life, Inc., 2000.

Carpenter, Kenneth. *Eggs, Nests, and Baby Dinosaurs—A Look at Dinosaur Reproduction*. Bloomington, Indiana: Indiana University Press, 2000.

Carpenter, Kenneth, and Philip J. Currie, editors. *Dinosaur Systematics: Approaches and Perspectives*. Cambridge, England: Cambridge University Press, 1990.

Carpenter, Kenneth, Karl F. Hirsch, and John R. Horner, editors. *Dinosaur Eggs and Babies*. Cambridge, England: Cambridge University Press, 1994.

Carroll, Robert L. *Vertebrate Paleontology and Evolution*. New York: W. H. Freeman and Co., 1988.

Case, Gerard R. *A Pictorial Guide to Fossils*. New York: Van Nostrand Reinhold Company, 1982.

Casier, Edgard. *Les Iguanodons de Bernissart*. Brussell,

Belgium: Editions du Patrimoine, Institute Royal des Sciences Naturelles de Belgique, 1978.
Charig, Alan. *A New Look at the Dinosaurs.* London: British Museum (Natural History), 1979. (Reprinted, New York: Facts on File, 1983.)
Charig, Alan, and Brenda Horsefield. *Before the Ark.* London: British Broadcasting Corporation, 1975.
Chiappe, Luis M., and Lowell Dingus. *Walking on Eggs: The Astonishing Discovery of Thousands of Dinosaur Eggs in the Badlands of Patagonia.* Illustrations by Nicholas Frankfurt. New York: Scribner, 2001.
Cohen, Daniel. *Monster Dinosaur.* New York: J. B. Lippincott, 1983.
Colbert, Edwin H. *The Age of Reptiles.* New York: W. W. Norton & Company, 1965. (Reprinted as a trade paperback in 1966 by The Norton Library [W. W. Norton & Company].)
_____. *Digging into the Past: An Autobiography.* New York: Dembner Books, 1989.
_____. *The Dinosaur Book: The Ruling Reptiles and Their Relatives.* New York: McGraw-Hill (published for the American Museum of Natural History), 1945.
_____. *Dinosaurs.* New York: The American Museum of Natural History, Man and Nature Publications, Science Guide No. 70, 1953.
_____. *Dinosaurs: An Illustrated History.* New York: A Dembner Book produced for Red Dembner Enterprises Corp. (Hammond), 1983.
_____. *Dinosaurs of the Colorado Plateau.* Flagstaff, Arizona: Museum of Northern Arizona, Volumer 54, Numbers 2 and 3 in the *Plateau* series, 1983.
_____. *Dinosaurs: Their Discovery and Their World.* New York: E. P. Dutton & Co., 1961.
_____. *Evolution of the Vertebrates.* New York: John Wiley & Sons, 1955.
_____. *A Fossil-Hunter's Notebook: My Life with Dinosaurs and Other Friends.* New York: E. P. Dutton, 1980.
_____. *The Little Dinosaurs of Ghost Ranch.* New York: Columbia University Press, 1995.
_____. *Men and Dinosaurs: The Search in Field and Laboratory.* New York: E. P. Dutton and Company, 1968. (Reprinted as a trade paperback in 1984 as *The Great Dinosaur Hunters and Their Discoveries* by Dover Publications, New York.)
_____. *Wandering Lands and Animals.* Foreword by Laurence M. Gould. London: Hutchinson & Co., 1974. (Reprinted as a trade paperback in 1985 by Dover Publications, New York.)
_____. *William Diller Matthew, Paleontologist: The Splendid Drama Observed.* New York: Columbia University Press, 1992.
_____. *The World of Dinosaurs.* Illustrated by George Geygan. New York: Home Library Press, 1961. For young readers.
_____. *The Year of the Dinosaur.* Illustrated by Margaret Colbert. New York: Charles Scribner's Sons, 1977.
Colbert, Edwin H., and William A. Burns. *Digging for Dinosaurs.* New York: Columbia Record Club, 1960. Includes record narrated by Walter Cronkite and slide set.
Colbert, Edwin H., and Michael Morales. *Evolution of the Vertebrates* [revised edition]. New York: Wiley-Liss, 1991.
Cole, Stephen, and Stephen Cole. *Walking with Dinosaurs.* London: Dorling Kindersley Publishing, 2000. Companion volume to the *Walking with Dinosaurs* television series and DVD.
Cooley, Brian, and Mary Ann Wilson. *Make-A-Saurus, My Life with Raptors and other Dinosaurs.* Forewords by Philip J. Currie and Christopher Sloan. Toronto and Vancouver: Annick Press, 2000.
Costa, V. *The Dinosaur Safari Guide: Tracking North America's Prehistoric Past.* Stillwater, New Mexico: Voyageur Press, 1994.
Cowen, Richard. *History of Life,* 3rd Edition. Malden, Massachusetts: Blackwell Science, 2000.
Cox, Berry. *Prehistoric Animals.* New York: Grosset and Dunlap, 1970.
Cox, Berry, R. J. G. Savage, Brian Gardiner, and Colin Harrison. *The Simon and Schuster Encyclopedia of Dinosaurs and Prehistoric Creatures: A Visual Who's Who of Prehistoric Life.* Revised and updated by Douglas Palmer. "Artist for Reptiles, Ruling Reptiles and Mammal-like Reptiles," Steve Kirk. New York: Simon and Schuster, 1999.
Crumly, Charles R., editor. *Dinosaur Imagery.* Foreword by Philip J. Currie, preface by Paul C. Sereno and John J. Lanzendorf, photography by Michael Tropea. Orlando, Florida: Academic Press, 2000. Reproductions of more than 100 paintings and drawings, 40 sculptures, and other art pieces representing dinosaurs by world renowned "paleoartists" in Lanzendorf's collection, with commentary by 20 noted paleontologists.
Currie, Philip J., and Kevin Padian, editors. *Encyclopedia of Dinosaurs.* Foreword by Michael Crichton. San Diego: Academic Press, 1997.
Czerkas, Stephen A., and Sylvia J. Czerkas. *My Life with the Dinosaurs.* New York: Byron Preiss and Pocket Books/Simon and Schuster (A Minstrel Book), 1989.
_____. *Dinosaurs: A Global View.* Illustrations by Douglas Henderson, Mark Hallett, and John Sibbick. New York: Mallard Press (An imprint of BDD Promotional Book Company, in association with Dragon's World Ltd., London), 1990.
Czerkas, Sylvia Massey, and Donald F. Glut. *Dinosaurs, Mammoths and Cavemen: The Art of Charles R. Knight.* Foreword by Edwin H. Colbert. New York: E. P. Dutton, 1982.
Czerkas, Sylvia J., and Everett C. Olson, editors. *Dinosaurs Past and Present,* Volumes 1 and 2. Introduction by John M. Harris, illustrations including reproductions of drawings, paintings, and sculptures by noted "paleoartists." Seattle: Natural History Museum of Los

Angeles County, in association with University of Washington Press, 1987.

Davidson, Jane Pierce. *The Bone Sharp: The Life of Edward Drinker Cope*. Philadelphia: The Academy of Natural Sciences of Philadelphia, Special Publication No. 17, 1997.

Dean, Dennis R. *Gideon Mantell and the Discovery of Dinosaurs*. Cambridge, England: Cambridge University Press, 1999.

Debus, Allen A., Bob Morales, and Diane Debus. *Dinosaur Sculpting: A Complete Beginner's Guide from Dragon Attack! and Hell Creek Creations*. Bartlett, Illinois: Hell Creek Creations, 1995.

de Camp, L. Sprague, and Catherine Crook de Camp. *The Day of the Dinosaur*. Doubleday & Company: Garden City, New York, 1968.

DeCourten, Frank. *Dinosaurs of Utah*. Illustrated by Carel Brest Van Kempen. Salt Lake City: University of Utah Press, 19998.

Desmond, Adrian J. *Archetypes and Ancestors*. London: Blond and Briggs, 1982.

_____, *The Hot-Blooded Dinosaurs: A Revolution in Paleontology*. Great Britain: Blond & Briggs, 1975. (Reprinted, New York: The Dial Press/James Wade, 1976.)

Dingus, Lowell. *Next of Kin*. New York: Rizzoli International Publications, 1996.

Dingus, Lowell, and Luis M. Chiappe. *The Tiniest Dinosaurs: Discovering Dinosaur Eggs*. New York: Random House (A Doubleday Book for Young Readers), 1999.

Dingus, Lowell, Eugene S. Gaffney, Mark A. Norell, and Scott D. Sampson. *The Halls of Dinosaurs: A Guide to Saurischians and Ornithischians*. New York: American Museum of Natural History, 1996.

Dingus, Lowell, and Mark A. Norell. *A Nest of Dinosaurs: The Story of Oviraptor*. Garden City, New York: Doubleday and Company, 2000. For young readers.

Ditmars, Raymond L. *The Book of Prehistoric Animals*. Illustrated by Helene Carter. Philadelphia and London: J. B. Lippincott Company, 1935. For young readers.

Dixon, Dougal. *Dougal Dixon's Dinosaurs*. Honesdale, Pennsylvania: Boyds Mills Press, 1993.

Dodson, Peter. *The Horned Dinosaurs*. Princeton, New Jersey: Princeton University Press, 1996.

Dong Zhiming [also known as Zhi-Ming]. *Dinosaurian Faunas from China*. Beijing: China Ocean Press, and Berlin: Springer Verlag, 1992.

_____. *Dinosaurs from China*. Beijing: China Ocean Press, 1987.

Dong Zhi-Ming, Yoshikagu Hasegawa, and Yoichi Azuma. *The Age of Dinosaurs in Japan and China*. Fukui, Japan: Fukui Prefectural Museum, 2000.

Dunkle, David H. *The World of Dinosaurs*. Washington, D.C.: Smithsonian Institution, 1957.

Epstein, Sam and Beryl. *Prehistoric Animals*. Illustrated by W. R. Lohse. New York: Franklin Watts, 1956.

Fackham, Margery. *Tracking Dinosaurs in the Gobi Desert*. New York: Twenty-First Century Books (A Division of Henry Holt and Company), 1997.

Farlow, James O. *The Dinosaurs of Dinosaur Valley State Park*. Austin: Texas Parks and Wildlife Department, 1993.

Farlow, James O., and Michael K. Brett-Surman, editors. *The Complete Dinosaur*. Art editor: Robert F. Walters. Bloomington and Indianapolis, Indiana: Indiana University Press, 1997.

Fastovsky, David E., and David B. Weishampel. *The Evolution and Extinction of the Dinosaurs*. Cambridge, United Kingdom: Cambridge University Press, 1996.

Fenton, Carroll Lane, and Mildred Adams Fenton. *The Fossil Book: A Record of Prehistoric Life*. Garden City, New York: Doubleday & Company, 1958.

Fiffer, Steve. Tyrannosaurus *Sue: The Extraordinary Saga of the Largest, Most Fought Over* T. rex *Ever Found*. New York: W. H. Freeman and Company, 2000.

Ford, Tracy Lee. *How to Draw Dinosaurs*. Illustrated by Ford. Folsom, California: Prehistoric Times, 1999.

Frankel, Charles. *The End of the Dinosaurs: Chicxulub Crater and Mass Extinctions*. Cambridge: University of Cambridge Press, 2000.

Fraser, Nicholas C., and Hans-Dieter Sues, editors, *In the Shadow of the Dinosaurs: Early Mesozoic Tetrapods*. Cambridge, England: University of Cambridge Press, 2000.

Gallagher, William B. *When Dinosaurs Ruled New Jersey*. New Brunswick, New Jersey: Rutgers University Press, 1997.

Gillette, David D. *Seismosaurus: The Earth Shaker*. Illustrated by Mark Hallett. New York, Columbia University Press, 1994.

Gillette, David D., and Martin G. Lockley, editors. *Dinosaur Tracks and Traces*. Cambridge, New York, and Melbourne: Cambridge University Press, 1989.

Glut, Donald F. *The Age of Dinosaurs*. Illustrated by Helen I. Driggs. Philadelphia: Running Press Book Publishers, 1994. Coloring book for young readers.

_____. *Amazing Dinosaurs: A Miniature Guide to the Dinosaur Kingdom*. Illustrations (from Dinocardz Corp.) by Dave Marrs. Berkeley, California: The Nature Company, 1993. For young readers.

_____. *Carbon Dates: A Day by Day Almanac of Paleo Anniversaries and Dino Events*. Jefferson, North Carolina: McFarland & Company, 1999.

_____. *The Dinosaur Dictionary*. Introductions by Alfred Sherwood Romer and David Techter. Secaucus, New Jersey: Citadel Press, 1972.

_____. *The New Dinosaur Dictionary*. Introductions by Robert Allen Long and R. E. Molnar. Secaucus, New Jersey: Citadel Press, 1982.

_____. *Dinosaurs: The Encyclopedia*. Foreword by Michael K. Brett-Surman. Jefferson, North Carolina: McFarland & Company, 1997.

_____. *Dinosaurs: The Encyclopedia, Supplement 1.* Foreword by Ralph E. Molnar. Jefferson, North Carolina: McFarland & Company, 1999.

Good, John M., Theodore E. White, and Gilbert F. Stucker. *The Dinosaur Quarry: Dinosaur National Monument.* Washington, D. C.: National Park Service, 1958.

Gross, Renie. *Dinosaur Country: Unearthing the Badlands' Prehistoric Past.* Saskatoon, Saskatchewan, Canada: Western Producer Prairie Books, 1985.

Hager, Michael W. *Fossils of Wyoming.* Laramie, Wyoming: Wyoming Geological Survey, Bulletin 54, 1970.

Hagood, Allen. *Dinosaur: The Story Behind the Scenery.* Photographs by Hagood, provided by the National Park Service. Las Vegas, Nevada: KC Publications, 1976.

Halstead, L. Beverly, and Jenny Halstead. *Dinosaurs.* Poole, Dorset, England: Blandford Press, 1981.

Hasegawa, Yoshikazu, and Y. Shiraki. *Dinosaurs Resurrected.* Tokyo: Mirai Bunkasha, 1994.

Haubold, Hartmut. *Saurierfährten.* Wittenburg, Lutherstadt, Germany: Die Neue Brehm-Bucherei, 1984.

Henderson, Douglas. *Asteroid Impact.* New York: Dial Books, 2000. For young readers, illustrated by Henderson.

Hitchcock, Edward. *Ichnology. A Report on the Sandstone of the Connecticut Valley, Especially Its Fossil Footmarks.* Boston: White, 1858. (Reprinted by Arno Press, Natural Sciences in America Series.)

Horner, John R., and Edwin Dobb. *Dinosaur Lives.* New York: Harcourt Brace, 1998.

Horner, John R., and James Gorman. *Maia: A Dinosaur Grows Up.* Illustrated by Doug Henderson. Bozeman, Montana: Museum of the Rockies, Montana State University, 1985.

_____. *Digging Dinosaurs.* New York: Workman Publishing, 1988.

Horner, John R., and Don Lessem. *The Complete* T. rex. New York: Simon and Schuster, 1993.

Hotton, Nicholas, III. *Dinosaurs.* New York: Pyramid Publications, 1963.

Howard, Robert West. *The Dawnseekers: The First History of American Paleontology.* Foreword by Gilbert F. Stucker. New York and London: Harcourt Brace Jovanovich, 1975.

Hutchinson, H. N. *Extinct Monsters and Creatures of Other Days.* London, 1892.

Jacobs, Louis. *Cretaceous Airport: The Surprising Story of Real Dinosaurs at the DFW.* Dallas: The Saurus Institute, 1993.

_____. *Lone Star Dinosaurs.* Original artwork by Karen Carr. Number Twenty-Two, Louise Merrick Natural Environment Series. College Station: Texas A&M University Press, 1995.

_____. *Quest for the African Dinosaurs: Ancient Roots of the Modern World.* New York: Willard Books, 1993.

Jenkins, John T., and Jannice L. Jenkins. *Colorado's Dinosaurs.* Cover painting by Donna Braginetz. Denver, Colorado: Colorado Geological Survey, 1993.

Jones, R. L., and Kathryn Gabriel. *Dinosaurs On-Line.* Nashville: Cumberland House, 2000. A "guide to the best dinosaur sites on the internet."

Kieran, Monique. *Discoveries in Paleontology,* Albertosaurus, *Death of a Predator.* Vancouver: Raincoast Books, 2000. Published for the Royal Tyrrell Museum of Palaeontology.

Kielan-Jaworowska, Zofia. *Hunting for Dinosaurs.* Translated from the Polish (originally published in Poland under the title *Polowanie na Dinozaury*). Cambridge, Massachusetts: MIT Press, 1968.

Knight, Charles R. *Before the Dawn of History.* Illustrated by Knight (mostly reproductions of the author's murals for the Field Museum of Natural History). New York: McGraw-Hill, 1935.

_____. *Life Through the Ages.* Illustrated by Knight. New York: Alfred A. Knopf, 1946.

Kohl, Michael F., and John S. McIntosh. *Discovering Dinosaurs in the Old West: The Field Journals of Arthur Lakes.* Washington, D.C.: Smithsonian Institution Press, 1997.

Krantz, Peter M. *Dinsoaurs in Maryland.* Maryland Geological Survey, Educational Series No. 6, 1989.

Kurtén, Björn, *The Age of Dinosaurs.* New York and Toronto: McGraw-Hill Book Company, 1968.

Lambert, David, *The Dinosaur Data Book.* Oxford: Facts on File; New York: Avon, 1990.

_____. *A Field Guide to Dinosaurs.* Illustrated by Joe Robinson, Graham Rosewarne, Sean Gilbert, Ashley Haddock, Brian Hewson, Richard Hummerstone, Janos Marffy, Eitetsu Nozawa, Max Rutherford, and Jerry Watkiss. New York: Avon (A Division of the Hearst Corporation), 1983.

_____. *Guide to Dinosaurs—A Thrilling Journey Through Prehistoric Times.* Illustrated by Luis Rey, Gary Staab, Frank DeNota, and others. London: Dorling Kindersley Publishing, 2000. For young readers.

Lankster, Ray E. *Extinct Animals.* New York: Holt & Company, 1905.

Leonardi, Giuseppe. *Annotated Atlas of South American Tetrapod Footprints (Devonian to Holocene).* Boston: White, 1985. (Reprinted by Arno Press, Natural Sciences in America Series.)

_____. *Glossary and Manual of Tetrapod Footprint Palaeoichnology.* Brazil: Departmento Nacional da Produção Mineral, 1987.

Lessem, Don. *Dinosaur Worlds.* Honesdale, Pennsylvania: Boyds Mills Press, 1996.

Lessem, Don, and Donald F. Glut. *The Dinosaur Society Dinosaur Encyclopedia.* Illustrated by Tracy L. Ford, Brian Franczak, Gregory S. Paul, John Sibbick, and Kenneth Carpenter. New York: Random House, 1993.

Lockley, Martin G. *Tracking Dinosaurs: A New Look at an Ancient World.* Cambridge, United Kingdom: Cambridge University Press, 1991.

Lockley, Martin G., and Adrian P. Hunt. *Dinosaur Tracks and Other Fossil Footprints of the Western United States.* New York: Columbia University Press, 1995.

Lockley, Martin G., and Christian Meyer. *Dinosaur Tracks and Other Fossil Footprints of Europe.* New York: Columbia University Press, 2000.

Long, John A. *Dinosaurs of Australia and New Zealand and Other Animals of the Mesozoic Era.* Cambridge, Massachusetts: Harvard University Press, 1998.

Long, Robert A., and Rose Houk. *Dawn of the Dinosaurs: The Triassic in Petrified Forest.* Illustrated by Doug Henderson. Petrified Forest, Arizona: Petrified Forest Association, 1988.

Long, Robert A., and Samuel P. Welles. *All New Dinosaurs and Their Friends.* Illustrated by Gregory Irons. San Francisco: Bellerophon Books, 1975. For young readers.

[_____, and _____; written anonymously]. *The Last of the Dinosaurs.* Illustrated by Gregory Irons. San Francisco: Bellerophon Books, 1980. For young readers.

Lucas, Frederic A. *Animals Before Man in North America.* New York: D. Appleton & Company, 1902.

_____. *Animals of the Past.* New York: McClure, Phillips & Company, 1901. (Reissued in a slightly revised paperback edition in 1916 by the American Museum of Natural History, as Handbook Series no. 4.)

Lucas, Spencer G. *Dinosaurs: The Textbook.* Dubuque, Iowa: William C. Brown Publishers, 1994.

Man, John. *The Day of the Dinosaur.* New York: Park South Books (Publishing Marketing Enterprises), 1978.

Markle, Sandra. *Outside and Inside Dinosaurs.* New York: Atheneum (a division of Simon and Schuster), 2000. For young readers.

Markman, Harvey C. *Fossils: A Story of the Rocks and Their Record of Prehistoric Life.* Denver: Denver Museum of Natural History, Popular Series No. 3, fourth edition, (reprinted) 1961.

McCarthy, Steve, and Mick Gilbert. *The Crystal Palace Dinosaurs: The Story of the World's First Prehistoric Sculptures.* London: The Crystal Palace Foundation, 1994.

McGinnis, Helen J. *Carnegie's Dinosaurs: A Comprehensive Guide to Dinosaur Hall at Carnegie Museum of Natural History, Carnegie Institute.* Foreward by Craig C. Black, Introduction by Mary R. Dawson. Pittsburgh, Pennsylvania: The Board of Trustees, Carnegie Institute, 1982.

McGowen, Christopher. *Dinosaurs, Spitfires and Sea Dragons.* Cambridge, Massaschusetts: Harvard University Press, 1991.

McGowen, Tom. *Album of Dinosaurs.* Illustrated by Rod Ruth. New York: Rand McNally & Company, 1972.

McLoughlin, John C. *Archosauria: A New Look at the Old Dinosaur.* Illustrated by McLoughlin. New York: The Viking Press, 1979.

McNeill, Alexander R. *Dynamics of Dinosaurs and Other Extinct Giants.* New York: Columbia University Press, 1989.

Michard, Jean-Guy. *The Reign of the Dinosaurs.* Translated from the French by I. Mark Paris. New York: Harry N. Abrams, Publishers, "Discoveries," and London: Thames and Hudson, Ltd., 1989.

Mitchell, W. J. T. *The Last Dinosaur Book: The Life and Times of a Cultural Icon.* Chicago and London: University of Chicago Press, 1998.

Moody, Ron, *A Natural History of Dinosaurs.* London: Hamlyn, 1977.

Moore, Ruth. *Man, Time, & Fossils.* Drawings by Sue Richert. New York: Alfred A. Knopf, 1953.

Norell, Mark A., Eugene S. Gaffney, and Lowell Dingus. *Discovering Dinosaurs in the American Museum of Natural History.* New York: Alfred A. Knoof, "A Peter N. Nevraumont Book," 1995.

Norman, David. *Dinosaur!* New York: McMillan, 1994.

_____. *The Illustrated Encyclopedia of Dinosaurs: An Original and Compelling Insight into Life in the Dinosaur Kingdom.* Color restorations by John Sibbick. London: Salemander Books/New York: Crescent Books (Crown Publishing), 1985. Note: A companion book, *The Illustrated History of Pterosaurs,* written by Peter Wellnhofer and with color restorations by Sibbick, was published in 1991.

_____. *Prehistoric Life.* New York: McMillan, 1994.

_____. *The Prehistoric World of the Dinosaur.* New York: Gallery Books (A Division of W. H. Smith Publishers), 1988.

Novacek, Michael. *Dinosaurs of the Flaming Cliffs.* New York: Doubleday and Company, 1996.

Officer, Charles, and Jake Page. *The Great Dinosaur Extinction Controversy.* Reading, Massachusetts: Helix Books (Addison-Wesley Publishing Company), 1996.

Osborn, Henry Fairfield. *Cope, Master Naturalist.* Princeton, New Jersey: Princeton University Press, 1931.

Ostrom, John H. *The Strange World of Dinosaurs.* Illustrated by Joseph Sibal. New York: G. P. Putnam's Sons, 1964.

Ostrom, John H., and John S. McIntosh. *Marsh's Dinosaurs: The Collections from Como Bluff.* New Haven, Connecticut: Yale University Press, 1966.

Owen, Ellis. *Prehistoric Animals: The Extraordinary Story of Life Before Man.* London: Octopus Books Limited, 1975.

Padian, Kevin, editor. *The Beginning of the Age of Dinosaurs: Faunal Change Across the Triassic–Jurassic Boundary.* Cambridge, United Kingdom: Cambridge University Press, 1986.

Padian, Kevin, and Daniel J. Chure, editors. *The Age of*

Dinosaurs, Short Courses in Paleontology (2). The Paleontological Society, 1989.

Parkinson, J. *The Dinosaur in East Africa: An Account of the Giant Reptile Beds of Tendaguru, Tanganyika Territory*. London: H. F. & G. Witherby, 1930.

Paul, Gregory S. *Predatory Dinosaurs of the World: A Complete Illustrated Guide*. Illustrated by Paul. New York: Simon and Schuster, 1988.

Peters, David. *Giants of Land, Sea & Air—Past & Present*. Illustrated by Peters. New York: Alfred A. Knopf (Sierra Club Books), 1986.

Piveteau, Jean, editor. *Traité de Paléontologie*. Paris: Masson et Cie., 1955.

Powell, James L. *Night Comes to the Cretaceous: Dinosaur Extinction and the Transformation of Modern Geology*. New York: W. H. Freeman & Company, 1999.

Preston, Douglas. *Dinosaurs in the Attic*. New York: St. Martin's Press, 1986.

Ratkevich, Ronald Paul. *Dinosaurs of the Southwest*. Illustrated by John C. McLoughlin. Albuquerque: University of New Mexico Press, 1976.

Relf, Pat. *A Dinosaur Named SUE: The Story of the Colossal Fossil*. New York: Scholastic Inc., 2000. "With the SUE Science Team of Christopher A. Brochu, Matthew T. Carrano, John J. Flynn, Olivier C. Rieppel, and William F. Simpson."

Rich, Thomas H., and Patricia Vickers-Rich. *Dinosaurs of Darkness*. With illustrations by Peter Trustler. Bloomington, Indiana: Indiana University Press, 2000. Hunting dinosaur fossils in Australia.

Romer, Alfred Sherwood. *Vertebrate Paleontology*. Chicago: University of Chicago Press, 1945 (third edition).

_____. Chicago: University of Chicago Press, 1967 (third edition).

_____. *Osteology of the Reptiles*. Chicago: University of Chicago Press, 1956.

Rudwick, Martin J. S. *Scenes from Deep Time: Early Pictorial Representations of the Prehistoric World*. Chicago: University of Chicago Press, 1992.

Rukauina, Darko. *Doba Dinosaura*. Illustrated by Berislov Krzić. Croatia: Croatian Natural History Museum, [no date; 1994].

Russell, Dale A. *An Odyssey in Time: The Dinosaurs of North America*. Minocqua: University of Toronto Press and Northwood Press, 1989.

_____. *A Vanished World: The Dinosaurs of Western Canada*. Photographs by Susanne M. Swibold, paintings by Eleanor M. Kish. Ottawa: National Museum of Natural Sciences, National Museums of Canada, 1977.

Sabino, Leghissa, and Giuseppe Leonardi. *Sauropodi in Istria*. Genoa, Italy: Centro di Cultura Giuliano Dalmata, 1990.

Saito, T. *Wonder of the World's Dinosaurs* [title translated from the Japanese]. Tokyo: Kodansha, 1979.

Sanz, José Luis L., and Angela D. Buscalioni, editors. *Los Dinosaurios y Su Entorno Biótico*. Madrid, Spain: Instituto "Juan Valdes," 1992.

Sattler, Helen Roney. *Dinosaurs of North America*. Introduction by John H. Ostrom, illustrated by Anthony Rao. New York: Lothrop, Lee & Shepard Books, 1981.

_____. *The Illustrated Dinosaur Dictionary*. Foreword by John H. Ostrom, illustrated by Pamela Carroll. New York: Lothrop, Lee & Shepard Books, 1983.

Scheele, William E. *Prehistoric Animals*. Illustrated by Seeley. Cleveland and New York: The World Publishing Company, 1954. Sampson Low, 1975.

Schuchert, Charles, and Clara Mae Le Vene. *O. C. Marsh, Pioneer in Paleontology*. New Haven, Connecticut: Yale University Press, 1940.

Scully, V., R. F. Zallinger, L. J. Hickey, and J. H. Ostrom. *The Great Dinosaur Mural at Yale: The Age of Reptiles*. New York: Harry N. Abrams, 1990.

Senkowsky, Sonya. *The Academy of Natural Sciences*. Philadelphia: Wyco Colour and The Academy of Natural Sciences of Philadelphia, 1988.

Serjeant, William A. *Vertebrate Fossils and the Evolution of Scientific Concepts*. New York: Gordon and Breach Publishers, 1995.

Silverberg, Robert, editor. *The Ultimate Dinosaur* (new edition). Chapters written by paleontologists, plus fiction by noted speculative writers, illustrated by William Stout, Doug Henderson, Brian Franczak, others. New York: Ibooks, Inc., 2000.

Sloan, Christopher. *Feathered Dinosaurs*. Introduction by Philip J. Currie. Washington, D.C.: National Geographic Press, 2000.

Špinar, Zdeněk V., *Life Before Man*. Illustrated by Zdeně Burian. London: Thames and Hudson, 1972. (Reprinted, 1972, New York: American Heritage Press (A Division of McGraw-Hill Book Company), 1972.

Steel, Rodney, and Anthony Harvey, editors. *The Encyclopedia of Prehistoric Life*. Foreword by W. E. Swinton. Art Editor: John Ridgeway. New York: McGraw-Hill Book Company, 1979.

Steel, Rodney, and Hartmut Haubold. *Die Dinosaurier*. Wittemburg Lutherstadt, Germany: A. Ziemsen Verlag, 1979.

Sternberg, Charles H. *Hunting Dinosaurs in the Bad Lands of the Red Deer River, Alberta, Canada*. Lawrence, Kansas: The World Company Press, 1917.

_____. *The Life of a Fossil Hunter*. New York: Henry Holt and Company, 1909.

Stout, William. *The Dinosaurs Sketchbook*. Illustrated by Stout. Self-published, 1999. Limited-edition collection of rough and preliminary sketches done for Stout and Service's *The Dinosaurs* and Glut's *New Dinosaur Dictionary*.

Stout, William [concept and art], and William Service

[narration]. *The Dinosaurs: A Fantastic New View of a Lost Era*. Introduction by Peter Dodson, edited by Byron Preiss. New York: A Byron Preiss Book, Bantam Books, 1981. (Reissued in 2000 by ibooks in a revised and updated edition as *The New Dinosaurs.*)

Strevell, Charles Nettleton. *Story of the Strevell Museum*. Salt Lake City: Board of Education, Department of Visual Education, Salt Lake City Public Schools, 1940.

Swinton, W. E. *Dinosaurs*. London: Trustees of the British Museum (Natural History), 1962.

_____. *The Dinosaurs*. London: George Allen & Unwin (New York: Wiley-Interscience, a Division of John Wiley & Sons), 1970.

_____. *Fossil Amphibians and Reptiles*. London: Trustees of the British Museum (Natural History), 1965.

_____. *Giants Past and Present*. London: Robert Hale, 1966.

_____. *The Wonderful World of Prehistoric Animals*. Paintings by Maurice Wilson. Garden City, New York: Garden City Books, 1971.

Tanimoto, Masahiro. *Restorations of Fossil Vertebrates*, No. 1. Illustrated by Tanimoto. Osaka, Japan [privately published by the author], 1983.

Taquet, Phillipe. *L'Empreinte des Dinosaures*. Paris: Editions Odile Jacob, 1994.

Thomas, Roger D. K., and Everett C. Olson, editors. *A Cold Look at the Warm-Blooded Dinosaurs*. Boulder, Colorado: AAAS Selected Symposium, 28, Westview Press, 1980.

Thulborn, Tony. *Dinosaur Tracks*. New York: Chapman and Hall, 1990.

Untermann, G. E. and Billie R. Untermann. *A Popular Guide to the Geology of Dinosaur National Monument*. Dinosaur National Monument, Utah and Colorado: Dinosaur Nature Association, 1969.

_____. *Dinosaur Land and the Unique Uinta Country*. Vernal, Utah: G. E. and B. R. Untermann, 1972.

Vickers-Rich, Patricia, and Thomas H. [Hewitt] Rich. *The ICI Australia Catalogue of The Great Russian Dinosaurs Exhibition 1993–1995*. Clayton, Australia: Monash Science Centre, 1993.

_____, *Wildlife of Gondwana—Dinosaurs and Other Vertebrates from the Ancient Supercontinent*. Bloomington, Indiana: Indiana University Press, 2000.

Wallace, David Rains. *The Bonehunters' Revenge: Dinosaurs, Greed, and the Greatest Scientific Feud of the Gilded Age*. New York: Houghton and Mifflin, 2000. The famous feud between paleontologists Othniel Charles Marsh and Edward Drinker Cope.

Wallace, J. *The American Museum of Natural History's Book of Dinosaurs and Other Ancient Creatures*. New York: Simon and Schuster, 1994.

Watson, Jane Werner. *The Giant Golden Book of Dinosaurs*. Illustrated by Rudolph F. Zallinger. Racine, Wisconsin: Golden Press (Western Publishing Company), 1960. For young readers.

Weishampel, David B., Peter Dodson, and Halszka Osmólska, editors. *The Dinosauria*. Berkeley and Los Angeles: University of California Press, 1990. Note: A revised and updated edition of this benchmark publication is in production to date of this writing.

Weishampel, David B., and Luther Young. *Dinosaurs of the East Coast*. Baltimore: Johns Hopkins University Press, 1996.

Wendt, Herbert. *Before the Deluge*. Garden City, New York: Doubleday & Company, 1968.

West, Linda, and Dan Chure. *Dinosaur: The Dinosaur National Monument Quarry*. Jensen, Utah: Dinosaur Nature Association.

Whitaker, George O., and Joan Meyers. *Dinosaur Hunt*. New York: Harcourt, Brace and World, 1965. Photographs by Whitaker, drawings by Michael Insinna.

White, Theodore E. *Dinosaurs—at Home*. New York: Vantage Press, 1980.

Whybrow, Peter J. *Travels with the Fossil Hunters*. Cambridge: Cambridge University Press, 2000.

Wilford, John Noble. *The Riddle of the Dinosaur*. New York: Alfred A. Knopf, 1985.

Will, Richard. *Dinosaur Digs*. Castine, Maine: County Roads Press, 1992.

Wolberg, Donald L., Edmund Stump, and Gary Rosenberg, editors. *Dinofest International: Proceedings of a Symposium Held at Arizona State University*. Philadelphia: Academy of Natural Sciences, 1997.

Zallinger, Peter. *Dinosaurs*. Illustrated by Zallinger. New York: Random House. A "Random House Picturebook" for children.

Zangerl, Rainer. *Dinosaurs, Predator and Prey*. Chicago: Chicago Natural History Museum Press, 1956.

Zimmerman, Howard. *Dinosaurs! The Biggest, Baddest, Strangest, Fastest* New York: Byron Preiss Visual Publications, 2000. Showcasing artwork by noted "paleartists" including John Sibbick, Gregory S. Paul, Mark Hallett, Doug Henderson, William Stout, Alex Ebel, Donna Braginetz, Patrick O'Brien, and Luis Rey, cover painting by James Gurney. For young readers.

Glossary

Included herein are technical and also some nontechnical terms that appear in this volume, but which are generally not defined anywhere else in the text. Definitions of terms were based in part upon those published in a number of earlier sources, these including various dictionaries of the English language, and also the following texts: A Dictionary of Scientific Terms *(Kenneth and Henderson 1960),* The Illustrated Encyclopedia of Dinosaurs: An Original and Compelling Insight Into Life in the Dinosaur Kingdom *(Norman 1985),* The Dinosauria *(Weishampel, Dodson and Osmólska 1990),* Encyclopedia of Dinosaurs *(Currie and Padian 1997),* The Complete Dinosaur *(Farlow and Brett-Surman 1998), and "Ceratosaurus (Dinosauria, Theropoda), a Revised Osteology" (Madsen and Welles 2000).*

ABDUCTION Movement of part of the body away from the midline axis of the body (opposite of adduction).

ACETABULUM Cup-shaped socket in the pelvic girdle for the head of the femur.

ACOELOUS Vertebrae having flattened centra.

ACROMIAL Artery, process, or ligament pertaining to the acromion.

ACROMION Ventral prolongation of the scapular spine.

ACUMINATE Tapering to a point.

ADDUCTION Movement of part of a body toward the midline of the axis of the body (opposite of abduction).

ADDUCTOR Muscle that brings one bony part towards another.

AEOLIAN (See Eolian.)

AEROBIC Thriving only in the presence of free oxygen.

ALA Wing-like projection or structure.

ALIFORM Wing-shaped.

ALIMENTARY Pertaining to nutritive functions.

ALLOMETRY Study of relative growth; change of proportions relating to growth.

ALLUVIAL Deposits having been formed by finely divided minerals laid down by running water.

ALTRICIAL BEHAVIOR Behavior in which a parent or parents care for the newly born.

ALVEOLI Pits or sockets on the surface of an organ or a bone.

AMBIENS Thigh muscle.

AMPHIBIAN Tetrapod adapted to live on both land and in water.

AMPHICOELOUS Concave on both surfaces of a vertebral centra.

AMPHYPLATYAN Flat on both ends of vertebral centra.

ANALAGOUS Describing structures in different kinds of organisms which serve the same function, without being derived from the same ancestral structure.

ANASTOMOSING Connecting in a branch-like manner.

ANGIOSPERM Seed plant in which its seed is enveloped by a seed vessel fruit; the flowering plants.

ANGULAR In most vertebrates, a dermal bone in the lower jaw, upon which rest the dentary and splenial bones.

ANISODACTYL Having digits of unequal length.

ANTERIOR Toward the front end, sometimes referred to as "cranial."

ANTITROCHANTER Articular surface of the ilium of birds, against which the trochanter of the femur plays.

ANTORBITAL In front of the orbits of a skull, sometimes referred to as "preorbital."

ANTORBITAL FENESTRA Opening in the skull, behind the external nares and in front of the orbit.

ANTORBITAL FOSSA Depression surrounding the antorbital fenestra.

ANTRUM Sinus or cavity.

APICAL At the summit or tip.

APOMORPHIC In cladistics, the derived state occurring only within members of an ingroup, when a character exhibits two states within that ingroup.

APOMORPHY In cladistics, a derived character.

APPENDICULAR SKELETON That part of the skeleton including the pectoral girdles, forelimbs, pelvic girdles and hindlimbs.

ARBOREAL Living mostly or exclusively in trees, bushes, or shrubs.

AQUATIC Living in the water.

ARCADE In anatomy, a bony bridge.

ARCHOSAURIA Diapsid group of reptiles including dinosaurs, pterosaurs, thecodontians, and crocodiles, defined primarily by the possession of an antorbital fenestra.

ARCHOSAUROMORPH One of a group of diapsids including dinosaurs, birds, pterosaurs, crocodilians, and their close relatives.

ARCTOMETATARSALIAN CONDITION Central metatarsal (III) pinched proximally, therefore obscured from view anteriorly, reduced or excluded from contact with the tibiotarsus.

ARMOR Bony scutes, plates, shields, horns, spikes and clubs possessed by some dinosaurs.

ARTICULAR In dinosaurs, the bone toward the rear of the mandible by which the lower jaw articulates with the quadrate bone.

ARTICULATED Jointed or joined together.

ASPIRATION Act of expelling breath.

ASSEMBLAGE Large group of fossils and other items found at the same location, considered to originate from the same time period.

ASTRAGALUS Larger tarsal bone which mostly articulates with the tibia dorsally and metatarsus ventrally.

ATLANTAL Pertaining to the atlas bone.

ATLAS First cervical vertebra.

ATTRITIONAL Bone accumulations resulting from recurring "normal"

death events of numerous individual animals over a long span of time (as opposed to a catastrophic, short-term mass mortality event).
AUTAPOMORPHY In cladistics, a character state unique to one taxon.
AVIAN— Pertaining to birds.
AXIAL SKELETON That part of the skeleton including the vertebral column and ribs.
AXIS Second cervical vertebra.

BADLANDS Area of barren land heavily roughly eroded by water and wind into ridges, mesas, and peaks.
BARB Delicate thread-like structure that extends obliquely from a feather rachis, forming the vane.
BARBULE Small hooked process fringing barbs of a feather.
BASAL Placed at or near the base; in cladistics, placed at or neat the base or "trunk" or a phylogenetic tree; a group outside a more derived clade; the earliest form of a lineage.
BASI- Prefix meaning "basis."
BASICRANIUM Base of the skull.
BASIOCCIPITAL Median bone in the occipital region of the skull, forming at least part of the occipital condyle.
BASIPTERYGOID Process of the basisphenoid contacting the pterygoid.
BASISPHENOID Cranial bone between the basioccipital and presphenoid.
BATTERY Distinctive tooth pattern wherein a number of small, slender teeth are tightly wedged together along the length of the jaw, with multiple teeth stacked in a single tooth position (as in hadrosaurs), forming a grinding or cutting surface.
BAUPLAN General body plan for a group of organisms; literally, a German word meaning an architect's or a building plan.
BED In geology, distinct layers of sedimentary rock.
BICIPITAL Groove on the upper part of the humerus; crests of the greater and lesser tubercles of the humerus; also, divided into two parts at one end.
BIFURCATED Forked; having two prongs or branches.
BILOBATE Having two lobes.
BIOMASS Total estimated body mass or weight of all the animals of a population combined; also, the total mass or weight of a single individual.
BIOMECHANICS Study of the motion of a body of a given organism in the context of mechanical laws and principles.
BIOSTRATIGRAPHY Study of the distribution of fossils in distinct strata.
BIOTA Flora and fauna of a region.
BIPEDAL Habitually walking on two feet.
BIPINNATE FEATHER Feather barbs on both sides of the shaft.
BIVARIATE Variable condition occurring simultaneously with another variable.
BODY FOSSIL Fossil consisting of an actual part of the organism.
BONEBED (also **BONE BED**) Sedimentary layer having a large concentration of fossil remains.
BOSS Raised ridge or rounded body part, such as the bony mass on the snout of some ceratopsians.
BRACKISH Containing some salt; briny.
BRADYMETABOLIC Having a metabolism that runs at a low rate.
BRAINCASE Part of the skull enclosing the brain.
BRANCH On a cladogram, a line connecting a taxon to a node that joins it to another taxon, representing the divergence of a taxon from its nearest relatives.
BREVIS SHELF Median shelf on the postacetabular section of the ilium for the origin of some of the caudifumoralis brevis muscle.
BRONCHUS Tube connecting the trachea with the lung.
BROWSER Animal that feeds on high foliage (*e.g.*, bushes, not grasses).
BUCCAL Pertaining to the cheek; the surface of a tooth toward the cheek or lip.
BURSA Sac-like cavity in a bone containing viscid fluid for preventing friction at joints.
BUTTRESS Bony structure for reinforcement.

CALAMUS Hollow quill of a feather.
CALCANEUM Smaller tarsal bone, lateral to the astragalus and distal to the fibula.
CALCAREOUS Composed of, containing, or characteristic of calceum carbonate, calcium, or limestone.
CANCELLOUS Made up of lamillae and slender fibres, joining to form a network-like structure.
CANCELLOUS BONE Spongy bone, having tissues that are not closely packed.
CANINIFORM Teeth of "canine" form.
CAPITULUM Knob-like swelling at the end of a bone.
CARAPACE Hard outer covering to the body, like the shell of a turtle.
CARCASS Dead body of an animal.
CARINA On some bones and teeth, a keel-like ridge or edge.
CARNIVORE Flesh-eater.
CARNOSAUR In the original (and abandoned) usage, an informal term generally referring to any large theropod; in the modern sense, a member of the Carnosauria, a restricted group of large theropods.
CAROTICUS Groove or process to which a muscle is attached.
CARPAL Pertaining to the wrist; also, a bone of the wrist.
CARTILAGE Transluscent firm and elastic tissue usually found in connection with bones and on the articular ends of limb bones.
CAT SCAN (See CT scan.)
CATASTROPHIC Pertaining to theories and beliefs that mass extinctions were the result of cataclysmic events.
CAUDAL Pertaining to the tail; toward the tail; more recently, used in place of "posterior."
CEMENTUM Substance investing parts of the teeth, chemically and physically allied to bone.
CENTROCOEL Cavity inside the centrum of a vertebra.
CENTRUM Main body of the vertebra (ventral to the neural chord) from which rise the neural and hemal arches.
CERVICAL Pertaining to the neck.
CHARACTER Distinctive feature or trait of an organism, or any difference among organisms, that can be used in classification or in estimating phylogeny.
CHARACTER STATE Range of expressions or conditions of a character.
CHELONIAN Member of the Chelonia, a reptilian group including turtles and tortoises.
CHEVRON Bone that hangs below a caudal vertebra.
CHOANA Funnel-shaped internal nasal opening.
CHONDRIFICATION Conversion into cartilage.
CINGULUM Girdle-like structure on teeth.
CLADE Monophyletic taxon as diagnosed by synapomorphies.
CLADISTICS Scientific approach in taxonomy to classify groups of organisms in terms of the recency of their last common ancestor.
CLADOGRAM Diagram representing the distribution of shared-derived characters for groupings of organisms.
CLASSIFICATION Process of organizing clades into groups related by common descent.
CLAVICLE Collar-bone forming the anterior portion of the shoulder-girdle.

CNEMIAL CREST Crest along the anterior dorsal margin of the tibia.
COLD-BLOODED Informal term for "ectothermic."
COELUROSAUR In the original (and abandoned) usage, an informal term generally referring to all small theropods; in the modern sense, a large group of theropods including both small and gigantic forms.
COLLAGEN Gelatinous protein present in all multicellular organisms, particularly in connective tissue.
COMMON ANCESTOR In cladistics, a taxon exhibiting all synapomorphies of that taxon but neither autapomorphies nor the synapomorphies at higher levels within that taxon.
COMMUNITY Ecological relationships between a local environment and all its fauna and flora.
COMPETITION Simultaneous use of a limited resource by more than one species, resulting in conflicting efforts by them for continued survival.
CONDYLE Process on a bone utilized in articulation.
CONGENERIC Belonging to the same genus.
CONIFER One of a group of gymnosperms including pines, spruces, larches, firs, and related plants.
CONSERVATIVE Tending to remain unchanged, as in being similar to an ancestral group.
CONSPECIFIC Belonging to the same species.
CONTINENTAL DRIFT Continents moving on the Earth.
CONVERGENCE (also **CONVERGENT EVOLUTION**) Organisms evolving similar appearances due to responses to similar lifestyle demands, though not sharing direct common ancestors.
COOSSIFIED Bones fused together.
COPROLITE Fossilized dung.
CORACOID Bone between the scapula and the sternum, participates in the shoulder joint.
CORNUAL Having to do with horns or horncores.
CORONOID PROCESS In reptiles, prong-shaped bony process on the lower jaw for the attachment of jaw-closing muscles.
CORTICAL BONE Bone tissue on the outer surface.
CORTICES (Plural of cortex) outer part of a bone.
COSMOPOLITAN Having a very wide or worldwide distribution.
COSTAL Involving the ribs.
COSTOVERTEBRAL Articulation between the ribs and vertebrae.
COTYLE Cup-like cavity in a bone.
COTYLUS Ball-shaped structure.
COTYPE Additional type specimen, usually collected at the same time and from the same locality as the holotype, or a specimen, along with others, from which the type is defined.
CRANIA (also **CRANIAL SKELETON**) Bones of the skull, excluding those of the lower jaws.
CRANIAL Toward the head; more recently, used in place of "anterior."
CRANIUM Skull, particularly the braincase, but excluding bones of the lower jaw.
CREST Ridge or rounded area of bone; in hadrosaurids, a rounded area of bone on the upper part of the skull, sometimes containing hollow passages.
CRETACEOUS PERIOD Third and latest division of the Mesozoic Era, 144 to 65 million years ago.
CRISTA Crest or ridge.
CROCODILIAN Member of the Crocodilia, a successful group of Mesozoic and extant archosaurs related to dinosaurs.
CROWN Exposed part of the tooth.
CROWN GROUP All descendants of the closest common ancestor of living forms.
CRUCIATE Cross-shaped.
CRUSTACEAN A member of the Crustacea, a group of mostly aquatic invertebrates having segmented bodies, chitinous skeletons, and paired, jointed limbs.
CT SCAN (also **CAT SCAN**) Process by which a computer is used to process data from a tomograph in order to display a reconstructed cross section of an organism's body without physically cutting into it.
CUIRASSAL Having a protective covering of bony plates or scales.
CULTRIFORM Sharp-edged and pointed.
CURSORIAL Running.
CYCAD Flowering gymnosperm prevalent from the Triassic to Early Cretaceous.

DELTOID Thick, triangular muscle covering the shoulder joint.
DELTOPECTORAL CREST Bony flange of the humerus for attachment of the deltoid and pectoralis muscles.
DENTARY Largest bone of the lower jaw, usually bearing teeth.
DENTICLE Small bump-like processes along the edges of teeth.
DENTICULATE Having denticles.
DENTIGEROUS Tooth-bearing.
DENTITION Teeth.
DEPOSIT Accumulation of a substance (*e.g.*, sediment, bones).
DERIVED CHARACTER More specialized character evolved from a simpler, more primitive condition.
DERMAL Pertaining to the skin.
DERMAL ARMOR Platelets or small plates of bone that grew in the flesh but were not connected to the skeleton.
DERMAL PLATE (See Plate.)
DESCRIPTION In paleontology, a detailed verbal representation of material.
DETRITIVOROUS Animal that eats food in the form of particles or grains.
DIAGNOSIS Concise statement enumerating the distinctive characters of a particular organism.
DIAPOPHYSIS Lateral or transverse process of the neural arch.
DIASTEMA Toothless space in a jaw, generally between two different kinds of teeth (such as the canine and postcanines in mammals).
DIGIT Toe or finger.
DIGITIGRADE Walking with only the digits touching ground.
DIMORPHISM State of having two different forms, usually according to sex.
DINOSAUR One of a diverse group (Dinosauria) of terrestrial archosaurian reptiles that flourished from the Late Triassic through the Late Cretaceous periods of the Mesozoic Era, with an erect gait, closely related to other archosaurian groups such as crocodilians and pterosaurs, one lineage (manuraptoran theropods) seemingly the direct descendants of birds.
DINOSAUROMORPHA Ornithodiran clade including "lagosuchids," dinosaurs, and birds.
DISARTICULATED Pulled apart.
DISPERSAL In biogeography, spreading out.
DISTAL End of any structure farthest from the midline of an organism, or from the point of attachment; away from the mass of the body; segments of a limb or of elements within a limb; the edge of a tooth away from the symphysis along the tooth row.
DIVERGENCE In evolution, moving away from a central group or changing in form.
DIVERTICULUM Sac or tube, "blind" at the distal end, that branches off from a cavity or canal.
DORSAL Relating to the back; toward the back.
DORSI- (also **DORSO**) Prefix meaning "back."

ECOSYSTEM Ecological system formed by interaction of organisms and their environment.

ECOLOGY Biological study of the relationship between organisms and their environment.
ECTEPICONDYLE Lateral projection of the distal end of the humerus.
ECTETHMOIDS Lateral ethmoid bone ["ethmoids" being bones that form much of the walls of the nasal cavity].
ECTOPTERYGOID Ventral membrane bone behind the palatine, extending to the quadrate.
ECTOTHERMIC Relying on external sources of heat to maintain body temperature; popularly, "cold-blooded."
EDENTULOUS Toothless.
EMBAYMENT A bay-like shape or depression in a bone.
EMBRYO Young organism in pre-birth stages of development.
ENAMEL Form of calceum phosphate forming the hard outer covering on teeth.
ENANTIOTHORNES Group of Mesozoic birds.
ENDEMIC Relating to an indigenous species or population occurring in a specific geographic range.
ENDOCAST Fill-in of the brain cavity by sediment, revealing the shape of the brain.
ENDOCHONDRAL Forming or beginning within the cartilage.
ENDOCRANIAL Pertaining to the brain cavity.
ENDOCRANIUM Brain cavity.
ENDOSTEAL Internal bone, or that lining the cavities of bones.
ENDOTHERMIC Able to generate body heat internally by means of chemical reactions; popularly, "warm-blooded."
ENDOCRANIUM The brain cavity of the skull.
ENTEPICONDYLE Lower end of the humerus.
ENVIRONMENT Surroundings in which organisms live.
EOLIAN (also spelled **AEOLIAN**) Caused by the wind.
EPAXIAL Above the axis; dorsal.
EPICONDYLE Medial/inner projection at the distal end of the humerus and femur.
EPICONTINENTAL SEAS Large bodies of water that invade large land masses.
EPIDERMIS Outer, nonvascular, and protective layer of the skin.
EPIJUGAL Horn-like projection off the jugal in ceratopsians.
EPIPHYSIS Part or process of a bone formed from a separate center of ossification, later fusing with the bone.
EPIPODIUM Region of the rear part of the foot, or the bones in that region.
EPOCCIPITAL Small bone located on the edge of the ceratopsian frill.
ERA Largest division of geologic time.
EROSION Result of weathering on exposed rocks.
ESTUARINE Found in an estuary.
EURAMERICA Land mass (which began to separate during the Early Cretaceous) including what is now the continents of Europe and North America.
EURYAPSID Member of the Euryapsidae, a diverse group of aquatic or amphibious reptiles common during the Permian period and Mesozoic Era.
EUSTACHIAN Bony or cartilaginous tube or canal connecting the tympanic cavity with the nasal part of the pharynx.
EVOLUTION Change in the characteristics of a population of organisms, caused by natural selection over time.
EXOCCIPITAL Bone of the skull on each side of the foramen magnum.
EXOSTIC BONE Bone that has grown as a result of partial parietal detachment.
EXPIRATION Act of emitting air from the lungs.
EXPOSURE In geology, where rock is exposed due to weathering.
EXTENSOR Muscle that extends a limb or part of a limb; also used to designate surfaces of a limb, manus, or pes.
EXTINCTION Termination of a species.
EXTRUSIVE Rock derived from magma.

FACIES In geology, one of different types of contemporaneous deposits in a lateral series of deposits; also, the paleontological and lithological makeup of a sedimentary deposit.
FACULTATIVE Having the ability to live and adapt to certain conditions, while not being restricted to those conditions.
FAMILY In Linnaean classification, a grouping of similar genera.
FAUNA All the animals of a particular place and time.
FEMUR Thigh-bone.
FENESTRA Opening in a bone or between bones.
FIBRO-LAMELLAR BONE Somewhat open hard tissue, filled with blood vessels, indicative of fast-growing bone.
FIBULA Smaller, outer shin bone.
FILIGREE Bone having thread-like outgrowths.
FLEXOR Muscle which bends a joint; also used to designate surfaces of a limb, manus, or pes.
FLOAT Fossil material collected on the surface, rather than being excavated.
FLORA All the plants of a particular place and time.
FLUVIATILE Growing in or inhabiting or developing in or near a stream.
FLUVIO- Pertaining to a river, stream, or sea.
FONTANELLE Opening on the frill in some ceratopsians.
FORAMEN Opening through a bone or membraneous structure.
FORAMEN MAGNUM Opening in the occipital area of the skull through which the spinal cord passes.
FORM GENUS Term referring to a nominal genus based on a morphology but that may not actually be diagnosable from other genera.
FORMATION In geology, a formally defined and mappable unit of sedimentary rock.
FOSSA Pit or trench-like depression.
FOSSIL Preserved remains of an animal or plant at least 10,000 years old, usually formed through burial and possibly involving a chemical change; evidence of life in the geologic past.
FOSSILIZED Having become a fossil.
FRACTURE Break in a bone.
FRONTAL Bone of the skull roof in front of the parietal.
FRONTOPARIETAL Frontal and parietal bones, usually referring to suture or fusion of both bones.
FUNCTIONAL MORPHOLOGY Study of the movements and patterns of locomotion of an organism, mostly relative to its form or structure.
FUSION In anatomy, the firm joining together of bones, either naturally or abnormally.
FUSED Firmly jointed together, usually when bones grow together; coossified.

GASTRALIA Belly ribs that help to support the viscera in some dinosaurs.
GASTROLITH Small "stomach" stone that is swallowed for ballast or to grind up already consumed food.
GASTROPOD Mollusc belonging to the class Gastropoda.
GENUS Group of closely related species.
GEOLOGIC TIME Period of time spanning the formation of the Earth to the beginning of recorded history.
GEOLOGY Science of the study of the Earth.

GHOST LINEAGE Missing sections of a clade, unknown from the fossil record, but implied by phylogeny; theorized geological extension of the range of a taxon before its earliest known occurrence.
GIGANTOTHERMY The ability of an organism with a low ectothermic metabolic rate to maintain constant high temperature, even in relatively cool environments by the use of large size, circulatory adjustments, and layers of body insulation.
GINGLYMOID Hinge joint, or constructed like one.
GINGLYMUS Articulation constructed permitting motion only in one plane.
GIRDLE Curved or circular structure, particularly one that encircles another.
GIZZARD Muscular portion of the stomach utilized in grinding up food.
GLENOID Socket in the pectoral girdle to which the head of the humerus attaches.
GONDWANA Southern continent including South America, Africa, India, Madagascar, Australia, and Antarctica.
GRACILE Having a graceful or slim build of form.
GRADE In cladistics, a paraphyletic taxon as diagnosed by the absence and presence of synapomorphies, delineated based upon morphologic distance; also, in a series of bones, the gradual changing of shape of those bones.
GRAVIPORTAL Slow-moving or lumbering.
GRAZING Feeding on low-lying vegetation.
GREGARIOUS Animals of the same species living in groups rather than in isolation.
GUILD Group of animals having a characteristic mode of existence.

HAEMAL (See HEMAL.)
HALF-RING In ankylosaurs, the unification of the first and second transverse rows of keeled plates to form a pair of yokes around the neck.
HALLUX First digit of the pes.
HAMULARIS Hooked or hook-like.
HATCHLING Organism newly hatched from an egg.
HAVERSIAN BONE Kind of secondary bone that replaces primary bone, forming a series of vascular canals called "Haversian canals."
HEAD-BUTTING Behavior in which two (usually male) individuals of the same species compete for dominance of their group by repeatedly colliding head to head.
HEMAL (or HAEMAL) Pertaining to blood or blood vessels.
HERBIVORE Plant-eater.
HERD Large group of (usually herbivorous) animals of the same species.
HETERO- Prefix meaning "other" or "different."
HETEROCHRONY Condition of having a different beginning and ending of growth, or a different growth rate for a different feature, relative to the beginning and end, or the rate of development, of the same feature in an ancestor; a kind of evolutionary mechanism.
HETEROGENEOUS Having dissimilar elements.
HISTO- Pertaining to tissue.
HISTOLOGY Study of the fine structure of body tissues.
HOLOTYPE Single specimen chosen to designate a new species.
HOMEOTHERMY Maintaining a fairly constant body temperature regardless of environmental temperature changes.
HOMEOTIC Having to do with the assumption of one part of likeness to another (*e.g.*, them modification of a dorsal vertebra into a sacral vertebra).
HOMO- Prefix meaning "same" or "alike."
HOMODONT Having similar teeth throughout.
HOMOLOGOUS Similar because of common ancestry; similarity.
HOMOPLASY In cladistics, a shared similarity between taxa explained by character reversal, convergence, or chance, and not a result of common ancestry.
HORIZON Soil layer formed at a definite time and characterized by definite fossil species.
HUMERUS Upper arm bone.
HYPANTRUM In some reptiles, a notch on a vertebra for articulation with the hyposphene.
HYPAPOPHYSIS Ventral process on a vertebral centrum.
HYPAXIAL Below the vertebral column; ventral.
HYPER- Prefix meaning "more than," "greater than," etc.
HYPEREXTENSION Atypical extension of a body part.
HYPERMORPHOSIS Evolutionary change wherein sexual maturity takes place later in the descendant than in the ancestor.
HYPOSPHENE In some reptiles, a wedge-shaped process on the neural arch of a vertebra, fitting into the hypantrum.

IBERIAN PENINSULA Region of southwestern Europe, consisting of Spain and Portugal, separated from France by the Pyrenees mountains.
ICHNITE Fossil footprint.
ICHNO- Prefix meaning "track" or "footprint."
ILIUM Dorsal bone of the pelvic arch; hipbone.
IN SITU Referring to specimens in place in the ground where they are discovered.
INCISIFORM Incisor-shaped.
INDEX FOSSIL Fossil restricted to a particular span of geologic time which can, therefore, be reliably utilized to date rocks in which other fossils are found.
INFRA- Prefix meaning "below."
INFRAORDER In Linnaean classification, category between family and suborder.
INFRAPREZYGAPOPHYSAL Below the prezygapophysis.
INGROUP In cladistics, a monophyletic grouping of taxa.
INSECTIVOROUS Insect-eating.
INSPIRATION Act of drawing air into the lungs.
INTEGUMENT Outer covering, usually pertaining to skin.
INTERCENTRUM Second central ring in a vertebra having two vertical rings in each centrum.
INTERMEDIUM Small bone of the carpus and tarsus.
INTERORBITAL Between the orbits.
INTRA- Prefix meaning "within."
INTRASPECIFIC Within the same species.
INTRUSIVES Igneous rocks forced into another stratum while in molten state.
INVAGINATED Enclosed, as if in a sheath.
INVERTEBRATE Animal without a backbone.
ISCHIUM Ventral and posterior bone of each half of the pelvic girdle.

JUGAL Skull bone between the maxilla and quadrate.
JUNIOR SYNONYM Taxon suppressed because another name, pertaining to the same fossil materials, was published previously.
JURASSIC PERIOD Second and middle division of the Mesozoic Era, 206 to 144 million years ago.

KERATIN Matter composed of fibrous protein, the main constituent in vertebrates of such epidermal structures as hair, nails, and horn.
KINETIC In zoology, bones joined together but capable of movement.

K-T BOUNDARY (also **KT BOUNDARY**) In geologic time, the transition from the end of the Cretaceous (K) period to the beginning of the Tertiary (T), approximately 65 million years ago.
K-T EXTINCTION (also **KT EXTINCTION**) The termination of numerous (but not all) groups of animals and plants at the end of the Cretaceous period.

LABIAL Near the lip.
LACERTILIA Reptilian suborder comprising lizards.
LACRIMAL (also **LACRIMAL BONE, LACHRIMAL**) Skull bone contributing to the anterior border of the orbit.
LACUNAE Cavities in bones; also, spaces between cells.
LACUSTRINE Living in or beside a lake.
LAGS (See Lines of Arrested Growth.)
LAMELLA Thin scale- or plate-like structure.
LAMELLAR Referring to a thin, scale- or plate-like tissue structure.
LAMINA Thin sheet or layer.
LANDMARK (Morphologically) certain homologous features that are recognizable between animals (*e.g.*, orbit, teeth, etc.).
LATERAL At the side externally; away from the midline.
LAURASIA Hypothetical northern supercontinent including North America, Europe, and parts of Asia.
LECTOTYPE Specimen chosen from syntypes to redesignate the type of a species.
LEPIDOSAUR Reptilians including lizards, snakes and their close relatives.
LIGAMENT Strong fibrous band of tissue that support joints between bones and joins muscles to bones.
LIMNIC Pertaining to lakes.
LINEAGE Continuous line of descent, over an evolutionary span of time, from a particular ancestor.
LINEAR In a line.
LINES OF ARRESTED GROWTH (also **LAGs**) Pattern of development wherein there are pauses in the deposition of bone and a related slower growth rate.
LINGUAL Pertaining to the tongue; the surface of a tooth toward the tongue.
LITORAL Living or growing at or near the sea-shore.
LOCALITY In geology, a named place where specimens have been found.
LOCOMOTION An organism's ability to move from place to place; also, the manner in which an organism moves.
LONG BONE Limb bone.
LUMBAR Pertaining to the region of the loins.
LUNATE Crescent-shaped.

M. Abbreviation identifying a muscle, preceding the formal name for that muscle.
MAMMALIA Group of vertebrate animals distinguished by self-regulating body temperature, hair, and in females, milk-producing mammae, almost all species giving live birth.
MAMMILLA (also **MAMILLA**) Nipple-shaped structure; lower part of an eggshell unit, with a characteristic shape, also called the "cone."
MANDIBLE Lower jaw.
MANDIBULAR Relating to the mandible.
MANUS Part of the forelimb corresponding to the hand, comprising metacarpals and phalanges.
MARINE Pertaining to the sea.
MARL Muddy limestone.
MASS EXTINCTION Death of all members of a number of diverse animal groups apparently due to a common cause.
MATRIX Fossil-embedded rock.
MAXILLA Usually tooth-bearing principal bone in the upper jaw.
MAXILLARY Relating to the maxilla.
MECKELIAN FENESTRA Opening in the lower jaw for Meckel's cartilage and associated vessels/nerves.
MEDIAL From the inside or inner; toward the midline.
MEDIAL MALLEOLUS Lower extermity prolongation of the tibia.
MEDULLA Central part of a bone or organ.
MEGA- Prefix meaning "large."
MESIAL In a middle longitudinal or vertical plane; the edge of a tooth toward the symphysis or premaxillary midline.
MESOZOIC ERA Geologic time span during which nonavian dinosaurs flourished, 248 to 65 million years ago.
METABOLISM Constructive and destructive chemical changes in the body for maintenance, growth, and repair of an organism.
METACARPAL Relating to the metacarpus; also, a bone of the metacarpus, generally one per digit.
METACARPUS Bones of the manus between the wrist and fingers.
METAPHYSEAL Having to do with growing bone.
METAPODIALS In tetrapods, bones of the metacarpus and metatarsus.
METATARSAL Relating to the metatarsus; also, a bone of the metatarsus, generally one per digit.
METATARSUS Part of the foot between the tarsus and toes.
MICACEOUS Containing mica, any of a group of physically and chemically related mineral silicates.
MICRO- Prefix meaning "very small."
MIDLINE Imaginary line extending dorsally along the length of an animal.
MIGRATION Behavior pattern whereby a group of animals of the same species move from one location to another on a regular or recurring basis.
MINERALIZED Formerly organic matter that has been transformed into mineral matter.
MOLLUSC (also **MOLLUSK**) Member of the Mollusca, a group of bilaterally symmetrical invertebrates, such as snails, clams, cephalopods, and other forms.
MONOPHYLETIC Group of taxa including a common ancestor and all of its descendants; derived from a single origin; having the condition of "monophyly."
MORPH Shape; also used as a suffix to denote a general shape, as for a group of organisms, in "archosauromorph" or "dinosauromorph."
MORPHOGENESIS Development of shape.
MORPHOLOGY Science of form.
MORPHOMETRIC Regarding the analysis or measurement of an organism's shape or form.
MORPHOMETRICS Quantitative analysis of shape.
MORPHOTYPE Type specimen of one form of a polymorphic species.
MUSCULATURE Arrangement of muscles.
MUZZLE Anterior part of the head containing the nostrils and jaws.
MYA Abbreviation for "million years ago."

NARIAL Pertaining to the nostrils.
NARIS Nostril opening.
NASAL Bone near the front of the skull, between the premaxilla and the frontal; also, that which pertains to the nostrils or nose.
NEO- Prefix meaning "new."
NEOCOMIAN Old term used to designate a subdivision of the Early Cretaceous period, equivalent to Hauterivian.
NEOGNATHOUS More advanced palate of modern birds (excluding ratites).
NEOPANGAEA (also **NEOPANGEA**)

Giant land mass comprising all major land masses apart from Central Asia.
NEORNITHES Avian crown group including modern birds.
NEURAL Closely connected with nerves or nervous tissues.
NEURAL ARCH Bony bridge over the passage of the spinal cord.
NEURAL CANAL Canal formed by the neural arch and centrum.
NEURAL SPINE Spine rising up from the neural arch.
NEUROCENTRAL Having to do with a neurocentrum, a type of centrum in primitive vertebrates.
NEUROCRANIUM Bony or cartilaginous case containing the brain and capsules of special sense organs.
NICHE Unique place occupied by a particular species within a larger ecological community.
NODE (In cladistic classification) point on a cladogram where two or more lines meet, this constituting a taxon including all descendant taxa that will meet at that point; (morphologically) a knob or swelling.
NODE-BASED Defining a taxonomic group as the descendants of the most recent common ancestor of two other groups and all descendants of that ancestor.
NOMEN DUBIUM Taxon founded upon material of questionable diagnostic value.
NOMEN NUDUM Taxon improperly founded without published material, diagnosis, type designation, and figure.
NOMENCLATURE Official naming or system of naming of taxa.
NONAVIAN (also **NON-AVIAN**) Pertaining to dinosaurs other than birds; also, not pertaining to birds in general.
NUCHAL Pertaining to the neck.

OBLIGATE Limited or restricted to a particular mode of behavior or environmental condition.
OBTURATOR Pertaining to any structure in the area of the obturator foramen.
OBTURATOR FORAMEN Oval foramen within the ischium for the passage of the obturator nerve/vessels.
OCCIPUT Back part of the skull.
OCCIPITAL CONDYLE Condyle with which the skull moves on the atlas and axis.
OCCLUSAL Where surfaces of upper and lower teeth touch when the jaws are closed.
OCCLUSION Surfaces of the upper and lower teeth making contact with each other when the jaws are closed in a bite.
ODONTOID Tooth-like process.
OEDEMA (also **EDEMA**) Accessive accumulation of serous (containing, secreting, or resembling serum) fluid in the tissues.
OLECRANON Process for insertion of the triceps muscle at the proximal end of the ulna.
OLFACTORY Pertaining to the sense of smell.
OMNIVORE Animal that eats both plant and animal food.
ONTOGENY Growth and development of an individual.
OOSPECIES Specific name given to a genus of fossil egg.
OPISTHOCOELOUS Having the centrum concave posteriorly.
OPISTHOPUBIC Pubis that is directed rearward.
OPISTHOTIC Inferior posterior bony element of the otic capsule.
OPTIC LOBE Part of the brain connected with vision.
ORBIT Bony cavity in which the eye is housed.
ORDER In Linnaean classification, a category including related families within a class.
ORGANIC Relating to things alive.
ORGANISM Any individual living being.
ORNAMENTATION Visible external body feature (*e.g.*, horn, frill, etc.) that primarily functions in social behavior.
ORNITH- Prefix meaning "bird" or "birdlike."
ORNITHODIRA Group including, among other taxa, pterosaurs and dinosauromorphs.
ORNITHOTHORACES Avian group including all birds except *Archaeopteryx*.
ORNITHURAE Group of modern birds.
OSSICLE Bony platelets set under the skin, serving as secondary armor.
OSSIFICATION The process by which bone forms.
OSSIFIED TENDONS In ornithischians, strand-like calcified tissues that connect and strengthen the vertebrae.
OSTEO- Prefix meaning "bone" or "relating to bones."
OSTEOCYTE Bone cell.
OSTEODERM Bony plates or scutes in the skin.
OSTEOGENESIS Formation of bone.
OSTEOLOGY Part of zoology dealing with the structure and development of bones.
OSTEON Haversian bone growth.
OSTRACOD Microscopic crustacean consisting of a hinged, bivalved shell.
OTIC Pertaining to the ear.
OTO Prefix meaning "ear."
OUTGROUP In cladistics, the character state occurring in the nearest relatives of an ingroup.
OTOSPHENOIDAL CREST Bony crest formed by fusion of the opisthotic and spendoid bones.

PAEDOMORPHOSIS Evolutionary change in which adults of a descendant species retain juveniles characteristics of the ancestral species.
PALATE Roof of the mouth.
PALATINE One of the bones of the palate, located near the front of the skull and to the side of the vomer; also, pertaining to the palate.
PALEO- (also **PALAEO-**) Prefix meaning "ancient" or "past," pertaining to something very old or prehistoric.
PALEOBIOLOGY Study of ancient extinct organisms.
PALEOBIOPROVINCE Very ancient land area viewed in terms of its indigenous life forms.
PALEOECOLOGY Study of the relationships between extinct organisms and their paleo-environments.
PALEOGEOGRAPHIC Pertaining to paleogeography, the study of the geographic distribution of life forms in the geologic past.
PALEOGNATHOUS BIRDS Primitive birds (including ratites).
PALEONTOLOGY Scientific study of past life, based on the study of fossil and fossil traces.
PALMER Surface of the manus in contact with the ground.
PALPEBRAL Small bone located on the rim of the eye socket, often forming a bony eyelid.
PALYNOMORPH Spores, pollen, and cysts of certain algae.
PANGAEA (also **PANGEA**) Hypothetical huge supercontinent formed by the collision of all Earth's continents during the Permian period.
PAPILLA Conical dermal structure constituting the beginning of a feather.
PARA- Prefix meaning "beside."
PARALLELISM (See Convergence.)
PARAPHYLETIC In cladistics, relating to a taxonomic group including a hypothetical common ancestor and only some of that ancestor's descendants.
PARASAGITTAL Parallel to the midline of an animal.
PARASPHENOID Membrane bone forming the floor of the braincase.

PARATYPE Specimen used along with the holotype in defining a new species.
PARIETAL Bone of the skull roof behind the frontal.
PAROCCIPITAL PROCESS Bony process at the back of the skull.
PARSIMONY In cladistic analysis, a subjective criterion for selecting taxa, usually that which proposes the least number of homoplasies.
PATHOLOGY The study of disease.
PATHOLOGIC Diseased.
PECTINEAL Pertaining to a ridgeline on the femur and the attached pectineus muscle.
PECTORAL Pertaining to the chest area of the skeleton.
PECTORAL GIRDLE Bones of the shoulder, including scapula, corocoid, sternum, and clavicle.
PEDAL Pertaining to the foot.
PEDICLE Backward-projecting vertebral process.
PEDUNCLE Stalk- or stem-like process of a bone.
PELAGE Furry or hairy coat of an animal, this term usually referring to mammals but also in some dinosaurs.
PELVIC GIRDLE Hip area of the skeleton, composed of the ilium, ischium, and pubis.
PELYCOSAUR One of a group of mammal-like reptiles of the Carboniferous and Permian periods, some of which having dorsal "sails" supported by elongated neural spines (not dinosaurs).
PENNACEOUS FEATHER A visible plumage feather (as opposed to down).
PERAMORPHOSIS Evolutionary change wherein juveniles of a descendant species exhibit some adult characteristics of the ancestral species.
PERINATAL Close to the time of hatching or birth.
PERIOD Division of geologic time, a subdivision of an Era.
PERIOSTEAL BONE Bone formed by the periosteum.
PERIOSTEUM Tissue that forms bone at the periphory or outermost region of a growing bone.
PERIOTIC Skull bone enclosing parts of the membranous labyrinth of the internal ear.
PERMINERALIZATION Fossil forming process wherein additional mineral materials are deposited in pore spaces of the originally hard parts of animals.
PES Foot.
PETRIFY Minerals replacing a fossilized organism's hard tissues so that it becomes stone-like.
PHALANGEAL FORMULA Formula giving the number of phalanges in the digits of the manus and pes.
PHALANX Segment of the digits, a bone of the fingers or toes.
PHARYNX Musculo-membranous tube extending from the nasal cavities of the skull to the larynx.
PHENETIC RESEMBLANCES Resemblances in form.
PHENON Clusters of things with similar shape.
PHYLOGENETIC Concerning the evolutionary relationships within and among groups of organisms.
PHYLOGENY Evolutionary tree-like diagram or "tree" showing the relationships between ancestors and descendants.
PHYSIOLOGY Biological study dealing with the functions and activities of organisms.
PHYTOSAUR crocodile-like, semi-aquatic "thecodontians" of the Triassic (not dinosaurs).
PISCIVOROUS Fish-eating.
PLATE (also **DERMAL PLATE**) In paleobiology, a piece of bone embedded in the skin.
PLATE TECTONICS Study of the plates making up the Earth's crust.
PLATYCOELOUS Condition in which the posterior articular end of a vertebral centrum is flat.
PLESIOMORPHIC In cladistics, the more primitive character state of two that are exhibited within members of an ingroup while also occurring in the nearest outgroup; a primitive feature.
PLESIOSAUR General term for a member of a group of Mesozoic marine reptiles, some with smaller heads and longer necks, others with longer heads and shorter necks (not dinosaurs).
PLEUROCOEL Cavity in the side of a vertebral centrum.
PLEUROKINESIS Adaptation of skull bones to move to the side.
PLEUROKINETIC HINGE Hinge between the maxilla and remainder of the skull, allowing the maxillae to swing outward when the mouth is closed.
PLEXIFORM Having interwoven blood vessels, like a network.
PNEUMATIC Bones penetrated by canals and air spaces.
POLLEX In the manus, the thumb or innermost digit of the normal five.
POLYPHYLETIC Associated groups that do not share a single common ancestor.
POPLITEAL Region behind and above the knee-joint.
POSTACETABULAR PROCESS Portion of the ilium posterior to the acetabulum.
POSTCOTYLOID PROCESS Portion of jugal posterior to cup-shaped acetabular cavity.
POSTCRANIA (or POSTCRANIAL SKELETON) Skeleton excluding the skull.
POSTER Presentation of data at a technical gathering (*e.g.*, a symposium or annual meeting of the Society of Vertebrate Paleontology) in the form of a poster, general including drawings, photographs, charts, cladograms, graphs, etc.
POSTERIOR Toward or at the rear end, sometimes referred to as "caudal."
POSTMORTEM Following the death of an organism.
POSTURE Walking or standing position.
POSTZYGAPOPHYSIS Process on the posterior face of the neural arch, for articulation with the vertebra behind it.
POSTURE Walking or standing position.
PREACETABULAR PROCESS Portion of the ilium anterior to the acetabulum.
PREARTICULAR Bone in the lower jaw of primitive tetrapods.
PRECOCIAL Species in which the young are relatively advanced upon hatching.
PRECURSOR Earlier form of life from which a later form is descended.
PREDATOR Organism that hunts and eats other organisms.
PREDENTARY In ornithischians, a small crescent-shaped bone located at the tip of the lower jaw.
PREHISTORIC Referring to an era before recorded history.
PREMAXILLA A usually paired bone at the front of the upper jaw.
PREOCCUPIED In zoological nomenclature, a taxonomic name identical to one published previously by another author.
PREORBITAL Anterior to the orbit, sometimes referred to as "antorbital."
PREPARATION One or more procedures applied to a fossil specimen so that the specimen can be strengthened, handled, preserved, studied, displayed, etc.
PREPARATOR Person who prepares fossils for study or display.
PRESERVATION General condition of a fossil specimen, referring to its quality and completeness.

PREY Creature hunted and caught for food.
PREZYGAPOPHYSIS Process on the anterior face of the neural arch, for articulation with the vertebra in front of it.
PRIMARY BONE Bone that is formed as an organism grows.
PRIMITIVE Characters or features found in the common ancestor of a taxonomic group, which are also found in all members of that group, also referred to as "plesiomorphic"; also (more generally), less developed, earlier.
PRIORITY Rule in scientific nomenclature stating that, in the case of different taxonomic names given to the same form or groupings of forms, the name published first is valid.
PROCESS Outgrowth or projection of bone.
PROCOELOUS Condition in which the anterior articular end of a vertebral centrum is concave and the posterior end strongly convex.
PROGENESIS Evolutionary change wherein sexual maturity takes place earlier in the descendant species than in the ancestor.
PROKINESIS Primitive avian kind of cranial kinesis derived from either akinetic or mesokinetic archosaurian skulls.
PRONATE Inclined.
PRONATION Act by which the palm of the manus is turned downwards by pronator muscles.
PROOTIC Anterior bone of the otic capsule.
PROTEROSUCHID Member of the Proterosuchidae, a family of Late Triassic archosaurs including both quadrupedal and bipedal forms.
PROTOFEATHER Incipient feather including branching barbs, but lacking the aerodynamic quality of the true avian feather.
PROVENANCE Place of origin.
PROVENTRICULUS Second stomach; in birds, the glandular stomach anterior to the gizzard.
PROXIMAL Nearest to the center of the body; toward the mass of the body; segment of a limb or of elements within a limb.
PTEROSAUR One of a group of flying reptile of the Mesozoic, related to (but not) dinosaurs, with batlike wings consisting of membrane stretched from an elongated finger to the area of the hips, one group generally having teeth and long tails, the other toothless and possessing short tails.
PTERYGOID Wing-like posterior bone of the palate.
PUBIC Relating to the pubis.
PUBIS Antero-ventral bone of the pelvic girdle.
PUBOISCHIAL Place where the pubis and ischium meet.
PULMONARY Pertaining to the lungs.
PYGOSTYLE In birds and some theropod, a structure at the tail end of the vertebral column consisting of fused vertebrae.
PYOGENIC Pus-forming material, involving bacteria.

QUADRATE In birds, reptiles and amphibians, the bone with which the lower jaw articulates.
QUADRATOJUGAL Bone bone connecting or overlying the quadrate and jugal.
QUADRUPED Animal that walks on all four feet.
QUADRUPEDALITY Habitually walking on four legs.

RACHIS (also **spelled RHACHIS**) Shaft of a feather.
RADIALE Carpal bone aligned with the radius.
RADIATION Process by which a group of species diverge from a common ancestor, thereby producing an increased biological diversity, usually over a relatively short span of time.
RADIOMETRIC DATING Dating method involving the measurement of decay, at a constant known rate, in various naturally occurring radioactive isotopes.
RADIUS Smaller forelimb bone between the humerus and carpals, lying next to the ulna.
RAMUS Branch-like structure.
RAPTOR One of various modern birds of prey, including falcons and hawks; also, a suffix used in the names of a number of sometimes rather diverse theropods; more recently, a popular term inaccurately used to designate any dromaeosaur.
RAPTORIAL Subsisting by or adapted for the seizure of prey.
RATITE One of a group of flightless birds having an unkeeled sternum; also, an eggshell morphotype in which the shells structure is discrete only in the inner one-sixth to one half of the shell thickness (mammillary layer); most of the eggshell formed of a single, continuous layer.
RECONSTRUCTION Drawn or modeled skeleton or partial skeleton, based upon the original fossil remains, often incorporating extrapolation and/or knowledge of the more complete remains of other taxa (sometimes used to mean "restoration").
RECTRICES Stiff tail feathers of a bird or some nonavian theropods, used in steering.
RECURVED Curved backward.
RED BEDS Sedimentary beds that are reddish in color.
REMIGES Large feathers or quills on a bird's wing, consisting of primaries and secondaries.
REMODELING Resorption and reprecipitation of bone for the purpose of maintaining its physiological and mechanical competence.
RENAL Pertaining to the kidneys.
RESPIRATION Breathing process, accomplished by an exchange of gases between an organism and its surrounding atmosphere.
RESPIRATORY TURBINATE (also **RT**) Thin, complex structure consisting of cartilage or bone in the nasal airway.
RESTORATION In paleontology, a drawn, sculpted, or other representation of a fossil organism as it may have appeared in life (sometimes used as synonymous with "reconstruction").
RETICULAR Bone possessing network-like interstices.
REVERSAL (In cladistic classification) transformation of a character in an advanced lineage back to its ancestral state.
"REVULETIAN" Infrequently used term, often employed by authors in the New Mexico area, referring to the early to middle Norian stage of the Late Triassic period.
"RHAETIAN" Basically obsolete term referring to a stage of the Upper Triassic of England.
"RHAETIC" Basically lithostratigraphic obsolete term referring to a sequence representing a part of "Rhaetian" time, the lower part of the Westbury Formation.
RHYNCHOSAUR A member of the Rhynchosauria, a group of large, squat, beaked, archosaur-like reptiles of the Triassic.
RIB Elongate and sometimes curved bone of the trunk articulating with vertebrae.
ROBUST Strongly formed or built; also, a method of study or analysis, verified by past results, which will probably result in a correct inference.
ROSTRAL (also **ROSTRUM**) In ceratopsians, median unpaired bone located at the tip of the upper jaw; also, toward the rostrum or tip of the head.
RUGOSE Possessing a rough surface (or "rugosity").

SACRAL Pertaining to the sacrum.
SACRAL RIB Rib that connects the sacral vertebrae to the pelvis.
SACRUM Structure formed by the sacral vertebrae and pelvic girdle.
SAGITTAL Pertaining to the midline on the dorsal aspect of the cranium.
SAUROPOD HIATUS Expanse of time (from the Cenomanian to the Maastrichtian) during which sauropods seem to have been absent from the North American continent.
SCANSORIAL Adapted to climbing.
SCAPULA Shoulder blade.
SCAVENGER Animal that feeds on dead animal flesh or other decomposing organic matter.
SCUTE Horny or bony plate embedded in the skin.
SEDIMENT Deposit of inorganic and/or organic particles.
SEDIMENTARY ROCKS (also **SEDIMENTS**) Rocks formed from sediment.
SELECTION Principle that organisms having a certain hereditary characteristic will have a tendency to reproduce at a more successful rate than those of the same population not having this characteristic, consequently increasing their numbers in later generations.
SEMILUNATE Having the approximate shape of a half-crescent.
SENIOR SYNONYM Taxon having priority over another identically named taxon and regarded as the valid name, because of the former's earlier publication.
SEPTUM Partition separating spaces.
SEXUAL DIMORPHISM Marked differences in shape, shape, color, structure, etc. between the male and female of the same species.
SIGMOID S-shaped.
SINUS Space within a body.
SINUSOIDALLY Having a small space for blood.
SIPHONIUM Membranous tube that connects air passages of quadrate with air space in the mandible.
SISTER GROUP (or SISTER TAXON, SISTER CLADE) Group of organisms descended from the same common ancestor as its closest group.
SPATULATE Spatula-shaped.
SPECIALIZATION Modification in a particular way.
SPECIES In paleontology, a group of animals with a unique shared morphology; in zoology, a group of naturally interbreeding organisms that do not naturally interbreed with another such group.
SPECIMEN Sample for study.
SPINAL Having to do with the backbone or tail.
SPLENIAL Dermal bone in the lower jaw, covering much of Meckel's groove.
SQUAMOSAL In the vertebrate skull, a bone that forms part of the posterior side wall.
STEM-BASED Pertaining to a taxonomic group defined as all those entities that share a more recent common ancestor with one group than with another.
STEREOSCOPIC Pertaining to the ability to see a three- dimensional image.
STERNAL Pertaining to the breastbone or chest.
STERNUM Breastbone.
STRATIGRAPHY Study of the pattern of deposition.
STRATUM Layer of sediment.
STREPTOSTYLY Adaptation of the skull in which the quadrate is in movable articulation with the squamosal.
SUBFAMILY In Linnaean classification, a category smaller than a family, including genus one or more.
SUBGENUS Subtle classification between a genus and a species; a group of related species within a genus.
SUBORDER In Linnaean classification, a category smaller than an order, larger than an infraorder, including one or more families.
SUITE Group of characters associated with a particular organism or species.
SUPER- Prefix meaning "greater" or "above."
SUPER-CONTINENT (also **SUPERCONTINENT**) Large structures formed by the joining of various continental areas.
SUPRA- Prefix meaning "above" or "over."
SUPRAORBITAL Small bone along the upper rim of the orbit of the skull; in ceratopsians, a horn above the eye or brow.
SUPRATEMPORAL FENESTRA Opening in the top of the skull, posterior to the orbit.
SURANGULAR Bone of the upper rear area of the lower jaw, contacting (and posterior to) the dentary, the angular, and the articular.
SUTURE Line where bones contact each other.
SYMPHYSIS Line of junction of two pieces of bone.
SYMPLESIOMORPY In cladistics, a character state shared by a member of one higher-level taxon with a member of a more primitive higher-level taxon.
SYN- Prefix meaning both "together" and "with"; also "united" or "fused."
SYNAPOMORPHY Shared/derived feature defining a monophyletic group; unique character shared by two or more taxa.
SYNAPSID Member of the Synapsida, a group of tetrapods having a skull with one opening behind the eye socket, including pelycosaurs, therapsids, and mammals.
SYNCLINE In bedrock, a low, trough-like area in which rocks incline together from opposite sides.
SYNONYM Different names for the same taxon.
SYNSACRUM Single-unit structure formed by the fusion of several vertebrae.
SYNTYPE When a holotype and paratypes have not been selected, one of a series of specimens used to designate a species.
SYSTEMATICS Scientific study that involves the classification and naming of organisms according to specific principles.

TABLE In a vertebra, a bony platform lateral to the base of a neural spine.
TAPHONOMY Study of the processes of burial and fossilization of organisms.
TARSAL Ankle bone.
TARSOMETATARSUS In birds and some dinosaurs, a bone formed by the fusion of the distal row of tarsals with the second to fourth metatarsals.
TARSUS Region where the leg and foot join; ankle bones.
TAXON Definite unite in the classification of animals and plants.
TAXONOMIC Pertaining to or according to the principles of taxonomy.
TEMPORAL Bone on either side of the skull that forms part of its lateral surface; also, pertaining to that area of the skull.
TERRESTRIAL Land-dwelling.
TERRITORIAL Displaying a pattern of behavior whereby an organism or group of organisms of one species inhabit a particular area, and defend that area against intrusion by other individuals of that species.
TETRAPOD Vertebrate with four limbs.
TEXTURE Sequence of horizontal ultrastructural zones of an eggshell, also called "eggshell unit macrostructure."
"THECODONTIAN" (also **"THECODONT"**) One of an obsolete and artificial "order" (Thecodontia) of early archosaurian reptiles of the Late Permian and Early Triassic, some of which may have been ancestral to dinosaurs, pterosaurs, and crocodiles.

THERMOREGULATION One of various processes by which the body of an organism maintains internal temperature.
THORACIC Pertaining to the thorax; in the chest region.
THORAX Part of the body between the neck and abdomen.
TIBIA Shin bone.
TIBIOTARSUS In birds and some dinosaurs, the tibial bone to which are fused the proximal tarsals.
TOMOGRAPHY Recording internal images in a body via X-rays; a CT (or CAT) scan.
TOOTH BATTERY (See Battery.)
TOPOLOGICAL Pertaining to the specific areas of a skeleton or body.
TOPOLOGY Study of the specific areas of a skeleton or body.
TOPOTYPE Specimen from the locality of the type specimen.
TRACE FOSSIL Not the actual remains of an extinct organism, but rather the fossilized record of something left behind by that organism; the fossil record of a living animal.
TRACKWAY Series of at least three successive footprints made by a moving animal.
TRANSVERSE PROCESS Laterally directed process of the vertebral centrum, for attachment of intervertebral muscles.
TRIASSIC PERIOD First and earliest division of the Mesozoic Era, 248 to 206 million years ago.
TRICIPITAL Having three "heads" or insertions.
TRIDACTYL Having three digits.
TRIGEMINAL Consisting of or pertaining to three structures.
TRIGEMINAL NERVE Fifth cranial nerve.
TRIPODAL Upright stance incorporating the hind feet and tail.
TROCHANTER Prominence or process on the femur to which muscles are attached.
TROCHLEAR Pulley-shaped.
TROPHIC Pertaining to food or the feeding process.
TROPISM Tendency of an organism to react in a specific way to a specific kind of stimulus.
TUBER (plural TUBERA) Rounded protuberance; an anterior projection of the tibia.
TUBERCLE Small, rounded protuberance.
TUBERCULATE Having or resembling tubercles.
TUBERCULOUS Having many tubercles.
TUBERCULUM One of the heads of the rib, attaches to the transverse process of the vertebral centrum.
TYMPANIC Pertaining to the ear or eardrum.
TYPE LOCALITY Geographic site at which a type specimen or type species was found and collected.
TYPE SPECIMEN Specimen used to diagnose a new species.

ULNA In the forearm, the larger long bone on the medial side, parallel with the radius.
ULNARE In the proximal row of carpals, the bone at the distal end of the ulna.
UNCINATE PROCESS In birds and some reptiles, a process on the ribs which overlaps other ribs.
UNGUAL Phalanx bearing a nail or claw.
UTRICULUS Membraneous sac of the ear-labyrinth.

VACUITY Open space.
VAGUS Tenth cranial nerve.
VARIATION Range of appearance within a group of organisms.
VARIETY In biology, a taxonomic category below the species level, comprising naturally occurring or selectively bred individuals having varying characteristics.
VASCULAR Of or pertaining to the circulatory system.
VASCULARIZED Possessing blood vessels.
VASCULARIZATION Formation or development of blood vessels.
VENTER Smooth concave surface; also, abdomen, or a lower abdominal surface.
VENTRAL From beneath, relating to the belly or venter [abdomen or lower abdominal surface]; toward the belly.
VENTRI- (also **VENTRO**) Prefix meaning "belly."
VERTEBRA Bony segment of the backbone.
VERTEBRATE Animal with a backbone.
VERTEBRATE PALEONTOLOGY Scientific study of fossil animals having backbones.
VISCERA Internal organs of the body, particularly those of the digestive tract.
VOLCANISM Volcanic activity or force.
VOMER Bone at the front of the palate.

ZONAL BONE Bone material resulting from a development pattern that involves slow to moderate growth during intermittent periods.
ZOOLOGY Science dealing with the structure, behavior, functions, classification, evolution and distribution of animals.
ZYGAPOPHYSIS Bony, usually peglike process on the neural arch of a vertebra, by which it articulates with other vertebrae.

Bibliography

[Anonymous], 1834, Discovery of saurian bones in the Magnesian Conglomerate near Bristol: *London, Edinburgh and Dublin Philosophical Magazine*, Series 3, 5, p. 463.

_____, 1935, Saurian remains in the Magnesian Conglomerate of Bristol: *West of England Journal of Science and Literature*, 1, pp. 84–85.

_____, 1997, Marginocephalia, *in*: Philip J. Currie and Kevin Padian, editors, *Encyclopedia of Dinosaurs*. San Diego: Academic Press, p. 415.

Abel, Othenio, 1927, *Lebensbilder der Vorzeit*. Jena, Germany: Fischer, p. 637.

Alcober, O., Paul C. Sereno, Hans C. E. Larsson, Ricardo Néstor Martinez, and David J. Varricchio, 1998, A Late Cretaceous carcharodontosaurid (Theropoda: Allosauroidea) from Argentina: *Journal of Vertebrate Paleontology*, 18 (Supplement to Number 3), Abstracts of Papers, Fifty-eighth Annual Meeting, p. 23A.

Alexander, R. McN., 1976, Estimates of speeds of dinosaurs: *Nature*, 261, pp. 129–130.

_____, 1985, Mechanics of posture and gait of some large dinosaurs: *Zoological Journal of the Linnean Society*, 83, pp. 1–25.

_____, 1989, *Dynamics of Dinosaurs and Other Extinct Giant*. New York: Columbia University Press, 167 pages.

_____, 1996, *Tyrannosaurus* on the run: *Ibid.*, 379, p. 121.

Allain, Ronan, and Philippe Taquet, 2000, A new genus of Dromaeosauridae (Dinosauria, Theropoda) from the Upper Cretaceous of France: *Journal of Vertebrate Paleontology*, 20 (2), pp. 404–407.

Allain, Ronan, Philippe Taquet, Bernard Battail, Jean Dejax, Philippe Richir, Monette Véran, Franck Limon-Duparcmeur, Renaud Vacant, Octavio Mateus, Phouvong Sayarath, Bounxou Khenthavong, and Sitha Phouyavong, 1999, Un nouveau genre de dinosaure sauropode de la formation des Grès supérieurs (Aptien–Albien) du Laos: *Comtes Rendu des Séances de l'Académie des Sciences, Paris, de la Terre et des Planètes*, 329, pp. 609–616.

Alvarez, Luis W., Walter Alvarez, Frank Asaro and Helen V. Michel, 1980, Extraterrestrial cause for the Cretaceous–Tertiary extinction: *Science*, 209 (4448), pp. 1095–1108.

Ancell, Carrie A., Robert Harmon, and John R. Horner, 1998, Gar in a duck-bill: preservation of a scavenger and its prey: *Journal of Vertebrate Paleontology*, 18 (Supplement to Number 3), Abstracts of Papers, Fifty-eighth Annual Meeting, p. 24.

Andersen, Arthur F., Ralph E. Chapman, K. Kenny, and Hans C. E. Larsson, 1999, Animation of 3-D digital data: the walking *Triceratops*: *Journal of Vertebrate Paleontology*, 19 (Supplement to Number 3), Abstracts of Papers, Fifty-ninth Annual Meeting, p. 29A.

Anderson, Brian G., Spencer G. Lucas, Reese E. Barrick, Andrew B. Heckert, and George T. Basabilvazo, 1998, Dinosaur skin impressions and associated skeletal remains from the upper Campanian of southwestern New Mexico: new data on the integument morphology of hadrosaurs: *Journal of Vertebrate Paleontology*, 18 (4), pp. 739–745.

Anderson, J. F., A. Hall-Martin, and Dale A. Russell, 1985, Long-bone circumferences and weight in mammals, birds and dinosaurs: *Journal of the Zoological Society of London*, 207, pp. 53–61.

Andreis, R. R., G. E. Bossi, and D. K. Montardo, 1980, O grupo Rosário do Sul (Triássico) no Rio Grande do Sul: *XXXI Congresso Brasileiro de Geologia, Camborrí-SC*, 2, pp. 659–673.

Andrews, Charles W., 1913, On some bird remains from the Upper Cretaceous of Transylvania: *Geological Magazine*, 5, pp. 193–196.

Anton, John, 2000, Ceratopsian crests as acoustic devices: *Journal of Vertebrate Paleontology*, 20 (Supplement to Number 3), Abstracts of Papers, Sixtieth Annual Meeting, p. 27A.

Apesteguía, S., and A. Cambiaso, 1999, Hallazgo de hadrosaurios en la Formacion Paso del Sapo (Campaniano–Maastrichtiano, Chubut): *Ameghiniana, Revista de la Asociación Paleontológica Argentina, (Resúmenes)*, 36 (Número 4-Suplemento), p. 5R.

Archibald, J. David, 1996, *Dinosaur Extinction and the End of an Era. What the Fossils Say*. New York: Columbia University Press, 237 pages.

Arcucci, Andrea Beatriz, and Rodolfo Anibal Coria, 1998, Skull features on a new primitive theropod from Argentina: *Journal of Vertebrate Paleontology*, 18 (Supplement to Number 3), Abstracts of Papers, Fifty-eighth Annual Meeting, pp. 24A–25A.

Arinobu, Tetsuya, Ryoshi Ishiwatari, Kunio Kaiho, and Marcos A. Lamolda, 1999, Spike of pyrosynthetic polycyclic aromatic hydrocarbons associated with an abrupt decrease in d13C of a terrestrial biomarker at the Cretaceous–Tertiary boundary at Caravaca, Spain: *Geology*, 27, pp. 723–726.

Ashworth, William B., Jr., 1996, *Paper Dinosaurs*. Kansas City, Missouri: Linda Hall Library, 50 pages.

Auffenberg, W. 1978, Social and feeding behavior in *Varanus komodoensis*, *in*: N. Greenberg and P. D. MacLean, editors, *Behavior and Neurology of Lizards*. Washington, D.C.: National Institute of Mental Health, (ADM) 77-491, pp. 301–331.

Augusta, Josef, 1960, *Prehistoric Animals*. London: Paul Hamlyn, 43 pages, plus unpaginated plates.

Austen, J. H., 1852, *Guide to the Geology of the Isle of Purbeck*. Blandford.

Azevedo, Sergio Alex Kugland de, 1999, Os dinossauros Triássicos do sul do Brasil: dado actualizados e novas perspectivas: *Paleontologia em Destaque*, 14 (26), p. 57.

Azevedo, Sergio Alex Kugland de, Átila Augusto Stock da Rosa, Ruben Alexandre Boelter, and Luciano Artemio Leal, 1999, A prosauropod dinosaur from the Late Triassic of Southern Brazil: *Paleontologia em Destaque*, 14 (26), p. 59.

Azuma, Yoichi, and Philip J. Currie, 1995, A new giant dromaeosaurid from Japan: *Journal of Vertebrate Paleontology*, 15 (Supplement to Number 3), Abstracts of Papers, Fifty-fifth Annual Meeting, p. 17A.

_____, 2000. A new carnosaur (Dinosauria: Theropoda) from the Lower Cretaceous of Japan: *Canadian Journal of Earth Sciences*, 37 (12), pp. 1735–1753.

Baez, A. M., and C. A. Marsicano, 1998, A heterodontosaurian ornithischian in the Upper Triassic of southern Patagonia?, *in*: J. Almond, J. Anderson, P. Booth, A. Chinsamy-Turan, D. Cole, M. de Wit, B. Rubridge, R. Smith, J. Van Bever Donker, and B. C. Storey, editors, *Journal of African Earth Sciences*, 27 (14), Special Abstracts Issue, Gondwana 10: Event Stratigraphy of Gondwana, unpaginated.

Bai, Zigi, Yang Jie, and Wang Guohui, 1990, *Yimenosaurus*, a new genus of prosauropod from Yimen County, Yunnan Province: *Yuxiwenbo (Yuxi Culture and Scholarship)* 1, pp. 14–23 (in Chinese, translated by Will Downs, Bilby Research Center, Northern Arizona University, 1999).

Baird, Donald, 1980, A prosauropod dinosaur trackway from the Navajo Sandstone (Lower Jurassic) of Arizona, *in*: Louis L. Jacobs, editor, *Aspects of Vertebrate History*. Flagstaff, Arizona: Museum of Northern Arizona Press, pp. 219–230.

Baird, Donald, and John R. Horner, 1977, A fresh look at the dinosaurs of New Jersey and Delaware: *New Jersey Academy of Sciences Bulletin*, 22 (2), p. 50.

Bakker, Robert T., 1971, Ecology of the brontosaurs: *Nature*, 229, pp. 172–174.

_____, 1978, Dinosaur feeding behaviour and

the origin of flowering plants: *Ibid.*, 274, pp. 661–663.

_____, 1986, *The Dinosaur Heresies: New Theories Unlocking the Mystery of the Dinosaurs and Their Extinction.* New York: William Morrow and Company, 481 pages.

_____, 1988, Review of the Late Cretaceous nodosaurid Dinosauria, *Denversaurus schlessmania*, a new armor-plated dinosaur from the Latest Cretaceous of South Dakota, the last survivor of the nodosaurians, with comments on stegosaur-nodosaur relationships: *Hunteria*, 1 (3), pp. 3–23.

_____, 1998, Channeling the thunder in the thunder lizards: cranial/cervical adaptations for infra-sound control in apatosaurine dinosaurs: *Journal of Vertebrate Paleontology*, 18 (Supplement to Number 3), Abstracts of Papers, Fifty-eighth Annual Meeting, p. 25A.

Bakker, Robert T., Michael Williams, and Philip J. Currie, 1988, *Nanotyrannus*, a new genus of pygmy tyrannosaur, from the Latest Cretaceous of Montana: *Hunteria*, 1 (5), pp. 1–30.

Ballent, S. C., 1980, Ostrácodos de ambiente salobre de la Formación Allen (Cretácico Superior) en la provicia de Río Negro (República Argentina): *Ameghiniana*, 17, pp. 67–82.

Bang, B. G., and S. Cobb, 1968, The size of the olfactory bulb in 108 species of birds: *The Auk*, 86, pp. 55–61.

Banks, John C., Shelley B. Fenton, and James L. Hayward, 2000, Skeletal development of an extant avian dinosaur: *Journal of Vertebrate Paleontology*, 20 (Supplement to Number 3), Abstracts of Papers, Sixtieth Annual Meeting, p. 28A.

Barrett, Paul M., 1998*a*, Feeding in thyreophoran dinosaurs: *Journal of Vertebrate Paleontology*, 18 (Supplement to Number 3), Abstracts of Papers, Fifty-eighth Annual Meeting, p. 26A.

_____, 1998*b*, Herbivory in the non-avian Dinosauria, Ph.D. dissertation, University of Cambridge, 308 pages (unpublished).

_____, 1999*a*, A reassessment of the enigmatic ornithischian *Echinodon*: *Journal of Vertebrate Paleontology*, 19 (Supplement to Number 3), Abstracts of Papers, Fifty-ninth Annual Meeting, p. 31A.

_____, 1999*b*, A sauropod dinosaur from the Lower Lufeng Formation (Lower Jurassic) of Yunnan Province, People's Republic of China: *Journal of Vertebrate Paleontology*, 19 (4), pp. 785–787.

Barrett, Paul M., and Makoto Manabe, 2000, The dinosaur fauna from the earliest Cretaceous Tetori Group of Central Honshu, Japan: *Journal of Vertebrate Paleontology*, 20 (Supplement to Number 3), Abstracts of Papers, Sixtieth Annual Meeting, pp. 28A–29A.

Barrett, Paul M., and Paul Upchurch, 1994, Feeding mechanisms of *Diplodocus*: *Gaia: Revista de Geociencias, Museuo Nacional de Historia Natural*, University of Lisbon, 10.

Barrett, Paul M., You Hailu, Paul Upchurch, and Alex C. Burton, 1998, A new ankylosaurian dinosaur (Ornithischia: Ankylosauria) from the Upper Cretaceous of Shanxi Province, People's Republic of China: *Journal of Vertebrate Paleontology*, 18 (2), pp. 376–384.

Barrick, Reese E., Dale A. Russell, and William Showers, 1998, How much did dinosaurs eat: metabolic evidence from oxygen isotopes: *Journal of Vertebrate Paleontology*, 18 (Supplement to Number 3), Abstracts of Papers, Fifty-eighth Annual Meeting, p. 26A.

Barrick, Reese E., Michael K. Stoskopf, Jonathon D. Marcot, Dale A. Russell, and William J. Showers, 1998, The thermoregulatory functions of the *Triceratops* frill and horns: heat flow measured with oxygen isotopes: *Journal of Vertebrate Paleontology*, 18 (4), pp. 746–750.

Barsbold, Rinchen, 1974, Saurornithoididae, a new family of small theropod dinosaurs from Central Asia and North America: *Palaeontologica Polonica*, 30, pp. 5–22.

_____, 1981, Predatory toothless dinosaurs from Mongolia: *Trudy Somestnaya Sovetsko-Mongol'skaya Paleontologischeskaya Ekspeditsiya* [*The Joint Soviet-Mongolian Palaeontological Expedition: Transactions*], 15. pp. 28–39.

_____, 1983*a*, Carnivorous dinosaurs from the Cretaceous of Mongolia: *Ibid.*, 19, 117 pages.

_____, 1983*b*, O "ptich'ikh" chertakh v stroyenii khischnykh dinozavrov [On the "avian" characters in the structure of predatory dinosaur]: *Ibid.*, 24, pp. 96-1-3.

_____, 1997, Oviraptorosauria, *in*: Philip J. Currie and Kevin Padian, editors, *Encyclopedia of Dinosaurs*. San Diego: Academic Press, pp. 505–509.

Barsbold, Rinchen, Philip J. Currie, Nathan R. Myhrvold, Halszka Osmólska, Khisigjaw Tsogtbaatar, and Mahito Watabe, 2000, A pygostyle from a non-avian theropod: *Nature*, 403, pp. 155–156.

Barsbold, Rinchen, Teresa Maryańska, and Halszka Osmólska, 1990, Oviraptorosauria, *in*: David B. Weishampel, Peter Dodson, and Osmólska, editors, *The Dinosauria*. Berkeley and Los Angeles: University of California Press, pp. 249–258.

Barsbold, Rinchen, and Halszka Osmólska, 1990, Ornithomimosauria, *in*: David B. Weishampel, Peter Dodson, and Osmólska, editors, *The Dinosauria*. Berkeley and Los Angeles: University of California Press, pp. 225–244.

_____, 1999, The skull of *Velociraptor* (Theropoda) from the Late Cretaceous of Mongolia: *Acta Palaeontologica Polonica*, 44 (2), pp. 189–219.

Barsbold, Rinchen, Halszka Osmólska, and Seriozha M. Kurzanov, 1987, On a new troodontid (Dinosauria, Theropoda) from the Early Cretaceous of Mongolia: *Acta Palaeontological Polonica*, 32, pp. 121–132.

Barsbold, Rinchen, Halszka Osmólska, Mahito Watabe, Philip J. Currie, and Khishigjaw Tsootbataar, 2000, A new oviraptorosaur (Dinosauria, Theropoda) from Mongolia: the first dinosaur with a pygostyle: *Acta Palaeontologica Polonica*, 45, pp. 97–106.

Bauer, C. M., 1916, Contributions to the geology and paleontology of San Juan County, New Mexico. 1. Stratigraphy of a part of the Chaco River Valley: *U.S. Geological Survey, Professional Paper*, 98, pp. 271–278.

Baumel, J. J., J. A. Wilson, and D. R. Bergen, 1990, The ventilatory movements of the avian pelvis and tail: function of the muscles of the tail region of the pigeon (*Columbia livia*): *The Journal of Experimental Biology*, 151, pp. 263–277.

Beneden, P.-J. van, 1881, Sur l'arc pelvien chez les dinosauriens de Bernissart: *Bulletins de l'Academie Royale des Sciences, des Lettres et des Beaux-Arts de Belgique*, Classe des Sciences, (3) 1 (5), pp. 600–608.

Benedetto, J. L., 1977, Herrerasauridae, neuva familia de sauris quios Triásicos: *Ameghiniana*, 10, pp. 89–102.

Bennett, S. Christopher, 1993, The ontogeny of *Pteranodon* and other pterosaurs: *Paleobiology*, 19, pp. 92–106.

Bennett, Geb E., and Robert M. Sullivan, 2000, A new juvenile *Parasaurolophus* from the Upper Cretaceous Fruitland Formation, San Juan Basin, New Mexico: *Journal of Vertebrate Paleontology*, 20 (Supplement to Number 3), Abstracts of Papers, Sixtieth Annual Meeting, p. 29A.

Benton, Michael J., 1990, Origin and Interrelationships of Dinosaurs, *in*: David B. Weishampel, Peter Dodson, and Halszka Osmólska, editors, *The Dinosauria*. Berkeley and Los Angeles: University of California Press, pp. 11–30.

_____, 1993, Reptilia, *in*: M. J. Benton, editor, *The Fossil Record 2*. London: Chapman and Hall, pp. 681–715.

Benton, Michael J., and James M. Clark, 1988, Archosaur phylogeny and the relationships of the Crocodylia: *The Systematics Association Special Volume*, 35A, pp. 295–338.

Benton, Michael J., Lars Juul, Glenn W. Storrs, and Peter M. Galton, 2000, Anatomy and systematics of the prosauropod dinosaur *Thecodontosaurus antiquus* from the upper Triassic of southwest England: *Journal of Vertebrate Paleontology*, 20 (1), pp. 77–108.

Benton, Michael J., and P. S. Spencer, 1995, *Fossil Reptiles of Great Britain*. London: Chapman and Hall, 389 pages.

Berman, David S, and John S. McIntosh, 1978, Skull and relationships of the Upper Jurassic sauropod *Apatosaurus* (Reptilia, Saurischia): *Bulletin of Carnegie Museum of Natural History*, 8, 35 pages.

Berner, R. A., 1997, The rise of plants and their effect on weathering and atmospheric CO^2: *Science*, 276, pp. 544–546.

Bertini, Reinaldo J., and Rodrigo M. Santucci, 2000, Titanosaurids (Saurischia, Sauropoda) from southeastern Brazil: *Journal of Vertebrate Paleontology*, 20 (Supplement to Number 3), Abstracts of Papers, Sixtieth Annual Meeting, p. 30A.

Beurlen, K., 1971, As condicões ecológicas e faciológicas da Formacão Santana na Chapada do Araripe (Nordeste do Brasil): *Anais de Academia Brasileira de Ciências*, 43 (supl.), pp. 411–415, Rio de Janeiro.

Bidar, Alain, Louis Demay, and Gérard Thomel, 1972, *Compsognathus corralestris*, nouvelle espèce de Dinosaurien Théropode du Portlandien de Canjuers (Sud-Est de la France): *Annales du Muséum d'Historie Naturelle de Nice*, 1 (1), pp. 1–34.

Bilbey, Sue Ann, J. Evan Hall, and Dee A. Hall, 2000, Preliminary results on a new haplocanthosaurid sauropod dinosaur from the Lower Morrison Formation of northeastern Utah: *Journal of Vertebrate Paleontology*, 20 (Supplement to Number 3), Abstracts of Papers, Sixtieth Annual Meeting, p. 30A.

Blanckenhorn, M., 1898, Saurierfunde im fränkischen Keuper: *Sber. Phys.-Med. Soz. Erlangen*, 29, pp. 67–91, Erlangen.

Blows, William T., 1987, The armoured dinosaur *Polacanthus foxii* from the Lower Cretaceous of the Isle of Wight: *Palaeontology*, 30, pp. 557–580.

Bohor, B. F., D. M. Triplehorn, D. J. Nichols, and H. T. Millard, 1987, Dinosaurs, spherules and the "magic" layer: a new K-T boundary clay site in Wyoming: *Geology*, 15, pp. 896–899.

Bohlin, Birger, 1953, Fossil reptiles from Mongolia and Kansu: *The Sino-Swedish Expedition Publication*, 37 (6), pp. 1–105.

Bolotsky, Y. L., and S. K. Kurzanov, 1991, The hadrosaurs of the Amur region, *in*: *Geology of the Pacific Coast Border*, pp. 94–103.

Bonaparte, José F., 1972, Los tetrapodos del sector superior de la Formación Los Colorados, La Rioja, Argentina (TriaH'sico Superior), I Parte: *Opera Lilloana*, 22, pp. 1–183.

_____, 1976, *Pisanosaurus mertii* Casamiquela and the origin of the Ornithischia: *Journal of Paleontology*, 50, pp. 808–820.

_____, 1986*a*, The early radiation and phylogenetic relationships of the Jurassic sauropod dinosaurs, based on vertebral anatomy, *in* Kevin Padian, editor, *The Beginning of the Age of Dinosaurs: Faunal Change across the Triassic–Jurassic Boundary*. New York: Cambridge University Press, pp. 247–258.

_____, 1986*b*, History of terrestrial Cretaceous vertebrates of Gondwana: *IV Congreso Argentino de Paleontología y Bioestratigrafía*, Revista del Museo Argentina de Ciencias Naturales "B. Rivadavia," Paleontología, 4, pp. 17–123.

_____, 1986*c*, Les dinosaures (carnosaures, allosauride's, sauropodes, cétiosauride's) de Jurassic moyen de cerro cóndor (Chubut, Argentine): *Annales de Paléontologie (Vert.-Invert.)*, 72 (3), pp. 247–289.

_____, 1991*a*, Los vertebrados fosiles de la Formacion Rio Colorado, de la ciudad de neuquen y cercanias, Cretacico Superior, Argentina: *Revista del Museo Argentino de Ciencias Naturales (Bernardino Rivadavia), el Instituto Nacional de Investigacion de las Ciencias Naturales*, 4 (3), pp. 68–101.

_____, 1991*b*, The Gondwanan theropod families Abelisauridae and Noasauridae: *Historical Biology*, 5, pp. 1–25.

_____, 1996*a*, The presacral vertebra of *Apatosaurus* suggest a different family: Apatosauridae: *Dinofest International Symposium, April 18–21, 1996, Program and Abstracts*, p. 30.

_____, 1996*b*, Cretaceous tetrapods of Argentina: *Müncher Geowissenschaftliche Abhandlung*, A (30), pp. 73–130.

_____, 1998, An armoured sauropod from the Aptian of northern Patagonia, Argentina, *in*: Yukimitsu Tomida, Thomas R. Rich, and Patricia Vickers-Rich, editors, *Second Symposium Gondwana Dinosaur, 12–13 July, 1998, National Science Museum, Tokyo, Abstracts with Program*, p. 10.

_____, 1999*a*, An armoured sauropod from the Aptian of Northern Patagonia, Argentina, *in*: Yukimitsu Tomida, Thomas H. Rich, and Patricia Vickers-Rich, *Proceedings of the Second Gondwanan Dinosaur Symposium*, National Science Museum Monographs, 15, pp. 1–12.

_____, 1999*b*, Evolución de las vértebras presacras en Sauropodomorpha: *Ameghiniana*, 36 (2), pp. 115–187.

Bonaparte, José F., and Rodolfo Anibal Coria, 1993, Un nuevo y gigantesco saurópodo titanosaurio de la Formacion Rio Limay (Albiano–Cenomaniano) de la Provincia del Neuquén, Argentina: *Ameghiniana*, 30 (3), pp. 271–282, Buenos Aires.

Bonaparte, José F., and Jorge Ferigolo, 1998, A new and primitive saurischian dinosaur, *Guaibasaurus dandelariai*, gen. et sp. nov., from the Late Triassic Caturrita Formation of southern Brazil: *in*: Yukimitsu Tomida, Thomas R. Rich, and Patricia Vickers-Rich, editors, *Second Symposium Gondwana Dinosaur, 12–13 July, 1998, National Science Museum, Tokyo, Abstracts with Program*, p. 11.

Bonaparte, José F., Jorge Ferigolo, and Ana Maria Ribeiro, 1999, A new early Late Triassic saurischian dinosaur from Rio Grande do Sul State, Brazil: *in*: Yukimitsu Tomida, Rich, and Vickers-Rich, editors, *Proceedings of the Second Gondwanan Dinosaur Symposium*, National Science Museum Monographs, 15, Tokyo, pp. 89–109.

Bonaparte, José F., M. R. Franchi, Jaime E. Powell, and E. G. Sepúlveda, 1984, La Formación Los Alamitos (Campaniano–Maastrichtiano) del sudeste de Rio Negro, con descripción de *Kritosaurus australia* n. sp. (Hadrosauridae). Significado paleontográfico de los vertebrados: *Rebista de la Asociación Geológica Argentina*, 39 (3–4), pp. 284–299.

Bonaparte, José F., Wolf-Dieter Heinrich, and Rupert Wild, 2000, Review of *Janenschia* Wild, with the description of a new sauropod from the Tendaguru beds of Tanzania and a discussion on the systematic value of procoelous caudal vertebrae in sauropods: *Palaeontographica*, Abt. A, 256 (1–3), pp. 25–76, Stuttgart.

Bonaparte, José F., and Zofia Kielan-Jaworowska, 1987, Late Cretaceous dinosaur and mammal faunas of Laurasia and Gondwana, *in*: P. J. Currie and E. H. Koster, editors, *Fourth Symposium on Mesozoic Terrestrial Ecosystems, Short Papers*, Tyrrell Museum of Palaeontology, Occasional Papers, 3, pp. 24–29.

Bonaparte, José F., and Octavio Mateus, 1999, A new diplodocid, *Dinheirosaurus lourinhanensis* gen. et sp., from the Late Jurassic beds of Portugal: *Revista del Museo Argentino di Ciencias Naturales "Bernardino Rivadavia" e Instituto Nacional de Investigaciones de las Ciencias Naturales*, 2, pp. 13–29.

Bonaparte, José F., and Fernando E. Novas, 1985, *Abelisaurus comahuensis*, N.G., N.SP., carnosauria del Cretacio tardo de Patagonia: *Ameghiniana*, 21 (2–4), pp. 259–265.

Bonaparte, José F., and Jaime E. Powell, 1980, A continental assemblage of tetrapods from the Upper Cretaceous beds of El Brete, northwestern Argentina (Sauropoda–Coelurosauria–Carnosauria–Aves): *Memoires de la Société Géologique de France, Nouvelle Serie*, pp. 19–28.

Bonaparte, José F., and M. Vince, 1979, El hallazgo del primer nido de Dinosaurios triásicos (Saurischia, Prosauropoda), Triásico Superior de Patagonia, Argentina: *Ameghiniana*, 16, pp. 173–182.

Bonnan, Matthew F., 1999*a*, Pedal locomotion in sauropod dinosaurs: *North-Central Section of the Geological Society of America 33rd Annual Meeting Abstracts*, 31 (5), p. 5.

_____, 1999*b*, The evolution of manus shape and the antebrachium in sauropods: *Journal of Vertebrate Paleontology*, 19 (Supplement to Number 3), Abstracts of Papers, Fifty-ninth Annual Meeting, p. 33A.

_____, 2000, The presence of a calcaneum in a diplodocid sauropod: *Ibid.*, 20 (2), pp. 317–323.

Boscarolli, D., M. Laprocina, M. Tentor, G. Tunis, and S. Venturini, 1993, Prima segnalazione di resti di dinosauro nei calcari hauteriviani di piattaforma dell'Istria meridionale (Croazia): *Natura Nascosta*, 7, pp. 1–20.

Breithaupt, Brent H., Daniel J. Chure, and Elizabeth H. Southwell, 1999, AMNH 5753: the world's first free-standing theropod skeleton: *Journal of Vertebrate Paleontology*, 19 (Supplement to Number 3), Abstracts of Papers, Fifty-ninth Annual Meeting, p. 33A.

Brenner, K., 1973, Stratigraphie und Palaeogeographie des oberen Mittelkeupers in Südwest-Deutschland: *Arb. Inst. Geol. Paläont. Univ. Stuttgart*, 68, pp. 101–222, Stuttgart.

Brett-Surman, Michael K. [Keith], 1979, Phylogeny and palaeobiogeography of hadrosaurian dinosaurs: *Nature*, 277, pp. 560–562.

_____, 1989, A revision of the Hadrosauriae (Reptilia: Ornithischia) and their evolution during the Campanian and Maastrichtian: Ph.D. dissertation, Graduate School of Arts and Sciences of The George Washington University, Washington, D.C., 272 pages (unpublished).

Brett-Surman, Michael K., Steven Jabo, Peter Kroehler, and Mary Parrish, 1999, The lifespan of specimens on exhibition: preventing a second extinction: *Journal of Vertebrate Paleontology*, 19 (Supplement to Number 3), Abstracts of Papers, Fifty-ninth Annual Meeting, p. 33A.

Briggs, Derek E. G., Philip R. Wilby, Bernardino P. Pérez-Moreno, José Luis Sanz, and Marian Fregenal Martínez, 1997, The mineralization of dinosaur soft tissue in the Lower Cretaceous of Las Hoyas, Spain: *Journal of the Geological Society, London*, 154, pp. 587–588.

Brinkman, D. B., and J.-H. Peng, 1996, A new species of *Zangerlia* (Testudines: Nanshiungchelyidae) from the Upper Cretaceous redbeds at Bayan Mandahu, Inner Mongolia, and the relation ships of the genus: *Canadian Journal of Earth Sciences*, 33 (4), pp. 526–540.

Brinkman, Daniel L., Richard Cifelli, and Nicholas J. Czaplewski, 1998, First occurrence of *Deinonychus antirrhopus* (Dinosauria: Theropoda) from the Antlers Formation (Lower Cretaceous: Aptian–Albian) of Oklahoma: *Oklahoma Geological Survey*, bulletin 146, pp. 1–27.

Brinkmann, W., 1988, Zur Fundgeschichte und Systematik der Ornithopoden (Ornithischia, Reptilia) aus der Ober-Kreide von Europe: *Documenta Naturae*, 45, pp. 1–157.

Britt, Brooks B., 1991, The theropods of Dry Mesa Quarry (Morrison Formation, Late Jurassic), Colorado: with an emphasis on the osteology of *Torvosaurus tanneri*: *Brigham Young University Geology Studies*, 37, pp. 1–72.

_____, 1993, Pneumatic postcranial bones in dinosaurs and other archosaurs, Ph.D. dissertation, University of Calvary (unpublished).

_____, 2000, Pneumatic vertebrae and partitioned lungs: osteological and physiological evidence of the theropod-bird connection: *Publications in Paleontology* [The Florida Symposium on Dinosaur Bird Evolution, abstracts of papers], 2, p. 10.

Britt, Brooks B., Daniel J. Chure, Thomas R. Holtz, Clifford A. Miles, and Kenneth L. Stadtman, 2000, A reanalysis of the phylogenetic affinities of *Ceratosaurus* (Theropoda, Dinosauria) based on new specimens from Utah, Colorado and Wyoming: *Journal of Vertebrate Paleontology*, 20 (Supplement to Number 3), Abstracts of Papers, Sixtieth Annual Meeting, p. 32A.

Britt, Brooks B., Clifford A. Miles, Karen C. Cloward, and James H. Madsen, Jr., 1999, A juvenile *Ceratosaurus* (Theropoda, Dinosauria) from Bone Cabin Quarry West (Upper Jurassic, Morrison Formation, Wyoming): *Journal of Vertebrate Paleontology*, 19 (Supplement to Number 3), Abstracts of Papers, Fifty-ninth Annual Meeting, p. 33A.

Britt, Brooks B., and Scott D. Sampson, 1999, Evidence for a paraphyletic "Ceratosauria" and its implications for theropod dinosaur evolution: *Journal of Vertebrate Paleontology*, 19 (Supplement to Number 3), Abstracts of Papers, Fifty-ninth Annual Meeting, p. 36A.

Britt, Brooks B., R. D. Scheetz, John S. McIntosh, and Kennet L. Stadtman, 1998, Osteological characters of an Early Cretaceous titanosaurid sauropod from the Cedar Mountain Formation of Utah: *Journal of Vertebrate Paleontology*, 18 (Supplement to Number 3), Abstracts of Papers, Fifty-eighth Annual Meeting, p. 29A.

Brochu, Christopher A., 1996, Closure of neurocentral sutures during crocodilian ontogeny: implications for maturity assessment in fossil archosaurs: *Journal of Vertebrate Paleontology*, 16 (1), pp. 49–62.

_____, 2000*a*, A digitally-rendered endocast for *Tyrannosaurus rex*: *Journal of Vertebrate Paleontology*, 20 (1), pp. 1–6.

_____, 2000*b*, Postcranial axial morphology of a large *Tyrannosaurus rex* skeleton: *Ibid.*, 20 (Supplement to Number 3), Abstracts of Papers, Sixtieth Annual Meeting, p. 32A.

Brochu, Christopher A., and Mark A. Norell, 2000, Temporal congruence and the origina of birds: *Journal of Vertebrate Paleontology*, 20 (1), pp. 197–200.

Brookes, A., 1997, A reassessment of the small theropod material from the Wealden of Southern England, MSc thesis, University of Bristol (unpublished).

Broom, Robert, 1911, On the dinosaurs of the Stormberg, South Africa: *Annals of the South African Museum*, 7, pp. 291–308, Cape Town.

Brown, Barnum, 1908, The Ankylosauridae, a new family of armored dinosaurs from the Upper Cretaceous: *Bulletin of the American Museum of Natural History*, 24, pp. 187–201.

_____, 1910, The Cretaceous Ojo Alamo Beds of New Mexico with description of the new dinosaur genus *Kritosaurus*: *Ibid.*, 28, pp. 267–274.

Brown, Barnum, and Erich Maren Schlaikjer, 1940, The structure and relationships of *Protoceratops*: *New York Academy of Sciences, Annals*, 40, pp. 133–66.

_____, 1942, The skeleton of *Leptoceratops* with the description of a new species: *American Museum Novitates*, 955, pp. 1–12.

_____, 1943, A study of the troödontid dinosaurs with the description of a new genus and four new species: *Bulletin of the American Museum of Natural History*, 82, pp. 121–149.

Buffetaut, Éric, 1983, *Isalorhynchus genovefae* n.g.n.s. (Reptil ia, Rhynchocephalia), un nouveau Rhynchosaure du Trias de Madagascar: *Neus Jahrbuch für Geologie und Paläontologie, Monatshefte*, 1983, 465, Stuttgart.

_____, 1989, New remains of the enigmatic dinosaur *Spinosaurus* from the Cretaceous of Morocco and the affinities between *Spinosaurus* and *Baryonyx*: *Ibid.*, 2, pp. 79–87, Stuttgart.

_____, 1992, Remarks on the Cretaceous theropodan dinosaurs *Spinosaurus* and *Baryonyx*: *Neuse Jahrbuch für Geologie und Paläontologie, Monatshefte*, 2, pp. 88–96.

_____, 1995, An ankylosaurid dinosaur from the Upper Cretaceous of Shandong (China): *Geological Magazine*, 132, pp. 683–692.

_______, 1999, Mantell, Cuvier, Buckland and the identification of *Iguanodon*: a contribution based on unpublished annotations by Mantell: *Oryctos*, 2, pp. 101–109.

Buffetaut, Éric, and W. Brinkmann, 1990, A new nonmarine vertebrate fauna in the Upper Cretaceous of northern Sudan: *Berline Geowissenschaftliche Abhandlung*, A, 120 (1), pp. 183–202.

Buffetaut, Éric, and Jean Le Loeuff, 1991, Late Cretaceous dinosaur faunas of Europe: some correlation problems: *Cretaceous Research*, 12, pp. 159–176.

Buffetaut, Éric, and Varavudh Suteethorn, 1998, Iguanodontid dinosaurs from the Lower Cretaceous of Thailand: *Journal of Vertebrate Paleontology*, 18 (Supplement to Number 3), Abstracts of Papers, Fifty-eighth Annual Meeting, p. 30A.

_____, 1999, The dinosaur fauna of the Sao Khua Formation of Thailand and the beginning of the Cretaceous radiation of dinosaurs in Asia: *Palaeogeography, Palaeoclimatology, and Palaeoecology*, 150, pp. 13–23.

Buffetaut, Éric, Varavudh Suteethorn, Gilles Cuny, Haiyan Tong, Jean Le Loeuff, Sasidhorn Khansubha, and Sutee Jongautchariyakul, 2000, The earliest known sauropod dinosaur: *Nature*, 407 (7), pp. 72–74.

Buffetaut, Éric, Varavudh Suteethorn, and Haiyan Tong, 1996, The earliest known tyrannosaur from the Lower Cretaceous of Thailand: *Nature*, 381, pp. 689–691.

Burge, Donald L., John H. Bird, Brooks B. Britt, Daniel J. Chure, and Rodney L. Scheetz, 2000, A brachiosaurid from the Ruby Ranch (Cedar Mountain Fm.) near Price, Utah, and sauropod faunal change across the Jurassic–Cretaceous boundary of North America: *Journal of Vertebrate Paleontology*, 20 (Supplement to Number 3), Abstracts of Papers, Sixtieth Annual Meeting, p. 32A.

Burge, Donald L., John H. Bird, Brian K. McClelland, and Maria A. Ciconnetti, 1999, Comparison of four armored dinosaurs from the Cedar Mountain Formation of eastern Utah: *Journal of Vertebrate Paleontology*, 19 (Supplement to Number 3), Abstracts of Papers, Fifty-ninth Annual Meeting, p. 34A.

Burge, Donald L., Robert Gaston, James I. Kirkland, John H. Bird, and Kenneth Carpenter, 1998, The first complete, freestanding mount of a polacanthid ankylosaur: *Journal of Vertebrate Paleontology*, 18 (Supplement to Number 3), Abstracts of Papers, Fifty-eighth Annual Meeting, p. 30A.

Burnham, David A., Philip J. Currie, Robert T. Bakker, Zhonghe Zhou, and John H. Ostrom, 2000, Remarkable new birdlike dinosaur (Theropoda: Maniraptora) from the Upper Cretaceous of Montana: *The University of Kansas Paleontological Contributions*, New Series, 13, pp. 1–14.

Burnham, David A., Kraig Derstler, and Cliff L. Linster, 1997, A new specimen of *Velociraptor* (Dinosauria: Theropoda) from the Two Medicine Formation of Montana, *in*: Donald L. Wolberg, Edmund Stump, and Gary Rosenberg, editors. *Dinofest International: Proceedings of a Symposium Held at Arizona State University*. Philadelphia: Academy of Natural Sciences, pp. 73–75.

Burnham, David A., and Zhonghe Zhou, 1999, Comparing the furcula in birds and dinosaurs: *Journal of Vertebrate Paleontology*, 19 (Supplement to Number 3), Abstracts of Papers, Fifty-ninth Annual Meeting, p. 34A.

Bybee, Paul, and David K. Smith, 1999, A large, unusual allosaurid skull from eastern Utah: *Journal of Vertebrate Paleontology*, 19 (Supplement to Number 3), Abstracts of Papers, Fifty-ninth Annual Meeting, p. 35A.

Calvo, Jorge O., 2000, Dinosaur remains from the coast of Los Barreales Lake (Upper Cretaceous), Neuquén, Patagonia, Argentina: *Journal of Vertebrate Paleontology*, 20 (Supplement to Number 3), Abstracts of Papers, Sixtieth Annual Meeting, p. 33A.

Calvo, Jorge O., Karen Moreno, and David Rubilar, 1999, A new record of a titanosaurid sauropod from Los Barreales lake, Neuquén Province, Argentina: *Ameghiniana, Revista de la Asociación Paleontológica Argentina, (Resúmenes)*, 36 (Número 4-Suplemento), p. 7R.

Calvo, Jorge O., David Rubilar, and Karen Moreno, 1999, Report of a new theropod dinosaur from northwestern Patagonia: *Ameghiniana, Revista de la Asociación Paleontológica Argentina, (Resúmenes)*, 36 (Número 4-Suplemento), p. 7R.

Calvo, Jorge O., and Leonardo Salgado, 1995, *Rebbachisaurus tessonei* sp. nov. a new Sauropoda from the Albian–Cenomanian of Argentina; new evidence on the origin of the Diplodocidae: Gaia, 11, pp. 13–33.

Campagna, Tony, 2000, The PT interview: Michael Triebold: *Prehistoric Times*, 40, pp. 18–19.

Caneer, William T., 1999, One of two *T. rex* tracks from the Raton Basin left traces of the forearms and hands in addition to the foot: *Journal of Vertebrate Paleontology*, 19 (Supplement to Number 3), Abstracts of Papers, Fifty-ninth Annual Meeting, p. 35B.

Carpenter, Kenneth, 1982, Baby dinosaurs from the Late Cretaceous Lance and Hell Creek formations and a description of a new species of theropod: *Contributions to Geology*, 20 (2), pp. 123–134.

_____, 1984, Skeletal reconstruction and life restoration of *Sauropelta* (Ankylosauria: Nodosauridae) from the Cretaceous of North America: *Canadian Journal of Earth Sciences*, 21, pp. 1491–1498.

_____, 1990, Ankylosaur systematics: example using *Panoplosaurus* and *Edmontonia* (Ankylosauria: Nodosauridae): *Ibid.*, pp. 281–299.

_____, 1992, Tyrannosaurids (Dinosauria) of Asia and North America, *in*: Niall J. Mateer and Chen Pei-Ji, editors: *International Symposium on Non-marine Cretaceous Correlation*. Beijing: China Ocean Press, pp. 250–268.

_____, 1997*a*, Ankylosauria, *in*: Philip J. Currie and Kevin Padian, editors, *Encyclopedia of Dinosaurs*. San Diego: Academic Press, pp. 16–20.

_____, 1997*b*, Tyrannosauridae, *Ibid.*, pp. 766–768.

_____, 1998*a*, Role of bacteria in the perminilazation of dinosaur bone: *Journal of Vertebrate Paleontology*, 18 (Supplement to Number 3), Abstracts of Papers, Fifty-eighth Annual Meeting, p. 31A.

_____, 1998*b*, Ankylosaur odds and ends: *Ibid.*, p. 31A.

Carpenter, Kenneth, David W. Dilkes, and David B. Weishampel, 1995, The dinosaurs of the Niobrara Chalk Formation (Upper Cretaceous, Kansas): *Journal of Vertebrate Paleontology*, 15, pp. 275–297.

Carpenter, Kenneth, and James I. Kirkland, 1998, Review of Lower and Middle Cretaceous ankylosaurs from North America, *in*: Spencer G. Lucas, Kirkland, and J. W. Estep, editors, *Lower and Middle Cretaceous Terrestrial Ecosystems*. Albuquerque: New Mexico Museum of Natural History and Science, Bulletin No. 14, pp. 249–270.

Carpenter, Kenneth, James I. Kirkland, C. A. Miles, K. Cloward, and Donald Burge, 1996, Evolutionary significance of new ankylosaurs (Dinosauria) from the Upper Jurassic and Lower Cretaceous, Western interior: *Journal of Vertebrate Paleontology*, 16 (Supplement to Number 3), Abstracts of Papers, Fifty-sixth Annual Meeting, p. 25A.

Carpenter, Kenneth, Dale A. Russell, Donald Baird, and Robert K. Denton, 1997, Redescription of the holotype of *Dryptosaurus aquilunguis* (Dinosauria: Theropoda) from the Upper Cretaceous of New Jersey: *Journal of Vertebrate Paleontology*, 17 (30), pp. 561–573.

Carr, Thomas D. [David], 1996, Craniofacial ontogeny in tyrannosaurids: taxonomic implications: *Journal of Vertebrate Paleontology*, 16 (Supplement to Number 3), Abstracts of Papers, Fifty-sixth Annual Meeting, p. 25A.

_____, 1999*a*, FMNH PR308: part I (or: analyzing an enigmatic tyrannosaurid specimen): *Dinosaur World*, 6, pp. 16–18.

_____, 1999*b*, Craniofacial ontogeny in Tyrannosauridae (Dinosauria: Coelurosauria): *Journal of Vertebrate Paleontology*, 19 (3), pp. 497–520.

_____, 2000*a*, FMNHPR308 part II: the cross-dressing *Daspletosaurus*: *Dinosaur World*, 7, pp. 21–24.

_____, 2000*b*, A little skull from Montana: *Ibid.*, 8, pp. 60–67.

Carr, Thomas D., and Richard L. Essner, Jr., 1997, A quantitative approach to cranial ontogeny in tyrannosaurids (Dinosauria: Theropoda): *Journal of Vertebrate Paleontology*, 17 (Supplement to Number 3), Abstracts of Papers, Fifty-seventh Annual Meeting, p. 35A.

Carr, Thomas D., and Thomas E. Williamson, 1999, A new tyrannosaurid (Theropoda: Coelurosauria) from the San Juan Basin of New Mexico: *Journal of Vertebrate Paleontology*, 19 (Supplement to Number 3), Abstracts of Papers, Fifty-ninth Annual Meeting, p. 36A.

_____, 2000, A review of Tyrannosauridae (Dinosauria, Coelurosauria) from New Mexico, *in*: Spencer G. Lucas and Andrew B. Heckert, editors, *Dinosaurs of New Mexico*. Albuquerque: New Mexico Museum of Natural History and Science, Bulletin 17, pp. 113–145.

Carrano, Matthew T., 1999*a*, Phylogenetic, temporal, and scaling perspectives on body-size evolution in the Dinosauria: *Journal of Vertebrate Paleontology*, 19 (Supplement to Number 3), Abstracts of Papers, Fifty-ninth Annual Meeting, p. 36A.

_____, 1999*b*, What, if anything, is a cursor? Categories versus continua for determining locomotor habit in mammals and dinosaurs: *Journal of the Zoological Society of London*, 247, pp. 29–42.

_____, 2000, Homoplasy and the evolution of dinosaur locomotion: *Paleobiology*, 26 (3), pp. 489–512.

Carrano, Matthew T., Christine M. Janis, and J. John Sepkoski, Jr., 1999, Hadrosaurs as ungulate parallels: lost lifestyles and deficient data: *Acta Palaeontologica Polonica*, 44 (3), pp. 237–261.

Carrier, David R., and Colleen G. Farmer, 2000, The evolution of pelvic aspiration in archosaurs: *Paleobiology*, 26 (2), pp. 271–293.

Casanovas, M. Lourdes, Xabier [Javier, Xavier] Pereda Suberbiola, Jose Vicente Santafé, and David B. Weishampel, 1999, A primitive euhadrosaurian dinosaur from the uppermost Cretaceous of the Ager syncline (southern Pyrenees, Catalonia): *Geologie en Mijnbouw*, 78, pp. 345–356.

Casanovas-Cladellas, M. Lourdes, Jose vicente Santafé-Llopis, Carles Santisteban Bové, and Xavier Pereda Suberbiola, 1999, Estegosaurios (Dinosauria) del Jurásico Superior-Cretácico Inferior de la Comarca de los Serranos (Valencia, España): *Homenaje al Prof. J. Truyols*, no extr., pp. 57–63.

Case, Ermin C., 1916, Further evidence bearing on the age of the red beds in the Rio Grande Valley, New Mexico: *Science*, 77, pp. 708–709.

_____, 1922, New reptiles and stegocephalians from the Upper Triassic of western Texas: *Carnegie Institution of Washington Publication*, 321, 84 pages.

_____, 1927, The vertebral column of *Coelophysis* Cope: *Contributions from the Museum of Paleontology, University of Michigan*, 2, pp. 209–222.

_____, 1932, On the caudal region of *Coelophysis* sp. and on some new or little known forms from the Upper Triassic of western Texas: *Ibid.*, 4, pp. 81–92.

Case, Judd A., James E. Martin, Dan S. Chaney, Marcelo Reguero, Sergio A. Marenssi, Sergio M. Santillana, and Michael O. Woodburne, 2000, The first duck-billed dinosaur (family Hadrosauridae) from Antarctica: *Journal of Vertebrate Pale ontology*, 20 (3), pp. 612–614.

Casinos, A., C. Quintana, and C. viladieu, 1993, Allometry and adaptation in the long bones of a digging group of rodents (Ctenomyinae): *Zoological Journal of the Linnean Society*, 107 (2), pp. 107–115.

Chabli, S., 1988, master's thesis, Université de Paris, (unpublished).

Chang, K. H., J. S. Seung, and S. O. Park, 1983, Occurrence of a dinosaur limb bone near Tabri, southern Korea: *The Journal of the Geological Society of Korea*, 18, pp. 195–202 (in Korean with English summary).

Chapman, Ralph E., 1997, Technology and the study of dinosaurs, *in*: James O. Farlow and Michael K. Brett-Surman, editors, *The Complete Dinosaur*. Bloomington and Indianapolis, Indiana University Press, pp. 112–135.

Chapman, Ralph E., Arthur F. Andersen, and Steven Jabo, 1999, Construction of the virtual *Triceratops*: procedures, results, and potentials: *Journal of Vertebrate Paleontology*, 19 (Supplement to Number 3), Abstracts of Papers, Fifty-ninth Annual Meeting, p. 37A.

Charig, Alan G., 1971, Faunal provinces on land: evidence based on the distribution of fossil tetrapods, with especial reference to the reptiles of the Permian and Mesozoic: *Geology Journal*, special issue 4, pp. 111–128, Liverpool.

_____, 1972, The evolution of the archosaur pelvis and hind-limb: an explanation in functional terms, *in*: K. A. Joysey and T. S. Kemp, editors, *Studies in Vertebrate Evolution*. New York: Winchester, pp. 121–153.

_____, 1979, *A New Look at the Dinosaurs*. London: Heinemann, 160 pages.

_____, 1984, Competition between therapsids and archosaurs during the Triassic period: a review and synthesis of current theories, *in*: Mark W. J. Ferguson, editor, *The Structure, Development and Evolution of Reptiles*, Symposia of the Zoological Society of London, 52. London: Acedemic Press Inc. (London) Ltd., pp. 597–628.

Charig, Alan G., John Attridge, and A. W. Crompton, 1965, On the origin of the sauropods and the classification of the Saurischia: *Proceedings of the Linnean Society of London*, 176 (2), pp. 197–221, London.

Charig, Alan J., and Angela C. Milner, 1986, *Baryonyx*, a remarkable new theropod dinosaur: *Nature*, 324, pp. 359–361.

_____, 1990, The systematic position of *Baryonyx walkeri*, in the light of Gauthier's reclassification of the Teropoda, *in*: Kenneth Carpenter and Philip J. Currie, editors, *Dinosaur Systematics: Approaches and Perspectives*. Cambridge, New York and Melbourne: Cambridge University Press, pp. 127–140.

_____, 1997, *Baryonyx walkeri*, a fish-eating dinosaur from the Wealden of Surrey: *Bulletin of The Natural History Museum, London*, (Geology) 53 (1), pp. 11–70.

Charig, Alan J., and Barney Newman, 1962, Footprints in the Purbeck: *New Scientist*, 285, pp. 234–235.

Chatterjee, Sankar, 1985, *Postosuchus*, a new thecodontian reptile from the Triassic of Texas, and the origin of tyrannosaurs: *Philosophical Transactions of the Royal Society of London*, B, 309, pp. 395–460.

_____, 1991, Cranial anatomy and relationships of a new Triassic bird from Texas: *Ibid.*, B, 332, pp. 277–346.

_____, 1993, *Shuvosaurus*, a new theropod: *Natural Geographic Research Exploration*, 9 (3), pp. 274–285.

_____, 1995, The Triassic Bird *Protoavis*: *Archaeopteryx*, 13, pp. 15–31, Eichstätt.

_____, 1997, The beginnings of avian flight, *in*: Donald L. Wolberg, Edmund Stump, and Gary Rosenberg, editors. *Dinofest International: Proceedings of a Symposium Held at Arizona State University*. Philadelphia: Academy of Natural Sciences, pp. 311–335.

_____, 1998*a*, Dinosaurs in the land of gonds: *Journal of Vertebrate Paleontology*, 18 (Supplement to Number 3), Abstracts of Papers, Fifty-eighth Annual Meeting, p. 33A.

_____, 1998*b*, The avian status of *Protoavis*: *Archaeopteryx*, 16, pp. 99–122, Eichstätt.

_____, 1999, *Protoavis* and the early evolution of birds: *Palaeontographica*, Abt. A, 254, pp. 1–100.

Chatterjee, Sankar, and R. J. Templin, 2000, Feathered coelurosaurs: new light on the arboreal origin of avian flight: *Publications in Paleontology* [The Florida Symposium on Dinosaur Bird Evolution, abstracts of papers], 2, pp. 11–12.

Chen, Pei-Ji, Zhi-ming Dong, and Shuo-nan Zhen, 1998, An exceptionally well-preserved theropod dinosaur from the Yixian Formation of China: *Nature*, 391, pp. 147–152.

Cheng Zhengwu and Pang Qiqing, 1996, A new dinosaurian fauna from Tianzhen, Shanxi Province with its stratigraphical significance: *Acta Geoscientia Sinica*, 17, p. 135.

Chiappe, Luis M., 1993, Enantiornithine (Aves) tarsometatarsi from the Cretaceous Lecho Formation of northwestern Argentina: *American Museum Novitates*, 308, pp. 1–27.

_____, 1995, The first 85 million years of avian evolution: *Nature*, 378, pp. 349–355.

_____, 1996*a*, Early avian evolution in the southern hemisphere: the fossil record of birds in the Mesozoic of Gondwana, *in*: Fernando S. Novas and Ralph E. Molnar, editors, *Proceedings of the Gondwanan Dinosaur Symposium*: *Memoirs of the Queensland Museum*, 39 (part 3), pp. 533–555.

_____, 1996*b*, Late Cretaceous birds of Southern South America: anatomy and systematics of Enanithornithes and *Patagopteryx deferralist*: *Münchner Geowissenschaftliche Abhandlung*, Abh (A), 30, pp. 203–244.

_____, 1997, Aves, *in*: Philip J. Currie and Kevin Padian, editors, *Encyclopedia of Dinosaurs*. San Diego: Academic Press, pp. 32–38.

_____, 1999, Auca Mahuevo: an extraordinary dinosaur nesting ground from the Late Cretaceous of Patagonia: *Journal of Vertebrate Paleontology*, 19 (Supplement to Number 3), Abstracts of Papers, Fifty-ninth Annual Meeting, p. 37A.

Chiappe, Luis M., and Jorge Orlando Calvo, 1994, *Neuquenornis volans*, a new Late Cretaceous bird (Enantiornithes: Avisauridae) from Patagonia, Argentina: *Journal of Vertebrate Paleontology*, 14 (2), pp. 230–246.

Chiappe, Luis M., Rodolfo A. Coria, Lowell Dingus, Frankie Jackson, Anusuya Chinsamy, and Marilyn Fox, 1998, Sauropod dinosaur embryos from the Late Cretaceous of Patagonia: *Nature*, 396, pp. 258–261.

Chiappe, Luis M., Frankie Jackson, Lowell Dingus, Gerald Grellet-Tinner, and Rodolfo A. Coria, 1999, Auca Mahuevo: an extraordinary dinosaur nesting ground from the Late Cretaceous of Patagonia: *Journal of Vertebrate Paleontology*, 19 (Supplement to Number 3), Abstracts of Papers, Fifty-ninth Annual Meeting, p. 37A.

Chiappe, Luis M., Mark A. Norell, and James M. Clark, 1996, Phylogenetic position of *Mononykus* (Aves: Alvarezsauridae) from the Late Cretaceous of the Gobi Desert, *in*: Fernando S. Novas and Ralph E. Molnar, editors, *Proceedings of the Gondwanan Dinosaur Symposium*: *Memoirs of the Queensland Museum*, 39 (part 3), pp. 557–582.

_____, 1998, The skull of a relative of the stem-group bird *Mononykus*: *Nature*, 392, pp. 275–278.

Chiappe, Luis M., and Gerald Grellet-Tinner, 2000, Dinosaur eggshells and the origin of birds, 2000: *Publications in Paleontology* [The Florida Symposium on Dinosaur Bird Evolution, abstracts of papers], 2, pp. 12–13.

Chin, Karen, David A. Eberth, and Wendy J. Sloboda, 1999, Exceptional soft-tissue preservation in a theropod coprolite from the Upper Cretaceous Dinosaur Park Formation of Alberta: *Journal of Vertebrate Paleontology*, 19 (Supplement to Number 3), Abstracts of Papers, Fifty-ninth Annual Meeting, pp. 37A–38A.

Chinnery, Brenda J., and Ralph E. Chapman, 1998, A morphometric study of the ceratopsid postcranial skeleton: *Journal of Vertebrate Paleontology*, 18 (Supplement to Number 3), Abstracts of Papers, Fifty-eighth Annual Meeting, p. 33A.

Chinnery, Brenda J., Thomas R. Lipka, James I. Kirkland, Mary I. Parrish, and Michael K. Brett-Surman, 1998, Neoceratopsian teeth from the Lower to Middle Cretaceous of North America *in*: Spencer G. Lucas, James I. Kirkland, and John W. Estep, editors, *Lower and Middle Cretaceous Ecosystems*, New Mexico Museum of Natural History Bulletin, 14.

Chinnery, Brenda J., and David Lee Trexler, 1999, The first bonebed occurrence of a basal neoceratopsian, with new information on the skull morphology of *Leptoceratops*: *Journal of Vertebrate Paleontology*, 19 (Supplement to Number 3), Abstracts of Papers, Fifty-ninth Annual Meeting, p. 38A.

Chinnery, Brenda J., and David B. Weishampel, 1996, *Montanoceratops cerorhynchus* (Dinosauria: Ceratopsia) and relationships among basal neoceratopsians: *Journal of Vertebrate Paleontology*, 18 (3), pp. 569–585.

Chinsamy, Anusuya, 1997, Assessing the biology of fossil vertebrates through bone histology: *Palaeontologia Africana*, 33, pp. 29–35.

Chinsamy, Anusuya, Luis M. Chiappe, and Peter Dodson, 1994, Growth rings in Mesozoic birds: *Nature*, 368, pp. 196–197.

Chinsamy, Anusuya, and Michael Raath, 1992, Preparation of fossil bone for histological examination: *Palaeontologia Africana*, 29, pp. 39–44.

Christian, Andreas, and Wold-Dieter Heinrich, 1998, The neck posture of *Brachiosaurus brancai*: *Mitteilungen Museum für Naturkunde der Humboldt-Universität zu Berlin, Geowissenschaftliche Reihe*, 1, pp. 73–80.

Christian, Andreas, Wolf-Dieter Heinrich, and

Werner Golder, 1999, Posture and mechanics of the forelimbs of *Brachiosaurus brancai*: *Mitteilungen Museum für Naturkunde der Humboldt-Universität zu Berlin, Geowissenschaftliche Reihe*, 2, pp. 63–67.

Christian, Andreas, Reinhart H. G. Müller, Gundula Christian, and Holger Preuschoft, 1999, Limb swinging in elephants and giraffes and implications for the reconstruction of limb movements and speed estimates in large dinosaurs: *Mitteilungen Museum für Naturkunde der Humboldt-Universität zu Berlin, Geowissenschaftliche Reihe*, 2, pp. 81–90.

Christiansen, Per, 1997, Locomotion in sauropod dinosaurs: *Gaia*, 14, pp. 45–75.

_____, 1999, Long bone scaling and limb posture in non-avian theropods: evidence for differential allometry: *Journal of Vertebrate Paleontology*, 19 (4), pp. 666–680.

Christiansen, Per, and Gregory S. Paul, (in press), Limb bone scaling, limb proportions, and bone strength in neoceratopsian dinosaurs: *Gala*.

Chure, Daniel J., 1995, A reassessment of the gigantic theropod *Saurophagus maximus* from the Morrison Formation (Upper Jurassic) of Oklahoma, USA, *in*: A. Sun and Y. Wang, editors, *Sixth Symposium on Mesozoic Terrestrial Ecosystems and Biota, Short Papers*. Bejing: China Ocean Press, pp. 103–106.

_____, 1997, Museums and displays, *in*: Philip J. Currie and Kevin Padian, editors, *Encyclopedia of Dinosaurs*. San Diego: Academic Press, pp. 460–468.

_____, 1998*a*, "*Chilantaisaurus*" *maortuensis*, a large maniraptoran theropod from the Early Cretaeous (Albian) of Nei Mongol, PRC: *Journal of Vertebrate Paleontology*, 18 (Supplement to Number 3), Abstracts of Papers, Fifty-eighth Annual Meeting, pp. 33A–34A.

_____, 1998*b*, A reassessment of the Australian *Allosaurus* and its implications for the Australian refugium concept: *Ibid.*, p. 34A.

_____, 1999, The wrist of *Allosaurus* and the evolution of the semilunate carpal: *Ibid.*, 19 (Supplement to Number 3), Abstracts of Papers, Fifty-ninth Annual Meeting, p. 38A.

_____, 2000*a*, New data on the gastral basket of *Allosaurus*: *Publications in Paleontology* [The Florida Symposium on Dinosaur Bird Evolution, abstracts of papers], 2, p. 13.

_____, 2000*b*, New records of *Ceratosaurus*, *Torvosaurus*, *Coelurus*, and *Stokesosaurus* and their implication for theropod community structure and biozonation in the Late Jurassic of the Western Interior: *Journal of Vertebrate Paleontology*, 20 (Supplement to Number 3), Abstracts of Papers, Sixtieth Annual Meeting, p. 35A.

Chure, Daniel J., and John S. McIntosh, 1989, *A Bibliography of the Dinosauria (Exclusive of the Aves), 1677–1986*. Grand Junction, Colorado: Museum of Western Colorado, Paleontology Series 1, 226 pages.

Claessens, Leon P. A. M., 1997, Gastralia, *in*: Philip J. Currie and Kevin Padian, editors, *Encyclopedia of Dinosaurs*. San Diego: Academic Press, pp. 269–270.

Claessens, Leon P. A. M., Steven F. Perry, and Philip J. Currie, 1998, Using comparative anatomy to reconstruct theropod respiration: *Journal of Vertebrate Paleontology*, 18 (Supplement to Number 3), Abstracts of Papers, Fifty-eighth Annual Meeting, p. 34A.

Clark, James M., Mark A. Norell, and Rinchen Barsbold, 2000, Two new oviraptorid theropod dinosaurs from the Late Cretaceous Djadokhta Formation, Ukhaa Tolgood, Mongolia: *Journal of Vertebrate Paleontology*, 20 (Supplement to Number 3), Abstracts of Papers, Sixtieth Annual Meeting, p. 36A.

Clark, James M., Mark A. Norell, and Luis M. Chiappe, 1999, An oviraptorid skeleton from the Late Cretaceous of Ukhaa Tolgod, Mongolia, preserved in an avian-like brooding position over an oviraptorid nest: *American Museum Novitates*, 3265, pp. 1–36.

Clark, James M., Mark A. Norell, and Peter J. Makovicky, 1999, An oviraptorid skeleton from the Late Cretaceous of Ukhaa Tolgod, Mongolia, preserved in an avian-like brooding position over an oviraptorid nest: *American Museum Novitates*, 3265, pp. 1–36.

Clark, James M., Altangerel Perle, and Mark A. Norell, 1994, The skull of *Erlikosaurus andrewsi*, a Late Cretaceous "segnosaur" (Theropoda: Therizinosauridae) from Mongolia: *American Museum Novitates*, 3115, pp. 1–39.

Clemens, William A., 1963, Fossil mammals of the type Lance Formation, Wyoming. Part 1. Introduction and Multituberculа: *University of California Publications in Geology and Science*, 48, pp. 1–105.

_____, 1973, The roles of fossil vertebrates in interpretation of Late Cretaceous stratigraphy of the San Juan Basin, New Mexico, *in*: James E. Fassett, editor, *Cretaceous and Tertiary rocks of the southern Colorado Plateau*. Durango, Colorado: Four Corners Society, pp. 154–167.

Clemens, William A., and J. David Archibald, 1980, Evolution of terrestrial faunas during the Cretaceous–Tertiary transition, *in*: *Ecosystèmes Continentaux du Mésozoïque: Société Géologique de France, Mémoir 139* (new series), pp. 67–74.

Clemens, William A., J. A. Lillegraven, E. H. Lindsay, and George Gaylord Simpson, 1979, Where, when, and what — a survey of known Mesozoic mammal distribution, *in*: J. A. Lillegraven, Zofia, Kielan-Jaworowska, and W. A. Clemens, editors, *Mesozoic Mammals: the first two-thirds of mammalian history*. Berkeley: University of California Press, pp. 7–58.

Clouse, V., and John R. Horner, 1993, Eggs and embryos from the Judith River Formation of Montana: *Journal of Vertebrate Paleontology*, 19 (Supplement to Number 3), Abstracts of Papers, Fifty-ninth Annual Meeting, p. 31A.

Cobb, S., 1960, Observations on the comparative anatomy of the avian brain: *Perspectives in Biology and Medicine*, 3, pp. 383–408.

Colbert, Edwin H. [Harris], 1961, *Dinosaurs: Their Discovery and Their World*. New York: E. P. Dutton & Co., xiv, 300 pages.

_____, 1962, The weights of dinosaurs: *American Museum Novitates*, 2076, pp. 1–16.

_____, 1964*a*, The Triassic genera *Podokesaurus* and *Coelophysis*: *Ibid.*, 2168, pp. 1–12.

_____, 1964*b*, Relationships of saurischian dinosaurs: *Ibid.*, 2181, pp. 1–24.

_____, 1968, *Men and Dinosaurs: The Search in Field and Laboratory*. New York: E. P. Dutton and Company, xviii, 238 pages.

_____, 1970, A saurischian dinosaur from the Triassic of Brazil: *Ibid.*, 2405, pp. 1–39.

_____, 1981, A primitive ornithischian dinosaur from the Kayenta Formation of Northern Arizona: *Museum of Northern Arizona Press Bulletin Series*, 53, 61 pages.

_____, 1983, *Dinosaurs: An Illustrated History*. Maplewood, New Jersey: Hammond Incorporated, 224 pages.

_____, 1989, The Triassic dinosaur *Coelophysis*: *Museum of Northern Arizona Bulletin*, 57, xv, 160 pages.

Connely, Melissa V., 1997, Analysis of head-neck functions and feeding ecology of common Jurassic sauropod dinosaurs based on a new find from Como Bluff, Wyoming: *Journal of Vertebrate Paleontology*, 17 (Supplement to Number 3), Abstracts of Papers, Fifty-seventh Annual Meeting, pp. 39A–40A.

Connely, Milissa Victoria, and Russell Hawley, 1998, A proposed reconstruction of the jaw musculature and other soft cranial tissues of *Apatosaurus*: *Journal of Vertebrate Paleontology*, 18 (Supplement to Number 3), Abstracts of Papers, Fifty-eighth Annual Meeting, p. 35A.

Coombs, Walter P., Jr., 1971, The Ankylosauria, Ph.D. thesis, Ann Arbor (Michigan) Microfilms Intern., Columbia University, New York, 487 pages (unpublished).

_____, 1975, Sauropod habits and habitats: *Palaeogeography, Palaeoclimatology, and Palaeoecology*, 17, pp. 1–33.

_____, 1978*a*, The families of the ornithischian dinosaur order Ankylosauria: *Paleontology*, 21, part 1, pp. 143–170.

_____, 1978*b*, Theoretical aspects of cursorial adaptations in dinosaurs: *Revisions in Biology*, 53 (4), pp. 393–410.

_____, 1979, Osteology and myology of the hindlimb in the Ankylosauria (Reptilia: Ornithischia): *Journal of Paleontology*, 53, pp. 666–684.

_____, 1986, A juvenile dinosaur referable to the genus *Euoplocephalus* (Reptilia, Ornithischia): *Journal of Vertebrate Paleontology*, 6, pp. 162–173.

_____, 1995, A new nodosaurid (Dinosauria: Ornithischia) from the Lower Cretaceous of Texas: *Ibid.*, 15 (2), pp. 298–312.

Coombs, Walter P., Jr., and Thomas A. Demere, 1996, A Late Cretaceous nodosaurid ankylosaur (Dinosauria: Ornithischia) from marine sediments of coastal California: *Journal of Paleontology*, 70, pp. 311–326.

Coombs, Walter P., Jr., and Teresa Maryańska, 1990, Ankylosauria, *in*: David B. Weishampel, Peter Dodson, and Halszka Osmólska, editors, *The Dinosauria*. Berkeley and Los Angeles: University of California Press, pp. 456–483.

Coombs, Walter P., Jr., David B. Weishampel, and Lawrence M. Witmer, 1990, Basal Thyreophora, *The Dinosauria, in*: David B. Weishampel, Dodson, and Halszka Osmólska, editors, *The Dinosauria.* Berkeley and Los Angeles: University of California Press, pp. 427–434.

Cooper, Lisa Noelle, and John R. Horner, 1999, Growth rate of *Hypacrosaurus stebingeri* as hypothesized from lines of arrested growth and whole femur circumference: *Journal of Vertebrate Paleontology*, 19 (Supplement to Number 3), Abstracts of Papers, Fifty-ninth Annual Meeting, p. 39A.

Cooper, Lisa Noelle, Ellen Lamm, and John R. Horner, 1998, Annual changes in mass of *Hypacrosaurus stebingeri* as hypothesized from lines of arrested growth: *Journal of Vertebrate Paleontology*, 18 (Supplement to Number 3), Abstracts of Papers, Fifty-eighth Annual Meeting, p. 35A.

Cooper, Michael R., 1980, The first record of the prosauropod dinosaur *Euskelosaurus* from Zimbabwe: *Arnoldia Zimbabwe*, 9 (3), pp. 1–17, Bulawayo.

_____, 1981, The prosauropod dinosaur *Massospondylus carinatus* Owen from Zimbabwe: its biology, mode of life and phylogenetic significance: *Occasional Papers of the National Museums of Rhodesia, Natural Science*, 6, pp. 689–840, Bulawayo.

_____, 1984, A reassessment of *Vulcanodon karibaensis* Raath (Dinosauria: Saurischia) and the origin of the Sauropoda: *Paleontologica Africana*, 25, pp. 203–231.

Cope, Edward Drinker, 1869, Synopsis of the extinct Batrachia, Reptilia, and Aves of North America: *Transactions of the American Philosophical Society*, 15, pp. 1–252.

_____, 1876, On some extinct reptiles and Barrachia from the Judith River Formation and Fox Hills Beds of Montana: *Proceedings of the Academy of Natural Sciences of Philadelphia*, 28, pp. 340–359.

_____, 1887, The dinosaurian genus *Coelurus*: *American Naturalist*, 21, pp. 367–369.

_____, 1889, On a new genus of Triassic Dinosauria: *Ibid.*, 23, p. 626.

Coria, Rodolfo A. [Anibal], 1999, Ornithopod dinosaurs from the Neuquén Group, Patagonia, Argentina: Phylogeny and Biostratigraphy: *in*: Yukimitsu Tomida, Rich, and Vickers-Rich, editors, *Proceedings of the Second Gondwanan Dinosaur Symposium*, National Science Museum Monographs, 15, Tokyo, pp. 47–60.

Coria, Rodolfo A., and Jorge Orlando Calvo, (in press), A new iguanodontian ornithopod from Newquén Basin, Patagonia, Argentina: *Journal of Vertebrate Paleontology*.

Coria, Rodolfo A., Luis M. Chiappe, and Lowell Dingus, 2000, A new abelisaur theropod from the Upper Cretaceous of Patagonia: *Journal of Vertebrate Paleontology*, 20 (Supplement to Number 3), Abstracts of Papers, Sixtieth Annual Meeting, pp. 36A–37A.

Coria, Rodolfo A., and Leonardo Salgado, 1995, A new giant carnivorous dinosaur from the Cretaceous of Patagonia: *Nature*, 377, pp. 224–226.

_____, 1996, A basal iguanodontian (Ornithischia: Ornithopoda) from the Late Cretaceous of South America: *Journal of Vertebrate Paleontology*, 16 (3), pp. 445–457.

_____, 1999, A primitive abelisaur theropod from the Rio Limay Formation (Upper Cretaceous) of Patagonia: *Ibid.*, 19 (Supplement to Number 3), Abstracts of Papers, Fifty-ninth Annual Meeting, p. 39A.

_____, A basal Abelisauria Novas 1992 (Theropoda–Ceratosauria) from the Cretaceous of Patagonia, Argentina, *in*: B. P. Perez-Moreno, T. R. Holtz, Jr., J. L. Sanz, and J. Mora talla, editors., *Aspects of Theropoda Paleobiology*, Gaia 15, pp. 89–102 [cover dated 1998, published in 2000].

Cott, H.-B., 1961, *Transactions of the Zoological Society of London*, 29, pp. 211–356.

Crouse, Kathy, and Kenneth Carpenter, 1999, A toddler ornithopod from the Morrison Formation of Garden Park, Colorado: *Journal of Vertebrate Paleontology*, 19 (Supplement to Number 3), Abstracts of Papers, Fifty-ninth Annual Meeting, p. 40A.

Csiki, Zoltan, 1995, Studui paleontologic si tafonomic al faunal Maastrichtian superioare din depozitele continentale ale Bazinului Hateg (Formatiunile de Densus-Ciula si de Sinpetru): implicatii paleoecologice. Thesis, University of Bucharest, 97 pages (unpublished).

_____, 1999, New evidence of armoured titanosaurids in the Late Cretaceous—*Magyarosaurus dacus* from the Hateg Basin d(Romania): *Oryctos*, 2, pp. 93–99.

Csiki, Zoltan, and Dan Grigorescu, 1998, Small theropods from the Late Cretaceous of the Hateg Basin (western Romania)—an unexpected diversity at the top of the food chain: *Oryctos*, 1, pp. 87–104.

Cunha, Fausto Luiz de Souza, D. Régo, and R. Capilla, 1987, Nova occurrència de rápteis cretácicos, no "Sito Myzobuchi," da Formacão Baurú, em Álvares Machado, S. P.: *Anais de 10º Congresso Brasileiro de Paleontologia*, pp. 155–171, Rio de Janeiro.

Cunha, Fausto Luiz de Souza, and J. M. Suarez, 1985, Restos de dinossauros na Formacão Bauru, Nunicipio de Ávares Machado, S. P.: *Anais da Academia Brasileira de Ciências*, 57 (1), p. 141.

Currie, Philip J., 1985, Cranial anatomy of *Stenonychosaurus inequalis* (Saurischia, Theropoda) and its bearing on the origin of birds: *Canadian Journal of Earth Science*, 22, pp. 1643–1658.

_____, 1987, Birdlike characteristics of the jaws and teeth of troodontid theropods (Dinosauria: Saurischia): *Journal of Vertebrate Paleontology*, 7, pp. 72–81.

_____, 1990, Elmisauridae, *in*: David B. Weishampel, Peter Dodson, and Halszka Osmólska, editors, *The Dinosauria.* Berkeley and Los Angeles: University of California Press, pp. 245–248.

_____, 1987*b*, Theropods of the Judith River Formation of Dinosaur Provincial Park, *in*: P. J. Currie and E. H. Koster, editors, *Fourth Symposium on Mesozoic Terrestrial Ecosystems, Short Papers*, Tyrrell Museum of Palaeontology, Occasional Papers, pp. 52–60.

_____, 1992, Saurischian dinosaurs of the Late Cretaceous of Asia and North America, *in*: Niall J. Mateer and Chen Pei-Ji, editors, *Aspects of Nonmarine Cretaceous Geology*. Beijing: China Ocean Press, pp. 237–249.

_____, 1995, New information on the anatomy and relationships of *Dromaeosaurus albertensis* (Dinosauria: Theropoda): *Journal of Vertebrate Paleontology*, 15 (3), pp. 237–249.

_____, 1997, Theropoda, *in*: Philip J. Currie and Kevin Padian, editors, *Encyclopedia of Dinosaurs*. San Diego: Academic Press, pp. 731–737.

_____, 1999, Skeletal anatomy of the feathered dinosaurs from China: *Ameghiniana, Revista de la Asociación Paleontológica Argentina, (Resúmenes)*, 36 (Número 4-Suplemento), p. 9R.

_____, 2000, On a pack of theropods from Argentina: *DinoPress*, 1, pp. 79–89 (in Japanese; "English Text," pp. 17–23)

Currie, Philip J., and Kenneth Carpenter, 2000, A new specimen of *Acrocanthosaurus atokensis* (Theropoda, Dinosauria) from the Lower Cretaceous Antlers Formation (Lower Cretaceous, Aptian) of Oklahoma, USA: *Geodiversitas*, 22 (2), pp. 207–246.

Currie, Philip J., Stephen J. Godfrey and Lev A. Nessov, 1993, New caenagnathid (Dinosauria: Theropoda) specimens from the Upper Cretaceous of North America and Asia: *Canadian Journal of Earth Sciences*, 30 (10–11), pp. 2255–2272.

Currie, Philip J., Mark A. Norell, Ji Qiang, and Ji Shūan, 1998, The anatomy of two feathered theropods from Liaoning, China: *Journal of Vertebrate Paleontology*, 18 (Supplement to Number 3), Abstracts of Papers, Fifty-eighth Annual Meeting, p. 36A.

Currie, Philip J., and Kevin Padian, editors, *Encyclopedia of Dinosaurs*. San Diego: Academic Press, xxx, 869 pages.

Currie, Philip J., and J.-H. Peng, 1993, A juvenile specimen of *Saurornithoides mongoliensis* from the Upper Cretaceous of northern China: *Canadian Journal of Earth Sciences*, 30, pp. 2037–2081.

Currie, Philip J., J. Keith Rigby, Jr., and Robert E. Sloan, 1990, *in*: Kenneth Carpenter and Currie, editors, *Dinosaur Systematics: Approaches and Perspectives*. Cambridge, New York and Melbourne: Cambridge University Press, pp. 107–125.

Currie, Philip J., and Dale A. Russell, 1988, Osteology and relationships of *Chirostenotes pergracilis* (Saurischia, Theropoda) from the Judith River (Oldman) Formation of Alberta, Canada: *Canadian Journal of Earth Sciences*, 25, pp. 972–986.

Currie, Philip J., and David J. Varricchio, 2000, New dromaeosaurids from the Late Cretaceous of western North America: *Publications in Paleontology* [The Florida Symposium on Dinosaur Bird Evolution, abstracts of papers], 2, p. 14.

Currie, Philip J., and Xi-Jin Zhao, 1993*a*, A new carnosaur (Dinosauria, Theropoda) from the Jurassic of Xinjiang, People's Republic of China: *Canadian Journal of Earth Sciences*, 30 (10–11), pp. 2037–2081.

_____, 1993*b*, A new troodontid (Dinosauria, Theropoda) braincase from the Dinosaur Park Formation (Campanian) of Alberta: *Ibid.*, 30, pp. 2231–2247.

_____, 1994, A new troodontid (Dinosauria, Theropoda) braincase from the Dinosaur Park Formation (Campanian) of Alberta: *Ibid.*, 30, pp. 2231–2247.

Curry, Kristina A., 1998, Histological quantification of growth rates in *Apatosaurus*: *Journal of Vertebrate Paleontology*, 19 (Supplement to Number 3), Abstracts of Papers, Fifty-ninth Annual Meeting, pp. 36–37A.

_____, 1999, Ontogenetic histology of *Apatosaurus* (Dinosauria: Sauropoda): new insights on growth rates and longevity: *Journal of Vertebrate Paleontology*, 19 (4), pp. 654–665.

Curry Rogers, Kristina A., J. Castanet, J. Cubo, and J. J. Boisard, 2000, How fast is fast? Bone growth rate quantifica tion in extant ratites and implications for dinosaurs: *Journal of Vertebrate Paleontology*, 20 (Supplement to Number 3), Abstracts of Papers, Sixtieth Annual Meeting, p. 37A.

Curtice, Brian David, 1996, Codex of diplodocid caudal vertebrae from the Dry Mesa dinosaur quarry, master's thesis, Brigham Young University, Provo, Utah, 250 pages.

_____, 1998, Sauropod vertebral variation and its phylogenetic impact: *Journal of Vertebrate Paleontology*, 18 (Supplement to Number 3), Abstracts of Papers, Fifty-eighth Annual Meeting, p. 37A.

_____, 2000, The axial skeleton of *Sonorosaurus thompsoni* Ratkevich 1998, *in*: R. D. McCord and D. Boaz, editors, *Southwest Paleontological Symposium, Proceedings 2000*, Mesa Southwest Museum Bulletin No. 7, pp. 83–87.

Czerkas, Stephen A., 1994, The history and interpretation of sauropod skin impressions: *Gaia*, 10, pp. 173–182.

_____, 1998, The lips, beaks, and cheeks of ornithischians: *Journal of Vertebrate Paleontology*, 18 (Supplement to Number 3), Abstracts of Papers, Fifty-eighth Annual Meeting, p. 37A.

_____, 2000, A new toothed bird from China: *Publications in Paleontology* [The Florida Symposium on Dinosaur Bird Evolution, abstracts of papers], 2, p. 14.

Czerkas, Sylvia, and Donald F. Glut, 1982, *Dinosaurs, Mammoths and Cavemen: The Art of Charles R. Knight*. New York: E. P. Dutton, vii, 120 pages.

Dal Sasso, Cristiano, and Marco Signore, 1998*a*, Exceptional soft-tissue preservation in a theropod dinosaur from Italy: *Nature*, 392, pp. 838–387.

_____, 1998*b*, *Scipionyx samniticus* (Theropoda: Coelurosauria) and its exceptionally preserved internal organs: *Journal of Vertebrate Paleontology*, 18 (Supplement to Number 3), Abstracts of Papers, Fifty-eighth Annual Meeting, p. 37A.

Dalla Vecchia, Fabio M., 1994*a*, I dinosauri dell'Istria, *in*: G. Ligabue, editor, *Il tempo dei dinosauri*. Le Scienze quaderni, pp. 76, 82–86.

_____, 1994*b*, Jurassic and Cretaceous sauropod evidence in the Mesozoic carbonate platforms of the Southern Alps and Dinarids, *in*: M. G. Lockley, V. F. dos Santos, C. A. Meyer, and A. Hunt, editors, *Aspects of Sauropod Paleobiology*. *Gaia*, 10, pp. 65–73.

_____, 1997*a*, Terrestrial tetrapod evidence on the Norian (Late Triassic) and Cretaceous carbonate platforms of northern Adriatic region (Italy, Slovenia and Croatia): *Sargetia*, 17, "Proceedings of the International Symposium 'Mesozoic Vertebrate Faunas of Central Europe' Deva, 22–24th August 1996," pp. 177–201.

_____, 1997*b*, Dinosauri Cretacei nella piattaforma carbonatica Adriatico-Dinarica: *Natura Nascosta*, 15, pp. 22–28 (in Italian, reprinted and translated as Dinosaurs in the Cretaceous Adriatic-Dinaric carbonate platform: *Geoitalia*, 1. Forum Italiano di Sciencze della Terra, Riassunti, 2, pp. 60–62).

_____, 1997*c*, Dinosauri cretacei nella piattaforma carbonatica Adriatico-Dinarica: *Natura Nascosta*, 15, pp. 22–28.

_____, 1998, Remains of Sauropoda (Reptilia, Saurischia) in the Lower Cretaceous (Upper Hauterivian/Lower Barremian) Limestone of SW Istria (Croatia): *Geologia Croatia*, 51 (2), pp. 105–134.

_____, 1999, Atlas of the sauropod bones from the upper Hauterivian–lower Barremian of Bale/Valle (SW Istria, Croatia): *Natura Nascosta*, 18, pp. 6–41.

Dalla Vecchio, Fabio M., and A. Tarlao, 1995, Dinosaur evidence in the Cretaceous of Istria (Croatia), *in*: L. Vlahović, I. Velić, and M. Sparica, editors: *I. hrvatski geoloski kongres (First Croatian Geological Congress), Zbornik radova (Proceedings)*, 1, pp. 151–154.

Dalla Vecchia, Fabio M., A. Tarlao, and G. Tunis, 1993, Theropod (Reptilia: Dinosauria) footprints in the Albian (Lower Cretaceous) of the Quieto/Mirna river mouth (NW Istria, Croatia) and dinosaur population of the Istrian region during the Cretaceous: *Mem. Sci. Geol. Padova*, 45, pp. 139–148.

Dalton, Rex, 2000, Doubts grow over discovery of fossilized "dinosaur heart": *Nature*, 407, pp. 275–276.

Dantas, P. M., J. L. Sanz, and Galopim de Carvalho, A. M., 1992, Dinossáurio da Praia De Porto Dinheiro (dados preliminares): *GAIA*, 5 (1), pp. 31–35.

Dantas, P. M., J. L. Sanz, C. M. Silva, F. Ortega, V. F. Santos, and M. Cachao, 1998, *Lourinhasaurus n. gen.* Novo dinossáurio saurópode do Jurassico superior (Kimmeridgian superior–Titoniano inferior) de Portugal: *Communicacóes do Instituto Geológico e Mineiro*, 84 (1), pp. 91–94. V Congresso Nacional de Geologia, Lisboa, 1998.

Darby, D. G., and R. W. Ojakangas, 1980, Gastroliths from an Upper Cretaceous plesiosaur: *Journal of Paleontology*, 54, pp. 548–556.

de Klerk, William J., Catherine A. Forster, Scott D. Sampson, Anusuya Chinsamy, and Callum F. Ross, 2000, A new coelurosaurian dinosaur from the Early Cretaceous of South Africa: *Journal of Vertebrate Paleontology*, 20 (2), pp. 324–332.

Dean, Dennis R., 1995, Gideon Mantell and the discovery of *Iguanodon*, *in*: Willam A. Sarjeant, editor, *Vertebrate Fossils and the Evolution of Scientific Concepts*. Amsterdam: Gordon and Breach, pp. 207–217.

_____, 1999, *Gideon Mantell and the Discovery of Dinosaurs*. Cambridge: University of Cambridge Press.

Debus, Allen A., and Diane E. Debus, 2000, "Joe Who? *Jobaria!*": *Fossil News*, 6 (6), pp. 5–8.

Debus, Allen A., and Steve Brusatte, 2000, Long live the Queen! *T. rex* reigns at the Field: *Dinosaur World*, 8, pp. 30–35.

Deck, Linda T., 1999, What's new in dinosaur exhibits at the National Museum of Natural History: *Journal of Vertebrate Paleontology*, 19 (Supplement to Number 3), Abstracts of Papers, Fifty-ninth Annual Meeting, pp. 41A–42A.

_____, 2000, Presenting how we did it: the virtual *Triceratops*: *Journal of Vertebrate Paleontology*, 20 (Supplement to Number 3), Abstracts of Papers, Sixtieth Annual Meeting, p. 38A.

Deeming, D. Charles, and David M. Unwin, 1993, Fossil embryos and neonates: are they what we want them to be?: *Journal of Vertebrate Paleontology*, 15 (Supplement to Number 3), Abstracts of Papers, Fifty-third Annual Meeting, p. 32A.

Dehm, R., 1935, Beobachtungen im oberen Bunten Keuper Mittelfran kens: *Neus Jarbuch für Mineralogie, Geologie, und Paläontologie*, 1935, B, pp. 97–109, Stuttgart.

Delair, Justin B., 1958, The Mesozoic reptiles of Dorset. Part 1. *Proceedings of the Dorset Natural History and Archaeological Society*, 79, pp. 47–72.

Depéret, Charles, 1896, Note sur les dinosauriens sauropodes et théropodes du Crétacé supérieur de Madagascar: *Bulletin, Societé Géologique de France, Series 3*, 24, pp. 176–194.

Derstler, Kraig, and David Burnham, 2000, Phylogenetic context of *Bambiraptor feinbergi*: *Publications in Paleontology* [The Florida Symposium on Dinosaur Bird Evolution, abstracts of papers], 2, p. 15.

Diegert, Carl F., and Thomas E. Williamson, 1998, A digital acoustic model of the lambeosaurine hadrosaur *Parasaurolophus tubicen*: *Journal of Vertebrate Paleontology*, 18 (Supplement to Number 3), Abstracts of Papers, Fifty-eighth Annual Meeting, p. 38A.

Dietrich, W. O., 1914, Die Gastropoden der Tendaguruschichten, der Aptstufe und der Oberkreide im sülichen Deutcsh-Ostafrika: *Archiv fü Biontologie*, 3 (4), pp. 97–153.

Dingus, Lowell, 1996, Eugene S. Gaffney, Mark A. Norell, and Scott D. Sampson, 1996, *The Halls of Dinosaurs: A Guide to Saurischians and Ornithischians*. New York: American Museum of Natural History, 100 pages.

Dodson, Peter, 1975, Taxonomic implications of relative growth in lambeosaurine hadrosaurs: *Systematic Zoology*, 24, pp. 37–54.

_____, 1986, *Avaceratops lammersi*, a new ceratopsid from the Judith River Formation of Montana: *Academy of Natural Sciences of Philadelphia, Proceedings*, 138 (2), pp. 305–317.

_____, 1990*a*, Sauropod paleoecology, *in*: David B. Weishampel, Peter Dodson and Halszka Osmólska, editors, *The Dinosauria*. Berkeley and Los Angeles: University of California Press, pp. 402–407.

_____, 1990*b*, Marginocephalia, *Ibid.*, pp. 562–563.

_____, 1991, Morphological and ecological trends in the evolution of ceratopsian dinosaurs, *in*: Zofia Kielan-Jaworowska, N. Heintz, and H. Arne Nakrem, editors, *Fifth Symposium on Mesozoic Terrestrial Ecosystems and Biota (extended abstracts), Contributions from the Paleontological Museum, University of Oslo*, pp. 17–18.

_____, 1993, Comparative craniology of the Ceratopsia, *in*: Dodson and Philip D. Gingerich, editors, *Functional Morphology and Evolution, American Journal of Science, 293-A (special volume)*, pp. 200–234.

_____, 1995, Reviews, *Dinosaur Eggs and Babies*: *Journal of Vertebrate Paleontology*, 15 (4), pp. 863–866.

_____, 1998, Ceratopsia, *in*: Philip J. and Kevin Padian, editors, *Encyclopedia of Dinosaurs*. San Diego: Academic Press, p. 106.

Dodson, Peter, and Philip J. Currie, 1988, The smallest ceratopsid skull — Judith River Formation of Alberta: *Canadian Journal of Earth Sciences*, 25, pp. 926–930.

_____, 1990, Neoceratopsia, *in*: David B. Weishampel, Dodson, and Halszka Osmólska, editors, *The Dinosauria*. Berkeley and Los Angeles: University of California Press, pp. 593–618.

Dodson, Peter, David W. Krause, Catherine A. Forster, Scott D. Sampson, and Florent Ravoavy, 1998*a*, Continuing discoveries of dinosaurs and associated fauna, Late Cretaceous of Madagascar, *in*: Donald L. Wolberg, K. Gittis, S. Miller, and A. Raynor, editors, *The Dinofest Symposium* [abstracts], Presented by The Academy of Natural Sciences of Philadelphia, Pennsylvania, pp. 11–12.

_____, 1998*b*, Titanosaurid (Sauropoda) osteoderms from the Late Cretaceous of Madagascar: *Journal of Vertebrate Paleontology*, 18 (3), pp. 563–568.

Dollo, Louis, 1883, Note sur les restes de dinosauriens rencon trés dans le Crétacé Supérieur de la Belgique: *Bulletin du Museé Royale d'Histoire Naturelle de Belgique*, 2, pp. 205–221.

Dong Zhiming [also known as Zhi-Ming], 1977, On the dinosaurian remains from Turpan, Xinjiang: *Vertebrata PalAsiatica*, 15 (1), 59–66.

_____, 1992, *Dinosaurian Faunas from China*. Beijing: China Ocean Press, and Berlin: Springer Verlag, 188 pages.

_____, 1993, The field activities of the Sino-Canadian Dinosaur Project in China: *Canadian Journal of Earth Sciences*, 30 (10/11), pp. 1987–1990.

Dong Zhi-Ming [formerly Zhiming], Philip J. Currie, and Dale A. Russell, 1989, The 1988 Field Program of The Dinosaur Project: *Vertebrata PalAsiatica*, 27 (3), pp. 233–236.

Dong Zhi-Ming [formerly Zhiming], Yoshikazu Hasegawa, and Yoichi Azuma, 2000, *The Age of Dinosaurs in Japan and China*. Fukui, Japan: Fukui Prefectural Museum, 65 pages.

Dong Zhi-Ming [formerly Zhiming], Zhou Shiwu and Zhang Zicheng, 1983, The dinosaurian remains from Sichuan Basin, China: *Paleontologia Sinica*, whole number 162, New Series C, 23, pp. 1–145.

Dong Zhi-Ming [formerly Zhiming], Tang Zilu, and Zhou Shiwu, 1982, Note on the new mid–Jurassic stegosaur from Sichuan Basin, China: *Vertebrata PalAsiatica*, 20, pp. 83–87.

Doré, A. C., 1991, The structural foundation and evolution of Mesozoic seaways between Europe and the Arctic: *Palaeogeography, Palaeoclimatology, and Palaeoecology*, 87, pp. 441–492.

Downs, Alex, 2000, *Coelophysis bauri* and *Syntarsus rhodesiensis* compared, with comments on the preparation and preservation of fossils from the Ghost Ranch *Coelophysis* quarry, *in*: Spencer G. Lucas and Andrew B. Heckert, editors, *Dinosaurs of New Mexico*. Albuquerque: New Mexico Museum of Natural History and Science, Bulletin 17, pp. 33–37.

Doyle, J. A., and L. J. Hickey, 1976, Pollen and leaves from the Mid-Cretaceous Potomac Group and their bearing on early angiosperm evolution, *in* C. B. Beck, editor: *Origin and Early Evolution of Angiosperms*. New York: Columbia University Press, pp. 139–206.

Dunbar, R. O., 1944, Como Bluff Anticline, Albany and Carbon Counties, Wyoming: *American Association of Petroleum Geologists Bulletin*, 28, pp. 1–24.

Dunham, A. E., K. L. Overall, W. P. Porter, and C. A. Forster, 1989, Implication of ecological energetics and biophysical and developmental constraints for life-history variation in dinosaurs, *in*: James A. Farlow, editor, *Paleobiology of the Dinosaurs, GSA Special Paper 238*. Boulder, Colorado: The Geological Society of America, pp. 1–21.

Dyke, Gareth J., and Joe Thorley, 1998, Reduced cladistic consensus methods and the inter-relationships of *Protoavis*, *Avimimus* and Mesozoic birds: *Archaeopteryx*, 16, pp. 123–129, Eichstätt.

Eaton, Theodore H., Jr., 1960, A new armored dinosaur from the Cretaceous of Kansas: *University of Kansas Paleontological Contribution*, 8, pp. 1–24.

Eaton, Jeffrey G., Steve Diem, J. David Archibald, C. Schierup and H. Monk 1999, Vertebrate paleontology of the Upper Cretaceous rocks of the Markagunt Plateau, southwestern Utah: *Utah Geological Survey, Miscellaneous Publication*, 99-1, pp. 323–353.

Eaton, Jeffrey H., Richard L. Cifelli, J. Howard Hutchison, James I. Kirkland, and J. Michael Parrish, 1999, Cretaceous vertebrate faunas from the Kaiparowits Plateau, south-central Utah: *Utah Geological Survey, Miscellaneous Publication*, 99-1, pp. 323–353.

Eberth, David A., Philip J. Currie, Rodolfo A. Coria, Alberto C. Garrido, and John-Paul Zonneveld, 2000, Large-theropod bonebed, Neuquen, Argentina: paleoecologocal importance: *Journal of Vertebrate Paleontology*, 20 (Supplement to Number 3), Abstracts of Papers, Sixtieth Annual Meeting, p. 39A.

Eberth, David A., and A. P. Hamblin, 1993, Tectonic, stratigraphic, and sedimentological significance of a regional discontinuity in the upper Judith River Group (Belly River wedge) of southern Alberta, Saskatchawan, and southern Montana: *Canadian Journal of Earth Science*, 30, pp. 174–200.

Efimov, M. B., 1988, The fossil crocodiles and Champsosauria of Mongolia and Russia: *Transactions of the Soviet-Mongolian Paleontological Expedition*, 36, pp. 1–109.

Ellenberger, F., and Ellenberger, P., 1958, Principaux types de pistes de vertébrés dans les couches du Stormberg au Basuto land (Afrique du Sud): *Comtes Rendu Somm. de la Société Géologique de France*, 4, pp. 65–67.

Ellenberger, Paul, 1970, Les niveaux paléontologiques de prémiere apparition des mammiferes primordiaux en Afrique du Sudat leur ichnologie: *Proceedings and Papers of the Second Gondwana Symposium, South Africa, 1970*, pp. 343–370.

Elźanowski, Andrzej, 1983, Birds in Cretaceous ecosystems: *Acta Palaeontologica Polonica*, 28, pp. 75–92.

_____, 1999, A comparison of the jaw skeleton in theropods and birds, with a description of the palate in the Oviraptoridae: *Smithsonian Contributions to Paleobiology*, 89, pp. 311–323.

Elźanowski, Andrzej, and Peter Wellnhofer, 1995, The skull of *Archaeopteryx* and the origin of birds: *Archaeopteryx*, 13, pp. 41–46.

Erickson, Gregory M., and Christopher A. Brochu, 1999, How the "terror crocodile" grew so big: *Nature*, 398, pp. 1–21.

Erickson, Gregory M., and Kristina A. Curry Rogers, 2000, Comparison of exponential growth rates between dinosaurs and extant vertebrates: *Journal of Vertebrate Paleontology*, 20 (Supplement to Number 3), Abstracts of Papers, Sixtieth Annual Meeting, p. 40A.

Erickson, Gregory M., and Tatyana A. Tumanova, 2000, Growth curve of *Psittacosaurus mongoliensis* Osborn (Ceratopsia: Psittacosauridae) inferred from long bone histology: *Zoological Journal of the Linnean Society*, 130, pp. 551–566.

Eun-Ju Park, Seong-Young Yang, and Philip J. Currie, 2000, Early Cretaceous dinosaur teeth of Korea: *Paleontological Society of Korea Special Publication*, 4, pp. 85–98.

Evans, S. E., A. R. Milner, and C. Werner, 1996, Sirenid salamanders and a gymnophionan amphibian from the Cretaceous of the Sudan: *Palaeontology*, 39 (1), pp. 77–95.

Fang, Pang, Lu, Zhang, Pan, Wang, Li and Cheng, 2000: The difinition of Lower, Middle and Upper Jurassic Series in Lufeng, Yunnan: *Proceedings of the Third National Stratigraphical Conference of China*. Beijing: Geological Publishing House, pp. 208–214.

Farke, Andrew A., 1997, The distribution and taxonomy of *Triceratops*, *in*: Donald L. Wolberg, Edmund Stump, and Gary Rosenberg, editors. *Dinofest International: Proceedings of Symposium Held at Arizona State University*. Philadelphia: Academy of Natural Sciences, pp. 47–49.

Farlow, James O., 1997, Dinosaurs and Geologic Time, *in*: Farlow and Michael K. Brett-Surman, editors, *The Complete Dinosaur*. Bloomington and Indianapolis: Indiana University Press, pp. 107–111.

Farlow, James O., and David L. Brinkman, 1994, Wear surfaces on the teeth of tyrannosaurs: *Paleontological Society*, special publication 7, pp. 165–175.

Farlow, James O., Matt B. Smith, and John M. Robinson, 1995, Body mass, bone "strength indicator," and cursorial potential of *Tyrannosaurus rex*: *Journal of Vertebrate Paleontology*, 15 (4), pp. 713–725.

Farmer, Colleen G., and David R. Carrier, 2000, Pelvic aspirations in the American alligator (*Alligator mississippiensis*): *Journal of Experimental Biology* (in press).

Farris, J. S., 1970, Methods for comparing Wagner tress: *Systematic Zoology*, 19, pp. 83–92.

Fassett, James E., 1982, Dinosaurs of the San Juan Basin, New Mexico, may have survived the event that resulted in creation of an iridium-enriched zone near the Cretaceous-Tertiary boundary: *Geological Society of America, Special Paper*, 190, pp. 435–447.

Fassett, James E., and Spencer G. Lucas, 2000, Evidence for Paleocene dinosaurs from the Ojo Alamo Sandstone, San Juan Basin, New Mexico, *in*: Spencer G. Lucas and Andrew B. Heckert, editors, *Dinosaurs of New Mexico*. Albuquerque: New Mexico Museum of Natural History and Science, Bulletin 17, pp. 221–230.

Fassett, James E., Spencer G. Lucas, and Michael O'Neill, 1987, Dinosaurs, pollen and spores, and the age of the Ojo Alamo Sandstone, San Juan Basin, New Mexico: *Geological Society of America, Special Paper*, 209, pp. 17–34.

Fassett, James E., Spencer G. Lucas, Robert A. Zielinski, and James R. Budahn, 2000, Compelling new evidence for Paleocene dinosaurs in the Ojo Alamo Sandstone, San Juan Basin, New Mexico and Colorado, USA; in catastrophic events and mass extinctions: impacts and beyond: *LPI Contribution*, 1053, Lunar and Planetary Institute, Houston, Texas, pp. 45–46.

Fassett, James E., and M. B. Steiner, 1997, Precise age of C33N-C32R magnetic polarity reversal, San Juan Basin, New Mexico and Colorado: *New Mexico Geological Society, Guidebook*, 48 pp. 239–247.

Fastovsky, David Eliot, 1987, Paleoenvironments of vertebrate-bearing strata during the Cretaceous–Tertiary transition, eastern Montana and western North Dakota: *Palaios*, 2, pp. 282–295.

Fastovsky, David Eliot, and Peter M. Sheehan, 1997, Demythicized dinosaur extinctions at the Cretaceous–Tertiary boundary, *in*: *Dinofest International: Proceedings of a Symposium Held at Arizona State University*. Philadelphia: Academy of Natural Sciences, pp. 527–531.

Fastovsky, David Eliot, and David B. Weishampel, 1996, *The Evolution and Extinction of the Dinosaurs*. New York and Cambridge: Cambridge University Press.

Fedde, M. R., 1976, Respiration, *in*: P. D. Sturke, editor, *Avian Physiology*. New York: Springer, pp. 122–145.

Fedde, M. R., P. D. DeWet, and R. L. Kitchell, 1969, Motor unit recruitment pattern and tonic activity in respiratory muscles of *Gallus domesticus*: *Journal of Neurophysiology*, 32, pp. 995–1004.

Feduccia, Alan, 1994, The great dinosaur debate: *Living Bird*, 13 (4), 28–33.

_____, 1996, *The Origin and Evolution of Birds*. New Haven, Connecticut: Yale University Press, 420 pages.

_____, 1999*a*, 1,2,3 = 2,3,4: accommodating the cladogram: *Proceedings of the National Academy of Sciences*, 96, pp. 4740–4742.

_____, 1999*b*, *The Origin and Evolution of Birds*, second edition. New Haven, Connecticut: Yale University Press, 466 pages.

_____, 2000, Persisting problems in the bird-theropod nexus: *Publications in Paleontology* [The Florida Symposium on Dinosaur Bird Evolution, abstracts of papers], 2, p. 15.

Fernandes, L. A., and A. M. Coimbra, 1996, A Bacia Bauru (Cretáceo Supereur, Brasil): *Anais da Academia Brasileira de Ciências*, 68 (2), pp. 195–205.

Fiffer, Steve, 2000, Tyrannosaurus *Sue: The Extraordinary Saga of the Largest, Most Fought Over* T. rex *Ever Found*. New York: W. H. Freeman and Company, 248 pages.

Fiorilla, Anthony R., 1998, Microwear on the teeth of theropod dinosaurs (Judith River Formation) of south-central Montana: inferences on diet: *Journal of Vertebrate Paleontology*, 18 (Supplement to Number 3), Abstracts of Papers, Fifty-eighth Annual Meeting, p. 41A.

Fiorilla, Anthony R., and Roland Gangloff, 2000, Theropod teeth from the Prince Creek Formation (Cretaceous) of northern Alaska, with speculations on Arctic dinosaur paleoecology: *Journal of Vertebrate Paleontology*, 20 (Supplement to Number 3), Abstracts of Papers, Sixtieth Annual Meeting, p. 41A.

Fischer, P., 1879, Recherches sur les reptiles fossiles de l'Afrique australe: *Nouvelles Archives du Muséum d'Historie Naturelle de Paris*, 6, pp. 163–200, Paris.

Fisher, Paul E., Dale A. Russell, Michael K. Stoskopf, Reese E. Barrick, Michael Hammer, and Andrew A. Kuzmitz, 2000, Cardiovascular evidence for an intermediate or higher metabolic rate in an ornithischian dinosaur: *Science*, 288, pp. 503–505.

Flynn, John J., J. Michael Parrish, Berthe Rakotosamimanana, William F. Simpson, Robin L. Whatley, and André R. Wyss, 1999, A Triassic fauna from Madagascar, including early dinosaurs: *Science*, 286, pp. 763–765.

Ford, Tracy L., A review of ankylosaur osteoderms from New Mexico and a preliminary review of ankylosaur armor, *in*: Spencer G. Lucas and Andrew B. Heckert, editors, *Dinosaurs of New Mexico*. Albuquerque: New Mexico Museum of Natural History and Science, Bulletin 17, pp. 157–176.

Forster, Catherine A., 1990, The postcranial skeleton of the ornithopod dinosaur *Tenontosaurus tilletti*: *Journal of Vertebrate Paleontology*, 10 (3), pp. 273–294.

_____, 1996, Species resolution in *Triceratops*: cladistic and morphometric approaches: *Ibid.*, 16 (20), pp. 259–270.

_____, 1997, Phylogeny of the Iguanodontia and Hadrosauridae: *Ibid.*, 17 (Supplement to Number 3), Abstracts of Papers, Fifty-seventh Annual Meeting, p. 47A.

Forster, Catherine A., Scott D. Sampson, Luis M. Chiappe, and David W. Krause, 1998, The theropod ancestry of birds: new evidence from the Late Cretaceous of Madagascar: *Science*, 279, pp. 1915–1919.

Forster, Cartherine A., and Paul C. Sereno, 1997, Marginocephalians, *in*: James O. Farlow and Michael K. Brett-Surman, editors, *The Complete Dinosaur*. Bloomington and Indianapolis: Indiana University Press, pp. 317–329.

Forster, Catherine A., Paul C. Sereno, Thomas W. Evans, and Timothy Rowe, 1993, A complete skull of *Chasmosaurus mariscalensis* (Dinosauria: Ceratopsidae) from the Aguja Formation (Late Campanian) of West Texas: *Journal of Vertebrate Paleontology*, 13, pp. 161–170.

Foster, John R., and Daniel J. Chure, 1999, Hindlimb proportion allometry in juvenile to adult *Allosaurus* (Dinosauria, Theropoda): *Journal of Vertebrate Paleontology*, 19 (Supplement to Number 3), Abstracts of Papers, Fifty-ninth Annual Meeting, p. 45A.

Foster, John R., Brian D. Curtice, and Darrin C. Pagnac, 1998, First reported occurrence of sauropod remains from the Lower Cretaceous (Lakota Formation), Black Hills, South Dakota: *Journal of Vertebrate Paleontology*, 18 (Supplement to Number 3), Abstracts of Papers, Fifty-eighth Annual Meeting, p. 43A.

Fox, William, 1866, Another new Wealden reptile: *The Athenaeum*, 2014, p. 740. (Reprinted in: *Geological Magazine*, 3, p. 383.)

Fraas, Eberhard, 1908, Ostafrikanische Dinosaurier: *Palaeontographica*, 55, pp. 105–144.

Frankfurt, Nicholas G., and Luis M. Chiappe, 1998, A possible oviraptorosaur from the Late Cretaceous of northwestern Argentina: *Journal of Vertebrate Paleontology*, 19 (1), pp. 101–105.

Frey, Eberhard, 1988, Das Tragsystem der Krokodile—eine biomechanische und phylogenetische Analyse: *Stuttgarter Beiträge zur Naturkunde*, 426, pp. 1–60.

Fritsch [Fric], 1878, *Die Reptilian und Fische der böhmischen Kreideformation*. Prague: ii, 46 pages.

_____, 1905, Synopsis der Saurier der böhm Kreiderformation: *Sitzungsberichte der Königl, Böhmischen Gesellschaft der Wissenschaften Mathematisch-Naturwissenschaftliche Classe*, 2 (101), pp. 1–7, Prague.

Galbrun, B., M. Feist, F. Colombo, R. Rocchia,

and Y. Tambareau, 1993, Magnetostratigraphy and biostratigraphy of Cretaceous–Teriary continental deposits, Ager Basin, Province of Lerida, Spain: *Palaeogeography, Palaeoclimatology, and Palaeoecology*, 104, pp. 41–52.

Galliano, Alessandra, and Marco Signore, 1999, Parental care in theropod dinosaurs: possible evidences from *Scipionyx samniticus*: *Journal of Vertebrate Paleontology*, 19 (Supplement to Number 3), Abstracts of Papers, Fifty-ninth Annual Meeting, p. 46A.

Gallup, Marc R., 1989, Functional morphology of the hindfoot of the Texas sauropod *Pleurocoelus* sp. indet., *in*: James O. Farlow, editor, *Paleobiology of the Dinosaurs*, Geological Society of America Special Paper 238, pp. 71–74.

Galton, Peter M., 1970, The posture of hadrosaurian dinosaurs: *Journal of Paleontology*, 44 (3), pp. 464–473.

_____, 1971, The prosauropod dinsoaur *Ammosaurus*, the crocodile *Protosuchus*, and their bearing on the age of the Navajo Sandstone of Northeastern Arizona: *Ibid.*, 45 (5), pp. 781–795.

_____, 1973*a*, The cheeks of ornithischian dinosaurs: *Lethaia*, 6, pp. 67–89.

_____, 1973*b*, On the anatomy and relationships of *Efraasia diagnostica* (Huene), n. gen., a prosauropod dinosaur (Reptilia: Saurischia) from the Upper Triassic of Germany: *Paläontographica Zeitschrift*, 47 (5), pp. 229–255.

_____, 1974, Notes on *Thescelosaurus*, a conservative ornithopod dinosaur from the Upper Cretaceous of North America, with comments on ornithopod classification: *Journal of Paleontology*, 48 (5), pp. 1048–1067.

_____, 1975, English hypsilophodontid dinosaurs (Reptilia: Ornithischia): *Palaeontology*, 18 (4), pp. 741–751.

_____, 1976, Prosauropod dinosaurs (Reptilia: Saurischia) of North America: *Postilla*, 169, pp. 1–98.

_____, 1980*a*, *Dryosaurus* and *Camptosaurus*, intercontinental genera of Upper Jurassic ornithopod dinosaurs: *Memoires de la Société Géologique de France, Nouvelle Serie*, 139, pp. 103–108.

_____, 1980*b*, European Jurassic ornithopod dinosaurs of the families Hypsilophodontidae and Camptosauridae: *Neus Jahrbuch für Geologie und Paläontologie, Abhandlungen*, 160 (1), pp. 73–95.

_____, 1982, Juveniles of the stegosaurian dinosaur *Stegosaurus* from the Upper Jurassic of North America: *Journal of Vertebrate Paleontology*, 2 (1), pp. 47–62.

_____, 1983, Armored dinosaurs (Ornithischia: Ankylosauria) from the Middle and Upper Jurassic of Europe: *Palaeontographica*, Abt. A., 182, pp. 1–25.

_____, 1984*a*, Cranial anatomy of the prosauropod dinosaur *Plateosaurus*, from the Knollenmergel (Middle Keuper, Upper Triassic) of Germany. 1. Two complete skulls from Trossingen/Württ with comments on the diet: *Geologica et Palaeontologica*, 18, pp. 139–171.

_____, 1984*b*, An early prosauropod from the Upper Triassic of Nordwürttemberg, West Germany: *Stuttgarter Beiträge zur Naturkunde*, (Serie B), (Geologie und Paläontologie), 196, pp. 1–25, Stuttgart.

_____, 1985*a*, the poposaurid thecodontian *Teratosaurus suevicus* v. Meyer, plus referred specimens mostly based on prosauropod dinosaurs, from the Middle Stubensandstein (Upper Triassic of Nordwürttemberg: *Ibid.*, 116, pp. 105–123, Oslo.

_____, 1985*b*, Cranial anatomy of the prosauropod dinosaur *Sellosaurus gracilis* from the Middle Stubensandstein (Upper Triassic) of Nordwürttemberg: *Ibid.*, (B), 118, pp. 1–39, Stuttgart.

_____, 1985*c*, Cranial anatomy of the prosauropod dinosaur *Plateosaurus* from the Knollenmergel (Middle Keuper) of Germany, II. all the cranial material and details of soft-part anatomy: *Geologica et Palaeontologica*, 19, pp. 119–159.

_____, 1985*d*, Notes on the Melanorosauridae, a family of large prosauropod dinosaurs (Saurischia: Sauropodomorpha): *Geobios*, 18 (5), pp. 671–676, Lyon.

_____, 1985*e*, Diet of prosauropod dinosaurs from the Late Triassic and Early Jurassic: *Lethaia*, 18, pp. 105–123, Oslo.

_____, 1985*f*, British plated dinosaurs (Ornithischia, Stegosauridae): *Journal of Vertebrate Paleontology*, 5 (3), pp. 211–254.

_____, 1986*a*, Prosauropod dinosaur *Plateosaurus* (=*Gresslyosaurus*) (Saurischia: Sauropodomorpha) from the Upper Triassic of Switzerland: *Geologica et Palaeontologica*, 20, pp. 167–183.

_____, 1986*b*, Herbivorous adaptations of Late Triassic dinosaurs, *in*: Kevin Padian, editor, *The Beginning of the Age of Dinosaurs: Faunal Change Across the Triassic–Jurassic Boundary*. Cambridge and New York: Cambridge University Press, pp. 203–221.

_____, 1990*a*, Basal Sauropodomorpha — Prosauropoda, *in*: David B. Weishampel, Peter Dodson and Halszka Osmólska, editors, *The Dinosauria*. Berkeley and Los Angeles: University of California Press, pp. 320–344.

_____, 1990*b*, Stegosauria, *Ibid.*, pp. 435–455.

_____, 1995, The species of the basal hypsilophodontid dinosaur *Thescelosaurus* Gilmore (Ornithischia: Ornithopoda) from the Upper Cretaceous of North America: *Neus Jahrbuch für Geologie und Paläontologie, Abhandlungen*, 198 (3), pp. 297–311.

_____, 1997*a*, Stegosauria, *in*: Philip J. Currie and Kevin Padian, editors, *Encyclopedia of Dinosaurs*. San Diego: Academic Press, pp. 701–703.

_____, 1997*b*, Comments on sexual dimorphism in the prosauropod dinosaur *Plateosaurus engelhardti* (Upper Cretaceous, Trossingen): *Neus Jahrbuch für Geologie und Paläontologie, Monatshefte*, 1997, H. 11, pp. 674–682, Stuttgart.

_____, 1997*c*, Cranial anatomy of the basal hypsilophodontid dinosaur *Thescelosaurus neglectus* Gilmore (Ornithischia: Ornithopoda) from the Upper Cretaceous of North America: *Revue Paléobiologie, Genève*, 16 (1), pp. 231–258.

_____, 1998*a*, The prosauropod dinosaur *Plateosaurus* (*Dimodosaurus*) *poligniensis* (Pidancet & Chopard, 1861) (Upper Triassic, Poligny, France): *Neus Jahrbuch für Geologie und Paläontologie, Abhandlungen*, 207 (2), pp. 255–288.

_____, 1998*b*, Saurischian dinosaurs from the Upper Triassic of England: *Camelotia* (Prosauropoda, Melanorosauridae) and *Avalonianus* (Theropoda, ?Carnosauria): *Palaeontographica*, Abt. A, 250, pp. 155–172, Stuttgart.

_____, 1999*a*, Sex, sacra and *Sellosaurus gracilis* (Saurischia, Sauropodomorpha, Upper Triassic, Germany) — or why the character "two sacral vertebrae" is plesiomorphic for Dinosauria: *Neus Jahrbuch für Geologie und Paläontologie, Abhandlungen*, 213 (1), pp. 19–55.

_____, 1999*b*, Cranial anatomy of the hypsilophodontid dinosaur *Bugenasaura infernalis* (Ornithischia: Ornithopoda) from the Upper Cretaceous of North America: *Revue Paléobiologie, Genève*, 18, pp. 517–534.

_____, 2000, The prosauropod dinosaur *Plateosaurus* Meyer 1837 (Saurischia: Sauropodomorpha). 1. the syntypes of *P. engel hardti* Meyer 1837 (Upper Triassic, Germany), with notes on other European prosauropods with "distally straight" femora: *Ibid.*, 216 (2), pp. 233–275.

_____, (in preparation), Postcranial anatomy of the prosauropod dinosaur *Sellosaurus gracilis* from the Middel Stubensandstein (Upper Triassic) of Nordwürttemberg, West Germany: *Stuttgarter Beiträge zur Naturkunde*, (B), Stuttgart.

_____, (in press a), Are *Spondylosoma* and *Staurikosaurus* (Santa Maria Formation, Middle–Upper Triassic, Brazil) the earliest saurischian dinosaurs?: *Paläontologische Zeitschrift*, Stuttgart.

_____, (in press b), The prosauropod dinosaur *Plateosaurus* Meyer, 1837 (Saurischia: Sauropodomorpha), II. Notes on referred species from Upper Triassic of western Europe: *Neus Jahrbuch für Geologie und Paläontologie, Monatshefte*, Stuttgart.

_____, (in press c), On the manus of saurischian dinosaurs and movements of the digits in representative forms, *in*: Donald L. Wolberg, editor, *Dinofest*. Philadelphia: The Academy of Natural Sciences of Philadelphia.

Galton, Peter M., and Gerard Boine, 1980, A stegosaurian dinosaur femur from the Kimmeridgian Beds (Upper Jurassic) of the Cap de la Hève, Normandy: *Bulletin Trimestriel de la Societe Geologique de Normandie et des Amis du Museum du Havre*, 67 (4), pp. 31–35.

Galton, Peter M., and M. A. Cluver, 1976, *Anchisaurus capensis* (Broom) and a revision of the Anchisauridae (Reptilia, Saur ischia): *Annals of the South African Museum*, 69 (6), pp. 121–159, Cape Town.

Galton, Peter M., and Jacques van Heerden, 1985, Partial hindlimb of *Blikanasaurus cromptoni* n. gen. and n. sp., representing a new family of prosauropod dinosaurs from the Upper Triassic of South Africa: *Géobios — Mémoires spéciaux*, 18 (4), pp. 509–516.

_____, 1998, Anatomy of the prosauropod dinosaur *Blikanasaurus cromptoni* (Upper

Triassic, South Africa), with notes on the other tetrapods from the lower Elliot Formation: *Paläontologische Zeitschrift*, 72 (1/2), pp. 163–177, Stuttgart.

Galton, Peter M., and H. Phillip Powell, 1980, The ornithischian dinosaur *Camptosaurus prestwichii* from the Upper Jurassic of England: *Paleontology*, 23, pp. 411–443.

Galton, Peter M., and Hans-Dieter Sues, 1983, New data on pachycephalosaurid dinosaurs (Reptilia: Ornithischia) from North America: *Canadian Journal of Earth Sciences*, 20, pp. 462–472.

Galton, Peter M., and Paul Upchurch, 2000, Prosauropod dinosaurs: homeotic transformations ("frame shifts") with third sacral as a caudosacral or a dorsosacral: *Journal of Vertebrate Paleontology*, 20 (Supplement to Number 3), Abstracts of Papers, Sixtieth Annual Meeting, p. 43A.

Galton, Peter M., and Alick D. Walker, 1996, Supposed prosauropod dinosaurs from Middle Triassic of England referred to Parasuchia and Dinosauriformes: *Neus Jahrbuch für Geologie und Paläontologie, Monatshefte 1996*, 12, pp. 727–738, Stuttgart.

Garcia, Géraldine, Marie Mincemaille, Monique Vianey-Liaud, Bernard Marandat, Edgar Lorenz, Gilles Cheylan, Henri Cap petta, Jacques Michaux, and Jean Sudre, 1999, Découverte du premier squelette presque complet de *Rhabdodon priscus* (Dinosauria, Ornithopoda) du Maastrichtian inférieur de Provence (Discovery of an almost complete skeleton of *Rhabdodon priscus* [Dinosauria, Ornithopoda] in the early Maastrichtian of Provence (southern France): *Comtes Rendu des Séances de l'Académie des Sciences, Paris, de la Terre et des Planètes*, 328, pp. 415–421 (in French with English abridgement).

Garstka, William R., Damien Marsic, Parker Carroll, Laura Heffelfinger, Tyler Lyson, and Joseph D. Ng, 2000, Analysis of the structure and articulation of the forelimb bones of the maniraptoran dinosaur, *Bambiraptor feinbergi*, support predation as the pre-adaptation of avian flight: *Publications in Paleontology* [The Florida Symposium on Dinosaur Bird Evolution, abstracts of papers], 2, p. 16.

Gatesy, Stephen M., 1990, Caudofemoral musculature and the evolution of theropod locomotion: *Paleobiology*, 16 (2), pp. 170–186.

_____, 1991, Hindlimb scaling in birds and other theropods: implications for terrestrial locomotion: *Journal of Morphology*, 209 (1), pp. 83 96.

_____, 2000, In search of the flight stroke: problems of shoulder motion homology: *Publications in Paleontology* [The Florida Symposium on Dinosaur Bird Evolution, abstracts of papers], 2, p. 17.

Gatesy, Stephen M., and Kevin M. Middleton, 1998, Reconstructing theropod foot function using 3-D computer-animated track simulation: *Journal of Vertebrate Paleontology*, 18 (Supplement to Number 3), Abstracts of Papers, Fifty-eighth Annual Meeting, p. 45A.

Gatesy, Stephen M., Kevin M. Middleton, Farish A. Jenkins, Jr., and Neil H. Shubin, 1999, Three-dimensional preservation of foot movements in Triassic theropod dinosaurs: *Nature*, 399, pp. 141–144.

Gaudry, A., 1890, *Les enchainements du monde animal dans les temps géologiques: Fossiles secondaires*. Paris: Masson, 323 pages.

Gauffre, Francis-Xavier, 1993*a*, The most recent melanorosaurid (Saurischia, Prosauropoda), Lower Jurassic of Lesotho, with remarks on the prosauropod phylogeny: *Neus Jahrbuch für Geologie und Paläontologie, Monatshefte*, 11, pp. 648–654, Stuttgart.

_____, 1993*b*, Biochronostratigraphy of the lower Elliot Formation (southern Africa) and preliminary results on the Maphutseng dinosaur (Saurischia: Prosauropoda) from the same formation of Lesotho: *Bulletin of the New Mexico Museum of Natural History and Science*, 3, pp. 147–149.

_____, 1993*c*, The prosauropod *Azendohsaurus laaroussi* from the Upper Triassic of Morocco: *Palaeontology*, 36, pp. 897–908.

_____, 1995, Phylogeny of prosauropod dinosaurs: *Journal of Vertebrate Paleontology*, 15 (Supplement to Number 3), Abstracts of Papers, Fifty-fifth Annual Meeting, p. 31A.

Gauthier, Jacques A., 1986, Saurischian monophyly and the origin of birds, *in*: Kevin Padian, editor, The Origin of Birds and the Evolution of Flight. *Memoirs of the California Academy of Sciences*, 8, pp. 1–55.

Gauthier, Jacques A., and Alan D. Gishlick, 2000, Re-examination of the manus of *Compsognathus* and its relevance to the original morphology of the coelurosaur manus: *Journal of Vertebrate Paleontology*, 20 (Supplement to Number 3), Abstracts of Papers, Sixtieth Annual Meeting, p. 43A.

Gauthier, Jacques A., and Kevin Padian, 1985, Phylogenetic, functional, and aerodynamic analyses of the origin of birds and their flight, *in*: Max K. Hecht, John H. Ostrom, G. Viohl and Peter Wellnhoffer, editors, *The Beginnings of Birds*. Eichstätt: Freunde des Jura-Museums, pp. 185–197.

Geist, Nicholas R., Arboreal Triassic archosaurs: re-evaluating the origin of birds: *Journal of Vertebrate Paleontology*, 19 (Supplement to Number 3), Abstracts of Papers, Fifty-ninth Annual Meeting, p. 47A.

Geist, Nicholas R., and Willem J. Hellenius, 1998, Information on dinosaur respiratory physiology and the structure of the theropod lung: *Journal of Vertebrate Paleontology*, 18 (Supplement to Number 3), Abstracts of Papers, Fifty-eighth Annual Meeting, p. 46A.

Geist, Nicholas R., Terry D. Jones, and John A. Ruben, 1997, Implications of soft tissue preservation in the compsognathid dinosaur, *Sinosauropteryx*: *Journal of Vertebrate Paleontology*, 17 (Supplement to Number 3), Abstracts of Papers, Fifty-seventh Annual Meeting, p. 48A.

Giffin, Emily B., 1989, Notes on pachycephalosaur (Ornithischia): *Journal of Paleontology*, 63 (4), pp. 525–529.

Gillette, David D., 1991, *Seismosaurus halli*, gen. et sp. nov., a new sauropod dinosaur from the Morrison Formation (Upper Jurassic/Lower Cretaceous) of New Mexico, USA: *Journal of Vertebrate Paleontology*, 11 (4), pp. 417–433.

_____, 1994, Seismosaurus, *the Earth Shaker*. New York: Columbia University Press, 205 pages.

Gillette, David D., J. W. Betchel, and P. Betchel, 1990, Gastroliths of a sauropod dinosaur from New Mexico: *Journal of Vertebrate Paleontology*, 9 (Supplement to Number 3), Abstracts of Papers, Forty-ninth Annual Meeting, p. 24A.

Gillette, David D., Donald L. Wolberg, and Adrian P. Hunt, 1986, *Tyrannosaurus rex* from the McRae Formation (Lancian, Upper Cretaceous), Elephant Butte Reservoir, Sierra County, New Mexico: *New Mexico Geological Society, Guidebook*, 37, pp. 235–238.

Gilmore, Charles Whitney, 1905, The mounted skeleton of *Triceratops prorsus*: *Proceedings of the United States National Museum*, 29 (1426), pp. 433–435.

_____, 1909, The mounted skeletons of *Camptosaurus* in the United States National Museum: *Ibid.*, 41 (1878), pp. 687–696.

_____, 1914, Osteology of the armored Dinosauria in the United States National Museum, with special reference to the genus *Stegosaurus*: *Memoirs of the United States National Museum*, 89, pp. 1–316.

_____, 1916*a*, Contributions to the geology and paleontology of San Juan County, New Mexico. 2. Vertebrate faunas of the Ojo Alamo, Kirtland, and Fruitland Formations: *United States Geological Survey Professional Paper*, 119, pp. 1–71.

_____, 1916*b*, Vertebrate faunas of the Ojo Alamo, Kirtland and Fruitland formations: *Ibid.*, 98Q, pp. 279–308.

_____, 1919, Reptilian faunas of the Torrejon, Puerco, and under lying Upper Cretaceous Formations of San Juan County, New Mexico: *Ibid.*, 119, pp. 1–68.

_____, 1920, Osteology of the carnivorous dinosauria in the United States National Museum, with special reference to the genera *Antrodemus* (*Allosaurus*) and *Ceratosaurus*: *Bulletin of the United States National Museum*, 110, pp. 1–154.

_____, 1922, A new sauropod dinosaur from the Ojo Alamo Formation of New Mexico: *Smithsonian Miscellaneous Collections*, 72 (2), pp. 1–9.

_____, 1925, A nearly complete articulated skeleton of *Camarasaurus*, a saurischian dinosaur from the Dinosaur National Monument: *Bulletin of the United States National Museum*, 81 (18), pp. 1–21.

_____, 1930, On dinosaurian reptiles from the Two Medicine Formation of Montana: *U.S. National Museum Proceedings*, 77, pp. 1–39.

_____, 1933*a*, Two new dinosaurian reptiles from Mongolia with notes on some fragmentary specimens: *American Museum Novitates*, 679, pp. 1–20.

_____, 1933*b*, On the dinosaurian fauna of the Iren Dabasu Formation: *Bulletin of the American Museum of Natural History*, 67, pp. 23–78.

_____, 1939, Ceratopsian dinosaurs from the

Two Medicine Formation, Upper Cretaceous of Montana: *Proceedings of the United States National Museum*, 87 (3066), pp. 1–18.

_____, 1946*a*, A new carnivorous dinosaur from the Lance Formation of Montana: *Smithsonian Miscellaneous Collections*, 106 (13), 19 pages.

_____, 1946*b*, Reptilian fauna of the North Horn Formation of Central Utah: *United States Geological Survey Professional Paper*, 210-C, pp. 29–53.

Gimenéz, Olga del Valle, 1996, Hallazgo de impronta de piel de un dinosaurio sauropodo en la provincia del Chubut: *Ameghiniana*, 33 (4), p. 465.

_____, 2000, The discovery of skin impressions of a sauropod dinosaur in Chubut Province, Argentina: *Journal of Vertebrate Paleontology*, 20 (Supplement to Number 3), Abstracts of Papers, Sixtieth Annual Meeting, p. 44A.

Gishlick, Alan D., 2000, An evaluation of the climbing capability of basal maniraptors: *Journal of Vertebrate Paleontology*, 20 (Supplement to Number 3), Abstracts of Papers, Sixtieth Annual Meeting, p. 44A.

Glut, Donald F., 1997, *Dinosaurs: The Encyclopedia*. Jefferson, North Carolina: McFarland & Company, xi, 1076 pages.

_____, 1999, *Dinosaurs: The Encyclopedia, Supplement 1*. Jefferson, North Carolina: McFarland & Company.

Godefroit, Pascal, Dong Zhi-Ming, Pierre Bultynck, Li Hong, and Feng Lu ["with the collaboration in the field of Shang Chang-Yong, Guo Dian-Yong, Dong Yu-Long, Sun Yan, Zhang Zhe-Min, Hugo De Potter, Georges Lenglet, Thierry Smith, and Eric Dermience"], 1998, Sino-Belgian Cooperation Program, "Cretaceous dinosaurs and mammals from Inner Mongolia," Part 1: New *Bactrosaurus* (Dinosauria: Euhadrosauria) material from Iren Dabasu (Inner Mongolia, P. R. China): *The Sino-Belgian Dinosaur Expedition in Inner Mongolia, Bulletin, Institut Royal des Sciences Naturelle de Belgique*, Suppl. 68, pp. 1–70.

Godefroit, Pascal, Xavier Pereda Suberbiola, Li Hong, and Dong Zhi-Ming, 1999, A new species of the ankylosaurid dinosaur *Pinacosaurus* from the Late Cretaceous of Inner Mongolia (P.R. China): *Bulletin van het Koninkluk Belgisch Instituut voor Natuurwetenschappen, Aardwetenschappen*, 69-Supp, B, pp. 17–36.

Godefroit, Pascal, Shuqin Zan, and Liyong Jin, 2000, *Charonosaurus jiayinensis* n.g., n.sp., a lambeosaurine dinosaur from the late Maastrichtian of northeastern China: *Comtes Rendu des Séances de l'Académie des Sciences, Paris, de la Terre et des Planètes*, 330, pp. 875–882.

Godfrey, Stephen J., and Robert B. Holmes, 1995, Cranial morphology and systematics of *Chasmosaurus* (Dinosauria: Ceratopsidae) from the Upper Cretaceous of western Canada: *Journal of Vertebrate Paleontology*, 15, pp. 726–742.

Golder, Werner, and Andreas Christain, 1999, CT findings in extremity bones of Tendaguru saurians: *Investigative Radiology*, in review.

Gomani, Elizabeth M., 1998, Sauropod cranial elements from Malawi: implications of titanosaurid cranial morphology, *Journal of Vertebrate Paleontology*, 18 (Supplement to Number 3), Abstracts of Papers, Fifty-eighth Annual Meeting, p. 46A.

Gomani, Elizabeth M., Louis Jacobs, and David Winkler, 1995, Comments on vertebral structure in African Titanosaurian sauropods: *Second Symposium on Gondwanan Dinosaurs*, National Science Museum, Tokyo, A, p. 21.

_____, 1999, Comparison of the African titanosaurian, *Malawi saurus*, with a North American Early Cretaceous sauropod, *in*: Yukimitsu Tomida, Thomas R. Rich, and Patricia Vickers-Rich, editors, *Proceedings of the Second Gondwanan Dinosaur Symposium*. Japan: National Science Museum Monographs, p. 15.

González Riga, Bernardo J., 1999, Descripción preliminar de un neuvo Titanosauria (Dinosauria: Saurischia) del Cretácico Superior de la provincia de Mendoza, Argentina: *Ameghiniana, Revista de la Asociación Paleontológica Argentina, (Resúmenes)*, 36 (Número 4-Suplemento), p. 13R.

González Riga, Bernardo J., and Silvio Casadóo, 2000, Primer registro de Dinosauria (Ornithischia, Hadrosauridae) en la provincia de la Pampa (Argentina) y sus implicancias paleo biogeográficas: *Ameghiniana*, 37 (3), pp. 341–351.

Good, John M., Theodore E. White, and Gilbert F. Stucker, 1958, *The Dinosaur Quarry*. Washington, D.C.: National Park Service, 47 pages.

Goodwin, Mark B., Emily A. Buchholtz, and Rolfe E. Johnson, 1998, Cranial anatomy and diagnosis of *Stygimoloch spinifer* (Ornithischia: Pachycephalosauria) with comments on cranial display structures in agonistic behavior: *Journal of Vertebrate Paleontology*, 18 (2), pp. 363–375.

Goodwin, Mark B., William A. Clemens, J. Howard Hutchinson, Craig B. Wood, Michael S. Zavada, Anne Kemp, Christopher J. Duffin, and Charles R. Schaff, 1999, Mesozoic continental vertebrates with associated palynostratigraphic dates from the northwestern Ethiopian plateau: *Journal of Vertebrate Paleontology*, 19 (4), pp. 728–741.

Gould, Stephen Jay, 1977, *Ontogeny and Phylogeny*. Cambridge, Massachusetts: Harvard University Press, 501 pages.

Gradstein, F. M., S. P. Jansa, S. P. Srivastava, M. A. Williamson, G. B. Carter, and B. Stam, 1990, Aspects of North Atlantic Paleooceanography, *in*: M. K. Keen and G. L. Williamson, editors, *Geology of the Continental Margin of Eastern Canada*, Geological Society of America, The Geology of North America, I-1, pp. 353–388.

Gradziński, R., Zofia Kielan-Jaworowska, and Teresa Maryańska, 1977, Upper Cretaceous Djadokhta, Barun Goyot and Nemegt formations of Mongolia, including remarks on previous subdivisions: *Acta Palaeontologica Polonica*, 27, pp. 281–318.

Gregory, William K., 1918, Note on the morphology and evolution of the femoral trochanters in reptiles and mammals: *American Museum of Natural History Bulletin*, 38, pp. 528–538.

Grigorescu, Dan, 1984*a*, New tetrapod groups in the Maastrichtian of the Hateg Basin: Coelurosaurians and multituberculates, *in*: W.-E. Reif and F. Westphal, editors, *Third Symposium on Mesozoic Terrestrial Ecosystems, Short Paper*, pp. 99–104, Tübingen.

_____, 1984*b*, New paleontological data on the dinosaur beds from the Hateg Basin: *Special Volume, 75 Years of the Laboratory of Paleontology, University of Bucharest*, pp. 111–118.

_____, J.-L. Hartenberger, C. Radulescu, Samson, and J. Sudre, 1985, Découverte de Mammifères et Dinosaures dans le Crétace supérieur de Pui (Roumanie): *Comptes Rendus des Séances de l'Academie des Sciences*, II, 301 (19), pp. 1365–1368.

Grigorescu, Dan, and E. Kessler, 1981, A new specimen of *Elopteryx nopcsai* from the dinosaurian beds of Hateg Basin: *Revue Roumaine de Géologie, Géophysique et Géographie, Géologie*, 24, pp. 171–175.

Greppin, J. B., 1870, Descríption géologiques du Jura bernois et de quelques districts adjacents: *Beitraege Geologie Karte Schweiz*, 5 (8), 357 pages.

Guan, Jian, J. Keith Rigby, and Zheng, Liang, 1999, A new dinosaur assemblage in the Chuanjie Basin of Yunnan, China: *The Album of Research Manuscripts for the 7th Conference of Chinese Society of Vertebrate Paleontology*, pp. 2–3.

Gümbel, C. W., 1865, Die geognostischen Verhältnisse des fränkischen Triasgebietes: *Bavaria*, 4 (11), pp. 3–77, Münnchen.

Gunga, Hanns-Christian, Karl Kirsch, F. Baartz, Lothar Röcker, Wolf-Dieter Heinrich, W. Lisowski, Albert Wiedemann, and Jörg Albertz, 1995, New data on the dimensions of *Brachiosaurus brancai* and their physiological implications: *Naturwissenschaften*, 82 (4), pp. 189–192.

Gunga, Hanns-Christian, Karl Kirsch, Jörn Rittweger, Lothar Röcker, Andrew Clarke, Jörg Albertz, Albert Wiedmann, Sascha Mokry, Tim Suthau, Aloys Wehr, Wolf-Dieter Heinrich, and Hans-Peter Schultze, 1999, Body size and body volume distribution in two sauropods from the Upper Jurassic of Tendaguru (Tanzania): *Mitteilungen Museum für Naturkunde der Humboldt-Universität zu Berlin, Geowissenschaftliche Reihe*, 2, pp. 91–102.

Gwinner, M. P., 1980, Eine einheitliche Gliederung des Keupers (Germanische Trias) in Südeutschland: *Neus Jahrbuch für Geologie und Paläontologie, Monatshefte*, 1980, pp. 229–234, Stuttgart.

Halstead, L. Beverly, and Halstead, Jenny, 1981, *Dinosaurs*. Poole, England: Blandford Press, 170 pages.

Hand, Jordon D., and Robert T. Bakker, 2000, Implications of the functional morphology of a new allosaurid forearm from the Como Bluffs: *Publications in Paleontology* [The Florida Symposium on Dinosaur Bird Evolution, abstracts of papers], 2, pp. 17–18.

Hanna, Rebecca R. (Lewis), Jeff W. LaRock, and John R. Horner, 1999, Pathological brachylophosaur bones from the Upper Cretaceous Judith River Formation, northeastern Montana: *Journal of Vertebrate Paleontology*, 19 (Supplement to Number 3), Abstracts of Papers, Fifty-ninth Annual Meeting, p. 49A.

Happ, John W., and Christopher M. Morrow, 1996, Separation of *Triceratops* (Dinosauria: Ceratopsidae) into two allopatric species by cranial morphology: *Journal of Vertebrate Paleontology*, 16 (Supplement to Number 3), Abstracts of Papers, Fifty-sixth Annual Meeting, p. 40A.

_____, 1999, A ceratopsian egg tooth as hatching strategy: *Journal of Vertebrate Paleontology*, 19 (Supplement to Number 3), Abstracts of Papers, Fifty-ninth Annual Meeting, p. 49A.

_____, 2000, Evidence of soft tissue associated with nasal and supraorbital horn cores, rostral and exoccipital of *Triceratops*: *Journal of Vertebrate Paleontology*, 20 (Supplement to Number 3), Abstracts of Papers, Sixtieth Annual Meeting, p. 47A.

Hargens, A. R., R. W. Millard, K. Petterson, and K. Johansen, 1987, Gravitational haemodynamics and oedema prevention in the giraffe: *Nature*, 329, pp. 59–60.

Harris, Jerald Davic, 1998, A reanalysis of *Acrocanthosaurus atokensis*, its phylogenetic status, and paleobiogeographic implications, based on a new specimen from Texas: *New Mexico Museum of Natural History and Science, Bulletin 13*, 75 pages.

Harrison, C. J. O., and Cyril A. Walker, 1975, The Bradycnemidae, a new family of owls from the Upper Cretaceous of Romania: *Palaeontology*, 18 (3), pp. 563–570.

Hartman, Scott, 2000, Primary and caudal feathers as locomotory adaptations of maniraptoran theropods: *Journal of Vertebrate Paleontology*, 20 (Supplement to Number 3), Abstracts of Papers, Sixtieth Annual Meeting, p. 47A.

Hasegawa, Yoshikazu, Makoto Manabe, Tomoki Kase, Shuichi Nakajima, and Yuji Takakuwa, 1999, An ornithomimid vertebra from the Early Cretaceous Sebayashi Formation, Sanchu Terrane, Gunma Prefecture, Japan: *Bulletin of the Gunma Museum of Natural History*, 3, pp. 1–6.

Hatcher, John Bell, 1901, *Diplodocus* (Marsh): its osteology, taxonomy, and probable habits, with a restoration of the skeleton: *Memoirs of the Carnegie Museum*, 1 (1), pp. 1–61.

_____, 1903, A new sauropod dinosaur from the Jurassic of Colorado: *Proceedings of the Biological Society of Washington*, 16, pp. 1–2.

Haubold, Hartmut, 1990, Ein neuer Dinosaurier (Ornithischia, Thyreophora) aus dem unteren Jura des Nördlichen Mitteleuropa: *Revue de Paléobiologie*, 9 (1), pp. 149–177.

Haughton, Sidney H., 1924, The fauna and stratigraphy of the Stormberg Series: *Annals of the South African Museum*, 12 (8), pp. 323–497, Cape Town.

Hay, Oliver P., 1902, *Bibliography and Catalogue of the Fossil Vertebrates of North America*, Bulletin USGS 179, 868 pages.

_____, 1908, On certain genera of carnosaurian dinosaurs, with special reference to *Ceratosaurus nasicornis* (Marsh): *Proceedings of the United States National Museum*, 35, pp. 351–366.

_____, 1930, *Second Bibliography and Catalogue of the Fossil Vertebrates of North America, Volume 2*. Washington, D.C.: Carnegie Institute Publication, 390, 1074 pages.

He Xinlu, Li Kui, Cai Kaiji, and Gao Yuhui, 1984, *Omeisaurus tianfuensis*— a new species of *Omeisaurus* from Dashanpu, Zigong, Sichuan: *Journal of Chengdu College of Geology*, 2, pp. 15–32.

He Xinlu, Wang Changsheng, Liu Shangzhong, Zhou Fengyun, Liu Tuqiang, Cai Kaiji, and Dai Bing, 1998, A new sauropod dinosaur from the Early Jurassic in Gongxian County, South Sichuan: *Acta Geologica Sichuan*, 18 (1), pp. 1–6 (in Chinese, reprinted and translated by Will Downs, Bilby Research Center, Northern Arizona University, 1999).

Head, Jason J., 1998, A new species of basal hadrosaurid (Dinosauria, Ornithischia) from the Cenomanian of Texas: *Journal of Vertebrate Paleontology*, 18 (4), pp. 718–738.

_____, 1999, Reassessment of the systematic position of *Eolambia caroljonesa* (Dinosauria, Iguanodontia) and the North American iguanodontian record: *Journal of Vertebrate Paleontology*, 19 (Supplement to Number 3), Abstracts of Papers, Fifty-ninth Annual Meeting, p. 50A.

Heaton, M. J., 1972, The palatal structure of some Canadian Hadrosauridae (Reptilia: Ornithischia): *Canadian Journal of Earth Science*, 9, pp. 185–205.

Heckert, Andrew B., and Spencer G. Lucas, 1999, Global correlation and chronology of Triassic tetrapods (Archosauria: Dinosauria): *Albertiana*, 23, pp. 22–35.

Heckert, Andrew B., Spencer G. Lucas, S. G. Rhinehart, and Jerry D. Harris, 2000, Preliminary description of coelophysoids from the Upper Triasic (Revueltian: early-mid Norian) Snyder quarry, north-central New Mexico: *New Mexico Museum of Natural History and Science, Bulletin 17*.

Heckert, Andrew B., Spencer G. Lucas, and Robert M. Sullivan, 2000, Triassic dinosaurs in New Mexico, *in*: Spencer G. Lucas and Andrew B. Heckert, editors, *Dinosaurs of New Mexico*. Albuquerque: New Mexico Museum of Natural History and Science, Bulletin 17, pp. 17–26.

Heckert, Andrew B., Kate E. Zeigler, Spencer G. Lucas, Larry F. Rinehart, and Jerald D. Harris, 2000, Preliminary description of soelophysoids (Dinosauria: Theropoda) from the Upper Triassic (Revueltian: Early–Mid Norian) Snyder quarry, north-central New Mexico, *in*: Spencer G. Lucas and Andrew B. Heckert, editors, *Dinosaurs of New Mexico*. Albuquerque: New Mexico Museum of Natural History and Science, Bulletin 17, pp. 27–32.

Heerden, Jacques van, and Peter M. Galton, 1997, The affinities of *Melanorosaurus*— a Late Triassic prosauropod dinosaur from South Africa: *Neus Jahrbuch für Geologie und Paläontologie, Monatshefte*, 1997, pp. 39–55, Stuttgart.

Heinrich, R. E., C. B. Ruff, and David B. Weishampel, 1993, Femoral ontogeny and locomotor biomechanics of *Dryosaurus lettowvorbecki* (Dinosauria, Iguanodontia): *Zoological Journal of the Linnean Society*, 108, pp. 179–196.

Heinrich, Wolf-Dieter, 1999, The taphonomy of dinosaurs from the Upper Jurassic of Tendaguru (Tanzania) based on field sketches of the German Tendaguru Expedition (1909–1913): *Mitteilungen Museum für Naturkunde der Humboldt-Universität zu Berlin, Geowissenschaftliche Reihe*, 2, pp. 25–61.

Helmdach, F.-F., 1973–74, A contribution to the stratigraphical subdivision on non-Marine sediments of the Portuguese Upper Jurassic: *Comunicacoes dos Servicos Geológicos de Portugal*, 57, pp. 1–21.

Henderson, Donald. M., 1999, Estimating the masses and centers of mass of extinct animals by mathematical slicing: *Palaeobiology*, 25 (1), pp. 88–106.

Hengst, Richard A., 1998, Testint the hepatic piston and costal models of theropod breathing: *Journal of Vertebrate Paleontology*, 18 (Supplement to Number 3), Abstracts of Papers, Fifty-eighth Annual Meeting, p. 50A.

Hennig, Edwin, 1912*a*, Die Entstehung der Dinosasaurier-Lager: *Sitzungsberichte der Gesellschaft naturforschender Freunde*, 1912, pp. 137–142, Berlin.

_____, 1912*b*, *Am Tendaguru: Leben und Wirken einer deutschen Forschungsexpedition zur Ausgrabung vorweltlicher Riesen saurier in Deutsch-Ostafrika*. Stuttgart: E. Schweitzerbart'sche Verlagsbuchhandlung, 151 pages.

_____, 1914*a*, Beiträge zur Geologie und Stratigraphie Deutsch-Ostrafrikas. I. Geologisch-stratigraphische Beobachtungen im Küstengebiet des südlichen Deutsch-Ostafrika. II. Geologisch-stratigraphische Beobachtungen im Gabiete der Jura-Ablagerungen an der Deutsch-Ostafrikanischen Zentralbahn: *Archiv für Biontologie*, 3 (3), pp. 1–72.

_____, 1914*b*, Die Invertebraten-Fauna der Saurierschichten am Tendaguru: *Ibid.*, 3 (4), pp. 155–185.

_____, 1915, *Kentrosaurus aethiopicus*. Die Stegosaurier-Funde von Tendaguru Deutsch-Ostavrika. II. Historisch-systematische Einführung: *Palaeontographica Supplement*, 7, pp. 103–253.

_____, 1925, *Kentrurosaurus aethiopicus*: die Stegosaurieurfunde vom Tendaguru, Deutsch-Ostafrika: *Palaeontographica, Supplement VII*, 1 Reihe, Teil 1, Lieferung 2, pp. 100–253.

Hernández-Rivera, René, and Carlos Delgado de Jesús, 1999, An endocast of a ceratopsian brain cavity from the Cerro del Pueblo Formation (Latest Campanian, Upper Cretaceous), Coahula, Mexico: *Journal of Vertebrate Paleontology*, 19 (Supplement to Number 3), Abstracts of Papers, Fifty-ninth Annual Meeting, pp. 50A–51A.

_____, 2000, Hadrosaurs skin impression and associated skeletal remains from Cerro del Pueblo Fm. (uppermost Campanian) southeastern Coahuila, Mexico: *Ibid.*, 20 (Supplement to Number 3), Abstracts of Papers, Sixtieth Annual Meeting, p. 48A.

Hickey, L. J., 1981, Land plant evidence compatible with gradual, not catastrophic, change at the end of the Cretaceous: *Nature*, 292, pp. 529–531.

Hicks, J. W., and C. G. Farmer, 1998, Gas exchange in vertebrate lungs: the dinosaur-avian: *Journal of Vertebrate Paleontology*, 18 (Supplement to Number 3), Abstracts of Papers, Fifty-eighth Annual Meeting, p. 50A.

Hill, Robert V., 1999, Phylogenetic relationships among Ankylosauria: an analysis of cranial characters: *Journal of Vertebrate Paleontology*, 19 (Supplement to Number 3), Abstracts of Papers, Fifty-ninth Annual Meeting, p. 51A.

Hitchcock, Edward, 1858, Ichnology: a report on the sandstone of the Connecticut Valley, especially its fossil footmarks, Boston 1858, 220 pages.

Hoffet, Josué-Heilmann, 1942, Description de quelques ossements de Titanosauriens du Sénonien du Bas-Laos: *Comtes Rendu des Séances Conseil Recherches des Sciences, Indochina*, pp. 51–57.

_____, 1944, Description des ossements les plus caractéristiques appartenant à des Avipelviens du Sénonien du Bas-Laos: *Bulletin, Conseil Recherches des Sciences, Indochina*, pp. 179–186.

Hohnke, L. A., 1973, Haemodynamics in the Sauropoda: *Nature*, 244, pp. 309–310.

Holden, Constance, 2000*a*, Florida meeting shows perils, promise of dealing for dinos: *Science*, 288, pp. 238–239.

_____, 2000*b*, Dinos and turkeys: connected by DNA?: *Ibid.*, p. 238.

Holl, Friedrich, 1829, *Handbuch der Petrefactenkunde*, part 1. Dresden, Germany: Hilscher, 115 pages.

Holland, William J., 1924, The skull of *Diplodocus*: *Memoirs of the Carnegie Museum*, 9, pp. 379–403.

Holmes, Robert, Kieren Shepherd, and Catherine A. Forster, 1999, An unusual chasmosaurine ceratopsid from Alberta: *Journal of Vertebrate Paleontology*, 19 (Supplement to Number 3), Abstracts of Papers, Fifty-ninth Annual Meeting, p. 52A.

Holtz, Thomas R., Jr., 1994, The phylogenetic position of the Tyrannosauridae: implications for theropod systematics: *Journal of Paleontology*, 68 (5). pp. 1100–1117.

_____, 1995, A new phylogeny of the Theropoda: *Journal of Vertebrate Paleontology*, 15 (Supplement to Number 3), Abstracts of Papers, Fifty-fifth Annual Meeting, p. 35A.

_____, 1996, Phylogenetic taxonomy of the Coelurosauria (Dinosauria: Theropoda): *Ibid.*, 70, pp. 536–538.

_____, 1997, Preliminary phylogenetic analysis of the Tyrannosauridae (Theropoda: Coelurosauria): *Journal of Vertebrate Paleontology*, 17 (Supplement to Number 3), Abstracts of Papers, Fifty-seventh Annual Meeting, p. 53A.

_____, 1998, Large theropod comparative cranial function: a new "twist" for tyrannosaurs: *Ibid.*, 18 (Supplement to Number 3), Abstracts of Papers, Fifty-eighth Annual Meeting, p. 51A.

_____, 1999, Consensus and conflict in current coelurosaur cladograms: *Ibid.*, 19 (Supplement to Number 3), Abstracts of Papers, Fifty-ninth Annual Meeting, p. 52A.

_____, A new phylogeny of carnivorous dinosaurs: *Gaia*, 15, pp. 5–62.

Holtz, Thomas R., Jr., and Kevin Padian, 1995, Definition and diagnosis of Theropoda and related taxa: *Journal of Vertebrate Paleontology*, 15 (Supplement to Number 3), Abstracts of Papers, Fifty-fifth Annual Meeting, p. 35A.

Hopson, James A., 1975, The evolution of cranial display structures in hadrosaurian dinosaurs: *Paleobiology*, 1, pp. 21–43.

Horner, John R., 1984, The nesting behavior of dinosaurs: *Scientific American*, 250 (4), pp. 130–137.

_____, 1990, Evidence of diphlyetic origination of the hadrosaurian (Reptilia: Ornithischia) dinosaurs, *in*: Kenneth Carpenter and Philip J. Currie, editors, *Dinosaur Systematics: Approaches and Perspectives*. Cambridge and New York: Cambridge University Press, pp. 179–182.

_____, 1992, Cranial morphology of *Prosaurolophus* (Ornithischia: Hadrosauridae) with descriptions of two new hadrosaurid species and an evaluation of hadrosaurid phylogenetic rela tionships: *Museum of the Rockies Occasional Paper*, 2, 120 pages.

_____, 1999, Egg clutches and embryos of two hadrosaurian dinosaurs: *Journal of Vertebrate Paleontology*, 19 (4), pp. 607–611.

Horner, John R., and V. Clouse, 1998, An undisturbed clutch of hadrosaur eggs from the Judith River Formation of Montana: *1st International Meeting on Dinosaur Paleobiology, Museu Nacional de Historia Natural, Universidade de Lisboa*, pp. 22–25.

Horner, John R., and Philip J. Currie, 1994, Embryonic and neonatal morphology and ontogeny of a new species of *Hypacrosaurus* (Ornithischia: Lambeosauridae) from Montana and Alberta, *in*: Kenneth Carpenter, Karl F. Hirsch, and John R. Horner, editors, *Dinosaur Eggs and Babies*. New York: Cambridge University Press, pp. 312–336.

Horner, John R., and James Gorman, 1988, *Digging Dinosaurs*. New York: Workman Publishing, 210 pages.

Horner, John R., Armand de Ricqlés, and Kevin Padian, 2000, Long bone histology of the hadrosaurid dinosaur *Maiasaura peeblesorum*: growth dynamics and physiology based on an ontogenetic series of skeletal elements: *Journal of Vertebrate Paleontology*, 20 (1), pp. 115–129.

Horner, John R., and Mark B. Goodwin, 1998, Did pachycephalosaurs really head-butt? An osteohistogenic cranial analysis: *Journal of Vertebrate Paleontology*, 18 (Supplement to Number 3), Abstracts of Papers, Fifty-eighth Annual Meeting, p. 52A.

Horner, John R., and Don Lessem, 1993, *The Complete* T. rex. New York: Simon and Schuster, 239 pages.

Horner, John R., and Robert Makela, 1979, Nest of juveniles provides evidence of family structure among dinosaurs: *Nature*, 282 (5736), pp. 296–298.

Horner, John R., Kevin Padian, and Armand de Ricqlès, 1999, Osteohistology of some embryonic and perinatal archosaurs: phylogenetic and behavioral implications for dinosaurs: *Journal of Vertebrate Paleontology*, 19 (Supplement to Number 3), Abstracts of Papers, Fifty-ninth Annual Meeting, p. 51A.

Horner, John R., and David B. Weishampel, 1988, A embryological study of two ornithischian dinosaurs: *Nature*, 332, pp. 256–257.

Hotton, Nicholas, III, 1980, An alternative to dinosaur endothermy: Roger D. K. Thomas and Everett C. Olson, editors. *A Cold Look at the Warm-Blooded Dinosaurs*. Boulder, Colorado: AAAS Selected Symposium, Westview Press, 28, pp. 311–350.

Hou L. H., Chao [Zhou] S. W., and Chu [Chao], 1976, New discovery of sauropod dinosaurs from Szechuan: *Vertebrata-Palasiatica*, XIV, 3, pp. 160–165.

Hou, L. H., and J. Y. Zhang, 1993, A new fossil bird from Lower Cretaceous of China: *Vertebrata PalAsiatica*, 31 (3), pp. 217–224 (in Chinese with English summary).

Howse, Straffor C. B., and Andrew R. Milner, 1993, *Ornithodesmus*— a maniraptoran theropod dinosaur from the Lower Cretaceous of the Isle of Wight, England: *Palaeontology*, 36 (2), pp. 425–437.

Hu Cheng-Chin, 1973, A new hadrosaur from the Cretaceous of Chuncheng, Shantung: *Acta Geologica Sinica*, 2, pp. 179–202.

Hu Show-yung, 1964, Carnosaurian remains from Alashan, Inner Mongolia: *Vertebrata PalAsiatica*, 8 (1), pp. 42–63.

Huene, Friedrich von, 1902, Überischt über die Reptilien der Trias: *Geologische und Paläontologische Abhandlungen, Neu Folge*, 8, pp. 97–156.

_____, 1905*a*, Über die Nomenklatur von *Zanclodon*: *Neus Jarbuch für Mineralogie, Geologie, und Paläontologie*, 1905, pp. 10–12, Stuttgart.

_____, 1905*b*, Über die Trias-Dinosaurier Europas: *Zeitschrift der Deutschen Geologischen Gesellschaft, Mh.*, 57, pp. 345–349, Berlin.

_____, 1906, Uber die Dinosaurier der aussereuropäischen Trias: *Geologische und Paläeontologische Abhandlungen*, 8, pp. 97–156, Jena.

_____, 1907–08, Die Dinosaurier der europäischen Triasformation mit Berucksichtigung der europaischen Vorkommisse: *Ibid.*, Supplement 1, 419 pages.

_____, 1908*a*, Die Dinosaurier der europäischen Triasformationen mit Berücksichtigung der aussereuropäischen Vorkommnisse: *Ibid.*, Supplement 1, pp. 1–419.

_____, 1908*b*, On phytosaurian remains from the Magnesian Conglom erate of Bristol (*Rileya platyodon*): *Annals and Magazine of Natural History*, Series 8, 1, pp. 229–230.

_____, Huene, 1914, Saurischia et Ornithischia Triadica ("Dinosauria" Triadica), *in*: *Fossilium Catalogus I. Animalia*, 4, pp. 1–21.

_____, 1926, Volstandige Osteologie eines Plateosauriden aus der schwäbischen Trias: *Neus Jahrbuch Fuer Geologie und Paläontologie Abhandlungen*, 15, pp. 129–179, Berlin.

_____, 1927, Short review of the present knowledge of the Sauropoda: *Memoirs of the Queensland Museum*, 9, pp. 121–126.

_____, 1929, Los Saurisquios y Ornitisquios del Cretaceo Argentina: *Annales Museo de La Plata*, 3, Serie 2a., 196 pages.

_____, 1932, Die fossile Reptil-Ordnung Saurischia, ihre Entwicklung und Geschichte: *Monographien zur Geologie und Palaeontologie*, series 1, 4, 361 pages.

_____, 1934, Ein neuer Coelurosaurier in der thüringischen Trias: *Paläeontologische Zeitschrift*, 1935, 16, pp. 10–170.

_____, 1942, *Die fossilen Reptilien des süamerikanischen. Ergebnissen der Sauriergrabungen in Südbrasilien 1928/29*. Munich: Beck'sche Verlagbuchhandlung, viii, 332 pages.

_____, 1956, *Paläontologie und Phylogenie der Niederen Tetrapoden*. Jena, Germany: Gustav Fisher, xii, pp. 103–253.

_____, 1958, Pre-Tertiary saurians of China: *Vertebrata PalAsiati ca*, 2 (4), pp. 201–207.

Huene, Friedrich von, and Charles Alfred Matley, 1933, The Cretaceous Saurischia and Ornithischia of the Central Provinces of India: *Paleontologica Indica*, 21 (1), pp. 1–74.

Hunt, Adrian P., 1994, Vertebrate paleontology and biostratigraphgy of the Bull Canyon Formation (Chinle Group, Norian), east-central New Mexico with revisions of the families Metoposauridae (Amphibia: Temnospondyli) and Parasuchidae (Reptilia: Archosauria), Ph.D. dissertation, University of New Mexico, Albuquerque, 403 pages (unpublished).

_____, 1995, The dinosaur fauna of the Bull Canyon Formation (Upper Triassic) of east-central New Mexico and west Texas and the biogeography of Late Triassic dinosaurs: *New Mexico Geology*, 17, p. 16.

_____, 1996, A new clade of herrerasaur-like theropods from the Late Triassic of western North America: *Journal of Vertebrate Paleontology*, 16 (Supplement to Number 3), Abstracts of Papers, Fifty-sixth Annual Meeting, p. 43A.

Hunt, Adrian P., and Spencer G. Lucas, 1990, Paleontology and biochronology of the Petrified Forest Member of the Upper Triassic Chinle Formation near San Ysidro, Sandoval County, New Mexico: *New Mexico Journal of Science*, 30, pp. 17–26.

_____, 1991, *Rioarribasaurus*, a new name for a Late Triassic dinosaur from New Mexico (USA): *Paläontraphica Zeitschrift*, 65 (1/2), pp. 191–198.

_____, 1992. Stratigraphy, paleontology and the age of the Fruitland and Kirtland formations (Upper Cretaceous), San Juan Basin, New Mexico: *New Mexico Geological Society Guidebook*, 43, pp. 217–239.

_____, 1993*a*, Jurassic vertebrates of New Mexico, *in*: S. G. Lucas and J. Zidek, editors: *Vertebrate Paleontology in New Mexico: New Mexico Museum of Natural History and Science, Bulletin*, 2, pp. 71–75, Albuquerque.

_____, 1993*b*, Cretaceous vertebrates of New Mexico: *Ibid.*, pp. 77–91.

_____, 1994, Ornithischian dinosaurs from the Upper Triassic of the United States, *in*: Nicholas C. Fraser and Hans-Dieter Sues, editors, *In the Shadow of the Dinosaurs: Early Mesozoic Tetrapods*. Cambridge: University of Cambridge Press, pp. 225–241.

_____, 1995, Two Late Triassic vertebrate faunas at Petrified Forest National Park: *National Park Service Paleontological Research*, Technical Report NPS/NRPO/NRTR-95/16, pp. 89–93.

Hunt, Adrian P., and Spencer G. Lucas, 1992, Stratigraphy, paleontology and the age of the Fruitland and Kirtland formations (Upper Cretaceous), San Juan Basin, New Mexico: *New Mexico Geological Society, Guidebook*, 43, pp. 217–239.

_____, 1993, Cretaceous vertebrates of New Mexico: *New Mexico Museum of Natural History and Science, Bulletin*, 2, pp. 77–91.

Hunt, Adrian P., Spencer G. Lucas, Andrew B. Heckert, Robert M. Sullivan, and Martin G. Lockley, 1998, Late Triassic dinosaurs from the western United States: *Geobios*, 31 (4), pp. 512–531.

Hurum, Jo/rn H., and Philip J. Currie, 2000, The crushing bite of tyrannosaurids: *Journal of Vertebrate Paleontology*, 20 (3), pp. 619–614.

Hutchinson, John R., 2000, Hindlimb function in extinct theropod dinosaurs: integrating osteological, soft tissue, and biomechanical data: *Journal of Vertebrate Paleontology*, 20 (Supplement to Number 3), Abstracts of Papers, Sixtieth Annual Meeting, p. 50A.

Hutchinson, John R., and Kevin Padian, 1997*a*, Carnosauria, *in*: Philip J. Currie and Kevin Padian, editors, *Encyclopedia of Dinosaurs*. San Diego: Academic Press, pp. 94–97.

_____, 1997*b*, Tetanurae, *Ibid.*, p. 727.

Huxley, Thomas H., 1866, On the remains of large dinosaurian reptiles from the Stormberg Mountains, South Africa: *Geological Magazine*, 3, p. 363, London.

_____, 1867, On *Acanthopholis horridus*, a new reptile from the Chalk Marl: *Ibid.*, 4, pp. 65–67.

_____, 1870, On the classification of the Dinosauria, with observations on the Dinosauria of the Trias: *Quarterly Journal of the Geological Society of London*, 26, pp. 32–50.

Ishigaki, Shinubo, 1999, Abundant dinosaur footprints from Upper Cretaceous of Gobi Desert, Mongolia: *Journal of Vertebrate Paleontology*, 19 (Supplement to Number 3), Abstracts of Papers, Fifty-ninth Annual Meeting, p. 54A.

Jabo, Steven J., Peter Allen Kroehler, Arthur F. Andersen, and Ralph E. Chapman, 1999, The use of three-dimensional computer imaging and scale-model prototypes in the mounting of a cast of *Triceratops*: *Journal of Vertebrate Paleontology*, 19 (Supplement to Number 3), Abstracts of Papers, Fifty-ninth Annual Meeting, p. 54A.

Jacobs, Louis L., 1995, *Lone Star Dinosaurs*. College Station, Texas: Texas A&M Press, Number 22: Louise Lindsey Merrick Natural Environment Series, xiv, 160 pages.

Jacobs, Louis L., Dale A. Winkler, William R. Downs, and Elizabeth M. Gomani, 1993, New material of an Early Cretaceous titanosaurid sauropod dinosaur from Malawi: *Palaeontology*, 36 (3), pp. 523–534.

Jacobs, Louis L., Dale A. Winkler, and Phillip A. Murry, 1991, On the age and correlation of the Trinity mammals, Early Cretaceous of Texas, USA: *Newsletters on Stratigraphy*, 24, pp. 35–43.

Jacobs, Louis L., Dale A. Winkler, Phillip A. Murry, and J. M. Maurice, 1994, A nodosaurid scutelling from the Texas shore of the Western Interior Sea, *in*: Kenneth Carpenter, Karl F. Hirsch, and John R. Horner, editors, *Dinosaur Eggs and Babies*. Cambridge and New York: Cambridge University Press, pp. 337–346.

Jacobsen, Aase Roland, 1999, Significance of a tooth-marked *Saurornitholestes* dentary on the feeding behavior of tyrannosaurids: *Journal of Vertebrate Paleontology*, 19 (Supplement to Number 3), Abstracts of Papers, Fifty-ninth Annual Meeting, p. 55A.

Jacobsen, Aase Roland, and Michael J. Ryan, 1999, Taphonomic aspects of theropod tooth-marked bones from an *Edmontosaurus* bone bed (lower Maastrichtian), Alberta, Canada: *Journal of Vertebrate Paleontology*, 19 (Supplement to Number 3), Abstracts of Papers, Fifty-ninth Annual Meeting, p. 55A.

Jaekel, O., 1914, Über die Wirbeltierfunde in der oberon Trias von Halberstadt: *Palaeont. Z.*, I, pp. 155–215, Berlin.

Jain, Sohan L., T. S. Kutty, Tapan Roy-Chowdbury, and Sankar Chatterjee, 1979, Some characteristics of *Barapasaurus tagorei*, a sauropod dinosaur from the Lower Jurassic of Deccan, India: *Proceedings of the IV International Gondwana Symposium, Calcutta*, 1, pp. 204–216.

Janensch, Werner, 1914*a*, Ubersicht uber die Wirbeltierfauna der Tendaguru-Schichten: *Archiv Für Biontologie*, 3, pp. 81–110.

_____, 1914*b*, Paläontologische Ausgrabungen in Deutsch-Ostafrika: *Blätter der Mitglieder des wissenschaftlichen Centralver eins*, 1914, pp. 1–16, Stuttgart.

_____, 1914*c*, Die invertebraten-Fauna der Saurierschichten am Tendaguru: *Archiv für Biontologie*, 3 (4), pp. 155–185.

_____, 1914*d*, Die Gliederung der Tendaguru-Schichten im Tenda guru-Gebiet und die Entstechung der Saurier-Lagerstätten: *Ibid.*, 3 (3), pp. 227–261.

_____, 1920, Über *Elaphrosaurus bambergi* und die Megalosaurier aus den Tendaguru-Schichten Deutsch-Ostafrikas: *Itzungsberichte der Gesellschaft Naturforschender Freunde zu Berlin 1920*, pp. 464–480.

_____, 1922, Das Handskellt von *Gigantosaurus robustus* und *Brachiosaurus brancai* aud den Tendaguru-Schichten Deutsch-Ostafrika: *Centralblatt fur Mineralogie, Geologie und Paläontologie*, 1922, pp. 225–235.

_____, 1925*a*, Die Coelurosauria und Theropoden der Tendaguru-Schichten Deutsch-Ostafrikas: *Palaeontographica, Supplement VII* (5), pp. 1–99.

_____, 1925*b*, Die Grabungsstellen der

Tendaguru-Gegend: *Ibid.,* (1), Teil 1, pp. 17–19.

_____, 1929*a*, Material und Formengehalt der Sauropoden in der ausbeute der Tendaguru-expedition: *Ibid.,* (1), teil 2, lieferung 1, pp. 1–34.

_____, 1929*b*, Die Wirbelsäule der Gattung *Dicraeosaurus*: *Ibid.,* (1), teil 2, lieferung 1, pp. 37–133.

_____, 1933, Eine *Estheria* aus den Tendaguruschichten: *Ibid.,* (2), teil 2, lieferung 1, pp. 95–98.

_____, 1935–1936, Die Sch del der Sauropoden *Brachiosaurus, Barosaurus,* and *Dicraeosaurus* aus den Tendaguruschiechten Deutsch-Ostafrikas: *Ibid.,* (1), teil 2, lieferung 3, pp. 145–298.

_____, 1950*a*, Die Wirbelsäule von *Brachiosaurus brancai*: *Ibid.,* (1), teil 3, lieferung 2, pp. 27–93.

_____, 1950*b*, Die aufgestelltes Skelett von *Dicraeosaurus hanse manni*: *Ibid.,* (1), teil 3, lieferung 2, pp. 95–102.

_____, 1950*c*, Die Skelettrekonstruktion von *Brachiosaurus brancai*: *Ibid.,* (1), teil 3, lieferung 2, pp. 95–103.

_____, 1961, Die Gliedmassen und Gliedmassengürtel der Sauropo den der Tendaguru-Schichten: *Ibid.,* (1), teil 3, lieferung 4, pp. 177–235.

Janis, Christine M., and Patricia Brady Wilhelm, 1993, Were there mammalian pursuit predators in the Tertiary? Dances with wolf avatars: *Journal of Mammalian Evolution*, 1 (2), pp. 103–125.

Jenny, J., A. Le Marrec, and M. Monbaron, 1981, Les Couches Rouges du Jurassique moyen du Haut Atlas central (Maroc): corrélations lithostratigraphiques, eléments de datation et cadre tectono-sédimentaire: *Bulletin de la Sociéte géologique de France*, 7, XXIII (23), pp. 627–639.

Jensen, James A., 1985, Three new sauropod dinosaurs from the Upper Jurassic of Colorado: *Great Basin Naturalist*, 45 (4), pp. 697–709.

Jensen, Marcia L., Zachary D. Sharp, and Spencer G. Lucas, 2000, Migrating Cretaceous theropods? Evidence from oxygen isotope geochemistry, Canada and New Mexico: *in*: Spencer G. Lucas and Andrew B. Heckert, editors, *Dinosaurs of New Mexico*. Albuquerque: New Mexico Museum of Natural History and Science, Bulletin 17, p. 91.

Jerzykiewicz, T., Philip J. Currie, David A. Eberth, P. A. Johnson, E. H. Koster and J. J. Zheng, 1993, Djadokhta Formation correlative strata in Chinese Inner Mongolia: an overview of the stratigraphy, sedimentary geology, and paleontology and comparisons with the type locality in the pre–Altai Gobi: *Canadian Journal of Earth Sciences*, 30, pp. 2180–2195.

Jerzykiewicz, T., and Dale A. Russell, 1991, Late Mesozoic stratigraphy and vertebrates of the Gobi Basin: *Cretaceous Research*, 12, pp. 345–377.

Ji Qiang, Philip J. Currie, Mark A. Norell, and Ji Shu-An, 1998, Two feathered dinosaurs from northeastern China: *Nature*, 393, pp. 753–761.

Ji Qiang and Ji S., 1996, On discovery of the earliest bird fossil in China and the origin of birds: *Chinese Geology*, 10 (233), pp. 30–34.

Ji Shúan, 1998, New pterosaurs (Reptilia: Pterosauria) from Northeastern China and the geological age problem of *Confuciuornis*: *Journal of Vertebrate Paleontology*, 18 (Supplement to Number 3), Abstracts of Papers, Fifty-eighth Annual Meeting, p. 54A.

Ji Qiang, Philip J. Currie, Mark A. Norell, and Ji Shu-An, 1998, Two feathered dinosaurs from northeastern China: *Nature*, 393, pp. 753–761.

Jiang Fu-Xing, Chen Jin-Hua, Cao Mei-Zhen, and Toshifumi Komatsu, 2000, A discussion on the age of the feathered dinosaurs-bearing beds of Liaoning, China: *Acta Palaeontologica Sinica*, 39 (2), pp. 307–311.

Jianu, Coralia-Maria, and David B. Weishampel, 1997, A new theropod dinosaur from the Hateg Basin, Western Romania, in the Hungarian Geological Survey collection: *Sargentia*, ser. Sci. Nat., 17, pp. 239–246.

_____, 1999, The smallest of the largest: a new look at possible dwarfing in sauropod dinosaurs: *Geologie en Mijnbouw*, 78, pp. 335–343.

Johnston, Christopher, 1859, Comments on *Astrodon*: *American Journal of Dental Science*, 9, p. 341.

Jones, Ramal D., and Donald L. Burge, 1995, Radiological surveying as a method for mapping dinosaur bone sites: *Journal of Vertebrate Paleontology*, 15 (Supplement to Number 3), Abstracts of Papers, Fifty-fifth Annual Meeting, p. 38A.

Jones, Terry D., James O. Farlow, John A. Ruben, Donald M. Henderson, and William J. Hillenius, 2000, Cursoriality in bipedal dinosaurs: *Nature*, 406, pp. 716–718.

Jones, Terry D., John A. Ruben, Larry D. Martin, Evgeny N. Kurochkin, Alan Feduccia, Paul F. A. Maderson, William J. Hillenius, Nicholas R. Geist, and Vladimir Alifanov, 2000, Nonavian feathers in a Late Triassic archosaur: *Science*, 288, pp. 2202–2205.

Jukes, J. Bette, 1847, *Narrative of the Surveying Voyage of the H.M.S. Fly*. London: Boone, vol. 1, pp. 1–423, vol. 2, pp. 1–362.

Kaneko, Ryuichi, 2000, Following dinosaur tracks in Thailand: *DinoPress*, 1, pp. 92–105 (in Japanese; "English Text," pp. 25–28)

Karhu, A. A., and A. S. Rautian, 1995, A new family of Maniraptora (Dinosauria: Saurischia) from the Late Cretaceous of Mongolia: *Paleontological Journal*, 30 (5), pp. 583–592.

Kase, Tomoki, 1984, Early Cretaceous Marine and Brackish-Water Gastropoda from Japan. Tokyo: National Science Museum, 263 pages.

Kaye, Fran Tannenbaum, and Kevian Padian, 1994, Microvertebrates from the *Placerias* quarry: a window on Late Triassic vertebrate diversity in the American Southwest, *in*: Nicholas C. Fraser and Hans-Dieter Sues, editors, *In the Shadow of the Dinosaurs: Early Mesozoic Tetrapods*. Cambridge: University of Cambridge Press, pp. 171–196.

Kellner, Alexander W. A., 1996, Remarks on Brazillian dinosaurs, *in*: Fernando S. Novas and Ralph E. Molnar, editors, *Proceedings of the Gondwanan Dinosaur Symposium*: *Memoirs of the Queensland Museum*, 39 (part 3), pp. 611–626.

_____, 1999, Short note on a new dinosaur (Theropoda, Coelurosauria) from the Santana Formation (Romualdo Member, Albian), northeastern Brazil: *Boletim do Museu Nacional*, Nova Série, 49, pp. 1–8, Rio de Janeiro.

_____, 2000, New information on a maniraptoriform theropod dinosaur from the Santana Formation (Aptian–Albian), Araripe Basin, Brazil: *Journal of Vertebrate Paleontology*, 20 (Supplement to Number 3), Abstracts of Papers, Sixtieth Annual Meeting, p. 51A.

Kellner, Alexander W. A., and Sergio A. K. de Azevedo, 1999, A new sauropod dinosaur (Titanosauria) from the Late Cretaceous of Brazil: *in*: Yukimitsu Tomida, Rich, and Vickers-Rich, editors, *Proceedings of the Second Gondwanan Dinosaur Symposium*, National Science Museum Monographs, 15, Tokyo, pp. 111–142.

Kellner, Alexander W. A., Sergio A. K. de Azevedo, Atila A. S. Rosa, Ruben A. Boelter, and Luciano A. Leal, 1999, The occurrence of Prosauropoda in the terrestrial Late Triassic Santa Maria Formation, southern Brazil: *Journal of Vertebrate Paleontology*, 19 (Supplement to Number 3), Abstracts of Papers, Fifty-ninth Annual Meeting, p. 57A.

Kellner, Alexander W. A., and Diogenese de A. Campos, 1996, First Early Cretaceous theropod dinosaur from Brazil with comments on Spinosauridae: *Neus Jahrbuch Fuer Geologie und Paläontologie Abhandlungen*, 199, pages 151–166.

Kennedy, W. J., and W. A. Cobban, 1990, Cenomanian ammonite faunas from Woodbine Formation and lower part of the Eagle Ford Group, Texas: *Palaeontology*, 33, pp. 75–154.

Kenneth, J. H., and W. D. Henderson, 1960, *A Dictionary of Scientific Terms*, seventh edition. Edinburgh and London: Oliver and Boyd, xv, 595 pages.

Kenworthy, Jason, Ralph E. Chapman, Thomas R. Holtz, Jr., and Rudyard W. Sadleir. 2000, On behavior and the shape of claws in dinosaurs, birds and mammals: *Journal of Vertebrate Paleontology*, 20 (Supplement to Number 3), Abstracts of Papers, Sixtieth Annual Meeting, p. 51A.

Kermack, D., 1984, New prosauropod material from South Wales: *Zoological Journal of the Linnean Society*, 82, pp. 101–117.

Kielan-Jaworowska, Zofia, and J. Hurum, 1997, Djadochtatheria — a new suborder of multituburculate mammals: *Acta Palaeontologica Polonica*, 42, pp. 201–242.

Kielan-Jaworowska, Zofia, and Rinchen Barsbold, 1972, Narrative of the Polish-Mongolian Palaeontological Expeditions 1967–1971: *Acta Palaeontologica Polonica*, 42, pp. 201–242.

Kim, Haang Mook, 1981, [Cretaceous dinosaur

fossils discovered from two dinosaur sites of Korea]: *The Journal of the Geological Society of Korea*, 17, p. 297.

_____, 1983, Cretaceous dinosaurs from Korea: *Ibid.*, 19, pp. 115–126 (in Korean with English summary).

Kirkland, James I. [Ian], 1996, Biogeography of western North America's mid–Cretaceous dinosaur faunas: losing European ties and the first great Asian–North American interchange: *Journal of Vertebrate Paleontology*, 16 (Supplement to Number 3), Abstracts of Papers, Sixtieth Annual Meeting, p. 45A.

_____, 1998*a*, A polacanthine ankylosaur (Ornithischia: Dinosauria) from the Early Cretaceous (Barremian) of eastern Utah, *in*: Spencer G. Lucas, James I. Kirkland, and John W. Estep, editors, *Lower and Middle Cretaceous Terrestrial Ecosystems*. Albuquerque: New Mexico Museum of Natural History and Science, Bulletin 14, pp. 271–282.

_______, 1998*b*, A new hadrosaurid from the Upper Cedar Mountain Formation (Albian–Cenomanian: Cretaceous) of eastern Utah — the oldest known hadrosaurid (lambeosaurine?), *Ibid.*, pages 283–296.

Kirkland, James I., Martha C. Aguillon-Martinez, René Hernández-Rivera, and Virginia A. Tidwell, 2000, A late Campanian brachiosaurid proxial caudal vertebra from Coahuila, Mexico: evidence against a Cretaceous North American sauropod hiatus: *Journal of Vertebrate Paleontology*, 20 (Supplement to Number 3), Abstracts of Papers, Sixtieth Annual Meeting, pp. 51A–52A.

Kirkland, James I., Brooks B. Britt, Donald L. Burge, Kenneth Carpenter, Richard L. Cifelli, Frank L. DeCourten, Jeffrey G. Eaton, Stephen Hasiotis, and Tim F. Lawton, 1997, Lower to Middle Cretaceous dinosaur faunas of the central Colorado Plateau: A Key to understanding 35 million years of tectonics, sedimentology, evolution, and biogeography: *Brigham Young University Geology Studies*, 42, part II, pp. 69–103.

Kirkland, James Ian, and Kenneth Carpenter, 1994, North America's first pre–Cretaceous ankylosaur (Dinosauria) from the Upper Jurassic Morrison Formation of Western Colorado: *BYU Geology Studies 1994*, 40, pp. 25–42.

Kirkland, James I., Robert Gaston, and Donald Burge, 1993, A large dromaeosaur (Theropoda) from the Lower Cretaceous of eastern Utah: *Hunteria*, 2 (10), pp. 1–16.

Kirkland, James I., and J. Michael Parrish, 1995, Theropod teeth from the Lower Cretaceous of Utah: *Journal of Vertebrate Paleontology*, 15 (Supplement to Number 3), Abstracts of Papers, Fifty-fifth Annual Meeting, p. 39A.

Kischlat, Edio-Ernst, 1999, A new dinosaurian "rescued" from the Brazillian Triassic: *Teyuawasu barberenai*, new taxon: *Paleontologia em Destaque*, 14 (26), p. 58.

Kischlat, Edio-Ernst, and Mário Costa Barberena, 1999, Triassic Brazilian dinosaurs: new data: *Paleontologia em Destaque*, 14 (26), p. 56.

Kitchin, F. L., 1929, On the age of the upper and middle deinosaur deposits at Tendaguru, Tanganyika Territory: *Geological Magazine*, 66 (5), pp. 193–220.

Kitching, J. W., and Michael A. Raath, 1984, Fossils from the Elliot and Clarens formations (Karoo sequence) of the Northeastern Cape, Orange Free State and Lesotho, and a suggested biozonation based on tetrapods: *Palaeontologia Africana*, 25, pp. 111–125, Johannesburg.

Kobayashi, T., 1963, On the Cretaceous Ban Na Yo Fauna of East Thailand with a note on the distribution of *Nippononaia*, *Trigonioides*, and *Plicatounio*: *Japanese Journal of Geology and Geography*, 34, pp. 35–43.

_____, 1968, The Cretaceous non-marine Pelecypods from the Nam Phung Dam site in the north-eastern part of the Khorat Plateau, Thailand with a note on the Trigonioididae: *Geology and Paleontology of southeast Asia*, 4, pp. 109–138.

Kobayashi, Yoshitsugu, and Yoichi Azuma, 1999, Cranial material of a new iguanodontian dinosaur from the Lower Cretaceous Kitadani Formation of Japan: *Journal of Vertebrate Paleontology*, 19 (Supplement to Number 3), Abstracts of Papers, Fifty-ninth Annual Meeting, p. 57A.

Koboyashi, Yoshitsugu, Jun-Chang Lu, and Dong Zhi-Ming, 1999, Herbivorous diet in an ornithischian dinosaur: *Nature*, 402, pp. 450–481.

Kobayashi, Yoshitsugu, Jun-Chang Lu, Zhi-Ming Dong, Rinchen Barsbold, Yoichi Azuma, and Yukimutsu Tomida, 1999, Herbivorous diet in an ornithomimid dinosaur: *Nature*, 402, pp. 480–481.

Krantz, Peter M., 1998, *Astrodon* rediscovered: America's first sauropod, *in*: Donald L. Wolberg, K. Gittis, S. Miller, and A. Raynor, editors, *The Dinofest Symposium* [abstracts], Presented by The Academy of Natural Sciences of Philadelphia, Pennsylvania, pp. 33–34.

Krause, David W., and Joseph H. Hartman, 1996, Late Cretaceous fossils from Madagascar and their implications for biogeographic relationships with the Indian subcontinent, *in*: A. Sahni, editor, *Cretaceous Stratigraphy and Palaeoenvironments*, Geological Society of India Memoir 37, pp. 135–154.

Kues, B. S., Thomas Lehman, and J. Keith Rigby, Jr., 1980, The teeth of *Alamosaurus sanjuanensis*, a Late Cretaceous sauropod: *Journal of Paleontology*, 54, pp. 864–869.

Kundrát, Martin, 1998, Comments on the significance of integumentary impression of the Early Jurassic theropod *Eubrontes minisculus*: *Journal of Vertebrate Paleontology*, 18 (Supplement to Number 3), Abstracts of Papers, Fifty-eighth Annual Meeting, p. 57A.

Kuhn, Oskar, 1936, Ornithischia (Stegosauria excluses), *in*: W. Quendstedt, editor, *Fossilium Catalogus, I. Animalia*, 78. Junk: 's-Gravenhage, 81 pages.

_____, 1959, Ein neuer Microsaurier aus dem deutschen Rotliegen den: *Neus Jahrbuch für Geologie und Paläontologie, Monatshefte, 1959*, pp. 424–426.

_____, 1961, *Die Familien der rezenten und fossilen Amphibien und Reptilien*. Bamberg, Germany: Verlaghaus Meisenbach KG, 79 pages.

_____, 1964, Ornithischia (Supplementum I), *in*: F. Westpal, editor, *Fossilium Catalogus I: Animalia Pars 105*. Uitgeverij, 80 pages.

Kurochkin, E. N., 1995, Synopsis of Mesozoic birds and early evolution of Class Aves: *Archaeopteryx*, 13, pp. 47–66.

Kurzanov, Seriozha M., 1981, Some unusual theropods from the Upper Cretaceous of Mongolia: *Iskopayemyye pozvonochnyye Mongolii (Fossil Vertebrates of Mongolia), Trudy-Sovmestnaya Sovetsko-Mongoliskaya Paleontologicheskaya Ekspeditsyia*, 15, pp. 39–49.

Kurzanov, Seriozha M., and A. F. Bannikov, 1983, [A new sauropod from the Upper Cretaceous of Mongolia], *Paleontologichesky Zhurnal 1983*, pp. 90–96 (reprinted in *Paleontological Journal*, 26 [3], pp. 91–97).

Kuznetsov, A. N., and A. G. Sennikov, 2000, On the function of a perforated acetabulum in archosaurs and birds: *Paleontological Journal*, 14 (4), pp. 439–448 (translated from *Paleontologicheskii Zhurnal*, 4, 2000, pp. 76–85).

Kyte, Frank T., 1998, A meteorite from the Cretaceous/Tertiary boundary: *Nature*, 396, pp. 237–239.

Lamanna, Matthew C., Joshua B. Smith, You Hailu, Thomas R. Holtz, Jr., and Peter Dodson, 1998, A reassessment of the Chinese theropod dinosaur *Dilophosaurus sinensis*: *Journal of Vertebrate Paleontology*, 18 (Supplement to Number 3), Abstracts of Papers, Fifty-eighth Annual Meeting, p. 57A.

Lamanna, Matthew C., Joshua B. Smith, Kenneth J. Lacovara, Peter Dodson, and Y. Attiya, 2000, New vertebrate discoveries from the Cretaceous of Egypt: *Journal of Vertebrate Paleontology*, 20 (Supplement to Number 3), Abstracts of Papers, Sixtieth Annual Meeting, p. 53A.

Lamb, James P., Jr., 1998, *Lophorhothon*, an iguanodontian, not a hadrosaur: *Journal of Vertebrate Paleontology*, 18 (Supplement to Number 3), Abstracts of Papers, Fifty-eighth Annual Meeting, p. 58A.

Lambe, Lawrence M., 1902, On vertebra of the Mid-Cretaceous of the North West Territory. 2. New genera and species from the Belly River Series (Mid-Cretaceous): *Geological Survey of Canada, Contributions to Canadian Paleontology*, 3, part 2, pp. 23–81.

_____, 1904, On the squamoso-parietal crest of two species of horned dinosaurs from the Cretaceous of Alberta: *Ottawa Naturalist*, 17, pp. 81–84.

_____, 1917, The Cretaceous theropodous dinosaur *Gorgosaurus*: *Canada Department of Mines, Geological Survey Memoir 100*, pp. 1–84.

_____, 1918, The Cretaceous genus *Stegoceras*, typifying a new family referred provisionally to the Stegosauria: *Transactions of the Royal Society of Canada*, series 3, 12, pp. 23–36.

Lambert, David, 1983, *A Field Guide to Dinosaurs*. New York: Avon, 256 pages.

_____, and the Diagram Group, 1990, *The Dinosaur Data Book*. Oxford: Facts on File; New York: Avon, 320 pages.

_____, 1993, *The Ultimate Dinosaur Book.* London: Dorling Kindersley Inc., 192 pages.

Langer, Max C., Fernando Abdala, Martha Richter, and Michael J. Benton, 1999, A sauropodomorph dinosaur from the Upper Triassic (Carnian) of southern Brazil: *Comtes Rendu des Séances de l'Académie des Sciences, Paris, de la Terre et des Planètes,* 329, pp. 511–517.

Langston, Wann, Jr., 1960, The vertebrate fauna of the Selma Formation of Alabama, Part VI, The dinosaurs: *Fieldiana: Geological Memoirs,* 3 (6), pp. 313–363.

_____, 1965, Pre-Cenozoic vertebrate paleontology in Alberta; its past and future: *Vertebrate Paleontology in Alberta,* Report on a Conference at the University of Alberta, Edmonton, pp. 9–31.

_____, 1974, Nonmammalian Comanchean tetrapods: *Geoscience and Man,* 8, pp. 77–102.

Lapparent, Albert F. de, and Georges Zbyszewski, 1957, Les dinosauriens du Portugal: *Memoires des Services géologiques du Portugal,* 2, pp. 1–63.

Larsson, Hans C. E., Paul C. Sereno, and Jeffrey A. Wilson, 2000, Forebrain enlargement among nonavian theropod dinosaurs: *Journal of Vertebrate Paleontology,* 20 (3), pp. 615–618.

Larson, Peter L., 1994, *Tyrannosaurus rex* sex, *in,* Donald L. Wolberg, Edmund Stump, and Gary Rosenberg, editors. *Dinofest International: Proceedings of a Symposium Held at Arizona State University.* Philadelphia: Academy of Natural Sciences, pp. 311–335.

_____, 1997, The King's new clothes: a fresh look at *Tyrannosaurus rex*: *in*: Donald L. Wolberg, Edmund Stump, and Gary Rosenberg, editors. *Dinofest International: Proceedings of a Symposium Held at Arizona State University.* Philadelphia: Academy of Natural Sciences, pp. 65–71.

_____, 1999, Guess who's coming to dinner: *Tyrannosaurus* vs. *Nanotyrannus*: variance in feeding habits: *Journal of Vertebrate Paleontology,* 19 (Supplement to Number 3), Abstracts of Papers, Fifty-ninth Annual Meeting, p. 58A.

Lawson, Douglas A., 1976, *Tyrannosaurus* and *Torosaurus*: Maastrichtian dinosaurs from Trans-Pecos Texas: *Journal of Paleontology,* 50, pp. 158–164.

Lawton, Tim. F., G. T. Basabilcazo, S. A. Hodgson, D. A. Wilson, G. H. Mack, W. C. McIntosh, S. G. Lucas, and K. K. Kietzke, 1993, Stratigraphy of the Little Hatchet Mountains, southwestern New Mexico: *New Mexico Geology,* 15, pp. 9–15.

Le Loeuff, Jean, 1991, The Campano–Maastrichtian vertebrate faunas from Southern Europe and their relationships with other faunas in the world: palaeobiogeographical implications: *Cretaceous Research,* 12, pp. 93–114.

_____, 1992, Les vertévés continentaux du Crétacé supérieur d'Europe: Paléoecologie, Biostratigraphie et Paléobiogeographie, Ph.D. thesis: *Memoires des Sciences de la Terre Université P. et M. Curie, Paris,* pp. 1–271.

_____, 1993, European titanosaurids: *Revue de Paléobiolie, Vol. Spéc,* 7, pp. 105–117.

_____, 1998, Evolution palébiogéographique des faunes de vertébre's continentaux du Jurassic supérieur à la fin du Crétacé. Mémoire d'habitation à dirger des recherches. Université Paul Sabatier, Toulouse, Tome 1, pp. 1–68.

Le Loeuff, Jean, and Éric Buffetaut, 1991, *Tarascosaurus salluvicus* nov.gen., nov. spéc., dinosaure théropode du Crétacé supéreur du sud de la France: *Géobios,* 25 (5), pp. 585–594.

_____, 1998, A new dromaeosaurid theropod from the Upper Cretaceous of southern France: *Oryctos,* 1, pp. 105–112.

Le Loeuff, Jean, Éric Buffetaut, L. Cavin, M. Martin, Valérie Mar tin, and H. Tong 1994, An armoured titanosaurid sauropod from the Late Cretaceous of Southern France and the occurrence of osteoderms in the Titanosauroidea: *Gaia,* 10, pp. 155–159.

Le Loeuff, Jean, Éric Buffetaut, Patrick Méchin, and Annie Méchin-Salessy, 1992, The first record of dromaeosaurid dinosaurs (Saurischia, Theropoda) in the Maastrichtian of Southern Europe: palaeobiogeographical implications: *Bulletin de la Sociéte géologique de France,* 16, pp. 337–343.

Leanza, H. A., and C. A. Hugo, 1996, Revisión estratigráfica del Cretácico Inferior continental en el ámbito sudroriental de la Cuenca Neuquina: *Revista Asociatión Geológica Argentina.*

Lee, Yuong-Nam, 1995, Mid-Cretaceous archosaurs from the Paw Paw and Woodbine formations, Texas: Ph.D. dissertation, Southern Methodist University, Dallas, Texas, 117 pages (unpublished).

_____, 1996, A new nodosaurid ankylosaur (Dinosauria: Or nithischia) from the Paw Paw Formation (Late Albian) of Texas: *Journal of Vertebrate Paleontology,* 16, pp. 323–345.

_____, 1997, The Archosauria from the Woodbine Formation (Cenomanian) in Texas: *Journal of Paleontology,* 71, pp. 1147–1156.

Lee, Yuong-Nam, Seong-Young Yang, and Eun-Jun Park, 1997, Sauropod dinosaur remains from the Gyeongsang Supergroup, Korea: *Paleontological Society of Korea, Special Publication,* 2, pp. 103–114.

Lehman, Thomas M., 1980, The Alamo Wash local fauna: A new look at the Ojo Alamo fauna, *in*: S. Lucas, K. Rigby, Jr., and B. Kues, editors: *Advances in San Juan Paleontology.* Albuquerque: University of New Mexico Press, pp. 189–192.

_____, 1981, The Alamo Wash Local Fauna, a new look at the old Ojo Alamo Fauna, *in*: Spencer G. Lucas, J. Keith Rigby, Jr., and B. S. Kues, editors, *Advances in San Juan Basin Paleontology.* Albuquerque: University of New Mexico Press, pp. 189–221.

_____, 1985, Depositional environments of the Naashoibito Member of the Kirtland Formation, Upper Cretaceous, San Juan Basin, New Mexico: *New Mexico Bureau of Mines and Mineral Resources, Circular,* 195, pp. 55–79.

_____, 1987, Late Maastrichtian paleoenvironments and dinosaur biogeography in the Western Interior of North America: *Palaeogeography, Palaeoclimatology, and Palaeoecology,* 60, pp. 189–217.

_____, 1989, *Chasmosaurus mariscalensis,* sp. nov., a new ceratopsian dinosaur from Texas: *Journal of Vertebrate Paleontology,* 9 (2), pp. 137–162.

_____, 1990, The ceratopsian subfamily Chasmosaurinae: sexual dimorphism and systematics, *in*: Kenneth Carpenter and Philip J. Currie, editors, *Dinosaur Systematics: Approaches and Perspectives.* Cambridge, New York and Melbourne: Cambridge University Press, pp. 211–229.

_____, 1996, A horned dinosaur from the El Picacho Formation of West Texas, and review of ceratopsian dinosaurs from the American Southwest: *Journal of Paleontology,* 70 (30), pp. 494–508.

_____, 1998, A gigantic skull and skeleton of the horned dinosaur *Pentaceratops sternbergi* from New Mexico: *Ibid.,* 72 (5), pp. 894–906.

Lehman, Thomas M., and Kenneth Carpenter, 1990, A partial skeleton of the tyrannosaurid dinosaur *Aublysodon* from the Upper Cretaceous of New Mexico: *Journal of Paleontology,* 64, pp. 1026–1032.

Leidy, Joseph, 1865, Memoir on the extinct reptiles of the Cretaceous Formations of the United States: *Smithsonian Contributions to Knowledge,* 14 (6), pp. 1–135.

_____, 1868, Remarks on a jaw fragment of *Megalosaurus*: *Academy of Natural Sciences of Philadelphia, Proceedings,* 20, pp. 197–200.

_____, 1870, (Proposal of *Poicilopleuron valens*): *Academy of Natural Sciences of Philadelphia, Proceedings,* 22, pp. 3–4.

_____, 1873, (Proposal of *Antrodemus valens*): Contributions to the extinct vertebrate fauna of the western territories: *Report of the U.S. Geological Survey of the Territories,* 1, pp. 14–358.

_____, 1868, Remarks on a jaw fragment of *Megalosaurus*: *Proceedings of the Academy of Natural Sciences of Philadelphia 1868,* pp. 197–200.

Lessem, Don, 1999, Creating the Jurassic Foundation: *Prehistoric Times,* 34, p. 31.

Lillegraven, Jason A., M. J. Kraus, and T. M. Brown, 1979, Paleogeography of the world of the Mesozoic, *in*: Lillegraven, Zofia Kielan-Jaworowska, and William A. Clemens, editors, *Mesozoic Mammals: The First Two-Thirds of Mammalian History.* Berkeley: University of California Press, pp. 277–308.

Lillywhite, Harvey B., 1987, Circulatory adaptations of snakes to gravity: *American Zoologist,* 27, pp. 81–95.

_____, 1991, Sauropods and gravity: *Natural History,* 12, p. 33.

Lockley, Martin G., 1986, A guide to dinosaur tracksites of the Colorado Plateau and American southwest: *University of Colorado Denver Geological Department Magazines Special Issue,* 1, pp. 1–56.

_____, 1991, *Tracking Dinosaurs.* Cambridge, England: Cambridge University Press, 238 pages.

Lockley, Martin G., and Adrian P. Hunt, 1995, Ceratopsid tracks and associated ichnofauna from the Laramie Formation (Upper Cretaceous: Maastrichtian) of Colorado: *Journal of Vertebrate Paleontology*, 15 (3), pp. 592–614.

Loewen, Mark, and Scott D. Sampson, 2000, Femoral ontogeny in *Allosaurus fragilis* (Theropoda: Allosauroidea) from the Late Jurassic Cleveland-Lloyd Dinosaur Quarry, Central Utah: *Journal of Vertebrate Paleontology*, 20 (Supplement to Number 3), Abstracts of Papers, Sixtieth Annual Meeting, p. 54A.

Long, John A., and Ralph E. Molnar, 1998, A new Jurassic theropod dinosaur from Western Australia: *Records of the Western Australian Museum*, 19, pp. 121–129.

Long, Robert A., and Phillip A. Murry, 1995, Late Triassic (Carnian and Norian) tetrapods from the southwestern United States: *New Mexico Museum of Natural History and Science Bulletin*, 4, 254 pages.

Longrich, Nick, 1999, On the semilunate carpal and trochanteric crest of maniraptoran theropods: *Journal of Vertebrate Paleontology*, 19 (Supplement to Number 3), Abstracts of Papers, Fifty-ninth Annual Meeting, p. 60A.

_____, 2000, Myrmecophagous Maniraptora? Alvarezsaurs as aardraptors: *Journal of Vertebrate Paleontology*, 20 (Supplement to Number 3), Abstracts of Papers, Sixtieth Annual Meeting, p. 55A.

López-Martinez, N., J. J. Moratalla, and J. L. Sanz, 2000, Dinosaurs nesting on tidal flats: *Palaeogeography, Palaeoclimatology, Palaeoecology*, 160, pp. 153–163.

Lorenz, J. C., and W. Gavin, 1984, Geology of the Two Medicine Formation and the sedimentology of a dinosaur nesting ground, *in*: J. D. Duncan and P. B. Garrison, editors, *Montana Geological Society, 1984 Field Conference and Symposium Guidebook, Northwest Montana and Adjacent Canada*. Billings, Montana: Montana Geological Society, pp. 175–186.

Lu Junchang, 1997, A new Iguanodontidae (*Probactrosaurus mazongshanensis* sp. nov.) from Mazongshan Area, Gansu Province, China, *in*: Dong Zhiming, editor, *Sino-Japanese Silk Road Dinosaur Expedition*. Beijing: China Ocean Press, pp. 27–47.

Lucas, Frederic A., 1902, Paleontological notes: the generic name *Omosaurus*: *Science* (new series), 16, p. 435.

Lucas, Spencer G., 2000, The gastromyths of "*Seismosaurus*," a Late Jurassic dinosaur from New Mexico: *in*: Spencer G. Lucas and Andrew B. Heckert, editors, *Dinosaurs of New Mexico*. Albuquerque: New Mexico Museum of Natural History and Science, Bulletin 17, pp. 61–67.

Lucas, Spencer G., O. J. Anderson, and C. Pigman, 1995, Jurassic stratigraphy in the Hagan basin, north-central New Mexico: *New Mexico Geological Society Guidebook* 46, pp. 317–326.

Lucas, Spencer G., and Andrew B. Heckert, 2000, Jurassic dinosaurs in New Mexico, *in*: Spencer G. Lucas and Andrew B. Heckert, editors, *Dinosaurs of New Mexico*. Albuquerque: New Mexico Museum of Natural History and Science, Bulletin 17, pp. 43–45.

Lucas, Spencer G., Andrew B. Heckert, and Robert M. Sullivan, 2000, Cretaceous dinosaurs in New Mexico, *in*: Spencer G. Lucas and Andrew B. Heckert, editors, *Dinosaurs of New Mexico*. Albuquerque: New Mexico Museum of Natural History and Science, Bulletin 17, pp. 83–90.

Lucas, Spencer G., and Adrian P. Hunt, 1989, *Alamosaurus* and the sauropod hiatus in the Cretaceous of North American Western Interior, *in*: James O. Farlow, editor, *Paleobiology of the Dinosaurs*, Geological Society of America Special Paper 238, Boulder, Colorado, pp. 75–85.

Lucas, Spencer G., Adrian P. Hunt, and Robert A. Long, 1992, The oldest dinosaurs: *Naturwissenschaften*, 79, pp. 171–172.

Lucas, Spencer G., B. S. Kues, and C. M. González-León, 1995, Paleontology of the Upper Cretaceous Cabullona Group, northeastern Senora, *in*: C. Jacques-Ayala, González-León, and J. Roldán-Quintana, editors, *Studies on the Mesozoic of Sonora and Adjacent Areas*, Geological Society of America Special Paper 301, Boulder, Colorado, pp. 143–165.

Lucas, Spencer G., Niall J. Mateer, Adrian P. Hunt, and F. Michael O'Neill, 1987, Dinosaurs, the age of the Fruitland and Kirtland formations, and the Cretaceous–Tertiary boundary in the San Juan basin, New Mexico, *in*: James E. Fassett and J. K. Rigby, Jr., editors, *The Cretaceous–Tertiary Boundary in the San Juan and Raton Basins, New Mexico and Colorado, Special Paper 209*. Boulder, Colorado: Geological Society of America, pp. 35–50.

Lucas, Spencer G., and J. Keith Rigby, Jr., 1979, Comment of "Biostratigraphy and magnetostratigraphy of Paleocene terrestrial deposits, San Juan Basin, New Mexico": *Geology*, 7, pp. 323–325.

Lucas, Spencer G., and Robert M. Sullivan, 2000*a*, Stratigraphy and vertebrate biostratigraphy across the Cretaceous–Tertiary boundary, Betonnie Tsosie Wash, San Juan Basin, New Mexico, *in*: Spencer G. Lucas and Andrew B. Heckert, editors, *Dinosaurs of New Mexico*. Albuquerque: New Mexico Museum of Natural History and Science, Bulletin 17, pp. 95–103.

_____, 2000*b*, The sauropod dinosaur *Alamosaurus* from the Upper Cretaceous of the San Juan Basin, New Mexico: *Ibid.*, pp. 147–156.

Lull, Richard Swann, 1908, The cranial musculature and the origin of the frill in the ceratopsian dinosaurs: *American Journal of Science*, 4 (25), pp. 387–399.

_____, 1911, The Reptilia of the Arundel Formation: Lower Cretaceous: *Maryland Geological Survey*, pp. 183–211.

_____, 1921, The Cretaceous armored dinosaur, *Nodosaurus textilis* Marsh: *Ibid.*, 1, pp. 97–125.

_____, 1933, A revision of the Ceratopsia or horned dinosaurs: *Peabody Museum of Natural History Memoirs*, 3 (3), pp. 1–175.

Lull, Richard Swann, and Nelda E. Wright, 1942, The hadrosaurian dinosaurs of North America: *Geological Society of America Special Paper* 40, xii, 242 pages.

Luo, Z., and X.-C. Wu, 1994, The smallest tetrapods of the Lower Lufeng Formation, Yunnan, China, *in*: Nicholas C. Fraser and Hans-Dieter Sues, editors, *In the Shadow of the Dinosaurs: Early Mesozoic Tetrapods*. Cambridge: University of Cambridge Press, pp. 51–270.

_____, 1995, Correlation of vertebrate assemblage of the Lower Lufeng Formation, Yunnan, China, *in*: A.-L. Sun and Y. Wang, editors, *6th Symposium on Mesozoic Terrestrial Ecosystems and Biotas, Short Papers*. Beijing: China Ocean Press, pp. 83–88.

Lydekker, Richard, 1890, Contributions to our knowledge of the Wealden and the Sauropterygians of the Purbeck and Oxford Clay: *Quarterly Journal of the Geological Society of London*, 46, pp. 36–53.

MacLeod, N., P. F. Rawson, P. L. Forey, F. T. Banner, M. K. BouDagher-Fadel, P. R. Brown, J. A. Burnett, P. Chambers, S. Culver, S. E. Evans, C. Jeffrey, M. A. Kaminski, A. R. Lord, A. C. Milner, A. R. Milner, N. Morris, E. Owen, B. R. Rosen, A. B. Smith, P. D. Taylor, E. Urquhardt, and J. R. Young, 1997, The Cretaceous–Tertiary biotic transition: *Journal of the Geological Society of London*, 154, pp. 265–292.

Madsen, James H., Jr., 1976, *Allosaurus fragilis*: a revised Osteology: *Utah Geological and Mineral Survey, a division of the Utah Department of Natural Resources*, Bulletin 109, xii, 163 pages.

Madsen, James H., Jr., and William L. Stokes, 1963, New information on the Jurassic dinosaur *Ceratosaurus*: *Geological Society of America, Special Paper*, 73, p. 90.

Madsen, James H., and Samuel P. Welles, 2000, *Ceratosaurus* (Dinosauria, Theropoda, a revised osteology: *Miscellaneous Publication*, 00-2, Utah Geological Survey, Utah Department of Resources, 80 pages.

Main, Russell P., Kevin Padian, and John R. Horner, 2000, Comparative histology, growth and evolution of archosaurian osteoderms: why did *Stegosaurus* have large dorsal plates?: *Journal of Vertebrate Paleontology*, 20 (Supplement to Number 3), Abstracts of Papers, Sixtieth Annual Meeting, p. 56A.

Makovicky, Peter J., 1997, A new small theropod from the Morrison Formation of Como Bluff, Wyoming: *Journal of Vertebrate Paleontology*, 17 (4), pp. 755–757.

Makovicky, Peter J., and Philip J. Currie, 1996, Discovery of furcula in tyrannosaurid theropods: *Journal of Vertebrate Paleontology*, 19 (Supplement to Number 3), Abstracts of Papers, Fifty-ninth Annual Meeting, p. 50A.

Makovicky, Peter J., and Mark A. Norell, 1998, A partial ornithomimid braincase from Ukhaa Tolgod (Upper Cretaceous, Mongolia): *American Museum Novitates*, p. 60A.

Makovicky, Peter J., Mark A. Norell, and James M. Clark, 1999, A new troodontid from the Upper Cretaceous of Mongolia: *Journal of Vertebrate Paleontology*, 19 (Supplement to Number 3), Abstracts of Papers, Fifty-ninth Annual Meeting, p. 61A.

Makovicky, Peter J., and Hans-Dieter Sues, 1997, A reappraisal of the phylogenetic affinities of *Microvenator celer* (Theropoda:

Dinosauria) from the Cloverly Formation: *Journal of Vertebrate Paleontology*, 17 (Supplement to Number 3), Abstracts of Papers, Fifty-seventh Annual Meeting, p. 62A.

_____, 1998, Anatomy and phyologenic relationships of the theropod dinosaur *Microvenator celer* from the Lower Cretaceous of Montana: *American Museum Novitates*, 3240, pp. 1–26.

Maleev, Evgeny [Eugene] Alexandrovich, 1952, [A new ankylosaur from the Upper Cretaceous of Mongolia]: *Dokladi Akademii Nauk S.S.S.R.*, 87 (2), pp. 273–276.

_____, 1955, Gigantskiye Khishchnye Dinosavri Mongoll [Giant carnivorous dinosaurs from the Upper Cretaceous of Mongolia]: *Ibid.*, 104 (4), pp. 779–782.

_____, 1965, [On the brain of carnivorous dinosaurs]: *Paleontologeskii Zhurnal*, 2, pp. 11–143.

_____, 1974, [Giant carnosaurs of the family Tyrannosauridae]: *Trudy Somestnaya Sovetsko-Mongol'skaya Paleontologischeskaya Ekspeditsiya* [*Transactions of the Joint Soviet-Mongolian Paleontological Expedition*], 1:32–191 (in Russian with English summary).

Manabe, Makoto, 1999, The early evolution of the Tyrannosauridae in Asia: *Journal of Paleontology*, 73 (6), pp. 1176–1178.

Manabe, Makoto, Paul M. Barrett, and Shinji Isaji, 2000, A refugium for relicts: *Nature*, 404, p. 953.

Manley, K., 1989, Technique for distinguishing *bona fide* and suspected gastroliths: *Abstracts of the Symposium on South western Geology and Paleontology 1989*, Museum of Northen Arizona, Flagstaff, p. 18.

_____, 1991*a*, Gastrolith identification and sauropod migration: *Geological Society of America, Abstracts with Program*, 23 (4), p. 45.

_____, 1991*b*, Two techniques for measuring surface polish as applied to gastroliths: *Ichnos*, 1, pp. 313–316.

_____, 1993, Surface polish measurements from *bona fide* and suspected dinosaur gastroliths: *Ibid.*, 2, pp. 167–169.

Mantell, Gideon Algernon, 1824, outlines of the natural history of the environs of Lewes, *in*: T. W. Horsefield, editor, *History and Antiquities of Lewes and its Vicinity*. Baxter, Lewes (also published separately at the same time and place), pp. iii–xxiv.

_____, 1825, Notice on the *Iguanodon*, a newly discovered fossil reptile, from the sandstone of Tilgate Forest, in Sussex: *Philosophical Transactions of the Royal Society*, 115 (1), pp. 179–186.

Manupella, G., 1996, Carta Geológica de Portugal 1/50.000. Falla da Lourinhã. (Coord.), Instituto Geológico e Mineiro de Portugal.

_____, 1998, Geologic data about the "Camadas de Alcobaca" (Upper Jurassic) North of Lourinhã and facies variation: *Classe de Ciências*, 37.

Markthaler, R., 1937, Die Feuerlettenkonglomerate und das Transgressionsproblem im Mittleren Keuper Franken: *Naturhist. Ges. Nürnberg, Abh.*, 26, pp. 51–164, Nürnberg.

Marsh, Othniel Charles, 1877, Notice on new dinosaurian reptiles from the Jurassic Formations: *American Journal of Science*, Third Edition, 14 (53), pp. 514–516.

_____, 1878, Notice of new dinosaurian reptiles: *Ibid.*, Third Series, 15 (87), pp. 241–244.

_____, 1879, Principle characters of American Jurassic dinosaurs, Part II: *Ibid.*, Third Series, 17 (97), pp. 86–92.

_____, 1880, Principle characters of American Jurassic dinosaurs, Part III: *Ibid.*, Third Series, 119, pp. 253–259.

_____, 1884*a*, Prinipal characters on American Jurassic dinosaurs, Part 8, The order of Theropoda: *Ibid.*, Third Series, 27 (38), pp. 329–340.

_____, 1884*b*, On the united metatarsal bones of *Ceratosaurus*: *Ibid.*, Third Series, 28, pp. 161–162.

_____, 1888*a*, Principle characters of American Jurassic dinosaurs, Part VII: *Ibid.*, 27, pp. 160–168.

_____, 1888*b*, Notice of a new genus of Sauropoda and other dinosaurs from the Patomac Formation: *Ibid.*, Third Series, 35, pp. 89–94.

_____, 1889, Notice of gigantic horned Dinosauria from the Cretaceous: *Ibid.*, Third Series, 38, pp. 173–175.

_____, 1891, The gigantic Ceratopsidae, or horned dinosaurs, of North America: *Ibid.*, Third Series, 41, pp. 167–178.

_____, 1892, Notes on some Mesozoic vertebrate fossils: *Ibid.*, Third Series, 44, pp. 171–176.

_____, 1896, The Dinosaurs of North America: *Sixteenth Annual Report of the U.S. Geological Survey*, 1, pp. 133–415.

Marsic, Damien, Parker Carroll, Laura Heffelfinger, Tyler Lyson, Joseph D. Ng, and William Garstka, 2000, DNA sequence of the mitochondrial 12S rRNA gene from *Triceratops* fossils: molecular evidence supports the evolutionary relationship between dinosaurs and birds: *Publications in Paleontology* [The Florida Symposium on Dinosaur Bird Evolution, abstracts of papers], 2, p. 19.

Martill, David M., 1993, Fossils of the Santana and Crato formations, Brazil, *Field Guides to Fossils*, No. 5. Oxford, United Kingdon: Palaeontological Association.

Martill, David M., A. R. I. Cruickshank, E. Frey, P. G. Small, and M. Clarke, 1996, A new crested maniraptoran dinosaur from the Santana Formation (Lower Cretaceous) of Brazil: *Journal of the Geological Society, London*, 153, pp. 5–8.

Martill, David M., Eberhard Frey, Hans-Dieter Sues, and Arthur R. I. Cruickshank, 2000, Skeletal remains of a small theropod dinosaur with associated soft structures from the Lower Cretaceous Santana Formation of northeastern Brazil: *Canadian Journal of Earth Sciences*, 37, pp. 891–900.

Martill, David M., and Paul G. Davis, 1998, Did dinosaurs come up to scratch?: *Nature*, 396, pp. 528–529.

Martin, John, Valérie Martin-Rolland, and Eberhard (Dino) Frey, 1998, Not cranes or masts, but beams: the biomechanics of sauropod necks: *Oryctos*, 1, pp. 113–120.

Martin, Larry D., 1991, Mesozoic birds and the origin of birds, *in*: H.-D. Schultz and L. Trueb, editors, *Origins of the Higher Groups of Tetrapods*. Ithaca, New York: Comstock Publishing Association, pp. 485–540.

_____, 1995, The relationship of *Mononykus* to ornithopod dinosaurs: *Journal of Vertebrate Paleontology*, 15 (Supplement to Number 3), Abstracts of Papers, Fifty-fifth Annual Meeting, p. 43A.

_____, 1997, The difference between dinosaurs and birds as applied to *Mononykus*, *in*: Donald L. Wolberg, Edmund Stump, and Gary Rosenberg, editors. *Dinofest International: Proceedings of a Symposium Held at Arizona State University*. Philadelphia: Academy of Natural Sciences, pp. 337–343.

Martin, Larry D., and C. Rinaldi, 1994, How to tell a bird from a dinosaur: *Maps Digest*, 17 (4), pp. 190–196.

Martin, Larry D., J. D. Stewart, and K. N. Whetstone, 1980, The origin of birds: structure of the tarsus and the teeth: *The Auk*, 97, pp. 86–93.

Martin, Valérie, Éric Buffetaut, and Varavudh Suteethorn, 1994, A new genus of sauropod dinosaur from the Sao Khua Formation (Late Jurassic to Early Cretaceous) of northeastern Thailand: *Comptes Rendus de l'Académie des Sciences de Paris*, 319 (II), pp. 125–132.

Martin-Rolland, Valérie, 1999, Les sauropodes chinois: *Revue Paléobiologie, Genève*, 18 (1), pp. 287–315.

Martin, Valérie, Varavudh Suteethorn, and Éric Buffetaut, 1999, Description of the type and referred material of *Phuwiangosaurus sirindhornae* Martin, Buffetaut and Suteethorn, 1994, a sauropod from the Lower Cretaceous of Thailand: *Oryctos*, 2, pp. 39–91.

Martinez, Ricardo Néstor, 1999, The first Southamerican record of *Massospondylus* (Dinosauria: Sauropodomorpha): *Journal of Vertebrate Paleontology*, 19 (Supplement to Number 3), Abstracts of Papers, Fifty-ninth Annual Meeting, p. 61A.

Martinez, Ricardo Néstor, and Matt C. Lamanna, H. B. Smith, Gabriel Casal, and Marcelo Luna, 1999, New Cretaceous theropod material from Patagonia: *Journal of Vertebrate Paleontology*, 19 (Supplement to Number 3), Abstracts of Papers, Fifty-ninth Annual Meeting, p. 62A.

Martínez, Rubén, 1998*a*, An articulated skull and neck of Sauropoda (Dinosauria: Saurischia) from the Upper Cretaceous of Central Patagonia, Argentina: *Journal of Vertebrate Paleontology*, 18 (Supplement to Number 3), Abstracts of Papers, Fifty-eighth Annual Meeting, p. 61A.

_____, 1998*b*, *Notohypsilophodon comodorensis* gen. et sp. nov. un Hypsilophodontidae (Ornithischia: Ornithopoda) del Cretaceo Superior de Chubut, Patagonia Central, Argentina: *Acta Geologica Leopoldensia*, 21 (46/47), pp. 119–135.

Maryańska, Teresa, 1971, New data on the skull of *Pinacosaurus grangeri* (Ankylosauria): *Palaeontologica Polonica*, 25, pp. 45–53.

_____, 1977, Ankylosauridae (Dinosauria) from Mongolia: *Ibid.*, 37, pp. 85–151.

_____, 1990, Pachycephalosauria, *in*: Philip J. Currie and Kevin Padian, editors, *Encyclopedia of Dinosaurs*. San Diego: Academic Press, pp. 564–577.

_____, 1997, New data on the skull of *Pinacosaurus grangeri* (Ankylosauria): *Palaeontologica Polonica*, 25, pp. 45–53.

Maryańska, Teresa, and Halszka Osmólska, 1974, Pachycephalosauria, a new suborder of ornithischian dinosaurs: *Acta Palaeontologica Polonica,* 26, pp. 133–182.

_____, 1977, The quadrate of oviraptorid dinosaurs: *Ibid.*, 42, pp. 377–387.

_____, 1981, Cranial anatomy of *Saurolophus angustirostris* with comments on the Asian Hadrosauridae (Dinosauria): *Ibid.*, 42, pp. 5–24.

_____, 1984, Phylogenetic classification of ornithischian dinosaurs: *27th International Geological Congress*, 1, pp. 286–287.

_____, 1985, On ornithischian phylogeny: *Acta Palaeontologica Polonica*, 46, pp. 137–150.

_____, 1997, The quadrate of oviraptorid dinosaurs: *Ibid.*, 42, pp. 361–371.

Mateer, Niall J., 1976, New topotypes of *Alamosaurus sanjuanensis* Gilmore (Reptilia: Sauropoda): *Bulletin of the Geological Institutions of the University of Uppsala*, New Series, 6, pp. 93–95.

_____, 1981, The reptilian megafauna from the Kirtland Shale (Late Cretaceous) of the San Juan Basin, New Mexico, *in*: S. Lucas, K. Rigby, Jr., and B. Kues, editors, *Advances in San Juan Basin Paleontology*. Albuquerque: University of New Mexico Press, pp. 49–75.

Mateus, Isabel, Horacio Mateus, Miguel Telles Antunes, Octavio Matteus, Philippe Taquet, Vasco Robeiro, Giuseppe Manuppella, 1998, Upper Jurassic theropod dinosaur embryos from Lourinha (Portugal): *Memorias da Academia de Ciencias de Lisboa*, 37, pp. 101–109.

Mateus, Octavio, 1998, *Lourinhanosaurus antunesi*, a new Upper Jurassic allosauroid (Dinosauria: Theropoda) from Lourinha, Portugal: *Classe de Ciências*, 37, pp. 111–124, erratum.

Mateus, Octavio, 1999, Upper Jurassic dinosaurs of Lourinhã (Portugal): *Journal of Vertebrate Paleontology*, 19 (Supplement to Number 3), Abstracts of Papers, Fifty-ninth Annual Meeting, p. 62A.

Mateus, Octavio, Philippe Taquet, Miguel Telles Antunes, Horatio Matteus, and Vasco Ribeiro, 1998, Theropod dinosaur nest from Lourinha, Portugal: *Journal of Vertebrate Paleontology*, 18 (Supplement to Number 3), Abstracts of Papers, Fifty-eighth Annual Meeting, p. 61A.

Matheron, Philippe, 1869, Notice sur les reptiles fossiles des dépôts fluvio-lacustres crétacés du bassin à lignite de Fuveau: *Mémoires de l'Académie des sciences, belles-lettres et arts de Marseille*, pp. 345–379,

Matsukawa, M., 1983, Stratigraphy and sedimentary environments of the Sanchu Cretaceous, Japan: *Memoirs of Ehime University, Natural Science*, Series D (Earth Science), 9 (4), pp. 1–50.

Matthew, William Diller, and Barnum Brown, 1922, The family Deinodontidae, with notice of a new genus from the Cretaceous of Alberta: *Bulletin of the American Museum of Natural History*, 46, pp. 367–385.

_____, 1923, Preliminary notices of skeletons and skulls of Deinodontidae from the Cretaceous of Alberta: *American Museum Novitates*, 89, pp. 1–10.

May, Kevin C., and Roland A. Gangloff, 1999, New dinosaur bonebed from the Prince Creek Formation, Colville River, National Petroleum Reserve — Alaska: *Journal of Vertebrate Paleontology*, 19 (Supplement to Number 3), Abstracts of Papers, Fifty-ninth Annual Meeting, p. 62A.

McCord, R., 1997, An Arizona titanosaurid sauropod and revision of the Late Cretaceous Adobe fauna: *Journal of Vertebrate Paleontology*, 17, pp. 620–622.

McGee, Elizabeth M., 1999, An analysis of bone orientation at the Cleveland-Lloyd Dinosaur Quarry using vector summation: a new spin on an old technique: *Journal of Vertebrate Paleontology*, 19 (Supplement to Number 3), Abstracts of Papers, Fifty-ninth Annual Meeting, p. 62A.

McGillavray, Edward Smith, 1852, *Narrative of the Voyage of the H.M.S. Rattlesnake*. London: Boone [facsimile edition: Libraries Board of South Australia, 1967], vol. 1, pp. 1–402, vol. 2, pp. 1–395.

McIntosh, John S., 1990, Sauropoda, *in*: David B. Weishampel, Peter Dodson, and Halszka Osmólska, editors, *The Dinosauria*. Berkeley and Los Angeles: University of California Press, pp. 345–401.

McIntosh, John S., and David B. Berman, 1975, Description of the palate and lower jaw of *Diplodocus* (Reptilia: Saurischia) with remarks on the nature of the skull of *Apatosaurus*: *Journal of Paleontology*, 49, pp. 187–199.

McIntosh, John S., Michael K. Brett-Surman, and James O. Farlow, 1997, Sauropods, *in*: Farlow and Brett-Surman, editors, *The Complete Dinosaur*. Bloomington and Indianapolis, Indiana University Press, pp. 264–290.

McIntosh, John S., Wade E. Miller, Kenneth L. Stadtman, and David D. Gillette, 1996, The osteology of *Camarasaurus lewisi* (Jensen, 1988): *BYU Geology Studies*, 41, pp. 73–115.

McKinney, M. L., and K. J. McNamara, 1991, *Heterochrony, The Evolution of Ontogeny*. New York: Plenum, 437 pages.

McMahon, T. A., and J. T. Bonner, 1983, *On Size and Life*. New York: Scientific American Books, W. H. Freeman, 255 pages.

McMillan, I. K., 1999, The foraminafera of the Late Valanginian to Hauterivian (Early Cretaceous) Sundays River Formation of the Algoa Basin, Eastern Cape Province, South Africa: *Annals of the South Africa Museum*, 106, pp. 1–120.

McNeil, Paul E., 1999, Histology of ossified tendons in ornithischian dinosaurs: *Journal of Vertebrate Paleontology*, 19 (Supplement to Number 3), Abstracts of Papers, Fifty-ninth Annual Meeting, p. 63A.

McWhinnie, Lorrie A., Rothschild, Bruce M., and Kenneth Carpenter, 1998, Post-traumatic chronic osteomyletis in *Stegosaurus* dermal spines: *Journal of Vertebrate Paleontology*, 18 (Supplement to Number 3), Abstracts of Papers, Fifty-eighth Annual Meeting, p. 62A.

McCarthy, Steve, and Mick Gilbert, 1994, *The Crystal Palace Dinosaurs: The Story of the World's First Prehistoric Sculptures*. London: The Crystal Palace Foundation, 100 pages.

Merewether, E. A., W. A. Cobban, and E. T. Cavanaugh, 1978, Frontier Formation and equivalent rocks in eastern Wyoming: *The Mountain Geologist*, 16, pp. 67–101.

Meyer, Hermann von, 1837, Mitteilung an Prof. Bronn (*Plateosaurus engelhardti*): *Neus Jarbuch für Mineralogie, Geologie, und Paläontologie*, 1837, p. 817, Stuttgart.

_____, 1839, (Letter on various fossil vertebrates): *Ibid.*, 1839, pp. 76–79, Stuttgart.

_____, 1855, *Zur Fauna der Vorwelt. Die Saurier des Muschel kalkes, mit Rücksicht auf die Saurier aus Buntem Sandstein und Keuper*. Frankfurt am Main, 167 pages.

Middleton, Kevin M., and Stephen M. Gatesy, 2000, Theropod forelimb design and evolution: *Zoological Journal of the Linnean Society*, 128, pp. 149–187.

Miles, C. A., Kenneth Carpenter, and K. Cloward, 1998, A new skeleton of *Coelurus fragilis* from the Morrison Formation of Wyoming: *Journal of Vertebrate Paleontology*, 18 (Supplement to Number 3), Abstracts of Papers, Fifty-eighth Annual Meeting, p. 64A.

Milner, Angela C., 1999, Theropod dinosaurs of the Purbeck Limestone Group: *Life and Environments in Purbeck Times*, 19th–22nd March 199, Dorset County Museum, abstracts of papers, pp. 39–40.

Molnar, Ralph E., 1973, The cranial morphology and mechanics of *Tyrannosaurus rex* (Reptilia: Saurischia): Ph.D. dissertation, University of California, Los Angeles, 451 pages.

_____, 1978, A new theropod dinosaur from the Upper Cretaceous of central Montana: *Journal of Paleontology*, 52 (1), pp. 73–82.

_____, 1980, An albertosaur from the Hell Creek Formation of Montana: *Journal of Paleontology*, 54 (1), pp. 102–108.

_____, 1985, Alternatives to *Archaeopteryx*: a survey of proposed early or ancestral birds, *in*: Max K. Hecht, John H. Ostrom, G. Viohl, and Peter Wellnhofer, editors, *The Beginnings of Birds*. Eichstätt: Freunde de Jura-Museums, pp. 209–217.

_____, 1989, Terrestrial tetrapods in Cretaceous Antarctica, *in*: J. A. Crane, editor, *Origins and Evolution of the Antarctic Biota*, Geological Society Special Publication, 47, pp. 131–140.

_____, 1996, Preliminary report on a new ankylosaur from the Early Cretaceous of Queensland, Australia: *Ibid.*, pp. 653–668.

Molnar, Ralph E., Alejandro Lopez Angriman, and Zulma Gasparini, 1996, An antarctic Cretaceous theropod, *in*: Fernando S. Novas and Ralph E. Molnar, editors, *Proceedings of the Gondwanan Dinosaur Symposium*: *Memoirs of the Queensland Museum*, 39 (part 3), pp. 669–674.

Molnar, Ralph E., and Kenneth Carpenter, 1989, The Jordan theropod (Maastrichtian, Montana, U.S.A.) referred to the genus *Aublysodon*: *Geobios*, 22 (4), pp. 445–454.

Molnar, Ralph E., and H. Trevor Clifford, 2000, Gut contents of a small ankylosaur: *Journal of Vertebrate Paleontology*, 20 (1), pp. 194–196.

Molnar, Ralph E., Timothy F. Flannery, and Thomas H. V. Rich, 1981, An allosaurid theropod dinosaur from the Early Cretaceous of Victoria, Australia: *Alcheringa*, pp. 141–146.

_____, 1985, Aussie *Allosaurus* after all: *Journal of Paleontology*, 59 (6), pp. 651–570.

Molnar, Ralph E., Seriozha M. Kurzanov, and Dong Zhi-Ming [Zhiming], 1990, Carnosauria, *in*: David B. Weishampel, Peter Dodson, and Halszka Osmólska, editors, *The Dinosauria*. Berkeley and Los Angeles: University of California Press, pp. 169–209.

Molnar, Ralph E., and N. S. Pledge, 1980, A new theropod dinosaur from South Australia: *Acheringa*, 4, pp. 281–287.

Monastersky, R., 2000, All mixed up over birds and dinosaurs: *Science News*, 157, p. 38.

Monbaron, Michel, 1983, Dinosauriens du Haut Atlas central (Meroc): état de recherche et précision sur la découverte d'un squelette complet de grand Cétiosaure: *Actes Socít' jurassienne d'Émulation (Porrentrury)*, pp. 203–234.

Monbaron, Michel, Dale A. Russell, and Philippe Taquet, 1999, *Atlasaurus imelakei* n.g., n.s., a brachiosaurid-like sauropod from the Middle Jurassic of Morocco: *Comtes Rendu des Séances de l'Académie des Sciences, Paris, de la Terre et des Planètes*, 329, pp. 519–526.

Monbaron, Michel, and Philippe Taquet, 1981, Decouverte du squelette complet d'un grand Cetiosaure (Dinosaure Sauropode) dans le hassin jurassique de Tilougguit Haut Atlas central Maroc: *Comtes Rendu des Séances de l'Académie des Sciences, Paris, Series II*, pp. 243–246.

Mongelli, Anthony, Jr., and David J. Varricchio, 1998, Theropod teeth of the Lower Two Medicine Formation (Campanian) of northwestern Montana: *Journal of Vertebrate Paleontology*, 18 (Supplement to Number 3), Abstracts of Papers, Fifty-eighth Annual Meeting, p. 65A.

Mongelli, Anthony, Jr., David J. Varrichio, and John J. Borkowski, 1999, Wear surfaces and breakage patterns of tyrannosaurid (Theropoda: Coelurosauria) teeth: *Journal of Vertebrate Paleontology*, 19 (Supplement to Number 3), Abstracts of Papers, Fifty-ninth Annual Meeting, p. 64A.

Moodie, Ron, 1910, An armoured dinosaur from the Upper Cretaceous of Wyoming: *University of Kansas Science Bulletin*, 5, pp. 257–273.

Moratalla, J. J., and Jaime E. Powell, 1994, Dinosaur nesting patterns, *in*: Kenneth Carpenter, Karl F. Hirsch, and John R. Horner, editors, *Dinosaur Eggs and Babies*. Cambridge: Cambridge University Press, pp. 37–46.

Morris, J., 1843, *A Catalogue of British Fossils*. London: British Museum, 222 pages.

Morris, William J., 1976, Hypsilophodontid dinosaurs: a new species and comments on their systematics, *in*:, C. S. Churcher, editor, *Athlon. Essays on Paleontology in Honour of Loris Shane Russell*. Toronto: University of Toronto Press, pp. 93–113.

Munter, R. C., Two theropod dinosaur specimens from Huizchal Canyon, Mexico: *Journal of Vertebrate Paleontology*, 19 (Supplement to Number 3), Abstracts of Papers, Fifty-ninth Annual Meeting, p. 65A.

Munyikwa, Darlington, and Michael A. Raath, 1999, Further material of the ceratosaurian dinosaur *Syntarsus* from the Elliot Formation (Early Jurassic) of South Africa: *Palaeontologia Africana*, 35, pp. 55–59.

Murry, Phillip A., and Robert A. Long, 1989, Geology and paleontology of the Chinle Formation, Petrified Forest National Park and vicinity, Arizona, and a discussion of vertebrate fossils of the southwestern Upper Triassic, *in*: Spencer G. Lucas and Adrian P. Hunt, editors, *The Dawn of the Age of Dinosaurs in the American Southwest*. Albuquerque: New Mexico Museum of Natural History, 414 pages.

_____, 1997, Dockum Group, *in*: Philip J. Currie and Kevin Padian, editors, *Encyclopedia of Dinosaurs*. San Diego: Academic Press, pp. 191–193.

Mussell, Jason, and David B. Weishampel, 2000, *Magyarosaurus* and the possible role of progenesis in sauropod development: *Journal of Vertebrate Paleontology*, 20 (Supplement to Number 3), Abstracts of Papers, Sixtieth Annual Meeting, pp. 59A–60A.

Myhrvold, Nathan P., and Philip J. Currie, 1997, Supersonic sauropods? Tail dynamics in the diplodocids: *Paleobiology*, 23 (4), pp. 393–409.

Nagao, Takumi, 1936, *Nipponosaurus sachalinensis*, a new genus and species of trachodont dinosaur from Japanese Saghalien: *Journal of the Faculty Science, University Hokkaido, Geology and Minerology*, 3 (2), pp. 185–220.

Naish, Darren, 1998*a*, Theropods eating theropods in the Wealden Group fauna of England: evidence from a previously undescribed tibia, *in*: John W. M. Jagt, Paul H. Lambers, Eric W. A. Mulder, and Anne S. Schulp, editors, *Third European Workshop on Vertebrate Palaeontology*, Maastricht, 6–9, May 1998, Programme and Abstracts Field Guide, [no pagination].

_____, 1998*b*, Predatory dinosaurs in England's Lower Cretaceous: crocodile heads, puffin snouts and the odd Lazarus taxon: *Newsletter of the European Paleontology Association*, 13, p. 15.

_____, 1999*a*, Fox, Owen and the small Wealden theropods *Calamos pondylus* and *Aristosuchus*: *Journal of Vertebrate Paleontology*, 19 (Supplement to Number 3), Abstracts of Papers, Fifty-ninth Annual Meeting, p. 66A.

_____, 1999*b*, Theropod dinosaur diversity and palaeobiology in the Wealden Group (Early Cretaceous) of England: evidence from a previously undescribed tibia: *Geologie en Mijnbouw*, 78, pp. 367–373.

_____, 2000, A small, unusual theropod (Dinosauria) femur from the Wealden Group (Lower Cretaceous) of the Isle of Wight, England: *Neus Jahrbuch für Geologie und Paläontologie, Monatshefte*, 2000 (4), pp. 217–234.

Nessov, Lev, A., 1995, *Dinosaurs of northern Eurasia: new information on group composition, ecology and paleobiogeographics*. St. Petersburg State University, Earth Science-research Institute: The Publishing House of the University of St. Petersburg, pp. 35–58.

Nessov, Lev A., and L. F. Kaznyshkina, 1989, Dinozavry-tseratopsii i krokodily mezozoya Sredney Asii, *in*: T. N. Bogdanova and L. I. Khozatskiy, editors, *Teoreticheskiye i prikladnyye aspekty sovremennoy paleontologii; trudy XXXIII sessii Vsesoyuznogo, Lenningrad*, pp. 144–154.

Newton, E. T., 1899, On a megalosauroid jaw from Rhaetic beds near Bridgend (Glamorgahshire): *Quarterly Journal of the Geological Society, London*, 55, pp. 89–96, London.

Nopcsa, Baron Franz [also Ferencz, Francis] (von Felsö-Szilvás), 1900, Dinosaurierreste aus Siebenbürgen. I. Schädel von *Limnosaurus transsylvanicus* nov. gen. et spec.—*Denkschriften der Akademie der Wissenschaften*, 68, pp. 555–591.

_____, 1914, Über das Vorkommen der Dinosaurier in Siebenbürgen: *Verh. Zool. Bot. Ges.*, 54, pp. 12–14.

_____, 1915, Erdély dinoszauruszai: *Magyar királyi Földtani Intézet Évkönyve*, 23, pp. 3–24.

_____, 1923, Notes on the British dinosaurs, Part IV: *Acanthopholis*: *Geological Magazine*, 60, pp. 193–199.

_____, 1929, Dinosauriereste aus Siebenbenburgen, V: *Geologica Hungarica Series Paleontology*, 4, pp. 1–76.

Norell, Mark A., James M. Clark, and Luis M. Chiappe, 1993, Naming names: *Nature*, 366, p. 518.

_____, 1996, Djadokhta series theropods: a summary review: *Dinofest International Symposium, April 18–21, 1996, Program and Abstracts*, p. 86.

Norell, Mark A., James M. Clark, Demberelynin Dashveg, Rinchen Barsbold, Luis M. Chiappe, Amy R. Davidson, Malcolm C. McKenna, Altangerel Perle, and Michael J. Novacek, 1994, A theropod dinosaur embryo and the affinities of the Flaming Cliffs dinosaur eggs: *Science*, 266, pp. 779–882.

Norell, Mark A., and Peter J. Makovicky, 1997, Important features of the dromaeosaur skeleton: information from a new specimen: *American Museum Novitates*, 3215, 28 pages.

_____, 1998, A revised look at the osteology of dromaeosaurs: evidence from new specimens of *Velociraptor*: *Journal of Vertebrate Paleontology*, 18 (Supplement to Number 3), Abstracts of Papers, Fifty-eighth Annual Meeting, p. 66A.

_____, 1999, Important features of the dromaeosaurid skeleton II: information from newly collected specimens of *Velociraptor mongoliensis*: *American Museum Novitates*, 3282, pp. 1–45.

Norell, Mark A., Peter J. Makovicky, and James M. Clark, 1997, A *Velociraptor* wishbone: *Nature*, 389, p. 247.

_____, 2000*a*, A new troodontid theropod

from Ukhaa Tolgod, Mongolia: *Journal of Vertebrate Paleontology*, 20 (1), pp. 7–11.
_____, 2000*b*, A review of the Dromaeosauridae: *Publications in Paleontology* [The Florida Symposium on Dinosaur Bird Evolution, abstracts of papers], 2, p. 20.
Norman, David B. [Bruce], 1980, On the ornithischian dinosaur *Iguanodon bernissartensis* from Belgium: *Mémoires de l'Institut Royal des Sciences Naturelles de Belgique*, 178, pp. 1–105.
_____, 1984*a*, On the cranial morphology and evolution of ornithopod dinosaurs: *Symposium of the Zoological Society of London*, 1984, pp. 521–547.
_____, 1984*b*, A systematic reappraisal of the reptile order Ornithischia, *in*: W. Reif and F. Westphal, editors, *Third Symposium on Mesozoic Terrestrial Ecosystems, Short Papers*, Attempto Verlag, Tübingen University Press, pp. 157–162.
_____, 1985, *The Illustrated Encyclopedia of Dinosaurs: An Original and Compelling Insight into Life in the Dinosaur Kingdom*. New York: Crescent Books (Crown Publishing, Inc.), 208 pages.
_____, 1986, On the anatomy of *Iguanodon atherfieldensis* (Ornithischia: Ornithopoda): *Bulletin, Institut Royale d'Histoire Naturelle de Belgique*, 56, pp. 281–372.
_____, 1990, Problematic Theropoda: "Coelurosaurs," *in*: David B. Weishampel, Peter Dodson, and Osmólska, editors, *The Dinosauria*. Berkeley and Los Angeles: University of California Press, pp. 280–305.
_____, 1991, *Dinosaurier*. München, Germany: C. Bertelsmann Verlag, 192 pages.
_____, 1998*a*, On Asian ornithopods (Dinosauria: Ornithischia). 3. A new species of iguanodontid dinosaur: *Zoological Journal of the Linnean Society*, 122, pp. 291–348.
_____, 1998*b*, *Probactrosaurus* from Asia and the origin of Hadrosaurs: *Journal of Vertebrate Paleontology*, 18 (Supplement to Number 3), Abstracts of Papers, Fifty-eighth Annual Meeting, p. 66A.
_____, 1999, *Iguanodon* the Early Cretaceous ornithopod?: *Ibid.*, 19 (Supplement to Number 3), Abstracts of Papers, Fifty-ninth Annual Meeting, p. 66A.
Norman, David B., and Paul M. Barrett, 1999, Ornithischian dinosaurs from the basal Cretaceous (Berriasian) of England: *Life and Environments in Purbeck Times*, 19th–22nd March 199, Dorset County Museum, abstracts of papers, pp. 45–46.
Norman, David B., and David B. Weishampel, 1990, Iguanodontidae and related ornithopods, *in*: Weishampel, Peter Dodson, and Halszka Osmolska, editors, *The Dinosauria*. Berkeley and Los Angeles: University of California Press, pp. 510–533.
Norton, James M., 1998, Eggs and nest place constraints on dinosaur lung morphology and function: *Journal of Vertebrate Paleontology*, 18 (Supplement to Number 3), Abstracts of Papers, Fifty-eighth Annual Meeting, p. 67A.
_____, 1999, A computer model for theropod rib motion: *Journal of Vertebrate Paleontology*, 19 (Supplement to Number 3), Abstracts of Papers, Fifty-ninth Annual Meeting, p. 66A.
Novas, Fernando E., 1989, The tibia and tarsus in Herrerasauridae (*Dinosauria, incertae sedis*) and the origin and evolution of the dinosaurian tarsus: *Journal of Paleontology*, 63, pp. 677–690.
_____, 1992*a*, Phylogenetic relationships of the basal dinosaurs, the Herrerasauridae: *Palaeontology*, 35, pp. 51–62, London.
_____, 1992*b*, La evolución de los dinosaurios carnívoros, *in*: José Luis Sanz and Ang'gela D. Buscalioni, editors, *Los Dinosaurios y su Entorno Biótico. Actas II Curso de Paleontologica en Cuenca*. Spain: Instituto "Juan de Valdés," Ayuntamiento de Cuenca, pp. 123–163.
_____, 1993, New information on the systematics and postcranial skeleton of *Herrerasaurus ischigualastensis* (Theropoda: Herrerasauridae) from the Ischigualasto Formation (Upper Triassic) of Argentina: *Journal of Vertebrate Paleontology*, 13 (4), pp. 400–423.
_____, 1996*a*, Dinosaur monophyly: *Journal of Vertebrate Paleontology*, 16 (4), pp. 723–741.
_____, 1996*b*, Alvarezsauridae, Cretaceous basal birds from Patagonia and Mongolia, *in*: Fernando S. Novas and Ralph E. Molnar, editors, *Proceedings of the Gondwanan Dinosaur Symposium*: *Memoirs of the Queensland Museum*, 39 (part 3), pp. 675–702.
_____, 1997*a*, Abelisauridae, *in*: Philip J. Currie and Kevin Padian, editors, *Encyclopedia of Dinosaurs*. San Diego: Academic Press, pp. 1–2.
_____, 1997*b*, Herrerasauridae, *Ibid.*, pp. 303–311.
_____, 1997*c*, New evidence concerning avian origins from the Late Cretaceous of Patagonia: *Nature*, 387, pp. 390–392.
_____, 2000, Avian-like traits in the ilium of *Unenlagia coma huensis*: *Publications in Paleontology* [The Florida Symposium on Dinosaur Bird Evolution, abstracts of papers], 2, p. 21.
Novas, Fernando E., S. Apesteguía, D. Pol, and A. Cambiaso, 1999, Un probable troodontido (Theropoda–Coelurosauria) del Cretácico Superior de Patagonia: *Ameghiniana, Revista de la Asociación Paleontológica Argentina, (Resúmenes)*, 36 (Número 4-Suplemento), p. 17R.
Novas, Fernando E., Rubén D. Martínez, S. de Valais, and A. Ambrosio, 1999, Nuevos registros de Charcharodontosauridae (Dinosauria, Theropoda) en el Cretácico de Patagonia: *Ameghiniana, Revista de la Asociación Paleontológica Argentina, (Resúmenes)*, 36 (Número 4-Suplemento), p. 17R.
Novas, Fernando E., and Pablo F. Puerta, 1997, New evidence concerning avian origins from the Late Cretaceous of Patagonia: *Nature*, 387, pp. 390–92.
Nowiński, Aleksander, 1971, *Nemegtosaurus mongoliensis* n. gen., n. sp. (Sauropoda) from the Uppermost Cretaceous of Mongolia. Results of the Polish-Mongolian Palaeontological Expeditions — Part III: *Palaeontological Polonica*, 25, pp. 57–81, Warsaw.
Obradovich, J. D., 1993, A Cretaceous time scale, *in*: W. Caldwell and E. Kauffman, editors, *Evolution of the Western Interior Basin*, Geological Association of Canada Special Papers, 39, pp. 379–396.
O'Connor, Patrick M., and Scott D. Sampson, 1998, The vertebral column of *Majungatholus atopus* (Theropoda: Abelisauridae) from the Late Cretaceous of Madagascar: *Journal of Vertebrate Paleontology*, 18 (Supplement to Number 3), Abstracts of Papers, Fifty-eighth Annual Meeting, p. 67A.
Olsen, Paul E., 1980, A comparison of the vertebrate assemblages from the Newark and Hartford Basins (early Mesozoic, Newark Supergroup) of Eastern North America, *in*: Louis L. Jacobs, editor, *Aspects of Verebrate History*, Museum of Northern Arizona, Flagstaff, pp. 35–54.
Olsen, Paul E., and Peter M. Galton, 1984, A review of the reptile and amphibian assemblages from the Stormberg Series of southern Africa, with special emphasis on the footprints and the age of the Stormberg: *Palaeontologia Africana*, 25, pp. 87–109, Johannesburg.
Olson, Storrs L., 2000, Birds — dino flap: *Backbone*, 13 (2), pp. 1–3.
Olshevsky, George, 1991, *A Revision of the Parainfraclass Archosauria Cope 1869, Excluding the Advanced Crocodylia*. Buffalo, New York: Publications Requiring Research, iv, 196 pages.
_____, 2000, An annotated checklist of dinosaur species by continent: *Mesozoic Meanderings*, 3, 157 pages.
Olshevsky, George, Tracy L. Ford, and S. Yamamoto, 1995*a*, [The origin and evolution of the tyrannosaurids, (part 1): *Kyoryugaku Saizensen* [*Dino-Frontline*], 9, pp. 92–119.
_____, (part 2): *Ibid.*, 10, pp. 75–99.
Organ, Chris, 2000, Ossified tendons in ornithischians: a locomotor adaptation that reduces tail deflection: *Journal of Vertebrate Paleontology*, 20 (Supplement to Number 3), Abstracts of Papers, Sixtieth Annual Meeting, p. 61A.
Osborn, Henry Fairfield, 1899, A skeleton of *Diplodocus*: Memoirs of the American Museum of Natural History, 1, pp. 191–214.
_____, 1904, Manus, sacrum, and caudals of Sauropoda: *Bulletin of the American Museum of Natural History*, 20, pp. 181–190.
_____, 1912, Crania of *Tyrannosaurus* and *Allosaurus*: *Memoirs of the American Museum of Natural History*, 1, pp. 33–54.
_____, 1923, A new genus and species of Ceratopsia from New Mexico, *Pentaceratops sternbergii*: *American Museum of Natural History Novitates*, 93, pp. 1–3.
_____, 1924, Three new Theropoda, *Protoceratops* zone, central Mongolia: *Ibid.*, 144, pp. 1–12.
Osborn, Henry Fairfield, and Charles Craig Mook, 1921, *Camarasaurus*, *Amphicoelias* and other sauropods of Cope: *Memoirs of the American Museum of Natural History*, (new series), 3, pp. 247–287.
Osmólska, Halszka, 1987, *Borogovia gracilicrus* gen. et sp.n., a new troodontid dinosaur from the Late Cretaceous of Mongolia:

Acta Palaeontological Polonica, 32 (1–2), pp. 133–150.

_____, 1990, Theropoda, *in*: David B. Weishampel, Peter Dodson and Halszka Osmólska, editors, *The Dinosauria*. Berkeley and Los Angeles: University of California Press, pp. 320–344.

_____, 1997, Ornithomimosauria, *in*: Philip J. Currie and Kevin Padian, editors, *Encyclopedia of Dinosaurs*. San Diego: Academic Press, pp. 499–505.

Osmólska, Halszka, and Rinchen Barsbold, 1990, Troodontidae, *in*: David B. Weishampel, Peter Dodson, and Osmólska, editors, *The Dinosauria*. Berkeley and Los Angeles: University of California Press, pp. 259–268.

Osmlólska, Halszka, Ewa Roniewicz, and Rinchen Barsbold, 1972, A new dinosaur, *Gallimimus bullatus* n. gen., n. sp. (Ornithomimidae) from the Upper Cretaceous of Mongolia: *Paleontologica Polonica*, 27, pp. 103–143.

Ostrom, John H., *Parasaurolophus cytocristatus*, a crested hadrosaurian dinosaur from New Mexico: *Fieldiana*, 14, pp. 143–168.

_____, 1969*a*, A new theropod dinosaur from the Lower Cretaceous of Montana: *Postilla, Peabody Museum of Natural History*, 128, pp. 1–17.

_____, 1969*b*, Osteology of *Deinonychus antirrhopus*, an unusual theropod from the Lower Cretaceous of Montana: *Bulletin of the Peabody Museum of Natural History*, 30, pp. 1–165.

_____, 1970, Stratigraphy and paleontology of the Cloverly Formation (Lower Cretaceous) of the Bighorn Basin area, Wyoming and Montana: *Ibid.*, 35, 234 pages.

_____, 1976*a*, Some hypothetical anatomical stages in the evolution of avian flight: *Smithsonian Contributions to Paleontology*, 27, pp. 1–21.

_____, 1976*b*, On a new specimen of the Lower Cretaceous theropod dinosaur *Deinonychus antirrhopus*: *Breviora*, 439, pp. 1–21.

_____, 1976*c*, *Archaeopteryx* and the origin of birds: *Biological Journal of the Linnean Society*, 8, pp. 91–182.

_____, 1978, The osteology of *Compsognathus longipes* Wagner: *Zitteliana*, 4, pp. 73–118.

_____, 1986, The cursorial origin of avian flight: *Memoires of the California Academy of Sciences*, 8 pp. 73–81.

_____, 1987, *Protoavis*, a Triassic bird?: *Archaeopteryx*, 14, pp. 39–42.

_____, 1990, The Dromaeosauria, *in*: David B. Weishampel, Peter Dodson, and Osmólska, editors, *The Dinosauria*. Berkeley and Los Angeles: University of California Press, pp. 269–279.

_____, 1991, The bird in the bush: *Nature*, 353, p. 212.

_____, 1994, On the origin of birds and avian flight, *in*: D. R. Prothero and R. M. Schoch (conv.), *Major Features of Vertebrate Evolution*. Short Courses in Paleontology 7, pp. 160–177.

_____, 1996, The questionable validity of *Protoavis*: *Archaeopteryx*, 14, pp. 39–42, Eichstätt.

Ostrom, John H., and John S. McIntosh, 1966, *Marsh's Dinosaurs. the Collections from Como Bluff*. New Haven, Connecticut: Yale University Press, 338 pages.

Ostrom, John H., and Wellnhofer, Peter, 1986, The Munich specimen of *Triceratops* with a revision of the genus: *Zitteliana Abandlungen der Bayerischen Staatssammlung für Paläontologie und historische Geologie*, 14, pp. 111–158.

_____, 1990, *Triceratops*, an example of flawed systematics, *in*: Kenneth Carpenter and Philip J. Currie, editors, *Dinosaur Systematics: Perspectives and Approaches*. Cambridge and New York: Cambridge University Press, pp. 245–254.

Owen, Richard, 1842, Report on British fossil reptiles. Part II. *Report of the British Association for the Advancement of Science 1842 (1841)*, pp. 60–204.

_____, 1854, On some fossil reptilian and mammalian remains from the Purbecks: *Quarterly Journal of the Geological Society*, 10, pp. 420–433.

_____, 1861, Monograph of the fossil Reptilia of the Wealden and Purbeck formations. Part V. Lacertilia (*Nuthetes* etc.): *Palaeontological Society Monographs*, 12, pp. 31–39.

_____, 1874, Monograph of the fossil Reptilia of the Wealden and Purbeck formations. Supplement V. Dinosauria (*Iguanodon*). *Ibid.*, 27, pp. 1–18.

_____, 1876, A monograph on the fossil Reptilia of the Wealden and Purbeck formations. Supplement VII. Crocodilia (*Poekipleuron*) and Dinosauria? (*Chondrosteosaurus*), [Wealden]: *Ibid.*, 30, pp. 1–7.

_____, 1879, Monograph of the fossil Reptilia of the Wealden and Purbeck formations. Supplement IX. Crocodilia (*Goniopholis, Brachydectes, Nannosuchus, Theropsuchus*, and *Nuthetes*), *Ibid.*, 33, pp. 1–19.

Owen-Smith, R. N., 1988, *Megaherbivores: The Influence of Very Large Body Size on Ecology*. Cambridge: University of Cambridge Press, 269 pages.

Padian, Kevin, 1986, On the type material of *Coelophysis* Cope (Saurischia: Theropoda) and a new specimen from the Petrified Forest of Arizona (Late Triassic: Chinle Formation), *in*: Kevin Padian, editor, *The Beginning of the Age of Dinosaurs: Faunal Change Across the Triassic–Jurassic Boundary*. Cambridge and New York: Cambridge University Press, pp. 45–60.

_____, 1996, Book review, *The Little Dinosaurs of Ghost Ranch*: *Journal of Vertebrate Paleontology*, 16 (2), pp. 365–366.

_____, 1997*a*, Ornithischia, *in*: Philip J. Currie and Padian, editors, *Encyclopedia of Dinosaurs*. San Diego: Academic Press, pp. 494–498.

_____, 1997*b*, Phylogeny of dinosaurs, *Ibid.*, pp. 546–551.

_____, 2000, Methods and standards of evidence: why the bird-dinosaur controversy is dead: *Publications in Paleontology* [The Florida Symposium on Dinosaur Bird Evolution, abstracts of papers], 2, p. 21.

Padian, Kevin, and Luis M. Chiappe, 1998, The origin and early evolution of birds: *Biology Review*, 73, pp. 1–42.

Padian, Kevin, John R. Horner, and Armand J. de Ricqlès, 1999, Dinosaurian growth rates and the evolution of life history strategies: *Journal of Vertebrate Paleontology*, 19 (Supplement to Number 3), Abstracts of Papers, Fifty-ninth Annual Meeting, pp. 67A–68A.

Padian, Kevin, John R. Hutchinson, and Thomas R. Holtz, Jr., 1997, Phylogenetic definitions and nomenclature of the major taxonomic categories of the theropod dinosaurs: *Journal of Vertebrate Paleontology*, 18 (Supplement to Number 3), Abstracts of Papers, Fifty-eighth Annual Meeting, p. 68A.

_____, 1999, Phylogenetic definitions and nomenclature of the major taxonomic categories of the carnivorous Dinosauria (Theropoda): *Ibid.*, 19 (1), pp. 69–80.

Padian, Kevin, and Cathleen L. May, 1993, The earliest dinosaurs: *New Mexico Museum of Natural History and Science Bulletin* 3, pp. 379–381.

Paik, Sung, 2000, Bone chip-filled burrows associated with bored dinosaur bone in floodplain paleosols of the Cretaceous Hasandong Formation, Korea: *Palaeogeography, Palaeoclima tology, and Palaeogeograohy*, 157, pp. 213–225.

Pang Qiqing, and Cheng Zhengwu, 1998, A new ankylosaur of Late Cretaceous from Tianzhen, Shanxi: *Progress in Natural Science*, 8 (3), pp. 326–334.

_____, 2000, A new family of sauropod dinosaur from the Upper Cretaceous of Tianzhen, Shanxi Province, China: *Acta Geologica Sinica*, 74 (2), pp. 117–125.

Pang Qiqing, Cheng Zhengwu, Yang Jianping, Xie Manze, Zhu Caifa, and Luo Junlin, 1996, The preliminary report on Late Cretaceous dinosaur fauna expedition, Tianzhen, Shanxi: *Journal of Hebei College Geology*, 19 (3–4), p. 227.

Papp, Michael J., and Lawrence Witmer M., 1998, Cheeks, beaks or freaks: a critical appraisal of buccal soft-tissue anatomy in ornithischian dinosaurs: *Journal of Vertebrate Paleontology*, 18 (Supplement to Number 3), Abstracts of Papers, Fifty-eighth Annual Meeting, p. 69A.

Parkinson, J., 1930, *The dinosaur in East Africa: An Account of the Giant Reptile Beds of Tendaguru, Tanganyika Territory*. London: H. F. & G. Witherby, 192 pages.

Parrish, J. Michael, 1998, The origin of high browsing: the paleoecological significance of the sauropodomorph bauplan: *Journal of Vertebrate Paleontology*, 18 (Supplement to Number 3), Abstracts of Papers, Fifty-eighth Annual Meeting, p. 69A.

_____, 1999, Dinosaur teeth from the Upper Cretaceous (Turonian–Judithian) of southern Utah: *Utah Geological Survey, Miscellaneous Publication*, 99-1, pp. 329–331.

Parrish, J. Michael, and Kent A. Stevens, 1998, Undoing the death pose: using computer imaging to restore the posture of articulated dinosaur skeletons: *Journal of Vertebrate Paleontology*, 18 (Supplement to Number 3), Abstracts of Papers, Fifty-eighth Annual Meeting, p. 69A.

Parsons, William L., and Kristen M. Parsons,

2000, A new skull of a nodosaurid from the Cloverly Formation of Montana and its comparison to *Sauropelta edwarsi*: *Journal of Vertebrate Paleontology*, 20 (Supplement to Number 3), Abstracts of Papers, Sixtieth Annual Meeting, p. 62A.

Patterson, Colin, 1993, Bird or dinosaur?: *Nature*, 362, pp. 21–22.

Paul, Gregory S., 1987, The science and art of restoring the life appearance of dinosaurs and their relatives, *in*: Sylvia J. Czerkas and Everett C. Olson, *Dinosaurs Past and Present*, Volume II. Los Angeles: Natural History Museum of Los Angeles County, pp. 5–49.

_____, 1988*a*, The brachiosaur giants of the Morrison and Tendaguru with a description of a new subgenus, *Giraffatitan*, and a comparison of the world's largest dinosaurs: *Hunteria*, 3 (3), pp. 1–14.

_____, 1988*b*, *Predatory Dinosaurs of the World: A Complete Illustrated Guide*. New York: Simon and Schuster, 403 pages.

_____, 1991, The many myths, some old, some new, of dinosaurology: *Modern Geology*, 16, pp. 69–99.

_____, 1993, Are *Syntarsus* and the Whitake quarry theropod the same genus?: *New Mexico Museum of Natural History and Science, Bulletin*, 3, pp. 397–402.

_____, 1994, Big sauropods — really really big sauropods: *The Dinosaur Report*, Fall, 1994, pp. 12–13.

_____, 1998, Differing bipedal and tripodal feeding modes in sauropods: *Journal of Vertebrate Paleontology*, 18 (Supplement to Number 3), Abstracts of Papers, Fifty-eighth Annual Meeting, p. 69A.

_____, 2000, Birds are dinosaurs, were some dinosaurs birds?: *Publications in Paleontology* [The Florida Symposium on Dinosaur Bird Evolution, abstracts of papers], 2, p. 22.

Paul, Gregory S., and Per Christiansen, 2000, Forelimb posture in neoceratopsian dinosaurs: implications for gait and locomotion: *Paleobiology*, 26 (3), pp. 450–465.

Peckzis, Jan, 1994, Implications of body-mass estimates for dinosaurs: *Journal of Vertebrate Paleontology*, 14 (4), pp. 520–533.

Peng Guangzhao, 1992, Jurassic ornithopod *Agilisaurus louderbacki* (Ornithopoda: Fabrosauridae) from Zigong, Sichuan, China: *Vertebrata PalAsiatica*, 30 (1), pp. 39–53.

_____, 1997, Fabrosauridae, *in*, Philip J. Currie and Kevin Padian, editors, *Encyclopedia of Dinosaurs*. San Diego: Academic Press, pp. 237–240.

Penkalski, Paul, 1993, The morphology of *Avaceratops limmersi*, a primitive ceratopsid from the Campanian of Montana: *Journal of Vertebrate Paleontology*, 13 (Supplement to Number 3), Abstracts of Papers, Fifty-third Annual Meeting, p. 52A.

_____, 1998, A preliminary systematic analysis of Ankylosauridae (Ornithischia: Thyreophora) from the Late Campanian of North America: *Ibid.*, 18 (Supplement to Number 3), Abstracts of Papers, Fifty-eighth Annual Meeting, pp. 69A–70A.

Penkalski, Paul, and Peter Dodson, 1999, The morphology and systematics of *Avaceratops*, a primitive horned dinosaur from the Judith River Formation (Late Campanian) of Montana, with the description of a second skull: *Journal of Vertebrate Paleontology*, 19 (4), pp. 692–711.

Pennycuick, C. J., 1986, *in*: Kevin Padian, editor, *The Origin of Birds and the Evolution of Flight*. San Francisco: California Academy of Sciences, pp. 83–98.

Pereda-Suberbiola, Xabier [Javier, Xavier], 1994, *Polacanthus* (Ornithischia, Ankylosauria), a Transatlantic armoured dinosaur from the Early Cretaceous of Europe and North America: *Palaeontographica*, A, 323, pp. 133–159.

Pereda-Suberbiola, Xabier [Javier, Exavier], 1994, *Polacanthus* (Ornithischia, Ankylosauria), a Transatlantic armoured dinosaur from the Early Cretaceous of Europe and North America: *Palaeontographica*, A, 323, pp. 133–159.

Pereda-Suberbiola, Xabier [Javier, Xavier], H. Astibia, and Éric Buffetaut, 1995, New remains of the armoured dinosaur *Struthiosaurus* from the Late Cretaceous of the Iberian peninsula (Laño locality, Basque-Cantabric basin): *Bulletin de la Sociéte Géologique de France*, 166, pp. 207–211.

Pérez-Moreno, Bernardino, P. Bernardo, José Luis bSanz, Ang'ela D. Buscalloni, José J. Moratalla, Francisco Ortega, and Diego Rasskin-Gutman, 1994, A unique multitoothed ornithomimosaur dinosaur from the Lower Cretaceous of Spain: *Nature*, 370, pp. 363–367.

Pérez-Moreno, Bernardino P. Daniel J. Chure, C. Pires, C. Marques da Silva, V. Dos Santos, P. Dantas, L. Póvoas, M. Cachão, José Luis Sanz, and A. M. Galopim de Carvalho (1999), 1999, On the presence of *Allosaurus fragilis* (Theropoda: Carnosauria) in the Upper Jurassic of Portugal: first evidence of an intercontinental dinosaur species: *Journal of the Geological Society, London*, 156, pp. 449–452.

Pérez-Moreno, B. P., José Luis Sanz, J. Sudre, and B. Sigé, 1993, A theropod dinosaur from the Lower Cretaceous of southern France: *Revue de Paléobiologie*, vol. spéc. 7, pp. 173–188.

Perle, Altangerel, Luis M. Chiappe, Rinchen Barsbold, James M. Clark, and Mark A. Norell, 1994, Skeletal morphology of *Mononykus olecrans* (Theropoda: Aviale) from the Late Cretaceous of Mongolia: *American Musuem Novitates*, 3105, pp. 1–29.

Perle, Altangerel, Mark A. Norell, Luis M. Chiappe, and James M. Clark, 1993, Flightless bird from the Cretaceous of Mongolia: *Nature*, 362, pp. 623–626.

Perle, Altangerel, Mark A. Norell, and James M. Clark, 1999, A new manuraptoran theropod —*Achillobator giganticus* (Dromaeosauridae)— from the Upper Cretaceous of Burkhant, Mongolia: *Mongolian-American Museum Palaeontological Project*, 101, National University of Mongolia, pp. 1–105.

Perry, Steven F., and Christian Reuter, 1999, Hypothetical lung structure of *Brachiosaurus* (Dinosauria: Sauropoda) based on functional constraints: *Mitteilungen Museum für Naturkunde der Humboldt-Universität zu Berlin, Geowissenschaftliche Reihe*, 2, pp. 75–79.

Peters, D. S., 1994, Die Entstehung der Vögel. Verändem die jüngsten Fossilifunde das Modell?: *Senckenberg-Buch*, 70, pp. 403–424.

Peters, R. H., 1983, *The Ecological Implications of Body Size*. Cambridge: Cambridge University Press, 329 pages.

Peterson, O. A., and Charles Whitney Gilmore, 1902, *Elosaurus parvus*, a new genus and species of Sauropoda: *Annals of the Carnegie Museum*, 1, pp. 490–499.

Pi, Ouyang and Ye, 1996, A new species of a sauropod from Zigong, Sichuan: *Papers on Geosciences*, 50, pp. 87–91.

Pidancet, J., and S. Chopard, 1862, Note sur un saurien gigantesque aux Marnes irisées: *Comptes Rendus des Séances de l'Academie des Sciences, Paris*, 54, pp. 1259–1262, Paris.

Pilmore, C. L., M. G. Lockley, R. F. Fleming, K. R. Johnson, 1994, Footprints in the rocks — new evidence from the Raton basin that dinosaurs flourished on land until the terminal Cretaceous impact event, *in*: New developments regarding the KT event and other catastrophes in earth history: *Lunar and Planetary Institute Contribution*, 825, pp. 89–90.

Plieninger, T., 1852, *Belodon plienigeri* H. V. Meyer. Ein Saurier der Keuperformation: *Jahreshect des Vereins für Vaterländische Naturkunde in Württemberg*, 8, pp. 389–524, Stuttgart.

Pomes, M. L., 1988, Stratigraphy, paleontology, and paleobiogeography of lower vertebrates from the Cedar Mountain Formation (Lower Cretaceous), Emery County, Utah, master's thesis, Fort Hays State University (Kansas), 87 pages (unpublished).

Pons, D., P.-Y. Berthou, and D. de A. Campos, 1990, Quelques observations sur la palynologie de l'Aptien supérieur et de l'Albien du Bassin d' Araripe, *in*: D. de A. Campos, M.S.S. Viana, P. M. Brito, and G. Beurlen, editors, *Atas do simposio sobre a Bacia do Araripe e Bacias Interiores do Nordeste, Crato, 14–16 June 1990*, pp. 241–516, Crato.

Poupa, O., and B. Ostadal, 1969, Experimental cardiomegalies and "cardiomegalies" in free-living animals: *Annals of the New York Academy of Sciences*, 156, pp. 445–468.

Powell, Jaime Eduardo, 1986, Revisión de los Titanosauridos de América del Sur. Tésis Doctoral inédita Fac. de Ciencias Exactas y Naturales, Universidad Nacional de Tucumán, Argentina, 472 pages (unpublished).

_____, 1987, Morfologia del esqueleto axial de los dinosaurios titanosáuridos (Saurischia–Sauropoda) del Estado de Minas Gerais, Brasil: *Anain X Congresso Brasileiro de Paleontologia*, pp. 155–171, Rio de Janeiro.

_____, 1992, Osteologia de *Saltasaurus loricatus* (Sauropoda–Titanosauridae) del Cretácico Superior del Noroeste Argenti no, *in*: José Luis Sanz and Ang'gela D.

Buscalioni, editors, *Actas 2o Curso de Paleontología en Cuenca*, Instituto "Juan de Valdés," Ayuntamiento de Cuenca, pp. 165–230.

Prieto-Marquez, Albert, 2000, On the postcrania of *Brachylophosaurus goodwini* (Dinosauria: Ornithopoda): implications for hadrosaur morphology: *Journal of Vertebrate Paleontology*, 20 (Supplement to Number 3), Abstracts of Papers, Sixtieth Annual Meeting, p. 63A.

Probst, C., and R. Windolf, 1993, *Dinosaurier in Deutschland*: München: Bertelsmann, 396 pages.

Prum, Richard O., 2000, A developmental model of the origin of feathers: *Publications in Paleontology* [The Florida Symposi um on Dinosaur Bird Evolution, abstracts of papers], 2, p. 22.

Quenstedt, F. A., 1856, Sonst und jetzt: *Populäre Vorträge über Geologie*, 288 pages, Tübingen (Laupp).

Raath, Michael R., 1969, A new coelurosaurian dinosaur from the Forest Sandstone of Rhodesia: *Arnoldia (Rhodesia)*, 4 (28), pp. 1–25.

_____, 1972, Fossil vertebrate studies in Rhodesia: a new dinosaur (reptilia: saurischia) from near the Trias–Jurassic boundary: *Ibid.*, 5 (30), pp. 1–37.

_____, 1977, The anatomy of the Triassic theropod *Syntarsus rhodesiensis* (Saurischia: Podokesauridae) and a consideration of its biology, Ph.D. thesis, Rhodes University, Grahamstown (unpublished).

_____, 1996, Extinction of dinosaurs from central Gondwana: *Proceedings of the Gondwanan Dinosaur Symposium*: *Memoirs of the Queensland Museum*, 39 (part 3), pp. 703–709.

Raath, Michael R., and John S. McIntosh, 1987, Sauropod dinosaurs from the Central Zambezi Valley, Zimbabwe, and the age of the Kadzi Formation: *South Africa Journal of Geology*, 90 (2), pp. 107–119.

Raath, Michael R., P. M. Oesterlen, and J. W. Kitching, 1992, First record of Triassic Rhynchosauria (Reptilia: Diapsida) from the Lower Zambezi Valley, Zimbabwe: *Palaeontologia Africana*, 29, pp. 1–10.

Racey, A., J. G. S. Goodall, M. A. Love, S. Polachan, and P. D. Jones, 1994, New age data for the Mesozoic Khorat Group of northeast Thailand, *in*: Angsuwathana *et al.*, editors, *Proceedings of the International Symposium on Stratigraphic correlation of southeast Asia*, pp. 245–256.

Racey, A., M. A. Love, and S. Polachan, 1996, Stratigraphy and reservoir potential of the Mesozoic Khorat Group, NE Thailand: *Journal of Petrology and Geology*, 19, pp. 5–40.

Rage, J. C., 1988, Gondwana, Tethys, and terrestrial vertebrates during the Mesozoic and Cenozoic, *in*: M. G. Audley Charles and A. Hallam, editors, *Gondwana and Tethys*. Oxford and New York: Oxford University Press, Geological Society Special Publication (London), 37, pp. 255–273.

Ralph, R., 1993, John MacGillivray — his life and work: *Archives of Natural History*, 20, pp. 185–195.

Randolph, S. E., 1994, The relative timing of the origin of flight and endothermy. Evidence from the comparative biology of birds and mammals: *Zoological Journal of the Linnean Society*, 112, pp. 389–397.

Ratkevich, Ronald P., 1997, Dinosaur remains of southern Arizona, *in*: Donald L. Wolberg, Edmund Stump, and Gary Rosenberg, editors. *Dinofest International: Proceedings of a Symposium Held at Arizona State University*. Philadelphia: Academy of Natural Sciences, pp. 213–221.

_____, 1998, New Cretaceous brachiosaurid dinosaur, *Sonorasaurus thompsoni* gen. et sp. nov, from Arizona: *Journal of the Arizona–Nevada Academy of Science*, 31 (1), pp. 71–82.

Rauhut, Oliver W. M., 1995, Zut systematischen Stellung der afrikanischen Theropoden *Carcharodontosaurus* Stromer 1931 und *Bahariasaurus* Stromer 1934: *Berliner geowissenschaftliche Abhandlung*, E16, pp. 357–375.

_____, 1998, *Elaphrosaurus bambergi* and the early evolution of theropod dinosaurs: *Journal of Vertebrate Paleontology*, 18 (Supplement to Number 3), Abstracts of Papers, Fifty-eighth Annual Meeting, p. 71A.

_____, 1999, A dinosaur fauna from the Late Cretaceous (Cenomanian) of northern Sudan: *Palaeontologia Africana*, 35, pp. 61–84.

Rayfield, J. Emily, 1999, A three dimensional model of the skull of *Allosaurus fragilis* analysed using finite element analysis: *Journal of Vertebrate Paleontology*, 19 (Supplement to Number 3), Abstracts of Papers, Fifty-ninth Annual Meeting, p. 69A.

_____, 2000, *Allosaurus fragilis*: mechanical behaviour of the skull and implications for feeding strategy: *Journal of Vertebrate Paleontology*, 20 (Supplement to Number 3), Abstracts of Papers, Sixtieth Annual Meeting, p. 63A.

Rawson, P. F., D. Currey, F. C. Dilley, J. M. Hancock, W. J. Kennedy, J. W. Neale, Wood, and B. C. Worssam 1978, A correlation of Cretaceous rocks in the British Isles: *Special Report of the Geological Society of London*, 9, pp. 1–70.

Reck, Hans, 1925, Grabungen auf fossile Wirbeltiere in Deutsch-Ostafrika: *Geologische Charakterbilder*, 31, pp. 1–36.

Reed, Christina, 2000, A dinosaur's tell-tale heart: *Geotimes*, June, pp. 8–9.

Reeside, J. B., 1924, Upper Cretaceous and Tertiary formations of the western part of the San Juan Basin of Colorado and New Mexico: *U.S. Geological Survey, Professional Paper*, 134, 70 pages.

Reid, Robin E. H., 1984*a*, Primary bone and dinosaurian physiology: *Geological Magazine*, 121, pp. 599–598.

_____, 1984*b*, The histology of dinosaurian bone, and its possible bearing on dinosaurian physiology: *Symposium of the Zoological Society of London*, 52, pp. 629–663.

_____, 1997, Dinosaurian physiology: the case for "intermediate" dinosaurs: *in*: Farlow and Brett-Surman, editors, *The Complete Dinosaur*. Bloomington and Indianapolis: Indiana University Press, pp. 449–473.

Rensberger, John M., Melinda Mielkey, Tammy Lee, and Basker Moorthy, 1999, Structure of lamellar bone in coelurosaurs and ornithischians parallels differences in birds and mammals: *Journal of Vertebrate Paleontology*, 19 (Supplement to Number 3), Abstracts of Papers, Fifty-ninth Annual Meeting, pp. 69A–70A.

Retellack, Gregory J., 2000, End Cretaceous acid rain as a selective extinction mechanisms between birds and dinosaurs: *Publications in Paleontology* [The Florida Symposium on Dinosaur Bird Evolution, abstracts of papers], 2, p. 23.

Rich, Thomas H., and Patricia [Vickers-] Rich, 1989, Polar dinosaurs and biotas of the Early Cretaceous of southeastern Australia: *National Geographic Society Research Reports*, 5, pp. 15–53.

_____, 1994, Neoceratopsians and ornithomimosaurs: dinosaurs of Gondwana origin?: *National Geographic Research*, 5 (1), pp. 15–53.

_____, 1999, The Hypsilophodontidae from southeastern Australia: *in*: Yukimitsu Tomida, Rich, and Vickers-Rich, editors, *Proceedings of the Second Gondwanan Dinosaur Symposium*, National Science Museum Monographs, 15, Tokyo, pp. 167–180.

Rich, Thomas H., Patricia Vickers-Rich, Olga Gimenéz, Rubén Cúneo, Pablo Puerta, and Raul Vacca, 1999, A new sauropod dinosaur from Chubut Province, Argentina, *in*: Yukimitsu Tomida, Rich, and Vickers-Rich, editors, *Proceedings of the Second Gondwanan Dinosaur Symposium*, National Science Museum Monographs, 15, Tokyo, pp. 61–84.

Richards, H. Robins, 1999, Is *Spinosuchus* a dinosaur?: *Journal of Vertebrate Paleontology*, 19 (Supplement to Number 3), Abstracts of Papers, Fifty-ninth Annual Meeting, p. 70A.

Ricklefs, R. E., and M. Stark, 1998, Embryonic growth and development, *in*: M. Stark and R. E. Ricklefs, editors, *Avian Growth and Development, Evolution within the Altricial-Precocial Spectrum*. New York: Oxford University Press, pp. 31–58.

Ricqlès, Armand J. de, 1983, Cyclical growth in the long limb bones of sauropod dinosaurs: *Acta Palaeontologia Palonica*, 18, pp. 225–232.

Rigby, J. Keith, Jr., 1981, A skeleton of *Gillisonchus gillianus* (Mammalia; Condylarthra) from the early Paleocene (Puercan) Ojo Alamo Sandstone, San Juan Basin, New Mexico, with comments on the local stratigraphy of Betonnie Tsosie Wash, *in*: Spencer G. Lucas, J. Keith Rigby, Jr., and B. S. Kues, editors, *Advances in San Juan Basin Paleontology*. Albuquerque: University of New Mexico Press, pp. 89–126.

Rigby, J. Keith, Jr., and Spencer G. Lucas, 1977, Fossil mammals from the Ojo Alamo Sandstone, New Mexico: *Geological Society Guidebook*, 28, supplement, pp. 55–56.

Rigby, J. Keith, and J. Keith Rigby, Jr., 1990, Geology of the Sand Arroyo and Bug Creek quadrangles, McCone County,

Montana: *Brigham Young University Geology Studies*, 36, pp. 69–
Riggs, Elmer S., 1903, *Brachiosaurus altithorax*, the largest known dinosaur: *American Journal of Science, Series 4*, 15, pp. 299–306.
Riley, Henry, and Samuel Stutchbury, 1836, A description of various fossil remains of three distinct saurian animals discovered in the autumn of 1834, in the Magnesian Conglomerate on Durnham Down, near Bristol: *Proceedings of the Geological Society of Lodon*, Series 2, 5, pp. 349–357.
_____, 1837, On an additional species of the newly-discovered saurian animals in the Magnesian Conglomerate of Durdham Down, near Bristol: *Annual Report of the British Association for the Advancement of Science, Transactions of the Section 1836*, pp. 90–94.
_____, 1840, A description of various fossil remains of three distinct saurian animals, recently discovered in the Magnesian Conglomerate near Bristol: *Transactions of the Geological Society of London*, 150, pp. 1025–1034.
Rodríguez de la Rosa, Rubén, and Francisco J. Aranda-Manteca, 2000, Were there venomous theropods?: *Journal of Vertebrate Paleontology*, 20 (Supplement to Number 3), Abstracts of Papers, Sixtieth Annual Meeting, p. 64A.
Romer, Alfred Sherwood, 1923*a*, The ilium in dinosaurs and birds: *Bulletin of the American Museum of Natural History*, 48, pp. 141–145.
_____, 123*b*, The pelvic musculature of saurischian dinosaurs: *Ibid.*, 48, pp. 605–617.
_____, 1927, The pelvic musculature of ornithischian dinosaurs: *Acta Zoologica*, 8, pp. 225–275.
_____, 1956, *Osteology of the Reptiles*. Chicago: University of Chicago Press, xxi, 772 pages.
Rosenbaum, Jason N., and Kevin Padian, 2000, New material of the basal thyreophoran *Scutellosaurus lawleri* from the Kayenta Formation (Lower Jurassic) of Arizona: *PaleoBios*, 20 (1), pp. 13–23.
Rothschild, Bruce M., and David S Berman, 1991, Fusion of caudal vertebrae in the Late Jurassic sauropods: *Journal of Vertebrate Paleontology*, 11 (1), pp. 29–36.
Rowe, Timothy, 1989, A new species of theropod dinosaur *Syntarsus* from the Early Jurassic Kayenta Formation of Arizona: *Journal of Vertebrate Paleontology*, 9 (2), pp. 125–136.
Rowe, Timothy, and Jacques A. Gauthier, 1990, Ceratosauria, *in*: David B. Weishampel, Peter Dodson and Halszka Osmólska, editors, *The Dinosauria*. Berkeley and Los Angeles: University of California Press, pp. 151–168.
Rowe, Timothy, Ron Tykoski, and John Hutchinson, 1997, Ceratosauria, *in*: Philip J. Currie and Kevin Padian, editors, *Encyclopedia of Dinosaurs*. San Diego: Academic Press, pp. 106–110.
Rozhestvensky, Anatoly Konstantinovich, 1965, Growth changes in Asian dinosaurs and some problems of their taxonomy: *Palaeontologicheskii Zhurnal*, 3, pp. 95–109.
_____, 1966, Novye iguanodonty iz Tsentral'nol Azii. Filogeneticheskie i taksonomicheskie vzaimootnosheniya poznik Iguanodontidae i rannikh Hadrosauridae: *Ibid.*, 3, pp. 103–116. (New iguanodonts from Central Asia. Phylogenetic and taxonomic relationships of late Iguanodontidae and early Hadrosauridae: *Palaeontological Journal*, 3, pp. 103–116.)
_____, 1967, New iguanodonts from Central Asia: *International Geology Review*, 9, pp. 556–566.
Ruben, John A., 1996, Evolution of endothermy in mammals, birds and their ancestors, *in*: I. A. Johnson and A. F. Bennett, editors, *Animals and Temperature Phenotypic and Evolutionary Adaptation*. Cambridge and New York: Cambridge University Press, pp. 347–376.
Ruben, John A., Cristiano Dal Sasso, Nicholas R. Geist, Willem J. Hillenius, Terry D. Jones, and Marco Signore, 1998, Pulmonary function and metabolic physiology of theropod dinosaurs: *Science*, 283, pp. 514–516.
Ruben, John A., and Terry D. Jones, 2000, Contrasting locomotory styles in bipedal dinosaurs and cursorial birds: *Publications in Paleontology* [The Florida Symposium on Dinosaur Bird Evolution, abstracts of papers], 2, pp. 23–24.
Ruben, John A., Terry D. Jones, and Nicholas R. Geist, 1998, Respiratory physiology of the dinosaurs: *BioEssays*, 20, pp. 852–859.
Ruben, John A., Terry D. Jones, Nicholas R. Geist, and W. Jaap Hillenius, 1997, Lung structure and ventilation in theropod dinosaurs and early birds: *Science*, 278, pp. 1267–1270.
Rubilar, David, Alexander Vargas, and David Lemus, 2000, The dinosaur-bird transition compared to the development of the chick (*Gallus gallus*): *Journal of Vertebrate Paleontology*, 20 (Supplement to Number 3), Abstracts of Papers, Sixtieth Annual Meeting, pp. 65A–66A.
Russell, Dale A., 1969, A new specimen of *Stenonychosaurus* from the Oldman Formation (Cretaceous) of Alberta: *Canadian Journal of Earth Sciences*, 6, pp. 595–612.
_____, 1970*a*, A skeletal reconstruction of *Leptoceratops gracilis* from the upper Edmonton Formation (Cretaceous) of Alberta: *Ibid.*, 7 (1), pp. 181–184.
_____, 1970*b*, Tyrannosaurs from the Late Cretaceous of western Canada: *National Museum of Natural Sciences, Publications in Paleontology*, 1, viii, 34 pages.
_____, 1993, The role of Central Asia in dinosaurian biogeography: *Canadian Journal of Earth Sciences*, 20, pp. 2002–2012.
_____, 1995, China and the lost worlds of the dinosaurian era: *Historical Biology*, 10, pp. 3–12.
_____, 1996, Isolated dinosaur bones from the Middle Cretaceous of the Tafilalt, Morocco: *Bulletin du Museum national d'Histoire naturelle, Paris*, sér. 4 (18), Section C, nos. 2–3, pp. 3409–402.
Russell, Dale A., P. Béland, and John S. McIntosh (1980), Paleoecology of the dinosaurs of Tendaguru (Tanzania): *Mémoires de la Sociéte Geologique de France*, Nouvelle Série, 59 (139), pp. 1–36.
Russell, Dale A., and Dong Zhi-Ming, 1993*a*, The affinities of a new theropod from the Alxa Desert, Inner Mongolia, People's Republic of China: *Canadian Journal of Earth Sciences*, 30 (10–11), pp. 2107–2127.
_____, 1993*b*, A nearly complete skeleton of a new troodontid dinosaur from the Early Cretaceous of the Ordos Basin, Inner Mongolia, People's Republic of China: *Ibid.*, 30 (10–11), pp. 2163–2173.
Russell, Dale A., and Zhong Zheng, 1993, A large mamenchisaurid from the Junggar Basin, Xinjiang, People's Republic of China: *Canadian Journal of Earth Sciences*, 30 (10–11), pp. 2082–2095.
Russell, Loris, 1940, *Edmontonia rugosidens* (Gilmore), an armoured dinosaur from the Belly River Series of Alberta: *University of Toronto Studies, Geological Series*, 43, pp. 3–27.
Ryan, Michael J., and Philip J. Currie, 1996, First report of Protoceratopsidae (Neoceratopsia) from the late Campanian, Judith River Group, Alberta, Canada: *Journal of Vertebrate Paleontology*, 16 (Supplement to Number 3), Abstracts of Papers, Fifty-sixth Annual Meeting, p. 61A.
Ryan, Michael J., David A. Eberth, Donald A. Brinkman, and Anthony P. Russell, 1998, *Journal of Vertebrate Paleontology*, 18 (Supplement to Number 3), Abstracts of Papers, Fifty-eighth Annual Meeting, p. 73A.
Ryan, Michael J., A. P. Russell, Mike Getty, David A. Eberth, and Donald B. Brinkman, 1999, A new centrosaurine ceratopsid from the Oldman Formation of Alberta, Canada: *Journal of Vertebrate Paleontology*, 19 (Supplement to Number 3), Abstracts of Papers, Fifty-ninth Annual Meeting, p. 72A.
Rybcznski, Natalia, and Matthew K. Vickaryous, 1998, Evidence of complex jaw movement in a Late Cretaceous ankylosaurid (Dinosauria: Thyreophora): *Journal of Vertebrate Paleontology*, 18 (Supplement to Number 3), Abstracts of Papers, Fifty-eighth Annual Meeting, p. 73A.
Sabath, Karol, and Gerard Gierlinski, 1998, Feathers of nonavian dinosaurs: *Journal of Vertebrate Paleontology*, 18 (Supplement to Number 3), Abstracts of Papers, Fifty-eighth Annual Meeting, pp. 73A–74A.
Sadleir, Rudyard W., 1998, Theropod teeth from the Cretaceous of Morocco: *Journal of Vertebrate Paleontology*, 18 (Supplement to Number 3), Abstracts of Papers, Fifty-eighth Annual Meeting, p. 74A.
Sadleir, Rudyard W., and Ralph E. Chapman, 1999, Theoretical morphology of theropod teeth: *Journal of Vertebrate Paleontology*, 19 (Supplement to Number 3), Abstracts of Papers, Fifty-ninth Annual Meeting, p. 72A.
Salgado, Leonardo, 1999, The macroevolution of the Diplodocimorpha (Dinosauria; Sauropoda): a developmental model: *Ameghiniana*, 36 (2), pp. 203–216.
Salgado, Leonardo, and C. Azpilicueta, 2000, Un nuevo saltasauri no (Sauropoda, Titanosauridae) de la provincia de Rio Negro (Formacion Allen, Cretacico supe-

rior), Patagonia, Argentina: *Ameghiniana*, 37 (3), pp. 259–264.

Salgado, Leonardo, and José F. Bonaparte, 1991, Un nuevo sauropodo Dicraeosauridae, *Amargasaurus cazui* gen. et sp. nov., de la Formacion la Amarga, Neocomiano de la Provincia del Neuquen, Argentina: *Ameghiniana*, 28 (3–4), pp. 333–346, Buenos Aires.

Salgado, Leonardo, and Jorge Orlando Calvo, 1997, Evolution of titanosaurid sauropods. II. The cranial evidence: *Ameghiniana*, 34 (1), pp. 33–48.

Salgado, Leonardo, Jorge Orlando Calvo, and Rodolfo Aníbal Coria, 1995, Relaciones filogenéticas de *Pleurocoelus* Marsh (Sauropoda): *Resúmes II, Journadas Argentinas de Paleontologia de Vertebrados*, Tucumán.

Salgado, Leonardo, Rodolfo Aníbal Coria, and Jorge Orlando Calvo, 1997, Evolution of titanosaurid sauropods. I: Phylogenetic analysis based on the postcranial evidence: *Ameghiniana*, 34 (1), pp. 3–32.

Sampson, Scott D., 1990, Ontogeny and variation in the crania of centrosaurine ceratopsians (Ornithischia: Ceratopsidae): phylogenetic implications: *Journal of Vertebrate Paleontology*, 15 (Supplement to Number 3), Abstracts of Papers, Fifty-fifth Annual Meeting, p. 40A.

_____, 1993, Cranial ornamentation in ceratopsid dinosaurs: systematic, behavioral, and evolutionary implications: Ph.D. dissertation, University of Toronto, Toronto, 299 pages.

Sampson, Scott D., Matthew T. Carrano, and Catherine A. Forster, 2000, A theropod dinosaur with bizarre dentition from the Late Cretaceous of Madagascar: *Journal of Vertebrate Paleontology*, 20 (Supplement to Number 3), Abstracts of Papers, Sixtieth Annual Meeting, p. 66A.

Sampson, Scott D., Catherine A. Forster, and David W. Krause, 1998, The Late Cretaceous dinosaurs of Madagascar: *Journal of Vertebrate Paleontology*, 18 (Supplement to Number 3), Abstracts of Papers, Fifty-eighth Annual Meeting, p. 74A.

Sampson, Scott D., Michael J. Ryan, and Darren H. Tanke, 1997, Cranofacial ontogeny in centrosaurine dinosaurs: taxonomic and behavorial implications: *Zoological Journal of the Linnean Society*, pp. 293–337.

Sampson, Scott D., and Lawrence M. Witmer, 1999, Novel narial anatomy in ceratopsid dinosaurs: *Journal of Vertebrate Paleontology*, 19 (Supplement to Number 3), Abstracts of Papers, Fifty-ninth Annual Meeting, pp. 72A–73A.

Sanberger, F., 1894, *Zanclodon* im obersten Keuper Unterfrankens: *Neus Jarbuch für Mineralogie, Geologie, und Paläontologie*, 1894 (1), pp. 203–204, Stuttgart.

Sander, P. Martin, 1999, Life history of Tendaguru sauropods as inferred from long bone histology: *Mitteilungen Museum für Naturkunde der Humboldt-Universität zu Berlin, Geowis senschaftliche Reihe*, 2, pp. 103–112.

_____, 2000, Long bone histology of the Tendaguru sauropods: implications for growth and biology: *Palaeobiology*, 26 (3), pp. 466–488.

Sander, P. Martin, and Kenneth Carpenter, 1998, Gastroliths from a camarasaurid in the Cedar Mountain Formation: *Journal of Vertebrate Paleontology*, 18 (Supplement to Number 3), Abstracts of Papers, Fifty-eighth Annual Meeting, p. 74A.

Sanders, Richard K., Matthew J. Wedel, Paul C. Sereno, and Gary T. Staab, 2000, A restoration of the cranio-cervical system in *Jobaria*: *Journal of Vertebrate Paleontology*, 20 (Supplement to Number 3), Abstracts of Papers, Sixtieth Annual Meeting, p. 67A.

Sanford, W. A., 1894, On bones of an animal resembling the Megalosaur, found in the Rhaetic formation at Wedmore: *Proceedings of the Somerset Archaeological and Natural History Society* (New Series), 20, pp. 227–235, Taunton.

Sankey, Julia T., and Donald B. Brinkman, 2000, New theropod and bird teeth from the Late Cretaceous (Campanian) Judith River Group, Alberta: *Journal of Vertebrate Paleontology*, 20 (Supplement to Number 3), Abstracts of Papers, Sixtieth Annual Meeting, p. 67A.

Sanz, José Luis, and Ang'ela D. Buscalioni, 1987, New evidence of armoured titanosaurs in the Upper Cretaceous of Spain, *in*: P. J. Currie and E. H. Koster, editors, *Fourth Symposium on Mesozoic Terrestrial Ecosystems, Short Papers*, Tyrrell Museum of Palaeontology, Occasional Papers, pp. 199–204.

Sanz, José Luis, Jaime E. Powell, Hean Le Loeuff, Rubén Martinez, and Xabier Suberbiola, 1999, Sauropod remains from the Upper Cretaceous of Laño (northcentral Spain), Titanosaur phylogenetic relationships: *Est. Museo de Ciencias Naturales de Alava*, 14 (1), pp. 235–255.

Sarjeant, William A. S., 1996, Dinosaur extinction: sudden or slow, cataclysmic or climatic?: *Geoscience Canada*, 23, pp. 161–164.

Scheetz, Rodney D., 1998, Phylogeny of basal ornithopod dinosaurs and the dissolution of the Hypsilophodontidae: *Journal of Vertebrate Paleontology*, 18 (Supplement to Number 3), Abstracts of Papers, Fifty-eighth Annual Meeting, p. 75A.

Scheid, P., and J. Piper, 1970, Analysis of gas exchanges in the avian lung: theory and experiments in the domestic fowl: *Respiration Physiology*, 1970 (9), pp. 246–262.

Schmidt-Nielsen, K., 1984, *Scaling: Why is Animal Size so Important?* Cambridge: Cambridge University Press, 241 pages.

Schmitt, J. G., John R. Horner, Rebecca R. Laws, and Frank Jackson, 1998, Debris-flow deposition of a hadrosaur-bearing bone bed, Upper Cretaceous Two Medicine Formation, northwest Montana: *Journal of Vertebrate Paleontology*, 18 (Supplement to Number 3), Abstracts of Papers, Fifty-eighth Annual Meeting, p. 76A.

Schudak, M. E., 1999, Charophytes from the Middle Dinosaur Member of the Tendaguru Formation (Upper Jurassic of Tanzania): *Mitteilungen Museum für Naturkunde der Humboldt-Universität zu Berlin, Geowissenschaftliche Reihe*, 2, pp. 201–205.

Schudak, M. E., U. Schudak, and E. Pietrzeniuk, 1999, Ostracods from the Middle Dinosaur Member of the Tendaguru Formation (Upper Jurassic of Tanania): (manuscript).

Scotese, C. R., and J. Golonka, 1992, PALEOMAP Paleogeographic Atlas: *PALEOMAP Progress Report #20*, Department of Geology, University of Texas at Arlington.

Seeley, Harry Govier, 1869, *Index to the Fossil Remains of Aves, Ornithosauria, and Reptilia from the Secondary Strata arranged in the Woodwardian Museum of the University of Cambridge*. Cambridge, England: Deighton, Bell and Co., x, 143 pages.

_____, 1876, On *Macrurosaurus* (Seeley), a long-tailed animal with procoelous vertebrae from the Cambridge Upper Greensand, preserved in the Woorwardian Museum of the University of Cambridge: *Quarterly Journal of the Geological Society, London*, 32, pp. 440–444.

_____, 1879, On the Dinosauria of the Cambridge Greensand: *Ibid.*, 37, pp. 620–707.

_____, 1887, On *Aristosuchus pusillus* (Owen), being further notes on the fossils descried by Sir R. Owen as *Poikilopleuron pusillus*, Owen: *Ibid.*, 43, pp. 221–228.

_____, 1891, On *Agrosaurus macgillivrayi* (Seeley), a saurischian reptile from the N.E. coast of Australia: *Ibid.*, 47, pp. 164–165.

_____, 1894, On *Euskelosaurus browni* (HUXLEY): *Annals and Magazine of the British Association for the Advancement of Science*, 57, pp. 698–699, London.

_____, 1895, On *Thecodontosaurus* and *Palaeosaurus*: *Annals and Magazine of Natural History*, Series 6, 15, pp. 102–132.

_____, 1898, On large terrestrial saurians from the Rhaetic beds of Wedmore Hill, described as *Avalonius sanfordi* and *Picrodon herveyi*: *Geological Magazine* (New Series), 4 (5), pp. 1–6, London.

Sereno, Paul C., 1984, The phylogeny of the Ornithischia: a reappraisal, *in*: W. Reif and F. Westphal, editors, *Third Symposium on Mesozoic Terrestrial Ecosystems, Short Papers*, Attempto Verlag, Tübingen University Press, pp. 219–226.

_____, 1986, Phylogeny of the bird-hipped dinosaurs (order Ornithischia): *National Geographic Research*, 2, pp. 234–256.

_____, 1997, The origin and evolution of dinosaurs: *Annual Review of Earth Planet Science*, 25, pp. 435–489.

_____, 1989, Prosauropod monophyly and basal sauropodomorph phylogeny: *Journal of Vertebrate Paleontology*, 10 (Supplement to Number 3), Abstracts of Papers, Fiftieth Annual Meeting, p. 38A.

_____, 1990, Psittacosauridae, *in*: Philip J. Currie and Kevin Padian, editors, *Encyclopedia of Dinosaurs*. San Diego: Academic Press, pp. 579–592.

_____, 1991, *Lesothosaurus*, "fabrosaurids," and the early evolution of Ornithischia: *Journal of Vertebrate Paleontology*, 11 (2), pp. 168–197.

_____, 1997, The origin and evolution of dinosaurs: *Annual Review of Earth and Planetary Sciences*, 25, pp. 435–490.

_____, 1998, A rationale for phylogenetic definitions, with application to the higher-level taxonomy of Dinosauria: *Neus Jahrbuch für Geologie und Paläontologie, Abhandlungen*, 1998 (189).

_____, 1999*a*, Alvarezsaurids (Dinosauria, Coelurosauria), birds or ornithomimosaurs?: *Journal of Vertebrate Paleontology*, 19 (Supplement to Number 3), Abstracts of Papers, Fifty-ninth Annual Meeting, p. 75A.

_____, 1999*b*, A rationale for dinosaurian taxonomy: *Ibid.*, 19 (4), pp. 788–790.

_____, 1999*c*, The evolution of dinosaurs: *Science*, 284, pp. 2137–2147.

_____, 2000, The fossil record, systematics and evolution of pachycephalosaurs and ceratopsians from Asia, *in*: M. J. Benton, M. A. Shishkin, D. M. Unwin and E. N. Kurochkin, editors, *The Age of Dinosaurs in Russia and Mongolia*. Cambridge: Cambridge University Press, pp. 480–516.

Sereno, Paul C., and Andrea Beatriz Arcucci, 1994, Dinosaurian precursors from the Middle Triassic of Argentina: *Marasuchus lilloensis*, gen. nov.: *Journal of Vertebrate Paleontology*, 14, pp. 53–73.

Sereno, Paul C., Allison L. Beck, Didier B. Dutheil, Boubacar Gado, Hans C. E. Larsson, Gabrielle H. Lyon, Jonathan D. Marcot, Oliver W. M. Rauhut, Rudyard W. Sadleir, Christian A. Sidor, David D. Varricchio, Gregory P. Wilson, and Jeffrey A. Wilson, 1998, A long-snouted predatory dinosaur from Africa and the evolution of spinosaurids: *Science*, 282 (5392), pp. 1298–1302.

Sereno, Paul C., Allison L. Beck, Didier B. Dutheil, M. Iarochene, Hans C. E. Larsson, Gabrielle H. Lyon, Paul M. Magwene, Christian A. Sidor, David J. Varracchio, and Jeffrey A. Wilson, 1996, Predatory dinosaurs from the Sahara and Late Cretaceous faunal differentiation: *Science*, 272, pp. 996–990.

Sereno, Paul C., Allison L. Beck, Didier B. Dutheil, Hans C. E. Larsson, Gabrielle H. Lyon, Bourahima Moussa, Rudyard W. Sadleir, Christian A. Sidor, David J. Varrichio, Gregory P. Wilson, and Jeffrey A. Wilson, 1999, Cretaceous sauropods from the Sahara and the uneven rate of skeletal evolution among dinosaurs: *Science*, 286, pp. 1342–1347.

Sereno, Paul C., and Dong Zhi-Ming, 1992, The skull of the basal stegosaur *Huayangosaurus taibaii* and a cladistic diagnosis of Stegosauria: *Journal of Vertebrate Paleontology*, 12 (3), pp. 318–343.

Sereno, Paul C., Didier B. Dutheil, M. Iarochene, Hans C. E. Larsson, Gabrielle H. Lyon, Paul M. Magwene, Christian A. Sidor, David J. Varracchio, and Jeffrey A. Wilson, 1996, Predatory dinosaurs from the Sahara and Late Cretaceous faunal differentiation: *Science*, 272, pp. 996–990.

Sereno, Paul C., Catherine A. Forster, Raymond R. Rogers, and Alfredo M. Monetta, 1993, Primitive dinosaur skeleton from Argentina and the early evolution of Dinosauria: *Nature*, 361, pp. 64–66.

Sereno, Paul C., and Fernando E. Novas, 1992, The complete skull and skeleton of an early dinosaur: *Science*, 258, pp. 1137–1140.

_____, 1993, The skull and neck of the basal theropod *Herrerasaurus ischigualastensis*: *Journal of Vertebrate Paleontology*, 13, pp. 451–476.

Sereno, Paul C., Jeffrey A. Wilson, Hans C. E. Larsson, Didier B. Dutheil, and Hans-Dieter Sues, 1994, Early Cretaceous dinosaurs from the Sahara: *Science*, 266, pp. 267–271.

Sereno, Paul C., Jeffrey A. Wilson, Hans C. E. Larsson, David J. Varricchio, and Oliver W. M. Rauhut, 1998, New theropods from Middle Cretaceous deposits of the Tenere Desert, Niger Republic: *Journal of Vertebrate Paleontology*, 18 (Supplement to Number 3), Abstracts of Papers, Fifty-eighth Annual Meeting, p. 77A.

Seymour, Roger S., 1976, Dinosaurs, endothermy and blood pressure: *Nature*, 262, 207–208.

Seymour, Roger S., and Harvey B. Lillywhite, 2000, Hearts, neck posture and metabolic intensity of sauropod dinosaurs: *Proceedings of the Royal Society of London*, B, 267, pp. 1883–1887.

Sharov, Alexander G., 1970, An unusual reptile from the Lower Triassic of Fergana: *Paleontologichesky Zhurnal*, 1, 127–130.

Sheehan, Peter M., David E. Fastovsky, Claudia Barreto, and Raymond G. Hoffman, 2000, Dinosaur abundance was not declining in a "3 m gap" at the top of the Hell Creek Formation, Montana and North Dakota: *Geology*, 28 (6), pp. 523–526.

Sheehan, Peter M., David E. Fastovsky, Raymond G. Hoffman, and D. L. Gabriel, 1991, Sudden extinction of the dinosaurs: latest Cretaceous, upper Great Plains, USA: *Science*, 254, pp. 835–839.

Shen, Y., 1981, [Fossil conchostracans from the Chijinpu Forma tion (Upper Cretaceous) and the Xinminpu Group (Lower Cretaceous) in Hexi Corridor, Gansu]: *Acta Palaeontologia Sinica*, 20, pp. 266–272 (in Chinese with English summary).

Shúan, Ji, and Be Haichen, 1998, [Discovery of the psittacosaur id skin impressions and its significance]: *Geological Review*, 44 (6), pp. 603–606 (in Chinese with English summary).

Shubin, Neil H., and H.-D. Sues, 1991, Biogeography of early Mesozoic continental tetrapods: patterns and implications: *Paleobiology*, 17 (3), pp. 214–230.

Shukolyukov, A., and G. W. Lugmair, 1998, Isotopic evidence for the Cretaceous–Tertiary impactor and its type: *Science*, 282, pp. 927–929.

Silva, Helder de Paula, and Alexander W. A. Kellner, 1999, Novos dentes de theropoda co Cretáceo continental do Brasil: *Paleontologia em Destaque*, 14 (26), p. 66.

Simmons, David J., 1965, The non-therapsid reptiles of the Lufeng Basin, Yunnan, China: *Fieldiana, Geology*, 15, pp. 1–93.

Simon, Lewis M., 2000, Report to members: *Archaeoraptor* fossil trail: *National Geographic*, 198 (4), pp. 128–132.

Sinclair, William J., and Walter Granger, 1914, Paleocene deposits of the San Juan Basin, New Mexico: *Bulletin of the American Museum of Natural History*, 33, pp. 297–316.

Sloan, Christopher P., 1999, Feathers for *T. rex*?: new birdlike fossils are missing links in dinosaur evolution: *National Geographic*, 196 (5), pp. 98–107.

Smith, David K., 1998, A morphometric analysis of *Allosaurus*: *Journal of Vertebrate Paleontology*, 18 (1), pp. 126–142.

_____, 1999, Patterns of size-related variation within *Allosaurus*: *Journal of Vertebrate Paleontology*, 19 (2), pp. 402–403.

Smith, David K., Arthur Anderson, Hans Larsson, and Paul Bybee, Results of a high resolution CT scan of *Coelophysis bauri* (Upper Triassic Chinle Formation, New Mexico): *Journal of Vertebrate Paleontology*, 20 (Supplement to Number 3), Abstracts of Papers, Sixtieth Annual Meeting, p. 70A.

Smith, David K., Paul Bybee, and Arthur Anderson, 1999, An updated analysis of cranial kenetics in *Allosaurus*: *Journal of Vertebrate Paleontology*, 19 (Supplement to Number 3), Abstracts of Papers, Fifty-ninth Annual Meeting, p. 76A.

Smith, Joshua B., You Hailu, and Peter Dodson, 1998, The age of the *Sinosauropteryx* quarry, northeastern China: *Journal of Vertebrate Paleontology*, 18 (Supplement to Number 3), Abstracts of Papers, Fifty-eighth Annual Meeting, p. 78A.

Spotila, James R., 1980, Constraints of body size and environment on the temperature regulation of dinosaurs, *in*: Roger D. K. Thomas and Everett C. Olson, editors. *A Cold Look at the Warm-Blooded Dinosaurs*. Boulder, Colorado: AAAS Selected Symposium, 28, Westview Press, pp. 233–252.

Spotila, James R., Michael P. O'Connor, and Frank V. Paladino, 1991, Hot and cold running dinosaurs. Size, metabolism and migration: *Modern Geology*, 16, pp. 203–227.

Starkov, Alexei I., 2000, Co-adaptive evolution: largest theropods and large sauropods: *Journal of Vertebrate Paleontology*, 20 (Supplement to Number 3), Abstracts of Papers, Sixtieth Annual Meeting, p. 71A.

Steel, Rodney, 1969, Ornithischia, *in*, Oskar Kuhn, editor, *Handbuch der Palaeoherpetologie*, part 15. Stuttgart: Gustav Fischer Verlag, 87 pages.

_____, 1970, Saurischia, *in*, Oskar Kuhn, editor, *Ibid.*, part 16. Stuttgart: Gustav Fischer Verlag, 87 pages.

Sternberg, Charles M., 1927, Horned dinosaur group in the National Museum of Canada: *Canadian Field Naturalist*, 41, pp. 67–73.

_____, 1940, *Thescelosaurus edmontonicus*, n. sp., and classification of the Hypsilophodontidae: *Journal of Paleontology*, 12, pp. 481–494.

_____, 1945, Pachycephalosauridae proposed for dome-headed dinosaurs, *Stegoceras lambei*, n. sp., described: *Journal of Paleontology*, 19, pp. 534–538.

_____, 1946, Canadian dinosaurs: *Bulletin of*

the National Museum of Canada, 103, pp. 1–20.
_____, 1951, Complete skeleton of *Leptoceratops gracilis* Brown from the Upper Edmonton Formation on Red Deer River, Alberta: *Annual Report of the National Museum for the Fiscal Year 1948–1949*, Bulletin 123, pp. 225–255.
Sternfield, R., 1911, Zur Nomenklatur der Gattung *Gigantosaurus* Fraas: *Sitzungherichte der Gesellischafti Naturfürschender Freunde zu Berlin*, 398 pages.
Stevens, Kent A., and J. Michael Parrish, 1999, Neck posture and feeding of two Jurassic sauropod dinosaurs: *Science*, 284, pp. 798–800.
Stokstad, Erik, 2000*a*, Learning to dissect dinosaurs — digitally: *Science*, 288, pp. 1728–1732.
_____, 2000*b*, Feathers, or flight of fancy?: *Ibid.*, 288, pp. 2124–2125.
Stovall, J. W., 1938, The Morrison of Oklahoma and its dinosaurs: *Journal of Geology*, 5 (46), pp. 583–600.
Stovall, J. Willis, and Wann Langston, Jr., 1950, *Acrocanthosaurus atokensis*, a new genus and species of Lower Cretaceous Theropoda from Oklahoma: *The American Midland Naturalist*, 43 (3), pp. 696–728.
Stromer, Ernst, 1909, Über Fossilfunde im Rhät und Unteren Lias bei Altdorf in Mittelfranken: *Naturhist. Ges. Nürnberg, Abh.*, 18, pp. 3–6, Nürnberg.
_____, 1915, Ergebnisse der Forschungsreisen Prof. E. Stromers in den Wüsten Ägyptens. II. Wirbeltierreste der Baharije-Stufe (unterstes Cenoman). III. Das Original des Theropoden *Spinosaurus aegyptiacus* n. g. n. sp.: *Abhandlungen der Bayerischen Akademie der Wissenschaften*, 18 (3), pp. 1–32.
_____, 1931, Wirbeltierreste de Baharijestufe (unterstes Cenoman). 10. Ein Skelettrest von *Carcharodontosaurus* nov. gen: *Ibid.*, 9, pp. 1–23.
_____, 1934, Ergebnisse der Forschungsreisen Prof. E. Stromer in den Wüsten Ägyptens. II: Wirbeltier-Reste der Baharije-Stufe unterestes (Cenoman). 13. Dinosauria: *Ibid.*, 22, pp. 1–79.
Suberbiola, Xabier [Javier, Xavier] Pereda, and Paul M. Barrett, 1999, A systematic review of ankylosaurian dinosaur remains from the Albian of England: *Special Papers in Palaeontology*, 60, pp. 177–208.
Sues, Hans-Dieter, 1978, A new small theropod dinosaur from the Judith River Formation (Campanian) of Alberta, Canada: *Zoological Journal of the Linnean Society*, 62, pp. 381–400.
_____, 1980, Anatomy and relationships of a new hypsilophodontid dinosaur from the Lower Cretaceous of North America: *Palaeontographica*, Abt. A, 169 (1–3), pp. 51–72.
_____, 1990, *Staurikosaurus* and Herrerasauridae, *in*: David B. Weishampel, Peter Dodson and Halszka Osmólska, editors, *The Dinosauria*. Berkeley and Los Angeles: University of California Press, pp. 143–147.
_____, 1997*a*, On *Chirostenotes*, a Late Cretaceous oviraptorosaur (Dinosauria: Theropoda) from Western North America: *Journal of Vertebrate Paleontology*, 17 (4), pp. 698–716.
_____, 1997*b*, Hysilophodontidae, *in*: Philip J. Currie and Kevin Padian, editors, *Encyclopedia of Dinosaurs*. San Diego: Academic Press, pp. 356–358.
Sues, Hans-Dieter, Eberhard Frey, and David M. Martill, 1999, The skull of *Irritator challengeri* (Dinosauria: Theropoda: Spinosauridae): *Journal of Vertebrate Paleontology*, 19 (Supplement to Number 3), Abstracts of Papers, Fifty-ninth Annual Meeting, p. 79A.
Sues, Hans-Dieter, and Peter M. Galton, 1987, Anatomy and classification of the North American Pachycephalosauria (Dinosauria: Ornithischia): *Palaeontografica*, Abstract A, 198, pp. 1–40.
Sues, Hans-Dieter, and David B. Norman, 1990, Hypsilophodontidae, *Tenontosauridae*, Dryosauridae, *in*: David B. Weishampel, Peter Dodson and Halszka Osmólska, editors, *The Dinosauria*. Berkeley and Los Angeles: University of California Press, pp. 498–509.
Sullivan, Robert M., 1987, A reassessment of reptilian diversity across the Cretaceous-Tertiary boundary: *Natural History Museum of Los Angeles County, Contributions in Science*, 391, 26 pages.
_____, 1993, Comments on the proposed designation of a neotype for *Coelophysis bauri* (Cope, 1887) (Reptilia, Saurischia): *Bulletin of Zoological Nomenclature*, 50, pp. 150–151.
_____, 1994, Topotypic material of *Coelophysis bauri* (Cope) and the *Coelophysis-Rioarribasaurus-Syntarsus* problem: *Journal of Vertebrate Paleontology*, 14 (Supplement to Number 3), Abstracts of Papers, Fifty-fourth Annual Meeting, p. 48A.
_____, 1995, Comment on the proposed designation of a neotype for *Coelophysis bauri* (Cope, 1887) (Reptilia, Saurischia): *Bulletin of Zoological Nomenclature*, 52, pp. 76–78.
_____, 1997, A juvenile *Ornithomimus antiquus* (Dinosauria: Theropoda: Ornithomimosauria), from the Upper Cretaceous Kirtland Formation (De-na-zin Member), San Juan Basin, New Mexico: *New Mexico Geological Society Guidebook, 48th Field Conference, Mesozoic Geology and Paleontology of the Four Corners Region*, pp. 249–253.
_____, 1999, *Nodocephalosaurus kirtlandensis*, gen. et sp. nov., a new ankylosaurid dinosaur (Ornithischia: Ankylosauria) from the Upper Cretaceous Kirtland Formation (Upper Campanian), San Juan Basin, New Mexico: *Journal of Vertebrate Paleontology*, 19 (1), pp. 126–139.
_____, 2000*a*, Reply to Upchurch and Barrett: *Ibid.*, 20 (1), pp. 218–219.
_____, 2000*b*, *Prenocephale edmontonensis* (Brown and Schlaikjer) new comb. and *P. brevis* new comb. (Dinosauria: Ornithischia: Pachycephalosauria) from the Upper Cretaceous of North America, *in*: Spencer G. Lucas and Andrew B. Heckert, editors, *Dinosaurs of New Mexico*. Albuquerque: New Mexico Museum of Natural History and Science, Bulletin 17, pp. 177–190.
_____, 2000*c*, *Stegoceras* revisited: *Journal of Vertebrate Paleontology*, 20 (Supplement to Number 3), Abstracts of Papers, Sixtieth Annual Meeting, p. 72A.
Sullivan, Robert M., and G. E. Bennett, III, A juvenile *Parasaur olophus* (Ornithischia: Hadrosauridae) from the Upper Cretaceous Fruitland Formation of New Mexico, *in*: Spencer G. Lucas and Andrew B. Heckert, editors, *Dinosaurs of New Mexico*. Albuquerque: New Mexico Museum of Natural History and Science, Bulletin 17, pp. 215–220.
Sullivan, Robert M., and Spencer G. Lucas, 1998, *Eucoelophysis baldwini*, a new theropod dinosaur from the Upper Triassic of New Mexico, and the status of the original types of *Coelophysis*: *Journal of Vertebrate Paleontology*, 19 (1), pp. 81–90.
_____, 2000*a*, *Alamosaurus* (Dinosauria: Sauropoda) from the late Campanian of New Mexico and its significance: *Ibid.*, 20 (2), pp. 400–403.
_____, 2000*b*, First occurrence of *Saurornitholestes* (Theropoda: Dromaeosauridae) from the Upper Cretaceous of New Mexico, *in*: Spencer G. Lucas and Andrew B. Heckert, editors, *Dinosaurs of New Mexico*. Albuquerque: New Mexico Museum of Natural History and Science, Bulletin 17, pp. 105–108.
Sullivan, Robert M., Spencer G. Lucas, Andrew B. Heckert, and Adrian P. Hunt, 1996, The type locality of *Coelophysis*, a Late Triassic dinosaur from north-central New Mexico (USA): *Paläontologische Zeitschrift*, 70, pp. 245–255.
Sullivan, Robert M., Darren H. Tanke, and Bruce M. Rothschild, 2000, An impact fracture in an ornithomimid (Ornithomimosauria: Dinosauria) metatarsal from the Upper Cretaceous (late Campanian) of New Mexico, *in*: Spencer G. Lucas and Andrew B. Heckert, editors, *Dinosaurs of New Mexico*. Albuquerque: New Mexico Museum of Natural History and Science, Bulletin 17, pp. 109–111.
Sullivan, Robert M., and Thomas E. Williamson, 1996, A new skull of *Parasaurolophus* (long-crested form) from New Mexico: external and internal (CT scans) features and their functional implications: *Journal of Vertebrate Paleontology*, 16 (Supplement to Number 3), Abstracts of Papers, Fifty-sixth Annual Meeting, p. 68A.
_____, 1999, A new skull of *Parasaurolophus* (Dinosauria: Hadrosauridae) from the Kirtland Formation of New Mexico and a revision of the genus: *New Mexico Museum of Natural History and Science*, Bulletin 15, 52 pages.
Sun, A., Li, J., Ye, X., Dong Zhiming, and Hou, L., 1992, *The Chinese Fossil Reptiles and Their Kin*. Beijing: Science Press, 260 pages.
Suteethorn, Varavudh, Valérie Martin, Eric Buffetaut, S. Triamwi Chanon, and Y. Chaimanee, 1995, A new dinosaur locality in the lower Cretaceous of northeastern Thailand: *Comptes Rendus de l'Académie des Sciences de Paris*, 321, pp. 1041–1047.

Suzuki, Daisuke, David B. Weishampel, and Nachio Minoura, 2000, *Nipponosaurus sachalinensis* Nagao, 1936 (Dinosauria: Hadrosauridae): anatomy and systematic position: *Journal of Vertebrate Paleontology*, 20 (Supplement to Number 3), Abstracts of Papers, Sixtieth Annual Meeting, p. 72A.

Sweeny, Frank, 2000, Built for speed — a biomechanical analysis of the distal phalanges of *Tyrannosaurus rex*, a theropod dinosaur: *Journal of Vertebrate Paleontology*, 20 (Supplement to Number 3), Abstracts of Papers, Sixtieth Annual Meeting, p. 72A.

Swinton, W. E., 1934, *The Dinosaurs: A Short History of a Great Group of Extinct Reptiles*. London: Thomas Murby and Co., xii, 233 pages (1st edition).

_____, 1970, *The Dinosaurs*. New York: Wiley — Interscience, A Division of John Wiley & Sons, Inc., 331 pages.

Swisher, C. C., Wang, Y.-q., Wang. X.-l., Xu, X., and Wang, Y., 1998, "Ar/Ar dating of the lower Yixian Fm., Liaoning Province, northeastern China: *Chinese Science Bulletin* (supplement), 43, 125 pages.

Tannenbaum, F., 1983, The microvertebrate fauna of the *Placerias* and Downs quarries, Chinle Formation (Upper Triassic) near St. Johns, Arizona, master's thesis, University of California, Berkeley, 111 pages (unpublished).

Taquet, Philippe, 1967, Deécouvertes paléontologiques récentes dans le nord du Niger, *in*: *Problémes actuels de Paléontologie*. Paris: Ediciones du Centre national de la recherche scientifique, pp. 415–418.

_____, 1976, Géologie et Paléontologie du Gisement de Gadoufaoua (Aptien du Niger): Cahiers de Paléontologie. Paris: *Editions du Centre National de la Recherche Scientifique*, pp. 1–191.

_____, 1984, Une curieuse spécialisation du crâne de certainss dinosauriens carnivores du Crétacé le meseau long et étroit des Spinosauridés: *Comptes Rendus des Séances de l'Academie des Sciences*, 299 II (5), pp. 217–222.

_____, 1994*a*, *L'empreinte des dinosaures*. Paris: Éditions Odile Jacob.

_____, 1994*b*, Chercheur d'os au Laos, *in*: Josué-Heilmann Hoffet, editor, *Commune d'Oberhausbergen, d'Oberhausbergen au Laos*, pp. 49–62.

Taquet, Philippe, B. Battail, and J. Dejax, 1992, New discoveries of sauropod and ornithopod dinosaurs in the Lower Cretaceous of Laos: *Journal of Vertebrate Paleontology*, 12 (3), p. 55.

Taquet, Philippe, B. Battail, J. Dejax, P. Richir, and M. Veran, 1995, First discovery of dinosaur footprints and new discoveries of dinosaur bones in the Lower Cretaceous of Savannakhet Province, Laos, *in*: *IGCP Symposium on Geology of SE Asia, Hanoi, Journal of Geology*, 5–6, p. 167.

Taquet, Philippe, and Dale A. Russell, 1998*a*, The ponderous iguanodont from Gadoufaoua, Niger: *Journal of Vertebrate Paleontology*, 18 (Supplement to Number 3), Abstracts of Papers, Fifty-eighth Annual Meeting, p. 81A.

_____, 1998*b*, New data on spinosaurid dinosaurs from the Early Cretaceous of the Sahara: *Comptes Rendus des Séances de l'Académie des Sciences*, 327, pp. 347–353.

_____, 1999, A massively-constructed iguanodont from Gadoufaoua, Lower Cretaceous of Niger: *Ann. Paléontol.* 85, 1, pp. 85–96.

Tarsitano, and M. K. Hecht, 1980, A reconsideration of the reptilian relationships of *Archaeopteryx*: *Zoological Journal of the Linnean Society*, 69, pp. 149–182.

Thayer, D. W., and R. P. Ratkevich, 1995, In progress excavation in the Mid-Cretaceous Turney Ranch Formation in southeastern Arizona: *Fossils of Arizona*, volume 3. Mesa, Arizona: Southwest Paleonteological Society.

_____, 1996, Dinosaur remains in southern Arizona, *in*: Donald L. Wolberg and Edmund Stump, editors, *Dinosaur International Symposium Program and Abstracts*. Tempe: Arizona State University, p. 108.

Thulborn, Richard A., 1972, The post-cranial skeleton of the Triassic ornithischian dinosaur *Fabrosaurus australis*: *Paleontology*, 15, pp. 29–60.

_____, 1984, The avian relationships of *Archaeopteryx*, and the origin of birds: *Zoological Journal of the Linnean Society*, 82, pp. 119–158.

_____, 1990, *Dinosaur Tracks*. London and New York: Chapman and Hall, xvii, 410 pages.

Thurmond, J. T., 1974, Lower vertebrate faunas of the Trinity Division in north-central Texas: *Geoscience and Man*, 8, pp. 103–129.

Tidwell, Virginia, Kenneth Carpenter, and William Brooks, 1999, New sauropod from the Lower Cretaceous of Utah, USA: *Oryctos*, 2, pp. 21–37.

Triebold, Michael, Frederick Nuss, and Candace Nuss, 2000, Initial report on a new North American oviraptor: *Publications in Paleontology* [The Florida Symposium on Dinosaur Bird Evolution, abstracts of papers], 2, p. 25.

Triebold, Michael, and Dale A. Russell, 1995, A new small dinosaur locality in the Hell Creek Formation: *Journal of Vertebrate Paleontology*, 15 (Supplement to Number 3), Abstracts of Papers, Fifty-seventh Annual Meeting, p. 57A.

Tumanova, Tatyana A., 1983, The first ankylosaurs from the Lower Cretaceous of Mongolia: *Transactions of the Joint Soviet-Mongolian Palaeontological Expedition*, 24, pp. 110–118.

_____, 1985, Cranial morphology of the ankylosaur *Shamosaurus scutatus* from the Lower Cretaceous of Mongolia, *in*: *Actes du Colloque "Les Dinosaures de la Chine á la France."* Muséum d'Histoire Naturelle, Tolouse, 2–6, pp. 73–79.

_____, 1998, Mongolian ankylosaurs: implications in morphological evolution: *Journal of Vertebrate Paleontology*, 18 (Supplement to Number 3), Abstracts of Papers, Fifty-eighth Annual Meeting, p. 83A.

Turner, J. C., 1983, Descripciòn geològica de la hoja 44d, Colòn Conhué Provincia del Chubut. Carta Geològico-Ecomòica de la Republica Argentina. Escala 1:200,000: *Servico GeolNgico Nacional, Buenos Aires*, 197, pp. 1–78.

Tykoski, Ronald S., 1998, The osteology of *Syntarsus kayentakatae* and its implications for ceratosaurid phylogeny, M.S. thesis, University of Texas, Austin.

Upchurch, Paul, 1993, The anatomy, phylogeny, and systematics of the sauropod dinosaurs, Ph.D. dissertation, University of Cambridge, 489 pages (unpublished).

_____, 1994, Sauropod phylogeny and palaeoecology: *Gaia: Revista de Geociencias, Museuo Nacional de Historia Natural*, University of Lisbon, 10, pp. 249–260.

_____, 1995, The evolutionary history of sauropod dinosaurs: *Philosophical Transactions of the Royal Society of London*, Series B 349, pp. 365–390.

_____, 1997*a*, Prosauropoda, *in*: Philip J. Currie and Kevin Padian, editors, *Encyclopedia of Dinosaurs*. San Diego: Academic Press, pp. 599–607.

_____, 1997*b*, Sauropodomorpha, *Ibid.*, pp. 658–660.

_____, 1999, The phylogenetic relationships of the Nemegtosauridae (Saurischia, Sauropoda): *Journal of Vertebrate Paleontology*, 19 (1), pp. 106–125.

Upchurch, Paul, and Paul M. Barrett, 2000, The taxonomic status of *Shanxia tianzhenensis* (Ornithischia, Ankylosauridae); a response to Sullivan: *Journal of Vertebrate Paleontology*, 20 (1), pp. 216–217.

Urlichs, B., 1966, Zur Fossilführung und Genese des Feuerlettens, der Rät-Lias-Grenzschichten und des unteren Lias bei Nürnberg: *Erlanger Geol. Abh.*, 64, pp. 1–42, Erlangen.

_____, 1968, Erläuterungen zur Geologischen Karte von Bayern 1:25000, Blatt Nr. 6533 Rothenbach a.d. Pegnitz: 126 pages, 1. geological map, Munchen (Bayer. Geol. Landesamt).

Van Valkenburgh, Blaire, and Ralph E. Molnar, 2000, Dinosaurian and mammalian predator guilds compared: *Journal of Vertebrate Paleontology*, 20 (Supplement to Number 3), Abstracts of Papers, Sixtieth Annual Meeting, p. 75A.

Vargus, Alexander, 1999, Evolution of arm size in theropods: *Ameghiniana, Revista de la Asociación Paleontológica Argentina, (Resúmenes)*, 36 (Número 4-Suplemento), p. 23R.

Vargas, Alexander, Alexander W. A. Kellner, Guillermo Chong Diaz, David Rubilar, and Mario Soares, 2000, New sauropod dinosaur from the Atacama Desert, Northern Chile: *Journal of Vertebrate Paleontology*, 20 (Supplement to Number 3), Abstracts of Papers, Sixtieth Annual Meeting, p. 75A.

Varrichio, David J., 1999, Gut contents for a Cretaceous tyrannosaur: implications for theropod dinosaur digestive tracts: *Journal of Vertebrate Paleontology*, 19 (Supplement to Number 3), Abstracts of Papers, Fifty-ninth Annual Meeting, p. 82A.

Varricchio, David J., and Frankie Jackson, 2000*a*, Egg pairs and parenting: reproduction in the theropod dinosaur *Troodon formosus*: *Publications in Paleontology* [The Florida Sympos ium on Dinosaur Bird Evolution, abstracts of papers], 2, p. 25.

_____, 2000*b*, Physiological implications of reproductive behavior in the dinosaur *Troodon formosus*: *Journal of Vertebrate Paleontology*, 20 (Supplement to Number 3), Abstracts of Papers, Sixtieth Annual Meeting, p. 75A.

Varricchio, David J., Frankie Jackson, J. J. Borkowski, and John R. Horner, 1997, Nest and egg clutches of the dinosaur *Troodon formosus* and the evolution of avian reproductive traits: *Nature*, 385, pp. 247–250.

Varricchio, David J., Frankie Jackson, and Clive N. Trueman, 1998, A nesting trace with eggs for the Cretaceous theropod dinosaur *Troodon formosus*: *Journal of Vertebrate Paleontology*, 19 (1), pp. 91–100.

Vene, Metello, 1998, Scoperte dinosauri d'Italia "Il nostro Jurassic Park" ("Our Jurassic Park"): *Airone*, 206.

Vickaryous, Matthew K., Allan P. Russell, Philip J. Currie, Kenneth Carpenter, and James I. Kirkland, 1998, The cranial sculpturing of ankylosaurs (Dinosauria: Ornithischia): reappraisal of developmental hypotheses: *Journal of Vertebrate Paleontology*, 18 (Supplement to Number 3), Abstracts of Papers, Fifty-eighth Annual Meeting, pp. 83A–84A.

Vickers-Rich, Patricia, Thomas H. Rich, Gregory C. McNamara, and Angela Milner, 1999, Is *Agrosaurus macgillivrayi* Australia's oldest dinosaur? *Records of the Western Australian Museum*, Supplement 57, pp. 191–200.

Viohl, Günter, 1999, Fund eines neuen kleinen Theropoden [Discovery of a new small theropod]: *Archaeopteryx*, 17, pp. 15–19, Eichstätt.

Wagner, Günter P., and Jacques A. Gauthier, 1999, 1, 2, 3 = 2, 3, 4: a solution to the problem of the homology of the digits of the avian hand: *Proceedings of the National Academy of Sciences*, 96, pp. 5111–5116.

Waldman, M., 1974, Megalosaurids from the Bajocian (Middle Jurassic) of Dorset: *Palaeontology*, 17 (2), pp. 325–339.

Walker, Alick D., 1969, The reptile fauna of the "Lower Keuper" Sandstone: *Geological Magazine*, 106, pp. 470–476, London.

_____, 1972, New light on the origin of birds and crocodiles: *Nature*, 237, pp. 257–263.

_____, 1977, Evolution of the pelvis in birds and dinosaurs, *in*: S. M. Andrews, R. S. Miles, and Walker, editors, *Problems in Vertebrate Evolution*. New York: Acedemic Press, pp. 319–357.

_____, 1990, A revision of *Sphenosuchus acutus* Haughton, a crocodylomorph reptile from the Elliot Formation (Late Triassic to Early Jurassic) of South Africa: *Philosophical Transactions of the Royal Society of London*, 8 (330), pp. 1–120.

Walker, C. A., 1981, New subclass of birds from the Cretaceous of South America: *Nature*, 292, 51–53.

Wall, William P., and Peter M. Galton, 1979, Notes on pachycephalosaurid dinosaurs (Reptilia: Ornithischia) from North America, with comments on their status as ornithopods: *Canadian Journal of Earth Sciences*, 16 (6), pp. 1176–1186.

Wang, S. E., 1998, Correlation of the Continental Jurassic in the north of China to the Pacific Jurassic in Northwest Scotland, United Kingdom. *Acta Geologica Sinica*, 72 (1), pp. 11–21 (in Chinese with English summary).

Wang Xiao-lin, Wang Yuanqing, Wang Yuan, Xu Xing, Tang Zhilu, Zhang Fucheng, and Hu Yaoming, 1998, Stratigraphic sequence and vertebrate-bearing beds of the lower part of the Yixian Formation in Sihetun and neighboring area, Western Liaoning, China: *Vertebrata PalAsiatica*, 36 (2), pp. 81–101.

Wang, Yuanqing, Wang, Xiaolin, Wang, Yuan, and Xu, Xing, 2000, Vertebrae assemblages and age of the Jehol Group: *Journal of Vertebrate Paleontology*, 20 (Supplement to Number 3), Abstracts of Papers, Sixtieth Annual Meeting, p. 76A.

Watabe, Mahito, Hideo Nakaya, and R. Nakahara, 2000, New skeleton of nodosaurine ankylosaur from Lance Formation (Upper Cretaceous) in Wyoming: taxonomy and morphology of the armor: *Journal of Vertebrate Paleontology*, 20 (Supplement to Number 3), Abstracts of Papers, Sixtieth Annual Meeting, p. 77A.

Watabe, Mahito, David B. Weishampel, Rinchen Barsbold, Khisigjaw Tsogtbataar, and S. Suzuki, 2000, New nearly complete skeleton of the bird-like theropod, *Avimimus*, from the Upper Cretaceous of the Gobi Desert, Mongolia: *Journal of Vertebrate Paleontology*, 20 (Supplement to Number 3), Abstracts of Papers, Sixtieth Annual Meeting, p. 77A.

Webster, Donovan, 2000, Debut Sue: *National Geographic*, 197 (6), pp. 24–37.

Wedel, Matthew J., 2000*a*, Reconstructing *Brachiosaurus*: *Prehistoric Times*, 42, p. 47.

_____, 2000*b*, New material of sauropod dinosaurs from the Cloverly Formation: *Journal of Vertebrate Paleontology*, 20 (Supplement to Number 3), Abstracts of Papers, Sixtieth Annual Meeting, p. 77A.

Wedel, Matthew J., Richard L. Cifelli, and R. Kent Sanders, 2000*a*, *Sauroposeiden proteles*, a new sauropod from the early Cretaceous of Oklahoma: *Journal of Vertebrate Paleontology*, 20 (1), pp. 109–114.

_____, 2000*b*, Osteology, paleobiology, and relationships of the sauropod dinosaur Sauroposeidon: *Acta Paleontologica Polonica*, 45 (4), pp. 343–390.

Weishampel, David B., 1981, Acoustic analyses of potential vocalization in lambeosaurine dinosaurs (Reptilia: Ornithischia): comparative anatomy and homolgies: *Journal of Paleontology*, 55, pp. 1046–1057.

_____, 1984, Evolution of jaw mechanisms in ornithopod dinosaurs: *Advances in Anatomy, Embryology, and Cell Biology*, 87, pp. 1–110.

_____, 1990, Dinosaur distributions, *in*: David B. Weishampel, Peter Dodson and Halszka Osmólska, editors, *The Dinosauria*. Berkeley and Los Angeles: University of California Press, pp. 63–139.

_____, 1997, Dinosaurian cacophony: *BioScience*, 47, pp. 150–159.

Weishampel, David B., and Ralph Chapman, 1990, Morphometric study of *Plateosaurus* from Trössingen (Baden-Wurttemberg, Federal Republic of Germany), *in*: Kenneth Carpenter and Philip J. Currie, editors, *Dinosaur Systematic: Approaches and Perspectives*. New York: Cambridge University Press, pp. 43–52.

Weishampel, David B., David Eliot Fastovsky, Mahito Watabe, Rinchen Barsbold, and Khisigjaw Tsogtbataar, 2000, New embryonic and hatchling dinosaur remains from the Late Cretaceous of Mongolia: *Journal of Vertebrate Paleontology*, 20 (Supplement to Number 3), Abstracts of Papers, Sixtieth Annual Meeting, p. 78A.

Weishampel, David B., Dan Grigorescu, and David B. Norman, 1991, The dinosaurs of Translyvania: island biogeography in the Late Cretaceous: *National Geographic Research & Exploration*, 7 (2), pp. 196–215.

Weishampel, David B., and R. E. Heinrich, 1992, Systematics of Hypsilophodontidae and basal Iguanodontia (Dinosauria: Ornithopoda), *Historical Biology*, 6, pp. 159–184.

Weishampel, David B., and John R. Horner, 1986, The hadrosaurid dinosaurs from the Iren Dabasu Fauna (People's Republic of China, late Cretaceous): *Journal of Vertebrate Paleontology*, 6 (1), pp. 38–45.

_____, 1990, Hadrosauridae, *in*: Weishampel, Peter Dodson, and Halszka Osmólska, editors, *The Dinosauria*. Berkeley and Los Angeles: University of California Press, pp. 534–561.

Weishampel, David B., Coralia-Maria Jianu, and Zoltan Csiki, Systematics and biogeography of the titanosaurid sauropod *Magyarosaurus* from the Late Cretaceous of Romania (in preparation).

_____, 1990, Hadrosauridae, *in*: David B. Weishampel, Peter Dodson and Halszka Osmólska, editors, *The Dinosauria*. Berkeley and Los Angeles: University of California Press, pp. 534–561.

Weishampel, David B., and James A. Jensen, 1979, *Parasaurolophus* (Reptilia: Hadrosauridae) from Utah: *Journal of Paleontology*, 53, pp. 1422–1427.

Weishampel, David B., and Coralia-Maria Jianu, 1996, New theropod dinosaur material from the Hateg Basin (Late Cretaceous, Western Romania): *Neus Jahrbuch für Geologie und Paläontologie, Abhandlungen*, 200 (3), pp. 387–404.

Weishampel, David B., Coralia-Maria Jianu, Z. Csiki, and David B. Norman, 1998, *Rhabdodon*, an unusual euornithopod dinosaur from the Late Cretaceous of western Romania: *Journal of Vertebrate Paleontology*, 18 (Supplement to Number 3), Abstracts of Papers, Fifty-eighth Annual Meeting, p. 85A.

Weishampel, David B., Eric W. A. Mulder, Rudi W. Dortangs, John W. M. Jagt, Coralia-Maria Jianu, Marcel M. M. Kuypers, Hans H. G. Peeters, and Anne S. Schulp, 1999, Dinosaur remains from the type Maastrichtian: and update: *Geologie en Mijn bouw*, 78, pp. 357–365.

Weishampel, David B., and F. Westphal, 1998, Die Plateosaurier von Trössingen: *Ausstellungskataloge der Universität Tübingen*, 19, pp. 1–27, Tübingen (Attempo).

Weishampel, David B., and Lawrence M.

Witmer, 1990*a*, *Lesothosaurus*, *Pisanosaurus*, and *Technosaurus*, *in*: Weishampel, Peter Dodson, and Halszka Osmólska, editors, *The Dinosauria*. Berkeley and Los Angeles: University of California Press, pp. 416–426.

_____, 1990*b*, Heterodontosauridae, *Ibid.*, pp. 486–497.

Weishampel, David B., and L. Young, 1996, *Dinosaurs of the East Coast*. Baltimore: Johns Hopkins University Press.

Welles, Samuel P., 1983, *Allosaurus* (Saurischia, Theropoda) not yet in Australia: *Journal of Paleontology*, 57 (2), pp. 196–197.

_____, 1984, *Dilophosaurus wetherilli* (Dinosauria, Theropoda)—osteology and comparisons: *Palaeontographica, Abteilung* A, 185, pp. 88–180.

Welles, Samuel P., and Robert A. Long, 1974, The tarsus of theropod dinosaurs: *Annals of the South African Museum*, 64, pp. 191–218.

Wellnhofer, Peter, 1974, Das fünte Skelettexemplar von *Archaeopteryx*: *Paläontographica*, Ser A, 147, pp. 169–216.

_____, 1992, A new specimen of *Archaeopteryx* from the Solnhofen Limestone, *in*: Kenneth E. Campbell, editor, *Papers in Avian Paleontology*. Los Angeles: Natural History Museum of Los Angeles County, pp. 3–23.

_____, 1993, Prosauropod dinosaurs from the Feuerletten (Middle Norian) of Ellingen near Weissenburg in Bavaria: *Revisions in Paleobiology*, special volume 7, pp. 263–271, Geneva.

_____, 1994, New data on the origin and early evolution of birds: *Comptes Rendus des Séances de l'Academie des Sciences, Paris*, 319 (11), pp. 299–308.

_____, 2000, The plumage of *Archaeopteryx*: feathers of a dinosaur: *Publications in Paleontology* [The Florida Symposium on Dinosaur Bird Evolution, abstracts of papers], 2, p. 26.

Welman, J., 1995*a*, The evolution of the braincase of early diapsids and its bearing on the origin of dinosaurs, crocodiles, and birds, Ph.D. thesis, University of the Orange Free State, Bloemfontein, South Africa, 198 pages.

_____, 1995*b*, *Euparkaria* and the origin of birds: *South African Journal of Science*, 91, pp. 533–537.

_____, 1998, *Euskelosaurus* and the origin of dinosaurs: *in*: J. Almond, J. Anderson, P. Booth, A. Chinsamy-Turan, D. Cole, M. de Wit, B. Rubridge, R. Smith, J. Van Bever Donker, and B. C. Storey, editors, *Journal of African Earth Sciences*, 27 (14), Special Abstracts Issue, Gondwana 10: Event Stratigraphy of Gondwana, unpaginated.

Werner, C., 1993, Eine neue Fundstelle terrestrischer Werbeltiere aus der Kreide des Sudan: *Berline Geowissenschaftliche Abhandlung*, E, 9, pp. 201–209.

_____, 1994, Die kontinentale Wirbeltierfauna aus der unteren Okerkreide des Sudan (Wadi Milk Formation): *Ibid.*, E, 13, pp. 221–249.

Werner, C., and J. C. Rage, 1994, Mid-Cretaceous snakes from Sudan. A preliminary report on an unexpectedly diverse snake fauna: *Comtes Rendu des Séances de l'Académie des Sciences, Paris II*, 319, pp. 241–252.

Westgate, James W., R. B. Brown, Dana Cope, and Jeffrey Pittman, 2000, A Late Cretaceous dinosaur-bearing community from coastal deposits in Chihuahua, Mexico: *Journal of Vertebrate Paleontology*, 20 (Supplement to Number 3), Abstracts of Papers, Sixtieth Annual Meeting, p. 78A.

Westphal, F., and J. P. Durand, 1990, Magnétostratigraphie des séries continentales fluvio-lacustres du Crétacé supérieur dans le synclinal de l'Arc (région d'Aix-en-Provence, France): *Bulletin, Societé Géologique de France*, (8), VI, 4, pp. 609–620.

Wheeler, P. E., 1978, Elaborate CNS cooling structure in large dinosaurs: *Nature*, 275, pp. 441–443.

White, P. D., David E. Fastovsky, and Peter M. Sheehan, 1998, Taphony and suggested structure of the dinosaurian assemblage of the Hell Creek Formation (Maastrichthian), eastern Montana and western North Dakota: *Palaios*, 13, pp. 41–51.

White, Theodore E., 1964, The Dinosaur Quarry, *in*: E. F. Sabatka, editor, *Guidebook to the Geology and Mineral Resources of the Unita Basin: Intermountain Association of Petroleum Geologists, 13th Annual Field Conference*, pp. 21–28.

_____, 1972, Catalogue of the genera of dinosaurs: *Annals of the Carnegie Museum*, 44 (9), pp. 117–155.

Whittle, Christopher H., and Laura Onorato, 2000, On the origins of gastroliths—determining the weathering environment of rounded and polished stones by scanning-electron-microscope examination, *in*: Spencer G. Lucas and Andrew B. Heckert, editors, *Dinosaurs of New Mexico*. Albuquerque: New Mexico Museum of Natural History and Science, Bulletin 17, pp. 69–73.

Wiedermann, Albert, and Aloys Wehr, 1998, Vermessung von Dino-saurierskeletten mit Stereophotogrammetrie und Laserscanner: *Publikationen der Deutschen Gesellschaft für Photogrammetrie und Fernerkundung*, 6. pp. 301–308.

Wiedermann, Albert, Tim Suthau, and Jörg Albertz, 1999, Survey of dinosaur skeletons: *Mitteilungen Museum für Naturkunde der Humboldt-Universität zu Berlin, Geowissenschaftliche Reihe*, 2, pp. 113–119.

Wild, Rupert, 1991, *Janenschia* n.g. *robusta* (E. Fraas 1908) pro *Tornieria robusta* (Reptilia, Saurischia, Sauropodomorpha): *Stuttgarter Beitrge zur Naturkunde*, Serie B (Geologie und Paleontologie), 174 (4), pp. 1–4.

Wilhite, Ray, and Brian Curtice, 1998, Ontogenetic variation in sauropod dinosaurs: *Journal of Vertebrate Paleontology*, 18 (Supplement to Number 3), Abstracts of Papers, Fifty-eighth Annual Meeting, p. 86A.

Williams, M. E., Catastrophic versus noncatastrophic extinction of the dinosaurs: testing falsifiability, and the burden of proof: *Journal of Paleontology*, 68, pp. 183–190.

Williamson, Thomas E., 1996, The beginning of the age of mammals in the San Juan Basin, New Mexico: Biostratigraphy and evolution of Paleocene mammals of the Nacimiento Formation: *New Mexico Museum of Natural History and Science Bulletin*, 8, 141 pages.

_____, 2000, Review of Hadrosauridae (Dinosauria, Ornithischia) from the San Juan Basin, New Mexico, *in*: Spencer G. Lucas and Andrew B. Heckert, editors, *Dinosaurs of New Mexico*. Albuquerque: New Mexico Museum of Natural History and Science, Bulletin 17, pp. 191–213.

Williamson, Thomas E., and Paul L. Sealey, 1999, A pachycephalosaurid (Ornithischia: Pachycephalosauria) skull from the Upper Cretaceous Kirtland Formation, San Juan Basin, New Mexico: *Journal of Vertebrate Paleontology*, 19 (Supplement to Number 3), Abstracts of Papers, Fifty-ninth Annual Meeting, p. 84A.

Williston, Samuel Wendell, 1905, A new armored dinosaur from the Upper Cretaceous of Wyoming: *Science*, 20, pp. 503–504.

_____, 1908, The Ankylosauridae: *American Naturalist*, 42, pp. 629–630.

Wilson, Jeffrey A., 1997, A reevaluation of Mongolian sauropods: implications for sauropod biogeography: *Journal of Vertebrate Paleontology*, 17 (Supplement to Number 3), Abstracts of Papers, Fifty-seventh Annual Meeting, pp. 84A–85A.

_____, 1998, Evolution and phylogeny of sauropod dinosaurs: *Journal of Vertebrate Paleontology*, 18 (Supplement to Number 3), Abstracts of Papers, Fifty-eighth Annual Meeting, pp. 86A–87A.

_____, 1999*a*, master's thesis, University of Chicago, (unpublished).

_____, 1999*b*, A nomenclature for vertebral laminae in sauropods and other saurischian dinosaurs: *Journal of Vertebrate Paleontology*, 19 (4), pp. 639–653.

Wilson, Jeffrey A., Ricardo N. Martinez, and Oscar Alcober, 1999, Distal tail segment of a titanosaur (Dinosauria: Sauropoda) from the Upper Cretaceous of Mendoza, Argentina: *Journal of Vertebrate Paleontology*, 19 (3), pp. 591–594.

Wilson, Jeffrey A., and Paul C. Sereno, 1998, Early evolution and higher-level phylogeny of sauropod dinosaurs: *Journal of Vertebrate Paleontology*, 15 (Supplement to Number 2), Society of Vertebrate Paleontology Memoir 5, 68 pages.

Wiman, Carl, 1929, Die Kreide-Dinosaurier aus Shantung: *Palaeontologia Sinica*, Series C, 6 (1), pp. 1–67.

_____, 1931, *Parasaurolophus tubicen*, n. sp. aus der Kreide in New Mexico: *Nova Acta Regia Societas Scientarum Upsaliensis*, Series 4, 7 (5), pp. 1–11.

_____, 1933, Über Ceratopsia aus der oberen Kreide in New Mexico: *Ibid.*, series 4 (7), pp. 1–19.

Winkler, Dale A., Elizabeth M. Gomani, and Louis L. Jacobs, 2000, Comparative taphonomy of an Early Cretaceous sauropod quarry, Malawi, Africa: *Paleontological Society of Korea Special Publication*, 4, pp. 99–114.

Winkler, Dale A., Louis L. Jacobs, and Phillip A. Murry, 1997, Jones Ranch: An Early

Cretaceous sauropod bonebed in Texas: *Journal of Vertebrate Paleontology*, 17 (Supplement to Number 3), Abstracts of Papers, Fifthy-seventh Annual Meeting, p. 85A.

Winkler, Dale A., Phillip A. Murry, and Louis L. Jacobs, 1998, The new ornithopod dinosaur from Proctor Lake, Texas, and the deconstruction of the family Hypsilophodontidae: *Journal of Vertebrate Paleontology*, 18 (Supplement to Number 3), Abstracts of Papers, Fifty-eighth Annual Meeting, p. 87A.

Winkler, Dale A., Phillip A. Murry, Louis L. Jacobs, William R. Downs, J. R. Branch, and Patrick Trudel, 1988, The Proctor Lake dinosaur locality, Lower Cretaceous of Texas: *Hunteria*, 2(5), pp. 1–8.

Winkler, T. C., 1865, Musée Teyler. Catalogue systématique de la collection paléontologique, 4: *Les Héritiers Loosjes (Haar lem)*, pp. 459–473.

Withers, P. C., 1992, *Comparative Animal Physiology*. Fort Worth: Saunders College Publishing, 949 pages.

Witmer, Lawrence M., 1987, The nature of the antorbital fossa of archosaurs: shifting the null hypothesis, *in*: P. J. Currie and E. M. Koster, editors, *Fourth Symposium on Mesozoic Terrestrial Ecosystems, Short Papers*, Revised Edition, Occasional Paper of the Tyrrell Museum of Palaeontology 3, pp. 234–239.

Witmer, Lawrence M., and W. Desmond Maxwell, 1996, The skull of *Deinonychus* (Dinosauria: Theropoda): new insights and implications: *Journal of Vertebrate Paleontology*, 16 (Supplement to Number 3), Abstracts of Papers, Fifty-sixth Annual Meeting, p. 73A.

Witmer, Lawrence M., and Scott D. Sampson, 1999, Nasal conchae and blood supply in some dinosaurs: physiological implications: *Journal of Vertebrate Paleontology*, 19 (Supplement to Number 3), Abstracts of Papers, Fifty-ninth Annual Meeting, p. 85A.

Wolf, Douglas G., 2000, New information on the skull of *Zuniceratops christopheri*, a neoceratopsian dinosaur from the Cretaceous Moreno Hill Formation, New Mexico: *in*: Spencer G. Lucas and Andrew B. Heckert, editors, *Dinosaurs of New Mexico*. Albuquerque: New Mexico Museum of Natural History and Science, Bulletin 17, pp. 93–94.

Wolfe, Douglas G., and James I. Kirkland, 1998, *Zuniceratops christopheri* n. gen. & n. sp., a ceratopsian dinosaur from the Moreno Hill Formation (Cretaceous, Turonian) of west-central New Mexico, *in*: Spencer G. Lucas, James I. Kirkland, and John W. Estep, editors, *Lower and Middle Cretaceous Terrestrial Ecosystems*. Albuquerque: New Mexico Museum of Natural History and Science, Bulletin 14, pages 303–318.

Wright, Joanna L., 1999, Ichnological evidence for the use of the forelimb in iguanodontid locomotion: *Special Papers in Palaeontology*, 60, pp. 209–219.

Xu, Xing, 2000, A new feathered maniraptoran dinosaur: *Publications in Paleontology* [The Florida Symposium on Dinosaur Bird Evolution, abstracts of papers], 2, p. 27.

Xu Xing, Wang Xiao-Lin, and You Hai-Lu, 2000, A primitive ornithopod from the Early Cretaceous Yikian Formation of Liaoning: *Vertebrata PalAsiatica*, 38 (4), pp. 318–325.

Xu, Xing, Zhi-lu Tang, and Xiao-lin Wang, 1999, A therizinosauroid dinosaur with integumentary structures from China: *Nature*, 399, pp. 350–354.

Xu, Xing, and Xiao-Lin Wang, 2000, Troodontid-like pes the dromaeosaurid *Sinornithosaurus*: *Paleontological Society of Korea Special Publication*, 4, pp. 179–188.

Xu, Xing, Xiao-Lin Wang, and Xiao-Chun Wu, 1999, A dromaeosaurid dinosaur with a filamentous integument from the Yixian Formation of China: *Nature*, 401 (16), pp. 262–265.

Xu, Xing, Zhonghe Zhou, and Xiaolin Wang, 2000, The smallest known non-avian theropod dinosaur: *Nature*, 408, pp. 705–708.

Xing, Zhao Xi-jin, Lu Jun-Chang, Dong Zhiming, and Huang Wan-bo, 1996, A new hadrosaur-like iguanodontian dinosaur from the Late Cretaceous Sangpin Formation of Henan Province, China: *Journal of Vertebrate Paleontology*, 19 (Supplement to Number 3), Abstracts of Papers, Fifty-ninth Annual Meeting, p. 86A.

Xu Xing, and Wang Xiao-lin, 1998, New psittacosaur (Ornithischia, Ceratopsia) occurrence from the Yixian Formation of Liaoning, China and its stratigraphical significance: *Vertebrata PalAsiatica*, 36 (2), pp. 148–158.

Xue Xiangu, Zhang Yunxiang, Bi Yan, Yue Leping, and Chen Danling, 1996, *The Development and Environmental Changes of the Intermontane Basins in the Eastern Part of Qinling Mountains*. Beijing: Geological Publishing House, 181 pages.

Yang, Zhungjian [Young Chung-Chien], 1935, On a new nodosaurid from Ninghsia: *Palaeontologica Sinica*, Series C, 11, pp. 5–27.

_____, 1937, A new dinosaur from Sinkiang: *Palaeontologica Sineca*, New Series C., 105 (2), pp. 1–23.

_____, 1939, On a new Sauropoda, with notes on other fragmentary reptiles from Szechuan: *Bulletin of the Geological Survey of China*, 19 (3), pp. 299–315.

_____, 1954, On a new sauropod from Yiping, Szechuan, China: *Scientia Sinica*, 3, pp. 491–504.

_____, 1958, New sauropods from China: *Vertebrata PalAsiatica*, 2 (4), pp. 231–236.

_____, 1959, On a new fossil egg from Laiyang, Shantung: *Vertebrata PalAsiatica*, 3, pp. 34–35.

Yang, Zhungjian [Young Chung-Chien] and Chao, H. C., 1972, *Mamenchisaurus hochuanensis* sp. nov.: *Institute of Vertebrate Paleontology and Paleoanthropology Monograph*, series A, 8, pp. 1–30.

You, Hailu, and Peter Dodson, 2000, A new phylogeny of basal Neoceratopsia: *Journal of Vertebrate Paleontology*, 20 (Supplement to Number 3), Abstracts of Papers, Sixtieth Annual Meeting, p. 80A.

You, Hailu, Peter Dodson, Dong Zhiming, and Yoichi Azuma, 1999, *Archaeoceratops* and other dinosaurs from the late Early Cretaceous of the Mazongshan area, northwest China: *Journal of Vertebrate Paleontology*, 19 (Supplement to Number 3), Abstracts of Papers, Fifty-ninth Annual Meeting, p. 86A.

Zbyszewski, Georges, 1946, Les ossements d'*Omosaurus* découverts près de Baleal (Peniche): *Communicacoes dos Servicos Geológicos de Portugal*, 27, pp. 3–12.

Zelenitsky, D. K., and L. V. Hills, 1997, Normal and pathological eggshells of *Spheroolithus albertensis*, oosp. nov., from the Oldman Formation (Judith River Group, late Campanian), southern Alberta: *Journal of Vertebrate Paleontology*, 17, pp. 167–171.

Zhang, F-K [Futeng], 1975, A new thecodont *Lotosaurus*, from Middle Triassic of Hunan: *Vertebrata PalAsiatica*, Vol. XIII, No. 3: 144–147.

Zhang, Yihong, 1988, *The Middle Jurassic Dinosaur Fauna from Dashanpu, Zigong, Sichuan*, vol. 1. Chengdu: Sichuan Publishing House of Science and Technology, pp. 1–89.

Zhang Yihong, and Chen Wei, 1996, Preliminary research on the classification of sauropods from Sichuan Basin, China: *in*: Michael Morales, editor, *Museum of Northern Arizona Bulletin 60*, pp. 97–108.

_____, 1985, Reptile, *in*: S.-E. Wang, editor, *The Jurassic in China*. Beijing: Geological Publishing House, pp. 286–291.

Zhao, Zi-Kui, 2000, Nesting behavior of dinosaurs as interpreted from the Chinese Cretaceous dinosaur eggs: *Paleontological Society of Korea Special Publication*, 4, pp. 115–126.

Zhao, Zi-Kui, and H. Z. Mao, 1997, Biomechanical properties of dinosaur eggshells from Bayan Manduhu, Inner Mongolia: *Vertebrata PalAsiatica*, 31, pp. 77–84.

Zhao, Xinjin [also spelled Xin Jin], Cheng Zhengwu, and Xu Xing, 1999, The earliest ceratopsian form the Tuchengzi Formation of Liaoning, China: *Journal of Vertebrate Paleontology*, 19 (4), pp. 681–691.

Zhao, Xin Jin [also spelled Xijin], and Philip J. Currie, 1993, A large crested theropod from the Jurassic of Xinjiang, People's Republic of China: *Canadian Journal of Earth Science*, 30, pp. 2027–2036.

Zhao, Xin Jin [also spelled Xijin], and Xu Xing, 1998, The oldest coelurosaurian: *Nature*, 394, pp. 234–235.

Zhou Zhong-He and Wang Xiao-Lin, 2000, A new species of *Caudipteryx* from the Yixian Formation of Liaoning, northeast China: *Vertebrata PalAsiatica*, 38 (2), pp. 111–127.

Zhou Zhong-He, Wang Xiao-Lin, Zhang Fu-Cheng, and Xu Xing, 2000, Important features *Caudipteryx*— evidence from two nearly complete new specimens: *Vertebrata PalAsiatica*, 38 (4), pp. 241–245.

Zittel, K. A., 1893, *Traité de Paléontologie III, Paléozoologie, Vertebrata (Pices, Amphibia, Reptilia, Aves)*. Paris: Doin, xii, 894 pages. [French edition by C. Barrois; original German edition published in 1890, Oldenbourg, Muchen and Lepizig].

Index

Included in this index are genera and species; selected authors, discoverers, and other persons; selected institutions, organizations, stratigraphic horizons, localities, sites, events, and miscellaneous places; and also relevant general topics that are discussed or appear in the text as well as in picture captions. Junior synonyms are cross-referenced to the currently and most widely accepted senior synonym. Page numbers in *italics* indicate subjects shown or implied in illustrations, or names and topics mentioned or implied in picture captions or credits. Not indexed here are dinosaurian taxa above the level of genus, nor institutions referred to only as abbreviations used to designate catalogue numbers for fossil specimens.

Abel, Othenio 253–4
Abelisaurus 33, 127
Abrictosaurus 166
Abrosaurus 155
Academic Press vi, 617
The Academy of Natural Sciences of Philadelphia *175*, 218, 229–30, 545, 605, 615, 618, 621
Acanthopholis 164, 187–8, *188*, 189, 214, 216; ?*A. eucercus* 188–9; *A. horridus* 187–8, *188*, 189; "A. hughesii" 189; "A. keepingi" 189; *A. macrocerus* 187–8; *A. platypus* 187–8; *A. stereocercus* 187–8, 216
Achelousaurus 182, 232, 528
Achillobator 144, 187, *190*, 190–1; *A. giganticus* *190*, 190–1
Acrocanthosaurus *6*, 132, 187, 191–2, *192–3*, 193, 611; *A. atokensis* *6* 191, *192–3*
Actiosaurus 595
Adamantina Formation 195
Adasaurus 144, 582; *A. mongoliensis* 582
Adolphe, Bruce 577
Aegyptosaurus 157
Aepisaurus 147
Aeolosaurus 61, 157–8, 187, *194*, 194–5, 343; *A. rionegrinus* 194
Aetosaurs 47, 52, 332, 524
Afrovenator 127, *130*, 372, *374*; *A. abakensis* *130*, *374*
Agathaumas 183
The Age of Dinosaurs in Japan and China 341
Agilisaurus 159, 258–9, 371; *A. louderbacki* 371
Agrosaurus see *Thecodontosaurus*; *A. macgillivrayi* see *Thecodontosaurus antiquus*
Água Negra locality 51
Aguja Formation 75
Agustinia 151, 187, 195, *195–6*, 196–7, *197–8*, 198, 120; *A. ligabuei* 195, *195–6*, 196, *197–8*
Air sacs 29
Airone 79
Alamosaurus 109, 157–8, 187, 198, *199*, 200–1, 178, 358; *A. sanjuanensis* 109, 198, *199*
Albany Museum 428
Albert (Burgersdorp)-Bernard's Spruit (Ezelsklip) locality 337
Albert Kersten Geocentere 547
Albertosaurus 33, 94, 135, 187, 200–2, *202*, 304–6, 495, 562, 570, 572–3; *Albertosaurus* cf. *A. lancensis* see *Tyrannosaurus rex*; *Albertosaurus* sp. 201; ?*Albertosaurus* sp. 202
Alcober, Oscar 58, 151, 157–8, 481, 483
Alectrosaurus 36, 135, 229, 527; *A. olseni* 36
Alexander, R. McN. 247, 249, 251, 254, 275
Alioramus 135
Aliwal North locality 337
Aliwal North-Kraai River locality 337
Aliwalia 120, 187, 202, *202–3*, 203; *A. rex* 202, *202–3*
All About Dinosaurs vi
Allain, Ronan 358, 476–7, 534–5, *535–6*, 536, 582–3
Allen Formation 79, 194, 481
Allometry 21, 72, 205–6, 209, 230, 235, 251–2, 286, 312, 359, 395, 498, 548, 577
Allosaurus *6*, *14*, 23–4, *87*, 98–9, *130*, 130–2, 187, 191, 203–4, *204–5*, 205–6, *206–7*, 207, *208*, 209, *209*, 210–2, 241, 294, 338–9, 385, 434, 501, 575, 606, 608, 610–1; *A. atrox* *208–9*, 218–9, 614; *A. fragilis* *130*, 191–2, 203–4, *204–5*, 205–6, *206*, *209*, 210–1, 218–9; *A. maximus* *6*, *207*, *223*; *Allosaurus* sp. *14*, 211
Alocodon 158
Altirhinus 73, 171, *172*; *A. kurzanovi* *172*
Altispinax 127
Alvarez, Luis W. 106
Alvarez, Walter W. 106, 616
Alvarezsaurus 104, 116, 137, 138, 187, 212–3, 412, 445–6, 509, 595; *A. calvoi* 212
Alwalkeria 19, 120
Alxasaurus 98, 105, 142
Amargasaurus 55, 152, *152*, 187, 213, *214*, 254; *A. cazaui* 55, *152*, 213, *214*; *A. sattleri* 55, 213, 254
Amber 96
Amblin Dreamworks Entertainment 4
American Association of Museums ix
American Museum Novitates 191
American Museum of Natural History *59*, *77*, *85*, *137*, 144, *162*, 201, *206*, 210, 260, 299, 304–5, 312, *330–4*, *359*, 377, 407, 411, 415, *491*, 408, *524*, *562*, 585, 616, 618, 620, 622
Ammonites 603
Ammosaurus 52, 147, 461–2
Ampelosaurus 157–8, 392
Amphibians 12, 109, 332, 474, 516, 622
Amphicoelias 117, 151, 198
Amtosaurus 164, 166
Amurosaurus 175, 187, 213–4; *A. riabini* 213–4
Amygdalodon 149
"Anabisetia" 73–4; "A. saldivai" 73
Anasazisaurus see *Kritosaurus*; *A. horneri* see *Kritosaurus navajovius*
Anatotitan 80, 174; *A. copei* 80
Ancell, Carrie A. 254, 256
Anchiceratops 183, 449
Anchisaurus 146, 461–2
"Ancient Life" *6*
Andersen, Arthur F. 206, 212, 298, 301, 556–60
Anderson, Brian G. 75–6
Anderson, J. F. 192–3, 228
Andesaurus 49, 61, 157–8, 187, 194, 214
Andreis, R. R. 347
Andrews, Charles W. *256*, 256–7, 325–6, *326*, 328
Andrews, Roy Chapman vi, 411, 616
Angaturama see *Irritator*; *A. limai* see *Irritator challengeri*
Animal Kingdom 575
Animantarx 164
Ankylosaurus 118, 160–1, *163*, 165, 166, 418–20, 503, 505, 522; *A. magniventris* *163*
Anoplosaurus 164, 187, 189, 214–5, *215*, 216–7; *A. curtonotus* 214–5, *215*, 216; *A. major* 214, 216
Anserimimus 139
Antarctosaurus 117, 158, 187, *217*, 217–8, 317, 416, 418; *A. wichmannianus* *217*, 217–8, 416
Antlers Formation 191, 310, *311*, 312, 491
Anton, John *85*, 85–6
"Antonio" 79–80, *80*
Antrodemus *7*, 130, 187, 211, *218*, 218–9; *A. valens* *218*, 218–9
Antunes, Miguel Telles 21
Anunciacão, Carlos 315
Anura 474
Apatosaurus *5*, 49, *54*, 54–5, 58, *59*, 63, 67, *68*, *137*, 152, *153*, 187, *206*, 210, 219–20, *220–1*, 221–2, *222–3*, 223–4, *224–5*, 225, 228, 317, 387–8, *492*, 602, *605*, 606–7, *608*, 608, 610, *612*; *A. alenquerensis* see *Lourinhasaurus alenquerensis*; *A. excelsus* *5*, *59*, *68*, *137*, *153*, 220, *220–3*, 223, *224*, 225, *608*; *A. louisae* 219, *224–5*; *Apatosaurus* sp. 222
Apesteguía, S. 45, 78–9
Aragosaurus 155
Aralosaurus 174
Arboreality 21, 98, 102, 167, 270, 275–6, 404
Archaeoceratops 23, 138, 180–1, 187, 225–6, 289, 509; *A. oshimai* 226
Archaeopteryx 14, 87, 94, 96–9, 101, 102–6, 234, 239, 241–2, 244, 260, 270–2, 275, 300, 327, 403–4, 412–3, 468–9, 512, 586–7; *A. lithographica* 103, 260, 327, 586–7
Archaeoraptor see *Microraptor*; *A. liaoningensis* see *Microraptor zhaoianus*
Archaeornithoides 121
Archaeornithomimus 139
Archibald, J. David 106–8, 528
Arctosaurus 146
Arcucci, Andrea Beatriz 19, 24
Argana Formation 49
Argana series 19
Argentinian National Commission of Atomic Energy 537
Argentinosaurus 49, 117, 157

Index

Argyrosaurus 157–8
Arinobu, Tetsuya 106
Aristosuchus 133, 187, 226; *A. pusillus* 226
Arizona-Sonora Desert Museum 517
Arrhinoceratops 183
Arroyo Seco locality 60, 331–2
Arstanosaurus 180
Art Technology 613
Arundel Formation 227, 466
Ash, Sid 110, 616
Ashland Community Hospital 549
Ashworth, William B., Jr. 260, 263
Asiaceratops 118, 180, 415
Asiamerica 121
Asiatosaurus 147, 335
Asmussia 253
Associated Press 579
Asteroids 108–9, 111, 619
Astibia, Humberto 216–7, 383
Astrodon 156, 187, 226–7, 492, 518; *A. johnstoni* 227; *A. pusillus* see *Dacentrurus armatus*
Atlantosaurus 147, 219
Atlasaurus 156, 187, 227–8, *228–9*, 229; *A. imelakei* 227–8, *228–9*
Atlascopcosaurus 72–3, 168, 477–8; *A. loadsi* 72, 477–8
Attridge, John 146, 202–3, 245–6, 401–2
Aublysodon 36, 135, 187, 201, 229–30, 309, 524–5, 527–8; ?*A. lateralis* 527; *A. mirandus* 229, 525; *Aublysodon* cf. *A. mirandus* see *Stygivenator molnari*; *A. amplus* see ?*Stygivenator mirandus*; *A. cristatus* see ?*Stygivenator mirandis*; ?*Aublysodon* sp. 229; cf. *Aublysodon* sp. 229
Auca Mahuevo site 61–2, 485–6
Aucasaurus garridoi 592
Augusta, Josef 247, 616
Augustia ligabuei see *Agustinia ligabuei*
Austrosaurus 147
Avaceratops 182–3, 187, 230, *230–1*, 231–2, *232–3*, 233; *A. lammersi* 230, *230–3*, 554
Avalonia sanfordi see ?*Megalosaurus cambrensis*
Avalonianus see *Megalosaurus*; *A. sanfordi* see ?*Megalosaurus cambrensis*
Aves (also Aviale, avian theropods) *see* Birds
Avimimus 134, 187, *234*, 234–5, 257, 327, 469; *A. portentosus* *234*, 234–5, 327
Azendohsaurus 19, 49–51, 146
Azevedo, Sergio Alex Kugland de 19, 51, *194*, 194–5, 343
Azuma, Yoichi 38, 73, 130, 212, 226, 337–8, *338–40*, 341, 618

"Back to the Future" 613
Backbone 406
Bacteria 9–10, 524
Bactrosaurus 75, 118, 171–2, 187, 235–6, *236–7*, 237–8, *238*, 539; *B. johnsoni* 75, 172, 235–6, *236–7*, 237–8, *238*, 539
Badlands National Monument 605, *605*, 614
Baez, A. M. 351–2
Bagaceratops 180–1, 415
Bagaraatan 44, 130; *B. ostromi* 44
Bahariasaurus 132
Baharija Formation 267
Bai Ziqi 587, *587–8*, 588, *588–9*, 589
Baird, Donald 134, 377–8, 462–3
Bajo Barreal Formation 46, 56, 426, 428
Bakker, Robert T. vi, 12, 21, 24, 34, 62, 220, 224–5, 238–9, *239–40*, 240, *241*, 242, 247, 285, 287, 305, 310, 320, 322, 485–6, 565, 567, 575, 579, 616
Baldwin, David 296, 330
Ballent, S. C. 481, 483
"Bambi" 238
Bambiraptor 46, 94, 99, 103, 144, 187, 238–9, *239–40*, 240–1, *241–2*, 242, 495; *B. feinbergi* 99, 238–9, *239–40*, 240–1, *241–2*, 242, 495
Ban Nong Mek site 451
Bang, B. G. 577, 579
Banks, John C. 11
Barapasaurus 60, 148, 317, 346, 365–6, 537, 548; *B. tarorei* 365
Barnard's Spruit locality 337
Barosaurus 152, 219, 246, 250, 254, 367, 606, 610; ?*B. africanus* 254, 367
Barrett, Paul M. 41, *43*, 53, 55, 70, 73, *188*, 188–9, 214–5, *215*, 216–7, 243, 320–1, 362–3, 503, 505–6
Barrick, Reese E. 75, 91–3, 551, 554–5, *557*, 560
Barsbold, Rinchen 30–1, 38–40, *101*, 104, 105–6, 138–40, 142–4, 235, 257, 260, 327, 422–6, 471, 476–7, *584*, 584–7
Barsboldia 175
Barun Goyot [Barungoiotian] Formation *61*, 419, 442, 582, 586
Barunbayan Formation *475*
Baryonyx 30, 129–30, 134, 187, 243, 303–4, 364, 530–2, *532*; *B. walkeri* 30, 243, 303, *532*
Basabilvazo, George T. 76
Baum, L. Frank 432
Baumel, J. J. 12
Bauru Group 25, 194
Bavarisaurus 303
Bayan Mandahu Formation 453, 455, 457
Bayan Shireh Formation 190–1
Bayn Dzak 586
Beckett, J. *359*
Beckles, Samuel H. 430
Becklespinax 116, 129, 131
Beds of Bugeen Tsav 422–3, 489
Beipiaosaurus 94, 97, 116, 142, 187, 243–4, *244*, 516; *B. inexpectus* 243, *244*
Bell, Claude Kenneth 607, *608*
Bellusaurus 156, 157
Belly River beds 439
Belodon 595
Beneden, P.-J., van 359, 363
Benitez, Aldo 537
Bennett, Christopher S. 563, 579
Bennett, Geb E. 440–2
Benton, Michael J. 44, 48, 51, 121, 123, 146–7, 158, 166, 245–6, 327, 346, 462–3, 488–9, *489*, 502, 533–4, 543, *543–5*, 545–6, 548–9, 595, 616
Beringia 415
Berman, David S. 54, 219, 225, 294
Berner, R. A. 228
Bertini, Reinaldo J. 194–5
Betasuchus 121; *B. bredai* 78
Betonnie Tsosie Wash locality 109
Beurlen, K. 487–8
Bidar, Alain 29
Bienosaurus lufengensis 592
Big Bend National Park 75
Bihariosaurus 171
Bilbey, Sue Ann 55–6
Bipedality (in sauropods) 219, 249
BIR Laboratories 298
Bird, John H. 254
Bird, Roland T. *84*, 616
Birds vi, 5, 10–2, 14–5, 17, 21, 23, 32, 38, 40, 42, 44, 66, 69, 71, 76, 84, 86–7, 89–95, *95*, 96–100, *100*, 101–5, *105*, 106, 109, 119–21, 124–5, 130, 133–5, 137–8, 140, 142–3, 145, 158, *176*, 179, 207, 210, 213, 222, 224, 234, 239–42, 244, 252, 257, 260, *270*, 270–6, 298, 300, 304, 313, 326–7, 359, 375, 386, 395, 404–6, *411*, 411–3, 423–5, 442, 444, 467–9, *469–70*, 470, 474–5, 506, 508–9, 512–3, 516, 551, 558–61, 577–8, 582, 584–7, 595, 613, 615–6
Biting and bite marks 34–5, 70, 211, 220, 304, 306, 323–4, *324*, 335, 375, 464, 566, 574–5, 578
Bivalves 332
Black, Craig C. 620
Black Hills Institute of Geological Research, Inc. 191, *572*, 578
Black Hills Petrified Forest 606
Blanckenhorn, M. 460, 463
"Blikana dinosaur" ("Blikana melanorosaurid") 245
Blikanasaurus 147, 187, 202, *245*, 244–6; *B. cromptoni* 245, *245*
Blagoveschensk locality 213
Blood and circulation 66–7, 86, 88, 134, 251–2, 550–1, 559
Blows, William T. 216–7, 354–5, 616
Bluewater Creek Formation 265, 294, 301
Boca Formation 23
Bohlin, Birger 292–4
Bohor, B. F. 106
Bolor's Hill locality 260
Bolotsky, Y. L. 213–4
Bolt quarry *see* Dinosaur Hill
Bonaparte, José F. 40, 48–9, 79, 120–1, 123, 127, 149, 152, *152*, 154–5, 157, 195, *195–6*, 196–7, *197–8*, 198, 212–3, 217–8, 230, 315–6, *316–7*, 317, 346, *346–7*, 347–8, *348–9*, 349–50, *350*, 351, 358–9, 365–8, *368*, 369, 379, *380–1*, 382, 388, 446–7, 489, 501–2, 537–40, *540–1*, 541–2, 616
Bonebeds 25–6, 33, 60, 69, 72, 82, 201, 213, 222, 235–6, *236*, 253, 255, 263, 289, 378–9, 393, 395, 397, 435, 591
Bone Cabin Quarry *280*, 285–6, 301, *524*
Bone histology 70, 81–2, 91, 221–4, 249–50, 254, 359, 395, 428, 475, 524
Bonnan, Matthew F. 67–8, *68*, 317–20
Bore holes 52–3, 393
Borogovia 143, 260; *B. gracilicrus* 260
Borsuk-Bialynicka, Magdalena 604
Boscarolli, Dario 351, 354
Boss, N. H. 438
Boston Museum of Science *107*
Bothriospondylus 156
Brachiosaurus 49, 56, 60, 63, *65*, 67, 155–6, 187, 198, 228, 246, *246–7*, 247–8, *248*, 249–54, 277, 317, 342, 367–8, 372, 491–2, *493*, 517–9, 548, 606, 610, 614; *B. altithorax* 63, *65*, 246, *246–7*, 606, 614; ?*B. atalaiensis* 63; *B. brancai* 49, 63, 246–7, *247–8*, 251, 253–4, 367, *493*, 518
Brachyceratops 182, 231
Brachylophosaurus 174, 187, 254–5, *255*, 256; *B. cabadensis* *255*; *B. goodwini* 255–6; *Brachylophosaurus* sp. 255
Brachypodosaurus 162
Brachytaenius 595
Bradycnema 44, 130, 187, *256*, 256–7, 260, 325–7, 350; *B. draculae* *256*, 256–7, 260, 327, 351
Braginetz, Donna 619, 622
Brains 21, 23, 66, 88, 98, 143, 147, 160, 292, 555, 559, *571*, 577
Breithaupt, Brent H. 210
Brett-Surman, Michael K. 18, 115, 120, *124*, 172, 174–5, 235–6, 238, *238*, 317, 320, 328, *351*, 377, 384, 438, 467, 539, 556, 560, 592, 623, 616, 618
Breviceratops 180–1, 415
Briggs, Derek E. G. 448
Brinkman, David L. 310, *311*, 312–3, 401–2
Brinkman, Donald B. 41–2, 82, 457
Brinkmann, W. 189, 214, 217
Bristol City Museum 545
Bristol Institution 543
British Geological Survey *400*
British-South African Expedition 245
Britt, Brooks B. 57, 99–100, 124–5, 254, 283, 285–7, 377, 409, 492, 494, 503, 517–8, 615
Brochu, Christopher A. 94–5, 250, 518–9, *571*, 577–9, 616, 621
Brongniart, Alexandre 360–1
Brooding *18*, 40, 90, 140, 561
Brookes, A. 30
Broom, Robert 337
Brown, Alfred 336–7
Brown, Barnum 68, *69*, 164, 201, *206*, 211–2, 304–5, 309–10, 377, 407–8, *408*, 414–6, 420, 463–5, 520, 522, 602, 616
Bruhathkayosaurus 127
Brusatte, Steve 576, 578–9
Buckland, William 360
Budhan, James R. 109
Buffetaut, Éric 11, 36, 49, 74, 151, 216–7, 257, 327–8, 360, 362–3,

365–6, *366*, 393, 448, 451–2, *452–3*, 453, 476–7, 520, 536, 581, *581–2*, 582–3, *583*
Bugenasaura 168, 187, 257–8, *258–9*, 259, 549; *B. infernalis* 257, *258–9*, 259
Bull Canyon Formation 294–7
Bulletin of Zoological Nomenclature 362
Bunurong Marine Park 477
Bureau of Land Management (BLM) 55, 110, 606, 613–4
Burge, Donald L. 30, 254, 342, 354–5, 503, 542
Burian, Zdenevk 247, 602, 616, 621
Burnham, Dave 99, 238–9, *239–40*, 240, *241*, 242
Burpee Museum of Natural History *22*
Bybee, Paul 205–6, 212, 298, 301
Byrd, Gary 471
Byronosaurus 143, 187, 259–60; *B. jaffei* 143, 259–60

Cactus Park bonebed 222
Caenagnathasia 141
Calamosaurus 121
Calamospondylus 121, 226; *C. oweni* 226
The Calgary Zoo Prehistoric Park 601, 603
Calgary Zoological Society *183*, 601
Callobatrachus 516
Callovosaurus 168
Calvo, Jorge Orlando 53, 55, 57–8, 61, 73, 151, 154, 194–5, 217–8, 279, 327, 416–7, 469, 471, 499–500
Camadas de Alcobaca Formation 315
Camarasaurus 49, 53–4, *54*, 56, 58, 60, 63, 66, 155, *155*, 187, 198, *221*, 225, 228, 260–1, *261–2*, 262–3, 294, 317–9, 367–9, 372, 387–8, 541, 607, *608–9*, 610, 614; *C. alenquerensis* see *Lourinhasaurus alenquerensis*; *C. grandis* 263; *C. lentus 54, 221*, 225, 260, *261*; *C. supremus 155, 262*, 294, 367; *Camarasaurus* sp. 262
Cambridge Greensand 189, 214, 216
Camelotia 147, 187, 263, *264*, 265, 399, 401, 460–1; *C. borealis* 263, *264*, 265, 460–1
Camelus domesticus 66
Camp, Charles 265
Camp des deux arbres 388
Campagna, Tony 323, 325, 579
Campos, Diogenese de A. 129, 303–4, 364–5
Camposaurus 126, 187, 265–6, *266*, 267, 296; *C. arizonensis* 265, *266*
Camptosaurus 73, 75, *170*, 170–1, 362, 606, 610; *C. browni* see *Camptosaurus dispar*; *C. dispar* 75, *170*
Campylodoniscus 157
Canadian Museum of Nature *184, 529*, 621
Cañadón Asfalto Formation 537–8
Caneer, William T. 579
Cannibalism 30
Cañon del Colorado Formation 51, 58, 60–1
Carcharodontosaurus 21, 23, 25, 32, 132, 187, 191–2, *267*, 267–8, 510, 520, 577; *C. saharicus 267*, 267–8
Cardiodon 149
Carnegie Museum of Natural History *54*, 66, *204*, 260, 299, 318, *319*, 320, 620
Carnegie Quarry 55, 260
Carnotaurus 24–6, 127, *128*, 207, 385, 396; *C. sastrei* 25–6, *128*
Carpenter, Kenneth 9–10, 36, 70–1, 132, 134, 161, 163–5, *175*, 191–2, *192*, 193, 201–2, 216–7, 229, 246, 276, *276–7*, 277–8, *278*, 279, 301, 309–10, 321–3, 335–6, 341, 343, 355, 362, *420–1*, 422, 432, 466, 481, 490–1, 502–3, 505–6, 510–11, 521–4, 527–8, 542, 562, 568–9, 574, 576, 579, 615–6, 619
Carr, Thomas D. [David] 33, *35*, 201–2, 229, 304–6, *307*, 309–10, 527–8, 562–3, *563–4*, 565–70, 573–6, 579
Carrano, Matthew T. 10, 17, 24–5, 80–1, 621
Carrier, David R. 12, *13–4*, 14–6, *17*, 89
Carroll, Robert L. 616
Casanovas [Casanovas-Cladellas], M. Lourdes 75, 304, 467, 473–4, 539
Case, Ermin C. 46–8, *268*, 268–9, 295, 301
Case, Judd A. *73*, 75
Caseosaurus 47, 123, 187, *268*, 268–9, 295; *C. crosbyensis 268*, 268–9, 295
Casper College 220
Catalogue of British Fossils 544
Catalogue of Casts of Fossils from the Principal Museums of Europe and America with Short Descriptions and Illustrations 3
Cathayornis 241
Caturrita Formation 51, 346–7
Caudipteryx 23, 38, 44, 98–9, 103, 143, 187, 243–4, 269, *269–70*, 270–1, *271–2*, 272–3, *273–4*, 274–6, 386, 424–5, 513, 515; *C. asianensis* 143; *C. dongi 269*, 269–70, 274; *C. zoui* 270, *272–3*; *Caudipteryx* sp. *273–4*, 275
CBR-Romontbos quarry 77
Cedar Mountain Formation 57, 254, 276–8, 282, 310, 341, 490, 503, 518, 542
Cedarosaurus 156, 187, 276, *276–7*, *277–8*, *278*, *279*; *C. weiskopfae* 276, *276–8*
Centemodon 595
Center for the Exploration of the Dinosaurian World 550
Central Asiatic Expedition of the American Museum of Natural History 411
Central Tethys Sea 495, 532–3
Centrosaurus 82, *83*, 182, 187, 232, 279, 328, 528; *C. apertus* 328; *C. nasicornis 83*; *Centrosaurus* sp see *Styracosaurus*
Ceratops 183
Ceratosaurus 23, 123–5, *125*, 126–7, 187, 210–1, 279–80, *280–1*, 281–2, *282–3*, 283, *284–5*, 285–6, *286*, 287, 313, 331, 533, *603*, 606–7, *609*, 610, 623; *C. dentisulcatus* 211, 279–82, *284*; *C. ingens* see *Megalosaurus ingens*; *C. magnicornis 125*, 279–82, *282*, 283, *283*, 287; *C. nasicornis* 279–80, *280–1*, 282, 285, *286*, 287, 533; *C. roechlingi* see *Ceratosaurus* sp.; *Ceratosaurus* sp. 211, 282–3, *285*, 285
Cerrito del Olmo locality 304
Cerro de los Dinosaurios locality 77
Cerro de Pueblo Formation 56, 77, 292
Cetiosauriscus 58, 151
Cetiosaurus 63, 117, 149
Chablis, S. 533
Chaney, Dan S. *74*, 75
Chang, K. H. 579
Changdusaurus 161
Chaomidizianzi Formation *100*
Chaoyangia 40, 270
Chaoyangosaurus 118, 179, 187, 287–8, *288*, 289; *C. youngi* 287–8, *288*
Chaoyoungosaurus see *Chaoyangosaurus*; *C. liaosiesis* see *Chaoyangosaurus youngi*
Chapman, Ralph E. 4, 32, 96, 181–2, 460–1, 463, 556–60
Chapman, Sandra D. 362
Charig, Alan J. 14–5, 17, 30, 127, 129, 134, 146, 202–3, 245–7, 303–4, 363, 364–5, 401–2, 432, 448, 617
Charonosaurus 175, 187, 213–4, 289–90, *290*; *C. jiayinensis* 213–4, 289, *290*
Chasmosaurus 183, *184*, 187, 231, 290–1, *291–2*, 292, 323, 448–50, 552–3; *C. belli* 183, *184*, 231, 290–1, *291–2*; *C. mariscalensis* 183, 290; *C. russelli* see *Chasmosaurus belli*
Chassternbergia see *Edmontonia*; *C. rugosidens* see *Edmontonia rugosidens*
Chatterjee, Sankar 48, 98–9, 234–5, 364, 365–6, 467–9, *469–70*, 470–1
Cheeks 68–9, *69, 83*, 147, 158–9, 166, 410
Chelonians 57, 69, 75, 109, 189, 213, 254, 268, 289, 302, 397, 432, 457, 476, 490, 560, 578
Chen Pei-Ji 29, 133
Chen Wei 149, 397, 399, 590
Cheng Zhengwu 287–8, *288*, 289, 355, *356–7*, 357–8, 503–4, *504*, 505–6, 552
Chialingosaurus 4, 161
Chiayüsaurus 292–3, *293*, 294; *C. asianensis* 292–3; *C. lacustris* 292–3, *293*
Chiappe, Luis M. 26, 30–1, 36, 40–1, *42*, 61–2, 67, 87, 89, 100, 103, 138, 213, 274, 276, 327, 410–1, *411*, 412–3, 444, 468–71, 484–6, 506, 508, *508–9*, 509, 512, 515, 586–7, 617–8
Chiayüsaurus 156, 187
Chicago Children's Museum *531–2*, 533
Chicago Sun-Times 375, 406, 494, 533, 578
Chicago Tribune 375
Chicxulub crater 108
Chilantaisaurus 127, 187, 294; *"C." maortuensis* 294; *C. tashuikouensis* 294
Chimney Buttes locality 586
Chin, Karen 35
"China's Feathered Dinosaurs" *100, 515–7*
Chindesaurus 24, 48, 123, 187, 266, 269, 294–5, *295*, 296; *C. bryansmalli* 48, 266, 269, 294–5, *295*, 296
Chinese Academy of Sciences 270
Chingkankousaurus 135
Chinle Formation 19, 294, 613
Chinle Group 265, 301, 330–1, 334
Chinnery, Brenda J. 181–2, 378–9, *413–4*, 414–6, 592
Chinsamy, Anusuya 67, 87, 428–9, *429–30*, 430, 484–6
Chinshakiangosaurus 147
Chirostenotes 41, 105–6, 139–41, 424; *C. elegans* 140; *C. pergracilis* 140
Chondrosteosaurus 156, 352
Christian, Andreas 247–8, *248*, 249, 251, 254
Christainsen, Per 21, *22, 84*, 84–5, 249, 254
Christman, E. S. 210
Chuandongocoelurus 121
Chuanjiesaurus 149, 187, 296; *C. anaensis* 296
Chubutisaurus 157
Chure, Daniel J. 98, 125, 191, 193, 203–5, 207, 209–12, 241–2, 254, 279, 285–7, 294, 328, 407, 409, 431–2, 575, 579, 596, 620, 622
Cionodon 174
Circulation *see* Blood and circulation
Citipati osmolskae 592
Claessens, Leon P. A. M. 14, 16, 88–9, *89*
Claorhynchus 179
Claosaurus 171
Clark, James M. 30–1, 36, 40, 44–5, 48, 89, 138, 143–4, *190*, 190–1, 213, 259–60, 270, 274, 276, 327, 410–1, *411*, 412–3, 444, 496, 508, *508–9*, 509, 586–7
Clasmodosaurus 147
Clemens, William A. 106–8, 310, 313, 402, 419
Cleveland-Lloyd Dinosaur Quarry 205–7, 209–10, 282, 606
Cleveland Museum of Natural History *563*, 606
Climate 87, 98, 250, 252, 358, 538, 561
Climbing *see* Scansoriality
Clouse, V. 394–5
Cloverly Formation 32, 56–7, 310, 312, *312*, 407, 490, 492, 517–8, 582
Cobb, S. 577, 579
Coelophysis 20, 47, 100, *124*, 125–6, 187, 207, 210, 265, *268*, 269, 296–7, *297–8*, 298–9, *299*, 300–1, 329–30, *330–1*, 331–2, *332–3*, 333–4, *334*, 533; *C.*

bauri 124, 265, 296–7, *297–8*, 298–9, *299*, 300, 329, *330*, 330–1, *331*, 349–50, 533–4; *C. longicollis* 296, *330–1*, 331–2, *332–4*; *C. willistoni* 296, 331; *Coelophysis* sp. see *Caseosaurus crosbyensis*
Coeluroides 121
Coelurus 29, 98, 127, 187, 210, 296, 301, 330, *330–2*, 332, *332–4*, 409; *C. fragilis* 301; *C. longicollis* see *Coelophysis longicollis*; *C. willistoni* see *Coelophysis willistoni*
Coffee Mine 466
Colalura Sandstone 432
Colbert, Edwin H. 126, 146, 251, 253–4, 265, 267, 296–7, *298*, 299–301, 330, 334, 349, 351, 495–6, *496*, 498, 533–4, 616–7
Colbert, Margaret 617
Coldwell, Robert 432
Cololites 409–10
Coloradisaurus 147
Colorado National Monument 606
"Colossal Claw" Dinosaur Discovery Exhibit *532*, 533
Comets 108–9
Como Bluff 41, *170, 206*, 210–1, 220, 224, 285
The Complete Dinosaur 115, 618, 623
Compsognathus 133, 187, 212, 302–2, *302*, 303, 409; *C. longipes* 29, 302–3
Compsosuchus 132
Computer imagery *64*, 64–6, 70, 95–6, 181–2, 206, 210–1, 219, *260*, 260–1, 297–8, 312–3, 441, 461, 463, 550, *551*, 552, 557–8, *571*, 576–7
Computer tomography (CT) 81, 86, 92, 211, 249, 298, 397, 405–6, 436, 549–50, *551*, 575, 577–8
Conchoraptor 40, 140
Confusciousornis 23, 99, *100*, 104–6, 138, 239, 241, 270–2, 474, 513, 515–6; *C. sanctus 100*, 474
Connochaetes 84
Connely, Melissa Victoria 220, 225
Cooley, Brian *517*, 617
Coombs, Walter P., Jr. 21, 62, 70–1, 162–3, *163*, 164, 188–9, 214, 216–7, 320–1, 323, 343, 431, 456–7, 505–6, 522–3, 542
Cooper, Lisa Noelle 359–60
Cooper, Michael R. 317, 320, 337, 366
Cope, Edward Drinker 10, *206*, 210, 296–7, 301, 330–1, *331*, 527–8, 539, 618, 620, 622
"Cope's Rule" 10
Coprolites 35, 302, 304
Cordylus cataphractus 486
Coria, Rodolfo Anibal 24–6, 33, 61, 67, 72–4, 154, 157, 168, 191, 193, 194–5, 279, 363–4, 428, 484–6
Corythosaurus 17, 76, *77*, 80, *92*, 175, *175*, 438, 441; *C. casuarius* 76, *77*, 80, *92, 175*; *C. intermedius* see *Corythosaurus casuarius*; *Corythosaurus* sp. *175*
Cosesaurus 102
Cotylosaurs 596
County Seat and Courthouse Tracks 606
Cowan, John (Jack) Pemberton *7–8*
Cranbrook Institute of Science 601, 606
Craspedodon 168
Crataeomus 164
Craterosaurus 161
Crato Formation 96–7
Creisler, Ben 279, 596
Crerar Library *132, 267*
Cretaceous Dinosaur Project of the State of Coahuila 292
Crichton, Michael vi, 617
Cristatusaurus 187, *303*, 303–4, 519–20; *C. lapparenti 303*, 303–4, 519–20
Crocodilians and crocodyliforms *4*, 12, 15–6, 21, 28–9, 32, 40, 44, 57, 63, 69, 84, 86, 89–95, 102, 129, 191, 213, 226, 250, 252, 268, 289, 298, 304, 306, 337, 359, 364, 393, 468, 476, 486, 490, 495, 502, 518, 524, 529, 533, 561, 579, 595, 606
Crocodylus niloticus 486
Crompton, A. W. 146, 202–3, 245–6, 401–2
Cronkite, Walter 617
Cross, Robert 329
Crouse, Kathy 432
Cruickshank, A. R. I. 26, *26–7*, 27–8, *28*, 29, 129, 448, 495
Crustaceans 19, 332
Cryolophosaurus 115–6, 131, 434; *C. elliotti* 434
Cryptodraco 164, 354
Crystal Gardens 375
The Crystal Palace 3, 601, 602, 620
Csiki, Zoltan 44, *45–6*, 256–7, 326–8, *392*, 392–3, 401–2, 479, 481
CT *see* Computer tomography
Cunha, Fausto Luiz de Souza 194
Currie, Philip J. vi, 16, 23, 25–6, 30, 33–5, 46, 71, 88–9, *89, 101*, 105, 121, 123–4, 126–7, 130–2, 140, 143–5, 181, 184, 191–2, *192*, 193, 201–2, 212, 216, 219–20, 225, 229–30, 232–3, 238–9, *239–40*, 240, *241*, 242, 257, 260, 270–1, 273, 275–6, 293–4, 305, 320, 327, 335–6, 337–8, *338–40*, 341, 385–6, 393–5, 401–2, 405–6, 409, 422, *422–3*, 423–4, *424–5*, 425–6, 456–7, 468, 471, 476–7, 495, 511, 515, 527–8, 552, 560, 565, 567, 579, 573–4, 586–7, 615–7, 623
Curry [Rogers], Kristina A. 11, 220–5
Cursoriality 89, 102, 138, 143, 159, 180, 210, 271, 275, 404, 512, 575
Curtice, Brian David 56, 58, 262–3, *263*, 320, 498, *499*, 500, 517–8, *518*, 519
Cuvier, Baron Georges 252, 362
Cynodonts 245, 332
Czerkas, Stephen A. 69, *95, 120, 394*, 405–6, 486, 617
Czerkas, Sylvia M. 62, *95, 206*, 210, 212, *394*, 405, 617
Dacentrurus 161, 187, 227, 304; *D. armatus* 227, 304
Dachungosaurus 145
Dakota Formation 510
Dal Sasso, Christiano 27–8, 101, 102, 495
Dalla Vecchia, Fabio M. 352, *352–3*, 354
Dallas Paleontological Society 471
Dalton, Rex 552
Dalton Wells quarry 57, 518
Damalasaurus 147
Danchungosaurus 145
Dandakosaurus 121
Dantas, P. M. 315–7, 386–7, *387*, 388
Danubiosaurus 162
Darby, D. G. 499–500
Dark Red Beds 53
Darling, Lois 265, 299
Daspletosaurus 5, 13, 35, 122, 135, *136*, 187, 201, 229, 304–5, *305–6*, 306–7, *307–8*, 309–10, 562, 565, 574; *D. torosus 5, 13, 35, 122, 136*, 304, *305–6*, 306, *307–8*; *Daspletosaurus* sp. 201, 229, 310
Datousaurus 156
Dawson, Mary R. 620
de Camp, L. Sprague and Catherine Crook 617
de Klerk, William J. 428–9, *429–30*, 430
Dean, Dennis R. 360, 363
Debus, Allen A. *22, 130, 341*, 375, *376–7*, 377, *570*, 576, 578–9, 618
Debus, Diane E. 375, 377, 618
Deck, Linda T. 556, 560, 558, 560
Deeming, Charles D. 393, 395
Dehm, R. 460, 463
Deinocheirus 121
Deinodon 135
Deinonychus 15, 30, 42, 46, 57, 90, *95*, 100, 103, 143–5, 187, 191, 239, 257, 270, 275, 310, *311–2*, 312–3, 476, 582–4, 586, 605; *D. antirrhopus 15*, 30, 57, *95*, 100, 103, 191, 239, 310, *311–2*, 312, 339, 476, 512, 582–4
Deinosuchus 250
Delair, Justin B. 431–2
Delgado, Ricardo *533*
Deltadromeus 133, *133*; *D. agilis 133*
Dendrorhynchoides 516
Densus-Ciula Formation 44
Denton, Robert K. 134
Denver Musuem of Natural History 246, 277, 620
Depéret, Charles 50
Derstler, Kraig 238–9, *239–40*, 240, *241*, 242
Desmatosuchus 47
Desmond, Adrian vi, 618
Detrich, Alan *574*, 578
Detrich Fossils *574*, 578
Dianchungosaurus 166
Diceratops 183–4
Diceros 84
Diclonius 174
Dicraeosaurus 55, 63, 152, *152*, 213, 246, 250–2, 262, 367–8; *D. hansemanni* 55, 367; *D. sattleri* see *Amargasaurus sattleri*
Dicynodonts 50
Diegert, Carl F. 441–2
Diet 12, 17, 25, 32–6, 38–9, 43 *43*, 51, 53, 66, 68, 80–1, 86, 91, *92*, 109, 120–1, 123, 133, 137–9, 141, 145–6, 158, 166–7, 173, 195, 202, 211–2, 252, 255, 274, 303–4, 310, 323–4, 335, 351, 360, 364, 372, 386, 401, 409–10, 430, 448, 473, 488, 495, 527, 561, 575, 578–9
Difunta Group 292
Dilophosaurus 126, 187, 300, 313, *313–4*, 314–5, *315*, 331, 434, 501, 533, 612; *"D." sinensis* 314, *314*; *D. wetherilli 313*, 313–4, *315*, 533
Dinamation International Corporation 601, 606
Dinamation International Society 201, 606
Dingus, Lowell 26, 61, 67, *142*, 484–6, 617–8, 620
Dinheirosaurus 152, 187, 315–6, *316–7*, 317, 386, 388; *D. lourinhanensis* 315, *316–7*
"Dinny" 602
Dinofest (also DinoFest and Dinofest International) *18, 76, 95, 125, 129, 132, 149–50, 154, 165, 180, 182, 232, 314, 373–4, 379, 413*, 500, *500*, 517, *572*, 601, 622
Dino-Frontline see *Kyoryugaki Saizensen*
DINOLAB *125, 129, 155, 160*
DinoLand U.S.A. 606
Dinornis 40
Dinosaur Beds 397
Dinosaur Caverns *see* Grand Canyon Caverns
Dinosaur Cove 72
The Dinosaur Data Book 339
Dinosaur Flats 606–7
Dinosaur Gardens 607–8, *608–9*
Dinosaur Gardens Prehistorical Zoo (Domke's Gardens) 608
"Dinosaur Giants" 375, *375, 376–7*
Dinosaur Hill 297, 608
Dinosaur Land 608, 613; *see also* Prehistoric Forest
Dinosaur Ledge 610
Dinosaur Museum *399*, 405
Dinosaur National Monument *54*, 55, 205–7, 260, 283, 317, *318*, 319, 608–11, 619, 622
Dinosaur Park *153*, 601–2, 605, *610*, 610–1, 614
Dinosaur Park Formation 35, 229, 233, 291, 304, 309, 439, 463–4, 494, 528
Dinosaur Park Miniature Golf 602
Dinosaur Provincial Park 82, 309, 528, 602, 615
Dinosaur Quarries 610
Dinosaur Quarry 610–1
The Dinosaur Quarry 219
"Dinosaur Renaissance" vi, 615
Dinosaur Ridge 611
The Dinosaur Society 4, 619
Dinosaur State Park 611
Dinosaur Track Site 611
Dinosaur Tracks 611
The Dinosaur Triangle 606–8, 611, 614, 616
Dinosaur Valley *282*, 602, 611
Dinosaur Valley State Park 611, *612*

Dinosaur Wall 611
Dinosaur World 602, 611
Dinosaur World 3, 304, 306, 562
The Dinosauria vi, 51, 489, 622–3
Dinosauria International *500*
Dinosaurierfrellichtmuseum (Dinosaurierpark) 602
The Dinosaurs 219
"The Dinosaurs of Jurassic Park—The Lost World" 4
Dinotyrannus see *Tyrannosaurus*; *D. megagracilis* see *Tyrannosaurus rex*
Diplodocus 49, 53–6, 58, 60, 62, *62–3*, 63, 67, 148, 151–3, 187, 198, 213, 217, *217*, 219, 249, 315, 317–8, *318–9*, 319–20, 367–8, 418, 492, 498, *500*, 606–7, 610; *D. carnegii* 62, *62–4*, 318, *318–9*, 367, 607; ?*D. hallorum* 317, 320, 498; *see also Seismosaurus halli*; *D. hayi 62, 318*; *Diplodocus* sp. 318
Diplotomodon 133
"Dippy" 607
Disease 254, 323, 523–4, 559
Discover 517
"Discovering Dinosaurs" *175*
Disneyland® Park 614
Display *see* Intraspecific display, rivalry, and combat
Ditmars, Raymond L. 618
Dixon, Dougal 618
Djadokhta Formation 37, 40, *234*, 410–1, 447, 456–7, 506, 582, 585–6
DNA 101, 559–60
Dockum Group 294
Dockum Formation 47, 268–9, 301, 467
Dodson, Peter vi, 16, 60, 62, 80, 87, 176, 178, 180–1, 183–4, 226, 230, *230–1*, 231–3, 268, 314, 319–20, 379, 393, 395, 439–42, 515–7, 552, 554, 560, 592, 618, 622–3
Dolicosuchus 121
Dollo, Louis *361*
Dong Zhi-Ming [Zhiming] 29, 38, 74–5, 105, 133, 141–2, 160–1, 171, 172, 174–5, 219, 226, 235–6, *236–7*, 237–8, *238*, 257, 274, 276, 293–4, 327, 339, 341, 365–6, 385–6, 400, 402, 409, 431–2, 453, *454*, 455–7, 476–7, 527, 539, 590, 595, 618
Donovan, Tim 537, 579
Doré, A. C. 216
Dortoka sp. 476
Douglass, Earl *204*, 317, 319
Downs, Alex 265, 267, 299–301, 310, 313, 534
Draconyx loureiroi 592
Dracopelta 164
Drepanosaurids 470
Drinker 166, 259
Dromaeosaurus 42–3, *43*, 144, *145*, 191, 401, 582–4, 586; *D. albertensis* 42, *43*, *145*, 191, 401, 582–4, 586
Dromiceiomimus 37, 139; *D. samueli* 37
Dry Mesa Dinosaur Quarry 205–7, 283
Dryosaurus 72–3, 169, *169*, 253, 606, 610; *D. altus 169*; *D. lettowverbecki* 72
Dryptosauroides 127, 187, 320; *D. grandis* 320
Dryptosaurus 115, 134, 211; ?*D. medius* 211
"Dueling Dinosaurs" *34*, *565*, 601, 611
Dunham, A. E. 250, 254
Durdham Down locality 543, *545*, 545–8
Dwarfism 391–3, 401, 561–2, 565, 567, 570
Dyke, Gareth J. 234–5, 469, 471
Dyoplosaurus see *Euoplocephalus*; *D. acutosquameus* see *Euoplocephalus tutus*
Dysganus 33, 181; *D. encaustus* 33
Dyslocosaurus 152
Dystrophaeus 151, 317
Dystylosaurus 156

Eaton, Jeffrey G. 528
Eaton, Theodore H., Jr. 229–30, 511
Eberth, David A. 33, 35, 82, 439, 456–7, 528
Echinodon 166, 187, 320–1, *321*, 431; *E. becklesii* 320, *321*
Echkar Formation 418
Ectothermy 67, 86–94, 253, 551
Edaphosaurus 555
Edelweiss farm 533
Edmarka 127
Edmonton Formation 463–4
Edmontonia 162, 163–4, 187, 189, 216, 321–2, *322*, 323, 466, 490, 510–1, 521, 542; *E. australis* 321–2, *322*; *E.* (*Chassternbergia*) *rugosidens* see *Edmontonia rugosidens*; *E. longiceps* 321, 510; *E. rugosidens 162*, 321, 323, 510
Edmontosaurus 33, 75, 80, *88*, 174, *174*, *177*, 187, 323, *323–4*, 324–5, 575, 606; *E. annectens* 80, *88*, *174*, *177*, 323, *323–4*; *E. edmontoni* see *Edmontosaurus annectens*; *E. regalis* 80; *E. saskatchewanensis* 80
Evimov, M. B. 191
Efraasia see *Sellosaurus*
Egg Mountain 560
Egg teeth (in ceratopsids) 558
Eggs 11, *18*, 18–21, 26, 40, *61*, 61–2, *69*, 74, *87*, 89–90, *90*, 100–1, 120, 139, *177*, 239, 304, 394–5, 476, 484–6, 502, *516*, 548, 558, 560–1, 616–8
Egyptian Geological Museum 268
Einiosaurus 182, 232, 528
El Gallo Formation 23
El Tranquilo Group 351
Elaphrosaurus 124, 127, 187, 325, *325*; *E. bambergi* 325, *325*
Ellenberger, Paul 337, 365–6
Elliot Formation 202, 245, 336, 337
Elmisaurus 138, 141
Elopteryx 44, 143, 187, 256–7, *325*, 325–8, 581–2, 595; *E. nopcsai* 325, *326*, 326–8, 582
Elosaurus parvus see *Haplocanthosaurus* sp.
Elrhaz Formation 303, 418
Elżanowski, Andrzej 104–5, *105*, 106, 257, 275–6
Emausaurus 117, 159, 496, 498
Embasaurus 121
Embryos 21, 30, 39–40, 54, 61, 67, 90, 92, 96–7, 99, 103, 393–5, 484–6, 561
The Encyclopedia of Dinosaurs vi, 617, 623
Endocasts 21, 23, 292, *571*, 577
Endothermy 67, 86–8, *88*, 89–93, *93*, 94, 102, 147, 252–3, 300, 475, 551, 561, 622
Enigmosaurus 142
Eobrontosaurus 152
Eoenantiornis 516
Eolambia 118, 171, 187, 290, 328, 474; *E. caroljonesa* 328
Eoraptor 17, 115, 118, 121, 123
Eosipterus 474, 515–6
Eotyrannus lengi 592
Epachthosaurus 157–8, 187, 328
Epcot Center 611
Erickson, Gregory M. 11 250, 475–6
Erlhaz Formation 388, 529
Erlikosaurus 104, 105, 142, 260
Ershijiazi locality 287
Erythrosuchids 543
Erythrosuchus 337
Estheria see *Asmussia*
Eubrontes minisculus 300
Eucentrosaurus see *Centrosaurus*; *E. apertus* see *Centrosaurus apertus*
Eucoelophysis 24, 124, 187, 297, 329, 328–9, *329–30*, 330–1, *331–2*, 332–3, *333–4*, 335; *E. baldwini* 24, 328–9, *329–30*, 330–1, *331–2*, 332, *333–4*, 334; ?*Eucoelophysis* sp. 333
Eumerella Formation 72, 478
Euycynodonts 50
Euhelopus 63, 156, 187, 317, 334–5, *335*, 552; *E. zdanskyi 335*
Euoplocephalus 70–1, 118, 164, 165, 166, 187, 335–6, *336*, 418–20, 505–6; *E. tutus* 70–1, 335–6, *336*
Euparkaria 95
Euronychodon 44, *45*, 144; *E. portucalensis* 44
Euskelosaurus 52, 146–7, 187, 202, 245, 336–7, 548; *E. browni* 202, 336–7
Eustreptospondylus 129, *131*; *E. oxoniensis 131*
Evans, S. E. 12
Evolution 10, 12, 16–7, 36, 48–9, 53–5, 58–9, 63, 67–8, *68*, 82–3, 87, 90–1, 94–103, 106, 124–5, 144, 148, 167–8, 174–5, 183, 207, 212–3, 218–9, 225, 237, 244, 271, 287, 326, 328, 334, 337, 345, 358, 364, 366, 372, 374, 390–2, 404, 406, 418–9, 441, 446–7, 449–50, 452, 468, 473, 477, 489, 491, 495, 502–3, 514, 516, 524, 530, 533, 551, 571, 576, 582, 621
Explorer's Hall 406
Extinction 5, 17–8, 20, 42, 106–7, *107*, 106–7, 200, 212, 290, 312, 358, 391, 395, 616, 618, 620–1

Fabrosaurus 159
Fantanele microvertebrate site 44
Farke, Andrew A. 552, 560
Farlow, James O. 21, 115, 246, 254, 275–6, 317, 320, 401–2, 575, 579, 616, 618, 623
Farmer, Colleen G. 12, *13–4*, 14–5, *15*, 16, *17*, 89, 91
Farrell, Mike 579
Farris, J. S. 392–3
The Farwell Dinosaur Park 611
Fassett, James E. *110*, 109–10, 200–1, 419, 439, 495
Fastovsky, David Eliot 39, 107–8, 166, 247, 254, 471, 502, 618
Faunal exchange 75, 79, 200, 205, 211–2, 317, 328, 358, 364, 366, 415–6, 429, 446–7, 465, 474, 533, 539, 570–1, 582, 620
Feather Quarry 430
"Feather-bed Marl" 430
Feathers 11, 94, *95*, 96–9, 101–3, 121, 133–4, 139, 142–4, *208–9*, 242, *242*, 243–4, *244*, 270–3, 276, *299*, 300, 403–6, 413, 423–5, 474, 488, 511–2, *512*, 513–4, *514*, 515, *585*
Fedde, M. R. 16
Feduccia, Alan 95, 97, 102–3, 239, 242, 260, 271, 276, 412–3
Felch's Quarry 211
Fengjahe Formation 587–8
Fenton, Carroll Land and Mildred Adams vi, 618
Fernandes, L. A. 194
Fernañdez Estancia locality 537–8
Field Columbian Museum *3*, *220*, *263*
The Field Museum *viii*, *3*, *5*, *13*, 49, *49–50*, 56, *65*, *69*, *83*, *100*, *120*, *122*, *136*, *153*, *176*, *220–2*, *224*, 225, *246–7*, *263*, 305, *305–8*, 309–10, *396*, *437*, *515–7*, *521–3*, *557*, *569–70*, 576–8, 608, 610, 615, 619, 622
Fiffer, Steve 576, 579
"Fighting Dinosaurs" 586
Fiorillo, Anthony R. 42–3, *43*, 561
Fires 107, 109, 201
Fischer, P. 336–7
Fish 11, 19, 37, 75, 109, 254–5, 268, 302, 332, 359, 364, 397, 487, 490, 516, 530, 615
Fisher, Patrick *207*
Fisher, Paul E. 92–3, 549–51, *551*
Flaming Cliffs 586, 620
Flamme de la Paix 533
Flannery, Timothy F. 211–2, 434
Flat Rocks site 477
Flight 10, 17, 96–9, *100*, 101–2, 191, 241–2, 244, 270, 275–6, 404, 406, 423, 512, 587
Flintstones Villages 601
Florida Institute of Paleontology 242
The Florida Symposium on Bird Evolution 94, 242, 406
Flynn, John J. *49*, 49–50, 615, 621
"Fontllonga hadrosaurid" 75, 118, 173
Footprints and trackways 5, 20, 67, 84, *84*, 107, 119, 201, 228, 235, 247–8, 251, 298, 300, 337, *362*, 363, 365, 462, 471, 489–90, 557–8, 579, 601–4, 606–7, 610–1, 613–4, 618–20, 622
Ford, Tracy L. 163–4, 229–30, 317, 321–2, *322*, 323, 328, 342–3, 371, 524–5, 527–8, 570–1, 579, 618–9

Forster, Catherine A. 24–5, 60, 103, 183–4, 233, 250, 254, 291–2, 393, 415–6, 428–9, *429–30*, 430, 481, 489, 502, 512, 515, 539, 552–3, 560, 615
Fort Crittenden Formation 200
Fossil Beach 603
The Fossil Book vi
Fossil Forest 438
Foster, John R. 56, 209–10, 212, 575, 579
Fox, William 226
Fraas, Eberhard 367, 369, 460, 463
Fraley, Phil 576–7
Franczak, Brian *152, 225, 336, 560*, 619, 621
Frankfurt, Nicholas G. 40–1, *42*
Fredericks, Mike *233, 242, 412, 514*
Frey, Eberhard 129, 251, 254, *262*, 364–5, 448, 495
Frithiof, Ron 579
Fritsch [Fric], Anton 596
Frontier Formation 422, 520–1
Fruita Paleontological Area 282
Fruitland Formation 200, 437, 440, 448–9, 495
Fudali, Henryck 603
Fukui Prefectural Museum 73
Fukuiraptor 132, 187, 212, 337–8, *338–40*, 341; *F. kitadanensis* 337–8, *338–40*
Fulengia 146
Fulgurotherium 72, 168; *F. australe* 72

Gaffney, Eugene S. 618, 620
Galbrun, B. 75
Gallagher, William B. 618
Galliano, Alessandra 495
Gallimimus 37, 139, 187, 212, 241, 341; *G. bullatus* 37
Galloping 85
Gallup, Marc R. 310, 313
Gallus gallus 99
Galopim de Carvalho, A. M. 315–7
Galton, Peter M. 16, 19, 51–3, 68, 106, 119–20, 146–7, 160–1, 168, 178, 202, *202–3*, 203, 205, 212, 214, 216–7, 227, *245*, 245–6, 257–8, *258–9*, 259, 263, 265, 304–5, 320–1, 336–7, 350–1, 365–6, 399–402, 431–2, 457, *458–9*, 459–65, 468–9, 500–1, *501–2*, 502, 520, 524, 543, *543–5*, 545–6, *546–7*, 547–9, 595
Galtonia 158
Garbani, Harley G. *174*, 525
Garcia, Géraldine 479, *479–80*, 480
Garden Park Fossil Area 611
Gargoyleosaurus 163, 165
Garlington, Joe 613
Garstka, William R. 99, 101, 241–2
Garudimimus 105, 138, 260; *G. brevipes* 260
Gasosaurus 127
Gasparinisaura 74, 118, 167–8
Gaston, Robert 30, 342
Gastonia 71, 163, 165, *341*, 341–2, 354–5, 510, 521; *G. burgei* 71, *342*, 343
Gastroliths 21, 38, 274, 276, 278, 315, 385–6, 409–10, 428, 498–500, 517
Gastropods 253
Gatesy, Stephen M. 21, 95–6, 99, 275, 297–8, 301
Gaudry, A. 461, 463
Gauffre, Francois-Xavier 49, 146–7, 202, 245, 336–7, 365–6, 548–9
Gauthier, Jacques A. 31, 40, 97, 119–21, 126, 134–5, 144–7, 158, 265, 267, 287, 299, 301–3, 313–4, 327, 409, 429–30, 442, 444, 462–3, 486, 489, 501–2, 512, 515, 533–4, 545, 548–9, 582–3
Geiser, Ray 436
Geissman, John 8
Geist, Nicholas R. 28, 86–8, 101–3, 495, 513, 515, 587
Genusaurus 124
Genyodectes 123
Geological Enterprises, Inc. 191
Geological Musuem 223
Geological Sciences Museum 110, 400, *400*
Geological Society (London) 360
Geological Society of America 8
Geranosaurus 166
German Tendaguru expedition 253, 367
Germs 55
Geygan, George 617
Ghost Ranch *124*, 265, 296–7, *297–8*, 298–9, 329–30, 332, 617
Gierliski, Gerard 300–1
Giffin, Emily B. 464–5
Gifford, Cameron *136*
Giganotosaurus 21, 25, 32, 132, 191–2, 577; *G. carolinii* 25, 191
Gigantosaurus 156, 367, 369; *G. africanus* see *Barosaurus africanus*; *G. robustus* see *Janenschia robusta*
Gigantothermy 93–4, 556
Gillette, David D. 317, 320, 377, 387–8, 498, 500, 576, 579, 615, 618
Gilmore, Charles Whitney 71, 84, *170*, 198, *199*, 200–2, 209–10, 212, 218–9, 225, 232–3, 235–6, 238, 260, 263, 282, 287, *342*, 343, 379, 420, 453, 455–7, 511, 556, 560, 562, 567, 579
Gilmoreosaurus 171–2, 235, 237; *G. mongoliensis* 235, 237
Gilpin, Orville L. *136, 221, 307*
Gimenéz, Olga 537, *537–8*, 538–9
"*Ginnareemimus*" 38
Giraffatitan see *Brachiosaurus*; *G. brancai* see *Brachiosaurus brancai*
Gishkin, Alan D. 42, 301–3
Glen Rose Formation 611
Glut, Donald F. vi, 62, *206*, 210, 212, 617–9, 621
Glyptodon 522
Glyptodontopelta 164, 166, 187, *342*, 342–3, 520; *G. mimus* *342*, 342–3
Glyptodonts 342–3
Gobipteryx 104, 105
Godefroit, Pascal 171–2, 174–5, 213–4, 235–6, *236–7*, 237–8, *238*, 289–90, *290*, 455, *456*, 456–7, 539
Godfrey, Stephen J. 105, 140, 291, 552, 560
Gojirasaurus 126, 295
Golder, Werner 11, 247–8, *248*, 249, 254
Gomani, Elizabeth M. 60, 157, 217–8, 279, 397
Gondwana (Gondwanaland) 25, 41, 58, 75, 195, 212, 325, 343, 364, 428–9, 446, 485, 532–3, 622
Gondwanatitan see *Aeolosaurus*; *G. faustoi* see *Aeolosaurus*
Gongbusaurus 158
Gongiopholis 57
Gongxianosaurus 147, 187, 343–4, *344–5*, 346, 365; *G. shibeinsis* 343–4, *344–5*, 346
González Riga, B. 60–1, 79, *79–80*
Good, John M. 219, 619
Goodwin, Mark B. 81–2, *82*, 178
Gorgosaurus *13*, 35, *35*, 135, *135–6*, 304–5, *305–6*, 306, *307–8*, 527, 562–3, 565, 567–9, 572, 575, 613; *G. lancensis* see *Tyrannosaurus rex*; *G. libratus* *13*, 35, *35*, *135–6*, 304–5, *305–6*, 306, *307–8*, 562–3, 565, 567–9, 613
"Gorgy" *307*
Gould, Stephen Jay 54
Gow, Christopher E. 245
Goyocephale 177
Graciliceratops 118–9, 180, 187, 346; *G. mongoliensis* 346
Gradstein, F. M. 229
Gradziński, R. 423, 426, 585–7
Graffham, Allen 191
Grand Canyon Caverns (Dinosaur Caverns) 611
Granger, Walter 109, 210, 411
Graves Museum of Archaeology and Natural History 242
Gravitholus 178, 520; *G. albertae* see *Gravitholus sternbergi*; *G. sternbergi* 520
"The Great Russian Dinosaurs Exhibition" *61, 141, 467, 472–3, 475, 566*, 622
Gregory, William King 71
Grés á Reptiles Formation 581
Grès Supérieur Formation 534–5
Greppin, J. B. 283, 287
Gresslyosaurus see *Plateosaurus*; *G. ingens* see *Plateosaurus engelhardti*; *G. robustus* see *Plateosaurus robustus*; *Gresslyosaurus* cf. *plieningeri* see *Plateosaurus polignien sis*
Grigorescu, Dan 44, *45–6*, 171–2, 174–5, 256–7, 326–8, 392–3, 401–2
Growth *see* Ontogeny
Grupo Malargüe 79
Gruss, Franz 602
Gryposaurus 76, 79–80, 174, 377; *G. incurvimanus* see "*Kritosaurus*" *incurvimanus*; *G. notabilis* 377
Guaibasaurus 120–1, 123, 187, 346–7, *347–8*, 348–9, *349–50*, 350–1; *G. candelariensis* 120, 346, *346–7*, 347, *348–50*; *G. dandelaraia* see *Guaibasaurus candelariensis*
Guan, Jian 397, 399
Gugyedong Formation 579
Gulpen Formation 77
Gümbel, C. W. 460, 463
Gunga, Hanns-Christian 67, 249–52, 254
Gurney, James 616, 622
Gwinner, M. P. 459, 463f
Gyeongsang Supergroup 293, 579
Gyposaurus see *Massospondylus*

Hadrosaurus *3*, 174, 377; *H. foulkii* *3*
Hagenbeck, Carl 602, *603*
Hall, Cephis 191
"Hall of Saurischian Dinosaurs" *136*, 305, *562*
Hallett, Mark 615, 618, 622
Halstead, L. Beverly 148, 619
Halstead, Jenny 148, 619
Halticosaurus 121, 145
Hammer, Jeff 549
Hammer, Michael 92–3, 549–52
Hand, Jordon D. 24
Hanna, Rebecca R. (Laws) 255–6
Haplocanthosaurus 49, 63, 154, 158, 198, 214, 225, 328, 518, 541; *H. priscus* 158; *Haplocanthosaurus* sp. 225
Happ, John W. 552, 560, 558–60
Harris, Jerald Davic 192–3
Harris, John M. 617
Harold Washington Library Center *130*
Harpymimus 138
Harrison, C. J. O. 256–7, 326–8
Hartman, Scott 102
Harvard University 312, *312*, 496, 620
Hasandong Formation 52, 293
Hasegawa, Yoshikazu 37, *39*, 293, 339, 341, 618–9
Hating Beds 30
Hatcher, John Bell 62, *63*, 227
Hatchlings 11, 72, *177*, 224, 391, 393, 395, 432, 440, 471, *472–3*, 495, 558, 575
Haubold, Hartmut 498, 619, 621
Haughton, Sidney H. 337
Haversian bone 70, 223, 250, 395
Hawkins, Benjamin Waterhouse 3, 602
Hay, Oliver P. 211–2, 282, 287, 300–1
Hayashi, Shigeo *340*
Hayashibara Museum of Natural Sciences 39, 235, 323, 471, 490
Hayashibara Museum of Natural Sciences-Mongolian Paleontological Center Joint Paleontological Expeditions 39, 235, 471, 490
Hayden, Ferdinand Vandiveer 218
He Xinlu 343–4, *344–5*, 365–6, 371, 432
Head, Jason J. 328, 471, *473*, 473–4
Head-butting 81–2, *82, 175*, 177
Hearts (in dinosaurs) 66–7, 92–3, *93*, 252, 549–50, *550–1*, 551–2
Heaton, M. J. 437, 441
Heck, Ed *552*
Heckert, Andrew B. 19–20, 24, 47–8, 76–7, 109, 262–3, 265–6, *266*, 267–9, 294–7, 301, 320, 329–34, 498, *499*, 500

Heerden, Jacques van 147, 202, *202–3*, 203, *245*, 245–6, 336–7, 501–2
Heinrich, R. E. 72, 167, 168, 257, 259, 477–8
Heinrich, Wolf-Dieter 247–8, *248*, 249–54, 367–8, *368*, 369, 539–40, *540–1*, 541–2
Heishansaurus *4*, 166
Helby Beds 546–7
Hell Creek Formation 40, *41*, 106–8, *174*, 229, 257, *434*, 435, 464, 525–7, 549, 558–9, 567, 571, *572*, 578–9
Helmdach, F.-F. 315
Heloderma 76
Henderson, Donald M. 251, 275–6, 621
Henderson, Douglas 619, 622
Hengst, Richard A. 86
Hennig, Edward 214, 217, 253–4
Heptasteornis see *Bradycneme*; *H. andrewsi* see *Bardycneme draculae*
Herbivory (in theropods) 121, 133, 137–9, 141, 143, 274–5
Hernández-Rivera, René 56, 77, 156, 292
"Herrerasaurids A, B and C" 295
Herrerasaurus 19, *120*, 123, 202, 207, 268–9, 348–50, 502, 543, 548; *H. ischigualastensis 120*
"Hesperisaurus" 12
Hesperornis 40
Heterochrony 54–5, 391, 570
Heterodontosaurus 166, 187, 287–8, *351*, 351–2; *H. tucki 351*
Heterosaurus see *Iguanodon*
Hickey, L. J. 106
Hicks, J. W. 91
Hidalgo Formation 200
Hierosaurus 164
Hikanodon see *Iguanodon*
Hill, Elizabeth A. 320
Hill, Robert V. 162–3
Hincliffe, Steven 432
Hinkle Family 205
Hirsch, Karl F. 616
Histology *see* Bone histology
Histriasaurus 151, 187, 352, *353–4*, 354; *H. boscarolli* 352, *353–4*, 354
Hitchcock, Edward 300–1, 619
H.M.S. *Fly* 546
Hoffet, Josué-Heilmann 74, 358, 535–6
Holden, Constance 406–7, 559–60
Holl, Friedrich 362–3
Holland, William J. 294
Holmes, Robert B. 291, 552, 560
Holtz, Thomas R., Jr. 20, 27, 30, 35, 36, 96, 103, 119–27, 130–5, 139, 142–4, 257, 285, 287, 314, 325–6, 328, 364, 385–6, 429–30, 470, 527, 533–4, 573–4, 616
Homalocephale 177
Homeothermy 91, 147, 252, 555
Hoplitosaurus 70–1, 163, 165, 187, 354–5, 422, 542, 543; *H. marshi* 70
Hoplitosuchus 543
Hopson, James A. 439–42
Horner, John R. 16, 18, 75, 81–2, 90–1, 174–5, 235, 238, 254–6, 359–60, 377–9, 393–5, 409, 437, 439, 441–2, 471, 474, 498, 524, 539, 560–1, 575, 579, 616, 619
Horseshoe Canyon Formation 33, 46, 579
The Hot-Blooded Dinosaurs: A Revolution in Paleontology vi, 618
Hotton, Nicholas, III 252, 254, 619
Hou, L. H. 270, 276, 293–4, 589–90, *590*
Houston Museum of Natural Science *62*, *318*
Howard McCleod Correctional Center 310
Howse, Straffor C. B. 143, 257
Hu Cheng-Chin 110
Hu Show-yung 294
Huabeisaurus 151, 187, 355, *356–7*, 357–9, 534–5; *H. allocotus* 355, *356–7*, 358–9
Huayangosaurus 161; *H. taibaii 161*
Hubbel, F. F. 210
Hudiesaurus 154, 156; *H. sinojapanorum 154*
Huene, Friedrich von 19, 46–7, 53, 126, 214, *217*, 217–8, 263, 265, 317, 320, 358–9, 391–3, 400–2, 418, 457, *458–9*, 459–63, 501–2, 543, *545*, 546–9, 596
Hughes Tool Company *7–8*
Huiquanpu Formation 355, 503
Hulsanpes 144, 582; *H. perlei* 582
Humboldt-Museum für Naturkunde 63, 247, 251, 253, 518
Hungarian Geological Survey 44
Hungerbühler, A. 595
Hunt, Adrian P. 19, 24, 47–8, 76, *84*, 198, 200–2, 248, 254, 262–3, 265–6, *266*, 267–9, 294–7, 301, 329–30, 334, 377–8, 419–20, 439–42, 576, 579, 620
Hunterian Museum 361–2
Huoyanshan Formation 293
Hurum, Jørn H. 34, 585–7
Hustead, Ted *605*, 614
Hutchinson, H. N. 619
Hutchinson, J. Howard 437, 528
Hutchinson, John R. 121–7, 130–2, 135, 139, 143–4, 576, 579
Hutt, Steve *167*
Huxley, Thomas H. 187–9, 336–7, 544, 549
Hylaeosaurus 70, 163, 165, 189, 216, 354; *H. armatus* 70
Hypacrosaurus 80, 175, 187, 359, *359*, 393–4, 418, 439, 441, 498; *H. altispinus* 80, 418; *H. stebingeri* 80, 359, 393–4; *Hypacrosaurus* sp. *359*
Hypselosaurus 157
Hypsibema 174
Hypsilophodon 73, 75, 166–7, *167*, 168, 259; *H. foxii* 75, *167*

Ichnites *see* Footprints and trackways
Ichthyosaur State Park 601
Ichthyosaurs 595, 601
ICZN *see* International Commission on Zoological Nomenclature
Iguanodon 12, 73–5, 166, 171, 174, 187, 216, 359–60, *360–1*, 361–2, *362*, 363, 388–9, 533, 596, 602–3, 616; *I. anglicus* see *Iguanodon bernissartensis*; *I. Atherfieldensis* 363, 389; *I. bernissartensis* 12, 359, *360–1*, 362–3, 388–9; *I. dawsoni* 362; *I. exogyrarum* see *Ponerosteus*; *I. fittoni* 362; *I. hoggi* 362; *Iguanodon* sp. 75; *?Iguanodon* sp. 12
"Iguanodontide trapu" *see Lurdosaurus*
Iliosuchus 127
Illustrated Encyclopedia of Dinosaurs: An Original and Compelling The Insight Into Life in the Dinosaur Kingdon 620, 623
The Illustrated London News 4
Ilokelesia 115, 126, 187, 363–4; *I. aguadagrandensis* 363–4
Impact (scenarios of extinction) *107*, 107–9, 111, 618–9
Imigration *see* Faunal exchanges
Indosaurus 127, 187, 654
Indosuchus 127
Infection (in "Sue") 578
Ingenia 106, 140, *141*, 424
Inner Mongolian Musuem 235, 255
Inosaurus 121
Insectivory 109, 133, 17, 430
Insects 36, 53, 96–7, 133, 137
Institut des sciences de l'évolution de Montepellier 481
Institut für Paläontologie of the Free University 11
Institut Royal des Sciences Naturelle de Belgique 235, 362
Institute of Geophysics and Planetary Sciences 108
Institute of Vertebrate Paleontology and Paleoanthropology 270, 405, 505, 547
Integument *see* Skin and skin impressions
International Code of Zoological Nomenclature 450
International Commission on Zoological Nomenclature (ICZN) 279, 295, 330, 362, 406, 463, 505
Intestinal tract 26, *26–7*, 27–8
Intraspecific display, rivalry, and combat 81, *82*, 160, 219, 441, 524, 554
Invertebrates 26, 36–7, 53, 75, 96–7, 109, 133, 137, 243, 253, 268, 287, 302–3, 310, 315, 332, 358, 397, 422, 475, 487, 490, 533, 535, 579, 602, 613, 615
Iren Dabasu Formation 36, 229, 235, 237–8, 410–1
Iridium 107–8
Irons, Gregory 620
Irritator 130, 187, 214, 304, 364–5, 487, 531; *I. Challengeri* 364, 487
Isalo II 49, *49–50*
Isanosaurus 147, 187, 365–6, *365*; *I. attavipachi* 365, *366*
Ischigualasto Formation 19, 50–1
Ishigaki, Shinobu 489–90
Ischyrosaurus 156
Itemerus 121
Itoshiro Subgroup 35
Iuticosaurus 157

Jabo, Steven J. 556, 558, 560
Jacobs, Louis L. 157, 167–8, 189, 216–7, 279, 310, 313, 397, 619
Jacobsen, Aase Roland 33
Jacobson, Arland 257
Jaekel, O. 457, 460, 463
Jain, Sohan L. 365–6
Jainosaurus 157
Janensch, Werner 12, 59, 63, 213, 245, 247, 251, 253–4, 282–3, *285*, 287, 293–4, *325*, 325–6, 367, 369, 539–42
Janenschia 155, 157, 187, 250, 367–8, *368*, 369, 539–42; *J. robusta* 367, *368*, 542
Janis, Christine M. 17, 80–1
Jason's Giant Site 359
Javelina Formation 576
Jaxartosaurus 175
Jehol Group 511, 516
Jeholodens 516
Jeholosaurus 187, 369, *369–70*, 370–1; *J. shangyuanensis* 369, *369–70*, 370
Jenghizkhan see *Tyrannosaurus*; *J. bataar* see *Tyrannosaurus bataar*
Jenkins, Farrish A. *312*, 496
Jensen, James A. 205, 437, 442, 580–1
Jensen, Marcia L. 93
Jenny, A. 228
Jerzykiewicz, T. 419, 423, 426, 455–7, 505–6
Ji Qiang 23, 143, 270–1, 273, 275–6, 511, 515, 517
Ji Shúan (Shu-An) 23, 143, 179, 271, 275–6, 475–6, 511, 515
Jiang Fu-Xing 243–4, 270, 276, 475–6
Jianu, Coralia-Maria 44, 77–8, *78*, 327–8, 391–3, 479, 481
Jingshanosaurus 52, 147, 346
Jiufutang Formation 403
Jobaria 66, 117, 148, *149–50*, 187, 371, *371–2*, 372, *373–4*, 374–5, *375–7*, 377, 418; *J. tiguidensis 149–50*, 371, *371–2*, 372, *373–4*, 374, *375–7*, 377
Jobu Formation 35
Johns Hopkins University 547
Johnston, Christopher 226–7, 456–7
Joint Russian-Mongolian Paleontological Expedition 442
Jonas, Louis Paul *92*, 602, 606, *607*, 611, *612*, 614
Jones, Ramal D. 542
Jones, Terry D. 86–8, 102–3, 271, 275–6, 495, 513, 515, 587
Jones Ranch 397
"Jordan theropod" *see Stygivenator molnari*
Jubbulpuria 121
Judith River Formation 33, 35, 42, 93, 230, 254–5, 394, 525, 527, 561, 582
Judith River Group 33, 42–3
Jukes, J. B. 546, 549
Jura-Museum 302
Jurassic Foundation 4
The Jurassic in China 287
Jurassic Park vi, 23, 612
"Jurassic Park — The Ride" 612–3
Juul, Lars 146–7, 462–3, 543, *543–5*, 545–6, 548–9, 595
Juveniles and subadults 11–2, 24, 26, 29–33, 35–9, 44, 54, 61, *62*, 71, 73–4, 82, 96, 98, *124*, 146,

171, *174*, *180*, 189, 201, 207, *208–9*, 209–10, 213, 216, 221–3, 225, 227, 229–33, 238–9, 250–1, 254, 260, 262, 279, *280*, 283, 285–6, 289, 292, 299, 301, 303–5, 309, 312, 330, 336, 337, 343–4, 346, *359*, 366, 371–2, *374*, 375, 378, 382, 391–3, 395, 408–9, 414, 418, 428, 431–2, 435, 437–8, 440–1, *454*, 455–7, 464, 475, 479, 481, 492, 495, 498, 500, 518, 525, *526*, 526–8, 545, *545*, 548, 558, 562–3, *563*, 567, *567–8*, 568, 570, 572–5, 579, 583, 586, 592

Kadzi Formation 367
"Kagasaurus" 341
Kaiji, Cai 370–1
Kaijiangosaurus 127
Kaiparowits Formation 437, 439
Kaisen, Peter 201, *524*
Kakuru 132
Kaneko, Ryuichi 37–8
Kangnasaurus 169
Kannemeyeriids 50
Karhu, A. A. 442, *442–4*, 444
Karoo Supergroup 245
Kase, Tomoki 37, *39*
"Katsuyamasaurus" 341
Kawinga Formation 48
Kaye, Fran Tannenbaum 256–7
Kayenta Formation 495–6, 613
Kellner, Alexander W. A. 25, 51, 57, 129, *194*, 194–5, 303–4, 343, 364–5, 486–7, *487*, 488
Kelmayisaurus 127
Kem Kem beds 32
Kemp, K. 552
Kenneth, J. H. 623
Kentrosaurus 161, 253
Kenworthy, Jason 96
Kermack, D. 545, *545*, 548–9
Khan mckennai 592
Khok Kruat Formation 74
Khongil locality 503
Khukhteska Svita 475
Kielan-Jaworowska, Zofia 423, 426, 446–7, 585–7, 604, 619
Kim, Haang Mook 579–80, *580*, 581
Kimmeridge Clay 259, 402
Kinesis 16, *17*, 89, *89*, 206, 211, 224, 298, 473, 508, *508*, 573–4
King Kong vi
Kingdom of the Dinosaurs 601, 613
Kinneer, Billy 277
Kirkland, James I. 8, 30, 56, 70–1, 156, 165, 174–5, 181, 229–30, 283, 319, 313, 328, 335–6, 342–3, 355, 419–20, *420–1*, 422, 466, 490–1, 503, 505, 510–1, 521–3, 528, 542, 590, 592, *604*
Kirtland Formation 33, 39, 81, 93, 109, 110, 111, 198, 200, 202, 229, 309, 321, 377, 418, 420, 436, 438–9, 463, 465, 494–5, 576
Kirtland Shale 342–3, 448–9, 527
Kirkwood Formation 428
Kischilat, Edio-Ernst 19, 542–3
Kish, Elenor M. 615, 621
Kitadani Formation 73, 338–9
Kitadani quarry 338
"Kitadanisaurus" 341
Kitchin, F. L. 253–4
Kitching, J. W. 337, 533
Klamelisaurus 156
Kleinwelka Dinosaur Park 602
Knight, Charles R. *83*, *153*, *206*, 210, 212, *557*, 610, *610*, 615, 619
Knott's Berry Farm 601, 613
Kobayashi, Yoshisugu 38, 73, 274, 276, 535
Kokoro Company, Ltd. 601
Koparian 143, 260
K. douglassi 260
Kotasaurus 147, 346
Koreanosaurus 145
Kraai River locality 337
Krantz, Peter M. 227, 466, 619
Krause, David W. 24, 60, 103, 393, 512, 515
Kritosaurus 79, *80*, 110, 174, *178*, 187, 214, 309, 377–8, 416; "*K.*" *australis* 79; "*K.*" *incurvimanus* 80, *178*; *K. navajovius* 309, 377–8
Krzic, Berislav *161*, *172*, *291*, *435*, *455*, 621
Kuchino, Krzyszstof 602
Kuehneodon sp. 21
Kues, B. S. 198, 200–1
Kuhn, Oskar 214, 217, 233, 401–2, 545, 549, 595–6
Kukhteskaya Svita 502
Kulceratops 180, 415
Kundrát, Martin 300–1
Kunmingosaurus 147
Kurochkin, E. N. 468, 471
Kurtén, Björn 619
Kurzanov, Seriozha M. 143, 213–4, 219, 234–5, 260, 400, 402, 416–7, 431–2
Kuszholia 595
Kutty, T. S. 365–6
Kuwajima Formation 41, *43*
Kuzmitz, Andrew 92–3, 549–52
Kuznetsov, A. N. 123
Kyoryugaki Saizensen 527, 570
Kyte, Frank T. 108

L. D. Engdahl Ranch 571
"La Cãrare" locality 393
Labocania 127, 294; *L. anomala* 294
Labrosaurus see *Allosaurus*; *L. ferox* see *Allosaurus fragilis*; *L. frailis* see *Allosaurus fragilis*; *L. huene*; *L. lucaris* see *Allosaurus fragilis*; *L. medius* see ?*Dryptosaurus medius*; *L. meriani* see *Ceratosaurus* sp.; *L.* (?) *stechowi* see *Ceratosaurus* sp.; *L. sulcatus* see *Ceratosaurus* sp.
Lacey Point *see* Dinosaur Hill
Laevisuchus 25, 127; *L. indicus* 25
LAGs *see* Lines of arrested growth
Lakota Formation 56, 355
Lamanna, Matthew C. 46, 268, 314
Lamb, James P., Jr. 384
Lambe, Lawrence M. 279, 328, 463–5, 520, 575, 579
Lambeosaurus *5*, 80, 175, *176*, 441; *L. lambei* *5*, 80, *176*
Lambert, David 251, 254, 339, 341, 619
Lameta Formation 320
Lametasaurus 164
Lamirata, Michael 440
Lanasaurus 166
Lancanjiangosaurus 147
Lance Formation 323, 402, 525
Land bridges and connections *see* Faunal exchanges
Langer, Max C. 51, 123, 488–9, *489*
Langston, Wann, Jr. 192–3, 279, *318*, 384, 448, 579
Lankster, Ray E. 619
Laño quarry 382–3
Lanzendorf, John *375*, 617
Lao Chang Quing 397
Laosaurus 168
Laornis 595
Laplatasaurus 58, 157, 198, 218; *L. araukanicus* 58, 157, 198
Lapparent, Albert F. de 63, 304, 387–8
Lapparentosaurus 156, 365; *L. madagascariensis* 365
Larainosaurus 484
Larson, Peter L. 502, 562, 574–5, 578–9
Larsson, Hans C. E. 21, 23, 32, 66, 129–31, 191–3, 243, 257, 267–8, 298, 303–4, 327, 364–5, 371, *371–2*, 374–5, 377, *417*, 417–8, 429–30, 520, 529–30, *530–1*, 531–3, 556–8, 560
Latchford, J. D. 227
Laurasia 212, 429, 446, 532–3
Lawson, Douglas A. 576, 579
Lawton, Tim F. 200–1, 503
Le Loeuff, Jean 157–8, 214, 257, 327–8, 365–6, *366*, 382, *382–3*, 383–4, *384*, 393, 476–7, 484, 486, 581, *581–2*, 582–3, *583*
Leanza, H. A. 196
Lecho Formation 40
Leaellynasaura 72–3, 168; *L. amicagraphica* 72
Lee, Yuong-Nam 292–3, *293*, 294, 471, 474, 579–80, *580*, 581
Lehman, Thomas M. 181–3, 198, 200–2, 229–30, 290–2, 309–10, 419–20, 448–9, *449–50*, 450, 527–8, 552–3, *553*, *556*, 560, 576, 579
Leidy, Joseph 218–9, 227, 229–30, 527–8
Leipsanosaurus 115, 164
Leonardi, Giuseppe 619, 621
Lepidosaurs 321, 595
Lepisosteus 254–5
Leptoceratops 119, 180–1, 187, 232, 289, 378, *378–9*, 379, 414–5; *L. cerorhynchus* see *Montanoceratops cerorhynchus*; *L. gracilis* 378, *378–9*, 379, 415; *Leptoceratops* sp. 379
Lesothosaurus 16, 117, 158, 258–9, 370, 498; *L. diagnosticus* 498
Lessem, Don 4, 379, 575, 579, 619
Lessemsaurus 147, 187, 379, *380–1–2*
L. sauropoides 379, *380–1*, 382
Lexovisaurus 161
Li Yinxian 243, 397, 399, 432, 595
Liaobatrachus 516
Liaoningornis 474, 516; *L. longiditrus* 474
Life Before Man 602, 615, 621
"Life Over Time" *5*, *222*, 578
Ligabueino 127
Ligaments *see* nuscles, ligaments, tendons, musculature and muscle scars
Likoelesaurus 121
Liliensternus 20, 126, 460, 533; *L. liliensterni* 460, 533
Lillegraven, J. A. 446–7
Limnosaurus see *Telmatosaurus*
Lines of arrested growth (LAGs) 87, 223, 250, 359, 395
Linster, Wes 238–9
Liolaemus gravenhorsti 99
Lipka, Tom R. 310, 466, 592
Lirainosaurus 117, 157–8, 187, 382, *382–3*, 383–4, *384*, 484; *L. astibae* 382, *382–3*, 383, *384*
Lisboasaurus 4
Lizards *3*, 12, 24, 36, 69, 76–7, 84, 88, 91, 99, *122*, 279, 320, 360, 363, 430, 432, 486, 601
Lockhart, Frank 610
Lockley, Martin G. 47–8, 84, *84*, 106, 248, 254, 265–6, *266*, 267–9, 294–7, 301, 618, 620
Locomotion 10, 21, *22*, 39, 70, 84–5, 99, 102, 138, 147, 180, 191, 247–9, 251, 275, 297–8, 319, 323, 363, 398, 434, 498, 557–8, 575, 579
Loewen, Mark 210, 212
Lohan Cura Formation 195–6
Loncosaurus 166
Long, John A. 432, *433*, 434, 616, 620
Long, Robert A. 24, *47*, 47–8, *184*, 265–7, *268*, 294–6, *298*, 301, *330–4*, 434, 572, 579, 618, 620
Longevity 224
Longisquama 102–3; *L. insignis* 102
Longrich, Nick 36, 98
Lopez de Bertodano Formation 75
López-Martinez, N. 19
Lophorothon 5, 168, 187, 384
Lorenz, Edgar 481
Lorenz, J. C. 395
Loricosaurus 157
Los Alamitos Formation 79
Los Angeles County Museum of Natural History *see* Natural History Museum of Los Angeles County
Los Colorados Formation 24, 382
Losilla locality 304
The Losillasaurus giganteus
Lost World: Jurassic Park 4
Louis Paul Jonas Studios 602, 606, *607*, 611, *612*, 614
Louisiana Purchase Exposition 556, *559*
Lourinhã Formation 203, 384–5, 387
Lourinhanosaurus 131, 187, 384–5, *385–6*, 386, *387*, 402; *L. antunesi* 384–5, *385–7*
Lourinhasaurus 152, 187, 315–6, 386–7, *387*, 388; *L. alenquerensis* 315–6, *387*, 387–8
Love, Sid 191
Love Site 359
Lower Stubensandstein 501
Lu Junchang 38, 466–7
Lucas, Frederic A. 543, 620
Lucas, Spencer G. 8, 19–20, 24, 47–8, 76–7, 93, 109–10, *110*,

198, 200–2, 262–3, 265–6, *266*, 267–9, 294–7, 301, 320, 328–30, *330*, 331–5, 377–8, 419–20, 439–42, *494*, 494–5, 498–9, *499*, 500, 620
Lucas, Timothy 552
Lucianosaurus 158, 266
Lufeng Formation 53, 146, 344, 397
Lufengosaurus 51–2, *52*, 146–7, 346, 461, 548; *L. hueni 52*
Lugmair, G. W. 108
Lujiatum Locality 370
Lukousaurus 121
Lull, Richard Swann 68, 75, 110, 377–8, 422, 438, 441–2, 466, 511, 553, 560
Lurdusaurus 171, 187, 388–9, *389–90*, 390–1, *391*, 418; *L. arenatus* 388–9, *389–91*
Lusitanosaurus 159
Lycorhinus 166, 352
Lydekker, Richard 432
Lyon, Gabrielle H. 129–30, 191–3, 243, 257, 267–8, 303–4, 327, 364–5, 371, *371–2*, 374–5, 377, 429–30, 520, 529–30, *530–1*, 531–3

Maastricht Formation 77
MacGillivray, John 546
MacLeod, N. 106–7
Macrodontophion 120
Macrurosaurus 147, 189
Madej, Waldemar 603
Madsen, James H., Jr. 125–6, 192–3, 205, 209, *209*, 211–2, 219, 279–83, *283–5*, 285, 287, 313–4, 385–6, 623
Maevarano Formation 24–5, 60
Magnosaurus 127
Magyarosaurus 157–8, 187, 391–2, *392*, 393, 402; *M. dacus 392*, 393, 402
Maiasaura 174, 187, 239, 256, 393–4, *394*, 395, 498, 613, 619; *M. peeblesorum* 174, 239, 256, 393–4, *394*, 395, 613
"Maidstone" *Iguanodon* 360
Main, Russell P. *303*, 303–4, 524
Majungasaurus see *Majungatholus*
Majungatholus 127, 187, *396*, 396–7; *M. atopus* 396, *396*
Makela, Robert 393, 395
Makovicky, Peter J. 15, 31–2, 36–7, *38*, 41, 44–5, 71, 103, 134, 139, 143–4, 259–60, 272, 274, 276, 407, *407–8*, 408–9, 476–7, 495, 512, 515, 584, 586–8
Malawisaurus 49, 60, 117, 157, 187, 194, 374, 397
Maleev, Evgeny [Eugene] Alexandrovich 164, 371, 568–9, 577, 579
Maleevosaurus novojilovi see *Tyrannosaurus bataar*
Maleevus 164, 166
Maleri Formation 19
Malin, Millard Fillmore 607, *609*
Mamenchisaurus 63, 156–7, 187, 219, 293, 317, 346, 367, 397, *398*, 399, 590; *M. constructus* 397; *M. hochuanensis* 367, 397, 399, 590; *M. jingyanensis* 399, 397, 590; *M. sinocanadorum* 397; *M. youngi* 397, *398*; *Mamenchisaurus* c.f. 397
Mammal-like reptiles *see* Therapsids
Mammals *3*, 10–1, 16–7, 21, 36, 54, 63, 66–9, 71, 76, 80–1, *84*, 84–7, 91, *92*, 93–4, 96, 99, 102, 109–10, 142, 160, 222, 224, 247, 250–2, 300, 310, 321, 342, 359–60, *361*, 362, 391, 395, 451, 475, 494–5, 516, 521–2, 524, 551, 555–6, 559, 596, 615
Mammut 556
Manabe, Makoto 35–6, *36*, 37, *39*, 41, *43*, 73, 243
Mandschurosaurus 171, 535; *M. laosensis* 535
Manley, K. 499–500
Manupella, G. 315, 385–6
Mantell, Gideon Algernon 359–63, 617
Mantell, Mary Ann 359–60
Marasuchus 48, 52, 349, 543
Marília Formation 195
Markthaler, R. 460, 463
Marsic, Damien 99, 101, 241–2, 559–60
Marsh, Othniel Charles 29, 71, *170*, 211, 218–9, 227, 282, 287, 525, 527–8, 552, 559–60, 595, 621–2
Marshosaurus 132, 218–9
Marsicano, C. A. 351–2
Martill, David M. 26, *26–7*, 27–8, *28*, 29, 129, 364–5, 448, 495
Martin, John *62*, 62–3, *63*, 64, 251, 254, *262*
Martin, Larry D. 95, 102–3, 272, 276, 412–3, 512, 515, 587
Martin, Valérie 451–2, *452–3*
Martin-Rolland [Martin], Valérie *62*, 62–3, *63*, 64, 151, 156, 251, 254, *262*, 334–5, 432, 590, 595
Martinelli, Agustin 195–6
Martinez, Ricardo Néstor 32, 46, 51, 56, 58, 151, 157–8, 481, 483
Martińez, Rubén 32, 157, 214, 328, 382, *382–3*, 383–4, *384*, 426, *426–7*, 428, 484, 486
Maryańska, Teresa 104, 142, 158, 162, 164, 176–7, 188–9, 216–7, 235, 238, 409, 419, 423, 426, 455–7, 463, 465, 505–6, 511, 585–7, 604
Maryland-National Capitol Park and Planning Commissions 227
Masiakasaurus knopfleri 592
Mass deaths 33, 72, 91, 201–2, 205, 223, 253, 516
Massospondylus 51–2, 146, 337, 461, 499, 501
Mata Amarilla Formation 32
Mateer, Niall J. 198, 200–2, 419, 437, 442
Mateus, Luis 21, 304, 385, 388
Mateus, Octavio 315–6, *316–7*, 317, 358, 384–5, *385–6*, 386, 402, 534–5, *535–6*, 536
Matheron, Philippe 481
Matley, Charles Alfred 358–9
Matthew, William Diller 211–2, 304, 310, 617
May, Kevin C. 158, 435
McCord, R. D. 200–1, 517
McDonald, J. Reed 571
McDonald's 577
McGee, Elizabeth M. 205, 212
McGinnis, Helen J. 620
McGowan, Christopher 620
McGraw, George 610
McIntosh, John S. 54–5, 57, 148–9, 152, 155, 200–1, 211–2, 217–8, 225, 227, 246, 253–4, 279, 294, 317–8, 320, 328, 335, 358–9, 367, 369, 387–8, 407, 409, 417, 431–2, 486, 517–9, 535–6, 538–9, 540, 542, 589–90, 596, 619
McKeever, F. S. 526
McKinney, M. L. 55
McLoughlin, John C. 620–1
McMahon, T. A. 248, 254
McNeil, Paul E. 69–70, *77*
McRae Formation 576
McWhinney, Lorrie A. 523–4
Mechin, Patrick 327–8, 581
Méchin-Salessy, Annie 327–8, 581
Megacervixosaurus 152
"Megadontosaurus ferox" *see Microvenator celer*
Megalancosaurus 102, 470
Megaloceras 3
Megaloolithus siruguei 19
Megalosaurus 127, 187, 233, 264, 283, 360, 399, *399–400*, 400–1, *401*, 402, 430–1, 453, 595, 602–3; *M. bredai* see *Betasuchus bredai*; ?*M. cambrensis* 399–400, *400–1*, 401; *M. cuvieri* see *Eustreptospondylus oxoniensis.*; *M. herperis* 400; ?*M. hungaricus* 401–2; *M. insignis* 402; *M. meriani* see *Ceratosaurus* sp.; ?*M. pannoniensis* 402; *Megalosaurus* sp. 402; *see also Magyarosaurus dacus*; *Megalosaurus* sp. 402
Megaraptor 46, 132
Melanorosaurus 52, 147, 245, 265, 501; *M. readi* 265
Mereck, John 103
Mesa Southwest Museum *61, 141, 234, 295*
Mesolimnadia 287
Mesozoic Meanderings 527, 570, 595
Metabolism 9, 66–7, 86–7, *87*, 91, 93–4, 251–3, 313, 395, 551, 556
Meteorites 108–9
Metriacanthosaurus 127
Meyer, Hermann von 457, 459, 595–6
Microceratops 118, 180, 346, 415; *M. gobiensis* 346
Microdontosaurus 147
Microhadrosaurus 174
Micropachycephalosaurus 177
Microraptor 144, 187, 226, 402, *402–3*, 403–4, *404*, 405–7
M. zhaoianus 402–3, 403, *404*, 405
Microvenator 31–2, 116, 139, 187, 327, 407, *407–8*, 408–9; *M. celer* 327, 407, *407–8*, 408–9
Middle Keuper 401
Middle Saurian Bed (Tendaguru) 253
Middle Stubensandstein 501
Middleton, Kevin M. 95–6, 297–8, 301
Miles, Clifford A. 125, 285–7, 301, 354–5
Mill Canyon Dinosaur Trail 613
Miller, Wade E. 387–8
Milner, Angela C. 30, 106–7, 127, 129, 134, 303–4, 362, 354–65, 431–2, 448, 456–7, 459
Milner, Andrew R. 12, 106–7, 143, 257
Minmi 70, 118, 162–3, 187, 409–10; *Minmi* sp. 409
Minott, Garfield *267*
Minter, K. 30
"Mr. T." 556
"Mr. Z" *574*, 578–9
Moenave Dinosaur Tracks 613
Mollusks 19, 109, 490
Molnar, Ralph E. 33, 75, 94, 102, 128–9, 132, 192, 211–2, 219, 229–30, 234–5, 314, 400, 402, 409–10, 431–2, 432, *433*, 434, 525–8, 546, 549, 566, 570, 572–3, 579, 618–9
Monash Science Center 546
Monash University 463, 547
Monastersky, R. 406–7
Monbaron, Michel 227–8, *228–9*, 229
Le Monde Prehistorique (Prehistoric World) 602
Mongelli, Anthony, Jr. 33–4, 43, *43*
Mongolian Academy of Sciences 36, 44, 144, 260, 408, 585–6
Mongolian Academy of Sciences-American Museum of Natural History Expedition 36, 44, 144, 260, 408, 585–6
Mongolian-Japanese Palaeontological Expeditions 423, 586
Mongolian Museum of Natural History 191
Mongolian-Russian Paleontological Expedition 191, 410
Mongolosaurus 147
Monkonosaurus 161
Monoclonius 182, 232, 279
Monolophosaurus 116, 131, 385
Mononychus see *Mononykus*; *M. olecranus* see *Mononykus olecrans*
Mononykus 31, 96, 104, 137–8, 187, 213, 327, 410–1, *411–2*, 412–3, 442, 444–6, 469, 508, 595; *M. olecrans* 31, 327, 410, *411–2*, 412, 444, 508
Montana State University 395
Montanoceratops 119, 180–1, 187, 232, 413, *413–4*, 414–6; *M. cerorhynchus 413–4*, 414
Moodie, Ron 521–3, 620
Mook, Charles Craig 293–4, 367, 369
Morales, Bob 618
Morales, Michael 617
Moratalla, José J. 18–9, 364, 447–8
Moreno Hill Formation 591
Morinosaurus 156
Morosaurus impar see *Camarasaurus grandis*
Morris, J. 544, 549
Morris, William J. 257
Morrison Formation 12, 55, *170*, 203, 205, 210, 218–9, 222, 246, 254, 279, 282–3, 285, 301, 432, 500, 519, 606, 608, 610–1, 614
Mosasaurs 78, 605
Mosasaurus camperi 78
Munyikwa, Darlington 533–4, *534*

Munter, R. C. 23
Murry, Phillip A. 24, 47–8, 167–8, 189, 216–7, 265–7, 294–6, 301, 310, 313, 397
Muscles, ligaments, tendons, musculature, and muscle scars 10, 12, *14–5*, 15–6, *17*, 19–21, 32, 35–6, 38, 55, 63, 66–71, 86, 96, 98–9, 104, 119–20, 130, 134, 137, 142, 166, 168–9, 210–1, 217, 219–20, 244, 248, 260, 272, 274, 302, 319, 326, 328, 345, 348, 363, 375, 407, 409, 446, 476, 493, 495, 498, 502, 541–2, 554, 558, 566, 576, 578, 588
Muscle scars *see* Muscles, ligaments, tendons, musculature, and muscle scars
Musée des Dinosaures 534–5
Museo Argentino de Ciencias Naturales "Bernardino Rivadaria" *152*, 196, 616
Museo di storia naturale di Trieste 79, *80*
Museo Nacional de Ciencias Naturales de Alvala 384
Museo Paleontològico Cittadino of Monfalcone (Gorizia) 352
Museo Paleontològico Egidio Feruglio 537, 539, 547
Museu de Ciência Naturais da Fundacão Zoobotañica do Rio Grande do Sul 347
Museu de Historia Natural da Universidade de Lisboa 315
Museu da Lourinhã 315
Museum Juan Cornelio Moyano de Mendoza 58
Museu Municipal "Carmen Funes" 25
Museu Nacional 51, 57, 194
Museu Nacional de Historia Natural 57
Museum für Naturkunde 540
Museum Isle of Wight *167*
Museum of Comparative Zoology 312, *312*
Museum of Geology and Paleontology 57
Museum of Natural History 103, *510*
Museum of Northern Arizona *451*, *496*, 616–7
Museum of Paleontology 402, *438*, 496
Museum of the Earth 604
Museum of the Municipality of Bale 352
Museum of the Rockies 613, 619
Museum of Western Colorado *87*, 282, *282*, 608, 611, 614, 616
Museum Paleontological Expedition of 1901 *220*
Museum Victoria 547
Mussaurus 54, 146–7; *M. patagonicus* 54
Mussell, Jason 392–3
Muttaburasaurus 118, 168, 479
Mygatt-Moore Quarry 283
Myhrvold, Nathan P. *101*, 219–20, 225, 423, 426
Mymooropelta 163, 165, 354, 521, 542
Mystery Hill and Prehistoric Forest 613
Myzobuchi, Yoshitoshi 194
Naashoibitosaurus see *Kritosaurus*; *N. ostromi* see *Kritosaurus navajovius*
Nacimiento Formation 109
Nagao, Takumi 418
Naish, Darren *29*, 29–31, *31*, 32, 226, 416
Nakajima, Shuichi 37, *39*
Nam Phong Formation 365
Nambango site 540
Nanosaurus 168
Nanotyrannus see *Tyrannosaurus*; *N. lancensis* see *Tyrannosaurus rex*
Nanshiungosaurus 142
Naomichleys 57
NASA 559
National Geographic 405–6, 423, 511, 577, 576
National Geographic Society *100*, 374, 405–6, *515–7*
National Museum of Natural History *124*, *139*, *170*, *199*, 209, *209*, *218*, 235, *280–1*, 299, 406, 556, 558, 614
National University of Comahue 57
Natural History Museum and Biodiversity Research Center 242
The Natural History Museum, London 30, 188, 249, 362, 432, 545, 547, 622
Natural History Museum of Aix-en-Provence 481
Natural History Museum of Los Angeles County *34*, *52*, *90*, *128*, *173*, *174*, *261*, *313*, *467*, *472–3*, *475*, 525, *525–6*, 526, *555–6*, *565–8*, 571, 601, 611, 617
Nature 406
Navahopus falcipollex 462
Navajo Sandstone 462
Nedcolbertia 134, 187, 416; cf. *Nedcolbertia* 416
Nelson, E. V. (Ernie) 614
Nemegt Formation 39, 410, 423, 447, 463, 568, 582
Nemegtosaurus 56, 151, 157, 187, 416–7, 452, 575; *N. mongoliensis* 416
Neopangaea 539
Neosodon 155
Neovenator 385; *N. salerii* 30, 131
Nerinia 253
Nessov, Lev A. 105, 140, 183, 416, 595
Nestlings 393–5, 471
Nests and nesting 11, *18*, 18–9, 21, 39–40, 62, 89–91, 395, 471, 560–1, 618
Neuquén Group 72, 428
Neuquensaurus 58, 158, 317, 481; *N. australis* 481
New Mexico Bureau of Mines and Mineral Resources 110
New Mexico Museum of Natural History and Science *155*, *160*, *436*, *499*, 611
New York Journal 320
New York World 320
Newman, Barney 363
Newton, E. T. 400, 402
Ngexisaurus 132
Nigersaurus 66, 152, 187, 374, *417*, 417–8; *N. taqueti* 417, *417*
Nikell, Michael W. *470*
"1999 Geologic Time Scale" 8
Niobrarasaurus 164, 354, 490, 510, 542
Nipponosaurus 174–5, 187, 418; *N. sachalinensis* 418
Niveau des inncents locality 417
Noasaurus 25, 40, 127; *N. leali* 25, 40
Nodocephalosaurus 118, 164–6, 187, 309, 343, 418–9, *419*, 420, 505; *N. kirtlandensis* 309, 418–9, *419*, 420
Nodosaurus 70, 164, 187, 216, 420, *421–2*, 423, 510–1, 521; *N. textilis* 70, 216, *421–3*, 423, 521–2
Nomingia *101*, 101–2, 139, 187, 422, *422–3*, 423–4, *424–5*, 425–6; *N. gobiensis* 101, *101*, 422, *422–3*, 423–4, *424–5*
Nopcsa, Baron Franz [also Ferencz, Francis] (von Felsö-Szilvás) 44, 163, 189, 214–7, 326, 391, 393, 401–2, 539
Norell, Mark A. [Allen] 15, 23, 30–1, 36–7, *38*, 40, 44–5, 89, 94–5, 103, 138, 143–4, *190*, 190–1, 213, 259–60, 270–4, 275–6, 327, 410–1, *411*, 412–3, 444, 476–7, 495, 508, *508–9*, 509, 511, 512, 515, 584–7, 615, 618, 620
Norman, David B. 16, *80*, 126, 158, 166, 167, 168–72, 174–5, 214, 216–7, 226, 234–5, 257, 259, 320–1, 327, 362–3, 392–3, 467, 471, 473–4, 479, 481, 539, 596, 620, 623
North Carolina State Museum of Natural Sciences *6*, *93*, *177*, 191, *193*, 325, *549–50*, 550–1, *551–2*
North Carolina State University *549–50*, 550–1, *551–2*
Norton, James M. 89–90, 312–3
"Nothronychus" 592
Notoceratops 158
Notohypsilophodon 168, 187, 426, *426–7*, 428; *N. comodorensis* 426, *426–7*, 428
Novacek, Michael J. 30, 620
Novas, Fernando E. 19, 31–2, 45, 48, 103–4, 119–20, 123, 137, 212–3, 241–2, 265, 269, 301, 328, 410, 413, 444–5, *445–6*, 446–7, 489, 502, 512, 515
Novey, Bill 613
Nowiński, Aleksander 416–7
Nqwebasaurus 132, 187, 428–9, *429–30*, 430; *N. thwazi* 428, *429–30*
Nurosaurus 155, 595; *N. qaganensis* 595
Nuss, Fred 40, *41*, 578
Nuss Fossils 40, *41*, 578
Nuthetes 4, 144, 187, 320, *321*, 430–1, *431*, 432, 595; *N. destructor* *321*, 430–1, *431*
Nyasasaurus 121
Obradovich, J. D. 200–2, 419, 495
O'Connor, Patrick M. 88, 396–7
Ohmdenosaurus 148
Ojo Alamo Formation *see* Kirtland Shale
Ojo Alamo Sandstone 109–10, *110*, 111, 201, 377
Oklahoma Museum of Natural History *see* Sam Noble Oklahoma Museum of Natural History
Okura, Masatoshi 35
Oldman Formation 82, 463–4
Olsen, Paul E. 126, 337
Olshevsky, George 103, 219, 227, 229–30, 246, 254, 279, 295, 317, 321, 323, 341–2, 371, 431–42, 524–5, 527–8, *566–8*, 570–1, 579, 595–6, 615
Olson, Everett C. 617, 622
Olson, Storrs L. 406–7
Omeisaurus *4*, 60, 63, 156, 187, 219, 228, 346, 432, 537, 539, 590; ?*O. changshouensis* see *O. junghsiensis*; *O. fuxiensis* see *Zigongosaurus fuxiensis*; *O. junghsiensis* 432; ?*O. luoquanensis* 432; *O. tianfuensis* 219, 539
Omnivory 43, 137
O'Neill, Michael 110, 111, 198, 200–2, 419
Onychosaurus 159
Ontogeny 11, 31, 38, 54–5, 82, 91, 96–9, 201, 205–7, 209–10, 216, 220–4, 230, 232, 235, 239, 249–51, 255, 259, 262, *263*, 286, 289, 299, 306, 312, 359, 379, 391, 394–5, 409, 428, 432, 437, 439–40, 450, 456–7, 467, 475, 486, 492, 498, 528, 539, 548, 552, 562–3, *564*, 565–7, 569–72, 575, 579, 592
Opisthocoelicaudia 58, 61, 158, 200
Oplosaurus 156, 246
Oregon State University 102
Organ, Chris 70
"Orphan Mesa theropod" *see Eucoelophysis*
Ornatotholus see *Stegoceras*; *O. browni* see *Stegoceras validum*
Ornithodesmus 143, 260; *O. cluniculus* 260
Ornitholestes 116, *142*, 143, 212, 301, 409; *O. hermanni* *142*, 301
Ornithomimoides 127
Ornithomimus 37, 39, 134–5, 139; *O. altus* see *Struthiomimus altus*; *O. edmontonicus* 37, 39
Ornithotarsus 174
Orodromeus 18, 168
Orthogoniosaurus 127
Orthomerus 174
Orzack, Steve *312*
Osborn, Henry Fairfield 59, 201, *206*, 210, 293–4, 309, 318, 320, 367, 369, 448, 450, 577, 579, 583, 587, 620
Oshanosaurus 166
Osmólska, Halszka vi, 30, *101*, 104, 105, 121, 138–9, 143–4, 158, 176, 235, 238, 257, 260, 409, 422–6, 463, 465, 476–7, 514–5, *584*, 584–7, 604, 622–3
Oson, Kenneth H. 257
Ossified tendons 56, 69–70, 76, *77*, 82, 158, 164, 166, 168–9, 177, 254, 286, 354, 456–7
Osteogenesis 11, 223, 524
Osteology of the Reptiles 219, 621
Ostrom, John H. 12, 14, 29–30, 32, 103, 211–2, 238–9, *239–40*, 240, *241*, 242, 310, 312–3, 327, 407–9, 412–3, 440, 442, 468, 470–1, 476–7, 490–2, 494, 511–2, 515, 518–9, 552–3, *553*, 560, 582–3, 586–7, 620–1

Othnielia 166, 168, 187, 259, 432
Ouranosaurus 12, 73, 75, *76*, 171–2, 175, 388–9, 418, 533; *O. nigeriensis* 75, *76*, 172, 175, 388–9, 533
"Outlines of the Natural History of the Environs of Lewes" 360
Oviraptor *18*, 40, 90, *90*, 98, 104, 106, 139–40, 240, 424, 618; *O. philoceratops* 40; *Oviraptor* sp. 104, *105*
Owen, Richard 3, 226, 320–1, *321*, 360, 363, 430–1, *431*, 432, 543, 549, 596, 602, 620
Owen-Smith, R. N. 252, 254
Ozraptor 121, 187, 432, *433*, 434; *O. subotaii* 432, *433*

Pachycephalosaurus 81, 178, 187, 434, *434–5*, 435, 463, 465; *P. wyomingensis* *434–5*, 435
Pachyrhinosaurus *182*, 182–3, 187, 232, 435; *P. canadensis* *182*
Packs and pack hunting 12, 26, 200, *208–9*, 575
Padian, Kevin vi, 48, 90–1, 94, 102, 103, 121–7, 130–2, 135, 139, 143–4, 158–9, 258–9, 265–7, 274, 297, 299, 301, 395, 441–2, 495–6, *496*, 497, 512, 515, 524, 617, 620, 623
Paedomorphosis 96, 392
Page, Jim *93*, *549–50*
Paik, Sung 52–3
Palaeosauriscus 543–4, 546, 549, 595–6; *P. cylindrodon* 543–4, 546, 549, 595; *P. diagnosticus* see *Sellosaurus gracilis*; *P. platyodon* 543–4, 546, 549, 595
Palaeosaurus see *Palaeosauriscus*
Palaeoscincus 164
Paleocene dinosaurs 33, 109–10, *110*, 111
Paleontological Institute, Academy of Science 102, 467
Paleontological Museum 420
Paleorhinus 19
Paleosaurus see *Palaeosauriscus*
Pallenberg, Josef Franz 602, *603*
Palmer, A. R. 8
Pampa de Agnia Formation 537
Pan-American Exposition 556, *558*
Pang Qiqing 355, *356–7*, 357–8, 503–4, *504*, 505–6, 552
Pangaea 18, 364, 366, 429, 533
Panoplosaurus 164, 216, 420, 466, 490, 510, 521, 542
Pant-y-ffynnon Quarry 548
Papp, Michael J. 68–9
Paranthodon 161
Paradys farm 533
Paralatitan stromeri 592
Pararhabdodon 175
Parasaurolophus 175, 187, 200, 290, 309, 419, 435–6, *436–7*, 437–8, *438–9*, 439–42, 495; *P. cyrtocristatus* see *Parasaurolophus tubicen*; *P. tubicen* 309, 435–6, *436–7*, 437, *439*, 438–41; *P. walkeri* 435–6, *438–9*, 439–40; *Parasaurolophus* sp. 437–8, 440; *see also Parasaurolophus tubicen*; cf. *Parasaurolophus* sp. 438
Parasites 96–7
Parco Zoo del Garda (Parco dei Dinosauri) 603
Pareiasaurids 596
Pareiasaurus 596
Parental care *61*, 61–2, 91, 495
Park, Eun-Jun 579–80, *580*, 581
Parker, Neave *4*
Parkinson, J. 253–4, 621
Parksosaurus 168
Paronychodon 42, 144
Parrish, J. Michael 49–51, *52*, 62, *62–3*, 63–4, *64*, 65–7, 229–30, 251, 254, *261*, 261–2, *262*, 263, 310, 313, 372, 377, 615
Parsons, William L. 490–1
Parvicursor 104, 138, 187, 413, 442, *442–4*, 444, 508; *P. remotus* 442, *442–4*, 444, 508
Paso del Sapo Formation 78–9
"Passage Beds" 245
PAST (Prehistoric Animal Structures, Inc.) *64*, *247*, *373–4*, 374, *531–2*, 533
Patagonykus 4, 104, 116, 137–8, 187, 412, 444–5, *445–6*, 446–7, 595; *P. puertai* 444–5, *445–6*
Patagopteryx 411
Patagosaurus 49, 55, 149, 158, 537; *P. fariasi* 55
Pathology 30–1, 39, 209, 211, 255, 323, 424, 523–4, 578
Patterson, Colin 412–3
Paul, Gregory S. 21, 62, 67, *84*, 84–5, 101, 126, 128, 145, *208–9*, 229–30, 234–5, *238*, 246–7, 251, 254, 257, 275–6, *286*, 298–9, *299*, 301, 305, 310, *315*, 317, 327–8, 434, 498, 500, 527–8, 567, 572–4, 579, *585*, 586–7, 619, 621–2
Paw Paw Formation 447
Pawpawsaurus 70, 164, 187, 216, 490, 510–1, 542; *P. campbelli* 70, 216
Peabody Museum of Natural History *68*, 310, 312, 407, 524, 545, 166
Peckzis, Jan 251, 254
Pebbly Arkose Formation 488
Peebles, Peter 432
"Peiropolis titanosaur" 157–8
Peishansaurus 162
Pekin Formation 19
Pekinosaurus 19
Pelecanimimus 135, 139, 187, 260, 447–8
Pellegrinisaurus 157
Pelorosaurus 156
Pelycosaurs 555
Peng Guangzhao 30, 159, 257–9, 327, 371, 476–7
Penkalski, Paul 71, 183–4, 230, *230–1*, 231–3, 554, 560, 592
Pentaceratops 183, 187, 200, 291, 309, 448–9, *449–50*, 450, *451*, 495, 552–3, 613; *P. sternbergi* 309, 448, *449–50*, 450, *451*, 613
Pérez-Moreno, Bernardo P. 203–5, 212, 327, 447–8
Pereda-Suberbiola, Xabier *see* Suberbiola, Xabier Pereda
Perle, Altangerel 30–1, *190*, 190–1, 327, 410–1, *411*, 412–3, 586–7
Permineralization 9–10, 47
Perry, Steven F. 88–9, *89*, 252–4
Peters, D. S. 251–2, 254, 468
Peterson, Brad 110
Peterson, O. A. 225
Petrified Creatures Museum of Natural History 613
Petrified Forest Formation 296–7, 328, 330, 332
Petrified Forest National Park 297, 613, 616, 620
Petrodus sp. 76
Phaedrolosaurus 144
Phalaecrocorax 326
Phil Fraley Productions, Inc. *569*, 576
Phrynosoma 279
Phu Pha Ngo site 451
Phuwiangosaurus 151, 187, 450–2, *452–3*, 453, 535–6, 605; *P. sirindhornae* 451, *452–3*, 535
Phyllodon 168
Phytosaurs 19, 47, 301, 330, 332, 524, 546, 595–6, 613
Piatnitzkysaurus 116, 129, 131, 385
Picrodon see *Megalosaurus*; *P. herveyi* see *?Megalosaurus herveyi*
Pidancet, J. 457, 463
Pinacosaurus 164, 165, 187, 216, 336, 418, 453, *454–5*, 455–7, 503, 505; *P. grangeri* 336, 453, *455*, 455–7; *P.* cf. *grangeri* see *Pinacosaurus* sp.; *P. mephistocephalus* 453, *454*, 455–7; *P. nighsiensis* see *Pinacosaurus* sp.; *Pinacosaurus* sp. 457
Pisanosaurus 19, 117, 158
Piveteau, Jean 621
Piveteausaurus 127–8
Placerias quarry 265–6, 294, 301
Planicoxa venenica 592
Plants 8, 12, 17, 19, 37, 42–3, 51, *52*, 66, 68, 81, 89–90, 96, 98, 102, 109–10, *110*, 166, 200, *208–9*, 228, 239, 252, 255, 268, 289, 302, 323, 409–10, 457, 473, 490, 515, 533, 538, 561, 579, 601, 606, 613, 615
Plate techtonics 364, 446, 352
Plateosaurus *7*, 48, 51–2, 146–7, 187, 263, 265, 336–7, 401, 457, *458–9*, 459–63, 548; *P. engelhardti* *7*, 263, 265, 401, 457, *458–9*, 459–63; *P. erlenbergiensis* 457, 460; *P. fraasianus* see *Plateosaurus longiceps*; *P. longiceps* 457, 463; *P. plieningeri* 457, 461–2; *?P. poligniensis* 457, 460–1; *P. reinigeri* 457, 461–2; *P. robustus* 457, 460
Pledge, N. S. 434
Plesiosaurs 499, 521, 596
Pleurocoelus *6*, 156, *193*, 226–7, 277, 279, 397, 492, 611; *P. nanus* *6*, *193*, 277, 279
Plieninger, T. 461, 463, 596
Pneumatoarthrus 174
Podokesaurus 121, 126, 265, 296; *P. holyokensis* 126
Poekilopleuron 127–8, 218
"*Poicilopleuron*" see *Poekilopleuron*; "*P.*" *valens* see *Antrodemus valens*
Polacanthoides see *Polacanthus*
Polacanthus *7*, 70, 163, 165, 216, 343, 354–5, 521–2, 542; *P. foxii* 70
Polish-Mongolian Palaeontological Expeditions 585, 603
Polyonax 183
Polysternon sp. 476
Pomes, M. L. 310, 313
Ponerosteus 596
Porter, Elbert H. 607, *609*
Portezuelo Formation 45
Portland Formation 320–1
Postosuchus 48, 52; *P. kirkpatricki* 48
Posture 14, 40, 51, 62–3, *63–4*, 64–7, 72–3, 81, *84*, 84–5, 94–6, 120–1, 123, 127, 129, 142, 145–8, 158–60, 162, 166, 168–9, *170*, 171, *175*, 176–81, 191, 210, 219, 235, *238*, 239, 245, 247–8, *248*, 249, 271, 275, 360, *361*, 362–3, 389–90, 413, *459*, 462, 481, 498, 512, 552, 557–8, 576–7
Potash Road Dinosaur Tracks 613
Potomac Group 310
Powder Hill Dinosaur Park 613
Powell, Jaime Eduardo 18, 40, 55, 79, 157–8, 198, 205, 212, 214, 217–8, 328, 358–9, 382, *382–3*, 383–4, *384*, 481, 483
Predatory Dinosaurs of the World: A Complete Illustrated Guide 126, *208–9*, 229, *286*, *299*, *315*, 527, 572, *585*
Prehistoric Animals 247
Prehistoric Forest (Dinosaur Land) 613
Prehistoric Gardens 613–4
Prehistoric North Carolina Exhibit 551
Prehistoric Parks 602
Prehistoric Times 3, 323, 579, 618
Prenocephale 81, 177–8, 187, 463–4, *464–5*, 465–6, *466*; *P. brevis* 463–5; *P. edmontonensis* 463–4, *464–5*, 465; *P. prenes* 463–4; *Prenocephale* sp. indet. 465, *466*
Price River II Quarry 254
Priconodon 70, 164, 187, 466; *P. crassus* 70, 466
Prieto-Marquez, Albert 255–6
Primeval World 614
Prince Creek Formation 435, 561
Princeton Museum of Natural History 378
Princeton University 378, 395, 616, 618
Priodontognathus 70, 164
Prismatoolithus 18, 561; *P. levis* 561
Probactrosaurus 75, 118, 171, 187, 328, 466–7, *467*; *?P. alashanicus* 466–7; *P. gobiensis* 75, 466–7, *467*; *P. mazongshanensis* 466
Probst, C. 459–60, 463
Proceratosaurus 126–7
Procerosaurus see *Ponerosteus*
Procompsognathus 126
Prodeinodon 121
Project Exploration 375
Prokinesis 507–9
Prosaurolophus 12, 80, 174, *179*; *P. maximus* 80, *179*
Protarchaeopteryx 98, 116, 143, 243–4, 271, 425, 474, 513, 515–6; *P. robusta* 474
Proterosuchids 48
Proterosuchus 337
Protoavis 98, 121, 234, 467–9, *469–70*, 470–1; *P. texensis* 467, *469–70*
Protoceratops 68, *69*, *90*, *180*, 180–1, 187, 231, 415, 456–7, 471,

472–3, 585, *609*; *P. andrewsi 69, 180*, 415, 471, *472–3*, 585
Protofeathers 98, 121, 133, 244, 405, 515, *517*
Protognathosaurus 149
Protohadros 118, 172, 187, 471, *473*, 473–4
P. byrdi 471, *473*
Protorosaurus 596
Prum, Richard O. 97, 103
Pseudograpta 287
Psittacosaurus 74, 179, 187, 289, 415, 474, *474–5*, 475–6, 499, 516–7; *P. mazongshanensis* 474–5; *P. meileyingensis* 474; *P. mongoliensis* 474–5, *475*; *P. neimongoliensis* 474; *P. ordosensis* 474; *P. sinensis* 474; *P. xinjiangensis* 474; *Psittacosaurus* sp. *474*, 475
Pteranodon 8; *P. ingens*
Pterosaurs *6*, 12, 48, 91, 101, 395, 468, 474, 515–6, 533, 606, 615, 620
Pterospondylus 121
Puerta, Pablo 103, 241–2, 512, 515, 537, 547
Purbeck Limestone Formation 320, 362, *362*, 430, 432
Purbeck Limestone Group 363
Pygostyle (in nonavian theropods) 99, *101*, 102, 235, 271, *423*, 423–4
Pyroraptor 144, 187, 476–7, 582; *P. olympius* 476

Quaesitosaurus 416; *Q. orientalis* 416
Qagganur Formation 595
Qantassaurus 168, 187, *477*, 477–8; *Q. intrepidus 477*, 477–8
Queensland Museum 409
Quenstedt, F. A. 457, 463
Quetzalcoatlus northropi 6
Quilmesaurus curriei 592
Qinlingosaurus 147, 187, *478*, 478–9; *Q. luonanensis 478*, 478–9

Raath, Michael A. 299–301, 317, 320, 334–6, 367, 369, 428, 502, 533–4, *534*, 540, 542
Rabbit Valley Trail Through Time 614
Racey, A. 365–6, 451
Rage, J. C. 539
Rahona see *Rahonavis*
Rahonavis 102, 104, 241, 404–5, 514
Ralph, R. 546, 549
Randolph, S. E. 87
Rapator 121
Rapetosaurus krausei 592
Ratkevich, Ronald P. 8, 202, 517–9, 621
Rauhut, Oliver W. M. *11*, 11–2, 124, 132, 191, 193, 243, 268, 303–4, 325–6, 364–5, 520, 529–30, *530–1*, 531–3
Rauisuchians 48, 52, 202, 245, 295, 301, 470
Rawson, P. F. 189, 214
Rayfield, J. Emily 210–2
Raymond M. Alf Museum *83*
Rayososaurus 152, 418
Rebbachisaurus 53, 151–2, 352, 416, 418, 499; *R. tessonei* 53
Reck, Hans 253–4
Red Beds of the Khermeen Tsav 419
Reed, Christina 551–2
Reeside, J. B. 109, 201, 343, 420
Regnosaurus 161
Regurgitates 302
Reid, Robin E. H. 251, 551
Rensberger, John M. 10–1
Reproduction *see* Sex
Reser, Pete 436
Respiration 12, *13–4*, 14–5, *15*, 16, 86–9, *89*, 91–2, 97, 99–100, 103, 241, 252–3, 313, 441, 448
Respiratory turbinates (RT) 86–8
Retallack, Gregory J. 108
Revueltosaurus 158, 266
Rey, Luis 619, 622
Rhabdodon 75, 118, 167, 187, 476, 479, *479–80*, 481; *R. priscus* 476, 479, *479–80*, 481; *R. robustus* 75
Rhea americana 375
Rhoetosaurus 149, 345
Rhynchocephalians 302, 468
Rhynchosaurs 17, 50
Rich, Leaellyn 547
Rich, Patricia *see* Patricia Vickers-Rich
Rich, Thomas H. V. 72, 138, 149, 168, 211–2, *477*, 477–8, 537, *537–8*, 538–9, 546–7, 549, 621–2
Richardoestesia 42–3, 141; *R. gilmorei* 42; *R. isosceles* 592
Richards, H. Robins 48
Ricklefs, R. E. 395
Ricqlés, Armand J. de 90–1, 395, 441–2, 498
Rieppel, Olivier C. 615, 621
Rigby, J. Keith, Jr. 106, 109, 198, 200–1, 229–30, 397, 399, 401–2, 528
Riggs, Elmer S. 63, 246, 254, 606, 608
Riggs Hill 614
Riley, Henry 457, 463, *543*, 543–5, 549, 596
Rileya bristolensis 546
Rileyasuchus 596
Ringbone Formation 75, 200
Río Colorado Formation 32, 57, 61, 447, 485–6
Río Limay Formation 25, 33, 73, 363
Río Neuquèn Formation 57, 72, 445, 447
Río Neuquèn Supergroup 45
Rioarribasaurus see *Coelophysis*
Riojasaurus 48, 52, 117, 146–7, 350, 382
Robinson, John M. 247, 254, 575, 579
Robinson, Trevor 432
Rocasaurus 158, 187, 481, *482–3*, 483, *483–5*; *R. muniozi* 482, *482–5*; *Rocasaurus* sp. 481, *485*
Rock Point Formation 296, 330
Rocques-Hautes locality 582
Rodríguez de Rosa, Rubén 23–4
Romer, Alfred Sherwood 14, 219, 227, 618, 621
Roosevelt, Frankylin Delano 605
Roosevelt, Theodore 613
Rosenbaum, Jason N. 159, 495–6, *496*, 497
Ross, Callum F. 428–9, *429–30*, 430
Rothschild, Bruce M. 39, 219, 225, 523–4
Rowe, Timothy 71, 126–7, 183, 265, 267, 287, 299, 301, 313–4, 327, 405, 533–4
Roy-Chowdbury, Tapan 365–6
Royal Ontario Museum 103, *179*
Royal Tyrrell Museum of Palaeontology 42, *88, 135, 140, 142, 145, 169, 175*, 201, *255, 262, 292*, 309, 405, *564*, 602, 619
Rozhestvensky, Anatoly Konstantinovich 466–7, 568, 570, 579
RT *see* Respiratory turbinates
Ruben, John A. 28, 86–8, 101–2, 271, 275–6, 495, 513, 515, 587
Rubilar, David 99
Ruehleia bedheimensis 592
Russell, Dale A. 8, 23, 82, 91–3, *93*, 105, 129, 133, 141–2, 192–3, 216, 227–8, *228–9*, 229, 251, 253–4, 257, 267–8, 293–4, *303*, 303–6, 309–10, 327, 335–6, 388–9, *389–90*, 391, *391*, 397, 399, 409, 420, 423, 426, 446–8, 455–7, 476–7, 505–6, *519*, 519–20, 528, 539, 554–5, *557*, 560, 562, 567, 573, 578–9, 582, 621
Russell, Loris 71, 163
Russian Academy of Sciences 559
Ryan, Michael J. 33, 39, 82, 181, 231–3, 528
Rybczynski, Natalia 70, 335–6
Ryder, Richard C. *558–9*

Sabath, Karol 300–1, *584*
Sacrison, Stan *572*
Sadleir, Rudyard W. 32, 66, 96, 192–3, 243, 303–4, 364–5, *417*, 417–8, 520, 529–30, *530–1*, 531–3
Saichania 118, 164–6, 419–20, 505, *604*; *S. chulsanensis 604*
St. Mary Formation 414
Salgado, Leonardo 25, 53–5, 74, 151, 153–4, 158, 168, 191, 193–5, 213, *214*, 217–8, 225, 279, 363–4, 416–7, 481, *482–3*, 483, *484–5*, 499–500
Salitral Moreno locality 481
Saltopus 121
Saltasaurus 40, 147–9, 151, 153, 155–8, 187, 198, 358, 367, 481, 483–6; *S. loricatus* 40, 158, 198, 367, 481, 486
Sam Noble Oklahoma Museum of Natural History *6*, 191, *207, 223*, 491
Sampson, Scott D. 24–5, 39, 60, 82–3, *83*, 88, *88*, 103, 124, *142*, 181, 210, 212, 231–3, 393, 396–7, 428–9, *429–30*, 430, 512, 515, 563, 579
San Pedro Arroyo Formation 47
Sander, P. Martin 249–52, 254, 278–9
Sanders, Richard K. 375, 377, 491–4
Sandy Quarry *41, 434–5*, 435
Sanford, W. A. 263, 265, 400–2
Sanford Maudlin Photography *6, 223*
Sangpin Formation 74
Sangonghesaurus 166
Sankey, Judith T. 41–2
Sano, Kazuhiko *193*
Sanpasaurus 4, 120
Sânpetru Formation 479
Santa Maria Formation 488
Santa Rosa Formation 24, 48
Santa Susana Field Laboratory 577
Santana Formation 26, 29, 364, 486–7
Santanaraptor 134, 187, 486–7, *487*, 488; *S. placidus* 486, *487*, 488
Sanz, José Luis 19, 157–8, 203–5, 212, 214, 315–7, 327, 328, 364, 382, *382–3*, 383–4, *384*, 386–7, *387*, 388, 393, 447–8, 484, 486, 621
Sao Khua Formation 38, 74, 451
Sarcolestes 164
Sarcos 612
Sarcosaurus 124
Sarjeant, William A. S. 106–7
Saturnalia 51–2, 116, 123, 146, 187, 488–9, *489*, 548; *S. tupiniquim* 488, *489*
Saurolophus 80, 110, 174, 187, 489–90; *S. osborni* 80
Sauropelta 70–1, 164, 187, 216, 354, 422, 466, 490–1, *491*, 510–1, 521, 542; *S. edwardsi* 70, 216, 490–1, *491*
Saurophaganax see *Allosaurus*
Sauroplites 162
"Sauropod hiatus" 56–7, 200
Sauropod Tracksite 614
Sauroposeidon 156, 187, 491–2, *492–3*, 493–4; *S. proteles* 491, *492–3*, 494
Saurornithoides 42, 45, 143, 260; *S. junior* 143, 260; *S. mongoliensis* 45, 260; *S. youngi* 143
Saurornitholestes 33, 42–4, 46, 93, 103, 104, 144, 187, 239, 327, 476, *494*, 494–5, 582–3, 586; *S. langstoni* 42, 44, 103, 104, 239, 476, *494*, 494–5, 582–3
Sauvage, E. 595
Savage, Donald E. 201, 310, 448
Sawicki, Josef 603
SBDE *see* Sino-Belgian Dinosaur Expedition
Scansoriality 42, 98, 101
Scavenging 30, 33, *206*, 210, 306, 375, 393, 575
Scelidosaurus 70–1, 117, 159, 498
Scheele, William E. 621
Scheetz, Rodney D. 57, 167, 254
Schlaijker, Erich Maren 68, *69*, 414, 416, 463–5, 520
Schmidt-Nielsen, K. 251–2, 254
Schmitt, J. G. 393, 395
Science 406
Science Museum of Minnesota 242
Science News 406
Scientific American vi, *193*
Scipionyx 27–8, 86, 98, 101, 134, 187, 302, 495; *S. samniticus* 495
Scleromochlus 95, 548
Scollard Formation 579
Scolosaurus see *Euoplocephalus*; *S. cutleri* see *Euoplocephalus tutus*
Scotch College 432
Scotese, C. R. 446–7
Scripps Institution of Oceanography 108

Scutellosaurus 91, 117, 159, 187, 495–6, *496–7*, 498, 524; *S. lawleri* 159, 495–6, *496–7*, 498
Sebayashi Formation 37
Secernosaurus 171–2
Sedgwick Museum 189
Seeley, Harry Govier 188–9, 202–3, 214–7, 226, 233, 263, *264*, 265, 336–7, 367, 369, 400–1, *401*, 402, 546–9
Segnosaurus 142
Seismosaurus 152, 187, 317, 320, 498–9, *499–500*, 500; *S. halli* 152, 320, 498, *499–500*, 618
Sellosaurus 50, 52, 120, 147, 187, 350, 461, 500–1, *501–2*, 502, 548, 595; *S. gracilis* 120, 500–1, *501–2*, 502
"Senonian lacustre" locality 79
Sentestheria 287
Sereno, Paul C. 16–9, 21, 23, 32, 48, 53, 66, 104, 115, 120–3, 129–30, *130*, 131, *133*, 137–8, 144, 146–9, 151, 153–61, 164, 168–70, 174–80, 183, 191–3, 213, 216–7, 232–3, 243, 257, 259, *267*, 267–8, 273–4, 275–6, 287–9, 303–4, 317, 320, 327, 346, 358–9, 364–5, 370–1, *371–2*, 374–5, 377, 388, 413, 415–7, *417*, 418, 429–30, 444, 447, 456–7, 470–1, 481, 486, 488–9, 502, 509, 520, 529–30, *530–1*, 531–2, *532*, 533, 535–6, 539, 548–9, 617
Serjeant, William A. 621
Sex 24, 85, 220–1, 249–50, 441, 502, 553, 561
Sexual dimorphism 24, 40, 71, 80–1, 120, *175*, 178, 219, 250, 439–41, 450, 457, 463–4, 498, 502, 520, 548, 552–3, *556*, 577–8, 592
Seymour, Roger S. 66–7
Shamosaurus 164, 166, 187, 418, 502–3, 505; *S. scutatus* 502–3
Shanshanosaurus 127, 229, 528; *S. houyanshanensis* 229, 528
Shantungosaurus 110, 174, 289; *S. giganteus* 111, 289
Shanyang Formation 478–9, 506
Shanyangosaurus 135, 187, 506, *507*
S. niupanggouensis *507*
Shanxia 164, 166, 187, 357, 503–4, *504*, 505–6, 552; *S. tianzhenensis* 503, *504*
Shar Tsav locality 235
Sharov, Alexander G. 102–3
Shaximiao Formation 589
Sheehan, Peter M. 106–8
Shen, Y. 293–4
Shendi Formation 11
Shubin, Neil H. 245, 298
Shukolyukov, A. 108
Shunosaurus 156, 219, 317, 346, 365–6, 548; *S. lii* 219, 365
Shuvosaurus inexpectatus 295
Shuvuuia 104, 138, 187, 411, 413, 506, 508, *508–9*, 509; *S. deserti* 506, 508, *508–9*
Siamosaurus 130
Siamotyrannus 135
Sibbick, John *545*, 619–20, 622
Sichuan Geological Survey 344
Sidwell, James Q. 613
Sigilmassasaurus see *Carcharodontosaurus*; *S. brevicollis* see *Carcharodontosaurus saharicus*
Sierra Sculpture *565*, 611
Signore, Marco 495
Sihetun locality 243, 270
Silesian Park of Culture and Recreation *604*
Siluosaurus 168
Silva, Helder de Paula 25, 315–7
Silverberg, Robert 621
Silvisaurus 70–1, 164, 187, 189, 216, 466, 490, *510*, 509–11, 542; *S. condrayi* 71, 189, 216
Simmons, David J. 53
Simpson, George Gaylord 310, 313
Simpson, William F. 521, 577, 615, 621
Simoliophis 268
Simon, Lewis M. 406
Simpson, William F. 49–50
Sinclair, William J. 109
"Sinclair Dinosaur Exhibit" 610, *610*
Sinclair Oil Corporation (Sinclair Refining Company) *92*, 606, *607*, *610*, 611, *612*, 614
"Sinclair's Dinoland" *92*, 606, 611, *612*, 614
Sino-Belgian Dinosaur Expedition 235, *236*, 456
Sino-Canadian Dinosaur Project 456
Sino-Swedish Expedition 457
Sinocoelurus *4*, 121
Sinornis 411
Sinornithoides youngi 260
Sinornithosaurus 94, 101, 144, 145, 187, 239–40, 274, 404, 511–2, *512–3*, 513–4, *514*, 514–6; *S. millenii* 239, 511, *512–4*
Sinosauropteryx 97, 101, 133, 187, 243–4, 271, 300, 302, 405, 474–5, 513, 515, *515–6*, 516–7, *517*
S. primus 300, 474, *515–7*
Sinosaurus 121
Sinotyrannus 35
Sinpetru Formation 44, 256, 325–6
Sinraptor 24, 100, 131–2, 339, 385
Skin and skin impressions 44, 61, 67, 75–7, 172–3, 219, 300, 375, 393, 403–4, 475, 485–6, 488, 511–2, *512*, 538
Slade, Warren 436, 438
Sloan, Christopher P. 226, 229–30, 401–2, 405, 407, 423, 426, 528
Smith, Charles 298
Smith, David K. 205–6, 212, 298, 301
Smith, Joshua B. 268, 314, 515–7
Smith, Matt B. 21, 247, 254, 575, 579
Smithsonian Institution *124*, *139*, *170*, *199*, *209*, *218*, 235, *280–1*, 556, 558, *558–9*, 614, 619
Snakes 24, 36, 252, 578
Snyder, Mark 331
Snyder quarry 331, 333–4
Social behavior 25–6, 33, 495
Society of Vertebrate Paleontology (SVP) ix, 4, 342, 405
Society of Vertebrate Paleontology News Bulletin 320
Solemys sp. 476
Solnhofen Limestone 302
Sonorosaurus 8, 187, 277, 517–8, *518*, 519; *S. thompsoni* 517, *518*
Sorenson, C., Jr. 299
Sound 85–6, 219–20, 225, 441–2
Soviet-Mongolian Palaeontological Expeditions 586
Sphenodontians 50
Sphenospondylus see *Iguanodon*
Spielberg, Steven vi
"Spike" 613
Spinar, Zdenek 602, 621
Spinosaurus 129–30, 187, 267–8, 303, *519*, 519–20, 530–2; *S. aegyptiacus* 303, 519–20; *S. ?maroccanus* 303, *519*, 519–20; "*Spinosaurus* B" 267, 520
Spinosuchus 46–7, *47*, 48, 595; *S. caseanus* 46–7, *47*, 48
Spondylosoma 19, 123
Spotila, James R. 88, 252, 254
Squalodon 596
Squamates 268, 468
Staab, Gary 375, *375*, 377, 619
Stadtman, Kenneth L. 57, 125, 223, 285–7, 387–8, 518–9
Staganolepis 52
"Stan" *572*
Stanley, Edward Smith 546
Starkov, Alexei I. 575, 579
State Museum of New York 428
State Museum of Pennsylvania 198
Staurikosaurus 19, 123, 202, 268–9, 348–50; *S. pricei* 19
Steel, Rodney 214, 217, 235, 238, 401–2, 431–2, 621
Stegocephalians 47
Stegoceras 81, *175*, 177–8, 187, 432, 463–5, 520; *S. breve* see *Prenocephale brevis*; *S. edmontonense* see *Prenocephale edmontonensis*; "*S.*" *lambei* 520; *S. validum* *175*, 463–5, 520
Stegopelta 70–1, 164, 166, 187, 342–3, 466, 520–1, *521–2*, 522–3, *523*; *S. landerensis* 70–1, 520–1, *521–3*
Stegosaurides 162
Stegosaurus 71, *160*, 160–1, 187, 320, 523–4, *524*, 555–6, 601, *603*, 606–7, *607*, *609*, 610–1, 614; *S. stenops* *524*; *S. ungulatus* *160*, 524
Stenopelix 118, 176
Sternberg, Charles H. 621
Sternberg, Charles M. 257, 259, 415–6, 464–5, 520, *564*
Sternberg family 602
Sternfield, R. 367, 369
Stevens, Kent A. 62, *62–3*, 63–4, *64*, 65–7, 251, 254, *261*, 261–2, *262*, 263, 372, 377
Stewart, J. D. 272, 276
Stokes, William L. 282, 287
Stokesosaurus 134, 210, 218–9
Stokstad, Erik 102, 577, 579
Stoneage *20*
Stormberg Group 245
Stoskopf, Michael 549–52, 554–5, *557*, 560
Stout, William 621–2
Stovall, J. Willis 192–3, 201, 287, 309, 448
Straight, William 551
Straight Cliffs Formation 229
Stromer, Ernst 129, 191, 193, 267–8, 460, 463, 519–20
Struthio camellus 375
Struthiomimus 37, 139, *139–40*, 212; *S. altus* 37, *139–40*
Struthiosaurus 164, 216, 354, 481, 490, 510, 521
Stucker, Gilbert F. 619
Stutchbury, Samuel 361, 467, 463, *543*, 543–5, 549, 596
Stygimoloch 81, *82*, 178, 229, *434*, 435, 463, 465; *S. spinifer* *82*, 178
Stygivenator 135, 187, 524–5, *525–6*, 526–8, 561; *?S. amplus* 525–6; *?S. cristatus* 525–6; *S. molnari* 525, *525–6*
Styracosaurus 82, 182, *183*, 187, 232, 528, *529*; *S. albertensis* *183*, 528, *529*
Subadults *see* Juveniles and subadults
Suberbiola, Xabier [Javier] Pereda 71, 75, 157–8, *189*, 188–9, 214–5, *215*, 216–7, 304, 328, 354–5, 382, *382–3*, 383–4, *384*, 453, *454*, 455–7, 467, 473–4, 484, 486, 539
Suchomimus x, 130, 187, 191, 243, 418, 520, 529–30, *530–1*, 531–2, *532–3*, 533; *S. tenerensis* 529–30, *530–3*, 533
"Sue" *viii*, 201, *569–71*, *573*, 576–9, 606, 618, 621
Sues, Hans-Dieter 23, 26, *26–7*, 27–8, *28*, 29, 32, 103–5, 131, 134, 139–40, 166, 167, 169, 178, 202–3, 245, 257–9, 362, 364–5, 402, 407, *407–8*, 408–9, 464–5, 476–7, 495, 520, 618
Sullivan, Emmett A. *153*, 309, *605*, 610, *610*, 614
Sullivan, Robert M. 24, 39, 47–8, 77, 109, 165, 177–8, 198, 200–1, 265–6, *266*, 267–9, 290, 294–7, 301, 328–9, *329–30*, 330–2, 334–5, 418–9, *419*, 420, 435–6, *436*, 437–9, *439*, 440–2, 463–4, *464–5*, 465, *466*, *494*, 494–5, 505–6, 520, *604*
Summerville Formation 262–3
Sun, A. 293–4
Supersaurus 152, 498, 580
Suteethorn, Varavudh 36, 74, 151, 365–6, *366*, 451–2, *452–3*, 536
Suzuki, Daisuke 174, 235, 418
SVP *see* Society of Vertebrate Paleontology
Sweeney, Frank 575–6, 579
Swinton, W. E. *4*, 78, 219, 431–2, 621–2
Swisher, C. C. 243
Syngonosaurus macrocercus see *Acanthopholis macrocercus*
Syntarsus 23, 71, 126, 187, 207, 265, 296–300, 314, 329, 332, 334, 337, 533–4, *534*; *S. kayentakatae* 297, 300, 314, 533; *S. rhodesiensis* 265, 298–300, 319, 334, 533–4, *534*
Szechuanosaurus *4*, 116, 131

Talarurus 164, 166, 418, 505
Tang Vat locality 534–5
Tangvayosaurus 157, 187, 355, 358, 534–5, *535–6*, 536; *T. hoffeti* 534–5, *535–6*; *Tangvayosaurus* sp. 535
Tanimoto, Masahiro 622

Index

Tanius 4, 171
Tanke, Darren H. 39, 181, 231–3
Tannenbaum, F. 266–7
Taphonomy 26, 33–4, 39, 89, 91, 111, 228, 253–5, 278, 289, 393, 397, 410, 468, 543, 592
Taquet, Philippe 21, 129, 227–8, *228–9*, 229, 358, 388–9, *389–90*, 391, *391*, 417–8, 447–8, 476–7, *519*, 519–20, 533–5, *535–6*, 536, 582–3, 622
Tarascosaurus 127
Tarbosaurus see *nTyrannosaurus*; *T. bataar* see *Tyrannosaurus bataar*; *T. efremovi* see *Tyrannosaurus bataar*
Tarchia 118, 164–6, 187, 419–20, 502, 505, 536–7, *604*; *T. gigantea* 536, *604*
Tarsitano (1st name?) 95
Tate Geological Museum 220
Tatisaurus 159
Taveirosaurus 158
Tawasaurus 146
Technosaurus 117, 158
Tecovas Formation 294
Tecovasaurus 158, 267; *T. murryi* 267
Tégama Group 417, 529
Tégama Series 388
Tehuelchesaurus 149, 187, 537, *537–8*, 538–9; *T. benitezii* 537, *537–8*, 538–9
Teinurosaurus 121
Telmatosaurus 44, 75, 118, 171–2, 187, 238, 539; *T. transsylvanicus* 75, 171, 238, 539
Tendaguria 155, 187, 366, 367, 539–40, *540–1*, 542; *T. tanzaniensis* 539–40, *540–1*, 541–2
Tendaguru Beds (Formation) 246, 249–50, 253, 282, 287, 367, 539–40, 621
Tendaguru Saurian Beds *see* Tendaguru Beds (Formation)
Tendons *see* Muscles, ligaments, tendons, musculature and muscle scars
Tenontosaurus 118, 167–8, 310, *311*, 312, 384, 479, 481; *T. tilleti* 481
?*Teratosaurus suevicus* see *Sellosaurus gracilis*
Termatosaurus 596
"Terror of the South" *193*
Tetori Group 35, 73, 338
Tetrapodosaurus borealis 490
Tetrasauropus unguiferus 337
Texasetes 70–1, 164, 187, 189, 216, 521–2, 542; *T. pleurohalio* 71, 216
Teylers Musem 77
Teyuwasu 120, 542–3; *T. barberenai* 542
Thayer, D. W. 310, 313, 517, 519
Thecoelurus 121
"Thecodontians" 123, 202, 245, 301, 399, 595, 615
Thecodontosaurus 50–2, 146, 187, 457, 461–2, 543, *543–4*, 544–5, *545–6*, 546–7, *547*, 548–9; *T. antiquus* 457, *543–4*, 544–5, *545–6*, 546–7, *547*, 548, 595; *Thecodontosaurus* sp. 545
Thecospondylus 121
Therapsids 615, 617
Therizinosaurus 98, 142
Theropod "A" 42
Therosaurus see *Iguanodon*
Thescelosaurus 93, *93*, 168, 187, 257–9, 549, *549–50*, 550–1, *551–2*, 552; *T. neglectus 93*, 549, *549–52*; ?*Thescelosaurus* sp. see *Bugenasaura infernalis*
Thespesius 174
Third Asiatic Expedition of the American Museum of Natural History *69*
Thomas, David A. 110, 377, 605, 607, 611, 613
Thotobolosaurus 145
Thulborn, Anthony A. [Tony] 234–5, 248–9, 254, 498, 622
Thurmond, J. T. 310
Tianchisaurus 164, 166, 177
Tianchungosaurus 117
Tianzhenosaurus see *Shanxia*
T. youngi see *Shanxia tianzhenensis*
Tichosteus 120
Tidwell, Virginia 56, 156, 276, *276–7*, 277–8, *278*, 279
Tienshanosaurus see *Euhelopus*
Tilougguit Formation 227
"Tinker" 575, 579
Tiourén Formation 371–2
Tipping, Michael 436
Titanosaurus 58, 60–1, 157–8, 317, 358, 535; *T. araukanicus* see *Laplatasaurus araukanicus*; *T. colberti* 157; *T. falloti see Tangvuayosaurus* sp.; *T. indicus* 157; *T. madagascarensis* 60
Tochisaurus 143, 260; *T. nemegtensis* 260
Tong, Haiyan 365–6, *366*, 393
Toolebuc Formation 409
Tornieria see *Barosaurus*; *T. africana* see *Barosaurus africanus*; *T. robusta* see *Janenschia robusta*; *Tornieria* sp. see *Janenschia*
Torosaurus 84, 183–4, 200, *557*, 575
Torvosaurus 127, *129*, 210, 385, 399; *T. tanneri 129*
Trace fossils *see* Eggs, Footprints and trackways
Trachodon 179
Tracks and trackways *see* Footprints and trackways
Trauma (in *Stegosaurus* dermal spines) 523–4
Traversodontids 245
Trcic, Michael *376–7*
Tree of Life Project 70
Tremp Formation 19, 75
Tribelesodon 596
Triceratops 8, 16, *17*, 68, *83*, *84–5*, 86, 101, *107*, 108, 119–20, 158, 183, 187, 232–3, 291–2, 448–50, 552–3, *553–4*, 554–5, *555–6*, 556–7, *557–8*, 558–9, *559–60*, 560, *565*, 574–5, *610*, 611, 614; *T. horridus 83*, *84–5*, *107*, 552–3, *553–6*, 556, *557–60*, *565*, *610*, 611, 614; *T. prorsus* see *Triceratops horridus*; *T. serratus* see *Triceratops horridus*; *Triceratops* sp. 86
Triceratops Rescue Project 556
Triebold, Michael 40, *41*, 323–4, *324*, *434*, 435
Triebold Paleontology, Inc. 40, *41*, *323–4*
Trigonia 253, 367
Trigonia smeei Bed (Tendaguru) 253, 367
Trilophosurus 48
Trimucrodon 158
Trinity Group 310
Troodon 23, 42, 45, 143, 187, 260, 520, 560–1; *T. edmontonensis* see *Prenocephale edmontonensis*; *T. formosus* 42, 45, 260, 560; *T. sternbergi* see *Gravitholus sternbergi*
Trössingen quarry 459–63
Tsagantegia 118, 164, 165, 418, 505
Tsintaosaurus 173, 175, 290; *T. spinorhinus 173*
Tuchengzi Formation 287, 516
Tucson Gem, Mineral, and Fossil Show 405
Tugrikin-Shireh 586
Tugrugeen Shireh 508, 586
Tugulusaurus 121
Tumanova, Tatyana A. 191, 475–6, 502–3, 537
Tuojiangosaurus 161
Turanoceratops 119, 181, 183–4, 416
Turner, J. C. 537
Turner Falls System 300
Turney Ranch Formation 517
Turtles *see* Cheloneons
Tustea locality 44
Twin Mountains Formation 397
Two Medicine Formation 33, 43, 46, 71, 238, 304, 309, 378–9, 393–5, 528, 560
Tykoski, Ron 300–1, 334–5
Tylocephale 178
Typhlops 36
Tyrannosaurus viii, *20*, 21, *22*, 23, 25, 33–4, *34*, *83*, *89*, 104, 135, *137*, 187, 191–2, 200, 317, 371, 524, 526, *526*, 527–9, *557*, 561–2, *562–3*, 563, *564–5*, 565–6, *566–7*, 567–8, *568–9*, 569–70, *570–1*, 571–2, *572–3*, 573–4, *574*, 575–9, 605–8, *610*, 611, 614, 618; *T. bataar* 34, 192, *566*, 567–71, 573–5; *T. rex viii*, *20*, *22*, 25, 33–4, *34*, *83*, *89*, *137*, 191, *526*, 526–9, *557*, 562, *562–5*, 565–7, *567–8*, 568, *569–72*, 572–3, *573–4*, 574–9, 605–6, *610*, 611, 614
Tyrannosaurus Sue: A Cretaceous Concerto 577
Tyson, Philip T. 227

U. S. Geological Survey 111
Udanoceratops 180–1, 415
Uncinate processes (in dinosaurs) 89, 101
Uncle Beazley 614
Uintatherium 3
Uitenhage Group 428
Ukhaa Tolghod locality 44, 259–60, 508
Ulansuhai Formation 38
Ultrasaurus 147, 187, 579–80, *580*, 581; *U. tabriensis* 579–80, *580*
Unenlagia 103–4, 241, 404, 587; *U. comahuensis* 103–4, 587
United States National Museum *see* National Museum of Natural History
Universal Studios 4, 612–3
Universät zu Berlin 253
Universidad Catolica del Norte 57
Universidad Federal do Rio de Janeiro 194
Universidade de Salamanca 316
University Museum *131*
University of Alabama 560
University of California 102, 108, 402, *438*, 496, 622
University of Cambridge 189
University of Chicago 57, *132*, *267*, 372, 521, 621
University of Colorado *553–4*
University of Colorado Museum *553–4*
University of Copenhagen 543
University of Glasgow 361
University of Kansas 102–3, 242, 510
University of Maryland 103
University of Melbourne 547
University of Michigan Museum of Paleontology 47, *47*, *268*
University of New Mexico 200, 621
University of North Carolina *93*, 102
University of Oklahoma *207*, 309, 448
University of Pennsylvania 268
University of Queensland 547
University of San Juan 57
University of South Wales 525
University of Texas 405, 496
University of Upsalla 420
University of Wyoming 223, 614
University of Wyoming Geological Museum 614
Unquillosaurus 127
Untermann, G. E. and Billie R. 622
Upchurch, Paul 51–3, 55, 145, 151, 158, 200, 217–8, 279, 317–8, 320, 365–6, 391, 393, 416–7, 452–3, 461, 463, 489, 492, 494, 503, 505–7
Upper Gault Clay 214
Upper Keuper 460
Upper Saurian Bed (Tendaguru) 282, 287, 368, 541
Upper Transitional Sands 367
Urlichs, B. 459–60, 463
Utah Field House of Natural History 607, *609*
Utahraptor 144, 607

Vacca, Raul 537, *537–8*, 538–9
Valdoraptor 132
Valdosaurus 169, 533; *V. nigeriensis* 533
Valle de la Prehistoria 602
Valley Anatomical Preparations 614
The Valley of Dinosaurs 603–4, *604*
Van Howd, Douglas *565*, 611,
Van Valkenburgh, Blaire 94
Van Wyngarden, Jerry *87*
Varanus 88, 430; *V. crocodilinus* 430
Vargus, Alexander 57, 99
Variraptor 144, 187, 327, 529–30, *530*, 581, *581–2*, 582–3, *583*; *V. mechinorum* 581, *581–2*, 582, *583*
Varrichio, David J. 32–4, 43, *43*, 46, 66, 129–30, 192–3, 243, 257, 267–8, 303–4, 310, 327,

364–5, 371, *371–2*, 374–5, 377, *417*, 417–8, 429–30, 520, 529–30, *530–1*, 531–3, 560–1
Vectisaurus see *Iguanodon*
Vegetation *see* Plants
Velocipes 121
Velociraptor 40, 46, 89, 99, 103–4, 144–5, 187, 239, 240, 260, 468, 495, 512, 582–4, *584–5*, 585–7; *V.* cf. *langstoni* see *Bambiraptor feinbergi*; *V. mongoliensis* 103, 144, 260, 495, 582–4, *584–5*, 585–7
Velocisaurus 127
Vene, Metello 79, *80*
Venenosaurus dicrocei 592
Venomous theropods 23
Vertebrate Paleontology 227, 616, 621
Vickaryous, Matthew K. 70, 335–6
Vickers-Rich, Patricia 72, 138, 149, 168, *477*, 477–8, 537, *537–8*, 538–9, 546–7, 549, 621–2
Victorian Hypsilophodontid Type 1 and 2 72
Viohl, Günter *302*, 302–3
Vision 66, 98, 585
Volcanoes and volcanics 107, 109, 393, 515–6
Volkeimeria 156, 365; *V. chubutensis* 365
Völkl, Pino 302
Vulcanodon 148, 317, 346, 366, 548

Wadi Milk Formation *11*, 11–2
Wagner, Günter P. 97
Wakinosaurus 127–8
Waldman, M. 400, 402
Walgettosuchus 127
Walker, Alick D. 14, 95, 245–6, 337
Walker, Cyril A. 40, 256–7, 326–8
Walker Museum 521
Walkeria see *Alwalkeria*
Wall, William P. 464–5, 520
Wall Drug Store *605*, 614
Wall Street Journal 578
Waller, Don *573*
Walt Disney World 611
Walters, Robert F. 618
Wang Xiao-lin 143, 243, 269, *269*, 270, *270–2*, 272–3, *273–4*, 274–6, 369, *369–70*, 370–1, 402, *402–3*, 403–4, *404*, 405–7, *474*, 474–6, 475–7, 511–2, *512–3*, 513–5, 515, 517
Wang, Yuanqing 475–6, 515, 517
Wannanosaurus 177
Ward, Henry A. *3*
Warner Valley Tracksite 614
Watabe, Mahito 39, *101*, 235, 322–3, 422–6, 471
Wawmda locality 227
Wealden Group 30–1, 226, 321, 359, 362, 416
Webster, Donovan 576–7, 579
Wedel, Matthew J. 56–7, 375, 377, 491–2, *492–3*, 493–4
WED Enterprises 614
Wedmore Hill quarry 399–400
Weinstein, John *viii*, *49–50*, *120*, *305*, *557*, *569*, 615
Weishampel, David B. vi, 16, 39, 44, 71–2, 75, 77–8, *78*, 166, 167, 168–9, 171–2, 174–5, 181, 205, 212, 214, 216–7, 235, 247, 254, 256–7, 259, 320–1, 326, 363, 370–1, 377–8, 391–5, 409, *413–4*, 414–6, 418, 431, 437, 439, 441–2, 448, 450, 460–3, 467, 471, 473–4, 477–9, 481, 502, 505–6, 539, 592, 596, 623, 615, 622
Weiss, Hans and Klaus-Dieter 302
Welbedacht farm 533
Welles, Samuel P. *47*, 125–6, *184*, 211–2, *268*, 279–83, *283–5*, 285, 287, *298*, 313–4, *330–4*, 402, 434, 533–4, 572, 579, 623
Wellnhofer, Peter 14, 103, 104, 263, 265, 271–2, 276, 302, 365–6, 412–3, 457, *459*, 460–3, 468, 471, 552–3, *553*, 560, 620
Welman, J. 95, 337
Werner, C. 11–2
Wesley College 547
Wessex Formation 30, 226
West, Linda 622
Western Australian Museum 432, 434
Western Interior Seaway 384, 474
Westgate, James W. 75
Westphal, F. 327, 462–3, 481
Wheeler, P. E. 441–2
"Whiplash" tails 58, *59*, 151–2, 157, 219–20
Whitaker, George O. 622
Whitaker Quarry 296–7, 330
White, P. D. 107–8
White, Theodore E. 219, 283, 287, 431–2, 619, 622
Whittle, Christopher H. 499
Widmann, Fred 440
Wiedermann, Albert 251, 254
Wilcox, Charles 430
Wild, Rupert 155, 157, 367–8, *368*, 369, 460, 539–40, *540–1*, 541–2
Wilhite, Ray 262–3, *263*
Williams, Larry *20*
Williams, Michael E. 34, 106, 107, 108, 305, 310, 565, 567, 579
Williamson, Thomas E. 33, 81, 200–2, 229–30, 290, 309–10, 377–8, 419–20, 435–9, *439*, 440–2, 527–8, 573–4, 576, 579, 592
Williston, Samuel Wendell 520–3
"Willo" *549*, 551, *552*
Wilson, Edward 545
Wilson, Jeffrey A. 21, 23, 53, 58–60, 66, 129–31, 146–9, 151, 153–8, 192–3, 243, 257, 267–8, 303–4, 317, 320, 327, 358–9, 364–5, 371, *371–2*, 374–5, 377, 388, 416–7, *417*, 418, 429–30, 481, 483, 486, 520, 529–30, *530–1*, 531–3, 535–6, 548–9
Wilson, Larry 606
Wilson, Woodrow 608
Wiman, Carl 293–4, 309, 334–5, *335*, 448, 450
Winkler, Dale A. 157, 167–8, 189, 216–7, 279, 310, 313, 397
Winkler, T. C. 77–8
Wirst, Bob 466
Withers, P. C. 251–2, 254
Witmer, Lawrence M. 16, 68–9, 82–3, *83*, 88, *88*, 166, 310, 320–1, 431, 566, 579, 584, 586–7
Wodehouseia spinata 289
Wolberg, Donald 576, 579, 622
Wold Creek quarry 56
Wolfe, Douglas G. 181, 590–1, *591*, 592
Wonthaggi Formation 72, 477
Woodbine Formation 471, 473
Works Progress Administration (WPA) 610
World's Fairs and Expositions *92*, *153*, 556, *558–9*, 602, 606, *607*, *610*, 610–1, *612*, 614
WPA see Works Progress Administration
Wright, Joanna L. *362*, 363
Wright, Nelda E. 75, 110, 377–8, 438, 441–2
Wuerhosaurus 161
Wyleia 121
The Wyoming Dinosaur Center 614

Xenotarsosaurus 127
Xiaosaurus 158, 371
Xinminbao Group 293
Xu Xing 43–4, 74–5, 243–4, *244*, 272, 274, 275–6, 287–8, *288*, 289, 369, *369–70*, 370–1, 402, *402–3*, 403–4, *404*, 405–7, 429–30, *474*, 474–6, 476–7, 511–2, *512–3*, 513–5, 517
Xuanhanosaurus 127–8, 338
Xue Xiangxu *478*, 478–9, 506, *507*

Yale University *68*, 310, 407, 524, 621
Yandusaurus 168, 371; *Y. multidens* 371
Yang, Seong-Young 292–3, *293*, 294, 579–80, *580*, 581
Yang, Zhungjian [Young Chung-Chien] 334–5, 394–5, 397, 399, 432, 457
Yangchuanosaurus 131, *132*, 385; *Y. shangyouensis 132*, 385
Yaverlandia 178
Yimenosaurus 147, 187, 587, *587–8*, 588–9, *589*; *Y. youngi* 587, *587–8*, 588, *589*
Yingshanosaurus 161
Yixian Formation 44, 243–4, 270, 369–70, 403, 474–5, 511, 515–7
You, Hailu 180–1, 226, 314, 369, *369–70*, 370–1, 379, 503, 505, 506, 515–7
Young Locality *87*
Yuliangze Formation 289
Yunnanosaurus 52–3, 146, 461, 548; *Y. robustus* 53

Zallinger, Jean 616
Zallinger, Peter 622
Zallinger, Rudolph F. 615, 621–2
Zanclodon 459; *Z. bavaricus* 460; *Z. cambrensis* see *?Megalosaurus cambrensis*
Zangerl, Rainer 622
Zatomus 121
Zbyszewsky, Georges 63, 304, 387–8
Zelenitsky, D. K. 561
Zephyrosaurus 168
Zhang Yihong 48, 149, 270, 275–6, 365–6, 385–6, 397, 399, 590
Zhangheotherium 516
Zhangijagou locality 270
Zhao, Xin Jin [Zhao Xijin] 18, 74–5, 131, 260, 287–8, *288*, 289, 385–6, 429–30, 468, 471
Zhen Shuo-nan 29, 133
Zheng, Zhong 293–4, 397, 399
Zhou, Fengyun 343–4, *344–5*, 346, 365–6
Zhou Xinjin 589
Zhou Zhong-He [Zhonghe] 99, 143, 238–9, *239–40*, 240, *241*, 242, 269, *269*, 270, *270–2*, 272–3, *273–4*, 274–6, 402, *402–3*, 403–4, *404*, 405–7
Zielinski, Robert 109
Zigongosaurus 156, 187, 589–90, *590*; *Z. fuxiensis* 589–90, *590*
Ziliujing Formation 343–4
Zimmerman, Howard 622
Zittel, K. A. 214, 217
Zizhongosaurus 156, 365; *Z. chuanchengensis* 366
Zuniceratops 119, 181, 184, 187, 590–1, *591*, 592; *Z. christopheri* 365, 590–1, *591*, 592